THE CIA WORLD FACTBOOK 2009

CENTRAL INTELLIGENCE AGENCY

Skyhorse Publishing

Skyhorse Publishing books may be purchased in bulk at special discounts for sales promotion, corporate gifts, fund-raising, or educational purposes. Special editions can also be created to specifications. For details, contact the Special Sales Department, Skyhorse Publishing, 555 Eighth Avenue, Suite 903, New York, NY 10018 or info@skyhorsepublishing.com.

www.skyhorsepublishing.com

ISBN: 978-1-60239-282-3

10 9 8 7 6 5 4 3 2 1

In general, information available as of June 2008 was used in the preparation of this edition.

Printed in the United States of America

TABLE OF CONTENTS

INTRODUCTION

The World Factbook is prepared by the Central Intelligence Agency for the use of US Government officials, and the style, format, coverage, and content are designed to meet their specific requirements. Information is provided by Antarctic Information Program (National Science Foundation), Armed Forces Medical Intelligence Center (Department of Defense), Bureau of the Census (Department of Commerce), Bureau of Labor Statistics (Department of Labor), Central Intelligence Agency, Council of Managers of National Antarctic Programs, Defense Intelligence Agency (Department of Defense), Department of Energy, Department of State, Fish and Wildlife Service (Department of the Interior), Maritime Administration (Department of Transportation), National Geospatial-Intelligence Agency (Department of Defense), Naval Facilities Engineering Command (Department of Defense), Office of Insular Affairs (Department of the Interior), Office of Naval Intelligence (Department of Defense), US Board on Geographic Names (Department of the Interior), US Transportation Command (Department of Defense), Oil & Gas Journal, and other public and private sources.

Comments and queries are welcome and may be addressed to:

Central Intelligence Agency
Attn.: Office of Public Affairs
Washington, DC 20505
Hours: Monday–Friday 8:00 AM–4:30 PM
  Eastern Standard Time
Telephone: [1] (703) 482-0623
FAX: [1] (703) 482-1739

The Intelligence Cycle is the process by which information is acquired, converted into intelligence, and made available to policymakers. **Information** *is raw data from any source, data that may be fragmentary, contradictory, unreliable, ambiguous, deceptive, or wrong.* **Intelligence** *is information that has been collected, integrated, evaluated, analyzed, and interpreted.* **Finished intelligence** *is the final product of the Intelligence Cycle ready to be delivered to the policymaker.*

The three types of finished intelligence are: basic, current, and estimative. Basic intelligence provides the fundamental and factual reference material on a country or issue. Current intelligence reports on new developments. Estimative intelligence judges probable outcomes. The three are mutually supportive: basic intelligence is the foundation on which the other two are constructed; current intelligence continually updates the inventory of knowledge; and estimative intelligence revises overall interpretations of country and issue prospects for guidance of basic and current intelligence. *The World Factbook, The President's Daily Brief*, and the *National Intelligence Estimates* are examples of the three types of finished intelligence.

The United States has carried on foreign intelligence activities since the days of George Washington but only since World War II have they been coordinated on a government-wide basis. Three programs have highlighted the development of coordinated basic intelligence since that time: (1) *the Joint Army Navy Intelligence Studies* (JANIS), (2) *the National Intelligence Survey* (NIS), and (3) *The World Factbook*.

During World War II, intelligence consumers realized that the production of basic intelligence by different components of the US Government resulted in a great duplication of effort and conflicting information. The Japanese attack on Pearl Harbor in 1941 brought home to leaders in Congress and the executive branch the need for integrating departmental reports to national policymakers. Detailed and coordinated information was needed not only on such major powers as Germany and Japan, but also on places of little previous interest. In the Pacific Theater, for example, the Navy and Marines had to launch amphibious operations against many islands about which information was unconfirmed or nonexistent. Intelligence authorities resolved that the United States should never again be caught unprepared.

In 1943, Gen. George B. Strong (G-2), Adm. H. C. Train (Office of Naval Intelligence—ONI), and Gen. William J. Donovan (Director of the Office of Strategic Services—OSS) decided that a joint effort should be initiated. A steering committee was appointed on 27 April 1943 that recommended the formation of a Joint Intelligence Study Publishing Board to assemble, edit, coordinate, and publish the *Joint Army Navy Intelligence Studies* (JANIS). JANIS was the first interdepartmental basic intelligence program to fulfill the needs of the US Government for an authoritative and coordinated appraisal of strategic basic intelligence. Between April 1943 and July 1947, the board published 34 JANIS studies. JANIS performed well in the war effort, and numerous letters of commendation were received, including a statement from Adm. Forrest Sherman, Chief of Staff, Pacific Ocean Areas, which said, "JANIS has become the indispensable reference work for the shore-based planners."

The need for more comprehensive basic intelligence in the postwar world was well expressed in 1946 by George S. Pettee, a noted author on national security. He wrote in *The Future of American Secret Intelligence* (Infantry Journal Press, 1946, page 46) that world leadership in peace requires even more elaborate intelligence than in war. "The conduct of peace involves all countries, all human activities—not just the enemy and his war production."

The Central Intelligence Agency was established on 26 July 1947 and officially began operating on 18 September 1947. Effective 1 October 1947, the Director of Central Intelligence assumed operational responsibility for JANIS. On 13 January 1948, the National Security Council issued Intelligence Directive (NSCID) No. 3, which authorized the *National Intelligence Survey* (NIS) program as a peacetime replacement for the wartime JANIS program. Before adequate NIS country sections could be produced, government agencies had to develop more comprehensive gazetteers and better maps. The US Board on Geographic Names (BGN) compiled the names; the Department of the Interior produced the gazetteers; and CIA produced the maps.

The Hoover Commission's Clark Committee, set up in 1954 to study the structure and administration of the CIA, reported to Congress in 1955 that: "The National Intelligence Survey is an invaluable publication which provides the essential elements of basic intelligence on all areas of the world. There will always be a continuing requirement for keeping the Survey up-to-date." The *Factbook* was created as an annual summary and update to the encyclopedic NIS studies. The first classified *Factbook* was published in August 1962, and the first unclassified version was published in June 1971. The NIS program was terminated in 1973 except for the *Factbook*, map, and gazetteer components. The 1975 *Factbook* was the first to be made available to the public with sales through the US Government Printing Office (GPO). The *Factbook* was first made available on the Internet in June 1997. The year 2008 marks the 61st anniversary of the establishment of the Central Intelligence Agency and the 65th year of continuous basic intelligence support to the US Government by *The World Factbook* and its two predecessor programs.

Abbreviations This information is included in **Appendix A: Abbreviations**, which includes all abbreviations and acronyms used in the *Factbook*, with their expansions.

Acronyms An acronym is an abbreviation coined from the initial letter of each successive word in a term or phrase. In general, an acronym made up solely from the first letter of the major words in the expanded form is rendered in all capital letters (NATO from North Atlantic Treaty Organization; an exception would be ASEAN for Association of Southeast Asian Nations). In general, an acronym made up of more than the first letter of the major words in the expanded form is rendered with only an initial capital letter (Comsat from Communications Satellite Corporation; an exception would be NAM from Nonaligned Movement). Hybrid forms are sometimes used to distinguish between initially identical terms (ICC for International Chamber of Commerce and ICCt for International Criminal Court).

Administrative divisions This entry generally gives the numbers, designatory terms, and first-order administrative divisions as approved by the US Board on Geographic Names (BGN). Changes that have been reported but not acted on by the BGN are noted.

Age structure This entry provides the distribution of the population according to age. Information is included by sex and age group (*0–14 years, 15–64 years, 65 years and over*). The age structure of a population affects a nation's key socioeconomic issues. Countries with young populations (high percentage under age 15) need to invest more in schools, while countries with older populations (high percentage ages 65 and over) need to invest more in the health sector. The age structure can also be used to help predict potential political issues. For example, the rapid growth of a young adult population unable to find employment can lead to unrest.

Agriculture—products This entry is an ordered listing of major crops and products starting with the most important.

Airports This entry gives the total number of airports or airfields recognizable from the air. The runway(s) may be paved (concrete or asphalt surfaces) or unpaved (grass, earth, sand, or gravel surfaces) but may include closed or abandoned installations. Airports or airfields that are no longer recognizable (overgrown, no facilities, etc.) are not included. Note that not all airports have accommodations for refueling, maintenance, or air traffic control.

Airports—with paved runways This entry gives the total number of airports with paved runways (concrete or asphalt surfaces) by length. For airports with more than one runway, only the longest runway is included according to the following five groups—(1) *over 3,047 m*, (2) *2,438 to 3,047 m*, (3) *1,524 to 2,437 m*, (4) *914 to 1,523 m*, and (5) *under 914 m*. Only airports with usable runways are included in this listing. Not all airports have facilities for refueling, maintenance, or air traffic control.

Airports—with unpaved runways This entry gives the total number of airports with unpaved runways (grass, dirt, sand, or gravel surfaces) by length. For airports with more than one runway, only the longest runway is included according to the following five groups—(1) *over 3,047 m*, (2) *2,438 to 3,047 m*, (3) *1,524 to 2,437 m*, (4) *914 to 1,523 m*, and (5) *under 914 m*. Only airports with usable runways are included in this listing. Not all airports have facilities for refueling, maintenance, or air traffic control.

Appendixes This section includes *Factbook*-related material by topic.

Area This entry includes three subfields. *Total area* is the sum of all land and water areas delimited by international boundaries and/or coastlines. *Land area* is the aggregate of all surfaces delimited by international boundaries and/or coastlines, excluding inland water bodies (lakes, reservoirs, rivers). *Water area* is the sum of the surfaces of all inland water bodies, such as lakes, reservoirs, or rivers, as delimited by international boundaries and/or coastlines.

Area—comparative This entry provides an area comparison based on total area equivalents. Most entities are compared with the entire US or one of the 50 states based on area measurements (1990 revised) provided by the US Bureau of the Census. The smaller entities are compared with Washington, DC (178 sq km, 69 sq mi) or The Mall in Washington, DC (0.59 sq km, 0.23 sq mi, 146 acres).

Background This entry usually highlights major historic events and current issues and may include a statement about one or two key future trends.

Birth rate This entry gives the average annual number of births during a year per 1,000 persons in the population at midyear; also known as crude birth rate. The birth rate is usually the dominant factor in determining the rate of population growth. It depends on both the level of fertility and the age structure of the population.

Budget This entry includes revenues, expenditures, and capital expenditures. These figures are calculated on an exchange rate basis, i.e., not in purchasing power parity (PPP) terms.

Capital This entry gives the *name* of the seat of government, its *geographic coordinates*, the *time difference* relative to **Coordinated Universal Time (UTC)** and the time observed in Washington, DC, and, if applicable, information on *daylight saving time* (**DST**). Where appropriate, a special *note* has been added to highlight those countries that have multiple time zones.

Climate This entry includes a brief description of typical weather regimes throughout the year.

Coastline This entry gives the total length of the boundary between the land area (including islands) and the sea.

Communications This category deals with the means of exchanging information and includes the telephone, radio, television, and Internet host entries.

Communications—note This entry includes miscellaneous communications information of significance not included elsewhere.

Constitution This entry includes the dates of adoption, revisions, and major amendments.

Coordinated Universal Time (UTC) UTC is the international atomic time scale that serves as the basis of timekeeping for most of the world. The hours, minutes, and seconds expressed by UTC represent the time of day at the Prime Meridian (0° longitude) located near Greenwich, England as reckoned from midnight. UTC is calculated by the Bureau International des Poids et Measures (BIPM) in Sevres, France. The BIPM averages data collected from more than 200 atomic time and frequency standards located at about 50 laboratories worldwide. UTC is the basis for all civil time with the Earth divided into time zones expressed as positive or negative differences from UTC. UTC is also referred to as "Zulu time." See the Standard Time Zones of the World map included with the **Reference Maps**.

Country data codes see **Data codes**.

Country map Most versions of the *Factbook* provide a country map in color. The maps were produced from the best information available at the time of preparation. Names and/or boundaries may have changed subsequently.

Country name This entry includes all forms of the country's name approved by the US Board on Geographic Names (Italy is used as an example): *conventional long form* (Italian Republic), *conventional short form* (Italy), *local long form* (Repubblica Italiana), *local short form* (Italia), *former* (Kingdom of Italy), as well as the *abbreviation*. Also see the **Terminology** note.

Crude oil See entry for oil.

Currency (code) This entry identifies the national medium of exchange and, in parenthesis, gives the International Organization for Standardization (ISO) 4217 alphabetic currency code for each country.

Current account balance This entry records a country's net trade in goods and services, plus net earnings from rents, interest, profits, and dividends, and net transfer payments (such as pension funds and worker remittances) to and from the rest of the world during the period specified. These figures are calculated on an exchange rate basis, i.e., not in purchasing power parity (PPP) terms.

Data codes This information is presented in **Appendix D: Cross-Reference List of Country Data Codes** and **Appendix E: Cross-Reference List of Hydrographic Data Codes**.

Date of information In general, information available as of 1 January 2007 was used in the preparation of this edition.

Daylight Saving Time (DST) This entry is included for those entities that have adopted a policy of adjusting the official local time forward, usually one hour, from Standard Time during summer months. Such policies are most common in mid-latitude regions.

Death rate This entry gives the average annual number of deaths during a year per 1,000 population at midyear; also known as crude death rate. The death rate, while only a rough indicator of the mortality situation in a country, accurately indicates the current mortality impact on population growth. This indicator is significantly affected by age distribution, and most countries will eventually show a rise in the overall death rate, in spite of continued decline in mortality at all ages, as declining fertility results in an aging population.

Debt—external This entry gives the total public and private debt owed to nonresidents repayable in foreign currency, goods, or services. These figures are calculated on an exchange rate basis, i.e., not in purchasing power parity (PPP) terms.

Dependency status This entry describes the formal relationship between a particular nonindependent entity and an independent state.

Dependent areas This entry contains an alphabetical listing of all nonindependent entities associated in some way with a particular independent state.

Diplomatic representation The US Government has diplomatic relations with 189 independent states, including 187 of the 192 UN members (excluded UN members are Bhutan, Cuba, Iran, North Korea, and the US itself). In addition, the US has diplomatic relations with 2 independent states that is not in the UN, the Holy See and Kosovo, as well as with the EU.

Diplomatic representation from the US This entry includes the *chief of mission, embassy* address, *mailing address, telephone* number, *FAX* number, *branch office* locations, *consulate general* locations, and *consulate* locations.

Diplomatic representation in the US This entry includes the *chief of mission, chancery, telephone, FAX, consulate general* locations, and *consulate* locations.

Disputes—international This entry includes a wide variety of situations that range from traditional bilateral boundary disputes to unilateral claims of one sort or another. Information regarding disputes over international terrestrial and maritime boundaries has been reviewed by the US Department of State. References to other situations involving borders or frontiers may also be included, such as resource disputes, geopolitical questions, or irredentist issues; however, inclusion does not necessarily constitute official acceptance or recognition by the US Government.

Distribution of family income—Gini index This index measures the degree of inequality in the distribution of family income in a country. The index is calculated from the Lorenz curve, in which cumulative family income is plotted against the number of families arranged from the poorest to the richest. The index is the ratio of (a) the area between a country's Lorenz curve and the 45 degree helping line to (b) the entire triangular area under the 45 degree line. The more nearly equal a country's income distribution, the closer its Lorenz curve to the 45 degree line and the lower its Gini index, e.g., a Scandinavian country with an index of 25. The more unequal a country's income distribution, the farther its Lorenz curve from the 45 degree line and the higher its Gini index, e.g., a Sub-Saharan country with an index of 50. If income were distributed with perfect equality, the Lorenz curve would coincide with the 45 degree line and the index would be zero; if income were distributed with perfect inequality, the Lorenz curve would coincide with the horizontal axis and the right vertical axis and the index would be 100.

Economic aid—donor This entry refers to net official development assistance (ODA) from Organization for Economic Cooperation and Development (OECD) nations to developing countries and multilateral organizations. ODA is defined as financial assistance that is concessional in character, has the main objective to promote economic development and welfare of the less developed countries (LDCs), and contains a grant element of at least 25%. The entry does not cover other official flows (OOF) or private flows. These figures are calculated on an exchange rate basis, i.e., not in purchasing power parity (PPP) terms.

Economic aid—recipient This entry, which is subject to major problems of definition and statistical coverage, refers to the net inflow of Official Development Finance (ODF) to recipient countries. The figure includes assistance from the World Bank, the IMF, and other international organizations and from individual nation donors. Formal commitments of aid are included in the data. Omitted from the data are grants by private organizations. Aid comes in various forms including outright grants and loans. The entry thus is the difference between new inflows and repayments. These figures are calculated on an exchange rate basis, i.e., not in purchasing power parity (PPP) terms.

Economy This category includes the entries dealing with the size, development, and management of productive resources, i.e., land, labor, and capital.

Economy—overview This entry briefly describes the type of economy, including the degree of market orientation, the level of economic development, the most important natural resources, and the unique areas of specialization. It also characterizes major economic events and policy changes in the most recent 12 months and may include a statement about one or two key future macroeconomic trends.

Electricity—consumption This entry consists of total electricity generated annually plus imports and minus exports, expressed in kilowatt-hours. The discrepancy between the amount of electricity generated and/or imported and the amount consumed and/or exported is accounted for as loss in transmission and distribution.

Electricity—exports This entry is the total exported electricity in kilowatt-hours.

Electricity—imports This entry is the total imported electricity in kilowatt-hours.

Electricity—production This entry is the annual electricity generated expressed in kilowatt-hours. The discrepancy between the amount of electricity generated and/or imported and the amount consumed and/or exported is accounted for as loss in transmission and distribution.

Elevation extremes This entry includes both the highest point and the lowest point.

Entities Some of the independent states, dependencies, areas of special sovereignty, and governments included in this publication are not independent, and others are not officially recognized by the US Government. "Independent state" refers to a people politically organized into a sovereign state with a definite territory. "Dependencies" and "areas of special sovereignty" refer to a broad category of political entities that are associated in some way with an independent state. "Country" names used in the table of contents or for page headings are usually the short-form names as approved by the US Board on Geographic Names and may include independent states, dependencies, and areas of special sovereignty, or other geographic entities. There are a total of 266 separate geographic entities in *The World Factbook* that may be categorized as follows:

INDEPENDENT STATES

194 Afghanistan, Albania, Algeria, Andorra, Angola, Antigua and Barbuda, Argentina, Armenia, Australia, Austria, Azerbaijan, The Bahamas, Bahrain, Bangladesh, Barbados, Belarus, Belgium, Belize, Benin, Bhutan, Bolivia, Bosnia and Herzegovina, Botswana, Brazil, Brunei, Bulgaria, Burkina Faso, Burma, Burundi, Cambodia, Cameroon, Canada, Cape Verde, Central African Republic, Chad, Chile, China, Colombia, Comoros, Democratic Republic of the Congo, Republic of the Congo, Costa Rica, Cote d'Ivoire, Croatia, Cuba, Cyprus, Czech Republic, Denmark, Djibouti, Dominica, Dominican Republic, Ecuador, Egypt, El Salvador, Equatorial Guinea, Eritrea, Estonia, Ethiopia, Fiji, Finland, France, Gabon, The Gambia, Georgia, Germany, Ghana, Greece, Grenada, Guatemala, Guinea, Guinea-Bissau, Guyana, Haiti, Holy See, Honduras, Hungary, Iceland, India, Indonesia, Iran, Iraq, Ireland, Israel, Italy, Jamaica, Japan, Jordan, Kazakhstan, Kenya, Kiribati, North Korea, South Korea, Kosovo, Kuwait, Kyrgyzstan, Laos, Latvia, Lebanon, Lesotho, Liberia, Libya, Liechtenstein, Lithuania, Luxembourg, Macedonia, Madagascar, Malawi, Malaysia, Maldives, Mali, Malta, Marshall Islands, Mauritania, Mauritius, Mexico, Federated States of Micronesia, Moldova, Monaco, Mongolia, Montenegro, Morocco, Mozambique, Namibia, Nauru, Nepal, Netherlands, NZ, Nicaragua, Niger, Nigeria, Norway, Oman, Pakistan, Palau, Panama, Papua New Guinea, Paraguay, Peru, Philippines, Poland, Portugal, Qatar, Romania, Russia, Rwanda, Saint Kitts and Nevis, Saint Lucia, Saint Vincent and the Grenadines, Samoa, San Marino, Sao Tome and Principe, Saudi Arabia, Senegal, Serbia, Seychelles, Sierra Leone,

Singapore, Slovakia, Slovenia, Solomon Islands, Somalia, South Africa, Spain, Sri Lanka, Sudan, Suriname, Swaziland, Sweden, Switzerland, Syria, Tajikistan, Tanzania, Thailand, Timor-Leste, Togo, Tonga, Trinidad and Tobago, Tunisia, Turkey, Turkmenistan, Tuvalu, Uganda, Ukraine, UAE, UK, US, Uruguay, Uzbekistan, Vanuatu, Venezuela, Vietnam, Yemen, Zambia, Zimbabwe

OTHER

2 Taiwan, European Union

DEPENDENCIES AND AREAS OF SPECIAL SOVEREIGNTY

6 Australia—Ashmore and Cartier Islands, Christmas Island, Cocos (Keeling) Islands, Coral Sea Islands, Heard Island and McDonald Islands, Norfolk Island
2 China—Hong Kong, Macau
2 Denmark—Faroe Islands, Greenland
9 France—Clipperton Island, French Polynesia, French Southern and Antarctic Lands, Mayotte, New Caledonia, Saint Barthelemy, Saint Martin, Saint Pierre and Miquelon, Wallis and Futuna
2 Netherlands—Aruba, Netherlands Antilles
3 New Zealand—Cook Islands, Niue, Tokelau
3 Norway—Bouvet Island, Jan Mayen, Svalbard
17 UK—Akrotiri, Anguilla, Bermuda, British Indian Ocean Territory, British Virgin Islands, Cayman Islands, Dhekelia, Falkland Islands, Gibraltar, Guernsey, Jersey, Isle of Man, Montserrat, Pitcairn Islands, Saint Helena, South Georgia and the South Sandwich Islands, Turks and Caicos Islands
14 US—American Samoa, Baker Island*, Guam, Howland Island*, Jarvis Island*, Johnston Atoll*, Kingman Reef*, Midway Islands*, Navassa Island, Northern Mariana Islands, Palmyra Atoll*, Puerto Rico, Virgin Islands, Wake Island (*consolidated in United States Pacific Island Wildlife Refuges entry)

MISCELLANEOUS

6 Antarctica, Gaza Strip, Paracel Islands, Spratly Islands, West Bank, Western Sahara

OTHER ENTITIES

5 oceans—Arctic Ocean, Atlantic Ocean, Indian Ocean, Pacific Ocean, Southern Ocean
1 World
266 total
Environment—current issues This entry lists the most pressing and important environmental problems. The following terms and abbreviations are used throughout the entry:

Acidification—the lowering of soil and water pH due to acid precipitation and deposition usually through precipitation; this process disrupts ecosystem nutrient flows and may kill freshwater fish and plants dependent on more neutral or alkaline conditions (see acid rain).

Acid rain—characterized as containing harmful levels of sulfur dioxide or nitrogen oxide; acid rain is damaging and potentially deadly to the earth's fragile ecosystems; acidity is measured using the pH scale where 7 is neutral, values greater than 7 are considered alkaline, and values below 5.6 are considered acid precipitation; note—a pH of 2.4 (the acidity of vinegar) has been measured in rainfall in New England.

Aerosol—a collection of airborne particles dispersed in a gas, smoke, or fog.

Afforestation—converting a bare or agricultural space by planting trees and plants; reforestation involves replanting trees on areas that have been cut or destroyed by fire.

Asbestos—a naturally occurring soft fibrous mineral commonly used in fireproofing materials and considered to be highly carcinogenic in particulate form.

Biodiversity—also biological diversity; the relative number of species, diverse in form and function, at the genetic, organism, community, and ecosystem level; loss of biodiversity reduces an ecosystem's ability to recover from natural or man-induced disruption.

Bio-indicators—a plant or animal species whose presence, abundance, and health reveal the general condition of its habitat.

Biomass—the total weight or volume of living matter in a given area or volume.

Carbon cycle—the term used to describe the exchange of carbon (in various forms, e.g., as carbon dioxide) between the atmosphere, ocean, terrestrial biosphere, and geological deposits.

Catchments—assemblages used to capture and retain rainwater and runoff; an important water management technique in areas with limited freshwater resources, such as Gibraltar.

DDT (dichloro-diphenyl-trichloro-ethane)—a colorless, odorless insecticide that has toxic effects on most animals; the use of DDT was banned in the US in 1972.

Defoliants—chemicals which cause plants to lose their leaves artificially; often used in agricultural practices for weed control, and may have detrimental impacts on human and ecosystem health.

Deforestation—the destruction of vast areas of forest (e.g., unsustainable forestry practices, agricultural and range land clearing, and the over exploitation of wood products for use as fuel) without planting new growth.

Desertification—the spread of desert-like conditions in arid or semi-arid areas, due to overgrazing, loss of agriculturally productive soils, or climate change.

Dredging—the practice of deepening an existing waterway; also, a technique used for collecting bottom-dwelling marine organisms (e.g., shellfish) or harvesting coral, often causing significant destruction of reef and ocean-floor ecosystems.

Drift-net fishing—done with a net, miles in extent, that is generally anchored to a boat and left to float with the tide; often results in an over harvesting and waste of large populations of non-commercial marine species (by-catch) by its effect of "sweeping the ocean clean."

Ecosystems—ecological units comprised of complex communities of organisms and their specific environments.

Effluents—waste materials, such as smoke, sewage, or industrial waste which are released into the environment, subsequently polluting it.

Endangered species—a species that is threatened with extinction either by direct hunting or habitat destruction.

Freshwater—water with very low soluble mineral content; sources include lakes, streams, rivers, glaciers, and underground aquifers.

Greenhouse gas—a gas that "traps" infrared radiation in the lower atmosphere causing surface warming; water vapor, carbon dioxide, nitrous oxide, methane, hydrofluorocarbons, and ozone are the primary greenhouse gases in the Earth's atmosphere.

Groundwater—water sources found below the surface of the earth often in naturally occurring reservoirs in permeable rock strata; the source for wells and natural springs.

Highlands Water Project—a series of dams constructed jointly by Lesotho and South Africa to redirect Lesotho's abundant water supply into a rapidly growing area in South Africa; while it is the largest infrastructure project in southern Africa, it is also the most costly and controversial; objections to the project include claims that it forces people from their homes, submerges farmlands, and squanders economic resources.

Inuit Circumpolar Conference (ICC)—represents the 145,000 Inuits of Russia, Alaska, Canada, and Greenland in international environmental issues; a General Assembly convenes every three years to determine the focus of the ICC; the most current concerns are long-range transport of pollutants, sustainable development, and climate change.

Metallurgical plants—industries which specialize in the science, technology, and processing of metals; these plants produce highly concentrated and toxic wastes which can contribute to pollution of ground water and air when not properly disposed.

Noxious substances—injurious, very harmful to living beings.

Overgrazing—the grazing of animals on plant material faster than it can naturally regrow leading to the permanent loss of plant cover, a common effect of too many animals grazing limited range land.

Ozone shield—a layer of the atmosphere composed of ozone gas (O3) that resides approximately 25 miles above the Earth's surface and absorbs solar ultraviolet radiation that can be harmful to living organisms.

Poaching—the illegal killing of animals or fish, a great concern with respect to endangered or threatened species.

Pollution—the contamination of a healthy environment by man-made waste.

Potable water—water that is drinkable, safe to be consumed.

Salination—the process through which fresh (drinkable) water becomes salt (undrinkable) water; hence, desalination is the reverse process; also involves the accumulation of salts in topsoil caused by evaporation of excessive irrigation water, a process that can eventually render soil incapable of supporting crops.

Siltation—occurs when water channels and reservoirs become clotted with silt and mud, a side effect of deforestation and soil erosion.

Slash-and-burn agriculture—a rotating cultivation technique in which trees are cut down and burned in order to clear land for temporary agriculture; the land is used until its productivity declines at which point a new plot is selected and the process repeats; this practice is sustainable while population levels are low and time is permitted for regrowth of natural vegetation; conversely, where these conditions do not exist, the practice can have disastrous consequences for the environment.

Soil degradation—damage to the land's productive capacity because of poor agricultural practices such as the excessive use of pesticides or fertilizers, soil compaction from heavy equipment, or erosion of topsoil, eventually resulting in reduced ability to produce agricultural products.

Soil erosion—the removal of soil by the action of water or wind, compounded by poor agricultural practices, deforestation, overgrazing, and desertification.

Ultraviolet (UV) radiation—a portion of the electromagnetic energy emitted by the sun and naturally filtered in the upper atmosphere by the ozone layer; UV radiation can be harmful to living organisms and has been linked to increasing rates of skin cancer in humans.

Water-born diseases—those in which bacteria survive in, and are transmitted through, water; always a serious threat in areas with an untreated water supply.

Environment—international agreements This entry separates country participation in international environmental agreements into two levels—*party to* and *signed, but not ratified*. Agreements are listed in alphabetical order by the abbreviated form of the full name.

Environmental agreements This information is presented in **Appendix C: Selected International Environmental Agreements**, which includes the name, abbreviation, date opened for signature, date entered into force, objective, and parties by category.

Ethnic groups This entry provides an ordered listing of ethnic groups starting with the largest and normally includes the percent of total population.

Exchange rates This entry provides the official value of a country's monetary unit at a given date or over a given period of time, as expressed in units of local currency per US dollar and as determined by international market forces or official fiat.

Executive branch This entry includes several subfields. *Chief of state* includes the name and title of the titular leader of the country who represents the state at official and ceremonial functions but may not be involved with the day-to-day activities of the government. *Head of* *government* includes the name and title of the top administrative leader who is designated to manage the day-to-day activities of the government. For example, in the UK, the monarch is the chief of state, and the prime minister is the head of government. In the US, the president is both the chief of state and the head of government. *Cabinet* includes the official name for this body of high-ranking advisers and the method for selection of members. *Elections* includes the nature of election process or accession to power, date of the last election, and date of the next election. *Election results* includes the percent of vote for each candidate in the last election.

Exports This entry provides the total US dollar amount of merchandise exports on an f.o.b. (free on board) basis. These figures are calculated on an exchange rate basis, i.e., not in purchasing power parity (PPP) terms.

Exports—commodities This entry provides a listing of the highest-valued exported products; it sometimes includes the percent of total dollar value.

Exports—partners This entry provides a rank ordering of trading partners starting with the most important; it sometimes includes the percent of total dollar value.

Fiscal year This entry identifies the beginning and ending months for a country's accounting period of 12 months, which often is the calendar year but which may begin in any month. All yearly references are for the calendar year (CY) unless indicated as a noncalendar fiscal year (FY).

Flag description This entry provides a written flag description produced from actual flags or the best information available at the time the entry was written. The flags of independent states are used by their dependencies unless there is an officially recognized local flag. Some disputed and other areas do not have flags.

Flag graphic Most versions of the *Factbook* include a color flag at the beginning of the country profile. The flag graphics were produced from actual flags or the best information available at the time of preparation. The flags of independent states are used by their dependencies unless there is an officially recognized local flag. Some disputed and other areas do not have flags.

Freshwater withdrawal (domestic/industrial/agricultural) This entry provides the annual quantity of water in cubic kilometers removed from available sources for use in any purpose. Water drawn-off is not necessarily entirely consumed and some portion may be returned for further use downstream. Domestic sector use refers to water supplied by public distribution systems. Note that some of this total may be used for small industrial and/or limited agricultural purposes. Industrial sector use is the quantity of water used by self-supplied industries not connected to a public distribution system. Agricultural sector use includes water used for irrigation and livestock watering, and does not account for agriculture directly dependent on rainfall. Included are figures for *total* annual water withdrawal and *per capita* water withdrawal.

GDP (official exchange rate) This entry gives the gross domestic product (GDP) or value of all final goods and services produced within a nation in a given year. A nation's GDP at offical exchange rates (OER) is the home-currency-denominated annual GDP figure divided by the bilateral average US exchange rate with that country in that year. The measure is simple to compute and gives a precise measure of the value of output. Many economists prefer this measure when gauging the economic power an economy maintains vis-à-vis its neighbors, judging that an exchange rate captures the purchasing power a nation enjoys in the international marketplace. Official exchange rates, however, can be artificially fixed and/or subject to manipulation—resulting in claims of the country having an under- or over-valued currency—and are not necessarily the equivalent of a market-determined exchange rate. Moreover, even if the official exchange rate is market-determined, market exchange rates are frequently established by a relatively small set of goods and services (the ones the country trades) and may not capture the value of the larger set of goods the country produces. Furthermore, OER-converted GDP is not well suited to comparing domestic GDP over time, since appreciation/depreciation from one year to the next will make the OER GDP value rise/fall regardless of whether home-currency-denominated GDP changed.

GDP (purchasing power parity) This entry gives the gross domestic product (GDP) or value of all final goods and services produced within a nation in a given year. A nation's GDP at purchasing power parity (PPP) exchange rates is the sum value of all goods and services produced in the country valued at prices prevailing in the United States. This is the measure most economists prefer when looking at per-capita welfare and when comparing living conditions or use of resources across countries. The measure is difficult to compute, as a US dollar value has to be assigned to all goods and services in the country regardless of whether these goods and services have a direct equivalent in the United States (for example, the value of an ox-cart or non-US military equipment); as a result, PPP estimates for some countries are based on a small and sometimes different set of goods and services. In addition, many countries do not formally participate in the World Bank's PPP project that calculates these measures, so the resulting GDP estimates for these countries may lack precision. For many developing countries, PPP-based GDP measures are multiples of the official exchange rate (OER) measure. The difference between the OER- and PPP-denominated GDP values for most of the wealthy industrialized countries are generally much smaller.

GDP—composition by sector This entry gives the percentage contribution of *agriculture*, *industry*, and *services* to total GDP. The distribution will total less than 100 percent if the data are incomplete.

GDP—per capita (PPP) This entry shows GDP on a purchasing power parity basis divided by population as of 1 July for the same year.

GDP—real growth rate This entry gives GDP growth on an annual basis adjusted for inflation and expressed as a percent.

GDP methodology In the **Economy** category, GDP dollar estimates for countries are reported both on an official exchange rate (OER) and a purchasing power parity (PPP) basis. Both measures contain information that is useful to the reader. The PPP method involves the use of standardized international dollar price weights, which are applied to the quantities of final goods and services produced in a given economy. The data derived from the PPP method probably provide the best available starting point for comparisons of economic strength and well-being between countries. In contrast, the currency exchange rate method involves a variety of international and domestic financial forces that may not capture the value of domestic output. Furthermore, exchange rates may suddenly go up or down by 10% or more because of market forces or official fiat whereas real output has remained unchanged. On 12 January 1994, for example, the 14 countries of the African Financial Community (whose currencies are tied to the French franc) devalued their currencies by 50%. This move, of course, did not cut the real output of these countries by half. Whereas PPP estimates for OECD countries are quite reliable, PPP estimates for developing countries are often rough approximations. In developing countries with weak currencies, the exchange rate estimate of GDP in dollars is typically one-fourth to one-half the PPP estimate. Most of the GDP estimates for developing countries are based on extrapolation of PPP numbers published by the UN International Comparison Program (UNICP) and by Professors Robert Summers and Alan Heston of the University of Pennsylvania and their colleagues. GDP derived using the OER method should be used for the purpose of calculating the share of items such as exports, imports, military expenditures, external debt, or the current account balance, because the dollar values presented in the *Factbook* for these items have been converted at official exchange rates, not at PPP. One should use the OER GDP figure to calculate the proportion of, say, Chinese defense expenditures in GDP, because that share will be the same as one calculated in local currency units. Comparison of OER GDP with PPP GDP may also indicate whether a currency is over- or under-valued. If OER GDP is smaller than PPP GDP, the official exchange rate may be undervalued, and vice versa. However, there is no strong historical evidence that market exchange rates move in the direction implied by the PPP rate, at least not in the short- or medium-term. Note: the numbers for GDP and other economic data should not be chained together from successive volumes of the *Factbook* because of changes in the US dollar measuring rod, revisions of data by statistical agencies, use of new or different sources of information, and changes in national statistical methods and practices.

GNP Gross national product (GNP) is the value of all final goods and services produced within a nation in a given year, plus income earned by its citizens abroad, minus income earned by foreigners from domestic production. The *Factbook*, following current practice, uses GDP rather than GNP to measure national production. However, the user must realize that in certain countries net remittances from citizens working abroad may be important to national well-being.

GWP This entry gives the gross world product (GWP) or aggregate value of all final goods and services produced worldwide in a given year.

Geographic coordinates This entry includes rounded latitude and longitude figures for the purpose of finding the approximate geographic center of an entity and is based on the locations provided in the Geographic Names Server (GNS), maintained by the National Geospatial-Intelligence Agency on behalf of the US Board on Geographic Names.

Geographic names This information is presented in **Appendix F: Cross-Reference List of Geographic Names**. It includes a listing of various alternate names, former names, local names, and regional names referenced to one or more related *Factbook* entries. Spellings are normally, but not always, those approved by the US Board on Geographic Names (BGN). Alternate names and additional information are included in parentheses.

Geography This category includes the entries dealing with the natural environment and the effects of human activity.

Geography—note This entry includes miscellaneous geographic information of significance not included elsewhere.

Gini index See entry for **Distribution of family income—Gini index**

Government This category includes the entries dealing with the system for the adoption and administration of public policy.

Government—note This entry includes miscellaneous government information of significance not included elsewhere.

Government type This entry gives the basic form of government. Definitions of the major governmental terms are as follows. (Note that for some countries more than one definition applies.):

Absolute monarchy—a form of government where the monarch rules unhindered, i.e., without any laws, constitution, or legally organized opposition.

Anarchy—a condition of lawlessness or political disorder brought about by the absence of governmental authority.

Authoritarian—a form of government in which state authority is imposed onto many aspects of citizens' lives.

Commonwealth—a nation, state, or other political entity founded on law and united by a compact of the people for the common good.

Communist—a system of government in which the state plans and controls the economy and a single—often authoritarian—party holds power; state controls are imposed with the elimination of private ownership of property or capital while claiming to make progress toward a higher social order in which all goods are equally shared by the people (i.e., a classless society).

Confederacy (Confederation)—a union by compact or treaty between states, provinces, or territories, that creates a central government with limited powers; the constituent entities retain supreme authority over all matters except those delegated to the central government.

Constitutional—a government by or operating under an authoritative document (constitution) that sets forth the system of fundamental laws and principles that determines the nature, functions, and limits of that government.

Constitutional democracy—a form of government in which the sovereign power of the people is spelled out in a governing constitution.

Constitutional monarchy—a system of government in which a monarch is guided by a constitution whereby his/her rights, duties, and responsibilities are spelled out in written law or by custom.

Democracy—a form of government in which the supreme power is retained by the people, but which is usually exercised indirectly through a system of representation and delegated authority periodically renewed.

Democratic republic—a state in which the supreme power rests in the body of citizens entitled to vote for officers and representatives responsible to them.

Dictatorship—a form of government in which a ruler or small clique wield absolute power (not restricted by a constitution or laws).

Ecclesiastical—a government administered by a church.

Emirate—similar to a monarchy or sultanate, but a government in which the supreme power is in the hands of an emir (the ruler of a Muslim state); the emir may be an absolute overlord or a sovereign with constitutionally limited authority.

Federal (Federation)—a form of government in which sovereign power is formally divided—usually by means of a constitution—between a central authority and a number of constituent regions (states, colonies, or provinces) so that each region retains some management of its internal affairs; differs from a confederacy in that the central government exerts influence directly upon both individuals as well as upon the regional units.

Federal republic—a state in which the powers of the central government are restricted and in which the component parts (states, colonies, or provinces) retain a degree of self-government; ultimate sovereign power rests with the voters who chose their governmental representatives.

Islamic republic—a particular form of government adopted by some Muslim states; although such a state is, in theory, a theocracy, it remains a republic, but its laws are required to be compatible with the laws of Islam.

Maoism—the theory and practice of Marxism-Leninism developed in China by Mao Zedong (Mao Tse-tung), which states that a continuous revolution is necessary if the leaders of a communist state are to keep in touch with the people.

Marxism—the political, economic, and social principles espoused by 19th century economist Karl Marx; he viewed the struggle of workers as a progression of historical forces that would proceed from a class struggle of the proletariat (workers) exploited by capitalists (business owners), to a socialist "dictatorship of the proletariat," to, finally, a classless society—Communism.

Marxism-Leninism—an expanded form of communism developed by Lenin from doctrines of Karl Marx; Lenin saw imperialism as the final stage of capitalism and shifted the focus of workers' struggle from developed to underdeveloped countries.

Monarchy—a government in which the supreme power is lodged in the hands of a monarch who reigns over a state or territory, usually for life and by hereditary right; the monarch may be either a sole absolute ruler or a sovereign—such as a king, queen, or prince—with constitutionally limited authority.

Oligarchy—a government in which control is exercised by a small group of individuals whose authority generally is based on wealth or power.

Parliamentary democracy—a political system in which the legislature (parliament) selects the government—a prime minister, premier, or chancellor along with the cabinet ministers—according to party strength as expressed in elections; by this system, the government acquires a dual responsibility: to the people as well as to the parliament.

Parliamentary government (Cabinet-Parliamentary government)—a government in which members of an executive branch (the cabinet and its leader—a prime minister, premier, or chancellor) are nominated to their positions by a legislature or parliament, and are directly responsible to it; this type of government can be dissolved at will by the parliament (legislature) by means of a no confidence vote or the leader of the cabinet may dissolve the parliament if it can no longer function.

Parliamentary monarchy—a state headed by a monarch who is not actively involved in policy formation or implementation (i.e., the exercise of sovereign powers by a monarch in a ceremonial capacity); true governmental leadership is carried out by a cabinet and its head—a prime minister, premier, or chancellor—who are drawn from a legislature (parliament).

Presidential—a system of government where the executive branch exists separately from a legislature (to which it is generally not accountable).

Republic—a representative democracy in which the people's elected deputies (representatives), not the people themselves, vote on legislation.

Socialism—a government in which the means of planning, producing, and distributing goods is controlled by a central government that theoretically seeks a more just and equitable distribution of property and labor; in actuality, most socialist governments have ended up being no more than dictatorships over workers by a ruling elite.

Sultanate—similar to a monarchy, but a government in which the supreme power is in the hands of a sultan (the head of a Muslim state); the sultan may be an absolute ruler or a sovereign with constitutionally limited authority.

Theocracy—a form of government in which a Deity is recognized as the supreme civil ruler, but the Deity's laws are interpreted by ecclesiastical authorities (bishops, mullahs, etc.); a government subject to religious authority.

Totalitarian—a government that seeks to subordinate the individual to the state by controlling not only all political and economic matters, but also the attitudes, values, and beliefs of its population.

Greenwich Mean Time (GMT) The mean solar time at the Greenwich Meridian, Greenwich, England, with the hours and days, since 1925, reckoned from midnight. GMT is now a historical term having been replaced by UTC on 1 January 1972. See **Coordinated Universal Time**.

Gross domestic product see **GDP**

Gross national product see **GNP**

Gross world product see **GWP**

HIV/AIDS—adult prevalence rate This entry gives an estimate of the percentage of adults (aged 15–49) living with HIV/AIDS. The adult prevalence rate is calculated by dividing the estimated number of adults living with HIV/AIDS at yearend by the total adult population at yearend.

HIV/AIDS—deaths This entry gives an estimate of the number of adults and children who died of AIDS during a given calendar year.

HIV/AIDS—people living with HIV/AIDS This entry gives an estimate of all people (adults and children) alive at yearend with HIV infection, whether or not they have developed symptoms of AIDS.

Heliports This entry gives the total number of heliports with hard-surface runways, helipads, or landing areas that support routine sustained helicopter operations exclusively and have support facilities including one or more of the following facilities: lighting, fuel, passenger handling, or maintenance. It includes former airports used exclusively for helicopter operations but excludes heliports limited to day operations and natural clearings that could support helicopter landings and takeoffs.

Household income or consumption by percentage share Data on household income or consumption come from household surveys, the results adjusted for household size. Nations use different standards and procedures in collecting and adjusting the data. Surveys based on income will normally show a more unequal distribution than surveys based on consumption. The quality of surveys is improving with time, yet caution is still necessary in making inter-country comparisons.

Hydrographic data codes see **Data codes**

Illicit drugs This entry gives information on the five categories of illicit drugs—narcotics, stimulants, depressants (sedatives), hallucinogens, and cannabis. These categories include many drugs legally produced and prescribed by doctors as well as those illegally produced and sold outside of medical channels.

Cannabis (Cannabis sativa) is the common hemp plant, which provides hallucinogens with some sedative properties, and includes marijuana (pot, Acapulco gold, grass, reefer), tetrahydrocannabinol (THC, Marinol), hashish (hash), and hashish oil (hash oil).

Coca (mostly *Erythroxylum coca*) is a bush with leaves that contain the stimulant used to make cocaine. Coca is not to be confused with cocoa, which comes from cacao seeds and is used in making chocolate, cocoa, and cocoa butter.

Cocaine is a stimulant derived from the leaves of the coca bush.

Depressants (sedatives) are drugs that reduce tension and anxiety and include chloral hydrate, barbiturates (Amytal, Nembutal, Seconal, phenobarbital), benzodiazepines (Librium, Valium), methaqualone (Quaalude), glutethimide (Doriden), and others (Equanil, Placidyl, Valmid).

Drugs are any chemical substances that effect a physical, mental, emotional, or behavioral change in an individual.

Drug abuse is the use of any licit or illicit chemical substance that results in physical, mental, emotional, or behavioral impairment in an individual.

Hallucinogens are drugs that affect sensation, thinking, self-awareness, and emotion. Hallucinogens include LSD (acid, microdot), mescaline and peyote (mexc, buttons, cactus), amphetamine variants (PMA, STP, DOB), phencyclidine (PCP, angel dust, hog), phencyclidine analogues (PCE, PCPy, TCP), and others (psilocybin, psilocyn).

Hashish is the resinous exudate of the cannabis or hemp plant (*Cannabis sativa*).

Heroin is a semisynthetic derivative of morphine.

Mandrax is a trade name for methaqualone, a pharmaceutical depressant.

Marijuana is the dried leaf of the cannabis or hemp plant (*Cannabis sativa*).

Methaqualone is a pharmaceutical depressant, referred to as mandrax in Southwest Asia and Africa.

Narcotics are drugs that relieve pain, often induce sleep, and refer to opium, opium derivatives, and synthetic substitutes. Natural narcotics include opium (paregoric, parepectolin), morphine (MS-Contin, Roxanol), codeine (Tylenol with codeine, Empirin with codeine, Robitussan AC), and thebaine. Semisynthetic narcotics include heroin (horse, smack), and hydromorphone (Dilaudid). Synthetic narcotics include meperidine or Pethidine (Demerol, Mepergan), methadone (Dolophine, Methadose), and others (Darvon, Lomotil).

Opium is the brown, gummy exudate of the incised, unripe seedpod of the opium poppy.

Opium poppy (Papaver somniferum) is the source for the natural and semisynthetic narcotics.

Poppy straw is the entire cut and dried opium poppy-plant material, other than the seeds. Opium is extracted from poppy straw in commercial operations that produce the drug for medical use.

Qat (kat, khat) is a stimulant from the buds or leaves of *Catha edulis* that is chewed or drunk as tea.

Quaaludes is the North American slang term for methaqualone, a pharmaceutical depressant.

Stimulants are drugs that relieve mild depression, increase energy and activity, and include cocaine (coke, snow, crack), amphetamines (Desoxyn, Dexedrine), ephedrine, ecstasy (clarity, essence, doctor, Adam), phenmetrazine (Preludin), methylphenidate (Ritalin), and others (Cylert, Sanorex, Tenuate).

Imports This entry provides the total US dollar amount of merchandise imports on a c.i.f. (cost, insurance, and freight) or f.o.b. (free on board) basis. These figures are calculated on an exchange rate basis, i.e., not in purchasing power parity (PPP) terms.

Imports—commodities This entry provides a listing of the highest-valued imported products; it sometimes includes the percent of total dollar value.

Imports—partners This entry provides a rank ordering of trading partners starting with the most important; it sometimes includes the percent of total dollar value.

Independence For most countries, this entry gives the date that sovereignty was achieved and from which nation, empire, or trusteeship. For the other countries, the date given may not represent "independence" in the strict sense, but rather some significant nationhood event such as the traditional founding date or the date of unification, federation, confederation, establishment, fundamental change in the form of government, or state succession. Dependent areas include the notation "none" followed by the nature of their dependency status. Also see the **Terminology** note.

Industrial production growth rate This entry gives the annual percentage increase in industrial production (includes manufacturing, mining, and construction).

Industries This entry provides a rank ordering of industries starting with the largest by value of annual output.

Infant mortality rate This entry gives the number of deaths of infants under one year old in a given year per 1,000 live births in the same year; included is the total death rate, and deaths by sex, *male* and *female*. This rate is often used as an indicator of the level of health in a country.

Inflation rate (consumer prices) This entry furnishes the annual percent change in consumer prices compared with the previous year's consumer prices.

International disputes see **Disputes—international**

International organization participation This entry lists in alphabetical order by abbreviation those international organizations in which the subject country is a member or participates in some other way.

International organizations This information is presented in **Appendix B: International Organizations and Groups** which includes the name, abbreviation, date established, aim, and members by category.

Internet country code This entry includes the two-letter codes maintained by the International Organization for Standardization (ISO) in the ISO 3166 Alpha-2 list and used by the Internet Assigned Numbers Authority (IANA) to establish country-coded top-level domains (ccTLDs).

Internet hosts This entry lists the number of Internet hosts available within a country. An Internet host is a computer connected directly to the Internet; normally an Internet Service Provider's (ISP) computer is a host. Internet users may use either a hard-wired terminal, at an institution with a mainframe computer connected directly to the Internet, or may connect remotely by way of a modem via telephone line, cable, or satellite to the Internet Service Provider's host computer. The number of hosts is one indicator of the extent of Internet connectivity.

Internet users This entry gives the number of users within a country that access the Internet. Statistics vary from country to country and may include users who access the Internet at least several times a week to those who access it only once within a period of several months.

Introduction This category includes one entry, **Background**.

Investment (gross fixed) This entry records total business spending on fixed assets, such as factories, machinery, equipment, dwellings, and inventories of raw materials, which provide the basis for future production. It is measured gross of the depreciation of the assets, i.e., it includes invesment that merely replaces worn-out or scrapped capital.

Irrigated land This entry gives the number of square kilometers of land area that is artificially supplied with water.

Judicial branch This entry contains the name(s) of the highest court(s) and a brief description of the selection process for members.

Labor force This entry contains the total labor force figure.

Labor force—by occupation This entry lists the percentage distribution of the labor force by occupation. The distribution will total less than 100 percent if the data are incomplete.

Land boundaries This entry contains the *total* length of all land boundaries and the individual lengths for each of the contiguous *border countries*. When available, official lengths published by national statistical agencies are used. Because surveying methods may differ, country border lengths reported by contiguous countries may differ.

Land use This entry contains the percentage shares of total land area for three different types of land use: *arable land*—land cultivated for crops like wheat, maize, and rice that are replanted after each harvest; *permanent crops*—land cultivated for crops like citrus, coffee, and rubber that are not replanted after each harvest; includes land under flowering shrubs, fruit trees, nut trees, and vines, but excludes land under trees grown for wood or timber; *other*—any land not arable or under permanent crops; includes permanent meadows and pastures, forests and woodlands, built-on areas, roads, barren land, etc.

Languages This entry provides a rank ordering of languages starting with the largest and sometimes includes the percent of total population speaking that language.

Legal system This entry contains a brief description of the legal system's historical roots, role in government, and acceptance of International Court of Justice (ICJ) jurisdiction.

Legislative branch This entry contains information on the structure (unicameral, bicameral, tricameral), formal name, number of seats, and term of office. *Elections* includes the nature of election process or accession to power, date of the last election, and date of the next election. *Election results* includes the percent of vote and/or number of seats held by each party in the last election.

Life expectancy at birth This entry contains the average number of years to be lived by a group of people born in the same year, if mortality at each age remains constant in the future. The entry includes *total population* as well as the *male* and *female* components. Life expectancy at birth is also a measure of overall quality of life in a country and summarizes the mortality at all ages. It can also be thought of as indicating the potential return on investment in human capital and is necessary for the calculation of various actuarial measures.

Literacy This entry includes a *definition* of literacy and Census Bureau percentages for the *total population*, *males*, and *females*. There are no universal definitions and standards of literacy. Unless otherwise specified, all rates are based on the most common definition—the ability to read and write at a specified age. Detailing the standards that individual countries use to assess the ability to read and write is beyond the scope of the *Factbook*. Information on literacy, while not a perfect measure of educational results, is probably the most easily available and valid for international comparisons. Low levels of literacy, and education in general, can impede the economic development of a country in the current rapidly changing, technology-driven world.

Location This entry identifies the country's regional location, neighboring countries, and adjacent bodies of water.

Major infectious diseases This entry lists major infectious diseases likely to be encountered in countries where the risk of such diseases is assessed to be very high as compared to the United States. These infectious diseases represent risks to US government personnel traveling to the specified country for a period of less than three years. The **degree of risk** is assessed by considering the foreign nature of these infectious diseases, their severity, and the probability of being affected by the diseases present. The diseases listed do not necessarily represent the total disease burden experienced by the local population. The risk to an individual traveler varies considerably by the specific location, visit duration, type of activities, type of accommodations, time of year, and other factors. Consultation with a travel medicine physician is needed to evaluate individual risk and recommend appropriate preventive measures such as vaccines. Diseases are organized into the following six exposure categories shown in italics *and listed in typical descending order of risk*. Note: The sequence of exposure categories listed in individual country entries may vary according to local conditions.

food or waterborne diseases acquired through eating or drinking on the local economy:

Hepatitis A—viral disease that interferes with the functioning of the liver; spread through consumption of food or water contaminated with fecal matter, principally in areas of poor sanitation; victims exhibit fever, jaundice, and diarrhea; 15% of victims will experience prolonged symptoms over 6–9 months; vaccine available.

Hepatitis E—water-borne viral disease that interferes with the functioning of the liver; most commonly spread through fecal contamination of drinking water; victims exhibit jaundice, fatigue, abdominal pain, and dark colored urine.

Typhoid fever—bacterial disease spread through contact with food or water contaminated by fecal matter or sewage; victims exhibit sustained high fevers; left untreated, mortality rates can reach 20%.

vectorborne diseases acquired through the bite of an infected arthropod:

Malaria—caused by single-cell parasitic protozoa *Plasmodium*; transmitted to humans via the bite of the female Anopheles mosquito; parasites multiply in the liver attacking red blood cells resulting in cycles of fever, chills, and sweats accompanied by anemia; death due to damage to vital organs and interruption of blood supply to the brain; endemic in 100, mostly tropical, countries with 90% of cases and the majority of 1.5–2.5 million estimated annual deaths occurring in sub-Saharan Africa.

Dengue fever—mosquito-borne (*Aedes aegypti*) viral disease associated with urban environments; manifests as sudden onset of fever and severe headache; occasionally produces shock and hemorrhage leading to death in 5% of cases.

Yellow fever—mosquito-borne viral disease; severity ranges from influenza-like symptoms to severe hepatitis and hemorrhagic fever; occurs only in tropical South America and sub-Saharan Africa, where most cases are reported; fatality rate is less than 20%.

Japanese Encephalitis—mosquito-borne (*Culex tritaeniorhynchus*) viral disease associated with rural areas in Asia; acute encephalitis can progress to paralysis, coma, and death; fatality rates 30%.

African Trypanosomiasis—caused by the parasitic protozoa *Trypanosoma*; transmitted to humans via the bite of bloodsucking Tsetse flies; infection leads to malaise and irregular fevers and, in advanced cases when the parasites invade the central nervous system, coma and death; endemic in 36 countries of sub-Saharan Africa; cattle and wild animals act as reservoir hosts for the parasites.

Cutaneous Leishmaniasis—caused by the parasitic protozoa *leishmania*; transmitted to humans via the bite of sandflies; results in skin lesions that may become chronic; endemic in 88 countries; 90% of cases occur in Iran, Afghanistan, Syria, Saudi Arabia, Brazil, and Peru; wild and domesticated animals as well as humans can act as reservoirs of infection.

Plague—bacterial disease transmitted by fleas normally associated with rats; person-to-person airborne transmission also possible; recent plague epidemics occurred in areas of Asia, Africa, and South America associated with rural areas or small towns and villages; manifests as fever, headache, and painfully swollen lymph nodes; disease progresses rapidly and without antibiotic treatment leads to pneumonic form with a death rate in excess of 50%.

Crimean-Congo hemorrhagic fever—tick-borne viral disease; infection may also result from exposure to infected animal blood or tissue; geographic distribution includes Africa, Asia, the Middle East, and Eastern Europe; sudden onset of fever, headache, and muscle aches followed by hemorrhaging in the bowels, urine, nose, and gums; mortality rate is approximately 30%.

Rift Valley fever—viral disease affecting domesticated animals and humans; transmission is by mosquito and other biting insects; infection may also occur through handling of infected meat or contact with blood; geographic distribution includes eastern and southern Africa where cattle and sheep are raised; symptoms are generally mild with fever and some liver abnormalities, but the disease may progress to hemorrhagic fever, encephalitis, or ocular disease; fatality rates are low at about 1% of cases.

Chikungunya—mosquito-borne (*Aedes aegypti*) viral disease associated with urban environments, similar to Dengue Fever; characterized by sudden onset of fever, rash, and severe joint pain usually lasting 3–7 days, some cases result in persistent arthritis.

water contact diseases acquired through swimming or wading in freshwater lakes, streams, and rivers:

Leptospirosis—bacterial disease that affects animals and humans; infection occurs through contact with water, food, or soil contaminated by animal urine; symptoms include high fever, severe headache, vomiting, jaundice, and diarrhea; untreated, the disease can result in kidney damage, liver failure, meningitis, or respiratory distress; fatality rates are low but left untreated recovery can take months.

Schistosomiasis—caused by parasitic trematode flatworm *Schistosoma*; fresh water snails act as intermediate host and release larval form of parasite that penetrates the skin of people exposed to contaminated water; worms mature and reproduce in the blood vessels, liver, kidneys, and intestines releasing eggs, which become trapped in tissues triggering an immune response; may manifest as either urinary or intestinal disease resulting in decreased work or learning capacity; mortality, while generally low, may occur in advanced cases usually due to bladder cancer; endemic in 74 developing countries with 80% of infected people living in sub-Saharan Africa; humans act as the reservoir for this parasite.

aerosolized dust or soil contact disease acquired through inhalation of aerosols contaminated with rodent urine:

Lassa fever—viral disease carried by rats of the genus *Mastomys*; endemic in portions of West Africa; infection occurs through direct contact with or consumption of food contaminated by rodent urine or fecal matter containing virus particles; fatality rate can reach 50% in epidemic outbreaks.

respiratory disease acquired through close contact with an infectious person:

Meningococcal meningitis—bacterial disease causing an inflammation of the lining of the brain and spinal cord; one of the most important bacterial pathogens is *Neisseria meningitidis* because of its potential to cause epidemics; symptoms include stiff neck, high fever, headaches, and vomiting; bacteria are transmitted from person to person by respiratory droplets and facilitated by close and prolonged contact resulting from crowded living conditions, often with a seasonal distribution; death occurs in 5–15% of cases, typically within 24–48 hours of onset of symptoms; highest burden of meningococcal disease occurs in the hyperendemic region of sub-Saharan Africa known as the "Meningitis Belt" which stretches from Senegal east to Ethiopia.

animal contact disease acquired through direct contact with local animals:

Rabies—viral disease of mammals usually transmitted through the bite of an infected animal, most commonly dogs; virus affects the central nervous system causing brain alteration and death; symptoms initially are non-specific fever and headache progressing to neurological symptoms; death occurs within days of the onset of symptoms.

Manpower available for military service This entry gives the number of males and females falling in the military age range for a country (defined as being ages 16–49) and assumes that every individual is fit to serve.

Manpower fit for military service This entry gives the number of males and females falling in the military age range for a country (defined as being ages 16–49) and who are not otherwise disqualified for health reasons; accounts for the health situation in the country and provides a more realistic estimate of the actual number fit to serve.

Manpower reaching militarily significant age annually This entry gives the number of males and females entering the military manpower pool (i.e., reaching age 16) in any given year and is a measure of the availability of military-age young adults.

Map references This entry includes the name of the *Factbook* reference map on which a country may be found. The entry on **Geographic coordinates** may be helpful in finding some smaller countries.

Maritime claims This entry includes the following claims, the definitions of which are excerpted from the United Nations Convention on the Law of the Sea (UNCLOS), which alone contains the full and definitive descriptions:

territorial sea—the sovereignty of a coastal state extends beyond its land territory and internal waters to an adjacent belt of sea, described as the territorial sea in the UNCLOS (Part II); this sovereignty extends to the air space over the territorial sea as well as its underlying seabed and subsoil; every state has the right to establish the breadth of its territorial sea up to a limit not exceeding 12 nautical miles; the normal baseline for measuring the breadth of the territorial sea is the mean low-water line along the coast as marked on large-scale charts officially recognized by the coastal state; the UNCLOS describes specific rules for archipelagic states.

contiguous zone—according to the UNCLOS (Article 33), this is a zone contiguous to a coastal state's territorial sea, over which it may exercise the control necessary to: prevent infringement of its customs, fiscal, immigration, or sanitary laws and regulations within its territory or territorial sea; punish infringement of the above laws and regulations committed within its territory or territorial sea; the contiguous zone may not extend beyond 24 nautical miles from the baselines from which the breadth of the territorial sea is measured (e.g. the US has claimed a 12-nautical mile contiguous zone in addition to its 12-nautical mile territorial sea).

exclusive economic zone (EEZ)—the UNCLOS (Part V) defines the EEZ as a zone beyond and adjacent to the territorial sea in which a coastal state has: sovereign rights for the purpose of exploring and exploiting, conserving and managing the natural resources, whether living or non-living, of the waters superjacent to the seabed and of the seabed and its subsoil, and with regard to other activities for the economic exploitation and exploration of the zone, such as the production of energy from the water, currents, and winds; jurisdiction with regard to the establishment and use of artificial islands, installations, and structures; marine scientific research; the protection and preservation of the marine environment; the outer limit of the exclusive economic zone shall not exceed 200 nautical miles from the baselines from which the breadth of the territorial sea is measured.

continental shelf—the UNCLOS (Article 76) defines the continental shelf of a coastal state as comprising the seabed and subsoil of the submarine areas that extend beyond its territorial sea throughout the natural prolongation of its land territory to the outer edge of the continental margin, or to a distance of 200 nautical miles from the baselines from which the breadth of the territorial sea is measured where the outer edge of the continental margin does not extend up to that distance; the continental margin comprises the submerged prolongation of the landmass of the coastal state, and consists of the seabed and subsoil of the shelf, the slope and the rise; wherever the continental margin extends beyond 200 nautical miles from the baseline, coastal states may extend their claim to a distance not to exceed 350 nautical miles from the baseline or 100 nautical miles from the 2500 meter isobath; it does not include the deep ocean floor with its oceanic ridges or the subsoil thereof.

exclusive fishing zone—while this term is not used in the UNCLOS, some states (e.g., the United Kingdom) have chosen not to claim an EEZ, but rather to claim jurisdiction over the living resources off their coast; in such cases, the term exclusive fishing zone is often used; the breadth of this zone is normally the same as the EEZ or 200 nautical miles.

Market value of publicly traded shares This entry gives the value of shares issued by publicly traded companies at a price determined in the national stock markets on the final day of the period indicated. It is simply the latest price per share multiplied by the total number of outstanding shares, cumulated over all companies listed on the particular exchange.

Median age This entry is the age that divides a population into two numerically equal groups; that is, half the people are younger than this age and half are older. It is a single index that summarizes the age distribution of a population. Currently, the median age ranges from a low of about 15 in Uganda and Gaza Strip to 40 or more in several European countries and Japan. See the entry for "Age structure" for the importance of a young versus an older age structure and, by implication, a low versus a higher median age.

Merchant marine Merchant marine may be defined as all ships engaged in the carriage of goods; or all commercial vessels (as opposed to all nonmilitary ships), which excludes tugs, fishing vessels, offshore oil rigs, etc. This entry contains information in four fields—*total*, *ships by type*, *foreign-owned*, and *registered in other countries*.

Total includes the number of ships (1,000 GRT or over), total DWT for those ships, and total GRT for those ships. DWT or dead weight tonnage is the total weight of cargo, plus bunkers, stores, etc., that a ship can carry when immersed to the appropriate load line. GRT or gross register tonnage is a figure obtained by measuring the entire sheltered volume of a ship available for cargo and passengers and converting it to tons on the basis of 100 cubic feet per ton; there is no stable relationship between GRT and DWT.

Ships by type includes a listing of barge carriers, bulk cargo ships, cargo ships, chemical tankers, combination bulk carriers, combination ore/oil carriers, container ships, liquefied gas tankers, livestock carriers, multifunctional large-load carriers, petroleum tankers, passenger ships, passenger/cargo ships, railcar carriers, refrigerated cargo ships, roll-on/roll-off cargo ships, short-sea passenger ships, specialized tankers, and vehicle carriers.

Foreign-owned are ships that fly the flag of one country but belong to owners in another.

Registered in other countries are ships that belong to owners in one country but fly the flag of another.

Military This category includes the entries dealing with a country's military structure, manpower, and expenditures.

Military—note This entry includes miscellaneous military information of significance not included elsewhere.

Military branches This entry lists the service branches subordinate to defense ministries or the equivalent (typically ground, naval, air, and marine forces).

Military expenditures—percent of GDP This entry gives spending on defense programs for the most recent year available as a percent of gross domestic product (GDP); the GDP is calculated on an exchange rate basis, i.e., not in terms of purchasing power parity (PPP).

Military service age and obligation This entry gives the required ages for voluntary or conscript military service and the length of service obligation.

Money figures All money figures are expressed in contemporaneous US dollars unless otherwise indicated.

National holiday This entry gives the primary national day of celebration—usually independence day.

Nationality This entry provides the identifying terms for citizens—*noun* and *adjective*.

Natural gas—consumption This entry is the total natural gas consumed in cubic meters (cu m). The discrepancy between the amount of natural gas produced and/or imported and the amount consumed and/or exported is due to the omission of stock changes and other complicating factors.

Natural gas—exports This entry is the total natural gas exported in cubic meters (cu m).

Natural gas—imports This entry is the total natural gas imported in cubic meters (cu m).

Natural gas—production This entry is the total natural gas produced in cubic meters (cu m). The discrepancy between the amount of natural gas produced and/or imported and the amount consumed and/or exported is due to the omission of stock changes and other complicating factors.

Natural gas—proved reserves This entry is the stock of proved reserves of natural gas in cubic meters (cu m). Proved reserves are those quantities of natural gas, which, by analysis of geological and engineering data, can be estimated with a high degree of confidence to be commercially recoverable from a given date forward, from known reservoirs and under current economic conditions.

Natural hazards This entry lists potential natural disasters.

Natural resources This entry lists a country's mineral, petroleum, hydropower, and other resources of commercial importance.

Net migration rate This entry includes the figure for the difference between the number of persons entering and leaving a country during the year per 1,000 persons (based on midyear population). An excess of persons entering the country is referred to as net immigration (e.g., 3.56 migrants/1,000 population); an excess of persons leaving the country as net emigration (e.g., -9.26 migrants/1,000 population). The net migration rate indicates the contribution of migration to the overall level of population change. High levels of migration can cause problems such as increasing unemployment and potential ethnic strife (if people are coming in) or a reduction in the labor force, perhaps in certain key sectors (if people are leaving).

Oil—consumption This entry is the total oil consumed in barrels per day (bbl/day). The discrepancy between the amount of oil produced and/or imported and the amount consumed and/or exported is due to the omission of stock changes, refinery gains, and other complicating factors.

Oil—exports This entry is the total oil exported in barrels per day (bbl/day), including both crude oil and oil products.

Oil—imports This entry is the total oil imported in barrels per day (bbl/day), including both crude oil and oil products.

Oil—production This entry is the total oil produced in barrels per day (bbl/day). The discrepancy between the amount of oil produced and/or imported and the amount consumed and/or exported is due to the omission of stock changes, refinery gains, and other complicating factors.

Oil—proved reserves This entry is the stock of proved reserves of crude oil in barrels (bbl). Proved reserves are those quantities of petroleum which, by analysis of geological and engineering data, can be estimated with a high degree of confidence to be commercially recoverable from a given date forward, from known reservoirs and under current economic conditions.

People This category includes the entries dealing with the characteristics of the people and their society.

People—note This entry includes miscellaneous demographic information of significance not included elsewhere.

Personal Names—Capitalization The *Factbook* capitalizes the surname or family name of individuals for the convenience of our users who are faced with a world of different cultures and naming conventions. The need for capitalization, bold type, underlining, italics, or some other indicator of the individual's surname is apparent in the following examples: MAO Zedong, Fidel CASTRO Ruz, George W. BUSH, and TUNKU SALAHUDDIN Abdul Aziz Shah ibni Al-Marhum Sultan Hisammuddin Alam Shah. By knowing the surname, a short form without all capital letters can be used with confidence as in President Castro, Chairman

Mao, President Bush, or Sultan Tunku Salahuddin. The same system of capitalization is extended to the names of leaders with surnames that are not commonly used such as Queen ELIZABETH II. For Vietnamese names, the given name is capitalized because officials are referred to by their given name rather than by their surname. For example, the president of Vietnam is Tran Duc LUONG. His surname is Tran, but he is referred to by his given name—President LUONG.

Personal Names—Spelling The romanization of personal names in the *Factbook* normally follows the same transliteration system used by the US Board on Geographic Names for spelling place names. At times, however, a foreign leader expressly indicates a preference for, or the media or official documents regularly use, a romanized spelling that differs from the transliteration derived from the US Government standard. In such cases, the *Factbook* uses the alternative spelling.

Personal Names—Titles The *Factbook* capitalizes any valid title (or short form of it) immediately preceding a person's name. A title standing alone is not capitalized. Examples: President PUTIN and President BUSH are chiefs of state. In Russia, the president is chief of state and the premier is the head of the government, while in the US, the president is both chief of state and head of government.

Petroleum See entries under **Oil**.

Petroleum products See entries under **Oil**.

Pipelines This entry gives the lengths and types of pipelines for transporting products like natural gas, crude oil, or petroleum products.

Political parties and leaders This entry includes a listing of significant political organizations and their leaders.

Political pressure groups and leaders This entry includes a listing of political, social, labor, or religious organizations with leaders involved in politics, but not standing for legislative election.

Population This entry gives an estimate from the US Bureau of the Census based on statistics from population censuses, vital statistics registration systems, or sample surveys pertaining to the recent past and on assumptions about future trends. The total population presents one overall measure of the potential impact of the country on the world and within its region. Note: Starting with the 1993 *Factbook*, demographic estimates for some countries (mostly African) have explicitly taken into account the effects of the growing impact of the HIV/AIDS epidemic. These countries are currently: The Bahamas, Benin, Botswana, Brazil, Burkina Faso, Burma, Burundi, Cambodia, Cameroon, Central African Republic, Democratic Republic of the Congo, Republic of the Congo, Cote d'Ivoire, Ethiopia, Gabon, Ghana, Guyana, Haiti, Honduras, Kenya, Lesotho, Malawi, Mozambique, Namibia, Nigeria, Rwanda, South Africa, Swaziland, Tanzania, Thailand, Togo, Uganda, Zambia, and Zimbabwe.

Population below poverty line National estimates of the percentage of the population falling below the poverty line are based on surveys of sub-groups, with the results weighted by the number of people in each group. Definitions of poverty vary considerably among nations. For example, rich nations generally employ more generous standards of poverty than poor nations.

Population growth rate The average annual percent change in the population, resulting from a surplus (or deficit) of births over deaths and the balance of migrants entering and leaving a country. The rate may be positive or negative. The growth rate is a factor in determining how great a burden would be imposed on a country by the changing needs of its people for infrastructure (e.g., schools, hospitals, housing, roads), resources (e.g., food, water, electricity), and jobs. Rapid population growth can be seen as threatening by neighboring countries.

Ports and terminals This entry lists major ports and terminals primarily on the basis of the amount of cargo tonnage shipped through the facilities on an annual basis. In some instances, the number of containers handled or ship visits were also considered.

Public debt This entry records the cumulative total of all government borrowings less repayments that are denominated in a country's home currency. Public debt should not be confused with external debt, which reflects the foreign currency liabilities of both the private and public sector and must be financed out of foreign exchange earnings.

Radio broadcast stations This entry includes the total number of AM, FM, and shortwave broadcast stations.

Railways This entry states the total route length of the railway network and of its component parts by gauge: *broad*, *standard*, *narrow*, and *dual*. Other gauges are listed under *note*.

Reference maps This section includes world and regional maps.

Refugees and internally displaced persons This entry includes those persons residing in a country as *refugees* or internally displaced persons (*IDPs*). The definition of a refugee according to a United Nations Convention is "a person who is outside his/her country of nationality or habitual residence; has a well-founded fear of persecution because of his/her race, religion, nationality, membership in a particular social group or political opinion; and is unable or unwilling to avail himself/herself of the protection of that country, or to return there, for fear of persecution." The UN established the Office of the UN High Commissioner for Refugees (UNHCR) in 1950 to handle refugee matters worldwide. The UN Relief and Works Agency for Palestine Refugees in the Near East (UNRWA) has a different operational definition for a Palestinian refugee: "a person whose normal place of residence was Palestine during the period 1 June 1946 to 15 May 1948 and who lost both home and means of livelihood as a result of the 1948 conflict." However, UNHCR also assists some 400,000 Palestinian refugees not covered under the UNRWA definition. The term "internally displaced person" is not specifically covered in the UN

Convention; it is used to describe people who have fled their homes for reasons similar to refugees, but who remain within their own national territory and are subject to the laws of that state.

Religions This entry is an ordered listing of religions by adherents starting with the largest group and sometimes includes the percent of total population. The core characteristics and beliefs of the world's major religions are described below.

Baha'i—Founded by Mirza Husayn-Ali (known as Baha'u'llah) in Iran in 1852, Baha'i faith emphasizes monotheism and believes in one eternal transcendent God. Its guiding focus is to encourage the unity of all peoples on the earth so that justice and peace may be achieved on earth. Baha'i revelation contends the prophets of major world religions reflect some truth or element of the divine, believes all were manifestations of God given to specific communities in specific times, and that Baha'u'llah is an additional prophet meant to call all humankind. Bahais are an open community, located worldwide, with the greatest concentration of believers in South Asia.

Buddhism—Religion or philosophy inspired by the 5th century B.C. teachings of Siddhartha Gautama (also known as Gautama Buddha "the enlightened one"). Buddhism focuses on the goal of spiritual enlightenment centered on an understanding of Gautama Buddha's Four Noble Truths on the nature of suffering, and on the Eightfold Path of spiritual and moral practice, to break the cycle of suffering of which we are a part. Buddhism ascribes to a karmic system of rebirth. Several schools and sects of Buddhism exist, differing often on the nature of the Buddha, the extent to which enlightenment can be achieved—for one or for all, and by whom—religious orders or laity.

Basic Groupings

Theravada Buddhism: The oldest Buddhist school, Theravada is practiced mostly in Sri Lanka, Cambodia, Laos, Burma, and Thailand, with minority representation elsewhere in Asia and the West. Theravadans follow the Pali Canon of Buddha's teachings, and believe that one may escape the cycle of rebirth, worldly attachment, and suffering for oneself; this process may take one or several lifetimes.

Mahayana Buddhism, including subsets Zen and Tibetan Buddhism: Forms of Mahayana Buddhism are common in East Asia and Tibet, and parts of the West. Mahayanas have additional scriptures beyond the Pali Canon and believe the Buddha is eternal and still teaching. Unlike Theravada Buddhism, Mahayana schools maintain the Buddha-nature is present in all beings and all will ultimately achieve enlightenment.

Christianity—Descending from Judaism, Christianity's central belief maintains Jesus of Nazareth is the promised messiah of the Hebrew Scriptures, and that his life, death, and resurrection are salvific for the world. Christianity is one of the three monotheistic Abrahamic faiths, along with Islam and Judaism, which traces its spiritual lineage to Abraham of the Hebrew Scriptures. Its sacred texts include the Hebrew Bible and the New Testament (or the Christian Gospels).

Basic Groupings

Catholicism (or Roman Catholicism): This is the oldest established western Christian church and the world's largest single religious body. It is supranational, and recognizes a hierarchical structure with the Pope, or Bishop of Rome, as its head, located at the Vatican. Catholics believe the Pope is the divinely ordered head of the Church from a direct spiritual legacy of Jesus' apostle Peter. Catholicism is comprised of 23 particular Churches, or Rites—one Western (Latin-Rite) and 22 Eastern. The Latin Rite is by far the largest, making up about 98% of Catholic membership. Eastern-Rite Churches, such as the Maronite Church and the Ukrainian Catholic Church, are in communion with Rome although they preserve their own worship traditions and their immediate hierarchy consists of clergy within their own rite. The Catholic Church has a comprehensive theological and moral doctrine specified for believers in its catechism, which makes it unique among most forms of Christianity.

Mormonism (including the Church of Jesus Christ of Latter-Day Saints): Originating in 1830 in the United States under Joseph Smith, Mormonism is not characterized as a form of Protestant Christianity because it claims additional revealed Christian scriptures after the Hebrew Bible and New Testament. The Book of Mormon maintains there was an appearance of Jesus in the New World following the Christian account of his resurrection, and that the Americas are uniquely blessed continents. Mormonism believes earlier Christian traditions, such as the Roman Catholic, Orthodox, and Protestant reform faiths, are apostasies and that Joseph Smith's revelation of the Book of Mormon is a restoration of true Christianity. Mormons have a hierarchical religious leadership structure, and actively proselytize their faith; they are located primarily in the Americas and in a number of other Western countries.

Orthodox Christianity: The oldest established eastern form of Christianity, the Holy Orthodox Church, has a ceremonial head in the Bishop of Constantinople (Istanbul), also known as a Patriarch, but its various regional forms (e.g., Greek Orthodox, Russian Orthodox, Serbian Orthodox, Ukrainian Orthodox) are autocephalous (independent of Constantinople's authority, and have their own Patriarchs). Orthodox churches are highly nationalist and ethnic. The Orthodox Christian faith shares many theological tenets with the Roman Catholic Church, but diverges on some key premises and does not recognize the governing authority of the Pope.

Protestant Christianity: Protestant Christianity originated in the 16th century as an attempt to reform

Roman Catholicism's practices, dogma, and theology. It encompasses several forms or denominations which are extremely varied in structure, beliefs, relationship to state, clergy, and governance. Many protestant theologies emphasize the primary role of scripture in their faith, advocating individual interpretation of Christian texts without the mediation of a final religious authority such as the Roman Pope. The oldest Protestant Christianities include Lutheranism, Calvinism (Presbyterians), and Anglican Christianity (Episcopalians), which have established liturgies, governing structure, and formal clergy. Other variants on Protestant Christianity, including Pentecostal movements and independent churches, may lack one or more of these elements, and their leadership and beliefs are individualized and dynamic.

Hinduism—Originating in the Vedic civilization of India (second and first millennium B.C.), Hinduism is an extremely diverse set of beliefs and practices with no single founder or religious authority. Hinduism has many scriptures; the Vedas, the Upanishads, and the Bhagavad-Gita are among some of the most important. Hindus may worship one or many deities, usually with prayer rituals within their own home. The most common figures of devotion are the gods Vishnu, Shiva, and a mother goddess, Devi. Most Hindus believe the soul, or *atman*, is eternal, and goes through a cycle of birth, death, and rebirth (*samsara*) determined by one's positive or negative karma, or the consequences of one's actions. The goal of religious life is to learn to act so as to finally achieve liberation (*moksha*) of one's soul, escaping the rebirth cycle.

Islam—The third of the monotheistic Abrahamic faiths, Islam originated with the teachings of Muhammad in the 7th century. Muslims believe Muhammad is the final of all religious prophets (beginning with Abraham) and that the Qu'ran, which is the Islamic scripture, was revealed to him by God. Islam derives from the word submission, and obedience to God is a primary theme in this religion. In order to live an Islamic life, believers must follow the five pillars, or tenets, of Islam, which are the testimony of faith (*shahada*), daily prayer (*salah*), giving alms (*zakah*), fasting during Ramadan (*sawm*), and the pilgrimage to Mecca (*hajj*).

Basic Groupings

The two primary branches of Islam are Sunni and Shia, which split from each other over a religio-political leadership dispute about the rightful successor to Muhammad. The Shia believe Muhammad's cousin and son-in-law, Ali, was the only divinely ordained Imam (religious leader), while the Sunni maintain the first three caliphs after Muhammad were also legitimate authorities. In modern Islam, Sunnis and Shia continue to have different views of acceptable schools of Islamic jurisprudence, and who is a proper Islamic religious authority. Islam also has an active mystical branch, Sufism, with various Sunni and Shia subsets.

Sunni Islam accounts for over 75% of the world's Muslim population. It recognizes the Abu Bakr as the first caliph after Muhammad. Sunni has four schools of Islamic doctrine and law—Hanafi, Maliki, Shafi'i, and Hanbali—which uniquely interpret the *Hadith*, or recorded oral traditions of Muhammad. A Sunni Muslim may elect to follow any one of these schools, as all are considered equally valid.

Shia Islam represents 10–20% of Muslims worldwide, and its distinguishing feature is its reverence for Ali as an infallible, divinely inspired leader, and as the first Imam of the Muslim community after Muhammad. A majority of Shia are known as "Twelvers," because they believe that the 11 familial successor imams after Muhammad culminate in a 12th Imam (al-Mahdi) who is hidden in the world and will reappear at its end to redeem the righteous.

Variants

Ismaili faith: A sect of Shia Islam, its adherents are also known as "Seveners," because they believe that the rightful seventh Imam in Islamic leadership was Isma'il, the elder son of Imam Jafar al-Sadiq. Ismaili tradition awaits the return of the seventh Imam as the Mahdi, or Islamic messianic figure. Ismailis are located in various parts of the world, particularly South Asia and the Levant.

Alawi faith: Another Shia sect of Islam, the name reflects followers' devotion to the religious authority of Ali. Alawites are a closed, secretive religious group who assert they are Shia Muslims, although outside scholars speculate their beliefs may have a syncretic mix with other faiths originating in the Middle East. Alawis live mostly in Syria, Lebanon, and Turkey.

Druze faith: A highly secretive tradition and a closed community that derives from the Ismaili sect of Islam; its core beliefs are thought to emphasize a combination of Gnostic principles believing that the Fatimid caliph, al-Hakin, is the one who embodies the key aspects of goodness of the universe, which are, the intellect, the word, the soul, the preceder, and the follower. The Druze have a key presence in Syria, Lebanon, and Israel.

Jainism—Originating in India, Jain spiritual philosophy believes in an eternal human soul, the eternal universe, and a principle of "the own nature of things." It emphasizes compassion for all living things, seeks liberation of the human soul from reincarnation through enlightenment, and values personal responsibility due to the belief in the immediate consequences of one's behavior. Jain philosophy teaches non-violence and prescribes vegetarianism for monks and laity alike; its adherents are a highly influential religious minority in Indian society.

Judaism—One of the first known monotheistic religions, likely dating to between 2000–1500 B.C., Judaism is the native faith of the Jewish people, based upon the belief in a covenant of responsibility between a sole omnipotent creator God and Abraham, the patriarch of Judaism's Hebrew Bible, or *Tanakh*. Divine revelation of principles and prohibitions in the Hebrew Scriptures form the basis of Jewish law, or *halakhah*, which is a key component of the faith. While there are extensive traditions of Jewish halakhic and theological discourse, there is no final dogmatic authority in the tradition. Local communities have their own religious leadership. Modern Judaism has three basic categories of faith: Orthodox, Conservative, and Reform/Liberal. These differ in their views and observance of Jewish law, with the Orthodox representing the most traditional practice, and Reform/Liberal communities the most accommodating of individualized interpretations of Jewish identity and faith.

Shintoism—A native animist tradition of Japan, Shinto practice is based upon the premise that every being and object has its own spirit or *kami*. Shinto practitioners worship several particular *kamis*, including the *kamis* of nature, and families often have shrines to their ancestors' *kamis*. Shintoism has no fixed tradition of prayers or prescribed dogma, but is characterized by individual ritual. Respect for the *kamis* in nature is a key Shinto value. Prior to the end of World War II, Shinto was the state religion of Japan, and bolstered the cult of the Japanese emperor.

Sikhism—Founded by the Guru Nanak (born 1469), Sikhism believes in a non-anthropomorphic, supreme, eternal, creator God; centering one's devotion to God is seen as a means of escaping the cycle of rebirth. Sikhs follow the teachings of Nanak and nine subsequent gurus. Their scripture, the Guru Granth Sahib—also known as the Adi Granth—is considered the living Guru, or final authority of Sikh faith and theology. Sikhism emphasizes equality of humankind and disavows caste, class, or gender discrimination.

Taoism—Chinese philosophy or religion based upon Lao Tzu's Tao Te Ching, which centers on belief in the Tao, or the way, as the flow of the universe and the nature of things. Taoism encourages a principle of non-force, or wu-wei, as the means to live harmoniously with the Tao. Taoists believe the esoteric world is made up of a perfect harmonious balance and nature, while in the manifest world—particularly in the body—balance is distorted. The Three Jewels of the Tao—compassion, simplicity, and humility—serve as the basis for Taoist ethics.

Zoroastrianism—Originating from the teachings of Zoroaster in about the 9th or 10th century B.C., Zoroastrianism may be the oldest continuing creedal religion. Its key beliefs center on a transcendent creator God, Ahura Mazda, and the concept of free will. The key ethical tenets of Zoroastrianism expressed in its scripture, the Avesta, are based on a dualistic worldview where one may prevent chaos if one chooses to serve God and exercises good thoughts, good words, and good deeds. Zoroastrianism is generally a closed religion and members are almost always born to Zoroastrian parents. Prior to the spread of Islam, Zoroastrianism dominated greater Iran. Today, though a minority, Zoroastrians remain primarily in Iran, India, and Pakistan.

Reserves of foreign exchange and gold This entry gives the dollar value for the stock of all financial assets that are available to the central monetary authority for use in meeting a country's balance of payments needs as of the end-date of the period specified. This category includes not only foreign currency and gold, but also a country's holdings of Special Drawing Rights in the International Monetary Fund, and its reserve position in the Fund.

Roadways This entry gives the total length of the road network and includes the length of the *paved* and *unpaved* portions.

Sex ratio This entry includes the number of males for each female in five age groups—*at birth, under 15 years, 15–64 years, 65 years and over*, and for the *total population*. Sex ratio at birth has recently emerged as an indicator of certain kinds of sex discrimination in some countries. For instance, high sex ratios at birth in some Asian countries are now attributed to sex-selective abortion and infanticide due to a strong preference for sons. This will affect future marriage patterns and fertility patterns. Eventually, it could cause unrest among young adult males who are unable to find partners.

Stock of direct foreign investment—abroad This entry gives the cumulative US dollar value of all investments in foreign countries made directly by residents—primarily companies—of the home country, as of the end of the time period indicated. Direct investment excludes investment through purchase of shares.

Stock of direct foreign investment—at home This entry gives the cumulative US dollar value of all investments in the home country made directly by residents—primarily companies—of other countries as of the end of the time period indicated. Direct investment excludes investment through purchase of shares.

Suffrage This entry gives the age at enfranchisement and whether the right to vote is universal or restricted.

Telephone numbers All telephone numbers in *The World Factbook* consist of the country code in brackets, the city or area code (where required) in parentheses, and the local number. The one component that is not presented is the international access code, which varies from country to country. For example, an international direct dial telephone call placed from the US to Madrid, Spain, would be as follows: 011 [34] (1) 577-xxxx, where 011 is the international access code for station-to-station calls; 01 is for calls other than station-to-station calls, [34] is the country code for Spain, (1) is the city code for Madrid, 577 is the local exchange, and xxxx is the local telephone number. An international direct dial telephone call placed from another country to the US would be as follows: international access code + [1] (202) 939-xxxx, where [1] is the country code for the US, (202) is the area code for Washington, DC, 939 is the local exchange, and xxxx is the local telephone number.

Telephone system This entry includes a brief general assessment of the system with details on the domestic and international components. The following terms and abbreviations are used throughout the entry:

Arabsat—Arab Satellite Communications Organization (Riyadh, Saudi Arabia).

Autodin—Automatic Digital Network (US Department of Defense).

CB—citizen's band mobile radio communications.

Cellular telephone system—the telephones in this system are radio transceivers, with each instrument having its own private radio frequency and sufficient radiated power to reach the booster station in its area (cell), from which the telephone signal is fed to a telephone exchange.

Central American Microwave System—a trunk microwave radio relay system that links the countries of Central America and Mexico with each other.

Coaxial cable—a multichannel communication cable consisting of a central conducting wire, surrounded by and insulated from a cylindrical conducting shell; a large number of telephone channels can be made available within the insulated space by the use of a large number of carrier frequencies.

Comsat—Communications Satellite Corporation (US).

DSN—Defense Switched Network (formerly Automatic Voice Network or Autovon); basic general-purpose, switched voice network of the Defense Communications System (US Department of Defense).

Eutelsat—European Telecommunications Satellite Organization (Paris).

Fiber-optic cable—a multichannel communications cable using a thread of optical glass fibers as a transmission medium in which the signal (voice, video, etc.) is in the form of a coded pulse of light.

GSM—a global system for mobile (cellular) communications devised by the Groupe Special Mobile of the pan-European standardization organization, Conference Europeanne des Posts et Telecommunications (CEPT) in 1982.

HF—high frequency; any radio frequency in the 3,000- to 30,000-kHz range.

Inmarsat—International Maritime Satellite Organization (London); provider of global mobile satellite communications for commercial, distress, and safety applications at sea, in the air, and on land.

Intelsat—International Telecommunications Satellite Organization (Washington, DC).

Intersputnik—International Organization of Space Communications (Moscow); first established in the former Soviet Union and the East European countries, it is now marketing its services worldwide with earth stations in North America, Africa, and East Asia.

Landline—communication wire or cable of any sort that is installed on poles or buried in the ground.

Marecs—Maritime European Communications Satellite used in the Inmarsat system on lease from the European Space Agency.

Marisat—satellites of the Comsat Corporation that participate in the Inmarsat system.

Medarabtel—the Middle East Telecommunications Project of the International Telecommunications Union (ITU) providing a modern telecommunications network, primarily by microwave radio relay, linking Algeria, Djibouti, Egypt, Jordan, Libya, Morocco, Saudi Arabia, Somalia, Sudan, Syria, Tunisia, and Yemen; it was initially started in Morocco in 1970 by the Arab Telecommunications Union (ATU) and was known at that time as the Middle East Mediterranean Telecommunications Network.

Microwave radio relay—transmission of long distance telephone calls and television programs by highly directional radio microwaves that are received and sent on from one booster station to another on an optical path.

NMT—Nordic Mobile Telephone; an analog cellular telephone system that was developed jointly by the national telecommunications authorities of the Nordic countries (Denmark, Finland, Iceland, Norway, and Sweden).

Orbita—a Russian television service; also the trade name of a packet-switched digital telephone network.

Radiotelephone communications—the two-way transmission and reception of sounds by broadcast radio on authorized frequencies using telephone handsets.

PanAmSat—PanAmSat Corporation (Greenwich, CT).

SAFE—South African Far East Cable.

Satellite communication system—a communication system consisting of two or more earth stations and at least one satellite that provide long distance transmission of voice, data, and television; the system usually serves as a trunk connection between telephone exchanges; if the earth stations are in the same country, it is a domestic system.

Satellite earth station—a communications facility with a microwave radio transmitting and receiving antenna and required receiving and transmitting equipment for communicating with satellites.

Satellite link—a radio connection between a satellite and an earth station permitting communication between them, either one-way (down link from satellite to earth station—television receive-only transmission) or two-way (telephone channels).

SHF—super high frequency; any radio frequency in the 3,000- to 30,000-MHz range.

Shortwave—radio frequencies (from 1.605 to 30 MHz) that fall above the commercial broadcast band and are used for communication over long distances.

Solidaridad—geosynchronous satellites in Mexico's system of international telecommunications in the Western Hemisphere.

Statsionar—Russia's geostationary system for satellite telecommunications.

Submarine cable—a cable designed for service under water.

TAT—Trans-Atlantic Telephone; any of a number of high-capacity submarine coaxial telephone cables linking Europe with North America.

Telefax—facsimile service between subscriber stations via the public switched telephone network or the international Datel network.

Telegraph—a telecommunications system designed for unmodulated electric impulse transmission.

Telex—a communication service involving teletypewriters connected by wire through automatic exchanges.

Tropospheric scatter—a form of microwave radio transmission in which the troposphere is used to scatter and reflect a fraction of the incident radio waves back to earth; powerful, highly directional antennas are used to transmit and receive the microwave signals; reliable over-the-horizon communications are realized for distances up to 600 miles in a single hop; additional hops can extend the range of this system for very long distances.

Trunk network—a network of switching centers, connected by multichannel trunk lines.

UHF—ultra high frequency; any radio frequency in the 300- to 3,000-MHz range.

VHF—very high frequency; any radio frequency in the 30- to 300-MHz range.

Telephones—main lines in use This entry gives the total number of main telephone lines in use.

Telephones—mobile cellular This entry gives the total number of mobile cellular telephone subscribers.

Television broadcast stations This entry gives the total number of separate broadcast stations plus any repeater stations.

Terminology Due to the highly structured nature of the *Factbook* database, some collective generic terms have to be used. For example, the word **Country** in the **Country name** entry refers to a wide variety of dependencies, areas of special sovereignty, uninhabited islands, and other entities in addition to the traditional countries or independent states. **Military** is also used as an umbrella term for various civil defense, security, and defense activities in many entries. The **Independence** entry includes the usual colonial independence dates and former ruling states as well as other significant nationhood dates such as the traditional founding date or the date of unification, federation, confederation, establishment, or state succession that are not strictly independence dates. Dependent areas have the nature of their dependency status noted in this same entry.

Terrain This entry contains a brief description of the topography.

Time Difference This entry is expressed in *The World Factbook* in two ways. First, it is stated as the difference in hours between the capital of an entity and **Coordinated Universal Time (UTC)** during Standard Time. Additionally, the difference in time between the capital of an entity and that observed in Washington,

D.C. is also provided. Note that the time difference assumes both locations are simultaneously observing Standard Time or Daylight Saving Time.

Time zones Ten countries (Australia, Brazil, Canada, Indonesia, Kazakhstan, Mexico, New Zealand, Russia, Spain, and the United States) and the island of Greenland observe more than one official time depending on the number of designated time zones within their boundaries. An illustration of time zones throughout the world and within countries can be seen in the Standard Time Zones of the World map included in the **Reference Maps** section of *The World Factbook*.

Total fertility rate This entry gives a figure for the average number of children that would be born per woman if all women lived to the end of their childbearing years and bore children according to a given fertility rate at each age. The total fertility rate (TFR) is a more direct measure of the level of fertility than the crude birth rate, since it refers to births per woman. This indicator shows the potential for population change in the country. A rate of two children per woman is considered the replacement rate for a population, resulting in relative stability in terms of total numbers. Rates above two children indicate populations growing in size and whose median age is declining. Higher rates may also indicate difficulties for families, in some situations, to feed and educate their children and for women to enter the labor force. Rates below two children indicate populations decreasing in size and growing older. Global fertility rates are in general decline and this trend is most pronounced in industrialized countries, especially Western Europe, where populations are projected to decline dramatically over the next 50 years.

Total renewable water resources This entry provides the long-term average water availability for a country in cubic kilometers of precipitation, recharged ground water, and surface inflows from surrounding countries. The values have been adjusted to account for overlap resulting from surface flow recharge of groundwater sources. Total renewable water resources provides the water total available to a country but does not include water resource totals that have been reserved for upstream or downstream countries through international agreements. Note that these values are averages and do not accurately reflect the total available in any given year. Annual available resources can vary greatly due to short-term and long-term climatic and weather variations.

Trafficking in persons Trafficking in persons is modern-day slavery, involving victims who are forced, defrauded, or coerced into labor or sexual exploitation. The International Labor Organization (ILO), the UN agency charged with addressing labor standards, employment, and social protection issues, estimates that 12.3 million people worldwide are enslaved in forced labor, bonded labor, forced child labor, sexual servitude, and involuntary servitude at any given time. Human trafficking is a multi-dimensional threat, depriving people of their human rights and freedoms, risking global health, promoting social breakdown, inhibiting development by depriving countries of their human capital, and helping fuel the growth of organized crime. In 2000, the US Congress passed the Trafficking Victims Protection Act (TVPA), reauthorized in 2003 and 2005, which provides tools for the US to combat trafficking in persons, both domestically and abroad. One of the law's key components is the creation of the US Department of State's annual *Trafficking in Persons Report*, which assesses the government response (i.e., the *current situation*) in some 150 countries with a significant number of victims trafficked across their borders who are recruited, harbored, transported, provided, or obtained for forced labor or sexual exploitation. Countries in the annual report are rated in three tiers, based on government efforts to combat trafficking. The countries identified in this entry are those listed in the 2007 *Trafficking in Persons Report* as *Tier 2 Watch List* or *Tier 3* based on the following tier rating *definitions*:

Tier 2 Watch List countries do not fully comply with the minimum standards for the elimination of trafficking but are making significant efforts to do so, and meet one of the following criteria:
1. they display a high or significantly increasing number of victims,
2. they have failed to provide evidence of increasing efforts to combat trafficking in persons, or,
3. they have committed to take action over the next year.
Tier 3 countries neither satisfy the minimum standards for the elimination of trafficking nor demonstrate a significant effort to do so. Countries in this tier are subject to potential non-humanitarian and non-trade sanctions.

Transnational issues This category includes four entries—**Disputes**—international, **Refugees and internally displaced persons**, **Trafficking in persons**, and **Illicit drugs**—that deal with current issues going beyond national boundaries.

Transportation This category includes the entries dealing with the means for movement of people and goods.

Transportation—note This entry includes miscellaneous transportation information of significance not included elsewhere.

UTC (Coordinated Universal Time) See entry for Coordinated Universal Time.

Unemployment rate This entry contains the percent of the labor force that is without jobs. Substantial underemployment might be noted.

Waterways This entry gives the total length of navigable rivers, canals, and other inland bodies of water.

Weights and Measures This information is presented in **Appendix G: Weights and Measures** and includes mathematical notations (mathematical powers and names), metric interrelationships (prefix; symbol; length, weight, or capacity; area; volume), and standard conversion factors.

Years All year references are for the calendar year (CY) unless indicated as fiscal year (FY). The calendar year is an accounting period of 12 months from 1 January to 31 December. The fiscal year is an accounting period of 12 months other than 1 January to 31 December.

Note: Information for the US and US dependencies was compiled from material in the public domain and does not represent Intelligence Community estimates.

GUIDE TO COUNTRY PROFILES

INTRODUCTION
Background

GEOGRAPHY
Location
Geographic coordinates
Map references
Area
total
land
water
Area—comparative
Land boundaries
total
border countries
Coastline
Maritime claims
territorial sea
contiguous zone
exclusive economic zone
exclusive fishing zone
Climate
Terrain
Elevation extremes
lowest point
highest point
Natural resources
Land use
arable land
permanent crops
other
Irrigated land
Total renewable water resources
**Freshwater withdrawal
(domestic/industrial/agricultural)**
total
per capita
Natural hazards
Environment—current issues
Environment—international agreements
party to
signed, but not ratified
Geography—note

PEOPLE
Population
Age structure
0–14 years
15–64 years
65 years and over
Median Age
total
male
female
Population growth rate
Birth rate
Death rate
Net migration rate
Sex ratio
at birth

under 15 years
15–64 years
65 years and over
total population
Infant mortality rate
total
male
female
Life expectancy at birth
total population
male
female
Total fertility rate
HIV/AIDS—adult prevalence rate
HIV/AIDS—people living with
HIV/AIDS
HIV/AIDS—deaths
Major infectious diseases
degree of risk
food or waterborne diseases
vectorborne diseases
water contact diseases
aerosolized dust or soil contact disease
respiratory disease
animal contact disease
Nationality
noun
adjective
Ethnic groups
Religions
Languages
Literacy
definition
total population
male
female
People—note

GOVERNMENT
Country name
conventional long form
conventional short form
local long form
local short form
former
abbreviation
Dependency status
Government type
Capital
name
geographic coordinates
time difference
daylight saving time
Administrative divisions
Dependent areas
Independence
National holiday
Constitution
Legal system
Suffrage
Executive branch

chief of state
head of government
cabinet
elections
election results
Legislative branch
elections
election results
Judicial branch
Political parties and leaders
Political pressure groups and leaders
International organization participation
Diplomatic representation in the US
chief of mission
chancery
telephone
FAX
consulate(s) general
consulate(s)
Diplomatic representation from the US
chief of mission
embassy
mailing address
telephone
FAX
consulate(s) general
consulate(s)
branch office(s)
Flag description
Government—note

ECONOMY
Economy—overview
GDP (purchasing power parity)
GDP (official exchange rate)
GDP—real growth rate
GDP—per capita
GDP—composition by sector
agriculture
industry
services
Labor force
Labor force—by occupation
agriculture
industry
services
Unemployment rate
Population below poverty line
**Household income or consumption by
percentage share**
lowest 10%
highest 10%
Distribution of family income—Gini index
Inflation rate (consumer prices)
Investment (gross fixed)
Budget
revenues
expenditures
Public debt
Agriculture—products
Industries

Industrial production growth rate
Electricity—production
Electricity—consumption
Electricity—exports
Electricity—imports
Oil—production
Oil—consumption
Oil—exports
Oil—imports
Oil—proved reserves
Natural Gas—production
Natural Gas—consumption
Natural Gas—exports
Natural Gas—imports
Natural Gas—proved reserves
Current account balance
Exports
Exports—commodities
Exports—partners
Imports
Imports—commodities
Imports—partners
Reserves of foreign exchange and gold
Debt—external
Stock of direct foreign investment—at home
Stock of direct foreign investment—abroad
Market value of publicly traded shares
Economic aid—donor
Economic aid—recipient
Currency (code)
Exchange rates
Fiscal year

COMMUNICATIONS

Telephones—main lines in use
Telephones—mobile cellular

Telephone system
general assessment
domestic
international
Radio broadcast stations
Television broadcast stations
Internet country code
Internet hosts
Internet users
Communications—note

TRANSPORTATION

Airports
Airports—with paved runways
total
over 3,047 m
2,438 to 3,047 m
1,524 to 2,437 m
914 to 1,523 m
under 914 m
Airports—with unpaved runways
total
over 3,047 m
2,438 to 3,047 m
1,524 to 2,437 m
914 to 1,523 m
under 914 m
Heliports
Pipelines
Railways
total
broad gauge
standard gauge
narrow gauge
dual gauge
Roadways
total
paved

unpaved
Waterways
Merchant marine
total
ships by type
foreign-owned
registered in other countries
Ports and terminals
Transportation—note

MILITARY

Military branches
Military service age and obligation
Manpower available for military service
males age 15–49
females age 15–49
Manpower fit for military service
males age 15–49
females age 15–49
Manpower reaching military age annually
males
females
Military expenditures—percent of GDP
Military—note

TRANSNATIONAL ISSUES

Disputes—international
Refugees and internally displaced persons
refugees
IDPs
Trafficking in persons
current situation
tier rating
Illicit drugs

AFGHANISTAN

Background: Ahmad Shah DURRANI unified the Pashtun tribes and founded Afghanistan in 1747. The country served as a buffer between the British and Russian empires until it won independence from notional British control in 1919. A brief experiment in democracy ended in a 1973 coup and a 1978 Communist counter-coup. The Soviet Union invaded in 1979 to support the tottering Afghan Communist regime, touching off a long and destructive war. The USSR withdrew in 1989 under relentless pressure by internationally supported anti-Communist mujahedin rebels. Subsequently, a series of civil wars saw Kabul finally fall in 1996 to the Taliban, a hardline Pakistani-sponsored movement that emerged in 1994 to end the country's civil war and anarchy. Following the 11 September 2001 terrorist attacks in New York City, a US, Allied, and anti-Taliban Northern Alliance military action toppled the Taliban for sheltering Osama BIN LADIN. The UN-sponsored Bonn Conference in 2001 established a process for political reconstruction that included the adoption of a new constitution and a presidential election in 2004, and National Assembly elections in 2005. On 7 December 2004, Hamid KARZAI became the first democratically elected president of Afghanistan. The National Assembly was inaugurated on 19 December 2005.

GEOGRAPHY

Location: Southern Asia, north and west of Pakistan, east of Iran
Geographic coordinates: 33 00 N, 65 00 E
Map references: Asia
Area:
total: 647,500 sq km

land: 647,500 sq km
water: 0 sq km
Area—comparative: slightly smaller than Texas
Land boundaries:
total: 5,529 km
border countries: China 76 km, Iran 936 km, Pakistan 2,430 km, Tajikistan 1,206 km, Turkmenistan 744 km, Uzbekistan 137 km
Coastline: 0 km (landlocked)
Maritime claims: none (landlocked)
Climate: arid to semiarid; cold winters and hot summers
Terrain: mostly rugged mountains; plains in north and southwest
Elevation extremes:
lowest point: Amu Darya 258 m
highest point: Nowshak 7,485 m
Natural resources: natural gas, petroleum, coal, copper, chromite, talc, barites, sulfur, lead, zinc, iron ore, salt, precious and semiprecious stones
Land use:
arable land: 12.13%
permanent crops: 0.21%
other: 87.66% (2005)
Irrigated land: 27,200 sq km (2003)
Total renewable water resources: 65 cu km (1997)
Freshwater withdrawal (domestic/industrial/agricultural): *total:* 23.26 cu km/yr (2%/0%/98%)
per capita: 779 cu m/yr (2000)
Natural hazards: damaging earthquakes occur in Hindu Kush mountains; flooding; droughts
Environment—current issues: limited natural fresh water resources; inadequate supplies of potable water; soil degradation; overgrazing; deforestation (much of the remaining forests are being cut down for fuel and building materials); desertification; air and water pollution
Environment—international agreements: *party to:* Biodiversity, Climate Change, Desertification, Endangered Species, Environmental Modification, Marine Dumping, Ozone Layer Protection
signed, but not ratified: Hazardous Wastes, Law of the Sea, Marine Life Conservation
Geography—note: landlocked; the Hindu Kush mountains that run northeast to southwest divide the northern provinces from the rest of the country; the highest peaks are in the northern Vakhan (Wakhan Corridor)

PEOPLE

Population: 32,738,376 (July 2008 est.)
Age structure:
0–14 years: 44.6% (male 7,474,394/female 7,121,145)

15–64 years: 53% (male 8,901,880/female 8,447,983)
65 years and over: 2.4% (male 383,830/female 409,144) (2008 est.)
Median age:
total: 17.6 years
male: 17.6 years
female: 17.6 years (2008 est.)
Population growth rate: 2.626% (2008 est.)
Birth rate: 45.82 births/1,000 population (2008 est.)
Death rate: 19.56 deaths/1,000 population (2008 est.)
Net migration rate: 21 migrant(s)/1,000 population (2005 est.)
Sex ratio:
at birth: 1.05 male(s)/female
under 15 years: 1.05 male(s)/female
15–64 years: 1.05 male(s)/female
65 years and over: 0.94 male(s)/female
total population: 1.05 male(s)/female (2008 est.)
Infant mortality rate:
total: 154.67 deaths/1,000 live births
male: 158.88 deaths/1,000 live births
female: 150.24 deaths/1,000 live births (2008 est.)
Life expectancy at birth:
total population: 44.21 years
male: 44.04 years
female: 44.39 years (2008 est.)
Total fertility rate: 6.58 children born/woman (2008 est.)
HIV/AIDS—adult prevalence rate: 0.01% (2001 est.)
HIV/AIDS—people living with HIV/AIDS: NA
HIV/AIDS—deaths: NA
Major infectious diseases:
degree of risk: high
food or waterborne diseases: bacterial and protozoal diarrhea, hepatitis A, and typhoid fever
vectorborne disease: malaria
animal contact disease: rabies
note: highly pathogenic H5N1 avian influenza has been identified in this country; it poses a negligible risk with extremely rare cases possible among US citizens who have close contact with birds (2008)
Nationality:
noun: Afghan(s)
adjective: Afghan
Ethnic groups: Pashtun 42%, Tajik 27%, Hazara 9%, Uzbek 9%, Aimak 4%, Turkmen 3%, Baloch 2%, other 4%
Religions: Sunni Muslim 80%, Shi'a Muslim 19%, other 1%
Languages: Afghan Persian or Dari (official) 50%, Pashto (official) 35%, Turkic languages (primarily Uzbek and

Turkmen) 11%, 30 minor languages (primarily Balochi and Pashai) 4%, much bilingualism

Literacy:
definition: age 15 and over can read and write
total population: 28.1%
male: 43.1%
female: 12.6% (2000 est.)

GOVERNMENT

Country name:
conventional long form: Islamic Republic of Afghanistan
conventional short form: Afghanistan
local long form: Jomhuri-ye Eslami-ye Afghanestan
local short form: Afghanestan
former: Republic of Afghanistan
Government type: Islamic republic
Capital: *name:* Kabul
geographic coordinates: 34 31 N, 69 11 E
time difference: UTC+4.5 (9.5 hours ahead of Washington, DC during Standard Time)
Administrative divisions: 34 provinces (velayat, singular—velayat); Badakhshan, Badghis, Baghlan, Balkh, Bamian, Daykondi, Farah, Faryab, Ghazni, Ghowr, Helmand, Herat, Jowzjan, Kabol, Kandahar, Kapisa, Khowst, Konar, Kondoz, Laghman, Lowgar, Nangarhar, Nimruz, Nurestan, Oruzgan, Paktia, Paktika, Panjshir, Parvan, Samangan, Sar-e Pol, Takhar, Vardak, Zabol
Independence: 19 August 1919 (from UK control over Afghan foreign affairs)
National holiday: Independence Day, 19 August (1919)
Constitution: new constitution drafted 14 December 2003–4 January 2004; signed 16 January 2004
Legal system: based on mixed civil and Shari'a law; has not accepted compulsory ICJ jurisdiction
Suffrage: 18 years of age; universal
Executive branch:
chief of state: President of the Islamic Republic of Afghanistan Hamid KARZAI (since 7 December 2004); Vice Presidents Ahmad Zia MASOOD and Abdul Karim KHALILI (since 7 December 2004); note—the president is both the chief of state and head of government; former King ZAHIR Shah held the honorific, "Father of the Country," and presided symbolically over certain occasions but lacked any governing authority; the honorific is not hereditary; King ZAHIR Shah died on 23 July 2007
head of government: President of the Islamic Republic of Afghanistan Hamid KARZAI (since 7 December 2004); Vice Presidents Ahmad Zia MASOOD and

Abdul Karim KHALILI (since 7 December 2004)
cabinet: 25 ministers; note—under the new constitution, ministers are appointed by the president and approved by the National Assembly
elections: the president and two vice presidents are elected by direct vote for a five-year term (eligible for a second term); if no candidate receives 50% or more of the vote in the first round of voting, the two candidates with the most votes will participate in a second round; a president can only be elected for two terms; election last held 9 October 2004 (next to be held in 2009)
election results: Hamid KARZAI elected president; percent of vote—Hamid KARZAI 55.4%, Yunus QANUNI 16.3%, Ustad Mohammad MOHAQQEQ 11.6%, Abdul Rashid DOSTAM 10.0%, Abdul Latif PEDRAM 1.4%, Masooda JALAL 1.2%
Legislative branch: the bicameral National Assembly consists of the Wolesi Jirga or House of People (no more than 249 seats), directly elected for five-year terms, and the Meshrano Jirga or House of Elders (102 seats, one-third elected from provincial councils for four-year terms, one-third elected from local district councils for three-year terms, and one-third nominated by the president for five-year terms)
note: on rare occasions the government may convene a Loya Jirga (Grand Council) on issues of independence, national sovereignty, and territorial integrity; it can amend the provisions of the constitution and prosecute the president; it is made up of members of the National Assembly and chairpersons of the provincial and district councils
elections: last held 18 September 2005 (next to be held for the Wolesi Jirga by September 2009; next to be held for the provincial councils to the Meshrano Jirga by September 2008)
election results: the single non-transferable vote (SNTV) system used in the election did not make use of political party slates; most candidates ran as independents
Judicial branch: the constitution establishes a nine-member Stera Mahkama or Supreme Court (its nine justices are appointed for 10-year terms by the president with approval of the Wolesi Jirga) and subordinate High Courts and Appeals Courts; there is also a minister of justice; a separate Afghan Independent Human Rights Commission established by the Bonn Agreement is charged with investigating human rights abuses and war crimes

Political parties and leaders:
Afghanistan Peoples' Treaty Party (Hizb-e-Wolesi Tarhun Afghanistan) [Sayyed Amir TAHSEEN]; Afghanistan's Islamic Mission Organization (Tanzim Daawat-e-Islami-e-Afghanistan) [Abdul Rasoul SAYYAF]; Afghanistan's Islamic Nation Party (Hezb-e-Umat-e-Islam-e-Afghanistan) [Toran Noor Aqa Ahmad ZAI]; Afghanistan's National Islamic Party (Hezb-e-Mili Islami-e-Afghanistan) [Rohullah LOUDIN]; Afghanistan's Welfare Party (Hezb-e-Refah-e-Afghanistan) [Meer Asef ZAEEFI]; Afghan Social Democratic Party (Hezb-e-Afghan Melat) [Anwarul Haq AHADI]; Afghan Society for the Call to the Koran and Sunna (Hezb-e-Jamahat-ul-Dawat ilal Quran-wa-Sunat-e-Afghanistan) [Mawlawee Samiullah NAJEEBEE]; Comprehensive Movement of Democracy and Development of Afghanistan Party (Hizb-e-Nahzat Faragir Democracy wa Taraqi-e-Afghanistan) [Sher Mohammad BAZGAR]; Democratic Party of Afghanistan (Hezb-e-Democracy Afghanistan) [Tawos ARAB]; Democratic Party of Afghanistan (Hezb-e-Domcrat-e-Afghanistan) [Abdul Kabir RANJBAR]; Elites People of Afghanistan Party (Hezb-e-Nakhbagan-e-Mardom-e-Afghanistan) [Abdul Hamid JAWAD]; Freedom and Democracy Movement of Afghanistan (Hezb-e-Nahzat-e-Aazadee Wa Democracy-e-Afghanistan) [Abdul Raqib Jawid KOHISTANEE]; Freedom Party of Afghanistan (Hezb-e-Azadee-e-Afghanistan) [Ilaj Abdul MALEK]; Freedom Party of Afghanistan (Hezb-e-Isteqlal-e-Afghanistan) [Dr. Ghulam Farooq NEJRABEE]; Hizullah-e-Afghanistan [Qari Ahmad ALI]; Human Rights Protection and Development Party of Afghanistan (Hezb-e-Ifazat Az Uqooq-e-Bashar Wa Inkishaf-e-Afghanistan) [Baryalai NASRATI]; Islamic Justice Party of Afghanistan (Hezb-e-Adalat-e-Islami Afghanistan) [Mohammad Kabir MARZBAN]; Islamic Movement of Afghanistan (Hezb-e Harakat-e-Islami-e-Afghanistan) [Mohammad Ali JAWID]; Islamic Movement of Afghanistan Party (Hizb-e-Nahzat-e-Melli Islami Afghanistan) [Mohammad Mukhtar MUFLEH]; Islamic Party of Afghanistan (Hizb-e-Islami Afghanistan) [Mohammad Khalid FAROOQI]; Islamic Party of the Afghan Land (De Afghan Watan Islami Gond) [Mohammad Hassan FEROZKHEL]; Islamic People's Movement of Afghanistan (Hezb-e-Harakat-e-Islami Mardom-e-Afghanistan) [Ilhaj Said

Hussain ANWARY]; Islamic Society of Afghanistan (Hezb-e Jamihat-e-Islami) [Ustad RABBANI]; Islamic Unity of the Nation of Afghanistan Party (Hezb-e-Wahdat-e-Islami-e-Melat-e-Afghanistan) [Qurban Ali URFANI]; Islamic Unity Party of Afghanistan (Hezb-e-Wahdat-e-Islami-e-Afghanistan) [Mohammad Karim KHALILI]; Islamic Unity Party of the People of Afghanistan (Hezb-e-Wahdat-e-Islami Mardom-e-Afghanistan) [Ustad Mohammad MOHAQQEQ]; Labor and Progress of Afghanistan Party (Hezb-e-Kar Wa Tawsiha-e-Afghanistan) [Zulfiqar OMID]; Muslim People of Afghanistan Party (Hezb-e-Mardom-e-Mosalman-e-Afghanistan) [Besmellah JOYAN]; Muslim Unity Movement Party of Afghanistan (Hezb-e-Tahreek Wahdat-ul-Musimeen Afghanistan) [Wazir Mohammad WAHDAT]; National and Islamic Sovereignty Movement Party of Afghanistan (Hizb-e-Eqtedar-e-Melli wa Islami Afghanistan) [Ahmad Shah AHMADZAI]; National Congress Party of Afghanistan (Hezb-e-Kangra-e-Mili-e-Afghanistan) [Abdul Latif PEDRAM]; National Country Party (Hezb-e-Mili Heward) [Ghulam MOHAMMAD]; National Development Party of Afghanistan (Hezb-e-Taraqee Mili Afghanistan) [Dr. Aref BAKTASH]; National Freedom Seekers Party (Hezb-e-Aazaadi Khwahan Maihan) [Abdul Hadi DABEER]; National Independence Party of Afghanistan (Hezb-e Esteqlal-e-Mili Afghanistan) [Taj Mohammad WARDAK]; National Islamic Fighters Party of Afghanistan (De Afghanistan De Mili Mubarizeeno Islami Gond) [Amanat NINGARHAREE]; National Islamic Front of Afghanistan (Mahaz-e-Mili Islami Afghanistan) [Pir Sayed Ahmad GAILANEE]; National Islamic Moderation Party of Afghanistan (Hezb-e-Eatedal-e-Mili Islami-e-Afghanistan) [Qara Bik Eized YAAR]; National Islamic Movement of Afghanistan (Hezb-e-Junbish Mili Islami-e-Afghanistan) [Sayed NOORULLAH]; National Islamic Unity Party of Afghanistan (Hezb-e-Wahdat-e-Mili Islami-e-Afghanistan) [Mohammad AKBAREE]; National Movement of Afghanistan (Nahzat-e-Mili Afghanistan) [Ahmad Wali MASOOUD], National Party of Afghanistan (Hezb-e-Mili Afghanistan) [Abdul Rashid ARYAN]; National Patch of Afghanistan Party (Hezb-e Paiwand Mihahani Afghanistan) [Sayed Kamal SADAT]; National Peace Islamic Party of Afghanistan (De Afghanistan De Solay Mili Islami Gond) [Shah Mohammood Popal ZAI]; National Peace & Islamic Party of the Tribes of Afghanistan (Hezb-e-Sulh-e-Mili Islami Aqwam-e-Afghanistan) [Abdul Qaher SHARIATEE]; National Peace & Unity Party of Afghanistan (Hezb-e-Sulh Wa Wahdat-e-Mili-e-Afghanistan) [Abdul Qader IMAMI]; National Prosperity and Islamic Party of Afghanistan (Hezb-e-Sahadat-e-Mili Islami-e-Afghanistan) [Mohammad Osman SALEKZADA]; National Prosperity Party (Hezb-e-Refah-e-Mili Afghanistan) [Mohammad Hassan JAHFAREE]; National Solidarity Movement of Afghanistan (Hezb-e-Nahzat-e-Hambastagee Mili-e-Afghanistan) [Pir Sayed Eshaq GAILANEE]; National Solidarity Party of Afghanistan (Hezb-e-Paiwand Mili Afghanistan) [Sayed Mansoor NADREEI]; National Sovereignty Party (Hezb-e-Eqtedar-e-Mili) [Sayed Mustafa KAZEMI]; National Stability Party (Hezb-e-Subat-e-Mili Islami-e-Afghanistan) [Mohammad Same KHAROTI]; National Stance Party (Hizb-e-Melli Dareez) [Habibullah JANEBDAR]; National Tribal Unity Islamic Party of Afghanistan (Hezb-e-Mili Wahdat-e-Aqwam-e-Islami-e-Afghanistan) [Mohammad Shah KHOGYANI]; National United Front (Jumbah-e Mutahed-e Milli) [Burhanuddin RABBANI] (a coalition); National Unity Movement (Hezb-e-Tahreek Wahdat-e-Mili-e-Afghanistan) [Sultan Mohammad GHAZI]; National Unity Movement of Afghanistan (Hezb-e-Harakat-e-Mili Wahdat-e-Afghanistan) [Mohammad Nadir AATASH]; National Unity Party of Afghanistan (Hezb-e-Wahdat-e-Mili Afghanistan) [Abdul Rashid JALILI]; New Afghanistan Party (Hezb-e-Afghanistan-e-Naween) [Mohammad Yunis QANUNI]; Peace and National Welfare Activists Society (Hezb-e-Majmeh Mili Faleen-Sulh-e-Afghanistan) [Shamsul Haq Noor SHAMS]; Peace Movement (De Afghanistan De Solay Ghorzang Gond) [Shahnawaz TANAI]; People's Aspirations Party of Afghanistan (Hezb-e-Aarman-e-Mardom-e-Afghanistan) [Ilhaj Saraj-u-din ZAFAREE]; People's Freedom Seekers Party of Afghanistan (Hezb-e-Aazadee Khwahan Mardom-e-Afghanistan) [Feda Mohammad EHSAS]; People's Liberal Freedom Seekers Party of Afghanistan (Hezb-e-Lebral-e-Aazadee Khwa-e-Afghanistan) [Ajmal SUHAIL]; People's Message Party of Afghanistan (Hezb-e-Resalat-e-Mardom-e-Afghanistan) [Noor Aqa WAINEE]; People's Movement of the National Unity of Afghanistan (De Afghanistan De Mili Wahdat Wolesi Tahreek) [Abdul Hakim NOORZAI]; People's Party of Afghanistan (Hezb-e-Mardom-e-Afghanistan) [Ahmad Shah ASAR]; People's Prosperity Party of Afghanistan (Hezb-e-Falah-e-Mardom-e-Afghanistan) [Ustad Mohammad ZAREEF]; People's Sovereignty Movement of Afghanistan (Nahzat-e-Hakemyat-e-Mardom-e-Afghanistan) [Hayatullah SUBHANEE]; People's Uprising Party of Afghanistan (Hezb-e-Rastakhaiz-e-Mardom-e-Afghanistan) [Sayed Zahir Qayed Omul BELADI]; People's Welfare Party of Afghanistan (Hezb-e-Refah-e-Mardom-e-Afghanistan) [Mia Gul WASIQ]; People's Welfare Party of Afghanistan (Hezb-e-Sahadat-e-Mardom-e-Afghanistan) [Mohammad Zubair PAIROZ]; Progressive Democratic Party of Afghanistan (Hezb-e-Taraqee Democrat Afghanistan) [Wali ARYA]; Republican Party (Hezb-e-Jamhoree Khwahane-Afghanistan) [Sebghatullah SANJAR]; Solidarity Party of Afghanistan (Hezb-e-Hambastagee-e-Afghanistan) [Abdul Khaleq NEMAT]; The Afghanistan's Mujahid Nation's Islamic Unity Movement (Da Afghanistan Mujahid Woles Yaowaali Islami Tahreek) [Saeedullah SAEED]; The People of Afghanistan's Democratic Movement (Hezb-e-Junbish Democracy Mardom-e-Afghanistan) [Sharif NAZARI]; Tribes Solidarity Party of Afghanistan (Hezb-e Hambastagee Mili Aqwam-e-Afghanistan) [Mohammad Zarif NASERI]; Understanding and Democracy Party of Afghanistan (Hezb-e-Tafahum Wa Democracy-e-Afghanistan) [Ahamad SHAHEEN]; United Afghanistan Party (Hezb-e-Afghanistan-e-Wahid) [Mohammad Wasil RAHIMEE]; United Islamic Party of Afghanistan (Hizb-e-Mutahed Islami Afghanistan) [Wahidullah SABAWOON]; Young Afghanistan's Islamic Organization (Hezb-e-Islami-e-Afghanistan-e-Jawan) [Sayed Jawad HUSSINEE]; Youth Solidarity Party of Afghanistan (Hezb-e-Hambastagee Mili Jawanan-e-Afghanistan) [Mohammad Jamil KARZAI]; note—includes only political parties approved by the Ministry of Justice

International organization participation: ADB, CP, ECO, FAO, G-77, IAEA, IBRD, ICAO, ICCt, IDA, IDB, IFAD, IFC, IFRCS, ILO, IMF, Interpol, IOC, IOM, ISO (correspondent), ITSO, ITU, MIGA, NAM, OIC, OPCW, OSCE (partner), SAARC, SACEP, UN,

UNCTAD, UNESCO, UNIDO, UNWTO, UPU, WCO, WFTU, WHO, WIPO, WMO, WTO (observer)

Diplomatic representation in the US:
chief of mission: Ambassador Said Tayeb JAWAD
chancery: 2341 Wyoming Avenue NW, Washington, DC 20008
telephone: [1] (202) 483-6410
FAX: [1] (202) 483-6488
consulate(s) general: Los Angeles, New York

Diplomatic representation from the US:
chief of mission: Ambassador William B. WOOD
embassy: The Great Masood Road, Kabul
mailing address: U.S. Embassy Kabul, APO, AE 09806
telephone: [93] 700 108 001
FAX: [00 93] (20) 230-1364

Flag description: three equal vertical bands of black (hoist side), red, and green, with the national emblem in white centered on the red band and slightly overlapping the other two bands; the center of the emblem features a mosque with pulpit and flags on either side, below the mosque are numerals for the solar year 1298 (1919 in the Gregorian calendar, the year of Afghan independence from the UK); this central image is circled by a border consisting of sheaves of wheat on the left and right, in the upper-center is an Arabic inscription of the Shahada (Muslim creed) below which are rays of the rising sun over the Takbir (Arabic expression meaning "God is great"), and at bottom center is a scroll bearing the name Afghanistan

ECONOMY

Economy—overview: Afghanistan's economy is recovering from decades of conflict. The economy has improved significantly since the fall of the Taliban regime in 2001 largely because of the infusion of international assistance, the recovery of the agricultural sector, and service sector growth. Real GDP growth exceeded 7% in 2007. Despite the progress of the past few years, Afghanistan is extremely poor, landlocked, and highly dependent on foreign aid, agriculture, and trade with neighboring countries. Much of the population continues to suffer from shortages of housing, clean water, electricity, medical care, and jobs. Criminality, insecurity, and the Afghan Government's inability to extend rule of law to all parts of the country pose challenges to future economic growth. It will probably take the remainder of the decade and continuing donor aid and attention to significantly raise Afghanistan's living standards from its current level, among the lowest in the world. International pledges made by more than 60 countries and international financial institutions at the Berlin Donors Conference for Afghan reconstruction in March 2004 reached $8.9 billion for 2004–09. While the international community remains committed to Afghanistan's development, pledging over $24 billion at three donors' conferences since 2002, Kabul will need to overcome a number of challenges. Expanding poppy cultivation and a growing opium trade generate roughly $4 billion in illicit economic activity and looms as one of Kabul's most serious policy concerns. Other long-term challenges include: budget sustainability, job creation, corruption, government capacity, and rebuilding war torn infrastructure.

GDP (purchasing power parity): $35 billion (2007 est.)

GDP (official exchange rate): $8.842 billion (2007 est.)

GDP—real growth rate: 12.4% (2007 est.)

GDP—per capita (PPP): $1,000 (2007 est.)

GDP—composition by sector:
agriculture: 38%
industry: 24%
services: 38%
note: data exclude opium production (2005 est.)

Labor force: 15 million (2004 est.)

Labor force—by occupation: *agriculture:* 80%
industry: 10%
services: 10% (2004 est.)

Unemployment rate: 40% (2005 est.)

Population below poverty line: 53% (2003)

Household income or consumption by percentage share:
lowest 10%: NA%
highest 10%: NA%

Inflation rate (consumer prices): 13% (2007 est.)

Budget:
revenues: $715 million
expenditures: $2.6 billion
note: Afghanistan has also received $273 million from the Reconstruction Trust Fund and $63 million from the Law and Order Trust Fund (2007 est.)

Agriculture—products: opium, wheat, fruits, nuts; wool, mutton, sheepskins, lambskins

Industries: small-scale production of textiles, soap, furniture, shoes, fertilizer, cement; handwoven carpets; natural gas, coal, copper

Industrial production growth rate: NA%

Electricity—production: 754.2 million kWh (2005)

Electricity—production by source:
fossil fuel: 36.3%
hydro: 63.7%
nuclear: 0%
other: 0% (2001)

Electricity—consumption: 801.4 million kWh (2005)

Electricity—exports: 0 kWh (2005)

Electricity—imports: 100 million kWh (2005)

Oil—production: 0 bbl/day (2005)

Oil—consumption: 5,000 bbl/day (2005 est.)

Oil—exports: 0 bbl/day (2004)

Oil—imports: 4,120 bbl/day (2004)

Oil—proved reserves: 0 bbl (1 January 2006 est.)

Natural gas—production: 19.18 million cu m (2005 est.)

Natural gas—consumption: 19.18 million cu m (2005 est.)

Natural gas—exports: 0 cu m (2005 est.)

Natural gas—imports: 0 cu m (2005)

Natural gas—proved reserves: 47.53 billion cu m (1 January 2006 est.)

Current account balance: NA

Exports: $274 million; note—not including illicit exports or reexports (2006)

Exports—commodities: opium, fruits and nuts, handwoven carpets, wool, cotton, hides and pelts, precious and semi-precious gems

Exports—partners: India 22.8%, Pakistan 21.7%, US 15.2%, UK 6.5%, Finland 4.4% (2006)

Imports: $3.823 billion (2006)

Imports—commodities: capital goods, food, textiles, petroleum products

Imports—partners: Pakistan 37.9%, US 12%, Germany 7.2%, India 5.1% (2006)

Economic aid—recipient: $2.775 billion (2005)

Debt—external: $8 billion in bilateral debt, mostly to Russia; Afghanistan has $500 million in debt to Multilateral Development Banks (2004)

Market value of publicly traded shares: publicly traded shares: $NA

Currency (code): afghani (AFA)

Currency code: AFA

Exchange rates: afghanis per US dollar—NA (2007), 46 (2006), 47.7 (2005), 48 (2004), 49 (2003)

Fiscal year: 21 March—20 March

COMMUNICATIONS

Telephones—main lines in use: 280,000 (2005)

Telephones—mobile cellular: 2.52 million (2006)

Telephone system:

general assessment: limited landline telephone service; an increasing number of Afghans utilize mobile-cellular phone networks in major cities
domestic: aided by the presence of multiple providers, mobile-cellular telephone service is improving rapidly
international: country code—93; five VSAT's installed in Kabul, Herat, Mazar-e-Sharif, Kandahar, and Jalalabad provide international and domestic voice and data connectivity (2007)
Radio broadcast stations: AM 21, FM 5, shortwave 1 (broadcasts in Pashto, Dari (Afghan Persian), Urdu, and English) (2006)
Radios: 167,000 (1999)
Television broadcast stations: at least 7 (1 government-run central television station in Kabul and regional stations in 6 of the 34 provinces) (2006)
Televisions: 100,000 (1999)
Internet country code: .af
Internet hosts: 21 (2007)
Internet Service Providers (ISPs): 1 (2000)
Internet users: 535,000 (2006)
Communications—note: Internet access is growing through Internet cafes as well as public "telekiosks" in Kabul (2005)

TRANSPORTATION

Airports: 46 (2007)
Airports—with paved runways:
total: 12
over 3,047 m: 4
2,438 to 3,047 m: 2
1,524 to 2,437 m: 4
914 to 1,523 m: 1
under 914 m: 1 (2007)

Airports—with unpaved runways:
total: 34
over 3,047 m: 1
2,438 to 3,047 m: 4
1,524 to 2,437 m: 16
914 to 1,523 m: 4
under 914 m: 9 (2007)
Heliports: 9 (2007)
Pipelines: gas 466 km (2007)
Roadways:
total: 34,782 km
paved: 8,229 km
unpaved: 26,553 km (2004)
Waterways: 1,200 km (chiefly Amu Darya, which handles vessels up to 500 DWT) (2007)
Ports and terminals: Kheyrabad, Shir Khan

MILITARY

Military branches: Afghan Armed Forces: Afghan National Army (ANA, includes Afghan National Army Air Corps) (2008)
Military service age and obligation: 22 years of age; inductees are contracted into service for a 4-year term (2005)
Manpower available for military service:
males age 16–49: 7,431,147
females age 16–49: 7,004,819 (2008 est.)
Manpower fit for military service:
males age 16–49: 4,234,180
females age 16–49: 3,946,685 (2008 est.)
Manpower reaching militarily significant age annually:
males age 16–49: 371,451
females age 16–49: 351,295 (2008 est.)
Military expenditures—percent of GDP: 1.9% (2006 est.)

TRANSNATIONAL ISSUES

Disputes—international:—international: Pakistan, with UN and other international assistance, repatriated 2.3 million Afghan refugees with less than a million still remaining, many at their own choosing; Pakistan has proposed and Afghanistan protests construction of a fence and laying of mines along portions of their border; Coalition and Pakistani forces continue to monitor remote tribal areas to control the border with Afghanistan and stem terrorist and other illegal activities
Refugees and internally displaced persons: internally displaced persons: *IDPs:* 132,246 (mostly Pashtuns and Kuchis displaced in south and west due to drought and instability) (2007)
Illicit drugs: world's largest producer of opium; cultivation dropped 48% to 107,400 hectares in 2005; better weather and lack of widespread disease returned opium yields to normal levels, meaning potential opium production declined by only 10% to 4,475 metric tons; if the entire poppy crop were processed, it is estimated that 526 metric tons of heroin could be processed; many narcotics-processing labs throughout the country; drug trade is a source of instability and some antigovernment groups profit from the trade; significant domestic use of opiates; 80–90% of the heroin consumed in Europe comes from Afghan opium; vulnerable to narcotics money laundering through informal financial networks; source of hashish

AKROTIRI

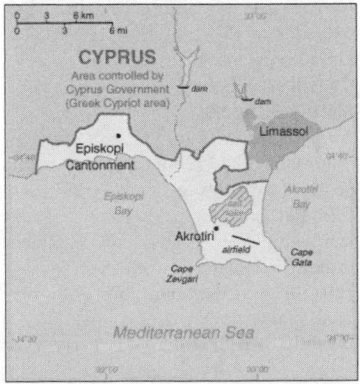

Background: By terms of the 1960 Treaty of Establishment that created the independent Republic of Cyprus, the UK retained full sovereignty and jurisdiction over two areas of almost 254 square kilometers—Akrotiri and Dhekelia. The southernmost and smallest of these is the Akrotiri Sovereign Base Area, which is also referred to as the Western Sovereign Base Area.

GEOGRAPHY

Location: Eastern Mediterranean, peninsula on the southwest coast of Cyprus
Geographic coordinates: 34 37 N, 32 58 E
Map references: Middle East
Area:
total: 123 sq km
note: includes a salt lake and wetlands
Area—comparative: about 0.7 times the size of Washington, DC

Land boundaries:
total: 47.4 km
border countries: Cyprus 47.4 km
Coastline: 56.3 km
Climate: temperate; Mediterranean with hot, dry summers and cool winters
Environment—current issues: shooting around the salt lake; note—breeding place for loggerhead and green turtles; only remaining colony of griffon vultures is on the base
Geography—note: British extraterritorial rights also extended to several small off-post sites scattered across Cyprus; of the Sovereign Base Area land, 60% is privately owned and farmed, 20% is owned by the Ministry of Defense, and 20% is SBA Crown land

INTRODUCTION

5

PEOPLE

Population: approximately 15,700 live on the Sovereign Base Areas of Akrotiri and Dhekelia including 7,700 Cypriots, 3,600 Service and UK-based contract personnel, and 4,400 dependents

Languages: English, Greek

GOVERNMENT

Country name:

conventional long form: Akrotiri Sovereign Base Area

conventional short form: Akrotiri

Dependency status: a special form of UK overseas territory; administered by an administrator who is also the Commander, British Forces Cyprus

Capital: *name:* Episkopi Cantonment (base administrative center for Akrotiri and Dhekelia)

geographic coordinates: 34 40 N, 32 51 E

time difference: UTC+2 (7 hours ahead of Washington, DC during Standard Time)

daylight saving time: +1hr, begins last Sunday in March; ends last Sunday in October

Constitution: Sovereign Base Areas of Akrotiri and Dhekelia Order in Council 1960, effective 16 August 1960, functions as a basic legal document

Legal system: the Sovereign Base Area Administration has its own court system to deal with civil and criminal matters; laws applicable to the Cypriot population are, as far as possible, the same as the laws of the Republic of Cyprus

Executive branch:

chief of state: Queen ELIZABETH II (since 6 February 1952)

head of government: Administrator Air Vice-Marshal Richard LACEY (since 26 April 2006); note—reports to the British Ministry of Defense

elections: none; the monarch is hereditary; the administrator is appointed by the monarch

Diplomatic representation in the US: none (overseas territory of the UK)

Diplomatic representation from the US: none (overseas territory of the UK)

Flag description: the flag of the UK is used

ECONOMY

Economy—overview: Economic activity is limited to providing services to the military and their families located in Akrotiri. All food and manufactured goods must be imported.

Currency (code): euro (EUR) adopted 1 January 2008; note—the Cypriot pound (CYP) formerly used

Exchange rates: Cypriot pounds per US dollar—0.4286 (2007), 0.46019 (2006), 0.4641 (2005), 0.4686 (2004), 0.5174 (2003)

COMMUNICATIONS

Radio broadcast stations: AM NA, FM 1, shortwave NA (British Forces Broadcasting Service (BFBS) provides Radio 1 and Radio 2 service to Akrotiri, Dhekelia, and Nicosia) (2006)

Television broadcast stations: 0 (British Forces Broadcasting Service (BFBS) provides multi-channel satellite service to Akrotiri, Dhekelia, and Nicosia) (2006)

MILITARY

Military—note: Akrotiri has a full RAF base, Headquarters for British Forces on Cyprus, and Episkopi Support Unit

ALBANIA

INTRODUCTION

Background: Albania declared its independence from the Ottoman Empire in 1912, but was conquered by Italy in 1939. Communist partisans took over the country in 1944. Albania allied itself first with the USSR (until 1960), and then with China (to 1978). In the early 1990s, Albania ended 46 years of xenophobic Communist rule and established a multiparty democracy. The transition has proven challenging as successive governments have tried to deal with high unemployment, widespread corruption, a dilapidated physical infrastructure, powerful organized crime networks, and combative political opponents. Albania has made progress in its democratic development since first holding multiparty elections in 1991, but deficiencies remain. International observers judged elections to be largely free and fair since the restoration of political stability following the collapse of pyramid schemes in 1997. In the 2005 general elections, the Democratic Party and its allies won a decisive victory on pledges of reducing crime and corruption, promoting economic growth, and decreasing the size of government. The election, and particularly the orderly transition of power, was considered an important step forward. Although Albania's economy continues to grow, the country is still one of the poorest in Europe, hampered by a large informal economy and an inadequate energy and transportation infrastructure. Albania has played a largely helpful role in managing inter-ethnic tensions in southeastern Europe, and is continuing to work toward joining NATO and the EU. Albania, with troops in Iraq and Afghanistan, has been a strong supporter of the global war on terrorism.

GEOGRAPHY

Location: Southeastern Europe, bordering the Adriatic Sea and Ionian Sea, between Greece in the south and Montenegro and Kosovo to the north

Geographic coordinates: 41 00 N, 20 00 E

Map references: Europe

Area:

total: 28,748 sq km

land: 27,398 sq km

water: 1,350 sq km

Area—comparative: slightly smaller than Maryland

Land boundaries:

total: 717 km

border countries: Greece 282 km, Macedonia 151 km, Montenegro 172 km, Kosovo 112 km

Coastline: 362 km

Maritime claims:

territorial sea: 12 nm

continental shelf: 200-m depth or to the depth of exploitation

Climate: mild temperate; cool, cloudy, wet winters; hot, clear, dry summers; interior is cooler and wetter

Terrain: mostly mountains and hills; small plains along coast

Elevation extremes:

lowest point: Adriatic Sea 0 m

highest point: Maja e Korabit (Golem Korab) 2,764 m

Natural resources: petroleum, natural gas, coal, bauxite, chromite, copper, iron ore, nickel, salt, timber, hydropower

Land use:
arable land: 20.1%
permanent crops: 4.21%
other: 75.69% (2005)
Irrigated land: 3,530 sq km (2003)
Total renewable water resources: 41.7 cu km (2001)
Freshwater withdrawal (domestic/industrial/agricultural): *total:* 1.71 cu km/yr (27%/11%/62%)
per capita: 546 cu m/yr (2000)
Natural hazards: destructive earthquakes; tsunamis occur along southwestern coast; floods; drought
Environment—current issues: deforestation; soil erosion; water pollution from industrial and domestic effluents
Environment—international agreements: *party to:* Biodiversity, Climate Change, Climate Change-Kyoto Protocol, Desertification, Endangered Species, Hazardous Wastes, Law of the Sea, Ozone Layer Protection, Wetlands
signed, but not ratified: none of the selected agreements
Geography—note: strategic location along Strait of Otranto (links Adriatic Sea to Ionian Sea and Mediterranean Sea)

PEOPLE

Population: 3,619,778 (July 2008 est.)
Age structure:
0–14 years: 23.6% (male 447,126/female 406,757)
15–64 years: 66.9% (male 1,239,819/female 1,180,720)
65 years and over: 9.5% (male 160,241/female 185,115) (2008 est.)
Median age:
total: 29.5 years
male: 28.9 years
female: 30.2 years (2008 est.)
Population growth rate: 0.538% (2008 est.)
Birth rate: 15.22 births/1,000 population (2008 est.)
Death rate: 5.44 deaths/1,000 population (2008 est.)
Net migration rate: -4.41 migrant(s)/1,000 population (2008 est.)
Sex ratio:
at birth: 1.1 male(s)/female
under 15 years: 1.1 male(s)/female
15–64 years: 1.05 male(s)/female
65 years and over: 0.87 male(s)/female
total population: 1.04 male(s)/female (2008 est.)
Infant mortality rate:
total: 19.31 deaths/1,000 live births
male: 19.74 deaths/1,000 live births
female: 18.83 deaths/1,000 live births (2008 est.)
Life expectancy at birth:
total population: 77.78 years
male: 75.12 years

female: 80.71 years (2008 est.)
Total fertility rate: 2.02 children born/woman (2008 est.)
HIV/AIDS—adult prevalence rate: NA
HIV/AIDS—people living with HIV/AIDS: NA
HIV/AIDS—deaths: NA
Nationality:
noun: Albanian(s)
adjective: Albanian
Ethnic groups: Albanian 95%, Greek 3%, other 2% (Vlach, Roma (Gypsy), Serb, Macedonian, Bulgarian) (1989 est.)
note: in 1989, other estimates of the Greek population ranged from 1% (official Albanian statistics) to 12% (from a Greek organization)
Religions: Muslim 70%, Albanian Orthodox 20%, Roman Catholic 10%
note: percentages are estimates; there are no available current statistics on religious affiliation; all mosques and churches were closed in 1967 and religious observances prohibited; in November 1990, Albania began allowing private religious practice
Languages: Albanian (official—derived from Tosk dialect), Greek, Vlach, Romani, Slavic dialects
Literacy:
definition: age 9 and over can read and write
total population: 98.7%
male: 99.2%
female: 98.3% (2001 census)

GOVERNMENT

Country name:
conventional long form: Republic of Albania
conventional short form: Albania
local long form: Republika e Shqiperise
local short form: Shqiperia
former: People's Socialist Republic of Albania
Government type: emerging democracy
Capital: *name:* Tirana (Tirane)
geographic coordinates: 41 19 N, 19 49 E
time difference: UTC+1 (6 hours ahead of Washington, DC during Standard Time)
daylight saving time: +1hr, begins last Sunday in March; ends last Sunday in October
Administrative divisions: 12 counties (qarqe, singular—qark); Berat, Diber, Durres, Elbasan, Fier, Gjirokaster, Korce, Kukes, Lezhe, Shkoder, Tirane, Vlore
Independence: 28 November 1912 (from the Ottoman Empire)
National holiday: Independence Day, 28 November (1912)
Constitution: adopted by popular referendum on 22 November 1998; promulgated 28 November 1998

Legal system: has a civil law system; has not accepted compulsory ICJ jurisdiction; has accepted jurisdiction of the International Criminal Court for its citizens
Suffrage: 18 years of age; universal
Executive branch:
chief of state: President of the Republic Bamir TOPI (since 24 July 2007)
head of government: Prime Minister Sali BERISHA (since 10 September 2005)
cabinet: Council of Ministers proposed by the prime minister, nominated by the president, and approved by parliament
elections: president elected by the People's Assembly for a five-year term (eligible for a second term); four election rounds held between 8 and 20 July 2007 (next election to be held in 2012); prime minister appointed by the president
election results: Bamir TOPI elected president; People's Assembly vote, fourth round (three-fifths majority (84 votes) required): Bamir TOPI 85 votes, Neritan CEKA 5 votes
Legislative branch: unicameral Assembly or Kuvendi (140 seats; 100 members are elected by direct popular vote and 40 by proportional vote to serve four-year terms)
elections: last held 3 July 2005 (next to be held in 2009)
election results: percent of vote by party—NA; seats by party—PD 56, PS 42, PR 11, PSD 7, LSI 5, other 19
Judicial branch: Constitutional Court, Supreme Court (chairman is elected by the People's Assembly for a four-year term), and multiple appeals and district courts
Political parties and leaders: Agrarian Environmentalist Party or PAA [Lufter XHUVELI]; Christian Democratic Party or PDK [Nard NDOKA]; Communist Party of Albania or PKSH [Hysni MILLOSHI]; Democratic Alliance Party or AD [Neritan CEKA]; Democratic Party or PD [Sali BERISHA]; Legality Movement Party or PLL [Ekrem SPAHIA]; Liberal Union Party or BLD [Arjan STAROVA]; Movement for National Development or LZhK [Dashamir SHEHI]; National Front Party (Balli Kombetar) or PBK [Adriatik ALIMADHI]; New Democratic Party or PDR [Genc POLLO]; Party of National Unity or PUK [Idajet BEQIRI]; Republican Party or PR [Fatmir MEDIU]; Social Democracy Party of Albania or PDSSh [Paskal MILO]; Social Democratic Party or PSD [Skender GJINUSHI]; Socialist Movement for Integration or LSI [Ilir META]; Socialist Party or PS [Edi RAMA]; Union for Human Rights Party or PBDNj [Vangjel DULE]

Political pressure groups and leaders: Citizens Advocacy Office [Kreshnik SPAHIU]; Confederation of Trade Unions of Albania or KSSH [Kastriot MUCO]; Front for Albanian National Unification or FBKSH [Gafur ADILI]; Mjaft Movement; Omonia [Jani JANI]; Union of Independent Trade Unions of Albania or BSPSH [Gezim KALAJA]

International organization participation: BSEC, CE, CEI, EAPC, EBRD, FAO, IAEA, IBRD, ICAO, ICCt, ICRM, IDA, IDB, IFAD, IFC, IFRCS, ILO, IMF, IMO, Interpol, IOC, IOM, IPU, ISO (correspondent), ITU, ITUC, MIGA, OIC, OIF, OPCW, OSCE, PFP, SECI, UN, UNCTAD, UNESCO, UNIDO, UNOMIG, UNWTO, UPU, WCO, WFTU, WHO, WIPO, WMO, WTO

Diplomatic representation in the US:
chief of mission: Ambassador Aleksander SALLABANDA
chancery: 2100 S Street NW, Washington, DC 20008
telephone: [1] (202) 223-4942
FAX: [1] (202) 628-7342

Diplomatic representation from the US:
chief of mission: Ambassador Dr. John L. WITHERS, II
embassy: Rruga e Elbasanit, Labinoti #103, Tirana
mailing address: US Department of State, 9510 Tirana Place, Dulles, VA 20189-9510
telephone: [355] (4) 247285
FAX: [355] (4) 232222

Flag description: red with a black two-headed eagle in the center

ECONOMY

Economy—overview: Lagging behind its Balkan neighbors, Albania is making the difficult transition to a more modern open-market economy. The government has taken measures to curb violent crime, and recently adopted a fiscal reform package aimed at reducing the large gray economy and attracting foreign investment. The economy is bolstered by annual remittances from abroad of $600–$800 million, mostly from Albanians residing in Greece and Italy; this helps offset the towering trade deficit. Agriculture, which accounts for more than one-fifth of GDP, is held back because of lack of modern equipment, unclear property rights, and the prevalence of small, inefficient plots of land. Energy shortages and antiquated and inadequate infrastructure contribute to Albania's poor business environment, which make it difficult to attract and sustain foreign investment. The completion of a new thermal power plant near Vlore

and improved transmission line between Albania and Montenegro will help relieve the energy shortages. Also, the government is moving slowly to improve the poor national road and rail network, a long-standing barrier to sustained economic growth. On the positive side, macroeconomic growth was strong in 2003–07 and inflation is low and stable.

GDP (purchasing power parity): $19.92 billion
note: Albania has a large gray economy that may be as large as 50% of official GDP (2007 est.)

GDP (official exchange rate): $10.62 billion (2007 est.)

GDP—real growth rate: 6% (2007 est.)

GDP—per capita (PPP): $6,300 (2007 est.)

GDP—composition by sector:
agriculture: 21.2%
industry: 20.1%
services: 58.7% (2007 est.)

Labor force: 1.09 million (not including 352,000 emigrant workers) (September 2006 est.)

Labor force—by occupation: *agriculture:* 58%
industry: 15%
services: 27% (September 2006 est.)

Unemployment rate: 13% official rate, but may exceed 30% due to preponderance of near-subsistence farming (2007 est.)

Population below poverty line: 25% (2004 est.)

Household income or consumption by percentage share:
lowest 10%: 3.4%
highest 10%: 24.4% (2004)

Distribution of family income—Gini index: 26.7 (2005)

Inflation rate (consumer prices): 2.9% (2007 est.)

Investment (gross fixed): 23.3% of GDP (2007 est.)

Budget:
revenues: $2.786 billion
expenditures: $3.159 billion (2007 est.)

Public debt: 52.5% of GDP (2007 est.)

Agriculture—products: wheat, corn, potatoes, vegetables, fruits, sugar beets, grapes; meat, dairy products

Industries: food processing, textiles and clothing; lumber, oil, cement, chemicals, mining, basic metals, hydropower

Industrial production growth rate: 2% (2007 est.)

Electricity—production: 5.385 billion kWh (2005)

Electricity—production by source:
fossil fuel: 2.9%
hydro: 97.1%
nuclear: 0%
other: 0% (2001)

Electricity—consumption: 3.323 billion kWh (2005)

Electricity—exports: 300 million kWh (2005)

Electricity—imports: 371 million kWh (2005)

Oil—production: 7,006 bbl/day (2005 est.)

Oil—consumption: 29,000 bbl/day (2005 est.)

Oil—exports: 1,240 bbl/day (2004 est.)

Oil—imports: 21,600 bbl/day (2005 est.)

Oil—proved reserves: 198.1 million bbl (1 January 2006 est.)

Natural gas—production: 28.77 million cu m (2005 est.)

Natural gas—consumption: 28.77 million cu m (2005 est.)

Natural gas—exports: 0 cu m (2005 est.)

Natural gas—imports: 0 cu m (2005)

Natural gas—proved reserves: 814.7 million cu m (1 January 2006 est.)

Current account balance: -$877 million (2007 est.)

Exports: $1.089 billion f.o.b. (2007 est.)

Exports—commodities: textiles and footwear; asphalt, metals and metallic ores, crude oil; vegetables, fruits, tobacco

Exports—partners: Italy 67.6%, Serbia and Montenegro 5.8%, Greece 5.4% (2006)

Imports: $3.891 billion f.o.b. (2007 est.)

Imports—commodities: machinery and equipment, foodstuffs, textiles, chemicals

Imports—partners: Italy 31.9%, Greece 17.7%, Turkey 8.1%, Germany 5.7% (2006)

Economic aid—recipient: ODA: $318.7 million
note: top donors were Italy, EU, Germany (2005 est.)

Reserves of foreign exchange and gold: $2.248 billion (31 December 2007 est.)

Debt—external: $1.55 billion (2004)

Market value of publicly traded shares: $NA

Currency (code): lek (ALL)
note: the plural of lek is leke

Currency code: ALL

Exchange rates: leke per US dollar— 92.668 (2007), 98.384 (2006), 102.649 (2005), 102.78 (2004), 121.863 (2003)

Fiscal year: calendar year

COMMUNICATIONS

Telephones—main lines in use: 353,600 (2005)

Telephones—mobile cellular: 1.53 million (2005)

Telephone system:
general assessment: despite new investment in fixed lines, the density of main lines remains low with roughly 10 lines per 100 people; cellular telephone use is

widespread and generally effective; combined fixed line and mobile telephone density is approximately 60 telephones per 100 persons

domestic: offsetting the shortage of fixed line capacity, mobile phone service has been available since 1996; by 2003 two companies were providing mobile services at a greater density than some of Albania's neighbors; Internet broadband services initiated in 2005; internet cafes are popular in Tirana and have started to spread outside the capital

international: country code—355; submarine cable provides connectivity to Italy, Croatia, and Greece; the Trans-Balkan Line, a combination submarine cable and land fiber-optic system, provides additional connectivity to Bulgaria, Macedonia, and Turkey; international traffic carried by fiber-optic cable and, when necessary, by microwave radio relay from the Tirana exchange to Italy and Greece (2007)

Radio broadcast stations: AM 13, FM 46, shortwave 1 (2005)

Radios: 1 million (2001)

Television broadcast stations: 65 (3 national, 62 local); 2 cable networks (2005)

Televisions: 700,000 (2001)

Internet country code: .al

Internet hosts: 852 (2007)

Internet Service Providers (ISPs): 10 (2001)

Internet users: 471,200 (2006)

TRANSPORTATION

Airports: 11 (2007)

Airports—with paved runways:

total: 3

2,438 to 3,047 m: 3 (2007)

Airports—with unpaved runways:

total: 8

over 3,047 m: 1

1,524 to 2,437 m: 2

914 to 1,523 m: 1

under 914 m: 4 (2007)

Heliports: 1 (2007)

Pipelines: gas 339 km; oil 207 km (2007)

Railways:

total: 447 km

standard gauge: 447 km 1.435-m gauge (2006)

Roadways:

total: 18,000 km

paved: 7,020 km

unpaved: 10,980 km (2002)

Waterways: 43 km (2007)

Merchant marine:

total: 24 ships (1000 GRT or over) 56,550 GRT/85,521 DWT

by type: cargo 23, roll on/roll off 1

foreign-owned: 1 (Turkey 1)

registered in other countries: 3 (Georgia 2, Panama 1) (2007)

Ports and terminals: Durres, Sarande, Shengjin, Vlore

MILITARY

Military branches: Land Forces Command (Army), Naval Forces Command, Air Defense Command, General Staff Headquarters (includes Logistics Command, Training and Doctrine Command) (2007)

Military service age and obligation: 19 years of age (2004)

Manpower available for military service:

males age 16–49: 944,592

females age 16–49: 908,527 (2008 est.)

Manpower fit for military service:

males age 16–49: 798,454

females age 16–49: 767,143 (2008 est.)

Manpower reaching militarily significant age annually:

males age 16–49: 36,340

females age 16–49: 33,077 (2008 est.)

Military expenditures—percent of GDP: 1.49% (2005 est.)

TRANSNATIONAL ISSUES

Disputes—international: the Albanian Government calls for the protection of the rights of ethnic Albanians in neighboring countries, and the peaceful resolution of interethnic disputes; some ethnic Albanian groups in neighboring countries advocate for a "greater Albania," but the idea has little appeal among Albanian nationals; the mass emigration of unemployed Albanians remains a problem for developed countries, chiefly Greece and Italy

Illicit drugs: increasingly active transshipment point for Southwest Asian opiates, hashish, and cannabis transiting the Balkan route and—to a lesser extent—cocaine from South America destined for Western Europe; limited opium and growing cannabis production; ethnic Albanian narcotrafficking organizations active and expanding in Europe; vulnerable to money laundering associated with regional trafficking in narcotics, arms, contraband, and illegal aliens

ALGERIA

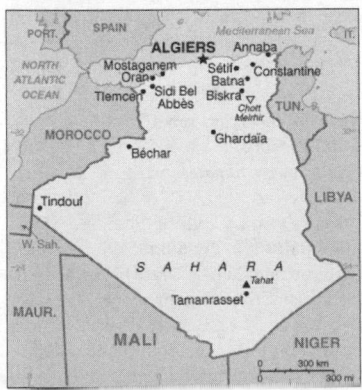

INTRODUCTION

Background: After more than a century of rule by France, Algerians fought through much of the 1950s to achieve independence in 1962. Algeria's primary political party, the National Liberation Front (FLN), has dominated politics ever since. Many Algerians in the subsequent generation were not satisfied, however, and moved to counter the FLN's centrality in Algerian politics. The surprising first round success of the Islamic Salvation Front (FIS) in the December 1991 balloting spurred the Algerian army to intervene and postpone the second round of elections to prevent what the secular elite feared would be an extremist-led government from assuming power. The army began a crackdown on the FIS that spurred FIS supporters to begin attacking government targets. The government later allowed elections featuring pro-government and moderate religious-based parties, but did not appease the activists who progressively widened their attacks. The fighting escalated into an insurgency, which saw intense fighting between 1992–98 and which resulted in over 100,000 deaths—many attributed to indiscriminate massacres of villagers by extremists. The government gained the upper hand by the late-1990s and FIS's armed wing, the Islamic Salvation Army, disbanded in January 2000. However, small numbers of armed militants persist in confronting government forces and conducting ambushes and occasional attacks on villages. The army placed Abdelaziz BOUTEFLIKA in the presidency in 1999 in a fraudulent election but claimed neutrality in his 2004 landslide reelection victory. Longstanding problems continue to face BOUTEFLIKA in his second term, including the ethnic minority Berbers' ongoing autonomy campaign, large-scale unemployment, a shortage of housing, unreliable electrical and water supplies, government inefficiencies and corruption, and the continuing activities of extremist militants. The 2006 merger of the Salafist Group for Preaching and Combat (GSPC) with al-Qaida (followed by a name change to al-Qaida in the Lands of the Islamic Maghreb) signaled an increase in bombings, including high-profile, mass-casualty suicide attacks targeted against the Algerian government and Western interests. Algeria must also diversify its petroleum-based economy, has yielded a large cash reserve but which has not been used to redress Algeria's many social and infrastructure problems.

GEOGRAPHY

Location: Northern Africa, bordering the Mediterranean Sea, between Morocco and Tunisia
Geographic coordinates: 28 00 N, 3 00 E
Map references: Africa
Area:
total: 2,381,740 sq km
land: 2,381,740 sq km
water: 0 sq km
Area—comparative: slightly less than 3.5 times the size of Texas
Land boundaries:
total: 6,343 km
border countries: Libya 982 km, Mali 1,376 km, Mauritania 463 km, Morocco 1,559 km, Niger 956 km, Tunisia 965 km, Western Sahara 42 km
Coastline: 998 km
Maritime claims:
territorial sea: 12 nm
exclusive fishing zone: 32–52 nm
Climate: arid to semiarid; mild, wet winters with hot, dry summers along coast; drier with cold winters and hot summers on high plateau; sirocco is a hot, dust/sand-laden wind especially common in summer
Terrain: mostly high plateau and desert; some mountains; narrow, discontinuous coastal plain
Elevation extremes:
lowest point: Chott Melrhir -40 m
highest point: Tahat 3,003 m
Natural resources: petroleum, natural gas, iron ore, phosphates, uranium, lead, zinc
Land use:
arable land: 3.17%
permanent crops: 0.28%
other: 96.55% (2005)
Irrigated land: 5,690 sq km (2003)
Total renewable water resources: 14.3 cu km (1997)
Freshwater withdrawal (domestic/ industrial/agricultural): *total:* 6.07 cu km/yr (22%/13%/65%)
per capita: 185 cu m/yr (2000)
Natural hazards: mountainous areas subject to severe earthquakes; mudslides and floods in rainy season
Environment—current issues: soil erosion from overgrazing and other poor farming practices; desertification; dumping of raw sewage, petroleum refining wastes, and other industrial effluents is leading to the pollution of rivers and coastal waters; Mediterranean Sea, in particular, becoming polluted from oil wastes, soil erosion, and fertilizer runoff; inadequate supplies of potable water
Environment—international agreements: *party to:* Biodiversity, Climate Change, Climate Change-Kyoto Protocol, Desertification, Endangered Species, Environmental Modification, Hazardous Wastes, Law of the Sea, Ozone Layer Protection, Ship Pollution, Wetlands
signed, but not ratified: none of the selected agreements
Geography—note: second-largest country in Africa (after Sudan)

PEOPLE

Population: 33,769,669 (July 2008 est.)
Age structure:
0–14 years: 26.3% (male 4,528,919/ female 4,349,746)
15–64 years: 68.7% (male 11,699,701/ female 11,509,619)
65 years and over: 5% (male 779,467/ female 902,217) (2008 est.)
Median age:
total: 26 years
male: 25.8 years
female: 26.2 years (2008 est.)
Population growth rate: 1.209% (2008 est.)
Birth rate: 17.03 births/1,000 population (2008 est.)
Death rate: 4.62 deaths/1,000 population (2008 est.)
Net migration rate: -0.31 migrant(s)/ 1,000 population (2008 est.)
Sex ratio:
at birth: 1.05 male(s)/female
under 15 years: 1.04 male(s)/female
15–64 years: 1.02 male(s)/female
65 years and over: 0.86 male(s)/female
total population: 1.01 male(s)/female (2008 est.)
Infant mortality rate:
total: 28.75 deaths/1,000 live births
male: 31.95 deaths/1,000 live births
female: 25.39 deaths/1,000 live births (2008 est.)
Life expectancy at birth:
total population: 73.77 years
male: 72.13 years
female: 75.49 years (2008 est.)
Total fertility rate: 1.82 children born/woman (2008 est.)
HIV/AIDS—adult prevalence rate: 0.1%; note—no country specific models provided (2001 est.)
HIV/AIDS—people living with HIV/AIDS: 9,100 (2003 est.)
HIV/AIDS—deaths: fewer than 500 (2003 est.)
Nationality:
noun: Algerian(s)
adjective: Algerian
Ethnic groups: Arab-Berber 99%, European less than 1%
note: almost all Algerians are Berber in origin, not Arab; the minority who identify themselves as Berber live mostly in the mountainous region of Kabylie east of Algiers; the Berbers are also Muslim but identify with their Berber rather than Arab cultural heritage; Berbers have long agitated, sometimes violently, for autonomy; the government is unlikely to grant autonomy but has offered to begin sponsoring teaching Berber language in schools
Religions: Sunni Muslim (state religion) 99%, Christian and Jewish 1%
Languages: Arabic (official), French, Berber dialects
Literacy:
definition: age 15 and over can read and write
total population: 69.9%
male: 79.6%
female: 60.1% (2002 est.)

GOVERNMENT

Country name:
conventional long form: People's Democratic Republic of Algeria
conventional short form: Algeria
local long form: Al Jumhuriyah al Jaza'iriyah ad Dimuqratiyah ash Sha'biyah
local short form: Al Jaza'ir
Government type: republic
Capital: *name:* Algiers
geographic coordinates: 36 45 N, 3 03 E
time difference: UTC+1 (6 hours ahead of Washington, DC during Standard Time)
Administrative divisions: 48 provinces (wilayat, singular—wilaya); Adrar, Ain Defla, Ain Temouchent, Alger, Annaba, Batna, Bechar, Bejaia, Biskra, Blida, Bordj Bou Arreridj, Bouira, Boumerdes, Chlef, Constantine, Djelfa, El Bayadh, El Oued, El Tarf, Ghardaia, Guelma, Illizi, Jijel, Khenchela, Laghouat, Mascara, Medea, Mila, Mostaganem, M'Sila, Naama, Oran, Ouargla, Oum el Bouaghi, Relizane, Saida, Setif, Sidi Bel Abbes, Skikda, Souk Ahras,

Tamanghasset, Tebessa, Tiaret, Tindouf, Tipaza, Tissemsilt, Tizi Ouzou, Tlemcen
Independence: 5 July 1962 (from France)
National holiday: Revolution Day, 1 November (1954)
Constitution: 8 September 1963; revised 19 November 1976, effective 22 November 1976; revised 3 November 1988, 23 February 1989, and 28 November 1996
Legal system: socialist, based on French and Islamic law; judicial review of legislative acts in ad hoc Constitutional Council composed of various public officials, including several Supreme Court justices; has not accepted compulsory ICJ jurisdiction
Suffrage: 18 years of age; universal
Executive branch:
chief of state: President Abdelaziz BOUTEFLIKA (since 28 April 1999)
head of government: Prime Minister Abdelaziz BELKHADEM
cabinet: Cabinet of Ministers appointed by the president
elections: president elected by popular vote for a five-year term (eligible for a second term); election last held 8 April 2004 (next to be held in April 2009); prime minister appointed by the president
election results: Abdelaziz BOUTEFLIKA reelected president for second term; percent of vote—Abdelaziz BOUTEFLIKA 85%, Ali BENFLIS 6.4%, Abdellah DJABALLAH 5%
Legislative branch: bicameral Parliament consists of the National People's Assembly or Al-Majlis Al-Shabi Al-Watani (389 seats; members elected by popular vote to serve five-year terms) and the Council of Nations (Senate) (144 seats; one-third of the members appointed by the president, two-thirds elected by indirect vote; to serve six-year terms; the constitution requires half the council to be renewed every three years)
elections: National People's Assembly—last held 17 May 2007 (next to be held in 2012); Council of Nations (Senate)—last held 28 December 2006 (next to be held in 2009)
election results: National People's Assembly—percent of vote by party—NA; seats by party—FLN 136, RND 61, MSP 52, PT 26, RCD 19, FNA 13, other 49, independents 33; Council of Nations—percent of vote by party—NA; seats by party—FLN 29, RND 12, MSP 3, RCD 1, independents 3, presidential appointees (unknown affiliation) 24; note—Council seating reflects the number of replaced council members rather than the whole Council

Judicial branch: Supreme Court
Political parties and leaders: Ahd 54 [Ali Fauzi REBAINE]; Algerian National Front or FNA [Moussa TOUATI]; Islamic Salvation Front or FIS (outlawed April 1992) [Ali BELHADJ, Dr. Abassi MADANI, Rabeh KEBIR]; National Democratic Rally (Rassemblement National Democratique) or RND [Ahmed OUYAHIA]; National Entente Movement or MEN [Ali BOUK-HAZNA]; National Liberation Front or FLN [Abdelaziz BELKHADEM, secretary general]; National Reform Movement or Islah (formerly MRN) [Mohamed BOULAHIA]; National Renewal Party or PRA [Mohamed BENSMAIL]; Rally for Culture and Democracy or RCD [Said SADI]; Renaissance Movement or EnNahda Movement [Fatah RABEI]; Socialist Forces Front or FFS [Hocine Ait AHMED]; Social Liberal Party or PSL [Ahmed KHELIL]; Society of Peace Movement or MSP [Boudjerra SOLTANI]; Workers Party or PT [Louisa HANOUNE]
note: a law banning political parties based on religion was enacted in March 1997
Political pressure groups and leaders: The Algerian Human Rights League or LADDH [Hocine ZEHOUANE]; SOS Disparus [Nacera DUTOUR]; Somoud [Ali MERABET]
International organization participation: ABEDA, AfDB, AFESD, AMF, AMU, AU, BIS, FAO, G-15, G-24, G-77, IAEA, IBRD, ICAO, ICC, ICCt (signatory), ICRM, IDA, IDB, IFAD, IFC, IFRCS, IHO, ILO, IMF, IMO, IMSO, Interpol, IOC, IOM, IPU, ISO, ITSO, ITU, ITUC, LAS, MIGA, NAM, OAPEC, OAS (observer), OIC, OPCW, OPEC, OSCE (partner), UN, UNCTAD, UNESCO, UNHCR, UNIDO, UNMEE, UNWTO, UPU, WCO, WHO, WIPO, WMO, WTO (observer)
Diplomatic representation in the US:
chief of mission: Ambassador Amine KHERBI
chancery: 2118 Kalorama Road NW, Washington, DC 20008
telephone: [1] (202) 265-2800
FAX: [1] (202) 667-2174
Diplomatic representation from the US:
chief of mission: Ambassador Robert S. FORD
embassy: 5 Chemin Cheikh Bachir, El-Ibrahimi, El-Biar 16000 Algiers
mailing address: B. P. 408, Alger-Gare, 16030 Algiers
telephone: [213] 70-08-2000
FAX: [213] 21-60-7355

Flag description: two equal vertical bands of green (hoist side) and white; a red, five-pointed star within a red crescent centered over the two-color boundary
note: the crescent, star, and color green are traditional symbols of Islam (the state religion)

ECONOMY

Economy—overview: The hydrocarbons sector is the backbone of the economy, accounting for roughly 60% of budget revenues, 30% of GDP, and over 95% of export earnings. Algeria has the eighth-largest reserves of natural gas in the world and is the fourth-largest gas exporter; it ranks 14th in oil reserves. Sustained high oil prices in recent years have helped improve Algeria's financial and macroeconomic indicators. Algeria is running substantial trade surpluses and building up record foreign exchange reserves. Algeria has decreased its external debt to less than 10% of GDP after repaying its Paris Club and London Club debt in 2006. Real GDP has risen due to higher oil output and increased government spending. The government's continued efforts to diversify the economy by attracting foreign and domestic investment outside the energy sector, however, has had little success in reducing high unemployment and improving living standards. Structural reform within the economy, such as development of the banking sector and the construction of infrastructure, moves ahead slowly hampered by corruption and bureaucratic resistance.
GDP (purchasing power parity): $224.7 billion (2007 est.)
GDP (official exchange rate): $131.6 billion (2007 est.)
GDP—real growth rate: 4.6% (2007 est.)
GDP—per capita (PPP): $6,500 (2007 est.)
GDP—composition by sector:
agriculture: 8.2%
industry: 61.4%
services: 30.4% (2007 est.)
Labor force: 9.38 million (2007 est.)
Labor force—by occupation: agriculture 14%, industry 13.4%, construction and public works 10%, trade 14.6%, government 32%, other 16% (2003 est.)
Unemployment rate: 13% (2007 est.)
Population below poverty line: 25% (2005 est.)
Household income or consumption by percentage share:
lowest 10%: 2.8%
highest 10%: 26.8% (1995)
Distribution of family income—Gini index: 35.3 (1995)

Inflation rate (consumer prices): 3.7% (2007 est.)

Investment (gross fixed): 24.6% of GDP (2007 est.)

Budget:
revenues: $56.36 billion
expenditures: $40.8 billion (2007 est.)

Public debt: 19% of GDP (2007 est.)

Agriculture—products: wheat, barley, oats, grapes, olives, citrus, fruits; sheep, cattle

Industries: petroleum, natural gas, light industries, mining, electrical, petrochemical, food processing

Industrial production growth rate: 5% (2007 est.)

Electricity—production: 31.91 billion kWh (2005 est.)

Electricity—production by source:
fossil fuel: 99.7%
hydro: 0.3%
nuclear: 0%
other: 0% (2001)

Electricity—consumption: 27.52 billion kWh (2005 est.)

Electricity—exports: 275 million kWh (2005 est.)

Electricity—imports: 359 million kWh (2005 est.)

Oil—production: 2.09 million bbl/day (2005 est.)

Oil—consumption: 250,000 bbl/day (2005 est.)

Oil—exports: 1.724 million bbl/day (2004 est.)

Oil—imports: 12,390 bbl/day (2004 est.)

Oil—proved reserves: 14.68 billion bbl (2007 est.)

Natural gas—production: 84.4 billion cu m (2005 est.)

Natural gas—consumption: 21.8 billion cu m (2005 est.)

Natural gas—exports: 62.6 billion cu m (2005 est.)

Natural gas—imports: 0 cu m (2005)

Natural gas—proved reserves: 4.359 trillion cu m (1 January 2006 est.)

Current account balance: $30.58 billion (2007 est.)

Exports: $59.52 billion f.o.b. (2007 est.)

Exports—commodities: petroleum, natural gas, and petroleum products 97%

Exports—partners: US 27.2%, Italy 17%, Spain 9.8%, France 8.8%, Canada 8.1%, Belgium 4.3% (2006)

Imports: $27.52 billion f.o.b. (2007 est.)

Imports—commodities: capital goods, foodstuffs, consumer goods

Imports—partners: France 22.1%, Italy 8.6%, China 8.6%, Germany 5.9%, Spain 5.9%, US 4.8%, Turkey 4.5% (2006)

Economic aid—recipient: $370.6 million (2005 est.)

Reserves of foreign exchange and gold: $110.6 billion (31 December 2007 est.)

Debt—external: $4.392 billion (31 December 2007 est.)

Stock of direct foreign investment—at home: $10.63 billion (2007 est.)

Stock of direct foreign investment—abroad: $851 million (2007 est.)

Market value of publicly traded shares: $NA

Currency (code): Algerian dinar (DZD)

Currency code: DZD

Exchange rates: Algerian dinars per US dollar—69.9 (2007), 72.647 (2006), 73.276 (2005), 72.061 (2004), 77.395 (2003)

Fiscal year: calendar year

COMMUNICATIONS

Telephones—main lines in use: 2.841 million (2006)

Telephones—mobile cellular: 20.998 million (2006)

Telephone system:
general assessment: a weak network of fixed-main lines, which remains low at less than 10 telephones per 100 persons, is partially offset by the rapid increase in mobile cellular subscribership; in 2006, combined fixed-line and mobile telephone density surpassed 70 telephones per 100 persons
domestic: privatization of Algeria's telecommunications sector began in 2000; three mobile cellular licenses have been issued and, in 2005, a consortium led by Egypt's Orascom Telecom won a 15-year license to build and operate a fixed-line network in Algeria; the license will allow Orascom to develop high-speed data and other specialized services and contribute to meeting the large unfulfilled demand for basic residential telephony; internet broadband services began in 2003 with approximately 200,000 subscribers in 2006
international: country code—213; landing point for the SEA-ME-WE-4 fiber-optic submarine cable system that provides links to Europe, the Middle East, and Asia; microwave radio relay to Italy, France, Spain, Morocco, and Tunisia; coaxial cable to Morocco and Tunisia; participant in Medarabtel; satellite earth stations—51 (Intelsat, Intersputnik, and Arabsat) (2007)

Radio broadcast stations: AM 25, FM 1, shortwave 8 (1999)

Radios: 7.1 million (1997)

Television broadcast stations: 46 (plus 216 repeaters) (1995)

Televisions: 3.1 million (1997)

Internet country code: .dz

Internet hosts: 2,077 (2007)

Internet Service Providers (ISPs): 2 (2000)

Internet users: 2.46 million (2006)

TRANSPORTATION

Airports: 150 (2007)

Airports—with paved runways:
total: 52
over 3,047 m: 10
2,438 to 3,047 m: 27
1,524 to 2,437 m: 10
914 to 1,523 m: 4
under 914 m: 1 (2007)

Airports—with unpaved runways:
total: 98
2,438 to 3,047 m: 3
1,524 to 2,437 m: 26
914 to 1,523 m: 44
under 914 m: 25 (2007)

Heliports: 2 (2007)

Pipelines: condensate 1,532 km; gas 13,861 km; liquid petroleum gas 2,408 km; oil 6,878 km (2007)

Railways:
total: 3,973 km
standard gauge: 2,888 km 1.435-m gauge (283 km electrified)
narrow gauge: 1,085 km 1.055-m gauge (2006)

Roadways:
total: 108,302 km
paved: 76,028 km
unpaved: 32,274 km (2004)

Merchant marine:
total: 35 ships (1000 GRT or over) 694,686 GRT/707,251 DWT
by type: bulk carrier 6, cargo 8, chemical tanker 2, liquefied gas 9, passenger/cargo 3, petroleum tanker 4, roll on/roll off 2, specialized tanker 1
foreign-owned: 12 (UK 12) (2007)

Ports and terminals: Algiers, Annaba, Arzew, Bejaia, Djendjene, Jijel, Mostaganem, Oran, Skikda

MILITARY

Military branches: National Popular Army (ANP; includes Land Forces), Algerian National Navy (MRA), Air Force (QJJ), Territorial Air Defense Force (2005)

Military service age and obligation: 19–30 years of age for compulsory military service; conscript service obligation—18 months (6 months basic training, 12 months civil projects) (2006)

Manpower available for military service:
males age 16–49: 9,736,757
females age 16–49: 9,590,978 (2008 est.)

Manpower fit for military service:
males age 16–49: 8,141,864
females age 16–49: 8,215,895 (2008 est.)

Manpower reaching militarily significant age annually:
males age 16–49: 374,365
females age 16–49: 360,942 (2008 est.)

Military expenditures—percent of GDP: 3.3% (2006)

TRANSNATIONAL ISSUES

Disputes—international: Algeria supports the Polisario Front exiled in Algeria and who represent the Sahrawi Arab Democratic Republic; Algeria rejects Moroccan administration of Western Sahara; most of the approximately 90,000 Western Saharan Sahrawi refugees are sheltered in camps in Tindouf, Algeria; Algeria's border with Morocco remains an irritant to bilateral relations, each nation accusing the other of harboring militants and arms smuggling; Algeria remains concerned about armed bandits operating throughout the Sahel who sometimes destabilize southern Algerian towns; dormant disputes include Libyan claims of about 32,000 sq km still reflected on its maps of southeastern Algeria and the FLN's assertions of a claim to Chirac Pastures in southeastern Morocco

Refugees and internally displaced persons: *refugees (country of origin):* 90,000 (Western Saharan Sahrawi, mostly living in Algerian-sponsored camps in the southwestern Algerian town of Tindouf) *IDPs:* undetermined (civil war during 1990s) (2007)

Trafficking in persons: *current situation:* Algeria is a transit and destination country for men, women, and children from sub-Saharan Africa and Asia trafficked for forced labor and sexual exploitation; many victims willingly migrate to Algeria en route to European countries with the help of smugglers, where they are often forced into prostitution, labor, and begging in order to pay off their smuggling debt; some Algerian children are reportedly trafficked within the country for domestic servitude *tier rating:* Tier 3—Algeria does not adequately identify trafficking victims among illegal immigrants; the government did not take serious law enforcement actions to punish traffickers who force women into commercial sexual exploitation or men into involuntary servitude; the government reported no investigations of trafficking of children for domestic servitude or improvements in protection services for victims of trafficking

AMERICAN SAMOA

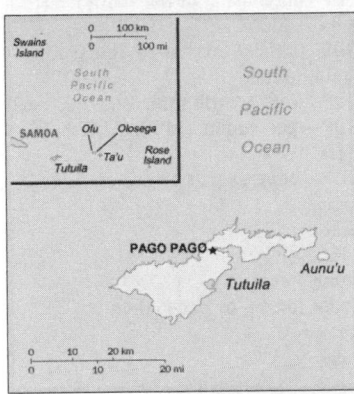

INTRODUCTION

Background: Settled as early as 1000 B.C., Samoa was "discovered" by European explorers in the 18th century. International rivalries in the latter half of the 19th century were settled by an 1899 treaty in which Germany and the US divided the Samoan archipelago. The US formally occupied its portion—a smaller group of eastern islands with the excellent harbor of Pago Pago—the following year.

GEOGRAPHY

Location: Oceania, group of islands in the South Pacific Ocean, about half way between Hawaii and New Zealand

Geographic coordinates: 14 20 S, 170 00 W

Map references: Oceania

Area:
total: 199 sq km
land: 199 sq km
water: 0 sq km

note: includes Rose Island and Swains Island

Area—comparative: slightly larger than Washington, DC

Land boundaries: 0 km

Coastline: 116 km

Maritime claims:
territorial sea: 12 nm
exclusive economic zone: 200 nm

Climate: tropical marine, moderated by southeast trade winds; annual rainfall averages about 3 m; rainy season (November to April), dry season (May to October); little seasonal temperature variation

Terrain: five volcanic islands with rugged peaks and limited coastal plains, two coral atolls (Rose Island, Swains Island)

Elevation extremes:
lowest point: Pacific Ocean 0 m
highest point: Lata Mountain 964 m

Natural resources: pumice, pumicite

Land use:
arable land: 10%
permanent crops: 15%
other: 75% (2005)

Irrigated land: NA

Natural hazards: typhoons common from December to March

Environment—current issues: limited natural fresh water resources; the water division of the government has spent substantial funds in the past few years to improve water catchments and pipelines

Geography—note: Pago Pago has one of the best natural deepwater harbors in the South Pacific Ocean, sheltered by shape from rough seas and protected by peripheral mountains from high winds; strategic location in the South Pacific Ocean

PEOPLE

Population: 57,496 (July 2008 est.)

Age structure:
0–14 years: 32.5% (male 9,684/female 9,006)
15–64 years: 64.6% (male 19,377/female 17,757)
65 years and over: 2.9% (male 591/female 1,081) (2008 est.)

Median age:
total: 24 years
male: 23.9 years
female: 24.2 years (2008 est.)

Population growth rate: -0.322% (2008 est.)

Birth rate: 21.24 births/1,000 population (2008 est.)

Death rate: 3.24 deaths/1,000 population (2008 est.)

Net migration rate: -21.22 migrant(s)/1,000 population (2008 est.)

Sex ratio:
at birth: 1.06 male(s)/female
under 15 years: 1.08 male(s)/female
15–64 years: 1.09 male(s)/female
65 years and over: 0.55 male(s)/female
total population: 1.06 male(s)/female (2008 est.)

Infant mortality rate:
total: 8.69 deaths/1,000 live births
male: 9.28 deaths/1,000 live births
female: 8.08 deaths/1,000 live births (2008 est.)

Life expectancy at birth:
total population: 76.45 years
male: 72.9 years
female: 80.21 years (2008 est.)

Total fertility rate: 2.98 children born/woman (2008 est.)

HIV/AIDS—adult prevalence rate: NA

HIV/AIDS—people living with HIV/AIDS: NA

HIV/AIDS—deaths: NA

Nationality:

noun: American Samoan(s) (US nationals)

adjective: American Samoan

Ethnic groups: native Pacific islander 92.9%, Asian 2.9%, white 1.2%, mixed 2.8%, other 0.2% (2000 census)

Religions: Christian Congregationalist 50%, Roman Catholic 20%, Protestant and other 30%

Languages: Samoan 90.6% (closely related to Hawaiian and other Polynesian languages), English 2.9%, Tongan 2.4%, other Pacific islander 2.1%, other 2%

note: most people are bilingual (2000 census)

Literacy:

definition: age 15 and over can read and write

total population: 97%

male: 98%

female: 97% (1980 est.)

GOVERNMENT

Country name:

conventional long form: Territory of American Samoa

conventional short form: American Samoa

abbreviation: AS

Dependency status: unincorporated and unorganized territory of the US; administered by the Office of Insular Affairs, US Department of the Interior

Government type: NA

Capital: *name:* Pago Pago

geographic coordinates: 14 16 S, 170 42 W

time difference: UTC-11 (6 hours behind Washington, DC during Standard Time)

Administrative divisions: none (territory of the US); there are no first-order administrative divisions as defined by the US Government, but there are three districts and two islands* at the second order; Eastern, Manu'a, Rose Island*, Swains Island*, Western

Independence: none (territory of the US)

National holiday: Flag Day, 17 April (1900)

Constitution: ratified 2 June 1966, effective 1 July 1967

Legal system: NA

Suffrage: 18 years of age; universal

Executive branch:

chief of state: President George W. BUSH of the US (since 20 January 2001); Vice President Richard B. CHENEY (since 20 January 2001)

head of government: Governor Togiola TULAFONO (since 7 April 2003)

cabinet: Cabinet made up of 12 department directors

elections: under the US Constitution, residents of unincorporated territories, such as American Samoa, do not vote in elections for US president and vice president; however, they may vote in Democratic and Republican presidential primary elections; governor and lieutenant governor elected on the same ticket by popular vote for four-year terms (eligible for a second term); election last held 2 and 16 November 2004 (next to be held in November 2008)

election results: Togiola TULAFONO elected governor; percent of vote—Togiola TULAFONO 55.7%, Afoa Moega LUTU 44.3%

Legislative branch: bicameral Fono or Legislative Assembly consists of the House of Representatives (21 seats; 20 members are elected by popular vote and 1 is an appointed, nonvoting delegate from Swains Island; members serve two-year terms) and the Senate (18 seats; members are elected from local chiefs to serve four-year terms)

elections: House of Representatives—last held 7 November 2006 (next to be held in November 2008); Senate—last held 2 November 2004 (next to be held in November 2008)

election results: House of Representatives—percent of vote by party—NA; seats by party—NA; Senate—percent of vote by party—NA; seats by party—independents 18

note: American Samoa elects one non-voting representative to the US House of Representatives; election last held on 7 November 2006 (next to be held in November 2008); results—Eni F. H. FALEOMAVAEGA reelected as delegate

Judicial branch: High Court (chief justice and associate justices are appointed by the US Secretary of the Interior)

Political parties and leaders: Democratic Party [Oreta M. TOGAFAU]; Republican Party [Tautai A. F. FAALEVAO]

Political pressure groups and leaders: NA

International organization participation: Interpol (subbureau), IOC, SPC, UPU

Diplomatic representation in the US: none (territory of the US)

Diplomatic representation from the US: none (territory of the US)

Flag description: blue, with a white triangle edged in red that is based on the outer side and extends to the hoist side; a brown and white American bald eagle flying toward the hoist side is carrying two traditional Samoan symbols of authority, a staff and a war club

ECONOMY

Economy—overview: American Samoa has a traditional Polynesian economy in which more than 90% of the land is communally owned. Economic activity is strongly linked to the US with which American Samoa conducts most of its commerce. Tuna fishing and tuna processing plants are the backbone of the private sector, with canned tuna the primary export. Transfers from the US Government add substantially to American Samoa's economic well being. Attempts by the government to develop a larger and broader economy are restrained by Samoa's remote location, its limited transportation, and its devastating hurricanes. Tourism is a promising developing sector.

note: as a territory of the US, American Samoa does not treat the US as an external trade partner

GDP (purchasing power parity): $510.1 million (2003 est.)

GDP (official exchange rate): $333.8 million (2005)

GDP—real growth rate: 3% (2003 est.)

GDP—per capita (PPP): $5,800 (2005 est.)

GDP—composition by sector:

agriculture: NA%

industry: NA%

services: NA%

Labor force: 17,630 (2005)

Labor force—by occupation:

agriculture: 34%

industry: 33%

services: 33% (1990)

Unemployment rate: 29.8% (2005)

Population below poverty line: NA%

Household income or consumption by percentage share:

lowest 10%: NA%

highest 10%: NA%

Inflation rate (consumer prices): NA%

Budget:

revenues: $121 million (37% in local revenue and 63% in US grants)

expenditures: $127 million (FY96/97)

Agriculture—products: bananas, coconuts, vegetables, taro, breadfruit, yams, copra, pineapples, papayas; dairy products, livestock

Industries: tuna canneries (largely supplied by foreign fishing vessels), handicrafts

Industrial production growth rate: NA%

Electricity—production: 180 million kWh (2005)

Electricity—production by source:

fossil fuel: 100%

hydro: 0%

nuclear: 0%

other: 0% (2001)

Electricity—consumption: 167.4 million kWh (2005)
Electricity—exports: 0 kWh (2005)
Electricity—imports: 0 kWh (2005)
Oil—production: 0 bbl/day (2005)
Oil—consumption: 4,000 bbl/day (2005 est.)
Oil—exports: 0 bbl/day (2004)
Oil—imports: 3,807 bbl/day (2004)
Oil—proved reserves: 0 bbl (1 January 2006 est.)
Natural gas—production: 0 cu m (2005 est.)
Natural gas—consumption: 0 cu m (2005 est.)
Natural gas—exports: 0 cu m (2005 est.)
Natural gas—imports: 0 cu m (2005)
Natural gas—proved reserves: 0 cu m (1 January 2006 est.)
Exports: $445.6 million (FY04 est.)
Exports—commodities: canned tuna 93% (2004 est.)
Exports—partners: Indonesia 28.2%, India 22.3%, Australia 15.3%, Japan 11.2%, NZ 7.1% (2006)
Imports: $308.8 million (FY04 est.)
Imports—commodities: materials for canneries 56%, food 8%, petroleum products 7%, machinery and parts 6% (2004 est.)
Imports—partners: Australia 66%, Samoa 13.8%, NZ 10.8% (2006)
Economic aid—recipient: important financial support from the US, more than $40 million in 1994
Debt—external: $NA
Currency (code): US dollar (USD)
Currency code: USD
Exchange rates: the US dollar is used
Fiscal year: 1 October—30 September

COMMUNICATIONS

Telephones—main lines in use: 10,400 (2004)
Telephones—mobile cellular: 2,200 (2004)
Telephone system:
general assessment: NA
domestic: good telex, telegraph, facsimile, and cellular telephone services; domestic satellite system with 1 Comsat earth station
international: country code—1-684; satellite earth station—1 (Intelsat-Pacific Ocean)
Radio broadcast stations: AM 2, FM 3, shortwave 0 (2005)

Radios: 57,000 (1997)
Television broadcast stations: 1 (2006)
Televisions: 14,000 (1997)
Internet country code: .as
Internet hosts: 1,824 (2007)
Internet Service Providers (ISPs): 1 (2000)
Internet users: NA

TRANSPORTATION

Airports: 3 (2007)
Airports—with paved runways:
total: 3
over 3,047 m: 1
914 to 1,523 m: 1
under 914 m: 1 (2007)
Roadways:
total: 185 km (2004)
Ports and terminals: Pago Pago

MILITARY

Military—note: defense is the responsibility of the US

TRANSNATIONAL ISSUES

Disputes—international: Tokelau included American Samoa's Swains Island (Olohega) in its 2006 draft constitution

ANDORRA

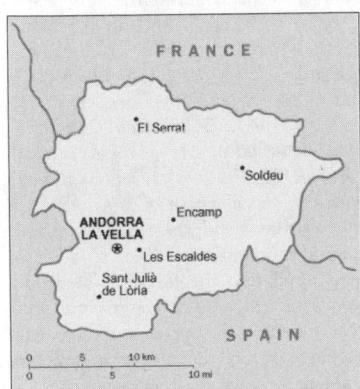

INTRODUCTION

Background: For 715 years, from 1278 to 1993, Andorrans lived under a unique co-principality, ruled by French and Spanish leaders (from 1607 onward, the French chief of state and the Spanish bishop of Urgel). In 1993, this feudal system was modified with the titular heads of state retained, but the government transformed into a parliamentary democracy. Long isolated and impoverished, mountainous Andorra achieved considerable prosperity since World War II through its tourist industry. Many immigrants (legal and illegal) are attracted to the thriving economy with its lack of income taxes.

GEOGRAPHY

Location: Southwestern Europe, between France and Spain
Geographic coordinates: 42 30 N, 1 30 E
Map references: Europe
Area:
total: 468 sq km
land: 468 sq km
water: 0 sq km
Area—comparative: 2.5 times the size of Washington, DC
Land boundaries:
total: 120.3 km
border countries: France 56.6 km, Spain 63.7 km
Coastline: 0 km (landlocked)
Maritime claims: none (landlocked)
Climate: temperate; snowy, cold winters and warm, dry summers
Terrain: rugged mountains dissected by narrow valleys
Elevation extremes:
lowest point: Riu Runer 840 m
highest point: Coma Pedrosa 2,946 m
Natural resources: hydropower, mineral water, timber, iron ore, lead

Land use:
arable land: 2.13%
permanent crops: 0%
other: 97.87% (2005)
Irrigated land: NA
Natural hazards: avalanches
Environment—current issues: deforestation; overgrazing of mountain meadows contributes to soil erosion; air pollution; wastewater treatment and solid waste disposal
Environment—international agreements: *party to:* Biodiversity, Desertification, Hazardous Wastes
signed, but not ratified: none of the selected agreements
Geography—note: landlocked; straddles a number of important crossroads in the Pyrenees

PEOPLE

Population: 72,413 (July 2008 est.)
Age structure:
0–14 years: 14.3% (male 5,404/female 4,966)
15–64 years: 71% (male 26,892/female 24,519)
65 years and over: 14.7% (male 5,246/female 5,386) (2008 est.)
Median age:
total: 42.1 years

male: 42.3 years
female: 41.9 years (2008 est.)
Population growth rate: 0.797% (2008 est.)
Birth rate: 8.23 births/1,000 population (2008 est.)
Death rate: 6.63 deaths/1,000 population (2008 est.)
Net migration rate: 6.37 migrant(s)/1,000 population (2008 est.)
Sex ratio:
at birth: 1.07 male(s)/female
under 15 years: 1.09 male(s)/female
15–64 years: 1.1 male(s)/female
65 years and over: 0.97 male(s)/female
total population: 1.08 male(s)/female (2008 est.)
Infant mortality rate:
total: 4.03 deaths/1,000 live births
male: 4.36 deaths/1,000 live births
female: 3.67 deaths/1,000 live births (2008 est.)
Life expectancy at birth:
total population: 83.53 years
male: 80.63 years
female: 86.63 years (2008 est.)
Total fertility rate: 1.32 children born/woman (2008 est.)
HIV/AIDS—adult prevalence rate: NA
HIV/AIDS—people living with HIV/AIDS: NA
HIV/AIDS—deaths: NA
Nationality:
noun: Andorran(s)
adjective: Andorran
Ethnic groups: Spanish 43%, Andorran 33%, Portuguese 11%, French 7%, other 6% (1998)
Religions: Roman Catholic (predominant)
Languages: Catalan (official), French, Castilian, Portuguese
Literacy:
definition: NA
total population: 100%
male: 100%
female: 100%

GOVERNMENT

Country name:
conventional long form: Principality of Andorra
conventional short form: Andorra
local long form: Principat d'Andorra
local short form: Andorra
Government type: parliamentary democracy (since March 1993) that retains as its chiefs of state a coprincipality; the two princes are the president of France and bishop of Seo de Urgel, Spain, who are represented locally by coprinces' representatives
Capital: *name:* Andorra la Vella
geographic coordinates: 42 30 N, 1 31 E
time difference: UTC+1 (6 hours ahead of

Washington, DC during Standard Time)
daylight saving time: +1hr, begins last Sunday in March; ends last Sunday in October
Administrative divisions: 7 parishes (parroquies, singular—parroquia); Andorra la Vella, Canillo, Encamp, Escaldes-Engordany, La Massana, Ordino, Sant Julia de Loria
Independence: 1278 (formed under the joint suzerainty of the French Count of Foix and the Spanish Bishop of Urgel)
National holiday: Our Lady of Meritxell Day, 8 September (1278)
Constitution: Andorra's first written constitution was drafted in 1991, approved by referendum 14 March 1993, effective 28 April 1993
Legal system: based on French and Spanish civil codes; no judicial review of legislative acts; has not accepted compulsory ICJ jurisdiction
Suffrage: 18 years of age; universal
Executive branch:
chief of state: French Coprince Nicolas SARKOZY (since 16 May 2007); represented by Philippe MASSONI (since 26 July 2002) and Spanish Coprince Bishop Joan Enric VIVES i SICILIA (since 12 May 2003); represented by Nemesi MARQUES i OSTE (since 30 July 2003)
head of government: Executive Council President Albert PINTAT SANTOLARIA (since 27 May 2005)
cabinet: Executive Council or Govern designated by the Executive Council president
elections: Executive Council president elected by the General Council and formally appointed by the coprinces for a four-year term; election last held 24 April 2005 (next to be held in April-May 2009)
election results: Albert PINTAT SANTOLARIA elected executive council president; percent of General Council vote—NA
Legislative branch: unicameral General Council of the Valleys or Consell General de las Valls (28 seats; members are elected by direct popular vote, 14 from a single national constituency and 14 to represent each of the seven parishes; to serve four-year terms)
elections: last held on 24 April 2005 (next to be held in March-April 2009)
election results: percent of vote by party—PLA 41.2%, PS 38.1%, CDA-S21 11%, other 9.7%; seats by party—PLA 14, PS 12, CDA-S21 2
Judicial branch: Tribunal of Judges or Tribunal de Batlles; Tribunal of the Courts or Tribunal de Corts; Supreme Court of Justice of Andorra or Tribunal Superior de Justicia d'Andorra; Supreme

Council of Justice or Consell Superior de la Justicia; Fiscal Ministry or Ministeri Fiscal; Constitutional Tribunal or Tribunal Constitucional
Political parties and leaders: Andorran Democratic Center Party (formerly Democratic Party or PD) and Century 21 or CDA and S21 [Enric TARRADO]; Liberal Party of Andorra or PLA [Albert PINTAT SANTOLARIA] (formerly Liberal Union or UL); Social Democratic Party or PS [Jaume BARTUMEU CASSANY] (formerly part of National Democratic Group or AND)
Political pressure groups and leaders: NA
International organization participation: CE, FAO, ICAO, ICCt, ICRM, IFRCS, Interpol, IOC, IPU, ITU, OIF, OPCW, OSCE, UN, UNCTAD, UNESCO, Union Latina, UNWTO, WCO, WHO, WIPO, WTO (observer)
Diplomatic representation in the US:
chief of mission: Ambassador Carles FONT-ROSSELL
chancery: 2 United Nations Plaza, 25th Floor, New York, NY 10017
telephone: [1] (212) 750-8064
FAX: [1] (212) 750-6630
Diplomatic representation from the US: the US does not have an embassy in Andorra; the US Ambassador to Spain is accredited to Andorra; US interests in Andorra are represented by the Consulate General's office in Barcelona (Spain); mailing address: Paseo Reina Elisenda de Montcada, 23, 08034 Barcelona, Spain; telephone: [34] (3) 280-2227; FAX: [34] (3) 205-5206
Flag description: three equal vertical bands of blue (hoist side), yellow, and red with the national coat of arms centered in the yellow band; the coat of arms features a quartered shield
note: similar to the flags of Chad and Romania, which do not have a national coat of arms in the center, and the flag of Moldova, which does bear a national emblem

ECONOMY

Economy—overview: Tourism, the mainstay of Andorra's tiny, well-to-do economy, accounts for more than 80% of GDP. An estimated 11.6 million tourists visit annually, attracted by Andorra's duty-free status and by its summer and winter resorts. Andorra's comparative advantage has recently eroded as the economies of neighboring France and Spain have been opened up, providing broader availability of goods and lower tariffs. The banking sector, with its partial "tax haven" status, also contributes substantially to the economy.

Agricultural production is limited—only 2% of the land is arable—and most food has to be imported. The principal live-stock activity is sheep raising. Manufacturing output consists mainly of cigarettes, cigars, and furniture. Andorra is a member of the EU Customs Union and is treated as an EU member for trade in manufactured goods (no tariffs) and as a non-EU member for agricultural products.

GDP (purchasing power parity): $2.77 billion (2005)

GDP (official exchange rate): $NA

GDP—real growth rate: 3.5% (2005 est.)

GDP—per capita (PPP): $38,800 (2005)

GDP—composition by sector:
agriculture: NA%
industry: NA%
services: NA%

Labor force: 42,420 (2005)

Labor force—by occupation: *agriculture:* 0.3%
industry: 20.3%
services: 79.4% (2005)

Unemployment rate: 0% (1996 est.)

Population below poverty line: NA%

Household income or consumption by percentage share:
lowest 10%: NA%
highest 10%: NA%

Inflation rate (consumer prices): 3.2% (2005)

Budget:
revenues: $333.5 million

expenditures: $386.6 million (2005)

Agriculture—products: small quantities of rye, wheat, barley, oats, vegetables; sheep

Industries: tourism (particularly skiing), cattle raising, timber, banking, tobacco, furniture

Industrial production growth rate: NA%

Electricity—production: NA kWh

Electricity—production by source: NA

Electricity—consumption: NA kWh

Electricity—exports: NA kWh

Electricity—imports: NA kWh; note—most electricity supplied by Spain and France; Andorra generates a small amount of hydropower

Exports: $148.7 million f.o.b. (2005)

Exports—commodities: tobacco products, furniture

Imports: $1.879 billion (2005)

Imports—commodities: consumer goods, food, electricity

Economic aid—recipient: $0

Debt—external: $NA

Currency (code): euro (EUR)

Currency code: EUR

Exchange rates: euros per US dollar—0.7345 (2007), 0.7964 (2006), 0.8041 (2005), 0.8054 (2004), 0.886 (2003)

Fiscal year: calendar year

COMMUNICATIONS

Telephones—main lines in use: 35,400 (2005)

Telephones—mobile cellular: 64,600 (2005)

Telephone system:
general assessment: NA
domestic: modern system with microwave radio relay connections between exchanges
international: country code—376; land-line circuits to France and Spain

Radio broadcast stations: AM 0, FM 15, shortwave 0 (1998)

Radios: 16,000 (1997)

Television broadcast stations: 0 (1997)

Televisions: 27,000 (1997)

Internet country code: .ad

Internet hosts: 15,486 (2007)

Internet Service Providers (ISPs): 1 (2000)

Internet users: 23,200 (2006)

TRANSPORTATION

Roadways:
total: 270 km

MILITARY

Military branches: no regular military forces, Police Service of Andorra

Manpower available for military service:
males age 16–49: 18,685 (2008 est.)

Manpower fit for military service:
males age 16–49: 14,976 (2008 est.)

Manpower reaching militarily significant age annually:
males age 16–49: 372 (2008 est.)

Military—note: defense is the responsibility of France and Spain

TRANSNATIONAL ISSUES

Disputes—international: none

ANGOLA

INTRODUCTION

Background: Angola is rebuilding its country after the end of a 27-year civil war in 2002. Fighting between the Popular Movement for the Liberation of Angola (MPLA), led by Jose Eduardo DOS SANTOS, and the National Union for the Total Independence of Angola (UNITA), led by Jonas SAV-IMBI, followed independence from Portugal in 1975. Peace seemed imminent in 1992 when Angola held national elections, but UNITA renewed fighting after being beaten by the MPLA at the polls. Up to 1.5 million lives may have been lost—and 4 million people dis-placed—in the quarter century of fighting. SAVIMBI's death in 2002 ended UNITA's insurgency and strength-ened the MPLA's hold on power. President DOS SANTOS has announced legislative elections will be held in September 2008, with presiden-tial elections planned for sometime in 2009.

GEOGRAPHY

Location: Southern Africa, bordering the South Atlantic Ocean, between Namibia and Democratic Republic of the Congo

Geographic coordinates: 12 30 S, 18 30 E

Map references: Africa

Area:
total: 1,246,700 sq km
land: 1,246,700 sq km
water: 0 sq km

Area—comparative: slightly less than twice the size of Texas

Land boundaries:
total: 5,198 km
border countries: Democratic Republic of the Congo 2,511 km (of which 225 km is the boundary of discontiguous Cabinda Province), Republic of the Congo 201 km, Namibia 1,376 km, Zambia 1,110 km

Coastline: 1,600 km

Maritime claims:
territorial sea: 12 nm

17

contiguous zone: 24 nm

exclusive economic zone: 200 nm

Climate: semiarid in south and along coast to Luanda; north has cool, dry season (May to October) and hot, rainy season (November to April)

Terrain: narrow coastal plain rises abruptly to vast interior plateau

Elevation extremes:

lowest point: Atlantic Ocean 0 m

highest point: Morro de Moco 2,620 m

Natural resources: petroleum, diamonds, iron ore, phosphates, copper, feldspar, gold, bauxite, uranium

Land use:

arable land: 2.65%

permanent crops: 0.23%

other: 97.12% (2005)

Irrigated land: 800 sq km (2003)

Total renewable water resources: 184 cu km (1987)

Freshwater withdrawal (domestic/industrial/agricultural): *total:* 0.35 cu km/yr (23%/17%/60%)

per capita: 22 cu m/yr (2000)

Natural hazards: locally heavy rainfall causes periodic flooding on the plateau

Environment—current issues: overuse of pastures and subsequent soil erosion attributable to population pressures; desertification; deforestation of tropical rain forest, in response to both international demand for tropical timber and to domestic use as fuel, resulting in loss of biodiversity; soil erosion contributing to water pollution and siltation of rivers and dams; inadequate supplies of potable water

Environment—international agreements: *party to:* Biodiversity, Climate Change, Climate Change-Kyoto Protocol, Desertification, Law of the Sea, Marine Dumping, Ozone Layer Protection, Ship Pollution

signed, but not ratified: none of the selected agreements

Geography—note: the province of Cabinda is an exclave, separated from the rest of the country by the Democratic Republic of the Congo

PEOPLE

Population: 12,531,357 (July 2008 est.)

Age structure:

0–14 years: 43.6% (male 2,760,264/female 2,707,665)

15–64 years: 53.6% (male 3,416,914/female 3,302,552)

65 years and over: 2.7% (male 151,609/female 192,353) (2008 est.)

Median age:

total: 18 years

male: 18 years

female: 18 years (2008 est.)

Population growth rate: 2.136% (2008 est.)

Birth rate: 44.09 births/1,000 population (2008 est.)

Death rate: 24.44 deaths/1,000 population (2008 est.)

Net migration rate: 1.72 migrant(s)/1,000 population (2008 est.)

Sex ratio:

at birth: 1.05 male(s)/female

under 15 years: 1.02 male(s)/female

15–64 years: 1.03 male(s)/female

65 years and over: 0.79 male(s)/female

total population: 1.02 male(s)/female (2008 est.)

Infant mortality rate:

total: 182.31 deaths/1,000 live births

male: 194.38 deaths/1,000 live births

female: 169.64 deaths/1,000 live births (2008 est.)

Life expectancy at birth:

total population: 37.92 years

male: 36.99 years

female: 38.9 years (2008 est.)

Total fertility rate: 6.2 children born/woman (2008 est.)

HIV/AIDS—adult prevalence rate: 3.9% (2003 est.)

HIV/AIDS—people living with HIV/AIDS: 240,000 (2003 est.)

HIV/AIDS—deaths: 21,000 (2003 est.)

Major infectious diseases:

degree of risk: very high

food or waterborne diseases: bacterial and protozoal diarrhea, hepatitis A, typhoid fever

vectorborne diseases: malaria, African trypanosomiasis (sleeping sickness)

water contact disease: schistosomiasis (2008)

Nationality:

noun: Angolan(s)

adjective: Angolan

Ethnic groups: Ovimbundu 37%, Kimbundu 25%, Bakongo 13%, mestico (mixed European and native African) 2%, European 1%, other 22%

Religions: indigenous beliefs 47%, Roman Catholic 38%, Protestant 15% (1998 est.)

Languages: Portuguese (official), Bantu and other African languages

Literacy:

definition: age 15 and over can read and write

total population: 67.4%

male: 82.9%

female: 54.2% (2001 est.)

GOVERNMENT

Country name:

conventional long form: Republic of Angola

conventional short form: Angola

local long form: Republica de Angola

local short form: Angola

former: People's Republic of Angola

Government type: republic; multiparty presidential regime

Capital: *name:* Luanda

geographic coordinates: 8 50 S, 13 14 E

time difference: UTC+1 (6 hours ahead of Washington, DC during Standard Time)

Administrative divisions: 18 provinces (provincias, singular—provincia); Bengo, Benguela, Bie, Cabinda, Cuando Cubango, Cuanza Norte, Cuanza Sul, Cunene, Huambo, Huila, Luanda, Lunda Norte, Lunda Sul, Malanje, Moxico, Namibe, Uige, Zaire

Independence: 11 November 1975 (from Portugal)

National holiday: Independence Day, 11 November (1975)

Constitution: adopted by People's Assembly 25 August 1992

Legal system: based on Portuguese civil law system and customary law; modified to accommodate political pluralism and increased use of free markets; has not accepted compulsory ICJ jurisdiction

Suffrage: 18 years of age; universal

Executive branch:

chief of state: President Jose Eduardo DOS SANTOS (since 21 September 1979); note—the president is both chief of state and head of government

head of government: President Jose Eduardo DOS SANTOS (since 21 September 1979); Fernando de Piedade Dias DOS SANTOS was appointed prime minister on 6 December 2002

cabinet: Council of Ministers appointed by the president

elections: president elected by universal ballot for a five-year term (eligible for a second consecutive or discontinuous term) under the 1992 constitution; President DOS SANTOS originally elected (in 1979) without opposition under a one-party system and stood for reelection in Angola's first multiparty elections 29–30 September 1992 (next to be held in 2009)

election results: Jose Eduardo DOS SANTOS 49.6%, Jonas SAVIMBI 40.1%, making a run-off election necessary; the run-off was not held because SAVIMBI's National Union for the Total Independence of Angola (UNITA) repudiated the results of the first election; the civil war resumed leaving DOS SANTOS in his current position as the president

Legislative branch: unicameral National Assembly or Assembleia Nacional (220 seats; members elected by proportional vote to serve four-year terms)

elections: last held 29–30 September 1992 (next to be held on 5–6 September 2008) *election results:* percent of vote by party—MPLA 54%, UNITA 34%, other 12%; seats by party—MPLA 129, UNITA 70, PRS 6, FNLA 5, PLD 3, other 7

Judicial branch: Supreme Court and separate provincial courts (judges are appointed by the president)

Political parties and leaders: Liberal Democratic Party or PLD [Analia de Victoria PEREIRA]; National Front for the Liberation of Angola or FNLA [disputed between Ngola KABANGU and Lucas NGONDA]; National Union for the Total Independence of Angola or UNITA (largest opposition party) [Isaias SAMAKUVA]; Popular Movement for the Liberation of Angola or MPLA (ruling party in power since 1975) [Jose Eduardo DOS SANTOS]; Social Renewal Party or PRS [Eduardo KUANGANA]

note: about a dozen minor parties participated in the 1992 elections but only won a few seats; they and more than 100 other smaller parties have little influence in the National Assembly

Political pressure groups and leaders: Front for the Liberation of the Enclave of Cabinda or FLEC [N'zita Henriques TIAGO, Antonio Bento BEMBE]

note: FLEC's small-scale, highly factionalized armed struggle for the independence of Cabinda Province ended after BEMBE's faction signed a peace accord in August 2006; other factions have since demobilized under provisions of the accord, although the two main faction leaders have not acceded to the accord

International organization participation: ACP, AfDB, AU, CPLP, FAO, G-77, IAEA, IBRD, ICAO, ICCt (signatory), ICRM, IDA, IFAD, IFC, IFRCS, ILO, IMF, IMO, Interpol, IOC, IOM, IPU, ISO (correspondent), ITSO, ITU, ITUC, MIGA, NAM, OAS (observer), OPEC, SADC, UN, UNCTAD, UNESCO, UNIDO, Union Latina, UNWTO, UPU, WCO, WFTU, WHO, WIPO, WMO, WTO

Diplomatic representation in the US:
chief of mission: Ambassador Josefina Perpetua Pitra DIAKITE
chancery: 2108 16th Street NW, Washington, DC 20009
telephone: [1] (202) 785-1156
FAX: [1] (202) 785-1258
consulate(s) general: Houston, New York

Diplomatic representation from the US:
chief of mission: Ambassador Dan MOZENA
embassy: number 32 Rua Houari Boumedienne (in the Miramar area of Luanda), Luanda
mailing address: international mail: Caixa Postal 6468, Luanda; pouch: US Embassy Luanda, US Department of State, 2550 Luanda Place, Washington, DC 20521-2550
telephone: [244] (222) 64-1000
FAX: [244] (222) 64-1232

Flag description: two equal horizontal bands of red (top) and black with a centered yellow emblem consisting of a five-pointed star within half a cogwheel crossed by a machete (in the style of a hammer and sickle)

ECONOMY

Economy—overview: Angola's high growth rate is driven by its oil sector, with record oil prices and rising petroleum production. Oil production and its supporting activities contribute about 85% of GDP. Increased oil production supported growth averaging more than 15% per year from 2004 to 2007. A postwar reconstruction boom and resettlement of displaced persons has led to high rates of growth in construction and agriculture as well. Much of the country's infrastructure is still damaged or undeveloped from the 27-year-long civil war. Remnants of the conflict such as widespread land mines still mar the countryside even though an apparently durable peace was established after the death of rebel leader Jonas SAVIMBI in February 2002. Subsistence agriculture provides the main livelihood for most of the people, but half of the country's food must still be imported. In 2005, the government started using a $2 billion line of credit, since increased to $7 billion, from China to rebuild Angola's public infrastructure, and several large-scale projects were completed in 2006. Angola also has large credit lines from Brazil, Portugal, Germany, Spain, and the EU. The central bank in 2003 implemented an exchange rate stabilization program using foreign exchange reserves to buy kwanzas out of circulation. This policy became more sustainable in 2005 because of strong oil export earnings; it has significantly reduced inflation. Although consumer inflation declined from 325% in 2000 to under 13% in 2007, the stabilization policy has put pressure on international net liquidity. Angola became a member of OPEC in late 2006 and in late 2007 was assigned a production quota of 1.9 million barrels a day, somewhat less than the 2–2.5 million bbl Angola's government had wanted. To fully take advantage of its rich national resources—gold, diamonds, extensive forests, Atlantic fisheries, and large oil deposits—Angola will need to implement government reforms, increase transparency, and reduce corruption. The government has rejected a formal IMF monitored program, although it continues Article IV consultations and ad hoc cooperation. Corruption, especially in the extractive sectors, and the negative effects of large inflows of foreign exchange, are major challenges facing Angola.

GDP (purchasing power parity): $91.29 billion (2007 est.)

GDP (official exchange rate): $61.36 billion (2007 est.)

GDP—real growth rate: 21.1% (2007 est.)

GDP—per capita (PPP): $5,600 (2007 est.)

GDP—composition by sector:
agriculture: 9.5%
industry: 65.8%
services: 24.6% (2007 est.)

Labor force: 6.64 million (2007 est.)

Labor force—by occupation: *agriculture:* 85%
industry and services: 15% (2003 est.)

Unemployment rate: extensive unemployment and underemployment affecting more than half the population (2001 est.)

Population below poverty line: 70% (2003 est.)

Household income or consumption by percentage share:
lowest 10%: NA%
highest 10%: NA%

Inflation rate (consumer prices): 12.2% (2007 est.)

Investment (gross fixed): 9.1% of GDP (2007 est.)

Budget:
revenues: $17.29 billion
expenditures: $15.78 billion (2007 est.)

Public debt: 12% of GDP (2007 est.)

Agriculture—products: bananas, sugarcane, coffee, sisal, corn, cotton, manioc (tapioca), tobacco, vegetables, plantains; livestock; forest products; fish

Industries: petroleum; diamonds, iron ore, phosphates, feldspar, bauxite, uranium, and gold; cement; basic metal products; fish processing; food processing, brewing, tobacco products, sugar; textiles; ship repair

Industrial production growth rate: 24.4% (2007 est.)

Electricity—production: 2.585 billion kWh (2005)

Electricity—production by source:
fossil fuel: 36.4%
hydro: 63.6%
nuclear: 0%
other: 0% (2001)

Electricity—consumption: 2.201 billion kWh (2005)

Electricity—exports: 0 kWh (2005)

19

Electricity—imports: 0 kWh (2005)

Oil—production: 1.26 million bbl/day (2005 est.)

Oil—consumption: 50,000 bbl/day (2005 est.)

Oil—exports: 1.021 million bbl/day (2004)

Oil—imports: 18,290 bbl/day (2004)

Oil—proved reserves: 25 billion bbl (2007 est.)

Natural gas—production: 767.3 million cu m (2005 est.)

Natural gas—consumption: 767.3 million cu m (2005 est.)

Natural gas—exports: 0 cu m (2005 est.)

Natural gas—imports: 0 cu m (2005)

Natural gas—proved reserves: 44 billion cu m (1 January 2006 est.)

Current account balance: $6.747 billion (2007 est.)

Exports: $44.32 billion f.o.b. (2007 est.)

Exports—commodities: crude oil, diamonds, refined petroleum products, gas, coffee, sisal, fish and fish products, timber, cotton

Exports—partners: US 38.1%, China 34.2%, Taiwan 5.8%, France 4.9%, Chile 4.1% (2006)

Imports: $12.29 billion f.o.b. (2007 est.)

Imports—commodities: machinery and electrical equipment, vehicles and spare parts; medicines, food, textiles, military goods

Imports—partners: US 15.3%, Portugal 15%, South Korea 10.1%, China 8.8%, Brazil 8.2%, South Africa 6.7%, France 6.2% (2006)

Economic aid—recipient: $441.8 million (2005)

Reserves of foreign exchange and gold: $11.33 billion (31 December 2007 est.)

Debt—external: $8.234 billion (31 December 2007 est.)

Stock of direct foreign investment—at home: $17.23 billion (2007 est.)

Stock of direct foreign investment—abroad: $227 million (2006 est.)

Currency (code): kwanza (AOA)

Currency code: AOA

Exchange rates: kwanza per US dollar—76.6 (2007), 80.4 (2006), 88.6 (2005), 83.541 (2004), 74.606 (2003)

Fiscal year: calendar year

COMMUNICATIONS

Telephones—main lines in use: 98,200 (2006)

Telephones—mobile cellular: 2.264 million (2006)

Telephone system:
general assessment: system inadequate; fewer than one fixed-line per 100 persons; combined fixed line and mobile telephone density approached 20 telephones per 100 persons in 2006
domestic: state-owned telecom had monopoly for fixed-lines until 2005; demand outstripped capacity, prices were high, and services poor; Telecom Namibia, through an Angolan company, became the first private licensed operator in Angola's fixed-line telephone network; Angola Telecom established mobile-cellular service in Luanda in 1993 and the network has been extended to larger towns; a privately-owned, mobile-cellular service provider began operations in 2001
international: country code—244; landing point for the SAT-3/WASC fiber-optic submarine cable that provides connectivity to Europe and Asia; satellite earth stations—29 (2007)

Radio broadcast stations: AM 21, FM 6, shortwave 7 (2001)

Radios: 815,000 (2000)

Television broadcast stations: 6 (2000)

Televisions: 196,000 (2000)

Internet country code: .ao

Internet hosts: 3,337 (2007)

Internet Service Providers (ISPs): 1 (2000)

Internet users: 85,000 (2005)

TRANSPORTATION

Airports: 232 (2007)

Airports—with paved runways:
total: 31
over 3,047 m: 5
2,438 to 3,047 m: 8
1,524 to 2,437 m: 12
914 to 1,523 m: 5
under 914 m: 1 (2007)

Airports—with unpaved runways:
total: 201
over 3,047 m: 2
2,438 to 3,047 m: 5
1,524 to 2,437 m: 30
914 to 1,523 m: 95
under 914 m: 69 (2007)

Pipelines: gas 234 km; liquid petroleum gas 85 km; oil 896 km; oil/gas/water 5 km (2007)

Railways:
total: 2,761 km
narrow gauge: 2,638 km 1.067-m gauge; 123 km 0.600-m gauge (2006)

Roadways:
total: 51,429 km
paved: 5,349 km
unpaved: 46,080 km (2001)

Waterways: 1,300 km (2007)

Merchant marine:
total: 5 ships (1000 GRT or over) 6,865 GRT/8,825 DWT
by type: cargo 1, passenger/cargo 2, petroleum tanker 2
foreign-owned: 1 (Spain 1)
registered in other countries: 6 (Bahamas 6) (2007)

Ports and terminals: Cabinda, Lobito, Luanda, Namibe

MILITARY

Military branches: Angolan Armed Forces (FAA): Army, Navy (Marinha de Guerra, MdG), Angolan National Air Force (FANA) (2007)

Military service age and obligation: 17 years of age for compulsory military service; conscript service obligation—2 years plus time for training (2001)

Manpower available for military service:
males age 16–49: 2,856,492
females age 16–49: 2,755,864 (2008 est.)

Manpower fit for military service:
males age 16–49: 1,430,658
females age 16–49: 1,371,689 (2008 est.)

Manpower reaching militarily significant age annually:
males age 16–49: 142,791
females age 16–49: 139,539 (2008 est.)

Military expenditures—percent of GDP: 5.7% (2006)

TRANSNATIONAL ISSUES

Disputes—international: many Cabindan separatists have returned to the province from exile since the 2006 ceasefire and peace agreement; concerns from international experts and local populations over the Okavango Delta ecology in Botswana and human displacement scuttled Namibian plans to construct a hydroelectric dam at Popavalle (Popa Falls) along the Angola-Namibia border

Refugees and internally displaced persons: *refugees (country of origin):* 12,615 (Democratic Republic of Congo)
IDPs: 61,700 (27-year civil war ending in 2002; 4 million IDPs already have returned) (2007)

Illicit drugs: used as a transshipment point for cocaine destined for Western Europe and other African states, particularly South Africa

ANGUILLA

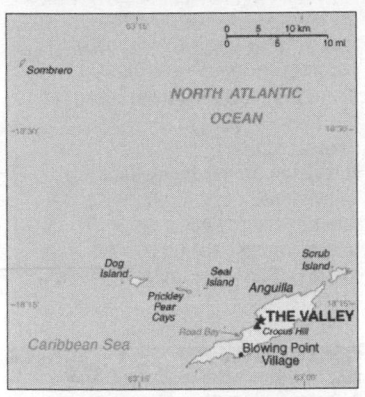

INTRODUCTION

Background: Colonized by English settlers from Saint Kitts in 1650, Anguilla was administered by Great Britain until the early 19th century, when the island—against the wishes of the inhabitants—was incorporated into a single British dependency, along with Saint Kitts and Nevis. Several attempts at separation failed. In 1971, two years after a revolt, Anguilla was finally allowed to secede; this arrangement was formally recognized in 1980, with Anguilla becoming a separate British dependency.

GEOGRAPHY

Location: Caribbean, islands between the Caribbean Sea and North Atlantic Ocean, east of Puerto Rico
Geographic coordinates: 18 15 N, 63 10 W
Map references: Central America and the Caribbean
Area:
total: 102 sq km
land: 102 sq km
water: 0 sq km
Area—comparative: about half the size of Washington, DC
Land boundaries: 0 km
Coastline: 61 km
Maritime claims:
territorial sea: 3 nm
exclusive fishing zone: 200 nm
Climate: tropical; moderated by northeast trade winds
Terrain: flat and low-lying island of coral and limestone
Elevation extremes:
lowest point: Caribbean Sea 0 m
highest point: Crocus Hill 65 m
Natural resources: salt, fish, lobster
Land use:
arable land: 0%
permanent crops: 0%

other: 100% (mostly rock with sparse scrub oak, few trees, some commercial salt ponds) (2005)
Irrigated land: NA
Natural hazards: frequent hurricanes and other tropical storms (July to October)
Environment—current issues: supplies of potable water sometimes cannot meet increasing demand largely because of poor distribution system
Geography—note: the most northerly of the Leeward Islands in the Lesser Antilles

PEOPLE

Population: 14,108 (July 2008 est.)
Age structure:
0–14 years: 24.8% (male 1,795/female 1,706)
15–64 years: 67.6% (male 4,569/female 4,970)
65 years and over: 7.6% (male 510/female 558) (2008 est.)
Median age:
total: 32.3 years
male: 31.3 years
female: 33.4 years (2008 est.)
Population growth rate: 2.332% (2008 est.)
Birth rate: 13.11 births/1,000 population (2008 est.)
Death rate: 4.39 deaths/1,000 population (2008 est.)
Net migration rate: 14.6 migrant(s)/1,000 population (2008 est.)
Sex ratio:
at birth: 1.03 male(s)/female
under 15 years: 1.05 male(s)/female
15–64 years: 0.92 male(s)/female
65 years and over: 0.91 male(s)/female
total population: 0.95 male(s)/female (2008 est.)
Infant mortality rate:
total: 3.54 deaths/1,000 live births
male: 4.01 deaths/1,000 live births
female: 3.06 deaths/1,000 live births (2008 est.)
Life expectancy at birth:
total population: 80.53 years
male: 78.01 years
female: 83.12 years (2008 est.)
Total fertility rate: 1.75 children born/woman (2008 est.)
HIV/AIDS—adult prevalence rate: NA
HIV/AIDS—people living with HIV/AIDS: NA
HIV/AIDS—deaths: NA
Nationality:
noun: Anguillan(s)
adjective: Anguillan
Ethnic groups: black (predominant)

90.1%, mixed, mulatto 4.6%, white 3.7%, other 1.5% (2001 census)
Religions: Anglican 29%, Methodist 23.9%, other Protestant 30.2%, Roman Catholic 5.7%, other Christian 1.7%, other 5.2%, none or unspecified 4.3% (2001 census)
Languages: English (official)
Literacy:
definition: age 12 and over can read and write
total population: 95%
male: 95%
female: 95% (1984 est.)

GOVERNMENT

Country name:
conventional long form: none
conventional short form: Anguilla
Dependency status: overseas territory of the UK
Government type: NA
Capital: *name:* The Valley
geographic coordinates: 18 13 N, 63 03 W
time difference: UTC-4 (1 hour ahead of Washington, DC during Standard Time)
Administrative divisions: none (overseas territory of the UK)
Independence: none (overseas territory of the UK)
National holiday: Anguilla Day, 30 May (1967)
Constitution: Anguilla Constitutional Order 1 April 1982; amended 1990
Legal system: based on English common law
Suffrage: 18 years of age; universal
Executive branch:
chief of state: Queen ELIZABETH II (since 6 February 1952); represented by Governor Andrew N. GEORGE (since 10 July 2006)
head of government: Chief Minister Osbourne FLEMING (since 3 March 2000)
cabinet: Executive Council appointed by the governor from among the elected members of the House of Assembly
elections: the monarch is hereditary; governor appointed by the monarch; following legislative elections, the leader of the majority party or the leader of the majority coalition is usually appointed chief minister by the governor
Legislative branch: unicameral House of Assembly (11 seats; 7 members elected by direct popular vote, 2 ex officio members, and 2 appointed; members serve five-year terms)
elections: last held 21 February 2005 (next to be held in 2010)
election results: percent of vote by party—

AUF 38.9%, AUM 19.4%, ANSA 19.2%, APP 9.5%, independents 13%; seats by party—AUF 4, ANSA 2, AUM 1

Judicial branch: High Court (judge provided by Eastern Caribbean Supreme Court)

Political parties and leaders: Anguilla United Front or AUF [Osbourne FLEMING, Victor BANKS] (a coalition of the Anguilla Democratic Party or ADP and the Anguilla National Alliance or ANA); Anguilla United Movement or AUM [Hubert HUGHES]; Anguilla Progressive Party or APP [Roy ROGERS]; Anguilla Strategic Alternative or ANSA [Edison BAIRD]

Political pressure groups and leaders: NA

International organization participation: Caricom (associate), CDB, Interpol (subbureau), OECS, UPU

Diplomatic representation in the US: none (overseas territory of the UK)

Diplomatic representation from the US: none (overseas territory of the UK)

Flag description: blue, with the flag of the UK in the upper hoist-side quadrant and the Anguillan coat of arms centered in the outer half of the flag; the coat of arms depicts three orange dolphins in an interlocking circular design on a white background with blue wavy water below

ECONOMY

Economy—overview: Anguilla has few natural resources, and the economy depends heavily on luxury tourism, offshore banking, lobster fishing, and remittances from emigrants. Increased activity in the tourism industry has spurred the growth of the construction sector, contributing to economic growth. Anguillan officials have put substantial effort into developing the offshore financial sector, which is small, but growing. In the medium term, prospects for the economy will depend largely on the tourism sector and, therefore, on revived income growth in the industrialized nations as well as on favorable weather conditions.

GDP (purchasing power parity): $108.9 million (2004 est.)

GDP (official exchange rate): $108.9 million (2004 est.)

GDP—real growth rate: 10.2% (2004 est.)

GDP—per capita (PPP): $8,800 (2004 est.)

GDP—composition by sector:
agriculture: 4%
industry: 18%
services: 78% (2002 est.)

Labor force: 6,049 (2001)

Labor force—by occupation: agriculture/fishing/forestry/mining 4%, manufacturing 3%, construction 18%, transportation and utilities 10%, commerce 36%, services 29% (2000 est.)

Unemployment rate: 8% (2002)

Population below poverty line: 23% (2002)

Household income or consumption by percentage share:
lowest 10%: NA%
highest 10%: NA%

Inflation rate (consumer prices): 5.3% (2006 est.)

Budget:
revenues: $22.8 million
expenditures: $22.5 million (2000 est.)

Agriculture—products: small quantities of tobacco, vegetables; cattle raising

Industries: tourism, boat building, offshore financial services

Industrial production growth rate: 3.1% (1997 est.)

Electricity—production: NA kWh

Electricity—production by source:
fossil fuel: NA
hydro: NA
nuclear: NA
other: NA

Current account balance: -$42.87 million (2003 est.)

Exports: $13 million (2006)

Exports—commodities: lobster, fish, livestock, salt, concrete blocks, rum

Exports—partners: UK, US, Puerto Rico, Saint-Martin (2006)

Imports: $143 million (2006)

Imports—commodities: fuels, foodstuffs, manufactures, chemicals, trucks, textiles

Imports—partners: US, Puerto Rico, UK (2006)

Economic aid—recipient: $9 million (2004 est.)

Debt—external: $8.8 million (1998)

Currency (code): East Caribbean dollar (XCD)

Currency code: XCD

Exchange rates: East Caribbean dollars per US dollar—2.7 (2007), 2.7 (2006), 2.7 (2005), 2.7 (2004), 2.7 (2003)
note: fixed rate since 1976

Fiscal year: 1 April—31 March

COMMUNICATIONS

Telephones—main lines in use: 6,200 (2002)

Telephones—mobile cellular: 1,800 (2002)

Telephone system:
general assessment: NA
domestic: modern internal telephone system
international: country code—1-264; landing point for the East Caribbean Fiber System (ECFS) submarine cable with links to 13 other islands in the eastern Caribbean extending from the British Virgin Islands to Trinidad; microwave radio relay to island of Saint Martin (Guadeloupe and Netherlands Antilles) (2007)

Radio broadcast stations: AM 2, FM 7, shortwave 0 (2004)

Radios: 3,000 (1997)

Television broadcast stations: 1 (1997)

Televisions: 1,000 (1997)

Internet country code: .ai

Internet hosts: 319 (2007)

Internet Service Providers (ISPs): 16 (2000)

Internet users: 3,000 (2002)

TRANSPORTATION

Airports: 3 (2007)

Airports—with paved runways:
total: 1
1,524 to 2,437 m: 1 (2007)

Airports—with unpaved runways:
total: 2
under 914 m: 2 (2007)

Roadways:
total: 175 km
paved: 82 km
unpaved: 93 km (2004)

Ports and terminals: Blowing Point, Road Bay

MILITARY

Manpower available for military service:
males age 16–49: 3,538 (2008 est.)

Manpower fit for military service:
males age 16–49: 2,929 (2008 est.)

Manpower reaching militarily significant age annually:
males age 16–49: 103 (2008 est.)

Military—note: defense is the responsibility of the UK

TRANSNATIONAL ISSUES

Disputes—international: none

Illicit drugs: transshipment point for South American narcotics destined for the US and Europe

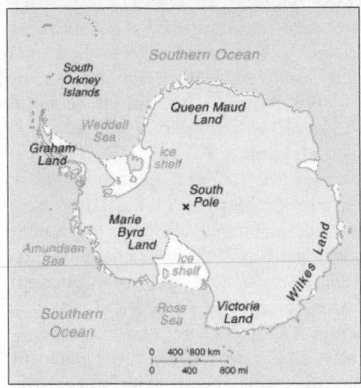

INTRODUCTION

Background: Speculation over the existence of a "southern land" was not confirmed until the early 1820s when British and American commercial operators and British and Russian national expeditions began exploring the Antarctic Peninsula region and other areas south of the Antarctic Circle. Not until 1840 was it established that Antarctica was indeed a continent and not just a group of islands. Several exploration "firsts" were achieved in the early 20th century. Following World War II, there was an upsurge in scientific research on the continent. A number of countries have set up a range of year-round and seasonal stations, camps, and refuges to support scientific research in Antarctica. Seven have made territorial claims, but not all countries recognize these claims. In order to form a legal framework for the activities of nations on the continent, an Antarctic Treaty was negotiated that neither denies nor gives recognition to existing territorial claims; signed in 1959, it entered into force in 1961.

GEOGRAPHY

Location: continent mostly south of the Antarctic Circle
Geographic coordinates: 90 00 S, 0 00 E
Map references: Antarctic Region
Area:
total: 14 million sq km
land: 14 million sq km (280,000 sq km ice-free, 13.72 million sq km ice-covered) (est.)
note: fifth-largest continent, following Asia, Africa, North America, and South America, but larger than Australia and the subcontinent of Europe
Area—comparative: slightly less than 1.5 times the size of the US
Land boundaries: 0 km

note: see entry on Disputes—international
Coastline: 17,968 km
Maritime claims: Australia, Chile, and Argentina claim Exclusive Economic Zone (EEZ) rights or similar over 200 nm extensions seaward from their continental claims, but like the claims themselves, these zones are not accepted by other countries; 21 of 28 Antarctic consultative nations have made no claims to Antarctic territory (although Russia and the US have reserved the right to do so) and do not recognize the claims of the other nations; also see the Disputes—international entry
Climate: severe low temperatures vary with latitude, elevation, and distance from the ocean; East Antarctica is colder than West Antarctica because of its higher elevation; Antarctic Peninsula has the most moderate climate; higher temperatures occur in January along the coast and average slightly below freezing
Terrain: about 98% thick continental ice sheet and 2% barren rock, with average elevations between 2,000 and 4,000 meters; mountain ranges up to nearly 5,000 meters; ice-free coastal areas include parts of southern Victoria Land, Wilkes Land, the Antarctic Peninsula area, and parts of Ross Island on McMurdo Sound; glaciers form ice shelves along about half of the coastline, and floating ice shelves constitute 11% of the area of the continent
Elevation extremes:
lowest point: Bentley Subglacial Trench -2,555 m
highest point: Vinson Massif 4,897 m
note: the lowest known land point in Antarctica is hidden in the Bentley Subglacial Trench; at its surface is the deepest ice yet discovered and the world's lowest elevation not under seawater
Natural resources: iron ore, chromium, copper, gold, nickel, platinum and other minerals, and coal and hydrocarbons have been found in small uncommercial quantities; none presently exploited; krill, finfish, and crab have been taken by commercial fisheries
Land use:
arable land: 0%
permanent crops: 0%
other: 100% (ice 98%, barren rock 2%) (2005)
Natural hazards: katabatic (gravity-driven) winds blow coastward from the high interior; frequent blizzards form near the foot of the plateau; cyclonic

storms form over the ocean and move clockwise along the coast; volcanism on Deception Island and isolated areas of West Antarctica; other seismic activity rare and weak; large icebergs may calve from ice shelf
Environment—current issues: in 1998, NASA satellite data showed that the Antarctic ozone hole was the largest on record, covering 27 million square kilometers; researchers in 1997 found that increased ultraviolet light passing through the hole damages the DNA of icefish, an Antarctic fish lacking hemoglobin; ozone depletion earlier was shown to harm one-celled Antarctic marine plants; in 2002, significant areas of ice shelves disintegrated in response to regional warming
Geography—note: the coldest, windiest, highest (on average), and driest continent; during summer, more solar radiation reaches the surface at the South Pole than is received at the Equator in an equivalent period; mostly uninhabitable

PEOPLE

Population: no indigenous inhabitants, but there are both permanent and summer-only staffed research stations
note: 28 nations, all signatory to the Antarctic Treaty, operate through their National Antarctic Program a number of seasonal-only (summer) and year-round research stations on the continent and its nearby islands south of 60 degrees south latitude (the region covered by the Antarctic Treaty); these stations' population of persons doing and supporting science or engaged in the management and protection of the Antarctic region varies from approximately 4,000 in summer to 1,000 in winter; in addition, approximately 1,000 personnel, including ship's crew and scientists doing onboard research, are present in the waters of the treaty region; peak summer (December-February) population—4,219 total; Argentina 667, Australia 200, Brazil 40, Bulgaria 15, Chile 237, China 70, Czech Republic 20, Ecuador 26, Finland 20, France 100, France and Italy jointly 45, Germany 90, India 65, Italy 90, Japan 125, South Korea 70, NZ 85, Norway 44, Peru 28, Poland 40, Romania 3, Russia 429, South Africa 80, Spain 28, Sweden 20, Ukraine 24, UK 205, US 1,293, Uruguay 60 (2007–2008); winter (June-August) station population—1,088 total; Argentina 176, Australia 62, Brazil 12, Chile 96, China 29, France 26, France and Italy jointly 13, Germany 9, India

25, Italy 2, Japan 40, South Korea 18, NZ 10, Norway 7, Poland 12, Russia 148, South Africa 10, Ukraine 12, UK 37, US 337, Uruguay 9 (2008); research stations operated within the Antarctic Treaty area (south of 60 degrees south latitude) by National Antarctic Programs: year-round stations—38 total; Argentina 6, Australia 3, Brazil 1, Chile 4, China 2, France 1, France and Italy jointly 1, Germany 1, India 1, Japan 1, South Korea 1, NZ 1, Norway 1, Poland 1, Russia 5, South Africa 1, Ukraine 1, UK 2, US 3, Uruguay 1 (2008); a range of seasonal-only (summer) stations, camps, and refuges—Argentina, Australia, Bulgaria, Brazil, Chile, China, Czech Republic, Ecuador, Finland, France, Germany, India, Italy, Japan, South Korea, New Zealand, Norway, Peru, Poland, Romania, Russia, Spain, Sweden, Ukraine, UK, US, and Uruguay (2007–2008); in addition, during the austral summer some nations have numerous occupied locations such as tent camps, summer-long temporary facilities, and mobile traverses in support of research (March 2008 est.)

GOVERNMENT

Country name:
conventional long form: none
conventional short form: Antarctica
Government type: Antarctic Treaty Summary—the Antarctic Treaty, signed on 1 December 1959 and entered into force on 23 June 1961, establishes the legal framework for the management of Antarctica; the 30th Antarctic Treaty Consultative Meeting was held in Delhi, India in April/May 2007; at these periodic meetings, decisions are made by consensus (not by vote) of all consultative member nations; at the end of 2007, there were 46 treaty member nations: 28 consultative and 18 non-consultative; consultative (decision-making) members include the seven nations that claim portions of Antarctica as national territory (some claims overlap) and 21 nonclaimant nations; the US and Russia have reserved the right to make claims; the US does not recognize the claims of others; Antarctica is administered through meetings of the consultative member nations; decisions from these meetings are carried out by these member nations (with respect to their own nationals and operations) in accordance with their own national laws; the years in parentheses indicate when a consultative member-nation acceded to the Treaty and when it was accepted as a consultative member, while no date indicates the country was an original 1959

treaty signatory; claimant nations are—Argentina, Australia, Chile, France, NZ, Norway, and the UK. Nonclaimant consultative nations are—Belgium, Brazil (1975/1983), Bulgaria (1978/1998) China (1983/1985), Ecuador (1987/1990), Finland (1984/1989), Germany (1979/1981), India (1983/1983), Italy (1981/1987), Japan, South Korea (1986/1989), Netherlands (1967/1990), Peru (1981/1989), Poland (1961/1977), Russia, South Africa, Spain (1982/1988), Sweden (1984/1988), Ukraine (1992/2004), Uruguay (1980/1985), and the US; non-consultative members, with year of accession in parentheses, are—Austria (1987), Belarus (2006), Canada (1988), Colombia (1989), Cuba (1984), Czech Republic (1962/1993), Denmark (1965), Estonia (2001), Greece (1987), Guatemala (1991), Hungary (1984), North Korea (1987), Papua New Guinea (1981), Romania (1971), Slovakia (1962/1993), Switzerland (1990), Turkey (1996), and Venezuela (1999); note—Czechoslovakia acceded to the Treaty in 1962 and separated into the Czech Republic and Slovakia in 1993; Article 1—area to be used for peaceful purposes only; military activity, such as weapons testing, is prohibited, but military personnel and equipment may be used for scientific research or any other peaceful purpose; Article 2—freedom of scientific investigation and cooperation shall continue; Article 3—free exchange of information and personnel, cooperation with the UN and other international agencies; Article 4—does not recognize, dispute, or establish territorial claims and no new claims shall be asserted while the treaty is in force; Article 5—prohibits nuclear explosions or disposal of radioactive wastes; Article 6—includes under the treaty all land and ice shelves south of 60 degrees 00 minutes south and reserves high seas rights; Article 7—treaty-state observers have free access, including aerial observation, to any area and may inspect all stations, installations, and equipment; advance notice of all expeditions and of the introduction of military personnel must be given; Article 8—allows for jurisdiction over observers and scientists by their own states; Article 9—frequent consultative meetings take place among member nations; Article 10—treaty states will discourage activities by any country in Antarctica that are contrary to the treaty; Article 11—disputes to be settled peacefully by the parties concerned or, ultimately, by the ICJ; Articles 12, 13, 14—deal with upholding, interpreting, and amending

the treaty among involved nations; other agreements—some 200 recommendations adopted at treaty consultative meetings and ratified by governments include—Agreed Measures for Fauna and Flora (1964) which were later incorporated into the Environmental Protocol; Convention for the Conservation of Antarctic Seals (1972); Convention on the Conservation of Antarctic Marine Living Resources (1980); a mineral resources agreement was signed in 1988 but remains unratified; the Protocol on Environmental Protection to the Antarctic Treaty was signed 4 October 1991 and entered into force 14 January 1998; this agreement provides for the protection of the Antarctic environment through six specific annexes: 1) environmental impact assessment, 2) conservation of Antarctic fauna and flora, 3) waste disposal and waste management, 4) prevention of marine pollution, 5) area protection and management and 6) liability arising from environmental emergencies; it prohibits all activities relating to mineral resources except scientific research; a permanent Antarctic Treaty Secretariat was established in 2004 in Buenos Aires, Argentina
Legal system: Antarctica is administered through meetings of the consultative member nations; decisions from these meetings are carried out by these member nations (with respect to their own nationals and operations) in accordance with their own national laws; US law, including certain criminal offenses by or against US nationals, such as murder, may apply extraterritorially; some US laws directly apply to Antarctica; for example, the Antarctic Conservation Act, 16 U.S.C. section 2401 et seq., provides civil and criminal penalties for the following activities, unless authorized by regulation of statute: the taking of native mammals or birds; the introduction of nonindigenous plants and animals; entry into specially protected areas; the discharge or disposal of pollutants; and the importation into the US of certain items from Antarctica; violation of the Antarctic Conservation Act carries penalties of up to $10,000 in fines and one year in prison; the National Science Foundation and Department of Justice share enforcement responsibilities; Public Law 95-541, the US Antarctic Conservation Act of 1978, as amended in 1996, requires expeditions from the US to Antarctica to notify, in advance, the Office of Oceans, Room 5805, Department of State, Washington, DC 20520, which reports such plans to

other nations as required by the Antarctic Treaty; for more information, contact Permit Office, Office of Polar Programs, National Science Foundation, Arlington, Virginia 22230; telephone: (703) 292-8030, or visit their website at www.nsf.gov; more generally, access to the Antarctic Treaty area, that is to all areas between 60 and 90 degrees south latitude, is subject to a number of relevant legal instruments and authorization procedures adopted by the states party to the Antarctic Treaty

ECONOMY

Economy—overview: Fishing off the coast and tourism, both based abroad, account for Antarctica's limited economic activity. Antarctic fisheries in 2005–06 (1 July–30 June) reported landing 128,081 metric tons (estimated fishing from the area covered by the Convention on the Conservation of Antarctic Marine Living Resources (CCAMLR), which extends slightly beyond the Antarctic Treaty area). Unregulated fishing, particularly of Patagonian toothfish (Dissostichus eleginoides), is a serious problem. The CCAMLR determines the recommended catch limits for marine species. A total of 36,460 tourists visited the Antarctic Treaty area in the 2006–07 Antarctic summer, up from the 30,877 visitors the previous year (estimates provided to the Antarctic Treaty by the International Association of Antarctica Tour Operators (IAATO); this does not include passengers on overflights). Nearly all of them were passengers on commercial (nongovernmental) ships and several yachts that make trips during the summer. Most tourist trips last approximately two weeks.

COMMUNICATIONS

Telephones—main lines in use: 0; note—information for US bases only (2001)
Telephone system:
general assessment: local systems at some research stations
domestic: commercial cellular networks

operating in a small number of locations
international: country code—none allocated; via satellite (including mobile Inmarsat and Iridium systems) to and from all research stations, ships, aircraft, and most field parties (2007)
Radio broadcast stations: FM 2, shortwave 1 (information for US bases only); note—many research stations have a local FM radio station (2007)
Radios: NA
Television broadcast stations: 1 (cable system with 6 channels; American Forces Antarctic Network-McMurdo—information for US bases only) (2002)
Televisions: several hundred at McMurdo Station (US)
note: information for US bases only (2001)
Internet country code: .aq
Internet hosts: 7,744 (2007)
Internet Service Providers (ISPs): NA

TRANSPORTATION

Airports: 27 (2008)
Airports—with unpaved runways:
total: 27
over 3,047 m: 6
2,438 to 3,047 m: 5
1,524 to 2,437 m: 1
914 to 1,523 m: 9
under 914 m: 6 (2008)
Heliports: 53
note: all year-round and seasonal stations operated by National Antarctic Programs stations have some kind of helicopter landing facilities, prepared (helipads) or unprepared (2007)
Ports and terminals: there are no developed ports and harbors in Antarctica; most coastal stations have offshore anchorages, and supplies are transferred from ship to shore by small boats, barges, and helicopters; a few stations have a basic wharf facility; US coastal stations include McMurdo (77 51 S, 166 40 E), and Palmer (64 43 S, 64 03 W); government use only except by permit (see Permit Office under "Legal System"); all ships at port are subject to inspection in accordance with Article 7, Antarctic Treaty; offshore anchorage is sparse and intermittent; relevant legal instruments

and authorization procedures adopted by the states parties to the Antarctic Treaty regulating access to the Antarctic Treaty area, to all areas between 60 and 90 degrees of latitude south, have to be complied with (see "Legal System"); The Hydrographic Committee on Antarctica (HCA), a special hydrographic commission of International Hydrographic Organization (IHO), is responsible for hydrographic surveying and nautical charting matters in Antarctic Treaty area; it coordinates and facilitates provision of accurate and appropriate charts and other aids to navigation in support of safety of navigation in region; membership of HCA is open to any IHO Member State whose government has acceded to the Antarctic Treaty and which contributes resources and/or data to IHO Chart coverage of the area; members of HCA are Argentina, Australia, Brazil, Chile, China, Ecuador, France, Germany, Greece, India, Italy, NZ, Norway, Russia, South Africa, Spain, UK, and US (2007)

MILITARY

Military—note: the Antarctic Treaty prohibits any measures of a military nature, such as the establishment of military bases and fortifications, the carrying out of military maneuvers, or the testing of any type of weapon; it permits the use of military personnel or equipment for scientific research or for any other peaceful purposes

TRANSNATIONAL ISSUES

Disputes—international: Antarctic Treaty freezes claims (see Antarctic Treaty Summary in Government type entry); Argentina, Australia, Chile, France, NZ, Norway, and UK claim land and maritime sectors (some overlapping) for a large portion of the continent; the US and many other states do not recognize these territorial claims and have made no claims themselves (the US and Russia reserve the right to do so); no claims have been made in the sector between 90 degrees west and 150 degrees west

ANTIGUA AND BARBUDA

INTRODUCTION

Background: The Siboney were the first to inhabit the islands of Antigua and Barbuda in 2400 B.C., but Arawak Indians populated the islands when COLUMBUS landed on his second voyage in 1493. Early

settlements by the Spanish and French were succeeded by the English who formed a colony in 1667. Slavery, established to run the sugar plantations on Antigua, was abolished in 1834. The islands became an independent state within the British Commonwealth of Nations in 1981.

GEOGRAPHY

Location: Caribbean, islands between the Caribbean Sea and the North Atlantic Ocean, east-southeast of Puerto Rico
Geographic coordinates: 17 03 N, 61 48 W

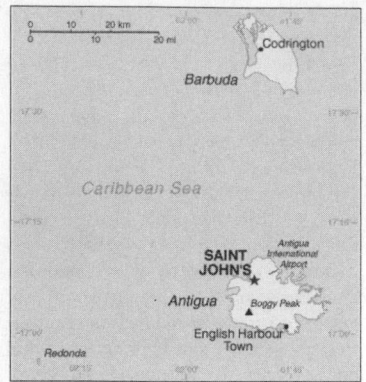

Map references: Central America and the Caribbean

Area:

total: 442.6 sq km (Antigua 280 sq km; Barbuda 161 sq km)

land: 442.6 sq km

water: 0 sq km

note: includes Redonda, 1.6 sq km

Area—comparative: 2.5 times the size of Washington, DC

Land boundaries: 0 km

Coastline: 153 km

Maritime claims:

territorial sea: 12 nm

contiguous zone: 24 nm

exclusive economic zone: 200 nm

continental shelf: 200 nm or to the edge of the continental margin

Climate: tropical maritime; little seasonal temperature variation

Terrain: mostly low-lying limestone and coral islands, with some higher volcanic areas

Elevation extremes:

lowest point: Caribbean Sea 0 m

highest point: Boggy Peak 402 m

Natural resources: NEGL; pleasant climate fosters tourism

Land use: *arable land:* 18.18%

permanent crops: 4.55%

other: 77.27% (2005)

Irrigated land: NA

Total renewable water resources: 0.1 cu km (2000)

Freshwater withdrawal (domestic/industrial/agricultural): *total:* 0.005 cu km/yr (60%/20%/20%)

per capita: 63 cu m/yr (1990)

Natural hazards: hurricanes and tropical storms (July to October); periodic droughts

Environment—current issues: water management—a major concern because of limited natural fresh water resources—is further hampered by the clearing of trees to increase crop production, causing rainfall to run off quickly

Environment—international agreements: *party to:* Biodiversity, Climate Change, Climate Change-Kyoto Protocol, Desertification, Endangered Species, Environmental Modification, Hazardous Wastes, Law of the Sea, Marine Dumping, Ozone Layer Protection, Ship Pollution, Wetlands, Whaling

signed, but not ratified: none of the selected agreements

Geography—note: Antigua has a deeply indented shoreline with many natural harbors and beaches; Barbuda has a large western harbor

PEOPLE

Population: 69,842 (July 2008 est.)

Age structure:

0–14 years: 26.9% (male 9,561/female 9,220)

15–64 years: 69.5% (male 24,467/female 24,076)

65 years and over: 3.6% (male 927/female 1,591) (2008 est.)

Median age:

total: 30.5 years

male: 30 years

female: 31.1 years (2008 est.)

Population growth rate: 0.508% (2008 est.)

Birth rate: 16.35 births/1,000 population (2008 est.)

Death rate: 5.25 deaths/1,000 population (2008 est.)

Net migration rate: -6.01 migrant(s)/1,000 population (2008 est.)

Sex ratio:

at birth: 1.05 male(s)/female

under 15 years: 1.04 male(s)/female

15–64 years: 1.02 male(s)/female

65 years and over: 0.58 male(s)/female

total population: 1 male(s)/female (2008 est.)

Infant mortality rate:

total: 17.67 deaths/1,000 live births

male: 21.26 deaths/1,000 live births

female: 13.89 deaths/1,000 live births (2008 est.)

Life expectancy at birth:

total population: 72.69 years

male: 70.28 years

female: 75.22 years (2008 est.)

Total fertility rate: 2.22 children born/woman (2008 est.)

HIV/AIDS—adult prevalence rate: NA

HIV/AIDS—people living with HIV/AIDS: NA

HIV/AIDS—deaths: NA

Nationality:

noun: Antiguan(s), Barbudan(s)

adjective: Antiguan, Barbudan

Ethnic groups: black 91%, mixed 4.4%, white 1.7%, other 2.9% (2001 census)

Religions: Anglican 25.7%, Seventh Day Adventist 12.3%, Pentecostal 10.6%, Moravian 10.5%, Roman Catholic 10.4%, Methodist 7.9%, Baptist 4.9%, Church of God 4.5%, other Christian 5.4%, other 2%, none or unspecified 5.8% (2001 census)

Languages: English (official), local dialects

Literacy:

definition: age 15 and over has completed five or more years of schooling

total population: 85.8%

male: NA%

female: NA% (2003 est.)

GOVERNMENT

Country name:

conventional long form: none

conventional short form: Antigua and Barbuda

Government type: constitutional monarchy with a parliamentary system of government

Capital: *name:* Saint John's

geographic coordinates: 17 07 N, 61 51 W

time difference: UTC-4 (1 hour ahead of Washington, DC during Standard Time)

Administrative divisions: 6 parishes and 2 dependencies*; Barbuda*, Redonda*, Saint George, Saint John, Saint Mary, Saint Paul, Saint Peter, Saint Philip

Independence: 1 November 1981 (from UK)

National holiday: Independence Day (National Day), 1 November (1981)

Constitution: 1 November 1981

Legal system: based on English common law

Suffrage: 18 years of age; universal

Executive branch:

chief of state: Queen ELIZABETH II (since 6 February 1952); represented by Governor General Louisse LAKE-TACK (since 17 July 2007)

head of government: Prime Minister Winston Baldwin SPENCER (since 24 March 2004)

cabinet: Council of Ministers appointed by the governor general on the advice of the prime minister

elections: the monarch is hereditary; governor general chosen by the monarch on the advice of the prime minister; following legislative elections, the leader of the majority party or the leader of the majority coalition is usually appointed prime minister by the governor general

Legislative branch: bicameral Parliament consists of the Senate (17 seats; members appointed by the governor general) and the House of Representatives (17 seats; members are elected by proportional representation to serve five-year terms)

elections: House of Representatives—last held 23 March 2004 (next to be held in 2009) *election results:* percent of vote by party—NA; seats by party—ALP 4, UPP 13

Judicial branch: Eastern Caribbean Supreme Court (based in Saint Lucia; one judge of the Supreme Court is a resident of the islands and presides over the Court of Summary Jurisdiction); member Caribbean Court of Justice

Political parties and leaders: Antigua Labor Party or ALP [Lester Bryant BIRD]; Barbudans for a Better Barbuda [Ordrick SAMUEL]; Barbuda People's Movement or BPM [Thomas H. FRANK]; Barbuda People's Movement for Change [Arthur NIBBS]; United Progressive Party or UPP [Baldwin SPENCER] (a coalition of three parties—Antigua Caribbean Liberation Movement or ACLM, Progressive Labor Movement or PLM, United National Democratic Party or UNDP)

Political pressure groups and leaders: Antigua Trades and Labor Union or ATLU [William ROBINSON]; People's Democratic Movement or PDM [Hugh MARSHALL]

International organization participation: ACP, C, Caricom, CDB, FAO, G-77, IBRD, ICAO, ICCt, ICRM, IDA, IFAD, IFC, IFRCS, ILO, IMF, IMO, Interpol, IOC, ISO (subscriber), ITU, ITUC, MIGA, NAM, OAS, OECS, OPANAL, OPCW, UN, UNCTAD, UNESCO, UPU, WCL, WFTU, WHO, WIPO, WMO, WTO

Diplomatic representation in the US:
chief of mission: Ambassador Deborah Mae LOVELL
chancery: 3216 New Mexico Avenue NW, Washington, DC 20016
telephone: [1] (202) 362-5122
FAX: [1] (202) 362-5225
consulate(s) general: Miami

Diplomatic representation from the US: the US does not have an embassy in Antigua and Barbuda; the US Ambassador to Barbados is accredited to Antigua and Barbuda

Flag description: red, with an inverted isosceles triangle based on the top edge of the flag; the triangle contains three horizontal bands of black (top), light blue, and white, with a yellow rising sun in the black band

ECONOMY

Economy—overview: Antigua has a relatively high GDP per capita in comparison to most other Caribbean nations. It has experienced solid growth since 2003, driven by a construction boom in hotels and housing that which should wind down in 2008. Tourism continues to dominate the economy, accounting for more than half of GDP. The dual-island nation's agricultural production is focused on the domestic market and constrained by a limited water supply and a labor shortage stemming from the lure of higher wages in tourism and construction. Manufacturing comprises enclave-type assembly for export with major products being bedding, handicrafts, and electronic components. Prospects for economic growth in the medium term will continue to depend on income growth in the industrialized world, especially in the US, which accounts for slightly more than one-third of tourist arrivals. Since taking office in 2004, the SPENCER government has adopted an ambitious fiscal reform program, but will continue to be saddled by its debt burden with a debt-to-GDP ratio exceeding 100%.

GDP (purchasing power parity): $1.526 billion (2007 est.)

GDP (official exchange rate): $1.089 billion (2007 est.)

GDP—real growth rate: 6.1% (2007 est.)

GDP—per capita (PPP): $18,300 (2007 est.)

GDP—composition by sector:
agriculture: 3.8%
industry: 22%
services: 74.3% (2002 est.)

Labor force: 30,000 (1991)

Labor force—by occupation: *agriculture:* 7%
industry: 11%
services: 82% (1983)

Unemployment rate: 11% (2001 est.)

Population below poverty line: NA%

Household income or consumption by percentage share:
lowest 10%: NA%
highest 10%: NA%

Inflation rate (consumer prices): 1.5% (2007 est.)

Budget:
revenues: $123.7 million
expenditures: $145.9 million (2000 est.)

Agriculture—products: cotton, fruits, vegetables, bananas, coconuts, cucumbers, mangoes, sugarcane; livestock

Industries: tourism, construction, light manufacturing (clothing, alcohol, household appliances)

Industrial production growth rate: NA%

Electricity—production: 105 million kWh (2005)

Electricity—production by source:
fossil fuel: 100%
hydro: 0%
nuclear: 0%
other: 0% (2001)

Electricity—consumption: 97.65 million kWh (2005)

Electricity—exports: 0 kWh (2005)

Electricity—imports: 0 kWh (2005)

Oil—production: 0 bbl/day (2005)

Oil—consumption: 4,000 bbl/day (2005 est.)

Oil—exports: 177.7 bbl/day (2004)

Oil—imports: 4,215 bbl/day (2004)

Oil—proved reserves: 0 bbl (1 January 2006 est.)

Natural gas—production: 0 cu m (2005 est.)

Natural gas—consumption: 0 cu m (2005 est.)

Natural gas—exports: 0 cu m (2005 est.)

Natural gas—imports: 0 cu m (2005)

Natural gas—proved reserves: 0 cu m (1 January 2006 est.)

Current account balance: -$211 million (2004)

Exports: $84.3 million (2007 est.)

Exports—commodities: petroleum products, bedding, handicrafts, electronic components, transport equipment, food and live animals

Exports—partners: Spain 34%, Germany 20.7%, Italy 7.7%, Singapore 5.8%, UK 4.9% (2006)

Imports: $522.8 million (2007 est.)

Imports—commodities: food and live animals, machinery and transport equipment, manufactures, chemicals, oil

Imports—partners: US 21.1%, China 16.4%, Germany 13.3%, Singapore 12.7%, Spain 6.5% (2006)

Economic aid—recipient: $7.23 million (2005)

Debt—external: $359.8 million (June 2006)

Currency (code): East Caribbean dollar (XCD)

Currency code: XCD

Exchange rates: East Caribbean dollars per US dollar—2.7 (2007), 2.7 (2006), 2.7 (2005), 2.7 (2004), 2.7 (2003)
note: fixed rate since 1976

Fiscal year: 1 April—31 March

COMMUNICATIONS

Telephones—main lines in use: 40,000 (2006)

Telephones—mobile cellular: 102,000 (2006)

Telephone system:
general assessment: NA
domestic: good automatic telephone system
international: country code—1-268; landing point for the East Caribbean Fiber System (ECFS) submarine cable with links to 13 other islands in the eastern Caribbean extending from the British Virgin Islands to Trinidad; satellite earth stations—2; tropospheric scatter to Saba (Netherlands Antilles)

and Guadeloupe (2007)
Radio broadcast stations: AM 4, FM 2, shortwave 0 (1998)
Radios: 36,000 (1997)
Television broadcast stations: 2 (1997)
Televisions: 31,000 (1997)
Internet country code: .ag
Internet hosts: 2,133 (2007)
Internet Service Providers (ISPs): 16 (2000)
Internet users: 32,000 (2006)

TRANSPORTATION

Airports: 3 (2007)
Airports—with paved runways:
total: 2
2,438 to 3,047 m: 1
under 914 m: 1 (2007)
Airports—with unpaved runways:
total: 1
under 914 m: 1 (2007)
Roadways:
total: 1,165 km
paved: 384 km
unpaved: 781 km (2002)
Merchant marine:
total: 1,059 ships (1000 GRT or over) 8,158,597 GRT/10,757,767 DWT
by type: bulk carrier 46, cargo 612, carrier 4, chemical tanker 6, container 350, liquefied gas 11, petroleum tanker 1, refrigerated cargo 9, roll on/roll off 20
foreign-owned: 1,021 (Australia 1, Colombia 1, Cyprus 2, Denmark 15, Estonia 15, France 1, Germany 891, Greece 3, Iceland 9, Latvia 9, Lebanon 1, Lithuania 6, Netherlands 19, Norway 7, NZ 2, Poland 2, Russia 5, Slovenia 6, Sweden 1, Switzerland 5, Turkey 7, UK 4, US 8, Vietnam 1) (2007)
Ports and terminals: Saint John's

MILITARY

Military branches: Royal Antigua and Barbuda Defense Force (2007)
Military service age and obligation: 18 years of age for voluntary military service; no conscription (2008)
Manpower available for military service:
males age 16–49: 19,560
females age 16–49: 18,977 (2008 est.)
Manpower fit for military service:
males age 16–49: 15,591
females age 16–49: 15,542 (2008 est.)
Manpower reaching militarily significant age annually:
males age 16–49: 627
females age 16–49: 609 (2008 est.)
Military expenditures—percent of GDP: NA

TRANSNATIONAL ISSUES

Disputes—international: none
Illicit drugs: considered a minor transshipment point for narcotics bound for the US and Europe; more significant as an offshore financial center

ARCTIC OCEAN

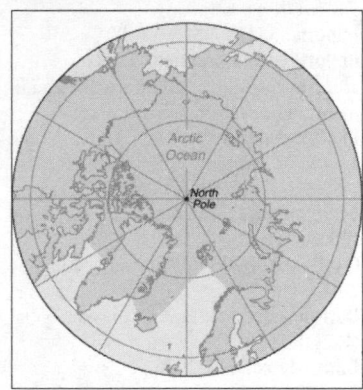

INTRODUCTION

Background: The Arctic Ocean is the smallest of the world's five oceans (after the Pacific Ocean, Atlantic Ocean, Indian Ocean, and the recently delimited Southern Ocean). The Northwest Passage (US and Canada) and Northern Sea Route (Norway and Russia) are two important seasonal waterways. A sparse network of air, ocean, river, and land routes circumscribes the Arctic Ocean.

GEOGRAPHY

Location: body of water between Europe, Asia, and North America, mostly north of the Arctic Circle
Geographic coordinates: 90 00 N, 0 00 E
Map references: Arctic Region
Area:
total: 14.056 million sq km
note: includes Baffin Bay, Barents Sea, Beaufort Sea, Chukchi Sea, East Siberian Sea, Greenland Sea, Hudson Bay, Hudson Strait, Kara Sea, Laptev Sea, Northwest Passage, and other tributary water bodies
Area—comparative: slightly less than 1.5 times the size of the US
Coastline: 45,389 km
Climate: polar climate characterized by persistent cold and relatively narrow annual temperature ranges; winters characterized by continuous darkness, cold and stable weather conditions, and clear skies; summers characterized by continuous daylight, damp and foggy weather, and weak cyclones with rain or snow
Terrain: central surface covered by a perennial drifting polar icepack that, on average, is about 3 meters thick, although pressure ridges may be three times that thickness; clockwise drift pattern in the Beaufort Gyral Stream, but nearly straight-line movement from the New Siberian Islands (Russia) to Denmark Strait (between Greenland and Iceland); the icepack is surrounded by open seas during the summer, but more than doubles in size during the winter and extends to the encircling landmasses; the ocean floor is about 50% continental shelf (highest percentage of any ocean) with the remainder a central basin interrupted by three submarine ridges (Alpha Cordillera, Nansen Cordillera, and Lomonosov Ridge)
Elevation extremes:
lowest point: Fram Basin -4,665 m
highest point: sea level 0 m
Natural resources: sand and gravel aggregates, placer deposits, polymetallic nodules, oil and gas fields, fish, marine mammals (seals and whales)
Natural hazards: ice islands occasionally break away from northern Ellesmere Island; icebergs calved from glaciers in western Greenland and extreme northeastern Canada; permafrost in islands; virtually ice locked from October to June; ships subject to superstructure icing from October to May
Environment—current issues: endangered marine species include walruses and whales; fragile ecosystem slow to change and slow to recover from disruptions or damage; thinning polar icepack
Geography—note: major chokepoint is the southern Chukchi Sea (northern access to the Pacific Ocean via the Bering Strait); strategic location between North America and Russia; shortest marine link between the extremes of eastern and western Russia; floating research stations operated by the US and Russia; maximum snow cover in March or April about 20 to 50 centimeters over the frozen ocean; snow cover lasts about 10 months

ECONOMY

Economy—overview: Economic activity is limited to the exploitation of natural resources, including petroleum, natural gas, fish, and seals.

Ports and terminals: Churchill (Canada), Murmansk (Russia), Prudhoe Bay (US)

Transportation—note: sparse network of air, ocean, river, and land routes; the Northwest Passage (North America) and Northern Sea Route (Eurasia) are important seasonal waterways

Disputes—international: some maritime disputes (see littoral states)

ARGENTINA

INTRODUCTION

Background: In 1816, the United Provinces of the Rio Plata declared their independence from Spain. After Bolivia, Paraguay, and Uruguay went their separate ways, the area that remained became Argentina. The country's population and culture were heavily shaped by immigrants from throughout Europe, but most particularly Italy and Spain, which provided the largest percentage of newcomers from 1860 to 1930. Up until about the mid-20th century, much of Argentina's history was dominated by periods of internal political conflict between Federalists and Unitarians and between civilian and military factions. After World War II, an era of Peronist authoritarian rule and interference in subsequent governments was followed by a military junta that took power in 1976. Democracy returned in 1983, and has persisted despite numerous challenges, the most formidable of which was a severe economic crisis in 2001–02 that led to violent public protests and the resignation of several interim presidents. The economy has recovered strongly since bottoming out in 2002.

GEOGRAPHY

Location: Southern South America, bordering the South Atlantic Ocean, between Chile and Uruguay
Geographic coordinates: 34 00 S, 64 00 W
Map references: South America
Area:
total: 2,766,890 sq km
land: 2,736,690 sq km
water: 30,200 sq km
Area—comparative: slightly less than three-tenths the size of the US
Land boundaries:
total: 9,861 km
border countries: Bolivia 832 km, Brazil 1,261 km, Chile 5,308 km, Paraguay 1,880 km, Uruguay 580 km
Coastline: 4,989 km
Maritime claims:
territorial sea: 12 nm
contiguous zone: 24 nm
exclusive economic zone: 200 nm
continental shelf: 200 nm or to the edge of the continental margin
Climate: mostly temperate; arid in southeast; subantarctic in southwest
Terrain: rich plains of the Pampas in northern half, flat to rolling plateau of Patagonia in south, rugged Andes along western border
Elevation extremes:
lowest point: Laguna del Carbon -105 m (located between Puerto San Julian and Comandante Luis Piedra Buena in the province of Santa Cruz)
highest point: Cerro Aconcagua 6,960 m (located in the northwestern corner of the province of Mendoza)
Natural resources: fertile plains of the pampas, lead, zinc, tin, copper, iron ore, manganese, petroleum, uranium
Land use:
arable land: 10.03%
permanent crops: 0.36%
other: 89.61% (2005)
Irrigated land: 15,500 sq km (2003)
Total renewable water resources: 814 cu km (2000)

Freshwater withdrawal (domestic/industrial/agricultural): *total:* 29.19 cu km/yr (17%/9%/74%)
per capita: 753 cu m/yr (2000)
Natural hazards: San Miguel de Tucuman and Mendoza areas in the Andes subject to earthquakes; pamperos are violent windstorms that can strike the pampas and northeast; heavy flooding
Environment—current issues: environmental problems (urban and rural) typical of an industrializing economy such as deforestation, soil degradation, desertification, air pollution, and water pollution
note: Argentina is a world leader in setting voluntary greenhouse gas targets
Environment—international agreements: *party to:* Antarctic-Environmental Protocol, Antarctic-Marine Living Resources, Antarctic Seals, Antarctic Treaty, Biodiversity, Climate Change, Climate Change-Kyoto Protocol, Desertification, Endangered Species, Environmental Modification, Hazardous Wastes, Law of the Sea, Marine Dumping, Ozone Layer Protection, Ship Pollution, Wetlands, Whaling
signed, but not ratified: Marine Life Conservation
Geography—note: second-largest country in South America (after Brazil); strategic location relative to sea lanes between the South Atlantic and the South Pacific Oceans (Strait of Magellan, Beagle Channel, Drake Passage); diverse geophysical landscapes range from tropical climates in the north to tundra in the far south; Cerro Aconcagua is the Western Hemisphere's tallest mountain, while Laguna del Carbon is the lowest point in the Western Hemisphere

PEOPLE

Population: 40,677,348 (July 2008 est.)
Age structure:
0–14 years: 24.6% (male 5,123,722/female 4,893,843)
15–64 years: 64.6% (male 13,143,693/female 13,127,372)
65 years and over: 10.8% (male 1,801,101/female 2,587,617) (2008 est.)
Median age:
total: 30.3 years

male: 29.3 years
female: 31.3 years (2008 est.)
Population growth rate: 0.917% (2008 est.)
Birth rate: 16.32 births/1,000 population (2008 est.)
Death rate: 7.54 deaths/1,000 population (2008 est.)
Net migration rate: 0.39 migrant(s)/1,000 population (2008 est.)
Sex ratio:
at birth: 1.05 male(s)/female
under 15 years: 1.05 male(s)/female
15–64 years: 1 male(s)/female
65 years and over: 0.7 male(s)/female
total population: 0.97 male(s)/female (2008 est.)
Infant mortality rate:
total: 13.87 deaths/1,000 live births
male: 15.65 deaths/1,000 live births
female: 11.99 deaths/1,000 live births (2008 est.)
Life expectancy at birth:
total population: 76.52 years
male: 72.81 years
female: 80.43 years (2008 est.)
Total fertility rate: 2.09 children born/woman (2008 est.)
HIV/AIDS—adult prevalence rate: 0.7% (2001 est.)
HIV/AIDS—people living with HIV/AIDS: 130,000 (2001 est.)
HIV/AIDS—deaths: 1,500 (2003 est.)
Major infectious diseases:
degree of risk: intermediate
food or waterborne diseases: bacterial diarrhea, hepatitis A
water contact disease: leptospirosis (2008)
Nationality:
noun: Argentine(s)
adjective: Argentine
Ethnic groups: white (mostly Spanish and Italian) 97%, mestizo (mixed white and Amerindian ancestry), Amerindian, or other non-white groups 3%
Religions: nominally Roman Catholic 92% (less than 20% practicing), Protestant 2%, Jewish 2%, other 4%
Languages: Spanish (official), Italian, English, German, French
Literacy:
definition: age 15 and over can read and write
total population: 97.2%
male: 97.2%
female: 97.2% (2001 census)

GOVERNMENT

Country name:
conventional long form: Argentine Republic
conventional short form: Argentina
local long form: Republica Argentina
local short form: Argentina
Government type: republic

Capital: *name:* Buenos Aires
geographic coordinates: 34 36 S, 58 40 W
time difference: UTC-3 (2 hours ahead of Washington, DC during Standard Time)
daylight saving time: +1hr, begins first Sunday in October; ends third Saturday in March; note—a new policy of daylight saving time was initiated by the government on 30 December 2007
Administrative divisions: 23 provinces (provincias, singular—provincia) and 1 autonomous city* (distrito federal); Buenos Aires, Buenos Aires Capital Federal*, Catamarca, Chaco, Chubut, Cordoba, Corrientes, Entre Rios, Formosa, Jujuy, La Pampa, La Rioja, Mendoza, Misiones, Neuquen, Rio Negro, Salta, San Juan, San Luis, Santa Cruz, Santa Fe, Santiago del Estero, Tierra del Fuego—Antartida e Islas del Atlantico Sur, Tucuman
note: the US does not recognize any claims to Antarctica
Independence: 9 July 1816 (from Spain)
National holiday: Revolution Day, 25 May (1810)
Constitution: 1 May 1853; amended many times starting in 1860
Legal system: mixture of US and West European legal systems; has not accepted compulsory ICJ jurisdiction
Suffrage: 18 years of age; universal and compulsory
Executive branch:
chief of state: President Cristina FERNANDEZ DE KIRCHNER (since 10 December 2007); Vice President Julio COBOS (since 10 December 2007); note—the president is both the chief of state and head of government
head of government: President Cristina FERNANDEZ DE KIRCHNER (since 10 December 2007); Vice President Julio COBOS (since 10 December 2007)
cabinet: Cabinet appointed by the president
elections: president and vice president elected on the same ticket by popular vote for four-year terms (eligible for a second term); election last held 28 October 2007 (next election to be held in 2011)
election results: Cristina FERNANDEZ DE KIRCHNER elected president; percent of vote—Cristina FERNANDEZ DE KIRCHNER 45%, Elisa CARRIO 23%, Roberto LAVAGNA 17%, Alberto Rodriguez SAA 8%
Legislative branch: bicameral National Congress or Congreso Nacional consists of the Senate (72 seats; members are elected by direct vote; presently one-third of the members elected every two years to serve six-year terms) and the Chamber of Deputies (257 seats; mem-

bers are elected by direct vote; one-half of the members elected every two years to serve four-year terms)
elections: Senate—last held 28 October 2007 (next to be held in 2009); Chamber of Deputies—last held last held 28 October 2007 (next to be held in 2009)
election results: Senate—percent of vote by bloc or party—NA; seats by bloc or party—FV 12, UCR 4, CC 4, other 4; Chamber of Deputies—percent of vote by bloc or party—NA; seats by bloc or party—FV 5, UCR 10, PJ 10, PRO 6, CC 16, FJ 2, other 31; note—Senate and Chamber of Deputies seating reflect the number of replaced senators and deputies, rather than the whole Senate and Chamber of Deputies
Judicial branch: Supreme Court or Corte Suprema (the nine Supreme Court judges are appointed by the president with approval of the Senate)
note: the Supreme Court currently has two unfilled vacancies, and the Argentine Congress is considering a bill to reduce the number of Supreme Court judges to five
Political parties and leaders: Coalicion Civica (a broad coalition loosely affiliated with Elisa CARRIO); Front for Victory or FV (a broad coalition, including elements of the UCR and numerous provincial parties) [Cristina FERNANDEZ DE KIRCHNER]; Interbloque Federal or IF (a broad coalition of approximately 12 parties including PRO); Justicialist Front or FJ; Justicialist Party or PJ (Peronist umbrella political organization); Radical Civic Union or UCR [Gerardo MORALES]; Republican Proposal or PRO (including Federal Recreate Movement or RECREAR [Ricardo LOPEZ MURPHY] and Commitment for Change or CPC [Mauricio MACRI]); Socialist Party or PS [Ruben GIUSTINIANI]; Union For All [Patricia BULLRICH]; several provincial parties
Political pressure groups and leaders: Argentine Association of Pharmaceutical Labs (CILFA); Argentine Industrial Union (manufacturers' association); Argentine Rural Confederation or CRA (small to medium landowners' association); Argentine Rural Society (large landowners' association); business organizations; Central of Argentine Workers or CTA (a radical union for employed and unemployed workers); General Confederation of Labor or CGT (Peronist-leaning umbrella labor organization); Peronist-dominated labor movement; Piquetero groups (popular protest organizations

that can be either pro or anti-government); Roman Catholic Church; students

International organization participation: ABEDA, AfDB, Australia Group, BCIE, BIS, CAN (associate), CPLP (associate), CSN, FAO, G-15, G-24, G-77, IADB, IAEA, IBRD, ICAO, ICC, ICCt, ICRM, IDA, IFAD, IFC, IFRCS, IHO, ILO, IMF, IMO, IMSO, Interpol, IOC, IOM, IPU, ISO, ITSO, ITU, ITUC, LAES, LAIA, Mercosur, MIGA, MINURSO, MINUSTAH, NSG, OAS, OPANAL, OPCW, PCA, RG, UN, UN Security Council (temporary), UNCTAD, UNESCO, UNFICYP, UNHCR, UNIDO, Union Latina (observer), UNTSO, UNWTO, UPU, WCL, WCO, WFTU, WHO, WIPO, WMO, WTO, ZC

Diplomatic representation in the US:
chief of mission: Ambassador Hector Marcos TIMERMAN
chancery: 1600 New Hampshire Avenue NW, Washington, DC 20009
telephone: [1] (202) 238-6400
FAX: [1] (202) 332-3171
consulate(s) general: Atlanta, Chicago, Houston, Los Angeles, Miami, New York

Diplomatic representation from the US:
chief of mission: Ambassador Earl Anthony WAYNE
embassy: Avenida Colombia 4300, C1425GMN Buenos Aires
mailing address: international mail: use embassy street address; APO address: Unit 4334, APO AA 34034
telephone: [54] (11) 5777-4533
FAX: [54] (11) 5777-4240

Flag description: three equal horizontal bands of light blue (top), white, and light blue; centered in the white band is a radiant yellow sun with a human face known as the Sun of May

ECONOMY

Economy—overview: Argentina benefits from rich natural resources, a highly literate population, an export-oriented agricultural sector, and a diversified industrial base. Although one of the world's wealthiest countries 100 years ago, Argentina suffered during most of the 20th century from recurring economic crises, persistent fiscal and current account deficits, high inflation, mounting external debt, and capital flight. A severe depression, growing public and external indebtedness, and a bank run culminated in 2001 in the most serious economic, social, and political crisis in the country's turbulent history. Interim President Adolfo RODRIGUEZ SAA declared a default—the largest in history—on the government's foreign debt in December of that year, and abruptly resigned only a few days after taking office. His successor, Eduardo DUHALDE, announced an end to the peso's decade-long 1-to-1 peg to the US dollar in early 2002. The economy bottomed out that year, with real GDP 18% smaller than in 1998 and almost 60% of Argentines under the poverty line. Real GDP rebounded to grow by an average 9% annually over the subsequent five years, taking advantage of previously idled industrial capacity and labor, an audacious debt restructuring and reduced debt burden, excellent international financial conditions, and expansionary monetary and fiscal policies. Inflation, however, reached double-digit levels in 2006 and the government of President Nestor KIRCHNER responded with "voluntary" price agreements with businesses, as well as export taxes and restraints. Multi-year price freezes on electricity and natural gas rates for residential users stoked consumption and kept private investment away, leading to restrictions on industrial use and blackouts in 2007.

GDP (purchasing power parity): $523.7 billion (2007 est.)

GDP (official exchange rate): $260 billion (2007 est.)

GDP—real growth rate: 8.7% (2007 est.)

GDP—per capita (PPP): $13,300 (2007 est.)

GDP—composition by sector:
agriculture: 9.5%
industry: 34%
services: 56.5% (2007 est.)

Labor force: 16.03 million
note: urban areas only (2007 est.)

Labor force—by occupation: *agriculture:* 1%
industry: 23%
services: 76% (2007 est.)

Unemployment rate: 14.1% (2007 est.)

Population below poverty line: 23.4% (January-June 2007)

Household income or consumption by percentage share:
lowest 10%: 1%
highest 10%: 35% (January-March 2007)

Distribution of family income—Gini index: 49 (2006)

Inflation rate (consumer prices): 8.8% official rate; actual rate may be double the official rate (2007 est.)

Investment (gross fixed): 22% of GDP (2007 est.)

Budget:
revenues: $48.99 billion
expenditures: $61.33 billion (2007 est.)

Public debt: 59% of GDP (June 2007 est.)

Agriculture—products: sunflower seeds, lemons, soybeans, grapes, corn, tobacco, peanuts, tea, wheat; livestock

Industries: food processing, motor vehicles, consumer durables, textiles, chemicals and petrochemicals, printing, metallurgy, steel

Industrial production growth rate: 7.5% (2007 est.)

Electricity—production: 101.1 billion kWh (2005)

Electricity—production by source:
fossil fuel: 52.2%
hydro: 40.8%
nuclear: 6.7%
other: 0.2% (2001)

Electricity—consumption: 88.98 billion kWh (2005)

Electricity—exports: 4.14 billion kWh (2005)

Electricity—imports: 8.017 billion kWh (2005)

Oil—production: 801,700 bbl/day (2005 est.)

Oil—consumption: 480,000 bbl/day (2005 est.)

Oil—exports: 367,600 bbl/day (2004)

Oil—imports: 21,650 bbl/day (2004)

Oil—proved reserves: 2.086 billion bbl (2007 est.)

Natural gas—production: 43.76 billion cu m (2005 est.)

Natural gas—consumption: 38.79 billion cu m (2005 est.)

Natural gas—exports: 6.646 billion cu m (2005 est.)

Natural gas—imports: 1.669 billion cu m (2005)

Natural gas—proved reserves: 512.4 billion cu m (1 January 2006 est.)

Current account balance: $7.438 billion (2007 est.)

Exports: $54.6 billion f.o.b. (2007 est.)

Exports—commodities: soybeans and derivatives, petroleum and gas, vehicles, corn, wheat

Exports—partners: Brazil 17.5%, Chile 9.5%, US 8.9%, China 7.5% (2006)

Imports: $42.59 billion f.o.b. (2007 est.)

Imports—commodities: machinery, motor vehicles, petroleum and natural gas, organic chemicals, plastics

Imports—partners: Brazil 34.8%, US 12.6%, China 9.1%, Germany 4.5% (2006)

Economic aid—recipient: $99.66 million (2005)

Reserves of foreign exchange and gold: $46.12 billion (31 December 2007 est.)

Debt—external: $135.4 billion (31 December 2007)

Stock of direct foreign investment—at home: $64.11 billion (2007 est.)

Stock of direct foreign investment—

abroad: $26.16 billion (2007 est.)
Market value of publicly traded shares: $79.73 billion (2006)
Currency (code): Argentine peso (ARS)
Currency code: ARS
Exchange rates: Argentine pesos per US dollar—3.1105 (2007), 3.0543 (2006), 2.9037 (2005), 2.9233 (2004), 2.9006 (2003)
Fiscal year: calendar year

COMMUNICATIONS

Telephones—main lines in use: 9.46 million (2006)
Telephones—mobile cellular: 31.51 million (2006)
Telephone system:
general assessment: by opening the telecommunications market to competition and foreign investment with the "Telecommunications Liberalization Plan of 1998," Argentina encouraged the growth of modern telecommunications technology; fiber-optic cable trunk lines are being installed between all major cities; major networks are entirely digital and the availability of telephone service is improving; fixed-line telephone density is gradually increasing reaching nearly 25 lines per 100 people in 2006; mobile telephone density has been increasing rapidly and has reached a level of 80 telephones per 100 persons
domestic: microwave radio relay, fiber-optic cable, and a domestic satellite system with 40 earth stations serve the trunk network; more than 110,000 pay telephones are installed and mobile telephone use is rapidly expanding; broadband services are gaining ground
international: country code—54; landing point for the Atlantis-2, UNISUR, and South America-1 optical submarine cable systems that provide links to Europe, Africa, South and Central America, and US; satellite earth stations—112; 2 international gateways near Buenos Aires (2007)
Radio broadcast stations: AM 260 (includes 10 inactive stations), FM (probably more than 1,000, mostly unlicensed), shortwave 6 (1998)
Radios: 24.3 million (1997)
Television broadcast stations: 42 (plus 444 repeaters) (1997)
Televisions: 7.95 million (1997)
Internet country code: .ar
Internet hosts: 2.159 million (2007)
Internet Service Providers (ISPs): 33 (2000)
Internet users: 8.184 million (2006)

TRANSPORTATION

Airports: 1,272 (2007)
Airports—with paved runways:
total: 154
over 3,047 m: 4
2,438 to 3,047 m: 26
1,524 to 2,437 m: 65
914 to 1,523 m: 50
under 914 m: 9 (2007)
Airports—with unpaved runways:
total: 1,118
over 3,047 m: 2
2,438 to 3,047 m: 1
1,524 to 2,437 m: 44
914 to 1,523 m: 515
under 914 m: 556 (2007)
Heliports: 1 (2007)
Pipelines: gas 28,657 km; liquid petroleum gas 41 km; oil 5,607 km; refined products 3,052 km; unknown (oil/water) 13 km (2007)
Railways:
total: 31,902 km
broad gauge: 20,858 km 1.676-m gauge (141 km electrified)
standard gauge: 2,885 km 1.435-m gauge (26 km electrified)
narrow gauge: 7,922 km 1.000-m gauge; 237 km 0.750-m gauge (2006)
Roadways:
total: 231,374 km
paved: 69,412 km (includes 734 km of expressways)
unpaved: 161,962 km (2004)
Waterways: 11,000 km (2006)
Merchant marine:
total: 47 ships (1000 GRT or over) 542,556 GRT/892,818 DWT
by type: bulk carrier 4, cargo 11, chemical tanker 1, container 1, passenger 1, passenger/cargo 3, petroleum tanker 23, refrigerated cargo 2, roll on/roll off 1
foreign-owned: 12 (Chile 7, UK 4, Uruguay 1)
registered in other countries: 19 (Bolivia 1, Chile 1, Liberia 3, Panama 8, Paraguay 3, Uruguay 3) (2007)
Ports and terminals: Arroyo Seco, Bahia Blanca, Buenos Aires, La Plata, Punta Colorada, Rosario, San Lorenzo-San Martin

MILITARY

Military branches: Argentine Army (Ejercito Argentino), Navy of the Argentine Republic (Armada Republica; includes naval aviation and naval infantry), Argentine Air Force (Fuerza Aerea Argentina, FAA) (2008)
Military service age and obligation: 18–24 years of age for voluntary military service (18–21 requires parental permission); no conscription (2001)
Manpower available for military service:
males age 16–49: 10,029,488
females age 16–49: 9,889,002 (2008 est.)
Manpower fit for military service:
males age 16–49: 8,352,147
females age 16–49: 8,366,781 (2008 est.)
Manpower reaching militarily significant age annually:
males age 16–49: 350,040
females age 16–49: 334,830 (2008 est.)
Military expenditures—percent of GDP: 1.3% (2005 est.)
Military—note: the Argentine military is a well-organized force constrained by the country's prolonged economic hardship; the country has recently experienced a strong recovery, and the military is implementing a modernization plan aimed at making the ground forces lighter and more responsive (2008)

TRANSNATIONAL ISSUES

Disputes—international: Argentina continues to assert its claims to the UK-administered Falkland Islands (Islas Malvinas) and South Georgia and the South Sandwich Islands in its constitution, forcibly occupying the Falklands in 1982, but in 1995 agreed no longer to seek settlement by force; territorial claim in Antarctica partially overlaps UK and Chilean claims (see Antarctic disputes); unruly region at convergence of Argentina-Brazil-Paraguay borders is locus of money laundering, smuggling, arms and illegal narcotics trafficking, and fundraising for extremist organizations; uncontested dispute between Brazil and Uruguay over Braziliera/Brasiliera Island in the Quarai/Cuareim River leaves the tripoint with Argentina in question; in January 2007, ICJ provisionally ruled Uruguay may begin construction of two paper mills on the Uruguay River, which forms the border with Argentina, while the court examines further whether Argentina has the legal right to stop such construction with potential environmental implications to both countries; the joint boundary commission, established by Chile and Argentina in 2001 has yet to map and demarcate the delimited boundary in the inhospitable Andean Southern Ice Field (Campo de Hielo Sur)
Trafficking in persons: *current situation:* Argentina is primarily a destination country for women and children trafficked for sexual and labor exploitation with most victims trafficked internally, from rural to urban areas, for exploitation in prostitution; foreign women and children trafficked for commercial sexual exploitation come primarily from Paraguay, but also from Bolivia, Brazil, the Dominican Republic, Colombia, and Chile; Bolivians are trafficked for forced labor; Argentine women and girls are also trafficked to neighboring countries for sexual exploitation

tier rating: Tier 2 Watch List—Argentina failed to show evidence of increasing efforts to combat trafficking particularly in the key area of prosecutions

Illicit drugs: used as a transshipment country for cocaine headed for Europe; some money-laundering activity, especially in the Tri-Border Area; domestic consumption of drugs in urban centers is increasing

ARMENIA

INTRODUCTION

Background: Armenia prides itself on being the first nation to formally adopt Christianity (early 4th century). Despite periods of autonomy, over the centuries Armenia came under the sway of various empires including the Roman, Byzantine, Arab, Persian, and Ottoman. During World War I in the western portion of Armenia, Ottoman Turkey instituted a policy of forced resettlement coupled with other harsh practices that resulted in an estimated 1 million Armenian deaths. The eastern area of Armenia was ceded by the Ottomans to Russia in 1828; this portion declared its independence in 1918, but was conquered by the Soviet Red Army in 1920. Armenian leaders remain preoccupied by the long conflict with Muslim Azerbaijan over Nagorno-Karabakh, a primarily Armenian-populated region, assigned to Soviet Azerbaijan in the 1920s by Moscow. Armenia and Azerbaijan began fighting over the area in 1988; the struggle escalated after both countries attained independence from the Soviet Union in 1991. By May 1994, when a cease-fire took hold, Armenian forces held not only Nagorno-Karabakh but also a significant portion of Azerbaijan proper. The economies of both sides have been hurt by their inability to make substantial progress toward a peaceful resolution. Turkey imposed an economic blockade on Armenia and closed the common border because of the Armenian separatists' control of Nagorno-Karabakh and surrounding areas.

GEOGRAPHY

Location: Southwestern Asia, east of Turkey
Geographic coordinates: 40 00 N, 45 00 E
Map references: Asia
Area:
total: 29,800 sq km
land: 28,400 sq km
water: 1,400 sq km
Area—comparative: slightly smaller than Maryland
Land boundaries:
total: 1,254 km
border countries: Azerbaijan-proper 566 km, Azerbaijan-Naxcivan exclave 221 km, Georgia 164 km, Iran 35 km, Turkey 268 km
Coastline: 0 km (landlocked)
Maritime claims: none (landlocked)
Climate: highland continental, hot summers, cold winters
Terrain: Armenian Highland with mountains; little forest land; fast flowing rivers; good soil in Aras River valley
Elevation extremes:
lowest point: Debed River 400 m
highest point: Aragats Lerrnagagat' 4,090 m
Natural resources: small deposits of gold, copper, molybdenum, zinc, bauxite
Land use:
arable land: 16.78%
permanent crops: 2.01%
other: 81.21% (2005)
Irrigated land: 2,860 sq km (2003)
Total renewable water resources: 10.5 cu km (1997)
Freshwater withdrawal (domestic/industrial/agricultural): *total:* 2.95 cu km/yr (30%/4%/66%)
per capita: 977 cu m/yr (2000)
Natural hazards: occasionally severe earthquakes; droughts
Environment—current issues: soil pollution from toxic chemicals such as DDT; the energy crisis of the 1990s led to deforestation when citizens scavenged for firewood; pollution of Hrazdan (Razdan) and Aras Rivers; the draining of Sevana Lich (Lake Sevan), a result of its use as a source for hydropower, threatens drinking water supplies; restart of Metsamor nuclear power plant in spite of its location in a seismically active zone

Environment—international agreements: *party to:* Air Pollution, Biodiversity, Climate Change, Climate Change-Kyoto Protocol, Desertification, Hazardous Wastes, Law of the Sea, Ozone Layer Protection, Wetlands
signed, but not ratified: Air Pollution-Persistent Organic Pollutants
Geography—note: landlocked in the Lesser Caucasus Mountains; Sevana Lich (Lake Sevan) is the largest lake in this mountain range

PEOPLE

Population: 2,968,586 (July 2008 est.)
Age structure:
0–14 years: 18.7% (male 296,401/female 259,594)
15–64 years: 70.3% (male 975,438/female 1,111,989)
65 years and over: 11% (male 128,398/female 196,766) (2008 est.)
Median age:
total: 31.1 years
male: 28.4 years
female: 34 years (2008 est.)
Population growth rate: -0.077% (2008 est.)
Birth rate: 12.53 births/1,000 population (2008 est.)
Death rate: 8.34 deaths/1,000 population (2008 est.)
Net migration rate: -4.95 migrant(s)/1,000 population (2008 est.)
Sex ratio:
at birth: 1.15 male(s)/female
under 15 years: 1.14 male(s)/female
15–64 years: 0.88 male(s)/female
65 years and over: 0.65 male(s)/female
total population: 0.89 male(s)/female (2008 est.)
Infant mortality rate:
total: 20.94 deaths/1,000 live births
male: 25.82 deaths/1,000 live births
female: 15.33 deaths/1,000 live births (2008 est.)
Life expectancy at birth:
total population: 72.4 years
male: 68.79 years
female: 76.55 years (2008 est.)
Total fertility rate: 1.35 children born/woman (2008 est.)
HIV/AIDS—adult prevalence rate: 0.1% (2003 est.)
HIV/AIDS—people living with HIV/AIDS: 2,600 (2003 est.)

33

HIV/AIDS—deaths: fewer than 200 (2003 est.)

Nationality:

noun: Armenian(s)

adjective: Armenian

Ethnic groups: Armenian 97.9%, Yezidi (Kurd) 1.3%, Russian 0.5%, other 0.3% (2001 census)

Religions: Armenian Apostolic 94.7%, other Christian 4%, Yezidi (monotheist with elements of nature worship) 1.3%

Languages: Armenian 97.7%, Yezidi 1%, Russian 0.9%, other 0.4% (2001 census)

Literacy:

definition: age 15 and over can read and write

total population: 99.4%

male: 99.7%

female: 99.2% (2001 census)

GOVERNMENT

Country name:

conventional long form: Republic of Armenia

conventional short form: Armenia

local long form: Hayastani Hanrapetut'yun

local short form: Hayastan

former: Armenian Soviet Socialist Republic, Armenian Republic

Government type: republic

Capital: *name:* Yerevan

geographic coordinates: 40 10 N, 44 30 E

time difference: UTC+4 (9 hours ahead of Washington, DC during Standard Time)

daylight saving time: +1hr, begins last Sunday in March; ends last Sunday in October

Administrative divisions: 11 provinces (marzer, singular—marz); Aragatsotn, Ararat, Armavir, Geghark'unik', Kotayk', Lorri, Shirak, Syunik', Tavush, Vayots' Dzor, Yerevan

Independence: 21 September 1991 (from Soviet Union)

National holiday: Independence Day, 21 September (1991)

Constitution: adopted by nationwide referendum 5 July 1995; amendments adopted through a nationwide referendum 27 November 2005

Legal system: based on civil law system; has not accepted compulsory ICJ jurisdiction

Suffrage: 18 years of age; universal

Executive branch:

chief of state: President Serzh SARGSIAN (since 9 April 2008)

head of government: Prime Minister Tigran SARGSIAN (since 9 April 2008)

cabinet: Council of Ministers appointed by the prime minister

elections: president elected by popular vote for a five-year term (eligible for a second term); election last held 19 February 2008 (next to be held February 2013); prime minister appointed by the president based on majority or plurality support in parliament; the prime minister and Council of Ministers must resign if the National Assembly refuses to accept their program

election results: Serzh SARGSIAN elected president; percent of vote—Serzh SARGSIAN 52.9%, Levon TER-PETROSSIAN 21.5%, Artur BAGHDASARIAN 16.7%

Legislative branch: unicameral National Assembly (Parliament) or Azgayin Zhoghov (131 seats; members elected by popular vote, 90 members elected by party list and 41 by direct vote; to serve four-year terms)

elections: last held 12 May 2007 (next to be held in the spring of 2012)

election results: percent of vote by party—HHK 33.9%, Prosperous Armenia 15.1%, ARF (Dashnak) 13.2%, Rule of Law 7.1%, Heritage Party 6%, other 24.7%; seats by party—HHK 64, Prosperous Armenia 18, ARF (Dashnak) 16, Rule of Law 9, Heritage Party 7, independent 17

Judicial branch: Constitutional Court; Court of Cassation (Appeals Court)

Political parties and leaders: Armenian National Movement or ANM [Ararat ZURABYAN]; Armenian People's Party [Tigran KARAPETYAN]; Armenian Ramkavar Azadagan Party Alliance or HRAK (includes former Dashink Party, National Revival Party, and Ramkavar Liberal Party); Armenian Revolutionary Federation ("Dashnak" Party) or ARF [Hrant MARKARYAN]; Heritage Party [Raffi HOVHANNISYAN]; National Democratic Party [Shavarsh KOCHARIAN]; National Democratic Union or NDU [Vazgen MANUKIAN]; National Unity Party [Artashes GEGHAMYAN]; People's Party of Armenia [Stepan DEMIRCHYAN]; Prosperous Armenia [Gagik TSAROUKYAN]; Republic Party [Aram SARKISYAN]; Republican Party of Armenia or HHK [Serzh SARGSIAN]; Rule of Law Party (Orinats Yerkir) [Artur BAGHDASARIAN]; Union of Constitutional Rights [Hrant KHACHATURYAN]; United Labor Party [Gurgen ARSENYAN]

Political pressure groups and leaders: Yerkrapah Union [Manvel GRIGORIAN], Aylentrank (Impeachment) [Nikol PASHINYAN]

International organization participation: ACCT (observer), ADB, BSEC, CE, CIS, CSTO, EAEC (observer), EAPC, EBRD, FAO, GCTU, IAEA, IBRD, ICAO, ICCt (signatory), ICRM, IDA, IFAD, IFC, IFRCS, ILO, IMF, Interpol, IOC, IOM, IPU, ISO, ITSO, ITU, MIGA, NAM (observer), OAS (observer), OIF (observer), OPCW, OSCE, PFP, UN, UNCTAD, UNESCO, UNIDO, UNWTO, UPU, WCO, WFTU, WHO, WIPO, WMO, WTO

Diplomatic representation in the US:

chief of mission: Ambassador Tatoul MARKARIAN

chancery: 2225 R Street NW, Washington, DC 20008

telephone: [1] (202) 319-1976

FAX: [1] (202) 319-2982

consulate(s) general: Los Angeles

Diplomatic representation from the US:

chief of mission: Ambassador Joseph S. PENNINGTON

embassy: 1 American Ave., Yerevan 0082

mailing address: American Embassy Yerevan, US Department of State, 7020 Yerevan Place, Washington, DC 20521-7020

telephone: [374](10) 464-700

FAX: [374](10) 464-742

Flag description: three equal horizontal bands of red (top), blue, and orange

ECONOMY

Economy—overview: Since the breakup of the Soviet Union in 1991, Armenia has made progress in implementing many economic reforms including privatization, price reforms, and prudent fiscal policies. The conflict with Azerbaijan over the ethnic Armenian-dominated region of Nagorno-Karabakh contributed to a severe economic decline in the early 1990s. By 1994, however, the Armenian Government launched an ambitious IMF-sponsored economic liberalization program that resulted in positive growth rates. Economic growth has averaged over 13% in recent years. Armenia has managed to reduce poverty, slash inflation, stabilize its currency, and privatize most small- and medium-sized enterprises. Under the old Soviet central planning system, Armenia developed a modern industrial sector, supplying machine tools, textiles, and other manufactured goods to sister republics, in exchange for raw materials and energy. Armenia has since switched to small-scale agriculture and away from the large agroindustrial complexes of the Soviet era. Nuclear power plants built at Metsamor in the 1970s were closed following the 1988 Spitak Earthquake, though they sustained no damage. One of the two reactors was re-opened in 1995, but the Armenian government is under international pressure to close it due to concerns that the Soviet era design lacks important safeguards. Metsamor provides 40 percent of the country's electricity—hydropower accounts for about one-fourth. Economic ties with Russia remain close, especially in the energy sector. The electricity distribution system was privatized in 2002 and bought by Russia's RAO-UES in 2005. Construction of a pipeline to deliver natural gas from Iran to Armenia is halfway completed and is scheduled to be commissioned by January 2009. Armenia has some mineral deposits (copper, gold, bauxite). Pig iron, unwrought copper, and other nonferrous metals are Armenia's highest valued exports. Armenia's severe trade imbalance has been offset somewhat by international aid, remittances from Armenians working abroad, and foreign direct investment. Armenia joined the WTO in January 2003. The government made some improvements in tax and customs administration in recent years, but anti-corruption measures will be more difficult to implement. Despite strong economic growth, Armenia's unemployment rate remains high. Armenia will need to pursue additional economic reforms in order to improve its economic competitiveness and to build on recent improvements in poverty and unemployment, especially given its economic isolation from two of its nearest neighbors, Turkey and Azerbaijan.

GDP (purchasing power parity): $17.15 billion (2007 est.)

GDP (official exchange rate): $7.974 billion (2007 est.)

GDP—real growth rate: 13.8% (2007 est.)

GDP—per capita (PPP): $4,900 (2007 est.)

GDP—composition by sector:
agriculture: 17.2%
industry: 36.4%
services: 46.4% (2007 est.)

Labor force: 1.2 million (2007 est.)

Labor force—by occupation: *agriculture:* 46.2%
industry: 15.6%
services: 38.2% (2006 est.)

Unemployment rate: 7.1% (2007 est.)

Population below poverty line: 26.5% (2006 est.)

Household income or consumption by percentage share:
lowest 10%: 1.6%
highest 10%: 41.3% (2004)

Distribution of family income—Gini index: 37 (2006)

Inflation rate (consumer prices): 4.4% (2007 est.)

Investment (gross fixed): 32.9% of GDP (2007 est.)

Budget:
revenues: $1.667 billion
expenditures: $1.654 billion; including capital expenditures of $NA (2007 est.)

Agriculture—products: fruit (especially grapes), vegetables; livestock

Industries: diamond-processing, metal-cutting machine tools, forging-pressing machines, electric motors, tires, knitted wear, hosiery, shoes, silk fabric, chemicals, trucks, instruments, microelectronics, jewelry manufacturing, software development, food processing, brandy

Industrial production growth rate: 2.6% (2007 est.)

Electricity—production: 5.941 billion kWh (2006)

Electricity—production by source:
fossil fuel: 42.3%
hydro: 27%
nuclear: 30.7%
other: 0% (2001)

Electricity—consumption: 5.454 billion kWh (2006)

Electricity—exports: 754.5 million kWh; note—exports an unknown quantity to Georgia; includes exports to Nagorno-Karabakh region in Azerbaijan (2006)

Electricity—imports: 354.9 million kWh; note—imports an unknown quantity from Iran (2006)

Oil—production: 0 bbl/day (2005)

Oil—consumption: 40,000 bbl/day (2005 est.)

Oil—exports: 0 bbl/day (2004)

Oil—imports: 41,240 bbl/day (2004)

Oil—proved reserves: 0 bbl (1 January 2006 est.)

Natural gas—production: 0 cu m (2007 est.)

Natural gas—consumption: 2.2 billion cu m (2007 est.)

Natural gas—exports: 0 cu m (2007 est.)

Natural gas—imports: 2.2 billion cu m (2007)

Natural gas—proved reserves: 0 cu m (1 January 2006)

Current account balance: -$518 million (2007 est.)

Exports: $1.157 billion f.o.b. (2007 est.)

Exports—commodities: pig iron, unwrought copper, nonferrous metals, diamonds, mineral products, foodstuffs, energy

Exports—partners: Germany 18.3%, Netherlands 14.1%, Belgium 13.3%, Russia 13.1%, Israel 7%, US 6.1%, Georgia 5.1%, Iran 4.9% (2006)

Imports: $2.714 billion f.o.b. (2007 est.)

Imports—commodities: natural gas, petroleum, tobacco products, foodstuffs, diamonds

Imports—partners: Russia 21.8%, Ukraine 7.8%, Belgium 7.6%, Turkmenistan 7.1%, Italy 6.1%, Germany 5.7%, Iran 5.7%, Israel 4.8%, US 4.5%, Georgia 4.1% (2006)

Economic aid—recipient: ODA, $180 million (2007)

Reserves of foreign exchange and gold: $1.646 billion (December 2007 est.)

Debt—external: $1.372 billion (31 December 2007 est.)

Market value of publicly traded shares: $42.8 million (2005)

Currency (code): dram (AMD)

Currency code: AMD

Exchange rates: drams per US dollar—344.06 (2007), 414.69 (2006), 457.69 (2005), 533.45 (2004), 578.76 (2003)

Fiscal year: calendar year

COMMUNICATIONS

Telephones—main lines in use: 594,400 (2005)

Telephones—mobile cellular: 318,000 (2005)

Telephone system:
general assessment: telecommunications investments have made major inroads in modernizing and upgrading the outdated telecommunications network inherited from the Soviet era; now 100% privately owned and undergoing modernization and expansion; mobile-cellular services monopoly terminated in late 2004 and a second provider began operations in mid-2005
domestic: reliable modern landline and mobile-cellular services are available across Yerevan in major cities and towns; significant but ever-shrinking gaps remain in mobile-cellular coverage in rural areas
international: country code—374; Yerevan is connected to the Trans-Asia-Europe fiber-optic cable through Iran; additional international service is available by microwave radio relay and landline connections to the other countries of the Commonwealth of Independent States, through the Moscow international switch, and by satellite to the rest of the world; satellite earth stations—3 (2007)

Radio broadcast stations: AM 9, FM 16, shortwave 1 (2006)

Radios: 850,000 (1997)

Television broadcast stations: 48 (private television stations alongside 2 public networks; major Russian channels widely available) (2006)

Televisions: 825,000 (1997)

Internet country code: .am

Internet hosts: 8,270 (2007)

Internet Service Providers (ISPs): 9 (2001)

Internet users: 172,800 (2006)

35

TRANSPORTATION

Airports: 12 (2007)
Airports—with paved runways:
total: 10
over 3,047 m: 2
2,438 to 3,047 m: 2
1,524 to 2,437 m: 4
914 to 1,523 m: 2 (2007)
Airports—with unpaved runways:
total: 2
1,524 to 2,437 m: 1
914 to 1,523 m: 1 (2007)
Pipelines: gas 2,036 km (2007)
Railways:
total: 839 km
broad gauge: 839 km 1.520-m gauge (828 km electrified)
note: some lines are out of service (2006)
Roadways:
total: 7,700 km
paved: 7,700 km (includes 1,561 km of expressways) (2006)

MILITARY

Military branches: Armed Forces: Ground Forces, Nagorno-Karabakh Self Defense Force (NKSDF), Air Force and Air Defense (2008)
Military service age and obligation: 18–27 years of age for compulsory military service; 18 years of age for voluntary military service; 2-year conscript service obligation (2006)
Manpower available for military service:
males age 16–49: 809,576
females age 16–49: 870,864 (2008 est.)
Manpower fit for military service:
males age 16–49: 637,776
females age 16–49: 729,846 (2008 est.)
Manpower reaching militarily significant age annually:
males age 16–49: 30,548
females age 16–49: 29,170 (2008 est.)
Military expenditures—percent of GDP: 6.5% (FY01)

TRANSNATIONAL ISSUES

Disputes—international: Armenia supports ethnic Armenian secessionists in Nagorno-Karabakh and since the early 1990s, has militarily occupied 16% of Azerbaijan—Organization for Security and Cooperation in Europe (OSCE) continues to mediate dispute; over 800,000 mostly ethnic Azerbaijanis were driven from the occupied lands and Armenia; about 230,000 ethnic Armenians were driven from their homes in Azerbaijan into Armenia; Azerbaijan seeks transit route through Armenia to connect to Naxcivan exclave; border with Turkey remains closed over Nagorno-Karabakh dispute; ethnic Armenian groups in Javakheti region of Georgia seek greater autonomy; Armenians continue to emigrate, primarily to Russia, seeking employment
Refugees and internally displaced persons: *refugees (country of origin):* 113,295 (Azerbaijan)
IDPs: 8,400 (conflict with Azerbaijan over Nagorno-Karabakh, majority have returned home since 1994 ceasefire) (2007)
Trafficking in persons: *current situation:* Armenia is a major source and, to a lesser extent, a transit and destination country for women and girls trafficked for sexual exploitation largely to the UAE and Turkey; traffickers, many of them women, route victims directly into Dubai or through Moscow; profits derived from the trafficking of Armenian victims reportedly have increased
tier rating: Tier 2 Watch List—Armenia has failed to show evidence of increasing efforts, particularly in the areas of enforcement, trafficking-related corruption, and victim protection
Illicit drugs: illicit cultivation of small amount of cannabis for domestic consumption; minor transit point for illicit drugs—mostly opium and hashish—moving from Southwest Asia to Russia and to a lesser extent the rest of Europe

ARUBA

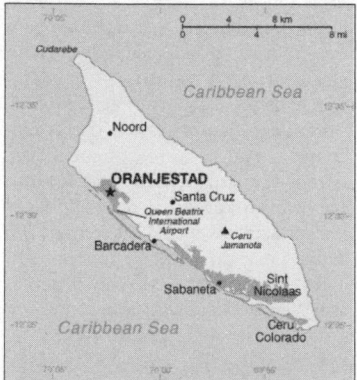

INTRODUCTION

Background: Discovered and claimed for Spain in 1499, Aruba was acquired by the Dutch in 1636. The island's economy has been dominated by three main industries. A 19th century gold rush was followed by prosperity brought on by the opening in 1924 of an oil refinery. The last decades of the 20th century saw a boom in the tourism industry. Aruba seceded from the Netherlands Antilles in 1986 and became a separate, autonomous member of the Kingdom of the Netherlands. Movement toward full independence was halted at Aruba's request in 1990.

GEOGRAPHY

Location: Caribbean, island in the Caribbean Sea, north of Venezuela
Geographic coordinates: 12 30 N, 69 58 W
Map references: Central America and the Caribbean
Area:
total: 193 sq km
land: 193 sq km
water: 0 sq km
Area—comparative: slightly larger than Washington, DC
Land boundaries: 0 km
Coastline: 68.5 km
Maritime claims: *territorial sea:* 12 nm
Climate: tropical marine; little seasonal temperature variation
Terrain: flat with a few hills; scant vegetation
Elevation extremes:
lowest point: Caribbean Sea 0 m
highest point: Mount Jamanota 188 m
Natural resources: NEGL; white sandy beaches
Land use:
arable land: 10.53%
permanent crops: 0%
other: 89.47% (2005)
Irrigated land: 0.01 sq km (1998 est.)
Natural hazards: hurricanes; lies outside the Caribbean hurricane belt and is rarely threatened
Environment—current issues: NA
Geography—note: a flat, riverless island renowned for its white sand beaches; its tropical climate is moderated by constant trade winds from the Atlantic Ocean; the temperature is almost constant at about 27 degrees Celsius (81 degrees Fahrenheit)

PEOPLE

Population: 101,541
note: estimate based on a revision of the base population, fertility, and mortality numbers, as well as a revision of

1985–1999 migration estimates from outmigration to inmigration, which is assumed to continue into the future; the new results are consistent with the 2000 census (July 2008 est.)

Age structure:
0–14 years: 19.4% (male 9,933/female 9,747)
15–64 years: 70.3% (male 34,123/female 37,228)
65 years and over: 10.4% (male 4,189/female 6,321) (2008 est.)

Median age:
total: 37.6 years
male: 35.8 years
female: 39.3 years (2008 est.)

Population growth rate: 1.501% (2008 est.)

Birth rate: 12.81 births/1,000 population (2008 est.)

Death rate: 7.65 deaths/1,000 population (2008 est.)

Net migration rate: 9.85 migrant(s)/1,000 population (2008 est.)

Sex ratio:
at birth: 1.02 male(s)/female
under 15 years: 1.02 male(s)/female
15–64 years: 0.92 male(s)/female
65 years and over: 0.66 male(s)/female
total population: 0.91 male(s)/female (2008 est.)

Infant mortality rate:
total: 14.26 deaths/1,000 live births
male: 18.92 deaths/1,000 live births
female: 9.51 deaths/1,000 live births (2008 est.)

Life expectancy at birth:
total population: 75.06 years
male: 72.03 years
female: 78.14 years (2008 est.)

Total fertility rate: 1.85 children born/woman (2008 est.)

HIV/AIDS—adult prevalence rate: NA
HIV/AIDS—people living with HIV/AIDS: NA
HIV/AIDS—deaths: NA

Nationality:
noun: Aruban(s)
adjective: Aruban; Dutch

Ethnic groups: mixed white/Caribbean Amerindian 80%, other 20%

Religions: Roman Catholic 82%, Protestant 8%, other (includes Hindu, Muslim, Confucian, Jewish) 10%

Languages: Papiamento (a Spanish-Portuguese-Dutch-English dialect) 66.3%, Spanish 12.6%, English (widely spoken) 7.7%, Dutch (official) 5.8%, other 2.2%, unspecified or unknown 5.3% (2000 census)

Literacy:
definition: NA
total population: 97.3%
male: 97.5%
female: 97.1% (2000 census)

GOVERNMENT

Country name:
conventional long form: none
conventional short form: Aruba

Dependency status: member country of the Kingdom of the Netherlands; full autonomy in internal affairs obtained in 1986 upon separation from the Netherlands Antilles; Dutch Government responsible for defense and foreign affairs

Government type: parliamentary democracy

Capital: *name:* Oranjestad
geographic coordinates: 12 31 N, 70 02 W
time difference: UTC-4 (1 hour ahead of Washington, DC during Standard Time)

Administrative divisions: none (part of the Kingdom of the Netherlands)

Independence: none (part of the Kingdom of the Netherlands)

National holiday: Flag Day, 18 March (1976)

Constitution: 1 January 1986

Legal system: based on Dutch civil law system, with some English common law influence

Suffrage: 18 years of age; universal

Executive branch:
chief of state: Queen BEATRIX of the Netherlands (since 30 April 1980); represented by Governor General Fredis REFUNJOL (since 11 May 2004)
head of government: Prime Minister Nelson O. ODUBER (since 30 October 2001)
cabinet: Council of Ministers elected by the Staten
elections: the monarch is hereditary; governor general appointed for a six-year term by the monarch; prime minister and deputy prime minister elected by the Staten for four-year terms; election last held in 2005 (next to be held by 2009)
election results: Nelson O. ODUBER elected prime minister; percent of legislative vote—NA

Legislative branch: unicameral Legislature or Staten (21 seats; members elected by direct popular vote to serve four-year terms)
elections: last held 23 September 2005 (next to be held in 2009)
election results: percent of vote by party—MEP 43%, AVP 32%, MPA 7%, RED 7%, PDR 6%, OLA 4%, PPA 2%; seats by party—MEP 11, AVP 8, MPA 1, RED 1

Judicial branch: Common Court of Justice of Aruba (judges are appointed by the monarch)

Political parties and leaders: Aliansa/Aruban Social Movement or MSA [Robert WEVER]; Aruban Liberal Organization or OLA [Glenbert CROES]; Aruban Patriotic Movement or MPA [Monica ARENDS-KOCK]; Aruban Patriotic Party or PPA [Benny NISBET]; Aruban People's Party or AVP [Mike EMAN]; People's Electoral Movement Party or MEP [Nelson O. ODUBER]; Real Democracy or PDR [Andin BIKKER]; RED [Rudy LAMPE]; Workers Political Platform or PTT [Gregorio WOLFF]

Political pressure groups and leaders: NA

International organization participation: Caricom (observer), ILO, IMF, Interpol, IOC, ITUC, UNESCO (associate), UNWTO (associate), UPU, WCL, WMO

Diplomatic representation in the US: none (represented by the Kingdom of the Netherlands); note—Mr. Henry BAARH, Minister Plenipotentiary for Aruba at the Embassy of the Kingdom of the Netherlands

Diplomatic representation from the US: the US does not have an embassy in Aruba; the Consul General to Netherlands Antilles is accredited to Aruba

Flag description: blue, with two narrow, horizontal, yellow stripes across the lower portion and a red, four-pointed star outlined in white in the upper hoist-side corner

ECONOMY

Economy—overview: Tourism is the mainstay of the small, open Aruban economy, with offshore banking and oil refining and storage also important. The rapid growth of the tourism sector over the last decade has resulted in a substantial expansion of other activities. Over 1.5 million tourists per year visit Aruba, with 75% of those from the US. Construction continues to boom, with hotel capacity five times the 1985 level. In addition, the country's oil refinery reopened in 1993, providing a major source of employment, foreign exchange earnings, and growth. Tourist arrivals have rebounded strongly following a dip after the 11 September 2001 attacks. The island experiences only a brief low season, and hotel occupancy in 2004 averaged 80%, compared to 68% throughout the rest of the Caribbean. The government has made cutting the budget and trade deficits a high priority.

GDP (purchasing power parity): $2.258 billion (2005 est.)

GDP (official exchange rate): $2.258 billion (2005 est.)

GDP—real growth rate: 2.4% (2005 est.)

GDP—per capita (PPP): $21,800 (2004 est.)
GDP—composition by sector:
agriculture: 0.4%
industry: 33.3%
services: 66.3% (2002 est.)
Labor force: 41,500 (2004 est.)
Labor force—by occupation: *agriculture:* NA%
industry: NA%
services: NA%
note: most employment is in wholesale and retail trade and repair, followed by hotels and restaurants; oil refining
Unemployment rate: 6.9% (2005 est.)
Population below poverty line: NA%
Household income or consumption by percentage share:
lowest 10%: NA%
highest 10%: NA%
Inflation rate (consumer prices): 3.4% (2005)
Budget: *revenues:* $507.9 million
expenditures: $577.9 million (2005 est.)
Public debt: 46.3% of GDP (2005)
Agriculture—products: aloes; livestock; fish
Industries: tourism, transshipment facilities, oil refining
Industrial production growth rate: NA%
Electricity—production: 770 million kWh (2005)
Electricity—production by source:
fossil fuel: 100%
hydro: 0%
nuclear: 0%
other: 0% (2001)
Electricity—consumption: 716.1 million kWh (2005)
Electricity—exports: 0 kWh (2005)
Electricity—imports: 0 kWh (2005)
Oil—production: 2,356 bbl/day (2005)
Oil—consumption: 7,000 bbl/day (2005 est.)
Oil—exports: 230,600 bbl/day (2004)
Oil—imports: 235,000 bbl/day (2004)
Oil—proved reserves: 0 bbl (1 January 2006 est.)
Natural gas—production: 0 cu m (2005 est.)

Natural gas—consumption: 0 cu m (2005 est.)
Natural gas—exports: 0 cu m (2005 est.)
Natural gas—imports: 0 cu m (2005)
Natural gas—proved reserves: 0 cu m (1 January 2006)
Exports: $124 million f.o.b.; note—includes oil reexports (2006)
Exports—commodities: live animals and animal products, art and collectibles, machinery and electrical equipment, transport equipment
Exports—partners: Netherlands 27.7%, Panama 25.5%, Colombia 12.8%, Venezuela 11.1%, US 9.4%, Netherlands Antilles 7.1% (2006)
Imports: $1.054 billion f.o.b. (2006)
Imports—commodities: machinery and electrical equipment, crude oil for refining and reexport, chemicals; foodstuffs
Imports—partners: US 53.6%, Netherlands 12.9%, UK 3.6% (2006)
Economic aid—recipient: $11.3 million (2004)
Debt—external: $478.6 million (2005 est.)
Currency (code): Aruban guilder/florin (AWG)
Currency code: AWG
Exchange rates: Aruban guilders/florins per US dollar—NA (2007), 1.79 (2006), 1.79 (2005), 1.79 (2004), 1.79 (2003)
Fiscal year: calendar year

COMMUNICATIONS

Telephones—main lines in use: 38,300 (2005)
Telephones—mobile cellular: 108,200 (2005)
Telephone system:
general assessment: modern fully automatic telecommunications system
domestic: increased competition through privatization; 3 wireless service providers are now licensed
international: country code—297; landing site for the PAN-AM submarine telecommunications cable system that extends from the US Virgin Islands through Aruba to Venezuela, Colombia,

Panama, and the west coast of South America; extensive interisland microwave radio relay links (2007)
Radio broadcast stations: AM 2, FM 16, shortwave 0 (2004)
Radios: 50,000 (1997)
Television broadcast stations: 1 (1997)
Televisions: 20,000 (1997)
Internet country code: .aw
Internet hosts: 16,914 (2007)
Internet Service Providers (ISPs): NA
Internet users: 24,000 (2005)

TRANSPORTATION

Airports: 1 (2007)
Airports—with paved runways:
total: 1
2,438 to 3,047 m: 1 (2007)
Roadways:
total: 800 km
Ports and terminals: Barcadera, Oranjestad, Sint Nicolaas

MILITARY

Military branches: no regular indigenous military forces; the Netherlands maintains a detachment of marines, a frigate, and an amphibious combat detachment in the neighboring Netherlands Antilles (2008)
Manpower available for military service:
males age 16–49: 24,585
females age 16–49: 25,742 (2008 est.)
Manpower fit for military service:
males age 16–49: 20,173
females age 16–49: 21,062 (2008 est.)
Manpower reaching militarily significant age annually:
males age 16–49: 705
females age 16–49: 719 (2008 est.)
Military—note: defense is the responsibility of the Kingdom of the Netherlands

TRANSNATIONAL ISSUES

Disputes—international: none
Illicit drugs: transit point for US- and Europe-bound narcotics with some accompanying money-laundering activity; relatively high percentage of population consumes cocaine

ASHMORE AND CARTIER ISLANDS

INTRODUCTION

Background: These uninhabited islands came under Australian authority in 1931; formal administration began two years later. Ashmore Reef supports a rich and diverse avian and marine habitat; in 1983, it became a National Nature Reserve. Cartier Island, a former bombing range, is now a marine reserve.

GEOGRAPHY

Location: Southeastern Asia, islands in the Indian Ocean, midway between northwestern Australia and Timor island
Geographic coordinates: 12 14 S, 123 05 E
Map references: Southeast Asia

Area:
total: 5 sq km
land: 5 sq km
water: 0 sq km
note: includes Ashmore Reef (West, Middle, and East Islets) and Cartier Island
Area—comparative: about eight times the size of The Mall in Washington, DC

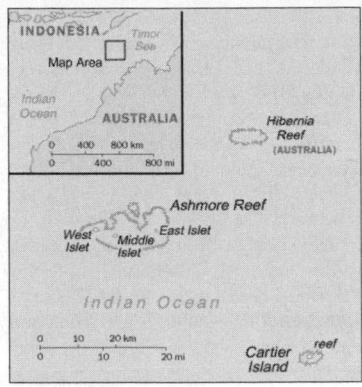

Land boundaries: 0 km
Coastline: 74.1 km
Maritime claims:
territorial sea: 12 nm
contiguous zone: 12 nm
exclusive fishing zone: 200 nm
continental shelf: 200-m depth or to the depth of exploitation
Climate: tropical
Terrain: low with sand and coral
Elevation extremes:
lowest point: Indian Ocean 0 m
highest point: unnamed location 3 m
Natural resources: fish
Land use: *arable land:* 0%

permanent crops: 0%
other: 100% (all grass and sand) (2005)
Irrigated land: 0 sq km
Natural hazards: surrounded by shoals and reefs that can pose maritime hazards
Environment—current issues: NA
Geography—note: Ashmore Reef National Nature Reserve established in August 1983

PEOPLE

Population: no indigenous inhabitants
note: Indonesian fishermen are allowed access to the lagoon and fresh water at Ashmore Reef's West Island
People—note: the landing of illegal immigrants from Indonesia's Rote Island has become an ongoing problem

GOVERNMENT

Country name:
conventional long form: Territory of Ashmore and Cartier Islands
conventional short form: Ashmore and Cartier Islands
Dependency status: territory of Australia; administered by the Australian Attorney-General's Department
Legal system: the laws of the Commonwealth of Australia and the laws of the Northern Territory of

Australia, where applicable, apply
Diplomatic representation in the US: none (territory of Australia)
Diplomatic representation from the US: none (territory of Australia)
Flag description: the flag of Australia is used

ECONOMY

Economy—overview: no economic activity

TRANSPORTATION

Ports and terminals: none; offshore anchorage only

MILITARY

Military—note: defense is the responsibility of Australia; periodic visits by the Royal Australian Navy and Royal Australian Air Force

TRANSNATIONAL ISSUES

Disputes—international: Indonesian groups challenge Australia's claim to these islands; Australia closed parts of the Ashmore and Cartier Reserve to Indonesian traditional fishing and placed restrictions on certain catches

ATLANTIC OCEAN

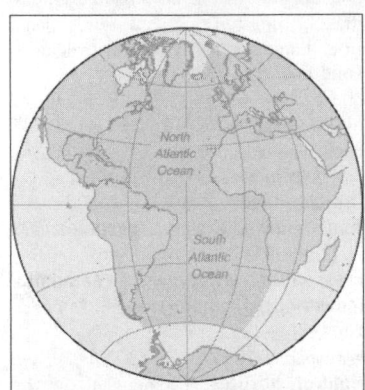

INTRODUCTION

Background: The Atlantic Ocean is the second largest of the world's five oceans (after the Pacific Ocean, but larger than the Indian Ocean, Southern Ocean, and Arctic Ocean). The Kiel Canal (Germany), Oresund (Denmark-Sweden), Bosporus (Turkey), Strait of Gibraltar (Morocco-Spain), and the Saint Lawrence Seaway (Canada-US) are important strategic access waterways. The decision by the International

Hydrographic Organization in the spring of 2000 to delimit a fifth world ocean, the Southern Ocean, removed the portion of the Atlantic Ocean south of 60 degrees south latitude.

GEOGRAPHY

Location: body of water between Africa, Europe, the Southern Ocean, and the Western Hemisphere
Geographic coordinates: 0 00 N, 25 00 W
Map references: Political Map of the World
Area:
total: 76.762 million sq km
note: includes Baltic Sea, Black Sea, Caribbean Sea, Davis Strait, Denmark Strait, part of the Drake Passage, Gulf of Mexico, Labrador Sea, Mediterranean Sea, North Sea, Norwegian Sea, almost all of the Scotia Sea, and other tributary water bodies
Area—comparative: slightly less than 6.5 times the size of the US
Coastline: 111,866 km
Climate: tropical cyclones (hurricanes) develop off the coast of Africa near Cape Verde and move westward into the

Caribbean Sea; hurricanes can occur from May to December, but are most frequent from August to November
Terrain: surface usually covered with sea ice in Labrador Sea, Denmark Strait, and coastal portions of the Baltic Sea from October to June; clockwise warm-water gyre (broad, circular system of currents) in the northern Atlantic, counterclockwise warm-water gyre in the southern Atlantic; the ocean floor is dominated by the Mid-Atlantic Ridge, a rugged north-south centerline for the entire Atlantic basin
Elevation extremes:
lowest point: Milwaukee Deep in the Puerto Rico Trench -8,605 m
highest point: sea level 0 m
Natural resources: oil and gas fields, fish, marine mammals (seals and whales), sand and gravel aggregates, placer deposits, polymetallic nodules, precious stones
Natural hazards: icebergs common in Davis Strait, Denmark Strait, and the northwestern Atlantic Ocean from February to August and have been spotted as far south as Bermuda and the Madeira Islands; ships subject to super-

structure icing in extreme northern Atlantic from October to May; persistent fog can be a maritime hazard from May to September; hurricanes (May to December)

Environment—current issues: endangered marine species include the manatee, seals, sea lions, turtles, and whales; drift net fishing is hastening the decline of fish stocks and contributing to international disputes; municipal sludge pollution off eastern US, southern Brazil, and eastern Argentina; oil pollution in Caribbean Sea, Gulf of Mexico, Lake Maracaibo, Mediterranean Sea, and North Sea; industrial waste and municipal sewage pollution in Baltic Sea, North Sea, and Mediterranean Sea

Geography—note: major chokepoints include the Dardanelles, Strait of Gibraltar, access to the Panama and Suez Canals; strategic straits include the Strait of Dover, Straits of Florida, Mona Passage, The Sound (Oresund), and Windward Passage; the Equator divides the Atlantic Ocean into the North Atlantic Ocean and South Atlantic Ocean

ECONOMY

Economy—overview: The Atlantic Ocean provides some of the world's most heavily trafficked sea routes, between and within the Eastern and Western Hemispheres. Other economic activity includes the exploitation of natural resources, e.g., fishing, dredging of aragonite sands (The Bahamas), and production of crude oil and natural gas (Caribbean Sea, Gulf of Mexico, and North Sea).

TRANSPORTATION

Ports and terminals: Alexandria (Egypt), Algiers (Algeria), Antwerp (Belgium), Barcelona (Spain), Buenos Aires (Argentina), Casablanca (Morocco), Colon (Panama), Copenhagen (Denmark), Dakar (Senegal), Gdansk (Poland), Hamburg (Germany), Helsinki (Finland), Las Palmas (Canary Islands, Spain), Le Havre (France), Lisbon (Portugal), London (UK), Marseille (France), Montevideo (Uruguay), Montreal (Canada), Naples (Italy), New Orleans (US), New York (US), Oran (Algeria), Oslo (Norway), Peiraiefs or Piraeus (Greece), Rio de Janeiro (Brazil), Rotterdam (Netherlands), Saint Petersburg (Russia), Stockholm (Sweden)

Transportation—note: Kiel Canal and Saint Lawrence Seaway are two important waterways; significant domestic commercial and recreational use of Intracoastal Waterway on central and south Atlantic seaboard and Gulf of Mexico coast of US

TRANSNATIONAL ISSUES

Disputes—international: some maritime disputes (see littoral states)

AUSTRALIA

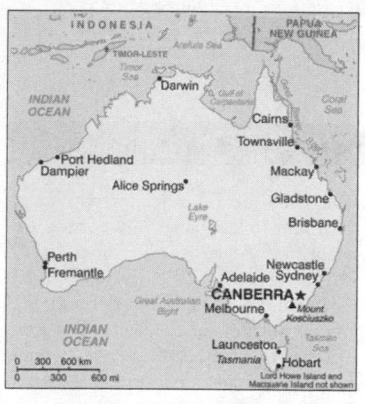

INTRODUCTION

Background: Aboriginal settlers arrived on the continent from Southeast Asia about 40,000 years before the first Europeans began exploration in the 17th century. No formal territorial claims were made until 1770, when Capt. James COOK took possession in the name of Great Britain. Six colonies were created in the late 18th and 19th centuries; they federated and became the Commonwealth of Australia in 1901. The new country took advantage of its natural resources to rapidly develop agricultural and manufacturing industries and to make a major contribution to the British effort in World Wars I and II. In recent decades, Australia has transformed itself into an internationally competitive, advanced market economy. It boasted one of the OECD's fastest growing economies during the 1990s, a performance due in large part to economic reforms adopted in the 1980s. Long-term concerns include climate-change issues such as the depletion of the ozone layer and more frequent droughts, and management and conservation of coastal areas, especially the Great Barrier Reef.

GEOGRAPHY

Location: Oceania, continent between the Indian Ocean and the South Pacific Ocean

Geographic coordinates: 27 00 S, 133 00 E

Map references: Oceania

Area:
total: 7,686,850 sq km
land: 7,617,930 sq km
water: 68,920 sq km
note: includes Lord Howe Island and Macquarie Island

Area—comparative: slightly smaller than the US contiguous 48 states

Land boundaries: 0 km

Coastline: 25,760 km

Maritime claims:
territorial sea: 12 nm
contiguous zone: 24 nm
exclusive economic zone: 200 nm
continental shelf: 200 nm or to the edge of the continental margin

Climate: generally arid to semiarid; temperate in south and east; tropical in north

Terrain: mostly low plateau with deserts; fertile plain in southeast

Elevation extremes:
lowest point: Lake Eyre -15 m
highest point: Mount Kosciuszko 2,229 m

Natural resources: bauxite, coal, iron ore, copper, tin, gold, silver, uranium, nickel, tungsten, mineral sands, lead, zinc, diamonds, natural gas, petroleum

Land use:
arable land: 6.15% (includes about 27 million hectares of cultivated grassland)
permanent crops: 0.04%
other: 93.81% (2005)

Irrigated land: 25,450 sq km (2003)

Total renewable water resources: 398 cu km (1995)

Freshwater withdrawal (domestic/industrial/agricultural): *total:* 24.06 cu km/yr (15%/10%/75%)
per capita: 1,193 cu m/yr (2000)

Natural hazards: cyclones along the coast; severe droughts; forest fires

Environment—current issues: soil erosion from overgrazing, industrial development, urbanization, and poor farming practices; soil salinity rising due to the use of poor quality water; desertification; clearing for agricultural purposes threatens the natural habitat of many unique animal and plant species; the Great Barrier Reef off the northeast coast, the largest coral reef in the world, is threatened by increased shipping and its popularity as a tourist site; limited natural fresh water resources

Environment—international agreements: *party to:* Antarctic-Environmental Protocol, Antarctic-Marine Living Resources, Antarctic Seals, Antarctic Treaty, Biodiversity, Climate Change, Climate Change-Kyoto Protocol, Desertification, Endangered Species, Environmental Modification, Hazardous Wastes, Law of the Sea, Marine Dumping, Marine Life Conservation, Ozone Layer Protection, Ship Pollution, Tropical Timber 83, Tropical Timber 94, Wetlands, Whaling *signed, but not ratified:* none of the selected agreements

Geography—note: world's smallest continent but sixth-largest country; population concentrated along the eastern and southeastern coasts; the invigorating sea breeze known as the "Fremantle Doctor" affects the city of Perth on the west coast, and is one of the most consistent winds in the world

PEOPLE

Population: 20,600,856 (July 2008 est.)

Age structure:

0–14 years: 19.1% (male 2,014,230/female 1,920,604)

15–64 years: 67.5% (male 7,005,588/female 6,895,817)

65 years and over: 13.4% (male 1,226,432/female 1,538,185) (2008 est.)

Median age:

total: 37.4 years

male: 36.6 years

female: 38.3 years (2008 est.)

Population growth rate: 0.801% (2008 est.)

Birth rate: 11.9 births/1,000 population (2008 est.)

Death rate: 7.62 deaths/1,000 population (2008 est.)

Net migration rate: 3.72 migrant(s)/1,000 population (2008 est.)

Sex ratio:

at birth: 1.05 male(s)/female

under 15 years: 1.05 male(s)/female

15–64 years: 1.02 male(s)/female

65 years and over: 0.8 male(s)/female

total population: 0.99 male(s)/female (2008 est.)

Infant mortality rate:

total: 4.51 deaths/1,000 live births

male: 4.89 deaths/1,000 live births

female: 4.11 deaths/1,000 live births (2008 est.)

Life expectancy at birth:

total population: 80.73 years

male: 77.86 years

female: 83.75 years (2008 est.)

Total fertility rate: 1.76 children born/woman (2008 est.)

HIV/AIDS—adult prevalence rate: 0.1% (2003 est.)

HIV/AIDS—people living with HIV/AIDS: 14,000 (2003 est.)

HIV/AIDS—deaths: fewer than 200 (2003 est.)

Nationality:

noun: Australian(s)

adjective: Australian

Ethnic groups: white 92%, Asian 7%, aboriginal and other 1%

Religions: Catholic 26.4%, Anglican 20.5%, other Christian 20.5%, Buddhist 1.9%, Muslim 1.5%, other 1.2%, unspecified 12.7%, none 15.3% (2001 Census)

Languages: English 79.1%, Chinese 2.1%, Italian 1.9%, other 11.1%, unspecified 5.8% (2001 Census)

Literacy:

definition: age 15 and over can read and write

total population: 99%

male: 99%

female: 99% (2003 est.)

GOVERNMENT

Country name:

conventional long form: Commonwealth of Australia

conventional short form: Australia

Government type: federal parliamentary democracy

Capital: *name:* Canberra

geographic coordinates: 35 17 S, 149 13 E

time difference: UTC+10 (15 hours ahead of Washington, DC during Standard Time)

daylight saving time: +1hr, begins last Sunday in October; ends last Sunday in March

note: Australia is divided into three time zones

Administrative divisions: 6 states and 2 territories*; Australian Capital Territory*, New South Wales, Northern Territory*, Queensland, South Australia, Tasmania, Victoria, Western Australia

Dependent areas: Ashmore and Cartier Islands, Christmas Island, Cocos (Keeling) Islands, Coral Sea Islands, Heard Island and McDonald Islands, Norfolk Island, Macquarie Island

Independence: 1 January 1901 (federation of UK colonies)

National holiday: Australia Day, 26 January (1788); ANZAC Day (commemorated as the anniversary of the landing of troops of the Australian and New Zealand Army Corps during World War I at Gallipoli, Turkey), 25 April (1915)

Constitution: 9 July 1900, effective 1 January 1901

Legal system: based on English common law; accepts compulsory ICJ jurisdiction, with reservations

Suffrage: 18 years of age; universal and compulsory

Executive branch:

chief of state: Queen of Australia ELIZABETH II (since 6 February 1952); represented by Governor General Maj. Gen. (Ret.) Michael JEFFERY (since 11 August 2003)

head of government: Prime Minister Kevin RUDD (since 3 December 2007); Deputy Prime Minister Julia GILLARD (since 3 December 2007)

cabinet: prime minister nominates, from among members of Parliament, candidates who are subsequently sworn in by the governor general to serve as government ministers

elections: the monarch is hereditary; governor general appointed by the monarch on the recommendation of the prime minister; following legislative elections, the leader of the majority party or leader of a majority coalition is sworn in as prime minister by the governor general

Legislative branch: bicameral Federal Parliament consists of the Senate (76 seats; 12 members from each of the six states and 2 from each of the two mainland territories; one-half of state members are elected every three years by popular vote to serve six-year terms while all territory members are elected every three years) and the House of Representatives (150 seats; members elected by popular preferential vote to serve terms of up to three-years; no state can have fewer than 5 representatives)

elections: Senate—last held 24 November 2007 (next to be held no later than 2010); House of Representatives—last held 24 November 2007 (next to be called no later than 2010)

election results: Senate—percent of vote by party—NA; seats by party—Liberal Party-National Party coalition 37, Australian Labor Party 32, Australian Greens 5, Family First Party 1, other 1; House of Representatives—percent of vote by party—NA; seats by party—Australian Labor Party 83, Liberal Party 55, National Party 10, independents 2

Judicial branch: High Court (the chief justice and six other justices are appointed by the governor general)

Political parties and leaders: Australian Democrats [Lyn ALLISON]; Australian Greens [Bob BROWN]; Australian Labor Party [Kevin RUDD]; Country Liberal Party [Jodeen CARNEY]; Family First Party [Steve FIELDING]; Liberal Party [Brendan NELSON]; The Nationals [Warren TRUSS]

International organization participation: ADB, ANZUS, APEC, ARF, ASEAN (dialogue partner), Australia Group, BIS, C, CP, EAS, EBRD, FAO,

IAEA, IBRD, ICAO, ICC, ICCt, ICRM, IDA, IEA, IFC, IFRCS, IHO, ILO, IMF, IMO, IMSO, Interpol, IOC, IOM, IPU, ISO, ITSO, ITU, ITUC, MIGA, NAM (guest), NEA, NSG, OECD, OPCW, Paris Club, PCA, PIF, Sparteca, SPC, UN, UNCTAD, UNESCO, UNFICYP, UNHCR, UNMIS, UNMIT, UNRWA, UNTSO, UNWTO, UPU, WCO, WFTU, WHO, WIPO, WMO, WTO, ZC

Diplomatic representation in the US:
chief of mission: Ambassador Dennis J. RICHARDSON
chancery: 1601 Massachusetts Avenue NW, Washington, DC 20036
telephone: [1] (202) 797-3000
FAX: [1] (202) 797-3168
consulate(s) general: Atlanta, Chicago, Honolulu, Los Angeles, New York, San Francisco

Diplomatic representation from the US:
chief of mission: Ambassador Robert D. McCALLUM, Jr.
embassy: Moonah Place, Yarralumla, Canberra, Australian Capital Territory 2600
mailing address: APO AP 96549
telephone: [61] (02) 6214-5600
FAX: [61] (02) 6214-5970
consulate(s) general: Melbourne, Perth, Sydney

Flag description: blue with the flag of the UK in the upper hoist-side quadrant and a large seven-pointed star in the lower hoist-side quadrant known as the Commonwealth or Federation Star, representing the federation of the colonies of Australia in 1901; the star depicts one point for each of the six original states and one representing all of Australia's internal and external territories; on the fly half is a representation of the Southern Cross constellation in white with one small five-pointed star and four larger, seven-pointed stars

ECONOMY

Economy—overview: Australia has an enviable, strong economy with a per capita GDP on par with the four dominant West European economies. Robust business and consumer confidence and high export prices for raw materials and agricultural products are fueling the economy, particularly in mining states. Australia's emphasis on reforms, low inflation, a housing market boom, and growing ties with China have been key factors behind the economy's 16 solid years of expansion. Drought, robust import demand, and a strong currency have pushed the trade deficit up in recent years, while infrastructure bottlenecks and a tight labor market are con-

straining growth in export volumes and stoking inflation. Australia's budget has been in surplus since 2002 due to strong revenue growth.

GDP (purchasing power parity): $760.8 billion (2007 est.)
GDP (official exchange rate): $908.8 billion (2007 est.)
GDP—real growth rate: 3.9% (2007 est.)
GDP—per capita (PPP): $36,300 (2007 est.)
GDP—composition by sector:
agriculture: 3%
industry: 26.4%
services: 70.6% (2007 est.)
Labor force: 10.95 million (2007 est.)
Labor force—by occupation: *agriculture:* 3.6%
industry: 21.2%
services: 75.2% (2004 est.)
Unemployment rate: 4.4% (2007 est.)
Population below poverty line: NA%
Household income or consumption by percentage share:
lowest 10%: 2%
highest 10%: 25.4% (1994)
Distribution of family income—Gini index: 30.5 (2006)
Inflation rate (consumer prices): 2.3% (2007 est.)
Investment (gross fixed): 27.3% of GDP (2007 est.)
Budget:
revenues: $321.3 billion
expenditures: $309.1 billion (2007 est.)
Public debt: 15.4% of GDP
note: The Commonwealth government eliminated its net debt in 2006, but continues a gross debt issue to support the market for risk-free securities. (2007 est.)
Agriculture—products: wheat, barley, sugarcane, fruits, cattle, sheep, poultry
Industries: mining, industrial and transportation equipment, food processing, chemicals, steel
Industrial production growth rate: 3.8% (2007 est.)
Electricity—production: 236.7 billion kWh (2005)
Electricity—production by source:
fossil fuel: 90.8%
hydro: 8.3%
nuclear: 0%
other: 0.9% (2001)
Electricity—consumption: 219.8 billion kWh (2005)
Electricity—exports: 0 kWh (2005)
Electricity—imports: 0 kWh (2005)
Oil—production: 572,400 bbl/day (2005 est.)
Oil—consumption: 903,200 bbl/day (2005 est.)
Oil—exports: 333,200 bbl/day (2004)
Oil—imports: 611,400 bbl/day (2004)

Oil—proved reserves: 1.437 billion bbl (1 January 2006 est.)
Natural gas—production: 38.62 billion cu m (2005 est.)
Natural gas—consumption: 25.72 billion cu m (2005 est.)
Natural gas—exports: 12.9 billion cu m (2005 est.)
Natural gas—imports: 0 cu m (2005)
Natural gas—proved reserves: 750.6 billion cu m (1 January 2006 est.)
Current account balance: -$56.2 billion (2007 est.)
Exports: $141.7 billion (2007 est.)
Exports—commodities: coal, iron ore, gold, meat, wool, alumina, wheat, machinery and transport equipment
Exports—partners: Japan 19.6%, China 12.3%, South Korea 7.5%, US 6.2%, India 5.5%, NZ 5.5%, UK 5% (2006)
Imports: $159.4 billion (2007 est.)
Imports—commodities: machinery and transport equipment, computers and office machines, telecommunication equipment and parts; crude oil and petroleum products
Imports—partners: China 14.4%, US 14.1%, Japan 9.6%, Singapore 6%, Germany 5.1% (2006)
Economic aid—donor: ODA, $2.123 billion (2006)
Reserves of foreign exchange and gold: $26.91 billion (31 December 2007 est.)
Debt—external: $824.9 billion (30 June 2007)
Stock of direct foreign investment—at home: $279 billion (2007 est.)
Stock of direct foreign investment—abroad: $257.9 billion (2007 est.)
Market value of publicly traded shares: $804.1 billion (2005)
Currency (code): Australian dollar (AUD)
Currency code: AUD
Exchange rates: Australian dollars per US dollar—1.2137 (2007), 1.3285 (2006), 1.3095 (2005), 1.3598 (2004), 1.5419 (2003)
Fiscal year: 1 July—30 June

COMMUNICATIONS

Telephones—main lines in use: 9.94 million (2006)
Telephones—mobile cellular: 19.76 million (2006)
Telephone system:
general assessment: excellent domestic and international service
domestic: domestic satellite system; significant use of radiotelephone in areas of low population density; rapid growth of mobile cellular telephones
international: country code—61; landing point for the SEA-ME-WE-3 optical

telecommunications submarine cable with links to Asia, the Middle East, and Europe; the Southern Cross fiber optic submarine cable provides links to New Zealand and the United States; satellite earth stations—19 (10 Intelsat—4 Indian Ocean and 6 Pacific Ocean, 2 Inmarsat—Indian and Pacific Ocean regions, 2 Globalstar, 5 other) (2007)

Radio broadcast stations: AM 262, FM 345, shortwave 1 (1998)

Radios: 25.5 million (1997)

Television broadcast stations: 104 (1997)

Televisions: 10.15 million (1997)

Internet country code: .au

Internet hosts: 9.458 million (2007)

Internet Service Providers (ISPs): 571 (2002)

Internet users: 15.3 million (2006)

TRANSPORTATION

Airports: 461 (2007)

Airports—with paved runways:
total: 317
over 3,047 m: 11
2,438 to 3,047 m: 12
1,524 to 2,437 m: 138
914 to 1,523 m: 143
under 914 m: 13 (2007)

Airports—with unpaved runways:
total: 144
1,524 to 2,437 m: 19
914 to 1,523 m: 109
under 914 m: 16 (2007)

Heliports: 1 (2007)

Pipelines: condensate/gas 469 km; gas 26,719 km; liquid petroleum gas 240 km; oil 3,720 km; oil/gas/water 110 km (2007)

Railways:
total: 38,550 km
broad gauge: 3,727 km 1.600-m gauge
standard gauge: 20,519 km 1.435-m gauge (1,877 km electrified)
narrow gauge: 14,074 km 1.067-m gauge (2,453 km electrified)
dual gauge: 230 km dual gauge (2006)

Roadways:
total: 812,972 km
paved: 341,448 km
unpaved: 471,524 km (2004)

Waterways: 2,000 km (mainly used for recreation on Murray and Murray-Darling river systems) (2006)

Merchant marine:
total: 52 ships (1000 GRT or over) 1,322,527 GRT/1,501,865 DWT
by type: bulk carrier 16, cargo 5, chemical tanker 1, container 1, liquefied gas 4, passenger 7, passenger/cargo 6, petroleum tanker 7, roll on/roll off 5
foreign-owned: 16 (Canada 2, France 1, Germany 2, Netherlands 2, Norway 1, Philippines 1, UK 2, US 5)
registered in other countries: 29 (Antigua and Barbuda 1, Bahamas 3, Bermuda 4, Fiji 1, The Gambia 1, Liberia 2, Marshall Islands 1, Panama 4, Singapore 6, Tonga 1, UK 1, US 2, Vanuatu 2, unknown 1) (2007)

Ports and terminals: Brisbane, Dampier, Fremantle, Gladstone, Hay Point, Melbourne, Newcastle, Port Hedland, Port Kembla, Port Walcott, Sydney

MILITARY

Military branches: Australian Defense Force (ADF): Australian Army, Royal Australian Navy, Royal Australian Air Force, Special Operations Command (2006)

Military service age and obligation: 17 years of age for voluntary military service (with parental consent); no conscription; women allowed to serve in Army combat units in non-combat support roles (2008)

Manpower available for military service:
males age 16–49: 4,999,988
females age 16–49: 4,870,043 (2008 est.)

Manpower fit for military service:
males age 16–49: 4,137,176
females age 16–49: 4,022,588 (2008 est.)

Manpower reaching militarily significant age annually:
males age 16–49: 146,248
females age 16–49: 139,697 (2008 est.)

Military expenditures—percent of GDP: 2.4% (2006)

TRANSNATIONAL ISSUES

Disputes—international: Timor-Leste and Australia agreed in 2005 to defer the disputed portion of the boundary for fifty years and to split hydrocarbon revenues evenly outside the Joint Petroleum Development Area covered by the 2002 Timor Sea Treaty; East Timor dispute hampers creation of a revised maritime boundary with Indonesia in the Timor Sea; Indonesian groups challenge Australia's claim to Ashmore and Cartier Islands; Australia closed parts of the Ashmore and Cartier Reserve to Indonesian traditional fishing and placed restrictions on certain catch; regional states continue to express concern over Australia's 2004 declaration of a 1,000-nautical mile-wide maritime identification zone; Australia asserts land and maritime claims to Antarctica (see Antarctica); in 2004 Australia submitted its claims to UN Commission on the Limits of the Continental Shelf (CLCS) to extend its continental margins covering over 3.37 million square kilometers or roughly thirty percent of its claimed exclusive economic zone; since 2003, Australian Defense Force leads the Regional Assistance Mission to the Solomon Islands (RAMSI) to maintain civil and political order and reinforce regional security

Illicit drugs: Tasmania is one of the world's major suppliers of licit opiate products; government maintains strict controls over areas of opium poppy cultivation and output of poppy straw concentrate; major consumer of cocaine and amphetamines

AUSTRIA

INTRODUCTION

Background: Once the center of power for the large Austro-Hungarian Empire, Austria was reduced to a small republic after its defeat in World War I. Following annexation by Nazi Germany in 1938 and subsequent occupation by the victorious Allies in 1945, Austria's status remained unclear for a decade. A State Treaty signed in 1955 ended the occupation, recognized Austria's independence, and forbade unification with Germany. A constitutional law that same year declared the country's "perpetual neutrality" as a condition for Soviet military withdrawal. The Soviet Union's collapse in 1991 and Austria's entry into the European Union in 1995 have altered the meaning of this neutrality. A prosperous, democratic country, Austria entered the EU Economic and Monetary Union in 1999.

GEOGRAPHY

Location: Central Europe, north of Italy and Slovenia

Geographic coordinates: 47 20 N, 13 20 E

Map references: Europe

Area:
total: 83,870 sq km
land: 82,444 sq km
water: 1,426 sq km
Area—comparative: slightly smaller than Maine
Land boundaries:
total: 2,562 km
border countries: Czech Republic 362 km, Germany 784 km, Hungary 366 km, Italy 430 km, Liechtenstein 35 km, Slovakia 91 km, Slovenia 330 km, Switzerland 164 km
Coastline: 0 km (landlocked)
Maritime claims: none (landlocked)
Climate: temperate; continental, cloudy; cold winters with frequent rain and some snow in lowlands and snow in mountains; moderate summers with occasional showers
Terrain: in the west and south mostly mountains (Alps); along the eastern and northern margins mostly flat or gently sloping
Elevation extremes:
lowest point: Neusiedler See 115 m
highest point: Grossglockner 3,798 m
Natural resources: oil, coal, lignite, timber, iron ore, copper, zinc, antimony, magnesite, tungsten, graphite, salt, hydropower
Land use:
arable land: 16.59%
permanent crops: 0.85%
other: 82.56% (2005)
Irrigated land: 40 sq km (2003)
Total renewable water resources: 84 cu km (2005)
Freshwater withdrawal (domestic/industrial/agricultural): *total:* 3.67 cu km/yr (35%/64%/1%)
per capita: 448 cu m/yr (1999)
Natural hazards: landslides; avalanches; earthquakes
Environment—current issues: some forest degradation caused by air and soil pollution; soil pollution results from the use of agricultural chemicals; air pollution results from emissions by coal- and oil-fired power stations and industrial plants and from trucks transiting Austria between northern and southern Europe
Environment—international agreements: *party to:* Air Pollution, Air Pollution-Nitrogen Oxides, Air Pollution-Persistent Organic Pollutants, Air Pollution-Sulfur 85, Air Pollution-Sulphur 94, Air Pollution-Volatile Organic Compounds, Antarctic Treaty, Biodiversity, Climate Change, Climate Change-Kyoto Protocol, Desertification, Endangered Species, Environmental Modification, Hazardous Wastes, Law of the Sea, Ozone Layer Protection, Ship Pollution, Tropical Timber 83, Tropical Timber 94, Wetlands, Whaling
signed, but not ratified: none of the selected agreements
Geography—note: landlocked; strategic location at the crossroads of central Europe with many easily traversable Alpine passes and valleys; major river is the Danube; population is concentrated on eastern lowlands because of steep slopes, poor soils, and low temperatures elsewhere

PEOPLE

Population: 8,205,533 (July 2008 est.)
Age structure:
0–14 years: 14.8% (male 621,326/female 592,131)
15–64 years: 67.5% (male 2,783,531/female 2,753,389)
65 years and over: 17.7% (male 599,415/female 855,741) (2008 est.)
Median age:
total: 41.7 years
male: 40.7 years
female: 42.8 years (2008 est.)
Population growth rate: 0.064% (2008 est.)
Birth rate: 8.66 births/1,000 population (2008 est.)
Death rate: 9.91 deaths/1,000 population (2008 est.)
Net migration rate: 1.88 migrant(s)/1,000 population (2008 est.)
Sex ratio:
at birth: 1.05 male(s)/female
under 15 years: 1.05 male(s)/female
15–64 years: 1.01 male(s)/female
65 years and over: 0.7 male(s)/female
total population: 0.95 male(s)/female (2008 est.)
Infant mortality rate:
total: 4.48 deaths/1,000 live births
male: 5.48 deaths/1,000 live births
female: 3.44 deaths/1,000 live births (2008 est.)
Life expectancy at birth:
total population: 79.36 years
male: 76.46 years
female: 82.41 years (2008 est.)
Total fertility rate: 1.38 children born/woman (2008 est.)
HIV/AIDS—adult prevalence rate: 0.3% (2003 est.)
HIV/AIDS—people living with HIV/AIDS: 10,000 (2003 est.)
HIV/AIDS—deaths: fewer than 100 (2003 est.)
Nationality:
noun: Austrian(s)
adjective: Austrian
Ethnic groups: Austrians 91.1%, former Yugoslavs 4% (includes Croatians, Slovenes, Serbs, and Bosniaks), Turks 1.6%, German 0.9%, other or unspecified 2.4% (2001 census)

Religions: Roman Catholic 73.6%, Protestant 4.7%, Muslim 4.2%, other 3.5%, unspecified 2%, none 12% (2001 census)
Languages: German (official nationwide) 88.6%, Turkish 2.3%, Serbian 2.2%, Croatian (official in Burgenland) 1.6%, other (includes Slovene, official in Carinthia, and Hungarian, official in Burgenland) 5.3% (2001 census)
Literacy:
definition: age 15 and over can read and write
total population: 98%
male: NA
female: NA

GOVERNMENT

Country name:
conventional long form: Republic of Austria
conventional short form: Austria
local long form: Republik Oesterreich
local short form: Oesterreich
Government type: federal republic
Capital: *name:* Vienna
geographic coordinates: 48 12 N, 16 22 E
time difference: UTC+1 (6 hours ahead of Washington, DC during Standard Time)
daylight saving time: +1hr, begins last Sunday in March; ends last Sunday in October
Administrative divisions: 9 states (Bundeslaender, singular—Bundesland); Burgenland, Kaernten (Carinthia), Niederoesterreich (Lower Austria), Oberoesterreich (Upper Austria), Salzburg, Steiermark (Styria), Tirol (Tyrol), Vorarlberg, Wien (Vienna)
Independence: 976 (Margravate of Austria established); 17 September 1156 (Duchy of Austria founded); 11 August 1804 (Austrian Empire proclaimed); 12 November 1918 (republic proclaimed)
National holiday: National Day, 26 October (1955); note—commemorates the passage of the law on permanent neutrality
Constitution: 1920; revised 1929; reinstated 1 May 1945; note—during the period 1 May 1934–1 May 1945 there was a fascist (corporative) constitution in place
Legal system: civil law system with Roman law origin; judicial review of legislative acts by the Constitutional Court; separate administrative and civil/penal supreme courts; accepts compulsory ICJ jurisdiction
Suffrage: 16 years of age; universal; note—reduced from 18 years of age in 2007
Executive branch:
chief of state: President Heinz FISCHER (SPOe) (since 8 July 2004)

head of government: Chancellor Alfred GUSENBAUER (SPOe) (since 11 January 2007); Vice Chancellor Wilhelm MOLTERER (OeVP) (since 11 January 2007)

cabinet: Council of Ministers chosen by the president on the advice of the chancellor

elections: president elected by direct popular vote for a six-year term (eligible for a second term); presidential election last held 25 April 2004 (next to be held in April 2010); chancellor formally chosen by the president but determined by the coalition parties forming a parliamentary majority; vice chancellor chosen by the president on the advice of the chancellor

election results: Heinz FISCHER elected president; percent of vote—Heinz FISCHER 52.4%, Benita FERRERO-WALDNER 47.6%

note: government coalition—SPOe and OeVP

Legislative branch: bicameral Federal Assembly or Bundesversammlung consists of Federal Council or Bundesrat (62 seats; members chosen by state parliaments with each state receiving 3 to 12 members according to its population; members serve a five- or six-year term) and the National Council or Nationalrat (183 seats; members elected by direct popular vote to serve four-year terms)

elections: National Council—last held 1 October 2006 (next scheduled for the fall of 2010)

election results: National Council—percent of vote by party—SPOe 35.3%, OeVP 34.3%, Greens 11.1%, FPOe 11.0%, BZOe 4.1%, other 4.2%; seats by party—SPOe 68, OeVP 66, Greens 21, FPOe 21, BZOe 7

Judicial branch: Supreme Judicial Court or Oberster Gerichtshof; Administrative Court or Verwaltungsgerichtshof; Constitutional Court or Verfassungsgerichtshof

Political parties and leaders: Alliance for the Future of Austria or BZOe [Peter WESTENTHALER]; Austrian People's Party or OeVP [Wilhelm MOLTERER]; Freedom Party of Austria or FPOe [Heinz Christian STRACHE]; Social Democratic Party of Austria or SPOe [Alfred GUSENBAUER]; The Greens [Alexander VAN DER BELLEN]

Political pressure groups and leaders: Austrian Trade Union Federation or OeGB (nominally independent but primarily Social Democratic); Federal Economic Chamber; OeVP-oriented Association of Austrian Industrialists or IV; Roman Catholic Church, including its chief lay organization, Catholic Action; three composite leagues of the Austrian People's Party or OeVP representing business, labor, farmers, and other nongovernment organizations in the areas of environment and human rights

International organization participation: ACCT (observer), ADB (nonregional members), AfDB, Australia Group, BIS, BSEC (observer), CE, CEI, CERN, EAPC, EBRD, EIB, EMU, ESA, EU, FAO, G-9, IADB, IAEA, IBRD, ICAO, ICC, ICCt, ICRM, IDA, IEA, IFAD, IFC, IFRCS, ILO, IMF, IMO, Interpol, IOC, IOM, IPU, ISO, ITSO, ITU, ITUC, MIGA, MINURSO, NAM (guest), NEA, NSG, OAS (observer), OECD, OIF (observer), OPCW, OSCE, Paris Club, PCA, PFP, Schengen Convention, SECI (observer), UN, UNCTAD, UNDOF, UNESCO, UNFICYP, UNHCR, UNIDO, UNMEE, UNOMIG, UNTSO, UNWTO, UPU, WCL, WCO, WEU (observer), WFTU, WHO, WIPO, WMO, WTO, ZC

Diplomatic representation in the US:

chief of mission: Ambassador Eva NOWOTNY

chancery: 3524 International Court NW, Washington, DC 20008-3035

telephone: [1] (202) 895-6700

FAX: [1] (202) 895-6750

consulate(s) general: Chicago, Los Angeles, New York

Diplomatic representation from the US:

chief of mission: Ambassador (vacant); Charge d'Affaires Scott KILNER

embassy: Boltzmanngasse 16, A-1090, Vienna

mailing address: use embassy street address

telephone: [43] (1) 31339-0

FAX: [43] (1) 3100682

Flag description: three equal horizontal bands of red (top), white, and red

ECONOMY

Economy—overview: Austria, with its well-developed market economy and high standard of living, is closely tied to other EU economies, especially Germany's. The Austrian economy also benefits greatly from strong commercial relations, especially in the banking and insurance sectors, with central, eastern, and southeastern Europe. The economy features a large service sector, a sound industrial sector, and a small, but highly developed agricultural sector. Membership in the EU has drawn an influx of foreign investors attracted by Austria's access to the single European market and proximity to the new EU economies. The outgoing government has successfully pursued a comprehensive economic reform program, aimed at streamlining government and creating a more competitive business environment, further strengthening Austria's attractiveness as an investment location. It has implemented effective pension reforms; however, lower taxes in 2005–06 led to a small budget deficit in 2006 and 2007. Boosted by strong exports, growth nevertheless reached 3.3% in both 2006 and 2007, although the economy may slow in 2008 because of the strong euro, high oil prices, and problems in international financial markets. To meet increased competition—especially from new EU members and Central European countries—Austria will need to continue restructuring, emphasizing knowledge-based sectors of the economy, and encouraging greater labor flexibility and greater labor participation by its aging population.

GDP (purchasing power parity): $317.8 billion (2007 est.)

GDP (official exchange rate): $373.9 billion (2007 est.)

GDP—real growth rate: 3.4% (2007 est.)

GDP—per capita (PPP): $38,400 (2007 est.)

GDP—composition by sector:

agriculture: 1.6%

industry: 30.3%

services: 68% (2007 est.)

Labor force: 3.566 million (2007 est.)

Labor force—by occupation: *agriculture:* 3%

industry: 27%

services: 70% (2005 est.)

Unemployment rate: 4.4% (2007 est.)

Population below poverty line: 5.9% (2004)

Household income or consumption by percentage share:

lowest 10%: 3.3%

highest 10%: 22.5% (2004)

Distribution of family income—Gini index: 26 (2005)

Inflation rate (consumer prices): 2.2% (2007 est.)

Investment (gross fixed): 21.1% of GDP (2007 est.)

Budget:

revenues: $179.3 billion

expenditures: $181.8 billion (2007 est.)

Public debt: 59.3% of GDP (2007 est.)

Agriculture—products: grains, potatoes, sugar beets, wine, fruit; dairy products, cattle, pigs, poultry; lumber

Industries: construction, machinery, vehicles and parts, food, metals, chemicals, lumber and wood processing, paper and paperboard, communications equipment, tourism

Industrial production growth rate: 2.5% (2007 est.)

Electricity—production: 61.02 billion kWh (2005 est.)

Electricity—production by source:
fossil fuel: 29.3%
hydro: 67.2%
nuclear: 0%
other: 3.5% (2001)
Electricity—consumption: 60.25 billion kWh (2005 est.)
Electricity—exports: 17.73 billion kWh (2005 est.)
Electricity—imports: 20.4 billion kWh (2005 est.)
Oil—production: 23,320 bbl/day (2005)
Oil—consumption: 295,100 bbl/day (2005 est.)
Oil—exports: 34,680 bbl/day (2004)
Oil—imports: 157,500 bbl/day (2005)
Oil—proved reserves: 62 million bbl (1 January 2006 est.)
Natural gas—production: 1.57 billion cu m (2005)
Natural gas—consumption: 9.217 billion cu m (2005)
Natural gas—exports: 936.1 million cu m (2005)
Natural gas—imports: 9.063 billion cu m (2005)
Natural gas—proved reserves: 14.39 billion cu m (1 January 2006 est.)
Current account balance: $10.04 billion (2007 est.)
Exports: $164.9 billion f.o.b. (2007 est.)
Exports—commodities: machinery and equipment, motor vehicles and parts, paper and paperboard, metal goods, chemicals, iron and steel, textiles, foodstuffs
Exports—partners: Germany 30.1%, Italy 9%, US 5.8%, Switzerland 4.7% (2006)
Imports: $162.7 billion f.o.b. (2007 est.)
Imports—commodities: machinery and equipment, motor vehicles, chemicals, metal goods, oil and oil products; foodstuffs
Imports—partners: Germany 45.4%, Italy 7%, Switzerland 4.5%, Netherlands 4.1% (2006)
Economic aid—donor: ODA, $1.498 billion (2006)
Reserves of foreign exchange and gold: $18.22 billion (2006 est.)
Debt—external: $752.5 billion (30 June 2007)
Stock of direct foreign investment—at home: $202.5 billion (2007 est.)
Stock of direct foreign investment—abroad: $182.2 billion (2007 est.)

Market value of publicly traded shares: $126.3 billion (2005)
Currency (code): euro (EUR)
Currency code: EUR
Exchange rates: euros per US dollar— 0.7345 (2007), 0.7964 (2006), 0.8041 (2005), 0.8054 (2004), 0.886 (2003)
Fiscal year: calendar year

COMMUNICATIONS

Telephones—main lines in use: 3.564 million (2006)
Telephones—mobile cellular: 9.255 million (2006)
Telephone system:
general assessment: highly developed and efficient
domestic: fixed-line subscribership has been in decline since the mid-1990s with mobile-cellular subscribership elipsing it by the late 1990s; the fiber-optic net is very extensive; all telephone applications and Internet services are available
international: country code—43; satellite earth stations—15; in addition, there are about 600 VSATs (very small aperture terminals) (2007)
Radio broadcast stations: AM 2, FM 65 (plus several hundred repeaters), shortwave 1 (2001)
Radios: 6.08 million (1997)
Television broadcast stations: 10 (plus more than 1,000 repeaters) (2001)
Televisions: 4.25 million (1997)
Internet country code: .at
Internet hosts: 2.427 million (2007)
Internet Service Providers (ISPs): 37 (2000)
Internet users: 4.2 million (2006)

TRANSPORTATION

Airports: 55 (2007)
Airports—with paved runways:
total: 25
over 3,047 m: 1
2,438 to 3,047 m: 5
1,524 to 2,437 m: 1
914 to 1,523 m: 3
under 914 m: 15 (2007)
Airports—with unpaved runways:
total: 30
1,524 to 2,437 m: 1
914 to 1,523 m: 3
under 914 m: 26 (2007)
Heliports: 1 (2007)
Pipelines: gas 2,722 km; oil 663 km; refined products 157 km (2007)

Railways:
total: 6,383 km
standard gauge: 5,924 km 1.435-m gauge (3,772 km electrified)
narrow gauge: 371 km 1.000-m gauge; 88 km 0.760-m gauge (25 km electrified) (2006)
Roadways:
total: 133,910 km
paved: 133,910 km (includes 2,050 km of expressways) (2003)
Waterways: 358 km (2007)
Merchant marine:
total: 7 ships (1000 GRT or over) 31,705 GRT/40,627 DWT
by type: cargo 5, container 2
foreign-owned: 2 (Netherlands 2)
registered in other countries: 4 (Cyprus 1, Malta 1, St Vincent and The Grenadines 2) (2007)
Ports and terminals: Enns, Krems, Linz, Vienna

MILITARY

Military branches: Land Forces (KdoLdSK), Air Forces (KdoLuSK)
Military service age and obligation: 18–35 years of age for compulsory military service; 16 years of age for male or female voluntary service; service obligation 7 months of training, followed by an 8-year reserve obligation (2006)
Manpower available for military service:
males age 16–49: 1,986,411
females age 16–49: 1,944,834 (2008 est.)
Manpower fit for military service:
males age 16–49: 1,617,385
females age 16–49: 1,583,886 (2008 est.)
Manpower reaching militarily significant age annually:
males age 16–49: 50,869
females age 16–49: 48,246 (2008 est.)
Military expenditures—percent of GDP: 0.9% (2005 est.)

TRANSNATIONAL ISSUES

Disputes—international: in 2006, Austrian public protests for the Czech Republic to close the Temelin nuclear power plant resulted in a parliamentary motion threatening international legal action
Illicit drugs: transshipment point for Southwest Asian heroin and South American cocaine destined for Western Europe; increasing consumption of European-produced synthetic drugs

AZERBAIJAN

ⓘ INTRODUCTION

Background: Azerbaijan—a nation with a majority-Turkic and majority-Muslim population—was briefly independent from 1918 to 1920; it regained its independence after the collapse of the Soviet Union in 1991. Despite a 1994 ceasefire, Azerbaijan has yet to resolve its conflict with Armenia over the Azerbaijani Nagorno-Karabakh enclave (largely Armenian populated). Azerbaijan has

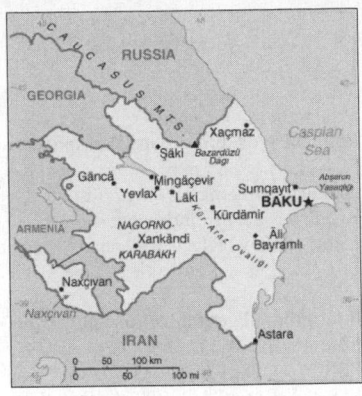

lost 16% of its territory and must support some 600,000 internally displaced persons as a result of the conflict. Corruption is ubiquitous, and the government has been accused of authoritarianism. Although the poverty rate has been reduced in recent years, the promise of widespread wealth from development of Azerbaijan's energy sector remains largely unfulfilled.

GEOGRAPHY

Location: Southwestern Asia, bordering the Caspian Sea, between Iran and Russia, with a small European portion north of the Caucasus range
Geographic coordinates: 40 30 N, 47 30 E
Map references: Asia
Area:
total: 86,600 sq km
land: 86,100 sq km
water: 500 sq km
note: includes the exclave of Naxcivan Autonomous Republic and the Nagorno-Karabakh region; the region's autonomy was abolished by Azerbaijani Supreme Soviet on 26 November 1991
Area—comparative: slightly smaller than Maine
Land boundaries:
total: 2,013 km
border countries: Armenia (with Azerbaijan-proper) 566 km, Armenia (with Azerbaijan-Naxcivan exclave) 221 km, Georgia 322 km, Iran (with Azerbaijan-proper) 432 km, Iran (with Azerbaijan-Naxcivan exclave) 179 km, Russia 284 km, Turkey 9 km
Coastline: 0 km (landlocked); note—Azerbaijan borders the Caspian Sea (800 km est.)
Maritime claims: none (landlocked)
Climate: dry, semiarid steppe
Terrain: large, flat Kur-Araz Ovaligi (Kura-Araks Lowland) (much of it below sea level) with Great Caucasus Mountains to the north, Qarabag Yaylasi (Karabakh Upland) in west; Baku lies on

Abseron Yasaqligi (Apsheron Peninsula) that juts into Caspian Sea
Elevation extremes:
lowest point: Caspian Sea -28 m
highest point: Bazarduzu Dagi 4,485 m
Natural resources: petroleum, natural gas, iron ore, nonferrous metals, bauxite
Land use: *arable land:* 20.62%
permanent crops: 2.61%
other: 76.77% (2005)
Irrigated land: 14,550 sq km (2003)
Total renewable water resources: 30.3 cu km (1997)
Freshwater withdrawal (domestic/industrial/agricultural): *total:* 17.25 cu km/yr (5%/28%/68%)
per capita: 2,051 cu m/yr (2000)
Natural hazards: droughts
Environment—current issues: local scientists consider the Abseron Yasaqligi (Apsheron Peninsula) (including Baku and Sumqayit) and the Caspian Sea to be the ecologically most devastated area in the world because of severe air, soil, and water pollution; soil pollution results from oil spills, from the use of DDT pesticide, and from toxic defoliants used in the production of cotton
Environment—international agreements: *party to:* Air Pollution, Biodiversity, Climate Change, Climate Change-Kyoto Protocol, Desertification, Endangered Species, Hazardous Wastes, Marine Dumping, Ozone Layer Protection, Ship Pollution, Wetlands
signed, but not ratified: none of the selected agreements
Geography—note: both the main area of the country and the Naxcivan exclave are landlocked

PEOPLE

Population: 8,177,717 (July 2008 est.)
Age structure:
0–14 years: 24.6% (male 1,061,318/female 947,607)
15–64 years: 68.6% (male 2,753,277/female 2,855,406)
65 years and over: 6.8% (male 208,293/female 351,816) (2008 est.)
Median age:
total: 27.9 years
male: 26.3 years
female: 29.7 years (2008 est.)
Population growth rate: 0.723% (2008 est.)
Birth rate: 17.52 births/1,000 population (2008 est.)
Death rate: 8.32 deaths/1,000 population (2008 est.)
Net migration rate: -1.97 migrant(s)/1,000 population (2008 est.)
Sex ratio:
at birth: 1.14 male(s)/female
under 15 years: 1.12 male(s)/female

15–64 years: 0.96 male(s)/female
65 years and over: 0.59 male(s)/female
total population: 0.97 male(s)/female (2008 est.)
Infant mortality rate:
total: 56.43 deaths/1,000 live births
male: 62.09 deaths/1,000 live births
female: 49.98 deaths/1,000 live births (2008 est.)
Life expectancy at birth:
total population: 66.31 years
male: 62.2 years
female: 71 years (2008 est.)
Total fertility rate: 2.05 children born/woman (2008 est.)
HIV/AIDS—adult prevalence rate: less than 0.1% (2003 est.)
HIV/AIDS—people living with HIV/AIDS: 1,400 (2003 est.)
HIV/AIDS—deaths: fewer than 100 (2001 est.)
Nationality:
noun: Azerbaijani(s)
adjective: Azerbaijani
Ethnic groups: Azeri 90.6%, Dagestani 2.2%, Russian 1.8%, Armenian 1.5%, other 3.9% (1999 census)
note: almost all Armenians live in the separatist Nagorno-Karabakh region
Religions: Muslim 93.4%, Russian Orthodox 2.5%, Armenian Orthodox 2.3%, other 1.8% (1995 est.)
note: religious affiliation is still nominal in Azerbaijan; percentages for actual practicing adherents are much lower
Languages: Azerbaijani (Azeri) 90.3%, Lezgi 2.2%, Russian 1.8%, Armenian 1.5%, other 3.3%, unspecified 1% (1999 census)
Literacy:
definition: age 15 and over can read and write
total population: 98.8%
male: 99.5%
female: 98.2% (1999 census)

GOVERNMENT

Country name:
conventional long form: Republic of Azerbaijan
conventional short form: Azerbaijan
local long form: Azarbaycan Respublikasi
local short form: Azarbaycan
former: Azerbaijan Soviet Socialist Republic
Government type: republic
Capital: *name:* Baku (Baki, Baky)
geographic coordinates: 40 23 N, 49 52 E
time difference: UTC+4 (9 hours ahead of Washington, DC during Standard Time)
daylight saving time: +1hr, begins last Sunday in March; ends last Sunday in October
Administrative divisions: 59 rayons (rayonlar; rayon—singular), 11 cities

(saharlar; sahar—singular), 1 autonomous republic (muxtar respublika)

rayons: Abseron Rayonu, Agcabadi Rayonu, Agdam Rayonu, Agdas Rayonu, Agstafa Rayonu, Agsu Rayonu, Astara Rayonu, Balakan Rayonu, Barda Rayonu, Beylaqan Rayonu, Bilasuvar Rayonu, Cabrayil Rayonu, Calilabad Rayonu, Daskasan Rayonu, Davaci Rayonu, Fuzuli Rayonu, Gadabay Rayonu, Goranboy Rayonu, Goycay Rayonu, Haciqabul Rayonu, Imisli Rayonu, Ismayilli Rayonu, Kalbacar Rayonu, Kurdamir Rayonu, Lacin Rayonu, Lankaran Rayonu, Lerik Rayonu, Masalli Rayonu, Neftcala Rayonu, Oguz Rayonu, Qabala Rayonu, Qax Rayonu, Qazax Rayonu, Qobustan Rayonu, Quba Rayonu, Qubadli Rayonu, Qusar Rayonu, Saatli Rayonu, Sabirabad Rayonu, Saki Rayonu, Salyan Rayonu, Samaxi Rayonu, Samkir Rayonu, Samux Rayonu, Siyazan Rayonu, Susa Rayonu, Tartar Rayonu, Tovuz Rayonu, Ucar Rayonu, Xacmaz Rayonu, Xanlar Rayonu, Xizi Rayonu, Xocali Rayonu, Xocavand Rayonu, Yardimli Rayonu, Yevlax Rayonu, Zangilan Rayonu, Zaqatala Rayonu, Zardab Rayonu

cities: Ali Bayramli Sahari, Baki Sahari, Ganca Sahari, Lankaran Sahari, Mingacevir Sahari, Naftalan Sahari, Saki Sahari, Sumqayit Sahari, Susa Sahari, Xankandi Sahari, Yevlax Sahari

autonomous republic: Naxcivan Muxtar Respublikasi

Independence: 30 August 1991 (from Soviet Union)

National holiday: Founding of the Democratic Republic of Azerbaijan, 28 May (1918)

Constitution: adopted 12 November 1995

Legal system: based on civil law system; has not accepted compulsory ICJ jurisdiction

Suffrage: 18 years of age; universal

Executive branch:

chief of state: President Ilham ALIYEV (since 31 October 2003)

head of government: Prime Minister Artur RASIZADE (since 4 November 2003); First Deputy Prime Minister Yaqub EYYUBOV (since June 2006)

cabinet: Council of Ministers appointed by the president and confirmed by the National Assembly

elections: president elected by popular vote to a five-year term (eligible for a second term); election last held 15 October 2003 (next to be held in October 2008); prime minister and first deputy prime minister appointed by the president and confirmed by the National Assembly

election results: Ilham ALIYEV elected president; percent of vote—Ilham ALIYEV 76.8%, Isa GAMBAR 14%

Legislative branch: unicameral National Assembly or Milli Mejlis (125 seats; members elected by popular vote to serve five-year terms)

elections: last held 6 November 2005 (next to be held in November 2010)

election results: percent of vote by party—NA; seats by party—Yeni 58, Azadliq coalition 8, CSP 2, Motherland 2, other parties with single seats 9, independents 42, undetermined 4

Judicial branch: Supreme Court

Political parties and leaders: Azadliq (Freedom) coalition (Popular Front Party, Liberal Party, Citizens' Development Party); Azerbaijan Democratic Party or ADP [Sardar JALALOGLU]; Azerbaijan Democratic Reforms Party (ADRP) Youth Movement [Ramin HAJILI]; Azerbaijan Popular Front or APF, now split in two [Ali KARIMLI, leader of "Reform" APF party; Mirmahmud MIRALI-OGLU, leader of "Classic" APF party]; Azerbaijan Public Forum [Eldar NAMAZOV]; Citizens' Development Party [Ali ALIYEV]; Civil Solidarity Party or CSP [Sabir RUSTAMKHANLY]; Dalga Youth Movement [Vafa JAFAROVA]; Green Party [Mais GULALIYEV and Tarana MAMMADOVA]; Hope (Umid) Party [Iqbal AGAZADE]; Ireli Youth Movement [Jeyhun OSMANLI, Roya TALIBOVA, Farhad MAMMADOV, Elnara GARIBOVA, Elnur MAMMADOV, Ziya ALIYEV]; Justice Party [Ilyas ISMAILOV]; Liberal Party of Azerbaijan [Lala Shovkat HACIYEVA]; Magam Youth Movement [Emin HUSEYNOV]; Motherland Party [Fazail AGAMALI]; Musavat (Equality) [Isa GAMBAR, chairman]; Musavat Party Youth Movement [Elnur MAMMADLI]; National Democratic Party or Grey Wolves (Nationalist, Pan-Turkic) [Iskender HAMIDOV]; Open Society Party [Rasul GULIYEV, in exile in the US]; Party for National Independence of Azerbaijan or PNIA [Ayaz RUSTAMOV]; Popular Front Party Youth Movement [Seymur KHAZIYEV]; Social Democratic Party of Azerbaijan or SDP [Araz ALIZADE and Ayaz MUTALIBOV (in exile)]; Turkish Nationalist Party [Vugar BAYTURAN]; United Azerbaijan Party [Karrar ABILOV]; United Azerbaijan National Unity Party [Hajibaba AZIMOV]; United Party [Tahir KARIMLI]; Yeni (New) Azerbaijan Party [President Ilham ALIYEV]; Yeni Azerbaijan Party Youth

Movement [Ramil HASANOV]; Yox (No) Youth Movement [Ali ISMAYILOV]

note: opposition parties regularly factionalize and form new parties;

Political pressure groups and leaders: Sadval, Lezgin movement; self-proclaimed Armenian Nagorno-Karabakh Republic; Talysh independence movement; Union of Pro-Azerbaijani Forces or UPAF; Karabakh Liberation Organization

International organization participation: ADB, BSEC, CE, CIS, EAPC, EBRD, ECO, FAO, GCTU, GUAM, IAEA, IBRD, ICAO, ICRM, IDA, IDB, IFAD, IFC, IFRCS, ILO, IMF, IMO, Interpol, IOC, IOM, IPU, ISO, ITSO, ITU, ITUC, MIGA, NAM (observer), OAS (observer), OIC, OPCW, OSCE, PFP, SECI (observer), UN, UNCTAD, UNESCO, UNIDO, UNWTO, UPU, WCO, WFTU, WHO, WIPO, WMO, WTO (observer)

Diplomatic representation in the US:

chief of mission: Ambassador Yashar ALIYEV

chancery: 2741 34th Street NW, Washington, DC 20008

telephone: [1] (202) 337-3500

FAX: [1] (202) 337-5911

Consulate(s) general: Los Angeles

Diplomatic representation from the US:

chief of mission: Ambassador Anne E. DERSE

embassy: 83 Azadliyg Prospecti, Baku AZ1007

mailing address: American Embassy Baku, US Department of State, 7050 Baku Place, Washington, DC 20521-7050

telephone: [994] (12) 4980-335 through 337

FAX: [994] (12) 4656-671

Flag description: three equal horizontal bands of blue (top), red, and green; a crescent and eight-pointed star in white are centered in red band

ECONOMY

Economy—overview: Azerbaijan's high economic growth in 2006 and 2007 is attributable to large and growing oil exports. Azerbaijan's oil production declined through 1997, but has registered an increase every year since. Negotiation of production-sharing arrangements (PSAs) with foreign firms, which have committed $60 billion to long-term oilfield development, should generate the funds needed to spur future industrial development. Oil production under the first of these PSAs, with the Azerbaijan International Operating Company, began in November 1997. A consortium of Western oil companies began pumping 1 million barrels a day from a large offshore field in early 2006, through a $4 billion pipeline it built from Baku to Turkey's Mediterranean port of Ceyhan. By 2010 revenues from this project will double the country's current GDP. Azerbaijan shares all the formidable problems of the former Soviet republics in making the transition from a command to a market economy, but its considerable energy resources brighten its long-term prospects. Baku has only recently begun making progress on economic reform, and old economic ties and structures are slowly being replaced. Several other obstacles impede Azerbaijan's economic progress: the need for stepped up foreign investment in the non-energy sector, the continuing conflict with Armenia over the Nagorno-Karabakh region, pervasive corruption, and elevated inflation. Trade with Russia and the other former Soviet republics is declining in importance, while trade is building with Turkey and the nations of Europe. Long-term prospects will depend on world oil prices, the location of new oil and gas pipelines in the region, and Azerbaijan's ability to manage its energy wealth.

GDP (purchasing power parity): $65.47 billion (2007 est.)
GDP (official exchange rate): $31.32 billion (2007 est.)
GDP—real growth rate: 23.4% (2007 est.)
GDP—per capita (PPP): $7,700 (2007 est.)
GDP—composition by sector:
agriculture: 6.2%
industry: 63.3%
services: 30.5% (2007 est.)
Labor force: 5.243 million (2007 est.)
Labor force—by occupation: *agriculture:* 41%
industry: 7%
services: 52% (2001)
Unemployment rate: 4.3% official rate (2007 est.)
Population below poverty line: 24% (2005 est.)
Household income or consumption by percentage share:
lowest 10%: 3.1%
highest 10%: 29.5% (2001)
Distribution of family income—Gini index: 36.5 (2001)
Inflation rate (consumer prices): 16.6% (2007 est.)
Investment (gross fixed): 28% of GDP (2007 est.)
Budget:
revenues: $7.67 billion
expenditures: $9.256 billion (2007 est.)
Public debt: 6.4% of GDP (2007 est.)

Agriculture—products: cotton, grain, rice, grapes, fruit, vegetables, tea, tobacco; cattle, pigs, sheep, goats
Industries: petroleum and natural gas, petroleum products, oilfield equipment; steel, iron ore; cement; chemicals and petrochemicals; textiles
Industrial production growth rate: 25% (2007 est.)
Electricity—production: 23.8 billion kWh (2007 est.)
Electricity—production by source:
fossil fuel: 89.7%
hydro: 10.3%
nuclear: 0%
other: 0% (2001)
Electricity—consumption: 27.5 billion kWh (2007 est.)
Electricity—exports: 880 million kWh (2005)
Electricity—imports: 2.082 billion kWh (2005)
Oil—production: 934,700 bbl/day (2007 est.)
Oil—consumption: 160,000 bbl/day (2007 est.)
Oil—exports: 795,600 bbl/day (2007 est.)
Oil—imports: 3,924 bbl/day (2004)
Oil—proved reserves: 7 billion bbl (17 April 2007 est.)
Natural gas—production: 6.3 billion cu m (2007 est.)
Natural gas—consumption: 9.8 billion cu m (2007 est.)
Natural gas—exports: 0 cu m (2005 est.)
Natural gas—imports: 4.373 billion cu m (2005)
Natural gas—proved reserves: 849.5 billion cu m (17 April 2007 est.)
Current account balance: $9.01 billion (2007 est.)
Exports: $21.27 billion f.o.b. (2007 est.)
Exports—commodities: oil and gas 90%, machinery, cotton, foodstuffs
Exports—partners: Italy 44.7%, Israel 10.7%, Turkey 6.1%, France 5.5%, Russia 5.4%, Iran 4.6%, Georgia 4.5% (2006)
Imports: $6.045 billion f.o.b. (2007 est.)
Imports—commodities: machinery and equipment, oil products, foodstuffs, metals, chemicals
Imports—partners: Russia 22.4%, UK 8.6%, Germany 7.7%, Turkey 7.3%, Turkmenistan 7%, Ukraine 6%, China 4.2% (2006)
Economic aid—recipient: ODA, $223.4 million (2005 est.)
Reserves of foreign exchange and gold: $4.273 billion (31 December 2007 est.)
Debt—external: $2.399 billion (31 December 2007 est.)

Stock of direct foreign investment—at home: $7.829 billion (2007 est.)
Stock of direct foreign investment—abroad: $4.912 billion (2007 est.)
Market value of publicly traded shares: $NA
Currency (code): Azerbaijani manat (AZN)
Currency code: AZM
Exchange rates: Azerbaijani manats per US dollar—0.8581 (2007), 0.8934 (2006), 4,727.1 (2005), 4,913.48 (2004), 4,910.73 (2003)
note: on 1 January 2006 Azerbaijan revalued its currency, with 5,000 old manats equal to 1 new manat
Fiscal year: calendar year

COMMUNICATIONS

Telephones—main lines in use: 1.189 million (2006)
Telephones—mobile cellular: 3.324 million (2006)
Telephone system:
general assessment: inadequate; requires considerable expansion and modernization; teledensity of 15 main lines per 100 persons is low; mobile-cellular penetration is increasing and is currently about 40 telephones per 100 persons
domestic: fixed-line telephony and a broad range of other telecom services are controlled by a state-owned telecommunications monopoly and growth has been stagnant; more competition exists in the mobile-cellular market with three providers in 2006; satellite service connects Baku to a modern switch in its exclave of Naxcivan
international: country code—994; the old Soviet system of cable and microwave is still serviceable; satellite earth stations—2 (2007)
Radio broadcast stations: AM 10, FM 17, shortwave 1 (1998)
Radios: 175,000 (1997)
Television broadcast stations: 2 (1997)
Televisions: 170,000 (1997)
Internet country code: .az
Internet hosts: 3,067 (2007)
Internet Service Providers (ISPs): 2 (2000)
Internet users: 829,100 (2006)

TRANSPORTATION

Airports: 35 (2007)
Airports—with paved runways:
total: 27
over 3,047 m: 2
2,438 to 3,047 m: 6
1,524 to 2,437 m: 13
914 to 1,523 m: 4
under 914 m: 2 (2007)
Airports—with unpaved runways:
total: 8

914 to 1,523 m: 1
under 914 m: 7 (2007)
Heliports: 1 (2007)
Pipelines: gas 3,857 km; oil 2,436 km (2007)
Railways:
total: 2,122 km
broad gauge: 2,122 km 1.520-m gauge (1,278 km electrified) (2006)
Roadways:
total: 59,141 km
paved: 29,210 km
unpaved: 29,931 km (2004)
Merchant marine:
total: 86 ships (1000 GRT or over) 421,061 GRT/460,968 DWT
by type: cargo 26, passenger 2, passenger/cargo 9, petroleum tanker 45, roll on/roll off 1, specialized tanker 3
registered in other countries: 4 (Georgia 1, Malta 3) (2007)
Ports and terminals: Baku (Baki)

MILITARY

Military branches: Army, Navy, Air and Air Defense Forces (2008)
Military service age and obligation: men between 18 and 35 are liable for military service; 18 years of age for voluntary military service; length of military service is 18 months and 12 months for university graduates (2006)
Manpower available for military service:
males age 16–49: 2,278,888
females age 16–49: 2,291,770 (2008 est.)
Manpower fit for military service:
males age 16–49: 1,696,167
females age 16–49: 1,923,556 (2008 est.)
Manpower reaching militarily significant age annually:
males age 16–49: 94,402
females age 16–49: 89,686 (2008 est.)
Military expenditures—percent of GDP: 2.6% (2005 est.)

TRANSNATIONAL ISSUES

Disputes—international: Armenia supports ethnic Armenian secessionists in Nagorno-Karabakh and since the early 1990s has militarily occupied 16% of Azerbaijan; over 800,000 mostly ethnic Azerbaijanis were driven from the occupied lands and Armenia; about 230,000 ethnic Armenians were driven from their homes in Azerbaijan into Armenia; Azerbaijan seeks transit route through Armenia to connect to Naxcivan exclave; Organization for Security and Cooperation in Europe (OSCE) continues to mediate dispute; Azerbaijan, Kazakhstan, and Russia have ratified Caspian seabed delimitation treaties based on equidistance, while Iran continues to insist on an even one-fifth allocation and challenges Azerbaijan's hydrocarbon exploration in disputed waters; bilateral talks continue with Turkmenistan on dividing the seabed and contested oilfields in the middle of the Caspian; Azerbaijan and Georgia continue to discuss the alignment of their boundary at certain crossing areas
Refugees and internally displaced persons: *refugees (country of origin):* 2,400 (Russia)
IDPs: 580,000–690,000 (conflict with Armenia over Nagorno-Karabakh) (2007)
Illicit drugs: limited illicit cultivation of cannabis and opium poppy, mostly for CIS consumption; small government eradication program; transit point for Southwest Asian opiates bound for Russia and to a lesser extent the rest of Europe

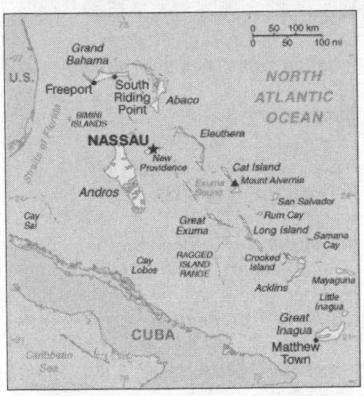

INTRODUCTION

Background: Lucayan Indians inhabited the islands when Christopher COLUMBUS first set foot in the New World on San Salvador in 1492. British settlement of the islands began in 1647; the islands became a colony in 1783. Since attaining independence from the UK in 1973, The Bahamas have prospered through tourism and international banking and investment management. Because of its geography, the country is a major transshipment point for illegal drugs, particularly shipments to the US and Europe, and its territory is used for smuggling illegal migrants into the US.

GEOGRAPHY

Location: Caribbean, chain of islands in the North Atlantic Ocean, southeast of Florida, northeast of Cuba
Geographic coordinates: 24 15 N, 76 00 W
Map references: Central America and the Caribbean
Area:
total: 13,940 sq km
land: 10,070 sq km
water: 3,870 sq km
Area—comparative: slightly smaller than Connecticut
Land boundaries: 0 km
Coastline: 3,542 km
Maritime claims:
territorial sea: 12 nm
exclusive economic zone: 200 nm
Climate: tropical marine; moderated by warm waters of Gulf Stream
Terrain: long, flat coral formations with some low rounded hills
Elevation extremes:
lowest point: Atlantic Ocean 0 m
highest point: Mount Alvernia, on Cat Island 63 m
Natural resources: salt, aragonite, timber, arable land

Land use:
arable land: 0.58%
permanent crops: 0.29%
other: 99.13% (2005)
Irrigated land: 10 sq km (2003)
Total renewable water resources: NA
Natural hazards: hurricanes and other tropical storms cause extensive flood and wind damage
Environment—current issues: coral reef decay; solid waste disposal
Environment—international agreements: *party to:* Biodiversity, Climate Change, Climate Change-Kyoto Protocol, Desertification, Endangered Species, Hazardous Wastes, Law of the Sea, Ozone Layer Protection, Ship Pollution, Wetlands
signed, but not ratified: none of the selected agreements
Geography—note: strategic location adjacent to US and Cuba; extensive island chain of which 30 are inhabited

PEOPLE

Population: 307,451
note: estimates for this country explicitly take into account the effects of excess mortality due to AIDS; this can result in lower life expectancy, higher infant mortality, higher death rates, lower population growth rates, and changes in the distribution of population by age and sex than would otherwise be expected (July 2008 est.)
Age structure:
0–14 years: 26.4% (male 40,608/female 40,506)
15–64 years: 66.9% (male 101,150/female 104,457)
65 years and over: 6.7% (male 8,472/female 12,258) (2008 est.)
Median age:
total: 28.4 years
male: 27.6 years
female: 29.2 years (2008 est.)
Population growth rate: 0.57% (2008 est.)
Birth rate: 17.06 births/1,000 population (2008 est.)
Death rate: 9.22 deaths/1,000 population (2008 est.)
Net migration rate: -2.14 migrant(s)/1,000 population (2008 est.)
Sex ratio:
at birth: 1.02 male(s)/female
under 15 years: 1 male(s)/female
15–64 years: 0.97 male(s)/female
65 years and over: 0.69 male(s)/female
total population: 0.96 male(s)/female (2008 est.)
Infant mortality rate:
total: 23.67 deaths/1,000 live births

male: 28.89 deaths/1,000 live births
female: 18.34 deaths/1,000 live births (2008 est.)
Life expectancy at birth:
total population: 65.72 years
male: 62.5 years
female: 69 years (2008 est.)
Total fertility rate: 2.13 children born/woman (2008 est.)
HIV/AIDS—adult prevalence rate: 3% (2003 est.)
HIV/AIDS—people living with HIV/AIDS: 5,600 (2003 est.)
HIV/AIDS—deaths: fewer than 200 (2003 est.)
Nationality:
noun: Bahamian(s)
adjective: Bahamian
Ethnic groups: black 85%, white 12%, Asian and Hispanic 3%
Religions: Baptist 35.4%, Anglican 15.1%, Roman Catholic 13.5%, Pentecostal 8.1%, Church of God 4.8%, Methodist 4.2%, other Christian 15.2%, none or unspecified 2.9%, other 0.8% (2000 census)
Languages: English (official), Creole (among Haitian immigrants)
Literacy:
definition: age 15 and over can read and write
total population: 95.6%
male: 94.7%
female: 96.5% (2003 est.)

GOVERNMENT

Country name:
conventional long form: Commonwealth of The Bahamas
conventional short form: The Bahamas
Government type: constitutional parliamentary democracy
Capital: *name:* Nassau
geographic coordinates: 25 05 N, 77 21 W
time difference: UTC-5 (same time as Washington, DC during Standard Time)
daylight saving time: +1hr, begins second Sunday in March; ends first Sunday in November
Administrative divisions: 21 districts; Acklins and Crooked Islands, Bimini, Cat Island, Exuma, Freeport, Fresh Creek, Governor's Harbour, Green Turtle Cay, Harbour Island, High Rock, Inagua, Kemps Bay, Long Island, Marsh Harbour, Mayaguana, New Providence, Nichollstown and Berry Islands, Ragged Island, Rock Sound, Sandy Point, San Salvador and Rum Cay
Independence: 10 July 1973 (from UK)
National holiday: Independence Day, 10 July (1973)
Constitution: 10 July 1973

Legal system: based on English common law

Suffrage: 18 years of age; universal

Executive branch:

chief of state: Queen ELIZABETH II (since 6 February 1952); represented by Governor General Arthur D. HANNA (since 1 February 2006)

head of government: Prime Minister Hubert A. INGRAHAM (since 4 May 2007)

cabinet: Cabinet appointed by the governor general on the prime minister's recommendation

elections: the monarch is hereditary; governor general appointed by the monarch; following legislative elections, the leader of the majority party or the leader of the majority coalition is usually appointed prime minister by the governor general; the prime minister recommends the deputy prime minister

Legislative branch: bicameral Parliament consists of the Senate (16 seats; members appointed by the governor general upon the advice of the prime minister and the opposition leader to serve five-year terms) and the House of Assembly (41 seats; members elected by direct popular vote to serve five-year terms); the government may dissolve the Parliament and call elections at any time

elections: last held 2 May 2007 (next to be held by May 2012)

election results: percent of vote by party—FNM 49.86%, PLP 47.02%; seats by party—FNM 23, PLP 18

Judicial branch: Privy Council in London; Courts of Appeal; Supreme (lower) Court; Magistrates' Courts

Political parties and leaders: Free National Movement or FNM [Hubert INGRAHAM]; Progressive Liberal Party or PLP [Perry CHRISTIE]

Political pressure groups and leaders: NA

International organization participation: ACP, C, Caricom, CDB, FAO, G-77, IADB, IBRD, ICAO, ICCt (signatory), ICFTU, ICRM, IDA, IFC, IFRCS, ILO, IMF, IMO, IMSO, Interpol, IOC, IOM, ITSO, ITU, LAES, MIGA, NAM, OAS, OPANAL, OPCW (signatory), UN, UNCTAD, UNESCO, UNIDO, UNWTO, UPU, WCO, WHO, WIPO, WMO, WTO (observer)

Diplomatic representation in the US:

chief of mission: Ambassador Cornelius A. SMITH

chancery: 2220 Massachusetts Avenue NW, Washington, DC 20008

telephone: [1] (202) 319-2660

FAX: [1] (202) 319-2668

consulate(s) general: Miami, New York

Diplomatic representation from the US:

chief of mission: Ambassador Ned L. SIEGEL

embassy: 42 Queen Street, Nassau

mailing address: local or express mail address: P. O. Box N-8197, Nassau; US Department of State, 3370 Nassau Place, Washington, DC 20521-3370

telephone: [1] (242) 322-1181, 356-3229 (after hours)

FAX: [1] (242) 356-0222

Flag description: three equal horizontal bands of aquamarine (top), gold, and aquamarine, with a black equilateral triangle based on the hoist side

ECONOMY

Economy—overview: The Bahamas is one of the wealthiest Caribbean countries with an economy heavily dependent on tourism and offshore banking. Tourism together with tourism-driven construction and manufacturing accounts for approximately 60% of GDP and directly or indirectly employs half of the archipelago's labor force. Steady growth in tourism receipts and a boom in construction of new hotels, resorts, and residences had led to solid GDP growth in recent years, but tourist arrivals have been on the decline since 2006. Financial services constitute the second-most important sector of the Bahamian economy and, when combined with business services, account for about 36% of GDP. However, since December 2000, when the government enacted new regulations on the financial sector, many international businesses have left The Bahamas. Manufacturing and agriculture combined contribute approximately a tenth of GDP and show little growth, despite government incentives aimed at those sectors. Overall growth prospects in the short run rest heavily on the fortunes of the tourism sector. Tourism, in turn, depends on growth in the US, the source of more than 80% of the visitors.

GDP (purchasing power parity): $8.332 billion (2007 est.)

GDP (official exchange rate): $6.586 billion (2007 est.)

GDP—real growth rate: 3.1% (2007 est.)

GDP—per capita (PPP): $25,000 (2007 est.)

GDP—composition by sector:

agriculture: 3%

industry: 7%

services: 90% (2001 est.)

Labor force: 181,900 (2006)

Labor force—by occupation: agriculture 5%, industry 5%, tourism 50%, other services 40% (2005 est.)

Unemployment rate: 7.6% (2006 est.)

Population below poverty line: 9.3% (2004)

Household income or consumption by percentage share:

lowest 10%: NA%

highest 10%: 27% (2000)

Inflation rate (consumer prices): 2.4% (2007 est.)

Budget:

revenues: $1.03 billion

expenditures: $1.03 billion (FY04/05)

Agriculture—products: citrus, vegetables; poultry

Industries: tourism, banking, cement, oil transshipment, salt, rum, aragonite, pharmaceuticals, spiral-welded steel pipe

Industrial production growth rate: NA%

Electricity—production: 1.894 billion kWh (2005)

Electricity—production by source:

fossil fuel: 100%

hydro: 0%

nuclear: 0%

other: 0% (2001)

Electricity—consumption: 1.762 billion kWh (2005)

Electricity—exports: 0 kWh (2005)

Electricity—imports: 0 kWh (2005)

Oil—production: 0 bbl/day (2005)

Oil—consumption: 26,000 bbl/day (2005 est.)

Oil—exports: transshipments of 41,290 bbl/day (2004)

Oil—imports: 68,250 bbl/day (2004)

Oil—proved reserves: 0 bbl (1 January 2006 est.)

Natural gas—production: 0 cu m (2005 est.)

Natural gas—consumption: 0 cu m (2005 est.)

Natural gas—exports: 0 cu m (2005 est.)

Natural gas—imports: 0 cu m (2005)

Natural gas—proved reserves: 0 cu m (1 January 2006 est.)

Current account balance: -$1.442 billion (2007 est.)

Exports: $674 million (2006)

Exports—commodities: mineral products and salt, animal products, rum, chemicals, fruit and vegetables

Exports—partners: Spain 22.8%, US 20.2%, Poland 13.8%, Germany 11.2%, UK 5.8%, Guatemala 5% (2006)

Imports: $2.401 billion (2006)

Imports—commodities: machinery and transport equipment, manufactures, chemicals, mineral fuels; food and live animals

Imports—partners: US 24.7%, Brazil 15.7%, Japan 13.1%, South Korea 7.8%, Spain 6.2% (2006)

Economic aid—recipient: $4.78 million (2004)

Debt—external: $342.6 million (2004 est.)

Market value of publicly traded shares: $NA

Currency (code): Bahamian dollar (BSD)

Currency code: BSD

Exchange rates: Bahamian dollars per US dollar—1 (2007), 1 (2006), 1 (2005), 1 (2004), 1 (2003)

Fiscal year: 1 July—30 June

COMMUNICATIONS

Telephones—main lines in use: 133,100 (2005)

Telephones—mobile cellular: 227,800 (2005)

Telephone system:

general assessment: modern facilities

domestic: totally automatic system; highly developed; the Bahamas Domestic Submarine Network links 14 of the islands and is designed to satisfy increasing demand for voice and broadband internet services

international: country code—1-242; landing point for the Americas Region Caribbean Ring System (ARCOS-1) fiber-optic submarine cable that provides links to South and Central America, parts of the Caribbean, and the US; satellite earth stations—2 (2007)

Radio broadcast stations: AM 3, FM 5, shortwave 0 (2006)

Radios: 215,000 (1997)

Television broadcast stations: 2 (2006)

Televisions: 67,000 (1997)

Internet country code: .bs

Internet hosts: 248 (2007)

Internet Service Providers (ISPs): 19 (2000)

Internet users: 103,000 (2005)

TRANSPORTATION

Airports: 62 (2007)

Airports—with paved runways:

total: 24

over 3,047 m: 2

2,438 to 3,047 m: 3

1,524 to 2,437 m: 12

914 to 1,523 m: 7 (2007)

Airports—with unpaved runways:

total: 38

1,524 to 2,437 m: 5

914 to 1,523 m: 11

under 914 m: 22 (2007)

Heliports: 1 (2007)

Roadways:

total: 2,693 km

paved: 1,546 km

unpaved: 1,147 km (2000)

Merchant marine:

total: 1,213 ships (1000 GRT or over) 40,403,455 GRT/54,276,183 DWT

by type: barge carrier 1, bulk carrier 225, cargo 240, chemical tanker 84, combination ore/oil 13, container 72, liquefied gas 49, livestock carrier 2, passenger 117, passenger/cargo 34, petroleum tanker 196, refrigerated cargo 118, roll on/roll off 18, specialized tanker 4, specialized tanker 1, vehicle carrier 39

foreign-owned: 1,134 (Angola 6, Australia 3, Belgium 15, Bermuda 12, Brazil 1, Canada 13, China 9, Croatia 1, Cuba 1, Cyprus 20, Denmark 66, Finland 8, France 43, Germany 40, Greece 214, Hong Kong 3, Iceland 1, Indonesia 3, Ireland 2, Italy 1, Japan 62, Jordan 2, Kenya 1, Malaysia 11, Monaco 11, Montenegro 2, Netherlands 24, Nigeria 2, Norway 232, Philippines 1, Poland 15, Russia 5, Saudi Arabia 15, Singapore 9, Slovenia 1, South Africa 1, Spain 11, Sweden 5, Switzerland 2, Taiwan 1, Thailand 1, Trinidad and Tobago 1, Turkey 5, UAE 20, UK 68, US 162, Uruguay 1, Venezuela 1)

registered in other countries: 3 (Barbados 1, Panama 2) (2007)

Ports and terminals: Freeport, Nassau, South Riding Point

MILITARY

Military branches: Royal Bahamian Defense Force: Land Force, Navy, Air Wing (2007)

Military service age and obligation: 18 years of age (est.); no conscription (2008)

Manpower available for military service:

males age 16–49: 80,200 (2008 est.)

Manpower fit for military service:

males age 16–49: 50,282 (2008 est.)

Manpower reaching militarily significant age annually:

males age 16–49: 3,016 (2008 est.)

Military expenditures—percent of GDP: 0.5% (2006)

TRANSNATIONAL ISSUES

Disputes—international: disagrees with the US on the alignment of a potential maritime boundary; continues to monitor and interdict drug dealers and Haitian refugees in Bahamian waters

Illicit drugs: transshipment point for cocaine and marijuana bound for US and Europe; offshore financial center

BAHRAIN

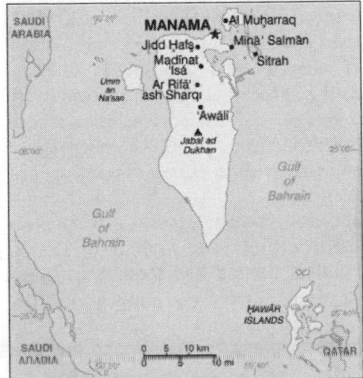

INTRODUCTION

Background: In 1783, the al-Khalifa family captured Bahrain from the Persians. In order to secure these hold-ings, it entered into a series of treaties with the UK during the 19th century that made Bahrain a British protectorate. The archipelago attained its independence in 1971. Bahrain's small size and central location among Persian Gulf countries require it to play a delicate balancing act in foreign affairs among its larger neighbors. Facing declining oil reserves, Bahrain has turned to petroleum processing and refining and has transformed itself into an international banking center. King HAMAD bin Isa al-Khalifa, after coming to power in 1999, pushed economic and political reforms to improve relations with the Shi'a community. Shi'a political societies participated in 2006 parliamentary and municipal elections. Al Wifaq, the largest Shi'a political society, won the largest number of seats in the elected chamber of the legislature. However, Shi'a discontent has resurfaced in recent years with street demonstrations and occasional low-level violence.

GEOGRAPHY

Location: Middle East, archipelago in the Persian Gulf, east of Saudi Arabia

Geographic coordinates: 26 00 N, 50 33 E

Map references: Middle East

Area:

total: 665 sq km

land: 665 sq km

water: 0 sq km

Area—comparative: 3.5 times the size of Washington, DC

Land boundaries: 0 km

Coastline: 161 km

Maritime claims:

territorial sea: 12 nm

contiguous zone: 24 nm
continental shelf: extending to boundaries to be determined
Climate: arid; mild, pleasant winters; very hot, humid summers
Terrain: mostly low desert plain rising gently to low central escarpment
Elevation extremes:
lowest point: Persian Gulf 0 m
highest point: Jabal ad Dukhan 122 m
Natural resources: oil, associated and nonassociated natural gas, fish, pearls
Land use: *arable land:* 2.82%
permanent crops: 5.63%
other: 91.55% (2005)
Irrigated land: 40 sq km (2003)
Total renewable water resources: 0.1 cu km (1997)
Freshwater withdrawal (domestic/industrial/agricultural): *total:* 0.3 cu km/yr (40%/3%/57%)
per capita: 411 cu m/yr (2000)
Natural hazards: periodic droughts; dust storms
Environment—current issues: desertification resulting from the degradation of limited arable land, periods of drought, and dust storms; coastal degradation (damage to coastlines, coral reefs, and sea vegetation) resulting from oil spills and other discharges from large tankers, oil refineries, and distribution stations; lack of freshwater resources (groundwater and seawater are the only sources for all water needs)
Environment—international agreements: *party to:* Biodiversity, Climate Change, Climate Change-Kyoto Protocol, Desertification, Hazardous Wastes, Law of the Sea, Ozone Layer Protection, Wetlands
signed, but not ratified: none of the selected agreements
Geography—note: close to primary Middle Eastern petroleum sources; strategic location in Persian Gulf, through which much of the Western world's petroleum must transit to reach open ocean

PEOPLE

Population: 718,306
note: includes 235,108 non-nationals (July 2008 est.)
Age structure:
0–14 years: 26.4% (male 95,709/female 93,747)
15–64 years: 69.8% (male 288,957/female 212,706)
65 years and over: 3.8% (male 14,224/female 12,963) (2008 est.)
Median age:
total: 29.9 years
male: 33 years
female: 26.4 years (2008 est.)

Population growth rate: 1.337% (2008 est.)
Birth rate: 17.26 births/1,000 population (2008 est.)
Death rate: 4.29 deaths/1,000 population (2008 est.)
Net migration rate: 0.4 migrant(s)/1,000 population (2008 est.)
Sex ratio:
at birth: 1.03 male(s)/female
under 15 years: 1.02 male(s)/female
15–64 years: 1.36 male(s)/female
65 years and over: 1.1 male(s)/female
total population: 1.25 male(s)/female (2008 est.)
Infant mortality rate:
total: 15.64 deaths/1,000 live births
male: 18.27 deaths/1,000 live births
female: 12.93 deaths/1,000 live births (2008 est.)
Life expectancy at birth:
total population: 74.92 years
male: 72.41 years
female: 77.5 years (2008 est.)
Total fertility rate: 2.53 children born/woman (2008 est.)
HIV/AIDS—adult prevalence rate: 0.2% (2001 est.)
HIV/AIDS—people living with HIV/AIDS: fewer than 600 (2003 est.)
HIV/AIDS—deaths: fewer than 200 (2003 est.)
Nationality:
noun: Bahraini(s)
adjective: Bahraini
Ethnic groups: Bahraini 62.4%, non-Bahraini 37.6% (2001 census)
Religions: Muslim (Shi'a and Sunni) 81.2%, Christian 9%, other 9.8% (2001 census)
Languages: Arabic, English, Farsi, Urdu
Literacy:
definition: age 15 and over can read and write
total population: 86.5%
male: 88.6%
female: 83.6% (2001 census)

GOVERNMENT

Country name:
conventional long form: Kingdom of Bahrain
conventional short form: Bahrain
local long form: Mamlakat al Bahrayn
local short form: Al Bahrayn
former: Dilmun
Government type: constitutional monarchy
Capital: *name:* Manama
geographic coordinates: 26 14 N, 50 34 E
time difference: UTC+3 (8 hours ahead of Washington, DC during Standard Time)
Administrative divisions: 5 governorates; Asamah, Janubiyah, Muharraq, Shamaliyah, Wasat

note: each governorate administered by an appointed governor
Independence: 15 August 1971 (from UK)
National holiday: National Day, 16 December (1971); note—15 August 1971 was the date of independence from the UK, 16 December 1971 was the date of independence from British protection
Constitution: adopted 14 February 2002
Legal system: based on Islamic law and English common law; has not accepted compulsory ICJ jurisdiction
Suffrage: 20 years of age; universal
Executive branch:
chief of state: King HAMAD bin Isa al-Khalifa (since 6 March 1999); Heir Apparent Crown Prince SALMAN bin Hamad (son of the monarch, born 21 October 1969)
head of government: Prime Minister KHALIFA bin Salman al-Khalifa (since 1971); Deputy Prime Ministers ALI bin Khalifa bin Salman al-Khalifa, MUHAMMAD bin Mubarak al-Khalifa, Jawad al-ARAIDH
cabinet: Cabinet appointed by the monarch
elections: the monarchy is hereditary; prime minister appointed by the monarch
Legislative branch: bicameral legislature consists of the Consultative Council (40 members appointed by the King) and the Council of Representatives or Chamber of Deputies (40 seats; members directly elected to serve four-year terms)
elections: Council of Representatives—last held November-December 2006 (next election to be held in 2010)
election results: Council of Representatives—percent of vote by party—NA; seats by party—al Wifaq (Shi'a) 17, al Asala (Sunni Salafi) 5, al Minbar (Sunni Muslim Brotherhood) 7, independents 11; note—seats by party as of February 2007—al Wifaq 17, al Asala 8, al Minbar 7, al Mustaqbal (Moderate Sunni pro-government) 4, unassociated independents (all Sunni) 3, independent affiliated with al Wifaq (Sunni opposing-tionist) 1
Judicial branch: High Civil Appeals Court
Political parties and leaders: political parties prohibited but political societies were legalized per a July 2005 law
Political pressure groups and leaders: Shi'a activists fomented unrest sporadically in 1994–97 and have recently engaged in protests with occasional low-level violence; protests related to a host of issues, including the 2002 constitution, elections, unemployment, and release of detainees; Sunni Islamist legis-

lators support a greater role for Shari'a in daily life; several small leftist and other groups are active

International organization participation: ABEDA, AFESD, AMF, FAO, G-77, GCC, IBRD, ICAO, ICC, ICCt (signatory), ICRM, IDA, IDB, IFC, IFRCS, IHO, ILO, IMF, IMO, IMSO, Interpol, IOC, IOM (observer), IPU, ISO, ITSO, ITU, ITUC, LAS, MIGA, NAM, OAPEC, OIC, OPCW, UN, UNCTAD, UNESCO, UNIDO, UNWTO, UPU, WCO, WFTU, WHO, WIPO, WMO, WTO

Diplomatic representation in the US:
chief of mission: Ambassador Nasir bin Muhammad al-BALUSHI
chancery: 3502 International Drive NW, Washington, DC 20008
telephone: [1] (202) 342-1111
FAX: [1] (202) 362-2192
consulate(s) general: New York

Diplomatic representation from the US:
chief of mission: Ambassador J. Adam ERELI
embassy: Building #979, Road 3119 (next to Al-Ahli Sports Club), Block 331, Zinj District, Manama
mailing address: PSC 451, Box 660, FPO AE 09834-5100; international mail: American Embassy, Box 26431, Manama
telephone: [973] 1724-2700
FAX: [973] 1727-0547

Flag description: red, the traditional color for flags of Persian Gulf states, with a white serrated band (five white points) on the hoist side; the five points represent the five pillars of Islam

ECONOMY

Economy—overview: With its highly developed communication and transport facilities, Bahrain is home to numerous multinational firms with business in the Gulf. Petroleum production and refining account for over 60% of Bahrain's export receipts, over 70% of government revenues, and 11% of GDP (exclusive of allied industries), underpinning Bahrain's strong economic growth in recent years. Aluminum is Bahrain's second major export after oil. Other major segments of Bahrain's economy are the financial and construction sectors. Bahrain is focused on Islamic banking and is competing on an international scale with Malaysia as a worldwide banking center. Bahrain is actively pursuing the diversification and privatization of its economy to reduce the country's dependence on oil. As part of this effort, in August 2006 Bahrain and the US implemented a Free Trade Agreement (FTA), the first FTA between the US and a Gulf state.

Continued strong growth hinges on Bahrain's ability to acquire new natural gas supplies as feedstock to support its expanding petrochemical and aluminum industries. Unemployment, especially among the young, and the depletion of oil and underground water resources are long-term economic problems.

GDP (purchasing power parity): $24.5 billion (2007 est.)
GDP (official exchange rate): $19.66 billion (2007 est.)
GDP—real growth rate: 6.6% (2007 est.)
GDP—per capita (PPP): $32,100 (2007 est.)
GDP—composition by sector:
agriculture: 0.3%
industry: 43.6%
services: 56% (2007 est.)
Labor force: 363,000
note: 44% of the population in the 15–64 age group is non-national (2007 est.)
Labor force—by occupation:
agriculture: 1%
industry: 79%
services: 20% (1997 est.)
Unemployment rate: 15% (2005 est.)
Population below poverty line: NA%
Household income or consumption by percentage share:
lowest 10%: NA%
highest 10%: NA%
Inflation rate (consumer prices): 3.4% (2007 est.)
Investment (gross fixed): 17.7% of GDP (2007 est.)
Budget:
revenues: $6.168 billion
expenditures: $5.205 billion (2007 est.)
Public debt: 29.4% of GDP (2007 est.)
Agriculture—products: fruit, vegetables; poultry, dairy products; shrimp, fish
Industries: petroleum processing and refining, aluminum smelting, iron pelletization, fertilizers, Islamic and offshore banking, insurance, ship repairing, tourism
Industrial production growth rate: 5.2% (2007 est.)
Electricity—production: 8.187 billion kWh (2005)
Electricity—production by source:
fossil fuel: 100%
hydro: 0%
nuclear: 0%
other: 0% (2001)
Electricity—consumption: 7.614 billion kWh (2005)
Electricity—exports: 0 kWh (2005)
Electricity—imports: 0 kWh (2005)
Oil—production: 184,000 bbl/day (2007 est.)
Oil—consumption: 31,000 bbl/day (2005 est.)
Oil—exports: 235,500 bbl/day (2004)

Oil—imports: 216,300 bbl/day (2004)
Oil—proved reserves: 118.6 million bbl (2007 est.)
Natural gas—production: 10.27 billion cu m (2005 est.)
Natural gas—consumption: 10.27 billion cu m (2005 est.)
Natural gas—exports: 0 cu m (2005 est.)
Natural gas—imports: 0 cu m (2005)
Natural gas—proved reserves: 88.26 billion cu m (1 January 2006 est.)
Current account balance: $3.913 billion (2007 est.)
Exports: $13.44 billion (2007 est.)
Exports—commodities: petroleum and petroleum products, aluminum, textiles
Exports—partners: Saudi Arabia 3.2%, US 3%, Japan 2.3% (2006)
Imports: $9.858 billion (2007 est.)
Imports—commodities: crude oil, machinery, chemicals
Imports—partners: Saudi Arabia 37.6%, Japan 6.8%, US 6.2%, UK 6.2%, Germany 5.1%, UAE 4.2% (2006)
Economic aid—recipient: $103.9 million (2004)
Reserves of foreign exchange and gold: $4.101 billion (31 December 2007 est.)
Debt—external: $7.895 billion (31 December 2007 est.)
Stock of direct foreign investment—at home: $14.61 billion (2007 est.)
Stock of direct foreign investment—abroad: $7.489 billion (2007 est.)
Market value of publicly traded shares: $21.12 billion (2006)
Currency (code): Bahraini dinar (BHD)
Currency code: BHD
Exchange rates: Bahraini dinars per US dollar—0.376 (2007), 0.376 (2006), 0.376 (2005), 0.376 (2004), 0.376 (2003)
Fiscal year: calendar year

COMMUNICATIONS

Telephones—main lines in use: 193,300 (2006)
Telephones—mobile cellular: 898,900 (2006)
Telephone system:
general assessment: modern system
domestic: modern fiber-optic integrated services; digital network with rapidly growing use of mobile-cellular telephones
international: country code—973; landing point for the Fiber-Optic Link Around the Globe (FLAG) submarine cable network that provides links to Asia, Middle East, Europe, and US; tropospheric scatter to Qatar and UAE; microwave radio relay to Saudi Arabia; satellite earth station—1 (2007)

Radio broadcast stations: AM 2, FM 3, shortwave 0 (1998)
Radios: 338,000 (1997)
Television broadcast stations: 4 (1997)
Televisions: 275,000 (1997)
Internet country code: .bh
Internet hosts: 2,413 (2007)
Internet Service Providers (ISPs): 1 (2000)
Internet users: 157,300 (2006)

TRANSPORTATION

Airports: 3 (2007)
Airports—with paved runways:
total: 3
over 3,047 m: 2
1,524 to 2,437 m: 1 (2007)
Heliports: 1 (2007)
Pipelines: gas 20 km; oil 52 km (2007)
Roadways:
total: 3,498 km
paved: 2,768 km
unpaved: 730 km (2003)
Merchant marine:
total: 7 ships (1000 GRT or over) 220,264 GRT/314,289 DWT
by type: bulk carrier 3, cargo 1, container 2, petroleum tanker 1

foreign-owned: 3 (Kuwait 3) (2007)
Ports and terminals: Mina' Salman, Sitrah

MILITARY

Military branches: Bahrain Defense Forces (BDF): Ground Force (includes Air Defense), Naval Force, Air Force, National Guard
Military service age and obligation: 17 years of age for voluntary military service; 15 years of age for NCOs, technicians, and cadets; no conscription (2008)
Manpower available for military service:
males age 16–49: 210,938
females age 16–49: 170,471 (2008 est.)
Manpower fit for military service:
males age 16–49: 171,536
females age 16–49: 142,714 (2008 est.)
Manpower reaching militarily significant age annually:
males age 16–49: 6,543
females age 16–49: 6,429 (2008 est.)
Military expenditures—percent of GDP: 4.5% (2006)

TRANSNATIONAL ISSUES

Disputes—international: none
Trafficking in persons: *current situation:* Bahrain is a destination country for men and women from South and Southeast Asia who migrate willingly to work as laborers or domestic servants, but may be subjected to conditions of involuntary servitude when faced with exorbitant recruitment and transportation fees, withholding of their passports, restrictions on their movement, non-payment of wages, and physical or sexual abuse; women from Eastern Europe, Central Asia, Morocco, and Thailand are also trafficked to Bahrain for the purpose of commercial sexual exploitation or forced labor
tier rating: Tier 3—Bahrain made no discernable progress in preventing trafficking in 2006; the government failed to enact a comprehensive anti-trafficking law and did not report any prosecutions or convictions for trafficking offenses, despite reports of a substantial problem of involutary servitude and trafficking for commercial sexual exploitation

BANGLADESH

INTRODUCTION

Background: Europeans began to set up trading posts in the area of Bangladesh in the 16th century; eventually the British came to dominate the region and it became part of British India. In 1947, West Pakistan and East Bengal (both primarily Muslim) separated from India (largely Hindu) and jointly became the new country of Pakistan. East Bengal became East Pakistan in 1955, but the awkward arrangement of a two-part country with its territorial units separated by 1,600 km left the Bengalis marginalized and dissatisfied. East Pakistan seceded from its union with West Pakistan in 1971 and was renamed Bangladesh. A military-backed caretaker regime suspended planned parliamentary elections in January 2007 in an effort to reform the political system and root out corruption; the regime has pledged new democratic elections by the end of 2008. About a third of this extremely poor country floods annually during the monsoon rainy season, hampering economic development.

GEOGRAPHY

Location: Southern Asia, bordering the Bay of Bengal, between Burma and India
Geographic coordinates: 24 00 N, 90 00 E
Map references: Asia
Area:
total: 144,000 sq km
land: 133,910 sq km
water: 10,090 sq km
Area—comparative: slightly smaller than Iowa
Land boundaries:
total: 4,246 km
border countries: Burma 193 km, India 4,053 km
Coastline: 580 km
Maritime claims:
territorial sea: 12 nm

contiguous zone: 18 nm
exclusive economic zone: 200 nm
continental shelf: up to the outer limits of the continental margin
Climate: tropical; mild winter (October to March); hot, humid summer (March to June); humid, warm rainy monsoon (June to October)
Terrain: mostly flat alluvial plain; hilly in southeast
Elevation extremes:
lowest point: Indian Ocean 0 m
highest point: Keokradong 1,230 m
Natural resources: natural gas, arable land, timber, coal
Land use:
arable land: 55.39%
permanent crops: 3.08%
other: 41.53% (2005)
Irrigated land: 47,250 sq km (2003)
Total renewable water resources: 1,210.6 cu km (1999)
Freshwater withdrawal (domestic/industrial/agricultural): *total:* 79.4 cu km/yr (3%/1%/96%)
per capita: 560 cu m/yr (2000)
Natural hazards: droughts, cyclones; much of the country routinely inundated during the summer monsoon season
Environment—current issues: many people are landless and forced to live on and cultivate flood-prone land; water-

borne diseases prevalent in surface water; water pollution, especially of fishing areas, results from the use of commercial pesticides; ground water contaminated by naturally occurring arsenic; intermittent water shortages because of falling water tables in the northern and central parts of the country; soil degradation and erosion; deforestation; severe overpopulation

Environment—international agreements: *party to:* Biodiversity, Climate Change, Climate Change-Kyoto Protocol, Desertification, Endangered Species, Environmental Modification, Hazardous Wastes, Law of the Sea, Ozone Layer Protection, Ship Pollution, Wetlands

signed, but not ratified: none of the selected agreements

Geography—note: most of the country is situated on deltas of large rivers flowing from the Himalayas: the Ganges unites with the Jamuna (main channel of the Brahmaputra) and later joins the Meghna to eventually empty into the Bay of Bengal

PEOPLE

Population: 153,546,901 (July 2008 est.)

Age structure:

0–14 years: 33.4% (male 26,364,370/ female 24,859,792)

15–64 years: 63.1% (male 49,412,903/ female 47,468,013)

65 years and over: 3.5% (male 2,912,321/female 2,529,502) (2008 est.)

Median age:

total: 22.8 years

male: 22.8 years

female: 22.9 years (2008 est.)

Population growth rate: 2.022% (2008 est.)

Birth rate: 28.86 births/1,000 population (2008 est.)

Death rate: 8 deaths/1,000 population (2008 est.)

Net migration rate: -0.65 migrant(s)/ 1,000 population (2008 est.)

Sex ratio:

at birth: 1.06 male(s)/female

under 15 years: 1.06 male(s)/female

15–64 years: 1.04 male(s)/female

65 years and over: 1.15 male(s)/female

total population: 1.05 male(s)/female (2008 est.)

Infant mortality rate:

total: 57.45 deaths/1,000 live births

male: 58.44 deaths/1,000 live births

female: 56.41 deaths/1,000 live births (2008 est.)

Life expectancy at birth:

total population: 63.21 years

male: 63.14 years

female: 63.28 years (2008 est.)

Total fertility rate: 3.08 children born/woman (2008 est.)

HIV/AIDS—adult prevalence rate: less than 0.1% (2001 est.)

HIV/AIDS—people living with HIV/AIDS: 13,000 (2001 est.)

HIV/AIDS—deaths: 650 (2001 est.)

Major infectious diseases:

degree of risk: high

food or waterborne diseases: bacterial and protozoal diarrhea, hepatitis A and E, and typhoid fever

vectorborne diseases: dengue fever and malaria are high risks in some locations

water contact disease: leptospirosis

animal contact disease: rabies

note: highly pathogenic H5N1 avian influenza has been identified in this country; it poses a negligible risk with extremely rare cases possible among US citizens who have close contact with birds (2008)

Nationality:

noun: Bangladeshi(s)

adjective: Bangladeshi

Ethnic groups: Bengali 98%, other 2% (includes tribal groups, non-Bengali Muslims) (1998)

Religions: Muslim 83%, Hindu 16%, other 1% (1998)

Languages: Bangla (official, also known as Bengali), English

Literacy:

definition: age 15 and over can read and write

total population: 43.1%

male: 53.9%

female: 31.8% (2003 est.)

GOVERNMENT

Country name:

conventional long form: People's Republic of Bangladesh

conventional short form: Bangladesh

local long form: Gana Prajatantri Banladesh

local short form: Banladesh

former: East Bengal, East Pakistan

Government type: parliamentary democracy

Capital: *name:* Dhaka

geographic coordinates: 23 43 N, 90 24 E

time difference: UTC+6 (11 hours ahead of Washington, DC during Standard Time)

Administrative divisions: 6 divisions; Barisal, Chittagong, Dhaka, Khulna, Rajshahi, Sylhet

Independence: 16 December 1971 (from West Pakistan); note—26 March 1971 is the date of independence from West Pakistan, 16 December 1971 is known as Victory Day and commemorates the official creation of the state of Bangladesh

National holiday: Independence Day, 26 March (1971); note—26 March 1971 is the date of independence from West Pakistan, 16 December 1971 is Victory Day and commemorates the official creation of the state of Bangladesh

Constitution: 4 November 1972, effective 16 December 1972; suspended following coup of 24 March 1982, restored 10 November 1986; amended many times

Legal system: based on English common law; has not accepted compulsory ICJ jurisdiction

Suffrage: 18 years of age; universal

Executive branch:

chief of state: President Iajuddin AHMED (since 6 September 2002)

note: the country has a caretaker government until a general election is held; Iajuddin AHMED remains as President and Minister of Defense, and all other Cabinet portfolios are held by Caretaker Advisers (CAs); the Chief CA, Fakhruddin AHMED, is roughly equivalent to a prime minister

elections: president elected by National Parliament for a five-year term (eligible for a second term); election scheduled for 16 September 2002 was not held since Iajuddin AHMED was the only presidential candidate; he was sworn in on 6 September 2002 (next election NA); following legislative elections, the leader of the party that wins the most seats is usually appointed prime minister by the president

election results: Iajuddin AHMED declared president-elect by the Election Commission; he ran unopposed as president; percent of National Parliament vote—NA

Legislative branch: unicameral National Parliament or Jatiya Sangsad; 300 seats elected by popular vote from single territorial constituencies; members serve five-year terms; note—parliament not in session during the extended caretaker regime

elections: last held 1 October 2001 (the scheduled January 2007 election has been postponed till late 2008)

election results: percent of vote by party— BNP and alliance partners 41%, AL 40%, other 19%; seats by party—BNP 193, AL 58, JI 17, JP (Ershad faction) 14, IOJ 2, JP (Manzur) 4, other 12; note— the election of October 2001 brought to power a majority BNP government aligned with three other smaller parties—JI, IOJ, and Jatiya Party (Manzur)

Judicial branch: Supreme Court (the chief justices and other judges are appointed by the president)

Political parties and leaders: Awami League or AL [Sheikh HASINA];

Bangladesh Communist Party or BCP [Manjurul A. KHAN]; Bangladesh Nationalist Party or BNP [Khaleda ZIA]; Islami Oikya Jote or IOJ [Mufti Fazlul Haq AMINI]; Jamaat-e-Islami Bangladesh or JIB [Matiur Rahman NIZAMI]; Jatiya Party or JP (Ershad faction) [Hussain Mohammad ERSHAD]; Jatiya Party (Manzur faction) [Naziur Rahman MANZUR]; Liberal Democratic Party or LDP [Badrudozza CHOWDHURY and Oli AHMED]

Political pressure groups and leaders: NA

International organization participation: ADB, ARF, BIMSTEC, C, CP, FAO, G-77, IAEA, IBRD, ICAO, ICC, ICCt (signatory), ICRM, IDA, IDB, IFAD, IFC, IFRCS, IHO, ILO, IMF, IMO, IMSO, Interpol, IOC, IOM, IPU, ISO, ITSO, ITU, ITUC, MIGA, MINURSO, MONUC, NAM, OIC, OPCW, SAARC, SACEP, UN, UNAMID, UNCTAD, UNESCO, UNHCR, UNIDO, UNMEE, UNMIL, UNMIS, UNMIT, UNOCI, UNOMIG, UNWTO, UPU, WCL, WCO, WFTU, WHO, WIPO, WMO, WTO

Diplomatic representation in the US:
chief of mission: Ambassador M. Humayun KABIR
chancery: 3510 International Drive NW, Washington, DC 20008
telephone: [1] (202) 244-0183
FAX: [1] (202) 244-7830/2771
consulate(s) general: Los Angeles, New York

Diplomatic representation from the US:
chief of mission: Ambassador James F. MORIARTY
embassy: Madani Avenue, Baridhara, Dhaka 1212
mailing address: G. P. O. Box 323, Dhaka 1000
telephone: [880] (2) 885-5500
FAX: [880] (2) 882-3744

Flag description: green field with a large red disk shifted slightly to the hoist side of center; the red disk represents the rising sun and the sacrifice to achieve independence; the green field symbolizes the lush vegetation of Bangladesh

ECONOMY

Economy—overview: The economy has grown 5–6% over the past few years despite inefficient state-owned enterprises, delays in exploiting natural gas resources, insufficient power supplies, and slow implementation of economic reforms. Bangladesh remains a poor, overpopulated, and inefficiently-governed nation. Although more than half of GDP is generated through the service sector, nearly two-thirds of Bangladeshis are employed in the agriculture sector, with rice as the single-most-important product. Garment exports and remittances from Bangladeshis working overseas, mainly in the Middle East and East Asia, fuel economic growth.

GDP (purchasing power parity): $206.7 billion (2007 est.)
GDP (official exchange rate): $72.42 billion (2007 est.)
GDP—real growth rate: 5.6% (2007 est.)
GDP—per capita (PPP): $1,300 (2007 est.)
GDP—composition by sector:
agriculture: 19%
industry: 28.7%
services: 52.3% (2007 est.)
Labor force: 69.4 million
note: extensive export of labor to Saudi Arabia, Kuwait, UAE, Oman, Qatar, and Malaysia; workers' remittances estimated at $4.8 billion in 2005–06. (2007 est.)
Labor force—by occupation: *agriculture:* 63%
industry: 11%
services: 26% (FY95/96)
Unemployment rate: 1% (includes underemployment) (2007 est.)
Population below poverty line: 45% (2004 est.)
Household income or consumption by percentage share:
lowest 10%: 3.7%
highest 10%: 27.9% (2000)
Distribution of family income—Gini index: 33.4 (2000)
Inflation rate (consumer prices): 8.4% (2007 est.)
Investment (gross fixed): 26% of GDP (2007 est.)
Budget:
revenues: $6.796 billion
expenditures: $9.794 billion (2007 est.)
Public debt: 37.4% of GDP (2007 est.)
Agriculture—products: rice, jute, tea, wheat, sugarcane, potatoes, tobacco, pulses, oilseeds, spices, fruit; beef, milk, poultry
Industries: cotton textiles, jute, garments, tea processing, paper newsprint, cement, chemical fertilizer, light engineering, sugar
Industrial production growth rate: 9.5% (2007 est.)
Electricity—production: 21.35 billion kWh (2005)
Electricity—production by source:
fossil fuel: 93.7%
hydro: 6.3%
nuclear: 0%
other: 0% (2001)
Electricity—consumption: 19.49 billion kWh (2005)
Electricity—exports: 0 kWh (2005)

Electricity—imports: 0 kWh (2005)
Oil—production: 6,746 bbl/day (2005)
Oil—consumption: 86,000 bbl/day (2005 est.)
Oil—exports: 1,100 bbl/day (2004)
Oil—imports: 81,010 bbl/day (2004)
Oil—proved reserves: 28 million bbl (1 January 2006 est.)
Natural gas—production: 13.43 billion cu m (2005 est.)
Natural gas—consumption: 13.43 billion cu m (2005 est.)
Natural gas—exports: 0 cu m (2005 est.)
Natural gas—imports: 0 cu m (2005)
Natural gas—proved reserves: 135.8 billion cu m (1 January 2006 est.)
Current account balance: $334 million (2007 est.)
Exports: $11.75 billion (2007 est.)
Exports—commodities: garments, jute and jute goods, leather, frozen fish and seafood
Exports—partners: US 25%, Germany 12.6%, UK 9.8%, France 5% (2006)
Imports: $16.03 billion (2007 est.)
Imports—commodities: machinery and equipment, chemicals, iron and steel, textiles, foodstuffs, petroleum products, cement
Imports—partners: China 17.7%, India 12.5%, Kuwait 7.9%, Singapore 5.5%, Hong Kong 4.1% (2006)
Economic aid—recipient: $1.321 billion (2005)
Reserves of foreign exchange and gold: $5.515 billion (31 December 2007 est.)
Debt—external: $21.23 billion (31 December 2007 est.)
Stock of direct foreign investment—at home: $4.938 billion (2007 est.)
Stock of direct foreign investment—abroad: $105 million (2007 est.)
Market value of publicly traded shares: $3.61 billion (2006)
Currency (code): taka (BDT)
Currency code: BDT
Exchange rates: taka per US dollar—69.893 (2007), 69.031 (2006), 64.328 (2005), 59.513 (2004), 58.15 (2003)
Fiscal year: 1 July—30 June

COMMUNICATIONS

Telephones—main lines in use: 1.134 million (2006)
Telephones—mobile cellular: 19.131 million (2006)
Telephone system:
general assessment: totally inadequate for a modern country; fixed-line telephone density of less than 1 per 100 persons; mobile-cellular telephone density of 13 per 100 persons
domestic: modernizing; introducing digital systems; trunk systems include VHF

and UHF microwave radio relay links, and some fiber-optic cable in cities
international: country code—880; landing point for the SEA-ME-WE-4 fiber-optic submarine cable system that provides links to Europe, the Middle East and Asia; satellite earth stations—6; international radiotelephone communications and landline service to neighboring countries (2007)
Radio broadcast stations: AM 15, FM 13, shortwave 2 (2006)
Radios: 6.15 million (1997)
Television broadcast stations: 15 (1999)
Televisions: 770,000 (1997)
Internet country code: .bd
Internet hosts: 376 (2007)
Internet Service Providers (ISPs): 10 (2000)
Internet users: 450,000 (2006)

TRANSPORTATION

Airports: 16 (2007)
Airports—with paved runways:
total: 15
over 3,047 m: 1
2,438 to 3,047 m: 4
1,524 to 2,437 m: 4
914 to 1,523 m: 1
under 914 m: 5 (2007)
Airports—with unpaved runways:
total: 1
1,524 to 2,437 m: 1 (2007)
Pipelines: gas 2,644 km (2007)
Railways: *total:* 2,768 km

broad gauge: 946 km 1.676-m gauge
narrow gauge: 1,822 km 1.000-m gauge (2006)
Roadways:
total: 239,226 km
paved: 22,726 km
unpaved: 216,500 km (2003)
Waterways: 8,370 km
note: includes up to 3,060 km main cargo routes; network reduced to 5,200 km in dry season (2006)
Merchant marine:
total: 41 ships (1000 GRT or over) 328,530 GRT/468,509 DWT
by type: bulk carrier 3, cargo 27, container 6, passenger/cargo 1, petroleum tanker 4
foreign-owned: 1 (China 1)
registered in other countries: 9 (Comoros 1, Honduras 1, Malta 3, Panama 1, Singapore 2, St Vincent and The Grenadines 1) (2007)
Ports and terminals: Chittagong, Mongla Port

MILITARY

Military branches: Bangladesh Defense Force: Bangladesh Army, Bangladesh Navy, Bangladesh Air Force (Bangladesh Biman Bahini, BAF) (2008)
Military service age and obligation: 16 years of age for voluntary military service; 17 years of age for officers (both with parental consent); conscription legally possible in emergency, but has never been implemented (2008)

Manpower available for military service:
males age 16–49: 41,199,340 (2008 est.)
Manpower fit for military service:
males age 16–49: 31,968,168 (2008 est.)
Military expenditures—percent of GDP: 1.5% (2006)

TRANSNATIONAL ISSUES

Disputes—international: discussions with India remain stalled to delimit a small section of river boundary, exchange territory for 51 small Bangladeshi exclaves in India and 111 small Indian exclaves in Bangladesh, allocate divided villages, and stop illegal cross-border trade, migration, violence, and transit of terrorists through the porous border; Bangladesh resists India's attempts to fence or wall off high-traffic sections of the porous boundary; a joint Bangladesh-India boundary inspection in 2005 revealed 92 pillars are missing; dispute with India over New Moore/South Talpatty/Purbasha Island in the Bay of Bengal deters maritime boundary delimitation; 21,000 Burmese Rohingya Muslim refugees reside in two camps in Bangladesh
Refugees and internally displaced persons: *refugees (country of origin):* 26,268 (Burma)
IDPs: 65,000 (land conflicts, religious persecution) (2007)
Illicit drugs: transit country for illegal drugs produced in neighboring countries

BARBADOS

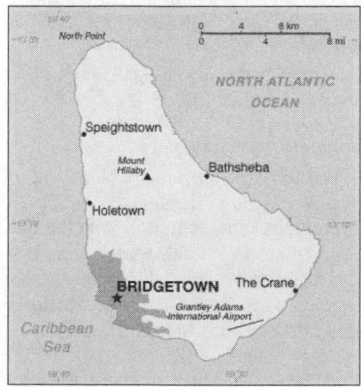

INTRODUCTION

Background: The island was uninhabited when first settled by the British in 1627. Slaves worked the sugar plantations established on the island until 1834 when slavery was abolished. The economy remained heavily dependent

on sugar, rum, and molasses production through most of the 20th century. The gradual introduction of social and political reforms in the 1940s and 1950s led to complete independence from the UK in 1966. In the 1990s, tourism and manufacturing surpassed the sugar industry in economic importance.

GEOGRAPHY

Location: Caribbean, island in the North Atlantic Ocean, northeast of Venezuela
Geographic coordinates: 13 10 N, 59 32 W
Map references: Central America and the Caribbean
Area:
total: 431 sq km
land: 431 sq km
water: 0 sq km
Area—comparative: 2.5 times the size of Washington, DC
Land boundaries: 0 km
Coastline: 97 km

Maritime claims:
territorial sea: 12 nm
exclusive economic zone: 200 nm
Climate: tropical; rainy season (June to October)
Terrain: relatively flat; rises gently to central highland region
Elevation extremes:
lowest point: Atlantic Ocean 0 m
highest point: Mount Hillaby 336 m
Natural resources: petroleum, fish, natural gas
Land use: *arable land:* 37.21%
permanent crops: 2.33%
other: 60.46% (2005)
Irrigated land: 50 sq km (2003)
Total renewable water resources: 0.1 cu km (2003)
Freshwater withdrawal (domestic/ industrial/agricultural): *total:* 0.09 cu km/yr (33%/44%/22%)
per capita: 333 cu m/yr (2000)
Natural hazards: infrequent hurricanes; periodic landslides

Environment—current issues: pollution of coastal waters from waste disposal by ships; soil erosion; illegal solid waste disposal threatens contamination of aquifers

Environment—international agreements: *party to:* Biodiversity, Climate Change, Climate Change-Kyoto Protocol, Desertification, Endangered Species, Hazardous Wastes, Law of the Sea, Marine Dumping, Ozone Layer Protection, Ship Pollution, Wetlands *signed, but not ratified:* none of the selected agreements

Geography—note: easternmost Caribbean island

PEOPLE

Population: 281,968 (July 2008 est.)

Age structure:

0–14 years: 19.3% (male 27,270/female 27,193)

15–64 years: 71.7% (male 99,357/female 102,683)

65 years and over: 9% (male 9,856/female 15,609) (2008 est.)

Median age:

total: 35.4 years

male: 34.2 years

female: 36.4 years (2008 est.)

Population growth rate: 0.36% (2008 est.)

Birth rate: 12.48 births/1,000 population (2008 est.)

Death rate: 8.58 deaths/1,000 population (2008 est.)

Net migration rate: -0.31 migrant(s)/1,000 population (2008 est.)

Sex ratio:

at birth: 1.01 male(s)/female

under 15 years: 1 male(s)/female

15–64 years: 0.97 male(s)/female

65 years and over: 0.63 male(s)/female

total population: 0.94 male(s)/female (2008 est.)

Infant mortality rate:

total: 11.05 deaths/1,000 live births

male: 12.4 deaths/1,000 live births

female: 9.69 deaths/1,000 live births (2008 est.)

Life expectancy at birth:

total population: 73.21 years

male: 71.2 years

female: 75.24 years (2008 est.)

Total fertility rate: 1.65 children born/woman (2008 est.)

HIV/AIDS—adult prevalence rate: 1.5% (2003 est.)

HIV/AIDS—people living with HIV/AIDS: 2,500 (2003 est.)

HIV/AIDS—deaths: fewer than 200 (2003 est.)

Nationality:

noun: Barbadian(s) or Bajan (colloquial)

adjective: Barbadian or Bajan (colloquial)

Ethnic groups: black 90%, white 4%, Asian and mixed 6%

Religions: Protestant 67% (Anglican 40%, Pentecostal 8%, Methodist 7%, other 12%), Roman Catholic 4%, none 17%, other 12%

Languages: English

Literacy:

definition: age 15 and over has ever attended school

total population: 99.7%

male: 99.7%

female: 99.7% (2002 est.)

GOVERNMENT

Country name:

conventional long form: none

conventional short form: Barbados

Government type: parliamentary democracy

Capital: *name:* Bridgetown

geographic coordinates: 13 06 N, 59 37 W

time difference: UTC-4 (1 hour ahead of Washington, DC during Standard Time)

Administrative divisions: 11 parishes and 1 city*; Bridgetown*, Christ Church, Saint Andrew, Saint George, Saint James, Saint John, Saint Joseph, Saint Lucy, Saint Michael, Saint Peter, Saint Philip, Saint Thomas

Independence: 30 November 1966 (from UK)

National holiday: Independence Day, 30 November (1966)

Constitution: 30 November 1966

Legal system: English common law; no judicial review of legislative acts; accepts compulsory ICJ jurisdiction, with reservations

Suffrage: 18 years of age; universal

Executive branch:

chief of state: Queen ELIZABETH II (since 6 February 1952); represented by Governor General Sir Clifford Straughn HUSBANDS (since 1 June 1996)

head of government: Prime Minister David THOMPSON (since 16 January 2008)

cabinet: Cabinet appointed by the governor general on the advice of the prime minister

elections: the monarch is hereditary; governor general appointed by the monarch; following legislative elections, the leader of the majority party or the leader of the majority coalition is usually appointed prime minister by the governor general; the prime minister recommends the deputy prime minister

Legislative branch: bicameral Parliament consists of the Senate (21 seats; members appointed by the governor general—12 on the advice of the Prime Minister, 2 on the advice of the opposition leader, and 7 at his discretion) and the House of Assembly (30

seats; members are elected by direct popular vote to serve five-year terms)

elections: House of Assembly—last held 15 January 2008 (next to be called in 2013)

election results: House of Assembly—percent of vote by party—DLP 52.5%, BLP 47.3%; seats by party—DLP 20, BLP 10

Judicial branch: Supreme Court of Judicature (judges are appointed by the Service Commissions for the Judicial and Legal Services); Caribbean Court of Justice is the highest court of appeal

Political parties and leaders: Barbados Labor Party or BLP [Mia MOTTLEY]; Democratic Labor Party or DLP [David THOMPSON]; People's Empowerment Party or PEP [David COMISSIONG]

Political pressure groups and leaders: Barbados Secondary Teachers' Union or BSTU [Patrick FROST]; Barbados Union of Teachers or BUT [Herbert GITTENS]; Congress of Trade Unions and Staff Associations of Barbados or CTUSAB, (includes the BWU, NUPW, BUT, and BSTU) [Leroy TROTMAN]; Barbados Workers Union or BWU [Leroy TROTMAN]; Clement Payne Labor Union [David COMISSIONG]; National Union of Public Workers [Joseph GODDARD]

International organization participation: ACP, C, Caricom, CDB, FAO, G-77, IADB, IBRD, ICAO, ICCt, ICRM, IDA, IFAD, IFC, IFRCS, ILO, IMF, IMO, Interpol, IOC, ISO, ITSO, ITU, ITUC, LAES, MIGA, NAM, OAS, OPANAL, OPCW, UN, UNCTAD, UNESCO, UNIDO, UPU, WCO, WFTU, WHO, WIPO, WMO, WTO

Diplomatic representation in the US:

chief of mission: Ambassador Michael Ian KING

chancery: 2144 Wyoming Avenue NW, Washington, DC 20008

telephone: [1] (202) 939-9200

FAX: [1] (202) 332-7467

consulate(s) general: Miami, New York

consulate(s): Los Angeles

Diplomatic representation from the US:

chief of mission: Ambassador Mary M. OURISMAN

embassy: U.S. Embassy, Wildey Business Park, Wildey, St. Michael

mailing address: P. O. Box 302, Bridgetown; CMR 1014, APO AA 34055

telephone: [1] (246) 436-4950

FAX: [1] (246) 429-5246, 429-3379

Flag description: three equal vertical bands of blue (hoist side), gold, and blue with the head of a black trident centered on the gold band; the trident head represents independence and a break with the past (the colonial coat of arms contained a complete trident)

ECONOMY

Economy—overview: Historically, the Barbadian economy was dependent on sugarcane cultivation and related activities. However, production in recent years has diversified into light industry and tourism, with about three-quarters of GDP and 80% of exports being attributed to services. Growth has rebounded since 2003, bolstered by increases in construction projects and tourism revenues—reflecting its success in the higher-end segment. The country enjoys one of the highest per capita incomes in the region and an investment grade rating which benefits from its political stability and stable institutions. Offshore finance and information services are important foreign exchange earners and thrive from having the same time zone as eastern US financial centers and a relatively highly educated workforce. The government continues its efforts to reduce unemployment, to encourage direct foreign investment, and to privatize remaining state-owned enterprises.

GDP (purchasing power parity): $5.317 billion (2007 est.)
GDP (official exchange rate): $3.739 billion (2007 est.)
GDP—real growth rate: 4.2% (2007 est.)
GDP—per capita (PPP): $19,300 (2007 est.)
GDP—composition by sector:
agriculture: 6%
industry: 16%
services: 78% (2000 est.)
Labor force: 128,500 (2001 est.)
Labor force—by occupation: agriculture: 10%
industry: 15%
services: 75% (1996 est.)
Unemployment rate: 10.7% (2003 est.)
Population below poverty line: NA%
Household income or consumption by percentage share:
lowest 10%: NA%
highest 10%: NA%
Inflation rate (consumer prices): 5.5% (2007 est.)
Budget:
revenues: $847 million (including grants)
expenditures: $886 million (2000 est.)
Agriculture—products: sugarcane, vegetables, cotton
Industries: tourism, sugar, light manufacturing, component assembly for export
Industrial production growth rate: -3.2% (2000 est.)
Electricity—production: 953 million kWh (2005)
Electricity—production by source:

fossil fuel: 100%
hydro: 0%
nuclear: 0%
other: 0% (2001)
Electricity—consumption: 886.3 million kWh (2005)
Electricity—exports: 0 kWh (2005)
Electricity—imports: 0 kWh (2005)
Oil—production: 1,002 bbl/day (2005)
Oil—consumption: 9,000 bbl/day (2005 est.)
Oil—exports: 1,666 bbl/day (2004)
Oil—imports: 7,071 bbl/day (2004)
Oil—proved reserves: 2.5 million bbl (1 January 2006 est.)
Natural gas—production: 27.97 million cu m (2005 est.)
Natural gas—consumption: 27.97 million cu m (2005 est.)
Natural gas—exports: 0 cu m (2005 est.)
Natural gas—imports: 0 cu m (2005)
Natural gas—proved reserves: 135.8 million cu m (1 January 2006 est.)
Current account balance: -$254 million (2007 est.)
Exports: $385 million (2006)
Exports—commodities: manufactures, sugar and molasses, rum, other foods and beverages, chemicals, electrical components
Exports—partners: US 27.6%, Trinidad and Tobago 15%, UK 10.2%, Saint Lucia 7%, Jamaica 6.5%, Saint Vincent and the Grenadines 4.3% (2006)
Imports: $1.586 billion (2006)
Imports—commodities: consumer goods, machinery, foodstuffs, construction materials, chemicals, fuel, electrical components
Imports—partners: US 37.7%, Trinidad and Tobago 22.6%, UK 5.9% (2006)
Economic aid—recipient: $2.07 million (2005)
Reserves of foreign exchange and gold: $620 million (2007)
Debt—external: $668 million (2003)
Market value of publicly traded shares: $5.513 billion (2005)
Currency (code): Barbadian dollar (BBD)
Currency code: BBD
Exchange rates: Barbadian dollars per US dollar—NA (2007), 2 (2006), 2 (2005), 2 (2004), 2 (2003)
Fiscal year: 1 April—31 March

COMMUNICATIONS

Telephones—main lines in use: 134,900 (2005)
Telephones—mobile cellular: 206,200 (2005)
Telephone system:
general assessment: fixed-line teledensity of roughly 50 per 100 persons; mobile-

cellular telephone density of 75 per 100 persons
domestic: island-wide automatic telephone system
international: country code—1-246; landing point for the East Caribbean Fiber System (ECFS) submarine cable with links to 13 other islands in the eastern Caribbean extending from the British Virgin Islands to Trinidad; satellite earth stations—1 (Intelsat -Atlantic Ocean); tropospheric scatter to Trinidad and Saint Lucia (2007)
Radio broadcast stations: AM 2, FM 6, shortwave 0 (2004)
Radios: 237,000 (1997)
Television broadcast stations: 1 (plus 2 cable channels) (2004)
Televisions: 76,000 (1997)
Internet country code: .bb
Internet hosts: 104 (2007)
Internet Service Providers (ISPs): 19 (2000)
Internet users: 160,000 (2005)

TRANSPORTATION

Airports: 1 (2007)
Airports—with paved runways:
total: 1
over 3,047 m: 1 (2007)
Roadways:
total: 1,600 km
paved: 1,600 km (2004)
Merchant marine:
total: 71 ships (1000 GRT or over) 539,579 GRT/793,899 DWT
by type: bulk carrier 13, cargo 39, chemical tanker 6, passenger 1, passenger/cargo 1, petroleum tanker 3, refrigerated cargo 5, roll on/roll off 2, specialized tanker 1
foreign-owned: 67 (Bahamas, The 1, Canada 9, Greece 11, India 1, Lebanon 1, Monaco 1, Norway 35, Sweden 5, UK 3)
registered in other countries: 1 (St Vincent and The Grenadines 1) (2007)
Ports and terminals: Bridgetown

MILITARY

Military branches: Royal Barbados Defense Force: Troops Command, Barbados Coast Guard (2007)
Military service age and obligation: 18 years of age for voluntary military service (younger requires parental consent); no conscription (2008)
Manpower available for military service:
males age 16–49: 75,265
females age 16–49: 75,389 (2008 est.)
Manpower fit for military service:
males age 16–49: 58,556
females age 16–49: 58,143 (2008 est.)
Military expenditures—percent of GDP: 0.5% (2006 est.)

Military—note: the Royal Barbados Defense Force includes a land-based Troop Command and a small Coast Guard; the primary role of the land element is to defend the island against external aggression; the Command consists of a single, part-time battalion with a small regular cadre that is deployed throughout the island; it increasingly supports the police in patrolling the coastline to prevent smuggling and other illicit activities (2007)

TRANSNATIONAL ISSUES

Disputes—international: in April 2006, the Permanent Court of Arbitration issued a decision that delimited a maritime boundary with Trinidad and Tobago and compelled Barbados to enter a fishing agreement limiting Barbadian fishermen's catches of flying fish in Trinidad and Tobago's exclusive economic zone; in 2005, Barbados and Trinidad and Tobago agreed to compulsory international arbitration under UNCLOS challenging whether the northern limit of Trinidad and Tobago's and Venezuela's maritime boundary extends into Barbadian waters; joins other Caribbean states to counter Venezuela's claim that Aves Island sustains human habitation, a criterion under the UN Convention on the Law of the Sea (UNCLOS), which permits Venezuela to extend its EEZ/continental shelf over a large portion of the eastern Caribbean Sea

Illicit drugs: one of many Caribbean transshipment points for narcotics bound for Europe and the US; offshore financial center

BELARUS

INTRODUCTION

Background: After seven decades as a constituent republic of the USSR, Belarus attained its independence in 1991. It has retained closer political and economic ties to Russia than any of the other former Soviet republics. Belarus and Russia signed a treaty on a two-state union on 8 December 1999 envisioning greater political and economic integration. Although Belarus agreed to a framework to carry out the accord, serious implementation has yet to take place. Since his election in July 1994 as the country's first president, Alexandr LUKASHENKO has steadily consolidated his power through authoritarian means. Government restrictions on freedom of speech and the press, peaceful assembly, and religion continue.

GEOGRAPHY

Location: Eastern Europe, east of Poland
Geographic coordinates: 53 00 N, 28 00 E
Map references: Europe
Area:
total: 207,600 sq km
land: 207,600 sq km
water: 0 sq km
Area—comparative: slightly smaller than Kansas
Land boundaries:
total: 3,098 km
border countries: Latvia 141 km, Lithuania 502 km, Poland 605 km, Russia 959 km, Ukraine 891 km
Coastline: 0 km (landlocked)
Maritime claims: none (landlocked)
Climate: cold winters, cool and moist summers; transitional between continental and maritime
Terrain: generally flat and contains much marshland
Elevation extremes:
lowest point: Nyoman River 90 m
highest point: Dzyarzhynskaya Hara 346 m
Natural resources: forests, peat deposits, small quantities of oil and natural gas, granite, dolomitic limestone, marl, chalk, sand, gravel, clay
Land use:
arable land: 26.77%
permanent crops: 0.6%
other: 72.63% (2005)
Irrigated land: 1,310 sq km (2003)
Total renewable water resources: 58 cu km (1997)
Freshwater withdrawal (domestic/industrial/agricultural): *total:* 2.79 cu km/yr (23%/47%/30%)
per capita: 286 cu m/yr (2000)
Natural hazards: NA
Environment—current issues: soil pollution from pesticide use; southern part of the country contaminated with fallout from 1986 nuclear reactor accident at Chornobyl' in northern Ukraine
Environment—international agreements: *party to:* Air Pollution, Air Pollution-Nitrogen Oxides, Air Pollution-Sulfur 85, Biodiversity, Climate Change, Climate Change-Kyoto Protocol, Desertification, Endangered Species, Environmental Modification, Hazardous Wastes, Law of the Sea, Marine Dumping, Ozone Layer Protection, Ship Pollution, Wetlands
signed, but not ratified: none of the selected agreements
Geography—note: landlocked; glacial scouring accounts for the flatness of Belarusian terrain and for its 11,000 lakes

PEOPLE

Population: 9,685,768 (July 2008 est.)
Age structure:
0–14 years: 14.4% (male 717,885/female 677,254)
15–64 years: 70.9% (male 3,333,699/female 3,531,920)
65 years and over: 14.7% (male 459,627/female 965,383) (2008 est.)
Median age:
total: 38.4 years
male: 35.4 years
female: 41.3 years (2008 est.)
Population growth rate: -0.393% (2008 est.)
Birth rate: 9.62 births/1,000 population (2008 est.)
Death rate: 13.92 deaths/1,000 population (2008 est.)
Net migration rate: 0.38 migrant(s)/1,000 population (2008 est.)
Sex ratio:
at birth: 1.06 male(s)/female
under 15 years: 1.06 male(s)/female
15–64 years: 0.94 male(s)/female
65 years and over: 0.48 male(s)/female
total population: 0.87 male(s)/female (2008 est.)
Infant mortality rate:
total: 6.53 deaths/1,000 live births
male: 7.56 deaths/1,000 live births
female: 5.44 deaths/1,000 live births (2008 est.)

Life expectancy at birth:
total population: 70.34 years
male: 64.63 years
female: 76.4 years (2008 est.)
Total fertility rate: 1.23 children born/woman (2008 est.)
HIV/AIDS—adult prevalence rate: 0.3% (2001 est.)
HIV/AIDS—people living with HIV/AIDS: 15,000 (2001 est.)
HIV/AIDS—deaths: 1,000 (2001 est.)
Nationality:
noun: Belarusian(s)
adjective: Belarusian
Ethnic groups: Belarusian 81.2%, Russian 11.4%, Polish 3.9%, Ukrainian 2.4%, other 1.1% (1999 census)
Religions: Eastern Orthodox 80%, other (including Roman Catholic, Protestant, Jewish, and Muslim) 20% (1997 est.)
Languages: Belarusian, Russian, other
Literacy:
definition: age 15 and over can read and write
total population: 99.6%
male: 99.8%
female: 99.4% (1999 census)

GOVERNMENT

Country name:
conventional long form: Republic of Belarus
conventional short form: Belarus
local long form: Respublika Byelarus'
local short form: Byelarus'
former: Belorussian (Byelorussian) Soviet Socialist Republic
Government type: republic in name, although in fact a dictatorship
Capital: *name:* Minsk
geographic coordinates: 53 54 N, 27 34 E
time difference: UTC+2 (7 hours ahead of Washington, DC during Standard Time)
daylight saving time: +1hr, begins last Sunday in March; ends last Sunday in October
Administrative divisions: 6 provinces (voblastsi, singular—voblasts') and 1 municipality* (horad); Brest, Homyel', Horad Minsk*, Hrodna, Mahilyow, Minsk, Vitsyebsk
note: administrative divisions have the same names as their administrative centers
Independence: 25 August 1991 (from Soviet Union)
National holiday: Independence Day, 3 July (1944); note—3 July 1944 was the date Minsk was liberated from German troops, 25 August 1991 was the date of independence from the Soviet Union
Constitution: 15 March 1994; revised by national referendum of 24 November 1996 giving the presidency greatly expanded powers and became effective

27 November 1996; revised again 17 October 2004 removing presidential term limits
Legal system: based on civil law system; has not accepted compulsory ICJ jurisdiction
Suffrage: 18 years of age; universal
Executive branch:
chief of state: President Aleksandr LUKASHENKO (since 20 July 1994)
head of government: Prime Minister Sergey SIDORSKIY (since 19 December 2003); First Deputy Prime Minister Vladimir SEMASHKO (since December 2003)
cabinet: Council of Ministers
elections: president elected by popular vote for a five-year term; first election took place 23 June and 10 July 1994; according to the 1994 constitution, the next election should have been held in 1999, however, Aleksandr LUKASHENKO extended his term to 2001 via a November 1996 referendum; subsequent election held 9 September 2001; an October 2004 referendum ended presidential term limits and allowed the president to run in a third election, which was held on 19 March 2006; prime minister and deputy prime ministers appointed by the president
election results: Aleksandr LUKASHENKO reelected president; percent of vote—Aleksandr LUKASHENKO 82.6%, Aleksandr MILINKEVICH 6%, Aleksandr KOZULIN 2.3%; note—election marred by electoral fraud
Legislative branch: bicameral National Assembly or Natsionalnoye Sobranie consists of the Council of the Republic or Soviet Respubliki (64 seats; 56 members elected by regional councils and eight members appointed by the president, to serve four-year terms) and the Chamber of Representatives or Palata Predstaviteley (110 seats; members elected by popular vote to serve four-year terms)
elections: last held 17 and 31 October 2004; international observers widely denounced the elections as flawed and undemocratic based on massive government falsification; pro-LUKASHENKO candidates won every seat after many opposition candidates were disqualified for technical reasons
election results: Soviet Respubliki—percent of vote by party—NA; seats by party—NA; Palata Predstaviteley—percent of vote by party—NA; seats by party—NA
Judicial branch: Supreme Court (judges are appointed by the president); Constitutional Court (half of the judges

appointed by the president and half appointed by the Chamber of Representatives)
Political parties and leaders: *pro-government parties:* Agrarian Party or AP [Mikhail SHIMANSKY]; Belarusian Communist Party or KPB; Belarusian Patriotic Movement (Belarusian Patriotic Party) or BPR [Nikolay ULAKHOVICH, chairman]; Liberal Democratic Party of Belarus [Sergey GAYDUKEVICH]; Party of Labor and Justice [Viktor SOKOLOV]; Social-Sports Party [Vladimir ALEXANDROVICH]
opposition parties: Belarusian Christian Democracy Party (unregistered) [Pavel SEVERINETS]; Belarusian Party of Communists or PKB [Sergey KALYAKIN]; Belarusian Party of Labor (unregistered) [Aleksandr BUKHVOSTOV, Leonid LEMESHONAK]; Belarusian Popular Front or BPF [Vintsyuk VYACHORKA]; Belarusian Social-Democratic Gramada [Stanislav SHUSHKEVICH]; Belarusian Social Democratic Party Hramada (People's Assembly) or BSDPH [Aleksandr KOZULIN; Anatoliy LEVKOVICH, acting]; Green Party [Oleg GROMYKO]; Party of Freedom and Progress (unregistered) [Vladimir NOVOSYAD]; United Civic Party or UCP [Anatoliy LEBEDKO]; Women's Party "Nadezhda" [Valentina MATUSEVICH, chairperson]
other opposition includes: Christian Conservative BPF [Zyanon PAZNIAK]; Ecological Party of Greens [Mikhail KARTASH]; Party of Popular Accord [Sergey YERMAKK]; Republican Party [Vladimir BELAZOR]
Political pressure groups and leaders: Assembly of Pro-Democratic NGOs [Sergey MATSKEVICH]; Belarusian Congress of Democratic Trade Unions [Aleksandr YAROSHUK]; Belarusian Helsinki Committee [Tatiana PROTKO]; Belarusian Organization of Working Women [Irina ZHIKHAR]; Charter 97 [Andrey SANNIKOV]; For Freedom (unregistered) [Aleksandr MILINKEVICH]; Lenin Communist Union of Youth (youth wing of the Belarusian Party of Communists or PKB); National Strike Committee of Entrepreneurs [Aleksandr VASILYEV, Valery LEVONEVSKY]; Partnership NGO [Nikolay ASTREYKA]; Perspektiva kiosk watchdog NGO [Anatol SHUMCHENKO]; Vyasna [Ales BYALATSKY]; Women's Independent Democratic Movement [Ludmila PETINA]; Youth Front (Malady Front) [Dmitriy DASHKE-

VICH, Sergey BAKHUN]; Zubr youth group [Vladimir KOBETS]

International organization participation: BSEC (observer), CEI, CIS, EAEC, EAPC, EBRD, GCTU, IAEA, IBRD, ICAO, ICRM, IDA, IFC, IFRCS, ILO, IMF, IMSO, Interpol, IOC, IOM, IPU, ISO, ITU, ITUC, MIGA, NAM, NSG, OPCW, OSCE, PCA, PFP, UN, UNCTAD, UNESCO, UNIDO, UNWTO, UPU, WCO, WFTU, WHO, WIPO, WMO, WTO (observer)

Diplomatic representation in the US:
chief of mission: Ambassador Mikhail KHVOSTOV
chancery: 1619 New Hampshire Avenue NW, Washington, DC 20009
telephone: [1] (202) 986-1604
FAX: [1] (202) 986-1805
consulate(s) general: New York

Diplomatic representation from the US:
chief of mission: Ambassador Karen B. STEWART
embassy: 46 Starovilenskaya Street, Minsk 220002
mailing address: PSC 78, Box B Minsk, APO 09723
telephone: [375] (17) 210-12-83, 217-7347, 217-7348
FAX: [375] (17) 234-7853

Flag description: red horizontal band (top) and green horizontal band one-half the width of the red band; a white vertical stripe on the hoist side bears Belarusian national ornamentation in red

ECONOMY

Economy—overview: Belarus has seen little structural reform since 1995, when President LUKASHENKO launched the country on the path of "market socialism." In keeping with this policy, LUKASHENKO reimposed administrative controls over prices and currency exchange rates and expanded the state's right to intervene in the management of private enterprises. Since 2005, the government has re-nationalized a number of private companies. In addition, businesses have been subject to pressure by central and local governments, e.g., arbitrary changes in regulations, numerous rigorous inspections, retroactive application of new business regulations, and arrests of "disruptive" businessmen and factory owners. A wide range of redistributive policies has helped those at the bottom of the ladder; the Gini coefficient is among the lowest in the world. Because of these restrictive economic policies, Belarus has had trouble attracting foreign investment. Nevertheless, GDP growth has been strong in recent years, reaching nearly

7% in 2007, despite the roadblocks of a tough, centrally directed economy with a high, but decreasing, rate of inflation. Belarus receives heavily discounted oil and natural gas from Russia and much of Belarus' growth can be attributed to the re-export of Russian oil at market prices. Trade with Russia—by far its largest single trade partner—decreased in 2007, largely as a result of a change in the way the Value Added Tax (VAT) on trade was collected. Russia has introduced an export duty on oil shipped to Belarus, which will increase gradually through 2009, and a requirement that Belarusian duties on re-exported Russian oil be shared with Russia—80% will go to Russia in 2008, and 85% in 2009. Russia also increased Belarusian natural gas prices from $47 per thousand cubic meters (tcm) to $100 per tcm in 2007, and plans to increase prices gradually to world levels by 2011. Russia's recent policy of bringing energy prices for Belarus to world market levels may result in a slowdown in economic growth in Belarus over the next few years. Some policy measures, including tightening of fiscal and monetary policies, improving energy efficiency, and diversifying exports, have been introduced, but external borrowing has been the main mechanism used to manage the growing pressures on the economy.

GDP (purchasing power parity): $105.2 billion (2007 est.)
GDP (official exchange rate): $44.77 billion (2007 est.)
GDP—real growth rate: 8.2% (2007 est.)
GDP—per capita (PPP): $10,900 (2007 est.)
GDP—composition by sector:
agriculture: 8.7%
industry: 40.6%
services: 50.6% (2007 est.)
Labor force: 4.3 million (31 December 2005)
Labor force—by occupation: *agriculture:* 14%
industry: 34.7%
services: 51.3% (2003 est.)
Unemployment rate: 1.6% officially registered unemployed; large number of underemployed workers (2005)
Population below poverty line: 27.1% (2003 est.)
Household income or consumption by percentage share:
lowest 10%: 3.4%
highest 10%: 23.5% (2002)
Distribution of family income—Gini index: 29.7 (2002)
Inflation rate (consumer prices): 8.4% (2007 est.)

Investment (gross fixed): 29.9% of GDP (2007 est.)
Budget:
revenues: $20.76 billion
expenditures: $21.18 billion (2007 est.)
Agriculture—products: grain, potatoes, vegetables, sugar beets, flax; beef, milk
Industries: metal-cutting machine tools, tractors, trucks, earthmovers, motorcycles, televisions, synthetic fibers, fertilizer, textiles, radios, refrigerators
Industrial production growth rate: 5% (2007 est.)
Electricity—production: 29.08 billion kWh (2005)
Electricity—production by source:
fossil fuel: 99.5%
hydro: 0.1%
nuclear: 0%
other: 0.4% (2001)
Electricity—consumption: 29.49 billion kWh (2005)
Electricity—exports: 5.053 billion kWh (2005)
Electricity—imports: 9.091 billion kWh (2005)
Oil—production: 33,700 bbl/day (2005 est.)
Oil—consumption: 156,000 bbl/day (2005 est.)
Oil—exports: 249,900 bbl/day (2004 est.)
Oil—imports: 378,200 bbl/day (2004 est.)
Oil—proved reserves: 198 million bbl (1 January 2006 est.)
Natural gas—production: 165 million cu m (2005 est.)
Natural gas—consumption: 19.47 billion cu m (2005 est.)
Natural gas—exports: 0 cu m (2005 est.)
Natural gas—imports: 19.31 billion cu m (2005)
Natural gas—proved reserves: 2.716 billion cu m (1 January 2006 est.)
Current account balance: -$2.944 billion (2007 est.)
Exports: $23.04 billion f.o.b. (2007 est.)
Exports—commodities: machinery and equipment, mineral products, chemicals, metals, textiles, foodstuffs
Exports—partners: Russia 34.7%, Netherlands 17.7%, UK 7.5%, Ukraine 6.3%, Poland 5.2% (2006)
Imports: $27.57 billion f.o.b. (2007 est.)
Imports—commodities: mineral products, machinery and equipment, chemicals, foodstuffs, metals
Imports—partners: Russia 58.6%, Germany 7.5%, Ukraine 5.5% (2006)
Economic aid—recipient: $53.76 million (2005)
Reserves of foreign exchange and gold: $2.469 billion (31 December 2007 est.)

Debt—external: $6.889 billion (31 December 2007)

Market value of publicly traded shares: $NA

Currency (code): Belarusian ruble (BYB/BYR)

Currency code: BYB/BYR

Exchange rates: Belarusian rubles per US dollar—2,145 (2007), 2,144.6 (2006), 2,150 (2005), 2,160.26 (2004), 2,051.27 (2003)

Fiscal year: calendar year

COMMUNICATIONS

Telephones—main lines in use: 3.368 million (2006)

Telephones—mobile cellular: 5.96 million (2006)

Telephone system:

general assessment: Belarus lags behind its neighbors in upgrading telecommunications infrastructure; state-owned Beltelcom is the sole provider of fixed-line local and long distance service; fixed-line teledensity of 33 per 100 persons; mobile-cellular telephone density of 58 per 100 persons; modernization of the network progressing with roughly two-thirds of switching equipment now digital

domestic: fixed-line penetration is improving although rural areas continue to be underserved; 4 GSM wireless networks are experiencing rapid growth; strict government controls on telecommunications technologies

international: country code—375; Belarus is a member of the Trans-European Line (TEL), Trans-Asia-Europe (TAE) fiber-optic line, and has access to the Trans-Siberia Line (TSL); 3 fiber-optic segments provide connectivity to Latvia, Poland, Russia, and Ukraine; worldwide service is available to Belarus through this infrastructure; additional analog lines to Russia; Intelsat, Eutelsat, and Intersputnik earth stations (2007)

Radio broadcast stations: AM 28, FM 37, shortwave 11 (1998)

Radios: 3.02 million (1997)

Television broadcast stations: 47 (plus 27 repeaters) (1995)

Televisions: 2.52 million (1997)

Internet country code: .by

Internet hosts: 20,685 (2007)

Internet Service Providers (ISPs): 23 (2002)

Internet users: 5.478 million (2006)

TRANSPORTATION

Airports: 67 (2007)

Airports—with paved runways:

total: 36

over 3,047 m: 2

2,438 to 3,047 m: 22

1,524 to 2,437 m: 4

914 to 1,523 m: 1

under 914 m: 7 (2007)

Airports—with unpaved runways:

total: 31

2,438 to 3,047 m: 1

1,524 to 2,437 m: 1

914 to 1,523 m: 2

under 914 m: 27 (2007)

Heliports: 1 (2007)

Pipelines: gas 5,250 km; oil 1,528 km; refined products 1,730 km (2007)

Railways:

total: 5,512 km

broad gauge: 5,497 km 1.520-m gauge (874 km electrified)

standard gauge: 15 km 1.435 m (2006)

Roadways:

total: 94,797 km

paved: 84,028 km

unpaved: 10,769 km (2005)

Waterways: 2,500 km (use limited by location on perimeter of country and by shallowness) (2003)

Ports and terminals: Mazyr

MILITARY

Military branches: Belarus Armed Forces: Land Force, Air and Air Defense Force (2008)

Military service age and obligation: 18–27 years of age for compulsory military service; conscript service obligation—18 months (2005)

Manpower available for military service:

males age 16–49: 2,491,643

females age 16–49: 2,528,779 (2008 est.)

Manpower fit for military service:

males age 16–49: 1,727,974

females age 16–49: 2,093,106 (2008 est.)

Manpower reaching militarily significant age annually:

males age 16–49: 64,232

females age 16–49: 60,788 (2008 est.)

Military expenditures—percent of GDP: 1.4% (2005 est.)

TRANSNATIONAL ISSUES

Disputes—international: as of January 2007, ground demarcations of the boundaries with Latvia and Lithuania were complete and mapped with final ratification documentation in preparation; 1997 boundary delimitation treaty with Ukraine remains unratified over unresolved financial claims, preventing demarcation and diminishing border security

Illicit drugs: limited cultivation of opium poppy and cannabis, mostly for the domestic market; transshipment point for illicit drugs to and via Russia, and to the Baltics and Western Europe; a small and lightly regulated financial center; new anti-money-laundering legislation does not meet international standards; few investigations or prosecutions of money-laundering activities

BELGIUM

INTRODUCTION

Background: Belgium became independent from the Netherlands in 1830; it was occupied by Germany during World Wars I and II. The country prospered in the past half century as a modern, technologically advanced European state and member of NATO and the EU. Tensions between the Dutch-speaking Flemings of the north and the French-speaking Walloons of the south have led in recent years to constitutional amendments granting these regions formal recognition and autonomy.

GEOGRAPHY

Location: Western Europe, bordering the North Sea, between France and the Netherlands

Geographic coordinates: 50 50 N, 4 00 E

Map references: Europe

Area:

total: 30,528 sq km

land: 30,278 sq km

water: 250 sq km

Area—comparative: about the size of Maryland

Land boundaries:

total: 1,385 km

border countries: France 620 km, Germany 167 km, Luxembourg 148 km, Netherlands 450 km

Coastline: 66.5 km

Maritime claims:
territorial sea: 12 nm
contiguous zone: 24 nm
exclusive economic zone: geographic coordinates define outer limit
continental shelf: median line with neighbors

Climate: temperate; mild winters, cool summers; rainy, humid, cloudy

Terrain: flat coastal plains in northwest, central rolling hills, rugged mountains of Ardennes Forest in southeast

Elevation extremes:
lowest point: North Sea 0 m
highest point: Signal de Botrange 694 m

Natural resources: construction materials, silica sand, carbonates

Land use:
arable land: 27.42%
permanent crops: 0.69%
other: 71.89%
note: includes Luxembourg (2005)

Irrigated land: 400 sq km (2003)

Total renewable water resources: 20.8 cu km (2005)

Freshwater withdrawal (domestic/industrial/agricultural): *total:* 7.44 cu km/yr (13%/85%/1%)
per capita: 714 cu m/yr (1998)

Natural hazards: flooding is a threat along rivers and in areas of reclaimed coastal land, protected from the sea by concrete dikes

Environment—current issues: the environment is exposed to intense pressures from human activities: urbanization, dense transportation network, industry, extensive animal breeding and crop cultivation; air and water pollution also have repercussions for neighboring countries; uncertainties regarding federal and regional responsibilities (now resolved) had slowed progress in tackling environmental challenges

Environment—international agreements: *party to:* Air Pollution, Air Pollution-Nitrogen Oxides, Air Pollution-Persistent Organic Pollutants, Air Pollution-Sulfur 85, Air Pollution-Sulfur 94, Air Pollution-Volatile Organic Compounds, Antarctic-Environmental Protocol, Antarctic-Marine Living Resources, Antarctic Seals, Antarctic Treaty, Biodiversity, Climate Change, Climate Change-Kyoto Protocol, Desertification, Endangered Species, Environmental Modification, Hazardous Wastes, Law of the Sea, Marine Dumping, Marine Life Conservation, Ozone Layer Protection, Ship Pollution, Tropical Timber 83, Tropical Timber 94, Wetlands, Whaling *signed, but not ratified:* none of the selected agreements

Geography—note: crossroads of Western Europe; most West European capitals within 1,000 km of Brussels, the seat of both the European Union and NATO

PEOPLE

Population: 10,403,951 (July 2008 est.)

Age structure:
0–14 years: 16.3% (male 864,287/female 828,435)
15–64 years: 66.3% (male 3,476,802/female 3,416,383)
65 years and over: 17.5% (male 751,745/female 1,066,299) (2008 est.)

Median age:
total: 41.4 years
male: 40.2 years
female: 42.7 years (2008 est.)

Population growth rate: 0.106% (2008 est.)

Birth rate: 10.22 births/1,000 population (2008 est.)

Death rate: 10.38 deaths/1,000 population (2008 est.)

Net migration rate: 1.22 migrant(s)/1,000 population (2008 est.)

Sex ratio:
at birth: 1.04 male(s)/female
under 15 years: 1.04 male(s)/female
15–64 years: 1.02 male(s)/female
65 years and over: 0.71 male(s)/female
total population: 0.96 male(s)/female (2008 est.)

Infant mortality rate:
total: 4.5 deaths/1,000 live births
male: 5.06 deaths/1,000 live births
female: 3.92 deaths/1,000 live births (2008 est.)

Life expectancy at birth:
total population: 79.07 years
male: 75.9 years
female: 82.38 years (2008 est.)

Total fertility rate: 1.65 children born/woman (2008 est.)

HIV/AIDS—adult prevalence rate: 0.2% (2003 est.)

HIV/AIDS—people living with HIV/AIDS: 10,000 (2003 est.)

HIV/AIDS—deaths: fewer than 100 (2003 est.)

Nationality:
noun: Belgian(s)
adjective: Belgian

Ethnic groups: Fleming 58%, Walloon 31%, mixed or other 11%

Religions: Roman Catholic 75%, other (includes Protestant) 25%

Languages: Dutch (official) 60%, French (official) 40%, German (official) less than 1%, legally bilingual (Dutch and French)

Literacy:
definition: age 15 and over can read and write
total population: 99%
male: 99%
female: 99% (2003 est.)

GOVERNMENT

Country name:
conventional long form: Kingdom of Belgium
conventional short form: Belgium
local long form: Royaume de Belgique/Koninkrijk Belgie
local short form: Belgique/Belgie

Government type: federal parliamentary democracy under a constitutional monarchy

Capital: *name:* Brussels
geographic coordinates: 50 50 N, 4 20 E
time difference: UTC+1 (6 hours ahead of Washington, DC during Standard Time)
daylight saving time: +1hr, begins last Sunday in March; ends last Sunday in October

Administrative divisions: 10 provinces (French: provinces, singular—province; Dutch: provincies, singular—provincie) and 3 regions* (French: regions; Dutch: gewesten); Brussels* (Bruxelles) capital region; Flanders* region (five provinces): Antwerpen (Antwerp), Limburg, Oost-Vlaanderen (East Flanders), Vlaams-Brabant (Flemish Brabant), West-Vlaanderen (West Flanders); Wallonia* region (five provinces): Brabant Wallon (Walloon Brabant), Hainaut, Liege, Luxembourg, Namur
note: as a result of the 1993 constitutional revision that furthered devolution into a federal state, there are now three levels of government (federal, regional, and linguistic community) with a complex division of responsibilities

Independence: 4 October 1830 (a provisional government declared independence from the Netherlands); 21 July 1831 (King LEOPOLD I ascended to the throne)

National holiday: 21 July (1831) ascension to the Throne of King LEOPOLD I

Constitution: 7 February 1831; amended many times; revised 14 July 1993 to create a federal state

Legal system: based on civil law system influenced by English constitutional theory; judicial review of legislative acts; accepts compulsory ICJ jurisdiction, with reservations

Suffrage: 18 years of age; universal and compulsory

Executive branch:
chief of state: King ALBERT II (since 9 August 1993); Heir Apparent Prince PHILIPPE, son of the monarch

head of government: Prime Minister Yves LETERME (20 March 2008)

cabinet: Council of Ministers are formally appointed by the monarch

elections: the monarchy is hereditary and constitutional; following legislative elections, the leader of the majority party or the leader of the majority coalition is usually appointed prime minister by the monarch and then approved by parliament

Legislative branch: bicameral Parliament consists of a Senate or Senaat in Dutch, Senat in French (71 seats; 40 members are directly elected by popular vote, 31 are indirectly elected; members serve four-year terms) and a Chamber of Deputies or Kamer van Volksvertegenwoordigers in Dutch, Chambre des Representants in French (150 seats; members are directly elected by popular vote on the basis of proportional representation to serve four-year terms)

elections: Senate and Chamber of Deputies—last held 10 June 2007 (next to be held no later than June 2011)

election results: Senate—percent of vote by party—CDV/N-VA 19.4%, Open VLD 12.4%, MR 12.3%, VB 11.9%, PS 10.2%, SP.A-Spirit 10%, CDH 5.9%, Ecolo 5.8%, Groen! 3.6%, Dedecker List 3.4%, FN 2.3%, other 2.8%; seats by party—CDV/N-VA 9, Open VLD 5, MR 6, VB 5, PS 4, SP.A-Spririt 4, CDH 2, Ecolo 2, Groen! 1, Dedecker List 1, FN 1 (note—there are also 31 indirectly elected senators); Chamber of Deputies—percent of vote by party— CDV/N-VA 18.5%, MR 12.5%, VB 12%, Open VLD 11.8%, PS 10.9%, SP.A-Spirit 10.3%, CDH 6.1%, Ecolo 5.1%, Dedecker List 4%, Groen! 4%, FN 2%, other 2.8%; seats by party— CDV/N-VA 30, MR 23, VB 17, Open VLD 18, PS 20, SP.A-Spirit 14, CDH 10, Ecolo 8, Dedecker List 5, Groen! 4, FN 1 *note:* as a result of the 1993 constitutional revision that furthered devolution into a federal state, there are now three levels of government (federal, regional, and linguistic community) with a complex division of responsibilities; this reality leaves six governments each with its own legislative assembly

Judicial branch: Supreme Court of Justice or Hof van Cassatie (in Dutch) or Cour de Cassation (in French) (judges are appointed for life by the government; candidacies have to be submitted by the High Justice Council)

Political parties and leaders: Flemish *parties:* Christian Democratic and Flemish or CDV [Etienne SCHOUPPE]; Dedecker List [Jean-Marie DEDECKER]; Flemish Liberals and Democrats or Open VLD [Bart SOMERS]; Groen! [Mieke

VOGELS] (formerly AGALEV, Flemish Greens); New Flemish Alliance or N-VA [Bart DE WEVER]; Social Progressive Alternative or SP.A [Caroline GENNEZ]; VlaamsProgressieven (Flemish Progressives) or VP [Bettina GEYSEN]—formerly Spirit; Vlaams Belang (Flemish Interest) or VB [Bruno VALKENIERS]

Francophone parties: Ecolo (Francophone Greens) [Jean-Michel JAVAUX, Isabelle DURANT, Claude BROUIR]; Humanist and Democratic Center or CDH [Joelle MILQUET]; National Front or FN [Michel BELACROIX]; Reform Movement or MR [Didier REYNDERS]; Socialist Party or PS [Elio DI RUPO]; other minor parties

Political pressure groups and leaders: Christian, Socialist, and Liberal Trade Unions; Federation of Belgian Industries; numerous other associations representing bankers, manufacturers, middle-class artisans, and the legal and medical professions; various organizations represent the cultural interests of Flanders and Wallonia; various peace groups such as Pax Christi and groups representing immigrants

International organization participation: ACCT, ADB (nonregional members), AfDB, Australia Group, Benelux, BIS, CE, CERN, EAPC, EBRD, EIB, EMU, ESA, EU, FAO, G-9, G-10, IADB, IAEA, IBRD, ICAO, ICC, ICCt, ICRM, IDA, IEA, IFAD, IFC, IFRCS, IHO, ILO, IMF, IMO, IMSO, Interpol, IOC, IOM, IPU, ISO, ITSO, ITU, ITUC, MIGA, NATO, NEA, NSG, OAS (observer), OECD, OIF, OPCW, OSCE, Paris Club, PCA, Schengen Convention, SECI (observer), UN, UN Security Council (temporary), UNCTAD, UNESCO, UNHCR, UNIDO, UNIFIL, UNRWA, UNTSO, UPU, WADB (nonregional), WCL, WCO, WEU, WHO, WIPO, WMO, WTO, ZC

Diplomatic representation in the US: *chief of mission:* Ambassador Dominique STRUYE DE SWIELANDE

chancery: 3330 Garfield Street NW, Washington, DC 20008

telephone: [1] (202) 333-6900

FAX: [1] (202) 338-4960

consulate(s) general: Los Angeles, New York

consulate(s): Atlanta

Diplomatic representation from the US: *chief of mission:* Ambassador Sam FOX

embassy: Regentlaan 27 Boulevard du Regent, B-1000 Brussels

mailing address: PSC 82, Box 002, APO AE 09710

telephone: [32] (2) 508-2111

FAX: [32] (2) 511-2725

Flag description: three equal vertical bands of black (hoist side), yellow, and red

note: the design was based on the flag of France

ECONOMY

Economy—overview: This modern, private-enterprise economy has capitalized on its central geographic location, highly developed transport network, and diversified industrial and commercial base. Industry is concentrated mainly in the populous Flemish area in the north. With few natural resources, Belgium must import substantial quantities of raw materials and export a large volume of manufactures, making its economy unusually dependent on the state of world markets. Roughly three-quarters of its trade is with other EU countries. Public debt is more than 85% of GDP. On the positive side, the government has succeeded in balancing its budget, and income distribution is relatively equal. Belgium began circulating the euro currency in January 2002. Economic growth in 2001–03 dropped sharply because of the global economic slowdown, with moderate recovery in 2004–07. Economic growth and foreign direct investment are expected to slow down in 2008, due to credit tightening, falling consumer and business confidence, and above average inflation. However, with the successful negotiation of the 2008 budget and devolution of power within the government, political tensions seem to be easing and could lead to an improvement in the economic outlook for 2008.

GDP (purchasing power parity): $376 billion (2007 est.)

GDP (official exchange rate): $453.6 billion (2007 est.)

GDP—real growth rate: 2.7% (2007 est.)

GDP—per capita (PPP): $35,300 (2007 est.)

GDP—composition by sector:
agriculture: 1.1%
industry: 24.5%
services: 74.4% (2007 est.)

Labor force: 4.96 million (2007 est.)

Labor force—by occupation:
agriculture: 2%
industry: 25%
services: 73% (2007 est.)

Unemployment rate: 7.5% (2007 est.)

Population below poverty line: 15.2% (2007 est.)

Household income or consumption by percentage share:
lowest 10%: 3.4%
highest 10%: 28.4% (2006)

Distribution of family income—Gini index: 28 (2005)

Inflation rate (consumer prices): 1.8% (2007 est.)

Investment (gross fixed): 21.3% of GDP (2007 est.)

Budget: *revenues:* $219.3 billion *expenditures:* $220.3 billion (2007 est.)

Public debt: 84.9% of GDP (2007 est.)

Agriculture—products: sugar beets, fresh vegetables, fruits, grain, tobacco; beef, veal, pork, milk

Industries: engineering and metal products, motor vehicle assembly, transportation equipment, scientific instruments, processed food and beverages, chemicals, basic metals, textiles, glass, petroleum

Industrial production growth rate: 3% (2007 est.)

Electricity—production: 80.84 billion kWh (2005)

Electricity—production by source: *fossil fuel:* 38.4% *hydro:* 0.6% *nuclear:* 59.3% *other:* 1.8% (2001)

Electricity—consumption: 82.99 billion kWh (2005)

Electricity—exports: 8.024 billion kWh (2005)

Electricity—imports: 14.33 billion kWh (2005)

Oil—production: 9,000 bbl/day (2006)

Oil—consumption: 591,000 bbl/day (2006 est.)

Oil—exports: 523,400 bbl/day (2004)

Oil—imports: 1.109 million bbl/day (2004)

Oil—proved reserves: 0 bbl (1 January 2006 est.)

Natural gas—production: 0 cu m (2005 est.)

Natural gas—consumption: 16.61 billion cu m (2005 est.)

Natural gas—exports: 0 cu m (2005 est.)

Natural gas—imports: 17.27 billion cu m (2005)

Natural gas—proved reserves: 0 cu m (1 January 2006)

Current account balance: $14.64 billion (2007 est.)

Exports: $322.1 billion f.o.b. (2007 est.)

Exports—commodities: machinery and equipment, chemicals, diamonds, metals and metal products, foodstuffs

Exports—partners: Germany 19.7%, France 16.9%, Netherlands 12%, UK 7.9%, US 6.2%, Italy 5.2% (2006)

Imports: $322.9 billion f.o.b. (2007 est.)

Imports—commodities: machinery and equipment, chemicals, diamonds, pharmaceuticals, foodstuffs, transportation equipment, oil products

Imports—partners: Netherlands 18.3%, Germany 17.3%, France 11.2%, UK 6.6%, Ireland 5.7%, US 5.4% (2006)

Economic aid—donor: ODA, $1.978 billion (2006)

Reserves of foreign exchange and gold: $16.51 billion (2007 est.)

Debt—external: $1.313 trillion (30 June 2007)

Stock of direct foreign investment—at home: $703.9 billion (2007 est.)

Stock of direct foreign investment—abroad: $537.6 billion (2007 est.)

Market value of publicly traded shares: $422.7 billion (2006)

Currency (code): euro (EUR)

Currency code: EUR

Exchange rates: euros per US dollar—0.7345 (2007), 0.7964 (2006), 0.8041 (2005), 0.8054 (2004), 0.886 (2003)

Fiscal year: calendar year

COMMUNICATIONS

Telephones—main lines in use: 4.719 million (2006)

Telephones—mobile cellular: 9.66 million (2006)

Telephone system: *general assessment:* highly developed, technologically advanced, and completely automated domestic and international telephone and telegraph facilities *domestic:* nationwide cellular telephone system; extensive cable network; limited microwave radio relay network *international:* country code—32; landing point for a number of submarine cables that provide links to Europe, the Middle East, and Asia; satellite earth stations—7 (Intelsat—3) (2007)

Radio broadcast stations: AM 7, FM 79, shortwave 1 (1998)

Radios: 8.075 million (1997)

Television broadcast stations: 25 (plus 10 repeaters) (1997)

Televisions: 4.72 million (1997)

Internet country code: .be

Internet hosts: 3.195 million (2007)

Internet Service Providers (ISPs): 61 (2000)

Internet users: 4.8 million (2005)

TRANSPORTATION

Airports: 43 (2007)

Airports—with paved runways: *total:* 27 *over 3,047 m:* 6 *2,438 to 3,047 m:* 7 *1,524 to 2,437 m:* 4 *914 to 1,523 m:* 1 *under 914 m:* 9 (2007)

Airports—with unpaved runways: *total:* 16 *914 to 1,523 m:* 1 *under 914 m:* 15 (2007)

Heliports: 1 (2007)

Pipelines: gas 1,562 km; oil 158 km; refined products 535 km (2007)

Railways: *total:* 3,536 km

standard gauge: 3,536 km 1.435-m gauge (2,950 km electrified) (2006)

Roadways: *total:* 150,567 km *paved:* 117,442 km (includes 1,747 km of expressways) *unpaved:* 33,125 km (2004)

Waterways: 2,043 km (1,528 km in regular commercial use) (2006)

Merchant marine: *total:* 68 ships (1000 GRT or over) 3,786,089 GRT/6,074,664 DWT *by type:* bulk carrier 20, cargo 5, chemical tanker 2, container 9, liquefied gas 16, passenger 1, petroleum tanker 10, roll on/roll off 5 *foreign-owned:* 9 (Denmark 3, France 1, Germany 1, Greece 4) *registered in other countries:* 123 (Bahamas 15, Bermuda 3, Cyprus 1, France 6, Gibraltar 3, Greece 16, Hong Kong 4, Liberia 1, Luxembourg 9, Malta 10, Marshall Islands 1, Mozambique 2, Netherlands 2, Netherlands Antilles 1, Panama 11, Portugal 9, Russia 6, Sierra Leone 1, Singapore 8, St Kitts and Nevis 1, St Vincent and The Grenadines 9, Vanuatu 4) (2007)

Ports and terminals: Antwerp, Gent, Liege, Zeebrugge

MILITARY

Military branches: Belgian Armed Forces: Land Operations Command, Naval Operations Command, Air Operations Command (2008)

Military service age and obligation: 18 years of age for voluntary military service; conscription suspended (2008)

Manpower available for military service: *males age 16–49:* 2,407,128 *females age 16–49:* 2,340,039 (2008 est.)

Manpower fit for military service: *males age 16–49:* 1,973,167 *females age 16–49:* 1,915,990 (2008 est.)

Manpower reaching militarily significant age annually: *males age 16–49:* 64,659 *females age 16–49:* 61,881 (2008 est.)

Military expenditures—percent of GDP: 1.3% (2005 est.)

TRANSNATIONAL ISSUES

Disputes—international: none

Illicit drugs: growing producer of synthetic drugs and cannabis; transit point for US-bound ecstasy; source of precursor chemicals for South American cocaine processors; transshipment point for cocaine, heroin, hashish, and marijuana entering Western Europe; despite a strengthening of legislation, the country remains vulnerable to money laundering related to narcotics, automobiles, alcohol, and tobacco; significant domestic consumption of ecstasy

INTRODUCTION

Background: Belize was the site of several Mayan city states until their decline at the end of the first millennium A.D. The British and Spanish disputed the region in the 17th and 18th centuries; it formally became the colony of British Honduras in 1854. Territorial disputes between the UK and Guatemala delayed the independence of Belize until 1981. Guatemala refused to recognize the new nation until 1992. Tourism has become the mainstay of the economy. Current concerns include an unsustainable foreign debt, high unemployment, growing involvement in the South American drug trade, growing urban crime, and increasing incidences of HIV/AIDS.

GEOGRAPHY

Location: Central America, bordering the Caribbean Sea, between Guatemala and Mexico
Geographic coordinates: 17 15 N, 88 45 W
Map references: Central America and the Caribbean
Area:
total: 22,966 sq km
land: 22,806 sq km
water: 160 sq km
Area—comparative: slightly smaller than Massachusetts
Land boundaries:
total: 516 km
border countries: Guatemala 266 km, Mexico 250 km
Coastline: 386 km
Maritime claims:
territorial sea: 12 nm in the north, 3 nm in the south; *note*—from the mouth of the Sarstoon River to Ranguana Cay, Belize's territorial sea is 3 nm; according to Belize's Maritime Areas Act, 1992, the purpose of this limitation is to provide a

framework for negotiating a definitive agreement on territorial differences with Guatemala
exclusive economic zone: 200 nm
Climate: tropical; very hot and humid; rainy season (May to November); dry season (February to May)
Terrain: flat, swampy coastal plain; low mountains in south
Elevation extremes:
lowest point: Caribbean Sea 0 m
highest point: Doyle's Delight 1,160 m
Natural resources: arable land potential, timber, fish, hydropower
Land use:
arable land: 3.05%
permanent crops: 1.39%
other: 95.56% (2005)
Irrigated land: 30 sq km (2003)
Total renewable water resources: 18.6 cu km (2000)
Freshwater withdrawal (domestic/industrial/agricultural): *total:* 0.15 cu km/yr (7%/73%/20%)
per capita: 556 cu m/yr (2000)
Natural hazards: frequent, devastating hurricanes (June to November) and coastal flooding (especially in south)
Environment—current issues: deforestation; water pollution from sewage, industrial effluents, agricultural runoff; solid and sewage waste disposal
Environment—international agreements: *party to:* Biodiversity, Climate Change, Climate Change-Kyoto Protocol, Desertification, Endangered Species, Hazardous Wastes, Law of the Sea, Ozone Layer Protection, Ship Pollution, Wetlands, Whaling
signed, but not ratified: none of the selected agreements
Geography—note: only country in Central America without a coastline on the North Pacific Ocean

PEOPLE

Population: 301,270 (July 2008 est.)
Age structure:
0–14 years: 38.4% (male 58,987/female 56,674)
15–64 years: 58.1% (male 88,521/female 86,450)
65 years and over: 3.5% (male 5,095/female 5,543) (2008 est.)
Median age:
total: 20.1 years
male: 20 years
female: 20.3 years (2008 est.)
Population growth rate: 2.207% (2008 est.)
Birth rate: 27.84 births/1,000 population (2008 est.)

Death rate: 5.77 deaths/1,000 population (2008 est.)
Net migration rate: NA
Sex ratio:
at birth: 1.05 male(s)/female
under 15 years: 1.04 male(s)/female
15–64 years: 1.02 male(s)/female
65 years and over: 0.92 male(s)/female
total population: 1.03 male(s)/female (2008 est.)
Infant mortality rate:
total: 23.65 deaths/1,000 live births
male: 26.35 deaths/1,000 live births
female: 20.81 deaths/1,000 live births (2008 est.)
Life expectancy at birth:
total population: 68.19 years
male: 66.39 years
female: 70.08 years (2008 est.)
Total fertility rate: 3.44 children born/woman (2008 est.)
HIV/AIDS—adult prevalence rate: 2.4% (2003 est.)
HIV/AIDS—people living with HIV/AIDS: 3,600 (2003 est.)
HIV/AIDS—deaths: fewer than 200 (2003 est.)
Major infectious diseases:
degree of risk: intermediate
food or waterborne diseases: bacterial diarrhea, hepatitis A, and typhoid fever
vectorborne diseases: dengue fever and malaria
water contact disease: leptospirosis (2008)
Nationality:
noun: Belizean(s)
adjective: Belizean
Ethnic groups: mestizo 48.7%, Creole 24.9%, Maya 10.6%, Garifuna 6.1%, other 9.7%
Religions: Roman Catholic 49.6%, Protestant 27% (Pentecostal 7.4%, Anglican 5.3%, Seventh-Day Adventist 5.2%, Mennonite 4.1%, Methodist 3.5%, Jehovah's Witnesses 1.5%), other 14%, none 9.4% (2000)
Languages: Spanish 46%, Creole 32.9%, Mayan dialects 8.9%, English 3.9% (official), Garifuna 3.4% (Carib), German 3.3%, other 1.4%, unknown 0.2% (2000 census)
Literacy: *definition:* age 15 and over can read and write
total population: 76.9%
male: 76.7%
female: 77.1% (2000 census)

GOVERNMENT

Country name:
conventional long form: none
conventional short form: Belize
former: British Honduras

Government type: parliamentary democracy

Capital: *name:* Belmopan
geographic coordinates: 17 15 N, 88 46 W
time difference: UTC-6 (1 hour behind Washington, DC during Standard Time)

Administrative divisions: 6 districts; Belize, Cayo, Corozal, Orange Walk, Stann Creek, Toledo

Independence: 21 September 1981 (from UK)

National holiday: Independence Day, 21 September (1981)

Constitution: 21 September 1981

Legal system: English law; has not accepted compulsory ICJ jurisdiction

Suffrage: 18 years of age; universal

Executive branch:
chief of state: Queen ELIZABETH II (since 6 February 1952); represented by Governor General Sir Colville YOUNG, Sr. (since 17 November 1993)
head of government: Prime Minister Dean BARROW (since 8 February 2008); Deputy Prime Minister Gaspar VEGA (since 12 February 2008)
cabinet: Cabinet appointed by the governor general on the advice of the prime minister
elections: the monarch is hereditary; governor general appointed by the monarch; following legislative elections, the leader of the majority party or the leader of the majority coalition is usually appointed prime minister by the governor general; prime minister recommends the deputy prime minister

Legislative branch: bicameral National Assembly consists of the Senate (12 seats; members appointed by the governor general—6 on the advice of the prime minister, 3 on the advice of the leader of the opposition, and 1 each on the advice of the Belize Council of Churches and Evangelical Association of Churches, the Belize Chamber of Commerce and Industry and the Belize Better Business Bureau, and the National Trade Union Congress and the Civil Society Steering Committee; to serve five-year terms) and the House of Representatives (31 seats; members are elected by direct popular vote to serve five-year terms)
elections: House of Representatives—last held 6 February 2008 (next to be held in 2013)
election results: percent of vote by party—NA; seats by party—UDP 25, PUP 6

Judicial branch: Supreme Court of Judicature (the chief justice is appointed by the governor general on the advice of the prime minister); Court of Appeal

Political parties and leaders: National Alliance for Belizean Rights or NABR;
National Reform Party or NRP [Cornelius DUECK]; People's National Party or PNP [Wil MAHEIA]; People's United Party or PUP [Said MUSA]; United Democratic Party or UDP [Dean BARROW]; Vision Inspired by the People or VIP [Paul MORGAN]; We the People Reform Movement or WTP [Hipolito BAUTISTA]

Political pressure groups and leaders: Society for the Promotion of Education and Research or SPEAR [Gustavo PERERA]; Association of Concerned Belizeans or ACB [David VASQUEZ]; National Trade Union Congress of Belize or NTUC/B [Rene GOMEZ]

International organization participation: ACP, C, Caricom, CDB, FAO, G-77, IADB, IAEA, IBRD, ICAO, ICCt, ICRM, IDA, IFAD, IFC, IFRCS, ILO, IMF, IMO, Interpol, IOC, IOM, ITU, ITUC, LAES, MIGA, NAM, OAS, OPANAL, OPCW, PCA, UN, UNCTAD, UNESCO, UNIDO, UPU, WCL, WHO, WIPO, WMO, WTO

Diplomatic representation in the US:
chief of mission: Ambassador (vacant)
chancery: 2535 Massachusetts Avenue NW, Washington, DC 20008
telephone: [1] (202) 332-9636
FAX: [1] (202) 332-6888
consulate(s) general: Los Angeles

Diplomatic representation from the US:
chief of mission: Ambassador Robert J. DIETER
embassy: Floral Park Road, Belmopan City, Cayo District
mailing address: P.O. Box 497, Belmopan City, Cayo District, Belize
telephone: [501] 822-4011
FAX: [501] 822-4012

Flag description: blue with a narrow red stripe along the top and the bottom edges; centered is a large white disk bearing the coat of arms; the coat of arms features a shield flanked by two workers in front of a mahogany tree with the related motto SUB UMBRA FLOREO (I Flourish in the Shade) on a scroll at the bottom, all encircled by a green garland

ECONOMY

Economy—overview: In this small, essentially private-enterprise economy, tourism is the number one foreign exchange earner followed by exports of marine products, citrus, cane sugar, bananas, and garments. The government's expansionary monetary and fiscal policies, initiated in September 1998, led to sturdy GDP growth averaging nearly 4% in 1999–2007. Oil discoveries in 2006 bolstered the economic growth in 2006 and 2007. Major concerns continue to be the sizable trade deficit and unsustainable foreign debt. In February 2007, the government restructured nearly all of its public external commercial debt, which will reduce interest payments and relieve liquidity concerns. A key short-term objective remains the reduction of poverty with the help of international donors.

GDP (purchasing power parity): $2.444 billion (2007 est.)

GDP (official exchange rate): $1.274 billion (2007 est.)

GDP—real growth rate: 2.2% (2007 est.)

GDP—per capita (PPP): $7,900 (2007 est.)

GDP—composition by sector:
agriculture: 21.3%
industry: 13.7%
services: 65% (2007 est.)

Labor force: 113,000
note: shortage of skilled labor and all types of technical personnel (2006 est.)

Labor force—by occupation:
agriculture: 22.5%
industry: 15.2%
services: 62.3% (2005 est.)

Unemployment rate: 9.4% (2006)

Population below poverty line: 33.5% (2002 est.)

Household income or consumption by percentage share:
lowest 10%: NA%
highest 10%: NA%

Inflation rate (consumer prices): 3% (2007 est.)

Investment (gross fixed): 22.6% of GDP (2007 est.)

Budget:
revenues: $307 million
expenditures: $344 million (2007 est.)

Agriculture—products: bananas, cacao, citrus, sugar; fish, cultured shrimp; lumber; garments

Industries: garment production, food processing, tourism, construction, oil

Industrial production growth rate: 0.5% (2007 est.)

Electricity—production: 200 million kWh (2007 est.)

Electricity—production by source:
fossil fuel: 59.9%
hydro: 40.1%
nuclear: 0%
other: 0% (2001)

Electricity—consumption: 162.8 million kWh (2005)

Electricity—exports: 0 kWh (2005)
Electricity—imports: 0 kWh (2005)
Oil—production: 2,413 bbl/day (2006)
Oil—consumption: 3,000 bbl/day (2006 est.)
Oil—exports: 1,960 bbl/day (2006)
Oil—imports: 6,754 bbl/day (2004)
Oil—proved reserves: 0 bbl (1 January 2006 est.)

Natural gas—production: 0 cu m (2005 est.)

Natural gas—consumption: 0 cu m (2005 est.)

Natural gas—exports: 0 cu m (2005 est.)

Natural gas—imports: 0 cu m (2005)

Natural gas—proved reserves: 0 cu m (1 January 2006 est.)

Current account balance: -$51 million (2007 est.)

Exports: $415 million f.o.b. (2007 est.)

Exports—commodities: sugar, bananas, citrus, clothing, fish products, molasses, wood

Exports—partners: US 33.9%, UK 33.6%, Cote d'Ivoire 3.7% (2006)

Imports: $641 million f.o.b. (2007 est.)

Imports—commodities: machinery and transport equipment, manufactured goods; fuels, chemicals, pharmaceuticals; food, beverages, tobacco

Imports—partners: US 35.7%, Mexico 13%, Cuba 7.7%, Guatemala 7.2%, China 4.3% (2006)

Economic aid—recipient: $12.91 million (2005)

Reserves of foreign exchange and gold: $109 million (31 December 2007 est.)

Debt—external: $1.2 billion (June 2005 est.)

Market value of publicly traded shares: $NA

Currency (code): Belizean dollar (BZD)

Currency code: BZD

Exchange rates: Belizean dollars per US dollar—2 (2007), 2 (2006), 2 (2005), 2 (2004), 2 (2003)

Fiscal year: 1 April—31 March

Telephones—main lines in use: 33,900 (2006)

Telephones—mobile cellular: 118,300 (2006)

Telephone system:

general assessment: above-average system; fixed-line teledensity of 12 per 100 persons; mobile-cellular telephone density of about 40 per 100 persons

domestic: trunk network depends primarily on microwave radio relay

international: country code—501; landing point for the Americas Region Caribbean Ring System (ARCOS-1) fiber-optic telecommunications submarine cable that provides links to South and Central America, parts of the Caribbean, and the US; satellite earth station—8 (Intelsat—2, unknown—6) (2007)

Radio broadcast stations: AM 1, FM 16, shortwave 0 (2006)

Radios: 133,000 (1997)

Television broadcast stations: 5 (2006)

Televisions: 41,000 (1997)

Internet country code: .bz

Internet hosts: 1,942 (2007)

Internet Service Providers (ISPs): 2 (2000)

Internet users: 34,000 (2006)

Airports: 44 (2007)

Airports—with paved runways:

total: 4

1,524 to 2,437 m: 1

914 to 1,523 m: 1

under 914 m: 2 (2007)

Airports—with unpaved runways:

total: 40

2,438 to 3,047 m: 1

914 to 1,523 m: 12

under 914 m: 27 (2007)

Roadways:

total: 2,872 km

paved: 488 km

unpaved: 2,384 km (2000)

Waterways: 825 km (navigable only by small craft) (2007)

Merchant marine:

total: 261 ships (1000 GRT or over) 940,852 GRT/1,275,111 DWT

by type: barge carrier 1, bulk carrier 36, cargo 190, chemical tanker 5, container 5, petroleum tanker 9, refrigerated cargo 8, roll on/roll off 6, specialized tanker 1

foreign-owned: 217 (China 107, Croatia 1, Cyprus 1, Estonia 1, Hong Kong 5, Iceland 1, Italy 4, Japan 2, South Korea 4, Latvia 14, Norway 3, Peru 1, Philippines 1, Russia 39, Singapore 3, Spain 2, Turkey 11, Ukraine 10, UAE 4, US 3) (2007)

Ports and terminals: Belize City, Big Creek

Military branches: Belize Defense Force (BDF): Army, BDF Air Wing, BDF Volunteer Guard (2007)

Military service age and obligation: 18 years of age for voluntary military service; laws allow for conscription only if volunteers are insufficient; conscription has never been implemented; volunteers typically outnumber available positions by 3:1 (2008)

Manpower available for military service:

males age 16–49: 74,605

females age 16–49: 72,926 (2008 est.)

Manpower fit for military service:

males age 16–49: 54,627

females age 16–49: 53,500 (2008 est.)

Manpower reaching militarily significant age annually:

males age 16–49: 3,580

females age 16–49: 3,449 (2008 est.)

Military expenditures—percent of GDP: 1.4% (2006)

Disputes—international: annual ministerial meetings under the OAS-initiated Agreement on the Framework for Negotiations and Confidence Building Measures continue to address Guatemalan land and maritime claims in Belize and Caribbean Sea; the Line of Adjacency created under the 2002 Differendum serves in lieu of the contiguous international boundary to control squatting in the sparsely inhabited rain forests of Belize's border region; Honduras claims Belizean-administered Sapodilla Cays in its constitution but agreed to a joint ecological park under the Differendum

Illicit drugs: transshipment point for cocaine; small-scale illicit producer of cannabis, primarily for local consumption; money-laundering activity related to narcotics trafficking and offshore sector

Background: Present day Benin was the site of Dahomey, a prominent West African kingdom that rose in the 15th century. The territory became a French Colony in 1872 and achieved independence on 1 August 1960, as the Republic of Benin. A succession of military governments ended in 1972 with the rise to power of Mathieu KEREKOU and the establishment of a government based on Marxist-Leninist principles. A move to representative government began in 1989. Two years later, free elections ushered in former Prime Minister Nicephore SOGLO as president, marking the first successful transfer of power in Africa from a dictatorship to a democracy. KEREKOU was returned to power by elections held in 1996 and 2001, though some irregularities were alleged. KEREKOU stepped down at the end of his second term in 2006 and was succeeded by Thomas YAYI Boni, a political outsider and independent. YAYI has begun a high profile fight against corrup-

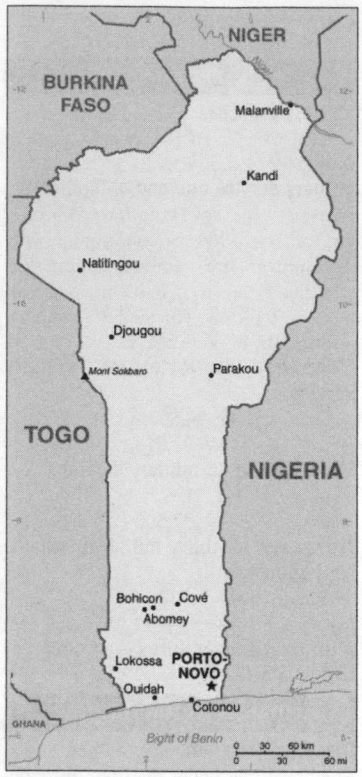

tion and has strongly promoted accelerating Benin's economic growth.

GEOGRAPHY

Location: Western Africa, bordering the Bight of Benin, between Nigeria and Togo
Geographic coordinates: 9 30 N, 2 15 E
Map references: Africa
Area:
total: 112,620 sq km
land: 110,620 sq km
water: 2,000 sq km
Area—comparative: slightly smaller than Pennsylvania
Land boundaries:
total: 1,989 km
border countries: Burkina Faso 306 km, Niger 266 km, Nigeria 773 km, Togo 644 km
Coastline: 121 km
Maritime claims: *territorial sea:* 200 nm
Climate: tropical; hot, humid in south; semiarid in north
Terrain: mostly flat to undulating plain; some hills and low mountains
Elevation extremes:
lowest point: Atlantic Ocean 0 m
highest point: Mont Sokbaro 658 m
Natural resources: small offshore oil deposits, limestone, marble, timber
Land use:
arable land: 23.53%

permanent crops: 2.37%
other: 74.1% (2005)
Irrigated land: 120 sq km (2003)
Total renewable water resources: 25.8 cu km (2001)
Freshwater withdrawal (domestic/industrial/agricultural): *total:* 0.13 cu km/yr (32%/23%/45%)
per capita: 15 cu m/yr (2001)
Natural hazards: hot, dry, dusty harmattan wind may affect north from December to March
Environment—current issues: inadequate supplies of potable water; poaching threatens wildlife populations; deforestation; desertification
Environment—international agreements: *party to:* Biodiversity, Climate Change, Climate Change-Kyoto Protocol, Desertification, Endangered Species, Environmental Modification, Hazardous Wastes, Law of the Sea, Ozone Layer Protection, Ship Pollution, Wetlands, Whaling
signed, but not ratified: none of the selected agreements
Geography—note: sandbanks create difficult access to a coast with no natural harbors, river mouths, or islands

PEOPLE

Population: 8,294,941
note: estimates for this country explicitly take into account the effects of excess mortality due to AIDS; this can result in lower life expectancy, higher infant mortality, higher death rates, lower population growth rates, and changes in the distribution of population by age and sex than would otherwise be expected (July 2008 est.)
Age structure:
0–14 years: 43.6% (male 1,824,803/female 1,790,723)
15–64 years: 54% (male 2,210,525/female 2,268,138)
65 years and over: 2.4% (male 80,081/female 120,671) (2008 est.)
Median age:
total: 17.9 years
male: 17.5 years
female: 18.3 years (2008 est.)
Population growth rate: 2.619% (2008 est.)
Birth rate: 37.36 births/1,000 population (2008 est.)
Death rate: 11.67 deaths/1,000 population (2008 est.)
Net migration rate: 0.5 migrant(s)/1,000 population (2008 est.)
Sex ratio:
at birth: 1.03 male(s)/female
under 15 years: 1.02 male(s)/female
15–64 years: 0.97 male(s)/female
65 years and over: 0.66 male(s)/female

total population: 0.98 male(s)/female (2008 est.)
Infant mortality rate:
total: 76.19 deaths/1,000 live births
male: 80.58 deaths/1,000 live births
female: 71.66 deaths/1,000 live births (2008 est.)
Life expectancy at birth:
total population: 53.85 years
male: 52.67 years
female: 55.06 years (2008 est.)
Total fertility rate: 4.96 children born/woman (2008 est.)
HIV/AIDS—adult prevalence rate: 1.9% (2003 est.)
HIV/AIDS—people living with HIV/AIDS: 68,000 (2003 est.)
HIV/AIDS—deaths: 5,800 (2003 est.)
Major infectious diseases:
degree of risk: very high
food or waterborne diseases: bacterial and protozoal diarrhea, hepatitis A, and typhoid fever
vectorborne diseases: malaria and yellow fever
respiratory disease: meningococcal meningitis (2008)
Nationality:
noun: Beninese (singular and plural)
adjective: Beninese
Ethnic groups: Fon and related 39.2%, Adja and related 15.2%, Yoruba and related 12.3%, Bariba and related 9.2%, Peulh and related 7%, Ottamari and related 6.1%, Yoa-Lokpa and related 4%, Dendi and related 2.5%, other 1.6% (includes Europeans), unspecified 2.9% (2002 census)
Religions: Christian 42.8% (Catholic 27.1%, Celestial 5%, Methodist 3.2%, other Protestant 2.2%, other 5.3%), Muslim 24.4%, Vodoun 17.3%, other 15.5% (2002 census)
Languages: French (official), Fon and Yoruba (most common vernaculars in south), tribal languages (at least six major ones in north)
Literacy:
definition: age 15 and over can read and write
total population: 34.7%
male: 47.9%
female: 23.3% (2002 census)

GOVERNMENT

Country name:
conventional long form: Republic of Benin
conventional short form: Benin
local long form: Republique du Benin
local short form: Benin
former: Dahomey
Government type: republic
Capital: *name:* Porto-Novo (official capital)
geographic coordinates: 6 29 N, 2 37 E

time difference: UTC+1 (6 hours ahead of Washington, DC during Standard Time)
note: Cotonou (seat of government)
Administrative divisions: 12 departments; Alibori, Atakora, Atlantique, Borgou, Collines, Kouffo, Donga, Littoral, Mono, Oueme, Plateau, Zou
Independence: 1 August 1960 (from France)
National holiday: National Day, 1 August (1960)
Constitution: adopted by referendum 2 December 1990
Legal system: based on French civil law and customary law; has not accepted compulsory ICJ jurisdiction
Suffrage: 18 years of age; universal
Executive branch:
chief of state: President Thomas YAYI Boni (since 6 April 2006); note—the president is both the chief of state and head of government
head of government: President Thomas YAYI Boni (since 6 April 2006)
cabinet: Council of Ministers appointed by the president
elections: president elected by popular vote for a five-year term (eligible for a second term); runoff election held 19 March 2006 (next to be held in March 2011)
election results: Thomas YAYI Boni elected president; percent of vote— Thomas YAYI Boni 74.5%, Adrien HOUNGBEDJI 25.5%
Legislative branch: unicameral National Assembly or Assemblee Nationale (83 seats; members are elected by direct popular vote to serve four-year terms)
elections: last held 31 March 2007 (next to be held by March 2011)
election results: percent of vote by party— NA; seats by party—FCBE 35, ADD 20, PRD 10, other and independents 18
Judicial branch: Constitutional Court or Cour Constitutionnelle; Supreme Court or Cour Supreme; High Court of Justice
Political parties and leaders: Alliance for Dynamic Democracy or ADD; Alliance of Progress Forces or AFP; African Movement for Democracy and Progress or MADEP [Sefou FAGBO-HOUN]; Benin Renaissance or RB [Rosine SOGLO]; Democratic Renewal Party or PRD [Adrien HOUNGBEDJI]; Force Cowrie for an Emerging Benin or FCBE; Impulse for Progress and Democracy or IPD [Theophile NATA]; Key Force or FC [Lazare SÈHOUÉTO]; Movement for the People's Alternative or MAP [Olivier CAPO-CHICHI]; Rally for Democracy and Progress or RDP [Dominique HOUNGNINOU];

Social Democrat Party or PSD [Bruno AMOUSSOU]; Union for the Relief or UPR [Issa SALIFOU]; Union for Democracy and National Solidarity or UDS [Sacca LAFIA]
note: approximately 20 additional minor parties
Political pressure groups and leaders: NA
International organization participation: ACCT, ACP, AfDB, AU, ECOWAS, Entente, FAO, FZ, G-77, IAEA, IBRD, ICAO, ICCt, ICRM, IDA, IDB, IFAD, IFC, IFRCS, ILO, IMF, IMO, Interpol, IOC, IOM, IPU, ISO (correspondent), ITSO, ITU, ITUC, MIGA, MONUC, NAM, OIC, OIF, OPCW, PCA, UN, UNCTAD, UNESCO, UNIDO, UNMIL, UNMIS, UNOCI, UNWTO, UPU, WADB (regional), WAEMU, WCL, WCO, WFTU, WHO, WIPO, WMO, WTO
Diplomatic representation in the US:
chief of mission: Ambassador Cyrille Segbe OGUIN
chancery: 2124 Kalorama Road NW, Washington, DC 20008
telephone: [1] (202) 232-6656
FAX: [1] (202) 265-1996
Diplomatic representation from the US:
chief of mission: Ambassador Gayleatha B. BROWN
embassy: Rue Caporal Bernard Anani, Cotonou
mailing address: 01 B. P. 2012, Cotonou
telephone: [229] 21-30-06-50
FAX: [229] 21-30-03-84
Flag description: two equal horizontal bands of yellow (top) and red (bottom) with a vertical green band on the hoist side

ECONOMY

Economy—overview: The economy of Benin remains underdeveloped and dependent on subsistence agriculture, cotton production, and regional trade. Growth in real output has averaged around 5% in the past seven years, but rapid population growth has offset much of this increase. Inflation has subsided over the past several years. In order to raise growth still further, Benin plans to attract more foreign investment, place more emphasis on tourism, facilitate the development of new food processing systems and agricultural products, and encourage new information and communication technology. Specific projects to improve the business climate by reforms to the land tenure system, the commercial justice system, and the financial sector were included in Benin's $307 million Millennium Challenge Account grant signed in February 2006. The 2001

privatization policy continues in telecommunications, water, electricity, and agriculture though the government annulled the privatization of Benin's state cotton company in November 2007 after the discovery of irregularities in the bidding process. The Paris Club and bilateral creditors have eased the external debt situation, with Benin benefiting from a G8 debt reduction announced in July 2005, while pressing for more rapid structural reforms. An insufficient electrical supply continues to adversely affect Benin's economic growth though the government recently has taken steps to increase domestic power production.
GDP (purchasing power parity): $12.1 billion (2007 est.)
GDP (official exchange rate): $5.433 billion (2007 est.)
GDP—real growth rate: 4.2% (2007 est.)
GDP—per capita (PPP): $1,500 (2007 est.)
GDP—composition by sector:
agriculture: 33.2%
industry: 14.5%
services: 52.3% (2007 est.)
Labor force: 5.38 million (2007 est.)
Unemployment rate: NA%
Population below poverty line: 37.4% (2007 est.)
Household income or consumption by percentage share:
lowest 10%: 3.1%
highest 10%: 29% (2003)
Distribution of family income—Gini Index: 36.5 (2003)
Inflation rate (consumer prices): 2% (2007 est.)
Investment (gross fixed): 19.4% of GDP (2007 est.)
Budget:
revenues: $959.2 million
expenditures: $1.211 billion (2007 est.)
Agriculture—products: cotton, corn, cassava (tapioca), yams, beans, palm oil, peanuts, cashews; livestock
Industries: textiles, food processing, construction materials, cement
Industrial production growth rate: 4.5% (2007 est.)
Electricity—production: 105 million kWh (2005)
Electricity—production by source:
fossil fuel: 14.2%
hydro: 85.8%
nuclear: 0%
other: 0% (2001)
Electricity—consumption: 587 million kWh (2005)
Electricity—exports: 0 kWh (2005)
Electricity—imports: 595 million kWh (2005)

Oil—production: 0 bbl/day (2007)

Oil—consumption: 9,232 bbl/day (2007 est.)

Oil—exports: 0 bbl/day (2007)

Oil—imports: 16,830 bbl/day (2007 est.)

Oil—proved reserves: 8.21 million bbl (1 January 2006 est.)

Natural gas—production: 0 cu m (2005 est.)

Natural gas—consumption: 0 cu m (2005 est.)

Natural gas—exports: 0 cu m (2005 est.)

Natural gas—imports: 0 cu m (2005)

Natural gas—proved reserves: 1.086 billion cu m (1 January 2006 est.)

Current account balance: -$310 million (2007 est.)

Exports: $585 million f.o.b. (2007 est.)

Exports—commodities: cotton, cashews, shea butter, textiles, palm products, seafood

Exports—partners: China 20.9%, Indonesia 7.7%, India 7%, Netherlands 6.2%, Niger 5.7%, Togo 4.6%, Nigeria 4.3% (2006)

Imports: $1.095 billion f.o.b. (2007 est.)

Imports—commodities: foodstuffs, capital goods, petroleum products

Imports—partners: China 46.6%, France 7.5%, Thailand 6% (2006)

Economic aid—recipient: $374.7 million (2006)

Reserves of foreign exchange and gold: $1.209 billion (31 December 2007 est.)

Debt—external: $1.2 billion (2007)

Market value of publicly traded shares: $NA

Currency (code): Communaute Financiere Africaine franc (XOF); note—responsible authority is the Central Bank of the West African States

Currency code: XOF

Exchange rates: Communaute Financiere Africaine francs (XOF) per US dollar—493.51 (2007), 522.59 (2006), 527.47 (2005), 528.29 (2004), 581.2 (2003)

note: since 1 January 1999, the XOF franc has been pegged to the euro at a rate of 655.957 XOF francs per euro

Fiscal year: calendar year

COMMUNICATIONS

Telephones—main lines in use: 77,300 (2006)

Telephones—mobile cellular: 1.056 million (2006)

Telephone system:

general assessment: inadequate; fixed-line network is almost saturated with fixed-line teledensity stuck at a meager 1 per 100 persons; mobile-cellular telephone subscribership is increasing

domestic: fair system of open-wire, microwave radio relay, and cellular connections; four mobile-cellular providers

international: country code—229; landing point for the SAT-3/WASC fiber-optic submarine cable that provides connectivity to Europe and Asia; satellite earth stations—7 (Intelsat-Atlantic Ocean) (2007)

Radio broadcast stations: AM 1, FM 34, shortwave 1 (2007)

Radios: 660,000 (2000)

Television broadcast stations: 6 (2007)

Televisions: 66,000 (2000)

Internet country code: .bj

Internet hosts: 798 (2007)

Internet Service Providers (ISPs): 4 (2002)

Internet users: 700,000 (2006)

TRANSPORTATION

Airports: 5 (2007)

Airports—with paved runways: *total:* 1 *1,524 to 2,437 m:* 1 (2007)

Airports—with unpaved runways:

total: 4

2,438 to 3,047 m: 1

1,524 to 2,437 m: 1

914 to 1,523 m: 2 (2007)

Railways: *total:* 758 km

narrow gauge: 758 km 1.000-m gauge (2006)

Roadways:

total: 16,000 km

paved: 1,400 km

unpaved: 14,600 km (2006)

Waterways: 150 km (on River Niger along northern border) (2005)

Ports and terminals: Cotonou

MILITARY

Military branches: Benin Armed Forces (FAB): Army (l'Arme de Terre), Benin Navy (Forces Navales Beninois, FNB), Benin People's Air Force (Force Aerienne Populaire de Benin, FAPB) (2008)

Military service age and obligation: 21 years of age for compulsory and voluntary military service; in practice, volunteers may be taken at the age of 18; both sexes are eligible for military service; conscript tour of duty—18 months (2006)

Manpower available for military service:

males age 16–49: 1,908,457

females age 16–49: 1,882,421 (2008 est.)

Manpower fit for military service:

males age 16–49: 1,173,742

females age 16–49: 1,162,113 (2008 est.)

Manpower reaching militarily significant age annually:

males age 16–49: 96,230

females age 16–49: 94,436 (2008 est.)

Military expenditures—percent of GDP: 1.7% (2006)

TRANSNATIONAL ISSUES

Disputes—international: two villages remain in dispute along the border with Burkina Faso; Benin accused Burkina Faso of moving boundary pillars; much of Benin-Niger boundary, including tripoint with Nigeria, remains undemarcated; in 2005, Nigeria ceded thirteen villages to Benin, but border relations remain strained by rival gang clashes; Benin and Togo announced plans in 2006 to construct a joint hydroelectric dam on the Mona River at the southern end of the border

Refugees and internally displaced persons: *refugees (country of origin):* 9,444 (Togo) (2007)

Illicit drugs: transshipment point used by Nigerian traffickers for narcotics destined for Western Europe; vulnerable to money laundering due to poorly enforced financial regulations

---

BERMUDA

INTRODUCTION

Background: Bermuda was first settled in 1609 by shipwrecked English colonists headed for Virginia. Tourism to the island to escape North American winters first developed in Victorian times. Tourism continues to be important to the island's economy, although international business has overtaken it in recent years.

Bermuda has developed into a highly successful offshore financial center. Although a referendum on independence from the UK was soundly defeated in 1995, the present government has reopened debate on the issue.

GEOGRAPHY

Location: North America, group of islands in the North Atlantic Ocean, east of South Carolina (US)

Geographic coordinates: 32 20 N, 64 45 W

Map references: North America

Area:

total: 53.3 sq km

land: 53.3 sq km

water: 0 sq km

Area—comparative: about one-third the size of Washington, DC

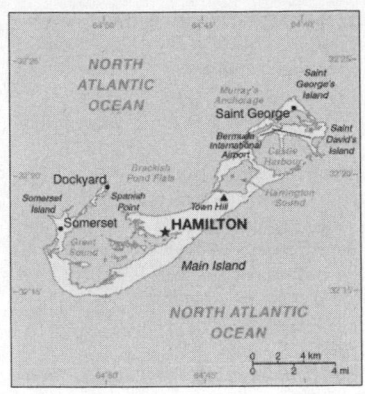

Land boundaries: 0 km
Coastline: 103 km
Maritime claims:
territorial sea: 12 nm
exclusive fishing zone: 200 nm
Climate: subtropical; mild, humid; gales, strong winds common in winter
Terrain: low hills separated by fertile depressions
Elevation extremes:
lowest point: Atlantic Ocean 0 m
highest point: Town Hill 76 m
Natural resources: limestone, pleasant climate fostering tourism
Land use:
arable land: 20%
permanent crops: 0%
other: 80% (55% developed, 45% rural/open space) (2005)
Irrigated land: NA
Natural hazards: hurricanes (June to November)
Environment—current issues: sustainable development
Geography—note: consists of about 138 coral islands and islets with ample rainfall, but no rivers or freshwater lakes; some land was leased by US Government from 1941 to 1995

PEOPLE

Population: 66,536 (July 2008 est.)
Age structure:
0–14 years: 18% (male 6,055/female 5,954)
15–64 years: 69.1% (male 22,795/female 23,189)
65 years and over: 12.8% (male 3,728/female 4,815) (2008 est.)
Median age:
total: 41 years
male: 40.1 years
female: 41.8 years (2008 est.)
Population growth rate: 0.546% (2008 est.)
Birth rate: 11.15 births/1,000 population (2008 est.)
Death rate: 7.98 deaths/1,000 population (2008 est.)

Net migration rate: 2.28 migrant(s)/1,000 population (2008 est.)
Sex ratio:
at birth: 1.02 male(s)/female
under 15 years: 1.02 male(s)/female
15–64 years: 0.98 male(s)/female
65 years and over: 0.77 male(s)/female
total population: 0.96 male(s)/female (2008 est.)
Infant mortality rate:
total: 7.87 deaths/1,000 live births
male: 9.31 deaths/1,000 live births
female: 6.4 deaths/1,000 live births (2008 est.)
Life expectancy at birth:
total population: 78.3 years
male: 76.15 years
female: 80.48 years (2008 est.)
Total fertility rate: 1.88 children born/woman (2008 est.)
HIV/AIDS—adult prevalence rate: 0.297% (2005)
HIV/AIDS—people living with HIV/AIDS: 163 (2005)
HIV/AIDS—deaths: 392 (2005)
Nationality:
noun: Bermudian(s)
adjective: Bermudian
Ethnic groups: black 54.8%, white 34.1%, mixed 6.4%, other races 4.3%, unspecified 0.4% (2000 census)
Religions: Anglican 23%, Roman Catholic 15%, African Methodist Episcopal 11%, other Protestant 18%, other 12%, unaffiliated 6%, unspecified 1%, none 14% (2000 census)
Languages: English (official), Portuguese
Literacy: *definition:* age 15 and over can read and write
total population: 98%
male: 98%
female: 99% (2005 est.)

GOVERNMENT

Country name:
conventional long form: none
conventional short form: Bermuda
former: Somers Islands
Dependency status: overseas territory of the UK
Government type: parliamentary; self-governing territory
Capital: *name:* Hamilton
geographic coordinates: 32 17 N, 64 47 W
time difference: UTC-4 (1 hour ahead of Washington, DC during Standard Time)
daylight saving time: +1hr, begins second Sunday in March; ends first Sunday in November
Administrative divisions: 9 parishes and 2 municipalities*; Devonshire, Hamilton, Hamilton*, Paget, Pembroke, Saint George*, Saint George's, Sandys, Smith's, Southampton, Warwick

Independence: none (overseas territory of the UK)
National holiday: Bermuda Day, 24 May
Constitution: 8 June 1968; amended 1989 and 2003
Legal system: English law
Suffrage: 18 years of age; universal
Executive branch:
chief of state: Queen ELIZABETH II (since 6 February 1952); represented by Governor Sir Richard GOZNEY (since 12 December 2007)
head of government: Premier Ewart BROWN (since 30 October 2006); Deputy Premier Paula COX
cabinet: Cabinet nominated by the premier, appointed by the governor
elections: the monarch is hereditary; governor appointed by the monarch; following legislative elections, the leader of the majority party or the leader of the majority coalition is usually appointed premier by the governor
Legislative branch: bicameral Parliament consists of the Senate (11 seats; members appointed by the governor, the premier, and the opposition) and the House of Assembly (36 seats; members are elected by popular vote to serve up to five-year terms)
elections: last general election held 18 December 2007 (next to be held not later than 2012)
election results: percent of vote by party—PLP 52.5%, UBP 47.3%; seats by party—PLP 22, UBP 14
Judicial branch: Supreme Court; Court of Appeal; Magistrate Courts
Political parties and leaders: Progressive Labor Party or PLP [Ewart BROWN]; United Bermuda Party or UBP [Kim SWAN]
Political pressure groups and leaders: Bermuda Employer's Union [Eddie SAINTS]; Bermuda Industrial Union or BIU [Derrick BURGESS]; Bermuda Public Services Union or BPSU [Ed BALL]; Bermuda Union of Teachers [Michael CHARLES]
International organization participation: Caricom (associate), Interpol (sub-bureau), IOC, ITUC, UPU, WCO
Diplomatic representation in the US: none (overseas territory of the UK)
Diplomatic representation from the US:
chief of mission: Consul General Gregory W. SLAYTON
consulate(s) general: Crown Hill, 16 Middle Road, Devonshire DVO3
mailing address: P. O. Box HM325, Hamilton HMBX; American Consulate General Hamilton, US Department of State, 5300 Hamilton Place, Washington, DC 20520-5300
telephone: [1] (441) 295-1342

FAX: [1] (441) 295-1592, [1] (441) 296-9233

Flag description: red, with the flag of the UK in the upper hoist-side quadrant and the Bermudian coat of arms (white and green shield with a red lion holding a scrolled shield showing the sinking of the ship Sea Venture off Bermuda in 1609) centered on the outer half of the flag

ECONOMY

Economy—overview: Bermuda enjoys the third highest per capita income in the world, more than 50% higher than that of the US. Its economy is primarily based on providing financial services for international business and luxury facilities for tourists. A number of reinsurance companies relocated to the island following the 11 September 2001 attacks and again after Hurricane Katrina in August 2005, contributing to the expansion of an already robust international business sector. Bermuda's tourism industry—which derives over 80% of its visitors from the US—continues to struggle but remains the island's number two industry. Most capital equipment and food must be imported. Bermuda's industrial sector is small, although construction continues to be important; the average cost of a house in June 2003 had risen to $976,000. Agriculture is limited with only 20% of the land being arable.

GDP (purchasing power parity): $4.5 billion (2004 est.)

GDP (official exchange rate): $NA

GDP—real growth rate: 4.6% (2004 est.)

GDP—per capita (PPP): $69,900 (2004 est.)

GDP—composition by sector:
agriculture: 1%
industry: 10%
services: 89% (2002 est.)

Labor force: 38,360 (2004)

Labor force—by occupation: agriculture and fishing 3%, laborers 17%, clerical 19%, professional and technical 21%, administrative and managerial 15%, sales 7%, services 19% (2004 est.)

Unemployment rate: 2.1% (2004 est.)

Population below poverty line: 19% (2000)

Household income or consumption by percentage share:
lowest 10%: NA%
highest 10%: NA%

Inflation rate (consumer prices): 2.8% (November 2005)

Budget: *revenues:* $738 million
expenditures: $665 million (FY04/05)

Agriculture—products: bananas, vegetables, citrus, flowers; dairy products, honey

Industries: international business, tourism, light manufacturing

Industrial production growth rate: NA%

Electricity—production: 618 million kWh (2005)

Electricity—production by source:
fossil fuel: 100%
hydro: 0%
nuclear: 0%
other: 0% (2001)

Electricity—consumption: 574.8 million kWh (2005)

Electricity—exports: 0 kWh (2005)

Electricity—imports: 0 kWh (2005)

Oil—production: 0 bbl/day (2005)

Oil—consumption: 4,400 bbl/day (2005 est.)

Oil—exports: 0 bbl/day (2005)

Oil—imports: 4,250 bbl/day (2004)

Oil—proved reserves: 0 bbl (1 January 2006 est.)

Natural gas—production: 0 cu m (2005 est.)

Natural gas—consumption: 0 cu m (2005 est.)

Natural gas—exports: 0 cu m (2005 est.)

Natural gas—imports: 0 cu m (2005)

Natural gas—proved reserves: 0 cu m (1 January 2006 est.)

Exports: $763 million (2006)

Exports—commodities: reexports of pharmaceuticals

Exports—partners: Spain 35.6%, UK 15.9%, Brazil 9.2%, Sweden 7.5% (2006)

Imports: $1.162 billion (2006)

Imports—commodities: clothing, fuels, machinery and transport equipment, construction materials, chemicals, food and live animals

Imports—partners: US 71.8%, Venezuela 6.9%, Canada 6.6% (2006)

Economic aid—recipient: $90,000 (2004)

Debt—external: $160 million (FY99/00)

Stock of direct foreign investment—at home: $NA

Stock of direct foreign investment—abroad: $NA

Market value of publicly traded shares: $2.125 billion (2005)

Currency (code): Bermudian dollar (BMD)

Currency code: BMD

Exchange rates: Bermudian dollar per US dollar—1.0000 (fixed rate pegged to the US dollar)

Fiscal year: 1 April—31 March

COMMUNICATIONS

Telephones—main lines in use: 57,700 (2006)

Telephones—mobile cellular: 60,100 (2006)

Telephone system:
general assessment: good
domestic: fully automatic digital telephone system; fiber optic trunk lines
international: country code—1-441; landing point for the Atlantica-1 telecommunications submarine cable that extends from the US to Brazil; satellite earth stations—3 (2007)

Radio broadcast stations: AM 5, FM 3, shortwave 0 (2005)

Radios: 82,000 (1997)

Television broadcast stations: 3 (2005)

Televisions: 66,000 (1997)

Internet country code: .bm

Internet hosts: 2,949 (2007)

Internet Service Providers (ISPs): 20 (2000)

Internet users: 42,000 (2005)

TRANSPORTATION

Airports: 1 (2007)

Airports—with paved runways:
total: 1
2,438 to 3,047 m: 1 (2007)

Roadways: *total:* 447 km
paved: 447 km
note: public roads—225 km; private roads—222 km (2002)

Merchant marine:
total: 133 ships (1000 GRT or over) 8,366,999 GRT/8,615,385 DWT
by type: bulk carrier 24, container 22, liquefied gas 30, passenger 23, passenger/cargo 5, petroleum tanker 15, refrigerated cargo 10, roll on/roll off 4
foreign-owned: 126 (Australia 4, Belgium 3, China 10, France 1, Germany 21, Greece 3, Hong Kong 4, Ireland 1, Israel 3, Japan 1, Nigeria 11, Norway 5, Singapore 1, Sweden 15, UK 20, US 23)
registered in other countries: 50 (Bahamas 12, Croatia 2, Marshall Islands 5, Philippines 31) (2007)

Ports and terminals: Hamilton, Saint George

MILITARY

Military branches: Bermuda Regiment (2008)

Military service age and obligation: 18–23 years of age; eligible men required to register for conscription as needed into the Bermuda Regiment, which is largely voluntary; term of service 39 months (2007)

Manpower available for military service:
males age 16–49: 15,623 (2008 est.)

Manpower fit for military service:
males age 16–49: 12,682 (2008 est.)

Manpower reaching militarily significant age annually:
males age 16–49: 426 (2008 est.)

Military expenditures—percent of GDP: 0.11% (2005 est.)
Military—note: defense is the responsibility of the UK

Disputes—international: none

BHUTAN

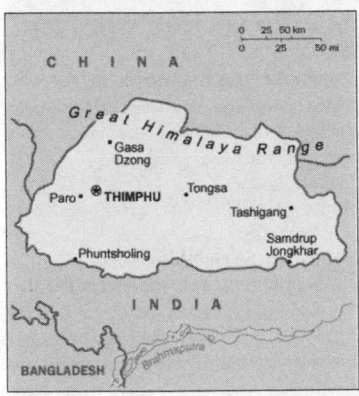

INTRODUCTION

Background: In 1865, Britain and Bhutan signed the Treaty of Sinchulu, under which Bhutan would receive an annual subsidy in exchange for ceding some border land to British India. Under British influence, a monarchy was set up in 1907; three years later, a treaty was signed whereby the British agreed not to interfere in Bhutanese internal affairs and Bhutan allowed Britain to direct its foreign affairs. This role was assumed by independent India after 1947. Two years later, a formal Indo-Bhutanese accord returned the areas of Bhutan annexed by the British, formalized the annual subsidies the country received, and defined India's responsibilities in defense and foreign relations. A refugee issue of over 100,000 Bhutanese in Nepal remains unresolved; 90% of the refugees are housed in seven United Nations Office of the High Commissioner for Refugees (UNHCR) camps. In March 2005, King Jigme Singye WANGCHUCK unveiled the government's draft constitution—which would introduce major democratic reforms—and pledged to hold a national referendum for its approval. In December 2006, the King abdicated the throne to his son, Jigme Khesar Namgyel WANGCHUCK, in order to give him experience as head of state before the democratic transition. In early 2007, India and Bhutan renegotiated their treaty to allow Bhutan greater autonomy in conducting its foreign policy, although Thimphu continues to coordinate policy decisions in this area with New Delhi. In July 2007, seven ministers of Bhutan's ten-member cabinet resigned to join the political process, leaving the remaining cabinet to act as a caretaker regime until a new government assumes power following parliamentary elections. Bhutan will complete its transition to full democracy in 2008, when its first fully democratic elections to a new parliament—expected to be completed by March 2008—and a concomitant referendum on the draft constitution will take place.

GEOGRAPHY

Location: Southern Asia, between China and India
Geographic coordinates: 27 30 N, 90 30 E
Map references: Asia
Area:
total: 47,000 sq km
land: 47,000 sq km
water: 0 sq km
Area—comparative: about one-half the size of Indiana
Land boundaries:
total: 1,075 km
border countries: China 470 km, India 605 km
Coastline: 0 km (landlocked)
Maritime claims: none (landlocked)
Climate: varies; tropical in southern plains; cool winters and hot summers in central valleys; severe winters and cool summers in Himalayas
Terrain: mostly mountainous with some fertile valleys and savanna
Elevation extremes:
lowest point: Drangme Chhu 97 m
highest point: Kula Kangri 7,553 m
Natural resources: timber, hydropower, gypsum, calcium carbonate
Land use:
arable land: 2.3%
permanent crops: 0.43%
other: 97.27% (2005)
Irrigated land: 400 sq km (2003)
Total renewable water resources: 95 cu km (1987)
Freshwater withdrawal (domestic/industrial/agricultural): *total:* 0.43 cu km/yr (5%/1%/94%)
per capita: 199 cu m/yr (2000)
Natural hazards: violent storms from the Himalayas are the source of the country's name, which translates as Land of the Thunder Dragon; frequent landslides during the rainy season
Environment—current issues: soil erosion; limited access to potable water
Environment—international agreements: *party to:* Biodiversity, Climate Change, Climate Change-Kyoto Protocol, Desertification, Endangered Species, Hazardous Wastes
signed, but not ratified: Law of the Sea
Geography—note: landlocked; strategic location between China and India; controls several key Himalayan mountain passes

PEOPLE

Population: 682,321
note: the Factbook population estimate is consistent with the first modern census of Bhutan, conducted in 2005; previous Factbook population estimates for this country, which were on the order of three times the total population reported here, were based on Bhutanese government publications that did not include the census (July 2008 est.)
Age structure:
0–14 years: 30.8% (male 107,360/female 103,093)
15–64 years: 63.7% (male 231,323/female 203,649)
65 years and over: 5.4% (male 19,561/female 17,335) (2008 est.)
Median age:
total: 23.5 years
male: 24.1 years
female: 22.8 years (2008 est.)
Population growth rate: 1.301% (2008 est.)
Birth rate: 20.56 births/1,000 population (2008 est.)
Death rate: 7.54 deaths/1,000 population (2008 est.)
Net migration rate: NA
Sex ratio:
at birth: 1.05 male(s)/female
under 15 years: 1.04 male(s)/female
15–64 years: 1.14 male(s)/female
65 years and over: 1.13 male(s)/female
total population: 1.11 male(s)/female (2008 est.)
Infant mortality rate:
total: 51.92 deaths/1,000 live births
male: 53.1 deaths/1,000 live births
female: 50.69 deaths/1,000 live births (2008 est.)

Life expectancy at birth:
total population: 65.53 years
male: 64.75 years
female: 66.35 years (2008 est.)
Total fertility rate: 2.48 children born/woman (2008 est.)
HIV/AIDS—adult prevalence rate: less than 0.1% (2001 est.)
HIV/AIDS—people living with HIV/AIDS: fewer than 100 (1999 est.)
HIV/AIDS—deaths: NA
Major infectious diseases:
degree of risk: intermediate
food or waterborne diseases: bacterial and protozoal diarrhea, hepatitis A, and typhoid fever
vectorborne diseases: malaria
water contact disease: leptospirosis (2008)
Nationality:
noun: Bhutanese (singular and plural)
adjective: Bhutanese
Ethnic groups: Bhote 50%, ethnic Nepalese 35% (includes Lhotsampas—one of several Nepalese ethnic groups), indigenous or migrant tribes 15%
Religions: Lamaistic Buddhist 75%, Indian- and Nepalese-influenced Hinduism 25%
Languages: Dzongkha (official), Bhotes speak various Tibetan dialects, Nepalese speak various Nepalese dialects
Literacy:
definition: age 15 and over can read and write
total population: 47%
male: 60%
female: 34% (2003 est.)

GOVERNMENT

Country name:
conventional long form: Kingdom of Bhutan
conventional short form: Bhutan
local long form: Druk Gyalkhap
local short form: Druk Yul
Government type: in transition to constitutional monarchy; special treaty relationship with India
Capital: *name:* Thimphu
geographic coordinates: 27 29 N, 89 36 E
time difference: UTC+6 (11 hours ahead of Washington, DC during Standard Time)
Administrative divisions: 20 districts (dzongkhag, singular and plural); Bumthang, Chhukha, Chirang, Daga, Gasa, Geylegphug, Ha, Lhuntshi, Mongar, Paro, Pemagatsel, Punakha, Samchi, Samdrup Jongkhar, Shemgang, Tashigang, Tashi Yangtse, Thimphu, Tongsa, Wangdi Phodrang
Independence: 1907 (became a unified kingdom under its first hereditary king)
National holiday: National Day (Ugyen WANGCHUCK became first hereditary king), 17 December (1907)

Constitution: none; note—a draft constitution was unveiled in March 2005 and is expected to be adopted by the new National Assembly in 2008
Legal system: based on Indian law and English common law; has not accepted compulsory ICJ jurisdiction
Suffrage: 18 years of age; universal
Executive branch:
chief of state: King Jigme Khesar Namgyel WANGCHUCK (since 14 December 2006); note—King Jigme Singye WANGCHUCK abdicated the throne on 14 December 2006 and his son immediately succeeded him
head of government: Prime Minister Jigme THINLEY (since 9 April 2008)
cabinet: Council of Ministers (Lhengye Shungtsog) nominated by the monarch, approved by the National Assembly; members serve fixed, five-year terms; note—there is also a Royal Advisory Council (Lodoi Tsokde), members nominated by the monarch
elections: the monarch is hereditary, but democratic reforms in July 1998 grant the National Assembly authority to remove the monarch with two-thirds vote; election of a new National Assembly occured in March 2008; the leader of the majority party is nominated as the prime minister
Legislative branch: new bicameral Parliament consists of the non-partisan National Council (25 seats; 20 members elected by each of the 20 electoral districts (dzongkhags) for four-year terms and 5 members nominated by the King); and the National Assembly (47 seats; members elected by direct, popular vote for five-year terms)
elections: National Council elections last held on 31 December 2007 and 29 January 2008 (next to be held by December 2012); National Assembly elections last held on 24 March 2008 (next to be held by March 2013)
election results: National Council—NA; National Assembly—percent of vote by party—DPT 67%, PDP 33%; seats by party—DPT 45, PDP 2
Judicial branch: Supreme Court of Appeal (the monarch); High Court (judges appointed by the monarch); note—the draft constitution establishes a Supreme Court, which will serve as chief court of appeal
Political parties and leaders: Bhutan Peace and Prosperity Party (Druk Phuensum Tshogpa) or DPT [Jigme THINLEY]; People's Democratic Party or PDP [Sangay NGEDUP]
Political pressure groups and leaders: Buddhist clergy; ethnic Nepalese organizations leading militant antigovernment

campaign; Indian merchant community; United Front for Democracy (exiled)
International organization participation: ADB, BIMSTEC, CP, FAO, G-77, IBRD, ICAO, IDA, IFAD, IFC, IMF, Interpol, IOC, IOM (observer), ISO (correspondent), ITSO, ITU, NAM, OPCW, SAARC, SACEP, UN, UNCTAD, UNESCO, UNIDO, UNWTO, UPU, WCO, WHO, WIPO, WMO, WTO (observer)
Diplomatic representation in the US: none; note—the Permanent Mission to the UN for Bhutan has consular jurisdiction in the US; address: 763 First Avenue, New York, NY 10017; telephone [1] (212) 682-2268; FAX [1] (212) 661-0551
consulate(s) general: New York
Diplomatic representation from the US: the US and Bhutan have no formal diplomatic relations, although informal contact is maintained between the Bhutanese and US Embassy in New Delhi (India)
Flag description: divided diagonally from the lower hoist-side corner; the upper triangle is yellow and the lower triangle is orange; centered along the dividing line is a large black and white dragon facing away from the hoist side

ECONOMY

Economy—overview: The economy, one of the world's smallest and least developed, is based on agriculture and forestry, which provide the main livelihood for more than 60% of the population. Agriculture consists largely of subsistence farming and animal husbandry. Rugged mountains dominate the terrain and make the building of roads and other infrastructure difficult and expensive. The economy is closely aligned with India's through strong trade and monetary links and dependence on India's financial assistance. The industrial sector is technologically backward, with most production of the cottage industry type. Most development projects, such as road construction, rely on Indian migrant labor. Model education, social, and environment programs are underway with support from multilateral development organizations. Each economic program takes into account the government's desire to protect the country's environment and cultural traditions. For example, the government, in its cautious expansion of the tourist sector, encourages visits by upscale, environmentally conscientious tourists. Detailed controls and uncertain policies in areas such as industrial licensing, trade, labor, and finance continue to hamper foreign investment. Hydropower exports to India had a major impact on growth in 2007.

GDP (purchasing power parity): $3.359 billion (2007 est.)

GDP (official exchange rate): $1.308 billion (2007 est.)

GDP—real growth rate: 22.4% (2007 est.)

GDP—per capita (PPP): $5,200 (2007 est.)

GDP—composition by sector:
agriculture: 24.7%
industry: 37.2%
services: 38.1% (2005)

Labor force: NA
note: major shortage of skilled labor

Labor force—by occupation: *agriculture:* 63%
industry: 6%
services: 31% (2004 est.)

Unemployment rate: 2.5% (2004)

Population below poverty line: 31.7% (2003)

Household income or consumption by percentage share:
lowest 10%: NA%
highest 10%: NA%

Inflation rate (consumer prices): 4.9% (2007 est.)

Budget:
revenues: $272 million
expenditures: $350 million
note: the government of India finances nearly three-fifths of Bhutan's budget expenditures (2005)

Public debt: 81.4% of GDP (2004)

Agriculture—products: rice, corn, root crops, citrus, foodgrains; dairy products, eggs

Industries: cement, wood products, processed fruits, alcoholic beverages, calcium carbide, tourism

Industrial production growth rate: 9.3% (1996 est.)

Electricity—production: 2 billion kWh (2005)

Electricity—production by source:
fossil fuel: 0.1%
hydro: 99.9%
nuclear: 0%
other: 0% (2001)

Electricity—consumption: 380 million kWh (2005)

Electricity—exports: 1.5 billion kWh (2005)

Electricity—imports: 20 million kWh (2005)

Oil—production: 0 bbl/day (2005)

Oil—consumption: 1,200 bbl/day (2005 est.)

Oil—exports: 0 bbl/day (2004)

Oil—imports: 1,138 bbl/day (2004)

Oil—proved reserves: 0 bbl (1 January 2006 est.)

Natural gas—production: 0 cu m (2005 est.)

Natural gas—consumption: 0 cu m (2005 est.)

Natural gas—exports: 0 cu m (2005 est.)

Natural gas—imports: 0 cu m (2005)

Natural gas—proved reserves: 0 cu m (1 January 2006 est.)

Current account balance: $116 million (2007 est.)

Exports: $350 million f.o.b. (2006)

Exports—commodities: electricity (to India), cardamom, gypsum, timber, handicrafts, cement, fruit, precious stones, spices

Exports—partners: India 67%, Hong Kong 29.8%, Thailand 1% (2006)

Imports: $320 million c.i.f. (2006)

Imports—commodities: fuel and lubricants, grain, aircraft, machinery and parts, vehicles, fabrics, rice

Imports—partners: India 60.7%, Japan 9.5%, Germany 5% (2006)

Economic aid—recipient: $90.02 million; note—substantial aid from India (2005)

Debt—external: $593 million (2004)

Market value of publicly traded shares: $NA

Currency (code): ngultrum (BTN); Indian rupee (INR)

Currency code: BTN; INR

Exchange rates: ngultrum per US dollar—41.487 (2007), 45.279 (2006), 44.101 (2005), 45.317 (2004), 46.583 (2003)
note: the ngultrum is pegged to the Indian rupee

Fiscal year: 1 July—30 June

COMMUNICATIONS

Telephones—main lines in use: 31,500 (2006)

Telephones—mobile cellular: 82,100 (2006)

Telephone system:
general assessment: urban towns and district headquarters have telecommunications services
domestic: very low teledensity; domestic service is very poor especially in rural areas; wireless service available since 2003
international: country code—975; international telephone and telegraph service via landline and microwave relay through India; satellite earth station—1 Intelsat (2007)

Radio broadcast stations: AM 0, FM 9, shortwave 1 (2007)

Radios: 37,000 (1997)

Television broadcast stations: 1 (2007)

Televisions: 11,000 (1997)

Internet country code: .bt

Internet hosts: 9,180 (2007)
Internet Service Providers (ISPs): NA
Internet users: 30,000 (2006)

TRANSPORTATION

Airports: 2 (2007)

Airports—with paved runways:
total: 1
1,524 to 2,437 m: 1 (2007)

Airports—with unpaved runways:
total: 1
914 to 1,523 m: 1 (2007)

Roadways:
total: 8,050 km
paved: 4,991 km
unpaved: 3,059 km (2003)

MILITARY

Military branches: Royal Bhutan Army (includes Royal Bodyguard and Royal Bhutan Police) (2008)

Military service age and obligation: 18 years of age for voluntary military service; no conscription (2008)

Manpower available for military service:
males age 16–49: 190,104
females age 16–49: 167,289 (2008 est.)

Manpower fit for military service:
males age 16–49: 146,063
females age 16–49: 131,193 (2008 est.)

Manpower reaching militarily significant age annually:
males age 16–49: 7,847
females age 16–49: 7,530 (2008 est.)

Military expenditures—percent of GDP: 1% (2005 est.)

TRANSNATIONAL ISSUES

Disputes—international: over 100,000 Bhutanese Lhotshampas (Hindus) have been confined in seven UN Office of the High Commissioner for Refugees camps since 1990; Bhutan cooperates with India to expel Indian Nagaland separatists; lacking any treaty describing the boundary, Bhutan and China continue negotiations to establish a boundary alignment to resolve substantial cartographic discrepancies, the largest of which lies in Bhutan's northwest

INTRODUCTION

Background: Bolivia, named after independence fighter Simon BOLIVAR, broke away from Spanish rule in 1825; much of its subsequent history has consisted of a series of nearly 200 coups and countercoups. Democratic civilian rule was established in 1982, but leaders have faced difficult problems of deep-seated poverty, social unrest, and illegal drug production. In December 2005, Bolivians elected Movement Toward Socialism leader Evo MORALES president—by the widest margin of any leader since the restoration of civilian rule in 1982—after he ran on a promise to change the country's traditional political class and empower the nation's poor majority. However, since taking office, his controversial strategies have exacerbated racial and economic tensions between the Amerindian populations of the Andean west and the non-indigenous communities of the eastern lowlands.

GEOGRAPHY

Location: Central South America, southwest of Brazil
Geographic coordinates: 17 00 S, 65 00 W
Map references: South America
Area:
total: 1,098,580 sq km
land: 1,084,390 sq km
water: 14,190 sq km
Area—comparative: slightly less than three times the size of Montana
Land boundaries:
total: 6,940 km
border countries: Argentina 832 km, Brazil 3,423 km, Chile 860 km, Paraguay 750 km, Peru 1,075 km
Coastline: 0 km (landlocked)
Maritime claims: none (landlocked)

Climate: varies with altitude; humid and tropical to cold and semiarid
Terrain: rugged Andes Mountains with a highland plateau (Altiplano), hills, lowland plains of the Amazon Basin
Elevation extremes:
lowest point: Rio Paraguay 90 m
highest point: Nevado Sajama 6,542 m
Natural resources: tin, natural gas, petroleum, zinc, tungsten, antimony, silver, iron, lead, gold, timber, hydropower
Land use:
arable land: 2.78%
permanent crops: 0.19%
other: 97.03% (2005)
Irrigated land: 1,320 sq km (2003)
Total renewable water resources: 622.5 cu km (2000)
Freshwater withdrawal (domestic/industrial/agricultural): *total:* 1.44 cu km/yr (13%/7%/81%)
per capita: 157 cu m/yr (2000)
Natural hazards: flooding in the northeast (March-April)
Environment—current issues: the clearing of land for agricultural purposes and the international demand for tropical timber are contributing to deforestation; soil erosion from overgrazing and poor cultivation methods (including slash-and-burn agriculture); desertification; loss of biodiversity; industrial pollution of water supplies used for drinking and irrigation
Environment—international agreements: *party to:* Biodiversity, Climate Change, Climate Change-Kyoto Protocol, Desertification, Endangered Species, Hazardous Wastes, Law of the Sea, Marine Dumping, Ozone Layer Protection, Ship Pollution, Tropical Timber 83, Tropical Timber 94, Wetlands
signed, but not ratified: Environmental Modification, Marine Life Conservation
Geography—note: landlocked; shares control of Lago Titicaca, world's highest navigable lake (elevation 3,805 m), with Peru

PEOPLE

Population: 9,247,816 (July 2008 est.)
Age structure:
0–14 years: 33.5% (male 1,580,887/female 1,519,960)
15–64 years: 61.8% (male 2,800,457/female 2,912,375)
65 years and over: 4.7% (male 192,701/female 241,436) (2008 est.)
Median age:
total: 22.6 years

male: 21.9 years
female: 23.3 years (2008 est.)
Population growth rate: 1.383% (2008 est.)
Birth rate: 22.31 births/1,000 population (2008 est.)
Death rate: 7.35 deaths/1,000 population (2008 est.)
Net migration rate: -1.14 migrant(s)/1,000 population (2008 est.)
Sex ratio:
at birth: 1.05 male(s)/female
under 15 years: 1.04 male(s)/female
15–64 years: 0.96 male(s)/female
65 years and over: 0.8 male(s)/female
total population: 0.98 male(s)/female (2008 est.)
Infant mortality rate:
total: 49.09 deaths/1,000 live births
male: 52.54 deaths/1,000 live births
female: 45.48 deaths/1,000 live births (2008 est.)
Life expectancy at birth:
total population: 66.53 years
male: 63.86 years
female: 69.33 years (2008 est.)
Total fertility rate: 2.67 children born/woman (2008 est.)
HIV/AIDS—adult prevalence rate: 0.1% (2003 est.)
HIV/AIDS—people living with HIV/AIDS: 4,900 (2003 est.)
HIV/AIDS—deaths: fewer than 500 (2003 est.)
Major infectious diseases:
degree of risk: high
food or waterborne diseases: bacterial diarrhea, hepatitis A, and typhoid fever
vectorborne diseases: dengue fever, malaria, and yellow fever
water contact disease: leptospirosis (2008)
Nationality:
noun: Bolivian(s)
adjective: Bolivian
Ethnic groups: Quechua 30%, mestizo (mixed white and Amerindian ancestry) 30%, Aymara 25%, white 15%
Religions: Roman Catholic 95%, Protestant (Evangelical Methodist) 5%
Languages: Spanish (official), Quechua (official), Aymara (official)
Literacy:
definition: age 15 and over can read and write
total population: 86.7%
male: 93.1%
female: 80.7% (2001 census)

GOVERNMENT

Country name:
conventional long form: Republic of Bolivia

conventional short form: Bolivia
local long form: Republica de Bolivia
local short form: Bolivia
Government type: republic
Capital: *name:* La Paz (administrative capital)
geographic coordinates: 16 30 S, 68 09 W
time difference: UTC-4 (1 hour ahead of Washington, DC during Standard Time)
note: Sucre (constitutional capital)
Administrative divisions: 9 departments (departamentos, singular—departamento); Beni, Chuquisaca, Cochabamba, La Paz, Oruro, Pando, Potosi, Santa Cruz, Tarija
Independence: 6 August 1825 (from Spain)
National holiday: Independence Day, 6 August (1825)
Constitution: 2 February 1967; revised in August 1994; possible referendum on new constitution to be held in 2008
Legal system: based on Spanish law and Napoleonic Code; has not accepted compulsory ICJ jurisdiction
Suffrage: 18 years of age, universal and compulsory (married); 21 years of age, universal and compulsory (single)
Executive branch:
chief of state: President Juan Evo MORALES Ayma (since 22 January 2006); Vice President Alvaro GARCIA Linera (since 22 January 2006); note—the president is both chief of state and head of government
head of government: President Juan Evo MORALES Ayma (since 22 January 2006); Vice President Alvaro GARCIA Linera (since 22 January 2006)
cabinet: Cabinet appointed by the president
elections: president and vice president elected on the same ticket by popular vote for a single five-year term; election last held 18 December 2005 (next to be held in 2010)
election results: Juan Evo MORALES Ayma elected president; percent of vote—Juan Evo MORALES Ayma 53.7%; Jorge Fernando QUIROGA Ramirez 28.6%; Samuel DORIA MEDINA Arana 7.8%; Michiaki NAGATANI Morishit 6.5%; Felipe QUISPE Huanca 2.2%; Guildo ANGULA Cabrera 0.7%
Legislative branch: bicameral National Congress or Congreso Nacional consists of Chamber of Senators or Camara de Senadores (27 seats; members are elected by proportional representation from party lists to serve five-year terms) and Chamber of Deputies or Camara de Diputados (130 seats; 70 members are directly elected from their districts and 60 are elected by proportional represen-

tation from party lists to serve five-year terms)
elections: Chamber of Senators and Chamber of Deputies—last held 18 December 2005 (next to be held in 2010)
election results: Chamber of Senators—percent of vote by party—NA; seats by party—PODEMOS 13, MAS 12, UN 1, MNR 1; Chamber of Deputies—percent of vote by party—NA; seats by party—MAS 73, PODEMOS 43, UN 8, MNR 6
Judicial branch: Supreme Court or Corte Suprema (judges appointed for 10-year terms by National Congress); District Courts (one in each department); provincial and local courts (to try minor cases); Constitutional Tribunal (five primary or titulares and five alternate or suplente magistrates appointed by Congress; to rule on constitutional issues); National Electoral Court (six members elected by Congress, Supreme Court, the President, and the political party with the highest vote in the last election for 4-year terms)
Political parties and leaders: Free Bolivia Movement or MBL [Franz BARRIOS]; Movement Toward Socialism or MAS [Juan Evo MORALES Ayma]; Movement Without Fear or MSM [Juan DEL GRANADO]; National Revolutionary Movement or MNR [Mirta QUEVEDO]; National Unity [Samuel DORIA MEDINA Arana]; Poder Democratico Nacional or PODEMOS [Jorge Fernando QUIROGA Ramirez]; Social Alliance [Rene JOAQUINO]
Political pressure groups and leaders: Cocalero groups; indigenous organizations; labor unions; Sole Confederation of Campesino Workers of Bolivia or CSUTCB
International organization participation: CAN, CSN, FAO, G-77, IADB, IAEA, IBRD, ICAO, ICCt, ICRM, IDA, IFAD, IFC, IFRCS, ILO, IMF, IMO, Interpol, IOC, IOM, IPU, ISO (correspondent), ITSO, ITU, LAES, LAIA, Mercosur (associate), MIGA, MINUSTAH, MONUC, NAM, OAS, OPANAL, OPCW, PCA, RG, UN, UNCTAD, UNESCO, UNFICYP, UNIDO, Union Latina, UNMEE, UNMIL, UNMIS, UNMISET, UNOCI, UNWTO, UPU, WCL, WCO, WFTU, WHO, WIPO, WMO, WTO
Diplomatic representation in the US:
chief of mission: Ambassador Gustavo GUZMAN Saldana
chancery: 3014 Massachusetts Avenue NW, Washington, DC 20008
telephone: [1] (202) 483-4410
FAX: [1] (202) 328-3712

consulate(s) general: Houston, Los Angeles, Miami, New York, Oklahoma City, San Francisco, Seattle, Washington, DC
Diplomatic representation from the US:
chief of mission: Ambassador Philip S. GOLDBERG
embassy: Avenida Arce 2780, Casilla 425, La Paz
mailing address: P. O. Box 425, La Paz; APO AA 34032
telephone: [591] (2) 216-8000
FAX: [591] (2) 216-8111
Flag description: three equal horizontal bands of red (top), yellow, and green with the coat of arms centered on the yellow band
note: similar to the flag of Ghana, which has a large black five-pointed star centered in the yellow band

ECONOMY

Economy—overview: Bolivia is one of the poorest and least developed countries in Latin America. Following a disastrous economic crisis during the early 1980s, reforms spurred private investment, stimulated economic growth, and cut poverty rates in the 1990s. The period 2003–05 was characterized by political instability, racial tensions, and violent protests against plans—subsequently abandoned—to export Bolivia's newly discovered natural gas reserves to large northern hemisphere markets. In 2005, the government passed a controversial hydrocarbons law that imposed significantly higher royalties and required foreign firms then operating under risk-sharing contracts to surrender all production to the state energy company, which was made the sole exporter of natural gas. The law also required that the state energy company regain control over the five companies that were privatized during the 1990s—a process that is still underway. In 2006, higher earnings for mining and hydrocarbons exports pushed the current account surplus to about 12% of GDP and the government's higher tax take produced a fiscal surplus after years of large deficits. Debt relief from the G8—announced in 2005—also has significantly reduced Bolivia's public sector debt burden. Private investment as a share of GDP, however, remains among the lowest in Latin America, and inflation reached double-digit levels in 2007.
GDP (purchasing power parity): $39.44 billion (2007 est.)
GDP (official exchange rate): $13.19 billion (2007 est.)
GDP—real growth rate: 4.2% (2007 est.)
GDP—per capita (PPP): $4,000 (2007 est.)

GDP—composition by sector:
agriculture: 14.5%
industry: 30.5%
services: 55% (2006 est.)
Labor force: 4.377 million (2007 est.)
Labor force—by occupation: *agriculture:* 40%
industry: 17%
services: 43% (2006 est.)
Unemployment rate: 7.5% in urban areas; widespread underemployment (2007 est.)
Population below poverty line: 60% (2006 est.)
Household income or consumption by percentage share:
lowest 10%: 0.3%
highest 10%: 47.2% (2002)
Distribution of family income—Gini index: 59.2 (2006)
Inflation rate (consumer prices): 8.7% (2007 est.)
Investment (gross fixed): 13.9% of GDP (2007 est.)
Budget:
revenues: $4.1 billion
expenditures: $5.495 billion (2007 est.)
Public debt: 46.6% of GDP (2007 est.)
Agriculture—products: soybeans, coffee, coca, cotton, corn, sugarcane, rice, potatoes; timber
Industries: mining, smelting, petroleum, food and beverages, tobacco, handicrafts, clothing
Industrial production growth rate: 1.1% (2007 est.)
Electricity—production: 5.293 billion kWh (2006)
Electricity—production by source:
fossil fuel: 44.4%
hydro: 54%
nuclear: 0%
other: 1.5% (2001)
Electricity—consumption: 3.385 billion kWh (2006)
Electricity—exports: 177,000 kWh (2005)
Electricity—imports: 18,000 kWh (2007)
Oil—production: 41,570 bbl/day (2007 est.)
Oil—consumption: 31,500 bbl/day (2007 est.)
Oil—exports: 18,500 bbl/day (2007 est.)
Oil—imports: 8,600 bbl/day (2007 est.)
Oil—proved reserves: 440.5 million bbl (1 January 2006 est.)
Natural gas—production: 12.74 billion cu m (2006 est.)
Natural gas—consumption: 1.486 billion cu m (2007 est.)
Natural gas—exports: 10.58 billion cu m (2006 est.)
Natural gas—imports: 0 cu m (2007 est.)

Natural gas—proved reserves: 651.8 billion cu m (1 January 2006 est.)
Current account balance: $1.758 billion (2007 est.)
Exports: $4.49 billion f.o.b. (2007 est.)
Exports—commodities: natural gas, soybeans and soy products, crude petroleum, zinc ore, tin
Exports—partners: Brazil 45.6%, US 10.8%, Argentina 9.2%, Colombia 6.8%, Japan 5.5%, South Korea 4.3% (2006)
Imports: $3.215 billion f.o.b. (2007 est.)
Imports—commodities: petroleum products, plastics, paper, aircraft and aircraft parts, prepared foods, automobiles, insecticides, soybeans
Imports—partners: Brazil 29.3%, Argentina 16%, Chile 12.1%, US 9.1%, Peru 8.1% (2006)
Economic aid—recipient: $582.9 million (2005 est.)
Reserves of foreign exchange and gold: $5.307 billion (31 October 2007)
Debt—external: $4.492 billion (31 December 2007 est.)
Stock of direct foreign investment—at home: $6.88 billion (31 December 2004)
Stock of direct foreign investment—abroad: $NA
Market value of publicly traded shares: $2.2 billion (2005)
Currency (code): boliviano (BOB)
Currency code: BOB
Exchange rates: bolivianos per US dollar—7.8616 (2007), 8.0159 (2006), 8.0661 (2005), 7.9363 (2004), 7.6592 (2003)
Fiscal year: calendar year

COMMUNICATIONS

Telephones—main lines in use: 646,300 (2005)
Telephones—mobile cellular: 2.421 million (2005)
Telephone system:
general assessment: privatization beginning in 1995; reliability has steadily improved; new subscribers face bureaucratic difficulties; most telephones are concentrated in La Paz and other cities; mobile-cellular telephone use expanding rapidly; fixed-line teledensity of 7 per 100 persons; mobile-cellular telephone density of 27 per 100 persons
domestic: primary trunk system, which is being expanded, employs digital microwave radio relay; some areas are served by fiber-optic cable; mobile cellular systems are being expanded
international: country code—591; satellite earth station—1 Intelsat (Atlantic Ocean) (2007)
Radio broadcast stations: AM 171, FM 73, shortwave 77 (1999)

Radios: 5.25 million (1997)
Television broadcast stations: 48 (1997)
Televisions: 900,000 (1997)
Internet country code: .bo
Internet hosts: 24,363 (2007)
Internet Service Providers (ISPs): 9 (2000)
Internet users: 580,000 (2006)

TRANSPORTATION

Airports: 1,061 (2007)
Airports—with paved runways:
total: 16
over 3,047 m: 4
2,438 to 3,047 m: 4
1,524 to 2,437 m: 5
914 to 1,523 m: 3 (2007)
Airports—with unpaved runways:
total: 1,045
over 3,047 m: 1
2,438 to 3,047 m: 4
1,524 to 2,437 m: 57
914 to 1,523 m: 183
under 914 m: 800 (2007)
Pipelines: gas 4,860 km; liquid petroleum gas 47 km; oil 2,475 km; refined products 1,589 km; unknown (oil/water) 247 km (2007)
Railways:
total: 3,504 km
narrow gauge: 3,504 km 1.000-m gauge (2006)
Roadways:
total: 62,479 km
paved: 3,749 km
unpaved: 58,730 km (2004)
Waterways: 10,000 km (commercially navigable) (2007)
Merchant marine:
total: 25 ships (1000 GRT or over) 73,877 GRT/110,148 DWT
by type: bulk carrier 1, cargo 12, carrier 1, passenger/cargo 2, petroleum tanker 9
foreign-owned: 9 (Argentina 1, China 1, Egypt 1, Iran 1, Italy 1, Singapore 1, Syria 1, Taiwan 1, Yemen 1) (2007)
Ports and terminals: Puerto Aguirre (inland port on the Paraguay/Parana waterway at the Bolivia/Brazil border); Bolivia has free port privileges in maritime ports in Argentina, Brazil, Chile, and Paraguay

MILITARY

Military branches: Bolivian Armed Forces: Bolivian Army (Ejercito Boliviano), Bolivian Navy (Armada Boliviana; includes marines), Bolivian Air Force (Fuerza Aerea Boliviana, FAB) (2008)
Military service age and obligation: 18 years of age for 12-month compulsory military service; when annual number of volunteers falls short of goal, compulsory

recruitment is effected, including conscription of boys as young as 14; 15–19 years of age for voluntary premilitary service, provides exemption from further military service (2008)

Manpower available for military service:
males age 16–49: 2,295,746
females age 16–49: 2,366,828 (2008 est.)
Manpower fit for military service:
males age 16–49: 1,600,219
females age 16–49: 1,815,514 (2008 est.)
Manpower reaching militarily significant age annually:
males age 16–49: 107,051
females age 16–49: 103,620 (2008 est.)
Military expenditures—percent of GDP: 1.9% (2006)

Disputes—international: Chile rebuffs Bolivia's reactivated claim to restore the Atacama corridor, ceded to Chile in 1884, offering instead unrestricted but not sovereign maritime access through Chile for Bolivian natural gas and other commodities

Illicit drugs: world's third-largest cultivator of coca (after Colombia and Peru) with an estimated 26,500 hectares under cultivation in August 2005, an 8% increase from 2004; transit country for Peruvian and Colombian cocaine destined for Brazil, Argentina, Chile, Paraguay, and Europe; cultivation steadily increasing despite eradication and alternative crop programs; money-laundering activity related to narcotics trade, especially along the borders with Brazil and Paraguay; major cocaine consumption

BOSNIA AND HERZEGOVINA

INTRODUCTION

Background: Bosnia and Herzegovina's declaration of sovereignty in October 1991 was followed by a declaration of independence from the former Yugoslavia on 3 March 1992 after a referendum boycotted by ethnic Serbs. The Bosnian Serbs—supported by neighboring Serbia and Montenegro—responded with armed resistance aimed at partitioning the republic along ethnic lines and joining Serb-held areas to form a "Greater Serbia." In March 1994, Bosniaks and Croats reduced the number of warring factions from three to two by signing an agreement creating a joint Bosniak/Croat Federation of Bosnia and Herzegovina. On 21 November 1995, in Dayton, Ohio, the warring parties initialed a peace agreement that brought to a halt three years of interethnic civil strife (the final agreement was signed in Paris on 14 December 1995). The Dayton Peace Accords retained Bosnia and Herzegovina's international boundaries and created a joint multi-ethnic and democratic government charged with conducting foreign, diplomatic, and fiscal policy. Also recognized was a second tier of government comprised of two entities roughly equal in size: the Bosniak/Croat Federation of Bosnia and Herzegovina and the Bosnian Serb-led Republika Srpska (RS). The Federation and RS governments were charged with overseeing most government functions. The Office of the High Representative (OHR) was established to oversee the implementation of the civilian aspects of the agreement. In 1995–96, a NATO-led international peacekeeping force (IFOR) of 60,000 troops served in Bosnia to implement and monitor the military aspects of the agreement. IFOR was succeeded by a smaller, NATO-led Stabilization Force (SFOR) whose mission was to deter renewed hostilities. European Union peacekeeping troops (EUFOR) replaced SFOR in December 2004; their mission is to maintain peace and stability throughout the country. EUFOR's mission changed from peacekeeping to civil policing in October 2007, with its presence reduced from nearly 7,000 to 2,500 troops.

GEOGRAPHY

Location: Southeastern Europe, bordering the Adriatic Sea and Croatia
Geographic coordinates: 44 00 N, 18 00 E
Map references: Europe
Area:
total: 51,129 sq km
land: 51,129 sq km
water: 0 sq km
Area—comparative: slightly smaller than West Virginia
Land boundaries:
total: 1,459 km
border countries: Croatia 932 km, Montenegro 225 km, Serbia 302 km
Coastline: 20 km
Maritime claims: no data available
Climate: hot summers and cold winters; areas of high elevation have short, cool summers and long, severe winters; mild, rainy winters along coast
Terrain: mountains and valleys
Elevation extremes:
lowest point: Adriatic Sea 0 m
highest point: Maglic 2,386 m
Natural resources: coal, iron ore, bauxite, copper, lead, zinc, chromite, cobalt, manganese, nickel, clay, gypsum, salt, sand, forests, hydropower
Land use:
arable land: 19.61%
permanent crops: 1.89%
other: 78.5% (2005)
Irrigated land: 30 sq km (2003)
Total renewable water resources: 37.5 cu km (2003)
Natural hazards: destructive earthquakes
Environment—current issues: air pollution from metallurgical plants; sites for disposing of urban waste are limited; water shortages and destruction of infrastructure because of the 1992–95 civil strife; deforestation
Environment—international agreements: *party to:* Air Pollution, Biodiversity, Climate Change, Climate Change-Kyoto Protocol, Desertification, Hazardous Wastes, Law of the Sea, Marine Life Conservation, Ozone Layer Protection, Wetlands
signed, but not ratified: none of the selected agreements
Geography—note: within Bosnia and Herzegovina's recognized borders, the country is divided into a joint Bosniak/Croat Federation (about 51% of the territory) and the Bosnian Serb-led Republika Srpska or RS (about 49% of the territory); the region called Herzegovina is contiguous to Croatia and Montenegro, and traditionally has been settled by an ethnic Croat majority in the west and an ethnic Serb majority in the east

PEOPLE

Population: 4,590,310 (July 2008 est.)

Age structure:

0–14 years: 14.7% (male 347,679/female 326,091)

15–64 years: 70.6% (male 1,634,053/female 1,606,341)

65 years and over: 14.7% (male 277,504/female 398,642) (2008 est.)

Median age:

total: 39.4 years

male: 38.2 years

female: 40.5 years (2008 est.)

Population growth rate: 0.666% (2008 est.)

Birth rate: 8.82 births/1,000 population (2008 est.)

Death rate: 8.54 deaths/1,000 population (2008 est.)

Net migration rate: 6.38 migrant(s)/1,000 population (2008 est.)

Sex ratio: *at birth:* 1.07 male(s)/female

under 15 years: 1.07 male(s)/female

15–64 years: 1.02 male(s)/female

65 years and over: 0.7 male(s)/female

total population: 0.97 male(s)/female (2008 est.)

Infant mortality rate:

total: 9.34 deaths/1,000 live births

male: 10.71 deaths/1,000 live births

female: 7.87 deaths/1,000 live births (2008 est.)

Life expectancy at birth:

total population: 78.33 years

male: 74.74 years

female: 82.19 years (2008 est.)

Total fertility rate: 1.24 children born/woman (2008 est.)

HIV/AIDS—adult prevalence rate: less than 0.1% (2001 est.)

HIV/AIDS—people living with HIV/AIDS: 900 (2003 est.)

HIV/AIDS—deaths: 100 (2001 est.)

Nationality:

noun: Bosnian(s), Herzegovinian(s)

adjective: Bosnian, Herzegovinian

Ethnic groups: Bosniak 48%, Serb 37.1%, Croat 14.3%, other 0.6% (2000)

note: Bosniak has replaced Muslim as an ethnic term in part to avoid confusion with the religious term Muslim—an adherent of Islam

Religions: Muslim 40%, Orthodox 31%, Roman Catholic 15%, other 14%

Languages: Bosnian, Croatian, Serbian

Literacy: *definition:* age 15 and over can read and write

total population: 96.7%

male: 99%

female: 94.4% (2000 est.)

GOVERNMENT

Country name:

conventional long form: none

conventional short form: Bosnia and Herzegovina

local long form: none

local short form: Bosna i Hercegovina

former: People's Republic of Bosnia and Herzegovina, Socialist Republic of Bosnia and Herzegovina

Government type: emerging federal democratic republic

Capital: *name:* Sarajevo

geographic coordinates: 43 52 N, 18 25 E

time difference: UTC+1 (6 hours ahead of Washington, DC during Standard Time)

daylight saving time: +1hr, begins last Sunday in March; ends last Sunday in October

Administrative divisions: 2 first-order administrative divisions and 1 internationally supervised district*—Brcko district (Brcko Distrikt)*, the Bosniak/Croat Federation of Bosnia and Herzegovina (Federacija Bosna i Hercegovina) and the Bosnian Serb-led Republika Srpska; *note*—Brcko district is in northeastern Bosnia and is an administrative unit under the sovereignty of Bosnia and Herzegovina; the district remains under international supervision

Independence: 1 March 1992 (from Yugoslavia; referendum for independence completed 1 March 1992; independence declared 3 March 1992)

National holiday: National Day, 25 November (1943)

Constitution: the Dayton Agreement, signed 14 December 1995 in Paris, included a new constitution now in force; *note*—each of the entities also has its own constitution

Legal system: based on civil law system; has not accepted compulsory ICJ jurisdiction

Suffrage: 18 years of age, universal

Executive branch:

chief of state: Chairman of the Presidency Haris SILAJDZIC (chairman since 6 March 2008; presidency member since 1 October 2006—Bosniak); other members of the three-member presidency rotating (every eight months): Nebojsa RADMANOVIC (presidency member since 1 October 2006—Serb); and Zeljko KOMSIC (presidency member since 1 October 2006—Croat)

head of government: Chairman of the Council of Ministers Nikola SPIRIC (since 11 January 2007)

cabinet: Council of Ministers nominated by the council chairman; approved by the National House of Representatives

elections: the three members of the presidency (one Bosniak, one Croat, one Serb) are elected by popular vote for a four-year term (eligible for a second term, but then ineligible for four years); the chairmanship rotates every eight months and resumes where it left off following each national election; election last held 1 October 2006 (next to be held in 2010); the chairman of the Council of Ministers is appointed by the presidency and confirmed by the National House of Representatives

election results: percent of vote—Nebojsa RADMANOVIC with 53.3% of the votes for the Serb seat; Zeljko KOMSIC received 39.6% of the votes for the Croat seat; Haris SILAJDZIC received 62.8% of the votes for the Bosniak seat

note: President of the Federation of Bosnia and Herzegovina: Borjana KRISTO (since 21 February 2007); Vice Presidents Spomenka MICIC (since NA 2007) and Mirsad KEBO (since NA 2007); President of the Republika Srpska: Rajko KUSMANOVIC (since 28 December 2007)

Legislative branch: bicameral Parliamentary Assembly or Skupstina consists of the national House of Representatives or Predstavnicki Dom (42 seats, 28 seats allocated for the Federation of Bosnia and Herzegovina and 14 seats for the Republika Srpska; members elected by popular vote on the basis of proportional representation, to serve four-year terms); and the House of Peoples or Dom Naroda (15 seats, 5 Bosniak, 5 Croat, 5 Serb; members elected by the Bosniak/Croat Federation's House of Representatives and the Republika Srpska's National Assembly to serve four-year terms); note—Bosnia's election law specifies four-year terms for the state and first-order administrative division entity legislatures

elections: national House of Representatives—elections last held 1 October 2006 (next to be held in 2010); House of Peoples—last constituted in January 2003 (next to be constituted in 2007)

election results: national House of Representatives—percent of vote by party/coalition—NA; seats by party/coalition—SDA 9, SBH 8, SNSD 7, SDP 5, SDS 3, HDZ-BH 3, HDZ1990 2, other 5; House of Peoples—percent of vote by party/coalition—NA; seats by party/coalition—NA

note: the Bosniak/Croat Federation has a bicameral legislature that consists of a House of Representatives (98 seats; members elected by popular vote to serve four-year terms); elections last held 1 October 2006 (next to be held in October 2010); percent of vote by party—NA; seats by party/coalition—

SDA 28, SBH 24, SDP 17, HDZ-BH 8, HDZ100 7, other 14; and a House of Peoples (58 seats—17 Bosniak, 17 Croat, 17 Serb, 7 other); last constituted December 2002; the Republika Srpska has a National Assembly (83 seats; members elected by popular vote to serve four-year terms); elections last held 1 October 2006 (next to be held in the fall of 2010); percent of vote by party—NA; seats by party/coalition—SNSD 41, SDS 17, PDP 8, DNS 4, SBH 4, SPRS 3, SDA 3, other 3; as a result of the 2002 constitutional reform process, a 28-member Republika Srpska Council of Peoples (COP) was established in the Republika Srpska National Assembly including eight Croats, eight Bosniaks, eight Serbs, and four members of the smaller communities

Judicial branch: BH Constitutional Court (consists of nine members: four members are selected by the Bosniak/Croat Federation's House of Representatives, two members by the Republika Srpska's National Assembly, and three non-Bosnian members by the president of the European Court of Human Rights); BH State Court (consists of nine judges and three divisions—Administrative, Appellate and Criminal—having jurisdiction over cases related to state-level law and appellate jurisdiction over cases initiated in the entities); a War Crimes Chamber opened in March 2005

note: the entities each have a Supreme Court; each entity also has a number of lower courts; there are 10 cantonal courts in the Federation, plus a number of municipal courts; the Republika Srpska has five municipal courts

Political parties and leaders: Alliance of Independent Social Democrats or SNSD [Milorad DODIK]; Bosnian Party or BOSS [Mirnes AJANOVIC]; Civic Democratic Party or GDS [Ibrahim SPAHIC]; Croat Christian Democratic Union of Bosnia and Herzegovina or HKDU [Marin TOPIC]; Croat Party of Rights or HSP [Zvonko JURISIC]; Croat Peasants Party or HSS [Marko TADIC]; Croatian Democratic Union of Bosnia and Herzegovina or HDZ-BH [Dragan COVIC]; Croatian Democratic Union 1990 or HDZ1990 [Bozo LJUBIC]; Croatian Democratic Union 100 or HDZ100; Croatian Peoples Union [Milenko BRKIC]; Democratic National Union or DNZ [Rifet DOLIC]; Democratic Peoples Alliance or DNS [Marko PAVIC]; Liberal Democratic Party or LDS [Rasim KADIC]; New Croat Initiative or NHI [Kresimir ZUBAK]; Party for Bosnia and Herzegovina or SBH [Haris SILAJDZIC]; Party for Democratic Action or SDA [Sulejman TIHIC]; Party of Democratic Progress or PDP [Mladen IVANIC]; Serb Democratic Party or SDS [Mladen BOSIC]; Serb Radical Party of the Republika Srpska or SRS-RS [Milanko MIHAJLICA]; Serb Radical Party-Dr. Vojislav Seselj or SRS-VS [Radislav KANJERIC]; Social Democratic Party of BIH or SDP [Zlatko LAGUMDZIJA]; Social Democratic Union or SDU [Sejfudin TOKIC]; Socialist Party of Republika Srpska or SPRS [Petar DJOKIC]

Political pressure groups and leaders: NA

International organization participation: BIS, CE, CEI, EAPC, EBRD, FAO, G-77, IAEA, IBRD, ICAO, ICCt, ICRM, IDA, IFAD, IFC, IFRCS, ILO, IMF, IMO, IMSO, Interpol, IOC, IOM, IPU, ISO, ITSO, ITU, ITUC, MIGA, NAM (observer), OAS (observer), OIC (observer), OPCW, OSCE, PFP, SECI, UN, UNCTAD, UNESCO, UNIDO, UNMEE, UNWTO, UPU, WHO, WIPO, WMO, WTO (observer)

Diplomatic representation in the US:
chief of mission: Ambassador Bisera TURKOVIC
chancery: 2109 E Street NW, Washington, DC 20037
telephone: [1] (202) 337-1500
FAX: [1] (202) 337-1502
consulate(s) general: Chicago, New York

Diplomatic representation from the US:
chief of mission: Ambassador Charles L. ENGLISH
embassy: Alipasina 43, 71000 Sarajevo
mailing address: use embassy street address
telephone: [387] (33) 445-700
FAX: [387] (33) 659-722
branch office(s): Banja Luka, Mostar

Flag description: a wide medium blue vertical band on the fly side with a yellow isosceles triangle abutting the band and the top of the flag; the remainder of the flag is medium blue with seven full five-pointed white stars and two half stars top and bottom along the hypotenuse of the triangle

ECONOMY

Economy—overview: Bosnia and Herzegovina ranked next to Macedonia as the poorest republic in the old Yugoslav federation. Although agriculture is almost all in private hands, farms are small and inefficient, and the republic traditionally is a net importer of food. The private sector is growing and foreign investment is slowly increasing, but government spending, at nearly 40% of adjusted GDP, remains unreasonably high. The interethnic warfare in Bosnia caused production to plummet by 80% from 1992 to 1995 and unemployment to soar. With an uneasy peace in place, output recovered in 1996–99 at high percentage rates from a low base; but output growth slowed in 2000–02. Part of the lag in output was made up in 2003–07 when GDP growth exceeded 5% per year. National-level statistics are limited and do not capture the large share of black market activity. The konvertibilna marka (convertible mark or BAM)—the national currency introduced in 1998—is pegged to the euro, and confidence in the currency and the banking sector has increased. Implementing privatization, however, has been slow, particularly in the Federation, although more successful in the Republika Srpska. Banking reform accelerated in 2001 as all the Communist-era payments bureaus were shut down; foreign banks, primarily from Western Europe, now control most of the banking sector. A sizeable current account deficit and high unemployment rate remain the two most serious macroeconomic problems. On 1 January 2006 a new value-added tax (VAT) went into effect. The VAT has been successful in capturing much of the gray market economy and has developed into a significant and predictable source of revenues for all layers of government. Bosnia and Herzegovina became a full member of the Central European Free Trade Agreement in September 2007. The country receives substantial reconstruction assistance and humanitarian aid from the international community but will have to prepare for an era of declining assistance.

GDP (purchasing power parity): $27.73 billion
note: Bosnia has a large informal sector that could also be as much as 50% of official GDP (2007 est.)

GDP (official exchange rate): $14.78 billion (2007 est.)

GDP—real growth rate: 5.8% (2007 est.)

GDP—per capita (PPP): $7,000 (2007 est.)

GDP—composition by sector:
agriculture: 10.2%
industry: 23.9%
services: 66% (2006 est.)

Labor force: 1.026 million (2001)

Labor force—by occupation: *agriculture:* NA%
industry: NA%
services: NA%

Unemployment rate: 45.5% official rate; grey economy may reduce actual unemployment to 25–30% (31 December 2004 est.)

Population below poverty line: 25% (2004 est.)

Household income or consumption by percentage share:
lowest 10%: 3.9%
highest 10%: 21.4% (2001)

Distribution of family income—Gini index: 26.2 (2001)

Inflation rate (consumer prices): 1.3% (2007 est.)

Budget:
revenues: $7.166 billion
expenditures: $7.094 billion (2007 est.)

Public debt: 34% of GDP (2007 est.)

Agriculture—products: wheat, corn, fruits, vegetables; livestock

Industries: steel, coal, iron ore, lead, zinc, manganese, bauxite, vehicle assembly, textiles, tobacco products, wooden furniture, tank and aircraft assembly, domestic appliances, oil refining

Industrial production growth rate: 6.7% (2007 est.)

Electricity—production: 12.22 billion kWh (2005)

Electricity—production by source:
fossil fuel: 53.5%
hydro: 46.5%
nuclear: 0%
other: 0% (2001)

Electricity—consumption: 8.574 billion kWh (2005)

Electricity—exports: 3.58 billion kWh (2005)

Electricity—imports: 2.174 billion kWh (2005)

Oil—production: 0 bbl/day (2005)

Oil—consumption: 26,000 bbl/day (2005 est.)

Oil—exports: 0 bbl/day (2004)

Oil—imports: 24,940 bbl/day (2004)

Oil—proved reserves: 0 bbl (1 January 2006 est.)

Natural gas—production: 0 cu m (2005 est.)

Natural gas—consumption: 383.6 million cu m (2005 est.)

Natural gas—exports: 0 cu m (2005 est.)

Natural gas—imports: 383.6 million cu m (2005)

Natural gas—proved reserves: 0 cu m (1 January 2006)

Current account balance: -$1.92 billion (2007 est.)

Exports: $4.243 billion f.o.b. (2007 est.)

Exports—commodities: metals, clothing, wood products

Exports—partners: Croatia 19.6%, Slovenia 16.8%, Italy 15.3%, Germany 12.3%, Austria 8.7%, Hungary 5.3% (2006)

Imports: $9.947 billion f.o.b. (2007 est.)

Imports—commodities: machinery and equipment, chemicals, fuels, foodstuffs

Imports—partners: Croatia 24%, Germany 14.5%, Slovenia 13.2%, Italy 10%, Austria 5.9%, Hungary 5.2% (2006)

Economic aid—recipient: $546.1 million (2005 est.)

Reserves of foreign exchange and gold: $4.525 billion (31 December 2007 est.)

Debt—external: $6.7 billion (31 December 2007 est.)

Market value of publicly traded shares: $NA

Currency (code): konvertibilna marka (convertible mark) (BAM)

Currency code: BAM

Exchange rates: konvertibilna maraka per US dollar—1.4419 (2007), 1.5576 (2006), 1.5727 (2005), 1.5752 (2004), 1.7329 (2003)
note: the convertible mark is pegged to the euro

Fiscal year: calendar year

COMMUNICATIONS

Telephones—main lines in use: 989,000 (2006)

Telephones—mobile cellular: 1.888 million (2006)

Telephone system:
general assessment: telephone and telegraph network needs modernization and expansion; many urban areas are below average as contrasted with services in other former Yugoslav republics
domestic: fixed-line teledensity is roughly 20 per 100 persons; mobile-cellular telephone density is about 22 per 100 persons
international: country code—387; no satellite earth stations (2006)

Radio broadcast stations: AM 8, FM 16, shortwave 1 (1998)

Radios: 940,000 (1997)

Television broadcast stations: 33 (plus 277 repeaters) (September 1995)

Televisions: NA

Internet country code: .ba

Internet hosts: 39,627 (2007)

Internet Service Providers (ISPs): 3 (2000)

Internet users: 950,000 (2006)

TRANSPORTATION

Airports: 28 (2007)

Airports—with paved runways:
total: 8
2,438 to 3,047 m: 4
1,524 to 2,437 m: 1
under 914 m: 3 (2007)

Airports—with unpaved runways:
total: 20
1,524 to 2,437 m: 1
914 to 1,523 m: 7
under 914 m: 12 (2007)

Heliports: 5 (2007)

Railways: *total:* 608 km
standard gauge: 608 km 1.435-m gauge (2006)

Roadways:
total: 21,846 km
paved: 11,425 km (4,714 km of interurban roads)
unpaved: 10,421 km (2006)

Waterways: Sava River (northern border) open to shipping but use limited (2006)

Ports and terminals: Bosanska Gradiska, Bosanski Brod, Bosanski Samac, and Brcko (all inland waterway ports on the Sava), Orasje

MILITARY

Military branches: Bosnia and Herzegovina Armed Forces (OSBiH): Army of Bosnia and Herzegovina, Air and Air Defense Forces of Bosnia and Herzegovina (Zrakoplovstvo i Protuzracna Obrana, ZPO) (2007)

Military service age and obligation: 17 years of age for voluntary military service in the Federation and in the Republika Srpska; conscription abolished January 2006; 4-month service obligation (2006)

Manpower available for military service: *males age 16–49:* 1,212,007
females age 16–49: 1,170,645 (2008 est.)

Manpower fit for military service:
males age 16–49: 996,225
females age 16–49: 962,927 (2008 est.)

Manpower reaching militarily significant age annually:
males age 16–49: 30,246
females age 16–49: 28,189 (2008 est.)

Military expenditures—percent of GDP: 4.5% (2005 est.)

TRANSNATIONAL ISSUES

Disputes—international: Bosnia and Herzegovina and Serbia have delimited most of their boundary, but sections along the Drina River remain in dispute; discussions continue with Croatia on several small disputed sections of the boundary related to maritime access that hinder final ratification of the 1999 border agreement

Refugees and internally displaced persons: *refugees (country of origin):* 7,269 (Croatia)
IDPs: 131,600 (Bosnian Croats, Serbs, and Muslims displaced in 1992–95 war) (2007)

Illicit drugs: increasingly a transit point for heroin being trafficked to Western Europe; minor transit point for marijuana; remains highly vulnerable to money-laundering activity given a primarily cash-based and unregulated economy, weak law enforcement, and instances of corruption

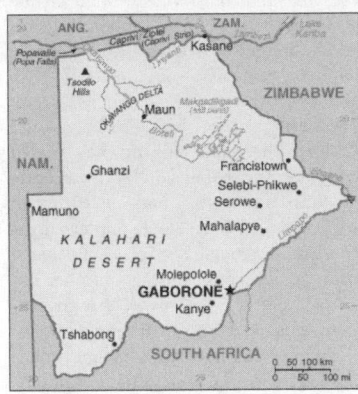

INTRODUCTION

Background: Formerly the British protectorate of Bechuanaland, Botswana adopted its new name upon independence in 1966. Four decades of uninterrupted civilian leadership, progressive social policies, and significant capital investment have created one of the most dynamic economies in Africa. Mineral extraction, principally diamond mining, dominates economic activity, though tourism is a growing sector due to the country's conservation practices and extensive nature preserves. Botswana has one of the world's highest known rates of HIV/AIDS infection, but also one of Africa's most progressive and comprehensive programs for dealing with the disease.

GEOGRAPHY

Location: Southern Africa, north of South Africa
Geographic coordinates: 22 00 S, 24 00 E
Map references: Africa
Area:
total: 600,370 sq km
land: 585,370 sq km
water: 15,000 sq km
Area—comparative: slightly smaller than Texas
Land boundaries:
total: 4,013 km
border countries: Namibia 1,360 km, South Africa 1,840 km, Zimbabwe 813 km
Coastline: 0 km (landlocked)
Maritime claims: none (landlocked)
Climate: semiarid; warm winters and hot summers
Terrain: predominantly flat to gently rolling tableland; Kalahari Desert in southwest
Elevation extremes:

lowest point: junction of the Limpopo and Shashe Rivers 513 m
highest point: Tsodilo Hills 1,489 m
Natural resources: diamonds, copper, nickel, salt, soda ash, potash, coal, iron ore, silver
Land use:
arable land: 0.65%
permanent crops: 0.01%
other: 99.34% (2005)
Irrigated land: 10 sq km (2003)
Total renewable water resources: 14.7 cu km (2001)
Freshwater withdrawal (domestic/industrial/agricultural): *total:* 0.19 cu km/yr (41%/18%/41%)
per capita: 107 cu m/yr (2000)
Natural hazards: periodic droughts; seasonal August winds blow from the west, carrying sand and dust across the country, which can obscure visibility
Environment—current issues: overgrazing; desertification; limited fresh water resources
Environment—international agreements: *party to:* Biodiversity, Climate Change, Climate Change-Kyoto Protocol, Desertification, Endangered Species, Hazardous Wastes, Law of the Sea, Ozone Layer Protection, Wetlands
signed, but not ratified: none of the selected agreements
Geography—note: landlocked; population concentrated in eastern part of the country

PEOPLE

Population: 1,842,323
note: estimates for this country explicitly take into account the effects of excess mortality due to AIDS; this can result in lower life expectancy, higher infant mortality, higher death rates, lower population growth rates, and changes in the distribution of population by age and sex than would otherwise be expected (July 2008 est.)
Age structure:
0–14 years: 35.2% (male 329,418/female 318,160)
15–64 years: 60.9% (male 566,239/female 556,286)
65 years and over: 3.9% (male 29,165/female 43,055) (2008 est.)
Median age:
total: 21.2 years
male: 21 years
female: 21.4 years (2008 est.)
Population growth rate: 1.434% (2008 est.)
Birth rate: 22.96 births/1,000 population (2008 est.)

Death rate: 14.02 deaths/1,000 population (2008 est.)
Net migration rate: 5.41 migrant(s)/1,000 population
note: there is an increasing flow of Zimbabweans into South Africa and Botswana in search of better economic opportunities (2008 est.)
Sex ratio:
at birth: 1.03 male(s)/female
under 15 years: 1.04 male(s)/female
15–64 years: 1.02 male(s)/female
65 years and over: 0.68 male(s)/female
total population: 1.01 male(s)/female (2008 est.)
Infant mortality rate:
total: 44.01 deaths/1,000 live births
male: 44.94 deaths/1,000 live births
female: 43.04 deaths/1,000 live births (2008 est.)
Life expectancy at birth:
total population: 50.16 years
male: 51.28 years
female: 49.02 years (2008 est.)
Total fertility rate: 2.66 children born/woman (2008 est.)
HIV/AIDS—adult prevalence rate: 37.3% (2003 est.)
HIV/AIDS—people living with HIV/AIDS: 350,000 (2003 est.)
HIV/AIDS—deaths: 33,000 (2003 est.)
Major infectious diseases:
degree of risk: high
food or waterborne diseases: bacterial diarrhea, hepatitis A, and typhoid fever
vectorborne disease: malaria (2008)
Nationality:
noun: Motswana (singular), Batswana (plural)
adjective: Motswana (singular), Batswana (plural)
Ethnic groups: Tswana (or Setswana) 79%, Kalanga 11%, Basarwa 3%, other, including Kgalagadi and white 7%
Religions: Christian 71.6%, Badimo 6%, other 1.4%, unspecified 0.4%, none 20.6% (2001 census)
Languages: Setswana 78.2%, Kalanga 7.9%, Sekgalagadi 2.8%, English 2.1% (official), other 8.6%, unspecified 0.4% (2001 census)
Literacy:
definition: age 15 and over can read and write
total population: 81.2%
male: 80.4%
female: 81.8% (2003 est.)

GOVERNMENT

Country name:
conventional long form: Republic of Botswana

conventional short form: Botswana
local long form: Republic of Botswana
local short form: Botswana
former: Bechuanaland

Government type: parliamentary republic

Capital: name: Gaborone
geographic coordinates: 24 45 S, 25 55 E
time difference: UTC+2 (7 hours ahead of Washington, DC during Standard Time)

Administrative divisions: 9 districts and 5 town councils*; Central, Francistown*, Gaborone*, Ghanzi, Jwaneng*, Kgalagadi, Kgatleng, Kweneng, Lobatse*, Northeast, Northwest, Selebi-Pikwe*, Southeast, Southern

Independence: 30 September 1966 (from UK)

National holiday: Independence Day (Botswana Day), 30 September (1966)

Constitution: March 1965, effective 30 September 1966

Legal system: based on Roman-Dutch law and local customary law; judicial review limited to matters of interpretation; accepts compulsory ICJ jurisdiction, with reservations

Suffrage: 18 years of age; universal

Executive branch:
chief of state: President Seretse Khama Ian KHAMA (since 1 April 2008); Vice President Mompati MERAFHE (since 1 April 2008); note—the president is both the chief of state and head of government
head of government: President Seretse Khama Ian KHAMA (since 1 April 2008); Vice President Mompati MERAFHE (since 1 April 2008)
cabinet: Cabinet appointed by the president
elections: president indirectly elected for a five-year term (eligible for a second term); election last held 20 October 2004 (next to be held in October 2009); vice president appointed by the president
election results: Festus G. MOGAE elected president; percent of National Assembly vote—52%

Legislative branch: bicameral Parliament consists of the House of Chiefs (a largely advisory 15-member body with 8 permanent members consisting of the chiefs of the principal tribes, and 7 non-permanent members serving 5-year terms, consisting of 4 elected subchiefs and 3 members selected by the other 12 members) and the National Assembly (63 seats, 57 members are directly elected by popular vote, 4 are appointed by the majority party, and 2, the President and Attorney-General, serve as ex-officio members; members serve five-year terms)

elections: National Assembly elections last held 30 October 2004 (next to be held in October 2009)
election results: percent of vote by party—BDP 51.7%, BNF 26.1%, BCP 16.6%, other 5%; seats by party—BDP 44, BNF 12, BCP 1

Judicial branch: High Court; Court of Appeal; Magistrates' Courts (one in each district)

Political parties and leaders: Botswana Alliance Movement or BAM [Ephraim Lepetu SETSHWAELO]; Botswana Congress Party or BCP [Otlaadisa KOOSALETSE]; Botswana Democratic Party or BDP [Festus G. MOGAE]; Botswana National Front or BNF [Otswoletse MOUPO]; Botswana Peoples Party or BPP; MELS Movement of Botswana or MELS; New Democratic Front or NDF
note: a number of minor parties joined forces in 1999 to form the BAM but did not capture any parliamentary seats—includes the United Action Party [Ephraim Lepetu SETSHWAELO]; the Independence Freedom Party or IFP [Motsamai MPHO]; the Botswana Progressive Union [D. K. KWELE]

Political pressure groups and leaders: NA

International organization participation: ACP, AfDB, AU, C, FAO, G-77, IAEA, IBRD, ICAO, ICCt, ICRM, IDA, IFAD, IFC, IFRCS, ILO, IMF, Interpol, IOC, IPU, ISO, ITSO, ITU, ITUC, MIGA, NAM, OPCW, SACU, SADC, UN, UNCTAD, UNESCO, UNIDO, UNMIS, UNWTO, UPU, WCO, WFTU, WHO, WIPO, WMO, WTO

Diplomatic representation in the US:
chief of mission: Ambassador Lapologang Caesar LEKOA
chancery: 1531-1533 New Hampshire Avenue NW, Washington, DC 20036
telephone: [1] (202) 244-4990
FAX: [1] (202) 244-4164

Diplomatic representation from the US:
chief of mission: Ambassador Katherine H. CANAVAN
embassy: address NA, Gaborone
mailing address: Embassy Enclave, P. O. Box 90, Gaborone
telephone: [267] 395-3982
FAX: [267] 395-6947

Flag description: light blue with a horizontal white-edged black stripe in the center

ECONOMY

Economy—overview: Botswana has maintained one of the world's highest economic growth rates since independence in 1966, though growth slowed to

4.7% annually in 2006–07. Through fiscal discipline and sound management, Botswana has transformed itself from one of the poorest countries in the world to a middle-income country with a per capita GDP of nearly $15,000 in 2007. Two major investment services rank Botswana as the best credit risk in Africa. Diamond mining has fueled much of the expansion and currently accounts for more than one-third of GDP and for 70–80% of export earnings. Tourism, financial services, subsistence farming, and cattle raising are other key sectors. On the downside, the government must deal with high rates of unemployment and poverty. Unemployment officially was 23.8% in 2004, but unofficial estimates place it closer to 40%. HIV/AIDS infection rates are the second highest in the world and threaten Botswana's impressive economic gains. An expected leveling off in diamond mining production overshadows long-term prospects.

GDP (purchasing power parity): $25.68 billion (2007 est.)

GDP (official exchange rate): $12.31 billion (2007 est.)

GDP—real growth rate: 5.4% (2007 est.)

GDP—per capita (PPP): $16,400 (2007 est.)

GDP—composition by sector:
agriculture: 1.6%
industry: 51.5% (including 36% mining)
services: 46.9% (2006 est.)

Labor force: 288,400 formal sector employees (2004)

Labor force—by occupation: agriculture: NA%
industry: NA%
services: NA%

Unemployment rate: 7.5% (2007 est.)

Population below poverty line: 30.3% (2003)

Household income or consumption by percentage share:
lowest 10%: NA%
highest 10%: NA%

Distribution of family income—Gini index: 63 (1993)

Inflation rate (consumer prices): 7.1% (2007 est.)

Investment (gross fixed): 18.2% of GDP (2007 est.)

Budget:
revenues: $4.741 billion
expenditures: $3.816 billion (2007 est.)

Public debt: 5% of GDP (2007 est.)

Agriculture—products: livestock, sorghum, maize, millet, beans, sunflowers, groundnuts

Industries: diamonds, copper, nickel, salt, soda ash, potash; livestock processing; textiles

Industrial production growth rate: 5% (2007 est.)

Electricity—production: 912 million kWh (2005)

Electricity—production by source:
fossil fuel: 100%
hydro: 0%
nuclear: 0%
other: 0% (2001)

Electricity—consumption: 2.602 billion kWh (2005)

Electricity—exports: 0 kWh (2005)

Electricity—imports: 1.754 billion kWh (2005)

Oil—production: 0 bbl/day (2005)

Oil—consumption: 12,000 bbl/day (2005 est.)

Oil—exports: 0 bbl/day (2004)

Oil—imports: 13,490 bbl/day (2004)

Oil—proved reserves: 0 bbl (1 January 2006 est.)

Natural gas—production: 0 cu m (2005 est.)

Natural gas—consumption: 0 cu m (2005 est.)

Natural gas—exports: 0 cu m (2005 est.)

Natural gas—imports: 0 cu m (2005)

Natural gas—proved reserves: 0 cu m (1 January 2006 est.)

Current account balance: $2.074 billion (2007 est.)

Exports: $5.025 billion f.o.b. (2007 est.)

Exports—commodities: diamonds, copper, nickel, soda ash, meat, textiles

Exports—partners: European Free Trade Association (EFTA) 87%, Southern African Customs Union (SACU) 7%, Zimbabwe 4% (2006)

Imports: $3.403 billion f.o.b. (2007 est.)

Imports—commodities: foodstuffs, machinery, electrical goods, transport equipment, textiles, fuel and petroleum products, wood and paper products, metal and metal products

Imports—partners: Southern African Customs Union (SACU) 74%, EFTA 17%, Zimbabwe 4% (2006)

Economic aid—recipient: $70.89 million (2005)

Reserves of foreign exchange and gold: $9.79 billion (31 December 2007 est.)

Debt—external: $408 million (31 December 2007 est.)

Market value of publicly traded shares: $3.947 billion (2006)

Currency (code): pula (BWP)
Currency code: BWP
Exchange rates: pulas per US dollar—6.2035 (2007), 5.8447 (2006), 5.1104 (2005), 4.6929 (2004), 4.9499 (2003)
Fiscal year: 1 April—31 March

COMMUNICATIONS

Telephones—main lines in use: 136,900 (2006)

Telephones—mobile cellular: 979,800 (2006)

Telephone system:
general assessment: the system is expanding with the growth of mobile-cellular service and participation in regional development; system is fully digital with fiber-optic cables linking the major population centers in the east; fixed-line connections declined in recent years and now stand at 8 per 100 persons; mobile-cellular telephone density currently is about 60 per 100 persons
domestic: small system of open-wire lines, microwave radio relay links, and a few radiotelephone communication stations; mobile-cellular service is growing fast
international: country code—267; international calls are made via satellite, using international direct dialing; 2 international exchanges; digital microwave radio relay links to Namibia, Zambia, Zimbabwe, and South Africa; satellite earth station—1 Intelsat (Indian Ocean) (2007)

Radio broadcast stations: AM 8, FM 13, shortwave 4 (2001)

Radios: 252,720 (2000)

Television broadcast stations: 2 (1 state-owned, 1 private)

Televisions: 31,000 (1997)

Internet country code: .bw

Internet hosts: 5,820 (2007)

Internet Service Providers (ISPs): 11 (2001)

Internet users: 60,000 (2005)

TRANSPORTATION

Airports: 85 (2007)

Airports—with paved runways:
total: 11
2,438 to 3,047 m: 2
1,524 to 2,437 m: 7
914 to 1,523 m: 2 (2007)

Airports—with unpaved runways:
total: 74

1,524 to 2,437 m: 3
914 to 1,523 m: 54
under 914 m: 17 (2007)

Railways:
total: 888 km
narrow gauge: 888 km 1.067-m gauge (2006)

Roadways:
total: 24,455 km
paved: 8,119 km
unpaved: 16,336 km (2004)

MILITARY

Military branches: Botswana Defense Force (includes an air wing) (2008)

Military service age and obligation: 18 is the apparent age of voluntary military service; the official qualifications for determining minimum age are unknown (2001)

Manpower available for military service:
males age 16–49: 487,853
females age 16–49: 464,278 (2008 est.)

Manpower fit for military service:
males age 16–49: 290,093
females age 16–49: 257,700 (2008 est.)

Manpower reaching militarily significant age annually:
males age 16–49: 23,007
females age 16–49: 22,551 (2008 est.)

Military expenditures—percent of GDP: 3.3% (2006)

TRANSNATIONAL ISSUES

Disputes—international: the alignment of the boundary with Namibia in the Kwando/Linyanti/Chobe River, including the Situngu marshlands, was resolved amicably in 2003; concerns from international experts and local populations over the ecology of the Okavango Delta in Botswana and human displacement scuttled Namibian plans to construct a hydroelectric dam at Popavalle (Popa Falls) along the Angola-Namibia border; Botswana has built electric fences to stem the thousands of Zimbabweans who flee to find work and escape political persecution; Namibia has long supported, and in 2004 Zimbabwe dropped objections to, plans between Botswana and Zambia to build a bridge over the Zambezi River, thereby de facto recognizing the short, but not clearly delimited, Botswana-Zambia boundary

BOUVET ISLAND

INTRODUCTION

Background: This uninhabited volcanic island is almost entirely covered by glaciers and is difficult to approach. It was discovered in 1739 by a French naval officer after whom the island was named. No claim was made until 1825, when the British flag was raised. In 1928, the UK waived its claim in favor of Norway, which had occupied the island the previous year. In 1971, Norway designated Bouvet Island and the adjacent territorial waters a nature reserve. Since 1977, it has run an automated meteorological station on the island.

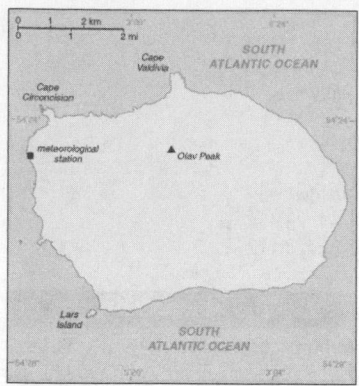

GEOGRAPHY

Location: island in the South Atlantic Ocean, southwest of the Cape of Good Hope (South Africa)
Geographic coordinates: 54 26 S, 3 24 E
Map references: Antarctic Region
Area:
total: 49 sq km
land: 49 sq km
water: 0 sq km
Area—comparative: about 0.3 times the size of Washington, DC

Land boundaries: 0 km
Coastline: 29.6 km
Maritime claims: *territorial sea:* 4 nm
Climate: antarctic
Terrain: volcanic; coast is mostly inaccessible
Elevation extremes:
lowest point: South Atlantic Ocean 0 m
highest point: Olav Peak 935 m
Natural resources: none
Land use:
arable land: 0%
permanent crops: 0%
other: 100% (93% ice) (2005)
Irrigated land: 0 sq km
Natural hazards: NA
Environment—current issues: NA
Geography—note: covered by glacial ice; declared a nature reserve Norway

PEOPLE

Population: uninhabited

GOVERNMENT

Country name:
conventional long form: none
conventional short form: Bouvet Island
Dependency status: territory of Norway; administered by the Polar Department of the Ministry of Justice and Police from Oslo
Legal system: the laws of Norway, where applicable, apply
Flag description: the flag of Norway is used

ECONOMY

Economy—overview: no economic activity; declared a nature reserve

COMMUNICATIONS

Internet country code: .bv
Internet hosts: 6 (2007)
Communications—note: automatic meteorological station

TRANSPORTATION

Ports and terminals: none; offshore anchorage only

MILITARY

Military—note: defense is the responsibility of Norway

TRANSNATIONAL ISSUES

Disputes—international: none

BRAZIL

INTRODUCTION

Background: Following three centuries under the rule of Portugal, Brazil became an independent nation in 1822 and a republic in 1889. By far the largest and most populous country in South America, Brazil overcame more than half a century of military intervention in the governance of the country when in 1985 the military regime peacefully ceded power to civilian rulers. Brazil continues to pursue industrial and agricultural growth and development of its interior. Exploiting vast natural resources and a large labor pool, it is today South America's leading economic power and a regional leader. Highly unequal income distribution and crime remain pressing problems.

GEOGRAPHY

Location: Eastern South America, bordering the Atlantic Ocean
Geographic coordinates: 10 00 S, 55 00 W
Map references: South America
Area:
total: 8,511,965 sq km
land: 8,456,510 sq km
water: 55,455 sq km
note: includes Arquipelago de Fernando de Noronha, Atol das Rocas, Ilha da Trindade, Ilhas Martin Vaz, and Penedos de Sao Pedro e Sao Paulo
Area—comparative: slightly smaller than the US
Land boundaries:
total: 16,885 km
border countries: Argentina 1,261 km, Bolivia 3,423 km, Colombia 1,644 km, French Guiana 730 km, Guyana 1,606 km, Paraguay 1,365 km, Peru 2,995 km, Suriname 593 km, Uruguay 1,068 km, Venezuela 2,200 km
Coastline: 7,491 km
Maritime claims:
territorial sea: 12 nm
contiguous zone: 24 nm
exclusive economic zone: 200 nm
continental shelf: 200 nm or to edge of the continental margin
Climate: mostly tropical, but temperate in south
Terrain: mostly flat to rolling lowlands in north; some plains, hills, mountains, and narrow coastal belt
Elevation extremes:
lowest point: Atlantic Ocean 0 m
highest point: Pico da Neblina 3,014 m
Natural resources: bauxite, gold, iron ore, manganese, nickel, phosphates, platinum, tin, uranium, petroleum, hydropower, timber
Land use:
arable land: 6.93%
permanent crops: 0.89%
other: 92.18% (2005)
Irrigated land: 29,200 sq km (2003)
Total renewable water resources: 8,233 cu km (2000)
Freshwater withdrawal (domestic/industrial/agricultural): *total:* 59.3 cu km/yr (20%/18%/62%)
per capita: 318 cu m/yr (2000)
Natural hazards: recurring droughts in northeast; floods and occasional frost in south
Environment—current issues: deforestation in Amazon Basin destroys the

habitat and endangers a multitude of plant and animal species indigenous to the area; there is a lucrative illegal wildlife trade; air and water pollution in Rio de Janeiro, Sao Paulo, and several other large cities; land degradation and water pollution caused by improper mining activities; wetland degradation; severe oil spills

Environment—international agreements: *party to:* Antarctic-Environmental Protocol, Antarctic-Marine Living Resources, Antarctic Seals, Antarctic Treaty, Biodiversity, Climate Change, Climate Change-Kyoto Protocol, Desertification, Endangered Species, Environmental Modification, Hazardous Wastes, Law of the Sea, Marine Dumping, Ozone Layer Protection, Ship Pollution, Tropical Timber 83, Tropical Timber 94, Wetlands, Whaling

signed, but not ratified: none of the selected agreements

Geography—note: largest country in South America; shares common boundaries with every South American country except Chile and Ecuador

PEOPLE

Population: 191,908,598

note: Brazil conducted a census in August 2000, which reported a population of 169,799,170; that figure was about 3.3% lower than projections by the US Census Bureau, and is close to the implied underenumeration of 4.6% for the 1991 census; estimates for this country explicitly take into account the effects of excess mortality due to AIDS; this can result in lower life expectancy, higher infant mortality, higher death rates, lower population growth rates, and changes in the distribution of population by age and sex than would otherwise be expected (July 2008 est.)

Age structure:

0–14 years: 24.9% (male 24,391,338/female 23,454,418)

15–64 years: 68.7% (male 65,330,427/female 66,431,982)

65 years and over: 6.4% (male 5,055,770/female 7,244,663) (2008 est.)

Median age:

total: 29 years

male: 28.3 years

female: 29.8 years (2008 est.)

Population growth rate: 0.98% (2008 est.)

Birth rate: 16.04 births/1,000 population (2008 est.)

Death rate: 6.22 deaths/1,000 population (2008 est.)

Net migration rate: -0.03 migrant(s)/1,000 population (2008 est.)

Sex ratio:

at birth: 1.05 male(s)/female

under 15 years: 1.04 male(s)/female

15–64 years: 0.98 male(s)/female

65 years and over: 0.7 male(s)/female

total population: 0.98 male(s)/female (2008 est.)

Infant mortality rate:

total: 26.67 deaths/1,000 live births

male: 30.28 deaths/1,000 live births

female: 22.89 deaths/1,000 live births (2008 est.)

Life expectancy at birth:

total population: 72.51 years

male: 68.57 years

female: 76.64 years (2008 est.)

Total fertility rate: 1.86 children born/woman (2008 est.)

HIV/AIDS—adult prevalence rate: 0.7% (2003 est.)

HIV/AIDS—people living with HIV/AIDS: 660,000 (2003 est.)

HIV/AIDS—deaths: 15,000 (2003 est.)

Nationality:

noun: Brazilian(s)

adjective: Brazilian

Ethnic groups: white 53.7%, mulatto (mixed white and black) 38.5%, black 6.2%, other (includes Japanese, Arab, Amerindian) 0.9%, unspecified 0.7% (2000 census)

Religions: Roman Catholic (nominal) 73.6%, Protestant 15.4%, Spiritualist 1.3%, Bantu/voodoo 0.3%, other 1.8%, unspecified 0.2%, none 7.4% (2000 census)

Languages: Portuguese (official and most widely spoken language); note—less common languages include Spanish (border areas and schools), German, Italian, Japanese, English, and a large number of minor Amerindian languages

Literacy:

definition: age 15 and over can read and write

total population: 88.6%

male: 88.4%

female: 88.8% (2004 est.)

GOVERNMENT

Country name:

conventional long form: Federative Republic of Brazil

conventional short form: Brazil

local long form: Republica Federativa do Brasil

local short form: Brasil

Government type: federal republic

Capital: *name:* Brasilia

geographic coordinates: 15 47 S, 47 55 W

time difference: UTC-3 (2 hours ahead of Washington, DC during Standard Time)

daylight saving time: +1hr, begins third Sunday in October; ends third Sunday in February

note: Brazil is divided into four time zones, including one for the Fernando de Noronha Islands

Administrative divisions: 26 states (estados, singular—estado) and 1 federal district* (distrito federal); Acre, Alagoas, Amapa, Amazonas, Bahia, Ceara, Distrito Federal*, Espirito Santo, Goias, Maranhao, Mato Grosso, Mato Grosso do Sul, Minas Gerais, Para, Paraiba, Parana, Pernambuco, Piaui, Rio de Janeiro, Rio Grande do Norte, Rio Grande do Sul, Rondonia, Roraima, Santa Catarina, Sao Paulo, Sergipe, Tocantins

Independence: 7 September 1822 (from Portugal)

National holiday: Independence Day, 7 September (1822)

Constitution: 5 October 1988

Legal system: based on Roman codes; has not accepted compulsory ICJ jurisdiction

Suffrage: voluntary between 16 and 18 years of age and over 70; compulsory over 18 and under 70 years of age; note—military conscripts do not vote

Executive branch:

chief of state: President Luiz Inacio LULA DA SILVA (since 1 January 2003); Vice President Jose ALENCAR (since 1 January 2003); note—the president is both the chief of state and head of government

head of government: President Luiz Inacio LULA DA SILVA (since 1 January 2003); Vice President Jose ALENCAR (since 1 January 2003)

cabinet: Cabinet appointed by the president

elections: president and vice president elected on the same ticket by popular vote for a single four-year term; election last held 1 October 2006 with runoff 29 October 2006 (next to be held 3 October 2010 and, if necessary, 31 October 2010)

election results: Luiz Inacio LULA DA SILVA (PT) reelected president—60.83%, Geraldo ALCKMIN (PSDB) 39.17%

Legislative branch: bicameral National Congress or Congresso Nacional consists of the Federal Senate or Senado Federal (81 seats; 3 members from each state and federal district elected according to the principle of majority to serve eight-year terms; one-third and two-thirds elected every four years, alternately) and the Chamber of Deputies or Camara dos Deputados (513 seats; members are elected by proportional representation to serve four-year terms)

elections: Federal Senate—last held 1 October 2006 for one-third of the Senate (next to be held in October 2010 for two-thirds of the Senate); Chamber of

Deputies—last held 1 October 2006 (next to be held in October 2010)

election results: Federal Senate—percent of vote by party—NA; seats by party—PFL 6, PSDB 5, PMDB 4, PTB 3, PT 2, PDT 1, PSB 1, PL 1, PPS 1, PRTB 1, PP 1, PCdoB 1; Chamber of Deputies—percent of vote by party—NA; seats by party—PMDB 89, PT 83, PFL 65, PSDB 65, PP 42, PSB 27, PDT 24, PL 23, PTB 22, PPS 21, PCdoB 13, PV 13, PSC 9, other 17; note—as of 1 January 2008: Federal Senate—seats by party—PMDB 20, DEM (formerly PFL) 14, PSDB 13, PT 12, PTB 6, PDT 5, PR 4, PRB 2, PSB 2, PCdoB 1, PP 1, PSOL 1; Chamber of Deputies—seats by party—PMDB 90, PT 83, PSDB 64, DEM (formerly PFL) 62, PP 41, PR 34, PSB 28, PDT 23, PTB 21, PPS 17, PV 13, PCdoB 13, PSC 7, PAN 4, PSOL 3, PMN 3, PTC 3, PHS 2, PTdoB 1, PRB 1

Judicial branch: Supreme Federal Tribunal or STF (11 ministers are appointed for life by the president and confirmed by the Senate); Higher Tribunal of Justice; Regional Federal Tribunals (judges are appointed for life); note—though appointed "for life," judges, like all federal employees, have a mandatory retirement age of 70

Political parties and leaders: Brazilian Democratic Movement Party or PMDB [Federal Deputy Michel TEMER]; Brazilian Labor Party or PTB [Roberto JEFFERSON]; Brazilian Renewal Labor Party or PRTB [Jose Levy FIDELIX da Cruz]; Brazilian Republican Party or PRB [Vitor Paulo Araujo DOS SANTOS]; Brazilian Social Democracy Party or PSDB [Senator Sergio GUERRA]; Brazilian Socialist Party or PSB [Governor Eduardo Henrique Accioly CAMPOS]; Christian Labor Party or PTC [Daniel TOURINHO]; Communist Party of Brazil or PCdoB [Jose Renato RABELO]; Democratic Labor Party or PDT [Carlos Roberto LUPI]; the Democrats or DEM (formerly Liberal Front Party or PFL) [Federal Deputy Rodrigo MAIA]; Freedom and Socialism Party or PSOL [Heloisa HELENA]; Green Party or PV [Jose Luiz de Franca PENNA]; Humanist Party of Solidarity or PHS [Paulo Roberto MATOS]; Labor Party of Brazil or PTdoB [Luis Henrique de Oliveira RESENDE]; Liberal Front Party or PFL (now known as the Democrats or DEM); National Mobilization Party or PMN [Oscar Noronha FILHO]; Party of the Republic or PR [Sergio TAMER]; Popular Socialist Party or PPS [Federal Deputy Fernando CORUJA]; Progressive Party or PP [Francisco DORNELLES]; Social

Christian Party or PSC [Vitor Jorge Abdala NOSSEIS]; Workers' Party or PT [Ricardo Jose Ribeiro BERZOINI]

Political pressure groups and leaders: Landless Workers' Movement or MST; labor unions and federations; large farmers' associations; religious groups including evangelical Christian churches and the Catholic Church

International organization participation: AfDB, BIS, CAN (associate), CPLP, CSN, FAO, G-15, G-24, G-77, IADB, IAEA, IBRD, ICAO, ICC, ICCt, ICRM, IDA, IFAD, IFC, IFRCS, IHO, ILO, IMF, IMO, IMSO, Interpol, IOC, IOM, IPU, ISO, ITSO, ITU, ITUC, LAES, LAIA, Mercosur, MIGA, MINURSO, MINUSTAH, NAM (observer), NSG, OAS, OPANAL, OPCW, PCA, RG, UN, UNCTAD, UNESCO, UNFICYP, UNHCR, UNIDO, Union Latina, UNITAR, UNMEE, UNMIL, UNMIS, UNMIT, UNOCI, UNWTO, UPU, WCL, WCO, WFTU, WHO, WIPO, WMO, WTO

Diplomatic representation in the US:
chief of mission: Ambassador Antonio de Aguiar PATRIOTA
chancery: 3006 Massachusetts Avenue NW, Washington, DC 20008
telephone: [1] (202) 238-2700
FAX: [1] (202) 238-2827
consulate(s) general: Boston, Chicago, Houston, Los Angeles, Miami, New York, San Francisco

Diplomatic representation from the US:
chief of mission: Ambassador Clifford M. SOBEL
embassy: Avenida das Nacoes, Quadra 801, Lote 3, Distrito Federal Cep 70403-900, Brasilia
mailing address: Unit 3500, APO AA 34030
telephone: [55] (61) 3312-7000
FAX: [55] (61) 3225-9136
consulate(s) general: Rio de Janeiro, Sao Paulo
consulate(s): Recife

Flag description: green with a large yellow diamond in the center bearing a blue celestial globe with 27 white five-pointed stars (one for each state and the Federal District) arranged in the same pattern as the night sky over Brazil; the globe has a white equatorial band with the motto ORDEM E PROGRESSO (Order and Progress)

ECONOMY

Economy—overview: Characterized by large and well-developed agricultural, mining, manufacturing, and service sectors, Brazil's economy outweighs that of all other South American countries and is expanding its presence in world mar-

kets. Having weathered 2001–03 financial turmoil, capital inflows are regaining strength and the currency has resumed appreciating. The appreciation has slowed export volume growth, but since 2004, Brazil's growth has yielded increases in employment and real wages. The resilience in the economy stems from commodity-driven current account surpluses, and sound macroeconomic policies that have bolstered international reserves to historically high levels, reduced public debt, and allowed a significant decline in real interest rates. A floating exchange rate, an inflation-targeting regime, and a tight fiscal policy are the three pillars of the economic program. From 2003 to 2007, Brazil ran record trade surpluses and recorded its first current account surpluses since 1992. Productivity gains coupled with high commodity prices contributed to the surge in exports. Brazil improved its debt profile in 2006 by shifting its debt burden toward real denominated and domestically held instruments. LULA DA SILVA restated his commitment to fiscal responsibility by maintaining the country's primary surplus during the 2006 election. Following his second inauguration, LULA DA SILVA announced a package of further economic reforms to reduce taxes and increase investment in infrastructure. The government's goal of achieving strong growth while reducing the debt burden is likely to create inflationary pressures.

GDP (purchasing power parity): $1.836 trillion (2007 est.)

GDP (official exchange rate): $1.314 trillion (2007 est.)

GDP—real growth rate: 5.4% (2007 est.)

GDP—per capita (PPP): $9,700 (2007 est.)

GDP—composition by sector:
agriculture: 5.5%
industry: 28.7%
services: 65.8% (2007 est.)

Labor force: 99.47 million (2007 est.)

Labor force—by occupation: *agriculture:* 20%
industry: 14%
services: 66% (2003 est.)

Unemployment rate: 9.3% (2007 est.)

Population below poverty line: 31% (2005)

Household income or consumption by percentage share:
lowest 10%: 0.9%
highest 10%: 44.8% (2004)

Distribution of family income—Gini index: 56.7 (2005)

Inflation rate (consumer prices): 3.6% (2007 est.)

Investment (gross fixed): 17.6% of GDP (2007 est.)

Budget:
revenues: $244 billion
expenditures: $219.9 billion (FY07)

Public debt: 45.1% of GDP (2007 est.)

Agriculture—products: coffee, soybeans, wheat, rice, corn, sugarcane, cocoa, citrus; beef

Industries: textiles, shoes, chemicals, cement, lumber, iron ore, tin, steel, aircraft, motor vehicles and parts, other machinery and equipment

Industrial production growth rate: 4.9% (2007 est.)

Electricity—production: 396.4 billion kWh (2005)

Electricity—production by source:
fossil fuel: 8.3%
hydro: 82.7%
nuclear: 4.4%
other: 4.6% (2001)

Electricity—consumption: 368.5 billion kWh (2005)

Electricity—exports: 160 million kWh (2005)

Electricity—imports: 39.2 billion kWh; note—supplied by Paraguay (2005)

Oil—production: 1.59 million bbl/day (2006 est.)

Oil—consumption: 2.1 million bbl/day (2006 est.)

Oil—exports: 278,400 bbl/day (2005)

Oil—imports: 674,500 bbl/day (2004)

Oil—proved reserves: 13.9 billion bbl (2007 est.)

Natural gas—production: 9.37 billion cu m (2005 est.)

Natural gas—consumption: 17.85 billion cu m (2005 est.)

Natural gas—exports: 0 cu m (2005 est.)

Natural gas—imports: 8.478 billion cu m (2005)

Natural gas—proved reserves: 312.7 billion cu m (1 January 2006 est.)

Current account balance: $3.555 billion (2007 est.)

Exports: $160.6 billion f.o.b. (2007 est.)

Exports—commodities: transport equipment, iron ore, soybeans, footwear, coffee, autos

Exports—partners: US 17.8%, Argentina 8.5%, China 6.1%, Netherlands 4.2%, Germany 4.1% (2006)

Imports: $120.6 billion f.o.b. (2007 est.)

Imports—commodities: machinery, electrical and transport equipment, chemical products, oil, automotive parts, electronics

Imports—partners: US 16.2%, Argentina 8.8%, China 8.7%, Germany 7.1%, Nigeria 4.3%, Japan 4.2% (2006)

Economic aid—recipient: $191.9 million (2005)

Reserves of foreign exchange and gold: $180.3 billion (31 December 2007)

Debt—external: $223.9 billion (31 December 2007)

Stock of direct foreign investment—at home: $249 billion (2007 est.)

Stock of direct foreign investment—abroad: $107 billion (2007 est.)

Market value of publicly traded shares: $711.1 billion (2006)

Currency (code): real (BRL)

Currency code: BRL

Exchange rates: reals per US dollar—1.85 (2007 est.), 2.1761 (2006), 2.4344 (2005), 2.9251 (2004), 3.0771 (2003)

Fiscal year: calendar year

COMMUNICATIONS

Telephones—main lines in use: 38.8 million (2006)

Telephones—mobile cellular: 99.919 million (2006)

Telephone system:
general assessment: good working system; fixed-line connections have remained relatively stable in recent years and stand at about 20 per 100 persons; mobile-cellular telephone density has risen to nearly 55 per 100 persons
domestic: extensive microwave radio relay system and a domestic satellite system with 64 earth stations; mobile-cellular usage has more than tripled in the past 5 years
international: country code—55; landing point for a number of submarine cables that provide direct links to South and Central America, the Caribbean, the US, Africa, and Europe; satellite earth stations—3 Intelsat (Atlantic Ocean), 1 Inmarsat (Atlantic Ocean region east), connected by microwave relay system to Mercosur Brazilsat B3 satellite earth station (2007)

Radio broadcast stations: AM 1,365, FM 296, shortwave 161 (of which 91 are collocated with AM stations) (1999)

Radios: 71 million (1997)

Television broadcast stations: 138 (1997)

Televisions: 36.5 million (1997)

Internet country code: .br

Internet hosts: 8.265 million (2007)

Internet Service Providers (ISPs): 50 (2000)

Internet users: 42.6 million (2006)

TRANSPORTATION

Airports: 4,263 (2007)

Airports—with paved runways:
total: 718
over 3,047 m: 7
2,438 to 3,047 m: 25
1,524 to 2,437 m: 167
914 to 1,523 m: 467

under 914 m: 52 (2007)

Airports—with unpaved runways:
total: 3,545
1,524 to 2,437 m: 83
914 to 1,523 m: 1,555
under 914 m: 1,907 (2007)

Heliports: 16 (2007)

Pipelines: condensate/gas 244 km; gas 12,070 km; liquid petroleum gas 351 km; oil 5,214 km; refined products 4,410 km (2007)

Railways:
total: 29,295 km
broad gauge: 4,932 km 1.600-m gauge (939 km electrified)
standard gauge: 194 km 1.440-m gauge
narrow gauge: 23,773 km 1.000-m gauge (581 km electrified)
dual gauge: 396 km 1.000 m and 1.600-m gauges (three rails) (78 km electrified) (2006)

Roadways:
total: 1,751,868 km
paved: 96,353 km
unpaved: 1,655,515 km (2004)

Waterways: 50,000 km (most in areas remote from industry and population) (2007)

Merchant marine:
total: 135 ships (1000 GRT or over) 2,020,182 GRT/3,039,015 DWT
by type: bulk carrier 20, cargo 21, carrier 1, chemical tanker 6, container 9, liquefied gas 12, passenger/cargo 12, petroleum tanker 47, roll on/roll off 7
foreign-owned: 16 (Chile 1, Denmark 2, Germany 7, Mexico 1, Norway 1, Spain 4)
registered in other countries: 5 (Bahamas 1, Ghana 1, Liberia 3) (2007)

Ports and terminals: Guaiba, Ilha Grande, Paranagua, Rio Grande, Santos, Sao Sebastiao, Tubarao

MILITARY

Military branches: Brazilian Army, Brazilian Navy (Marinha do Brasil (MB), includes Naval Air and Marine Corps (Corpo de Fuzileiros Navais)), Brazilian Air Force (Forca Aerea Brasileira, FAB) (2008)

Military service age and obligation: 21–45 years of age for compulsory military service; conscript service obligation—9 to 12 months; 17–45 years of age for voluntary service; an increasing percentage of the ranks are "long-service" volunteer professionals; women were allowed to serve in the armed forces beginning in early 1980s when the Brazilian Army became the first army in South America to accept women into career ranks; women serve in Navy and Air Force only in Women's Reserve Corps (2001)

93

Manpower available for military service:
males age 16–49: 52,449,957
females age 16–49: 52,375,921 (2008 est.)
Manpower fit for military service:
males age 16–49: 39,263,710
females age 16–49: 44,109,056 (2008 est.)
Manpower reaching militarily significant age annually:
males age 16–49: 1,668,722
females age 16–49: 1,609,437 (2008 est.)
Military expenditures—percent of GDP: 2.6% (2006 est.)

Disputes—international: unruly region at convergence of Argentina-Brazil-Paraguay borders is locus of money laundering, smuggling, arms and illegal narcotics trafficking, and fundraising for extremist organizations; uncontested dispute with Uruguay over certain islands in the Quarai/Cuareim and Invernada boundary streams and the resulting tripoint with Argentina
Illicit drugs: illicit producer of cannabis; trace amounts of coca cultivation in the Amazon region, used for domestic consumption; government has a large-scale eradication program to control cannabis; important transshipment country for Bolivian, Colombian, and Peruvian cocaine headed for Europe; also used by traffickers as a way station for narcotics air transshipments between Peru and Colombia; upsurge in drug-related violence and weapons smuggling; important market for Colombian, Bolivian, and Peruvian cocaine; illicit narcotics proceeds earned in Brazil are often laundered through the financial system; significant illicit financial activity in the Tri-Border Area

BRITISH INDIAN OCEAN TERRITORY

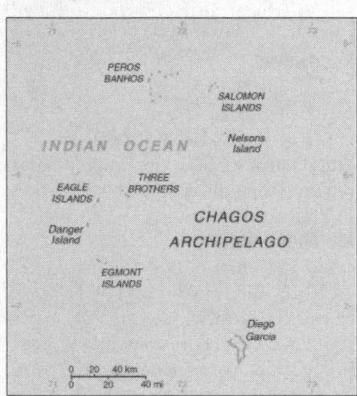

INTRODUCTION

Background: Established as a territory of the UK in 1965, a number of the British Indian Ocean Territory (BIOT) islands were transferred to the Seychelles when it attained independence in 1976. Subsequently, BIOT has consisted only of the six main island groups comprising the Chagos Archipelago. The largest and most southerly of the islands, Diego Garcia, contains a joint UK-US naval support facility. All of the remaining islands are uninhabited. Former agricultural workers, earlier residents in the islands, were relocated primarily to Mauritius but also to the Seychelles, between 1967 and 1973. In 2000, a British High Court ruling invalidated the local immigration order that had excluded them from the archipelago, but upheld the special military status of Diego Garcia.

GEOGRAPHY

Location: archipelago in the Indian Ocean, south of India, about halfway between Africa and Indonesia
Geographic coordinates: 6 00 S, 71 30 E; note—Diego Garcia 7 20 S, 72 25 E

Map references: Political Map of the World
Area:
total: 54,400 sq km
land: 60 sq km; Diego Garcia 44 sq km
water: 54,340 sq km
note: includes the entire Chagos Archipelago of 55 islands
Area—comparative: land area is about 0.3 times the size of Washington, DC
Land boundaries: 0 km
Coastline: 698 km
Maritime claims:
territorial sea: 3 nm
exclusive fishing zone: 200 nm
Climate: tropical marine; hot, humid, moderated by trade winds
Terrain: flat and low (most areas do not exceed two meters in elevation)
Elevation extremes:
lowest point: Indian Ocean 0 m
highest point: unnamed location on Diego Garcia 15 m
Natural resources: coconuts, fish, sugarcane
Land use:
arable land: 0%
permanent crops: 0%
other: 100% (2005)
Irrigated land: 0 sq km
Natural hazards: NA
Environment—current issues: NA
Geography—note: archipelago of 55 islands; Diego Garcia, largest and southernmost island, occupies strategic location in central Indian Ocean; island is site of joint US-UK military facility

PEOPLE

Population: no indigenous inhabitants
note: approximately 1,200 former agricultural workers resident in the Chagos Archipelago, often referred to as Chagossians or Ilois, were relocated to Mauritius and the Seychelles in the 1960s and 1970s; in November 2000 they were granted the right of return by a British High Court ruling, though no timetable has been set; in November 2004, approximately 4,000 UK and US military personnel and civilian contractors were living on the island of Diego Garcia

GOVERNMENT

Country name:
conventional long form: British Indian Ocean Territory
conventional short form: none
abbreviation: BIOT
Dependency status: overseas territory of the UK; administered by a commissioner, resident in the Foreign and Commonwealth Office in London
Legal system: the laws of the UK, where applicable, apply
Executive branch:
chief of state: Queen ELIZABETH II (since 6 February 1952)
head of government: Commissioner Leigh TURNER (since July 2006); Administrator Tony HUMPHRIES (since February 2005); note—both reside in the UK and are represented by the officer commanding British Forces on Diego Garcia
cabinet: NA
elections: none; the monarch is hereditary; commissioner and administrator appointed by the monarch
Diplomatic representation in the US: none (overseas territory of the UK)
Diplomatic representation from the US: none (overseas territory of the UK)
Flag description: white with six blue wavy horizontal stripes; the flag of the UK is in the upper hoist-side quadrant; the striped section bears a palm tree and yellow crown centered on the outer half of the flag

ECONOMY

Economy—overview: All economic activity is concentrated on the largest island of Diego Garcia, where a joint UK-US military facility is located. Construction projects and various services needed to support the military installation are performed by military and contract employees from the UK, Mauritius, the Philippines, and the US. There are no industrial or agricultural activities on the islands. When the native Ilois return, they plan to reestablish sugarcane production and fishing. The territory earns foreign exchange by selling fishing licenses and postage stamps.

Electricity—production: NA kWh; note—electricity supplied by the US military

Electricity—consumption: NA kWh

Currency (code): both the British Pound (GBP) and the US Dollar (USD) are accepted

COMMUNICATIONS

Telephones—main lines in use: NA

Telephone system:
general assessment: separate facilities for military and public needs are available
domestic: all commercial telephone services are available, including connection to the Internet
international: country code (Diego Garcia)—246; international telephone service is carried by satellite (2000)

Radio broadcast stations: AM 1, FM 2, shortwave 0 (1998)

Radios: NA

Television broadcast stations: 1 (1997)

Televisions: NA

Internet country code: .io

Internet hosts: 61 (2007)

Internet Service Providers (ISPs): 1 (2000)

TRANSPORTATION

Airports: 1 (2007)

Airports—with paved runways:
total: 1
over 3,047 m: 1 (2007)

Roadways:
note: short section of paved road between port and airfield on Diego Garcia

Ports and terminals: Diego Garcia

MILITARY

Military—note: defense is the responsibility of the UK; the US lease on Diego Garcia expires in 2016

TRANSNATIONAL ISSUES

Disputes—international: Mauritius and Seychelles claim the Chagos Archipelago including Diego Garcia; in 2001, the former inhabitants of the Chagos Archipelago, evicted in 1967 and 1973 and now residing chiefly in Mauritius, were granted UK citizenship and the right to repatriation; in May 2006, the High Court of London reversed U.K. Government's 2004 orders of council that banned habitation on the islands; a small group of Chagossians visited Diego Garcia in April 2006; repatriation is complicated by the exclusive US military lease of Diego Garcia that restricts access to the largest viable island in the chain

BRITISH VIRGIN ISLANDS

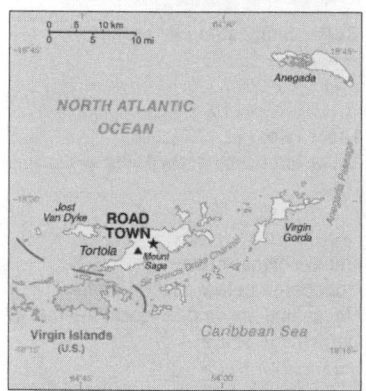

INTRODUCTION

Background: First inhabited by Arawak and later by Carib Indians, the Virgin Islands were settled by the Dutch in 1648 and then annexed by the English in 1672. The islands were part of the British colony of the Leeward Islands from 1872–1960; they were granted autonomy in 1967. The economy is closely tied to the larger and more populous US Virgin Islands to the west; the US dollar is the legal currency.

GEOGRAPHY

Location: Caribbean, between the Caribbean Sea and the North Atlantic Ocean, east of Puerto Rico

Geographic coordinates: 18 30 N, 64 30 W

Map references: Central America and the Caribbean

Area:
total: 153 sq km
land: 153 sq km
water: 0 sq km
note: comprised of 16 inhabited and more than 20 uninhabited islands; includes the islands of Tortola, Anegada, Virgin Gorda, Jost van Dyke

Area—comparative: about 0.9 times the size of Washington, DC

Land boundaries: 0 km

Coastline: 80 km

Maritime claims:
territorial sea: 3 nm
exclusive fishing zone: 200 nm

Climate: subtropical; humid; temperatures moderated by trade winds

Terrain: coral islands relatively flat; volcanic islands steep, hilly

Elevation extremes.
lowest point: Caribbean Sea 0 m
highest point: Mount Sage 521 m

Natural resources: NEGL

Land use:
arable land: 20%
permanent crops: 6.67%

other: 73.33% (2005)

Irrigated land: NA

Natural hazards: hurricanes and tropical storms (July to October)

Environment—current issues: limited natural fresh water resources (except for a few seasonal streams and springs on Tortola, most of the islands' water supply comes from wells and rainwater catchments)

Geography—note: strong ties to nearby US Virgin Islands and Puerto Rico

PEOPLE

Population: 24,004 (July 2008 est.)

Age structure:
0–14 years: 19.9% (male 2,431/female 2,356)
15–64 years: 74.5% (male 9,176/female 8,708)
65 years and over: 5.6% (male 700/female 633) (2008 est.)

Median age:
total: 32 years
male: 32.1 years
female: 31.8 years (2008 est.)

Population growth rate: 1.875% (2008 est.)

Birth rate: 14.75 births/1,000 population (2008 est.)

Death rate: 4.46 deaths/1,000 population (2008 est.)

Net migration rate: 8.46 migrant(s)/ 1,000 population (2008 est.)
Sex ratio: *at birth:* 1.05 male(s)/female
under 15 years: 1.03 male(s)/female
15–64 years: 1.05 male(s)/female
65 years and over: 1.11 male(s)/female
total population: 1.05 male(s)/female (2008 est.)
Infant mortality rate:
total: 15.54 deaths/1,000 live births
male: 18.15 deaths/1,000 live births
female: 12.8 deaths/1,000 live births (2008 est.)
Life expectancy at birth:
total population: 77.06 years
male: 75.87 years
female: 78.3 years (2008 est.)
Total fertility rate: 1.72 children born/woman (2008 est.)
HIV/AIDS—adult prevalence rate: NA
HIV/AIDS—people living with HIV/AIDS: NA
HIV/AIDS—deaths: NA
Nationality:
noun: British Virgin Islander(s)
adjective: British Virgin Islander
Ethnic groups: black 83%, other 17% (includes white, Indian, Asian and mixed)
Religions: Protestant 86% (Methodist 33%, Anglican 17%, Church of God 9%, Seventh-Day Adventist 6%, Baptist 4%, Jehovah's Witnesses 2%, other 15%), Roman Catholic 10%, other 2%, none 2% (1991)
Languages: English (official)
Literacy:
definition: age 15 and over can read and write
total population: 97.8% (1991 est.)
male: NA%
female: NA%

GOVERNMENT

Country name:
conventional long form: none
conventional short form: British Virgin Islands
abbreviation: BVI
Dependency status: overseas territory of the UK; internal self-governing
Government type: NA
Capital: *name:* Road Town
geographic coordinates: 18 27 N, 64 37 W
time difference: UTC-4 (1 hour ahead of Washington, DC during Standard Time)
Administrative divisions: none (overseas territory of the UK)
Independence: none (overseas territory of the UK)
National holiday: Territory Day, 1 July (1956)
Constitution: 13 June 2007
Legal system: English law
Suffrage: 18 years of age; universal

Executive branch:
chief of state: Queen ELIZABETH II (since 6 February 1952); represented by Governor David PEAREY (since 18 April 2006)
head of government: Premier Ralph T. O'NEAL (since 23 August 2007)
cabinet: Executive Council appointed by the governor from members of the House of Assembly
elections: the monarch is hereditary; governor appointed by the monarch; following legislative elections, the leader of the majority party or the leader of the majority coalition is usually appointed premier by the governor
Legislative branch: unicameral House of Assembly (13 elected seats and 1 nonvoting ex officio member in the attorney general; members are elected by direct popular vote, 1 member from each of nine electoral districts, 4 at-large members; members serve four-year terms)
elections: last held 20 August 2007 (next to be held in 2011)
election results: percent of vote by party— VIP 45.2%, NDP 39.6%, independent 15.2%; seats by party—VIP 10, NDP 2, independent 1
Judicial branch: Eastern Caribbean Supreme Court, consisting of the High Court of Justice and the Court of Appeal (one judge of the Supreme Court is a resident of the islands and presides over the High Court); Magistrate's Court; Juvenile Court; Court of Summary Jurisdiction
Political parties and leaders: Concerned Citizens Movement or CCM [Ethlyn SMITH]; National Democratic Party or NDP [Orlando SMITH]; United Party or UP [Gregory MADURO]; Virgin Islands Party or VIP [Ralph T. O'NEAL]
Political pressure groups and leaders: NA
International organization participation: Caricom (associate), CDB, Interpol (subbureau), IOC, OECS, UNESCO (associate), UPU
Diplomatic representation in the US: none (overseas territory of the UK)
Diplomatic representation from the US: none (overseas territory of the UK)
Flag description: blue, with the flag of the UK in the upper hoist-side quadrant and the Virgin Islander coat of arms centered in the outer half of the flag; the coat of arms depicts a woman flanked on either side by a vertical column of six oil lamps above a scroll bearing the Latin word VIGILATE (Be Watchful)

ECONOMY

Economy—overview: The economy, one of the most stable and prosperous in the Caribbean, is highly dependent on tourism, generating an estimated 45% of the national income. An estimated 820,000 tourists, mainly from the US, visited the islands in 2005. In the mid-1980s, the government began offering offshore registration to companies wishing to incorporate in the islands, and incorporation fees now generate substantial revenues. Roughly 400,000 companies were on the offshore registry by yearend 2000. The adoption of a comprehensive insurance law in late 1994, which provides a blanket of confidentiality with regulated statutory gateways for investigation of criminal offenses, made the British Virgin Islands even more attractive to international business. Livestock raising is the most important agricultural activity; poor soils limit the islands' ability to meet domestic food requirements. Because of traditionally close links with the US Virgin Islands, the British Virgin Islands has used the US dollar as its currency since 1959.
GDP (purchasing power parity): $853.4 million (2004 est.)
GDP (official exchange rate): $839.7 million (2003)
GDP—real growth rate: 1% (2002 est.)
GDP—per capita (PPP): $38,500 (2004 est.)
GDP—composition by sector:
agriculture: 1.8%
industry: 6.2%
services: 92% (1996 est.)
Labor force: 12,770 (2004)
Labor force—by occupation: *agriculture:* 0.6%
industry: 40%
services: 59.4% (2005)
Unemployment rate: 3.6% (1997)
Population below poverty line: NA%
Household income or consumption by percentage share:
lowest 10%: NA%
highest 10%: NA%
Inflation rate (consumer prices): 2% (2005)
Budget:
revenues: $204.7 million
expenditures: $180.4 million (2004)
Agriculture—products: fruits, vegetables; livestock, poultry; fish
Industries: tourism, light industry, construction, rum, concrete block, offshore financial center
Industrial production growth rate: NA%
Electricity—production: 45 million kWh (2005)
Electricity—production by source:
fossil fuel: 100%
hydro: 0%

nuclear: 0%
other: 0% (2001)
Electricity—consumption: 41.85 million kWh (2005)
Electricity—exports: 0 kWh (2005)
Electricity—imports: 0 kWh (2005)
Oil—production: 0 bbl/day (2005)
Oil—consumption: 600 bbl/day (2005 est.)
Oil—exports: 0 bbl/day (2004)
Oil—imports: 604.3 bbl/day (2004)
Oil—proved reserves: 0 bbl (1 January 2006 est.)
Natural gas—production: 0 cu m (2005 est.)
Natural gas—consumption: 0 cu m (2005 est.)
Natural gas—exports: 0 cu m (2005 est.)
Natural gas—imports: 0 cu m (2005)
Natural gas—proved reserves: 0 cu m (1 January 2006 est.)
Current account balance: $134.3 million (1999)
Exports: $25.3 million (2002)
Exports—commodities: rum, fresh fish, fruits, animals; gravel, sand
Exports—partners: Virgin Islands (US), Puerto Rico, US (2006)
Imports: $187 million (2002 est.)
Imports—commodities: building materials, automobiles, foodstuffs, machinery
Imports—partners: Virgin Islands (US), Puerto Rico, US (2006)

Economic aid—recipient: $NA
Debt—external: $36.1 million (1997)
Currency (code): US dollar (USD)
Currency code: USD
Exchange rates: the US dollar is used
Fiscal year: 1 April—31 March

Telephones—main lines in use: 11,700 (2002)
Telephones—mobile cellular: 8,000 (2002)
Telephone system:
general assessment: worldwide telephone service
domestic: NA
international: country code—1-284; connected via submarine cable to Bermuda; the East Caribbean Fiber System (ECFS) submarine cable provides connectivity to 13 other islands in the eastern Caribbean (2007)
Radio broadcast stations: AM 1, FM 5, shortwave 0 (2004)
Radios: 9,000 (1997)
Television broadcast stations: 1 (plus 1 cable company) (1997)
Televisions: 4,000 (1997)
Internet country code: .vg
Internet hosts: 490 (2007)
Internet Service Providers (ISPs): 16 (2000)
Internet users: 4,000 (2002)

Airports: 3 (2007)
Airports—with paved runways:
total: 2
914 to 1,523 m: 1
under 914 m: 1 (2007)
Airports—with unpaved runways:
total: 1
914 to 1,523 m: 1 (2007)
Roadways:
total: 177 km
paved: 177 km (2002)
Ports and terminals: Road Town

Manpower available for military service:
males age 16–49: 7,101 (2008 est.)
Manpower fit for military service:
males age 16–49: 5,921 (2008 est.)
Manpower reaching militarily significant age annually:
males age 16–49: 185 (2008 est.)
Military—note: defense is the responsibility of the UK

Disputes—international: none
Illicit drugs: transshipment point for South American narcotics destined for the US and Europe; large offshore financial center makes it vulnerable to money laundering

BRUNEI

Background: The Sultanate of Brunei's influence peaked between the 15th and 17th centuries when its control extended over coastal areas of northwest Borneo and the southern Philippines. Brunei subsequently entered a period of decline brought on by internal strife over royal succession, colonial expansion of European powers, and piracy. In 1888, Brunei became a British protectorate; independence was achieved in 1984. The same family has ruled Brunei for over six centuries. Brunei benefits from extensive petroleum and natural gas fields, the source of one of the highest per capita GDPs in Asia.

Location: Southeastern Asia, bordering the South China Sea and Malaysia
Geographic coordinates: 4 30 N, 114 40 E
Map references: Southeast Asia
Area:
total: 5,770 sq km
land: 5,270 sq km
water: 500 sq km
Area—comparative: slightly smaller than Delaware
Land boundaries:
total: 381 km
border countries: Malaysia 381 km
Coastline: 161 km
Maritime claims:
territorial sea: 12 nm

exclusive economic zone: 200 nm or to median line
Climate: tropical; hot, humid, rainy
Terrain: flat coastal plain rises to mountains in east; hilly lowland in west
Elevation extremes:
lowest point: South China Sea 0 m
highest point: Bukit Pagon 1,850 m
Natural resources: petroleum, natural gas, timber
Land use:
arable land: 2.08%
permanent crops: 0.87%
other: 97.05% (2005)
Irrigated land: 10 sq km (2003)
Total renewable water resources: 8.5 cu km (1999)
Freshwater withdrawal (domestic/industrial/agricultural): *total:* 0.09
per capita: 243 cu m/yr (1994)
Natural hazards: typhoons, earthquakes, and severe flooding are rare
Environment—current issues: seasonal smoke/haze resulting from forest fires in Indonesia

97

Environment—international agreements: *party to:* Biodiversity, Climate Change, Desertification, Endangered Species, Environmental Modification, Hazardous Wastes, Law of the Sea, Ozone Layer Protection, Ship Pollution *signed, but not ratified:* none of the selected agreements

Geography—note: close to vital sea lanes through South China Sea linking Indian and Pacific Oceans; two parts physically separated by Malaysia; almost an enclave within Malaysia

PEOPLE

Population: 381,371 (July 2008 est.)

Age structure:
0–14 years: 27.2% (male 53,400/female 50,333)
15–64 years: 69.6% (male 132,895/female 132,391)
65 years and over: 3.2% (male 5,927/female 6,425) (2008 est.)

Median age: *total:* 27.5 years
male: 27.5 years
female: 27.5 years (2008 est.)

Population growth rate: 1.785% (2008 est.)

Birth rate: 18.39 births/1,000 population (2008 est.)

Death rate: 3.28 deaths/1,000 population (2008 est.)

Net migration rate: 2.74 migrant(s)/1,000 population (2008 est.)

Sex ratio:
at birth: 1.05 male(s)/female
under 15 years: 1.06 male(s)/female
15–64 years: 1 male(s)/female
65 years and over: 0.92 male(s)/female
total population: 1.02 male(s)/female (2008 est.)

Infant mortality rate:
total: 12.69 deaths/1,000 live births
male: 15.19 deaths/1,000 live births
female: 10.07 deaths/1,000 live births (2008 est.)

Life expectancy at birth:
total population: 75.52 years
male: 73.32 years
female: 77.83 years (2008 est.)

Total fertility rate: 1.94 children born/woman (2008 est.)

HIV/AIDS—adult prevalence rate: less than 0.1% (2003 est.)

HIV/AIDS—people living with HIV/AIDS: fewer than 200 (2003 est.)

HIV/AIDS—deaths: fewer than 200 (2003 est.)

Nationality: *noun:* Bruneian(s)
adjective: Bruneian

Ethnic groups: Malay 67%, Chinese 15%, indigenous 6%, other 12%

Religions: Muslim (official) 67%, Buddhist 13%, Christian 10%, other (includes indigenous beliefs) 10%

Languages: Malay (official), English, Chinese

Literacy:
definition: age 15 and over can read and write
total population: 92.7%
male: 95.2%
female: 90.2% (2001 census)

GOVERNMENT

Country name:
conventional long form: Brunei Darussalam
conventional short form: Brunei
local long form: Negara Brunei Darussalam
local short form: Brunei

Government type: constitutional sultanate

Capital: *name:* Bandar Seri Begawan
geographic coordinates: 4 53 N, 114 56 E
time difference: UTC+8 (13 hours ahead of Washington, DC during Standard Time)

Administrative divisions: 4 districts (daerah-daerah, singular—daerah); Belait, Brunei and Muara, Temburong, Tutong

Independence: 1 January 1984 (from UK)

National holiday: National Day, 23 February (1984); note—1 January 1984 was the date of independence from the UK, 23 February 1984 was the date of independence from British protection

Constitution: 29 September 1959 (some provisions suspended under a State of Emergency since December 1962, others since independence on 1 January 1984)

Legal system: based on English common law; for Muslims, Islamic Shari'a law supersedes civil law in a number of areas; has not accepted compulsory ICJ jurisdiction

Suffrage: 18 years of age for village elections; universal

Executive branch:
chief of state: Sultan and Prime Minister Sir HASSANAL Bolkiah (since 5 October 1967); note—the monarch is both the chief of state and head of government
head of government: Sultan and Prime Minister Sir HASSANAL Bolkiah (since 5 October 1967)
cabinet: Council of Cabinet Ministers appointed and presided over by the monarch; deals with executive matters; note—there is also a Religious Council (members appointed by the monarch) that advises on religious matters, a Privy Council (members appointed by the monarch) that deals with constitutional matters, and the Council of Succession (members appointed by the monarch)

that determines the succession to the throne if the need arises
elections: none; the monarch is hereditary

Legislative branch: Legislative Council met on 25 September 2004 for first time in 20 years with 21 members appointed by the Sultan; passed constitutional amendments calling for a 45-seat council with 15 elected members; Sultan dissolved council on 1 September 2005 and appointed a new council with 29 members as of 2 September 2005; council met in March 2006 and in March 2007
elections: last held in March 1962 (date of next election NA)

Judicial branch: Supreme Court—chief justice and judges are sworn in by monarch for three-year terms; Judicial Committee of Privy Council in London is final court of appeal for civil cases; Shariah courts deal with Islamic laws (2006)

Political parties and leaders: National Development Party or NDP [YASSIN Affendi]
note: Brunei National Solidarity Party or PPKB [Abdul LATIF bin Chuchu] and People's Awareness Party or PAKAR [Awang Haji MAIDIN bin Haji Ahmad] were deregistered; parties are small and have limited activity

Political pressure groups and leaders: NA

International organization participation: ADB, APEC, APT, ARF, ASEAN, C, EAS, G-77, IBRD, ICAO, ICRM, IDB, IFRCS, ILO, IMF, IMO, IMSO, Interpol, IOC, ISO (correspondent), ITSO, ITU, NAM, OIC, OPCW, UN, UNCTAD, UNESCO, UNWTO, UPU, WCO, WHO, WIPO, WMO, WTO

Diplomatic representation in the US:
chief of mission: Ambassador Pengiran Anak Dato PUTEH
chancery: 3520 International Court NW, Washington, DC 20008
telephone: [1] (202) 237-1838
FAX: [1] (202) 885-0560

Diplomatic representation from the US:
chief of mission: Ambassador Emil SKODON
embassy: Third Floor, Teck Guan Plaza, Jalan Sultan, Bandar Seri Begawan, BS8811
mailing address: PSC 470 (BSB), FPO AP 96507; P.O. Box 2991, Bandar Seri Begawan BS8675, Negara Brunei Darussalam
telephone: [673] 222-0384
FAX: [673] 222-5293

Flag description: yellow with two diagonal bands of white (top, almost double width) and black starting from the upper hoist side; the national emblem in red is superimposed at the center; the emblem includes a swallow-tailed flag on top of a winged column within an upturned crescent above a scroll and flanked by two upraised hands

ECONOMY

Economy—overview: Brunei has a small well-to-do economy that encompasses a mixture of foreign and domestic entrepreneurship, government regulation, welfare measures, and village tradition. Crude oil and natural gas production account for just over half of GDP and more than 90% of exports. Per capita GDP is among the highest in Asia, and substantial income from overseas investment supplements income from domestic production. The government provides for all medical services and free education through the university level and subsidizes rice and housing. Brunei's leaders are concerned that steadily increased integration in the world economy will undermine internal social cohesion. Plans for the future include upgrading the labor force, reducing unemployment, strengthening the banking and tourist sectors, and, in general, further widening the economic base beyond oil and gas.

GDP (purchasing power parity): $19.64 billion (2007 est.)

GDP (official exchange rate): $12.39 billion (2007 est.)

GDP—real growth rate: 0.4% (2007 est.)

GDP—per capita (PPP): $51,000 (2007 est.)

GDP—composition by sector:
agriculture: 0.9%
industry: 71.6%
services: 27.5% (2005 est.)

Labor force: 180,400 (2006 est.)

Labor force—by occupation: *agriculture:* 2.9%
industry: 61.1%
services: 36% (2003 est.)

Unemployment rate: 4% (2006)

Population below poverty line: NA%

Household income or consumption by percentage share:
lowest 10%: NA%
highest 10%: NA%

Inflation rate (consumer prices): 0.4% (2007 est.)

Budget:
revenues: $3.765 billion
expenditures: $4.815 billion (2004 est.)

Agriculture—products: rice, vegetables, fruits; chickens, water buffalo, cattle, goats, eggs

Industries: petroleum, petroleum refining, liquefied natural gas, construction

Industrial production growth rate: 1.8% (2005 est.)

Electricity—production: 2.735 billion kWh (2005)

Electricity—production by source:
fossil fuel: 100%
hydro: 0%
nuclear: 0%
other: 0% (2001)

Electricity—consumption: 2.625 billion kWh (2005 est.)

Electricity—exports: 0 kWh (2005)

Electricity—imports: 0 kWh (2005)

Oil—production: 219,300 bbl/day (2006)

Oil—consumption: 14,900 bbl/day (2006 est.)

Oil—exports: 205,600 bbl/day (2006)

Oil—imports: 660.1 bbl/day (2004)

Oil—proved reserves: 1.35 billion bbl (1 January 2006 est.)

Natural gas—production: 11.03 billion cu m (2005 est.)

Natural gas—consumption: 2.254 billion cu m (2005 est.)

Natural gas—exports: 8.776 billion cu m (2005 est.)

Natural gas—imports: 0 cu m (2005)

Natural gas—proved reserves: 374.8 billion cu m (1 January 2006 est.)

Current account balance: $7.101 billion (2007 est.)

Exports: $6.767 billion f.o.b. (2006)

Exports—commodities: crude oil, natural gas, refined products, clothing

Exports—partners: Japan 30.8%, Indonesia 20.1%, South Korea 15%, Australia 11.6%, US 7.8% (2006)

Imports: $2 billion c.i.f. (2006)

Imports—commodities: machinery and transport equipment, manufactured goods, food, chemicals

Imports—partners: Singapore 31.6%, Malaysia 19%, UK 8.1%, Japan 5.6%, China 5.5%, Thailand 4.6% (2006)

Economic aid—recipient: $770,000 (2004)

Debt—external: $0 (2005)

Market value of publicly traded shares: $NA

Currency (code): Bruneian dollar (BND)

Currency code: BND

Exchange rates: Bruneian dollars per US dollar—NA (2007), 1.5886 (2006), 1.6644 (2005), 1.6902 (2004), 1.7422 (2003)

Fiscal year: 1 April—31 March

COMMUNICATIONS

Telephones—main lines in use: 80,200 (2006)

Telephones—mobile cellular: 254,000 (2006)

Telephone system:
general assessment: service throughout the country is excellent; international service is good to Southeast Asia, Middle East, Western Europe, and the US
domestic: every service available
international: country code—673; landing point for the SEA-ME-WE-3 optical telecommunications submarine cable that provides links to Asia, the Middle East, and Europe; the Asia-America Gateway submarine cable network, scheduled for completion by late 2008, will provide new links to Asia and the US; satellite earth stations—2 Intelsat (1 Indian Ocean and 1 Pacific Ocean) (2007)

Radio broadcast stations: AM 1, FM 2 (transmitting on 18 different frequencies), shortwave 0 (British Forces Broadcasting Service (BFBS) station transmits two FM signals with English and Nepali service) (2006)

Radios: 329,000 (1998)

Television broadcast stations: 4 (includes 2 UHF stations broadcasting a subscription service) (2006)

Televisions: 201,900 (1998)

Internet country code: .bn

Internet hosts: 15,347 (2007)

Internet Service Providers (ISPs): 2 (2000)

Internet users: 165,600 (2006)

TRANSPORTATION

Airports: 2 (2007)

Airports—with paved runways: *total:* 1
over 3,047 m: 1 (2007)

Airports—with unpaved runways:
total: 1
914 to 1,523 m: 1 (2007)

Heliports: 3 (2007)

Pipelines: gas 672 km; oil 463 km (2007)

Roadways:
total: 3,650 km
paved: 2,819 km
unpaved: 831 km (2005)

Waterways: 209 km (navigable by craft drawing less than 1.2 m) (2007)

Merchant marine:
total: 8 ships (1000 GRT or over) 465,937 GRT/413,393 DWT
by type: liquefied gas 8
foreign-owned: 8 (UK 8) (2007)

Ports and terminals: Lumut, Muara, Seria

MILITARY

Military branches: Royal Brunei Armed Forces (RBAF): Royal Brunei Land Forces, Royal Brunei Navy, Royal Brunei Air Force (Tentera Udara Diraja Brunei) (2008)

Military service age and obligation: 18 years of age (est.) for voluntary military service; non-Malays are ineligible to serve (2007)

Manpower available for military service:
males age 16–49: 108,356
females age 16–49: 110,153 (2008 est.)
Manpower fit for military service:
males age 16–49: 91,297
females age 16–49: 93,228 (2008 est.)
Manpower reaching militarily significant age annually:
males age 16–49: 3,223
females age 16–49: 3,182 (2008 est.)
Military expenditures—percent of GDP: 4.5% (2006)

Disputes—international: Brunei and Malaysia are still considering international adjudication over their disputed offshore and deepwater seabeds, where hydrocarbon exploration was terminated in 2003 international legal adjudication; Malaysia's land boundary with Brunei around Limbang is in dispute; Brunei established an exclusive economic fishing zone encompassing Louisa Reef in the southern Spratly Islands in 1984, but makes no public territorial claim to the offshore reefs; the 2002 "Declaration on the Conduct of Parties in the South China Sea" has eased tensions in the Spratly Islands but falls short of a legally binding "code of conduct" desired by several of the disputants

Illicit drugs: drug trafficking and illegally importing controlled substances are serious offenses in Brunei and carry a mandatory death penalty

BULGARIA

INTRODUCTION

Background: The Bulgars, a Central Asian Turkic tribe, merged with the local Slavic inhabitants in the late 7th century to form the first Bulgarian state. In succeeding centuries, Bulgaria struggled with the Byzantine Empire to assert its place in the Balkans, but by the end of the 14th century the country was overrun by the Ottoman Turks. Northern Bulgaria attained autonomy in 1878 and all of Bulgaria became independent from the Ottoman Empire in 1908. Having fought on the losing side in both World Wars, Bulgaria fell within the Soviet sphere of influence and became a People's Republic in 1946. Communist domination ended in 1990, when Bulgaria held its first multiparty election since World War II and began the contentious process of moving toward political democracy and a market economy while combating inflation, unemployment, corruption, and crime. The country joined NATO in 2004 and the EU in 2007.

GEOGRAPHY

Location: Southeastern Europe, bordering the Black Sea, between Romania and Turkey

Geographic coordinates: 43 00 N, 25 00 E
Map references: Europe
Area: *total:* 110,910 sq km
land: 110,550 sq km
water: 360 sq km
Area—comparative: slightly larger than Tennessee
Land boundaries:
total: 1,808 km
border countries: Greece 494 km, Macedonia 148 km, Romania 608 km, Serbia 318 km, Turkey 240 km
Coastline: 354 km
Maritime claims:
territorial sea: 12 nm
contiguous zone: 24 nm
exclusive economic zone: 200 nm
Climate: temperate; cold, damp winters; hot, dry summers
Terrain: mostly mountains with lowlands in north and southeast
Elevation extremes:
lowest point: Black Sea 0 m
highest point: Musala 2,925 m
Natural resources: bauxite, copper, lead, zinc, coal, timber, arable land
Land use:
arable land: 29.94%
permanent crops: 1.9%
other: 68.16% (2005)
Irrigated land: 5,880 sq km (2003)
Total renewable water resources: 19.4 cu km (2005)
Freshwater withdrawal (domestic/industrial/agricultural): *total:* 6.92 cu km/yr (3%/78%/19%)
per capita: 895 cu m/yr (2003)
Natural hazards: earthquakes, landslides
Environment—current issues: air pollution from industrial emissions; rivers polluted from raw sewage, heavy metals, detergents; deforestation; forest damage from air pollution and resulting acid rain; soil contamination from heavy metals from metallurgical plants and industrial wastes

Environment—international agreements: *party to:* Air Pollution, Air Pollution-Nitrogen Oxides, Air Pollution-Persistent Organic Pollutants, Air Pollution-Sulfur 85, Air Pollution-Sulfur 94, Air Pollution-Volatile Organic Compounds, Antarctic-Environmental Protocol, Antarctic-Marine Living Resources, Antarctic Treaty, Biodiversity, Climate Change, Climate Change-Kyoto Protocol, Desertification, Endangered Species, Environmental Modification, Hazardous Wastes, Law of the Sea, Marine Dumping, Ozone Layer Protection, Ship Pollution, Wetlands
signed, but not ratified: none of the selected agreements
Geography—note: strategic location near Turkish Straits; controls key land routes from Europe to Middle East and Asia

PEOPLE

Population: 7,262,675 (July 2008 est.)
Age structure:
0–14 years: 13.8% (male 514,238/female 489,608)
15–64 years: 68.6% (male 2,449,812/female 2,532,845)
65 years and over: 17.6% (male 520,962/female 755,210) (2008 est.)
Median age:
total: 41.1 years
male: 38.9 years
female: 43.4 years (2008 est.)
Population growth rate: -0.813% (2008 est.)
Birth rate: 9.58 births/1,000 population (2008 est.)
Death rate: 14.3 deaths/1,000 population (2008 est.)
Net migration rate: -3.41 migrant(s)/1,000 population (2008 est.)
Sex ratio:
at birth: 1.06 male(s)/female
under 15 years: 1.05 male(s)/female

15–64 years: 0.97 male(s)/female
65 years and over: 0.69 male(s)/female
total population: 0.92 male(s)/female
(2008 est.)
Infant mortality rate:
total: 18.51 deaths/1,000 live births
male: 22 deaths/1,000 live births
female: 14.8 deaths/1,000 live births
(2008 est.)
Life expectancy at birth:
total population: 72.83 years
male: 69.22 years
female: 76.66 years (2008 est.)
Total fertility rate: 1.4 children
born/woman (2008 est.)
HIV/AIDS—adult prevalence rate: less
than 0.1% (2001 est.)
**HIV/AIDS—people living with
HIV/AIDS:** 346 (2001 est.)
HIV/AIDS—deaths: 100 (2001 est.)
Nationality:
noun: Bulgarian(s)
adjective: Bulgarian
Ethnic groups: Bulgarian 83.9%, Turk
9.4%, Roma 4.7%, other 2% (including
Macedonian, Armenian, Tatar,
Circassian) (2001 census)
Religions: Bulgarian Orthodox 82.6%,
Muslim 12.2%, other Christian 1.2%,
other 4% (2001 census)
Languages: Bulgarian 84.5%, Turkish
9.6%, Roma 4.1%, other and unspecified
1.8% (2001 census)
Literacy:
definition: age 15 and over can read and
write
total population: 98.2%
male: 98.7%
female: 97.7% (2001 census)

GOVERNMENT

Country name:
conventional long form: Republic of
Bulgaria
conventional short form: Bulgaria
local long form: Republika Balgariya
local short form: Balgariya
Government type: parliamentary democracy
Capital: *name:* Sofia
geographic coordinates: 42 41 N, 23 19 E
time difference: UTC+2 (7 hours ahead of
Washington, DC during Standard Time)
daylight saving time: +1hr, begins last
Sunday in March; ends last Sunday in
October
Administrative divisions: 28 provinces
(oblasti, singular—oblast); Blagoevgrad,
Burgas, Dobrich, Gabrovo, Khaskovo,
Kurdzhali, Kyustendil, Lovech,
Montana, Pazardzhik, Pernik, Pleven,
Plovdiv, Razgrad, Ruse, Shumen, Silistra,
Sliven, Smolyan, Sofiya, Sofiya-Grad,
Stara Zagora, Turgovishte, Varna, Veliko
Turnovo, Vidin, Vratsa, Yambol

Independence: 3 March 1878 (as an
autonomous principality within the
Ottoman Empire); 22 September 1908
(complete independence from the
Ottoman Empire)
National holiday: Liberation Day, 3
March (1878)
Constitution: adopted 12 July 1991
Legal system: civil and criminal law
based on Roman law; accepts compulsory
ICJ jurisdiction with reservations
Suffrage: 18 years of age; universal
Executive branch:
chief of state: President Georgi PAR-
VANOV (since 22 January 2002); Vice
President Angel MARIN (since 22
January 2002)
head of government: Prime Minister
Sergei STANISHEV (since 16 August
2005); Deputy Prime Ministers Ivaylo
KALFIN, Daniel VULCHEV, and Emel
ETEM (since 16 August 2005) and
Meglena PLUGCHIEVA (since 25 April
2008)
cabinet: Council of Ministers nominated
by the prime minister and elected by the
National Assembly
elections: president and vice president
elected on the same ticket by popular
vote for a five-year term (eligible for a
second term); election last held 22 and
29 October 2006 (next to be held in
2011); chairman of the Council of
Ministers (prime minister) nominated by
the president and elected by the
National Assembly; deputy prime ministers
nominated by the prime minister
and elected by the National Assembly
election results: Georgi PARVANOV
reelected president; percent of vote—
Georgi PARVANOV 77.3%, Volen
SIDEROV 22.7%; Sergei STANISHEV
elected prime minister, result of legislative
vote—168 to 67
Legislative branch: unicameral
National Assembly or Narodno Sobranie
(240 seats; members elected by popular
vote to serve four-year terms)
elections: last held 25 June 2005 (next to
be held in June 2009)
election results: percent of vote by party—
CfB 31.1%, NMS2 19.9%, MRF 12.7%,
ATAKA 8.2%, UDF 7.7%, DSB 6.5%,
BPU 5.2%, other 8.7%; seats by party—
CfB 83, NMS2 53, MRF 33, UDF 20,
ATAKA 17, DSB 17, BPU 13, independents
4; note—seats by party as of January
2008—CfB 82, NMS2 36, MRF 34,
Bulgarian New Democracy 16, DSB 16,
UDF 16, BPU 13, ATAKA 11, independents 16
Judicial branch: Supreme Administrative
Court; Supreme Court of Cassation;
Constitutional Court (12 justices appointed
or elected for nine-year terms);

Supreme Judicial Council (consists of the
chairmen of the two Supreme Courts, the
Chief Prosecutor, and 22 other members;
responsible for appointing the justices, prosecutors,
and investigating magistrates in the
justice system; members of the Supreme
Judicial Council elected for five-year terms,
11 elected by the National Assembly and
11 by bodies of the judiciary)
Political parties and leaders: ATAKA
(Attack Coalition) (coalition of parties
headed by the Attack National Union);
Attack National Union [Volen
SIDEROV]; Bulgarian Agrarian
National Union-People's Union or
BANU [Anastasia MOZER]; Bulgarian
New Democracy [Borislav RALCHEV];
Bulgarian People's Union or BPU (coalition
of UFD, IMRO, and BANU);
Bulgarian Socialist Party or BSP [Sergei
STANISHEV]; Citizens for the
European Development of Bulgaria or
GERB [Tsvetan TSVETANOV];
Coalition for Bulgaria or CfB (coalition
of parties dominated by BSP) [Sergei
STANISHEV]; Democrats for a Strong
Bulgaria or DSB [Ivan KOSTOV];
Internal Macedonian Revolutionary
Organization or IMRO [Krasimir
KARAKACHANOV]; Movement for
Rights and Freedoms or MRF [Ahmed
DOGAN]; National Movement for
Stability and Progress or NMSS [Simeon
SAXE-COBURG-GOTHA] (formerly
National Movement Simeon II or
NMS2); New Time [Emil
KOSHLUKOV]; Union of Democratic
Forces or UDF [Petar STOYANOV];
Union of Free Democrats or UFD [Stefan
SOFIYANSKI]; United Democratic
Forces or UtDF (a coalition of center-
right parties dominated by UDF)
Political pressure groups and leaders:
Confederation of Independent Trade
Unions of Bulgaria or CITUB; Podkrepa
Labor Confederation; numerous
regional, ethnic, and national interest
groups with various agendas
International organization participation:
ACCT, Australia Group, BIS, BSEC, CE,
CEI, CERN, EAPC, EBRD, EIB, EU
(new member), FAO, G-9, IAEA, IBRD,
ICAO, ICC, ICCt, ICRM, IFC, IFRCS,
ILO, IMF, IMO, IMSO, Interpol, IOC,
IOM, IPU, ISO, ITSO, ITU, ITUC,
MIGA, NAM (guest), NATO, NSG,
OAS (observer), OIF, OPCW, OSCE,
PCA, SECI, UN, UNCTAD, UNESCO,
UNIDO, UNMEE, UNMIL, UNMIS,
UNWTO, UPU, WCL, WCO, WEU
(associate affiliate), WFTU, WHO,
WIPO, WMO, WTO, ZC
Diplomatic representation in the US:
chief of mission: Ambassador Elena B.
POPTODOROVA

chancery: 1621 22nd Street NW, Washington, DC 20008
telephone: [1] (202) 387-0174
FAX: [1] (202) 234-7973
consulate(s) general: Chicago, Los Angeles, New York
Diplomatic representation from the US:
chief of mission: Ambassador John Ross BEYRLE
embassy: 16 Kozyak Street, Sofia 1407
mailing address: American Embassy Sofia, US Department of State, 5740 Sofia Place, Washington, DC 20521-5740
telephone: [359] (2) 937-5100
FAX: [359] (2) 937-5320
Flag description: three equal horizontal bands of white (top), green, and red
note: the national emblem, formerly on the hoist side of the white stripe, has been removed

ECONOMY

Economy—overview: Bulgaria, a former communist country that entered the EU on 1 January 2007, has experienced strong growth since a major economic downturn in 1996. Successive governments have demonstrated commitment to economic reforms and responsible fiscal planning, but have failed so far to rein in rising inflation and large current account deficits. Bulgaria has averaged more than 6% growth since 2004, attracting significant amounts of foreign direct investment, but corruption in the public administration, a weak judiciary, and the presence of organized crime remain significant challenges.
GDP (purchasing power parity): $86.32 billion (2007 est.)
GDP (official exchange rate): $39.61 billion (2007 est.)
GDP—real growth rate: 6.2% (2007 est.)
GDP—per capita (PPP): $11,300 (2007 est.)
GDP—composition by sector:
agriculture: 6.3%
industry: 32.3%
services: 61.4% (2007 est.)
Labor force: 2.59 million (2007 est.)
Labor force—by occupation: *agriculture:* 8.5%
industry: 33.6%
services: 57.9% (2nd qtr. 2006 est.)
Unemployment rate: 7.7% (2007 est.)
Population below poverty line: 14.1% (2003 est.)
Household income or consumption by percentage share:
lowest 10%: 2.9%
highest 10%: 25.4% (2005)
Distribution of family income—Gini index: 31.6 (2005)
Inflation rate (consumer prices): 7.6% (2007 est.)
Investment (gross fixed): 29.8% of GDP (2007 est.)
Budget:
revenues: $16.84 billion
expenditures: $15.35 billion (2007 est.)
Public debt: 12.7% of GDP (2007 est.)
Agriculture—products: vegetables, fruits, tobacco, wine, wheat, barley, sunflowers, sugar beets; livestock
Industries: electricity, gas, water; food, beverages, tobacco; machinery and equipment, base metals, chemical products, coke, refined petroleum, nuclear fuel
Industrial production growth rate: 14% (2007 est.)
Electricity—production: 45.7 billion kWh (2006)
Electricity—production by source:
fossil fuel: 47.8%
hydro: 8.1%
nuclear: 44.1%
other: 0% (2001)
Electricity—consumption: 37.4 billion kWh (2006)
Electricity—exports: 7.8 billion kWh (2006)
Electricity—imports: 0 kWh (2006)
Oil—production: 3,661 bbl/day (2005 est.)
Oil—consumption: 108,000 bbl/day (2005 est.)
Oil—exports: 51,000 bbl/day (2005 est.)
Oil—imports: 138,800 bbl/day (2004 est.)
Oil—proved reserves: 15 million bbl (1 January 2006 est.)
Natural gas—production: 407,000 cu m (2005 est.)
Natural gas—consumption: 5.179 billion cu m (2005 est.)
Natural gas—exports: 0 cu m (2005 est.)
Natural gas—imports: 5.179 billion cu m (2005)
Natural gas—proved reserves: 5.703 billion cu m (1 January 2006 est.)
Current account balance: -$8.464 billion (2007 est.)
Exports: $18.44 billion f.o.b. (2007 est.)
Exports—commodities: clothing, footwear, iron and steel, machinery and equipment, fuels
Exports—partners: Turkey 11.6%, Italy 10.1%, Germany 9.6%, Greece 8%, Belgium 6.5%, France 4.2% (2006)
Imports: $28.67 billion f.o.b. (2007 est.)
Imports—commodities: machinery and equipment; metals and ores; chemicals and plastics; fuels, minerals, and raw materials
Imports—partners: Germany 14.8%, Italy 10.5%, Turkey 7.1%, Greece 6.2%, China 5%, France 4.9%, Romania 4.5% (2006)
Economic aid—recipient: $742 million (2005–06 est.)
Reserves of foreign exchange and gold: $17.38 billion (31 December 2007 est.)
Debt—external: $34.44 billion (30 June 2007)
Stock of direct foreign investment—at home: $28.84 billion (2007 est.)
Stock of direct foreign investment—abroad: $607 million (2007 est.)
Market value of publicly traded shares: $10.32 billion (2006)
Currency (code): lev (BGL)
Currency code: BGN
Exchange rates: leva per US dollar—1.4366 (2007), 1.5576 (2006), 1.5741 (2005), 1.5751 (2004), 1.7327 (2003)
Fiscal year: calendar year

COMMUNICATIONS

Telephones—main lines in use: 2.399 million (2006)
Telephones—mobile cellular: 8.253 million (2006)
Telephone system:
general assessment: an extensive but antiquated telecommunications network inherited from the Soviet era; quality has improved; the Bulgaria Telecommunications Company's fixed-line monopoly terminated in 2005 when alternative fixed-line operators were given access to its network; a drop in fixed-line connections in recent years has been offset by a sharp increase in mobile-cellular telephone use fostered by multiple service providers
domestic: a fairly modern digital cable trunk line now connects switching centers in most of the regions; the others are connected by digital microwave radio relay
international: country code—359; submarine cable provides connectivity to Ukraine and Russia; a combination submarine cable and land fiber-optic system provides connectivity to Italy, Albania, and Macedonia; satellite earth stations—3 (1 Intersputnik in the Atlantic Ocean region, 2 Intelsat in the Atlantic and Indian Ocean regions) (2007)
Radio broadcast stations: AM 31, FM 63, shortwave 2 (2001)
Radios: 4.51 million (1997)
Television broadcast stations: 39 (plus 1,242 repeaters) (2001)
Televisions: 3.31 million (1997)
Internet country code: .bg
Internet hosts: 298,781 (2007)
Internet Service Providers (ISPs): 200 (2001)
Internet users: 1.87 million (2006)

TRANSPORTATION

Airports: 214 (2007)
Airports—with paved runways:

total: 131
over 3,047 m: 2
2,438 to 3,047 m: 18
1,524 to 2,437 m: 15
914 to 1,523 m: 1
under 914 m: 95 (2007)
Airports—with unpaved runways:
total: 83
1,524 to 2,437 m: 2
914 to 1,523 m: 9
under 914 m: 72 (2007)
Heliports: 4 (2007)
Pipelines: gas 2,500 km; oil 339 km; refined products 156 km (2007)
Railways:
total: 4,294 km
standard gauge: 4,049 km 1.435-m gauge (2,710 km electrified)
narrow gauge: 245 km 0.760-m gauge (2006)
Roadways:
total: 44,033 km
paved: 43,593 km (includes 333 km of expressways)
unpaved: 440 km (2004)
Waterways: 470 km (2007)

Merchant marine:
total: 71 ships (1000 GRT or over) 833,153 GRT/1,194,660 DWT
by type: bulk carrier 37, cargo 14, chemical tanker 4, container 6, liquefied gas 1, passenger/cargo 2, petroleum tanker 3, roll on/roll off 4
foreign-owned: 3 (Germany 1, Ireland 1, Russia 1)
registered in other countries: 39 (Comoros 1, Malta 15, Mongolia 2, Panama 1, Slovakia 7, St Vincent and The Grenadines 13) (2007)
Ports and terminals: Burgas, Varna

MILITARY

Military branches: Bulgarian Armed Forces: Ground Forces, Naval Forces, Bulgarian Air Forces (Bulgarski Voennovazdyshni Sily, BVVS) (2008)
Military service age and obligation: 18–27 years of age for voluntary military service; conscript service obligation—9 months; as of May 2006, 67% of the Bulgarian Army comprised of professional soldiers; conscription ended as of 1 January 2008; Air and Air Defense Forces and Naval Forces became fully professional at the end of 2006 (2008)
Manpower available for military service:
males age 16–49: 1,701,979
females age 16–49: 1,691,092 (2008 est.)
Manpower fit for military service:
males age 16–49: 1,364,029
females age 16–49: 1,401,348 (2008 est.)
Manpower reaching militarily significant age annually:
males age 16–49: 39,477
females age 16–49: 37,339 (2008 est.)
Military expenditures—percent of GDP: 2.6% (2005 est.)

TRANSNATIONAL ISSUES

Disputes—international: none
Illicit drugs: major European transshipment point for Southwest Asian heroin and, to a lesser degree, South American cocaine for the European market; limited producer of precursor chemicals; some money laundering of drug-related proceeds through financial institutions

BURKINA FASO

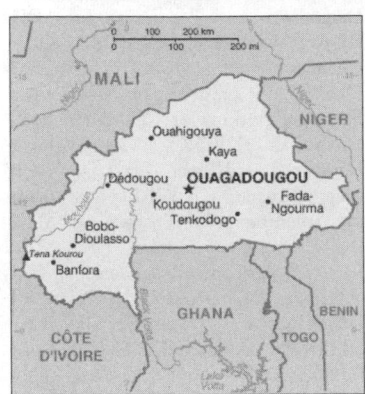

INTRODUCTION

Background: Burkina Faso (formerly Upper Volta) achieved independence from France in 1960. Repeated military coups during the 1970s and 1980s were followed by multiparty elections in the early 1990s. Current President Blaise COMPAORE came to power in a 1987 military coup and has won every election since then. Burkina Faso's high population density and limited natural resources result in poor economic prospects for the majority of its citizens. Recent unrest in Cote d'Ivoire and northern Ghana has hindered the ability of several hundred thousand seasonal Burkinabe farm workers to find employment in neighboring countries.

GEOGRAPHY

Location: Western Africa, north of Ghana
Geographic coordinates: 13 00 N, 2 00 W
Map references: Africa
Area:
total: 274,200 sq km
land: 273,800 sq km
water: 400 sq km
Area—comparative: slightly larger than Colorado
Land boundaries:
total: 3,193 km
border countries: Benin 306 km, Cote d'Ivoire 584 km, Ghana 549 km, Mali 1,000 km, Niger 628 km, Togo 126 km
Coastline: 0 km (landlocked)
Maritime claims: none (landlocked)
Climate: tropical; warm, dry winters; hot, wet summers
Terrain: mostly flat to dissected, undulating plains; hills in west and southeast
Elevation extremes:
lowest point: Mouhoun (Black Volta) River 200 m
highest point: Tena Kourou 749 m
Natural resources: manganese, limestone, marble; small deposits of gold, phosphates, pumice, salt

Land use:
arable land: 17.66%
permanent crops: 0.22%
other: 82.12% (2005)
Irrigated land: 250 sq km (2003)
Total renewable water resources: 17.5 cu km (2001)
Freshwater withdrawal (domestic/industrial/agricultural): *total*: 0.8 cu km/yr (13%/1%/86%)
per capita: 60 cu m/yr (2000)
Natural hazards: recurring droughts
Environment—current issues: recent droughts and desertification severely affecting agricultural activities, population distribution, and the economy; overgrazing; soil degradation; deforestation
Environment—international agreements: *party to*: Biodiversity, Climate Change, Climate Change-Kyoto Protocol, Desertification, Endangered Species, Hazardous Wastes, Law of the Sea, Marine Life Conservation, Ozone Layer Protection, Wetlands
signed, but not ratified: none of the selected agreements
Geography—note: landlocked savanna cut by the three principal rivers of the Black, Red, and White Voltas

PEOPLE

Population: 15,264,735

note: estimates for this country explicitly take into account the effects of excess mortality due to AIDS; this can result in lower life expectancy, higher infant mortality, higher death rates, lower population growth rates, and changes in the distribution of population by age and sex than would otherwise be expected (July 2008 est.)

Age structure:

0–14 years: 46.3% (male 3,549,034/female 3,521,684)

15–64 years: 51.1% (male 3,885,124/female 3,922,198)

65 years and over: 2.5% (male 154,476/female 232,219) (2008 est.)

Median age:

total: 16.7 years

male: 16.5 years

female: 16.9 years (2008 est.)

Population growth rate: 3.109% (2008 est.)

Birth rate: 44.68 births/1,000 population (2008 est.)

Death rate: 13.59 deaths/1,000 population (2008 est.)

Net migration rate: NA

Sex ratio:

at birth: 1.03 male(s)/female

under 15 years: 1.01 male(s)/female

15–64 years: 0.99 male(s)/female

65 years and over: 0.67 male(s)/female

total population: 0.99 male(s)/female (2008 est.)

Infant mortality rate:

total: 86.02 deaths/1,000 live births

male: 93.68 deaths/1,000 live births

female: 78.12 deaths/1,000 live births (2008 est.)

Life expectancy at birth:

total population: 52.55 years

male: 50.67 years

female: 54.49 years (2008 est.)

Total fertility rate: 6.34 children born/woman (2008 est.)

HIV/AIDS—adult prevalence rate: 4.2% (2003 est.)

HIV/AIDS—people living with HIV/AIDS: 300,000 (2003 est.)

HIV/AIDS—deaths: 29,000 (2003 est.)

Major infectious diseases:

degree of risk: very high

food or waterborne diseases: bacterial and protozoal diarrhea, hepatitis A, and typhoid fever

vectorborne disease: malaria

water contact disease: schistosomiasis

respiratory disease: meningococcal meningitis

note: highly pathogenic H5N1 avian influenza has been identified in this country; it poses a negligible risk with extremely rare cases possible among US citizens who have close contact with birds (2008)

Nationality:

noun: Burkinabe (singular and plural)

adjective: Burkinabe

Ethnic groups: Mossi over 40%, other approximately 60% (includes Gurunsi, Senufo, Lobi, Bobo, Mande, and Fulani)

Religions: Muslim 50%, indigenous beliefs 40%, Christian (mainly Roman Catholic) 10%

Languages: French (official), native African languages belonging to Sudanic family spoken by 90% of the population

Literacy:

definition: age 15 and over can read and write

total population: 21.8%

male: 29.4%

female: 15.2% (2003 est.)

GOVERNMENT

Country name:

conventional long form: none

conventional short form: Burkina Faso

local long form: none

local short form: Burkina Faso

former: Upper Volta, Republic of Upper Volta

Government type: parliamentary republic

Capital: *name:* Ouagadougou

geographic coordinates: 12 22 N, 1 31 W

time difference: UTC 0 (5 hours ahead of Washington, DC during Standard Time)

Administrative divisions: 45 provinces; Bale, Bam, Banwa, Bazega, Bougouriba, Boulgou, Boulkiemde, Comoe, Ganzourgou, Gnagna, Gourma, Houet, Ioba, Kadiogo, Kenedougou, Komondjari, Kompienga, Kossi, Koulpelogo, Kouritenga, Kourweogo, Leraba, Loroum, Mouhoun, Nahouri, Namentenga, Nayala, Noumbiel, Oubritenga, Oudalan, Passore, Poni, Sanguie, Sanmatenga, Seno, Sissili, Soum, Sourou, Tapoa, Tuy, Yagha, Yatenga, Ziro, Zondoma, Zoundweogo

Independence: 5 August 1960 (from France)

National holiday: Republic Day, 11 December (1958)

Constitution: 2 June 1991 approved by referendum, 11 June 1991 formally adopted; last amended January 2002

Legal system: based on French civil law system and customary law; has not accepted compulsory ICJ jurisdiction

Suffrage: universal

Executive branch:

chief of state: President Blaise COMPAORE (since 15 October 1987)

head of government: Prime Minister Tertius ZONGO (since 4 June 2007)

cabinet: Council of Ministers appointed by the president on the recommendation of the prime minister

elections: president elected by popular vote for a five-year term (eligible for a second term); election last held 13 November 2005 (next to be held in 2010); in April 2000, the constitution was amended reducing the presidential term from seven to five years, enforceable as of 2005; prime minister appointed by the president with the consent of the legislature

election results: Blaise COMPAORE reelected president; percent of popular vote—Blaise COMPAORE 80.3%, Benewende Stanislas SANKARA 4.9%

Legislative branch: unicameral National Assembly or Assemblee Nationale (111 seats; members are elected by popular vote to serve five-year terms)

elections: National Assembly election last held 6 May 2007 (next to be held in May 2012)

election results: percent of vote by party—NA; seats by party—CDP 73, ADF-RDA 14, UPR 5, UNIR-MS 4, CFD-B 3, UPS 2, PDP-PS 2, RDB 2, PDS 2, PAREN 1, PAI 1, RPC 1, UDPS 1

Judicial branch: Supreme Court; Appeals Court

Political parties and leaders: African Democratic Rally-Alliance for Democracy and Federation or ADF-RDA [Gilbert OUEDRAOGO]; Citizen's Popular Rally or RPC [Antoine QUARE]; Coalition of Democratic Forces of Burkina or CFD-B [Amadou Diemdioda DICKO]; Congress for Democracy and Progress or CDP [Roch Marc-Christian KABORE]; Movement for Tolerance and Progress or MTP [Nayabtigungou Congo KABORE]; Party for African Independence or PAI [Philippe OUEDRAOGO]; Party for Democracy and Progress/Socialist Party or PDP/PS [Ali LANKOANDE]; Party for Democracy and Socialism or PDS [Felix SOUBEIGA]; Party for National Rebirth or PAREN [Oumar DJIGU-IMDE]; Rally for the Development of Burkina or RDB [Antoine KAR-GOUGOU]; Rally of Ecologists of Burkina Faso or RDEB [Ram OUE-DRAGO]; Republican Party for Integration and Solidarity or PARIS [Cyril GOUNGOUNGA]; Union for Democracy and Social Progress or UDPS [Fidele HIEN]; Union for Rebirth—Sankarist Movement or UNIR-MS [Benewende STANISLAS]; Union for the Republic or UPR [Toussaint Abel COULIBALY]; Union of Sankarist Parties or UPS [Ernest Nongma OUE-DRAOGO]

Political pressure groups and leaders: Burkinabe General Confederation of

Labor or CGTB [Tole SAGNON]; Burkinabe Movement for Human Rights or MBDHP [Chrysigone ZOUGMORE]; Group of 14 February [Benewende STANISLAS]; National Confederation of Burkinabe Workers or CNTB [Laurent OUEDRAOGO]; National Organization of Free Unions or ONSL [Paul KABORE]; watchdog/political action groups throughout the country in both organizations and communities

International organization participation: ACCT, ACP, AfDB, AU, ECOWAS, Entente, FAO, FZ, G-77, IAEA, IBRD, ICAO, ICC, ICCt, ICRM, IDA, IDB, IFAD, IFC, IFRCS, ILO, IMF, Interpol, IOC, IOM, IPU, ISO (correspondent), ITSO, ITU, ITUC, MIGA, NAM, OIC, OIF, OPCW, PCA, UN, UN Security Council (temporary), UNAMID, UNCTAD, UNESCO, UNIDO, UNITAR, UNOCI, UNWTO, UPU, WADB (regional), WAEMU, WCL, WCO, WFTU, WHO, WIPO, WMO, WTO

Diplomatic representation in the US:
chief of mission: Ambassador Paramanga Ernest YONLI
chancery: 2340 Massachusetts Avenue NW, Washington, DC 20008
telephone: [1] (202) 332-5577
FAX: [1] (202) 667-1882

Diplomatic representation from the US:
chief of mission: Ambassador Jeanine E. JACKSON
embassy: 602 Avenue Raoul Follereau, Koulouba, Secteur 4
mailing address: 01 B. P. 35, Ouagadougou 01; pouch mail—US Department of State, 2440 Ouagadougou Place, Washington, DC 20521-2440
telephone: [226] 50-30-67-23
FAX: [226] 50-30-38-90, 50-31-23-68

Flag description: two equal horizontal bands of red (top) and green with a yellow five-pointed star in the center
note: uses the popular pan-African colors of Ethiopia

ECONOMY

Economy—overview: One of the poorest countries in the world, landlocked Burkina Faso has few natural resources and a weak industrial base. About 90% of the population is engaged in subsistence agriculture, which is vulnerable to periodic drought. Cotton is the main cash crop and the government has joined with three other cotton producing countries in the region—Mali, Niger, and Chad—to lobby in the World Trade Organization for fewer subsidies to producers in other competing countries. Since 1998, Burkina Faso has embarked upon a gradual but successful privatiza-

tion of state-owned enterprises. Having revised its investment code in 2004, Burkina Faso hopes to attract foreign investors. Thanks to this new code and other legislation favoring the mining sector, the country has seen an upswing in gold exploration and production. While the bitter internal crisis in neighboring Cote d'Ivoire is beginning to be resolved, it is still having a negative effect on Burkina Faso's trade and employment. In 2007 higher costs for energy and imported foodstuffs, as well as low cotton prices, dampened a GDP growth rate that had averaged 6% in the last 10 years. Burkina Faso received a Millennium Challenge Account threshold grant to improve girls' education at the primary school level, and appears likely to receive a grant in the areas of infrastructure, agriculture, and land reform.

GDP (purchasing power parity): $17.2 billion (2007 est.)
GDP (official exchange rate): $6.977 billion (2007 est.)
GDP—real growth rate: 4.2% (2007 est.)
GDP—per capita (PPP): $1,300 (2007 est.)
GDP—composition by sector:
agriculture: 29.7%
industry: 19.4%
services: 50.9% (2007 est.)
Labor force: 5 million
note: a large part of the male labor force migrates annually to neighboring countries for seasonal employment (2003)
Labor force—by occupation: *agriculture:* 90%
industry and services: 10% (2000 est.)
Unemployment rate: 77% (2004)
Population below poverty line: 46.4% (2004)
Household income or consumption by percentage share:
lowest 10%: 2.8%
highest 10%: 32.2% (2004)
Distribution of family income—Gini index: 39.5 (2007)
Inflation rate (consumer prices): -0.2% (2007 est.)
Investment (gross fixed): 17.6% of GDP (2007 est.)
Budget:
revenues: $1.49 billion
expenditures: $1.95 billion (2007 est.)
Agriculture—products: cotton, peanuts, shea nuts, sesame, sorghum, millet, corn, rice; livestock
Industries: cotton lint, beverages, agricultural processing, soap, cigarettes, textiles, gold
Industrial production growth rate: 5.2% (2007 est.)

Electricity—production: 516.2 million kWh (2005)
Electricity—production by source:
fossil fuel: 69.9%
hydro: 30.1%
nuclear: 0%
other: 0% (2001)
Electricity—consumption: 480.1 million kWh (2005)
Electricity—exports: 0 kWh (2005)
Electricity—imports: 0 kWh (2005)
Oil—production: 0 bbl/day (2005)
Oil—consumption: 8,300 bbl/day (2005 est.)
Oil—exports: 0 bbl/day (2004)
Oil—imports: 8,158 bbl/day (2004)
Oil—proved reserves: 0 bbl (1 January 2006 est.)
Natural gas—production: 0 cu m (2005 est.)
Natural gas—consumption: 0 cu m (2005 est.)
Natural gas—exports: 0 cu m (2005 est.)
Natural gas—imports: 0 cu m (2005)
Natural gas—proved reserves: 0 cu m (1 January 2006 est.)
Current account balance: -$688 million (2007)
Exports: $676 million f.o.b. (2007 est.)
Exports—commodities: cotton, livestock, gold
Exports—partners: China 41.9%, Singapore 14.4%, Ghana 5.9%, Thailand 4.9%, Niger 4.4% (2006)
Imports: $1.332 billion f.o.b. (2007 est.)
Imports—commodities: capital goods, foodstuffs, petroleum
Imports—partners: Cote d'Ivoire 25.9%, France 22.8%, Togo 7.2% (2006)
Economic aid—recipient: $659.6 million (2005)
Reserves of foreign exchange and gold: $1.029 billion (31 December 2007 est.)
Debt—external: $1.33 billion (2007)
Market value of publicly traded shares: $NA
Currency (code): Communaute Financiere Africaine franc (XOF); note—responsible authority is the Central Bank of the West African States
Currency code: XOF
Exchange rates: Communaute Financiere Africaine francs (XOF) per US dollar—493.51 (2007), 522.59 (2006), 527.47 (2005), 528.29 (2004), 581.2 (2003)
note: since 1 January 1999, the XOF franc has been pegged to the euro at a rate of 655.957 XOF francs per euro
Fiscal year: calendar year

COMMUNICATIONS

Telephones—main lines in use: 94,800 (2006)

105

Telephones—mobile cellular: 1.017 million (2006)

Telephone system:

general assessment: services only fair; in 2006 the government sold a 51 percent stake in the national telephone company and ultimately plans to retain only a 23 percent stake in the company; fixed-line connections stand at less than 1 per 100 persons; mobile-cellular usage, fostered by multiple providers, is increasing rapidly from a low base

domestic: microwave radio relay, open-wire, and radiotelephone communication stations

international: country code—226; satellite earth station—1 Intelsat (Atlantic Ocean) (2007)

Radio broadcast stations: AM 2, FM 26, shortwave 3

Radios: 394,020 (2000)

Television broadcast stations: 3 (1 national, 2 private)

Televisions: 131,340 (2002)

Internet country code: .bf

Internet hosts: 193 (2007)

Internet Service Providers (ISPs): 1 (2002)

Internet users: 80,000 (2006)

TRANSPORTATION

Airports: 33 (2007)

Airports—with paved runways:
total: 2
over 3,047 m: 1
2,438 to 3,047 m: 1 (2007)

Airports—with unpaved runways:
total: 31
1,524 to 2,437 m: 3
914 to 1,523 m: 11
under 914 m: 17 (2007)

Railways:
total: 622 km
narrow gauge: 622 km 1.000-m gauge
note: another 660 km of this railway extends into Cote D'Ivoire (2006)

Roadways:
total: 92,495 km
paved: 3,857 km
unpaved: 88,638 km (2004)

MILITARY

Military branches: Army, Air Force of Burkina Faso (Force Aerienne de Burkina Faso, FABF), National Gendarmerie (2008)

Military service age and obligation: 18 years of age for compulsory military service; 20 years of age for voluntary military service (2001)

Manpower available for military service:
males age 16–49: 3,364,288 (2008 est.)

Manpower fit for military service:
males age 16–49: 2,115,948 (2008 est.)

Military expenditures—percent of GDP: 1.2% (2006)

TRANSNATIONAL ISSUES

Disputes—international: two villages remain in dispute along the border with Benin; Benin accuses Burkina Faso of moving boundary pillars; in recent years citizens and rogue security forces rob and harass local populations on both sides of the poorly-defined Burkina Faso-Niger border; despite the presence of over 9,000 UN forces (UNOCI) in Cote d'Ivoire since 2004, ethnic conflict continues to spread into neighboring states who can no longer send their migrant workers to work in Ivorian cocoa plantations

BURMA

INTRODUCTION

Background: Britain conquered Burma over a period of 62 years (1824–1886) and incorporated it into its Indian Empire. Burma was administered as a province of India until 1937 when it became a separate, self-governing colony; independence from the Commonwealth was attained in 1948. Gen. NE WIN dominated the government from 1962 to 1988, first as military ruler, then as self-appointed president, and later as political kingpin. Despite multiparty legislative elections in 1990 that resulted in the main opposition party—the National League for Democracy (NLD)—winning a landslide victory, the ruling junta refused to hand over power. NLD leader and Nobel Peace Prize recipient AUNG SAN SUU KYI, who was under house arrest from 1989 to 1995 and 2000 to 2002, was imprisoned in May 2003 and subsequently transferred to house arrest. After Burma's ruling junta in August 2007 unexpectedly increased fuel prices, tens of thousands of Burmese marched in protest, led by prodemocracy activists and Buddhist monks. In late September 2007, the government brutally suppressed the protests, killing at least 13 people and arresting thousands for participating in the demonstrations. Since then, the regime has continued to raid homes and monasteries and arrest persons suspected of participating in the pro-democracy protests. The junta appointed Labor Minister AUNG KYI in October 2007 as liaison to AUNG SAN SUU KYI, who remains under house arrest and virtually incommunicado with her party and supporters.

GEOGRAPHY

Location: Southeastern Asia, bordering the Andaman Sea and the Bay of Bengal, between Bangladesh and Thailand

Geographic coordinates: 22 00 N, 98 00 E

Map references: Southeast Asia

Area:
total: 678,500 sq km
land: 657,740 sq km
water: 20,760 sq km

Area—comparative: slightly smaller than Texas

Land boundaries:
total: 5,876 km
border countries: Bangladesh 193 km, China 2,185 km, India 1,463 km, Laos 235 km, Thailand 1,800 km

Coastline: 1,930 km

Maritime claims:
territorial sea: 12 nm

contiguous zone: 24 nm

exclusive economic zone: 200 nm

continental shelf: 200 nm or to the edge of the continental margin

Climate: tropical monsoon; cloudy, rainy, hot, humid summers (southwest monsoon, June to September); less cloudy, scant rainfall, mild temperatures, lower humidity during winter (northeast monsoon, December to April)

Terrain: central lowlands ringed by steep, rugged highlands

Elevation extremes:

lowest point: Andaman Sea 0 m

highest point: Hkakabo Razi 5,881 m

Natural resources: petroleum, timber, tin, antimony, zinc, copper, tungsten, lead, coal, marble, limestone, precious stones, natural gas, hydropower

Land use: arable land: 14.92%

permanent crops: 1.31%

other: 83.77% (2005)

Irrigated land: 18,700 sq km (2003)

Total renewable water resources: 1,045.6 cu km (1999)

Freshwater withdrawal (domestic/industrial/agricultural): total: 33.23 cu km/yr (1%/1%/98%)

per capita: 658 cu m/yr (2000)

Natural hazards: destructive earthquakes and cyclones; flooding and landslides common during rainy season (June to September); periodic droughts

Environment—current issues: deforestation; industrial pollution of air, soil, and water; inadequate sanitation and water treatment contribute to disease

Environment—international agreements: party to: Biodiversity, Climate Change, Climate Change-Kyoto Protocol, Desertification, Endangered Species, Law of the Sea, Ozone Layer Protection, Ship Pollution, Tropical Timber 83, Tropical Timber 94

signed, but not ratified: none of the selected agreements

Geography—note: strategic location near major Indian Ocean shipping lanes

PEOPLE

Population: 47,758,181

note: estimates for this country take into account the effects of excess mortality due to AIDS; this can result in lower life expectancy, higher infant mortality, higher death rates, lower population growth rates, and changes in the distribution of population by age and sex than would otherwise be expected (July 2008 est.)

Age structure:

0–14 years: 25.7% (male 6,236,484/female 6,038,576)

15–64 years: 68.9% (male 16,300,380/female 16,627,045)

65 years and over: 5.4% (male 1,098,344/female 1,457,352) (2008 est.)

Median age:

total: 27.8 years

male: 27.2 years

female: 28.4 years (2008 est.)

Population growth rate: 0.8% (2008 est.)

Birth rate: 17.23 births/1,000 population (2008 est.)

Death rate: 9.23 deaths/1,000 population (2008 est.)

Net migration rate: NA

Sex ratio:

at birth: 1.06 male(s)/female

under 15 years: 1.03 male(s)/female

15–64 years: 0.98 male(s)/female

65 years and over: 0.75 male(s)/female

total population: 0.98 male(s)/female (2008 est.)

Infant mortality rate:

total: 49.12 deaths/1,000 live births

male: 55.53 deaths/1,000 live births

female: 42.33 deaths/1,000 live births (2008 est.)

Life expectancy at birth:

total population: 62.94 years

male: 60.73 years

female: 65.28 years (2008 est.)

Total fertility rate: 1.92 children born/woman (2008 est.)

HIV/AIDS—adult prevalence rate: 1.2% (2003 est.)

HIV/AIDS—people living with HIV/AIDS: 330,000 (2003 est.)

HIV/AIDS—deaths: 20,000 (2003 est.)

Major infectious diseases:

degree of risk: very high

food or waterborne diseases: bacterial and protozoal diarrhea, hepatitis A, and typhoid fever

vectorborne diseases: dengue fever and malaria

water contact disease: leptospirosis

animal contact disease: rabies

note: highly pathogenic H5N1 avian influenza has been identified in this country; it poses a negligible risk with extremely rare cases possible among US citizens who have close contact with birds (2008)

Nationality:

noun: Burmese (singular and plural)

adjective: Burmese

Ethnic groups: Burman 68%, Shan 9%, Karen 7%, Rakhine 4%, Chinese 3%, Indian 2%, Mon 2%, other 5%

Religions: Buddhist 89%, Christian 4% (Baptist 3%, Roman Catholic 1%), Muslim 4%, animist 1%, other 2%

Languages: Burmese, minority ethnic groups have their own languages

Literacy:

definition: age 15 and over can read and write

total population: 89.9%

male: 93.9%

female: 86.4% (2000 est.)

GOVERNMENT

Country name:

conventional long form: Union of Burma

conventional short form: Burma

local long form: Pyidaungzu Myanma Nainngngandaw (translated by the US Government as Union of Myanma and by the Burmese as Union of Myanmar)

local short form: Myanma Nainngngandaw

former: Socialist Republic of the Union of Burma

note: since 1989 the military authorities in Burma have promoted the name Myanmar as a conventional name for their state; this decision was not approved by any sitting legislature in Burma, and the US Government did not adopt the name, which is a derivative of the Burmese short-form name Myanma Nainngngandaw

Government type: military junta

Capital: name: Rangoon (Yangon)

geographic coordinates: 16 48 N, 96 09 E

time difference: UTC+6.5 (11.5 hours ahead of Washington, DC during Standard Time)

note: Nay Pyi Taw is administrative capital

Administrative divisions: 7 divisions (taing-myar, singular—taing) and 7 states (pyi ne-myar, singular—pyi ne)

divisions: Ayeyarwady, Bago, Magway, Mandalay, Sagaing, Tanintharyi, Yangon

states: Chin, Kachin, Kayah, Kayin, Mon, Rakhine, Shan

Independence: 4 January 1948 (from UK)

National holiday: Independence Day, 4 January (1948); Union Day, 12 February (1947)

Constitution: 30 May 2008

Legal system: based on English common law; has not accepted compulsory ICJ jurisdiction

Suffrage: 18 years of age; universal

Executive branch:

chief of state: Chairman of the State Peace and Development Council (SPDC) Sr. Gen. THAN SHWE (since 23 April 1992)

head of government: Prime Minister, Lt. Gen THEIN SEIN (since 24 October 2007)

cabinet: Cabinet is overseen by SPDC; military junta assumed power 18 September 1988 under name State Law and Order Restoration Council (SLORC)

elections: none

Legislative branch: unicameral People's Assembly or Pyithu Hluttaw (485 seats;

members elected by popular vote to serve four-year terms)

elections: last held 27 May 1990, but Assembly never allowed by junta to convene (junta has anounced plans to hold elections in 2010)

election results: percent of vote by party—NA; seats by party—NLD 392 (opposition), SNLD 23 (opposition), NUP 10 (pro-government), other 60

Judicial branch: remnants of the British-era legal system are in place, but there is no guarantee of a fair public trial; the judiciary is not independent of the executive

Political parties and leaders: National League for Democracy or NLD [AUNG SHWE, AUNG SAN SUU KYI]; National Unity Party or NUP (pro-regime) [TUN YE]; Shan Nationalities League for Democracy or SNLD [HKUN HTUN OO]; and other smaller parties

Political pressure groups and leaders: Ethnic Nationalities Council or ENC (based in Thailand); Federation of Trade Unions-Burma or FTUB (exile trade union and labor advocates); National Coalition Government of the Union of Burma or NCGUB (self-proclaimed government in exile) ["Prime Minister" Dr. SEIN WIN] consists of individuals, some legitimately elected to the People's Assembly in 1990 (the group fled to a border area and joined insurgents in December 1990 to form parallel government in exile); Kachin Independence Organization or KIO; Karen National Union or KNU; Karenni National People's Party or KNPP; National Council-Union of Burma or NCUB (exile coalition of opposition groups); several Shan factions; United Wa State Army or UWSA; Union Solidarity and Development Association or USDA (pro-regime, a social and political mass-member organization) [HTAY OO, general secretary]; 88 Generation Students (pro-democracy movement) [MIN KO NAING]

International organization participation: ADB, APT, ARF, ASEAN, BIMSTEC, CP, EAS, FAO, G-77, IAEA, IBRD, ICAO, ICRM, IDA, IFAD, IFC, IFRCS, IHO, ILO, IMF, IMO, Interpol, IOC, ISO (correspondent), ITU, NAM, OPCW (signatory), UN, UNCTAD, UNESCO, UNIDO, UPU, WCO, WHO, WIPO, WMO, WTO

Diplomatic representation in the US:
chief of mission: Ambassador (vacant); Charge d'Affaires MYINT LWIN
chancery: 2300 S Street NW, Washington, DC 20008
telephone: [1] (202) 332-3344
FAX: [1] (202) 332-4351
consulate(s) general: New York
Diplomatic representation from the US:

chief of mission: Ambassador (vacant); Charge d'Affaires Shari VILLAROSA
embassy: 110 University Avenue, Kamayut Township, Rangoon
mailing address: Box B, APO AP 96546
telephone: [95] (1) 556-509, 535-756
FAX: [95] (1) 650-306

Flag description: red with a blue rectangle in the upper hoist-side corner bearing 14, white, five-pointed stars encircling a cogwheel containing a stalk of rice; the 14 stars represent the seven administrative divisions and seven states

ECONOMY

Economy—overview: Burma, a resource-rich country, suffers from pervasive government controls, inefficient economic policies, and rural poverty. The junta took steps in the early 1990s to liberalize the economy after decades of failure under the "Burmese Way to Socialism," but those efforts stalled, and some of the liberalization measures were rescinded. Despite Burma's increasing oil and gas revenue, socio-economic conditions have deteriorated due to the regime's mismanagement of the economy. Lacking monetary or fiscal stability, the economy suffers from serious macroeconomic imbalances—including rising inflation, fiscal deficits, multiple official exchange rates that overvalue the Burmese kyat, a distorted interest rate regime, unreliable statistics, and an inability to reconcile national accounts to determine a realistic GDP figure. Most overseas development assistance ceased after the junta began to suppress the democracy movement in 1988 and subsequently refused to honor the results of the 1990 legislative elections. In response to the government of Burma's attack in May 2003 on AUNG SAN SUU KYI and her convoy, the US imposed new economic sanctions in August 2003 including a ban on imports of Burmese products and a ban on provision of financial services by US persons. Further, a poor investment climate hampers attracting outside investment slowing the inflow of foreign exchange. The most productive sectors will continue to be in extractive industries, especially oil and gas, mining, and timber with the latter especially causing environmental degradation. Other areas, such as manufacturing and services, are struggling with inadequate infrastructure, unpredictable import/export policies, deteriorating health and education systems, and endemic corruption. A major banking crisis in 2003 shuttered the country's 20 private banks and disrupted the economy. As of 2007, the largest private banks operated under tight restrictions limiting the private sector's access to formal credit. Moreover, the September 2007 crackdown on prodemocracy demonstrators, including thousands of monks, further strained the economy as the tourism industry, which directly employs about 500,000 people, suffered dramatic declines in foreign visitor levels. In November 2007, the European Union announced new sanctions banning investment and trade in Burmese gems, timber and precious stones, while the United States expanded its sanctions list to include more Burmese government and military officials and their family members, as well as prominent regime business cronies, their family members, and associated companies. Official statistics are inaccurate. Published statistics on foreign trade are greatly understated because of the size of the black market and unofficial border trade—often estimated to be as large as the official economy. Though the Burmese government has good economic relations with its neighbors, better investment and business climates and an improved political situation are needed to promote serious foreign investment, exports, and tourism.

GDP (purchasing power parity): $91.13 billion (2007 est.)
GDP (official exchange rate): $13.53 billion (2007 est.)
GDP—real growth rate: 5.5% (2007 est.)
GDP—per capita (PPP): $1,900 (2007 est.)
GDP—composition by sector:
agriculture: 50%
industry: 17.6%
services: 32.4% (2007 est.)
Labor force: 29.26 million (2007 est.)
Labor force—by occupation: *agriculture:* 70%
industry: 7%
services: 23% (2001)
Unemployment rate: 10.2% (2007 est.)
Population below poverty line: 32.7% (2007 est.)
Household income or consumption by percentage share:
lowest 10%: 2.8%
highest 10%: 32.4% (1998)
Inflation rate (consumer prices): 34.4% (2007 est.)
Investment (gross fixed): 12.7% of GDP (2007 est.)
Budget:
revenues: NA
expenditures: NA (2007 est.)
Agriculture—products: rice, pulses, beans, sesame, groundnuts, sugarcane; hardwood; fish and fish products

Industries: agricultural processing; wood and wood products; copper, tin, tungsten, iron; cement, construction materials; pharmaceuticals; fertilizer; natural gas; garments, jade and gems

Industrial production growth rate: 9.6% (2007 est.)

Electricity—production: 6.154 billion kWh (FY06)

Electricity—production by source:
fossil fuel: 44.5%
hydro: 43.4%
nuclear: 0%
other: 12.1% (2002)

Electricity—consumption: 3.744 billion kWh (FY06)

Electricity—exports: 0 kWh (2005)

Electricity—imports: 0 kWh (2005)

Oil—production: 7,700 bbl/day (2006 est.)

Oil—consumption: 20,460 bbl/day (2006 est.)

Oil—exports: 5,000 bbl/day (2006 est.)

Oil—imports: 19,180 bbl/day (2004 est.)

Oil—proved reserves: 1.963 billion bbl (2007 est.)

Natural gas—production: 12.47 billion cu m (2005)

Natural gas—consumption: 3.971 billion cu m (2005)

Natural gas—exports: 8.497 billion cu m (2005)

Natural gas—imports: 0 cu m (2005)

Natural gas—proved reserves: 271.6 billion cu m (1 January 2006 est.)

Current account balance: $541 million (2007 est.)

Exports: $6.346 billion f.o.b.
note: official export figures are grossly underestimated due to the value of timber, gems, narcotics, rice, and other products smuggled to Thailand, China, and Bangladesh (2007 est.)

Exports—commodities: natural gas, wood products, pulses, beans, fish, rice, clothing, jade and gems

Exports—partners: Thailand 48.8%, India 12.7%, China 5.2%, Japan 5.2% (2006)

Imports: $3.624 billion f.o.b.
note: import figures are grossly underestimated due to the value of consumer goods, diesel fuel, and other products smuggled in from Thailand, China, Malaysia, and India (2007 est.)

Imports—commodities: fabric, petroleum products, fertilizer, plastics, machinery, transport equipment; cement, construction materials, crude oil; food products, edible oil

Imports—partners: China 35.1%, Thailand 22.1%, Singapore 16.4%, Malaysia 4.8% (2006)

Economic aid—recipient: $144.7 million (2005 est.)

Reserves of foreign exchange and gold: $2.268 billion (31 December 2007 est.)

Debt—external: $7.133 billion (31 December 2007 est.)

Market value of publicly traded shares: $NA

Currency (code): kyat (MMK)

Currency code: MMK

Exchange rates: kyats per US dollar—1,296 (2007), 1,280 (2006), 5.761 (2005), 5.7459 (2004), 6.0764 (2003)
note: unofficial exchange rates ranged in 2004 from 815 kyat/US dollar to nearly 970 kyat/US dollar, and by yearend 2005, the unofficial exchange rate was 1,075 kyat/US dollar; data shown for 2003–05 are official exchange rates

Fiscal year: 1 April—31 March

COMMUNICATIONS

Telephones—main lines in use: 503,900 (2005)

Telephones—mobile cellular: 214,200 (2006)

Telephone system:
general assessment: meets minimum requirements for local and intercity service for business and government; international service is good
domestic: system capable of providing basic service; cellular mobile phone system functions more efficiently than traditional lines
international: country code—95; landing point for the SEA-ME-WE-3 optical telecommunications submarine cable that provides links to Asia, the Middle East, and Europe; satellite earth stations—2, Intelsat (Indian Ocean), and ShinSat (2007)

Radio broadcast stations: AM 1, FM 2, shortwave 3 (2007)

Radios: 4.2 million (1997)

Television broadcast stations: 4 (2008)

Televisions: 320,000 (2000)

Internet country code: .mm

Internet hosts: 101 (2007)

Internet Service Providers (ISPs): 1
note: as of September 2000, Internet connections were legal only for the government, tourist offices, and a few large businesses (2000)

Internet users: 31,500 (2005)

TRANSPORTATION

Airports: 86 (2007)

Airports—with paved runways:
total: 25
over 3,047 m: 8
2,438 to 3,047 m: 10
1,524 to 2,437 m: 5
914 to 1,523 m: 1
under 914 m: 1 (2007)

Airports—with unpaved runways:

total: 61
over 3,047 m: 1
1,524 to 2,437 m: 14
914 to 1,523 m: 14
under 914 m: 32 (2007)

Heliports: 4 (2007)

Pipelines: gas 2,790 km; oil 558 km (2007)

Railways:
total: 3,955 km
narrow gauge: 3,955 km 1.000-m gauge (2006)

Roadways:
total: 27,000 km
paved: 3,200 km
unpaved: 23,800 km (2005)

Waterways: 12,800 km (2007)

Merchant marine:
total: 33 ships (1000 GRT or over) 364,447 GRT/549,310 DWT
by type: bulk carrier 7, cargo 20, passenger 2, passenger/cargo 3, specialized tanker 1
foreign-owned: 8 (Germany 5, Japan 3) (2007)

Ports and terminals: Moulmein, Rangoon, Sittwe

MILITARY

Military branches: Myanmar Armed Forces (Tatmadaw): Army, Navy, Air Force (Tatmadaw Lay) (2008)

Military service age and obligation: 18 years of age for voluntary military service for both sexes; forced conscription of children, although officially prohibited, reportedly continues (2007)

Manpower available for military service:
males age 16–49: 13,402,788
females age 16–49: 13,437,042 (2008 est.)

Manpower fit for military service:
males age 16–49: 9,031,046
females age 16–49: 9,396,547 (2008 est.)

Manpower reaching militarily significant age annually:
males age 16–49: 423,809
females age 16–49: 415,843 (2008 est.)

Military expenditures—percent of GDP: 2.1% (2005 est.)

TRANSNATIONAL ISSUES

Disputes—international: over half of Burma's population consists of diverse ethnic groups who have substantial numbers of kin in neighboring countries; Thailand must deal with Karen and other ethnic rebels, illegal cross-border activities, Karen and other refugees, and asylum seekers from Burma; Thailand is studying the feasibility of jointly constructing the Hatgyi Dam on the Salween River near the border with Burma; in 2004, international environmentalist pressure prompted China to halt construction of 13 dams on the Salween River which flows through China, Burma, and Thailand; India seeks cooperation from Burma to keep Indian Nagaland separatists, such as the United Liberation Front of Assam, from hiding in remote Burmese Uplands; Burmese Rohingya Muslim refugees reside in two camps in Bangladesh

Refugees and internally displaced persons: *IDPs:* 503,000 (government offensives against ethnic insurgent groups near the eastern borders; most IDPs are ethnic Karen, Karenni, Shan, Tavoyan, and Mon) (2007)

Trafficking in persons: *current situation:* Burma is a source country for men, women, and children trafficked to East and Southeast Asia for sexual exploitation, domestic service, and forced com-

mercial labor; a significant number of victims are economic migrants who wind up in forced or bonded labor and forced prostitution; to a lesser extent, Burma is a country of transit and destination for women trafficked from China for sexual exploitation; internal trafficking of persons occurs primarily for labor in industrial zones and agricultural estates; internal trafficking of women and girls for sexual exploitation occurs from villages to urban centers and other areas; the military junta's economic mismanagement, human rights abuses, and policy of using forced labor are driving factors behind Burma's large trafficking problem

tier rating: Tier 3—Burma does not fully comply with the minimum standards for the elimination of trafficking and is not making significant efforts to do so

Illicit drugs: remains world's second largest producer of illicit opium with an estimated production in 2005 of 380 metric tons, up 13% from 2004 and cultivation in 2005 was 40,000 hectares, a 10% increase from 2004; the decline in opium production in the United Wa State Army's areas of greatest control was more than offset by increases in south and east Shan state; lack of government will to take on major narcotrafficking groups and lack of serious commitment against money laundering continues to hinder the overall antidrug effort; major source of methamphetamine and heroin for regional consumption; currently under Financial Action Task Force countermeasures due to continued failure to address its inadequate money-laundering controls (2005)

BURUNDI

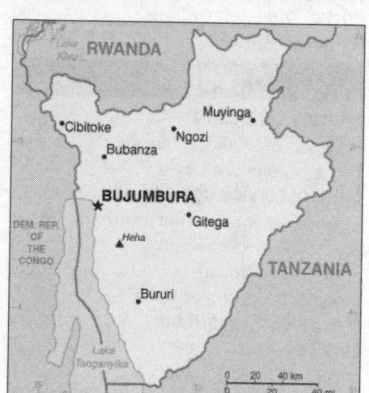

INTRODUCTION

Background: Burundi's first democratically elected president was assassinated in October 1993 after only 100 days in office, triggering widespread ethnic violence between Hutu and Tutsi factions. More than 200,000 Burundians perished during the conflict that spanned almost a dozen years. Hundreds of thousands of Burundians were internally displaced or became refugees in neighboring countries. An internationally brokered power-sharing agreement between the Tutsi-dominated government and the Hutu rebels in 2003 paved the way for a transition process that led to an integrated defense force, established a new constitution in 2005, and elected a majority Hutu government in 2005. The new government, led by President Pierre NKURUNZIZA, signed a South African brokered ceasefire with the country's last

rebel group in September of 2006 but still faces many challenges.

GEOGRAPHY

Location: Central Africa, east of Democratic Republic of the Congo
Geographic coordinates: 3 30 S, 30 00 E
Map references: Africa
Area:
total: 27,830 sq km
land: 25,650 sq km
water: 2,180 sq km
Area—comparative: slightly smaller than Maryland
Land boundaries:
total: 974 km
border countries: Democratic Republic of the Congo 233 km, Rwanda 290 km, Tanzania 451 km
Coastline: 0 km (landlocked)
Maritime claims: none (landlocked)
Climate: equatorial; high plateau with considerable altitude variation (772 m to 2,670 m above sea level); average annual temperature varies with altitude from 23 to 17 degrees centigrade but is generally moderate as the average altitude is about 1,700 m; average annual rainfall is about 150 cm; two wet seasons (February to May and September to November), and two dry seasons (June to August and December to January)
Terrain: hilly and mountainous, dropping to a plateau in east, some plains
Elevation extremes:
lowest point: Lake Tanganyika 772 m
highest point: Heha 2,670 m
Natural resources: nickel, uranium, rare earth oxides, peat, cobalt, copper, plat-

inum, vanadium, arable land, hydropower, niobium, tantalum, gold, tin, tungsten, kaolin, limestone
Land use:
arable land: 35.57%
permanent crops: 13.12%
other: 51.31% (2005)
Irrigated land: 210 sq km (2003)
Total renewable water resources: 3.6 cu km (1987)
Freshwater withdrawal (domestic/industrial/agricultural): *total:* 0.29 cu km/yr (17%/6%/77%)
per capita: 38 cu m/yr (2000)
Natural hazards: flooding, landslides, drought
Environment—current issues: soil erosion as a result of overgrazing and the expansion of agriculture into marginal lands; deforestation (little forested land remains because of uncontrolled cutting of trees for fuel); habitat loss threatens wildlife populations
Environment—international agreements: *party to:* Biodiversity, Climate Change, Climate Change-Kyoto Protocol, Desertification, Endangered Species, Hazardous Wastes, Ozone Layer Protection, Wetlands
signed, but not ratified: Law of the Sea
Geography—note: landlocked; straddles crest of the Nile-Congo watershed; the Kagera, which drains into Lake Victoria, is the most remote headstream of the White Nile

PEOPLE

Population: 8,691,005
note: estimates for this country explicitly

take into account the effects of excess mortality due to AIDS; this can result in lower life expectancy, higher infant mortality, higher death rates, lower population growth rates, and changes in the distribution of population by age and sex than would otherwise be expected (July 2008 est.)

Age structure:
0–14 years: 46.3% (male 2,021,320/ female 1,998,502)
15–64 years: 51.2% (male 2,210,157/ female 2,240,921)
65 years and over: 2.5% (male 87,600/female 132,505) (2008 est.)

Median age:
total: 16.7 years
male: 16.4 years
female: 17 years (2008 est.)

Population growth rate: 3.443% (2008 est.)

Birth rate: 41.72 births/1,000 population (2008 est.)

Death rate: 12.91 deaths/1,000 population (2008 est.)

Net migration rate: 5.62 migrant(s)/ 1,000 population (2008 est.)

Sex ratio:
at birth: 1.03 male(s)/female
under 15 years: 1.01 male(s)/female
15–64 years: 0.99 male(s)/female
65 years and over: 0.66 male(s)/female
total population: 0.99 male(s)/female (2008 est.)

Infant mortality rate:
total: 60.77 deaths/1,000 live births
male: 67.6 deaths/1,000 live births
female: 53.73 deaths/1,000 live births (2008 est.)

Life expectancy at birth:
total population: 51.71 years
male: 50.86 years
female: 52.6 years (2008 est.)

Total fertility rate: 6.4 children born/woman (2008 est.)

HIV/AIDS—adult prevalence rate: 6% (2003 est.)

HIV/AIDS—people living with HIV/AIDS: 250,000 (2003 est.)

HIV/AIDS—deaths: 25,000 (2003 est.)

Major infectious diseases:
degree of risk: very high
food or waterborne diseases: bacterial and protozoal diarrhea, hepatitis A, and typhoid fever
vectorborne disease: malaria (2008)

Nationality:
noun: Burundian(s)
adjective: Burundian

Ethnic groups: Hutu (Bantu) 85%, Tutsi (Hamitic) 14%, Twa (Pygmy) 1%, Europeans 3,000, South Asians 2,000

Religions: Christian 67% (Roman Catholic 62%, Protestant 5%), indigenous beliefs 23%, Muslim 10%

Languages: Kirundi (official), French (official), Swahili (along Lake Tanganyika and in the Bujumbura area)

Literacy:
definition: age 15 and over can read and write
total population: 59.3%
male: 67.3%
female: 52.2% (2000 est.)

GOVERNMENT

Country name:
conventional long form: Republic of Burundi
conventional short form: Burundi
local long form: Republique du Burundi/Republika y'u Burundi
local short form: Burundi
former: Urundi

Government type: republic

Capital: *name:* Bujumbura
geographic coordinates: 3 22 S, 29 21 E
time difference: UTC+2 (7 hours ahead of Washington, DC during Standard Time)

Administrative divisions: 17 provinces; Bubanza, Bujumbura Mairie, Bujumbura Rurale, Bururi, Cankuzo, Cibitoke, Gitega, Karuzi, Kayanza, Kirundo, Makamba, Muramvya, Muyinga, Mwaro, Ngozi, Rutana, Ruyigi

Independence: 1 July 1962 (from UN trusteeship under Belgian administration)

National holiday: Independence Day, 1 July (1962)

Constitution: 28 February 2005; ratified by popular referendum

Legal system: based on German and Belgian civil codes and customary law; has not accepted compulsory ICJ jurisdiction

Suffrage: NA years of age; universal (adult)

Executive branch:
chief of state: President Pierre NKURUNZIZA (since 26 August 2005); First Vice President Yves SAVINGUVU—Tutsi (since 9 November 2007); Second Vice President Gabriel NTISEZERANA—Hutu (since 9 February 2007); note—the president is both the chief of state and head of government
head of government: President Pierre NKURUNZIZA (since 26 August 2005); First Vice President Yves SAVINGUVU—Tutsi (since 9 November 2007); Second Vice President Gabriel NTISEZERANA—Hutu (since 9 February 2007)
cabinet: Council of Ministers appointed by president
elections: the president is elected by popular vote to a five-year term (eligible for a second term); note—the constitution adopted in February 2005 permits the

post-transition president to be elected by a two-thirds majority of the parliament; vice presidents nominated by the president, endorsed by parliament
election results: Pierre NKURUNZIZA was elected president by the parliament by a vote of 151 to 9; note—the constitution adopted in February 2005 permits the post-transition president to be elected by a two-thirds majority of the legislature

Legislative branch: bicameral Parliament or Parlement, consists of a National Assembly or Assemblee Nationale (minimum 100 seats, 60% Hutu and 40% Tutsi with at least 30% being women; additional seats appointed by a National Independent Electoral Commission to ensure ethnic representation; members are elected by popular vote to serve five-year terms) and a Senate (54 seats; 34 members elected by indirect vote to serve five-year terms, with remaining seats assigned to ethnic groups and former chiefs of state)
elections: National Assembly—last held 4 July 2005 (next to be held in 2010); Senate—last held 29 July 2005 (next to be held in 2010)
election results: National Assembly—percent of vote by party—CNDD-FDD 58.6%, FRODEBU 21.7%, UPRONA 7.2%, CNDD 4.1%, MRC-Rurenzangemero 2.1%, others 6.2%; seats by party—CNDD-FDD 59, FRODEBU 25, UPRONA 10, CNDD 4, MRC-Rurenzangemero 2; Senate—percent of vote by party—NA%; seats by party—CNDD-FDD 30, FRODEBU 3, CNDD 1

Judicial branch: Supreme Court or Cour Supreme; Constitutional Court; Courts of Appeal (there are three in separate locations); Tribunals of First Instance (17 at the province level and 123 small local tribunals)

Political parties and leaders: *governing parties:* Burundi Democratic Front or FRODEBU [Leonce NGENDAKUMANA]; National Council for the Defense of Democracy—Front for the Defense of Democracy or CNDD-FDD [Jeremie NGENDAKUMANA]; Unity for National Progress or UPRONA [Aloys RUBUKA]
note: a multiparty system was introduced after 1998, included are: National Council for the Defense of Democracy or CNDD [Leonard NYANGOMA]; National Resistance Movement for the Rehabilitation of the Citizen or MRC-Rurenzangemero [Epitace BANYAGANAKANDI]; Party for National Redress or PARENA [Jean-Baptiste BAGAZA]

Political pressure groups and leaders: none

International organization participation: ACCT, ACP, AfDB, AU, CEPGL, COMESA, EAC, FAO, G-77, IBRD, ICAO, ICCt, ICRM, IDA, IFAD, IFC, IFRCS, ILO, IMF, Interpol, IOC, IOM, IPU, ISO (subscriber), ITU, ITUC, MIGA, NAM, OIF, OPCW, UN, UNCTAD, UNESCO, UNIDO, UNWTO, UPU, WCO, WHO, WIPO, WMO, WTO

Diplomatic representation in the US:
chief of mission: Ambassador Celestin NIYONGABO
chancery: Suite 212, 2233 Wisconsin Avenue NW, Washington, DC 20007
telephone: [1] (202) 342-2574
FAX: [1] (202) 342-2578

Diplomatic representation from the US:
chief of mission: Ambassador Patricia Newton MOLLER
embassy: Avenue des Etats-Unis, Bujumbura
mailing address: B. P. 1720, Bujumbura
telephone: [257] 223454
FAX: [257] 222926

Flag description: divided by a white diagonal cross into red panels (top and bottom) and green panels (hoist side and fly side) with a white disk superimposed at the center bearing three red six-pointed stars outlined in green arranged in a triangular design (one star above, two stars below)

ECONOMY

Economy—overview: Burundi is a landlocked, resource-poor country with an underdeveloped manufacturing sector. The economy is predominantly agricultural with more than 90% of the population dependent on subsistence agriculture. Economic growth depends on coffee and tea exports, which account for 90% of foreign exchange earnings. The ability to pay for imports, therefore, rests primarily on weather conditions and international coffee and tea prices. The Tutsi minority, 14% of the population, dominates the government and the coffee trade at the expense of the Hutu majority, 85% of the population. An ethnic-based war that lasted for over a decade resulted in more than 200,000 deaths, forced more than 48,000 refugees into Tanzania, and displaced 140,000 others internally. Only one in two children go to school, and approximately one in 15 adults has HIV/AIDS. Food, medicine, and electricity remain in short supply. Burundi's GDP grew around 5% annually in 2006–07. Political stability and the end of the civil war have improved aid flows and economic

activity has increased, but underlying weaknesses—a high poverty rate, poor education rates, a weak legal system, and low administrative capacity—risk undermining planned economic reforms. Burundi will continue to remain heavily dependent on aid from bilateral and multilateral donors; the delay of funds after a corruption scandal cut off bilateral aid in 2007 reduced government's revenues and its ability to pay salaries.

GDP (purchasing power parity): $2.896 billion (2007 est.)

GDP (official exchange rate): $1.001 billion (2007 est.)

GDP—real growth rate: 3.6% (2007 est.)

GDP—per capita (PPP): $400 (2007 est.)

GDP—composition by sector:
agriculture: 33.7%
industry: 20.9%
services: 45.4% (2007 est.)

Labor force: 2.99 million (2002)

Labor force—by occupation: *agriculture:* 93.6%
industry: 2.3%
services: 4.1% (2002 est.)

Unemployment rate: NA%

Population below poverty line: 68% (2002 est.)

Household income or consumption by percentage share:
lowest 10%: 1.7%
highest 10%: 32.8% (1998)

Distribution of family income—Gini index: 42.4 (1998)

Inflation rate (consumer prices): 8.4% (2007 est.)

Investment (gross fixed): 24.4% of GDP (2007 est.)

Budget:
revenues: $264.2 million
expenditures: $335.4 million; including capital expenditures of $NA (2007 est.)

Agriculture—products: coffee, cotton, tea, corn, sorghum, sweet potatoes, bananas, manioc (tapioca); beef, milk, hides

Industries: light consumer goods such as blankets, shoes, soap; assembly of imported components; public works construction; food processing

Industrial production growth rate: 6.4% (2007 est.)

Electricity—production: 137 million kWh (2005)

Electricity—production by source:
fossil fuel: 0.6%
hydro: 99.4%
nuclear: 0%
other: 0% (2001)

Electricity—consumption: 161.4 million kWh (2005)

Electricity—exports: 0 kWh (2005)

Electricity—imports: 34 million kWh; note—supplied by the Democratic Republic of the Congo (2005)

Oil—production: 0 bbl/day (2005)

Oil—consumption: 2,900 bbl/day (2005 est.)

Oil—exports: 0 bbl/day (2004)

Oil—imports: 2,687 bbl/day (2004)

Oil—proved reserves: 0 bbl (1 January 2006 est.)

Natural gas—production: 0 cu m (2005 est.)

Natural gas—consumption: 0 cu m (2005 est.)

Natural gas—exports: 0 cu m (2005 est.)

Natural gas—imports: 0 cu m (2005)

Natural gas—proved reserves: 0 cu m (1 January 2006 est.)

Current account balance: -$124 million (2007 est.)

Exports: $44.5 million f.o.b. (2007 est.)

Exports—commodities: coffee, tea, sugar, cotton, hides

Exports—partners: Switzerland 33.7%, UK 12.2%, Pakistan 8.5%, Rwanda 5.3%, Egypt 4.2% (2006)

Imports: $272 million f.o.b. (2007 est.)

Imports—commodities: capital goods, petroleum products, foodstuffs

Imports—partners: Saudi Arabia 12.6%, Kenya 8.2%, Japan 7.8%, Russia 4.7%, UK 4.6%, France 4.4%, China 4.4% (2006)

Economic aid—recipient: $365 million (2005)

Reserves of foreign exchange and gold: $176.2 million (31 December 2007 est.)

Debt—external: $1.2 billion (2003)

Market value of publicly traded shares: $NA

Currency (code): Burundi franc (BIF)

Currency code: BIF

Exchange rates: Burundi francs per US dollar—1,065 (2007), 1,030 (2006), 1,138 (2005), 1,100.91 (2004), 1,082.62 (2003)

Fiscal year: calendar year

COMMUNICATIONS

Telephones—main lines in use: 31,100 (2005)

Telephones—mobile cellular: 153,000 (2005)

Telephone system:
general assessment: primitive system; telephone density one of the lowest in the world; fixed-line connections stand at well less than 1 per 100 persons; mobile-cellular usage is increasing but remains at a meager 2 per 100 persons
domestic: sparse system of open-wire, radiotelephone communications, and low-capacity microwave radio relay

international: country code—257; satellite earth station—1 Intelsat (Indian Ocean) (2007)
Radio broadcast stations: AM 0, FM 4, shortwave 1 (2001)
Radios: 440,000 (2001)
Television broadcast stations: 1 (2001)
Televisions: 25,000 (1997)
Internet country code: .bi
Internet hosts: 163 (2007)
Internet Service Providers (ISPs): 1 (2000)
Internet users: 60,000 (2006)

TRANSPORTATION

Airports: 8 (2007)
Airports—with paved runways:
total: 1
over 3,047 m: 1 (2007)
Airports—with unpaved runways:
total: 7
914 to 1,523 m: 4
under 914 m: 3 (2007)
Heliports: 1 (2007)
Roadways: *total:* 12,322 km

paved: 1,286 km
unpaved: 11,036 km (2004)
Waterways: mainly on Lake Tanganyika (2005)
Ports and terminals: Bujumbura

MILITARY

Military branches: National Defense Force (Forces de Defense Nationales, FDN): Army (includes Naval Detachment and Air Wing), Gendarmerie (2008)
Military service age and obligation: 16 years of age for compulsory and voluntary military service; children as young as 10 years of age have been conscripted into the armed forces; the enrollment of children is still not prohibited (2007)
Manpower available for military service:
males age 16–49: 1,878,544
females age 16–49: 1,851,676 (2008 est.)
Manpower fit for military service:
males age 16–49: 1,083,899
females age 16–49: 1,062,488 (2008 est.)

Manpower reaching militarily significant age annually:
males age 16–49: 98,105
females age 16–49: 98,533 (2008 est.)
Military expenditures—percent of GDP: 5.9% (2006 est.)

TRANSNATIONAL ISSUES

Disputes—international: conflicts among Tutsi, Hutu, other ethnic groups, associated political rebels, armed gangs, and various government forces have abated somewhat in the Great Lakes region; UN Operation in Burundi (ONUB) completed its mandate in December 2006 after a three-year peace-keeping mission
Refugees and internally displaced persons: *refugees (country of origin):* 9,849 (Democratic Republic of the Congo)
IDPs: 100,000 (armed conflict between government and rebels; most IDPs in northern and western Burundi) (2007)

CAMBODIA

INTRODUCTION

Background: Most Cambodians consider themselves to be Khmers, descendants of the Angkor Empire that extended over much of Southeast Asia and reached its zenith between the 10th and 13th centuries. Attacks by the Thai and Cham (from present-day Vietnam) weakened the empire, ushering in a long period of decline. The king placed the country under French protection in 1863 and it became part of French Indochina in 1887. Following Japanese occupation in World War II, Cambodia gained full independence from France in 1953. In April 1975, after a five-year struggle, Communist Khmer Rouge forces captured Phnom Penh and evacuated all cities and towns. At least 1.5 million Cambodians died from execution, forced hardships, or starvation during the Khmer Rouge regime under POL POT. A December 1978 Vietnamese invasion drove the Khmer Rouge into the countryside, began a 10-year Vietnamese occupation, and touched off almost 13 years of civil war. The 1991 Paris Peace Accords mandated democratic elections and a ceasefire, which was not fully respected by the Khmer Rouge. UN-sponsored elections in 1993 helped restore some semblance of normalcy under a coalition government. Factional fighting in 1997 ended the first coalition government, but a second round of national elections in 1998 led to the formation of another coalition government and renewed political stability. The remaining elements of the Khmer Rouge surrendered in early 1999. Some of the remaining Khmer Rouge leaders are awaiting trial by a UN-sponsored tribunal for crimes against humanity. Elections in July 2003 were relatively peaceful, but it took one year of negotiations between contending political par-

ties before a coalition government was formed. In October 2004, King SIHANOUK abdicated the throne due to illness and his son, Prince Norodom SIHAMONI, was selected to succeed him. Local elections were held in Cambodia in April 2007, and there was little in the way of pre-election violence that preceded prior elections. National elections are scheduled for July 2008.

GEOGRAPHY

Location: Southeastern Asia, bordering the Gulf of Thailand, between Thailand, Vietnam, and Laos

Geographic coordinates: 13 00 N, 105 00 E

Map references: Southeast Asia

Area:
total: 181,040 sq km
land: 176,520 sq km
water: 4,520 sq km

Area—comparative: slightly smaller than Oklahoma

Land boundaries:
total: 2,572 km
border countries: Laos 541 km, Thailand 803 km, Vietnam 1,228 km

Coastline: 443 km

Maritime claims:
territorial sea: 12 nm
contiguous zone: 24 nm
exclusive economic zone: 200 nm
continental shelf: 200 nm

Climate: tropical; rainy, monsoon season (May to November); dry season (December to April); little seasonal temperature variation

Terrain: mostly low, flat plains; mountains in southwest and north

Elevation extremes:
lowest point: Gulf of Thailand 0 m
highest point: Phnum Aoral 1,810 m

Natural resources: oil and gas, timber, gemstones, iron ore, manganese, phosphates, hydropower potential

Land use:
arable land: 20.44%
permanent crops: 0.59%
other: 78.97% (2005)

Irrigated land: 2,700 sq km (2003)

Total renewable water resources: 476.1 cu km (1999)

Freshwater withdrawal (domestic/industrial/agricultural): *total:* 4.08 cu km/yr (1%/0%/98%)
per capita: 290 cu m/yr (2000)

Natural hazards: monsoonal rains (June to November); flooding; occasional droughts

Environment—current issues: illegal logging activities throughout the country and strip mining for gems in the western

region along the border with Thailand have resulted in habitat loss and declining biodiversity (in particular, destruction of mangrove swamps threatens natural fisheries); soil erosion; in rural areas, most of the population does not have access to potable water; declining fish stocks because of illegal fishing and overfishing

Environment—international agreements: *party to:* Biodiversity, Climate Change, Climate Change-Kyoto Protocol, Desertification, Endangered Species, Hazardous Wastes, Marine Life Conservation, Ozone Layer Protection, Ship Pollution, Tropical Timber 94, Wetlands, Whaling
signed, but not ratified: Law of the Sea

Geography—note: a land of paddies and forests dominated by the Mekong River and Tonle Sap

PEOPLE

Population: 14,241,640
note: estimates for this country take into account the effects of excess mortality due to AIDS; this can result in lower life expectancy, higher infant mortality, higher death rates, lower population growth rates, and changes in the distribution of population by age and sex than would otherwise be expected (July 2008 est.)

Age structure:
0–14 years: 33.2% (male 2,389,668/female 2,338,838)
15–64 years: 63.2% (male 4,372,480/female 4,627,895)
65 years and over: 3.6% (male 193,338/female 319,421) (2008 est.)

Median age:
total: 21.7 years
male: 21 years
female: 22.5 years (2008 est.)

Population growth rate: 1.752% (2008 est.)

Birth rate: 25.68 births/1,000 population (2008 est.)

Death rate: 8.16 deaths/1,000 population (2008 est.)

Net migration rate: NA

Sex ratio:
at birth: 1.04 male(s)/female
under 15 years: 1.02 male(s)/female
15–64 years: 0.94 male(s)/female
65 years and over: 0.61 male(s)/female
total population: 0.95 male(s)/female (2008 est.)

Infant mortality rate:
total: 56.59 deaths/1,000 live births
male: 63.76 deaths/1,000 live births
female: 49.1 deaths/1,000 live births (2008 est.)

Life expectancy at birth:
total population: 61.69 years
male: 59.65 years
female: 63.83 years (2008 est.)

Total fertility rate: 3.08 children born/woman (2008 est.)

HIV/AIDS—adult prevalence rate: 2.6% (2003 est.)

HIV/AIDS—people living with HIV/AIDS: 170,000 (2003 est.)

HIV/AIDS—deaths: 15,000 (2003 est.)

Major infectious diseases:
degree of risk: very high
food or waterborne diseases: bacterial and protozoal diarrhea, hepatitis A, and typhoid fever
vectorborne diseases: dengue fever, Japanese encephalitis, and malaria
note: highly pathogenic H5N1 avian influenza has been identified in this country; it poses a negligible risk with extremely rare cases possible among US citizens who have close contact with birds (2008)

Nationality:
noun: Cambodian(s)
adjective: Cambodian

Ethnic groups: Khmer 90%, Vietnamese 5%, Chinese 1%, other 4%

Religions: Theravada Buddhist 95%, other 5%

Languages: Khmer (official) 95%, French, English

Literacy:
definition: age 15 and over can read and write
total population: 73.6%
male: 84.7%
female: 64.1% (2004 est.)

GOVERNMENT

Country name:
conventional long form: Kingdom of Cambodia
conventional short form: Cambodia
local long form: Preahreacheanachakr Kampuchea (phonetic pronunciation)
local short form: Kampuchea
former: Khmer Republic, Democratic Kampuchea, People's Republic of Kampuchea, State of Cambodia

Government type: multiparty democracy under a constitutional monarchy

Capital: *name:* Phnom Penh
geographic coordinates: 11 33 N, 104 55 E
time difference: UTC+7 (12 hours ahead of Washington, DC during Standard Time)

Administrative divisions. 20 provinces (khaitt, singular and plural) and 4 municipalities* (krong, singular and plural)
provinces: Banteay Mean Cheay, Batdambang, Kampong Cham, Kampong Chhnang, Kampong Spoe, Kampong Thum, Kampot, Kandal, Kaoh Kong, Krachen, Mondol Kiri, Otdar Mean Cheay, Pouthisat, Preah Vihear, Prey Veng, Rotanah Kiri, Siem Reab, Stoeng Treng, Svay Rieng, Takev
municipalities: Keb, Pailin, Phnum Penh (Phnom Penh), Preah Seihanu (Sihanoukville)

Independence: 9 November 1953 (from France)

National holiday: Independence Day, 9 November (1953)

Constitution: promulgated 21 September 1993

Legal system: primarily a civil law mixture of French-influenced codes from the United Nations Transitional Authority in Cambodia (UNTAC) period, royal decrees, and acts of the legislature, with influences of customary law and remnants of communist legal theory; increasing influence of common law; accepts compulsory ICJ jurisdiction with reservations

Suffrage: 18 years of age; universal

Executive branch:
chief of state: King Norodom SIHAMONI (since 29 October 2004)
head of government: Prime Minister HUN SEN (since 14 January 1985) [co-prime minister from 1993 to 1997]; Deputy Prime Ministers SAR KHENG (since 3 February 1992); SOK AN, LU LAY SRENG, TEA BANH, HOR NAMHONG, NHEK BUNCHHAY (since 16 July 2004); KEV PUT REAKSMEI (since 24 October 2006), BIN CHHIN (since 5 September 2007)
cabinet: Council of Ministers appointed by the monarch; in practice named by the prime minister
elections: the monarch is chosen by a Royal Throne Council; following legislative elections, a member of the majority party or majority coalition is named prime minister by the Chairman of the National Assembly and appointed by the king

Legislative branch: bicameral, consists of the National Assembly (123 seats); members elected by popular vote to serve five-year terms) and the Senate (61 seats; 2 members appointed by the monarch, 2 elected by the National Assembly, and 57 elected by parliamentarians and commune councils; members serve five-year terms)
elections: National Assembly—last held 27 July 2003 (next to be held on 27 July 2008); Senate—last held 22 January 2006 (next to be held in January 2011)
election results: National Assembly—percent of vote by party—CPP 47%, SRP 22%, FUNCINPEC 21%, other 10%; seats by party—CPP 73, FUNCINPEC 26, SRP 24; Senate—percent of vote by party—CPP 69%, FUNCINPEC 21%, SRP 10%; seats by party—CPP 45, FUNCINPEC 10, SRP 2 (January 2006)

Judicial branch: Supreme Council of the Magistracy (provided for in the constitution and formed in December 1997); Supreme Court (and lower courts) exercises judicial authority

Political parties and leaders: Cambodian People's Party or CPP [CHEA SIM]; National United Front for an Independent, Neutral, Peaceful, and Cooperative Cambodia or FUNCINPEC [KEV PUT REAKSMEI]; Norodom Ranariddh Party or NRP [Norodom RANARIDDH]; Sam Rangsi Party or SRP [SAM RANGSI]

Political pressure groups and leaders: NA

International organization participation: ACCT, ADB, APT, ARF, ASEAN, EAS, FAO, G-77, IBRD, ICAO, ICCt, ICRM, IDA, IFAD, IFC, IFRCS, ILO, IMF, IMO, Interpol, IOC, IOM, IPU, ISO (correspondent), ISO (subscriber), ITU, MIGA, NAM, OIF, OPCW, PCA, UN, UNCTAD, UNESCO, UNIDO, UNMIS, UNWTO, UPU, WCO, WFTU, WHO, WIPO, WMO, WTO

Diplomatic representation in the US:
chief of mission: Ambassador EK SEREYWATH
chancery: 4530 16th Street NW, Washington, DC 20011
telephone: [1] (202) 726-7742
FAX: [1] (202) 726-8381

Diplomatic representation from the US:
chief of mission: Ambassador Joseph A. MUSSOMELI
embassy: #1, Street 96, Sangkat Wat Phnom, Khan Daun Penh, Phnom Penh
mailing address: Box P, APO AP 96546
telephone: [855] (23) 728-000
FAX: [855] (23) 728-600

Flag description: three horizontal bands of blue (top), red (double width), and blue with a white three-towered temple representing Angkor Wat outlined in black in the center of the red band
note: only national flag to incorporate an actual building in its design

ECONOMY

Economy—overview: From 2001 to 2004, the economy grew at an average rate of 6.4%, driven largely by an expansion in the garment sector and tourism. The US and Cambodia signed a Bilateral Textile Agreement, which gave Cambodia a guaranteed quota of US textile imports and established a bonus for improving working conditions and enforcing Cambodian labor laws and international labor standards in the industry. With the January 2005 expiration of a WTO Agreement on Textiles and Clothing, Cambodia-based textile producers were forced to compete directly with lower-priced producing countries such as China and India. Better-than-expected garment sector performance led to more than 9% growth in 2007. Its vibrant garment industry employs more than 350,000 people and contributes more than 70% of Cambodia's exports. The Cambodian government has committed itself to a policy supporting high labor standards in an attempt to maintain buyer interest. In 2005, exploitable oil and natural gas deposits were found beneath Cambodia's territorial waters, representing a new revenue stream for the government if commercial extraction begins. Mining also is attracting significant investor interest, particularly in the northeastern parts of the country, and the government has said opportunities exist for mining bauxite, gold, iron and gems. In 2006, a US-Cambodia bilateral Trade and Investment Framework Agreement (TIFA) was signed and the first round of discussions took place in early 2007. The tourism industry continues to grow rapidly, with foreign arrivals reaching 2 million in 2007. In 2007 the government signed a joint venture agreement with two companies to form a new national airline. The long-term development of the economy remains a daunting challenge. The Cambodian government is working with bilateral and multilateral donors, including the World Bank and IMF, to address the country's many pressing needs. The major economic challenge for Cambodia over the next decade will be fashioning an economic environment in which the private sector can create enough jobs to handle Cambodia's demographic imbalance. More than 50% of the population is less than 21 years old. The population lacks education and productive skills, particularly in the poverty-ridden countryside, which suffers from an almost total lack of basic infrastructure.

GDP (purchasing power parity): $25.9 billion (2007 est.)

GDP (official exchange rate): $8.604 billion (2007 est.)

GDP—real growth rate: 9.6% (2007 est.)

GDP—per capita (PPP): $1,800 (2007 est.)

GDP—composition by sector:
agriculture: 31%
industry: 26%
services: 43% (2007 est.)

Labor force: 7 million (2003 est.)

Labor force—by occupation: *agriculture:* 75%
industry: NA%
services: NA%

Unemployment rate: 2.5% (2000 est.)

Population below poverty line: 35% (2004)

Household income or consumption by percentage share:
lowest 10%: 2.9%
highest 10%: 34.8% (2004)

Distribution of family income—Gini index: 41.7 (2004 est.)

Inflation rate (consumer prices): 5.9% (2007 est.)

Investment (gross fixed): 21.7% of GDP (2007 est.)

Budget:
revenues: $972.8 million
expenditures: $1.04 billion (2007 est.)

Agriculture—products: rice, rubber, corn, vegetables, cashews, tapioca

Industries: tourism, garments, rice milling, fishing, wood and wood products, rubber, cement, gem mining, textiles

Industrial production growth rate: 9% (2007 est.)

Electricity—production: 134 million kWh (2005)

Electricity—production by source:
fossil fuel: 65%
hydro: 35%
nuclear: 0%
other: 0% (2001)

Electricity—consumption: 206.6 million kWh (2005)

Electricity—exports: 0 kWh (2005)

Electricity—imports: 82 million kWh (2005)

Oil—production: 0 bbl/day (2005)

Oil—consumption: 3,700 bbl/day (2005 est.)

Oil—exports: 0 bbl/day (2004)

Oil—imports: 3,585 bbl/day (2004)

Oil—proved reserves: 0 bbl (1 January 2006 est.)

Natural gas—production: 0 cu m (2005 est.)

Natural gas—consumption: 0 cu m (2005 est.)

Natural gas—exports: 0 cu m (2005 est.)

Natural gas—imports: 0 cu m (2005)

Natural gas—proved reserves: NA

Current account balance: -$77 million (2007 est.)

Exports: $4.071 billion f.o.b. (2007 est.)

Exports—commodities: clothing, timber, rubber, rice, fish, tobacco, footwear

Exports—partners: US 53.3%, Hong Kong 15.2%, Germany 6.6%, UK 4.3% (2006)

Imports: $5.413 billion f.o.b. (2007 est.)

Imports—commodities: petroleum products, cigarettes, gold, construction materials, machinery, motor vehicles, pharmaceutical products

Imports—partners: Hong Kong 18.1%, China 17.5%, Thailand 13.9%, Taiwan 12.7%, Vietnam 9%, Singapore 5.3%, South Korea 4.9%, Japan 4.3% (2006)

Economic aid—recipient: $698.2 million pledged in grants and concession loans for 2007 by international donors (2007)

Reserves of foreign exchange and gold: $2.143 billion (31 December 2007 est.)

Debt—external: $3.891 billion (31 December 2007 est.)

Market value of publicly traded shares: $NA

Currency (code): riel (KHR)

Currency code: KHR

Exchange rates: riels per US dollar—4,006 (2007), 4,103 (2006), 4,092.5 (2005), 4,016.25 (2004), 3,973.33 (2003)

Fiscal year: calendar year

COMMUNICATIONS

Telephones—main lines in use: 32,800 (2006)

Telephones—mobile cellular: 1.14 million (2006)

Telephone system:
general assessment: mobile-phone systems are widely used in urban areas to bypass deficiencies in the fixed-line network; fixed-line connections stand at well less than 1 per 100 persons; mobile-cellular usage, aided by increasing competition among service providers, is increasing and stands at about 8 per 100 persons
domestic: adequate landline and/or cellular service in Phnom Penh and other provincial cities; mobile-phone coverage is rapidly expanding in rural areas
international: country code—855; adequate but expensive landline and cellular service available to all countries from Phnom Penh and major provincial cities; satellite earth station—1 Intersputnik (Indian Ocean region) (2007)

Radio broadcast stations: AM 2, FM 17, shortwave NA (2003)

Radios: 1.34 million (1997)

Television broadcast stations: 9 (including 2 TV relay stations with French and Vietnamese broadcasts); excludes 18 regional relay stations (2006)

Televisions: 94,000 (1997)

Internet country code: .kh

Internet hosts: 941 (2007)

Internet Service Providers (ISPs): 2 (2000)

Internet users: 44,000 (2005)

TRANSPORTATION

Airports: 17 (2007)
Airports—with paved runways:
total: 6
2,438 to 3,047 m: 2
1,524 to 2,437 m: 2
914 to 1,523 m: 2 (2007)
Airports—with unpaved runways:
total: 11
1,524 to 2,437 m: 1
914 to 1,523 m: 9
under 914 m: 1 (2007)
Heliports: 1 (2007)
Railways: total: 602 km
narrow gauge: 602 km 1.000-m gauge (2006)
Roadways:
total: 38,257 km
paved: 2,406 km
unpaved: 35,851 km (2004)
Waterways: 2,400 km (mainly on Mekong River) (2005)
Merchant marine:
total: 586 ships (1000 GRT or over) 1,889,909 GRT/2,682,881 DWT
by type: bulk carrier 40, cargo 487, chemical tanker 10, container 9, livestock carrier 3, passenger/cargo 5, petroleum tanker 11, refrigerated cargo 18, roll on/roll off 1, specialized tanker 1, vehicle carrier 1
foreign-owned: 463 (Canada 6, China 166, Cyprus 9, Egypt 14, Estonia 1, Gabon 1, Greece 5, Hong Kong 11, Indonesia 1, Japan 3, South Korea 29, Latvia 2, Lebanon 7, Nigeria 2, Romania 1, Russia 112, Singapore 2, Syria 32, Taiwan 1, Turkey 20, Ukraine 27, UAE 2, US 6, Yemen 3) (2007)
Ports and terminals: Phnom Penh, Kampong Saom (Sihanoukville)

MILITARY

Military branches: Royal Cambodian Armed Forces: Royal Cambodian Army, Royal Khmer Navy, Royal Cambodian Air Force (2008)
Military service age and obligation: conscription law of October 2006 requires all males between 18–30 to register for military service; 18-month service obligation (2006)
Manpower available for military service:
males age 16–49: 3,759,034
females age 16–49: 3,784,333 (2008 est.)
Manpower fit for military service:
males age 16–49: 2,581,045
females age 16–49: 2,676,075 (2008 est.)
Manpower reaching militarily significant age annually:
males age 16–49: 185,959
females age 16–49: 182,558 (2008 est.)
Military expenditures—percent of GDP: 3% (2005 est.)

TRANSNATIONAL ISSUES

Disputes—international: Southeast Asian states must maintain border surveillance to check the spread of avian flu; Cambodia and Thailand dispute sections of boundary with missing boundary markers and claims of Thai encroachments into Cambodian territory; maritime boundary with Vietnam is hampered by unresolved dispute over sovereignty of offshore islands; Cambodia accuses Thailand of obstructing access to Preah Vihear temple ruins awarded to Cambodia by ICJ decision in 1962
Trafficking in persons: current situation: Cambodia is a source, destination, and transit country for men, women, and children trafficked for the purposes of sexual exploitation and forced labor; a significant number of women and children are trafficked to Thailand and Malaysia for commercial sexual exploitation and forced labor; men are trafficked primarily to Thailand for forced labor in the construction and agricultural sectors, particularly the fishing industry, while women and girls are trafficked for factory and domestic work; children are trafficked to Vietnam and Thailand for the purpose of forced begging; Cambodia is a transit and destination point for women from Vietnam trafficked for sexual exploitation; trafficking for sexual exploitation also occurs within Cambodia's borders, from rural areas to the cities
tier rating: Tier 2 Watch List—Cambodia does not fully comply with the minimum standards for the elimination of trafficking; however, it is committed to making significant efforts to sustain progress over the coming year
Illicit drugs: narcotics-related corruption reportedly involving some in the government, military, and police; limited methamphetamine production; vulnerable to money laundering due to its cash-based economy and porous borders

CAMEROON

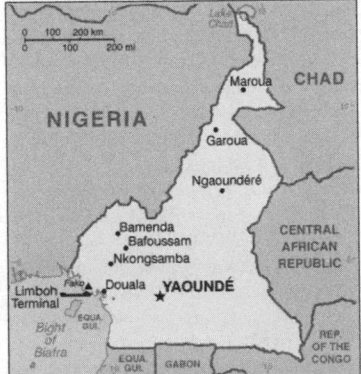

INTRODUCTION

Background: The former French Cameroon and part of British Cameroon merged in 1961 to form the present country. Cameroon has generally enjoyed stability, which has permitted the development of agriculture, roads, and railways, as well as a petroleum industry. Despite a slow movement toward democratic reform, political power remains firmly in the hands of President Paul BIYA.

GEOGRAPHY

Location: Western Africa, bordering the Bight of Biafra, between Equatorial Guinea and Nigeria
Geographic coordinates: 6 00 N, 12 00 E
Map references: Africa
Area:
total: 475,440 sq km
land: 469,440 sq km
water: 6,000 sq km
Area—comparative: slightly larger than California
Land boundaries:
total: 4,591 km
border countries: Central African Republic 797 km, Chad 1,094 km, Republic of the Congo 523 km, Equatorial Guinea 189 km, Gabon 298 km, Nigeria 1,690 km
Coastline: 402 km
Maritime claims:
territorial sea: 12 nm
contiguous zone: 24 nm
Climate: varies with terrain, from tropical along coast to semiarid and hot in north
Terrain: diverse, with coastal plain in southwest, dissected plateau in center, mountains in west, plains in north
Elevation extremes:
lowest point: Atlantic Ocean 0 m
highest point: Fako 4,095 m (on Mt. Cameroon)
Natural resources: petroleum, bauxite, iron ore, timber, hydropower

Land use:
arable land: 12.54%
permanent crops: 2.52%
other: 84.94% (2005)
Irrigated land: 260 sq km (2003)
Total renewable water resources: 285.5 cu km (2003)
Freshwater withdrawal (domestic/industrial/agricultural): *total:* 0.99 cu km/yr (18%/8%/74%)
per capita: 61 cu m/yr (2000)
Natural hazards: volcanic activity with periodic releases of poisonous gases from Lake Nyos and Lake Monoun volcanoes
Environment—current issues: waterborne diseases are prevalent; deforestation; overgrazing; desertification; poaching; overfishing
Environment—international agreements: *party to:* Biodiversity, Climate Change, Climate Change-Kyoto Protocol, Desertification, Endangered Species, Hazardous Wastes, Law of the Sea, Ozone Layer Protection, Tropical Timber 83, Tropical Timber 94, Wetlands, Whaling
signed, but not ratified: none of the selected agreements
Geography—note: sometimes referred to as the hinge of Africa; throughout the country there are areas of thermal springs and indications of current or prior volcanic activity; Mount Cameroon, the highest mountain in Sub-Saharan west Africa, is an active volcano

PEOPLE

Population: 18,467,692
note: estimates for this country explicitly take into account the effects of excess mortality due to AIDS; this can result in lower life expectancy, higher infant mortality, higher death rates, lower population growth rates, and changes in the distribution of population by age and sex than would otherwise be expected (July 2008 est.)
Age structure:
0–14 years: 41.1% (male 3,826,232/female 3,757,859)
15–64 years: 55.7% (male 5,164,338/female 5,122,817)
65 years and over: 3.2% (male 274,821/female 321,625) (2008 est.)
Median age:
total: 19 years
male: 18.9 years
female: 19.2 years (2008 est.)
Population growth rate: 2.218% (2008 est.)
Birth rate: 34.59 births/1,000 population (2008 est.)
Death rate: 12.41 deaths/1,000 population (2008 est.)
Net migration rate: NA

Sex ratio:
at birth: 1.03 male(s)/female
under 15 years: 1.02 male(s)/female
15–64 years: 1.01 male(s)/female
65 years and over: 0.85 male(s)/female
total population: 1.01 male(s)/female (2008 est.)
Infant mortality rate:
total: 64.57 deaths/1,000 live births
male: 69.39 deaths/1,000 live births
female: 59.62 deaths/1,000 live births (2008 est.)
Life expectancy at birth:
total population: 53.3 years
male: 52.54 years
female: 54.08 years (2008 est.)
Total fertility rate: 4.41 children born/woman (2008 est.)
HIV/AIDS—adult prevalence rate: 6.9% (2003 est.)
HIV/AIDS—people living with HIV/AIDS: 560,000 (2003 est.)
HIV/AIDS—deaths: 49,000 (2003 est.)
Major infectious diseases:
degree of risk: very high
food or waterborne diseases: bacterial and protozoal diarrhea, hepatitis A and E, and typhoid fever
vectorborne diseases: malaria and yellow fever
water contact disease: schistosomiasis
respiratory disease: meningococcal meningitis
animal contact disease: rabies (2008)
Nationality:
noun: Cameroonian(s)
adjective: Cameroonian
Ethnic groups: Cameroon Highlanders 31%, Equatorial Bantu 19%, Kirdi 11%, Fulani 10%, Northwestern Bantu 8%, Eastern Nigritic 7%, other African 13%, non-African less than 1%
Religions: indigenous beliefs 40%, Christian 40%, Muslim 20%
Languages: 24 major African language groups, English (official), French (official)
Literacy:
definition: age 15 and over can read and write
total population: 67.9%
male: 77%
female: 59.8% (2001 est.)

GOVERNMENT

Country name:
conventional long form: Republic of Cameroon
conventional short form: Cameroon
local long form: Republique du Cameroun/Republic of Cameroon
local short form: Cameroun/Cameroon
former: French Cameroon, British Cameroon, Federal Republic of Cameroon, United Republic of Cameroon

Government type: republic; multiparty presidential regime
Capital: *name:* Yaounde
geographic coordinates: 3 52 N, 11 31 E
time difference: UTC+1 (6 hours ahead of Washington, DC during Standard Time)
Administrative divisions: 10 provinces; Adamaoua, Centre, Est, Extreme-Nord, Littoral, Nord, Nord-Ouest, Ouest, Sud, Sud-Ouest
Independence: 1 January 1960 (from French-administered UN trusteeship)
National holiday: Republic Day (National Day), 20 May (1972)
Constitution: 20 May 1972 approved by referendum, adopted 2 June 1972; revised January 1996
Legal system: based on French civil law system, with common law influence; accepts compulsory ICJ jurisdiction
Suffrage: 20 years of age; universal
Executive branch:
chief of state: President Paul BIYA (since 6 November 1982)
head of government: Prime Minister Ephraim INONI (since 8 December 2004)
cabinet: Cabinet appointed by the president from proposals submitted by the prime minister
elections: president elected by popular vote for a seven-year term (eligible for a second term); election last held 11 October 2004 (next to be held by October 2011); prime minister appointed by the president
election results: President Paul BIYA reelected; percent of vote—Paul BIYA 70.9%, John FRU NDI 17.4%, Adamou Ndam NJOYA 4.5%, Garga Haman ADJI 3.7%
Legislative branch: unicameral National Assembly or Assemblee Nationale (180 seats; members are elected by direct popular vote to serve five-year terms); note—the president can either lengthen or shorten the term of the legislature
elections: last held 22 July 2007 (next to be held in 2012)
election results: percent of vote by party—NA; seats by party—RDCP 140, SDF 14, UDC 4, UNDP 4, MP 1, vacant 17; note—vacant seats will be determined in a yet to be scheduled by-election after the Supreme Court nullified results in five districts
note: the constitution calls for an upper chamber for the legislature, to be called a Senate, but it has yet to be established
Judicial branch: Supreme Court (judges are appointed by the president); High Court of Justice (consists of nine judges and six substitute judges, elected by the National Assembly)

Political parties and leaders: Cameroonian Democratic Union or UDC [Adamou Ndam NJOYA]; Cameroon People's Democratic Movement or RDPC [Paul BIYA]; Movement for the Defense of the Republic or MDR [Dakole DAISSALA]; Movement for the Liberation and Development of Cameroon or MLDC [Marcel YONDO]; National Union for Democracy and Progress or UNDP [Maigari BELLO BOUBA]; Progressive Movement or MP; Social Democratic Front or SDF [John FRU NDI]; Union of Peoples of Cameroon or UPC [Augustin Frederic KODOCK]

Political pressure groups and leaders: Southern Cameroon National Council [Ayamba Ette OTUN]; Human Rights Defense Group [Albert MUKONG, president]

International organization participation: ACCT, ACP, AfDB, AU, BDEAC, C, CEMAC, FAO, FZ, G-77, IAEA, IBRD, ICAO, ICC, ICCt (signatory), ICRM, IDA, IDB, IFAD, IFC, IFRCS, ILO, IMF, IMO, IMSO, Interpol, IOC, IOM, IPU, ISO (correspondent), ITSO, ITU, ITUC, MIGA, NAM, OIC, OIF, OPCW, PCA, UN, UNCTAD, UNESCO, UNIDO, UNWTO, UPU, WCL, WCO, WFTU, WHO, WIPO, WMO, WTO

Diplomatic representation in the US:
chief of mission: Ambassador Jerome MENDOUGA
chancery: 2349 Massachusetts Avenue NW, Washington, DC 20008
telephone: [1] (202) 265-8790
FAX: [1] (202) 387-3826

Diplomatic representation from the US:
chief of mission: Ambassador Janet E. GARVEY
embassy: Avenue Rosa Parks, Yaounde
mailing address: P. O. Box 817, Yaounde; pouch: American Embassy, US Department of State, Washington, DC 20521-2520
telephone: [237] 2220 15 00; Consular: [237] 2220 16 03
FAX: [237] 2220 16 00 Ext. 4531; Consular FAX: [237] 2220 17 52
branch office(s): Douala

Flag description: three equal vertical bands of green (hoist side), red, and yellow with a yellow five-pointed star centered in the red band
note: uses the popular pan-African colors of Ethiopia

ECONOMY

Economy—overview: Because of its modest oil resources and favorable agricultural conditions, Cameroon has one of the best-endowed primary commodity economies in sub-Saharan Africa. Still, it faces many of the serious problems facing other underdeveloped countries, such as a top-heavy civil service and a generally unfavorable climate for business enterprise. Since 1990, the government has embarked on various IMF and World Bank programs designed to spur business investment, increase efficiency in agriculture, improve trade, and recapitalize the nation's banks. In June 2000, the government completed an IMF-sponsored, three-year structural adjustment program; however, the IMF is pressing for more reforms, including increased budget transparency, privatization, and poverty reduction programs. In January 2001, the Paris Club agreed to reduce Cameroon's debt of $1.3 billion by $900 million; debt relief now totals $1.26 billion. International oil and cocoa prices have a significant impact on the economy.

GDP (purchasing power parity): $39.37 billion (2007 est.)

GDP (official exchange rate): $20.65 billion (2007 est.)

GDP—real growth rate: 3.3% (2007 est.)

GDP—per capita (PPP): $2,100 (2007 est.)

GDP—composition by sector:
agriculture: 43.9%
industry: 15.8%
services: 40.3% (2007 est.)

Labor force: 6.674 million (2007 est.)

Labor force—by occupation: *agriculture:* 70%
industry: 13%
services: 17% (2001 est.)

Unemployment rate: 30% (2001 est.)

Population below poverty line: 48% (2000 est.)

Household income or consumption by percentage share:
lowest 10%: 2.3%
highest 10%: 35.4% (2001)

Distribution of family income—Gini index: 44.6 (2001)

Inflation rate (consumer prices): 0.9% (2007 est.)

Investment (gross fixed): 17.3% of GDP (2007 est.)

Budget:
revenues: $4.178 billion
expenditures: $3.297 billion (2007 est.)

Public debt: 15.4% of GDP (2007 est.)

Agriculture—products: coffee, cocoa, cotton, rubber, bananas, oilseed, grains, root starches; livestock; timber

Industries: petroleum production and refining, aluminum production, food processing, light consumer goods, textiles, lumber, ship repair

Industrial production growth rate: 3.5% (2007 est.)

Electricity—production: 4.09 billion kWh (2005)

Electricity—production by source:
fossil fuel: 2.7%
hydro: 97.3%
nuclear: 0%
other: 0% (2001)

Electricity—consumption: 3.435 billion kWh (2005)

Electricity—exports: 0 kWh (2005)

Electricity—imports: 0 kWh (2005)

Oil—production: 82,670 bbl/day (2005 est.)

Oil—consumption: 24,200 bbl/day (2005 est.)

Oil—exports: 107,400 bbl/day (2004)

Oil—imports: 63,710 bbl/day (2004)

Oil—proved reserves: 95 million bbl (2007 est.)

Natural gas—production: 0 cu m (2005 est.)

Natural gas—consumption: 0 cu m (2005 est.)

Natural gas—exports: 0 cu m (2005 est.)

Natural gas—imports: 0 cu m (2005)

Natural gas—proved reserves: 105.9 billion cu m (1 January 2006 est.)

Current account balance: $85 million (2007 est.)

Exports: $3.82 billion f.o.b. (2007 est.)

Exports—commodities: crude oil and petroleum products, lumber, cocoa beans, aluminum, coffee, cotton

Exports—partners: Spain 21.4%, Italy 15.4%, France 11.6%, South Korea 7.3%, Netherlands 7.2%, US 5.7%, Belgium 4.2% (2006)

Imports: $3.714 billion f.o.b. (2007 est.)

Imports—commodities: machinery, electrical equipment, transport equipment, fuel, food

Imports—partners: France 23.6%, Nigeria 13.2%, China 7.2%, Belgium 6.1%, US 4.5% (2006)

Economic aid—recipient: $413.8 million (2005)

Reserves of foreign exchange and gold: $2.934 billion (31 December 2007 est.)

Debt—external: $2.555 billion (31 December 2007 est.)

Market value of publicly traded shares: $NA

Currency (code): Communaute Financiere Africaine franc (XAF); note—responsible authority is the Bank of the Central African States

Currency code: XAF

Exchange rates: Communaute Financiere Africaine francs (XAF) per US dollar—493.51 (2007), 522.59 (2006), 527.47 (2005), 528.29 (2004), 581.2 (2003)

Fiscal year: 1 July–30 June

COMMUNICATIONS

Telephones—main lines in use: 100,300 (2005)

Telephones—mobile cellular: 2.253 million (2005)

Telephone system:

general assessment: fixed-line connections stand at less than 1 per 100 persons; equipment is old and outdated, and connections with many parts of the country are unreliable; mobile-cellular usage, in part a reflection of the poor condition and general inadequacy of the fixed-line network, has been increasing steadily and currently stands at 14 per 100 persons

domestic: cable, microwave radio relay, and tropospheric scatter

international: country code—237; landing point for the SAT-3/WASC fiber-optic submarine cable that provides connectivity to Europe and Asia; satellite earth stations—2 Intelsat (Atlantic Ocean) (2007)

Radio broadcast stations: AM 2, FM 9, shortwave 3 (2001)

Radios: 2.27 million (1997)

Television broadcast stations: 1 (2001)

Televisions: 450,000 (1997)

Internet country code: .cm

Internet hosts: 512 (2007)

Internet Service Providers (ISPs): 1 (2002)

Internet users: 370,000 (2006)

TRANSPORTATION

Airports: 45 (2007)

Airports—with paved runways:

total: 11

over 3,047 m: 2

2,438 to 3,047 m: 4

1,524 to 2,437 m: 3

914 to 1,523 m: 1

under 914 m: 1 (2007)

Airports—with unpaved runways:

total: 34

1,524 to 2,437 m: 6

914 to 1,523 m: 20

under 914 m: 8 (2007)

Pipelines: gas 27 km; liquid petroleum gas 5 km; oil 1,110 km (2007)

Railways:

total: 987 km

narrow gauge: 987 km 1.000-m gauge (2006)

Roadways:

total: 50,000 km

paved: 5,000 km

unpaved: 45,000 km (2004)

Waterways: navigation mainly on Benue River; limited during rainy season (2005)

Merchant marine:

total: 1 ship (1000 GRT or over) 38,613 GRT/68,820 DWT

by type: petroleum tanker 1

foreign-owned: 1 (France 1) (2007)

Ports and terminals: Douala, Limboh Terminal

MILITARY

Military branches: Cameroon Armed Forces: Army, Navy (includes naval infantry), Air Force (Armee de l'Air du Cameroun, AAC) (2008)

Military service age and obligation: 18 years of age for voluntary military service; no conscription; the government makes periodic calls for volunteers (2006)

Manpower available for military service:

males age 16–49: 4,321,175

females age 16–49: 4,228,625 (2008 est.)

Manpower fit for military service:

males age 16–49: 2,567,428

females age 16–49: 2,498,990 (2008 est.)

Manpower reaching militarily significant age annually:

males age 16–49: 212,205

females age 16–49: 207,545 (2008 est.)

Military expenditures—percent of GDP: 1.3% (2006)

TRANSNATIONAL ISSUES

Disputes—international: Joint Border Commission with Nigeria reviewed 2002 ICJ ruling on the entire boundary and bilaterally resolved differences, including June 2006 Greentree Agreement that immediately cedes sovereignty of the Bakassi Peninsula to Cameroon with a phase-out of Nigerian control within two years while resolving patriation issues; implementation of the ICJ ruling on the Cameroon-Equatorial Guinea-Nigeria maritime boundary in the Gulf of Guinea is pending due to imprecisely defined coordinates and a sovereignty dispute between Equatorial Guinea and Cameroon over an island at the mouth of the Ntem River; only Nigeria and Cameroon have heeded the Lake Chad Commission's admonition to ratify the delimitation treaty, which also includes the Chad-Niger and Niger-Nigeria boundaries

Refugees and internally displaced persons: *refugees (country of origin):* 20,000–30,000 (Chad); 3,000 (Nigeria); 24,000 (Central African Republic) (2007)

CANADA

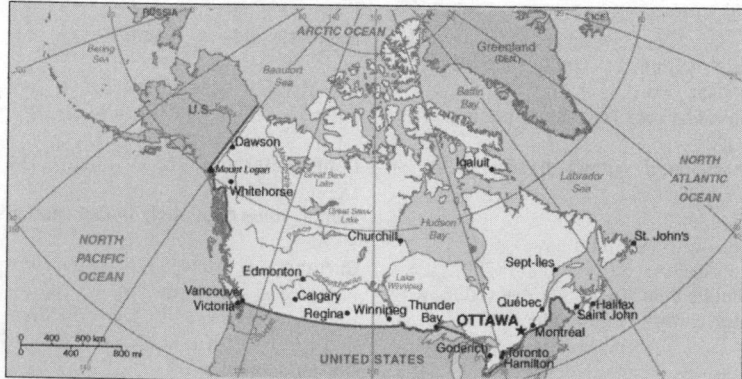

INTRODUCTION

Background: A land of vast distances and rich natural resources, Canada became a self-governing dominion in 1867 while retaining ties to the British crown. Economically and technologically the nation has developed in parallel with the US, its neighbor to the south across an unfortified border. Canada faces the political challenges of meeting public demands for quality improvements in health care and education services, as well as responding to separatist concerns in predominantly francophone Quebec. Canada also aims to develop its diverse energy resources while maintaining its commitment to the environment.

GEOGRAPHY

Location: Northern North America, bordering the North Atlantic Ocean on the east, North Pacific Ocean on the west, and the Arctic Ocean on the north, north of the conterminous US

Geographic coordinates: 60 00 N, 95 00 W

Map references: North America

Area:

total: 9,984,670 sq km

land: 9,093,507 sq km

water: 891,163 sq km

Area—comparative: somewhat larger than the US

Land boundaries:

total: 8,893 km

border countries: US 8,893 km (includes 2,477 km with Alaska)

Coastline: 202,080 km

Maritime claims:

territorial sea: 12 nm

contiguous zone: 24 nm

exclusive economic zone: 200 nm

continental shelf: 200 nm or to the edge of the continental margin

Climate: varies from temperate in south to subarctic and arctic in north

Terrain: mostly plains with mountains in west and lowlands in southeast

Elevation extremes:

lowest point: Atlantic Ocean 0 m

highest point: Mount Logan 5,959 m

Natural resources: iron ore, nickel, zinc, copper, gold, lead, molybdenum, potash, diamonds, silver, fish, timber, wildlife, coal, petroleum, natural gas, hydropower

Land use:

arable land: 4.57%

permanent crops: 0.65%

other: 94.78% (2005)

Irrigated land: 7,850 sq km (2003)

Total renewable water resources: 3,300 cu km (1985)

Freshwater withdrawal (domestic/industrial/agricultural): *total:* 44.72 cu km/yr (20%/69%/12%)

per capita: 1,386 cu m/yr (1996)

Natural hazards: continuous permafrost in north is a serious obstacle to development; cyclonic storms form east of the Rocky Mountains, a result of the mixing of air masses from the Arctic, Pacific, and North American interior, and produce most of the country's rain and snow east of the mountains

Environment—current issues: air pollution and resulting acid rain severely affecting lakes and damaging forests; metal smelting, coal-burning utilities, and vehicle emissions impacting on agricultural and forest productivity; ocean waters becoming contaminated due to agricultural, industrial, mining, and forestry activities

Environment—international agreements: *party to:* Air Pollution, Air Pollution-Nitrogen Oxides, Air Pollution-Persistent Organic Pollutants, Air Pollution-Sulfur 85, Air Pollution-Sulfur 94, Antarctic-Environmental Protocol, Antarctic-Marine Living Resources, Antarctic Seals, Antarctic Treaty, Biodiversity, Climate Change, Climate Change-Kyoto Protocol, Desertification, Endangered Species, Environmental Modification, Hazardous Wastes, Law of the Sea, Marine Dumping, Ozone Layer Protection, Ship Pollution, Tropical Timber 83, Tropical Timber 94, Wetlands

signed, but not ratified: Air Pollution-Volatile Organic Compounds, Marine Life Conservation

Geography—note: second-largest country in world (after Russia); strategic location between Russia and US via north polar route; approximately 90% of the population is concentrated within 160 km of the US border

PEOPLE

Population: 33,212,696 (July 2008 est.)

Age structure:

0–14 years: 16.3% (male 2,780,491/female 2,644,276)

15–64 years: 68.8% (male 11,547,354/female 11,300,639)

65 years and over: 14.9% (male 2,150,991/female 2,788,945) (2008 est.)

Median age:

total: 40.1 years

male: 39 years

female: 41.2 years (2008 est.)

Population growth rate: 0.83% (2008 est.)

Birth rate: 10.29 births/1,000 population (2008 est.)

Death rate: 7.61 deaths/1,000 population (2008 est.)

Net migration rate: 5.62 migrant(s)/1,000 population (2008 est.)

Sex ratio:

at birth: 1.06 male(s)/female

under 15 years: 1.05 male(s)/female

15–64 years: 1.02 male(s)/female

65 years and over: 0.77 male(s)/female

total population: 0.98 male(s)/female (2008 est.)

Infant mortality rate:

total: 5.08 deaths/1,000 live births

male: 5.4 deaths/1,000 live births

female: 4.75 deaths/1,000 live births (2008 est.)

Life expectancy at birth:

total population: 81.16 years

male: 78.65 years

female: 83.81 years (2008 est.)

Total fertility rate: 1.57 children born/woman (2008 est.)

HIV/AIDS—adult prevalence rate: 0.3% (2003 est.)

HIV/AIDS—people living with HIV/AIDS: 56,000 (2003 est.)

HIV/AIDS—deaths: 1,500 (2003 est.)

Nationality:

noun: Canadian(s)

adjective: Canadian

Ethnic groups: British Isles origin 28%, French origin 23%, other European 15%, Amerindian 2%, other, mostly Asian, African, Arab 6%, mixed background 26%

Religions: Roman Catholic 42.6%, Protestant 23.3% (including United Church 9.5%, Anglican 6.8%, Baptist 2.4%, Lutheran 2%), other Christian 4.4%, Muslim 1.9%, other and unspecified 11.8%, none 16% (2001 census)

Languages: English (official) 59.3%, French (official) 23.2%, other 17.5%

Literacy:

definition: age 15 and over can read and write

total population: 99%

male: 99%

female: 99% (2003 est.)

GOVERNMENT

Country name:

conventional long form: none

conventional short form: Canada

Government type: constitutional monarchy that is also a parliamentary democracy and a federation

Capital: *name:* Ottawa

geographic coordinates: 45 25 N, 75 42 W

time difference: UTC-5 (same time as Washington, DC during Standard Time)

daylight saving time: +1hr, begins second Sunday in March; ends first Sunday in November

note: Canada is divided into six time zones

Administrative divisions: 10 provinces and 3 territories*; Alberta, British Columbia, Manitoba, New Brunswick, Newfoundland and Labrador, Northwest Territories*, Nova Scotia, Nunavut*, Ontario, Prince Edward Island, Quebec, Saskatchewan, Yukon Territory*

Independence: 1 July 1867 (union of British North American colonies); 11 December 1931 (recognized by UK)

National holiday: Canada Day, 1 July (1867)

Constitution: made up of unwritten and written acts, customs, judicial decisions, and traditions; the written part of the constitution consists of the Constitution Act of 29 March 1867, which created a federation of four provinces, and the Constitution Act of 17 April 1982, which transferred formal control over the constitution from Britain to Canada, and added a Canadian Charter of Rights and Freedoms as well as procedures for constitutional amendments

Legal system: based on English common law, except in Quebec, where civil law system based on French law prevails; accepts compulsory ICJ jurisdiction, with reservations

Suffrage: 18 years of age; universal

Executive branch:

chief of state: Queen ELIZABETH II (since 6 February 1952); represented by Governor General Michaelle JEAN

(since 27 September 2005)

head of government: Prime Minister Stephen HARPER (since 6 February 2006)

cabinet: Federal Ministry chosen by the prime minister usually from among the members of his own party sitting in Parliament

elections: the monarchy is hereditary; governor general appointed by the monarch on the advice of the prime minister for a five-year term; following legislative elections, the leader of the majority party or the leader of the majority coalition in the House of Commons is automatically designated prime minister by the governor general

Legislative branch: bicameral Parliament or Parlement consists of the Senate or Senat (105 seats; members appointed by the governor general with the advice of the prime minister and serve until reaching 75 years of age) and the House of Commons or Chambre des Communes (308 seats; members elected by direct, popular vote to serve a maximum of five-year terms starting in 2009 elections)

elections: House of Commons—last held 23 January 2006 (next to be held 19 October 2009)

election results: House of Commons—percent of vote by party—Conservative Party 36.3%, Liberal Party 30.2%, New Democratic Party 17.5%, Bloc Quebecois 10.5%, Greens 4.5%, other 1%; seats by party—Conservative Party 124, Liberal Party 102, New Democratic Party 29, Bloc Quebecois 51, other 2; seats by party as of November 2007—Conservative Party 125, Liberal Party 96, New Democratic Party 30, Bloc Quebecois 49, other 4, vacant 4

Judicial branch: Supreme Court of Canada (judges are appointed by the prime minister through the governor general); Federal Court of Canada; Federal Court of Appeal; Provincial Courts (these are named variously Court of Appeal, Court of Queens Bench, Superior Court, Supreme Court, and Court of Justice)

Political parties and leaders: Bloc Quebecois [Gilles DUCEPPE]; Conservative Party of Canada [Stephen HARPER] (a merger of the Canadian Alliance and the Progressive Conservative Party); Green Party [Elizabeth MAY]; Liberal Party [Stephane DION]; New Democratic Party [Jack LAYTON]

Political pressure groups and leaders: NA

International organization participation: ACCT, ADB (nonregional members), AfDB, APEC, Arctic Council, ARF, ASEAN (dialogue partner), Australia Group, BIS, C, CDB, CE (observer), EAPC, EBRD, ESA (cooperating state), FAO, G-7, G-8, G-10, IADB, IAEA, IBRD, ICAO, ICC, ICCt, ICRM, IDA, IEA, IFAD, IFC, IFRCS, IHO, ILO, IMF, IMO, IMSO, Interpol, IOC, IOM, IPU, ISO, ITSO, ITU, ITUC, MIGA, MINUSTAH, NAFTA, NAM (guest), NATO, NEA, NSG, OAS, OECD, OIF, OPCW, OSCE, Paris Club, PCA, PIF (partner), SECI (observer), UN, UNAMSIL, UNCTAD, UNDOF, UNESCO, UNFICYP, UNHCR, UNRWA, UNTSO, UNWTO, UPU, WCL, WCO, WFTU, WHO, WIPO, WMO, WTO, ZC

Diplomatic representation in the US:
chief of mission: Ambassador Michael WILSON
chancery: 501 Pennsylvania Avenue NW, Washington, DC 20001
telephone: [1] (202) 682-1740
FAX: [1] (202) 682-7701
consulate(s) general: Atlanta, Boston, Buffalo, Chicago, Dallas, Denver, Detroit, Los Angeles, Miami, Minneapolis, New York, Phoenix, San Diego, San Francisco, Seattle, Tucson
consulate(s): Anchorage, Houston, Philadelphia, Princeton (New Jersey), Raleigh, San Jose (California)

Diplomatic representation from the US:
chief of mission: Ambassador David H. WILKINS
embassy: 490 Sussex Drive, Ottawa, Ontario K1N 1G8
mailing address: P. O. Box 5000, Ogdensburgh, NY 13669-0430; P.O. Box 866, Station B, Ottawa, Ontario K1P 5T1
telephone: [1] (613) 688-5335
FAX: [1] (613) 688-3082
consulate(s) general: Calgary, Halifax, Montreal, Quebec, Toronto, Vancouver, Winnipeg

Flag description: two vertical bands of red (hoist and fly side, half width), with white square between them; an 11-pointed red maple leaf is centered in the white square; the official colors of Canada are red and white

ECONOMY

Economy—overview: As an affluent, high-tech industrial society in the trillion-dollar class, Canada resembles the US in its market-oriented economic system, pattern of production, and affluent living standards. Since World War II, the impressive growth of the manufacturing, mining, and service sectors has transformed the nation from a largely rural economy into one primarily industrial and urban. The 1989 US-Canada Free Trade Agreement (FTA) and the 1994 North American Free Trade Agreement (NAFTA) (which includes Mexico) touched off a dramatic increase in trade and economic integration with the US. Given its great natural resources, skilled labor force, and modern capital plant, Canada enjoys solid economic prospects. Top-notch fiscal management has produced consecutive balanced budgets since 1997, although public debate continues over the equitable distribution of federal funds to the Canadian provinces. Exports account for roughly a third of GDP. Canada enjoys a substantial trade surplus with its principal trading partner, the US, which absorbs 80% of Canadian exports each year. Canada is the US's largest foreign supplier of energy, including oil, gas, uranium, and electric power. During 2007, Canada enjoyed good economic growth, moderate inflation, and the lowest unemployment rate in more than three decades.

GDP (purchasing power parity): $1.266 trillion (2007 est.)

GDP (official exchange rate): $1.432 trillion (2007 est.)

GDP—real growth rate: 2.7% (2007 est.)

GDP—per capita (PPP): $38,400 (2007 est.)

GDP—composition by sector:
agriculture: 2.1%
industry: 28.8%
services: 69.1% (2007 est.)

Labor force: 17.95 million (2007 est.)

Labor force—by occupation: agriculture 2%, manufacturing 13%, construction 6%, services 76%, other 3% (2006)

Unemployment rate: 6% (2007 est.)

Population below poverty line: 10.8%; note—this figure is the Low Income Cut-Off (LICO), a calculation that results in higher figures than found in many comparable economies; Canada does not have an official poverty line (2005)

Household income or consumption by percentage share:
lowest 10%: 2.6%
highest 10%: 24.8% (2000)

Distribution of family income—Gini index: 32.1 (2005)

Inflation rate (consumer prices): 2.1% (2007 est.)

Investment (gross fixed): 22.8% of GDP (2007 est.)

Budget:
revenues: $569.3 billion
expenditures: $555.2 billion (2007 est.)

Public debt: 68.5% of GDP (2007 est.)

Agriculture—products: wheat, barley, oilseed, tobacco, fruits, vegetables; dairy products; forest products; fish

Industries: transportation equipment, chemicals, processed and unprocessed minerals, food products, wood and paper products, fish products, petroleum and natural gas

Industrial production growth rate: 0.3% (2007 est.)

Electricity—production: 609.6 billion kWh (2005)

Electricity—production by source:
fossil fuel: 28%
hydro: 57.9%
nuclear: 12.9%
other: 1.3% (2001)

Electricity—consumption: 540.2 billion kWh (2005)

Electricity—exports: 42.93 billion kWh (2005)

Electricity—imports: 19.33 billion kWh (2005)

Oil—production: 3.092 million bbl/day (2005)

Oil—consumption: 2.29 million bbl/day (2005)

Oil—exports: 2.274 million bbl/day (2004)

Oil—imports: 1.185 million bbl/day (2004)

Oil—proved reserves: 178.8 billion bbl
note: includes oil sands (1 January 2006 est.)

Natural gas—production: 178.2 billion cu m (2005 est.)

Natural gas—consumption: 92.76 billion cu m (2005 est.)

Natural gas—exports: 101.9 billion cu m (2005 est.)

Natural gas—imports: 9.403 billion cu m (2005)

Natural gas—proved reserves: 1.537 trillion cu m (1 January 2006 est.)

Current account balance: $13.26 billion (2007 est.)

Exports: $433.1 billion f.o.b. (2007 est.)

Exports—commodities: motor vehicles and parts, industrial machinery, aircraft, telecommunications equipment; chemicals, plastics, fertilizers; wood pulp, timber, crude petroleum, natural gas, electricity, aluminum

Exports—partners: US 81.6%, UK 2.3%, Japan 2.1% (2006)

Imports: $386.9 billion f.o.b. (2007 est.)

Imports—commodities: machinery and equipment, motor vehicles and parts, crude oil, chemicals, electricity, durable consumer goods

Imports—partners: US 54.9%, China 8.7%, Mexico 4% (2006)

Economic aid—donor: ODA, $3.9 billion (2007)

Reserves of foreign exchange and gold: $39.31 billion (2007 est.)

Debt—external: $758.6 billion (30 June 2007)

Stock of direct foreign investment—at home: $527.4 billion (2007 est.)

Stock of direct foreign investment—abroad: $514.7 billion (2007 est.)

Market value of publicly traded shares: $1.481 trillion (2005)

Currency (code): Canadian dollar (CAD)

Currency code: CAD

Exchange rates: Canadian dollars per US dollar—1.0724 (2007), 1.1334 (2006), 1.2118 (2005), 1.301 (2004), 1.4011 (2003)

Fiscal year: 1 April—31 March

COMMUNICATIONS

Telephones—main lines in use: 21 million (2006)

Telephones—mobile cellular: 18.749 million (2006)

Telephone system:
general assessment: excellent service provided by modern technology
domestic: domestic satellite system with about 300 earth stations
international: country code—1; submarine cables provide links to the US and Europe; satellite earth stations—7 (5 Intelsat—4 Atlantic Ocean and 1 Pacific Ocean, and 2 Intersputnik—Atlantic Ocean region) (2007)

Radio broadcast stations: AM 245, FM 582, shortwave 6 (2004)

Radios: 32.3 million (1997)

Television broadcast stations: 80 (plus many repeaters) (1997)

Televisions: 21.5 million (1997)

Internet country code: .ca

Internet hosts: 4.196 million (2007)

Internet Service Providers (ISPs): 760 (2000 est.)

Internet users: 22 million (2005)

TRANSPORTATION

Airports: 1,343 (2007)

Airports—with paved runways:
total: 509
over 3,047 m: 18
2,438 to 3,047 m: 16
1,524 to 2,437 m: 149
914 to 1,523 m: 248
under 914 m: 78 (2007)

Airports—with unpaved runways:
total: 834
1,524 to 2,437 m: 68
914 to 1,523 m: 356
under 914 m: 410 (2007)

Heliports: 11 (2007)

Pipelines: crude and refined oil 23,564 km; liquid petroleum gas 74,980 km (2006)

Railways:
total: 48,068 km
standard gauge: 48,068 km 1.435-m gauge (2006)

Roadways:
total: 1,042,300 km
paved: 415,600 km (includes 17,000 km of expressways)
unpaved: 626,700 km (2006)

Waterways: 636 km
note: Saint Lawrence Seaway of 3,769 km, including the Saint Lawrence River of 3,058 km, shared with United States (2007)

Merchant marine:
total: 171 ships (1000 GRT or over) 2,191,099 GRT/2,815,416 DWT
by type: bulk carrier 60, cargo 10, carrier 1, chemical tanker 9, combination ore/oil 1, container 2, passenger 6, passenger/cargo 64, petroleum tanker 12, roll on/roll off 6
foreign-owned: 8 (Germany 3, Netherlands 1, Norway 1, US 3)
registered in other countries: 130 (Australia 2, Bahamas 13, Barbados 9, Cambodia 6, Cyprus 2, Denmark 1, Honduras 1, Hong Kong 39, Liberia 3, Malta 15, Marshall Islands 4, Panama 17, St Vincent and The Grenadines 6, Taiwan 3, US 4, Vanuatu 5) (2007)

Ports and terminals: Fraser River Port, Halifax, Hamilton, Montreal, Port-Cartier, Quebec City, Saint John (New Brunswick), Sept-Isles, Vancouver

MILITARY

Military branches: Canadian Forces: Land Forces Command (LFC), Maritime Command (MARCOM), Air Command (AIRCOM), Canada Command (homeland security) (2008)

Military service age and obligation: 16–34 years of age for voluntary military service; women comprise approximately 11% of Canada's armed forces (2006)

Manpower available for military service:
males age 16–49: 8,072,010
females age 16–49: 7,813,462 (2008 est.)

Manpower fit for military service:
males age 16–49: 6,646,281
females age 16–49: 6,417,924 (2008 est.)

Manpower reaching militarily significant age annually:
males age 16–49: 227,435
females age 16–49: 215,556 (2008 est.)

Military expenditures—percent of GDP: 1.1% (2005 est.)

TRANSNATIONAL ISSUES

Disputes—international: managed maritime boundary disputes with the US at Dixon Entrance, Beaufort Sea, Strait of Juan de Fuca, and around the disputed Machias Seal Island and North Rock; US works closely with Canada to intensify security measures to monitor and control legal and illegal personnel, transport, and commodities across the international border; sovereignty dispute with Denmark over Hans Island in the Kennedy Channel between Ellesmere Island and Greenland

Illicit drugs: illicit producer of cannabis for the domestic drug market and export to US; use of hydroponics technology permits growers to plant large quantities of high-quality marijuana indoors; increasing ecstasy production, some of which is destined for the US; vulnerable to narcotics money laundering because of its mature financial services sector

CAPE VERDE

INTRODUCTION

Background: The uninhabited islands were discovered and colonized by the Portuguese in the 15th century; Cape Verde subsequently became a trading center for African slaves and later an important coaling and resupply stop for whaling and transatlantic shipping. Following independence in 1975, and a tentative interest in unification with Guinea-Bissau, a one-party system was established and maintained until multiparty elections were held in 1990. Cape Verde continues to exhibit one of Africa's most stable democratic governments. Repeated droughts during the second half of the 20th century caused significant hardship and prompted heavy emigration. As a result, Cape Verde's expatriate population is greater than its domestic one. Most Cape Verdeans have both African and Portuguese antecedents.

GEOGRAPHY

Location: Western Africa, group of islands in the North Atlantic Ocean, west of Senegal
Geographic coordinates: 16 00 N, 24 00 W
Map references: Political Map of the World
Area:
total: 4,033 sq km
land: 4,033 sq km
water: 0 sq km
Area—comparative: slightly larger than Rhode Island

Land boundaries: 0 km
Coastline: 965 km
Maritime claims: measured from claimed archipelagic baselines
territorial sea: 12 nm
contiguous zone: 24 nm
exclusive economic zone: 200 nm
Climate: temperate; warm, dry summer; precipitation meager and very erratic
Terrain: steep, rugged, rocky, volcanic
Elevation extremes:
lowest point: Atlantic Ocean 0 m
highest point: Mt. Fogo 2,829 m (a volcano on Fogo Island)
Natural resources: salt, basalt rock, limestone, kaolin, fish, clay, gypsum
Land use: *arable land:* 11.41%
permanent crops: 0.74%
other: 87.85% (2005)
Irrigated land: 30 sq km (2003)
Total renewable water resources: 0.3 cu km (1990)
Freshwater withdrawal (domestic/industrial/agricultural): *total:* 0.02 cu km/yr (7%/2%/91%)
per capita: 39 cu m/yr (2000)
Natural hazards: prolonged droughts; seasonal harmattan wind produces obscuring dust; volcanically and seismically active
Environment—current issues: soil erosion; deforestation due to demand for wood used as fuel; water shortages; desertification; environmental damage has threatened several species of birds and reptiles; illegal beach sand extraction; overfishing
Environment—international agreements: *party to:* Biodiversity, Climate Change, Climate Change-Kyoto Protocol, Desertification, Endangered Species, Environmental Modification, Hazardous Wastes, Law of the Sea, Marine Dumping, Ozone Layer Protection, Ship Pollution, Wetlands
signed, but not ratified: none of the selected agreements
Geography—note: strategic location 500 km from west coast of Africa near major north-south sea routes; important communications station; important sea and air refueling site

PEOPLE

Population: 426,998 (July 2008 est.)

Age structure:
0–14 years: 36.1% (male 77,533/female 76,489)
15–64 years: 57.4% (male 120,208/female 125,009)
65 years and over: 6.5% (male 10,226/female 17,533) (2008 est.)
Median age:
total: 20.6 years
male: 19.9 years
female: 21.5 years (2008 est.)
Population growth rate: 0.595% (2008 est.)
Birth rate: 23.95 births/1,000 population (2008 est.)
Death rate: 6.26 deaths/1,000 population (2008 est.)
Net migration rate: -11.74 migrant(s)/1,000 population (2008 est.)
Sex ratio:
at birth: 1.03 male(s)/female
under 15 years: 1.01 male(s)/female
15–64 years: 0.96 male(s)/female
65 years and over: 0.58 male(s)/female
total population: 0.95 male(s)/female (2008 est.)
Infant mortality rate:
total: 42.55 deaths/1,000 live births
male: 48.66 deaths/1,000 live births
female: 36.25 deaths/1,000 live births (2008 est.)
Life expectancy at birth:
total population: 71.33 years
male: 67.99 years
female: 74.76 years (2008 est.)
Total fertility rate: 3.17 children born/woman (2008 est.)
HIV/AIDS—adult prevalence rate: 0.035% (2001 est.)
HIV/AIDS—people living with HIV/AIDS: 775 (2001)
HIV/AIDS—deaths: 225 (as of 2001)
Nationality:
noun: Cape Verdean(s)
adjective: Cape Verdean
Ethnic groups: Creole (mulatto) 71%, African 28%, European 1%
Religions: Roman Catholic (infused with indigenous beliefs), Protestant (mostly Church of the Nazarene)
Languages: Portuguese, Crioulo (a blend of Portuguese and West African words)
Literacy:
definition: age 15 and over can read and write

total population: 76.6%
male: 85.8%
female: 69.2% (2003 est.)

GOVERNMENT

Country name:
conventional long form: Republic of Cape Verde
conventional short form: Cape Verde
local long form: Republica de Cabo Verde
local short form: Cabo Verde
Government type: republic
Capital: *name:* Praia
geographic coordinates: 14 55 N, 23 31 W
time difference: UTC-1 (4 hours ahead of Washington, DC during Standard Time)
Administrative divisions: 17 municipalities (concelhos, singular—concelho); Boa Vista, Brava, Maio, Mosteiros, Paul, Praia, Porto Novo, Ribeira Grande, Sal, Santa Catarina, Santa Cruz, Sao Domingos, Sao Filipe, Sao Miguel, Sao Nicolau, Sao Vicente, Tarrafal
Independence: 5 July 1975 (from Portugal)
National holiday: Independence Day, 5 July (1975)
Constitution: 25 September 1992; a major revision on 23 November 1995 substantially increased the powers of the president; a 1999 revision created the position of national ombudsman (Provedor de Justica)
Legal system: based on the legal system of Portugal; has not accepted compulsory ICJ jurisdiction
Suffrage: 18 years of age; universal
Executive branch:
chief of state: President Pedro Verona PIRES (since 22 March 2001)
head of government: Prime Minister Jose Maria Pereira NEVES (since 1 February 2001)
cabinet: Council of Ministers appointed by the president on the recommendation of the prime minister
elections: president elected by popular vote for a five-year term (eligible for a second term); election last held 12 February 2006 (next to be held in February 2011); prime minister nominated by the National Assembly and appointed by the president
election results: Pedro PIRES reelected president; percent of vote—Pedro PIRES (PAICV) 51.2%, Carlos VIEGA (MPD) 48.8%
Legislative branch: unicameral National Assembly or Assembleia Nacional (72 seats; members are elected by popular vote to serve five-year terms)
elections: last held 22 January 2006 (next to be held in January 2011)
election results: percent of vote by party—PAICV 52.3%, MPD 44%, UCID 2.7%;

seats by party—PAICV 41, MPD 29, UCID 2
Judicial branch: Supreme Tribunal of Justice or Supremo Tribunal de Justia
Political parties and leaders: African Party for Independence of Cape Verde or PAICV [Jose Maria Pereira NEVES, chairman]; Democratic Alliance for Change or ADM [Dr. Eurico MONTEIRO] (a coalition of PCD, PTS, and UCID); Democratic Christian Party or PDC [Manuel RODRIGUES]; Democratic Renovation Party or PRD [Victor FIDALGO]; Democratic and Independent Cape Verdean Union or UCID [Antonio MONTEIRO]; Movement for Democracy or MPD [Agostinho LOPES]; Party for Democratic Convergence or PCD [Dr. Eurico MONTEIRO]; Party of Work and Solidarity or PTS [Isaias RODRIGUES]; Social Democratic Party or PSD [Joao ALEM]
Political pressure groups and leaders: NA
International organization participation: ACCT, ACP, AfDB, AU, CPLP, ECOWAS, FAO, G-77, IBRD, ICAO, ICCt (signatory), ICRM, IDA, IFAD, IFC, IFRCS, ILO, IMF, IMO, Interpol, IOC, IOM, IPU, ITSO, ITU, ITUC, MIGA, NAM, OIF, OPCW, UN, UNCTAD, UNESCO, UNIDO, Union Latina, UNWTO, UPU, WCO, WHO, WIPO, WMO, WTO (observer)
Diplomatic representation in the US:
chief of mission: Ambassador Fatima Lima VEIGA
chancery: 3415 Massachusetts Avenue NW, Washington, DC 20007
telephone: [1] (202) 965-6820
FAX: [1] (202) 965-1207
consulate(s) general: Boston
Diplomatic representation from the US:
chief of mission: Ambassador Roger D. PIERCE
embassy: Rua Abilio Macedo n6, Praia
mailing address: C. P. 201, Praia
telephone: [238] 2-60-89-00
FAX: [238] 2-61-13-55
Flag description: five unequal horizontal bands; the top-most band of blue—equal to one half the width of the flag—is followed by three bands of white, red, and white, each equal to 1/12 of the width, and a bottom stripe of blue equal to one quarter of the flag width; a circle of 10, yellow, five-pointed stars, each representing one of the islands, is centered on the red stripe and positioned 3/8 of the length of the flag from the hoist side

ECONOMY

Economy—overview: This island economy suffers from a poor natural resource base, including serious water shortages exacerbated by cycles of long-term drought. The economy is service-oriented, with commerce, transport, tourism, and public services accounting for about three-fourths of GDP. Although nearly 70% of the population lives in rural areas, the share of food production in GDP is low. About 82% of food must be imported. The fishing potential, mostly lobster and tuna, is not fully exploited. Cape Verde annually runs a high trade deficit, financed by foreign aid and remittances from emigrants; remittances supplement GDP by more than 20%. Economic reforms are aimed at developing the private sector and attracting foreign investment to diversify the economy. Future prospects depend heavily on the maintenance of aid flows, the encouragement of tourism, remittances, and the momentum of the government's development program.
GDP (purchasing power parity): $1.603 billion (2007 est.)
GDP (official exchange rate): $1.428 billion (2007 est.)
GDP—real growth rate: 6.9% (2007 est.)
GDP—per capita (PPP): $3,200 (2007 est.)
GDP—composition by sector:
agriculture: 9.3%
industry: 16.7%
services: 74% (2007 est.)
Labor force: 120,600 (1990)
Unemployment rate: 21% (2000 est.)
Population below poverty line: 30% (2000)
Household income or consumption by percentage share:
lowest 10%: NA%
highest 10%: NA%
Inflation rate (consumer prices): 4.4% (2007 est.)
Investment (gross fixed): 37.1% of GDP (2007 est.)
Budget:
revenues: $436.1 million
expenditures: $449.7 million (2007 est.)
Agriculture—products: bananas, corn, beans, sweet potatoes, sugarcane, coffee, peanuts; fish
Industries: food and beverages, fish processing, shoes and garments, salt mining, ship repair
Industrial production growth rate: 7.5% (2007 est.)
Electricity—production: 45 million kWh (2005)
Electricity—production by source:
fossil fuel: 100%
hydro: 0%
nuclear: 0%
other: 0% (2001)

Electricity—consumption: 41.85 million kWh (2005)
Electricity—exports: 0 kWh (2005)
Electricity—imports: 0 kWh (2005)
Oil—production: 0 bbl/day (2005)
Oil—consumption: 2,000 bbl/day (2005 est.)
Oil—exports: 0 bbl/day (2004)
Oil—imports: 2,080 bbl/day (2004)
Oil—proved reserves: 0 bbl (1 January 2006 est.)
Natural gas—production: 0 cu m (2005 est.)
Natural gas—consumption: 0 cu m (2005 est.)
Natural gas—exports: 0 cu m (2005 est.)
Natural gas—imports: 0 cu m (2005)
Natural gas—proved reserves: 0 cu m (1 January 2006 est.)
Current account balance: -$144 million (2007 est.)
Exports: $80.36 million f.o.b. (2007 est.)
Exports—commodities: fuel, shoes, garments, fish, hides
Exports—partners: Spain 44.2%, Portugal 21.7%, Netherlands 12.6%, Morocco 4.6% (2006)
Imports: $768 million f.o.b. (2007 est.)
Imports—commodities: foodstuffs, industrial products, transport equipment, fuels
Imports—partners: Portugal 41.1%, Netherlands 10.6%, Spain 6.5%, Italy 5.4%, Côte d'Ivoire 5.2%, Brazil 4.8% (2006)
Economic aid—recipient: $160.6 million (2005)
Reserves of foreign exchange and gold: $348 million (31 December 2007 est.)
Debt—external: $325 million (2002)
Currency (code): Cape Verdean escudo (CVE)
Currency code: CVE

Exchange rates: Cape Verdean escudos (CVE) per US dollar—81.235 (2007), 87.946 (2006), 88.67 (2005), 88.808 (2004), 97.703 (2003)
Fiscal year: calendar year

COMMUNICATIONS

Telephones—main lines in use: 71,600 (2006)
Telephones—mobile cellular: 108,900 (2006)
Telephone system:
general assessment: effective system, extensive modernization from 1996–2000 following partial privatization in 1995
domestic: major service provider is Cabo Verde Telecom (CVT); fiber-optic ring, completed in 2001, links all islands providing Internet access and ISDN services; cellular service introduced in 1998; broadband services launched in 2004
international: country code—238; landing point for the Atlantis-2 fiber-optic transatlantic telephone cable that provides links to South America, Senegal, and Europe; HF radiotelephone to Senegal and Guinea-Bissau; satellite earth station—1 Intelsat (Atlantic Ocean) (2007)
Radio broadcast stations: AM 0, FM 22 (plus 12 repeaters), shortwave 0 (2001)
Radios: 100,000 (2002 est.)
Television broadcast stations: 1 (plus 7 repeaters) (2001)
Televisions: 15,000 (2002 est.)
Internet country code: .cv
Internet hosts: 344 (2007)
Internet Service Providers (ISPs): 1 (2002)
Internet users: 29,000 (2005)

TRANSPORTATION

Airports: 8 (2007)
Airports—with paved runways:

total: 8
over 3,047 m: 1
1,524 to 2,437 m: 2
914 to 1,523 m: 4
under 914 m: 1 (2007)
Roadways:
total: 1,350 km
paved: 932 km
unpaved: 418 km (2000)
Merchant marine:
total: 8 ships (1000 GRT or over) 13,922 GRT/7,726 DWT
by type: cargo 2, chemical tanker 1, passenger/cargo 5
foreign-owned: 2 (Spain 1, UK 1) (2007)
Ports and terminals: Porto Grande

MILITARY

Military branches: People's Revolutionary Armed Forces (FARP): Army, Coast Guard (includes maritime air wing) (2007)
Military service age and obligation: 18 years of age (est.) for selective compulsory military service; 14-month conscript service obligation (2006)
Manpower available for military service:
males age 16–49: 103,650
females age 16–49: 103,553 (2008 est.)
Manpower fit for military service:
males age 16–49: 83,082
females age 16–49: 88,832 (2008 est.)
Military expenditures—percent of GDP: 0.7% (2005)

TRANSNATIONAL ISSUES

Disputes—international: none
Illicit drugs: used as a transshipment point for Latin American cocaine destined for Western Europe; the lack of a well-developed financial system limits the country's utility as a money-laundering center

CAYMAN ISLANDS

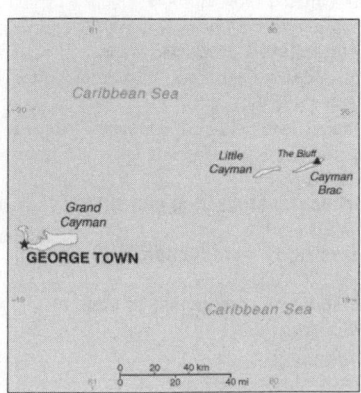

INTRODUCTION

Background: The Cayman Islands were colonized from Jamaica by the British during the 18th and 19th centuries, and were administered by Jamaica after 1863. In 1959, the islands became a territory within the Federation of the West Indies, but when the Federation dissolved in 1962, the Cayman Islands chose to remain a British dependency.

GEOGRAPHY

Location: Caribbean, three-island group (Grand Cayman, Cayman Brac, Little Cayman) in Caribbean Sea, 240 km

south of Cuba and 268 km northwest of Jamaica
Geographic coordinates: 19 30 N, 80 30 W
Map references: Central America and the Caribbean
Area: *total:* 262 sq km
land: 262 sq km
water: 0 sq km
Area—comparative: 1.5 times the size of Washington, DC
Land boundaries: 0 km
Coastline: 160 km
Maritime claims:
territorial sea: 12 nm
exclusive fishing zone: 200 nm

Climate: tropical marine; warm, rainy summers (May to October) and cool, relatively dry winters (November to April)
Terrain: low-lying limestone base surrounded by coral reefs
Elevation extremes:
lowest point: Caribbean Sea 0 m
highest point: The Bluff (Cayman Brac) 43 m
Natural resources: fish, climate and beaches that foster tourism
Land use:
arable land: 3.85%
permanent crops: 0%
other: 96.15% (2005)
Irrigated land: NA
Natural hazards: hurricanes (July to November)
Environment—current issues: no natural fresh water resources; drinking water supplies must be met by rainwater catchments
Geography—note: important location between Cuba and Central America

PEOPLE

Population: 47,862
note: most of the population lives on Grand Cayman (July 2008 est.)
Age structure:
0–14 years: 19.9% (male 4,774/female 4,759)
15–64 years: 71.1% (male 16,594/female 17,434)
65 years and over: 9% (male 2,022/female 2,279) (2008 est.)
Median age: *total:* 37.8 years
male: 37.4 years
female: 38.2 years (2008 est.)
Population growth rate: 2.449% (2008 est.)
Birth rate: 12.43 births/1,000 population (2008 est.)
Death rate: 4.83 deaths/1,000 population (2008 est.)
Net migration rate: 16.88 migrant(s)/1,000 population
note: major destination for Cubans trying to migrate to the US (2008 est.)
Sex ratio:
at birth: 1.02 male(s)/female
under 15 years: 1 male(s)/female
15–64 years: 0.95 male(s)/female
65 years and over: 0.89 male(s)/female
total population: 0.96 male(s)/female (2008 est.)
Infant mortality rate:
total: 7.1 deaths/1,000 live births
male: 8.16 deaths/1,000 live births
female: 6.03 deaths/1,000 live births (2008 est.)
Life expectancy at birth:
total population: 80.32 years
male: 77.68 years
female: 83 years (2008 est.)

Total fertility rate: 1.89 children born/woman (2008 est.)
HIV/AIDS—adult prevalence rate: NA
HIV/AIDS—people living with HIV/AIDS: NA
HIV/AIDS—deaths: NA
Nationality: *noun:* Caymanian(s)
adjective: Caymanian
Ethnic groups: mixed 40%, white 20%, black 20%, expatriates of various ethnic groups 20%
Religions: United Church (Presbyterian and Congregational), Anglican, Baptist, Church of God, other Protestant, Roman Catholic
Languages: English
Literacy:
definition: age 15 and over has ever attended school
total population: 98%
male: 98%
female: 98% (1970 est.)

GOVERNMENT

Country name:
conventional long form: none
conventional short form: Cayman Islands
Dependency status: overseas territory of the UK
Government type: British crown colony
Capital: *name:* George Town (on Grand Cayman)
geographic coordinates: 19 18 N, 81 23 W
time difference: UTC-5 (same time as Washington, DC during Standard Time)
Administrative divisions: 8 districts; Creek, Eastern, Midland, South Town, Spot Bay, Stake Bay, West End, Western
Independence: none (overseas territory of the UK)
National holiday: Constitution Day, first Monday in July
Constitution: 1959; revised 1962, 1972, and 1994
Legal system: British common law and local statutes
Suffrage: 18 years of age; universal
Executive branch:
chief of state: Queen ELIZABETH II (since 6 February 1952); represented by Governor Stuart JACK (since 23 November 2005)
head of government: Leader of Government Business Kurt TIBBETTS (since 18 May 2005)
cabinet: Executive Council (three members appointed by the governor, four members elected by the Legislative Assembly)
elections: the monarch is hereditary; the governor is appointed by the monarch; following legislative elections, the leader of the majority party or coalition is appointed by the governor Leader of Government Business

Legislative branch: unicameral Legislative Assembly (18 seats; 3 appointed members from the Executive Council and 15 elected by popular vote; to serve four-year terms)
elections: last held 11 May 2005 (next to be held in 2009)
election results: percent of vote by party—NA; seats by party—PPM 9, UDP 5, independent 1
Judicial branch: Summary Court; Grand Court; Cayman Islands Court of Appeal
Political parties and leaders: United Democratic Party or UDP [McKeeva BUSH]; People's Progressive Movement or PPM [Kurt TIBBETTS]; note—no national teams (loose groupings of political organizations) were formed for the 2000 elections
Political pressure groups and leaders: NA
International organization participation: Caricom (associate), CDB, Interpol (subbureau), IOC, UNESCO (associate), UPU
Diplomatic representation in the US: none (overseas territory of the UK)
Diplomatic representation from the US: none (overseas territory of the UK)
Flag description: blue, with the flag of the UK in the upper hoist-side quadrant and the Caymanian coat of arms centered on the outer half of the flag; the coat of arms includes a pineapple and turtle above a shield with three stars (representing the three islands) and a scroll at the bottom bearing the motto HE HATH FOUNDED IT UPON THE SEAS

ECONOMY

Economy—overview: With no direct taxation, the islands are a thriving offshore financial center. More than 68,000 companies were registered in the Cayman Islands as of 2003, including almost 500 banks, 800 insurers, and 5,000 mutual funds. A stock exchange was opened in 1997. Tourism is also a mainstay, accounting for about 70% of GDP and 75% of foreign currency earnings. The tourist industry is aimed at the luxury market and caters mainly to visitors from North America. Total tourist arrivals exceeded 2.1 million in 2003, with about half from the US. About 90% of the islands' food and consumer goods must be imported. The Caymanians enjoy one of the highest outputs per capita and one of the highest standards of living in the world.
GDP (purchasing power parity): $1.939 billion (2004 est.)
GDP (official exchange rate): $NA
GDP—real growth rate: 0.9% (2004 est.)

GDP—per capita (PPP): $43,800 (2004 est.)
GDP—composition by sector:
agriculture: 1.4%
industry: 3.2%
services: 95.4% (1994 est.)
Labor force: 23,450 (2004)
Labor force—by occupation: *agriculture:* 1.4%
industry: 12.6%
services: 86% (1995)
Unemployment rate: 4.4% (2004)
Population below poverty line: NA%
Household income or consumption by percentage share:
lowest 10%: NA%
highest 10%: NA%
Inflation rate (consumer prices): 4.4% (2004)
Budget:
revenues: $423.8 million
expenditures: $392.6 million (2004)
Agriculture—products: vegetables, fruit; livestock; turtle farming
Industries: tourism, banking, insurance and finance, construction, construction materials, furniture
Industrial production growth rate: NA%
Electricity—production: 400 million kWh (2005)
Electricity—production by source:
fossil fuel: 100%
hydro: 0%
nuclear: 0%
other: 0% (2001)
Electricity—consumption: 372 million kWh (2005)
Electricity—exports: 0 kWh (2005)
Electricity—imports: 0 kWh (2005)
Oil—production: 0 bbl/day (2005)
Oil—consumption: 2,700 bbl/day (2005 est.)
Oil—exports: 0 bbl/day (2004)
Oil—imports: 2,698 bbl/day (2004)
Oil—proved reserves: 0 bbl (1 January 2006 est.)
Natural gas—production: 0 cu m (2005 est.)
Natural gas—consumption: 0 cu m (2005 est.)
Natural gas—exports: 0 cu m (2005 est.)

Natural gas—imports: 0 cu m (2005)
Natural gas—proved reserves: 0 cu m (1 January 2006 est.)
Exports: $2.52 million (2004)
Exports—commodities: turtle products, manufactured consumer goods
Exports—partners: mostly US (2006)
Imports: $866.9 million (2004)
Imports—commodities: foodstuffs, manufactured goods
Imports—partners: US, Netherlands Antilles, Japan (2006)
Economic aid—recipient: $390,000 (2004)
Debt—external: $70 million (1996)
Stock of direct foreign investment—at home: $NA
Stock of direct foreign investment—abroad: $NA
Market value of publicly traded shares: $130 million (2005)
Currency (code): Caymanian dollar (KYD)
Currency code: KYD
Exchange rates: Caymanian dollars per US dollar—NA (2007), 0.8496 (2006)
Fiscal year: 1 April—31 March

COMMUNICATIONS

Telephones—main lines in use: 38,000 (2002)
Telephones—mobile cellular: 17,000 (2002)
Telephone system:
general assessment: reasonably good system
domestic: liberalization of telecom market in 2003; introduction of competition in the mobile-cellular market in 2004
international: country code—1-345; landing point for the MAYA-1 submarine telephone cable network that provides links to the US and parts of Central and South America; submarine cable provides connectivity to Jamaica; satellite earth station—1 Intelsat (Atlantic Ocean) (2007)
Radio broadcast stations: AM 1, FM 4, shortwave 0 (2004)
Radios: 36,000 (1997)
Television broadcast stations: 4 with cable system (2004)

Televisions: 7,000 (1997)
Internet country code: .ky
Internet hosts: 4,888 (2007)
Internet Service Providers (ISPs): 16 (2000)
Internet users: 9,909 (2003)

TRANSPORTATION

Airports: 3 (2007)
Airports—with paved runways:
total: 2
1,524 to 2,437 m: 2 (2007)
Airports—with unpaved runways:
total: 1
914 to 1,523 m: 1 (2007)
Roadways:
total: 785 km
paved: 785 km (2002)
Merchant marine:
total: 124 ships (1000 GRT or over) 2,953,923 GRT/4,597,716 DWT
by type: bulk carrier 33, cargo 11, chemical tanker 41, liquefied gas 1, passenger 1, petroleum tanker 17, refrigerated cargo 13, roll on/roll off 3, vehicle carrier 4
foreign-owned: 122 (Denmark 3, Germany 17, Greece 23, Italy 10, Japan 6, Norway 2, Singapore 10, Sweden 1, UK 9, US 41) (2007)
Ports and terminals: Cayman Brac, George Town

MILITARY

Military branches: no regular military forces; Royal Cayman Islands Police Force (2007)
Manpower available for military service:
males age 16–49: 11,790 (2008 est.)
Manpower fit for military service:
males age 16–49: 9,577 (2008 est.)
Manpower reaching militarily significant age annually:
males age 16–49: 336 (2008 est.)
Military—note: defense is the responsibility of the UK

TRANSNATIONAL ISSUES

Disputes—international: none
Illicit drugs: offshore financial center; vulnerable to drug transshipment to the US and Europe

CENTRAL AFRICAN REPUBLIC

INTRODUCTION

Background: The former French colony of Ubangi-Shari became the Central African Republic upon independence in 1960. After three tumultuous decades of misrule—mostly by military governments—civilian rule was established in 1993 and lasted for one decade. President Ange-Felix PATASSE's civilian government was plagued by unrest, and in March 2003 he was deposed in a military coup led by General Francois BOZIZE, who established a transitional government. Though the government has the tacit support of civil society groups and the main parties, a wide field of candidates contested the municipal, legislative, and presidential elections held in March and May of 2005 in which General BOZIZE was affirmed as president. The government still does not fully control the countryside, where pockets of lawlessness

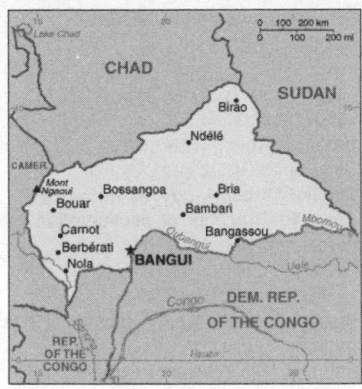

persist. Unrest in neighboring nations, Chad, Sudan, and the DRC, continues to affect stability in the Central African Republic as well.

GEOGRAPHY

Location: Central Africa, north of Democratic Republic of the Congo
Geographic coordinates: 7 00 N, 21 00 E
Map references: Africa
Area:
total: 622,984 sq km
land: 622,984 sq km
water: 0 sq km
Area—comparative: slightly smaller than Texas
Land boundaries:
total: 5,203 km
border countries: Cameroon 797 km, Chad 1,197 km, Democratic Republic of the Congo 1,577 km, Republic of the Congo 467 km, Sudan 1,165 km
Coastline: 0 km (landlocked)
Maritime claims: none (landlocked)
Climate: tropical; hot, dry winters; mild to hot, wet summers
Terrain: vast, flat to rolling, monotonous plateau; scattered hills in northeast and southwest
Elevation extremes:
lowest point: Oubangui River 335 m
highest point: Mont Ngaoui 1,420 m
Natural resources: diamonds, uranium, timber, gold, oil, hydropower
Land use:
arable land: 3.1%
permanent crops: 0.15%
other: 96.75% (2005)
Irrigated land: 20 sq km (2003)
Total renewable water resources: 144.4 cu km (2003)
Freshwater withdrawal (domestic/industrial/agricultural): *total:* 0.03 cu km/yr (80%/16%/4%)
per capita: 7 cu m/yr (2000)
Natural hazards: hot, dry, dusty harmattan winds affect northern areas; floods are common

Environment—current issues: tap water is not potable; poaching has diminished the country's reputation as one of the last great wildlife refuges; desertification; deforestation
Environment—international agreements: *party to:* Biodiversity, Climate Change, Climate Change-Kyoto Protocol, Desertification, Endangered Species, Hazardous Wastes, Ozone Layer Protection, Tropical Timber 94, Wetlands
signed, but not ratified: Law of the Sea
Geography—note: landlocked; almost the precise center of Africa

PEOPLE

Population: 4,434,873
note: estimates for this country explicitly take into account the effects of excess mortality due to AIDS; this can result in lower life expectancy, higher infant mortality, higher death rates, lower population growth rates, and changes in the distribution of population by age and sex than would otherwise be expected (July 2008 est.)
Age structure:
0–14 years: 41.3% (male 921,115/female 910,267)
15–64 years: 54.6% (male 1,203,280/female 1,217,956)
65 years and over: 4.1% (male 71,316/female 110,939) (2008 est.)
Median age:
total: 18.7 years
male: 18.3 years
female: 19 years (2008 est.)
Population growth rate: 1.487% (2008 est.)
Birth rate: 33.13 births/1,000 population (2008 est.)
Death rate: 18.26 deaths/1,000 population (2008 est.)
Net migration rate: NA
Sex ratio:
at birth: 1.03 male(s)/female
under 15 years: 1.01 male(s)/female
15–64 years: 0.99 male(s)/female
65 years and over: 0.64 male(s)/female
total population: 0.98 male(s)/female (2008 est.)
Infant mortality rate:
total: 82.36 deaths/1,000 live births
male: 88.98 deaths/1,000 live births
female: 75.53 deaths/1,000 live births (2008 est.)
Life expectancy at birth:
total population: 43.97 years
male: 43.94 years
female: 44 years (2008 est.)
Total fertility rate: 4.23 children born/woman (2008 est.)
HIV/AIDS—adult prevalence rate: 13.5% (2003 est.)

HIV/AIDS—people living with HIV/AIDS: 260,000 (2003 est.)
HIV/AIDS—deaths: 23,000 (2003 est.)
Major infectious diseases:
degree of risk: very high
food or waterborne diseases: bacterial and protozoal diarrhea, hepatitis A, and typhoid fever
vectorborne disease: malaria
respiratory disease: meningococcal meningitis (2008)
Nationality:
noun: Central African(s)
adjective: Central African
Ethnic groups: Baya 33%, Banda 27%, Mandjia 13%, Sara 10%, Mboum 7%, M'Baka 4%, Yakoma 4%, other 2%
Religions: indigenous beliefs 35%, Protestant 25%, Roman Catholic 25%, Muslim 15%
note: animistic beliefs and practices strongly influence the Christian majority
Languages: French (official), Sangho (lingua franca and national language), tribal languages
Literacy:
definition: age 15 and over can read and write
total population: 51%
male: 63.3%
female: 39.9% (2003 est.)

GOVERNMENT

Country name:
conventional long form: Central African Republic
conventional short form: none
local long form: Republique Centrafricaine
local short form: none
former: Ubangi-Shari, Central African Empire
abbreviation: CAR
Government type: republic
Capital: *name:* Bangui
geographic coordinates: 4 22 N, 18 35 E
time difference: UTC+1 (6 hours ahead of Washington, DC during Standard Time)
Administrative divisions—14 prefectures (prefectures, singular—prefecture), 2 economic prefectures* (prefectures economiques, singular—prefecture economique), and 1 commune**; Bamingui-Bangoran, Bangui**, Basse-Kotto, Haute-Kotto, Haut-Mbomou, Kemo, Lobaye, Mambere-Kadei, Mbomou, Nana-Grebizi*, Nana-Mambere, Ombella-Mpoko, Ouaka, Ouham, Ouham-Pende, Sangha-Mbaere*, Vakaga
Independence: 13 August 1960 (from France)
National holiday: Republic Day, 1 December (1958)
Constitution: ratified by popular referendum 5 December 2004; effective 27 December 2004

Legal system: based on French law; has not accepted compulsory ICJ jurisdiction

Suffrage: 21 years of age; universal

Executive branch:

chief of state: President Francois BOZIZE (since 15 March 2003 coup)

head of government: Prime Minister Faustin-Archange TOUADERA (since 22 January 2008)

cabinet: Council of Ministers

elections: under the new constitution, the president elected to a five-year term (eligible for a second term); elections last held 13 March and 8 May 2005 (next to be held in 2010); prime minister appointed by the political party with a parliamentary majority

election results: Francois BOZIZE elected president; percent of second round balloting—Francois BOZIZE (KNK) 64.6%, Martin ZIGUELE (MLPC) 35.4%

Legislative branch: unicameral National Assembly or Assemblee Nationale (109 seats; members are elected by popular vote to serve five-year terms)

elections: last held 13 March 2005 and 8 May 2005 (next to be held in 2010)

election results: percent of vote by party—MLPC 43%, RDC 18%, MDD 9%, FPP 6%, PSD 5%, ADP 4%, PUN 3%, FODEM 2%, PLD 2%, UPR 1%, FC 1%, independents 6%; seats by party—MLPC 47, RDC 20, MDD 8, FPP 7, PSD 6, ADP 5, PUN 3, FODEM 2, PLD 2, UPR 1, FC 1, independents 7

Judicial branch: Supreme Court or Cour Supreme; Constitutional Court (3 judges appointed by the president, 3 by the president of the National Assembly, and 3 by fellow judges); Court of Appeal; Criminal Courts; Inferior Courts

Political parties and leaders: Alliance for Democracy and Progress or ADP [Jacques MBOLIEDAS]; Central African Democratic Assembly or RDC [Andre KOLINGBA]; Civic Forum or FC [Gen. Timothee MALENDOMA]; Democratic Forum for Modernity or FODEM [Charles MASSI]; Liberal Democratic Party or PLD [Nestor KOMBO-NAGUEMON]; Movement for Democracy and Development or MDD [David DACKO]; Movement for the Liberation of the Central African People or MLPC [Ange-Felix PATASSE] (the party of deposed president); National Convergence or KNK; Patriotic Front for Progress or FPP [Abel GOUMBA]; People's Union for the Republic or UPR [Pierre Sammy MAKFOY]; National Unity Party or PUN [Jean-Paul NGOUPANDE]; Social Democratic Party or PSD [Enoch LAKOUE]

Political pressure groups and leaders: NA

International organization participation: ACCT, ACP, AfDB, AU, BDEAC, CEMAC, FAO, FZ, G-77, IAEA, IBRD, ICAO, ICCt, ICRM, IDA, IFAD, IFC, IFRCS, ILO, IMF, Interpol, IOC, ITSO, ITU, ITUC, MIGA, NAM, OIC (observer), OIF, OPCW, UN, UNCTAD, UNESCO, UNIDO, UNWTO, UPU, WCL, WCO, WHO, WIPO, WMO, WTO

Diplomatic representation in the US:

chief of mission: Ambassador Emmanuel TOUABOY

chancery: 1618 22nd Street NW, Washington, DC 20008

telephone: [1] (202) 483-7800

FAX: [1] (202) 332-9893

Diplomatic representation from the US:

chief of mission: Ambassador (vacant); Charge d'Affaires James PANOS

embassy: Avenue David Dacko, Bangui

mailing address: B. P. 924, Bangui

telephone: [236] 61 02 00

FAX: [236] 61 44 94

note: the embassy is currently operating with a minimal staff

Flag description: four equal horizontal bands of blue (top), white, green, and yellow with a vertical red band in center; a yellow five-pointed star to the hoist side of the blue band

ECONOMY

Economy—overview: Subsistence agriculture, together with forestry, remains the backbone of the economy of the Central African Republic (CAR), with more than 70% of the population living in outlying areas. The agricultural sector generates more than half of GDP. Timber has accounted for about 16% of export earnings and the diamond industry, for 40%. Important constraints to economic development include the CAR's landlocked position, a poor transportation system, a largely unskilled work force, and a legacy of misdirected macroeconomic policies. Factional fighting between the government and its opponents remains a drag on economic revitalization. Distribution of income is extraordinarily unequal. Grants from France and the international community can only partially meet humanitarian needs.

GDP (purchasing power parity): $3.099 billion (2007 est.)

GDP (official exchange rate): $1.714 billion (2007 est.)

GDP—real growth rate: 4.2% (2007 est.)

GDP—per capita (PPP): $700 (2007 est.)

GDP—composition by sector:

agriculture: 55%

industry: 20%

services: 25% (2001 est.)

Labor force: 1.857 million (2006)

Unemployment rate: 8% (23% for Bangui) (2001 est.)

Population below poverty line: NA%

Household income or consumption by percentage share:

lowest 10%: 0.7%

highest 10%: 47.7% (1993)

Distribution of family income—Gini index: 61.3 (1993)

Inflation rate (consumer prices): 0.9% (2007 est.)

Budget:

revenues: $250 million

expenditures: $273 million (2007 est.)

Agriculture—products: timber, cotton, coffee, tobacco, manioc (tapioca), yams, millet, corn, bananas; timber

Industries: gold and diamond mining, logging, brewing, textiles, footwear, assembly of bicycles and motorcycles

Industrial production growth rate: 3% (2002)

Electricity—production: 109 million kWh (2005)

Electricity—production by source:

fossil fuel: 19.8%

hydro: 80.2%

nuclear: 0%

other: 0% (2001)

Electricity—consumption: 101.4 million kWh (2005)

Electricity—exports: 0 kWh (2005)

Electricity—imports: 0 kWh (2005)

Oil—production: 0 bbl/day (2005)

Oil—consumption: 2,300 bbl/day (2005 est.)

Oil—exports: 0 bbl/day (2004)

Oil—imports: 2,201 bbl/day (2004)

Oil—proved reserves: 0 bbl (1 January 2006 est.)

Natural gas—production: 0 cu m (2005 est.)

Natural gas—consumption: 0 cu m (2005 est.)

Natural gas—exports: 0 cu m (2005 est.)

Natural gas—imports: 0 cu m (2005)

Natural gas—proved reserves: 0 cu m (1 January 2006)

Current account balance: -$77 million (2007 est.)

Exports: $146.7 million f.o.b. (2007 est.)

Exports—commodities: diamonds, timber, cotton, coffee, tobacco

Exports—partners: Belgium 30.7%, Spain 10.7%, Indonesia 8%, France 7.8%, China 6.9%, Democratic Republic of the Congo 6%, Turkey 5%, Italy 4.7% (2006)

Imports: $237.3 million f.o.b. (2007 est.)

Imports—commodities: food, textiles, petroleum products, machinery, electrical equipment, motor vehicles, chemicals, pharmaceuticals

Imports—partners: France 15.4%, Netherlands 15.1%, US 9.2%, Cameroon 8.9% (2006)

Economic aid—recipient: ODA, $95.29 million; note—traditional budget subsidies from France (2005 est.)

Debt—external: $1.153 billion (2007 est.)

Market value of publicly traded shares: $NA

Currency (code): Communaute Financiere Africaine franc (XAF); note—responsible authority is the Bank of the Central African States

Currency code: XAF

Exchange rates: Communaute Financiere Africaine francs (XAF) per US dollar—481.8 (2007), 522.59 (2006), 527.47 (2005), 528.29 (2004), 581.2 (2003)

Fiscal year: calendar year

COMMUNICATIONS

Telephones—main lines in use: 12,000 (2006)

Telephones—mobile cellular: 110,000 (2006)

Telephone system:

general assessment: limited telephone service; fixed-line connections for well less than 1 per 100 persons coupled with mobile-cellular usage of only about 3 per 100 persons

domestic: network consists principally of microwave radio relay and low-capacity, low-powered radiotelephone communication

international: country code—236; satel-lite earth station—1 Intelsat (Atlantic Ocean) (2007)

Radio broadcast stations: AM 1, FM 5, shortwave 1 (2001)

Radios: 283,000 (1997)

Television broadcast stations: 1 (2001)

Televisions: 18,000 (1997)

Internet country code: .cf

Internet hosts: 15 (2007)

Internet Service Providers (ISPs): 1 (2002)

Internet users: 13,000 (2006)

TRANSPORTATION

Airports: 51 (2007)

Airports—with paved runways:
total: 3
2,438 to 3,047 m: 1
1,524 to 2,437 m: 2 (2007)

Airports—with unpaved runways:
total: 48
2,438 to 3,047 m: 1
1,524 to 2,437 m: 10
914 to 1,523 m: 24
under 914 m: 13 (2007)

Roadways:
total: 24,307 km (2000)

Waterways: 2,800 km (primarily on the Oubangui and Sangha rivers) (2006)

Ports and terminals: Bangui, Nola, Salo, Nzinga

MILITARY

Military branches: Central African Armed Forces (Forces Armees Centrafricaines, FACA): Ground Forces, General Directorate of Gendarmerie Inspection (DGIG), Military Air Service, National Police (2008)

Military service age and obligation: 18 years of age for compulsory and voluntary military service; 2-year conscript service obligation (2006)

Manpower available for military service: *males age 16–49:* 1,032,828 *females age 16–49:* 999,330 (2008 est.)

Manpower fit for military service: *males age 16–49:* 534,141 *females age 16–49:* 495,303 (2008 est.)

Military expenditures—percent of GDP: 1.1% (2006 est.)

TRANSNATIONAL ISSUES

Disputes—international: periodic skirmishes over water and grazing rights among related pastoral populations along the border with southern Sudan persist

Refugees and internally displaced persons: *refugees (country of origin):* 7,900 (Sudan); 3,700 (Democratic Republic of the Congo); note—UNHCR resumed repatriation of Southern Sudanese refugees in 2006

IDPs: 197,000 (ongoing unrest following coup in 2003) (2007)

Trafficking in persons: *current situation:* Central African Republic is a source and destination country for children trafficked for domestic servitude, sexual exploitation, and forced labor in shops and commercial labor activities; while the majority of child victims are trafficked within the country, some are also trafficked to and from Cameroon and Nigeria

tier rating: Tier 2 Watch List—the Central African Republic failed to provide evidence of increasing efforts to combat trafficking in persons during 2005, specifically its inadequate law enforcement response to trafficking crimes

CHAD

INTRODUCTION

Background: Chad, part of France's African holdings until 1960, endured three decades of civil warfare as well as invasions by Libya before a semblance of peace was finally restored in 1990. The government eventually drafted a democratic constitution, and held flawed presidential elections in 1996 and 2001. In 1998, a rebellion broke out in northern Chad, which has sporadically flared up despite several peace agreements between the government and the rebels. In 2005, new rebel groups emerged in western Sudan and made probing attacks into eastern Chad, despite signing peace agreements in December 2006 and October 2007. Power remains in the hands of an ethnic minority. In June 2005, President Idriss DEBY held a referendum successfully removing constitutional term limits and won another controversial election in 2006. Sporadic rebel campaigns continued throughout 2006 and 2007, and the capital experienced a significant rebel threat in early 2008.

GEOGRAPHY

Location: Central Africa, south of Libya

Geographic coordinates: 15 00 N, 19 00 E

Map references: Africa

Area:
total: 1.284 million sq km
land: 1,259,200 sq km
water: 24,800 sq km

Area—comparative: slightly more than three times the size of California

Land boundaries: *total:* 5,968 km

border countries: Cameroon 1,094 km, Central African Republic 1,197 km, Libya 1,055 km, Niger 1,175 km, Nigeria 87 km, Sudan 1,360 km

Coastline: 0 km (landlocked)

Maritime claims: none (landlocked)

Climate: tropical in south, desert in north

Terrain: broad, arid plains in center, desert in north, mountains in northwest, lowlands in south

Elevation extremes:
lowest point: Djourab Depression 160 m
highest point: Emi Koussi 3,415 m

Natural resources: petroleum, uranium, natron, kaolin, fish (Lake Chad), gold, limestone, sand and gravel, salt

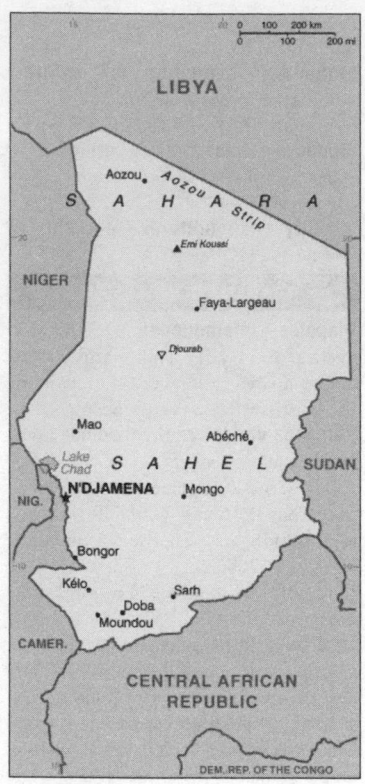

Land use: *arable land:* 2.8%
permanent crops: 0.02%
other: 97.18% (2005)
Irrigated land: 300 sq km (2003)
Total renewable water resources: 43 cu km (1987)
Freshwater withdrawal (domestic/industrial/agricultural): *total:* 0.23 cu km/yr (17%/0%/83%)
per capita: 24 cu m/yr (2000)
Natural hazards: hot, dry, dusty harmattan winds occur in north; periodic droughts; locust plagues
Environment—current issues: inadequate supplies of potable water; improper waste disposal in rural areas contributes to soil and water pollution; desertification
Environment—international agreements: *party to:* Biodiversity, Climate Change, Desertification, Endangered Species, Hazardous Wastes, Ozone Layer Protection, Wetlands
signed, but not ratified: Law of the Sea, Marine Dumping
Geography—note: landlocked; Lake Chad is the most significant water body in the Sahel

PEOPLE

Population: 10,111,337 (July 2008 est.)
Age structure:
0–14 years: 47% (male 2,408,638/female 2,346,984)

15–64 years: 50.1% (male 2,317,406/female 2,746,104)
65 years and over: 2.9% (male 123,561/female 168,644) (2008 est.)
Median age:
total: 16.4 years
male: 15.2 years
female: 17.5 years (2008 est.)
Population growth rate: 2.195% (2008 est.)
Birth rate: 41.61 births/1,000 population (2008 est.)
Death rate: 16.39 deaths/1,000 population (2008 est.)
Net migration rate: -3.27 migrant(s)/1,000 population (2008 est.)
Sex ratio:
at birth: 1.04 male(s)/female
under 15 years: 1.03 male(s)/female
15–64 years: 0.84 male(s)/female
65 years and over: 0.73 male(s)/female
total population: 0.92 male(s)/female (2008 est.)
Infant mortality rate:
total: 100.36 deaths/1,000 live births
male: 106.48 deaths/1,000 live births
female: 94 deaths/1,000 live births (2008 est.)
Life expectancy at birth:
total population: 47.43 years
male: 46.4 years
female: 48.5 years (2008 est.)
Total fertility rate: 5.43 children born/woman (2008 est.)
HIV/AIDS—adult prevalence rate: 4.8% (2003 est.)
HIV/AIDS—people living with HIV/AIDS: 200,000 (2003 est.)
HIV/AIDS—deaths: 18,000 (2003 est.)
Major infectious diseases:
degree of risk: very high
food or waterborne diseases: bacterial and protozoal diarrhea, hepatitis A, and typhoid fever
vectorborne disease: malaria
water contact disease: schistosomiasis
respiratory disease: meningococcal meningitis (2008)
Nationality:
noun: Chadian(s)
adjective: Chadian
Ethnic groups: Sara 27.7%, Arab 12.3%, Mayo-Kebbi 11.5%, Kanem-Bornou 9%, Ouaddai 8.7%, Hadjarai 6.7%, Tandjile 6.5%, Gorane 6.3%, Fitri-Batha 4.7%, other 6.4%, unknown 0.3% (1993 census)
Religions: Muslim 53.1%, Catholic 20.1%, Protestant 14.2%, animist 7.3%, other 0.5%, unknown 1.7%, atheist 3.1% (1993 census)
Languages: French (official), Arabic (official), Sara (in south), more than 120 different languages and dialects
Literacy: *definition:* age 15 and over can read and write French or Arabic

total population: 47.5%
male: 56%
female: 39.3% (2003 est.)

GOVERNMENT

Country name:
conventional long form: Republic of Chad
conventional short form: Chad
local long form: Republique du Tchad/Jumhuriyat Tshad
local short form: Tchad/Tshad
Government type: republic
Capital: *name:* N'Djamena
geographic coordinates: 12 06 N, 15 02 E
time difference: UTC+1 (6 hours ahead of Washington, DC during Standard Time)
Administrative divisions: 18 regions (regions, singular—region); Batha, Borkou-Ennedi-Tibesti, Chari-Baguirmi, Guera, Hadjer-Lamis, Kanem, Lac, Logone Occidental, Logone Oriental, Mandoul, Mayo-Kebbi Est, Mayo-Kebbi Ouest, Moyen-Chari, Ouaddai, Salamat, Tandjile, Ville de N'Djamena, Wadi Fira
Independence: 11 August 1960 (from France)
National holiday: Independence Day, 11 August (1960)
Constitution: passed by referendum 31 March 1996; a June 2005 referendum removed constitutional term limits
Legal system: based on French civil law system and Chadian customary law; has not accepted compulsory ICJ jurisdiction
Suffrage: 18 years of age; universal
Executive branch:
chief of state: President Lt. Gen. Idriss DEBY Itno (since 4 December 1990)
head of government: Prime Minister Youssof Saleh ABBAS (since 16 April 2008)
cabinet: Council of State, members appointed by the president on the recommendation of the prime minister
elections: president elected by popular vote to serve five-year term; if no candidate receives at least 50% of the total vote, the two candidates receiving the most votes must stand for a second round of voting; last held 3 May 2006 (next to be held by May 2011); prime minister appointed by the president
election results: Lt. Gen. Idriss DEBY Itno reelected president; percent of vote—Lt. Gen. Idriss DEBY 64.7%, Delwa Kassire KOUMAKOYE 15.1%, Albert Pahimi PADACKE 7.8%, Mahamat ABDOULAYE 7.1%, Brahim KOULA-MALLAH 5.3%; note—a June 2005 national referendum altered the constitution removing presidential term limits and permitting Lt. Gen. Idriss DEBY Itno to run for reelection
Legislative branch: unicameral National Assembly (155 seats; members

elected by popular vote to serve four-year terms); note—the 1996 constitution called for a Senate that has never been formed

elections: National Assembly—last held 21 April 2002 (next to be held by 2009); note—legislative elections, originally scheduled for 2006, were first delayed by National Assembly action and subsequently by an accord, signed in August 2007, between government and opposition parties

election results: percent of vote by party—NA; seats by party—MPS 110, RDP 12, FAR 9, RNDP 5, UNDR 5, URD 3, other 11

Judicial branch: Supreme Court; Court of Appeal; Criminal Courts; Magistrate Courts

Political parties and leaders: Federation Action for the Republic or FAR [Ngarledjy YORONGAR]; National Rally for Development and Progress or RNDP [Delwa Kassire KOUMAKOYE]; National Union for Democracy and Renewal or UNDR [Saleh KEBZABO]; Party for Liberty and Development or PLD [Ibni Oumar Mahamat SALEH]; Patriotic Salvation Movement or MPS [Mahamat Saleh AHMAT, chairman]; Rally for Democracy and Progress or RDP [Lol Mahamat CHOUA]; Union for Renewal and Democracy or URD [Gen. Wadal Abdelkader KAMOUGUE]

Political pressure groups and leaders: NA

International organization participation: ACCT, ACP, AfDB, AU, BDEAC, CEMAC, FAO, FZ, G-77, IAEA, IBRD, ICAO, ICCt, ICRM, IDA, IDB, IFAD, IFC, IFRCS, ILO, IMF, Interpol, IOC, ITSO, ITU, ITUC, MIGA, NAM, OIC, OIF, OPCW, UN, UNCTAD, UNESCO, UNIDO, UNOCI, UNWTO, UPU, WCL, WCO, WHO, WIPO, WMO, WTO

Diplomatic representation in the US:
chief of mission: Ambassador Mahamat Adam BECHIR
chancery: 2002 R Street NW, Washington, DC 20009
telephone: [1] (202) 462-4009
FAX: [1] (202) 265-1937

Diplomatic representation from the US:
chief of mission: Ambassador Louis NIGRO
embassy: Avenue Felix Eboue, N'Djamena
mailing address: B. P. 413, N'Djamena
telephone: [235] 251-62-11, [235] 251-70-09, [235] 251-77-59
FAX: [235] 251-56-54

Flag description: three equal vertical bands of blue (hoist side), yellow, and red

note: similar to the flag of Romania; also similar to the flags of Andorra and Moldova, both of which have a national coat of arms centered in the yellow band; design was based on the flag of France

ECONOMY

Economy—overview: Chad's primarily agricultural economy will continue to be boosted by major foreign direct investment projects in the oil sector that began in 2000. At least 80% of Chad's population relies on subsistence farming and livestock raising for its livelihood. Chad's economy has long been handicapped by its landlocked position, high energy costs, and a history of instability. Chad relies on foreign assistance and foreign capital for most public and private sector investment projects. A consortium led by two US companies has been investing $3.7 billion to develop oil reserves—estimated at 1 billion barrels—in southern Chad. Chinese companies are also expanding exploration efforts and plan to build a refinery. The nation's total oil reserves have been estimated to be 1.5 billion barrels. Oil production came on stream in late 2003. Chad began to export oil in 2004. Cotton, cattle, and gum arabic provide the bulk of Chad's non-oil export earnings.

GDP (purchasing power parity): $15.9 billion (2007 est.)

GDP (official exchange rate): $7.095 billion (2007 est.)

GDP—real growth rate: 0.6% (2007 est.)

GDP—per capita (PPP): $1,700 (2007 est.)

GDP—composition by sector:
agriculture: 21.5%
industry: 47.8%
services: 30.6% (2007 est.)

Labor force: 3.747 million (2006)

Labor force—by occupation: *agriculture:* 80% (subsistence farming, herding, and fishing)
industry and services: 20% (2006 est.)

Unemployment rate: NA%

Population below poverty line: 80% (2001 est.)

Household income or consumption by percentage share:
lowest 10%: NA%
highest 10%: NA%

Inflation rate (consumer prices): -8.8% (2007 est.)

Investment (gross fixed): 11.2% of GDP (2007 est.)

Budget:
revenues: $1.864 billion
expenditures: $1.749 billion (2007 est.)

Agriculture—products: cotton, sorghum, millet, peanuts, rice, potatoes, manioc (tapioca); cattle, sheep, goats, camels

Industries: oil, cotton textiles, meatpacking, brewing, natron (sodium carbonate), soap, cigarettes, construction materials

Industrial production growth rate: 2% (2007 est.)

Electricity—production: 95 million kWh (2005)

Electricity—production by source:
fossil fuel: 100%
hydro: 0%
nuclear: 0%
other: 0% (2001)

Electricity—consumption: 88.35 million kWh (2005)

Electricity—exports: 0 kWh (2005)

Electricity—imports: 0 kWh (2005)

Oil—production: 176,700 bbl/day (2005 est.)

Oil—consumption: 1,350 bbl/day (2005 est.)

Oil—exports: 170,000 bbl/day (2004)

Oil—imports: 1,316 bbl/day (2004)

Oil—proved reserves: 1.5 billion bbl (1 January 2006 est.)

Natural gas—production: 0 cu m (2005 est.)

Natural gas—consumption: 0 cu m (2005 est.)

Natural gas—exports: 0 cu m (2005 est.)

Natural gas—imports: 0 cu m (2005)

Natural gas—proved reserves: 0 cu m (1 January 2006 est.)

Current account balance: -$302 million (2007 est.)

Exports: $4.198 billion f.o.b. (2007 est.)

Exports—commodities: oil, cattle, cotton, gum arabic

Exports—partners: US 80.6%, China 10.4%, South Korea 2.3% (2006)

Imports: $1.158 billion f.o.b. (2007 est.)

Imports—commodities: machinery and transportation equipment, industrial goods, foodstuffs, textiles

Imports—partners: France 18.6%, Cameroon 17.6%, US 12.5%, Germany 7.3%, Saudi Arabia 5%, Belgium 4.9% (2006)

Economic aid—recipient: ODA, $379.8 million (2005)

Reserves of foreign exchange and gold: $997.3 million (31 December 2007 est.)

Debt—external: $1.6 billion (2005 est.)

Stock of direct foreign investment—at home: $4.5 billion (2006 est.)

Stock of direct foreign investment—abroad: $NA

Market value of publicly traded shares: $NA

Currency (code): Communaute Financiere Africaine franc (XAF);

note—responsible authority is the Bank of the Central African States

Currency code: XAF

Exchange rates: Communaute Financiere Africaine francs (XAF) per US dollar—480.1 (2007), 522.59 (2006), 527.47 (2005), 528.29 (2004), 581.2 (2003)

Fiscal year: calendar year

COMMUNICATIONS

Telephones—main lines in use: 13,000 (2006)

Telephones—mobile cellular: 466,100 (2006)

Telephone system:

general assessment: primitive system with high costs and low telephone density

domestic: fair system of radiotelephone communication stations

international: country code—235; satellite earth station—1 Intelsat (Atlantic Ocean) (2007)

Radio broadcast stations: AM 2, FM 4, shortwave 5 (2001)

Radios: 1.67 million (1997)

Television broadcast stations: 1 (2001)

Televisions: 10,000 (1997)

Internet country code: .td

Internet hosts: 72 (2007)

Internet Service Providers (ISPs): 1 (2002)

Internet users: 60,000 (2006)

TRANSPORTATION

Airports: 55 (2007)

Airports—with paved runways: *total:* 7

over 3,047 m: 2

2,438 to 3,047 m: 3

1,524 to 2,437 m: 1

under 914 m: 1 (2007)

Airports—with unpaved runways: *total:* 48

1,524 to 2,437 m: 16

914 to 1,523 m: 21

under 914 m: 11 (2007)

Pipelines: oil 250 km (2007)

Roadways:

total: 33,400 km

paved: 267 km

unpaved: 33,133 km (2000)

Waterways: Chari and Legone rivers are navigable only in wet season (2006)

MILITARY

Military branches: Armed Forces: Chadian National Army (Armee Nationale du Tchad, ANT), Chadian Air Force (Force Aerienne Tchadienne, FAT), Gendarmerie (2008)

Military service age and obligation: 20 years of age for conscripts, with 3-year service obligation; 18 years of age for volunteers; no minimum age restriction for volunteers with consent from a guardian; women are subject to 1 year of compulsory military or civic service at age of 21 (2004)

Manpower available for military service:

males age 16–49: 1,906,545

females age 16–49: 2,258,758 (2008 est.)

Manpower fit for military service:

males age 16–49: 1,066,565

females age 16–49: 1,279,318 (2008 est.)

Manpower reaching militarily significant age annually:

males age 16–49: 116,824

females age 16–49: 117,831 (2008 est.)

Military expenditures—percent of GDP: 4.2% (2006)

TRANSNATIONAL ISSUES

Disputes—international: since 2003, Janjawid armed militia and the Sudanese military have driven hundreds of thousands of Darfur residents into Chad; Chad remains an important mediator in the Sudanese civil conflict, reducing tensions with Sudan arising from cross-border banditry; Chadian Aozou rebels reside in southern Libya; only Nigeria and Cameroon have heeded the Lake Chad Commission's admonition to ratify the delimitation treaty, which also includes the Chad-Niger and Niger-Nigeria boundaries

Refugees and internally displaced persons: *refugees (country of origin):* 234,000 (Sudan); 54,200 (Central African Republic)

IDPs: 178,918 (2007)

CHILE

INTRODUCTION

Background: Prior to the coming of the Spanish in the 16th century, northern Chile was under Inca rule while Araucanian Indians (also known as Mapuches) inhabited central and southern Chile. Although Chile declared its independence in 1810, decisive victory over the Spanish was not achieved until 1818. In the War of the Pacific (1879–83), Chile defeated Peru and Bolivia and won its present northern regions. It was not until the 1880s that the Araucanian Indians were completely subjugated. A three-year-old Marxist government of Salvador ALLENDE was overthrown in 1973 by a military coup led by Augusto PINOCHET, who ruled until a freely elected president was installed in 1990. Sound economic policies, maintained consistently since the 1980s, have contributed to steady growth, reduced poverty rates by over half, and have helped secure the country's commitment to democratic and representative government. Chile has increasingly assumed regional and international leadership roles befitting its status as a stable, democratic nation.

GEOGRAPHY

Location: Southern South America, bordering the South Pacific Ocean, between Argentina and Peru

Geographic coordinates: 30 00 S, 71 00 W

Map references: South America

Area:

total: 756,950 sq km

land: 748,800 sq km

water: 8,150 sq km

note: includes Easter Island (Isla de Pascua) and Isla Sala y Gomez

Area—comparative: slightly smaller than twice the size of Montana

Land boundaries:

total: 6,339 km

border countries: Argentina 5,308 km, Bolivia 860 km, Peru 171 km

Coastline: 6,435 km

Maritime claims: *territorial sea:* 12 nm

contiguous zone: 24 nm

exclusive economic zone: 200 nm

continental shelf: 200/350 nm

Climate: temperate; desert in north; Mediterranean in central region; cool and damp in south

Terrain: low coastal mountains; fertile central valley; rugged Andes in east

Elevation extremes:

lowest point: Pacific Ocean 0 m

highest point: Nevado Ojos del Salado 6,880 m

Natural resources: copper, timber, iron ore, nitrates, precious metals, molybdenum, hydropower

Land use:

arable land: 2.62%

permanent crops: 0.43%

other: 96.95% (2005)

Irrigated land: 19,000 sq km (2003)

Total renewable water resources: 922 cu km (2000)

Freshwater withdrawal (domestic/industrial/agricultural): *total:* 12.55 cu km/yr (11%/25%/64%)

per capita: 770 cu m/yr (2000)

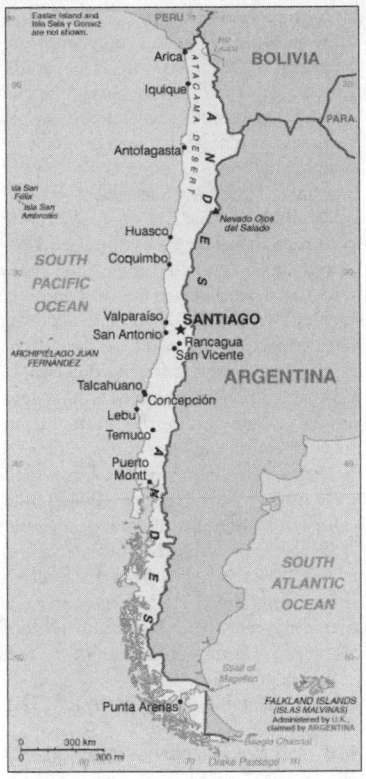

Natural hazards: severe earthquakes; active volcanism; tsunamis

Environment—current issues: widespread deforestation and mining threaten natural resources; air pollution from industrial and vehicle emissions; water pollution from raw sewage

Environment—international agreements: *party to:* Antarctic-Environmental Protocol, Antarctic-Marine Living Resources, Antarctic Seals, Antarctic Treaty, Biodiversity, Climate Change, Climate Change-Kyoto Protocol, Desertification, Endangered Species, Environmental Modification, Hazardous Wastes, Law of the Sea, Marine Dumping, Ozone Layer Protection, Ship Pollution, Wetlands, Whaling
signed, but not ratified: none of the selected agreements

Geography—note: strategic location relative to sea lanes between Atlantic and Pacific Oceans (Strait of Magellan, Beagle Channel, Drake Passage); Atacama Desert is one of world's driest regions

PEOPLE

Population: 16,454,143 (July 2008 est.)
Age structure:
0–14 years: 23.6% (male 1,987,962/female 1,899,489)

15–64 years: 67.6% (male 5,556,867/female 5,563,666)
65 years and over: 8.8% (male 602,789/female 843,370) (2008 est.)
Median age:
total: 31.1 years
male: 30.1 years
female: 32.1 years (2008 est.)
Population growth rate: 0.905% (2008 est.)
Birth rate: 14.82 births/1,000 population (2008 est.)
Death rate: 5.77 deaths/1,000 population (2008 est.)
Net migration rate: NA
Sex ratio:
at birth: 1.05 male(s)/female
under 15 years: 1.05 male(s)/female
15–64 years: 1 male(s)/female
65 years and over: 0.71 male(s)/female
total population: 0.98 male(s)/female (2008 est.)
Infant mortality rate:
total: 7.9 deaths/1,000 live births
male: 8.7 deaths/1,000 live births
female: 7.06 deaths/1,000 live births (2008 est.)
Life expectancy at birth:
total population: 77.15 years
male: 73.88 years
female: 80.59 years (2008 est.)
Total fertility rate: 1.95 children born/woman (2008 est.)
HIV/AIDS—adult prevalence rate: 0.3% (2003 est.)
HIV/AIDS—people living with HIV/AIDS: 26,000 (2003 est.)
HIV/AIDS—deaths: 1,400 (2003 est.)
Nationality:
noun: Chilean(s)
adjective: Chilean
Ethnic groups: white and white-Amerindian 95%, Amerindian 3%, other 2%
Religions: Roman Catholic 70%, Evangelical 15.1%, Jehovah's Witness 1.1%, other Christian 1%, other 4.6%, none 8.3% (2002 census)
Languages: Spanish (official), Mapudungun, German, English
Literacy:
definition: age 15 and over can read and write
total population: 95.7%
male: 95.8%
female: 95.6% (2002 census)

GOVERNMENT

Country name:
conventional long form: Republic of Chile
conventional short form: Chile
local long form: Republica de Chile
local short form: Chile
Government type: republic
Capital: *name:* Santiago

geographic coordinates: 33 27 S, 70 40 W
time difference: UTC-4 (1 hour ahead of Washington, DC during Standard Time)
daylight saving time: +1hr, begins second Sunday in October; ends second Sunday in March
Administrative divisions: 15 regions (regiones, singular—region); Aisen del General Carlos Ibanez del Campo, Antofagasta, Araucania, Arica y Parinacota, Atacama, Biobio, Coquimbo, Libertador General Bernardo O'Higgins, Los Lagos, Los Rios, Magallanes y de la Antartica Chilena, Maule, Region Metropolitana (Santiago), Tarapaca, Valparaiso
note: the US does not recognize claims to Antarctica
Independence: 18 September 1810 (from Spain)
National holiday: Independence Day, 18 September (1810)
Constitution: 11 September 1980, effective 11 March 1981; amended 1989, 1991, 1997, 1999, 2000, 2003, and 2005
Legal system: based on Code of 1857 derived from Spanish law and subsequent codes influenced by French and Austrian law; judicial review of legislative acts in the Supreme Court; has not accepted compulsory ICJ jurisdiction; note—in June 2005, Chile completed overhaul of its criminal justice system to a new, US-style adversarial system
Suffrage: 18 years of age; universal and compulsory
Executive branch:
chief of state: President Michelle BACHELET Jeria (since 11 March 2006); note—the president is both the chief of state and head of government
head of government: President Michelle BACHELET Jeria (since 11 March 2006)
cabinet: Cabinet appointed by the president
elections: president elected by popular vote for a single four-year term; election last held 11 December 2005, with runoff election held 15 January 2006 (next to be held in December 2009)
election results: Michelle BACHELET Jeria elected president; percent of vote—Michelle BACHELET Jeria 53.5%; Sebastian PINERA Echenique 46.5%
Legislative branch: bicameral National Congress or Congreso Nacional consists of the Senate or Senado (38 seats; members elected by popular vote to serve eight-year terms; one-half elected every four years) and the Chamber of Deputies or Camara de Diputados (120 seats; members are elected by popular vote to serve four-year terms)
elections: Senate—last held 11 December 2005 (next to be held in December

2009); Chamber of Deputies—last held 11 December 2005 (next to be held in December 2009)

election results: Senate—percent of vote by party—NA; seats by party—CPD 20 (PDC 6, PS 8, PPD 3, PRSD 3), APC 17 (UDI 9, RN 8), independent 1; Chamber of Deputies—percent of vote by party—NA; seats by party—CPD 65 (PDC 21, PPD 22, PS 15, PRSD 7), APC 54 (UDI 34, RN 20), independent 1; note—as of 8 January 2008: Senate—seats by party—CPD 18, (PDC 5, PS 8, PPD 2, PRSD 3), APC 16 (UDI 9, RN 7), independent 4; Chamber of Deputies—seats by party—CPD 57 (PDC 16, PPD 19, PS 15, PRSD 7), APC 53 (UDI 33, RN 20), independent 10.

Judicial branch: Supreme Court or Corte Suprema (judges are appointed by the president and ratified by the Senate from lists of candidates provided by the court itself; the president of the Supreme Court is elected every three years by the 20-member court); Constitutional Tribunal

Political parties and leaders: Alliance for Chile (Alianza) or APC (including National Renewal or RN [Carlos LARRAIN Pena] and Independent Democratic Union or UDI [Hernan LARRAIN Fernandez]); Coalition of Parties for Democracy (Concertacion) or CPD (including Christian Democratic Party or PDC [Soledad ALVEAR], Socialist Party or PS [Camilo ESCALONA Medina], Party for Democracy or PPD [Sergio BITAR Chacra], Radical Social Democratic Party or PRSD [Jose Antonio GOMEZ Urrutia]); Communist Party or PC [Guillermo TEILLIER]; Humanist Party [Marilen CABRERA Olmos]

Political pressure groups and leaders: revitalized university student federations at all major universities; Roman Catholic Church; United Labor Central or CUT includes trade unionists from the country's five largest labor confederations

International organization participation: ABEDA, APEC, BIS, CAN (associate), CSN, FAO, G-15, G-77, IADB, IAEA, IBRD, ICAO, ICC, ICCt (signatory), ICRM, IDA, IFAD, IFC, IFRCS, IHO, ILO, IMF, IMO, IMSO, Interpol, IOC, IOM, IPU, ISO, ITSO, ITU, ITUC, LAES, LAIA, Mercosur (associate), MIGA, MINUSTAH, NAM, OAS, OPANAL, OPCW, PCA, RG, UN, UNCTAD, UNESCO, UNFICYP, UNHCR, UNIDO, Union Latina, UNMOGIP, UNTSO, UNWTO, UPU, WCL, WCO, WFTU, WHO, WIPO, WMO, WTO

Diplomatic representation in the US:
chief of mission: Ambassador Mariano FERNANDEZ
chancery: 1732 Massachusetts Avenue NW, Washington, DC 20036
telephone: [1] (202) 785-1746
FAX: [1] (202) 887-5579
consulate(s) general: Chicago, Houston, Los Angeles, Miami, New York, Philadelphia, San Francisco, San Juan (Puerto Rico)

Diplomatic representation from the US:
chief of mission: Ambassador Paul E. SIMONS
embassy: Avenida Andres Bello 2800, Las Condes, Santiago
mailing address: APO AA 34033
telephone: [56] (2) 232-2600
FAX: [56] (2) 330-3710, 330-3160

Flag description: two equal horizontal bands of white (top) and red; a blue square the same height as the white band at the hoist-side end of the white band; the square bears a white five-pointed star in the center representing a guide to progress and honor; blue symbolizes the sky, white is for the snow-covered Andes, and red represents the blood spilled to achieve independence
note: design was influenced by the US flag

ECONOMY

Economy—overview: Chile has a market-oriented economy characterized by a high level of foreign trade. During the early 1990s, Chile's reputation as a role model for economic reform was strengthened when the democratic government of Patricio AYLWIN—which took over from the military in 1990—deepened the economic reform initiated by the military government. Growth in real GDP averaged 8% during 1991–97, but fell to half that level in 1998 because of tight monetary policies implemented to keep the current account deficit in check and because of lower export earnings—the latter a product of the global financial crisis. A severe drought exacerbated the recession in 1999, reducing crop yields and causing hydroelectric shortfalls and electricity rationing, and Chile experienced negative economic growth for the first time in more than 15 years. Despite the effects of the recession, Chile maintained its reputation for strong financial institutions and sound policy that have given it the strongest sovereign bond rating in South America. Between 2000 and 2007 growth ranged between 2%-6%. Throughout these years Chile maintained a low rate of inflation with GDP growth coming from high copper prices, solid export earnings (par-

ticularly forestry, fishing, and mining), and growing domestic consumption. President BACHELET in 2006 established an Economic and Social Stabilization Fund to hold excess copper revenues so that social spending can be maintained during periods of copper shortfalls. This fund probably surpassed $20 billion at the end of 2007. Chile continues to attract foreign direct investment, but most foreign investment goes into gas, water, electricity and mining. Unemployment has exhibited a downward trend over the past two years, dropping to 7.8% and 7.0% at the end of 2006 and 2007, respectively. Chile deepened its longstanding commitment to trade liberalization with the signing of a free trade agreement with the US, which took effect on 1 January 2004. Chile claims to have more bilateral or regional trade agreements than any other country. It has 57 such agreements (not all of them full free trade agreements), including with the European Union, Mercosur, China, India, South Korea, and Mexico.

GDP (purchasing power parity): $231.1 billion (2007 est.)
GDP (official exchange rate): $163.8 billion (2007 est.)
GDP—real growth rate: 5% (2007 est.)
GDP—per capita (PPP): $13,900 (2007 est.)
GDP—composition by sector:
agriculture: 4.8%
industry: 51.2%
services: 44% (2007 est.)
Labor force: 6.97 million (2007 est.)
Labor force—by occupation: *agriculture:* 13.6%
industry: 23.4%
services: 63% (2003)
Unemployment rate: 7% (2007 est.)
Population below poverty line: 18.2% (2005)
Household income or consumption by percentage share:
lowest 10%: 1.4%
highest 10%: 45% (2003)
Distribution of family income—Gini index: 54.9 (2003)
Inflation rate (consumer prices): 4.4% (2007 est.)
Investment (gross fixed): 20.6% of GDP (2007 est.)
Budget:
revenues: $44.96 billion
expenditures: $30.51 billion (2007 est.)
Public debt: 4.1% of GDP (2007 est.)
Agriculture—products: grapes, apples, pears, onions, wheat, corn, oats, peaches, garlic, asparagus, beans; beef, poultry, wool; fish; timber
Industries: copper, other minerals, food-

stuffs, fish processing, iron and steel, wood and wood products, transport equipment, cement, textiles

Industrial production growth rate: 11.1% (2007 est.)

Electricity—production: 47.6 billion kWh (2006)

Electricity—production by source:
fossil fuel: 47%
hydro: 51.5%
nuclear: 0%
other: 1.4% (2001)

Electricity—consumption: 48.31 billion kWh (2005)

Electricity—exports: 0 kWh (2005)

Electricity—imports: 2.152 billion kWh (2005)

Oil—production: 15,100 bbl/day (2006 est.)

Oil—consumption: 238,000 bbl/day (2006 est.)

Oil—exports: 31,510 bbl/day (2004)

Oil—imports: 222,900 bbl/day (2006 est.)

Oil—proved reserves: 150 million bbl (1 January 2006 est.)

Natural gas—production: 1.957 billion cu m (2005 est.)

Natural gas—consumption: 8.191 billion cu m (2005 est.)

Natural gas—exports: 0 cu m (2005 est.)

Natural gas—imports: 6.234 billion cu m (2005)

Natural gas—proved reserves: 93.97 billion cu m (1 January 2006 est.)

Current account balance: $6.05 billion (2007 est.)

Exports: $67.64 billion f.o.b. (2007 est.)

Exports—commodities: copper, fruit, fish products, paper and pulp, chemicals, wine

Exports—partners: US 15.6%, Japan 10.5%, China 8.6%, Netherlands 6.7%, South Korea 5.9%, Italy 4.9%, Brazil 4.8%, France 4.2% (2006)

Imports: $43.99 billion f.o.b. (2007 est.)

Imports—commodities: petroleum and petroleum products, chemicals, electrical and telecommunications equipment, industrial machinery, vehicles, natural gas

Imports—partners: US 15.6%, Argentina 12.6%, Brazil 11.8%, China 9.7% (2006)

Economic aid—recipient: $0 (2006)

Reserves of foreign exchange and gold: $16.84 billion (31 December 2007 est.)

Debt—external: $49.65 billion (30 June 2007)

Stock of direct foreign investment—at home: $98.53 billion (2007 est.)

Stock of direct foreign investment—abroad: $32.33 billion (2007 est.)

Market value of publicly traded shares: $174.6 billion (2006)

Currency (code): Chilean peso (CLP)

Currency code: CLP

Exchange rates: Chilean pesos per US dollar—526.25 (2007), 530.29 (2006), 560.09 (2005), 609.37 (2004), 691.43 (2003)

Fiscal year: calendar year

COMMUNICATIONS

Telephones—main lines in use: 3.326 million (2006)

Telephones—mobile cellular: 12.451 million (2006)

Telephone system:
general assessment: privatization began in 1988; advanced telecommunications infrastructure; modern system based on extensive microwave radio relay facilities; fixed-line connections have dropped in recent years as mobile-cellular usage continues to increase, reaching a level of 75 telephones per 100 persons
domestic: extensive microwave radio relay links; domestic satellite system with 3 earth stations
international: country code—56; submarine cables provide links to the US and to Central and South America; satellite earth stations—2 Intelsat (Atlantic Ocean) (2007)

Radio broadcast stations: AM 180 (8 inactive), FM 64, shortwave 17 (1 inactive) (1998)

Radios: 5.18 million (1997)

Television broadcast stations: 63 (plus 121 repeaters) (1997)

Televisions: 3.15 million (1997)

Internet country code: .cl

Internet hosts: 745,375 (2007)

Internet Service Providers (ISPs): 7 (2000)

Internet users: 4.156 million (2006)

TRANSPORTATION

Airports: 358 (2007)

Airports—with paved runways:
total: 79
over 3,047 m: 5
2,438 to 3,047 m: 8
1,524 to 2,437 m: 22
914 to 1,523 m: 25
under 914 m: 19 (2007)

Airports—with unpaved runways:
total: 279
2,438 to 3,047 m: 2
1,524 to 2,437 m: 12
914 to 1,523 m: 49
under 914 m: 216 (2007)

Pipelines: gas 2,550 km; gas/liquid petroleum gas 42 km; liquid petroleum gas 539 km; oil 1,002 km; refined products 757 km; unknown (oil/water) 97 km (2007)

Railways:
total: 6,585 km
broad gauge: 2,831 km 1.676-m gauge (1,317 km electrified)
narrow gauge: 3,754 km 1.000-m gauge (2006)

Roadways:
total: 79,605 km
paved: 16,080 km (includes 407 km of expressways)
unpaved: 63,525 km (2001)

Merchant marine:
total: 48 ships (1000 GRT or over) 719,668 GRT/1,016,892 DWT
by type: bulk carrier 10, cargo 6, chemical tanker 11, container 1, liquefied gas 2, passenger 4, passenger/cargo 2, petroleum tanker 8, roll on/roll off 1, vehicle carrier 3
foreign-owned: 1 (Argentina 1)
registered in other countries: 20 (Argentina 7, Brazil 1, Marshall Islands 4, Panama 8) (2007)

Ports and terminals: Coronel, Huasco, Lirquen, Puerto Ventanas, San Antonio, San Vicente, Valparaiso

MILITARY

Military branches: Army of the Nation, Chilean Navy (Armada de Chile, includes naval air, marine corps, and Maritime Territory and Merchant Marine Directorate (Directemar)), Chilean Air Force (Fuerza Aerea de Chile, FACh), Carabineros Corps (Cuerpo de Carabineros) (2008)

Military service age and obligation: 18–45 years of age for voluntary male and female military service, although the right to compulsory recruitment is retained; service obligation—12 months for Army, 22 months for Navy and Air Force (2008)

Manpower available for military service:
males age 16–49: 4,242,912
females age 16–49: 4,182,509 (2008 est.)

Manpower fit for military service:
males age 16–49: 3,542,448
females age 16–49: 3,500,059 (2008 est.)

Manpower reaching militarily significant age annually:
males age 16–49: 147,518
females age 16–49: 141,139 (2008 est.)

Military expenditures—percent of GDP: 2.7% (2006)

TRANSNATIONAL ISSUES

Disputes—international: Chile rebuffs Bolivia's reactivated claim to restore the Atacama corridor, ceded to Chile in 1884, offering instead unrestricted but not sovereign maritime access through Chile to Bolivian gas and other commodities; Chile rejects Peru's unilateral legislation to change its latitudinal maritime boundary with Chile to an equidistance line with a southwestern axis favoring Peru; territorial claim in Antarctica (Chilean Antarctic Territory) partially overlaps Argentine and British claims; the joint boundary commission, established by Chile and Argentina in 2001, has yet to map and demarcate the delimited boundary in the inhospitable Andean Southern Ice Field (Campo de Hielo Sur)

Illicit drugs: important transshipment country for cocaine destined for Europe; economic prosperity and increasing trade have made Chile more attractive to traffickers seeking to launder drug profits, especially through the Iquique Free Trade Zone, but a new anti-money-laundering law improves controls; imported precursors passed on to Bolivia; domestic cocaine consumption is rising; significant consumer of cocaine

CHINA

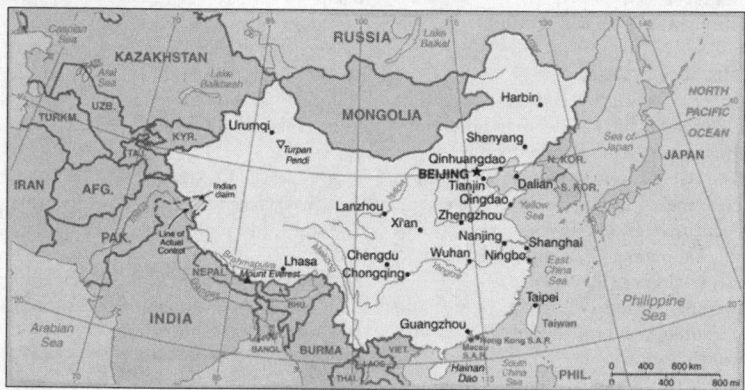

INTRODUCTION

Background: For centuries China stood as a leading civilization, outpacing the rest of the world in the arts and sciences, but in the 19th and early 20th centuries, the country was beset by civil unrest, major famines, military defeats, and foreign occupation. After World War II, the Communists under MAO Zedong established an autocratic socialist system that, while ensuring China's sovereignty, imposed strict controls over everyday life and cost the lives of tens of millions of people. After 1978, his successor DENG Xiaoping and other leaders focused on market-oriented economic development and by 2000 output had quadrupled. For much of the population, living standards have improved dramatically and the room for personal choice has expanded, yet political controls remain tight.

GEOGRAPHY

Location: Eastern Asia, bordering the East China Sea, Korea Bay, Yellow Sea, and South China Sea, between North Korea and Vietnam

Geographic coordinates: 35 00 N, 105 00 E

Map references: Asia

Area: *total:* 9,596,960 sq km

land: 9,326,410 sq km

water: 270,550 sq km

Area—comparative: slightly smaller than the US

Land boundaries:

total: 22,117 km

border countries: Afghanistan 76 km, Bhutan 470 km, Burma 2,185 km, India 3,380 km, Kazakhstan 1,533 km, North Korea 1,416 km, Kyrgyzstan 858 km, Laos 423 km, Mongolia 4,677 km, Nepal 1,236 km, Pakistan 523 km, Russia (northeast) 3,605 km, Russia (northwest) 40 km, Tajikistan 414 km, Vietnam 1,281 km

regional borders: Hong Kong 30 km, Macau 0.34 km

Coastline: 14,500 km

Maritime claims:

territorial sea: 12 nm

contiguous zone: 24 nm

exclusive economic zone: 200 nm

continental shelf: 200 nm or to the edge of the continental margin

Climate: extremely diverse; tropical in south to subarctic in north

Terrain: mostly mountains, high plateaus, deserts in west; plains, deltas, and hills in east

Elevation extremes:

lowest point: Turpan Pendi -154 m

highest point: Mount Everest 8,850 m

Natural resources: coal, iron ore, petroleum, natural gas, mercury, tin, tungsten, antimony, manganese, molybdenum, vanadium, magnetite, aluminum, lead, zinc, uranium, hydropower potential (world's largest)

Land use:

arable land: 14.86%

permanent crops: 1.27%

other: 83.87% (2005)

Irrigated land: 545,960 sq km (2003)

Total renewable water resources: 2,829.6 cu km (1999)

Freshwater withdrawal (domestic/industrial/agricultural): *total:* 549.76 cu km/yr (7%/26%/68%)

per capita: 415 cu m/yr (2000)

Natural hazards: frequent typhoons (about five per year along southern and eastern coasts); damaging floods; tsunamis; earthquakes; droughts; land subsidence

Environment—current issues: air pollution (greenhouse gases, sulfur dioxide particulates) from reliance on coal produces acid rain; water shortages, particularly in the north; water pollution from untreated wastes; deforestation; estimated loss of one-fifth of agricultural land since 1949 to soil erosion and economic development; desertification; trade in endangered species

Environment—international agreements: *party to:* Antarctic-Environmental Protocol, Antarctic Treaty, Biodiversity, Climate Change, Climate Change-Kyoto Protocol, Desertification, Endangered Species, Hazardous Wastes, Law of the Sea, Marine Dumping, Ozone Layer Protection, Ship Pollution, Tropical Timber 83, Tropical Timber 94, Wetlands, Whaling

signed, but not ratified: none of the selected agreements

Geography—note: world's fourth largest country (after Russia, Canada, and US); Mount Everest on the border with Nepal is the world's tallest peak

PEOPLE

Population: 1,330,044,605 (July 2008 est.)

Age structure:

0–14 years: 20.1% (male 142,085,665/female 125,300,391)

15–64 years: 71.9% (male 491,513,378/female 465,020,030)

65 years and over: 8% (male 50,652,480/female 55,472,661) (2008 est.)

Median age:
total: 33.6 years
male: 33.1 years
female: 34.2 years (2008 est.)
Population growth rate: 0.629% (2008 est.)
Birth rate: 13.71 births/1,000 population (2008 est.)
Death rate: 7.03 deaths/1,000 population (2008 est.)
Net migration rate: -0.39 migrant(s)/1,000 population (2008 est.)
Sex ratio:
at birth: 1.11 male(s)/female
under 15 years: 1.13 male(s)/female
15–64 years: 1.06 male(s)/female
65 years and over: 0.91 male(s)/female
total population: 1.06 male(s)/female (2008 est.)
Infant mortality rate:
total: 21.16 deaths/1,000 live births
male: 19.43 deaths/1,000 live births
female: · 23.08 deaths/1,000 live births (2008 est.)
Life expectancy at birth:
total population: 73.18 years
male: 71.37 years
female: 75.18 years (2008 est.)
Total fertility rate: 1.77 children born/woman (2008 est.)
HIV/AIDS—adult prevalence rate: 0.1% (2003 est.)
HIV/AIDS—people living with HIV/AIDS: 840,000 (2003 est.)
HIV/AIDS—deaths: 44,000 (2003 est.)
Major infectious diseases:
degree of risk: intermediate
food or waterborne diseases: bacterial diarrhea, hepatitis A, and typhoid fever
vectorborne diseases: Crimean Congo hemorrhagic fever, Japanese encephalitis, and malaria
water contact disease: leptospirosis
animal contact disease: rabies
note: highly pathogenic H5N1 avian influenza has been identified in this country; it poses a negligible risk to extremely rare cases possible among US citizens who have close contact with birds (2008)
Nationality:
noun: Chinese (singular and plural)
adjective: Chinese
Ethnic groups: Han Chinese 91.9%, Zhuang, Uyghur, Hui, Yi, Tibetan, Miao, Manchu, Mongol, Buyi, Korean, and other nationalities 8.1%
Religions: Daoist (Taoist), Buddhist, Christian 3%-4%, Muslim 1%-2%
note: officially atheist (2002 est.)
Languages: Standard Chinese or Mandarin (Putonghua, based on the Beijing dialect), Yue (Cantonese), Wu (Shanghainese), Minbei (Fuzhou), Minnan (Hokkien-Taiwanese), Xiang,

Gan, Hakka dialects, minority languages (see Ethnic groups entry)
Literacy:
definition: age 15 and over can read and write
total population: 90.9%
male: 95.1%
female: 86.5% (2000 census)

GOVERNMENT

Country name:
conventional long form: People's Republic of China
conventional short form: China
local long form: Zhonghua Renmin Gongheguo
local short form: Zhongguo
abbreviation: PRC
Government type: Communist state
Capital: *name:* Beijing
geographic coordinates: 39 55 N, 116 23 E
time difference: UTC+8 (13 hours ahead of Washington, DC during Standard Time)
note: despite its size, all of China falls within one time zone
Administrative divisions: 23 provinces (sheng, singular and plural), 5 autonomous regions (zizhiqu, singular and plural), and 4 municipalities (shi, singular and plural)
provinces: Anhui, Fujian, Gansu, Guangdong, Guizhou, Hainan, Hebei, Heilongjiang, Henan, Hubei, Hunan, Jiangsu, Jiangxi, Jilin, Liaoning, Qinghai, Shaanxi, Shandong, Shanxi, Sichuan, Yunnan, Zhejiang; (see note on Taiwan)
autonomous regions: Guangxi, Nei Mongol, Ningxia, Xinjiang Uygur, Xizang (Tibet)
municipalities: Beijing, Chongqing, Shanghai, Tianjin
note: China considers Taiwan its 23rd province; see separate entries for the special administrative regions of Hong Kong and Macau
Independence: 221 BC (unification under the Qin or Ch'in Dynasty); 1 January 1912 (Manchu Dynasty replaced by a Republic); 1 October 1949 (People's Republic established)
National holiday: Anniversary of the Founding of the People's Republic of China, 1 October (1949)
Constitution: most recent promulgation 4 December 1982
Legal system: based on civil law system; derived from Soviet and continental civil code legal principles; legislature retains power to interpret statutes; constitution ambiguous on judicial review of legislation; has not accepted compulsory ICJ jurisdiction
Suffrage: 18 years of age; universal
Executive branch:

chief of state: President HU Jintao (since 15 March 2003); Vice President XI Jinping (since 15 March 2008)
head of government: Premier WEN Jiabao (since 16 March 2003); Executive Vice Premier LI Keqiang (17 March 2008), Vice Premier HUI Liangyu (since 17 March 2003), Vice Premier ZHANG Deijiang (since 17 March 2008), and Vice Premier WANG Qishan (since 17 March 2008)
cabinet: State Council appointed by National People's Congress (NPC)
elections: president and vice president elected by National People's Congress for a five-year term (eligible for a second term); elections last held 15–17 March 2008 (next to be held in mid-March 2013); premier nominated by president, confirmed by National People's Congress
election results: HU Jintao elected president by National People's Congress with a total of 2,963 votes; XI Jinping elected vice president with a total of 2,919 votes
Legislative branch: unicameral National People's Congress or Quanguo Renmin Daibiao Dahui (2,987 seats; members elected by municipal, regional, and provincial people's congresses, and People's Liberation Army to serve five-year terms)
elections: last held December 2007–February 2008; date of next election—NA
election results: percent of vote—NA; seats—2,987
Judicial branch: Supreme People's Court (judges appointed by the National People's Congress); Local People's Courts (comprise higher, intermediate, and basic courts); Special People's Courts (primarily military, maritime, railway transportation, and forestry courts)
Political parties and leaders: Chinese Communist Party or CCP [HU Jintao]; eight registered small parties controlled by CCP
Political pressure groups and leaders: no substantial political opposition groups exist, although the government has identified the Falungong spiritual movement and the China Democracy Party as subversive groups
International organization participation: ADB, AfDB, APEC, APT, Arctic Council (observer), ARF, ASEAN (dialogue partner), BIS, CDB, EAS, FAO, G-24 (observer), G-77, IAEA, IBRD, ICAO, ICC, ICRM, IDA, IFAD, IFC, IFRCS, IHO, ILO, IMF, IMO, IMSO, Interpol, IOC, IOM (observer), IPU, ISO, ITSO, ITU, LAIA (observer), MIGA, MINURSO, MONUC, NAM (observer), NSG, OAS (observer),

OPCW, PCA, PIF (partner), SAARC (observer), SCO, UN, UN Security Council, UNAMID, UNCTAD, UNESCO, UNHCR, UNIDO, UNIFIL, UNMEE, UNMIL, UNMIS, UNMIT, UNOCI, UNTSO, UNWTO, UPU, WCO, WHO, WIPO, WMO, WTO, ZC

Diplomatic representation in the US:
chief of mission: Ambassador ZHOU Wenzhong
chancery: 2300 Connecticut Avenue NW, Washington, DC 20008
telephone: [1] (202) 328-2500
FAX: [1] (202) 328-2582
consulate(s) general: Chicago, Houston, Los Angeles, New York, San Francisco

Diplomatic representation from the US:
chief of mission: Ambassador Clark T. RANDT, Jr.
embassy: Xiu Shui Bei Jie 3, 100600 Beijing
mailing address: PSC 461, Box 50, FPO AP 96521-0002
telephone: [86] (10) 6532-3831
FAX: [86] (10) 6532-3178
consulate(s) general: Chengdu, Guangzhou, Hong Kong and Macau, Shanghai, Shenyang

Flag description: red with a large yellow five-pointed star and four smaller yellow five-pointed stars (arranged in a vertical arc toward the middle of the flag) in the upper hoist-side corner

ECONOMY

Economy—overview: China's economy during the last quarter century has changed from a centrally planned system that was largely closed to international trade to a more market-oriented economy that has a rapidly growing private sector and is a major player in the global economy. Reforms started in the late 1970s with the phasing out of collectivized agriculture, and expanded to include the gradual liberalization of prices, fiscal decentralization, increased autonomy for state enterprises, the foundation of a diversified banking system, the development of stock markets, the rapid growth of the non-state sector, and the opening to foreign trade and investment. China has generally implemented reforms in a gradualist or piecemeal fashion, including the sale of minority shares in four of China's largest state banks to foreign investors and refinements in foreign exchange and bond markets in 2005. After keeping its currency tightly linked to the US dollar for years, China in July 2005 revalued its currency by 2.1% against the US dollar and moved to an exchange rate system that references a basket of currencies. Cumulative appreciation of the ren-

minbi against the US dollar since the end of the dollar peg reached 15% in January 2008. The restructuring of the economy and resulting efficiency gains have contributed to a more than tenfold increase in GDP since 1978. Measured on a purchasing power parity (PPP) basis, China in 2007 stood as the second-largest economy in the world after the US, although in per capita terms the country is still lower middle-income. Annual inflows of foreign direct investment in 2007 rose to $75 billion. By the end of 2007, more than 5,000 domestic Chinese enterprises had established direct investments in 172 countries and regions around the world. The Chinese government faces several economic development challenges: (a) to sustain adequate job growth for tens of millions of workers laid off from state-owned enterprises, migrants, and new entrants to the work force; (b) to reduce corruption and other economic crimes; and (c) to contain environmental damage and social strife related to the economy's rapid transformation. Economic development has been more rapid in coastal provinces than in the interior, and approximately 200 million rural laborers have relocated to urban areas to find work. One demographic consequence of the "one child" policy is that China is now one of the most rapidly aging countries in the world. Deterioration in the environment—notably air pollution, soil erosion, and the steady fall of the water table, especially in the north—is another long-term problem. China continues to lose arable land because of erosion and economic development. In 2007 China intensified government efforts to improve environmental conditions, tying the evaluation of local officials to environmental targets, publishing a national climate change policy, and establishing a high level leading group on climate change, headed by Premier WEN Jiabao. The Chinese government seeks to add energy production capacity from sources other than coal and oil as its double-digit economic growth increases demand. Chinese energy officials in 2007 agreed to purchase five third generation nuclear reactors from Western companies. More power generating capacity came on line in 2006 as large scale investments—including the Three Gorges Dam across the Yangtze River—were completed.

GDP (purchasing power parity): $6.991 trillion (2007 est.)
GDP (official exchange rate): $3.251 trillion (2007 est.)
GDP—real growth rate: 11.4% (2007 est.)

GDP—per capita (PPP): $5,300 (2007 est.)
GDP—composition by sector:
agriculture: 11.3%
industry: 48.6%
services: 40.1%
note: industry includes construction (2007 est.)
Labor force: 803.3 million (2007 est.)
Labor force—by occupation:
agriculture: 43%
industry: 25%
services: 32% (2006 est.)
Unemployment rate: 4% unemployment in urban areas; substantial unemployment and underemployment in rural areas (2007 est.)
Population below poverty line: 8%
note: 21.5 million rural population live below the official "absolute poverty" line (approximately $90 per year); and an additional 35.5 million rural population above that but below the official "low income" line (approximately $125 per year) (2006 est.)
Household income or consumption by percentage share:
lowest 10%: 1.6%
highest 10%: 34.9% (2004)
Distribution of family income—Gini index: 47 (2007)
Inflation rate (consumer prices): 4.8% (2007 est.)
Investment (gross fixed): 40.4% of GDP (2007 est.)
Budget: *revenues:* $674.3 billion
expenditures: $651.6 billion (2007 est.)
Public debt: 18.4% of GDP (2007 est.)
Agriculture—products: rice, wheat, potatoes, corn, peanuts, tea, millet, barley, apples, cotton, oilseed; pork; fish
Industries: mining and ore processing, iron, steel, aluminum, and other metals, coal; machine building; armaments; textiles and apparel; petroleum; cement; chemicals; fertilizers; consumer products, including footwear, toys, and electronics; food processing; transportation equipment, including automobiles, rail cars and locomotives, ships, and aircraft; telecommunications equipment, commercial space launch vehicles, satellites
Industrial production growth rate: 13.4% (2007 est.)
Electricity—production: 3.256 trillion kWh (2007)
Electricity—production by source:
fossil fuel: 80.2%
hydro: 18.5%
nuclear: 1.2%
other: 0.1% (2001)
Electricity—consumption: 2.859 trillion kWh (2006)
Electricity—exports: 11.27 billion kWh (2006)

Electricity—imports: 5.39 billion kWh (2006)

Oil—production: 3.73 million bbl/day (2007 est.)

Oil—consumption: 6.93 million bbl/day (2007 est.)

Oil—exports: 79,060 bbl/day (2007)

Oil—imports: 3.19 million bbl/day (2007)

Oil—proved reserves: 12.8 billion bbl (2007 est.)

Natural gas—production: 58.6 billion cu m (2006 est.)

Natural gas—consumption: 55.6 billion cu m (2006 est.)

Natural gas—exports: 2.874 billion cu m (2006)

Natural gas—imports: 976 million cu m (2006)

Natural gas—proved reserves: 2.45 trillion cu m (2006 est.)

Current account balance: $360.7 billion (2007 est.)

Exports: $1.217 trillion f.o.b. (2007 est.)

Exports—commodities: machinery, electrical products, data processing equipment, apparel, textile, steel, mobile phones

Exports—partners: US 21%, Hong Kong 16%, Japan 9.5%, South Korea 4.6%, Germany 4.2% (2006)

Imports: $901.3 billion f.o.b. (2007 est.)

Imports—commodities: machinery and equipment, oil and mineral fuels, plastics, LED screens, data processing equipment, optical and medical equipment, organic chemicals, steel, copper

Imports—partners: Japan 14.6%, South Korea 11.3%, Taiwan 10.9%, US 7.5%, Germany 4.8% (2006)

Economic aid—recipient: $1.641 billion (FY07)

Reserves of foreign exchange and gold: $1.534 trillion (31 December 2007 est.)

Debt—external: $363 billion (31 December 2007 est.)

Stock of direct foreign investment—at home: $758.9 billion (2007 est.)

Stock of direct foreign investment—abroad: $93.75 billion (2007 est.)

Market value of publicly traded shares: $4.477 trillion (31 December 2007 est.)

Currency (code): Renminbi (RMB); note—also referred to by the unit yuan (CNY)

Currency code: CNY

Exchange rates: yuan per US dollar— 7.61 (2007), 7.97 (2006), 8.1943 (2005), 8.2768 (2004), 8.277 (2003)

Fiscal year: calendar year

COMMUNICATIONS

Telephones—main lines in use: 368 million (2006)

Telephones—mobile cellular: 461.1 million (2006)

Telephone system:

general assessment: domestic and international services are increasingly available for private use; unevenly distributed domestic system serves principal cities, industrial centers, and many towns; China continues to develop its telecommunications infrastructure, and is partnering with foreign providers to expand its global reach; 3 of China's 6 major telecommunications operators are part of an international consortium which, in December 2006, signed an agreement with Verizon Business to build the first next-generation fiber optic submarine cable system directly linking the US mainland and China

domestic: interprovincial fiber-optic trunk lines and cellular telephone systems have been installed; mobile-cellular subscribership is increasing rapidly; the number of internet users reached 162 million in 2007; a domestic satellite system with 55 earth stations is in place

international: country code—86; a number of submarine cables provide connectivity to Asia, the Middle East, Europe, and the US; satellite earth stations—7 (5 Intelsat—4 Pacific Ocean and 1 Indian Ocean, 1 Intersputnik—Indian Ocean region, and 1 Inmarsat—Pacific and Indian Ocean regions) (2007)

Radio broadcast stations: AM 369, FM 259, shortwave 45 (1998)

Radios: 417 million (1997)

Television broadcast stations: 3,240 (of which 209 are operated by China Central Television, 31 are provincial TV stations, and nearly 3,000 are local city stations) (1997)

Televisions: 400 million (1997)

Internet country code: .cn

Internet hosts: 10.637 million (2007)

Internet Service Providers (ISPs): 3 (2000)

Internet users: 162 million (2007)

TRANSPORTATION

Airports: 467 (2007)

Airports—with paved runways: *total:* 403
over 3,047 m: 58
2,438 to 3,047 m: 128
1,524 to 2,437 m: 130
914 to 1,523 m: 20
under 914 m: 67 (2007)

Airports—with unpaved runways:
total: 64
over 3,047 m: 4
2,438 to 3,047 m: 4
1,524 to 2,437 m: 13
914 to 1,523 m: 17
under 914 m: 26 (2007)

Heliports: 35 (2007)

Pipelines: gas 26,344 km; oil 17,240 km; refined products 6,106 km (2007)

Railways:
total: 75,438 km
standard gauge: 75,438 km 1.435-m gauge (20,151 km electrified) (2005)

Roadways:
total: 1,930,544 km
paved: 1,575,571 km (includes 41,005 km of expressways)
unpaved: 354,973 km (2005)

Waterways: 124,000 km navigable (2006)

Merchant marine:
total: 1,775 ships (1000 GRT or over) 22,219,786 GRT/33,819,636 DWT
by type: barge carrier 3, bulk carrier 415, cargo 689, carrier 3, chemical tanker 62, combination ore/oil 2, container 157, liquefied gas 35, passenger 8, passenger/cargo 84, petroleum tanker 250, refrigerated cargo 33, roll on/roll off 9, specialized tanker 8, vehicle carrier 17
foreign-owned: 12 (Ecuador 1, Greece 1, Hong Kong 6, Japan 2, South Korea 1, Norway 1)
registered in other countries: 1,366 (Bahamas 9, Bangladesh 1, Belize 107, Bermuda 10, Bolivia 1, Cambodia 166, Cyprus 10, France 5, Georgia 4, Germany 2, Honduras 3, Hong Kong 309, India 1, Indonesia 2, Liberia 32, Malaysia 1, Malta 13, Marshall Islands 3, Mongolia 3, Norway 47, Panama 473, Philippines 2, Sierra Leone 8, Singapore 19, St Vincent and The Grenadines 106, Thailand 1, Turkey 1, Tuvalu 25, unknown 33) (2007)

Ports and terminals: Dalian, Guangzhou, Ningbo, Qingdao, Qinhuangdao, Shanghai, Shenzhen, Tianjin

MILITARY

Military branches: People's Liberation Army (PLA): Ground Forces, Navy (includes marines and naval aviation), Air Force (includes airborne forces), and Second Artillery Corps (strategic missile force); People's Armed Police (PAP); Reserve and Militia Forces (2008)

Military service age and obligation: 18–22 years of age for selective compulsory military service, with 24-month service obligation; no minimum age for voluntary service (all officers are volunteers); 18–19 years of age for women high school graduates who meet requirements for specific military jobs (2007)

Manpower available for military service:
males age 16–49: 375,009,345
females age 16–49: 354,314,328 (2008 est.)

Manpower fit for military service:
males age 16–49: 313,321,639
females age 16–49: 295,951,438 (2008 est.)

Manpower reaching militarily significant age annually:
males age 16–49: 10,760,380
females age 16–49: 9,710,032 (2008 est.)

Military expenditures—percent of GDP: 4.3% (2006)

TRANSNATIONAL ISSUES

Disputes—international: based on principles drafted in 2005, China and India continue discussions to resolve all aspects of their extensive boundary and territorial disputes together with a security and foreign policy dialogue to consolidate discussions related to the boundary, regional nuclear proliferation, and other matters; recent talks and confidence-building measures have begun to defuse tensions over Kashmir, site of the world's largest and most militarized territorial dispute with portions under the de facto administration of China (Aksai Chin), India (Jammu and Kashmir), and Pakistan (Azad Kashmir and Northern Areas); India does not recognize Pakistan's ceding historic Kashmir lands to China in 1964; lacking any treaty describing the boundary, Bhutan and China continue negotiations to establish a boundary alignment to resolve substantial cartographic discrepancies, the largest of which lies in Bhutan's northwest; China asserts sovereignty over the Spratly Islands together with Malaysia, Philippines, Taiwan, Vietnam, and possibly Brunei; the 2002 "Declaration on the Conduct of Parties in the South China Sea" eased tensions in the Spratly's but is not the legally binding "code of conduct" sought by some parties; Vietnam and China continue to expand construction of facilities in the Spratly's and in March 2005, the national oil companies of China, the Philippines, and Vietnam signed a joint accord on marine seismic activities in the Spratly Islands; China occupies some of the Paracel Islands also claimed by Vietnam and Taiwan; China and Taiwan continue to reject both Japan's claims to the uninhabited islands of Senkaku-shoto (Diaoyu Tai) and Japan's unilaterally declared equidistance line in the East China Sea, the site of intensive hydrocarbon prospecting; certain islands in the Yalu and Tumen rivers are in dispute with North Korea; China seeks to stem illegal migration of North Koreans; China and Russia have demarcated the once disputed islands at the Amur and Ussuri confluence and in the Argun River in accordance with their 2004 Agreement; in 2006, China and Tajikistan pledged to commence demarcation of the revised boundary agreed to in the delimitation of 2002; demarcation of the China-Vietnam land boundary proceeds slowly and although the maritime boundary delimitation and fisheries agreements were ratified in June 2004, implementation remains stalled; in 2004, international environmentalist and political pressure from Burma and Thailand prompted China to halt construction of 13 dams on the Salween River

Refugees and internally displaced persons: *refugees (country of origin):* 300,897 (Vietnam); estimated 30,000–50,000 (North Korea)
IDPs: 90,000 (2007)

Trafficking in persons: *current situation:* China is a source, transit, and destination country for women, men, and children trafficked for purposes of sexual exploitation and forced labor; the majority of trafficking in China is internal, but there is also international trafficking of Chinese citizens; women are lured through false promises of legitimate employment into commercial sexual exploitation in Taiwan, Thailand, Malaysia, and Japan; Chinese men and women are smuggled to countries throughout the world at enormous personal expense and then forced into commercial sexual exploitation or exploitative labor to repay debts to traffickers; women and children are trafficked into China from Mongolia, Burma, North Korea, Russia, and Vietnam for forced labor, marriage, and sexual slavery; most North Koreans enter northeastern China voluntarily, but others reportedly are trafficked into China from North Korea; domestic trafficking remains the most significant problem in China, with an estimated minimum of 10,000–20,000 victims trafficked each year; the actual number of victims could be much greater; some experts believe that the serious and prolonged imbalance in the male-female birth ratio may now be contributing to Chinese and foreign girls and women being trafficked as potential brides
tier rating: Tier 2 Watch List—China failed to show evidence of increasing efforts to address transnational trafficking; while the government provides reasonable protection to internal victims of trafficking, protection for Chinese and foreign victims of transnational trafficking remain inadequate

Illicit drugs: major transshipment point for heroin produced in the Golden Triangle region of Southeast Asia; growing domestic drug abuse problem; source country for chemical precursors, despite new regulations on its large chemical industry

CHRISTMAS ISLAND

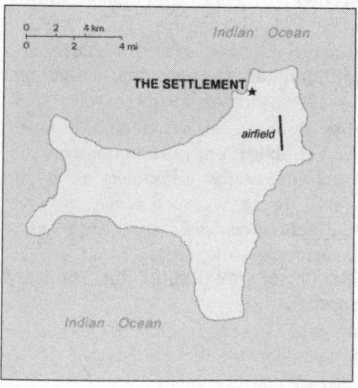

INTRODUCTION

Background: Named in 1643 for the day of its discovery, the island was annexed and settlement began by the UK in 1888. Phosphate mining began in the 1890s. The UK transferred sovereignty to Australia in 1958. Almost two-thirds of the island has been declared a national park.

GEOGRAPHY

Location: Southeastern Asia, island in the Indian Ocean, south of Indonesia
Geographic coordinates: 10 30 S, 105 40 E
Map references: Southeast Asia
Area:
total: 135 sq km
land: 135 sq km
water: 0 sq km
Area—comparative: about three-quarters the size of Washington, DC
Land boundaries: 0 km
Coastline: 138.9 km
Maritime claims:
territorial sea: 12 nm
contiguous zone: 12 nm
exclusive fishing zone: 200 nm
Climate: tropical with a wet season

(December to April) and dry season; heat and humidity moderated by trade winds

Terrain: steep cliffs along coast rise abruptly to central plateau

Elevation extremes:
lowest point: Indian Ocean 0 m
highest point: Murray Hill 361 m

Natural resources: phosphate, beaches

Land use:
arable land: 0%
permanent crops: 0%
other: 100% (mainly tropical rainforest; 63% of the island is a national park) (2005)

Irrigated land: NA

Natural hazards: the narrow fringing reef surrounding the island can be a maritime hazard

Environment—current issues: loss of rainforest; impact of phosphate mining

Geography—note: located along major sea lanes of Indian Ocean

PEOPLE

Population: 1,402 (July 2007 est.)

Age structure:
0–14 years: NA
15–64 years: NA
65 years and over: NA

Population growth rate: 0% (2008 est.)

Birth rate: NA

Death rate: NA (2008 est.)

Net migration rate: NA (2008 est.)

Sex ratio: NA

Infant mortality rate:
total: NA
male: NA
female: NA (2008 est.)

Life expectancy at birth:
total population: NA
male: NA
female: NA (2008 est.)

Total fertility rate: NA (2008 est.)

HIV/AIDS—adult prevalence rate: NA

HIV/AIDS—people living with HIV/AIDS: NA

HIV/AIDS—deaths: NA

Nationality:
noun: Christmas Islander(s)
adjective: Christmas Island

Ethnic groups: Chinese 70%, European 20%, Malay 10%
note: no indigenous population (2001)

Religions: Buddhist 36%, Muslim 25%, Christian 18%, other 21% (1997)

Languages: English (official), Chinese, Malay

Literacy: NA

GOVERNMENT

Country name:
conventional long form: Territory of Christmas Island
conventional short form: Christmas Island

Dependency status: non-self governing territory of Australia; administered from Canberra by the Australian Attorney-General's Department

Government type: NA

Capital: *name:* The Settlement
geographic coordinates: 10 25 S, 105 43 E
time difference: UTC+7 (12 hours ahead of Washington, DC during Standard Time)

Administrative divisions: none (territory of Australia)

Independence: none (territory of Australia)

National holiday: Australia Day, 26 January (1788)

Constitution: Christmas Island Act of 1958–59 (1 October 1958) as amended by the Territories Law Reform Act of 1992

Legal system: under the authority of the governor general of Australia and Australian law

Executive branch:
chief of state: Queen ELIZABETH II (since 6 February 1952), represented by the Australian governor general
head of government: Administrator Neil LUCAS (since 30 January 2006)
elections: the monarch is hereditary; administrator appointed by the governor general of Australia and represents the monarch and Australia

Legislative branch: unicameral Christmas Island Shire Council (9 seats; members elected by popular vote to serve four-year terms)
elections: held every two years with half the members standing for election; last held 20 October 2007 (next to be held in 2009)
election results: percent of vote—NA; seats—independents 9

Judicial branch: Supreme Court; District Court; Magistrate's Court

Political parties and leaders: none

Political pressure groups and leaders: none

International organization participation: none

Diplomatic representation in the US: none (territory of Australia)

Diplomatic representation from the US: none (territory of Australia)

Flag description: territorial flag; divided diagonally from upper hoist to lower fly; the upper triangle is green with a yellow image of the Golden Bosun Bird superimposed, while the lower triangle is blue with the Southern Cross constellation, representing Australia, superimposed; a centered yellow disk displays a green map of the island
note: the flag of Australia is used for official purposes

ECONOMY

Economy—overview: Phosphate mining had been the only significant economic activity, but in December 1987 the Australian Government closed the mine. In 1991, the mine was reopened. With the support of the government, a $34 million casino opened in 1993, but closed in 1998. The Australian Government in 2001 agreed to support the creation of a commercial space-launching site on the island, expected to begin operations in the near future.

GDP (purchasing power parity): $NA

Labor force: NA

Budget:
revenues: $NA
expenditures: $NA

Agriculture—products: NA

Industries: tourism, phosphate extraction (near depletion)

Electricity—production by source:
fossil fuel: NA
hydro: NA
nuclear: NA
other: NA

Exports: $NA

Exports—commodities: phosphate

Exports—partners: Australia, NZ (2006)

Imports: $NA

Imports—commodities: consumer goods

Imports—partners: principally Australia (2006)

Economic aid—recipient: $NA

Currency (code): Australian dollar (AUD)

Currency code: AUD

Exchange rates: Australian dollars per US dollar—1.2137 (2007), 1.3285 (2006), 1.3095 (2005), 1.3598 (2004), 1.5419 (2003)

Fiscal year: 1 July—30 June

COMMUNICATIONS

Telephones—main lines in use: NA

Telephone system:
general assessment: service provided by the Australian network
domestic: GSM mobile telephone service replaced older analog system in February 2005
international: country code—61-8; satellite earth station—1 (Intelsat provides telephone and telex service) (2005)

Radio broadcast stations: AM 1, FM 2, shortwave 0 (2006)

Radios: 1,000 (1997)

Television broadcast stations: 0 (TV broadcasts received via satellite from mainland Australia) (2006)

Televisions: 600 (1997)

Internet country code: .cx

Internet hosts: 1,826 (2007)
Internet Service Providers (ISPs): 2 (2000)
Internet users: 464 (2001)

TRANSPORTATION

Airports: 1 (2007)

Airports—with paved runways:
total: 1
1,524 to 2,437 m: 1 (2007)
Roadways:
total: 142 km
paved: 32 km
unpaved: 110 km (2006)

Ports and terminals: Flying Fish Cove

MILITARY

Military—note: defense is the responsibility of Australia

TRANSNATIONAL ISSUES

Disputes—international: none

CLIPPERTON ISLAND

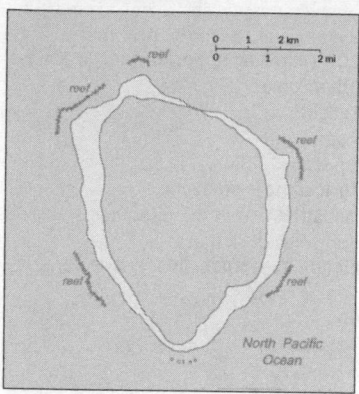

INTRODUCTION

Background: This isolated island was named for John CLIPPERTON, a pirate who made it his hideout early in the 18th century. Annexed by France in 1855, it was seized by Mexico in 1897. Arbitration eventually awarded the island to France, which took possession in 1935.

GEOGRAPHY

Location: Middle America, atoll in the North Pacific Ocean, 1,120 km southwest of Mexico
Geographic coordinates: 10 17 N, 109 13 W

Map references: Political Map of the World
Area:
total: 6 sq km
land: 6 sq km
water: 0 sq km
Area—comparative: about 12 times the size of The Mall in Washington, DC
Land boundaries: 0 km
Coastline: 11.1 km
Maritime claims:
territorial sea: 12 nm
exclusive economic zone: 200 nm
Climate: tropical; humid, average temperature 20–32 degrees C, wet season (May to October)
Terrain: coral atoll
Elevation extremes:
lowest point: Pacific Ocean 0 m
highest point: Rocher Clipperton 29 m
Natural resources: fish
Land use:
arable land: 0%
permanent crops: 0%
other: 100% (all coral) (2005)
Irrigated land: 0 sq km
Natural hazards: NA
Environment—current issues: NA
Geography—note: reef 12 km in circumference

PEOPLE

Population: uninhabited

GOVERNMENT

Country name:
conventional long form: none
conventional short form: Clipperton Island
local long form: none
local short form: Ile Clipperton
former: sometimes called Ile de la Passion
Dependency status: possession of France; administered directly by the Minister of Overseas France
Legal system: the laws of France, where applicable, apply
Flag description: the flag of France is used

ECONOMY

Economy—overview: Although 115 species of fish have been identified in the territorial waters of Clipperton Island, the only economic activity is tuna fishing.

TRANSPORTATION

Ports and terminals: none; offshore anchorage only

MILITARY

Military—note: defense is the responsibility of France

TRANSNATIONAL ISSUES

Disputes—international: none

COCOS (KEELING) ISLANDS

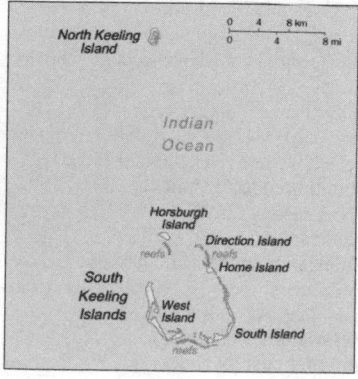

INTRODUCTION

Background: There are 27 coral islands in the group. Captain William KEELING discovered the islands in 1609, but they remained uninhabited until the 19th century. From the 1820s to 1978, members of the CLUNIE-ROSS family controlled the islands and the copra produced from local coconuts. Annexed by the UK in 1857, the Cocos Islands were transferred to the Australian Government in 1955. The population on the two inhabited islands generally is split between the ethnic Europeans on West Island and the ethnic Malays on Home Island.

GEOGRAPHY

Location: Southeastern Asia, group of islands in the Indian Ocean, southwest of Indonesia, about halfway from Australia to Sri Lanka
Geographic coordinates: 12 30 S, 96 50 E
Map references: Southeast Asia
Area:
total: 14 sq km
land: 14 sq km
water: 0 sq km

note: includes the two main islands of West Island and Home Island
Area—comparative: about 24 times the size of The Mall in Washington, DC
Land boundaries: 0 km
Coastline: 26 km
Maritime claims:
territorial sea: 12 nm
exclusive fishing zone: 200 nm
Climate: tropical with high humidity, moderated by the southeast trade winds for about nine months of the year
Terrain: flat, low-lying coral atolls
Elevation extremes:
lowest point: Indian Ocean 0 m
highest point: unnamed location 5 m
Natural resources: fish
Land use:
arable land: 0%
permanent crops: 0%
other: 100% (2005)
Irrigated land: NA
Natural hazards: cyclone season is October to April
Environment—current issues: fresh water resources are limited to rainwater accumulations in natural underground reservoirs
Geography—note: islands are thickly covered with coconut palms and other vegetation

PEOPLE

Population: 596 (July 2007 est.)
Age structure:
0–14 years: NA
15–64 years: NA
65 years and over: NA
Population growth rate: 0% (2008 est.)
Birth rate: NA
Death rate: NA (2008 est.)
Net migration rate: NA
Infant mortality rate:
total: NA
male: NA
female: NA (2008 est.)
Life expectancy at birth:
total population: NA
male: NA
female: NA (2008 est.)
Total fertility rate: NA (2008 est.)
HIV/AIDS—adult prevalence rate: NA
HIV/AIDS—people living with HIV/AIDS: NA
HIV/AIDS—deaths: NA
Nationality:
noun: Cocos Islander(s)
adjective: Cocos Islander
Ethnic groups: Europeans, Cocos Malays
Religions: Sunni Muslim 80%, other 20% (2002 est.)
Languages: Malay (Cocos dialect), English
Literacy:
NA

GOVERNMENT

Country name:
conventional long form: Territory of Cocos (Keeling) Islands
conventional short form: Cocos (Keeling) Islands
Dependency status: non-self governing territory of Australia; administered from Canberra by the Australian Attorney-General's Department
Government type: NA
Capital: *name:* West Island
geographic coordinates: 12 10 S, 96 50 E
time difference: UTC+6.5 (11.5 hours ahead of Washington, DC during Standard Time)
Administrative divisions: none (territory of Australia)
Independence: none (territory of Australia)
National holiday: Australia Day, 26 January (1788)
Constitution: Cocos (Keeling) Islands Act of 1955 (23 November 1955) as amended by the Territories Law Reform Act of 1992
Legal system: based upon the laws of Australia and local laws
Suffrage: NA
Executive branch:
chief of state: Queen ELIZABETH II (since 6 February 1952); represented by the Australian governor general
head of government: Administrator (non-resident) Neil LUCAS (since 30 January 2006)
cabinet: NA
elections: the monarch is hereditary; administrator appointed by the governor general of Australia and represents the monarch and Australia
Legislative branch: unicameral Cocos (Keeling) Islands Shire Council (7 seats)
elections: held every two years with half the members standing for election; last held in May 2007 (next to be held in May 2009)
Judicial branch: Supreme Court; Magistrate's Court
Political parties and leaders: none
Political pressure groups and leaders: none
International organization participation: none
Diplomatic representation in the US: none (territory of Australia)
Diplomatic representation from the US: none (territory of Australia)
Flag description: the flag of Australia is used

ECONOMY

Economy—overview: Grown throughout the islands, coconuts are the sole cash crop. Small local gardens and fishing contribute to the food supply, but additional food and most other necessities must be imported from Australia. There is a small tourist industry.
GDP (purchasing power parity): $NA
Labor force: NA
Labor force—by occupation: *note:* the Cocos Islands Cooperative Society Ltd. employs construction workers, stevedores, and lighterage workers; tourism employs others
Unemployment rate: 60% (2000 est.)
Budget:
revenues: $NA
expenditures: $NA
Agriculture—products: vegetables, bananas, pawpaws, coconuts
Industries: copra products and tourism
Electricity—production by source:
fossil fuel: NA
hydro: NA
nuclear: NA
other: NA
Exports: $NA
Exports—commodities: copra
Exports—partners: Australia (2006)
Imports: $NA
Imports—commodities: foodstuffs
Imports—partners: Australia (2006)
Economic aid—recipient: $NA
Currency (code): Australian dollar (AUD)
Currency code: AUD
Exchange rates: Australian dollars per US dollar—1.2137 (2007), 1.3285 (2006), 1.3095 (2005), 1.3598 (2004), 1.5419 (2003)
Fiscal year: 1 July—30 June

COMMUNICATIONS

Telephones—main lines in use: 287 (1992)
Telephone system:
general assessment: connected within Australia's telecommunication system; a local mobile-cellular network is in operation
domestic: NA
international: country code—61; telephone, telex, and facsimile communications with Australia and elsewhere via satellite; satellite earth station—1 (Intelsat) (2001)
Radio broadcast stations: AM 1, FM 2, shortwave 0 (2004)
Radios: 300 (1992)
Television broadcast stations: 4 (2007)
Televisions: NA
Internet country code: .cc
Internet Service Providers (ISPs): 2 (2000)
Internet users: NA

TRANSPORTATION

Airports: 1 (2007)

145

Airports—with paved runways:
total: 1
1,524 to 2,437 m: 1 (2007)
Roadways: *total:* 22 km
paved: 10 km
unpaved: 12 km (2006)

Ports and terminals: Port Refuge

MILITARY

Military—note: defense is the responsibility of Australia; the territory has a five-person police force

TRANSNATIONAL ISSUES

Disputes—international: none

COLOMBIA

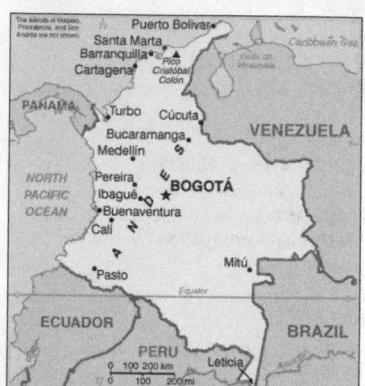

INTRODUCTION

Background: Colombia was one of the three countries that emerged from the collapse of Gran Colombia in 1830 (the others are Ecuador and Venezuela). A 40-year conflict between government forces and anti-government insurgent groups and illegal paramilitary groups—both heavily funded by the drug trade—escalated during the 1990s. The insurgents lack the military or popular support necessary to overthrow the government, and violence has been decreasing since about 2002, but insurgents continue attacks against civilians and large swaths of the countryside are under guerrilla influence. More than 32,000 former paramilitaries had demobilized by the end of 2006 and the United Self Defense Forces of Colombia (AUC) as a formal organization had ceased to function. Still, some renegades continued to engage in criminal activities. The Colombian Government has stepped up efforts to reassert government control throughout the country, and now has a presence in every one of its administrative departments. However, neighboring countries worry about the violence spilling over their borders.

GEOGRAPHY

Location: Northern South America, bordering the Caribbean Sea, between Panama and Venezuela, and bordering the North Pacific Ocean, between Ecuador and Panama

Geographic coordinates: 4 00 N, 72 00 W

Map references: South America

Area:
total: 1,138,910 sq km
land: 1,038,700 sq km
water: 100,210 sq km
note: includes Isla de Malpelo, Roncador Cay, and Serrana Bank

Area—comparative: slightly less than twice the size of Texas

Land boundaries:
total: 6,309 km
border countries: Brazil 1,644 km, Ecuador 590 km, Panama 225 km, Peru 1,800 km, Venezuela 2,050 km

Coastline: 3,208 km (Caribbean Sea 1,760 km, North Pacific Ocean 1,448 km)

Maritime claims:
territorial sea: 12 nm
exclusive economic zone: 200 nm
continental shelf: 200-m depth or to the depth of exploitation

Climate: tropical along coast and eastern plains; cooler in highlands

Terrain: flat coastal lowlands, central highlands, high Andes Mountains, eastern lowland plains

Elevation extremes:
lowest point: Pacific Ocean 0 m
highest point: Pico Cristobal Colon 5,775 m
note: nearby Pico Simon Bolivar also has the same elevation

Natural resources: petroleum, natural gas, coal, iron ore, nickel, gold, copper, emeralds, hydropower

Land use:
arable land: 2.01%
permanent crops: 1.37%
other: 96.62% (2005)

Irrigated land: 9,000 sq km (2003)

Total renewable water resources: 2,132 cu km (2000)

Freshwater withdrawal (domestic/industrial/agricultural): *total:* 10.71 cu km/yr (50%/4%/46%)
per capita: 235 cu m/yr (2000)

Natural hazards: highlands subject to volcanic eruptions; occasional earthquakes; periodic droughts

Environment—current issues: deforestation; soil and water quality damage from overuse of pesticides; air pollution, especially in Bogota, from vehicle emissions

Environment—international agreements: *party to:* Antarctic Treaty, Biodiversity, Climate Change, Climate Change-Kyoto Protocol, Desertification, Endangered Species, Hazardous Wastes, Marine Life Conservation, Ozone Layer Protection, Ship Pollution, Tropical Timber 83, Tropical Timber 94, Wetlands
signed, but not ratified: Law of the Sea

Geography—note: only South American country with coastlines on both the North Pacific Ocean and Caribbean Sea

PEOPLE

Population: 45,013,674 (July 2008 est.)

Age structure:
0–14 years: 29.4% (male 6,688,530/female 6,531,768)
15–64 years: 65.1% (male 14,292,647/female 15,017,204)
65 years and over: 5.5% (male 1,072,644/female 1,410,881) (2008 est.)

Median age:
total: 26.8 years
male: 25.9 years
female: 27.8 years (2008 est.)

Population growth rate: 1.405% (2008 est.)

Birth rate: 19.86 births/1,000 population (2008 est.)

Death rate: 5.54 deaths/1,000 population (2008 est.)

Net migration rate: -0.28 migrant(s)/1,000 population (2008 est.)

Sex ratio:
at birth: 1.03 male(s)/female
under 15 years: 1.02 male(s)/female
15–64 years: 0.95 male(s)/female
65 years and over: 0.76 male(s)/female
total population: 0.96 male(s)/female (2008 est.)

Infant mortality rate:
total: 19.51 deaths/1,000 live births
male: 23.18 deaths/1,000 live births
female: 15.7 deaths/1,000 live births (2008 est.)

Life expectancy at birth:
total population: 72.54 years
male: 68.71 years
female: 76.5 years (2008 est.)

Total fertility rate: 2.49 children born/woman (2008 est.)

HIV/AIDS—adult prevalence rate: 0.7% (2003 est.)

HIV/AIDS—people living with HIV/AIDS: 190,000 (2003 est.)

HIV/AIDS—deaths: 3,600 (2003 est.)

Major infectious diseases:

degree of risk: high

food or waterborne diseases: bacterial diarrhea, hepatitis A

vectorborne diseases: dengue fever, malaria, and yellow fever

water contact disease: leptospirosis (2008)

Nationality:

noun: Colombian(s)

adjective: Colombian

Ethnic groups: mestizo 58%, white 20%, mulatto 14%, black 4%, mixed black-Amerindian 3%, Amerindian 1%

Religions: Roman Catholic 90%, other 10%

Languages: Spanish

Literacy:

definition: age 15 and over can read and write

total population: 92.8%

male: 92.9%

female: 92.7% (2004 est.)

GOVERNMENT

Country name:

conventional long form: Republic of Colombia

conventional short form: Colombia

local long form: Republica de Colombia

local short form: Colombia

Government type: republic; executive branch dominates government structure

Capital: *name:* Bogota

geographic coordinates: 4 36 N, 74 05 W

time difference: UTC-5 (same time as Washington, DC during Standard Time)

Administrative divisions: 32 departments (departamentos, singular—departamento) and 1 capital district* (distrito capital); Amazonas, Antioquia, Arauca, Atlantico, Bogota*, Bolivar, Boyaca, Caldas, Caqueta, Casanare, Cauca, Cesar, Choco, Cordoba, Cundinamarca, Guainia, Guaviare, Huila, La Guajira, Magdalena, Meta, Narino, Norte de Santander, Putumayo, Quindio, Risaralda, San Andres y Providencia, Santander, Sucre, Tolima, Valle del Cauca, Vaupes, Vichada

Independence: 20 July 1810 (from Spain)

National holiday: Independence Day, 20 July (1810)

Constitution: 5 July 1991; amended many times

Legal system: based on Spanish law; a new criminal code modeled after US procedures was enacted into law in 2004 and reached full implemention in January 2008; judicial review of executive and legislative acts; has not accepted compulsory ICJ jurisdiction

Suffrage: 18 years of age; universal

Executive branch:

chief of state: President Alvaro URIBE Velez (since 7 August 2002); Vice President Francisco SANTOS (since 7 August 2002); note—the president is both the chief of state and head of government

head of government: President Alvaro URIBE Velez (since 7 August 2002); Vice President Francisco SANTOS (since 7 August 2002)

cabinet: Cabinet consists of a coalition of the three largest parties that supported President URIBE's reelection—the PSUN, PC, and CR—and independents

elections: president and vice president elected by popular vote for a four-year term (eligible for a second term); election last held 28 May 2006 (next to be held in May 2010)

election results: President Alvaro URIBE Velez reelected president; percent of vote—Alvaro URIBE Velez 62%, Carlos GAVIRIA Diaz 22%, Horacio SERPA Uribe 12%, other 4%

Legislative branch: bicameral Congress or Congreso consists of the Senate or Senado (102 seats; members are elected by popular vote to serve four-year terms) and the House of Representatives or Camara de Representantes (166 seats; members are elected by popular vote to serve four-year terms)

elections: Senate—last held 12 March 2006 (next to be held in March 2010); House of Representatives—last held 12 March 2006 (next to be held in March 2010)

election results: Senate—percent of vote by party—NA; seats by party—PSUN 20, PC 18, PL 18, CR 15, PDI 10, other parties 21; House of Representatives—percent of vote by party—NA; seats by party—PL 35, PSUN 33, PC 29, CR 20, PDA 8, other parties 41

Judicial branch: four roughly coequal, supreme judicial organs; Supreme Court of Justice or Corte Suprema de Justicia (highest court of criminal law; judges are selected by their peers from the nominees of the Superior Judicial Council for eight-year terms); Council of State (highest court of administrative law; judges are selected from the nominees of the Superior Judicial Council for eight-year terms); Constitutional Court (guards integrity and supremacy of the constitution; rules on constitutionality of laws, amendments to the constitution, and international treaties); Superior Judicial Council (administers and disciplines the civilian judiciary; resolves jurisdictional conflicts arising between other courts; members are elected by three sister courts and Congress for eight-year terms)

Political parties and leaders: Colombian Conservative Party or PC [Efrain Jose CEPEDA Sarabia]; Alternative Democratic Pole or PDA [Carlos GAVIRIA Diaz]; Liberal Party or PL [Cesar GAVIRIA Trujillo]; Radical Change or CR [German VARGAS Lleras]; Social National Unity Party or U Party [Carlos GARCIA Orjuela]

note: Colombia has 15 formally recognized political parties, and numerous unofficial parties that did not meet the vote threshold in the March 2006 legislative elections required for recognition

Political pressure groups and leaders: two largest insurgent groups active in Colombia—Revolutionary Armed Forces of Colombia or FARC and National Liberation Army or ELN

International organization participation: BCIE, CAN, Caricom (observer), CDB, CSN, FAO, G-3, G-24, G-77, IADB, IAEA, IBRD, ICAO, ICC, ICCt, ICRM, IDA, IFAD, IFC, IFRCS, IHO, ILO, IMF, IMO, IMSO, Interpol, IOC, IOM, IPU, ISO, ITSO, ITU, ITUC, LAES, LAIA, Mercosur (associate), MIGA, NAM, OAS, OPANAL, OPCW, PCA, RG, UN, UNCTAD, UNESCO, UNHCR, UNIDO, Union Latina, UNWTO, UPU, WCL, WCO, WFTU, WHO, WIPO, WMO, WTO

Diplomatic representation in the US:

chief of mission: Ambassador Carolina BARCO Isakson

chancery: 2118 Leroy Place NW, Washington, DC 20008

telephone: [1] (202) 387-8338

FAX: [1] (202) 232-8643

consulate(s) general: Atlanta, Boston, Chicago, Houston, Los Angeles, Miami, New York, San Francisco, San Juan (Puerto Rico), Washington, DC

Diplomatic representation from the US:

chief of mission: Ambassador William R. BROWNFIELD

embassy: Calle 24 Bis No. 48-50, Bogota, D.C.

mailing address: Carrera 45 No. 24B-27, Bogota, D.C.

telephone: [57] (1) 315-0811

FAX: [57] (1) 315-2197

Flag description: three horizontal bands of yellow (top, double-width), blue, and red

note: similar to the flag of Ecuador, which is longer and bears the Ecuadorian coat of arms superimposed in the center

ECONOMY

Economy—overview: Colombia's economy has experienced positive growth over the past five years despite a serious armed conflict. In fact, 2007 is regarded by policy makers and the private sector as one of the best economic years in recent history, after 2005. The economy continues to improve in part because of austere government budgets, focused efforts to reduce public debt levels, an export-oriented growth strategy, improved domestic security, and high commodity prices. Ongoing economic problems facing President URIBE include reforming the pension system, reducing high unemployment, and funding new exploration to offset declining oil production. The government's economic reforms and democratic security strategy, coupled with increased investment, have engendered a growing sense of confidence in the economy. However, the business sector continues to be concerned about failure of the US Congress to approve the signed FTA.

GDP (purchasing power parity): $319.5 billion (2007 est.)

GDP (official exchange rate): $171.6 billion (2007 est.)

GDP—real growth rate: 7% (2007 est.)

GDP—per capita (PPP): $6,700 (2007 est.)

GDP—composition by sector:
agriculture: 11.5%
industry: 36%
services: 52.5% (2007 est.)

Labor force: 20.5 million (2007 est.)

Labor force—by occupation: *agriculture:* 22.7%
industry: 18.7%
services: 58.5% (2000 est.)

Unemployment rate: 11.2% (2007 est.)

Population below poverty line: 49.2% (2005)

Household income or consumption by percentage share:
lowest 10%: 7.9%
highest 10%: 34.3% (2004)

Distribution of family income—Gini index: 53.8 (2005)

Inflation rate (consumer prices): 5.5% (2007 est.)

Investment (gross fixed): 22.5% of GDP (2007 est.)

Budget:
revenues: $63.69 billion
expenditures: $64.96 billion; including capital expenditures of $NA (2007 est.)

Public debt: 53.5% of GDP (2007 est.)

Agriculture—products: coffee, cut flowers, bananas, rice, tobacco, corn, sugarcane, cocoa beans, oilseed, vegetables; forest products; shrimp

Industries: textiles, food processing, oil, clothing and footwear, beverages, chemicals, cement; gold, coal, emeralds

Industrial production growth rate: 9.4% (2007 est.)

Electricity—production: 50.47 billion kWh (2005)

Electricity—production by source:
fossil fuel: 26%
hydro: 72.7%
nuclear: 0%
other: 1.3% (2001)

Electricity—consumption: 38.91 billion kWh (2005)

Electricity—exports: 1.758 billion kWh (2005)

Electricity—imports: 16 million kWh (2005)

Oil—production: 539,000 bbl/day (2005 est.)

Oil—consumption: 264,000 bbl/day (2005 est.)

Oil—exports: 289,700 bbl/day (2004)

Oil—imports: 6,453 bbl/day (2004)

Oil—proved reserves: 1.387 billion bbl (2007 est.)

Natural gas—production: 6.397 billion cu m (2005 est.)

Natural gas—consumption: 6.397 billion cu m (2005 est.)

Natural gas—exports: 0 cu m (2005 est.)

Natural gas—imports: 0 cu m (2005)

Natural gas—proved reserves: 109.7 billion cu m (1 January 2006 est.)

Current account balance: -$6.465 billion (2007 est.)

Exports: $30.58 billion f.o.b. (2007 est.)

Exports—commodities: petroleum, coffee, coal, nickel, emeralds, apparel, bananas, cut flowers

Exports—partners: US 35.8%, Venezuela 11.4%, Ecuador 5.4% (2006)

Imports: $31.17 billion f.o.b. (2007 est.)

Imports—commodities: industrial equipment, transportation equipment, consumer goods, chemicals, paper products, fuels, electricity

Imports—partners: US 26.8%, Brazil 8.6%, Mexico 8.5%, China 6%, Venezuela 5.6%, Japan 4.1% (2006)

Economic aid—recipient: $511.1 million (2005)

Reserves of foreign exchange and gold: $20.95 billion (31 December 2007 est.)

Debt—external: $41.16 billion (30 June 2007)

Stock of direct foreign investment—at home: $54.04 billion (2007 est.)

Stock of direct foreign investment—abroad: $10.38 billion (2007 est.)

Market value of publicly traded shares: $56.2 billion (2006)

Currency (code): Colombian peso (COP)

Currency code: COP

Exchange rates: Colombian pesos per US dollar—2,013.8 (2007), 2,358.6 (2006), 2,320.75 (2005), 2,628.61 (2004), 2,877.65 (2003)

Fiscal year: calendar year

COMMUNICATIONS

Telephones—main lines in use: 7.865 million (2006)

Telephones—mobile cellular: 29.763 million (2006)

Telephone system:
general assessment: modern system in many respects; telecommunications sector liberalized during the 1990s; multiple providers of both fixed-line and mobile-cellular services; fixed-line connections stand at about 18 per 100 persons; mobile cellular usage is about 70 per 100 persons
domestic: nationwide microwave radio relay system; domestic satellite system with 41 earth stations; fiber-optic network linking 50 cities
international: country code—57; submarine cables provide links to the US, parts of the Caribbean, and Central and South America; satellite earth stations—10 (6 Intelsat, 1 Inmarsat, 3 fully digitalized international switching centers) (2007)

Radio broadcast stations: AM 454, FM 34, shortwave 27 (1999)

Radios: 21 million (1997)

Television broadcast stations: 60 (1997)

Televisions: 4.59 million (1997)

Internet country code: .co

Internet hosts: 1.014 million (2007)

Internet Service Providers (ISPs): 18 (2000)

Internet users: 6.705 million (2006)

TRANSPORTATION

Airports: 934 (2007)

Airports—with paved runways:
total: 103
over 3,047 m: 2
2,438 to 3,047 m: 8
1,524 to 2,437 m: 39
914 to 1,523 m: 42
under 914 m: 12 (2007)

Airports—with unpaved runways:
total: 831
over 3,047 m: 1
1,524 to 2,437 m: 34
914 to 1,523 m: 216
under 914 m: 580 (2007)

Heliports: 2 (2007)

Pipelines: gas 4,329 km; oil 6,140 km; refined products 3,145 km (2007)

Railways:
total: 3,304 km
standard gauge: 150 km 1.435-m gauge
narrow gauge: 3,154 km 0.914-m gauge (2006)

Roadways: *total:* 164,257 km (2005)
Waterways: 18,000 km (2006)
Merchant marine:
total: 15 ships (1000 GRT or over) 35,949 GRT/49,161 DWT
by type: cargo 11, liquefied gas 1, petroleum tanker 3
registered in other countries: 5 (Antigua and Barbuda 1, Panama 4) (2007)
Ports and terminals: Barranquilla, Buenaventura, Cartagena, Santa Marta, Turbo

MILITARY

Military branches: National Army (Ejercito Nacional), National Navy (Armada Nacional, includes Naval Aviation, Naval Infantry (Infanteria de Marina, Colmar), and Coast Guard), Colombian Air Force (Fuerza Aerea de Colombia, FAC) (2008)
Military service age and obligation: 18–24 years of age for compulsory and voluntary military service; service obligation—18 months (2004)
Manpower available for military service: *males age 16–49:* 11,478,109
females age 16–49: 11,809,279 (2008 est.)

Manpower fit for military service:
males age 16–49: 8,056,336
females age 16–49: 9,919,952 (2008 est.)
Manpower reaching militarily significant age annually:
males age 16–49: 442,403
females age 16–49: 433,192 (2008 est.)
Military expenditures—percent of GDP: 3.4% (2005 est.)

TRANSNATIONAL ISSUES

Disputes—international: memorials and countermemorials were filed by the parties in Nicaragua's 1999 and 2001 proceedings against Honduras and Colombia at the ICJ over the maritime boundary and territorial claims in the western Caribbean Sea—final public hearings are scheduled for 2007; dispute with Venezuela over maritime boundary and Venezuelan-administered Los Monjes Islands near the Gulf of Venezuela; Colombian-organized illegal narcotics, guerrilla, and paramilitary activities penetrate all of its neighbors' borders and have caused over 300,000 persons to flee the country, mostly into neighboring states
Refugees and internally displaced

persons: *IDPs:* 1.8–3.5 million (conflict between government and illegal armed groups and drug traffickers) (2007)
Illicit drugs: illicit producer of coca, opium poppy, and cannabis; world's leading coca cultivator with 144,000 hectares in coca cultivation in 2005, a 26% increase over 2004, producing a potential of 545 mt of pure cocaine; the world's largest producer of coca derivatives; supplies cocaine to most of the US market and the great majority of other international drug markets; in 2005, aerial eradication dispensed herbicide to treat over 130,000 hectares but aggressive replanting on the part of coca growers means Colombia remains a key producer; a significant portion of non-US narcotics proceeds are either laundered or invested in Colombia through the black market peso exchange; important supplier of heroin to the US market; opium poppy cultivation fell 50% between 2003 and 2004 to 2,100 hectares yielding a potential 3.8 metric tons of pure heroin, mostly for the US market; no poppy estimate was conducted in 2005

COMOROS

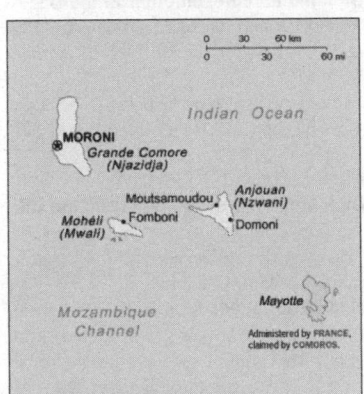

INTRODUCTION

Background: Comoros has endured more than 20 coups or attempted coups since gaining independence from France in 1975. In 1997, the islands of Anjouan and Moheli declared independence from Comoros. In 1999, military chief Col. AZALI seized power in a bloodless coup, and helped negotiate the 2000 Fomboni Accords power-sharing agreement in which the federal presidency rotates among the three islands, and each island maintains its own local government. AZALI won the 2002 Presidential election, and each island in the archipelago elected its own president. AZALI

stepped down in 2006 and President SAMBI took office. Since 2006, Anjouan's President Mohamed BACAR has refused to work effectively with the Union presidency. In 2007, BACAR effected Anjouan's de-facto secession from the Union, refusing to step down in favor of fresh Anjouanais elections when Comoros' other islands held legitimate elections in July. The African Union (AU) initially attempted to resolve the political crisis by applying sanctions and a naval blockade on Anjouan, but in March 2008, AU and Comoran soldiers seized the island. The move was generally welcomed by the island's inhabitants.

GEOGRAPHY

Location: Southern Africa, group of islands at the northern mouth of the Mozambique Channel, about two-thirds of the way between northern Madagascar and northern Mozambique
Geographic coordinates: 12 10 S, 44 15 E
Map references: Africa
Area:
total: 2,170 sq km
land: 2,170 sq km
water: 0 sq km
Area—comparative: slightly more than 12 times the size of Washington, DC

Land boundaries: 0 km
Coastline: 340 km
Maritime claims: *territorial sea:* 12 nm
exclusive economic zone: 200 nm
Climate: tropical marine; rainy season (November to May)
Terrain: volcanic islands, interiors vary from steep mountains to low hills
Elevation extremes:
lowest point: Indian Ocean 0 m
highest point: Le Kartala 2,360 m
Natural resources: NEGL
Land use:
arable land: 35.87%
permanent crops: 23.32%
other: 40.81% (2005)
Irrigated land: NA
Total renewable water resources: 1.2 cu km (2003)
Freshwater withdrawal (domestic/industrial/agricultural): *total:* 0.01 cu km/yr (48%/5%/47%)
per capita: 13 cu m/yr (1999)
Natural hazards: cyclones possible during rainy season (December to April); Le Kartala on Grand Comore is an active volcano
Environment—current issues: soil degradation and erosion results from crop cultivation on slopes without proper terracing; deforestation
Environment—international agreements: *party to:* Biodiversity, Climate

149

Change, Climate Change-Kyoto Protocol, Desertification, Endangered Species, Hazardous Wastes, Law of the Sea, Ozone Layer Protection, Ship Pollution, Wetlands

signed, but not ratified: none of the selected agreements

Geography—note: important location at northern end of Mozambique Channel

PEOPLE

Population: 731,775 (July 2008 est.)

Age structure:
0–14 years: 42.4% (male 155,662/female 154,520)
15–64 years: 54.6% (male 197,178/female 202,231)
65 years and over: 3% (male 10,203/female 11,981) (2008 est.)

Median age:
total: 18.7 years
male: 18.5 years
female: 19 years (2008 est.)

Population growth rate: 2.803% (2008 est.)

Birth rate: 35.78 births/1,000 population (2008 est.)

Death rate: 7.76 deaths/1,000 population (2008 est.)

Net migration rate: NA

Sex ratio:
at birth: 1.03 male(s)/female
under 15 years: 1.01 male(s)/female
15–64 years: 0.98 male(s)/female
65 years and over: 0.85 male(s)/female
total population: 0.98 male(s)/female (2008 est.)

Infant mortality rate:
total: 68.58 deaths/1,000 live births
male: 76.65 deaths/1,000 live births
female: 60.28 deaths/1,000 live births (2008 est.)

Life expectancy at birth:
total population: 63.1 years
male: 60.72 years
female: 65.55 years (2008 est.)

Total fertility rate: 4.9 children born/woman (2008 est.)

HIV/AIDS—adult prevalence rate: 0.12% (2001 est.)

HIV/AIDS—people living with HIV/AIDS: NA

HIV/AIDS—deaths: NA

Nationality:
noun: Comoran(s)
adjective: Comoran

Ethnic groups: Antalote, Cafre, Makoa, Oimatsaha, Sakalava

Religions: Sunni Muslim 98%, Roman Catholic 2%

Languages: Arabic (official), French (official), Shikomoro (a blend of Swahili and Arabic)

Literacy: *definition:* age 15 and over can read and write

total population: 56.5%
male: 63.6%
female: 49.3% (2003 est.)

GOVERNMENT

Country name:
conventional long form: Union of the Comoros
conventional short form: Comoros
local long form: Union des Comores
local short form: Comores

Government type: republic

Capital: *name:* Moroni
geographic coordinates: 11 42 S, 43 14 E
time difference: UTC+3 (8 hours ahead of Washington, DC during Standard Time)

Administrative divisions: 3 islands and 4 municipalities*; Grande Comore, Anjouan, Domoni*, Fomboni*, Moheli, Moroni*, Mutsamudu*

Independence: 6 July 1975 (from France)

National holiday: Independence Day, 6 July (1975)

Constitution: 23 December 2001

Legal system: French and Islamic law in a new consolidated code; has not accepted compulsory ICJ jurisdiction

Suffrage: 18 years of age; universal

Executive branch:
chief of state: President Ahmed Abdallah SAMBI (since 26 May 2006)
head of government: President Ahmed Abdallah SAMBI (since 26 May 2006)
cabinet: Council of Ministers appointed by the president
elections: as defined by the 2001 constitution, the presidency rotates every four years among the elected presidents from the three main islands in the Union; election last held 14 May 2006 (next to be held by May 2010); prime minister appointed by the president; note—the post of prime minister has been vacant since May 2002
election results: Ahmed Abdallah SAMBI elected president; percent of vote—Ahmed Abdallah SAMBI 58.0%, Ibrahim HALIDI 28.3%, Mohamed DJAANFAMI 13.7%

Legislative branch: unicameral Assembly of the Union (33 seats; 15 deputies are selected by the individual islands' local assemblies and 18 by universal suffrage; to serve for five years);
elections: last held 18 and 25 April 2004 (next to be held in 2009)
election results: percent of vote by party—NA; seats by party—CdIA 12, CRC 6; note—15 additional seats are filled by deputies from local island assemblies

Judicial branch: Supreme Court or Cour Supremes (two members appointed by the president, two members elected by the Federal Assembly, one elected by the

Council of each island, and others are former presidents of the republic)

Political parties and leaders: Convention for the Renewal of the Comoros or CRC [AZALI Assowmani]; Camp of the Autonomous Islands or CdIA (a coalition of parties organized by the islands' presidents in opposition to the Union President); Front National pour la Justice or FNJ [Ahmed RACHID] (Islamic party in opposition); Mouvement pour la Democratie et le Progress or MDP-NGDC [Abbas DJOUSSOUF]; Parti Comorien pour la Democratie et le Progress or PCDP [Ali MROUDJAE]; Rassemblement National pour le Development or RND [Omar TAMOU, Abdoulhamid AFFRAITANE]

Political pressure groups and leaders: NA

International organization participation: ACCT, ACP, AfDB, AMF, AU, COMESA, FAO, FZ, G-77, IBRD, ICAO, ICCt, ICRM, IDA, IDB, IFAD, IFC, IFRCS, ILO, IMF, IMO, InOC, Interpol, IOC, ITSO, ITU, ITUC, LAS, NAM, OIC, OIF, OPCW (signatory), UN, UNCTAD, UNESCO, UNIDO, UPU, WCO, WHO, WIPO, WMO, WTO (observer)

Diplomatic representation in the US:
chief of mission: Representative to the US and Ambassador to the UN Mohamed TOIHIRI
chancery: Mission to the US, 336 East 45th Street (2nd floor), New York, NY 10017
telephone: [1] (212) 750-1637

Diplomatic representation from the US: the US does not have an embassy in Comoros; the ambassador to Madagascar is accredited to Comoros

Flag description: four equal horizontal bands of yellow (top), white, red, and blue with a green isosceles triangle based on the hoist; centered within the triangle is a white crescent with the convex side facing the hoist and four white, five-pointed stars placed vertically in a line between the points of the crescent; the horizontal bands and the four stars represent the four main islands of the archipelago—Mwali, Njazidja, Nzwani, and Mahore (Mayotte—territorial collectivity of France, but claimed by Comoros)
note: the crescent, stars, and color green are traditional symbols of Islam

ECONOMY

Economy—overview: One of the world's poorest countries, Comoros is made up of three islands that have inadequate transportation links, a young and rapidly increasing population, and few natural resources. The low educational level of the labor force contributes to a subsistence level of economic activity, high unemployment, and a heavy dependence on foreign grants and technical assistance. Agriculture, including fishing, hunting, and forestry, contributes 40% to GDP, employs 80% of the labor force, and provides most of the exports. The country is not self-sufficient in food production; rice, the main staple, accounts for the bulk of imports. The government—which is hampered by internal political disputes—is struggling to upgrade education and technical training, privatize commercial and industrial enterprises, improve health services, diversify exports, promote tourism, and reduce the high population growth rate. The political problems caused the economy to contract in 2007. Remittances from 150,000 Comorans abroad help supplement GDP.

GDP (purchasing power parity): $1.262 billion (2007 est.)

GDP (official exchange rate): $442 million (2007 est.)

GDP—real growth rate: -1% (2007 est.)

GDP—per capita (PPP): $1,100 (2007 est.)

GDP—composition by sector:
agriculture: 40%
industry: 4%
services: 56% (2001 est.)

Labor force: 144,500 (1996 est.)

Labor force—by occupation: *agriculture:* 80%
industry and services: 20% (1996 est.)

Unemployment rate: 20% (1996 est.)

Population below poverty line: 60% (2002 est.)

Household income or consumption by percentage share:
lowest 10%: NA%
highest 10%: NA%

Inflation rate (consumer prices): 3% (2007 est.)

Budget:
revenues: $27.6 million
expenditures: $NA (2001 est.)

Agriculture—products: vanilla, cloves, ylang-ylang, perfume essences, copra, coconuts, bananas, cassava (tapioca)

Industries: fishing, tourism, perfume distillation

Industrial production growth rate: -2% (1999 est.)

Electricity—production: 20 million kWh (2005)

Electricity—production by source:
fossil fuel: 90.6%
hydro: 9.4%
nuclear: 0%
other: 0% (2001)

Electricity—consumption: 18.6 million kWh (2005)

Electricity—exports: 0 kWh (2005)

Electricity—imports: 0 kWh (2005)

Oil—production: 0 bbl/day (2005)

Oil—consumption: 700 bbl/day (2005 est.)

Oil—exports: 0 bbl/day (2004)

Oil—imports: 709.1 bbl/day (2004)

Oil—proved reserves: 0 bbl (1 January 2006 est.)

Natural gas—production: 0 cu m (2005 est.)

Natural gas—consumption: 0 cu m (2005 est.)

Natural gas—exports: 0 cu m (2005 est.)

Natural gas—imports: 0 cu m (2005)

Natural gas—proved reserves: 0 cu m (1 January 2006 est.)

Current account balance: $8 million (2007 est.)

Exports: $32 million f.o.b. (2006)

Exports—commodities: vanilla, ylang-ylang (perfume essence), cloves, copra

Exports—partners: Netherlands 35.8%, France 18.3%, Italy 12.8%, Singapore 7.8%, Turkey 5%, US 4.6% (2006)

Imports: $143 million f.o.b. (2006)

Imports—commodities: rice and other foodstuffs, consumer goods, petroleum products, cement, transport equipment

Imports—partners: France 24.8%, UAE 9.9%, South Africa 6.4%, Pakistan 6.3%, Kenya 5%, China 4.8%, India 4.4%, Italy 4.2% (2006)

Economic aid—recipient: $25.23 million (2005 est.)

Debt—external: $232 million (2000 est.)

Currency (code): Comoran franc (KMF)

Currency code: KMF

Exchange rates: Comoran francs (KMF) per US dollar—361.4 (2007), 391.8 (2006), 395.6 (2005), 396.21 (2004), 435.9 (2003)
note: the Comoran franc is pegged to the euro at a rate of 491.9677 Comoran francs per euro

Fiscal year: calendar year

COMMUNICATIONS

Telephones—main lines in use: 16,900 (2005)

Telephones—mobile cellular: 16,100 (2005)

Telephone system:
general assessment: sparse system of microwave radio relay and HF radiotelephone communication stations; fixed-line connections only about 2 per 100 persons; mobile cellular usage about 2 per 100 persons
domestic: HF radiotelephone communications and microwave radio relay

international: country code—269; HF radiotelephone communications to Madagascar and Reunion

Radio broadcast stations: AM 1, FM 4, shortwave 1 (2001)

Radios: 90,000 (1997)

Television broadcast stations: NA

Televisions: 1,000 (1997)

Internet country code: .km

Internet hosts: 6 (2007)

Internet Service Providers (ISPs): 1 (2000)

Internet users: 21,000 (2006)

TRANSPORTATION

Airports: 4 (2007)

Airports—with paved runways:
total: 4
2,438 to 3,047 m: 1
914 to 1,523 m: 3 (2007)

Roadways:
total: 880 km
paved: 673 km
unpaved: 207 km (2000)

Merchant marine:
total: 144 ships (1000 GRT or over) 657,755 GRT/954,498 DWT
by type: bulk carrier 11, cargo 101, chemical tanker 3, container 1, livestock carrier 4, passenger 1, passenger/cargo 1, petroleum tanker 9, refrigerated cargo 6, roll on/roll off 6, specialized tanker 1
foreign-owned: 70 (Bangladesh 1, Bulgaria 1, Cyprus 1, Greece 8, India 2, Kenya 1, Kuwait 1, Lebanon 5, Norway 1, Pakistan 2, Philippines 1, Russia 9, Saudi Arabia 1, Syria 8, Turkey 8, Ukraine 13, UAE 5, US 2) (2007)

Ports and terminals: Mayotte, Mutsamudu

MILITARY

Military branches: National Development Army (AND): Comoran Security Force; Comoran Federal Police (2008)

Manpower available for military service:
males age 16–49: 167,850
females age 16–49: 167,362 (2008 est.)

Manpower fit for military service:
males age 16–49: 121,550
females age 16–49: 131,015 (2008 est.)

Military expenditures—percent of GDP: 2.8% (2006)

TRANSNATIONAL ISSUES

Disputes—international: claims French-administered Mayotte

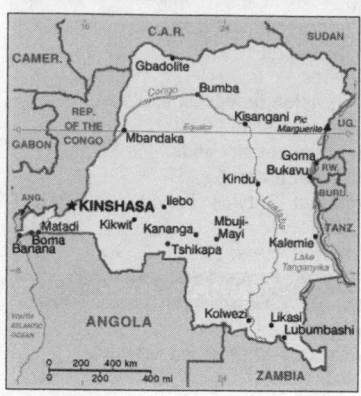

INTRODUCTION

Background: Established as a Belgian colony in 1908, the Republic of the Congo gained its independence in 1960, but its early years were marred by political and social instability. Col. Joseph MOBUTU seized power and declared himself president in a November 1965 coup. He subsequently changed his name—to MOBUTU Sese Seko—as well as that of the country—to Zaire. MOBUTU retained his position for 32 years through several sham elections, as well as through the use of brutal force. Ethnic strife and civil war, touched off by a massive inflow of refugees in 1994 from fighting in Rwanda and Burundi, led in May 1997 to the toppling of the MOBUTU regime by a rebellion backed by Rwanda and Uganda and fronted by Laurent KABILA. He renamed the country the Democratic Republic of the Congo (DRC), but in August 1998 his regime was itself challenged by a second insurrection again backed by Rwanda and Uganda. Troops from Angola, Chad, Namibia, Sudan, and Zimbabwe intervened to support KABILA's regime. A cease-fire was signed in July 1999 by the DRC, Congolese armed rebel groups, Angola, Namibia, Rwanda, Uganda, and Zimbabwe but sporadic fighting continued. Laurent KABILA was assassinated in January 2001 and his son, Joseph KABILA, was named head of state. In October 2002, the new president was successful in negotiating the withdrawal of Rwandan forces occupying eastern Congo; two months later, the Pretoria Accord was signed by all remaining warring parties to end the fighting and establish a government of national unity. A transitional government was set up in July 2003. Joseph KABILA as president and four vice presidents represented the former government, former rebel groups, the political opposition, and civil society. The transitional government held a successful constitutional referendum in December 2005 and elections for the presidency, National Assembly, and provincial legislatures in 2006. KABILA was inaugurated president in December 2006. The National Assembly was installed in September 2006. Its president, Vital KAMERHE, was chosen in December. Provincial assemblies were constituted in early 2007, and elected governors and national senators in January 2007.

GEOGRAPHY

Location: Central Africa, northeast of Angola

Geographic coordinates: 0 00 N, 25 00 E

Map references: Africa

Area:
total: 2,345,410 sq km
land: 2,267,600 sq km
water: 77,810 sq km

Area—comparative: slightly less than one-fourth the size of the US

Land boundaries:
total: 10,730 km
border countries: Angola 2,511 km (of which 225 km is the boundary of Angola's discontiguous Cabinda Province), Burundi 233 km, Central African Republic 1,577 km, Republic of the Congo 2,410 km, Rwanda 217 km, Sudan 628 km, Tanzania 459 km, Uganda 765 km, Zambia 1,930 km

Coastline: 37 km

Maritime claims:
territorial sea: 12 nm
exclusive economic zone: boundaries with neighbors

Climate: tropical; hot and humid in equatorial river basin; cooler and drier in southern highlands; cooler and wetter in eastern highlands; north of Equator—wet season (April to October), dry season (December to February); south of Equator—wet season (November to March), dry season (April to October)

Terrain: vast central basin is a low-lying plateau; mountains in east

Elevation extremes:
lowest point: Atlantic Ocean 0 m
highest point: Pic Marguerite on Mont Ngaliema (Mount Stanley) 5,110 m

Natural resources: cobalt, copper, niobium, tantalum, petroleum, industrial and gem diamonds, gold, silver, zinc, manganese, tin, uranium, coal, hydropower, timber

Land use:
arable land: 2.86%
permanent crops: 0.47%
other: 96.67% (2005)

Irrigated land: 110 sq km (2003)

Total renewable water resources: 1,283 cu km (2001)

Freshwater withdrawal (domestic/industrial/agricultural): *total:* 0.36 cu km/yr (53%/17%/31%)
per capita: 6 cu m/yr (2000)

Natural hazards: periodic droughts in south; Congo River floods (seasonal); in the east, in the Great Rift Valley, there are active volcanoes

Environment—current issues: poaching threatens wildlife populations; water pollution; deforestation; refugees responsible for significant deforestation, soil erosion, and wildlife poaching; mining of minerals (coltan—a mineral used in creating capacitors, diamonds, and gold) causing environmental damage

Environment—international agreements: *party to:* Biodiversity, Climate Change, Climate Change-Kyoto Protocol, Desertification, Endangered Species, Hazardous Wastes, Law of the Sea, Marine Dumping, Ozone Layer Protection, Tropical Timber 83, Tropical Timber 94, Wetlands
signed, but not ratified: Environmental Modification

Geography—note: straddles equator; has narrow strip of land that controls the lower Congo River and is only outlet to South Atlantic Ocean; dense tropical rain forest in central river basin and eastern highlands

PEOPLE

Population: 66,514,506
note: estimates for this country explicitly take into account the effects of excess mortality due to AIDS; this can result in lower life expectancy, higher infant mortality, higher death rates, lower population growth rates, and changes in the distribution of population by age and sex than would otherwise be expected (July 2008 est.)

Age structure:
0–14 years: 47.1% (male 15,711,817/female 15,594,449)
15–64 years: 50.4% (male 16,672,399/female 16,875,468)
65 years and over: 2.5% (male 674,766/female 985,607) (2008 est.)

Median age:
total: 16.3 years
male: 16.1 years
female: 16.5 years (2008 est.)

Population growth rate: 3.236% (2008 est.)

Birth rate: 43 births/1,000 population (2008 est.)

Death rate: 11.88 deaths/1,000 population (2008 est.)

Net migration rate: 1.24 migrant(s)/1,000 population (2008 est.)

Sex ratio:

at birth: 1.03 male(s)/female

under 15 years: 1.01 male(s)/female

15–64 years: 0.99 male(s)/female

65 years and over: 0.68 male(s)/female

total population: 0.99 male(s)/female (2008 est.)

Infant mortality rate:

total: 83.11 deaths/1,000 live births

male: 91.14 deaths/1,000 live births

female: 74.83 deaths/1,000 live births (2008 est.)

Life expectancy at birth:

total population: 53.98 years

male: 52.22 years

female: 55.8 years (2008 est.)

Total fertility rate: 6.28 children born/woman (2008 est.)

HIV/AIDS—adult prevalence rate: 4.2% (2003 est.)

HIV/AIDS—people living with HIV/AIDS: 1.1 million (2003 est.)

HIV/AIDS—deaths: 100,000 (2003 est.)

Major infectious diseases:

degree of risk: very high

food or waterborne diseases: bacterial and protozoal diarrhea, hepatitis A, and typhoid fever

vectorborne diseases: malaria, plague, and African trypanosomiasis (sleeping sickness)

water contact disease: schistosomiasis

animal contact disease: rabies (2008)

Nationality:

noun: Congolese (singular and plural)

adjective: Congolese or Congo

Ethnic groups: over 200 African ethnic groups of which the majority are Bantu; the four largest tribes—Mongo, Luba, Kongo (all Bantu), and the Mangbetu-Azande (Hamitic) make up about 45% of the population

Religions: Roman Catholic 50%, Protestant 20%, Kimbanguist 10%, Muslim 10%, other (includes syncretic sects and indigenous beliefs) 10%

Languages: French (official), Lingala (a lingua franca trade language), Kingwana (a dialect of Kiswahili or Swahili), Kikongo, Tshiluba

Literacy:

definition: age 15 and over can read and write French, Lingala, Kingwana, or Tshiluba

total population: 65.5%

male: 76.2%

female: 55.1% (2003 est.)

GOVERNMENT

Country name:

conventional long form: Democratic Republic of the Congo

conventional short form: none

local long form: Republique Democratique du Congo

local short form: none

former: Congo Free State, Belgian Congo, Congo/Leopoldville, Congo/Kinshasa, Zaire

abbreviation: DRC

Government type: republic

Capital: *name:* Kinshasa

geographic coordinates: 4 19 S, 15 18 E

time difference: UTC+1 (six hours ahead of Washington, DC during Standard Time)

Administrative divisions: 10 provinces (provinces, singular—province) and 1 city* (ville); Bandundu, Bas-Congo, Equateur, Kasai-Occidental, Kasai-Oriental, Katanga, Kinshasa*, Maniema, Nord-Kivu, Orientale, Sud-Kivu

note: according to the Constitution adopted in December 2005, the current administrative divisions will be subdivided into 26 new provinces by 2009

Independence: 30 June 1960 (from Belgium)

National holiday: Independence Day, 30 June (1960)

Constitution: 18 February 2006

Legal system: a new constitution was adopted by referendum 18 December 2005; accepts compulsory ICJ jurisdiction, with reservations

Suffrage: 18 years of age; universal and compulsory

Executive branch:

chief of state: President Joseph KABILA (since 17 January 2001); note—following the assassination of his father, Joseph KABILA succeeded to the presidency which he retained through the 2003–06 transition; he was subsequently elected president in October 2006

head of government: Prime Minister Antoine GIZENGA (since 30 December 2006);

cabinet: Ministers of State appointed by the president

elections: under the new constitution the president is elected by popular vote to a five-year term (eligible for a second term); elections last held 30 July 2006 with a second round held on 29 October 2006 (next to be held in 2011); prime minister appointed by the president

election results: results of 29 October 2006 elections (second round); Joseph KABILA 58%, Jean-Pierre BEMBA Gombo 42%

note: Joseph KABILA succeeded his father, Laurent Desire KABILA, following the latter's assassination in January 2001; negotiations with rebel leaders led to the establishment of a transitional government in July 2003 with free elections held on 30 July 2006 and 29 October 2006 confirming Joseph KABILA as president

Legislative branch: bicameral legislature consists of a National Assembly (500 seats; 61 members elected by majority vote in single-member constituencies, 439 members elected by open list proportional-representation in multi-member constituencies; to serve five-year terms) and a Senate (108 seats; members elected by provincial assemblies to serve five-year terms)

elections: National Assembly—last held 30 July 2006 (next to be held in 2011); Senate—last held 19 January 2007 (next to be held by 2012)

election results: National Assembly—percent of vote by party—NA; seats by party—PPRD 111, MLC 64, PALU 34, MSR 27, FR 26, RCD 15, independents 63, others 160 (includes 63 political parties that won 10 or fewer seats); Senate—percent of vote by party—NA; seats by party—PPRD 22, MLC 14, FR 7, RCD 7, PDC 6, CDC 3, MSR 3, PALU 2, independents 26, others 18 (political parties that won a single seat)

Judicial branch: Constitutional Court; Appeals Court or Cour de Cassation; Council of State; High Military Court; plus civil and military courts and tribunals

Political parties and leaders: Christian Democrat Party or PDC [Jose ENDUNDO]; Congolese Rally for Democracy or RCD [Azarias RUBERWA]; Convention of Christian Democrats or CDC; Forces of Renewal or FR [Mbusa NYAMWISI]; Movement for the Liberation of the Congo or MLC [Jean-Pierre BEMBA]; People's Party for Reconstruction and Democracy or PPRD [Joseph KABILA]; Social Movement for Renewal or MSR [Pierre LUMBI]; Unified Lumumbist Party or PALU [Antoine GIZENGA]; Union for Democracy and Social Progress or UDPS [Etienne TSHISEKEDI]; Union of Mobutuist Democrats or UDEMO [MOBUTU Nzanga]

Political pressure groups and leaders: NA

International organization participation: ACCT, ACP, AfDB, AU, CEPGL, COMESA, FAO, G-24, G-77, IAEA, IBRD, ICAO, ICCt, ICRM, IDA, IFAD, IFC, IFRCS, IHO (suspended), ILO, IMF, IMO, Interpol, IOC, IOM, IPU, ISO, ITSO, ITU, ITUC, MIGA, NAM,

OIF, OPCW, PCA, SADC, UN, UNCTAD, UNESCO, UNHCR, UNIDO, UNWTO, UPU, WCL, WCO, WFTU, WHO, WIPO, WMO, WTO
Diplomatic representation in the US:
chief of mission: Ambassador Faida MITIFU
chancery: 1800 New Hampshire Avenue NW, Washington, DC 20009: note—Consular Office at 1726 M Street, NW, Washington, DC, 20036
telephone: [1] (202) 234-7690, 7691
FAX: [1] (202) 234-2609
Diplomatic representation from the US:
chief of mission: Ambassador William GARVELINK
embassy: 310 Avenue des Aviateurs, Kinshasa
mailing address: Unit 31550, APO AE 09828
telephone: [243] (81) 556-0151
FAX: [243] (81) 556-0175
Flag description: sky blue field divided diagonally from the lower hoist corner to upper fly corner by a red stripe bordered by two narrow yellow stripes; a yellow, five-pointed star appears in the upper hoist corner

ECONOMY

Economy—overview: The economy of the Democratic Republic of the Congo—a nation endowed with vast potential wealth—is slowly recovering from two decades of decline. Conflict, which began in August 1998, dramatically reduced national output and government revenue, increased external debt, and resulted in the deaths of more than 3.5 million people from violence, famine, and disease. Foreign businesses curtailed operations due to uncertainty about the outcome of the conflict, lack of infrastructure, and the difficult operating environment. Conditions began to improve in late 2002 with the withdrawal of a large portion of the invading foreign troops. The transitional government reopened relations with international financial institutions and international donors, and President KABILA has begun implementing reforms, although progress is slow and the International Monetary Fund curtailed their program for the DRC at the end of March 2006 because of fiscal overruns. Much economic activity still occurs in the informal sector, and is not reflected in GDP data. Renewed activity in the mining sector, the source of most export income, boosted Kinshasa's fiscal position and GDP growth. Government reforms and improved security may lead to increased government revenues, outside budget assistance, and foreign direct investment, although an uncertain legal framework, corruption, and a lack of transparency in government policy are continuing long-term problems.
GDP (purchasing power parity): $18.84 billion (2007 est.)
GDP (official exchange rate): $10.14 billion (2007 est.)
GDP—real growth rate: 6.3% (2007 est.)
GDP—per capita (PPP): $300 (2007 est.)
GDP—composition by sector:
agriculture: 55%
industry: 11%
services: 34% (2000 est.)
Labor force: 15 million (2006 est.)
Labor force—by occupation: *agriculture:* NA%
industry: NA%
services: NA%
Unemployment rate: NA%
Population below poverty line: NA%
Household income or consumption by percentage share:
lowest 10%: NA%
highest 10%: NA%
Inflation rate (consumer prices): 16.7% (2007 est.)
Budget:
revenues: $700 million
expenditures: $2 billion (2006 est.)
Agriculture—products: coffee, sugar, palm oil, rubber, tea, quinine, cassava (tapioca), palm oil, bananas, root crops, corn, fruits; wood products
Industries: mining (diamonds, gold, copper, cobalt, coltan zinc), mineral processing, consumer products (including textiles, footwear, cigarettes, processed foods and beverages), cement, commercial ship repair
Industrial production growth rate: NA%
Electricity—production: 7.341 billion kWh (2005)
Electricity—production by source:
fossil fuel: 1.8%
hydro: 98.2%
nuclear: 0%
other: 0% (2001)
Electricity—consumption: 5.272 billion kWh (2005)
Electricity—exports: 1.8 billion kWh (2005)
Electricity—imports: 6 million kWh (2005)
Oil—production: 19,750 bbl/day (2005)
Oil—consumption: 11,000 bbl/day (2005 est.)
Oil—exports: 20,750 bbl/day (2004 est.)
Oil—imports: 8,220 bbl/day (2006 est.)
Oil—proved reserves: 187 million bbl (1 January 2006 est.)
Natural gas—production: 0 cu m (2005 est.)
Natural gas—consumption: 0 cu m (2005 est.)
Natural gas—exports: 0 cu m (2005 est.)
Natural gas—imports: 0 cu m (2005)
Natural gas—proved reserves: 950.5 million cu m (1 January 2006 est.)
Current account balance: -$402 million (2007 est.)
Exports: $1.587 billion f.o.b. (2006)
Exports—commodities: diamonds, copper, crude oil, coffee, cobalt
Exports—partners: Belgium 29.3%, China 21%, Brazil 12.3%, Chile 7.8%, Finland 7.2%, US 4.9% (2006)
Imports: $2.263 billion f.o.b. (2006)
Imports—commodities: foodstuffs, mining and other machinery, transport equipment, fuels
Imports—partners: South Africa 17.7%, Belgium 10.9%, France 8.5%, Zimbabwe 8.1%, Zambia 6.9%, Kenya 6.8%, Côte d'Ivoire 4.4% (2006)
Economic aid—recipient: $1.828 billion (2005)
Debt—external: $10 billion (2006 est.)
Market value of publicly traded shares: $NA
Currency (code): Congolese franc (CDF)
Currency code: CDF
Exchange rates: Congolese francs per US dollar—NA (2007), 464.69 (2006), 437.86 (2005), 401.04 (2004), 405.34 (2003)
Fiscal year: calendar year

COMMUNICATIONS

Telephones—main lines in use: 9,700 (2006)
Telephones—mobile cellular: 4.415 million (2006)
Telephone system:
general assessment: inadequate; state-owned fixed-line operator has been unable to expand fixed-line connections and there are now fewer than 10,000 connections; given the backdrop of a wholly inadequate fixed-line infrastructure, the use of cellular services has surged and subscribership now exceeds 4 million—roughly 7 per 100 persons
domestic: barely adequate wire and microwave radio relay service in and between urban areas; domestic satellite system with 14 earth stations
international: country code—243; satellite earth station—1 Intelsat (Atlantic Ocean) (2007)
Radio broadcast stations: AM 3, FM 11, shortwave 2 (2001)
Radios: 18.03 million (1997)
Television broadcast stations: 4 (2001)
Televisions: 6.478 million (1997)
Internet country code: .cd

Internet hosts: 2,209 (2007)
Internet Service Providers (ISPs): 1 (2001)
Internet users: 180,000 (2006)

TRANSPORTATION

Airports: 237 (2007)
Airports—with paved runways:
total: 26
over 3,047 m: 4
2,438 to 3,047 m: 2
1,524 to 2,437 m: 17
914 to 1,523 m: 2
under 914 m: 1 (2007)
Airports—with unpaved runways:
total: 211
1,524 to 2,437 m: 17
914 to 1,523 m: 95
under 914 m: 99 (2007)
Pipelines: gas 62 km; oil 71 km (2007)
Railways:
total: 5,138 km
narrow gauge: 3,987 km 1.067-m gauge (858 km electrified); 125 km 1.000-m gauge; 1,026 km 0.600-m gauge (2006)
Roadways:
total: 153,497 km
paved: 2,794 km
unpaved: 150,703 km (2004)
Waterways: 15,000 km (2005)
Merchant marine:
total: 1 ship (1000 GRT or over) 1,004 GRT/1,640 DWT
by type: petroleum tanker 1
foreign-owned: 1 (Congo, Republic of the 1) (2007)
Ports and terminals: Banana, Boma, Bukavu, Bumba, Goma, Kalemie, Kindu, Kinshasa, Kisangani, Matadi, Mbandaka Military Congo, Democratic Republic of the
Military branches: Armed Forces of the Democratic Republic of the Congo (FARDC): Army, Navy, Congolese Air Force (Force Aerienne Congolaise, FAC) (2008)
Military service age and obligation: 18–45 years of age for military service
Manpower available for military service:
males age 16–49: 14,101,263 (2008 est.)
Manpower fit for military service:
males age 16–49: 8,562,989 (2008 est.)
Military expenditures—percent of GDP: 2.5% (2006)

TRANSNATIONAL ISSUES

Disputes—international: heads of the Great Lakes states and UN pledge to abate tribal, rebel, and militia fighting in the northeastern region of the Democratic Republic of the Congo (DROC); in 2006, the UN Organization Mission in the Democratic Republic of the Congo (MONUC) maintained over 18,000 uniformed peacekeepers in the region, first deployed in 1999; despite significant repatriation efforts by governments and international organizations, in 2006, Angolans, Rwandans, Sudanese, and residents of other neighboring states reside as refugees in the DROC; members of Uganda's Lords Resistance Army forces take refuge in DROC's Garamba National Park; the location of the boundary in the broad Congo River with the Republic of the Congo is indefinite except in the Pool Malebo/Stanley Pool area
Refugees and internally displaced persons: refugees (country of origin): 132,295 (Angola); 37,313 (Rwanda); 17,777 (Burundi); 13,904 (Uganda); 6,181 (Sudan); 5,243 (Republic of Congo)
IDPs: 1.4 million (fighting between government forces and rebels since mid-1990s; most IDPs are in eastern provinces) (2007)
Illicit drugs: one of Africa's biggest producers of cannabis, but mostly for domestic consumption; while rampant corruption and inadequate supervision leaves the banking system vulnerable to money laundering, the lack of a well-developed financial system limits the country's utility as a money-laundering center

CONGO, REPUBLIC OF THE

and ushered in a period of ethnic and political unrest. Southern-based rebel groups agreed to a final peace accord in March 2003, but the calm is tenuous and refugees continue to present a humanitarian crisis. The Republic of Congo was once one of Africa's largest petroleum producers, but with declining production it will need new offshore oil finds to sustain its oil earnings over the long term.

GEOGRAPHY

Location: Western Africa, bordering the South Atlantic Ocean, between Angola and Gabon
Geographic coordinates: 1 00 S, 15 00 E
Map references: Africa
Area: total: 342,000 sq km
land: 341,500 sq km
water: 500 sq km
Area—comparative: slightly smaller than Montana
Land boundaries: total: 5,504 km
border countries: Angola 201 km, Cameroon 523 km, Central African Republic 467 km, Democratic Republic of the Congo 2,410 km, Gabon 1,903 km

INTRODUCTION

Background: Upon independence in 1960, the former French region of Middle Congo became the Republic of the Congo. A quarter century of experimentation with Marxism was abandoned in 1990 and a democratically elected government took office in 1992. A brief civil war in 1997 restored former Marxist President Denis SASSOU-NGUESSO,

Coastline: 169 km
Maritime claims: territorial sea: 200 nm
Climate: tropical; rainy season (March to June); dry season (June to October); persistent high temperatures and humidity; particularly enervating climate astride the Equator
Terrain: coastal plain, southern basin, central plateau, northern basin
Elevation extremes:
lowest point: Atlantic Ocean 0 m
highest point: Mount Berongou 903 m
Natural resources: petroleum, timber, potash, lead, zinc, uranium, copper, phosphates, gold, magnesium, natural gas, hydropower
Land use: arable land: 1.45%
permanent crops: 0.15%
other: 98.4% (2005)
Irrigated land: 20 sq km (2003)
Total renewable water resources: 832 cu km (1987)
Freshwater withdrawal (domestic/industrial/agricultural): total: 0.03 cu km/yr (59%/29%/12%)
per capita: 8 cu m/yr (2000)
Natural hazards: seasonal flooding

155

Environment—current issues: air pollution from vehicle emissions; water pollution from the dumping of raw sewage; tap water is not potable; deforestation
Environment—international agreements: *party to:* Biodiversity, Climate Change, Climate Change-Kyoto Protocol, Desertification, Endangered Species, Ozone Layer Protection, Ship Pollution, Tropical Timber 83, Tropical Timber 94, Wetlands
signed, but not ratified: Law of the Sea
Geography—note: about 70% of the population lives in Brazzaville, Pointe-Noire, or along the railroad between them

PEOPLE

Population: 3,903,318
note: estimates for this country explicitly take into account the effects of excess mortality due to AIDS; this can result in lower life expectancy, higher infant mortality, higher death rates, lower population growth rates, and changes in the distribution of population by age and sex than would otherwise be expected (July 2008 est.)
Age structure:
0–14 years: 46.1% (male 906,345/female 894,568)
15–64 years: 51% (male 989,126/female 1,002,682)
65 years and over: 2.8% (male 45,560/female 65,037) (2008 est.)
Median age:
total: 16.7 years
male: 16.5 years
female: 17 years (2008 est.)
Population growth rate: 2.696% (2008 est.)
Birth rate: 41.76 births/1,000 population (2008 est.)
Death rate: 12.28 deaths/1,000 population (2008 est.)
Net migration rate: -2.52 migrant(s)/1,000 population (2008 est.)
Sex ratio:
at birth: 1.03 male(s)/female
under 15 years: 1.01 male(s)/female
15–64 years: 0.99 male(s)/female
65 years and over: 0.7 male(s)/female
total population: 0.99 male(s)/female (2008 est.)
Infant mortality rate:
total: 81.29 deaths/1,000 live births
male: 86.9 deaths/1,000 live births
female: 75.51 deaths/1,000 live births (2008 est.)
Life expectancy at birth:
total population: 53.74 years
male: 52.52 years
female: 55 years (2008 est.)
Total fertility rate: 5.92 children born/woman (2008 est.)

HIV/AIDS—adult prevalence rate: 4.9% (2003 est.)
HIV/AIDS—people living with HIV/AIDS: 90,000 (2003 est.)
HIV/AIDS—deaths: 9,700 (2003 est.)
Major infectious diseases:
degree of risk: very high
food or waterborne diseases: bacterial and protozoal diarrhea, hepatitis A, and typhoid fever
vectorborne disease: malaria and African trypanosomiasis (sleeping sickness)
animal contact disease: rabies (2008)
Nationality:
noun: Congolese (singular and plural)
adjective: Congolese or Congo
Ethnic groups: Kongo 48%, Sangha 20%, M'Bochi 12%, Teke 17%, Europeans and other 3%
Religions: Christian 50%, animist 48%, Muslim 2%
Languages: French (official), Lingala and Monokutuba (lingua franca trade languages), many local languages and dialects (of which Kikongo is the most widespread)
Literacy: *definition:* age 15 and over can read and write
total population: 83.8%
male: 89.6%
female: 78.4% (2003 est.)

GOVERNMENT

Country name:
conventional long form: Republic of the Congo
conventional short form: Congo (Brazzaville)
local long form: Republique du Congo
local short form: none
former: Middle Congo, Congo/Brazzaville, Congo
Government type: republic
Capital: *name:* Brazzaville
geographic coordinates: 4 15 S, 15 17 E
time difference: UTC+1 (six hours ahead of Washington, DC during Standard Time)
Administrative divisions: 10 regions (regions, singular—region) and 1 commune*; Bouenza, Brazzaville*, Cuvette, Cuvette-Ouest, Kouilou, Lekoumou, Likouala, Niari, Plateaux, Pool, Sangha
Independence: 15 August 1960 (from France)
National holiday: Independence Day, 15 August (1960)
Constitution: approved by referendum 20 January 2002
Legal system: based on French civil law system and customary law; has not accepted compulsory ICJ jurisdiction
Suffrage: 18 years of age; universal
Executive branch:
chief of state: President Denis SASSOU-

NGUESSO (since 25 October 1997, following the civil war in which he toppled elected president Pascal LISSOUBA);
head of government: Prime Minister Isidore MVOUBA (since 7 January 2005)
cabinet: Council of Ministers appointed by the president
elections: president elected by popular vote for a seven-year term (eligible for a second term); election last held 10 March 2002 (next to be held in 2009)
election results: Denis SASSOU-NGUESSO reelected president; percent of vote—Denis SASSOU-NGUESSO 89.4%, Joseph Kignoumbi Kia MBOUNGOU 2.7%
Legislative branch: bicameral Parliament consists of the Senate (66 seats; members are elected by indirect vote to serve five-year terms) and the National Assembly (137 seats; members are elected by popular vote to serve five-year terms)
elections: Senate—last held 11 July 2002 (next to be held in July 2008); National Assembly—last held 24 June and 5 August 2007 (next to be held in 2012)
election results: Senate—percent of vote by party—NA; seats by party—FDU 56, other 10; National Assembly—percent of vote by party—NA; seats by party—PCT 46, MCDDI 11, UPADS 11, MAR 5, MSD 5, independents 37, other 22
Judicial branch: Supreme Court or Cour Supreme
Political parties and leaders: Action Movement for Renewal or MAR; Congolese Movement for Democracy and Integral Development or MCDDI [Michel MAMPOUYA]; Congolese Labour Party or PCT; Movement for Solidarity and Development or MSD; Pan-African Union for Social Development or UPADS [Martin MBERI]; Rally for Democracy and Social Progress or RDPS [Jean-Pierre Thystere TCHICAYA, president]; Rally for Democracy and the Republic or RDR [Raymond Damasge NGOLLO]; Union for Democracy and Republic or UDR; United Democratic Forces or FDU [Sebastian EBAO]; many less important parties
Political pressure groups and leaders: Congolese Trade Union Congress or CSC; General Union of Congolese Pupils and Students or UGEEC; Revolutionary Union of Congolese Women or URFC; Union of Congolese Socialist Youth or UJSC
International organization participation: ACCT, ACP, AfDB, AU, BDEAC, CEMAC, FAO, FZ, G-77, IBRD, ICAO, ICCt, ICRM, IDA, IFAD, IFC, IFRCS,

ILO, IMF, IMO, Interpol, IOC, IOM, IPU, ITSO, ITU, ITUC, MIGA, NAM, OIF, OPCW, UN, UNCTAD, UNESCO, UNIDO, UNWTO, UPU, WCL, WCO, WFTU, WHO, WIPO, WMO, WTO

Diplomatic representation in the US:
chief of mission: Ambassador Serge MOMBOULI
chancery: 4891 Colorado Avenue NW, Washington, DC 20011
telephone: [1] (202) 726-5500
FAX: [1] (202) 726-1860

Diplomatic representation from the US:
chief of mission: Ambassador Robert WEISBERG
embassy: BDEAC Building, 4th Floor, Brazzaville note—a new Embassy is expected to open in 2009
mailing address: NA
telephone: [242] 81-1480
FAX:: [243] 81-5324; note—until the new embassy in Brazzaville is operational, some duties are still handled in the US embassy in Kinshasha, Democratic Republic of the Congo

Flag description: divided diagonally from the lower hoist side by a yellow band; the upper triangle (hoist side) is green and the lower triangle is red
note: uses the popular pan-African colors of Ethiopia

ECONOMY

Economy—overview: The economy is a mixture of subsistance agriculture, an industrial sector based largely on oil, and support services, and a government characterized by budget problems and over-staffing. Oil has supplanted forestry as the mainstay of the economy, providing a major share of government revenues and exports. In the early 1980s, rapidly rising oil revenues enabled the government to finance large-scale development projects with GDP growth averaging 5% annually, one of the highest rates in Africa. The government has mortgaged a substantial portion of its oil earnings through oil-backed loans that have contributed to a growing debt burden and chronic revenue shortfalls. Economic reform efforts have been undertaken with the support of international organizations, notably the World Bank and the IMF. However, the reform program came to a halt in June 1997 when civil war erupted. Denis SASSOU-NGUESSO, who returned to power when the war ended in October 1997, publicly expressed interest in moving forward on economic reforms and privatization and in renewing cooperation with international financial institutions. Economic progress was badly hurt by slumping oil

prices and the resumption of armed conflict in December 1998, which worsened the republic's budget deficit. The current administration presides over an uneasy internal peace and faces difficult economic challenges of stimulating recovery and reducing poverty. Recovery of oil prices has boosted the economy's GDP and near-term prospects. In March 2006, the World Bank and the International Monetary Fund (IMF) approved Heavily Indebted Poor Countries (HIPC) treatment for Congo.

GDP (purchasing power parity): $13.23 billion (2007 est.)
GDP (official exchange rate): $7.657 billion (2007 est.)
GDP—real growth rate: -1.6% (2007 est.)
GDP—per capita (PPP): $3,700 (2007 est.)
GDP—composition by sector:
agriculture: 5.6%
industry: 57.1%
services: 37.3% (2006 est.)
Labor force: NA
Unemployment rate: NA%
Population below poverty line: NA%
Household income or consumption by percentage share:
lowest 10%: NA%
highest 10%: NA%
Inflation rate (consumer prices): 2.6% (2007 est.)
Investment (gross fixed): 30.8% of GDP (2007 est.)
Budget:
revenues: $3.522 billion
expenditures: $2.377 billion (2007 est.)
Agriculture—products: cassava (tapioca), sugar, rice, corn, peanuts, vegetables, coffee, cocoa; forest products
Industries: petroleum extraction, cement, lumber, brewing, sugar, palm oil, soap, flour, cigarettes
Industrial production growth rate: -6% (2007 est.)
Electricity—production: 352 million kWh (2005)
Electricity—production by source:
fossil fuel: 0.3%
hydro: 99.7%
nuclear: 0%
other: 0% (2001)
Electricity—consumption: 572 million kWh (2005)
Electricity—exports: 0 kWh (2005)
Electricity—imports: 418 million kWh (2005)
Oil—production: 235,900 bbl/day
Oil—consumption: 7,000 bbl/day (2005 est.)
Oil—exports: 229,700 bbl/day (2004 est.)
Oil—imports: 11,410 bbl/day (2004)

Oil—proved reserves: 1.506 billion bbl (1 January 2006 est.)
Natural gas—production: 115.1 million cu m (2005 est.)
Natural gas—consumption: 115.1 million cu m (2005 est.)
Natural gas—exports: 0 cu m (2005 est.)
Natural gas—imports: 0 cu m (2005)
Natural gas—proved reserves: 86.9 billion cu m (1 January 2006 est.)
Current account balance: -$1.49 billion (2007 est.)
Exports: $6.251 billion f.o.b. (2007 est.)
Exports—commodities: petroleum, lumber, plywood, sugar, cocoa, coffee, diamonds
Exports—partners: US 35.9%, China 31.4%, Taiwan 9.9%, South Korea 8% (2006)
Imports: $1.762 billion f.o.b. (2007 est.)
Imports—commodities: capital equipment, construction materials, foodstuffs
Imports—partners: France 23.5%, China 13.2%, US 7.6%, India 7%, Italy 5.6%, Belgium 5.3% (2006)
Economic aid—recipient: $1.449 billion (2005)
Reserves of foreign exchange and gold: $2.197 billion (31 December 2007 est.)
Debt—external: $5 billion (2000 est.)
Market value of publicly traded shares: $NA
Currency (code): Communaute Financiere Africaine franc (XAF); note—responsible authority is the Bank of the Central African States
Currency code: XAF
Exchange rates: Communaute Financiere Africaine francs (XAF) per US dollar—483.6 (2007), 522.59 (2006), 527.47 (2005), 528.29 (2004), 581.2 (2003)
Fiscal year: calendar year

COMMUNICATIONS

Telephones—main lines in use: 15,900 (2005)
Telephones—mobile cellular: 490,000 (2005)
Telephone system:
general assessment: services barely adequate for government use; key exchanges are in Brazzaville, Pointe-Noire, and Loubomo; intercity lines frequently out of order; fixed-line infrastructure inadequate providing less than 1 connection per 100 persons; mobile cellular subscribership has surged reaching 16 per 100 persons
domestic: primary network consists of microwave radio relay and coaxial cable
international: country code—242; satellite earth station—1 Intelsat (Atlantic Ocean)

Radio broadcast stations: AM 1, FM 5, shortwave 3 (2001)
Radios: 341,000 (1997)
Television broadcast stations: 1 (2001)
Televisions: 33,000 (1997)
Internet country code: .cg
Internet hosts: 3 (2007)
Internet Service Providers (ISPs): 1 (2000)
Internet users: 70,000 (2006)

Airports: 31 (2007)
Airports—with paved runways:
total: 5
over 3,047 m: 2
2,438 to 3,047 m: 1
1,524 to 2,437 m: 2 (2007)
Airports—with unpaved runways:
total: 26
1,524 to 2,437 m: 7
914 to 1,523 m: 10
under 914 m: 9 (2007)
Pipelines: gas 89 km; liquid petroleum gas 4 km; oil 758 km (2007)
Railways:
total: 894 km

narrow gauge: 894 km 1.067-m gauge (2006)
Roadways:
total: 17,289 km
paved: 864 km
unpaved: 16,425 km (2004)
Waterways: 1,125 km (commercially navigable on Congo and Oubanqui rivers) (2006)
Merchant marine:
registered in other countries: 1 (Congo, Democratic Republic of the 1) (2007)
Ports and terminals: Brazzaville, Djeno, Impfondo, Ouesso, Oyo, Pointe-Noire

Military branches: Congolese Armed Forces (Forces Armees Congolaises, FAC): Army, Navy, Congolese Air Force (Armee de l'Air Congolaise), Gendarmerie, Special Presidential Security Guard (GSSP) (2008)
Military service age and obligation: 18 years of age for voluntary military service; women allowed to serve (2007)
Manpower available for military service: *males age 16–49:* 842,771

females age 16–49: 833,624 (2008 est.)
Manpower fit for military service:
males age 16–49: 519,296
females age 16–49: 509,564 (2008 est.)
Manpower reaching militarily significant age annually:
males age 16–49: 45,671
females age 16–49: 45,248 (2008 est.)
Military expenditures—percent of GDP: 3.1% (2006)

Disputes—international: Congo hosts about 63,000 refugees from neighboring states, primarily from the Pool border area of the Democratic Republic of the Congo; the location of the boundary in the broad Congo River with the Democratic Republic of the Congo is indefinite except in the Pool Malebo/Stanley Pool area
Refugees and internally displaced persons: *refugees (country of origin):* 46,341 (Democratic Republic of Congo); 6,564 (Rwanda)
IDPs: 48,000 (multiple civil wars since 1992; most IDPs are ethnic Lari) (2007)

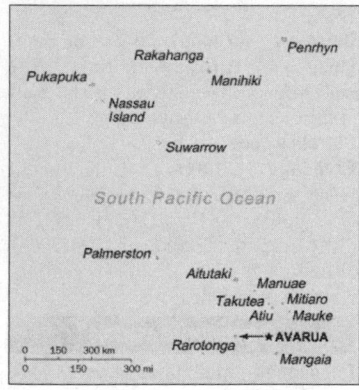

Background: Named after Captain COOK, who sighted them in 1770, the islands became a British protectorate in 1888. By 1900, administrative control was transferred to New Zealand; in 1965, residents chose self-government in free association with New Zealand. The emigration of skilled workers to New Zealand and government deficits are continuing problems.

Location: Oceania, group of islands in the South Pacific Ocean, about half way between Hawaii and New Zealand

Geographic coordinates: 21 14 S, 159 46 W
Map references: Oceania
Area:
total: 236.7 sq km
land: 236.7 sq km
water: 0 sq km
Area—comparative: 1.3 times the size of Washington, DC
Land boundaries: 0 km
Coastline: 120 km
Maritime claims:
territorial sea: 12 nm
exclusive economic zone: 200 nm
continental shelf: 200 nm or to the edge of the continental margin
Climate: tropical oceanic; moderated by trade winds; a dry season from April to November and a more humid season from December to March
Terrain: low coral atolls in north; volcanic, hilly islands in south
Elevation extremes:
lowest point: Pacific Ocean 0 m
highest point: Te Manga 652 m
Natural resources: NEGL
Land use:
arable land: 16.67%
permanent crops: 8.33%
other: 75% (2005)
Irrigated land: NA
Natural hazards: typhoons (November to March)

Environment—current issues: NA
Environment—international agreements: *party to:* Biodiversity, Climate Change, Climate Change-Kyoto Protocol, Desertification, Hazardous Wastes, Law of the Sea, Ozone Layer Protection
signed, but not ratified: none of the selected agreements
Geography—note: the northern Cook Islands are seven low-lying, sparsely populated, coral atolls; the southern Cook Islands, where most of the population lives, consist of eight elevated, fertile, volcanic isles, including the largest, Rarotonga, at 67 sq km

Population: 21,923 (July 2008 est.)
Age structure:
0–14 years: NA
15–64 years: NA
65 years and over: NA
Median age:
total: 25.3 years
male: 24.7 years
female: 25.9 years (2001 census)
Population growth rate: NA
Birth rate: 21 births/1,000 population (2001 census)
Death rate: NA (2008 est.)
Sex ratio: *total population:* 1.07 male(s)/female (2001 census)

Infant mortality rate:
total: NA
male: NA
female: NA (2008 est.)
Life expectancy at birth:
total population: NA
male: NA
female: NA (2008 est.)
Total fertility rate: 3.1 children born/woman (2008 est.)
HIV/AIDS—adult prevalence rate: NA
HIV/AIDS—people living with HIV/AIDS: NA
HIV/AIDS—deaths: NA
Nationality:
noun: Cook Islander(s)
adjective: Cook Islander
Ethnic groups: Cook Island Maori (Polynesian) 87.7%, part Cook Island Maori 5.8%, other 6.5% (2001 census)
Religions: Cook Islands Christian Church 55.9%, Roman Catholic 16.8%, Seventh-Day Adventists 7.9%, Church of Latter Day Saints 3.8%, other Protestant 5.8%, other 4.2%, unspecified 2.6%, none 3% (2001 census)
Languages: English (official), Maori
Literacy:
definition: NA
total population: 95%
male: NA%
female: NA%
People—note: 2001 census counted a resident population of 15,017

GOVERNMENT

Country name:
conventional long form: none
conventional short form: Cook Islands
former: Harvey Islands
Dependency status: self-governing in free association with New Zealand; Cook Islands is fully responsible for internal affairs; New Zealand retains responsibility for external affairs and defense, in consultation with the Cook Islands
Government type: self-governing parliamentary democracy
Capital: *name:* Avarua
geographic coordinates: 21 12 S, 159 46 W
time difference: UTC-10 (5 hours behind Washington, DC during Standard Time)
Administrative divisions: none
Independence: none (became self-governing in free association with New Zealand on 4 August 1965 and has the right at any time to move to full independence by unilateral action)
National holiday: Constitution Day, first Monday in August (1965)
Constitution: 4 August 1965
Legal system: based on New Zealand law and English common law
Suffrage: NA years of age; universal (adult)

Executive branch:
chief of state: Queen ELIZABETH II (since 6 February 1952); represented by Frederick GOODWIN (since 9 February 2001); New Zealand High Commissioner Brian DONNELLY (since 21 February 2008), representative of New Zealand
head of government: Prime Minister Jim MARURAI (since 14 December 2004); Deputy Prime Minister Terepai MAOATE (since 9 August 2005)
cabinet: Cabinet chosen by the prime minister; collectively responsible to Parliament
elections: the monarch is hereditary; the UK representative is appointed by the monarch; the New Zealand high commissioner is appointed by the New Zealand Government; following legislative elections, the leader of the majority party or the leader of the majority coalition usually becomes prime minister
Legislative branch: bicameral Parliament consisting of a Legislative Assembly (or lower house) (24 seats; members elected by popular vote to serve four-year terms) and a House of Ariki (or upper house) made up of traditional leaders
note: the House of Ariki advises on traditional matters and maintains considerable influence but has no legislative powers
elections: last held 26 September 2006 (next to be held by 2011)
election results: percent of vote by party—Demo 51.9%, CIP 45.5%, independent 2.7%; seats by party—Demo 15, CIP 8, independent 1
Judicial branch: High Court
Political parties and leaders: Cook Islands Party or CIP [Henry PUNA]; Democratic Party or Demo [Dr. Terepai MAOATE]
Political pressure groups and leaders: NA
International organization participation: ACP, ADB, FAO, ICAO, ICRM, IFAD, IFRCS, IOC, ITUC, OPCW, PIF, Sparteca, SPC, UNESCO, UPU, WHO, WMO
Diplomatic representation in the US: none (self-governing in free association with New Zealand)
Diplomatic representation from the US: none (self-governing in free association with New Zealand)
Flag description: blue, with the flag of the UK in the upper hoist-side quadrant and a large circle of 15 white five-pointed stars (one for every island) centered in the outer half of the flag

ECONOMY

Economy—overview: Like many other South Pacific island nations, the Cook Islands' economic development is hindered by the isolation of the country from foreign markets, the limited size of domestic markets, lack of natural resources, periodic devastation from natural disasters, and inadequate infrastructure. Agriculture, employing about one-third of the working population, provides the economic base with major exports made up of copra and citrus fruit. Black pearls are the Cook Islands' leading export. Manufacturing activities are limited to fruit processing, clothing, and handicrafts. Trade deficits are offset by remittances from emigrants and by foreign aid, overwhelmingly from New Zealand. In the 1980s and 1990s, the country lived beyond its means, maintaining a bloated public service and accumulating a large foreign debt. Subsequent reforms, including the sale of state assets, the strengthening of economic management, the encouragement of tourism, and a debt restructuring agreement, have rekindled investment and growth.
GDP (purchasing power parity): $183.2 million (2005 est.)
GDP (official exchange rate): $183.2 million (2005 est.)
GDP—real growth rate: 0.1% (2005 est.)
GDP—per capita (PPP): $9,100 (2005 est.)
GDP—composition by sector:
agriculture: 15.1%
industry: 9.6%
services: 75.3% (2004)
Labor force: 6,820 (2001)
Labor force—by occupation: *agriculture:* 29%
industry: 15%
services: 56% (1995)
Unemployment rate: 13.1% (2005)
Population below poverty line: NA%
Household income or consumption by percentage share:
lowest 10%: NA%
highest 10%: NA%
Inflation rate (consumer prices): 2.1% (2005 est.)
Budget:
revenues: $70.95 million
expenditures: $69.05 million (FY05/06)
Agriculture—products: copra, citrus, pineapples, tomatoes, beans, pawpaws, bananas, yams, taro, coffee; pigs; poultry
Industries: fruit processing, tourism, fishing, clothing, handicrafts
Industrial production growth rate: 1% (2002)
Electricity—production: 30 million kWh (2005)
Electricity—production by source:
fossil fuel: 100%

hydro: 0%
nuclear: 0%
other: 0% (2001)
Electricity—consumption: 27.9 million kWh (2005 est.)
Electricity—exports: 0 kWh (2005)
Electricity—imports: 0 kWh (2005)
Oil—production: 0 bbl/day (2005)
Oil—consumption: 450 bbl/day (2005 est.)
Oil—exports: 0 bbl/day (2004)
Oil—imports: 429.3 bbl/day (2004)
Oil—proved reserves: 0 bbl (1 January 2006 est.)
Natural gas—production: 0 cu m (2005 est.)
Natural gas—consumption: 0 cu m (2005 est.)
Natural gas—exports: 0 cu m (2005 est.)
Natural gas—imports: 0 cu m (2005)
Natural gas—proved reserves: 0 cu m (1 January 2006 est.)
Current account balance: $26.67 million (2005)
Exports: $5.222 million (2005)
Exports—commodities: copra, papayas, fresh and canned citrus fruit, coffee; fish; pearls and pearl shells; clothing
Exports—partners: Australia 34%, Japan 27%, NZ 25%, US 8% (2006)
Imports: $81.04 million (2005)
Imports—commodities: foodstuffs, textiles, fuels, timber, capital goods
Imports—partners: NZ 61%, Fiji 19%, US 9%, Australia 6% (2006)
Economic aid—recipient: $13.1 million; note—New Zealand continues to furnish the greater part (1995)

Debt—external: $141 million (1996 est.)
Currency (code): NZ dollar (NZD)
Currency code: NZD
Exchange rates: NZ dollars per US dollar—1.3811 (2007), 1.5408 (2006), 1.4203 (2005), 1.5087 (2004), 1.7221 (2003)
Fiscal year: 1 April—31 March

COMMUNICATIONS

Telephones—main lines in use: 6,200 (2002)
Telephones—mobile cellular: 1,500 (2002)
Telephone system:
general assessment: Telecom Cook Islands offers international direct dialing, Internet, email, fax, and Telex
domestic: individual islands are connected by a combination of satellite earth stations, microwave systems, and VHF and HF radiotelephone; within the islands, service is provided by small exchanges connected to subscribers by open-wire, cable, and fiber-optic cable
international: country code—682; satellite earth station—1 Intelsat (Pacific Ocean)
Radio broadcast stations: AM 1, FM 1, shortwave 0 (2004)
Radios: 14,000 (1997)
Television broadcast stations: 1 (outer islands receive satellite broadcasts) (2004)
Televisions: 4,000 (1997)
Internet country code: .ck
Internet hosts: 1,479 (2007)

Internet Service Providers (ISPs): 3 (2000)
Internet users: 3,600 (2002)

TRANSPORTATION

Airports: 9 (2007)
Airports—with paved runways:
total: 2
1,524 to 2,437 m: 2 (2007)
Airports—with unpaved runways:
total: 7
1,524 to 2,437 m: 2
914 to 1,523 m: 4
under 914 m: 1 (2007)
Roadways:
total: 320 km
paved: 33 km
unpaved: 287 km (2003)
Merchant marine:
total: 16 ships (1000 GRT or over) 112,129 GRT/126,160 DWT
by type: cargo 5, petroleum tanker 1, refrigerated cargo 9, roll on/roll off 1
foreign-owned: 11 (Norway 1, NZ 1, Sweden 9) (2007)
Ports and terminals: Avatiu

MILITARY

Military branches: no regular military forces; National Police Department (2007)
Military—note: defense is the responsibility of New Zealand, in consultation with the Cook Islands and at its request

TRANSNATIONAL ISSUES

Disputes—international: none

CORAL SEA ISLANDS

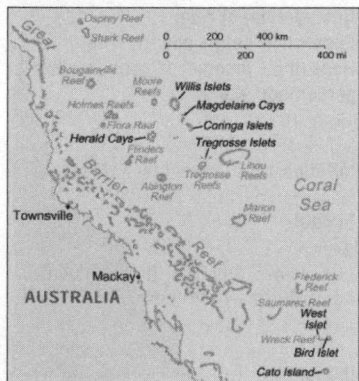

INTRODUCTION

Background: Scattered over more than three-quarters of a million square kilometers of ocean, the Coral Sea Islands were declared a territory of Australia in 1969. They are uninhabited except for a small meteorological staff on the Willis Islets. Automated weather stations, beacons, and a lighthouse occupy many other islands and reefs.

GEOGRAPHY

Location: Oceania, islands in the Coral Sea, northeast of Australia
Geographic coordinates: 18 00 S, 152 00 E
Map references: Oceania
Area:
total: less than 3 sq km
land: less than 3 sq km
water: 0 sq km
note: includes numerous small islands and reefs scattered over a sea area of about 780,000 sq km, with the Willis Islets the most important
Area—comparative: NA
Land boundaries: 0 km

Coastline: 3,095 km
Maritime claims:
territorial sea: 3 nm
exclusive fishing zone: 200 nm
Climate: tropical
Terrain: sand and coral reefs and islands (or cays)
Elevation extremes:
lowest point: Pacific Ocean 0 m
highest point: unnamed location on Cato Island 6 m
Natural resources: NEGL
Land use:
arable land: 0%
permanent crops: 0%
other: 100% (mostly grass or scrub cover) (2005)
Irrigated land: 0 sq km
Natural hazards: occasional tropical cyclones
Environment—current issues: no permanent fresh water resources

Geography—note: important nesting area for birds and turtles

PEOPLE

Population: no indigenous inhabitants *note:* there is a staff of three to four at the meteorological station on Willis Island (July 2007 est.)

GOVERNMENT

Country name:
conventional long form: Coral Sea Islands Territory
conventional short form: Coral Sea Islands
Dependency status: territory of Australia; administered from Canberra by the Australian Attorney-General's Department

Legal system: the laws of Australia, where applicable, apply
Executive branch: administered from Canberra by the Australian Attorney-General's Department
Diplomatic representation in the US: none (territory of Australia)
Diplomatic representation from the US: none (territory of Australia)
Flag description: the flag of Australia is used

ECONOMY

Economy—overview: no economic activity

COMMUNICATIONS

Communications—note: there are automatic weather stations on many of the isles and reefs relaying data to the mainland

TRANSPORTATION

Ports and terminals: none; offshore anchorage only

MILITARY

Military—note: defense is the responsibility of Australia

TRANSNATIONAL ISSUES

Disputes—international: none

COSTA RICA

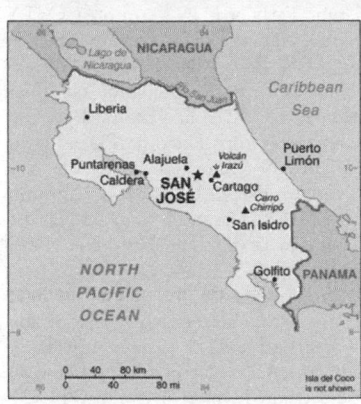

INTRODUCTION

Background: Although explored by the Spanish early in the 16th century, initial attempts at colonizing Costa Rica proved unsuccessful due to a combination of factors, including: disease from mosquito-infested swamps, brutal heat, resistance by natives, and pirate raids. It was not until 1563 that a permanent settlement of Cartago was established in the cooler, fertile central highlands. The area remained a colony for some two and a half centuries. In 1821, Costa Rica became one of several Central American provinces that jointly declared their independence from Spain. Two years later it joined the United Provinces of Central America, but this federation disintegrated in 1838, at which time Costa Rica proclaimed its sovereignty and independence. Since the late 19th century, only two brief periods of violence have marred the country's democratic development. Although it still maintains a large agricultural sector, Costa Rica has

expanded its economy to include strong technology and tourism industries. The standard of living is relatively high. Land ownership is widespread.

GEOGRAPHY

Location: Central America, bordering both the Caribbean Sea and the North Pacific Ocean, between Nicaragua and Panama
Geographic coordinates: 10 00 N, 84 00 W
Map references: Central America and the Caribbean
Area:
total: 51,100 sq km
land: 50,660 sq km
water: 440 sq km
note: includes Isla del Coco
Area—comparative: slightly smaller than West Virginia
Land boundaries:
total: 639 km
border countries: Nicaragua 309 km, Panama 330 km
Coastline: 1,290 km
Maritime claims:
territorial sea: 12 nm
exclusive economic zone: 200 nm
continental shelf: 200 nm
Climate: tropical and subtropical; dry season (December to April); rainy season (May to November); cooler in highlands
Terrain: coastal plains separated by rugged mountains including over 100 volcanic cones, of which several are major volcanoes
Elevation extremes:
lowest point: Pacific Ocean 0 m
highest point: Cerro Chirripo 3,810 m
Natural resources: hydropower
Land use: *arable land:* 4.4%

permanent crops: 5.87%
other: 89.73% (2005)
Irrigated land: 1,080 sq km (2003)
Total renewable water resources: 112.4 cu km (2000)
Freshwater withdrawal (domestic/industrial/agricultural): *total:* 2.68 cu km/yr (29%/17%/53%)
per capita: 619 cu m/yr (2000)
Natural hazards: occasional earthquakes, hurricanes along Atlantic coast; frequent flooding of lowlands at onset of rainy season and landslides; active volcanoes
Environment—current issues: deforestation and land use change, largely a result of the clearing of land for cattle ranching and agriculture; soil erosion; coastal marine pollution; fisheries protection; solid waste management; air pollution
Environment—international agreements: *party to:* Biodiversity, Climate Change, Climate Change-Kyoto Protocol, Desertification, Endangered Species, Environmental Modification, Hazardous Wastes, Law of the Sea, Marine Dumping, Ozone Layer Protection, Wetlands, Whaling
signed, but not ratified: Marine Life Conservation
Geography—note: four volcanoes, two of them active, rise near the capital of San Jose in the center of the country; one of the volcanoes, Irazu, erupted destructively in 1963–65

PEOPLE

Population: 4,195,914 (July 2008 est.)
Age structure:
0–14 years: 27.2% (male 584,782/female 557,952)
15–64 years: 66.8% (male 1,416,456/female 1,384,692)

65 years and over: 6% (male 116,461/female 135,571) (2008 est.)
Median age:
total: 27.1 years
male: 26.7 years
female: 27.6 years (2008 est.)
Population growth rate: 1.388% (2008 est.)
Birth rate: 17.71 births/1,000 population (2008 est.)
Death rate: 4.31 deaths/1,000 population (2008 est.)
Net migration rate: 0.48 migrant(s)/1,000 population (2008 est.)
Sex ratio:
at birth: 1.05 male(s)/female
under 15 years: 1.05 male(s)/female
15–64 years: 1.02 male(s)/female
65 years and over: 0.86 male(s)/female
total population: 1.02 male(s)/female (2008 est.)
Infant mortality rate:
total: 9.01 deaths/1,000 live births
male: 9.92 deaths/1,000 live births
female: 8.05 deaths/1,000 live births (2008 est.)
Life expectancy at birth:
total population: 77.4 years
male: 74.79 years
female: 80.14 years (2008 est.)
Total fertility rate: 2.17 children born/woman (2008 est.)
HIV/AIDS—adult prevalence rate: 0.6% (2003 est.)
HIV/AIDS—people living with HIV/AIDS: 12,000 (2003 est.)
HIV/AIDS—deaths: 900 (2003 est.)
Major infectious diseases:
degree of risk: intermediate
food or waterborne diseases: bacterial diarrhea and hepatitis A
vectorborne diseases: dengue fever (2008)
Nationality:
noun: Costa Rican(s)
adjective: Costa Rican
Ethnic groups: white (including mestizo) 94%, black 3%, Amerindian 1%, Chinese 1%, other 1%
Religions: Roman Catholic 76.3%, Evangelical 13.7%, Jehovah's Witnesses 1.3%, other Protestant 0.7%, other 4.8%, none 3.2%
Languages: Spanish (official), English
Literacy:
definition: age 15 and over can read and write
total population: 96%
male: 95.9%
female: 96.1% (2003 est.)

GOVERNMENT

Country name:
conventional long form: Republic of Costa Rica
conventional short form: Costa Rica
local long form: Republica de Costa Rica
local short form: Costa Rica
Government type: democratic republic
Capital: *name:* San Jose
geographic coordinates: 9 56 N, 84 05 W
time difference: UTC-6 (1 hour behind Washington, DC during Standard Time)
Administrative divisions: 7 provinces (provincias, singular—provincia); Alajuela, Cartago, Guanacaste, Heredia, Limon, Puntarenas, San Jose
Independence: 15 September 1821 (from Spain)
National holiday: Independence Day, 15 September (1821)
Constitution: 7 November 1949
Legal system: based on Spanish civil law system; judicial review of legislative acts in the Supreme Court; has accepted compulsory ICJ jurisdiction
Suffrage: 18 years of age; universal and compulsory
Executive branch:
chief of state: President Oscar ARIAS Sanchez (since 8 May 2006); First Vice President Laura CHINCHILLA (since 8 May 2006); Second Vice President (vacant); note—the president is both the chief of state and head of government
head of government: President Oscar ARIAS Sanchez (since 8 May 2006); First Vice President Laura CHINCHILLA (since 8 May 2006); Second Vice President (vacant)
cabinet: Cabinet selected by the president
elections: president and vice presidents elected on the same ticket by popular vote for a single four-year term; election last held 5 February 2006 (next to be held in February 2010)
election results: Oscar ARIAS Sanchez elected president; percent of vote—Oscar ARIAS Sanchez (PLN) 40.9%; Otton SOLIS (PAC) 39.8%, Otto GUEVARA Guth (PML) 8%, Ricardo TOLEDO (PUSC) 3%
Legislative branch: unicameral Legislative Assembly or Asamblea Legislativa (57 seats; members are elected by direct, popular vote to serve four-year terms)
elections: last held 5 February 2006 (next to be held in February 2010)
election results: percent of vote by party—NA; seats by party—PLN 25, PAC 17, PML 6, PUSC 5, PASE 1, PFA 1, PRN 1, PUN 1
Judicial branch: Supreme Court or Corte Suprema (22 justices are elected for renewable eight-year terms by the Legislative Assembly)
Political parties and leaders: Authentic Member from Heredia [Jose SALAS]; Citizen Action Party or PAC [Epsy CAMPBELL Barr]; Costa Rican Renovation Party or PRC [Gerardo Justo OROZCO Alvarez]; Democratic Force Party or PFD [Marco NUNEZ Gonzalez]; General Union Party or PUGEN [Carlos Alberto FERNANDEZ Vega]; Homeland First or PP [Juan Jose VARGAS Fallas]; Independent Worker Party or PIO [Jose Alberto CUBERO Carmona]; Libertarian Movement Party or PML [Otto GUEVARA Guth]; National Christian Alliance Party or ANC [Juan Carlos CHAVEZ Mora]; National Integration Party or PIN [Walter MUNOZ Cespedes]; National Liberation Party or PLN [Francisco Antonio PACHECO Fernandez]; National Patriotic Party or PPN [Daniel Enrique REYNOLDS Vargas]; National Restoration Party or PRN [Fabio Enrique DELGADO Hernandez]; National Union Party or PUN [Arturo ACOSTA Mora]; Nationalist Democratic Alliance or ADN [Jose Miguel VILLALOBOS Umana]; Patriotic Union or UP [Jose Miguel CORRALES Bolanos]; Social Christian Unity Party or PUSC [Luis FISHMAN Zonzinski]; Union for Change Party or UPC [Antonio ALVAREZ Desanti]; United Leftist Coalition or IU [Humberto VARGAS Carbonel]
Political pressure groups and leaders: Authentic Confederation of Democratic Workers or CATD (Communist Party affiliate); Chamber of Coffee Growers; Confederated Union of Workers or CUT (Communist Party affiliate); Costa Rican Confederation of Democratic Workers or CCTD (Liberation Party affiliate); Costa Rican Exporter's Chamber or CADEXCO; Costa Rican Solidarity Movement; Costa Rican Union of Private Sector Enterprises or UCCAEP [Rafael CARRILLO]; Federation of Public Service Workers or FTSP; National Association for Economic Development or ANFE; National Association of Educators or ANDE; National Association of Public and Private Employees or ANEP [Albino VARGAS]; Rerum Novarum or CTRN (PLN affiliate) [Gilbert BROWN]
International organization participation: BCIE, CACM, FAO, G-77, IADB, IAEA, IBRD, ICAO, ICC, ICCt, ICRM, IDA, IFAD, IFC, IFRCS, ILO, IMF, IMO, IMSO, Interpol, IOC, IOM, IPU, ISO, ITSO, ITU, ITUC, LAES, LAIA (observer), MIGA, NAM (observer), OAS, OPANAL, OPCW, PCA, RG, UN, UN Security Council (temporary), UNCTAD, UNESCO, UNHCR, UNIDO, Union Latina, UNWTO,

UPU, WCL, WCO, WFTU, WHO, WIPO, WMO, WTO

Diplomatic representation in the US:
chief of mission: Ambassador Tomas DUENAS
chancery: 2114 S Street NW, Washington, DC 20008
telephone: [1] (202) 234-2945
FAX: [1] (202) 265-4795
consulate(s) general: Atlanta, Chicago, Houston, Los Angeles, Miami, New Orleans, New York, San Juan (Puerto Rico), Tampa (temporarily closed), Washington, DC
consulate(s): San Francisco

Diplomatic representation from the US:
chief of mission: Ambassador Mark LANGDALE
embassy: Calle 120 Avenida O, Pavas, San Jose
mailing address: APO AA 34020
telephone: [506] 519-2000
FAX: [506] 519-2305

Flag description: five horizontal bands of blue (top), white, red (double width), white, and blue, with the coat of arms in a white elliptical disk on the hoist side of the red band; above the coat of arms a light blue ribbon contains the words, AMERICA CENTRAL, and just below it near the top of the coat of arms is a white ribbon with the words, REPUBLICA COSTA RICA

ECONOMY

Economy—overview: Costa Rica's basically stable economy depends on tourism, agriculture, and electronics exports. Poverty has remained around 20% for nearly 20 years, and the strong social safety net that had been put into place by the government has eroded due to increased financial constraints on government expenditures. Immigration from Nicaragua has increasingly become a concern for the government. The estimated 300,000–500,000 Nicaraguans estimated to be in Costa Rica legally and illegally are an important source of (mostly unskilled) labor, but also place heavy demands on the social welfare system. Foreign investors remain attracted by the country's political stability and high education levels, as well as the fiscal incentives offered in the free-trade zones. Exports have become more diversified in the past 10 years due to the growth of the high-tech manufacturing sector, which is dominated by the microprocessor industry. Tourism continues to bring in foreign exchange, as Costa Rica's impressive biodiversity makes it a key destination for eco-tourism. The government continues to grapple with its large internal and external deficits and sizable internal debt. Reducing inflation remains a difficult problem because of rising import prices, labor market rigidities, and fiscal deficits. Tax and public expenditure reforms will be necessary to close the budget gap. In October 2007, a national referendum voted in favor of the US-Central American Free Trade Agreement (CAFTA).

GDP (purchasing power parity): $45.77 billion (2007 est.)
GDP (official exchange rate): $26.24 billion (2007 est.)
GDP—real growth rate: 6.8% (2007 est.)
GDP—per capita (PPP): $10,300 (2007 est.)
GDP—composition by sector:
agriculture: 8.6%
industry: 29.4%
services: 62.1% (2007 est.)
Labor force: 1.92 million
note: this official estimate excludes Nicaraguans living in Costa Rica (2007 est.)
Labor force—by occupation: *agriculture:* 14%
industry: 22%
services: 64% (2006 est.)
Unemployment rate: 4.6% (2007 est.)
Population below poverty line: 16% (2006 est.)
Household income or consumption by percentage share:
lowest 10%: 1%
highest 10%: 37.4% (2003)
Distribution of family income—Gini index: 49.8 (2003)
Inflation rate (consumer prices): 9.4% (2007 est.)
Investment (gross fixed): 21.6% of GDP (2007 est.)
Budget:
revenues: $3.976 billion
expenditures: $3.808 billion (2007 est.)
Public debt: 46.6% of GDP (2007 est.)
Agriculture—products: bananas, pineapples, coffee, melons, ornamental plants, sugar, corn, rice, beans, potatoes; beef; timber
Industries: microprocessors, food processing, medical equipment, textiles and clothing, construction materials, fertilizer, plastic products
Industrial production growth rate: 7.3% (2007 est.)
Electricity—production: 8.349 billion kWh (2005)
Electricity—production by source:
fossil fuel: 1.5%
hydro: 81.9%
nuclear: 0%
other: 16.6% (2001)
Electricity—consumption: 7.776 billion kWh (2005)

Electricity—exports: 70 million kWh (2005)
Electricity—imports: 81 million kWh (2005)
Oil—production: 0 bbl/day (2004)
Oil—consumption: 43,000 bbl/day (2005 est.)
Oil—exports: 2,998 bbl/day (2004)
Oil—imports: 43,640 bbl/day (2004)
Oil—proved reserves: 0 bbl (1 January 2006 est.)
Natural gas—production: 0 cu m (2005 est.)
Natural gas—consumption: 0 cu m (2005 est.)
Natural gas—exports: 0 cu m (2005 est.)
Natural gas—imports: 0 cu m (2005)
Natural gas—proved reserves: 0 cu m (1 January 2006 est.)
Current account balance: -$1.519 billion (2007 est.)
Exports: $9.268 billion (2007 est.)
Exports—commodities: bananas, pineapples, coffee, melons, ornamental plants, sugar; seafood; electronic components, medical equipment
Exports—partners: US 27.5%, Netherlands 12.2%, China 11.7%, UK 6.2%, Mexico 5.8% (2006)
Imports: $12.26 billion (2007 est.)
Imports—commodities: raw materials, consumer goods, capital equipment, petroleum, construction materials
Imports—partners: US 41.2%, Venezuela 5.4%, Mexico 5.2%, Ireland 5%, Japan 4.9%, Brazil 4.3%, China 4.1% (2006)
Economic aid—recipient: $29.51 million (2005)
Reserves of foreign exchange and gold: $4.114 billion (31 December 2007 est.)
Debt—external: $7.422 billion (30 June 2007)
Stock of direct foreign investment—at home: $8.669 billion (2007 est.)
Stock of direct foreign investment—abroad: $491 million (2007 est.)
Market value of publicly traded shares: $1.478 billion (2005)
Currency (code): Costa Rican colon (CRC)
Currency code: CRC
Exchange rates: Costa Rican colones per US dollar—519.53 (2007), 511.3 (2006), 477.79 (2005), 437.91 (2004), 398.66 (2003)
Fiscal year: calendar year

COMMUNICATIONS

Telephones—main lines in use: 1.351 million (2006)
Telephones—mobile cellular: 1.444 million (2006)

Telephone system:

general assessment: good domestic telephone service in terms of breadth of coverage; restricted cellular telephone service; state-run monopoly provider is struggling with the demand for new lines, resulting in long waiting times

domestic: point-to-point and point-to-multi-point microwave, fiber-optic, and coaxial cable link rural areas; Internet service is available

international: country code—506; landing point for the Americas Region Caribbean Ring System (ARCOS-1) fiber-optic telecommunications submarine cable and the MAYA-1 submarine cable that provide links to South and Central America, parts of the Caribbean, and the US; connected to Central American Microwave System; satellite earth stations—2 Intelsat (Atlantic Ocean) (2007)

Radio broadcast stations: AM 65, FM 51, shortwave 19 (2002)

Radios: 980,000 (1997)

Television broadcast stations: 20 (plus 43 repeaters) (2002)

Televisions: 525,000 (1997)

Internet country code: .cr

Internet hosts: 13,792 (2007)

Internet Service Providers (ISPs): 3 (of which only one is legal) (2000)

Internet users: 1.214 million (2006)

TRANSPORTATION

Airports: 151 (2007)

Airports—with paved runways: *total:* 36
2,438 to 3,047 m: 2
1,524 to 2,437 m: 2
914 to 1,523 m: 21
under 914 m: 11 (2007)

Airports—with unpaved runways: *total:* 115
914 to 1,523 m: 19
under 914 m: 96 (2007)

Pipelines: refined products 242 km (2007)

Railways:

total: 278 km

narrow gauge: 278 km 1.067-m gauge

note: none of the railway network is in use (2007)

Roadways:

total: 35,330 km

paved: 8,621 km

unpaved: 26,709 km (2004)

Waterways: 730 km (seasonally navigable by small craft) (2007)

Merchant marine:

total: 1 ship (1000 GRT or over) 1,058 GRT/255 DWT

by type: passenger/cargo 1 (2007)

Ports and terminals: Caldera, Puerto Limon

MILITARY

Military branches: no regular military forces; Ministry of Public Security, Government, and Police (2008)

Manpower available for military service: *males age 16–49:* 1,134,205
females age 16–49: 1,095,763 (2008 est.)

Manpower fit for military service:
males age 16–49: 958,013
females age 16–49: 925,727 (2008 est.)

Manpower reaching militarily significant age annually:
males age 16–49: 40,767
females age 16–49: 38,899 (2008 est.)

Military expenditures—percent of GDP: 0.4% (2006)

TRANSNATIONAL ISSUES

Disputes—international: in September 2005, Costa Rica took its case before the ICJ to advocate the navigation, security, and commercial rights of Costa Rican vessels using the Río San Juan over which Nicaragua retains sovereignty

Refugees and internally displaced persons: *refugees (country of origin):* 9,699–11,500 (Colombia) (2007)

Illicit drugs: transshipment country for cocaine and heroin from South America; illicit production of cannabis in remote areas; domestic cocaine consumption, particularly crack cocaine, is rising; significant consumption of amphetamines

CÔTE D'IVOIRE

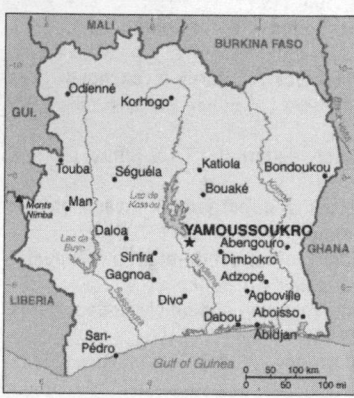

INTRODUCTION

Background: Close ties to France since independence in 1960, the development of cocoa production for export, and foreign investment made Côte d'Ivoire one of the most prosperous of the West African states, but did not protect it from political turmoil. In December 1999, a military coup—the first ever in Côte d'Ivoire's history—overthrew the government. Junta leader Robert GUEI blatantly rigged elections held in late 2000 and declared himself the winner. Popular protest forced him to step aside and brought Laurent GBAGBO into power. Ivorian dissidents and disaffected members of the military launched a failed coup attempt in September 2002. Rebel forces claimed the northern half of the country, and in January 2003 were granted ministerial positions in a unity government under the auspices of the Linas-Marcoussis Peace Accord. President GBAGBO and rebel forces resumed implementation of the peace accord in December 2003 after a three-month stalemate, but issues that sparked the civil war, such as land reform and grounds for citizenship, remained unresolved. In March 2007 President GBAGBO and former New Force rebel leader Guillaume SORO signed the Ouagadougou Political Agreement. As a result of the agreement, SORO joined GBAGBO's government as Prime Minister and the two agreed to reunite the country by dismantling the zone of confidence separating North from South, integrate rebel forces into the national armed forces, and hold elections. Several thousand French and UN troops remain in Côte d'Ivoire to help the parties implement their commitments and to support the peace process.

GEOGRAPHY

Location: Western Africa, bordering the North Atlantic Ocean, between Ghana and Liberia

Geographic coordinates: 8 00 N, 5 00 W

Map references: Africa

Area:

total: 322,460 sq km

land: 318,000 sq km

water: 4,460 sq km

Area—comparative: slightly larger than New Mexico

Land boundaries:

total: 3,110 km

border countries: Burkina Faso 584 km, Ghana 668 km, Guinea 610 km, Liberia 716 km, Mali 532 km

Coastline: 515 km

Maritime claims:
territorial sea: 12 nm
exclusive economic zone: 200 nm
continental shelf: 200 nm
Climate: tropical along coast, semiarid in far north; three seasons—warm and dry (November to March), hot and dry (March to May), hot and wet (June to October)
Terrain: mostly flat to undulating plains; mountains in northwest
Elevation extremes:
lowest point: Gulf of Guinea 0 m
highest point: Mont Nimba 1,752 m
Natural resources: petroleum, natural gas, diamonds, manganese, iron ore, cobalt, bauxite, copper, gold, nickel, tantalum, silica sand, clay, cocoa beans, coffee, palm oil, hydropower
Land use: *arable land:* 10.23%
permanent crops: 11.16%
other: 78.61% (2005)
Irrigated land: 730 sq km (2003)
Total renewable water resources: 81 cu km (2001)
Freshwater withdrawal (domestic/industrial/agricultural): *total:* 0.93 cu km/yr (24%/12%/65%)
per capita: 51 cu m/yr (2000)
Natural hazards: coast has heavy surf and no natural harbors; during the rainy season torrential flooding is possible
Environment—current issues: deforestation (most of the country's forests—once the largest in West Africa—have been heavily logged); water pollution from sewage and industrial and agricultural effluents
Environment—international agreements: *party to:* Biodiversity, Climate Change, Climate Change-Kyoto Protocol, Desertification, Endangered Species, Hazardous Wastes, Law of the Sea, Marine Dumping, Ozone Layer Protection, Ship Pollution, Tropical Timber 83, Tropical Timber 94, Wetlands, Whaling
signed, but not ratified: none of the selected agreements
Geography—note: most of the inhabitants live along the sandy coastal region; apart from the capital area, the forested interior is sparsely populated

Population: 18,373,060
note: estimates for this country explicitly take into account the effects of excess mortality due to AIDS; this can result in lower life expectancy, higher infant mortality, higher death rates, lower population growth rates, and changes in the distribution of population by age and sex than would otherwise be expected (July 2008 est.)

Age structure:
0–14 years: 40.4% (male 3,660,057/female 3,767,893)
15–64 years: 56.7% (male 5,233,772/female 5,180,841)
65 years and over: 2.9% (male 253,573/female 276,924) (2008 est.)
Median age:
total: 19.4 years
male: 19.6 years
female: 19.1 years (2008 est.)
Population growth rate: 1.96% (2008 est.)
Birth rate: 34.26 births/1,000 population (2008 est.)
Death rate: 14.65 deaths/1,000 population (2008 est.)
Net migration rate: NA
Sex ratio:
at birth: 1.03 male(s)/female
under 15 years: 0.97 male(s)/female
15–64 years: 1.01 male(s)/female
65 years and over: 0.92 male(s)/female
total population: 0.99 male(s)/female (2008 est.)
Infant mortality rate:
total: 85.71 deaths/1,000 live births
male: 101.96 deaths/1,000 live births
female: 68.98 deaths/1,000 live births (2008 est.)
Life expectancy at birth:
total population: 49.18 years
male: 46.62 years
female: 51.82 years (2008 est.)
Total fertility rate: 4.35 children born/woman (2008 est.)
HIV/AIDS—adult prevalence rate: 7% (2003 est.)
HIV/AIDS—people living with HIV/AIDS: 570,000 (2003 est.)
HIV/AIDS—deaths: 47,000 (2003 est.)
Major infectious diseases:
degree of risk: very high
food or waterborne diseases: bacterial diarrhea, hepatitis A, and typhoid fever
vectorborne diseases: malaria and yellow fever
water contact: schistosomiasis
note: highly pathogenic H5N1 avian influenza has been identified in this country; it poses a negligible risk with extremely rare cases possible among US citizens who have close contact with birds (2008)
Nationality:
noun: Ivoirian(s)
adjective: Ivoirian
Ethnic groups: Akan 42.1%, Voltaiques or Gur 17.6%, Northern Mandes 16.5%, Krous 11%, Southern Mandes 10%, other 2.8% (includes 130,000 Lebanese and 14,000 French) (1998)
Religions: Muslim 35–40%, indigenous 25–40%, Christian 20–30% (2001)
note: the majority of foreigners (migra-

tory workers) are Muslim (70%) and Christian (20%)
Languages: French (official), 60 native dialects with Dioula the most widely spoken
Literacy: *definition:* age 15 and over can read and write
total population: 50.9%
male: 57.9%
female: 43.6% (2003 est.)

Country name:
conventional long form: Republic of Côte d'Ivoire
conventional short form: Côte d'Ivoire
local long form: Republique de Côte d'Ivoire
local short form: Côte d'Ivoire
note: pronounced coat-div-whar
former: Ivory Coast
Government type: republic; multiparty presidential regime established 1960
note: the government is currently operating under a power-sharing agreement mandated by international mediators
Capital: *name:* Yamoussoukro
geographic coordinates: 6 49 N, 5 17 W
time difference: UTC 0 (5 hours ahead of Washington, DC during Standard Time)
note: although Yamoussoukro has been the official capital since 1983, Abidjan remains the commercial and administrative center; the US, like other countries, maintains its Embassy in Abidjan
Administrative divisions: 19 regions; Agneby, Bafing, Bas-Sassandra, Denguele, Dix-Huit Montagnes, Fromager, Haut-Sassandra, Lacs, Lagunes, Marahoue, Moyen-Cavally, Moyen-Comoe, N'zi-Comoe, Savanes, Sud-Bandama, Sud-Comoe, Vallee du Bandama, Worodougou, Zanzan
Independence: 7 August 1960 (from France)
National holiday: Independence Day, 7 August (1960)
Constitution: approved by referendum 23 July 2000
Legal system: based on French civil law system and customary law; judicial review in the Constitutional Chamber of the Supreme Court; accepts compulsory ICJ jurisdiction, with reservations
Suffrage: 18 years of age; universal
Executive branch:
chief of state: President Laurent GBAGBO (since 26 October 2000)
head of government: Prime Minister Guillaume SORO (since 4 April 2007)
cabinet: Council of Ministers appointed by the president; note—under the current power-sharing agreement the prime minister and the president share the authority to appoint ministers

elections: president elected by popular vote for a five-year term (no term limits); election last held 26 October 2000 (next to be held 30 November 2008; elections were to be held in 2005 but have been repeatedly postponed by the government; the UN Security Council has extended the government's mandate); prime minister appointed by the president

election results: Laurent GBAGBO elected president; percent of vote—Laurent GBAGBO 59.4%, Robert GUEI 32.7%, Francis WODIE 5.7%, other 2.2%

Legislative branch: unicameral National Assembly or Assemblee Nationale (225 seats; members are elected in single- and multi-district elections by direct popular vote to serve five-year terms)

elections: elections last held 10 December 2000 with by-elections on 14 January 2001 (next to be held by June 2008 after the government postponed the elections in 2005 and 2006)

election results: percent of vote by party—NA; seats by party—FPI 96, PDCI-RDA 94, RDR 5, PIT 4, other 2, independents 22, vacant 2

note: a Senate that was scheduled to be created in the October 2006 elections never took place

Judicial branch: Supreme Court or Cour Supreme consists of four chambers: Judicial Chamber for criminal cases, Audit Chamber for financial cases, Constitutional Chamber for judicial review cases, and Administrative Chamber for civil cases; there is no legal limit to the number of members

Political parties and leaders: Citizen's Democratic Union or UDCY [Theodore MEL EG]; Democratic Party of Côte d'Ivoire-African Democratic Rally or PDCI-RDA [Henri Konan BEDIE]; Ivorian Popular Front or FPI [Pascale Affi N'GUESSAN]; Ivorian Worker's Party or PIT [Francis WODIE]; Opposition Movement of the Future or MFA [Innocent Augustin ANAKY]; Rally of the Republicans or RDR [Alassane OUATTARA]; Union for Democracy and Peace in Côte d'Ivoire or UDPCI [Toikeuse MABRI]; over 144 smaller registered parties

Political pressure groups and leaders: Federation of University and High School Students of Côte d'Ivoire or FESCI [Serges KOFFI]; Rally of Houphouetists for Democracy and Peace or RHDP [Alphonse DJEDJE MADY]; Young Patriots [Charles BLE GOUDE]

International organization participation: ACCT, ACP, AfDB, AU, ECOWAS, Entente, FAO, FZ, G-24, G-77, IAEA, IBRD, ICAO, ICCt (signatory), ICRM, IDA, IDB, IFAD, IFC, IFRCS, ILO, IMF, IMO, Interpol, IOC, IOM, IPU, ITSO, ITU, ITUC, MIGA, NAM, OIC, OIF, OPCW, UN, UNCTAD, UNESCO, UNHCR, UNIDO, Union Latina, UNWTO, UPU, WADB (regional), WAEMU, WCL, WCO, WFTU, WHO, WIPO, WMO, WTO

Diplomatic representation in the US:
chief of mission: Ambassador Yao Charles KOFFI
chancery: 3421 Massachusetts Avenue NW, Washington, DC 20007
telephone: [1] (202) 797-0300
FAX: [1] (202) 244-3088

Diplomatic representation from the US:
chief of mission: Ambassador Wanda L. NESBITT
embassy: Riviera Golf 01, Abidjan
mailing address: B. P. 1712, Abidjan 01
telephone: [225] 22 49 40 00
FAX: [225] 22 49 43 23

Flag description: three equal vertical bands of orange (hoist side), white, and green
note: similar to the flag of Ireland, which is longer and has the colors reversed—green (hoist side), white, and orange; also similar to the flag of Italy, which is green (hoist side), white, and red; design was based on the flag of France

ECONOMY

Economy—overview: Côte d'Ivoire is the world's largest producer and exporter of cocoa beans and a significant producer and exporter of coffee and palm oil. Consequently, the economy is highly sensitive to fluctuations in international prices for these products, and, to a lesser extent, in climatic conditions. Despite government attempts to diversify the economy, it is still heavily dependent on agriculture and related activities, engaging roughly 68% of the population. Since 2006, oil and gas production have become more important engines of economic activity than cocoa. According to IMF statistics, earnings from oil and refined products were $1.3 billion in 2006, while cocoa-related revenues were $1 billion during the same period. Côte d'Ivoire's offshore oil and gas production has resulted in substantial crude oil exports and provides sufficient natural gas to fuel electricity exports to Ghana, Togo, Benin, Mali and Burkina Faso. Oil exploration by a number of consortiums of private companies continues offshore, and President GBAGBO has expressed hope that daily crude output could reach 200,000 barrels per day (b/d) by the end of the decade. Since the end of the civil war in 2003, political turmoil has continued to damage the economy, resulting in the loss of foreign investment and slow economic growth. GDP grew by 1.8% in 2006 and 1.7% in 2007. Per capita income has declined by 15% since 1999.

GDP (purchasing power parity): $32.18 billion (2007 est.)

GDP (official exchange rate): $19.6 billion (2007 est.)

GDP—real growth rate: 1.6% (2007 est.)

GDP—per capita (PPP): $1,700 (2007 est.)

GDP—composition by sector:
agriculture: 27.5%
industry: 22.2%
services: 50.2% (2007 est.)

Labor force: 6.907 million (68% agricultural) (2007 est.)

Labor force—by occupation: *agriculture:* 68%
industry and services: NA (2007 est.)

Unemployment rate: unemployment may have climbed to 40–50% as a result of the civil war

Population below poverty line: 42% (2006 est.)

Household income or consumption by percentage share:
lowest 10%: 2%
highest 10%: 34% (2002)

Distribution of family income—Gini index: 44.6 (2002)

Inflation rate (consumer prices): 2.1% (2007 est.)

Investment (gross fixed): 9.6% of GDP (2007 est.)

Budget:
revenues: $3.213 billion
expenditures: $3.826 billion (2007 est.)

Public debt: 81.1% of GDP (2007 est.)

Agriculture—products: coffee, cocoa beans, bananas, palm kernels, corn, rice, manioc (tapioca), sweet potatoes, sugar, cotton, rubber; timber

Industries: foodstuffs, beverages; wood products, oil refining, truck and bus assembly, textiles, fertilizer, building materials, electricity, ship construction and repair

Industrial production growth rate: 1.5% (2007 est.)

Electricity—production: 5.305 billion kWh (2005)

Electricity—production by source:
fossil fuel: 61.9%
hydro: 38.1%
nuclear: 0%
other: 0% (2001)

Electricity—consumption: 2.9 billion kWh (2005)

Electricity—exports: 1.397 billion kWh (2005)

Electricity—imports: 0 kWh (2005)
Oil—production: 57,700 bbl/day (2005 est.)
Oil—consumption: 27,000 bbl/day (2005 est.)
Oil—exports: 85,780 bbl/day (2004)
Oil—imports: 76,730 bbl/day (2004)
Oil—proved reserves: 250 million bbl (2007 est.)
Natural gas—production: 1.247 billion cu m (2005 est.)
Natural gas—consumption: 1.247 billion cu m (2005 est.)
Natural gas—exports: 0 cu m (2005 est.)
Natural gas—imports: 0 cu m (2005)
Natural gas—proved reserves: 27.16 billion cu m (1 January 2006 est.)
Current account balance: $265 million (2007 est.)
Exports: $18.5 billion f.o.b. (2007 est.)
Exports—commodities: cocoa, coffee, timber, petroleum, cotton, bananas, pineapples, palm oil, fish
Exports—partners: France 18.3%, Netherlands 9.7%, US 9.1%, Nigeria 7.2%, Germany 4.2% (2006)
Imports: $6.137 billion f.o.b. (2007 est.)
Imports—commodities: fuel, capital equipment, foodstuffs
Imports—partners: Nigeria 27.6%, France 25.4%, China 4.3% (2006)
Economic aid—recipient: ODA, $60 million (2007 est.)
Reserves of foreign exchange and gold: $2.519 billion (31 December 2007 est.)
Debt—external: $13.99 billion (31 December 2007 est.)
Stock of direct foreign investment—at home: $NA
Stock of direct foreign investment—abroad: $NA
Market value of publicly traded shares: $4.155 billion (2006)
Currency (code): Communaute Financiere Africaine franc (XOF); note—responsible authority is the Central Bank of the West African States
Currency code: XOF
Exchange rates: Communaute Financiere Africaine francs (CFA) per US dollar—481.83 (2007), 522.89 (2006), 527.47 (2005), 528.29 (2004), 581.2 (2003)
note: since 1 January 1999, the XOF franc has been pegged to the euro at a rate of 655.957 XOF francs per euro

Fiscal year: calendar year

COMMUNICATIONS

Telephones—main lines in use: 260,900 (2006)
Telephones—mobile cellular: 4.065 million (2006)
Telephone system:
general assessment: well developed by African standards; telecommunications sector privatized in late 1990s; mobile cellular usage has increased to 23 per 100 persons; fixed-line connections stand at about 2 per 100 persons
domestic: open-wire lines and microwave radio relay; 90% digitalized
international: country code—225; landing point for the SAT-3/WASC fiber-optic submarine cable that provides connectivity to Europe and Asia; satellite earth stations—2 Intelsat (1 Atlantic Ocean and 1 Indian Ocean) (2007)
Radio broadcast stations: AM 2, FM 9, shortwave 3 (1998)
Radios: 2.26 million (1997)
Television broadcast stations: 14 (1998)
Televisions: 1.09 million (2000)
Internet country code: .ci
Internet hosts: 1,373 (2007)
Internet Service Providers (ISPs): 5 (2001)
Internet users: 300,000 (2006)

TRANSPORTATION

Airports: 34 (2007)
Airports—with paved runways:
total: 7
over 3,047 m: 1
2,438 to 3,047 m: 2
1,524 to 2,437 m: 4 (2007)
Airports—with unpaved runways:
total: 27
1,524 to 2,437 m: 8
914 to 1,523 m: 14
under 914 m: 5 (2007)
Pipelines: condensate 102 km; gas 245 km; oil 112 km (2007)
Railways:
total: 660 km
narrow gauge: 660 km 1.000 meter gauge
note: an additional 622 km of this railroad extends into Burkina Faso (2006)
Roadways:
total: 80,000 km
paved: 6,500 km
unpaved: 73,500 km

note: includes intercity and urban roads; another 20,000 km of dirt roads are in poor condition and 150,000 km of dirt roads are impassable (2006)
Waterways: 980 km (navigable rivers, canals, and numerous coastal lagoons) (2006)
Ports and terminals: Abidjan, Espoir, San-Pedro

MILITARY

Military branches: Côte d'Ivoire Defense and Security Forces (FDSC): Army, Navy, Air Force (2006)
Military service age and obligation: 18 years of age for compulsory and voluntary military service (2008)
Manpower available for military service:
males age 16–49: 4,369,735
females age 16–49: 4,287,042 (2008 est.)
Manpower fit for military service:
males age 16–49: 2,393,104
females age 16–49: 2,381,607 (2008 est.)
Manpower reaching militarily significant age annually:
males age 16–49: 202,545
females age 16–49: 211,601 (2008 est.)
Military expenditures—percent of GDP: 1.6% (2005 est)

TRANSNATIONAL ISSUES

Disputes—international: despite the presence of over 9,000 UN forces (UNOCI) in Côte d'Ivoire since 2004, ethnic conflict there has displaced hundreds of thousands of Ivorians in and out of the country as well as driven out migrants from neighboring states who worked in Ivorian cocoa plantations; Ivorian rebels reportedly hide along the borders of neighboring states
Refugees and internally displaced persons: *refugees (country of origin):* 25,615 (Liberia)
IDPs: 709,000 (2002 coup; most IDPs are in western regions) (2007)
Illicit drugs: illicit producer of cannabis, mostly for local consumption; utility as a narcotic transshipment point to Europe reduced by ongoing political instability; while rampant corruption and inadequate supervision leave the banking system vulnerable to money laundering, the lack of a developed financial system limits the country's utility as a major money-laundering center

CROATIA

INTRODUCTION

Background: The lands that today comprise Croatia were part of the Austro-Hungarian Empire until the close of World War I. In 1918, the Croats, Serbs, and Slovenes formed a kingdom known after 1929 as Yugoslavia. Following World War II, Yugoslavia became a federal independent Communist state under the strong hand of Marshal TITO. Although Croatia declared its independence from Yugoslavia in 1991, it took four years of sporadic, but often bitter, fighting before occupying Serb armies

were mostly cleared from Croatian lands. Under UN supervision, the last Serb-held enclave in eastern Slavonia was returned to Croatia in 1998.

GEOGRAPHY

Location: Southeastern Europe, bordering the Adriatic Sea, between Bosnia and Herzegovina and Slovenia
Geographic coordinates: 45 10 N, 15 30 E
Map references: Europe
Area:
total: 56,542 sq km
land: 56,414 sq km
water: 128 sq km
Area—comparative: slightly smaller than West Virginia
Land boundaries:
total: 2,197 km
border countries: Bosnia and Herzegovina 932 km, Hungary 329 km, Serbia 241 km, Montenegro 25 km, Slovenia 670 km
Coastline: 5,835 km (mainland 1,777 km, islands 4,058 km)
Maritime claims:
territorial sea: 12 nm
continental shelf: 200-m depth or to the depth of exploitation
Climate: Mediterranean and continental; continental climate predominant with hot summers and cold winters; mild winters, dry summers along coast
Terrain: geographically diverse; flat plains along Hungarian border, low mountains and highlands near Adriatic coastline and islands
Elevation extremes:
lowest point: Adriatic Sea 0 m
highest point: Dinara 1,830 m
Natural resources: oil, some coal, bauxite, low-grade iron ore, calcium, gypsum, natural asphalt, silica, mica, clays, salt, hydropower
Land use:
arable land: 25.82%
permanent crops: 2.19%
other: 71.99% (2005)

Irrigated land: 110 sq km (2003)
Total renewable water resources: 105.5 cu km (1998)
Natural hazards: destructive earthquakes
Environment—current issues: air pollution (from metallurgical plants) and resulting acid rain is damaging the forests; coastal pollution from industrial and domestic waste; landmine removal and reconstruction of infrastructure consequent to 1992–95 civil strife
Environment—international agreements: *party to:* Air Pollution, Air Pollution-Sulfur 94, Biodiversity, Climate Change, Climate Change-Kyoto Protocol, Desertification, Endangered Species, Hazardous Wastes, Law of the Sea, Marine Dumping, Ozone Layer Protection, Ship Pollution, Wetlands, Whaling
signed, but not ratified: Air Pollution-Persistent Organic Pollutants
Geography—note: controls most land routes from Western Europe to Aegean Sea and Turkish Straits; most Adriatic Sea islands lie off the coast of Croatia—some 1,200 islands, islets, ridges, and rocks

PEOPLE

Population: 4,491,543 (July 2008 est.)
Age structure:
0–14 years: 15.8% (male 363,551/female 345,132)
15–64 years: 67.2% (male 1,501,949/female 1,517,962)
65 years and over: 17% (male 295,229/female 467,720) (2008 est.)
Median age:
total: 40.8 years
male: 38.9 years
female: 42.6 years (2008 est.)
Population growth rate: -0.043% (2008 est.)
Birth rate: 9.64 births/1,000 population (2008 est.)
Death rate: 11.66 deaths/1,000 population (2008 est.)
Net migration rate: 1.58 migrant(s)/1,000 population (2008 est.)
Sex ratio:
at birth: 1.06 male(s)/female
under 15 years: 1.05 male(s)/female
15–64 years: 0.99 male(s)/female
65 years and over: 0.63 male(s)/female
total population: 0.93 male(s)/female (2008 est.)
Infant mortality rate:
total: 6.49 deaths/1,000 live births
male: 6.51 deaths/1,000 live births
female: 6.46 deaths/1,000 live births (2008 est.)
Life expectancy at birth:
total population: 75.13 years

male: 71.49 years
female: 78.97 years (2008 est.)
Total fertility rate: 1.41 children born/woman (2008 est.)
HIV/AIDS—adult prevalence rate: less than 0.1% (2001 est.)
HIV/AIDS—people living with HIV/AIDS: 200 (2001 est.)
HIV/AIDS—deaths: fewer than 10 (2001 est.)
Major infectious diseases:
degree of risk: intermediate
food or waterborne diseases: bacterial diarrhea and hepatitis A
vectorborne diseases: tickborne encephalitis
note: highly pathogenic H5N1 avian influenza has been identified in this country; it poses a negligible risk with extremely rare cases possible among US citizens who have close contact with birds (2008)
Nationality:
noun: Croat(s), Croatian(s)
adjective: Croatian
Ethnic groups: Croat 89.6%, Serb 4.5%, other 5.9% (including Bosniak, Hungarian, Slovene, Czech, and Roma) (2001 census)
Religions: Roman Catholic 87.8%, Orthodox 4.4%, other Christian 0.4%, Muslim 1.3%, other and unspecified 0.9%, none 5.2% (2001 census)
Languages: Croatian 96.1%, Serbian 1%, other and undesignated 2.9% (including Italian, Hungarian, Czech, Slovak, and German) (2001 census)
Literacy: *definition:* age 15 and over can read and write
total population: 98.1%
male: 99.3%
female: 97.1% (2001 census)

GOVERNMENT

Country name:
conventional long form: Republic of Croatia
conventional short form: Croatia
local long form: Republika Hrvatska
local short form: Hrvatska
former: People's Republic of Croatia, Socialist Republic of Croatia
Government type: presidential/parliamentary democracy
Capital: *name:* Zagreb
geographic coordinates: 45 48 N, 16 00 E
time difference: UTC+1 (6 hours ahead of Washington, DC during Standard Time)
daylight saving time: +1hr, begins last Sunday in March; ends last Sunday in October
Administrative divisions: 20 counties (zupanije, zupanija—singular) and 1 city* (grad—singular); Bjelovarsko-Bilogorska Zupanija, Brodsko-Posavska

Zupanija, Dubrovacko-Neretvanska Zupanija, Istarska Zupanija, Karlovacka Zupanija, Koprivnicko-Krizevacka Zupanija, Krapinsko-Zagorska Zupanija, Licko-Senjska Zupanija, Medimurska Zupanija, Osjecko-Baranjska Zupanija, Pozesko-Slavonska Zupanija, Primorsko-Goranska Zupanija, Sibensko-Kninska Zupanija, Sisacko-Moslavacka Zupanija, Splitsko-Dalmatinska Zupanija, Varazdinska Zupanija, Viroviticko-Podravska Zupanija, Vukovarsko-Srijemska Zupanija, Zadarska Zupanija, Zagreb*, Zagrebacka Zupanija

Independence: 25 June 1991 (from Yugoslavia)

National holiday: Independence Day, 8 October (1991); note—25 June 1991 was the day the Croatian Parliament voted for independence; following a three-month moratorium to allow the European Community to solve the Yugoslav crisis peacefully, Parliament adopted a decision on 8 October 1991 to sever constitutional relations with Yugoslavia

Constitution: adopted on 22 December 1990; revised 2000, 2001

Legal system: based on Austro-Hungarian law system with Communist law influences; has not accepted compulsory ICJ jurisdiction

Suffrage: 18 years of age; universal (16 years of age, if employed)

Executive branch:
chief of state: President Stjepan (Stipe) MESIC (since 18 February 2000)
head of government: Prime Minister Ivo SANADER (since 9 December 2003); Deputy Prime Ministers Jadranka KOSOR (since 23 December 2003) and Damir POLANCEC (since 15 February 2005), Djurdja ADLESIC (since 12 January 2008), Slobodan UZELAC (since 12 January 2008)
cabinet: Council of Ministers named by the prime minister and approved by the parliamentary Assembly
elections: president elected by popular vote for a five-year term (eligible for a second term); election last held 16 January 2005 (next to be held in January 2010); the leader of the majority party or the leader of the majority coalition is usually appointed prime minister by the president and then approved by the Assembly
election results: Stjepan MESIC reelected president; percent of vote—Stjepan MESIC 66%, Jadranka KOSOR 34% in the second round

Legislative branch: unicameral Assembly or Sabor (153 seats; members elected from party lists by popular vote to serve four-year terms)

elections: last held 25 November 2007 (next to be held in November 2011)
election results: percent of vote by party—NA; number of seats by party—HDZ 66, SDP 56, HNS 7, HSS 6, HDSSB 3, IDS 3, SDSS 3, other 9

Judicial branch: Supreme Court; Constitutional Court; judges for both courts appointed for eight-year terms by the Judicial Council of the Republic, which is elected by the Assembly

Political parties and leaders: Croatian Democratic Congress of Slavonia and Baranja or HDSSB [Vladimir SISLJAGIC]; Croatian Democratic Union or HDZ [Ivo SANADER]; Croatian Party of the Right or HSP [Anto DJAPIC]; Croatian Peasant Party or HSS [Josip FRISCIC]; Croatian Pensioner Party or HSU [Vladimir JORDAN]; Croatian People's Party or HNS [Vesna PUSIC]; Croatian Social Liberal Party or HSLS [Djurdja ADLESIC]; Independent Democratic Serb Party or SDSS [Vojislav STANIMIROVIC]; Istrian Democratic Assembly or IDS [Ivan JAKOVCIC]; Social Democratic Party of Croatia or SDP [Zoran MILANOVIC]

Political pressure groups and leaders: NA

International organization participation: ACCT (observer), Australia Group, BIS, BSEC (observer), CE, CEI, EAPC, EBRD, FAO, IADB, IAEA, IBRD, ICAO, ICC, ICCt, ICRM, IDA, IFAD, IFC, IFRCS, IHO, ILO, IMF, IMO, IMSO, Interpol, IOC, IOM, IPU, ISO, ITSO, ITU, ITUC, MIGA, MINURSO, MINUSTAH, NAM (observer), NSG, OAS (observer), OIF (observer), OPCW, OSCE, PCA, PFP, SECI, UN, UN Security Council (temporary), UNCTAD, UNESCO, UNIDO, UNIFIL, UNMEE, UNMIL, UNMIS, UNMOGIP, UNOCI, UNOMIG, UNWTO, UPU, WCO, WHO, WIPO, WMO, WTO, ZC

Diplomatic representation in the US:
chief of mission: Ambassador Kolinda GRABAR-KITAROVIC
chancery: 2343 Massachusetts Avenue NW, Washington, DC 20008
telephone: [1] (202) 588-5899
FAX: [1] (202) 588-8936
consulate(s) general: Chicago, Los Angeles, New York

Diplomatic representation from the US:
chief of mission: Ambassador Robert A. BRADTKE
embassy: 2 Thomas Jefferson Street, 10010 Zagreb
mailing address: use street address
telephone: [385] (1) 661-2200
FAX: [385] (1) 661-2373

Flag description: three equal horizontal bands of red (top), white, and blue superimposed by the Croatian coat of arms (red and white checkered)

Economy—overview: Once one of the wealthiest of the Yugoslav republics, Croatia's economy suffered badly during the 1991–95 war as output collapsed and the country missed the early waves of investment in Central and Eastern Europe that followed the fall of the Berlin Wall. Since 2000, however, Croatia's economic fortunes have begun to improve slowly, with moderate but steady GDP growth between 4% and 6% led by a rebound in tourism and credit-driven consumer spending. Inflation over the same period has remained tame and the currency, the kuna, stable. Nevertheless, difficult problems still remain, including a stubbornly high unemployment rate, a growing trade deficit and uneven regional development. The state retains a large role in the economy, as privatization efforts often meet stiff public and political resistance. While macroeconomic stabilization has largely been achieved, structural reforms lag because of deep resistance on the part of the public and lack of strong support from politicians. The EU accession process should accelerate fiscal and structural reform.

GDP (purchasing power parity): $68.98 billion (2007 est.)

GDP (official exchange rate): $51.36 billion (2007 est.)

GDP—real growth rate: 5.8% (2007 est.)

GDP—per capita (PPP): $15,500 (2007 est.)

GDP—composition by sector:
agriculture: 7.2%
industry: 31.6%
services: 61.2% (2007 est.)

Labor force: 1.749 million (2007 est.)

Labor force—by occupation: *agriculture:* 2.7%
industry: 32.8%
services: 64.5% (2004)

Unemployment rate: 11.8% (2007 est.)

Population below poverty line: 11% (2003)

Household income or consumption by percentage share:
lowest 10%: 3.4%
highest 10%: 24.5% (2003 est.)

Distribution of family income—Gini index: 29 (2001)

Inflation rate (consumer prices): 2.9% (2007 est.)

Investment (gross fixed): 30.9% of GDP (2007 est.)

Budget: *revenues:* $22.57 billion

expenditures: $23.92 billion (2007 est.)
Public debt: 47.3% of GDP (2007 est.)
Agriculture—products: wheat, corn, sugar beets, sunflower seed, barley, alfalfa, clover, olives, citrus, grapes, soybeans, potatoes; livestock, dairy products
Industries: chemicals and plastics, machine tools, fabricated metal, electronics, pig iron and rolled steel products, aluminum, paper, wood products, construction materials, textiles, shipbuilding, petroleum and petroleum refining, food and beverages, tourism
Industrial production growth rate: 5.2% (2007 est.)
Electricity—production: 11.99 billion kWh (2005)
Electricity—production by source:
fossil fuel: 33.6%
hydro: 66%
nuclear: 0%
other: 0.4% (2001)
Electricity—consumption: 14.97 billion kWh (2005)
Electricity—exports: 3.634 billion kWh (2005)
Electricity—imports: 8.746 billion kWh (2005)
Oil—production: 27,190 bbl/day (2005 est.)
Oil—consumption: 99,000 bbl/day (2005 est.)
Oil—exports: 40,930 bbl/day (2004)
Oil—imports: 109,800 bbl/day (2004)
Oil—proved reserves: 69.14 million bbl (1 January 2006 est.)
Natural gas—production: 1.477 billion cu m (2005 est.)
Natural gas—consumption: 2.58 billion cu m (2005 est.)
Natural gas—exports: 0 cu m (2005 est.)
Natural gas—imports: 1.103 billion cu m (2005)
Natural gas—proved reserves: 27.16 billion cu m (1 January 2006 est.)
Current account balance: -$4.385 billion (2007 est.)
Exports: $12.02 billion f.o.b. (2007 est.)
Exports—commodities: transport equipment, textiles, chemicals, foodstuffs, fuels
Exports—partners: Italy 23.1%, Bosnia and Herzegovina 12.7%, Germany 10.4%, Slovenia 8.3%, Austria 6.1% (2006)
Imports: $26.54 billion f.o.b. (2007 est.)
Imports—commodities: machinery, transport and electrical equipment; chemicals, fuels and lubricants; foodstuffs
Imports—partners: Italy 16.7%, Germany 14.5%, Russia 9.7%, Slovenia 6.8%, Austria 5.4%, China 5.3% (2006)
Economic aid—recipient: ODA, $125.4 million (2005)

Reserves of foreign exchange and gold: $13.67 billion (31 December 2007 est.)
Debt—external: $45.29 billion (30 June 2007)
Stock of direct foreign investment—at home: $23.06 billion (2007 est.)
Stock of direct foreign investment—abroad: $3.07 billion (2007 est.)
Market value of publicly traded shares: $29.01 billion (2006)
Currency (code): kuna (HRK)
Currency code: HRK
Exchange rates: kuna per US dollar—5.3735 (2007), 5.8625 (2006), 5.9473 (2005), 6.0358 (2004), 6.7035 (2003)
Fiscal year: calendar year

COMMUNICATIONS

Telephones—main lines in use: 1.832 million (2006)
Telephones—mobile cellular: 4.47 million (2006)
Telephone system:
general assessment: the telecommunications network has improved steadily since the mid-1990s; the number of fixed telephone lines has increased to about 40 per 100 persons; virtually 100 mobile cellular telephones per 100 persons
domestic: more than 90 percent of local lines are digital
international: country code—385; digital international service is provided through the main switch in Zagreb; Croatia participates in the Trans-Asia-Europe (TEL) fiber-optic project, which consists of 2 fiber-optic trunk connections with Slovenia and a fiber-optic trunk line from Rijeka to Split and Dubrovnik; the ADRIA-1 submarine cable provides connectivity to Albania and Greece (2007)
Radio broadcast stations: AM 16, FM 98, shortwave 5 (1999)
Radios: 1.51 million (1997)
Television broadcast stations: 36 (plus 321 repeaters) (1995)
Televisions: 1.22 million (1997)
Internet country code: .hr
Internet hosts: 261,954 (2007)
Internet Service Providers (ISPs): 9 (2000)
Internet users: 1.576 million (2006)

TRANSPORTATION

Airports: 68 (2007)
Airports—with paved runways:
total: 23
over 3,047 m: 2
2,438 to 3,047 m: 6
1,524 to 2,437 m: 2
914 to 1,523 m: 4
under 914 m: 9 (2007)
Airports—with unpaved runways:
total: 45

1,524 to 2,437 m: 1
914 to 1,523 m: 7
under 914 m: 37 (2007)
Heliports: 2 (2007)
Pipelines: gas 1,556 km; oil 583 km (2007)
Railways: *total:* 2,726 km
standard gauge: 2,726 km 1.435-m gauge (1,199 km electrified) (2006)
Roadways:
total: 28,788 km (includes 877 km of expressways) (2006)
Waterways: 785 km (2007)
Merchant marine:
total: 75 ships (1000 GRT or over) 1,165,409 GRT/1,867,160 DWT
by type: bulk carrier 21, cargo 12, chemical tanker 3, passenger/cargo 28, petroleum tanker 7, refrigerated cargo 1, roll on/roll off 3
foreign-owned: 2 (Bermuda 2)
registered in other countries: 36 (Bahamas 1, Belize 1, Liberia 5, Malta 12, Marshall Islands 4, Panama 6, St Vincent and The Grenadines 7) (2007)
Ports and terminals: Omisalj, Ploce, Rijeka, Sibenik, Vukovar (on Danube)

MILITARY

Military branches: Armed Forces of the Republic of Croatia (Oruzane Snage Republike Hrvatske, OSRH), consists of five major commands directly subordinate to a General Staff: Ground Forces (Hrvatska Kopnena Vojska, HKoV), Naval Forces (Hrvatska Ratna Mornarica, HRM), Air Force, Joint Education and Training Command, Logistics Command; Military Police Force supports each of the three Croatian military forces (2007)
Military service age and obligation: 18–27 years of age for compulsory military service; 16 years of age with consent for voluntary service; 6-month conscript service obligation; full conversion to professional military service by 2010 (2006)
Manpower available for military service:
males age 16–49: 1,035,712
females age 16–49: 1,037,896 (2008 est.)
Manpower fit for military service:
males age 16–49: 771,323
females age 16–49: 855,937 (2008 est.)
Manpower reaching militarily significant age annually:
males age 16–49: 27,500
females age 16–49: 25,893 (2008 est.)
Military expenditures—percent of GDP: 2.39% (2005 est.)

TRANSNATIONAL ISSUES

Disputes—international: dispute remains with Bosnia and Herzegovina over several small disputed sections of the boundary related to maritime access that hinders ratification of the 1999 border agreement; the Croatia-Slovenia land and maritime boundary agreement, which would have ceded most of Pirin Bay and maritime access to Slovenia and several villages to Croatia, remains unratified and in dispute; Slovenia also protests Croatia's 2003 claim to an exclusive economic zone in the Adriatic; as a European Union peripheral state, neighboring Slovenia must conform to the strict Schengen border rules to curb illegal migration and commerce through southeastern Europe while encouraging close cross-border ties with Croatia

Refugees and internally displaced persons: IDPs: 2,900–7,000 (Croats and Serbs displaced in 1992–95 war) (2007)

Illicit drugs: transit point along the Balkan route for Southwest Asian heroin to Western Europe; has been used as a transit point for maritime shipments of South American cocaine bound for Western Europe

CUBA

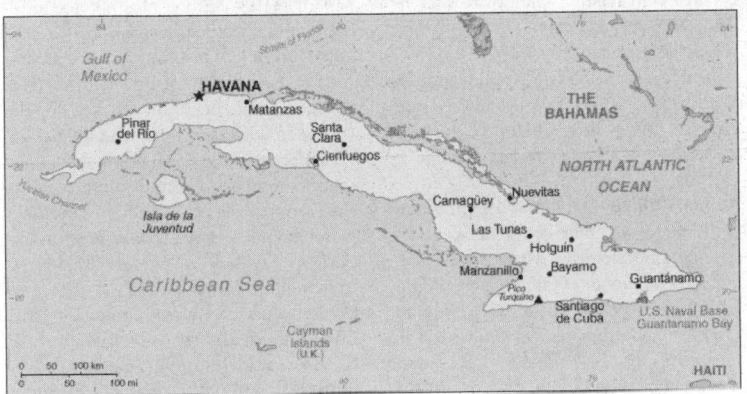

INTRODUCTION

Background: The native Amerindian population of Cuba began to decline after the European discovery of the island by Christopher COLUMBUS in 1492 and following its development as a Spanish colony during the next several centuries. Large numbers of African slaves were imported to work the coffee and sugar plantations, and Havana became the launching point for the annual treasure fleets bound for Spain from Mexico and Peru. Spanish rule, marked initially by neglect, became increasingly repressive, provoking an independence movement and occasional rebellions that were harshly suppressed. It was US intervention during the Spanish-American War in 1898 that finally overthrew Spanish rule. The subsequent Treaty of Paris established Cuban independence, which was granted in 1902 after a three-year transition period. Fidel CASTRO led a rebel army to victory in 1959; his iron rule held the subsequent regime together for nearly five decades. He stepped down as president in February 2008 in favor of his younger brother Raul CASTRO. Cuba's Communist revolution, with Soviet support, was exported throughout Latin America and Africa during the 1960s, 1970s, and 1980s. The country is now slowly recovering from a severe economic downturn in 1990, following the withdrawal of former Soviet subsidies, worth $4 billion to $6 billion annually. Cuba portrays its difficulties as the result of the US embargo in place since 1961. Illicit migration to the US—using homemade rafts, alien smugglers, air flights, or via the southwest border—is a continuing problem. The US Coast Guard intercepted 2,864 individuals attempting to cross the Straits of Florida in fiscal year 2006.

GEOGRAPHY

Location: Caribbean, island between the Caribbean Sea and the North Atlantic Ocean, 150 km south of Key West, Florida

Geographic coordinates: 21 30 N, 80 00 W

Map references: Central America and the Caribbean

Area:
total: 110,860 sq km
land: 110,860 sq km
water: 0 sq km

Area—comparative: slightly smaller than Pennsylvania

Land boundaries:
total: 29 km
border countries: US Naval Base at Guantanamo Bay 29 km
note: Guantanamo Naval Base is leased by the US and remains part of Cuba

Coastline: 3,735 km

Maritime claims:
territorial sea: 12 nm
contiguous zone: 24 nm
exclusive economic zone: 200 nm

Climate: tropical; moderated by trade winds; dry season (November to April); rainy season (May to October)

Terrain: mostly flat to rolling plains, with rugged hills and mountains in the southeast

Elevation extremes:
lowest point: Caribbean Sea 0 m
highest point: Pico Turquino 2,005 m

Natural resources: cobalt, nickel, iron ore, chromium, copper, salt, timber, silica, petroleum, arable land

Land use:
arable land: 27.63%
permanent crops: 6.54%
other: 65.83% (2005)

Irrigated land: 8,700 sq km (2003)

Total renewable water resources: 38.1 cu km (2000)

Freshwater withdrawal (domestic/industrial/agricultural): total: 8.2 cu km/yr (19%/12%/69%)
per capita: 728 cu m/yr (2000)

Natural hazards: the east coast is subject to hurricanes from August to November (in general, the country averages about one hurricane every other year); droughts are common

Environment—current issues: air and water pollution; biodiversity loss; deforestation

Environment—international agreements: party to: Antarctic Treaty, Biodiversity, Climate Change, Climate Change-Kyoto Protocol, Desertification, Endangered Species, Environmental Modification, Hazardous Wastes, Law of the Sea, Marine Dumping, Ozone Layer Protection, Ship Pollution, Wetlands
signed, but not ratified: Marine Life Conservation

Geography—note: largest country in Caribbean and westernmost island of the Greater Antilles

PEOPLE

Population: 11,423,952 (July 2008 est.)

Age structure:

0–14 years: 18.5% (male 1,088,311/female 1,030,499)
15–64 years: 70.5% (male 4,029,381/female 4,025,154)
65 years and over: 10.9% (male 569,002/female 681,605) (2008 est.)

Median age:
total: 36.8 years
male: 36.1 years
female: 37.5 years (2008 est.)

Population growth rate: 0.251% (2008 est.)

Birth rate: 11.27 births/1,000 population (2008 est.)

Death rate: 7.19 deaths/1,000 population (2008 est.)

Net migration rate: -1.57 migrant(s)/1,000 population (2008 est.)

Sex ratio:
at birth: 1.06 male(s)/female
under 15 years: 1.06 male(s)/female
15–64 years: 1 male(s)/female
65 years and over: 0.83 male(s)/female
total population: 0.99 male(s)/female (2008 est.)

Infant mortality rate:
total: 5.93 deaths/1,000 live births
male: 6.64 deaths/1,000 live births
female: 5.17 deaths/1,000 live births (2008 est.)

Life expectancy at birth:
total population: 77.27 years
male: 75.02 years
female: 79.64 years (2008 est.)

Total fertility rate: 1.6 children born/woman (2008 est.)

HIV/AIDS—adult prevalence rate: less than 0.1% (2003 est.)

HIV/AIDS—people living with HIV/AIDS: 3,300 (2003 est.)

HIV/AIDS—deaths: fewer than 200 (2003 est.)

Major infectious diseases:
degree of risk: intermediate
food or waterborne diseases: bacterial diarrhea and hepatitis A
vectorborne diseases: dengue fever (2008)

Nationality:
noun: Cuban(s)
adjective: Cuban

Ethnic groups: mulatto 51%, white 37%, black 11%, Chinese 1%

Religions: nominally 85% Roman Catholic prior to CASTRO assuming power; Protestants, Jehovah's Witnesses, Jews, and Santeria are also represented

Languages: Spanish

Literacy: *definition:* age 15 and over can read and write
total population: 99.8%
male: 99.8%
female: 99.8% (2002 census)

People—note: illicit emigration is a continuing problem; Cubans attempt to depart the island and enter the US using homemade rafts, alien smugglers, direct flights, or falsified visas; Cubans also use non-maritime routes to enter the US including direct flights to Miami and over-land via the southwest border

GOVERNMENT

Country name:
conventional long form: Republic of Cuba
conventional short form: Cuba
local long form: Republica de Cuba
local short form: Cuba

Government type: Communist state

Capital: *name:* Havana
geographic coordinates: 23 07 N, 82 21 W
time difference: UTC-5 (same time as Washington, DC during Standard Time)
daylight saving time: +1hr, begins last Sunday in March; ends last Sunday in October

Administrative divisions: 14 provinces (provincias, singular—provincia) and 1 special municipality* (municipio especial); Camaguey, Ciego de Avila, Cienfuegos, Ciudad de La Habana, Granma, Guantanamo, Holguin, Isla de la Juventud*, La Habana, Las Tunas, Matanzas, Pinar del Rio, Sancti Spiritus, Santiago de Cuba, Villa Clara

Independence: 20 May 1902 (from Spain 10 December 1898; administered by the US from 1898 to 1902); not acknowledged by the Cuban Government as a day of independence

National holiday: Triumph of the Revolution, 1 January (1959)

Constitution: 24 February 1976; amended July 1992 and June 2002

Legal system: based on Spanish civil law and influenced by American legal concepts, with large elements of Communist legal theory; has not accepted compulsory ICJ jurisdiction

Suffrage: 16 years of age; universal

Executive branch:
chief of state: President of the Council of State and President of the Council of Ministers Gen. Raul CASTRO Ruz (president since 24 February 2008); First Vice President of the Council of State and First Vice President of the Council of Ministers Gen. Jose Ramon MACHADO Ventura (since 24 February 2008); note—the president is both the chief of state and head of government
head of government: President of the Council of State and President of the Council of Ministers Gen. Raul CASTRO Ruz (president since 24 February 2008); First Vice President of the Council of State and First Vice President of the Council of Ministers Gen. Jose Ramon MACHADO Ventura (since 24 February 2008)

cabinet: Council of Ministers proposed by the president of the Council of State and appointed by the National Assembly or the 31-member Council of State, elected by the Assembly to act on its behalf when it is not in session
elections: president and vice presidents elected by the National Assembly for a term of five years; election last held 24 February 2008 (next to be held in 2013)
election results: Gen. Raul CASTRO Ruz elected president; percent of legislative vote—100%; Gen. Jose Ramon MACHADO Ventura elected vice president; percent of legislative vote—100%

Legislative branch: unicameral National Assembly of People's Power or Asemblea Nacional del Poder Popular (number of seats in the National Assembly is based on population; 614 seats; members elected directly from slates approved by special candidacy commissions to serve five-year terms)
elections: last held 20 January 2008 (next to be held in January 2013)
election results: Cuba's Communist Party is the only legal party, and officially sanctioned candidates run unopposed

Judicial branch: People's Supreme Court or Tribunal Supremo Popular (president, vice president, and other judges are elected by the National Assembly)

Political parties and leaders: Cuban Communist Party or PCC [Fidel CASTRO Ruz, first secretary]

Political pressure groups and leaders: NA

International organization participation: ACP, FAO, G-77, IAEA, ICAO, ICC, ICRM, IFAD, IFRCS, IHO, ILO, IMO, IMSO, Interpol, IOC, IOM (observer), IPU, ISO, ITSO, ITU, LAES, LAIA, NAM, OAS (excluded from formal participation since 1962), OPANAL, OPCW, PCA, UN, UNCTAD, UNESCO, UNIDO, Union Latina, UNWTO, UPU, WCL, WCO, WFTU, WHO, WIPO, WMO, WTO

Diplomatic representation in the US: none; note—Cuba has an Interests Section in the Swiss Embassy, headed by Principal Officer Jorge BOLANOS Suarez; address: Cuban Interests Section, Swiss Embassy, 2630 16th Street NW, Washington, DC 20009; telephone: [1] (202) 797-8518; FAX: [1] (202) 797-8521

Diplomatic representation from the US: none; note—the US has an Interests Section in the Swiss Embassy, headed by Principal Officer Michael E. PARMLY; address: USINT, Swiss Embassy, Calzada between L and M Streets, Vedado, Havana; telephone: [53] (7) 833-3551

through 3559 (operator assistance required); FAX: [53] (7) 833-3700; protecting power in Cuba is Switzerland

Flag description: five equal horizontal bands of blue (top, center, and bottom) alternating with white; a red equilateral triangle based on the hoist side bears a white, five-pointed star in the center

ECONOMY

Economy—overview: The government continues to balance the need for economic loosening against a desire for firm political control. It has rolled back limited reforms undertaken in the 1990s to increase enterprise efficiency and alleviate serious shortages of food, consumer goods, and services. The average Cuban's standard of living remains at a lower level than before the downturn of the 1990s, which was caused by the loss of Soviet aid and domestic inefficiencies. Since late 2000, Venezuela has been providing oil on preferential terms, and it currently supplies about 100,000 barrels per day of petroleum products. Cuba has been paying for the oil, in part, with the services of Cuban personnel in Venezuela, including some 20,000 medical professionals. In 2007, high metals prices continued to boost Cuban earnings from nickel and cobalt production. Havana continued to invest in the country's energy sector to mitigate electrical blackouts that had plagued the country since 2004.

GDP (purchasing power parity): $51.11 billion (2007 est.)

GDP (official exchange rate): $45.1 billion (2007 est.)

GDP—real growth rate: 7% (2007 est.)

GDP—per capita (PPP): $4,500 (2007 est.)

GDP—composition by sector:
agriculture: 5.2%
industry: 25%
services: 69.8% (2007 est.)

Labor force: 4.956 million
note: state sector 78%, non-state sector 22% (2007 est.)

Labor force—by occupation: *agriculture:* 20%
industry: 19.4%
services: 60.6% (2005)

Unemployment rate: 1.8% (2007 est.)

Population below poverty line: NA%

Household income or consumption by percentage share:
lowest 10%: NA%
highest 10%: NA%

Inflation rate (consumer prices): 3.1% (2007 est.)

Investment (gross fixed): 13.1% of GDP (2007 est.)

Budget: *revenues:* $41.84 billion
expenditures: $43.9 billion (2007 est.)

Public debt: 36.8% of GDP (2007 est.)

Agriculture—products: sugar, tobacco, citrus, coffee, rice, potatoes, beans; livestock

Industries: sugar, petroleum, tobacco, construction, nickel, steel, cement, agricultural machinery, pharmaceuticals

Industrial production growth rate: 2.5% (2007 est.)

Electricity—production: 16.45 billion kWh (2006)

Electricity—production by source:
fossil fuel: 93.9%
hydro: 0.6%
nuclear: 0%
other: 5.4% (2001)

Electricity—consumption: 13.87 billion kWh (2006)

Electricity—exports: 0 kWh (2006)

Electricity—imports: 0 kWh (2006)

Oil—production: 50,850 bbl/day (2006 est.)

Oil—consumption: 150,000 bbl/day (2006 est.)

Oil—exports: 0 bbl/day (2006)

Oil—imports: 98,100 bbl/day (2005)

Oil—proved reserves: 243.6 million bbl (1 January 2006 est.)

Natural gas—production: 1.058 billion cu m (2006)

Natural gas—consumption: 1.058 billion cu m (2006)

Natural gas—exports: 0 cu m (2006)

Natural gas—imports: 0 cu m (2006)

Natural gas—proved reserves: 67.89 billion cu m (1 January 2006 est.)

Current account balance: -$750 million (2007 est.)

Exports: $3.702 billion f.o.b. (2007 est.)

Exports—commodities: sugar, nickel, tobacco, fish, medical products, citrus, coffee

Exports—partners: Netherlands 21.8%, Canada 21.6%, China 18.7%, Spain 5.9% (2006)

Imports: $10.08 billion f.o.b. (2007 est.)

Imports—commodities: petroleum, food, machinery and equipment, chemicals

Imports—partners: Venezuela 26.6%, China 15.6%, Spain 9.8%, Germany 6.4%, Canada 5.6%, Italy 4.4%, US 4.3%, Brazil 4.2% (2006)

Economic aid—recipient: $87.8 million (2005 est.)

Reserves of foreign exchange and gold: $4.247 billion (31 December 2007 est.)

Debt—external: $16.79 billion (convertible currency); another $15–20 billion owed to Russia (31 December 2007 est.)

Stock of direct foreign investment—at home: $11.24 billion (2006 est.)

Stock of direct foreign investment—abroad: $4.138 billion (2006 est.)

Currency (code): Cuban peso (CUP) and Convertible peso (CUC)

Currency code: CUP (nonconvertible Cuban peso) and CUC (convertible Cuban peso)

Exchange rates: Convertible pesos per US dollar—0.9259 (2007), 0.9231 (2006)
note: Cuba has two currencies in circulation: the Cuban peso (CUP) and the convertible peso (CUC); in April 2005 the official exchange rate changed from $1 per CUC to $1.08 per CUC (0.93 CUC per $1), both for individuals and enterprises; individuals can buy 24 Cuban pesos (CUP) for each CUC sold, or sell 25 Cuban pesos for each CUC bought; enterprises, however, must exchange CUP and CUC at a 1:1 ratio.

Fiscal year: calendar year

COMMUNICATIONS

Telephones—main lines in use: 972,900 (2006)

Telephones—mobile cellular: 152,700 (2006)

Telephone system:
general assessment: greater investment beginning in 1994 and the establishment of a new Ministry of Information Technology and Communications in 2000 has resulted in improvements in the system; wireless service is expensive and must be paid in convertible pesos which effectively limits mobile cellular subscribership
domestic: national fiber-optic system under development; 95% of switches digitized by end of 2006; fixed telephone line density remains low, at less than 10 per 100 inhabitants; domestic cellular service expanding but remains at only about 2 per 100 persons
international: country code—53; fiber-optic cable laid to but not linked to US network; satellite earth station—1 Intersputnik (Atlantic Ocean region)

Radio broadcast stations: AM 169, FM 55, shortwave 1 (1998)

Radios: 3.9 million (1997)

Television broadcast stations: 58 (1997)

Televisions: 2.64 million (1997)

Internet country code: .cu

Internet hosts: 3,388 (2007)

Internet Service Providers (ISPs): 5 (2001)

Internet users: 240,000
note: private citizens are prohibited from buying computers or accessing the Internet without special authorization; foreigners may access the Internet in large hotels but are subject to firewalls; some Cubans buy illegal passwords on the black market or take advantage of

public outlets to access limited email and the government-controlled "intranet" (2006)

TRANSPORTATION

Airports: 165 (2007)
Airports—with paved runways:
total: 70
over 3,047 m: 7
2,438 to 3,047 m: 9
1,524 to 2,437 m: 18
914 to 1,523 m: 5
under 914 m: 31 (2007)
Airports—with unpaved runways:
total: 95
1,524 to 2,437 m: 1
914 to 1,523 m: 23
under 914 m: 71 (2007)
Pipelines: gas 49 km; oil 230 km (2007)
Railways:
total: 4,226 km
standard gauge: 4,226 km 1.435-m gauge (140 km electrified)
note: an additional 7,742 km of track is used by sugar plantations; about 65% of this track is standard gauge; the rest is narrow gauge (2006)
Roadways:
total: 60,858 km
paved: 29,820 km (includes 638 km of expressway)
unpaved: 31,038 km (2000)
Waterways: 240 km (2007)
Merchant marine:
total: 12 ships (1000 GRT or over) 35,030 GRT/51,388 DWT
by type: bulk carrier 2, cargo 3, chemical tanker 1, passenger 1, petroleum tanker 3, refrigerated cargo 2

foreign-owned: 1 (Spain 1)
registered in other countries: 16 (Bahamas 1, Cyprus 2, Netherlands Antilles 1, Panama 11, Spain 1) (2007)
Ports and terminals: Cienfuegos, Havana, Matanzas

MILITARY

Military branches: Revolutionary Armed Forces (Fuerzas Armadas Revolucionarias, FAR): Revolutionary Army (ER; includes Territorial Militia Troops, MTT), Revolutionary Navy (Marina de Guerra Revolucionaria, MGR; includes Marine Corps), Revolutionary Air and Air Defense Force (DAAFAR), Youth Labor Army (EJT) (2008)
Military service age and obligation: 17–28 years of age for compulsory military service; 2-year service obligation; both sexes subject to military service (2006)
Manpower available for military service: *males age 16–49:* 3,094,388
females age 16–49: 3,024,876 (2008 est.)
Manpower fit for military service:
males age 16–49: 2,543,044
females age 16–49: 2,481,823 (2008 est.)
Manpower reaching militarily significant age annually:
males age 16–49: 79,945
females age 16–49: 76,014 (2008 est.)
Military expenditures—percent of GDP: 3.8% (2006 est.)
Military—note: the collapse of the Soviet Union deprived the Cuban Army of its major economic and logistic support, and had a significant impact on

equipment numbers and serviceability; the army remains well trained and professional in nature; while the lack of replacement parts for its existing equipment and the current severe shortage of fuel have increasingly affected operational capabilities, Cuba remains able to offer considerable resistance to any regional power (2008)

TRANSNATIONAL ISSUES

Disputes—international: US Naval Base at Guantanamo Bay is leased to US and only mutual agreement or US abandonment of the area can terminate the lease
Trafficking in persons: *current situation:* Cuba is a source country for women and children trafficked for the purposes of sexual exploitation and forced child labor; Cuba is a major destination for sex tourism, which largely caters to European, Canadian, and Latin American tourists and involves large numbers of minors; there are reports that Cuban women have been trafficked to Mexico for sexual exploitation; forced labor victims also include children coerced into working in commercial agriculture
tier rating: Tier 3—Cuba does not fully comply with the minimum standards for the elimination of trafficking and is not making significant efforts to do so
Illicit drugs: territorial waters and air space serve as transshipment zone for US- and European-bound drugs; established the death penalty for certain drug-related crimes in 1999

CYPRUS

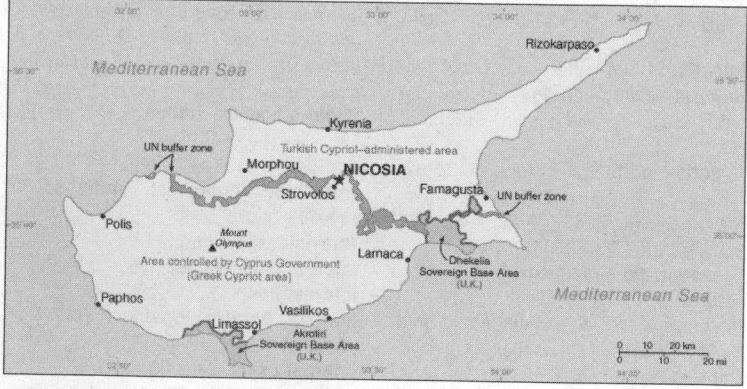

INTRODUCTION

Background: A former British colony, Cyprus became independent in 1960 following years of resistance to British rule. Tensions between the Greek Cypriot

majority and Turkish Cypriot minority came to a head in December 1963, when violence broke out in the capital of Nicosia. Despite the deployment of UN peacekeepers in 1964, sporadic intercommunal violence continued forcing

most Turkish Cypriots into enclaves throughout the island. In 1974, a Greek Government-sponsored attempt to seize control of Cyprus was met by military intervention from Turkey, which soon controlled more than a third of the island. In 1983, the Turkish-held area declared itself the "Turkish Republic of Northern Cyprus" (TRNC), but it is recognized only by Turkey. The latest two-year round of UN-brokered talks—between the leaders of the Greek Cypriot and Turkish Cypriot communities to reach an agreement to reunite the divided island—ended when the Greek Cypriots rejected the UN settlement plan in an April 2004 referendum. The entire island entered the EU on 1 May 2004, although the EU acquis—the body of common rights and obligations—applies only to the areas under direct government control, and is suspended in the areas administered by Turkish Cypriots. However, individual Turkish Cypriots able to document their eligibility for Republic of Cyprus citizenship legally enjoy the same rights accorded to other citizens of European Union states. The election of a new Cypriot president in 2008 served as the impetus for the UN to encourage both the Turkish and Cypriot Governments to reopen unification negotiations.

GEOGRAPHY

Location: Middle East, island in the Mediterranean Sea, south of Turkey
Geographic coordinates: 35 00 N, 33 00 E
Map references: Middle East
Area:
total: 9,250 sq km (of which 3,355 sq km are in north Cyprus)
land: 9,240 sq km
water: 10 sq km
Area—comparative: about 0.6 times the size of Connecticut
Land boundaries:
total: 150.4 km (approximately)
border sovereign base areas: Akrotiri 47.4 km, Dhekelia 103 km (approximately)
Coastline: 648 km
Maritime claims:
territorial sea: 12 nm
contiguous zone: 24 nm
continental shelf: 200-m depth or to the depth of exploitation
Climate: temperate; Mediterranean with hot, dry summers and cool winters
Terrain: central plain with mountains to north and south; scattered but significant plains along southern coast
Elevation extremes:
lowest point: Mediterranean Sea 0 m
highest point: Mount Olympus 1,951 m
Natural resources: copper, pyrites, asbestos, gypsum, timber, salt, marble, clay earth pigment
Land use:
arable land: 10.81%
permanent crops: 4.32%
other: 84.87% (2005)
Irrigated land: 400 sq km (2003)
Total renewable water resources: 0.4 cu km (2005)
Freshwater withdrawal (domestic/industrial/agricultural): *total:* 0.21 cu km/yr (27%/1%/71%)
per capita: 250 cu m/yr (2000)
Natural hazards: moderate earthquake activity; droughts
Environment—current issues: water resource problems (no natural reservoir catchments, seasonal disparity in rainfall, sea water intrusion to island's largest aquifer, increased salination in the north); water pollution from sewage and industrial wastes; coastal degradation; loss of wildlife habitats from urbanization
Environment—international agreements: *party to:* Air Pollution, Air Pollution-Nitrogen Oxides, Air Pollution-Persistent Organic Pollutants, Air Pollution-Sulfur 94, Biodiversity, Climate Change, Climate Change-Kyoto Protocol, Desertification, Endangered Species, Environmental Modification, Hazardous Wastes, Law of the Sea, Marine Dumping, Ozone Layer Protection, Ship Pollution, Wetlands
signed, but not ratified: none of the selected agreements
Geography—note: the third largest island in the Mediterranean Sea (after Sicily and Sardinia)

PEOPLE

Population: 792,604 (July 2008 est.)
Age structure:
0–14 years: 19.5% (male 78,922/female 75,523)
15–64 years: 68.5% (male 275,223/female 267,798)
65 years and over: 12% (male 41,592/female 53,546) (2008 est.)
Median age:
total: 35.3 years
male: 34.3 years
female: 36.4 years (2008 est.)
Population growth rate: 0.522% (2008 est.)
Birth rate: 12.56 births/1,000 population (2008 est.)
Death rate: 7.76 deaths/1,000 population (2008 est.)
Net migration rate: 0.42 migrant(s)/1,000 population (2008 est.)
Sex ratio:
at birth: 1.05 male(s)/female
under 15 years: 1.05 male(s)/female
15–64 years: 1.03 male(s)/female
65 years and over: 0.78 male(s)/female
total population: 1 male(s)/female (2008 est.)
Infant mortality rate:
total: 6.75 deaths/1,000 live births
male: 8.34 deaths/1,000 live births
female: 5.07 deaths/1,000 live births (2008 est.)
Life expectancy at birth:
total population: 78.15 years
male: 75.75 years
female: 80.67 years (2008 est.)
Total fertility rate: 1.79 children born/woman (2008 est.)
HIV/AIDS—adult prevalence rate: 0.1% (2003 est.)
HIV/AIDS—people living with HIV/AIDS: fewer than 1,000 (1999 est.)
HIV/AIDS—deaths: NA
Nationality: *noun:* Cypriot(s)
adjective: Cypriot

Ethnic groups: Greek 77%, Turkish 18%, other 5% (2001)
Religions: Greek Orthodox 78%, Muslim 18%, other (includes Maronite and Armenian Apostolic) 4%
Languages: Greek, Turkish, English
Literacy: *definition:* age 15 and over can read and write
total population: 97.6%
male: 98.9%
female: 96.3% (2003 est.)

GOVERNMENT

Country name:
conventional long form: Republic of Cyprus
conventional short form: Cyprus
local long form: Kypriaki Dimokratia/Kibris Cumhuriyeti
local short form: Kypros/Kibris
note: the Turkish Cypriot community, which administers the northern part of the island, refers to itself as the "Turkish Republic of Northern Cyprus" (TRNC)
Government type: republic
note: a separation of the two ethnic communities inhabiting the island began following the outbreak of communal strife in 1963; this separation was further solidified after the Turkish intervention in July 1974 that followed a Greek junta-supported coup attempt gave the Turkish Cypriots de facto control in the north; Greek Cypriots control the only internationally recognized government; on 15 November 1983 Turkish Cypriot "President" Rauf DENKTASH declared independence and the formation of a "Turkish Republic of Northern Cyprus" (TRNC), which is recognized only by Turkey
Capital: *name:* Nicosia (Lefkosia)
geographic coordinates: 35 10 N, 33 22 E
time difference: UTC+2 (7 hours ahead of Washington, DC during Standard Time)
daylight saving time: +1hr, begins last Sunday in March; ends last Sunday in October
Administrative divisions: 6 districts; Famagusta, Kyrenia, Larnaca, Limassol, Nicosia, Paphos; note—Turkish Cypriot area's administrative divisions include Kyrenia, all but a small part of Famagusta, and small parts of Lefkosia (Nicosia)
Independence: 16 August 1960 (from UK); note—Turkish Cypriots proclaimed self-rule on 13 February 1975 and independence in 1983, but these proclamations are only recognized by Turkey
National holiday: Independence Day, 1 October (1960); note—Turkish Cypriots celebrate 15 November (1983) as Independence Day

175

Constitution: 16 August 1960

note: from December 1963, the Turkish Cypriots no longer participated in the government; negotiations to create the basis for a new or revised constitution to govern the island and for better relations between Greek and Turkish Cypriots have been held intermittently since the mid-1960s; in 1975, following the 1974 Turkish intervention, Turkish Cypriots created their own constitution and governing bodies within the "Turkish Federated State of Cyprus," which became the "Turkish Republic of Northern Cyprus (TRNC)" when the Turkish Cypriots declared their independence in 1983; a new constitution for the "TRNC" passed by referendum on 5 May 1985, although the "TRNC" remains unrecognized by any country other than Turkey

Legal system: based on English common law, with civil law modifications; accepts compulsory ICJ jurisdiction, with reservations

Suffrage: 18 years of age; universal

Executive branch:

chief of state: President Dimitris CHRISTOFIAS (since 28 February 2008); note—the president is both the chief of state and head of government; post of vice president is currently vacant; under the 1960 constitution, the post is reserved for a Turkish Cypriot

head of government: President Dimitris CHRISTOFIAS (since 28 February 2008)

cabinet: Council of Ministers appointed jointly by the president and vice president

elections: president elected by popular vote for a five-year term; election last held 17 and 24 February 2008 (next to be held in February 2013)

election results: Dimitris CHRISTOFIAS elected president; percent of vote (first round)—Ioannis KASOULIDIS 33.5%, Dimitris CHRISTOFIAS 33.3%, Tassos PAPADOPOULOS 31.8%; (second round) Dimitris CHRISTOFIAS 53.4%, Ioannis KASOULIDIS 46.6%

note: Mehmet Ali TALAT became "president" of the "TRNC", 24 April 2005, after "presidential" elections on 17 April 2005; results—Mehmet Ali TALAT 55.6%, Dervis EROGLU 22.7%; Ferdi Sabit SOYER is "TRNC prime minister" and heads the Council of Ministers (cabinet) in coalition with "Foreign Minister and Deputy Prime Minister" Turgay AVCI

Legislative branch: unicameral—area under government control: House of Representatives or Vouli Antiprosopon (80 seats, 56 assigned to the Greek

Cypriots, 24 to Turkish Cypriots; note—only those assigned to Greek Cypriots are filled; members are elected by popular vote to serve five-year terms); area administered by Turkish Cypriots: Assembly of the Republic or Cumhuriyet Meclisi (50 seats; members are elected by popular vote to serve five-year terms)

elections: area under government control: last held 21 May 2006 (next to be held 2011); area administered by Turkish Cypriots: last held 14 December 2003 (next to be held in 2008)

election results: area under government control: House of Representatives—percent of vote by party—AKEL 31.1%, DISY 30.3%, DIKO 17.9%, EDEK 8.9%, EURO.KO 5.8%, Greens 2.0%; seats by party—AKEL (Communist) 18, DISY 18, DIKO 11, EDEK 5, EURO.KO 3, Greens 1; area administered by Turkish Cypriots: Assembly of the Republic—percent of vote by party—CTP 35.8%, UBP 32.3%, Peace and Democratic Movement 13.4%, DP 12.3%; seats by party—CTP 19, UBP 18, Peace and Democratic Movement 6, DP 7; note—"TRNC" seats by party as of September 2006—CTP 25, OP 3, UBP 13, DP 6, BDH 1, independents 2

Judicial branch: Supreme Court (judges are appointed jointly by the president and vice president)

note: there is also a Supreme Court in the area administered by Turkish Cypriots

Political parties and leaders: *area under government control:* Democratic Party or DIKO [Marios KAROYIAN]; Democratic Rally or DISY [Nikos ANASTASIADHIS]; European Democracy or EURO.DI [Prodromos PRODROMOU] (evolved from For Europe which merged with New Horizons); European Party or EURO.KO [Demetris SYLLOURIS]; Fighting Democratic Movement or ADIK [Dinos MIKHAILIDIS]; Green Party of Cyprus [George PERDIKIS]; Movement for Social Democrats or EDEK [Yannakis OMIROU]; Political Movement of Hunters [Michalis PAFITANIS]; Progressive Party of the Working People or AKEL (Communist Party) [Dimitris CHRISTOFIAS]; United Democrats or EDI [Michalis PAPAPETROU]

area administered by Turkish Cypriots: Communal Liberation Party or TKP [Huseyin ANGOLEMLI]; Cyprus Socialist Party or KSP [Kazim ONGEN]; Democratic Party or DP [Serder DENKTASH]; Freedom and Reform Party or OP [Turgay AVCI]; National Unity Party or UBP [Tahsin ERTUGRULOGLU]; Nationalist Justice Party or MAP [Ata TEPE]; New Party or YP [Huseyin

TURAN]; Our Party or BP [Okkay SADIKOGLU]; Patriotic Unity Movement or YBH [Oguz OZEN]; Peace and Democratic Movement or BDH [Mustafa AKINCI]; Renewal Progress Party or YAP [Ertugrul HASIPOGLU]; Republican Turkish Party or CTP [Ferdi Sabit SOYER]; United Cyprus Party or BKP [Isset IZCAN]

Political pressure groups and leaders: Confederation of Cypriot Workers or SEK (pro-West); Confederation of Revolutionary Labor Unions or Dev-Is; Federation of Turkish Cypriot Labor Unions or Turk-Sen; Pan-Cyprian Labor Federation or PEO (Communist controlled)

International organization participation: Australia Group, C, CE, EBRD, EIB, EMU, EU, FAO, IAEA, IBRD, ICAO, ICC, ICCt, IFAD, IFC, IHO, ILO, IMF, IMO, IMSO, Interpol, IOC, IOM, IPU, ISO, ITSO, ITU, ITUC, MIGA, NAM (guest), NSG, OAS (observer), OIF, OPCW, OSCE, PCA, UN, UNCTAD, UNESCO, UNHCR, UNIDO, UNIFIL, UNWTO, UPU, WCL, WCO, WFTU, WHO, WIPO, WMO, WTO

Diplomatic representation in the US:

chief of mission: Ambassador Andreas KAKOURIS

chancery: 2211 R Street NW, Washington, DC 20008

telephone: [1] (202) 462-5772, 462-0873

FAX: [1] (202) 483-6710

consulate(s) general: New York

note: representative of the Turkish Cypriot community in the US is Hilmi AKIL; office at 1667 K Street NW, Washington, DC; telephone [1] (202) 887-6198

Diplomatic representation from the US:

chief of mission: Ambassador Ronald L. SCHLICHER

embassy: corner of Metochiou and Ploutarchou Streets, 2407 Engomi, Nicosia

mailing address: P. O. Box 24536, 1385 Nicosia

telephone: [357] (22) 393939

FAX: [357] (22) 780944

Flag description: white with a copper-colored silhouette of the island (the name Cyprus is derived from the Greek word for copper) above two green crossed olive branches in the center of the flag; the branches symbolize the hope for peace and reconciliation between the Greek and Turkish communities

note: the "Turkish Republic of Northern Cyprus" flag has a white field with narrow horizontal red stripes positioned a small distance from the top and bottom edges between which is centered a red crescent and a red five-pointed star

ECONOMY

Economy—overview: The area of the Republic of Cyprus under government control has a market economy dominated by the service sector, which accounts for 78% of GDP. Tourism, financial services, and real estate are the most important sectors. Erratic growth rates over the past decade reflect the economy's reliance on tourism, which often fluctuates with political instability in the region and economic conditions in Western Europe. Nevertheless, the economy in the area under government control grew by an average of 3.6% per year during the period of 2000–06, well above the EU average. Cyprus joined the European Exchange Rate Mechanism (ERM2) in May 2005 and adopted the euro as its national currency on 1 January 2008. An aggressive austerity program in the preceeding years, aimed at paving the way for the euro, helped turn a soaring fiscal deficit (6.3% in 2003) into a surplus of 1.5% in 2007. As in the area administered by Turkish Cypriots, water shortages are a perennial problem; a few desalination plants are now on line. After 10 years of drought, the country received substantial rainfall from 2001–04 alleviating immediate concerns. Rainfall in 2005 and 2006, however, was well below average, making water rationing a necessity in 2007.

GDP (purchasing power parity): $36.53 billion (2007 est.)

GDP (official exchange rate): $21.3 billion (2007 est.)

GDP—real growth rate: 4.4% (2007 est.)

GDP—per capita (PPP): $46,900 (2007 est.)

GDP—composition by sector:
agriculture: 2.7%
industry: 19.2%
services: 78% (2007 est.)

Labor force: 393,000 (2007 est.)

Labor force—by occupation:
agriculture: 8.5%
industry: 20.5%
services: 71% (2006 est.)

Unemployment rate: 3.9% (2007 est.)

Population below poverty line: NA%

Household income or consumption by percentage share:
lowest 10%: NA%
highest 10%: NA%

Distribution of family income—Gini index: 29 (2005)

Inflation rate (consumer prices): 2.2% (2007 est.)

Investment (gross fixed): 20.8% of GDP (2007 est.)

Budget:
revenues: $9.996 billion

expenditures: $9.304 billion (2007 est.)

Public debt: 59.6% of GDP (2007 est.)

Agriculture—products: citrus, vegetables, barley, grapes, olives, vegetables; poultry, pork, lamb; dairy, cheese

Industries: tourism, food and beverage processing, cement and gypsum production, ship repair and refurbishment, textiles, light chemicals, metal products, wood, paper, stone, and clay products

Industrial production growth rate: 3.8% (2007 est.)

Electricity—production: 4.618 billion kWh (2006)

Electricity—production by source:
fossil fuel: 100%
hydro: 0%
nuclear: 0%
other: 0% (2001)

Electricity—consumption: 4.135 billion kWh (2006)

Electricity—exports: 0 kWh (2005)

Electricity—imports: 0 kWh (2005)

Oil—production: 300 bbl/day (2005 est.)

Oil—consumption: 56,000 bbl/day (2005 est.)

Oil—exports: 0 bbl/day (2004)

Oil—imports: 51,640 bbl/day (2004)

Oil—proved reserves: NA

Natural gas—production: 0 cu m (2007 est.)

Natural gas—consumption: NA

Natural gas—exports: 0 cu m (2007 est.)

Natural gas—imports: NA

Natural gas—proved reserves: 0 cu m (1 January 2006)

Current account balance: -$1.514 billion (2007 est.)

Exports: $1.489 billion f.o.b. (2007 est.)

Exports—commodities: citrus, potatoes, pharmaceuticals, cement, and clothing

Exports—partners: UK 15.1%, Greece 14.2%, France 7.6%, Germany 4.9%, UAE 4.1% (2006)

Imports: $7.786 billion f.o.b. (2007 est.)

Imports—commodities: consumer goods, petroleum and lubricants, intermediate goods, machinery, transport equipment

Imports—partners: Greece 17.6%, Italy 11.4%, Germany 9%, UK 8.9%, Israel 6.2%, France 4.3%, Netherlands 4.3%, China 4.2% (2006)

Economic aid—donor: $25.9 million (2006)

Economic aid—recipient: $15 million (2006)

Reserves of foreign exchange and gold: $6.507 billion (31 December 2007 est.)

Debt—external: $27.69 billion (31 December 2007 est.)

Stock of direct foreign investment—at home: $13.18 billion (2007 est.)

Stock of direct foreign investment—abroad: $5.324 billion (2007 est.)

Market value of publicly traded shares: $48.2 billion (December 2007)

Currency (code): Cypriot pound (CYP); euro (EUR) after 1 January 2008

Currency code: CYP; TRL

Exchange rates: *Cypriot pounds per US dollar:* 0.4286 (2007), 0.4586 (2006), 0.4641 (2005), 0.4686 (2004), 0.5174 (2003)

Fiscal year: calendar year

Economy of the area administered by Turkish Cypriots: *Economy—overview:* The Turkish Cypriot economy has roughly 30% of the per capita GDP of the south, and economic growth tends to be volatile, given the north's relative isolation, bloated public sector, reliance on the Turkish lira, and small market size. Agriculture and services, together, employ more than half of the work force. The Turkish Cypriot economy grew around 10.6% in 2006, fueled by growth in the construction and education sectors, as well as increased employment of Turkish Cypriots in the area under government control. GDP declined about 2.0% in 2007. The Turkish Cypriots are heavily dependent on transfers from the Turkish Government. Ankara directly finances around one-third of the "TRNC's" budget. Aid from Turkey has exceeded $400 million annually in recent years.

GDP (purchasing power parity): $1.865 billion (2006 est.)

GDP—real growth rate: -2% (2007 est.)

GDP—per capita: $11,800 (2006 est.)

GDP—composition by sector:
agriculture: 8.6%, industry: 22.5%, services: 69.1% (2006 est.)

Labor force: 95,030 (2007 est.)

Labor force—by occupation: agriculture: 14.5%, industry: 29%, services: 56.5% (2004)

Unemployment rate: 9.4% (2005 est.)

Population below poverty line: %NA

Inflation rate: 11.4% (2006)

Budget:
revenues: $2.5 billion, expenditures: $2.5 billion (2006)

Agriculture—products: citrus fruit, dairy, potatoes, grapes, olives, poultry, lamb

Industries: foodstuffs, textiles, clothing, ship repair, clay, gypsum, copper, furniture

Industrial production growth rate: -0.3% (2007 est.)

Electricity production: 998.9 million kWh (2005)

Electricity consumption: 797.9 million kWh (2005)

Exports: $68.1 million, f.o.b. (2007 est.)

Export—commodities: citrus, dairy, potatoes, textiles

Export—partners: Turkey 40%; direct trade between the area administered by Turkish Cypriots and the area under government control remains limited

Imports: $1.2 billion, f.o.b. (2007 est.)

Import—commodities: vehicles, fuel, cigarettes, food, minerals, chemicals, machinery

Import—partners: Turkey 60%; direct trade between the area administered by Turkish Cypriots and the area under government control remains limited

Economic aid—recipient: under a July 2006 agreement, Turkey plans to provide the area administered by Turkish Cypriots 1.875 billion YTL ($1.3 billion) over three years (600 million YTL in 2006, 625 million YTL in 2007 and 650 million YTL in 2008); Turkey has forgiven most past aid; additionally, the EU pledged financial assistance of Euro 259 million ($388 million) in 2004, which is yet to be disbursed.

Reserves of foreign exchange and gold: $NA

Debt—external: $NA

Currency (code): Turkish new lira (YTL)

Exchange rates: Turkish new lira per US dollar: 1.319 (2007) 1.4286 (2006) 1.3436 (2005) 1.4255 (2004) 1.5009 (2003)

COMMUNICATIONS

Telephones—main lines in use: area under government control: 408,300 (2006); area administered by Turkish Cypriots: 86,228 (2002)

Telephones—mobile cellular: area under government control: 777,500 (2006); area administered by Turkish Cypriots: 143,178 (2002)

Telephone system:

general assessment: excellent in both area under government control and area administered by Turkish Cypriots

domestic: open-wire, fiber-optic cable, and microwave radio relay

international: country code—357 (area administered by Turkish Cypriots uses the country code of Turkey—90); a number of submarine cables, including the SEA-ME-WE-3, combine to provide connectivity to Western Europe, the Middle East, and Asia; tropospheric scatter; satellite earth stations—8 (3 Intelsat—1 Atlantic Ocean and 2 Indian Ocean, 2 Eutelsat, 2 Intersputnik, and 1 Arabsat)

Radio broadcast stations: *area under government control:* AM 5, FM 76, shortwave 0

area administered by Turkish Cypriots: AM 1, FM 20, shortwave 1 (2004)

Radios: Greek Cypriot **Area:** 310,000 (1997); Turkish Cypriot **Area:** 56,450 (1994)

Television broadcast stations: *area under government control:* 8

area administered by Turkish Cypriots: 2 (plus 4 relay) (2004)

Televisions: Greek Cypriot **Area:** 248,000 (1997); Turkish Cypriot **Area:** 52,300 (1994)

Internet country code: .cy

Internet hosts: 36,964 (2007)

Internet Service Providers (ISPs): 6 (2000)

Internet users: 356,600 (2006)

TRANSPORTATION

Airports: 16 (2007)

Airports—with paved runways:

total: 13

2,438 to 3,047 m: 7

1,524 to 2,437 m: 2

914 to 1,523 m: 3

under 914 m: 1 (2007)

Airports—with unpaved runways:

total: 3

1,524 to 2,437 m: 1

under 914 m: 2 (2007)

Heliports: 10 (2007)

Roadways:

total: 14,630 km (area under government control: 12,280 km; area administered by Turkish Cypriots: 2,350 km)

paved: area under government control: 7,979 km (includes 257 km of expressways); area administered by Turkish Cypriots: 1,370 km

unpaved: area under government control: 4,301 km; area administered by Turkish Cypriots: 980 km (2006)

Merchant marine:

total: 868 ships (1000 GRT or over) 19,408,418 GRT/30,843,848 DWT

by type: bulk carrier 311, cargo 197, chemical tanker 58, container 163, liquefied gas 7, passenger 6, passenger/cargo 24, petroleum tanker 64, refrigerated cargo 17, roll on/roll off 16, vehicle carrier 5

foreign-owned: 724 (Austria 1, Belgium 1, Canada 2, China 10, Cuba 2, Denmark 1, Estonia 5, Germany 197, Greece 292, Hong Kong 2, India 1, Iran 2, Ireland 1, Israel 4, Italy 5, Japan 19, South Korea 2, Latvia 1, Lebanon 1, Netherlands 23, Norway 17, Philippines 1, Poland 18, Portugal 1, Russia 50, Singapore 1, Slovenia 4, Spain 7, Sweden 2, Switzerland 3, Syria 2, Turkey 1, Ukraine 6, UAE 10, UK 21, US 8)

registered in other countries: 133 (Antigua and Barbuda 2, Bahamas 20, Belize 1, Cambodia 9, Comoros 1, Georgia 1, Gibraltar 5, Greece 5, Isle of Man 4, Liberia 5, Malta 15, Marshall Islands 39, Norway 2, Panama 15, Russia 2, Samoa 1, St Vincent and The Grenadines 3, Turkey 2, UK 1, unknown 1) (2007)

Ports and terminals: area under government control: Larnaca, Limassol, Vasilikos; area administered by Turkish Cypriots: Famagusta, Kyrenia

MILITARY

Military branches: Republic of Cyprus: Greek Cypriot National Guard (Ethniki Forea, EF; includes air and naval elements); northern Cyprus: Turkish Cypriot Security Force (GKK) (2007)

Military service age and obligation: Greek Cypriot National Guard (GCNG): 18–50 years of age for compulsory military service for all Greek Cypriot males; 17 years of age for voluntary service; females are not conscripted; age of military eligibility 17 to 50; length of normal service is 25 months with a minimum of 3 months (2006)

Manpower available for military service: *Greek Cypriot National Guard (GCNG): males age 16–49:* 199,767 *females age 16–49:* 190,665 (2008 est.)

Manpower fit for military service: *Greek Cypriot National Guard (GCNG): males age 16–49:* 165,042 *females age 16–49:* 158,869 (2008 est.)

Manpower reaching militarily significant age annually: *Greek Cypriot National Guard (GCNG): males age 16–49:* 6,482 *females age 16–49:* 6,208 (2008 est.)

Military expenditures—percent of GDP: 3.8% (2005 est.)

TRANSNATIONAL ISSUES

Disputes—international: hostilities in 1974 divided the island into two de facto autonomous entities, the internationally recognized Cypriot Government and a Turkish-Cypriot community (north Cyprus); the 1,000-strong UN Peacekeeping Force in Cyprus (UNFICYP) has served in Cyprus since 1964 and maintains the buffer zone between north and south; on 1 May 2004, Cyprus entered the European Union still divided, with the EU's body of legislation and standards (acquis communitaire) suspended in the north

Refugees and internally displaced persons: *IDPs:* 210,000 (both Turkish and Greek Cypriots; many displaced for over 30 years) (2007)

Trafficking in persons: *current situation:* Cyprus is primarily a destination country for a large number of women trafficked from Eastern and Central Europe, the Philippines, and the Dominican Republic for the purpose of sexual exploitation; traffickers continued to fraudulently recruit victims for work as dancers in cabarets and nightclubs on short-term "artiste" visas, for work in pubs and bars on employment visas, or for illegal work on tourist or student visas; there were credible reports of female domestic workers from India, Sri Lanka, and the Philippines forced to work excessively long hours and denied proper compensation

tier rating: Tier 2 Watch List—Cyprus does not fully comply with the minimum standards for the elimination of trafficking and failed to show evidence of increasing efforts to address its serious trafficking for sexual exploitation problem; however, it is making significant efforts to do so

Illicit drugs: minor transit point for heroin and hashish via air routes and container traffic to Europe, especially from Lebanon and Turkey; some cocaine transits as well; despite a strengthening of anti-money-laundering legislation, remains vulnerable to money laundering; reporting of suspicious transactions in offshore sector remains weak

CZECH REPUBLIC

INTRODUCTION

Background: Following the First World War, the closely related Czechs and Slovaks of the former Austro-Hungarian Empire merged to form Czechoslovakia. During the interwar years, the new country's leaders were frequently preoccupied with meeting the demands of other ethnic minorities within the republic, most notably the Sudeten Germans and the Ruthenians (Ukrainians). After World War II, a truncated Czechoslovakia fell within the Soviet sphere of influence. In 1968, an invasion by Warsaw Pact troops ended the efforts of the country's leaders to liberalize Communist party rule and create "socialism with a human face." Anti-Soviet demonstrations the following year ushered in a period of harsh repression. With the collapse of Soviet authority in 1989, Czechoslovakia regained its freedom through a peaceful "Velvet Revolution." On 1 January 1993, the country underwent a "velvet divorce" into its two national components, the Czech Republic and Slovakia. The Czech Republic joined NATO in 1999 and the European Union in 2004.

GEOGRAPHY

Location: Central Europe, southeast of Germany

Geographic coordinates: 49 45 N, 15 30 E

Map references: Europe

Area:
total: 78,866 sq km
land: 77,276 sq km
water: 1,590 sq km

Area—comparative: slightly smaller than South Carolina

Land boundaries:
total: 2,290.2 km
border countries: Austria 466.3 km, Germany 810.3 km, Poland 761.8 km, Slovakia 251.8 km

Coastline: 0 km (landlocked)

Maritime claims: none (landlocked)

Climate: temperate; cool summers; cold, cloudy, humid winters

Terrain: Bohemia in the west consists of rolling plains, hills, and plateaus surrounded by low mountains; Moravia in the east consists of very hilly country

Elevation extremes:
lowest point: Elbe River 115 m
highest point: Snezka 1,602 m

Natural resources: hard coal, soft coal, kaolin, clay, graphite, timber

Land use:
arable land: 38.82%
permanent crops: 3%
other: 58.18% (2005)

Irrigated land: 240 sq km (2003)

Total renewable water resources: 16 cu km (2005)

Freshwater withdrawal (domestic/industrial/agricultural): *total:* 1.91 cu km/yr (41%/57%/2%)
per capita: 187 cu m/yr (2002)

Natural hazards: flooding

Environment—current issues: air and water pollution in areas of northwest Bohemia and in northern Moravia around Ostrava present health risks; acid rain damaging forests; efforts to bring industry up to EU code should improve domestic pollution

Environment—international agreements: *party to:* Air Pollution, Air Pollution-Nitrogen Oxides, Air Pollution-Persistent Organic Pollutants, Air Pollution-Sulfur 85, Air Pollution-Sulfur 94, Air Pollution-Volatile Organic Compounds, Antarctic-Environmental Protocol, Antarctic Treaty, Biodiversity, Climate Change, Climate Change-Kyoto Protocol, Desertification, Endangered Species, Environmental Modification, Hazardous Wastes, Law of the Sea, Ozone Layer Protection, Ship Pollution, Wetlands, Whaling

signed, but not ratified: none of the selected agreements

Geography—note: landlocked; strategically located astride some of oldest and most significant land routes in Europe; Moravian Gate is a traditional military corridor between the North European Plain and the Danube in central Europe

PEOPLE

Population: 10,220,911 (July 2008 est.)

Age structure:
0–14 years: 13.8% (male 723,521/female 684,786)
15–64 years: 71.2% (male 3,653,679/female 3,619,872)
65 years and over: 15.1% (male 604,419/female 934,634) (2008 est.)

Median age:
total: 39.8 years
male: 38.2 years
female: 41.6 years (2008 est.)

Population growth rate: -0.082% (2008 est.)

Birth rate: 8.89 births/1,000 population (2008 est.)

Death rate: 10.69 deaths/1,000 population (2008 est.)

Net migration rate: 0.97 migrant(s)/1,000 population (2008 est.)

Sex ratio:
at birth: 1.06 male(s)/female
under 15 years: 1.06 male(s)/female
15–64 years: 1.01 male(s)/female
65 years and over: 0.65 male(s)/female
total population: 0.95 male(s)/female (2008 est.)

Infant mortality rate:
total: 3.83 deaths/1,000 live births
male: 4.17 deaths/1,000 live births
female: 3.46 deaths/1,000 live births (2008 est.)

Life expectancy at birth:
total population: 76.62 years
male: 73.34 years

179

female: 80.08 years (2008 est.)
Total fertility rate: 1.23 children born/woman (2008 est.)
HIV/AIDS—adult prevalence rate: less than 0.1% (2001 est.)
HIV/AIDS—people living with HIV/AIDS: 2,500 (2001 est.)
HIV/AIDS—deaths: fewer than 10 (2001 est.)
Nationality: *noun:* Czech(s)
adjective: Czech
Ethnic groups: Czech 90.4%, Moravian 3.7%, Slovak 1.9%, other 4% (2001 census)
Religions: Roman Catholic 26.8%, Protestant 2.1%, other 3.3%, unspecified 8.8%, unaffiliated 59% (2001 census)
Languages: Czech 94.9%, Slovak 2%, other 2.3%, unidentified 0.8% (2001 census)
Literacy:
definition: NA
total population: 99%
male: 99%
female: 99% (2003 est.)

GOVERNMENT

Country name:
conventional long form: Czech Republic
conventional short form: Czech Republic
local long form: Ceska Republika
local short form: Cesko
Government type: parliamentary democracy
Capital: *name:* Prague
geographic coordinates: 50 05 N, 14 28 E
time difference: UTC+1 (6 hours ahead of Washington, DC during Standard Time)
daylight saving time: +1hr, begins last Sunday in March; ends last Sunday in October
Administrative divisions: 13 regions (kraje, singular—kraj) and 1 capital city* (hlavni mesto); Jihocesky Kraj, Jihomoravsky Kraj, Karlovarsky Kraj, Kralovehradecky Kraj, Liberecky Kraj, Moravskoslezsky Kraj, Olomoucky Kraj, Pardubicky Kraj, Plzensky Kraj, Praha (Prague)*, Stredocesky Kraj, Ustecky Kraj, Vysocina, Zlinsky Kraj
Independence: 1 January 1993 (Czechoslovakia split into the Czech Republic and Slovakia)
National holiday: Czech Founding Day, 28 October (1918)
Constitution: ratified 16 December 1992, effective 1 January 1993
Legal system: civil law system based on Austro-Hungarian codes; has not accepted compulsory ICJ jurisdiction; legal code modified to bring it in line with Organization on Security and Cooperation in Europe (OSCE) obligations and to expunge Marxist-Leninist legal theory

Suffrage: 18 years of age; universal
Executive branch:
chief of state: President Vaclav KLAUS (since 7 March 2003)
head of government: Prime Minister Mirek TOPOLANEK (since 9 January 2007); Deputy Prime Ministers Petr NECAS (since 9 January 2007), Martin BURSIK (since 9 January 2007), and Alexandr VONDRA (since 9 January 2007)
cabinet: Cabinet appointed by the president on the recommendation of the prime minister
elections: president elected by Parliament for a five-year term (eligible for a second term); last successful election held 15 February 2008 (after earlier elections held 8 and 9 February 2008 were inconclusive; next election to be held in February 2013); prime minister appointed by the president
election results: Vaclav KLAUS reelected president on 15 February 2008; Vaclav KLAUS 141 votes, Jan SVEJNAR 111 votes (third round; combined votes of both chambers of parliament)
Legislative branch: bicameral Parliament or Parlament consists of the Senate or Senat (81 seats; members are elected by popular vote to serve six-year terms; one-third elected every two years) and the Chamber of Deputies or Poslanecka Snemovna (200 seats; members are elected by popular vote to serve four-year terms)
elections: Senate—last held in two rounds 20–21 and 27–28 October 2006 (next to be held in October 2008); Chamber of Deputies—last held 2–3 June 2006 (next to be held by June 2010)
election results: Senate—percent of vote by party—NA; seats by party—ODS 41, CSSD 12, KDU-CSL 11, others 15, independents 2; Chamber of Deputies—percent of vote by party—ODS 35.4%, CSSD 32.3%, KSCM 12.8%, KDU-CSL 7.2%, Greens 6.3%, other 6%; seats by party—ODS 81, CSSD 74, KSCM 26, KDU-CSL 13, Greens 6; note—seats by party as of December 2007—ODS 81, CSSD 72, KSCM 26, KDU-CSL 13, Greens 6, unaffiliated 2 (former CSSD members)
Judicial branch: Supreme Court; Constitutional Court; chairman and deputy chairmen are appointed by the president for a 10-year term
Political parties and leaders: Association of Independent Candidates-European Democrats or SNK-ED [Helmut DOHNALEK]; Christian Democratic Union-Czechoslovak People's Party or KDU-CSL [Jiri CUNEK]; Civic Democratic Party or ODS [Mirek TOPOLANEK];

Communist Party of Bohemia and Moravia or KSCM [Vojtech FILIP]; Czech Social Democratic Party or CSSD [Jiri PAROUBEK]; Union of Freedom-Democratic Union or US-DEU [Jan CERNY]; Green Party [Martin BURSIK]; Independent Democrats (NEZDEM) [Vladimir ZELEZNY]; Party of Open Society (SOS) [Pavel NOVACEK]; Path of Change [Jiri LOBKOWITZ]
Political pressure groups and leaders: Czech-Moravian Confederation of Trade Unions or CMKOS [Milan STECH]
International organization participation: ACCT (observer), Australia Group, BIS, BSEC (observer), CE, CEI, CERN, EAPC, EBRD, EIB, ESA (cooperating state), EU, FAO, IAEA, IBRD, ICAO, ICC, ICCt (signatory), ICRM, IDA, IEA, IFC, IFRCS, ILO, IMF, IMO, IMSO, Interpol, IOC, IOM, IPU, ISO, ITSO, ITU, ITUC, MIGA, NAM (guest), NATO, NEA, NSG, OAS (observer), OECD, OIF (observer), OPCW, OSCE, PCA, Schengen Convention, UN, UNCTAD, UNESCO, UNIDO, UNMEE, UNMIL, UNOMIG, UNWTO, UPU, WCL, WCO, WEU (associate), WFTU, WHO, WIPO, WMO, WTO, ZC
Diplomatic representation in the US:
chief of mission: Ambassador Petr KOLAR
chancery: 3900 Spring of Freedom Street NW, Washington, DC 20008
telephone: [1] (202) 274-9100
FAX: [1] (202) 966-8540
consulate(s) general: Chicago, Los Angeles, New York
Diplomatic representation from the US:
chief of mission: Ambassador Richard W. GRABER
embassy: Trziste 15, 11801 Prague 1
mailing address: use embassy street address
telephone: [420] 257 022 000
FAX: [420] 257 022 809
Flag description: two equal horizontal bands of white (top) and red with a blue isosceles triangle based on the hoist side
note: identical to the flag of the former Czechoslovakia

ECONOMY

Economy—overview: The Czech Republic is one of the most stable and prosperous of the post-Communist states of Central and Eastern Europe. Growth in 2000–07 was supported by exports to the EU, primarily to Germany, and a strong recovery of foreign and domestic investment. Domestic demand is playing an ever more important role in underpinning growth as the availability of credit cards and mortgages increases. The current account deficit has declined to around 3.3% of GDP as demand for automotive and other products from the Czech Republic remains strong in the European Union. Rising inflation from higher food and energy prices are a risk to balanced economic growth. Significant increases in social spending in the run-up to June 2006 elections prevented, the government from meeting its goal of reducing its budget deficit to 3% of GDP in 2007. Negotiations on pension and additional healthcare reforms are continuing without clear prospects for agreement and implementation. Intensified restructuring among large enterprises, improvements in the financial sector, and effective use of available EU funds should strengthen output growth. The pro-business Civic Democratic Party-led government approved reforms in 2007 designed to cut spending on some social welfare benefits and reform the tax system with the aim of eventually reducing the budget deficit to 2.3% of GDP by 2010. Parliamentary approval for any additional reforms could prove difficult, however, because of the parliament's even split. The government withdrew a 2010 target date for euro adoption and instead aims to meet the eurozone criteria around 2012.

GDP (purchasing power parity): $248.9 billion (2007 est.)
GDP (official exchange rate): $175.3 billion (2007 est.)
GDP—real growth rate: 6.5% (2007 est.)
GDP—per capita (PPP): $24,200 (2007 est.)
GDP—composition by sector:
agriculture: 2.8%
industry: 38.4%
services: 58.8% (2007 est.)
Labor force: 5.36 million (2007 est.)
Labor force—by occupation: *agriculture:* 4.1%
industry: 37.6%
services: 58.3% (2003)
Unemployment rate: 6.6% (2007 est.)
Population below poverty line: NA%
Household income or consumption by percentage share:
lowest 10%: 4.3%
highest 10%: 22.4% (1996)
Distribution of family income—Gini index: 26 (2005)
Inflation rate (consumer prices): 2.8% (2007 est.)
Investment (gross fixed): 24.1% of GDP (2007 est.)
Budget:
revenues: $72.2 billion
expenditures: $76.29 billion (2007 est.)
Public debt: 26.6% of GDP (2007 est.)

Agriculture—products: wheat, potatoes, sugar beets, hops, fruit; pigs, poultry
Industries: metallurgy, machinery and equipment, motor vehicles, glass, armaments
Industrial production growth rate: 7.2% (2007 est.)
Electricity—production: 77.38 billion kWh (2005)
Electricity—production by source:
fossil fuel: 76.1%
hydro: 2.9%
nuclear: 20%
other: 1% (2001)
Electricity—consumption: 59.72 billion kWh (2005)
Electricity—exports: 24.99 billion kWh (2005)
Electricity—imports: 12.35 billion kWh (2005)
Oil—production: 18,030 bbl/day (2005)
Oil—consumption: 213,000 bbl/day (2005 est.)
Oil—exports: 20,930 bbl/day (2004)
Oil—imports: 203,700 bbl/day (2004)
Oil—proved reserves: 15 million bbl (1 January 2006 est.)
Natural gas—production: 165 million cu m (2005 est.)
Natural gas—consumption: 9.076 billion cu m (2005 est.)
Natural gas—exports: 81.52 million cu m (2005 est.)
Natural gas—imports: 8.976 billion cu m (2005)
Natural gas—proved reserves: 3.802 billion cu m (1 January 2006 est.)
Current account balance: -$4.384 billion (2007 est.)
Exports: $122.3 billion f.o.b. (2007 est.)
Exports—commodities: machinery and transport equipment 52%, raw materials and fuel 9%, chemicals 5% (2003)
Exports—partners: Germany 31.9%, Slovakia 8.5%, Poland 5.7%, France 5.5%, Austria 5.1%, UK 4.8%, Italy 4.6% (2006)
Imports: $116.6 billion f.o.b. (2007 est.)
Imports—commodities: machinery and transport equipment 46%, raw materials and fuels 15%, chemicals 10% (2003)
Imports—partners: Germany 32.5%, Netherlands 6.8%, Slovakia 6.2%, Poland 6.1%, Russia 5.7%, Austria 5%, Italy 4.4%, France 4.3% (2006)
Economic aid—recipient: $278.7 million in available EU structural adjustment and cohesion funds (2004)
Reserves of foreign exchange and gold: $34.59 billion (31 December 2007 est.)
Debt—external: $76.76 billion (31 December 2007)
Stock of direct foreign investment—at home: $86.75 billion (2007 est.)

Stock of direct foreign investment—abroad: $6.413 billion (2007 est.)
Market value of publicly traded shares: $48.6 billion (2006)
Currency (code): Czech koruna (CZK)
Currency code: CZK
Exchange rates: koruny per US dollar—20.53 (2007), 22.596 (2006), 23.957 (2005), 25.7 (2004), 28.209 (2003)
Fiscal year: calendar year

COMMUNICATIONS

Telephones—main lines in use: 2.888 million (2006)
Telephones—mobile cellular: 12.408 million (2006)
Telephone system:
general assessment: privatization and modernization of the Czech telecommunication system got a late start but is advancing steadily; access to the fixed-line telephone network expanded throughout the 1990s; mobile telephone usage increased sharply beginning in the mid-1990s and there are now about 120 mobile telephones per 100 persons
domestic: 93% of exchanges now digital; existing copper subscriber systems enhanced with Asymmetric Digital Subscriber Line (ADSL) equipment to accommodate Internet and other digital signals; trunk systems include fiber-optic cable and microwave radio relay
international: country code—420; satellite earth stations—6 (2 Intersputnik—Atlantic and Indian Ocean regions, 1 Intelsat, 1 Eutelsat, 1 Inmarsat, 1 Globalstar) (2007)
Radio broadcast stations: AM 31, FM 304, shortwave 17 (2000)
Radios: 3,159,134 (December 2000)
Television broadcast stations: 150 (plus 1,434 repeaters) (2000)
Televisions: 3,405,834 (December 2000)
Internet country code: .cz
Internet hosts: 1.668 million (2007)
Internet Service Providers (ISPs): more than 300 (2000)
Internet users: 3.541 million (2006)

TRANSPORTATION

Airports: 122 (2007)
Airports—with paved runways:
total: 45
over 3,047 m: 2
2,438 to 3,047 m: 10
1,524 to 2,437 m: 13
914 to 1,523 m: 2
under 914 m: 18 (2007)
Airports—with unpaved runways:
total: 77
1,524 to 2,437 m: 1
914 to 1,523 m: 26
under 914 m: 50 (2007)

181

Heliports: 1 (2007)
Pipelines: gas 7,010 km; oil 547 km; refined products 94 km (2007)
Railways:
total: 9,597 km
standard gauge: 9,597 km 1.435-m gauge (3,041 km electrified) (2006)
Roadways:
total: 127,865 km
paved: 127,865 km (includes 633 km of expressways) (2006)
Waterways: 664 km (principally on Elbe, Vltava, Oder, and other navigable rivers, lakes, and canals) (2006)
Merchant marine:
registered in other countries: 1 (St Vincent and The Grenadines 1) (2007)
Ports and terminals: Decin, Prague, Usti nad Labem

MILITARY

Military branches: Army of the Czech Republic (ACR): Joint Forces Command (includes Army and Air Forces), Support and Training Forces Command (2007)
Military service age and obligation: 18–28 years of age for voluntary and 19–28 for compulsory military service (2008)
Manpower available for military service: *males age 16–49:* 2,522,383
females age 16–49: 2,425,095 (2008 est.)
Manpower fit for military service:
males age 16–49: 2,100,789
females age 16–49: 2,018,101 (2008 est.)
Manpower reaching militarily significant age annually:
males age 16–49: 63,124

females age 16–49: 59,786 (2008 est.)
Military expenditures—percent of GDP: 1.46% (2007 est.)

TRANSNATIONAL ISSUES

Disputes—international: in 2006, Austrian public protests for the Czech Republic to close the Temelin nuclear power plant resulted in an Austrian parliamentary motion threatening international legal action
Illicit drugs: transshipment point for Southwest Asian heroin and minor transit point for Latin American cocaine to Western Europe; producer of synthetic drugs for local and regional markets; susceptible to money laundering related to drug trafficking, organized crime; significant consumer of ecstasy

DENMARK

INTRODUCTION

Background: Once the seat of Viking raiders and later a major north European power, Denmark has evolved into a modern, prosperous nation that is participating in the general political and economic integration of Europe. It joined NATO in 1949 and the EEC (now the EU) in 1973. However, the country has opted out of certain elements of the European Union's Maastricht Treaty, including the European Economic and Monetary Union (EMU), European defense cooperation, and issues concerning certain justice and home affairs.

GEOGRAPHY

Location: Northern Europe, bordering the Baltic Sea and the North Sea, on a peninsula north of Germany (Jutland); also includes two major islands (Sjaelland and Fyn)
Geographic coordinates: 56 00 N, 10 00 E
Map references: Europe
Area:
total: 43,094 sq km
land: 42,394 sq km
water: 700 sq km
note: includes the island of Bornholm in the Baltic Sea and the rest of metropolitan Denmark (the Jutland Peninsula, and the major islands of Sjaelland and Fyn), but excludes the Faroe Islands and Greenland
Area—comparative: slightly less than twice the size of Massachusetts
Land boundaries:
total: 68 km
border countries: Germany 68 km
Coastline: 7,314 km
Maritime claims:
territorial sea: 12 nm
contiguous zone: 24 nm
exclusive economic zone: 200 nm
continental shelf: 200-m depth or to the depth of exploitation

Climate: temperate; humid and overcast; mild, windy winters and cool summers
Terrain: low and flat to gently rolling plains
Elevation extremes:
lowest point: Lammefjord -7 m
highest point: Yding Skovhoej 173 m
Natural resources: petroleum, natural gas, fish, salt, limestone, chalk, stone, gravel and sand
Land use:
arable land: 52.59%
permanent crops: 0.19%
other: 47.22% (2005)
Irrigated land: 4,490 sq km (2003)
Total renewable water resources: 6.1 cu km (2003)
Freshwater withdrawal (domestic/industrial/agricultural): *total:* 0.67 cu km/yr (32%/26%/42%)
per capita: 123 cu m/yr (2002)
Natural hazards: flooding is a threat in some areas of the country (e.g., parts of Jutland, along the southern coast of the island of Lolland) that are protected from the sea by a system of dikes
Environment—current issues: air pollution, principally from vehicle and power plant emissions; nitrogen and phosphorus pollution of the North Sea; drinking and surface water becoming polluted from animal wastes and pesticides
Environment—international agreements: *party to:* Air Pollution, Air Pollution-Nitrogen Oxides, Air Pollution-Persistent Organic Pollutants, Air Pollution-Sulfur 85, Air Pollution-Sulfur 94, Air Pollution-Volatile Organic Compounds, Antarctic Treaty, Biodiversity, Climate Change, Climate Change-Kyoto Protocol, Desertification, Endangered Species, Environmental Modification, Hazardous Wastes, Law of the Sea, Marine Dumping, Marine Life Conservation, Ozone Layer Protection, Ship Pollution, Tropical Timber 83, Tropical Timber 94, Wetlands, Whaling *signed, but not ratified:* none of the selected agreements
Geography—note: controls Danish Straits (Skagerrak and Kattegat) linking Baltic and North Seas; about one-quarter of the population lives in greater Copenhagen

PEOPLE

Population: 5,484,723 (July 2008 est.)
Age structure:
0–14 years: 18.4% (male 516,735/female 490,532)
15–64 years: 65.9% (male 1,818,681/female 1,796,753)

65 years and over: 15.7% (male 374,388/female 487,634) (2008 est.)
Median age:
total: 40.3 years
male: 39.4 years
female: 41.2 years (2008 est.)
Population growth rate: 0.295% (2008 est.)
Birth rate: 10.71 births/1,000 population (2008 est.)
Death rate: 10.25 deaths/1,000 population (2008 est.)
Net migration rate: 2.49 migrant(s)/1,000 population (2008 est.)
Sex ratio:
at birth: 1.06 male(s)/female
under 15 years: 1.05 male(s)/female
15–64 years: 1.01 male(s)/female
65 years and over: 0.77 male(s)/female
total population: 0.98 male(s)/female (2008 est.)
Infant mortality rate:
total: 4.4 deaths/1,000 live births
male: 4.44 deaths/1,000 live births
female: 4.35 deaths/1,000 live births (2008 est.)
Life expectancy at birth:
total population: 78.13 years
male: 75.8 years
female: 80.59 years (2008 est.)
Total fertility rate: 1.74 children born/woman (2008 est.)
HIV/AIDS—adult prevalence rate: 0.2% (2003 est.)
HIV/AIDS—people living with HIV/AIDS: 5,000 (2003 est.)
HIV/AIDS—deaths: fewer than 100 (2003 est.)
Nationality:
noun: Dane(s)
adjective: Danish
Ethnic groups: Scandinavian, Inuit, Faroese, German, Turkish, Iranian, Somali
Religions: Evangelical Lutheran 95%, other Christian (includes Protestant and Roman Catholic) 3%, Muslim 2%
Languages: Danish, Faroese, Greenlandic (an Inuit dialect), German (small minority)
note: English is the predominant second language
Literacy:
definition: age 15 and over can read and write
total population: 99%
male: 99%
female: 99% (2003 est.)

GOVERNMENT

Country name:
conventional long form: Kingdom of Denmark

conventional short form: Denmark
local long form: Kongeriget Danmark
local short form: Danmark
Government type: constitutional monarchy
Capital: *name:* Copenhagen
geographic coordinates: 55 40 N, 12 35 E
time difference: UTC+1 (6 hours ahead of Washington, DC during Standard Time)
daylight saving time: +1hr, begins last Sunday in March; ends last Sunday in October
Administrative divisions: metropolitan Denmark—5 regions (regioner, singular—region); Hovedstaden, Midtjylland, Nordjylland, Sjaelland, Syddanmark
note: an extensive local government reform merged 271 municipalities into 98 and 13 counties into five regions, effective 1 January 2007
Independence: first organized as a unified state in 10th century; in 1849 became a constitutional monarchy
National holiday: none designated; Constitution Day, 5 June (1849) is generally viewed as the National Day
Constitution: 5 June 1953 constitution allowed for a unicameral legislature and a female chief of state
Legal system: civil law system; judicial review of legislative acts; accepts compulsory ICJ jurisdiction, with reservations
Suffrage: 18 years of age; universal
Executive branch:
chief of state: Queen MARGRETHE II (since 14 January 1972); Heir Apparent Crown Prince FREDERIK, elder son of the monarch (born 26 May 1968)
head of government: Prime Minister Anders Fogh RASMUSSEN (since 27 November 2001)
cabinet: Council of State appointed by the monarch
elections: the monarch is hereditary; following legislative elections, the leader of the majority party or the leader of the majority coalition is usually appointed prime minister by the monarch
Legislative branch: unicameral People's Assembly or Folketinget (179 seats, including 2 from Greenland and 2 from the Faroe Islands; members are elected by popular vote on the basis of proportional representation to serve four-year terms)
elections: last held 13 November 2007 (next to be held in 2011)
election results: percent of vote by party—Liberal Party 26.2%, Social Democrats 25.5%, Danish People's Party 13.9%, Socialist People's Party 13.0%, Conservative People's Party 10.4%, Social Liberal Party 5.1%, New Alliance 2.8%, Red-Green Alliance 2.2%, other 0.9%; seats by party—Liberal Party 46,

Social Democrats 45, Danish People's Party 25, Socialist People's Party 23, Conservative People's Party 18, Social Liberal Party 9, New Alliance 5, Red-Green Alliance 4; note—does not include the two seats from Greenland and the two seats from the Faroe Islands
Judicial branch: Supreme Court (judges are appointed by the monarch for life)
Political parties and leaders: Christian Democrats [Bodil KORNBEK] (was Christian People's Party); Conservative Party [Bendt BENDTSEN] (sometimes known as Conservative People's Party); Danish People's Party [Pia KJAERSGAARD]; Liberal Party [Anders Fogh RASMUSSEN]; New Alliance [Naser KHADER]; Red-Green Unity List (Alliance) [collective leadership] (bloc includes Left Socialist Party, Communist Party of Denmark, Socialist Workers' Party); Social Democratic Party [Helle THORNING-SCHMIDT]; Social Liberal Party [Margrethe VESTAGER]; Socialist People's Party [Villy SOEVNDAL]
Political pressure groups and leaders: NA
International organization participation: ADB (nonregional members), AfDB, Arctic Council, Australia Group, BIS, CBSS, CE, CERN, EAPC, EBRD, EIB, ESA, EU, FAO, G-9, IADB, IAEA, IBRD, ICAO, ICC, ICCt, ICRM, IDA, IEA, IFAD, IFC, IFRCS, IHO, ILO, IMF, IMO, IMSO, Interpol, IOC, IOM, IPU, ISO, ITSO, ITU, ITUC, MIGA, NATO, NC, NEA, NIB, NSG, OAS (observer), OECD, OPCW, OSCE, Paris Club, PCA, Schengen Convention, UN, UN Security Council (temporary), UNCTAD, UNESCO, UNHCR, UNIDO, UNMEE, UNMIL, UNMOGIP, UNOMIG, UNRWA, UNTSO, UPU, WCO, WEU (observer), WHO, WIPO, WMO, WTO, ZC
Diplomatic representation in the US:
chief of mission: Ambassador Friis Arne PETERSEN
chancery: 3200 Whitehaven Street NW, Washington, DC 20008
telephone: [1] (202) 234-4300
FAX: [1] (202) 328-1470
consulate(s) general: Chicago, New York
Diplomatic representation from the US:
chief of mission: Ambassador James P. CAIN
embassy: Dag Hammarskjolds Alle 24, 2100 Copenhagen
mailing address: PSC 73, APO AE 09716
telephone: [45] 33 41 71 00
FAX: [45] 35 43 02 23
Flag description: red with a white cross that extends to the edges of the flag; the

vertical part of the cross is shifted to the hoist side; the banner is referred to as the Dannebrog (Danish flag)
note: the shifted design element was subsequently adopted by the other Nordic countries of Finland, Iceland, Norway, and Sweden

ECONOMY

Economy—overview: The Danish economy has in recent years undergone strong expansion fueled primarily by private consumption growth, but also supported by exports and investments. This thoroughly modern market economy features high-tech agriculture, up-to-date small-scale and corporate industry, extensive government welfare measures, comfortable living standards, a stable currency, and high dependence on foreign trade. Unemployment is low and capacity constraints are limiting growth potential. Denmark is a net exporter of food and energy and enjoys a comfortable balance of payments surplus. Government objectives include streamlining the bureaucracy and further privatization of state assets. The government has been successful in meeting, and even exceeding, the economic convergence criteria for participating in the third phase (a common European currency) of the European Economic and Monetary Union (EMU), but so far Denmark has decided not to join 15 other EU members in the euro. Nonetheless, the Danish krone remains pegged to the euro. Economic growth gained momentum in 2004 and the upturn continued through 2007. The controversy over caricatures of the Prophet Muhammad printed in a Danish newspaper in September 2005 led to boycotts of some Danish exports to the Muslim world, especially exports of dairy products, but the boycotts did not have a significant impact on the overall Danish economy. Because of high GDP per capita, welfare benefits, a low Gini index, and political stability, the Danish living standards are among the highest in the world. A major long-term issue will be the sharp decline in the ratio of workers to retirees.
GDP (purchasing power parity): $203.7 billion (2007 est.)
GDP (official exchange rate): $311.9 billion (2007 est.)
GDP—real growth rate: 1.8% (2007 est.)
GDP—per capita (PPP): $37,400 (2007 est.)
GDP—composition by sector:
agriculture: 1.5%
industry: 26%
services: 72.4% (2007 est.)

Labor force: 2.86 million (2007 est.)
Labor force—by occupation: *agriculture:* 3%
industry: 21%
services: 76% (2004 est.)
Unemployment rate: 2.8% (2007 est.)
Population below poverty line: NA%
Household income or consumption by percentage share:
lowest 10%: 2%
highest 10%: 24% (2000 est.)
Distribution of family income—Gini index: 24 (2005)
Inflation rate (consumer prices): 1.7% (2007 est.)
Investment (gross fixed): 22.9% of GDP (2007 est.)
Budget:
revenues: $172.6 billion
expenditures: $158.8 billion (2007 est.)
Public debt: 26% of GDP (2007 est.)
Agriculture—products: barley, wheat, potatoes, sugar beets; pork, dairy products; fish
Industries: iron, steel, nonferrous metals, chemicals, food processing, machinery and transportation equipment, textiles and clothing, electronics, construction, furniture and other wood products, shipbuilding and refurbishment, windmills, pharmaceuticals, medical equipment
Industrial production growth rate: 1.5% (2007 est.)
Electricity—production: 43.35 billion kWh (2006)
Electricity—production by source:
fossil fuel: 82.7%
hydro: 0.1%
nuclear: 0%
other: 17.3% (2001)
Electricity—consumption: 34.02 billion kWh (2005)
Electricity—exports: 13.72 billion kWh (2006)
Electricity—imports: 6.77 billion kWh (2006)
Oil—production: 342,000 bbl/day (2006 est.)
Oil—consumption: 171,000 bbl/day (2006 est.)
Oil—exports: 320,000 bbl/day (2006)
Oil—imports: 164,000 bbl/day (2006 est.)
Oil—proved reserves: 1.328 billion bbl (1 January 2006 est.)
Natural gas—production: 9.87 billion cu m (2006 est.)
Natural gas—consumption: 4.775 billion cu m (2005 est.)
Natural gas—exports: 5.35 billion cu m (2005 est.)
Natural gas—imports: 0 cu m (2005)
Natural gas—proved reserves: 75.66 billion cu m (1 January 2006 est.)

Current account balance: $3.454 billion (2007 est.)
Exports: $101.2 billion f.o.b. (2007 est.)
Exports—commodities: machinery and instruments, meat and meat products, dairy products, fish, pharmaceuticals, furniture, windmills
Exports—partners: Germany 17.3%, Sweden 14.1%, UK 8.7%, US 6.2%, Netherlands 5.4%, Norway 5.4%, France 4.9% (2006)
Imports: $102.3 billion f.o.b. (2007 est.)
Imports—commodities: machinery and equipment, raw materials and semimanufactures for industry, chemicals, grain and foodstuffs, consumer goods
Imports—partners: Germany 21.4%, Sweden 14.2%, Norway 6.5%, Netherlands 6.3%, UK 5.7%, China 5%, France 4.4% (2006)
Economic aid—donor: ODA, $2.236 billion (2006)
Reserves of foreign exchange and gold: $34.32 billion (2006 est.)
Debt—external: $492.6 billion (30 June 2007)
Stock of direct foreign investment—at home: $149.8 billion (2007 est.)
Stock of direct foreign investment—abroad: $168.2 billion (2007 est.)
Market value of publicly traded shares: $178 billion (2005)
Currency (code): Danish krone (DKK)
Currency code: DKK
Exchange rates: Danish kroner per US dollar—5.4797 (2007), 5.9468 (2006), 5.9969 (2005), 5.9911 (2004), 6.5877 (2003)
Fiscal year: calendar year

COMMUNICATIONS

Telephones—main lines in use: 3.098 million (2006)
Telephones—mobile cellular: 5.841 million (2006)
Telephone system:
general assessment: excellent telephone and telegraph services
domestic: buried and submarine cables and microwave radio relay form trunk network, 4 cellular mobile communications systems
international: country code—45; a series of fiber-optic submarine cables link Denmark with Canada, Faroe Islands, Germany, Iceland, Netherlands, Norway, Poland, Russia, Sweden, and UK; satellite earth stations—18 (6 Intelsat, 10 Eutelsat, 1 Orion, 1 Inmarsat (Blaavand-Atlantic-East)); note—the Nordic countries (Denmark, Finland, Iceland, Norway, and Sweden) share the Danish earth station and the Eik, Norway, station for worldwide Inmarsat access
Radio broadcast stations: AM 2, FM 355, shortwave 0 (1998)
Radios: 6.02 million (1997)
Television broadcast stations: 26 (plus 51 repeaters) (1998)
Televisions: 3.121 million (1997)
Internet country code: .dk
Internet hosts: 3.114 million (2007)
Internet Service Providers (ISPs): 13 (2000)
Internet users: 3.171 million (2006)

TRANSPORTATION

Airports: 91 (2007)
Airports—with paved runways:
total: 28
over 3,047 m: 2
2,438 to 3,047 m: 7
1,524 to 2,437 m: 4
914 to 1,523 m: 12
under 914 m: 3 (2007)
Airports—with unpaved runways:
total: 63
914 to 1,523 m: 3
under 914 m: 60 (2007)
Pipelines: condensate 11 km; gas 4,073 km; oil 617 km; oil/gas/water 2 km (2007)
Railways:
total: 2,644 km
standard gauge: 2,644 km 1.435-m gauge (636 km electrified) (2007)
Roadways:
total: 72,362 km
paved: 72,362 km (includes 1,032 km of expressways) (2006)
Waterways: 400 km (2007)
Merchant marine:
total: 299 ships (1000 GRT or over) 8,767,265 GRT/10,604,081 DWT
by type: bulk carrier 7, cargo 64, chemical tanker 57, container 84, liquefied gas 2, livestock carrier 2, passenger 1, passenger/cargo 41, petroleum tanker 22, refrigerated cargo 7, roll on/roll off 8, specialized tanker 4
foreign-owned: 25 (Canada 1, Germany 13, Greece 4, Greenland 1, Norway 1, Sweden 4, UK 1)
registered in other countries: 468 (Antigua and Barbuda 15, Bahamas 66, Belgium 3, Brazil 2, Cayman Islands 3, Cyprus 1, Egypt 1, Estonia 2, France 3, Gibraltar 9, Hong Kong 12, Isle of Man 41, Italy 2, Jamaica 1, Liberia 12, Lithuania 9, Malta 10, Marshall Islands 9, Mexico 2, Netherlands 19, Netherlands Antilles 1, Norway 26, Panama 32, Portugal 3, Singapore 68, South Africa 1, Spain 2, St Vincent and The Grenadines 16, Sweden 4, UK 61, US 29, Venezuela 3) (2007)
Ports and terminals: Aalborg, Aarhus, Copenhagen, Ensted, Esbjerg, Fredericia, Kalundborg

MILITARY

Military branches: Defense Command: Army Operational Command, Admiral Danish Fleet, Island Command Greenland, Tactical Air Command, Home Guard (2008)

Military service age and obligation: 18 years of age for compulsory and voluntary military service; conscripts serve an initial training period that varies from 4 to 12 months according to specialization; reservists are assigned to mobilization units following completion of their con-script service; women eligible to volunteer for military service (2004)

Manpower available for military service:
males age 16–49: 1,235,067
females age 16–49: 1,215,418 (2008 est.)

Manpower fit for military service:
males age 16–49: 1,012,716
females age 16–49: 996,436 (2008 est.)

Manpower reaching militarily significant age annually:
males age 16–49: 36,561
females age 16–49: 34,603 (2008 est.)

Military expenditures—percent of GDP: 1.5% (2006; 1.28% 2007 est.)

TRANSNATIONAL ISSUES

Disputes—international: Iceland, the UK, and Ireland dispute Denmark's claim that the Faroe Islands' continental shelf extends beyond 200 nm; Faroese continue to study proposals for full independence; sovereignty dispute with Canada over Hans Island in the Kennedy Channel between Ellesmere Island and Greenland

DHEKELIA

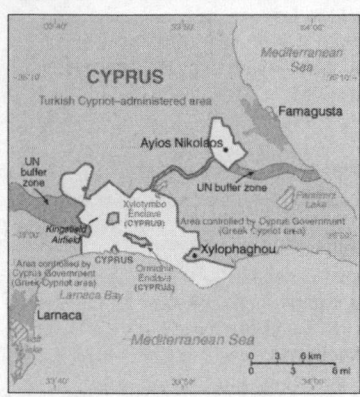

INTRODUCTION

Background: By terms of the 1960 Treaty of Establishment that created the independent Republic of Cyprus, the UK retained full sovereignty and jurisdiction over two areas of almost 254 square kilometers—Akrotiri and Dhekelia. The larger of these is the Dhekelia Sovereign Base Area, which is also referred to as the Eastern Sovereign Base Area.

GEOGRAPHY

Location: Eastern Mediterranean, on the southeast coast of Cyprus near Famagusta
Geographic coordinates: 34 59 N, 33 45 E
Map references: Middle East
Area: *total:* 130.8 sq km
note: area surrounds three Cypriot enclaves
Area—comparative: about three-quarters the size of Washington, DC
Land boundaries:
total: 103 km (approximately)
border countries: Cyprus 103 km (approximately)
Coastline: 27.5 km
Climate: temperate; Mediterranean with hot, dry summers and cool winters

Environment—current issues: netting and trapping of small migrant songbirds in the spring and autumn
Geography—note: British extraterritorial rights also extended to several small off-post sites scattered across Cyprus; of the Sovereign Base Area land 60% is privately owned and farmed, 20% is owned by the Ministry of Defense, and 20% is SBA Crown land

PEOPLE

Population: approximately 15,700 live on the Sovereign Base Areas of Akrotiri and Dhekelia including 7,700 Cypriots, 3,600 Service and UK Based Contract personnel, and 4,400 dependents
Languages: English, Greek

GOVERNMENT

Country name:
conventional long form: Dhekelia Sovereign Base Area
conventional short form: Dhekelia
Dependency status: a special form of UK overseas territory; administered by an administrator who is also the Commander, British Forces Cyprus
Capital: *name:* Episkopi Cantonment (base administrative center for Akrotiri and Dhekelia); located in Akrotiri
geographic coordinates: 34 40 N, 32 51 E
time difference: UTC+2 (7 hours ahead of Washington, DC during Standard Time)
daylight saving time: +1hr, begins last Sunday in March; ends last Sunday in October
Constitution: Sovereign Base Areas of Akrotiri and Dhekelia Order in Council 1960, effective 16 August 1960, functions as a basic legal document
Legal system: the Sovereign Base Area Administration has its own court system to deal with civil and criminal matters; laws applicable to the Cypriot population are, as far as possible, the same as the laws of the Republic of Cyprus

Executive branch:
chief of state: Queen ELIZABETH II (since 6 February 1952)
head of government: Administrator Air Vice-Marshal Richard LACEY (since 26 April 2006); note—reports to the British Ministry of Defense
elections: none; the monarch is hereditary; the administrator is appointed by the monarch
Diplomatic representation in the US: none (overseas territory of the UK)
Diplomatic representation from the US: none (overseas territory of the UK)
Flag description: the flag of the UK is used

ECONOMY

Economy—overview: Economic activity is limited to providing services to the military and their families located in Dhekelia. All food and manufactured goods must be imported.
Industries: none
Currency (code): euro (EUR) adopted 1 January 2008; note—the Cypriot pound (CYP) formerly used
Exchange rates: Cypriot pounds per US dollar—0.4286 (2007), 0.46019 (2006), 0.4641 (2005), 0.4686 (2004), 0.5174 (2003)

COMMUNICATIONS

Radio broadcast stations: AM NA, FM 1 (located in Akrotiri), shortwave NA (British Forces Broadcasting Service (BFBS) provides Radio 1 and Radio 2 service to Akrotiri, Dhekelia, and Nicosia) (2006)
Television broadcast stations: 0 (British Forces Broadcasting Service (BFBS) provides multi-channel satellite service to Akrotiri, Dhekelia, and Nicosia) (2006)

MILITARY

Military—note: includes Dhekelia Garrison and Ayios Nikolaos Station connected by a roadway

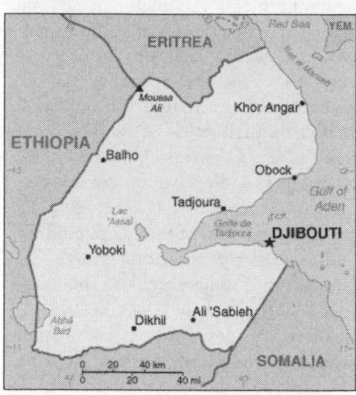

INTRODUCTION

Background: The French Territory of the Afars and the Issas became Djibouti in 1977. Hassan Gouled APTIDON installed an authoritarian one-party state and proceeded to serve as president until 1999. Unrest among the Afars minority during the 1990s led to a civil war that ended in 2001 following the conclusion of a peace accord between Afar rebels and the Issa-dominated government. In 1999, Djibouti's first multi-party presidential elections resulted in the election of Ismail Omar GUELLEH; he was re-elected to a second and final term in 2005. Djibouti occupies a strategic geographic location at the mouth of the Red Sea and serves as an important transshipment location for goods entering and leaving the east African highlands. The present leadership favors close ties to France, which maintains a significant military presence in the country, but also has strong ties with the US. Djibouti hosts the only US military base in sub-Saharan Africa and is a front-line state in the global war on terrorism.

GEOGRAPHY

Location: Eastern Africa, bordering the Gulf of Aden and the Red Sea, between Eritrea and Somalia
Geographic coordinates: 11 30 N, 43 00 E
Map references: Africa
Area:
total: 23,000 sq km
land: 22,980 sq km
water: 20 sq km
Area—comparative: slightly smaller than Massachusetts
Land boundaries:
total: 516 km
border countries: Eritrea 109 km, Ethiopia 349 km, Somalia 58 km
Coastline: 314 km

Maritime claims:
territorial sea: 12 nm
contiguous zone: 24 nm
exclusive economic zone: 200 nm
Climate: desert; torrid, dry
Terrain: coastal plain and plateau separated by central mountains
Elevation extremes:
lowest point: Lac Assal -155 m
highest point: Moussa Ali 2,028 m
Natural resources: geothermal areas, gold, clay, granite, limestone, marble, salt, diatomite, gypsum, pumice, petroleum
Land use:
arable land: 0.04%
permanent crops: 0%
other: 99.96% (2005)
Irrigated land: 10 sq km (2003)
Total renewable water resources: 0.3 cu km (1997)
Freshwater withdrawal (domestic/industrial/agricultural): *total:* 0.02 cu km/yr (84%/0%/16%)
per capita: 25 cu m/yr (2000)
Natural hazards: earthquakes; droughts; occasional cyclonic disturbances from the Indian Ocean bring heavy rains and flash floods
Environment—current issues: inadequate supplies of potable water; limited arable land; desertification; endangered species
Environment—international agreements: *party to:* Biodiversity, Climate Change, Climate Change-Kyoto Protocol, Desertification, Endangered Species, Hazardous Wastes, Law of the Sea, Ozone Layer Protection, Ship Pollution, Wetlands
signed, but not ratified: none of the selected agreements
Geography—note: strategic location near world's busiest shipping lanes and close to Arabian oilfields; terminus of rail traffic into Ethiopia; mostly wasteland; Lac Assal (Lake Assal) is the lowest point in Africa

PEOPLE

Population: 506,221 (July 2008 est.)
Age structure:
0–14 years: 43.3% (male 110,089/female 109,331)
15–64 years: 53.1% (male 139,164/female 129,614)
65 years and over: 3.6% (male 9,068/female 8,955) (2008 est.)
Median age:
total: 18.2 years
male: 18.6 years
female: 17.7 years (2008 est.)

Population growth rate: 1.945% (2008 est.)
Birth rate: 38.61 births/1,000 population (2008 est.)
Death rate: 19.16 deaths/1,000 population (2008 est.)
Net migration rate: NA
Sex ratio:
at birth: 1.03 male(s)/female
under 15 years: 1.01 male(s)/female
15–64 years: 1.07 male(s)/female
65 years and over: 1.01 male(s)/female
total population: 1.04 male(s)/female (2008 est.)
Infant mortality rate:
total: 99.13 deaths/1,000 live births
male: 106.65 deaths/1,000 live births
female: 91.38 deaths/1,000 live births (2008 est.)
Life expectancy at birth:
total population: 43.31 years
male: 41.89 years
female: 44.77 years (2008 est.)
Total fertility rate: 5.14 children born/woman (2008 est.)
HIV/AIDS—adult prevalence rate: 2.9% (2003 est.)
HIV/AIDS—people living with HIV/AIDS: 9,100 (2003 est.)
HIV/AIDS—deaths: 690 (2003 est.)
Major infectious diseases:
degree of risk: high
food or waterborne diseases: bacterial and protozoal diarrhea, hepatitis A and E, and typhoid fever
vectorborne disease: malaria
note: highly pathogenic H5N1 avian influenza has been identified in this country; it poses a negligible risk with extremely rare cases possible among US citizens who have close contact with birds (2008)
Nationality:
noun: Djiboutian(s)
adjective: Djiboutian
Ethnic groups: Somali 60%, Afar 35%, other 5% (includes French, Arab, Ethiopian, and Italian)
Religions: Muslim 94%, Christian 6%
Languages: French (official), Arabic (official), Somali, Afar
Literacy:
definition: age 15 and over can read and write
total population: 67.9%
male: 78%
female: 58.4% (2003 est.)

GOVERNMENT

Country name:
conventional long form: Republic of Djibouti

conventional short form: Djibouti
local long form: Republique de Djibouti/Jumhuriyat Jibuti
local short form: Djibouti/Jibuti
former: French Territory of the Afars and Issas, French Somaliland
Government type: republic
Capital: *name:* Djibouti
geographic coordinates: 11 35 N, 43 09 E
time difference: UTC+3 (8 hours ahead of Washington, DC during Standard Time)
Administrative divisions: 6 districts (cercles, singular—cercle); Ali Sabieh, Arta, Dikhil, Djibouti, Obock, Tadjourah
Independence: 27 June 1977 (from France)
National holiday: Independence Day, 27 June (1977)
Constitution: multiparty constitution approved by referendum 4 September 1992
Legal system: based on French civil law system, traditional practices, and Islamic law; accepts ICJ jurisdiction, with reservations
Suffrage: 18 years of age; universal
Executive branch:
chief of state: President Ismail Omar GUELLEH (since 8 May 1999)
head of government: Prime Minister Mohamed Dileita DILEITA (since 4 March 2001)
cabinet: Council of Ministers responsible to the president
elections: president elected by popular vote for a six-year term (eligible for a second term); election last held 8 April 2005 (next to be held by April 2011); prime minister appointed by the president
election results: Ismail Omar GUELLEH reelected president; percent of vote—Ismail Omar GUELLEH 100%
Legislative branch: unicameral Chamber of Deputies or Chambre des Deputes (65 seats; members elected by popular vote for five-year terms)
elections: last held 8 February 2008 (next to be held 2013)
election results: percent of vote by party—NA; seats—UMP (coalition of parties associated with President Ismail Omar GUELLAH) 65
Judicial branch: Supreme Court or Cour Supreme
Political parties and leaders: Democratic National Party or PND [ADEN Robleh Awaleh]; Democratic Renewal Party or PRD [Abdillahi HAMARITEH]; Djibouti Development Party or PDD [Mohamed Daoud CHEHEM]; Front pour la Restauration de l'Unite Democratique or FRUD [Ali Mohamed DAOUD]; People's Progress

Assembly or RPP [Ismail Omar GUELLEH] (governing party); Peoples Social Democratic Party or PPSD [Moumin Bahdon FARAH]; Republican Alliance for Democracy or ARD; Union for a Presidential Majority or UMP (a coalition of parties including RPP, FRUD, PND, and PPSD); Union for Democracy and Justice or UDJ
Political pressure groups and leaders: Union for Presidential Majority UMP (coalition includes RPP, FRUD, PPSD and PND); Union for Democratic Changeover or UAD (opposition coalition includes ARD, MRDD, and UDJ)
International organization participation: ACCT, ACP, AfDB, AFESD, AMF, AU, COMESA, FAO, G-77, IBRD, ICAO, ICCt, ICRM, IDA, IDB, IFAD, IFC, IFRCS, IGAD, ILO, IMF, IMO, Interpol, IOC, IPU, ITU, ITUC, LAS, MIGA, MINURSO, NAM, OIC, OIF, OPCW (signatory), UN, UNCTAD, UNESCO, UNIDO, UNWTO, UPU, WFTU, WHO, WIPO, WMO, WTO
Diplomatic representation in the US:
chief of mission: Ambassador Roble OLHAYE Oudine
chancery: Suite 515, 1156 15th Street NW, Washington, DC 20005
telephone: [1] (202) 331-0270
FAX: [1] (202) 331-0302
Diplomatic representation from the US:
chief of mission: Ambassador W. Stuart SYMINGTON
embassy: Plateau du Serpent, Boulevard Marechal Joffre, Djibouti
mailing address: B. P. 185, Djibouti
telephone: [253] 35 39 95
FAX: [253] 35 39 40
Flag description: two equal horizontal bands of light blue (top) and light green with a white isosceles triangle based on the hoist side bearing a red five-pointed star in the center

ECONOMY

Economy—overview: The economy is based on service activities connected with the country's strategic location and status as a free trade zone in the Horn of Africa. Two-thirds of Djibouti's inhabitants live in the capital city; the remainder are mostly nomadic herders. Scanty rainfall limits crop production to fruits and vegetables, and most food must be imported. Djibouti provides services as both a transit port for the region and an international transshipment and refueling center. Imports and exports from landlocked neighbor Ethiopia represent 85% of port activity at Djibouti's container terminal. Djibouti has few natural resources and little industry. The nation is, therefore, heavily dependent on for-

eign assistance to help support its balance of payments and to finance development projects. An unemployment rate of nearly 60% continues to be a major problem. While inflation is not a concern, due to the fixed tie of the Djiboutian franc to the US dollar, the artificially high value of the Djiboutian franc adversely affects Djibouti's balance of payments. Per capita consumption dropped an estimated 35% between 1999 and 2006 because of recession, civil war, and a high population growth rate (including immigrants and refugees). Faced with a multitude of economic difficulties, the government has fallen in arrears on long-term external debt and has been struggling to meet the stipulations of foreign aid donors.
GDP (purchasing power parity): $1.738 billion (2007 est.)
GDP (official exchange rate): $841 million (2007 est.)
GDP—real growth rate: 5.2% (2007 est.)
GDP—per capita (PPP): $2,300 (2007 est.)
GDP—composition by sector:
agriculture: 3.2%
industry: 14.9%
services: 81.9% (2006)
Labor force: 282,000 (2000)
Labor force—by occupation: *agriculture:* NA%
industry: NA%
services: NA%
Unemployment rate: 59% in urban areas, 83% in rural areas (2007 est.)
Population below poverty line: 42% (2007 est.)
Household income or consumption by percentage share:
lowest 10%: NA%
highest 10%: NA%
Inflation rate (consumer prices): 5% (2007 est.)
Budget:
revenues: $135 million
expenditures: $182 million (1999 est.)
Agriculture—products: fruits, vegetables; goats, sheep, camels, animal hides
Industries: construction, agricultural processing
Industrial production growth rate: 3% (1996 est.)
Electricity—production: 306 million kWh (2006)
Electricity—production by source:
fossil fuel: 100%
hydro: 0%
nuclear: 0%
other: 0% (2001)
Electricity—consumption: 226.9 million kWh (2006)
Electricity—exports: 0 kWh (2006)

Electricity—imports: 0 kWh (2006)
Oil—production: 0 bbl/day (2005)
Oil—consumption: 5,066 bbl/day (2007)
Oil—exports: 19.13 bbl/day (2004)
Oil—imports: 11,860 bbl/day (2004)
Oil—proved reserves: 0 bbl (1 January 2006 est.)
Natural gas—production: 0 cu m (2005 est.)
Natural gas—consumption: 0 cu m (2005 est.)
Natural gas—exports: 0 cu m (2005 est.)
Natural gas—imports: 0 cu m (2005)
Natural gas—proved reserves: 0 cu m (1 January 2006 est.)
Current account balance: -$212 million (2007 est.)
Exports: $340 million f.o.b. (2006)
Exports—commodities: reexports, hides and skins, coffee (in transit)
Exports—partners: Somalia 66.3%, Ethiopia 21.4%, Yemen 3.4% (2006)
Imports: $1.555 billion f.o.b. (2006)
Imports—commodities: foods, beverages, transport equipment, chemicals, petroleum products
Imports—partners: Saudi Arabia 21.4%, India 17.9%, China 11%, Ethiopia 4.6% (2006)
Economic aid—recipient: $78.6 million (2005)
Debt—external: $428 million (2006)
Currency (code): Djiboutian franc (DJF)
Currency code: DJF
Exchange rates: Djiboutian francs per US dollar—177.71 (2007), 174.75 (2006), 177.72 (2005), 177.72 (2004), 177.72 (2003)
Fiscal year: calendar year

COMMUNICATIONS

Telephones—main lines in use: 10,800 (2005)
Telephones—mobile cellular: 44,100 (2005)
Telephone system:
general assessment: telephone facilities in the city of Djibouti are adequate, as are the microwave radio relay connections to outlying areas of the country
domestic: microwave radio relay network; mobile cellular coverage is limited to the area in and around Djibouti city
international: country code—253; landing point for the SEA-ME-WE-3 optical telecommunications submarine cable with links to Asia, the Middle East, and Europe; satellite earth stations—2 (1 Intelsat—Indian Ocean and 1 Arabsat); Medarabtel regional microwave radio relay telephone network
Radio broadcast stations: AM 1, FM 2, shortwave 0 (2001)
Radios: 52,000 (1997)
Television broadcast stations: 1 (2001)
Televisions: 28,000 (1997)
Internet country code: .dj
Internet hosts: 168 (2007)
Internet Service Providers (ISPs): 1 (2000)
Internet users: 11,000 (2006)

TRANSPORTATION

Airports: 13 (2007)
Airports—with paved runways:
total: 3
over 3,047 m: 1
2,438 to 3,047 m: 1
1,524 to 2,437 m: 1 (2007)
Airports—with unpaved runways:
total: 10
1,524 to 2,437 m: 2
914 to 1,523 m: 5
under 914 m: 3 (2007)
Railways:
total: 100 km (Djibouti segment of the Addis Ababa-Djibouti railway)
narrow gauge: 100 km 1.000-m gauge
note: railway under joint control of Djibouti and Ethiopia but remains largely inoperable (2006)
Roadways:
total: 3,065 km
paved: 1,226 km
unpaved: 1,839 km (2000)
Merchant marine:
total: 1 ship (1000 GRT or over) 1,369 GRT/3,030 DWT
by type: cargo 1 (2007)

Ports and terminals: Djibouti

MILITARY

Military branches: Djibouti National Army (includes Navy and Air Force)
Military service age and obligation: 18 years of age for voluntary military service; 16–25 years of age for voluntary military training; no conscription (2008)
Manpower available for military service:
males age 16–49: 111,274
females age 16–49: 105,168 (2008 est.)
Manpower fit for military service:
males age 16–49: 54,460
females age 16–49: 51,684 (2008 est.)
Military expenditures—percent of GDP: 3.8% (2006)

TRANSNATIONAL ISSUES

Disputes—international: Djibouti maintains economic ties and border accords with "Somaliland" leadership while maintaining some political ties to various factions in Somalia; thousands of Somali refugees await repatriation in UNHCR camps in Djibouti
Refugees and internally displaced persons: *refugees (country of origin):* 8,642 (Somalia) (2007)
Trafficking in persons: *current situation:* Djibouti is a source, transit, and destination country for women and children trafficked for the purposes of sexual exploitation and possibly forced labor; small numbers are trafficked from Ethiopia and Somalia for sexual exploitation; economic migrants from these countries also fall victim to trafficking upon reaching Djibouti City or the Ethiopia-Djibouti trucking corridor; women and children from neighboring countries reportedly transit Djibouti to Arab countries and Somalia for ultimate use in forced labor or sexual exploitation *tier rating:* Tier 2 Watch List—Djibouti does not fully comply with the minimum standards for the elimination of trafficking; however, it is making significant efforts to do so based partly on the government's commitments to undertake future action

DOMINICA

INTRODUCTION

Background: Dominica was the last of the Caribbean islands to be colonized by Europeans due chiefly to the fierce resistance of the native Caribs. France ceded possession to Great Britain in 1763, which made the island a colony in 1805. In 1980, two years after independence, Dominica's fortunes improved when a corrupt and tyrannical administration was replaced by that of Mary Eugenia CHARLES, the first female prime minister in the Caribbean, who remained in office for 15 years. Some 3,000 Carib Indians still living on Dominica are the only pre-Columbian population remaining in the eastern Caribbean.

GEOGRAPHY

Location: Caribbean, island between the Caribbean Sea and the North Atlantic Ocean, about half way between Puerto Rico and Trinidad and Tobago

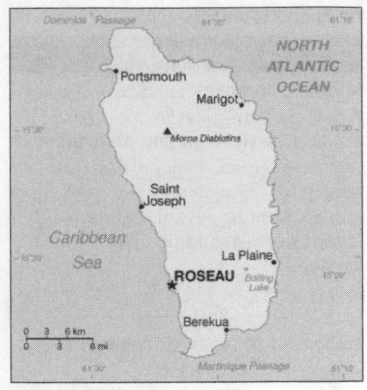

Geographic coordinates: 15 25 N, 61 20 W

Map references: Central America and the Caribbean

Area:
total: 754 sq km
land: 754 sq km
water: 0 sq km

Area—comparative: slightly more than four times the size of Washington, DC

Land boundaries: 0 km

Coastline: 148 km

Maritime claims:
territorial sea: 12 nm
contiguous zone: 24 nm
exclusive economic zone: 200 nm

Climate: tropical; moderated by northeast trade winds; heavy rainfall

Terrain: rugged mountains of volcanic origin

Elevation extremes:
lowest point: Caribbean Sea 0 m
highest point: Morne Diablatins 1,447 m

Natural resources: timber, hydropower, arable land

Land use:
arable land: 6.67%
permanent crops: 21.33%
other: 72% (2005)

Irrigated land: NA

Total renewable water resources: NA

Freshwater withdrawal (domestic/industrial/agricultural): *total:* 0.02 cu km/yr
per capita: 213 cu m/yr (1996)

Natural hazards: flash floods are a constant threat; destructive hurricanes can be expected during the late summer months

Environment—current issues: NA

Environment—international agreements: *party to:* Biodiversity, Climate Change, Climate Change-Kyoto Protocol, Desertification, Endangered Species, Environmental Modification, Hazardous Wastes, Law of the Sea, Ozone Layer Protection, Ship Pollution, Whaling
signed, but not ratified: none of the selected agreements

Geography—note: known as "The Nature Island of the Caribbean" due to its spectacular, lush, and varied flora and fauna, which are protected by an extensive natural park system; the most mountainous of the Lesser Antilles, its volcanic peaks are cones of lava craters and include Boiling Lake, the second-largest, thermally active lake in the world

PEOPLE

Population: 72,514 (July 2008 est.)

Age structure:
0–14 years: 24.7% (male 9,175/female 8,762)
15–64 years: 65.1% (male 24,192/female 22,995)
65 years and over: 10.2% (male 3,178/female 4,212) (2008 est.)

Median age:
total: 29.4 years
male: 29 years
female: 29.8 years (2008 est.)

Population growth rate: 0.196% (2008 est.)

Birth rate: 15.73 births/1,000 population (2008 est.)

Death rate: 8.32 deaths/1,000 population (2008 est.)

Net migration rate: -5.46 migrant(s)/1,000 population (2008 est.)

Sex ratio:
at birth: 1.05 male(s)/female
under 15 years: 1.05 male(s)/female
15–64 years: 1.05 male(s)/female
65 years and over: 0.75 male(s)/female
total population: 1.02 male(s)/female (2008 est.)

Infant mortality rate:
total: 14.12 deaths/1,000 live births
male: 19 deaths/1,000 live births
female: 9.01 deaths/1,000 live births (2008 est.)

Life expectancy at birth:
total population: 75.33 years
male: 72.39 years
female: 78.41 years (2008 est.)

Total fertility rate: 2.1 children born/woman (2008 est.)

HIV/AIDS—adult prevalence rate: NA

HIV/AIDS—people living with HIV/AIDS: NA

HIV/AIDS—deaths: NA

Nationality:
noun: Dominican(s)
adjective: Dominican

Ethnic groups: black 86.8%, mixed 8.9%, Carib Amerindian 2.9%, white 0.8%, other 0.7% (2001 census)

Religions: Roman Catholic 61.4%, Seventh Day Adventist 6%, Pentecostal 5.6%, Baptist 4.1%, Methodist 3.7%, Church of God 1.2%, Jehovah's Witnesses 1.2%, other Christian 7.7%, Rastafarian 1.3%, other or unspecified 1.6%, none 6.1% (2001 census)

Languages: English (official), French patois

Literacy: *definition:* age 15 and over has ever attended school
total population: 94%
male: 94%
female: 94% (2003 est.)

GOVERNMENT

Country name:
conventional long form: Commonwealth of Dominica
conventional short form: Dominica

Government type: parliamentary democracy

Capital: *name:* Roseau
geographic coordinates: 15 18 N, 61 24 W
time difference: UTC-4 (1 hour ahead of Washington, DC during Standard Time)

Administrative divisions: 10 parishes; Saint Andrew, Saint David, Saint George, Saint John, Saint Joseph, Saint Luke, Saint Mark, Saint Patrick, Saint Paul, Saint Peter

Independence: 3 November 1978 (from UK)

National holiday: Independence Day, 3 November (1978)

Constitution: 3 November 1978

Legal system: based on English common law; accepts ICJ jurisdiction

Suffrage: 18 years of age; universal

Executive branch:
chief of state: President Nicholas J. O. LIVERPOOL (since October 2003)
head of government: Prime Minister Roosevelt SKERRIT (since 8 January 2004)
cabinet: Cabinet appointed by the president on the advice of the prime minister
elections: president elected by the House of Assembly for a five-year term; election last held 1 October 2003 (next to be held in October 2008); prime minister appointed by the president
election results: Nicholas LIVERPOOL elected president; percent of legislative vote—NA%

Legislative branch: unicameral House of Assembly (30 seats; 9 members appointed, 21 elected by popular vote; to serve five-year terms)
elections: last held 5 May 2005 (next to be held by 5 August 2010); note—tradition dictates that the election will be held within five years of the last election, but technically it is five years from the first seating of parliament (12 May 2005) plus a 90-day grace period
election results: percent of vote by party—DLP 52.1%, UWP 43.6%, DFP 3.2%, other 1.1%; seats by party—DLP 12, UWP 8, independent 1

Judicial branch: Eastern Caribbean Supreme Court, consisting of the Court of Appeal and the High Court (located in Saint Lucia; one of the six judges must reside in Dominica and preside over the Court of Summary Jurisdiction)

Political parties and leaders: Dominica Freedom Party or DFP [Charles SAVARIN]; Dominica Labor Party or DLP [Roosevelt SKERRIT]; Dominica United Workers Party or UWP [Earl WILLIAMS]

Political pressure groups and leaders: Dominica Liberation Movement or DLM (a small leftist party)

International organization participation: ACCT, ACP, C, Caricom, CDB, FAO, G-77, IBRD, ICCt, ICRM, IDA, IFAD, IFC, IFRCS, ILO, IMF, IMO, Interpol, IOC, ISO (subscriber), ITU, ITUC, MIGA, NAM, OAS, OECS, OIF, OPANAL, OPCW, UN, UNCTAD, UNESCO, UNIDO, UPU, WCL, WHO, WIPO, WMO, WTO

Diplomatic representation in the US:
chief of mission: vacant
chancery: 3216 New Mexico Avenue NW, Washington, DC 20016
telephone: [1] (202) 364-6781
FAX: [1] (202) 364-6791
consulate(s) general: New York

Diplomatic representation from the US: the US does not have an embassy in Dominica; the US Ambassador to Barbados is accredited to Dominica

Flag description: green, with a centered cross of three equal bands—the vertical part is yellow (hoist side), black, and white and the horizontal part is yellow (top), black, and white; superimposed in the center of the cross is a red disk bearing a sisserou parrot encircled by 10 green, five-pointed stars edged in yellow; the 10 stars represent the 10 administrative divisions (parishes)

ECONOMY

Economy—overview: The Dominican economy depends on agriculture, primarily bananas, and remains highly vulnerable to climatic conditions and international economic developments. Tourism has increased as the government seeks to promote Dominica as an "eco-tourism" destination. In 2003, the government began a comprehensive restructuring of the economy—including elimination of price controls, privatization of the state banana company, and tax increases—to address Dominica's economic and financial crisis of 2001–02 and to meet IMF targets. This restructuring paved the way for the current economic recovery—real growth for 2006 reached a two-decade high—and will

help to reduce the debt burden, which remains at about 100% of GDP. In order to diversify the island's production base, the government is attempting to develop an offshore financial sector and is researching Dominica's capability to export geothermal energy.

GDP (purchasing power parity): $648 million (2007 est.)

GDP (official exchange rate): $311 million (2007 est.)

GDP—real growth rate: 0.9% (2007 est.)

GDP—per capita (PPP): $9,000 (2007 est.)

GDP—composition by sector:
agriculture: 17.7%
industry: 32.8%
services: 49.5% (2004 est.)

Labor force: 25,000 (2000 est.)

Labor force—by occupation: *agriculture:* 40%
industry: 32%
services: 28% (2000 est.)

Unemployment rate: 23% (2000 est.)

Population below poverty line: 30% (2002 est.)

Household income or consumption by percentage share:
lowest 10%: NA%
highest 10%: NA%

Inflation rate (consumer prices): 2.7% (2007 est.)

Budget:
revenues: $73.9 million
expenditures: $84.4 million (2001)

Agriculture—products: bananas, citrus, mangoes, root crops, coconuts, cocoa; forest and fishery potential not exploited

Industries: soap, coconut oil, tourism, copra, furniture, cement blocks, shoes

Industrial production growth rate: -10% (1997 est.)

Electricity—production: 80 million kWh (2005)

Electricity—production by source:
fossil fuel: 47.1%
hydro: 52.9%
nuclear: 0%
other: 0% (2001)

Electricity—consumption: 74.4 million kWh (2005)

Electricity—exports: 0 kWh (2005)

Electricity—imports: 0 kWh (2005)

Oil—production: 0 bbl/day (2005)

Oil—consumption: 800 bbl/day (2005 est.)

Oil—exports: 0 bbl/day (2004)

Oil—imports: 771.8 bbl/day (2004)

Oil—proved reserves: 0 bbl (1 January 2006 est.)

Natural gas—production: 0 cu m (2005 est.)

Natural gas—consumption: 0 cu m (2005 est.)

Natural gas—exports: 0 cu m (2005 est.)

Natural gas—imports: 0 cu m (2005)

Natural gas—proved reserves: 0 cu m (1 January 2006 est.)

Current account balance: -$72 million (2007 est.)

Exports: $94 million f.o.b. (2006)

Exports—commodities: bananas, soap, bay oil, vegetables, grapefruit, oranges

Exports—partners: UK 24.8%, Jamaica 12.3%, Antigua and Barbuda 9.8%, Guyana 8.3%, China 7.9%, Trinidad and Tobago 5.4%, Saint Lucia 4.5% (2006)

Imports: $296 million f.o.b. (2006)

Imports—commodities: manufactured goods, machinery and equipment, food, chemicals

Imports—partners: US 25.3%, China 22.7%, Trinidad and Tobago 13.8%, South Korea 4.8% (2006)

Economic aid—recipient: $15.17 million (2005 est.)

Debt—external: $213 million (2004)

Currency (code): East Caribbean dollar (XCD)

Currency code: XCD

Exchange rates: East Caribbean dollars per US dollar—2.7 (2007), 2.7 (2006), 2.7 (2005), 2.7 (2004), 2.7 (2003)

Fiscal year: 1 July–30 June

COMMUNICATIONS

Telephones—main lines in use: 21,000 (2004)

Telephones—mobile cellular: 41,800 (2004)

Telephone system:
general assessment: NA
domestic: fully automatic network
international: country code—1-767; landing point for the East Caribbean Fiber Optic System (ECFS) submarine cable with links to 13 other islands in the eastern Caribbean extending from the British Virgin Islands to Trinidad; microwave radio relay and SHF radiotelephone links to Martinique and Guadeloupe; VHF and UHF radiotelephone links to Saint Lucia

Radio broadcast stations: AM 2, FM 4, shortwave 0 (2003)

Radios: 46,000 (1997)

Television broadcast stations: 1 (2004)

Televisions: 6,000 (1997)

Internet country code: .dm

Internet hosts: 257 (2007)

Internet Service Providers (ISPs): 16 (2000)

Internet users: 26,000 (2005)

TRANSPORTATION

Airports: 2 (2007)

Airports—with paved runways: *total:* 2
914 to 1,523 m: 2 (2007)

Roadways:

total: 780 km

paved: 393 km

unpaved: 387 km (2000)

Merchant marine:

total: 53 ships (1000 GRT or over) 716,435 GRT/1,252,537 DWT

by type: bulk carrier 9, cargo 30, chemical tanker 2, container 1, petroleum tanker 7, refrigerated cargo 2, roll on/roll off 1, vehicle carrier 1

foreign-owned: 50 (Estonia 8, Greece 8, India 2, Latvia 2, Lebanon 1, Norway 1, NZ 3, Russia 2, Saudi Arabia 1, Singapore 8, Syria 2, Turkey 9, Ukraine 3) (2007)

Ports and terminals: Portsmouth, Roseau

MILITARY

Military branches: no regular military forces; Commonwealth of Dominica Police Force (includes Coast Guard) (2008)

Manpower available for military service:

males age 16–49: 18,584 (2008 est.)

Manpower fit for military service:

males age 16–49: 15,648 (2008 est.)

Manpower reaching militarily significant age annually:

males age 16–49: 756 (2008 est.)

Military expenditures—percent of GDP: NA (2006)

TRANSNATIONAL ISSUES

Disputes—international: Dominica is the only Caribbean state to challenge Venezuela's sovereignty claim over Aves Island and joins the other island nations in challenging whether the feature sustains human habitation, a criterion under the UN Convention on the Law of the Sea (UNCLOS), which permits Venezuela to extend its Exclusive Economic Zone (EEZ) and continental shelf claims over a large portion of the eastern Caribbean Sea

Illicit drugs: transshipment point for narcotics bound for the US and Europe; minor cannabis producer; anti-money-laundering enforcement is weak, making the country particularly vulnerable to money laundering

DOMINICAN REPUBLIC

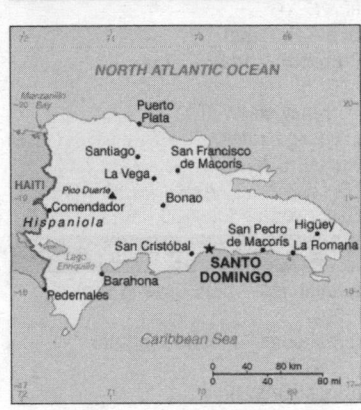

INTRODUCTION

Background: Explored and claimed by Christopher COLUMBUS on his first voyage in 1492, the island of Hispaniola became a springboard for Spanish conquest of the Caribbean and the American mainland. In 1697, Spain recognized French dominion over the western third of the island, which in 1804 became Haiti. The remainder of the island, by then known as Santo Domingo, sought to gain its own independence in 1821, but was conquered and ruled by the Haitians for 22 years; it finally attained independence as the Dominican Republic in 1844. In 1861, the Dominicans voluntarily returned to the Spanish Empire, but two years later they launched a war that restored independence in 1865. A legacy of unsettled, mostly non-representative rule followed, capped by the dictatorship of Rafael Leonidas TRUJILLO from 1930–61. Juan BOSCH was elected president in 1962, but was deposed in a military coup in 1963. In 1965, the United States led an intervention in the midst of a civil war sparked by an uprising to restore BOSCH. In 1966, Joaquin BALAGUER defeated BOSCH in an election to become president. BALAGUER maintained a tight grip on power for most of the next 30 years when international reaction to flawed elections forced him to curtail his term in 1996. Since then, regular competitive elections have been held in which opposition candidates have won the presidency. Former President (1996–2000) Leonel FERNANDEZ Reyna won election to a second term in 2004 following a constitutional amendment allowing presidents to serve more than one term.

GEOGRAPHY

Location: Caribbean, eastern two-thirds of the island of Hispaniola, between the Caribbean Sea and the North Atlantic Ocean, east of Haiti

Geographic coordinates: 19 00 N, 70 40 W

Map references: Central America and the Caribbean

Area:

total: 48,730 sq km

land: 48,380 sq km

water: 350 sq km

Area—comparative: slightly more than twice the size of New Hampshire

Land boundaries:

total: 360 km

border countries: Haiti 360 km

Coastline: 1,288 km

Maritime claims: measured from claimed archipelagic straight baselines

territorial sea: 6 nm

contiguous zone: 24 nm

exclusive economic zone: 200 nm

continental shelf: 200 nm or to the edge of the continental margin

Climate: tropical maritime; little seasonal temperature variation; seasonal variation in rainfall

Terrain: rugged highlands and mountains with fertile valleys interspersed

Elevation extremes:

lowest point: Lago Enriquillo -46 m

highest point: Pico Duarte 3,175 m

Natural resources: nickel, bauxite, gold, silver

Land use:

arable land: 22.49%

permanent crops: 10.26%

other: 67.25% (2005)

Irrigated land: 2,750 sq km (2003)

Total renewable water resources: 21 cu km (2000)

Freshwater withdrawal (domestic/industrial/agricultural): *total:* 3.39 cu km/yr (32%/2%/66%)

per capita: 381 cu m/yr (2000)

Natural hazards: lies in the middle of the hurricane belt and subject to severe storms from June to October; occasional flooding; periodic droughts

Environment—current issues: water shortages; soil eroding into the sea damages coral reefs; deforestation

Environment—international agreements: *party to:* Biodiversity, Climate Change, Climate Change-Kyoto Protocol, Desertification, Endangered Species, Hazardous Wastes, Marine Dumping, Marine Life Conservation, Ozone Layer Protection, Ship Pollution, Wetlands

signed, but not ratified: Law of the Sea

Geography—note: shares island of Hispaniola with Haiti

PEOPLE

Population: 9,507,133 (July 2008 est.)
Age structure:
0–14 years: 31.8% (male 1,537,981/female 1,482,546)
15–64 years: 62.4% (male 3,029,349/female 2,905,471)
65 years and over: 5.8% (male 255,898/female 295,888) (2008 est.)
Median age:
total: 24.7 years
male: 24.6 years
female: 24.8 years (2008 est.)
Population growth rate: 1.495% (2008 est.)
Birth rate: 22.65 births/1,000 population (2008 est.)
Death rate: 5.3 deaths/1,000 population (2008 est.)
Net migration rate: -2.4 migrant(s)/1,000 population (2008 est.)
Sex ratio:
at birth: 1.04 male(s)/female
under 15 years: 1.04 male(s)/female
15–64 years: 1.04 male(s)/female
65 years and over: 0.86 male(s)/female
total population: 1.03 male(s)/female (2008 est.)
Infant mortality rate:
total: 26.93 deaths/1,000 live births
male: 29.01 deaths/1,000 live births
female: 24.78 deaths/1,000 live births (2008 est.)
Life expectancy at birth:
total population: 73.39 years
male: 71.61 years
female: 75.24 years (2008 est.)
Total fertility rate: 2.78 children born/woman (2008 est.)
HIV/AIDS—adult prevalence rate: 1.7% (2003 est.)
HIV/AIDS—people living with HIV/AIDS: 88,000 (2003 est.)
HIV/AIDS—deaths: 7,900 (2003 est.)
Major infectious diseases:
degree of risk: high
food or waterborne diseases: bacterial diarrhea, hepatitis A, and typhoid fever
vectorborne diseases: dengue fever and malaria
water contact disease: leptospirosis (2008)
Nationality:
noun: Dominican(s)
adjective: Dominican
Ethnic groups: mixed 73%, white 16%, black 11%
Religions: Roman Catholic 95%, other 5%
Languages: Spanish
Literacy: *definition:* age 15 and over can read and write
total population: 87%
male: 86.8%
female: 87.2% (2002 census)

GOVERNMENT

Country name:
conventional long form: Dominican Republic
conventional short form: The Dominican
local long form: Republica Dominicana
local short form: La Dominicana
Government type: democratic republic
Capital: *name:* Santo Domingo
geographic coordinates: 18 28 N, 69 54 W
time difference: UTC-4 (1 hour ahead of Washington, DC during Standard Time)
Administrative divisions: 31 provinces (provincias, singular—provincia) and 1 district* (distrito); Azua, Bahoruco, Barahona, Dajabon, Distrito Nacional*, Duarte, El Seibo, Elias Pina, Espaillat, Hato Mayor, Independencia, La Altagracia, La Romana, La Vega, Maria Trinidad Sanchez, Monsenor Nouel, Monte Cristi, Monte Plata, Pedernales, Peravia, Puerto Plata, Salcedo, Samana, San Cristobal, San Jose de Ocoa, San Juan, San Pedro de Macoris, Sanchez Ramirez, Santiago, Santiago Rodriguez, Santo Domingo, Valverde
Independence: 27 February 1844 (from Haiti)
National holiday: Independence Day, 27 February (1844)
Constitution: 28 November 1966; amended 25 July 2002
Legal system: based on French civil codes; Criminal Procedures Code modified in 2004 to include important elements of an accusatory system; accepts compulsory ICJ jurisdiction
Suffrage: 18 years of age, universal and compulsory; married persons regardless of age; note—members of the armed forces and national police cannot vote
Executive branch:
chief of state: President Leonel FERNANDEZ Reyna (since 16 August 2004); Vice President Rafael ALBURQUERQUE de Castro (since 16 August 2004); note—the president is both the chief of state and head of government
head of government: President Leonel FERNANDEZ Reyna (since 16 August 2004); Vice President Rafael ALBURQUERQUE de Castro (since 16 August 2004)
cabinet: Cabinet nominated by the president
elections: president and vice president elected on the same ticket by popular vote for four-year terms (eligible for a second consecutive term); election last held 16 May 2008 (next to be held in May 2012)
election results: Leonel FERNANDEZ reelected president; percent of vote—Leonel FERNANDEZ 53.6%, Miguel

VARGAS 41%, Amable ARISTY less than 5%
Legislative branch: bicameral National Congress or Congreso Nacional consists of the Senate or Senado (32 seats; members are elected by popular vote to serve four-year terms) and the House of Representatives or Camara de Diputados (178 seats; members are elected by popular vote to serve four-year terms)
elections: Senate—last held 16 May 2006 (next to be held in May 2010); House of Representatives—last held 16 May 2006 (next to be held in May 2010)
election results: Senate—percent of vote by party—NA; seats by party—PLD 22, PRD 6, PRSC 4; House of Representatives—percent of vote by party—NA; seats by party—PLD 96, PRD 60, PRSC 22
Judicial branch: Supreme Court or Corte Suprema (judges are appointed by the National Judicial Council comprised of the president, the leaders of both chambers of congress, the president of the Supreme Court, and an additional non-governing party congressional representative)
Political parties and leaders: Dominican Liberation Party or PLD [Leonel FERNANDEZ Reyna]; Dominican Revolutionary Party or PRD [Ramon ALBURQUERQUE]; National Progressive Front [Vincent CASTILLO, Pelegrin CASTILLO]; Social Christian Reformist Party or PRSC [Enrique ANTUN]
Political pressure groups and leaders: Citizen Participation Group (Participacion Ciudadania); Collective of Popular Organizations or COP; Foundation for Institution-Building and Justice (FINJUS)
International organization participation: ACP, Caricom (observer), FAO, G-77, IADB, IAEA, IBRD, ICAO, ICC, ICCt (signatory), ICRM, IDA, IFAD, IFC, IFRCS, IHO (suspended), ILO, IMF, IMO, Interpol, IOC, IOM, IPU, ISO (correspondent), ITSO, ITU, ITUC, LAES, LAIA (observer), MIGA, NAM, OAS, OPANAL, OPCW (signatory), PCA, RG, UN, UNCTAD, UNESCO, UNIDO, Union Latina, UNOCI, UNWTO, UPU, WCL, WCO, WFTU, WHO, WIPO, WMO, WTO
Diplomatic representation in the US:
chief of mission: Ambassador Flavio Dario ESPINAL Jacobo
chancery: 1715 22nd Street NW, Washington, DC 20008
telephone: [1] (202) 332-6280
FAX: [1] (202) 265-8057
consulate(s) general: Anchorage, Boston, Chicago, Mayaguez (Puerto Rico),

193

Miami, New Orleans, New York, San Francisco, San Juan (Puerto Rico)
Diplomatic representation from the US:
chief of mission: Ambassador P. Robert FANNIN
embassy: corner of Calle Cesar Nicolas Penson and Calle Leopoldo Navarro, Santo Domingo
mailing address: Unit 5500, APO AA 34041-5500
telephone: [1] (809) 221-2171
FAX: [1] (809) 686-7437
Flag description: a centered white cross that extends to the edges divides the flag into four rectangles—the top ones are blue (hoist side) and red, and the bottom ones are red (hoist side) and blue; a small coat of arms featuring a shield supported by an olive branch (left) and a palm branch (right) is at the center of the cross; above the shield a blue ribbon displays the motto, DIOS, PATRIA, LIBERTAD (God, Fatherland, Liberty), and below the shield, REPUBLICA DOMINICANA appears on a red ribbon

ECONOMY

Economy—overview: The Dominican Republic has enjoyed strong GDP growth since 2005, with double digit growth in 2006. In 2007, exports were bolstered by the nearly 50% increase in nickel prices; however, prices are expected to fall in 2008, contributing to a slowdown in GDP growth for the year. Although the country has long been viewed primarily as an exporter of sugar, coffee, and tobacco, in recent years the service sector has overtaken agriculture as the economy's largest employer due to growth in tourism and free trade zones. The economy is highly dependent upon the US, the source of nearly three-fourths of exports, and remittances represent about a tenth of GDP, equivalent to almost half of exports and three-quarters of tourism receipts. With the help of strict fiscal targets agreed to in the 2004 renegotiation of an IMF standby loan, President FERNANDEZ has stabilized the country's financial situation, lowering inflation to less than 6%. A fiscal expansion is expected for 2008 prior to the elections in May and for Tropical Storm Noel reconstruction. Although the economy is growing at a respectable rate, high unemployment and underemployment remains an important challenge. The country suffers from marked income inequality; the poorest half of the population receives less than one-fifth of GNP, while the richest 10% enjoys nearly 40% of national income. The Central America-Dominican Republic Free Trade Agreement (CAFTA-DR) came into force in March 2007, which should boost investment and exports and diminishs losses to the Asian garment industry.
GDP (purchasing power parity): $61.79 billion (2007 est.)
GDP (official exchange rate): $36.4 billion (2007 est.)
GDP—real growth rate: 8.5% (2007 est.)
GDP—per capita (PPP): $7,000 (2007 est.)
GDP—composition by sector:
agriculture: 11.7%
industry: 23.8%
services: 64.4% (2007 est.)
Labor force: 4.027 million (2007 est.)
Labor force—by occupation: *agriculture:* 17%
industry: 24.3%
services: 58.7% (1998 est.)
Unemployment rate: 15.6% (2007 est.)
Population below poverty line: 42.2% (2004)
Household income or consumption by percentage share:
lowest 10%: 1.4%
highest 10%: 41.1% (2004)
Distribution of family income—Gini index: 51.6 (2004)
Inflation rate (consumer prices): 6.1% (2007 est.)
Investment (gross fixed): 17.1% of GDP (2007 est.)
Budget:
revenues: $7.942 billion
expenditures: $7.766 billion (2007 est.)
Public debt: 40.6% of GDP (2007 est.)
Agriculture—products: sugarcane, coffee, cotton, cocoa, tobacco, rice, beans, potatoes, corn, bananas; cattle, pigs, dairy products, beef, eggs
Industries: tourism, sugar processing, ferronickel and gold mining, textiles, cement, tobacco
Industrial production growth rate: 2.4% (2007 est.)
Electricity—production: 12.22 billion kWh (2005)
Electricity—production by source:
fossil fuel: 92%
hydro: 7.6%
nuclear: 0%
other: 0.4% (2001)
Electricity—consumption: 8.791 billion kWh (2005)
Electricity—exports: 0 kWh (2005)
Electricity—imports: 0 kWh (2005)
Oil—production: 12 bbl/day (2004)
Oil—consumption: 116,000 bbl/day (2005 est.)
Oil—exports: 0 bbl/day (2004)
Oil—imports: 116,700 bbl/day (2004)
Oil—proved reserves: 0 bbl (1 January 2006 est.)
Natural gas—production: 0 cu m (2005 est.)
Natural gas—consumption: 239.8 million cu m (2005 est.)
Natural gas—exports: 0 cu m (2005 est.)
Natural gas—imports: 239.8 million cu m (2005)
Natural gas—proved reserves: 0 cu m (1 January 2006 est.)
Current account balance: -$2.041 billion (2007 est.)
Exports: $7.237 billion f.o.b. (2007 est.)
Exports—commodities: ferronickel, sugar, gold, silver, coffee, cocoa, tobacco, meats, consumer goods
Exports—partners: US 72.7%, UK 3.2%, Belgium 2.4% (2006)
Imports: $13.82 billion f.o.b. (2007 est.)
Imports—commodities: foodstuffs, petroleum, cotton and fabrics, chemicals and pharmaceuticals
Imports—partners: US 46.9%, Venezuela 8.4%, Colombia 6.3%, Mexico 5.7% (2006)
Economic aid—recipient: $76.99 million (2005)
Reserves of foreign exchange and gold: $2.562 billion (31 December 2007 est.)
Debt—external: $10.21 billion (31 December 2007 est.)
Stock of direct foreign investment—at home: $12.37 billion (2007 est.)
Stock of direct foreign investment—abroad: $59 million (2007 est.)
Market value of publicly traded shares: $NA
Currency (code): Dominican peso (DOP)
Currency code: DOP
Exchange rates: Dominican pesos per US dollar—33.113 (2007), 33.406 (2006), 30.409 (2005), 42.12 (2004), 30.831 (2003)
Fiscal year: calendar year

COMMUNICATIONS

Telephones—main lines in use: 897,000 (2006)
Telephones—mobile cellular: 4.606 million (2006)
Telephone system:
general assessment: relatively efficient system based on island-wide microwave radio relay network
domestic: fixed telephone line density is about 10 per 100 persons; multiple providers of mobile cellular service with a subscribership of roughly 50 per 100 persons
international: country code—1-809; landing point for the Americas Region Caribbean Ring System (ARCOS-1) fiber-optic telecommunications subma-

rine cable that provides links to South and Central America, parts of the Caribbean, and US; satellite earth station—1 Intelsat (Atlantic Ocean)

Radio broadcast stations: AM 120, FM 56, shortwave 4 (1998)

Radios: 1.44 million (1997)

Television broadcast stations: 25 (2003)

Televisions: 770,000 (1997)

Internet country code: .do

Internet hosts: 81,218 (2007)

Internet Service Providers (ISPs): 24 (2000)

Internet users: 1.232 million (2006)

TRANSPORTATION

Airports: 34 (2007)

Airports—with paved runways:
total: 15
over 3,047 m: 3
2,438 to 3,047 m: 4
1,524 to 2,437 m: 4
914 to 1,523 m: 3
under 914 m: 1 (2007)

Airports—with unpaved runways:
total: 19
1,524 to 2,437 m: 3
914 to 1,523 m: 5
under 914 m: 11 (2007)

Railways:
total: 517 km
standard gauge: 375 km 1.435-m gauge
narrow gauge: 142 km 0.762-m gauge
note: additional 1,226 km operated by sugar companies in 1.076 m, 0.889 m, and 0.762-m gauges (2006)

Roadways:
total: 12,600 km
paved: 6,224 km
unpaved: 6,376 km (2000)

Merchant marine:
total: 1 ship (1000 GRT or over) 1,587 GRT/1,165 DWT
by type: cargo 1
registered in other countries: 1 (Panama 1) (2007)

Ports and terminals: Boca Chica, Caucedo, Puerto Plata, Rio Haina, Santo Domingo

MILITARY

Military branches: Army, Navy, Air Force (Fuerza Aerea Dominicana, FAD) (2007)

Military service age and obligation: 18 years of age for voluntary military service (2007)

Manpower available for military service: *males age 16–49:* 2,440,203

females age 16–49: 2,326,694 (2008 est.)

Manpower fit for military service:
males age 16–49: 2,020,490
females age 16–49: 1,883,875 (2008 est.)

Manpower reaching militarily significant age annually:
males age 16–49: 96,971
females age 16–49: 93,116 (2008 est.)

Military expenditures—percent of GDP: 0.8% (2006)

TRANSNATIONAL ISSUES

Disputes—international: Haitian migrants cross the porous border into the Dominican Republic to find work; illegal migrants from the Dominican Republic cross the Mona Passage each year to Puerto Rico to find better work

Illicit drugs: transshipment point for South American drugs destined for the US and Europe; has become a transshipment point for ecstasy from the Netherlands and Belgium destined for US and Canada; substantial money laundering activity; Colombian narcotics traffickers favor the Dominican Republic for illicit financial transactions; significant amphetamine consumption

INTRODUCTION

Background: What is now Ecuador formed part of the northern Inca Empire until the Spanish conquest in 1533. Quito became a seat of Spanish colonial government in 1563 and part of the Viceroyalty of New Granada in 1717. The territories of the Viceroyalty—New Granada (Colombia), Venezuela, and Quito—gained their independence between 1819 and 1822 and formed a federation known as Gran Colombia. When Quito withdrew in 1830, the traditional name was changed in favor of the "Republic of the Equator." Between 1904 and 1942, Ecuador lost territories in a series of conflicts with its neighbors. A border war with Peru that flared in 1995 was resolved in 1999. Although Ecuador marked 25 years of civilian governance in 2004, the period has been marred by political instability. Protests in Quito have contributed to the mid-term ouster of Ecuador's last three democratically elected Presidents. In 2007, a Constituent Assembly was elected to draft a new constitution; Ecuador's twentieth since gaining independence.

GEOGRAPHY

Location: Western South America, bordering the Pacific Ocean at the Equator, between Colombia and Peru
Geographic coordinates: 2 00 S, 77 30 W
Map references: South America
Area:
total: 283,560 sq km
land: 276,840 sq km
water: 6,720 sq km
note: includes Galapagos Islands
Area—comparative: slightly smaller than Nevada
Land boundaries:
total: 2,010 km
border countries: Colombia 590 km, Peru 1,420 km

Coastline: 2,237 km
Maritime claims:
territorial sea: 200 nm
continental shelf: 100 nm from 2,500-m isobath
Climate: tropical along coast, becoming cooler inland at higher elevations; tropical in Amazonian jungle lowlands
Terrain: coastal plain (costa), inter-Andean central highlands (sierra), and flat to rolling eastern jungle (oriente)
Elevation extremes:
lowest point: Pacific Ocean 0 m
highest point: Chimborazo 6,267 m
Natural resources: petroleum, fish, timber, hydropower
Land use:
arable land: 5.71%
permanent crops: 4.81%
other: 89.48% (2005)
Irrigated land: 8,650 sq km (2003)
Total renewable water resources: 432 cu km (2000)
Freshwater withdrawal (domestic/industrial/agricultural): *total:* 16.98 cu km/yr (12%/5%/82%)
per capita: 1,283 cu m/yr (2000)
Natural hazards: frequent earthquakes, landslides, volcanic activity; floods; periodic droughts
Environment—current issues: deforestation; soil erosion; desertification; water pollution; pollution from oil production wastes in ecologically sensitive areas of the Amazon Basin and Galapagos Islands
Environment—international agreements: *party to:* Antarctic-Environmental Protocol, Antarctic Treaty, Biodiversity, Climate Change, Climate Change-Kyoto Protocol, Desertification, Endangered Species, Hazardous Wastes, Ozone Layer Protection, Ship Pollution, Tropical Timber 83, Tropical Timber 94, Wetlands
signed, but not ratified: none of the selected agreements
Geography—note: Cotopaxi in Andes is highest active volcano in world

PEOPLE

Population: 13,927,650 (July 2008 est.)
Age structure:
0–14 years: 32.1% (male 2,274,986/female 2,189,437)
15–64 years: 62.7% (male 4,355,909/female 4,381,141)
65 years and over: 5.2% (male 340,861/female 385,316) (2008 est.)
Median age:
total: 24.2 years
male: 23.7 years

female: 24.7 years (2008 est.)
Population growth rate: 0.935% (2008 est.)
Birth rate: 21.54 births/1,000 population (2008 est.)
Death rate: 4.21 deaths/1,000 population (2008 est.)
Net migration rate: -7.98 migrant(s)/1,000 population (2008 est.)
Sex ratio:
at birth: 1.05 male(s)/female
under 15 years: 1.04 male(s)/female
15–64 years: 0.99 male(s)/female
65 years and over: 0.88 male(s)/female
total population: 1 male(s)/female (2008 est.)
Infant mortality rate:
total: 21.35 deaths/1,000 live births
male: 25.61 deaths/1,000 live births
female: 16.88 deaths/1,000 live births (2008 est.)
Life expectancy at birth:
total population: 76.81 years
male: 73.94 years
female: 79.84 years (2008 est.)
Total fertility rate: 2.59 children born/woman (2008 est.)
HIV/AIDS—adult prevalence rate: 0.3% (2003 est.)
HIV/AIDS—people living with HIV/AIDS: 21,000 (2003 est.)
HIV/AIDS—deaths: 1,700 (2003 est.)
Major infectious diseases:
degree of risk: high
food or waterborne diseases: bacterial diarrhea, hepatitis A, and typhoid fever
vectorborne diseases: dengue fever, malaria, and yellow fever
water contact disease: leptospirosis (2008)
Nationality:
noun: Ecuadorian(s)
adjective: Ecuadorian
Ethnic groups: mestizo (mixed Amerindian and white) 65%, Amerindian 25%, Spanish and others 7%, black 3%
Religions: Roman Catholic 95%, other 5%
Languages: Spanish (official), Amerindian languages (especially Quechua)
Literacy:
definition: age 15 and over can read and write
total population: 91%
male: 92.3%
female: 89.7% (2001 census)

GOVERNMENT

Country name:
conventional long form: Republic of Ecuador
conventional short form: Ecuador

local long form: Republica del Ecuador

local short form: Ecuador

Government type: republic

Capital: name: Quito

geographic coordinates: 0 13 S, 78 30 W

time difference: UTC-5 (same time as Washington, DC during Standard Time)

Administrative divisions: 24 provinces (provincias, singular—provincia); Azuay, Bolivar, Canar, Carchi, Chimborazo, Cotopaxi, El Oro, Esmeraldas, Galapagos, Guayas, Imbabura, Loja, Los Rios, Manabi, Morona-Santiago, Napo, Orellana, Pastaza, Pichincha, Santa Elena, Santo Domingo de los Tsachilas, Sucumbios, Tungurahua, Zamora-Chinchipe

Independence: 24 May 1822 (from Spain)

National holiday: Independence Day (independence of Quito), 10 August (1809)

Constitution: 10 August 1998

Legal system: based on civil law system; has not accepted compulsory ICJ jurisdiction

Suffrage: 18 years of age; universal, compulsory for literate persons ages 18–65, optional for other eligible voters

Executive branch:

chief of state: President Rafael CORREA Delgado (since 15 January 2007); Vice President Lenin MORENO Garces (since 15 January 2007); note—the president is both the chief of state and head of government

head of government: President Rafael CORREA Delgado (since 15 January 2007); Vice President Lenin MORENO Garces (since 15 January 2007)

cabinet: Cabinet appointed by the president

elections: the president and vice president are elected on the same ticket by popular vote for a four-year term (may not serve consecutive terms); election last held 15 October 2006 with a runoff election on 26 November 2006 (next to be held in October 2010)

election results: Rafael CORREA Delgado elected president; percent of vote— Rafael CORREA Delgado 56.7%; Alvaro NOBOA 43.3%

Legislative branch: unicameral National Congress or Congreso Nacional (100 seats; members are elected through a party-list proportional representation system to serve four-year terms)

elections: last held 15 October 2006 (next to be held in October 2010)

election results: percent of vote by party— NA; seats by party—PRIAN 28; PSP 24; PSC 13; ID 7; PRE 6; MUPP-NP 6; RED 5; UDC 5; other 6; note—defections by

members of National Congress are commonplace, resulting in frequent changes in the numbers of seats held by the various parties; as of 29 November 2007, Congress is on indefinite recess while a Constituent Assembly is convened

Judicial branch: Supreme Court or Corte Suprema (according to the Constitution, new justices are elected by the full Supreme Court; in December 2004, however, Congress successfully replaced the entire court via a simple-majority resolution)

Political parties and leaders: Alianza PAIS Movement [Rafael Vicente CORREA Delgado]; Christian Democratic Union or UDC [Diego ORDONEZ Guerrero]; Democratic Left or ID [Andres PAEZ Benalcazar]; Ethical and Democratic Network or RED [Leon ROLDOS]; Institutional Renewal and National Action Party or PRIAN [Alvaro NOBOA]; Pachakutik Plurinational Unity Movement—New Country or MUPP-NP [Jorge GUAMAN]; Patriotic Society Party or PSP [Lucio GUTIERREZ Borbua]; Popular Democratic Movement or MPD [Ciro GUZMAN Aldaz]; Roldosist Party or PRE [Abdala BUCARAM Ortiz, director]; Social Christian Party or PSC [Pascual DEL CIOPPO]; Socialist Party—Broad Front or PS-FA [Gustavo AYALA Cruz]

Political pressure groups and leaders: Confederation of Indigenous Nationalities of Ecuador or CONAIE [Marlon SANTI, president]; Coordinator of Social Movements or CMS [F. Napoleon SANTOS]; Federation of Indigenous Evangelists of Ecuador or FEINE [Marco MURILLO, president]; National Federation of Indigenous Afro-Ecuatorianos and Peasants or FENOCIN [Pedro DE LA CRUZ, president]

International organization participation: CAN, CSN, FAO, G-77, IADB, IAEA, IBRD, ICAO, ICC, ICCt, ICRM, IDA, IFAD, IFC, IFRCS, IHO, ILO, IMF, IMO, Interpol, IOC, IOM, IPU, ISO, ITSO, ITU, ITUC, LAES, LAIA, Mercosur (associate), MIGA, MINUSTAH, NAM, OAS, OPANAL, OPCW, OPEC, PCA, RG, UN, UNCTAD, UNESCO, UNHCR, UNIDO, Union Latina, UNMIL, UNMIS, UNOCI, UNWTO, UPU, WCL, WCO, WFTU, WHO, WIPO, WMO, WTO

Diplomatic representation in the US:

chief of mission: Ambassador Luis Benigno GALLEGOS Chiriboga

chancery: 2535 15th Street NW, Washington, DC 20009

telephone: [1] (202) 234-7200

FAX: [1] (202) 667-3482

consulate(s) general: Atlanta, Boston, Chicago, Dallas, Denver, Houston, Jersey City (New Jersey), Los Angeles, Miami, New Orleans, New York, San Francisco, San Juan (Puerto Rico), Washington, DC

Diplomatic representation from the US:

chief of mission: Ambassador Linda L. JEWELL

embassy: Avenida 12 de Octubre y Avenida Patria, Quito

mailing address: APO AA 34039

telephone: [593] (2) 256-2890

FAX: [593] (2) 250-2052

consulate(s) general: Guayaquil

Flag description: three horizontal bands of yellow (top, double width), blue, and red with the coat of arms superimposed at the center of the flag; similar to the flag of Colombia, which is shorter and does not bear a coat of arms

ECONOMY

Economy—overview: Ecuador is substantially dependent on its petroleum resources, which have accounted for more than half of the country's export earnings and one-fourth of public sector revenues in recent years. In 1999/2000, Ecuador suffered a severe economic crisis, with GDP contracted by more than 6%, with a significant increase in poverty. The banking system also collapsed, and Ecuador defaulted on its external debt later that year. In March 2000, Congress approved a series of structural reforms that also provided for the adoption of the US dollar as legal tender. Dollarization stabilized the economy, and positive growth returned in the years that followed, helped by high oil prices, remittances, and increased non-traditional exports. From 2002–06 the economy grew 5.5%, the highest five-year average in 25 years. The poverty rate declined but remained high at 38% in 2006. In 2006 the government of Alfredo PALACIO (2005–07) seized the assets of Occidental Petroleum for alleged contract violations and imposed a windfall revenue tax on foreign oil companies, leading to the suspension of free trade negotiations with the US. These measures, combined with chronic underinvestment in the state oil company, Petroecuador, led to a drop in petroleum production in 2007. PALACIO's successor, Rafael CORREA, raised the specter of debt default—but Ecuador has paid its debt on time. He also decreed a higher windfall revenue tax on private oil companies, then sought to renegotiate their contracts to overcome the debilitating effect of the tax. This generated economic uncertainty; private investment has dropped and economic growth has slowed significantly.

GDP (purchasing power parity): $98.79 billion (2007 est.)

GDP (official exchange rate): $44.18 billion (2007 est.)

GDP—real growth rate: 1.9% (2007 est.)

GDP—per capita (PPP): $7,200 (2007 est.)

GDP—composition by sector:
agriculture: 6.7%
industry: 35.1%
services: 58.2% (2007 est.)

Labor force: 4.51 million (urban) (2007 est.)

Labor force—by occupation: *agriculture:* 8%
industry: 24%
services: 68% (2001)

Unemployment rate: 9.3% (2007 est.)

Population below poverty line: 38.3% (2006)

Household income or consumption by percentage share:
lowest 10%: 2%
highest 10%: 35%
note: data for urban households only (October 2006)

Distribution of family income—Gini index: 46
note: data are for urban households (2006)

Inflation rate (consumer prices): 2.2% (2007 est.)

Investment (gross fixed): 22.2% of GDP (2007 est.)

Budget:
revenues: $13.46 billion
expenditures: planned $11.96 billion (2007 est.)

Public debt: 33.6% of GDP (2007 est.)

Agriculture—products: bananas, coffee, cocoa, rice, potatoes, manioc (tapioca), plantains, sugarcane; cattle, sheep, pigs, beef, pork, dairy products; balsa wood; fish, shrimp

Industries: petroleum, food processing, textiles, wood products, chemicals

Industrial production growth rate: 1.4% (2007 est.)

Electricity—production: 12.94 billion kWh (2005)

Electricity—production by source:
fossil fuel: 81%
hydro: 19%
nuclear: 0%
other: 0% (2001)

Electricity—consumption: 8.855 billion kWh (2005)

Electricity—exports: 16 million kWh (2005)

Electricity—imports: 1.723 billion kWh (2005)

Oil—production: 538,000 bbl/day (2005)

Oil—consumption: 162,000 bbl/day (2005)

Oil—exports: 420,600 bbl/day (2004 est.)

Oil—imports: 44,680 bbl/day (2004)

Oil—proved reserves: 4.727 billion bbl (2007 est.)

Natural gas—production: 249.4 million cu m (2005 est.)

Natural gas—consumption: 249.4 million cu m (2005 est.)

Natural gas—exports: 0 cu m (2005 est.)

Natural gas—imports: 0 cu m (2005)

Natural gas—proved reserves: 9.369 billion cu m (1 January 2006 est.)

Current account balance: $1.464 billion (2007 est.)

Exports: $14.37 billion (2007 est.)

Exports—commodities: petroleum, bananas, cut flowers, shrimp, cacao, coffee, hemp, wood, fish

Exports—partners: US 53.6%, Peru 8.2%, Colombia 5.6%, Chile 4.4% (2006)

Imports: $12.76 billion (2007 est.)

Imports—commodities: industrial materials, fuels and lubricants, nondurable consumer goods

Imports—partners: US 23.1%, Colombia 13.3%, Brazil 7.3%, Panama 4% (2006)

Economic aid—recipient: $209.5 million (2005)

Reserves of foreign exchange and gold: $3.521 billion (30 November 2007 est.)

Debt—external: $16.93 billion (31 December 2007)

Stock of direct foreign investment—at home: $16.31 billion (2007 est.)

Stock of direct foreign investment—abroad: $1.749 billion (2007 est.)

Market value of publicly traded shares: $4.04 billion (2006)

Currency (code): US dollar (USD)

Currency code: USD

Exchange rates: the US dollar is used; the sucre was eliminated in 2000

Fiscal year: calendar year

Telephones—main lines in use: 1.754 million (2006)

Telephones—mobile cellular: 8.485 million (2006)

Telephone system:
general assessment: generally elementary but being expanded
domestic: fixed-line services provided by three state-owned enterprises; plans to transfer the state-owned operators to private ownership have repeatedly failed; fixed-line density stands at about 13 per 100 persons; mobile cellular use has surged and has a subscribership of nearly 65 per 100 persons

international: country code—593; landing point for the PAN-AM submarine telecommunications cable that provides links to the west coast of South America, Panama, Colombia, Venezuela, and extending onward to Aruba and the US Virgin Islands in the Caribbean; satellite earth station—1 Intelsat (Atlantic Ocean) (2007)

Radio broadcast stations: AM 392, FM 35, shortwave 29 (2001)

Radios: 5 million (2001)

Television broadcast stations: 7 (plus 14 repeaters) (2000)

Televisions: 2.5 million (2001)

Internet country code: .ec

Internet hosts: 28,420 (2007)

Internet Service Providers (ISPs): 31 (2001)

Internet users: 1.549 million (2006)

Airports: 406 (2007)

Airports—with paved runways:
total: 104
over 3,047 m: 4
2,438 to 3,047 m: 3
1,524 to 2,437 m: 17
914 to 1,523 m: 26
under 914 m: 54 (2007)

Airports—with unpaved runways:
total: 302
914 to 1,523 m: 34
under 914 m: 268 (2007)

Heliports: 1 (2007)

Pipelines: extra heavy crude oil 578 km; gas 71 km; oil 1,389 km; refined products 1,185 km (2007)

Railways:
total: 966 km
narrow gauge: 966 km 1.067-m gauge (2006)

Roadways:
total: 43,197 km
paved: 6,467 km
unpaved: 36,730 km (2004)

Waterways: 1,500 km (most inaccessible) (2006)

Merchant marine:
total: 33 ships (1000 GRT or over) 190,931 GRT/306,280 DWT
by type: chemical tanker 1, liquefied gas 1, passenger 8, petroleum tanker 22, specialized tanker 1
foreign-owned: 2 (Philippines 1, US 1)
registered in other countries: 3 (China 1, Panama 2) (2007)

Ports and terminals: Esmeraldas, Guayaquil, Manta, Puerto Bolivar

MILITARY

Military branches: Army, Navy (includes Naval Infantry, Naval Aviation, Coast Guard), Air Force (Fuerza Aerea Ecuatoriana, FAE) (2007)
Military service age and obligation: 20 years of age for selective conscript military service; 12-month service obligation (2006)
Manpower available for military service:
males age 16–49: 3,536,602
females age 16–49: 3,559,188 (2008 est.)
Manpower fit for military service:
males age 16–49: 3,030,664
females age 16–49: 3,037,892 (2008 est.)

Manpower reaching militarily significant age annually:
males age 16–49: 144,821
females age 16–49: 139,091 (2008 est.)
Military expenditures—percent of GDP: 2.8% (2006)

TRANSNATIONAL ISSUES

Disputes—international: organized illegal narcotics operations in Colombia penetrate across Ecuador's shared border, which thousands of Colombians also cross to escape the violence in their home country
Refugees and internally displaced persons: *refugees (country of origin):* 11,526 (Colombia); note—UNHCR estimates

as many as 250,000 Columbians are seeking asylum in Ecuador, many of whom do not register as refugees for fear of deportation (2007)
Illicit drugs: significant transit country for cocaine originating in Colombia and Peru, with over half of the US-bound cocaine passing through Ecuadorian Pacific waters; importer of precursor chemicals used in production of illicit narcotics; attractive location for cash-placement by drug traffickers laundering money because of dollarization and weak anti-money-laundering regime; increased activity on the northern frontier by trafficking groups and Colombian insurgents

EGYPT

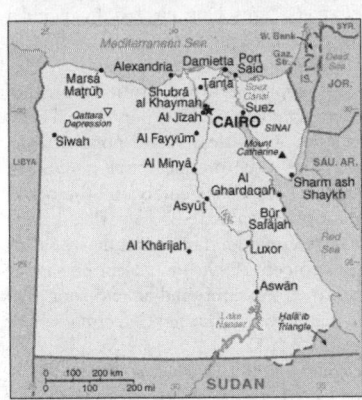

INTRODUCTION

Background: The regularity and richness of the annual Nile River flood, coupled with semi-isolation provided by deserts to the east and west, allowed for the development of one of the world's great civilizations. A unified kingdom arose circa 3200 B.C., and a series of dynasties ruled in Egypt for the next three millennia. The last native dynasty fell to the Persians in 341 B.C., who in turn were replaced by the Greeks, Romans, and Byzantines. It was the Arabs who introduced Islam and the Arabic language in the 7th century and who ruled for the next six centuries. A local military caste, the Mamluks took control about 1250 and continued to govern after the conquest of Egypt by the Ottoman Turks in 1517. Following the completion of the Suez Canal in 1869, Egypt became an important world transportation hub, but also fell heavily into debt. Ostensibly to protect its investments, Britain seized control of Egypt's government in 1882, but nominal allegiance to the Ottoman

Empire continued until 1914. Partially independent from the UK in 1922, Egypt acquired full sovereignty with the overthrow of the British-backed monarchy in 1952. The completion of the Aswan High Dam in 1971 and the resultant Lake Nasser have altered the time-honored place of the Nile River in the agriculture and ecology of Egypt. A rapidly growing population (the largest in the Arab world), limited arable land, and dependence on the Nile all continue to overtax resources and stress society. The government has struggled to meet the demands of Egypt's growing population through economic reform and massive investment in communications and physical infrastructure.

GEOGRAPHY

Location: Northern Africa, bordering the Mediterranean Sea, between Libya and the Gaza Strip, and the Red Sea north of Sudan, and includes the Asian Sinai Peninsula
Geographic coordinates: 27 00 N, 30 00 E
Map references: Africa
Area:
total: 1,001,450 sq km
land: 995,450 sq km
water: 6,000 sq km
Area—comparative: slightly more than three times the size of New Mexico
Land boundaries:
total: 2,665 km
border countries: Gaza Strip 11 km, Israel 266 km, Libya 1,115 km, Sudan 1,273 km
Coastline: 2,450 km
Maritime claims:
territorial sea: 12 nm
contiguous zone: 24 nm

exclusive economic zone: 200 nm
continental shelf: 200 m depth or to the depth of exploitation
Climate: desert; hot, dry summers with moderate winters
Terrain: vast desert plateau interrupted by Nile valley and delta
Elevation extremes:
lowest point: Qattara Depression -133 m
highest point: Mount Catherine 2,629 m
Natural resources: petroleum, natural gas, iron ore, phosphates, manganese, limestone, gypsum, talc, asbestos, lead, zinc
Land use:
arable land: 2.92%
permanent crops: 0.5%
other: 96.58% (2005)
Irrigated land: 34,220 sq km (2003)
Total renewable water resources: 86.8 cu km (1997)
Freshwater withdrawal (domestic/industrial/agricultural): *total:* 68.3 cu km/yr (8%/6%/86%)
per capita: 923 cu m/yr (2000)
Natural hazards: periodic droughts; frequent earthquakes, flash floods, landslides; hot, driving windstorm called khamsin occurs in spring; dust storms, sandstorms
Environment—current issues: agricultural land being lost to urbanization and windblown sands; increasing soil salination below Aswan High Dam; desertification; oil pollution threatening coral reefs, beaches, and marine habitats; other water pollution from agricultural pesticides, raw sewage, and industrial effluents; limited natural fresh water resources away from the Nile, which is the only perennial water source; rapid growth in population overstraining the Nile and natural resources

Environment—international agreements: *party to:* Biodiversity, Climate Change, Climate Change-Kyoto Protocol, Desertification, Endangered Species, Environmental Modification, Hazardous Wastes, Law of the Sea, Marine Dumping, Ozone Layer Protection, Ship Pollution, Tropical Timber 83, Tropical Timber 94, Wetlands
signed, but not ratified: none of the selected agreements

Geography—note: controls Sinai Peninsula, only land bridge between Africa and remainder of Eastern Hemisphere; controls Suez Canal, a sea link between Indian Ocean and Mediterranean Sea; size, and juxtaposition to Israel, establish its major role in Middle Eastern geopolitics; dependence on upstream neighbors; dominance of Nile basin issues; prone to influxes of refugees

PEOPLE

Population: 81,713,517 (July 2008 est.)

Age structure:
0–14 years: 31.8% (male 13,292,961/female 12,690,711)
15–64 years: 63.5% (male 26,257,440/female 25,627,390)
65 years and over: 4.7% (male 1,636,560/female 2,208,455) (2008 est.)

Median age:
total: 24.5 years
male: 24.1 years
female: 24.9 years (2008 est.)

Population growth rate: 1.682% (2008 est.)

Birth rate: 22.12 births/1,000 population (2008 est.)

Death rate: 5.09 deaths/1,000 population (2008 est.)

Net migration rate: -0.21 migrant(s)/1,000 population (2008 est.)

Sex ratio:
at birth: 1.05 male(s)/female
under 15 years: 1.05 male(s)/female
15–64 years: 1.02 male(s)/female
65 years and over: 0.74 male(s)/female
total population: 1.02 male(s)/female (2008 est.)

Infant mortality rate:
total: 28.36 deaths/1,000 live births
male: 30.06 deaths/1,000 live births
female: 26.57 deaths/1,000 live births (2008 est.)

Life expectancy at birth:
total population: 71.85 years
male: 69.3 years
female: 74.52 years (2008 est.)

Total fertility rate: 2.72 children born/woman (2008 est.)

HIV/AIDS—adult prevalence rate: less than 0.1% (2001 est.)

HIV/AIDS—people living with HIV/AIDS: 12,000 (2001 est.)

HIV/AIDS—deaths: 700 (2003 est.)

Major infectious diseases:
degree of risk: intermediate
food or waterborne diseases: bacterial diarrhea, hepatitis A, and typhoid fever
water contact disease: schistosomiasis
note: highly pathogenic H5N1 avian influenza has been identified in this country; it poses a negligible risk with extremely rare cases possible among US citizens who have close contact with birds (2008)

Nationality:
noun: Egyptian(s)
adjective: Egyptian

Ethnic groups: Egyptian 98%, Berber, Nubian, Bedouin, and Beja 1%, Greek, Armenian, other European (primarily Italian and French) 1%

Religions: Muslim (mostly Sunni) 90%, Coptic 9%, other Christian 1%

Languages: Arabic (official), English and French widely understood by educated classes

Literacy:
definition: age 15 and over can read and write
total population: 71.4%
male: 83%
female: 59.4% (2005 est.)

GOVERNMENT

Country name:
conventional long form: Arab Republic of Egypt
conventional short form: Egypt
local long form: Jumhuriyat Misr al-Arabiyah
local short form: Misr
former: United Arab Republic (with Syria)

Government type: republic

Capital: *name:* Cairo
geographic coordinates: 30 03 N, 31 15 E
time difference: UTC+2 (7 hours ahead of Washington, DC during Standard Time)
daylight saving time: +1hr, begins last Friday in April; ends last Thursday in September

Administrative divisions: 26 governorates (muhafazat, singular—muhafazah); Ad Daqahliyah, Al Bahr al Ahmar, Al Buhayrah, Al Fayyum, Al Gharbiyah, Al Iskandariyah, Al Isma'iliyah, Al Jizah, Al Minufiyah, Al Minya, Al Qahirah, Al Qalyubiyah, Al Wadi al Jadid, As Suways, Ash Sharqiyah, Aswan, Asyut, Bani Suwayf, Bur Sa'id, Dumyat, Janub Sina', Kafr ash Shaykh, Matruh, Qina, Shamal Sina', Suhaj

Independence: 28 February 1922 (from UK)

National holiday: Revolution Day, 23 July (1952)

Constitution: 11 September 1971; amended 22 May 1980, 25 May 2005, and 26 March 2007

Legal system: based on Islamic and civil law (particularly Napoleonic codes); judicial review by Supreme Court and Council of State (oversees validity of administrative decisions); accepts compulsory ICJ jurisdiction with reservations

Suffrage: 18 years of age; universal and compulsory

Executive branch:
chief of state: President Mohamed Hosni MUBARAK (since 14 October 1981)
head of government: Prime Minister Ahmed Mohamed NAZIF (since 9 July 2004)
cabinet: Cabinet appointed by the president
elections: president elected by popular vote for six-year term (no term limits); note—a national referendum in May 2005 approved a constitutional amendment that changed the presidential election to a multicandidate popular vote; previously the president was nominated by the People's Assembly and the nomination was validated by a national, popular referendum; last referendum held 26 September 1999; first election under terms of constitutional amendment held 7 September 2005; next election scheduled for 2011
election results: Hosni MUBARAK reelected president; percent of vote—Hosni MUBARAK 88.6%, Ayman NOUR 7.6%, Noman GOMAA 2.9%

Legislative branch: bicameral system consists of the People's Assembly or Majlis al-Sha'b (454 seats; 444 elected by popular vote, 10 appointed by the president; members serve five-year terms) and the Advisory Council or Majlis al-Shura that traditionally functions only in a consultative role but 2007 constitutional amendments could grant the Council new powers (264 seats; 176 elected by popular vote, 88 appointed by the president; members serve six-year terms; midterm elections for half of the elected members)
elections: People's Assembly—three-phase voting—last held 7 and 20 November, 1 December 2005;(next to be held November-December 2010); Advisory Council—last held June 2007 (next to be held May-June 2010)
election results: People's Assembly—percent of vote by party—NA; seats by party—NDP 311, NWP 6, Tagammu 2, Tomorrow Party 1, independents 112 (12 seats to be determined by rerun elections, 10 seats appointed by President);

Advisory Council—percent of vote by party—NA; seats by party—NDP 84, Tagammu 1, independents 3

Judicial branch: Supreme Constitutional Court

Political parties and leaders: National Democratic Party or NDP (governing party) [Mohamed Hosni MUBARAK]; National Progressive Unionist Grouping or Tagammu [Rifaat EL-SAID]; New Wafd Party or NWP [Mahmoud ABAZA]; Tomorrow Party [Moussa Mustafa MOUSSA]

note: formation of political parties must be approved by the government; only parties with representation in elected bodies are listed

Political pressure groups and leaders: despite a constitutional ban against religious-based parties and political activity, the technically illegal Muslim Brotherhood constitutes Hosni MUBARAK's potentially most significant political opposition; MUBARAK has alternated between tolerating limited political activity by the Brotherhood and blocking its influence; civic society groups are sanctioned, but constrained in practical terms; only trade unions and professional associations affiliated with the government are officially sanctioned

International organization participation: ABEDA, ACCT, AfDB, AFESD, AMF, AU, BSEC (observer), CAEU, COMESA, EBRD, FAO, G-15, G-24, G-77, IAEA, IBRD, ICAO, ICC, ICCt (signatory), ICRM, IDA, IDB, IFAD, IFC, IFRCS, IHO, ILO, IMF, IMO, IMSO, Interpol, IOC, IOM, IPU, ISO, ITSO, ITU, LAS, MIGA, MINURSO, NAM, OAPEC, OAS (observer), OIC, OIF, OSCE (partner), PCA, UN, UNAMID, UNCTAD, UNESCO, UNHCR, UNIDO, UNITAR, UNMIL, UNMIS, UNOMIG, UNRWA, UNWTO, UPU, WCO, WFTU, WHO, WIPO, WMO, WTO

Diplomatic representation in the US:
chief of mission: Ambassador Nabil FAHMY
chancery: 3521 International Court NW, Washington, DC 20008
telephone: [1] (202) 895-5400
FAX: [1] (202) 244-4319
consulate(s) general: Chicago, Houston, New York, San Francisco

Diplomatic representation from the US:
chief of mission: Ambassador Francis J. RICCIARDONE, Jr.
embassy: 8 Kamal El Din Salah St., Garden City, Cairo
mailing address: Unit 64900, Box 15, APO AE 09839-4900
telephone: [20] (2) 2797-3300
FAX: [20] (2) 2797-3200

Flag description: three equal horizontal bands of red (top), white, and black; the national emblem (a gold Eagle of Saladin facing the hoist side with a shield superimposed on its chest above a scroll bearing the name of the country in Arabic) centered in the white band; design is based on the Arab Liberation flag and similar to the flag of Syria, which has two green stars in the white band, Iraq, which has an Arabic inscription centered in the white band, and Yemen, which has a plain white band

ECONOMY

Economy—overview: Occupying the northeast corner of the African continent, Egypt is bisected by the highly fertile Nile valley, where most economic activity takes place. In the last 30 years, the government has reformed the highly centralized economy it inherited from President Gamel Abdel NASSER. In 2005, Prime Minister Ahmed NAZIF's government reduced personal and corporate tax rates, reduced energy subsidies, and privatized several enterprises. The stock market boomed, and GDP grew about 5% per year in 2005–06, and topped 7% in 2007. Despite these achievements, the government has failed to raise living standards for the average Egyptian, and has had to continue providing subsidies for basic necessities. The subsidies have contributed to a sizeable budget deficit—roughly 7.5% of GDP in 2007—and represent a significant drain on the economy. Foreign direct investment has increased significantly in the past two years, but the NAZIF government will need to continue its aggressive pursuit of reforms in order to sustain the spike in investment and growth and begin to improve economic conditions for the broader population. Egypt's export sectors—particularly natural gas—have bright prospects.

GDP (purchasing power parity): $404 billion (2007 est.)

GDP (official exchange rate): $127.9 billion (2007 est.)

GDP—real growth rate: 7.1% (2007 est.)

GDP—per capita (PPP): $5,500 (2007 est.)

GDP—composition by sector:
agriculture: 13.8%
industry: 41.1%
services: 45.1% (2007 est.)

Labor force: 22.1 million (2007 est.)

Labor force—by occupation: *agriculture:* 32%
industry: 17%
services: 51% (2001 est.)

Unemployment rate: 9.1% (2007 est.)

Population below poverty line: 20% (2005 est.)

Household income or consumption by percentage share:
lowest 10%: 3.7%
highest 10%: 29.5% (2000)

Distribution of family income—Gini index: 34.4 (2001)

Inflation rate (consumer prices): 11% (2007 est.)

Investment (gross fixed): 21.2% of GDP (2007 est.)

Budget:
revenues: $35.12 billion
expenditures: $44.86 billion (2007 est.)

Public debt: 105.8% of GDP (2007 est.)

Agriculture—products: cotton, rice, corn, wheat, beans, fruits, vegetables; cattle, water buffalo, sheep, goats

Industries: textiles, food processing, tourism, chemicals, pharmaceuticals, hydrocarbons, construction, cement, metals, light manufactures

Industrial production growth rate: 7.5% (2007 est.)

Electricity—production: 102.5 billion kWh (2005)

Electricity—production by source:
fossil fuel: 81%
hydro: 19%
nuclear: 0%
other: 0% (2001)

Electricity—consumption: 84.49 billion kWh (2005)

Electricity—exports: 946 million kWh (2005)

Electricity—imports: 168 million kWh (2005)

Oil—production: 688,100 bbl/day (2005 est.)

Oil—consumption: 635,000 bbl/day (2005 est.)

Oil—exports: 152,600 bbl/day (2004 est.)

Oil—imports: 69,860 bbl/day (2004)

Oil—proved reserves: 3.75 billion bbl (2007 est.)

Natural gas—production: 40.76 billion cu m (2005 est.)

Natural gas—consumption: 32.81 billion cu m (2005 est.)

Natural gas—exports: 7.951 billion cu m (2005 est.)

Natural gas—imports: 0 cu m (2005)

Natural gas—proved reserves: 1.589 trillion cu m (1 January 2006 est.)

Current account balance: $1.862 billion (2007 est.)

Exports: $25.72 billion f.o.b. (2007 est.)

Exports—commodities: crude oil and petroleum products, cotton, textiles, metal products, chemicals

Exports—partners: Italy 12%, US 11.3%, Spain 8.7%, UK 5.5%, France 5.4%, Syria 5.1%, Saudi Arabia 4.3%, Germany 4.2% (2006)

Imports: $43.43 billion f.o.b. (2007 est.)
Imports—commodities: machinery and equipment, foodstuffs, chemicals, wood products, fuels
Imports—partners: US 11.4%, China 8.3%, Germany 6.5%, Italy 5.4%, Saudi Arabia 5%, France 4.6% (2006)
Economic aid—recipient: ODA, $925.9 million (2005)
Reserves of foreign exchange and gold: $31.37 billion (31 December 2007 est.)
Debt—external: $30.2 billion (30 June 2007)
Stock of direct foreign investment—at home: $47.16 billion (2007 est.)
Stock of direct foreign investment—abroad: $1.295 billion (2007 est.)
Market value of publicly traded shares: $93.48 billion (2006)
Currency (code): Egyptian pound (EGP)
Currency code: EGP
Exchange rates: Egyptian pounds per US dollar—5.67 (2007), 5.725 (2006), 5.78 (2005), 6.1962 (2004), 5.8509 (2003)
Fiscal year: 1 July—30 June

COMMUNICATIONS

Telephones—main lines in use: 10.808 million (2006)
Telephones—mobile cellular: 18.001 million (2006)
Telephone system:
general assessment: large system; underwent extensive upgrading during 1990s and is reasonably modern; Telecom Egypt, the landline monopoly, has been increasing service availability and in 2006 fixed-line density stood at 14 per 100 persons; as of 2007 there were three mobile-cellular networks and service is expanding rapidly
domestic: principal centers at Alexandria, Cairo, Al Mansurah, Ismailia, Suez, and Tanta are connected by coaxial cable and microwave radio relay
international: country code—20; landing point for both the SEA-ME-WE-3 AND SEA-ME-WE-4 submarine cable networks; linked to the international submarine cable FLAG (Fiber-Optic Link Around the Globe); satellite earth stations—4 (2 Intelsat—Atlantic Ocean and Indian Ocean, 1 Arabsat, and 1 Inmarsat); tropospheric scatter to Sudan; microwave radio relay to Israel; a participant in Medarabtel
Radio broadcast stations: AM 42 (plus 15 repeaters), FM 14, shortwave 3 (1999)
Radios: 20.5 million (1997)
Television broadcast stations: 98 (September 1995)

Televisions: 7.7 million (1997)
Internet country code: .eg
Internet hosts: 5,363 (2007)
Internet Service Providers (ISPs): 50 (2000)
Internet users: 6 million (2006)

TRANSPORTATION

Airports: 88 (2007)
Airports—with paved runways: *total:* 72
over 3,047 m: 15
2,438 to 3,047 m: 36
1,524 to 2,437 m: 16
under 914 m: 5 (2007)
Airports—with unpaved runways:
total: 16
2,438 to 3,047 m: 1
1,524 to 2,437 m: 3
914 to 1,523 m: 5
under 914 m: 7 (2007)
Heliports: 3 (2007)
Pipelines: condensate 483 km; condensate/gas 74 km; gas 6,466 km; liquid petroleum gas 957 km; oil 5,518 km; oil/gas/water 37 km; refined products 895 km (2007)
Railways: *total:* 5,063 km
standard gauge: 5,063 km 1.435-m gauge (62 km electrified) (2006)
Roadways: *total:* 92,370 km
paved: 74,820 km
unpaved: 17,550 km (2004)
Waterways: 3,500 km
note: includes Nile River, Lake Nasser, Alexandria-Cairo Waterway, and numerous smaller canals in delta; Suez Canal (193.5 km including approaches) navigable by oceangoing vessels drawing up to 17.68 m (2006)
Merchant marine:
total: 77 ships (1000 GRT or over) 1,032,116 GRT/1,553,065 DWT
by type: bulk carrier 13, cargo 33, container 2, passenger/cargo 5, petroleum tanker 14, roll on/roll off 10
foreign-owned: 10 (Denmark 1, Greece 8, Lebanon 1)
registered in other countries: 55 (Bolivia 1, Cambodia 14, Georgia 14, Honduras 4, North Korea 1, Panama 13, Sao Tome and Principe 1, Saudi Arabia 1, St Kitts and Nevis 2, St Vincent and The Grenadines 4) (2007)
Ports and terminals: Ayn Sukhnah, Alexandria, Damietta, El Dekheila, Sidi Kurayr, Suez

MILITARY

Military branches: Army, Navy, Air Force, Air Defense Command
Military service age and obligation: 18–30 years of age for male conscript military service; service obligation 12–36 months, followed by a 9-year reserve obligation (2008)

Manpower available for military service:
males age 16–49: 21,247,777
females age 16–49: 20,406,408 (2008 est.)
Manpower fit for military service:
males age 16–49: 18,153,158
females age 16–49: 17,405,837 (2008 est.)
Manpower reaching militarily significant age annually:
males age 16–49: 825,300
females age 16–49: 786,590 (2008 est.)
Military expenditures—percent of GDP: 3.4% (2005 est.)

TRANSNATIONAL ISSUES

Disputes—international: while Sudan retains claim to the Hala'ib Triangle north of the 1899 Treaty boundary along the 22nd Parallel, both states withdrew their military presence in the 1990s and Egypt has invested in and effectively administers the area; Egypt vigilantly monitors the Sinai and borders with Israel and the Gaza Strip to deter terrorist, smuggling, and other illegal activities; Egypt does not extend domestic asylum to some 70,000 persons who identify themselves as Palestinians but who largely lack UNRWA assistance and, until recently, UNHCR recognition as refugees
Refugees and internally displaced persons: *refugees (country of origin):* 60,000—80,000 (Iraq); 70,198 (Palestinian Territories); 12,157 (Sudan) (2007)
Trafficking in persons: *current situation:* Egypt is a transit country for women trafficked from Eastern Europe to Israel for the purpose of sexual exploitation; these women generally arrive as tourists and are subsequently trafficked through the Sinai Desert by Bedouin tribes; men and women from sub-Saharan Africa and Asia are believed to be trafficked through the Sinai Desert to Israel and Europe for labor exploitation; some Egyptian children from rural areas are trafficked within the country to work as domestic servants or laborers in the agriculture industry
tier rating: Tier 2 Watch List—Egypt is placed on the Tier 2 Watch List for its failure to show evidence of increasing efforts to address trafficking over the past year, particularly in the area of law enforcement
Illicit drugs: transit point for cannabis, heroin, and opium moving to Europe, Israel, and North Africa; transit stop for Nigerian drug couriers; concern as money laundering site due to lax enforcement of financial regulations

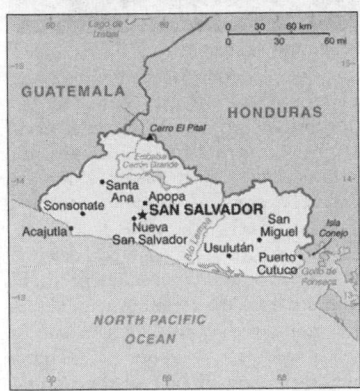

INTRODUCTION

Background: El Salvador achieved independence from Spain in 1821 and from the Central American Federation in 1839. A 12-year civil war, which cost about 75,000 lives, was brought to a close in 1992 when the government and leftist rebels signed a treaty that provided for military and political reforms.

GEOGRAPHY

Location: Central America, bordering the North Pacific Ocean, between Guatemala and Honduras

Geographic coordinates: 13 50 N, 88 55 W

Map references: Central America and the Caribbean

Area:
total: 21,040 sq km
land: 20,720 sq km
water: 320 sq km

Area—comparative: slightly smaller than Massachusetts

Land boundaries:
total: 545 km
border countries: Guatemala 203 km, Honduras 342 km

Coastline: 307 km

Maritime claims:
territorial sea: 12 nm
contiguous zone: 24 nm
exclusive economic zone: 200 nm

Climate: tropical; rainy season (May to October); dry season (November to April); tropical on coast; temperate in uplands

Terrain: mostly mountains with narrow coastal belt and central plateau

Elevation extremes:
lowest point: Pacific Ocean 0 m
highest point: Cerro El Pital 2,730 m

Natural resources: hydropower, geothermal power, petroleum, arable land

Land use:
arable land: 31.37%

permanent crops: 11.88%
other: 56.75% (2005)

Irrigated land: 450 sq km (2003)

Total renewable water resources: 25.2 cu km (2001)

Freshwater withdrawal (domestic/industrial/agricultural): *total:* 1.28 cu km/yr (25%/16%/59%)
per capita: 186 cu m/yr (2000)

Natural hazards: known as the Land of Volcanoes; frequent and sometimes destructive earthquakes and volcanic activity; extremely susceptible to hurricanes

Environment—current issues: deforestation; soil erosion; water pollution; contamination of soils from disposal of toxic wastes

Environment—international agreements: *party to:* Biodiversity, Climate Change, Climate Change-Kyoto Protocol, Desertification, Endangered Species, Hazardous Wastes, Ozone Layer Protection, Wetlands
signed, but not ratified: Law of the Sea

Geography—note: smallest Central American country and only one without a coastline on Caribbean Sea

PEOPLE

Population: 7,066,403 (July 2008 est.)

Age structure:
0–14 years: 35.8% (male 1,291,147/female 1,237,453)
15–64 years: 59% (male 1,987,671/female 2,179,620)
65 years and over: 5.2% (male 162,100/female 208,412) (2008 est.)

Median age:
total: 22.2 years
male: 21.1 years
female: 23.4 years (2008 est.)

Population growth rate: 1.679% (2008 est.)

Birth rate: 25.72 births/1,000 population (2008 est.)

Death rate: 5.53 deaths/1,000 population (2008 est.)

Net migration rate: -3.4 migrant(s)/1,000 population (2008 est.)

Sex ratio:
at birth: 1.05 male(s)/female
under 15 years: 1.04 male(s)/female
15–64 years: 0.91 male(s)/female
65 years and over: 0.78 male(s)/female
total population: 0.95 male(s)/female (2008 est.)

Infant mortality rate:
total: 22.19 deaths/1,000 live births
male: 25.06 deaths/1,000 live births
female: 19.18 deaths/1,000 live births (2008 est.)

Life expectancy at birth:
total population: 72.06 years
male: 68.45 years
female: 75.84 years (2008 est.)

Total fertility rate: 3.04 children born/woman (2008 est.)

HIV/AIDS—adult prevalence rate: 0.7% (2003 est.)

HIV/AIDS—people living with HIV/AIDS: 29,000 (2003 est.)

HIV/AIDS—deaths: 2,200 (2003 est.)

Major infectious diseases:
degree of risk: high
food or waterborne diseases: bacterial diarrhea, hepatitis A, and typhoid fever
vectorborne diseases: dengue fever
water contact disease: leptospirosis (2008)

Nationality:
noun: Salvadoran(s)
adjective: Salvadoran

Ethnic groups: mestizo 90%, white 9%, Amerindian 1%

Religions: Roman Catholic 83%, other 17%
note: there is extensive activity by Protestant groups throughout the country; by the end of 1992, there were an estimated 1 million Protestant evangelicals in El Salvador

Languages: Spanish, Nahua (among some Amerindians)

Literacy:
definition: age 10 and over can read and write
total population: 80.2%
male: 82.8%
female: 77.7% (2003 est.)

GOVERNMENT

Country name:
conventional long form: Republic of El Salvador
conventional short form: El Salvador
local long form: Republica de El Salvador
local short form: El Salvador

Government type: republic

Capital: *name:* San Salvador
geographic coordinates: 13 42 N, 89 12 W
time difference: UTC-6 (1 hour behind Washington, DC during Standard Time)

Administrative divisions: 14 departments (departamentos, singular—departamento); Ahuachapan, Cabanas, Chalatenango, Cuscatlan, La Libertad, La Paz, La Union, Morazan, San Miguel, San Salvador, San Vicente, Santa Ana, Sonsonate, Usulutan

Independence: 15 September 1821 (from Spain)

National holiday: Independence Day, 15 September (1821)

Constitution: 20 December 1983

Legal system: based on civil and Roman law with traces of common law; judicial review of legislative acts in the Supreme Court; has not accepted compulsory ICJ jurisdiction

Suffrage: 18 years of age; universal

Executive branch:

chief of state: President Elias Antonio SACA Gonzalez (since 1 June 2004); Vice President Ana Vilma Albanez DE ESCOBAR (since 1 June 2004); note—the president is both the chief of state and head of government

head of government: President Elias Antonio SACA Gonzalez (since 1 June 2004); Vice President Ana Vilma Albanez DE ESCOBAR (since 1 June 2004)

cabinet: Council of Ministers selected by the president

elections: president and vice president elected on the same ticket by popular vote for a single five-year term; election last held 21 March 2004 (next to be held in March 2009)

election results: Elias Antonio SACA Gonzalez elected president; percent of vote—Elias Antonio SACA Gonzalez 57.7%, Schafik HANDAL 35.6%, Hector SILVA 3.9%, other 2.8%

Legislative branch: unicameral Legislative Assembly or Asamblea Legislativa (84 seats; members are elected by direct, popular vote to serve three-year terms)

elections: last held 12 March 2006 (next to be held in March 2009)

election results: percent of vote by party—NA; seats by party—ARENA 34, FMLN 32, PCN 10, PDC 6, CD 2

Judicial branch: Supreme Court or Corte Suprema (15 judges are selected by the Legislative Assembly; the 15 judges are assigned to four Supreme Court chambers—constitutional, civil, penal, and administrative conflict)

Political parties and leaders: Christian Democratic Party or PDC [Rodolfo PARKER]; Democratic Convergence or CD [Ruben ZAMORA] (formerly United Democratic Center or CDU); Farabundo Marti National Liberation Front or FMLN [Medardo GONZALEZ]; National Conciliation Party or PCN [Ciro CRUZ ZEPEDA]; National Republican Alliance or ARENA [Elias Antonio SACA Gonzalez]; Popular Social Christian Party or PPSC [Rene AGUILUZ]; Revolutionary Democratic Front or FDR [Julio Cesar HERNANDEZ Carcamo]

Political pressure groups and leaders: labor organizations—Electrical Industry Union of El Salvador or SIES; Federation of the Construction Industry, Similar Transport and other activities, or FESINCONTRANS; National Confederation of Salvadoran Workers or CNTS; National Union of Salvadoran Workers or UNTS; Port Industry Union of El Salvador or SIPES; Salvadoran Union of Ex-Petrolleros and Peasant Workers or USEPOC; Salvadoran Workers Central or CTS; Workers Union of Electrical Corporation or STCEL; business organizations—National Association of Small Enterprise or ANEP; Salvadoran Assembly Industry Association or ASIC; Salvadoran Industrial Association or ASI

International organization participation: BCIE, CACM, FAO, G-77, IADB, IAEA, IBRD, ICAO, ICC, ICRM, IDA, IFAD, IFC, IFRCS, ILO, IMF, IMO, Interpol, IOC, IOM, IPU, ISO (correspondent), ITSO, ITU, ITUC, MIGA, MINURSO, NAM (observer), OAS, OPANAL, OPCW, PCA, RG, UN, UNCTAD, UNESCO, UNIDO, Union Latina, UNMIL, UNMIS, UNOCI, UNWTO, UPU, WCL, WCO, WFTU, WHO, WIPO, WMO, WTO

Diplomatic representation in the US:

chief of mission: Ambassador Rene Antonio LEON Rodriguez

chancery: 1400 16th Street, Washington, DC 20036

telephone: [1] (202) 265-9671

FAX: [1] (202) 234-3763

consulate(s) general: Boston, Chicago, Dallas, Elizabeth (New Jersey), Houston, Las Vegas, Los Angeles, Miami, New York (2), Nogales (Arizona), Santa Ana (California), San Francisco, Washington (DC), Woodbridge (Virginia), Woodstock (Georgia)

Diplomatic representation from the US:

chief of mission: Ambassador Charles L. GLAZER

embassy: Final Boulevard Santa Elena Sur, Antiguo Cuscatlan, La Libertad, San Salvador

mailing address: Unit 3116, APO AA 34023

telephone: [503] 2278-4444

FAX: [503] 2278-6011

Flag description: three equal horizontal bands of blue (top), white, and blue with the national coat of arms centered in the white band; the coat of arms features a round emblem encircled by the words REPUBLICA DE EL SALVADOR EN LA AMERICA CENTRAL; similar to the flag of Nicaragua, which has a different coat of arms centered in the white band—it features a triangle encircled by the words REPUBLICA DE NICARAGUA on top and AMERICA CENTRAL on the bottom; also similar to the flag of Honduras, which has five blue stars arranged in an X pattern centered in the white band

ECONOMY

Economy—overview: The smallest country in Central America, El Salvador has the third largest economy, but growth has been modest in recent years. Robust growth in non-traditional exports have offset declines in the maquila exports, while remittances and external aid offset the trade deficit from high oil prices and strong import demand for consumer and intermediate goods. El Salvador leads the region in remittances per capita with inflows equivalent to nearly all export income. Implementation in 2006 of the Central America-Dominican Republic Free Trade Agreement (CAFTA), which El Salvador was the first to ratify, has strengthened an already positive export trend. With the adoption of the US dollar as its currency in 2001, El Salvador lost control over monetary policy and must concentrate on maintaining a disciplined fiscal policy. The current government has pursued economic diversification, with some success in promoting textile production, international port services, and tourism through tax incentives. It is committed to opening the economy to trade and investment, and has embarked on a wave of privatizations extending to telecom, electricity distribution, banking, and pension funds. In late 2006, the government and the Millennium Challenge Corporation signed a five-year, $461 million compact to stimulate economic growth and reduce poverty in the country's northern region through investments in education, public services, enterprise development, and transportation infrastructure.

GDP (purchasing power parity): $41.65 billion (2007 est.)

GDP (official exchange rate): $20.37 billion (2007 est.)

GDP—real growth rate: 4.7% (2007 est.)

GDP—per capita (PPP): $5,800 (2007 est.)

GDP—composition by sector:

agriculture: 9.7%

industry: 27.6%

services: 62.7% (2007 est.)

Labor force: 2.913 million (2007 est.)

Labor force—by occupation: *agriculture:* 19%

industry: 23%

services: 58% (2006 est.)

Unemployment rate: 6.2% official rate; but the economy has much underemployment (2007 est.)

Population below poverty line: 30.7% (2006 est.)

Household income or consumption by percentage share:
lowest 10%: 0.7%
highest 10%: 38.8% (2002)

Distribution of family income—Gini index: 52.4 (2002)

Inflation rate (consumer prices): 3.9% (2007 est.)

Investment (gross fixed): 14.7% of GDP (2007 est.)

Budget:
revenues: $3.659 billion
expenditures: $3.716 billion (2007 est.)

Public debt: 37.9% of GDP (2007 est.)

Agriculture—products: coffee, sugar, corn, rice, beans, oilseed, cotton, sorghum; beef, dairy products; shrimp

Industries: food processing, beverages, petroleum, chemicals, fertilizer, textiles, furniture, light metals

Industrial production growth rate: 3.4% (2007 est.)

Electricity—production: 5.316 billion kWh (2006)

Electricity—production by source:
fossil fuel: 44%
hydro: 30.9%
nuclear: 0%
other: 25.1% (2001)

Electricity—consumption: 5.319 billion kWh (2006)

Electricity—exports: 111.1 million kWh (2007)

Electricity—imports: 38.6 million kWh (2007)

Oil—production: 0 bbl/day (2005)

Oil—consumption: 43,200 bbl/day (2005 est.)

Oil—exports: 4,963 bbl/day (2006)

Oil—imports: 45,210 bbl/day (2006)

Oil—proved reserves: 0 bbl (1 January 2006 est.)

Natural gas—production: 0 cu m (2005 est.)

Natural gas—consumption: 0 cu m (2005 est.)

Natural gas—exports: 0 cu m (2005 est.)

Natural gas—imports: 0 cu m (2005)

Natural gas—proved reserves: 0 cu m (1 January 2006 est.)

Current account balance: -$985 million (2007 est.)

Exports: $3.83 billion (2007 est.)

Exports—commodities: offshore assembly exports, coffee, sugar, shrimp, textiles, chemicals, electricity

Exports—partners: US 49.6%, Guatemala 14.4%, Honduras 8.8%, Nicaragua 5% (2006)

Imports: $8.208 billion (2007 est.)

Imports—commodities: raw materials, consumer goods, capital goods, fuels, foodstuffs, petroleum, electricity

Imports—partners: US 32.2%, Guatemala 9.3%, Mexico 7.4%, Germany 6.3%, China 4.7% (2006)

Economic aid—recipient: $267.6 million of which $55 million from US (2005)

Reserves of foreign exchange and gold: $2.198 billion (31 December 2007 est.)

Debt—external: $9.574 billion (December 2007)

Stock of direct foreign investment—at home: $5.352 billion (2007 est.)

Stock of direct foreign investment—abroad: $239 million (2007 est.)

Market value of publicly traded shares: $3.623 billion (2005)

Currency (code): US dollar (USD)

Currency code: USD

Exchange rates: the US dollar became El Salvador's currency in 2001

Fiscal year: calendar year

COMMUNICATIONS

Telephones—main lines in use: 1.037 million (2006)

Telephones—mobile cellular: 3.852 million (2006)

Telephone system:
general assessment: the four mobile-cellular service providers are expanding services rapidly and in 2006 mobile-cellular density stood at roughly 55 per 100 persons; growth in fixed-line services has slowed in the face of mobile-cellular competition
domestic: nationwide microwave radio relay system
international: country code—503; satellite earth station—1 Intelsat (Atlantic Ocean); connected to Central American Microwave System

Radio broadcast stations: AM 52, FM 144, shortwave 0 (2005)

Radios: 2.75 million (1997)

Television broadcast stations: 5 (1997)

Televisions: 600,000 (1990)

Internet country code: .sv

Internet hosts: 12,519 (2007)

Internet Service Providers (ISPs): 4 (2000)

Internet users: 637,000 (2005)

TRANSPORTATION

Airports: 65 (2007)

Airports—with paved runways:
total: 4
over 3,047 m: 1
1,524 to 2,437 m: 1
914 to 1,523 m: 2 (2007)

Airports—with unpaved runways:
total: 61
1,524 to 2,437 m: 1
914 to 1,523 m: 12
under 914 m: 48 (2007)

Heliports: 1 (2007)

Railways:
total: 562 km
narrow gauge: 562 km 0.914-m gauge
note: railways not in operation since 2005 because of disuse and high costs that led to a lack of maintenance (2007)

Roadways:
total: 10,886 km
paved: 2,827 km
unpaved: 8,059 km (2000)

Waterways: Rio Lempa partially navigable for small craft (2007)

Ports and terminals: Acajutla, Puerto Cutuco

MILITARY

Military branches: Salvadoran Army (ES), Salvadoran Navy (FNES), Salvadoran Air Force (Fuerza Aerea Salvadorena, FAS) (2008)

Military service age and obligation: 18 years of age for selective compulsory military service; 16 years of age for voluntary service; 12-month service obligation (2006)

Manpower available for military service:
males age 16–49: 1,634,816
females age 16–49: 1,775,474 (2008 est.)

Manpower fit for military service:
males age 16–49: 1,168,406
females age 16–49: 1,519,375 (2008 est.)

Manpower reaching militarily significant age annually:
males age 16–49: 73,915
females age 16–49: 71,252 (2008 est.)

Military expenditures—percent of GDP: 5% (2006)

TRANSNATIONAL ISSUES

Disputes—international: International Court of Justice (ICJ) ruled on the delimitation of "bolsones" (disputed areas) along the El Salvador-Honduras boundary, in 1992, with final agreement by the parties in 2006 after an Organization of American States (OAS) survey and a further ICJ ruling in 2003; the 1992 ICJ ruling advised a tripartite resolution to a maritime boundary in the Gulf of Fonseca advocating Honduran access to the Pacific; El Salvador continues to claim tiny Conejo Island, not identified in the ICJ decision, off Honduras in the Gulf of Fonseca

Illicit drugs: transshipment point for cocaine; small amounts of marijuana produced for local consumption; significant use of cocaine

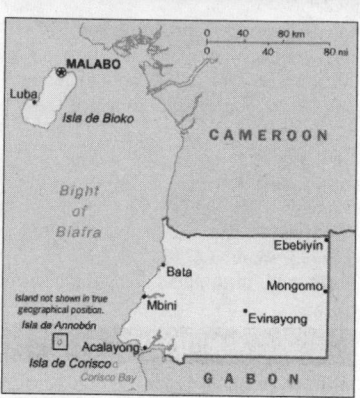

INTRODUCTION

Background: Equatorial Guinea gained independence in 1968 after 190 years of Spanish rule. This tiny country, composed of a mainland portion plus five inhabited islands, is one of the smallest on the African continent. President Teodoro OBIANG NGUEMA MBA-SOGO has ruled the country since 1979 when he seized power in a coup. Although nominally a constitutional democracy since 1991, the 1996 and 2002 presidential elections—as well as the 1999 and 2004 legislative elections—were widely seen as flawed. The president exerts almost total control over the political system and has discouraged political opposition. Equatorial Guinea has experienced rapid economic growth due to the discovery of large offshore oil reserves, and in the last decade has become Sub-Saharan Africa's third largest oil exporter. Despite the country's economic windfall from oil production resulting in a massive increase in government revenue in recent years, there have been few improvements in the population's living standards.

GEOGRAPHY

Location: Western Africa, bordering the Bight of Biafra, between Cameroon and Gabon
Geographic coordinates: 2 00 N, 10 00 E
Map references: Africa
Area:
total: 28,051 sq km
land: 28,051 sq km
water: 0 sq km
Area—comparative: slightly smaller than Maryland
Land boundaries:
total: 539 km
border countries: Cameroon 189 km, Gabon 350 km

Coastline: 296 km
Maritime claims:
territorial sea: 12 nm
exclusive economic zone: 200 nm
Climate: tropical; always hot, humid
Terrain: coastal plains rise to interior hills; islands are volcanic
Elevation extremes:
lowest point: Atlantic Ocean 0 m
highest point: Pico Basile 3,008 m
Natural resources: petroleum, natural gas, timber, gold, bauxite, diamonds, tantalum, sand and gravel, clay
Land use:
arable land: 4.63%
permanent crops: 3.57%
other: 91.8% (2005)
Irrigated land: NA
Total renewable water resources: 26 cu km (2001)
Freshwater withdrawal (domestic/industrial/agricultural): *total:* 0.11 cu km/yr (83%/16%/1%)
per capita: 220 cu m/yr (2000)
Natural hazards: violent windstorms, flash floods
Environment—current issues: tap water is not potable; deforestation
Environment—international agreements: *party to:* Biodiversity, Climate Change, Climate Change-Kyoto Protocol, Desertification, Endangered Species, Hazardous Wastes, Law of the Sea, Marine Dumping, Ozone Layer Protection, Ship Pollution, Wetlands
signed, but not ratified: none of the selected agreements
Geography—note: insular and continental regions widely separated

PEOPLE

Population: 616,459 (July 2008 est.)
Age structure:
0–14 years: 42% (male 131,696/female 127,253)
15–64 years: 53.8% (male 162,458/female 169,445)
65 years and over: 4.2% (male 11,394/female 14,213) (2008 est.)
Median age:
total: 18.9 years
male: 18.3 years
female: 19.5 years (2008 est.)
Population growth rate: 2.732% (2008 est.)
Birth rate: 37.04 births/1,000 population (2008 est.)
Death rate: 9.72 deaths/1,000 population (2008 est.)
Net migration rate: NA
Sex ratio:
at birth: 1.03 male(s)/female

under 15 years: 1.03 male(s)/female
15–64 years: 0.96 male(s)/female
65 years and over: 0.8 male(s)/female
total population: 0.98 male(s)/female (2008 est.)
Infant mortality rate:
total: 83.75 deaths/1,000 live births
male: 84.85 deaths/1,000 live births
female: 82.61 deaths/1,000 live births (2008 est.)
Life expectancy at birth:
total population: 61.23 years
male: 60.36 years
female: 62.13 years (2008 est.)
Total fertility rate: 5.16 children born/woman (2008 est.)
HIV/AIDS—adult prevalence rate: 3.4% (2001 est.)
HIV/AIDS—people living with HIV/AIDS: 5,900 (2001 est.)
HIV/AIDS—deaths: 370 (2001 est.)
Major infectious diseases:
degree of risk: very high
food or waterborne diseases: bacterial and protozoal diarrhea, hepatitis A, and typhoid fever
vectorborne disease: malaria (2008)
Nationality:
noun: Equatorial Guinean(s) or Equatoguinean(s)
adjective: Equatorial Guinean or Equatoguinean
Ethnic groups: Fang 85.7%, Bubi 6.5%, Mdowe 3.6%, Annobon 1.6%, Bujeba 1.1%, other 1.4% (1994 census)
Religions: nominally Christian and predominantly Roman Catholic, pagan practices
Languages: Spanish 67.6% (official), other 32.4% (includes French (official), Fang, Bubi) (1994 census)
Literacy:
definition: age 15 and over can read and write
total population: 85.7%
male: 93.3%
female: 78.4% (2003 est.)

GOVERNMENT

Country name:
conventional long form: Republic of Equatorial Guinea
conventional short form: Equatorial Guinea
local long form: Republica de Guinea Ecuatorial/Republique de Guinee equatoriale
local short form: Guinea Ecuatorial/Guinee equatoriale
former: Spanish Guinea
Government type: republic
Capital: *name:* Malabo

geographic coordinates: 3 45 N, 8 47 E

time difference: UTC+1 (6 hours ahead of Washington, DC during Standard Time)

Administrative divisions: 7 provinces (provincias, singular—provincia); Annobon, Bioko Norte, Bioko Sur, Centro Sur, Kie-Ntem, Litoral, Wele-Nzas

Independence: 12 October 1968 (from Spain)

National holiday: Independence Day, 12 October (1968)

Constitution: approved by national referendum 17 November 1991; amended January 1995

Legal system: partly based on Spanish civil law and tribal custom; has not accepted compulsory ICJ jurisdiction

Suffrage: 18 years of age; universal

Executive branch:

chief of state: President Brig. Gen. (Ret.) Teodoro OBIANG NGUEMA MBASOGO (since 3 August 1979 when he seized power in a military coup)

head of government: Prime Minister Ricardo Mangue Obama NFUBEA (since 14 August 2006); First Deputy Prime Minister Mercelino Oyono NTUTUMU (since 15 June 2004)

cabinet: Council of Ministers appointed by the president

elections: president elected by popular vote for a seven-year term (no term limits); election last held 15 December 2002 (next to be held in December 2009); prime minister and deputy prime ministers appointed by the president

election results: Teodoro OBIANG NGUEMA MBASOGO reelected president; percent of vote—Teodoro OBIANG NGUEMA MBASOGO 97.1%, Celestino Bonifacio BACALE 2.2%; elections marred by widespread fraud

Legislative branch: unicameral House of People's Representatives or Camara de Representantes del Pueblo (100 seats; members directly elected by popular vote to serve five-year terms)

elections: last held 25 April 2004 (next to be held 4 May 2008)

election results: percent of vote by party—NA; seats by party—PDGE 98, CPDS 2

note: Parliament has little power since the constitution vests all executive authority in the president

Judicial branch: Supreme Tribunal

Political parties and leaders: Convergence Party for Social Democracy or CPDS [Placido MICO Abogo]; Democratic Party for Equatorial Guinea or PDGE (ruling party) [Teodoro OBIANG NGUEMA MBASOGO]; Party for Progress of Equatorial Guinea or PPGE [Severo MOTO]; Popular

Action of Equatorial Guinea or APGE [Avelino MOCACHE]; Popular Union or UP

Political pressure groups and leaders: NA

International organization participation: ACCT, ACP, AfDB, AU, BDEAC, CEMAC, FAO, FZ, G-77, IBRD, ICAO, ICRM, IDA, IFAD, IFC, IFRCS, ILO, IMF, IMO, Interpol, IOC, ITSO, ITU, MIGA, NAM, OAS (observer), OIF, OPCW, UN, UNCTAD, UNESCO, UNIDO, UNWTO, UPU, WHO, WIPO, WTO (observer)

Diplomatic representation in the US:

chief of mission: Ambassador Purificacion ANGUE ONDO

chancery: 2020 16th Street NW, Washington, DC 20009

telephone: [1] (202) 518-5700

FAX: [1] (202) 518-5252

Diplomatic representation from the US:

chief of mission: Ambassador Donald C. JOHNSON

embassy: adjacent to the golf course at the base of Mont Febe; note—relocated embassy is opened for limited functions; inquiries should continue to be directed to the US Embassy in Yaounde, Cameroon

mailing address: B.P. 817, Yaounde, Cameroon; US Embassy Yaounde, US Department of State, Washington, DC 20521-2520

telephone: [237] 220 15 00

FAX: [237] 220 16 20

Flag description: three equal horizontal bands of green (top), white, and red with a blue isosceles triangle based on the hoist side and the coat of arms centered in the white band; the coat of arms has six yellow six-pointed stars (representing the mainland and five offshore islands) above a gray shield bearing a silk-cotton tree and below which is a scroll with the motto UNIDAD, PAZ, JUSTICIA (Unity, Peace, Justice)

ECONOMY

Economy—overview: The discovery and exploitation of large oil reserves have contributed to dramatic economic growth in recent years. Forestry, farming, and fishing are also major components of GDP. Subsistence farming predominates. Although pre-independence Equatorial Guinea counted on cocoa production for hard currency earnings, the neglect of the rural economy under successive regimes has diminished potential for agriculture-led growth (the government has stated its intention to reinvest some oil revenue into agriculture). A number of aid programs sponsored by the World Bank and the IMF have been cut off

since 1993, because of corruption and mismanagement. No longer eligible for concessional financing because of large oil revenues, the government has been trying to agree on a "shadow" fiscal management program with the World Bank and IMF. Government officials and their family members own most businesses. Undeveloped natural resources include titanium, iron ore, manganese, uranium, and alluvial gold. Growth remained strong in 2007, led by oil. Equatorial Guinea now has the fourth highest per capita income in the world, after Luxembourg, Bermuda, and Jersey.

GDP (purchasing power parity): $15.54 billion (2007 est.)

GDP (official exchange rate): $10.49 billion (2007 est.)

GDP—real growth rate: 12.4% (2007 est.)

GDP—per capita (PPP): $12,900 (2007 est.)

GDP—composition by sector:

agriculture: 2.9%

industry: 92.2%

services: 4.8% (2007 est.)

Labor force: NA

Unemployment rate: 30% (1998 est.)

Population below poverty line: NA%

Household income or consumption by percentage share:

lowest 10%: NA%

highest 10%: NA%

Inflation rate (consumer prices): 4.6% (2007 est.)

Investment (gross fixed): 38% of GDP (2007 est.)

Budget:

revenues: $4.963 billion

expenditures: $2.494 billion (2007 est.)

Public debt: 3.7% of GDP (2007 est.)

Agriculture—products: coffee, cocoa, rice, yams, cassava (tapioca), bananas, palm oil nuts; livestock; timber

Industries: petroleum, fishing, sawmilling, natural gas

Industrial production growth rate: 10.1% (2007 est.)

Electricity—production: 28 million kWh (2005)

Electricity—production by source:

fossil fuel: 94.3%

hydro: 5.7%

nuclear: 0%

other: 0% (2001)

Electricity—consumption: 26.04 million kWh (2005)

Electricity—exports: 0 kWh (2005)

Electricity—imports: 0 kWh (2005)

Oil—production: 396,100 bbl/day (2005 est.)

Oil—consumption: 1,000 bbl/day (2005 est.)

Oil—exports: 371,700 bbl/day (2004)

Oil—imports: 1,026 bbl/day (2004)
Oil—proved reserves: 563.5 million bbl (1 January 2002 est.)
Natural gas—production: 1.247 billion cu m (2005 est.)
Natural gas—consumption: 1.247 billion cu m (2005 est.)
Natural gas—exports: 0 cu m (2005 est.)
Natural gas—imports: 0 cu m (2005)
Natural gas—proved reserves: 35.31 billion cu m (1 January 2006 est.)
Current account balance: $188 million (2007 est.)
Exports: $9.915 billion f.o.b. (2007 est.)
Exports—commodities: petroleum, methanol, timber, cocoa
Exports—partners: China 30.9%, US 22.2%, Spain 12.6%, Taiwan 10.6%, Portugal 6.1% (2006)
Imports: $3.098 billion f.o.b. (2007 est.)
Imports—commodities: petroleum sector equipment, other equipment
Imports—partners: US 37.7%, Spain 9.8%, Cote d'Ivoire 7.9%, France 6.1%, South Korea 6.1%, UK 5.8%, Italy 5% (2006)
Economic aid—recipient: $39 million (2005)
Reserves of foreign exchange and gold: $3.837 billion (31 December 2007 est.)
Debt—external: $341 million (31 December 2007 est.)
Currency (code): Communaute Financiere Africaine franc (XAF); note—responsible authority is the Bank of the Central African States
Currency code: XAF
Exchange rates: Communaute Financiere Africaine francs (XAF) per US dollar—481.83 (2007), 522.4 (2006), 527.47 (2005), 528.29 (2004), 581.2 (2003)
Fiscal year: calendar year

COMMUNICATIONS

Telephones—main lines in use: 10,000 (2005)
Telephones—mobile cellular: 96,900 (2005)

Telephone system:
general assessment: digital fixed-line network in most major urban areas and good mobile coverage
domestic: fixed-line density is about 2 per 100 persons; mobile-cellular subscribership has been increasing and in 2005 stood at about 20 percent of the population
international: country code—240; international communications from Bata and Malabo to African and European countries; satellite earth station—1 Intelsat (Indian Ocean)
Radio broadcast stations: AM 0, FM 3, shortwave 5 (2001)
Radios: 180,000 (1997)
Television broadcast stations: 1 (2001)
Televisions: 4,000 (1997)
Internet country code: .gq
Internet hosts: 81 (2007)
Internet Service Providers (ISPs): 1 (2002)
Internet users: 8,000 (2006)

TRANSPORTATION

Airports: 5 (2007)
Airports—with paved runways: *total:* 5
2,438 to 3,047 m: 1
1,524 to 2,437 m: 1
914 to 1,523 m: 1
under 914 m: 2 (2007)
Pipelines: condensate 42 km; condensate/gas 5 km; gas 80 km; oil 54 km (2007)
Roadways:
total: 2,880 km (2000)
Merchant marine:
total: 1 ship (1000 GRT or over) 1,745 GRT/3,434 DWT
by type: cargo 1 (2007)
Ports and terminals: Bata, Malabo

MILITARY

Military branches: National Guard (Guardia Nacional (Army), with Coast Guard (Navy) and Air Wing) (2008)
Military service age and obligation: 18 years of age (est.) for compulsory military service (2008)

Manpower available for military service:
males age 16–49: 136,725
females age 16–49: 138,018 (2008 est.)
Manpower fit for military service:
males age 16–49: 101,712
females age 16–49: 104,381 (2008 est.)
Military expenditures—percent of GDP: 0.1% (2006 est.)

TRANSNATIONAL ISSUES

Disputes—international: in 2002, ICJ ruled on an equidistance settlement of Cameroon-Equatorial Guinea-Nigeria maritime boundary in the Gulf of Guinea, but a dispute between Equatorial Guinea and Cameroon over an island at the mouth of the Ntem River and imprecisely defined maritime coordinates in the ICJ decision delay final delimitation; UN urges Equatorial Guinea and Gabon to resolve the sovereignty dispute over Gabon-occupied Mbane and lesser islands and to create a maritime boundary in the hydrocarbon-rich Corisco Bay
Trafficking in persons: *current situation:* Equatorial Guinea is mainly a destination country for children trafficked for forced labor, involuntary domestic servitude, and commercial sexual exploitation from surrounding countries—primarily Benin, Nigeria, Gabon, and Cameroon; victims work in the agricultural and commercial sectors of Malabo and Bata, where demand is high due to a booming oil sector and a flourishing expatriate business community; children work as farmhands, street vendors, or household servants; girls are trafficked for commercial sexual exploitation
tier rating: Tier 3—failed to demonstrate the political commitment to address its human trafficking problem; despite efforts to raise awareness of trafficking problems, in 2006 the government failed to investigate and prosecute traffickers or protect victims

ERITREA

INTRODUCTION

Background: Eritrea was awarded to Ethiopia in 1952 as part of a federation. Ethiopia's annexation of Eritrea as a province 10 years later sparked a 30-year struggle for independence that ended in 1991 with Eritrean rebels defeating governmental forces; independence was overwhelmingly approved in a 1993 referendum. A two-and-a-half-year border war with Ethiopia that erupted in 1998 ended under UN auspices in December 2000. Eritrea currently hosts a UN peacekeeping operation that is monitoring a 25 km-wide Temporary Security Zone (TSZ) on the border with Ethiopia. An international commission, organized to resolve the border dispute, posted its findings in 2002. However, both parties have been unable to reach agreement on implementing the decision. On 30 November 2007, the Eritrea-Ethiopia Boundary Commission remotely demarcated the border by coordinates and dissolved itself, leaving Ethiopia still occupying several tracts of disputed territory, including the town of Badme. Eritrea accepted the EEBC's "virtual demarcation" decision and called on Ethiopia to remove its troops from the TSZ which it states is Eritrean territory. Ethiopia has not accepted the virtual demarcation decision.

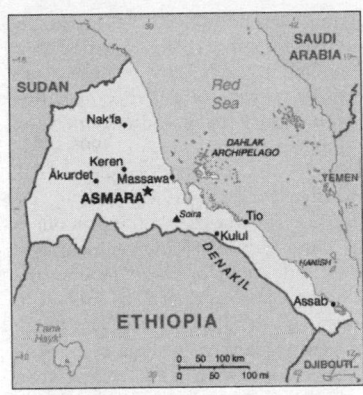

GEOGRAPHY

Location: Eastern Africa, bordering the Red Sea, between Djibouti and Sudan
Geographic coordinates: 15 00 N, 39 00 E
Map references: Africa
Area:
total: 121,320 sq km
land: 121,320 sq km
water: 0 sq km
Area—comparative: slightly larger than Pennsylvania
Land boundaries:
total: 1,626 km
border countries: Djibouti 109 km, Ethiopia 912 km, Sudan 605 km
Coastline: 2,234 km (mainland on Red Sea 1,151 km, islands in Red Sea 1,083 km)
Maritime claims: *territorial sea:* 12 nm
Climate: hot, dry desert strip along Red Sea coast; cooler and wetter in the central highlands (up to 61 cm of rainfall annually, heaviest June to September); semiarid in western hills and lowlands
Terrain: dominated by extension of Ethiopian north-south trending highlands, descending on the east to a coastal desert plain, on the northwest to hilly terrain and on the southwest to flat-to-rolling plains
Elevation extremes:
lowest point: near Kulul within the Denakil depression -75 m
highest point: Soira 3,018 m
Natural resources: gold, potash, zinc, copper, salt, possibly oil and natural gas, fish
Land use:
arable land: 4.78%
permanent crops: 0.03%
other: 95.19% (2005)
Irrigated land: 210 sq km (2003)
Total renewable water resources: 6.3 cu km (2001)
Freshwater withdrawal (domestic/industrial/agricultural): *total:* 0.3 cu km/yr (3%/0%/97%)
per capita: 68 cu m/yr (2000)

Natural hazards: frequent droughts; locust swarms
Environment—current issues: deforestation; desertification; soil erosion; overgrazing; loss of infrastructure from civil warfare
Environment—international agreements: *party to:* Biodiversity, Climate Change, Climate Change-Kyoto Protocol, Desertification, Endangered Species, Hazardous Wastes, Ozone Layer Protection
signed, but not ratified: none of the selected agreements
Geography—note: strategic geopolitical position along world's busiest shipping lanes; Eritrea retained the entire coastline of Ethiopia along the Red Sea upon de jure independence from Ethiopia on 24 May 1993

PEOPLE

Population: 5,028,475 (July 2008 est.)
Age structure:
0–14 years: 42.9% (male 1,085,116/female 1,072,262)
15–64 years: 53.5% (male 1,332,349/female 1,355,494)
65 years and over: 3.6% (male 88,068/female 95,186) (2008 est.)
Median age: *total:* 18.1 years
male: 17.9 years
female: 18.3 years (2008 est.)
Population growth rate: 2.447% (2008 est.)
Birth rate: 33.62 births/1,000 population (2008 est.)
Death rate: 9.15 deaths/1,000 population (2008 est.)
Net migration rate: NA
Sex ratio:
at birth: 1.03 male(s)/female
under 15 years: 1.01 male(s)/female
15–64 years: 0.98 male(s)/female
65 years and over: 0.93 male(s)/female
total population: 0.99 male(s)/female (2008 est.)
Infant mortality rate:
total: 44.22 deaths/1,000 live births
male: 49.93 deaths/1,000 live births
female: 38.35 deaths/1,000 live births (2008 est.)
Life expectancy at birth:
total population: 60.05 years
male: 58.29 years
female: 61.87 years (2008 est.)
Total fertility rate: 4.84 children born/woman (2008 est.)
HIV/AIDS—adult prevalence rate: 2.7% (2003 est.)
HIV/AIDS—people living with HIV/AIDS: 60,000 (2003 est.)
HIV/AIDS—deaths: 6,300 (2003 est.)
Major infectious diseases:
degree of risk: high

food or waterborne diseases: bacterial diarrhea, hepatitis A, and typhoid fever
vectorborne disease: malaria (2008)
Nationality:
noun: Eritrean(s)
adjective: Eritrean
Ethnic groups: Tigrinya 50%, Tigre and Kunama 40%, Afar 4%, Saho (Red Sea coast dwellers) 3%, other 3%
Religions: Muslim, Coptic Christian, Roman Catholic, Protestant
Languages: Afar, Arabic, Tigre and Kunama, Tigrinya, other Cushitic languages
Literacy:
definition: age 15 and over can read and write
total population: 58.6%
male: 69.9%
female: 47.6% (2003 est.)

GOVERNMENT

Country name:
conventional long form: State of Eritrea
conventional short form: Eritrea
local long form: Hagere Ertra
local short form: Ertra
former: Eritrea Autonomous Region in Ethiopia
Government type: transitional government
note: following a successful referendum on independence for the Autonomous Region of Eritrea on 23–25 April 1993, a National Assembly, composed entirely of the People's Front for Democracy and Justice or PFDJ, was established as a transitional legislature; a Constitutional Commission was also established to draft a constitution; ISAIAS Afworki was elected president by the transitional legislature; the constitution, ratified in May 1997, did not enter into effect, pending parliamentary and presidential elections; parliamentary elections were scheduled in December 2001, but were postponed indefinitely; currently the sole legal party is the People's Front for Democracy and Justice (PFDJ)
Capital: *name:* Asmara (Asmera)
geographic coordinates: 15 20 N, 38 56 E
time difference: UTC+3 (8 hours ahead of Washington, DC during Standard Time)
Administrative divisions: 6 regions (zobatat, singular—zoba); Anseba, Debub (Southern), Debubawi K'eyih Bahri (Southern Red Sea), Gash Barka, Ma'akel (Central), Semenawi Keyih Bahri (Northern Red Sea)
Independence: 24 May 1993 (from Ethiopia)
National holiday: Independence Day, 24 May (1993)
Constitution: a transitional constitution, decreed on 19 May 1993, was replaced by

a new constitution adopted on 23 May 1997, but not yet implemented

Legal system: primary basis is the Ethiopian legal code of 1957, with revisions; new civil, commercial, and penal codes have not yet been promulgated; government also issues unilateral proclamations setting laws and policies; also relies on customary and post-independence-enacted laws and, for civil cases involving Muslims, Islamic law; does not accept compulsory ICJ jurisdiction

Suffrage: 18 years of age; universal

Executive branch:

chief of state: President ISAIAS Afworki (since 8 June 1993); note—the president is both the chief of state and head of government and is head of the State Council and National Assembly

head of government: President ISAIAS Afworki (since 8 June 1993)

cabinet: State Council is the collective executive authority; members appointed by the president

elections: president elected by the National Assembly for a five-year term (eligible for a second term); the most recent and only election held 8 June 1993 (next election date uncertain as the National Assembly did not hold a presidential election in December 2001 as anticipated)

election results: ISAIAS Afworki elected president; percent of National Assembly vote—ISAIAS Afworki 95%, other 5%

Legislative branch: unicameral National Assembly (150 seats; members elected by direct popular vote to serve five-year terms)

elections: in May 1997, following the adoption of the new constitution, 75 members of the PFDJ Central Committee (the old Central Committee of the EPLF), 60 members of the 527-member Constituent Assembly, which had been established in 1997 to discuss and ratify the new constitution, and 15 representatives of Eritreans living abroad were formed into a Transitional National Assembly to serve as the country's legislative body until countrywide elections to a National Assembly were held; although only 75 of 150 members of the Transitional National Assembly were elected, the constitution stipulates that once past the transition stage, all members of the National Assembly will be elected by secret ballot of all eligible voters; National Assembly elections scheduled for December 2001 were postponed indefinitely

Judicial branch: High Court—regional, subregional, and village courts; also have military and special courts

Political parties and leaders: People's

Front for Democracy and Justice or PFDJ [ISAIAS Afworki] (the only party recognized by the government); note—a National Assembly committee drafted a law on political parties in January 2001, but the full National Assembly has yet to debate or vote on it

Political pressure groups and leaders: Eritrean Islamic Jihad or EIJ (also including Eritrean Islamic Jihad Movement or EIJM (also known as the Abu Sihel Movement)); Eritrean Islamic Salvation or EIS (also known as the Arafa Movement); Eritrean Liberation Front or ELF [ABDULLAH Muhammed]; Eritrean National Alliance or ENA (a coalition including EIJ, EIS, ELF, and a number of ELF factions) [HERUY Tedla Biru]; Eritrean Public Forum or EPF [ARADOM Iyob]; Eritrean Democratic Party (EDP) [HAGOS, Mesfin]

International organization participation: ACP, AfDB, AU, COMESA, FAO, G-77, IAEA, IBRD, ICAO, ICCt (signatory), IDA, IFAD, IFC, IFRCS (observer), ILO, IMF, IMO, Interpol, IOC, ISO (correspondent), ITU, ITUC, LAS (observer), MIGA, NAM, OPCW, PCA, UN, UNCTAD, UNESCO, UNIDO, UNWTO, UPU, WCO, WFTU, WHO, WIPO, WMO

Diplomatic representation in the US:

chief of mission: Ambassador GHIRMAI Ghebremariam

chancery: 1708 New Hampshire Avenue NW, Washington, DC 20009

telephone: [1] (202) 319-1991

FAX: [1] (202) 319-1304

consulate(s) general: Oakland (California)

Diplomatic representation from the US:

chief of mission: Ambassador Ronald MCMULLEN

embassy: 179 Alaa Street, Asmara

mailing address: P. O. Box 211, Asmara

telephone: [291] (1) 120004

FAX: [291] (1) 127584

Flag description: red isosceles triangle (based on the hoist side) dividing the flag into two right triangles; the upper triangle is green, the lower one is blue; a gold wreath encircling a gold olive branch is centered on the hoist side of the red triangle

ECONOMY

Economy—overview: Since independence from Ethiopia in 1993, Eritrea has faced the economic problems of a small, desperately poor country, accentuated by the recent implementation of restrictive economic policies. Eritrea has a command economy under the control of the sole political party, the People's Front for

Democracy and Justice (PFDJ). Like the economies of many African nations, the economy is largely based on subsistence agriculture, with 80% of the population involved in farming and herding. The Ethiopian-Eritrea war in 1998–2000 severely hurt Eritrea's economy. GDP growth fell to zero in 1999 and to -12.1% in 2000. The May 2000 Ethiopian offensive into northern Eritrea caused some $600 million in property damage and loss, including losses of $225 million in livestock and 55,000 homes. The attack prevented planting of crops in Eritrea's most productive region, causing food production to drop by 62%. Even during the war, Eritrea developed its transportation infrastructure, asphalting new roads, improving its ports, and repairing war-damaged roads and bridges. Since the war ended, the government has maintained a firm grip on the economy, expanding the use of the military and party-owned businesses to complete Eritrea's development agenda. The government strictly controls the use of foreign currency, limiting access and availability. Few private enterprises remain in Eritrea. Eritrea's economy is heavily dependent on taxes paid by members of the diaspora. Erratic rainfall and the delayed demobilization of agriculturalists from the military continue to interfere with agricultural production, and Eritrea's recent harvests have not been able to meet the food needs of the country. The government continues to place its hope for additional revenue on the development of several international mining projects. Despite difficulties for international companies in working with the Eritrean government, a Canadian mining company signed a contract with the GSE in 2007 and plans to begin mineral extraction in 2010. Eritrea also anticipates opening a free trade zone at the port of Massawa in 2008. Eritrea's economic future depends upon its ability to master social problems such as illiteracy, unemployment, and low skills, and more importantly, on the government's willingness to support a true market economy.

GDP (purchasing power parity): $3.619 billion (2007 est.)

GDP (official exchange rate): $1.316 billion (2007 est.)

GDP—real growth rate: 1.3% (2007 est.)

GDP—per capita (PPP): $800 (2007 est.)

GDP—composition by sector:

agriculture: 21.7%

industry: 22.6%

services: 55.7% (2007 est.)

Labor force: NA
Labor force—by occupation: *agriculture:* 80%
industry and services: 20% (2004 est.)
Unemployment rate: NA%
Population below poverty line: 50% (2004 est.)
Household income or consumption by percentage share:
lowest 10%: NA%
highest 10%: NA%
Inflation rate (consumer prices): 9.3% (2007 est.)
Investment (gross fixed): 21.1% of GDP (2007 est.)
Budget:
revenues: $234.6 million
expenditures: $471.4 million (2007 est.)
Agriculture—products: sorghum, lentils, vegetables, corn, cotton, tobacco, sisal; livestock, goats; fish
Industries: food processing, beverages, clothing and textiles, light manufacturing, salt, cement
Industrial production growth rate: 2% (2007 est.)
Electricity—production: 274 million kWh (2005)
Electricity—production by source:
fossil fuel: 100%
hydro: 0%
nuclear: 0%
other: 0% (2001)
Electricity—consumption: 228 million kWh (2005)
Electricity—exports: 0 kWh (2005)
Electricity—imports: 0 kWh (2005)
Oil—production: 0 bbl/day (2005 est.)
Oil—consumption: 5,000 bbl/day (2005 est.)
Oil—exports: 54.59 bbl/day (2004)
Oil—imports: 4,924 bbl/day (2004)
Oil—proved reserves: 0 bbl (1 January 2006 est.)
Natural gas—production: 0 cu m (2005 est.)
Natural gas—consumption: 0 cu m (2005 est.)
Natural gas—exports: 0 cu m (2005 est.)
Natural gas—imports: 0 cu m (2005)
Natural gas—proved reserves: 0 cu m (1 January 2006 est.)
Current account balance: -$62 million (2007 est.)
Exports: $16.95 million f.o.b. (2007 est.)
Exports—commodities: livestock, sorghum, textiles, food, small manufactures
Exports—partners: Italy 19%, Austria 11.6%, France 11.5%, US 5.6%, Ethiopia 4.7%, Taiwan 4.6%, China 4.4% (2006)
Imports: $577 million f.o.b. (2007 est.)

Imports—commodities: machinery, petroleum products, food, manufactured goods
Imports—partners: Italy 16.8%, China 16.7%, Netherlands 6.9%, Saudi Arabia 6.8%, Turkey 6.6%, Sudan 6.3%, Germany 5.7% (2006)
Economic aid—recipient: $355.2 million (2005)
Reserves of foreign exchange and gold: $22.66 million (31 December 2007 est.)
Debt—external: $311 million (2000 est.)
Currency (code): nakfa (ERN)
Currency code: ERN
Exchange rates: nakfa (ERN) per US dollar—15.5 (2007), 15.4 (2006), 14.5 (2005), 13.788 (2004), 13.878 (2003)
note: the official exchange rate is 15 nakfa to the dollar
Fiscal year: calendar year

COMMUNICATIONS

Telephones—main lines in use: 37,700 (2006)
Telephones—mobile cellular: 62,000 (2006)
Telephone system:
general assessment: inadequate
domestic: inadequate; most telephones are in Asmara; government is seeking international tenders to improve the system (2002)
international: country code—291; note—international connections exist
Radio broadcast stations: AM 2, FM NA, shortwave 2 (2000)
Radios: 345,000 (1997)
Television broadcast stations: 2 (2006)
Televisions: 1,000 (1997)
Internet country code: .er
Internet hosts: 1,446 (2007)
Internet Service Providers (ISPs): 5 (2001)
Internet users: 100,000 (2006)

TRANSPORTATION

Airports: 18 (2007)
Airports—with paved runways:
total: 4
over 3,047 m: 2
2,438 to 3,047 m: 2 (2007)
Airports—with unpaved runways:
total: 14
over 3,047 m: 1
2,438 to 3,047 m: 1
1,524 to 2,437 m: 6
914 to 1,523 m: 4
under 914 m: 2 (2007)
Heliports: 1 (2007)
Railways:
total: 306 km
narrow gauge: 306 km 0.950-m gauge (2006)

Roadways:
total: 4,010 km
paved: 874 km
unpaved: 3,136 km (2000)
Merchant marine:
total: 5 ships (1000 GRT or over) 12,529 GRT/15,023 DWT
by type: cargo 2, liquefied gas 1, petroleum tanker 1, roll on/roll off 1 (2007)
Ports and terminals: Assab, Massawa

MILITARY

Military branches: Eritrean Armed Forces: Ground Forces, Navy, Air Force (2008)
Military service age and obligation: 18–40 years of age for male and female voluntary and compulsory military service; 16-month conscript service obligation (2006)
Manpower available for military service:
males age 16–49: 1,108,836
females age 16–49: 1,096,120 (2008 est.)
Manpower fit for military service:
males age 16–49: 715,531
females age 16–49: 731,511 (2008 est.)
Manpower reaching militarily significant age annually:
males age 16–49: 62,504
females age 16–49: 62,153 (2008 est.)
Military expenditures—percent of GDP: 6.3% (2006 est.)

TRANSNATIONAL ISSUES

Disputes—international: Eritrea and Ethiopia agreed to abide by 2002 Ethiopia-Eritrea Boundary Commission's (EEBC) delimitation decision but, neither party responded to the revised line detailed in the November 2006 EEBC Demarcation Statement; UN Peacekeeping Mission to Ethiopia and Eritrea (UNMEE), which has monitored the 25-km-wide Temporary Security Zone in Eritrea since 2000, is extended for six months in 2007 despite Eritrean restrictions on its operations and reduced force of 17,000; Sudan accuses Eritrea of supporting eastern Sudanese rebel groups
Refugees and internally displaced persons: *IDPs:* 32,000 (border war with Ethiopia from 1998–2000; most IDPs are near the central border region) (2007)

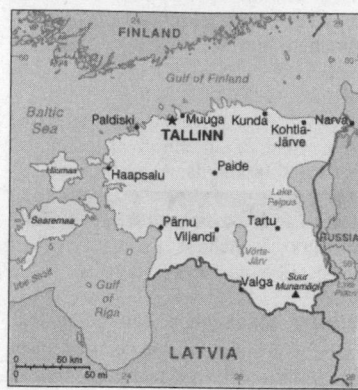

INTRODUCTION

Background: After centuries of Danish, Swedish, German, and Russian rule, Estonia attained independence in 1918. Forcibly incorporated into the USSR in 1940—an action never recognized by the US—it regained its freedom in 1991, with the collapse of the Soviet Union. Since the last Russian troops left in 1994, Estonia has been free to promote economic and political ties with Western Europe. It joined both NATO and the EU in the spring of 2004.

GEOGRAPHY

Location: Eastern Europe, bordering the Baltic Sea and Gulf of Finland, between Latvia and Russia
Geographic coordinates: 59 00 N, 26 00 E
Map references: Europe
Area:
total: 45,226 sq km
land: 43,211 sq km
water: 2,015 sq km
note: includes 1,520 islands in the Baltic Sea
Area—comparative: slightly smaller than New Hampshire and Vermont combined
Land boundaries:
total: 633 km
border countries: Latvia 339 km, Russia 294 km
Coastline: 3,794 km
Maritime claims:
territorial sea: 12 nm
exclusive economic zone: limits fixed in coordination with neighboring states
Climate: maritime, wet, moderate winters, cool summers
Terrain: marshy, lowlands; flat in the north, hilly in the south
Elevation extremes:
lowest point: Baltic Sea 0 m

highest point: Suur Munamagi 318 m
Natural resources: oil shale, peat, phosphorite, clay, limestone, sand, dolomite, arable land, sea mud
Land use:
arable land: 12.05%
permanent crops: 0.35%
other: 87.6% (2005)
Irrigated land: 40 sq km (2003)
Total renewable water resources: 21.1 cu km (2005)
Freshwater withdrawal (domestic/industrial/agricultural): *total:* 1.41 cu km/yr (56%/39%/5%)
per capita: 1,060 cu m/yr (2002)
Natural hazards: sometimes flooding occurs in the spring
Environment—current issues: air polluted with sulfur dioxide from oil-shale burning power plants in northeast; however, the amount of pollutants emitted to the air have fallen steadily, the emissions of 2000 were 80% less than in 1980; the amount of unpurified wastewater discharged to water bodies in 2000 was one-20th the level of 1980; in connection with the start-up of new water purification plants, the pollution load of wastewater decreased; Estonia has more than 1,400 natural and manmade lakes, the smaller of which in agricultural areas need to be monitored; coastal seawater is polluted in certain locations
Environment—international agreements: *party to:* Air Pollution, Air Pollution-Nitrogen Oxides, Air Pollution-Persistent Organic Pollutants, Air Pollution-Sulfur 85, Air Pollution-Volatile Organic Compounds, Antarctic Treaty, Biodiversity, Climate Change, Climate Change-Kyoto Protocol, Endangered Species, Hazardous Wastes, Law of the Sea, Ozone Layer Protection, Ship Pollution, Wetlands
signed, but not ratified: none of the selected agreements
Geography—note: the mainland terrain is flat, boggy, and partly wooded; offshore lie more than 1,500 islands

PEOPLE

Population: 1,307,605 (July 2008 est.)
Age structure:
0–14 years: 14.9% (male 100,143/female 94,450)
15–64 years: 67.5% (male 420,896/female 462,072)
65 years and over: 17.6% (male 76,171/female 153,873) (2008 est.)
Median age:
total: 39.6 years
male: 36.2 years

female: 43.2 years (2008 est.)
Population growth rate: -0.632% (2008 est.)
Birth rate: 10.28 births/1,000 population (2008 est.)
Death rate: 13.35 deaths/1,000 population (2008 est.)
Net migration rate: -3.24 migrant(s)/1,000 population (2008 est.)
Sex ratio:
at birth: 1.06 male(s)/female
under 15 years: 1.06 male(s)/female
15–64 years: 0.91 male(s)/female
65 years and over: 0.5 male(s)/female
total population: 0.84 male(s)/female (2008 est.)
Infant mortality rate:
total: 7.45 deaths/1,000 live births
male: 8.62 deaths/1,000 live births
female: 6.21 deaths/1,000 live births (2008 est.)
Life expectancy at birth:
total population: 72.56 years
male: 67.16 years
female: 78.3 years (2008 est.)
Total fertility rate: 1.42 children born/woman (2008 est.)
HIV/AIDS—adult prevalence rate: 1.1% (2001 est.)
HIV/AIDS—people living with HIV/AIDS: 7,800 (2003 est.)
HIV/AIDS—deaths: fewer than 200 (2003 est.)
Major infectious diseases:
degree of risk: intermediate
food or waterborne diseases: bacterial diarrhea and hepatitis A
vectorborne disease: tickborne encephalitis (2008)
Nationality:
noun: Estonian(s)
adjective: Estonian
Ethnic groups: Estonian 67.9%, Russian 25.6%, Ukrainian 2.1%, Belarusian 1.3%, Finn 0.9%, other 2.2% (2000 census)
Religions: Evangelical Lutheran 13.6%, Orthodox 12.8%, other Christian (including Methodist, Seventh-Day Adventist, Roman Catholic, Pentecostal) 1.4%, unaffiliated 34.1%, other and unspecified 32%, none 6.1% (2000 census)
Languages: Estonian (official) 67.3%, Russian 29.7%, other 2.3%, unknown 0.7% (2000 census)
Literacy:
definition: age 15 and over can read and write
total population: 99.8%
male: 99.8%
female: 99.8% (2000 census)

GOVERNMENT

Country name:
conventional long form: Republic of Estonia
conventional short form: Estonia
local long form: Eesti Vabariik
local short form: Eesti
former: Estonian Soviet Socialist Republic
Government type: parliamentary republic
Capital: *name:* Tallinn
geographic coordinates: 59 26 N, 24 43 E
time difference: UTC+2 (7 hours ahead of Washington, DC during Standard Time)
daylight saving time: +1hr, begins last Sunday in March; ends last Sunday in October
Administrative divisions: 15 counties (maakonnad, singular—maakond): Harjumaa (Tallinn), Hiiumaa (Kardla), Ida-Virumaa (Johvi), Jarvamaa (Paide), Jogevamaa (Jogeva), Laanemaa (Haapsalu), Laane-Virumaa (Rakvere), Parnumaa (Parnu), Polvamaa (Polva), Raplamaa (Rapla), Saaremaa (Kuressaare), Tartumaa (Tartu), Valgamaa (Valga), Viljandimaa (Viljandi), Vorumaa (Voru)
note: counties have the administrative center name following in parentheses
Independence: 20 August 1991 (from Soviet Union)
National holiday: Independence Day, 24 February (1918); note—24 February 1918 was the date Estonia declared its independence from Soviet Russia; 20 August 1991 was the date it declared its independence from the Soviet Union
Constitution: adopted 28 June 1992
Legal system: based on civil law system; accepts compulsory ICJ jurisdiction with reservations
Suffrage: 18 years of age; universal for all Estonian citizens
Executive branch:
chief of state: President Toomas Hendrik ILVES (since 9 October 2006)
head of government: Prime Minister Andrus ANSIP (since 12 April 2005)
cabinet: Council of Ministers appointed by the prime minister, approved by Parliament
elections: president elected by Parliament for a five-year term (eligible for a second term); if a candidate does not secure two-thirds of the votes after three rounds of balloting in the Parliament, then an electoral assembly (made up of Parliament plus members of local governments) elects the president, choosing between the two candidates with the largest percentage of votes; election last held 23 September 2006 (next to be held

in the fall of 2011); prime minister nominated by the president and approved by Parliament
election results: Toomas Hendrik ILVES elected president on 23 September 2006 by a 345-member electoral assembly; ILVES received 174 votes to incumbent Arnold RUUTEL's 162; remaining 9 ballots left blank or invalid
Legislative branch: unicameral Parliament or Riigikogu (101 seats; members are elected by popular vote to serve four-year terms)
elections: last held 4 March 2007 (next to be held in March 2011)
election results: percent of vote by party—Estonian Reform Party 27.8%, Center Party of Estonia 26.1%, Union of Pro Patria and Res Publica 17.9%, Social Democratic Party 10.6%, Estonian Greens 7.1%, Estonian People's Union 7.1%, other 5%; seats by party—Estonian Reform Party 31, Center Party 29, Union of Pro Patria and Res Publica 19, Social Democratic Party 10, Estonian Greens 6, Estonian People's Union 6
Judicial branch: National Court (chairman appointed by Parliament for life)
Political parties and leaders: Center Party of Estonia (Keskerakond) [Edgar SAVISAAR]; Estonian Greens (Rohelised) [Marek STRANDBERG]; Estonian People's Union (Rahvaliit) [Villu REILJAN]; Estonian Reform Party (Reformierakond) [Andrus ANSIP]; Estonian United Russian People's Party or EUVRP [Yevgeniy TOMBERG]; Social Democratic Party (formerly People's Party Moodukad or Moderates) [Ivari PADAR]; Union of Pro Patria and Res Publica (Isamaa je Res Publica Liit) [Mart LAAR]
Political pressure groups and leaders: Nochnoy Dozor/Night Watch anti-fascist movement (leader Alexander KOROBOV)
International organization participation: Australia Group, BA, BIS, CBSS, CE, EAPC, EBRD, EIB, EU, FAO, IAEA, IBRD, ICAO, ICCt, ICRM, IDA, IFC, IFRCS, IHO, ILO, IMF, IMO, Interpol, IOC, IOM, IPU, ISO (correspondent), ITU, ITUC, MIGA, NATO, NIB, NSG, OAS (observer), OPCW, OSCE, PCA, Schengen Convention, UN, UNCTAD, UNESCO, UNHCR, UNITAR, UNTSO, UPU, WCO, WEU (associate partner), WHO, WIPO, WMO, WTO
Diplomatic representation in the US:
chief of mission: Ambassador Vaino REINART
chancery: 2131 Massachusetts Avenue NW, Washington, DC 20008

telephone: [1] (202) 588-0101
FAX: [1] (202) 588-0108
consulate(s) general: New York
Diplomatic representation from the US:
chief of mission: Ambassador Stanley Davis PHILLIPS
embassy: Kentmanni 20, 15099 Tallinn
mailing address: use embassy street address
telephone: [372] 668-8100
FAX: [372] 668-8134
Flag description: pre-1940 flag restored by Supreme Soviet in May 1990—three equal horizontal bands of blue (top), black, and white

ECONOMY

Economy—overview: Estonia, a 2004 European Union entrant, has a modern market-based economy and one of the highest per capita income levels in Central Europe. The economy benefits from strong electronics and telecommunications sectors and strong trade ties with Finland, Sweden, and Germany. The current government has pursued relatively sound fiscal policies, resulting in balanced budgets and low public debt. In 2007, however, a large current account deficit and rising inflation put pressure on Estonia's currency, which is pegged to the euro, highlighting the need for growth in export-generating industries.
GDP (purchasing power parity): $29.35 billion (2007 est.)
GDP (official exchange rate): $21.28 billion (2007 est.)
GDP—real growth rate: 7.1% (2007 est.)
GDP—per capita (PPP): $21,100 (2007 est.)
GDP—composition by sector:
agriculture: 3%
industry: 28.5%
services: 68.5% (2007 est.)
Labor force: 687,000 (2007 est.)
Labor force—by occupation: *agriculture:* 11%
industry: 20%
services: 69% (1999 est.)
Unemployment rate: 4.7% (2007 est.)
Population below poverty line: 5% (2003)
Household income or consumption by percentage share:
lowest 10%: 2.5%
highest 10%: 27.6% (2003)
Distribution of family income—Gini index: 34 (2005)
Inflation rate (consumer prices): 6.6% (2007 est.)
Investment (gross fixed): 31.9% of GDP (2007 est.)
Budget:
revenues: $7.854 billion
expenditures: $7.171 billion (2007 est.)

213

Public debt: 3.4% of GDP (2007 est.)

Agriculture—products: potatoes, vegetables; livestock and dairy products; fish

Industries: engineering, electronics, wood and wood products, textiles; information technology, telecommunications

Industrial production growth rate: 7.7% (2007 est.)

Electricity—production: 9.599 billion kWh (2005)

Electricity—production by source:
fossil fuel: 99.8%
hydro: 0.1%
nuclear: 0%
other: 0.2% (2001)

Electricity—consumption: 6.888 billion kWh (2005)

Electricity—exports: 1.953 billion kWh (2005)

Electricity—imports: 345 million kWh (2005)

Oil—production: 6,930 bbl/day (2005 est.)

Oil—consumption: 29,000 bbl/day (2005 est.)

Oil—exports: 3,958 bbl/day (2004)

Oil—imports: 54,000 bbl/day (2004)

Oil—proved reserves: 0 bbl (1 January 2006 est.)

Natural gas—production: 0 cu m (2005 est.)

Natural gas—consumption: 1.458 billion cu m (2005 est.)

Natural gas—exports: 0 cu m (2005 est.)

Natural gas—imports: 1.458 billion cu m (2005)

Natural gas—proved reserves: 0 cu m (1 January 2006 est.)

Current account balance: -$3.402 billion (2007 est.)

Exports: $11.08 billion f.o.b. (2007 est.)

Exports—commodities: machinery and equipment 33%, wood and paper 15%, textiles 14%, food products 8%, furniture 7%, metals, chemical products (2001)

Exports—partners: Finland 18.2%, Sweden 12.2%, Latvia 9.1%, Russia 7.9%, US 6.6%, Germany 5%, Lithuania 4.8%, Gibraltar 4.5% (2006)

Imports: $14.69 billion f.o.b. (2007 est.)

Imports—commodities: machinery and equipment 33.5%, chemical products 11.6%, textiles 10.3%, foodstuffs 9.4%, transportation equipment 8.9% (2001)

Imports—partners: Finland 18.4%, Russia 12.9%, Germany 12.3%, Sweden 9.2%, Lithuania 6.4%, Latvia 5.8% (2006)

Economic aid—recipient: $135.5 million (2004)

Reserves of foreign exchange and gold: $3.27 billion (31 December 2007 est.)

Debt—external: $23.08 billion (30 June 2007)

Stock of direct foreign investment—at home: $16.59 billion (2007 est.)

Stock of direct foreign investment—abroad: $5.873 billion (2007 est.)

Market value of publicly traded shares: $5.963 billion (2006)

Currency (code): Estonian kroon (EEK)

Currency code: EEK

Exchange rates: krooni per US dollar—11.535 (2007), 12.473 (2006), 12.584 (2005), 12.596 (2004), 13.856 (2003)
note: the krooni is pegged to the euro

Fiscal year: calendar year

COMMUNICATIONS

Telephones—main lines in use: 541,900 (2006)

Telephones—mobile cellular: 1.659 million (2006)

Telephone system:
general assessment: foreign investment in the form of joint business ventures greatly improved telephone service; substantial fiber-optic cable systems carry telephone, TV, and radio traffic in the digital mode; Internet services are widely available; schools and libraries are connected to the Internet, a large percentage of the population files income-tax returns online, and online voting was used for the first time in the 2005 local elections
domestic: a wide range of high quality voice, data, and Internet services is available throughout the country
international: country code—372; fiber-optic cables to Finland, Sweden, Latvia, and Russia provide worldwide packet-switched service; 2 international switches are located in Tallinn (2001)

Radio broadcast stations: AM 0, FM 98, shortwave 0 (2001)

Radios: 1.01 million (1997)

Television broadcast stations: 3 (2001)

Televisions: 605,000 (1997)

Internet country code: .ee

Internet hosts: 387,336 (2007)

Internet Service Providers (ISPs): 38 (2001)

Internet users: 760,000 (2006)

TRANSPORTATION

Airports: 19 (2007)

Airports—with paved runways:
total: 12
over 3,047 m: 1
2,438 to 3,047 m: 7
1,524 to 2,437 m: 1
914 to 1,523 m: 3 (2007)

Airports—with unpaved runways:
total: 7
over 3,047 m: 1
1,524 to 2,437 m: 2
914 to 1,523 m: 1
under 914 m: 3 (2007)

Heliports: 1 (2007)

Pipelines: gas 859 km (2007)

Railways:
total: 968 km
broad gauge: 968 km 1.520 m/1.524-m gauge (2006)

Roadways: total: 57,016 km
paved: 12,926 km (includes 99 km of expressways)
unpaved: 44,090 km (2005)

Waterways: 320 km (2006)

Merchant marine:
total: 33 ships (1000 GRT or over) 393,655 GRT/93,245 DWT
by type: cargo 7, chemical tanker 1, passenger/cargo 23, petroleum tanker 2
foreign-owned: 4 (Denmark 2, Norway 2)
registered in other countries: 67 (Antigua and Barbuda 15, Belize 1, Cambodia 1, Cyprus 5, Dominica 8, Latvia 1, Liberia 1, Malta 7, Norway 1, Panama 3, Slovakia 2, St Kitts and Nevis 1, St Vincent and The Grenadines 20, Vanuatu 1) (2007)

Ports and terminals: Kuivastu, Kunda, Muuga, Tallinn, Virtsu

MILITARY

Military branches: Estonian Defense Forces: Land Force, Navy, Air Force (Eesti Ohuvagi), Volunteer Defense League (Kaitseliit, KL) (2008)

Military service age and obligation: compulsory military service for men between 19 and 28; conscription lasts 11 months for junior NCOs and reserve platoon leaders; reserve officers and designated specialists have a different conscript service obligation; Estonia has committed to retaining conscription for men up to 2010 and, unlike Latvia and Lithuania, has no plan to transition to a contract armed forces; 17 years of age for volunteers; reserve commitment up to the age of 60 (2006)

Manpower available for military service: males age 16–49: 306,273
females age 16–49: 317,852 (2008 est.)

Manpower fit for military service:
males age 16–49: 218,448 (in 2004, 51% of the young men called up for service were determined to be unfit; main obstacles to conscription were psychiatric and behavioral)
females age 16–49: 264,187 (2008 est.)

Manpower reaching militarily significant age annually:
males age 16–49: 8,322
females age 16–49: 7,846 (2008 est.)

Military expenditures—percent of GDP: 2% (2005 est.)

TRANSNATIONAL ISSUES

Disputes—international: Russia recalled its signature to the 1996 technical border agreement with Estonia in 2005, rather than concede to Estonia's appending prepared a unilateral declaration referencing Soviet occupation and territorial losses; Russia demands better accommodation of Russian-speaking population in Estonia; Estonian citizen groups continue to press for realignment of the boundary based on the 1920 Tartu Peace Treaty that would bring the now divided ethnic Setu people and parts of the Narva region within Estonia; as a member state that forms part of the EU's external border, Estonia must implement the strict Schengen border rules with Russia

Illicit drugs: growing producer of synthetic drugs; increasingly important transshipment zone for cannabis, cocaine, opiates, and synthetic drugs since joining the European Union and the Schengen Accord; potential money laundering related to organized crime and drug trafficking is a concern, as is possible use of the gambling sector to launder funds; major use of opiates and ecstasy

ETHIOPIA

INTRODUCTION

Background: Unique among African countries, the ancient Ethiopian monarchy maintained its freedom from colonial rule with the exception of the 1936–41 Italian occupation during World War II. In 1974, a military junta, the Derg, deposed Emperor Haile SELASSIE (who had ruled since 1930) and established a socialist state. Torn by bloody coups, uprisings, wide-scale drought, and massive refugee problems, the regime was finally toppled in 1991 by a coalition of rebel forces, the Ethiopian People's Revolutionary Democratic Front (EPRDF). A constitution was adopted in 1994, and Ethiopia's first multiparty elections were held in 1995. A border war with Eritrea late in the 1990s ended with a peace treaty in December 2000. The Eritrea-Ethiopia Border Commission in November 2007 remotely demarcated the border by geographical coordinates, but final demarcation of the boundary on the ground is currently on hold because of Ethiopian objections to an international commission's finding requiring it to surrender territory considered sensitive to Ethiopia.

GEOGRAPHY

Location: Eastern Africa, west of Somalia

Geographic coordinates: 8 00 N, 38 00 E

Map references: Africa

Area:
total: 1,127,127 sq km
land: 1,119,683 sq km
water: 7,444 sq km

Area—comparative: slightly less than twice the size of Texas

Land boundaries:
total: 5,328 km
border countries: Djibouti 349 km, Eritrea 912 km, Kenya 861 km, Somalia 1,600 km, Sudan 1,606 km

Coastline: 0 km (landlocked)

Maritime claims: none (landlocked)

Climate: tropical monsoon with wide topographic-induced variation

Terrain: high plateau with central mountain range divided by Great Rift Valley

Elevation extremes:
lowest point: Denakil Depression -125 m
highest point: Ras Dejen 4,533 m

Natural resources: small reserves of gold, platinum, copper, potash, natural gas, hydropower

Land use:
arable land: 10.01%
permanent crops: 0.65%
other: 89.34% (2005)

Irrigated land: 2,900 sq km (2003)

Total renewable water resources: 110 cu km (1987)

Freshwater withdrawal (domestic/industrial/agricultural): *total:* 5.56 cu km/yr (6%/0%/94%)
per capita: 72 cu m/yr (2002)

Natural hazards: geologically active Great Rift Valley susceptible to earthquakes, volcanic eruptions; frequent droughts

Environment—current issues: deforestation; overgrazing; soil erosion; desertification; water shortages in some areas from water-intensive farming and poor management

Environment—international agreements: *party to:* Biodiversity, Climate Change, Climate Change-Kyoto Protocol, Desertification, Endangered Species, Hazardous Wastes, Ozone Layer Protection

signed, but not ratified: Environmental Modification, Law of the Sea

Geography—note: landlocked—entire coastline along the Red Sea was lost with the de jure independence of Eritrea on 24 May 1993; the Blue Nile, the chief headstream of the Nile by water volume, rises in T'ana Hayk (Lake Tana) in northwest Ethiopia; three major crops are believed to have originated in Ethiopia: coffee, grain sorghum, and castor bean

PEOPLE

Population: 78,254,090
note: estimates for this country explicitly take into account the effects of excess mortality due to AIDS; this can result in lower life expectancy, higher infant mortality, higher death rates, lower population growth rates, and changes in the distribution of population by age and sex than would otherwise be expected (July 2008 est.)

Age structure:
0–14 years: 43.1% (male 16,932,540/female 16,818,931)
15–64 years: 54.1% (male 21,128,196/female 21,211,755)
65 years and over: 2.8% (male 979,166/female 1,183,502) (2008 est.)

Median age:
total: 18.1 years
male: 18 years
female: 18.2 years (2008 est.)

Population growth rate: 2.231% (2008 est.)

Birth rate: 36.8 births/1,000 population (2008 est.)

Death rate: 14.49 deaths/1,000 population (2008 est.)

Net migration rate: NA
note: repatriation of Ethiopian refugees residing in Sudan is expected to continue for several years; some Sudanese, Somali, and Eritrean refugees, who fled to Ethiopia from the fighting or famine in their own countries, continue to return to their homes (2008 est.)

Sex ratio:
at birth: 1.03 male(s)/female

under 15 years: 1.01 male(s)/female
15–64 years: 1 male(s)/female
65 years and over: 0.83 male(s)/female
total population: 1 male(s)/female (2008 est.)

Infant mortality rate:
total: 90.24 deaths/1,000 live births
male: 99.72 deaths/1,000 live births
female: 80.47 deaths/1,000 live births (2008 est.)

Life expectancy at birth:
total population: 49.43 years
male: 48.26 years
female: 50.64 years (2008 est.)

Total fertility rate: 4.99 children born/woman (2008 est.)

HIV/AIDS—adult prevalence rate: 4.4% (2003 est.)

HIV/AIDS—people living with HIV/AIDS: 1.5 million (2003 est.)

HIV/AIDS—deaths: 120,000 (2003 est.)

Major infectious diseases:
degree of risk: high
food or waterborne diseases: bacterial and protozoal diarrhea, hepatitis A and E, and typhoid fever
vectorborne diseases: malaria
respiratory disease: meningococcal meningitis
animal contact disease: rabies
water contact disease: schistosomiasis (2008)

Nationality: *noun:* Ethiopian(s)
adjective: Ethiopian

Ethnic groups: Oromo 32.1%, Amara 30.1%, Tigraway 6.2%, Somalie 5.9%, Guragie 4.3%, Sidama 3.5%, Welaita 2.4%, other 15.4% (1994 census)

Religions: Christian 60.8% (Orthodox 50.6%, Protestant 10.2%), Muslim 32.8%, traditional 4.6%, other 1.8% (1994 census)

Languages: Amarigna 32.7%, Oromigna 31.6%, Tigrigna 6.1%, Somaligna 6%, Guaragigna 3.5%, Sidamigna 3.5%, Hadiyigna 1.7%, other 14.8%, English (major foreign language taught in schools) (1994 census)

Literacy: *definition:* age 15 and over can read and write
total population: 42.7%
male: 50.3%
female: 35.1% (2003 est.)

GOVERNMENT

Country name:
conventional long form: Federal Democratic Republic of Ethiopia
conventional short form: Ethiopia
local long form: Ityop'iya Federalawi Demokrasiyawi Ripeblik
local short form: Ityop'iya
former: Abyssinia, Italian East Africa
abbreviation: FDRE

Government type: federal republic

Capital: *name:* Addis Ababa
geographic coordinates: 9 02 N, 38 42 E
time difference: UTC+3 (8 hours ahead of Washington, DC during Standard Time)

Administrative divisions: 9 ethnically based states (kililoch, singular—kilil) and 2 self-governing administrations* (astedaderoch, singular—astedader); Adis Abeba* (Addis Ababa), Afar, Amara (Amhara), Binshangul Gumuz, Dire Dawa*, Gambela Hizboch (Gambela Peoples), Hareri Hizb (Harari People), Oromiya (Oromia), Sumale (Somali), Tigray, Ye Debub Biheroch Bihereseboch na Hizboch (Southern Nations, Nationalities and Peoples)

Independence: oldest independent country in Africa and one of the oldest in the world—at least 2,000 years

National holiday: National Day (defeat of MENGISTU regime), 28 May (1991)

Constitution: ratified 8 December 1994, effective 22 August 1995

Legal system: based on civil law; currently transitional mix of national and regional courts; has not accepted compulsory ICJ jurisdiction

Suffrage: 18 years of age; universal

Executive branch:
chief of state: President GIRMA Woldegiorgis (since 8 October 2001)
head of government: Prime Minister MELES Zenawi (since August 1995)
cabinet: Council of Ministers as provided for in the December 1994 constitution; ministers are selected by the prime minister and approved by the House of People's Representatives
elections: president elected by the House of People's Representatives for a six-year term (eligible for a second term); election last held 9 October 2007 (next to be held in October 2013); prime minister designated by the party in power following legislative elections
election results: GIRMA Woldegiorgis elected president; percent of vote by the House of People's Representatives—79%

Legislative branch: bicameral Parliament consists of the House of Federation (or upper chamber responsible for interpreting the constitution and federal-regional issues) (108 seats; members are chosen by state assemblies to serve five-year terms) and the House of People's Representatives (or lower chamber responsible for passing legislation) (547 seats; members are directly elected by popular vote from single-member districts to serve five-year terms)
elections: last held 15 May 2005 (next to be held in 2010)
election results: percent of vote—NA; seats by party—EPRDF 327, CUD 109, UEDF 52, SPDP 23, OFDM 11,

BGPDUF 8, ANDP 8, independent 1, others 6, undeclared 2
note: some seats still remain vacant as detained opposition MPs did not take their seats

Judicial branch: Federal Supreme Court (the president and vice president of the Federal Supreme Court are recommended by the prime minister and appointed by the House of People's Representatives; for other federal judges, the prime minister submits to the House of People's Representatives for appointment candidates selected by the Federal Judicial Administrative Council)

Political parties and leaders: Afar National Democratic Party or ANDP; Benishangul Gumuz People's Democratic Unity Front or BGPDUF [Mulualem BESSE]; Coalition for Unity and Democratic Party or CUDP [AYELE Chamisso] (awarded to AYELE by the National Electoral Board on 11 January 2008, but AYELE has virtually no support among former CUD MPs, other CUD MPs must now be affiliated with their original CUD-precursor parties); Ethiopian People's Revolutionary Democratic Front or EPRDF [MELES Zenawi] (an alliance of Amhara National Democratic Movement or ANDM, Oromo People's Democratic Organization or OPDO, the South Ethiopian People's Democratic Front or SEPDF, and Tigrayan Peoples' Liberation Front or TPLF); Gurage Nationalities' Democratic Movement or GNDM; Oromo Federalist Democratic Movement or OFDM [BULCHA Demeksa]; Omoro People's Congress or OPC [IMERERA Gudina]; Somali People's Democratic Party or SPDP; United Ethiopian Democratic Forces or UEDF [BEYENE Petros]

Political pressure groups and leaders: Ethiopian People's Patriotic Front or EPPF; Ogaden National Liberation Front or ONLF; Oromo Liberation Front or OLF [DAOUD Ibsa]

International organization participation: ACP, AfDB, AU, COMESA, FAO, G-24, G-77, IAEA, IBRD, ICAO, ICRM, IDA, IFAD, IFC, IFRCS, IGAD, ILO, IMF, IMO, Interpol, IOC, IOM (observer), IPU, ISO, ITSO, ITU, ITUC, MIGA, NAM, OPCW, PCA, UN, UNAMID, UNCTAD, UNESCO, UNHCR, UNIDO, UNMIL, UNOCI, UNWTO, UPU, WCO, WFTU, WHO, WIPO, WMO, WTO (observer)

Diplomatic representation in the US:
chief of mission: Ambassador Samuel ASSEFA
chancery: 3506 International Drive NW, Washington, DC 20008

telephone: [1] (202) 364-1200
FAX: [1] (202) 587-0195
consulate(s) general: Los Angeles
consulate(s): New York
Diplomatic representation from the US:
chief of mission: Ambassador Donald Y. YAMAMOTO
embassy: Entoto Street, Addis Ababa
mailing address: P. O. Box 1014, Addis Ababa
telephone: [251] 11-517-40-00
FAX: [251] 11-517-40-01
Flag description: three equal horizontal bands of green (top), yellow, and red with a yellow pentagram and single yellow rays emanating from the angles between the points on a light blue disk centered on the three bands; Ethiopia is the oldest independent country in Africa, and the three main colors of her flag were so often adopted by other African countries upon independence that they became known as the pan-African colors

ECONOMY

Economy—overview: Ethiopia's poverty-stricken economy is based on agriculture, accounting for almost half of GDP, 60% of exports, and 80% of total employment. The agricultural sector suffers from frequent drought and poor cultivation practices. Coffee is critical to the Ethiopian economy with exports of some $350 million in 2006, but historically low prices have seen many farmers switching to qat to supplement income. The war with Eritrea in 1998–2000 and recurrent drought have buffeted the economy, in particular coffee production. In November 2001, Ethiopia qualified for debt relief from the Highly Indebted Poor Countries (HIPC) initiative, and in December 2005 the IMF voted to forgive Ethiopia's debt to the body. Under Ethiopia's constitution, the state owns all land and provides long-term leases to the tenants; the system continues to hamper growth in the industrial sector as entrepreneurs are unable to use land as collateral for loans. Drought struck again late in 2002, leading to a 3.3% decline in GDP in 2003. Normal weather patterns helped agricultural and GDP growth recover during 2004–07.
GDP (purchasing power parity): $62.19 billion (2007 est.)
GDP (official exchange rate): $19.43 billion (2007 est.)
GDP—real growth rate: 11.4% (2007 est.)
GDP—per capita (PPP): $800 (2007 est.)
GDP—composition by sector:

agriculture: 47%
industry: 13.2%
services: 39.8% (2007 est.)
Labor force: 27.27 million (1999)
Labor force—by occupation: *agriculture:* 80%
industry: 8%
services: 12% (1985)
Unemployment rate: NA%
Population below poverty line: 38.7% (FY05/06 est.)
Household income or consumption by percentage share:
lowest 10%: 3.9%
highest 10%: 25.5% (2000)
Distribution of family income—Gini index: 30 (2000)
Inflation rate (consumer prices): 17% (2007 est.)
Investment (gross fixed): 26% of GDP (2007 est.)
Budget:
revenues: $2.947 billion
expenditures: $3.687 billion (2007 est.)
Public debt: 45.6% of GDP (2007 est.)
Agriculture—products: cereals, pulses, coffee, oilseed, cotton, sugarcane, potatoes, qat, cut flowers; hides, cattle, sheep, goats; fish
Industries: food processing, beverages, textiles, leather, chemicals, metals processing, cement
Industrial production growth rate: 11% (2007 est.)
Electricity—production: 2.864 billion kWh (2005)
Electricity—production by source:
fossil fuel: 1.3%
hydro: 97.6%
nuclear: 0%
other: 1.2% (2001)
Electricity—consumption: 2.577 billion kWh (2005)
Electricity—exports: 0 kWh (2005)
Electricity—imports: 0 kWh (2005)
Oil—production: 7.334 bbl/day (2005 est.)
Oil—consumption: 29,000 bbl/day (2005 est.)
Oil—exports: 0 bbl/day (2004)
Oil—imports: 28,460 bbl/day (2004)
Oil—proved reserves: 428,000 bbl (1 January 2006 est.)
Natural gas—production: 0 cu m (2005 est.)
Natural gas—consumption: 0 cu m (2005 est.)
Natural gas—exports: 0 cu m (2005 est.)
Natural gas—imports: 0 cu m (2005)
Natural gas—proved reserves: 23.9 billion cu m (1 January 2006 est.)
Current account balance: -$881 million (2007 est.)
Exports: $1.216 billion f.o.b. (2007 est.)

Exports—commodities: coffee, qat, gold, leather products, live animals, oilseeds
Exports—partners: China 11%, Germany 9.1%, Japan 7.8%, US 7.2%, Saudi Arabia 6.1%, Djibouti 6%, Italy 5.2% (2006)
Imports: $4.783 billion f.o.b. (2007 est.)
Imports—commodities: food and live animals, petroleum and petroleum products, chemicals, machinery, motor vehicles, cereals, textiles
Imports—partners: Saudi Arabia 18.1%, China 11.4%, India 8.1%, Italy 5.1% (2006)
Economic aid—recipient: $1.6 billion (FY05/06)
Reserves of foreign exchange and gold: $1.261 billion (31 December 2007 est.)
Debt—external: $2.622 billion (31 December 2007 est.)
Market value of publicly traded shares: $NA
Currency (code): birr (ETB)
Currency code: ETB
Exchange rates: birr per US dollar— 8.96 (2007), 8.69 (2006), 8.68 (2005), 8.6356 (2004), 8.5997 (2003)
note: since 24 October 2001, exchange rates are determined on a daily basis via interbank transactions regulated by the Central Bank
Fiscal year: 8 July—7 July

COMMUNICATIONS

Telephones—main lines in use: 725,000 (2006)
Telephones—mobile cellular: 866,700 (2006)
Telephone system:
general assessment: inadequate telephone system; the number of fixed lines and mobile telephones is increasing from a very small base; combined fixed and mobile-cellular teledensity is only about 2 per 100 persons
domestic: open-wire; microwave radio relay; radio communication in the HF, VHF, and UHF frequencies; 2 domestic satellites provide the national trunk service
international: country code—251; open-wire to Sudan and Djibouti; microwave radio relay to Kenya and Djibouti; satellite earth stations—3 Intelsat (1 Atlantic Ocean and 2 Pacific Ocean)
Radio broadcast stations: AM 8, FM 0, shortwave 1 (2001)
Radios: 15.2 million (2002)
Television broadcast stations: 1 (plus 24 repeaters) (2001)
Televisions: 682,000 (2002)
Internet country code: .et
Internet hosts: 89 (2007)

217

Internet Service Providers (ISPs): 1 (2002)
Internet users: 164,000 (2005)

TRANSPORTATION

Airports: 84 (2007)
Airports—with paved runways:
total: 15
over 3,047 m: 3
2,438 to 3,047 m: 5
1,524 to 2,437 m: 5
914 to 1,523 m: 1
under 914 m: 1 (2007)
Airports—with unpaved runways:
total: 69
over 3,047 m: 3
2,438 to 3,047 m: 5
1,524 to 2,437 m: 11
914 to 1,523 m: 29
under 914 m: 21 (2007)
Railways:
total: 699 km (Ethiopian segment of the Addis Ababa-Djibouti railroad)
narrow gauge: 699 km 1.000-m gauge
note: railway under joint control of Djibouti and Ethiopia but remains largely inoperable (2006)
Roadways:
total: 36,469 km
paved: 6,980 km
unpaved: 29,489 km (2004)
Merchant marine:
total: 10 ships (1000 GRT or over) 120,383 GRT/152,418 DWT
by type: cargo 8, roll on/roll off 2 (2007)
Ports and terminals: Ethiopia is land-locked and uses ports of Djibouti in Djibouti and Berbera in Somalia

MILITARY

Military branches: Ethiopian National Defense Force (ENDF): Ground Forces, Ethiopian Air Force (ETAF) (2008)
note: Ethiopia is landlocked and has no navy; following the secession of Eritrea, Ethiopian naval facilities remained in Eritrean possession
Military service age and obligation: 18 years of age for compulsory and voluntary military service; theoretically, no compulsory military service, but the military can conduct call-ups when necessary and compliance is compulsory (2008)
Manpower available for military service: *males age 16–49:* 17,666,967 *females age 16–49:* 17,530,211 (2008 est.)
Manpower fit for military service:
males age 16–49: 10,060,775
females age 16–49: 9,854,710 (2008 est.)
Manpower reaching militarily significant age annually:
males age 16–49: 910,602
females age 16–49: 911,081 (2008 est.)
Military expenditures—percent of GDP: 3% (2006)

TRANSNATIONAL ISSUES

Disputes—international: Eritrea and Ethiopia agreed to abide by the 2002 Eritrea-Ethiopia Boundary Commission's (EEBC) delimitation decision, but neither party responded to the revised line detailed in the November 2006 EEBC Demarcation Statement; UN Peacekeeping Mission to Ethiopia and Eritrea (UNMEE), which has monitored the 25-km-wide Temporary Security Zone in Eritrea since 2000, is extended for six months in 2007 despite Eritrean restrictions on its operations and reduced force of 17,000; the undemarcated former British administrative line has little meaning as a political separation to rival clans within Ethiopia's Ogaden and southern Somalia's Oromo region; Ethiopian forces invaded southern Somalia and routed Islamist Courts from Mogadishu in January 2007; "Somaliland" secessionists provide port facilities in Berbera and trade ties to land-locked Ethiopia; civil unrest in eastern Sudan has hampered efforts to demarcate the porous boundary with Ethiopia
Refugees and internally displaced persons: *refugees (country of origin):* 66,980 (Sudan); 16,576 (Somalia); 13,078 (Eritrea)
IDPs: 200,000 (border war with Eritrea from 1998–2000, ethnic clashes in Gambela, and ongoing Ethiopian military counterinsurgency in Somali region; most IDPs are in Tigray and Gambela Provinces) (2007)
Illicit drugs: transit hub for heroin originating in Southwest and Southeast Asia and destined for Europe, as well as cocaine destined for markets in southern Africa; cultivates qat (khat) for local use and regional export, principally to Djibouti and Somalia (legal in all three countries); the lack of a well-developed financial system limits the country's utility as a money laundering center

FALKLAND ISLANDS (ISLAS MALVINAS)

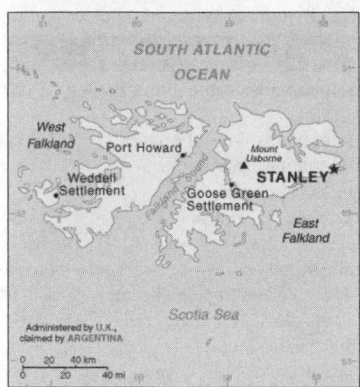

INTRODUCTION

Background: Although first sighted by an English navigator in 1592, the first landing (English) did not occur until almost a century later in 1690, and the first settlement (French) was not established until 1764. The colony was turned over to Spain two years later and the islands have since been the subject of a territorial dispute, first between Britain and Spain, then between Britain and Argentina. The UK asserted its claim to the islands by establishing a naval garrison there in 1833. Argentina invaded the islands on 2 April 1982. The British responded with an expeditionary force that landed seven weeks later and after fierce fighting forced an Argentine surrender on 14 June 1982.

GEOGRAPHY

Location: Southern South America, islands in the South Atlantic Ocean, east of southern Argentina
Geographic coordinates: 51 45 S, 59 00 W
Map references: South America
Area:
total: 12,173 sq km
land: 12,173 sq km
water: 0 sq km
note: includes the two main islands of East and West Falkland and about 200 small islands
Area—comparative: slightly smaller than Connecticut
Land boundaries: 0 km
Coastline: 1,288 km
Maritime claims:
territorial sea: 12 nm
continental shelf: 200 nm
exclusive fishing zone: 200 nm
Climate: cold marine; strong westerly winds, cloudy, humid; rain occurs on more than half of days in year; average annual rainfall is 24 inches in Stanley;

occasional snow all year, except in January and February, but does not accumulate
Terrain: rocky, hilly, mountainous with some boggy, undulating plains
Elevation extremes:
lowest point: Atlantic Ocean 0 m
highest point: Mount Usborne 705 m
Natural resources: fish, squid, wildlife, calcified seaweed, sphagnum moss
Land use:
arable land: 0%
permanent crops: 0%
other: 100% (99% permanent pastures, 1% other) (2005)
Irrigated land: NA
Natural hazards: strong winds persist throughout the year
Environment—current issues: overfishing by unlicensed vessels is a problem; reindeer were introduced to the islands in 2001 for commercial reasons; this is the only commercial reindeer herd in the world unaffected by the 1986 Chornobyl disaster
Geography—note: deeply indented coast provides good natural harbors; short growing season

PEOPLE

Population: 3,140 (July 2008 est.)
Age structure:
0–14 years: NA
15–64 years: NA
65 years and over: NA
Population growth rate: 0.011% (2008 est.)
Birth rate: NA
Death rate: NA
Net migration rate: NA
Infant mortality rate:
total: NA
male: NA
female: NA
Life expectancy at birth:
total population: NA
male: NA
female: NA
Total fertility rate: NA
HIV/AIDS—adult prevalence rate: NA
HIV/AIDS—people living with HIV/AIDS: NA
HIV/AIDS—deaths: NA
Nationality:
noun: Falkland Islander(s)
adjective: Falkland Island
Ethnic groups: British
Religions: primarily Anglican, Roman Catholic, United Free Church, Evangelist Church, Jehovah's Witnesses, Lutheran, Seventh-Day Adventist
Languages: English
Literacy: NA

GOVERNMENT

Country name:
conventional long form: none
conventional short form: Falkland Islands (Islas Malvinas)
Dependency status: overseas territory of the UK; also claimed by Argentina
Government type: NA
Capital: *name:* Stanley
geographic coordinates: 51 42 S, 57 51 W
time difference: UTC-4 (1 hour ahead of Washington, DC during Standard Time)
daylight saving time: +1hr, begins first Sunday in September; ends third Sunday in April
Administrative divisions: none (overseas territory of the UK; also claimed by Argentina)
Independence: none (overseas territory of the UK; also claimed by Argentina)
National holiday: Liberation Day, 14 June (1982)
Constitution: 3 October 1985; amended 1997 and 1998
Legal system: English common law
Suffrage: 18 years of age; universal
Executive branch:
chief of state: Queen ELIZABETH II (since 6 February 1952)
head of government: Governor Alan HUCKLE (since 25 August 2006); Chief Executive Dr. Tim THOROGOOD (since 3 January 2008)
cabinet: Executive Council; three members elected by the Legislative Council, two ex officio members (chief executive and the financial secretary), and the governor
elections: the monarchy is hereditary; governor appointed by the monarch
Legislative branch: unicameral Legislative Council (10 seats; 2 members are ex officio and 8 are elected by popular vote; to serve four-year terms); presided over by the governor
elections: last held 17 November 2005 (next to be held in November 2009)
election results: percent of vote—NA; seats—independents 8
Judicial branch: Supreme Court (chief justice is a nonresident); Magistrates Court (senior magistrate presides over civil and criminal divisions); Court of Summary Jurisdiction
Political parties and leaders: none; all independents
Political pressure groups and leaders: none
International organization participation: ICFTU, UPU
Diplomatic representation in the US: none (overseas territory of the UK; also claimed by Argentina)

Diplomatic representation from the US: none (overseas territory of the UK; also claimed by Argentina)

Flag description: blue with the flag of the UK in the upper hoist-side quadrant and the Falkland Island coat of arms centered on the outer half of the flag; the coat of arms contains a white ram (sheep raising was once the major economic activity) above the sailing ship Desire (whose crew discovered the islands) with a scroll at the bottom bearing the motto DESIRE THE RIGHT

ECONOMY

Economy—overview: The economy was formerly based on agriculture, mainly sheep farming, but today fishing contributes the bulk of economic activity. In 1987, the government began selling fishing licenses to foreign trawlers operating within the Falkland Islands' exclusive fishing zone. These license fees total more than $40 million per year, which help support the island's health, education, and welfare system. Squid accounts for 75% of the fish taken. Dairy farming supports domestic consumption; crops furnish winter fodder. Exports feature shipments of high-grade wool to the UK and the sale of postage stamps and coins. The islands are now self-financing except for defense. The British Geological Survey announced a 200-mile oil exploration zone around the islands in 1993, and early seismic surveys suggest substantial reserves capable of producing 500,000 barrels per day; to date, no exploitable site has been identified. An agreement between Argentina and the UK in 1995 seeks to defuse licensing and sovereignty conflicts that would dampen foreign interest in exploiting potential oil reserves. Tourism, especially eco-tourism, is increasing rapidly, with about 30,000 visitors in 2001. Another large source of income is interest paid on money the government has in the bank. The British military presence also provides a sizeable economic boost.

GDP (purchasing power parity): $75 million (2002 est.)
GDP (official exchange rate): $NA
GDP—real growth rate: NA%
GDP—per capita (PPP): $25,000 (2002 est.)
GDP—composition by sector:
agriculture: 95%
industry: NA%
services: NA%
Labor force: 1,724 (est.) (1996)
Labor force—by occupation: *agriculture:* 95% (mostly sheepherding and fishing) *industry and services:* 5% (1996)

Unemployment rate: full employment; labor shortage (2001)
Population below poverty line: NA%
Household income or consumption by percentage share:
lowest 10%: NA%
highest 10%: NA%
Inflation rate (consumer prices): 3.6% (1998)
Budget:
revenues: $66.2 million
expenditures: $67.9 million (FY98/99 est.)
Agriculture—products: fodder and vegetable crops; sheep, dairy products; fish, squid
Industries: fish and wool processing; tourism
Industrial production growth rate: NA%
Electricity—production: 16 million kWh (2005)
Electricity—production by source:
fossil fuel: 100%
hydro: 0%
nuclear: 0%
other: 0% (2001)
Electricity—consumption: 14.88 million kWh (2005)
Electricity—exports: 0 kWh (2005)
Electricity—imports: 0 kWh (2005)
Oil—production: 0 bbl/day (2005 est.)
Oil—consumption: 240 bbl/day (2005 est.)
Oil—exports: 0 bbl/day (2004)
Oil—imports: 227.9 bbl/day (2004)
Oil—proved reserves: 0 bbl (1 January 2006 est.)
Natural gas—production: 0 cu m (2005 est.)
Natural gas—consumption: 0 cu m (2005 est.)
Natural gas—exports: 0 cu m (2005 est.)
Natural gas—imports: 0 cu m (2005)
Natural gas—proved reserves: 0 cu m (1 January 2006 est.)
Exports: $125 million (2004 est.)
Exports—commodities: wool, hides, meat, fish, squid
Exports—partners: Spain 81.9%, US 6%, UK 4.5% (2006)
Imports: $90 million (2004 est.)
Imports—commodities: fuel, food and drink, building materials, clothing
Imports—partners: UK 72.5%, US 15.1%, Netherlands 8.5% (2006)
Economic aid—recipient: $0 (1997 est.)
Debt—external: $NA
Currency (code): Falkland pound (FKP)
Currency code: FKP
Exchange rates: Falkland pounds per US dollar—0.4993 (2007), 0.5434 (2006), 0.5504 (2005), 0.5462 (2004), 0.6125 (2003)

note: the Falkland pound is at par with the British pound
Fiscal year: 1 April—31 March

COMMUNICATIONS

Telephones—main lines in use: 2,400 (2002)
Telephones—mobile cellular: 0 (2001)
Telephone system:
general assessment: NA
domestic: government-operated radio-telephone and private VHF/CB radio-telephone networks provide effective service to almost all points on both islands
international: country code—500; satellite earth station—1 Intelsat (Atlantic Ocean) with links through London to other countries
Radio broadcast stations: AM 1, FM 7, shortwave 0 (British Forces Broadcasting Service (BFBS) provides Radio 1 and Radio 2 service) (2006)
Radios: 1,000 (1997)
Television broadcast stations: 2 (British Forces Broadcasting Service (BFBS) provides multi-channel satellite service to members of UK Forces as well as islanders); cable television is available in Stanley (2006)
Televisions: 1,000 (1997)
Internet country code: .fk
Internet hosts: 104 (2007)
Internet Service Providers (ISPs): 2 (2000)
Internet users: 1,900 (2002)

TRANSPORTATION

Airports: 6 (2007)
Airports—with paved runways: *total:* 2 *2,438 to 3,047 m:* 1 *914 to 1,523 m:* 1 (2007)
Airports—with unpaved runways: *total:* 4 *under 914 m:* 4 (2007)
Roadways:
total: 440 km
paved: 50 km
unpaved: 390 km (2003)
Ports and terminals: Stanley

MILITARY

Military branches: no regular military forces
Military expenditures—percent of GDP: NA
Military—note: defense is the responsibility of the UK

TRANSNATIONAL ISSUES

Disputes—international: Argentina, which claims the islands in its constitution and briefly occupied them by force in 1982, agreed in 1995 to no longer seek settlement by force; UK continues to reject Argentine requests for sovereignty talks

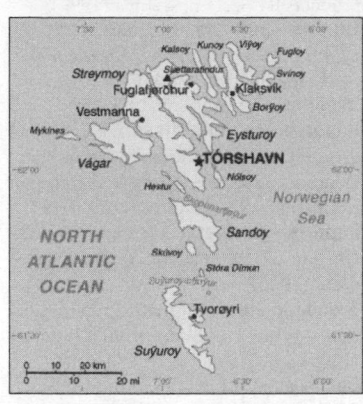

TÓRSHAVN

INTRODUCTION

Background: The population of the Faroe Islands is largely descended from Viking settlers who arrived in the 9th century. The islands have been connected politically to Denmark since the 14th century. A high degree of self government was attained in 1948.

GEOGRAPHY

Location: Northern Europe, island group between the Norwegian Sea and the North Atlantic Ocean, about half way between Iceland and Norway
Geographic coordinates: 62 00 N, 7 00 W
Map references: Europe
Area:
total: 1,399 sq km
land: 1,399 sq km
water: 0 sq km (some lakes and streams)
Area—comparative: eight times the size of Washington, DC
Land boundaries: 0 km
Coastline: 1,117 km
Maritime claims:
territorial sea: 3 nm
continental shelf: 200 nm or agreed boundaries or median line
exclusive fishing zone: 200 nm or agreed boundaries or median line
Climate: mild winters, cool summers; usually overcast; foggy, windy
Terrain: rugged, rocky, some low peaks; cliffs along most of coast
Elevation extremes:
lowest point: Atlantic Ocean 0 m
highest point: Slaettaratindur 882 m
Natural resources: fish, whales, hydropower, possible oil and gas
Land use:
arable land: 2.14%
permanent crops: 0%
other: 97.86% (2005)
Irrigated land: 0 sq km

Natural hazards: NA
Environment—current issues: NA
Environment—international agreements: *party to:* Marine Dumping -associate member to the London Convention and Ship Pollution
Geography—note: archipelago of 17 inhabited islands and one uninhabited island, and a few uninhabited islets; strategically located along important sea lanes in northeastern Atlantic; precipitous terrain limits habitation to small coastal lowlands

PEOPLE

Population: 48,668 (July 2008 est.)
Age structure:
0–14 years: 21.9% (male 5,489/female 5,166)
15–64 years: 64% (male 16,650/female 14,482)
65 years and over: 14.1% (male 3,233/female 3,648) (2008 est.)
Median age:
total: 36.7 years
male: 36 years
female: 37.5 years (2008 est.)
Population growth rate: 0.376% (2008 est.)
Birth rate: 13.25 births/1,000 population (2008 est.)
Death rate: 8.67 deaths/1,000 population (2008 est.)
Net migration rate: -0.82 migrant(s)/1,000 population (2008 est.)
Sex ratio:
at birth: 1.07 male(s)/female
under 15 years: 1.06 male(s)/female
15–64 years: 1.15 male(s)/female
65 years and over: 0.89 male(s)/female
total population: 1.09 male(s)/female (2008 est.)
Infant mortality rate:
total: 6.46 deaths/1,000 live births
male: 6.69 deaths/1,000 live births
female: 6.2 deaths/1,000 live births (2008 est.)
Life expectancy at birth:
total population: 79.29 years
male: 76.86 years
female: 81.89 years (2008 est.)
Total fertility rate: 2.45 children born/woman (2008 est.)
HIV/AIDS—adult prevalence rate: NA
HIV/AIDS—people living with HIV/AIDS: NA
HIV/AIDS—deaths: NA
Nationality:
noun: Faroese (singular and plural)
adjective: Faroese
Ethnic groups: Scandinavian
Religions: Evangelical Lutheran

Languages: Faroese (derived from Old Norse), Danish
Literacy:
NA; note—probably 100%, the same as Denmark proper

GOVERNMENT

Country name:
conventional long form: none
conventional short form: Faroe Islands
local long form: none
local short form: Foroyar
Dependency status: part of the Kingdom of Denmark; self-governing overseas administrative division of Denmark since 1948
Government type: NA
Capital: *name:* Torshavn
geographic coordinates: 62 01 N, 6 46 W
time difference: UTC 0 (5 hours ahead of Washington, DC during Standard Time)
daylight saving time: +1hr, begins last Sunday in March; ends last Sunday in October
Administrative divisions: none (part of the Kingdom of Denmark; self-governing overseas administrative division of Denmark; there are no first-order administrative divisions as defined by the US Government, but there are 34 municipalities
Independence: none (part of the Kingdom of Denmark; self-governing overseas administrative division of Denmark)
National holiday: Olaifest (Olavasoka), 29 July
Constitution: 5 June 1953 (Danish constitution)
Legal system: the laws of Denmark, where applicable, apply
Suffrage: 18 years of age; universal
Executive branch:
chief of state: Queen MARGRETHE II of Denmark (since 14 January 1972), represented by High Commissioner Birgit KLEIS, chief administrative officer (since 1 November 2001)
head of government: Prime Minister Joannes EIDESGAARD (since 3 February 2004)
cabinet: Landsstyri appointed by the prime minister
elections: the monarch is hereditary; high commissioner appointed by the monarch; following legislative elections, the leader of the majority party or the leader of the majority coalition is usually elected prime minister by the Faroese Parliament; election last held 20 January 2004 (next to be held no later than January 2008)

election results: Joannes EIDESGAARD elected prime minister; percent of parliamentary vote—NA

Legislative branch: unicameral Faroese Parliament or Logting (33 seats; members are elected by popular vote on a proportional basis from the seven constituencies to serve four-year terms)
elections: last held 19 January 2008 (next to be held no later than January 2012)
election results: percent of vote by party—Union Party 21%, Social Democratic Party 19.4%, Republican Party 23.3%, People's Party 20.1%, Center Party 8.4%, Self-Government Party 7.2%, other 0.6%; seats by party—Republican Party 8, Union Party 7, Social Democratic Party 6, People's Party 7, Center Party 3, Independence Party 2
note: election of two seats to the Danish Parliament was last held on 13 November 2007 (next to be held no later than November 2011); results—percent of vote by party—NA; seats by party—Republican Party 1, Union Party 1

Judicial branch: none

Political parties and leaders: Center Party [Jenis A. RANA]; Independence Party [Kari P. HOJGAARD]; People's Party [Jorgen NICLASEN]; Republican Party [Hogni HOYDAL]; Social Democratic Party [Joannes EIDESGAARD]; Union Party [Kaj Leo JOHANNESEN]

Political pressure groups and leaders: NA

International organization participation: Arctic Council, IMO (associate), NC, NIB, UPU

Diplomatic representation in the US: none (self-governing overseas administrative division of Denmark)

Diplomatic representation from the US: none (self-governing overseas administrative division of Denmark)

Flag description: white with a red cross outlined in blue extending to the edges of the flag; the vertical part of the cross is shifted toward the hoist side in the style of the Dannebrog (Danish flag)

ECONOMY

Economy—overview: The Faroese economy is dependent on fishing, which makes the economy vulnerable to price swings. Since 2003 the Faroese economy has picked up as a result of higher prices for fish and for housing. Unemployment is minimal and government finances are relatively sound. Oil finds close to the Islands give hope for economically recoverable deposits, which could eventually lay the basis for a more diversified economy and lessen dependence on Danish economic assistance. Aided by a

substantial annual subsidy (about 15% of GDP) from Denmark, the Faroese have a standard of living not far below the Danes and other Scandinavians.

GDP (purchasing power parity): $1 billion (2001 est.)

GDP (official exchange rate): $1.7 billion (2005 est.)

GDP—real growth rate: 2.4% (2005 est.)

GDP—per capita (PPP): $31,000 (2001 est.)

GDP—composition by sector:
agriculture: 27%
industry: 11%
services: 62% (1999)

Labor force: 24,250 (October 2000)

Labor force—by occupation: *agriculture:* 33%
industry: 33%
services: 34% (October 2000)

Unemployment rate: 2.1% (2006)

Population below poverty line: NA%

Household income or consumption by percentage share:
lowest 10%: NA%
highest 10%: NA%

Inflation rate (consumer prices): 1.8% (2005)

Budget:
revenues: $588 million
expenditures: $623 million (2005)

Agriculture—products: milk, potatoes, vegetables; sheep; salmon, other fish

Industries: fishing, fish processing, small ship repair and refurbishment, handicrafts

Industrial production growth rate: 8% (1999 est.)

Electricity—production: 290 million kWh (2005)

Electricity—production by source:
fossil fuel: 62.4%
hydro: 37.6%
nuclear: 0%
other: 0% (2001)

Electricity—consumption: 269.7 million kWh (2005)

Electricity—exports: 0 kWh (2005)

Electricity—imports: 0 kWh (2005)

Oil—production: 0 bbl/day (2005 est.)

Oil—consumption: 4,600 bbl/day (2005 est.)

Oil—exports: 0 bbl/day (2004)

Oil—imports: 4,580 bbl/day (2004)

Oil—proved reserves: 0 bbl (1 January 2006 est.)

Natural gas—production: 0 cu m (2005 est.)

Natural gas—consumption: 0 cu m (2005 est.)

Natural gas—exports: 0 cu m (2005 est.)

Natural gas—imports: 0 cu m (2005)

Natural gas—proved reserves: 0 cu m (1 January 2006 est.)

Exports: $634 million f.o.b. (2006)

Exports—commodities: fish and fish products 94%, stamps, ships (1999)

Exports—partners: Denmark 31%, UK 27.4%, Norway 10.3%, Nigeria 9.5%, Netherlands 5.6% (2006)

Imports: $751 million c.i.f. (2006)

Imports—commodities: consumer goods 36%, raw materials and semi-manufactures 32%, machinery and transport equipment 29%, fuels, fish, salt (1999)

Imports—partners: Denmark 52.6%, Norway 20.7%, Iceland 6.1%, Sweden 4.3% (2006)

Economic aid—recipient: $105 million; note—annual subsidy from Denmark (2005)

Debt—external: $64 million (1999)

Currency (code): Danish krone (DKK)

Currency code: DKK

Exchange rates: Danish kroner per US dollar—5.4797 (2007), 5.9468 (2006), 5.9969 (2005), 5.9911 (2004), 6.5877 (2003)

Fiscal year: calendar year

COMMUNICATIONS

Telephones—main lines in use: 23,000 (2006)

Telephones—mobile cellular: 50,000 (2006)

Telephone system:
general assessment: good international communications; good domestic facilities
domestic: digitalization was completed in 1998; both NMT (analog) and GSM (digital) mobile telephone systems are installed
international: country code—298; satellite earth stations—1 Orion; 1 fiber-optic submarine cable to the Shetland Islands, linking the Faroe Islands with Denmark and Iceland; fiber-optic submarine cable connection to Canada-Europe cable

Radio broadcast stations: AM 1, FM 13, shortwave 0 (1998)

Radios: 26,000 (1997)

Television broadcast stations: 3 (plus 43 repeaters) (September 1995)

Televisions: 15,000 (1997)

Internet country code: .fo

Internet hosts: 8,490 (2007)

Internet Service Providers (ISPs): 2 (2000)

Internet users: 34,000 (2006)

TRANSPORTATION

Airports: 1 (2007)

Airports—with paved runways: *total:* 1
914 to 1,523 m: 1 (2007)

Roadways: *total:* 463 km (2006)

Merchant marine:
total: 16 ships (1000 GRT or over) 92,454 GRT/63,291 DWT

by type: cargo 10, container 2, passenger/cargo 3, petroleum tanker 1
foreign-owned: 8 (Iceland 4, Norway 4) (2007)
Ports and terminals: Torshavn, Vagur

MILITARY

Military branches: no regular military forces
Manpower available for military service: *males age 16–49:* 11,725 (2008 est.)

Manpower fit for military service: *males age 16–49:* 9,735 (2008 est.)
Manpower reaching militarily significant age annually: *males age 16–49:* 400 (2008 est.)
Military expenditures—percent of GDP: NA
Military—note: defense is the responsibility of Denmark

TRANSNATIONAL ISSUES

Disputes—international: because anticipated offshore hydrocarbon resources have not been realized, earlier Faroese proposals for full independence have been deferred; Iceland, the UK, and Ireland dispute Denmark's claim that the Faroe Islands' continental shelf extends beyond 200 nm

FIJI

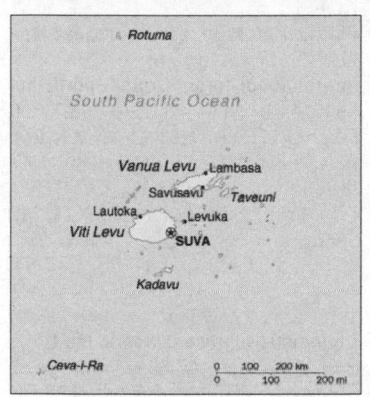

INTRODUCTION

Background: Fiji became independent in 1970, after nearly a century as a British colony. Democratic rule was interrupted by two military coups in 1987, caused by concern over a government perceived as dominated by the Indian community (descendants of contract laborers brought to the islands by the British in the 19th century). The coups and a 1990 constitution that cemented native Melanesian control of Fiji, led to heavy Indian emigration; the population loss resulted in economic difficulties, but ensured that Melanesians became the majority. A new constitution enacted in 1997 was more equitable. Free and peaceful elections in 1999 resulted in a government led by an Indo-Fijian, but a civilian-led coup in May 2000 ushered in a prolonged period of political turmoil. Parliamentary elections held in August 2001 provided Fiji with a democratically elected government led by Prime Minister Laisenia QARASE. Re-elected in May 2006, QARASE was ousted in a December 2006 military coup led by Commodore Voreqe BAINIMARAMA, who initially appointed himself acting president. In January 2007, BAINIMARAMA was appointed interim prime minister.

GEOGRAPHY

Location: Oceania, island group in the South Pacific Ocean, about two-thirds of the way from Hawaii to New Zealand
Geographic coordinates: 18 00 S, 175 00 E
Map references: Oceania
Area: *total:* 18,270 sq km
land: 18,270 sq km
water: 0 sq km
Area—comparative: slightly smaller than New Jersey
Land boundaries: 0 km
Coastline: 1,129 km
Maritime claims: measured from claimed archipelagic straight baselines
territorial sea: 12 nm
exclusive economic zone: 200 nm
continental shelf: 200-m depth or to the depth of exploitation; rectilinear shelf claim added
Climate: tropical marine; only slight seasonal temperature variation
Terrain: mostly mountains of volcanic origin
Elevation extremes:
lowest point: Pacific Ocean 0 m
highest point: Tomanivi 1,324 m
Natural resources: timber, fish, gold, copper, offshore oil potential, hydropower
Land use:
arable land: 10.95%
permanent crops: 4.65%
other: 84.4% (2005)
Irrigated land: 30 sq km (2003)
Total renewable water resources: 28.6 cu km (1987)
Freshwater withdrawal (domestic/industrial/agricultural): *total:* 0.07 cu km/yr (14%/14%/71%)
per capita: 82 cu m/yr (2000)
Natural hazards: cyclonic storms can occur from November to January
Environment—current issues: deforestation; soil erosion
Environment—international agreements: *party to:* Biodiversity, Climate Change, Climate Change-Kyoto

Protocol, Desertification, Endangered Species, Law of the Sea, Marine Life Conservation, Ozone Layer Protection, Tropical Timber 83, Tropical Timber 94, Wetlands
signed, but not ratified: none of the selected agreements
Geography—note: includes 332 islands; approximately 110 are inhabited

PEOPLE

Population: 931,741 (July 2008 est.)
Age structure:
0–14 years: 30.6% (male 145,430/female 139,498)
15–64 years: 64.8% (male 302,460/female 301,344)
65 years and over: 4.6% (male 19,413/female 23,596) (2008 est.)
Median age:
total: 25.2 years
male: 24.7 years
female: 25.7 years (2008 est.)
Population growth rate: 1.388% (2008 est.)
Birth rate: 22.15 births/1,000 population (2008 est.)
Death rate: 5.66 deaths/1,000 population (2008 est.)
Net migration rate: -2.62 migrant(s)/1,000 population (2008 est.)
Sex ratio:
at birth: 1.05 male(s)/female
under 15 years: 1.04 male(s)/female
15–64 years: 1 male(s)/female
65 years and over: 0.82 male(s)/female
total population: 1.01 male(s)/female (2008 est.)
Infant mortality rate:
total: 11.88 deaths/1,000 live births
male: 13.07 deaths/1,000 live births
female: 10.63 deaths/1,000 live births (2008 est.)
Life expectancy at birth:
total population: 70.44 years
male: 67.9 years
female: 73.1 years (2008 est.)
Total fertility rate: 2.68 children born/woman (2008 est.)

HIV/AIDS—adult prevalence rate: 0.1% (2003 est.)

HIV/AIDS—people living with HIV/AIDS: 600 (2003 est.)

HIV/AIDS—deaths: fewer than 200 (2003 est.)

Nationality:

noun: Fijian(s)

adjective: Fijian

Ethnic groups: Fijian 54.8% (predominantly Melanesian with a Polynesian admixture), Indian 37.4%, other 7.9% (European, other Pacific Islanders, Chinese) (2005 estimate)

Religions: Christian 53% (Methodist 34.5%, Roman Catholic 7.2%, Assembly of God 3.8%, Seventh Day Adventist 2.6%, other 4.9%), Hindu 34% (Sanatan 25%, Arya Samaj 1.2%, other 7.8%), Muslim 7% (Sunni 4.2%. other 2.8%), other or unspecified 5.6%, none 0.3% (1996 census)

Languages: English (official), Fijian (official), Hindustani

Literacy:

definition: age 15 and over can read and write

total population: 93.7%

male: 95.5%

female: 91.9% (2003 est.)

GOVERNMENT

Country name:

conventional long form: Republic of the Fiji Islands

conventional short form: Fiji

local long form: Republic of the Fiji Islands/Matanitu ko Viti

local short form: Fiji/Viti

Government type: republic

Capital: *name:* Suva (on Viti Levu)

geographic coordinates: 18 08 S, 178 25 E

time difference: UTC+12 (17 hours ahead of Washington, DC during Standard Time)

Administrative divisions: 4 divisions and 1 dependency*; Central, Eastern, Northern, Rotuma*, Western

Independence: 10 October 1970 (from UK)

National holiday: Independence Day, second Monday of October (1970)

Constitution: enacted on 25 July 1997 to encourage multiculturalism and make multiparty government mandatory; effective 28 July 1998

Legal system: based on British system; has not accepted compulsory ICJ jurisdiction

Suffrage: 21 years of age; universal

Executive branch:

chief of state: President Ratu Josefa ILOILOVATU Uluivuda (since 18 July 2000); note—ILOILOVATU was reaffirmed as president by the Great Council

of Chiefs in a statement issued on 22 December, and reappointed by the coup leader Commodore Voreqe BAINIMARAMA in January 2007

head of government: Prime Minister Laisenia QARASE (since 10 September 2000); note—although QARASE is still the legal prime minister, he has been confined to his home island; the president appointed Commodore Voreqe BAINIMARAMA interim prime minister under the military regime

cabinet: Cabinet appointed by the prime minister from among the members of Parliament and is responsible to Parliament; note—coup leader Commodore Voreqe BAINIMARAMA has appointed an interim cabinet

elections: president elected by the Great Council of Chiefs for a five-year term (eligible for a second term); prime minister appointed by the president; election last held 8 March 2006

election results: Ratu Josefa ILOILO-VATU Uluivuda elected president by the Great Council of Chiefs; percent of vote—NA

Legislative branch: bicameral Parliament consists of the Senate (32 seats; 14 appointed by the president on the advice of the Great Council of Chiefs, 9 appointed by the president on the advice of the Prime Minister, 8 on the advice of the Opposition Leader, and 1 appointed on the advice of the council of Rotuma) and the House of Representatives (71 seats; 23 reserved for ethnic Fijians, 19 reserved for ethnic Indians, 3 reserved for other ethnic groups, 1 reserved for the council of Rotuma constituency encompassing the whole of Fiji, and 25 open seats; members serve five-year terms)

elections: House of Representatives—last held 6–13 May 2006 (next to be held in 2011)

election results: House of Representatives—percent of vote by party—SDL 44.6%, FLP 39.2%, UPP 0.8%, independents 4.9%, other 10.5%; seats by party—SDL 36, FLP 31, UPP 2, independents 2

Judicial branch: Supreme Court (judges are appointed by the president); Court of Appeal; High Court; Magistrates' Courts

Political parties and leaders: Dodonu Ni Taukei Party or DNT [Fereti S. DEWA]; Fiji Democratic Party or FDP [Filipe BOLE] (a merger of the Christian Democrat Alliance or VLV [Poesci Waqalevu BUNE], Fijian Association Party or FAP, Fijian Political Party or SVT (primarily Fijian) [Sitiveni RABUKA], and New Labor Unity Party or NLUP [Ofa SWANN]); Fiji Labor

Party or FLP [Mahendra CHAUDHRY]; General Voters Party or GVP (became part of United General Party); Girmit Heritage Party or GHP; Justice and Freedom Party or AIM; Lio 'On Famor Rotuma Party or LFR; National Federation Party or NFP (primarily Indian) [Pramond RAE]; Nationalist Vanua Takolavo Party or NVTLP [Saula TELAWA]; Party of National Unity or PANU [Ponipate LESAVUA]; Party of the Truth or POTT; United Fiji Party/Sogosogo Duavata ni Lewenivanua or SDL [Laisenia QARASE]; United Peoples Party or UPP [Millis Mick BEDDOES]

Political pressure groups and leaders: NA

International organization participation: ACP, ADB, C (suspended), CP, FAO, G-77, IBRD, ICAO, ICCt, ICRM, IDA, IFAD, IFC, IFRCS, IHO, ILO, IMF, IMO, Interpol, IOC, ISO, ITSO, ITU, ITUC, MIGA, OPCW, PCA, PIF, Sparteca, SPC, UN, UNCTAD, UNESCO, UNIDO, UNMIS, UNMIT, UNWTO, UPU, WCO, WFTU, WHO, WIPO, WMO, WTO

Diplomatic representation in the US:

chief of mission: Ambassador (vacant); Charge d'Affaires Penijamini R. LOMA-LOMA

chancery: 2000 M Street, NW, Suite 710, Washington, DC 20036

telephone: [1] (202) 466-8320

FAX: [1] (202) 466-8325

Diplomatic representation from the US:

chief of mission: Ambassador Larry Miles DINGER

embassy: 31 Loftus Street, Suva

mailing address: P. O. Box 218, Suva

telephone: [679] 331-4466

FAX: [679] 330-0081

Flag description: light blue with the flag of the UK in the upper hoist-side quadrant and the Fijian shield centered on the outer half of the flag; the shield depicts a yellow lion above a white field quartered by the cross of Saint George featuring stalks of sugarcane, a palm tree, bananas, and a white dove

ECONOMY

Economy—overview: Fiji, endowed with forest, mineral, and fish resources, is one of the most developed of the Pacific island economies, though still with a large subsistence sector. Sugar exports, remittances from Fijians working abroad, and a growing tourist industry—with 400,000 to 500,000 tourists annually—are the major sources of foreign exchange. Fiji's sugar has special access to European Union markets, but will be harmed by the EU's decision to cut sugar subsidies. Sugar processing makes up one-third of industrial activity but is not efficient. Fiji's tourism industry was damaged by the December 2006 coup and is facing an uncertain recovery time. The coup has created a difficult business climate. Tourist arrivals for 2007 are estimated to be down almost 6%, with substantial job losses in the service sector. In July 2007 the Reserve Bank of Fiji announced the economy was expected to contract by 3.1% in 2007. Fiji's current account deficit reached 23% of GDP in 2006. The EU has suspended all aid until the interim government takes steps toward new elections. Long-term problems include low investment, uncertain land ownership rights, and the government's inability to manage its budget. Overseas remittances from Fijians working in Kuwait and Iraq have decreased significantly.

GDP (purchasing power parity): $5.079 billion (2007 est.)
GDP (official exchange rate): $4.969 billion (2007 est.)
GDP—real growth rate: 3.9% (2007 est.)
GDP—per capita (PPP): $5,500 (2007 est.)
GDP—composition by sector:
agriculture: 8.9%
industry: 13.5%
services: 77.6% (2004 est.)
Labor force: 117,500 (2006 est.)
Labor force—by occupation: *agriculture:* 70%
industry and services: 30% (2001 est.)
Unemployment rate: 7.6% (1999)
Population below poverty line: 25.5% (FY90/91)
Household income or consumption by percentage share:
lowest 10%: NA%
highest 10%: NA%
Inflation rate (consumer prices): 4.8% (2007)
Budget:
revenues: $1.363 billion
expenditures: $1.376 billion (2006)
Agriculture—products: sugarcane, coconuts, cassava (tapioca), rice, sweet potatoes, bananas; cattle, pigs, horses, goats; fish
Industries: tourism, sugar, clothing, copra, gold, silver, lumber, small cottage industries
Industrial production growth rate: NA%
Electricity—production: 1.046 billion kWh (2005)
Electricity—production by source:
fossil fuel: 18.5%
hydro: 81.5%
nuclear: 0%
other: 0% (2001)

Electricity—consumption: 735.6 million kWh (2006)
Electricity—exports: 0 kWh (2005)
Electricity—imports: 0 kWh (2005)
Oil—production: 0 bbl/day (2005 est.)
Oil—consumption: 9,000 bbl/day (2005 est.)
Oil—exports: 2,268 bbl/day (2004)
Oil—imports: 10,870 bbl/day (2004)
Oil—proved reserves: 0 bbl (1 January 2006 est.)
Natural gas—production: 0 cu m (2005 est.)
Natural gas—consumption: 0 cu m (2005 est.)
Natural gas—exports: 0 cu m (2005 est.)
Natural gas—imports: 0 cu m (2005)
Natural gas—proved reserves: 0 cu m (1 January 2006 est.)
Current account balance: -$507 million (2006 est.)
Exports: $1.202 billion f.o.b. (2006)
Exports—commodities: sugar, garments, gold, timber, fish, molasses, coconut oil
Exports—partners: US 16.8%, Australia 13.9%, UK 13.5%, Japan 5.3%, Samoa 4.7%, Tonga 4.1% (2006)
Imports: $3.12 billion c.i.f. (2006)
Imports—commodities: manufactured goods, machinery and transport equipment, petroleum products, food, chemicals
Imports—partners: Singapore 28.8%, Australia 23.3%, NZ 16.8%, China 4.7% (2006)
Economic aid—recipient: $63.96 million (2005)
Debt—external: $127 million (2004 est.)
Stock of direct foreign investment—at home: $NA
Stock of direct foreign investment—abroad: $NA
Market value of publicly traded shares: $586.7 million (2005)
Currency (code): Fijian dollar (FJD)
Currency code: FJD
Exchange rates: Fijian dollars per US dollar—NA (2007), 1.7313 (2006), 1.691 (2005), 1.7331 (2004), 1.8958 (2003)
Fiscal year: calendar year

COMMUNICATIONS

Telephones—main lines in use: 112,500 (2005)
Telephones—mobile cellular: 205,000 (2005)
Telephone system:
general assessment: modern local, interisland, and international (wire/radio integrated) public and special-purpose telephone, telegraph, and teleprinter facilities; regional radio communications center

domestic: telephone or radio telephone links to almost all inhabited islands; most towns and large villages have automatic telephone exchanges and direct dialing; combined fixed and mobile-cellular density is about 35 per 100 persons
international: country code—679; access to important cable links between US and Canada as well as between NZ and Australia; satellite earth stations—2 Inmarsat (Pacific Ocean)
Radio broadcast stations: AM 13, FM 40, shortwave 0 (1998)
Radios: 541,476 (1999)
Television broadcast stations: NA
Televisions: 88,110 (1999)
Internet country code: .fj
Internet hosts: 12,137 (2007)
Internet Service Providers (ISPs): 2 (2000)
Internet users: 80,000 (2006)

TRANSPORTATION

Airports: 28 (2007)
Airports—with paved runways: *total:* 3
over 3,047 m: 1
1,524 to 2,437 m: 1
914 to 1,523 m: 1 (2007)
Airports—with unpaved runways:
total: 25
914 to 1,523 m: 7
under 914 m: 18 (2007)
Railways:
total: 597 km
narrow gauge: 597 km 0.600-m gauge
note: belongs to the government-owned Fiji Sugar Corporation; used to haul sugarcane during harvest season (May to December) (2006)
Roadways:
total: 3,440 km
paved: 1,692 km
unpaved: 1,748 km (2000)
Waterways: 203 km
note: 122 km navigable by motorized craft and 200-metric-ton barges (2006)
Merchant marine:
total: 8 ships (1000 GRT or over) 17,376 GRT/8,788 DWT
by type: passenger 3, passenger/cargo 3, roll on/roll off 2
foreign-owned: 1 (Australia 1) (2007)
Ports and terminals: Lautoka, Suva

MILITARY

Military branches: Republic of Fiji Military Forces (RFMF): Land Forces, Naval Forces (2008)
Military service age and obligation: 18 years of age for voluntary military service; reserve obligation to age 45 (2006)
Manpower available for military service: *males age 16–49:* 242,567
females age 16–49: 238,556 (2008 est.)

Manpower fit for military service:
males age 16–49: 189,282
females age 16–49: 202,350 (2008 est.)
Manpower reaching militarily signifi-

cant age annually: *males age 16–49:* 9,077
females age 16–49: 8,728 (2008 est.)
Military expenditures—percent of

GDP: 2.2% (2005 est.)

Disputes—international: none

FINLAND

INTRODUCTION

Background: Finland was a province and then a grand duchy under Sweden from the 12th to the 19th centuries, and an autonomous grand duchy of Russia after 1809. It won its complete independence in 1917. During World War II, it was able to successfully defend its freedom and resist invasions by the Soviet Union—albeit with some loss of territory. In the subsequent half century, the Finns made a remarkable transformation from a farm/forest economy to a diversified modern industrial economy; per capita income is now among the highest in Western Europe. A member of the European Union since 1995, Finland was the only Nordic state to join the euro system at its initiation in January 1999.

GEOGRAPHY

Location: Northern Europe, bordering the Baltic Sea, Gulf of Bothnia, and Gulf

of Finland, between Sweden and Russia
Geographic coordinates: 64 00 N, 26 00 E
Map references: Europe
Area: *total:* 338,145 sq km
land: 304,473 sq km
water: 33,672 sq km
Area—comparative: slightly smaller than Montana
Land boundaries:
total: 2,681 km
border countries: Norway 727 km, Sweden 614 km, Russia 1,340 km
Coastline: 1,250 km
Maritime claims:
territorial sea: 12 nm (in the Gulf of Finland—3 nm)
contiguous zone: 24 nm
exclusive fishing zone: 12 nm; extends to continental shelf boundary with Sweden
continental shelf: 200-m depth or to the depth of exploitation
Climate: cold temperate; potentially subarctic but comparatively mild because of moderating influence of the North Atlantic Current, Baltic Sea, and more than 60,000 lakes
Terrain: mostly low, flat to rolling plains interspersed with lakes and low hills
Elevation extremes:
lowest point: Baltic Sea 0 m
highest point: Haltiatunturi 1,328 m
Natural resources: timber, iron ore, copper, lead, zinc, chromite, nickel, gold, silver, limestone
Land use:
arable land: 6.54%
permanent crops: 0.02%
other: 93.44% (2005)
Irrigated land: 640 sq km (2003)
Total renewable water resources: 110 cu km (2005)
Freshwater withdrawal (domestic/industrial/agricultural): *total:* 2.33 cu km/yr (14%/84%/3%)
per capita: 444 cu m/yr (1999)
Natural hazards: NA
Environment—current issues: air pollution from manufacturing and power plants contributing to acid rain; water pollution from industrial wastes, agricultural chemicals; habitat loss threatens wildlife populations
Environment—international agreements: *party to:* Air Pollution, Air

Pollution-Nitrogen Oxides, Air Pollution-Persistent Organic Pollutants, Air Pollution-Sulfur 85, Air Pollution-Sulfur 94, Air Pollution-Volatile Organic Compounds, Antarctic-Environmental Protocol, Antarctic-Marine Living Resources, Antarctic Treaty, Biodiversity, Climate Change, Climate Change-Kyoto Protocol, Desertification, Endangered Species, Environmental Modification, Hazardous Wastes, Law of the Sea, Marine Dumping, Marine Life Conservation, Ozone Layer Protection, Ship Pollution, Tropical Timber 83, Tropical Timber 94, Wetlands, Whaling
signed, but not ratified: none of the selected agreements
Geography—note: long boundary with Russia; Helsinki is northernmost national capital on European continent; population concentrated on small southwestern coastal plain

PEOPLE

Population: 5,244,749 (July 2008 est.)
Age structure:
0–14 years: 16.6% (male 443,738/female 427,875)
15–64 years: 66.8% (male 1,773,232/female 1,731,808)
65 years and over: 16.6% (male 349,826/female 518,270) (2008 est.)
Median age:
total: 41.8 years
male: 40.3 years
female: 43.4 years (2008 est.)
Population growth rate: 0.112% (2008 est.)
Birth rate: 10.39 births/1,000 population (2008 est.)
Death rate: 10 deaths/1,000 population (2008 est.)
Net migration rate: 0.73 migrant(s)/1,000 population (2008 est.)
Sex ratio:
at birth: 1.04 male(s)/female
under 15 years: 1.04 male(s)/female
15–64 years: 1.02 male(s)/female
65 years and over: 0.67 male(s)/female
total population: 0.96 male(s)/female (2008 est.)
Infant mortality rate:
total: 3.5 deaths/1,000 live births
male: 3.81 deaths/1,000 live births

female: 3.17 deaths/1,000 live births (2008 est.)

Life expectancy at birth:
total population: 78.82 years
male: 75.31 years
female: 82.46 years (2008 est.)

Total fertility rate: 1.73 children born/woman (2008 est.)

HIV/AIDS—adult prevalence rate: less than 0.1% (2003 est.)

HIV/AIDS—people living with HIV/AIDS: 1,500 (2003 est.)

HIV/AIDS—deaths: fewer than 100 (2003 est.)

Nationality:
noun: Finn(s)
adjective: Finnish

Ethnic groups: Finn 93.4%, Swede 5.6%, Russian 0.5%, Estonian 0.3%, Roma (Gypsy) 0.1%, Sami 0.1% (2006)

Religions: Lutheran Church of Finland 82.5%, Orthodox Church 1.1%, other Christian 1.1%, other 0.1%, none 15.1% (2006)

Languages: Finnish 91.5% (official), Swedish 5.5% (official), other 3% (small Sami- and Russian-speaking minorities) (2006)

Literacy:
definition: age 15 and over can read and write
total population: 100%
male: 100%
female: 100% (2000 est.)

GOVERNMENT

Country name:
conventional long form: Republic of Finland
conventional short form: Finland
local long form: Suomen tasavalta/ Republiken Finland
local short form: Suomi/Finland

Government type: republic

Capital: *name*: Helsinki
geographic coordinates: 60 10 N, 24 56 E
time difference: UTC+2 (7 hours ahead of Washington, DC during Standard Time)
daylight saving time: +1hr, begins last Sunday in March; ends last Sunday in October

Administrative divisions: 6 provinces (laanit, singular—laani); Aland, Etela-Suomen Laani, Ita-Suomen Laani, Lansi-Suomen Laani, Lappi, Oulun Laani

Independence: 6 December 1917 (from Russia)

National holiday: Independence Day, 6 December (1917)

Constitution: 1 March 2000

Legal system: civil law system based on Swedish law; the president may request the Supreme Court to review laws; accepts compulsory ICJ jurisdiction with reservations

Suffrage: 18 years of age; universal

Executive branch:
chief of state: President Tarja HALONEN (since 1 March 2000)
head of government: Prime Minister Matti VANHANEN (since 24 June 2003); Deputy Prime Minister Jyrki KATAINEN (since 19 April 2007)
cabinet: Council of State or Valtioneuvosto appointed by the president, responsible to parliament
elections: president elected by popular vote for a six-year term (eligible for a second term); election last held 15 January 2006 (next to be held in January 2012); the president appoints the prime minister and deputy prime minister from the majority party or the majority coalition after parliamentary elections and the parliament must approve the appointment; Prime Minister VANHANEN reelected 17 April 2007
election results: percent of vote—Tarja HALONEN (SDP) 46.3%, Sauli NIINISTO (Kok) 24.1%, Matti Vanhanen (Kesk) 18.6%, Heidi HAUTALA (VIHR) 3.5%; a runoff election between HALONEN and NIINISTO was held 29 January 2006—HALONEN 51.8%, NIINISTO 48.2%; Matti VANHANEN reelected prime minister; election results 121-71
note: government coalition—Kesk, KOK, VIHR, and SFP

Legislative branch: unicameral Parliament or Eduskunta (200 seats); members are elected by popular vote on a proportional basis to serve four-year terms)
elections: last held 18 March 2007 (next to be held March 2011)
election results: percent of vote by party—Kesk 23.1%, Kok 22.3%, SDP 21.4%, VAS 8.8%, VIHR 8.5%, KD 4.9%, SFP 4.5%, True Finns 4.1%, other 3.4%; seats by party—Kesk 51, Kok 50, SDP 45, VAS 17, VIHR 15, SFP 9, KD 7, True Finns 5, other 1

Judicial branch: Supreme Court or Korkein Oikeus (judges appointed by the president)

Political parties and leaders: Center Party or Kesk [Matti VANHANEN]; Christian Democrats or KD [Paivi RASANEN]; Green Party or VIHR [Tarja CRONBERG]; Left Alliance or VAS [Martti KORHONEN] (composed of People's Democratic League and Democratic Alternative); National Coalition (conservative) Party or Kok [Jyrki KATAINEN]; Social Democratic Party or SDP [Eero HEINALUOMA]; Swedish People's Party or SFP [Stefan WALLIN]; True Finns [Timo SOINI]

International organization participation: ADB (nonregional members), AfDB, Arctic Council, Australia Group, BIS, CBSS, CE, CERN, EAPC, EBRD, EIB, EMU, ESA, EU, FAO, G-9, IADB, IAEA, IBRD, ICAO, ICC, ICCt, ICRM, IDA, IEA, IFAD, IFC, IFRCS, IHO, ILO, IMF, IMO, IMSO, Interpol, IOC, IOM, IPU, ISO, ITSO, ITU, ITUC, MIGA, NAM (guest), NC, NEA, NIB, NSG, OAS (observer), OECD, OPCW, OSCE, Paris Club, PCA, PFP, Schengen Convention, UN, UNCTAD, UNESCO, UNFICYP, UNHCR, UNIDO, UNIFIL, UNMEE, UNMIL, UNMIS, UNMOGIP, UNTSO, UPU, WCO, WEU (observer), WFTU, WHO, WIPO, WMO, WTO, ZC

Diplomatic representation in the US:
chief of mission: Ambassador Pekka LINTU
chancery: 3301 Massachusetts Avenue NW, Washington, DC 20008
telephone: [1] (202) 298-5800
FAX: [1] (202) 298-6030
consulate(s) general: Los Angeles, New York

Diplomatic representation from the US:
chief of mission: Ambassador Marilyn WARE
embassy: Itainen Puistotie 14B, 00140 Helsinki
mailing address: APO AE 09723
telephone: [358] (9) 616250
FAX: [358] (9) 6162 5800

Flag description: white with a blue cross extending to the edges of the flag; the vertical part of the cross is shifted to the hoist side in the style of the Dannebrog (Danish flag)

ECONOMY

Economy—overview: Finland has a highly industrialized, largely free-market economy with per capita output roughly that of the UK, France, Germany, and Italy. Its key economic sector is manufacturing—principally the wood, metals, engineering, telecommunications, and electronics industries. Trade is important; exports equal nearly two-fifths of GDP. Finland excels in high-tech exports, e.g., mobile phones. Except for timber and several minerals, Finland depends on imports of raw materials, energy, and some components for manufactured goods. Because of the climate, agricultural development is limited to maintaining self-sufficiency in basic products. Forestry, an important export earner, provides a secondary occupation for the rural population. High unemployment remains a persistent problem. In 2007 Russia announced plans to impose high tariffs on raw timber exported to Finland. The Finnish pulp and paper industry will be threatened if these duties are put into place in 2008 and 2009, and the matter is now being handled by the European Union.

GDP (purchasing power parity): $185.5 billion (2007 est.)

GDP (official exchange rate): $245 billion (2007)

GDP—real growth rate: 4.4% (2007)

GDP—per capita (PPP): $35,300 (2007 est.)

GDP—composition by sector:
agriculture: 2.6%
industry: 31.9%
services: 65.6% (2007 est.)

Labor force: 2.675 million (2007 est.)

Labor force—by occupation: agriculture and forestry 4.4%, industry 18.6%, construction 6%, commerce 16.3%, finance, insurance, and business services 13.9%, transport and communications 7.6%, public services 33.2% (2004)

Unemployment rate: 6.8% (2007 est.)

Population below poverty line: NA%

Household income or consumption by percentage share:
lowest 10%: 4%
highest 10%: 22.6% (2000)

Distribution of family income—Gini index: 26 (2005)

Inflation rate (consumer prices): 1.6% (2007 est.)

Investment (gross fixed): 20.3% of GDP (2007 est.)

Budget:
revenues: $62.02 billion
expenditures: $58.16 billion (2007)

Public debt: 31.3% of GDP (2007)

Agriculture—products: barley, wheat, sugar beets, potatoes; dairy cattle; fish

Industries: metals and metal products, electronics, machinery and scientific instruments, shipbuilding, pulp and paper, foodstuffs, chemicals, textiles, clothing

Industrial production growth rate: 4.6% (2007)

Electricity—production: 73.47 billion kWh (2007 est.)

Electricity—production by source:
fossil fuel: 39%
hydro: 18.7%
nuclear: 30.4%
other: 11.8% (2001)

Electricity—consumption: 88.27 billion kWh (2007 est.)

Electricity—exports: 1.059 billion kWh (2007 est.)

Electricity—imports: 15.85 billion kWh (2007 est.)

Oil—production: 8,951 bbl/day (2005 est.)

Oil—consumption: 219,700 bbl/day (2005 est.)

Oil—exports: 126,300 bbl/day (January-September 2007 est.)

Oil—imports: 281,300 bbl/day (January-September 2007 est.)

Oil—proved reserves: NA bbl

Natural gas—production: 0 cu m (2007 est.)

Natural gas—consumption: 3.909 billion cu m (2007 est.)

Natural gas—exports: 0 cu m (2007 est.)

Natural gas—imports: 4.123 billion cu m (2007 est.)

Natural gas—proved reserves: 0 cu m (1 January 2006)

Current account balance: $11.27 billion (2007)

Exports: $104.9 billion f.o.b. (2007)

Exports—commodities: machinery and equipment, chemicals, metals; timber, paper, pulp

Exports—partners: Germany 11.3%, Sweden 10.5%, Russia 10.1%, UK 6.5%, US 6.5%, Netherlands 5.1% (2006)

Imports: $81.54 billion f.o.b. (2007)

Imports—commodities: foodstuffs, petroleum and petroleum products, chemicals, transport equipment, iron and steel, machinery, textile yarn and fabrics, grains

Imports—partners: Germany 15.6%, Russia 14%, Sweden 13.7%, Netherlands 6.6%, China 5.4%, UK 4.7%, Denmark 4.5% (2006)

Economic aid—donor: ODA, $1.023 billion (2007)

Reserves of foreign exchange and gold: $7.804 billion (2007)

Debt—external: $271.2 billion (30 June 2007)

Stock of direct foreign investment—at home: $72.94 billion (2007 est.)

Stock of direct foreign investment—abroad: $98.89 billion (2007 est.)

Market value of publicly traded shares: $1.095 trillion (January 2008)

Currency (code): euro (EUR)

Currency code: EUR

Exchange rates: euros per US dollar—0.7345 (2007), 0.7964 (2006), 0.8041 (2005), 0.8054 (2004), 0.886 (2003)

Fiscal year: calendar year

COMMUNICATIONS

Telephones—main lines in use: 1.92 million (2006)

Telephones—mobile cellular: 5.67 million (2006)

Telephone system:
general assessment: modern system with excellent service
domestic: digital fiber-optic fixed-line network and an extensive cellular network provide domestic needs
international: country code—358; submarine cables provide links to Estonia and Sweden; satellite earth stations—access to Intelsat transmission service via a Swedish satellite earth station, 1 Inmarsat (Atlantic and Indian Ocean regions); note—Finland shares the Inmarsat earth station with the other Nordic countries (Denmark, Iceland, Norway, and Sweden)

Radio broadcast stations: AM 2, FM 186, shortwave 1 (1998)

Radios: 7.7 million (1997)

Television broadcast stations: 120 (plus 431 repeaters) (1999); note—On 1 September 2007, Finland became one of the first countries in the world to broadcast all television signals digitally

Televisions: 3.2 million (1997)

Internet country code: .fi; note—Aland Islands assigned .ax

Internet hosts: 2.323 million (2007)

Internet Service Providers (ISPs): 3 (2002)

Internet users: 2.925 million (2006)

TRANSPORTATION

Airports: 148 (2007)

Airports—with paved runways:
total: 76
over 3,047 m: 2
2,438 to 3,047 m: 27
1,524 to 2,437 m: 10
914 to 1,523 m: 22
under 914 m: 15 (2007)

Airports—with unpaved runways:
total: 72
914 to 1,523 m: 4
under 914 m: 68 (2007)

Pipelines: gas 694 km (2007)

Railways:
total: 5,741 km
broad gauge: 5,741 km 1.524-m gauge (2,619 km electrified) (2006)

Roadways: *total:* 78,821 km
paved: 50,854 km (includes 700 km of expressways)
unpaved: 27,967 km (2008)

Waterways: 7,842 km
note: includes Saimaa Canal system of 3,577 km; southern part leased from Russia (2006)

Merchant marine:
total: 92 ships (1000 GRT or over) 1,362,014 GRT/1,002,280 DWT
by type: bulk carrier 3, cargo 26, chemical tanker 6, container 3, passenger 5, passenger/cargo 20, petroleum tanker 4, roll on/roll off 23, vehicle carrier 2
foreign-owned: 5 (Germany 2, Norway 1, Sweden 2)
registered in other countries: 43 (Bahamas 8, Germany 4, Gibraltar 3, Marshall Islands 2, Netherlands 14, Norway 1, Sweden 10, UK 1) (2007)

Ports and terminals: Hamina, Helsinki, Kokkola, Kotka, Naantali, Pori, Raahe, Rauma, Turku

MILITARY

Military branches: Finnish Defense Forces (FDF): Army, Navy (includes Coastal Defense Forces), Air Force (Suomen Ilmavoimat) (2007)
Military service age and obligation: 18 years of age for male voluntary and compulsory national military and nonmilitary service; service obligation 6–12 months (2007)

Manpower available for military service:
males age 16–49: 1,169,910
females age 16–49: 1,121,187 (2008 est.)
Manpower fit for military service:
males age 16–49: 965,131
females age 16–49: 923,224 (2008 est.)
Manpower reaching militarily significant age annually:
males age 16–49: 34,152
females age 16–49: 32,870 (2008 est.)

Military expenditures—percent of GDP: 2% (2005 est.)

TRANSNATIONAL ISSUES

Disputes—international: various groups in Finland advocate restoration of Karelia and other areas ceded to the Soviet Union, but the Finnish Government asserts no territorial demands

FRANCE

INTRODUCTION

Background: Although ultimately a victor in World Wars I and II, France suffered extensive losses in its empire, wealth, manpower, and rank as a dominant nation-state. Nevertheless, France today is one of the most modern countries in the world and is a leader among European nations. Since 1958, it has constructed a hybrid presidential-parliamentary governing system resistant to the instabilities experienced in earlier more purely parliamentary administrations. In recent years, its reconciliation and cooperation with Germany have proved central to the economic integration of Europe, including the introduction of a common exchange currency, the euro, in January 1999. At present, France is at the forefront of efforts to develop the EU's military capabilities to supplement progress toward an EU foreign policy.

GEOGRAPHY

Location: *metropolitan France:* Western Europe, bordering the Bay of Biscay and English Channel, between Belgium and Spain, southeast of the UK; bordering the Mediterranean Sea, between Italy and Spain
French Guiana: Northern South America, bordering the North Atlantic

Ocean, between Brazil and Suriname
Guadeloupe: Caribbean, islands between the Caribbean Sea and the North Atlantic Ocean, southeast of Puerto Rico
Martinique: Caribbean, island between the Caribbean Sea and North Atlantic Ocean, north of Trinidad and Tobago
Reunion: Southern Africa, island in the Indian Ocean, east of Madagascar
Geographic coordinates: *metropolitan France:* 46 00 N, 2 00 E
French Guiana: 4 00 N, 53 00 W
Guadeloupe: 16 15 N, 61 35 W
Martinique: 14 40 N, 61 00 W
Reunion: 21 06 S, 55 36 E
Map references: *metropolitan France:* Europe
French Guiana: South America
Guadeloupe: Central America and the Caribbean
Martinique: Central America and the Caribbean
Reunion: World
Area:
total: 643,427 sq km; 547,030 sq km (metropolitan France)
land: 640,053 sq km; 545,630 sq km (metropolitan France)
water: 3,374 sq km; 1,400 sq km (metropolitan France)
note: the first numbers include the overseas regions of French Guiana, Guadeloupe, Martinique, and Reunion
Area—comparative: slightly less than the size of Texas
Land boundaries:
metropolitan France—total: 2,889 km
border countries: Andorra 56.6 km, Belgium 620 km, Germany 451 km, Italy 488 km, Luxembourg 73 km, Monaco 4.4 km, Spain 623 km, Switzerland 573 km
French Guiana—total: 1,183 km
border countries: Brazil 673 km, Suriname 510 km
Coastline: *total:* 4,668 km
metropolitan France: 3,427 km
Maritime claims: *territorial sea:* 12 nm

contiguous zone: 24 nm
exclusive economic zone: 200 nm (does not apply to the Mediterranean)
continental shelf: 200-m depth or to the depth of exploitation
Climate: *metropolitan France:* generally cool winters and mild summers, but mild winters and hot summers along the Mediterranean; occasional strong, cold, dry, north-to-northwesterly wind known as mistral
French Guiana: tropical; hot, humid; little seasonal temperature variation
Guadeloupe and Martinique: subtropical tempered by trade winds; moderately high humidity; rainy season (June to October); vulnerable to devastating cyclones (hurricanes) every eight years on average
Reunion: tropical, but temperature moderates with elevation; cool and dry (May to November), hot and rainy (November to April)
Terrain: *metropolitan France:* mostly flat plains or gently rolling hills in north and west; remainder is mountainous, especially Pyrenees in south, Alps in east
French Guiana: low-lying coastal plains rising to hills and small mountains
Guadeloupe: Basse-Terre is volcanic in origin with interior mountains; Grande-Terre is low limestone formation; most of the seven other islands are volcanic in origin
Martinique: mountainous with indented coastline; dormant volcano
Reunion: mostly rugged and mountainous; fertile lowlands along coast
Elevation extremes:
lowest point: Rhone River delta -2 m
highest point: Mont Blanc 4,807 m
Natural resources: *metropolitan France:* coal, iron ore, bauxite, zinc, uranium, antimony, arsenic, potash, feldspar, fluorspar, gypsum, timber, fish
French Guiana: gold deposits, petroleum, kaolin, niobium, tantalum, clay
Land use: *arable land:* 33.46%

permanent crops: 2.03%

other: 64.51%

note: French Guiana—arable land 0.13%, permanent crops 0.04%, other 99.83% (90% forest, 10% other); Guadeloupe—arable land 11.70%, permanent crops 2.92%, other 85.38%; Martinique—arable land 9.09%, permanent crops 10.0%, other 80.91%; Reunion—arable land 13.94%, permanent crops 1.59%, other 84.47% (2005)

Irrigated land: *total:* 26,190 sq km;

metropolitan France: 26,000 sq km (2003)

Total renewable water resources: 189 cu km (2005)

Freshwater withdrawal (domestic/industrial/agricultural): *total:* 33.16 cu km/yr (16%/74%/10%)

per capita: 548 cu m/yr (2000)

Natural hazards: *metropolitan France:* flooding; avalanches; midwinter windstorms; drought; forest fires in south near the Mediterranean

overseas departments: hurricanes (cyclones), flooding, volcanic activity (Guadeloupe, Martinique, Reunion)

Environment—current issues: some forest damage from acid rain; air pollution from industrial and vehicle emissions; water pollution from urban wastes, agricultural runoff

Environment—international agreements: *party to:* Air Pollution, Air Pollution-Nitrogen Oxides, Air Pollution-Persistent Organic Pollutants, Air Pollution-Sulfur 85, Air Pollution-Sulfur 94, Air Pollution-Volatile Organic Compounds, Antarctic-Environmental Protocol, Antarctic-Marine Living Resources, Antarctic Seals, Antarctic Treaty, Biodiversity, Climate Change, Climate Change-Kyoto Protocol, Desertification, Endangered Species, Hazardous Wastes, Law of the Sea, Marine Dumping, Marine Life Conservation, Ozone Layer Protection, Ship Pollution, Tropical Timber 83, Tropical Timber 94, Wetlands, Whaling

signed, but not ratified: none of the selected agreements

Geography—note: largest West European nation

PEOPLE

Population: *total:* 64,057,790

note: 60,876,136 in metropolitan France (July 2008 est.)

Age structure:

0–14 years: 18.6% (male 6,091,571/female 5,803,127)

15–64 years: 65.2% (male 20,884,919/female 20,849,988)

65 years and over: 16.3% (male 4,335,996/female 6,092,189) (2008 est.)

Median age:

total: 39.2 years

male: 37.7 years

female: 40.7 years (2008 est.)

Population growth rate: 0.574% (2008 est.)

Birth rate: 12.73 births/1,000 population (2008 est.)

Death rate: 8.48 deaths/1,000 population (2008 est.)

Net migration rate: 1.48 migrant(s)/1,000 population (2008 est.)

Sex ratio:

at birth: 1.05 male(s)/female

under 15 years: 1.05 male(s)/female

15–64 years: 1 male(s)/female

65 years and over: 0.71 male(s)/female

total population: 0.96 male(s)/female (2008 est.)

Infant mortality rate:

total: 3.36 deaths/1,000 live births

male: 3.69 deaths/1,000 live births

female: 3.02 deaths/1,000 live births (2008 est.)

Life expectancy at birth:

total population: 80.87 years

male: 77.68 years

female: 84.23 years (2008 est.)

Total fertility rate: 1.98 children born/woman (2008 est.)

HIV/AIDS—adult prevalence rate: 0.4% (2003 est.)

HIV/AIDS—people living with HIV/AIDS: 120,000 (2003 est.)

HIV/AIDS—deaths: fewer than 1,000 (2003 est.)

Nationality:

noun: Frenchman(men), Frenchwoman(women)

adjective: French

Ethnic groups: Celtic and Latin with Teutonic, Slavic, North African, Indochinese, Basque minorities

overseas departments: black, white, mulatto, East Indian, Chinese, Amerindian

Religions: Roman Catholic 83%-88%, Protestant 2%, Jewish 1%, Muslim 5%-10%, unaffiliated 4%

overseas departments: Roman Catholic, Protestant, Hindu, Muslim, Buddhist, pagan

Languages: French 100%, rapidly declining regional dialects and languages (Provencal, Breton, Alsatian, Corsican, Catalan, Basque, Flemish)

overseas departments: French, Creole patois

Literacy:

definition: age 15 and over can read and write

total population: 99%

male: 99%

female: 99% (2003 est.)

GOVERNMENT

Country name:

conventional long form: French Republic

conventional short form: France

local long form: Republique francaise

local short form: France

Government type: republic

Capital: *name:* Paris

geographic coordinates: 48 52 N, 2 20 E

time difference: UTC+1 (6 hours ahead of Washington, DC during Standard Time)

daylight saving time: +1hr, begins last Sunday in March; ends last Sunday in October

Administrative divisions: 26 regions (regions, singular—region); Alsace, Aquitaine, Auvergne, Basse-Normandie (Lower Normandy), Bourgogne, Bretagne (Brittany), Centre, Champagne-Ardenne, Corse (Corsica), Franche-Comte, Guadeloupe, Guyane (French Guiana), Haute-Normandie (Upper Normandy), Ile-de-France, Languedoc-Roussillon, Limousin, Lorraine, Martinique, Midi-Pyrenees, Nord-Pas-de-Calais, Pays de la Loire, Picardie, Poitou-Charentes, Provence-Alpes-Cote d'Azur, Reunion, Rhone-Alpes

note: France is divided into 22 metropolitan regions (including the "territorial collectivity" of Corse or Corsica) and 4 overseas regions (including French Guiana, Guadeloupe, Martinique, and Reunion) and is subdivided into 96 metropolitan departments and 4 overseas departments (which are the same as the overseas regions)

Dependent areas: Clipperton Island, French Polynesia, French Southern and Antarctic Lands, Mayotte, New Caledonia, Saint Barthelemy, Saint Martin, Wallis and Futuna

note: the US does not recognize claims to Antarctica; New Caledonia has been considered a "sui generis" collectivity of France since 1999, a unique status falling between that of an independent country and a French overseas department

Independence: 486 (Frankish tribes unified); 843 (Western Francia established from the division of the Carolingian Empire)

National holiday: Fete de la Federation, 14 July (1790); note—although often incorrectly referred to as Bastille Day, the celebration actually commemorates the holiday held on the first anniversary of the storming of the Bastille (on 14 July 1789) and the establishment of a constitutional monarchy; other names for the holiday are Fete Nationale (National Holiday) and quatorze juillet (14th of July)

Constitution: adopted by referendum 28 September 1958, effective 4 October 1958

note: amended concerning election of president in 1962; amended to comply with provisions of 1992 EC Maastricht Treaty, 1997 Amsterdam Treaty, 2003 Treaty of Nice; amended to tighten immigration laws in 1993; amended in 2000 to change the seven-year presidential term to a five-year term; amended in 2005 to make the EU constitutional treaty compatible with the Constitution of France and to ensure that the decision to ratify EU accession treaties would be made by referendum

Legal system: civil law system with indigenous concepts; review of administrative but not legislative acts; has not accepted compulsory ICJ jurisdiction

Suffrage: 18 years of age; universal

Executive branch:

chief of state: President Nicolas SARKOZY (since 16 May 2007)

head of government: Prime Minister Francois FILLON (since 17 May 2007)

cabinet: Council of Ministers appointed by the president at the suggestion of the prime minister

elections: president elected by popular vote for a five-year term (changed from seven-year term in October 2000); election last held 22 April and 6 May 2007 (next to be held spring 2012); prime minister nominated by the National Assembly majority and appointed by the president

election results: Nicolas SARKOZY wins the election; First Round: percent of vote—Nicolas SARKOZY 31.18%, Segolene ROYAL 25.87%, Francois BAYROU 18.57%, Jean-Marie LE PEN 10.44%, others 13.94%; Second Round: SARKOZY 53.1% and ROYAL 46.9%

Legislative branch: bicameral Parliament or Parlement consists of the Senate or Senat (331 seats, 305 for metropolitan France, 9 for overseas departments, 5 for dependencies, and 12 for French nationals abroad; members are indirectly elected by an electoral college to serve six-year terms; one third elected every three years); note—between 2006 and 2011, 15 new seats will be added to the Senate for a total of 348 seats—326 for metropolitan France and overseas departments, 2 for New Caledonia, 2 for Mayotte, 1 for Saint-Pierre and Miquelon, 1 for Saint-Barthelemy, 1 for Saint-Martin, 3 for overseas territories, and 12 for French nationals abroad; starting in 2008, members will be indirectly elected by an electoral college to serve six-year terms, with one-half elected every three years; and the National Assembly or Assemblee Nationale (577 seats, 555 for metropolitan France, 15 for overseas departments, 7 for dependencies; members are elected by popular vote under a single-member majority system to serve five-year terms)

elections: Senate—last held 26 September 2004 (next to be held in September 2008); National Assembly—last held 10 and 17 June 2007 (next to be held in June 2012)

election results: Senate—percent of vote by party—NA; seats by party—UMP 156, PS 97, UDF (now MoDem) 33, PCF 23, RDSE 15, other 7; National Assembly—percent of vote by party—UMP 46.37%, PS 42.25%, miscellaneous left wing parties 2.47%, PCF 2.28%, NC 2.12%, PRG 1.65%, miscellaneous right wing parties 1.17%, the Greens 0.45, other 1.24%; seats by party—UMP 313, PS 186, NC 22, miscellaneous left wing parties 15, PCF 15, miscellaneous right wing parties 9, PRG 7, the Greens 4, other 6

Judicial branch: Supreme Court of Appeals or Cour de Cassation (judges are appointed by the president from nominations of the High Council of the Judiciary); Constitutional Council or Conseil Constitutionnel (three members appointed by the president, three appointed by the president of the National Assembly, and three appointed by the president of the Senate); Council of State or Conseil d'Etat

Political parties and leaders: Democratic Movement or MoDem [Francois BAYROU] (previously Union for French Democracy or UDF); Democratic and Social European Rally or RDSE [Pierre LAFFITTE] (mainly Radical Republican and Socialist Parties, and PRG); French Communist Party or PCF [Marie-George BUFFET]; Greens [Cecile DUFLOT]; Left Radical Party or PRG [Jean-Michel BAYLET] (previously Radical Socialist Party or PRS and the Left Radical Movement or MRG); Movement for France or MPF [Philippe DE VILLIERS]; National Front or FN [Jean-Marie LE PEN]; New Center or NC [Herve MORIN]; Rally for France or RPF [Charles PASQUA]; Republican and Citizen Movement or MRC [Jean Pierre CHEVENEMENT and Georges SARRE]; Socialist Party or PS [Francois HOLLANDE]; Union for a Popular Movement or UMP [Patrick DEVEDJIAN, Jean-Claude GAUDIN, Jean-Pierre RAFFARIN, Pierre MEHAIGNERIE]; Radical Party [Jean-Louis BORLOO]

Political pressure groups and leaders: historically Communist labor union (Confederation Generale du Travail) or CGT, approximately 700,000 members (claimed); left-leaning labor union (Confederation Francaise Democratique du Travail) or CFDT, approximately 803,000 members (claimed); independent labor union (Confederation Generale du Travail—Force Ouvriere) or FO, 300,000 members (est.); independent white-collar union (Confederation Generale des Cadres) or CGC, 196,000 members (claimed); employers' union (Mouvement des Entreprises de France) or MEDEF, 750,000 companies as members (claimed)

French Guiana: NA

Guadeloupe: Christian Movement for the Liberation of Guadeloupe or KLPG; General Federation of Guadeloupe Workers or CGT-G; General Union of Guadeloupe Workers or UGTG; Movement for an Independent Guadeloupe or MPGI; The Socialist Renewal Movement

Martinique: Caribbean Revolutionary Alliance or ARC; Central Union for Martinique Workers or CSTM; Frantz Fanon Circle; League of Workers and Peasants; Proletarian Action Group or GAP

Reunion: NA

International organization participation: ABEDA, ACCT, ADB (nonregional members), AfDB, Arctic Council (observer), Australia Group, BDEAC, BIS, BSEC (observer), CBSS (observer), CE, CERN, EAPC, EBRD, EIB, EMU, ESA, EU, FAO, FZ, G-5, G-7, G-8, G-10, IADB, IAEA, IBRD, ICAO, ICC, ICCt, ICRM, IDA, IEA, IFAD, IFC, IFRCS, IFTU, IHO, ILO, IMF, IMO, IMSO, InOC, Interpol, IOC, IOM, IPU, ISO, ITSO, ITU, MIGA, MINURCAT, MINURSO, MINUSTAH, NATO, NEA, NSG, OAS (observer), OECD, OIF, OPCW, OSCE, Paris Club, PCA, PIF (partner), Schengen Convention, SECI (observer), SPC, UN, UN Security Council, UNCTAD, UNESCO, UNHCR, UNIDO, UNIFIL, Union Latina, UNITAR, UNMEE, UNMIL, UNOCI, UNOMIG, UNRWA, UNTSO, UNWTO, UPU, WADB (nonregional), WCL, WCO, WEU, WFTU, WHO, WIPO, WMO, WTO, ZC

Diplomatic representation in the US:

chief of mission: Ambassador Pierre VIMONT

chancery: 4101 Reservoir Road NW, Washington, DC 20007

telephone: [1] (202) 944-6000

FAX: [1] (202) 944-6166

consulate(s) general: Atlanta, Boston, Chicago, Houston, Los Angeles, Miami, New Orleans, New York, San Francisco, Washington, DC

Diplomatic representation from the US:
chief of mission: Ambassador Craig R. STAPLETON
embassy: 2 Avenue Gabriel, 75382 Paris Cedex 08
mailing address: PSC 116, APO AE 09777
telephone: [33] (1) 43-12-22-22
FAX: [33] (1) 42 66 97 83
consulate(s) general: Marseille, Strasbourg
Flag description: three equal vertical bands of blue (hoist side), white, and red; known as the "Le drapeau tricolore" (French Tricolor), the origin of the flag dates to 1790 and the French Revolution; the design and/or colors are similar to a number of other flags, including those of Belgium, Chad, Ireland, Cote d'Ivoire, Luxembourg, and Netherlands; the official flag for all French dependent areas

ECONOMY

Economy—overview: France is in the midst of transition from a well-to-do modern economy that has featured extensive government ownership and intervention to one that relies more on market mechanisms. The government has partially or fully privatized many large companies, banks, and insurers, and has ceded stakes in such leading firms as Air France, France Telecom, Renault, and Thales. It maintains a strong presence in some sectors, particularly power, public transport, and defense industries. The telecommunications sector is gradually being opened to competition. France's leaders remain committed to a capitalism in which they maintain social equity by means of laws, tax policies, and social spending that reduce income disparity and the impact of free markets on public health and welfare. Widespread opposition to labor reform has in recent years hampered the government's ability to revitalize the economy. In 2007, the government launched divisive labor reform efforts that will continue into 2008. France's tax burden remains one of the highest in Europe (nearly 50% of GDP in 2005). France brought the budget deficit within the eurozone's 3%-of-GDP limit for the first time in 2007 and has reduced unemployment to roughly 8%. With at least 75 million foreign tourists per year, France is the most visited country in the world and maintains the third largest income in the world from tourism.
GDP (purchasing power parity): $2.047 trillion (2007 est.)
GDP (official exchange rate): $2.56 trillion (2007 est.)
GDP—real growth rate: 1.9% (2007 est.)

GDP—per capita (PPP): $33,200 (2007 est.)
GDP—composition by sector:
agriculture: 2.2%
industry: 21%
services: 76.7% (2007 est.)
Labor force: 27.81 million (2007 est.)
Labor force—by occupation:
agriculture: 4.1%
industry: 24.4%
services: 71.5% (1999)
Unemployment rate: 8.3% (2007 est.)
Population below poverty line: 6.2% (2004)
Household income or consumption by percentage share: *lowest 10%:* 3%
highest 10%: 24.8% (2004)
Distribution of family income—Gini index: 28 (2005)
Inflation rate (consumer prices): 1.6% (2007 est.)
Investment (gross fixed): 21.7% of GDP (2007 est.)
Budget:
revenues: $1.288 trillion
expenditures: $1.358 trillion (2007 est.)
Public debt: 64% of GDP (2007 est.)
Agriculture—products: wheat, cereals, sugar beets, potatoes, wine grapes; beef, dairy products; fish
Industries: machinery, chemicals, automobiles, metallurgy, aircraft, electronics; textiles, food processing; tourism
Industrial production growth rate: 2% (2007 est.)
Electricity—production: 543.6 billion kWh (2005)
Electricity—production by source:
fossil fuel: 8.2%
hydro: 14%
nuclear: 77.1%
other: 0.7% (2001)
Electricity—consumption: 451.5 billion kWh (2005)
Electricity—exports: 68.33 billion kWh (2005)
Electricity—imports: 8.035 billion kWh (2005)
Oil—production: 73,180 bbl/day (2005 est.)
Oil—consumption: 1.999 million bbl/day (2005 est.)
Oil—exports: 474,200 bbl/day (2005)
Oil—imports: 1.89 million bbl/day (2005)
Oil—proved reserves: 158.4 million bbl (1 January 2006 est.)
Natural gas—production: 1.4 billion cu m (2004 est.)
Natural gas—consumption: 47.26 billion cu m (2005 est.)
Natural gas—exports: 863.2 million cu m (2005 est.)
Natural gas—imports: 47.02 billion cu m (2005 est.)

Natural gas—proved reserves: 341 billion cu m (1 January 2006 est.)
Current account balance: -$33.39 billion (2007 est.)
Exports: $548 billion f.o.b. (2007 est.)
Exports—commodities: machinery and transportation equipment, aircraft, plastics, chemicals, pharmaceutical products, iron and steel, beverages
Exports—partners: Germany 15.6%, Spain 9.6%, Italy 8.9%, UK 8.3%, Belgium 7.3%, US 6.6%, Netherlands 4% (2006)
Imports: $600.1 billion f.o.b. (2007 est.)
Imports—commodities: machinery and equipment, vehicles, crude oil, aircraft, plastics, chemicals
Imports—partners: Germany 18.9%, Belgium 11.1%, Italy 8.4%, Spain 7%, Netherlands 6.8%, UK 6.6%, US 4.6% (2006)
Economic aid—donor: ODA, $10.6 billion (2006)
Reserves of foreign exchange and gold: $98.24 billion (2006 est.)
Debt—external: $4.396 trillion (30 June 2007)
Stock of direct foreign investment—at home: $872.4 billion (2007 est.)
Stock of direct foreign investment—abroad: $1.211 trillion (2007 est.)
Market value of publicly traded shares: $1.71 trillion (2005)
Currency (code): euro (EUR)
Currency code: EUR
Exchange rates: euros per US dollar—0.7345 (2007), 0.7964 (2006), 0.8041 (2005), 0.8054 (2004), 0.886 (2003)
Fiscal year: calendar year

COMMUNICATIONS

Telephones—main lines in use: 34.63 million; 33.897 million (metropolitan France) (2006)
Telephones—mobile cellular: 53.023 million; 51.662 million (metropolitan France) (2006)
Telephone system:
general assessment: highly developed
domestic: extensive cable and microwave radio relay; extensive introduction of fiber-optic cable; domestic satellite system
international: country code—33; numerous submarine cables provide links throughout Europe, Asia, Australia, the Middle East, and US; satellite earth stations—more than 3 (2 Intelsat (with total of 5 antennas—2 for Indian Ocean and 3 for Atlantic Ocean), NA Eutelsat, 1 Inmarsat—Atlantic Ocean region); HF radiotelephone communications with more than 20 countries
overseas departments: country codes: French Guiana—594; Guadeloupe—590; Martinique—596; Reunion—262

Radio broadcast stations: AM 41, FM about 3,500 (this figure is an approximation and includes many repeaters), shortwave 2 (1998)
Radios: 55.3 million (1997)
Television broadcast stations: 584 (plus 9,676 repeaters) (1995)
Televisions: 34.8 million (1997)
Internet country code: metropolitan France—.fr; French Guiana—.gf; Guadeloupe—.gp; Martinique—.mq; Reunion—.re
Internet hosts: 12.556 million; 12,555,000 (metropolitan France) (2007)
Internet Service Providers (ISPs): 62 (2000)
Internet users: 31.295 million; 30.838 million (metropolitan France) (2007)

TRANSPORTATION

Airports: 476 (2007)
Airports—with paved runways:
total: 292
over 3,047 m: 14
2,438 to 3,047 m: 27
1,524 to 2,437 m: 97
914 to 1,523 m: 80
under 914 m: 74 (2007)
Airports—with unpaved runways:
total: 184
1,524 to 2,437 m: 4
914 to 1,523 m: 72
under 914 m: 108 (2007)
Heliports: 3 (2007)
Pipelines: gas 14,665 km; oil 3,032 km; refined products 4,947 km (2007)
Railways:
total: 29,370 km
standard gauge: 29,203 km 1.435-m gauge (14,778 km electrified)
narrow gauge: 167 km 1.000-m gauge (2006)
Roadways:
total: 950,985 km

paved: 950,985 km (metropolitan France; includes 10,490 km of expressways)
note: there are another 5,083 km of roadways in overseas departments (2005)
Waterways: *metropolitan France:* 8,500 km (1,686 km accessible to craft of 3,000 metric tons)
French Guiana: 3,760 km (460 km navigable by small oceangoing vessels and coastal and river steamers, 3,300 km by native craft) (2006)
Merchant marine:
total: 141 ships (1000 GRT or over) 5,777,107 GRT/7,533,631 DWT
by type: bulk carrier 2, cargo 1, chemical tanker 31, container 25, liquefied gas 14, passenger 3, passenger/cargo 32, petroleum tanker 22, roll on/roll off 7, vehicle carrier 4
foreign-owned: 56 (Belgium 6, China 5, Denmark 3, Germany 1, Italy 2, Japan 5, Norway 17, NZ 1, Saudi Arabia 1, Singapore 2, Sweden 10, Switzerland 3)
registered in other countries: 145 (Antigua and Barbuda 1, Australia 1, Bahamas 43, Belgium 1, Bermuda 1, Cameroon 1, Gibraltar 1, Hong Kong 1, Indonesia 1, Isle of Man 2, Italy 5, South Korea 8, Liberia 5, Luxembourg 14, Malta 4, Morocco 13, Netherlands 1, Norway 3, Panama 15, Singapore 1, St Vincent and The Grenadines 7, Taiwan 1, UK 9, Wallis and Futuna 6) (2007)
Ports and terminals: Bordeaux, Calais, Dunkerque, Le Havre, Marseille, Nantes, Paris, Rouen, Strasbourg

MILITARY

Military branches: Army (includes Marines, Foreign Legion, Army Light Aviation), Navy (Marine Nationale, includes Naval Air), Air Force (Armee de l'Air, includes air defense), National Gendarmerie (2008)

Military service age and obligation: 17–40 years of age for male or female voluntary military service); no conscription; 12-month service obligation; women serve in noncombat military posts (2005)
Manpower available for military service:
males age 16–49: 14,646,427
females age 16–49: 14,379,630 (2008 est.)
Manpower fit for military service:
males age 16–49: 12,110,718
females age 16–49: 11,849,988 (2008 est.)
Manpower reaching militarily significant age annually:
males age 16–49: 401,379
females age 16–49: 382,409 (2008 est.)
Military expenditures—percent of GDP: 2.6% (2005 est.)

TRANSNATIONAL ISSUES

Disputes—international: Madagascar claims the French territories of Bassas da India, Europa Island, Glorioso Islands, and Juan de Nova Island; Comoros claims Mayotte; Mauritius claims Tromelin Island; territorial dispute between Suriname and the French overseas department of French Guiana; France asserts a territorial claim in Antarctica (Adelie Land); France and Vanuatu claim Matthew and Hunter Islands, east of New Caledonia
Illicit drugs: *metropolitan France:* transshipment point for South American cocaine, Southwest Asian heroin, and European synthetics
French Guiana: small amount of marijuana grown for local consumption; minor transshipment point to Europe
Martinique: transshipment point for cocaine and marijuana bound for the US and Europe

FRENCH POLYNESIA

INTRODUCTION

Background: The French annexed various Polynesian island groups during the 19th century. In September 1995, France stirred up widespread protests by resuming nuclear testing on the Mururoa atoll after a three-year moratorium. The tests were suspended in January 1996. In recent years, French Polynesia's autonomy has been considerably expanded.

GEOGRAPHY

Location: Oceania, archipelagoes in the South Pacific Ocean about half way between South America and Australia

Geographic coordinates: 15 00 S, 140 00 W
Map references: Oceania
Area:
total: 4,167 sq km (118 islands and atolls)
land: 3,660 sq km
water: 507 sq km
Area—comparative: slightly less than one-third the size of Connecticut
Land boundaries: 0 km
Coastline: 2,525 km
Maritime claims:
territorial sea: 12 nm
exclusive economic zone: 200 nm
Climate: tropical, but moderate

Terrain: mixture of rugged high islands and low islands with reefs

Elevation extremes:
lowest point: Pacific Ocean 0 m
highest point: Mont Orohena 2,241 m

Natural resources: timber, fish, cobalt, hydropower

Land use:
arable land: 0.75%
permanent crops: 5.5%
other: 93.75% (2005)

Irrigated land: 10 sq km (2003)

Natural hazards: occasional cyclonic storms in January

Environment—current issues: NA

Geography—note: includes five archipelagoes (four volcanic, one coral); Makatea in French Polynesia is one of the three great phosphate rock islands in the Pacific Ocean—the others are Banaba (Ocean Island) in Kiribati and Nauru

PEOPLE

Population: 283,019 (July 2008 est.)

Age structure:
0–14 years: 24.8% (male 35,903/female 34,364)
15–64 years: 68.6% (male 100,700/female 93,492)
65 years and over: 6.6% (male 9,374/female 9,186) (2008 est.)

Median age:
total: 28.7 years
male: 29 years
female: 28.4 years (2008 est.)

Population growth rate: 1.425% (2008 est.)

Birth rate: 16.16 births/1,000 population (2008 est.)

Death rate: 4.67 deaths/1,000 population (2008 est.)

Net migration rate: 2.77 migrant(s)/1,000 population (2008 est.)

Sex ratio:
at birth: 1.05 male(s)/female
under 15 years: 1.04 male(s)/female
15–64 years: 1.08 male(s)/female
65 years and over: 1.02 male(s)/female
total population: 1.07 male(s)/female (2008 est.)

Infant mortality rate:
total: 7.7 deaths/1,000 live births
male: 8.84 deaths/1,000 live births
female: 6.5 deaths/1,000 live births (2008 est.)

Life expectancy at birth:
total population: 76.51 years
male: 74.07 years
female: 79.08 years (2008 est.)

Total fertility rate: 1.95 children born/woman (2008 est.)

HIV/AIDS—adult prevalence rate: NA

HIV/AIDS—people living with HIV/AIDS: NA

HIV/AIDS—deaths: NA

Nationality:
noun: French Polynesian(s)
adjective: French Polynesian

Ethnic groups: Polynesian 78%, Chinese 12%, local French 6%, metropolitan French 4%

Religions: Protestant 54%, Roman Catholic 30%, other 10%, no religion 6%

Languages: French 61.1% (official), Polynesian 31.4% (official), Asian languages 1.2%, other 0.3%, unspecified 6% (2002 census)

Literacy: *definition:* age 14 and over can read and write
total population: 98%
male: 98%
female: 98% (1977 est.)

GOVERNMENT

Country name:
conventional long form: Overseas Lands of French Polynesia
conventional short form: French Polynesia
local long form: Pays d'outre-mer de la Polynesie Francaise
local short form: Polynesie Francaise
former: French Colony of Oceania

Dependency status: overseas lands of France; overseas territory of France from 1946–2004

Government type: NA

Capital: *name:* Papeete
geographic coordinates: 17 32 S, 149 34 W
time difference: UTC-10 (5 hours behind Washington, DC during Standard Time)

Administrative divisions: none (overseas lands of France); there are no first-order administrative divisions as defined by the US Government, but there are five archipelagic divisions named Archipel des Marquises, Archipel des Tuamotu, Archipel des Tubuai, Iles du Vent, Iles Sous-le-Vent

Independence: none (overseas lands of France)

National holiday: Bastille Day, 14 July (1789)

Constitution: 4 October 1958 (French Constitution)

Legal system: the laws of France, where applicable, apply

Suffrage: 18 years of age; universal

Executive branch:
chief of state: President Nicolas SARKOZY (since 16 May 2007), represented by High Commissioner of the Republic Anne BOQUET (since September 2005)
head of government: President of French Polynesia Gaston TONG SANG (since 15 April 2008); President of the Territorial Assembly Antony GEROS (since 9 May 2004)

cabinet: Council of Ministers; president submits a list of members of the Territorial Assembly for approval by them to serve as ministers
elections: French president elected by popular vote for a five-year term; high commissioner appointed by the French president on the advice of the French Ministry of Interior; president of the territorial government and the president of the Territorial Assembly are elected by the members of the assembly for five-year terms (no term limits)

Legislative branch: unicameral Territorial Assembly or Assemblee Territoriale (57 seats; members are elected by popular vote to serve five-year terms)
elections: last held 27 January 2008 (first round) and 10 February 2008 (second round) (next to be held NA 2013)
election results: percent of vote by party—Our Home alliance 45.2%, Union for Democracy alliance 37.2%, Popular Rally (Tahoeraa Huiraatira) 17.2% other 0.5%; seats by party—Our Home alliance 27, Union for Democracy alliance 20, Popular Rally 10
note: one seat was elected to the French Senate on 27 September 1998 (next to be held in September 2007); results—percent of vote by party—NA; seats by party—NA; two seats were elected to the French National Assembly on 9 June-16 June 2002 (next to be held in 2007); results—percent of vote by party—NA; seats by party—UMP/RPR 1, UMP 1

Judicial branch: Court of Appeal or Cour d'Appel; Court of the First Instance or Tribunal de Premiere Instance; Court of Administrative Law or Tribunal Administratif

Political parties and leaders: Alliance for a New Democracy or ADN [Nicole BOUTEAU and Philip SCHYLE](includes the parties The New Star and This Country is Yours); Independent Front for the Liberation of Polynesia (Tavini Huiraatira) [Oscar TEMARU]; New Fatherland Party (Ai'a Api) [Emile VERNAUDON]; Our Home alliance; People's Rally for the Republic of Polynesia or RPR (Tahoeraa Huiraatira) [Gaston FLOSSE]; Union for Democracy alliance or UPD [Oscar TEMARU]

Political pressure groups and leaders: NA

International organization participation: FZ, ITUC, PIF (associate member), SPC, UPU, WMO

Diplomatic representation in the US: none (overseas lands of France)

Diplomatic representation from the US: none (overseas lands of France)

Flag description: two narrow red horizontal bands encase a wide white band; centered on the white band is a disk with a blue and white wave pattern on the lower half and a gold and white ray pattern on the upper half; a stylized red, blue, and white ship rides on the wave pattern; the French flag is used for official occasions

Government—note: under certain acts of France, French Polynesia has acquired autonomy in all areas except those relating to police and justice, monetary policy, tertiary education, immigration, and defense and foreign affairs; the duties of its president are fashioned after those of the French prime minister

ECONOMY

Economy—overview: Since 1962, when France stationed military personnel in the region, French Polynesia has changed from a subsistence agricultural economy to one in which a high proportion of the work force is either employed by the military or supports the tourist industry. With the halt of French nuclear testing in 1996, the military contribution to the economy fell sharply. Tourism accounts for about one-fourth of GDP and is a primary source of hard currency earnings. Other sources of income are pearl farming and deep-sea commercial fishing. The small manufacturing sector primarily processes agricultural products. The territory benefits substantially from development agreements with France aimed principally at creating new businesses and strengthening social services.

GDP (purchasing power parity): $4.58 billion (2003 est.)

GDP (official exchange rate): $3.8 billion (2002)

GDP—real growth rate: 5.1% (2002)

GDP—per capita (PPP): $17,500 (2003 est.)

GDP—composition by sector:
agriculture: 3.1%
industry: 19%
services: 77.8% (2005)

Labor force: 65,930 (December 2005)

Labor force—by occupation: agriculture: 13%
industry: 19%
services: 68% (2002)

Unemployment rate: 11.7% (2005)

Population below poverty line: NA%

Household income or consumption by percentage share:
lowest 10%: NA%
highest 10%: NA%

Inflation rate (consumer prices): 1.1% (2006 est.)

Budget:
revenues: $865 million

expenditures: $644.1 million (1999)

Agriculture—products: fish; coconuts, vanilla, vegetables, fruits, coffee; poultry, beef, dairy products

Industries: tourism, pearls, agricultural processing, handicrafts, phosphates

Industrial production growth rate: NA%

Electricity—production: 462 million kWh (2005)

Electricity—production by source:
fossil fuel: 60.7%
hydro: 39.3%
nuclear: 0%
other: 0% (2001)

Electricity—consumption: 429.7 million kWh (2005)

Electricity—exports: 0 kWh (2005)

Electricity—imports: 0 kWh (2005)

Oil—production: 0 bbl/day (2005 est.)

Oil—consumption: 5,800 bbl/day (2005 est.)

Oil—exports: 0 bbl/day (2004)

Oil—imports: 5,678 bbl/day (2004)

Oil—proved reserves: 0 bbl (1 January 2006 est.)

Natural gas—production: 0 cu m (2005 est.)

Natural gas—consumption: 0 cu m (2005 est.)

Natural gas—exports: 0 cu m (2005 est.)

Natural gas—imports: 0 cu m (2005 est.)

Natural gas—proved reserves: 0 cu m (1 January 2006 est.)

Exports: $211 million f.o.b. (2005 est.)

Exports—commodities: cultured pearls, coconut products, mother-of-pearl, vanilla, shark meat

Exports—partners: France 46.3%, Japan 20.8%, Niger 12.8%, US 12.5% (2006)

Imports: $1.706 billion f.o.b. (2005 est.)

Imports—commodities: fuels, foodstuffs, machinery and equipment

Imports—partners: France 52.7%, Singapore 14.9%, NZ 6.8%, US 6.6% (2006)

Economic aid—recipient: $579.8 million (2004)

Debt—external: $NA

Market value of publicly traded shares: $NA

Currency (code): Comptoirs Francais du Pacifique franc (XPF)

Currency code: XPF

Exchange rates: Comptoirs Francais du Pacifique francs (XPF) per US dollar—87.59 (2007), 94.97 (2006), 95.89 (2005), 96.04 (2004), 105.66 (2003)
note: pegged at the rate of 119.25 XPF to the euro

Fiscal year: calendar year

COMMUNICATIONS

Telephones—main lines in use: 53,600 (2006)

Telephones—mobile cellular: 152,000 (2006)

Telephone system:
general assessment: NA
domestic: NA
international: country code—689; satellite earth station—1 Intelsat (Pacific Ocean)

Radio broadcast stations: AM 2, FM 14, shortwave 2 (1998)

Radios: 128,000 (1997)

Television broadcast stations: 7 (plus 17 repeaters) (1997)

Televisions: 40,000 (1997)

Internet country code: .pf

Internet hosts: 14,059 (2007)

Internet Service Providers (ISPs): 2 (2000)

Internet users: 65,000 (2006)

TRANSPORTATION

Airports: 54 (2007)

Airports—with paved runways:
total: 37
over 3,047 m: 2
1,524 to 2,437 m: 5
914 to 1,523 m: 27
under 914 m: 3 (2007)

Airports—with unpaved runways:
total: 17
914 to 1,523 m: 9
under 914 m: 8 (2007)

Heliports: 1 (2007)

Roadways:
total: 2,590 km
paved: 1,735 km
unpaved: 855 km (1999)

Merchant marine:
total: 13 ships (1000 GRT or over) 23,684 GRT/17,291 DWT
by type: cargo 4, passenger 2, passenger/cargo 5, refrigerated cargo 1, roll on/roll off 1
registered in other countries: 2 (Wallis and Futuna 2) (2007)

Ports and terminals: Papeete

MILITARY

Military branches: no regular military forces; Gendarmerie and National Police Force (2007)

Manpower available for military service:
males age 16–49: 79,540 (2008 est.)

Manpower fit for military service:
males age 16–49: 64,287 (2008 est.)

Manpower reaching militarily significant age annually:
males age 16–49: 2,699 (2008 est.)

Military—note: defense is the responsibility of France

TRANSNATIONAL ISSUES

Disputes—international: none

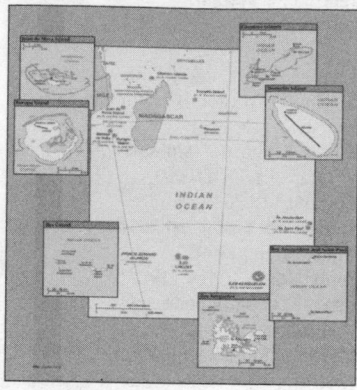

INTRODUCTION

Background: In February 2007, the Iles Eparses became an integral part of the French Southern and Antarctic Lands (TAAF). The Southern Lands are now divided into five administrative districts, two of which are archipelagos, Iles Crozet and Iles Kerguelen; the third is a district composed of two volcanic islands, Ile Saint-Paul and Ile Amsterdam; the fourth, Iles Eparses, consists of five scattered tropical islands around Madagascar. They contain no permanent inhabitants and are visited only by researchers studying the native fauna, scientists at the various scientific stations, fishermen, and military personnel. The fifth district is the Antarctic portion, which consists of "Adelie Land," a thin slice of the Antarctic continent discovered and claimed by the French in 1840.

Ile Amsterdam: Discovered but not named in 1522 by the Spanish, the island subsequently received the appellation of Nieuw Amsterdam from a Dutchman; it was claimed by France in 1843. A short-lived attempt at cattle farming began in 1871. A French meteorological station established on the island in 1949 is still in use.

Ile Saint Paul: Claimed by France since 1893, the island was a fishing industry center from 1843 to 1914. In 1928, a spiny lobster cannery was established, but when the company went bankrupt in 1931, seven workers were abandoned. Only two survived until 1934 when rescue finally arrived.

Iles Crozet: A large archipelago formed from the Crozet Plateau, Iles Crozet is divided into two main groups: L'Occidental (the West), which includes Ile aux Cochons, Ilots des Apotres, Ile des Pingouins, and the reefs Brisants de l'Heroine; and L'Oriental (the east), which includes Ile d'Est and Ile de la Possession (the largest island of the Crozets). Discovered and claimed by France in 1772, the islands were used for seal hunting and as a base for whaling. Originally administered as a dependency of Madagascar, they became part of the TAAF in 1955.

Iles Kerguelen: This island group, discovered in 1772, is made up of one large island (Ile Kerguelen) and about 300 smaller islands. A permanent group of 50 to 100 scientists resides at the main base at Port-aux-Francais.

Adelie Land: The only non-insular district of the TAAF is the Antarctic claim known as "Adelie Land." The US Government does not recognize it as a French dependency.

Bassas da India: A French possession since 1897, this atoll is a volcanic rock surrounded by reefs and is awash at high tide.

Europa Island: This heavily wooded island has been a French possession since 1897; it is the site of a small military garrison that staffs a weather station.

Glorioso Islands: A French possession since 1892, the Glorioso Islands are composed of two lushly vegetated coral islands (Ile Glorieuse and Ile du Lys) and three rock islets. A military garrison operates a weather and radio station on Ile Glorieuse.

Juan de Nova Island: Named after a famous 15th century Spanish navigator and explorer, the island has been a French possession since 1897. It has been exploited for its guano and phosphate. Presently a small military garrison oversees a meteorological station.

Tromelin Island: First explored by the French in 1776, the island came under the jurisdiction of Reunion in 1814. At present, it serves as a sea turtle sanctuary and is the site of an important meteorological station.

GEOGRAPHY

Location: southeast and east of Africa, islands in the southern Indian Ocean, some near Madagascar and others about equidistant between Africa, Antarctica, and Australia; note—French Southern and Antarctic Lands include Ile Amsterdam, Ile Saint-Paul, Iles Crozet, Iles Kerguelen, Bassas da India, Europa Island, Glorioso Islands, Juan de Nova Island, and Tromelin Island in the southern Indian Ocean, along with the French-claimed sector of Antarctica, "Adelie Land"; the US does not recognize the French claim to "Adelie Land"

Geographic coordinates: *Ile Amsterdam (Ile Amsterdam et Ile Saint-Paul):* 37 50 S, 77 32 E

Ile Saint-Paul (Ile Amsterdam et Ile Saint-Paul): 38 72 S, 77 53 E

Iles Crozet: 46 25 S, 51 00 E

Iles Kerguelen: 49 15 S, 69 35 E

Bassas da India (Iles Eparses): 21 30 S, 39 50 E

Europa Island (Iles Eparses): 22 20 S, 40 22 E

Glorioso Islands (Iles Eparses): 11 30 S, 47 20 E

Juan de Nova Island (Iles Eparses): 17 03 S, 42 45 E

Tromelin Island (Iles Eparses): 15 52 S, 54 25 E

Map references: Antarctic Region, Africa

Area:

Ile Amsterdam (Ile Amsterdam et Ile Saint-Paul): total—55 sq km; land—55 sq km; water—0 sq km

Ile Saint-Paul (Ile Amsterdam et Ile Saint-Paul): total—7 sq km; land—7 sq km; water—0 sq km

Iles Crozet: total—352 sq km; land—352 sq km; water—0 sq km

Iles Kerguelen: total—7,215 sq km; land—7,215 sq km; water—0 sq km

Bassas da India (Iles Eparses): total—80 sq km; land—0.2 sq km; water—79.8 sq km (lagoon)

Europa Island (Iles Eparses): total—28 sq km; land—28 sq km; water—0 sq km

Glorioso Islands (Iles Eparses): total—5 sq km; land—5 sq km; water—0 sq km

Juan de Nova Island (Iles Eparses): total—4.4 sq km; land—4.4 sq km; water—0 sq km

Tromelin Island (Iles Eparses): total—1 sq km; land—1 sq km; water—0 sq km

note: excludes "Adelie Land" claim of about 500,000 sq km in Antarctica that is not recognized by the US

Area—comparative: *Ile Amsterdam (Ile Amsterdam et Ile Saint-Paul):* less than one-half the size of Washington, DC

Ile Saint-Paul (Ile Amsterdam et Ile Saint-Paul): more than 10 times the size of The Mall in Washington, DC

Iles Crozet: about twice the size of Washington, DC

Iles Kerguelen: a little larger than Delaware

Bassas da India (Iles Eparses): land area about one-third the size of The Mall in Washington, DC

Europa Island (Iles Eparses): about one-sixth the size of Washington, DC

Glorioso Islands (Iles Eparses): about eight times the size of The Mall in Washington, DC

Juan de Nova Island (Iles Eparses): about seven times the size of The Mall in Washington, DC

Tromelin Island (Iles Eparses): about 1.7 times the size of The Mall in Washington, DC

Land boundaries: 0 km

Coastline: *Ile Amsterdam (Ile Amsterdam et Ile Saint-Paul)*: 28 km

Ile Saint-Paul (Ile Amsterdam et Ile Saint-Paul):

Iles Kerguelen: 2,800 km

Bassas da India (Iles Eparses): 35.2 km

Europa Island (Iles Eparses): 22.2 km

Glorioso Islands (Iles Eparses): 35.2 km

Juan de Nova Island (Iles Eparses): 24.1 km

Tromelin Island (Iles Eparses): 3.7 km

Maritime claims:

territorial sea: 12 nm

exclusive economic zone: 200 nm from Iles Kerguelen and Iles Eparses (does not include the rest of French Southern and Antarctic Lands); Juan de Nova Island and Tromelin Island claim a continental shelf of 200-m depth or to the depth of exploitation

Climate: *Ile Amsterdam et Ile Saint-Paul*: oceanic with persistent westerly winds and high humidity

Iles Crozet: windy, cold, wet, and cloudy

Iles Kerguelen: oceanic, cold, overcast, windy

Iles Eparses: tropical

Terrain: *Ile Amsterdam (Ile Amsterdam et Ile Saint-Paul)*: a volcanic island with steep coastal cliffs; the center floor of the volcano is a large plateau

Ile Saint-Paul (Ile Amsterdam et Ile Saint-Paul): triangular in shape, the island is the top of a volcano, rocky with steep cliffs on the eastern side; has active thermal springs

Iles Crozet: a large archipelago formed from the Crozet Plateau is divided into two groups of islands

Iles Kerguelen: the interior of the large island of Ile Kerguelen is composed of rugged terrain of high mountains, hills, valleys, and plains with a number of peninsulas stretching off its coasts

Bassas da India (Iles Eparses): atoll, awash at high tide; shallow (15 m) lagoon

Europa Island, Glorioso Islands, Juan de Nova Island: low, flat, and sandy

Tromelin Island (Iles Eparses): low, flat, sandy; likely volcanic seamount

Elevation extremes:

lowest point: Indian Ocean 0 m

highest point: Mont de la Dives on Ile Amsterdam (Ile Amsterdam et Ile Saint-Paul) 867 m; unnamed location on Ile

Saint-Paul (Ile Amsterdam et Ile Saint-Paul) 272 m; Pic Marion-Dufresne in Iles Crozet 1,090 m; Mont Ross in Iles Kerguelen 1,850 m; unnamed location on Bassas de India (Iles Eparses) 2.4 m; unnamed location on Europa Island (Iles Eparses) 24 m; unnamed location on Glorioso Islands (Iles Eparses) 12 m; unnamed location on Juan de Nova Island (Iles Eparses) 10 m; unnamed location on Tromelin Island (Iles Eparses) 7 m

Natural resources: fish, crayfish

note: Glorioso Islands and Tromelin Island (Iles Eparses) have guano, phosphates, and coconuts

Land use:

Ile Amsterdam (Ile Amsterdam et Ile Saint-Paul)—100% trees, grasses, ferns, and moss; Ile Saint-Paul (Ile Amsterdam et Ile Saint-Paul)—100% grass, ferns, and moss; Iles Crozet—100% tossock grass, heath, and fern; Iles Kerguelen—100% tossock grass and Kerguelen cabbage; Bassas da India (Iles Eparses)—100% rock, coral reef, and sand; Europa Island (Iles Eparses)—100% mangrove swamp and dry woodlands; Glorioso Islands (Iles Eparses)—100% lush vegetation and coconut palms; Juan de Nova Island (Iles Eparses)—90% forest, 10% other; Tromelin Island (Iles Eparses)—100% grasses and scattered brush (2005)

Irrigated land: 0 sq km

Natural hazards: Ile Amsterdam and Ile Saint-Paul are inactive volcanoes; Iles Eparses subject to periodic cyclones; Bassas da India is a maritime hazard since it is under water for a period of three hours prior to and following the high tide and surrounded by reefs

Environment—current issues: introduction of foreign species on Iles Crozet has caused severe damage to the original ecosystem; overfishing of Patagonian Toothfish around Iles Crozet and Iles Kerguelen

Geography—note: islands component is widely scattered across remote locations in the southern Indian Ocean

Bassas da India (Iles Eparses): the atoll is a circular reef that sits atop a long-extinct, submerged volcano

Europa Island and Juan de Nova Island (Iles Eparses): wildlife sanctuary for seabirds and sea turtles

Glorioso Island (Iles Eparses): the islands and rocks are surrounded by an extensive reef system

Tromelin Island (Iles Eparses): climatologically important location for forecasting cyclones in the western Indian Ocean; wildlife sanctuary (seabirds, tortoises)

PEOPLE

Population: no indigenous inhabitants

Ile Amsterdam (Ile Amsterdam et Ile Saint-Paul): has no permanent residents but has a meteorological station

Ile Saint-Paul (Ile Amsterdam et Ile Saint-Paul): is uninhabited but is frequently visited by fishermen and has a scientific research cabin for short stays

Iles Crozet: are uninhabited except for 18 to 30 people staffing the Alfred Faure research station on Ile del la Possession

Iles Kerguelen: 50 to 100 scientists are located at the main base at Port-aux-Francais on Ile Kerguelen

Bassas da India (Iles Eparses): uninhabitable

Europa Island, Glorioso Islands, Juan de Nova Island (Iles Eparses): a small French military garrison and a few meteorologists on each possession; visited by scientists

Tromelin Island (Iles Eparses): uninhabited, except for visits by scientists

GOVERNMENT

Country name:

conventional long form: Territory of the French Southern and Antarctic Lands

conventional short form: French Southern and Antarctic Lands

local long form: Territoire des Terres Australes et Antarctiques Francaises

local short form: Terres Australes et Antarctiques Francaises

abbreviation: TAAF

Dependency status: overseas territory of France since 1955; administered from Paris by Administrateur Superieur Eric PILLOTON (since 10 April 2007)

Administrative divisions: none (overseas territory of France); there are no first-order administrative divisions as defined by the US Government, but there are five administrative districts named Iles Crozet, Iles Eparses, Iles Kerguelen, Ile Saint-Paul et Ile Amsterdam; the fifth district is the "Adelie Land" claim in Antarctica that is not recognized by the US

Legal system: the laws of France, where applicable, apply

Executive branch:

chief of state: President Nicolas SARKOZY (since 16 May 2007), represented by Senior Administrator Eric PILLOTON (10 April 2007)

International organization participation: UPU

Diplomatic representation in the US: none (overseas territory of France)

Diplomatic representation from the US: none (overseas territory of France)

Flag description: the flag of France is used

ECONOMY

Economy—overview: Economic activity is limited to servicing meteorological and geophysical research stations, military bases, and French and other fishing fleets. The fish catches landed on Iles Kerguelen by foreign ships are exported to France and Reunion.

COMMUNICATIONS

Internet country code: .tf
Internet hosts: 33 (2007)
Communications—note: one or more meteorological stations on each possession; note—meteorological station on Tromelin Island (Iles Eparses) is important for forecasting cyclones

TRANSPORTATION

Airports: 4 (one each on Europa Island, Glorioso Islands, Juan de Nova Island, and Tromelin Island in the Iles Eparses district) (2006)
Ports and terminals: none; offshore anchorage only
Transportation—note: *aids to navigation—lighthouses:* Europa Island 18m; Juan de Nova Island (W side) 37m; Tromelin Island (NW point) 11m (all in the Iles Eparses district)

MILITARY

Military—note: defense is the responsibility of France

TRANSNATIONAL ISSUES

Disputes—international: French claim to "Adelie Land" in Antarctica is not recognized by the US
Bassas da India, Europa Island, Glorioso Islands, Juan de Nova Island (Iles Eparses): claimed by Madagascar
Tromelin Island (Iles Eparses): claimed by Mauritius

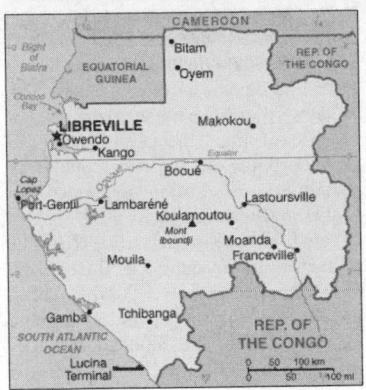

INTRODUCTION

Background: Only two autocratic presidents have ruled Gabon since independence from France in 1960. The current president of Gabon, El Hadj Omar BONGO Ondimba—one of the longest-serving heads of state in the world—has dominated the country's political scene for four decades. President BONGO introduced a nominal multiparty system and a new constitution in the early 1990s. However, allegations of electoral fraud during local elections in 2002–03 and the presidential elections in 2005 have exposed the weaknesses of formal political structures in Gabon. Gabon's political opposition remains weak, divided, and financially dependent on the current regime. Despite political conditions, a small population, abundant natural resources, and considerable foreign support have helped make Gabon one of the more prosperous and stable African countries.

GEOGRAPHY

Location: Western Africa, bordering the Atlantic Ocean at the Equator, between Republic of the Congo and Equatorial Guinea

Geographic coordinates: 1 00 S, 11 45 E

Map references: Africa

Area:
total: 267,667 sq km
land: 257,667 sq km
water: 10,000 sq km

Area—comparative: slightly smaller than Colorado

Land boundaries:
total: 2,551 km
border countries: Cameroon 298 km, Republic of the Congo 1,903 km, Equatorial Guinea 350 km

Coastline: 885 km

Maritime claims:
territorial sea: 12 nm

contiguous zone: 24 nm

exclusive economic zone: 200 nm

Climate: tropical; always hot, humid

Terrain: narrow coastal plain; hilly interior; savanna in east and south

Elevation extremes:
lowest point: Atlantic Ocean 0 m
highest point: Mont Iboundji 1,575 m

Natural resources: petroleum, natural gas, diamond, niobium, manganese, uranium, gold, timber, iron ore, hydropower

Land use:
arable land: 1.21%
permanent crops: 0.64%
other: 98.15% (2005)

Irrigated land: 70 sq km (2003)

Total renewable water resources: 164 cu km (1987)

Freshwater withdrawal (domestic/industrial/agricultural): *total:* 0.12 cu km/yr (50%/8%/42%)
per capita: 87 cu m/yr (2000)

Natural hazards: NA

Environment—current issues: deforestation; poaching

Environment—international agreements: *party to:* Biodiversity, Climate Change, Climate Change-Kyoto Protocol, Desertification, Endangered Species, Law of the Sea, Marine Dumping, Ozone Layer Protection, Ship Pollution, Tropical Timber 83, Tropical Timber 94, Wetlands, Whaling
signed, but not ratified: none of the selected agreements

Geography—note: a small population and oil and mineral reserves have helped Gabon become one of Africa's wealthier countries; in general, these circumstances have allowed the country to maintain and conserve its pristine rain forest and rich biodiversity

PEOPLE

Population: 1,485,832

note: estimates for this country explicitly take into account the effects of excess mortality due to AIDS; this can result in lower life expectancy, higher infant mortality, higher death rates, lower population growth rates, and changes in the distribution of population by age and sex than would otherwise be expected (July 2008 est.)

Age structure:
0–14 years: 42.1% (male 314,078/female 311,900)
15–64 years: 53.9% (male 399,586/female 401,602)
65 years and over: 3.9% (male 24,388/female 34,278) (2008 est.)

Median age:
total: 18.6 years

male: 18.4 years
female: 18.9 years (2008 est.)

Population growth rate: 1.954% (2008 est.)

Birth rate: 35.75 births/1,000 population (2008 est.)

Death rate: 12.59 deaths/1,000 population (2008 est.)

Net migration rate: -3.62 migrant(s)/1,000 population (2008 est.)

Sex ratio:
at birth: 1.03 male(s)/female
under 15 years: 1.01 male(s)/female
15–64 years: 0.99 male(s)/female
65 years and over: 0.71 male(s)/female
total population: 0.99 male(s)/female (2008 est.)

Infant mortality rate:
total: 52.65 deaths/1,000 live births
male: 61.27 deaths/1,000 live births
female: 43.77 deaths/1,000 live births (2008 est.)

Life expectancy at birth:
total population: 53.52 years
male: 52.5 years
female: 54.57 years (2008 est.)

Total fertility rate: 4.68 children born/woman (2008 est.)

HIV/AIDS—adult prevalence rate: 8.1% (2003 est.)

HIV/AIDS—people living with HIV/AIDS: 48,000 (2003 est.)

HIV/AIDS—deaths: 3,000 (2003 est.)

Major infectious diseases:
degree of risk: very high
food or waterborne diseases: bacterial diarrhea, hepatitis A, and typhoid fever
vectorborne disease: malaria (2008)

Nationality:
noun: Gabonese (singular and plural)
adjective: Gabonese

Ethnic groups: Bantu tribes, including four major tribal groupings (Fang, Bapounou, Nzebi, Obamba); other Africans and Europeans, 154,000, including 10,700 French and 11,000 persons of dual nationality

Religions: Christian 55%-75%, animist, Muslim less than 1%

Languages: French (official), Fang, Myene, Nzebi, Bapounou/Eschira, Bandjabi

Literacy:
definition: age 15 and over can read and write
total population: 63.2%
male: 73.7%
female: 53.3% (1995 est.)

GOVERNMENT

Country name:
conventional long form: Gabonese Republic

conventional short form: Gabon
local long form: Republique gabonaise
local short form: Gabon
Government type: republic; multiparty presidential regime
Capital: *name:* Libreville
geographic coordinates: 0 23 N, 9 27 E
time difference: UTC+1 (6 hours ahead of Washington, DC during Standard Time)
Administrative divisions: 9 provinces; Estuaire, Haut-Ogooue, Moyen-Ogooue, Ngounie, Nyanga, Ogooue-Ivindo, Ogooue-Lolo, Ogooue-Maritime, Woleu-Ntem
Independence: 17 August 1960 (from France)
National holiday: Independence Day, 17 August (1960)
Constitution: adopted 14 March 1991
Legal system: based on French civil law system and customary law; judicial review of legislative acts in Constitutional Chamber of the Supreme Court; has not accepted compulsory ICJ jurisdiction
Suffrage: 21 years of age; universal
Executive branch:
chief of state: President El Hadj Omar BONGO Ondimba (since 2 December 1967)
head of government: Prime Minister Jean Eyeghe NDONG (since 20 January 2006)
cabinet: Council of Ministers appointed by the prime minister in consultation with the president
elections: president elected by popular vote for a seven-year term (no term limits); election last held 27 November 2005 (next to be held in 2012); prime minister appointed by the president
election results: President El Hadj Omar BONGO Ondimba reelected; percent of vote—El Hadj Omar BONGO Ondimba 79.2%, Pierre MAMBOUNDOU 13.6%, Zacharie MYBOTO 6.6%
Legislative branch: bicameral legislature consists of the Senate (91 seats; members elected by members of municipal councils and departmental assemblies to serve six-year terms) and the National Assembly or Assemblee Nationale (120 seats; members are elected by direct, popular vote to serve five-year terms)
elections: Senate—last held 26 January and 9 February 2003 (next to be held by January 2009); National Assembly—last held 17 and 24 December 2006 (next to be held in December 2011)
election results: Senate—percent of vote by party—NA; seats by party—PDG 53, RNB 20, PGP 4, ADERE 3, RDP 1, CLR 1, independents 9; National Assembly—percent of vote by party—NA; seats by party—PDG 82, RPG 8, UPG 8, UGDD

4, ADERE 3, CLR 2, PGP-Ndaot 2, PSD 2, independents 4, others 5
Judicial branch: Supreme Court or Cour Supreme consisting of three chambers—Judicial, Administrative, and Accounts; Constitutional Court; Courts of Appeal; Court of State Security; County Courts
Political parties and leaders: Circle of Liberal Reformers or CLR [General Jean Boniface ASSELE]; Congress for Democracy and Justice or CDJ [Jules Aristide Bourdes OGOULIGUENDE]; Democratic and Republican Alliance or ADERE [Divungui-di-Ndinge DIDJOB]; Gabonese Democratic Party or PDG (former sole party) [Simplice Nguedet MANZELA]; Gabonese Party for Progress or PGP [Benoit Mouity NZAMBA]; Gabonese Union for Democracy and Development or UGDD [Zacherie MYBOTO]; National Rally of Woodcutters or RNB; National Rally of Woodcutters-Rally for Gabon or RNB-RPG (Bucherons) [Fr. Paul M'BA-ABESSOLE]; Party of Development and Social Solidarity or PDS [Seraphin Ndoat REMBOGO]; People's Unity Party or PUP [Louis Gaston MAYILA]; Social Democratic Party or PSD [Pierre Claver MAGANGA-MOUSSAVOU]; Union for Democracy and Social Integration or UDIS; Union of Gabonese Patriots or UPG [Pierre MAMBOUNDOU]
Political pressure groups and leaders: NA
International organization participation: ACCT, ACP, AfDB, AU, BDEAC, CEMAC, FAO, FZ, G-24, G-77, IAEA, IBRD, ICAO, ICCt, ICRM, IDA, IDB, IFAD, IFC, IFRCS, ILO, IMF, IMO, IMSO, Interpol, IOC, IOM, IPU, ISO (correspondent), ITSO, ITU, ITUC, MIGA, NAM, OIC, OIF, OPCW, UN, UNCTAD, UNESCO, UNIDO, UNMIS, UNWTO, UPU, WCL, WCO, WHO, WIPO, WMO, WTO
Diplomatic representation in the US:
chief of mission: Ambassador Carlos BOUNGOU
chancery: Suite 200, 2034 20th Street NW, Washington, DC 20009
telephone: [1] (202) 797-1000
FAX: [1] (202) 332-0668
consulate(s): New York
Diplomatic representation from the US:
chief of mission: Ambassador Eunice S. REDDICK
embassy: Boulevard du Bord de Mer, Libreville
mailing address: Centre Ville, B. P. 4000, Libreville
telephone: [241] 76 20 03 through 76 20 04, after hours—74 34 92
FAX: [241] 74 55 07

Flag description: three equal horizontal bands of green (top), yellow, and blue

ECONOMY

Economy—overview: Gabon enjoys a per capita income four times that of most of sub-Saharan African nations. but because of high income inequality, a large proportion of the population remains poor. Gabon depended on timber and manganese until oil was discovered offshore in the early 1970s. The oil sector now accounts for 50% of GDP. Gabon continues to face fluctuating prices for its oil, timber, and manganese exports. Despite the abundance of natural wealth, poor fiscal management hobbles the economy. The devaluation of the CFA franc—its currency—by 50% in January 1994 sparked a one-time inflationary surge, to 35%; the rate dropped to 6% in 1996. The IMF provided a one-year standby arrangement in 1994–95, a three-year Enhanced Financing Facility (EFF) at near commercial rates beginning in late 1995, and stand-by credit of $119 million in October 2000. Those agreements mandated progress in privatization and fiscal discipline. France provided additional financial support in January 1997 after Gabon met IMF targets for mid-1996. In 1997, an IMF mission to Gabon criticized the government for overspending on off-budget items, overborrowing from the central bank, and slipping on its schedule for privatization and administrative reform. The rebound of oil prices since 1999 have helped growth, but drops in production have hampered Gabon from fully realizing potential gains, and will continue to temper the gains for most of this decade. In December 2000, Gabon signed a new agreement with the Paris Club to reschedule its official debt. A follow-up bilateral repayment agreement with the US was signed in December 2001. Gabon signed a 14-month Stand-By Arrangement with the IMF in May 2004, and received Paris Club debt rescheduling later that year. Short-term progress depends on an upbeat world economy and fiscal and other adjustments in line with IMF policies.
GDP (purchasing power parity): $20.18 billion (2007 est.)
GDP (official exchange rate): $11.3 billion (2007 est.)
GDP—real growth rate: 5.6% (2007 est.)
GDP—per capita (PPP): $14,100 (2007 est.)
GDP—composition by sector:
agriculture: 5.8%
industry: 58.2%

services: 36% (2007 est.)
Labor force: 582,000 (2007 est.)
Labor force—by occupation: *agriculture:* 60%
industry: 15%
services: 25%
Unemployment rate: 21% (2006 est.)
Population below poverty line: NA%
Household income or consumption by percentage share:
lowest 10%: NA%
highest 10%: NA%
Inflation rate (consumer prices): 5% (2007 est.)
Investment (gross fixed): 24.2% of GDP (2007 est.)
Budget:
revenues: $3.534 billion
expenditures: $2.347 billion (2007 est.)
Public debt: 50% of GDP (2007 est.)
Agriculture—products: cocoa, coffee, sugar, palm oil, rubber; cattle; okoume (a tropical softwood); fish
Industries: petroleum extraction and refining; manganese, gold; chemicals, ship repair, food and beverages, textiles, lumbering and plywood, cement
Industrial production growth rate: 5.2% (2007 est.)
Electricity—production: 1.52 billion kWh (2005)
Electricity—production by source:
fossil fuel: 34.5%
hydro: 65.5%
nuclear: 0%
other: 0% (2001)
Electricity—consumption: 1.241 billion kWh (2005)
Electricity—exports: 0 kWh (2005)
Electricity—imports: 0 kWh (2005)
Oil—production: 266,000 bbl/day (2005 est.)
Oil—consumption: 13,000 bbl/day (2005 est.)
Oil—exports: 228,000 bbl/day (2004)
Oil—imports: 2,436 bbl/day (2004)
Oil—proved reserves: 1.748 billion bbl (2007 est.)
Natural gas—production: 95.91 million cu m (2005 est.)
Natural gas—consumption: 95.91 million cu m (2005 est.)
Natural gas—exports: 0 cu m (2005 est.)
Natural gas—imports: 0 cu m (2005)
Natural gas—proved reserves: 32.59 billion cu m (1 January 2006 est.)
Current account balance: $1.45 billion (2007 est.)
Exports: $6.952 billion f.o.b. (2007 est.)
Exports—commodities: crude oil 77%, timber, manganese, uranium (2001)
Exports—partners: US 27.6%, China 15.9%, France 7.8%, Trinidad and Tobago 5.4%, Thailand 4.3% (2006)

Imports: $2.107 billion f.o.b. (2007 est.)
Imports—commodities: machinery and equipment, foodstuffs, chemicals, construction materials
Imports—partners: France 35.4%, US 7.6%, Netherlands 5.5%, Cameroon 4.5%, Belgium 4.3% (2006)
Economic aid—recipient: $53.87 million (2005)
Reserves of foreign exchange and gold: $1.236 billion (31 December 2007 est.)
Debt—external: $4.895 billion (31 December 2007 est.)
Market value of publicly traded shares: $NA
Currency (code): Communaute Financiere Africaine franc (XAF); note—responsible authority is the Bank of the Central African States
Currency code: XAF
Exchange rates: Communaute Financiere Africaine francs (XAF) per US dollar—481.83 (2007), 522.89 (2006), 527.47 (2005), 528.29 (2004), 581.2 (2003)
Fiscal year: calendar year

COMMUNICATIONS

Telephones—main lines in use: 36,500 (2006)
Telephones—mobile cellular: 764,700 (2006)
Telephone system:
general assessment: adequate service by African standards and improving with the help of a growing mobile cell network system with three providers; mobile-cellular subscribership exceeded 50 per 100 persons in 2006
domestic: adequate system of cable, microwave radio relay, tropospheric scatter, radiotelephone communication stations, and a domestic satellite system with 12 earth stations
international: country code—241; landing point for the SAT-3/WASC fiber-optic submarine cable that provides connectivity to Europe and Asia; satellite earth stations—3 Intelsat (Atlantic Ocean)
Radio broadcast stations: AM 6, FM 7 (plus 11 repeaters), shortwave 4 (2001)
Radios: 208,000 (1997)
Television broadcast stations: 4 (plus 4 repeaters) (2001)
Televisions: 63,000 (1997)
Internet country code: .ga
Internet hosts: 288 (2007)
Internet Service Providers (ISPs): 1 (2001)
Internet users: 81,000 (2006)

TRANSPORTATION

Airports: 53 (2007)

Airports—with paved runways:
total: 10
over 3,047 m: 1
2,438 to 3,047 m: 1
1,524 to 2,437 m: 7
914 to 1,523 m: 1 (2007)
Airports—with unpaved runways:
total: 43
1,524 to 2,437 m: 7
914 to 1,523 m: 13
under 914 m: 23 (2007)
Pipelines: gas 384 km; oil 1,427 km (2007)
Railways:
total: 814 km
standard gauge: 814 km 1.435-m gauge (2006)
Roadways:
total: 9,170 km
paved: 937 km
unpaved: 8,233 km (2004)
Waterways: 1,600 km (310 km on Ogooue River) (2007)
Merchant marine:
registered in other countries: 2 (Cambodia 1, Panama 1) (2007)
Ports and terminals: Gamba, Libreville, Lucinda, Port-Gentil

MILITARY

Military branches: Army, Navy, Air Force, National Gendarmerie, National Police
Military service age and obligation: 20 years of age for compulsory and voluntary military service (2007)
Manpower available for military service:
males age 16–49: 331,181
females age 16–49: 332,498 (2008 est.)
Manpower fit for military service:
males age 16–49: 192,717
females age 16–49: 188,539 (2008 est.)
Manpower reaching militarily significant age annually:
males age 16–49: 16,558
females age 16–49: 16,577 (2008 est.)
Military expenditures—percent of GDP: 3.4% (2005 est.)

TRANSNATIONAL ISSUES

Disputes—international: UN urges Equatorial Guinea and Gabon to resolve the sovereignty dispute over Gabon-occupied Mbane Island and lesser islands and to establish a maritime boundary in hydrocarbon-rich Corisco Bay
Refugees and internally displaced persons: *refugees (country of origin):* 7,178 (Republic of Congo) (2007)

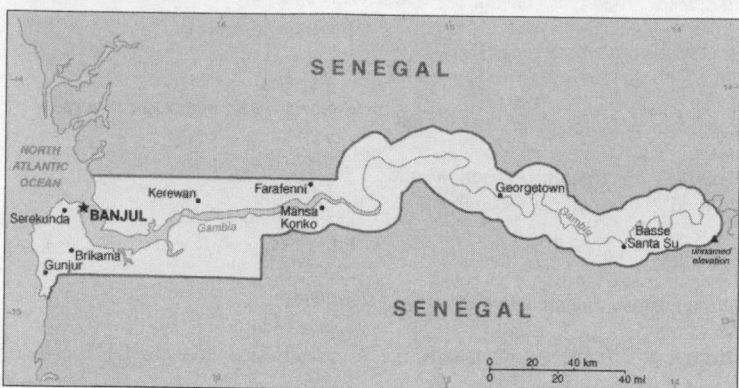

INTRODUCTION

Background: The Gambia gained its independence from the UK in 1965. Geographically surrounded by Senegal, it formed a short-lived federation of Senegambia between 1982 and 1989. In 1991 the two nations signed a friendship and cooperation treaty, but tensions have flared up intermittently since then. Yahya A. J. J. JAMMEH led a military coup in 1994 that overthrew the president and banned political activity. A new constitution and presidential elections in 1996, followed by parliamentary balloting in 1997, completed a nominal return to civilian rule. JAMMEH has been elected president in all subsequent elections, including most recently in late 2006.

GEOGRAPHY

Location: Western Africa, bordering the North Atlantic Ocean and Senegal
Geographic coordinates: 13 28 N, 16 34 W
Map references: Africa
Area: *total:* 11,300 sq km
land: 10,000 sq km
water: 1,300 sq km
Area—comparative: slightly less than twice the size of Delaware
Land boundaries:
total: 740 km
border countries: Senegal 740 km
Coastline: 80 km
Maritime claims:
territorial sea: 12 nm
contiguous zone: 18 nm
exclusive fishing zone: 200 nm
continental shelf: extent not specified
Climate: tropical; hot, rainy season (June to November); cooler, dry season (November to May)
Terrain: flood plain of the Gambia River flanked by some low hills

Elevation extremes:
lowest point: Atlantic Ocean 0 m
highest point: unnamed location 53 m
Natural resources: fish, titanium (rutile and ilmenite), tin, zircon, silica sand, clay, petroleum
Land use:
arable land: 27.88%
permanent crops: 0.44%
other: 71.68% (2005)
Irrigated land: 20 sq km (2003)
Total renewable water resources: 8 cu km (1982)
Freshwater withdrawal (domestic/industrial/agricultural): *total:* 0.03 cu km/yr (23%/12%/65%)
per capita: 20 cu m/yr (2000)
Natural hazards: drought (rainfall has dropped by 30% in the last 30 years)
Environment—current issues: deforestation; desertification; water-borne diseases prevalent
Environment—international agreements: *party to:* Biodiversity, Climate Change, Climate Change-Kyoto Protocol, Desertification, Endangered Species, Hazardous Wastes, Law of the Sea, Ozone Layer Protection, Ship Pollution, Wetlands, Whaling
signed, but not ratified: none of the selected agreements
Geography—note: almost an enclave of Senegal; smallest country on the continent of Africa

PEOPLE

Population: 1,735,464 (July 2008 est.)
Age structure:
0–14 years: 43.9% (male 382,385/female 378,853)
15–64 years: 53.4% (male 459,315/female 466,689)
65 years and over: 2.8% (male 24,303/female 23,919) (2008 est.)
Median age:
total: 17.9 years

male: 17.7 years
female: 18 years (2008 est.)
Population growth rate: 2.724% (2008 est.)
Birth rate: 38.36 births/1,000 population (2008 est.)
Death rate: 11.74 deaths/1,000 population (2008 est.)
Net migration rate: 0.61 migrant(s)/1,000 population (2008 est.)
Sex ratio:
at birth: 1.03 male(s)/female
under 15 years: 1.01 male(s)/female
15–64 years: 0.98 male(s)/female
65 years and over: 1.02 male(s)/female
total population: 1 male(s)/female (2008 est.)
Infant mortality rate:
total: 68.72 deaths/1,000 live births
male: 75.07 deaths/1,000 live births
female: 62.18 deaths/1,000 live births (2008 est.)
Life expectancy at birth:
total population: 54.95 years
male: 53.06 years
female: 56.9 years (2008 est.)
Total fertility rate: 5.13 children born/woman (2008 est.)
HIV/AIDS—adult prevalence rate: 1.2% (2003 est.)
HIV/AIDS—people living with HIV/AIDS: 6,800 (2003 est.)
HIV/AIDS—deaths: 600 (2003 est.)
Major infectious diseases:
degree of risk: very high
food or waterborne diseases: bacterial and protozoal diarrhea, hepatitis A, and typhoid fever
vectorborne diseases: dengue fever, malaria, Crimean-Congo hemorrhagic fever, and yellow fever
water contact disease: schistosomiasis
respiratory disease: meningococcal meningitis (2008)
Nationality:
noun: Gambian(s)
adjective: Gambian
Ethnic groups: African 99% (Mandinka 42%, Fula 18%, Wolof 16%, Jola 10%, Serahuli 9%, other 4%), non-African 1%
Religions: Muslim 90%, Christian 9%, indigenous beliefs 1%
Languages: English (official), Mandinka, Wolof, Fula, other indigenous vernaculars
Literacy:
definition: age 15 and over can read and write
total population: 40.1%
male: 47.8%
female: 32.8% (2003 est.)

GOVERNMENT

Country name:
conventional long form: Republic of The Gambia
conventional short form: The Gambia
Government type: republic
Capital: *name:* Banjul
geographic coordinates: 13 27 N, 16 34 W
time difference: UTC 0 (5 hours ahead of Washington, DC during Standard Time)
Administrative divisions: 5 divisions and 1 city*; Banjul*, Central River, Lower River, North Bank, Upper River, Western
Independence: 18 February 1965 (from UK)
National holiday: Independence Day, 18 February (1965)
Constitution: approved by national referendum 8 August 1996; effective 16 January 1997
Legal system: based on a composite of English common law, Islamic law, and customary law; accepts compulsory ICJ jurisdiction with reservations
Suffrage: 18 years of age; universal
Executive branch:
chief of state: President Yahya A. J. J. JAMMEH (since 18 October 1996); note—from 1994 to 1996 he was chairman of the Junta; Vice President Isatou NJIE-SAIDY (since 20 March 1997); note—the president is both the chief of state and head of government
head of government: President Yahya A. J. J. JAMMEH (since 18 October 1996); Vice President Isatou NJIE-SAIDY (since 20 March 1997)
cabinet: Cabinet appointed by the president
elections: president elected by popular vote for a five-year term (no term limits); election last held 22 September 2006 (next to be held in 2011)
election results: Yahya A. J. J. JAMMEH reelected president; percent of vote— Yahya A. J. J. JAMMEH 67.3%, Ousainou DARBOE 26.6%, Halifa SALLAH 6.0%
Legislative branch: unicameral National Assembly (53 seats; 48 members elected by popular vote, 5 appointed by the president; to serve five-year terms)
elections: last held 25 January 2007 (next to be held in 2012)
election results: percent of vote by party— NA; seats by party—APRC 47, UDP 4, NADD 1, independent 1
Judicial branch: Supreme Court
Political parties and leaders: Alliance for Patriotic Reorientation and Construction or APRC (the ruling party) [Yahya A. J. J. JAMMEH];

Gambia People's Democratic Party or GPDP [Henry GOMEZ]; National Alliance for Democracy and Development or NADD [Halifa SALLAH]; National Convention Party or NCP [Sheriff DIBBA]; National Reconciliation Party or NRP [Hamat N. K. BAH]; People's Democratic Organization for Independence and Socialism or PDOIS [Sidia JATTA]; United Democratic Party or UDP [Ousainou DARBOE]
Political pressure groups and leaders: NA
International organization participation: ACP, AfDB, AU, C, ECOWAS, FAO, G-77, IBRD, ICAO, ICCt, ICRM, IDA, IDB, IFAD, IFC, IFRCS, ILO, IMF, IMO, Interpol, IOC, IOM, IPU, ITSO, ITU, ITUC, MIGA, NAM, OIC, OPCW, UN, UNAMID, UNCTAD, UNESCO, UNIDO, UNMEE, UNMIL, UNOCI, UNWTO, UPU, WCL, WCO, WFTU, WHO, WIPO, WMO, WTO
Diplomatic representation in the US:
chief of mission: Ambassador (vacant); Charge d'Affaires Abdul Rahman COLE (since 24 December 2007)
chancery: Suite 905, 1156 15th Street NW, Washington, DC 20005
telephone: [1] (202) 785-1379
FAX: [1] (202) 785-1430
Diplomatic representation from the US:
chief of mission: Ambassador Barry L. WELLS
embassy: Kairaba Avenue, Fajara, Banjul
mailing address: P. M. B. No. 19, Banjul
telephone: [220] 439-2856, 437-6169, 437-6170
FAX: [220] 439-2475
Flag description: three equal horizontal bands of red (top), blue with white edges, and green

ECONOMY

Economy—overview: The Gambia has no confirmed mineral or natural resource deposits and has a limited agricultural base. About 75% of the population depends on crops and livestock for its livelihood. Small-scale manufacturing activity features the processing of peanuts, fish, and hides. Reexport trade normally constitutes a major segment of economic activity, but a 1999 government-imposed preshipment inspection plan, and instability of the Gambian dalasi (currency) have drawn some of the reexport trade away from The Gambia. The Gambia's natural beauty and proximity to Europe has made it one of the larger markets for tourism in West Africa. The government's 1998 seizure of the private peanut firm Alimenta eliminated the largest purchaser of Gambian

groundnuts. Despite an announced program to begin privatizing key parastatals, no plans have been made public that would indicate that the government intends to follow through on its promises. Unemployment and underemployment rates remain extremely high; short-run economic progress depends on sustained bilateral and multilateral aid, on responsible government economic management, on continued technical assistance from the IMF and bilateral donors, and on expected growth in the construction sector.
GDP (purchasing power parity): $2.106 billion (2007 est.)
GDP (official exchange rate): $653 million (2007 est.)
GDP—real growth rate: 7% (2007 est.)
GDP—per capita (PPP): $1,300 (2007 est.)
GDP—composition by sector:
agriculture: 32.8%
industry: 8.7%
services: 58.5% (2007 est.)
Labor force: 400,000 (1996)
Labor force—by occupation: *agriculture:* 75%
industry: 19%
services: 6% (1996)
Unemployment rate: NA%
Population below poverty line: NA%
Household income or consumption by percentage share:
lowest 10%: 1.8%
highest 10%: 37% (1998)
Distribution of family income—Gini index: 50.2 (1998)
Inflation rate (consumer prices): 5% (2007 est.)
Investment (gross fixed): 25.4% of GDP (2007 est.)
Budget:
revenues: $181.1 million
expenditures: $163.4 million (2007 est.)
Agriculture—products: rice, millet, sorghum, peanuts, corn, sesame, cassava (tapioca), palm kernels; cattle, sheep, goats
Industries: processing peanuts, fish, and hides; tourism, beverages, agricultural machinery assembly, woodworking, metalworking, clothing
Industrial production growth rate: -2.3% (2007 est.)
Electricity—production: 145 million kWh (2005)
Electricity—production by source:
fossil fuel: 100%
hydro: 0%
nuclear: 0%
other: 0% (2001)
Electricity—consumption: 134.9 million kWh (2005)
Electricity—exports: 0 kWh (2005)

243

Electricity—imports: 0 kWh (2005)

Oil—production: 0 bbl/day (2005 est.)

Oil—consumption: 2,030 bbl/day (2005 est.)

Oil—exports: 41.5 bbl/day (2004)

Oil—imports: 2,050 bbl/day (2004)

Oil—proved reserves: 0 bbl (1 January 2006 est.)

Natural gas—production: 0 cu m (2005 est.)

Natural gas—consumption: 0 cu m (2005 est.)

Natural gas—exports: 0 cu m (2005 est.)

Natural gas—imports: 0 cu m (2005)

Natural gas—proved reserves: 0 cu m (1 January 2006 est.)

Current account balance: -$70 million (2007 est.)

Exports: $93 million f.o.b. (2007 est.)

Exports—commodities: peanut products, fish, cotton lint, palm kernels, re-exports

Exports—partners: India 38.6%, UK 15.9%, Indonesia 7.9%, France 7%, Italy 4.6% (2006)

Imports: $271 million f.o.b. (2007 est.)

Imports—commodities: foodstuffs, manufactures, fuel, machinery and transport equipment

Imports—partners: China 25.2%, Senegal 11.3%, Cote d'Ivoire 8.1%, Brazil 6.6%, Netherlands 4.5%, UK 4% (2006)

Economic aid—recipient: $58.15 million (2005)

Reserves of foreign exchange and gold: $142.8 million (31 December 2007 est.)

Debt—external: $628.8 million (2003 est.)

Market value of publicly traded shares: $NA

Currency (code): dalasi (GMD)

Currency code: GMD

Exchange rates: dalasi per US dollar— 27.79 (2007), 28.066 (2006), 28.575 (2005), 30.03 (2004), 27.306 (2003)

Fiscal year: calendar year

COMMUNICATIONS

Telephones—main lines in use: 52,900 (2006)

Telephones—mobile cellular: 404,300 (2006)

Telephone system:

general assessment: adequate; a packet switched data network is available; two mobile-cellular service providers

domestic: adequate network of microwave radio relay and open-wire; combined fixed-line and mobile-cellular teledensity approaching 30 telephones per 100 persons

international: country code—220; microwave radio relay links to Senegal and Guinea-Bissau; satellite earth station—1 Intelsat (Atlantic Ocean) (1997)

Radio broadcast stations: AM 3, FM 2, shortwave 0 (2001)

Radios: 196,000 (1997)

Television broadcast stations: 1 (government-owned) (1997)

Televisions: 5,000 (2000)

Internet country code: .gm

Internet hosts: 6 (2007)

Internet Service Providers (ISPs): 2 (2001)

Internet users: 58,000 (2005)

TRANSPORTATION

Airports: 1 (2007)

Airports—with paved runways:

total: 1

over 3,047 m: 1 (2007)

Roadways:

total: 3,742 km

paved: 723 km

unpaved: 3,019 km (2004)

Waterways: 390 km (on River Gambia; small ocean-going vessels can reach 190 km) (2006)

Merchant marine:

total: 5 ships (1000 GRT or over) 32,064 GRT/9,751 DWT

by type: passenger/cargo 4, petroleum tanker 1

foreign-owned: 1 (Australia 1) (2007)

Ports and terminals: Banjul

MILITARY

Military branches: Office of the Chief of Defense: Gambian National Army (National Guard, GNA), Gambian Navy (GN) (2008)

Military service age and obligation: 18 years of age for voluntary military service; no conscription (2008)

Manpower available for military service:

males age 16–49: 379,668

females age 16–49: 384,438 (2008 est.)

Manpower fit for military service:

males age 16–49: 230,202

females age 16–49: 244,480 (2008 est.)

Military expenditures—percent of GDP: 0.5% (2006)

TRANSNATIONAL ISSUES

Disputes—international: attempts to stem refugees, cross-border raids, arms smuggling, and other illegal activities by separatists from southern Senegal's Casamance region, as well as from conflicts in other west African states

Refugees and internally displaced persons: *refugees (country of origin):* 5,955 (Sierra Leone) (2007)

GAZA STRIP

INTRODUCTION

Background: The September 1993 Israel-PLO Declaration of Principles on Interim Self-Government Arrangements provided for a transitional period of Palestinian self-rule in the West Bank and Gaza Strip. Under a series of agreements signed between May 1994 and September 1999, Israel transferred to the Palestinian Authority (PA) security and civilian responsibility for Palestinian-populated areas of the West Bank and Gaza. Negotiations to determine the permanent status of the West Bank and Gaza stalled following the outbreak of an intifada in September 2000, as Israeli forces reoccupied most Palestinian-controlled areas. In April 2003, the Quartet (US, EU, UN, and Russia) presented a roadmap to a final settlement of the conflict by 2005 based on reciprocal steps by the two parties leading to two states, Israel and a democratic Palestine. The proposed date for a permanent status agreement was postponed indefinitely due to violence and accusations that both sides had not followed through on their commitments. Following Palestinian leader Yasir ARAFAT's death in late 2004, Mahmud ABBAS was elected PA president in January 2005. A month later, Israel and the PA agreed to the Sharm el-Sheikh

Commitments in an effort to move the peace process forward. In September 2005, Israel unilaterally withdrew all its settlers and soldiers and dismantled its military facilities in the Gaza Strip and withdrew settlers and redeployed soldiers from four small northern West Bank settlements. Nonetheless, Israel controls maritime, airspace, and most access to the Gaza Strip. A November 2005 PA-Israeli agreement authorized the reopening of the Rafah border crossing between the Gaza Strip and Egypt under joint PA and Egyptian control. In January 2006, the Islamic Resistance Movement, HAMAS, won control of the Palestinian Legislative Council (PLC). The international community refused to accept the HAMAS-led government because it did not recognize Israel, would not renounce violence, and refused to honor previous peace agreements between Israel and the PA. HAMAS took control of the PA government in March 2006, but President ABBAS had little success negotiating with HAMAS to present a political platform acceptable to the international community so as to lift economic sanctions on Palestinians. The PLC was unable to convene throughout most of 2006 as a result of Israel's detention of many HAMAS PLC members and Israeli-imposed travel restrictions on other PLC members. Violent clashes took place between Fatah and HAMAS supporters in the Gaza Strip in 2006 and early 2007, resulting in numerous Palestinian deaths and injuries. ABBAS and HAMAS Political Bureau Chief MISHAL in February 2007 signed the Mecca Agreement in Saudi Arabia that resulted in the formation of a Palestinian National Unity Government (NUG) headed by HAMAS member Ismail HANIYA. However, fighting continued in the Gaza Strip, and in June, HAMAS militants succeeded in a violent takeover of all military and governmental institutions in the Gaza Strip. ABBAS dismissed the NUG and through a series of Presidential decrees formed a PA government in the West Bank led by independent Salam FAYYAD. HAMAS rejected the NUG's dismissal and has called for resuming talks with Fatah, but ABBAS has ruled out negotiations until HAMAS agrees to a return of PA control over the Gaza Strip and recognizes the FAYYAD-led government. FAYYAD and his PA government initiated a series of security and economic reforms to improve conditions in the West Bank. ABBAS participated in talks with Israel's Prime Minister OLMERT and secured the release of some Palestinian prisoners and previously withheld customs revenue. During a November 2007 international meeting in Annapolis Maryland, ABBAS and OLMERT agreed to resume peace negotiations with the goal of reaching a final peace settlement by the end of 2008.

GEOGRAPHY

Location: Middle East, bordering the Mediterranean Sea, between Egypt and Israel
Geographic coordinates: 31 25 N, 34 20 E
Map references: Middle East
Area:
total: 360 sq km
land: 360 sq km
water: 0 sq km
Area—comparative: slightly more than twice the size of Washington, DC
Land boundaries:
total: 62 km
border countries: Egypt 11 km, Israel 51 km
Coastline: 40 km
Maritime claims: Israeli-occupied with current status subject to the Israeli-Palestinian Interim Agreement—permanent status to be determined through further negotiation
Climate: temperate, mild winters, dry and warm to hot summers
Terrain: flat to rolling, sand- and dune-covered coastal plain
Elevation extremes:
lowest point: Mediterranean Sea 0 m
highest point: Abu 'Awdah (Joz Abu 'Auda) 105 m
Natural resources: arable land, natural gas
Land use:
arable land: 29%
permanent crops: 21%
other: 50% (2002)
Irrigated land: 150 sq km; note—includes West Bank (2003)
Natural hazards: droughts
Environment—current issues: desertification; salination of fresh water; sewage treatment; water-borne disease; soil degradation; depletion and contamination of underground water resources
Geography—note: strategic strip of land along Mideast-North African trade routes has experienced an incredibly turbulent history; the town of Gaza itself has been besieged countless times in its history

PEOPLE

Population: 1,537,269 (July 2008 est.)
Age structure:
0–14 years: 47.1% (male 370,767/female 353,571)
15–64 years: 50.4% (male 396,110/female 378,743)
65 years and over: 2.5% (male 15,716/female 22,362) (2008 est.)
Median age:
total: 16.2 years
male: 16.1 years
female: 16.4 years (2008 est.)
Population growth rate: 3.609% (2008 est.)
Birth rate: 38.38 births/1,000 population (2008 est.)
Death rate: 3.67 deaths/1,000 population (2008 est.)
Net migration rate: 1.38 migrant(s)/1,000 population (2008 est.)
Sex ratio:
at birth: 1.05 male(s)/female
under 15 years: 1.05 male(s)/female
15–64 years: 1.05 male(s)/female
65 years and over: 0.7 male(s)/female
total population: 1.04 male(s)/female (2008 est.)
Infant mortality rate:
total: 21.35 deaths/1,000 live births
male: 22.34 deaths/1,000 live births
female: 20.3 deaths/1,000 live births (2008 est.)
Life expectancy at birth:
total population: 72.34 years
male: 71.01 years
female: 73.73 years (2008 est.)
Total fertility rate: 5.51 children born/woman (2008 est.)
HIV/AIDS—adult prevalence rate: NA
HIV/AIDS—people living with HIV/AIDS: NA
HIV/AIDS—deaths: NA
Nationality:
noun: NA
adjective: NA
Ethnic groups: Palestinian Arab
Religions: Muslim (predominantly Sunni) 99.3%, Christian 0.7%
Languages: Arabic, Hebrew (spoken by many Palestinians), English (widely understood)
Literacy:
definition: age 15 and over can read and write
total population: 92.4%
male: 96.7%
female: 88% (2004 est.)

GOVERNMENT

Country name:
conventional long form: none
conventional short form: Gaza Strip
local long form: none
local short form: Qita Ghazzah

ECONOMY

Economy—overview: High population density, limited land access, and strict internal and external security controls have kept economic conditions in the Gaza Strip—the smaller of the two areas under the Palestinian Authority (PA)-even more degraded than in the West Bank. The beginning of the second intifadah in September 2000 sparked an economic downturn, largely the result of Israeli closure policies; these policies, which were imposed to address security concerns in Israel, disrupted labor and trade access to and from the Gaza Strip. In 2001, and even more severely in 2003, Israeli military measures in PA areas resulted in the destruction of capital, the disruption of administrative structures, and widespread business closures. The Israeli withdrawal from the Gaza Strip in September 2005 offered some medium-term opportunities for economic growth, but continued Israeli-imposed crossings closures, which became more restrictive after Hamas violently took over the territory in June 2007, have resulted in widespread private sector layoffs and shortages of most goods.

GDP (purchasing power parity): $5.034 billion (includes West Bank) (2006 est.)

GDP (official exchange rate): $5.328 billion (includes West Bank) (2006 est.)

GDP—real growth rate: -8% (includes West Bank) (2006 est.)

GDP—per capita (PPP): $1,100 (includes West Bank) (2006 est.)

GDP—composition by sector:
agriculture: 8%
industry: 13%
services: 79% (includes West Bank) (2006 est.)

Labor force: 267,000 (2006)

Labor force—by occupation:
agriculture: 12%
industry: 18%
services: 70% (2005)

Unemployment rate: 34.8% (2006)

Population below poverty line: 80% (2007 est.)

Household income or consumption by percentage share:
lowest 10%: NA%
highest 10%: NA%

Inflation rate (consumer prices): 3.6% (includes West Bank) (2006)

Budget: *revenues:* $1.149 billion
expenditures: $2.31 billion
note: includes West Bank (2006)

Agriculture—products: olives, citrus, vegetables; beef, dairy products

Industries: generally small family businesses that produce textiles, soap, olivewood carvings, and mother-of-pearl souvenirs; the Israelis had established some small-scale modern industries in an industrial center, but operations ceased prior to Israel's evacuation of Gaza Strip settlements

Industrial production growth rate: 2.4% (includes West Bank) (2005)

Electricity—production: 140,000 kWh (2005)

Electricity—consumption: 230,000 kWh (2005)

Electricity—exports: 0 kWh (2005)

Electricity—imports: 90,000 kWh; note—from Israeli Electric Company (2005)

Exports: $301 million f.o.b.; (includes West Bank) (2005)

Exports—commodities: citrus, flowers, textiles

Exports—partners: Israel, Egypt, West Bank (2006)

Imports: $2.44 billion c.i.f.; (includes West Bank) (2005)

Imports—commodities: food, consumer goods, construction materials

Imports—partners: Israel, Egypt, West Bank (2006)

Economic aid—recipient: $1.4 billion; (includes West Bank) (2006 est.)

Debt—external: $NA

Currency (code): new Israeli shekel (ILS)

Currency code: ILS

Exchange rates: new Israeli shekels per US dollar—4.14 (2007), 4.4565 (2006), 4.4877 (2005), 4.482 (2004), 4.5541 (2003)

Fiscal year: calendar year

COMMUNICATIONS

Telephones—main lines in use: 349,000 (includes West Bank) (2005)

Telephones—mobile cellular: 1.095 million (includes West Bank) (2005)

Telephone system:
general assessment: NA
domestic: Israeli company BEZEK and the Palestinian company PALTEL are responsible for fixed line services; the Palestinian JAWAL company provides cellular services
international: country code—970 (2004)

Radio broadcast stations: AM 0, FM 10, shortwave 0 (2008)

Radios: NA; note—most Palestinian households have radios (1999)

Television broadcast stations: 1 (2008)

Televisions: NA; note—most Palestinian households have televisions (1997)

Internet country code: .ps; note—same as West Bank

Internet Service Providers (ISPs): 3 (1999)

Internet users: 243,000 (includes West Bank) (2005)

TRANSPORTATION

Airports: 2 (2007)

Airports—with paved runways:
total: 1
over 3,047 m: 1 (2007)

Airports—with unpaved runways:
total: 1
under 914 m: 1 (2007)

Heliports: 1 (2007)

Roadways:
note: see entry for West Bank

Ports and terminals: Gaza

MILITARY

Military branches: in accordance with the peace agreement, the Palestinian Authority is not permitted conventional military forces; there are, however, public security forces (2008)

Manpower available for military service:
males age 16–49: 337,670 (2008 est.)

Manpower fit for military service:
males age 16–49: 291,467 (2008 est.)
Manpower reaching militarily significant age annually: *males age 16–49:* 19,567 (2008 est.)

Military expenditures—percent of GDP: NA

TRANSNATIONAL ISSUES

Disputes—international: West Bank and Gaza Strip are Israeli-occupied with current status subject to the Israeli-Palestinian Interim Agreement—permanent status to be determined through further negotiation; Israel removed settlers and military personnel from the Gaza Strip in August 2005

Refugees and internally displaced persons: *refugees (country of origin):* 1.017 million (Palestinian Refugees (UNRWA)) (2007)

GEORGIA

INTRODUCTION

Background: The region of present-day Georgia contained the ancient kingdoms of Colchis and Kartli-Iberia. The area came under Roman influence in the first centuries A.D. and Christianity became the state religion in the 330s. Domination by Persians, Arabs, and Turks was followed by a Georgian golden age (11th-13th centuries) that was cut short by the Mongol invasion of 1236. Subsequently, the Ottoman and Persian empires competed for influence in the region. Georgia was absorbed into the

Russian Empire in the 19th century. Independent for three years (1918–1921) following the Russian revolution, it was forcibly incorporated into the USSR until the Soviet Union dissolved in 1991. An attempt by the incumbent Georgian government to manipulate national legislative elections in November 2003 touched off widespread protests that led to the resignation of Eduard SHEVARDNADZE, president since 1995. New elections in early 2004 swept Mikheil SAAKASHVILI into power along with his National Movement party. Progress on market reforms and democratization has been made in the years since independence, but this progress has been complicated by two ethnic conflicts in the breakaway regions of Abkhazia and South Ossetia. These two territories remain outside the control of the central government and are ruled by de facto, unrecognized governments, supported by Russia. Russian-led peacekeeping operations continue in both regions.

GEOGRAPHY

Location: Southwestern Asia, bordering the Black Sea, between Turkey and Russia
Geographic coordinates: 42 00 N, 43 30 E
Map references: Asia
Area:
total: 69,700 sq km
land: 69,700 sq km
water: 0 sq km
Area—comparative: slightly smaller than South Carolina
Land boundaries:
total: 1,461 km
border countries: Armenia 164 km, Azerbaijan 322 km, Russia 723 km, Turkey 252 km
Coastline: 310 km
Maritime claims:
territorial sea: 12 nm
exclusive economic zone: 200 nm

Climate: warm and pleasant; Mediterranean-like on Black Sea coast
Terrain: largely mountainous with Great Caucasus Mountains in the north and Lesser Caucasus Mountains in the south; Kolkhet'is Dablobi (Kolkhida Lowland) opens to the Black Sea in the west; Mtkvari River Basin in the east; good soils in river valley flood plains, foothills of Kolkhida Lowland
Elevation extremes:
lowest point: Black Sea 0 m
highest point: Mt'a Shkhara 5,201 m
Natural resources: forests, hydropower, manganese deposits, iron ore, copper, minor coal and oil deposits; coastal climate and soils allow for important tea and citrus growth
Land use:
arable land: 11.51%
permanent crops: 3.79%
other: 84.7% (2005)
Irrigated land: 4,690 sq km (2003)
Total renewable water resources: 63.3 cu km (1997)
Freshwater withdrawal (domestic/industrial/agricultural): *total:* 3.61 cu km/yr (20%/21%/59%)
per capita: 808 cu m/yr (2000)
Natural hazards: earthquakes
Environment—current issues: air pollution, particularly in Rust'avi; heavy pollution of Mtkvari River and the Black Sea; inadequate supplies of potable water; soil pollution from toxic chemicals
Environment—international agreements: *party to:* Air Pollution, Biodiversity, Climate Change, Climate Change-Kyoto Protocol, Desertification, Endangered Species, Hazardous Wastes, Law of the Sea, Ozone Layer Protection, Ship Pollution, Wetlands
signed, but not ratified: none of the selected agreements
Geography—note: strategically located east of the Black Sea; Georgia controls much of the Caucasus Mountains and the routes through them

PEOPLE

Population: 4,630,841 (July 2008 est.)
Age structure:
0–14 years: 16.3% (male 402,961/female 352,735)
15–64 years: 67.1% (male 1,496,802/female 1,610,725)
65 years and over: 16.6% (male 307,795/female 459,823) (2008 est.)
Median age: *total:* 38.3 years
male: 35.8 years
female: 40.7 years (2008 est.)
Population growth rate: -0.325% (2008 est.)
Birth rate: 10.62 births/1,000 population (2008 est.)
Death rate: 9.51 deaths/1,000 population (2008 est.)
Net migration rate: -4.36 migrant(s)/1,000 population (2008 est.)
Sex ratio:
at birth: 1.13 male(s)/female
under 15 years: 1.14 male(s)/female
15–64 years: 0.93 male(s)/female
65 years and over: 0.67 male(s)/female
total population: 0.91 male(s)/female (2008 est.)
Infant mortality rate:
total: 16.78 deaths/1,000 live births
male: 18.81 deaths/1,000 live births
female: 14.48 deaths/1,000 live births (2008 est.)
Life expectancy at birth:
total population: 76.51 years
male: 73.21 years
female: 80.26 years (2008 est.)
Total fertility rate: 1.43 children born/woman (2008 est.)
HIV/AIDS—adult prevalence rate: less than 0.1% (2001 est.)
HIV/AIDS—people living with HIV/AIDS: 3,000 (2003 est.)
HIV/AIDS—deaths: fewer than 200 (2003 est.)
Nationality:
noun: Georgian(s)
adjective: Georgian
Ethnic groups: Georgian 83.8%, Azeri 6.5%, Armenian 5.7%, Russian 1.5%, other 2.5% (2002 census)
Religions: Orthodox Christian 83.9%, Muslim 9.9%, Armenian-Gregorian 3.9%, Catholic 0.8%, other 0.8%, none 0.7% (2002 census)
Languages: Georgian 71% (official), Russian 9%, Armenian 7%, Azeri 6%, other 7%
note: Abkhaz is the official language in Abkhazia
Literacy: *definition:* age 15 and over can read and write
total population: 100%
male: 100%
female: 100% (2004 est.)

247

GOVERNMENT

Country name:
conventional long form: none
conventional short form: Georgia
local long form: none
local short form: Sak'art'velo
former: Georgian Soviet Socialist Republic
Government type: republic
Capital: *name:* T'bilisi
geographic coordinates: 41 43 N, 44 47 E
time difference: UTC+4 (9 hours ahead of Washington, DC during Standard Time)
Administrative divisions: 9 regions (mkharebi, singular—mkhare), 1 city (k'alak'i), and 2 autonomous republics (avtomnoy respubliki, singular—avtom respublika)
regions: Guria, Imereti, Kakheti, Kvemo Kartli, Mtskheta-Mtianeti, Racha-Lechkhumi and Kvemo Svaneti, Samegrelo and Zemo Svaneti, Samtskhe-Javakheti, Shida Kartli
city: Tbilisi
autonomous republics: Abkhazia or Ap'khazet'is Avtonomiuri Respublika (Sokhumi), Ajaria or Acharis Avtonomiuri Respublika (Bat'umi)
note: the administrative centers of the two autonomous republics are shown in parentheses
Independence: 9 April 1991 (from Soviet Union)
National holiday: Independence Day, 26 May (1918); note—26 May 1918 was the date of independence from Soviet Russia, 9 April 1991 was the date of independence from the Soviet Union
Constitution: adopted 24 August 1995
Legal system: based on civil law system; accepts compulsory ICJ jurisdiction
Suffrage: 18 years of age; universal
Executive branch:
chief of state: President Mikheil SAAKASHVILI (since 25 January 2004); the president is both the chief of state and head of government for the power ministries: state security (includes interior) and defense
head of government: President Mikheil SAAKASHVILI (since 25 January 2004); Prime Minister Vladimir "Lado" GURGENIDZE (since 19 November 2007); the president is both the chief of state and head of government for the power ministries: state security (includes interior) and defense; the prime minister is head of the remaining ministries of government
cabinet: Cabinet of Ministers
elections: president elected by popular vote for a five-year term (eligible for a second term); election last held 5 January 2008 (next to be held January 2013)

election results: Mikheil SAAKASHVILI reelected president; percent of vote— Mikheil SAAKASHVILI 53.5%, Levan GACHECHILADZE 25.7%, Badri PATARKATSISHVILI 7.1%
Legislative branch: unicameral Parliament or Parlamenti (also known as Supreme Council or Umaghlesi Sabcho) (235 seats; 150 members elected by proportional representation, 75 from single-seat constituencies, and 10 represent displaced persons from Abkhazia; to serve five-year terms)
elections: last held 21 May 2008 (next to be held in spring 2012)
election results: percent of vote by party— National Movement-Democratic Front 59.2%, National Council-New Rights 17.7%, other parties 23.1%; seats by party—National Movement-Democratic Front 120, National Council-New Rights 16
Judicial branch: Supreme Court (judges elected by the Supreme Council on the president's or chairman of the Supreme Court's recommendation); Constitutional Court; first and second instance courts
Political parties and leaders: Burjanadze-Democrats [Nino BURJANADZE]; Georgian People's Front [Nodar NATADZE]; Georgian United Communist Party or UCPG [Panteleimon GIORGADZE]; Georgia's Way Party [Salome ZOURABICHVILI]; Greens [Giorgi GACHECHILADZE]; Industry Will Save Georgia (Industrialists) or IWSG [Georgi TOPADZE]; Labor Party [Shalva NATELASHVILI]; National Council-New Rights (a bloc uniting a nine-party alliance with New Rights); National Democratic Party or NDP [Bachuki KARDAVA]; National Movement-Democratic Front [Mikheil SAAKASHVILI] (bloc composed of National Movement and Burjanadze-Democrats); National Movement [Mikheil SAAKASHVILI]; New Rights [David GAMKRELIDZE]; Republican Party [David USUPASHVILI]; Socialist Party or SPG [Irakli MINDELI]; Traditionalists [Akaki ASATIANI]; Union of National Forces-Conservatives [Koba DAVITASHVILI and Zviad DZIDZIGURI]
Political pressure groups and leaders: Georgian independent deputies from Abkhaz government in exile; separatists in the breakaway regions of Abkhazia and South Ossetia
International organization participation: ACCT (observer), ADB, BSEC, CE, CIS, EAPC, EBRD, FAO, GCTU, GUAM, IAEA, IBRD, ICAO, ICC,

ICCt, ICRM, IDA, IFAD, IFC, IFRCS, ILO, IMF, IMO, Interpol, IOC, IOM, IPU, ISO (correspondent), ITSO, ITU, ITUC, MIGA, OAS (observer), OIF (observer), OPCW, OSCE, PFP, SECI (observer), UN, UNCTAD, UNESCO, UNIDO, UNWTO, UPU, WCO, WHO, WIPO, WMO, WTO
Diplomatic representation in the US:
chief of mission: Ambassador Vasil SIKHARULIDZE
chancery: 2209 Massachusetts Avenue NW, Washington, DC 20008
telephone: [1] (202) 387-2390
FAX: [1] (202) 393-4537
Diplomatic representation from the US:
chief of mission: Ambassador John F. TEFFT
embassy: 11 George Balanchine Street, T'bilisi 0131
mailing address: 7060 T'bilisi Place, Washington, DC 20521-7060
telephone: [995] (32) 27-70-00
FAX: [995] (32) 53-23-10
Flag description: white rectangle, in its central portion a red cross connecting all four sides of the flag; in each of the four corners is a small red bolnur-katskhuri cross; the five-cross flag appears to date back to the 14th century

ECONOMY

Economy—overview: Georgia's economy has sustained robust GDP growth of close to 10% in 2006 and 12% in 2007, based on strong inflows of foreign investment and robust government spending. However, a widening trade deficit and higher inflation are emerging risks to the economy. Areas of recent improvement include increasing foreign direct investment as well as growth in the construction, banking services and mining sectors. Georgia's main economic activities include the cultivation of agricultural products such as grapes, citrus fruits, and hazelnuts; mining of manganese and copper; and output of a small industrial sector producing alcoholic and nonalcoholic beverages, metals, machinery, aircraft and chemicals. The country imports nearly all its needed supplies of natural gas and oil products. It has sizeable hydropower capacity, a growing component of its energy supplies. Despite the severe damage the economy suffered due to civil strife in the 1990s, Georgia, with the help of the IMF and World Bank, has made substantial economic gains since 2000, achieving positive GDP growth and curtailing inflation. Georgia's GDP growth neared 10% in 2006 and 2007 despite restrictions on commerce with Russia. Areas of recent improvement include increased foreign direct investment as well as growth in the construction, banking services, and mining sectors. In addition, the reinvigorated privatization process has met with success. However, a widening trade deficit and higher inflation are emerging risks to the economy. Georgia has suffered from a chronic failure to collect tax revenues; however, the new government is making progress and has reformed the tax code, improved tax administration, increased tax enforcement, and cracked down on corruption. Government revenues have increased nearly four fold since 2003. Due to improvements in customs and financial (tax) enforcement, smuggling is a declining problem. Georgia has overcome the chronic energy shortages of the past by renovating hydropower plants and by bringing newly available natural gas supplies from Azerbaijan. It also has an increased ability to pay for more expensive gas imports from Russia. The country is pinning its hopes for long-term growth on a determined effort to reduce regulation, taxes and corruption in order to attract foreign investment. The construction on the Baku-T'bilisi-Ceyhan oil pipeline, the Baku-T'bilisi-Erzerum gas pipeline, and the Kars-Akhalkalaki Railroad are part of a strategy to capitalize on Georgia's strategic location between Europe and Asia and develop its role as a transit point for gas, oil and other goods.

GDP (purchasing power parity): $20.5 billion (2007 est.)

GDP (official exchange rate): $10.29 billion (2007 est.)

GDP—real growth rate: 12.4% (2007 est.)

GDP—per capita (PPP): $4,700 (2007 est.)

GDP—composition by sector:
agriculture: 13.1%
industry: 29.3%
services: 57.6% (2007 est.)

Labor force: 2.02 million (2007 est.)

Labor force—by occupation: *agriculture:* 55.6%
industry: 8.9%
services: 35.5% (2006 est.)

Unemployment rate: 13.6% (2006 est.)

Population below poverty line: 31% (2006)

Household income or consumption by percentage share:
lowest 10%: 2.4%
highest 10%: 27% (2005)

Distribution of family income—Gini index: 40.4 (2003)

Inflation rate (consumer prices): 9.2% (2007 est.)

Investment (gross fixed): 29.5% of GDP (2007 est.)

Budget:
revenues: $3.68 billion
expenditures: $3.08 billion (2007 est.)

Agriculture—products: citrus, grapes, tea, hazelnuts, vegetables; livestock

Industries: steel, aircraft, machine tools, electrical appliances, mining (manganese and copper), chemicals, wood products, wine

Industrial production growth rate: 13% (2007 est.)

Electricity—production: 8.338 billion kWh (2007)

Electricity—production by source:
fossil fuel: 19.7%
hydro: 80.3%
nuclear: 0%
other: 0% (2001)

Electricity—consumption: 8.146 billion kWh (2007)

Electricity—exports: 625 million kWh (2007)

Electricity—imports: 433 million kWh (2007)

Oil—production: 1,979 bbl/day (2005 est.)

Oil—consumption: 13,400 bbl/day (2005 est.)

Oil—exports: 2,400 bbl/day (2004)

Oil—imports: 13,530 bbl/day (2004)

Oil—proved reserves: 35 million bbl (1 January 2006 est.)

Natural gas—production: 14.39 million cu m (2005 est.)

Natural gas—consumption: 1.8 billion cu m (2007 est.)

Natural gas—exports: 0 cu m (2005)

Natural gas—imports: 1.264 billion cu m (2005)

Natural gas—proved reserves: 8.147 billion cu m (1 January 2006 est.)

Current account balance: -$2.028 billion (2007 est.)

Exports: $1.97 billion (2007 est.)

Exports—commodities: scrap metal, wine, mineral water, ores, vehicles, fruits and nuts

Exports—partners: Turkey 12.7%, Azerbaijan 9.4%, Russia 7.7%, Armenia 7.5%, Turkmenistan 7.3%, Bulgaria 6.4%, US 6%, Ukraine 5.8%, Canada 5%, Germany 4.6% (2006)

Imports: $4.79 billion (2007 est.)

Imports—commodities: fuels, vehicles, machinery and parts, grain and other foods, pharmaceuticals

Imports—partners: Russia 15.2%, Turkey 14.2%, Germany 9.5%, Ukraine 8.7%, Azerbaijan 8.7% (2006)

Economic aid—recipient: ODA, $309.8 million (2005 est.)

Reserves of foreign exchange and gold: $1.361 billion (31 December 2007 est.)

Debt—external: $4.5 billion (2007)

Market value of publicly traded shares: $1.39 billion (2007)

Currency (code): lari (GEL)

Currency code: GEL

Exchange rates: lari per US dollar—1.7 (2007), 1.78 (2006), 1.8127 (2005), 1.9167 (2004), 2.1457 (2003)

Fiscal year: calendar year

COMMUNICATIONS

Telephones—main lines in use: 544,000 (2007)

Telephones—mobile cellular: 2.4 million (2007)

Telephone system:
general assessment: fixed-line telecommunications network has only limited coverage outside Tbilisi; multiple mobile-cellular providers provide services to an increasing subscribership throughout the country
domestic: cellular telephone networks now cover the entire country; urban telephone density is about 20 per 100 people; rural telephone density is about 4 per 100 people; intercity facilities include a fiber-optic line between T'bilisi and K'ut'aisi; nationwide pager service is available
international: country code—995; the Georgia-Russia fiber optic submarine cable provides connectivity to Russia; international service is available by microwave, landline, and satellite through the Moscow switch; international electronic mail and telex service are available

Radio broadcast stations: AM 7, FM 12, shortwave 4 (1998)

Radios: 3.02 million (1997)

Television broadcast stations: 12 (plus repeaters) (1998)

Televisions: 2.57 million (1997)

Internet country code: .ge

Internet hosts: 30,193 (2007)

Internet Service Providers (ISPs): 6 (2000)

Internet users: 332,000 (2006)

TRANSPORTATION

Airports: 23 (2007)

Airports—with paved runways: *total:* 19
over 3,047 m: 1
2,438 to 3,047 m: 7
1,524 to 2,437 m: 5
914 to 1,523 m: 4
under 914 m: 2 (2007)

Airports—with unpaved runways:
total: 4
1,524 to 2,437 m: 1
914 to 1,523 m: 2
under 914 m: 1 (2007)

Heliports: 3 (2007)

Pipelines: gas 1,591 km; oil 1,253 km (2007)

Railways:

total: 1,612 km

broad gauge: 1,575 km 1.520-m gauge (1,575 electrified)

narrow gauge: 37 km 0.912-m gauge (37 electrified) (2006)

Roadways:

total: 20,247 km

paved: 7,973 km

unpaved: 12,274 km (2004)

Merchant marine:

total: 209 ships (1000 GRT or over) 958,504 GRT/1,408,540 DWT

by type: bulk carrier 25, cargo 159, carrier 2, chemical tanker 1, container 5, liquefied gas 2, passenger/cargo 3, petroleum tanker 4, refrigerated cargo 4, roll on/roll off 3, vehicle carrier 1

foreign-owned: 180 (Albania 2, Azerbaijan 1, China 4, Cyprus 1, Egypt 14, Germany 2, Greece 7, Lebanon 3, Monaco 10, Romania 15, Russia 17, Slovenia 2, Syria 54, Turkey 23, Ukraine 24, UAE 1) (2007)

Ports and terminals: Bat'umi, P'ot'i

Transportation—note: large parts of transportation network are in poor condition because of lack of maintenance and repair

MILITARY

Military branches: Georgian Armed Forces: Land Forces, Navy, Air and Air Defense Forces, National Guard (2008)

Military service age and obligation: 18 to 34 years of age for compulsory and voluntary active duty military service; conscript service obligation—18 months (2005)

Manpower available for military service: *males age 16–49:* 1,113,251

females age 16–49: 1,168,021 (2008 est.)

Manpower fit for military service:

males age 16–49: 910,720

females age 16–49: 967,566 (2008 est.)

Manpower reaching militarily significant age annually:

males age 16–49: 35,917

females age 16–49: 34,566 (2008 est.)

Military expenditures—percent of GDP: 0.59% (2005 est.)

Military—note: a CIS peacekeeping force of Russian troops is deployed in the Abkhazia region of Georgia together with a UN military observer group; a Russian peacekeeping battalion is deployed in South Ossetia

TRANSNATIONAL ISSUES

Disputes—international: Russia and Georgia agree on delimiting 80% of their common border, leaving certain small, strategic segments and the maritime boundary unresolved; OSCE observers monitor volatile areas such as the Pankisi Gorge in the Akhmeti region and the Argun Gorge in Abkhazia; UN Observer Mission in Georgia has maintained a peacekeeping force in Georgia since 1993; Meshkheti Turks scattered throughout the former Soviet Union seek to return to Georgia; boundary with Armenia remains undemarcated; ethnic Armenian groups in Javakheti region of Georgia seek greater autonomy from the Georgian government; Azerbaijan and Georgia continue to discuss the alignment of their boundary at certain crossing areas

Refugees and internally displaced persons: *refugees (country of origin):* 1,100 (Russia)

IDPs: 220,000–240,000 (displaced from Abkhazia and South Ossetia) (2007)

Illicit drugs: limited cultivation of cannabis and opium poppy, mostly for domestic consumption; used as transshipment point for opiates via Central Asia to Western Europe and Russia

GERMANY

INTRODUCTION

Background: As Europe's largest economy and second most populous nation, Germany is a key member of the continent's economic, political, and defense organizations. European power struggles immersed Germany in two devastating World Wars in the first half of the 20th century and left the country occupied by the victorious Allied powers of the US, UK, France, and the Soviet Union in 1945. With the advent of the Cold War, two German states were formed in 1949: the western Federal Republic of Germany (FRG) and the eastern German Democratic Republic (GDR). The democratic FRG embedded itself in key Western economic and security organizations, the EC, which became the EU, and NATO, while the Communist GDR was on the front line of the Soviet-led Warsaw Pact. The decline of the USSR and the end of the Cold War allowed for German unification in 1990. Since then, Germany has expended considerable funds to bring Eastern productivity and wages up to Western standards. In January 1999, Germany and 10 other EU countries introduced a common European exchange currency, the euro.

GEOGRAPHY

Location: Central Europe, bordering the Baltic Sea and the North Sea, between the Netherlands and Poland, south of Denmark

Geographic coordinates: 51 00 N, 9 00 E

Map references: Europe

Area: *total:* 357,021 sq km

land: 349,223 sq km

water: 7,798 sq km

Area—comparative: slightly smaller than Montana

Land boundaries:

total: 3,621 km

border countries: Austria 784 km, Belgium 167 km, Czech Republic 646 km, Denmark 68 km, France 451 km, Luxembourg 138 km, Netherlands 577 km, Poland 456 km, Switzerland 334 km

Coastline: 2,389 km

Maritime claims:

territorial sea: 12 nm

exclusive economic zone: 200 nm

continental shelf: 200-m depth or to the depth of exploitation

Climate: temperate and marine; cool, cloudy, wet winters and summers; occasional warm mountain (foehn) wind

Terrain: lowlands in north, uplands in center, Bavarian Alps in south

Elevation extremes:

lowest point: Neuendorf bei Wilster -3.54 m

highest point: Zugspitze 2,963 m

Natural resources: coal, lignite, natural gas, iron ore, copper, nickel, uranium, potash, salt, construction materials, timber, arable land

Land use:
arable land: 33.13%
permanent crops: 0.6%
other: 66.27% (2005)
Irrigated land: 4,850 sq km (2003)
Total renewable water resources: 188 cu km (2005)
Freshwater withdrawal (domestic/industrial/agricultural): *total:* 38.01 cu km/yr (12%/68%/20%)
per capita: 460 cu m/yr (2001)
Natural hazards: flooding
Environment—current issues: emissions from coal-burning utilities and industries contribute to air pollution; acid rain, resulting from sulfur dioxide emissions, is damaging forests; pollution in the Baltic Sea from raw sewage and industrial effluents from rivers in eastern Germany; hazardous waste disposal; government established a mechanism for ending the use of nuclear power over the next 15 years; government working to meet EU commitment to identify nature preservation areas in line with the EU's Flora, Fauna, and Habitat directive
Environment—international agreements: *party to:* Air Pollution, Air Pollution-Nitrogen Oxides, Air Pollution-Persistent Organic Pollutants, Air Pollution-Sulfur 85, Air Pollution-Sulfur 94, Air Pollution-Volatile Organic Compounds, Antarctic-Environmental Protocol, Antarctic-Marine Living Resources, Antarctic Seals, Antarctic Treaty, Biodiversity, Climate Change, Climate Change-Kyoto Protocol, Desertification, Endangered Species, Environmental Modification, Hazardous Wastes, Law of the Sea, Marine Dumping, Ozone Layer Protection, Ship Pollution, Tropical Timber 83, Tropical Timber 94, Wetlands, Whaling
signed, but not ratified: none of the selected agreements
Geography—note: strategic location on North European Plain and along the entrance to the Baltic Sea

PEOPLE

Population: 82,369,548 (July 2008 est.)
Age structure:
0–14 years: 13.8% (male 5,826,066/female 5,524,568)
15–64 years: 66.2% (male 27,763,917/female 26,739,934)
65 years and over: 20% (male 6,892,743/female 9,622,320) (2008 est.)
Median age:
total: 43.4 years
male: 42.2 years
female: 44.7 years (2008 est.)
Population growth rate: -0.044% (2008 est.)

Birth rate: 8.18 births/1,000 population (2008 est.)
Death rate: 10.8 deaths/1,000 population (2008 est.)
Net migration rate: 2.19 migrant(s)/1,000 population (2008 est.)
Sex ratio:
at birth: 1.06 male(s)/female
under 15 years: 1.05 male(s)/female
15–64 years: 1.04 male(s)/female
65 years and over: 0.72 male(s)/female
total population: 0.97 male(s)/female (2008 est.)
Infant mortality rate:
total: 4.03 deaths/1,000 live births
male: 4.46 deaths/1,000 live births
female: 3.58 deaths/1,000 live births (2008 est.)
Life expectancy at birth:
total population: 79.1 years
male: 76.11 years
female: 82.26 years (2008 est.)
Total fertility rate: 1.41 children born/woman (2008 est.)
HIV/AIDS—adult prevalence rate: 0.1% (2001 est.)
HIV/AIDS—people living with HIV/AIDS: 43,000 (2001 est.)
HIV/AIDS—deaths: fewer than 1,000 (2003 est.)
Nationality:
noun: German(s)
adjective: German
Ethnic groups: German 91.5%, Turkish 2.4%, other 6.1% (made up largely of Greek, Italian, Polish, Russian, Serbo-Croatian, Spanish)
Religions: Protestant 34%, Roman Catholic 34%, Muslim 3.7%, unaffiliated or other 28.3%
Languages: German
Literacy:
definition: age 15 and over can read and write
total population: 99%
male: 99%
female: 99% (2003 est.)
People—note: second most populous country in Europe after Russia

GOVERNMENT

Country name:
conventional long form: Federal Republic of Germany
conventional short form: Germany
local long form: Bundesrepublik Deutschland
local short form: Deutschland
former: German Empire, German Republic, German Reich
Government type: federal republic
Capital: *name:* Berlin
geographic coordinates: 52 31 N, 13 24 E
time difference: UTC+1 (6 hours ahead of Washington, DC during Standard Time)

daylight saving time: +1hr, begins last Sunday in March; ends last Sunday in October
Administrative divisions: 16 states (Laender, singular—Land); Baden-Wuerttemberg, Bayern (Bavaria), Berlin, Brandenburg, Bremen, Hamburg, Hessen, Mecklenburg-Vorpommern (Mecklenburg-Western Pomerania), Niedersachsen (Lower Saxony), Nordrhein-Westfalen (North Rhine-Westphalia), Rheinland-Pfalz (Rhineland-Palatinate), Saarland, Sachsen (Saxony), Sachsen-Anhalt (Saxony-Anhalt), Schleswig-Holstein, Thueringen (Thuringia); note—Bayern, Sachsen, and Thueringen refer to themselves as free states (Freistaaten, singular—Freistaat)
Independence: 18 January 1871 (German Empire unification); divided into four zones of occupation (UK, US, USSR, and later, France) in 1945 following World War II; Federal Republic of Germany (FRG or West Germany) proclaimed 23 May 1949 and included the former UK, US, and French zones; German Democratic Republic (GDR or East Germany) proclaimed 7 October 1949 and included the former USSR zone; unification of West Germany and East Germany took place 3 October 1990; all four powers formally relinquished rights 15 March 1991
National holiday: Unity Day, 3 October (1990)
Constitution: 23 May 1949, known as Basic Law; became constitution of the united Germany 3 October 1990
Legal system: civil law system with indigenous concepts; judicial review of legislative acts in the Federal Constitutional Court; has not accepted compulsory ICJ jurisdiction
Suffrage: 18 years of age; universal
Executive branch:
chief of state: President Horst KOEHLER (since 1 July 2004)
head of government: Chancellor Angela MERKEL (since 22 November 2005)
cabinet: Cabinet or Bundesminister (Federal Ministers) appointed by the president on the recommendation of the chancellor
elections: president elected for a five-year term (eligible for a second term) by a Federal Convention, including all members of the Federal Assembly and an equal number of delegates elected by the state parliaments; election last held 23 May 2004 (next scheduled for 23 May 2009); chancellor elected by an absolute majority of the Federal Assembly for a four-year term; Bundestag vote for Chancellor last held 22 November 2005

251

(next will follow the national elections to be held by autumn 2009)

election results: Horst KOEHLER elected president; received 604 votes of the Federal Convention against 589 for Gesine SCHWAN; Angela MERKEL elected chancellor; vote by Federal Assembly 397 to 202 with 12 abstentions

Legislative branch: bicameral Parliament or Parlament consists of the Federal Assembly or Bundestag (614 seats; elected by popular vote under a system combining direct and proportional representation; a party must win 5% of the national vote or three direct mandates to gain proportional representation and caucus recognition; to serve four-year terms) and the Federal Council or Bundesrat (69 votes; state governments are directly represented by votes; each has three to six votes depending on population and are required to vote as a block)

elections: Bundestag—last held on 18 September 2005 (next to be held no later than autumn 2009); note—there are no elections for the Bundesrat; composition is determined by the composition of the state-level governments; the composition of the Bundesrat has the potential to change any time one of the 16 states holds an election

election results: Bundestag—percent of vote by party—CDU/CSU 35.2%, SPD 34.3%, FDP 9.8%, Left 8.7%, Greens 8.1%, other 3.9%; seats by party—CDU/CSU 225, SPD 222, FDP 61, Left 53, Greens 51, independents 2

Judicial branch: Federal Constitutional Court or Bundesverfassungsgericht (half the judges are elected by the Bundestag and half by the Bundesrat)

Political parties and leaders: Alliance '90/Greens [Claudia ROTH and Reinhard BUETIKOFER]; Christian Democratic Union or CDU [Angela MERKEL]; Christian Social Union or CSU [Erwin HUBER]; Free Democratic Party or FDP [Guido WESTERWELLE]; Left Party or Die Linke [Lothar BISKY and Oskar LAFONTAINE]; Social Democratic Party or SPD [Kurt BECK]

Political pressure groups and leaders: business associations and employers' organizations; religious, trade unions, immigrant, expellee, and veterans groups

International organization participation: ADB (nonregional members), AfDB, Arctic Council (observer), Australia Group, BIS, BSEC (observer), CBSS, CDB, CE, CERN, EAPC, EBRD, EIB, EMU, ESA, EU, FAO, G-5, G-7, G-8, G-10, IADB, IAEA, IBRD, ICAO, ICC, ICCt, ICRM, IDA, IEA, IFAD, IFC, IFRCS, IHO, ILO, IMF, IMO, IMSO, Interpol, IOC, IOM, IPU, ISO, ITSO, ITU, ITUC, MIGA, NAM (guest), NATO, NEA, NSG, OAS (observer), OECD, OPCW, OSCE, Paris Club, PCA, Schengen Convention, SECI (observer), UN, UNCTAD, UNESCO, UNHCR, UNIDO, UNIFIL, UNMEE, UNMIL, UNMIS, UNOMIG, UNRWA, UNWTO, UPU, WADB (nonregional), WCO, WEU, WHO, WIPO, WMO, WTO, ZC

Diplomatic representation in the US:
chief of mission: Ambassador Klaus SCHARIOTH
chancery: 4645 Reservoir Road NW, Washington, DC 20007
telephone: [1] (202) 298-4000
FAX: [1] (202) 298-4249
consulate(s) general: Atlanta, Boston, Chicago, Houston, Los Angeles, Miami, New York, San Francisco

Diplomatic representation from the US:
chief of mission: Ambassador William R. TIMKEN, Jr.
embassy: Neustaedtische Kirchstrasse 4-5, 10117 Berlin; note—a new embassy is being built near the Brandenburg Gate in Berlin; ground was broken in October 2004 and completion is scheduled for 2008
mailing address: PSC 120, Box 1000, APO AE 09265
telephone: [49] (030) 2375174
FAX: [49] (030) 8305-1215
consulate(s) general: Duesseldorf, Frankfurt am Main, Hamburg, Leipzig, Munich

Flag description: three equal horizontal bands of black (top), red, and gold

ECONOMY

Economy—overview: Germany's affluent and technologically powerful economy—the fifth largest in the world in PPP terms—showed considerable improvement in 2007 with 2.6% growth. After a long period of stagnation with an average growth rate of 0.7% between 2001–05 and chronically high unemployment, stronger growth led to a considerable fall in unemployment to about 8% near the end of 2007. Among the most important reasons for Germany's high unemployment during the past decade were macroeconomic stagnation, the declining level of investment in plant and equipment, company restructuring, flat domestic consumption, structural rigidities in the labor market, lack of competition in the service sector, and high interest rates. The modernization and integration of the eastern German economy continues to be a costly long-term process, with annual transfers from west to east amounting to roughly $80 billion. The former government of Chancellor Gerhard SCHROEDER launched a comprehensive set of reforms of labor market and welfare-related institutions. The current government of Chancellor Angela MERKEL has initiated other reform measures, such as a gradual increase in the mandatory retirement age from 65 to 67 and measures to increase female participation in the labor market. Germany's aging population, combined with high chronic unemployment, has pushed social security outlays to a level exceeding contributions, but higher government revenues from the cyclical upturn in 2006–07 and a 3% rise in the value-added tax pushed Germany's budget deficit well below the EU's 3% debt limit. Corporate restructuring and growing capital markets are setting the foundations that could help Germany meet the long-term challenges of European economic integration and globalization, although some economists continue to argue the need for change in inflexible labor and services markets. Growth may fall below 2% in 2008 as the strong euro, high oil prices, tighter credit markets, and slowing growth abroad take their toll.

GDP (purchasing power parity): $2.81 trillion (2007 est.)
GDP (official exchange rate): $3.322 trillion (2007 est.)
GDP—real growth rate: 2.5% (2007 est.)
GDP—per capita (PPP): $34,200 (2007 est.)
GDP—composition by sector:
agriculture: 0.8%
industry: 29%
services: 70.1% (2007 est.)
Labor force: 43.51 million (2007 est.)
Labor force—by occupation: *agriculture:* 2.8%
industry: 33.4%
services: 63.8% (1999)
Unemployment rate: 8.4%
note: this is the International Labor Organization's estimated rate for international comparisons; Germany's Federal Employment Office estimated a seasonally adjusted rate of 10.8% (2007 est.)
Population below poverty line: 11% (2001 est.)
Household income or consumption by percentage share:
lowest 10%: 3.2%
highest 10%: 22.1% (2000)
Distribution of family income—Gini index: 28 (2005)
Inflation rate (consumer prices): 2.3% (2007 est.)
Investment (gross fixed): 18.3% of GDP (2007 est.)

Budget:

revenues: $1.476 trillion

expenditures: $1.473 trillion (2007 est.)

Public debt: 63.2% of GDP (2007 est.)

Agriculture—products: potatoes, wheat, barley, sugar beets, fruit, cabbages; cattle, pigs, poultry

Industries: among the world's largest and most technologically advanced producers of iron, steel, coal, cement, chemicals, machinery, vehicles, machine tools, electronics, food and beverages, shipbuilding, textiles

Industrial production growth rate: 2.1% (2007 est.)

Electricity—production: 579.4 billion kWh (2005)

Electricity—production by source:

fossil fuel: 61.8%

hydro: 4.2%

nuclear: 29.9%

other: 4.1% (2001)

Electricity—consumption: 545.5 billion kWh (2005)

Electricity—exports: 61.43 billion kWh (2005)

Electricity—imports: 56.86 billion kWh (2005)

Oil—production: 141,700 bbl/day (2005)

Oil—consumption: 2.618 million bbl/day (2005 est.)

Oil—exports: 518,700 bbl/day (2004)

Oil—imports: 2.953 million bbl/day (2004)

Oil—proved reserves: 367.2 million bbl (1 January 2006 est.)

Natural gas—production: 19.9 billion cu m (2005 est.)

Natural gas—consumption: 96.84 billion cu m (2005 est.)

Natural gas—exports: 9.42 billion cu m (2005 est.)

Natural gas—imports: 86.99 billion cu m (2005)

Natural gas—proved reserves: 246.5 billion cu m (1 January 2006 est.)

Current account balance: $185 billion (2007 est.)

Exports: $1.334 trillion f.o.b. (2007 est.)

Exports—commodities: machinery, vehicles, chemicals, metals and manufactures, foodstuffs, textiles

Exports—partners: France 9.6%, US 8.6%, UK 7.3%, Italy 6.7%, Netherlands 6.3%, Austria 5.6%, Belgium 5.3%, Spain 4.7% (2006)

Imports: $1.089 trillion f.o.b. (2007 est.)

Imports—commodities: machinery, vehicles, chemicals, foodstuffs, textiles, metals

Imports—partners: Netherlands 11.9%, France 8.6%, Belgium 7.3%, China 6%, UK 5.7%, Italy 5.7%, US 4.7%, Austria 4.3% (2006)

Economic aid—donor: ODA, $10.44 billion (2006)

Reserves of foreign exchange and gold: $136.2 billion (31 December 2007 est.)

Debt—external: $4.489 trillion (30 June 2007)

Stock of direct foreign investment—at home: $811 billion (2007 est.)

Stock of direct foreign investment—abroad: $1.123 trillion (2007 est.)

Market value of publicly traded shares: $1.221 trillion (2005)

Currency (code): euro (EUR)

Currency code: EUR

Exchange rates: euros per US dollar—0.7345 (2007), 0.7964 (2006), 0.8041 (2005), 0.8054 (2004), 0.886 (2003)

Fiscal year: calendar year

COMMUNICATIONS

Telephones—main lines in use: 54.2 million (2006)

Telephones—mobile cellular: 84.3 million (2006)

Telephone system:

general assessment: Germany has one of the world's most technologically advanced telecommunications systems; as a result of intensive capital expenditures since reunification, the formerly backward system of the eastern part of the country, dating back to World War II, has been modernized and integrated with that of the western part

domestic: Germany is served by an extensive system of automatic telephone exchanges connected by modern networks of fiber-optic cable, coaxial cable, microwave radio relay, and a domestic satellite system; cellular telephone service is widely available, expanding rapidly, and includes roaming service to many foreign countries

international: country code—49; Germany's international service is excellent worldwide, consisting of extensive land and undersea cable facilities as well as earth stations in the Inmarsat, Intelsat, Eutelsat, and Intersputnik satellite systems (2001)

Radio broadcast stations: AM 51, FM 787, shortwave 4 (1998)

Radios: 77.8 million (1997)

Television broadcast stations: 373 (plus 8,042 repeaters) (1995)

Televisions: 51.4 million (1998)

Internet country code: .de

Internet hosts: 16.494 million (2007)

Internet Service Providers (ISPs): 200 (2001)

Internet users: 38.6 million (2006)

TRANSPORTATION

Airports: 550 (2007)

Airports—with paved runways:

total: 331

over 3,047 m: 14

2,438 to 3,047 m: 52

1,524 to 2,437 m: 58

914 to 1,523 m: 72

under 914 m: 135 (2007)

Airports—with unpaved runways:

total: 219

2,438 to 3,047 m: 1

1,524 to 2,437 m: 3

914 to 1,523 m: 34

under 914 m: 181 (2007)

Heliports: 28 (2007)

Pipelines: condensate 37 km; gas 25,094 km; oil 3,546 km; refined products 3,828 km (2007)

Railways:

total: 48,215 km

standard gauge: 47,962 km 1.435-m gauge (20,278 km electrified)

narrow gauge: 229 km 1.000-m gauge (16 km electrified); 24 km 0.750-m gauge (2006)

Roadways:

total: 231,500 km

paved: 231,500 km (includes 12,400 km of expressways) (2006)

Waterways: 7,467 km

note: Rhine River carries most goods; Main-Danube Canal links North Sea and Black Sea (2006)

Merchant marine:

total: 382 ships (1000 GRT or over) 12,085,484 GRT/14,261,476 DWT

by type: bulk carrier 1, cargo 50, chemical tanker 11, container 269, liquefied gas 5, passenger 5, passenger/cargo 26, petroleum tanker 12, roll on/roll off 3

foreign-owned: 7 (China 2, Finland 4, Ireland 1)

registered in other countries: 2,716 (Antigua and Barbuda 891, Australia 2, Bahamas 40, Belgium 1, Bermuda 21, Brazil 7, Bulgaria 1, Burma 5, Canada 3, Cayman Islands 17, Cyprus 197, Denmark 12, Faroe Islands 1, Finland 2, France 1, Georgia 2, Gibraltar 117, Hong Kong 10, Isle of Man 61, Italy 1, Jamaica 1, Liberia 728, Luxembourg 10, Malaysia 2, Malta 67, Marshall Islands 214, Morocco 1, Netherlands 70, Netherlands Antilles 48, Norway 2, NZ 1, Panama 38, Portugal 22, Russia 2, Singapore 18, Spain 9, Sri Lanka 6, St Vincent and The Grenadines 3, Sweden 4, Turkey 1, UK 71, US 6) (2007)

Ports and terminals: Bremen, Bremerhaven, Duisburg, Hamburg, Karlsruhe, Lubeck, Rostock, Wilhemshaven

MILITARY

Military branches: Federal Armed Forces (Bundeswehr): Army (Heer),

Navy (Deutsche Marine, includes naval air arm), Air Force (Luftwaffe), Joint Service Support Command (Streitkraeftebasis), Central Medical Service (Zentraler Sanitaetsdienst) (2006)

Military service age and obligation: 18 years of age (conscripts serve a 9-month tour of compulsory military service) (2004)

Manpower available for military service:
males age 16–49: 19,594,118

females age 16–49: 18,543,955 (2008 est.)

Manpower fit for military service:
males age 16–49: 15,906,930
females age 16–49: 15,051,183 (2008 est.)

Manpower reaching militarily significant age annually:
males age 16–49: 442,972
females age 16–49: 420,801 (2008 est.)

Military expenditures—percent of GDP: 1.5% (2005 est.)

TRANSNATIONAL ISSUES

Disputes—international: none
Illicit drugs: source of precursor chemicals for South American cocaine processors; transshipment point for and consumer of Southwest Asian heroin, Latin American cocaine, and European-produced synthetic drugs; major financial center

GHANA

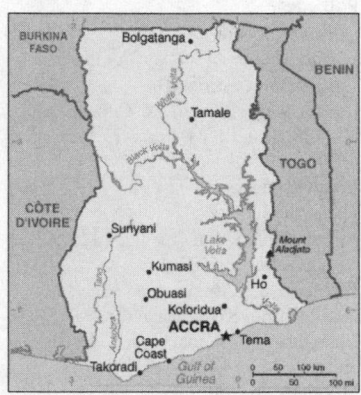

INTRODUCTION

Background: Formed from the merger of the British colony of the Gold Coast and the Togoland trust territory, Ghana in 1957 became the first sub-Saharan country in colonial Africa to gain its independence. Ghana endured a long series of coups before Lt. Jerry RAWLINGS took power in 1981 and banned political parties. After approving a new constitution and restoring multiparty politics in 1992, RAWLINGS won presidential elections in 1992 and 1996, but was constitutionally prevented from running for a third term in 2000. John KUFUOR succeeded him and was reelected in 2004. Kufuor is constitutionally barred from running for a third term in upcoming Presidential elections, which are scheduled for December 2008.

GEOGRAPHY

Location: Western Africa, bordering the Gulf of Guinea, between Cote d'Ivoire and Togo
Geographic coordinates: 8 00 N, 2 00 W
Map references: Africa
Area:
total: 239,460 sq km

land: 230,940 sq km
water: 8,520 sq km
Area—comparative: slightly smaller than Oregon
Land boundaries:
total: 2,094 km
border countries: Burkina Faso 549 km, Cote d'Ivoire 668 km, Togo 877 km
Coastline: 539 km
Maritime claims:
territorial sea: 12 nm
contiguous zone: 24 nm
exclusive economic zone: 200 nm
continental shelf: 200 nm
Climate: tropical; warm and comparatively dry along southeast coast; hot and humid in southwest; hot and dry in north
Terrain: mostly low plains with dissected plateau in south-central area
Elevation extremes:
lowest point: Atlantic Ocean 0 m
highest point: Mount Afadjato 880 m
Natural resources: gold, timber, industrial diamonds, bauxite, manganese, fish, rubber, hydropower, petroleum, silver, salt, limestone
Land use:
arable land: 17.54%
permanent crops: 9.22%
other: 73.24% (2005)
Irrigated land: 310 sq km (2003)
Total renewable water resources: 53.2 cu km (2001)
Freshwater withdrawal (domestic/industrial/agricultural): *total:* 0.98 cu km/yr (24%/10%/66%)
per capita: 44 cu m/yr (2000)
Natural hazards: dry, dusty, northeastern harmattan winds occur from January to March; droughts
Environment—current issues: recurrent drought in north severely affects agricultural activities; deforestation; overgrazing; soil erosion; poaching and habitat destruction threatens wildlife populations; water pollution; inadequate supplies of potable water

Environment—international agreements: *party to:* Biodiversity, Climate Change, Climate Change-Kyoto Protocol, Desertification, Endangered Species, Environmental Modification, Hazardous Wastes, Law of the Sea, Ozone Layer Protection, Ship Pollution, Tropical Timber 83, Tropical Timber 94, Wetlands
signed, but not ratified: Marine Life Conservation
Geography—note: Lake Volta is the world's largest artificial lake

PEOPLE

Population: 23,382,848
note: estimates for this country explicitly take into account the effects of excess mortality due to AIDS; this can result in lower life expectancy, higher infant mortality, higher death rates, lower population growth rates, and changes in the distribution of population by age and sex than would otherwise be expected (July 2008 est.)

Age structure:
0–14 years: 37.8% (male 4,470,382/ female 4,360,359)
15–64 years: 58.7% (male 6,852,363/ female 6,866,470)
65 years and over: 3.6% (male 386,150/female 447,124) (2008 est.)
Median age:
total: 20.4 years
male: 20.2 years
female: 20.7 years (2008 est.)
Population growth rate: 1.928% (2008 est.)
Birth rate: 29.22 births/1,000 population (2008 est.)
Death rate: 9.39 deaths/1,000 population (2008 est.)
Net migration rate: -0.55 migrant(s)/ 1,000 population (2008 est.)
Sex ratio:
at birth: 1.03 male(s)/female
under 15 years: 1.03 male(s)/female

15–64 years: 1 male(s)/female
65 years and over: 0.86 male(s)/female
total population: 1 male(s)/female (2008 est.)

Infant mortality rate:
total: 52.31 deaths/1,000 live births
male: 56.64 deaths/1,000 live births
female: 47.85 deaths/1,000 live births (2008 est.)

Life expectancy at birth:
total population: 59.49 years
male: 58.65 years
female: 60.35 years (2008 est.)

Total fertility rate: 3.78 children born/woman (2008 est.)

HIV/AIDS—adult prevalence rate: 3.1% (2003 est.)

HIV/AIDS—people living with HIV/AIDS: 350,000 (2003 est.)

HIV/AIDS—deaths: 30,000 (2003 est.)

Major infectious diseases:
degree of risk: very high
food or waterborne diseases: bacterial and protozoal diarrhea, hepatitis A, and typhoid fever
vectorborne diseases: malaria and yellow fever
water contact disease: schistosomiasis
respiratory disease: meningococcal meningitis
note: highly pathogenic H5N1 avian influenza has been identified in this country; it poses a negligible risk with extremely rare cases possible among US citizens who have close contact with birds (2008)

Nationality:
noun: Ghanaian(s)
adjective: Ghanaian

Ethnic groups: Akan 45.3%, Mole-Dagbon 15.2%, Ewe 11.7%, Ga-Dangme 7.3%, Guan 4%, Gurma 3.6%, Grusi 2.6%, Mande-Busanga 1%, other tribes 1.4%, other 7.8% (2000 census)

Religions: Christian 68.8% (Pentecostal/Charismatic 24.1%, Protestant 18.6%, Catholic 15.1%, other 11%), Muslim 15.9%, traditional 8.5%, other 0.7%, none 6.1% (2000 census)

Languages: Asante 14.8%, Ewe 12.7%, Fante 9.9%, Boron (Brong) 4.6%, Dagomba 4.3%, Dangme 4.3%, Dagarte (Dagaba) 3.7%, Akyem 3.4%, Ga 3.4%, Akuapem 2.9%, other 36.1% (includes English (official)) (2000 census)

Literacy:
definition: age 15 and over can read and write
total population: 57.9%
male: 66.4%
female: 49.8% (2000 census)

GOVERNMENT

Country name:
conventional long form: Republic of Ghana
conventional short form: Ghana
former: Gold Coast

Government type: constitutional democracy

Capital: *name:* Accra
geographic coordinates: 5 33 N, 0 13 W
time difference: UTC 0 (5 hours ahead of Washington, DC during Standard Time)

Administrative divisions: 10 regions; Ashanti, Brong-Ahafo, Central, Eastern, Greater Accra, Northern, Upper East, Upper West, Volta, Western

Independence: 6 March 1957 (from UK)

National holiday: Independence Day, 6 March (1957)

Constitution: approved 28 April 1992

Legal system: based on English common law and customary law; has not accepted compulsory ICJ jurisdiction

Suffrage: 18 years of age; universal

Executive branch:
chief of state: President John Agyekum KUFUOR (since 7 January 2001); Vice President Alhaji Aliu MAHAMA (since 7 January 2001); note—the president is both the chief of state and head of government
head of government: President John Agyekum KUFUOR (since 7 January 2001); Vice President Alhaji Aliu MAHAMA (since 7 January 2001)
cabinet: Council of Ministers; president nominates members subject to approval by Parliament
elections: president and vice president elected on the same ticket by popular vote for four-year terms (eligible for a second term); election last held 7 December 2004 (next to be held in December 2008)
election results: John Agyekum KUFUOR reelected president in election; percent of vote—John KUFUOR 52.4%, John ATTA-MILLS 44.6%

Legislative branch: unicameral Parliament (230 seats; members are elected by direct, popular vote to serve four-year terms)
elections: last held 7 December 2004 (next to be held in December 2008)
election results: percent of vote by party—NA; seats by party—NPP 128, NDC 94, PNC 4, CPP 3, independent 1

Judicial branch: Supreme Court

Political parties and leaders: Convention People's Party or CPP [Ladi NYLANDER]; Democratic Freedom Party or DFP [Alhaji Abudu Rahman ISSAKAH]; Every Ghanaian Living Everywhere or EGLE; Great Consolidated Popular Party or GCPP [Dan LARTEY]; National Democratic Congress or NDC [Dr. Kwabena ADJEI]; New Patriotic Party or NPP [Peter MAC-MANU]; People's National Convention or PNC [Alhaji Amed RAMADAN]; Reform Party [Kyeretwie OPUKU]; United Renaissance Party or URP [Charles WAYO]

Political pressure groups and leaders: NA

International organization participation: ACP, AfDB, AU, C, ECOWAS, FAO, G-24, G-77, IAEA, IBRD, ICAO, ICC, ICCt, ICRM, IDA, IFAD, IFC, IFRCS, ILO, IMF, IMO, IMSO, Interpol, IOC, IOM, IPU, ISO, ITSO, ITU, ITUC, MIGA, MINURSO, MONUC, NAM, OAS (observer), OIF, OPCW, UN, UNAMID, UNCTAD, UNESCO, UNHCR, UNIDO, UNIFIL, UNITAR, UNMEE, UNMIL, UNOCI, UNOMIG, UNWTO, UPU, WCL, WCO, WFTU, WHO, WIPO, WMO, WTO

Diplomatic representation in the US:
chief of mission: Ambassador Dr. Kwame BAWUAH-EDUSEI
chancery: 1156 15th St. NW #905, Washington, DC 20005
telephone: [1] (202) 785-1379
FAX: [1] (202) 785-1430
consulate(s) general: New York

Diplomatic representation from the US:
chief of mission: Ambassador Pamela E. BRIDGEWATER
embassy: 24 4th Circular Rd. Cantonments, Accra
mailing address: P. O. Box 194, Accra
telephone: [233] (21) 741-000
FAX: [233] (21) 741-389

Flag description: three equal horizontal bands of red (top), yellow, and green with a large black five-pointed star centered in the yellow band; uses the popular pan-African colors of Ethiopia; similar to the flag of Bolivia, which has a coat of arms centered in the yellow band

ECONOMY

Economy—overview: Well endowed with natural resources, Ghana has roughly twice the per capita output of the poorest countries in West Africa. Even so, Ghana remains heavily dependent on international financial and technical assistance. Gold and cocoa production, and individual remittances, are major sources of foreign exchange. The domestic economy continues to revolve around agriculture, which accounts for about 35% of GDP and employs about 55% of the work force, mainly small landholders. Ghana opted for debt relief under the Heavily Indebted Poor Country (HIPC) program in 2002, and is also benefiting from the Multilateral Debt Relief Initiative that took effect in 2006. Thematic priorities under its current Growth and Poverty

Reduction Strategy, which also provides the framework for development partner assistance, are: macroeconomic stability; private sector competitiveness; human resource development; and good governance and civic responsibility. Sound macro-economic management along with high prices for gold and cocoa helped sustain GDP growth in 2007. Ghana signed a Millennium Challenge Corporation (MCC) Compact in 2006, which aims to assist in transforming Ghana's agricultural sector.

GDP (purchasing power parity): $31.33 billion (2007 est.)

GDP (official exchange rate): $14.86 billion (2007 est.)

GDP—real growth rate: 6.4% (2007 est.)

GDP—per capita (PPP): $1,400 (2007 est.)

GDP—composition by sector:
agriculture: 37.3%
industry: 25.3%
services: 37.5% (2006 est.)

Labor force: 11.29 million (2007 est.)

Labor force—by occupation: *agriculture:* 56%
industry: 15%
services: 29% (2005 est.)

Unemployment rate: 11% (2000 est.)

Population below poverty line: 28.5% (2007 est.)

Household income or consumption by percentage share:
lowest 10%: 2.2%
highest 10%: 30.1% (1999)

Distribution of family income—Gini index: 39.4 (2005–06)

Inflation rate (consumer prices): 9.6% (2007 est.)

Investment (gross fixed): 31.6% of GDP (2007 est.)

Budget:
revenues: $4.262 billion
expenditures: $5.481 billion (2007 est.)

Public debt: 48.4% of GDP (2007 est.)

Agriculture—products: cocoa, rice, cassava (tapioca), peanuts, corn, shea nuts, bananas; timber

Industries: mining, lumbering, light manufacturing, aluminum smelting, food processing, cement, small commercial ship building

Industrial production growth rate: 7.4% (2007 est.)

Electricity—production: 7.042 billion kWh (2007 est.)

Electricity—production by source:
fossil fuel: 5%
hydro: 95%
nuclear: 0%
other: 0% (2001)

Electricity—consumption: 6.906 billion kWh (2007 est.)

Electricity—exports: 256 million kWh (2007 est.)

Electricity—imports: 461 million kWh (2007 est.)

Oil—production: 700 bbl/day (2007 est.)

Oil—consumption: 47,000 bbl/day (2005 est.)

Oil—exports: 8,041 bbl/day (2004)

Oil—imports: 45,010 bbl/day (2004)

Oil—proved reserves: 16.5 million bbl (1 January 2006 est.)

Natural gas—production: 0 cu m (2005 est.)

Natural gas—consumption: 0 cu m (2005 est.)

Natural gas—exports: 0 cu m (2005 est.)

Natural gas—imports: 0 cu m (2005)

Natural gas—proved reserves: 22.81 billion cu m (1 January 2006 est.)

Current account balance: -$1.896 billion (2007 est.)

Exports: $4.194 billion f.o.b. (2007 est.)

Exports—commodities: gold, cocoa, timber, tuna, bauxite, aluminum, manganese ore, diamonds, horticulture

Exports—partners: Netherlands 11.3%, UK 8.7%, US 6.7%, Spain 5.7%, Belgium 5.2%, France 4.4% (2006)

Imports: $8.073 billion f.o.b. (2007 est.)

Imports—commodities: capital equipment, petroleum, foodstuffs

Imports—partners: Nigeria 16.7%, China 13%, UK 5.7%, Belgium 4.7%, US 4.7%, South Africa 4.1%, France 4.1% (2006)

Economic aid—recipient: $1.316 billion in loans and grants (2007)

Reserves of foreign exchange and gold: $2.837 billion (31 December 2007 est.)

Debt—external: $4.898 billion (31 December 2007 est.)

Stock of direct foreign investment—at home: $NA

Stock of direct foreign investment—abroad: $NA

Market value of publicly traded shares: $13.01 billion (2007)

Currency (code): Ghana cedi (GHC)

Currency code: GHC

Exchange rates: cedis per US dollar—0.95 (2007), 9,174.8 (2006), 9,072.5 (2005), 9,004.6 (2004), 8,677.4 (2003)
note: in 2007 Ghana revalued its currency with 10,000 old cedis equal to 1 new cedis

Fiscal year: calendar year

COMMUNICATIONS

Telephones—main lines in use: 356,400 (2006)

Telephones—mobile cellular: 5.207 million (2006)

Telephone system:
general assessment: fixed-line infrastructure outdated and unreliable; competition among multiple mobile-cellular providers has spurred growth with subscribership about 25 per 100 persons and rising
domestic: primarily microwave radio relay; wireless local loop has been installed
international: country code—233; landing point for the SAT-3/WASC fiber-optic submarine cable that provides connectivity to Europe and Asia; satellite earth stations—4 Intelsat (Atlantic Ocean); microwave radio relay link to Panaftel system connects Ghana to its neighbors

Radio broadcast stations: AM 0, FM 86, shortwave 3 (2007)

Radios: 12.5 million (2001)

Television broadcast stations: 7 (2007)

Televisions: 1.9 million (2001)

Internet country code: .gh

Internet hosts: 2,899 (2007)

Internet Service Providers (ISPs): 12 (2000)

Internet users: 609,800 (2006)

TRANSPORTATION

Airports: 12 (2007)

Airports—with paved runways:
total: 7
over 3,047 m: 1
1,524 to 2,437 m: 4
914 to 1,523 m: 2 (2007)

Airports—with unpaved runways:
total: 5
914 to 1,523 m: 3
under 914 m: 2 (2007)

Pipelines: oil 13 km; refined products 316 km (2007)

Railways:
total: 953 km
narrow gauge: 953 km 1.067-m gauge (2006)

Roadways:
total: 62,221 km
paved: 9,955 km
unpaved: 52,266 km (2006)

Waterways: 1,293 km
note: 168 km for launches and lighters on Volta, Ankobra, and Tano rivers; 1,125 km of arterial and feeder waterways on Lake Volta (2007)

Merchant marine:
total: 3 ships (1000 GRT or over) 5,032 GRT/7,282 DWT
by type: petroleum tanker 1, refrigerated cargo 2
foreign-owned: 1 (Brazil 1) (2007)

Ports and terminals: Tema

MILITARY

Military branches: Ghanaian Army, Ghanaian Navy, Ghanaian Air Force (2007)

Military service age and obligation: 18 years of age for voluntary military service; no conscription (2008)
Manpower available for military service: *males age 16–49:* 5,802,096
females age 16–49: 5,729,939 (2008 est.)
Manpower fit for military service:
males age 16–49: 3,737,481
females age 16–49: 3,729,699 (2008 est.)
Manpower reaching militarily significant age annually:
males age 16–49: 273,265

females age 16–49: 267,204 (2008 est.)
Military expenditures—percent of GDP: 0.8% (2006 est.)

TRANSNATIONAL ISSUES

Disputes—international: Ghana struggles to accommodate returning nationals who worked in the cocoa plantations and escaped fighting in Cote d'Ivoire
Refugees and internally displaced persons: *refugees (country of origin):* 35,653 (Liberia); 8,517 (Togo) (2007)

Illicit drugs: illicit producer of cannabis for the international drug trade; major transit hub for Southwest and Southeast Asian heroin and, to a lesser extent, South American cocaine destined for Europe and the US; widespread crime and money laundering problem, but the lack of a well developed financial infrastructure limits the country's utility as a money laundering center; significant domestic cocaine and cannabis use

GIBRALTAR

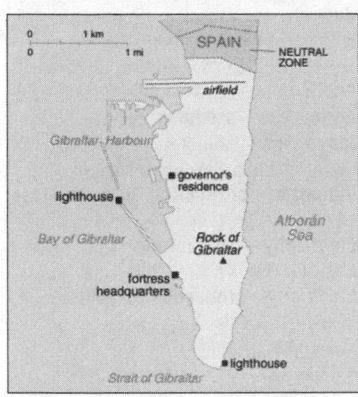

INTRODUCTION

Background: Strategically important, Gibraltar was reluctantly ceded to Great Britain by Spain in the 1713 Treaty of Utrecht; the British garrison was formally declared a colony in 1830. In a referendum held in 1967, Gibraltarians voted overwhelmingly to remain a British dependency. The subsequent granting of autonomy in 1969 by the UK led to Spain closing the border and severing all communication links. A series of talks were held by the UK and Spain between 1997 and 2002 on establishing temporary joint sovereignty over Gibraltar. In response to these talks, the Gibraltar Government called a referendum in late 2002 in which the majority of citizens voted overwhelmingly against any sharing of sovereignty with Spain. Since the referendum, tripartite talks on other issues have been held with Spain, the UK, and Gibraltar, and in September 2006 a three-way agreement was signed. Spain agreed to remove restrictions on air movements, to speed up customs procedures, to implement international telephone dialing, and to allow mobile roaming agreements. Britain agreed to pay increased pensions to Spaniards who had been employed in

Gibraltar before the border closed. Spain will be allowed to open a cultural institute from which the Spanish flag will fly. A new noncolonial constitution came into effect in 2007, but the UK retains responsibility for defense, foreign relations, internal security, and financial stability.

GEOGRAPHY

Location: Southwestern Europe, bordering the Strait of Gibraltar, which links the Mediterranean Sea and the North Atlantic Ocean, on the southern coast of Spain
Geographic coordinates: 36 08 N, 5 21 W
Map references: Europe
Area:
total: 6.5 sq km
land: 6.5 sq km
water: 0 sq km
Area—comparative: a little less than one half the size of Rhode Island
Land boundaries:
total: 1.2 km
border countries: Spain 1.2 km
Coastline: 12 km
Maritime claims: *territorial sea:* 3 nm
Climate: Mediterranean with mild winters and warm summers
Terrain: a narrow coastal lowland borders the Rock of Gibraltar
Elevation extremes:
lowest point: Mediterranean Sea 0 m
highest point: Rock of Gibraltar 426 m
Natural resources: none
Land use:
arable land: 0%
permanent crops: 0%
other: 100% (2005)
Irrigated land: NA
Natural hazards: NA
Environment—current issues: limited natural freshwater resources: large concrete or natural rock water catchments collect rainwater (no longer used for drinking water) and adequate desalination plant

Geography—note: strategic location on Strait of Gibraltar that links the North Atlantic Ocean and Mediterranean Sea

PEOPLE

Population: 28,002 (July 2008 est.)
Age structure:
0–14 years: 16.9% (male 2,426/female 2,309)
15–64 years: 66.6% (male 9,507/female 9,153)
65 years and over: 16.5% (male 2,103/female 2,504) (2008 est.)
Median age:
total: 40.3 years
male: 39.8 years
female: 40.7 years (2008 est.)
Population growth rate: 0.125% (2008 est.)
Birth rate: 10.71 births/1,000 population (2008 est.)
Death rate: 9.46 deaths/1,000 population (2008 est.)
Net migration rate: 0 migrant(s)/1,000 population (2008 est.)
Sex ratio:
at birth: 1.06 male(s)/female
under 15 years: 1.05 male(s)/female
15–64 years: 1.04 male(s)/female
65 years and over: 0.84 male(s)/female
total population: 1.01 male(s)/female (2008 est.)
Infant mortality rate:
total: 4.91 deaths/1,000 live births
male: 5.46 deaths/1,000 live births
female: 4.33 deaths/1,000 live births (2008 est.)
Life expectancy at birth:
total population: 80.06 years
male: 77.17 years
female: 83.09 years (2008 est.)
Total fertility rate: 1.65 children born/woman (2008 est.)
HIV/AIDS—adult prevalence rate: NA
HIV/AIDS—people living with HIV/AIDS: NA
HIV/AIDS—deaths: NA
Nationality: *noun:* Gibraltarian(s)
adjective: Gibraltar

Ethnic groups: Spanish, Italian, English, Maltese, Portuguese, German, North Africans

Religions: Roman Catholic 78.1%, Church of England 7%, other Christian 3.2%, Muslim 4%, Jewish 2.1%, Hindu 1.8%, other or unspecified 0.9%, none 2.9% (2001 census)

Languages: English (used in schools and for official purposes), Spanish, Italian, Portuguese

Literacy:
definition: NA
total population: above 80%
male: NA
female: NA

GOVERNMENT

Country name:
conventional long form: none
conventional short form: Gibraltar

Dependency status: overseas territory of the UK

Government type: NA

Capital: *name:* Gibraltar
geographic coordinates: 36 08 N, 5 21 W
time difference: UTC+1 (6 hours ahead of Washington, DC during Standard Time)
daylight saving time: +1hr, begins last Sunday in March; ends last Sunday in October

Administrative divisions: none (overseas territory of the UK)

Independence: none (overseas territory of the UK)

National holiday: National Day, 10 September (1967); note—day of the national referendum to decide whether to remain with the UK or go with Spain

Constitution: 5 June 2006; came into force 2 January 2007

Legal system: the laws of the UK, where applicable, apply

Suffrage: 18 years of age; universal; and British citizens who have been residents six months or more

Executive branch:
chief of state: Queen ELIZABETH II (since 6 February 1952); represented by Governor Sir Robert FULTON (since 27 October 2006)
head of government: Chief Minister Peter CARUANA (since 17 May 1996)
cabinet: Council of Ministers appointed from among the 17 elected members of the Parliament by the governor in consultation with the chief minister
elections: the monarch is hereditary; governor appointed by the monarch; following legislative elections, the leader of the majority party or the leader of the majority coalition is usually appointed chief minister by the governor

Legislative branch: unicameral Parliament (18 seats: 17 members elected by popular vote, 1 for the Speaker appointed by Parliament; to serve four-year terms)
elections: last held 11 October 2007 (next to be held not later than October 2011)
election results: percent of vote by party—GSD 49.3%, GSLP 31.8%, Gibraltar Liberal Party 13.6%; seats by party—GSD 10, GSLP 4, Gibraltar Liberal Party 3

Judicial branch: Supreme Court; Court of Appeal

Political parties and leaders: Gibraltar Liberal Party [Joseph GARCIA]; Gibraltar Social Democrats or GSD [Peter CARUANA]; Gibraltar Socialist Labor Party or GSLP [Joseph John BOSSANO]

Political pressure groups and leaders: Chamber of Commerce; Gibraltar Representatives Organization; Women's Association

International organization participation: Interpol (subbureau), UPU

Diplomatic representation in the US: none (overseas territory of the UK)

Diplomatic representation from the US: none (overseas territory of the UK)

Flag description: two horizontal bands of white (top, double width) and red with a three-towered red castle in the center of the white band; hanging from the castle gate is a gold key centered in the red band

ECONOMY

Economy—overview: Self-sufficient Gibraltar benefits from an extensive shipping trade, offshore banking, and its position as an international conference center. The British military presence has been sharply reduced and now contributes about 7% to the local economy, compared with 60% in 1984. The financial sector, tourism (almost 5 million visitors in 1998), shipping services fees, and duties on consumer goods also generate revenue. The financial sector, the shipping sector, and tourism each contribute 25%-30% of GDP. Telecommunications accounts for another 10%. In recent years, Gibraltar has seen major structural change from a public to a private sector economy, but changes in government spending still have a major impact on the level of employment.

GDP (purchasing power parity): $1.066 billion (2005 est.)

GDP (official exchange rate): $1.066 billion (2005 est.)

GDP—real growth rate: 7% (2005 est.)

GDP—per capita (PPP): $38,200 (2005 est.)

GDP—composition by sector:
agriculture: NA%
industry: NA%
services: NA%

Labor force: 12,690 (including non-Gibraltar laborers) (2001)

Labor force—by occupation: *agriculture:* negligible
industry: 40%
services: 60% (2001)

Unemployment rate: 3% (2005 est.)

Population below poverty line: NA%

Household income or consumption by percentage share:
lowest 10%: NA%
highest 10%: NA%

Inflation rate (consumer prices): 2.9% (2005)

Budget:
revenues: $455.1 million
expenditures: $423.6 million (2005 est.)

Public debt: 15.7% of GDP (2005 est.)

Agriculture—products: none

Industries: tourism, banking and finance, ship repairing, tobacco

Industrial production growth rate: NA%

Electricity—production: 141 million kWh (2005)

Electricity—production by source:
fossil fuel: 100%
hydro: 0%
nuclear: 0%
other: 0% (2001)

Electricity—consumption: 141 million kWh (2005)

Electricity—exports: 0 kWh (2005)

Electricity—imports: 0 kWh (2005)

Oil—production: 0 bbl/day (2005 est.)

Oil—consumption: 25,000 bbl/day (2005 est.)

Oil—exports: 0 bbl/day (2004)

Oil—imports: 24,350 bbl/day (2004)

Oil—proved reserves: 0 bbl (1 January 2006 est.)

Natural gas—production: 0 cu m (2005 est.)

Natural gas—consumption: 0 cu m (2005 est.)

Natural gas—exports: 0 cu m (2005 est.)

Natural gas—imports: 0 cu m (2005)

Natural gas—proved reserves: 0 cu m (1 January 2006 est.)

Exports: $271 million f.o.b. (2004 est.)

Exports—commodities: (principally reexports) petroleum 51%, manufactured goods 41%, other 8%

Exports—partners: UK 30.8%, Spain 22.7%, Germany 13.7%, Turkmenistan 10.4%, Switzerland 8.3%, Italy 6.7% (2006)

Imports: $2.967 billion c.i.f. (2004 est.)

Imports—commodities: fuels, manufactured goods, and foodstuffs

Imports—partners: Spain 23.4%, Russia 12.3%, Italy 12%, UK 9%, France 8.9%, Netherlands 6.8%, US 4.7% (2006)

Economic aid—recipient: $NA
Debt—external: $NA
Currency (code): Gibraltar pound (GIP)
Currency code: GIP
Exchange rates: Gibraltar pounds per US dollar—0.4993 (2007), 0.5434 (2006), 0.5504 (2005), 0.5462 (2004), 0.6125 (2003)
note: the Gibraltar pound is at par with the British pound
Fiscal year: 1 July—30 June

COMMUNICATIONS

Telephones—main lines in use: 24,512 (2002)
Telephones—mobile cellular: 9,797 (2002)
Telephone system:
general assessment: adequate, automatic domestic system and adequate international facilities
domestic: automatic exchange facilities
international: country code—350; radiotelephone; microwave radio relay; satellite earth station—1 Intelsat (Atlantic Ocean)
Radio broadcast stations: AM 1, FM 5, shortwave 0 (1998)
Radios: 37,000 (1997)

Television broadcast stations: 1 (plus 3 repeaters) (1997)
Televisions: 10,000 (1997)
Internet country code: .gi
Internet hosts: 380 (2007)
Internet Service Providers (ISPs): 2 (2000)
Internet users: 6,200 (2002)

TRANSPORTATION

Airports: 1 (2007)
Airports—with paved runways:
total: 1
1,524 to 2,437 m: 1 (2007)
Roadways:
total: 29 km
paved: 29 km (2002)
Merchant marine:
total: 216 ships (1000 GRT or over) 1,422,155 GRT/1,866,572 DWT
by type: barge carrier 2, bulk carrier 5, cargo 117, chemical tanker 39, container 31, passenger 1, petroleum tanker 13, roll on/roll off 7, specialized tanker 1
foreign-owned: 201 (Belgium 3, Cyprus 5, Denmark 9, Finland 3, France 1, Germany 117, Greece 8, Iceland 1, Italy 1, Netherlands 11, Norway 27, Sweden 10, UAE 2, UK 3)

registered in other countries: 7 (Liberia 7) (2007)
Ports and terminals: Gibraltar

MILITARY

Military branches: Royal Gibraltar Regiment
Manpower available for military service:
males age 16–49: 6,308 (2008 est.)
Manpower fit for military service:
males age 16–49: 5,244 (2008 est.)
Manpower reaching militarily significant age annually:
males age 16–49: 190 (2008 est.)
Military—note: defense is the responsibility of the UK; the Royal Gibraltar Regiment replaced the last British regular infantry forces in 1992

TRANSNATIONAL ISSUES

Disputes—international: in 2002, Gibraltar residents voted overwhelmingly by referendum to reject any "shared sovereignty" arrangement; the government of Gibraltar insists on equal participation in talks between the UK and Spain; Spain disapproves of UK plans to grant Gibraltar even greater autonomy

GREECE

INTRODUCTION

Background: Greece achieved independence from the Ottoman Empire in 1829. During the second half of the 19th century and the first half of the 20th century, it gradually added neighboring islands and territories, most with Greek-speaking populations. In World War II, Greece was first invaded by Italy (1940) and subsequently occupied by Germany (1941–44); fighting endured in a protracted civil war between supporters of the king and Communist rebels.

Following the latter's defeat in 1949, Greece joined NATO in 1952. A military dictatorship, which in 1967 suspended many political liberties and forced the king to flee the country, lasted seven years. The 1974 democratic elections and a referendum created a parliamentary republic and abolished the monarchy. In 1981, Greece joined the EC (now the EU); it became the 12th member of the European Economic and Monetary Union in 2001.

GEOGRAPHY

Location: Southern Europe, bordering the Aegean Sea, Ionian Sea, and the Mediterranean Sea, between Albania and Turkey
Geographic coordinates: 39 00 N, 22 00 E
Map references: Europe
Area:
total: 131,940 sq km
land: 130,800 sq km
water: 1,140 sq km
Area—comparative: slightly smaller than Alabama
Land boundaries:
total: 1,228 km
border countries: Albania 282 km,

Bulgaria 494 km, Turkey 206 km, Macedonia 246 km
Coastline: 13,676 km
Maritime claims:
territorial sea: 12 nm
continental shelf: 200-m depth or to the depth of exploitation
Climate: temperate; mild, wet winters; hot, dry summers
Terrain: mostly mountains with ranges extending into the sea as peninsulas or chains of islands
Elevation extremes:
lowest point: Mediterranean Sea 0 m
highest point: Mount Olympus 2,917 m
Natural resources: lignite, petroleum, iron ore, bauxite, lead, zinc, nickel, magnesite, marble, salt, hydropower potential
Land use:
arable land: 20.45%
permanent crops: 8.59%
other: 70.96% (2005)
Irrigated land: 14,530 sq km (2003)
Total renewable water resources: 72 cu km (2005)
Freshwater withdrawal (domestic/industrial/agricultural): total: 8.7 cu km/yr (16%/3%/81%)
per capita: 782 cu m/yr (1997)

Natural hazards: severe earthquakes
Environment—current issues: air pollution; water pollution
Environment—international agreements: *party to:* Air Pollution, Air Pollution-Nitrogen Oxides, Air Pollution-Sulfur 94, Antarctic-Environmental Protocol, Antarctic-Marine Living Resources, Antarctic Treaty, Biodiversity, Climate Change, Climate Change-Kyoto Protocol, Desertification, Endangered Species, Environmental Modification, Hazardous Wastes, Law of the Sea, Marine Dumping, Ozone Layer Protection, Ship Pollution, Tropical Timber 83, Tropical Timber 94, Wetlands
signed, but not ratified: Air Pollution-Persistent Organic Pollutants, Air Pollution-Volatile Organic Compounds
Geography—note: strategic location dominating the Aegean Sea and southern approach to Turkish Straits; a peninsular country, possessing an archipelago of about 2,000 islands

PEOPLE

Population: 10,722,816 (July 2008 est.)
Age structure:
0–14 years: 14.3% (male 789,137/female 742,469)
15–64 years: 66.6% (male 3,568,101/female 3,575,572)
65 years and over: 19.1% (male 898,337/female 1,149,200) (2008 est.)
Median age:
total: 41.5 years
male: 40.4 years
female: 42.6 years (2008 est.)
Population growth rate: 0.146% (2008 est.)
Birth rate: 9.54 births/1,000 population (2008 est.)
Death rate: 10.42 deaths/1,000 population (2008 est.)
Net migration rate: 2.33 migrant(s)/1,000 population (2008 est.)
Sex ratio:
at birth: 1.06 male(s)/female
under 15 years: 1.06 male(s)/female
15–64 years: 1 male(s)/female
65 years and over: 0.78 male(s)/female
total population: 0.96 male(s)/female (2008 est.)
Infant mortality rate:
total: 5.25 deaths/1,000 live births
male: 5.77 deaths/1,000 live births
female: 4.7 deaths/1,000 live births (2008 est.)
Life expectancy at birth:
total population: 79.52 years
male: 76.98 years
female: 82.21 years (2008 est.)
Total fertility rate: 1.36 children born/woman (2008 est.)

HIV/AIDS—adult prevalence rate: 0.2% (2001 est.)
HIV/AIDS—people living with HIV/AIDS: 9,100 (2001 est.)
HIV/AIDS—deaths: fewer than 100 (2003 est.)
Nationality:
noun: Greek(s)
adjective: Greek
Ethnic groups: Population: Greek 93%, other (foreign citizens) 7% (2001 census)
note: percents represent citizenship, since Greece does not collect data on ethnicity
Religions: Greek Orthodox 98%, Muslim 1.3%, other 0.7%
Languages: Greek 99% (official), other 1% (includes English and French)
Literacy:
definition: age 15 and over can read and write
total population: 96%
male: 97.8%
female: 94.2% (2001 census)

GOVERNMENT

Country name:
conventional long form: Hellenic Republic
conventional short form: Greece
local long form: Elliniki Dhimokratia
local short form: Ellas or Ellada
former: Kingdom of Greece
Government type: parliamentary republic
Capital: *name:* Athens
geographic coordinates: 37 59 N, 23 44 E
time difference: UTC+2 (7 hours ahead of Washington, DC during Standard Time)
daylight saving time: +1hr, begins last Sunday in March; ends last Sunday in October
Administrative divisions: 51 prefectures (nomoi, singular—nomos) and 1 autonomous region*; Achaia, Agion Oros* (Mt. Athos), Aitolia kai Akarnania, Argolis, Arkadia, Arta, Attiki, Chalkidiki, Chanion, Chios, Dodekanisos, Drama, Evros, Evrytania, Evvoia, Florina, Fokidos, Fthiotis, Grevena, Ileia, Imathia, Ioannina, Irakleion, Karditsa, Kastoria, Kavala, Kefallinia, Kerkyra, Kilkis, Korinthia, Kozani, Kyklades, Lakonia, Larisa, Lasithi, Lefkas, Lesvos, Magnisia, Messinia, Pella, Pieria, Preveza, Rethynnis, Rodopi, Samos, Serrai, Thesprotia, Thessaloniki, Trikala, Voiotia, Xanthi, Zakynthos
Independence: 1829 (from the Ottoman Empire)
National holiday: Independence Day, 25 March (1821)
Constitution: 11 June 1975; amended March 1986 and April 2001

Legal system: based on codified Roman law; judiciary divided into civil, criminal, and administrative courts; accepts compulsory ICJ jurisdiction with reservations
Suffrage: 18 years of age; universal and compulsory
Executive branch:
chief of state: President Karolos PAPOULIAS (since 12 March 2005)
head of government: Prime Minister Konstandinos (Kostas) KARAMANLIS (since 7 March 2004)
cabinet: Cabinet appointed by the president on the recommendation of the prime minister
elections: president elected by parliament for a five-year term (eligible for a second term); election last held 8 February 2005 (next to be held by February 2010); according to the Greek Constitution, presidents may only serve two terms; president appoints leader of the party securing plurality of vote in election to become prime minister and form a government
election results: Karolos PAPOULIAS elected president; number of parliamentary votes, 279 out of 300
Legislative branch: unicameral Parliament or Vouli ton Ellinon (300 seats; members are elected by direct popular vote to serve four-year terms)
elections: elections last held 16 September 2007 (next to be held by 2011)
election results: percent of vote by party—ND 41.8%, PASOK 38.1%, KKE 8.2%, Synaspismos 5%, LAOS 3.8%, other 3.1%; seats by party—ND 152, PASOK 102, KKE 22, Synaspismos 14, LAOS 10
Judicial branch: Supreme Judicial Court; Special Supreme Tribunal; all judges appointed for life by the president after consultation with a judicial council
Political parties and leaders: Coalition of the Left and Progress (Synaspismos) [Alekos ALAVANOS]; Communist Party of Greece or KKE [Aleka PAPARIGA]; New Democracy or ND (conservative) [Konstandinos KARAMANLIS]; Panhellenic Socialist Movement or PASOK [Yiorgos PAPANDREOU]; Popular Orthodox Rally or LAOS [Yeoryios KARATZAFERIS]
Political pressure groups and leaders: General Confederation of Greek Workers or GSEE [Ioannis PANAGOPOULOS]; Federation of Greek Industries or SEV [Dimitris DASKALOPOULOS]; Civil Servants Confederation or ADEDY [Spyros PAPASPYROS]
International organization participation: Australia Group, BIS, BSEC, CE,

CERN, EAPC, EBRD, EIB, EMU, EU, FAO, IAEA, IBRD, ICAO, ICC, ICCt, ICRM, IDA, IEA, IFAD, IFC, IFRCS, IHO, ILO, IMF, IMO, IMSO, Interpol, IOC, IOM, IPU, ISO, ITSO, ITU, ITUC, MIGA, MINURSO, NAM (guest), NATO, NEA, NSG, OAS (observer), OECD, OIF, OPCW, OSCE, PCA, Schengen Convention, SECI, UN, UN Security Council (temporary), UNCTAD, UNESCO, UNHCR, UNIDO, UNIFIL, UNMEE, UNMIS, UNOMIG, UNWTO, UPU, WCO, WEU, WFTU, WHO, WIPO, WMO, WTO, ZC

Diplomatic representation in the US:
chief of mission: Ambassador Alexandros P. MALLIAS
chancery: 2221 Massachusetts Avenue NW, Washington, DC 20008
telephone: [1] (202) 939-1300
FAX: [1] (202) 939-1324
consulate(s) general: Boston, Chicago, Los Angeles, New York, San Francisco, Tampa
consulate(s): Atlanta, Houston

Diplomatic representation from the US:
chief of mission: Ambassador Daniel V. SPECKHARD
embassy: 91 Vasilisis Sophias Avenue, 10160 Athens
mailing address: PSC 108, APO AE 09842-0108
telephone: [30] (210) 721-2951
FAX: [30] (210) 645-6282
consulate(s) general: Thessaloniki

Flag description: nine equal horizontal stripes of blue alternating with white; there is a blue square in the upper hoist-side corner bearing a white cross; the cross symbolizes Greek Orthodoxy, the established religion of the country

ECONOMY

Economy—overview: Greece has a capitalist economy with the public sector accounting for about 40% of GDP and with per capita GDP at least 75% of the leading euro-zone economies. Tourism provides 15% of GDP. Immigrants make up nearly one-fifth of the work force, mainly in agricultural and unskilled jobs. Greece is a major beneficiary of EU aid, equal to about 3.3% of annual GDP. The Greek economy grew by nearly 4.0% per year between 2003 and 2007, due partly to infrastructural spending related to the 2004 Athens Olympic Games, and in part to an increased availability of credit, which has sustained record levels of consumer spending. Greece violated the EU's Growth and Stability Pact budget deficit criteria of no more than 3% of GDP from 2001 to 2006, but finally met that criteria in 2007. Public debt, infla-

tion, and unemployment are above the euro-zone average, but are falling. The Greek Government continues to grapple with cutting government spending, reducing the size of the public sector, and reforming the labor and pension systems, in the face of often vocal opposition from the country's powerful labor unions and the general public. The economy remains an important domestic political issue in Greece and, while the ruling New Democracy government has had some success in improving economic growth and reducing the budget deficit, Athens faces long-term challenges in its effort to continue its economic reforms, especially social security reform and privatization.

GDP (purchasing power parity): $324.6 billion (2007 est.)
GDP (official exchange rate): $314.6 billion (2007 est.)
GDP—real growth rate: 4% (2007 est.)
GDP—per capita (PPP): $29,200 (2007 est.)
GDP—composition by sector:
agriculture: 3.6%
industry: 24.8%
services: 71.6% (2007 est.)
Labor force: 4.92 million (2007 est.)
Labor force—by occupation: *agriculture:* 12%
industry: 20%
services: 68% (2004 est.)
Unemployment rate: 8.3% (2007 est.)
Population below poverty line: NA%
Household income or consumption by percentage share:
lowest 10%: 2.5%
highest 10%: 26% (2000 est.)
Distribution of family income—Gini index: 33 (2005)
Inflation rate (consumer prices): 3% (2007 est.)
Investment (gross fixed): 26.2% of GDP (2007 est.)
Budget:
revenues: $111.8 billion
expenditures: $120.6 billion (2007 est.)
Public debt: 89.7% of GDP (2007 est.)
Agriculture—products: wheat, corn, barley, sugar beets, olives, tomatoes, wine, tobacco, potatoes; beef, dairy products
Industries: tourism, food and tobacco processing, textiles, chemicals, metal products; mining, petroleum
Industrial production growth rate: 3.2% (2007 est.)
Electricity—production: 56.13 billion kWh (2005)
Electricity—production by source:
fossil fuel: 94.5%
hydro: 3.8%
nuclear: 0%
other: 1.7% (2001)

Electricity—consumption: 54.31 billion kWh (2005)
Electricity—exports: 1.836 billion kWh (2005)
Electricity—imports: 5.616 billion kWh (2005)
Oil—production: 5,687 bbl/day (2005 est.)
Oil—consumption: 415,700 bbl/day (2005 est.)
Oil—exports: 119,200 bbl/day (2004)
Oil—imports: 550,400 bbl/day (2004)
Oil—proved reserves: 7 million bbl (1 January 2006 est.)
Natural gas—production: 15.35 million cu m (2005 est.)
Natural gas—consumption: 2.724 billion cu m (2005 est.)
Natural gas—exports: 0 cu m (2005 est.)
Natural gas—imports: 2.707 billion cu m (2005)
Natural gas—proved reserves: 950.5 million cu m (1 January 2006 est.)
Current account balance: -$43.7 billion (2007 est.)
Exports: $23.91 billion f.o.b. (2007 est.)
Exports—commodities: food and beverages, manufactured goods, petroleum products, chemicals, textiles
Exports—partners: Germany 11.4%, Italy 11.3%, Bulgaria 6.4%, UK 6%, Cyprus 5.4%, Turkey 5.1%, France 4.4%, US 4.4%, Spain 4% (2006)
Imports: $80.79 billion f.o.b. (2007 est.)
Imports—commodities: machinery, transport equipment, fuels, chemicals
Imports—partners: Germany 12.6%, Italy 11.5%, Russia 7.1%, France 5.9%, Netherlands 5.1%, South Korea 4.2% (2006)
Economic aid—donor: $424 million (2006)
Economic aid—recipient: $8 billion annually from EU (2000–06); Greece will receive about $3.8 billion per year between 2007–13 under the EU's Community Support Funds IV
Reserves of foreign exchange and gold: $3.658 billion (31 December 2007 est.)
Debt—external: $86.72 billion (31 December 2007)
Stock of direct foreign investment—at home: $43.18 billion (2007 est.)
Stock of direct foreign investment—abroad: $18.02 billion (2007 est.)
Market value of publicly traded shares: $145 billion (2005)
Currency (code): euro (EUR)
Currency code: EUR
Exchange rates: euros per US dollar—0.7345 (2007), 0.7964 (2006), 0.8041 (2005), 0.8054 (2004), 0.886 (2003)
Fiscal year: calendar year

261

COMMUNICATIONS

Telephones—main lines in use: 6.185 million (2006)
Telephones—mobile cellular: 11.098 million (2006)
Telephone system:
general assessment: adequate, modern networks reach all areas; good mobile telephone and international service
domestic: microwave radio relay trunk system; extensive open-wire connections; submarine cable to offshore islands
international: country code—30; landing point for the SEA-ME-WE-3 optical telecommunications submarine cable that provides links to Europe, Middle East, and Asia; a number of smaller submarine cables provide connectivity to various parts of Europe, the Middle East, and Cyprus; tropospheric scatter; satellite earth stations—4 (2 Intelsat—1 Atlantic Ocean and 1 Indian Ocean, 1 Eutelsat, and 1 Inmarsat—Indian Ocean region)
Radio broadcast stations: AM 26, FM 88, shortwave 4 (1998)
Radios: 5.02 million (1997)
Television broadcast stations: 36 (plus 1,341 repeaters); also 2 stations in the US Armed Forces Radio and Television Service (1995)
Televisions: 2.54 million (1997)
Internet country code: .gr
Internet hosts: 905,824 (2007)
Internet Service Providers (ISPs): 27 (2000)
Internet users: 2.048 million (2006)

TRANSPORTATION

Airports: 81 (2007)
Airports—with paved runways: *total:* 66
over 3,047 m: 5
2,438 to 3,047 m: 15
1,524 to 2,437 m: 20
914 to 1,523 m: 17
under 914 m: 9 (2007)
Airports—with unpaved runways:
total: 15
914 to 1,523 m: 3
under 914 m: 12 (2007)
Heliports: 9 (2007)

Pipelines: gas 1,166 km; oil 94 km (2007)
Railways: *total:* 2,571 km
standard gauge: 1,565 km 1.435-m gauge (764 km electrified)
narrow gauge: 961 km 1.000-m gauge; 22 km 0.750-m gauge
dual gauge: 23 km combined 1.435 m and 1.000-m gauges (three rail system) (2006)
Roadways:
total: 117,533 km
paved: 107,895 km (includes 880 km of expressways)
unpaved: 9,638 km (2005)
Waterways: 6 km
note: Corinth Canal (6 km) crosses the Isthmus of Corinth; shortens sea voyage by 325 km (2007)
Merchant marine:
total: 824 ships (1000 GRT or over) 33,654,384 GRT/57,898,789 DWT
by type: bulk carrier 246, cargo 66, carrier 1, chemical tanker 52, combination ore/oil 1, container 43, liquefied gas 6, passenger 11, passenger/cargo 109, petroleum tanker 269, roll on/roll off 19, specialized tanker 1
foreign-owned: 49 (Belgium 16, Cyprus 5, Italy 1, South Korea 2, UK 15, US 10)
registered in other countries: 2,324 (Antigua and Barbuda 3, Bahamas 214, Barbados 11, Belgium 4, Bermuda 3, Cambodia 5, Cayman Islands 23, China 1, Comoros 8, Cyprus 292, Denmark 4, Dominica 8, Egypt 8, Georgia 7, Gibraltar 8, Honduras 1, Hong Kong 30, Isle of Man 48, Italy 13, Jamaica 8, Lebanon 2, Liberia 311, Maldives 1, Malta 448, Marshall Islands 226, Norway 6, Panama 505, Philippines 3, Portugal 4, Russia 1, Sao Tome and Principe 1, Saudi Arabia 2, Sierra Leone 1, Singapore 14, Slovakia 4, St Kitts and Nevis 2, St Vincent and The Grenadines 81, UAE 3, UK 6, Uruguay 1, Venezuela 3, unknown 8) (2007)
Ports and terminals: Agioitheodoroi, Aspropyrgos, Pachi, Piraeus, Thessaloniki

MILITARY

Military branches: Hellenic Army (Ellinikos Stratos, ES), Hellenic Navy (Ellinikos Polemiko Navtiko, EPN), Hellenic Air Force (Elliniki Polimiki Aeroporia, EPA) (2007)
Military service age and obligation: 18 years of age for compulsory military service; during wartime the law allows for recruitment beginning January of the year of inductee's 18th birthday, thus including 17 year olds; 17 years of age for volunteers; conscript service obligation—12 months for the Army, Air Force; 15 months for Navy; women are eligible for voluntary military service (2007)
Manpower available for military service:
males age 16–49: 2,535,174
females age 16–49: 2,517,273 (2008 est.)
Manpower fit for military service:
males age 16–49: 2,084,469
females age 16–49: 2,065,956 (2008 est.)
Manpower reaching militarily significant age annually:
males age 16–49: 53,858
females age 16–49: 50,488 (2008 est.)
Military expenditures—percent of GDP: 4.3% (2005 est.)

TRANSNATIONAL ISSUES

Disputes—international: Greece and Turkey continue discussions to resolve their complex maritime, air, territorial, and boundary disputes in the Aegean Sea; Cyprus question with Turkey; Greece rejects the use of the name Macedonia or Republic of Macedonia; the mass migration of unemployed Albanians still remains a problem for developed countries, chiefly Greece and Italy
Illicit drugs: a gateway to Europe for traffickers smuggling cannabis and heroin from the Middle East and Southwest Asia to the West and precursor chemicals to the East; some South American cocaine transits or is consumed in Greece; money laundering related to drug trafficking and organized crime

GREENLAND

INTRODUCTION

Background: Greenland, the world's largest island, is about 81% ice-capped. Vikings reached the island in the 10th century from Iceland; Danish colonization began in the 18th century, and Greenland was made an integral part of Denmark in 1953. It joined the European Community (now the EU) with Denmark in 1973, but withdrew in 1985 over a dispute centered on stringent fishing quotas. Greenland was granted self-government in 1979 by the Danish parliament; the law went into effect the following year. Denmark continues to exercise control of Greenland's foreign affairs in consultation with Greenland's Home Rule Government.

GEOGRAPHY

Location: Northern North America, island between the Arctic Ocean and the North Atlantic Ocean, northeast of Canada

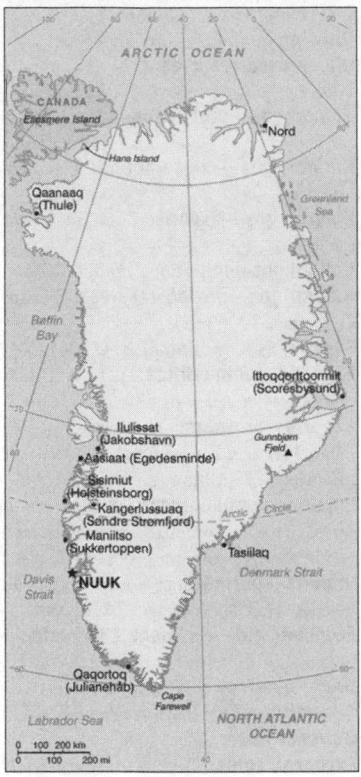

Geographic coordinates: 72 00 N, 40 00 W

Map references: Arctic Region

Area:

total: 2,166,086 sq km

land: 2,166,086 sq km (410,449 sq km ice-free, 1,755,637 sq km ice-covered) (2000 est.)

Area—comparative: slightly more than three times the size of Texas

Land boundaries: 0 km

Coastline: 44,087 km

Maritime claims:

territorial sea: 3 nm

exclusive fishing zone: 200 nm or agreed boundaries or median line

continental shelf: 200 nm or agreed boundaries or median line

Climate: arctic to subarctic; cool summers, cold winters

Terrain: flat to gradually sloping icecap covers all but a narrow, mountainous, barren, rocky coast

Elevation extremes:

lowest point: Atlantic Ocean 0 m

highest point: Gunnbjorn 3,700 m

Natural resources: coal, iron ore, lead, zinc, molybdenum, diamonds, gold, platinum, niobium, tantalite, uranium, fish, seals, whales, hydropower, possible oil and gas

Land use:

arable land: 0%

permanent crops: 0%

other: 100% (2005)

Irrigated land: NA

Natural hazards: continuous permafrost over northern two-thirds of the island

Environment—current issues: protection of the arctic environment; preservation of the Inuit traditional way of life, including whaling and seal hunting

Geography—note: dominates North Atlantic Ocean between North America and Europe; sparse population confined to small settlements along coast; close to one-quarter of the population lives in the capital, Nuuk; world's second largest ice cap

PEOPLE

Population: 56,326 (July 2008 est.)

Age structure:

0–14 years: 23.5% (male 6,788/female 6,464)

15–64 years: 69.2% (male 20,890/female 18,096)

65 years and over: 7.3% (male 1,977/female 2,111) (2008 est.)

Median age:

total: 34 years

male: 35.4 years

female: 32.3 years (2008 est.)

Population growth rate: -0.034% (2008 est.)

Birth rate: 16.08 births/1,000 population (2008 est.)

Death rate: 8.04 deaths/1,000 population (2008 est.)

Net migration rate: -8.38 migrant(s)/1,000 population (2008 est.)

Sex ratio:

at birth: 1.02 male(s)/female

under 15 years: 1.05 male(s)/female

15–64 years: 1.15 male(s)/female

65 years and over: 0.94 male(s)/female

total population: 1.11 male(s)/female (2008 est.)

Infant mortality rate:

total: 14.56 deaths/1,000 live births

male: 15.9 deaths/1,000 live births

female: 13.19 deaths/1,000 live births (2008 est.)

Life expectancy at birth:

total population: 70.53 years

male: 66.95 years

female: 74.2 years (2008 est.)

Total fertility rate: 2.39 children born/woman (2008 est.)

HIV/AIDS—adult prevalence rate: NA

HIV/AIDS—people living with HIV/AIDS: 100 (1999)

HIV/AIDS—deaths: NA

Nationality:

noun: Greenlander(s)

adjective: Greenlandic

Ethnic groups: Greenlander 88% (Inuit and Greenland-born whites), Danish and others 12% (2000)

Religions: Evangelical Lutheran

Languages: Greenlandic (East Inuit), Danish, English

Literacy: *definition:* age 15 and over can read and write

total population: 100%

male: 100%

female: 100% (2001 est.)

GOVERNMENT

Country name:

conventional long form: none

conventional short form: Greenland

local long form: none

local short form: Kalaallit Nunaat

Dependency status: part of the Kingdom of Denmark; self-governing overseas administrative division of Denmark since 1979

Government type: parliamentary democracy within a constitutional monarchy

Capital: *name:* Nuuk (Godthab)

geographic coordinates: 64 11 N, 51 45 W

time difference: UTC-3 (2 hours ahead of Washington, DC during Standard Time)

daylight saving time: +1hr, begins last Sunday in March; ends last Sunday in October

note: Greenland is divided into four time zones

Administrative divisions: 3 districts (landsdele); Avannaa (Nordgronland), Tunu (Ostgronland), Kitaa (Vestgronland)

note: there are 18 municipalities in Greenland

Independence: none (extensive self-rule as part of the Kingdom of Denmark; foreign affairs is the responsibility of Denmark, but Greenland actively participates in international agreements relating to Greenland)

National holiday: June 21 (longest day)

Constitution: 5 June 1953 (Danish constitution)

Legal system: the laws of Denmark, where applicable, apply

Suffrage: 18 years of age; universal

Executive branch:

chief of state: Queen MARGRETHE II of Denmark (since 14 January 1972), represented by High Commissioner Soren MOLLER (since April 2005)

head of government: Prime Minister Hans ENOKSEN (since 14 December 2002)

cabinet: Home Rule Government is elected by the parliament (Landstinget) on the basis of the strength of parties

elections: the monarchy is hereditary; high commissioner appointed by the monarch; prime minister is elected by parliament (usually the leader of the majority party);

election results: Hans ENOKSEN reelected prime minister

note: government coalition—Siumut and Inuit Ataqatigiit

Legislative branch: unicameral Parliament or Landstinget (31 seats; members are elected by popular vote on the basis of proportional representation to serve four-year terms)

elections: last held on 15 November 2005 (next to be held by December 2009)

election results: percent of vote by party—Siumut 30.7%, Demokratiit 22.8%, IA 22.6%, Atassut Party 19.1%; Katusseqatigiit 4.1%, other 0.7%; seats by party—Siumut 10, Demokratiit 7, IA 7, Atassut 6, Katuseqatigiit 1

note: two representatives were elected to the Danish Parliament or Folketing on 13 November 2007 (next to be held in November 2011); percent of vote by party—NA; seats by party—Siumut 1, Inuit Ataqatigiit 1

Judicial branch: High Court or Landsret (appeals can be made to the Ostre Landsret or Eastern Division of the High Court or Supreme Court in Copenhagen)

Political parties and leaders: Atassut Party (Solidarity) [Finn KARLSEN] (a conservative party favoring continuing close relations with Denmark); Demokratiit [Per BERTHELSEN]; Inuit Ataqatigiit or IA (Eskimo Brotherhood) [Josef MOTZFELDT] (a leftist party favoring complete independence from Denmark rather than home rule); Kattusseqatigiit (Candidate List) (an independent right-of-center party with no official platform); Siumut (Forward Party) [Hans ENOKSEN] (a social democratic party advocating more distinct Greenlandic identity and greater autonomy from Denmark)

Political pressure groups and leaders: NA

International organization participation: Arctic Council, NC, NIB, UPU

Diplomatic representation in the US: none (self-governing overseas administrative division of Denmark)

Diplomatic representation from the US: none (self-governing overseas administrative division of Denmark)

Flag description: two equal horizontal bands of white (top) and red with a large disk slightly to the hoist side of center—the top half of the disk is red, the bottom half is white

ECONOMY

Economy—overview: The economy remains critically dependent on exports of fish and a substantial subsidy from the Danish Government, which supplies about half of government revenues. The public sector, including publicly owned enterprises and the municipalities, plays the dominant role in the economy. Several interesting hydrocarbon and mineral exploration activities are ongoing. Press reports in early 2007 indicated that two international aluminum companies were considering building smelters in Greenland to take advantage of local hydropower potential. Tourism is the only sector offering any near-term potential, and even this is limited due to a short season and high costs. Air Greenland began summer-season direct flights to the US east coast in May 2007, potentially opening a major new tourism market.

GDP (purchasing power parity): $1.1 billion (2001 est.)

GDP (official exchange rate): $1.7 billion (2005)

GDP—real growth rate: 2% (2005 est.)

GDP—per capita (PPP): $20,000 (2001 est.)

GDP—composition by sector:
agriculture: NA%
industry: NA%
services: NA%

Labor force: 32,120 (2004)

Unemployment rate: 9.3% (2005 est.)

Population below poverty line: NA%

Household income or consumption by percentage share:
lowest 10%: NA%
highest 10%: NA%

Inflation rate (consumer prices): 1% (2005 est.)

Budget:
revenues: $1.36 billion
expenditures: $1.27 billion (2005)

Agriculture—products: forage crops, garden and greenhouse vegetables; sheep, reindeer; fish

Industries: fish processing (mainly shrimp and Greenland halibut); gold, niobium, tantalite, uranium, iron and diamond mining; handicrafts, hides and skins, small shipyards

Industrial production growth rate: NA%

Electricity—production: 300 million kWh (2005)

Electricity—production by source:
fossil fuel: 100%
hydro: 0%
nuclear: 0%
other: 0%
note: Greenland is shifting its electricity production from fossil fuel to hydropower production (2001)

Electricity—consumption: 279 million kWh (2005)

Electricity—exports: 0 kWh (2005)

Electricity—imports: 0 kWh (2005)

Oil—production: 0 bbl/day (2005 est.)

Oil—consumption: 3,880 bbl/day (2005 est.)

Oil—exports: 149.1 bbl/day (2004)

Oil—imports: 4,013 bbl/day (2004)

Oil—proved reserves: 0 bbl (1 January 2006 est.)

Natural gas—production: 0 cu m (2005 est.)

Natural gas—consumption: 0 cu m (2005 est.)

Natural gas—exports: 0 cu m (2005 est.)

Natural gas—imports: 0 cu m (2005)

Natural gas—proved reserves: 0 cu m (1 January 2006 est.)

Exports: $480 million f.o.b. (2006)

Exports—commodities: fish and fish products 94% (prawns 63%) (2001 est.)

Exports—partners: Denmark 67.1%, Japan 12.1%, China 5.6% (2006)

Imports: $712 million c.i.f. (2006)

Imports—commodities: machinery and transport equipment, manufactured goods, food, petroleum products

Imports—partners: Denmark 69.9%, Sweden 16.3%, Norway 3.7% (2006)

Economic aid—recipient: $512 million; note—subsidy from Denmark (2005)

Debt—external: $25 million (1999)

Currency (code): Danish krone (DKK)

Currency code: DKK

Exchange rates: Danish kroner per US dollar—5.4797 (2007), 5.9468 (2006), 5.9969 (2005), 5.9911 (2004), 6.5877 (2003)

Fiscal year: calendar year

COMMUNICATIONS

Telephones—main lines in use: 25,300 (2002)

Telephones—mobile cellular: 32,200 (2004)

Telephone system:
general assessment: adequate domestic and international service provided by satellite, cables and microwave radio relay; totally digitalized in 1995
domestic: microwave radio relay and satellite
international: country code—299; satellite earth stations—15 (12 Intelsat, 1 Eutelsat, 2 Americom GE-2 (all Atlantic Ocean)) (2000)

Radio broadcast stations: AM 5, FM 12, shortwave 0 (1998)

Radios: 30,000 (1998 est.)

Television broadcast stations: 1 (plus some local low-power stations, and 3 Armed Forces Radio and Television Service (AFRTS) stations (1997)

Televisions: 30,000 (1998 est.)

Internet country code: .gl

Internet hosts: 15,329 (2007)

Internet Service Providers (ISPs): 1 (2000)

Internet users: 38,000 (2005)

TRANSPORTATION

Airports: 14 (2007)
Airports—with paved runways: *total:* 9
2,438 to 3,047 m: 2
1,524 to 2,437 m: 1
914 to 1,523 m: 1
under 914 m: 5 (2007)
Airports—with unpaved runways:
total: 5
1,524 to 2,437 m: 1
914 to 1,523 m: 2
under 914 m: 2 (2007)
Roadways:
note: although there are short roads in towns, there are no roads between towns;
inter-urban transport takes place either by sea or air (2005)
Merchant marine:
total: 2 ships (1000 GRT or over) 3,422 GRT/2,340 DWT
by type: cargo 1, passenger 1
registered in other countries: 1 (Denmark 1) (2007)
Ports and terminals: Sisimiut

MILITARY

Military branches: no regular military forces
Manpower available for military service:
males age 16–49: 15,221 (2008 est.)
Manpower fit for military service:
males age 16–49: 10,739 (2008 est.)
Manpower reaching militarily significant age annually:
males age 16–49: 509 (2008 est.)
Military—note: defense is the responsibility of Denmark

TRANSNATIONAL ISSUES

Disputes—international: managed dispute between Canada and Denmark over Hans Island in the Kennedy Channel between Canada's Ellesmere Island and Greenland

GRENADA

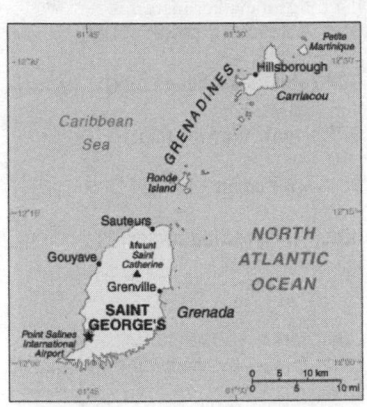

INTRODUCTION

Background: Carib Indians inhabited Grenada when COLUMBUS discovered the island in 1498, but it remained uncolonized for more than a century. The French settled Grenada in the 17th century, established sugar estates, and imported large numbers of African slaves. Britain took the island in 1762 and vigorously expanded sugar production. In the 19th century, cacao eventually surpassed sugar as the main export crop; in the 20th century, nutmeg became the leading export. In 1967, Britain gave Grenada autonomy over its internal affairs. Full independence was attained in 1974, making Grenada one of the smallest independent countries in the Western Hemisphere. Grenada was seized by a Marxist military council on 19 October 1983. Six days later the island was invaded by US forces and those of six other Caribbean nations, which quickly captured the ringleaders and their hundreds of Cuban advisers. Free elections were reinstituted the following year and have continued since that time. Hurricane Ivan struck Grenada in September of 2004 causing severe damage.

GEOGRAPHY

Location: Caribbean, island between the Caribbean Sea and Atlantic Ocean, north of Trinidad and Tobago
Geographic coordinates: 12 07 N, 61 40 W
Map references: Central America and the Caribbean
Area: *total:* 344 sq km
land: 344 sq km
water: 0 sq km
Area—comparative: twice the size of Washington, DC
Land boundaries: 0 km
Coastline: 121 km
Maritime claims:
territorial sea: 12 nm
exclusive economic zone: 200 nm
Climate: tropical; tempered by northeast trade winds
Terrain: volcanic in origin with central mountains
Elevation extremes:
lowest point: Caribbean Sea 0 m
highest point: Mount Saint Catherine 840 m
Natural resources: timber, tropical fruit, deepwater harbors
Land use:
arable land: 5.88%
permanent crops: 29.41%
other: 64.71% (2005)
Irrigated land: NA
Total renewable water resources: NA
Natural hazards: lies on edge of hurricane belt; hurricane season lasts from June to November
Environment—current issues: NA
Environment—international agreements: *party to:* Biodiversity, Climate Change, Climate Change-Kyoto Protocol, Desertification, Endangered Species, Law of the Sea, Ozone Layer Protection, Whaling
signed, but not ratified: none of the selected agreements
Geography—note: the administration of the islands of the Grenadines group is divided between Saint Vincent and the Grenadines and Grenada

PEOPLE

Population: 90,343 (July 2008 est.)
Age structure:
0–14 years: 32.4% (male 14,725/female 14,524)
15–64 years: 64.7% (male 30,911/female 27,502)
65 years and over: 3% (male 1,310/female 1,371) (2008 est.)
Median age:
total: 22.4 years
male: 22.9 years
female: 21.9 years (2008 est.)
Population growth rate: 0.406% (2008 est.)
Birth rate: 21.61 births/1,000 population (2008 est.)
Death rate: 6.31 deaths/1,000 population (2008 est.)
Net migration rate: -11.23 migrant(s)/1,000 population (2008 est.)
Sex ratio: *at birth:* 1 male(s)/female
under 15 years: 1.01 male(s)/female
15–64 years: 1.12 male(s)/female
65 years and over: 0.96 male(s)/female
total population: 1.08 male(s)/female (2008 est.)
Infant mortality rate:
total: 13.58 deaths/1,000 live births
male: 13.25 deaths/1,000 live births
female: 13.91 deaths/1,000 live births (2008 est.)
Life expectancy at birth:
total population: 65.6 years

male: 63.74 years
female: 67.47 years (2008 est.)
Total fertility rate: 2.27 children born/woman (2008 est.)
HIV/AIDS—adult prevalence rate: NA
HIV/AIDS—people living with HIV/AIDS: NA
HIV/AIDS—deaths: NA
Nationality:
noun: Grenadian(s)
adjective: Grenadian
Ethnic groups: black 82%, mixed black and European 13%, European and East Indian 5%, and trace of Arawak/Carib Amerindian
Religions: Roman Catholic 53%, Anglican 13.8%, other Protestant 33.2%
Languages: English (official), French patois
Literacy: *definition:* age 15 and over can read and write
total population: 96%
male: NA
female: NA (2003 est.)

GOVERNMENT

Country name:
conventional long form: none
conventional short form: Grenada
Government type: parliamentary democracy
Capital: *name:* Saint George's
geographic coordinates: 12 03 N, 61 45 W
time difference: UTC-4 (1 hour ahead of Washington, DC during Standard Time)
Administrative divisions: 6 parishes and 1 dependency*; Carriacou and Petite Martinique*, Saint Andrew, Saint David, Saint George, Saint John, Saint Mark, Saint Patrick
Independence: 7 February 1974 (from UK)
National holiday: Independence Day, 7 February (1974)
Constitution: 19 December 1973
Legal system: based on English common law; has not accepted compulsory ICJ jurisdiction
Suffrage: 18 years of age; universal
Executive branch:
chief of state: Queen ELIZABETH II (since 6 February 1952); represented by Governor General Daniel WILLIAMS (since 9 August 1996)
head of government: Prime Minister Keith MITCHELL (since 22 June 1995)
cabinet: Cabinet appointed by the governor general on the advice of the prime minister
elections: the monarch is hereditary; governor general appointed by the monarch; following legislative elections, the leader of the majority party or the leader of the majority coalition is usually appointed prime minister by the governor general

Legislative branch: bicameral Parliament consists of the Senate (13 seats, 10 appointed by the government and 3 by the leader of the opposition) and the House of Representatives (15 seats; members are elected by popular vote to serve five-year terms)
elections: last held on 27 November 2003 (next to be held by early 2009)
election results: House of Representatives—percent of vote by party—NNP 46.6%, NDC 44.1%, other 9.3%; seats by party—NNP 8, NDC 7
Judicial branch: Eastern Caribbean Supreme Court, consisting of a court of Appeal and a High Court of Justice (a High Court judge is assigned to and resides in Grenada)
Political parties and leaders: Grenada United Labor Party or GULP [Gloria Payne BANFIELD]; National Democratic Congress or NDC [Tillman THOMAS]; New National Party or NNP [Keith MITCHELL]
Political pressure groups and leaders: NA
International organization participation: ACP, C, Caricom, CDB, FAO, G-77, IBRD, ICAO, ICRM, IDA, IFAD, IFC, IFRCS, ILO, IMF, IMO, Interpol, IOC, ITU, ITUC, MIGA, NAM, OAS, OECS, OPANAL, OPCW, UN, UNCTAD, UNESCO, UNIDO, UPU, WHO, WIPO, WTO
Diplomatic representation in the US:
chief of mission: Ambassador Denis G. ANTOINE
chancery: 1701 New Hampshire Avenue NW, Washington, DC 20009
telephone: [1] (202) 265-2561
FAX: [1] (202) 265-2468
consulate(s) general: New York
Diplomatic representation from the US:
chief of mission: the US Ambassador to Barbados is accredited to Grenada
embassy: Lance-aux-Epines Stretch, Saint George's
mailing address: P. O. Box 54, Saint George's
telephone: [1] (473) 444-1173 through 1177
FAX: [1] (473) 444-4820
Flag description: a rectangle divided diagonally into yellow triangles (top and bottom) and green triangles (hoist side and outer side), with a red border around the flag; there are seven yellow, five-pointed stars with three centered in the top red border, three centered in the bottom red border, and one on a red disk superimposed at the center of the flag; there is also a symbolic nutmeg pod on the hoist-side triangle (Grenada is the world's second-largest producer of nutmeg, after Indonesia); the seven stars represent the seven administrative divisions

ECONOMY

Economy—overview: Grenada relies on tourism as its main source of foreign exchange, especially since the construction of an international airport in 1985. Strong performances in construction and manufacturing, together with the development of an offshore financial industry, have also contributed to growth in national output. Grenada has rebounded from the devastating effects of Hurricanes Ivan (2004) and Emily (2005), but is now saddled with the debt burden from the rebuilding process. The agricultural sector, particularly nutmeg and cocoa cultivation, has gradually recovered, and the tourism sector has seen substantial increases in foreign direct investment as the regional share of the tourism market increases.
GDP (purchasing power parity): $1.108 billion (2007 est.)
GDP (official exchange rate): $590 million (2007 est.)
GDP—real growth rate: 3.1% (2007 est.)
GDP—per capita (PPP): $10,500 (2007 est.)
GDP—composition by sector:
agriculture: 5.4%
industry: 18%
services: 76.6% (2003)
Labor force: 42,300 (1996)
Labor force—by occupation: *agriculture:* 24%
industry: 14%
services: 62% (1999 est.)
Unemployment rate: 12.5% (2000)
Population below poverty line: 32% (2000)
Household income or consumption by percentage share:
lowest 10%: NA%
highest 10%: NA%
Inflation rate (consumer prices): 3.7% (2007 est.)
Budget:
revenues: $85.8 million
expenditures: $102.1 million (1997)
Agriculture—products: bananas, cocoa, nutmeg, mace, citrus, avocados, root crops, sugarcane, corn, vegetables
Industries: food and beverages, textiles, light assembly operations, tourism, construction
Industrial production growth rate: 0.7% (1997 est.)
Electricity—production: 150 million kWh (2005)
Electricity—production by source:
fossil fuel: 100%
hydro: 0%
nuclear: 0%
other: 0% (2001)

Electricity—consumption: 139.5 million kWh (2005)
Electricity—exports: 0 kWh (2005)
Electricity—imports: 0 kWh (2005)
Oil—production: 0 bbl/day (2005 est.)
Oil—consumption: 1,800 bbl/day (2005 est.)
Oil—exports: 0 bbl/day (2004)
Oil—imports: 1,776 bbl/day (2004)
Oil—proved reserves: 0 bbl (1 January 2006 est.)
Natural gas—production: 0 cu m (2005 est.)
Natural gas—consumption: 0 cu m (2005 est.)
Natural gas—exports: 0 cu m (2005 est.)
Natural gas—imports: 0 cu m (2005)
Natural gas—proved reserves: 0 cu m (1 January 2006 est.)
Current account balance: -$138 million (2007 est.)
Exports: $38 million (2006)
Exports—commodities: bananas, cocoa, nutmeg, fruit and vegetables, clothing, mace
Exports—partners: Saint Lucia 18.8%, Antigua and Barbuda 12.8%, Saint Kitts & Nevis 11.5%, Dominica 11.4%, US 11.4% (2006)
Imports: $343 million (2006)
Imports—commodities: food, manufactured goods, machinery, chemicals, fuel
Imports—partners: Trinidad and Tobago 33.7%, US 24.2%, UK 4.3% (2006)
Economic aid—recipient: $44.87 million (2005)

Debt—external: $347 million (2004)
Market value of publicly traded shares: $NA
Currency (code): East Caribbean dollar (XCD)
Currency code: XCD
Exchange rates: East Caribbean dollars per US dollar—2.7 (2007), 2.7 (2006), 2.7 (2005), 2.7 (2004), 2.7 (2003)
Fiscal year: calendar year

COMMUNICATIONS

Telephones—main lines in use: 27,700 (2006)
Telephones—mobile cellular: 46,200 (2006)
Telephone system:
general assessment: automatic, islandwide telephone system
domestic: interisland VHF and UHF radiotelephone links
international: country code—1-473; landing point for the East Caribbean Fiber Optic System (ECFS) submarine cable with links to 13 other islands in the eastern Caribbean extending from the British Virgin Islands to Trinidad and SHF radiotelephone links to Trinidad and Tobago and Saint Vincent; VHF and UHF radio links to Trinidad
Radio broadcast stations: AM 2, FM 13, shortwave 0 (1998)
Radios: 57,000 (1997)
Television broadcast stations: 2 (1997)
Televisions: 33,000 (1997)
Internet country code: .gd
Internet hosts: 7 (2007)

Internet Service Providers (ISPs): 14 (2000)
Internet users: 19,000 (2003)

TRANSPORTATION

Airports: 3 (2007)
Airports—with paved runways:
total: 3
2,438 to 3,047 m: 1
1,524 to 2,437 m: 1
under 914 m: 1 (2007)
Roadways:
total: 1,127 km
paved: 687 km
unpaved: 440 km (2000)
Ports and terminals: Saint George's

MILITARY

Military branches: no regular military forces; Royal Grenada Police Force (includes Coast Guard) (2007)
Manpower available for military service:
males age 16–49: 27,309 (2008 est.)
Manpower fit for military service:
males age 16–49: 20,249 (2008 est.)
Manpower reaching militarily significant age annually:
males age 16–49: 1,034 (2008 est.)
Military expenditures—percent of GDP: NA

TRANSNATIONAL ISSUES

Disputes—international: none
Illicit drugs: small-scale cannabis cultivation; lesser transshipment point for marijuana and cocaine to US

GUAM

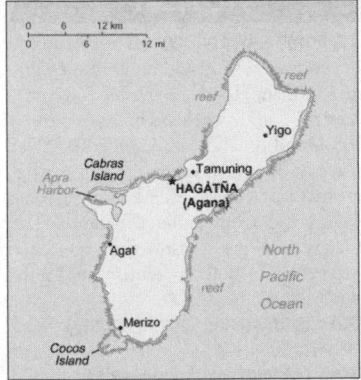

INTRODUCTION

Background: Guam was ceded to the US by Spain in 1898. Captured by the Japanese in 1941, it was retaken by the US three years later. The military installation on the island is one of the most strategically important US bases in the Pacific.

GEOGRAPHY

Location: Oceania, island in the North Pacific Ocean, about three-quarters of the way from Hawaii to the Philippines
Geographic coordinates: 13 28 N, 144 47 E
Map references: Oceania
Area: *total:* 541.3 sq km
land: 541.3 sq km
water: 0 sq km
Area—comparative: three times the size of Washington, DC
Land boundaries: 0 km
Coastline: 125.5 km
Maritime claims:
territorial sea: 12 nm
exclusive economic zone: 200 nm
Climate: tropical marine; generally warm and humid, moderated by northeast trade winds; dry season (January to June), rainy season (July to December); little seasonal temperature variation
Terrain: volcanic origin, surrounded by coral reefs; relatively flat coralline limestone plateau (source of most fresh water), with steep coastal cliffs and narrow coastal plains in north, low hills in center, mountains in south
Elevation extremes:
lowest point: Pacific Ocean 0 m
highest point: Mount Lamlam 406 m
Natural resources: aquatic wildlife (supporting tourism), fishing (largely undeveloped)
Land use:
arable land: 3.64%
permanent crops: 18.18%
other: 78.18% (2005)
Irrigated land: NA
Natural hazards: frequent squalls during rainy season; relatively rare, but potentially very destructive typhoons (June—December)

Environment—current issues: extirpation of native bird population by the rapid proliferation of the brown tree snake, an exotic, invasive species

Geography—note: largest and southernmost island in the Mariana Islands archipelago; strategic location in western North Pacific Ocean

PEOPLE

Population: 175,877 (July 2008 est.)

Age structure:

0–14 years: 28.2% (male 25,644/female 23,910)

15–64 years: 64.8% (male 58,034/female 55,900)

65 years and over: 7% (male 5,801/female 6,588) (2008 est.)

Median age:

total: 28.9 years

male: 28.7 years

female: 29.2 years (2008 est.)

Population growth rate: 1.373% (2008 est.)

Birth rate: 18.37 births/1,000 population (2008 est.)

Death rate: 4.65 deaths/1,000 population (2008 est.)

Net migration rate: NA

Sex ratio: at birth: 1.06 male(s)/female

under 15 years: 1.07 male(s)/female

15–64 years: 1.04 male(s)/female

65 years and over: 0.88 male(s)/female

total population: 1.04 male(s)/female (2008 est.)

Infant mortality rate:

total: 6.55 deaths/1,000 live births

male: 7.22 deaths/1,000 live births

female: 5.84 deaths/1,000 live births (2008 est.)

Life expectancy at birth:

total population: 78.93 years

male: 75.86 years

female: 82.19 years (2008 est.)

Total fertility rate: 2.55 children born/woman (2008 est.)

HIV/AIDS—adult prevalence rate: NA

HIV/AIDS—people living with HIV/AIDS: NA

HIV/AIDS—deaths: NA

Nationality:

noun: Guamanian(s) (US citizens)

adjective: Guamanian

Ethnic groups: Chamorro 37.1%, Filipino 26.3%, other Pacific islander 11.3%, white 6.9%, other Asian 6.3%, other ethnic origin or race 2.3%, mixed 9.8% (2000 census)

Religions: Roman Catholic 85%, other 15% (1999 est.)

Languages: English 38.3%, Chamorro 22.2%, Philippine languages 22.2%, other Pacific island languages 6.8%, Asian languages 7%, other languages 3.5% (2000 census)

Literacy:

definition: age 15 and over can read and write

total population: 99%

male: 99%

female: 99% (1990 est.)

GOVERNMENT

Country name:

conventional long form: Territory of Guam

conventional short form: Guam

local long form: Guahan

local short form: Guahan

Dependency status: organized, unincorporated territory of the US with policy relations between Guam and the US under the jurisdiction of the Office of Insular Affairs, US Department of the Interior

Government type: NA

Capital: name: Hagatna (Agana)

geographic coordinates: 13 28 N, 144 44 E

time difference: UTC+10 (15 hours ahead of Washington, DC during Standard Time)

Administrative divisions: none (territory of the US)

Independence: none (territory of the US)

National holiday: Discovery Day, first Monday in March (1521)

Constitution: Organic Act of Guam, 1 August 1950

Legal system: modeled on US; US federal laws apply

Suffrage: 18 years of age; universal; US citizens, but do not vote in US presidential elections

Executive branch:

chief of state: President George W. BUSH of the US (since 20 January 2001); Vice President Richard B. CHENEY (since 20 January 2001)

head of government: Governor Felix P. CAMACHO (since 6 January 2003); Lieutenant Governor Dr. Michael W. CRUZ (since 1 January 2007)

cabinet: heads of executive departments; appointed by the governor with the consent of the Guam legislature

elections: under the US Constitution, residents of unincorporated territories, such as Guam, do not vote in elections for US president and vice president; however, they may vote in Democratic and Republican presidential primary elections; governor and lieutenant governor elected on the same ticket by popular vote for four-year term (can serve two consecutive terms, then must wait a full term before running again); election last held 7 November 2006 (next to be held in November 2010)

election results: Felix P. CAMACHO reelected governor; Dr. Michael W.

CRUZ elected lieutenant governor; percent of vote—NA

Legislative branch: unicameral Legislature (15 seats; members are elected by popular vote to serve two-year terms)

elections: last held 7 November 2006 (next to be held in November 2008)

election results: percent of vote by party—NA; seats by party—Republican Party 8, Democratic Party 7

note: Guam elects one nonvoting delegate to the US House of Representatives; election last held 7 November 2006 (next to be held in November 2008); results—percent of vote by party—NA; seats by party—Democratic Party 1

Judicial branch: Federal District Court (judge is appointed by the president); Territorial Superior Court (judges appointed for eight-year terms by the governor)

Political parties and leaders: Democratic Party [leader Michael PHILLIPS]; Republican Party [Philip J. FLORES] (controls the legislature)

Political pressure groups and leaders: NA

International organization participation: IOC, SPC, UPU

Diplomatic representation in the US: none (territory of the US)

Diplomatic representation from the US: none (territory of the US)

Flag description: territorial flag is dark blue with a narrow red border on all four sides; centered is a red-bordered, pointed, vertical ellipse containing a beach scene, outrigger canoe with sail, and a palm tree with the word GUAM superimposed in bold red letters; US flag is the national flag

ECONOMY

Economy—overview: The economy depends largely on US military spending and tourism. Total US grants, wage payments, and procurement outlays amounted to $1.3 billion in 2004. Over the past 30 years, the tourist industry has grown to become the largest income source following national defense. The Guam economy continues to experience expansion in both its tourism and military sectors.

GDP (purchasing power parity): $2.5 billion (2005 est.)

GDP (official exchange rate): $2.773 billion (2001)

GDP—real growth rate: NA%

GDP—per capita (PPP): $15,000 (2005 est.)

GDP—composition by sector:

agriculture: NA%

industry: NA%

services: NA%

Labor force: 62,050 (2002 est.)

Labor force—by occupation:

agriculture: 26%

industry: 10%

services: 64% (2004 est.)

Unemployment rate: 11.4% (2002 est.)

Population below poverty line: 23% (2001 est.)

Household income or consumption by percentage share:

lowest 10%: NA%

highest 10%: NA%

Inflation rate (consumer prices): 2.5% (2005 est.)

Budget: *revenues:* $319.6 million

expenditures: $427.8 million (2002 est.)

Agriculture—products: fruits, copra, vegetables; eggs, pork, poultry, beef

Industries: US military, tourism, construction, transshipment services, concrete products, printing and publishing, food processing, textiles

Industrial production growth rate: NA%

Electricity—production: 1.793 billion kWh (2005)

Electricity—production by source:

fossil fuel: 100%

hydro: 0%

nuclear: 0%

other: 0% (2001)

Electricity—consumption: 1.667 billion kWh (2005)

Electricity—exports: 0 kWh (2005)

Electricity—imports: 0 kWh (2005)

Oil—production: 0 bbl/day (2005 est.)

Oil—consumption: 13,530 bbl/day (2005 est.)

Oil—exports: 0 bbl/day (2004)

Oil—imports: 12,130 bbl/day (2004)

Oil—proved reserves: 0 bbl (1 January 2006 est.)

Natural gas—production: 0 cu m (2005 est.)

Natural gas—consumption: 0 cu m (2005 est.)

Natural gas—exports: 0 cu m (2005 est.)

Natural gas—imports: 0 cu m (2005)

Natural gas—proved reserves: 0 cu m (1 January 2006 est.)

Exports: $45 million f.o.b. (2004 est.)

Exports—commodities: transshipments of refined petroleum products, construction materials, fish, food and beverage products

Exports—partners: Japan 67.2%, Singapore 11.6%, UK 4.8% (2006)

Imports: $701 million f.o.b. (2004 est.)

Imports—commodities: petroleum and petroleum products, food, manufactured goods

Imports—partners: Singapore 50%, South Korea 21.4%, Japan 14%, Hong Kong 4.6% (2006)

Economic aid—recipient: Guam receives large transfer payments from the US Federal Treasury into which Guamanians pay no income or excise taxes; under the provisions of a special law of Congress, the Guam Treasury, rather than the US Treasury, receives federal income taxes paid by military and civilian Federal employees stationed in Guam (2001 est.)

Debt—external: $NA

Currency (code): US dollar (USD)

Currency code: USD

Exchange rates: the US dollar is used

Fiscal year: 1 October—30 September

COMMUNICATIONS

Telephones—main lines in use: 80,000 (2001)

Telephones—mobile cellular: 98,000 (2004)

Telephone system:

general assessment: modern system, integrated with US facilities for direct dialing, including free use of 800 numbers

domestic: modern digital system, including cellular mobile service and local access to the Internet

international: country code—1-671; major landing point for submarine cables between Asia and the US (Guam is a trans-Pacific communications hub for major carriers linking the US and Asia); satellite earth stations—2 Intelsat (Pacific Ocean)

Radio broadcast stations: AM 3, FM 11, shortwave 2 (2005)

Radios: 221,000 (1997)

Television broadcast stations: 3 (2006)

Televisions: 106,000 (1997)

Internet country code: .gu

Internet hosts: 36 (2007)

Internet Service Providers (ISPs): 20 (2000)

Internet users: 65,000 (2005)

TRANSPORTATION

Airports: 5 (2007)

Airports—with paved runways:

total: 4

over 3,047 m: 2

2,438 to 3,047 m: 1

914 to 1,523 m: 1 (2007)

Airports—with unpaved runways:

total: 1

under 914 m: 1 (2007)

Roadways:

total: 977 km (2004)

Ports and terminals: Apra Harbor

MILITARY

Military—note: defense is the responsibility of the US

TRANSNATIONAL ISSUES

Disputes—international: none

GUATEMALA

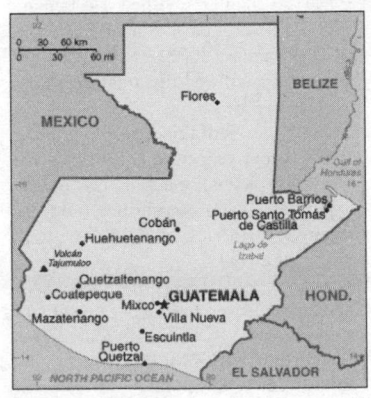

INTRODUCTION

Background: The Mayan civilization flourished in Guatemala and surrounding regions during the first millennium A.D. After almost three centuries as a Spanish colony, Guatemala won its independence in 1821. During the second half of the 20th century, it experienced a variety of military and civilian governments, as well as a 36-year guerrilla war. In 1996, the government signed a peace agreement formally ending the conflict, which had left more than 100,000 people dead and had created, by some estimates, some 1 million refugees.

GEOGRAPHY

Location: Central America, bordering the North Pacific Ocean, between El Salvador and Mexico, and bordering the Gulf of Honduras (Caribbean Sea) between Honduras and Belize

Geographic coordinates: 15 30 N, 90 15 W

Map references: Central America and the Caribbean

Area:

total: 108,890 sq km

land: 108,430 sq km

water: 460 sq km

Area—comparative: slightly smaller than Tennessee

Land boundaries:

total: 1,687 km

border countries: Belize 266 km, El Salvador 203 km, Honduras 256 km, Mexico 962 km

Coastline: 400 km

Maritime claims:

territorial sea: 12 nm

exclusive economic zone: 200 nm

continental shelf: 200-m depth or to the depth of exploitation

Climate: tropical; hot, humid in low-lands; cooler in highlands

Terrain: mostly mountains with narrow coastal plains and rolling limestone plateau

Elevation extremes:

lowest point: Pacific Ocean 0 m

highest point: Volcan Tajumulco 4,211 m

Natural resources: petroleum, nickel, rare woods, fish, chicle, hydropower

Land use:

arable land: 13.22%

permanent crops: 5.6%

other: 81.18% (2005)

Irrigated land: 1,300 sq km (2003)

Total renewable water resources: 111.3 cu km (2000)

Freshwater withdrawal (domestic/industrial/agricultural): *total:* 2.01 cu km/yr (6%/13%/80%)

per capita: 160 cu m/yr (2000)

Natural hazards: numerous volcanoes in mountains, with occasional violent earthquakes; Caribbean coast extremely susceptible to hurricanes and other tropical storms

Environment—current issues: defor-estation in the Peten rainforest; soil ero-sion; water pollution

Environment—international agreements: *party to:* Antarctic Treaty, Biodiversity, Climate Change, Climate Change-Kyoto Protocol, Desertification, Endangered Species, Environmental Modification, Hazardous Wastes, Law of the Sea, Marine Dumping, Ozone Layer Protection, Ship Pollution, Wetlands, Whaling

signed, but not ratified: none of the selected agreements

Geography—note: no natural harbors on west coast

PEOPLE

Population: 13,002,206 (July 2008 est.)

Age structure:

0–14 years: 40.1% (male 2,653,915/female 2,565,841)

15–64 years: 56.2% (male 3,539,874/female 3,762,471)

65 years and over: 3.7% (male 222,303/female 257,802) (2008 est.)

Median age: *total:* 19.2 years

male: 18.6 years

female: 19.7 years (2008 est.)

Population growth rate: 2.11% (2008 est.)

Birth rate: 28.55 births/1,000 population (2008 est.)

Death rate: 5.19 deaths/1,000 population (2008 est.)

Net migration rate: -2.26 migrant(s)/1,000 population (2008 est.)

Sex ratio:

at birth: 1.05 male(s)/female

under 15 years: 1.03 male(s)/female

15–64 years: 0.94 male(s)/female

65 years and over: 0.86 male(s)/female

total population: 0.97 male(s)/female (2008 est.)

Infant mortality rate:

total: 28.79 deaths/1,000 live births

male: 31.21 deaths/1,000 live births

female: 26.24 deaths/1,000 live births (2008 est.)

Life expectancy at birth:

total population: 69.99 years

male: 68.22 years

female: 71.86 years (2008 est.)

Total fertility rate: 3.59 children born/woman (2008 est.)

HIV/AIDS—adult prevalence rate: 1.1% (2003 est.)

HIV/AIDS—people living with HIV/AIDS: 78,000 (2003 est.)

HIV/AIDS—deaths: 5,800 (2003 est.)

Major infectious diseases:

degree of risk: intermediate

food or waterborne diseases: bacterial diar-rhea, hepatitis A, and typhoid fever

vectorborne disease: dengue fever and malaria (2008)

Nationality:

noun: Guatemalan(s)

adjective: Guatemalan

Ethnic groups: Mestizo (mixed Amerindian-Spanish—in local Spanish called Ladino) and European 59.4%, K'iche 9.1%, Kaqchikel 8.4%, Mam 7.9%, Q'eqchi 6.3%, other Mayan 8.6%, indigenous non-Mayan 0.2%, other 0.1% (2001 census)

Religions: Roman Catholic, Protestant, indigenous Mayan beliefs

Languages: Spanish 60%, Amerindian languages 40% (23 officially recognized Amerindian languages, including Quiche, Cakchiquel, Kekchi, Mam, Garifuna, and Xinca)

Literacy: *definition:* age 15 and over can read and write

total population: 69.1%

male: 75.4%

female: 63.3% (2002 census)

GOVERNMENT

Country name:

conventional long form: Republic of Guatemala

conventional short form: Guatemala

local long form: Republica de Guatemala

local short form: Guatemala

Government type: constitutional demo-cratic republic

Capital: *name:* Guatemala

geographic coordinates: 14 37 N, 90 31 W

time difference: UTC-6 (1 hour behind Washington, DC during Standard Time)

daylight saving time: +1hr, begins last Sunday in April; ends last Friday in September; note—there is no DST planned for 2007–2009

Administrative divisions: 22 depart-ments (departamentos, singular—depar-tamento); Alta Verapaz, Baja Verapaz, Chimaltenango, Chiquimula, El Progreso, Escuintla, Guatemala, Huehuetenango, Izabal, Jalapa, Jutiapa, Peten, Quetzaltenango, Quiche, Retalhuleu, Sacatepequez, San Marcos, Santa Rosa, Solola, Suchitepequez, Totonicapan, Zacapa

Independence: 15 September 1821 (from Spain)

National holiday: Independence Day, 15 September (1821)

Constitution: 31 May 1985, effective 14 January 1986; note—suspended 25 May 1993 by former President Jorge SER-RANO; reinstated 5 June 1993 following ouster of president; amended November 1993

Legal system: civil law system; judicial review of legislative acts; has not accepted compulsory ICJ jurisdiction

Suffrage: 18 years of age; universal; note—active duty members of the armed forces may not vote and are restricted to their barracks on election day

Executive branch:

chief of state: President Alvaro COLOM Caballeros (since 14 January 2008); Vice President Rafael ESPADA (since 14 January 2008); note—the president is both the chief of state and head of gov-ernment

head of government: President Alvaro COLOM Caballeros (since 14 January 2008); Vice President Rafael ESPADA (since 14 January 2008)

cabinet: Council of Ministers appointed by the president

elections: president elected by popular vote for a four-year term (may not serve consecutive terms); election last held 9 September 2007; runoff held 4 November 2007 (next to be held September 2011)

election results: Alvaro COLOM Caballeros elected president; percent of vote—Alvaro COLOM Caballeros 52.8%, Otto PEREZ Molina 47.2%

Legislative branch: unicameral Congress of the Republic or Congreso de la Republica (158 seats; members are elected by popular vote to serve four-year terms)

elections: last held 9 September 2007 (next to be held in September 2011)

election results: percent of vote by party—UNE 30.4%, GANA 23.4%, PP 18.9%, FRG 9.5%, PU 5.1%, other 12.7%; seats by party—UNE 48, GANA 37, PP 30, FRG 15, PU 8, CASA 5, EG 4, PAN 4, UCN 4, URNG 2, UD 1

Judicial branch: Constitutional Court or Corte de Constitucionalidad is Guatemala's highest court (five judges are elected for concurrent five-year terms); Supreme Court of Justice or Corte Suprema de Justicia (13 members serve concurrent five-year terms and elect a president of the Court each year from among their number; the president of the Supreme Court of Justice also supervises trial judges around the country, who are named to five-year terms)

Political parties and leaders: Center of Social Action or CASA [Eduardo SUGER]; Democracy Front or FRENTE [Alfonso CABRERA]; Democratic Union or UD [Manuel CONDE Orellana]; Encounter for Guatemala or EG [Nineth MONTENGRO]; Grand National Alliance or GANA [Alfredo VILLA]; Guatemalan National Revolutionary Unity or URNG [Hector NUILA]; Guatemalan Republican Front or FRG [Efrain RIOS Montt]; National Advancement Party or PAN [Ruben Dario MORALES]; National Unity for Hope or UNE [Alvaro COLOM Caballeros]; Patriot Party or PP [Ret. Gen. Otto PEREZ Molina]; Unionista Party or PU [Fritz GARCIA]; Unity of National Change or UCN [Sidney SHAW]

Political pressure groups and leaders: Agrarian Owners Group or UNAGRO; Alliance Against Impunity or AAI; Committee for Campesino Unity or CUC; Coordinating Committee of Agricultural, Commercial, Industrial, and Financial Associations or CACIF; Mutual Support Group or GAM

International organization participation: BCIE, CACM, FAO, G-24, G-77, IADB, IAEA, IBRD, ICAO, ICC, ICRM, IDA, IFAD, IFC, IFRCS, IHO, ILO, IMF, IMO, Interpol, IOC, IOM, IPU, ISO (correspondent), ITSO, ITU, ITUC, LAES, LAIA (observer), MIGA, MINUSTAH, MONUC, NAM, OAS, OPANAL, OPCW, PCA, RG, UN, UNCTAD, UNESCO, UNIDO, UNIFIL, Union Latina, UNMEE, UNMIS, UNOCI, UNWTO, UPU, WCL, WCO, WFTU, WHO, WIPO, WMO, WTO

Diplomatic representation in the US:
chief of mission: Ambassador Francisco VILLAGRAN de Leon
chancery: 2220 R Street NW, Washington, DC 20008
telephone: [1] (202) 745-4952
FAX: [1] (202) 745-1908
consulate(s) general: Chicago, Denver, Houston, Los Angeles, Miami, New York, Providence, San Francisco

Diplomatic representation from the US:
chief of mission: Ambassador James M. DERHAM
embassy: 7-01 Avenida Reforma, Zone 10, Guatemala City
mailing address: APO AA 34024
telephone: [502] 2326-4000
FAX: [502] 2326-4654

Flag description: three equal vertical bands of light blue (hoist side), white, and light blue with the coat of arms centered in the white band; the coat of arms includes a green and red quetzal (the national bird) and a scroll bearing the inscription LIBERTAD 15 DE SEPTIEMBRE DE 1821 (the original date of independence from Spain) all superimposed on a pair of crossed rifles and a pair of crossed swords and framed by a wreath

ECONOMY

Economy—overview: Guatemala is the most populous of the Central American countries with a GDP per capita roughly one-half that of Argentina, Brazil, and Chile. The agricultural sector accounts for about one-tenth of GDP, two-fifths of exports, and half of the labor force. Coffee, sugar, and bananas are the main products, with sugar exports benefiting from increased global demand for ethanol. The 1996 signing of peace accords, which ended 36 years of civil war, removed a major obstacle to foreign investment, and Guatemala since then has pursued important reforms and macroeconomic stabilization. On 1 July 2006, the Central American Free Trade Agreement (CAFTA) entered into force between the US and Guatemala and has since spurred increased investment in the export sector. The distribution of income remains highly unequal with about 56% of the population below the poverty line. Other ongoing challenges include increasing government revenues, negotiating further assistance from international donors, upgrading both government and private financial operations, curtailing drug trafficking and rampant crime, and narrowing the trade deficit. Given Guatemala's large expatriate community in the United States, it is the top remittance recipient in Central America, with inflows serving as a primary source of foreign income equivalent to nearly two-thirds of exports.

GDP (purchasing power parity): $62.53 billion (2007 est.)
GDP (official exchange rate): $33.69 billion (2007 est.)
GDP—real growth rate: 5.7% (2007 est.)
GDP—per capita (PPP): $4,700 (2007 est.)
GDP—composition by sector:
agriculture: 13.2%
industry: 25.9%
services: 60.8% (2007 est.)
Labor force: 3.958 million (2007 est.)
Labor force—by occupation: *agriculture:* 50%
industry: 15%
services: 35% (1999 est.)
Unemployment rate: 3.2% (2005 est.)
Population below poverty line: 56.2% (2004 est.)
Household income or consumption by percentage share:
lowest 10%: 0.9%
highest 10%: 43.4% (2002)
Distribution of family income—Gini index: 55.1 (2007)
Inflation rate (consumer prices): 6.8% (2007 est.)
Investment (gross fixed): 17.4% of GDP (2007 est.)
Budget:
revenues: $4.38 billion
expenditures: $4.872 billion (2007 est.)
Public debt: 20.8% of GDP (2007 est.)
Agriculture—products: sugarcane, corn, bananas, coffee, beans, cardamom; cattle, sheep, pigs, chickens
Industries: sugar, textiles and clothing, furniture, chemicals, petroleum, metals, rubber, tourism
Industrial production growth rate: 5% (2007 est.)
Electricity—production: 7.281 billion kWh (2005)
Electricity—production by source:
fossil fuel: 51.9%
hydro: 35.2%
nuclear: 0%
other: 12.9% (2001)
Electricity—consumption: 6.361 billion kWh (2005)
Electricity—exports: 339 million kWh (2005)
Electricity—imports: 23 million kWh (2005)
Oil—production: 20,100 bbl/day (2006 est.)
Oil—consumption: 73,510 bbl/day (2006 est.)
Oil—exports: 15,560 bbl/day (2006 est.)
Oil—imports: 72,960 bbl/day (2006 est.)
Oil—proved reserves: 526 million bbl (1 January 2006 est.)

Natural gas—production: 0 cu m (2005 est.)

Natural gas—consumption: 0 cu m (2005 est.)

Natural gas—exports: 0 cu m (2005 est.)

Natural gas—imports: 0 cu m (2005)

Natural gas—proved reserves: 2.96 billion cu m (1 January 2006 est.)

Current account balance: -$1.685 billion (2007 est.)

Exports: $6.94 billion f.o.b. (2007 est.)

Exports—commodities: coffee, sugar, petroleum, apparel, bananas, fruits and vegetables, cardamom

Exports—partners: US 44.6%, El Salvador 11.9%, Honduras 7.2%, Mexico 5.2% (2006)

Imports: $12.62 billion f.o.b. (2007 est.)

Imports—commodities: fuels, machinery and transport equipment, construction materials, grain, fertilizers, electricity

Imports—partners: US 33.2%, Mexico 8.8%, China 6.5%, El Salvador 5.3%, South Korea 4.9% (2006)

Economic aid—recipient: $253.6 million (2005 est.)

Reserves of foreign exchange and gold: $4.139 billion (31 December 2007 est.)

Debt—external: $5.908 billion (31 December 2007 est.)

Market value of publicly traded shares: $NA

Currency (code): quetzal (GTQ), US dollar (USD), others allowed

Currency code: GTQ; USD

Exchange rates: quetzales per US dollar—7.6833 (2007), 7.6026 (2006), 7.6339 (2005), 7.9465 (2004), 7.9409 (2003)

Fiscal year: calendar year

COMMUNICATIONS

Telephones—main lines in use: 1.355 million (2006)

Telephones—mobile cellular: 7.179 million (2006)

Telephone system:
general assessment: fairly modern network centered in the city of Guatemala
domestic: state-owned telecommunications company privatized in the late 1990s opening the way for competition; fixed-line teledensity 11 per 100 persons; mobile-cellular teledensity approaching 60 per 100 persons

international: country code—502; landing point for both the Americas Region Caribbean Ring System (ARCOS-1) and the SAM-1 fiber optic submarine cable system that together provide connectivity to South and Central America, parts of the Caribbean, and the US; connected to Central American Microwave System; satellite earth station—1 Intelsat (Atlantic Ocean)

Radio broadcast stations: AM 130, FM 487, shortwave 15 (2000)

Radios: 835,000 (1997)

Television broadcast stations: 26 (plus 27 repeaters) (1997)

Televisions: 1.323 million (1997)

Internet country code: .gt

Internet hosts: 40,927 (2007)

Internet Service Providers (ISPs): 5 (2000)

Internet users: 1.32 million (2006)

TRANSPORTATION

Airports: 402 (2007)

Airports—with paved runways:
total: 12
2,438 to 3,047 m: 3
1,524 to 2,437 m: 2
914 to 1,523 m: 4
under 914 m: 3 (2007)

Airports—with unpaved runways:
total: 390
2,438 to 3,047 m: 1
1,524 to 2,437 m: 6
914 to 1,523 m: 82
under 914 m: 301 (2007)

Pipelines: oil 480 km (2007)

Railways:
total: 886 km
narrow gauge: 886 km 0.914-m gauge (2006)

Roadways:
total: 14,095 km
paved: 4,863 km (includes 75 km of expressways)
unpaved: 9,232 km (2000)

Waterways: 990 km
note: 260 km navigable year round; additional 730 km navigable during high-water season (2007)

Ports and terminals: Puerto Quetzal, Santo Tomas de Castilla

MILITARY

Military branches: Army, Navy (includes Marines), Air Force

Military service age and obligation: all male citizens between the ages of 18 and 50 are liable for military service; conscript service obligation varies from 12 to 24 months; women can serve as officers (2007)

Manpower available for military service:
males age 16–49: 2,861,696
females age 16–49: 3,062,967 (2008 est.)

Manpower fit for military service:
males age 16–49: 2,310,272
females age 16–49: 2,622,450 (2008 est.)

Manpower reaching militarily significant age annually:
males age 16–49: 161,550
females age 16–49: 159,760 (2008 est.)

Military expenditures—percent of GDP: 0.4% (2006)

TRANSNATIONAL ISSUES

Disputes—international: annual ministerial meetings under the OAS-initiated Agreement on the Framework for Negotiations and Confidence Building Measures continue to address Guatemalan land and maritime claims in Belize and the Caribbean Sea; the Line of Adjacency created under the 2002 Differendum serves in lieu of the contiguous international boundary to control squatting in the sparsely inhabited rain forests of Belize's border region; Mexico must deal with thousands of impoverished Guatemalans and other Central Americans who cross the porous border looking for work in Mexico and the United States

Refugees and internally displaced persons: *IDPs:* undetermined (the UN does not estimate there are any IDPs, although some NGOs estimate over 200,000 IDPs as a result of over three decades of internal conflict that ended in 1996) (2007)

Illicit drugs: major transit country for cocaine and heroin; in 2005, cultivated 100 hectares of opium poppy after reemerging as a potential source of opium in 2004; potential production of less than 1 metric ton of pure heroin; marijuana cultivation for mostly domestic consumption; proximity to Mexico makes Guatemala a major staging area for drugs (particularly for cocaine); money laundering is a serious problem; corruption is a major problem

GUERNSEY

INTRODUCTION

Background: Guernsey and the other Channel Islands represent the last remnants of the medieval Dukedom of Normandy, which held sway in both France and England. The islands were the only British soil occupied by German troops in World War II. Guernsey is a British crown dependency, but is not part of the UK. However, the UK Government is constitutionally respon-

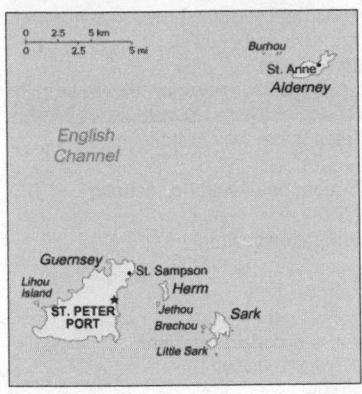

sible for its defense and international representation.

GEOGRAPHY

Location: Western Europe, islands in the English Channel, northwest of France
Geographic coordinates: 49 28 N, 2 35 W
Map references: Europe
Area:
total: 78 sq km
land: 78 sq km
water: 0 sq km
note: includes Alderney, Guernsey, Herm, Sark, and some other smaller islands
Area—comparative: about one-half the size of Washington, DC
Land boundaries: 0 km
Coastline: 50 km
Maritime claims:
territorial sea: 3 nm
exclusive fishing zone: 12 nm
Climate: temperate with mild winters and cool summers; about 50% of days are overcast
Terrain: mostly level with low hills in southwest
Elevation extremes:
lowest point: Atlantic Ocean 0 m
highest point: unnamed location on Sark 114 m
Natural resources: cropland
Land use: *arable land:* NA
permanent crops: NA
other: NA
Irrigated land: NA
Natural hazards: NA
Environment—current issues: NA
Geography—note: large, deepwater harbor at Saint Peter Port

PEOPLE

Population: 65,726 (July 2008 est.)
Age structure:
0–14 years: 14.6% (male 4,849/female 4,727)
15–64 years: 67.5% (male 22,013/female 22,380)
65 years and over: 17.9% (male 4,988/female 6,769) (2008 est.)
Median age:
total: 42.1 years
male: 41 years
female: 43 years (2008 est.)
Population growth rate: 0.228% (2008 est.)
Birth rate: 8.57 births/1,000 population (2008 est.)
Death rate: 10.09 deaths/1,000 population (2008 est.)
Net migration rate: 3.8 migrant(s)/1,000 population (2008 est.)
Sex ratio:
at birth: 1.03 male(s)/female
under 15 years: 1.03 male(s)/female
15–64 years: 0.98 male(s)/female
65 years and over: 0.74 male(s)/female
total population: 0.94 male(s)/female (2008 est.)
Infant mortality rate:
total: 4.53 deaths/1,000 live births
male: 5.05 deaths/1,000 live births
female: 3.98 deaths/1,000 live births (2008 est.)
Life expectancy at birth:
total population: 80.65 years
male: 77.64 years
female: 83.76 years (2008 est.)
Total fertility rate: 1.4 children born/woman (2008 est.)
HIV/AIDS—adult prevalence rate: NA
HIV/AIDS—people living with HIV/AIDS: NA
HIV/AIDS—deaths: NA
Nationality:
noun: Channel Islander(s)
adjective: Channel Islander
Ethnic groups: UK and Norman-French descent with small percentages from other European countries
Religions: Anglican, Roman Catholic, Presbyterian, Baptist, Congregational, Methodist
Languages: English, French, Norman-French dialect spoken in country districts
Literacy:
NA

GOVERNMENT

Country name:
conventional long form: Bailiwick of Guernsey
conventional short form: Guernsey
Dependency status: British crown dependency
Government type: parliamentary democracy
Capital: *name:* Saint Peter Port
geographic coordinates: 49 27 N, 2 32 W
time difference: UTC 0 (5 hours ahead of Washington, DC during Standard Time)
daylight saving time: +1hr, begins last Sunday in March; ends last Sunday in October
Administrative divisions: none (British crown dependency); there are no first-order administrative divisions as defined by the US Government, but there are 10 parishes including Castel, Forest, Saint Andrew, Saint Martin, Saint Peter Port, Saint Pierre du Bois, Saint Sampson, Saint Saviour, Torteval, Vale
Independence: none (British crown dependency)
National holiday: Liberation Day, 9 May (1945)
Constitution: unwritten; partly statutes, partly common law and practice
Legal system: the laws of the UK, where applicable, apply; justice is administered by the Royal Court
Suffrage: 18 years of age; universal
Executive branch:
chief of state: Queen ELIZABETH II (since 6 February 1952), represented by Lieutenant Governor Sir Fabian MALBON (since 28 October 2005)
head of government: Chief Minister Lyndon TROTT (since 1 May 2008)
cabinet: Policy Council elected by the States of Deliberation
elections: the monarch is hereditary; lieutenant governor appointed by the monarch; chief minister is elected by States of Deliberation
election results: Lyndon TROTT elected chief minister, percent of vote of the States of Deliberation NA
Legislative branch: unicameral States of Deliberation (45 seats; members are elected by popular vote for four years); note—Alderney and Sark have parliaments
elections: last held 23 April 2008 (next to be held in 2012)
election results: percent of vote—NA; seats—all independents
Judicial branch: Royal Court (judges elected by an electoral college and the bailiff)
Political parties and leaders: none; all independents
Political pressure groups and leaders: none
International organization participation: UPU
Diplomatic representation in the US: none (British crown dependency)
Diplomatic representation from the US: none (British crown dependency)
Flag description: white with the red cross of Saint George (patron saint of England) extending to the edges of the flag and a yellow equal-armed cross of William the Conqueror superimposed on the Saint George cross

ECONOMY

Economy—overview: Financial services—banking, fund management, insurance—account for about 23% of employment and about 55% of total income in this tiny, prosperous Channel Island economy. Tourism, manufacturing, and horticulture, mainly tomatoes and cut flowers, have been declining. Financial services, construction, retail, and the public sector have been growing. Light tax and death duties make Guernsey a popular tax haven. The evolving economic integration of the EU nations is changing the environment under which Guernsey operates.

GDP (purchasing power parity): $2.742 billion (2005)

GDP (official exchange rate): $2.742 billion (2005)

GDP—real growth rate: 3% (2005 est.)

GDP—per capita (PPP): $44,600 (2005)

GDP—composition by sector:
agriculture: 3%
industry: 10%
services: 87% (2000)

Labor force: 31,470 (March 2006)

Unemployment rate: 0.9% (March 2006 est.)

Population below poverty line: NA%

Household income or consumption by percentage share:
lowest 10%: NA%
highest 10%: NA%

Inflation rate (consumer prices): 3.4% (June 2006)

Budget:
revenues: $563.6 million
expenditures: $530.9 million (2005)

Agriculture—products: tomatoes, greenhouse flowers, sweet peppers, eggplant, fruit; Guernsey cattle

Industries: tourism, banking

Industrial production growth rate: NA%

Electricity—production: NA kWh

Electricity—production by source:
fossil fuel: NA
hydro: NA
nuclear: NA
other: NA

Electricity—consumption: NA kWh

Electricity—exports: 0 kWh (2002)

Electricity—imports: 0 kWh (2002)

Exports: $NA

Exports—commodities: tomatoes, flowers and ferns, sweet peppers, eggplant, other vegetables

Exports—partners: UK; note—regarded as internal trade (2006)

Imports: $NA

Imports—commodities: coal, gasoline, oil, machinery and equipment

Imports—partners: UK; note—regarded as internal trade (2006)

Economic aid—recipient: $NA

Debt—external: $NA

Currency (code): Guernsey pound
note: the British pound is also legal tender

Currency code: GBP

Exchange rates: Guernsey pounds per US dollar—0.4993 (2007), 0.5418 (2006), 0.5493 (2005), 0.5462 (2004), 0.6125 (2003)

note: the Guernsey pound is at par with the British pound

Fiscal year: calendar year

COMMUNICATIONS

Telephones—main lines in use: 45,100 (2005)

Telephones—mobile cellular: 43,800 (2004)

Telephone system:
general assessment: NA
domestic: NA
international: 1 submarine cable

Radio broadcast stations: AM 1, FM 1, shortwave 0 (1998)

Radios: NA

Television broadcast stations: 1 (1997)

Televisions: NA

Internet country code: .gg

Internet hosts: 2,734 (2007)

Internet Service Providers (ISPs): NA

Internet users: 36,000 (2005)

TRANSPORTATION

Airports: 2 (2007)

Airports—with paved runways:
total: 2
914 to 1,523 m: 1
under 914 m: 1 (2007)

Ports and terminals: Saint Peter Port, Saint Sampson

MILITARY

Military—note: defense is the responsibility of the UK

TRANSNATIONAL ISSUES

Disputes—international: none

GUINEA

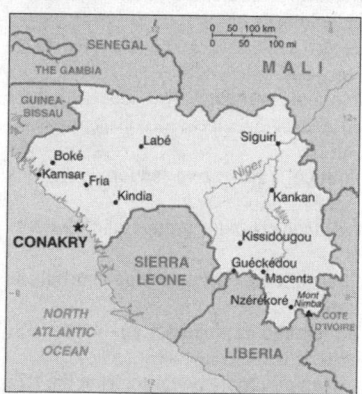

INTRODUCTION

Background: Guinea has had only two presidents since gaining its independence from France in 1958. Lansana CONTE came to power in 1984 when the military seized the government after the death of the first president, Sekou TOURE. Guinea did not hold democratic elections until 1993 when Gen. CONTE (head of the military government) was elected president of the civilian government. He was reelected in 1998 and again in 2003, though all the polls have been marred by irregularities. Guinea has maintained its internal stability despite spillover effects from conflict in Sierra Leone and Liberia. As those countries have rebuilt, Guinea's own vulnerability to political and economic crisis has increased. Declining economic conditions and popular dissatisfaction with corruption and bad governance prompted two massive strikes in 2006; a third nationwide strike in early 2007 sparked violent protests in many Guinean cities and prompted two weeks of martial law. To appease the unions and end the unrest, CONTE named a new prime minister in March 2007.

GEOGRAPHY

Location: Western Africa, bordering the North Atlantic Ocean, between Guinea-Bissau and Sierra Leone

Geographic coordinates: 11 00 N, 10 00 W

Map references: Africa

Area:
total: 245,857 sq km
land: 245,857 sq km
water: 0 sq km

Area—comparative: slightly smaller than Oregon

Land boundaries:
total: 3,399 km
border countries: Cote d'Ivoire 610 km, Guinea-Bissau 386 km, Liberia 563 km, Mali 858 km, Senegal 330 km, Sierra Leone 652 km

Coastline: 320 km

Maritime claims: *territorial sea:* 12 nm *exclusive economic zone:* 200 nm

Climate: generally hot and humid; monsoonal-type rainy season (June to November) with southwesterly winds; dry season (December to May) with northeasterly harmattan winds

Terrain: generally flat coastal plain, hilly to mountainous interior

Elevation extremes:
lowest point: Atlantic Ocean 0 m
highest point: Mont Nimba 1,752 m

Natural resources: bauxite, iron ore, diamonds, gold, uranium, hydropower, fish, salt

Land use: *arable land:* 4.47%
permanent crops: 2.64%
other: 92.89% (2005)

Irrigated land: 950 sq km (2003)

Total renewable water resources: 226 cu km (1987)

Freshwater withdrawal (domestic/industrial/agricultural): *total:* 1.51 cu km/yr (8%/2%/90%)
per capita: 161 cu m/yr (2000)

Natural hazards: hot, dry, dusty harmattan haze may reduce visibility during dry season

Environment—current issues: deforestation; inadequate supplies of potable water; desertification; soil contamination and erosion; overfishing, overpopulation in forest region; poor mining practices have led to environmental damage

Environment—international agreements: *party to:* Biodiversity, Climate Change, Climate Change-Kyoto Protocol, Desertification, Endangered Species, Hazardous Wastes, Law of the Sea, Ozone Layer Protection, Ship Pollution, Wetlands, Whaling
signed, but not ratified: none of the selected agreements

Geography—note: the Niger and its important tributary the Milo have their sources in the Guinean highlands

PEOPLE

Population: 10,211,437 (July 2008 est.)
Age structure:
0–14 years: 44.3% (male 2,282,453/female 2,239,611)
15–64 years: 52.5% (male 2,684,444/female 2,680,472)
65 years and over: 3.2% (male 142,327/female 182,130) (2008 est.)

Median age: *total:* 17.7 years
male: 17.5 years
female: 17.9 years (2008 est.)

Population growth rate: 2.612% (2008 est.)

Birth rate: 41.31 births/1,000 population (2008 est.)

Death rate: 15.19 deaths/1,000 population (2008 est.)

Net migration rate: -3 migrant(s)/1,000 population (2005 est.)

Sex ratio:
at birth: 1.03 male(s)/female
under 15 years: 1.02 male(s)/female
15–64 years: 1 male(s)/female
65 years and over: 0.78 male(s)/female
total population: 1 male(s)/female (2008 est.)

Infant mortality rate:
total: 87.17 deaths/1,000 live births
male: 92.23 deaths/1,000 live births
female: 81.96 deaths/1,000 live births (2008 est.)

Life expectancy at birth:
total population: 49.8 years
male: 48.66 years
female: 50.97 years (2008 est.)

Total fertility rate: 5.71 children born/woman (2008 est.)

HIV/AIDS—adult prevalence rate: 3.2% (2003 est.)

HIV/AIDS—people living with HIV/AIDS: 140,000 (2003 est.)

HIV/AIDS—deaths: 9,000 (2003 est.)

Major infectious diseases:
degree of risk: very high
food or waterborne diseases: bacterial and protozoal diarrhea, hepatitis A, and typhoid fever
vectorborne diseases: malaria and yellow fever
water contact disease: schistosomiasis
respiratory disease: meningococcal meningitis
aerosolized dust or soil contact disease: Lassa fever (2008)

Nationality:
noun: Guinean(s)
adjective: Guinean

Ethnic groups: Peuhl 40%, Malinke 30%, Soussou 20%, smaller ethnic groups 10%

Religions: Muslim 85%, Christian 8%, indigenous beliefs 7%

Languages: French (official); note—each ethnic group has its own language

Literacy: *definition:* age 15 and over can read and write
total population: 29.5%
male: 42.6%
female: 18.1% (2003 est.)

GOVERNMENT

Country name:
conventional long form: Republic of Guinea
conventional short form: Guinea
local long form: Republique de Guinee
local short form: Guinee
former: French Guinea

Government type: republic

Capital: *name:* Conakry
geographic coordinates: 9 33 N, 13 42 W
time difference: UTC 0 (5 hours ahead of Washington, DC during Standard Time)

Administrative divisions: 33 prefectures and 1 special zone (zone special)*; Beyla, Boffa, Boke, Conakry*, Coyah, Dabola, Dalaba, Dinguiraye, Dubreka, Faranah, Forecariah, Fria, Gaoual, Gueckedou, Kankan, Kerouane, Kindia, Kissidougou, Koubia, Koundara, Kouroussa, Labe, Lelouma, Lola, Macenta, Mali, Mamou, Mandiana, Nzerekore, Pita, Siguiri, Telimele, Tougue, Yomou

Independence: 2 October 1958 (from France)

National holiday: Independence Day, 2 October (1958)

Constitution: 23 December 1990 (Loi Fundamentale)

Legal system: based on French civil law system, customary law, and decree; accepts compulsory ICJ jurisdiction with reservations

Suffrage: 18 years of age; universal

Executive branch:
chief of state: President Lansana CONTE (head of military government since 5 April 1984, elected president 19 December 1993)
head of government: Prime Minister Ahmed Tidiane SOUARE (since 23 May 2008)
cabinet: Council of Ministers appointed by the president
elections: president elected by popular vote for a seven-year term (no term limits); candidate must receive a majority of the votes cast to be elected president; election last held 21 December 2003 (next to be held in December 2010); the prime minister is appointed by the president
election results: Lansana CONTE reelected president; percent of vote—Lansana CONTE 95.3%, Mamadou Bhoye BARRY 4.6%

Legislative branch: unicameral People's National Assembly or Assemblee Nationale Populaire (114 seats; members are elected by a mixed system of direct popular vote and proportional party lists)
elections: last held 30 June 2002 (next to be held in 2008)
election results: percent of vote by party—PUP 61.6%, UPR 26.6%, other 11.8%; seats by party—PUP 85, UPR 20, other 9
note: legislative elections were due in 2007 but have been postponed

Judicial branch: Court of First Instance or Tribunal de Premiere Instance; Court of Appeal or Cour d'Appel; Supreme Court or Cour Supreme

Political parties and leaders: National Union for Progress or UPN [Mamadou Bhoye BARRY]; Party for Unity and Progress or PUP (the governing party) [Lansana CONTE]; People's Party of

Guinea or PPG [Charles Pascal TOLNO]; Rally for the Guinean People or RPG [Alpha CONDE]; Union of Democratic Forces of Guinea or UFDG [Cellou Dalein DIALLO]; Union of Republican Forces or UFR [Sidya TOURE]; Union for Progress of Guinea or UPG [Jean-Marie DORE, secretary-general]; Union for Progress and Renewal or UPR [Ousmane BAH]

Political pressure groups and leaders: National Confederation of Guinean Workers—Labor Union of Guinean Workers or CNTG-USTG Alliance: National Confederation of Guinean Workers [Rabiatou Sarah DIALLO] and Labor Union of Guinean Workers [Dr. Ibrahima FOFANA]; Syndicate of Guinean Teachers and Researchers or SLECG [Dr. Louis M'Bemba SOUMAH]; National Council of Civil Society Organizations of Guinea CNOSCG [Ben Sekou SYLLA]

International organization participation: ACCT, ACP, AfDB, AU, ECOWAS, FAO, G-77, IBRD, ICAO, ICCt, ICRM, IDA, IDB, IFAD, IFC, IFRCS, ILO, IMF, IMO, Interpol, IOC, IOM, ITSO, ITU, ITUC, MIGA, MINURSO, NAM, OIC, OIF, OPCW, UN, UNCTAD, UNESCO, UNHCR, UNIDO, UNMIS, UNOCI, UNWTO, UPU, WCL, WCO, WFTU, WHO, WIPO, WMO, WTO

Diplomatic representation in the US:
chief of mission: Ambassador Mory Karamoko KABA
chancery: 2112 Leroy Place NW, Washington, DC 20008
telephone: [1] (202) 483-9420
FAX: [1] (202) 483-8688

Diplomatic representation from the US:
chief of mission: Ambassador Phillip CARTER III
embassy: Koloma, Conakry, east of Hamdallaye Circle
mailing address: B. P. 603, Transversale No. 2, Centre Administratif de Koloma, Commune de Ratoma, Conakry
telephone: [224] 30-42-08-61 through 68
FAX: [224] 30-42-08-73

Flag description: three equal vertical bands of red (hoist side), yellow, and green; uses the popular pan-African colors of Ethiopia

ECONOMY

Economy—overview: Guinea possesses major mineral, hydropower, and agricultural resources, yet remains an underdeveloped nation. The country has almost half of the world's bauxite reserves and is the second-largest bauxite producer. The mining sector accounts for over 70% of exports. Long-run improvements in government fiscal arrangements, literacy, and the legal framework are needed if the country is to move out of poverty. Investor confidence has been sapped by rampant corruption, a lack of electricity and other infrastructure, a lack of skilled workers, and the political uncertainty due to the failing health of President Lansana CONTE. Guinea is trying to reengage with the IMF and World Bank, which cut off most assistance in 2003, and is working closely with technical advisors from the U.S. Treasury Department, the World Bank and IMF, seeking to return to a fully funded program. Growth rose slightly in 2006–07, primarily due to increases in global demand and commodity prices on world markets, but the standard of living fell. The Guinea franc depreciated sharply as the prices for basic necessities like food and fuel rose beyond the reach of most Guineans. Dissatisfaction with economic conditions prompted nationwide strikes in February and June 2006.

GDP (purchasing power parity): $10.69 billion (2007 est.)
GDP (official exchange rate): $4.714 billion (2007 est.)
GDP—real growth rate: 1.5% (2007 est.)
GDP—per capita (PPP): $1,100 (2007 est.)
GDP—composition by sector:
agriculture: 22%
industry: 40.5%
services: 37.6% (2007 est.)
Labor force: 3.7 million (2006 est.)
Labor force—by occupation: *agriculture:* 76%
industry and services: 24% (2006 est.)
Unemployment rate: NA%
Population below poverty line: 47% (2006 est.)
Household income or consumption by percentage share:
lowest 10%: 1.9%
highest 10%: 41% (2006)
Distribution of family income—Gini index: 38.1 (2006)
Inflation rate (consumer prices): 22.9% (2007 est.)
Investment (gross fixed): 11.3% of GDP (2007 est.)
Budget:
revenues: $375 million
expenditures: $802.3 million (2007 est.)
Agriculture—products: rice, coffee, pineapples, palm kernels, cassava (tapioca), bananas, sweet potatoes; cattle, sheep, goats; timber
Industries: bauxite, gold, diamonds, iron; alumina refining; light manufacturing, and agricultural processing
Industrial production growth rate: 7.6% (2007 est.)

Electricity—production: 840 million kWh
note: excludes electricity generated at interior mining sites (2006)
Electricity—production by source:
fossil fuel: 45.5%
hydro: 54.5%
nuclear: 0%
other: 0% (2001)
Electricity—consumption: 832.9 million kWh (2006)
Electricity—exports: 0 kWh (2006)
Electricity—imports: 0 kWh (2006)
Oil—production: 0 bbl/day (2006 est.)
Oil—consumption: 9,650 bbl/day (2006 est.)
Oil—exports: 0 bbl/day (2004)
Oil—imports: 8,481 bbl/day (2004)
Oil—proved reserves: 0 bbl (1 January 2006 est.)
Natural gas—production: 0 cu m (2005 est.)
Natural gas—consumption: 0 cu m (2005 est.)
Natural gas—exports: 0 cu m (2005 est.)
Natural gas—imports: 0 cu m (2005)
Natural gas—proved reserves: 0 cu m (1 January 2006 est.)
Current account balance: -$433 million (2007 est.)
Exports: $894 million f.o.b. (2007 est.)
Exports—commodities: bauxite, alumina, gold, diamonds, coffee, fish, agricultural products
Exports—partners: Russia 11.6%, Ukraine 9.6%, Spain 9%, South Korea 8.8%, France 7.7%, US 7.7%, Germany 5.4%, Ireland 5.1% (2006)
Imports: $936 million f.o.b. (2007 est.)
Imports—commodities: petroleum products, metals, machinery, transport equipment, textiles, grain and other foodstuffs
Imports—partners: China 8.6%, France 8%, Netherlands 4.8%, Belgium 4.4% (2006)
Economic aid—recipient: $182.1 million (2005)
Reserves of foreign exchange and gold: $122 million (31 December 2007 est.)
Debt—external: $3.307 billion (31 December 2007 est.)
Market value of publicly traded shares: $NA
Currency (code): Guinean franc (GNF)
Currency code: GNF
Exchange rates: Guinean francs per US dollar—4,122.8 (2007), 5,350 (2006), 3,644.3 (2005), 2,225 (2004), 1,984.9 (2003)
Fiscal year: calendar year

COMMUNICATIONS

Telephones—main lines in use: 26,300 (2005)

Telephones—mobile cellular: 189,000 (2005)

Telephone system:

general assessment: inadequate system of open-wire lines, small radiotelephone communication stations, and new microwave radio relay system

domestic: Conakry reasonably well served; coverage elsewhere remains inadequate and large companies tend to rely on their own systems for nationwide links; combined fixed and mobile-cellular teledensity is about 2 per 100 persons

international: country code—224; satellite earth station—1 Intelsat (Atlantic Ocean)

Radio broadcast stations: AM 0, FM 5, shortwave 3 (2006)

Radios: 357,000 (1997)

Television broadcast stations: 6 (2001)

Televisions: 85,000 (1997)

Internet country code: .gn

Internet hosts: 173 (2007)

Internet Service Providers (ISPs): 4 (2001)

Internet users: 50,000 (2006)

TRANSPORTATION

Airports: 16 (2007)

Airports—with paved runways:

total: 5
over 3,047 m: 1
2,438 to 3,047 m: 1
1,524 to 2,437 m: 3 (2007)

Airports—with unpaved runways:

total: 11
1,524 to 2,437 m: 6
914 to 1,523 m: 3
under 914 m: 2 (2007)

Railways:

total: 837 km
standard gauge: 175 km 1.435-m gauge
narrow gauge: 662 km 1.000-m gauge (2006)

Roadways:

total: 44,348 km
paved: 4,342 km
unpaved: 40,006 km (2003)

Waterways: 1,300 km (navigable by shallow-draft native craft) (2005)

Ports and terminals: Conakry, Kamsar

MILITARY

Military branches: Armed Forces: Army, Navy (Marine Guineenne, includes Marines), Air Force, Presidential Guard (2008)

Military service age and obligation: 18 years of age for compulsory military service; 2-year conscript service obligation (2006)

Manpower available for military service:

males age 16–49: 2,230,049
females age 16–49: 2,193,236 (2008 est.)

Manpower fit for military service:

males age 16–49: 1,268,193
females age 16–49: 1,259,913 (2008 est.)

Military expenditures—percent of GDP: 1.7% (2006)

TRANSNATIONAL ISSUES

Disputes—international: conflicts among rebel groups, warlords, and youth gangs in neighboring states have spilled over into Guinea, resulting in domestic instability; Sierra Leone considers Guinea's definition of the flood plain limits to define the left bank boundary of the Makona and Moa rivers excessive and protests Guinea's continued occupation of these lands, including the hamlet of Yenga, occupied since 1998

Refugees and internally displaced persons: *refugees (country of origin):* 21,856 (Liberia); 5,259 (Sierra Leone); 3,900 (Cote d'Ivoire)

IDPs: 19,000 (cross-border incursions from Cote d'Ivoire, Liberia, Sierra Leone) (2007)

GUINEA-BISSAU

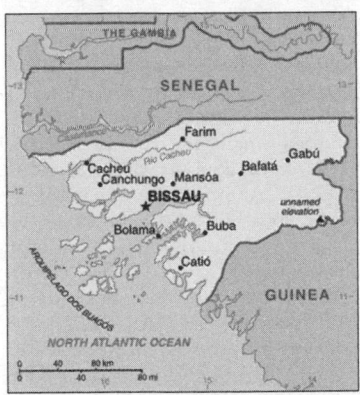

INTRODUCTION

Background: Since independence from Portugal in 1974, Guinea-Bissau has experienced considerable political and military upheaval. In 1980, a military coup established authoritarian dictator Joao Bernardo 'Nino' VIEIRA as president. Despite setting a path to a market economy and multiparty system, VIEIRA's regime was characterized by the suppression of political opposition and the purging of political rivals. Several coup attempts through the 1980s and early 1990s failed to unseat him. In 1994 VIEIRA was elected president in the country's first free elections. A military mutiny and resulting civil war in 1998 eventually led to VIEIRA's ouster in May 1999. In February 2000, a transitional government turned over power to opposition leader Kumba YALA, after he was elected president in transparent polling. In September 2003, after only three years in office, YALA was ousted by the military in a bloodless coup, and businessman Henrique ROSA was sworn in as interim president. In 2005, former President VIEIRA was re-elected president pledging to pursue economic development and national reconciliation.

GEOGRAPHY

Location: Western Africa, bordering the North Atlantic Ocean, between Guinea and Senegal

Geographic coordinates: 12 00 N, 15 00 W

Map references: Africa

Area:

total: 36,120 sq km
land: 28,000 sq km
water: 8,120 sq km

Area—comparative: slightly less than three times the size of Connecticut

Land boundaries: *total:* 724 km

border countries: Guinea 386 km, Senegal 338 km

Coastline: 350 km

Maritime claims:

territorial sea: 12 nm
exclusive economic zone: 200 nm

Climate: tropical; generally hot and humid; monsoonal-type rainy season (June to November) with southwesterly winds; dry season (December to May) with northeasterly harmattan winds

Terrain: mostly low coastal plain rising to savanna in east

Elevation extremes:

lowest point: Atlantic Ocean 0 m
highest point: unnamed location in the northeast corner of the country 300 m

Natural resources: fish, timber, phosphates, bauxite, clay, granite, limestone, unexploited deposits of petroleum

Land use: *arable land:* 8.31%

permanent crops: 6.92%
other: 84.77% (2005)

Irrigated land: 250 sq km (2003)

Total renewable water resources: 31 cu km (2003)

Freshwater withdrawal (domestic/industrial/agricultural): *total:* 0.18 cu km/yr (13%/5%/82%)

per capita: 113 cu m/yr (2000)

Natural hazards: hot, dry, dusty harmattan haze may reduce visibility during dry season; brush fires

Environment—current issues: deforestation; soil erosion; overgrazing; overfishing

Environment—international agreements: *party to:* Biodiversity, Climate Change, Climate Change-Kyoto Protocol, Desertification, Endangered Species, Law of the Sea, Ozone Layer Protection, Wetlands

signed, but not ratified: none of the selected agreements

Geography—note: this small country is swampy along its western coast and low-lying inland

PEOPLE

Population: 1,503,182 (July 2008 est.)

Age structure:

0–14 years: 41% (male 307,353/female 308,726)

15–64 years: 55.9% (male 404,747/female 436,245)

65 years and over: 3.1% (male 18,819/female 27,292) (2008 est.)

Median age:

total: 19.2 years

male: 18.6 years

female: 19.8 years (2008 est.)

Population growth rate: 2.035% (2008 est.)

Birth rate: 36.4 births/1,000 population (2008 est.)

Death rate: 16.05 deaths/1,000 population (2008 est.)

Net migration rate: -2 migrant(s)/1,000 population (2005 est.)

Sex ratio:

at birth: 1.03 male(s)/female

under 15 years: 1 male(s)/female

15–64 years: 0.93 male(s)/female

65 years and over: 0.69 male(s)/female

total population: 0.95 male(s)/female (2008 est.)

Infant mortality rate:

total: 101.64 deaths/1,000 live births

male: 111.74 deaths/1,000 live births

female: 91.25 deaths/1,000 live births (2008 est.)

Life expectancy at birth:

total population: 47.52 years

male: 45.71 years

female: 49.39 years (2008 est.)

Total fertility rate: 4.72 children born/woman (2008 est.)

HIV/AIDS—adult prevalence rate: 10% (2003 est.)

HIV/AIDS—people living with HIV/AIDS: 17,000 (2001 est.)

HIV/AIDS—deaths: 1,200 (2001 est.)

Major infectious diseases:

degree of risk: very high

food or waterborne diseases: bacterial and protozoal diarrhea, hepatitis A, and typhoid fever

vectorborne diseases: malaria and yellow fever

water contact disease: schistosomiasis

respiratory disease: meningococcal meningitis (2008)

Nationality:

noun: Guinean(s)

adjective: Guinean

Ethnic groups: African 99% (includes Balanta 30%, Fula 20%, Manjaca 14%, Mandinga 13%, Papel 7%), European and mulatto less than 1%

Religions: indigenous beliefs 50%, Muslim 45%, Christian 5%

Languages: Portuguese (official), Crioulo, African languages

Literacy:

definition: age 15 and over can read and write

total population: 42.4%

male: 58.1%

female: 27.4% (2003 est.)

GOVERNMENT

Country name:

conventional long form: Republic of Guinea-Bissau

conventional short form: Guinea-Bissau

local long form: Republica da Guine-Bissau

local short form: Guine-Bissau

former: Portuguese Guinea

Government type: republic

Capital: *name:* Bissau

geographic coordinates: 11 51 N, 15 35 W

time difference: UTC 0 (5 hours ahead of Washington, DC during Standard Time)

Administrative divisions: 9 regions (regioes, singular—regiao): Bafata, Biombo, Bissau, Bolama, Cacheu, Gabu, Oio, Quinara, Tombali; note—Bolama may have been renamed Bolama/Bijagos

Independence: 24 September 1973 (declared); 10 September 1974 (from Portugal)

National holiday: Independence Day, 24 September (1973)

Constitution: 16 May 1984; amended 4 May 1991, 4 December 1991, 26 February 1993, 9 June 1993, and in 1996

Legal system: based on French civil law; accepts compulsory ICJ jurisdiction

Suffrage: 18 years of age; universal

Executive branch:

chief of state: President Joao Bernardo 'Nino' VIEIRA (since 1 October 2005)

head of government: Prime Minister Martinho N'Dafa CABI (since 9 April 2007)

cabinet: NA

elections: president elected by popular vote for a five-year term (no term limits); election last held 24 July 2005 (next to be held in 2010); prime minister appointed by the president after consultation with party leaders in the legislature

election results: Joao Bernardo VIEIRA elected president; percent of vote, second ballot—Joao Bernardo VIEIRA 52.4%, Malam Bacai SANHA 47.6%

Legislative branch: unicameral National People's Assembly or Assembleia Nacional Popular (100 seats; members are elected by popular vote to serve four-year terms)

elections: last held 28 March 2004 (next to be held 16 November 2008)

election results: percent of vote by party—PAIGC 31.5%, PRS 24.8%, PUSD 16.1%, UE 4.1%, APU 1.3%, 13 other parties 22.2%; seats by party—PAIGC 45, PRS 35, PUSD 17, UE 2, APU 1

Judicial branch: Supreme Court or Supremo Tribunal da Justica (consists of nine justices appointed by the president and serve at his pleasure; final court of appeals in criminal and civil cases); Regional Courts (one in each of nine regions; first court of appeals for Sectoral Court decisions; hear all felony cases and civil cases valued at more than $1,000); 24 Sectoral Courts (judges are not necessarily trained lawyers; they hear civil cases valued at less than $1,000 and misdemeanor criminal cases)

Political parties and leaders: African Party for the Independence of Guinea-Bissau and Cape Verde or PAIGC [Carlos GOMES Junior]; Party for Social Renewal or PRS [Kumba YALA]; Democratic Social Front or FDS; Electoral Union or UE; Guinea-Bissau Civic Forum/Social Democracy or FCGSD [Antonieta Rosa GOMES]; Guinea-Bissau Democratic Party or PDG; Guinea-Bissau Socialist Democratic Party or PDSG [Serifo BALDE]; Labor and Solidarity Party or PST [Iancuba INDJAI]; Party for Democratic Convergence or PCD [Victor MANDINGA]; Party for Renewal and Progress or PRP; Progress Party or PP [Ibrahima SOW]; Union for Change or UM [Amine SAAD]; Union of Guinean Patriots or UPG [Francisca VAZ]; United Platform or UP (coalition formed by PCD, FDS, FLING, and RGB-MB); United Popular Alliance or APU; United Social Democratic Party or PUSD

Political pressure groups and leaders: NA

International organization participation: ACCT, ACP, AfDB, AU, CPLP, ECOWAS, FAO, FZ, G-77, IBRD, ICAO, ICCt (signatory), ICRM, IDA,

IDB, IFAD, IFC, IFRCS, ILO, IMF, IMO, Interpol, IOC, IOM, ITSO, ITU, ITUC, MIGA, NAM, OIC, OIF, OPCW (signatory), UN, UNCTAD, UNESCO, UNIDO, Union Latina, UNWTO, UPU, WADB (regional), WAEMU, WFTU, WHO, WIPO, WMO, WTO
Diplomatic representation in the US:
chief of mission: none; note—Guinea-Bissau does not have official representation in Washington, DC
Diplomatic representation from the US: the US Embassy suspended operations on 14 June 1998 in the midst of violent conflict between forces loyal to then President VIEIRA and military-led junta; the US Ambassador to Senegal is accredited to Guinea-Bissau
Flag description: two equal horizontal bands of yellow (top) and green with a vertical red band on the hoist side; there is a black five-pointed star centered in the red band; uses the popular pan-African colors of Ethiopia

ECONOMY

Economy—overview: One of the five poorest countries in the world, Guinea-Bissau depends mainly on farming and fishing. Cashew crops have increased remarkably in recent years, and the country now ranks sixth in cashew production. Guinea-Bissau exports fish and seafood along with small amounts of peanuts, palm kernels, and timber. Rice is the major crop and staple food. However, intermittent fighting between Senegalese-backed government troops and a military junta destroyed much of the country's infrastructure and caused widespread damage to the economy in 1998; the civil war led to a 28% drop in GDP that year, with partial recovery in 1999–2002. Before the war, trade reform and price liberalization were the most successful part of the country's structural adjustment program under IMF sponsorship. The tightening of monetary policy and the development of the private sector had also begun to reinvigorate the economy. Because of high costs, the development of petroleum, phosphate, and other mineral resources is not a near-term prospect. Offshore oil prospecting is underway in several sectors but has not yet led to commercially viable crude deposits. The inequality of income distribution is one of the most extreme in the world. The government and international donors continue to work out plans to forward economic development from a lamentably low base. In December 2003, the World Bank, IMF, and UNDP were forced to step in to provide emergency budgetary support in the amount of $107 million for 2004, representing over 80% of the total national budget. Government drift and indecision, however, resulted in continued low growth in 2002–06. Higher raw material prices boosted growth to 3.7% in 2007.
GDP (purchasing power parity): $808 million (2007 est.)
GDP (official exchange rate): $343 million (2007 est.)
GDP—real growth rate: 2.5% (2007 est.)
GDP—per capita (PPP): $500 (2007 est.)
GDP—composition by sector:
agriculture: 62%
industry: 12%
services: 26% (1999 est.)
Labor force: 480,000 (1999)
Labor force—by occupation: *agriculture:* 82%
industry and services: 18% (2000 est.)
Unemployment rate: NA%
Population below poverty line: NA%
Household income or consumption by percentage share:
lowest 10%: 0.5%
highest 10%: 42.4% (1991)
Inflation rate (consumer prices): 3.8% (2007 est.)
Budget:
revenues: $NA
expenditures: $NA
Agriculture—products: rice, corn, beans, cassava (tapioca), cashew nuts, peanuts, palm kernels, cotton; timber; fish
Industries: agricultural products processing, beer, soft drinks
Industrial production growth rate: 4.7% (2003 est.)
Electricity—production: 60 million kWh (2005)
Electricity—production by source:
fossil fuel: 100%
hydro: 0%
nuclear: 0%
other: 0% (2001)
Electricity—consumption: 55.8 million kWh (2005)
Electricity—exports: 0 kWh (2005)
Electricity—imports: 0 kWh (2005)
Oil—production: 0 bbl/day (2005 est.)
Oil—consumption: 2,480 bbl/day (2005 est.)
Oil—exports: 0 bbl/day (2004)
Oil—imports: 2,463 bbl/day (2004)
Oil—proved reserves: 0 bbl (1 January 2006 est.)
Natural gas—production: 0 cu m (2005 est.)
Natural gas—consumption: 0 cu m (2005 est.)
Natural gas—exports: 0 cu m (2005 est.)
Natural gas—imports: 0 cu m (2005)
Natural gas—proved reserves: 0 cu m (1 January 2006 est.)
Current account balance: -$6 million (2007 est.)
Exports: $133 million f.o.b. (2006)
Exports—commodities: cashew nuts, shrimp, peanuts, palm kernels, sawn lumber
Exports—partners: India 76.1%, Nigeria 18.1%, Italy 1.4% (2006)
Imports: $200 million f.o.b. (2006)
Imports—commodities: foodstuffs, machinery and transport equipment, petroleum products
Imports—partners: Portugal 18.7%, Senegal 16.3%, Italy 13%, Pakistan 4.5% (2006)
Economic aid—recipient: $79.12 million (2005)
Debt—external: $941.5 million (2000 est.)
Market value of publicly traded shares: $NA
Currency (code): Communaute Financiere Africaine franc (XOF); note—responsible authority is the Central Bank of the West African States
Currency code: XOF; GWP
Exchange rates: Communaute Financiere Africaine francs (XOF) per US dollar—493.51 (2007), 522.59 (2006), 527.47 (2005), 528.29 (2004), 581.2 (2003)
note: since 1 January 1999, the XOF franc has been pegged to the euro at a rate of 655.957 XOF francs per euro
Fiscal year: calendar year

COMMUNICATIONS

Telephones—main lines in use: 10,200 (2005)
Telephones—mobile cellular: 95,000 (2005)
Telephone system:
general assessment: small system
domestic: combination of microwave radio relay, open-wire lines, radiotelephone, and cellular communications; fixed-line teledensity less than 1 per 100 persons; mobile-cellular teledensity reached 7 per 100 in 2005
international: country code—245
Radio broadcast stations: AM 1 (transmitter out of service), FM 4, shortwave 0 (2001)
Radios: 49,000 (1997)
Television broadcast stations: NA (2005)
Televisions: NA
Internet country code: .gw
Internet hosts: 0 (2007)
Internet Service Providers (ISPs): 2 (2002)
Internet users: 37,000 (2006)

TRANSPORTATION

Airports: 27 (2007)
Airports—with paved runways: *total:* 3
over 3,047 m: 1
1,524 to 2,437 m: 1
914 to 1,523 m: 1 (2007)
Airports—with unpaved runways:
total: 24
1,524 to 2,437 m: 1
914 to 1,523 m: 4
under 914 m: 19 (2007)
Roadways:
total: 3,455 km
paved: 965 km
unpaved: 2,490 km (2002)
Waterways: rivers are navigable for some distance; many inlets and creeks give shallow-water access to much of interior (2007)

Ports and terminals: Bissau, Buba, Cacheu, Farim

MILITARY

Military branches: People's Revolutionary Armed Force (FARP): Army, Navy, Air Force; paramilitary force
Military service age and obligation: 18 years of age for selective compulsory military service (2006)
Manpower available for military service:
males age 16–49: 344,087
females age 16–49: 347,886 (2008 est.)
Manpower fit for military service:
males age 16–49: 188,605
females age 16–49: 195,429 (2008 est.)
Military expenditures—percent of GDP: 3.1% (2005 est.)

TRANSNATIONAL ISSUES

Disputes—international: in 2006, political instability within Senegal's Casamance region resulted in thousands of Senegalese refugees, cross-border raids, and arms smuggling into Guinea-Bissau
Refugees and internally displaced persons: *refugees (country of origin):* 7,454 (Senegal) (2007)
Illicit drugs: increasingly important transit country for South American cocaine enroute to Europe; enabling environment for trafficker operations thanks to pervasive corruption; archipelago-like geography around the capital facilitates drug smuggling

GUYANA

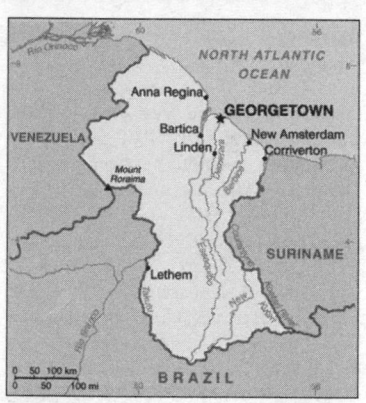

INTRODUCTION

Background: Originally a Dutch colony in the 17th century, by 1815 Guyana had become a British possession. The abolition of slavery led to black settlement of urban areas and the importation of indentured servants from India to work the sugar plantations. This ethnocultural divide has persisted and has led to turbulent politics. Guyana achieved independence from the UK in 1966, and since then it has been ruled mostly by socialist-oriented governments. In 1992, Cheddi JAGAN was elected president in what is considered the country's first free and fair election since independence. After his death five years later, his wife, Janet JAGAN, became president but resigned in 1999 due to poor health. Her successor, Bharrat JAGDEO, was reelected in 2001 and again in 2006.

GEOGRAPHY

Location: Northern South America, bordering the North Atlantic Ocean, between Suriname and Venezuela
Geographic coordinates: 5 00 N, 59 00 W
Map references: South America
Area: *total:* 214,970 sq km
land: 196,850 sq km
water: 18,120 sq km
Area—comparative: slightly smaller than Idaho
Land boundaries: *total:* 2,949 km
border countries: Brazil 1,606 km, Suriname 600 km, Venezuela 743 km
Coastline: 459 km
Maritime claims: *territorial sea:* 12 nm
exclusive economic zone: 200 nm
continental shelf: 200 nm or to the outer edge of the continental margin
Climate: tropical; hot, humid, moderated by northeast trade winds; two rainy seasons (May to August, November to January)
Terrain: mostly rolling highlands; low coastal plain; savanna in south
Elevation extremes:
lowest point: Atlantic Ocean 0 m
highest point: Mount Roraima 2,835 m
Natural resources: bauxite, gold, diamonds, hardwood timber, shrimp, fish
Land use: *arable land:* 2.23%
permanent crops: 0.14%
other: 97.63% (2005)
Irrigated land: 1,500 sq km (2003)
Total renewable water resources: 241 cu km (2000)
Freshwater withdrawal (domestic /industrial/agricultural): *total:* 1.64 cu km/yr (2%/1%/98%)
per capita: 2,187 cu m/yr (2000)
Natural hazards: flash floods are a constant threat during rainy seasons
Environment—current issues: water pollution from sewage and agricultural and industrial chemicals; deforestation
Environment—international agreements: *party to:* Biodiversity, Climate Change, Climate Change-Kyoto Protocol, Desertification, Endangered Species, Hazardous Wastes, Law of the Sea, Ozone Layer Protection, Ship Pollution, Tropical Timber 83, Tropical Timber 94
signed, but not ratified: none of the selected agreements
Geography—note: the third-smallest country in South America after Suriname and Uruguay; substantial portions of its western and eastern territories are claimed by Venezuela and Suriname respectively

PEOPLE

Population: 770,794
note: estimates for this country explicitly take into account the effects of excess mortality due to AIDS; this can result in lower life expectancy, higher infant mortality, higher death rates, lower population growth rates, and changes in the distribution of population by age and sex than would otherwise be expected (July 2008 est.)
Age structure:
0–14 years: 25.9% (male 101,712/female 97,907)
15–64 years: 68.7% (male 267,239/female 262,188)
65 years and over: 5.4% (male 17,610/female 24,138) (2008 est.)
Median age: *total:* 28.2 years
male: 27.7 years
female: 28.7 years (2008 est.)
Population growth rate: 0.211% (2008 est.)
Birth rate: 17.85 births/1,000 population (2008 est.)
Death rate: 8.29 deaths/1,000 population (2008 est.)

Net migration rate: -7.45 migrant(s)/ 1,000 population (2008 est.)
Sex ratio: *at birth:* 1.05 male(s)/female
under 15 years: 1.04 male(s)/female
15–64 years: 1.02 male(s)/female
65 years and over: 0.73 male(s)/female
total population: 1.01 male(s)/female (2008 est.)
Infant mortality rate:
total: 30.43 deaths/1,000 live births
male: 33.87 deaths/1,000 live births
female: 26.82 deaths/1,000 live births (2008 est.)
Life expectancy at birth:
total population: 66.43 years
male: 63.81 years
female: 69.18 years (2008 est.)
Total fertility rate: 2.03 children born/woman (2008 est.)
HIV/AIDS—adult prevalence rate: 2.5% (2003 est.)
HIV/AIDS—people living with HIV/AIDS: 11,000 (2003 est.)
HIV/AIDS—deaths: 1,100 (2003 est.)
Major infectious diseases:
degree of risk: high
food or waterborne diseases: bacterial and protozoal diarrhea, hepatitis A, and typhoid fever
vectorborne diseases: dengue fever and malaria
water contact disease: leptospirosis (2008)
Nationality:
noun: Guyanese (singular and plural)
adjective: Guyanese
Ethnic groups: East Indian 50%, black 36%, Amerindian 7%, white, Chinese, and mixed 7%
Religions: Christian 50%, Hindu 35%, Muslim 10%, other 5%
Languages: English, Amerindian dialects, Creole, Caribbean Hindustani (a dialect of Hindi), Urdu
Literacy: *definition:* age 15 and over has ever attended school
total population: 98.8%
male: 99.1%
female: 98.5% (2003 est.)

GOVERNMENT

Country name:
conventional long form: Cooperative Republic of Guyana
conventional short form: Guyana
former: British Guiana
Government type: republic
Capital: *name:* Georgetown
geographic coordinates: 6 48 N, 58 10 W
time difference: UTC-4 (1 hour ahead of Washington, DC during Standard Time)
Administrative divisions: 10 regions; Barima-Waini, Cuyuni-Mazaruni, Demerara-Mahaica, East Berbice-Corentyne, Essequibo Islands-West Demerara, Mahaica-Berbice, Pomeroon-Supenaam, Potaro-Siparuni, Upper Demerara-Berbice, Upper Takutu-Upper Essequibo

Independence: 26 May 1966 (from UK)
National holiday: Republic Day, 23 February (1970)
Constitution: 6 October 1980
Legal system: based on English common law with certain admixtures of Roman-Dutch law; has not accepted compulsory ICJ jurisdiction
Suffrage: 18 years of age; universal
Executive branch:
chief of state: President Bharrat JAGDEO (since 11 August 1999); note—assumed presidency after resignation of President Janet JAGAN and was reelected in 2001, and again in 2006
head of government: Prime Minister Samuel HINDS (since October 1992, except for a period as chief of state after the death of President Cheddi JAGAN on 6 March 1997)
cabinet: Cabinet of Ministers appointed by the president, responsible to the legislature
elections: president elected by popular vote as leader of a party list in parliamentary elections, which must be held at least every five years (no term limits); elections last held 28 August 2006 (next to be held by August 2011); prime minister appointed by the president
election results: President Bharrat JAGDEO reelected; percent of vote 54.6%
Legislative branch: unicameral National Assembly (65 seats; members elected by popular vote, also not more than 4 non-elected non-voting ministers and 2 non-elected non-voting parliamentary secretaries appointed by the president; to serve five-year terms)
elections: last held 28 August 2006 (next to be held by August 2011)
election results: percent of vote by party—PPP/C 54.6%, PNC/R 34%, AFC 8.1%, other 3.3%; seats by party—PPP/C 36, PNC/R 22, AFC 5, other 2
Judicial branch: Supreme Court of Judicature, consisting of the High Court and the Judicial Court of Appeal, with right of final appeal to the Caribbean Court of Justice
Political parties and leaders: Alliance for Change or AFC [Raphael TROTMAN and Khemraj RAMJATTAN]; Guyana Action Party or GAP [Paul HARDY]; Justice for All Party [C.N. SHARMA]; People's National Congress/Reform or PNC/R [Robert Herman Orlando CORBIN]; People's Progressive Party/Civic or PPP/C [Bharrat JAGDEO]; Rise, Organize, and Rebuild or ROAR [Ravi DEV]; The United Force or TUF [Manzoor NADIR]; The Unity Party [Joey JAGAN]; Vision Guyana [Peter RAMSAROOP]; Working People's Alliance or WPA [Rupert ROOPNARAINE]
Political pressure groups and leaders: Amerindian People's Association;

Guyana Citizens Initiative; Guyana Bar Association; Guyana Human Rights Association; Guyana Public Service Union or GPSU; Private Sector Commission; Trades Union Congress
International organization participation: ACP, C, Caricom, CDB, CSN, FAO, G-77, IADB, IBRD, ICAO, ICCt, ICRM, IDA, IFAD, IFC, IFRCS, ILO, IMF, IMO, Interpol, IOC, IOM (observer), ISO (subscriber), ITU, ITUC, LAES, MIGA, NAM, OAS, OIC, OPANAL, OPCW, PCA, RG, UN, UNCTAD, UNESCO, UNIDO, UPU, WCL, WCO, WFTU, WHO, WIPO, WMO, WTO
Diplomatic representation in the US:
chief of mission: Ambassador Bayney KARRAN
chancery: 2490 Tracy Place NW, Washington, DC 20008
telephone: [1] (202) 265-6900
FAX: [1] (202) 232-1297
consulate(s) general: New York
Diplomatic representation from the US:
chief of mission: Ambassador David M. ROBINSON
embassy: 100 Young and Duke Streets, Kingston, Georgetown
mailing address: P. O. Box 10507, Georgetown; US Embassy, 3170 Georgetown Place, Washington DC 20521-3170
telephone: [592] 225-4900 through 4909
FAX: [592] 225-8497
Flag description: green, with a red isosceles triangle (based on the hoist side) superimposed on a long, yellow arrowhead; there is a narrow, black border between the red and yellow, and a narrow, white border between the yellow and the green

ECONOMY

Economy—overview: The Guyanese economy exhibited moderate economic growth in 2001–07, based on expansion in the agricultural and mining sectors, a more favorable atmosphere for business initiatives, a more realistic exchange rate, fairly low inflation, and the continued support of international organizations. Economic recovery since the 2005 flood-related contraction has been buoyed by increases in remittances and foreign direct investment. Chronic problems include a shortage of skilled labor and a deficient infrastructure. The government is juggling a sizable external debt against the urgent need for expanded public investment. In March 2007, the Inter-American Development Bank, Guyana's principal donor, canceled Guyana's nearly $470 million debt, equivalent to nearly 48% of GDP. The bauxite mining sector should benefit in the near term from restructuring and partial privatization, and the state-owned sugar industry will conduct efficiency increasing modernizations. Export earnings from agriculture and mining have fallen sharply, while the import bill has risen, driven by higher energy prices. Guyana's entrance into the Caricom Single Market and Economy (CSME) in January 2006 will broaden the country's export market, primarily in the raw materials sector.

GDP (purchasing power parity): $2.92 billion (2007 est.)

GDP (official exchange rate): $1.039 billion (2007 est.)

GDP—real growth rate: 5.4% (2007 est.)

GDP—per capita (PPP): $3,800 (2007 est.)

GDP—composition by sector:
agriculture: 31.1%
industry: 21.7%
services: 47.1% (2007 est.)

Labor force: 418,000 (2001 est.)

Labor force—by occupation:
agriculture: NA%
industry: NA%
services: NA%

Unemployment rate: 9.1% (understated) (2000)

Population below poverty line: NA%

Household income or consumption by percentage share: *lowest 10%:* 1.3% *highest 10%:* 33.8% (1999)

Distribution of family income—Gini index: 43.2 (1999)

Inflation rate (consumer prices): 12.2% (2007 est.)

Investment (gross fixed): 41.1% of GDP (2007 est.)

Budget: *revenues:* $446.2 million *expenditures:* $531.2 million (2007 est.)

Agriculture—products: sugarcane, rice, shrimp, fish, vegetable oils; beef, pork, poultry, dairy products

Industries: bauxite, sugar, rice milling, timber, textiles, gold mining

Industrial production growth rate: -26.4% (2007 est.)

Electricity—production: 807.3 million kWh (2005)

Electricity—production by source:
fossil fuel: 99.4%
hydro: 0.6%
nuclear: 0%
other: 0% (2001)

Electricity—consumption: 750.7 million kWh (2005)

Electricity—exports: 0 kWh (2005)

Electricity—imports: 0 kWh (2005)

Oil—production: 0 bbl/day (2005 est.)

Oil—consumption: 10,500 bbl/day (2005 est.)

Oil—exports: 0 bbl/day (2004)

Oil—imports: 10,070 bbl/day (2004)

Oil—proved reserves: 0 bbl (1 January 2006 est.)

Natural gas—production: 0 cu m (2005 est.)

Natural gas—consumption: 0 cu m (2005 est.)

Natural gas—exports: 0 cu m (2005 est.)

Natural gas—imports: 0 cu m (2005)

Natural gas—proved reserves: 0 cu m (1 January 2006 est.)

Current account balance: -$189 million (2007 est.)

Exports: $684 million f.o.b. (2007 est.)

Exports—commodities: sugar, gold, bauxite, alumina, rice, shrimp, molasses, rum, timber

Exports—partners: US 18.8%, Canada 18.4%, UK 8.7%, Portugal 6.5%, Trinidad and Tobago 4.9%, Netherlands 4.3%, Belgium 4.3%, Jamaica 4.1% (2006)

Imports: $950 million f.o.b. (2007 est.)

Imports—commodities: manufactures, machinery, petroleum, food

Imports—partners: Trinidad and Tobago 23%, US 21.3%, China 9.7%, Cuba 6.3%, UK 4.5% (2006)

Economic aid—recipient: $136.8 million (2005)

Reserves of foreign exchange and gold: $324 million (31 December 2007 est.)

Debt—external: $1.2 billion (2002)

Market value of publicly traded shares: $187.3 million (2005)

Currency (code): Guyanese dollar (GYD)

Currency code: GYD

Exchange rates: Guyanese dollars per US dollar—201.89 (2007), 200.28 (2006), 200.79 (2005), 198.31 (2004), 193.88 (2003)

Fiscal year: calendar year

COMMUNICATIONS

Telephones—main lines in use: 110,100 (2005)

Telephones—mobile cellular: 281,400 (2005)

Telephone system:
general assessment: fair system for long-distance service
domestic: microwave radio relay network for trunk lines; fixed-line teledensity is about 15 per 100 persons; many areas still lack fixed-line telephone services; mobile-cellular teledensity reached 37 per 100 persons in 2005
international: country code—592; tropospheric scatter to Trinidad; satellite earth station—1 Intelsat (Atlantic Ocean)

Radio broadcast stations: AM 3, FM 3, shortwave 1 (1998)

Radios: 420,000 (1997)

Television broadcast stations: 3 (1 public station; 2 private stations which relay US satellite services) (1997)

Televisions: 46,000 (1997)

Internet country code: .gy

Internet hosts: 3,000 (2007)

Internet Service Providers (ISPs): 3 (2000)

Internet users: 160,000 (2005)

TRANSPORTATION

Airports: 93 (2007)

Airports—with paved runways: *total:* 9
1,524 to 2,437 m: 3
under 914 m: 6 (2007)

Airports—with unpaved runways:
total: 84
1,524 to 2,437 m: 1
914 to 1,523 m: 14
under 914 m: 69 (2007)

Roadways: *total:* 7,970 km
paved: 590 km
unpaved: 7,380 km (2000)

Waterways: Berbice, Demerara, and Essequibo rivers are navigable by ocean-going vessels for 150 km, 100 km, and 80 km respectively (2006)

Merchant marine:
total: 7 ships (1000 GRT or over) 12,516 GRT/14,193 DWT
by type: cargo 5, petroleum tanker 1, refrigerated cargo 1
registered in other countries: 2 (St Vincent and The Grenadines 2, unknown 1) (2007)

Ports and terminals: Georgetown

MILITARY

Military branches: Guyana Defense Force: Army (includes Coast Guard, Air Corps) (2007)

Military service age and obligation: 18 years of age for voluntary military service; no conscription (2008)

Manpower available for military service:
males age 16–49: 220,797 (2008 est.)

Manpower fit for military service:
males age 16–49: 150,623 (2008 est.)

Military expenditures—percent of GDP: 1.8% (2006)

TRANSNATIONAL ISSUES

Disputes—international: all of the area west of the Essequibo River is claimed by Venezuela preventing any discussion of a maritime boundary; Guyana has expressed its intention to join Barbados in asserting claims before UNCLOS that Trinidad and Tobago's maritime boundary with Venezuela extends into their waters; Suriname claims a triangle of land between the New and Kutari/Koetari rivers in a historic dispute over the headwaters of the Courantyne; Guyana seeks arbitration under provisions of the UN Convention on the Law of the Sea (UNCLOS) to resolve the long-standing dispute with Suriname over the axis of the territorial sea boundary in potentially oil-rich waters

Illicit drugs: transshipment point for narcotics from South America—primarily Venezuela—to Europe and the US; producer of cannabis; rising money laundering related to drug trafficking and human smuggling

HAITI

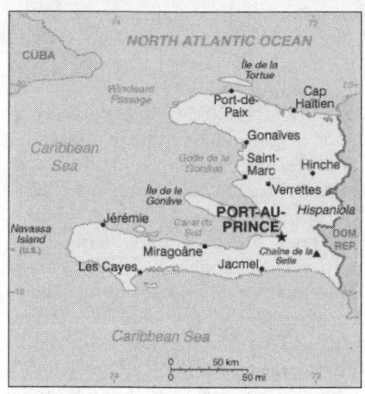

INTRODUCTION

Background: The native Taino Amerindians—who inhabited the island of Hispaniola when it was discovered by COLUMBUS in 1492—were virtually annihilated by Spanish settlers within 25 years. In the early 17th century, the French established a presence on Hispaniola, and in 1697, Spain ceded to the French the western third of the island, which later became Haiti. The French colony, based on forestry and sugar-related industries, became one of the wealthiest in the Caribbean, but only through the heavy importation of African slaves and considerable environmental degradation. In the late 18th century, Haiti's nearly half million slaves revolted under Toussaint L'OUVERTURE. After a prolonged struggle, Haiti became the first black republic to declare its independence in 1804. The poorest country in the Western Hemisphere, Haiti has been plagued by political violence for most of its history. After an armed rebellion led to the forced resignation and exile of President Jean-Bertrand ARISTIDE in February 2004, an interim government took office to organize new elections under the auspices of the United Nations Stabilization Mission in Haiti (MINUSTAH). Continued violence and technical delays prompted repeated postponements, but Haiti finally did inaugurate a democratically elected president and parliament in May of 2006.

GEOGRAPHY

Location: Caribbean, western one-third of the island of Hispaniola, between the Caribbean Sea and the North Atlantic Ocean, west of the Dominican Republic
Geographic coordinates: 19 00 N, 72 25 W
Map references: Central America and the Caribbean

Area:
total: 27,750 sq km
land: 27,560 sq km
water: 190 sq km
Area—comparative: slightly smaller than Maryland
Land boundaries:
total: 360 km
border countries: Dominican Republic 360 km
Coastline: 1,771 km
Maritime claims:
territorial sea: 12 nm
contiguous zone: 24 nm
exclusive economic zone: 200 nm
continental shelf: to depth of exploitation
Climate: tropical; semiarid where mountains in east cut off trade winds
Terrain: mostly rough and mountainous
Elevation extremes:
lowest point: Caribbean Sea 0 m
highest point: Chaine de la Selle 2,680 m
Natural resources: bauxite, copper, calcium carbonate, gold, marble, hydropower
Land use:
arable land: 28.11%
permanent crops: 11.53%
other: 60.36% (2005)
Irrigated land: 920 sq km (2003)
Total renewable water resources: 14 cu km (2000)
Freshwater withdrawal (domestic/industrial/agricultural): *total:* 0.99 cu km/yr (5%/1%/94%)
per capita: 116 cu m/yr (2000)
Natural hazards: lies in the middle of the hurricane belt and subject to severe storms from June to October; occasional flooding and earthquakes; periodic droughts
Environment—current issues: extensive deforestation (much of the remaining forested land is being cleared for agriculture and used as fuel); soil erosion; inadequate supplies of potable water
Environment—international agreements: *party to:* Biodiversity, Climate Change, Climate Change-Kyoto Protocol, Desertification, Law of the Sea, Marine Dumping, Marine Life Conservation, Ozone Layer Protection
signed, but not ratified: Hazardous Wastes
Geography—note: shares island of Hispaniola with Dominican Republic (western one-third is Haiti, eastern two-thirds is the Dominican Republic)

PEOPLE

Population: 8,924,553
note: estimates for this country explicitly take into account the effects of excess mortality due to AIDS; this can result in lower life expectancy, higher infant mortality, higher death rates, lower population growth rates, and changes in the distribution of population by age and sex than would otherwise be expected (July 2008 est.)
Age structure:
0–14 years: 41.8% (male 1,881,509/female 1,851,591)
15–64 years: 54.7% (male 2,386,761/female 2,495,233)
65 years and over: 3.5% (male 135,695/female 173,764) (2008 est.)
Median age: *total:* 18.5 years
male: 18.1 years
female: 19 years (2008 est.)
Population growth rate: 2.493% (2008 est.)
Birth rate: 35.69 births/1,000 population (2008 est.)
Death rate: 10.15 deaths/1,000 population (2008 est.)
Net migration rate: -0.61 migrant(s)/1,000 population (2008 est.)
Sex ratio:
at birth: 1.03 male(s)/female
under 15 years: 1.02 male(s)/female
15–64 years: 0.96 male(s)/female
65 years and over: 0.78 male(s)/female
total population: 0.97 male(s)/female (2008 est.)
Infant mortality rate:
total: 62.33 deaths/1,000 live births
male: 66.88 deaths/1,000 live births
female: 57.64 deaths/1,000 live births (2008 est.)
Life expectancy at birth:
total population: 57.56 years
male: 55.83 years
female: 59.35 years (2008 est.)
Total fertility rate: 4.79 children born/woman (2008 est.)
HIV/AIDS—adult prevalence rate: 5.6% (2003 est.)
HIV/AIDS—people living with HIV/AIDS: 280,000 (2003 est.)
HIV/AIDS—deaths: 24,000 (2003 est.)
Major infectious diseases:
degree of risk: high
food or waterborne diseases: bacterial and protozoal diarrhea, hepatitis A and E, and typhoid fever
vectorborne diseases: dengue fever and malaria
water contact disease: leptospirosis (2008)
Nationality:
noun: Haitian(s)
adjective: Haitian
Ethnic groups: black 95%, mulatto and white 5%
Religions: Roman Catholic 80%, Protestant 16% (Baptist 10%,

Pentecostal 4%, Adventist 1%, other 1%), none 1%, other 3%
note: roughly half of the population practices voodoo
Languages: French (official), Creole (official)
Literacy: *definition:* age 15 and over can read and write
total population: 52.9%
male: 54.8%
female: 51.2% (2003 est.)

GOVERNMENT

Country name:
conventional long form: Republic of Haiti
conventional short form: Haiti
local long form: Republique d'Haiti/Repiblik d' Ayiti
local short form: Haiti/Ayiti
Government type: republic
Capital: *name:* Port-au-Prince
geographic coordinates: 18 32 N, 72 20 W
time difference: UTC-5 (same time as Washington, DC during Standard Time)
daylight saving time: +1hr, begins first Sunday in April; ends last Sunday in October
Administrative divisions: 10 departments (departements, singular—departement); Artibonite, Centre, Grand 'Anse, Nippes, Nord, Nord-Est, Nord-Ouest, Ouest, Sud, Sud-Est
Independence: 1 January 1804 (from France)
National holiday: Independence Day, 1 January (1804)
Constitution: approved March 1987; suspended June 1988 with most articles reinstated March 1989; constitutional government ousted in a military coup in September 1991, although in October 1991, military government claimed to be observing the constitution; returned to constitutional rule in October 1994; constitution, while technically in force between 2004–2006, was not enforced; returned to constitutional rule in May 2006
Legal system: based on Roman civil law system; accepts compulsory ICJ jurisdiction
Suffrage: 18 years of age; universal
Executive branch:
chief of state: President Rene PREVAL (since 14 May 2006)
head of government: Prime Minister Jacques-Edouard ALEXIS (since 30 May 2006); note—on 12 April 2008 ALEXIS was ousted in a no-confidence vote by the Senate over his inability to quell violence associated with the ongoing food crisis
cabinet: Cabinet chosen by the prime minister in consultation with the president
elections: president elected by popular vote for a five-year term (may not serve consecutive terms); election last held 7 February 2006 (next to be held in 2011);

prime minister appointed by the president, ratified by the National Assembly
election results: Rene PREVAL elected president; percent of vote—Rene PREVAL 51%
Legislative branch: bicameral National Assembly or Assemblee Nationale consists of the Senate (30 seats; members elected by popular vote to serve six-year terms; one-third elected every two years) and the Chamber of Deputies (99 seats; members are elected by popular vote to serve four-year terms); note—in reestablishing the Senate, the candidate in each department receiving the most votes in the last election serves six years, the candidate with the second most votes serves four years, and the candidate with the third most votes serves two years
elections: Senate—last held 21 April 2006 with run-off elections on 3 December 2006 (next regular election, for one third of seats, to be held in 2008); Chamber of Deputies—last held 21 April 2006 with run-off elections on 3 December 2006 and 29 April 2007 (next regular election to be held in 2010)
election results: Senate—percent of vote by party—NA; seats by party—L'ESPWA 11, FUSION 5, OPL 4, FL 3, LAAA 2, UNCRH 2, PONT 2, ALYANS 1; Chamber of Deputies—percent of vote by party—NA; seats by party—L'ESPWA 23, FUSION 17, FRN 12, OPL 10, ALYANS 10, LAAA 5, MPH 3, MOCHRENA 3, other 10; results for six other seats contested on 3 December 2006 remain unknown
Judicial branch: Supreme Court or Cour de Cassation
Political parties and leaders: Artibonite in Action or LAAA [Youri LATORTUE]; Assembly of Progressive National Democrats or RDNP [Leslie MANIGAT]; Convention for Democratic Unity or KID [Evans PAUL]; Cooperative Action to Build Haiti or KONBA [Evans LESCOUFALIR]; Democratic Alliance or ALYANS [Evans PAUL] (coalition composed of KID and PPRH); Effort and Solidarity to Create an Alternative for the People or ESKAMP [Joseph JASME]; For Us All or PONT [Jean-Marie CHERESTAL]; Front for Hope or L'ESPWA [Rene PREVAL] (alliance of ESKAMP, PLB, and grass-roots organizations Grand-Anse Resistance Committee, the Central Plateau Peasants' Group, and Kombit Sudest); Haitian Christian Democratic Party or PDCH [Osner FEVRY and Marie-Denise CLAUDE]; Haitian Democratic and Reform Movement or MODEREH [Dany TOUSSAINT and Pierre Soncon PRINCE];

Heads Together or Tet-Ansanm [Dr. Gerard BLOT]; Independent Movement for National Reconciliation or MIRN [Luc FLEURINORD]; Justice for Peace and National Development or JPDN [Rigaud DUPLAN]; Fanmi Lavalas or FL [Rudy HERIVEAUX]; Liberal Party of Haiti or PLH [Gehy MICHEL]; Merging of Haitian Social Democratic Parties or FUSION or FPSDH [Serge GILLES] (coalition of Ayiti Capable, Haitian National Revolutionary Party, and National Congress of Democratic Movements); Mobilization for Haiti's Development or MPH [Samir MOURRA]; Mobilization for National Development or MDN [Hubert de RONCERAY]; Movement for National Reconstruction or MRN [Jean Henold BUTEAU]; Movement for the Installation of Democracy in Haiti or MIDH [Marc BAZIN]; National Christian Union for the Reconstruction of Haiti or UNCRH [Marie Claude GERMAIN]; National Front for the Reconstruction of Haiti or FRN [Guy PHILIPPE]; New Christian Movement for a New Haiti or MOCHRENA [Luc MESADIEU]; Open the Gate Party or PLB [Anes LUBIN]; Popular Party for the Renewal of Haiti or PPRH [Claude ROMAIN]; Struggling People's Organization or OPL [Edgard LEBLANC]; Union of Nationalist and Progressive Haitians or UNITE [Edouard FRANCISQUE]
Political pressure groups and leaders: Autonomous Organizations of Haitian Workers or CATH [Fignole ST-CYR]; Confederation of Haitian Workers or CTH; Federation of Workers Trade Unions or FOS; General Organization of Independent Haitian Workers [Patrick NUMAS]; Grand-Anse Resistance Committee, or KOREGA; National Popular Assembly or APN; Papaye Peasants Movement or MPP [Chavannes JEAN-BAPTISTE]; Popular Organizations Gathering Power or PROP; Roman Catholic Church; Protestant Federation of Haiti
International organization participation: ACCT, ACP, Caricom, CDB, FAO, G-77, IADB, IAEA, IBRD, ICAO, ICCt (signatory), ICRM, IDA, IFAD, IFC, IFRCS, ILO, IMF, IMO, Interpol, IOC, IOM, ITSO, ITU, ITUC, LAES, MIGA, NAM, OAS, OIF, OPANAL, OPCW (signatory), PCA, UN, UNCTAD, UNESCO, UNIDO, Union Latina, UNWTO, UPU, WCL, WCO, WFTU, WHO, WIPO, WMO, WTO
Diplomatic representation in the US:
chief of mission: Ambassador Raymond JOSEPH

chancery: 2311 Massachusetts Avenue NW, Washington, DC 20008
telephone: [1] (202) 332-4090
FAX: [1] (202) 745-7215
consulate(s) general: Boston, Chicago, Miami, New York, San Juan (Puerto Rico)
Diplomatic representation from the US:
chief of mission: Ambassador Janet A. SANDERSON
embassy: 5 Harry S Truman Boulevard, Bicentenaire-Port-au-Prince
mailing address: P. O. Box 1761, Port-au-Prince
telephone: [509] 222-0200
FAX: [509] 223-9038
Flag description: two equal horizontal bands of blue (top) and red with a centered white rectangle bearing the coat of arms, which contains a palm tree flanked by flags and two cannons above a scroll bearing the motto L'UNION FAIT LA FORCE (Union Makes Strength)

ECONOMY

Economy—overview: Haiti is the poorest country in the Western Hemisphere, with 80% of the population living under the poverty line and 54% in abject poverty. Two-thirds of all Haitians depend on the agricultural sector, mainly small-scale subsistence farming, and remain vulnerable to damage from frequent natural disasters, exacerbated by the country's widespread deforestation. A macroeconomic program developed in 2005 with the help of the International Monetary Fund helped the economy grow 3.5% in 2007, the highest growth rate since 1999. US economic engagement under the Haitian Hemispheric Opportunity through Partnership Encouragement (HOPE) Act, passed in December 2006, has boosted the garment and automotive parts exports and investment by providing tariff-free access to the US. Haiti suffers from high inflation, a lack of investment because of insecurity and limited infrastructure, and a severe trade deficit. In 2005, Haiti paid its arrears to the World Bank, paving the way for reengagement with the Bank. The government relies on formal international economic assistance for fiscal sustainability. Remittances are the primary source of foreign exchange, equaling nearly a quarter of GDP and more than twice the earnings from exports.
GDP (purchasing power parity): $11.14 billion (2007 est.)
GDP (official exchange rate): $5.435 billion (2007 est.)
GDP—real growth rate: 3.2% (2007 est.)

GDP—per capita (PPP): $1,300 (2007 est.)
GDP—composition by sector:
agriculture: 28%
industry: 20%
services: 52% (2004 est.)
Labor force: 3.6 million
note: shortage of skilled labor, unskilled labor abundant (1995)
Labor force—by occupation: *agriculture:* 66%
industry: 9%
services: 25% (1995)
Unemployment rate: widespread unemployment and underemployment; more than two-thirds of the labor force do not have formal jobs (2002 est.)
Population below poverty line: 80% (2003 est.)
Household income or consumption by percentage share:
lowest 10%: 0.7%
highest 10%: 47.7% (2001)
Distribution of family income—Gini index: 59.2 (2001)
Inflation rate (consumer prices): 9% (2007 est.)
Investment (gross fixed): 28.9% of GDP (2006 est.)
Budget: *revenues:* $926.3 million
expenditures: $1.045 billion (2007 est.)
Agriculture—products: coffee, mangoes, sugarcane, rice, corn, sorghum; wood
Industries: sugar refining, flour milling, textiles, cement, light assembly based on imported parts
Industrial production growth rate: 2.5% (2007 est.)
Electricity—production: 535 million kWh (2005)
Electricity—production by source:
fossil fuel: 60.3%
hydro: 39.7%
nuclear: 0%
other: 0% (2001)
Electricity—consumption: 322 million kWh (2005)
Electricity—exports: 0 kWh (2005)
Electricity—imports: 0 kWh (2005)
Oil—production: 0 bbl/day (2005 est.)
Oil—consumption: 12,000 bbl/day (2005 est.)
Oil—exports: 0 bbl/day (2004)
Oil—imports: 11,840 bbl/day (2004)
Oil—proved reserves: 0 bbl (1 January 2006 est.)
Natural gas—production: 0 cu m (2005 est.)
Natural gas—consumption: 0 cu m (2005 est.)
Natural gas—exports: 0 cu m (2005 est.)
Natural gas—imports: 0 cu m (2005)
Natural gas—proved reserves: 0 cu m (1 January 2006 est.)

Current account balance: $11 million (2007 est.)
Exports: $524 million f.o.b. (2007 est.)
Exports—commodities: apparel, manufactures, oils, cocoa, mangoes, coffee
Exports—partners: US 79.8%, Dominican Republic 7.6%, Canada 3% (2006)
Imports: $1.614 billion f.o.b. (2007 est.)
Imports—commodities: food, manufactured goods, machinery and transport equipment, fuels, raw materials
Imports—partners: US 46.6%, Netherlands Antilles 11.9%, Brazil 3.8% (2006)
Economic aid—recipient: $515 million (2005 est.)
Reserves of foreign exchange and gold: $370 million (31 December 2007 est.)
Debt—external: $1.463 billion (31 December 2007 est.)
Market value of publicly traded shares: $NA
Currency (code): gourde (HTG)
Currency code: HTG
Exchange rates: gourdes per US dollar—37.138 (2007), 40.232 (2006), 40.449 (2005), 38.352 (2004), 42.367 (2003)
Fiscal year: 1 October—30 September

COMMUNICATIONS

Telephones—main lines in use: 145,300 (2005)
Telephones—mobile cellular: 500,200 (2005)
Telephone system:
general assessment: domestic facilities barely adequate; international facilities slightly better; telephone density in Haiti remains the lowest in the Latin American and Caribbean region
domestic: coaxial cable and microwave radio relay trunk service; combined fixed and mobile-cellular teledensity is about 8 per 100 persons
international: country code—509; satellite earth station—1 Intelsat (Atlantic Ocean)
Radio broadcast stations: AM 41, FM 26, shortwave 0 (1999)
Radios: 415,000 (1997)
Television broadcast stations: 2 (plus a cable TV service) (1997)
Televisions: 38,000 (1997)
Internet country code: .ht
Internet hosts: 7 (2007)
Internet Service Providers (ISPs): 3 (2000)
Internet users: 650,000 (2006)

TRANSPORTATION

Airports: 14 (2007)
Airports—with paved runways: *total:* 4

2,438 to 3,047 m: 1
914 to 1,523 m: 3 (2007)
Airports—with unpaved runways:
total: 10
914 to 1,523 m: 1
under 914 m: 9 (2007)
Roadways:
total: 4,160 km
paved: 1,011 km
unpaved: 3,149 km (2000)
Ports and terminals: Cap-Haitien

MILITARY

Military branches: no regular military forces—small Coast Guard; the regular Haitian Armed Forces (FAdH)—Army, Navy, and Air Force—have been demo-

bilized but still exist on paper unless they are constitutionally abolished (2007)
Manpower available for military service: males age 16–49: 2,047,083
females age 16–49: 2,047,953 (2008 est.)
Manpower fit for military service:
males age 16–49: 1,303,743
females age 16–49: 1,332,316 (2008 est.)
Manpower reaching militarily significant age annually:
males age 16–49: 105,655
females age 16–49: 104,376 (2008 est.)
Military expenditures—percent of GDP: 0.4% (2006)

TRANSNATIONAL ISSUES

Disputes—international: since 2004,

about 8,000 peacekeepers from the UN Stabilization Mission in Haiti (MINUSTAH) maintain civil order in Haiti; despite efforts to control illegal migration, Haitians cross into the Dominican Republic and sail to neighboring countries; Haiti claims US-administered Navassa Island
Illicit drugs: Caribbean transshipment point for cocaine en route to the US and Europe; substantial bulk cash smuggling activity; Colombian narcotics traffickers favor Haiti for illicit financial transactions; pervasive corruption; significant consumer of cannabis

HEARD ISLAND AND MCDONALD ISLANDS

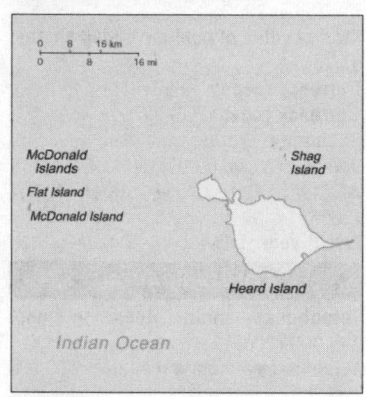

INTRODUCTION

Background: These uninhabited, barren, sub-Antarctic islands were transferred from the UK to Australia in 1947. Populated by large numbers of seal and bird species, the islands have been designated a nature preserve.

GEOGRAPHY

Location: islands in the Indian Ocean, about two-thirds of the way from Madagascar to Antarctica
Geographic coordinates: 53 06 S, 72 31 E
Map references: Antarctic Region
Area:
total: 412 sq km
land: 412 sq km
water: 0 sq km
Area—comparative: slightly more than two times the size of Washington, DC
Land boundaries: 0 km
Coastline: 101.9 km

Maritime claims:
territorial sea: 12 nm
exclusive fishing zone: 200 nm
Climate: antarctic
Terrain: Heard Island—80% ice-covered, bleak and mountainous, dominated by a large massif (Big Ben) and an active volcano (Mawson Peak); McDonald Islands—small and rocky
Elevation extremes:
lowest point: Indian Ocean 0 m
highest point: Mawson Peak, on Big Ben 2,745 m
Natural resources: fish
Land use:
arable land: 0%
permanent crops: 0%
other: 100% (2005)
Irrigated land: 0 sq km
Natural hazards: Mawson Peak, an active volcano, is on Heard Island
Environment—current issues: NA
Geography—note: Mawson Peak on Heard Island is the highest Australian mountain (at 2,745 meters, it is taller than Mt. Kosciuszko in Australia proper), and one of only two active volcanoes located in Australian territory, the other being McDonald Island; in 1992, McDonald Island broke its dormancy and began erupting; it has erupted several times since, the most recent being in 2005

PEOPLE

Population: uninhabited

GOVERNMENT

Country name:
conventional long form: Territory of Heard

Island and McDonald Islands
conventional short form: Heard Island and McDonald Islands
abbreviation: HIMI
Dependency status: territory of Australia; administered from Canberra by the Australian Antarctic Division of the Department of the Environment, Water, Heritage and the Arts
Legal system: the laws of Australia, where applicable, apply
Diplomatic representation in the US: none (territory of Australia)
Diplomatic representation from the US: none (territory of Australia)
Flag description: the flag of Australia is used

ECONOMY

Economy—overview: The islands have no indigenous economic activity, but the Australian Government allows limited fishing in the surrounding waters.

COMMUNICATIONS

Internet country code: .hm

TRANSPORTATION

Ports and terminals: none; offshore anchorage only

MILITARY

Military—note: defense is the responsibility of Australia; Australia conducts fisheries patrols

TRANSNATIONAL ISSUES

Disputes—international: none

HOLY SEE (VATICAN CITY)

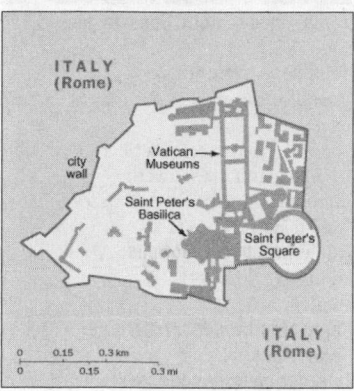

INTRODUCTION

Background: Popes in their secular role ruled portions of the Italian peninsula for more than a thousand years until the mid 19th century, when many of the Papal States were seized by the newly united Kingdom of Italy. In 1870, the pope's holdings were further circumscribed when Rome itself was annexed. Disputes between a series of "prisoner" popes and Italy were resolved in 1929 by three Lateran Treaties, which established the independent state of Vatican City and granted Roman Catholicism special status in Italy. In 1984, a concordat between the Holy See and Italy modified certain of the earlier treaty provisions, including the primacy of Roman Catholicism as the Italian state religion. Present concerns of the Holy See include religious freedom, international development, the environment, the Middle East, China, the decline of religion in Europe, terrorism, interreligious dialogue and reconciliation, and the application of church doctrine in an era of rapid change and globalization. About one billion people worldwide profess the Catholic faith.

GEOGRAPHY

Location: Southern Europe, an enclave of Rome (Italy)
Geographic coordinates: 41 54 N, 12 27 E
Map references: Europe
Area:
total: 0.44 sq km
land: 0.44 sq km
water: 0 sq km
Area—comparative: about 0.7 times the size of The Mall in Washington, DC
Land boundaries:
total: 3.2 km
border countries: Italy 3.2 km
Coastline: 0 km (landlocked)

Maritime claims: none (landlocked)
Climate: temperate; mild, rainy winters (September to May) with hot, dry summers (May to September)
Terrain: urban; low hill
Elevation extremes:
lowest point: unnamed location 19 m
highest point: unnamed location 75 m
Natural resources: none
Land use:
arable land: 0%
permanent crops: 0%
other: 100% (urban area) (2005)
Irrigated land: 0 sq km
Natural hazards: NA
Environment—current issues: NA
Environment—international agreements: *party to:* Climate Change
signed, but not ratified: Air Pollution, Environmental Modification
Geography—note: landlocked; enclave in Rome, Italy; world's smallest state; beyond the territorial boundary of Vatican City, the Lateran Treaty of 1929 grants the Holy See extraterritorial authority over 23 sites in Rome and five outside of Rome, including the Pontifical Palace at Castel Gandolfo (the Pope's summer residence)

PEOPLE

Population: 824 (July 2008 est.)
Population growth rate: 0.003% (2008 est.)
HIV/AIDS—adult prevalence rate: NA
HIV/AIDS—people living with HIV/AIDS: NA
HIV/AIDS—deaths: NA
Nationality:
noun: none
adjective: none
Ethnic groups: Italians, Swiss, other
Religions: Roman Catholic
Languages: Italian, Latin, French, various other languages
Literacy:
definition: NA
total population: 100%
male: 100%
female: 100%

GOVERNMENT

Country name:
conventional long form: The Holy See (State of the Vatican City)
conventional short form: Holy See (Vatican City)
local long form: Santa Sede (Stato della Citta del Vaticano)
local short form: Santa Sede (Citta del Vaticano)
Government type: ecclesiastical

Capital: *name:* Vatican City
geographic coordinates: 41 54 N, 12 27 E
time difference: UTC+1 (6 hours ahead of Washington, DC during Standard Time)
daylight saving time: +1hr, begins last Sunday in March; ends last Sunday in October
Administrative divisions: none
Independence: 11 February 1929 (from Italy); note—the three treaties signed with Italy on 11 February 1929 acknowledged, among other things, the full sovereignty of the Vatican and established its territorial extent; however, the origin of the Papal States, which over the years have varied considerably in extent, may be traced back to the 8th century
National holiday: Coronation Day of Pope BENEDICT XVI, 24 April (2005)
Constitution: new Fundamental Law promulgated by Pope JOHN PAUL II on 26 November 2000, effective 22 February 2001 (replaces the first Fundamental Law of 1929)
Legal system: based on Code of Canon Law and revisions to it
Suffrage: limited to cardinals less than 80 years old
Executive branch:
chief of state: Pope BENEDICT XVI (since 19 April 2005)
head of government: Secretary of State Cardinal Tarcisio BERTONE (since 15 September 2006)
cabinet: Pontifical Commission for the State of Vatican City appointed by the pope
elections: pope elected for life by the College of Cardinals; election last held 19 April 2005 (next to be held after the death of the current pope); secretary of state appointed by the pope
election results: Joseph RATZINGER elected Pope BENEDICT XVI
Legislative branch: unicameral Pontifical Commission for the State of Vatican City
Judicial branch: there are three tribunals responsible for civil and criminal matters within Vatican City; three other tribunals rule on issues pertaining to the Holy See
note: judicial duties were established by the Motu Proprio of Pope PIUS XII on 1 May 1946
Political parties and leaders: none
Political pressure groups and leaders: none (exclusive of influence exercised by church officers)
International organization participation: CE (observer), CPLP (associate), IAEA, IOM (observer), ITSO, ITU,

ITUC, NAM (guest), OAS (observer), OPCW, OSCE, UN (observer), UNCTAD, UNHCR, Union Latina (observer), UNWTO (observer), UPU, WIPO, WTO (observer)

Diplomatic representation in the US:
chief of mission: Apostolic Nuncio Archbishop Pietro SAMBI
chancery: 3339 Massachusetts Avenue NW, Washington, DC 20008
telephone: [1] (202) 333-7121
FAX: [1] (202) 337-4036

Diplomatic representation from the US:
chief of mission: Ambassador Mary Ann GLENDON
embassy: Villa Domiziana, Via delle Terme Deciane 26, 00153 Rome
mailing address: PSC 59, Box 66, APO AE 09624
telephone: [39] (06) 4674-3428
FAX: [39] (06) 575-8346

Flag description: two vertical bands of yellow (hoist side) and white with the arms of the Holy See, consisting of the crossed keys of Saint Peter surmounted by the three-tiered papal tiara, centered in the white band

ECONOMY

Economy—overview: This unique, non-commercial economy is supported financially by an annual contribution (known as Peter's Pence) from Roman Catholic dioceses throughout the world; by the sale of postage stamps, coins, medals, and tourist mementos; by fees for admission to museums; and by the sale of publications. Investments and real estate income also account for a sizable portion of revenue. The incomes and living standards of lay workers are comparable to those of counterparts who work in the city of Rome.

GDP (purchasing power parity): $NA
Labor force: NA
Labor force—by occupation: *note:* essentially services with a small amount of industry; nearly all dignitaries, priests, nuns, guards, and the approximately 3,000 lay workers live outside the Vatican
Population below poverty line: NA%
Budget:
revenues: $310 million
expenditures: $307 million (2006)
Industries: printing; production of coins, medals, postage stamps; a small amount of mosaics and staff uniforms; worldwide banking and financial activities
Electricity—production: NA kWh
Electricity—consumption: NA kWh
Electricity—imports: NA kWh; note—electricity supplied by Italy
Currency (code): euro (EUR)
Currency code: EUR
Exchange rates: euros per US dollar—0.7345 (2007), 0.7964 (2006), 0.8041 (2005), 0.8054 (2004), 0.886 (2003)

Fiscal year: calendar year

COMMUNICATIONS

Telephones—main lines in use: 5,120 (2005)
Telephone system:
general assessment: automatic digital exchange
domestic: connected via fiber optic cable to Telecom Italia network
international: country code—39; uses Italian system
Radio broadcast stations: AM 4, FM 3, shortwave 2 (2004)
Radios: NA
Television broadcast stations: 1 (2005)
Televisions: NA
Internet country code: .va
Internet hosts: 20 (2007)
Internet Service Providers (ISPs): NA
Internet users: 93 (2000)

MILITARY

Military branches: Pontifical Swiss Guard (Corpo della Guardia Svizzera Pontificia) (2007)
Military—note: defense is the responsibility of Italy; ceremonial and limited security duties performed by Pontifical Swiss Guard

TRANSNATIONAL ISSUES

Disputes—international: none

HONDURAS

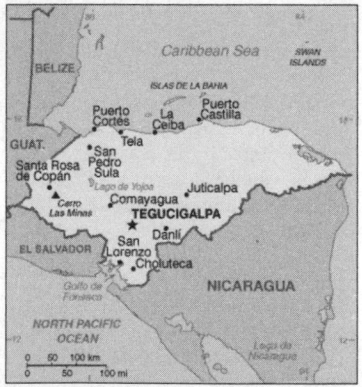

INTRODUCTION

Background: Once part of Spain's vast empire in the New World, Honduras became an independent nation in 1821. After two and a half decades of mostly military rule, a freely elected civilian government came to power in 1982. During the 1980s, Honduras proved a haven for anti-Sandinista contras fighting the Marxist Nicaraguan Government and an ally to Salvadoran Government forces fighting leftist guerrillas. The country was devastated by Hurricane Mitch in 1998, which killed about 5,600 people and caused approximately $2 billion in damage.

GEOGRAPHY

Location: Central America, bordering the Caribbean Sea, between Guatemala and Nicaragua and bordering the Gulf of Fonseca (North Pacific Ocean), between El Salvador and Nicaragua
Geographic coordinates: 15 00 N, 86 30 W
Map references: Central America and the Caribbean
Area:
total: 112,090 sq km
land: 111,890 sq km
water: 200 sq km
Area—comparative: slightly larger than Tennessee
Land boundaries:
total: 1,520 km
border countries: Guatemala 256 km, El Salvador 342 km, Nicaragua 922 km
Coastline: 820 km
Maritime claims:
territorial sea: 12 nm
contiguous zone: 24 nm
exclusive economic zone: 200 nm
continental shelf: natural extension of territory or to 200 nm
Climate: subtropical in lowlands, temperate in mountains
Terrain: mostly mountains in interior, narrow coastal plains
Elevation extremes:
lowest point: Caribbean Sea 0 m
highest point: Cerro Las Minas 2,870 m
Natural resources: timber, gold, silver, copper, lead, zinc, iron ore, antimony, coal, fish, hydropower
Land use:
arable land: 9.53%
permanent crops: 3.21%
other: 87.26% (2005)
Irrigated land: 800 sq km (2003)
Total renewable water resources: 95.9 cu km (2000)

Freshwater withdrawal (domestic/industrial/agricultural): *total:* 0.86 cu km/yr (8%/12%/80%)
per capita: 119 cu m/yr (2000)
Natural hazards: frequent, but generally mild, earthquakes; extremely susceptible to damaging hurricanes and floods along the Caribbean coast
Environment—current issues: urban population expanding; deforestation results from logging and the clearing of land for agricultural purposes; further land degradation and soil erosion hastened by uncontrolled development and improper land use practices such as farming of marginal lands; mining activities polluting Lago de Yojoa (the country's largest source of fresh water), as well as several rivers and streams, with heavy metals
Environment—international agreements: *party to:* Biodiversity, Climate Change, Climate Change-Kyoto Protocol, Desertification, Endangered Species, Hazardous Wastes, Law of the Sea, Marine Dumping, Ozone Layer Protection, Ship Pollution, Tropical Timber 83, Tropical Timber 94, Wetlands
signed, but not ratified: none of the selected agreements
Geography—note: has only a short Pacific coast but a long Caribbean shoreline, including the virtually uninhabited eastern Mosquito Coast

PEOPLE

Population: 7,639,327
note: estimates for this country explicitly take into account the effects of excess mortality due to AIDS; this can result in lower life expectancy, higher infant mortality, higher death rates, lower population growth rates, and changes in the distribution of population by age and sex than would otherwise be expected (July 2008 est.)
Age structure:
0–14 years: 38.7% (male 1,508,835/female 1,446,530)
15–64 years: 57.8% (male 2,210,187/female 2,203,620)
65 years and over: 3.5% (male 121,839/female 148,316) (2008 est.)
Median age: *total:* 20 years
male: 19.7 years
female: 20.4 years (2008 est.)
Population growth rate: 2.024% (2008 est.)
Birth rate: 26.93 births/1,000 population (2008 est.)
Death rate: 5.36 deaths/1,000 population (2008 est.)
Net migration rate: -1.33 migrant(s)/1,000 population (2008 est.)

Sex ratio:
at birth: 1.05 male(s)/female
under 15 years: 1.04 male(s)/female
15–64 years: 1 male(s)/female
65 years and over: 0.82 male(s)/female
total population: 1.01 male(s)/female (2008 est.)
Infant mortality rate:
total: 24.61 deaths/1,000 live births
male: 27.63 deaths/1,000 live births
female: 21.43 deaths/1,000 live births (2008 est.)
Life expectancy at birth:
total population: 69.37 years
male: 67.81 years
female: 71.01 years (2008 est.)
Total fertility rate: 3.38 children born/woman (2008 est.)
HIV/AIDS—adult prevalence rate: 1.8% (2003 est.)
HIV/AIDS—people living with HIV/AIDS: 63,000 (2003 est.)
HIV/AIDS—deaths: 4,100 (2003 est.)
Major infectious diseases:
degree of risk: high
food or waterborne diseases: bacterial diarrhea, hepatitis A, and typhoid fever
vectorborne diseases: dengue fever and malaria
water contact disease: leptospirosis (2008)
Nationality:
noun: Honduran(s)
adjective: Honduran
Ethnic groups: mestizo (mixed Amerindian and European) 90%, Amerindian 7%, black 2%, white 1%
Religions: Roman Catholic 97%, Protestant 3%
Languages: Spanish, Amerindian dialects
Literacy:
definition: age 15 and over can read and write
total population: 80%
male: 79.8%
female: 80.2% (2001 census)

GOVERNMENT

Country name:
conventional long form: Republic of Honduras
conventional short form: Honduras
local long form: Republica de Honduras
local short form: Honduras
Government type: democratic constitutional republic
Capital: *name:* Tegucigalpa
geographic coordinates: 14 06 N, 87 13 W
time difference: UTC-6 (1 hour behind Washington, DC during Standard Time)
daylight saving time: +1hr, begins second Sunday in March; ends first Sunday in November
Administrative divisions: 18 departments (departamentos, singular—depar-

tamento); Atlantida, Choluteca, Colon, Comayagua, Copan, Cortes, El Paraiso, Francisco Morazan, Gracias a Dios, Intibuca, Islas de la Bahia, La Paz, Lempira, Ocotepeque, Olancho, Santa Barbara, Valle, Yoro
Independence: 15 September 1821 (from Spain)
National holiday: Independence Day, 15 September (1821)
Constitution: 11 January 1982, effective 20 January 1982; amended many times
Legal system: rooted in Roman and Spanish civil law with increasing influence of English common law; recent judicial reforms include abandoning Napoleonic legal codes in favor of the oral adversarial system; accepts ICJ jurisdiction with reservations
Suffrage: 18 years of age; universal and compulsory
Executive branch:
chief of state: President Manuel ZELAYA Rosales (since 27 January 2006); Vice President Elvin Ernesto SANTOS Ordonez (since 27 January 2006); note—the president is both the chief of state and head of government
head of government: President Manuel ZELAYA Rosales (since 27 January 2006); Vice President Elvin Ernesto SANTOS Ordonez (since 27 January 2006)
cabinet: Cabinet appointed by president
elections: president elected by popular vote for a four-year term; election last held 27 November 2005 (next to be held in November 2009)
election results: Manuel ZELAYA Rosales elected president—49.8%, Porfirio "Pepe" LOBO Sosa 46.1%, other 4.1%
Legislative branch: unicameral National Congress or Congreso Nacional (128 seats; members are elected proportionally to the number of votes their party's presidential candidate receives to serve four-year terms)
elections: last held 27 November 2005 (next to be held in November 2009)
election results: percent of vote by party—NA; seats by party—PL 62, PN 55, PUD 5, PDC 4, PINU 2
Judicial branch: Supreme Court of Justice or Corte Suprema de Justicia (15 judges are elected for seven-year terms by the National Congress)
Political parties and leaders: Christian Democratic Party or PDC [Felicito AVILA]; Democratic Unification Party or PUD [Cesar HAM]; Liberal Party or PL [Patricia RODAS]; National Innovation and Unity Party or PINU [Jorge AQUILAR Paredes]; National Party of Honduras or PN [Porfirio LOBO]

Political pressure groups and leaders: Committee for the Defense of Human Rights in Honduras or CODEH; Confederation of Honduran Workers or CTH; Coordinating Committee of Popular Organizations or CCOP; General Workers Confederation or CGT; Honduran Council of Private Enterprise or COHEP; National Association of Honduran Campesinos or ANACH; National Union of Campesinos or UNC; Popular Bloc or BP; United Confederation of Honduran Workers or CUTH

International organization participation: BCIE, CACM, FAO, G-77, IADB, IAEA, IBRD, ICAO, ICCt, ICRM, IDA, IFAD, IFC, IFRCS, ILO, IMF, IMO, Interpol, IOC, IOM, ISO (subscriber), ITSO, ITU, ITUC, LAES, LAIA (observer), MIGA, MINURSO, NAM, OAS, OPANAL, OPCW, PCA, RG, UN, UNCTAD, UNESCO, UNIDO, Union Latina, UNWTO, UPU, WCL, WCO, WFTU, WHO, WIPO, WMO, WTO

Diplomatic representation in the US:
chief of mission: Ambassador Roberto FLORES BERMUDEZ
chancery: Suite 4-M, 3007 Tilden Street NW, Washington, DC 20008
telephone: [1] (202) 966-7702
FAX: [1] (202) 966-9751
consulate(s) general: Atlanta, Chicago, Houston, Los Angeles, Miami, New Orleans, New York, Phoenix, San Francisco
honorary consulate(s):. Boston, Detroit, Jacksonville

Diplomatic representation from the US:
chief of mission: Ambassador Charles A. FORD
embassy: Avenida La Paz, Apartado Postal No. 3453, Tegucigalpa
mailing address: American Embassy, APO AA 34022, Tegucigalpa
telephone: [504] 236-9320, 238-5114
FAX: [504] 236-9037

Flag description: three equal horizontal bands of blue (top), white, and blue with five blue, five-pointed stars arranged in an X pattern centered in the white band; the stars represent the members of the former Federal Republic of Central America—Costa Rica, El Salvador, Guatemala, Honduras, and Nicaragua; similar to the flag of El Salvador, which features a round emblem encircled by the words REPUBLICA DE EL SALVADOR EN LA AMERICA CENTRAL centered in the white band; also similar to the flag of Nicaragua, which features a triangle encircled by the word REPUBLICA DE NICARAGUA on top and AMERICA CENTRAL on the bottom, centered in the white band

ECONOMY

Economy—overview: Honduras, the second poorest country in Central America and one of the poorest countries in the Western Hemisphere, with an extraordinarily unequal distribution of income and massive unemployment, is banking on expanded trade under the US-Central America Free Trade Agreement (CAFTA) and on debt relief under the Heavily Indebted Poor Countries (HIPC) initiative. Despite improvements in tax collections, the government's fiscal deficit is growing due to increases in current expenditures and financial losses from the state energy and telephone companies. Honduras is the fastest growing remittance destination in the region with inflows representing over a quarter of GDP, equivalent to nearly three-quarters of exports. The economy relies heavily on a narrow range of exports, notably bananas and coffee, making it vulnerable to natural disasters and shifts in commodity prices, however, investments in the maquila and non-traditional export sectors are slowly diversifying the economy. Growth remains dependent on the economy of the US, its largest trading partner, and on reduction of the high crime rate, as a means of attracting and maintaining investment.

GDP (purchasing power parity): $30.65 billion (2007 est.)
GDP (official exchange rate): $12.28 billion (2007 est.)
GDP—real growth rate: 6.3% (2007 est.)
GDP—per capita (PPP): $4,100 (2007 est.)
GDP—composition by sector:
agriculture: 13.4%
industry: 28.1%
services: 58.6% (2007 est.)
Labor force: 2.78 million (2007 est.)
Labor force—by occupation: *agriculture:* 34%
industry: 23%
services: 43% (2003 est.)
Unemployment rate: 27.8% (2007 est.)
Population below poverty line: 50.7% (2004)
Household income or consumption by percentage share:
lowest 10%: 1.2%
highest 10%: 42.2% (2003)
Distribution of family income—Gini index: 53.8 (2003)
Inflation rate (consumer prices): 6.9% (2007 est.)
Investment (gross fixed): 30.4% of GDP (2007 est.)
Budget:
revenues: $2.344 billion

expenditures: $2.631 billion; including capital expenditures of $106 million (2007 est.)
Public debt: 24.1% of GDP (2007 est.)
Agriculture—products: bananas, coffee, citrus; beef; timber; shrimp, tilapia, lobster; corn, African palm
Industries: sugar, coffee, textiles, clothing, wood products
Industrial production growth rate: 4.4% (2007 est.)
Electricity—production: 5.339 billion kWh (2005)
Electricity—production by source:
fossil fuel: 50.2%
hydro: 49.8%
nuclear: 0%
other: 0% (2001)
Electricity—consumption: 4.036 billion kWh (2005)
Electricity—exports: 0 kWh (2005)
Electricity—imports: 57 million kWh (2005)
Oil—production: 0 bbl/day (2005 est.)
Oil—consumption: 43,000 bbl/day (2005 est.)
Oil—exports: 765.4 bbl/day (2004)
Oil—imports: 42,620 bbl/day (2004)
Oil—proved reserves: 0 bbl (1 January 2006 est.)
Natural gas—production: 0 cu m (2005 est.)
Natural gas—consumption: 0 cu m (2005 est.)
Natural gas—exports: 0 cu m (2005 est.)
Natural gas—imports: 0 cu m (2005)
Natural gas—proved reserves: 0 cu m (1 January 2006 est.)
Current account balance: -$1.225 billion (2007 est.)
Exports: $5.594 billion f.o.b. (2007 est.)
Exports—commodities: coffee, shrimp, bananas, gold, palm oil, fruit, lobster, lumber
Exports—partners: US 70.5%, Guatemala 3.5%, El Salvador 3.4% (2006)
Imports: $8.556 billion f.o.b. (2007 est.)
Imports—commodities: machinery and transport equipment, industrial raw materials, chemical products, fuels, foodstuffs
Imports—partners: US 53%, Guatemala 7%, El Salvador 4.5%, Costa Rica 4.1%, Mexico 4.1% (2006)
Economic aid—recipient: $680.8 million (2005)
Reserves of foreign exchange and gold: $2.545 billion (31 December 2007 est.)
Debt—external: $3.41 billion (31 December 2007 est.)
Market value of publicly traded shares: $NA

Currency (code): lempira (HNL)
Currency code: HNL
Exchange rates: lempiras per US dollar—18.9 (2007), 18.895 (2006), 18.92 (2005), 18.206 (2004), 17.345 (2003)
Fiscal year: calendar year

COMMUNICATIONS

Telephones—main lines in use: 708,400 (2006)
Telephones—mobile cellular: 2.241 million (2006)
Telephone system:
general assessment: inadequate system
domestic: beginning in 2003, private sub-operators allowed to provide fixed-lines in order to expand telephone coverage; fixed-line teledensity has increased to about 10 per 100 persons; mobile-cellular telephone service has been increasing rapidly and subscribership in 2006 exceeded 30 per 100 persons
international: country code—504; landing point for both the Americas Region Caribbean Ring System (ARCOS-1) and the MAYA-1 fiber optic submarine cable system that together provide connectivity to South and Central America, parts of the Caribbean, and the US; satellite earth stations—2 Intelsat (Atlantic Ocean); connected to Central American Microwave System
Radio broadcast stations: AM 241, FM 53, shortwave 12 (1998)
Radios: 2.45 million (1997)
Television broadcast stations: 11 (plus 17 repeaters) (1997)
Televisions: 570,000 (1997)
Internet country code: .hn
Internet hosts: 4,672 (2007)
Internet Service Providers (ISPs): 8 (2000)
Internet users: 337,300 (2006)

TRANSPORTATION

Airports: 112 (2007)
Airports—with paved runways:

total: 12
2,438 to 3,047 m: 3
1,524 to 2,437 m: 2
914 to 1,523 m: 4
under 914 m: 3 (2007)
Airports—with unpaved runways:
total: 100
1,524 to 2,437 m: 2
914 to 1,523 m: 15
under 914 m: 83 (2007)
Railways:
total: 699 km
narrow gauge: 279 km 1.067-m gauge; 420 km 0.914-m gauge (2006)
Roadways:
total: 13,600 km
paved: 2,775 km
unpaved: 10,825 km (2000)
Waterways: 465 km (most navigable only by small craft) (2007)
Merchant marine:
total: 126 ships (1000 GRT or over) 352,534 GRT/481,217 DWT
by type: bulk carrier 9, cargo 58, chemical tanker 5, container 1, liquefied gas 1, livestock carrier 1, passenger 4, passenger/cargo 7, petroleum tanker 27, refrigerated cargo 8, roll on/roll off 4, specialized tanker 1
foreign-owned: 40 (Bangladesh 1, Canada 1, China 3, Egypt 4, Greece 1, Hong Kong 1, Israel 1, Japan 4, South Korea 6, Lebanon 2, Mexico 1, Singapore 10, Taiwan 2, Tanzania 1, US 1, Vietnam 1) (2007)
Ports and terminals: La Ceiba, Puerto Cortes, San Lorenzo, Tela

MILITARY

Military branches: Army, Navy (includes Naval Infantry), Honduran Air Force (Fuerza Aerea Hondurena, FAH) (2008)
Military service age and obligation: 18 years of age for voluntary 2 to 3-year military service (2004)
Manpower available for military service:
males age 16–49: 1,868,940

females age 16–49: 1,825,770 (2008 est.)
Manpower fit for military service:
males age 16–49: 1,359,406
females age 16–49: 1,371,418 (2008 est.)
Manpower reaching militarily significant age annually:
males age 16–49: 90,876
females age 16–49: 87,292 (2008 est.)
Military expenditures—percent of GDP: 0.6% (2006 est.)

TRANSNATIONAL ISSUES

Disputes—international: International Court of Justice (ICJ) ruled on the delimitation of "bolsones" (disputed areas) along the El Salvador-Honduras border in 1992 with final settlement by the parties in 2006 after an Organization of American States (OAS) survey and a further ICJ ruling in 2003; the 1992 ICJ ruling advised a tripartite resolution to a maritime boundary in the Gulf of Fonseca with consideration of Honduran access to the Pacific; El Salvador continues to claim tiny Conejo Island, not mentioned in the ICJ ruling, off Honduras in the Gulf of Fonseca; Honduras claims the Belizean-administered Sapodilla Cays off the coast of Belize in its constitution, but agreed to a joint ecological park around the cays should Guatemala consent to a maritime corridor in the Caribbean under the OAS-sponsored 2002 Belize-Guatemala Differendum; memorials and counter-memorials were filed by the parties in Nicaragua's 1999 and 2001 proceedings against Honduras and Colombia at the ICJ over the maritime boundary and territorial claims in the western Caribbean Sea—final public hearings are scheduled for 2007
Illicit drugs: transshipment point for drugs and narcotics; illicit producer of cannabis, cultivated on small plots and used principally for local consumption; corruption is a major problem; some money-laundering activity

HONG KONG

INTRODUCTION

Background: Occupied by the UK in 1841, Hong Kong was formally ceded by China the following year; various adjacent lands were added later in the 19th century. Pursuant to an agreement signed by China and the UK on 19 December 1984, Hong Kong became the Hong Kong Special Administrative Region (SAR) of China on 1 July 1997. In this

agreement, China promised that, under its "one country, two systems" formula, China's socialist economic system would not be imposed on Hong Kong and that Hong Kong would enjoy a high degree of autonomy in all matters except foreign and defense affairs for the next 50 years.

GEOGRAPHY

Location: Eastern Asia, bordering the South China Sea and China

Geographic coordinates: 22 15 N, 114 10 E
Map references: Southeast Asia
Area: *total:* 1,092 sq km
land: 1,042 sq km
water: 50 sq km
Area—comparative: six times the size of Washington, DC
Land boundaries:
total: 30 km
regional border: China 30 km

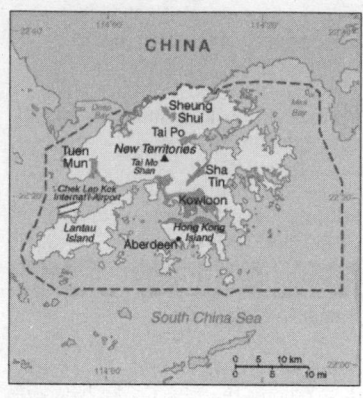

Coastline: 733 km
Maritime claims: *territorial sea:* 3 nm
Climate: subtropical monsoon; cool and humid in winter, hot and rainy from spring through summer, warm and sunny in fall
Terrain: hilly to mountainous with steep slopes; lowlands in north
Elevation extremes:
lowest point: South China Sea 0 m
highest point: Tai Mo Shan 958 m
Natural resources: outstanding deep-water harbor, feldspar
Land use:
arable land: 5.05%
permanent crops: 1.01%
other: 93.94% (2001)
Irrigated land: 20 sq km (1998 est.)
Natural hazards: occasional typhoons
Environment—current issues: air and water pollution from rapid urbanization
Environment—international agreements: *party to:* Marine Dumping (associate member), Ship Pollution (associate member)
Geography—note: more than 200 islands

PEOPLE

Population: 7,018,636 (July 2008 est.)
Age structure:
0–14 years: 12.6% (male 463,300/female 422,945)
15–64 years: 74.4% (male 2,535,246/female 2,684,495)
65 years and over: 13% (male 425,500/female 487,150) (2008 est.)
Median age:
total: 41.7 years
male: 41.4 years
female: 42 years (2008 est.)
Population growth rate: 0.532% (2008 est.)
Birth rate: 7.37 births/1,000 population (2008 est.)
Death rate: 6.6 deaths/1,000 population (2008 est.)
Net migration rate: 4.55 migrant(s)/1,000 population (2008 est.)

Sex ratio: *at birth:* 1.08 male(s)/female
under 15 years: 1.1 male(s)/female
15–64 years: 0.94 male(s)/female
65 years and over: 0.87 male(s)/female
total population: 0.95 male(s)/female (2008 est.)
Infant mortality rate:
total: 2.93 deaths/1,000 live births
male: 3.11 deaths/1,000 live births
female: 2.73 deaths/1,000 live births (2008 est.)
Life expectancy at birth:
total population: 81.77 years
male: 79.07 years
female: 84.69 years (2008 est.)
Total fertility rate: 1 children born/woman (2008 est.)
HIV/AIDS—adult prevalence rate: 0.1% (2003 est.)
HIV/AIDS—people living with HIV/AIDS: 2,600 (2003 est.)
HIV/AIDS—deaths: fewer than 200 (2003 est.)
Nationality:
noun: Chinese/Hong Konger
adjective: Chinese/Hong Kong
Ethnic groups: Chinese 94.9%, Filipino 2.1%, other 3% (2001 census)
Religions: eclectic mixture of local religions 90%, Christian 10%
Languages: Chinese (Cantonese) 89.2% (official), other Chinese dialects 6.4%, English 3.2% (official), other 1.2% (2001 census)
Literacy:
definition: age 15 and over has ever attended school
total population: 93.5%
male: 96.9%
female: 89.6% (2002)

GOVERNMENT

Country name:
conventional long form: Hong Kong Special Administrative Region
conventional short form: Hong Kong
local long form: Xianggang Tebie Xingzhengqu
local short form: Xianggang
abbreviation: HK
Dependency status: special administrative region of China
Government type: limited democracy
Administrative divisions: none (special administrative region of China)
Independence: none (special administrative region of China)
National holiday: National Day (Anniversary of the Founding of the People's Republic of China), 1 October (1949); note—1 July 1997 is celebrated as Hong Kong Special Administrative Region Establishment Day
Constitution: Basic Law, approved in March 1990 by China's National

People's Congress, is Hong Kong's "mini-constitution"
Legal system: based on English common law
Suffrage: direct election—18 years of age for a number of non-executive positions; universal for permanent residents living in the territory of Hong Kong for the past seven years; indirect election—limited to about 220,000 members of functional constituencies and an 800-member election committee drawn from broad regional groupings, central government bodies, and municipal organizations
Executive branch:
chief of state: President of China HU Jintao (since 15 March 2003)
head of government: Chief Executive Donald TSANG (since 24 June 2005)
cabinet: Executive Council consists of 15 official members and 16 non-official members
elections: chief executive elected for five-year term by 800-member electoral committee; last held on 25 March 2007 (next to be held in 2012)
election results: Donald TSANG elected chief executive receiving 84.1% of the vote of the election committee; Alan LEONG received 15.9%
Legislative branch: unicameral Legislative Council or LEGCO (60 seats; in 2004, 30 seats indirectly elected by functional constituencies, 30 elected by popular vote; members serve four-year terms)
elections: last held 12 September 2004 (next to be held in September 2008)
election results: percent of vote by party—pro-democracy 63%, pro-Beijing 37%; seats by party—(pro-Beijing 34) DAB 12, Liberal Party 10, FTU 1, independents 11; (pro-democracy 25) Democratic Party 9, CTU 2, ADPL 1, Frontier Party 1, NWSC 1, independents 11; non-voting LEGCO president 1
Judicial branch: Court of Final Appeal in the Hong Kong Special Administrative Region
Political parties and leaders: Association for Democracy and People's Livelihood or ADPL [Frederick FUNG Kin-kee]; Citizens Party [Alex CHAN Kai-chung]; Civic Party [KUAN Hsin-chi]; Democratic Alliance for the Betterment and Progress of Hong Kong or DAB [TAM Yiu Cheng]; Democratic Party [Albert HO]; Frontier Party [Emily LAU Wai-hing]; League of Social Democrats [Raymond WONG]; Liberal Party [James TIEN Pei-chun]
note: political blocs include: pro-democracy—ADPL, Democratic Party, Frontier Party, League of Social Democrats; pro-Beijing—DAB, Liberal Party, The Alliance (a group of five generally pro-

government and pro-business Legco members from functional constituencies); there is no political party ordinance, so there are no registered political parties; politically active groups register as societies or companies

Political pressure groups and leaders: Chinese General Chamber of Commerce (pro-China); Chinese Manufacturers' Association of Hong Kong; Confederation of Trade Unions or CTU (pro-democracy) [LAU Chin-shek, president; LEE Cheuk-yan, general secretary]; Federation of Hong Kong Industries; Federation of Trade Unions or FTU (pro-China) [CHENG Yiu-tong, executive councilor]; Hong Kong Alliance in Support of the Patriotic Democratic Movement in China [Szeto WAH, chairman]; Hong Kong and Kowloon Trade Union Council (pro-Taiwan); Hong Kong General Chamber of Commerce; Hong Kong Professional Teachers' Union [CHEUNG Man-kwong, president]; Neighborhood and Workers' Service Center or NWSC (pro-democracy); The Alliance [Bernard CHARNWUT, exco member]

International organization participation: ADB, APEC, BIS, ICC, IHO, IMF, IMO (associate), IOC, ISO (correspondent), ITUC, UNWTO (associate), UPU, WCL, WCO, WMO, WTO

Diplomatic representation in the US: none (special administrative region of China); Hong Kong Economic and Trade Office in Washington and two other cities carries out normal liaison and communication with the US Government and other US entities

Diplomatic representation from the US: *chief of mission:* Consul General James B. CUNNINGHAM

consulate(s) general: 26 Garden Road, Hong Kong

mailing address: PSC 461, Box 1, FPO AP 96521-0006

telephone: [852] 2523-9011

FAX: [852] 2845-1598

Flag description: red with a stylized, white, five-petal bauhinia flower in the center

ECONOMY

Economy—overview: Hong Kong has a free market economy highly dependent on international trade. In 2006, the total value of goods and services trade, including the sizable share of reexports, was equivalent to 400% of GDP. The territory has become increasingly integrated with mainland China over the past few years through trade, tourism, and financial links. The mainland has long been Hong Kong's largest trading partner, accounting for 46% of Hong Kong's total trade by value in 2006. As a result of China's easing of travel restrictions, the number of mainland tourists to the territory has surged from 4.5 million in 2001 to 13.6 million in 2006, when they outnumbered visitors from all other countries combined. Hong Kong has also established itself as the premier stock market for Chinese firms seeking to list abroad. Bolstered by several successful initial public offerings in early 2007, by September 2007 mainland companies accounted for one-third of the firms listed on the Hong Kong Stock Exchange, and more than half of the Exchange's market capitalization. During the past decade, as Hong Kong's manufacturing industry moved to the mainland, its service industry has grown rapidly and now accounts for 91% of the territory's GDP. Hong Kong's natural resources are limited, and food and raw materials must be imported. GDP growth averaged a strong 5% from 1989 to 2007, despite the economy suffering two recessions during the Asian financial crisis in 1997–98 and the global downturn in 2001–02. Hong Kong continues to link its currency closely to the US dollar, maintaining an arrangement established in 1983.

GDP (purchasing power parity): $292.8 billion (2007 est.)

GDP (official exchange rate): $206.7 billion (2007 est.)

GDP—real growth rate: 6.3% (2007 est.)

GDP—per capita (PPP): $42,000 (2007 est.)

GDP—composition by sector: *agriculture:* 0.1%

industry: 8.1%

services: 91.7% (2007 est.)

Labor force: 3.62 million (2007 est.)

Labor force—by occupation: manufacturing 6.5%, construction 2.1%, wholesale and retail trade, restaurants, and hotels 43.3%, financing, insurance, and real estate 20.7%, transport and communications 7.8%, community and social services 19.5%

note: above data exclude public sector (2007 est.)

Unemployment rate: 4.1% (2007 est.)

Population below poverty line: NA%

Household income or consumption by percentage share: *lowest 10%:* NA%

highest 10%: NA%

Distribution of family income—Gini index: 53.3 (2007)

Inflation rate (consumer prices): 2% (2007 est.)

Investment (gross fixed): 20.3% of GDP (2007 est.)

Budget: *revenues:* $36.9 billion

expenditures: $29.4 billion (FY07-08 est.)

Public debt: 7.1% of GDP (2007 est.)

Agriculture—products: fresh vegetables; poultry, pork; fish

Industries: textiles, clothing, tourism, banking, shipping, electronics, plastics, toys, watches, clocks

Industrial production growth rate: -0.9% (2007 est.)

Electricity—production: 38.6 billion kWh (2006)

Electricity—production by source: *fossil fuel:* 100%

hydro: 0%

nuclear: 0%

other: 0% (2001)

Electricity—consumption: 40.3 billion kWh (2006)

Electricity—exports: 4.5 billion kWh (2006)

Electricity—imports: 10.9 billion kWh (2006)

Oil—production: 0 bbl/day (2006 est.)

Oil—consumption: 292,000 bbl/day (2006 est.)

Oil—exports: 22,420 bbl/day (2006)

Oil—imports: 314,700 bbl/day (2006)

Oil—proved reserves: 0 bbl (1 January 2006 est.)

Natural gas—production: 0 cu m (2006 est.)

Natural gas—consumption: 2.944 billion cu m (2006 est.)

Natural gas—exports: 0 cu m (2006 est.)

Natural gas—imports: 2.944 billion cu m (2006)

Natural gas—proved reserves: 0 cu m (1 January 2006 est.)

Current account balance: $25.46 billion (2007 est.)

Exports: $345.9 billion f.o.b., including reexports (2007 est.)

Exports—commodities: electrical machinery and appliances, textiles, apparel, footwear, watches and clocks, toys, plastics, precious stones, printed material

Exports—partners: China 47%, US 15.1%, Japan 4.9% (2006)

Imports: $365.6 billion (2007 est.)

Imports—commodities: raw materials and semi-manufactures, consumer goods, capital goods, foodstuffs, fuel (most is reexported)

Imports—partners: China 45.9%, Japan 10.3%, Taiwan 7.5%, Singapore 6.3%, US 4.8%, South Korea 4.6% (2006)

Economic aid—recipient: $6.95 million (2004)

Reserves of foreign exchange and gold: $152.7 billion (31 December 2007 est.)

Debt—external: $588 billion (2007 est.)
Stock of direct foreign investment—at home: $780.4 billion (2007 est.)
Stock of direct foreign investment—abroad: $716.2 billion (2007 est.)
Market value of publicly traded shares: $2.97 trillion (2007 est.)
Currency (code): Hong Kong dollar (HKD)
Currency code: HKD
Exchange rates: Hong Kong dollars per US dollar—7.802 (2007), 7.7678 (2006), 7.7773 (2005), 7.788 (2004), 7.7868 (2003)
Fiscal year: 1 April—31 March

COMMUNICATIONS

Telephones—main lines in use: 3.87 million (2007)
Telephones—mobile cellular: 9.913 million (2007)
Telephone system:
general assessment: modern facilities provide excellent domestic and international services
domestic: microwave radio relay links and extensive fiber-optic network
international: country code—852; multiple international submarine cables provide connections to Asia, US, Australia, the Middle East, and Western Europe; satellite earth stations—3 Intelsat (1 Pacific Ocean and 2 Indian Ocean); coaxial cable to Guangzhou, China
Radio broadcast stations: AM 5, FM 9, shortwave 0 (2004)
Radios: 4.45 million (1997)
Television broadcast stations: 55 (2 TV networks, each broadcasting on 2 channels) (2007)
Televisions: 1.84 million (1997)

Internet country code: .hk
Internet hosts: 812,137 (2007)
Internet Service Providers (ISPs): 17 (2000)
Internet users: 3.77 million (2006)

TRANSPORTATION

Airports: 2 (2007)
Airports—with paved runways:
total: 2
over 3,047 m: 1
1,524 to 2,437 m: 1 (2007)
Heliports: 5 (2007)
Roadways:
total: 2,009 km
paved: 2,009 km (2007)
Merchant marine:
total: 1,009 ships (1000 GRT or over) 34,556,075 GRT/57,423,309 DWT
by type: barge carrier 2, bulk carrier 499, cargo 135, chemical tanker 51, combination ore/oil 3, container 173, liquefied gas 24, passenger 6, passenger/cargo 5, petroleum tanker 91, roll on/roll off 4, specialized tanker 8, vehicle carrier 8
foreign-owned: 617 (Belgium 4, Canada 39, China 309, Denmark 12, France 1, Germany 10, Greece 30, Indonesia 7, Japan 78, South Korea 6, Lebanon 1, Norway 30, Pakistan 1, Philippines 10, Portugal 1, Singapore 11, Syria 1, Taiwan 11, UAE 1, UK 32, US 22)
registered in other countries: 275 (Bahamas 3, Belize 5, Bermuda 4, Cambodia 11, China 6, Cyprus 2, Honduras 1, India 1, Liberia 21, Malaysia 14, Malta 1, Marshall Islands 4, Mongolia 1, Norway 5, Panama 137, Philippines 2, Seychelles 1, Singapore 37, St Vincent and The Grenadines 7, Tuvalu 10, UK 2, unknown 7) (2007)

Ports and terminals: Hong Kong

MILITARY

Military branches: no regular indigenous military forces; Hong Kong garrison of China's People's Liberation Army (PLA) includes elements of the PLA Ground Forces, PLA Navy, and PLA Air Force; these forces are under the direct leadership of the Central Military Commission in Beijing and under administrative control of the adjacent Guangzhou Military Region (2007)
Manpower available for military service:
males age 16–49: 1,772,820
females age 16–49: 1,941,448 (2008 est.)
Manpower fit for military service:
males age 16–49: 1,438,165
females age 16–49: 1,561,252 (2008 est.)
Manpower reaching militarily significant age annually:
males age 16–49: 42,173
females age 16–49: 38,753 (2008 est.)
Military expenditures—percent of GDP: NA
Military—note: defense is the responsibility of China

TRANSNATIONAL ISSUES

Disputes—international: none
Illicit drugs: despite strenuous law enforcement efforts, faces difficult challenges in controlling transit of heroin and methamphetamine to regional and world markets; modern banking system provides conduit for money laundering; rising indigenous use of synthetic drugs, especially among young people

HUNGARY

INTRODUCTION

Background: Hungary became a Christian kingdom in A.D. 1000 and for many centuries served as a bulwark against Ottoman Turkish expansion in Europe. The kingdom eventually became part of the polyglot Austro-Hungarian Empire, which collapsed during World War I. The country fell under Communist rule following World War II. In 1956, a revolt and an announced withdrawal from the Warsaw Pact were met with a massive military intervention by Moscow. Under the leadership of Janos KADAR in 1968, Hungary began liberalizing its economy, introducing so-called "Goulash Communism." Hungary held its first multiparty elections in 1990 and initiated a free market economy. It joined NATO in 1999 and the EU in 2004.

GEOGRAPHY

Location: Central Europe, northwest of Romania
Geographic coordinates: 47 00 N, 20 00 E

Map references: Europe
Area: *total:* 93,030 sq km
land: 92,340 sq km
water: 690 sq km
Area—comparative: slightly smaller than Indiana
Land boundaries:
total: 2,171 km
border countries: Austria 366 km, Croatia 329 km, Romania 443 km, Serbia 151 km, Slovakia 677 km, Slovenia 102 km, Ukraine 103 km
Coastline: 0 km (landlocked)
Maritime claims: none (landlocked)
Climate: temperate; cold, cloudy, humid winters; warm summers
Terrain: mostly flat to rolling plains; hills and low mountains on the Slovakian border
Elevation extremes:
lowest point: Tisza River 78 m
highest point: Kekes 1,014 m
Natural resources: bauxite, coal, natural gas, fertile soils, arable land
Land use:
arable land: 49.58%
permanent crops: 2.06%
other: 48.36% (2005)
Irrigated land: 2,300 sq km (2003)
Total renewable water resources: 120 cu km (2005)
Freshwater withdrawal (domestic/ industrial/agricultural): *total:* 21.03 cu km/yr (9%/59%/32%)
per capita: 2,082 cu m/yr (2001)
Environment—current issues: the upgrading of Hungary's standards in waste management, energy efficiency, and air, soil, and water pollution to meet EU requirements will require large investments
Environment—international agreements: *party to:* Air Pollution, Air Pollution-Nitrogen Oxides, Air Pollution-Persistent Organic Pollutants, Air Pollution-Sulfur 85, Air Pollution-Sulfur 94, Air Pollution-Volatile Organic Compounds, Antarctic Treaty, Biodiversity, Climate Change, Climate Change-Kyoto Protocol, Desertification, Endangered Species, Environmental Modification, Hazardous Wastes, Law of the Sea, Marine Dumping, Ozone Layer Protection, Ship Pollution, Wetlands, Whaling
signed, but not ratified: none of the selected agreements
Geography—note: landlocked; strategic location astride main land routes between Western Europe and Balkan Peninsula as well as between Ukraine and Mediterranean basin; the north-south flowing Duna (Danube) and Tisza Rivers divide the country into three large regions

PEOPLE

Population: 9,930,915 (July 2008 est.)
Age structure:
0–14 years: 15.2% (male 774,092/female 730,485)
15–64 years: 69.3% (male 3,393,630/female 3,488,011)
65 years and over: 15.6% (male 559,483/female 985,214) (2008 est.)
Median age:
total: 39.1 years
male: 36.8 years
female: 41.8 years (2008 est.)
Population growth rate: -0.254% (2008 est.)
Birth rate: 9.59 births/1,000 population (2008 est.)
Death rate: 12.99 deaths/1,000 population (2008 est.)
Net migration rate: 0.86 migrant(s)/1,000 population (2008 est.)
Sex ratio:
at birth: 1.06 male(s)/female
under 15 years: 1.06 male(s)/female
15–64 years: 0.97 male(s)/female
65 years and over: 0.57 male(s)/female
total population: 0.91 male(s)/female (2008 est.)
Infant mortality rate:
total: 8.03 deaths/1,000 live births
male: 8.74 deaths/1,000 live births
female: 7.29 deaths/1,000 live births (2008 est.)
Life expectancy at birth:
total population: 73.18 years
male: 69 years
female: 77.62 years (2008 est.)
Total fertility rate: 1.34 children born/woman (2008 est.)
HIV/AIDS—adult prevalence rate: 0.1% (2001 est.)
HIV/AIDS—people living with HIV/AIDS: 2,800 (2001 est.)
HIV/AIDS—deaths: fewer than 100 (2001 est.)
Major infectious diseases:
degree of risk: intermediate
food or waterborne diseases: bacterial diarrhea and hepatitis A
vectorborne diseases: tickborne encephalitis (2008)
Nationality:
noun: Hungarian(s)
adjective: Hungarian
Ethnic groups: Hungarian 92.3%, Roma 1.9%, other or unknown 5.8% (2001 census)
Religions: Roman Catholic 51.9%, Calvinist 15.9%, Lutheran 3%, Greek Catholic 2.6%, other Christian 1%, other or unspecified 11.1%, unaffiliated 14.5% (2001 census)
Languages: Hungarian 93.6%, other or unspecified 6.4% (2001 census)

Literacy: *definition:* age 15 and over can read and write
total population: 99.4%
male: 99.5%
female: 99.3% (2003 est.)

GOVERNMENT

Country name:
conventional long form: Republic of Hungary
conventional short form: Hungary
local long form: Magyar Koztarsasag
local short form: Magyarorszag
Government type: parliamentary democracy
Capital: *name:* Budapest
geographic coordinates: 47 30 N, 19 05 E
time difference: UTC+1 (6 hours ahead of Washington, DC during Standard Time)
daylight saving time: +1hr, begins last Sunday in March; ends last Sunday in October
Administrative divisions: 19 counties (megyek, singular—megye), 23 urban counties (singular—megyei varos), and 1 capital city (fovaros)
counties: Bacs-Kiskun, Baranya, Bekes, Borsod-Abauj-Zemplen, Csongrad, Fejer, Gyor-Moson-Sopron, Hajdu-Bihar, Heves, Jasz-Nagykun-Szolnok, Komarom-Esztergom, Nograd, Pest, Somogy, Szabolcs-Szatmar-Bereg, Tolna, Vas, Veszprem, Zala
urban counties: Bekescsaba, Debrecen, Dunaujvaros, Eger, Erd, Gyor, Hodmezovasarhely, Kaposvar, Kecskemet, Miskolc, Nagykanizsa, Nyiregyhaza, Pecs, Salgotarjan, Sopron, Szeged, Szekesfehervar, Szekszard, Szolnok, Szombathely, Tatabanya, Veszprem, Zalaegerszeg
capital city: Budapest
Independence: 25 December 1000 (crowning of King STEPHEN I, traditional founding date)
National holiday: Saint Stephen's Day, 20 August
Constitution: 18 August 1949, effective 20 August 1949; revised 19 April 1972; 18 October 1989 revision ensured legal rights for individuals and constitutional checks on the authority of the prime minister and also established the principle of parliamentary oversight; 1997 amendment streamlined the judicial system
Legal system: based on the German-Austrian legal system; accepts compulsory ICJ jurisdiction with reservations
Suffrage: 18 years of age; universal
Executive branch:
chief of state: President Laszlo SOLYOM (since 5 August 2005)
head of government: Prime Minister Ferenc GYURCSANY (since 29 September 2004)

THE CIA WORLD FACTBOOK

cabinet: Council of Ministers prime minister elected by the National Assembly on the recommendation of the president; other ministers proposed by the prime minister and appointed and relieved of their duties by the president

elections: president elected by the National Assembly for a five-year term (eligible for a second term); election last held 6–7 June 2005 (next to be held by June 2010); prime minister elected by the National Assembly on the recommendation of the president; election last held 29 September 2004

election results: Laszlo SOLYOM elected president by a simple majority in the third round of voting, 185 to 182; Ferenc GYURCSANY elected prime minister; result of legislative vote—197 to 12

note: to be elected, the president must win two-thirds of legislative vote in the first two rounds or a simple majority in the third round

Legislative branch: unicameral National Assembly or Orszaggyules (386 seats; members are elected by popular vote under a system of proportional and direct representation to serve four-year terms)

elections: last held 9 and 23 April 2006 (next to be held in April 2010)

election results: percent of vote by party (5% or more of the vote required for parliamentary representation in the first round)—MSzP 43.2%, Fidesz-KDNP 42%, SzDSz 6.5%, MDF 5%, other 3.3%; seats by party—MSzP 190, Fidesz-KDNP 164, SzDSz 20, MDF 11, independent 1

Judicial branch: Constitutional Court (judges are elected by the National Assembly for nine-year terms)

Political parties and leaders: Alliance of Free Democrats or SzDSz [Janos KOKA]; Christian Democratic People's Party or KDNP [Zsolt SEMJEN]; Hungarian Civic Alliance or Fidesz [Viktor ORBAN, chairman]; Hungarian Democratic Forum or MDF [Ibolya DAVID]; Hungarian Socialist Party or MSzP [Ferenc GYURCSANY]

Political pressure groups and leaders: NA

International organization participation: ACCT (observer), Australia Group, BIS, CE, CEI, CERN, EAPC, EBRD, EIB, ESA (cooperating state), EU, FAO, G-9, IAEA, IBRD, ICAO, ICC, ICCt, ICRM, IDA, IEA, IFC, IFRCS, ILO, IMF, IMO, IMSO, Interpol, IOC, IOM, IPU, ISO, ITSO, ITU, ITUC, MIGA, MINURSO, NAM (guest), NATO, NEA, NSG, OAS (observer), OECD, OIF (observer), OPCW, OSCE, PCA, Schengen Convention, SECI, UN, UNCTAD,

UNESCO, UNFICYP, UNHCR, UNIDO, UNOMIG, UNWTO, UPU, WCL, WCO, WEU (associate), WFTU, WHO, WIPO, WMO, WTO, ZC

Diplomatic representation in the US:
chief of mission: Ambassador Ferenc SOMOGYI
chancery: 3910 Shoemaker Street NW, Washington, DC 20008
telephone: [1] (202) 362-6730
FAX: [1] (202) 966-8135
consulate(s) general: Chicago, Los Angeles, New York

Diplomatic representation from the US:
chief of mission: Ambassador April H. FOLEY
embassy: Szabadsag ter 12, H-1054 Budapest
mailing address: pouch: American Embassy Budapest, 5270 Budapest Place, US Department of State, Washington, DC 20521-5270
telephone: [36] (1) 475-4400
FAX: [36] (1) 475-4764

Flag description: three equal horizontal bands of red (top), white, and green

ECONOMY

Economy—overview: Hungary has made the transition from a centrally planned to a market economy, with a per capita income nearly two-thirds that of the EU-25 average. The private sector accounts for more than 80% of GDP. Foreign ownership of and investment in Hungarian firms are widespread, with cumulative foreign direct investment totaling more than $60 billion since 1989. Hungary issues investment-grade sovereign debt. International observers, however, have expressed concerns over Hungary's fiscal and current account deficits. In 2007, Hungary eliminated a trade deficit that had persisted for several years. Inflation declined from 14% in 1998 to a low of 3.7% in 2006, but jumped to 7.8% in 2007. Unemployment has persisted above 6%. Hungary's labor force participation rate of 57% is one of the lowest in the Organization for Economic Cooperation and Development (OECD). Germany is by far Hungary's largest economic partner. Policy challenges include cutting the public sector deficit to 4% of GDP by 2008, from about 6% in 2007. The government's austerity program of tax hikes and subsidy cuts has reduced Hungary's large budget deficit, but the reforms have dampened domestic consumption, slowing GDP growth to about 2% in 2007. The government will need to pass additional reforms to ensure the long-term stability of public finances. The government plans to eventually lower its

public sector deficit to below 3% of GDP to adopt the euro.

GDP (purchasing power parity): $191.3 billion (2007 est.)

GDP (official exchange rate): $138.4 billion (2007 est.)

GDP—real growth rate: 1.3% (2007 est.)

GDP—per capita (PPP): $19,000 (2007 est.)

GDP—composition by sector:
agriculture: 2.8%
industry: 31.5%
services: 65.7% (2007 est.)

Labor force: 4.19 million (2007 est.)

Labor force—by occupation: *agriculture:* 5.5%
industry: 33.3%
services: 61.2% (2003)

Unemployment rate: 7.3% (2007 est.)

Population below poverty line: 8.6% (1993 est.)

Household income or consumption by percentage share:
lowest 10%: 4%
highest 10%: 22.2% (2002)

Distribution of family income—Gini index: 28 (2005)

Inflation rate (consumer prices): 7.9% (2007 est.)

Investment (gross fixed): 20.9% of GDP (2007 est.)

Budget:
revenues: $63.97 billion
expenditures: $71.69 billion (2007 est.)

Public debt: 67% of GDP (2007 est.)

Agriculture—products: wheat, corn, sunflower seed, potatoes, sugar beets; pigs, cattle, poultry, dairy products

Industries: mining, metallurgy, construction materials, processed foods, textiles, chemicals (especially pharmaceuticals), motor vehicles

Industrial production growth rate: 3.4% (2007 est.)

Electricity—production: 33.69 billion kWh (2005)

Electricity—production by source:
fossil fuel: 60.1%
hydro: 0.5%
nuclear: 39%
other: 0.3% (2001)

Electricity—consumption: 35.98 billion kWh (2005)

Electricity—exports: 9.41 billion kWh (2005)

Electricity—imports: 15.64 billion kWh (2005)

Oil—production: 42,180 bbl/day (2005 est.)

Oil—consumption: 152,200 bbl/day (2005 est.)

Oil—exports: 58,380 bbl/day (2004)

Oil—imports: 150,000 bbl/day (2004)

Oil—proved reserves: 102.5 million bbl (1 January 2006 est.)

296

Natural gas—production: 2.904 billion cu m (2005 est.)

Natural gas—consumption: 14.37 billion cu m (2005 est.)

Natural gas—exports: 0 cu m (2005 est.)

Natural gas—imports: 11.51 billion cu m (2005)

Natural gas—proved reserves: 32.86 billion cu m (1 January 2006 est.)

Current account balance: -$7.75 billion (2007 est.)

Exports: $88.77 billion f.o.b. (2007 est.)

Exports—commodities: machinery and equipment 61.1%, other manufactures 28.7%, food products 6.5%, raw materials 2%, fuels and electricity 1.6% (2003)

Exports—partners: Germany 29.2%, Italy 5.6%, France 4.9%, Austria 4.9%, UK 4.4%, Romania 4.1%, Poland 4% (2006)

Imports: $87.88 billion f.o.b. (2007 est.)

Imports—commodities: machinery and equipment 51.6%, other manufactures 35.7%, fuels and electricity 7.7%, food products 3.1%, raw materials 2.0% (2003)

Imports—partners: Germany 27%, Russia 8.2%, China 6.9%, Austria 6.2%, France 4.7%, Italy 4.6%, Netherlands 4.3%, Poland 4.2% (2006)

Economic aid—recipient: $302.6 million (2004)

Reserves of foreign exchange and gold: $24.05 billion (31 December 2007 est.)

Debt—external: $125.9 billion (31 December 2007)

Stock of direct foreign investment—at home: $108.6 billion (2007 est.)

Stock of direct foreign investment—abroad: $45.54 billion (2007 est.)

Market value of publicly traded shares: $41.93 billion (2006)

Currency (code): forint (HUF)

Currency code: HUF

Exchange rates: forints per US dollar— 186.16 (2007), 210.39 (2006), 199.58 (2005), 202.75 (2004), 224.31 (2003)

Fiscal year: calendar year

COMMUNICATIONS

Telephones—main lines in use: 3.35 million (2006)

Telephones—mobile cellular: 9.965 million (2006)

Telephone system:
general assessment: the telephone system has been modernized and is capable of satisfying all requests for telecommunication service

domestic: the system is digitalized and highly automated; trunk services are carried by fiber-optic cable and digital microwave radio relay; a program for fiber-optic subscriber connections was initiated in 1996; competition among mobile-cellular service providers has led to a sharp increase in the use of mobile cellular phones since 2000 and a decrease in the number of fixed-line connections

international: country code—36; Hungary has fiber-optic cable connections with all neighboring countries; the international switch is in Budapest; satellite earth stations—2 Intelsat (Atlantic Ocean and Indian Ocean regions), 1 Inmarsat, 1 very small aperture terminal (VSAT) system of ground terminals

Radio broadcast stations: AM 17, FM 57, shortwave 3 (1998)

Radios: 7.01 million (1997)

Television broadcast stations: 35 (plus 161 repeaters) (1995)

Televisions: 4.42 million (1997)

Internet country code: .hu

Internet hosts: 2.313 million (2007)

Internet Service Providers (ISPs): 16 (2000)

Internet users: 3.5 million (2006)

TRANSPORTATION

Airports: 46 (2007)

Airports—with paved runways:
total: 20
over 3,047 m: 2
2,438 to 3,047 m: 8
1,524 to 2,437 m: 4
914 to 1,523 m: 4
under 914 m: 2 (2007)

Airports—with unpaved runways:
total: 26
2,438 to 3,047 m: 2
1,524 to 2,437 m: 3
914 to 1,523 m: 11
under 914 m: 10 (2007)

Heliports: 5 (2007)

Pipelines: gas 4,397 km; oil 990 km; refined products 335 km (2007)

Railways:
total: 8,057 km
broad gauge: 36 km 1.524-m gauge
standard gauge: 7,802 km 1.435-m gauge (2,628 km electrified)
narrow gauge: 219 km 0.760-m gauge (2006)

Roadways:
total: 159,568 km
paved: 70,050 km (30,874 km of interurban roads including 626 km of expressways)
unpaved: 89,518 km (2005)

Waterways: 1,622 km (most on Danube River) (2007)

Ports and terminals: Budapest, Dunaujvaros, Gyor-Gonyu, Csepel, Baja, Mohacs (2003)

MILITARY

Military branches: Ground Forces, Hungarian Air Force (Magyar Legiero, ML) (2008)

Military service age and obligation: 18 years of age for voluntary military service; conscription abolished in June 2004; 6-month service obligation, with reserve obligation to age 50 (2006)

Manpower available for military service:
males age 16–49: 2,391,400
females age 16–49: 2,337,240 (2008 est.)

Manpower fit for military service:
males age 16–49: 1,890,105
females age 16–49: 1,943,422 (2008 est.)

Manpower reaching militarily significant age annually:
males age 16–49: 62,197
females age 16–49: 59,267 (2008 est.)

Military expenditures—percent of GDP: 1.75% (2005 est.)

TRANSNATIONAL ISSUES

Disputes—international: bilateral government, legal, technical and economic working group negotiations continue in 2006 with Slovakia over Hungary's failure to complete its portion of the Gabcikovo-Nagymaros hydroelectric dam project along the Danube; as a member state that forms part of the EU's external border, Hungary has implemented the strict Schengen border rules

Illicit drugs: transshipment point for Southwest Asian heroin and cannabis and for South American cocaine destined for Western Europe; limited producer of precursor chemicals, particularly for amphetamine and methamphetamine; efforts to counter money laundering, related to organized crime and drug trafficking, are improving, but remain vulnerable; significant consumer of ecstasy

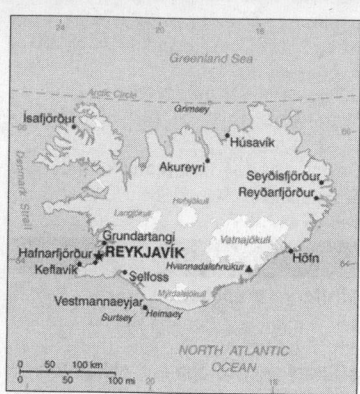

INTRODUCTION

Background: Settled by Norwegian and Celtic (Scottish and Irish) immigrants during the late 9th and 10th centuries A.D., Iceland boasts the world's oldest functioning legislative assembly, the Althing, established in 930. Independent for over 300 years, Iceland was subsequently ruled by Norway and Denmark. Fallout from the Askja volcano of 1875 devastated the Icelandic economy and caused widespread famine. Over the next quarter century, 20% of the island's population emigrated, mostly to Canada and the US. Limited home rule from Denmark was granted in 1874 and complete independence attained in 1944. Literacy, longevity, income, and social cohesion are first-rate by world standards.

GEOGRAPHY

Location: Northern Europe, island between the Greenland Sea and the North Atlantic Ocean, northwest of the UK

Geographic coordinates: 65 00 N, 18 00 W

Map references: Arctic Region

Area: *total:* 103,000 sq km
land: 100,250 sq km
water: 2,750 sq km

Area—comparative: slightly smaller than Kentucky

Land boundaries: 0 km

Coastline: 4,970 km

Maritime claims: *territorial sea:* 12 nm
exclusive economic zone: 200 nm
continental shelf: 200 nm or to the edge of the continental margin

Climate: temperate; moderated by North Atlantic Current; mild, windy winters; damp, cool summers

Terrain: mostly plateau interspersed with mountain peaks, icefields; coast deeply indented by bays and fiords

Elevation extremes:
lowest point: Atlantic Ocean 0 m
highest point: Hvannadalshnukur 2,110 m (at Vatnajokull glacier)

Natural resources: fish, hydropower, geothermal power, diatomite

Land use:
arable land: 0.07%
permanent crops: 0%
other: 99.93% (2005)

Irrigated land: NA

Total renewable water resources: 170 cu km (2005)

Freshwater withdrawal (domestic/industrial/agricultural): *total:* 0.17 cu km/yr (34%/66%/0%)
per capita: 567 cu m/yr (2003)

Natural hazards: earthquakes and volcanic activity

Environment—current issues: water pollution from fertilizer runoff; inadequate wastewater treatment

Environment—international agreements: *party to:* Air Pollution, Air Pollution-Persistent Organic Pollutants, Biodiversity, Climate Change, Climate Change-Kyoto Protocol, Desertification, Endangered Species, Hazardous Wastes, Kyoto Protocol, Law of the Sea, Marine Dumping, Ozone Layer Protection, Ship Pollution, Transboundary Air Pollution, Wetlands, Whaling
signed, but not ratified: Environmental Modification, Marine Life Conservation

Geography—note: strategic location between Greenland and Europe; westernmost European country; Reykjavik is the northernmost national capital in the world; more land covered by glaciers than in all of continental Europe

PEOPLE

Population: 304,367 (July 2008 est.)

Age structure:
0–14 years: 21% (male 32,500/female 31,566)
15–64 years: 67% (male 103,231/female 100,545)
65 years and over: 12% (male 16,530/female 19,995) (2008 est.)

Median age:
total: 34.8 years
male: 34.4 years
female: 35.3 years (2008 est.)

Population growth rate: 0.783% (2008 est.)

Birth rate: 13.5 births/1,000 population (2008 est.)

Death rate: 6.81 deaths/1,000 population (2008 est.)

Net migration rate: 1.13 migrant(s)/1,000 population (2008 est.)

Sex ratio: *at birth:* 1.04 male(s)/female

under 15 years: 1.03 male(s)/female
15–64 years: 1.03 male(s)/female
65 years and over: 0.83 male(s)/female
total population: 1 male(s)/female (2008 est.)

Infant mortality rate:
total: 3.25 deaths/1,000 live births
male: 3.39 deaths/1,000 live births
female: 3.1 deaths/1,000 live births (2008 est.)

Life expectancy at birth:
total population: 80.55 years
male: 78.43 years
female: 82.76 years (2008 est.)

Total fertility rate: 1.91 children born/woman (2008 est.)

HIV/AIDS—adult prevalence rate: 0.2% (2001 est.)

HIV/AIDS—people living with HIV/AIDS: 220 (2001 est.)

HIV/AIDS—deaths: fewer than 100 (2003 est.)

Nationality: *noun:* Icelander(s)
adjective: Icelandic

Ethnic groups: homogeneous mixture of descendants of Norse and Celts 94%, population of foreign origin 6%

Religions: Lutheran Church of Iceland 85.5%, Reykjavik Free Church 2.1%, Roman Catholic Church 2%, Hafnarfjorour Free Church 1.5%, other Christian 2.7%, other or unspecified 3.8%, unaffiliated 2.4% (2004)

Languages: Icelandic, English, Nordic languages, German widely spoken

Literacy: *definition:* age 15 and over can read and write
total population: 99%
male: 99%
female: 99% (2003 est.)

GOVERNMENT

Country name:
conventional long form: Republic of Iceland
conventional short form: Iceland
local long form: Lydveldid Island
local short form: Island

Government type: constitutional republic

Capital: *name:* Reykjavik
geographic coordinates: 64 09 N, 21 57 W
time difference: UTC (5 hours ahead of Washington, DC during Standard Time)

Administrative divisions: 8 regions; Austurland, Hofudhborgarsvaedhi, Nordhurland Eystra, Nordhurland Vestra, Sudhurland, Sudhurnes, Vestfirdhir, Vesturland

Independence: 1 December 1918 (became a sovereign state under the Danish Crown); 17 June 1944 (from Denmark)

National holiday: Independence Day, 17 June (1944)

Constitution: 16 June 1944, effective 17 June 1944; amended many times

Legal system: civil law system based on Danish law; has not accepted compulsory ICJ jurisdiction

Suffrage: 18 years of age; universal

Executive branch:

chief of state: President Olafur Ragnar GRIMSSON (since 1 August 1996)

head of government: Prime Minister Geir H. HAARDE (since 7 June 2006)

cabinet: Cabinet appointed by the prime minister

elections: president, largely a ceremonial post, is elected by popular vote for a four-year term (no term limits); election last held 26 June 2004 (next to be held in June 2008); following legislative elections, the leader of the majority party or the leader of the majority coalition is usually the prime minister

election results: Olafur Ragnar GRIMSSON 85.6%, Baldur AGUSTSSON 12.5%, Astthor MAGNUSSON 1.9%

Legislative branch: unicameral Parliament or Althing (63 seats; members are elected by popular vote to serve four-year terms)

elections: last held 12 May 2007 (next to be held by May 2011)

election results: percent of vote by party—Independence Party 36.6%, Social Democratic Alliance 26.8%, Progressive Party 11.7%, Left-Green Movement 14.3%, Liberal Party 7.3%, other 3.3%; seats by party—Independence Party 25, Social Democratic Alliance 18, Progressive Party 7, Left-Green Alliance 9, Liberal Party 4

Judicial branch: Supreme Court or Haestirettur (justices are appointed for life by the Minister of Justice); eight district courts (justices are appointed for life by the Minister of Justice)

Political parties and leaders: Independence Party or IP [Geir H. HAARDE]; Left-Green Movement or LGM [Steingrimur SIGFUSSON]; Liberal Party or LP [Gudjon KRISTJANSSON]; Progressive Party or PP [Gudni AGUSTSSON]; Social Democratic Alliance or SDA [Ingibjorg Solrun GISLADOTTIR] (includes People's Alliance or PA, Social Democratic Party or SDP, Women's List)

Political pressure groups and leaders: NA

International organization participation: Arctic Council, Australia Group, BIS, CBSS, CE, EAPC, EBRD, EFTA, FAO, IAEA, IBRD, ICAO, ICC, ICCt, ICRM, IDA, IFAD, IFC, IFRCS, IHO, ILO, IMF, IMO, IMSO, Interpol, IOC, IPU, ISO, ITSO, ITU, ITUC, MIGA, NATO, NC, NEA, NIB, OECD, OPCW, OSCE, PCA, Schengen Convention, UN, UNCTAD, UNESCO, UPU, WCO, WEU (associate), WHO, WIPO, WMO, WTO

Diplomatic representation in the US:

chief of mission: Ambassador Albert JONSSON

chancery: Suite 1200, 1156 15th Street NW, Washington, DC 20005-1704

telephone: [1] (202) 265-6653

FAX: [1] (202) 265-6656

consulate(s) general: New York

Diplomatic representation from the US:

chief of mission: Ambassador Carol VAN VOORST

embassy: Laufasvegur 21, 101 Reykjavik

mailing address: US Department of State, 5640 Reykjavik Place, Washington, D.C. 20521-5640

telephone: [354] 562-9100

FAX: [354] 562-9118

Flag description: blue with a red cross outlined in white extending to the edges of the flag; the vertical part of the cross is shifted to the hoist side in the style of the Dannebrog (Danish flag)

ECONOMY

Economy—overview: Iceland's Scandinavian-type economy is basically capitalistic, yet with an extensive welfare system (including generous housing subsidies), low unemployment, and remarkably even distribution of income. In the absence of other natural resources (except for abundant geothermal power), the economy depends heavily on the fishing industry, which provides 70% of export earnings and employs 6% of the work force. The economy remains sensitive to declining fish stocks as well as to fluctuations in world prices for its main **Exports:** fish and fish products, aluminum, and ferrosilicon. Substantial foreign investment in the aluminum and hydropower sectors has boosted economic growth which, nevertheless, has been volatile and characterized by recurrent imbalances. Government policies include reducing the current account deficit, limiting foreign borrowing, containing inflation, revising agricultural and fishing policies, and diversifying the economy. The government remains opposed to EU membership, primarily because of Icelanders' concern about losing control over their fishing resources. Iceland's economy has been diversifying into manufacturing and service industries in the last decade, and new developments in software production, biotechnology, and financial services are taking place. The tourism sector is also expanding, with the recent trends in ecotourism and whale watching. The 2006 closure of the US military base at Keflavik had very little impact on the national economy; Iceland's low unemployment rate aided former base employees in finding alternate employment.

GDP (purchasing power parity): $12.14 billion (2007 est.)

GDP (official exchange rate): $20 billion (2007 est.)

GDP—real growth rate: 3.8% (2007 est.)

GDP—per capita (PPP): $38,800 (2007 est.)

GDP—composition by sector:

agriculture: 5.2%

industry: 25.7%

services: 69.1% (2007 est.)

Labor force: 181,000 (2007 est.)

Labor force—by occupation: *agriculture:* 5.1%

industry: 23%

services: 71.8% (2005)

Unemployment rate: 1% (2007 est.)

Population below poverty line: NA%

Household income or consumption by percentage share:

lowest 10%: NA%

highest 10%: NA%

Distribution of family income—Gini index: 25 (2005)

Inflation rate (consumer prices): 5% (2007 est.)

Investment (gross fixed): 27.5% of GDP (2007 est.)

Budget:

revenues: $9.635 billion

expenditures: $8.597 billion (2007 est.)

Public debt: 27.6% of GDP (2007 est.)

Agriculture—products: potatoes, green vegetables; mutton, dairy products; fish

Industries: fish processing; aluminum smelting, ferrosilicon production; geothermal power, tourism

Industrial production growth rate: 9% (2007 est.)

Electricity—production: 8.533 billion kWh (2005)

Electricity—production by source:

fossil fuel: 0.1%

hydro: 82.5%

nuclear: 0%

other: 17.5% (geothermal) (2001)

Electricity—consumption: 8.152 billion kWh (2005)

Electricity—exports: 0 kWh (2005)

Electricity—imports: 0 kWh (2005)

Oil—production: 0 bbl/day (2005 est.)

Oil—consumption: 18,460 bbl/day (2005 est.)

Oil—exports: 0 bbl/day (2004)

Oil—imports: 17,450 bbl/day (2004)

Oil—proved reserves: 0 bbl (1 January 2006 est.)
Natural gas—production: 0 cu m (2005 est.)
Natural gas—consumption: 0 cu m (2005 est.)
Natural gas—exports: 0 cu m (2005 est.)
Natural gas—imports: 0 cu m (2005)
Natural gas—proved reserves: 0 cu m (1 January 2006 est.)
Current account balance: -$3.125 billion (2007 est.)
Exports: $4.793 billion f.o.b. (2007 est.)
Exports—commodities: fish and fish products 70%, aluminum, animal products, ferrosilicon, diatomite
Exports—partners: Netherlands 16.5%, UK 15.7%, Germany 15%, US 10.8%, Spain 6.4% (2006)
Imports: $6.181 billion (2007 est.)
Imports—commodities: machinery and equipment, petroleum products, foodstuffs, textiles
Imports—partners: US 12.8%, Germany 12.3%, Norway 7.1%, Sweden 6.9%, Denmark 6.1%, UK 5.3%, China 5.3%, Netherlands 4.8%, Japan 4.1% (2006)
Economic aid—donor: $6.7 million (2004)
Reserves of foreign exchange and gold: $2.436 billion (31 December 2007 est.)
Debt—external: $3.073 billion (2002)
Stock of direct foreign investment—at home: $NA
Stock of direct foreign investment—abroad: $NA
Market value of publicly traded shares: $27.8 billion (2005)
Currency (code): Icelandic krona (ISK)
Currency code: ISK
Exchange rates: Icelandic kronur per US dollar—63.391 (2007), 70.195 (2006), 62.982 (2005), 70.192 (2004), 76.709 (2003)
Fiscal year: calendar year

COMMUNICATIONS

Telephones—main lines in use: 193,700 (2006)

Telephones—mobile cellular: 328,500 (2006)
Telephone system:
general assessment: telecommunications infrastructure is modern and fully digitized, with satellite-earth stations, fiber-optic cables, and an extensive broadband network
domestic: liberalization of the telecommunications sector beginning in the late 1990s has led to increased competition especially in the mobile services segment of the market
international: country code—354; the CANTAT-3 and FARICE-1 submarine cable systems provide connectivity to Canada, the Faroe Islands, UK, Denmark, and Germany; a planned new section of the Hibernia-Atlantic submarine cable will provide additional connectivity to Canada, US, and Ireland; satellite earth stations—2 Intelsat (Atlantic Ocean), 1 Inmarsat (Atlantic and Indian Ocean regions); note—Iceland shares the Inmarsat earth station with the other Nordic countries (Denmark, Finland, Norway, and Sweden)
Radio broadcast stations: AM 3, FM about 70 (including repeaters), shortwave 1 (1998)
Radios: 260,000 (1997)
Television broadcast stations: 14 (plus 156 repeaters) (1997)
Televisions: 98,000 (1997)
Internet country code: .is
Internet hosts: 270,942 (2007)
Internet Service Providers (ISPs): 20 (2001)
Internet users: 194,000 (2006)

TRANSPORTATION

Airports: 99 (2007)
Airports—with paved runways:
total: 5
over 3,047 m: 1
1,524 to 2,437 m: 3
914 to 1,523 m: 1 (2007)
Airports—with unpaved runways:
total: 94
1,524 to 2,437 m: 3

914 to 1,523 m: 28
under 914 m: 63 (2007)
Roadways:
total: 13,058 km
paved/oiled gravel: 4,397 km (does not include urban roads)
unpaved: 8,661 km (2007)
Merchant marine:
total: 2 ships (1000 GRT or over) 4,704 GRT/729 DWT
by type: passenger/cargo 2
registered in other countries: 41 (Antigua and Barbuda 9, Bahamas 1, Belize 1, Faroe Islands 4, Gibraltar 1, Malta 7, Norway 3, St Vincent and The Grenadines 15) (2007)
Ports and terminals: Grundartangi, Hafnarfjordur, Reykjavik

MILITARY

Military branches: no regular military forces; Icelandic National Police (2008)
Manpower available for military service:
males age 16–49: 74,896 (2008 est.)
Manpower fit for military service:
males age 16–49: 62,342 (2008 est.)
Military expenditures—percent of GDP: 0% (2005 est.)
Military—note: Iceland has no standing military force; under a 1951 bilateral agreement—still valid—its defense was provided by the US-manned Icelandic Defense Force (IDF) headquartered at Keflavik; however, all US military forces in Iceland were withdrawn as of October 2006; although wartime defense of Iceland remains a NATO commitment, in April 2007, Iceland and Norway signed a bilateral agreement providing for Norwegian aerial surveillance and defense of Icelandic airspace (2008)

TRANSNATIONAL ISSUES

Disputes—international: Iceland, the UK, and Ireland dispute Denmark's claim that the Faroe Islands' continental shelf extends beyond 200 nm

INDIA

INTRODUCTION

Background: Aryan tribes from the northwest infiltrated onto the Indian subcontinent about 1500 B.C.; their merger with the earlier Dravidian inhabitants created the classical Indian culture. The Maurya Empire of the 4th and 3rd centuries B.C.—which reached its zenith under ASHOKA—united much of South Asia. The Golden Age ushered in by the Gupta dynasty (4th to 6th centuries A.D.) saw a flowering of Indian science, art, and culture. Arab incursions starting in the 8th century and Turkic in the 12th were followed by those of European traders, beginning in the late 15th century. By the 19th century, Britain had assumed political control of virtually all Indian lands. Indian armed forces in the British army played a vital role in both World Wars. Nonviolent resistance to British colonialism led by Mohandas GANDHI and Jawaharlal NEHRU brought independence in 1947. The subcontinent was divided into the secular state of India and the smaller Muslim state of Pakistan. A third war between the two countries in 1971 resulted in East Pakistan becoming the

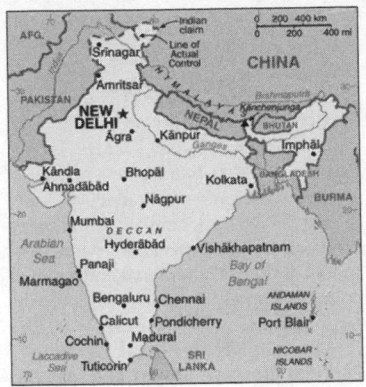

separate nation of Bangladesh. India's nuclear weapons testing in 1998 caused Pakistan to conduct its own tests that same year. The dispute between the countries over the state of Kashmir is ongoing, but discussions and confidence-building measures have led to decreased tensions since 2002. Despite impressive gains in economic investment and output, India faces pressing problems such as significant overpopulation, environmental degradation, extensive poverty, and ethnic and religious strife.

GEOGRAPHY

Location: Southern Asia, bordering the Arabian Sea and the Bay of Bengal, between Burma and Pakistan
Geographic coordinates: 20 00 N, 77 00 E
Map references: Asia
Area:
total: 3,287,590 sq km
land: 2,973,190 sq km
water: 314,400 sq km
Area—comparative: slightly more than one-third the size of the US
Land boundaries:
total: 14,103 km
border countries: Bangladesh 4,053 km, Bhutan 605 km, Burma 1,463 km, China 3,380 km, Nepal 1,690 km, Pakistan 2,912 km
Coastline: 7,000 km
Maritime claims:
territorial sea: 12 nm
contiguous zone: 24 nm
exclusive economic zone: 200 nm
continental shelf: 200 nm or to the edge of the continental margin
Climate: varies from tropical monsoon in south to temperate in north
Terrain: upland plain (Deccan Plateau) in south, flat to rolling plain along the Ganges, deserts in west, Himalayas in north
Elevation extremes:
lowest point: Indian Ocean 0 m
highest point: Kanchenjunga 8,598 m

Natural resources: coal (fourth-largest reserves in the world), iron ore, manganese, mica, bauxite, titanium ore, chromite, natural gas, diamonds, petroleum, limestone, arable land
Land use:
arable land: 48.83%
permanent crops: 2.8%
other: 48.37% (2005)
Irrigated land: 558,080 sq km (2003)
Total renewable water resources: 1,907.8 cu km (1999)
Freshwater withdrawal (domestic/industrial/agricultural): *total:* 645.84 cu km/yr (8%/5%/86%)
per capita: 585 cu m/yr (2000)
Natural hazards: droughts; flash floods, as well as widespread and destructive flooding from monsoonal rains; severe thunderstorms; earthquakes
Environment—current issues: deforestation; soil erosion; overgrazing; desertification; air pollution from industrial effluents and vehicle emissions; water pollution from raw sewage and runoff of agricultural pesticides; tap water is not potable throughout the country; huge and growing population is overstraining natural resources
Environment—international agreements: *party to:* Antarctic-Environmental Protocol, Antarctic-Marine Living Resources, Antarctic Treaty, Biodiversity, Climate Change, Climate Change-Kyoto Protocol, Desertification, Endangered Species, Environmental Modification, Hazardous Wastes, Law of the Sea, Ozone Layer Protection, Ship Pollution, Tropical Timber 83, Tropical Timber 94, Wetlands, Whaling
signed, but not ratified: none of the selected agreements
Geography—note: dominates South Asian subcontinent; near important Indian Ocean trade routes; Kanchenjunga, third tallest mountain in the world, lies on the border with Nepal

PEOPLE

Population: 1,147,995,898 (July 2008 est.)
Age structure:
0–14 years: 31.5% (male 189,238,487/female 172,168,306)
15–64 years: 63.3% (male 374,157,581/female 352,868,003)
65 years and over: 5.2% (male 28,285,796/female 31,277,725) (2008 est.)
Median age:
total: 25.1 years
male: 24.7 years
female: 25.5 years (2008 est.)
Population growth rate: 1.578% (2008 est.)

Birth rate: 22.22 births/1,000 population (2008 est.)
Death rate: 6.4 deaths/1,000 population (2008 est.)
Net migration rate: -0.05 migrant(s)/1,000 population (2008 est.)
Sex ratio:
at birth: 1.12 male(s)/female
under 15 years: 1.1 male(s)/female
15–64 years: 1.06 male(s)/female
65 years and over: 0.9 male(s)/female
total population: 1.06 male(s)/female (2008 est.)
Infant mortality rate:
total: 32.31 deaths/1,000 live births
male: 36.94 deaths/1,000 live births
female: 27.12 deaths/1,000 live births (2008 est.)
Life expectancy at birth:
total population: 69.25 years
male: 66.87 years
female: 71.9 years (2008 est.)
Total fertility rate: 2.76 children born/woman (2008 est.)
HIV/AIDS—adult prevalence rate: 0.9% (2001 est.)
HIV/AIDS—people living with HIV/AIDS: 5.1 million (2001 est.)
HIV/AIDS—deaths: 310,000 (2001 est.)
Major infectious diseases:
degree of risk: high
food or waterborne diseases: bacterial diarrhea, hepatitis A and E, and typhoid fever
vectorborne diseases: chikungunya, dengue fever, Japanese encephalitis, and malaria
animal contact disease: rabies
note: highly pathogenic H5N1 avian influenza has been identified in this country; it poses a negligible risk with extremely rare cases possible among US citizens who have close contact with birds (2008)
Nationality:
noun: Indian(s)
adjective: Indian
Ethnic groups: Indo-Aryan 72%, Dravidian 25%, Mongoloid and other 3% (2000)
Religions: Hindu 80.5%, Muslim 13.4%, Christian 2.3%, Sikh 1.9%, other 1.8%, unspecified 0.1% (2001 census)
Languages: English enjoys associate status but is the most important language for national, political, and commercial communication; Hindi is the national language and primary tongue of 30% of the people; there are 21 other official
Languages: Assamese, Bengali, Bodo, Dogri, Gujarati, Kannada, Kashmiri, Konkani, Maithili, Malayalam, Manipuri, Marathi, Nepali, Oriya, Punjabi, Sanscrit, Santhali, Sindhi, Tamil, Telugu, and Urdu; Hindustani is a

THE CIA WORLD FACTBOOK

popular variant of Hindi/Urdu spoken widely throughout northern India but is not an official language

Literacy:

definition: age 15 and over can read and write

total population: 61%

male: 73.4%

female: 47.8% (2001 census)

GOVERNMENT

Country name:

conventional long form: Republic of India

conventional short form: India

local long form: Republic of India/Bharatiya Ganarajya

local short form: India/Bharat

Government type: federal republic

Capital: *name:* New Delhi

geographic coordinates: 28 36 N, 77 12 E

time difference: UTC+5.5 (10.5 hours ahead of Washington, DC during Standard Time)

Administrative divisions: 28 states and 7 union territories*; Andaman and Nicobar Islands*, Andhra Pradesh, Arunachal Pradesh, Assam, Bihar, Chandigarh*, Chhattisgarh, Dadra and Nagar Haveli*, Daman and Diu*, Delhi*, Goa, Gujarat, Haryana, Himachal Pradesh, Jammu and Kashmir, Jharkhand, Karnataka, Kerala, Lakshadweep*, Madhya Pradesh, Maharashtra, Manipur, Meghalaya, Mizoram, Nagaland, Orissa, Puducherry*, Punjab, Rajasthan, Sikkim, Tamil Nadu, Tripura, Uttar Pradesh, Uttarakhand, West Bengal

Independence: 15 August 1947 (from UK)

National holiday: Republic Day, 26 January (1950)

Constitution: 26 January 1950; amended many times

Legal system: based on English common law; judicial review of legislative acts; accepts compulsory ICJ jurisdiction with reservations; separate personal law codes apply to Muslims, Christians, and Hindus

Suffrage: 18 years of age; universal

Executive branch:

chief of state: President Pratibha PATIL (since 25 July 2007); Vice President Hamid ANSARI (since 11 August 2007)

head of government: Prime Minister Manmohan SINGH (since 22 May 2004)

cabinet: Cabinet appointed by the president on the recommendation of the prime minister

elections: president elected by an electoral college consisting of elected members of both houses of Parliament and the legislatures of the states for a five-year

term (no term limits); election last held 21 July 2007 (next to be held in July 2012); vice president elected by both houses of Parliament for a five-year term; election last held 12 August 2002 (next to be held August 2007); prime minister chosen by parliamentary members of the majority party following legislative elections; election last held April—May 2004 (next to be held May 2009)

election results: Pratibha PATIL elected president; percent of vote—65.8%; Bhairon Singh SHEKHAWAT—34.2%

Legislative branch: bicameral Parliament or Sansad consists of the Council of States or Rajya Sabha (a body consisting of not more than 250 members up to 12 of whom are appointed by the president, the remainder are chosen by the elected members of the state and territorial assemblies; members serve six-year terms) and the People's Assembly or Lok Sabha (545 seats; 543 elected by popular vote, 2 appointed by the president; members serve five-year terms)

elections: People's Assembly—last held 20 April through 10 May 2004 (next must be held by May 2009)

election results: People's Assembly—percent of vote by party—NA; seats by party—INC 147, BJP 129, CPI (M) 43, SP 38, RJD 23, DMK 16, BSP 15, SS 12, BJD 11, CPI 10, NCP 10, JD (U) 8, SAD 8, PMK 6, JMM 5, LJSP 4, MDMK 4, TDP 4, TRS 4, independent 6, other 29, vacant 13; note—seats by party as of December 2006

Judicial branch: Supreme Court (one chief justice and 25 associate justices are appointed by the president and remain in office until they reach the age of 65 or are removed for "proved misbehavior")

Political parties and leaders: Bahujan Samaj Party or BSP [MAYAWATI]; Bharatiya Janata Party or BJP [Rajnath SINGH]; Biju Janata Dal or BJD [Naveen PATNAIK]; Communist Party of India or CPI [Ardhendu Bhushan BARDHAN]; Communist Party of India-Marxist or CPI-M [Prakash KARAT]; Dravida Munnetra Kazagham or DMK [M. KARUNANIDHI]; Indian National Congress or INC [Sonia GANDHI]; Janata Dal (United) or JD(U) [Sharad YADAV]; Jharkhand Mukti Morcha or JMM [Shibu SOREN]; Left Front (an alliance of Indian leftist parties); Lok Jan Shakti Party or LJSP [Ram Vilas PASWAN]; Marumalarchi Dravida Munnetra Kazhagam or MDMK [V. Gopalswamy VAIKO]; Nationalist Congress Party or NCP [Sharad PAWAR]; Pattali Makkal Katchi or PMK [S. RAMADOSS]; Rashtriya Janata Dal or RJD [Laloo Prasad YADAV];

Samajwadi Party or SP [Mulayam Singh YADAV]; Shiromani Akali Dal or SAD [Parkash Singh BADAL]; Shiv Sena or SS [Bal THACKERAY]; Telangana Rashtriya Samithi or TRS [K. Chandrashekhar RAO]; Telugu Desam Party or TDP [Chandrababu NAIDU]; United Progressive Alliance or UPA [Sonia GANDHI] (India's ruling party coalition of 12 political parties); note—India has dozens of national and regional political parties; only parties or coalitions with four or more seats in the People's Assembly are listed

Political pressure groups and leaders: numerous religious or militant/chauvinistic organizations, including Vishwa Hindu Parishad, Bajrang Dal, and Rashtriya Swayamsevak Sangh; various separatist groups seeking greater communal and/or regional autonomy, including the All Parties Hurriyat Conference in the Kashmir Valley and the National Socialist Council of Nagaland in the Northeast

International organization participation: ADB, AfDB, ARF, ASEAN (dialogue partner), BIMSTEC, BIS, C, CERN (observer), CP, EAS, FAO, G-15, G-24, G-77, IAEA, IBRD, ICAO, ICC, ICRM, IDA, IFAD, IFC, IFRCS, IHO, ILO, IMF, IMO, IMSO, Interpol, IOC, IOM (observer), IPU, ISO, ITSO, ITU, ITUC, LAS (observer), MIGA, MONUC, NAM, OAS (observer), OPCW, PCA, PIF (partner), SAARC, SACEP, SCO (observer), UN, UNCTAD, UNDOF, UNESCO, UNHCR, UNIDO, UNIFIL, UNMEE, UNMIS, UNOCI, UNWTO, UPU, WCL, WCO, WFTU, WHO, WIPO, WMO, WTO

Diplomatic representation in the US:

chief of mission: Ambassador Ranendra SEN

chancery: 2107 Massachusetts Avenue NW, Washington, DC 20008; note— Consular Wing located at 2536 Massachusetts Avenue NW, Washington, DC 20008

telephone: [1] (202) 939-7000

FAX: [1] (202) 265-4351

consulate(s) general: Chicago, Houston, New York, San Francisco

Diplomatic representation from the US:

chief of mission: Ambassador David C. MULFORD

embassy: Shantipath, Chanakyapuri, New Delhi 110021

mailing address: use embassy street address

telephone: [91] (011) 2419-8000

FAX: [91] (11) 2419-0017

consulate(s) general: Chennai (Madras), Kolkata (Calcutta), Mumbai (Bombay)

Flag description: three equal horizontal bands of saffron (subdued orange) (top), white, and green with a blue chakra (24-spoked wheel) centered in the white band; similar to the flag of Niger, which has a small orange disk centered in the white band

ECONOMY

Economy—overview: India's diverse economy encompasses traditional village farming, modern agriculture, handicrafts, a wide range of modern industries, and a multitude of services. Services are the major source of economic growth, accounting for more than half of India's output with less than one third of its labor force. About three-fifths of the work force is in agriculture, leading the United Progressive Alliance (UPA) government to articulate an economic reform program that includes developing basic infrastructure to improve the lives of the rural poor and boost economic performance. The government has reduced controls on foreign trade and investment. Higher limits on foreign direct investment were permitted in a few key sectors, such as telecommunications. However, tariff spikes in sensitive categories, including agriculture, and incremental progress on economic reforms still hinder foreign access to India's vast and growing market. Privatization of government-owned industries remains stalled and continues to generate political debate; populist pressure from within the UPA government and from its Left Front allies continues to restrain needed initiatives. The economy has posted an average growth rate of more than 7% in the decade since 1997, reducing poverty by about 10 percentage points. India achieved 8.5% GDP growth in 2006, and again in 2007, significantly expanding production of manufactures. India is capitalizing on its large numbers of well-educated people skilled in the English language to become a major exporter of software services and software workers. Economic expansion has helped New Delhi continue to make progress in reducing its federal fiscal deficit. However, strong growth combined with easy consumer credit and a real estate boom fueled inflation concerns in 2006 and 2007, leading to a series of central bank interest rate hikes that have slowed credit growth and eased inflation concerns. The huge and growing population is the fundamental social, economic, and environmental problem.

GDP (purchasing power parity): $2.989 trillion (2007 est.)

GDP (official exchange rate): $1.099 trillion (2007 est.)

GDP—real growth rate: 9.2% (2007 est.)

GDP—per capita (PPP): $2,700 (2007 est.)

GDP—composition by sector:
agriculture: 17.6%
industry: 29.4%
services: 52.9% (2007 est.)

Labor force: 516.4 million (2007 est.)

Labor force—by occupation: *agriculture:* 60%
industry: 12%
services: 28% (2003)

Unemployment rate: 7.2% (2007 est.)

Population below poverty line: 25% (2007 est.)

Household income or consumption by percentage share:
lowest 10%: 3.6%
highest 10%: 31.1% (2004)

Distribution of family income—Gini index: 36.8 (2004)

Inflation rate (consumer prices): 6.4% (2007 est.)

Investment (gross fixed): 34.6% of GDP (2007 est.)

Budget:
revenues: $141.8 billion
expenditures: $178.3 billion (2007 est.)

Public debt: 58% of GDP (federal and state debt combined) (2007 est.)

Agriculture—products: rice, wheat, oilseed, cotton, jute, tea, sugarcane, potatoes; cattle, water buffalo, sheep, goats, poultry; fish

Industries: textiles, chemicals, food processing, steel, transportation equipment, cement, mining, petroleum, machinery, software

Industrial production growth rate: 8.9% (2007 est.)

Electricity—production: 661.6 billion kWh (2005)

Electricity—production by source:
fossil fuel: 81.7%
hydro: 14.5%
nuclear: 3.4%
other: 0.3% (2001)

Electricity—consumption: 488.5 billion kWh (2005)

Electricity—exports: 67 million kWh (2005)

Electricity—imports: 1.764 billion kWh (2005)

Oil—production: 834,600 bbl/day (2005 est.)

Oil—consumption: 2.438 million bbl/day (2005 est.)

Oil—exports: 350,000 bbl/day (2005 est.)

Oil—imports: 2.098 million bbl/day (2004 est.)

Oil—proved reserves: 5.7 billion bbl (2007 est.)

Natural gas—production: 28.68 billion cu m (2005 est.)

Natural gas—consumption: 34.47 billion cu m (2005 est.)

Natural gas—exports: 0 cu m (2005 est.)

Natural gas—imports: 5.793 billion cu m (2005)

Natural gas—proved reserves: 1.056 trillion cu m (1 January 2006 est.)

Current account balance: -$19.35 billion (2007 est.)

Exports: $150.8 billion f.o.b. (2007 est.)

Exports—commodities: petroleum products, textile goods, gems and jewelry, engineering goods, chemicals, leather manufactures

Exports—partners: US 17%, UAE 8.3%, China 7.7%, UK 4.3% (2006)

Imports: $230.2 billion f.o.b. (2007 est.)

Imports—commodities: crude oil, machinery, gems, fertilizer, chemicals

Imports—partners: China 8.7%, US 6%, Germany 4.6%, Singapore 4.6%, Australia 4% (2006)

Economic aid—recipient: $1.724 billion (2005)

Reserves of foreign exchange and gold: $275 billion (31 December 2007 est.)

Debt—external: $148.1 billion (31 December 2007)

Stock of direct foreign investment—at home: $95.28 billion (2007 est.)

Stock of direct foreign investment—abroad: $37.62 billion (2007 est.)

Market value of publicly traded shares: $818.9 billion (2006)

Currency (code): Indian rupee (INR)

Currency code: INR

Exchange rates: Indian rupees per US dollar—41.487 (2007), 45.3 (2006), 44.101 (2005), 45.317 (2004), 46.583 (2003)

Fiscal year: 1 April—31 March

COMMUNICATIONS

Telephones—main lines in use: 49.75 million (2005)

Telephones—mobile cellular: 166.1 million (2006)

Telephone system:
general assessment: recent deregulation and liberalization of telecommunications laws and policies have prompted rapid growth; local and long distance service provided throughout all regions of the country, with services primarily concentrated in the urban areas; steady improvement is taking place with the recent admission of private and private-public investors, but combined fixed and mobile telephone density remains low at about 20 for each 100 persons nationwide and much lower for persons in rural

areas; fastest growth is in cellular service with modest growth in fixed lines

domestic: mobile cellular service introduced in 1994 and organized nationwide into four metropolitan areas and 19 telecom circles each with about three private service providers and one state-owned service provider; in recent years significant trunk capacity added in the form of fiber-optic cable and one of the world's largest domestic satellite systems, the Indian National Satellite system (INSAT), with 6 satellites supporting 33,000 very small aperture terminals (VSAT)

international: country code—91; a number of major international submarine cable systems, including Sea-Me-We-3 with landing sites at Cochin and Mumbai (Bombay), Sea-Me-We-4 with a landing site at Chennai, Fiber-Optic Link Around the Globe (FLAG) with a landing site at Mumbai (Bombay), South Africa—Far East (SAFE) with a landing site at Cochin, the i2i cable network linking to Singapore with landing sites at Mumbai (Bombay) and Chennai (Madras), and Tata Indicom linking Singapore and Chennai (Madras), provide a significant increase in the bandwidth available for both voice and data traffic; satellite earth stations—8 Intelsat (Indian Ocean) and 1 Inmarsat (Indian Ocean region); 9 gateway exchanges operating from Mumbai (Bombay), New Delhi, Kolkata (Calcutta), Chennai (Madras), Jalandhar, Kanpur, Gandhinagar, Hyderabad, and Ernakulam

Radio broadcast stations: AM 153, FM 91, shortwave 68 (1998)

Radios: 116 million (1997)

Television broadcast stations: 562 (1997)

Televisions: 63 million (1997)

Internet country code: .in

Internet hosts: 2.306 million (2007)

Internet Service Providers (ISPs): 43 (2000)

Internet users: 60 million (2005)

TRANSPORTATION

Airports: 346 (2007)

Airports—with paved runways:
total: 250
over 3,047 m: 18
2,438 to 3,047 m: 52
1,524 to 2,437 m: 75
914 to 1,523 m: 84
under 914 m: 21 (2007)

Airports—with unpaved runways:
total: 96
over 3,047 m: 1
2,438 to 3,047 m: 1
1,524 to 2,437 m: 7

914 to 1,523 m: 40
under 914 m: 47 (2007)

Heliports: 30 (2007)

Pipelines: condensate/gas 9 km; gas 7,488 km; liquid petroleum gas 1,861 km; oil 7,883 km; refined products 6,422 km (2007)

Railways:
total: 63,221 km
broad gauge: 46,807 km 1.676-m gauge (17,343 km electrified)
narrow gauge: 13,290 km 1.000-m gauge (165 km electrified); 3,124 km 0.762-m gauge and 0.610-m gauge (2006)

Roadways:
total: 3,383,344 km
paved: 1,603,705 km
unpaved: 1,779,639 km (2002)

Waterways: 14,500 km
note: 5,200 km on major rivers and 485 km on canals suitable for mechanized vessels (2006)

Merchant marine:
total: 477 ships (1000 GRT or over) 8,350,093 GRT/14,339,440 DWT
by type: bulk carrier 101, cargo 220, chemical tanker 18, combination ore/oil 1, container 9, liquefied gas 19, passenger 3, passenger/cargo 10, petroleum tanker 95, roll on/roll off 1
foreign-owned: 5 (China 1, Hong Kong 1, UAE 2, UK 1)
registered in other countries: 54 (Barbados 1, Comoros 2, Cyprus 1, Dominica 2, North Korea 1, Liberia 2, Malta 3, Mauritius 2, Panama 25, Singapore 9, St Kitts and Nevis 1, St Vincent and The Grenadines 5, unknown 2) (2007)

Ports and terminals: Chennai, Haldia, Jawaharal Nehru, Kandla, Kolkata (Calcutta), Mormugao, Mumbai (Bombay), New Mangalore, Vishakhapatnam

MILITARY

Military branches: Army, Navy (includes naval air arm), Air Force, Coast Guard (2008)

Military service age and obligation: 16 years of age for voluntary military service; no conscription (2008)

Manpower available for military service:
males age 16–49: 301,094,084
females age 16–49: 283,047,141 (2008 est.)

Manpower fit for military service:
males age 16–49: 231,161,111
females age 16–49: 236,633,962 (2008 est.)

Manpower reaching militarily significant age annually:
males age 16–49: 11,592,516
females age 16–49: 10,636,857 (2008 est.)

Military expenditures—percent of GDP: 2.5% (2006)

TRANSNATIONAL ISSUES

Disputes—international: since China and India launched a security and foreign policy dialogue in 2005, consolidated discussions related to the dispute over most of their rugged, militarized boundary, regional nuclear proliferation, Indian claims that China transferred missiles to Pakistan, and other matters continue; various talks and confidence-building measures have cautiously begun to defuse tensions over Kashmir, particularly since the October 2005 earthquake in the region; Kashmir nevertheless remains the site of the world's largest and most militarized territorial dispute with portions under the de facto administration of China (Aksai Chin), India (Jammu and Kashmir), and Pakistan (Azad Kashmir and Northern Areas); India and Pakistan have maintained the 2004 cease fire in Kashmir and initiated discussions on defusing the armed stand-off in the Siachen glacier region; Pakistan protests India's fencing the highly militarized Line of Control and construction of the Baglihar Dam on the Chenab River in Jammu and Kashmir, which is part of the larger dispute on water sharing of the Indus River and its tributaries; UN Military Observer Group in India and Pakistan (UNMOGIP) has maintained a small group of peace-keepers since 1949; India does not recognize Pakistan's ceding historic Kashmir lands to China in 1964; to defuse tensions and prepare for discussions on a maritime boundary, India and Pakistan seek technical resolution of the disputed boundary in Sir Creek estuary at the mouth of the Rann of Kutch in the Arabian Sea; Pakistani maps continue to show its Junagadh claim in Indian Gujarat State; discussions with Bangladesh remain stalled to delimit a small section of river boundary, to exchange territory for 51 Bangladeshi exclaves in India and 111 Indian exclaves in Bangladesh, to allocate divided villages, and to stop illegal cross-border trade, migration, violence, and transit of terrorists through the porous border; Bangladesh protests India's attempts to fence off high-traffic sections of the border; dispute with Bangladesh over New Moore/South Talpatty/Purbasha Island in the Bay of Bengal deters maritime boundary delimitation; India seeks cooperation from Bhutan and Burma to keep Indian Nagaland and Assam separatists from hiding in remote areas along the borders;

Joint Border Committee with Nepal continues to examine contested boundary sections, including the 400 square kilometer dispute over the source of the Kalapani River; India maintains a strict border regime to keep out Maoist insurgents and control illegal cross-border activities from Nepal

Refugees and internally displaced persons: *refugees (country of origin):* 77,200 (Tibet/China); 69,609 (Sri Lanka); 9,472 (Afghanistan)

IDPs: at least 600,000 (about half are Kashmiri Pandits from Jammu and Kashmir) (2007)

Trafficking in persons: *current situation:* India is a source, destination, and transit country for men, women, and children trafficked for the purposes of forced or bonded labor and commercial sexual exploitation; the large population of men, women, and children—numbering in the millions—in debt bondage face involuntary servitude in brick kilns, rice mills, and embroidery factories, while some children endure involuntary servitude as domestic servants; internal trafficking of women and girls for the purposes of commercial sexual exploitation and forced marriage also occurs; the government estimates that 90 percent of India's sex trafficking is internal; India is also a destination for women and girls from Nepal and Bangladesh trafficked for the purpose of commercial sexual exploitation; boys from Afghanistan, Pakistan, and Bangladesh are trafficked through India to the Gulf states for involuntary servitude as child camel jockeys; Indian men and women migrate willingly to the Persian Gulf region for work as domestic servants and low-skilled laborers, but some later find themselves in situations of involuntary servitude including extended working hours, nonpayment of wages, restrictions on their movement by withholding of their passports or confinement to the home, and physical or sexual abuse

tier rating: Tier 2 Watch List—India has been on the Tier 2 Watch List since 2004 for its failure to show evidence of increasing efforts to address trafficking in persons

Illicit drugs: world's largest producer of licit opium for the pharmaceutical trade, but an undetermined quantity of opium is diverted to illicit international drug markets; transit point for illicit narcotics produced in neighboring countries; illicit producer of methaqualone; vulnerable to narcotics money laundering through the hawala system; licit ketamine and precursor production

INDIAN OCEAN

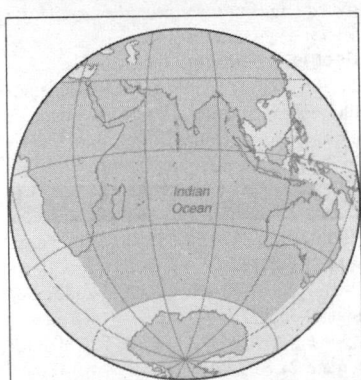

Indian Ocean

INTRODUCTION

Background: The Indian Ocean is the third largest of the world's five oceans (after the Pacific Ocean and Atlantic Ocean, but larger than the Southern Ocean and Arctic Ocean). Four critically important access waterways are the Suez Canal (Egypt), Bab el Mandeb (Djibouti-Yemen), Strait of Hormuz (Iran-Oman), and Strait of Malacca (Indonesia-Malaysia). The decision by the International Hydrographic Organization in the spring of 2000 to delimit a fifth ocean, the Southern Ocean, removed the portion of the Indian Ocean south of 60 degrees south latitude.

GEOGRAPHY

Location: body of water between Africa, the Southern Ocean, Asia, and Australia

Geographic coordinates: 20 00 S, 80 00 E

Map references: Political Map of the World

Area:
total: 68.556 million sq km
note: includes Andaman Sea, Arabian Sea, Bay of Bengal, Flores Sea, Great Australian Bight, Gulf of Aden, Gulf of Oman, Java Sea, Mozambique Channel, Persian Gulf, Red Sea, Savu Sea, Strait of Malacca, Timor Sea, and other tributary water bodies

Area—comparative: about 5.5 times the size of the US

Coastline: 66,526 km

Climate: northeast monsoon (December to April), southwest monsoon (June to October); tropical cyclones occur during May/June and October/November in the northern Indian Ocean and January/February in the southern Indian Ocean

Terrain: surface dominated by counterclockwise gyre (broad, circular system of currents) in the southern Indian Ocean; unique reversal of surface currents in the northern Indian Ocean; low atmospheric pressure over southwest Asia from hot, rising, summer air results in the southwest monsoon and southwest-to-northeast winds and currents, while high pressure over northern Asia from cold, falling, winter air results in the northeast monsoon and northeast-to-southwest winds and currents; ocean floor is dominated by the Mid-Indian Ocean Ridge and subdivided by the Southeast Indian Ocean Ridge, Southwest Indian Ocean Ridge, and Ninetyeast Ridge

Elevation extremes:
lowest point: Java Trench -7,258 m
highest point: sea level 0 m

Natural resources: oil and gas fields, fish, shrimp, sand and gravel aggregates, placer deposits, polymetallic nodules

Natural hazards: occasional icebergs pose navigational hazard in southern reaches

Environment—current issues: endangered marine species include the dugong, seals, turtles, and whales; oil pollution in the Arabian Sea, Persian Gulf, and Red Sea

Geography—note: major chokepoints include Bab el Mandeb, Strait of Hormuz, Strait of Malacca, southern access to the Suez Canal, and the Lombok Strait

ECONOMY

Economy—overview: The Indian Ocean provides major sea routes connecting the Middle East, Africa, and East Asia with Europe and the Americas. It carries a particularly heavy traffic of petroleum and petroleum products from the oilfields of the Persian Gulf and Indonesia. Its fish are of great and growing importance to the bordering countries for domestic consumption and export. Fishing fleets from Russia, Japan, South Korea, and Taiwan also exploit the Indian Ocean, mainly for shrimp and tuna. Large reserves of hydrocarbons are being tapped in the offshore areas of

Saudi Arabia, Iran, India, and western Australia. An estimated 40% of the world's offshore oil production comes from the Indian Ocean. Beach sands rich in heavy minerals and offshore placer deposits are actively exploited by bordering countries, particularly India, South Africa, Indonesia, Sri Lanka, and Thailand.

TRANSPORTATION

Ports and terminals: Chennai (Madras; India), Colombo (Sri Lanka), Durban (South Africa), Jakarta (Indonesia), Kolkata (Calcutta; India) Melbourne (Australia), Mumbai (Bombay; India), Richards Bay (South Africa)

TRANSNATIONAL ISSUES

Disputes—international: some maritime disputes (see littoral states)

INDONESIA

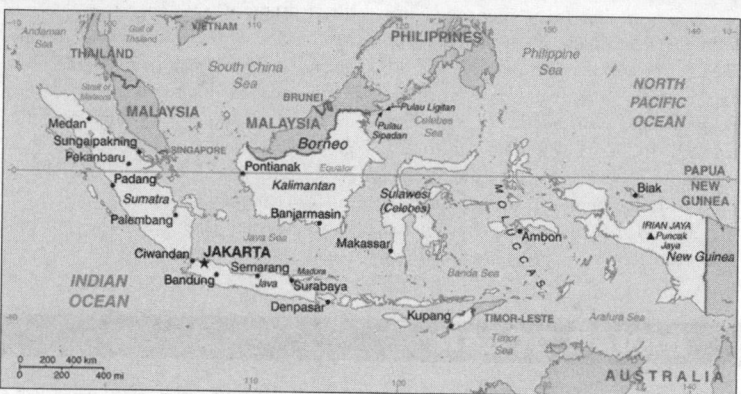

INTRODUCTION

Background: The Dutch began to colonize Indonesia in the early 17th century; the islands were occupied by Japan from 1942 to 1945. Indonesia declared its independence after Japan's surrender, but it required four years of intermittent negotiations, recurring hostilities, and UN mediation before the Netherlands agreed to relinquish its colony. Indonesia is the world's largest archipelagic state and home to the world's largest Muslim population. Current issues include: alleviating poverty, preventing terrorism, consolidating democracy after four decades of authoritarianism, implementing financial sector reforms, stemming corruption, holding the military and police accountable for human rights violations, and controlling avian influenza. In 2005, Indonesia reached a historic peace agreement with armed separatists in Aceh, which led to democratic elections in December 2006. Indonesia continues to face a low intensity separatist movement in Papua.

GEOGRAPHY

Location: Southeastern Asia, archipelago between the Indian Ocean and the Pacific Ocean
Geographic coordinates: 5 00 S, 120 00 E
Map references: Southeast Asia
Area:
total: 1,919,440 sq km
land: 1,826,440 sq km
water: 93,000 sq km
Area—comparative: slightly less than three times the size of Texas
Land boundaries:
total: 2,830 km
border countries: Timor-Leste 228 km, Malaysia 1,782 km, Papua New Guinea 820 km
Coastline: 54,716 km
Maritime claims: measured from claimed archipelagic straight baselines
territorial sea: 12 nm
exclusive economic zone: 200 nm
Climate: tropical; hot, humid; more moderate in highlands
Terrain: mostly coastal lowlands; larger islands have interior mountains
Elevation extremes:
lowest point: Indian Ocean 0 m
highest point: Puncak Jaya 5,030 m
Natural resources: petroleum, tin, natural gas, nickel, timber, bauxite, copper, fertile soils, coal, gold, silver
Land use:
arable land: 11.03%
permanent crops: 7.04%
other: 81.93% (2005)
Irrigated land: 45,000 sq km (2003)
Total renewable water resources: 2,838 cu km (1999)
Freshwater withdrawal (domestic/industrial/agricultural): *total:* 82.78 cu km/yr (8%/1%/91%)
per capita: 372 cu m/yr (2000)
Natural hazards: occasional floods, severe droughts, tsunamis, earthquakes, volcanoes, forest fires

Environment—current issues: deforestation; water pollution from industrial wastes, sewage; air pollution in urban areas; smoke and haze from forest fires
Environment—international agreements: *party to:* Biodiversity, Climate Change, Climate Change-Kyoto Protocol, Desertification, Endangered Species, Hazardous Wastes, Law of the Sea, Marine Life Conservation, Ozone Layer Protection, Ship Pollution, Tropical Timber 83, Tropical Timber 94, Wetlands
signed, but not ratified: none of the selected agreements
Geography—note: archipelago of 17,508 islands (6,000 inhabited); straddles equator; strategic location astride or along major sea lanes from Indian Ocean to Pacific Ocean

PEOPLE

Population: 237,512,355 (July 2008 est.)
Age structure:
0–14 years: 28.4% (male 34,343,198/female 33,175,135)
15–64 years: 65.7% (male 78,330,830/female 77,812,339)
65 years and over: 5.8% (male 6,151,305/female 7,699,548) (2008 est.)
Median age: *total:* 27.2 years
male: 26.7 years
female: 27.7 years (2008 est.)
Population growth rate: 1.175% (2008 est.)
Birth rate: 19.24 births/1,000 population (2008 est.)
Death rate: 6.24 deaths/1,000 population (2008 est.)
Net migration rate: -1.25 migrant(s)/1,000 population (2008 est.)
Sex ratio: *at birth:* 1.05 male(s)/female
under 15 years: 1.04 male(s)/female
15–64 years: 1.01 male(s)/female
65 years and over: 0.8 male(s)/female
total population: 1 male(s)/female (2008 est.)
Infant mortality rate:
total: 31.04 deaths/1,000 live births
male: 36.14 deaths/1,000 live births
female: 25.68 deaths/1,000 live births (2008 est.)

Life expectancy at birth:
total population: 70.46 years
male: 67.98 years
female: 73.07 years (2008 est.)
Total fertility rate: 2.34 children born/woman (2008 est.)
HIV/AIDS—adult prevalence rate: 0.1% (2003 est.)
HIV/AIDS—people living with HIV/AIDS: 110,000 (2003 est.)
HIV/AIDS—deaths: 2,400 (2003 est.)
Major infectious diseases:
degree of risk: high
food or waterborne diseases: bacterial diarrhea, hepatitis A and E, and typhoid fever
vectorborne diseases: chikungunya, dengue fever, and malaria
note: highly pathogenic H5N1 avian influenza has been identified in this country; it poses a negligible risk with extremely rare cases possible among US citizens who have close contact with birds (2008)
Nationality: *noun:* Indonesian(s)
adjective: Indonesian
Ethnic groups: Javanese 40.6%, Sundanese 15%, Madurese 3.3%, Minangkabau 2.7%, Betawi 2.4%, Bugis 2.4%, Banten 2%, Banjar 1.7%, other or unspecified 29.9% (2000 census)
Religions: Muslim 86.1%, Protestant 5.7%, Roman Catholic 3%, Hindu 1.8%, other or unspecified 3.4% (2000 census)
Languages: Bahasa Indonesia (official, modified form of Malay), English, Dutch, local dialects (the most widely spoken of which is Javanese)
Literacy: *definition:* age 15 and over can read and write
total population: 90.4%
male: 94%
female: 86.8% (2004 est.)

GOVERNMENT

Country name:
conventional long form: Republic of Indonesia
conventional short form: Indonesia
local long form: Republik Indonesia
local short form: Indonesia
former: Netherlands East Indies, Dutch East Indies
Government type: republic
Capital: *name:* Jakarta
geographic coordinates: 6 10 S, 106 49 E
time difference: UTC+7 (12 hours ahead of Washington, DC during Standard Time)
note: Indonesia is divided into three time zones
Administrative divisions: 30 provinces (propinsi-propinsi, singular—propinsi), 2 special regions* (daerah-daerah istimewa, singular—daerah istimewa), and 1 special

capital city district** (daerah khusus ibukota); Aceh*, Bali, Banten, Bengkulu, Gorontalo, Jakarta Raya**, Jambi, Jawa Barat, Jawa Tengah, Jawa Timur, Kalimantan Barat, Kalimantan Selatan, Kalimantan Tengah, Kalimantan Timur, Kepulauan Bangka Belitung, Kepulauan Riau, Lampung, Maluku, Maluku Utara, Nusa Tenggara Barat, Nusa Tenggara Timur, Papua, Papua Barat (Irian Jaya Barat), Riau, Sulawesi Barat, Sulawesi Selatan, Sulawesi Tengah, Sulawesi Tenggara, Sulawesi Utara, Sumatera Barat, Sumatera Selatan, Sumatera Utara, Yogyakarta*
note: following the implementation of decentralization beginning on 1 January 2001, the 440 districts or regencies have become the key administrative units responsible for providing most government services
Independence: 17 August 1945 (declared)
note: recognized by the Netherlands on 27 December 1949; in August 2005, the Netherlands announced it recognized de facto Indonesian independence on 17 August 1945
National holiday: Independence Day, 17 August (1945)
Constitution: August 1945; abrogated by Federal Constitution of 1949 and Provisional Constitution of 1950, restored 5 July 1959; series of amendments concluded in 2002
Legal system: based on Roman-Dutch law, substantially modified by indigenous concepts and by new criminal procedures and election codes; has not accepted compulsory ICJ jurisdiction
Suffrage: 17 years of age; universal and married persons regardless of age
Executive branch:
chief of state: President Susilo Bambang YUDHOYONO (since 20 October 2004); Vice President Muhammad Yusuf KALLA (since 20 October 2004); note—the president is both the chief of state and head of government
head of government: President Susilo Bambang YUDHOYONO (since 20 October 2004); Vice President Muhammad Yusuf KALLA (since 20 October 2004)
cabinet: Cabinet appointed by the president
elections: president and vice president were elected for five-year terms (eligible for a second term) by direct vote of the citizenry; last held 20 September 2004 (next to be held in 2009)
election results: Susilo Bambang YUDHOYONO elected president receiving 60.6% of vote; MEGAWATI Sukarnoputri received 39.4%

Legislative branch: House of Representatives or Dewan Perwakilan Rakyat (DPR) (550 seats; members elected to serve five-year terms); House of Regional Representatives (Dewan Perwakilan Daerah or DPD), constitutionally mandated role includes providing legislative input to DPR on issues affecting regions; People's Consultative Assembly (Majelis Permusyawaratan Rakyat or MPR) has role in inaugurating and impeaching president and in amending constitution; consists of popularly elected members in DPR and DPD; MPR does not formulate national policy
elections: last held 5 April 2004 (next to be held in April 2009)
election results: percent of vote by party—Golkar 21.6%, PDI-P 18.5%, PKB 10.6%, PPP 8.2%, PD 7.5%, PKS 7.3%, PAN 6.4%, others 19.9%; seats by party—Golkar 128, PDI-P 109, PPP 58, PD 55, PAN 53, PKB 52, PKS 45, others 50
note: because of election rules, the number of seats won does not always follow the percentage of votes received by parties
Judicial branch: Supreme Court or Mahkamah Agung (justices appointed by the president from a list of candidates selected by the legislature); a separate Constitutional Court or Mahkamah Konstitusi was invested by the president on 16 August 2003; in March 2004 the Supreme Court assumed administrative and financial responsibility for the lower court system from the Ministry of Justice and Human Rights; Labor Court under supervision of Supreme Court began functioning in January 2006
Political parties and leaders: Crescent Moon and Star Party or PBB [MS KABAN]; Democratic Party or PD [Hadi UTOMO]; Functional Groups Party or Golkar [Yusuf KALLA]; Indonesia Democratic Party-Struggle or PDI-P [MEGAWATI Sukarnoputri]; National Awakening Party or PKB; National Mandate Party or PAN [Sutrisno BACHIR]; Prosperous Justice Party or PKS [Tifatul SEMBIRING]; United Development Party or PPP [Suryadharma ALI]
Political pressure groups and leaders: NA
International organization participation: ADB, APEC, APT, ARF, ASEAN, BIS, CP, EAS, FAO, G-15, G-77, IAEA, IBRD, ICAO, ICC, ICRM, IDA, IDB, IFAD, IFC, IFRCS, IHO, ILO, IMF, IMO, IMSO, Interpol, IOC, IOM (observer), IPU, ISO, ITSO, ITU, ITUC, MIGA, MONUC, NAM, OIC, OPCW, OPEC, PIF (partner), UN, UN Security Council (temporary),

307

UNCTAD, UNESCO, UNIDO, UNIFIL, UNMIL, UNOMIG, UNWTO, UPU, WCL, WCO, WFTU, WHO, WIPO, WMO, WTO

Diplomatic representation in the US:
chief of mission: Ambassador SUD-JADNAN Parnohadiningrat
chancery: 2020 Massachusetts Avenue NW, Washington, DC 20036
telephone: [1] (202) 775-5200
FAX: [1] (202) 775-5365
consulate(s) general: Chicago, Houston, Los Angeles, New York, San Francisco

Diplomatic representation from the US:
chief of mission: Ambassador Cameron R. HUME
embassy: Jalan 1 Medan Merdeka Selatan 4-5, Jakarta 10110
mailing address: Unit 8129, Box 1, FPO AP 96520
telephone: [62] (21) 3435-9000
FAX: [62] (21) 3435-9922
consulate(s) general: Surabaya

Flag description: two equal horizontal bands of red (top) and white; similar to the flag of Monaco, which is shorter; also similar to the flag of Poland, which is white (top) and red

ECONOMY

Economy—overview: Indonesia, a vast polyglot nation, has been undergoing significant economic reforms under President YUDHOYONO. Indonesia's debt-to-GDP ratio has been declining steadily, its foreign exchange reserves are at an all-time high of over $50 billion, and its stock market has been one of the three best performers in the world in 2006 and 2007, as global investors sought out higher returns in emerging markets. The government has introduced significant reforms in the financial sector, including tax and customs reforms, the introduction of Treasury bills, and improved capital market supervision. Indonesia's new investment law, passed in March 2007, seeks to address some of the concerns of foreign and domestic investors. Indonesia still struggles with poverty and unemployment, inadequate infrastructure, corruption, a complex regulatory environment, and unequal resource distribution among regions. Indonesia has been slow to privatize over 100 state-owned enterprises, several of which have monopolies in key sectors. The non-bank financial sector, including pension funds and insurance, remains weak. Capital markets are underdeveloped. The high global price of oil in 2007 increased the cost of domestic fuel and electricity subsidies, and are contributing to concerns about higher food prices. Located on the Pacific "Ring of Fire" Indonesia remains vulnerable to volcanic and tectonic disasters. Significant progress has been made in rebuilding Aceh after the devastating December 2004 tsunami, and the province now shows more economic activity than before the disaster. Unfortunately, Indonesia suffered new disasters in 2006 and early 2007 including: a major earthquake near Yogyakarta, an industrial accident in Sidoarjo, East Java that created a "mud volcano," a tsunami in South Java, and major flooding in Jakarta, all of which caused additional damages in the billions of dollars. Donors are assisting Indonesia with its disaster mitigation and early warning efforts.

GDP (purchasing power parity): $837.8 billion (2007 est.)

GDP (official exchange rate): $432.9 billion (2007 est.)

GDP—real growth rate: 6.3% (2007 est.)

GDP—per capita (PPP): $3,700 (2007 est.)

GDP—composition by sector:
agriculture: 13.8%
industry: 46.7%
services: 39.4% (2007 est.)

Labor force: 109.9 million (2007 est.)

Labor force—by occupation: *agriculture:* 43.3%
industry: 18%
services: 38.7% (2004 est.)

Unemployment rate: 9.6% (2007 est.)

Population below poverty line: 17.8% (2006)

Household income or consumption by percentage share:
lowest 10%: 3.6%
highest 10%: 28.5% (2002)

Distribution of family income—Gini index: 36.3 (2005)

Inflation rate (consumer prices): 6.4% (2007 est.)

Investment (gross fixed): 24.9% of GDP (2007 est.)

Budget: *revenues:* $79.25 billion
expenditures: $84.85 billion (2007 est.)

Public debt: 34.1% of GDP (2007 est.)

Agriculture—products: rice, cassava (tapioca), peanuts, rubber, cocoa, coffee, palm oil, copra; poultry, beef, pork, eggs

Industries: petroleum and natural gas, textiles, apparel, footwear, mining, cement, chemical fertilizers, plywood, rubber, food, tourism

Industrial production growth rate: 4.7% (2007 est.)

Electricity—production: 125.9 billion kWh (2006 est.)

Electricity—production by source:
fossil fuel: 86.9%
hydro: 10.5%
nuclear: 0%
other: 2.6% (2001)

Electricity—consumption: 108 billion kWh (2006 est.)

Electricity—exports: 0 kWh (2006 est.)

Electricity—imports: 0 kWh (2006 est.)

Oil—production: 1.07 million bbl/day (2006 est.)

Oil—consumption: 1.1 million bbl/day (2006 est.)

Oil—exports: 470,000 bbl/day (2006 est.)

Oil—imports: 500,000 bbl/day (2006 est.)

Oil—proved reserves: 4.43 billion bbl (2007 est.)

Natural gas—production: 74 billion cu m (2006 est.)

Natural gas—consumption: 37.5 billion cu m (2006 est.)

Natural gas—exports: 29.6 billion cu m (2006 est.)

Natural gas—imports: 0 cu m (2006)

Natural gas—proved reserves: 2.63 trillion cu m (1 January 2007 est.)

Current account balance: $11.01 billion (2007 est.)

Exports: $118 billion f.o.b. (2007 est.)

Exports—commodities: oil and gas, electrical appliances, plywood, textiles, rubber

Exports—partners: Japan 19.4%, Singapore 11.8%, US 11.5%, China 7.7%, South Korea 6.4%, Taiwan 4.2% (2006)

Imports: $84.93 billion f.o.b. (2007 est.)

Imports—commodities: machinery and equipment, chemicals, fuels, foodstuffs

Imports—partners: Singapore 29.6%, China 11.2%, Japan 8.8%, South Korea 5.3%, Malaysia 4.8% (2006)

Economic aid—recipient: ODA, $2.524 billion (2006 est.)
note: Indonesia ended 2006 with $67 billion in official foreign debt (about 25% of GDP), with Japan ($25 billion), the World Bank ($8.5 billion) and the Asian Development Bank ($8.4 billion) as the largest creditors; about $6 billion in grant assistance was pledged to rebuild Aceh after the December 2004 tsunami; President YUDHOYONO disbanded the Consultative Group on Indonesia (CGI) donor forum in January 2007

Reserves of foreign exchange and gold: $56.92 billion (31 December 2007 est.)

Debt—external: $140.7 billion (31 December 2007)

Stock of direct foreign investment—at home: $58.13 billion (2007 est.)

Stock of direct foreign investment—abroad: $9.225 billion (2006 est.)

Market value of publicly traded shares: $138.9 billion (2006)

Currency (code): Indonesian rupiah (IDR)

Currency code: IDR

Exchange rates: Indonesian rupiah per US dollar—9,056 (2007 est.), 9,159.3

(2006), 9,704.7 (2005), 8,938.9 (2004), 8,577.1 (2003)

Fiscal year: calendar year

COMMUNICATIONS

Telephones—main lines in use: 14.821 million (2006)

Telephones—mobile cellular: 63.803 million (2006)

Telephone system:

general assessment: domestic service fair, international service good

domestic: interisland microwave system and HF radio police net; domestic satellite communications system; coverage provided by existing network has been expanded by use of over 200,000 telephone kiosks many located in remote areas

international: country code—62; landing point for both the SEA-ME-WE-3 AND SEA-ME-WE-4 submarine cable networks that provide links throughout Asia, the Middle East, and Europe; satellite earth stations—2 Intelsat (1 Indian Ocean and 1 Pacific Ocean)

Radio broadcast stations: AM 678, FM 43, shortwave 82 (1998)

Radios: 31.5 million (1997)

Television broadcast stations: 54 local TV stations (11 national TV networks; each with its group of local transmitters) (2006)

Televisions: 13.75 million (1997)

Internet country code: .id

Internet hosts: 559,359 (2007)

Internet Service Providers (ISPs): 24 (2000)

Internet users: 16 million (2005)

TRANSPORTATION

Airports: 652 (2007)

Airports—with paved runways: *total:* 158

over 3,047 m: 4

2,438 to 3,047 m: 15

1,524 to 2,437 m: 51

914 to 1,523 m: 49

under 914 m: 39 (2007)

Airports—with unpaved runways:

total: 494

1,524 to 2,437 m: 5

914 to 1,523 m: 27

under 914 m: 462 (2007)

Heliports: 17 (2007)

Pipelines: condensate 963 km; condensate/gas 81 km; gas 9,003 km; oil 7,471 km; oil/gas/water 77 km; refined products 1,365 km (2007)

Railways:

total: 6,458 km

narrow gauge: 5,961 km 1.067-m gauge (125 km electrified); 497 km 0.750-m gauge (2006)

Roadways:

total: 391,009 km

paved: 216,714 km

unpaved: 174,295 km (2005)

Waterways: 21,579 km (2007)

Merchant marine:

total: 965 ships (1000 GRT or over) 4,409,198 GRT/5,825,591 DWT

by type: bulk carrier 53, cargo 522, chemical tanker 25, container 66, liquefied gas 7, livestock carrier 1, passenger 44, passenger/cargo 67, petroleum tanker 155, refrigerated cargo 2, roll on/roll off 11, specialized tanker 8, vehicle carrier 4

foreign-owned: 45 (China 2, France 1, Japan 5, South Korea 1, Philippines 1, Singapore 26, Switzerland 3, Taiwan 2, Thailand 1, UK 3)

registered in other countries: 105 (Bahamas 3, Cambodia 1, Hong Kong 7, Liberia 1, Panama 37, Singapore 56, unknown 5) (2007)

Ports and terminals: Banjarmasin, Belawan, Ciwandan, Kotabaru, Krueg Geukueh, Palembang, Panjang, Sungai Pakning, Tanjung Perak, Tanjung Priok

MILITARY

Military branches: Indonesian Armed Forces (Tentara Nasional Indonesia, TNI): Army (TNI-Angkatan Darat (TNI-AD)), Navy (TNI-Angkatan Laut (TNI-AL); includes marines, naval air arm), Air Force (TNI-Angkatan Udara (TNI-AU)), National Air Defense Command (Kommando Pertahanan Udara Nasional (Kohanudnas)) (2008)

Military service age and obligation: 18 years of age for selective compulsory and voluntary military service; 2-year conscript service obligation, with reserve obligation to age 45 (2006)

Manpower available for military service:

males age 16–49: 63,800,825

females age 16–49: 61,729,717 (2008 est.)

Manpower fit for military service:

males age 16–49: 52,367,788

females age 16–49: 52,129,123 (2008 est.)

Manpower reaching militarily significant age annually:

males age 16–49: 2,181,303

females age 16–49: 2,110,397 (2008 est.)

Military expenditures—percent of GDP: 3% (2005 est.)

TRANSNATIONAL ISSUES

Disputes—international: Indonesia has a stated foreign policy objective of establishing stable fixed land and maritime boundaries with all of its neighbors; Timor-Leste-Indonesia Boundary Committee has resolved all but a small portion of the land boundary, but discussions on maritime boundaries are stalemated over sovereignty of the uninhabited coral island of Pulau Batek/Fatu Sinai in the north and alignment with Australian claims in the south; many refugees from Timor-Leste who left in 2003 still reside in Indonesia and refuse repatriation; a 1997 treaty between Indonesia and Australia settled some parts of their maritime boundary but outstanding issues remain; ICJ's award of Sipadan and Ligitan islands to Malaysia in 2002 left the sovereignty of Unarang rock and the maritime boundary in the Ambalat oil block in the Celebes Sea in dispute; the ICJ decision has prompted Indonesia to assert claims to and to establish a presence on its smaller outer islands; Indonesia and Singapore continue to work on finalization of their 1973 maritime boundary agreement by defining unresolved areas north of Indonesia's Batam Island; Indonesian secessionists, squatters, and illegal migrants create repatriation problems for Papua New Guinea; piracy remains a problem in the Malacca Strait; maritime delimitation talks continue with Palau; Indonesian groups challenge Australia's claim to Ashmore Reef; Australia has closed parts of the Ashmore and Cartier Reserve to Indonesian traditional fishing and placed restrictions on certain catches

Refugees and internally displaced persons: *IDPs:* 200,000–350,000 (government offensives against rebels in Aceh; most IDPs in Aceh, Central Kalimantan, Central Sulawesi Provinces, and Maluku) (2007)

Illicit drugs: illicit producer of cannabis largely for domestic use; producer of methamphetamine and ecstasy

IRAN

INTRODUCTION

Background: Known as Persia until 1935, Iran became an Islamic republic in 1979 after the ruling monarchy was overthrown and the shah was forced into exile. Conservative clerical forces established a theocratic system of government with ultimate political authority vested in a learned religious scholar referred to commonly as the Supreme Leader who, according to the constitution, is

accountable only to the Assembly of Experts. US-Iranian relations have been strained since a group of Iranian students seized the US Embassy in Tehran on 4 November 1979 and held it until 20 January 1981. During 1980–88, Iran fought a bloody, indecisive war with Iraq that eventually expanded into the Persian Gulf and led to clashes between US Navy and Iranian military forces between 1987 and 1988. Iran has been designated a state sponsor of terrorism for its activities in Lebanon and elsewhere in the world and remains subject to US and UN economic sanctions and export controls because of its continued involvement in terrorism and conventional weapons proliferation. Following the election of reformer Hojjat ol-Eslam Mohammad KHATAMI as president in 1997 and similarly a reformer Majles (parliament) in 2000, a campaign to foster political reform in response to popular dissatisfaction was initiated. The movement floundered as conservative politicians, through the control of unelected institutions, prevented reform measures from being enacted and increased repressive measures. Starting with nationwide municipal elections in 2003 and continuing through Majles elections in 2004, conservatives reestablished control over Iran's elected government institutions, which culminated with the August 2005 inauguration of hardliner Mahmud AHMADI-NEJAD as president. In December 2006 and March 2007, the international community passed resolutions 1737 and 1747 respectively after Iran failed to comply with UN demands to halt the enrichment of uranium or to agree to full IAEA oversight of its nuclear program. In October 2007, Iranian entities were also subject to US sanctions under EO 13382 designations for proliferation activities and EO 13224 designations for providing material support to the Taliban and other terrorist organizations.

GEOGRAPHY

Location: Middle East, bordering the Gulf of Oman, the Persian Gulf, and the Caspian Sea, between Iraq and Pakistan

Geographic coordinates: 32 00 N, 53 00 E

Map references: Middle East

Area:

total: 1.648 million sq km

land: 1.636 million sq km

water: 12,000 sq km

Area—comparative: slightly larger than Alaska

Land boundaries: *total:* 5,440 km

border countries: Afghanistan 936 km, Armenia 35 km, Azerbaijan-proper 432 km, Azerbaijan-Naxcivan exclave 179 km, Iraq 1,458 km, Pakistan 909 km, Turkey 499 km, Turkmenistan 992 km

Coastline: 2,440 km; note—Iran also borders the Caspian Sea (740 km)

Maritime claims:

territorial sea: 12 nm

contiguous zone: 24 nm

exclusive economic zone: bilateral agreements or median lines in the Persian Gulf

continental shelf: natural prolongation

Climate: mostly arid or semiarid, subtropical along Caspian coast

Terrain: rugged, mountainous rim; high, central basin with deserts, mountains; small, discontinuous plains along both coasts

Elevation extremes:

lowest point: Caspian Sea -28 m

highest point: Kuh-e Damavand 5,671 m

Natural resources: petroleum, natural gas, coal, chromium, copper, iron ore, lead, manganese, zinc, sulfur

Land use:

arable land: 9.78%

permanent crops: 1.29%

other: 88.93% (2005)

Irrigated land: 76,500 sq km (2003)

Total renewable water resources: 137.5 cu km (1997)

Freshwater withdrawal (domestic/industrial/agricultural): *total:* 72.88 cu km/yr (7%/2%/91%)

per capita: 1,048 cu m/yr (2000)

Natural hazards: periodic droughts, floods; dust storms, sandstorms; earthquakes

Environment—current issues: air pollution, especially in urban areas, from vehicle emissions, refinery operations, and industrial effluents; deforestation; overgrazing; desertification; oil pollution in the Persian Gulf; wetland losses from drought; soil degradation (salination); inadequate supplies of potable water; water pollution from raw sewage and industrial waste; urbanization

Environment—international agreements: *party to:* Biodiversity, Climate Change, Climate Change-Kyoto Protocol, Desertification, Endangered Species, Hazardous Wastes, Marine Dumping, Ozone Layer Protection, Ship Pollution, Wetlands

signed, but not ratified: Environmental Modification, Law of the Sea, Marine Life Conservation

Geography—note: strategic location on the Persian Gulf and Strait of Hormuz, which are vital maritime pathways for crude oil transport

PEOPLE

Population: 65,875,223 (July 2008 est.)

Age structure:

0–14 years: 22.3% (male 7,548,116/female 7,164,921)

15–64 years: 72.3% (male 24,090,976/female 23,522,861)

65 years and over: 5.4% (male 1,713,533/female 1,834,816) (2008 est.)

Median age:

total: 26.4 years

male: 26.2 years

female: 26.7 years (2008 est.)

Population growth rate: 0.792% (2008 est.)

Birth rate: 16.89 births/1,000 population (2008 est.)

Death rate: 5.69 deaths/1,000 population (2008 est.)

Net migration rate: -3.28 migrant(s)/1,000 population (2008 est.)

Sex ratio:

at birth: 1.05 male(s)/female

under 15 years: 1.05 male(s)/female

15–64 years: 1.02 male(s)/female

65 years and over: 0.93 male(s)/female

total population: 1.03 male(s)/female (2008 est.)

Infant mortality rate:

total: 36.93 deaths/1,000 live births

male: 37.12 deaths/1,000 live births

female: 36.73 deaths/1,000 live births (2008 est.)

Life expectancy at birth:

total population: 70.86 years

male: 69.39 years

female: 72.4 years (2008 est.)

Total fertility rate: 1.71 children born/woman (2008 est.)

HIV/AIDS—adult prevalence rate: 0.2% (2005 est.)

HIV/AIDS—people living with HIV/AIDS: 66,000 (2005 est.)

HIV/AIDS—deaths: 1,600 (2005 est.)

Major infectious diseases:

degree of risk: intermediate

food or waterborne diseases: bacterial diarrhea and hepatitis A

vectorborne diseases: Crimean Congo hemorrhagic fever and malaria *note:* highly pathogenic H5N1 avian influenza has been identified in this country; it poses a negligible risk with extremely rare cases possible among US citizens who have close contact with birds (2008)

Nationality:

noun: Iranian(s)

adjective: Iranian

Ethnic groups: Persian 51%, Azeri 24%, Gilaki and Mazandarani 8%, Kurd 7%, Arab 3%, Lur 2%, Baloch 2%, Turkmen 2%, other 1%

Religions: Muslim 98% (Shi'a 89%, Sunni 9%), other (includes Zoroastrian, Jewish, Christian, and Baha'i) 2%

Languages: Persian and Persian dialects 58%, Turkic and Turkic dialects 26%, Kurdish 9%, Luri 2%, Balochi 1%, Arabic 1%, Turkish 1%, other 2%

Literacy:

definition: age 15 and over can read and write

total population: 77%

male: 83.5%

female: 70.4% (2002 est.)

GOVERNMENT

Country name:

conventional long form: Islamic Republic of Iran

conventional short form: Iran

local long form: Jomhuri-ye Eslami-ye Iran

local short form: Iran

former: Persia

Government type: theocratic republic

Capital: *name:* Tehran

geographic coordinates: 35 40 N, 51 25 E

time difference: UTC+3.5 (8.5 hours ahead of Washington, DC during Standard Time)

Administrative divisions: 30 provinces (ostanha, singular—ostan); Ardabil, Azarbayjan-e Gharbi, Azarbayjan-e Sharqi, Bushehr, Chahar Mahall va Bakhtiari, Esfahan, Fars, Gilan, Golestan, Hamadan, Hormozgan, Ilam, Kerman, Kermanshah, Khorasan-e Janubi, Khorasan-e Razavi, Khorasan-e Shemali, Khuzestan, Kohgiluyeh va Buyer Ahmad, Kordestan, Lorestan, Markazi, Mazandaran, Qazvin, Qom, Semnan, Sistan va Baluchestan, Tehran, Yazd, Zanjan

Independence: 1 April 1979 (Islamic Republic of Iran proclaimed)

National holiday: Republic Day, 1 April (1979)

Constitution: 2–3 December 1979; revised 1989 to expand powers of the presidency and eliminate the prime ministership

Legal system: based on Sharia law system; has not accepted compulsory ICJ jurisdiction

Suffrage: 16 years of age; universal

Executive branch:

chief of state: Supreme Leader Ali Hoseini-KHAMENEI (since 4 June 1989)

head of government: President Mahmud AHMADI-NEJAD (since 3 August 2005); First Vice President Parviz DAVUDI (since 11 September 2005)

cabinet: Council of Ministers selected by the president with legislative approval; the Supreme Leader has some control over appointments to the more sensitive ministries

note: also considered part of the Executive branch of government are three oversight bodies: 1) Assembly of Experts (Majles-Khebregan), a popularly elected body charged with determining the succession of the Supreme Leader, reviewing his performance, and deposing him if deemed necessary; 2) Expediency Council or the Council for the Discernment of Expediency (Majma-e-Tashkise-Maslahat-e-Nezam) exerts supervisory authority over the executive, judicial, and legislative branches and resolves legislative issues on which the Majles and the Council of Guardians disagree and since 1989 has been used to advise national religious leaders on matters of national policy; in 2005 the Council's powers were expanded to act as a supervisory body for the government; 3) Council of Guardians of the Constitution or Council of Guardians or Guardians Council (Shora-ye Negaban-e Qanun-e Assasi) determines whether proposed legislation is both constitutional and faithful to Islamic law, vets candidates for suitability, and supervises national elections

elections: Supreme Leader is appointed for life by the Assembly of Experts; president is elected by popular vote for a four-year term (eligible for a second term and third nonconsecutive term); last held 17 June 2005 with a two-candidate runoff on 24 June 2005 (next presidential election slated for 2009)

election results: Mahmud AHMADI-NEJAD elected president; percent of vote—Mahmud AHMADI-NEJAD 62%, Ali Akbar Hashemi-RAFSAN-JANI 36%

Legislative branch: unicameral Islamic Consultative Assembly or Majles-e-Shura-ye-Eslami or Majles (290 seats; members elected by popular vote to serve four-year terms)

elections: last held 14 March 2008 with a runoff held 25 April 2008 (next to be held in 2012)

election results: percent of vote—NA; seats by party—conservatives/Islamists

170, reformers 46, independents 71, religious minorities 3

Judicial branch: The Supreme Court (Qeveh Qazaieh) and the four-member High Council of the Judiciary have a single head and overlapping responsibilities; together they supervise the enforcement of all laws and establish judicial and legal policies; lower courts include a special clerical court, a revolutionary court, and a special administrative court

Political parties and leaders: formal political parties are a relatively new phenomenon in Iran and most conservatives still prefer to work through political pressure groups rather than parties, and often political parties or coalitions are formed prior to elections and disbanded soon thereafter; a loose pro-reform coalition called the 2nd Khordad Front, which includes political parties as well as less formal groups and organizations, achieved considerable success at elections to the sixth Majles in early 2000; groups in the coalition include: Islamic Iran Participation Front (IIPF), Executives of Construction Party (Kargozaran), Solidarity Party, Islamic Labor Party, Mardom Salari, Mojahedin of the Islamic Revolution Organization (MIRO), and Militant Clerics Society (Ruhaniyun); the coalition participated in the seventh Majles elections in early 2004; following his defeat in the 2005 presidential elections, former MCS Secretary General and sixth Majles Speaker Mehdi KARUBI formed the National Trust Party; a new conservative group, Islamic Iran Developers Coalition (Abadgaran), took a leading position in the new Majles after winning a majority of the seats in February 2004; following the 2004 Majles elections, traditional and hardline conservatives have attempted to close ranks under the United Front of Principlists; the IIPF has repeatedly complained that the overwhelming majority of its candidates have been unfairly disqualified from the 2008 elections

Political pressure groups and leaders: the Islamic Republic Party (IRP) was Iran's sole political party until its dissolution in 1987; Iran now has a variety of groups engaged in political activity; some are oriented along political lines or based on an identity group; others are more akin to professional political parties seeking members and recommending candidates for office; some are active participants in the Revolution's political life while others reject the state; political pressure groups conduct most of Iran's political activities; groups that generally support the Islamic Republic include

Ansar-e Hizballah, Followers of the Line of the Imam and the Leader, Islamic Coalition Party (Motalefeh), Islamic Engineers Society, and Tehran Militant Clergy Association (Ruhaniyat); active pro-reform student groups include the Office of Strengthening Unity (OSU); opposition groups include Freedom Movement of Iran, the National Front, Marz-e Por Gohar, Baluchistan People's Party (BPP), and various ethnic and Monarchist organizations; armed political groups that have been repressed by the government include Democratic Party of Iranian Kurdistan (KDPI), Komala, Mujahidin-e Khalq Organization (MEK or MKO), People's Fedayeen, Jundallah, and the People's Free Life Party of Kurdistan (PJAK)

International organization participation: ABEDA, CP, ECO, FAO, G-24, G-77, IAEA, IBRD, ICAO, ICC, ICCt (signatory), ICRM, IDA, IDB, IFAD, IFC, IFRCS, IHO, ILO, IMF, IMO, IMSO, Interpol, IOC, IOM, IPU, ISO, ITSO, ITU, MIGA, NAM, OIC, OPCW, OPEC, PCA, SAARC, SCO (observer), UN, UNCTAD, UNESCO, UNHCR, UNIDO, UNMEE, UNWTO, UPU, WCL, WCO, WFTU, WHO, WIPO, WMO, WTO (observer)

Diplomatic representation in the US: none; note—Iran has an Interests Section in the Pakistani Embassy; address: Iranian Interests Section, Pakistani Embassy, 2209 Wisconsin Avenue NW, Washington, DC 20007; telephone: [1] (202) 965-4990; FAX [1] (202) 965-1073

Diplomatic representation from the US: none; note—the American Interests Section is located in the Swiss Embassy compound at Africa Avenue, West Farzan Street, number 59, Tehran, Iran; telephone 021 8878 2964 or 021 8879 2364; FAX 021 8877 3265

Flag description: three equal horizontal bands of green (top), white, and red; the national emblem (a stylized representation of the word Allah in the shape of a tulip, a symbol of martyrdom) in red is centered in the white band; ALLAH AKBAR (God is Great) in white Arabic script is repeated 11 times along the bottom edge of the green band and 11 times along the top edge of the red band

ECONOMY

Economy—overview: Iran's economy is marked by an inefficient state sector, reliance on the oil sector (which provides 85% of government revenues), and statist policies that create major distortions throughout. Most economic activity is controlled by the state. Private sector activity is typically small-scale workshops, farming, and services. President Mahmud AHMADI-NEJAD failed to make any notable progress in fulfilling the goals of the nation's latest five-year plan. A combination of price controls and subsidies, particularly on food and energy, continue to weigh down the economy, and administrative controls, widespread corruption, and other rigidities undermine the potential for private-sector-led growth. As a result of these inefficiencies, significant informal market activity flourishes and shortages are common. High oil prices in recent years have enabled Iran to amass nearly $70 billion in foreign exchange reserves. Yet this increased revenue has not eased economic hardships, which include double-digit unemployment and inflation. The economy has seen only moderate growth. Iran's educated population, economic inefficiency and insufficient investment—both foreign and domestic—have prompted an increasing number of Iranians to seek employment overseas, resulting in significant "brain drain."

GDP (purchasing power parity): $753 billion (2007 est.)

GDP (official exchange rate): $294.1 billion (2007 est.)

GDP—real growth rate: 5.8% (2007 est.)

GDP—per capita (PPP): $10,600 (2007 est.)

GDP—composition by sector:
agriculture: 10.7%
industry: 42.9%
services: 46.4% (2007 est.)

Labor force: 28.7 million
note: shortage of skilled labor (2006 est.)

Labor force—by occupation:
agriculture: 25%
industry: 31%
services: 45% (June 2007)

Unemployment rate: 12% according to the Iranian government (2007 est.)

Population below poverty line: 18% (2007 est.)

Household income or consumption by percentage share: *lowest 10%:* 2%
highest 10%: 33.7% (1998)

Distribution of family income—Gini index: 43 (1998)

Inflation rate (consumer prices): 17.5% (2007 est.)

Investment (gross fixed): 27.6% of GDP (2007 est.)

Budget: *revenues:* $64 billion
expenditures: $64 billion (2007 est.)

Public debt: 25.2% of GDP (2007 est.)

Agriculture—products: wheat, rice, other grains, sugar beets, sugar cane, fruits, nuts, cotton; dairy products, wool; caviar

Industries: petroleum, petrochemicals, fertilizers, caustic soda, textiles, cement and other construction materials, food processing (particularly sugar refining and vegetable oil production), ferrous and non-ferrous metal fabrication, armaments

Industrial production growth rate: 4.8% excluding oil (2007 est.)

Electricity—production: 170.4 billion kWh (2005)

Electricity—production by source:
fossil fuel: 97.1%
hydro: 2.9%
nuclear: 0%
other: 0% (2001)

Electricity—consumption: 136.2 billion kWh (2005)

Electricity—exports: 2.761 billion kWh (2005)

Electricity—imports: 2.074 billion kWh (2005)

Oil—production: 4.15 million bbl/day (2006 est.)

Oil—consumption: 1.63 million bbl/day (2006 est.)

Oil—exports: 2.52 million bbl/day (2006 est.)

Oil—imports: 153,600 bbl/day (2004)

Oil—proved reserves: 138.4 billion bbl based on Iranian claims (2007 est.)

Natural gas—production: 101 billion cu m (2005 est.)

Natural gas—consumption: 98.19 billion cu m (2005 est.)

Natural gas—exports: 4.33 billion cu m (2005 est.)

Natural gas—imports: 5.8 billion cu m (2005)

Natural gas—proved reserves: 26.37 trillion cu m (1 January 2006 est.)

Current account balance: $30.47 billion (2007 est.)

Exports: $83.99 billion f.o.b. (2007 est.)

Exports—commodities: petroleum 80%, chemical and petrochemical products, fruits and nuts, carpets

Exports—partners: Japan 14%, China 12.8%, Turkey 7.2%, Italy 6.3%, South Korea 6%, Netherlands 4.6% (2006)

Imports: $53.73 billion f.o.b. (2007 est.)

Imports—commodities: industrial raw materials and intermediate goods, capital goods, foodstuffs and other consumer goods, technical services

Imports—partners: Germany 12.1%, China 10.5%, UAE 9.4%, France 5.6%, South Korea 5.4%, Italy 5.4%, Russia 4.5% (2006)

Economic aid—recipient: $104 million (2005 est.)

Reserves of foreign exchange and gold: $64.46 billion (2007 est.)

Debt—external: $20.65 billion (31 December 2007 est.)

Stock of direct foreign investment—at home: $6.026 billion (2007 est.)

Stock of direct foreign investment—abroad: $903 million (2007 est.)

Market value of publicly traded shares: $45.2 billion (December 2007)

Currency (code): Iranian rial (IRR)

Currency code: IRR

Exchange rates: rials per US dollar—9,407.5 (2007), 9,227.1 (2006), 8,964 (2005), 8,614 (2004), 8,193.9 (2003)

note: Iran has been using a managed floating exchange rate regime since unifying multiple exchange rates in March 2002

Fiscal year: 21 March—20 March

COMMUNICATIONS

Telephones—main lines in use: 21.981 million (2006)

Telephones—mobile cellular: 13.659 million (2006)

Telephone system:

general assessment: currently being modernized and expanded with the goal of not only improving the efficiency and increasing the volume of the urban service but also bringing telephone service to several thousand villages, not presently connected

domestic: the addition of new fiber cables and modern switching and exchange systems installed by Iran's state-owned telecom company have improved and expanded the main line network greatly; main line availability has more than doubled to 22 million lines since 2000; additionally, mobile service has increased dramatically serving nearly 13.7 million subscribers in 2006

international: country code—98; submarine fiber-optic cable to UAE with access to Fiber-Optic Link Around the Globe (FLAG); Trans-Asia-Europe (TAE) fiber-optic line runs from Azerbaijan through the northern portion of Iran to Turkmenistan with expansion to Georgia and Azerbaijan; HF radio and microwave radio relay to Turkey, Azerbaijan, Pakistan, Afghanistan, Turkmenistan, Syria, Kuwait, Tajikistan, and Uzbekistan; satellite earth stations—13 (9 Intelsat and 4 Inmarsat) (2006)

Radio broadcast stations: AM 72, FM 5, shortwave 5 (1998)

Radios: 17 million (1997)

Television broadcast stations: 28 (plus 450 repeaters) (1997)

Televisions: 4.61 million (1997)

Internet country code: .ir

Internet hosts: 6,111 (2007)

Internet Service Providers (ISPs): 100 (2002)

Internet users: 18 million (2006)

TRANSPORTATION

Airports: 331 (2007)

Airports—with paved runways:
total: 129

over 3,047 m: 40
2,438 to 3,047 m: 28
1,524 to 2,437 m: 24
914 to 1,523 m: 32
under 914 m: 5 (2007)

Airports—with unpaved runways:
total: 202
over 3,047 m: 1
1,524 to 2,437 m: 10
914 to 1,523 m: 145
under 914 m: 46 (2007)

Heliports: 14 (2007)

Pipelines: condensate 7 km; condensate/gas 397 km; gas 19,161 km; liquid petroleum gas 570 km; oil 8,438 km; refined products 7,936 km (2007)

Railways:
total: 8,367 km
broad gauge: 94 km 1.676-m gauge
standard gauge: 8,273 km 1.435-m gauge (146 km electrified) (2006)

Roadways:
total: 179,388 km
paved: 120,782 km (includes 878 km of expressways)
unpaved: 58,606 km (2003)

Waterways: 850 km (on Karun River; additional service on Lake Urmia) (2006)

Merchant marine:
total: 131 ships (1000 GRT or over) 4,721,202 GRT/8,309,580 DWT
by type: bulk carrier 35, cargo 45, chemical tanker 4, container 9, liquefied gas 1, passenger/cargo 4, petroleum tanker 29, roll on/roll off 4
foreign-owned: 1 (UAE 1)
registered in other countries: 33 (Bolivia 1, Cyprus 2, Malta 24, Panama 4, St Kitts and Nevis 1, St Vincent and The Grenadines 1) (2007)

Ports and terminals: Assaluyeh, Bandar Abbas, Bandar-e-Eman Khomeyni

MILITARY

Military branches: Islamic Republic of Iran Regular Forces (Artesh): Ground Forces, Navy, Air Force of the Military of the Islamic Republic of Iran (Niru-ye Hava'i-ye Artesh-e Jomhuri-ye Eslami-ye Iran; includes air defense); Islamic Revolutionary Guard Corps (Sepah-e Pasdaran-e Enqelab-e Eslami, IRGC): Ground Forces, Navy, Air Force, Qods Force (special operations), and Basij Force (Popular Mobilization Army); Law Enforcement Forces (2008)

Military service age and obligation: 19 years of age for compulsory military service; 16 years of age for volunteers; 17 years of age for Law Enforcement Forces; 15 years of age for Basij Forces (Popular Mobilization Army); conscript military service obligation—18 months; women exempt from military service (2008)

Manpower available for military service:
males age 16–49: 20,212,275
females age 16–49: 19,638,751 (2008 est.)

Manpower fit for military service:
males age 16–49: 17,416,126
females age 16–49: 16,928,226 (2008 est.)

Manpower reaching militarily significant age annually:
males age 16–49: 766,668
females age 16–49: 727,654 (2008 est.)

Military expenditures—percent of GDP: 2.5% (2006)

TRANSNATIONAL ISSUES

Disputes—international: Iran protests Afghanistan's limiting flow of dammed tributaries to the Helmand River in periods of drought; Iraq's lack of a maritime boundary with Iran prompts jurisdiction disputes beyond the mouth of the Shatt al Arab in the Persian Gulf; Iran and UAE dispute Tunb Islands and Abu Musa Island, which are occupied by Iran; Iran stands alone among littoral states in insisting upon a division of the Caspian Sea into five equal sectors

Refugees and internally displaced persons: *refugees (country of origin):* 914,268 (Afghanistan); 54,024 (Iraq) (2007)

Trafficking in persons: *current situation:* Iran is a source, transit, and destination country for women and girls trafficked for the purposes of sexual exploitation and involuntary servitude; according to foreign observers, women and girls are trafficked to Pakistan, Turkey, the Persian Gulf, and Europe for sexual exploitation, while boys from Bangladesh, Pakistan, and Afghanistan are trafficked through Iran en route to Persian Gulf states where they are ultimately forced to work as camel jockeys, beggars, or laborers; Afghan women and girls are trafficked to the country for forced marriages and sexual exploitation; women and children are also trafficked internally for the purposes of forced marriage, sexual exploitation, and involuntary servitude

tier rating: Tier 3—Iran is downgraded to Tier 3 after persistent, credible reports of Iranian authorities punishing victims of trafficking with beatings, imprisonment, and execution

Illicit drugs: despite substantial interdiction efforts, Iran remains a key transshipment point for Southwest Asian heroin to Europe; highest percentage of the population in the world using opiates; lacks anti-money-laundering laws

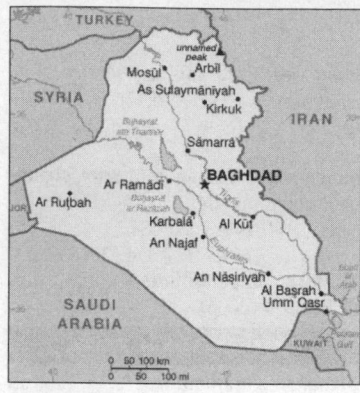

INTRODUCTION

Background: Formerly part of the Ottoman Empire, Iraq was occupied by Britain during the course of World War I; in 1920, it was declared a League of Nations mandate under UK administration. In stages over the next dozen years, Iraq attained its independence as a kingdom in 1932. A "republic" was proclaimed in 1958, but in actuality a series of military strongmen ruled the country until 2003. The last was SADDAM Husayn. Territorial disputes with Iran led to an inconclusive and costly eight-year war (1980–88). In August 1990, Iraq seized Kuwait but was expelled by US-led, UN coalition forces during the Gulf War of January-February 1991. Following Kuwait's liberation, the UN Security Council (UNSC) required Iraq to scrap all weapons of mass destruction and long-range missiles and to allow UN verification inspections. Continued Iraqi noncompliance with UNSC resolutions over a period of 12 years led to the US-led invasion of Iraq in March 2003 and the ouster of the SADDAM Husayn regime. Coalition forces remain in Iraq under a UNSC mandate, helping to provide security and to support the freely elected government. The Coalition Provisional Authority, which temporarily administered Iraq after the invasion, transferred full governmental authority on 28 June 2004 to the Iraqi Interim Government, which governed under the Transitional Administrative Law for Iraq (TAL). Under the TAL, elections for a 275-member Transitional National Assembly (TNA) were held in Iraq on 30 January 2005. Following these elections, the Iraqi Transitional Government (ITG) assumed office. The TNA was charged with drafting Iraq's permanent constitution, which was approved in a 15 October 2005 constitutional referendum. An election under the constitution for a 275-member Council of Representatives (CoR) was held on 15 December 2005. The CoR approval in the selection of most of the cabinet ministers on 20 May 2006 marked the transition from the ITG to Iraq's first constitutional government in nearly a half-century.

GEOGRAPHY

Location: Middle East, bordering the Persian Gulf, between Iran and Kuwait
Geographic coordinates: 33 00 N, 44 00 E
Map references: Middle East
Area:
total: 437,072 sq km
land: 432,162 sq km
water: 4,910 sq km
Area—comparative: slightly more than twice the size of Idaho
Land boundaries:
total: 3,650 km
border countries: Iran 1,458 km, Jordan 181 km, Kuwait 240 km, Saudi Arabia 814 km, Syria 605 km, Turkey 352 km
Coastline: 58 km
Maritime claims:
territorial sea: 12 nm
continental shelf: not specified
Climate: mostly desert; mild to cool winters with dry, hot, cloudless summers; northern mountainous regions along Iranian and Turkish borders experience cold winters with occasionally heavy snows that melt in early spring, sometimes causing extensive flooding in central and southern Iraq
Terrain: mostly broad plains; reedy marshes along Iranian border in south with large flooded areas; mountains along borders with Iran and Turkey
Elevation extremes:
lowest point: Persian Gulf 0 m
highest point: unnamed peak; 3,611 m; note—this peak is neither Gundah Zhur 3,607 m nor Kuh-e Hajji-Ebrahim 3,595 m
Natural resources: petroleum, natural gas, phosphates, sulfur
Land use: *arable land:* 13.12%
permanent crops: 0.61%
other: 86.27% (2005)
Irrigated land: 35,250 sq km (2003)
Total renewable water resources: 96.4 cu km (1997)
Freshwater withdrawal (domestic/industrial/agricultural): *total:* 42.7 cu km/yr (3%/5%/92%)
per capita: 1,482 cu m/yr (2000)

Natural hazards: dust storms, sandstorms, floods
Environment—current issues: government water control projects have drained most of the inhabited marsh areas east of An Nasiriyah by drying up or diverting the feeder streams and rivers; a once sizable population of Marsh Arabs, who inhabited these areas for thousands of years, has been displaced; furthermore, the destruction of the natural habitat poses serious threats to the area's wildlife populations; inadequate supplies of potable water; development of the Tigris and Euphrates rivers system contingent upon agreements with upstream riparian Turkey; air and water pollution; soil degradation (salination) and erosion; desertification
Environment—international agreements: *party to:* Biodiversity, Climate Change, Law of the Sea
signed, but not ratified: Environmental Modification
Geography—note: strategic location on Shatt al Arab waterway and at the head of the Persian Gulf

PEOPLE

Population: 28,221,181 (July 2008 est.)
Age structure:
0–14 years: 39.2% (male 5,613,420/female 5,438,770)
15–64 years: 57.9% (male 8,270,573/female 8,057,423)
65 years and over: 3% (male 396,751/female 444,244) (2008 est.)
Median age: *total:* 20.2 years
male: 20.1 years
female: 20.2 years (2008 est.)
Population growth rate: 2.562% (2008 est.)
Birth rate: 30.77 births/1,000 population (2008 est.)
Death rate: 5.14 deaths/1,000 population (2008 est.)
Net migration rate: NA
Sex ratio: *at birth:* 1.05 male(s)/female
under 15 years: 1.03 male(s)/female
15–64 years: 1.03 male(s)/female
65 years and over: 0.89 male(s)/female
total population: 1.02 male(s)/female (2008 est.)
Infant mortality rate:
total: 45.43 deaths/1,000 live births
male: 51.06 deaths/1,000 live births
female: 39.53 deaths/1,000 live births (2008 est.)
Life expectancy at birth:
total population: 69.62 years
male: 68.32 years
female: 70.99 years (2008 est.)

Total fertility rate: 3.97 children born/woman (2008 est.)

HIV/AIDS—adult prevalence rate: less than 0.1% (2001 est.)

HIV/AIDS—people living with HIV/AIDS: fewer than 500 (2003 est.)

HIV/AIDS—deaths: NA

Major infectious diseases:

degree of risk: intermediate

food or waterborne diseases: bacterial diarrhea, hepatitis A, and typhoid fever

note: highly pathogenic H5N1 avian influenza has been identified in this country; it poses a negligible risk with extremely rare cases possible among US citizens who have close contact with birds (2008)

Nationality: *noun:* Iraqi(s)

adjective: Iraqi

Ethnic groups: Arab 75%-80%, Kurdish 15%-20%, Turkoman, Assyrian, or other 5%

Religions: Muslim 97% (Shi'a 60%-65%, Sunni 32%-37%), Christian or other 3%

Languages: Arabic, Kurdish (official in Kurdish regions), Turkoman (a Turkish dialect), Assyrian (Neo-Aramaic), Armenian

Literacy: *definition:* age 15 and over can read and write

total population: 74.1%

male: 84.1%

female: 64.2% (2000 est.)

GOVERNMENT

Country name:

conventional long form: Republic of Iraq

conventional short form: Iraq

local long form: Al Jumhuriyah al-Iraqiyah

local short form: Al Iraq

Government type: parliamentary democracy

Capital: *name:* Baghdad

geographic coordinates: 33 20 N, 44 23 E

time difference: UTC+3 (8 hours ahead of Washington, DC during Standard Time)

daylight saving time: +1hr, begins 1 April; ends 1 October

Administrative divisions: 18 governorates (muhafazat, singular—muhafazah) and 1 region*; Al Anbar, Al Basrah, Al Muthanna, Al Qadisiyah, An Najaf, Arbil, As Sulaymaniyah, At Ta'mim, Babil, Baghdad, Dahuk, Dhi Qar, Diyala, Karbala', Kurdistan Regional Government*, Maysan, Ninawa, Salah ad Din, Wasit

Independence: 3 October 1932 (from League of Nations mandate under British administration); note—on 28 June 2004 the Coalition Provisional Authority transferred sovereignty to the Iraqi-controlled Government

National holiday: Revolution Day, 17 July (1968); note—this holiday was celebrated under the SADDAM Husayn regime; the Government of Iraq has yet to declare a new national holiday

Constitution: ratified on 15 October 2005 (subject to review by the Constitutional Review Committee and a possible public referendum)

Legal system: based on European civil and Islamic law under the framework outlined in the Iraqi Constitution; has not accepted compulsory ICJ jurisdiction

Suffrage: 18 years of age; universal

Executive branch:

chief of state: President Jalal TALABANI (since 6 April 2005); Vice Presidents Adil ABD AL-MAHDI and Tariq al-HASHIMI (since 22 April 2006); note—the president and vice presidents comprise the Presidency Council)

head of government: Prime Minister Nuri al-MALIKI (since 20 May 2006); Deputy Prime Minister Barham SALIH (since 20 May 2006); second deputy prime minister positon vacant

cabinet: 34 ministers appointed by the Presidency Council, plus Prime Minister Nuri al-MALIKI, and Deputy Prime Minister Barham SALIH; second deputy prime minister position vacant

elections: held 15 December 2005 to elect a 275-member Council of Representatives

Legislative branch: Council of Representatives (consisting of 275 members elected by a closed-list, proportional representation system)

elections: held 15 December 2005 to elect a 275-member Council of Representatives; the Council of Representatives elected the Presidency Council and approved the prime minister and two deputy prime ministers

election results: Council of Representatives—percent of vote by party—Unified Iraqi Alliance 41%, Kurdistan Alliance 22%, Tawafuq Coalition 15%, Iraqi National List 8%, Iraqi Front for National Dialogue 4%, other 10%; number of seats by party (as of November 2007)—Unified Iraqi Alliance (including the Sadrist bloc with 30 and Fadilah with 15) 130, Kurdistan Alliance 53, Tawafuq Front 44, Iraqi National List 25, Fadilah 15, Iraqi Front for National Dialogue 11, other 12

Judicial branch: the Iraq Constitution calls for the federal judicial power to be comprised of the Higher Juridical Council, Federal Supreme Court, Federal Court of Cassation, Public Prosecution Department, Judiciary Oversight Commission and other federal courts that are regulated in accordance with the law

Political parties and leaders: Assyrian Democratic Movement [Yunadim KANNA]; Badr Organization [Hadi al-AMIRI]; Constitutional Monarchy Movement or CMM [Sharif Ali Bin al-HUSAYN]; Da'wa al-Islamiya Party [Ibrahim al-JA'FARI]; General Conference of Iraqi People [Adnan al-DULAYMI]; Independent Iraqi Alliance or IIA [Falah al-NAQIB]; Iraqi Communist Party [Hamid MAJEED]; Iraqi Front for National Dialogue [Salih al-MUTLAQ]; Iraqi Hizballah [Karim Mahmud al-MUHAMMADAWI]; Iraqi Independent Democrats or IID [Adnan PACHACHI, Mahdi al-HAFIZ]; Iraqi Islamic Party or IIP [Tariq al-HASHIMI]; Iraqi National Accord or INA [Ayad ALLAWI]; Iraqi National Congress or INC [Ahmad CHALABI]; Iraqi National Council for Dialogue or INCD [Khalaf Ulayan al-Khalifawi al-DULAYMI]; Iraqi National Unity Movement or INUM [Ahmad al-KUBAYSI]; Islamic Action Organization or IAO [Ayatollah Muhammad al-MUDARRISI]; Islamic Supreme Council of Iraq or ISCI [Abd al-Aziz al-HAKIM]; Jama'at al Fadilah or JAF [Muhammad Ali al-YAQUBI]; Kurdistan Democratic Party or KDP [Masud BARZANI]; Kurdistan Islamic Union [Salah ad-Din Muhammad BAHA al-DIN]; National Reconciliation and Liberation Party [Mishan al-JABBURI]; Patriotic Union of Kurdistan or PUK [Jalal TALABANI]; Sadrist Trend [Muqtada al-SADR] (not an organized political party, but it fields independent candidates affiliated with Muqtada al-SADR); Sahawa al-Iraq [Ahmed al-RISHAWI]

note: the Kurdistan Alliance, Iraqi National List, Tawafuq Front, Iraqi Front for National Dialogue, and Unified Iraqi Alliance were only electoral slates consisting of the representatives from the various Iraqi political parties

Political pressure groups and leaders: an insurgency against the Government of Iraq and Coalition forces is primarily concentrated in Baghdad and in areas north, northeast, and west of the capital; the diverse, multigroup insurgency consists principally of Sunni Arabs with a shared desire to oust the Coalition, end US influence in Iraq, and reassert Sunni Arab dominance; a number of predominantly Shia militias, some associated with political parties, challenge governmental authority in Baghdad and southern Iraq

International organization participation: ABEDA, AFESD, AMF, CAEU, FAO, G-77, IAEA, IBRD, ICAO,

ICRM, IDA, IDB, IFAD, IFC, IFRCS, ILO, IMF, IMO, IMSO, Interpol, IOC, ISO, ITSO, ITU, LAS, NAM, OAPEC, OIC, OPEC, PCA, UN, UNCTAD, UNESCO, UNIDO, UNWTO, UPU, WCO, WFTU, WHO, WIPO, WMO, WTO (observer)

Diplomatic representation in the US:
chief of mission: Ambassador Samir Shakir al-SUMAYDI
chancery: 3421 Massachusetts Ave, NW, Washington, DC 20007
telephone: [1] (202) 483-7500 (Consular section)
FAX: [1] (202) 333-1129

Diplomatic representation from the US:
chief of mission: Ambassador Ryan C. CROCKER
embassy: Baghdad
mailing address: APO AE 09316
telephone: 1-240-553-0589 ext. 5340 or 5635; note—Consular Section
FAX: NA

Flag description: three equal horizontal bands of red (top), white, and black; the Takbir (Arabic expression meaning "God is great") in green Arabic script is centered in the white band; similar to the flag of Syria, which has two stars but no script, Yemen, which has a plain white band, and that of Egypt, which has a gold Eagle of Saladin centered in the white band; design is based upon the Arab Liberation colors; Council of Representatives approved this flag as a compromise temporary replacement for Ba'athist Saddam-era flag

ECONOMY

Economy—overview: Iraq's economy is dominated by the oil sector, which has traditionally provided about 95% of foreign exchange earnings. Although looting, insurgent attacks, and sabotage have undermined economy rebuilding efforts, economic activity is beginning to pick up in areas recently secured by the US military surge. Oil exports are around levels seen before Operation Iraqi Freedom, and total government revenues have benefited from high oil prices. Despite political uncertainty, Iraq is making some progress in building the institutions needed to implement economic policy and has negotiated a debt reduction agreement with the Paris Club and a new Stand-By Arrangement with the IMF. Iraq has received pledges for $13.5 billion in foreign aid for 2004-07 from outside of the US, more than $33 billion in total pledges. The International Compact with Iraq was established in May 2007 to integrate Iraq into the regional and global economy, and the Iraqi government is seeking to pass laws to strengthen its economy. This legislation includes a hydrocarbon law to establish a modern legal framework to allow Iraq to develop its resources and a revenue sharing law to equitably divide oil revenues within the nation, although both are still bogged down in discussions. The Central Bank has been successful in controlling inflation through appreciation of the dinar against the US dollar. Reducing corruption and implementing structural reforms, such as bank restructuring and developing the private sector, will be key to Iraq's economic success.

GDP (purchasing power parity): $102.3 billion (2007 est.)
GDP (official exchange rate): $55.44 billion (2007 est.)
GDP—real growth rate: 5% (2007 est.)
GDP—per capita (PPP): $3,600 (2007 est.)
GDP—composition by sector:
agriculture: 5%
industry: 68%
services: 27% (2006 est.)
Labor force: 7.4 million (2004 est.)
Labor force—by occupation: *agriculture:* NA%
industry: NA%
services: NA%
Unemployment rate: 18% to 30% (2006 est.)
Population below poverty line: NA%
Household income or consumption by percentage share:
lowest 10%: NA%
highest 10%: NA%
Inflation rate (consumer prices): 4.7% (2007 est.)
Budget:
revenues: $42.3 billion
expenditures: $48.4 billion (FY08 est.)
Agriculture—products: wheat, barley, rice, vegetables, dates, cotton; cattle, sheep, poultry
Industries: petroleum, chemicals, textiles, leather, construction materials, food processing, fertilizer, metal fabrication/processing
Industrial production growth rate: 7.9% (2007 est.)
Electricity—production: 33.53 billion kWh (2007 est.)
Electricity—production by source:
fossil fuel: 98.4%
hydro: 1.6%
nuclear: 0%
other: 0% (2001)
Electricity—consumption: 35.84 billion kWh (2007 est.)
Electricity—exports: 0 kWh (2007)
Electricity—imports: 2.315 billion kWh (2007 est.)
Oil—production: 2.11 million bbl/day (2007 est.)
Oil—consumption: 295,000 bbl/day (2007 est.)
Oil—exports: 1.67 million bbl/day (2007 est.)
Oil—imports: NA bbl/day
Oil—proved reserves: 115 billion bbl (1 January 2007 est.)
Natural gas—production: 3.5 billion cu m (2007 est.)
Natural gas—consumption: 980 million cu m
note: 1.48 billion cu m were flared (2005 est.)
Natural gas—exports: 0 cu m (2005 est.)
Natural gas—imports: 0 cu m (2005)
Natural gas—proved reserves: 3.17 trillion cu m (1 January 2007 est.)
Current account balance: $7.802 billion (2007 est.)
Exports: $38.11 billion f.o.b. (2007 est.)
Exports—commodities: crude oil 84%, crude materials excluding fuels 8%, food and live animals 5%
Exports—partners: US 46.8%, Italy 10.7%, Spain 6.2%, Canada 6.2% (2006)
Imports: $24.81 billion f.o.b. (2007 est.)
Imports—commodities: food, medicine, manufactures
Imports—partners: Syria 27.1%, Turkey 21%, US 12.1%, Jordan 5.1% (2006)
Economic aid—recipient: $21.65 billion (2005)
Reserves of foreign exchange and gold: $25.66 billion (31 December 2007 est.)
Debt—external: $100.9 billion (31 December 2007 est.)
Market value of publicly traded shares: $NA
Currency (code): New Iraqi dinar (NID) as of 22 January 2004
Currency code: NID, IQD prior to 22 January 2004
Exchange rates: New Iraqi dinars per US dollar—1,255 (2007), 1,466 (2006), 1,475 (2005), 1,890 (second half, 2003)
Fiscal year: calendar year

COMMUNICATIONS

Telephones—main lines in use: 1.547 million (2005)
Telephones—mobile cellular: 10.9 million (2007)
Telephone system:
general assessment: the 2003 liberation of Iraq severely disrupted telecommunications throughout Iraq including international connections; widespread government efforts to rebuild domestic and international communications through fiber optic links are in progress; the mobile cellular market has expanded rapidly with an estimated 10.9 million current users

domestic: repairs to switches and lines destroyed during 2003 continue; additional switching capacity is improving access; cellular service is available and centered on 3 GSM networks which are being expanded beyond their regional roots, improving country-wide connectivity; wireless local loop licences have been issued with the hope of overcoming the lack of fixed-line infrastructure
international: country code—964; satellite earth stations—4 (2 Intelsat—1 Atlantic Ocean and 1 Indian Ocean, 1 Intersputnik—Atlantic Ocean region, and 1 Arabsat (inoperative)); local microwave radio relay connects border regions to Jordan, Kuwait, Syria, and Turkey; planned international fiber-optic connections to Iran (terrestrial) with a link to the Fiber-Optic Link Around the Globe (FLAG) submarine fiber-optic cable (2007)
Radio broadcast stations: after 17 months of unregulated media growth, there are approximately 80 radio stations (types NA) on the air inside Iraq (2004)
Radios: 4.85 million (1997)
Television broadcast stations: 21 (2004)
Televisions: 1.75 million (1997)
Internet country code: .iq
Internet hosts: 3 (2007)
Internet Service Providers (ISPs): 1 (2000)
Internet users: 36,000 (2004)

TRANSPORTATION

Airports: 110 (2007)
Airports—with paved runways:
total: 76

over 3,047 m: 19
2,438 to 3,047 m: 37
1,524 to 2,437 m: 5
914 to 1,523 m: 6
under 914 m: 9 (2007)
Airports—with unpaved runways:
total: 34
over 3,047 m: 3
2,438 to 3,047 m: 4
1,524 to 2,437 m: 4
914 to 1,523 m: 13
under 914 m: 10 (2007)
Heliports: 17 (2007)
Pipelines: gas 2,250 km; liquid petroleum gas 918 km; oil 5,509 km; refined products 1,637 km (2007)
Railways:
total: 2,272 km
standard gauge: 2,272 km 1.435-m gauge (2006)
Roadways:
total: 45,550 km
paved: 38,399 km
unpaved: 7,151 km (2000)
Waterways: 5,279 km
note: Euphrates River (2,815 km), Tigris River (1,899 km), and Third River (565 km) are principal waterways (2006)
Merchant marine:
total: 13 ships (1000 GRT or over) 67,796 GRT/101,317 DWT
by type: cargo 11, petroleum tanker 2 (2007)
Ports and terminals: Al Basrah, Khawr az Zubayr, Umm Qasr

MILITARY

Military branches: Iraqi Armed Forces: Iraqi Army (includes Iraqi Special Operations Force, Iraqi Intervention

Force), Iraqi Navy (former Iraqi Coastal Defense Force), Iraqi Air Force (former Iraqi Army Air Corps) (2005)
Military service age and obligation: 18–40 years of age for voluntary military service (2006)
Manpower available for military service:
males age 16–49: 7,086,200
females age 16–49: 6,808,954 (2008 est.)
Manpower fit for military service:
males age 16–49: 6,019,795
females age 16–49: 5,878,905 (2008 est.)
Manpower reaching militarily significant age annually:
males age 16–49: 302,926
females age 16–49: 294,747 (2008 est.)
Military expenditures—percent of GDP: 8.6% (2006)

TRANSNATIONAL ISSUES

Disputes—international: coalition forces assist Iraqis in monitoring internal and cross-border security; approximately two million Iraqis have fled the conflict in Iraq, with the majority taking refuge in Syria and Jordan, and lesser numbers to Egypt, Lebanon, Iran, and Turkey; Iraq's lack of a maritime boundary with Iran prompts jurisdiction disputes beyond the mouth of the Shatt al Arab in the Persian Gulf; Turkey has expressed concern over the autonomous status of Kurds in Iraq
Refugees and internally displaced persons: *refugees (country of origin):* 10,000–15,000 (Palestinian Territories); 11,773 (Iran); 16,832 (Turkey)
IDPs: 2.4 million (ongoing US-led war and ethno-sectarian violence) (2007)

IRELAND

by Norsemen that began in the late 8th century were finally ended when King Brian BORU defeated the Danes in 1014. English invasions began in the 12th century and set off more than seven centuries of Anglo-Irish struggle marked by fierce rebellions and harsh repressions. A failed 1916 Easter Monday Rebellion touched off several years of guerrilla warfare that in 1921 resulted in independence from the UK for 26 southern counties; six northern (Ulster) counties remained part of the UK. In 1948 Ireland withdrew from the British Commonwealth; it joined the European Community in 1973. Irish governments have sought the peaceful unification of Ireland and have cooperated with Britain against terrorist groups. A peace settlement for Northern Ireland is being

implemented with some difficulties. In 2006, the Irish and British governments developed and began to implement the St. Andrews Agreement, building on the Good Friday Agreement approved in 1998.

GEOGRAPHY

Location: Western Europe, occupying five-sixths of the island of Ireland in the North Atlantic Ocean, west of Great Britain
Geographic coordinates: 53 00 N, 8 00 W
Map references: Europe
Area: *total:* 70,280 sq km
land: 68,890 sq km
water: 1,390 sq km
Area—comparative: slightly larger than West Virginia

INTRODUCTION

Background: Celtic tribes arrived on the island between 600–150 B.C. Invasions

Land boundaries:
total: 360 km
border countries: UK 360 km
Coastline: 1,448 km
Maritime claims:
territorial sea: 12 nm
exclusive fishing zone: 200 nm
Climate: temperate maritime; modified by North Atlantic Current; mild winters, cool summers; consistently humid; overcast about half the time
Terrain: mostly level to rolling interior plain surrounded by rugged hills and low mountains; sea cliffs on west coast
Elevation extremes:
lowest point: Atlantic Ocean 0 m
highest point: Carrauntoohil 1,041 m
Natural resources: natural gas, peat, copper, lead, zinc, silver, barite, gypsum, limestone, dolomite
Land use:
arable land: 16.82%
permanent crops: 0.03%
other: 83.15% (2005)
Irrigated land: NA
Total renewable water resources: 46.8 cu km (2003)
Freshwater withdrawal (domestic/ industrial/agricultural): *total:* 1.18 cu km/yr (23%/77%/0%)
per capita: 284 cu m/yr (1994)
Natural hazards: NA
Environment—current issues: water pollution, especially of lakes, from agricultural runoff
Environment—international agreements: *party to:* Air Pollution, Air Pollution-Nitrogen Oxides, Air Pollution-Sulfur 94, Biodiversity, Climate Change, Climate Change-Kyoto Protocol, Desertification, Endangered Species, Environmental Modification, Hazardous Wastes, Law of the Sea, Marine Dumping, Ozone Layer Protection, Ship Pollution, Tropical Timber 83, Tropical Timber 94, Wetlands, Whaling
signed, but not ratified: Air Pollution-Persistent Organic Pollutants, Marine Life Conservation
Geography—note: strategic location on major air and sea routes between North America and northern Europe; over 40% of the population resides within 100 km of Dublin

PEOPLE

Population: 4,156,119 (July 2008 est.)
Age structure:
0–14 years: 20.9% (male 448,333/female 418,476)
15–64 years: 67.3% (male 1,400,222/female 1,398,194)
65 years and over: 11.8% (male 218,459/female 272,435) (2008 est.)

Median age:
total: 34.6 years
male: 33.9 years
female: 35.4 years (2008 est.)
Population growth rate: 1.133% (2008 est.)
Birth rate: 14.33 births/1,000 population (2008 est.)
Death rate: 7.77 deaths/1,000 population (2008 est.)
Net migration rate: 4.76 migrant(s)/ 1,000 population (2008 est.)
Sex ratio:
at birth: 1.07 male(s)/female
under 15 years: 1.07 male(s)/female
15–64 years: 1 male(s)/female
65 years and over: 0.8 male(s)/female
total population: 0.99 male(s)/female (2008 est.)
Infant mortality rate:
total: 5.14 deaths/1,000 live births
male: 5.63 deaths/1,000 live births
female: 4.61 deaths/1,000 live births (2008 est.)
Life expectancy at birth:
total population: 78.07 years
male: 75.44 years
female: 80.88 years (2008 est.)
Total fertility rate: 1.85 children born/woman (2008 est.)
HIV/AIDS—adult prevalence rate: 0.1% (2001 est.)
HIV/AIDS—people living with HIV/AIDS: 2,800 (2001 est.)
HIV/AIDS—deaths: fewer than 100 (2003 est.)
Nationality:
noun: Irishman(men), Irishwoman(women), Irish (collective plural)
adjective: Irish
Ethnic groups: Celtic, English
Religions: Roman Catholic 88.4%, Church of Ireland 3%, other Christian 1.6%, other 1.5%, unspecified 2%, none 3.5% (2002 census)
Languages: English (official) is the language generally used, Irish (Gaelic or Gaeilge) (official) spoken mainly in areas located along the western seaboard
Literacy:
definition: age 15 and over can read and write
total population: 99%
male: 99%
female: 99% (2003 est.)

GOVERNMENT

Country name:
conventional long form: none
conventional short form: Ireland
local long form: none
local short form: Eire
Government type: republic, parliamentary democracy

Capital: *name:* Dublin
geographic coordinates: 53 19 N, 6 14 W
time difference: UTC 0 (5 hours ahead of Washington, DC during Standard Time)
daylight saving time: +1hr, begins last Sunday in March; ends last Sunday in October
Administrative divisions: 26 counties; Carlow, Cavan, Clare, Cork, Donegal, Dublin, Galway, Kerry, Kildare, Kilkenny, Laois, Leitrim, Limerick, Longford, Louth, Mayo, Meath, Monaghan, Offaly, Roscommon, Sligo, Tipperary, Waterford, Westmeath, Wexford, Wicklow
note: Cavan, Donegal, and Monaghan are part of Ulster Province
Independence: 6 December 1921 (from UK by treaty)
National holiday: Saint Patrick's Day, 17 March
Constitution: adopted 1 July 1937 by plebiscite; effective 29 December 1937
Legal system: based on English common law, substantially modified by indigenous concepts; judicial review of legislative acts in Supreme Court; has not accepted compulsory ICJ jurisdiction
Suffrage: 18 years of age; universal
Executive branch:
chief of state: President Mary MCALEESE (since 11 November 1997)
head of government: Prime Minister Brian COWEN (since 7 May 2008)
cabinet: Cabinet appointed by the president with previous nomination by the prime minister and approval of the House of Representatives
elections: president elected by popular vote for a seven-year term (eligible for a second term); election last held 31 October 1997 (next scheduled for October 2011); note—Mary MCALEESE appointed to a second term when no other candidate qualified for the 2004 presidential election; prime minister (taoiseach) nominated by the House of Representatives and appointed by the president
election results: Mary MCALEESE elected president; percent of vote—Mary MCALEESE 44.8%, Mary BANOTTI 29.6%
note: government coalition—Fianna Fail, the Green Party, the Progressive Democrats, and independent members of Parliament
Legislative branch: bicameral Parliament or Oireachtas consists of the Senate or Seanad Eireann (60 seats; 49 members elected by the universities and from candidates put forward by five vocational panels, 11 are nominated by the prime minister; to serve five-year terms) and the House of Representatives or Dail

Eireann (166 seats; members are elected by popular vote on the basis of proportional representation to serve five-year terms)

elections: Senate—last held in July 2007 (next to be held by July 2012); House of Representatives—last held 24 May 2007 (next to be held by May 2012)

election results: Senate—percent of vote by party—NA; seats by party—Fianna Fail 28, Fine Gael 14, Labor Party 6, Progressive Democrats 2, Green Party 2, Sein Fein 1, independents 7; House of Representatives—percent of vote by party—Fianna Fail 41.6%, Fine Gael 27.3%, Labor Party 10.1%, Sinn Fein 6.9%, Green Party 4.7%, Progressive Democrats 2.7%, other 6.7%; seats by party—Fianna Fail 78, Fine Gael 51, Labor Party 20, Sinn Fein 4, Green Party 6, Progressive Democrats 2, other 5

Judicial branch: Supreme Court (judges appointed by the president on the advice of the prime minister and cabinet)

Political parties and leaders: Fianna Fail [Brian COWEN]; Fine Gael [Enda KENNY]; Green Party [John GORMLEY]; Labor Party [Eamon GILMORE]; Progressive Democrats [Mary HARNEY, acting leader]; Sinn Fein [Gerry ADAMS]; Socialist Party [Joe HIGGINS]; The Workers' Party [Sean GARLAND]

Political pressure groups and leaders: NA

International organization participation: ADB (nonregional members), Australia Group, BIS, CE, EAPC, EBRD, EIB, EMU, ESA, EU, FAO, IAEA, IBRD, ICAO, ICC, ICCt, ICRM, IDA, IEA, IFAD, IFC, IFRCS, IHO, ILO, IMF, IMO, Interpol, IOC, IOM, IPU, ISO, ITSO, ITU, ITUC, MIGA, MINURSO, NEA, NSG, OAS (observer), OECD, OPCW, OSCE, Paris Club, PCA, PFP, UN, UNCTAD, UNESCO, UNFICYP, UNHCR, UNIDO, UNIFIL, UNMIL, UNOCI, UNTSO, UPU, WCO, WEU (observer), WHO, WIPO, WMO, WTO, ZC

Diplomatic representation in the US:
chief of mission: Ambassador Michael COLLINS
chancery: 2234 Massachusetts Avenue NW, Washington, DC 20008
telephone: [1] (202) 462-3939
FAX: [1] (202) 232-5993
consulate(s) general: Boston, Chicago, New York, San Francisco

Diplomatic representation from the US:
chief of mission: Ambassador Thomas C. FOLEY
embassy: 42 Elgin Road, Ballsbridge, Dublin 4

mailing address: use embassy street address
telephone: [353] (1) 668-8777
FAX: [353] (1) 668-9946

Flag description: three equal vertical bands of green (hoist side), white, and orange; similar to the flag of Cote d'Ivoire, which is shorter and has the colors reversed—orange (hoist side), white, and green; also similar to the flag of Italy, which is shorter and has colors of green (hoist side), white, and red

ECONOMY

Economy—overview: Ireland is a small, modern, trade-dependent economy with growth averaging 6% in 1995–2007. Agriculture, once the most important sector, is now dwarfed by industry and services. Although the exports sector, dominated by foreign multinationals, remains a key component of Ireland's economy, construction has most recently fueled economic growth along with strong consumer spending and business investment. Property prices have risen more rapidly in Ireland in the decade up to 2006 than in any other developed world economy. Per capita GDP is 40% above that of the four big European economies and the second highest in the EU behind Luxembourg, and in 2007 surpassed that of the United States. The Irish Government has implemented a series of national economic programs designed to curb price and wage inflation, invest in infrastructure, increase labor force skills, and promote foreign investment. A slowdown in the property market, more intense global competition, and increased costs, however, have compelled government economists to lower Ireland's growth forecast slightly for 2008. Ireland joined in circulating the euro on 1 January 2002 along with 11 other EU nations.

GDP (purchasing power parity): $186.2 billion (2007 est.)

GDP (official exchange rate): $258.6 billion (2007 est.)

GDP—real growth rate: 5.3% (2007 est.)

GDP—per capita (PPP): $43,100 (2007 est.)

GDP—composition by sector:
agriculture: 5%
industry: 46%
services: 49% (2002 est.)

Labor force: 2.22 million (2007 est.)

Labor force—by occupation: *agriculture:* 6%
industry: 27%
services: 67% (2006 est.)

Unemployment rate: 4.6% (2007 est.)

Population below poverty line: 7% (2005 est.)

Household income or consumption by percentage share:
lowest 10%: 2.9%
highest 10%: 27.2% (2000)

Distribution of family income—Gini index: 32 (2005)

Inflation rate (consumer prices): 3% (2007 est.)

Investment (gross fixed): 25.3% of GDP (2007 est.)

Budget:
revenues: $93.97 billion
expenditures: $88.27 billion (2007 est.)

Public debt: 24.7% of GDP (2007 est.)

Agriculture—products: turnips, barley, potatoes, sugar beets, wheat; beef, dairy products

Industries: steel, lead, zinc, silver, aluminum, barite, and gypsum mining processing; food products, brewing, textiles, clothing; chemicals, pharmaceuticals; machinery, rail transportation equipment, passenger and commercial vehicles, ship construction and refurbishment; glass and crystal; software, tourism

Industrial production growth rate: 5% (2006 est.)

Electricity—production: 24.13 billion kWh (2005)

Electricity—production by source:
fossil fuel: 95.9%
hydro: 2.3%
nuclear: 0%
other: 1.7% (2001)

Electricity—consumption: 24.09 billion kWh (2005)

Electricity—exports: 1 million kWh (2005)

Electricity—imports: 2.045 billion kWh (2005)

Oil—production: 0 bbl/day (2005 est.)

Oil—consumption: 192,000 bbl/day (2005 est.)

Oil—exports: 23,360 bbl/day (2004)

Oil—imports: 204,400 bbl/day (2004)

Oil—proved reserves: 0 bbl (1 January 2006 est.)

Natural gas—production: 546.7 million cu m (2005 est.)

Natural gas—consumption: 3.895 billion cu m (2005 est.)

Natural gas—exports: 0 cu m (2005 est.)

Natural gas—imports: 3.348 billion cu m (2005)

Natural gas—proved reserves: 9.505 billion cu m (1 January 2006 est.)

Current account balance: -$11.69 billion (2007 est.)

Exports: $115.6 billion f.o.b. (2007 est.)

Exports—commodities: machinery and equipment, computers, chemicals, pharmaceuticals; live animals, animal products

Exports—partners: US 18.7%, UK 17.9%, Belgium 14.4%, Germany 7.8%, France 5.8%, Italy 4.2% (2006)
Imports: $84.2 billion f.o.b. (2007 est.)
Imports—commodities: data processing equipment, other machinery and equipment, chemicals, petroleum and petroleum products, textiles, clothing
Imports—partners: UK 37.5%, US 11.5%, Germany 9.6%, Netherlands 4.6% (2006)
Economic aid—donor: ODA, $1.022 billion (2006)
Reserves of foreign exchange and gold: $926.2 million (2006 est.)
Debt—external: $1.841 trillion (30 June 2007)
Stock of direct foreign investment—at home: $173.9 billion (2007 est.)
Stock of direct foreign investment—abroad: $135.6 billion (2007 est.)
Market value of publicly traded shares: $114.1 billion (2005)
Currency (code): euro (EUR)
Currency code: EUR
Exchange rates: euros per US dollar— 0.7345 (2007), 0.7964 (2006), 0.8041 (2005), 0.8054 (2004), 0.886 (2003)
Fiscal year: calendar year

COMMUNICATIONS

Telephones—main lines in use: 2.097 million (2006)
Telephones—mobile cellular: 4.69 million (2006)
Telephone system:
general assessment: modern digital system using cable and microwave radio relay
domestic: microwave radio relay
international: country code—353; landing point for the Hibernia-Atlantic submarine cable with links to the US, Canada, and UK; satellite earth station—1 Intelsat (Atlantic Ocean)
Radio broadcast stations: AM 9, FM 106, shortwave 0 (1998)
Radios: 2.55 million (1997)

Television broadcast stations: 4 (many repeaters) (2001)
Televisions: 1.82 million (2001)
Internet country code: .ie
Internet hosts: 429,487 (2007)
Internet Service Providers (ISPs): 22 (2000)
Internet users: 1.437 million (2006)

TRANSPORTATION

Airports: 34 (2007)
Airports—with paved runways:
total: 15
over 3,047 m: 1
2,438 to 3,047 m: 1
1,524 to 2,437 m: 4
914 to 1,523 m: 4
under 914 m: 5 (2007)
Airports—with unpaved runways:
total: 19
914 to 1,523 m: 3
under 914 m: 16 (2007)
Pipelines: gas 1,855 km (2007)
Railways:
total: 3,237 km
broad gauge: 1,872 km 1.600-m gauge (37 km electrified)
narrow gauge: 1,365 km 0.914-m gauge (operated by the Irish Peat Board to transport peat to power stations and briquetting plants) (2006)
Roadways:
total: 96,602 km
paved: 96,602 km (includes 200 km of expressways) (2003)
Waterways: 956 km (pleasure craft only) (2007)
Merchant marine:
total: 27 ships (1000 GRT or over) 116,091 GRT/161,808 DWT
by type: cargo 23, chemical tanker 2, container 1, roll on/roll off 1
foreign-owned: 3 (Spain 1, US 2)
registered in other countries: 18 (Bahamas 2, Bermuda 1, Bulgaria 1, Cyprus 1, Germany 1, Isle of Man 1, Netherlands 9, Panama 1, UK 1, unknown 1) (2007)

Ports and terminals: Cork, Dublin, Shannon Foynes

MILITARY

Military branches: Irish Defense Forces (Oglaigh na h-Eireann): Army (includes Naval Service and Air Corps) (2008)
Military service age and obligation: 17–25 years of age for voluntary military service; 16 years of age can be recruited for apprentice specialist positions; maximum obligation 12 years; 17–35 years of age for the Reserve Defense Forces (2008)
Manpower available for military service:
males age 16–49: 1,024,635
females age 16–49: 1,024,276 (2008 est.)
Manpower fit for military service:
males age 16–49: 854,982
females age 16–49: 852,592 (2008 est.)
Manpower reaching militarily significant age annually:
males age 16–49: 28,610
females age 16–49: 27,095 (2008 est.)
Military expenditures—percent of GDP: 0.9% (2005 est.)

TRANSNATIONAL ISSUES

Disputes—international: Ireland, Iceland, and the UK dispute Denmark's claim that the Faroe Islands' continental shelf extends beyond 200 nm
Illicit drugs: transshipment point for and consumer of hashish from North Africa to the UK and Netherlands and of European-produced synthetic drugs; increasing consumption of South American cocaine; minor transshipment point for heroin and cocaine destined for Western Europe; despite recent legislation, narcotics-related money laundering—using bureaux de change, trusts, and shell companies involving the offshore financial community—remains a concern

ISLE OF MAN

INTRODUCTION

Background: Part of the Norwegian Kingdom of the Hebrides until the 13th century when it was ceded to Scotland, the isle came under the British crown in 1765. Current concerns include reviving the almost extinct Manx Gaelic language. Isle of Man is a British crown dependency but is not part of the UK. However, the UK Government remains constitutionally responsible for its defense and international representation.

GEOGRAPHY

Location: Western Europe, island in the Irish Sea, between Great Britain and Ireland
Geographic coordinates: 54 15 N, 4 30 W
Map references: Europe
Area:
total: 572 sq km
land: 572 sq km
water: 0 sq km
Area—comparative: slightly more than three times the size of Washington, DC

Land boundaries: 0 km
Coastline: 160 km
Maritime claims:
territorial sea: 12 nm
exclusive fishing zone: 12 nm
Climate: temperate; cool summers and mild winters; overcast about one-third of the time
Terrain: hills in north and south bisected by central valley
Elevation extremes:
lowest point: Irish Sea 0 m
highest point: Snaefell 621 m
Natural resources: none

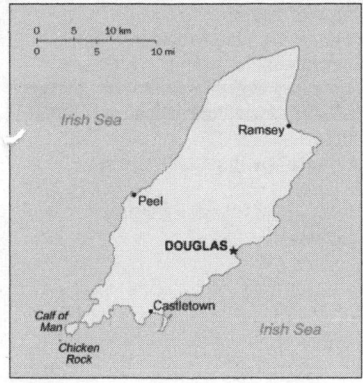

Land use:

arable land: 9%

permanent crops: 0%

other: 91% (permanent pastures, forests, mountain, and heathland) (2002)

Irrigated land: 0 sq km

Natural hazards: NA

Environment—current issues: waste disposal (both household and industrial); transboundary air pollution

Geography—note: one small islet, the Calf of Man, lies to the southwest and is a bird sanctuary

PEOPLE

Population: 76,220 (July 2008 est.)

Age structure:

0–14 years: 17% (male 6,629/female 6,318)

15–64 years: 65.9% (male 25,251/female 24,959)

65 years and over: 17.1% (male 5,294/female 7,769) (2008 est.)

Median age:

total: 40 years

male: 38.8 years

female: 41.3 years (2008 est.)

Population growth rate: 0.509% (2008 est.)

Birth rate: 10.86 births/1,000 population (2008 est.)

Death rate: 11.02 deaths/1,000 population (2008 est.)

Net migration rate: 5.25 migrant(s)/1,000 population (2008 est.)

Sex ratio:

at birth: 1.06 male(s)/female

under 15 years: 1.05 male(s)/female

15–64 years: 1.01 male(s)/female

65 years and over: 0.68 male(s)/female

total population: 0.95 male(s)/female (2008 est.)

Infant mortality rate:

total: 5.62 deaths/1,000 live births

male: 6.55 deaths/1,000 live births

female: 4.64 deaths/1,000 live births (2008 est.)

Life expectancy at birth:

total population: 78.8 years

male: 75.46 years

female: 82.32 years (2008 est.)

Total fertility rate: 1.65 children born/woman (2008 est.)

HIV/AIDS—adult prevalence rate: NA

HIV/AIDS—people living with HIV/AIDS: NA

HIV/AIDS—deaths: NA

Nationality:

noun: Manxman (men), Manxwoman (women)

adjective: Manx

Ethnic groups: Manx (Norse-Celtic descent), Britons

Religions: Anglican, Roman Catholic, Methodist, Baptist, Presbyterian, Society of Friends

Languages: English, Manx Gaelic

Literacy:

NA

GOVERNMENT

Country name:

conventional long form: none

conventional short form: Isle of Man

abbreviation: I.O.M.

Dependency status: British crown dependency

Government type: parliamentary democracy

Capital: *name:* Douglas

geographic coordinates: 54 09 N, 4 29 W

time difference: UTC 0 (five hours ahead of Washington, DC during Standard Time)

daylight saving time: +1hr, begins last Sunday in March; ends last Sunday in October

Administrative divisions: none; there are no first-order administrative divisions as defined by the US Government, but there are 24 local authorities each with its own elections

Independence: none (British crown dependency)

National holiday: Tynwald Day, 5 July

Constitution: unwritten; note—The Isle of Man Constitution Act of 1961 does not embody the unwritten Manx Constitution

Legal system: the laws of the UK, where applicable, apply and Manx statutes

Suffrage: 16 years of age; universal

Executive branch:

chief of state: Lord of Mann Queen ELIZABETH II (since 6 February 1952); represented by Lieutenant Governor Sir Paul K. HADDACKS (since 17 October 2005)

head of government: Chief Minister Tony BROWN (since 14 December 2006)

cabinet: Council of Ministers

elections: the monarch is hereditary; lieutenant governor appointed by the monarch for a five-year term; the chief minister is elected by the Tynwald; election last held 14 December 2006 (next to be held in December 2008)

election results: House of Keys speaker Tony BROWN elected chief minister by the Tynwald

Legislative branch: bicameral Tynwald consists of the Legislative Council (11 seats; members composed of the President of Tynwald, the Lord Bishop of Sodor and Man, a nonvoting attorney general, and 8 others named by the House of Keys) and the House of Keys (24 seats; members are elected by popular vote to serve five-year terms)

elections: House of Keys—last held 23 November 2006 (next to be held in November 2011)

election results: House of Keys—percent of vote by party—NA; seats by party—Liberal Vannin Party 2, Man Labor Party 1, independents 21

Judicial branch: High Court of Justice (justices are appointed by the Lord Chancellor of England on the nomination of the lieutenant governor)

Political parties and leaders: Alliance for Progressive Government; Liberal Vannin Party [Peter KARRAN]; Man Labor Party; Man Nationalist Party (Mec Vannin) [Bernard MOFFATT]

note: most members sit as independents

Political pressure groups and leaders: none

International organization participation: UPU

Diplomatic representation in the US: none (British crown dependency)

Diplomatic representation from the US: none (British crown dependency)

Flag description: red with the Three Legs of Man emblem (Trinacria), in the center; the three legs are joined at the thigh and bent at the knee; in order to have the toes pointing clockwise on both sides of the flag, a two-sided emblem is used

ECONOMY

Economy—overview: Offshore banking, manufacturing, and tourism are key sectors of the economy. The government offers incentives to high-technology companies and financial institutions to locate on the island; this has paid off in expanding employment opportunities in high-income industries. As a result, agriculture and fishing, once the mainstays of the economy, have declined in their shares of GDP. The Isle of Man also attracts online gambling sites and the film industry. Trade is mostly with the UK. The Isle of Man enjoys free access to EU markets.

GDP (purchasing power parity): $2.719 billion (2005 est.)

GDP (official exchange rate): $2.719 billion (2005 est.)
GDP—real growth rate: 5.2% (2005)
GDP—per capita (PPP): $35,000 (2005 est.)
GDP—composition by sector:
agriculture: 1%
industry: 13%
services: 86% (2000 est.)
Labor force: 39,690 (2001)
Labor force—by occupation: agriculture, forestry, and fishing 3%, manufacturing 11%, construction 10%, transport and communication 8%, wholesale and retail distribution 11%, professional and scientific services 18%, public administration 6%, banking and finance 18%, tourism 2%, entertainment and catering 3%, miscellaneous services 10% (2001)
Unemployment rate: 1.5% (December 2006 est.)
Population below poverty line: NA%
Household income or consumption by percentage share:
lowest 10%: NA%
highest 10%: NA%
Inflation rate (consumer prices): 3.1% (December 2006 est.)
Budget:
revenues: $965 million
expenditures: $943 million (FY05/06 est.)
Agriculture—products: cereals, vegetables; cattle, sheep, pigs, poultry
Industries: financial services, light manufacturing, tourism
Industrial production growth rate: 3.2% (FY96/97)
Exports: $NA
Exports—commodities: tweeds, herring, processed shellfish, beef, lamb

Exports—partners: UK (2006)
Imports: $NA
Imports—commodities: timber, fertilizers, fish
Imports—partners: UK (2006)
Economic aid—recipient: $NA
Debt—external: $NA
Market value of publicly traded shares: $NA
Currency (code): Isle of Man pound (IMP), also known as the Manx pound
note: the British pound is also legal tender, but change is given in IMP
Currency code: GBP
Exchange rates: Manx pounds per US dollar—0.4993 (2007), 0.5418 (2006), 0.5493 (2005), 0.5462 (2004), 0.6125 (2003)
note: the Manx pound is at par with the British pound
Fiscal year: 1 April—31 March

COMMUNICATIONS

Telephones—main lines in use: 51,000 (1999)
Telephone system:
general assessment: NA
domestic: landline, telefax, mobile cellular telephone system
international: fiber-optic cable, microwave radio relay, satellite earth station, submarine cable
Radio broadcast stations: AM 1, FM 1, shortwave 0 (1998)
Radios: NA
Television broadcast stations: 0 (receives broadcasts from the UK and satellite) (1999)
Televisions: 27,490 (1999)
Internet country code: .im

Internet hosts: 159 (2007)
Internet Service Providers (ISPs): NA
Internet users: NA

TRANSPORTATION

Airports: 1 (2007)
Airports—with paved runways:
total: 1
1,524 to 2,437 m: 1 (2007)
Railways:
total: 65 km
standard gauge: 7 km 1.067-m gauge (7 km electrified)
narrow gauge: 58 km 0.914-m gauge (29 km electrified)
note: primarily summer tourist attractions (2006)
Roadways:
total: 800 km
Merchant marine:
total: 297 ships (1000 GRT or over) 8,377,775 GRT/13,890,881 DWT
by type: bulk carrier 33, cargo 65, chemical tanker 54, combination ore/oil 1, container 17, liquefied gas 34, passenger/cargo 1, petroleum tanker 74, refrigerated cargo 5, roll on/roll off 8, vehicle carrier 5
foreign-owned: 210 (Cyprus 4, Denmark 41, France 2, Germany 61, Greece 48, Ireland 1, Italy 1, Japan 4, Monaco 3, Netherlands 1, Norway 33, Singapore 2, Sweden 3, Turkey 2, US 4) (2007)
Ports and terminals: Douglas, Ramsey

MILITARY

Military—note: defense is the responsibility of the UK

TRANSNATIONAL ISSUES

Disputes—international: none

ISRAEL

INTRODUCTION

Background: Following World War II, the British withdrew from their mandate of Palestine, and the UN partitioned the area into Arab and Jewish states, an arrangement rejected by the Arabs. Subsequently, the Israelis defeated the Arabs in a series of wars without ending the deep tensions between the two sides. The territories Israel occupied since the 1967 war are not included in the Israel country profile, unless otherwise noted. On 25 April 1982, Israel withdrew from the Sinai pursuant to the 1979 Israel-Egypt Peace Treaty. In keeping with the framework established at the Madrid Conference in October 1991, bilateral negotiations were conducted between Israel and Palestinian representatives and Syria to achieve a permanent settlement. Israel and Palestinian officials signed on 13 September 1993 a Declaration of Principles (also known as the "Oslo Accords") guiding an interim period of Palestinian self-rule. Outstanding territorial and other disputes with Jordan were resolved in the 26 October 1994 Israel-Jordan Treaty of Peace. In addition, on 25 May 2000, Israel withdrew unilaterally from southern Lebanon, which it had occupied since 1982. In April 2003, US President BUSH, working in conjunction with the EU, UN, and Russia—the "Quartet"—took the lead in laying out a roadmap to a final settlement of the conflict by 2005, based on reciprocal steps by the two parties leading to two states, Israel and a democratic Palestine. However, progress toward a permanent status agreement was undermined by Israeli-Palestinian violence between September 2003 and February 2005. An Israeli-Palestinian agreement reached at Sharm al-Sheikh in February 2005, along with an internally-brokered Palestinian ceasefire, significantly reduced the violence. In the summer of 2005, Israel unilaterally disengaged from the Gaza Strip, evacuating settlers and its military while retaining control over most points of entry into the Gaza Strip. The election of HAMAS in January 2006 to head the Palestinian Legislative Council froze relations between Israel and the Palestinian Authority (PA). Ehud

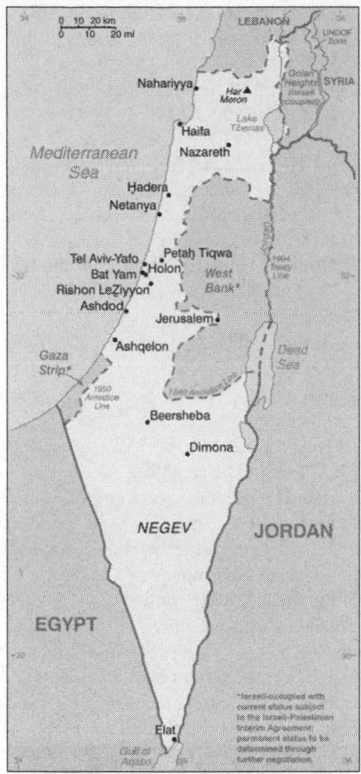

OLMERT became prime minister in March 2006; following an Israeli military operation in Gaza in June-July 2006 and a 34-day conflict with Hizballah in Lebanon in June-August 2006, he shelved plans to unilaterally evacuate from most of the West Bank. OLMERT in June 2007 resumed talks with the PA after HAMAS seized control of the Gaza Strip and PA President Mahmoud ABBAS formed a new government without HAMAS.

GEOGRAPHY

Location: Middle East, bordering the Mediterranean Sea, between Egypt and Lebanon

Geographic coordinates: 31 30 N, 34 45 E

Map references: Middle East

Area: *total:* 20,770 sq km
land: 20,330 sq km
water: 440 sq km

Area—comparative: slightly smaller than New Jersey

Land boundaries:
total: 1,017 km
border countries: Egypt 266 km, Gaza Strip 51 km, Jordan 238 km, Lebanon 79 km, Syria 76 km, West Bank 307 km

Coastline: 273 km

Maritime claims: *territorial sea:* 12 nm
continental shelf: to depth of exploitation

Climate: temperate; hot and dry in southern and eastern desert areas

Terrain: Negev desert in the south; low coastal plain; central mountains; Jordan Rift Valley

Elevation extremes:
lowest point: Dead Sea -408 m
highest point: Har Meron 1,208 m

Natural resources: timber, potash, copper ore, natural gas, phosphate rock, magnesium bromide, clays, sand

Land use:
arable land: 15.45%
permanent crops: 3.88%
other: 80.67% (2005)

Irrigated land: 1,940 sq km (2003)

Total renewable water resources: 1.7 cu km (2001)

Freshwater withdrawal (domestic/industrial/agricultural): *total:* 2.05 cu km/yr (31%/7%/62%)
per capita: 305 cu m/yr (2000)

Natural hazards: sandstorms may occur during spring and summer; droughts; periodic earthquakes

Environment—current issues: limited arable land and natural fresh water resources pose serious constraints; desertification; air pollution from industrial and vehicle emissions; groundwater pollution from industrial and domestic waste, chemical fertilizers, and pesticides

Environment—international agreements: *party to:* Biodiversity, Climate Change, Climate Change-Kyoto Protocol, Desertification, Endangered Species, Hazardous Wastes, Ozone Layer Protection, Ship Pollution, Wetlands, Whaling
signed, but not ratified: Marine Life Conservation

Geography—note: there are 242 Israeli settlements and civilian land use sites in the West Bank, 42 in the Israeli-occupied Golan Heights, 0 in the Gaza Strip, and 29 in East Jerusalem (August 2005 est.); Lake Tiberias (Sea of Galilee) is an important freshwater source

PEOPLE

Population: 7,112,359
note: includes about 187,000 Israeli settlers in the West Bank, about 20,000 in the Israeli-occupied Golan Heights, and fewer than 177,000 in East Jerusalem (July 2008 est.)

Age structure:
0–14 years: 28% (male 1,018,229/female 971,083)
15–64 years: 62.2% (male 2,242,928/female 2,183,688)
65 years and over: 9.8% (male 303,289/female 393,142) (2008 est.)

Median age:
total: 28.9 years

male: 28.2 years
female: 29.7 years (2008 est.)

Population growth rate: 1.713% (2008 est.)

Birth rate: 20.02 births/1,000 population (2008 est.)

Death rate: 5.41 deaths/1,000 population (2008 est.)

Net migration rate: 2.52 migrant(s)/1,000 population (2008 est.)

Sex ratio:
at birth: 1.05 male(s)/female
under 15 years: 1.05 male(s)/female
15–64 years: 1.03 male(s)/female
65 years and over: 0.77 male(s)/female
total population: 1 male(s)/female (2008 est.)

Infant mortality rate:
total: 4.28 deaths/1,000 live births
male: 4.43 deaths/1,000 live births
female: 4.12 deaths/1,000 live births (2008 est.)

Life expectancy at birth:
total population: 80.61 years
male: 78.54 years
female: 82.79 years (2008 est.)

Total fertility rate: 2.77 children born/woman (2008 est.)

HIV/AIDS—adult prevalence rate: 0.1% (2001 est.)

HIV/AIDS—people living with HIV/AIDS: 3,000 (1999 est.)

HIV/AIDS—deaths: 100 (2001 est.)

Nationality:
noun: Israeli(s)
adjective: Israeli

Ethnic groups: Jewish 76.4% (of which Israel-born 67.1%, Europe/America-born 22.6%, Africa-born 5.9%, Asia-born 4.2%), non-Jewish 23.6% (mostly Arab) (2004)

Religions: Jewish 76.4%, Muslim 16%, Arab Christians 1.7%, other Christian 0.4%, Druze 1.6%, unspecified 3.9% (2004)

Languages: Hebrew (official), Arabic used officially for Arab minority, English most commonly used foreign language

Literacy:
definition: age 15 and over can read and write
total population: 97.1%
male: 98.5%
female: 95.9% (2004 est.)

GOVERNMENT

Country name:
conventional long form: State of Israel
conventional short form: Israel
local long form: Medinat Yisra'el
local short form: Yisra'el

Government type: parliamentary democracy

Capital: *name:* Jerusalem
geographic coordinates: 31 46 N, 35 14 E

time difference: UTC+2 (7 hours ahead of Washington, DC during Standard Time)
daylight saving time: +1hr, begins last Friday in March; ends the Sunday between the holidays of Rosh Hashana and Yom Kippur
note: Israel proclaimed Jerusalem as its capital in 1950, but the US, like nearly all other countries, maintains its Embassy in Tel Aviv

Administrative divisions: 6 districts (mehozot, singular—mehoz); Central, Haifa, Jerusalem, Northern, Southern, Tel Aviv

Independence: 14 May 1948 (from League of Nations mandate under British administration)

National holiday: Independence Day, 14 May (1948); note—Israel declared independence on 14 May 1948, but the Jewish calendar is lunar and the holiday may occur in April or May

Constitution: no formal constitution; some of the functions of a constitution are filled by the Declaration of Establishment (1948), the Basic Laws of the parliament (Knesset), and the Israeli citizenship law; note—since May 2003 the Constitution, Law, and Justice Committee of the Knesset has been working on a draft constitution

Legal system: mixture of English common law, British Mandate regulations, and, in personal matters, Jewish, Christian, and Muslim legal systems; in December 1985, Israel informed the UN Secretariat that it would no longer accept compulsory ICJ jurisdiction

Suffrage: 18 years of age; universal

Executive branch:
chief of state: President Shimon PERES (since 15 July 2007)
head of government: Prime Minister Ehud OLMERT (since May 2006); Deputy Prime Minister Tzipora "Tzipi" LIVNI (since May 2006); Ehud OLMERT won the right to lead the government when his Kadima Party won 29 seats in elections held on 28 March 2006
cabinet: Cabinet selected by prime minister and approved by the Knesset
elections: president is largely a ceremonial role and is elected by the Knesset for a seven-year term (one-term limit); election last held 13 June 2007 (next to be held in 2014 but can be called earlier); following legislative elections, the president assigns a Knesset member—traditionally the leader of the largest party—the task of forming a governing coalition
note: government coalition—Kadima, Labor Party, GIL (Pensioners), and SHAS
election results: Shimon PERES elected president; number of votes in first round—Shimon PERES 58, Reuven RIVLIN 37, Colette AVITAL 21; PERES elected president in second round with 86 votes (unopposed)

Legislative branch: unicameral Knesset (120 seats; members elected by popular vote to serve four-year terms)
elections: last held 28 March 2006 (next scheduled to be held in 2010 but can be called earlier)
election results: percent of vote by party—Kadima 22%, Labor 15.1%, SHAS 9.5%, Likud 9%, Yisrael Beiteinu 9%, NU/NRP 7.1%, GIL 5.9%, Torah and Shabbat Judaism 4.7%, Meretz-YAHAD 3.8%, United Arab List 3%, Balad 2.3%, HADASH 2.7%, other 5.9%; seats by party—Kadima 29, Labor 19, Likud 12, SHAS 12, Yisrael Beiteinu 11, NU/NRP 9, GIL 7, Torah and Shabbat Judaism 6, Meretz-YAHAD 5, United Arab List 4, Balad 3, HADASH 3

Judicial branch: Supreme Court (justices appointed by Judicial Selection Committee—made up of all three branches of the government; mandatory retirement age is 70)

Political parties and leaders: Democratic Front for Peace and Equality (HADASH) [Muhammad BARAKA]; GIL (Pensioners) [Rafael "Rafi" EITAN]; Kadima [Ehud OLMERT]; Labor Party [Ehud BARAK]; Likud [Binyamin NETANYAHU]; Meretz-YAHAD [Yossi BEILIN]; National Democratic Assembly (Balad) [Jamal ZAHALKA]; National Union (NU)/National Religious Party (NRP) [Binyamin ELON]; SHAS [Eliyahu YISHAI]; Torah and Shabbat Judaism [Yaakov LITZMAN]; United Arab List [Ibrahim SARSOUR]; Yisrael Beiteinu [Avigdor LIEBERMAN]

Political pressure groups and leaders: Peace Now [Yariv OPPENHEIMER, Secretary General] supports territorial concessions in the West Bank and Gaza Strip; YESHA Council of Settlements [Danny DAYAN, Chairman] promotes settler interests and opposes territorial compromise; B'Tselem [Jessica MONTELL, Executive Director] monitors human rights abuses

International organization participation: BIS, BSEC (observer), CERN (observer), EBRD, FAO, IADB, IAEA, IBRD, ICAO, ICC, ICCt (signatory), ICRM, IDA, IFAD, IFC, IFRCS, ILO, IMF, IMO, IMSO, Interpol, IOC, IOM, IPU, ISO, ITSO, ITU, ITUC, MIGA, OAS (observer), OPCW (signatory), OSCE (partner), PCA, SECI (observer), UN, UNCTAD, UNESCO, UNHCR, UNIDO, UNWTO, UPU, WCO, WHO, WIPO, WMO, WTO

Diplomatic representation in the US:
chief of mission: Ambassador Salai MERIDOR
chancery: 3514 International Drive NW, Washington, DC 20008
telephone: [1] (202) 364-5500
FAX: [1] (202) 364-5607
consulate(s) general: Atlanta, Boston, Chicago, Houston, Los Angeles, Miami, New York, Philadelphia, San Francisco

Diplomatic representation from the US:
chief of mission: Ambassador Richard H. JONES
embassy: 71 Hayarkon Street, Tel Aviv 63903
mailing address: PSC 98, Box 29, APO AE 09830
telephone: [972] (3) 519-7575
FAX: [972] (3) 516-4390
consulate(s) general: Jerusalem; note—an independent US mission, established in 1928, whose members are not accredited to a foreign government

Flag description: white with a blue hexagram (six-pointed linear star) known as the Magen David (Shield of David) centered between two equal horizontal blue bands near the top and bottom edges of the flag

ECONOMY

Economy—overview: Israel has a technologically advanced market economy with substantial, though diminishing, government participation. It depends on imports of crude oil, grains, raw materials, and military equipment. Despite limited natural resources, Israel has intensively developed its agricultural and industrial sectors over the past 20 years. Israel imports substantial quantities of grain but is largely self-sufficient in other agricultural products. Cut diamonds, high-technology equipment, and agricultural products (fruits and vegetables) are the leading exports. Israel usually posts sizable trade deficits, which are covered by large transfer payments from abroad and by foreign loans. Roughly half of the government's external debt is owed to the US, its major source of economic and military aid. Israel's GDP, after contracting slightly in 2001 and 2002 due to the Palestinian conflict and troubles in the high-technology sector, has grown by about 5% per year since 2003. The economy grew an estimated 5.4% in 2007, the fastest pace since 2000. The government's prudent fiscal policy and structural reforms over the past few years have helped to induce strong foreign investment, tax revenues, and private consumption, setting the economy on a solid growth path.

GDP (purchasing power parity): $185.9 billion (2007 est.)

GDP (official exchange rate): $161.9 billion (2007 est.)
GDP—real growth rate: 5.3% (2007 est.)
GDP—per capita (PPP): $25,800 (2007 est.)
GDP—composition by sector:
agriculture: 2.7%
industry: 30.2%
services: 67.1% (2007 est.)
Labor force: 2.894 million (2007 est.)
Labor force—by occupation: agriculture 18.5%, industry 23.7%, services 50%, other 7.8% (2002)
Unemployment rate: 7.3% (2007 est.)
Population below poverty line: 21.6%
note: Israel's poverty line is $7.30 per person per day (2005)
Household income or consumption by percentage share:
lowest 10%: 2.4%
highest 10%: 28.3% (2005)
Distribution of family income—Gini index: 38.6 (2005)
Inflation rate (consumer prices): 0.5% (2007 est.)
Investment (gross fixed): 18.5% of GDP (2007 est.)
Budget:
revenues: $53.6 billion
expenditures: $53.63 billion (2007 est.)
Public debt: 80.6% of GDP (2007 est.)
Agriculture—products: citrus, vegetables, cotton; beef, poultry, dairy products
Industries: high-technology projects (including aviation, communications, computer-aided design and manufactures, medical electronics, fiber optics), wood and paper products, potash and phosphates, food, beverages, and tobacco, caustic soda, cement, construction, metals products, chemical products, plastics, diamond cutting, textiles, footwear
Industrial production growth rate: 4.1% (2007 est.)
Electricity—production: 46.85 billion kWh (2005)
Electricity—production by source:
fossil fuel: 99.9%
hydro: 0.1%
nuclear: 0%
other: 0% (2001)
Electricity—consumption: 43.28 billion kWh (2005)
Electricity—exports: 1.663 billion kWh (2005)
Electricity—imports: 0 kWh (2005)
Oil—production: 100 bbl/day (2006 est.)
Oil—consumption: 249,500 bbl/day (2006 est.)
Oil—exports: 75,980 bbl/day (2004)
Oil—imports: 315,200 bbl/day (2004)
Oil—proved reserves: 2 million bbl (1 January 2006 est.)

Natural gas—production: 709.7 million cu m (2005 est.)
Natural gas—consumption: 709.7 million cu m (2005 est.)
Natural gas—exports: 0 cu m (2005 est.)
Natural gas—imports: 0 cu m (2005)
Natural gas—proved reserves: 37.34 billion cu m (1 January 2006 est.)
Current account balance: $4.993 billion (2007 est.)
Exports: $50.24 billion f.o.b. (2007 est.)
Exports—commodities: machinery and equipment, software, cut diamonds, agricultural products, chemicals, textiles and apparel
Exports—partners: US 38.4%, Belgium 6.5%, Hong Kong 5.9% (2006)
Imports: $55.76 billion f.o.b. (2007 est.)
Imports—commodities: raw materials, military equipment, investment goods, rough diamonds, fuels, grain, consumer goods
Imports—partners: US 12.4%, Belgium 8.2%, Germany 6.7%, Switzerland 5.9%, UK 5.1%, China 5.1% (2006)
Economic aid—recipient: $240 million from US (FY06)
Reserves of foreign exchange and gold: $28.52 billion (31 December 2007 est.)
Debt—external: $89.95 billion (31 December 2007)
Stock of direct foreign investment—at home: $57.43 billion (2007 est.)
Stock of direct foreign investment—abroad: $41.98 billion (2007 est.)
Market value of publicly traded shares: $173.3 billion (2006)
Currency (code): new Israeli shekel (ILS); note—NIS is the currency abbreviation; ILS is the International Organization for Standardization (ISO) code for the NIS
Currency code: ILS
Exchange rates: new Israeli shekels per US dollar—4.14 (2007), 4.4565 (2006), 4.4877 (2005), 4.482 (2004), 4.5541 (2003)
Fiscal year: calendar year

COMMUNICATIONS

Telephones—main lines in use: 3.005 million (2006)
Telephones—mobile cellular: 8.404 million (2006)
Telephone system:
general assessment: most highly developed system in the Middle East although not the largest
domestic: good system of coaxial cable and microwave radio relay; all systems are digital; four privately-owned mobile-cellular service providers with countrywide coverage; mobile-cellular teledensity is more than 130 per 100 persons

international: country code—972; submarine cables provide links to Europe, Cyprus, and parts of the Middle East; satellite earth stations—3 Intelsat (2 Atlantic Ocean and 1 Indian Ocean)
Radio broadcast stations: AM 23, FM 15, shortwave 2 (1998)
Radios: 3.07 million (1997)
Television broadcast stations: 17 (plus 36 repeaters) (1995)
Televisions: 1.69 million (1997)
Internet country code: .il
Internet hosts: 671,030 (2007)
Internet Service Providers (ISPs): 21 (2000)
Internet users: 1.899 million (2006)

TRANSPORTATION

Airports: 53 (2007)
Airports—with paved runways:
total: 30
over 3,047 m: 2
2,438 to 3,047 m: 5
1,524 to 2,437 m: 7
914 to 1,523 m: 10
under 914 m: 6 (2007)
Airports—with unpaved runways:
total: 23
1,524 to 2,437 m: 1
914 to 1,523 m: 2
under 914 m: 20 (2007)
Heliports: 3 (2007)
Pipelines: gas 160 km; oil 442 km; refined products 261 km (2007)
Railways:
total: 853 km
standard gauge: 853 km 1.435-m gauge (2006)
Roadways:
total: 17,686 km
paved: 17,686 km (includes 146 km of expressways) (2006)
Merchant marine:
total: 18 ships (1000 GRT or over) 716,382 GRT/845,053 DWT
by type: cargo 2, container 16
registered in other countries: 51 (Bermuda 3, Cyprus 4, Honduras 1, North Korea 1, Liberia 9, Malta 21, Panama 2, Slovakia 6, St Vincent and The Grenadines 4) (2007)
Ports and terminals: Ashdod, Elat (Eilat), Hadera, Haifa

MILITARY

Military branches: Israel Defense Forces (IDF), Israel Naval Forces (INF), Israel Air Force (IAF) (2007)
Military service age and obligation: 18 years of age for compulsory (Jews, Druzes) and voluntary (Christians, Muslims, Circassians) military service; both sexes are obligated to military service; conscript service obligation—36 months for enlisted men, 21 months for

enlisted women, 48 months for officers; reserve obligation to age 41–51 (men), 24 (women) (2008)

Manpower available for military service:
males age 16–49: 1,717,362
females age 16–49: 1,636,574 (2008 est.)

Manpower fit for military service:
males age 16–49: 1,452,926
females age 16–49: 1,383,796 (2008 est.)

Manpower reaching militarily significant age annually:
males age 16–49: 60,602
females age 16–49: 57,532 (2008 est.)

Military expenditures—percent of GDP: 7.3% (2006)

TRANSNATIONAL ISSUES

Disputes—international: West Bank and Gaza Strip are Israeli-occupied with current status subject to the Israeli-Palestinian Interim Agreement—permanent status to be determined through further negotiation; Israel continues construction of a "seam line" separation barrier along parts of the Green Line and within the West Bank; Israel withdrew its settlers and military from the Gaza Strip and from four settlements in the West Bank in August 2005; Golan Heights is Israeli-occupied (Lebanon claims the Shab'a Farms area of Golan Heights); since 1948, about 350 peace-keepers from the UN Truce Supervision Organization (UNTSO) headquartered in Jerusalem monitor ceasefires, supervise armistice agreements, prevent isolated incidents from escalating, and assist other UN personnel in the region

Refugees and internally displaced persons: *IDPs:* 150,000–420,000 (Arab villagers displaced from homes in northern Israel) (2007)

Illicit drugs: increasingly concerned about ecstasy, cocaine, and heroin abuse; drugs arrive in country from Lebanon and, increasingly, from Jordan; money-laundering center

ITALY

INTRODUCTION

Background: Italy became a nation-state in 1861 when the regional states of the peninsula, along with Sardinia and Sicily, were united under King Victor EMMANUEL II. An era of parliamentary government came to a close in the early 1920s when Benito MUSSOLINI established a Fascist dictatorship. His alliance with Nazi Germany led to Italy's defeat in World War II. A democratic republic replaced the monarchy in 1946 and economic revival followed. Italy was a charter member of NATO and the European Economic Community (EEC). It has been at the forefront of European economic and political unification, joining the Economic and Monetary Union in 1999. Persistent problems include illegal immigration, organized crime, corruption, high unemployment, sluggish economic growth, and the low incomes and technical standards of southern Italy compared with the prosperous north.

GEOGRAPHY

Location: Southern Europe, a peninsula extending into the central Mediterranean Sea, northeast of Tunisia

Geographic coordinates: 42 50 N, 12 50 E

Map references: Europe

Area:
total: 301,230 sq km
land: 294,020 sq km
water: 7,210 sq km
note: includes Sardinia and Sicily

Area—comparative: slightly larger than Arizona

Land boundaries:
total: 1,932.2 km
border countries: Austria 430 km, France 488 km, Holy See (Vatican City) 3.2 km, San Marino 39 km, Slovenia 232 km, Switzerland 740 km

Coastline: 7,600 km

Maritime claims:
territorial sea: 12 nm
continental shelf: 200 m depth or to the depth of exploitation

Climate: predominantly Mediterranean; Alpine in far north; hot, dry in south

Terrain: mostly rugged and mountainous; some plains, coastal lowlands

Elevation extremes:
lowest point: Mediterranean Sea 0 m
highest point: Mont Blanc (Monte Bianco) de Courmayeur 4,748 m (a secondary peak of Mont Blanc)

Natural resources: coal, mercury, zinc, potash, marble, barite, asbestos, pumice, fluorspar, feldspar, pyrite (sulfur), natural gas and crude oil reserves, fish, arable land

Land use:
arable land: 26.41%
permanent crops: 9.09%
other: 64.5% (2005)

Irrigated land: 27,500 sq km (2003)

Total renewable water resources: 175 cu km (2005)

Freshwater withdrawal (domestic/industrial/agricultural): *total:* 41.98 cu km/yr (18%/37%/45%)
per capita: 723 cu m/yr (1998)

Natural hazards: regional risks include landslides, mudflows, avalanches, earthquakes, volcanic eruptions, flooding; land subsidence in Venice

Environment—current issues: air pollution from industrial emissions such as sulfur dioxide; coastal and inland rivers polluted from industrial and agricultural effluents; acid rain damaging lakes; inadequate industrial waste treatment and disposal facilities

Environment—international agreements: *party to:* Air Pollution, Air Pollution-Nitrogen Oxides, Air Pollution-Persistent Organic Pollutants, Air Pollution-Sulfur 85, Air Pollution-Sulfur 94, Air Pollution-Volatile Organic Compounds, Antarctic-Environmental Protocol, Antarctic-Marine Living Resources, Antarctic Seals, Antarctic Treaty, Biodiversity, Climate Change, Climate Change-Kyoto Protocol, Desertification, Endangered Species, Environmental Modification, Hazardous Wastes, Law of the Sea, Marine Dumping, Ozone Layer Protection, Ship Pollution, Tropical Timber 83, Tropical Timber 94, Wetlands, Whaling
signed, but not ratified: none of the selected agreements

Geography—note: strategic location dominating central Mediterranean as well as southern sea and air approaches to Western Europe

PEOPLE

Population: 58,145,321 (July 2008 est.)
Age structure: *0–14 years:* 13.6% (male 4,086,951/female 3,842,765)

15–64 years: 66.3% (male 19,534,247/female 19,024,776)

65 years and over: 20% (male 4,864,189/female 6,792,393) (2008 est.)

Median age:

total: 42.9 years

male: 41.4 years

female: 44.4 years (2008 est.)

Population growth rate: -0.019% (2008 est.)

Birth rate: 8.36 births/1,000 population (2008 est.)

Death rate: 10.61 deaths/1,000 population (2008 est.)

Net migration rate: 2.06 migrant(s)/1,000 population (2008 est.)

Sex ratio:

at birth: 1.07 male(s)/female

under 15 years: 1.06 male(s)/female

15–64 years: 1.03 male(s)/female

65 years and over: 0.72 male(s)/female

total population: 0.96 male(s)/female (2008 est.)

Infant mortality rate:

total: 5.61 deaths/1,000 live births

male: 6.19 deaths/1,000 live births

female: 5 deaths/1,000 live births (2008 est.)

Life expectancy at birth:

total population: 80.07 years

male: 77.13 years

female: 83.2 years (2008 est.)

Total fertility rate: 1.3 children born/woman (2008 est.)

HIV/AIDS—adult prevalence rate: 0.5% (2001 est.)

HIV/AIDS—people living with HIV/AIDS: 140,000 (2001 est.)

HIV/AIDS—deaths: fewer than 1,000 (2003 est.)

Nationality:

noun: Italian(s)

adjective: Italian

Ethnic groups: Italian (includes small clusters of German-, French-, and Slovene-Italians in the north and Albanian-Italians and Greek-Italians in the south)

Religions: Roman Catholic 90% (approximately; about one-third practicing), other 10% (includes mature Protestant and Jewish communities and a growing Muslim immigrant community)

Languages: Italian (official), German (parts of Trentino-Alto Adige region are predominantly German speaking), French (small French-speaking minority in Valle d'Aosta region), Slovene (Slovene-speaking minority in the Trieste-Gorizia area)

Literacy: definition: age 15 and over can read and write

total population: 98.4%

male: 98.8%

female: 98% (2001 census)

GOVERNMENT

Country name:

conventional long form: Italian Republic

conventional short form: Italy

local long form: Repubblica Italiana

local short form: Italia

former: Kingdom of Italy

Government type: republic

Capital: name: Rome

geographic coordinates: 41 54 N, 12 29 E

time difference: UTC+1 (6 hours ahead of Washington, DC during Standard Time)

daylight saving time: +1hr, begins last Sunday in March; ends last Sunday in October

Administrative divisions: 15 regions (regioni, singular—regione) and 5 autonomous regions* (regioni autonome, singular—regione autonoma); Abruzzo, Basilicata, Calabria, Campania, Emilia-Romagna, Friuli-Venezia Giulia*, Lazio (Latium), Liguria, Lombardia, Marche, Molise, Piemonte (Piedmont), Puglia (Apulia), Sardegna* (Sardinia), Sicilia*, Toscana (Tuscany), Trentino-Alto Adige* (Trentino-South Tyrol), Umbria, Valle d'Aosta* (Aosta Valley), Veneto

Independence: 17 March 1861 (Kingdom of Italy proclaimed; Italy was not finally unified until 1870)

National holiday: Republic Day, 2 June (1946)

Constitution: passed 11 December 1947, effective 1 January 1948; amended many times

Legal system: based on civil law system; appeals treated as new trials; judicial review under certain conditions in Constitutional Court; has not accepted compulsory ICJ jurisdiction

Suffrage: 18 years of age; universal (except in senatorial elections, where minimum age is 25)

Executive branch:

chief of state: President Giorgio NAPOLITANO (since 15 May 2006)

head of government: Prime Minister Silvio BERLUSCONI (referred to in Italy as the president of the Council of Ministers) (since 8 May 2008) note—in Italy the prime minister is referred to as the president of the Council of Ministers

cabinet: Council of Ministers nominated by the prime minister and approved by the president

elections: president elected by an electoral college consisting of both houses of parliament and 58 regional representatives for a seven-year term (no term limits); election last held 10 May 2006 (next to be held in May 2013); prime minister appointed by the president and confirmed by parliament

election results: Giorgio NAPOLITANO elected president on the fourth round of voting; electoral college vote—543

Legislative branch: bicameral Parliament or Parlamento consists of the Senate or Senato della Repubblica (315 seats; members elected by proportional vote with the winning coalition in each region receiving 55% of seats from that region; to serve five-year terms) and the Chamber of Deputies or Camera dei Deputati (630 seats; members elected by popular vote with the winning national coalition receiving 54% of chamber seats; to serve five-year terms)

elections: Senate—last held 13–14 April 2008 (next to be held April 2010); Chamber of Deputies—last held 13–14 April 2008 (next to be held in April 2010)

election results: Senate—percent of vote by party—NA; seats by party—S. BERLUSCONI coalition 174 (PdL 147, LN 25, MpA 2), W. VELTRONI coalition 132 (PD 118, IdV 3), UdC 3, other 6; Chamber of Deputies—percent of vote by party—NA; seats by party—S. BERLUSCONI coalition 344 (PdL 276, LN 60, MpA 8), W. VELTRONI coalition 246 (PD 217, IdV 29), UdC 36, other 4

Judicial branch: Constitutional Court or Corte Costituzionale (composed of 15 judges: one-third appointed by the president, one-third elected by parliament, one-third elected by the ordinary and administrative Supreme Courts)

Political parties and leaders: Silvio BERLUSCONI coalition: People of Freedom or PdL [Silvio BERLUSCONI]; Lega Nord or LN [Umberto BOSSI]; Movement for Autonomy or MpA [Raffaele LOMBARDO]

Walter VELTRONI coalition: Democratic Party or PD [Walter VELTRONI]; Italy of Values or IdV [Antonio DI PIETRO]

other non-allied parties: Union of the Centre or UdC [Savino PEZZOTTA]

Political pressure groups and leaders: Italian manufacturers and merchants associations (Confindustria, Confcommercio); organized farm groups (Confcoltivatori, Confagricoltura); Roman Catholic Church; three major trade union confederations (Confederazione Generale Italiana del Lavoro or CGIL [Guglielmo EPIFANI] which is left wing, Confederazione Italiana dei Sindacati Lavoratori or CISL [Raffaele BONANNO], which is Roman Catholic centrist, and Unione Italiana del Lavoro or UIL [Luigi ANGELETTI] which is lay centrist)

International organization participation: ADB (nonregional members), AfDB, Arctic Council (observer),

Australia Group, BIS, BSEC (observer), CBSS (observer), CDB, CE, CEI, CERN, EAPC, EBRD, EIB, EMU, ESA, EU, FAO, G-7, G-8, G-10, IADB, IAEA, IBRD, ICAO, ICC, ICCt, ICRM, IDA, IEA, IFAD, IFC, IFRCS, IHO, ILO, IMF, IMO, IMSO, Interpol, IOC, IOM, IPU, ISO, ITSO, ITU, ITUC, LAIA (observer), MIGA, MINURSO, NAM (guest), NATO, NEA, NSG, OAS (observer), OECD, OPCW, OSCE, Paris Club, PCA, Schengen Convention, SECI (observer), UN, UN Security Council (temporary), UNCTAD, UNESCO, UNHCR, UNIDO, UNIFIL, Union Latina, UNMOGIP, UNRWA, UNTSO, UNWTO, UPU, WCL, WCO, WEU, WHO, WIPO, WMO, WTO, ZC

Diplomatic representation in the US:
chief of mission: Ambassador Giovanni CASTELLANETA
chancery: 3000 Whitehaven Street NW, Washington, DC 20008
telephone: [1] (202) 612-4400
FAX: [1] (202) 518-2151
consulate(s) general: Boston, Chicago, Houston, Miami, New York, Los Angeles, Philadelphia, San Francisco
consulate(s): Detroit

Diplomatic representation from the US:
chief of mission: Ambassador Ronald P. SPOGLI
embassy: Via Vittorio Veneto 121, 00187-Rome
mailing address: PSC 59, Box 100, APO AE 09624
telephone: [39] (06) 46741
FAX: [39] (06) 488-2672, 4674-2356
consulate(s) general: Florence, Milan, Naples

Flag description: three equal vertical bands of green (hoist side), white, and red; similar to the flag of Ireland, which is longer and is green (hoist side), white, and orange; also similar to the flag of the Cote d'Ivoire, which has the colors reversed—orange (hoist side), white, and green; inspired by the French flag brought to Italy by Napoleon in 1797

ECONOMY

Economy—overview: Italy has a diversified industrial economy with roughly the same total and per capita output as France and the UK. This capitalistic economy remains divided into a developed industrial north, dominated by private companies, and a less-developed, welfare-dependent, agricultural south, with 20% unemployment. Most raw materials needed by industry and more than 75% of energy requirements are imported. Over the past decade, Italy has pursued a tight fiscal policy in order to meet the requirements of the Economic and Monetary Unions and has benefited from lower interest and inflation rates. The current government has enacted numerous short-term reforms aimed at improving competitiveness and long-term growth. Italy has moved slowly, however, on implementing needed structural reforms, such as lightening the high tax burden and overhauling Italy's rigid labor market and over-generous pension system, because of the current economic slowdown and opposition from labor unions. But the leadership faces a severe economic constraint: Italy's official debt remains above 100% of GDP, and the government has found it difficult to bring the budget deficit down to a level that would allow a rapid decrease in that debt. The economy continues to grow by less than the euro-zone average and growth is expected to decelerate from 1.9% in 2006 and 2007 to under 1.5% in 2008 as the euro-zone and world economies slow.

GDP (purchasing power parity): $1.786 trillion (2007 est.)
GDP (official exchange rate): $2.105 trillion (2007 est.)
GDP—real growth rate: 1.5% (2007 est.)
GDP—per capita (PPP): $30,400 (2007 est.)
GDP—composition by sector:
agriculture: 1.9%
industry: 28.9%
services: 69.2% (2007 est.)
Labor force: 24.71 million (2007 est.)
Labor force—by occupation: *agriculture:* 5%
industry: 32%
services: 63% (2001)
Unemployment rate: 6% (2007 est.)
Population below poverty line: NA%
Household income or consumption by percentage share:
lowest 10%: 2.3%
highest 10%: 26.8% (2000)
Distribution of family income—Gini index: 33 (2005)
Inflation rate (consumer prices): 2% (2007 est.)
Investment (gross fixed): 21.6% of GDP (2007 est.)
Budget:
revenues: $981.8 billion
expenditures: $1.021 trillion (2007 est.)
Public debt: 104% of GDP (2007 est.)
Agriculture—products: fruits, vegetables, grapes, potatoes, sugar beets, soybeans, grain, olives; beef, dairy products; fish
Industries: tourism, machinery, iron and steel, chemicals, food processing, textiles, motor vehicles, clothing, footwear, ceramics

Industrial production growth rate: 1.5% (2007 est.)
Electricity—production: 278.5 billion kWh (2005)
Electricity—production by source:
fossil fuel: 78.6%
hydro: 18.4%
nuclear: 0%
other: 3% (2001)
Electricity—consumption: 307.1 billion kWh (2005)
Electricity—exports: 1.109 billion kWh (2005)
Electricity—imports: 50.26 billion kWh (2005)
Oil—production: 164,800 bbl/day (2005 est.)
Oil—consumption: 1.732 million bbl/day (2005 est.)
Oil—exports: 521,400 bbl/day (2004)
Oil—imports: 2.182 million bbl/day (2004)
Oil—proved reserves: 621.7 million bbl (1 January 2006 est.)
Natural gas—production: 11.49 billion cu m (2005 est.)
Natural gas—consumption: 82.64 billion cu m (2005 est.)
Natural gas—exports: 379.8 million cu m (2005 est.)
Natural gas—imports: 70.45 billion cu m (2005)
Natural gas—proved reserves: 217.3 billion cu m (1 January 2006 est.)
Current account balance: -$47.25 billion (2007 est.)
Exports: $501.4 billion f.o.b. (2007 est.)
Exports—commodities: engineering products, textiles and clothing, production machinery, motor vehicles, transport equipment, chemicals; food, beverages and tobacco; minerals, and nonferrous metals
Exports—partners: Germany 13.2%, France 11.8%, US 7.4%, Spain 7.4%, UK 6.1% (2006)
Imports: $498.6 billion f.o.b. (2007 est.)
Imports—commodities: engineering products, chemicals, transport equipment, energy products, minerals and nonferrous metals, textiles and clothing; food, beverages, and tobacco
Imports—partners: Germany 16.8%, France 9.3%, Netherlands 5.6%, China 5.1%, Spain 4.3%, Belgium 4.2% (2006)
Economic aid—donor: ODA, $3.641 billion (2006)
Reserves of foreign exchange and gold: $94.33 billion (31 December 2007 est.)
Debt—external: $2.345 trillion (30 June 2007)
Stock of direct foreign investment—at home: $324.3 billion (2007 est.)
Stock of direct foreign investment—abroad: $460.7 billion (2007 est.)

Market value of publicly traded shares: $798.2 billion (2005)
Currency (code): euro (EUR)
Currency code: EUR
Exchange rates: euros per US dollar— 0.7345 (2007), 0.7964 (2006), 0.8041 (2005), 0.8054 (2004), 0.886 (2003)
Fiscal year: calendar year

COMMUNICATIONS

Telephones—main lines in use: 25.049 million (2005)
Telephones—mobile cellular: 71.5 million (2005)
Telephone system:
general assessment: modern, well developed, fast; fully automated telephone, telex, and data services
domestic: high-capacity cable and microwave radio relay trunks
international: country code—39; a series of submarine cables provide links to Asia, Middle East, Europe, North Africa, and US; satellite earth stations—3 Intelsat (with a total of 5 antennas—3 for Atlantic Ocean and 2 for Indian Ocean), 1 Inmarsat (Atlantic Ocean region), and NA Eutelsat
Radio broadcast stations: AM about 100, FM about 4,600, shortwave 9 (1998)
Radios: 50.5 million (1997)
Television broadcast stations: 358 (plus 4,728 repeaters) (1995)
Televisions: 30.3 million (1997)
Internet country code: .it
Internet hosts: 4.117 million (2007)
Internet Service Providers (ISPs): 93 (Italy and Holy See) (2000)
Internet users: 28.855 million (2006)

TRANSPORTATION

Airports: 132 (2007)
Airports—with paved runways:
total: 101
over 3,047 m: 7
2,438 to 3,047 m: 32
1,524 to 2,437 m: 15
914 to 1,523 m: 34
under 914 m: 13 (2007)
Airports—with unpaved runways:
total: 31
1,524 to 2,437 m: 1
914 to 1,523 m: 11
under 914 m: 19 (2007)
Heliports: 5 (2007)
Pipelines: gas 18,863 km; oil 1,258 km (2007)
Railways:
total: 19,460 km
standard gauge: 18,038 km 1.435-m gauge (11,354 km electrified)
narrow gauge: 123 km 1.000-m gauge (123 km electrified); 1,299 km 0.950-m gauge (161 km electrified) (2006)
Roadways:
total: 484,688 km
paved: 484,688 km (includes 6,529 km of expressways) (2004)
Waterways: 2,400 km
note: used for commercial traffic; of limited overall value compared to road and rail (2006)
Merchant marine:
total: 604 ships (1000 GRT or over) 12,529,192 GRT/13,150,989 DWT
by type: bulk carrier 53, cargo 46, carrier 1, chemical tanker 141, combination ore/oil 1, container 32, liquefied gas 33, livestock carrier 3, passenger 17, passenger/cargo 156, petroleum tanker 40, refrigerated cargo 4, roll on/roll off 35, specialized tanker 14, vehicle carrier 28
foreign-owned: 62 (Denmark 2, France 5, Germany 1, Greece 13, Sweden 1, Switzerland 5, Taiwan 11, Turkey 1, UK 7, US 16)
registered in other countries: 169 (Bahamas 1, Belize 4, Bolivia 1, Cayman Islands 10, Cyprus 5, France 2, Gibraltar 1, Greece 1, Isle of Man 1, Liberia 31, Malta 45, Marshall Islands 3, Norway 4, Panama 10, Portugal 11, Singapore 4, Slovakia 1, Spain 1, St Vincent and The Grenadines 19, Sweden 7, Turkey 3, UK 4) (2007)
Ports and terminals: Augusta, Genoa, Livorno, Ravenna, Sarroch, Taranto, Trieste, Venice

MILITARY

Military branches: Italian Army (Esercito Italiano, EI), Italian Navy (Marina Militare Italiana, MMI), Italian Air Force (Aeronautica Militare Italiana, AMI), Carabinieri Corps (Arma dei Carabinieri, CC) (2008)
Military service age and obligation: 18–27 year of age for voluntary military service; conscription abolished January 2005; women may serve in any military branch; 10-month service obligation, with a reserve obligation to age 45 (Army and Air Force) or 39 (Navy) (2006)
Manpower available for military service:
males age 16–49: 13,884,079
females age 16–49: 13,158,378 (2008 est.)
Manpower fit for military service:
males age 16–49: 11,285,488
females age 16–49: 10,680,672 (2008 est.)
Manpower reaching militarily significant age annually:
males age 16–49: 290,740
females age 16–49: 273,569 (2008 est.)
Military expenditures—percent of GDP: 1.8% (2005 est.)

TRANSNATIONAL ISSUES

Disputes—international: Italy's long coastline and developed economy entices tens of thousands of illegal immigrants from southeastern Europe and northern Africa
Illicit drugs: important gateway for and consumer of Latin American cocaine and Southwest Asian heroin entering the European market; money laundering by organized crime and from smuggling

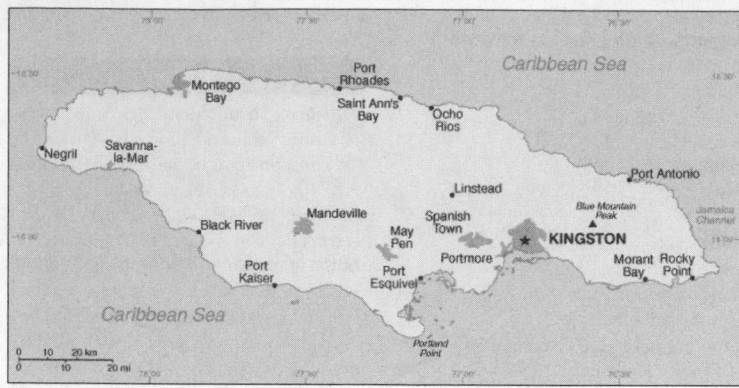

INTRODUCTION

Background: The island—discovered by Christopher COLUMBUS in 1494—was settled by the Spanish early in the 16th century. The native Taino Indians, who had inhabited Jamaica for centuries, were gradually exterminated and replaced by African slaves. England seized the island in 1655 and established a plantation economy based on sugar, cocoa, and coffee. The abolition of slavery in 1834 freed a quarter million slaves, many of whom became small farmers. Jamaica gradually obtained increasing independence from Britain, and in 1958 it joined other British Caribbean colonies in forming the Federation of the West Indies. Jamaica gained full independence when it withdrew from the Federation in 1962. Deteriorating economic conditions during the 1970s led to recurrent violence as rival gangs affiliated with the major political parties evolved into powerful organized crime networks involved in international drug smuggling and money laundering. Violent crime, drug trafficking, and poverty pose significant challenges to the government today. Nonetheless, many rural and resort areas remain relatively safe and contribute substantially to the economy.

GEOGRAPHY

Location: Caribbean, island in the Caribbean Sea, south of Cuba

Geographic coordinates: 18 15 N, 77 30 W

Map references: Central America and the Caribbean

Area:
total: 10,991 sq km
land: 10,831 sq km
water: 160 sq km

Area—comparative: slightly smaller than Connecticut

Land boundaries: 0 km

Coastline: 1,022 km

Maritime claims: measured from claimed archipelagic straight baselines
territorial sea: 12 nm
contiguous zone: 24 nm
exclusive economic zone: 200 nm
continental shelf: 200 nm or to edge of the continental margin

Climate: tropical; hot, humid; temperate interior

Terrain: mostly mountains, with narrow, discontinuous coastal plain

Elevation extremes:
lowest point: Caribbean Sea 0 m
highest point: Blue Mountain Peak 2,256 m

Natural resources: bauxite, gypsum, limestone

Land use:
arable land: 15.83%
permanent crops: 10.01%
other: 74.16% (2005)

Irrigated land: 250 sq km (2002)

Total renewable water resources: 9.4 cu km (2000)

Freshwater withdrawal (domestic/industrial/agricultural): *total:* 0.41 cu km/yr (34%/17%/49%)
per capita: 155 cu m/yr (2000)

Natural hazards: hurricanes (especially July to November)

Environment—current issues: heavy rates of deforestation; coastal waters polluted by industrial waste, sewage, and oil spills; damage to coral reefs; air pollution in Kingston results from vehicle emissions

Environment—international agreements: *party to:* Biodiversity, Climate Change, Climate Change-Kyoto Protocol, Desertification, Endangered Species, Hazardous Wastes, Law of the Sea, Marine Dumping, Marine Life Conservation, Ozone Layer Protection, Ship Pollution, Wetlands

signed, but not ratified: none of the selected agreements

Geography—note: strategic location between Cayman Trench and Jamaica Channel, the main sea lanes for the Panama Canal

PEOPLE

Population: 2,804,332 (July 2008 est.)

Age structure:
0–14 years: 32% (male 455,871/female 440,928)
15–64 years: 60.6% (male 837,241/female 861,906)
65 years and over: 7.4% (male 93,415/female 114,971) (2008 est.)

Median age:
total: 23.4 years
male: 22.9 years
female: 24 years (2008 est.)

Population growth rate: 0.779% (2008 est.)

Birth rate: 20.04 births/1,000 population (2008 est.)

Death rate: 6.37 deaths/1,000 population (2008 est.)

Net migration rate: -5.88 migrant(s)/1,000 population (2008 est.)

Sex ratio:
at birth: 1.05 male(s)/female
under 15 years: 1.03 male(s)/female
15–64 years: 0.97 male(s)/female
65 years and over: 0.81 male(s)/female
total population: 0.98 male(s)/female (2008 est.)

Infant mortality rate:
total: 15.57 deaths/1,000 live births
male: 16.19 deaths/1,000 live births
female: 14.92 deaths/1,000 live births (2008 est.)

Life expectancy at birth:
total population: 73.59 years
male: 71.88 years
female: 75.38 years (2008 est.)

Total fertility rate: 2.3 children born/woman (2008 est.)

HIV/AIDS—adult prevalence rate: 1.2% (2003 est.)

HIV/AIDS—people living with HIV/AIDS: 22,000 (2003 est.)

HIV/AIDS—deaths: 900 (2003 est.)

Nationality:
noun: Jamaican(s)
adjective: Jamaican

Ethnic groups: black 91.2%, mixed 6.2%, other or unknown 2.6% (2001 census)

Religions: Protestant 62.5% (Seventh-Day Adventist 10.8%, Pentecostal 9.5%, Other Church of God 8.3%, Baptist 7.2%, New Testament Church of God 6.3%, Church of God in Jamaica 4.8%, Church of God of Prophecy 4.3%,

Anglican 3.6%, other Christian 7.7%), Roman Catholic 2.6%, other or unspecified 14.2%, none 20.9%, (2001 census)
Languages: English, English patois
Literacy:
definition: age 15 and over has ever attended school
total population: 87.9%
male: 84.1%
female: 91.6% (2003 est.)

GOVERNMENT

Country name:
conventional long form: none
conventional short form: Jamaica
Government type: constitutional parliamentary democracy
Capital: *name:* Kingston
geographic coordinates: 18 00 N, 76 48 W
time difference: UTC-5 (same time as Washington, DC during Standard Time)
Administrative divisions: 14 parishes; Clarendon, Hanover, Kingston, Manchester, Portland, Saint Andrew, Saint Ann, Saint Catherine, Saint Elizabeth, Saint James, Saint Mary, Saint Thomas, Trelawny, Westmoreland
note: for local government purposes, Kingston and Saint Andrew were amalgamated in 1923 into the present single corporate body known as the Kingston and Saint Andrew Corporation
Independence: 6 August 1962 (from UK)
National holiday: Independence Day, 6 August (1962)
Constitution: 6 August 1962
Legal system: based on English common law; has not accepted compulsory ICJ jurisdiction
Suffrage: 18 years of age; universal
Executive branch:
chief of state: Queen ELIZABETH II (since 6 February 1952); represented by Governor General Kenneth O. HALL (since 15 February 2006)
head of government: Prime Minister Bruce GOLDING (since 11 September 2007)
cabinet: Cabinet appointed by the governor general on the advice of the prime minister
elections: the monarch is hereditary; governor general appointed by the monarch on the recommendation of the prime minister; following legislative elections, the leader of the majority party or the leader of the majority coalition in the House of Representatives is appointed prime minister by the governor general; the deputy prime minister is recommended by the prime minister
Legislative branch: bicameral Parliament consists of the Senate (a 21-member body appointed by the governor general on the recommendations of the

prime minister and the leader of the opposition; ruling party is allocated 13 seats, and the opposition is allocated 8 seats) and the House of Representatives (60 seats; members are elected by popular vote to serve five-year terms)
elections: last held 3 September 2007 (next to be held no later than October 2012)
election results: percent of vote by party—JLP 50.1%, PNP 49.8%; seats by party—JLP 33, PNP 27
Judicial branch: Supreme Court (judges appointed by the governor general on the advice of the prime minister); Court of Appeal
Political parties and leaders: Jamaica Labor Party or JLP [Bruce GOLDING]; People's National Party or PNP [Portia SIMPSON-MILLER]; National Democratic Movement or NDM [Michael WILLIAMS]
Political pressure groups and leaders: New Beginnings Movement or NBM; Rastafarians (black religious/racial cultists, pan-Africanists)
International organization participation: ACP, C, Caricom, CDB, FAO, G-15, G-77, IADB, IAEA, IBRD, ICAO, ICCt (signatory), ICRM, IDA, IFAD, IFC, IFRCS, IHO, ILO, IMF, IMO, Interpol, IOC, IOM, ISO, ITSO, ITU, LAES, MIGA, NAM, OAS, OPANAL, OPCW, UN, UNCTAD, UNESCO, UNIDO, UNWTO, UPU, WCO, WFTU, WHO, WIPO, WMO, WTO
Diplomatic representation in the US:
chief of mission: Ambassador Anthony JOHNSON
chancery: 1520 New Hampshire Avenue NW, Washington, DC 20036
telephone: [1] (202) 452-0660
FAX: [1] (202) 452-0081
consulate(s) general: Miami, New York
Diplomatic representation from the US:
chief of mission: Ambassador Brenda LaGrange JOHNSON
embassy: 142 Old Hope Road, Kingston 6
mailing address: P.O. Box 541, Kingston 5
telephone: [1] (876) 702-6000
FAX: [1] (876) 702-6348
Flag description: diagonal yellow cross divides the flag into four triangles—green (top and bottom) and black (hoist side and outer side)

ECONOMY

Economy—overview: The Jamaican economy is heavily dependent on services, which now account for more than 60% of GDP. The country continues to derive most of its foreign exchange from tourism, remittances, and bauxite/alumina. Remittances account for nearly

20% of GDP and are equivalent to tourism revenues. Jamaica's economy, already saddled with a record of sluggish growth, will suffer an economic setback from damages caused by Hurricane Dean in August 2007. The economy faces serious long-term problems: high but declining interest rates, increased foreign competition, exchange rate instability, a sizable merchandise trade deficit, large-scale unemployment and underemployment, and a debt-to-GDP ratio of 135%. Jamaica's onerous debt burden—the fourth highest per capita—is the result of government bailouts to ailing sectors of the economy, most notably the financial sector in the mid-to-late 1990s. Inflation also has declined, standing at about 7% at the end of 2007. High unemployment exacerbates the serious crime problem, including gang violence that is fueled by the drug trade. The GOLDING administration faces the difficult prospect of having to achieve fiscal discipline in order to maintain debt payments while simultaneously attacking a serious and growing crime problem that is hampering economic growth.
GDP (purchasing power parity): $20.67 billion (2007 est.)
GDP (official exchange rate): $11.21 billion (2007 est.)
GDP—real growth rate: 1.4% (2007 est.)
GDP—per capita (PPP): $7,700 (2007 est.)
GDP—composition by sector:
agriculture: 5.1%
industry: 32.7%
services: 62.2% (2007 est.)
Labor force: 1.255 million (2007 est.)
Labor force—by occupation: *agriculture:* 17%
industry: 19%
services: 64% (2006)
Unemployment rate: 9.9% (2007 est.)
Population below poverty line: 14.8% (2003 est.)
Household income or consumption by percentage share:
lowest 10%: 2.1%
highest 10%: 35.8% (2004)
Distribution of family income—Gini index: 45.5 (2004)
Inflation rate (consumer prices): 9.3% (2007 est.)
Investment (gross fixed): 34.2% of GDP (2007 est.)
Budget:
revenues: $3.707 billion
expenditures: $4.251 billion (2007 est.)
Public debt: 127.2% of GDP (2007 est.)
Agriculture—products: sugarcane, bananas, coffee, citrus, yams, ackees, vegetables; poultry, goats, milk; crustaceans, mollusks

Industries: tourism, bauxite/alumina, agro processing, light manufactures, rum, cement, metal, paper, chemical products, telecommunications

Industrial production growth rate: 1.2% (2007 est.)

Electricity—production: 6.985 billion kWh (2005)

Electricity—production by source:
fossil fuel: 96.8%
hydro: 1.8%
nuclear: 0%
other: 1.4% (2001)

Electricity—consumption: 6.131 billion kWh (2005)

Electricity—exports: 0 kWh (2005)

Electricity—imports: 0 kWh (2005)

Oil—production: 0 bbl/day (2005 est.)

Oil—consumption: 72,000 bbl/day (2005 est.)

Oil—exports: 1,531 bbl/day (2004)

Oil—imports: 71,420 bbl/day (2004)

Oil—proved reserves: 0 bbl (1 January 2006 est.)

Natural gas—production: 0 cu m (2005 est.)

Natural gas—consumption: 0 cu m (2005 est.)

Natural gas—exports: 0 cu m (2005 est.)

Natural gas—imports: 0 cu m (2005)

Natural gas—proved reserves: 0 cu m (1 January 2006 est.)

Current account balance: -$1.623 billion (2007 est.)

Exports: $2.331 billion f.o.b. (2007 est.)

Exports—commodities: alumina, bauxite, sugar, bananas, rum, coffee, yams, beverages, chemicals, wearing apparel, mineral fuels

Exports—partners: US 30.2%, Canada 15.6%, China 15.2%, UK 10.3%, Netherlands 7%, Norway 4.6% (2006)

Imports: $5.784 billion f.o.b. (2007 est.)

Imports—commodities: food and other consumer goods, industrial supplies, fuel, parts and accessories of capital goods, machinery and transport equipment, construction materials

Imports—partners: US 39.3%, Trinidad and Tobago 13.6%, Venezuela 9.5% (2006)

Economic aid—recipient: $35.74 million (2005)

Reserves of foreign exchange and gold: $1.905 billion (31 December 2007 est.)

Debt—external: $9.657 billion (31 December 2007 est.)

Market value of publicly traded shares: $12.28 billion (2006)

Currency (code): Jamaican dollar (JMD)

Currency code: JMD

Exchange rates: Jamaican dollars per US dollar—69.034 (2007), 65.768 (2006), 62.51 (2005), 61.197 (2004), 57.741 (2003)

Fiscal year: 1 April—31 March

COMMUNICATIONS

Telephones—main lines in use: 319,000 (2005)

Telephones—mobile cellular: 2.804 million (2005)

Telephone system:
general assessment: fully automatic domestic telephone network
domestic: the 1999 agreement to open the market for telecommunications services resulted in rapid growth in mobile-cellular telephone usage; mobile-cellular teledensity now exceeds 100 per 100 persons; the number of fixed-lines in use has been declining
international: country code—1-876; the Fibralink submarine cable network provides enhanced delivery of business and broadband traffic and is linked to the Americas Region Caribbean Ring System (ARCOS-1) submarine cable in the Dominican Republic; the link to ARCOS-1 provides seamless connectivity to US, parts of the Caribbean, Central America, and South America; satellite earth stations—2 Intelsat (Atlantic Ocean)

Radio broadcast stations: AM 10, FM 13, shortwave 0 (1998)

Radios: 1.215 million (1997)

Television broadcast stations: 7 (1997)

Televisions: 460,000 (1997)

Internet country code: .jm

Internet hosts: 1,213 (2007)

Internet Service Providers (ISPs): 21 (2000)

Internet users: 1.232 million (2005)

TRANSPORTATION

Airports: 34 (2007)

Airports—with paved runways:
total: 11
2,438 to 3,047 m: 2
914 to 1,523 m: 4
under 914 m: 5 (2007)

Airports—with unpaved runways:
total: 23
914 to 1,523 m: 2
under 914 m: 21 (2007)

Roadways:
total: 21,552 km
paved: 15,937 km (includes 33 km of expressways)
unpaved: 5,615 km (2005)

Merchant marine:
total: 13 ships (1000 GRT or over) 161,700 GRT/241,663 DWT
by type: bulk carrier 6, cargo 2, carrier 1, petroleum tanker 1, roll on/roll off 3
foreign-owned: 12 (Denmark 1, Germany 1, Greece 8, Latvia 2)
registered in other countries: 1 (Panama 1) (2007)

Ports and terminals: Kingston, Port Esquivel, Port Kaiser, Port Rhoades, Rocky Point

MILITARY

Military branches: Jamaica Defense Force: Ground Forces, Coast Guard, Air Wing (2007)

Military service age and obligation: 18 years of age for voluntary military service; younger recruits may be conscripted with parental consent (2001)

Manpower available for military service:
males age 16–49: 688,480
females age 16–49: 709,548 (2008 est.)

Manpower fit for military service:
males age 16–49: 566,477
females age 16–49: 583,075 (2008 est.)

Manpower reaching militarily significant age annually:
males age 16–49: 32,000
females age 16–49: 31,428 (2008 est.)

Military expenditures—percent of GDP: 0.6% (2006 est.)

TRANSNATIONAL ISSUES

Disputes—international: none

Illicit drugs: transshipment point for cocaine from South America to North America and Europe; illicit cultivation and consumption of cannabis; government has an active manual cannabis eradication program; corruption is a major concern; substantial money-laundering activity; Colombian narcotics traffickers favor Jamaica for illicit financial transactions

JAN MAYEN

INTRODUCTION

Background: This desolate, arctic, mountainous island was named after a Dutch whaling captain who indisputably discovered it in 1614 (earlier claims are inconclusive). Visited only occasionally by seal hunters and trappers over the following centuries, the island came under Norwegian sovereignty in 1929. The long dormant Haakon VII Toppen/Beerenberg volcano resumed

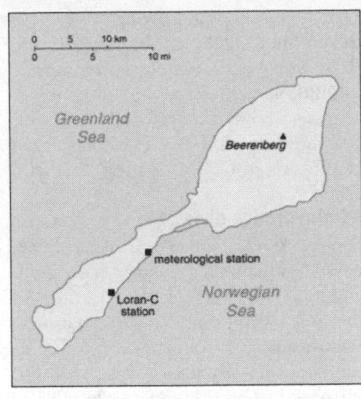

activity in 1970; the most recent eruption occurred in 1985. It is the northernmost active volcano on earth.

GEOGRAPHY

Location: Northern Europe, island between the Greenland Sea and the Norwegian Sea, northeast of Iceland
Geographic coordinates: 71 00 N, 8 00 W
Map references: Arctic Region
Area:
total: 377 sq km
land: 377 sq km
water: 0 sq km
Area—comparative: slightly more than twice the size of Washington, DC
Land boundaries: 0 km
Coastline: 124.1 km
Maritime claims:
territorial sea: 4 nm
contiguous zone: 10 nm

exclusive economic zone: 200 nm
continental shelf: 200 m depth or to the depth of exploitation
Climate: arctic maritime with frequent storms and persistent fog
Terrain: volcanic island, partly covered by glaciers
Elevation extremes:
lowest point: Norwegian Sea 0 m
highest point: Haakon VII Toppen/Beerenberg 2,277 m
Natural resources: none
Land use:
arable land: 0%
permanent crops: 0%
other: 100% (2005)
Irrigated land: 0 sq km
Natural hazards: dominated by the volcano Haakon VII Toppen/Beerenberg; volcanic activity resumed in 1970; the most recent eruption occurred in 1985
Environment—current issues: NA
Geography—note: barren volcanic island with some moss and grass

PEOPLE

Population: no indigenous inhabitants
note: personnel operate the Long Range Navigation (Loran-C) base and the weather and coastal services radio station

GOVERNMENT

Country name:
conventional long form: none
conventional short form: Jan Mayen
Dependency status: territory of Norway; since August 1994, administered from Oslo through the county governor

(fylkesmann) of Nordland; however, authority has been delegated to a station commander of the Norwegian Defense Communication Service
Legal system: the laws of Norway, where applicable, apply
Flag description: the flag of Norway is used

ECONOMY

Economy—overview: Jan Mayen is a volcanic island with no exploitable natural resources. Economic activity is limited to providing services for employees of Norway's radio and meteorological stations on the island.

COMMUNICATIONS

Radio broadcast stations: NA; note—there is one radio and meteorological station (1998)
Internet Service Providers (ISPs): 13 (Jan Mayen and Svalbard) (2000)

TRANSPORTATION

Airports: 1 (2007)
Airports—with unpaved runways:
total: 1
1,524 to 2,437 m: 1 (2007)
Ports and terminals: none; offshore anchorage only

MILITARY

Military—note: defense is the responsibility of Norway

TRANSNATIONAL ISSUES

Disputes—international: none

JAPAN

INTRODUCTION

Background: In 1603, a Tokugawa shogunate (military dictatorship) ushered in a long period of isolation from foreign influence in order to secure its

power. For more than two centuries this policy enabled Japan to enjoy stability and a flowering of its indigenous culture. Following the Treaty of Kanagawa with the US in 1854, Japan opened its ports and began to intensively modernize and industrialize. During the late 19th and early 20th centuries, Japan became a regional power that was able to defeat the forces of both China and Russia. It occupied Korea, Formosa (Taiwan), and southern Sakhalin Island. In 1931–32 Japan occupied Manchuria, and in 1937 it launched a full-scale invasion of China. Japan attacked US forces in 1941—triggering America's entry into World War II—and soon occupied much of East and Southeast Asia. After its defeat in World War II, Japan recovered to become an economic power and a staunch ally of the US. While the emperor retains his throne as a symbol of

national unity, elected politicians—with heavy input from bureaucrats and business executives—wield actual decision-making power. The economy experienced a major slowdown starting in the 1990s following three decades of unprecedented growth, but Japan still remains a major economic power, both in Asia and globally.

GEOGRAPHY

Location: Eastern Asia, island chain between the North Pacific Ocean and the Sea of Japan, east of the Korean Peninsula
Geographic coordinates: 36 00 N, 138 00 E
Map references: Asia
Area:
total: 377,835 sq km
land: 374,744 sq km
water: 3,091 sq km

note: includes Bonin Islands (Ogasawara-gunto), Daito-shoto, Minami-jima, Okino-tori-shima, Ryukyu Islands (Nansei-shoto), and Volcano Islands (Kazan-retto)

Area—comparative: slightly smaller than California

Land boundaries: 0 km

Coastline: 29,751 km

Maritime claims:

territorial sea: 12 nm; between 3 nm and 12 nm in the international straits—La Perouse or Soya, Tsugaru, Osumi, and Eastern and Western Channels of the Korea or Tsushima Strait

contiguous zone: 24 nm

exclusive economic zone: 200 nm

Climate: varies from tropical in south to cool temperate in north

Terrain: mostly rugged and mountainous

Elevation extremes:

lowest point: Hachiro-gata -4 m

highest point: Mount Fuji 3,776 m

Natural resources: negligible mineral resources, fish

Land use: _arable land:_ 11.64%

permanent crops: 0.9%

other: 87.46% (2005)

Irrigated land: 25,920 sq km (2003)

Total renewable water resources: 430 cu km (1999)

Freshwater withdrawal (domestic/industrial/agricultural): _total:_ 88.43 cu km/yr (20%/18%/62%)

per capita: 690 cu m/yr (2000)

Natural hazards: many dormant and some active volcanoes; about 1,500 seismic occurrences (mostly tremors) every year; tsunamis; typhoons

Environment—current issues: air pollution from power plant emissions results in acid rain; acidification of lakes and reservoirs degrading water quality and threatening aquatic life; Japan is one of the largest consumers of fish and tropical timber, contributing to the depletion of these resources in Asia and elsewhere

Environment—international agreements: _party to:_ Antarctic-Environmental Protocol, Antarctic-Marine Living Resources, Antarctic Seals, Antarctic Treaty, Biodiversity, Climate Change, Climate Change-Kyoto Protocol, Desertification, Endangered Species, Environmental Modification, Hazardous Wastes, Law of the Sea, Marine Dumping, Ozone Layer Protection, Ship Pollution, Tropical Timber 83, Tropical Timber 94, Wetlands, Whaling

Geography—note: strategic location in northeast Asia

PEOPLE

Population: 127,288,419 (July 2008 est.)

Age structure:

0–14 years: 13.7% (male 8,926,439/female 8,460,629)

15–64 years: 64.7% (male 41,513,061/female 40,894,057)

65 years and over: 21.6% (male 11,643,845/female 15,850,388) (2008 est.)

Median age: _total:_ 43.8 years

male: 42.1 years

female: 45.7 years (2008 est.)

Population growth rate: -0.139% (2008 est.)

Birth rate: 7.87 births/1,000 population (2008 est.)

Death rate: 9.26 deaths/1,000 population (2008 est.)

Net migration rate: NA

Sex ratio: _at birth:_ 1.06 male(s)/female

under 15 years: 1.06 male(s)/female

15–64 years: 1.02 male(s)/female

65 years and over: 0.73 male(s)/female

total population: 0.95 male(s)/female (2008 est.)

Infant mortality rate:

total: 2.8 deaths/1,000 live births

male: 3 deaths/1,000 live births

female: 2.58 deaths/1,000 live births (2008 est.)

Life expectancy at birth:

total population: 82.07 years

male: 78.73 years

female: 85.59 years (2008 est.)

Total fertility rate: 1.22 children born/woman (2008 est.)

HIV/AIDS—adult prevalence rate: less than 0.1% (2003 est.)

HIV/AIDS—people living with HIV/AIDS: 12,000 (2003 est.)

HIV/AIDS—deaths: 500 (2003 est.)

Nationality:

noun: Japanese (singular and plural)

adjective: Japanese

Ethnic groups: Japanese 98.5%, Koreans 0.5%, Chinese 0.4%, other 0.6%

note: up to 230,000 Brazilians of Japanese origin migrated to Japan in the 1990s to work in industries; some have returned to Brazil (2004)

Religions: observe both Shinto and Buddhist 84%, other 16% (including Christian 0.7%)

Languages: Japanese

Literacy: _definition:_ age 15 and over can read and write

total population: 99%

male: 99%

female: 99% (2002)

GOVERNMENT

Country name:

conventional long form: none

conventional short form: Japan

local long form: Nihon-koku/Nippon-koku

local short form: Nihon/Nippon

Government type: constitutional monarchy with a parliamentary government

Capital: _name:_ Tokyo

geographic coordinates: 35 41 N, 139 45 E

time difference: UTC+9 (14 hours ahead of Washington, DC during Standard Time)

Administrative divisions: 47 prefectures; Aichi, Akita, Aomori, Chiba, Ehime, Fukui, Fukuoka, Fukushima, Gifu, Gunma, Hiroshima, Hokkaido, Hyogo, Ibaraki, Ishikawa, Iwate, Kagawa, Kagoshima, Kanagawa, Kochi, Kumamoto, Kyoto, Mie, Miyagi, Miyazaki, Nagano, Nagasaki, Nara, Niigata, Oita, Okayama, Okinawa, Osaka, Saga, Saitama, Shiga, Shimane, Shizuoka, Tochigi, Tokushima, Tokyo, Tottori, Toyama, Wakayama, Yamagata, Yamaguchi, Yamanashi

Independence: 660 B.C. (traditional founding by Emperor JIMMU)

National holiday: Birthday of Emperor AKIHITO, 23 December (1933)

Constitution: 3 May 1947

Legal system: modeled after German civil law system with English-American influence; judicial review of legislative acts in the Supreme Court; accepts compulsory ICJ jurisdiction with reservations

Suffrage: 20 years of age; universal

Executive branch:

chief of state: Emperor AKIHITO (since 7 January 1989)

head of government: Prime Minister Yasuo FUKUDA (since 26 September 2007)

cabinet: Cabinet appointed by the prime minister

elections: Diet designates prime minister; constitution requires that prime minister commands parliamentary majority; following legislative elections, leader of majority party or leader of majority coalition in House of Representatives usually becomes prime minister; monarch is hereditary

election results: FUKUDA elected prime minister with 338 of 477 votes cast in the House of Representatives; he received 106 of 240 votes cast in the House of Councillors; vote of House of Representatives prevailed

Legislative branch: bicameral Diet or Kokkai consists of the House of Councillors or Sangi-in (242 seats—members elected for six-year terms; half reelected every three years; 146 members in multi-seat constituencies and 96 by proportional representation) and the House of Representatives or Shugi-in (480 seats—members elected for four-year terms; 300 in single-seat constituencies; 180 members by proportional representation in 11 regional blocs) *elections:* House of Councillors—last held 29 July 2007 (next to be held in July 2010); House of Representatives—last held 11 September 2005 (next election by September 2009)

election results: House of Councillors—percent of vote by party—NA; seats by party—DPJ 109, LDP 83, Komeito 20, JCP 7, SDP 5, others 18

: House of Representatives—percent of vote by party (in single-seat constituencies)—LDP 47.8%, DPJ 36.4%, others 15.8%; seats by party—LDP 296, DPJ 113, Komeito 31, JCP 9, SDP 7, others 24 (2007)

Judicial branch: Supreme Court (chief justice is appointed by the monarch after designation by the cabinet; all other justices are appointed by the cabinet)

Political parties and leaders: Democratic Party of Japan or DPJ [Ichiro OZAWA]; Japan Communist Party or JCP [Kazuo SHII]; Komeito [Akihiro OTA]; Liberal Democratic Party or LDP [Yasuo FUKUDA]; Social Democratic Party or SDP [Mizuho FUKUSHIMA]

Political pressure groups and leaders: NA

International organization participation: ADB, AfDB, APEC, APT, ARF, ASEAN (dialogue partner), Australia Group, BIS, CE (observer), CERN (observer), CP, EAS, EBRD, FAO, G-5, G-7, G-8, G-10, IADB, IAEA, IBRD, ICAO, ICC, ICCt, ICRM, IDA, IEA, IFAD, IFC, IFRCS, IHO, ILO, IMF, IMO, IMSO, Interpol, IOC, IOM, IPU, ISO, ITSO, ITU, ITUC, LAIA, MIGA, NEA, NSG, OAS (observer), OECD, OPCW, OSCE (partner), Paris Club, PCA, PIF (partner), SAARC (observer), SECI (observer), UN, UN Security Council (temporary), UNCTAD, UNDOF, UNESCO, UNHCR, UNIDO, UNITAR, UNRWA, UNWTO, UPU, WCL, WCO, WFTU, WHO, WIPO, WMO, WTO, ZC

Diplomatic representation in the US:
chief of mission: Ambassador Ryozo KATO
chancery: 2520 Massachusetts Avenue NW, Washington, DC 20008
telephone: [1] (202) 238-6700
FAX: [1] (202) 328-2187
consulate(s) general: Anchorage, Atlanta, Boston, Chicago, Denver, Detroit, Agana (Guam), Honolulu, Houston, Los Angeles, Miami, New Orleans, New York, Portland (Oregon), San Francisco, Seattle

Diplomatic representation from the US:
chief of mission: Ambassador J. Thomas SCHIEFFER
embassy: 1-10-5 Akasaka, Minato-ku, Tokyo 107-8420
mailing address: APO AP 96337-5004

telephone: [81] (03) 3224-5000
FAX: [81] (03) 3505-1862
consulate(s) general: Naha (Okinawa), Osaka-Kobe, Sapporo
consulate(s): Fukuoka, Nagoya

Flag description: white with a large red disk (representing the sun without rays) in the center

ECONOMY

Economy—overview: Government-industry cooperation, a strong work ethic, mastery of high technology, and a comparatively small defense allocation (1% of GDP) helped Japan advance with extraordinary rapidity to the rank of second most technologically powerful economy in the world after the US and the third-largest economy in the world after the US and China, measured on a purchasing power parity (PPP) basis. One notable characteristic of the economy has been how manufacturers, suppliers, and distributors have worked together in closely-knit groups called keiretsu. A second basic feature has been the guarantee of lifetime employment for a substantial portion of the urban labor force. Both features have now eroded. Japan's industrial sector is heavily dependent on imported raw materials and fuels. The tiny agricultural sector is highly subsidized and protected, with crop yields among the highest in the world. Usually self sufficient in rice, Japan must import about 55% of its food on a caloric basis. Japan maintains one of the world's largest fishing fleets and accounts for nearly 15% of the global catch. For three decades, overall real economic growth had been spectacular—a 10% average in the 1960s, a 5% average in the 1970s, and a 4% average in the 1980s. Growth slowed markedly in the 1990s, averaging just 1.7%, largely because of the after effects of overinvestment and an asset price bubble during the late 1980s that required a protracted period of time for firms to reduce excess debt, capital, and labor. From 2000 to 2001, government efforts to revive economic growth proved short lived and were hampered by the slowing of the US, European, and Asian economies. In 2002–07, growth improved and the lingering fears of deflation in prices and economic activity lessened, leading the central bank to raise interest rates to 0.25% in July 2006, up from the near 0% rate of the six years prior, and to 0.50% in February 2007. In addition, the 10-year privatization of Japan Post, which has functioned not only as the national postal delivery system but also, through its banking and insurance facilities as

Japan's largest financial institution, was completed in October 2007, marking a major milestone in the process of structural reform. Nevertheless, Japan's huge government debt, which totals 182% of GDP, and the aging of the population are two major long-run problems. Some fear that a rise in taxes could endanger the current economic recovery. Debate also continues on the role of and effects of reform in restructuring the economy, particularly with respect to increasing income disparities.

GDP (purchasing power parity): $4.29 trillion (2007 est.)

GDP (official exchange rate): $4.384 trillion (2007 est.)

GDP—real growth rate: 2.1% (2007 est.)

GDP—per capita (PPP): $33,600 (2007 est.)

GDP—composition by sector:
agriculture: 1.4%
industry: 26.5%
services: 72% (2007 est.)

Labor force: 66.69 million (2007 est.)

Labor force—by occupation: *agriculture:* 4.6%
industry: 27.8%
services: 67.7% (2004)

Unemployment rate: 3.9% (2007 est.)

Population below poverty line: NA%

Household income or consumption by percentage share:
lowest 10%: 4.8%
highest 10%: 21.7% (1993)

Distribution of family income—Gini index: 38.1 (2002)

Inflation rate (consumer prices): 0% (2007 est.)

Investment (gross fixed): 23.8% of GDP (2007 est.)

Budget:
revenues: $1.448 trillion
expenditures: $1.597 trillion (2007 est.)

Public debt: 195.5% of GDP (2007 est.)

Agriculture—products: rice, sugar beets, vegetables, fruit; pork, poultry, dairy products, eggs; fish

Industries: among world's largest and technologically advanced producers of motor vehicles, electronic equipment, machine tools, steel and nonferrous metals, ships, chemicals, textiles, processed foods

Industrial production growth rate: 1.3% (2007 est.)

Electricity—production: 1.025 trillion kWh (2005)

Electricity—production by source:
fossil fuel: 60%
hydro: 8.4%
nuclear: 29.8%
other: 1.8% (2001)

Electricity—consumption: 974.2 billion kWh (2005)

335

Electricity—exports: 0 kWh (2005)
Electricity—imports: 0 kWh (2005)
Oil—production: 125,000 bbl/day (2006)
Oil—consumption: 5.353 million bbl/day (2005)
Oil—exports: 94,830 bbl/day (2004)
Oil—imports: 5.425 million bbl/day (2004)
Oil—proved reserves: 58.5 million bbl (1 January 2006 est.)
Natural gas—production: 4.85 billion cu m (2005 est.)
Natural gas—consumption: 83.67 billion cu m (2005 est.)
Natural gas—exports: 0 cu m (2005 est.)
Natural gas—imports: 77.6 billion cu m (2005)
Natural gas—proved reserves: 38.02 billion cu m (1 January 2006 est.)
Current account balance: $212.8 billion (2007 est.)
Exports: $676.9 billion f.o.b. (2007 est.)
Exports—commodities: transport equipment, motor vehicles, semiconductors, electrical machinery, chemicals
Exports—partners: US 22.8%, China 14.3%, South Korea 7.8%, Taiwan 6.8%, Hong Kong 5.6% (2006)
Imports: $572.4 billion f.o.b. (2007 est.)
Imports—commodities: machinery and equipment, fuels, foodstuffs, chemicals, textiles, raw materials
Imports—partners: China 20.5%, US 12%, Saudi Arabia 6.4%, UAE 5.5%, Australia 4.8%, South Korea 4.7%, Indonesia 4.2% (2006)
Economic aid—donor: ODA, $11.19 billion (2006)
Reserves of foreign exchange and gold: $954.1 billion (31 December 2007 est.)
Debt—external: $1.492 trillion (30 June 2007)
Stock of direct foreign investment—at home: $109 billion (2007 est.)
Stock of direct foreign investment—abroad: $527.8 billion (2007 est.)
Market value of publicly traded shares: $4.737 trillion (2005)
Currency (code): yen (JPY)
Currency code: JPY
Exchange rates: yen per US dollar— 117.99 (2007), 116.18 (2006), 110.22 (2005), 108.19 (2004), 115.93 (2003)
Fiscal year: 1 April—31 March

COMMUNICATIONS

Telephones—main lines in use: 55.155 million (2006)
Telephones—mobile cellular: 101.7 million (2006)
Telephone system:
general assessment: excellent domestic and international service
domestic: high level of modern technology and excellent service of every kind
international: country code—81; numerous submarine cables provide links throughout Asia, Australia, the Middle East, Europe, and US; satellite earth stations—5 Intelsat (4 Pacific Ocean and 1 Indian Ocean), 1 Intersputnik (Indian Ocean region), and 1 Inmarsat (Pacific and Indian Ocean regions
Radio broadcast stations: AM 215 (plus 370 repeaters), FM 89 (plus 485 repeaters), shortwave 21 (2001)
Radios: 120.5 million (1997)
Television broadcast stations: 211 (plus 7,341 repeaters); in addition, US Forces are served by 3 TV stations and 2 TV cable services (1999)
Televisions: 86.5 million (1997)
Internet country code: .jp
Internet hosts: 33.333 million (2007)
Internet Service Providers (ISPs): 73 (2000)
Internet users: 87.54 million (2006)

TRANSPORTATION

Airports: 176 (2007)
Airports—with paved runways:
total: 145
over 3,047 m: 7
2,438 to 3,047 m: 41
1,524 to 2,437 m: 40
914 to 1,523 m: 28
under 914 m: 29 (2007)
Airports—with unpaved runways:
total: 31
914 to 1,523 m: 4
under 914 m: 27 (2007)
Heliports: 14 (2007)
Pipelines: gas 3,939 km; oil 170 km; oil/gas/water 104 km (2007)
Railways:
total: 23,474 km
standard gauge: 3,204 km 1.435-m gauge (3,204 km electrified)
narrow gauge: 77 km 1.372-m gauge (77 km electrified); 20,182 km 1.067-m gauge (13,334 km electrified); 11 km 0.762-m gauge (11 km electrified) (2006)
Roadways:
total: 1.193 million km
paved: 942,000 km (includes 7,383 km of expressways)
unpaved: 251,000 km (2005)
Waterways: 1,770 km (seagoing vessels use inland seas) (2007)
Merchant marine:
total: 676 ships (1000 GRT or over) 10,386,894 GRT/11,689,142 DWT
by type: bulk carrier 131, cargo 29, carrier 3, chemical tanker 23, container 10, liquefied gas 58, passenger 14, passenger/cargo 142, petroleum tanker 157, refrigerated cargo 2, roll on/roll off 52, vehicle carrier 55
registered in other countries: 2,692 (Bahamas 62, Belize 2, Bermuda 1, Burma 3, Cambodia 3, Cayman Islands 6, China 2, Cyprus 19, France 5, Honduras 4, Hong Kong 78, Indonesia 5, Isle of Man 4, South Korea 1, Liberia 111, Malaysia 4, Malta 3, Marshall Islands 5, Mongolia 1, Norway 1, Panama 2,151, Philippines 69, Portugal 10, Singapore 108, Sweden 1, Thailand 4, UK 1, Vanuatu 28, unknown 2) (2007)
Ports and terminals: Chiba, Kawasaki, Kobe, Mizushima, Moji, Nagoya, Osaka, Tokyo, Tomakomai, Yohohama

MILITARY

Military branches: Japanese Ministry of Defense (MOD): Ground Self-Defense Force (Rikujou Jietai, GSDF), Maritime Self-Defense Force (Kaijou Jietai, MSDF), Air Self-Defense Force (Koku Jieitai, ASDF) (2008)
Military service age and obligation: 18 years of age for voluntary military service (2001)
Manpower available for military service:
males age 16–49: 27,819,804
females age 16–49: 26,863,794 (2008 est.)
Manpower fit for military service:
males age 16–49: 22.963 million
females age 16–49: 22,134,127 (2008 est.)
Manpower reaching militarily significant age annually:
males age 16–49: 622,168
females age 16–49: 590,153 (2008 est.)
Military expenditures—percent of GDP: 0.8% (2006)

TRANSNATIONAL ISSUES

Disputes—international: the sovereignty dispute over the islands of Etorofu, Kunashiri, and Shikotan, and the Habomai group, known in Japan as the "Northern Territories" and in Russia as the "Southern Kuril Islands," occupied by the Soviet Union in 1945, now administered by Russia and claimed by Japan, remains the primary sticking point to signing a peace treaty formally ending World War II hostilities; Japan and South Korea claim Liancourt Rocks (Take-shima/Tok-do) occupied by South Korea since 1954; China and Taiwan dispute both Japan's claims to the uninhabited islands of the Senkaku-shoto (Diaoyu Tai) and Japan's unilaterally declared exclusive economic zone in the East China Sea, the site of intensive hydrocarbon prospecting

JERSEY

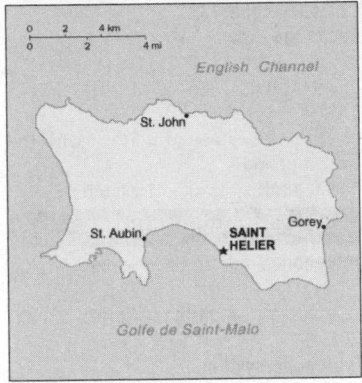

INTRODUCTION

Background: Jersey and the other Channel Islands represent the last remnants of the medieval Dukedom of Normandy that held sway in both France and England. These islands were the only British soil occupied by German troops in World War II. Jersey is a British crown dependency but is not part of the UK. However, the UK Government is constitutionally responsible for its defense and international representation.

GEOGRAPHY

Location: Western Europe, island in the English Channel, northwest of France
Geographic coordinates: 49 15 N, 2 10 W
Map references: Europe
Area:
total: 116 sq km
land: 116 sq km
water: 0 sq km
Area—comparative: about two-thirds the size of Washington, DC
Land boundaries: 0 km
Coastline: 70 km
Maritime claims:
territorial sea: 3 nm
exclusive fishing zone: 12 nm
Climate: temperate; mild winters and cool summers
Terrain: gently rolling plain with low, rugged hills along north coast
Elevation extremes:
lowest point: Atlantic Ocean 0 m
highest point: unnamed location 143 m
Natural resources: arable land
Land use:
arable land: 0%
permanent crops: 0%
other: 100% (2005)
Irrigated land: NA
Natural hazards: NA

Environment—current issues: NA
Geography—note: largest and southernmost of Channel Islands; about 30% of population concentrated in Saint Helier

PEOPLE

Population: 91,533 (July 2008 est.)
Age structure:
0–14 years: 16.6% (male 7,851/female 7,298)
15–64 years: 67.5% (male 30,744/female 30,997)
65 years and over: 16% (male 6,499/female 8,144) (2008 est.)
Median age:
total: 42.3 years
male: 41.6 years
female: 43.1 years (2008 est.)
Population growth rate: 0.221% (2008 est.)
Birth rate: 8.84 births/1,000 population (2008 est.)
Death rate: 9.36 deaths/1,000 population (2008 est.)
Net migration rate: 2.73 migrant(s)/1,000 population (2008 est.)
Sex ratio:
at birth: 1.08 male(s)/female
under 15 years: 1.08 male(s)/female
15–64 years: 0.99 male(s)/female
65 years and over: 0.8 male(s)/female
total population: 0.97 male(s)/female (2008 est.)
Infant mortality rate:
total: 5.01 deaths/1,000 live births
male: 5.36 deaths/1,000 live births
female: 4.63 deaths/1,000 live births (2008 est.)
Life expectancy at birth:
total population: 79.65 years
male: 77.15 years
female: 82.35 years (2008 est.)
Total fertility rate: 1.58 children born/woman (2008 est.)
HIV/AIDS—adult prevalence rate: NA
HIV/AIDS—people living with HIV/AIDS: NA
HIV/AIDS—deaths: NA
Nationality:
noun: Channel Islander(s)
adjective: Channel Islander
Ethnic groups: Jersey 51.1%, Britons 34.8%, Irish, French, and other white 6.6%, Portuguese/Madeiran 6.4%, other 1.1% (2001 census)
Religions: Anglican, Roman Catholic, Baptist, Congregational New Church, Methodist, Presbyterian
Languages: English 94.5% (official), Portuguese 4.6%, other 0.9% (2001 census)
Literacy: NA

GOVERNMENT

Country name:
conventional long form: Bailiwick of Jersey
conventional short form: Jersey
Dependency status: British crown dependency
Government type: parliamentary democracy
Capital: *name:* Saint Helier
geographic coordinates: 49 11 N, 2 06 W
time difference: UTC 0 (5 hours ahead of Washington, DC during Standard Time)
daylight saving time: +1hr, begins last Sunday in March; ends last Sunday in October
Administrative divisions: none (British crown dependency); there are no first-order administrative divisions as defined by the US Government, but there are 12 parishes including Grouville, Saint Brelade, Saint Clement, Saint Helier, Saint John, Saint Lawrence, Saint Martin, Saint Mary, Saint Quen, Saint Peter, Saint Saviour, and Trinity
Independence: none (British crown dependency)
National holiday: Liberation Day, 9 May (1945)
Constitution: unwritten; partly statutes, partly common law and practice
Legal system: the laws of the UK, where applicable, apply and local statutes; justice is administered by the Royal Court
Suffrage: 16 years of age; universal
Executive branch:
chief of state: Queen ELIZABETH II (since 6 February 1952); represented by Lieutenant Governor Andrew RIDGEWAY (since 14 June 2006)
head of government: Chief Minister Frank WALKER (since December 2005); Bailiff Philip Martin BAILHACHE (since February 1995)
cabinet: Cabinet (since December 2005)
elections: ministers of the Cabinet including the chief minister are elected by the Assembly of States; the monarch is hereditary; lieutenant governor and bailiff appointed by the monarch
Legislative branch: unicameral Assembly of the States of Jersey (58 seats; 55 are voting members, of which 12 are senators elected for six-year terms, 12 are constables or heads of parishes elected for three-year terms, 29 are deputies elected for three-year terms, the bailiff and the deputy bailiff, and 3 nonvoting members includes the Dean of Jersey, the Attorney General, and the Solicitor General appointed by the monarch)

elections: last held 19 October 2005 for senators and 23 November 2005 for deputies (next to be held in 2008)

election results: percent of vote—NA; seats—independents 55

Judicial branch: Royal Court (judges elected by an electoral college and the bailiff)

Political parties and leaders: two declared parties: Centre Party; Jersey Democratic Alliance

note: all senators and deputies elected in 2005 were independents

Political pressure groups and leaders: none

Diplomatic representation in the US: none (British crown dependency)

Diplomatic representation from the US: none (British crown dependency)

Flag description: white with a diagonal red cross extending to the corners of the flag; in the upper quadrant, surmounted by a yellow crown, a red shield with the three lions of England in yellow

ECONOMY

Economy—overview: Jersey's economy is based on international financial services, agriculture, and tourism. In 2005 the finance sector accounted for about 50% of the island's output. Potatoes, cauliflower, tomatoes, and especially flowers are important export crops, shipped mostly to the UK. The Jersey breed of dairy cattle is known worldwide and represents an important export income earner. Milk products go to the UK and other EU countries. Tourism accounts for one-quarter of GDP. In recent years, the government has encouraged light industry to locate in Jersey, with the result that an electronics industry has developed alongside the traditional manufacturing of knitwear. All raw material and energy requirements are imported, as well as a large share of Jersey's food needs. Light taxes and death duties make the island a popular tax haven. Living standards come close to those of the UK.

GDP (purchasing power parity): $5.1 billion (2005 est.)

GDP (official exchange rate): $5.1 billion (2005 est.)

GDP—real growth rate: NA%

GDP—per capita (PPP): $57,000 (2005 est.)

GDP—composition by sector:

agriculture: 1%

industry: 2%

services: 97% (2005)

Labor force: 53,560 (June 2006)

Unemployment rate: 2.2% (2006 est.)

Population below poverty line: NA%

Household income or consumption by percentage share:

lowest 10%: NA%

highest 10%: NA%

Inflation rate (consumer prices): 3.7% (December 2006)

Budget:

revenues: $829 million

expenditures: $851 million (2005)

Agriculture—products: potatoes, cauliflower, tomatoes; beef, dairy products

Industries: tourism, banking and finance, dairy, electronics

Industrial production growth rate: NA%

Electricity—consumption: 630.1 million kWh (2004 est.)

Electricity—imports: NA kWh; note—electricity supplied by France

Exports: $NA

Exports—commodities: light industrial and electrical goods, dairy cattle, foodstuffs, textiles

Exports—partners: UK (2006)

Imports: $NA

Imports—commodities: machinery and transport equipment, manufactured goods, foodstuffs, mineral fuels, chemicals

Imports—partners: UK (2006)

Debt—external: $NA

Market value of publicly traded shares: $NA

Currency (code): Jersey pound

note: the British pound is also legal tender

Currency code: GBP

Exchange rates: Jersey pounds per US dollar—0.4993 (2007), 0.5418 (2006), 0.5493 (2005), 0.5462 (2004), 0.6125 (2003)

note: the Jersey pound is at par with the British pound

Fiscal year: 1 April—31 March

COMMUNICATIONS

Telephones—main lines in use: 73,900 (2001)

Telephones—mobile cellular: 83,900 (2004)

Telephone system:

general assessment: NA

domestic: NA

international: submarine cable connectivity to Guernsey and UK

Radio broadcast stations: AM NA, FM 1, shortwave 0 (1998)

Radios: NA

Television broadcast stations: 2 (1997)

Televisions: NA

Internet country code: .je

Internet hosts: 2,219 (2007)

Internet Service Providers (ISPs): NA

Internet users: 27,000 (2005)

TRANSPORTATION

Airports: 1 (2007)

Airports—with paved runways:

total: 1

1,524 to 2,437 m: 1 (2007)

Roadways:

total: 577 km (2002)

Ports and terminals: Gorey, Saint Aubin, Saint Helier

MILITARY

Military—note: defense is the responsibility of the UK

TRANSNATIONAL ISSUES

Disputes—international: none

JORDAN

INTRODUCTION

Background: Following World War I and the dissolution of the Ottoman Empire, the UK received a mandate to govern much of the Middle East. Britain separated out a semi-autonomous region of Transjordan from Palestine in the early 1920s, and the area gained its independence in 1946; it adopted the name of Jordan in 1950. The country's long-time ruler was King HUSSEIN (1953–99). A pragmatic leader, he successfully navigated competing pressures from the major powers (US, USSR, and UK), various Arab states, Israel, and a large internal Palestinian population, despite several wars and coup attempts. In 1989 he reinstituted parliamentary elections and gradual political liberalization; in 1994 he signed a peace treaty with Israel. King ABDALLAH II, the son of King HUSSEIN, assumed the throne following his father's death in February 1999. Since then, he has consolidated his power and undertaken an aggressive economic reform program. Jordan acceded to the World Trade Organization in 2000, and began to participate in the European Free Trade Association in 2001. Municipal elections were held in July 2007 under a system in which 20% of seats in all municipal councils were reserved by quota for women. Parliamentary elections were held in November 2007 and saw independent pro-government candidates win the vast majority of seats. In November 2007, King Abdallah instructed his new

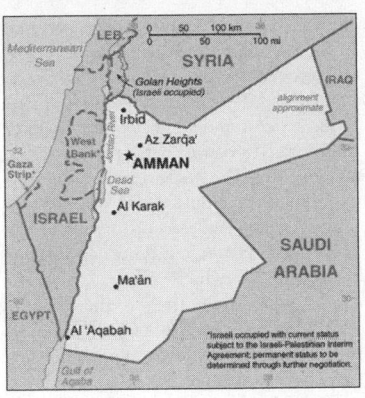

prime minister to focus on socioeconomic reform, developing a healthcare and housing network for civilians and military personnel, and improving the educational system.

GEOGRAPHY

Location: Middle East, northwest of Saudi Arabia
Geographic coordinates: 31 00 N, 36 00 E
Map references: Middle East
Area: *total:* 92,300 sq km
land: 91,971 sq km
water: 329 sq km
Area—comparative: slightly smaller than Indiana
Land boundaries:
total: 1,635 km
border countries: Iraq 181 km, Israel 238 km, Saudi Arabia 744 km, Syria 375 km, West Bank 97 km
Coastline: 26 km
Maritime claims: *territorial sea:* 3 nm
Climate: mostly arid desert; rainy season in west (November to April)
Terrain: mostly desert plateau in east, highland area in west; Great Rift Valley separates East and West Banks of the Jordan River
Elevation extremes:
lowest point: Dead Sea -408 m
highest point: Jabal Ram 1,734 m
Natural resources: phosphates, potash, shale oil
Land use: *arable land:* 3.32%
permanent crops: 1.18%
other: 95.5% (2005)
Irrigated land: 750 sq km (2003)
Total renewable water resources: 0.9 cu km (1997)
Freshwater withdrawal (domestic/industrial/agricultural): *total:* 1.01 cu km/yr (21%/4%/75%)
per capita: 177 cu m/yr (2000)
Natural hazards: droughts; periodic earthquakes
Environment—current issues: limited natural fresh water resources; deforestation; overgrazing; soil erosion; desertification

Environment—international agreements: *party to:* Biodiversity, Climate Change, Climate Change-Kyoto Protocol, Desertification, Endangered Species, Hazardous Wastes, Law of the Sea, Marine Dumping, Ozone Layer Protection, Wetlands
signed, but not ratified: none of the selected agreements
Geography—note: strategic location at the head of the Gulf of Aqaba and as the Arab country that shares the longest border with Israel and the occupied West Bank

PEOPLE

Population: 6,198,677 (July 2008 est.)
Age structure:
0–14 years: 32.2% (male 1,017,233/female 976,284)
15–64 years: 63.7% (male 2,110,293/female 1,840,531)
65 years and over: 4.1% (male 122,975/female 131,361) (2008 est.)
Median age:
total: 23.9 years
male: 24.6 years
female: 23.2 years (2008 est.)
Population growth rate: 2.338% (2008 est.)
Birth rate: 20.13 births/1,000 population (2008 est.)
Death rate: 2.72 deaths/1,000 population (2008 est.)
Net migration rate: 5.97 migrant(s)/1,000 population (2008 est.)
Sex ratio:
at birth: 1.06 male(s)/female
under 15 years: 1.04 male(s)/female
15–64 years: 1.15 male(s)/female
65 years and over: 0.94 male(s)/female
total population: 1.1 male(s)/female (2008 est.)
Infant mortality rate:
total: 15.57 deaths/1,000 live births
male: 18.62 deaths/1,000 live births
female: 12.34 deaths/1,000 live births (2008 est.)
Life expectancy at birth:
total population: 78.71 years
male: 76.19 years
female: 81.39 years (2008 est.)
Total fertility rate: 2.47 children born/woman (2008 est.)
HIV/AIDS—adult prevalence rate: less than 0.1% (2001 est.)
HIV/AIDS—people living with HIV/AIDS: 600 (2003 est.)
HIV/AIDS—deaths: fewer than 500 (2003 est.)
Nationality:
noun: Jordanian(s)
adjective: Jordanian
Ethnic groups: Arab 98%, Circassian 1%, Armenian 1%

Religions: Sunni Muslim 92%, Christian 6% (majority Greek Orthodox, but some Greek and Roman Catholics, Syrian Orthodox, Coptic Orthodox, Armenian Orthodox, and Protestant denominations), other 2% (several small Shi'a Muslim and Druze populations) (2001 est.)
Languages: Arabic (official), English widely understood among upper and middle classes
Literacy: *definition:* age 15 and over can read and write
total population: 89.9%
male: 95.1%
female: 84.7% (2003 est.)

GOVERNMENT

Country name:
conventional long form: Hashemite Kingdom of Jordan
conventional short form: Jordan
local long form: Al Mamlakah al Urduniyah al Hashimiyah
local short form: Al Urdun
former: Transjordan
Government type: constitutional monarchy
Capital: *name:* Amman
geographic coordinates: 31 57 N, 35 56 E
time difference: UTC+2 (7 hours ahead of Washington, DC during Standard Time)
daylight saving time: +1hr, begins last Thursday in March; ends last Friday in September
Administrative divisions: 12 governorates (muhafazat, singular—muhafazah); Ajlun, Al 'Aqabah, Al Balqa', Al Karak, Al Mafraq, 'Amman, At Tafilah, Az Zarqa', Irbid, Jarash, Ma'an, Madaba
Independence: 25 May 1946 (from League of Nations mandate under British administration)
National holiday: Independence Day, 25 May (1946)
Constitution: 1 January 1952; amended many times
Legal system: based on Islamic law and French codes; judicial review of legislative acts in a specially provided High Tribunal; has not accepted compulsory ICJ jurisdiction
Suffrage: 18 years of age; universal
Executive branch:
chief of state: King ABDALLAH II (since 7 February 1999); Prince HUSSEIN (born 1994), eldest son of King ABDALLAH II, is considered to be first in line to inherit the throne
head of government: Prime Minister Nader al-DAHABI (since 25 November 2007)
cabinet: Cabinet appointed by the prime minister in consultation with the monarch

339

elections: the monarch is hereditary; prime minister appointed by the monarch

Legislative branch: bicameral National Assembly or Majlis al-'Umma consists of the Senate, also called the House of Notables or Majlis al-Ayan (55 seats; members appointed by the monarch from designated categories of public figures to serve four-year terms) and the Chamber of Deputies, also called the House of Representatives or Majlis al-Nuwaab (110 seats; members elected by popular vote on the basis of proportional representation to serve four-year terms; note—six seats are reserved for women and are allocated by a special electoral panel if no women are elected)

elections: Chamber of Deputies—last held 20 November 2007 (next to be held in 2011)

election results: Chamber of Deputies—percent of vote by party—IAF 5.5 %, independents and other 94.5%; seats by party—IAF 6, independents and other 104; note—seven women will serve in the next Assembly—six of whom filled women's quota seats and one was directly elected

Judicial branch: Court of Cassation; Supreme Court (court of final appeal)

Political parties and leaders: al-Ahd Party; Arab Islamic Democratic Movement [Yusuf ABU BAKR]; Arab Land Party [Dr. Ayishah Salih HIJAZAYN]; Arab Socialist Ba'th Party [Taysir al-HIMSI]; Ba'th Arab Progressive Party [Fu'ad DABBUR]; Freedom Party; Future Party; Islamic Action Front or IAF [Zaki Sa'ed BANI IRSHEID]; Islamic Center Party [Marwan al-FAURI]; Jordanian Arab Ansar Party; Jordanian Arab New Dawn Party; Jordanian Arab Party; Jordanian Citizens' Rights Movement; Jordanian Communist Party [Munir HAMARINAH]; Jordanian Communist Workers Party; Jordanian Democratic Left Party [Musa MA'AYTEH]; Jordanian Democratic Popular Unity Party [Sa'id Dhiyab Ali MUSTAFA]; Jordanian Generations Party [Muhammad KHALAYLEH]; Jordanian Green Party [Muhammad BATAYNEH]; Jordanian Labor Party [Dr. Mazin Sulayman Jiryis HANNA]; Jordanian Peace Party; Jordanian People's Committees Movement; Jordanian People's Democratic Party (Hashd) [Ahmad YUSUF]; Jordanian Rafah Party; Jordanian Renaissance Party; Mission Party; Nation Party [Ahmad al-HANANDEH]; National Action Party (Haqq) [Tariq al-KAYYALI]; National Constitutional Party [Abdul Hadi MAJALI]; National Popular Democratic Movement [Mahmud al-NUWAYHI]; Progressive Party [Fawwaz al-ZUBI]

Political pressure groups and leaders: Anti-Normalization Committee [Ali Abu SUKKAR, president vice chairman]; Jordan Bar Association [Hussein Mujalli, chairman]; Jordanian Press Association [Sayf al-SHARIF, president]; Muslim Brotherhood [Salem AL-FALAHAT, controller general]

International organization participation: ABEDA, AFESD, AMF, CAEU, FAO, G-77, IAEA, IBRD, ICAO, ICC, ICCt, ICRM, IDA, IDB, IFAD, IFC, IFRCS, ILO, IMF, IMO, Interpol, IOC, IOM, IPU, ISO, ITSO, ITU, ITUC, LAS, MIGA, MINUSTAH, MONUC, NAM, OIC, OPCW, OSCE (partner), PCA, UN, UNCTAD, UNESCO, UNHCR, UNIDO, UNMEE, UNMIL, UNMIS, UNOCI, UNOMIG, UNRWA, UNWTO, UPU, WCO, WFTU, WHO, WIPO, WMO, WTO

Diplomatic representation in the US:
chief of mission: Ambassador ZEID Ra'ad Zeid al-Hussein, Prince
chancery: 3504 International Drive NW, Washington, DC 20008
telephone: [1] (202) 966-2664
FAX: [1] (202) 966-3110

Diplomatic representation from the US:
chief of mission: Ambassador David HALE
embassy: Abdun, Amman
mailing address: P. O. Box 354, Amman 11118 Jordan; Unit 70200, Box 5, APO AE 09892-0200
telephone: [962] (6) 590-6000
FAX: [962] (6) 592-0121

Flag description: three equal horizontal bands of black (top), representing the Abbassid Caliphate, white, representing the Ummayyad Caliphate, and green, representing the Fatimid Caliphate; a red isosceles triangle on the hoist side, representing the Great Arab Revolt of 1916, and bearing a small white seven-pointed star symbolizing the seven verses of the opening Sura (Al-Fatiha) of the Holy Koran; the seven points on the star represent faith in One God, humanity, national spirit, humility, social justice, virtue, and aspirations; design is based on the Arab Revolt flag of World War I

ECONOMY

Economy—overview: Jordan is a small Arab country with insufficient supplies of water, oil, and other natural resources. Poverty, unemployment, and inflation are fundamental problems, but King ABDALLAH II, since assuming the throne in 1999, has undertaken some broad economic reforms in a long-term effort to improve living standards. Since Jordan's graduation from its most recent IMF program in 2002, Amman has continued to follow IMF guidelines, practicing careful monetary policy, making substantial headway with privatization, and opening the trade regime. Jordan's exports have significantly increased under the free trade accord with the US and Jordanian Qualifying Industrial Zones (QIZ), which allow Jordan to export goods duty free to the US. In 2006, Jordan reduced its debt-to-GDP ratio significantly. These measures have helped improve productivity and have made Jordan more attractive for foreign investment. Before the US-led war in Iraq, Jordan imported most of its oil from Iraq. Since 2003, however, Jordan has been more dependent on oil from other Gulf nations. The government ended subsidies for petroleum and other consumer goods in 2008 in an effort to control the budget. The main challenges facing Jordan are reducing dependence on foreign grants, reducing the budget deficit, attracting investments, and creating jobs.

GDP (purchasing power parity): $27.99 billion (2007 est.)

GDP (official exchange rate): $16.01 billion (2007 est.)

GDP—real growth rate: 5.7% (2007 est.)

GDP—per capita (PPP): $4,900 (2007 est.)

GDP—composition by sector:
agriculture: 3.5%
industry: 10.3%
services: 86.2% (2007 est.)

Labor force: 1.563 million (2007 est.)

Labor force—by occupation: *agriculture:* 5%
industry: 12.5%
services: 82.5% (2001 est.)

Unemployment rate: 13.5% official rate; unofficial rate is approximately 30% (2007 est.)

Population below poverty line: 14.2% (2002)

Household income or consumption by percentage share:
lowest 10%: 2.7%
highest 10%: 30.6% (2003)

Distribution of family income—Gini index: 38.8 (2003)

Inflation rate (consumer prices): 5.4% (2007 est.)

Investment (gross fixed): 27.8% of GDP (2007 est.)

Budget:
revenues: $5.117 billion
expenditures: $6.468 billion (2007 est.)

Public debt: 72.7% of GDP (2007 est.)

Agriculture—products: citrus, tomatoes, cucumbers, olives; sheep, poultry, stone fruits, strawberries, dairy

Industries: clothing, phosphate mining, fertilizers, pharmaceuticals, petroleum refining, cement, potash, inorganic chemicals, light manufacturing, tourism
Industrial production growth rate: 7.7% (2007 est.)
Electricity—production: 9.074 billion kWh (2005)
Electricity—production by source:
fossil fuel: 99.4%
hydro: 0.6%
nuclear: 0%
other: 0% (2001)
Electricity—consumption: 8.49 billion kWh (2005)
Electricity—exports: 4 million kWh (2005)
Electricity—imports: 741 million kWh (2005)
Oil—production: 0 bbl/day (2005 est.)
Oil—consumption: 109,000 bbl/day (2005 est.)
Oil—exports: 0 bbl/day (2004 est.)
Oil—imports: 106,400 bbl/day (2004 est.)
Oil—proved reserves: 1 million bbl (1 January 2006 est.)
Natural gas—production: 268.5 million cu m (2005 est.)
Natural gas—consumption: 1.496 billion cu m (2005 est.)
Natural gas—exports: 0 cu m (2005 est.)
Natural gas—imports: 1.228 billion cu m (2005)
Natural gas—proved reserves: 5.975 billion cu m (1 January 2006 est.)
Current account balance: -$2.769 billion (2007 est.)
Exports: $5.7 billion f.o.b. (2007 est.)
Exports—commodities: clothing, pharmaceuticals, potash, phosphates, fertilizers, vegetables, manufactures;
Exports—partners: US 25%, Iraq 12.2%, India 7.6%, Saudi Arabia 7.6%, UAE 5.6%, Switzerland 5.3%, Syria 5.2% (2006)
Imports: $12.02 billion f.o.b. (2007 est.)
Imports—commodities: crude oil, textile fabrics, machinery, transport equipment, manufactured goods
Imports—partners: Saudi Arabia 25.4%, China 10.5%, Germany 7.9%, US 4.8%, Egypt 4.2% (2006)
Economic aid—recipient: ODA, $752 million (2005 est.)
Reserves of foreign exchange and gold: $7.929 billion (31 December 2007 est.)
Debt—external: $8.206 billion (31 December 2007 est.)
Stock of direct foreign investment—at home: $10.24 billion (2007 est.)
Market value of publicly traded shares: $29.73 billion (2006)
Currency (code): Jordanian dinar (JOD)

Currency code: JOD
Exchange rates: Jordanian dinars per US dollar—0.709 (2007), 0.709 (2006), 0.709 (2005), 0.709 (2004), 0.709 (2003)
Fiscal year: calendar year

COMMUNICATIONS

Telephones—main lines in use: 614,000 (2006)
Telephones—mobile cellular: 4.343 million (2006)
Telephone system:
general assessment: service has improved recently with increased use of digital switching equipment; microwave radio relay transmission and coaxial and fiber-optic cable are employed on trunk lines; growing mobile-cellular usage in both urban and rural areas is reducing use of fixed-line services; Internet penetration remains modest and slow-growing
domestic: 1995 telecommunications law opened all non-fixed-line services to private competition; in 2005, monopoly over fixed-line services terminated and the entire telecommunications sector was opened to competition; mobile-cellular usage is increasing rapidly and tele-density is approaching 75 per 100 persons
international: country code—962; landing point for the Fiber-Optic Link Around the Globe (FLAG) submarine cable network that provides links to Asia, Middle East, Europe; satellite earth stations—33 (3 Intelsat, 1 Arabsat, and 29 land and maritime Inmarsat terminals); fiber-optic cable to Saudi Arabia and microwave radio relay link with Egypt and Syria; participant in Medarabtel
Radio broadcast stations: FM 31 (2007)
Radios: 1.66 million (1997)
Television broadcast stations: 22 (2007)
Televisions: 500,000 (1997)
Internet country code: .jo
Internet hosts: 2,500 (2007)
Internet Service Providers (ISPs): 5 (2000)
Internet users: 796,900 (2006)

TRANSPORTATION

Airports: 17 (2007)
Airports—with paved runways:
total: 15
over 3,047 m: 7
2,438 to 3,047 m: 6
914 to 1,523 m: 1
under 914 m: 1 (2007)
Airports—with unpaved runways:
total: 2
under 914 m: 2 (2007)
Heliports: 1 (2007)

Pipelines: gas 426 km; oil 49 km (2007)
Railways:
total: 505 km
narrow gauge: 505 km 1.050-m gauge (2006)
Roadways:
total: 7,601 km
paved: 7,601 km (2005)
Merchant marine:
total: 30 ships (1000 GRT or over) 410,472 GRT/564,643 DWT
by type: bulk carrier 2, cargo 11, container 3, passenger/cargo 8, petroleum tanker 2, roll on/roll off 4
foreign-owned: 15 (UAE 15)
registered in other countries: 15 (Bahamas 2, Panama 11, Syria 2) (2007)
Ports and terminals: Al 'Aqabah

MILITARY

Military branches: Jordanian Armed Forces (JAF): Royal Jordanian Land Force, Royal Jordanian Navy, Royal Jordanian Air Force (Al-Quwwat al-Jawwiya al-Malakiya al-Urduniya), Special Operations Command (Socom); Public Security Directorate (normally falls under Ministry of Interior, but comes under JAF in wartime or crisis situations) (2006)
Military service age and obligation: 17 years of age for voluntary military service; conscription at age 18 was suspended in 1999, although all males under age 37 are required to register; women not subject to conscription, but can volunteer to serve in non-combat military positions (2004)
Manpower available for military service:
males age 16–49: 1,812,551
females age 16–49: 1,559,155 (2008 est.)
Manpower fit for military service:
males age 16–49: 1,546,766
females age 16–49: 1,339,366 (2008 est.)
Manpower reaching militarily significant age annually:
males age 16–49: 68,067
females age 16–49: 65,512 (2008 est.)
Military expenditures—percent of GDP: 8.6% (2006)

TRANSNATIONAL ISSUES

Disputes—international: approximately two million Iraqis have fled the conflict in Iraq, with the majority taking refuge in Syria and Jordan; 2004 Agreement settles border dispute with Syria pending demarcation
Refugees and internally displaced persons: *refugees (country of origin):* 1,835,704 (Palestinian Refugees (UNRWA)); 500,000 (Iraq)
IDPs: 160,000 (1967 Arab-Israeli War) (2007)

KAZAKHSTAN

INTRODUCTION

Background: Native Kazakhs, a mix of Turkic and Mongol nomadic tribes who migrated into the region in the 13th century, were rarely united as a single nation. The area was conquered by Russia in the 18th century, and Kazakhstan became a Soviet Republic in 1936. During the 1950s and 1960s agricultural "Virgin Lands" program, Soviet citizens were encouraged to help cultivate Kazakhstan's northern pastures. This influx of immigrants (mostly Russians, but also some other deported nationalities) skewed the ethnic mixture and enabled non-Kazakhs to outnumber natives. Independence in 1991 caused many of these newcomers to emigrate. Kazakhstan's economy is larger than those of all the other Central Asian states combined, largely due to the country's vast natural resources and a recent history of political stability. Current issues include: developing a cohesive national identity; expanding the development of the country's vast energy resources and exporting them to world markets; achieving a sustainable economic growth; diversifying the economy outside the oil, gas, and mining sectors; enhancing Kazakhstan's competitiveness; and strengthening relations with neighboring states and other foreign powers.

GEOGRAPHY

Location: Central Asia, northwest of China; a small portion west of the Ural River in eastern-most Europe
Geographic coordinates: 48 00 N, 68 00 E
Map references: Asia
Area:
total: 2,717,300 sq km
land: 2,669,800 sq km
water: 47,500 sq km

Area—comparative: slightly less than four times the size of Texas
Land boundaries:
total: 12,012 km
border countries: China 1,533 km, Kyrgyzstan 1,051 km, Russia 6,846 km, Turkmenistan 379 km, Uzbekistan 2,203 km
Coastline: 0 km (landlocked); note—Kazakhstan borders the Aral Sea, now split into two bodies of water (1,070 km), and the Caspian Sea (1,894 km)
Maritime claims: none (landlocked)
Climate: continental, cold winters and hot summers, arid and semiarid
Terrain: extends from the Volga to the Altai Mountains and from the plains in western Siberia to oases and desert in Central Asia
Elevation extremes:
lowest point: Vpadina Kaundy -132 m
highest point: Khan Tangiri Shyngy (Pik Khan-Tengri) 6,995 m
Natural resources: major deposits of petroleum, natural gas, coal, iron ore, manganese, chrome ore, nickel, cobalt, copper, molybdenum, lead, zinc, bauxite, gold, uranium
Land use:
arable land: 8.28%
permanent crops: 0.05%
other: 91.67% (2005)
Irrigated land: 35,560 sq km (2003)
Total renewable water resources: 109.6 cu km (1997)
Freshwater withdrawal (domestic/industrial/agricultural): *total:* 35 cu km/yr (2%/17%/82%)
per capita: 2,360 cu m/yr (2000)
Natural hazards: earthquakes in the south, mudslides around Almaty
Environment—current issues: radioactive or toxic chemical sites associated with former defense industries and test ranges scattered throughout the country pose health risks for humans and ani-

mals; industrial pollution is severe in some cities; because the two main rivers that flowed into the Aral Sea have been diverted for irrigation, it is drying up and leaving behind a harmful layer of chemical pesticides and natural salts; these substances are then picked up by the wind and blown into noxious dust storms; pollution in the Caspian Sea; soil pollution from overuse of agricultural chemicals and salination from poor infrastructure and wasteful irrigation practices
Environment—international agreements: *party to:* Air Pollution, Biodiversity, Climate Change, Desertification, Endangered Species, Hazardous Wastes, Ozone Layer Protection, Ship Pollution, Wetlands
signed, but not ratified: Climate Change-Kyoto Protocol
Geography—note: landlocked; Russia leases approximately 6,000 sq km of territory enclosing the Baykonur Cosmodrome; in January 2004, Kazakhstan and Russia extended the lease to 2050

PEOPLE

Population: 15,340,533 (July 2008 est.)
Age structure:
0–14 years: 22.1% (male 1,734,622/female 1,659,723)
15–64 years: 69.6% (male 5,219,983/female 5,463,468)
65 years and over: 8.2% (male 443,483/female 819,254) (2008 est.)
Median age:
total: 29.3 years
male: 27.8 years
female: 31.1 years (2008 est.)
Population growth rate: 0.374% (2008 est.)
Birth rate: 16.44 births/1,000 population (2008 est.)
Death rate: 9.39 deaths/1,000 population (2008 est.)
Net migration rate: -3.31 migrant(s)/1,000 population (2008 est.)
Sex ratio:
at birth: 1.06 male(s)/female
under 15 years: 1.05 male(s)/female
15–64 years: 0.96 male(s)/female
65 years and over: 0.54 male(s)/female
total population: 0.93 male(s)/female (2008 est.)
Infant mortality rate:
total: 26.56 deaths/1,000 live births
male: 31.03 deaths/1,000 live births
female: 21.83 deaths/1,000 live births (2008 est.)
Life expectancy at birth:
total population: 67.55 years

male: 62.24 years

female: 73.16 years (2008 est.)

Total fertility rate: 1.88 children born/woman (2008 est.)

HIV/AIDS—adult prevalence rate: 0.2% (2001 est.)

HIV/AIDS—people living with HIV/AIDS: 16,500 (2001 est.)

HIV/AIDS—deaths: fewer than 200 (2003 est.)

Nationality:

noun: Kazakhstani(s)

adjective: Kazakhstani

Ethnic groups: Kazakh (Qazaq) 53.4%, Russian 30%, Ukrainian 3.7%, Uzbek 2.5%, German 2.4%, Tatar 1.7%, Uygur 1.4%, other 4.9% (1999 census)

Religions: Muslim 47%, Russian Orthodox 44%, Protestant 2%, other 7%

Languages: Kazakh (Qazaq, state language) 64.4%, Russian (official, used in everyday business, designated the "language of interethnic communication") 95% (2001 est.)

Literacy:

definition: age 15 and over can read and write

total population: 99.5%

male: 99.8%

female: 99.3% (1999 est.)

GOVERNMENT

Country name:

conventional long form: Republic of Kazakhstan

conventional short form: Kazakhstan

local long form: Qazaqstan Respublikasy

local short form: Qazaqstan

former: Kazakh Soviet Socialist Republic

Government type: republic; authoritarian presidential rule, with little power outside the executive branch

Capital: *name*: Astana

geographic coordinates: 51 10 N, 71 25 E

time difference: UTC+6 (11 hours ahead of Washington, DC during Standard Time)

note: Kazakhstan is divided into two time zones

Administrative divisions: 14 provinces (oblystar, singular—oblys) and 3 cities* (qala, singular—qalasy); Almaty Oblysy, Almaty Qalasy*, Aqmola Oblysy (Astana), Aqtobe Oblysy, Astana Qalasy*, Atyrau Oblysy, Batys Qazaqstan Oblysy (Oral), Bayqongyr Qalasy*, Mangghystau Oblysy (Aqtau), Ongtustik Qazaqstan Oblysy (Shymkent), Pavlodar Oblysy, Qaraghandy Oblysy, Qostanay Oblysy, Qyzylorda Oblysy, Shyghys Qazaqstan Oblysy (Oskemen), Soltustik Qazaqstan Oblysy (Petropavlovsk), Zhambyl Oblysy (Taraz)

note: administrative divisions have the same names as their administrative cen-

ters (exceptions have the administrative center name following in parentheses); in 1995, the Governments of Kazakhstan and Russia entered into an agreement whereby Russia would lease for a period of 20 years an area of 6,000 sq km enclosing the Baykonur space launch facilities and the city of Bayqongyr (Baykonur, formerly Leninsk); in 2004, a new agreement extended the lease to 2050

Independence: 16 December 1991 (from Soviet Union)

National holiday: Independence Day, 16 December (1991)

Constitution: first post-independence constitution adopted 28 January 1993; new constitution adopted by national referendum 30 August 1995

Legal system: based on Islamic law and Roman law; has not accepted compulsory ICJ jurisdiction

Suffrage: 18 years of age; universal

Executive branch:

chief of state: President Nursultan A. NAZARBAYEV (chairman of the Supreme Soviet from 22 February 1990, elected president 1 December 1991)

head of government: Prime Minister Karim MASIMOV (since 10 January 2007); Deputy Prime Ministers Umirzak SHUKEYEV (since 27 August 2007) and Yerbol ORYNBAYEV (since 29 October 2007)

cabinet: Council of Ministers appointed by the president

elections: president elected by popular vote for a seven-year term (no term limits); election last held 4 December 2005 (next to be held in 2012); prime minister and first deputy prime minister appointed by the president

election results: Nursultan A. NAZARBAYEV reelected president; percent of vote—Nursultan A. NAZARBAYEV 91.1%, Zharmakhan A. TUYAKBAI 6.6%, Alikhan M. BAIMENOV 1.6%

note: President NAZARBAYEV arranged a referendum in 1995 that extended his term of office and expanded his presidential powers: only he can initiate constitutional amendments, appoint and dismiss the government, dissolve Parliament, call referenda at his discretion, and appoint administrative heads of regions and cities

Legislative branch: bicameral Parliament consists of the Senate (47 seats; 7 members are appointed by the president; other members are elected by local assemblies; to serve six-year terms) and the Mazhilis (107 seats; 9 out of the 107 Mazhilis members are elected from the Assembly of the People of

Kazakhstan, which represents the country's ethnic minorities; members are popularly elected to serve five-year terms)

elections: Senate—(indirect) last held December 2005; next to be held in 2011; Mazhilis—last held 18 August 2007 (next to be held in 2012)

election results: Senate—percent of vote by party—NA; seats by party—NA; Mazhilis—percent of vote by party— Nur-Otan 88.1%, NSDP 4.6%, Ak Zhol 3.3%, Auyl 1.6%, Communist People's Party 1.3%, Patriots Party .8% Ruhaniyat .4%; seats by party—Nur-Otan 98; note—parties must achieve a threshold of 7% of the electorate to qualify for seats in the Mazhilis

Judicial branch: Supreme Court (44 members); Constitutional Council (7 members)

Political parties and leaders: Adilet (Justice) [Maksut NARIKBAYEV, Zeynulla ALSHIMBAYEV, Bakhytbek AKHMETZHAN, Yerkin ONGARBAYEV, Tolegan SYDYKOV] (formerly Democratic Party of Kazakhstan); Agrarian and Industrial Union of Workers Block or AIST (Agrarian Party and Civic Party); Ak Zhol Party (Bright Path) [Alikhan BAIMENOV]; Auyl (Village) [Gani KALIYEV]; Communist Party of Kazakhstan or KPK [Serikbolsyn ABDILDIN]; Communist People's Party of Kazakhstan [Vladislav KOSAREV]; National Social Democratic Party (NSDP)[Zharmakhan TUYAKBAY]; Nur-Otan [Bakhytzhan ZHUMAGULOV] (the Agrarian, Asar, and Civic parties merged with Otan); Patriots' Party [Gani KASYMOV]; Rukhaniyat (Spirituality) [Altynshash ZHAGANOVA]

Political pressure groups and leaders: Adil-Soz [Tamara KALEYEVA]; Almaty Helsinki Group [Ninel FOKINA]; Confederation of Free Trade Unions [Sergei BELKIN]; For a Just Kazakhstan [Bolat ABILOV]; For Fair Elections [Yevgeniy ZHOVTIS, Sabit ZHUSUPOV, Sergey DUVANOV, Ibrash NUSUPBAYEV]; Kazakhstan International Bureau on Human Rights [Yevgeniy ZHOVTIS, executive director]; Pan-National Social Democratic Party of Kazakhstan [Zharmakhan TUYAKBAI]; Pensioners Movement or Pokoleniye [Irina SAVOSTINA, chairwoman]; Republican Network of International Monitors [Dos KUSHIM]; Transparency International [Sergei ZLOTNIKOV]

International organization participation: ADB, CIS, CSTO, EAEC, EAPC, EBRD, ECO, FAO, GCTU, IAEA,

IBRD, ICAO, ICRM, IDA, IDB, IFAD, IFC, IFRCS, ILO, IMF, IMO, Interpol, IOC, IOM, IPU, ISO, ITSO, ITU, MIGA, NAM (observer), NSG, OAS (observer), OIC, OPCW, OSCE, PFP, SCO, UN, UNCTAD, UNESCO, UNIDO, UNWTO, UPU, WCL, WCO, WFTU, WHO, WIPO, WMO, WTO (observer)

Diplomatic representation in the US:
chief of mission: Ambassador Yerlan IDRISOV
chancery: 1401 16th Street NW, Washington, DC 20036
telephone: [1] (202) 232-5488
FAX: [1] (202) 232-5845
consulate(s): New York

Diplomatic representation from the US:
chief of mission: Ambassador John M. ORDWAY
embassy: Ak Bulak 4, Str. 23-22, Building #3, Astana 010010
mailing address: use embassy street address
telephone: [7] (7172) 70-21-00
FAX: [7] (7172) 34-08-90

Flag description: sky blue background representing the endless sky and a gold sun with 32 rays above a soaring golden steppe eagle in the center; on the hoist side is a "national ornamentation" in gold

ECONOMY

Economy—overview: Kazakhstan, the largest of the former Soviet republics in territory, excluding Russia, possesses enormous fossil fuel reserves and plentiful supplies of other minerals and metals. It also has a large agricultural sector featuring livestock and grain. Kazakhstan's industrial sector rests on the extraction and processing of these natural resources. The breakup of the USSR in December 1991 and the collapse in demand for Kazakhstan's traditional heavy industry products resulted in a short-term contraction of the economy, with the steepest annual decline occurring in 1994. In 1995–97, the pace of the government program of economic reform and privatization quickened, resulting in a substantial shifting of assets into the private sector. Kazakhstan enjoyed double-digit growth in 2000–01—8% or more per year in 2002–07—thanks largely to its booming energy sector, but also to economic reform, good harvests, and foreign investment. Inflation, however, jumped to more than 10% in 2007. In the energy sector, the opening of the Caspian Consortium pipeline in 2001, from western Kazakhstan's Tengiz oilfield to the Black Sea, substantially raised export capacity. In 2006 Kazakhstan completed

the Atasu-Alashankou portion of an oil pipeline to China that is planned in future construction to extend from the country's Caspian coast eastward to the Chinese border. The country has embarked upon an industrial policy designed to diversify the economy away from overdependence on the oil sector by developing its manufacturing potential. The policy aims to reduce the influence of foreign investment and foreign personnel. The government has engaged in several disputes with foreign oil companies over the terms of production agreements; tensions continue. Upward pressure on the local currency continued in 2007 due to massive oil-related foreign-exchange inflows. Aided by strong growth and foreign exchange earnings, Kazakhstan aspires to become a regional financial center and has created a banking system comparable to those in Central Europe.

GDP (purchasing power parity): $167.6 billion (2007 est.)

GDP (official exchange rate): $103.8 billion (2007 est.)

GDP—real growth rate: 8.5% (2007 est.)

GDP—per capita (PPP): $11,100 (2007 est.)

GDP—composition by sector:
agriculture: 5.8%
industry: 39.4%
services: 54.8% (2007 est.)

Labor force: 8.229 million (2007 est.)

Labor force—by occupation: *agriculture:* 32.2%
industry: 18%
services: 49.8% (2005)

Unemployment rate: 7.3% (2007 est.)

Population below poverty line: 13.8% (2007)

Household income or consumption by percentage share:
lowest 10%: 3.3%
highest 10%: 26.5% (2004 est.)

Distribution of family income—Gini index: 30.4 (2005)

Inflation rate (consumer prices): 10.8% (2007)

Investment (gross fixed): 29.7% of GDP (2007 est.)

Budget:
revenues: $23.58 billion
expenditures: $25.33 billion (2007 est.)

Public debt: 7.7% of GDP (2007 est.)

Agriculture—products: grain (mostly spring wheat), cotton; livestock

Industries: oil, coal, iron ore, manganese, chromite, lead, zinc, copper, titanium, bauxite, gold, silver, phosphates, sulfur, iron and steel; tractors and other agricultural machinery, electric motors, construction materials

Industrial production growth rate: 4.5% (2007)

Electricity—production: 76.34 billion kWh (2007)

Electricity—production by source:
fossil fuel: 84.3%
hydro: 15.7%
nuclear: 0%
other: 0% (2001)

Electricity—consumption: 76.43 billion kWh (2007)

Electricity—exports: 3.7 billion kWh (2007)

Electricity—imports: 4 billion kWh (2007)

Oil—production: 1.338 million bbl/day (2005 est.)

Oil—consumption: 234,000 bbl/day (2005 est.)

Oil—exports: 1 million bbl/day (2005 est.)

Oil—imports: 113,600 bbl/day (2004)

Oil—proved reserves: 9 billion bbl (1 January 2006 est.)

Natural gas—production: 16.69 billion cu m (2007)

Natural gas—consumption: 8.4 billion cu m (2007)

Natural gas—exports: 10.27 billion cu m (2007)

Natural gas—imports: 3.901 billion cu m (2007)

Natural gas—proved reserves: 1.765 trillion cu m (1 January 2006 est.)

Current account balance: -$6.851 billion (2007 est.)

Exports: $48.35 billion f.o.b. (2007 est.)

Exports—commodities: oil and oil products 59%, ferrous metals 19%, chemicals 5%, machinery 3%, grain, wool, meat, coal (2001)

Exports—partners: Germany 12.4%, Russia 11.6%, China 10.9%, Italy 10.5%, France 7.6%, Romania 5% (2006)

Imports: $33.21 billion f.o.b. (2007 est.)

Imports—commodities: machinery and equipment, metal products, foodstuffs

Imports—partners: Russia 36.4%, China 19.3%, Germany 7.4% (2006)

Economic aid—recipient: $229.2 million (2005)

Reserves of foreign exchange and gold: $17.39 billion (31 December 2007 est.)

Debt—external: $96.37 billion (31 December 2007)

Stock of direct foreign investment—at home: $40.16 billion (2007 est.)

Stock of direct foreign investment—abroad: $3.97 billion (September 2007)

Market value of publicly traded shares: $10.52 billion (2005)

Currency (code): tenge (KZT)

Currency code: KZT

Exchange rates: tenge per US dollar—122.55 (2007), 126.09 (2006), 132.88 (2005), 136.04 (2004), 149.58 (2003)
Fiscal year: calendar year

COMMUNICATIONS

Telephones—main lines in use: 2.928 million (2006)
Telephones—mobile cellular: 7.83 million (2006)
Telephone system:
general assessment: inherited an outdated telecommunications network from the Soviet era requiring modernization
domestic: intercity by landline and microwave radio relay; number of fixed-line connections is gradually increasing and fixed-line teledensity is about 20 per 100 persons; mobile-cellular usage is increasing rapidly and subscriptions now exceed 50 per 100 persons
international: country code—7; international traffic with other former Soviet republics and China carried by landline and microwave radio relay and with other countries by satellite and by the Trans-Asia-Europe (TAE) fiber-optic cable; satellite earth stations—2 Intelsat
Radio broadcast stations: AM 60, FM 17, shortwave 9 (1998)
Radios: 6.47 million (1997)
Television broadcast stations: 12 (plus 9 repeaters) (1998)
Televisions: 3.88 million (1997)
Internet country code: .kz
Internet hosts: 33,217 (2007)
Internet Service Providers (ISPs): 10 (with their own international channels) (2001)
Internet users: 1.247 million (2006)

TRANSPORTATION

Airports: 97 (2007)
Airports—with paved runways:
total: 65

over 3,047 m: 9
2,438 to 3,047 m: 27
1,524 to 2,437 m: 17
914 to 1,523 m: 4
under 914 m: 8 (2007)
Airports—with unpaved runways:
total: 32
over 3,047 m: 4
2,438 to 3,047 m: 6
1,524 to 2,437 m: 6
914 to 1,523 m: 4
under 914 m: 12 (2007)
Heliports: 5 (2007)
Pipelines: condensate 658 km; gas 11,082 km; oil 10,376 km; refined products 1,095 km (2007)
Railways: *total:* 13,700 km
broad gauge: 13,700 km 1.520-m gauge (3,700 km electrified) (2006)
Roadways:
total: 90,018 km
paved: 84,104 km
unpaved: 5,914 km (2004)
Waterways: 4,000 km (on the Ertis ((Irtysh)) River (80%) and Syr Darya ((Syrdariya)) River) (2006)
Merchant marine:
total: 5 ships (1000 GRT or over) 30,011 GRT/49,223 DWT
by type: petroleum tanker 4, refrigerated cargo 1 (2007)
Ports and terminals: Aqtau (Shevchenko), Atyrau (Gur'yev), Oskemen (Ust-Kamenogorsk), Pavlodar, Semey (Semipalatinsk)

MILITARY

Military branches: Ground Forces, Naval Force, Air and Air Defense Forces, Republican Guard
Military service age and obligation: 18 years of age for compulsory military service; conscript service obligation—2 years; minimum age for volunteers NA (2004)

Manpower available for military service:
males age 16–49: 4,176,731
females age 16–49: 4,219,636 (2008 est.)
Manpower fit for military service:
males age 16–49: 2,871,205
females age 16–49: 3,551,032 (2008 est.)
Manpower reaching militarily significant age annually:
males age 16–49: 145,495
females age 16–49: 140,149 (2008 est.)
Military expenditures—percent of GDP: 0.9% (Ministry of Defense expenditures) (FY02)

TRANSNATIONAL ISSUES

Disputes—international: Kyrgyzstan has yet to ratify the 2001 boundary delimitation with Kazakhstan; field demarcation of the boundaries with Turkmenistan commenced in 2005, and with Uzbekistan in 2004; demarcation is scheduled to get underway with Russia in 2007; demarcation with China was completed in 2002; creation of a seabed boundary with Turkmenistan in the Caspian Sea remains under discussion; equidistant seabed treaties have been ratified with Azerbaijan and Russia in the Caspian Sea, but no resolution has been made on dividing the water column among any of the littoral states
Refugees and internally displaced persons: *refugees (country of origin):* 3,700 (Russia); 508 (Afghanistan) (2007)
Illicit drugs: significant illicit cultivation of cannabis for CIS markets, as well as limited cultivation of opium poppy and ephedra (for the drug ephedrine); limited government eradication of illicit crops; transit point for Southwest Asian narcotics bound for Russia and the rest of Europe; significant consumer of opiates

KENYA

INTRODUCTION

Background: Founding president and liberation struggle icon Jomo KENYATTA led Kenya from independence in 1963 until his death in 1978, when President Daniel Toroitich arap MOI took power in a constitutional succession. The country was a de facto one-party state from 1969 until 1982 when the ruling Kenya African National Union (KANU) made itself the sole legal party in Kenya. MOI acceded to internal and external pressure for political liberalization in late 1991. The ethnically fractured opposition failed to dislodge KANU from power in elections in 1992 and 1997, which were marred by violence and fraud, but were viewed as having generally reflected the will of the Kenyan people. President MOI stepped down in December 2002 following fair and peaceful elections. Mwai KIBAKI, running as the candidate of the multi-ethnic, united opposition group, the National Rainbow Coalition (NARC), defeated KANU candidate Uhuru KENYATTA and assumed the presidency following a campaign centered on an anticorruption platform. KIBAKI's NARC coalition splintered in 2005 over the constitutional review process.

Government defectors joined with KANU to form a new opposition coalition, the Orange Democratic Movement, which defeated the government's draft constitution in a popular referendum in November 2005. KIBAKI¿s reelection in December 2007 brought charges of vote rigging from ODM candidate Raila ODINGA and unleashed two months of violence in which as many as 1,500 people died. UN-sponsored talks in late February produced a powersharing accord bringing ODINGA into the government in the restored position of prime minister.

GEOGRAPHY

Location: Eastern Africa, bordering the Indian Ocean, between Somalia and Tanzania
Geographic coordinates: 1 00 N, 38 00 E
Map references: Africa
Area:
total: 582,650 sq km
land: 569,250 sq km
water: 13,400 sq km
Area—comparative: slightly more than twice the size of Nevada
Land boundaries:
total: 3,477 km
border countries: Ethiopia 861 km, Somalia 682 km, Sudan 232 km, Tanzania 769 km, Uganda 933 km
Coastline: 536 km
Maritime claims:
territorial sea: 12 nm
exclusive economic zone: 200 nm
continental shelf: 200 m depth or to the depth of exploitation
Climate: varies from tropical along coast to arid in interior
Terrain: low plains rise to central highlands bisected by Great Rift Valley; fertile plateau in west
Elevation extremes:
lowest point: Indian Ocean 0 m
highest point: Mount Kenya 5,199 m
Natural resources: limestone, soda ash, salt, gemstones, fluorspar, zinc, diatomite, gypsum, wildlife, hydropower
Land use:
arable land: 8.01%
permanent crops: 0.97%
other: 91.02% (2005)
Irrigated land: 1,030 sq km (2003)
Total renewable water resources: 30.2 cu km (1990)
Freshwater withdrawal (domestic/industrial/agricultural): *total:* 1.58 cu km/yr (30%/6%/64%)
per capita: 46 cu m/yr (2000)
Natural hazards: recurring drought; flooding during rainy seasons
Environment—current issues: water pollution from urban and industrial wastes; degradation of water quality from increased use of pesticides and fertilizers; water hyacinth infestation in Lake Victoria; deforestation; soil erosion; desertification; poaching
Environment—international agreements: *party to:* Biodiversity, Climate Change, Climate Change-Kyoto Protocol, Desertification, Endangered Species, Hazardous Wastes, Law of the Sea, Marine Dumping, Marine Life Conservation, Ozone Layer Protection, Ship Pollution, Wetlands, Whaling
signed, but not ratified: none of the selected agreements
Geography—note: the Kenyan Highlands comprise one of the most successful agricultural production regions in Africa; glaciers are found on Mount Kenya, Africa's second highest peak; unique physiography supports abundant and varied wildlife of scientific and economic value

PEOPLE

Population: 37,953,838
note: estimates for this country explicitly take into account the effects of excess mortality due to AIDS; this can result in lower life expectancy, higher infant mortality, higher death rates, lower population growth rates, and changes in the distribution of population by age and sex than would otherwise be expected (July 2008 est.)
Age structure:
0–14 years: 42.2% (male 8,065,789/female 7,953,077)
15–64 years: 55.2% (male 10,498,468/female 10,434,764)
65 years and over: 2.6% (male 457,886/female 543,854) (2008 est.)
Median age: *total:* 18.6 years
male: 18.5 years
female: 18.8 years (2008 est.)
Population growth rate: 2.758% (2008 est.)
Birth rate: 37.89 births/1,000 population (2008 est.)
Death rate: 10.3 deaths/1,000 population (2008 est.)
Net migration rate: -1 migrant(s)/1,000 population (2005 est.)
Sex ratio:
at birth: 1.02 male(s)/female
under 15 years: 1.01 male(s)/female
15–64 years: 1.01 male(s)/female
65 years and over: 0.84 male(s)/female
total population: 1 male(s)/female (2008 est.)
Infant mortality rate:
total: 56.01 deaths/1,000 live births
male: 58.95 deaths/1,000 live births
female: 53.02 deaths/1,000 live births (2008 est.)
Life expectancy at birth:
total population: 56.64 years
male: 56.42 years
female: 56.87 years (2008 est.)
Total fertility rate: 4.7 children born/woman (2008 est.)
HIV/AIDS—adult prevalence rate: 6.7% (2003 est.)
HIV/AIDS—people living with HIV/AIDS: 1.2 million (2003 est.)
HIV/AIDS—deaths: 150,000 (2003 est.)
Major infectious diseases:
degree of risk: high
food or waterborne diseases: bacterial and protozoal diarrhea, hepatitis A, and typhoid fever
vectorborne disease: malaria
water contact disease: schistosomiasis (2008)
Nationality:
noun: Kenyan(s)
adjective: Kenyan
Ethnic groups: Kikuyu 22%, Luhya 14%, Luo 13%, Kalenjin 12%, Kamba 11%, Kisii 6%, Meru 6%, other African 15%, non-African (Asian, European, and Arab) 1%
Religions: Protestant 45%, Roman Catholic 33%, Muslim 10%, indigenous beliefs 10%, other 2%
note: a large majority of Kenyans are Christian, but estimates for the percentage of the population that adheres to Islam or indigenous beliefs vary widely
Languages: English (official), Kiswahili (official), numerous indigenous languages
Literacy:
definition: age 15 and over can read and write
total population: 85.1%
male: 90.6%
female: 79.7% (2003 est.)

GOVERNMENT

Country name:
conventional long form: Republic of Kenya
conventional short form: Kenya
local long form: Republic of Kenya/Jamhuri y Kenya
local short form: Kenya
former: British East Africa
Government type: republic
Capital: *name:* Nairobi
geographic coordinates: 1 17 S, 36 49 E
time difference: UTC+3 (8 hours ahead of Washington, DC during Standard Time)
Administrative divisions: 7 provinces and 1 area*; Central, Coast, Eastern, Nairobi Area*, North Eastern, Nyanza, Rift Valley, Western
Independence: 12 December 1963 (from UK)
National holiday: Independence Day, 12 December (1963)

Constitution: 12 December 1963; amended as a republic 1964; reissued with amendments 1979, 1982, 1986, 1988, 1991, 1992, 1997, 2001; note—a new draft constitution was defeated by popular referendum in 2005

Legal system: based on Kenyan statutory law, Kenyan and English common law, tribal law, and Islamic law; judicial review in High Court; accepts compulsory ICJ jurisdiction with reservations; constitutional amendment of 1982 making Kenya a de jure one-party state repealed in 1991

Suffrage: 18 years of age; universal

Executive branch:

chief of state: President Mwai KIBAKI (since 30 December 2002); Vice President Stephene Kalonzo MUSYOKA (since 10 January 2008);

head of government: Prime Minister Raila Amolo ODINGA (since 17 April 2008)

cabinet: Cabinet appointed by the president

elections: president elected by popular vote for a five-year term (eligible for a second term); in addition to receiving the largest number of votes in absolute terms, the presidential candidate must also win 25% or more of the vote in at least five of Kenya's seven provinces and one area to avoid a runoff; election last held 27 December 2007 (next to be held in December 2012); vice president appointed by the president

election results: President Mwai KIBAKI reelected; percent of vote—Mwai KIBAKI 46%, Raila ODINGA 44%, Kalonzo MUSYOKA 9%

Legislative branch: unicameral National Assembly or Bunge (224 seats; 210 members elected by popular vote to serve five-year terms, 12 so-called "nominated" members who are appointed by the president but selected by the parties in proportion to their parliamentary vote totals, 2 ex-officio members)

elections: last held 27 December 2007 (next to be held in December 2012)

election results: percent of vote by party— NA; seats by party—ODM 99, PNU 43, ODM-K 16, KANU 14 other 38; ex-officio 2; seats appointed by the president—TBD

Judicial branch: Court of Appeal (chief justice is appointed by the president); High Court

Political parties and leaders: Forum for the Restoration of Democracy-Kenya or FORD-Kenya [Musikari KOMBO]; Forum for the Restoration of Democracy-People or FORD-People [Simeon NYACHAE]; Kenya African National Union or KANU [Uhuru KENYATTA]; National Rainbow Coalition-Kenya or NARC-Kenya [Raphael TUJU]; Orange Democratic Movement or ODM [Raila ODINGA]; Orange Democratic Movement-Kenya or ODM-K [Kalonzo MUSYOKA]; Party of National Unity or PNU [Mwai KIBAKI]; Shirikisho Party of Kenya or SPK [Chirau Ali MWAK-WERE]

Political pressure groups and leaders: Council of Islamic Preachers of Kenya or CIPK [Sheikh Idris MOHAMMED]; Kenya Human Rights Commission [L. Muthoni WANYEKI]; labor unions; Muslim Human Rights Forum [Ali-Amin KIMATHI]; National Convention Executive Council or NCEC, a prore-form coalition of political parties and nongovernment organizations [Ndung'u WAINANA]; Protestant National Council of Churches of Kenya or NCCK [Canon Peter Karanja MWANGI]; Roman Catholic and other Christian churches; Supreme Council of Kenya Muslims or SUPKEM [Shaykh Abdul Gafur al-BUSAIDY]

International organization participation: ACP, AfDB, AU, C, COMESA, EAC, EADB, FAO, G-15, G-77, IAEA, IBRD, ICAO, ICCt, ICRM, IDA, IFAD, IFC, IFRCS, IGAD, ILO, IMF, IMO, IMSO, Interpol, IOC, IOM, IPU, ISO, ITSO, ITU, ITUC, MIGA, MINURSO, NAM, OPCW, PCA, UN, UNAMID, UNCTAD, UNESCO, UNHCR, UNIDO, UNMEE, UNMIL, UNMIS, UNOCI, UNWTO, UPU, WCO, WHO, WIPO, WMO, WTO

Diplomatic representation in the US:

chief of mission: Ambassador Peter Rateng Oginga OGEGO

chancery: 2249 R Street NW, Washington, DC 20008

telephone: [1] (202) 387-6101

FAX: [1] (202) 462-3829

consulate(s) general: Los Angeles

Diplomatic representation from the US:

chief of mission: Ambassador Michael RANNEBERGER

embassy: US Embassy, United Nations Avenue, Gigiri; P. O. Box 606 Village Market Nairobi

mailing address: Box 21A, Unit 64100, APO AE 09831

telephone: [254] (20) 537-800

FAX: [254] (20) 537-810

Flag description: three equal horizontal bands of black (top), red, and green; the red band is edged in white; a large warrior's shield covering crossed spears is superimposed at the center

ECONOMY

Economy—overview: The regional hub for trade and finance in East Africa, Kenya has been hampered by corruption and by reliance upon several primary goods whose prices have remained low. In 1997, the IMF suspended Kenya's Enhanced Structural Adjustment Program due to the government's failure to maintain reforms and curb corruption. A severe drought from 1999 to 2000 compounded Kenya's problems, causing water and energy rationing and reducing agricultural output. As a result, GDP contracted by 0.2% in 2000. The IMF, which had resumed loans in 2000 to help Kenya through the drought, again halted lending in 2001 when the government failed to institute several anticorruption measures. Despite the return of strong rains in 2001, weak commodity prices, endemic corruption, and low investment limited Kenya's economic growth to 1.2%. Growth lagged at 1.1% in 2002 because of erratic rains, low investor confidence, meager donor support, and political infighting up to the elections. In the key December 2002 elections, Daniel Arap MOI's 24-year-old reign ended, and a new opposition government took on the formidable economic problems facing the nation. After some early progress in rooting out corruption and encouraging donor support, the KIBAKI government was rocked by high-level graft scandals in 2005 and 2006. In 2006 the World Bank and IMF delayed loans pending action by the government on corruption. The international financial institutions and donors have since resumed lending, despite little action on the government's part to deal with corruption. The scandals have not weighed down growth, with estimated real GDP growth at more than 6 percent in 2007.

GDP (purchasing power parity): $58.88 billion (2007 est.)

GDP (official exchange rate): $29.3 billion (2007 est.)

GDP—real growth rate: 7% (2007 est.)

GDP—per capita (PPP): $1,700 (2007 est.)

GDP—composition by sector:

agriculture: 23.8%

industry: 16.7%

services: 59.5% (2007 est.)

Labor force: 11.85 million (2005 est.)

Labor force—by occupation: *agriculture:* 75%

industry and services: 25% (2003 est.)

Unemployment rate: 40% (2001 est.)

Population below poverty line: 50% (2000 est.)

Household income or consumption by percentage share:

lowest 10%: 2%

highest 10%: 37.2% (2000)

Distribution of family income—Gini index: 44.5 (1997)

Inflation rate (consumer prices): 9.8% (2007 est.)

Investment (gross fixed): 20.4% of GDP (2007 est.)

Budget:

revenues: $5.525 billion

expenditures: $6.493 billion (2007 est.)

Public debt: 40.5% of GDP (2007 est.)

Agriculture—products: tea, coffee, corn, wheat, sugarcane, fruit, vegetables; dairy products, beef, pork, poultry, eggs

Industries: small-scale consumer goods (plastic, furniture, batteries, textiles, clothing, soap, cigarettes, flour), agricultural products, horticulture, oil refining; aluminum, steel, lead; cement, commercial ship repair, tourism

Industrial production growth rate: 6.1% (2007 est.)

Electricity—production: 5.502 billion kWh (2005)

Electricity—production by source:

fossil fuel: 17.7%

hydro: 71%

nuclear: 0%

other: 11.3% (2001)

Electricity—consumption: 4.464 billion kWh (2005)

Electricity—exports: 0 kWh (2005)

Electricity—imports: 28 million kWh (2005)

Oil—production: 0 bbl/day (2005 est.)

Oil—consumption: 64,000 bbl/day (2005 est.)

Oil—exports: 8,563 bbl/day (2004)

Oil—imports: 70,540 bbl/day (2004)

Oil—proved reserves: 0 bbl (1 January 2006 est.)

Natural gas—production: 0 cu m (2005 est.)

Natural gas—consumption: 0 cu m (2005 est.)

Natural gas—exports: 0 cu m (2005 est.)

Natural gas—imports: 0 cu m (2005)

Natural gas—proved reserves: 0 cu m (1 January 2006 est.)

Current account balance: -$1.014 billion (2007 est.)

Exports: $4.054 billion f.o.b. (2007 est.)

Exports—commodities: tea, horticultural products, coffee, petroleum products, fish, cement

Exports—partners: Uganda 16.1%, UK 10.4%, US 8.3%, Netherlands 8%, Tanzania 7.8%, Pakistan 5% (2006)

Imports: $8.54 billion f.o.b. (2007 est.)

Imports—commodities: machinery and transportation equipment, petroleum products, motor vehicles, iron and steel, resins and plastics

Imports—partners: UAE 11.8%, India 8.8%, China 8.3%, Saudi Arabia 8.3%, US 7%, South Africa 6.4%, UK 5.3%, Japan 4.7% (2006)

Economic aid—recipient: $768.3 million (2005)

Reserves of foreign exchange and gold: $3.355 billion (31 December 2007 est.)

Debt—external: $6.713 billion (31 December 2007 est.)

Stock of direct foreign investment—at home: $1.249 billion (2007 est.)

Stock of direct foreign investment—abroad: $47.4 million (2007 est.)

Market value of publicly traded shares: $11.38 billion (2006)

Currency (code): Kenyan shilling (KES)

Currency code: KES

Exchange rates: Kenyan shillings per US dollar—68.309 (2007), 72.101 (2006), 75.554 (2005), 79.174 (2004), 75.936 (2003)

Fiscal year: 1 July—30 June

COMMUNICATIONS

Telephones—main lines in use: 293,400 (2006)

Telephones—mobile cellular: 6.485 million (2006)

Telephone system:

general assessment: inadequate; fixed-line telephone system is small and inefficient; trunks are primarily microwave radio relay; business data commonly transferred by a very small aperture terminal (VSAT) system

domestic: no recent growth in fixed-line infrastructure and the sole provider, Telkom Kenya, is slated for privatization; multiple providers in the mobile-cellular segment of the market fostering a boom in mobile-cellular telephone usage

international: country code—254; satellite earth stations—4 Intelsat

Radio broadcast stations: AM 24, FM 18, shortwave 6 (2001)

Radios: 3.07 million (1997)

Television broadcast stations: 8 (2001)

Televisions: 730,000 (1997)

Internet country code: .ke

Internet hosts: 2,120 (2007)

Internet Service Providers (ISPs): 65 (2001)

Internet users: 2.77 million (2006)

TRANSPORTATION

Airports: 225 (2007)

Airports—with paved runways: *total:* 15

over 3,047 m: 4

2,438 to 3,047 m: 1

1,524 to 2,437 m: 4

914 to 1,523 m: 5

under 914 m: 1 (2007)

Airports—with unpaved runways:

total: 210

1,524 to 2,437 m: 12

914 to 1,523 m: 113

under 914 m: 85 (2007)

Pipelines: refined products 900 km (2007)

Railways:

total: 2,778 km

narrow gauge: 2,778 km 1.000-m gauge (2006)

Roadways:

total: 63,265 km (interurban roads)

paved: 8,933 km

unpaved: 54,332 km

note: there also are 100,000 km of rural roads and 14,500 km of urban roads for a national total of 177,765 km (2004)

Waterways: part of Lake Victoria system is within boundaries of Kenya (2006)

Merchant marine:

total: 1 ship (1000 GRT or over) 3,737 GRT/5,558 DWT

by type: petroleum tanker 1

registered in other countries: 5 (Bahamas 1, Comoros 1, St Vincent and The Grenadines 2, Tuvalu 1, unknown 1) (2007)

Ports and terminals: Mombasa

MILITARY

Military branches: Kenyan Army, Kenyan Navy, Kenyan Air Force (2007)

Military service age and obligation: 18 years of age (est.) for voluntary service, with a 9-year obligation (2007)

Manpower available for military service:

males age 16–49: 9,044,685

females age 16–49: 8,805,736 (2008 est.)

Manpower fit for military service:

males age 16–49: 5,688,259

females age 16–49: 5,396,166 (2008 est.)

Military expenditures—percent of GDP: 2.8% (2006)

TRANSNATIONAL ISSUES

Disputes—international: Kenya served as an important mediator in brokering Sudan's north-south separation in February 2005; Kenya provides shelter to almost a quarter of a million refugees, including Ugandans who flee across the border periodically to seek protection from Lord's Resistance Army (LRA) rebels; Kenya works hard to prevent the clan and militia fighting in Somalia from spreading across the border, which has long been open to nomadic pastoralists; the boundary that separates Kenya's and Sudan's sovereignty is unclear in the "Ilemi Triangle," which Kenya has administered since colonial times

Refugees and internally displaced persons: *refugees (country of origin):* 173,702 (Somalia); 73,004 (Sudan); 16,428 (Ethiopia)

IDPs: 250,000–400,000 (2007 post-election violence; KANU attacks on opposition tribal groups in 1990s) (2007)

Trafficking in persons: *current situation:* Kenya is a source, transit, and destination country for men, women, and children trafficked for forced labor and sexual exploitation; children are trafficked within the country for domestic servitude, street vending, agricultural labor, and sexual exploitation; men, women, and girls are trafficked to the Middle East, other African nations, Western Europe, and North America for domestic servitude, enslavement in massage parlors and brothels, and manual labor; Chinese women trafficked for sexual exploitation reportedly transit Nairobi and Bangladeshis may transit Kenya for forced labor in other countries *tier rating:* Tier 2 Watch List—Kenya is placed on the Tier 2 Watch List due to a lack of evidence of increasing efforts to combat severe forms of trafficking

Illicit drugs: widespread harvesting of small plots of marijuana; transit country for South Asian heroin destined for Europe and North America; Indian methaqualone also transits on way to South Africa; significant potential for money-laundering activity given the country's status as a regional financial center; massive corruption, and relatively high levels of narcotics-associated activities

KIRIBATI

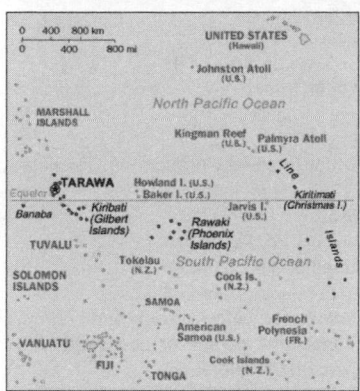

INTRODUCTION

Background: The Gilbert Islands were granted self-rule by the UK in 1971 and complete independence in 1979 under the new name of Kiribati. The US relinquished all claims to the sparsely inhabited Phoenix and Line Island groups in a 1979 treaty of friendship with Kiribati.

GEOGRAPHY

Location: Oceania, group of 33 coral atolls in the Pacific Ocean, straddling the Equator; the capital Tarawa is about half way between Hawaii and Australia; note—on 1 January 1995, Kiribati proclaimed that all of its territory lies in the same time zone as its Gilbert Islands group (UTC +12) even though the Phoenix Islands and the Line Islands under its jurisdiction lie on the other side of the International Date Line

Geographic coordinates: 1 25 N, 173 00 E

Map references: Oceania

Area: *total:* 811 sq km

land: 811 sq km

water: 0 sq km

note: includes three island groups—Gilbert Islands, Line Islands, Phoenix Islands

Area—comparative: four times the size of Washington, DC

Land boundaries: 0 km

Coastline: 1,143 km

Maritime claims:

territorial sea: 12 nm

exclusive economic zone: 200 nm

Climate: tropical; marine, hot and humid, moderated by trade winds

Terrain: mostly low-lying coral atolls surrounded by extensive reefs

Elevation extremes:

lowest point: Pacific Ocean 0 m

highest point: unnamed location on Banaba 81 m

Natural resources: phosphate (production discontinued in 1979)

Land use:

arable land: 2.74%

permanent crops: 47.95%

other: 49.31% (2005)

Irrigated land: NA

Natural hazards: typhoons can occur any time, but usually November to March; occasional tornadoes; low level of some of the islands make them sensitive to changes in sea level

Environment—current issues: heavy pollution in lagoon of south Tarawa atoll due to heavy migration mixed with traditional practices such as lagoon latrines and open-pit dumping; ground water at risk

Environment—international agreements: *party to:* Biodiversity, Climate Change, Climate Change-Kyoto Protocol, Desertification, Hazardous Wastes, Law of the Sea, Marine Dumping, Ozone Layer Protection, Whaling

signed, but not ratified: none of the selected agreements

Geography—note: 21 of the 33 islands are inhabited; Banaba (Ocean Island) in Kiribati is one of the three great phosphate rock islands in the Pacific Ocean—the others are Makatea in French Polynesia, and Nauru

PEOPLE

Population: 110,356 (July 2008 est.)

Age structure:

0–14 years: 37.9% (male 21,180/female 20,604)

15–64 years: 58.7% (male 31,993/female 32,797)

65 years and over: 3.4% (male 1,606/female 2,176) (2008 est.)

Median age: *total:* 20.6 years

male: 20.1 years

female: 21.1 years (2008 est.)

Population growth rate: 2.235% (2008 est.)

Birth rate: 30.31 births/1,000 population (2008 est.)

Death rate: 7.97 deaths/1,000 population (2008 est.)

Net migration rate: NA

Sex ratio:

at birth: 1.05 male(s)/female

under 15 years: 1.03 male(s)/female

15–64 years: 0.98 male(s)/female

65 years and over: 0.74 male(s)/female

total population: 0.99 male(s)/female (2008 est.)

Infant mortality rate:

total: 44.69 deaths/1,000 live births

male: 49.61 deaths/1,000 live births

female: 39.53 deaths/1,000 live births (2008 est.)

Life expectancy at birth:

total population: 62.85 years

male: 59.79 years

female: 66.06 years (2008 est.)

Total fertility rate: 4.08 children born/woman (2008 est.)

HIV/AIDS—adult prevalence rate: NA

HIV/AIDS—people living with HIV/AIDS: NA

HIV/AIDS—deaths: NA

Nationality:

noun: I-Kiribati (singular and plural)

adjective: I-Kiribati

Ethnic groups: Micronesian 98.8%, other 1.2% (2000 census)

Religions: Roman Catholic 52%, Protestant (Congregational) 40%, other (includes Seventh-Day Adventist, Muslim, Baha'i, Latter-day Saints, Church of God) 8% (1999)

Languages: I-Kiribati, English (official)
Literacy: NA

GOVERNMENT

Country name:
conventional long form: Republic of Kiribati
conventional short form: Kiribati
local long form: Republic of Kiribati
local short form: Kiribati
note: pronounced keer-ree-bahss
former: Gilbert Islands
Government type: republic
Capital: name: Tarawa
geographic coordinates: 1 19 N, 172 58 E
time difference: UTC+12 (17 hours ahead of Washington, DC during Standard Time)
Administrative divisions: 3 units; Gilbert Islands, Line Islands, Phoenix Islands; note—in addition, there are 6 districts (Banaba, Central Gilberts, Line Islands, Northern Gilberts, Southern Gilberts, Tarawa) and 21 island councils—one for each of the inhabited islands (Abaiang, Abemama, Aranuka, Arorae, Banaba, Beru, Butaritari, Kanton, Kiritimati, Kuria, Maiana, Makin, Marakei, Nikunau, Nonouti, Onotoa, Tabiteuea, Tabuaeran, Tamana, Tarawa, Teraina)
Independence: 12 July 1979 (from UK)
National holiday: Independence Day, 12 July (1979)
Constitution: 12 July 1979
Legal system: NA
Suffrage: 18 years of age; universal
Executive branch:
chief of state: President Anote TONG (since 10 July 2003); Vice President Teima ONORIO; note—the president is both the chief of state and head of government
head of government: President Anote TONG (since 10 July 2003); Vice President Teima ONORIO
cabinet: 12-member cabinet appointed by the president from among the members of the House of Parliament
elections: the House of Parliament chooses the presidential candidates from among its members and then those candidates compete in a general election; president is elected by popular vote for a four-year term (eligible for two more terms); election last held 17 October 2007 (next to be held in 2011); vice president appointed by the president
election results: Anote TONG 63.7%, Nabuti MWEMWENIKARAWA 32.9%
Legislative branch: unicameral House of Parliament or Maneaba Ni Maungatabu (46 seats; 44 members elected by popular vote, 1 ex officio member—the attorney general, 1 nominated by the Rabi Council of Leaders (representing Banaba Island); to serve four-year terms)
elections: legislative elections were held in two rounds—the first round on 22 August 2007 and the second round on 30 August 2007 (next to be held in 2011)
election results: percent of vote by party—NA; seats by party—NA, other 2 (includes attorney general)
Judicial branch: Court of Appeal; High Court; 26 Magistrates' courts; judges at all levels are appointed by the president
Political parties and leaders: Boutokaan Te Koaua Party or BTK [Taberannang TIMEON]; Maneaban Te Mauri Party or MTM [Teburoro TITO]; Maurin Kiribati Pati or MKP; National Progressive Party or NPP [Dr. Harry TONG]
note: there is no tradition of formally organized political parties in Kiribati; they more closely resemble factions or interest groups because they have no party headquarters, formal platforms, or party structures
Political pressure groups and leaders: NA
International organization participation: ACP, ADB, C, FAO, IBRD, ICAO, ICRM, IDA, IFAD, IFC, IFRCS, ILO, IMF, IMO, IOC, ITU, ITUC, OPCW, PIF, Sparteca, SPC, UN, UNCTAD, UNESCO, UPU, WHO, WMO
Diplomatic representation in the US: Kiribati does not have an embassy in the US; there is an honorary consulate in Honolulu
Diplomatic representation from the US: the US does not have an embassy in Kiribati; the ambassador to Fiji is accredited to Kiribati
Flag description: the upper half is red with a yellow frigate bird flying over a yellow rising sun, and the lower half is blue with three horizontal wavy white stripes to represent the ocean

ECONOMY

Economy—overview: A remote country of 33 scattered coral atolls, Kiribati has few natural resources. Commercially viable phosphate deposits were exhausted at the time of independence from the UK in 1979. Copra and fish now represent the bulk of production and exports. The economy has fluctuated widely in recent years. Economic development is constrained by a shortage of skilled workers, weak infrastructure, and remoteness from international markets. Tourism provides more than one-fifth of GDP. Private sector initiatives and a financial sector are in the early stages of development. Foreign financial aid from UK, Japan, Australia, New Zealand, and China equals more than 10% of GDP. Remittances from seamen on merchant ships abroad account for more than $5 million each year. Kiribati receives around $15 million annually for the government budget from an Australian trust fund.
GDP (purchasing power parity): $348 million (2007 est.)
GDP (official exchange rate): $67 million (2007 est.)
GDP—real growth rate: 2% (2007 est.)
GDP—per capita (PPP): $3,600 (2007 est.)
GDP—composition by sector:
agriculture: 8.9%
industry: 24.2%
services: 66.8% (2004)
Labor force: 7,870 economically active, not including subsistence farmers (2001 est.)
Labor force—by occupation: agriculture: 2.7%
industry: 32%
services: 65.3% (2000)
Unemployment rate: 2% official rate; underemployment 70% (1992 est.)
Population below poverty line: NA%
Household income or consumption by percentage share:
lowest 10%: NA%
highest 10%: NA%
Inflation rate (consumer prices): 0.2% (2007 est.)
Budget:
revenues: $55.52 million
expenditures: $59.71 million (FY05)
Agriculture—products: copra, taro, breadfruit, sweet potatoes, vegetables; fish
Industries: fishing, handicrafts
Industrial production growth rate: 0.7% (1991 est.)
Electricity—production: 9 million kWh (2005)
Electricity—production by source:
fossil fuel: 100%
hydro: 0%
nuclear: 0%
other: 0% (2001)
Electricity—consumption: 8.37 million kWh (2005)
Electricity—exports: 0 kWh (2005)
Electricity—imports: 0 kWh (2005)
Oil—production: 0 bbl/day (2005 est.)
Oil—consumption: 220 bbl/day (2005 est.)
Oil—exports: 0 bbl/day (2004)
Oil—imports: 216.4 bbl/day (2004)
Oil—proved reserves: 0 bbl (1 January 2006 est.)
Natural gas—production: 0 cu m (2005 est.)
Natural gas—consumption: 0 cu m (2005 est.)

Natural gas—exports: 0 cu m (2005 est.)

Natural gas—imports: 0 cu m (2005)

Natural gas—proved reserves: 0 cu m (1 January 2006 est.)

Current account balance: -$21 million (2004)

Exports: $17 million f.o.b. (2004 est.)

Exports—commodities: copra 62%, coconuts, seaweed, fish

Exports—partners: US 22.8%, Belgium 21.5%, Japan 14.3%, Samoa 7.8%, Australia 7.5%, Malaysia 6.7%, Taiwan 5.6%, Denmark 4.6% (2006)

Imports: $62 million c.i.f. (2004 est.)

Imports—commodities: foodstuffs, machinery and equipment, miscellaneous manufactured goods, fuel

Imports—partners: Australia 33%, Fiji 27.1%, Japan 18.1%, NZ 6.9% (2006)

Economic aid—recipient: $27.84 million largely from UK and Japan (2005)

Debt—external: $10 million (1999 est.)

Market value of publicly traded shares: $NA

Currency (code): Australian dollar (AUD)

Currency code: AUD

Exchange rates: Australian dollars per US dollar—1.2137 (2007), 1.3285 (2006), 1.3095 (2005), 1.3598 (2004), 1.5419 (2003)

Fiscal year: NA

COMMUNICATIONS

Telephones—main lines in use: 4,500 (2002)

Telephones—mobile cellular: 600 (2004)

Telephone system:
general assessment: generally good quality national and international service
domestic: wire line service available on Tarawa and Kiritimati (Christmas Island); connections to outer islands by HF/VHF radiotelephone; wireless service available in Tarawa since 1999
international: country code—686; Kiribati is being linked to the Pacific Ocean Cooperative Telecommunications Network, which should improve telephone service; satellite earth station—1 Intelsat (Pacific Ocean)

Radio broadcast stations: AM 1, FM 2, shortwave 1 (may be inactive) (2002)

Radios: 17,000 (1997)

Television broadcast stations: 1 (possibly inactive) (2002)

Televisions: 1,000 (1997)

Internet country code: .ki

Internet hosts: 41 (2007)

Internet Service Providers (ISPs): 1 (2000)

Internet users: 2,000 (2006)

TRANSPORTATION

Airports: 19 (2007)

Airports—with paved runways:
total: 4
1,524 to 2,437 m: 4 (2007)

Airports—with unpaved runways:
total: 15
914 to 1,523 m: 11
under 914 m: 4 (2007)

Roadways:
total: 670 km (2000)

Waterways: 5 km (small network of canals in Line Islands) (2007)

Merchant marine:
total: 7 ships (1000 GRT or over) 28,435 GRT/42,682 DWT
by type: bulk carrier 1, cargo 3, passenger/cargo 1, refrigerated cargo 2
foreign-owned: 3 (Malaysia 1, Singapore 1, Turkey 1) (2007)

Ports and terminals: Betio

MILITARY

Military branches: no regular military forces (constitutionally prohibited); Police Force (2008)

Manpower available for military service:
males age 16–49: 26,377 (2008 est.)

Manpower fit for military service:
males age 16–49: 17,577 (2008 est.)

Manpower reaching militarily significant age annually:
males age 16–49: 1,247 (2008 est.)

Military expenditures—percent of GDP: NA

Military—note: Kiribati does not have military forces; defense assistance is provided by Australia and NZ

TRANSNATIONAL ISSUES

Disputes—international: none

KOREA, NORTH

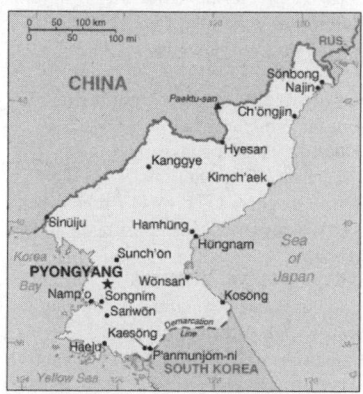

INTRODUCTION

Background: An independent kingdom for much of its long history, Korea was occupied by Japan in 1905 following the Russo-Japanese War. Five years later, Japan formally annexed the entire peninsula. Following World War II, Korea was split with the northern half coming under Soviet-sponsored Communist domination. After failing in the Korean War (1950–53) to conquer the US-backed Republic of Korea (ROK) in the southern portion by force, North Korea (DPRK), under its founder President KIM Il Sung, adopted a policy of ostensible diplomatic and economic "self-reliance" as a check against excessive Soviet or Communist Chinese influence. The DPRK demonized the US as the ultimate threat to its social system through state-funded propaganda, and molded political, economic, and military policies around the core ideological objective of eventual unification of Korea under Pyongyang's control. KIM's son, the current ruler KIM Jong Il, was officially designated as his father's successor in 1980, assuming a growing political and managerial role until the elder KIM's death in 1994. After decades of economic mismanagement and resource misallocation, the DPRK since the mid-1990s has relied heavily on international aid to feed its population while continuing to expend resources to maintain an army of approximately 1 million. North Korea's history of regional military provocations, proliferation of military-related items, and long-range missile development—as well as its nuclear, chemical, and biological weapons programs and massive conventional armed forces—are of major concern to the international community. In December 2002, following revelations that the DPRK was pursuing a nuclear weapons program based on enriched uranium in violation of a 1994 agreement with the US to freeze and ultimately dismantle its existing plutonium-based program, North Korea expelled monitors from the International Atomic Energy Agency (IAEA). In January 2003, it declared its withdrawal from the international Non-Proliferation Treaty. In mid-2003

351

Pyongyang announced it had completed the reprocessing of spent nuclear fuel rods (to extract weapons-grade plutonium) and was developing a "nuclear deterrent." Beginning in August 2003, North Korea, China, Japan, Russia, South Korea, and the US have participated in the Six-Party Talks aimed at resolving the stalemate over the DPRK's nuclear programs. North Korea pulled out of the talks in November 2005. It test-fired ballistic missiles in July 2006 and conducted a nuclear test in October 2006. North Korea returned to the Six-Party Talks in December 2006 and subsequently signed two agreements on denuclearization. The 13 February 2007 Initial Actions Agreement shut down the North's nuclear facilities at Yongbyon in July 2007. In the 3 October 2007 Second Phase Actions Agreement, Pyongyang pledged to disable those facilities and provide a correct and complete declaration of its nuclear programs. Under the supervision of US nuclear experts, North Korean personnel completed a number of agreed-upon disablement actions at the three core facilities at the Yongbyon nuclear complex by the end of 2007. North Korea also began the discharge of spent fuel rods in December 2007, but it did not provide a declaration of its nuclear programs by the end of the year.

GEOGRAPHY

Location: Eastern Asia, northern half of the Korean Peninsula bordering the Korea Bay and the Sea of Japan, between China and South Korea
Geographic coordinates: 40 00 N, 127 00 E
Map references: Asia
Area: *total:* 120,540 sq km
land: 120,410 sq km
water: 130 sq km
Area—comparative: slightly smaller than Mississippi
Land boundaries:
total: 1,673 km
border countries: China 1,416 km, South Korea 238 km, Russia 19 km
Coastline: 2,495 km
Maritime claims: *territorial sea:* 12 nm
exclusive economic zone: 200 nm
note: military boundary line 50 nm in the Sea of Japan and the exclusive economic zone limit in the Yellow Sea where all foreign vessels and aircraft without permission are banned
Climate: temperate with rainfall concentrated in summer
Terrain: mostly hills and mountains separated by deep, narrow valleys; coastal plains wide in west, discontinuous in east

Elevation extremes:
lowest point: Sea of Japan 0 m
highest point: Paektu-san 2,744 m
Natural resources: coal, lead, tungsten, zinc, graphite, magnesite, iron ore, copper, gold, pyrites, salt, fluorspar, hydropower
Land use: *arable land:* 22.4%
permanent crops: 1.66%
other: 75.94% (2005)
Irrigated land: 14,600 sq km (2003)
Total renewable water resources: 77.1 cu km (1999)
Freshwater withdrawal (domestic/industrial/agricultural): *total:* 9.02 cu km/yr (20%/25%/55%)
per capita: 401 cu m/yr (2000)
Natural hazards: late spring droughts often followed by severe flooding; occasional typhoons during the early fall
Environment—current issues: water pollution; inadequate supplies of potable water; waterborne disease; deforestation; soil erosion and degradation
Environment—international agreements: *party to:* Antarctic Treaty, Biodiversity, Climate Change, Climate Change-Kyoto Protocol, Environmental Modification, Ozone Layer Protection, Ship Pollution
signed, but not ratified: Law of the Sea
Geography—note: strategic location bordering China, South Korea, and Russia; mountainous interior is isolated and sparsely populated

PEOPLE

Population: 23,479,089 (July 2008 est.)
Age structure:
0–14 years: 22.9% (male 2,733,352/female 2,654,186)
15–64 years: 68.2% (male 7,931,484/female 8,083,626)
65 years and over: 8.8% (male 751,401/female 1,325,040) (2008 est.)
Median age:
total: 32.7 years
male: 31.2 years
female: 34.2 years (2008 est.)
Population growth rate: 0.732% (2008 est.)
Birth rate: 14.61 births/1,000 population (2008 est.)
Death rate: 7.29 deaths/1,000 population (2008 est.)
Net migration rate: NA
Sex ratio:
at birth: 1.05 male(s)/female
under 15 years: 1.03 male(s)/female
15–64 years: 0.98 male(s)/female
65 years and over: 0.57 male(s)/female
total population: 0.95 male(s)/female (2008 est.)
Infant mortality rate:
total: 21.86 deaths/1,000 live births

male: 23.46 deaths/1,000 live births
female: 20.18 deaths/1,000 live births (2008 est.)
Life expectancy at birth:
total population: 72.2 years
male: 69.45 years
female: 75.08 years (2008 est.)
Total fertility rate: 2 children born/woman (2008 est.)
HIV/AIDS—adult prevalence rate: NA
HIV/AIDS—people living with HIV/AIDS: NA
HIV/AIDS—deaths: NA
Nationality:
noun: Korean(s)
adjective: Korean
Ethnic groups: racially homogeneous; there is a small Chinese community and a few ethnic Japanese
Religions: traditionally Buddhist and Confucianist, some Christian and syncretic Chondogyo (Religion of the Heavenly Way)
note: autonomous religious activities now almost nonexistent; government-sponsored religious groups exist to provide illusion of religious freedom
Languages: Korean
Literacy:
definition: age 15 and over can read and write
total population: 99%
male: 99%
female: 99%

GOVERNMENT

Country name:
conventional long form: Democratic People's Republic of Korea
conventional short form: North Korea
local long form: Choson-minjujuui-inmin-konghwaguk
local short form: Choson
abbreviation: DPRK
Government type: Communist state one-man dictatorship
Capital: *name:* Pyongyang
geographic coordinates: 39 01 N, 125 45 E
time difference: UTC+9 (14 hours ahead of Washington, DC during Standard Time)
Administrative divisions: 9 provinces (do, singular and plural) and 4 municipalities (si, singular and plural)
provinces: Chagang-do (Chagang), Hamgyong-bukto (North Hamgyong), Hamgyong-namdo (South Hamgyong), Hwanghae-bukto (North Hwanghae), Hwanghae-namdo (South Hwanghae), Kangwon-do (Kangwon), P'yongan-bukto (North P'yongan), P'yongan-namdo (South P'yongan), Yanggang-do (Yanggang)
municipalities: Kaesong-si (Kaesong), Najin Sonbong-si (Najin-Sonbong),

Namp'o-si (Namp'o), P'yongyang-si (Pyongyang)

Independence: 15 August 1945 (from Japan)

National holiday: Founding of the Democratic People's Republic of Korea (DPRK), 9 September (1948)

Constitution: adopted 1948; completely revised 27 December 1972, revised again in April 1992, and September 1998

Legal system: based on Prussian civil law system with Japanese influences and Communist legal theory; no judicial review of legislative acts; has not accepted compulsory ICJ jurisdiction

Suffrage: 17 years of age; universal

Executive branch:
chief of state: KIM Jong Il (since July 1994); note—on 3 September 2003, rubberstamp Supreme People's Assembly (SPA) reelected KIM Jong Il chairman of the National Defense Commission, a position accorded nation's "highest administrative authority"; SPA reelected KIM Yong Nam president of its Presidium also with responsibility of representing state and receiving diplomatic credentials
head of government: Premier KIM Yong Il (since 11 April 2007); Vice Premiers KWAK Pom Gi (since 5 September 1998), JON Sung Hun (since 3 September 2003), RO Tu Chol (since 3 September 2003), THAE Jong Su (since 16 October 2007)
cabinet: Naegak (cabinet) members, except for Minister of People's Armed Forces, are appointed by SPA
elections: last held in September 2003 (next to be held in September 2008)
election results: KIM Jong Il and KIM Yong Nam were only nominees for positions and ran unopposed

Legislative branch: unicameral Supreme People's Assembly or Ch'oego Inmin Hoeui (687 seats; members elected by popular vote to serve five-year terms)
elections: last held 3 August 2003 (next to be held in August 2008)
election results: percent of vote by party—NA; seats by party—NA; ruling party approves a list of candidates who are elected without opposition; some seats are held by minor parties

Judicial branch: Central Court (judges are elected by the Supreme People's Assembly)

Political parties and leaders: major party—Korean Workers' Party or KWP [KIM Jong Il]; minor parties—Chondoist Chongu Party [RYU Mi Yong] (under KWP control), Social Democratic Party [KIM Yong Dae] (under KWP control)

Political pressure groups and leaders: none

International organization participation: ARF, FAO, G-77, ICAO, ICRM, IFAD, IFRCS, IHO, IMO, IOC, IPU, ISO, ITSO, ITU, NAM, UN, UNCTAD, UNESCO, UNIDO, UNWTO, UPU, WFTU, WHO, WIPO, WMO

Diplomatic representation in the US: none; North Korea has a Permanent Mission to the UN in New York

Diplomatic representation from the US: none; note—Swedish Embassy in Pyongyang represents the US as consular protecting power

Flag description: three horizontal bands of blue (top), red (triple width), and blue; the red band is edged in white; on the hoist side of the red band is a white disk with a red five-pointed star

ECONOMY

Economy—overview: North Korea, one of the world's most centrally directed and least open economies, faces chronic economic problems. Industrial capital stock is nearly beyond repair as a result of years of underinvestment and shortages of spare parts. Industrial and power output have declined in parallel from pre-1990 levels. Due in part to severe summer flooding followed by dry weather conditions in the fall of 2006, the nation suffered its 13th year of food shortages because of on-going systemic problems including a lack of arable land, collective farming practices, and persistent shortages of tractors and fuel. During the summer of 2007, severe flooding again occurred. Large-scale international food aid deliveries have allowed the people of North Korea to escape widespread starvation since famine threatened in 1995, but the population continues to suffer from prolonged malnutrition and poor living conditions. Large-scale military spending draws off resources needed for investment and civilian consumption. Since 2002, the government has formalized an arrangement whereby private "farmers' markets" were allowed to begin selling a wider range of goods. It also permitted some private farming on an experimental basis in an effort to boost agricultural output. In October 2005, the government tried to reverse some of these policies by forbidding private sales of grains and reinstituting a centralized food rationing system. By December 2005, the government terminated most international humanitarian assistance operations in North Korea (calling instead for developmental assistance only) and restricted the activities of remaining international and non-governmental aid organizations such as the World Food Program. External food aid now comes primarily from China and South Korea in the form of grants and long-term concessional loans. During the October 2007 summit, South Korea also agreed to develop some of North Korea's infrastructure and natural resources and light industry. Firm political control remains the Communist government's overriding concern, which will likely inhibit the loosening of economic regulations.

GDP (purchasing power parity): $40 billion
note: North Korea does not publish any reliable National Income Accounts data; the datum shown here is derived from purchasing power parity (PPP) GDP estimates for North Korea that were made by Angus MADDISON in a study conducted for the OECD; his figure for 1999 was extrapolated to 2007 using estimated real growth rates for North Korea's GDP and an inflation factor based on the US GDP deflator; the result was rounded to the nearest $10 billion (2007 est.)

GDP (official exchange rate): $2.22 billion (2006 est.)

GDP—real growth rate: -1.1% (2006 est.)

GDP—per capita (PPP): $1,900 (2007 est.)

GDP—composition by sector:
agriculture: 23.3%
industry: 43.1%
services: 33.6% (2002 est.)

Labor force: 20 million
note: estimates vary widely (2004 est.)

Labor force—by occupation: *agriculture:* 37%
industry and services: 63% (2004 est.)

Unemployment rate: NA%

Population below poverty line: NA%

Household income or consumption by percentage share:
lowest 10%: NA%
highest 10%: NA%

Inflation rate (consumer prices): NA%

Budget: *revenues:* $2.88 billion $NA
expenditures: $2.98 billion $NA

Agriculture—products: rice, corn, potatoes, soybeans, pulses; cattle, pigs, pork, eggs

Industries: military products; machine building, electric power, chemicals; mining (coal, iron ore, limestone, magnesite, graphite, copper, zinc, lead, and precious metals), metallurgy; textiles, food processing; tourism

Industrial production growth rate: NA%

Electricity—production: 22.5 billion kWh (2006 est.)

Electricity—production by source:
fossil fuel: 29%
hydro: 71%
nuclear: 0%
other: 0% (2001)

Electricity—consumption: 18.57 billion kWh (2005)
Electricity—exports: 0 kWh (2007)
Electricity—imports: 0 kWh (2007)
Oil—production: 141 bbl/day (2005 est.)
Oil—consumption: 10,520 bbl/day (2006)
Oil—exports: 0 bbl/day (2006)
Oil—imports: 10,520 bbl/day (2006 est.)
Oil—proved reserves: NA bbl
Natural gas—production: 0 cu m (2007 est.)
Natural gas—consumption: 0 cu m (2007 est.)
Natural gas—exports: 0 cu m (2007 est.)
Natural gas—imports: 0 cu m (2007)
Natural gas—proved reserves: 0 cu m (1 January 2007)
Exports: $1.466 billion f.o.b. (2006)
Exports—commodities: minerals, metallurgical products, manufactures (including armaments), textiles, agricultural and fishery products
Exports—partners: South Korea 32%, China 29%, Thailand 9% (2006)
Imports: $2.879 billion c.i.f. (2006)
Imports—commodities: petroleum, coking coal, machinery and equipment, textiles, grain
Imports—partners: China 27%, South Korea 16%, Thailand 9%, Russia 7% (2006)
Economic aid—recipient: $372 million
note: approximately 65,000 metric tons in food aid through the World Food Program appeals in 2007, plus additional aid from bilateral donors and non-governmental organizations (2007 est.)
Debt—external: $12.5 billion (2001 est.)
Currency (code): North Korean won (KPW)
Currency code: KPW
Exchange rates: official: North Korean won per US dollar—140 (2007), 141 (2006), 170 (December 2004), market: North Korean won per US dollar—2,500–3,000 (December 2006)
Fiscal year: calendar year

COMMUNICATIONS

Telephones—main lines in use: 980,000 (2003)
Telephone system:
general assessment: NA
domestic: NA
international: country code—850; satellite earth stations—2 (1 Intelsat—Indian Ocean, 1 Russian—Indian Ocean region); other international connections through Moscow and Beijing
Radio broadcast stations: AM 17 (including 11 stations of Korean Central Broadcasting Station; North Korea has a "national intercom" cable radio station wired throughout the country that is a significant source of information for the average North Korean citizen; it is wired into most residences and workplaces and carries news and commentary), FM 14, shortwave 14 (2006)
Radios: 3.36 million (1997)
Television broadcast stations: 4 (includes Korean Central Television, Mansudae Television, Korean Educational and Cultural Network, and Kaesong Television targeting South Korea) (2003)
Televisions: 1.2 million (1997)
Internet country code: .kp
Internet Service Providers (ISPs): 1 (2000)
Internet users: NA

TRANSPORTATION

Airports: 77 (2007)
Airports—with paved runways:
total: 36
over 3,047 m: 2
2,438 to 3,047 m: 22
1,524 to 2,437 m: 8
914 to 1,523 m: 1
under 914 m: 3 (2007)
Airports—with unpaved runways:
total: 41
2,438 to 3,047 m: 2
1,524 to 2,437 m: 19
914 to 1,523 m: 13
under 914 m: 7 (2007)
Heliports: 23 (2007)
Pipelines: oil 154 km (2007)
Railways:
total: 5,235 km
standard gauge: 5,235 km 1.435-m gauge (3,500 km electrified) (2006)
Roadways:
total: 25,554 km
paved: 724 km
unpaved: 24,830 km (2006)
Waterways: 2,250 km (most navigable only by small craft) (2007)
Merchant marine:
total: 171 ships (1000 GRT or over) 854,268 GRT/1,225,453 DWT
by type: bulk carrier 12, cargo 131, chemical tanker 1, container 1, livestock carrier 1, passenger/cargo 4, petroleum tanker 14, refrigerated cargo 4, roll on/roll off 3
foreign-owned: 29 (Egypt 1, India 1, Israel 1, Lebanon 3, Lithuania 1, Pakistan 1, Romania 6, Russia 1, Syria 7, Turkey 1, UAE 4, Yemen 2)
registered in other countries: (unknown 1) (2007)
Ports and terminals: Ch'ongjin, Haeju, Hungnam (Hamhung), Kimch'aek, Kosong, Najin, Namp'o, Sinuiju, Songnim, Sonbong (formerly Unggi), Ungsang, Wonsan

MILITARY

Military branches: North Korean People's Army: Ground Forces, Navy, Air Force; civil security forces (2005)
Military service age and obligation: 17 years of age (2004)

Manpower available for military service: *males age 16–49:* 6,225,747 *females age 16–49:* 6,188,270 (2008 est.)
Manpower fit for military service:
males age 16–49: 5,141,240
females age 16–49: 5,139,447 (2008 est.)
Manpower reaching militarily significant age annually:
males age 16–49: 199,628
females age 16–49: 192,388 (2008 est.)
Military expenditures—percent of GDP: NA

TRANSNATIONAL ISSUES

Disputes—international: risking arrest, imprisonment, and deportation, tens of thousands of North Koreans cross into China to escape famine, economic privation, and political oppression; North Korea and China dispute the sovereignty of certain islands in Yalu and Tumen rivers; Military Demarcation Line within the 4-km wide Demilitarized Zone has separated North from South Korea since 1953; periodic incidents in the Yellow Sea with South Korea which claims the Northern Limiting Line as a maritime boundary; North Korea supports South Korea in rejecting Japan's claim to Liancourt Rocks (Tok-do/Take-shima)
Refugees and internally displaced persons: *IDPs:* undetermined (flooding in mid-2007 and famine during mid-1990s) (2007)
Trafficking in persons: *current situation:* North Korea is a source country for men, women, and children trafficked for the purposes of forced labor and sexual exploitation; North Korea's own system of political repression includes forced labor in a network of prison camps where an estimated 150,000 to 200,000 persons are incarcerated; the illegal status of North Koreans in China and other countries increases their vulnerability to trafficking schemes and sexual and physical abuse; North Koreans forcibly returned from China may be subject to hard labor in prison camps operated by the government *tier rating:* Tier 3—North Korea does not fully comply with minimum standards for the elimination of trafficking and is not making significant efforts to do so
Illicit drugs: for years, from the 1970s into the 2000s, citizens of the Democratic People's Republic of (North) Korea (DPRK), many of them diplomatic employees of the government, were apprehended abroad while trafficking in narcotics, including two in Turkey in December 2004; police investigations in Taiwan and Japan in recent years have linked North Korea to large illicit shipments of heroin and methamphetamine, including an attempt by the North Korean merchant ship Pong Su to deliver 150 kg of heroin to Australia in April 2003

Background: An independent Korean state or collection of states has existed almost continuously for several millennia. Between its initial unification in the 7th century—from three predecessor Korean states—until the 20th century, Korea existed as a single independent country. In 1905, following the Russo-Japanese War, Korea became a protectorate of imperial Japan, and in 1910 it was annexed as a colony. Korea regained its independence following Japan's surrender to the United States in 1945. After World War II, a Republic of Korea (ROK) was set up in the southern half of the Korean Peninsula while a Communist-style government was installed in the north (the DPRK). During the Korean War (1950–53), US troops and UN forces fought alongside soldiers from the ROK to defend South Korea from DPRK attacks supported by China and the Soviet Union. An armistice was signed in 1953, splitting the peninsula along a demilitarized zone at about the 38th parallel. Thereafter, South Korea achieved rapid economic growth with per capita income rising to roughly 14 times the level of North Korea. In 1993, KIM Young-sam became South Korea's first civilian president following 32 years of military rule. South Korea today is a fully functioning modern democracy. In June 2000, a historic first North-South summit took place between the South's President KIM Dae-jung and the North's leader KIM Jong Il. In October 2007, a second North-South summit took place between the South's President ROH Moo-hyun and the North Korean leader.

GEOGRAPHY

Location: Eastern Asia, southern half of the Korean Peninsula bordering the Sea of Japan and the Yellow Sea

Geographic coordinates: 37 00 N, 127 30 E

Map references: Asia

Area:

total: 98,480 sq km

land: 98,190 sq km

water: 290 sq km

Area—comparative: slightly larger than Indiana

Land boundaries:

total: 238 km

border countries: North Korea 238 km

Coastline: 2,413 km

Maritime claims:

territorial sea: 12 nm; between 3 nm and 12 nm in the Korea Strait

contiguous zone: 24 nm

exclusive economic zone: 200 nm

continental shelf: not specified

Climate: temperate, with rainfall heavier in summer than winter

Terrain: mostly hills and mountains; wide coastal plains in west and south

Elevation extremes:

lowest point: Sea of Japan 0 m

highest point: Halla-san 1,950 m

Natural resources: coal, tungsten, graphite, molybdenum, lead, hydropower potential

Land use:

arable land: 16.58%

permanent crops: 2.01%

other: 81.41% (2005)

Irrigated land: 8,780 sq km (2003)

Total renewable water resources: 69.7 cu km (1999)

Freshwater withdrawal (domestic/industrial/agricultural): *total:* 18.59 cu km/yr (36%/16%/48%)

per capita: 389 cu m/yr (2000)

Natural hazards: occasional typhoons bring high winds and floods; low-level seismic activity common in southwest

Environment—current issues: air pollution in large cities; acid rain; water pollution from the discharge of sewage and industrial effluents; drift net fishing

Environment—international agreements: *party to:* Antarctic-Environmental Protocol, Antarctic-Marine Living Resources, Antarctic Treaty, Biodiversity, Climate Change, Climate Change-Kyoto Protocol, Desertification, Endangered Species, Environmental Modification, Hazardous Wastes, Law of the Sea, Marine Dumping, Ozone Layer Protection, Ship Pollution, Tropical Timber 83, Tropical Timber 94, Wetlands, Whaling

signed, but not ratified: none of the selected agreements

Geography—note: strategic location on Korea Strait

Population: 49,232,844 (July 2008 est.)

Age structure:

0–14 years: 17.7% (male 4,579,018/female 4,157,631)

15–64 years: 72.3% (male 18,150,771/female 17,464,610)

65 years and over: 9.9% (male 1,997,032/female 2,883,782) (2008 est.)

Median age:

total: 36.4 years

male: 35.3 years

female: 37.4 years (2008 est.)

Population growth rate: 0.371% (2008 est.)

Birth rate: 9.83 births/1,000 population (2008 est.)

Death rate: 6.12 deaths/1,000 population (2008 est.)

Net migration rate: NA

Sex ratio:

at birth: 1.08 male(s)/female

under 15 years: 1.1 male(s)/female

15–64 years: 1.04 male(s)/female

65 years and over: 0.69 male(s)/female

total population: 1.01 male(s)/female (2008 est.)

Infant mortality rate:

total: 5.94 deaths/1,000 live births

male: 6.33 deaths/1,000 live births

female: 5.53 deaths/1,000 live births (2008 est.)

Life expectancy at birth:

total population: 77.42 years

male: 74 years

female: 81.1 years (2008 est.)

Total fertility rate: 1.29 children born/woman (2008 est.)

HIV/AIDS—adult prevalence rate: less than 0.1% (2003 est.)

HIV/AIDS—people living with HIV/AIDS: 8,300 (2003 est.)

HIV/AIDS—deaths: fewer than 200 (2003 est.)

Nationality:

noun: Korean(s)

adjective: Korean

Ethnic groups: homogeneous (except for about 20,000 Chinese)

Religions: Christian 26.3% (Protestant 19.7%, Roman Catholic 6.6%), Buddhist 23.2%, other or unknown 1.3%, none 49.3% (1995 census)

Languages: Korean, English widely taught in junior high and high school

Literacy: *definition:* age 15 and over can read and write

total population: 97.9%
male: 99.2%
female: 96.6% (2002)

GOVERNMENT

Country name:
conventional long form: Republic of Korea
conventional short form: South Korea
local long form: Taehan-min'guk
local short form: Han'guk
abbreviation: ROK
Government type: republic
Capital: *name:* Seoul
geographic coordinates: 37 33 N, 126 59 E
time difference: UTC+9 (14 hours ahead of Washington, DC during Standard Time)
Administrative divisions: 9 provinces (do, singular and plural) and 7 metropolitan cities (gwangyoksi, singular and plural)
provinces: Cheju-do, Cholla-bukto (North Cholla), Cholla-namdo (South Cholla), Ch'ungch'ong-bukto (North Ch'ungch'ong), Ch'ungch'ong-namdo (South Ch'ungch'ong), Kangwon-do, Kyonggi-do, Kyongsang-bukto (North Kyongsang), Kyongsang-namdo (South Kyongsang)
metropolitan cities: Inch'on-gwangyoksi (Inch'on), Kwangju-gwangyoksi (Kwangju), Pusan-gwangyoksi (Pusan), Soul-t'ukpyolsi (Seoul), Taegu-gwangyoksi (Taegu), Taejon-gwangyoksi (Taejon), Ulsan-gwangyoksi (Ulsan)
Independence: 15 August 1945 (from Japan)
National holiday: Liberation Day, 15 August (1945)
Constitution: 17 July 1948; note— amended or rewritten nine times; current constitution approved on 29 October 1987
Legal system: combines elements of continental European civil law systems, Anglo-American law, and Chinese classical thought; has not accepted compulsory ICJ jurisdiction
Suffrage: 19 years of age; universal
Executive branch:
chief of state: President LEE Myung-bak (since 25 February 2008)
head of government: Prime Minister HAN Seung-soo (since 29 February 2008)
cabinet: State Council appointed by the president on the prime minister's recommendation
elections: president elected by popular vote for a single five-year term; election last held 19 December 2007 (next to be held on in December 2012); prime minister appointed by president with consent of National Assembly; deputy prime ministers appointed by president on prime minister's recommendation

election results: ROH Moo-hyun elected president on 19 December 2002; percent of vote—ROH Moo-hyun (MDP) 48.9%; LEE Hoi-chang (GNP) 46.6%; others 4.5%; LEE Myung-bak elected president on 19 December 2007; percent of vote—LEE Myung-bak (GNP) 48.7%; CHUNG Dong-young (UNDP) 26.1%); LEE Hoi-chang (independent) 15.1; others 10.1%
Legislative branch: unicameral National Assembly or Kukhoe (299 seats; 243 members elected in single-seat constituencies, 56 elected by proportional representation; to serve four-year terms)
elections: last held 9 April 2008 (next to be held in April 2012)
election results: percent of vote by party— NA; seats by party—GNP 153, UDP 81, LFP 18, Pro-Park Alliance 14, DLP 5, CKP 3, independents 25
Judicial branch: Supreme Court (justices appointed by the president with consent of National Assembly); Constitutional Court (justices appointed by the president based partly on nominations by National Assembly and Chief Justice of the court)
Political parties and leaders: Creative Korea Party or CKP [MOON Kook-hyun]; Democratic Labor Party or DLP [CHUN Young-se]; Grand National Party or GNP [KANG Jae-sup]; Liberty Forward Party or LFP [SIM Dae-pyung]; United Democratic Party or UDP [SOHN Hak-kyu]
Political pressure groups and leaders: Federation of Korean Industries; Federation of Korean Trade Unions; Korean Confederation of Trade Unions; Korean National Council of Churches; Korean Traders Association; Korean Veterans' Association; National Council of Labor Unions; National Democratic Alliance of Korea; National Federation of Farmers' Associations; National Federation of Student Associations
International organization participation: ADB, AfDB, APEC, APT, ARF, ASEAN (dialogue partner), Australia Group, BIS, CP, EAS, EBRD, FAO, IADB, IAEA, IBRD, ICAO, ICC, ICCt, ICRM, IDA, IEA, IFAD, IFC, IFRCS, IHO, ILO, IMF, IMO, IMSO, Interpol, IOC, IOM, IPU, ISO, ITSO, ITU, ITUC, LAIA, MIGA, NEA, NSG, OAS (observer), OECD, OPCW, OSCE (partner), PCA, PIF (partner), SAARC (observer), UN, UNCTAD, UNESCO, UNFICYP, UNHCR, UNIDO, UNIFIL, UNMIL, UNMIS, UNMOGIP, UNOMIG, UNWTO, UPU, WCL, WCO, WHO, WIPO, WMO, WTO, ZC
Diplomatic representation in the US:
chief of mission: Ambassador LEE Tae-sik

chancery: 2450 Massachusetts Avenue NW, Washington, DC 20008
telephone: [1] (202) 939-5600
FAX: [1] (202) 387-0205
consulate(s) general: Agana (Guam), Atlanta, Boston, Chicago, Honolulu, Houston, Los Angeles, New York, San Francisco, Seattle
Diplomatic representation from the US:
chief of mission: Ambassador Alexander VERSHBOW
embassy: 32 Sejong-no, Jongno-gu, Seoul 110-710
mailing address: US Embassy Seoul, APO AP 96205-5550
telephone: [82] (2) 397-4114
FAX: [82] (2) 738-8845
Flag description: white with a red (top) and blue yin-yang symbol in the center; there is a different black trigram from the ancient I Ching (Book of Changes) in each corner of the white field

ECONOMY

Economy—overview: Since the 1960s, South Korea has achieved an incredible record of growth and integration into the high-tech modern world economy. Four decades ago, GDP per capita was comparable with levels in the poorer countries of Africa and Asia. In 2004, South Korea joined the trillion dollar club of world economies. Today its GDP per capita is roughly the same as that of Greece and Spain. This success was achieved by a system of close government/business ties including directed credit, import restrictions, sponsorship of specific industries, and a strong labor effort. The government promoted the import of raw materials and technology at the expense of consumer goods and encouraged savings and investment over consumption. The Asian financial crisis of 1997–98 exposed longstanding weaknesses in South Korea's development model including high debt/equity ratios, massive foreign borrowing, and an undisciplined financial sector. GDP plunged by 6.9% in 1998, then recovered by 9.5% in 1999 and 8.5% in 2000. Growth fell back to 3.3% in 2001 because of the slowing global economy, falling exports, and the perception that much-needed corporate and financial reforms had stalled. Led by consumer spending and exports, growth in 2002 was an impressive 7%, despite anemic global growth. Between 2003 and 2007, growth moderated to about 4–5% annually. A downturn in consumer spending was offset by rapid export growth. Moderate inflation, low unemployment, and an export surplus in 2007 characterize this solid economy, but inflation and unemployment are increasing in the face of rising oil prices.

GDP (purchasing power parity): $1.201 trillion (2007 est.)

GDP (official exchange rate): $957.1 billion (2007 est.)

GDP—real growth rate: 5% (2007 est.)

GDP—per capita (PPP): $24,800 (2007 est.)

GDP—composition by sector:
agriculture: 3%
industry: 39.4%
services: 57.6% (2007 est.)

Labor force: 24.22 million (2007 est.)

Labor force—by occupation:
agriculture: 7.5%
industry: 17.3%
services: 75.2% (2007)

Unemployment rate: 3.3% (2007 est.)

Population below poverty line: 15% (2003 est.)

Household income or consumption by percentage share:
lowest 10%: 2.9%
highest 10%: 25% (2005 est.)

Distribution of family income—Gini index: 35.1 (2006)

Inflation rate (consumer prices): 2.5% (2007)

Investment (gross fixed): 28.8% of GDP (2007 est.)

Budget:
revenues: $248 billion
expenditures: $246.5 billion (2007 est.)

Public debt: 33.4% of GDP (2007 est.)

Agriculture—products: rice, root crops, barley, vegetables, fruit; cattle, pigs, chickens, milk, eggs; fish

Industries: electronics, telecommunications, automobile production, chemicals, shipbuilding, steel

Industrial production growth rate: 7.6% (2007 est.)

Electricity—production: 403.2 billion kWh (2007)

Electricity—production by source:
fossil fuel: 62.4%
hydro: 0.8%
nuclear: 36.6%
other: 0.2% (2001)

Electricity—consumption: 368.6 billion kWh (2007)

Electricity—exports: 0 kWh (2005)

Electricity—imports: 0 kWh (2005)

Oil—production: 17,050 bbl/day (2005)

Oil—consumption: 2.13 million bbl/day (2006)

Oil—exports: NA bbl/day

Oil—imports: 2.41 million bbl/day (2006)

Oil—proved reserves: 0 bbl (1 January 2006 est.)

Natural gas—production: 1.66 billion cu m (2006)

Natural gas—consumption: 34.2 billion cu m (2006)

Natural gas—exports: 2,450 cu m (2006)

Natural gas—imports: 35.86 billion cu m (2006)

Natural gas—proved reserves: 0 cu m (1 January 2006 est.)

Current account balance: $5.954 billion (2007 est.)

Exports: $371.5 billion f.o.b. (2007)

Exports—commodities: semiconductors, wireless telecommunications equipment, motor vehicles, computers, steel, ships, petrochemicals

Exports—partners: China 22%, US 12.5%, Japan 7.1%, Hong Kong 5% (2007)

Imports: $356.8 billion f.o.b. (2007)

Imports—commodities: machinery, electronics and electronic equipment, oil, steel, transport equipment, organic chemicals, plastics

Imports—partners: China 17.7%, Japan 16%, US 10.7%, Saudi Arabia 5.9%, UAE 4.2% (2007)

Economic aid—donor: ODA, $455.3 million (2006)

Economic aid—recipient: $68.07 million (2004)

Reserves of foreign exchange and gold: $262.2 billion (31 December 2007)

Debt—external: $220.1 billion (31 December 2007)

Stock of direct foreign investment—at home: $120.7 billion (2007 est.)

Stock of direct foreign investment—abroad: $82.1 billion (2006)

Market value of publicly traded shares: $1.051 trillion (2007)

Currency (code): South Korean won (KRW)

Currency code: KRW

Exchange rates: South Korean won per US dollar—929.2 (2007), 954.8 (2006), 1,024.1 (2005), 1,145.3 (2004), 1,191.6 (2003)

Fiscal year: calendar year

COMMUNICATIONS

Telephones—main lines in use: 26.866 million (2006)

Telephones—mobile cellular: 40.197 million (2006)

Telephone system:
general assessment: excellent domestic and international services
domestic: NA
international: country code—82; numerous submarine cables provide links throughout Asia, Australia, the Middle East, Europe, and US; satellite earth stations—6 (3 Intelsat—1 Pacific Ocean and 2 Indian Ocean, 3 Inmarsat—1 Pacific Ocean and 2 Indian Ocean)

Radio broadcast stations: AM 61, FM 150, shortwave 2 (2005)

Radios: 47.5 million (2000)

Television broadcast stations: 43 (plus 59 cable operators and 190 relay cable operators) (2005)

Televisions: 15.9 million (1997)

Internet country code: .kr

Internet hosts: 315,537 (2007)

Internet Service Providers (ISPs): 11 (2000)

Internet users: 34.12 million (2006)

TRANSPORTATION

Airports: 105 (2007)

Airports—with paved runways:
total: 68
over 3,047 m: 3
2,438 to 3,047 m: 21
1,524 to 2,437 m: 14
914 to 1,523 m: 11
under 914 m: 19 (2007)

Airports—with unpaved runways:
total: 37
914 to 1,523 m: 3
under 914 m: 34 (2007)

Heliports: 536 (2007)

Pipelines: gas 1,482 km; refined products 827 km (2007)

Railways:
total: 3,472 km
standard gauge: 3,472 km 1.435-m gauge (1,342 km electrified) (2006)

Roadways:
total: 102,293 km
paved: 78,581 km (includes 3,060 km of expressways)
unpaved: 23,712 km (2005)

Waterways: 1,608 km (most navigable only by small craft) (2007)

Merchant marine:
total: 738 ships (1000 GRT or over) 10,636,466 GRT/17,371,943 DWT
by type: bulk carrier 187, cargo 202, carrier 1, chemical tanker 119, container 81, liquefied gas 26, passenger 5, passenger/cargo 21, petroleum tanker 57, refrigerated cargo 19, roll on/roll off 8, specialized tanker 4, vehicle carrier 8
foreign-owned: 22 (China 2, France 8, Japan 1, Sweden 2, UK 1, US 7, Vietnam 1)
registered in other countries: 386 (Belize 4, Cambodia 29, China 1, Cyprus 2, Greece 2, Honduras 6, Hong Kong 6, Indonesia 1, Liberia 4, Malta 3, Marshall Islands 3, Netherlands 1, Panama 316, Russia 1, Singapore 7, unknown 4) (2007)

Ports and terminals: Inch'on, P'ohang, Pusan, Ulsan

MILITARY

Military branches: Republic of Korea Army, Navy (includes Marine Corps), Air Force (2008)

Military service age and obligation: 20–30 years of age for compulsory military service; conscript service obligation—24–28 months, depending on the military branch involved (to be reduced to 18 months beginning 2016); 18 years of age for voluntary military service; women, in service since 1950, admitted to 7 service branches, including infantry, but excluded from artillery, armor, anti-air, and chaplaincy corps; some 4,000 women serve as commissioned and non-commissioned officers, approx. 2.3% of all officers (2007)

Manpower available for military service:
males age 16–49: 13,691,809

females age 16–49: 13,029,859 (2008 est.)

Manpower fit for military service:
males age 16–49: 11,282,699
females age 16–49: 10,683,668 (2008 est.)

Manpower reaching militarily significant age annually:
males age 16–49: 382,305
females age 16–49: 336,998 (2008 est.)

Military expenditures—percent of GDP: 2.7% (2006)

TRANSNATIONAL ISSUES

Disputes—international: Military Demarcation Line within the 4-km wide Demilitarized Zone has separated North from South Korea since 1953; periodic incidents with North Korea in the Yellow Sea over the Northern Limiting Line, which South Korea claims as a maritime boundary; South Korea and Japan claim Liancourt Rocks (Tok-do/Take-shima), occupied by South Korea since 1954

KOSOVO

INTRODUCTION

Background: Serbs migrated to the territories of modern Kosovo in the 7th century but did not fully incorporate them into the Serbian realm until the early 13th century. The Serbian defeat at the Battle of Kosovo in 1389 led to five centuries of Ottoman rule during which large numbers of Turks and Albanians moved to Kosovo. By the end of the 19th century, Albanians replaced the Serbs as the dominant ethnic group in Kosovo. Serbia reacquired control over Kosovo from the Ottoman Empire during the First Balkan War (1912). After World War II (1945), the government of the Socialist Federal Republic of Yugoslavia led by Josip TITO reorganized Kosovo as an autonomous province within the constituent republic of Serbia. Over the next four decades, Kosovo Albanians lobbied for greater autonomy, and Kosovo was granted the status almost equal to that of a republic in the 1974 Yugoslav Constitution. Despite the legislative concessions, Albanian nationalism increased in the 1980s leading to nationalist riots and calls for Kosovo's independence. Serbs in Kosovo complained of mistreatment and Serb nationalist leaders, such as Slobodan MILOSEVIC,

exploited those charges to win support among Serbian voters many of whom viewed Kosovo as their cultural heartland. Under MILOSEVIC's leadership, Serbia instituted a new constitution in 1989 that drastically curtailed Kosovo's autonomy. Kosovo Albanian leaders responded in 1991 by organizing a referendum that declared Kosovo independent from Serbia. The MILOSEVIC regime carried out repressive measures against the Albanians in the early 1990s as the unofficial government of Kosovo, led by Ibrahim RUGOVA, tried to use passive resistance to gain international assistance and recognition of its demands for independence. In 1995, Albanians dissatisfied with RUGOVA's nonviolent strategy created the Kosovo Liberation Army and launched an insurgency. In 1998, MILOSEVIC authorized a counterinsurgency campaign that resulted in massacres and massive expulsions of ethnic Albanians by Serbian military, police, and paramilitary forces. The international community tried to resolve the conflict peacefully, but MILOSEVIC rejected the proposed international settlement—the Rambouillet Accords—leading to a three-month NATO bombing of Serbia beginning in March 1999, which forced Serbia to withdraw its military and police forces from Kosovo in June 1999. UN Security Council Resolution 1244 (1999) placed Kosovo under a transitional administration, the UN Interim Administration Mission in Kosovo (UNMIK), pending a determination of Kosovo's future status. Under the resolution, Serbia's territorial integrity was protected, but it was UNMIK that assumed responsibility for governing Kosovo. In 2001, UNMIK promulgated a Constitutional Framework, which established Kosovo's Provisional Institutions of Self-Government (PISG). In succeeding years UNMIK increasingly devolved

responsibilities to the PISG. A UN-led process began in late 2005 to determine Kosovo's future status. Negotiations held intermittently between 2006 and 2007 on issues related to decentralization, religious heritage, and minority rights failed to yield a resolution between Serbia's willingness to grant a high degree of autonomy and the Albanians' call for full independence for Kosovo. On 17 February 2008, the Kosovo Assembly declared its independence from Serbia.

GEOGRAPHY

Location: Southeast Europe, between Serbia and Macedonia

Geographic coordinates: 42 35 N, 21 00 E

Map references: Europe

Area:
total: 10,887 sq km
land: 10,887 sq km
water: 0 sq km

Area—comparative: slightly larger than Delaware

Land boundaries:
total: 702 km
border countries: Albania 112 km, Macedonia 159 km, Montenegro 79 km, Serbia 352 km

Coastline: 0 km (landlocked)

Maritime claims: none (landlocked)

Climate: influenced by continental air masses resulting in relatively cold winters with heavy snowfall and hot, dry summers and autumns; Mediterranean and alpine influences create regional variation; maximum rainfall between October and December

Terrain: flat fluvial basin with an elevation of 400–700 m above sea level surrounded by several high mountain ranges with elevations of 2,000 to 2,500 m

Elevation extremes:
lowest point: Drini i Bardhe/Beli Drim 297 m (located on the border with Albania)
highest point: Gjeravica/Deravica 2,565 m

Natural resources: nickel, lead, zinc, magnesium, lignite, kaolin, chrome, bauxite

PEOPLE

Population: 2,126,708 (2007 est.)

Nationality:

noun: Kosovar (Albanian), Kosovac (Serbian)

adjective: Kosovar (Albanian), Kosovski (Serbian)

note: Kosovan, a neutral term, is sometimes also used as a noun or adjective

Ethnic groups: Albanians 88%, Serbs 7%, other 5% (Bosniak, Gorani, Roma, Turk, Ashkali, Egyptian)

Religions: Muslim, Serbian Orthodox, Roman Catholic

Languages: Albanian (official), Serbian (official), Bosniak, Turkish, Roma

GOVERNMENT

Country name:

conventional long form: Republic of Kosovo

conventional short form: Kosovo

local long form: Republika e Kosoves (Republika Kosova)

local short form: Kosova (Kosovo)

former: Kosovo and Metohija Autonomous Province

Government type: republic

Capital: *name:* Pristina (Prishtine)

geographic coordinates: 42 40 N, 21 10 E

time difference: UTC+1 (6 hours ahead of Washington, DC during Standard Time)

daylight saving time: +1hr, begins last Sunday in March; ends last Sunday in October

Administrative divisions: 30 municipalities (komunat, singular—komuna in Albanian; opstine, singular—opstina in Serbian); Decan (Decani), Dragash (Dragas), Ferizaj (Urosevac), Fushe Kosove (Kosovo Polje), Gjakove (Dakovica), Gllogoc/Drenas (Glogovac), Gjilan (Gnjilane), Istog (Istok), Kacanik, Kamenice/Dardana (Kamenica), Kline (Klina), Leposaviq (Leposavic), Lipjan (Lipljan), Malisheve (Malisevo), Mitrovice (Mitrovica), Novoberde (Novo Brdo), Obiliq (Obilic), Peje (Pec), Podujeve (Podujevo), Prishtine (Pristina), Prizren, Rahovec (Orahovac), Shtime (Stimlje), Shterpce (Strpce), Skenderaj (Srbica), Suhareke (Suva Reka), Viti (Vitina), Vushtrri (Vucitrn), Zubin Potok, Zvecan

Independence: 17 February 2008 (from Serbia)

Constitution: Constitutional Framework of 2001; note—the Kosovo Government is charged with putting forward an AHTISAARI (UN Special Envoy) Plan-compliant draft of a new constitution soon after independence

Legal system: evolving legal system based on terms of UN Special Envoy Martti AHTISAARI's Plan for Kosovo's supervised independence

Suffrage: 18 years of age; universal

Executive branch:

chief of state: President Fatmir SEJDIU (since 10 February 2006)

head of government: Prime Minister Hashim THACI (since 9 January 2008)

cabinet: ministers; elected by the Kosovo Assembly

elections: the president is elected for a five-year term by the Kosovo Assembly; election last held 9 January 2008 (next to be held by in 2013); the prime minister is elected by the Kosovo Assembly

election results: Fatmir SEJDIU reelected president; first round: Fatmir SEDIU 62, Naim MALOKU 37; second round: Fatmir SEDIU 61, Naim MALOKU 37; and Hashim THACI elected to be prime minister by the Assembly

Legislative branch: unicameral Kosovo Assembly of the Provisional Government (120 seats; 100 seats directly elected, 10 seats for Serbs, 10 seats for other minorities; to serve three-year terms)

elections: last held 17 November 2007 (next to be held in 2011)

election results: percent of vote by party—PDK 34.3%, LDK 22.6%, AKR 12.3%, LDD 10.0%, AAK 9.6%, other 11.2%; seats by party—PDK 37, LDK 25, AKR 13, LDD 11, AAK 10, other 4

Judicial branch: Supreme Court judges are appointed by the Special Representative of the Secretary-General (SRSG); district courts judges are appointed by the SRSG; municipal courts judges are appointed by the SRSG note: after the termination of UNMIK's mandate, the Kosovo Judicial Council (KJC) will propose to the president candidates for appointment or reappointment as judges and prosecutors; the KJC is also responsible for decisions on the promotion and transfer of judges and disciplinary proceedings against judges; at least 15% of Supreme Court and district court judges shall be from nonmajority communities

Political parties and leaders: Albanian Christian Democratic Party of Kosovo or PShDK [Mark KRASNIQI]; Alliance for the Future of Kosovo or AAK [Ramush HARADINAJ]; Alliance of Independent Social Democrats of Kososvo and Metohija or SDSKIM [Slavisa PETKOVIC]; Autonomous Liberal Party of SLS [Slobodan PETROVIC]; Bosniak Vakat Coalition [Dzezair MURATI]; Citizens' Initiative of Gora or GIG [Murselj HALILI]; Council of Independent Social Democrats of Kosovo or SNSDKIM [Ljubisa ZIVIC]; Democratic League of Dardania or LDD [Nexhat DACI]; Democratic League of Kosovo or LDK [Fatmir SEJDIU]; Democratic Party of Ashkali of Kosovo or PDAK [Sabit RAHMANI]; Democratic Party of Kosovo or PDK [Hashim THACI]; Kosovo Democratic Turkish Party of KDTP [Mahir YAGCILAR]; New Democratic Initiative of Kosovo or IRDK [Xhevdet NEZIRAJ]; New Democratic Party or ND [Branislav GRBIC]; New Kosovo Alliance or AKR [Behxhet PACOLLI]; Popular Movement of Kosovo or LPK [Emrush XHEMAJLI]; Reform Party Ora; Serb National Party or SNS [Mihailo SCEPANOVIC]; Serbian Kosovo and Metohija Party or SKMS [Dragisa MIRIC]; United Roma Party of Kosovo or PREBK [Haxhi Zylfi MERXHA]; Democratic Action Party or SDA [Numan BALIC]; Serbian List for Kosovo and Metohija [Oliver IVANOVIC]; Serbian National Council of Northern Kosovo and Metohija or SNV [Milan IVANOVIC]; Democratic Party of Bosniaks [Dzezair MURAIT]; Democratic Party Vatan [Sadik IDRIZI]; Gorani Citizens Initiative [Mursel HALJILJI]; Serbian People Party [Mihailo SCEPANOVIC]; Serbian Democratic Party of Kosovo and Metohija [Slavisa PETKOVIC]; Serb Liberal Party [Slobodan PETROVIC]; Independent League of Social-Democrats of Kosovo and Metohija [Ljubisa ZIVIC]

International organization participation: ITUC

Diplomatic representation from the US:

chief of mission: Ambassador (vacant); Charge d'Affaires Tina KAIDANOW

embassy: Arberia/Dragodan, Nazim Hikmet 30, Pristina, Kosovo

mailing address: use embassy street address

telephone: 381 38 59 59 3000

FAX: 381 38 549 890

Flag description: centered on a dark blue field is the geographical shape of Kosovo in a gold color surmounted by six white, five-pointed stars—each representing one of the major ethnic groups of Kosovo—arrayed in a slight arc

ECONOMY

Economy—overview: Over the past few years Kosovo's economy has shown significant progress in transitioning to a market-based system, but it is still highly dependent on the international community and the diaspora for financial and technical assistance. Remittances from the diaspora—located mainly in Germany and Switzerland—account for about 30% of GDP. Kosovo's citizens are the poorest in Europe with an average annual per capita income of only $1800—about one-third the level of neighboring Albania. Unemployment—at more than 40% of the population—is a severe problem that encourages outward migration. Most of Kosovo's population lives in rural towns outside of the capital, Pristina. Inefficient, near-subsistence farming is common—the result of small plots, limited mechanization, and lack of technical expertise. Economic growth is largely driven by the private sector—mostly small-scale retail businesses. With international assistance, Kosovo has been able to privatize 50% of its state-owned enterprises (SOEs) by number, and over 90% of SOEs by value. Minerals and metals—including lignite, lead, zinc, nickel, chrome, aluminum, magnesium, and a wide variety of construction materials—once formed the backbone of industry, but output has declined because investment has been insufficient to replace ageing Eastern Bloc equipment. Technical and financial problems in the power sector also impedes industrial development. The US has worked with the World Bank to prepare a commercial tender for the development of new power generating and mining capacity. The official currency of Kosovo is the euro, but the Serbian dinar is also used in the Serb enclaves. Kosovo's tie to the euro has helped keep inflation low. Kosovo has maintained a budget surplus as a result of efficient tax collection and inefficient budget execution. While maintaining ultimate oversight, UNMIK continues to work with the EU and with Kosovo's government to accelerate economic growth, lower unemployment, and attract foreign investment. In order to help integrate Kosovo into regional economic structures, UNMIK signed (on behalf of Kosovo) its accession to the Central Europe Free Trade Area (CEFTA) in 2006. In February 2008, UNMIK also represented Kosovo at the newly established Regional Cooperation Council (RCC).

GDP (purchasing power parity): $4 billion (2007 est.)

GDP (official exchange rate): $3.237 billion (2007 est.)

GDP—real growth rate: 2.6% (2007 est.)

GDP—per capita (PPP): $1,800 (2007 est.)

GDP—composition by sector:
agriculture: 20%
industry: 20%
services: 60% (2007 est.)

Labor force: 832,000 (June 2007 est.)

Labor force—by occupation: *agriculture*: 21.4%
industry: NA
services: NA (2006 est,)

Unemployment rate: 43% (2007 est.)

Population below poverty line: 37% (2007 est.)

Distribution of family income—Gini index: 30 (FY05/06)

Inflation rate (consumer prices): 2% (2007 est.)

Investment (gross fixed): 29% of GDP (2006 est.)

Budget:
revenues: $1.364 billion
expenditures: $1.008 billion (2007 est.)

Public debt: NA

Agriculture—products: NA

Industries: mineral mining, construction materials, base metals, leather, machinery, appliances

Electricity—production: 3.996 billion kWh (2006)

Electricity—consumption: 4.281 billion kWh (2006)

Oil—production: 0 bbl/day (2007)

Oil—consumption: NA bbl

Oil—proved reserves: NA bbl

Natural gas—production: 0 cu m (2007)

Natural gas—consumption: 0 cu m (2007)

Natural gas—proved reserves: NA cu m

Current account balance: -$58.3 million (2007)

Exports: $148.4 million (2007)

Exports—commodities: mining and processed metal products, scrap metals, leather products, machinery, appliances

Exports—partners: Central Europe Free Trade Area (CFTA) 56% (2006)

Imports: $NA

Imports—commodities: foodstuffs, wood, petroleum, chemicals, machinery and electrical equipment

Imports—partners: EU 35%, Macedonia 15%, Serbia 13%, Turkey 8% (2006)

Economic aid—recipient: $324 million (2007)

Reserves of foreign exchange and gold: $NA

Debt—external: according to the national bank of Serbia, Kosovo's external debt was around $1.2 billion; Kosovo was willing to accept around $900 million (2007)

Currency (code): euro (EUR); Serbian Dinar (RSD) is also in circulation

Exchange rates: euros per US dollar—0.7345 (2007)

COMMUNICATIONS

Telephones—main lines in use: 106,300 (2006)

Telephones—mobile cellular: 562,000 (2006)

TRANSPORTATION

Airports: 10 (2008)

Airports—with paved runways:
total: 6
2,438 to 3,047 m: 1
1,524 to 2,437 m: 1
under 914 m: 4 (2008)

Airports—with unpaved runways:
total: 4
under 914 m: 4 (2008)

Heliports: 2 (2008)

Railways:
total: 430 km (2005)

Roadways:
total: 1,924 km
paved: 1,666 km
unpaved: 258 km (2006)

TRANSNATIONAL ISSUES

Disputes—international: Serbia with several other states protest the US and other states' recognition of Kosovo's declaring itself as a sovereign and independent state in February 2008; ethnic Serbian municipalities along Kosovo's northern border challenge final status of Kosovo-Serbia boundary; several thousand NATO-led KFOR peacekeepers under UNMIK authority continue to keep the peace within Kosovo between the ethnic Albanian majority and the Serb minority in Kosovo; Kosovo authorities object to alignment of the Kosovo boundary with Macedonia in accordance with the 2000 Macedonia-Serbia and Montenegro delimitation agreement

Refugees and internally displaced persons: *IDP's*: 21,000 (2007)

KUWAIT

INTRODUCTION

Background: Britain oversaw foreign relations and defense for the ruling Kuwaiti AL-SABAH dynasty from 1899 until independence in 1961. Kuwait was attacked and overrun by Iraq on 2 August 1990. Following several weeks of aerial bombardment, a US-led, UN coalition began a ground assault on 23 February 1991 that liberated Kuwait in four days. Kuwait spent more than $5 bil-

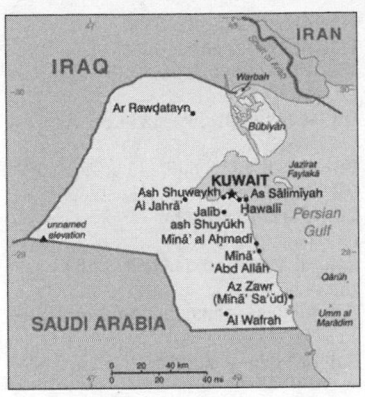

lion to repair oil infrastructure damaged during 1990–91. The AL-SABAH family has ruled since returning to power in 1991 and reestablished an elected legislature that in recent years has become increasingly assertive.

GEOGRAPHY

Location: Middle East, bordering the Persian Gulf, between Iraq and Saudi Arabia
Geographic coordinates: 29 30 N, 45 45 E
Map references: Middle East
Area: *total:* 17,820 sq km
land: 17,820 sq km
water: 0 sq km
Area—comparative: slightly smaller than New Jersey
Land boundaries:
total: 462 km
border countries: Iraq 240 km, Saudi Arabia 222 km
Coastline: 499 km
Maritime claims: *territorial sea:* 12 nm
Climate: dry desert; intensely hot summers; short, cool winters
Terrain: flat to slightly undulating desert plain
Elevation extremes:
lowest point: Persian Gulf 0 m
highest point: unnamed location 306 m
Natural resources: petroleum, fish, shrimp, natural gas
Land use:
arable land: 0.84%
permanent crops: 0.17%
other: 98.99% (2005)
Irrigated land: 130 sq km (2003)
Total renewable water resources: 0.02 cu km (1997)
Freshwater withdrawal (domestic/industrial/agricultural): *total:* 0.44 cu km/yr (45%/2%/52%)
per capita: 164 cu m/yr (2000)
Natural hazards: sudden cloudbursts are common from October to April and bring heavy rain, which can damage roads and houses; sandstorms and dust storms occur throughout the year but are most common between March and August

Environment—current issues: limited natural fresh water resources; some of world's largest and most sophisticated desalination facilities provide much of the water; air and water pollution; desertification
Environment—international agreements: *party to:* Biodiversity, Climate Change, Climate Change-Kyoto Protocol, Desertification, Endangered Species, Environmental Modification, Hazardous Wastes, Law of the Sea, Ozone Layer Protection
signed, but not ratified: Marine Dumping
Geography—note: strategic location at head of Persian Gulf

PEOPLE

Population: 2,596,799
note: includes 1,291,354 non-nationals (July 2008 est.)
Age structure:
0–14 years: 26.6% (male 351,057/female 338,634)
15–64 years: 70.6% (male 1,172,460/female 659,927)
65 years and over: 2.9% (male 46,770/female 27,951) (2008 est.)
Median age:
total: 26.1 years
male: 28 years
female: 22.6 years (2008 est.)
Population growth rate: 3.591%
note: this rate reflects a return to pre-Gulf crisis immigration of expatriates (2008 est.)
Birth rate: 21.9 births/1,000 population (2008 est.)
Death rate: 2.37 deaths/1,000 population (2008 est.)
Net migration rate: 16.39 migrant(s)/1,000 population (2008 est.)
Sex ratio:
at birth: 1.04 male(s)/female
under 15 years: 1.04 male(s)/female
15–64 years: 1.78 male(s)/female
65 years and over: 1.67 male(s)/female
total population: 1.53 male(s)/female (2008 est.)
Infant mortality rate:
total: 9.22 deaths/1,000 live births
male: 10.2 deaths/1,000 live births
female: 8.21 deaths/1,000 live births (2008 est.)
Life expectancy at birth:
total population: 77.53 years
male: 76.38 years
female: 78.73 years (2008 est.)
Total fertility rate: 2.81 children born/woman (2008 est.)
HIV/AIDS—adult prevalence rate: 0.12% (2001 est.)
HIV/AIDS—people living with HIV/AIDS: NA
HIV/AIDS—deaths: NA
Nationality:
noun: Kuwaiti(s)
adjective: Kuwaiti

Ethnic groups: Kuwaiti 45%, other Arab 35%, South Asian 9%, Iranian 4%, other 7%
Religions: Muslim 85% (Sunni 70%, Shi'a 30%), other (includes Christian, Hindu, Parsi) 15%
Languages: Arabic (official), English widely spoken
Literacy: *definition:* age 15 and over can read and write
total population: 93.3%
male: 94.4%
female: 91% (2005 census)

GOVERNMENT

Country name:
conventional long form: State of Kuwait
conventional short form: Kuwait
local long form: Dawlat al Kuwayt
local short form: Al Kuwayt
Government type: constitutional emirate
Capital: *name:* Kuwait
geographic coordinates: 29 22 N, 47 58 E
time difference: UTC+3 (8 hours ahead of Washington, DC during Standard Time)
Administrative divisions: 6 governorates (muhafazat, singular—muhafazah); Al Ahmadi, Al 'Asimah, Al Farwaniyah, Al Jahra', Hawalli, Mubarak Al Kabir
Independence: 19 June 1961 (from UK)
National holiday: National Day, 25 February (1950)
Constitution: approved and promulgated 11 November 1962
Legal system: civil law system with Islamic law significant in personal matters; has not accepted compulsory ICJ jurisdiction
Suffrage: NA years of age; universal (adult); note—males in the military or police are not allowed to vote; adult females were allowed to vote as of 16 May 2005; all voters must have been citizens for 20 years
Executive branch:
chief of state: Amir SABAH al-Ahmad al-Jabir al-Sabah (since 29 January 2006); Crown Prince NAWAF al-Ahmad al-Jabir al-Sabah
head of government: Prime Minister NASIR MUHAMMAD al-Ahmad al-Sabah (since 3 April 2007); First Deputy Prime Minister JABIR Mubarak al-Hamad al-Sabah (since 9 February 2006); Deputy Prime Ministers MUHAMMAD al-Sabah al-Salim al-Sabah (since 9 February 2006) and Faysal al-HAJJI (since 5 April 2007)
cabinet: Council of Ministers appointed by the prime minister and approved by the Amir
elections: none; the amir is hereditary; the amir appoints the prime minister and deputy prime ministers
Legislative branch: unicameral National Assembly or Majlis al-Umma (50 seats; members elected by popular

361

vote to serve four-year terms; all cabinet ministers are also ex officio voting members of the National Assembly)

elections: last held 17 May 2008 (next election to be held in 2012)

election results: percent of vote by bloc—NA; seats by bloc—Sunni 21, Islamic Salafi Alliance 10, Liberals 7, Shiites 5, Popular Action Bloc 4, Islamic Constitutional Movement 3

Judicial branch: High Court of Appeal

Political parties and leaders: none; formation of political parties is in practice illegal but is not forbidden by law

Political pressure groups and leaders: a number of political groups act as de facto parties; several legislative blocs operate in the National Assembly: tribal groups, merchants, Shia activists, Islamists, secular liberals and pro-government deputies; in mid-2006, a coalition of Islamists, liberals, and Shia campaigned successfully for electoral reform to reduce corruption

International organization participation: ABEDA, AfDB, AFESD, AMF, BDEAC, CAEU, FAO, G-77, GCC, IAEA, IBRD, ICAO, ICC, ICCt (signatory), ICRM, IDA, IDB, IFAD, IFC, IFRCS, IHO, ILO, IMF, IMO, IMSO, Interpol, IOC, IPU, ISO, ITSO, ITU, ITUC, LAS, MIGA, NAM, OAPEC, OIC, OPCW, OPEC, PCA, UN, UNCTAD, UNESCO, UNIDO, UNITAR, UNWTO, UPU, WCO, WFTU, WHO, WIPO, WMO, WTO

Diplomatic representation in the US:

chief of mission: Ambassador SALIM al-Abdallah al-Jabir al-Sabah

chancery: 2940 Tilden Street NW, Washington, DC 20008

telephone: [1] (202) 966-0702

FAX: [1] (202) 966-0517

Diplomatic representation from the US:

chief of mission: Ambassador Deborah K. JONES

embassy: Bayan 36302, Area 14, Al-Masjed Al-Aqsa Street (near the Bayan palace), Kuwait City

mailing address: P. O. Box 77 Safat 13001 Kuwait; or PSC 1280 APO AE 09880-9000

telephone: [965] 259-1001

FAX: [965] 538-0282

Flag description: three equal horizontal bands of green (top), white, and red with a black trapezoid based on the hoist side; design, which dates to 1961, based on the Arab revolt flag of World War I

ECONOMY

Economy—overview: Kuwait is a small, rich, relatively open economy with self-reported crude oil reserves of about 104 billion barrels—10% of world reserves. Petroleum accounts for nearly half of GDP, 95% of export revenues, and 80% of government income. High oil prices in recent years have helped build Kuwait's budget and trade surpluses and foreign reserves. As a result of this positive fiscal situation, the need for economic reforms is less urgent and the government has not earnestly pushed through new initiatives. Despite its vast oil reserves, Kuwait experienced power outages during the summer months in 2006 and 2007 because demand exceeded power generating capacity. Power outages are likely to worsen, given its high population growth rates, unless the government can increase generating capacity. In May 2007 Kuwait changed its currency peg from the US dollar to a basket of currencies in order to curb inflation and to reduce its vulnerability to external shocks.

GDP (purchasing power parity): $130.1 billion (2007 est.)

GDP (official exchange rate): $111.3 billion (2007 est.)

GDP—real growth rate: 4.6% (2007 est.)

GDP—per capita (PPP): $39,300 (2007 est.)

GDP—composition by sector:

agriculture: 0.3%

industry: 51.5%

services: 48.1% (2007 est.)

Labor force: 2.093 million

note: non-Kuwaitis represent about 80% of the labor force (2007 est.)

Labor force—by occupation: *agriculture:* NA%

industry: NA%

services: NA%

Unemployment rate: 2.2% (2004 est.)

Population below poverty line: NA%

Household income or consumption by percentage share:

lowest 10%: NA%

highest 10%: NA%

Inflation rate (consumer prices): 5% (2007 est.)

Investment (gross fixed): 21.4% of GDP (2007 est.)

Budget: *revenues:* $84.76 billion

expenditures: $36.8 billion (2007 est.)

Public debt: 7.8% of GDP (2007 est.)

Agriculture—products: practically no crops; fish

Industries: petroleum, petrochemicals, cement, shipbuilding and repair, water desalination, food processing, construction materials

Industrial production growth rate: 2.2% (2007 est.)

Electricity—production: 41.11 billion kWh (2005)

Electricity—production by source:

fossil fuel: 100%

hydro: 0%

nuclear: 0%

other: 0% (2001)

Electricity—consumption: 36.28 billion kWh (2005)

Electricity—exports: 0 kWh (2005)

Electricity—imports: 0 kWh (2005)

Oil—production: 2.669 million bbl/day (2005 est.)

Oil—consumption: 333,000 bbl/day (2005 est.)

Oil—exports: 2.2 million bbl/day (2004)

Oil—imports: 2,611 bbl/day (2004)

Oil—proved reserves: 101.5 billion bbl (2007 est.)

Natural gas—production: 11.8 billion cu m (2005 est.)

Natural gas—consumption: 11.8 billion cu m (2005 est.)

Natural gas—exports: 0 cu m (2005 est.)

Natural gas—imports: 0 cu m (2005)

Natural gas—proved reserves: 1.521 trillion cu m (1 January 2006 est.)

Current account balance: $52.73 billion (2007 est.)

Exports: $61.43 billion f.o.b. (2007 est.)

Exports—commodities: oil and refined products, fertilizers

Exports—partners: Japan 20.4%, South Korea 16.2%, Taiwan 10.8%, Singapore 9.7%, US 9%, Netherlands 5.3%, China 4.1% (2006)

Imports: $19.4 billion f.o.b. (2007 est.)

Imports—commodities: food, construction materials, vehicles and parts, clothing

Imports—partners: US 14.1%, Japan 7.8%, Germany 7.7%, Saudi Arabia 6.8%, China 5.7%, UK 5.4%, Italy 4.6% (2006)

Economic aid—recipient: $2.6 million (2004)

Reserves of foreign exchange and gold: $16.78 billion (31 December 2007 est.)

Debt—external: $34.67 billion (31 December 2007 est.)

Stock of direct foreign investment—at home: $963 million (2007 est.)

Stock of direct foreign investment—abroad: $19.88 billion (2007 est.)

Market value of publicly traded shares: $128.9 billion (2006)

Currency (code): Kuwaiti dinar (KD)

Currency code: KWD

Exchange rates: Kuwaiti dinars per US dollar—0.2844 (2007), 0.29 (2006), 0.292 (2005), 0.2947 (2004), 0.298 (2003)

Fiscal year: 1 April—31 March

COMMUNICATIONS

Telephones—main lines in use: 510,300 (2005)

Telephones—mobile cellular: 2.536 million (2006)

Telephone system:

general assessment: the quality of service is excellent

domestic: new telephone exchanges provide a large capacity for new subscribers; trunk traffic is carried by microwave

radio relay, coaxial cable, and open-wire and fiber-optic cable; a cellular telephone system operates throughout Kuwait, and the country is well supplied with pay telephones
international: country code—965; linked to international submarine cable Fiber-Optic Link Around the Globe (FLAG); linked to Bahrain, Qatar, UAE via the Fiber-Optic Gulf (FOG) cable; coaxial cable and microwave radio relay to Saudi Arabia; satellite earth stations—6 (3 Intelsat—1 Atlantic Ocean and 2 Indian Ocean, 1 Inmarsat—Atlantic Ocean, and 2 Arabsat)
Radio broadcast stations: AM 6, FM 11, shortwave 1 (1998)
Radios: 1.175 million (1997)
Television broadcast stations: 13 (plus several satellite channels) (1997)
Televisions: 875,000 (1997)
Internet country code: .kw
Internet hosts: 2,013 (2007)
Internet Service Providers (ISPs): 3 (2000)
Internet users: 816,700 (2006)

TRANSPORTATION

Airports: 7 (2007)
Airports—with paved runways:
total: 4
over 3,047 m: 1
2,438 to 3,047 m: 2
1,524 to 2,437 m: 1 (2007)
Airports—with unpaved runways:
total: 3
1,524 to 2,437 m: 1
under 914 m: 2 (2007)
Heliports: 4 (2007)
Pipelines: gas 269 km; oil 540 km; refined products 57 km (2007)

Roadways:
total: 5,749 km
paved: 4,887 km
unpaved: 862 km (2004)
Merchant marine:
total: 38 ships (1000 GRT or over) 2,195,831 GRT/3,566,308 DWT
by type: bulk carrier 2, cargo 1, container 6, liquefied gas 5, livestock carrier 3, petroleum tanker 21
registered in other countries: 28 (Bahrain 3, Comoros 1, Liberia 1, Libya 1, Panama 1, Qatar 7, Saudi Arabia 6, UAE 8) (2007)
Ports and terminals: Ash Shu'aybah, Ash Shuwaykh, Az Zawr (Mina' Sa'ud), Mina' 'Abd Allah, Mina' al Ahmadi

MILITARY

Military branches: Land Forces, Kuwaiti Navy, Kuwaiti Air Force (Al-Quwwat al-Jawwiya al-Kuwaitiya), National Guard (2007)
Military service age and obligation: 18 years of age for compulsory and voluntary military service; reserve obligation to age 40 with 1 month annual training; women have served in police forces since 1999 (2006)
Manpower available for military service:
males age 16–49: 1,032,408
females age 16–49: 568,657 (2008 est.)
Manpower fit for military service:
males age 16–49: 892,816
females age 16–49: 500,540 (2008 est.)
Manpower reaching militarily significant age annually:
males age 16–49: 17,737
females age 16–49: 18,519 (2008 est.)
Military expenditures—percent of GDP: 5.3% (2006)

TRANSNATIONAL ISSUES

Disputes—international: Kuwait and Saudi Arabia continue negotiating a joint maritime boundary with Iran; no maritime boundary exists with Iraq in the Persian Gulf
Trafficking in persons: *current situation*: Kuwait is a destination country for men and women who migrate legally from South and Southeast Asia for domestic or low-skilled labor, but are subjected to conditions of involuntary servitude by employers in Kuwait including conditions of physical and sexual abuse, non-payment of wages, confinement to the home, and withholding of passports to restrict their freedom of movement; Kuwait is reportedly a transit point for South and East Asian workers recruited for low-skilled work in Iraq; some of these workers are deceived as to the true location and nature of this work, and others are subjected to conditions of involuntary servitude in Iraq; in past years, Kuwait was also a destination country for children exploited as camel jockeys, but this form of trafficking appears to have ceased
tier rating: Tier 3—insufficient efforts in 2006 to prosecute and punish abusive employers and those who traffic women for sexual exploitation; the government failed for the third year in a row to live up to promises to provide shelter and protective services for victims of involuntary domestic servitude and other forms of trafficking

KYRGYZSTAN

INTRODUCTION

Background: A Central Asian country of incredible natural beauty and proud nomadic traditions, most of Kyrgyzstan was formally annexed to Russia in 1876. The Kyrgyz staged a major revolt against the Tsarist Empire in 1916 in which almost one-sixth of the Kyrgyz population was killed. Kyrgyzstan became a Soviet republic in 1936 and achieved independence in 1991 when the USSR dissolved. Nationwide demonstrations in the spring of 2005 resulted in the ouster of President Askar AKAYEV, who had run the country since 1990. Subsequent presidential elections in July 2005 were won overwhelmingly by former prime minister Kurmanbek BAKIYEV. The political opposition organized demonstrations in Bishkek in April, May, and November 2006 resulting in the adoption of a new constitution that transferred some of the president's powers to parliament and the government. In December 2006, the Kyrgyz parliament voted to adopt new amendments, restoring some of the presidential powers lost in the November 2006 constitutional change. By late-September 2007, both previous versions of the constitution were declared illegal, and the country reverted to the AKAYEV-era 2003 constitution, which was subsequently modified in a flawed referendum initiated by BAKIYEV. The president then dissolved parliament, called for early elections, and gained control of the new parliament through his newly-created political party, Ak Jol, in December 2007 elections. Current concerns include: privatization of state-owned enterprises, negative trends in democracy and political freedoms, reduction of corruption, improving interethnic relations, and combating terrorism.

GEOGRAPHY

Location: Central Asia, west of China
Geographic coordinates: 41 00 N, 75 00 E
Map references: Asia
Area: *total:* 198,500 sq km
land: 191,300 sq km
water: 7,200 sq km
Area—comparative: slightly smaller than South Dakota
Land boundaries:
total: 3,878 km
border countries: China 858 km, Kazakhstan 1,051 km, Tajikistan 870 km, Uzbekistan 1,099 km
Coastline: 0 km (landlocked)
Maritime claims: none (landlocked)
Climate: dry continental to polar in high Tien Shan; subtropical in southwest (Fergana Valley); temperate in northern foothill zone
Terrain: peaks of Tien Shan and associated valleys and basins encompass entire nation
Elevation extremes:
lowest point: Kara-Daryya (Karadar'ya) 132 m
highest point: Jengish Chokusu (Pik Pobedy) 7,439 m
Natural resources: abundant hydropower; significant deposits of gold and rare earth metals; locally exploitable coal, oil, and natural gas; other deposits of nepheline, mercury, bismuth, lead, and zinc
Land use:
arable land: 6.55%
permanent crops: 0.28%
other: 93.17%
note: Kyrgyzstan has the world's largest natural-growth walnut forest (2005)
Irrigated land: 10,720 sq km (2003)
Total renewable water resources: 46.5 cu km (1997)
Freshwater withdrawal (domestic/industrial/agricultural): *total:* 10.08 cu km/yr (3%/3%/94%)
per capita: 1,916 cu m/yr (2000)

Natural hazards: NA
Environment—current issues: water pollution; many people get their water directly from contaminated streams and wells; as a result, water-borne diseases are prevalent; increasing soil salinity from faulty irrigation practices
Environment—international agreements: *party to:* Air Pollution, Biodiversity, Climate Change, Climate Change-Kyoto Protocol, Desertification, Hazardous Wastes, Ozone Layer Protection, Wetlands
signed, but not ratified: none of the selected agreements
Geography—note: landlocked; entirely mountainous, dominated by the Tien Shan range; many tall peaks, glaciers, and high-altitude lakes

PEOPLE

Population: 5,356,869 (July 2008 est.)
Age structure:
0–14 years: 29.9% (male 817,369/female 784,782)
15–64 years: 64% (male 1,681,440/female 1,748,222)
65 years and over: 6.1% (male 127,263/female 197,793) (2008 est.)
Median age:
total: 24.2 years
male: 23.3 years
female: 25 years (2008 est.)
Population growth rate: 1.38% (2008 est.)
Birth rate: 23.31 births/1,000 population (2008 est.)
Death rate: 6.97 deaths/1,000 population (2008 est.)
Net migration rate: -2.55 migrant(s)/1,000 population (2008 est.)
Sex ratio:
at birth: 1.05 male(s)/female
under 15 years: 1.04 male(s)/female
15–64 years: 0.96 male(s)/female
65 years and over: 0.64 male(s)/female
total population: 0.96 male(s)/female (2008 est.)
Infant mortality rate:
total: 32.3 deaths/1,000 live births
male: 37.33 deaths/1,000 live births
female: 27 deaths/1,000 live births (2008 est.)
Life expectancy at birth:
total population: 69.12 years
male: 65.12 years
female: 73.33 years (2008 est.)
Total fertility rate: 2.67 children born/woman (2008 est.)
HIV/AIDS—adult prevalence rate: less than 0.1% (2001 est.)
HIV/AIDS—people living with HIV/AIDS: 3,900 (2003 est.)
HIV/AIDS—deaths: fewer than 200 (2003 est.)

Nationality:
noun: Kyrgyzstani(s)
adjective: Kyrgyzstani
Ethnic groups: Kyrgyz 64.9%, Uzbek 13.8%, Russian 12.5%, Dungan 1.1%, Ukrainian 1%, Uygur 1%, other 5.7% (1999 census)
Religions: Muslim 75%, Russian Orthodox 20%, other 5%
Languages: Kyrgyz 64.7% (official), Uzbek 13.6%, Russian 12.5% (official), Dungun 1%, other 8.2% (1999 census)
Literacy:
definition: age 15 and over can read and write
total population: 98.7%
male: 99.3%
female: 98.1% (1999 census)

GOVERNMENT

Country name:
conventional long form: Kyrgyz Republic
conventional short form: Kyrgyzstan
local long form: Kyrgyz Respublikasy
local short form: Kyrgyzstan
former: Kirghiz Soviet Socialist Republic
Government type: republic
Capital: *name:* Bishkek
geographic coordinates: 42 52 N, 74 36 E
time difference: UTC+6 (11 hours ahead of Washington, DC during Standard Time)
Administrative divisions: 7 provinces (oblastlar, singular—oblasty) and 1 city* (shaar); Batken Oblasty, Bishkek Shaary*, Chuy Oblasty (Bishkek), Jalal-Abad Oblasty, Naryn Oblasty, Osh Oblasty, Talas Oblasty, Ysyk-Kol Oblasty (Karakol)
note: administrative divisions have the same names as their administrative centers (exceptions have the administrative center name following in parentheses)
Independence: 31 August 1991 (from Soviet Union)
National holiday: Independence Day, 31 August (1991)
Constitution: adopted 5 May 1993; *note*—amendment proposed by President Askar AKAYEV and passed in a national referendum on 2 February 2003 significantly expanded the powers of the president at the expense of the legislature; during large-scale demonstrations in November 2006, President BAKIYEV and the opposition negotiated a new constitution granting greater powers to the parliament and the government; amendments added on 30 December 2006 redistributed some power back to the president, but both November and December 2006 versions were annulled in September 2007, and a new version was approved by referendum on 21 October 2007; the BAKIYEV-ini-

tiated referendum was criticized by Western observers for voting irregularities, particularly ballot stuffing

Legal system: based on French and Russian laws; has not accepted compulsory ICJ jurisdiction

Suffrage: 18 years of age; universal

Executive branch:

chief of state: President Kurmanbek BAKIYEV (since 14 August 2005)

head of government: Prime Minister Igor CHUDINOV (since 24 December 2007)

cabinet: Cabinet of Ministers proposed by the prime minister, appointed by the president; ministers in charge of defense and security, appointed solely by the president

elections: Kurmanbek BAKIYEV elected by popular vote for a five-year term (eligible for a second term); election last held 10 July 2005 (next scheduled for 2010); prime minister nominated by the parliamentary party holding more than 50% of the seats; if no such party exists, the president selects the party that will nominate a prime minister

election results: Kurmanbek BAKIYEV elected president; percent of vote— Kurmanbek BAKIYEV 88.6%, Tursunbai BAKIR-UULU 3.9%, other candidates 7.5%

Legislative branch: unicameral Supreme Council or Jorgorku Kenesh (90 seats; members are elected by popular vote to serve five-year terms)

elections: last held 16 December 2007 (next to be held in 2012)

election results: Supreme Council—percent of vote by party—NA; seats by party—Ak Jol 71, Social Democratic Party 11, KCP 8

Judicial branch: Supreme Court; Constitutional Court (judges of both the Supreme and Constitutional Courts are appointed for 10-year terms by the Jorgorku Kenesh on the recommendation of the president; their mandatory retirement age is 70 years); Higher Court of Arbitration; Local Courts (judges appointed by the president on the recommendation of the National Council on Legal Affairs for a probationary period of five years, then 10 years)

Political parties and leaders: Ak Jol [Avtandil ARABAYEV, Elmira IBRAIMOVA, Vladimir NIFADYEV, co-chairs]; Ar-Namys (Dignity) Party [Emil ALIYEV]; Asaba (Banner National Revival Party) [Azimbek BEKNAZAROV]; Ata-Meken (Fatherland) [Omurbek TEKEBAYEV]; Democratic Movement of Kyrgyzstan or DDK [Viktor TCHETRNOMORETS]; Erkindik (Freedom) Party [Topchubek TURGUNALIYEV]; Moya Strana (My Country Party of Action) [Medet SADYRKULOV]; Party of Communists of Kyrgyzstan or KCP [Ishak MASALIYEV]; Party of Justice and Progress [Muratbek IMANALIEV]; Party of Peasants [Esengul ISAKOV]; Republican Party of Labor and Unity [Tabaldy OROZALIYEV]; Sanjira (Tree of Life) [Ednan KARABAYEV]; Social Democratic Party [Almaz ATAMBAYEV]; Union of Democratic Forces [Kubatbek BAIBOLOV]

Political pressure groups and leaders: Adilet Legal Clinic [Cholpon JAKUPOVA]; Coalition for Democracy and Civil Society [Dinara OSHURAKHUNOVA]; Interbilim [Asiya SASYKBAYEVA]

International organization participation: ADB, CIS, CSTO, EAEC, EAPC, EBRD, ECO, FAO, GCTU, IAEA, IBRD, ICAO, ICCt (signatory), ICRM, IDA, IDB, IFAD, IFC, IFRCS, ILO, IMF, Interpol, IOC, IOM, IPU, ISO (correspondent), ITSO, ITU, MIGA, NAM (observer), OIC, OPCW, OSCE, PCA, PFP, SCO, UN, UNCTAD, UNESCO, UNIDO, UNMEE, UNMIL, UNMIS, UNWTO, UPU, WCO, WFTU, WHO, WIPO, WMO, WTO

Diplomatic representation in the US:

chief of mission: Ambassador Zamira SYDYKOVA

chancery: 2360 Massachusetts Ave. NW, Washington, DC 20008

telephone: [1] (202) 338-5141

FAX: [1] (202) 386-7550

consulate(s): New York

Diplomatic representation from the US:

chief of mission: Ambassador Marie L. YOVANOVITCH

embassy: 171 Prospect Mira, Bishkek 720016

mailing address: use embassy street address

telephone: [996] (312) 551-241, (517) 777-217

FAX: [996] (312) 551-264

Flag description: red field with a yellow sun in the center having 40 rays representing the 40 Kyrgyz tribes; on the obverse side the rays run counterclockwise, on the reverse, clockwise; in the center of the sun is a red ring crossed by two sets of three lines, a stylized representation of the roof of the traditional Kyrgyz yurt

ECONOMY

Economy—overview: Kyrgyzstan is a poor, mountainous country with a predominantly agricultural economy. Cotton, tobacco, wool, and meat are the main agricultural products, although only tobacco and cotton are exported in any quantity. Industrial exports include gold, mercury, uranium, natural gas, and electricity. Following independence, Kyrgyzstan was progressive in carrying out market reforms such as an improved regulatory system and land reform. Kyrgyzstan was the first Commonwealth of Independent States (CIS) country to be accepted into the World Trade Organization. Much of the government's stock in enterprises has been sold. Drops in production had been severe after the breakup of the Soviet Union in December 1991, but by mid-1995, production began to recover and exports began to increase. The economy is heavily weighted toward gold export and a drop in output at the main Kumtor gold mine sparked a 0.5% decline in GDP in 2002 and a 0.6% decline in 2005. GDP grew more than 6% in 2007, partly due to higher gold prices internationally. The government made steady strides in controlling its substantial fiscal deficit, nearly closing the gap between revenues and expenditures in 2006, before boosting expenditures more than 20% in 2007. The government and international financial institutions have been engaged in a comprehensive medium-term poverty reduction and economic growth strategy. In 2005, Bishkek agreed to pursue much-needed tax reform and, in 2006, became eligible for the heavily indebted poor countries (HIPC) initiative. Progress fighting corruption, further restructuring of domestic industry, and success in attracting foreign investment are keys to future growth.

GDP (purchasing power parity): $10.5 billion (2007 est.)

GDP (official exchange rate): $3.748 billion (2007 est.)

GDP—real growth rate: 8.2% (2007 est.)

GDP—per capita (PPP): $2,000 (2007 est.)

GDP—composition by sector:

agriculture: 30.9%

industry: 19.9%

services: 49.1% (2007 est.)

Labor force: 2.7 million (2000)

Labor force—by occupation: *agriculture:* 55%

industry: 15%

services: 30% (2000 est.)

Unemployment rate: 18% (2004 est.)

Population below poverty line: 40% (2004 est.)

Household income or consumption by percentage share:

lowest 10%: 3.8%

highest 10%: 24.3% (2003)

Distribution of family income—Gini index: 30.3 (2003)

Inflation rate (consumer prices): 10.2% (2007 est.)

Investment (gross fixed): 15.4% of GDP (2007 est.)

Budget:

revenues: $920.8 million

expenditures: $993.3 million (2007 est.)

Agriculture—products: tobacco, cotton, potatoes, vegetables, grapes, fruits and berries; sheep, goats, cattle, wool

Industries: small machinery, textiles, food processing, cement, shoes, sawn logs, refrigerators, furniture, electric motors, gold, rare earth metals

Industrial production growth rate: 7.3% (2007 est.)

Electricity—production: 15.15 billion kWh (2005)

Electricity—production by source:

fossil fuel: 7.6%

hydro: 92.4%

nuclear: 0%

other: 0% (2001)

Electricity—consumption: 8.206 billion kWh (2005)

Electricity—exports: 2.684 billion kWh (2005)

Electricity—imports: 0 kWh (2005)

Oil—production: 1,965 bbl/day (2005)

Oil—consumption: 12,000 bbl/day (2005 est.)

Oil—exports: 3,221 bbl/day (2004)

Oil—imports: 13,770 bbl/day (2004)

Oil—proved reserves: 40 million bbl (1 January 2006 est.)

Natural gas—production: 28.77 million cu m (2005 est.)

Natural gas—consumption: 709.7 million cu m (2005 est.)

Natural gas—exports: 0 cu m (2005 est.)

Natural gas—imports: 680.9 million cu m (2005)

Natural gas—proved reserves: 5.432 billion cu m (1 January 2006 est.)

Current account balance: -$244 million (2007 est.)

Exports: $1.04 billion f.o.b. (2007 est.)

Exports—commodities: cotton, wool, meat, tobacco; gold, mercury, uranium, natural gas, hydropower; machinery; shoes

Exports—partners: Switzerland 26.1%, Kazakhstan 20.4%, Russia 19.3%, Afghanistan 9.4%, China 4.8% (2006)

Imports: $2.509 billion f.o.b. (2007 est.)

Imports—commodities: oil and gas, machinery and equipment, chemicals, foodstuffs

Imports—partners: Russia 38.1%, China 14.4%, Kazakhstan 11.7%, US 5.7% (2006)

Economic aid—recipient: $268.5 million from the US (2005)

Reserves of foreign exchange and gold: $1.177 billion (31 December 2007 est.)

Debt—external: $2.966 billion (30 June 2007)

Stock of direct foreign investment—at home: $NA

Stock of direct foreign investment—abroad: $NA

Market value of publicly traded shares: $41.99 million (2005)

Currency (code): som (KGS)

Currency code: KGS

Exchange rates: soms per US dollar—37.746 (2007), 40.149 (2006), 41.012 (2005), 42.65 (2004), 43.648 (2003)

Fiscal year: calendar year

COMMUNICATIONS

Telephones—main lines in use: 458,900 (2006)

Telephones—mobile cellular: 1,261,800 (2006)

Telephone system:

general assessment: telecommunications infrastructure is growing; fixed line penetration remains low and concentrated in urban areas

domestic: 4 mobile cellular service providers with growing coverage

international: country code—996; connections with other CIS countries by landline or microwave radio relay and with other countries by leased connections with Moscow international gateway switch and by satellite; satellite earth stations—2 (1 Intersputnik, 1 Intelsat); connected internationally by the Trans-Asia-Europe (TAE) fiber-optic line (2006)

Radio broadcast stations: AM 3 (plus 10 repeater stations), FM 23, shortwave NA (2007)

Radios: 520,000 (1997)

Television broadcast stations: 8 (2 countrywide and 6 regional stations; state-owned); note—there are about 20 private TV stations, most of which rebroadcast other channels (2007)

Televisions: 210,000 (1997)

Internet country code: .kg

Internet hosts: 80,990 (2007)

Internet Service Providers (ISPs): NA

Internet users: 298,100 (2006)

TRANSPORTATION

Airports: 30 (2007)

Airports—with paved runways: *total:* 18

over 3,047 m: 1

2,438 to 3,047 m: 3

1,524 to 2,437 m: 11

under 914 m: 3 (2007)

Airports—with unpaved runways:

total: 12

1,524 to 2,437 m: 1

914 to 1,523 m: 1

under 914 m: 10 (2007)

Pipelines: gas 254 km; oil 16 km (2007)

Railways:

total: 470 km

broad gauge: 470 km 1.520-m gauge (2006)

Roadways:

total: 18,500 km

paved: 16,854 km

unpaved: 1,646 km (2000)

Waterways: 600 km (2007)

Ports and terminals: Balykchy (Ysyk-Kol or Rybach'ye)

MILITARY

Military branches: Army, Air Force, National Guard (2005)

Military service age and obligation: 18 years of age for compulsory military service (2001)

Manpower available for military service:

males age 16–49: 1,398,878

females age 16–49: 1,419,374 (2008 est.)

Manpower fit for military service:

males age 16–49: 1,061,942

females age 16–49: 1,211,249 (2008 est.)

Manpower reaching militarily significant age annually:

males age 16–49: 60,706

females age 16–49: 58,721 (2008 est.)

Military expenditures—percent of GDP: 1.4% (2005 est.)

TRANSNATIONAL ISSUES

Disputes—international: Kyrgyzstan has yet to ratify the 2001 boundary delimitation with Kazakhstan; disputes in Isfara Valley delay completion of delimitation with Tajikistan; delimitation of 130 km of border with Uzbekistan is hampered by serious disputes around enclaves and other areas

Illicit drugs: limited illicit cultivation of cannabis and opium poppy for CIS markets; limited government eradication of illicit crops; transit point for Southwest Asian narcotics bound for Russia and the rest of Europe; major consumer of opiates

LAOS

INTRODUCTION

Background: Modern-day Laos has its roots in the ancient Lao kingdom of Lan Xang, established in the 14th Century under King FA NGUM. For 300 years Lan Xang had influence reaching into present-day Cambodia and Thailand, as well as over all of what is now Laos. After centuries of gradual decline, Laos came under the domination of Siam (Thailand) from the late 18th century until the late 19th century when it became part of French Indochina. The Franco-Siamese Treaty of 1907 defined the current Lao border with Thailand. In 1975, the Communist Pathet Lao took control of the government ending a six-century-old monarchy and instituting a strict socialist regime closely aligned to Vietnam. A gradual return to private enterprise and the liberalization of foreign investment laws began in 1986. Laos became a member of ASEAN in 1997.

GEOGRAPHY

Location: Southeastern Asia, northeast of Thailand, west of Vietnam
Geographic coordinates: 18 00 N, 105 00 E
Map references: Southeast Asia
Area: *total:* 236,800 sq km
land: 230,800 sq km
water: 6,000 sq km
Area—comparative: slightly larger than Utah
Land boundaries:
total: 5,083 km
border countries: Burma 235 km, Cambodia 541 km, China 423 km, Thailand 1,754 km, Vietnam 2,130 km
Coastline: 0 km (landlocked)
Maritime claims: none (landlocked)
Climate: tropical monsoon; rainy season (May to November); dry season (December to April)

Terrain: mostly rugged mountains; some plains and plateaus
Elevation extremes:
lowest point: Mekong River 70 m
highest point: Phou Bia 2,817 m
Natural resources: timber, hydropower, gypsum, tin, gold, gemstones
Land use:
arable land: 4.01%
permanent crops: 0.34%
other: 95.65% (2005)
Irrigated land: 1,750 sq km (2003)
Total renewable water resources: 333.6 cu km (2003)
Freshwater withdrawal (domestic/industrial/agricultural): *total:* 3 cu km/yr (4%/6%/90%)
per capita: 507 cu m/yr (2000)
Natural hazards: floods, droughts
Environment—current issues: unexploded ordnance; deforestation; soil erosion; most of the population does not have access to potable water
Environment—international agreements: *party to:* Biodiversity, Climate Change, Climate Change-Kyoto Protocol, Desertification, Endangered Species, Environmental Modification, Law of the Sea, Ozone Layer Protection *signed, but not ratified:* none of the selected agreements
Geography—note: landlocked; most of the country is mountainous and thickly forested; the Mekong River forms a large part of the western boundary with Thailand

PEOPLE

Population: 6,677,534 (July 2008 est.)
Age structure:
0–14 years: 41% (male 1,374,966/female 1,362,945)
15–64 years: 55.9% (male 1,846,375/female 1,885,029)
65 years and over: 3.1% (male 91,028/female 117,191) (2008 est.)
Median age: *total:* 19.2 years
male: 18.9 years
female: 19.5 years (2008 est.)
Population growth rate: 2.344% (2008 est.)
Birth rate: 34.46 births/1,000 population (2008 est.)
Death rate: 11.02 deaths/1,000 population (2008 est.)
Net migration rate: NA
Sex ratio:
at birth: 1.05 male(s)/female
under 15 years: 1.01 male(s)/female
15–64 years: 0.98 male(s)/female
65 years and over: 0.78 male(s)/female
total population: 0.98 male(s)/female (2008 est.)

Infant mortality rate:
total: 79.61 deaths/1,000 live births
male: 88.9 deaths/1,000 live births
female: 69.88 deaths/1,000 live births (2008 est.)
Life expectancy at birth:
total population: 56.29 years
male: 54.19 years
female: 58.47 years (2008 est.)
Total fertility rate: 4.5 children born/woman (2008 est.)
HIV/AIDS—adult prevalence rate: 0.1% (2003 est.)
HIV/AIDS—people living with HIV/AIDS: 1,700 (2003 est.)
HIV/AIDS—deaths: fewer than 200 (2003 est.)
Major infectious diseases:
degree of risk: very high
food or waterborne diseases: bacterial and protozoal diarrhea, hepatitis A, and typhoid fever
vectorborne diseases: dengue fever, Japanese encephalitis, and malaria
note: highly pathogenic H5N1 avian influenza has been identified in this country; it poses a negligible risk with extremely rare cases possible among US citizens who have close contact with birds (2008)
Nationality:
noun: Lao(s) or Laotian(s)
adjective: Lao or Laotian
Ethnic groups: Lao Loum (lowland) 68%, Lao Theung (upland) 22%, Lao Soung (highland) including the Hmong and the Yao 9%, ethnic Vietnamese/Chinese 1%
Religions: Buddhist 65%, animist 32.9%, Christian 1.3%, other and unspecified 0.8% (1995 census)
Languages: Lao (official), French, English, and various ethnic languages
Literacy: *definition:* age 15 and over can read and write
total population: 68.7%
male: 77%
female: 60.9% (2001 est.)

GOVERNMENT

Country name:
conventional long form: Lao People's Democratic Republic
conventional short form: Laos
local long form: Sathalanalat Paxathipatai Paxaxon Lao
local short form: none
Government type: Communist state
Capital: *name:* Vientiane
geographic coordinates: 17 58 N, 102 36 E
time difference: UTC+7 (12 hours ahead of Washington, DC during Standard Time)

Administrative divisions: 16 provinces (khoueng, singular and plural) and 1 capital city* (nakhon luang, singular and plural); Attapu, Bokeo, Bolikhamxai, Champasak, Houaphan, Khammouan, Louangnamtha, Louangphrabang, Oudomxai, Phongsali, Salavan, Savannakhet, Viangchan (Vientiane)*, Viangchan, Xaignabouli, Xekong, Xiangkhoang
Independence: 19 July 1949 (from France)
National holiday: Republic Day, 2 December (1975)
Constitution: promulgated 14 August 1991
Legal system: based on traditional customs, French legal norms and procedures, and socialist practice; has not accepted compulsory ICJ jurisdiction
Suffrage: 18 years of age; universal
Executive branch:
chief of state: President Lt. Gen. CHOUMMALI Saignason (since 8 June 2006); Vice President BOUN-GNANG Volachit (since 8 June 2006)
head of government: Prime Minister BOUASONE Bouphavanh (since 8 June 2006); Deputy Prime Ministers Maj. Gen. ASANG Laoli (since May 2002), Maj. Gen. DOUANGCHAI Phichit (since 8 June 2006), SOMSAVAT Lengsavat (since 26 February 1998), and THONGLOUN Sisoulit (since 27 March 2001)
cabinet: Ministers appointed by president, approved by National Assembly
elections: president and vice president elected by National Assembly for five-year terms; election last held 8 June 2006 (next to be held in 2011); prime minister nominated by president and elected by National Assembly for five-year term
election results: CHOUMMALI Saignason elected president; BOUN-GNANG Volachit elected vice president; percent of National Assembly vote—100%; BOUASONE Bouphavanh elected prime minister; percent of National Assembly vote—97%
Legislative branch: unicameral National Assembly (115 seats; members elected by popular vote from a list of candidates selected by the Lao People's Revolutionary Party to serve five-year terms)
elections: last held 30 April 2006 (next to be held in 2011)
election results: percent of vote by party—NA; seats by party—LPRP 113, independents 2
Judicial branch: People's Supreme Court (the president of the People's Supreme Court is elected by the National Assembly on the recommenda-

tion of the National Assembly Standing Committee; the vice president of the People's Supreme Court and the judges are appointed by the National Assembly Standing Committee)
Political parties and leaders: Lao People's Revolutionary Party or LPRP [CHOUMMALI Saignason]; other parties proscribed
Political pressure groups and leaders: political parties and groups other than LPRP are proscribed
International organization participation: ACCT, ADB, APT, ARF, ASEAN, CP, EAS, FAO, G-77, IBRD, ICAO, ICRM, IDA, IFAD, IFC, IFRCS, ILO, IMF, Interpol, IOC, IPU, ISO (subscriber), ITU, MIGA, NAM, OIF, OPCW, PCA, UN, UNCTAD, UNESCO, UNIDO, UNWTO, UPU, WCO, WFTU, WHO, WIPO, WMO, WTO (observer)
Diplomatic representation in the US:
chief of mission: Ambassador PHIANE Philakone
chancery: 2222 S Street NW, Washington, DC 20008
telephone: [1] (202) 332-6416
FAX: [1] (202) 332-4923
Diplomatic representation from the US:
chief of mission: Ambassador Ravic R. HUSO
embassy: 19 Rue Bartholonie, That Dam, Vientiane
mailing address: American Embassy Vientiane, APO AP 96546
telephone: [856] 21-26-7000
FAX: [856] 21-26-7190
Flag description: three horizontal bands of red (top), blue (double width), and red with a large white disk centered in the blue band

ECONOMY

Economy—overview: The government of Laos, one of the few remaining one-party Communist states, began decentralizing control and encouraging private enterprise in 1986. The results, starting from an extremely low base, were striking—growth averaged 6% per year in 1988–2007 except during the short-lived drop caused by the Asian financial crisis beginning in 1997. Despite this high growth rate, Laos remains a country with a underdeveloped infrastructure, particularly in rural areas. It has no railroads, a rudimentary road system, and limited external and internal telecommunications, though the government is sponsoring major improvements in the road system with support from Japan and China. Electricity is available in urban areas and in most rural districts. Subsistence agriculture, dominated by

rice, accounts for about 40% of GDP and provides 80% of total employment. The economy will continue to benefit from aid from international donors and from foreign investment in hydropower and mining. Construction will be another strong economic driver, especially as hydroelectric dam and road projects gain steam. Several policy changes since 2004 may help spur growth. In late 2004, Laos gained Normal Trade Relations status with the US, allowing Laos-based producers to benefit from lower tariffs on exports. Laos is taking steps to join the World Trade Organization in the next few years; the resulting trade policy reforms will improve the business environment. On the fiscal side, a value-added tax (VAT) regime, slated to begin in 2008, should help streamline the government's inefficient tax system.
GDP (purchasing power parity): $12.65 billion (2007 est.)
GDP (official exchange rate): $4.028 billion (2007 est.)
GDP—real growth rate: 7.5% (2007 est.)
GDP—per capita (PPP): $2,100 (2007 est.)
GDP—composition by sector:
agriculture: 41.3%
industry: 32.2%
services: 26.5% (2007 est.)
Labor force: 2.1 million (2006 est.)
Labor force—by occupation: *agriculture:* 80%
industry and services: 20% (2005 est.)
Unemployment rate: 2.4% (2005 est.)
Population below poverty line: 30.7% (2005 est.)
Household income or consumption by percentage share:
lowest 10%: 3.4%
highest 10%: 28.5% (2002)
Distribution of family income—Gini index: 34.6 (2002)
Inflation rate (consumer prices): 4.5% (2007 est.)
Budget:
revenues: $472.3 million
expenditures: $646.1 million (2007 est.)
Agriculture—products: sweet potatoes, vegetables, corn, coffee, sugarcane, tobacco, cotton, tea, peanuts, rice; water buffalo, pigs, cattle, poultry
Industries: copper, tin, gold, and gypsum mining; timber, electric power, agricultural processing, construction, garments, tourism, cement
Industrial production growth rate: 12% (2007 est.)
Electricity—production: 1.715 billion kWh (2005)
Electricity—production by source:
fossil fuel: 1.4%

hydro: 98.6%
nuclear: 0%
other: 0% (2001)
Electricity—consumption: 1.193 billion kWh (2005)
Electricity—exports: 728 million kWh (2005)
Electricity—imports: 326 million kWh (2005)
Oil—production: 0 bbl/day (2005 est.)
Oil—consumption: 2,950 bbl/day (2005 est.)
Oil—exports: 0 bbl/day (2004)
Oil—imports: 2,898 bbl/day (2004)
Oil—proved reserves: 0 bbl (1 January 2006 est.)
Natural gas—production: 0 cu m (2005 est.)
Natural gas—consumption: 0 cu m (2005 est.)
Natural gas—exports: 0 cu m (2005 est.)
Natural gas—imports: 0 cu m (2005)
Natural gas—proved reserves: 0 cu m (1 January 2006 est.)
Current account balance: -$930 million (2007 est.)
Exports: $970 million (2007 est.)
Exports—commodities: wood products, coffee, electricity, tin, copper, gold
Exports—partners: Thailand 42.1%, Vietnam 9.5%, China 4% (2006)
Imports: $1.376 billion f.o.b. (2007 est.)
Imports—commodities: machinery and equipment, vehicles, fuel, consumer goods
Imports—partners: Thailand 68.8%, China 11.3%, Vietnam 5.6% (2006)
Economic aid—recipient: $379 million (2006 est.)
Reserves of foreign exchange and gold: $541 million (31 December 2007 est.)
Debt—external: $3.179 billion (2006)
Currency (code): kip (LAK)
Currency code: LAK
Exchange rates: kips per US dollar—9,658 (2007), 10,235 (2006), 10,820 (2005), 10,585.5 (2004), 10,569 (2003)
Fiscal year: 1 October—30 September

COMMUNICATIONS

Telephones—main lines in use: 90,067 (2006)
Telephones—mobile cellular: 638,200 (2006)

Telephone system:
general assessment: service to general public is poor but improving; the government relies on a radiotelephone network to communicate with remote areas
domestic: multiple service providers; combined fixed-line and mobile-cellular subscribership about 10 per 100 persons
international: country code—856; satellite earth station—1 Intersputnik (Indian Ocean region)
Radio broadcast stations: AM 7, FM 14, shortwave 2 (2006)
Radios: 730,000 (1997)
Television broadcast stations: 7 (includes 1 station relaying Vietnam Television from Hanoi) (2006)
Televisions: 52,000 (1997)
Internet country code: .la
Internet hosts: 935 (2007)
Internet Service Providers (ISPs): 1 (2000)
Internet users: 25,000 (2005)

TRANSPORTATION

Airports: 42 (2007)
Airports—with paved runways:
total: 9
2,438 to 3,047 m: 2
1,524 to 2,437 m: 4
914 to 1,523 m: 3 (2007)
Airports—with unpaved runways:
total: 33
1,524 to 2,437 m: 1
914 to 1,523 m: 9
under 914 m: 23 (2007)
Pipelines: refined products 540 km (2007)
Roadways:
total: 31,210 km
paved: 4,494 km
unpaved: 26,716 km (2003)
Waterways: 4,600 km
note: primarily Mekong and tributaries; 2,900 additional km are intermittently navigable by craft drawing less than 0.5 m (2007)
Merchant marine:
total: 1 ship (1000 GRT or over) 2,370 GRT/3,110 DWT
by type: cargo 1 (2007)

MILITARY

Military branches: Lao People's Armed Forces (LPAF): Lao People's Army (LPA; includes Riverine Force), Air Force (2008)

Military service age and obligation: 15 years of age for compulsory military service; minimum 18-month conscript service obligation (2006)
Manpower available for military service: *males age 16–49:* 1,549,774
females age 16–49: 1,570,702 (2008 est.)
Manpower fit for military service:
males age 16–49: 993,162
females age 16–49: 1,052,053 (2008 est.)
Manpower reaching militarily significant age annually:
males age 16–49: 73,973
females age 16–49: 72,758 (2008 est.)
Military expenditures—percent of GDP: 0.5% (2006)
Military—note: serving one of the world's least developed countries, the Lao People's Armed Forces (LPAF) is small, poorly funded, and ineffectively resourced; its mission focus is border and internal security, primarily in countering ethnic Hmong insurgent groups; together with the Lao People's Revolutionary Party and the government, the Lao People's Army (LPA) is the third pillar of state machinery, and as such is expected to suppress political and civil unrest and similar national emergencies, but the LPA also has upgraded skills to respond to avian influenza outbreaks; there is no perceived external threat to the state and the LPA maintains strong ties with the neighboring Vietnamese military (2008)

TRANSNATIONAL ISSUES

Disputes—international: Southeast Asian states have enhanced border surveillance to check the spread of avian flu; talks continue on completion of demarcation with Thailand but disputes remain over islands in the Mekong River; concern among Mekong Commission members that China's construction of dams on the Mekong River will affect water levels
Illicit drugs: estimated opium poppy cultivation in 2005 was 5,600 hectares, about a 45% decrease from 2004; estimated potential opium production in 2005 was 28 metric tons, a significant decrease from 200 metric tons in 2003; unsubstantiated reports of domestic methamphetamine production; growing domestic methamphetamine problem

LATVIA

INTRODUCTION

Background: The name "Latvia" originates from the ancient Latgalians, one of four eastern Baltic tribes that formed the ethnic core of the Latvian people (ca. 8th–12th centuries A.D.). The region subsequently came under the control of Germans, Poles, Swedes, and finally, Russians. A Latvian republic emerged following World War I, but it was annexed by the USSR in 1940—an

action never recognized by the US and many other countries. Latvia reestablished its independence in 1991 following the breakup of the Soviet Union. Although the last Russian troops left in 1994, the status of the Russian minority (some 30% of the population) remains of concern to Moscow. Latvia joined both NATO and the EU in the spring of 2004.

GEOGRAPHY

Location: Eastern Europe, bordering the Baltic Sea, between Estonia and Lithuania

Geographic coordinates: 57 00 N, 25 00 E

Map references: Europe

Area: total: 64,589 sq km
land: 63,589 sq km
water: 1,000 sq km

Area—comparative: slightly larger than West Virginia

Land boundaries:
total: 1,348 km
border countries: Belarus 141 km, Estonia 343 km, Lithuania 588 km, Russia 276 km

Coastline: 498 km

Maritime claims:
territorial sea: 12 nm
exclusive economic zone: 200 nm
continental shelf: 200 m depth or to the depth of exploitation

Climate: maritime; wet, moderate winters

Terrain: low plain

Elevation extremes:
lowest point: Baltic Sea 0 m
highest point: Galzina Kalns 312 m

Natural resources: peat, limestone, dolomite, amber, hydropower, wood, arable land

Land use:
arable land: 28.19%
permanent crops: 0.45%
other: 71.36% (2005)

Irrigated land: 200 sq km
note: land in Latvia is often too wet, and in need of drainage, not irrigation;

approximately 16,000 sq km or 85% of agricultural land has been improved by drainage (2003)

Total renewable water resources: 49.9 cu km (2005)

Freshwater withdrawal (domestic/industrial/agricultural): total: 0.25 cu km/yr (55%/33%/12%)
per capita: 108 cu m/yr (2003)

Natural hazards: NA

Environment—current issues: Latvia's environment has benefited from a shift to service industries after the country regained independence; the main environmental priorities are improvement of drinking water quality and sewage system, household, and hazardous waste management, as well as reduction of air pollution; in 2001, Latvia closed the EU accession negotiation chapter on environment committing to full enforcement of EU environmental directives by 2010

Environment—international agreements: party to: Air Pollution, Air Pollution-Persistent Organic Pollutants, Biodiversity, Climate Change, Climate Change-Kyoto Protocol, Endangered Species, Hazardous Wastes, Law of the Sea, Ozone Layer Protection, Ship Pollution, Wetlands
signed, but not ratified: none of the selected agreements

Geography—note: most of the country is composed of fertile, low-lying plains, with some hills in the east

PEOPLE

Population: 2,245,423 (July 2008 est.)

Age structure:
0–14 years: 13.4% (male 154,077/female 146,825)
15–64 years: 69.7% (male 760,976/female 803,106)
65 years and over: 16.9% (male 124,658/female 255,781) (2008 est.)

Median age:
total: 39.9 years
male: 36.9 years
female: 43 years (2008 est.)

Population growth rate: -0.629% (2008 est.)

Birth rate: 9.62 births/1,000 population (2008 est.)

Death rate: 13.63 deaths/1,000 population (2008 est.)

Net migration rate: -2.29 migrant(s)/1,000 population (2008 est.)

Sex ratio:
at birth: 1.05 male(s)/female
under 15 years: 1.05 male(s)/female
15–64 years: 0.95 male(s)/female
65 years and over: 0.49 male(s)/female
total population: 0.86 male(s)/female (2008 est.)

Infant mortality rate:
total: 8.96 deaths/1,000 live births
male: 10.85 deaths/1,000 live births
female: 6.97 deaths/1,000 live births (2008 est.)

Life expectancy at birth:
total population: 71.88 years
male: 66.68 years
female: 77.35 years (2008 est.)

Total fertility rate: 1.29 children born/woman (2008 est.)

HIV/AIDS—adult prevalence rate: 0.6% (2001 est.)

HIV/AIDS—people living with HIV/AIDS: 7,600 (2001 est.)

HIV/AIDS—deaths: fewer than 500 (2003 est.)

Major infectious diseases:
degree of risk: intermediate
food or waterborne diseases: bacterial diarrhea and hepatitis A
vectorborne diseases: tickborne encephalitis (2008)

Nationality:
noun: Latvian(s)
adjective: Latvian

Ethnic groups: Latvian 57.7%, Russian 29.6%, Belarusian 4.1%, Ukrainian 2.7%, Polish 2.5%, Lithuanian 1.4%, other 2% (2002)

Religions: Lutheran, Roman Catholic, Russian Orthodox

Languages: Latvian (official) 58.2%, Russian 37.5%, Lithuanian and other 4.3% (2000 census)

Literacy: definition: age 15 and over can read and write
total population: 99.7%
male: 99.8%
female: 99.7% (2000 census)

GOVERNMENT

Country name:
conventional long form: Republic of Latvia
conventional short form: Latvia
local long form: Latvijas Republika
local short form: Latvija
former: Latvian Soviet Socialist Republic

Government type: parliamentary democracy

Capital: *name:* Riga
geographic coordinates: 56 57 N, 24 06 E
time difference: UTC+2 (7 hours ahead of Washington, DC during Standard Time)
daylight saving time: +1hr, begins last Sunday in March; ends last Sunday in October

Administrative divisions: 26 counties (singular—rajons) and 7 municipalities*: Aizkraukles Rajons, Aluksnes Rajons, Balvu Rajons, Bauskas Rajons, Cesu Rajons, Daugavpils*, Daugavpils Rajons, Dobeles Rajons, Gulbenes Rajons, Jekabpils Rajons, Jelgava*, Jelgavas Rajons, Jurmala*, Kraslavas Rajons, Kuldigas Rajons, Liepaja*, Liepajas Rajons, Limbazu Rajons, Ludzas Rajons, Madonas Rajons, Ogres Rajons, Preilu Rajons, Rezekne*, Rezeknes Rajons, Riga*, Rigas Rajons, Saldus Rajons, Talsu Rajons, Tukuma Rajons, Valkas Rajons, Valmieras Rajons, Ventspils*, Ventspils Rajons

Independence: 18 November 1918 (from Soviet Russia)

National holiday: Independence Day, 18 November (1918); note—18 November 1918 was the date Latvia declared itself independent from Soviet Russia; 4 May 1990 is when it declared the renewal of independence; 21 August 1991 was the date of de facto independence from the Soviet Union

Constitution: 15 February 1922; restored to force by the Constitutional Law of the Republic of Latvia adopted by the Supreme Council on 21 August 1991; multiple amendments

Legal system: based on civil law system with traces of Socialist legal traditions and practices; has not accepted compulsory ICJ jurisdiction

Suffrage: 18 years of age; universal for Latvian citizens

Executive branch:
chief of state: President Valdis ZATLERS (since 8 July 2007)
head of government: Prime Minister Ivars GODMANIS (since 20 December 2007)
cabinet: Council of Ministers nominated by the prime minister and appointed by Parliament
elections: president elected by Parliament for a four-year term (eligible for a second term); election last held 31 May 2007 (next to be held in 2011); prime minister appointed by the president, confirmed by Parliament
election results: Valdis ZATLERS elected president; parliamentary vote—Valdis ZATLERS 58, Aivars ENDZINS 39

Legislative branch: unicameral Parliament or Saeima (100 seats; members are elected by proportional representation from party lists by popular vote to serve four-year terms)

elections: last held 7 October 2006 (next to be held in October 2010)
election results: percent of vote by party—TP 19.5%, ZZS 16.7%, JL 16.4%, SC 14.4%; LPP/LC 8.6%; TB/LNNK 7%; PCTVL 6%; seats by party—TP 23, ZZS 18, JL 18, SC 17, LPP/LC 10, TB/LNNK 8, PCTVL 6; note—seats by party as of February 2008—TP 21, ZZS 17, SC 17, JL 14, LPP/LC 10, TB/LNNK 5, PCTVL 6, independents 10

Judicial branch: Supreme Court (judges' appointments are confirmed by Parliament); Constitutional Court (judges' appointments are confirmed by Parliament)

Political parties and leaders: First Party of Latvia/Latvia's Way or LPP/LC [Ainars SLESERS, Ivars GODMANIS]; For Human Rights in a United Latvia or PCTVL [Jakovs PLINERS]; For the Fatherland and Freedom/Latvian National Independence Movement or TB/LNNK [Roberts ZILE, Maris GRINBLATS]; Harmony Center or SC [Janis URBANOVICS, Nils USAKOVS]; Latvian Social Democratic Workers Party (Social Democrats) or LSDSP [Juris BOJARS]; Latvian Socialist Party or LSP [Alfreds RUBIKS]; New Democrats or JD [Maris GULBIS]; New Era Party or JL [Einars REPSE, Krisjanis KARINS]; People's Party or TP [Aigars KALVITIS]; The Union of Latvian Greens and Farmers Party or ZZS [Augusts BRIGMANIS]

Political pressure groups and leaders: Headquarters for the Protection of Russian Schools (SHTAB) [Aleksandr KAZAKOV]

International organization participation: Australia Group, BA, BIS, CBSS, CE, EAPC, EBRD, EIB, EU, FAO, IAEA, IBRD, ICAO, ICCt, ICRM, IDA, IFC, IFRCS, IHO, ILO, IMF, IMO, IMSO, Interpol, IOC, IOM, IPU, ISO (correspondent), ITU, ITUC, MIGA, NATO, NIB, NSG, OAS (observer), OPCW, OSCE, PCA, Schengen Convention, UN, UNCTAD, UNESCO, UNWTO, UPU, WCO, WEU (associate partner), WHO, WIPO, WMO, WTO

Diplomatic representation in the US:
chief of mission: Ambassador Andrejs PILDEGOVICS
chancery: 2306 Massachusetts Ave. NW, Washington, DC 20008
telephone: [1] (202) 328-2840
FAX: [1] (202) 328-2860

Diplomatic representation from the US:
chief of mission: Ambassador Charles LARSON Jr.
embassy: 7 Raina Boulevard, Riga LV-1510

mailing address: American Embassy Riga, PSC 78, Box Riga, APO AE 09723
telephone: [371] 703-6200
FAX: [371] 782-0047

Flag description: three horizontal bands of maroon (top), white (half-width), and maroon

ECONOMY

Economy—overview: Latvia's economy experienced GDP growth of more than 10% per year during 2006–07. The majority of companies, banks, and real estate have been privatized, although the state still holds sizable stakes in a few large enterprises. Latvia officially joined the World Trade Organization in February 1999. EU membership, a top foreign policy goal, came in May 2004. The current account deficit—more than 22% of GDP in 2007—and inflation—at nearly 10% per year—remain major concerns.

GDP (purchasing power parity): $39.73 billion (2007 est.)
GDP (official exchange rate): $27.34 billion (2007 est.)
GDP—real growth rate: 10.2% (2007 est.)
GDP—per capita (PPP): $17,400 (2007 est.)
GDP—composition by sector:
agriculture: 3.3%
industry: 22%
services: 74.7% (2007 est.)
Labor force: 1.167 million (2007 est.)
Labor force—by occupation: *agriculture:* 13%
industry: 19%
services: 68% (2005 est.)
Unemployment rate: 5.7% (2007 est.)
Population below poverty line: NA%
Household income or consumption by percentage share:
lowest 10%: 2.5%
highest 10%: 29.1% (2003)
Distribution of family income—Gini index: 37.7 (2003)
Inflation rate (consumer prices): 10.1% (2007 est.)
Investment (gross fixed): 31.6% of GDP (2007 est.)
Budget:
revenues: $10.47 billion
expenditures: $10.29 billion (2007 est.)
Public debt: 7.4% of GDP (2007 est.)
Agriculture—products: grain, sugar beets, potatoes, vegetables; beef, pork, milk, eggs; fish
Industries: buses, vans, street and railroad cars; synthetic fibers, agricultural machinery, fertilizers, washing machines, radios, electronics, pharmaceuticals, processed foods, textiles; note—dependent on imports for energy and raw materials

371

Industrial production growth rate: 5.4% (2007 est.)

Electricity—production: 4.778 billion kWh (2005)

Electricity—production by source:
fossil fuel: 29.1%
hydro: 70.9%
nuclear: 0%
other: 0% (2001)

Electricity—consumption: 6.09 billion kWh (2005)

Electricity—exports: 707 million kWh (2005)

Electricity—imports: 2.855 billion kWh (2005)

Oil—production: 0 bbl/day (2005 est.)

Oil—consumption: 34,000 bbl/day (2005 est.)

Oil—exports: 6,765 bbl/day (2004)

Oil—imports: 39,190 bbl/day (2004)

Oil—proved reserves: 0 bbl (1 January 2006 est.)

Natural gas—production: 0 cu m (2005 est.)

Natural gas—consumption: 1.861 billion cu m (2005 est.)

Natural gas—exports: 0 cu m (2005 est.)

Natural gas—imports: 1.861 billion cu m (2005)

Current account balance: -$6.381 billion (2007 est.)

Exports: $8.143 billion f.o.b. (2007 est.)

Exports—commodities: wood and wood products, machinery and equipment, metals, textiles, foodstuffs

Exports—partners: Lithuania 14.2%, Estonia 12.3%, Russia 11.5%, Germany 9.8%, UK 7.6%, Sweden 6.3%, Denmark 4.8% (2006)

Imports: $14.82 billion f.o.b. (2007 est.)

Imports—commodities: machinery and equipment, chemicals, fuels, vehicles

Imports—partners: Germany 15.5%, Lithuania 12.9%, Russia 8%, Estonia 7.7%, Poland 7.2%, Finland 5.7%, Sweden 5%, Belarus 4.7% (2006)

Economic aid—recipient: $162 million (2004)

Reserves of foreign exchange and gold: $5.758 billion (31 December 2007 est.)

Debt—external: $33.53 billion (31 December 2007)

Stock of direct foreign investment—at home: $8.62 billion (2007 est.)

Stock of direct foreign investment—abroad: $699 million (2007 est.)

Market value of publicly traded shares: $2.705 billion (2006)

Currency (code): lat (LVL)

Currency code: LVL

Exchange rates: lati per US dollar—0.5162 (2007), 0.5597 (2006), 0.5647 (2005), 0.5402 (2004), 0.5715 (2003)

Fiscal year: calendar year

Telephones—main lines in use: 657,400 (2006)

Telephones—mobile cellular: 2.184 million (2006)

Telephone system:
general assessment: recent efforts focused on bringing competition to the telecommunications sector; the number of fixed lines is decreasing as wireless telephony expands
domestic: number of telecommunications operators has grown rapidly since the fixed-line market opened to competition in 2003; combined fixed-line and mobile-cellular subscribership is roughly 125 per 100 persons
international: country code—371; the Latvian network is now connected via fiber optic cable to Estonia, Finland, and Sweden

Radio broadcast stations: AM 8, FM 56, shortwave 1 (1998)

Radios: 1.76 million (1997)

Television broadcast stations: 44 (plus 31 repeaters) (1995)

Televisions: 1.22 million (1997)

Internet country code: .lv

Internet hosts: 234,014 (2007)

Internet Service Providers (ISPs): 41 (2001)

Internet users: 1.071 million (2006)

Airports: 42 (2007)

Airports—with paved runways:
total: 21
2,438 to 3,047 m: 7
1,524 to 2,437 m: 3
914 to 1,523 m: 2
under 914 m: 9 (2007)

Airports—with unpaved runways:
total: 21
914 to 1,523 m: 1
under 914 m: 20 (2007)

Pipelines: gas 948 km; oil 82 km; refined products 415 km (2007)

Railways: *total:* 2,303 km
broad gauge: 2,270 km 1.520-m gauge (257 km electrified)
narrow gauge: 33 km 0.750-m gauge (2006)

Roadways: *total:* 69,829 km
paved: 69,829 km (2005)

Waterways: 300 km (2006)

Merchant marine:
total: 22 ships (1000 GRT or over) 201,684 GRT/221,186 DWT
by type: cargo 9, liquefied gas 2, passenger/cargo 4, petroleum tanker 5, roll on/roll off 2
foreign-owned: 1 (Estonia 1)
registered in other countries: 122 (Antigua and Barbuda 9, Belize 14, Cambodia 2, Cyprus 1, Dominica 2, Jamaica 2, Liberia 15, Malta 36, Marshall Islands 10, Panama 5, Russia 2, St Kitts and Nevis 4, St Vincent and The Grenadines 20) (2007)

Ports and terminals: Riga, Ventspils

Military branches: Latvian Republic Defense Force: Ground Forces, Navy (Latvijas Juras Speki; includes Coast Guard (Latvijas Kara Flotes)), Air Force (Latvijas Gaisa Spelki), Border Guard, Latvian Home Guard (Latvijas Zemessardze) (2008)

Military service age and obligation: 18 years of age for voluntary military service; conscription abolished January 2007; under current law, every citizen is entitled to serve in the armed forces for life (2006)

Manpower available for military service:
males age 16–49: 568,683
females age 16–49: 565,826 (2008 est.)

Manpower fit for military service:
males age 16–49: 412,849
females age 16–49: 468,827 (2008 est.)

Manpower reaching militarily significant age annually:
males age 16–49: 14,506
females age 16–49: 13,982 (2008 est.)

Military expenditures—percent of GDP: 1.2% (2005 est.)

Disputes—international: Russia refuses to sign the 1997 boundary treaty due to Latvian insistence on a unilateral clarificatory declaration referencing Soviet occupation of Latvia and territorial losses; Russia demands better Latvian treatment of ethnic Russians in Latvia; as of January 2007, ground demarcation of the boundary with Belarus was complete and mapped with final ratification documentation in preparation; the Latvian parliament has not ratified its 1998 maritime boundary treaty with Lithuania, primarily due to concerns over oil exploration rights; as a member state that forms part of the EU's external border, Latvia has implemented the strict Schengen border rules with Russia

Illicit drugs: transshipment and destination point for cocaine, synthetic drugs, opiates, and cannabis from Southwest Asia, Western Europe, Latin America, and neighboring Balkan countries; despite improved legislation, vulnerable to money laundering due to nascent enforcement capabilities and comparatively weak regulation of offshore companies and the gaming industry; CIS organized crime (including counterfeiting, corruption, extortion, stolen cars, and prostitution) accounts for most laundered proceeds

LEBANON

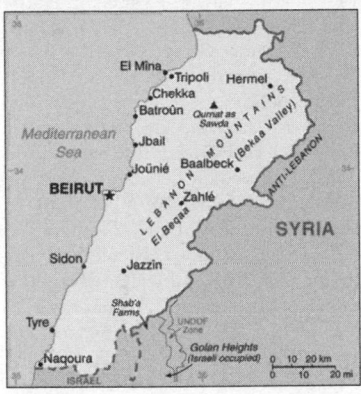

INTRODUCTION

Background: Following the capture of Syria from the Ottoman Empire by Anglo-French forces in 1918, France received a mandate over this territory and separated out the region of Lebanon in 1920. France granted this area independence in 1943. A lengthy civil war (1975–1990) devastated the country, but Lebanon has since made progress toward rebuilding its political institutions. Under the Ta'if Accord—the blueprint for national reconciliation—the Lebanese established a more equitable political system, particularly by giving Muslims a greater voice in the political process while institutionalizing sectarian divisions in the government. Since the end of the war, Lebanon has conducted several successful elections. Most militias have been disbanded, and the Lebanese Armed Forces (LAF) have extended authority over about two-thirds of the country. Hizballah, a radical Shi'a organization listed by the US State Department as a Foreign Terrorist Organization, retains its weapons. During Lebanon's civil war, the Arab League legitimized in the Ta'if Accord Syria's troop deployment, numbering about 16,000 based mainly east of Beirut and in the Bekaa Valley. Israel's withdrawal from southern Lebanon in May 2000 and the passage in October 2004 of UNSCR 1559—a resolution calling for Syria to withdraw from Lebanon and end its interference in Lebanese affairs—encouraged some Lebanese groups to demand that Syria withdraw its forces as well. The assassination of former Prime Minister Rafiq HARIRI and 20 others in February 2005 led to massive demonstrations in Beirut against the Syrian presence ("the Cedar Revolution"), and Syria withdrew the remainder of its mili-

tary forces in April 2005. In May-June 2005, Lebanon held its first legislative elections since the end of the civil war free of foreign interference, handing a majority to the bloc led by Saad HARIRI, the slain prime minister's son. Lebanon continues to be plagued by violence—Hizballah kidnapped two Israeli soldiers in July 2006 leading to a 34-day conflict with Israel. The LAF in May-September 2007 battled Sunni extremist group Fatah al-Islam in the Nahr al-Barid Palestinian refugee camp; and the country has witnessed a string of politically motivated assassinations since the death of Rafiq HARIRI. Lebanese politicians in November 2007 were unable to agree on a successor to Emile LAHUD when he stepped down as president, creating a political vacuum.

GEOGRAPHY

Location: Middle East, bordering the Mediterranean Sea, between Israel and Syria

Geographic coordinates: 33 50 N, 35 50 E

Map references: Middle East

Area:
total: 10,400 sq km
land: 10,230 sq km
water: 170 sq km

Area—comparative: about 0.7 times the size of Connecticut

Land boundaries: *total:* 454 km
border countries: Israel 79 km, Syria 375 km

Coastline: 225 km

Maritime claims: *territorial sea:* 12 nm

Climate: Mediterranean; mild to cool, wet winters with hot, dry summers; Lebanon mountains experience heavy winter snows

Terrain: narrow coastal plain; El Beqaa (Bekaa Valley) separates Lebanon and Anti-Lebanon Mountains

Elevation extremes:
lowest point: Mediterranean Sea 0 m
highest point: Qurnat as Sawda' 3,088 m

Natural resources: limestone, iron ore, salt, water-surplus state in a water-deficit region, arable land

Land use: *arable land:* 16.35%
permanent crops: 13.75%
other: 69.9% (2005)

Irrigated land: 1,040 sq km (2003)

Total renewable water resources: 4.8 cu km (1997)

Freshwater withdrawal (domestic/industrial/agricultural): *total:* 1.38 cu km/yr (33%/1%/67%)
per capita: 385 cu m/yr (2000)

Natural hazards: dust storms, sandstorms

Environment—current issues: deforestation; soil erosion; desertification; air pollution in Beirut from vehicular traffic and the burning of industrial wastes; pollution of coastal waters from raw sewage and oil spills

Environment—international agreements: *party to:* Biodiversity, Climate Change, Climate Change-Kyoto Protocol, Desertification, Hazardous Wastes, Law of the Sea, Ozone Layer Protection, Ship Pollution, Wetlands
signed, but not ratified: Environmental Modification, Marine Life Conservation

Geography—note: Nahr el Litani is the only major river in Near East not crossing an international boundary; rugged terrain historically helped isolate, protect, and develop numerous factional groups based on religion, clan, and ethnicity

PEOPLE

Population: 3,971,941 (July 2008 est.)

Age structure:
0–14 years: 26% (male 526,994/female 505,894)
15–64 years: 66.8% (male 1,275,021/female 1,380,131)
65 years and over: 7.1% (male 128,002/female 155,899) (2008 est.)

Median age:
total: 28.8 years
male: 27.6 years
female: 30 years (2008 est.)

Population growth rate: 1.154% (2008 est.)

Birth rate: 17.61 births/1,000 population (2008 est.)

Death rate: 6.06 deaths/1,000 population (2008 est.)

Net migration rate: NA

Sex ratio:
at birth: 1.05 male(s)/female
under 15 years: 1.04 male(s)/female
15–64 years: 0.92 male(s)/female
65 years and over: 0.82 male(s)/female
total population: 0.95 male(s)/female (2008 est.)

Infant mortality rate:
total: 22.59 deaths/1,000 live births
male: 25.08 deaths/1,000 live births
female: 19.97 deaths/1,000 live births (2008 est.)

Life expectancy at birth:
total population: 73.41 years
male: 70.91 years
female: 76.04 years (2008 est.)

Total fertility rate: 1.87 children born/woman (2008 est.)

HIV/AIDS—adult prevalence rate: 0.1% (2001 est.)

HIV/AIDS—people living with HIV/AIDS: 2,800 (2003 est.)

HIV/AIDS—deaths: fewer than 200 (2003 est.)

Nationality:

noun: Lebanese (singular and plural)

adjective: Lebanese

Ethnic groups: Arab 95%, Armenian 4%, other 1%

note: many Christian Lebanese do not identify themselves as Arab but rather as descendents of the ancient Canaanites and prefer to be called Phoenicians

Religions: Muslim 59.7% (Shi'a, Sunni, Druze, Isma'ilite, Alawite or Nusayri), Christian 39% (Maronite Catholic, Greek Orthodox, Melkite Catholic, Armenian Orthodox, Syrian Catholic, Armenian Catholic, Syrian Orthodox, Roman Catholic, Chaldean, Assyrian, Copt, Protestant), other 1.3%

note: 17 religious sects recognized

Languages: Arabic (official), French, English, Armenian

Literacy:

definition: age 15 and over can read and write

total population: 87.4%

male: 93.1%

female: 82.2% (2003 est.)

GOVERNMENT

Country name:

conventional long form: Lebanese Republic

conventional short form: Lebanon

local long form: Al Jumhuriyah al Lubnaniyah

local short form: Lubnan

former: Greater Lebanon

Government type: republic

Capital: *name:* Beirut

geographic coordinates: 33 52 N, 35 30 E

time difference: UTC+2 (7 hours ahead of Washington, DC during Standard Time)

daylight saving time: +1hr, begins last Sunday in March; ends last Sunday in October

Administrative divisions: 8 governorates (mohafazat, singular—mohafazah); Aakar, Baalbek-Hermel, Beqaa, Beyrouth, Liban-Nord, Liban-Sud, Mont-Liban, Nabatiye

Independence: 22 November 1943 (from League of Nations mandate under French administration)

National holiday: Independence Day, 22 November (1943)

Constitution: 23 May 1926; amended a number of times, most recently Charter of Lebanese National Reconciliation (Ta'if Accord) of October 1989

Legal system: mixture of Ottoman law, canon law, Napoleonic code, and civil law; no judicial review of legislative acts; has not accepted compulsory ICJ jurisdiction

Suffrage: 21 years of age; compulsory for all males; authorized for women at age 21 with elementary education

Executive branch:

chief of state: President Michel SULAYMAN (as of 25 May 2008)

head of government: Prime Minister Fuad SINIORA (since 30 June 2005); Deputy Prime Minister Elias MURR (since April 2005)

cabinet: Cabinet chosen by the prime minister in consultation with the president and members of the National Assembly

elections: president elected by the National Assembly for a six-year term (may not serve consecutive terms); election last held 25 May 2008 (next to be held in 2014); the prime minister and deputy prime minister appointed by the president in consultation with the National Assembly

election results: Michel SULAYMAN elected president; National Assembly vote—118 for, 6 abstentions, 3 invalidated

Legislative branch: unicameral National Assembly or Majlis Alnuwab (Arabic) or Assemblee Nationale (French) (128 seats; members elected by popular vote on the basis of sectarian proportional representation to serve four-year terms)

elections: last held in four rounds on 29 May, 5, 12, 19 June 2005 (next to be held in 2009)

election results: percent of vote by group—NA; seats by group—Future Movement Bloc 36; Democratic Gathering 15; Development and Resistance Bloc 15; Free Patriotic Movement 15; Loyalty to the Resistance 14; Qornet Shehwan 6; Lebanese Forces 5; Popular Bloc 4; Tripoli Independent Bloc 3; Kataeb Reform Movement 2; Syrian National Socialist Party 2; Tashnaq 2; Syrian Ba'th Party 1; Democratic Left 1; Democratic Renewal Movement 1; Kataeb Party 1; Nasserite Popular Movement 1; independent 4

Judicial branch: four Courts of Cassation (three courts for civil and commercial cases and one court for criminal cases); Constitutional Council (called for in Ta'if Accord—rules on constitutionality of laws); Supreme Council (hears charges against the president and prime minister as needed)

Political parties and leaders: *14 March Coalition:* Democratic Gathering Bloc [Walid JUNBLATT, leader of Progressive Socialist Party]; Democratic Left [Ilyas ATALLAH]; Democratic Renewal Movement [Nassib LAHUD]; Future Movement Bloc [Sa'ad HARIRI]; Kataeb Party [Amine GEMAYEL]; Lebanese Forces [Samir JA'JA]; Tripoli Independent Bloc

8 March Coalition: Development and Resistance Bloc [Nabih BERRI, leader of Amal Movement]; Free Patriotic Movement [Michel AWN]; Loyalty to the Resistance Bloc [Mohammad RA'AD] (includes Hizballah Party [Hassan NASRALLAH]); Nasserite Popular Movement [Ussama SAAD]; Popular Bloc [Elias SKAFF]; Syrian Ba'th Party [Sayez SHUKR]; Syrian Social Nationalist Party [Ali QANSO]

Independent: Metn Bloc [Michel MURR]; Tashnaq

Political pressure groups and leaders: none

International organization participation: ABEDA, ACCT, AFESD, AMF, FAO, G-24, G-77, IAEA, IBRD, ICAO, ICC, ICRM, IDA, IDB, IFAD, IFC, IFRCS, ILO, IMF, IMO, IMSO, Interpol, IOC, IPU, ISO, ITSO, ITU, LAS, MIGA, NAM, OAS (observer), OIC, OIF, PCA, UN, UNCTAD, UNESCO, UNHCR, UNIDO, UNRWA, UNWTO, UPU, WCO, WFTU, WHO, WIPO, WMO, WTO (observer)

Diplomatic representation in the US:

chief of mission: Ambassador (vacant); Charge d'Affaires Antoine CHEDID

chancery: 2560 28th Street NW, Washington, DC 20008

telephone: [1] (202) 939-6300

FAX: [1] (202) 939-6324

consulate(s) general: Detroit, New York, Los Angeles

Diplomatic representation from the US:

chief of mission: Ambassador (vacant); Charge d'Affaires Michele J. SISON

embassy: Awkar, Lebanon; (Awkar facing the Municipality)

mailing address: P. O. Box 70-840, Antelias, Lebanon; from US: US Embassy Beirut, 6070 Beirut Place, Washington, DC 20521-6070

telephone: [961] (4) 542600, 543600

FAX: [961] (4) 544136

Flag description: three horizontal bands consisting of red (top), white (middle, double width), and red (bottom) with a green cedar tree centered in the white band

ECONOMY

Economy—overview: The 1975–90 civil war seriously damaged Lebanon's economic infrastructure, cut national output by half, and all but ended Lebanon's position as a Middle Eastern entrepot and banking hub. In the years since, Lebanon has rebuilt much of its war-torn physical and financial infrastructure by borrowing heavily—mostly from domestic banks. In an attempt to reduce the ballooning national debt, the Rafiq HARIRI government began an austerity program, reining in government expenditures, increasing revenue collection, and privatizing state enterprises, but economic and financial reform initiatives stalled and public debt continued to grow despite receipt of more than $2 billion in bilateral assistance at the Paris II Donors Conference. The Israeli-Hizballah conflict in July-August 2006 caused an estimated $3.6 billion in infrastructure damage, and prompted international donors to pledge nearly $1 billion in recovery and reconstruction assistance. Donors met again in January 2007 and pledged over $7.5 billion to Lebanon for development projects and budget support, conditioned on progress on Beirut's fiscal reform and privatization program. Internal Lebanese political tension continues to hamper economic activity, particularly in the tourism and retail sectors.

GDP (purchasing power parity): $42.27 billion (2007 est.)

GDP (official exchange rate): $24.64 billion (2007 est.)

GDP—real growth rate: 4% (2007 est.)

GDP—per capita (PPP): $11,300 (2007 est.)

GDP—composition by sector:
agriculture: 5.1%
industry: 19%
services: 75.9% (2007 est.)

Labor force: 1.5 million
note: in addition, there are as many as 1 million foreign workers (2005 est.)

Labor force—by occupation: *agriculture:* NA%
industry: NA%
services: NA%

Unemployment rate: 20% (2006 est.)

Population below poverty line: 28% (1999 est.)

Household income or consumption by percentage share:
lowest 10%: NA%
highest 10%: NA%

Inflation rate (consumer prices): 4.1% (2007 est.)

Investment (gross fixed): 22% of GDP (2007 est.)

Budget:
revenues: $6.472 billion
expenditures: $8.35 billion (2007 est.)

Public debt: 186.6% of GDP (2007 est.)

Agriculture—products: citrus, grapes, tomatoes, apples, vegetables, potatoes, olives, tobacco; sheep, goats

Industries: banking, tourism, food pro-

cessing, wine, jewelry, cement, textiles, mineral and chemical products, wood and furniture products, oil refining, metal fabricating

Industrial production growth rate: NA%

Electricity—production: 9.183 billion kWh (2005)

Electricity—production by source:
fossil fuel: 97.2%
hydro: 2.8%
nuclear: 0%
other: 0% (2001)

Electricity—consumption: 10.58 billion kWh (2005)

Electricity—exports: 0 kWh (2005)

Electricity—imports: 455 million kWh (2005)

Oil—production: 0 bbl/day (2005 est.)

Oil—consumption: 106,000 bbl/day (2005 est.)

Oil—exports: 0 bbl/day (2004)

Oil—imports: 102,300 bbl/day (2004)

Oil—proved reserves: 0 bbl (1 January 2006 est.)

Natural gas—production: 0 cu m (2005 est.)

Natural gas—consumption: 0 cu m (2005 est.)

Natural gas—exports: 0 cu m (2005 est.)

Natural gas—imports: 0 cu m (2005)

Natural gas—proved reserves: 0 cu m (1 January 2006 est.)

Current account balance: -$2.634 billion (2007 est.)

Exports: $3.445 billion f.o.b. (2007 est.)

Exports—commodities: authentic jewelry, inorganic chemicals, miscellaneous consumer goods, fruit and vegetables, tobacco, construction minerals, electric power machinery and switchgear, textile fibers, paper

Exports—partners: Syria 27.1%, UAE 12.2%, Switzerland 6.1%, Saudi Arabia 5.8%, Turkey 4.6% (2006)

Imports: $10.75 billion f.o.b. (2007 est.)

Imports—commodities: petroleum products, cars, medicinal products, clothing, meat and live animals, consumer goods, paper, textile fabrics, tobacco, electrical machinery

Imports—partners: Syria 11.6%, Italy 9.7%, US 9.3%, France 7.7%, Germany 6%, China 5%, Saudi Arabia 4.7% (2006)

Economic aid—recipient: of the $7.6 billion in grants and loans pledged to Lebanon at the Paris III conference in January 2007, Beirut as of mid-December 2007 had signed agreements for $3 billion, including $1 billion in project financing, $750 million in direct budget support, $750 million in private sector credit, and $285 million in in-kind aid;

about $500 million of the $1.7 billion pledged for direct budget support has been disbursed to Lebanon; donors in August 2006 also pledged nearly $1.8 billion in aid to help Lebanon recover from the 2006 Israel-Hizballah war; during the conflict, Saudi Arabia and Kuwait provided $1.5 billion in concessional loans to the Lebanese central bank to maintain confidence in the Lebanese currency. (2005)

Reserves of foreign exchange and gold: $20.55 billion (31 December 2007 est.)

Debt—external: $31.52 billion (31 December 2007 est.)

Stock of direct foreign investment—at home: $NA

Stock of direct foreign investment—abroad: $NA

Market value of publicly traded shares: $8.279 billion (2006)

Currency (code): Lebanese pound (LBP)

Currency code: LBP

Exchange rates: Lebanese pounds per US dollar—1,507.5 (2007), 1,507.5 (2006), 1,507.5 (2005), 1,507.5 (2004), 1,507.5 (2003)

Fiscal year: calendar year

COMMUNICATIONS

Telephones—main lines in use: 681,400 (2006)

Telephones—mobile cellular: 1.103 million (2006)

Telephone system:
general assessment: repair of the telecommunications system, severely damaged during the civil war, now complete
domestic: two wireless networks provide good service; political instability hampers privatization and deployment of new technologies; combined fixed-line and mobile-cellular subscribership approaching 50 per 100 persons
international: country code—961; submarine cable link to Cyprus; satellite earth stations—2 Intelsat (1 Indian Ocean and 1 Atlantic Ocean); coaxial cable to Syria

Radio broadcast stations: AM 20, FM 22, shortwave 4 (1998)

Radios: 2.85 million (1997)

Television broadcast stations: 15 (plus 5 repeaters) (1995)

Televisions: 1.18 million (1997)

Internet country code: .lb

Internet hosts: 5,635 (2007)

Internet Service Providers (ISPs): 22 (2000)

Internet users: 950,000 (2006)

TRANSPORTATION

Airports: 7 (2007)

Airports—with paved runways: *total:* 5

over 3,047 m: 1
2,438 to 3,047 m: 2
914 to 1,523 m: 1
under 914 m: 1 (2007)
Airports—with unpaved runways:
total: 2
914 to 1,523 m: 2 (2007)
Pipelines: gas 43 km (2007)
Railways:
total: 401 km
standard gauge: 319 km 1.435 m
narrow gauge: 82 km 1.050 m
note: rail system became unusable because of damage done during fighting in the 1980s and in 2006 (2006)
Roadways:
total: 6,970 km (includes 170 km of expressways) (2005)
Merchant marine:
total: 35 ships (1000 GRT or over) 132,871 GRT/140,011 DWT
by type: bulk carrier 3, cargo 14, livestock carrier 12, passenger/cargo 1, refrigerated cargo 1, roll on/roll off 2, vehicle carrier 2
foreign-owned: 3 (Greece 2, Syria 1)
registered in other countries: 55 (Antigua and Barbuda 1, Barbados 1, Cambodia 7,

Comoros 5, Cyprus 1, Dominica 1, Egypt 1, Georgia 3, Honduras 2, Hong Kong 1, North Korea 3, Liberia 2, Malta 12, Mongolia 1, Panama 3, St Vincent and The Grenadines 7, Syria 4, unknown 2) (2007)
Ports and terminals: Beirut, Tripoli

MILITARY

Military branches: Lebanese Armed Forces (LAF): Army (includes Navy), Air Force (2008)
Military service age and obligation: 18–30 years of age for voluntary military service; no conscription (2007)
Manpower available for military service:
males age 16–49: 1,106,879
females age 16–49: 1,122,595 (2008 est.)
Manpower fit for military service:
males age 16–49: 934,828
females age 16–49: 948,327 (2008 est.)
Military expenditures—percent of GDP: 3.1% (2005 est.)

TRANSNATIONAL ISSUES

Disputes—international: lacking a treaty or other documentation describing

the boundary, portions of the Lebanon-Syria boundary are unclear with several sections in dispute; since 2000, Lebanon has claimed Shab'a Farms area in the Israeli-occupied Golan Heights; the roughly 2,000-strong UN Interim Force in Lebanon (UNIFIL) has been in place since 1978
Refugees and internally displaced persons: *refugees (country of origin):* 405,425 (Palestinian Refugees (UNRWA)); 50,000–60,000 (Iraq)
IDPs: 17,000 (1975–90 civil war, Israeli invasions); 200,000 (July–August 2006 war) (2007)
Illicit drugs: cannabis cultivation dramatically reduced to 2,500 hectares in 2002 despite continued significant cannabis consumption; opium poppy cultivation minimal; small amounts of Latin American cocaine and Southwest Asian heroin transit country on way to European markets and for Middle Eastern consumption; money laundering of drug proceeds fuels concern that extremists are benefiting from drug trafficking

LESOTHO

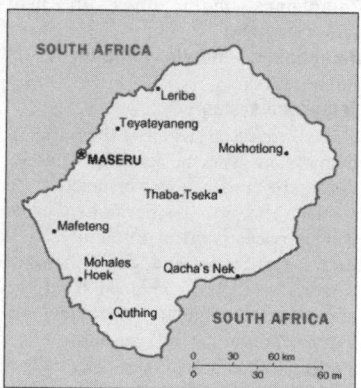

INTRODUCTION

Background: Basutoland was renamed the Kingdom of Lesotho upon independence from the UK in 1966. The Basuto National Party ruled for the first two decades. King MOSHOESHOE was exiled in 1990, but returned to Lesotho in 1992 and was reinstated in 1995. Constitutional government was restored in 1993 after seven years of military rule. In 1998, violent protests and a military mutiny following a contentious election prompted a brief but bloody intervention by South African and Botswanan military forces under the aegis of the

Southern African Development Community. Subsequent constitutional reforms restored relative political stability. Peaceful parliamentary elections were held in 2002, but the National Assembly elections of February 2007 were hotly contested and aggrieved parties continue to periodically demonstrate their distrust of the results.

GEOGRAPHY

Location: Southern Africa, an enclave of South Africa
Geographic coordinates: 29 30 S, 28 30 E
Map references: Africa
Area: *total:* 30,355 sq km
land: 30,355 sq km
water: 0 sq km
Area—comparative: slightly smaller than Maryland
Land boundaries:
total: 909 km
border countries: South Africa 909 km
Coastline: 0 km (landlocked)
Maritime claims: none (landlocked)
Climate: temperate; cool to cold, dry winters; hot, wet summers
Terrain: mostly highland with plateaus, hills, and mountains
Elevation extremes:
lowest point: junction of the Orange and Makhaleng Rivers 1,400 m

highest point: Thabana Ntlenyana 3,482 m
Natural resources: water, agricultural and grazing land, diamonds, sand, clay, building stone
Land use:
arable land: 10.87%
permanent crops: 0.13%
other: 89% (2005)
Irrigated land: 30 sq km (2003)
Total renewable water resources: 5.2 cu km (1987)
Freshwater withdrawal (domestic/industrial/agricultural): *total:* 0.05 cu km/yr (40%/40%/20%)
per capita: 28 cu m/yr (2000)
Natural hazards: periodic droughts
Environment—current issues: population pressure forcing settlement in marginal areas results in overgrazing, severe soil erosion, and soil exhaustion; desertification; Highlands Water Project controls, stores, and redirects water to South Africa
Environment—international agreements: *party to:* Biodiversity, Climate Change, Climate Change-Kyoto Protocol, Desertification, Endangered Species, Hazardous Wastes, Law of the Sea, Marine Life Conservation, Ozone Layer Protection, Wetlands
signed, but not ratified: none of the selected agreements

Geography—note: landlocked, completely surrounded by South Africa; mountainous, more than 80% of the country is 1,800 m above sea level

PEOPLE

Population: 2,128,180

note: estimates for this country explicitly take into account the effects of excess mortality due to AIDS; this can result in lower life expectancy, higher infant mortality, higher death rates, lower population growth rates, and changes in the distribution of population by age and sex than would otherwise be expected (July 2008 est.)

Age structure:

0–14 years: 35.3% (male 377,784/female 372,840)

15–64 years: 59.8% (male 621,687/female 649,981)

65 years and over: 5% (male 42,348/female 63,540) (2008 est.)

Median age:

total: 21.2 years

male: 20.6 years

female: 21.8 years (2008 est.)

Population growth rate: 0.129% (2008 est.)

Birth rate: 24.41 births/1,000 population (2008 est.)

Death rate: 22.33 deaths/1,000 population (2008 est.)

Net migration rate: -0.78 migrant(s)/1,000 population (2008 est.)

Sex ratio:

at birth: 1.03 male(s)/female

under 15 years: 1.01 male(s)/female

15–64 years: 0.96 male(s)/female

65 years and over: 0.67 male(s)/female

total population: 0.96 male(s)/female (2008 est.)

Infant mortality rate:

total: 78.59 deaths/1,000 live births

male: 83.01 deaths/1,000 live births

female: 74.03 deaths/1,000 live births (2008 est.)

Life expectancy at birth:

total population: 40.17 years

male: 40.97 years

female: 39.34 years (2008 est.)

Total fertility rate: 3.13 children born/woman (2008 est.)

HIV/AIDS—adult prevalence rate: 28.9% (2003 est.)

HIV/AIDS—people living with HIV/AIDS: 320,000 (2003 est.)

HIV/AIDS—deaths: 29,000 (2003 est.)

Nationality:

noun: Mosotho (singular), Basotho (plural)

adjective: Basotho

Ethnic groups: Sotho 99.7%, Europeans, Asians, and other 0.3%,

Religions: Christian 80%, indigenous beliefs 20%

Languages: Sesotho (southern Sotho), English (official), Zulu, Xhosa

Literacy: *definition:* age 15 and over can read and write

total population: 84.8%

male: 74.5%

female: 94.5% (2003 est.)

GOVERNMENT

Country name:

conventional long form: Kingdom of Lesotho

conventional short form: Lesotho

local long form: Kingdom of Lesotho

local short form: Lesotho

former: Basutoland

Government type: parliamentary constitutional monarchy

Capital: *name:* Maseru

geographic coordinates: 29 19 S, 27 29 E

time difference: UTC+2 (7 hours ahead of Washington, DC during Standard Time)

Administrative divisions: 10 districts; Berea, Butha-Buthe, Leribe, Mafeteng, Maseru, Mohale's Hoek, Mokhotlong, Qacha's Nek, Quthing, Thaba-Tseka

Independence: 4 October 1966 (from UK)

National holiday: Independence Day, 4 October (1966)

Constitution: 2 April 1993

Legal system: based on English common law and Roman-Dutch law; judicial review of legislative acts in High Court and Court of Appeal; accepts compulsory ICJ jurisdiction with reservations

Suffrage: 18 years of age; universal

Executive branch:

chief of state: King LETSIE III (since 7 February 1996); note—King LETSIE III formerly occupied the throne from November 1990 to February 1995 while his father was in exile

head of government: Prime Minister Pakalitha MOSISILI (since 23 May 1998)

cabinet: Cabinet

elections: according to the constitution, the leader of the majority party in the Assembly automatically becomes prime minister; the monarch is hereditary, but, under the terms of the constitution that came into effect after the March 1993 election, the monarch is a "living symbol of national unity" with no executive or legislative powers; under traditional law the college of chiefs has the power to depose the monarch, determine who is next in the line of succession, or who shall serve as regent in the event that the successor is not of mature age

Legislative branch: bicameral Parliament consists of the Senate (33 members—22 principal chiefs and 11 other members appointed by the ruling

party) and the Assembly (120 seats, 80 by popular vote and 40 by proportional vote; members elected by popular vote for five-year terms)

elections: last held 17 February 2007 (next to be held in 2012)

election results: percent of vote by party—NA; seats by party—LCD 61, NIP 21, ABC 17, LWP 10, ACP 4, BNP 3, other 4

Judicial branch: High Court (chief justice appointed by the monarch acting on the advice of the Prime Minister); Court of Appeal; Magistrate Courts; customary or traditional court

Political parties and leaders: Alliance of Congress Parties or ACP; All Basotho Convention or ABC [Thomas THABANE]; Basotholand African Congress or BAC [Khauhelo RALITAPOLE]; Basotho Congress Party or BCP [Ntsukunyane MPHANYA]; Basotho National Party or BNP [Maj. Gen. Justin Metsing LEKHANYA]; Kopanang Basotho Party or KPB [Pheelo MOSALA]; Lesotho Congress for Democracy or LCD (the governing party) [Pakalitha MOSISILI]; Lesotho Education Party or LEP [Thabo PITSO]; Lesotho Workers Party or LWP [Macaefa BILLY]; Maremaltou Freedom Party or MFP [Vincent MALEBO]; National Independent Party or NIP [Anthony MANYELI]; New Lesotho Freedom Party or NLFP [Manapo MAJARA]; Popular Front for Democracy or PFD [Lekhetho RAKUOANE]; Sefate Democratic Union or SDU [Bofihla NKUEBE]; Social Democratic Party of SDP [Masitise SELESO]

Political pressure groups and leaders: NA

International organization participation: ACP, AfDB, AU, C, FAO, G-77, IBRD, ICAO, ICCt, ICRM, IDA, IFAD, IFC, IFRCS, ILO, IMF, Interpol, IOC, ISO (subscriber), ITU, MIGA, NAM, OPCW, SACU, SADC, UN, UNCTAD, UNESCO, UNHCR, UNIDO, UNWTO, UPU, WCO, WFTU, WHO, WIPO, WMO, WTO

Diplomatic representation in the US:

chief of mission: Ambassador (vacant); Charge d'Affaires Mabasia MOHOBANE

chancery: 2511 Massachusetts Avenue NW, Washington, DC 20008

telephone: [1] (202) 797-5533 through 5536

FAX: [1] (202) 234-6815

Diplomatic representation from the US:

chief of mission: Ambassador Robert NOLAN

embassy: 254 Kingsway, Maseru West (Consular Section)

mailing address: P. O. Box 333, Maseru 100, Lesotho
telephone: [266] 22 312666
FAX: [266] 22 310116

Flag description: three horizontal stripes of blue (top), white, and green in the proportions of 3:4:3; the colors represent rain, peace, and prosperity respectively; centered in the white stripe is a black Basotho hat representing the indigenous people; the flag was unfurled in October 2006 to celebrate 40 years of independence

ECONOMY

Economy—overview: Small, land-locked, and mountainous, Lesotho relies on remittances from miners employed in South Africa and customs duties from the Southern Africa Customs Union for the majority of government revenue. However, the government has recently strengthened its tax system to reduce dependency on customs duties. Completion of a major hydropower facility in January 1998 permitted the sale of water to South Africa and generated royalties for Lesotho. Lesotho produces about 90% of its own electrical power needs. As the number of mineworkers has declined steadily over the past several years, a small manufacturing base has developed based on farm products that support the milling, canning, leather, and jute industries, as well as a rapidly expanding apparel-assembly sector. The latter has grown significantly mainly due to Lesotho qualifying for the trade benefits contained in the Africa Growth and Opportunity Act. The economy is still primarily based on subsistence agriculture, especially livestock, although drought has decreased agricultural activity. The extreme inequality in the distribution of income remains a major drawback. Lesotho has signed an Interim Poverty Reduction and Growth Facility with the IMF. In July 2007, Lesotho signed a Millennium Challenge Account Compact with the US worth $362.5 million.

GDP (purchasing power parity): $3.092 billion (2007 est.)

GDP (official exchange rate): $1.6 billion (2007 est.)

GDP—real growth rate: 4.9% (2007 est.)

GDP—per capita (PPP): $1,300 (2007 est.)

GDP—composition by sector:
agriculture: 15.2%
industry: 45%
services: 39.7% (2007 est.)

Labor force: 838,000 (2000 est.)

Labor force—by occupation: *agriculture:* 86% of resident population engaged in subsistence agriculture; roughly 35% of the active male wage earners work in South Africa
industry and services: 14% (2002 est.)

Unemployment rate: 45% (2002)

Population below poverty line: 49% (1999)

Household income or consumption by percentage share:
lowest 10%: 0.9%
highest 10%: 43.4% (2002 est.)

Distribution of family income—Gini index: 63.2 (1995)

Inflation rate (consumer prices): 8% (2007 est.)

Investment (gross fixed): 51.9% of GDP (2007 est.)

Budget: *revenues:* $779.9 million
expenditures: $696.9 million (2007 est.)

Agriculture—products: corn, wheat, pulses, sorghum, barley; livestock

Industries: food, beverages, textiles, apparel assembly, handicrafts, construction, tourism

Industrial production growth rate: 12% (2007 est.)

Electricity—production: 350 million kWh; note—electricity supplied by South Africa (2005)

Electricity—consumption: 338.5 million kWh (2005)

Electricity—exports: 0 kWh (2005)

Electricity—imports: 13 million kWh; note—electricity supplied by South Africa (2005)

Oil—production: 0 bbl/day (2005 est.)

Oil—consumption: 1,400 bbl/day (2005)

Oil—exports: 0 bbl/day (2004)

Oil—imports: 1,400 bbl/day (2004)

Oil—proved reserves: 0 bbl (1 January 2006 est.)

Natural gas—production: 0 cu m (2005 est.)

Natural gas—consumption: 0 cu m (2005 est.)

Natural gas—exports: 0 cu m (2005 est.)

Natural gas—imports: 0 cu m (2005)

Natural gas—proved reserves: 0 cu m (1 January 2006 est.)

Current account balance: $77 million (2007 est.)

Exports: $853 million f.o.b. (2007 est.)

Exports—commodities: manufactures 75% (clothing, footwear, road vehicles), wool and mohair, food and live animals (2000)

Exports—partners: US 79.8%, Belgium 14.5%, Canada 1.7% (2006)

Imports: $1.604 billion f.o.b. (2007 est.)

Imports—commodities: food; building materials, vehicles, machinery, medicines, petroleum products

Imports—partners: Hong Kong 26.2%, Taiwan 24.7%, China 24.5%, Germany 6.2%, South Korea 4.1% (2006)

Economic aid—recipient: $68.82 million (2005)

Reserves of foreign exchange and gold: $879 million (31 December 2007 est.)

Debt—external: $689 million (31 December 2007 est.)

Currency (code): loti (LSL); South African rand (ZAR)

Currency code: LSL; ZAR

Exchange rates: maloti per US dollar—7.25 (2007), 6.85 (2006), 6.3593 (2005), 6.4597 (2004), 7.5648 (2003)

Fiscal year: 1 April—31 March

COMMUNICATIONS

Telephones—main lines in use: 48,000 (2005)

Telephones—mobile cellular: 249,800 (2005)

Telephone system:
general assessment: rudimentary system consisting of a modest but growing number of landlines, a small microwave radio relay system, and a small radiotelephone communication system; mobile-cellular telephone system is expanding
domestic: privatized in 2001, Telecom Lesotho tasked with providing an additional 50,000 fixed-line connections within five years, a target not met; mobile-cellular service is expanding with a subscribership approaching 15 per 100 persons; rural services are scant
international: country code—266; satellite earth station—1 Intelsat (Atlantic Ocean)

Radio broadcast stations: AM 1, FM 2, shortwave 1 (1998)

Radios: NA (2002)

Television broadcast stations: 1 (2000)

Televisions: NA

Internet country code: .ls

Internet hosts: 66 (2007)

Internet Service Providers (ISPs): 1 (2000)

Internet users: 51,500 (2005)

TRANSPORTATION

Airports: 28 (2007)

Airports—with paved runways:
total: 3
over 3,047 m: 1
914 to 1,523 m: 1
under 914 m: 1 (2007)

Airports—with unpaved runways:
total: 25
914 to 1,523 m: 4
under 914 m: 21 (2007)

Roadways:
total: 5,940 km
paved: 1,087 km
unpaved: 4,853 km (1999)

MILITARY

Military branches: Lesotho Defense Force (LDF): Army (includes Air Wing) (2008)
Military service age and obligation: 18 years of age for voluntary military service; no conscription (2008)
Manpower available for military service: *males age 16–49:* 525,203
females age 16–49: 522,485 (2008 est.)
Manpower fit for military service:
males age 16–49: 262,101

females age 16–49: 238,350 (2008 est.)
Military expenditures—percent of GDP: 2.6% (2006)
Military—note: Lesotho's declared policy is maintenance of its independent sovereignty and preservation of internal security; in practice, external security is guaranteed by South Africa; restructuring of the Lesotho Defense Force (LDF) and Ministry of Defense and Public Service over the past five years has focused on subordinating the defense apparatus to civilian control and

restoring the LDF's cohesion; the restructuring has considerably improved capabilities and professionalism, but the LDF is disproportionately large for a small, poor country; the government has outlined a reduction to a planned 1,500-man strength, but these plans have met with vociferous resistance from the political opposition and from inside the LDF (2008)

TRANSNATIONAL ISSUES

Disputes—international: none

LIBERIA

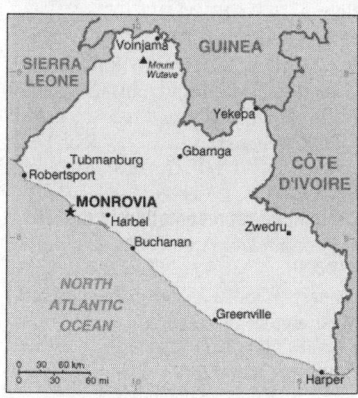

INTRODUCTION

Background: Settlement of freed slaves from the US in what is today Liberia began in 1822; by 1847, the Americo-Liberians were able to establish a republic. William TUBMAN, president from 1944–71, did much to promote foreign investment and to bridge the economic, social, and political gaps between the descendents of the original settlers and the inhabitants of the interior. In 1980, a military coup led by Samuel DOE ushered in a decade of authoritarian rule. In December 1989, Charles TAYLOR launched a rebellion against DOE's regime that led to a prolonged civil war in which DOE himself was killed. A period of relative peace in 1997 allowed for elections that brought TAYLOR to power, but major fighting resumed in 2000. An August 2003 peace agreement ended the war and prompted the resignation of former president Charles TAYLOR, who faces war crimes charges in The Hague related to his involvement in Sierra Leone's civil war. After two years of rule by a transitional government, democratic elections in late 2005 brought President Ellen JOHNSON SIRLEAF to power. The UN

Mission in Liberia (UNMIL) maintains a strong presence throughout the country, but the security situation is still fragile and the process of rebuilding the social and economic structure of this war-torn country will take many years.

GEOGRAPHY

Location: Western Africa, bordering the North Atlantic Ocean, between Cote d'Ivoire and Sierra Leone
Geographic coordinates: 6 30 N, 9 30 W
Map references: Africa
Area: *total:* 111,370 sq km
land: 96,320 sq km
water: 15,050 sq km
Area—comparative: slightly larger than Tennessee
Land boundaries:
total: 1,585 km
border countries: Guinea 563 km, Cote d'Ivoire 716 km, Sierra Leone 306 km
Coastline: 579 km
Maritime claims: *territorial sea:* 200 nm
Climate: tropical; hot, humid; dry winters with hot days and cool to cold nights; wet, cloudy summers with frequent heavy showers
Terrain: mostly flat to rolling coastal plains rising to rolling plateau and low mountains in northeast
Elevation extremes:
lowest point: Atlantic Ocean 0 m
highest point: Mount Wuteve 1,380 m
Natural resources: iron ore, timber, diamonds, gold, hydropower
Land use: *arable land:* 3.43%
permanent crops: 1.98%
other: 94.59% (2005)
Irrigated land: 30 sq km (2003)
Total renewable water resources: 232 cu km (1987)
Freshwater withdrawal (domestic/industrial/agricultural): *total:* 0.11 cu km/yr (27%/18%/55%)
per capita: 34 cu m/yr (2000)

Natural hazards: dust-laden harmattan winds blow from the Sahara (December to March)
Environment—current issues: tropical rain forest deforestation; soil erosion; loss of biodiversity; pollution of coastal waters from oil residue and raw sewage
Environment—international agreements: *party to:* Biodiversity, Climate Change, Climate Change-Kyoto Protocol, Desertification, Endangered Species, Hazardous Wastes, Ozone Layer Protection, Ship Pollution, Tropical Timber 83, Tropical Timber 94, Wetlands
signed, but not ratified: Environmental Modification, Law of the Sea, Marine Life Conservation
Geography—note: facing the Atlantic Ocean, the coastline is characterized by lagoons, mangrove swamps, and river-deposited sandbars; the inland grassy plateau supports limited agriculture

PEOPLE

Population: 3,334,587 (July 2008 est.)
Age structure:
0–14 years: 44% (male 734,375/female 731,287)
15–64 years: 53.3% (male 879,848/female 896,319)
65 years and over: 2.8% (male 45,175/female 47,583) (2008 est.)
Median age:
total: 18 years
male: 17.8 years
female: 18.2 years (2008 est.)
Population growth rate: 3.661% (2008 est.)
Birth rate: 42.92 births/1,000 population (2008 est.)
Death rate: 21.45 deaths/1,000 population (2008 est.)
Net migration rate: 15.14 migrant(s)/1,000 population (2008 est.)
Sex ratio:
at birth: 1.03 male(s)/female

under 15 years: 1 male(s)/female
15–64 years: 0.98 male(s)/female
65 years and over: 0.95 male(s)/female
total population: 0.99 male(s)/female
(2008 est.)
Infant mortality rate:
total: 143.89 deaths/1,000 live births
male: 159.5 deaths/1,000 live births
female: 127.81 deaths/1,000 live births
(2008 est.)
Life expectancy at birth:
total population: 41.13 years
male: 39.85 years
female: 42.46 years (2008 est.)
Total fertility rate: 5.87 children
born/woman (2008 est.)
HIV/AIDS—adult prevalence rate: 5.9%
(2003 est.)
**HIV/AIDS—people living with
HIV/AIDS:** 100,000 (2003 est.)
HIV/AIDS—deaths: 7,200 (2003 est.)
Major infectious diseases:
degree of risk: very high
food or waterborne diseases: bacterial and
protozoal diarrhea, hepatitis A, and
typhoid fever
vectorborne diseases: malaria and yellow
fever
water contact disease: schistosomiasis
aerosolized dust or soil contact disease:
Lassa fever
animal contact disease: rabies (2008)
Nationality:
noun: Liberian(s)
adjective: Liberian
Ethnic groups: indigenous African 95%
(including Kpelle, Bassa, Gio, Kru,
Grebo, Mano, Krahn, Gola, Gbandi,
Loma, Kissi, Vai, Dei, Bella, Mandingo,
and Mende), Americo-Liberians 2.5%
(descendants of immigrants from the US
who had been slaves), Congo People
2.5% (descendants of immigrants from
the Caribbean who had been slaves)
Religions: Christian 40%, Muslim 20%,
indigenous beliefs 40%
Languages: English 20% (official), some
20 ethnic group languages, of which a
few can be written and are used in corre-
spondence
Literacy:
definition: age 15 and over can read and
write
total population: 57.5%
male: 73.3%
female: 41.6% (2003 est.)

GOVERNMENT

Country name:
conventional long form: Republic of
Liberia
conventional short form: Liberia
Government type: republic
Capital: *name:* Monrovia
geographic coordinates: 6 18 N, 10 48 W

time difference: UTC 0 (5 hours ahead of
Washington, DC during Standard Time)
Administrative divisions: 15 counties;
Bomi, Bong, Gbarpolu, Grand Bassa,
Grand Cape Mount, Grand Gedeh,
Grand Kru, Lofa, Margibi, Maryland,
Montserrado, Nimba, River Cess, River
Gee, Sinoe
Independence: 26 July 1847
National holiday: Independence Day, 26
July (1847)
Constitution: 6 January 1986
Legal system: dual system of statutory
law based on Anglo-American common
law for the modern sector and customary
law based on unwritten tribal practices
for indigenous sector; accepts compulsory ICJ jurisdiction with reservations
Suffrage: 18 years of age; universal
Executive branch:
chief of state: President Ellen JOHNSON
SIRLEAF (since 16 January 2006);
note—the President is both the chief of
state and head of government
head of government: President Ellen
JOHNSON SIRLEAF (since 16 January
2006)
cabinet: Cabinet appointed by the president and confirmed by the Senate
elections: president elected by popular
vote for a six-year term (eligible for a
second term); election last held 8
November 2005 (next to be held in
2011)
election results: Ellen JOHNSON SIR-
LEAF elected president; percent of vote,
second round—Ellen JOHNSON SIR-
LEAF 59.6%, George WEAH 40.4%
Legislative branch: bicameral National
Assembly consists of the Senate (30
seats; note—number of seats changed in
11 October 2005 elections; members
elected by popular vote to serve nine-
year terms) and the House of
Representatives (64 seats; members
elected by popular vote to serve six-year
terms)
elections: Senate—last held 11 October
2005 (next to be held in 2011); House of
Representatives—last held 11 October
2005 (next to be held in 2011)
election results: Senate—percent of vote
by party—NA; seats by party—COTOL
7, NPP 4, CDC 3, LP 3, UP 3, APD 3,
other 7; House of Representatives—percent of vote by party—NA; seats by
party—CDC 15, LP 9, COTOL 8, UP 8,
APD 5, NPP 4, other 15
note: junior senators—those who
received the second most votes in each
county in the 11 October 2005 election—will only serve a six-year first term
because the Liberian constitution mandates staggered Senate elections to
ensure continuity of government; all sen-

ators will be eligible for nine-year terms
thereafter
Judicial branch: Supreme Court
Political parties and leaders: Alliance
for Peace and Democracy or APD
[Togba-na TIPOTEH]; Coalition for the
Transformation of Liberia or COTOL
[H. Varney SHERMAN]; Congress for
Democratic Change or CDC [George
WEAH]; Liberty Party or LP [Charles
BRUMSKINE]; National Patriotic Party
or NPP [Roland MASSAQUOI]; Unity
Party or UP [Ellen JOHNSON SIR-
LEAF]
Political pressure groups and leaders:
Demobilized former military officers
**International organization participa-
tion:** ACP, AfDB, AU, ECOWAS, FAO,
G-77, IAEA, IBRD, ICAO, ICCt,
ICRM, IDA, IFAD, IFC, IFRCS, ILO,
IMF, IMO, IMSO, Interpol, IOC, IOM,
IPU, ITU, ITUC, MIGA, NAM,
OPCW (signatory), UN, UNCTAD,
UNESCO, UNIDO, UPU, WCL,
WCO, WFTU, WHO, WIPO, WMO
Diplomatic representation in the US:
chief of mission: Ambassador Charles A.
MINOR
chancery: 5201 16th Street NW,
Washington, DC 20011
telephone: [1] (202) 723-0437
FAX: [1] (202) 723-0436
consulate(s) general: New York
Diplomatic representation from the US:
chief of mission: Ambassador Donald E.
BOOTH
embassy: 111 United Nations Drive, P.
O. Box 10-0098, Mamba Point, 1000
Monrovia, 10
mailing address: use embassy street address
telephone: [231] 7-705-4825 or 4826
FAX: [231] 7-701-0370
Flag description: 11 equal horizontal
stripes of red (top and bottom) alternating with white; there is a white five-
pointed star on a blue square in the upper
hoist-side corner; the design was based
on the US flag

ECONOMY

Economy—overview: Civil war and
government mismanagement destroyed
much of Liberia's economy, especially
the infrastructure in and around the capital, Monrovia. Many businesses fled the
country, taking capital and expertise
with them, but with the conclusion of
fighting and the installation of a demo-
cratically-elected government in 2006,
some have returned. Richly endowed
with water, mineral resources, forests,
and a climate favorable to agriculture,
Liberia had been a producer and exporter
of basic products—primarily raw timber
and rubber. Local manufacturing, mainly

foreign owned, had been small in scope. President JOHNSON SIRLEAF, a Harvard-trained banker and administrator, has taken steps to reduce corruption, build support from international donors, and encourage private investment. Embargos on timber and diamond exports have been lifted, opening new sources of revenue for the government. The reconstruction of infrastructure and the raising of incomes in this ravaged economy will largely depend on generous financial and technical assistance from donor countries and foreign investment in key sectors, such as infrastructure and power generation.

GDP (purchasing power parity): $1.34 billion (2007 est.)

GDP (official exchange rate): $730 million (2007 est.)

GDP—real growth rate: 9.4% (2007 est.)

GDP—per capita (PPP): $400 (2007 est.)

GDP—composition by sector:
agriculture: 76.9%
industry: 5.4%
services: 17.7% (2002 est.)

Labor force—by occupation: *agriculture:* 70%
industry: 8%
services: 22% (2000 est.)

Unemployment rate: 85% (2003 est.)

Population below poverty line: 80% (2000 est.)

Household income or consumption by percentage share:
lowest 10%: NA%
highest 10%: NA%

Inflation rate (consumer prices): 11.2% (2007 est.)

Budget:
revenues: NA
expenditures: NA

Agriculture—products: rubber, coffee, cocoa, rice, cassava (tapioca), palm oil, sugarcane, bananas; sheep, goats; timber

Industries: rubber processing, palm oil processing, timber, diamonds

Industrial production growth rate: NA%

Electricity—production: 319.3 million kWh (2005)

Electricity—production by source:
fossil fuel: 100%
hydro: 0%
nuclear: 0%
other: 0% (2001)

Electricity—consumption: 296.9 million kWh (2005)

Electricity—exports: 0 kWh (2005)

Electricity—imports: 0 kWh (2005)

Oil—production: 0 bbl/day (2005 est.)

Oil—consumption: 3,550 bbl/day (2005 est.)

Oil—exports: 23.31 bbl/day (2004)

Oil—imports: 3,532 bbl/day (2004)

Oil—proved reserves: 0 bbl (1 January 2006 est.)

Natural gas—production: 0 cu m (2005 est.)

Natural gas—consumption: 0 cu m (2005 est.)

Natural gas—exports: 0 cu m (2005 est.)

Natural gas—imports: 0 cu m (2005)

Natural gas—proved reserves: 0 cu m (1 January 2006 est.)

Current account balance: -$224 million (2007)

Exports: $1.197 billion f.o.b. (2006)

Exports—commodities: rubber, timber, iron, diamonds, cocoa, coffee

Exports—partners: Germany 40.1%, South Africa 12%, Poland 11.7%, US 8.5%, Spain 8.2% (2006)

Imports: $7.143 billion f.o.b. (2006)

Imports—commodities: fuels, chemicals, machinery, transportation equipment, manufactured goods; foodstuffs

Imports—partners: South Korea 43.2%, Singapore 15%, Japan 12.8%, China 8.2% (2006)

Economic aid—recipient: $236.2 million (2005)

Debt—external: $3.2 billion (2005 est.)

Stock of direct foreign investment—at home: $NA

Stock of direct foreign investment—abroad: $NA

Market value of publicly traded shares: $NA

Currency (code): Liberian dollar (LRD)

Currency code: LRD

Exchange rates: Liberian dollars per US dollar—NA (2007), 59.43 (2006), 53.098 (2005), 54.906 (2004), 59.379 (2003)

Fiscal year: calendar year

COMMUNICATIONS

Telephones—main lines in use: 6,900 (2002)

Telephones—mobile cellular: 160,000 (2005)

Telephone system:
general assessment: the limited services available are found almost exclusively in the capital Monrovia; coverage extended to a number of other towns and rural areas by four mobile-cellular network operators
domestic: combined fixed and mobile-cellular teledensity only about 5 per 100 persons
international: country code—231; satellite earth station—1 Intelsat (Atlantic Ocean)

Radio broadcast stations: AM 0, FM 10, shortwave 2 (2007)

Radios: 790,000 (1997)

Television broadcast stations: 4 (plus 4 repeaters) (2007)

Televisions: 70,000 (1997)

Internet country code: .lr

Internet hosts: 38 (2007)

Internet Service Providers (ISPs): 2 (2001)

Internet users: 1,000 (2002)

TRANSPORTATION

Airports: 53 (2007)

Airports—with paved runways:
total: 2
over 3,047 m: 1
1,524 to 2,437 m: 1 (2007)

Airports—with unpaved runways:
total: 51
1,524 to 2,437 m: 5
914 to 1,523 m: 8
under 914 m: 38 (2007)

Railways:
total: 490 km
standard gauge: 345 km 1.435-m gauge
narrow gauge: 145 km 1.067-m gauge
note: sections of railway are inoperable because of damage suffered during the civil war (2008)

Roadways: *total:* 10,600 km
paved: 657 km
unpaved: 9,943 km (1999)

Merchant marine:
total: 1,948 ships (1000 GRT or over) 71,387,243 GRT/109,450,945 DWT
by type: barge carrier 3, bulk carrier 338, cargo 91, chemical tanker 211, combination ore/oil 9, container 614, liquefied gas 81, passenger 2, passenger/cargo 1, petroleum tanker 455, refrigerated cargo 91, roll on/roll off 6, specialized tanker 11, vehicle carrier 35
foreign-owned: 1,904 (Argentina 3, Australia 2, Belgium 1, Brazil 3, Canada 3, China 32, Croatia 5, Cyprus 5, Denmark 12, Estonia 1, France 5, Germany 728, Gibraltar 7, Greece 311, Hong Kong 21, India 2, Indonesia 1, Israel 9, Italy 31, Japan 111, South Korea 4, Kuwait 1, Latvia 15, Lebanon 2, Mexico 1, Monaco 8, Netherlands 28, Norway 42, Poland 14, Qatar 2, Russia 87, Saudi Arabia 24, Singapore 42, Slovenia 1, Sweden 11, Switzerland 11, Taiwan 82, Turkey 7, Ukraine 24, UAE 22, UK 74, US 103, Uruguay 3, Vietnam 3) (2007)

Ports and terminals: Buchanan, Monrovia

MILITARY

Military branches: Armed Forces of Liberia (AFL): Army, Navy, Air Force

Military service age and obligation: 16 years of age for voluntary military service; no conscription (2008)

Manpower available for military service: *males age 16–49*: 729,813
females age 16–49: 741,223 (2008 est.)
Manpower fit for military service:
males age 16–49: 371,287
females age 16–49: 373,265 (2008 est.)
Military expenditures—percent of GDP: 1.3% (2006 est.)

TRANSNATIONAL ISSUES

Disputes—international: although civil unrest continues to abate with the assistance of 18,000 UN Mission in Liberia (UNMIL) peacekeepers, as of January 2007, Liberian refugees still remain in Guinea, Cote d'Ivoire, Sierra Leone, and Ghana; Liberia, in turn, shelters refugees fleeing turmoil in Cote d'Ivoire; despite the presence of over 9,000 UN forces (UNOCI) in Cote d'Ivoire since 2004, ethnic conflict continues to spread into neighboring states who can no longer send their migrant workers to Ivorian cocoa plantations; UN sanctions ban Liberia from exporting diamonds and timber

Refugees and internally displaced persons: *refugees (country of origin)*: 12,600 (Cote d'Ivoire)

IDPs: 13,000 (civil war from 1990–2004; IDP resettlement began in November 2004) (2007)

Illicit drugs: transshipment point for Southeast and Southwest Asian heroin and South American cocaine for the European and US markets; corruption, criminal activity, arms-dealing, and diamond trade provide significant potential for money laundering, but the lack of well-developed financial system limits the country's utility as a major money-laundering center

LIBYA

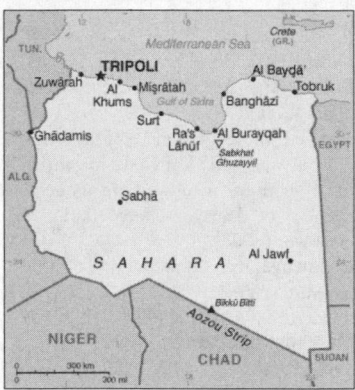

INTRODUCTION

Background: The Italians supplanted the Ottoman Turks in the area around Tripoli in 1911 and did not relinquish their hold until 1943 when defeated in World War II. Libya then passed to UN administration and achieved independence in 1951. Following a 1969 military coup, Col. Muammar Abu Minyar al-QADHAFI began to espouse his own political system, the Third Universal Theory. The system is a combination of socialism and Islam derived in part from tribal practices and is supposed to be implemented by the Libyan people themselves in a unique form of "direct democracy." QADHAFI has always seen himself as a revolutionary and visionary leader. He used oil funds during the 1970s and 1980s to promote his ideology outside Libya, supporting subversives and terrorists abroad to hasten the end of Marxism and capitalism. In addition, beginning in 1973, he engaged in military operations in northern Chad's Aozou Strip—to gain access to minerals and to use as a base of influence in Chadian politics—but was forced to retreat in 1987. UN sanctions in 1992

isolated QADHAFI politically following the downing of Pan AM Flight 103 over Lockerbie, Scotland. During the 1990s, QADHAFI began to rebuild his relationships with Europe. UN sanctions were suspended in April 1999 and finally lifted in September 2003 after Libya accepted responsibility for the Lockerbie bombing. In December 2003, Libya announced that it had agreed to reveal and end its programs to develop weapons of mass destruction and to renounce terrorism. QADHAFI has made significant strides in normalizing relations with Western nations since then. He has received various Western European leaders as well as many working-level and commercial delegations, and made his first trip to Western Europe in 15 years when he traveled to Brussels in April 2004. Libya has responded in good faith to legal cases brought against it in US courts for terrorist acts that predate its renunciation of violence. Claims for compensation in the Lockerbie bombing, LaBelle disco bombing, and UTA 772 bombing cases are ongoing. The US rescinded Libya's designation as a state sponsor of terrorism in June 2006. In late 2007, Libya was elected by the General Assembly to a nonpermanent seat on the United Nations Security Council for the 2008–09 term.

GEOGRAPHY

Location: Northern Africa, bordering the Mediterranean Sea, between Egypt and Tunisia
Geographic coordinates: 25 00 N, 17 00 E
Map references: Africa
Area: *total*: 1,759,540 sq km
land: 1,759,540 sq km
water: 0 sq km
Area—comparative: slightly larger than Alaska

Land boundaries:
total: 4,348 km
border countries: Algeria 982 km, Chad 1,055 km, Egypt 1,115 km, Niger 354 km, Sudan 383 km, Tunisia 459 km
Coastline: 1,770 km
Maritime claims:
territorial sea: 12 nm
note: Gulf of Sidra closing line—32 degrees, 30 minutes north
exclusive fishing zone: 62 nm
Climate: Mediterranean along coast; dry, extreme desert interior
Terrain: mostly barren, flat to undulating plains, plateaus, depressions
Elevation extremes:
lowest point: Sabkhat Ghuzayyil -47 m
highest point: Bikku Bitti 2,267 m
Natural resources: petroleum, natural gas, gypsum
Land use:
arable land: 1.03%
permanent crops: 0.19%
other: 98.78% (2005)
Irrigated land: 4,700 sq km (2003)
Total renewable water resources: 0.6 cu km (1997)
Freshwater withdrawal (domestic/industrial/agricultural): *total*: 4.27 cu km/yr (14%/3%/83%)
per capita: 730 cu m/yr (2000)
Natural hazards: hot, dry, dust-laden ghibli is a southern wind lasting one to four days in spring and fall; dust storms, sandstorms
Environment—current issues: desertification; limited natural fresh water resources; the Great Manmade River Project, the largest water development scheme in the world, is being built to bring water from large aquifers under the Sahara to coastal cities
Environment—international agreements: *party to*: Biodiversity, Climate Change, Climate Change-Kyoto Protocol, Desertification, Endangered

Species, Hazardous Wastes, Marine Dumping, Ozone Layer Protection, Ship Pollution, Wetlands

signed, but not ratified: Law of the Sea

Geography—note: more than 90% of the country is desert or semidesert

PEOPLE

Population: 6,173,579

note: includes 166,510 non-nationals (July 2008 est.)

Age structure:

0–14 years: 33.2% (male 1,046,400/female 1,002,148)

15–64 years: 62.6% (male 1,988,038/female 1,875,034)

65 years and over: 4.2% (male 128,386/female 133,573) (2008 est.)

Median age:

total: 23.6 years

male: 23.7 years

female: 23.5 years (2008 est.)

Population growth rate: 2.216% (2008 est.)

Birth rate: 25.62 births/1,000 population (2008 est.)

Death rate: 3.46 deaths/1,000 population (2008 est.)

Net migration rate: NA

Sex ratio:

at birth: 1.05 male(s)/female

under 15 years: 1.04 male(s)/female

15–64 years: 1.06 male(s)/female

65 years and over: 0.96 male(s)/female

total population: 1.05 male(s)/female (2008 est.)

Infant mortality rate:

total: 21.94 deaths/1,000 live births

male: 24.14 deaths/1,000 live births

female: 19.63 deaths/1,000 live births (2008 est.)

Life expectancy at birth:

total population: 77.07 years

male: 74.81 years

female: 79.44 years (2008 est.)

Total fertility rate: 3.15 children born/woman (2008 est.)

HIV/AIDS—adult prevalence rate: 0.3% (2001 est.)

HIV/AIDS—people living with HIV/AIDS: 10,000 (2001 est.)

HIV/AIDS—deaths: NA

Nationality:

noun: Libyan(s)

adjective: Libyan

Ethnic groups: Berber and Arab 97%, other 3% (includes Greeks, Maltese, Italians, Egyptians, Pakistanis, Turks, Indians, and Tunisians)

Religions: Sunni Muslim 97%, other 3%

Languages: Arabic, Italian, English, all are widely understood in the major cities

Literacy: *definition:* age 15 and over can read and write

total population: 82.6%

male: 92.4%

female: 72% (2003 est.)

GOVERNMENT

Country name:

conventional long form: Great Socialist People's Libyan Arab Jamahiriya

conventional short form: Libya

local long form: Al Jumahiriyah al Arabiyah al Libiyah ash Shabiyah al Ishtirakiyah al Uzma

local short form: none

Government type: Jamahiriya (a state of the masses) in theory, governed by the populace through local councils; in practice, an authoritarian state

Capital: *name:* Tripoli

geographic coordinates: 32 53 N, 13 10 E

time difference: UTC+2 (7 hours ahead of Washington, DC during Standard Time)

Administrative divisions: 25 municipalities (baladiyat, singular—baladiyah); Ajdabiya, Al 'Aziziyah, Al Fatih, Al Jabal al Akhdar, Al Jufrah, Al Khums, Al Kufrah, An Nuqat al Khams, Ash Shati', Awbari, Az Zawiyah, Banghazi, Darnah, Ghadamis, Gharyan, Misratah, Murzuq, Sabha, Sawfajjin, Surt, Tarabulus, Tarhunah, Tubruq, Yafran, Zlitan; note—the 25 municipalities may have been replaced by 13 regions

Independence: 24 December 1951 (from UN trusteeship)

National holiday: Revolution Day, 1 September (1969)

Constitution: none; note—following the September 1969 military overthrow of the Libyan government, the Revolutionary Command Council replaced the existing constitution with the Constitutional Proclamation in December 1969; in March 1977, Libya adopted the Declaration of the Establishment of the People's Authority

Legal system: based on Italian and French civil law systems and Islamic law; separate religious courts; no constitutional provision for judicial review of legislative acts; has not accepted compulsory ICJ jurisdiction

Suffrage: 18 years of age; universal and compulsory

Executive branch:

chief of state: Revolutionary Leader Col. Muammar Abu Minyar al-QADHAFI (since 1 September 1969); note—holds no official title, but is de facto chief of state

head of government: Secretary of the General People's Committee (Prime Minister) al-Baghdadi Ali al-MAH-MUDI (since 5 March 2006)

cabinet: General People's Committee established by the General People's Congress

elections: national elections are indirect through a hierarchy of people's committees; head of government elected by the General People's Congress; election last held March 2006 (next to be held NA)

election results: NA

Legislative branch: unicameral General People's Congress (approximately 2,700 seats; members elected indirectly through a hierarchy of people's committees)

Judicial branch: Supreme Court

Political parties and leaders: none

Political pressure groups and leaders: various Arab nationalist movements with almost negligible memberships may be functioning clandestinely, as well as some Islamic elements; an anti-QADHAFI Libyan exile movement exists, primarily based in London, but has little influence

International organization participation: ABEDA, AfDB, AFESD, AMF, AMU, AU, CAEU, COMESA, FAO, G-77, IAEA, IBRD, ICAO, ICRM, IDA, IDB, IFAD, IFC, IFRCS, ILO, IMF, IMO, IMSO, Interpol, IOC, IOM, IPU, ISO, ITSO, ITU, LAS, MIGA, NAM, OAPEC, OIC, OPCW, OPEC, PCA, UN, UN Security Council (temporary), UNCTAD, UNESCO, UNIDO, UNWTO, UPU, WCL, WCO, WFTU, WHO, WIPO, WMO, WTO (observer)

Diplomatic representation in the US:

chief of mission: ambassador (vacant); Charge d'Affaires Ali Suleiman AUJALI

chancery: 2600 Virginia Avenue NW, Suite 705, Washington, DC 20037

telephone: [1] (202) 944-9601

FAX: [1] (202) 944-9060

Diplomatic representation from the US:

chief of mission: Ambassador (vacant); Charge d'Affaires J. Christopher Stevens

embassy: Corinthia Bab Africa Hotel, Souq At-Tlat Al-Qadim, Tripoli

mailing address: US Embassy, 8850 Tripoli Place, Washington, DC 20521-8850

telephone: [218] 21-335-1848

Flag description: plain green; green is the traditional color of Islam (the state religion)

ECONOMY

Economy—overview: The Libyan economy depends primarily upon revenues from the oil sector, which contribute about 95% of export earnings, about one-quarter of GDP, and 60% of public sector wages. Substantial revenues from the energy sector coupled with a small population give Libya one of the highest per capita GDPs in Africa, but little of this income flows down to the lower orders of society. Libyan officials in the past five years have made progress on economic reforms as part of a broader campaign to reintegrate the country into the international fold. This effort picked up steam after UN sanctions were lifted in September 2003 and as Libya announced in December 2003 that it would abandon programs to build weapons of mass destruction. Almost all US unilateral sanctions against Libya were removed in April 2004, helping Libya attract more foreign direct investment, mostly in the energy sector. Libyan oil and gas licensing rounds continue to draw high international interest; the National Oil Company set a goal of nearly doubling oil production to 3 million bbl/day by 2015. Libya faces a long road ahead in liberalizing the socialist-oriented economy, but initial steps— including applying for WTO membership, reducing some subsidies, and announcing plans for privatization—are laying the groundwork for a transition to a more market-based economy. The non-oil manufacturing and construction sectors, which account for more than 20% of GDP, have expanded from processing mostly agricultural products to include the production of petrochemicals, iron, steel, and aluminum. Climatic conditions and poor soils severely limit agricultural output, and Libya imports about 75% of its food. Libya's primary agricultural water source remains the Great Manmade River Project, but significant resources are being invested in desalinization research to meet growing water demands.

GDP (purchasing power parity): $74.75 billion (2007 est.)
GDP (official exchange rate): $57.06 billion (2007 est.)
GDP—real growth rate: 6.8% (2007 est.)
GDP—per capita (PPP): $12,300 (2007 est.)
GDP—composition by sector:
agriculture: 2.1%
industry: 83.1%
services: 14.9% (2007 est.)
Labor force: 1.83 million (2007 est.)
Labor force—by occupation: *agriculture:* 17%
industry: 23%
services: 59% (2004 est.)
Unemployment rate: 30% (2004 est.)
Population below poverty line: 7.4% (2005 est.)
Household income or consumption by percentage share:
lowest 10%: NA%
highest 10%: NA%
Inflation rate (consumer prices): 6.7% (2007 est.)
Investment (gross fixed): 8.9% of GDP (2007 est.)

Budget:
revenues: $39.85 billion
expenditures: $19.47 billion (2007 est.)
Public debt: 4.7% of GDP (2007 est.)
Agriculture—products: wheat, barley, olives, dates, citrus, vegetables, peanuts, soybeans; cattle
Industries: petroleum, iron and steel, food processing, textiles, handicrafts, cement
Industrial production growth rate: 5.6% (2007 est.)
Electricity—production: 21.15 billion kWh (2005)
Electricity—production by source:
fossil fuel: 100%
hydro: 0%
nuclear: 0%
other: 0% (2001)
Electricity—consumption: 18.18 billion kWh (2005)
Electricity—exports: 0 kWh (2005)
Electricity—imports: 0 kWh (2005)
Oil—production: 1.72 million bbl/day (2006 est.)
Oil—consumption: 266,000 bbl/day (2005 est.)
Oil—exports: 1.326 million bbl/day (2004)
Oil—imports: 1,233 bbl/day (2004)
Oil—proved reserves: 45 billion bbl (2007 est.)
Natural gas—production: 10.84 billion cu m (2005 est.)
Natural gas—consumption: 5.591 billion cu m (2005 est.)
Natural gas—exports: 5.246 billion cu m (2005 est.)
Natural gas—imports: 0 cu m (2005)
Natural gas—proved reserves: 1.43 trillion cu m (1 January 2006 est.)
Current account balance: $24.28 billion (2007 est.)
Exports: $40.47 billion f.o.b. (2007 est.)
Exports—commodities: crude oil, refined petroleum products, natural gas, chemicals
Exports—partners: Italy 36.8%, Germany 14.3%, Spain 8.7%, US 6.1%, France 5.6%, Turkey 5.3% (2006)
Imports: $14.47 billion f.o.b. (2007 est.)
Imports—commodities: machinery, semi-finished goods, food, transport equipment, consumer products
Imports—partners: Italy 18.9%, Germany 7.8%, China 7.6%, Tunisia 6.3%, France 5.8%, Turkey 5.3%, US 4.7%, South Korea 4.3%, UK 4% (2006)
Economic aid—recipient: ODA, $24.44 million (2005 est.)
Reserves of foreign exchange and gold: $79.6 billion (31 December 2007 est.)
Debt—external: $4.837 billion (31 December 2007 est.)

Stock of direct foreign investment—at home: $6.286 billion (2007 est.)
Stock of direct foreign investment—abroad: $3.333 billion (2007 est.)
Market value of publicly traded shares: $NA
Currency (code): Libyan dinar (LYD)
Currency code: LYD
Exchange rates: Libyan dinars per US dollar—1.2604 (2007), 1.3108 (2006), 1.3084 (2005), 1.305 (2004), 1.2929 (2003)
Fiscal year: calendar year

COMMUNICATIONS

Telephones—main lines in use: 483,000 (2006)
Telephones—mobile cellular: 3.928 million (2006)
Telephone system:
general assessment: telecommunications system is being modernized; mobile cellular telephone system became operational in 1996; combined fixed line and mobile telephone density reached 75 telephones per 100 persons in 2006
domestic: microwave radio relay, coaxial cable, cellular, tropospheric scatter, and a domestic satellite system with 14 earth stations
international: country code—218; satellite earth stations—4 Intelsat, NA Arabsat, and NA Intersputnik; submarine cables to France and Italy; microwave radio relay to Tunisia and Egypt; tropospheric scatter to Greece; participant in Medarabtel (1999)
Radio broadcast stations: AM 16, FM 3, shortwave 3 (2001)
Radios: 1.35 million (1997)
Television broadcast stations: 12 (plus 1 repeater) (1999)
Televisions: 730,000 (1997)
Internet country code: .ly
Internet hosts: 24 (2007)
Internet Service Providers (ISPs): 1 (2002)
Internet users: 232,000 (2005)

TRANSPORTATION

Airports: 141 (2007)
Airports—with paved runways:
total: 60
over 3,047 m: 23
2,438 to 3,047 m: 6
1,524 to 2,437 m: 23
914 to 1,523 m: 6
under 914 m: 2 (2007)
Airports—with unpaved runways:
total: 81
over 3,047 m: 5
2,438 to 3,047 m: 2
1,524 to 2,437 m: 15
914 to 1,523 m: 41
under 914 m: 18 (2007)

Heliports: 2 (2007)
Pipelines: condensate 882 km; gas 3,425 km; oil 6,956 km (2007)
Railways:
0 km
note: Libya has announced plans to build seven lines totaling 2,757 km of 1.435-m gauge track (2006)
Roadways:
total: 83,200 km
paved: 47,590 km
unpaved: 35,610 km (1999)
Merchant marine:
total: 17 ships (1000 GRT or over) 67,200 GRT/85,931 DWT
by type: cargo 11, liquefied gas 3, petroleum tanker 2, roll on/roll off 1
foreign-owned: 3 (Kuwait 1, Norway 1, Syria 1)
registered in other countries: 4 (Malta 3, Tunisia 1) (2007)
Ports and terminals: As Sidrah, Az Zuwaytinah, Marsa al Burayqah, Ra's Lanuf, Tripoli, Zawiyah

MILITARY

Military branches: Armed Peoples on Duty (APOD, Army), Libyan Arab Navy, Libyan Arab Air Force (Al-Quwwat al-Jawwiya al-Jamahiriya al-Arabia al-Libyya, LAAF) (2008)
Military service age and obligation: 17 years of age (2004)
Manpower available for military service: *males age 16–49:* 1,682,183
females age 16–49: 1,611,001 (2008 est.)
Manpower fit for military service:
males age 16–49: 1,439,941
females age 16–49: 1,381,914 (2008 est.)
Manpower reaching militarily significant age annually:
males age 16–49: 61,305
females age 16–49: 58,788 (2008 est.)
Military expenditures—percent of GDP: 3.9% (2005 est.)

TRANSNATIONAL ISSUES

Disputes—international: Libya has claimed more than 32,000 sq km in southeastern Algeria and about 25,000 sq km in the Tommo region of Niger in a currently dormant dispute; various Chadian rebels from the Aozou region reside in southern Libya
Refugees and internally displaced persons: *refugees (country of origin):* 8,000 (Palestinian Territories) (2007)
Trafficking in persons: *current situation:* Libya is a transit and destination country for men, women, and children from sub-Saharan Africa and Asia trafficked for forced labor and sexual exploitation; many victims willingly migrate to Libya en route to Europe with the help of smugglers, but may be forced into prostitution or work as laborers and beggars to pay off their $800–$1,200 smuggling debt; laborers from Egypt, Sudan, and Ethiopia are reportedly trafficked to Libya for the purpose of labor exploitation
tier rating: Tier 2 Watch List—Libya is placed on the Tier 2 Watch List for its lack of evidence of increasing efforts to address trafficking since 2004

LIECHTENSTEIN

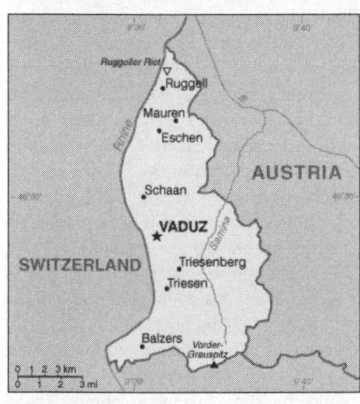

INTRODUCTION

Background: The Principality of Liechtenstein was established within the Holy Roman Empire in 1719. Occupied by both French and Russian troops during the Napoleanic wars, it became a sovereign state in 1806 and joined the Germanic Confederation in 1815. Liechtenstein became fully independent in 1866 when the Confederation dissolved. Until the end of World War I, it was closely tied to Austria, but the economic devastation caused by that conflict forced Liechtenstein to enter into a customs and monetary union with Switzerland. Since World War II (in which Liechtenstein remained neutral), the country's low taxes have spurred outstanding economic growth. In 2000, shortcomings in banking regulatory oversight resulted in concerns about the use of financial institutions for money laundering. However, Liechtenstein implemented anti-money-laundering legislation and a Mutual Legal Assistance Treaty with the US went into effect in 2003.

GEOGRAPHY

Location: Central Europe, between Austria and Switzerland
Geographic coordinates: 47 16 N, 9 32 E
Map references: Europe
Area:
total: 160 sq km
land: 160 sq km
water: 0 sq km
Area—comparative: about 0.9 times the size of Washington, DC
Land boundaries:
total: 76 km
border countries: Austria 34.9 km, Switzerland 41.1 km
Coastline: 0 km (doubly landlocked)
Maritime claims: none (landlocked)
Climate: continental; cold, cloudy winters with frequent snow or rain; cool to moderately warm, cloudy, humid summers
Terrain: mostly mountainous (Alps) with Rhine Valley in western third
Elevation extremes:
lowest point: Ruggeller Riet 430 m
highest point: Vorder-Grauspitz 2,599 m
Natural resources: hydroelectric potential, arable land
Land use: *arable land:* 25%
permanent crops: 0%
other: 75% (2005)
Irrigated land: NA
Natural hazards: NA
Environment—current issues: NA
Environment—international agreements: *party to:* Air Pollution, Air Pollution-Nitrogen Oxides, Air Pollution-Persistent Organic Pollutants, Air Pollution-Sulfur 85, Air Pollution-Sulfur 94, Air Pollution-Volatile Organic Compounds, Biodiversity, Climate Change, Climate Change-Kyoto Protocol, Desertification, Endangered Species, Hazardous Wastes, Ozone Layer Protection, Wetlands
signed, but not ratified: Law of the Sea
Geography—note: along with Uzbekistan, one of only two doubly landlocked countries in the world; variety of microclimatic variations based on elevation

PEOPLE

Population: 34,498 (July 2008 est.)
Age structure:
0–14 years: 16.9% (male 2,892/female 2,927)
15–64 years: 69.8% (male 11,905/female 12,180)
65 years and over: 13.3% (male 1,964/female 2,630) (2008 est.)

Median age:

total: 40.5 years

male: 40 years

female: 41 years (2008 est.)

Population growth rate: 0.713% (2008 est.)

Birth rate: 9.86 births/1,000 population (2008 est.)

Death rate: 7.42 deaths/1,000 population (2008 est.)

Net migration rate: 4.7 migrant(s)/1,000 population (2008 est.)

Sex ratio:

at birth: 1 male(s)/female

under 15 years: 0.99 male(s)/female

15–64 years: 0.98 male(s)/female

65 years and over: 0.75 male(s)/female

total population: 0.94 male(s)/female (2008 est.)

Infant mortality rate:

total: 4.52 deaths/1,000 live births

male: 6.03 deaths/1,000 live births

female: 3 deaths/1,000 live births (2008 est.)

Life expectancy at birth:

total population: 79.95 years

male: 76.38 years

female: 83.52 years (2008 est.)

Total fertility rate: 1.51 children born/woman (2008 est.)

HIV/AIDS—adult prevalence rate: NA

HIV/AIDS—people living with HIV/AIDS: NA

HIV/AIDS—deaths: NA

Nationality:

noun: Liechtensteiner(s)

adjective: Liechtenstein

Ethnic groups: Alemannic 86%, Italian, Turkish, and other 14%

Religions: Roman Catholic 76.2%, Protestant 7%, unknown 10.6%, other 6.2% (June 2002)

Languages: German (official), Alemannic dialect

Literacy: *definition:* age 10 and over can read and write

total population: 100%

male: 100%

female: 100%

GOVERNMENT

Country name:

conventional long form: Principality of Liechtenstein

conventional short form: Liechtenstein

local long form: Fuerstentum Liechtenstein

local short form: Liechtenstein

Government type: constitutional monarchy

Capital: *name:* Vaduz

geographic coordinates: 47 08 N, 9 31 E

time difference: UTC+1 (6 hours ahead of Washington, DC during Standard Time)

daylight saving time: +1hr, begins last Sunday in March; ends last Sunday in October

Administrative divisions: 11 communes (Gemeinden, singular—Gemeinde); Balzers, Eschen, Gamprin, Mauren, Planken, Ruggell, Schaan, Schellenberg, Triesen, Triesenberg, Vaduz

Independence: 23 January 1719 (Principality of Liechtenstein established); 12 July 1806 (independence from the Holy Roman Empire)

National holiday: Assumption Day, 15 August

Constitution: 5 October 1921

Legal system: local civil and penal codes based on civil law system; accepts compulsory ICJ jurisdiction with reservations

Suffrage: 18 years of age; universal

Executive branch:

chief of state: Prince HANS ADAM II (since 13 November 1989, assumed executive powers 26 August 1984); Heir Apparent Prince ALOIS, son of the monarch (born 11 June 1968); note—on 15 August 2004, HANS ADAM transferred the official duties of the ruling prince to ALOIS, but HANS ADAM retains status of chief of state

head of government: Head of Government Otmar HASLER (since 5 April 2001); Deputy Head of Government Klaus TSCHUETSCHER (since 21 April 2005)

cabinet: Cabinet elected by the Parliament, confirmed by the monarch

elections: the monarch is hereditary; following legislative elections, the leader of the majority party in the Landtag is usually appointed the head of government by the monarch and the leader of the largest minority party in the Landtag is usually appointed the deputy head of government by the monarch if there is a coalition government

Legislative branch: unicameral Parliament or Landtag (25 seats; members are elected by popular vote under proportional representation to serve four-year terms)

elections: last held 11 and 13 March 2005 (next to be held by 2009)

election results: percent of vote by party—FBP 48.7%, VU 38.2%, FL 13%; seats by party—FBP 12, VU 10, FL 3

Judicial branch: Supreme Court or Oberster Gerichtshof; Court of Appeal or Obergericht

Political parties and leaders: Patriotic Union or VU [Adolf HEEB] (was Fatherland Union); Progressive Citizens' Party or FBP [Marcus VOGT]; The Free List or FL [Claudia HEEB-FLECK and Egon MATT]

Political pressure groups and leaders: NA

International organization participation: CE, EBRD, EFTA, IAEA, ICCt, ICRM, IFRCS, Interpol, IOC, IPU, ITSO, ITU, OPCW, OSCE, PCA, UN, UNCTAD, UPU, WCL, WIPO, WTO

Diplomatic representation in the US:

chief of mission: Ambassador Claudia FRITSCHE

chancery: 888 17th Street NW, Suite 1250, Washington, DC 20006

telephone: [1] (202) 331-0590

FAX: [1] (202) 331-3221

Diplomatic representation from the US: the US does not have an embassy in Liechtenstein; the US Ambassador to Switzerland is accredited to Liechtenstein

Flag description: two equal horizontal bands of blue (top) and red with a gold crown on the hoist side of the blue band

ECONOMY

Economy—overview: Despite its small size and limited natural resources, Liechtenstein has developed into a prosperous, highly industrialized, free-enterprise economy with a vital financial service sector and living standards on a par with its large European neighbors. The Liechtenstein economy is widely diversified with a large number of small businesses. Low business taxes—the maximum tax rate is 20%—and easy incorporation rules have induced many holding or so-called letter box companies to establish nominal offices in Liechtenstein, providing 30% of state revenues. The country participates in a customs union with Switzerland and uses the Swiss franc as its national currency. It imports more than 90% of its energy requirements. Liechtenstein has been a member of the European Economic Area (an organization serving as a bridge between the European Free Trade Association (EFTA) and the EU) since May 1995. The government is working to harmonize its economic policies with those of an integrated Europe.

GDP (purchasing power parity): $1.786 billion (2001 est.)

GDP (official exchange rate): $36.33 billion (2007 est.)

GDP—real growth rate: 11% (1999 est.)

GDP—per capita (PPP): $25,000 (1999 est.)

GDP—composition by sector:

agriculture: 6%

industry: 39%

services: 55% (2001)

Labor force: 29,500 of whom 13,900 commute from Austria, Switzerland, and Germany to work each day (31 December 2001)

Labor force—by occupation:

agriculture: 2%
industry: 47%
services: 51% (31 December 2001)
Unemployment rate: 1.3% (September 2002)
Population below poverty line: NA%
Household income or consumption by percentage share:
lowest 10%: NA%
highest 10%: NA%
Inflation rate (consumer prices): 1% (2001)
Budget:
revenues: $424.2 million
expenditures: $414.1 million (1998 est.)
Agriculture—products: wheat, barley, corn, potatoes; livestock, dairy products
Industries: electronics, metal manufacturing, dental products, ceramics, pharmaceuticals, food products, precision instruments, tourism, optical instruments
Industrial production growth rate: NA%
Exports: $2.47 billion (1996)
Exports—commodities: small specialty machinery, connectors for audio and video, parts for motor vehicles, dental products, hardware, prepared foodstuffs, electronic equipment, optical products
Exports—partners: EU 62.6% (Germany 24.3%, Austria 9.5%, France 8.9%, Italy 6.6%, UK 4.6%), US 18.9%, Switzerland 15.7% (2006)
Imports: $917.3 million (1996)
Imports—commodities: agricultural products, raw materials, energy products, machinery, metal goods, textiles, foodstuffs, motor vehicles

Imports—partners: EU, Switzerland (2006)
Debt—external: $0 (2001)
Market value of publicly traded shares: $NA
Currency (code): Swiss franc (CHF)
Currency code: CHF
Exchange rates: Swiss francs per US dollar—1.1973 (2007), 1.2539 (2006), 1.2452 (2005), 1.2435 (2004), 1.3467 (2003)
Fiscal year: calendar year

COMMUNICATIONS

Telephones—main lines in use: 20,000 (2005)
Telephones—mobile cellular: 27,500 (2005)
Telephone system:
general assessment: automatic telephone system
domestic: NA
international: country code—423; linked to Swiss networks by cable and microwave radio relay
Radio broadcast stations: AM 0, FM 4, shortwave 0 (1998)
Radios: 21,000 (1997)
Television broadcast stations: NA (linked to Swiss networks) (1997)
Televisions: 12,000 (1997)
Internet country code: .li
Internet hosts: 4,753 (2007)
Internet Service Providers (ISPs): 44 (Liechtenstein and Switzerland) (2000)
Internet users: 22,000 (2006)

TRANSPORTATION

Pipelines: gas 20 km (2007)

Railways:
9 km 1.435-m gauge (electrified)
note: belongs to the Austrian Railway System connecting Austria and Switzerland (2006)
Roadways:
total: 380 km
paved: 380 km (2006)
Waterways: 28 km (2006)

MILITARY

Military branches: no regular military forces (constitutionally prohibited); Principality of Liechtenstein National Police (Landespolizei, LP) (2008)
Manpower available for military service:
males age 16–49: 8,102 (2008 est.)
Manpower fit for military service:
males age 16–49: 6,584 (2008 est.)
Manpower reaching militarily significant age annually:
males age 16–49: 202 (2008 est.)
Military—note: Liechtenstein has no military forces, but is interested in European security policy and is an active member of the Organization for Security and Cooperation in Europe (OSCE)

TRANSNATIONAL ISSUES

Disputes—international: none
Illicit drugs: has strengthened money laundering controls, but money laundering remains a concern due to Liechtenstein's sophisticated offshore financial services sector

LITHUANIA

INTRODUCTION

Background: Lithuanian lands were united under MINDAUGAS in 1236; over the next century, through alliances and conquest, Lithuania extended its territory to include most of present-day Belarus and Ukraine. By the end of the 14th century Lithuania was the largest state in Europe. An alliance with Poland in 1386 led the two countries into a union through the person of a common ruler. In 1569, Lithuania and Poland formally united into a single dual state, the Polish-Lithuanian Commonwealth. This entity survived until 1795, when its remnants were partitioned by surrounding countries. Lithuania regained its independence following World War I but was annexed by the USSR in 1940—an action never recognized by the US and many other countries. On 11 March 1990, Lithuania became the first of the Soviet republics to declare its independence, but Moscow did not recognize this proclamation until September of 1991 (following the abortive coup in Moscow). The last Russian troops withdrew in 1993. Lithuania subsequently restructured its economy for integration into Western European institutions; it joined both NATO and the EU in the spring of 2004.

GEOGRAPHY

Location: Eastern Europe, bordering the Baltic Sea, between Latvia and Russia
Geographic coordinates: 56 00 N, 24 00 E
Map references: Europe
Area: *total:* 65,200 sq km
land: NA sq km
water: NA sq km
Area—comparative: slightly larger than West Virginia
Land boundaries:
total: 1,613 km

border countries: Belarus 653.5 km, Latvia 588 km, Poland 103.7 km, Russia (Kaliningrad) 267.8 km
Coastline: 99 km
Maritime claims: *territorial sea:* 12 nm
Climate: transitional, between maritime and continental; wet, moderate winters and summers
Terrain: lowland, many scattered small lakes, fertile soil
Elevation extremes:
lowest point: Baltic Sea 0 m
highest point: Juozapines Kalnas 293.6 m
Natural resources: peat, arable land, amber
Land use:
arable land: 44.81%
permanent crops: 0.9%
other: 54.29% (2005)
Irrigated land: 70 sq km (2003)
Total renewable water resources: 24.5 cu km (2005)
Freshwater withdrawal (domestic/industrial/agricultural): *total:* 3.33 cu km/yr (78%/15%/7%)
per capita: 971 cu m/yr (2003)
Natural hazards: NA
Environment—current issues: contamination of soil and groundwater with petroleum products and chemicals at military bases
Environment—international agreements: *party to:* Air Pollution, Air Pollution-Nitrogen Oxides, Air Pollution-Persistent Organic Pollutants, Biodiversity, Climate Change, Climate Change-Kyoto Protocol, Endangered Species, Hazardous Wastes, Law of the Sea, Ozone Layer Protection, Ship Pollution, Wetlands
signed, but not ratified: none of the selected agreements
Geography—note: fertile central plains are separated by hilly uplands that are ancient glacial deposits

PEOPLE

Population: 3,565,205 (July 2008 est.)
Age structure:
0–14 years: 14.5% (male 264,668/female 250,997)
15–64 years: 69.5% (male 1,214,236/female 1,263,198)
65 years and over: 16% (male 197,498/female 374,608) (2008 est.)
Median age:
total: 39 years
male: 36.4 years
female: 41.6 years (2008 est.)
Population growth rate: -0.284% (2008 est.)
Birth rate: 9 births/1,000 population (2008 est.)
Death rate: 11.12 deaths/1,000 population (2008 est.)

Net migration rate: -0.72 migrant(s)/1,000 population (2008 est.)
Sex ratio:
at birth: 1.06 male(s)/female
under 15 years: 1.05 male(s)/female
15–64 years: 0.96 male(s)/female
65 years and over: 0.53 male(s)/female
total population: 0.89 male(s)/female (2008 est.)
Infant mortality rate:
total: 6.57 deaths/1,000 live births
male: 7.86 deaths/1,000 live births
female: 5.21 deaths/1,000 live births (2008 est.)
Life expectancy at birth:
total population: 74.67 years
male: 69.72 years
female: 79.89 years (2008 est.)
Total fertility rate: 1.22 children born/woman (2008 est.)
HIV/AIDS—adult prevalence rate: 0.1% (2001 est.)
HIV/AIDS—people living with HIV/AIDS: 1,300 (2003 est.)
HIV/AIDS—deaths: fewer than 200 (2003 est.)
Major infectious diseases:
degree of risk: intermediate
food or waterborne diseases: bacterial diarrhea and hepatitis A
vectorborne diseases: tickborne encephalitis (2008)
Nationality:
noun: Lithuanian(s)
adjective: Lithuanian
Ethnic groups: Lithuanian 83.4%, Polish 6.7%, Russian 6.3%, other or unspecified 3.6% (2001 census)
Religions: Roman Catholic 79%, Russian Orthodox 4.1%, Protestant (including Lutheran and Evangelical Christian Baptist) 1.9%, other or unspecified 5.5%, none 9.5% (2001 census)
Languages: Lithuanian (official) 82%, Russian 8%, Polish 5.6%, other and unspecified 4.4% (2001 census)
Literacy: *definition:* age 15 and over can read and write
total population: 99.6%
male: 99.6%
female: 99.6% (2001 census)

GOVERNMENT

Country name:
conventional long form: Republic of Lithuania
conventional short form: Lithuania
local long form: Lietuvos Respublika
local short form: Lietuva
former: Lithuanian Soviet Socialist Republic
Government type: parliamentary democracy
Capital: *name:* Vilnius

geographic coordinates: 54 41 N, 25 19 E
time difference: UTC+2 (7 hours ahead of Washington, DC during Standard Time)
daylight saving time: +1hr, begins last Sunday in March; ends last Sunday in October
Administrative divisions: 10 counties (apskritys, singular—apskritis); Alytaus, Kauno, Klaipedos, Marijampoles, Panevezio, Siauliu, Taurages, Telsiu, Utenos, Vilniaus
Independence: 11 March 1990 (declared); 6 September 1991 (recognized by Soviet Union)
National holiday: Independence Day, 16 February (1918); note—16 February 1918 was the date Lithuania declared its independence from Soviet Russia and established its statehood; 11 March 1990 was the date it declared its independence from the Soviet Union
Constitution: adopted 25 October 1992
Legal system: based on civil law system; legislative acts can be appealed to the constitutional court; has not accepted compulsory ICJ jurisdiction
Suffrage: 18 years of age; universal
Executive branch:
chief of state: President Valdas ADAMKUS (since 12 July 2004)
head of government: Prime Minister Gediminas KIRKILAS (since 4 July 2006)
cabinet: Council of Ministers appointed by the president on the nomination of the prime minister
elections: president elected by popular vote for a five-year term (eligible for a second term); election last held 13 and 27 June 2004 (next to be held in June 2009); prime minister appointed by the president on the approval of the Parliament
election results: Valdas ADAMKUS elected president; percent of vote—Valdas ADAMKUS 52.2%, Kazimiera PRUNSKIENE 47.8%; Gediminas KIRKILAS approved by Parliament 85-13 with five abstentions
Legislative branch: unicameral Parliament or Seimas (141 seats; 71 members are elected by popular vote, 70 are elected by proportional representation; serve four-year terms)
elections: last held 10 and 24 October 2004 (next to be held in October 2008)
election results: percent of vote by party—Labor 28.6%, Working for Lithuania (Social Democrats and Social Liberals) 20.7%, TS 14.6%, For Order and Justice (Liberal Democrats and Lithuanian People's Union) 11.4%, Liberal and Center Union 9.1%, Farmers and New Democracy Union 6.6%, other 9%; seats by faction—Social Democrats 38, TS 25,

Labor 23, Farmers National Union 13 (combined with Civil Democracy), Liberal Democrats/Order and Justice 11, New Union Social Liberals 10, Liberal and Center Union 10, Liberal Movement 9, independent 2 (as of April 2008)

Judicial branch: Constitutional Court; Supreme Court; Court of Appeal; judges for all courts appointed by the president

Political parties and leaders: Civil Democracy Party or PDP [Viktor MUNTIANAS]; Electoral Action of Lithuanian Poles [Valdemar TOMASZEVSKI]; National Farmer's Union or VLS [Kazimiera PRUN-SKIENE]; Homeland Union/Conservative Party or TS [Andrius KUBILIUS]; Labor Party or DP [Viktor USPASKICH]; Liberal and Center Union [Arturas ZUOKAS]; Liberal Democrats/Order and Justice Party or TT [Rolandas PAKSAS]; Liberal Movement or LLS [Petras AUS-TREVICIUS]; Social Democratic Party or LSDP [Gediminas KIRKILAS]; Social Liberal/New Union [Arturas PAULAUSKAS]; Young Lithuania and New Nationalists [Stanislovas BUSKE-VICIUS]

Political pressure groups and leaders: NA

International organization participation: ACCT (observer), Australia Group, BA, BIS, CBSS, CE, EAPC, EBRD, EIB, EU, FAO, IAEA, IBRD, ICAO, ICC, ICCt, ICRM, IDA, IFC, IFRCS, ILO, IMF, IMO, Interpol, IOC, IOM, IPU, ISO, ITU, ITUC, MIGA, NATO, NIB, NSG, OIF (observer), OPCW, OSCE, PCA, Schengen Convention, UN, UNCTAD, UNESCO, UNIDO, UNOMIG, UNWTO, UPU, WCL, WCO, WEU (associate partner), WHO, WIPO, WMO, WTO

Diplomatic representation in the US:
chief of mission: Ambassador Audrius BRUZGA
chancery: 4590 MacArthur Blvd. NW, Suite 200, Washington, DC 20007
telephone: [1] (202) 234-5860
FAX: [1] (202) 328-0466
consulate(s) general: Chicago, New York

Diplomatic representation from the US:
chief of mission: Ambassador John A. CLOUD
embassy: Akmenu Gatve 6, Vilnius, LT-03106
mailing address: American Embassy, Akmenu Gatve 6, Vilnius LT-03106
telephone: [370] (5) 266 5500
FAX: [370] (5) 266 5510

Flag description: three equal horizontal bands of yellow (top), green, and red

ECONOMY

Economy—overview: Lithuania, the Baltic state that has conducted the most trade with Russia, has grown rapidly since rebounding from the 1998 Russian financial crisis. Unemployment fell to 3.2% in 2007 while wages continued to grow at double digit rates, contributing to rising inflation. Exports and imports also grew strongly, and the current account deficit rose to nearly 15% of GDP in 2007. Trade has been increasingly oriented toward the West. Lithuania has gained membership in the World Trade Organization and joined the EU in May 2004. Privatization of the large, state-owned utilities is nearly complete. Foreign government and business support have helped in the transition from the old command economy to a market economy.

GDP (purchasing power parity): $59.64 billion (2007 est.)

GDP (official exchange rate): $38.35 billion (2007 est.)

GDP—real growth rate: 8.8% (2007 est.)

GDP—per capita (PPP): $17,700 (2007 est.)

GDP—composition by sector:
agriculture: 5.3%
industry: 33.3%
services: 61.4% (2007 est.)

Labor force: 1.587 million (2007 est.)

Labor force—by occupation: *agriculture:* 15.8%
industry: 28.2%
services: 56% (2004)

Unemployment rate: 3.5%
note: based on survey data, official registered unemployment of 5.7% (2007 est.)

Population below poverty line: 4% (2003)

Household income or consumption by percentage share:
lowest 10%: 2.7%
highest 10%: 27.7% (2003)

Distribution of family income—Gini index: 36 (2005)

Inflation rate (consumer prices): 5.8% (2007 est.)

Investment (gross fixed): 26.6% of GDP (2007 est.)

Budget:
revenues: $13.14 billion
expenditures: $13.6 billion (2007 est.)

Public debt: 14.5% of GDP (2007 est.)

Agriculture—products: grain, potatoes, sugar beets, flax, vegetables; beef, milk, eggs; fish

Industries: metal-cutting machine tools, electric motors, television sets, refrigerators and freezers, petroleum refining, shipbuilding (small ships), furniture making, textiles, food processing, fertilizers, agricultural machinery, optical equipment, electronic components, computers, amber jewelry

Industrial production growth rate: 7.4% (2007 est.)

Electricity—production: 13.48 billion kWh (2005)

Electricity—production by source:
fossil fuel: 16.5%
hydro: 5.7%
nuclear: 77.7%
other: 0% (2001)

Electricity—consumption: 9.296 billion kWh (2005)

Electricity—exports: 8.607 billion kWh (2005)

Electricity—imports: 5.641 billion kWh (2005)

Oil—production: 13,160 bbl/day (2005 est.)

Oil—consumption: 57,000 bbl/day (2005 est.)

Oil—exports: 145,100 bbl/day (2004)

Oil—imports: 187,800 bbl/day (2004)

Oil—proved reserves: 12 million bbl (1 January 2006 est.)

Natural gas—production: 0 cu m (2005)

Natural gas—consumption: 2.916 billion cu m (2005 est.)

Natural gas—exports: 0 cu m (2005 est.)

Natural gas—imports: 2.916 billion cu m (2005)

Natural gas—proved reserves: 0 cu m (1 January 2006 est.)

Current account balance: -$4.988 billion (2007 est.)

Exports: $17.18 billion f.o.b. (2007 est.)

Exports—commodities: mineral products 23%, textiles and clothing 16%, machinery and equipment 11%, chemicals 6%, wood and wood products 5%, foodstuffs 5% (2001)

Exports—partners: Russia 12.8%, Latvia 11.1%, Germany 8.6%, Estonia 6.5%, Poland 6.1%, Netherlands 4.8%, Sweden 4.5%, UK 4.4%, US 4.3%, Denmark 4.2%, France 4.2% (2006)

Imports: $22.8 billion f.o.b. (2007 est.)

Imports—commodities: mineral products, machinery and equipment, transport equipment, chemicals, textiles and clothing, metals

Imports—partners: Russia 24.3%, Germany 14.9%, Poland 9.5%, Latvia 4.8% (2006)

Economic aid—recipient: $249.7 million (2004)

Reserves of foreign exchange and gold: $7.721 billion (31 December 2007 est.)

Debt—external: $27.19 billion (31 December 2007)

Stock of direct foreign investment—at home: $14.63 billion (2007 est.)

Stock of direct foreign investment—abroad: $1.783 billion (2007 est.)
Market value of publicly traded shares: $10.19 billion (2006)
Currency (code): litas (LTL)
Currency code: LTL
Exchange rates: litai per US dollar— 2.5362 (2007), 2.7498 (2006), 2.774 (2005), 2.7806 (2004), 3.0609 (2003)
Fiscal year: calendar year

COMMUNICATIONS

Telephones—main lines in use: 792,400 (2006)
Telephones—mobile cellular: 4.718 million (2006)
Telephone system:
general assessment: adequate; being modernized to provide improved international capability and better residential access
domestic: rapid expansion of mobile-cellular services has resulted in a steady decline in the number of main line subscriptions; mobile-cellular teledensity has increased to about 135 per 100 persons while fixed-line teledensity has dropped to 22 per 100 persons
international: country code—370; major international connections to Denmark, Sweden, and Norway by submarine cable for further transmission by satellite; landline connections to Latvia and Poland
Radio broadcast stations: AM 29, FM 142, shortwave 1 (2001)
Radios: 1.9 million (1997)
Television broadcast stations: 27 (may have as many as 100 transmitters, including repeater stations) (2001)
Televisions: 1.7 million (1997)
Internet country code: .lt
Internet hosts: 1.301 million (2007)
Internet Service Providers (ISPs): 32 (2001)
Internet users: 1.083 million (2006)

TRANSPORTATION

Airports: 87 (2007)

Airports—with paved runways:
total: 30
over 3,047 m: 3
2,438 to 3,047 m: 1
1,524 to 2,437 m: 7
914 to 1,523 m: 2
under 914 m: 17 (2007)
Airports—with unpaved runways:
total: 57
over 3,047 m: 1
914 to 1,523 m: 3
under 914 m: 53 (2007)
Pipelines: gas 1,695 km; oil 228 km; refined products 121 km (2007)
Railways:
total: 1,771 km
broad gauge: 1,749 km 1.524-m gauge (122 km electrified)
standard gauge: 22 km 1.435-m gauge (2006)
Roadways: *total:* 79,497 km
paved: 70,549 km (includes 417 km of expressways)
unpaved: 8,948 km (2005)
Waterways: 425 km (2005)
Merchant marine:
total: 50 ships (1000 GRT or over) 363,795 GRT/366,624 DWT
by type: bulk carrier 4, cargo 22, chemical tanker 1, container 1, passenger/cargo 5, petroleum tanker 1, refrigerated cargo 16
foreign-owned: 9 (Denmark 9)
registered in other countries: 20 (Antigua and Barbuda 6, North Korea 1, Norway 1, Panama 5, St Vincent and The Grenadines 7, unknown 3) (2007)
Ports and terminals: Klaipeda

MILITARY

Military branches: Ground Forces, Naval Force, Lithuanian Military Air Forces, National Defense Volunteer Forces (2005)
Military service age and obligation: 19–45 years of age for compulsory military service; 18 years of age for volun-

teers; 12-month conscript service obligation (2006)
Manpower available for military service: males age 16–49: 915,187
females age 16–49: 906,097 (2008 est.)
Manpower fit for military service:
males age 16–49: 678,434
females age 16–49: 749,483 (2008 est.)
Manpower reaching militarily significant age annually:
males age 16–49: 25,907
females age 16–49: 24,735 (2008 est.)
Military expenditures—percent of GDP: 1.2% (2006; 1.23% 2007 est.)

TRANSNATIONAL ISSUES

Disputes—international: Lithuania and Russia committed to demarcating their boundary in 2006 in accordance with the land and maritime treaty ratified by Russia in May 2003 and by Lithuania in 1999; Lithuania operates a simplified transit regime for Russian nationals traveling from the Kaliningrad coastal exclave into Russia, while still conforming, as a EU member state having an external border with a non-EU member, to strict Schengen border rules; the Latvian parliament has not ratified its 1998 maritime boundary treaty with Lithuania, primarily due to concerns over potential hydrocarbons; as of January 2007, ground demarcation of the boundary with Belarus was complete and mapped with final ratification documents in preparation
Illicit drugs: transshipment and destination point for cannabis, cocaine, ecstasy, and opiates from Southwest Asia, Latin America, Western Europe, and neighboring Baltic countries; growing production of high-quality amphetamines, but limited production of cannabis, methamphetamines; susceptible to money laundering despite changes to banking legislation

LUXEMBOURG

INTRODUCTION

Background: Founded in 963, Luxembourg became a grand duchy in 1815 and an independent state under the Netherlands. It lost more than half of its territory to Belgium in 1839, but gained a larger measure of autonomy. Full independence was attained in 1867. Overrun by Germany in both World Wars, it ended its neutrality in 1948 when it entered into the Benelux Customs Union and when it joined NATO the following year. In 1957, Luxembourg became one of the six founding countries

of the European Economic Community (later the European Union), and in 1999 it joined the euro currency area.

GEOGRAPHY

Location: Western Europe, between France and Germany
Geographic coordinates: 49 45 N, 6 10 E
Map references: Europe
Area: *total:* 2,586 sq km
land: 2,586 sq km
water: 0 sq km
Area—comparative: slightly smaller than Rhode Island

Land boundaries:
total: 359 km
border countries: Belgium 148 km, France 73 km, Germany 138 km
Coastline: 0 km (landlocked)
Maritime claims: none (landlocked)
Climate: modified continental with mild winters, cool summers
Terrain: mostly gently rolling uplands with broad, shallow valleys; uplands to slightly mountainous in the north; steep slope down to Moselle flood plain in the southeast
Elevation extremes:
lowest point: Moselle River 133 m
highest point: Buurgplaatz 559 m
Natural resources: iron ore (no longer exploited), arable land
Land use:
arable land: 27.42%
permanent crops: 0.69%
other: 71.89% (includes Belgium) (2005)
Irrigated land: NA
Total renewable water resources: 1.6 cu km (2005)
Freshwater withdrawal (domestic/industrial/agricultural): *total:* 0.06 cu km/yr (42%/45%/13%)
per capita: 121 cu m/yr (1999)
Natural hazards: NA
Environment—current issues: air and water pollution in urban areas, soil pollution of farmland
Environment—international agreements: *party to:* Air Pollution, Air Pollution-Nitrogen Oxides, Air Pollution-Persistent Organic Pollutants, Air Pollution-Sulfur 85, Air Pollution-Sulfur 94, Air Pollution-Volatile Organic Compounds, Biodiversity, Climate Change, Climate Change-Kyoto Protocol, Desertification, Endangered Species, Hazardous Wastes, Law of the Sea, Marine Dumping, Ozone Layer Protection, Ship Pollution, Tropical Timber 83, Tropical Timber 94, Wetlands
signed, but not ratified: Environmental Modification
Geography—note: landlocked; the only Grand Duchy in the world

PEOPLE

Population: 486,006 (July 2008 est.)
Age structure:
0–14 years: 18.6% (male 46,729/female 43,889)
15–64 years: 66.6% (male 163,356/female 160,425)
65 years and over: 14.7% (male 29,206/female 42,401) (2008 est.)
Median age:
total: 39 years
male: 38 years
female: 40 years (2008 est.)

Population growth rate: 1.188% (2008 est.)
Birth rate: 11.77 births/1,000 population (2008 est.)
Death rate: 8.43 deaths/1,000 population (2008 est.)
Net migration rate: 8.54 migrant(s)/1,000 population (2008 est.)
Sex ratio:
at birth: 1.07 male(s)/female
under 15 years: 1.06 male(s)/female
15–64 years: 1.02 male(s)/female
65 years and over: 0.69 male(s)/female
total population: 0.97 male(s)/female (2008 est.)
Infant mortality rate:
total: 4.62 deaths/1,000 live births
male: 4.62 deaths/1,000 live births
female: 4.62 deaths/1,000 live births (2008 est.)
Life expectancy at birth:
total population: 79.18 years
male: 75.91 years
female: 82.67 years (2008 est.)
Total fertility rate: 1.78 children born/woman (2008 est.)
HIV/AIDS—adult prevalence rate: 0.2% (2001 est.)
HIV/AIDS—people living with HIV/AIDS: fewer than 500 (2003 est.)
HIV/AIDS—deaths: fewer than 100 (2003 est.)
Nationality:
noun: Luxembourger(s)
adjective: Luxembourg
Ethnic groups: Celtic base (with French and German blend), Portuguese, Italian, Slavs (from Montenegro, Albania, and Kosovo) and European (guest and resident workers)
Religions: Roman Catholic 87%, other (includes Protestant, Jewish, and Muslim) 13% (2000)
Languages: Luxembourgish (national language), German (administrative language), French (administrative language)
Literacy:
definition: age 15 and over can read and write
total population: 100%
male: 100%
female: 100% (2000 est.)

GOVERNMENT

Country name:
conventional long form: Grand Duchy of Luxembourg
conventional short form: Luxembourg
local long form: Grand Duche de Luxembourg
local short form: Luxembourg
Government type: constitutional monarchy
Capital: *name:* Luxembourg

geographic coordinates: 49 36 N, 6 07 E
time difference: UTC+1 (6 hours ahead of Washington, DC during Standard Time)
daylight saving time: +1hr, begins last Sunday in March; ends last Sunday in October
Administrative divisions: 3 districts; Diekirch, Grevenmacher, Luxembourg
Independence: 1839 (from the Netherlands)
National holiday: National Day (Birthday of Grand Duchess Charlotte) 23 June; note—the actual date of birth was 23 January 1896, but the festivities were shifted by five months to allow observance during a more favorable time of year
Constitution: 17 October 1868; occasional revisions
Legal system: based on civil law system; accepts compulsory ICJ jurisdiction
Suffrage: 18 years of age; universal and compulsory
Executive branch:
chief of state: Grand Duke HENRI (since 7 October 2000); Heir Apparent Prince GUILLAUME (son of the monarch)
head of government: Prime Minister Jean-Claude JUNCKER (since 20 January 1995); Deputy Prime Minister Jean ASSELBORN (since 31 July 2004)
cabinet: Council of Ministers recommended by the prime minister and appointed by the monarch
elections: the monarch is hereditary; following popular elections to the Chamber of Deputies, the leader of the majority party or the leader of the majority coalition is usually appointed prime minister by the monarch; the deputy prime minister is appointed by the monarch; they are responsible to the Chamber of Deputies
note: government coalition—CSV and LSAP
Legislative branch: unicameral Chamber of Deputies or Chambre des Deputes (60 seats; members are elected by popular vote to serve five-year terms)
elections: last held 13 June 2004 (next to be held by June 2009)
election results: percent of vote by party—CSV 36.1%, LSAP 23.4%, DP 16.1%, Green Party 11.6%, ADR 10%, other 2.8%; seats by party—CSV 24, LSAP 14, DP 10, Green Party 7, ADR 5
note: there is also a Council of State that serves as an advisory body to the Chamber of Deputies; the Council of State has 21 members appointed by the Grand Duke on the advice of the prime minister
Judicial branch: judicial courts and tribunals (3 Justices of the Peace, 2 district courts, and 1 Supreme Court of

Appeals); administrative courts and tribunals (State Prosecutor's Office, administrative courts and tribunals, and the Constitutional Court); judges for all courts are appointed for life by the monarch

Political parties and leaders: Alternative Democratic Reform Party or ADR [Robert MENLEN]; Christian Social People's Party or CSV [Francois BILTGEN] (also known as Christian Social Party or PCS); Democratic Party or DP [Claude MEISCH]; Green Party [Francois BAUSCH]; Luxembourg Socialist Workers' Party or LSAP [Alex BODRY]; dei Lenk/la Gauche (the Left); other minor parties

Political pressure groups and leaders: ABBL (bankers' association); ALEBA (financial sector trade union); Centrale Paysanne (federation of agricultural producers); CEP (professional sector chamber); CGFP (trade union representing civil service); Chambre de Commerce (Chamber of Commerce); Chambre des Metiers (Chamber of Artisans); FEDIL (federation of industrialists); Greenpeace (environment protection); LCGP (center-right trade union); Mouvement Ecologique (protection of ecology); OGBL (center-left trade union)

International organization participation: ACCT, ADB (nonregional members), Australia Group, Benelux, CE, EAPC, EBRD, EIB, EMU, ESA, EU, FAO, IAEA, IBRD, ICAO, ICC, ICCt, ICRM, IDA, IEA, IFAD, IFC, IFRCS, ILO, IMF, IMO, Interpol, IOC, IOM, IPU, ISO, ITSO, ITU, ITUC, MIGA, NATO, NEA, NSG, OAS (observer), OECD, OIF, OPCW, OSCE, PCA, Schengen Convention, UN, UNCTAD, UNESCO, UNHCR, UNIDO, UNIFIL, UNRWA, UPU, WCL, WCO, WEU, WHO, WIPO, WMO, WTO, ZC

Diplomatic representation in the US: *chief of mission:* Ambassador Joseph WEYLAND
chancery: 2200 Massachusetts Avenue NW, Washington, DC 20008
telephone: [1] (202) 265-4171/72
FAX: [1] (202) 328-8270
consulate(s) general: New York, San Francisco

Diplomatic representation from the US: *chief of mission:* Ambassador Ann WAGNER
embassy: 22 Boulevard Emmanuel Servais, L-2535 Luxembourg City
mailing address: American Embassy Luxembourg, Unit 1410, APO AE 09126-1410 (official mail); American Embassy Luxembourg, PSC 9, Box 9500, APO AE 09123 (personal mail)

telephone: [352] 46 01 23
FAX: [352] 46 14 01

Flag description: three equal horizontal bands of red (top), white, and light blue; similar to the flag of the Netherlands, which uses a darker blue and is shorter; design was based on the flag of France

ECONOMY

Economy—overview: This stable, high-income economy—benefiting from its proximity to France, Belgium, and Germany—features solid growth, low inflation, and low unemployment. The industrial sector, initially dominated by steel, has become increasingly diversified to include chemicals, rubber, and other products. Growth in the financial sector, which now accounts for about 28% of GDP, has more than compensated for the decline in steel. Most banks are foreign owned and have extensive foreign dealings. Agriculture is based on small family-owned farms. The economy depends on foreign and cross-border workers for about 60% of its labor force. Although Luxembourg, like all EU members, suffered from the global economic slump in the early part of this decade, the country continues to enjoy an extraordinarily high standard of living—GDP per capita ranks first in the world. After two years of strong economic growth in 2006-07, turmoil in the world financial markets will slow Luxembourg's economy in 2008, but growth will remain above the European average.

GDP (purchasing power parity): $38.56 billion (2007 est.)

GDP (official exchange rate): $50.16 billion (2007 est.)

GDP—real growth rate: 5.4% (2007 est.)

GDP—per capita (PPP): $80,500 (2007 est.)

GDP—composition by sector:
agriculture: 1%
industry: 13%
services: 86% (2005 est.)

Labor force: 205,000 of whom 121,600 are foreign cross-border workers commuting primarily from France, Belgium, and Germany (2007 est.)

Labor force—by occupation: *agriculture:* 1%
industry: 13%
services: 86% (2004 est.)

Unemployment rate: 4.4% (2007 est.)

Population below poverty line: NA%

Household income or consumption by percentage share:
lowest 10%: 3.5%
highest 10%: 23.8% (2000)

Distribution of family income—Gini index: 26 (2005)

Inflation rate (consumer prices): 2.3% (2007 est.)

Investment (gross fixed): 17.4% of GDP (2007 est.)

Budget:
revenues: $20.53 billion
expenditures: $19.27 billion (2007 est.)

Public debt: 5.4% of GDP (2007 est.)

Agriculture—products: wine, grapes, barley, oats, potatoes, wheat, fruits; dairy products, livestock products

Industries: banking and financial services, iron and steel, information technology, telecommunications, cargo transportation, food processing, chemicals, metal products, engineering, tires, glass, aluminum, tourism

Industrial production growth rate: 1.7% (2007 est.)

Electricity—production: 3.156 billion kWh (2005 est.)

Electricity—production by source:
fossil fuel: 57.3%
hydro: 25.2%
nuclear: 0%
other: 17.5% (2001)

Electricity—consumption: 6.315 billion kWh (2005 est.)

Electricity—exports: 3.131 billion kWh (2005 est.)

Electricity—imports: 6.392 billion kWh (2005 est.)

Oil—production: 0 bbl/day (2005 est.)

Oil—consumption: 64,020 bbl/day (2005 est.)

Oil—exports: 283 bbl/day (2004)

Oil—imports: 61,070 bbl/day (2004)

Oil—proved reserves: 0 bbl (1 January 2006 est.)

Natural gas—production: 0 cu m (2005 est.)

Natural gas—consumption: 1.356 billion cu m (2005 est.)

Natural gas—exports: 0 cu m (2005 est.)

Natural gas—imports: 1.356 billion cu m (2005)

Natural gas—proved reserves: 0 cu m (1 January 2006 est.)

Current account balance: $4.746 billion (2007 est.)

Exports: $19.85 billion f.o.b. (2007 est.)

Exports—commodities: machinery and equipment, steel products, chemicals, rubber products, glass

Exports—partners: Germany 19.3%, France 15.5%, Italy 9.5%, UK 9.4%, Belgium 8.8%, Spain 5.2%, Netherlands 4.4% (2006)

Imports: $24.75 billion c.i.f. (2007 est.)

Imports—commodities: minerals, metals, foodstuffs, quality consumer goods

Imports—partners: Belgium 26.4%, Germany 20.2%, China 16.8%, France

8.5%, UK 5.5%, Netherlands 4.2% (2006)

Economic aid—donor: ODA, $291 million (2006)

Reserves of foreign exchange and gold: $205.5 million (2006 est.)

Debt—external: $NA

Stock of direct foreign investment—at home: $NA

Stock of direct foreign investment—abroad: $NA

Market value of publicly traded shares: $79.4 billion (2006)

Currency (code): euro (EUR)

Currency code: EUR

Exchange rates: euros per US dollar—0.7345 (2007), 0.7964 (2006), 0.8041 (2005), 0.8054 (2004), 0.886 (2003)

Fiscal year: calendar year

COMMUNICATIONS

Telephones—main lines in use: 246,700 (2006)

Telephones—mobile cellular: 713,800 (2006)

Telephone system:
general assessment: highly developed, completely automated and efficient system, mainly buried cables
domestic: nationwide cellular telephone system; market for mobile-cellular phones is virtually saturated with roughly 150 cellular phones per 100 persons
international: country code—352

Radio broadcast stations: AM 2, FM 9, shortwave 2 (1999)

Radios: 285,000 (1997)

Television broadcast stations: 5 (1999)

Televisions: 285,000 (1998 est.)

Internet country code: .lu

Internet hosts: 132,090 (2007)

Internet Service Providers (ISPs): 8 (2000)

Internet users: 339,000 (2006)

TRANSPORTATION

Airports: 2 (2007)

Airports—with paved runways:
total: 1
over 3,047 m: 1 (2007)

Airports—with unpaved runways:
total: 1
under 914 m: 1 (2007)

Heliports: 1 (2007)

Pipelines: gas 155 km (2007)

Railways:
total: 275 km
standard gauge: 275 km 1.435-m gauge (243 km electrified) (2006)

Roadways:
total: 5,227 km
paved: 5,227 km (includes 147 km of expressways) (2004)

Waterways: 37 km (on Moselle River) (2007)

Merchant marine:
total: 45 ships (1000 GRT or over) 682,955 GRT/858,985 DWT

by type: bulk carrier 7, chemical tanker 14, container 7, liquefied gas 2, passenger 3, passenger/cargo 1, petroleum tanker 4, roll on/roll off 7
foreign-owned: 44 (Belgium 9, France 14, Germany 10, Netherlands 1, UK 7, US 3) (2007)

Ports and terminals: Mertert

MILITARY

Military branches: Army (2007)

Military service age and obligation: 17–25 years of age for male and female voluntary military service; soldiers under 18 are not deployed into combat or with peacekeeping missions; no conscription (2008)

Manpower available for military service:
males age 16–49: 116,305
females age 16–49: 114,566 (2008 est.)

Manpower fit for military service:
males age 16–49: 95,152
females age 16–49: 93,792 (2008 est.)

Manpower reaching militarily significant age annually:
males age 16–49: 3,066
females age 16–49: 2,909 (2008 est.)

Military expenditures—percent of GDP: 0.9% (2005 est.)

TRANSNATIONAL ISSUES

Disputes—international: none

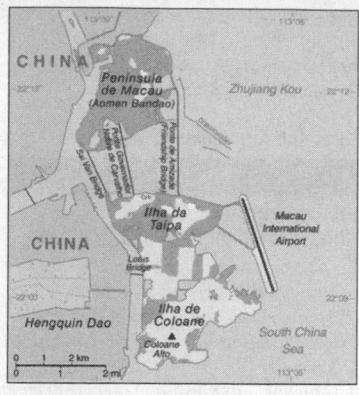

INTRODUCTION

Background: Colonized by the Portuguese in the 16th century, Macau was the first European settlement in the Far East. Pursuant to an agreement signed by China and Portugal on 13 April 1987, Macau became the Macau Special Administrative Region (SAR) of China on 20 December 1999. In this agreement, China promised that, under its "one country, two systems" formula, China's socialist economic system would not be practiced in Macau, and that Macau would enjoy a high degree of autonomy in all matters except foreign and defense affairs for the next 50 years.

GEOGRAPHY

Location: Eastern Asia, bordering the South China Sea and China
Geographic coordinates: 22 10 N, 113 33 E
Map references: Southeast Asia
Area:
total: 28.2 sq km
land: 28.2 sq km
water: 0 sq km
Area—comparative: less than one-sixth the size of Washington, DC
Land boundaries:
total: 0.34 km
regional border: China 0.34 km
Coastline: 41 km
Maritime claims: not specified
Climate: subtropical; marine with cool winters, warm summers
Terrain: generally flat
Elevation extremes:
lowest point: South China Sea 0 m
highest point: Coloane Alto 172.4 m
Natural resources: NEGL
Land use:
arable land: 0%
permanent crops: 0%
other: 100% (2005)
Irrigated land: NA

Natural hazards: typhoons
Environment—current issues: NA
Environment—international agreements: *party to:* Marine Dumping (associate member), Ship Pollution (associate member)
Geography—note: essentially urban; an area of land reclaimed from the sea measuring 5.2 sq km and known as Cotai now connects the islands of Coloane and Taipa; the island area is connected to the mainland peninsula by three bridges

PEOPLE

Population: 460,823 (July 2008 est.)
Age structure:
0–14 years: 14.7% (male 35,107/female 32,756)
15–64 years: 77.1% (male 169,317/female 186,069)
65 years and over: 8.2% (male 16,053/female 21,521) (2008 est.)
Median age:
total: 37 years
male: 36.4 years
female: 37.5 years (2008 est.)
Population growth rate: 0.83% (2008 est.)
Birth rate: 8.73 births/1,000 population (2008 est.)
Death rate: 4.72 deaths/1,000 population (2008 est.)
Net migration rate: 4.28 migrant(s)/1,000 population (2008 est.)
Sex ratio:
at birth: 1.05 male(s)/female
under 15 years: 1.07 male(s)/female
15–64 years: 0.91 male(s)/female
65 years and over: 0.75 male(s)/female
total population: 0.92 male(s)/female (2008 est.)
Infant mortality rate:
total: 4.3 deaths/1,000 live births
male: 4.49 deaths/1,000 live births
female: 4.11 deaths/1,000 live births (2008 est.)
Life expectancy at birth:
total population: 82.35 years
male: 79.52 years
female: 85.33 years (2008 est.)
Total fertility rate: 1.05 children born/woman (2008 est.)
HIV/AIDS—adult prevalence rate: NA
HIV/AIDS—people living with HIV/AIDS: NA
HIV/AIDS—deaths: NA
Nationality:
noun: Chinese
adjective: Chinese
Ethnic groups: Chinese 95.7%, Macanese (mixed Portuguese and Asian ancestry) 1%, other 3.3% (2001 census)
Religions: Buddhist 50%, Roman

Catholic 15%, none and other 35% (1997 est.)
Languages: Cantonese 87.9%, Hokkien 4.4%, Mandarin 1.6%, other Chinese dialects 3.1%, other 3% (2001 census)
Literacy:
definition: age 15 and over can read and write
total population: 91.3%
male: 95.3%
female: 87.8% (2001 census)

GOVERNMENT

Country name:
conventional long form: Macau Special Administrative Region
conventional short form: Macau
local long form: Aomen Tebie Xingzhengqu (Chinese); Regiao Administrativa Especial de Macau (Portuguese)
local short form: Aomen (Chinese); Macau (Portuguese)
Dependency status: special administrative region of China
Government type: limited democracy
Administrative divisions: none (special administrative region of China)
Independence: none (special administrative region of China)
National holiday: National Day (Anniversary of the Founding of the People's Republic of China), 1 October (1949); note—20 December 1999 is celebrated as Macau Special Administrative Region Establishment Day
Constitution: Basic Law, approved on 31 March 1993 by China's National People's Congress, is Macau's "mini-constitution"
Legal system: based on Portuguese civil law system
Suffrage: direct election 18 years of age for some non-executive positions, universal for permanent residents living in Macau for the past seven years; indirect election limited to organizations registered as "corporate voters" (257 are currently registered) and a 300-member Election Committee drawn from broad regional groupings, municipal organizations, and central government bodies
Executive branch:
chief of state: President of China HU Jintao (since 15 March 2003)
head of government: Chief Executive Edmund HO Hau-wah (since 20 December 1999)
cabinet: Executive Council consists of one government secretary, three legislators, four businessmen, one pro-Beijing unionist, and one pro-Beijing educator
elections: chief executive chosen by a

300-member Election Committee for a five-year term (eligible for a second term); election last held 29 August 2004 (next to be held in 2009)

election results: Edmund HO Hau-wah reelected received 296 votes; three members submitted blank ballots; one member was absent

Legislative branch: unicameral Legislative Assembly (29 seats; 12 members elected by popular vote, 10 by indirect vote, and 7 appointed by the chief executive; to serve four-year terms)

elections: last held 25 September 2005 (next in September 2009)

election results: percent of vote—New Democratic Macau Association 18.8%, Macau United Citizens' Association 16.6%, Union for Development 13.3%, Union for Promoting Progress 9.6%, Macau Development Alliance 9.3%, others 32.4%; seats by political group—New Democratic Macau Association 2, Macau United Citizens' Association 2, Union for Development 2, Union for Promoting Progress 2, Macau Development Alliance 1, others 3; 10 seats filled by professional and business groups; seven members appointed by chief executive

Judicial branch: Court of Final Appeal in Macau Special Administrative Region

Political parties and leaders: Civil Service Union [Jose Maria Pereira COUTINHO]; Development Union [KWAN Tsui-hang]; Macau Development Alliance [Angela LEONG On-kei]; Macau United Citizens' Association [CHAN Meng-kam]; New Democratic Macau Association [Antonio NG Kuok-cheong]; United Forces

note: there is no political party ordinance, so there are no registered political parties; politically active groups register as societies or companies

Political pressure groups and leaders: NA

International organization participation: IHO, IMF, IMO (associate), ISO (correspondent), UNESCO (associate), UNWTO (associate), UPU, WCO, WMO, WTO

Diplomatic representation in the US: none (special administrative region of China)

Diplomatic representation from the US: the US has no offices in Macau; US Consulate General in Hong Kong is accredited to Macau

Flag description: light green with a lotus flower above a stylized bridge and water in white, beneath an arc of five gold, five-pointed stars: one large in center of arc and four smaller

ECONOMY

Economy—overview: Macau's economy has enjoyed strong growth in recent years on the back of its expanding tourism and gaming sectors. Since opening up its locally-controlled casino industry to foreign competition in 2001, the territory has attracted tens of billions of dollars in foreign investment that have helped transform it into the world's largest gaming center. In 2006, Macau's gaming revenue surpassed that of the Las Vegas strip, and gaming-related taxes accounted for 75% of total government revenue. The expanding casino sector, and China's decision beginning in 2002 to relax travel restrictions, have reenergized Macau's tourism industry, which saw total visitors grow to 27 million in 2007, up 62% in three years. Macau's strong economic growth has put pressure its labor market prompting businesses to look abroad to meet their staffing needs. The resulting influx of non-resident workers, who totaled one-fifth of the workforce in 2006, has fueled tensions among some segments of the population. Macau's traditional manufacturing industry has been in a slow decline. In 2006, exports of textiles and garments generated only $1.8 billion compared to $6.9 billion in gross gaming receipts. Macau's textile industry will continue to move to the mainland because of the termination in 2005 of the Multi-Fiber Agreement, which provided a near guarantee of export markets, leaving the territory more dependent on gambling and trade-related services to generate growth. However, the Closer Economic Partnership Agreement (CEPA) between Macau and mainland China that came into effect on 1 January 2004 offers many Macau-made products tariff-free access to the mainland. Macau's currency, the Pataca, is closely tied to the Hong Kong dollar, which is also freely accepted in the territory.

GDP (purchasing power parity): $12.5 billion (2006)

GDP (official exchange rate): $14.3 billion (2006)

GDP—real growth rate: 16.6% (2006)

GDP—per capita (PPP): $28,400 (2006)

GDP—composition by sector:

agriculture: 0.1%

industry: 3.9%

services: 96% (2006 est.)

Labor force: 275,000 (2006)

Labor force—by occupation: manufacturing 11.1%, construction 11.7%, transport and communications 6.3%, wholesale and retail trade 13.7%, restau-rants and hotels 11.3%, gambling 19.8%, public sector 7.7%, financial services 2.6%, other services and agriculture 15.7% (2006)

Unemployment rate: 3.1% (2006)

Population below poverty line: NA%

Household income or consumption by percentage share:

lowest 10%: NA%

highest 10%: NA%

Inflation rate (consumer prices): 7.2% (2006)

Budget:

revenues: $4.6 billion

expenditures: $3.4 billion (2006)

Agriculture—products: only 2% of land area is cultivated, mainly by vegetable growers; fishing, mostly for crustaceans, is important; some of the catch is exported to Hong Kong

Industries: tourism, gambling, clothing, textiles, electronics, footwear, toys

Industrial production growth rate: 3.8% (3rd quarter)

Electricity—production: 1.67 billion kWh (2006)

Electricity—production by source:

fossil fuel: 100%

hydro: 0%

nuclear: 0%

other: 0% (2001)

Electricity—consumption: 2.37 billion kWh (2006)

Electricity—exports: 0 kWh (2006)

Electricity—imports: 964.4 million kWh (2006)

Oil—production: 0 bbl/day (2006 est.)

Oil—consumption: 13,920 bbl/day (2006 est.)

Oil—exports: 21 bbl/day (2005)

Oil—imports: 13,870 bbl/day (2006)

Oil—proved reserves: 0 bbl (1 January 2006 est.)

Natural gas—production: 0 cu m (2006 est.)

Natural gas—consumption: 0 cu m (2006 est.)

Natural gas—exports: 0 cu m (2006 est.)

Natural gas—imports: 0 cu m (2006)

Natural gas—proved reserves: 0 cu m (1 January 2006 est.)

Exports: $2.557 billion f.o.b.; note—includes reexports (2006)

Exports—commodities: clothing, textiles, footwear, toys, electronics, machinery and parts

Exports—partners: US 44.1%, China 14.8%, Hong Kong 11.3%, Germany 7.3%, UK 4.1% (2006)

Imports: $4.559 billion c.i.f. (2006)

Imports—commodities: raw materials and semi-manufactured goods, consumer goods (foodstuffs, beverages, tobacco), capital goods, mineral fuels and oils

Imports—partners: China 45.2%, Hong Kong 10.2%, Japan 8.4%, US 5.5%, Singapore 4.1%, France 4% (2006)
Economic aid—recipient: $13.7 million (2004)
Debt—external: $0 (2006)
Stock of direct foreign investment—at home: $6.5 billion (2006)
Stock of direct foreign investment—abroad: $1.1 billion (2006)
Market value of publicly traded shares: $413.1 million (2004)
Currency (code): pataca (MOP)
Currency code: MOP
Exchange rates: patacas per US dollar—8.011 (2007), 8.0015 (2006), 8.011 (2005), 8.022 (2004), 8.021 (2003)
Fiscal year: calendar year

Telephones—main lines in use: 178,013 (2007)
Telephones—mobile cellular: 794,323 (2007)
Telephone system:
general assessment: fairly modern communication facilities maintained for domestic and international services
domestic: termination of monopoly over mobile-cellular telephone services in 2001 spurred sharp increase in subscriptions with mobile-cellular teledensity approaching 140 per 100 persons in 2006; fixed-line teledensity about 40 per 100 persons

international: country code—853; landing point for the SEA-ME-WE-3 submarine cable network that provides links to Asia, the Middle East, and Europe; HF radiotelephone communication facility; satellite earth station—1 Intelsat (Indian Ocean)
Radio broadcast stations: AM 0, FM 2, shortwave 0 (1998)
Radios: 160,000 (1997)
Television broadcast stations: 1 (2006)
Televisions: 49,000 (1997)
Internet country code: .mo
Internet hosts: 232 (2007)
Internet Service Providers (ISPs): 1 (2000)
Internet users: 300,000 (2007)

Airports: 1 (2007)
Airports—with paved runways:
total: 1
over 3,047 m: 1 (2007)
Heliports: 1 (2007)
Roadways:
total: 384 km
paved: 384 km (2006)
Ports and terminals: Macau

Military branches: no regular military forces
Manpower available for military service:
males age 16–49: 121,825 (2008 est.)

Manpower fit for military service:
males age 16–49: 100,826 (2008 est.)
Military—note: defense is the responsibility of China

Disputes—international: none
Trafficking in persons: *current situation:* Macau is a transit and destination territory for women trafficked for the purpose of commercial sexual exploitation; most females in Macau's sizeable sex industry come from the interior regions of China or Mongolia, though a significant number also come from Russia, Eastern Europe, Thailand, and Vietnam; the majority of women in Macau's prostitution trade appear to have entered Macau and the sex trade voluntarily, though there is evidence that some are deceived or coerced into sexual servitude, often through the use of debt bondage; organized criminal syndicates are reportedly involved in bringing women to Macau, and fear of reprisals from these groups may prevent some women from seeking help
tier rating: Tier 2 Watch List—Macau is placed on the Tier 2 Watch List for failing to show evidence of increasing efforts to address trafficking since 2004
Illicit drugs: transshipment point for drugs going into mainland China; consumer of opiates and amphetamines

MACEDONIA

Background: Macedonia gained its independence peacefully from Yugoslavia in 1991, but Greece's objection to the new state's use of what it considered a Hellenic name and symbols delayed international recognition, which occurred under the provisional designa-

tion of "the Former Yugoslav Republic of Macedonia." In 1995, Greece lifted a 20-month trade embargo and the two countries agreed to normalize relations. The United States began referring to Macedonia by its constitutional name, Republic of Macedonia, in 2004 and negotiations continue between Greece and Macedonia to resolve the name issue. Some ethnic Albanians, angered by perceived political and economic inequities, launched an insurgency in 2001 that eventually won the support of the majority of Macedonia's Albanian population and led to the internationally-brokered Framework Agreement, which ended the fighting by establishing a set of new laws enhancing the rights of minorities. Fully implementating the Framework Agreement and stimulating economic growth and development continue to be challenges for Macedonia, although progress has been made on both fronts over the past several years.

Location: Southeastern Europe, north of Greece
Geographic coordinates: 41 50 N, 22 00 E
Map references: Europe
Area:
total: 25,333 sq km
land: 24,856 sq km
water: 477 sq km
Area—comparative: slightly larger than Vermont
Land boundaries:
total: 766 km
border countries: Albania 151 km, Bulgaria 148 km, Greece 246 km, Kosovo 159 km, Serbia 62 km
Coastline: 0 km (landlocked)
Maritime claims: none (landlocked)
Climate: warm, dry summers and autumns; relatively cold winters with heavy snowfall
Terrain: mountainous territory covered with deep basins and valleys; three large

lakes, each divided by a frontier line; country bisected by the Vardar River

Elevation extremes:
lowest point: Vardar River 50 m
highest point: Golem Korab (Maja e Korabit) 2,764 m

Natural resources: low-grade iron ore, copper, lead, zinc, chromite, manganese, nickel, tungsten, gold, silver, asbestos, gypsum, timber, arable land

Land use: *arable land:* 22.01%
permanent crops: 1.79%
other: 76.2% (2005)

Irrigated land: 550 sq km (2003)

Total renewable water resources: 6.4 cu km (2001)

Freshwater withdrawal (domestic/industrial/agricultural): *total:* 2.27
per capita: 1,118 cu m/yr (2000)

Natural hazards: high seismic risks

Environment—current issues: air pollution from metallurgical plants

Environment—international agreements: *party to:* Air Pollution, Biodiversity, Climate Change, Climate Change-Kyoto Protocol, Desertification, Endangered Species, Hazardous Wastes, Law of the Sea, Ozone Layer Protection, Wetlands
signed, but not ratified: none of the selected agreements

Geography—note: landlocked; major transportation corridor from Western and Central Europe to Aegean Sea and Southern Europe to Western Europe

PEOPLE

Population: 2,061,315 (July 2008 est.)

Age structure:
0–14 years: 19.5% (male 207,954/female 193,428)
15–64 years: 69.3% (male 719,708/female 708,033)
65 years and over: 11.3% (male 101,036/female 131,156) (2008 est.)

Median age: *total:* 34.8 years
male: 33.8 years
female: 35.8 years (2008 est.)

Population growth rate: 0.262% (2008 est.)

Birth rate: 12 births/1,000 population (2008 est.)

Death rate: 8.81 deaths/1,000 population (2008 est.)

Net migration rate: -0.57 migrant(s)/1,000 population (2008 est.)

Sex ratio:
at birth: 1.08 male(s)/female
under 15 years: 1.08 male(s)/female
15–64 years: 1.02 male(s)/female
65 years and over: 0.77 male(s)/female
total population: 1 male(s)/female (2008 est.)

Infant mortality rate:
total: 9.27 deaths/1,000 live births
male: 9.45 deaths/1,000 live births
female: 9.08 deaths/1,000 live births (2008 est.)

Life expectancy at birth:
total population: 74.45 years
male: 71.95 years
female: 77.13 years (2008 est.)

Total fertility rate: 1.58 children born/woman (2008 est.)

HIV/AIDS—adult prevalence rate: less than 0.1% (2001 est.)

HIV/AIDS—people living with HIV/AIDS: fewer than 200 (2003 est.)

HIV/AIDS—deaths: fewer than 100 (2003 est.)

Nationality:
noun: Macedonian(s)
adjective: Macedonian

Ethnic groups: Macedonian 64.2%, Albanian 25.2%, Turkish 3.9%, Roma (Gypsy) 2.7%, Serb 1.8%, other 2.2% (2002 census)

Religions: Macedonian Orthodox 64.7%, Muslim 33.3%, other Christian 0.37%, other and unspecified 1.63% (2002 census)

Languages: Macedonian 66.5%, Albanian 25.1%, Turkish 3.5%, Roma 1.9%, Serbian 1.2%, other 1.8% (2002 census)

Literacy: *definition:* age 15 and over can read and write
total population: 96.1%
male: 98.2%
female: 94.1% (2002 census)

GOVERNMENT

Country name:
conventional long form: Republic of Macedonia
conventional short form: Macedonia
local long form: Republika Makedonija
local short form: Makedonija
note: the provisional designation used by the UN, EU, and NATO is the Former Yugoslav Republic of Macedonia (FYROM)
former: People's Republic of Macedonia, Socialist Republic of Macedonia

Government type: parliamentary democracy

Capital: *name:* Skopje
geographic coordinates: 42 00 N, 21 26 E
time difference: UTC+1 (6 hours ahead of Washington, DC during Standard Time)
daylight saving time: +1hr, begins last Sunday in March; ends last Sunday in October

Administrative divisions: 84 municipalities (opstini, singular—opstina); Aerodrom (Skopje), Aracinovo, Berovo, Bitola, Bogdanci, Bogovinje, Bosilovo, Brvenica, Butel (Skopje), Cair (Skopje), Caska, Centar (Skopje), Centar Zupa, Cesinovo-Oblesevo, Cucer-Sandevo,

Debar, Debarca, Delcevo, Demir Hisar, Demir Kapija, Dojran, Dolneni, Drugovo, Gazi Baba (Skopje), Gevgelija, Gjorce Petrov (Skopje), Gostivar, Gradsko, Ilinden, Jegunovce, Karbinci, Karpos (Skopje), Kavadarci, Kicevo, Kisela Voda (Skopje), Kocani, Konce, Kratovo, Kriva Palanka, Krivogastani, Krusevo, Kumanovo, Lipkovo, Lozovo, Makedonska Kamenica, Makedonski Brod, Mavrovo i Rostusa, Mogila, Negotino, Novaci, Novo Selo, Ohrid, Oslomej, Pehcevo, Petrovec, Plasnica, Prilep, Probistip, Radovis, Rankovce, Resen, Rosoman, Saraj (Skopje), Sopiste, Staro Nagoricane, Stip, Struga, Strumica, Studenicani, Suto Orizari (Skopje), Sveti Nikole, Tearce, Tetovo, Valandovo, Vasilevo, Veles, Vevcani, Vinica, Vranestica, Vrapciste, Zajas, Zelenikovo, Zelino, Zrnovci
note: the 10 municipalities followed by Skopje in parentheses collectively constitute the larger Skopje Municipality

Independence: 8 September 1991 (referendum by registered voters endorsed independence from Yugoslavia)

National holiday: Ilinden Uprising Day, 2 August (1903); note—also known as Saint Elijah's Day

Constitution: adopted 17 November 1991, effective 20 November 1991; amended November 2001 by a series of new constitutional amendments strengthening minority rights and in 2005 with amendments related to the judiciary

Legal system: based on civil law system; judicial review of legislative acts; has not accepted compulsory ICJ jurisdiction

Suffrage: 18 years of age; universal

Executive branch:
chief of state: President Branko CRVENKOVSKI (since 12 May 2004)
head of government: Prime Minister Nikola GRUEVSKI (since 26 August 2006)
cabinet: Council of Ministers elected by the majority vote of all the deputies in the Assembly; note—current cabinet formed by the government coalition parties VMRO/DPMNE, NSDP, PDSh/DPA, and several small parties
elections: president elected by popular vote for a five-year term (eligible for a second term); two-round election last held 14 April and 28 April 2004 (next to be held by April 2009); prime minister elected by the Assembly following legislative elections
election results: Branko CRVENKOVSKI elected president on second-round ballot; percent of vote—Branko CRVENKOVSKI 62.7%, Sasko KEDEV 37.3%

Legislative branch: unicameral Assembly or Sobranie (120 seats; members elected by popular vote from party lists based on the percentage of the overall vote the parties gain in each of six electoral districts; serve four-year terms)

elections: last held 5 July 2006 (next to be held by July 2010)

election results: percent of vote by party—VMRO-DPMNE 33%, SDSM 22%, BDI/DUI 12%, PDSh/DPA 7%, NSDP 6%, VMRO-Narodna 6%, other 14%; seats by party—VMRO-DPMNE 45, SDSM 32, BDI/DUI 17, PDSh/DPA 11, NSDP 7, VMRO-Narodna 6, other 2

Judicial branch: Supreme Court—the Assembly appoints the judges; Constitutional Court—the Assembly appoints the judges; Republican Judicial Council—the Assembly appoints the judges

Political parties and leaders: Democratic Alliance [Pavle TRAJANOV]; Democratic League of Bosniaks [Rafet MUMINOVIC]; Democratic Party of Albanians or PDSh/DPA [Menduh THACI]; Democratic Party of Serbs [Ivan STOILJKOVIC]; Democratic Party of Turks [Kenan HASIPI]; Democratic Renewal of Macedonia [Liljana POPOVSKA]; Democratic Union of Albanians or BDSh [BardYL MAHMUTI]; Democratic Union of Vlachs for Macedonia [Mitko KOSTOV]; Democratic Union for Integration or BDI/DUI [Ali AHMETI]; Internal Macedonian Revolutionary Organization-Democratic Party for Macedonian National Unity or VMRO-DPMNE [Nikola GRUEVSKI]; Internal Macedonian Revolutionary Organization-People's Party or VMRO-Narodna [Gjorgji TRENDAFILOV]; League for Democracy [Gjorgi MARJANOVIC]; Liberal Democratic Party or LDP [Jovan MANSIEVSKI]; Liberal Party [Stojan ANDOV]; National Alternative [Harun ALIU]; National Democratic Union or BDK [Hysni SHAQIR]; New Social Democratic Party or NSDP [Tito PETKOVSKI]; Party for Democratic Prosperity or PPD/PDP [Abduljhadi VEJSELI]; Party for European Future or PEI [Fijat CANOSKI]; Party of Free Democrats or PSD [Ljubco JORDANOVSKI]; Social Democratic Alliance of Macedonia or SDSM [Radmila SEKERINSKA]; Socialist Party of Macedonia or SP [Ljubisav IVANOV-ZINGO]; Union of Romas or SR [Shaban SALIU]; United Party for Emancipation or OPE [Nezdet MUSTAFA]

Political pressure groups and leaders: Federation of Free Trade Unions [Svetlana PETROVIC]; Federation of Trade Unions [Vanco MURATOVSKI]; Trade Union of Education, Science and Culture [Dojcin CVETANOSKI]; World Macedonian Congress [Todor PETROV]

International organization participation: BIS, CE, CEI, EAPC, EBRD, FAO, IAEA, IBRD, ICAO, ICCt, ICRM, IDA, IFAD, IFC, IFRCS, ILO, IMF, IMO, Interpol, IOC, IOM (observer), IPU, ISO, ITU, ITUC, MIGA, OIF, OPCW, OSCE, PCA, PFP, SECI, UN, UNCTAD, UNESCO, UNIDO, UNIFIL, UNWTO, UPU, WCL, WCO, WHO, WIPO, WMO, WTO

Diplomatic representation in the US:
chief of mission: Ambassador Zoran JOLEVSKI
chancery: 2129 Wyoming Avenue NW, Washington, DC 20008
telephone: [1] (202) 667-0501
FAX: [1] (202) 667-2131
consulate(s) general: New York, Southfield (Michigan); note—consulate general in Chicago is due to open in 2008

Diplomatic representation from the US:
chief of mission: Ambassador Gillian A. MILOVANOVIC
embassy: Bul. Ilindenska bb, 1000 Skopje
mailing address: American Embassy Skopje, US Department of State, 7120 Skopje Place, Washington, DC 20521-7120 (pouch)
telephone: [389] 2 311-6180
FAX: [389] 2 311-7103

Flag description: a yellow sun with eight broadening rays extending to the edges of the red field

ECONOMY

Economy—overview: At independence in September 1991, Macedonia was the least developed of the Yugoslav republics, producing a mere 5% of the total federal output of goods and services. The collapse of Yugoslavia ended transfer payments from the central government and eliminated advantages from inclusion in a de facto free trade area. An absence of infrastructure, UN sanctions on the downsized Yugoslavia, and a Greek economic embargo over a dispute about the country's constitutional name and flag hindered economic growth until 1996. GDP subsequently rose each year through 2000. In 2001, during a civil conflict, the economy shrank 4.5% because of decreased trade, intermittent border closures, increased deficit spending on security needs, and investor uncertainty. Growth barely recovered in 2002 to 0.9%, then averaged 4% per year

during 2003–07, expanding to 5.1% in 2007. Macedonia has maintained macroeconomic stability with low inflation, but it has so far lagged the region in attracting foreign investment and creating jobs, despite making extensive fiscal and business sector reforms. Official unemployment remains high at nearly 35%, but may be overstated based on the existence of an extensive gray market, estimated to be more than 20 percent of GDP, that is not captured by official statistics.

GDP (purchasing power parity): $17.35 billion
note: Macedonia has a large informal sector (2007 est.)

GDP (official exchange rate): $7.497 billion (2007 est.)

GDP—real growth rate: 5% (2007 est.)

GDP—per capita (PPP): $8,500 (2007 est.)

GDP—composition by sector:
agriculture: 11.9%
industry: 28.2%
services: 59.9% (2007 est.)

Labor force: 890,000 (2007 est.)

Labor force—by occupation: *agriculture:* 19.6%
industry: 30.4%
services: 50% (September 2007)

Unemployment rate: 35% (2007 est.)

Population below poverty line: 29.8% (2006)

Household income or consumption by percentage share:
lowest 10%: 2.4%
highest 10%: 29.6% (2003)

Distribution of family income—Gini index: 39 (2003)

Inflation rate (consumer prices): 2.3% (2007 est.)

Investment (gross fixed): 17.7% of GDP (2007 est.)

Budget:
revenues: $2.508 billion
expenditures: $2.487 billion (2007 est.)

Public debt: 30.8% of GDP (2007 est.)

Agriculture—products: grapes, wine, tobacco, vegetables, fruits; milk, eggs

Industries: food processing, beverages, textiles, chemicals, iron, steel, cement, energy, pharmaceuticals

Industrial production growth rate: 3.7% (2007 est.)

Electricity—production: 6.051 billion kWh (2007)

Electricity—production by source:
fossil fuel: 83.7%
hydro: 16.3%
nuclear: 0%
other: 0% (2001)

Electricity—consumption: 8.651 billion kWh (2007)

Electricity—exports: 0 kWh (2007)

Electricity—imports: 2.6 billion kWh (2007)

Oil—production: 0 bbl/day (2007)

Oil—consumption: 21,700 bbl/day (2007)

Oil—exports: 4,134 bbl/day (2004)

Oil—imports: 23,150 bbl/day (2004)

Oil—proved reserves: 0 bbl (1 January 2008 est.)

Natural gas—production: 0 cu m (2007)

Natural gas—consumption: 102.8 million cu m (2007)

Natural gas—exports: 0 cu m (2007)

Natural gas—imports: 102.8 million cu m (2007)

Natural gas—proved reserves: 0 cu m (1 January 2008 est.)

Current account balance: -$202 million (November 200

Exports: $3.35 billion f.o.b. (2007 est.)

Exports—commodities: food, beverages, tobacco; textiles, miscellaneous manufactures, iron and steel

Exports—partners: Serbia and Montenegro 23.2%, Germany 15.6%, Greece 15.1%, Italy 9.9%, Bulgaria 5.4%, Croatia 5.2% (2006)

Imports: $4.977 billion f.o.b. (2007 est.)

Imports—commodities: machinery and equipment, automobiles, chemicals, fuels, food products

Imports—partners: Russia 15.1%, Germany 9.8%, Greece 8.5%, Serbia and Montenegro 7.5%, Bulgaria 6.7%, Italy 6% (2006)

Economic aid—recipient: $230.3 million (2005)

Reserves of foreign exchange and gold: $2.219 billion (31 December 2007)

Debt—external: $3.967 billion (31 December 2007)

Stock of direct foreign investment—at home: $2.405 billion (2007 est.)

Stock of direct foreign investment—abroad: $NA

Market value of publicly traded shares: $646 million (2005)

Currency (code): Macedonian denar (MKD)

Currency code: MKD

Exchange rates: Macedonian denars per US dollar—44.732 (2007), 48.978 (2006), 48.92 (2005), 49.41 (2004), 54.322 (2003)

Fiscal year: calendar year

COMMUNICATIONS

Telephones—main lines in use: 490,900 (2006)

Telephones—mobile cellular: 1.417 million (2006)

Telephone system:

general assessment: competition from the mobile-cellular segment of the telecommunications market has led to a drop in fixed-line telephone subscriptions

domestic: combined fixed line and mobile telephone density exceeds 90 per 100 persons

international: country code—389

Radio broadcast stations: AM 29, FM 63, shortwave 0 (2007)

Radios: 410,000 (1997)

Television broadcast stations: 52 (2007)

Televisions: 510,000 (1997)

Internet country code: .mk

Internet hosts: 6,001 (2007)

Internet Service Providers (ISPs): 6 (2000)

Internet users: 268,000 (2006)

TRANSPORTATION

Airports: 17 (2007)

Airports—with paved runways:

total: 10

2,438 to 3,047 m: 2

under 914 m: 8 (2007)

Airports—with unpaved runways:

total: 7

914 to 1,523 m: 3

under 914 m: 4 (2007)

Pipelines: gas 268 km; oil 120 km (2007)

Railways:

total: 699 km

standard gauge: 699 km 1.435-m gauge (223 km electrified) (2006)

Roadways: *total:* 8,684 km

paved: 5,540 km

unpaved: 3,144 km (1999)

MILITARY

Military branches: Army of the Republic of Macedonia (ARM): Joint Operational Command, with subordinate Air Wing (Makedonsko Voeno Vozduhoplovstvo, MVV), Special Operations Regiment (2007)

Military service age and obligation: 18 years of age for voluntary military service (2007)

Manpower available for military service:

males age 16–49: 532,856

females age 16–49: 513,684 (2008 est.)

Manpower fit for military service:

males age 16–49: 444,693

females age 16–49: 428,341 (2008 est.)

Manpower reaching militarily significant age annually:

males age 16–49: 15,141

females age 16–49: 14,434 (2008 est.)

Military expenditures—percent of GDP: 6% (2005 est.)

TRANSNATIONAL ISSUES

Disputes—international: ethnic Albanians in Kosovo object to demarcation of the boundary with Serbia in accordance with the 2000 Macedonia-Serbia and Montenegro delimitation agreement; Greece continues to reject the use of the name Macedonia or Republic of Macedonia

Refugees and internally displaced persons: IDPs: fewer than 1,000 (ethnic conflict in 2001) (2007)

Illicit drugs: major transshipment point for Southwest Asian heroin and hashish; minor transit point for South American cocaine destined for Europe; although not a financial center and most criminal activity is thought to be domestic, money laundering is a problem due to a mostly cash-based economy and weak enforcement

MADAGASCAR

INTRODUCTION

Background: Formerly an independent kingdom, Madagascar became a French colony in 1896 but regained independence in 1960. During 1992–93, free presidential and National Assembly elections were held ending 17 years of single-party rule. In 1997, in the second presidential race, Didier RATSIRAKA, the leader during the 1970s and 1980s, was returned to the presidency. The 2001 presidential election was contested between the followers of Didier RATSIRAKA and Marc RAVALOMANANA, nearly causing secession of half of the country. In April 2002, the High Constitutional Court announced RAVALOMANANA the winner. RAVALOMANANA is now in his second term following a landslide victory in the generally free and fair presidential elections of 2006.

GEOGRAPHY

Location: Southern Africa, island in the Indian Ocean, east of Mozambique

Geographic coordinates: 20 00 S, 47 00 E

Map references: Africa

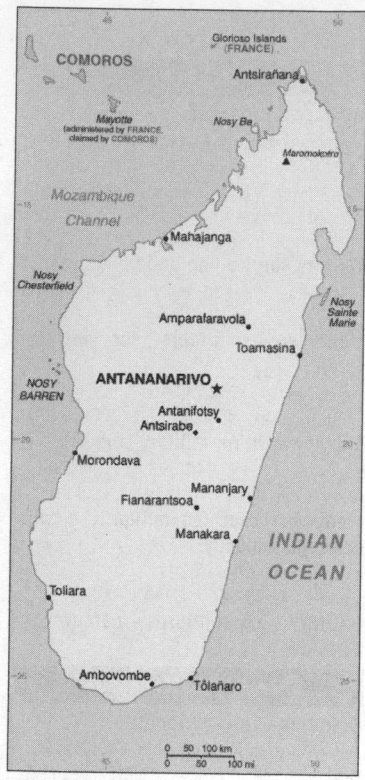

Area:
total: 587,040 sq km
land: 581,540 sq km
water: 5,500 sq km
Area—comparative: slightly less than twice the size of Arizona
Land boundaries: 0 km
Coastline: 4,828 km
Maritime claims:
territorial sea: 12 nm
contiguous zone: 24 nm
exclusive economic zone: 200 nm
continental shelf: 200 nm or 100 nm from the 2,500-m isobath
Climate: tropical along coast, temperate inland, arid in south
Terrain: narrow coastal plain, high plateau and mountains in center
Elevation extremes:
lowest point: Indian Ocean 0 m
highest point: Maromokotro 2,876 m
Natural resources: graphite, chromite, coal, bauxite, salt, quartz, tar sands, semi-precious stones, mica, fish, hydropower
Land use:
arable land: 5.03%
permanent crops: 1.02%
other: 93.95% (2005)
Irrigated land: 10,860 sq km (2003)
Total renewable water resources: 337 cu km (1984)
Freshwater withdrawal (domestic/industrial/agricultural): *total:* 14.96 cu

km/yr (3%/2%/96%)
per capita: 804 cu m/yr (2000)
Natural hazards: periodic cyclones, drought, and locust infestation
Environment—current issues: soil erosion results from deforestation and overgrazing; desertification; surface water contaminated with raw sewage and other organic wastes; several endangered species of flora and fauna unique to the island
Environment—international agreements: *party to:* Biodiversity, Climate Change, Climate Change-Kyoto Protocol, Desertification, Endangered Species, Hazardous Wastes, Law of the Sea, Marine Life Conservation, Ozone Layer Protection, Ship Pollution, Wetlands
signed, but not ratified: none of the selected agreements
Geography—note: world's fourth-largest island; strategic location along Mozambique Channel

PEOPLE

Population: 20,042,551 (July 2008 est.)
Age structure:
0–14 years: 43.7% (male 4,408,615/female 4,349,862)
15–64 years: 53.2% (male 5,298,805/female 5,371,764)
65 years and over: 3.1% (male 275,087/female 338,418) (2008 est.)
Median age:
total: 17.9 years
male: 17.7 years
female: 18.1 years (2008 est.)
Population growth rate: 3.005% (2008 est.)
Birth rate: 38.38 births/1,000 population (2008 est.)
Death rate: 8.32 deaths/1,000 population (2008 est.)
Net migration rate: NA
Sex ratio:
at birth: 1.03 male(s)/female
under 15 years: 1.01 male(s)/female
15–64 years: 0.99 male(s)/female
65 years and over: 0.81 male(s)/female
total population: 0.99 male(s)/female (2008 est.)
Infant mortality rate:
total: 55.59 deaths/1,000 live births
male: 60.59 deaths/1,000 live births
female: 50.45 deaths/1,000 live births (2008 est.)
Life expectancy at birth:
total population: 62.52 years
male: 60.58 years
female: 64.51 years (2008 est.)
Total fertility rate: 5.19 children born/woman (2008 est.)
HIV/AIDS—adult prevalence rate: 1.7% (2003 est.)

HIV/AIDS—people living with HIV/AIDS: 140,000 (2003 est.)
HIV/AIDS—deaths: 7,500 (2003 est.)
Major infectious diseases:
degree of risk: very high
food or waterborne diseases: bacterial and protozoal diarrhea, hepatitis A, and typhoid fever
vectorborne diseases: chikungunya, malaria, and plague
water contact disease: schistosomiasis (2008)
Nationality:
noun: Malagasy (singular and plural)
adjective: Malagasy
Ethnic groups: Malayo-Indonesian (Merina and related Betsileo), Cotiers (mixed African, Malayo-Indonesian, and Arab ancestry—Betsimisaraka, Tsimihety, Antaisaka, Sakalava), French, Indian, Creole, Comoran
Religions: indigenous beliefs 52%, Christian 41%, Muslim 7%
Languages: English (official), French (official), Malagasy (official)
Literacy: *definition:* age 15 and over can read and write
total population: 68.9%
male: 75.5%
female: 62.5% (2003 est.)

GOVERNMENT

Country name:
conventional long form: Republic of Madagascar
conventional short form: Madagascar
local long form: Republique de Madagascar/Repoblikan'i Madagasikara
local short form: Madagascar/Madagasikara
former: Malagasy Republic
Government type: republic
Capital: *name:* Antananarivo
geographic coordinates: 18 55 S, 47 31 E
time difference: UTC+3 (8 hours ahead of Washington, DC during Standard Time)
Administrative divisions: 6 provinces (faritany); Antananarivo, Antsiranana, Fianarantsoa, Mahajanga, Toamasina, Toliara
Independence: 26 June 1960 (from France)
National holiday: Independence Day, 26 June (1960)
Constitution: 19 August 1992 by national referendum
Legal system: based on French civil law system and traditional Malagasy law; accepts compulsory ICJ jurisdiction with reservations
Suffrage: 18 years of age; universal
Executive branch:
chief of state: President Marc RAVALO-MANANA (since 6 May 2002)
head of government: Prime Minister

Charles RABEMANANJARA (25 January 2007)

cabinet: Council of Ministers appointed by the prime minister

elections: president elected by popular vote for a five-year term (eligible for a second term); election last held 3 December 2006 (next to be held in December 2011); prime minister appointed by the president

election results: percent of vote—Marc RAVALOMANANA 54.8%, Jean LAHINIRIKO 11.7%, Roland RATSIRAKA 10.1%, Herizo RAZAFIMAHALEO 9.1%, Norbert RATSIRAHONANA 4.2%, Ny Hasina ANDRIAMANJATO 4.2%, Elia RAVELOMANANTSOA 2.6%, Pety RAKOTONIAINA 1.7%, other 1.6%

Legislative branch: bicameral legislature consists of a National Assembly or Assemblee Nationale (127 seats—reduced from 160 seats by an April 2007 national referendum; members are elected by popular vote to serve four-year terms) and a Senate or Senat (100 seats; two-thirds of the seats filled by regional assemblies; the remaining one-third of seats appointed by the president; to serve four-year terms)

elections: National Assembly—last held 23 September 2007 (next to be held in 2011)

election results: National Assembly—percent of vote by party—NA; seats by party—TIM 106, LEADER/Fanilo 1, independents 20

Judicial branch: Supreme Court or Cour Supreme; High Constitutional Court or Haute Cour Constitutionnelle

Political parties and leaders: Association for the Rebirth of Madagascar or AREMA [Pierrot RAJAONARIVELO]; Democratic Party for Union in Madagascar or PSDUM [Jean LAHINIRIKO]; Economic Liberalism and Democratic Action for National Recovery or LEADER/Fanilo [Herizo RAZAFIMAHALEO]; Fihaonana Party or FP [Guy-Willy RAZANAMASY]; I Love Madagascar or TIM [Marc RAVALOMANANA]; Renewal of the Social Democratic Party or RPSD [Evariste MARSON]

Political pressure groups and leaders: Committee for the Defense of Truth and Justice or KMMR; Committee for National Reconciliation or CRN [Albert Zafy]; National Council of Christian Churches or FFKM

International organization participation: ACCT, ACP, AfDB, AU, COMESA, FAO, G-77, IAEA, IBRD, ICAO, ICC, ICCt (signatory), ICRM, IDA, IFAD, IFC, IFRCS, ILO, IMF, IMO, InOC, Interpol, IOC, IOM, IPU, ISO (correspondent), ITSO, ITU, ITUC, MIGA, NAM, OIF, OPCW, SADC, UN, UNCTAD, UNESCO, UNHCR, UNIDO, UNWTO, UPU, WCL, WCO, WFTU, WHO, WIPO, WMO, WTO

Diplomatic representation in the US:
chief of mission: Ambassador Jocelyn Bertin RADIFERA
chancery: 2374 Massachusetts Avenue NW, Washington, DC 20008
telephone: [1] (202) 265-5525, 5526
FAX: [1] (202) 265-3034
consulate(s) general: New York

Diplomatic representation from the US:
chief of mission: Ambassador R. Niels MARQUARDT
embassy: 14-16 Rue Rainitovo, Antsahavola, Antananarivo 101
mailing address: B. P. 620, Antsahavola, Antananarivo
telephone: [261] (20) 22-212-57, 22-212-73, 22-209-56
FAX: [261] (20) 22-345-39

Flag description: two equal horizontal bands of red (top) and green with a vertical white band of the same width on hoist side

ECONOMY

Economy—overview: Having discarded past socialist economic policies, Madagascar has since the mid 1990s followed a World Bank- and IMF-led policy of privatization and liberalization. This strategy placed the country on a slow and steady growth path from an extremely low level. Agriculture, including fishing and forestry, is a mainstay of the economy, accounting for more than one-fourth of GDP and employing 80% of the population. Exports of apparel have boomed in recent years primarily due to duty-free access to the US. Deforestation and erosion, aggravated by the use of firewood as the primary source of fuel, are serious concerns. President RAVALOMANANA has worked aggressively to revive the economy following the 2002 political crisis, which triggered a 12% drop in GDP that year. Poverty reduction and combating corruption will be the centerpieces of economic policy for the next few years.

GDP (purchasing power parity): $18.12 billion (2007 est.)
GDP (official exchange rate): $7.322 billion (2007 est.)
GDP—real growth rate: 6.3% (2007 est.)
GDP—per capita (PPP): $1,100 (2007 est.)
GDP—composition by sector:
agriculture: 26.8%
industry: 15.8%
services: 57.4% (2007 est.)
Labor force: 7.3 million (2000)
Population below poverty line: 50% (2004 est.)
Household income or consumption by percentage share:
lowest 10%: 1.9%
highest 10%: 36.6% (2001)
Distribution of family income—Gini index: 47.5 (2001)
Inflation rate (consumer prices): 10.3% (2007 est.)
Investment (gross fixed): 25.9% of GDP (2007 est.)
Budget:
revenues: $1.227 billion
expenditures: $1.629 billion (2007 est.)
Agriculture—products: coffee, vanilla, sugarcane, cloves, cocoa, rice, cassava (tapioca), beans, bananas, peanuts; livestock products
Industries: meat processing, seafood, soap, breweries, tanneries, sugar, textiles, glassware, cement, automobile assembly plant, paper, petroleum, tourism
Industrial production growth rate: 6% (2007 est.)
Electricity—production: 1.046 billion kWh (2005)
Electricity—production by source:
fossil fuel: 36.1%
hydro: 63.9%
nuclear: 0%
other: 0% (2001)
Electricity—consumption: 973.2 million kWh (2005)
Electricity—exports: 0 kWh (2005)
Electricity—imports: 0 kWh (2005)
Oil—production: 92.18 bbl/day (2005 est.)
Oil—consumption: 17,000 bbl/day (2005 est.)
Oil—exports: 363.9 bbl/day (2004)
Oil—imports: 17,830 bbl/day (2004)
Oil—proved reserves: 0 bbl (1 January 2006 est.)
Natural gas—production: 0 cu m (2005 est.)
Natural gas—consumption: 0 cu m (2005 est.)
Natural gas—exports: 0 cu m (2005 est.)
Natural gas—imports: 0 cu m (2005)
Natural gas—proved reserves: 0 cu m (1 January 2006 est.)
Current account balance: -$1.106 billion (2007 est.)
Exports: $989 million f.o.b. (2007 est.)
Exports—commodities: coffee, vanilla, shellfish, sugar, cotton cloth, chromite, petroleum products
Exports—partners: France 32.2%, US 25.3%, Germany 6%, Italy 5%, UK 4.1% (2006)

Imports: $1.933 billion f.o.b. (2007 est.)

Imports—commodities: capital goods, petroleum, consumer goods, food

Imports—partners: France 14.5%, China 12%, Iran 9.3%, Mauritius 5.6%, Hong Kong 4.7% (2006)

Economic aid—recipient: $929.2 million (2005)

Reserves of foreign exchange and gold: $821 million (31 December 2007 est.)

Debt—external: $4.6 billion (2002)

Stock of direct foreign investment—at home: $NA

Stock of direct foreign investment—abroad: $NA

Market value of publicly traded shares: $NA

Currency (code): ariary (MGA)

Currency code: MGF

Exchange rates: Malagasy ariary per US dollar—1,880 (2007), 2,161.4 (2006), 2,003 (2005), 1,868.9 (2004), 1,238.3 (2003)

Fiscal year: calendar year

COMMUNICATIONS

Telephones—main lines in use: 129,800 (2006)

Telephones—mobile cellular: 1.046 million (2006)

Telephone system:

general assessment: system is above average for the region; Antananarivo's main telephone exchange modernized, but the rest of the analogue-based telephone system is poorly developed; planning to add 50,000 new private-subscriber fixed lines beginning in 2005

domestic: combined fixed-line and mobile telephone density only about 7 per 100 persons

international: country code—261; submarine cable to Bahrain; satellite earth stations—2 (1 Intelsat—Indian Ocean, 1 Intersputnik—Atlantic Ocean region)

Radio broadcast stations: AM 2 (plus a number of repeater stations), FM 9, shortwave 6 (2001)

Radios: 3.05 million (1997)

Television broadcast stations: 1 (plus 36 repeaters) (2001)

Televisions: 325,000 (1997)

Internet country code: .mg

Internet hosts: 9,734 (2007)

Internet Service Providers (ISPs): 2 (2000)

Internet users: 110,000 (2006)

TRANSPORTATION

Airports: 104 (2007)

Airports—with paved runways:

total: 27

over 3,047 m: 1

2,438 to 3,047 m: 2

1,524 to 2,437 m: 6

914 to 1,523 m: 17

under 914 m: 1 (2007)

Airports—with unpaved runways:

total: 77

1,524 to 2,437 m: 2

914 to 1,523 m: 41

under 914 m: 34 (2007)

Railways: *total:* 854 km

narrow gauge: 854 km 1.000-m gauge (2006)

Roadways: *total:* 49,827 km

paved: 5,780 km

unpaved: 44,047 km (1999)

Waterways: 600 km (2006)

Merchant marine:

total: 9 ships (1000 GRT or over) 13,896 GRT/18,466 DWT

by type: cargo 5, passenger/cargo 2, petroleum tanker 2 (2007)

Ports and terminals: Antsiranana, Mahajanga, Toamasina, Toliara

MILITARY

Military branches: People's Armed Forces: Intervention Force, Development Force, and Aeronaval Force (navy and air); National Gendarmerie

Military service age and obligation: 18–50 years of age for compulsory military service; 18-month conscript service obligation (either military or equivalent civil service) (2006)

Manpower available for military service:

males age 16–49: 4,443,341

females age 16–49: 4,441,124 (2008 est.)

Manpower fit for military service:

males age 16–49: 3,034,600

females age 16–49: 3,271,732 (2008 est.)

Manpower reaching militarily significant age annually:

males age 16–49: 230,088

females age 16–49: 229,932 (2008 est.)

Military expenditures—percent of GDP: 1% (2006)

TRANSNATIONAL ISSUES

Disputes—international: claims Bassas da India, Europa Island, Glorioso Islands, and Juan de Nova Island (all administered by France)

Illicit drugs: illicit producer of cannabis (cultivated and wild varieties) used mostly for domestic consumption; transshipment point for heroin

MALAWI

INTRODUCTION

Background: Established in 1891, the British protectorate of Nyasaland became the independent nation of Malawi in 1964. After three decades of one-party rule under President Hastings Kamuzu BANDA the country held multiparty elections in 1994, under a provisional constitution that came into full effect the following year. Current President Bingu wa MUTHARIKA, elected in May 2004 after a failed attempt by the previous president to amend the constitution to permit another term, struggled to assert his authority against his predecessor and subsequently started his own party, the Democratic Progressive Party (DPP) in 2005. As president, MUTHARIKA has overseen substantial economic improvement but because of political deadlock in the legislature, his minority party has been unable to pass significant legislation, and anti-corruption measures have stalled. Population growth, increasing pressure on agricultural lands, corruption, and the spread of HIV/AIDS pose major problems for Malawi.

GEOGRAPHY

Location: Southern Africa, east of Zambia

Geographic coordinates: 13 30 S, 34 00 E

Map references: Africa

Area: *total:* 118,480 sq km

land: 94,080 sq km

water: 24,400 sq km

Area—comparative: slightly smaller than Pennsylvania

Land boundaries:

total: 2,881 km

border countries: Mozambique 1,569 km, Tanzania 475 km, Zambia 837 km

Coastline: 0 km (landlocked)

Maritime claims: none (landlocked)

Climate: sub-tropical; rainy season (November to May); dry season (May to November)

Terrain: narrow elongated plateau with rolling plains, rounded hills, some mountains

Elevation extremes:

lowest point: junction of the Shire River

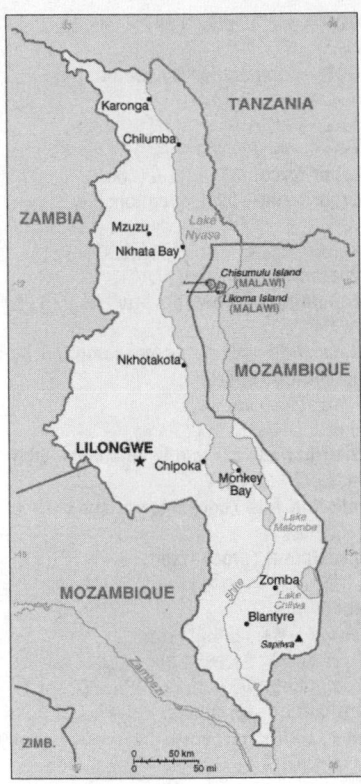

and international boundary with Mozambique 37 m

highest point: Sapitwa (Mount Mlanje) 3,002 m

Natural resources: limestone, arable land, hydropower, unexploited deposits of uranium, coal, and bauxite

Land use:
arable land: 20.68%
permanent crops: 1.18%
other: 78.14% (2005)

Irrigated land: 560 sq km (2003)

Total renewable water resources: 17.3 cu km (2001)

Freshwater withdrawal (domestic/industrial/agricultural): *total:* 1.01 cu km/yr (15%/5%/80%)
per capita: 78 cu m/yr (2000)

Natural hazards: NA

Environment—current issues: deforestation; land degradation; water pollution from agricultural runoff, sewage, industrial wastes; siltation of spawning grounds endangers fish populations

Environment—international agreements: *party to:* Biodiversity, Climate Change, Climate Change-Kyoto Protocol, Desertification, Endangered Species, Environmental Modification, Hazardous Wastes, Marine Life Conservation, Ozone Layer Protection, Ship Pollution, Wetlands
signed, but not ratified: Law of the Sea

Geography—note: landlocked; Lake Nyasa, some 580 km long, is the country's most prominent physical feature

PEOPLE

Population: 13,931,831

note: estimates for this country explicitly take into account the effects of excess mortality due to AIDS; this can result in lower life expectancy, higher infant mortality, higher death rates, lower population growth rates, and changes in the distribution of population by age and sex than would otherwise be expected (July 2008 est.)

Age structure:
0–14 years: 46% (male 3,208,112/female 3,194,600)
15–64 years: 51.4% (male 3,592,073/female 3,563,840)
65 years and over: 2.7% (male 159,450/female 213,756) (2008 est.)

Median age: *total:* 16.8 years
male: 16.7 years
female: 16.8 years (2008 est.)

Population growth rate: 2.39% (2008 est.)

Birth rate: 41.79 births/1,000 population (2008 est.)

Death rate: 17.89 deaths/1,000 population (2008 est.)

Net migration rate: NA

Sex ratio:
at birth: 1.01 male(s)/female
under 15 years: 1 male(s)/female
15–64 years: 1.01 male(s)/female
65 years and over: 0.75 male(s)/female
total population: 1 male(s)/female (2008 est.)

Infant mortality rate:
total: 90.55 deaths/1,000 live births
male: 94.69 deaths/1,000 live births
female: 86.35 deaths/1,000 live births (2008 est.)

Life expectancy at birth:
total population: 43.45 years
male: 43.74 years
female: 43.15 years (2008 est.)

Total fertility rate: 5.67 children born/woman (2008 est.)

HIV/AIDS—adult prevalence rate: 14.2% (2003 est.)

HIV/AIDS—people living with HIV/AIDS: 900,000 (2003 est.)

HIV/AIDS—deaths: 84,000 (2003 est.)

Major infectious diseases:
degree of risk: very high
food or waterborne diseases: bacterial and protozoal diarrhea, hepatitis A, and typhoid fever
vectorborne diseases: malaria and plague
water contact disease: schistosomiasis (2008)

Nationality: *noun:* Malawian(s)
adjective: Malawian

Ethnic groups: Chewa, Nyanja, Tumbuka, Yao, Lomwe, Sena, Tonga, Ngoni, Ngonde, Asian, European

Religions: Christian 79.9%, Muslim 12.8%, other 3%, none 4.3% (1998 census)

Languages: Chichewa 57.2% (official), Chinyanja 12.8%, Chiyao 10.1%, Chitumbuka 9.5%, Chisena 2.7%, Chilomwe 2.4%, Chitonga 1.7%, other 3.6% (1998 census)

Literacy:
definition: age 15 and over can read and write
total population: 62.7%
male: 76.1%
female: 49.8% (2003 est.)

GOVERNMENT

Country name:
conventional long form: Republic of Malawi
conventional short form: Malawi
local long form: Dziko la Malawi
local short form: Malawi
former: British Central African Protectorate, Nyasaland Protectorate, Nyasaland

Government type: multiparty democracy

Capital: *name:* Lilongwe
geographic coordinates: 13 59 S, 33 47 E
time difference: UTC+2 (7 hours ahead of Washington, DC during Standard Time)

Administrative divisions: 28 districts; Balaka, Blantyre, Chikwawa, Chiradzulu, Chitipa, Dedza, Dowa, Karonga, Kasungu, Likoma, Lilongwe, Machinga (Kasupe), Mangochi, Mchinji, Mulanje, Mwanza, Mzimba, Neno, Ntcheu, Nkhata Bay, Nkhotakota, Nsanje, Ntchisi, Phalombe, Rumphi, Salima, Thyolo, Zomba

Independence: 6 July 1964 (from UK)

National holiday: Independence Day (Republic Day), 6 July (1964)

Constitution: 18 May 1994

Legal system: based on English common law and customary law; judicial review of legislative acts in the Supreme Court of Appeal; accepts compulsory ICJ jurisdiction with reservations

Suffrage: 18 years of age; universal

Executive branch:
chief of state: President Bingu wa MUTHARIKA (since 24 May 2004); note—the president is both the chief of state and head of government
head of government: President Bingu wa MUTHARIKA (since 24 May 2004)
cabinet: 46-member Cabinet named by the president
elections: president elected by popular vote for a five-year term (eligible for a second term); election last held 20 May 2004 (next to be held in May 2009)

403

election results: Bingu wa MUTHARIKA elected president; percent of vote—Bingu wa MUTHARIKA 35.9%, John TEMBO 27.1%, Gwandaguluwe CHAKUAMBA 25.7%, Brown MPINGANJIRA 8.7%, Justin MALEWEZI 2.5%

Legislative branch: unicameral National Assembly (193 seats; members elected by popular vote to serve five-year terms)
elections: last held 20 May 2004 (next to be held in May 2009)
election results: percent of vote by party—NA; seats by party—MCP 56, UDF 49, independents 39, RP 15, others 25, vacancies 8

Judicial branch: Supreme Court of Appeal; High Court (chief justice appointed by the president, puisne judges appointed on the advice of the Judicial Service Commission); magistrate's courts

Political parties and leaders: Alliance for Democracy or AFORD; Congress for National Unity or CONU; Democratic Progressive Party or DPP [Bingu wa MUTHARIKA]; Malawi Congress Party or MCP [John TEMBO]; Malawi Democratic Party or MDP [Kampelo KALUA]; Malawi Forum for Unity and Development or MAFUNDE [George MNESA]; Mgwirizano Coalition or MC [Gwandaguluwe CHAKUAMBA] (coalition of MAFUNDE, MDP, MGODE, NUP, PETRA, PPM, RP); Movement for Genuine Democratic Change or MGODE [Sam Kandodo BANDA]; National Democratic Alliance or NDA [Brown MPINGANJIRA]; National Unity Party or NUP [Harry CHIUME]; People's Progressive Movement or PPM [Aleke BANDA]; People's Transformation Movement or PETRA [Kamuzu CHIBAMBO]; Republican Party or RP [Gwandaguluwe CHAKUAMBA]; United Democratic Front or UDF [Bakili MULUZI]

Political pressure groups and leaders: NA

International organization participation: ACP, AfDB, AU, C, COMESA, FAO, G-77, IAEA, IBRD, ICAO, ICCt, ICRM, IDA, IFAD, IFC, IFRCS, ILO, IMF, IMO, Interpol, IOC, ISO (correspondent), ITSO, ITU, ITUC, MIGA, NAM, OPCW, SADC, UN, UNAMID, UNCTAD, UNESCO, UNIDO, UNMIL, UNMIS, UNWTO, UPU, WCL, WCO, WFTU, WHO, WIPO, WMO, WTO

Diplomatic representation in the US:
chief of mission: Ambassador Hawa NDILOWE
chancery: 1156 15th Street, NW, Suite 320, Washington, DC 20005
telephone: [1] (202) 721-0270
FAX: [1] (202) 721-0288

Diplomatic representation from the US:
chief of mission: Ambassador Alan EASTHAM
embassy: Area 40, Plot 24, Kenyatta Road
mailing address: P. O. Box 30016, Lilongwe 3, Malawi
telephone: [265] (1) 773 166
FAX: [265] (1) 770 471

Flag description: three equal horizontal bands of black (top), red, and green with a radiant, rising, red sun centered in the black band
Government—note: no party has a majority in the fractured legislature

ECONOMY

Economy—overview: Landlocked Malawi ranks among the world's most densely populated and least developed countries. The economy is predominately agricultural with about 85% of the population living in rural areas. Agriculture accounts for more than one-third of GDP and 90% of export revenues. The performance of the tobacco sector is key to short-term growth as tobacco accounts for more than half of exports. The economy depends on substantial inflows of economic assistance from the IMF, the World Bank, and individual donor nations. In December 2007, the US granted Malawi eligibility status to receive financial support within the Millennium Challenge Corporation (MCC) initiative. Malawi will now begin a consultative process to develop a five-year program before funding can begin. In 2006, Malawi was approved for relief under the Heavily Indebted Poor Countries (HIPC) program. The government faces many challenges including developing a market economy, improving educational facilities, facing up to environmental problems, dealing with the rapidly growing problem of HIV/AIDS, and satisfying foreign donors that fiscal discipline is being tightened. In 2005, President MUTHARIKA championed an anticorruption campaign. Since 2005 President MUTHARIKA'S government has exhibited improved financial discipline under the guidance of Finance Minister Goodall GONDWE and signed a three year Poverty Reduction and Growth Facility worth $56 million with the IMF. Improved relations with the IMF lead other international donors to resume aid as well.

GDP (purchasing power parity): $10.51 billion (2007 est.)
GDP (official exchange rate): $3.538 billion (2007 est.)
GDP—real growth rate: 7.4% (2007 est.)
GDP—per capita (PPP): $800 (2007 est.)
GDP—composition by sector:
agriculture: 37.8%
industry: 18.1%
services: 44.1% (2007 est.)
Labor force: 4.5 million (2001 est.)
Labor force—by occupation: *agriculture:* 90%
industry and services: 10% (2003 est.)
Unemployment rate: NA%
Population below poverty line: 53% (2004)
Household income or consumption by percentage share:
lowest 10%: 2.9%
highest 10%: 31.8% (2004)
Distribution of family income—Gini index: 39 (2004)
Inflation rate (consumer prices): 8.1% (2007 est.)
Investment (gross fixed): 8.4% of GDP (2007 est.)
Budget:
revenues: $1.128 billion
expenditures: $1.185 billion (2007 est.)
Public debt: 50.6% of GDP (2007 est.)
Agriculture—products: tobacco, sugarcane, cotton, tea, corn, potatoes, cassava (tapioca), sorghum, pulses, groundnuts, Macadamia nuts; cattle, goats
Industries: tobacco, tea, sugar, sawmill products, cement, consumer goods
Industrial production growth rate: 4.4% (2007 est.)
Electricity—production: 1.397 billion kWh (2005)
Electricity—production by source:
fossil fuel: 3.3%
hydro: 96.7%
nuclear: 0%
other: 0% (2001)
Electricity—consumption: 1.299 billion kWh (2005)
Electricity—exports: 0 kWh (2005)
Electricity—imports: 0 kWh (2005)
Oil—production: 0 bbl/day (2005 est.)
Oil—consumption: 6,000 bbl/day (2005 est.)
Oil—exports: 0 bbl/day (2004)
Oil—imports: 6,263 bbl/day (2004)
Oil—proved reserves: 0 bbl (1 January 2006 est.)
Natural gas—production: 0 cu m (2005 est.)
Natural gas—consumption: 0 cu m (2005 est.)
Natural gas—exports: 0 cu m (2005 est.)
Natural gas—imports: 0 cu m (2005)
Natural gas—proved reserves: 0 cu m (1 January 2006 est.)
Current account balance: -$113 million (2007 est.)
Exports: $604 million f.o.b. (2007 est.)

Exports—commodities: tobacco 53%, tea, sugar, cotton, coffee, peanuts, wood products, apparel

Exports—partners: South Africa 12.6%, Germany 9.7%, Egypt 9.6%, US 9.5%, Zimbabwe 8.5%, Russia 5.4%, Netherlands 4.4% (2006)

Imports: $866 million f.o.b. (2007 est.)

Imports—commodities: food, petroleum products, semimanufactures, consumer goods, transportation equipment

Imports—partners: South Africa 34.1%, India 8%, Zambia 7.6%, US 6.3%, Tanzania 5.8%, Germany 4.5%, China 4.2% (2006)

Economic aid—recipient: $575.3 million (2005)

Reserves of foreign exchange and gold: $217 million (31 December 2007 est.)

Debt—external: $894 million (31 December 2007 est.)

Stock of direct foreign investment—at home: $NA

Stock of direct foreign investment—abroad: $NA

Market value of publicly traded shares: $NA

Currency (code): Malawian kwacha (MWK)

Currency code: MWK

Exchange rates: Malawian kwachas per US dollar—141.12 (2007), 135.96 (2006), 108.894 (2005), 108.898 (2004), 97.433 (2003)

Fiscal year: 1 July—30 June

COMMUNICATIONS

Telephones—main lines in use: 102,700 (2005)

Telephones—mobile cellular: 429,300 (2005)

Telephone system:

general assessment: rudimentary

domestic: fixed-line subscribership remains less than 1 per 100 persons; privatization of Malawi Telecommunications (MTL), a necessary step in bringing improvement to telecommunications services, completed in 2006; mobile-cellular services are expanding but cellular network coverage is limited and is based around the main urban areas

international: country code—265; satellite earth stations—2 Intelsat (1 Indian Ocean, 1 Atlantic Ocean)

Radio broadcast stations: AM 9, FM 5 (plus 15 repeater stations), shortwave 2 (plus one shortwave station on standby) (2001)

Radios: 2.6 million (1997)

Television broadcast stations: 1 (2001)

Televisions: NA

Internet country code: .mw

Internet hosts: 347 (2007)

Internet Service Providers (ISPs): 3 (2002)

Internet users: 59,700 (2006)

TRANSPORTATION

Airports: 39 (2007)

Airports—with paved runways:

total: 6

over 3,047 m: 1

1,524 to 2,437 m: 1

914 to 1,523 m: 4 (2007)

Airports—with unpaved runways:

total: 33

1,524 to 2,437 m: 1

914 to 1,523 m: 16

under 914 m: 16 (2007)

Railways:

total: 797 km

narrow gauge: 797 km 1.067-m gauge (2006)

Roadways:

total: 15,451 km

paved: 6,956 km

unpaved: 8,495 km (2003)

Waterways: 700 km (on Lake Nyasa (Lake Malawi) and Shire River) (2007)

Ports and terminals: Chipoka, Monkey Bay, Nkhata Bay, Nkhotakota, Chilumba

MILITARY

Military branches: Malawi Armed Forces: Army (includes Air Wing and Naval Detachment) (2007)

Military service age and obligation: 18 years of age for voluntary military service; standard obligation is 2 years of active duty and 5 years of reserve service (2007)

Manpower available for military service:

males age 16–49: 3,050,444 (2008 est.)

Manpower fit for military service:

males age 16–49: 1,676,117 (2008 est.)

Military expenditures—percent of GDP: 1.3% (2006)

TRANSNATIONAL ISSUES

Disputes—international: disputes with Tanzania over the boundary in Lake Nyasa (Lake Malawi) and the meandering Songwe River remain dormant

MALAYSIA

INTRODUCTION

Background: During the late 18th and 19th centuries, Great Britain established colonies and protectorates in the area of current Malaysia; these were occupied by Japan from 1942 to 1945. In 1948, the British-ruled territories on the Malay Peninsula formed the Federation of Malaya, which became independent in 1957. Malaysia was formed in 1963 when the former British colonies of Singapore and the East Malaysian states of Sabah and Sarawak on the northern coast of Borneo joined the Federation. The first several years of the country's history were marred by a Communist insurgency, Indonesian confrontation with Malaysia, Philippine claims to Sabah, and Singapore's secession from the Federation in 1965. During the 22-year term of Prime Minister MAHATHIR bin Mohamad (1981–2003), Malaysia was successful in diversifying its economy from dependence on exports of raw materials to expansion in manufacturing, services, and tourism.

GEOGRAPHY

Location: Southeastern Asia, peninsula bordering Thailand and northern one-third of the island of Borneo, bordering Indonesia, Brunei, and the South China Sea, south of Vietnam

Geographic coordinates: 2 30 N, 112 30 E

Map references: Southeast Asia

Area:

total: 329,750 sq km

land: 328,550 sq km

water: 1,200 sq km

Area—comparative: slightly larger than New Mexico

Land boundaries:

total: 2,669 km

border countries: Brunei 381 km, Indonesia 1,782 km, Thailand 506 km

Coastline: 4,675 km (Peninsular Malaysia 2,068 km, East Malaysia 2,607 km)

Maritime claims:

territorial sea: 12 nm

exclusive economic zone: 200 nm

continental shelf: 200 m depth or to the depth of exploitation; specified boundary in the South China Sea

Climate: tropical; annual southwest (April to October) and northeast (October to February) monsoons

Terrain: coastal plains rising to hills and mountains

Elevation extremes:

lowest point: Indian Ocean 0 m

highest point: Gunung Kinabalu 4,100 m

Natural resources: tin, petroleum, timber, copper, iron ore, natural gas, bauxite

Land use: *arable land:* 5.46%

permanent crops: 17.54%

other: 77% (2005)

Irrigated land: 3,650 sq km (2003)

Total renewable water resources: 580 cu km (1999)

Freshwater withdrawal (domestic/industrial/agricultural): *total:* 9.02 cu km/yr (17%/21%/62%)

per capita: 356 cu m/yr (2000)

Natural hazards: flooding, landslides, forest fires

Environment—current issues: air pollution from industrial and vehicular emissions; water pollution from raw sewage; deforestation; smoke/haze from Indonesian forest fires

Environment—international agreements: *party to:* Biodiversity, Climate Change, Climate Change-Kyoto Protocol, Desertification, Endangered Species, Hazardous Wastes, Law of the Sea, Marine Life Conservation, Ozone Layer Protection, Ship Pollution, Tropical Timber 83, Tropical Timber 94, Wetlands

Geography—note: strategic location along Strait of Malacca and southern South China Sea

PEOPLE

Population: 25,274,133 (July 2008 est.)

Age structure:

0–14 years: 31.8% (male 4,135,013/female 3,898,761)

15–64 years: 63.3% (male 8,026,755/female 7,965,332)

65 years and over: 4.9% (male 548,970/female 699,302) (2008 est.)

Median age:

total: 24.6 years

male: 24 years

female: 25.3 years (2008 est.)

Population growth rate: 1.742% (2008 est.)

Birth rate: 22.44 births/1,000 population (2008 est.)

Death rate: 5.02 deaths/1,000 population (2008 est.)

Net migration rate: NA

note: does not reflect net flow of an unknown number of illegal immigrants from other countries in the region (2008 est.)

Sex ratio: *at birth:* 1.07 male(s)/female

under 15 years: 1.06 male(s)/female

15–64 years: 1.01 male(s)/female

65 years and over: 0.79 male(s)/female

total population: 1.01 male(s)/female (2008 est.)

Infant mortality rate:

total: 16.39 deaths/1,000 live births

male: 18.92 deaths/1,000 live births

female: 13.68 deaths/1,000 live births (2008 est.)

Life expectancy at birth:

total population: 73.03 years

male: 70.32 years

female: 75.94 years (2008 est.)

Total fertility rate: 2.98 children born/woman (2008 est.)

HIV/AIDS—adult prevalence rate: 0.4% (2003 est.)

HIV/AIDS—people living with HIV/AIDS: 52,000 (2003 est.)

HIV/AIDS—deaths: 2,000 (2003 est.)

Major infectious diseases:

degree of risk: high

food or waterborne diseases: bacterial diarrhea, hepatitis A, and typhoid fever

vectorborne diseases: dengue fever and malaria

note: highly pathogenic H5N1 avian influenza has been identified in this country; it poses a negligible risk with extremely rare cases possible among US citizens who have close contact with birds (2008)

Nationality:

noun: Malaysian(s)

adjective: Malaysian

Ethnic groups: Malay 50.4%, Chinese 23.7%, indigenous 11%, Indian 7.1%, others 7.8% (2004 est.)

Religions: Muslim 60.4%, Buddhist 19.2%, Christian 9.1%, Hindu 6.3%, Confucianism, Taoism, other traditional Chinese religions 2.6%, other or unknown 1.5%, none 0.8% (2000 census)

Languages: Bahasa Malaysia (official), English, Chinese (Cantonese, Mandarin, Hokkien, Hakka, Hainan, Foochow), Tamil, Telugu, Malayalam, Panjabi, Thai

note: in East Malaysia there are several indigenous languages; most widely spoken are Iban and Kadazan

Literacy:

definition: age 15 and over can read and write

total population: 88.7%

male: 92%

female: 85.4% (2000 census)

GOVERNMENT

Country name:

conventional long form: none

conventional short form: Malaysia

local long form: none

local short form: Malaysia

former: Federation of Malaya

Government type: constitutional monarchy

note: nominally headed by paramount ruler and a bicameral Parliament consisting of a nonelected upper house and an elected lower house; all Peninsular Malaysian states have hereditary rulers except Melaka and Pulau Pinang (Penang); those two states along with Sabah and Sarawak in East Malaysia have governors appointed by government; powers of state governments are limited by federal constitution; under terms of federation, Sabah and Sarawak retain certain constitutional prerogatives (e.g., right to maintain their own immigration controls); Sabah holds 25 seats in House of Representatives; Sarawak has 31 seats

Capital: *name:* Kuala Lumpur

geographic coordinates: 3 10 N, 101 42 E

time difference: UTC+8 (13 hours ahead of Washington, DC during Standard Time)

note: Putrajaya is referred to as administrative center not capital; Parliament meets in Kuala Lumpur

Administrative divisions: 13 states (negeri-negeri, singular—negeri) Johor, Kedah, Kelantan, Melaka, Negeri Sembilan, Pahang, Perak, Perlis, Pulau Pinang, Sabah, Sarawak, Selangor, and Terengganu; and one federal territory (wilayah persekutuan) with three components, city of Kuala Lumpur, Labuan, and Putrajaya

Independence: 31 August 1957 (from UK)

National holiday: Independence Day/Malaysia Day, 31 August (1957)

Constitution: 31 August 1957 (amended many times, latest in 2007)

Legal system: based on English common law; judicial review of legislative acts in the Supreme Court at request of supreme head of the federation; Islamic law is applied to Muslims in matters of family law and religion; has not accepted compulsory ICJ jurisdiction

Suffrage: 21 years of age; universal

Executive branch:

chief of state: Paramount Ruler Sultan MIZAN Zainal Abidin (since 13 December 2006)

head of government: Prime Minister ABDULLAH bin Ahmad Badawi (since 31 October 2003); Deputy Prime Minister Mohamed NAJIB bin Abdul Razak (since 7 January 2004)

cabinet: Cabinet appointed by the prime minister from among the members of Parliament with consent of the paramount ruler

elections: paramount ruler elected by and from the hereditary rulers of nine of the states for five-year terms; election last held on 3 November 2006 (next to be held in 2011); prime minister designated from among the members of the House of Representatives; following legislative elections, the leader of the party that wins a plurality of seats in the House of Representatives becomes prime minister

election results: Sultan MIZAN Zainal Abidin elected paramount ruler

note: position of paramount ruler is primarily ceremonial; in practice, selection is based on principle of rotation among rulers of states

Legislative branch: bicameral Parliament or Parlimen consists of Senate or Dewan Negara (70 seats; 44 appointed by paramount ruler, 26 elected by 13 state legislatures; to serve three-year terms with limit of two terms) and House of Representatives or Dewan Rakyat (222 seats; members elected by popular vote to serve five-year terms)

elections: House of Representatives—last held on 8 March 2008 (next to be held by March 2013)

election results: House of Representatives—percent of vote—BN coalition 50.3%, opposition parties 46.8%, others 2.9%; seats—BN coalition 140, opposition parties 82

Judicial branch: Civil Courts include Federal Court, Court of Appeal, High Court of Malaya on peninsula Malaysia, and High Court of Sabah and Sarawak in states of Borneo (judges appointed by the paramount ruler on the advice of the prime minister); Sharia Courts include Sharia Appeal Court, Sharia High Court, and Sharia Subordinate Courts at state-level and deal with religious and family matters such as custody, divorce, and inheritance, only for Muslims; decisions of Sharia courts cannot be appealed to civil courts

Political parties and leaders: *National Front (Barisan Nasional) or BN (ruling coalition) consists of the following parties:* Gerakan Rakyat Malaysia Party or PGRM [KOH Tsu Koon—acting]; Liberal Democratic Party (Parti Liberal Demokratik—Sabah) or LDP [LIEW Vui Keong]; Malaysian Chinese Association (Persatuan China Malaysia) or MCA [ONG Ka Ting]; Malaysian Indian Congress (Kongres India Malaysia) or MIC [S. Samy VELLU]; Parti Bersatu Rakyat Sabah or PBRS [Joseph KURUP]; Parti Bersatu Sabah or PBS [Joseph PAIRIN Kitingan]; Parti Pesaka Bumiputera Bersatu or PBB [Abdul TAIB Mahmud]; Parti Rakyat Sarawak or PRS [James MASING]; Sabah Progressive Party (Parti Progresif Sabah) or SAPP [YONG Teck Lee]; Sarawak United People's Party (Parti Bersatu Rakyat Sarawak) or SUPP [George CHAN Hong Nam]; United Malays National Organization or UMNO [ABDULLAH bin Ahmad Badawi]; United Pasokmomogun Kadazandusun Murut Organization (Pertubuhan Pasko Momogun Kadazan Dusun Bersatu) or UPKO [Bernard DOMPOK]; People's Progressive Party (Parti Progresif Penduduk Malaysia) or PPP [M.Kayveas]; Sarawak Progressive Democratic Party or SPDP [William MAWAN])

opposition parties—DAP, PAS, and PKR are members of the People's Alliance (Pakatan Rakyat) or PR: Democratic Action Party (Parti Tindakan Demokratik) or DAP [KARPAL Singh]; Islamic Party of Malaysia (Parti Islam se Malaysia) or PAS [Abdul HADI Awang]; People's Justice Party (Parti Keadilan Rakyat) or PKR [WAN AZIZAH Wan Ismael]; Sarawak National Party or SNAP [Edwin DUNDANG]

Political pressure groups and leaders: NA

International organization participation: ADB, APEC, APT, ARF, ASEAN, BIS, C, CP, EAS, FAO, G-15, G-77, IAEA, IBRD, ICAO, ICC, ICRM, IDA, IDB, IFAD, IFC, IFRCS, IHO, ILO, IMF, IMO, IMSO, Interpol, IOC, IPU, ISO, ITSO, ITU, ITUC, MIGA, MINURSO, MONUC, NAM, OIC, OPCW, PCA, PIF (partner), UN, UNCTAD, UNESCO, UNIDO, UNIFIL, UNMEE, UNMIL, UNMIS, UNMIT, UNWTO, UPU, WCL, WCO, WFTU, WHO, WIPO, WMO, WTO

Diplomatic representation in the US:

chief of mission: Ambassador RAJMAH binti Hussain

chancery: 3516 International Court NW, Washington, DC 20008

telephone: [1] (202) 572-9700

FAX: [1] (202) 572-9882

consulate(s) general: Los Angeles, New York

Diplomatic representation from the US:

chief of mission: Ambassador James KEITH

embassy: 376 Jalan Tun Razak, Kuala Lumpur 50400

mailing address: US Embassy Kuala Lumpur, APO AP 96535-8152

telephone: [60] (3) 2168-5000

FAX: [60] (3) 2142-2207

Flag description: 14 equal horizontal stripes of red (top) alternating with white (bottom); there is a blue rectangle in the upper hoist-side corner bearing a yellow crescent and a yellow 14-pointed star; the crescent and the star are traditional symbols of Islam; the design was based on the flag of the US

ECONOMY

Economy—overview: Malaysia, a middle-income country, has transformed itself since the 1970s from a producer of raw materials into an emerging multi-sector economy. Since coming to office in 2003, Prime Minister ABDULLAH has tried to move the economy farther up the value-added production chain by attracting investments in high technology industries, medical technology, and pharmaceuticals. The Government of Malaysia is continuing efforts to boost domestic demand to wean the economy off of its dependence on exports. Nevertheless, exports—particularly of electronics—remain a significant driver of the economy. As an oil and gas exporter, Malaysia has profited from higher world energy prices, although the rising cost of domestic gasoline and diesel fuel forced Kuala Lumpur to reduce government subsidies. Malaysia "unpegged" the ringgit from the US dollar in 2005 and the currency appreciated 6% per year against the dollar in 2006–07. Although this has helped to hold down the price of imports, inflationary pressures began to build in 2007. Healthy foreign exchange reserves and a small external debt greatly reduce the risk that Malaysia will experience a financial crisis over the near term similar to the one in 1997. The government presented its five-year national development agenda in April 2006 through the

Ninth Malaysia Plan, a comprehensive blueprint for the allocation of the national budget from 2006–10. With national elections expected within the year, ABDULLAH has unveiled a series of ambitious development schemes for several regions that have had trouble attracting business investment. Real GDP growth has averaged about 6% per year under ABDULLAH, but regions outside of Kuala Lumpur and the manufacturing hub Penang have not fared as well.

GDP (purchasing power parity): $357.4 billion (2007 est.)

GDP (official exchange rate): $186.5 billion (2007 est.)

GDP—real growth rate: 6.3% (2007 est.)

GDP—per capita (PPP): $13,300 (2007 est.)

GDP—composition by sector:
agriculture: 9.9%
industry: 45.3%
services: 44.8% (2007 est.)

Labor force: 10.94 million (2007 est.)

Labor force—by occupation: agriculture: 13%
industry: 36%
services: 51% (2005 est.)

Unemployment rate: 3.2% (2007 est.)

Population below poverty line: 5.1% (2002 est.)

Household income or consumption by percentage share:
lowest 10%: 1.4%
highest 10%: 39.2% (2003 est.)

Distribution of family income—Gini index: 46.1 (2002)

Inflation rate (consumer prices): 2.1%
note: approximately 30% of goods are price-controlled (2007 est.)

Investment (gross fixed): 21.8% of GDP (2007 est.)

Budget:
revenues: $40.69 billion
expenditures: $46.7 billion (2007 est.)

Public debt: 41.6% of GDP (2007 est.)

Agriculture—products: Peninsular Malaysia—rubber, palm oil, cocoa, rice; Sabah—subsistence crops, rubber, timber, coconuts, rice; Sarawak—rubber, pepper, timber

Industries: Peninsular Malaysia—rubber and oil palm processing and manufacturing, light manufacturing, electronics, tin mining and smelting, logging, timber processing; Sabah—logging, petroleum production; Sarawak—agriculture processing, petroleum production and refining, logging

Industrial production growth rate: 3.2% (2007 est.)

Electricity—production: 82.36 billion kWh (2005)

Electricity—production by source:
fossil fuel: 89.5%
hydro: 10.5%
nuclear: 0%
other: 0% (2001)

Electricity—consumption: 78.72 billion kWh (2005)

Electricity—exports: 0 kWh (2005)

Electricity—imports: 0 kWh (2005)

Oil—production: 751,800 bbl/day (2005 est.)

Oil—consumption: 501,000 bbl/day (2005 est.)

Oil—exports: 611,200 bbl/day (2004)

Oil—imports: 278,600 bbl/day (2004)

Oil—proved reserves: 3 billion bbl (2007 est.)

Natural gas—production: 60.9 billion cu m (2005 est.)

Natural gas—consumption: 31.84 billion cu m (2005 est.)

Natural gas—exports: 29.06 billion cu m (2005 est.)

Natural gas—imports: 0 cu m (2005)

Natural gas—proved reserves: 2.037 trillion cu m (1 January 2006 est.)

Current account balance: $26.05 billion (2007 est.)

Exports: $181.2 billion f.o.b. (2007 est.)

Exports—commodities: electronic equipment, petroleum and liquefied natural gas, wood and wood products, palm oil, rubber, textiles, chemicals

Exports—partners: US 18.8%, Singapore 15.4%, Japan 8.9%, China 7.2%, Thailand 5.3%, Hong Kong 4.9% (2006)

Imports: $145.7 billion f.o.b. (2007 est.)

Imports—commodities: electronics, machinery, petroleum products, plastics, vehicles, iron and steel products, chemicals

Imports—partners: Japan 13.3%, US 12.6%, China 12.2%, Singapore 11.7%, Thailand 5.5%, Taiwan 5.5%, South Korea 5.4%, Germany 4.4% (2006)

Economic aid—recipient: $31.6 million (2005)

Reserves of foreign exchange and gold: $101.1 billion (31 December 2007 est.)

Debt—external: $53.45 billion (31 December 2007)

Stock of direct foreign investment—at home: $86.31 billion (2007 est.)

Stock of direct foreign investment—abroad: $44.15 billion (2007 est.)

Market value of publicly traded shares: $235.4 billion (2006)

Currency (code): ringgit (MYR)

Currency code: MYR

Exchange rates: ringgits per US dollar—3.46 (2007), 3.6683 (2006), 3.8 (2005), 3.8 (2004), 3.8 (2003)

Fiscal year: calendar year

COMMUNICATIONS

Telephones—main lines in use: 4.342 million (2006)

Telephones—mobile cellular: 19.464 million (2006)

Telephone system:
general assessment: modern system; international service excellent
domestic: good intercity service provided on Peninsular Malaysia mainly by microwave radio relay; adequate intercity microwave radio relay network between Sabah and Sarawak via Brunei; domestic satellite system with 2 earth stations; combined fixed-line and mobile cellular teledensity approaching 100 per 100 persons
international: country code—60; landing point for several major international submarine cable networks that provide connectivity to Asia, Middle East, and Europe; satellite earth stations—2 Intelsat (1 Indian Ocean, 1 Pacific Ocean) (2001)

Radio broadcast stations: AM 35, FM 391, shortwave 15 (2001)

Radios: 10.9 million (1999)

Television broadcast stations: 88 (mainland Malaysia 51, Sabah 16, and Sarawak 21) (2006)

Televisions: 10.8 million (1999)

Internet country code: .my

Internet hosts: 337,674 (2007)

Internet Service Providers (ISPs): 7 (2000)

Internet users: 11.292 million (2006)

TRANSPORTATION

Airports: 116 (2007)

Airports—with paved runways: total: 36
over 3,047 m: 5
2,438 to 3,047 m: 9
1,524 to 2,437 m: 8
914 to 1,523 m: 8
under 914 m: 6 (2007)

Airports—with unpaved runways:
total: 80
1,524 to 2,437 m: 1
914 to 1,523 m: 7
under 914 m: 72 (2007)

Heliports: 2 (2007)

Pipelines: condensate 282 km; gas 5,273 km; oil 1,750 km; oil/gas/water 19 km; refined products 114 km (2007)

Railways:
total: 1,890 km
standard gauge: 57 km 1.435-m gauge (57 km electrified)
narrow gauge: 1,833 km 1.000-m gauge (150 km electrified) (2006)

Roadways: total: 98,721 km
paved: 80,280 km (includes 1,821 km of expressways)
unpaved: 18,441 km (2004)

Waterways: 7,200 km

note: Peninsular Malaysia 3,200 km; Sabah 1,500 km; Sarawak 2,500 km (2005)

Merchant marine:

total: 304 ships (1000 GRT or over) 6,154,877 GRT/8,364,578 DWT

by type: bulk carrier 16, cargo 98, chemical tanker 30, container 47, liquefied gas 30, livestock carrier 1, passenger/cargo 5, petroleum tanker 68, roll on/roll off 5, vehicle carrier 4

foreign-owned: 43 (China 1, Germany 2, Hong Kong 14, Japan 4, Singapore 22)

registered in other countries: 67 (Bahamas 11, Kiribati 1, Marshall Islands 3, Mongolia 1, Panama 14, Philippines 2, Singapore 28, Thailand 3, US 4, unknown 1) (2007)

Ports and terminals: Bintulu, Johor Bahru, Kuantan, Labuan, George Town (Penang), Port Kelang, Tanjung Pelepas

MILITARY

Military branches: Malaysian Armed Forces (Angkatan Tentera Malaysia, ATM): Malaysian Army (Tentera Darat Malaysia), Royal Malaysian Navy (Tentera Laut Diraja Malaysia, TLDM), Royal Malaysian Air Force (Tentera Udara Diraja Malaysia, TUDM) (2008)

Military service age and obligation: 18 years of age for voluntary military service (2005)

Manpower available for military service: *males age 16–49:* 6,440,338 *females age 16–49:* 6,280,826 (2008 est.)

Manpower fit for military service: *males age 16–49:* 5,374,006 *females age 16–49:* 5,316,865 (2008 est.)

Manpower reaching militarily significant age annually: *males age 16–49:* 260,725

females age 16–49: 247,309 (2008 est.)

Military expenditures—percent of GDP: 2.03% (2005 est.)

TRANSNATIONAL ISSUES

Disputes—international: Malaysia has asserted sovereignty over the Spratly Islands together with China, Philippines, Taiwan, Vietnam, and possibly Brunei; while the 2002 "Declaration on the Conduct of Parties in the South China Sea" has eased tensions over the Spratly Islands, it is not the legally binding "code of conduct" sought by some parties; Malaysia was not party to the March 2005 joint accord among the national oil companies of China, the Philippines, and Vietnam on conducting marine seismic activities in the Spratly Islands; disputes continue over deliveries of fresh water to Singapore, Singapore's land reclamation, bridge construction, and maritime boundaries in the Johor and Singapore Straits; in November 2007, the ICJ will hold public hearings in response to the Memorials and Countermemorials filed by the parties in 2003 and 2005 over sovereignty of Pedra Branca Island/Pulau Batu Puteh, Middle Rocks and South Ledge; ICJ awarded Ligitan and Sipadan islands, also claimed by Indonesia and Philippines, to Malaysia but left maritime boundary and sovereignty of Unarang rock in the hydrocarbon-rich Celebes Sea in dispute; separatist violence in Thailand's predominantly Muslim southern provinces prompts measures to close and monitor border with Malaysia to stem terrorist activities; Philippines retains a dormant claim to Malaysia's Sabah State in northern Borneo; Brunei and Malaysia are still considering international adjudi-

cation over their disputed offshore and deepwater seabeds, where hydrocarbon exploration was terminated in 2003; Malaysia's land boundary with Brunei around Limbang is in dispute; piracy remains a problem in the Malacca Strait

Refugees and internally displaced persons: *refugees (country of origin):* 15,174 (Indonesia); 21,544 (Burma) (2007)

Trafficking in persons: *current situation:* Malaysia is a destination and, to a lesser extent, a source and transit country for women and children trafficked for the purposes of sexual exploitation; foreign victims, mostly women and girls from Burma, Cambodia, China, Indonesia, the Philippines, Thailand, and Vietnam, are trafficked to Malaysia for commercial sexual exploitation; economic migrants from countries in the region who work as domestic servants or laborers in the construction and agricultural sectors face exploitative conditions in Malaysia that meet the definition of involuntary servitude; some Malaysian women, primarily of Chinese ethnicity, are trafficked abroad for sexual exploitation

tier rating: Tier 3—lack of satisfactory progress in combating trafficking in 2006; the government failed to prosecute traffickers arrested and detained under existing law and failed to provide adequate shelters and services to victims of trafficking

Illicit drugs: drug trafficking prosecuted vigorously and carries severe penalties; heroin still primary drug of abuse, but synthetic drug demand remains strong; continued ecstasy and methamphetamine producer for domestic users and, to a lesser extent, the regional drug market

MALDIVES

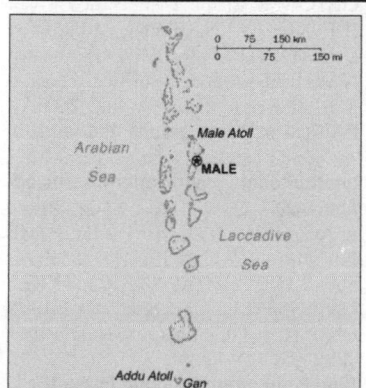

INTRODUCTION

Background: The Maldives was long a

sultanate, first under Dutch and then under British protection. It became a republic in 1968, three years after independence. Since 1978, President Maumoon Abdul GAYOOM—currently in his sixth term in office—has dominated the islands' political scene. Following riots in the capital Male in August 2004, the president and his government pledged to embark upon democratic reforms including a more representative political system and expanded political freedoms. Progress has been slow, however, and many promised reforms have been slow to come to fruition. Nonetheless, political parties were legalized in 2005. A constituent assembly—termed the "special majlis"—has pledged to complete the drafting of a

new constitution by the end of 2007 and first-ever presidential elections under a multi-candidate, multi-party system are slated for November 2008. Tourism and fishing are being developed on the archipelago.

GEOGRAPHY

Location: Southern Asia, group of atolls in the Indian Ocean, south-southwest of India

Geographic coordinates: 3 15 N, 73 00 E

Map references: Asia

Area:

total: 300 sq km

land: 300 sq km

water: 0 sq km

Area—comparative: about 1.7 times the size of Washington, DC

Land boundaries: 0 km
Coastline: 644 km
Maritime claims: measured from claimed archipelagic straight baselines
territorial sea: 12 nm
contiguous zone: 24 nm
exclusive economic zone: 200 nm
Climate: tropical; hot, humid; dry, northeast monsoon (November to March); rainy, southwest monsoon (June to August)
Terrain: flat, with white sandy beaches
Elevation extremes:
lowest point: Indian Ocean 0 m
highest point: unnamed location on Wilingili island in the Addu Atoll 2.4 m
Natural resources: fish
Land use:
arable land: 13.33%
permanent crops: 30%
other: 56.67% (2005)
Irrigated land: NA
Total renewable water resources: 0.03 cu km (1999)
Freshwater withdrawal (domestic/industrial/agricultural): *total:* 0.003 cu km/yr (98%/2%/0%)
per capita: 9 cu m/yr (1987)
Natural hazards: low level of islands makes them sensitive to sea level rise
Environment—current issues: depletion of freshwater aquifers threatens water supplies; global warming and sea level rise; coral reef bleaching
Environment—international agreements: *party to:* Biodiversity, Climate Change, Climate Change-Kyoto Protocol, Desertification, Hazardous Wastes, Law of the Sea, Ozone Layer Protection, Ship Pollution
signed, but not ratified: none of the selected agreements
Geography—note: 1,190 coral islands grouped into 26 atolls (200 inhabited islands, plus 80 islands with tourist resorts); archipelago with strategic location astride and along major sea lanes in Indian Ocean

PEOPLE

Population: 379,174 (July 2008 est.)
Age structure:
0–14 years: 42.4% (male 82,616/female 78,165)
15–64 years: 54.5% (male 105,465/female 101,115)
65 years and over: 3.1% (male 5,753/female 6,060) (2008 est.)
Median age:
total: 18.3 years
male: 18.2 years
female: 18.4 years (2008 est.)
Population growth rate: 2.69% (2008 est.)
Birth rate: 33.61 births/1,000 population (2008 est.)

Death rate: 6.71 deaths/1,000 population (2008 est.)
Net migration rate: NA
Sex ratio:
at birth: 1.05 male(s)/female
under 15 years: 1.06 male(s)/female
15–64 years: 1.04 male(s)/female
65 years and over: 0.95 male(s)/female
total population: 1.05 male(s)/female (2008 est.)
Infant mortality rate:
total: 51.62 deaths/1,000 live births
male: 50.78 deaths/1,000 live births
female: 52.5 deaths/1,000 live births (2008 est.)
Life expectancy at birth:
total population: 65.12 years
male: 63.73 years
female: 66.58 years (2008 est.)
Total fertility rate: 4.66 children born/woman (2008 est.)
HIV/AIDS—adult prevalence rate: 0.1% (2001 est.)
HIV/AIDS—people living with HIV/AIDS: fewer than 100 (2001 est.)
HIV/AIDS—deaths: NA
Nationality:
noun: Maldivian(s)
adjective: Maldivian
Ethnic groups: South Indians, Sinhalese, Arabs
Religions: Sunni Muslim
Languages: Maldivian Dhivehi (dialect of Sinhala, script derived from Arabic), English spoken by most government officials
Literacy:
definition: age 15 and over can read and write
total population: 96.3%
male: 96.2%
female: 96.4% (2000 census)

GOVERNMENT

Country name:
conventional long form: Republic of Maldives
conventional short form: Maldives
local long form: Dhivehi Raajjeyge Jumhooriyyaa
local short form: Dhivehi Raajje
Government type: republic
Capital: *name:* Male
geographic coordinates: 4 10 N, 73 30 E
time difference: UTC+5 (10 hours ahead of Washington, DC during Standard Time)
Administrative divisions: 19 atolls (atholhu, singular and plural) and the capital city*; Alifu, Baa, Dhaalu, Faafu, Gaafu Alifu, Gaafu Dhaalu, Gnaviyani, Haa Alifu, Haa Dhaalu, Kaafu, Laamu, Lhaviyani, Maale* (Male), Meemu, Noonu, Raa, Seenu, Shaviyani, Thaa, Vaavu

Independence: 26 July 1965 (from UK)
National holiday: Independence Day, 26 July (1965)
Constitution: adopted 1 January 1998
Legal system: based on Islamic law with admixtures of English common law primarily in commercial matters; has not accepted compulsory ICJ jurisdiction
Suffrage: 21 years of age; universal
Executive branch:
chief of state: President Maumoon Abdul GAYOOM (since 11 November 1978); note—the president is both the chief of state and head of government
head of government: President Maumoon Abdul GAYOOM (since 11 November 1978)
cabinet: Cabinet of Ministers appointed by the president
elections: president nominated by the Majlis; nomination must be ratified by a national referendum (at least a 51% approval margin is required); president elected for a five-year term; election last held 17 October 2003 (next to be held in 2008)
election results: President Maumoon Abdul GAYOOM reelected in referendum held 17 October 2003; percent of popular vote—Maumoon Abdul GAYOOM 90.3%
Legislative branch: unicameral People's Council or Majlis (50 seats; 42 members elected by popular vote, 8 appointed by the president; to serve five-year terms)
elections: last held 22 January 2005 (next to be held in 2010)
election results: percent of vote—NA; seats—independents 50
Judicial branch: High Court
Political parties and leaders: Adhaalath (Justice) Party or AP [Abdul Majeed Abdul BARI]; Dhivehi Rayyithunge Party (Maldivian People's Party) or DRP [Maumoon Abdul GAYOOM]; Islamic Democratic Party or IDP [Omar NASEER]; Maldivian Democratic Party or MDP [Mohamed NASHEED]; note—political parties were allowed to register in June 2005
Political pressure groups and leaders: various unregistered political parties
International organization participation: ADB, C, CP, FAO, G-77, IBRD, ICAO, IDA, IDB, IFAD, IFC, IMF, IMO, Interpol, IOC, IPU, ITU, MIGA, NAM, OIC, OPCW, SAARC, SACEP, UN, UNCTAD, UNESCO, UNIDO, UNWTO, UPU, WCO, WHO, WIPO, WMO, WTO
Diplomatic representation in the US:
chief of mission: Ambassador Mohamed Hussain MANIKU
chancery: 800 2nd Avenue, Suite 400E, New York, NY 10017

telephone: [1] (212) 599-6194
FAX: [1] (212) 599-6195
Diplomatic representation from the US: the US does not have an embassy in Maldives; the US Ambassador to Sri Lanka is accredited to Maldives and makes periodic visits
Flag description: red with a large green rectangle in the center bearing a vertical white crescent; the closed side of the crescent is on the hoist side of the flag

ECONOMY

Economy—overview: Tourism, Maldives' largest industry, accounts for 28% of GDP and more than 60% of the Maldives' foreign exchange receipts. Over 90% of government tax revenue comes from import duties and tourism-related taxes. Fishing is the second leading sector. Agriculture and manufacturing continue to play a lesser role in the economy, constrained by the limited availability of cultivable land and the shortage of domestic labor. Most staple foods must be imported. Industry, which consists mainly of garment production, boat building, and handicrafts, accounts for about 7% of GDP. The Maldivian Government began an economic reform program in 1989 initially by lifting import quotas and opening some exports to the private sector. Subsequently, it has liberalized regulations to allow more foreign investment. Real GDP growth averaged over 7.5% per year for more than a decade. In late December 2004, a major tsunami left more than 100 dead, 12,000 displaced, and property damage exceeding $300 million. As a result of the tsunami, the GDP contracted by about 3.6% in 2005. A rebound in tourism, post-tsunami reconstruction, and development of new resorts helped the economy recover quickly. The trade deficit has expanded sharply as a result of high oil prices and imports of construction material. Diversifying beyond tourism and fishing and increasing employment are the major challenges facing the government. Over the longer term Maldivian authorities worry about the impact of erosion and possible global warming on their low-lying country; 80% of the area is 1 meter or less above sea level.
GDP (purchasing power parity): $1.588 billion (2007 est.)
GDP (official exchange rate): $1.049 billion (2007 est.)
GDP—real growth rate: 6.6% (2007 est.)
GDP—per capita (PPP): $4,600 (2007 est.)
GDP—composition by sector:
agriculture: 16%
industry: 7%
services: 77% (2006 est.)
Labor force: 101,300 (2004)
Labor force—by occupation: *agriculture:* 22%
industry: 18%
services: 60% (1995)
Unemployment rate: NEGL% (2003 est.)
Population below poverty line: 21% (2004)
Household income or consumption by percentage share:
lowest 10%: NA%
highest 10%: NA%
Inflation rate (consumer prices): 5% (2007 est.)
Budget:
revenues: $508 million (including foreign grants)
expenditures: $671 million (2006 est.)
Agriculture—products: coconuts, corn, sweet potatoes; fish
Industries: tourism, fish processing, shipping, boat building, coconut processing, garments, woven mats, rope, handicrafts, coral and sand mining
Industrial production growth rate: -0.9% (2004 est.)
Electricity—production: 169 million kWh (2005)
Electricity—production by source:
fossil fuel: 100%
hydro: 0%
nuclear: 0%
other: 0% (2001)
Electricity—consumption: 157.1 million kWh (2005)
Electricity—exports: 0 kWh (2005)
Electricity—imports: 0 kWh (2005)
Oil—production: 0 bbl/day (2005 est.)
Oil—consumption: 5,000 bbl/day (2005 est.)
Oil—exports: 1,517 bbl/day (2004)
Oil—imports: 6,390 bbl/day (2004)
Oil—proved reserves: 0 bbl (1 January 2006 est.)
Natural gas—production: 0 cu m (2005 est.)
Natural gas—consumption: 0 cu m (2005 est.)
Natural gas—exports: 0 cu m (2005 est.)
Natural gas—imports: 0 cu m (2005)
Natural gas—proved reserves: 0 cu m (1 January 2006 est.)
Current account balance: -$472 million (2007)
Exports: $167 million f.o.b. (2006)
Exports—commodities: fish
Exports—partners: Thailand 33.1%, UK 14.3%, Sri Lanka 11.9%, Japan 10.3%, France 6.9%, Algeria 6.1% (2006)
Imports: $930 million f.o.b. (2006)
Imports—commodities: petroleum products, ships, foodstuffs, clothing, intermediate and capital goods
Imports—partners: Singapore 23.2%, UAE 15.8%, India 11.1%, Malaysia 7.9%, Thailand 6.9%, Sri Lanka 5.7% (2006)
Economic aid—recipient: $66.83 million (2005)
Debt—external: $482 million (2006 est.)
Market value of publicly traded shares: $NA
Currency (code): rufiyaa (MVR)
Currency code: MVR
Exchange rates: rufiyaa per US dollar— NA (2007), 12.8 (2006), 12.8 (2005), 12.8 (2004), 12.8 (2003)
Fiscal year: calendar year

COMMUNICATIONS

Telephones—main lines in use: 32,500 (2006)
Telephones—mobile cellular: 262,600 (2006)
Telephone system:
general assessment: telephone services have improved; each island now has at least 1 public telephone, and there are mobile cellular networks with rapidly expanding subscribership
domestic: interatoll communication through microwave links; all inhabited islands and resorts are connected with telephone and fax service
international: country code—960; linked to international submarine cable Fiber-Optic Link Around the Globe (FLAG); satellite earth station—3 Intelsat (Indian Ocean)
Radio broadcast stations: AM 1, FM 1, shortwave 1 (1998)
Radios: 35,000 (1999)
Television broadcast stations: 1 (2006)
Televisions: 10,000 (1999)
Internet country code: .mv
Internet hosts: 1,082 (2007)
Internet Service Providers (ISPs): 1 (2000)
Internet users: 20,100 (2005)

TRANSPORTATION

Airports: 5 (2007)
Airports—with paved runways:
total: 3
over 3,047 m: 1
2,438 to 3,047 m: 1
914 to 1,523 m: 1 (2007)
Airports—with unpaved runways:
total: 2
914 to 1,523 m: 2 (2007)
Roadways:
total: 88 km
paved roads: 88 km—60 km in Male; 14 km on Addu Atolis; 14 km on Laamu

note: village roads are mainly compacted coral (2006)

Merchant marine:
total: 22 ships (1000 GRT or over) 85,935 GRT/114,054 DWT
by type: cargo 17, petroleum tanker 3, refrigerated cargo 2
foreign-owned: 1 (Greece 1)
registered in other countries: 2 (Panama 1, Tuvalu 1) (2007)
Ports and terminals: Male

MILITARY

Military branches: Maldives National Defense Force (MNDF): Quick Reaction Force, Security Protection Group, Coast Guard (2007)

Military service age and obligation: 18 years of age for voluntary military service; no conscription (2008)

Manpower available for military service:
males age 16–49: 89,505
females age 16–49: 85,745 (2008 est.)

Manpower fit for military service:
males age 16–49: 72,150
females age 16–49: 69,058 (2008 est.)

Military expenditures—percent of GDP: 5.5% (2005 est.)

Military—note: the Maldives National Defense Force (MNDF), with its small size and with little serviceable equipment, is inadequate to prevent external aggression and is primarily tasked to reinforce the Maldives Police Service (MPS) and ensure security in the exclusive economic zone (2008)

TRANSNATIONAL ISSUES

Disputes—international: none
Refugees and internally displaced persons: *IDPs:* 1,000–10,000 (December 2004 tsunami victims) (2007)

MALI

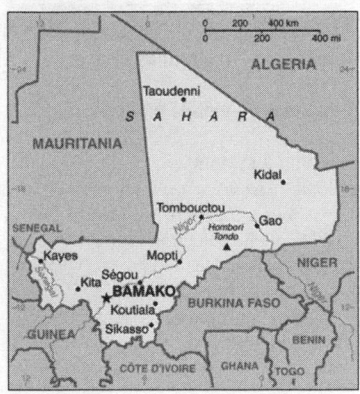

INTRODUCTION

Background: The Sudanese Republic and Senegal became independent of France in 1960 as the Mali Federation. When Senegal withdrew after only a few months, what formerly made up the Sudanese Republic was renamed Mali. Rule by dictatorship was brought to a close in 1991 by a military coup—led by the current president Amadou TOURE—enabling Mali's emergence as one of the strongest democracies on the continent. President Alpha KONARE won Mali's first democratic presidential election in 1992 and was reelected in 1997. In keeping with Mali's two-term constitutional limit, KONARE stepped down in 2002 and was succeeded by Amadou TOURE, who was subsequently elected to a second term in 2007. The elections were widely judged to be free and fair.

GEOGRAPHY

Location: Western Africa, southwest of Algeria
Geographic coordinates: 17 00 N, 4 00 W

Map references: Africa
Area:
total: 1.24 million sq km
land: 1.22 million sq km
water: 20,000 sq km
Area—comparative: slightly less than twice the size of Texas
Land boundaries:
total: 7,243 km
border countries: Algeria 1,376 km, Burkina Faso 1,000 km, Guinea 858 km, Cote d'Ivoire 532 km, Mauritania 2,237 km, Niger 821 km, Senegal 419 km
Coastline: 0 km (landlocked)
Maritime claims: none (landlocked)
Climate: subtropical to arid; hot and dry (February to June); rainy, humid, and mild (June to November); cool and dry (November to February)
Terrain: mostly flat to rolling northern plains covered by sand; savanna in south, rugged hills in northeast
Elevation extremes:
lowest point: Senegal River 23 m
highest point: Hombori Tondo 1,155 m
Natural resources: gold, phosphates, kaolin, salt, limestone, uranium, gypsum, granite, hydropower
note: bauxite, iron ore, manganese, tin, and copper deposits are known but not exploited
Land use:
arable land: 3.76%
permanent crops: 0.03%
other: 96.21% (2005)
Irrigated land: 2,360 sq km (2003)
Total renewable water resources: 100 cu km (2001)
Freshwater withdrawal (domestic/industrial/agricultural): *total:* 6.55 cu km/yr (9%/1%/90%)
per capita: 484 cu m/yr (2000)
Natural hazards: hot, dust-laden harmattan haze common during dry seasons; recurring droughts; occasional Niger River flooding

Environment—current issues: deforestation; soil erosion; desertification; inadequate supplies of potable water; poaching
Environment—international agreements: *party to:* Biodiversity, Climate Change, Climate Change-Kyoto Protocol, Desertification, Endangered Species, Hazardous Wastes, Law of the Sea, Ozone Layer Protection, Wetlands, Whaling
signed, but not ratified: none of the selected agreements
Geography—note: landlocked; divided into three natural zones: the southern, cultivated Sudanese; the central, semi-arid Sahelian; and the northern, arid Saharan

PEOPLE

Population: 12,324,029 (July 2008 est.)
Age structure:
0–14 years: 48.2% (male 3,004,003/female 2,937,138)
15–64 years: 48.7% (male 2,976,314/female 3,028,433)
65 years and over: 3.1% (male 150,597/female 227,544) (2008 est.)
Median age:
total: 15.8 years
male: 15.4 years
female: 16.2 years (2008 est.)
Population growth rate: 2.725% (2008 est.)
Birth rate: 49.38 births/1,000 population (2008 est.)
Death rate: 16.16 deaths/1,000 population (2008 est.)
Net migration rate: -5.97 migrant(s)/1,000 population (2008 est.)
Sex ratio:
at birth: 1.03 male(s)/female
under 15 years: 1.02 male(s)/female
15–64 years: 0.98 male(s)/female

65 years and over: 0.66 male(s)/female
total population: 0.99 male(s)/female
(2008 est.)
Infant mortality rate:
total: 103.83 deaths/1,000 live births
male: 113.41 deaths/1,000 live births
female: 93.97 deaths/1,000 live births
(2008 est.)
Life expectancy at birth:
total population: 49.94 years
male: 48 years
female: 51.94 years (2008 est.)
Total fertility rate: 7.34 children
born/woman (2008 est.)
HIV/AIDS—adult prevalence rate: 1.9%
(2003 est.)
**HIV/AIDS—people living with
HIV/AIDS:** 140,000 (2003 est.)
HIV/AIDS—deaths: 12,000 (2003 est.)
Major infectious diseases:
degree of risk: very high
food or waterborne diseases: bacterial and
protozoal diarrhea, hepatitis A, and
typhoid fever
vectorborne disease: malaria
water contact disease: schistosomiasis
respiratory disease: meningococcal
meningitis (2008)
Nationality:
noun: Malian(s)
adjective: Malian
Ethnic groups: Mande 50% (Bambara,
Malinke, Soninke), Peul 17%, Voltaic
12%, Songhai 6%, Tuareg and Moor
10%, other 5%
Religions: Muslim 90%, Christian 1%,
indigenous beliefs 9%
Languages: French (official), Bambara
80%, numerous African languages
Literacy:
definition: age 15 and over can read and
write
total population: 46.4%
male: 53.5%
female: 39.6% (2003 est.)

GOVERNMENT

Country name:
conventional long form: Republic of Mali
conventional short form: Mali
local long form: Republique de Mali
local short form: Mali
former: French Sudan and Sudanese
Republic
Government type: republic
Capital: *name:* Bamako
geographic coordinates: 12 39 N, 8 00 W
time difference: UTC 0 (5 hours ahead of
Washington, DC during Standard Time)
Administrative divisions: 8 regions
(regions, singular—region); Gao, Kayes,
Kidal, Koulikoro, Mopti, Segou, Sikasso,
Tombouctou
Independence: 22 September 1960
(from France)

National holiday: Independence Day, 22
September (1960)
Constitution: adopted 12 January 1992
Legal system: based on French civil law
system and customary law; judicial
review of legislative acts in
Constitutional Court; has not accepted
compulsory ICJ jurisdiction
Suffrage: 18 years of age; universal
Executive branch:
chief of state: President Amadou Toumani
TOURE (since 8 June 2002)
head of government: Prime Minister
Modibo SIDIBE (since 28 September
2007)
cabinet: Council of Ministers appointed
by the prime minister
elections: president elected by popular
vote for a five-year term (eligible for a
second term); election last held on 29
April 2007 (next to be held in April
2012); prime minister appointed by the
president
election results: Amadou Toumani
TOURE reelected president; percent of
vote—Amadou Toumani TOURE
71.2%, Ibrahim Boubacar KEITA 19.2%,
other 9.6%
Legislative branch: unicameral
National Assembly or Assemblee
Nationale (147 seats; members elected
by popular vote to serve five-year terms)
elections: last held on 1 and 22 July 2007
(next to be held in July 2012)
election results: percent of vote by party—
NA; seats by party—ADP coalition 113
(including ADEMA 51, URD 34, MPR
8, CNID 7, UDD 3, and other 10), FDR
coalition 15 (including RPM 11,
PARENA 4), SADI 4, independent 15
Judicial branch: Supreme Court or Cour
Supreme
Political parties and leaders: Alliance
for Democratic Change (political group
comprised mainly of Tuareg from Mali's
northern region); African Solidarity for
Democracy and Independence or SADI
[Oumar MARIKO, secretary general];
Alliance for Democracy and Progress or
ADP (a coalition of political parties
including ADEMA and URD formed in
December 2006 to support the presiden-
tial candidacy of Amadou TOURE);
Alliance for Democracy or ADEMA
[Diouncondo TRAORE]; Convergence
2007 [Soumeylou Boubeye MAIGA];
Front for Democracy and the Republic or
FDR (a coalition of political parties
including RPM and PARENA formed to
oppose the presidential candidacy of
Amadou TOURE); National Congress
for Democratic Initiative or CNID
[Mountaga TALL]; Party for Democracy
and Progress or PDP [Me Idrissa
TRAORE]; Party for National Renewal

or PARENA [Tiebile DRAME]; Patriotic
Movement for Renewal or MPR
[Choguel MAIGA]; Rally for Democracy
and Labor or RDT; Rally for Mali or
RPM [Ibrahim Boubacar KEITA];
Sudanese Union/African Democratic
Rally or US/RDA [Mamadou Bamou
TOURE]; Union for Democracy and
Development or UDD [Moussa Balla
COULIBALY]; Union for Republic and
Democracy or URD [Soumaila CISSE]
**International organization participa-
tion:** ACCT, ACP, AfDB, AU,
ECOWAS, FAO, FZ, G-77, IAEA,
IBRD, ICAO, ICCt, ICRM, IDA, IDB,
IFAD, IFC, IFRCS, ILO, IMF, Interpol,
IOC, IOM, IPU, ITSO, ITU, ITUC,
MIGA, NAM, OIC, OIF, OPCW, UN,
UNAMID, UNCTAD, UNESCO,
UNIDO, UNMIL, UNMIS, UNWTO,
UPU, WADB (regional), WAEMU,
WCO, WFTU, WHO, WIPO, WMO,
WTO
Diplomatic representation in the US:
chief of mission: Ambassador Abdoulaye
DIOP
chancery: 2130 R Street NW,
Washington, DC 20008
telephone: [1] (202) 332-2249, 939-8950
FAX: [1] (202) 332-6603
Diplomatic representation from the US:
chief of mission: Ambassador Terrence P.
MCCULLEY
embassy: located just off the Roi Bin
Fahad Aziz Bridge just west of the
Bamako central district
mailing address: ACI 2000, Rue 243,
Porte 297, Bamako
telephone: [223] 270-2300
FAX: [223] 270-2479
Flag description: three equal vertical
bands of green (hoist side), yellow, and
red; uses the popular pan-African colors
of Ethiopia

ECONOMY

Economy—overview: Mali is among the
poorest countries in the world, with 65%
of its land area desert or semidesert and
with a highly unequal distribution of
income. Economic activity is largely
confined to the riverine area irrigated by
the Niger. About 10% of the population
is nomadic and some 80% of the labor
force is engaged in farming and fishing.
Industrial activity is concentrated on
processing farm commodities. Mali is
heavily dependent on foreign aid and
vulnerable to fluctuations in world prices
for cotton, its main export, along with
gold. The government has continued its
successful implementation of an IMF-
recommended structural adjustment program that is helping the economy grow,
diversify, and attract foreign investment.

Mali's adherence to economic reform and the 50% devaluation of the CFA franc in January 1994 have pushed up economic growth to a 5% average in 1996–2007. Worker remittances and external trade routes for the landlocked country have been jeopardized by continued unrest in neighboring Cote d'Ivoire.

GDP (purchasing power parity): $13.47 billion (2007 est.)

GDP (official exchange rate): $6.745 billion (2007 est.)

GDP—real growth rate: 2.5% (2007 est.)

GDP—per capita (PPP): $1,000 (2007 est.)

GDP—composition by sector:
agriculture: 45%
industry: 17%
services: 38% (2001 est.)

Labor force: 5.4 million (2007 est.)

Labor force—by occupation: *agriculture:* 80%
industry and services: 20% (2005 est.)

Unemployment rate: 30% (2004 est.)

Population below poverty line: 36.1% (2005 est.)

Household income or consumption by percentage share:
lowest 10%: 2.4%
highest 10%: 30.2% (2001)

Distribution of family income—Gini index: 40.1 (2001)

Inflation rate (consumer prices): 2.5% (2007 est.)

Budget:
revenues: $1.5 billion
expenditures: $1.8 billion (2006 est.)

Agriculture—products: cotton, millet, rice, corn, vegetables, peanuts; cattle, sheep, goats

Industries: food processing; construction; phosphate and gold mining

Industrial production growth rate: NA%

Electricity—production: 804 million kWh (2006)

Electricity—production by source:
fossil fuel: 41.7%
hydro: 58.3%
nuclear: 0%
other: 0% (2001)

Electricity—consumption: 804 million kWh (2006 est.)

Electricity—exports: 0 kWh; note—recent hydropower developments may be providing electricity to Senegal and Mauritania (2007 est.)

Electricity—imports: 0 kWh (2007)

Oil—production: 0 bbl/day (2006 est.)

Oil—consumption: 5,600 bbl/day (2006 est.)

Oil—exports: 0 bbl/day (2006)

Oil—imports: 5,600 bbl/day (2006 est.)

Oil—proved reserves: 0 bbl (1 January 2006 est.)

Natural gas—production: 0 cu m (2005 est.)

Natural gas—consumption: 0 cu m (2005 est.)

Natural gas—exports: 0 cu m (2005 est.)

Natural gas—imports: 0 cu m (2005)

Natural gas—proved reserves: 0 cu m (1 January 2006 est.)

Current account balance: -$446 million (2007 est.)

Exports: $294 million f.o.b. (2006)

Exports—commodities: cotton, gold, livestock

Exports—partners: China 26.8%, Germany 24.9%, Thailand 7.1%, Taiwan 4.9%, Bangladesh 4% (2006)

Imports: $2.358 billion f.o.b. (2006)

Imports—commodities: petroleum, machinery and equipment, construction materials, foodstuffs, textiles

Imports—partners: France 12.8%, Senegal 12.2%, Cote d'Ivoire 10.5% (2006)

Economic aid—recipient: $691.5 million (2005)

Debt—external: $2.8 billion (2002)

Market value of publicly traded shares: $NA

Currency (code): Communaute Financiere Africaine franc (XOF); note—responsible authority is the Central Bank of the West African States

Currency code: XOF

Exchange rates: Communaute Financiere Africaine francs (XOF) per US dollar—493.51 (2007), 522.59 (2006), 527.47 (2005), 528.29 (2004), 581.2 (2003)
note: since 1 January 1999, the XOF franc has been pegged to the euro at a rate of 655.957 XOF francs per euro

Fiscal year: calendar year

COMMUNICATIONS

Telephones—main lines in use: 82,500 (2006)

Telephones—mobile cellular: 1.513 million (2006)

Telephone system:
general assessment: domestic system unreliable but improving; provides only minimal service
domestic: fixed-line availability is gradually increasing, but subscribership remains less than 1 per 100 persons; mobile-cellular subscribership has increased sharply to 13 per 100 persons
international: country code—223; satellite earth stations—2 Intelsat (1 Atlantic Ocean, 1 Indian Ocean)

Radio broadcast stations: AM 1, FM 230 (27 regional and government stations, and 203 private stations), shortwave 1 (2001)

Radios: 570,000 (1997)

Television broadcast stations: 2 (plus repeaters) (2007)

Televisions: 45,000 (1997)

Internet country code: .ml

Internet hosts: 28 (2007)

Internet Service Providers (ISPs): 13 (2001)

Internet users: 70,000 (2006)

TRANSPORTATION

Airports: 29 (2007)

Airports—with paved runways:
total: 8
2,438 to 3,047 m: 4
1,524 to 2,437 m: 4 (2007)

Airports—with unpaved runways:
total: 21
2,438 to 3,047 m: 1
1,524 to 2,437 m: 5
914 to 1,523 m: 7
under 914 m: 8 (2007)

Railways:
total: 729 km
narrow gauge: 729 km 1.000-m gauge (2006)

Roadways:
total: 18,709 km
paved: 3,368 km
unpaved: 15,341 km (2004)

Waterways: 1,800 km (2007)

Ports and terminals: Koulikoro

MILITARY

Military branches: Malian Armed Forces: Army, Republic of Mali Air Force (Force Aerienne de la Republique du Mali, FARM), National Guard (2008)

Military service age and obligation: 18 years of age for compulsory and voluntary military service; conscript service obligation—2 years (2008)

Manpower available for military service:
males age 16–49: 2,603,700
females age 16–49: 2,441,776 (2008 est.)

Manpower fit for military service:
males age 16–49: 1,594,184
females age 16–49: 1,529,871 (2008 est.)

Military expenditures—percent of GDP: 1.9% (2006)

TRANSNATIONAL ISSUES

Disputes—international: none

Refugees and internally displaced persons: *refugees (country of origin):* 6,300 (Mauritania) (2007)

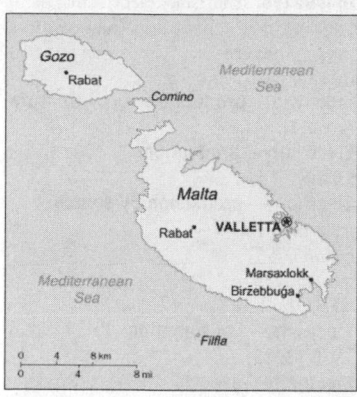

INTRODUCTION

Background: Great Britain formally acquired possession of Malta in 1814. The island staunchly supported the UK through both World Wars and remained in the Commonwealth when it became independent in 1964. A decade later Malta became a republic. Since about the mid-1980s, the island has transformed itself into a freight transshipment point, a financial center, and a tourist destination. Malta became an EU member in May 2004 and began to use the euro as currency in 2008.

GEOGRAPHY

Location: Southern Europe, islands in the Mediterranean Sea, south of Sicily (Italy)
Geographic coordinates: 35 50 N, 14 35 E
Map references: Europe
Area:
total: 316 sq km
land: 316 sq km
water: 0 sq km
Area—comparative: slightly less than twice the size of Washington, DC
Land boundaries: 0 km
Coastline: 196.8 km (does not include 56.01 km for the island of Gozo)
Maritime claims:
territorial sea: 12 nm
contiguous zone: 24 nm
continental shelf: 200 m depth or to the depth of exploitation
exclusive fishing zone: 25 nm
Climate: Mediterranean; mild, rainy winters; hot, dry summers
Terrain: mostly low, rocky, flat to dissected plains; many coastal cliffs
Elevation extremes:
lowest point: Mediterranean Sea 0 m
highest point: Ta'Dmejrek 253 m (near Dingli)

Natural resources: limestone, salt, arable land
Land use:
arable land: 31.25%
permanent crops: 3.13%
other: 65.62% (2005)
Irrigated land: 20 sq km (2003)
Total renewable water resources: 0.07 cu km (2005)
Freshwater withdrawal (domestic/industrial/agricultural): *total:* 0.02 cu km/yr (74%/1%/25%)
per capita: 50 cu m/yr (2000)
Natural hazards: NA
Environment—current issues: limited natural fresh water resources; increasing reliance on desalination
Environment—international agreements: *party to:* Air Pollution, Biodiversity, Climate Change, Climate Change-Kyoto Protocol, Desertification, Endangered Species, Hazardous Wastes, Law of the Sea, Marine Dumping, Ozone Layer Protection, Ship Pollution, Wetlands
signed, but not ratified: none of the selected agreements
Geography—note: the country comprises an archipelago, with only the three largest islands (Malta, Ghawdex or Gozo, and Kemmuna or Comino) being inhabited; numerous bays provide good harbors; Malta and Tunisia are discussing the commercial exploitation of the continental shelf between their countries, particularly for oil exploration

PEOPLE

Population: 403,532 (July 2008 est.)
Age structure:
0–14 years: 16.4% (male 33,954/female 32,158)
15–64 years: 69.7% (male 142,338/female 138,792)
65 years and over: 13.9% (male 24,240/female 32,050) (2008 est.)
Median age: *total:* 39.2 years
male: 37.9 years
female: 40.6 years (2008 est.)
Population growth rate: 0.407% (2008 est.)
Birth rate: 10.33 births/1,000 population (2008 est.)
Death rate: 8.29 deaths/1,000 population (2008 est.)
Net migration rate: 2.03 migrant(s)/1,000 population (2008 est.)
Sex ratio:
at birth: 1.06 male(s)/female
under 15 years: 1.06 male(s)/female
15–64 years: 1.03 male(s)/female
65 years and over: 0.76 male(s)/female

total population: 0.99 male(s)/female (2008 est.)
Infant mortality rate:
total: 3.79 deaths/1,000 live births
male: 4.25 deaths/1,000 live births
female: 3.3 deaths/1,000 live births (2008 est.)
Life expectancy at birth:
total population: 79.3 years
male: 77.08 years
female: 81.64 years (2008 est.)
Total fertility rate: 1.51 children born/woman (2008 est.)
HIV/AIDS—adult prevalence rate: 0.2% (2001 est.)
HIV/AIDS—people living with HIV/AIDS: fewer than 500 (2003 est.)
HIV/AIDS—deaths: fewer than 100 (2003 est.)
Nationality:
noun: Maltese (singular and plural)
adjective: Maltese
Ethnic groups: Maltese (descendants of ancient Carthaginians and Phoenicians with strong elements of Italian and other Mediterranean stock)
Religions: Roman Catholic 98%
Languages: Maltese (official), English (official)
Literacy: *definition:* age 10 and over can read and write
total population: 92.8%
male: 92%
female: 93.6% (2003 est.)

GOVERNMENT

Country name:
conventional long form: Republic of Malta
conventional short form: Malta
local long form: Repubblika ta' Malta
local short form: Malta
Government type: republic
Capital: *name:* Valletta
geographic coordinates: 35 53 N, 14 30 E
time difference: UTC+1 (6 hours ahead of Washington, DC during Standard Time)
daylight saving time: +1hr, begins last Sunday in March; ends last Sunday in October
Administrative divisions: none (administered directly from Valletta); note—local councils carry out administrative orders
Independence: 21 September 1964 (from UK)
National holiday: Independence Day, 21 September (1964)
Constitution: 1964 constitution; amended many times
Legal system: based on English common law and Roman civil law; accepts compulsory ICJ jurisdiction with reservations

Suffrage: 18 years of age; universal

Executive branch:

chief of state: President Edward FENECH ADAMI (since 4 April 2004)

head of government: Prime Minister Lawrence GONZI (since 23 March 2004)

cabinet: Cabinet appointed by the president on the advice of the prime minister

elections: president elected by the House of Representatives for a five-year term (eligible for a second term); election last held on 29 March 2004 (next to be held by April 2009); following legislative elections, the leader of the majority party or leader of a majority coalition is usually appointed prime minister by the president for a five-year term; the deputy prime minister is appointed by the president on the advice of the prime minister

election results: Eddie FENECH ADAMI elected president; House of Representatives vote—33 out of 65 votes

Legislative branch: unicameral House of Representatives (usually 65 seats; members are elected by popular vote on the basis of proportional representation to serve five-year terms; note—additional seats are given to the party with the largest popular vote to ensure a legislative majority)

elections: last held on 8 March 2008 (next to be held by March 2013)

election results: percent of vote by party—PN 49.3%, MLP 48.8%, other 1.9%; seats by party—PN 35, MLP 34

Judicial branch: Constitutional Court; Court of Appeal; judges for both courts are appointed by the president on the advice of the prime minister

Political parties and leaders: Alternativa Demokratika/Alliance for Social Justice or AD [Harry VASSALLO]; Malta Labor Party or MLP [acting leader Charles MANGION]; Nationalist Party or PN [Lawrence GONZI]

Political pressure groups and leaders: NA

International organization participation: Australia Group, C, CE, CPLP (associate), EBRD, EIB, EMU, EU, FAO, IAEA, IBRD, ICAO, ICCt, ICRM, IFAD, IFC, IFRCS, ILO, IMF, IMO, IMSO, Interpol, IOC, IOM, IPU, ISO, ITSO, ITU, ITUC, MIGA, NSG, OPCW, OSCE, PCA, Schengen Convention, UN, UNCTAD, UNESCO, UNIDO, UNWTO, UPU, WCL, WCO, WHO, WIPO, WMO, WTO

Diplomatic representation in the US:

chief of mission: Ambassador Mark MICELI-FARRUGIA

chancery: 2017 Connecticut Avenue NW, Washington, DC 20008

telephone: [1] (202) 462-3611, 3612

FAX: [1] (202) 387-5470

consulate(s): New York

Diplomatic representation from the US:

chief of mission: Ambassador Molly BORDONARO

embassy: 3rd Floor, Development House, Saint Anne Street, Floriana, VLT 01

mailing address: P. O. Box 535, Valletta, CMR01

telephone: [356] 2561 4000

FAX: [356] 21 243229

Flag description: two equal vertical bands of white (hoist side) and red; in the upper hoist-side corner is a representation of the George Cross, edged in red

ECONOMY

Economy—overview: Major resources are limestone, a favorable geographic location, and a productive labor force. Malta produces only about 20% of its food needs, has limited fresh water supplies, and has few domestic energy sources. The economy is dependent on foreign trade, manufacturing (especially electronics and pharmaceuticals), and tourism. Economic recovery of the European economy has lifted exports, tourism, and overall growth. Malta adopted the euro on 1 January 2008.

GDP (purchasing power parity): $21.89 billion (2007 est.)

GDP (official exchange rate): $7.419 billion (2007 est.)

GDP—real growth rate: 3.8% (2007 est.)

GDP—per capita (PPP): $53,400 (2007 est.)

GDP—composition by sector:

agriculture: 2.7%

industry: 22.3%

services: 74.9% (2003 est.)

Labor force: 164,000 (2006 est.)

Labor force—by occupation: *agriculture:* 3%

industry: 22%

services: 75% (2005 est.)

Unemployment rate: 6.3% (2007 est.)

Population below poverty line: NA%

Household income or consumption by percentage share:

lowest 10%: NA%

highest 10%: NA%

Distribution of family income—Gini index: 28 (2005)

Inflation rate (consumer prices): 0.7% (2007 est.)

Investment (gross fixed): 16.9% of GDP (2007 est.)

Budget:

revenues: $3.479 billion

expenditures: $3.549 billion (2007 est.)

Agriculture—products: potatoes, cauliflower, grapes, wheat, barley, tomatoes, citrus, cut flowers, green peppers; pork, milk, poultry, eggs

Industries: tourism, electronics, ship building and repair, construction, food and beverages, pharmaceuticals, footwear, clothing, tobacco

Industrial production growth rate: NA%

Electricity—production: 2.106 billion kWh (2005)

Electricity—production by source:

fossil fuel: 100%

hydro: 0%

nuclear: 0%

other: 0% (2001)

Electricity—consumption: 1.959 billion kWh (2005)

Electricity—exports: 0 kWh (2005)

Electricity—imports: 0 kWh (2005)

Oil—production: 0 bbl/day (2005 est.)

Oil—consumption: 18,600 bbl/day (2005 est.)

Oil—exports: 0 bbl/day (2004)

Oil—imports: 18,210 bbl/day (2004)

Oil—proved reserves: 0 bbl (1 January 2006 est.)

Natural gas—production: 0 cu m (2005 est.)

Natural gas—consumption: 0 cu m (2005 est.)

Natural gas—exports: 0 cu m (2005 est.)

Natural gas—imports: 0 cu m (2005)

Natural gas—proved reserves: 0 cu m (1 January 2006 est.)

Current account balance: -$459 million (2007 est.)

Exports: $3.278 billion f.o.b. (2007 est.)

Exports—commodities: machinery and transport equipment, manufactures

Exports—partners: France 15.2%, Singapore 13.2%, US 13%, Germany 12.5%, UK 9.5%, Japan 4.9%, Hong Kong 4.2% (2006)

Imports: $4.113 billion f.o.b. (2007 est.)

Imports—commodities: machinery and transport equipment, manufactured and semi-manufactured goods; food, drink, tobacco

Imports—partners: Italy 28%, UK 10.5%, France 8.7%, Germany 7.6%, Singapore 6.8%, US 5.6% (2006)

Economic aid—recipient: $6.19 million (2004)

Reserves of foreign exchange and gold: $3.694 billion (31 December 2007 est.)

Debt—external: $188.8 million (2005)

Stock of direct foreign investment—at home: $NA

Stock of direct foreign investment—abroad: $NA

Market value of publicly traded shares: $4.097 billion (2005)

Currency (code): euro (EUR) as of 1 January 2008; Maltese lira (MTL) before then

Currency code: MTL
Exchange rates: euros per US dollar—0.6795 (January 2008), Maltese liri per US dollar—0.3106 (2007), 0.37 (2006), 0.34578 (2005), 0.34466 (2004), 0.37723 (2003)
Fiscal year: calendar year

COMMUNICATIONS

Telephones—main lines in use: 202,300 (2006)
Telephones—mobile cellular: 346,800 (2006)
Telephone system:
general assessment: automatic system satisfies normal requirements; fixed-line teledensity 50 per 100 persons; mobile-cellular teledensity about 90 per 100 persons
domestic: submarine cable and microwave radio relay between islands
international: country code—356; submarine cable connects to Italy; satellite earth station—1 Intelsat (Atlantic Ocean)
Radio broadcast stations: AM 1, FM 18, shortwave 6 (1999)
Radios: 255,000 (1997)
Television broadcast stations: 5 (2006)
Televisions: 280,000 (1997)
Internet country code: .mt
Internet hosts: 21,386 (2007)

Internet Service Providers (ISPs): 6 (2002)
Internet users: 127,200 (2005)

TRANSPORTATION

Airports: 1 (2007)
Airports—with paved runways:
total: 1
over 3,047 m: 1 (2007)
Roadways:
total: 2,227 km
paved: 2,014 km
unpaved: 213 km (2004)
Merchant marine:
total: 1,281 ships (1000 GRT or over) 25,213,650 GRT/41,033,203 DWT
by type: bulk carrier 439, cargo 382, chemical tanker 125, combination ore/oil 2, container 65, liquefied gas 15, livestock carrier 1, passenger 15, passenger/cargo 14, petroleum tanker 132, refrigerated cargo 41, roll on/roll off 31, specialized tanker 2, vehicle carrier 17
foreign-owned: 1,197 (Austria 1, Azerbaijan 3, Bangladesh 3, Belgium 10, Bulgaria 15, Canada 15, China 13, Croatia 12, Cyprus 15, Denmark 10, Estonia 7, France 4, Germany 67, Greece 448, Hong Kong 1, Iceland 7, India 3, Iran 24, Israel 21, Italy 45, Japan 3, South Korea 3, Latvia 36, Lebanon 12, Libya 3, Monaco 1, Netherlands 3, Norway 71, Pakistan 2, Poland 25, Portugal 3, Romania 10, Russia 66, Slovenia 3, Spain 1, Sweden 1, Switzerland 22, Syria 4, Turkey 143, Ukraine 28, UAE 10, UK 12, US 11)
registered in other countries: 4 (Panama 2, Portugal 1, St Vincent and The Grenadines 1) (2007)
Ports and terminals: Marsaxlokk (Malta Freeport), Valletta

MILITARY

Military branches: Armed Forces of Malta (AFM; includes air and maritime elements) (2007)
Military service age and obligation: 17 years 6 months of age for voluntary military service; no conscription (2008)
Manpower available for military service:
males age 16–49: 96,309
females age 16–49: 92,242 (2008 est.)
Manpower fit for military service:
males age 16–49: 80,227
females age 16–49: 76,623 (2008 est.)
Military expenditures—percent of GDP: 0.7% (2006 est.)

TRANSNATIONAL ISSUES

Disputes—international: none
Illicit drugs: minor transshipment point for hashish from North Africa to Western Europe

MARSHALL ISLANDS

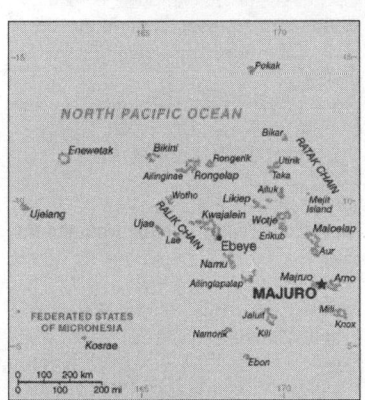

INTRODUCTION

Background: After almost four decades under US administration as the easternmost part of the UN Trust Territory of the Pacific Islands, the Marshall Islands attained independence in 1986 under a Compact of Free Association. Compensation claims continue as a result of US nuclear testing on some of the atolls between 1947 and 1962. The Marshall Islands hosts the US Army Kwajalein Atoll (USAKA) Reagan Missile Test Site, a key installation in the US missile defense network.

GEOGRAPHY

Location: Oceania, two archipelagic island chains of 29 atolls, each made up of many small islets, and five single islands in the North Pacific Ocean, about half way between Hawaii and Australia
Geographic coordinates: 9 00 N, 168 00 E
Map references: Oceania
Area:
total: 181.3 sq km
land: 181.3 sq km
water: 0 sq km
note: the archipelago includes 11,673 sq km of lagoon waters and includes the atolls of Bikini, Enewetak, Kwajalein, Majuro, Rongelap, and Utirik
Area—comparative: about the size of Washington, DC
Land boundaries: 0 km
Coastline: 370.4 km
Maritime claims:
territorial sea: 12 nm

contiguous zone: 24 nm
exclusive economic zone: 200 nm
Climate: tropical; hot and humid; wet season May to November; islands border typhoon belt
Terrain: low coral limestone and sand islands
Elevation extremes:
lowest point: Pacific Ocean 0 m
highest point: unnamed location on Likiep 10 m
Natural resources: coconut products, marine products, deep seabed minerals
Land use:
arable land: 11.11%
permanent crops: 44.44%
other: 44.45% (2005)
Irrigated land: 0 sq km
Natural hazards: infrequent typhoons
Environment—current issues: inadequate supplies of potable water; pollution of Majuro lagoon from household waste and discharges from fishing vessels
Environment—international agreements: *party to:* Biodiversity, Climate Change, Climate Change-Kyoto Protocol, Desertification, Hazardous Wastes, Law of the Sea, Ozone Layer

417

Protection, Ship Pollution, Wetlands, Whaling

signed, but not ratified: none of the selected agreements

Geography—note: the Marshall Islands Bikini and Enewetak are former US nuclear test sites; Kwajalein, the famous World War II battleground, is used as a US missile test range; island city of Ebeye is the second largest settlement in the Marshall Islands, after the capital of Majuro, and one of the most densely populated locations in the Pacific

PEOPLE

Population: 63,174 (July 2008 est.)

Age structure:

0–14 years: 38.5% (male 12,404/female 11,946)

15–64 years: 58.6% (male 18,937/female 18,095)

65 years and over: 2.8% (male 869/female 923) (2008 est.)

Median age: *total:* 21 years

male: 21 years

female: 20.9 years (2008 est.)

Population growth rate: 2.142% (2008 est.)

Birth rate: 31.52 births/1,000 population (2008 est.)

Death rate: 4.57 deaths/1,000 population (2008 est.)

Net migration rate: -5.52 migrant(s)/1,000 population (2008 est.)

Sex ratio:

at birth: 1.05 male(s)/female

under 15 years: 1.04 male(s)/female

15–64 years: 1.05 male(s)/female

65 years and over: 0.94 male(s)/female

total population: 1.04 male(s)/female (2008 est.)

Infant mortality rate:

total: 26.36 deaths/1,000 live births

male: 29.58 deaths/1,000 live births

female: 22.98 deaths/1,000 live births (2008 est.)

Life expectancy at birth:

total population: 70.9 years

male: 68.88 years

female: 73.03 years (2008 est.)

Total fertility rate: 3.68 children born/woman (2008 est.)

HIV/AIDS—adult prevalence rate: NA

HIV/AIDS—people living with HIV/AIDS: NA

HIV/AIDS—deaths: NA

Nationality:

noun: Marshallese (singular and plural)

adjective: Marshallese

Ethnic groups: Micronesian

Religions: Protestant 54.8%, Assembly of God 25.8%, Roman Catholic 8.4%, Bukot nan Jesus 2.8%, Mormon 2.1%, other Christian 3.6%, other 1%, none 1.5% (1999 census)

Languages: Marshallese (official) 98.2%, other languages 1.8% (1999 census)

note: English (official), widely spoken as a second language

Literacy: *definition:* age 15 and over can read and write

total population: 93.7%

male: 93.6%

female: 93.7% (1999)

GOVERNMENT

Country name:

conventional long form: Republic of the Marshall Islands

conventional short form: Marshall Islands

local long form: Republic of the Marshall Islands

local short form: Marshall Islands

abbreviation: RMI

former: Trust Territory of the Pacific Islands, Marshall Islands District

Government type: constitutional government in free association with the US; the Compact of Free Association entered into force 21 October 1986 and the Amended Compact entered into force in May 2004

Capital: *name:* Majuro

geographic coordinates: 7 06 N, 171 23 E

time difference: UTC+12 (17 hours ahead of Washington, DC during Standard Time)

Administrative divisions: 33 municipalities; Ailinginae, Ailinglaplap, Ailuk, Arno, Aur, Bikar, Bikini, Bokak, Ebon, Enewetak, Erikub, Jabat, Jaluit, Jemo, Kili, Kwajalein, Lae, Lib, Likiep, Majuro, Maloelap, Mejit, Mili, Namorik, Namu, Rongelap, Rongrik, Toke, Ujae, Ujelang, Utirik, Wotho, Wotje

Independence: 21 October 1986 (from the US-administered UN trusteeship)

National holiday: Constitution Day, 1 May (1979)

Constitution: 1 May 1979

Legal system: based on adapted Trust Territory laws, acts of the legislature, municipal, common, and customary laws; has not accepted compulsory ICJ jurisdiction

Suffrage: 18 years of age; universal

Executive branch:

chief of state: President Litokwa TOMEING (since 7 January 2008); note—the president is both the chief of state and head of government

head of government: President Litokwa TOMEING (since 7 January 2008)

cabinet: Cabinet selected by the president from among the members of the legislature

elections: president elected by Parliament from among its members for a four-year term; election last held 7 January 2008 (next to be held in 2012)

election results: Litokwa TOMEING elected president; TOMEING received 18 votes to 15 for incumbent NOTE

Legislative branch: unicameral legislature or Nitijela (33 seats; members elected by popular vote to serve four-year terms)

elections: last held 19 November 2007 (next to be held by November 2011)

election results: percent of vote by party—NA; seats by party—independents 4

note: the Council of Chiefs or Ironij is a 12-member body comprised of tribal chiefs that advises on matters affecting customary law and practice

Judicial branch: Supreme Court; High Court; Traditional Rights Court

Political parties and leaders: traditionally there have been no formally organized political parties; what has existed more closely resembles factions or interest groups because they do not have party headquarters, formal platforms, or party structures; the following two "groupings" have competed in legislative balloting in recent years—Aelon Kein Ad Party [Michael KABUA] and United Democratic Party or UDP [Litokwa TOMEING]

Political pressure groups and leaders: NA

International organization participation: ACP, ADB, FAO, G-77, IAEA, IBRD, ICAO, ICCt, IDA, IFC, ILO, IMF, IMO, IMSO, Interpol, IOC, ITU, OPCW, PIF, Sparteca, SPC, UN, UNCTAD, UNESCO, WHO

Diplomatic representation in the US:

chief of mission: Ambassador Banny DE BRUM

chancery: 2433 Massachusetts Avenue NW, Washington, DC 20008

telephone: [1] (202) 234-5414

FAX: [1] (202) 232-3236

consulate(s) general: Honolulu

Diplomatic representation from the US:

chief of mission: Ambassador Clyde BISHOP

embassy: Oceanside, Mejen Weto, Long Island, Majuro

mailing address: P. O. Box 1379, Majuro, Republic of the Marshall Islands 96960-1379

telephone: [692] 247-4011

FAX: [692] 247-4012

Flag description: blue with two stripes radiating from the lower hoist-side corner—orange (top) and white; there is a white star with four large rays and 20 small rays on the hoist side above the two stripes

ECONOMY

Economy—overview: US Government assistance is the mainstay of this tiny island economy. The Marshall Islands received more than $1 billion in aid from the US from 1986–2002. Agricultural production, primarily subsistence, is concentrated on small farms; the most important commercial crops are coconuts and breadfruit. Small-scale industry is limited to handicrafts, tuna processing, and copra. The tourist industry, now a small source of foreign exchange employing less than 10% of the labor force, remains the best hope for future added income. The islands have few natural resources, and imports far exceed exports. Under the terms of the Amended Compact of Free Association, the US will provide millions of dollars per year to the Marshall Islands (RMI) through 2023, at which time a Trust Fund made up of US and RMI contributions will begin perpetual annual payouts. Government downsizing, drought, a drop in construction, the decline in tourism, and less income from the renewal of fishing vessel licenses have held GDP growth to an average of 1% over the past decade.

GDP (purchasing power parity): $115 million (2001 est.)

GDP (official exchange rate): $144 million (2005)

GDP—real growth rate: 3.5% (2005 est.)

GDP—per capita (PPP): $2,900 (2005 est.)

GDP—composition by sector:
agriculture: 31.7%
industry: 14.9%
services: 53.4% (2004 est.)

Labor force: 14,680 (2000)

Labor force—by occupation: *agriculture:* 21.4%
industry: 20.9%
services: 57.7% (2000)

Unemployment rate: 30.9% (2000 est.)

Population below poverty line: NA%

Household income or consumption by percentage share:
lowest 10%: NA%
highest 10%: NA%

Inflation rate (consumer prices): 3% (2005 est.)

Budget:
revenues: $42 million
expenditures: $40 million (1999)

Agriculture—products: coconuts, tomatoes, melons, taro, breadfruit, fruits; pigs, chickens

Industries: copra, tuna processing, tourism, craft items (from seashells, wood, and pearls)

Industrial production growth rate: NA%

Electricity—production by source:
fossil fuel: 99%

hydro: 0%
nuclear: 0%
other: 1% (solar)

Exports: $9.1 million f.o.b. (2000)

Exports—commodities: copra cake, coconut oil, handicrafts, fish

Exports—partners: US, Japan, Australia, China (2006)

Imports: $54.7 million f.o.b. (2000)

Imports—commodities: foodstuffs, machinery and equipment, fuels, beverages and tobacco

Imports—partners: US, Japan, Australia, NZ, Singapore, Fiji, China, Philippines (2006)

Economic aid—recipient: $56.56 million (2005)

Debt—external: $86.5 million (FY99/00 est.)

Currency (code): US dollar (USD)

Currency code: USD

Exchange rates: the US dollar is used

Fiscal year: 1 October—30 September

COMMUNICATIONS

Telephones—main lines in use: 4,500 (2004)

Telephones—mobile cellular: 600 (2004)

Telephone system:
general assessment: digital switching equipment; modern services include telex, cellular, Internet, international calling, caller ID, and leased data circuits
domestic: Majuro Atoll and Ebeye and Kwajalein islands have regular, seven-digit, direct-dial telephones; other islands interconnected by high frequency radiotelephone (used mostly for government purposes) and mini-satellite telephones
international: country code—692; satellite earth stations—2 Intelsat (Pacific Ocean); US Government satellite communications system on Kwajalein (2001)

Radio broadcast stations: AM 1, FM 3, shortwave 0 (additionally, the US Armed Forces Radio and Television Services (Central Pacific Network) operate one FM and one AM station on Kwajalein) (2005)

Radios: NA

Television broadcast stations: 2 (both are US military stations; Marshalls Broadcasting Service, a cable company, operates on Majuro) (2005)

Televisions: NA

Internet country code: .mh

Internet hosts: 3 (2007)

Internet Service Providers (ISPs): 1 (2002)

Internet users: 2,200 (2006)

TRANSPORTATION

Airports: 15 (2007)

Airports—with paved runways:
total: 4
1,524 to 2,437 m: 3
914 to 1,523 m: 1 (2007)

Airports—with unpaved runways:
total: 11
914 to 1,523 m: 10
under 914 m: 1 (2007)

Roadways:
total: 64.5 km
paved: 64.5 km
note: paved roads on major islands (Majuro, Kwajalein), otherwise stone-, coral-, or laterite-surfaced roads and tracks (2002)

Merchant marine:
total: 902 ships (1000 GRT or over) 33,260,440 GRT/55,644,008 DWT
by type: barge carrier 2, bulk carrier 215, cargo 61, carrier 1, chemical tanker 165, combination ore/oil 6, container 171, liquefied gas 28, passenger 6, petroleum tanker 228, refrigerated cargo 2, roll on/roll off 10, specialized tanker 2, vehicle carrier 5
foreign-owned: 857 (Australia 1, Belgium 1, Bermuda 5, Canada 4, Chile 4, China 3, Croatia 4, Cyprus 39, Denmark 9, Finland 2, Germany 214, Greece 226, Hong Kong 4, Italy 3, Japan 5, South Korea 3, Latvia 10, Malaysia 3, Monaco 7, Netherlands 5, Norway 62, Romania 1, Russia 4, Saudi Arabia 4, Singapore 12, Slovenia 3, Spain 3, Sweden 1, Switzerland 14, Turkey 41, UAE 14, UK 17, US 129) (2007)

Ports and terminals: Majuro

MILITARY

Military branches: no regular military forces; under the 1983 Compact of Free Association, the US has full authority and responsibility for security and defense of the Marshall Islands; Marshall Islands Police (2008)

Manpower available for military service:
males age 16–49: 15,708 (2008 est.)

Manpower fit for military service:
males age 16–49: 12,864 (2008 est.)

Manpower reaching militarily significant age annually:
males age 16–49: 512 (2008 est.)

Military expenditures—percent of GDP: NA

Military—note: defense is the responsibility of the US

TRANSNATIONAL ISSUES

Disputes—international: claims US territory of Wake Island

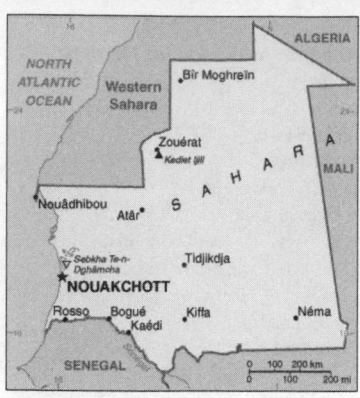

INTRODUCTION

Background: Independent from France in 1960, Mauritania annexed the southern third of the former Spanish Sahara (now Western Sahara) in 1976, but relinquished it after three years of raids by the Polisario guerrilla front seeking independence for the territory. Maaouya Ould Sid Ahmed TAYA seized power in a coup in 1984 and ruled Mauritania with a heavy hand for over two decades. A series of presidential elections that he held were widely seen as flawed. A bloodless coup in August 2005 deposed President TAYA and ushered in a military council that oversaw a transition to democratic rule. Independent candidate Sidi Ould Cheikh ABDALLAHI was inaugurated in April 2007 as Mauritania's first freely and fairly elected president. The country continues to experience ethnic tensions among its black population (Afro-Mauritanians) and White and Black Moor (Arab-Berber) communities, although the new government is attempting to ameliorate some of these tensions.

GEOGRAPHY

Location: Northern Africa, bordering the North Atlantic Ocean, between Senegal and Western Sahara
Geographic coordinates: 20 00 N, 12 00 W
Map references: Africa
Area: *total:* 1,030,700 sq km
land: 1,030,400 sq km
water: 300 sq km
Area—comparative: slightly larger than three times the size of New Mexico
Land boundaries: *total:* 5,074 km
border countries: Algeria 463 km, Mali 2,237 km, Senegal 813 km, Western Sahara 1,561 km

Coastline: 754 km
Maritime claims:
territorial sea: 12 nm
contiguous zone: 24 nm
exclusive economic zone: 200 nm
continental shelf: 200 nm or to the edge of the continental margin
Climate: desert; constantly hot, dry, dusty
Terrain: mostly barren, flat plains of the Sahara; some central hills
Elevation extremes:
lowest point: Sebkhet Te-n-Dghamcha -5 m
highest point: Kediet Ijill 915 m
Natural resources: iron ore, gypsum, copper, phosphate, diamonds, gold, oil, fish
Land use:
arable land: 0.2%
permanent crops: 0.01%
other: 99.79% (2005)
Irrigated land: 490 sq km (2002)
Total renewable water resources: 11.4 cu km (1997)
Freshwater withdrawal (domestic/industrial/agricultural): *total:* 1.7 cu km/yr (9%/3%/88%)
per capita: 554 cu m/yr (2000)
Natural hazards: hot, dry, dust/sand-laden sirocco wind blows primarily in March and April; periodic droughts
Environment—current issues: overgrazing, deforestation, and soil erosion aggravated by drought are contributing to desertification; limited natural fresh water resources away from the Senegal, which is the only perennial river; locust infestation
Environment—international agreements: *party to:* Biodiversity, Climate Change, Climate Change-Kyoto Protocol, Desertification, Endangered Species, Hazardous Wastes, Law of the Sea, Ozone Layer Protection, Ship Pollution, Wetlands, Whaling
signed, but not ratified: none of the selected agreements
Geography—note: most of the population concentrated in the cities of Nouakchott and Nouadhibou and along the Senegal River in the southern part of the country

PEOPLE

Population: 3,364,940 (July 2008 est.)
Age structure:
0–14 years: 45.3% (male 763,845/female 759,957)
15–64 years: 52.5% (male 872,924/female 894,980)
65 years and over: 2.2% (male 29,147/female 44,087) (2008 est.)

Median age:
total: 17.2 years
male: 16.9 years
female: 17.4 years (2008 est.)
Population growth rate: 2.852% (2008 est.)
Birth rate: 40.14 births/1,000 population (2008 est.)
Death rate: 11.61 deaths/1,000 population (2008 est.)
Net migration rate: NA
Sex ratio:
at birth: 1.03 male(s)/female
under 15 years: 1.01 male(s)/female
15–64 years: 0.98 male(s)/female
65 years and over: 0.66 male(s)/female
total population: 0.98 male(s)/female (2008 est.)
Infant mortality rate:
total: 66.65 deaths/1,000 live births
male: 69.69 deaths/1,000 live births
female: 63.52 deaths/1,000 live births (2008 est.)
Life expectancy at birth:
total population: 53.91 years
male: 51.61 years
female: 56.28 years (2008 est.)
Total fertility rate: 5.69 children born/woman (2008 est.)
HIV/AIDS—adult prevalence rate: 0.6% (2003 est.)
HIV/AIDS—people living with HIV/AIDS: 9,500 (2003 est.)
HIV/AIDS—deaths: fewer than 500 (2003 est.)
Major infectious diseases:
degree of risk: high
food or waterborne diseases: bacterial and protozoal diarrhea, hepatitis A, and typhoid fever
vectorborne diseases: malaria and Rift Valley fever (2008)
Nationality:
noun: Mauritanian(s)
adjective: Mauritanian
Ethnic groups: mixed Moor/black 40%, Moor 30%, black 30%
Religions: Muslim 100%
Languages: Arabic (official), Pulaar, Soninke, French, Hassaniya, Wolof
Literacy:
definition: age 15 and over can read and write
total population: 51.2%
male: 59.5%
female: 43.4% (2000 census)

GOVERNMENT

Country name:
conventional long form: Islamic Republic of Mauritania
conventional short form: Mauritania

local long form: Al Jumhuriyah al Islamiyah al Muritaniyah
local short form: Muritaniyah
Government type: Democratic Republic
Capital: *name:* Nouakchott
geographic coordinates: 18 07 N, 16 02 W
time difference: UTC 0 (5 hours ahead of Washington, DC during Standard Time)
Administrative divisions: 12 regions (regions, singular—region) and 1 capital district*; Adrar, Assaba, Brakna, Dakhlet Nouadhibou, Gorgol, Guidimaka, Hodh Ech Chargui, Hodh El Gharbi, Inchiri, Nouakchott*, Tagant, Tiris Zemmour, Trarza
Independence: 28 November 1960 (from France)
National holiday: Independence Day, 28 November (1960)
Constitution: 12 July 1991
Legal system: a combination of Islamic law and French civil law; has not accepted compulsory ICJ jurisdiction
Suffrage: 18 years of age; universal
Executive branch:
chief of state: Sidi Ould Cheikh ABDEL-LAHI (since 19 April 2007)
head of government: Prime Minister Yahya Ould Ahmed El WAGHEF (since 6 May 2008)
cabinet: Council of Ministers
elections: president elected by popular vote for a five-year term (eligible for a second consecutive term); election last held 11 March 2007 with a runoff between the two leading candidates held on 25 March 2007 (next to be held 2012); prime minister appointed by the president
election results: percent of vote—(second round) Sidi Ould Cheikh ABDELLAHI 52.8%, Ahmed Ould DADDAH 47.2%
Legislative branch: bicameral legislature consists of the Senate or Majlis al-Shuyukh (56 seats; 53 members elected by municipal leaders and 3 members elected by Mauritanians abroad to serve six-year terms; a portion of seats up for election every two years) and the National Assembly or Majlis al-Watani (95 seats; members elected by popular vote to serve five-year terms)
elections: Senate—last held 21 January and 4 February 2007 (next to be held 2009); National Assembly—last held 19 November and 3 December 2006 (next to be held in 2011)
election results: Senate—percent of vote by party—NA; seats by party—Mithaq (coalition of independents and parties associated with the former regime) 37, CFCD (coalition of political parties) 15, representatives of the diaspora 3, undecided 1; National Assembly—percent of vote by party—NA; seats by party—

Mithaq 51 (independents 37, PRDR 7, UDP 3, RDU 3, Alternative (El-Badil) 1), CFCD 41 (RFD 16, UFP 9, APP 6, Centrist Reformists 4, HATEM-PMUC 3, RD 2, PUDS 1), RNDLE 1, UCD 1, FP 1
Judicial branch: Supreme Court or Cour Supreme; Court of Appeals; lower courts
Political parties and leaders: Alternative or El-Badil; Centrist Reformists (independent moderate Islamists); Coalition for Forces for Democratic Change or CFCD (coalition of political parties including APP, Centrist Reformists (independent moderate Islamists), HATEM-PMUC, PUDS, RD, RFD, UFP); Democratic and Social Republican Party or PRDS; Democratic Renewal or RD; Mauritanian Party for Unity and Change or HATEM-PMUC; Mithaq (coalition of independents and parties associated with the former regime including Alternative or El-Badil, PRDR, UDP, RDU); National Rally for Freedom, Democracy and Equality or RNDLE; Popular Front or FP [Ch'bih Ould CHEIKH MALAININE]; Popular Progressive Alliance or APP [Messoud Ould BOULKHEIR]; Rally of Democratic Forces or RFD [Ahmed Ould DADDAH]; Rally for Democracy and Unity or RDU [Ahmed Ould SIDI BABA]; Republican Party for Democracy and Renewal or PRDR [Boullah Ould MOGUEYA] (formerly ruling Democratic and Social Republican Party or PRDS); Socialist and Democratic Unity Party or PUDS; Union for Democracy and Progress or UDP [Naha Mint MOUKNASS]; Union of Democratic Centre or UCD; Union of the Forces for Progress or UFP
Political pressure groups and leaders: Arab nationalists; Ba'thists; General Confederation of Mauritanian Workers or CGTM [Abdallahi Ould MOHAMED, secretary general]; Independent Confederation of Mauritanian Workers or CLTM [Samory Ould BEYE]; Islamists; Mauritanian Workers Union or UTM [Mohamed Ely Ould BRAHIM, secretary general]
International organization participation: ABEDA, ACCT, ACP, AfDB, AFESD, AMF, AMU, AU, CAEU, FAO, G-77, IAEA, IBRD, ICAO, ICRM, IDA, IDB, IFAD, IFC, IFRCS, ILO, IMF, IMO, Interpol, IOC, IOM, ITSO, ITU, ITUC, LAS, MIGA, NAM, OIC, OIF, OPCW, UN, UNCTAD, UNESCO, UNIDO, UNWTO, UPU, WCL, WCO, WHO, WIPO, WMO, WTO
Diplomatic representation in the US: *chief of mission:* Ambassador Ibrahima DIA

chancery: 2129 Leroy Place NW, Washington, DC 20008
telephone: [1] (202) 232-5700, 5701
FAX: [1] (202) 319-2623
Diplomatic representation from the US: *chief of mission:* Ambassador Mark M. BOULWARE
embassy: 288 Rue Abdallaye (between Presidency building and Spanish Embassy), Nouakchott
mailing address: BP 222, Nouakchott
telephone: [222] 525-2660/525-2663
FAX: [222] 525-1592
Flag description: green with a yellow five-pointed star above a yellow, horizontal crescent; the closed side of the crescent is down; the crescent, star, and color green are traditional symbols of Islam

ECONOMY

Economy—overview: Half the population still depends on agriculture and livestock for a livelihood, even though many of the nomads and subsistence farmers were forced into the cities by recurrent droughts in the 1970s and 1980s. Mauritania has extensive deposits of iron ore, which account for nearly 40% of total exports. The nation's coastal waters are among the richest fishing areas in the world, but overexploitation by foreigners threatens this key source of revenue. The country's first deepwater port opened near Nouakchott in 1986. In the past, drought and economic mismanagement resulted in a buildup of foreign debt, which now stands at more than three times the level of annual exports. In February 2000, Mauritania qualified for debt relief under the Heavily Indebted Poor Countries (HIPC) initiative and in December 2001 received strong support from donor and lending countries at a triennial Consultative Group review. A new investment code approved in December 2001 improved the opportunities for direct foreign investment. Ongoing negotiations with the IMF involve problems of economic reforms and fiscal discipline. In 2001, exploratory oil wells in tracts 80 km offshore indicated potential extraction at current world oil prices. Oil prospects, while initially promising, have failed to materialize. Meantime the government emphasizes reduction of poverty, improvement of health and education, and promoting privatization of the economy.
GDP (purchasing power parity): $5.947 billion (2007 est.)
GDP (official exchange rate): $2.756 billion (2007 est.)
GDP—real growth rate: 0.9% (2007 est.)
GDP—per capita (PPP): $2,000 (2007 est.)

GDP—composition by sector:
agriculture: 25%
industry: 29%
services: 46% (2001 est.)
Labor force: 786,000 (2001)
Labor force—by occupation: *agriculture:* 50%
industry: 10%
services: 40% (2001 est.)
Unemployment rate: 20% (2004 est.)
Population below poverty line: 40% (2004 est.)
Household income or consumption by percentage share:
lowest 10%: 2.5%
highest 10%: 29.5% (2000)
Distribution of family income—Gini index: 39 (2000)
Inflation rate (consumer prices): 7.3% (2007 est.)
Budget: *revenues:* $421 million
expenditures: $378 million (2002 est.)
Agriculture—products: dates, millet, sorghum, rice, corn; cattle, sheep
Industries: fish processing, mining of iron ore and gypsum
Industrial production growth rate: 2% (2000 est.)
Electricity—production: 248 million kWh (2005)
Electricity—production by source:
fossil fuel: 85.9%
hydro: 14.1%
nuclear: 0%
other: 0% (2001)
Electricity—consumption: 230.6 million kWh (2005)
Electricity—exports: 0 kWh (2005)
Electricity—imports: 0 kWh (2005)
Oil—production: 75,000 bbl/day (2006 est.)
Oil—consumption: 20,000 bbl/day (2005 est.)
Oil—exports: 0 bbl/day (2004)
Oil—imports: 19,960 bbl/day (2004)
Oil—proved reserves: 0 bbl (1 January 2006 est.)
Natural gas—production: 0 cu m (2005 est.)
Natural gas—consumption: 0 cu m (2005 est.)
Natural gas—exports: 0 cu m (2005 est.)
Natural gas—imports: 0 cu m (2005)
Natural gas—proved reserves: 0 cu m (1 January 2006 est.)
Current account balance: -$184 million (2007 est.)
Exports: $1.395 billion f.o.b. (2006)
Exports—commodities: iron ore, fish and fish products, gold

Exports—partners: China 26.1%, Italy 11.7%, France 10.5%, Spain 6.9%, Belgium 6.8%, Japan 5.4%, Cote d'Ivoire 4.6% (2006)
Imports: $1.475 billion f.o.b. (2006)
Imports—commodities: machinery and equipment, petroleum products, capital goods, foodstuffs, consumer goods
Imports—partners: France 11.9%, China 8.1%, Belgium 6.8%, US 6.7%, Italy 5.9%, Spain 5.7%, Brazil 5.5% (2006)
Economic aid—recipient: $190.4 million (2005)
Debt—external: NA
Market value of publicly traded shares: $NA
Currency (code): ouguiya (MRO)
Currency code: MRO
Exchange rates: ouguiyas per US dollar—NA (2007), 271.3 (2006), 267.04 (2005), 265.8 (2004), 263.03 (2003)
Fiscal year: calendar year

COMMUNICATIONS

Telephones—main lines in use: 34,900 (2006)
Telephones—mobile cellular: 1.06 million (2006)
Telephone system:
general assessment: limited system of cable and open-wire lines, minor microwave radio relay links, and radiotelephone communications stations; mobile-cellular services expanding rapidly
domestic: Mauritel, the national telecommunications company, was privatized in 2001 but remains the monopoly provider of fixed-line services; fixed-line teledensity 1 per 100 persons; mobile-cellular network coverage extends mainly to urban areas with a teledensity approaching 35 per 100 persons; mostly cable and open-wire lines; a domestic satellite telecommunications system links Nouakchott with regional capitals
international: country code—222; satellite earth stations—3 (1 Intelsat—Atlantic Ocean, 2 Arabsat)
Radio broadcast stations: AM 1, FM 14, shortwave 1 (2001)
Radios: 410,000 (2001)
Television broadcast stations: 1 (2002)
Televisions: 98,000 (2001)
Internet country code: .mr
Internet hosts: 14 (2007)
Internet Service Providers (ISPs): 5 (2001)
Internet users: 100,000 (2006)

TRANSPORTATION

Airports: 25 (2007)
Airports—with paved runways:
total: 8
2,438 to 3,047 m: 3
1,524 to 2,437 m: 5 (2007)
Airports—with unpaved runways:
total: 17
1,524 to 2,437 m: 9
914 to 1,523 m: 7
under 914 m: 1 (2007)
Railways:
717 km
standard gauge: 717 km 1.435-m gauge (2006)
Roadways:
total: 7,660 km
paved: 866 km
unpaved: 6,794 km (1999)
Ports and terminals: Nouadhibou, Nouakchott

MILITARY

Military branches: Mauritanian Armed Forces: Army, Mauritanian Navy (Marine Mauritanienne; includes naval infantry), Air Force (Force Aerienne Islamique de Mauritanie, FAIM) (2008)
Military service age and obligation: 18 years of age (est.); conscript service obligation—2 years; majority of servicemen believed to be volunteers; service in Air Force and Navy is voluntary (2006)
Manpower available for military service:
males age 16–49: 740,675
females age 16–49: 744,709 (2008 est.)
Manpower fit for military service:
males age 16–49: 463,305
females age 16–49: 484,777 (2008 est.)
Military expenditures—percent of GDP: 5.5% (2006)

TRANSNATIONAL ISSUES

Disputes—international: Mauritanian claims to Western Sahara remain dormant
Trafficking in persons: *current situation:* Mauritania is a source and destination country for children trafficked for the purpose of forced labor, begging, and domestic servitude; adults and children are subjected to slavery-related practices rooted in ancestral master-slave relationships in isolated parts of the country where a barter economy exists
tier rating: Tier 2 Watch List—Mauritania is placed on the Tier 2 Watch List for its failure to show evidence of increased efforts to combat trafficking, particularly in the area of law enforcement

MAURITIUS

INTRODUCTION

Background: Although known to Arab and Malay sailors as early as the 10th century, Mauritius was first explored by the Portuguese in 1505; it was subsequently held by the Dutch, French, and British before independence was attained in 1968. A stable democracy

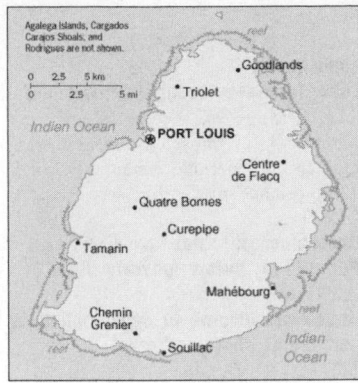

with regular free elections and a positive human rights record, the country has attracted considerable foreign investment and has earned one of Africa's highest per capita incomes. Recent poor weather, declining sugar prices, and declining textile and apparel production, have slowed economic growth, leading to some protests over standards of living in the Creole community.

GEOGRAPHY

Location: Southern Africa, island in the Indian Ocean, east of Madagascar
Geographic coordinates: 20 17 S, 57 33 E
Map references: Political Map of the World
Area:
total: 2,040 sq km
land: 2,030 sq km
water: 10 sq km
note: includes Agalega Islands, Cargados Carajos Shoals (Saint Brandon), and Rodrigues
Area—comparative: almost 11 times the size of Washington, DC
Land boundaries: 0 km
Coastline: 177 km
Maritime claims: measured from claimed archipelagic straight baselines
territorial sea: 12 nm
exclusive economic zone: 200 nm
continental shelf: 200 nm or to the edge of the continental margin
Climate: tropical, modified by southeast trade winds; warm, dry winter (May to November); hot, wet, humid summer (November to May)
Terrain: small coastal plain rising to discontinuous mountains encircling central plateau
Elevation extremes:
lowest point: Indian Ocean 0 m
highest point: Mont Piton 828 m
Natural resources: arable land, fish
Land use: *arable land:* 49.02%
permanent crops: 2.94%
other: 48.04% (2005)

Irrigated land: 220 sq km (2003)
Total renewable water resources: 2.2 cu km (2001)
Freshwater withdrawal (domestic/industrial/agricultural): *total:* 0.61 cu km/yr (25%/14%/60%)
per capita: 488 cu m/yr (2000)
Natural hazards: cyclones (November to April); almost completely surrounded by reefs that may pose maritime hazards
Environment—current issues: water pollution, degradation of coral reefs
Environment—international agreements: *party to:* Antarctic-Marine Living Resources, Biodiversity, Climate Change, Climate Change-Kyoto Protocol, Desertification, Endangered Species, Environmental Modification, Hazardous Wastes, Law of the Sea, Marine Life Conservation, Ozone Layer Protection, Ship Pollution, Wetlands
signed, but not ratified: none of the selected agreements
Geography—note: the main island, from which the country derives its name, is of volcanic origin and is almost entirely surrounded by coral reefs

PEOPLE

Population: 1,274,189 (July 2008 est.)
Age structure:
0–14 years: 23% (male 148,573/female 143,859)
15–64 years: 70.1% (male 443,968/female 449,670)
65 years and over: 6.9% (male 35,269/female 52,850) (2008 est.)
Median age:
total: 31.5 years
male: 30.6 years
female: 32.3 years (2008 est.)
Population growth rate: 0.8% (2008 est.)
Birth rate: 14.64 births/1,000 population (2008 est.)
Death rate: 6.55 deaths/1,000 population (2008 est.)
Net migration rate: -0.09 migrant(s)/1,000 population (2008 est.)
Sex ratio:
at birth: 1.05 male(s)/female
under 15 years: 1.03 male(s)/female
15–64 years: 0.99 male(s)/female
65 years and over: 0.67 male(s)/female
total population: 0.97 male(s)/female (2008 est.)
Infant mortality rate:
total: 12.56 deaths/1,000 live births
male: 14.94 deaths/1,000 live births
female: 10.06 deaths/1,000 live births (2008 est.)
Life expectancy at birth:
total population: 73.75 years
male: 70.28 years
female: 77.4 years (2008 est.)

Total fertility rate: 1.83 children born/woman (2008 est.)
HIV/AIDS—adult prevalence rate: 0.1% (2001 est.)
HIV/AIDS—people living with HIV/AIDS: 700 (2001 est.)
HIV/AIDS—deaths: fewer than 100 (2001 est.)
Nationality:
noun: Mauritian(s)
adjective: Mauritian
Ethnic groups: Indo-Mauritian 68%, Creole 27%, Sino-Mauritian 3%, Franco-Mauritian 2%
Religions: Hindu 48%, Roman Catholic 23.6%, Muslim 16.6%, other Christian 8.6%, other 2.5%, unspecified 0.3%, none 0.4% (2000 census)
Languages: Creole 80.5%, Bhojpuri 12.1%, French 3.4%, English (official; spoken by less than 1% of the population), other 3.7%, unspecified 0.3% (2000 census)
Literacy:
definition: age 15 and over can read and write
total population: 84.4%
male: 88.4%
female: 80.5% (2000 census)

GOVERNMENT

Country name:
conventional long form: Republic of Mauritius
conventional short form: Mauritius
local long form: Republic of Mauritius
local short form: Mauritius
Government type: parliamentary democracy
Capital: *name:* Port Louis
geographic coordinates: 20 09 S, 57 29 E
time difference: UTC+4 (9 hours ahead of Washington, DC during Standard Time)
Administrative divisions: 9 districts and 3 dependencies*; Agalega Islands*, Black River, Cargados Carajos Shoals*, Flacq, Grand Port, Moka, Pamplemousses, Plaines Wilhems, Port Louis, Riviere du Rempart, Rodrigues*, Savanne
Independence: 12 March 1968 (from UK)
National holiday: Independence Day, 12 March (1968)
Constitution: 12 March 1968; amended 12 March 1992
Legal system: based on French civil law system with elements of English common law in certain areas; accepts compulsory ICJ jurisdiction with reservations
Suffrage: 18 years of age; universal
Executive branch:
chief of state: President Sir Anerood JUGNAUTH (since 7 October 2003);

Vice President Abdool Raouf BUNDHUN (since 25 February 2002)
head of government: Prime Minister Navinchandra RAMGOOLAM (since 5 July 2005)
cabinet: Council of Ministers appointed by the president on the recommendation of the prime minister
elections: president and vice president elected by the National Assembly for five-year terms (eligible for a second term); election last held 25 February 2002 (next to be held in May 2008); prime minister and deputy prime minister appointed by the president, responsible to the National Assembly
election results: Karl OFFMANN elected president and Raouf BUNDHUN elected vice president; percent of vote by the National Assembly—NA%; note— Karl OFFMANN stepped down on 30 September 2003
Legislative branch: unicameral National Assembly (70 seats; 62 members elected by popular vote, 8 appointed by the election commission to give representation to various ethnic minorities; to serve five-year terms)
elections: last held on 3 July 2005 (next to be held in 2010)
election results: percent of vote by party— NA; seats by party—AS 38, MSM/MMM 22, OPR 2; appointed seats—AS 4, MSM/MMM 2, OPR 2
Judicial branch: Supreme Court
Political parties and leaders: Alliance Sociale or AS [Navinchandra RAMGOOLAM] (governing coalition— includes MLD, MMSM, MR, MSD, PMXD); Mauritian Labor Party or MLP [Navinchandra RAMGOOLAM]; Mauritian Militant Movement or MMM [Paul BERENGER]; Mauritian Socialist Militant Movement or MMSM [Madan DOLLOO]; Militant Socialist Movement or MSM [Nando BODHA]; Mouvement Republicain or MR [Jayarama VALAYDEN]; Parti Mauricien Xavier Duval or PMXD [Xavier Luc DUVAL]; Rodrigues Movement or MR [Joseph (Nicholas) Von MALLY]; Rodrigues Peoples Organization or OPR [Serge CLAIR]
Political pressure groups and leaders: various labor unions
International organization participation: ACCT, ACP, AfDB, AU, C, COMESA, FAO, G-77, IAEA, IBRD, ICAO, ICCt, ICRM, IDA, IFAD, IFC, IFRCS, IHO, ILO, IMF, IMO, IMSO, InOC, Interpol, IOC, IOM, IPU, ISO, ITSO, ITU, ITUC, MIGA, NAM, OIF, OPCW, PCA, SADC, UN, UNCTAD, UNESCO, UNIDO, UNWTO, UPU, WCL, WCO, WFTU, WHO, WIPO, WMO, WTO

Diplomatic representation in the US:
chief of mission: Ambassador Usha JEETAH
chancery: 4301 Connecticut Avenue NW, Suite 441, Washington, DC 20008
telephone: [1] (202) 244-1491, 1492
FAX: [1] (202) 966-0983
Diplomatic representation from the US:
chief of mission: Ambassador Cesar CABRERA
embassy: 4th Floor, Rogers House, John Kennedy Street, Port Louis
mailing address: international mail: P. O. Box 544, Port Louis; US mail: American Embassy, Port Louis, US Department of State, Washington, DC 20521-2450
telephone: [230] 202-4400
FAX: [230] 208-9534
Flag description: four equal horizontal bands of red (top), blue, yellow, and green

ECONOMY

Economy—overview: Since independence in 1968, Mauritius has developed from a low-income, agriculturally based economy to a middle-income diversified economy with growing industrial, financial, and tourist sectors. For most of the period, annual growth has been in the order of 5% to 6%. This remarkable achievement has been reflected in more equitable income distribution, increased life expectancy, lowered infant mortality, and a much-improved infrastructure. The economy rests on sugar, tourism, textiles and apparel, and financial services, and is expanding into fish processing, information and communications technology, and hospitality and property development. Sugarcane is grown on about 90% of the cultivated land area and accounts for 15% of export earnings. The government's development strategy centers on creating vertical and horizontal clusters of development in these sectors. Mauritius has attracted more than 32,000 offshore entities, many aimed at commerce in India, South Africa, and China. Investment in the banking sector alone has reached over $1 billion. Mauritius, with its strong textile sector, has been well poised to take advantage of the Africa Growth and Opportunity Act (AGOA).
GDP (purchasing power parity): $14.06 billion (2007 est.)
GDP (official exchange rate): $6.959 billion (2007 est.)
GDP—real growth rate: 4.6% (2007 est.)
GDP—per capita (PPP): $11,200 (2007 est.)
GDP—composition by sector:
agriculture: 4.8%

industry: 25%
services: 70.1% (2007 est.)
Labor force: 574,000 (2007 est.)
Labor force—by occupation: agriculture and fishing 9%, construction and industry 30%, transportation and communication 7%, trade, restaurants, hotels 22%, finance 6%, other services 25% (2007)
Unemployment rate: 8.8% (2007 est.)
Population below poverty line: 8% (2006 est.)
Household income or consumption by percentage share:
lowest 10%: NA%
highest 10%: NA%
Distribution of family income—Gini index: 39 (2006 est.)
Inflation rate (consumer prices): 8.8% (2007 est.)
Investment (gross fixed): 25.1% of GDP (2007 est.)
Budget:
revenues: $1.331 billion
expenditures: $1.632 billion; including capital expenditures of $NA (2007 est.)
Public debt: 58.8% of GDP (2007 est.)
Agriculture—products: sugarcane, tea, corn, potatoes, bananas, pulses; cattle, goats; fish
Industries: food processing (largely sugar milling), textiles, clothing, mining, chemicals, metal products, transport equipment, nonelectrical machinery, tourism
Industrial production growth rate: 4.7% (2007 est.)
Electricity—production: 2.35 billion kWh (2006)
Electricity—production by source:
fossil fuel: 90.8%
hydro: 9.2%
nuclear: 0%
other: 0% (2001)
Electricity—consumption: 2.068 billion kWh (2006)
Electricity—exports: 0 kWh (2006)
Electricity—imports: 0 kWh (2006)
Oil—production: 0 bbl/day (2006 est.)
Oil—consumption: 23,650 bbl/day (2006 est.)
Oil—exports: 0 bbl/day (2006)
Oil—imports: 23,650 bbl/day (2006)
Oil—proved reserves: 0 bbl (1 January 2006 est.)
Natural gas—production: 0 cu m (2005 est.)
Natural gas—consumption: 0 cu m (2005 est.)
Natural gas—exports: 0 cu m (2005 est.)
Natural gas—imports: 0 cu m (2005)
Natural gas—proved reserves: 0 cu m (1 January 2006 est.)
Current account balance: -$549 million (2007 est.)

Exports: $2.218 billion f.o.b. (2007 est.)
Exports—commodities: clothing and textiles, sugar, cut flowers, molasses, fish
Exports—partners: UK 32.5%, France 15.1%, UAE 11.4%, US 8.3%, Madagascar 4.8% (2006)
Imports: $3.628 billion f.o.b. (2007 est.)
Imports—commodities: manufactured goods, capital equipment, foodstuffs, petroleum products, chemicals
Imports—partners: France 14.3%, India 13.6%, China 8.6%, South Africa 7.3% (2006)
Economic aid—recipient: $31.93 million (2005)
Reserves of foreign exchange and gold: $1.822 billion (31 December 2007 est.)
Debt—external: $2.136 billion (31 December 2007 est.)
Stock of direct foreign investment—at home: $NA
Stock of direct foreign investment—abroad: $NA
Market value of publicly traded shares: $5.7 billion (2007)
Currency (code): Mauritian rupee (MUR)
Currency code: MUR
Exchange rates: Mauritian rupees per US dollar—31.798 (2007), 31.656 (2006), 29.496 (2005), 27.499 (2004), 27.902 (2003)
Fiscal year: 1 July—30 June

COMMUNICATIONS

Telephones—main lines in use: 357,300 (2006)
Telephones—mobile cellular: 772,400 (2006)

Telephone system:
general assessment: small system with good service
domestic: monopoly over fixed-line services terminated in 2005; fixed-line teledensity roughly 30 per 100 persons; mobile-cellular services launched in 1989 with teledensity in 2006 exceeding 60 per 100 persons
international: country code—230; landing point for the SAFE submarine cable that provides links to Asia and South Africa where it connects to the SAT-3/WASC submarine cable that provides further links to parts of East Africa, and Europe; satellite earth station—1 Intelsat (Indian Ocean); new microwave link to Reunion; HF radiotelephone links to several countries
Radio broadcast stations: AM 4, FM 9, shortwave 0 (2001)
Radios: 420,000 (1997)
Television broadcast stations: 2 (plus several repeaters) (1997)
Televisions: 258,000 (1997)
Internet country code: .mu
Internet hosts: 9,792 (2007)
Internet Service Providers (ISPs): 2 (2000)
Internet users: 182,000 (2006)

TRANSPORTATION

Airports: 5 (2007)
Airports—with paved runways:
total: 2
over 3,047 m: 1
914 to 1,523 m: 1 (2007)
Airports—with unpaved runways:
total: 3
914 to 1,523 m: 2

under 914 m: 1 (2007)
Roadways:
total: 2,020 km
paved: 2,020 km (includes 75 km of expressways) (2005)
Merchant marine:
total: 5 ships (1000 GRT or over) 19,417 GRT/19,700 DWT
by type: bulk carrier 2, passenger/cargo 2, refrigerated cargo 1
foreign-owned: 2 (India 2) (2007)
Ports and terminals: Port Louis

MILITARY

Military branches: no regular military forces; National Police Force, Special Mobile Force, National Coast Guard (2007)
Manpower available for military service:
males age 16–49: 341,018 (2008 est.)
Military expenditures—percent of GDP: 0.3% (2006 est.)

TRANSNATIONAL ISSUES

Disputes—international: Mauritius claims the Chagos Archipelago (UK-administered British Indian Ocean Territory), and its former inhabitants, who reside chiefly in Mauritius; claims French-administered Tromelin Island
Illicit drugs: consumer and transshipment point for heroin from South Asia; small amounts of cannabis produced and consumed locally; significant offshore financial industry creates potential for money laundering, but corruption levels are relatively low and the government appears generally to be committed to regulating its banking industry

MAYOTTE

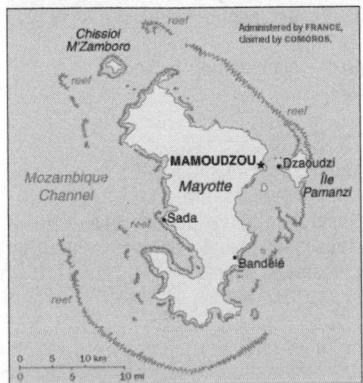

INTRODUCTION

Background: Mayotte was ceded to France along with the other islands of the Comoros group in 1843. It was the only island in the archipelago that voted in 1974 to retain its link with France and forego independence.

GEOGRAPHY

Location: Southern Indian Ocean, island in the Mozambique Channel, about half way between northern Madagascar and northern Mozambique
Geographic coordinates: 12 50 S, 45 10 E
Map references: Africa
Area:
total: 374 sq km
land: 374 sq km
water: 0 sq km
Area—comparative: slightly more than twice the size of Washington, DC
Land boundaries: 0 km
Coastline: 185.2 km
Maritime claims: *territorial sea:* 12 nm

exclusive economic zone: 200 nm
Climate: tropical; marine; hot, humid, rainy season during northeastern monsoon (November to May); dry season is cooler (May to November)
Terrain: generally undulating, with deep ravines and ancient volcanic peaks
Elevation extremes:
lowest point: Indian Ocean 0 m
highest point: Benara 660 m
Natural resources: NEGL
Land use:
arable land: NA%
permanent crops: NA%
other: NA%
Irrigated land: NA
Natural hazards: cyclones during rainy season
Environment—current issues: NA
Geography—note: part of Comoro Archipelago (18 islands)

PEOPLE

Population: 216,306 (July 2008 est.)

Age structure:

0–14 years: 45.5% (male 49,521/female 48,996)

15–64 years: 52.7% (male 61,267/female 52,641)

65 years and over: 1.8% (male 1,971/female 1,910) (2008 est.)

Median age:

total: 17.2 years

male: 18.1 years

female: 16.2 years (2008 est.)

Population growth rate: 3.465% (2008 est.)

Birth rate: 39.79 births/1,000 population (2008 est.)

Death rate: 7.36 deaths/1,000 population (2008 est.)

Net migration rate: 2.22 migrant(s)/1,000 population (2008 est.)

Sex ratio:

at birth: 1.03 male(s)/female

under 15 years: 1.01 male(s)/female

15–64 years: 1.16 male(s)/female

65 years and over: 1.03 male(s)/female

total population: 1.09 male(s)/female (2008 est.)

Infant mortality rate:

total: 57.88 deaths/1,000 live births

male: 63.59 deaths/1,000 live births

female: 52 deaths/1,000 live births (2008 est.)

Life expectancy at birth:

total population: 62.54 years

male: 60.3 years

female: 64.85 years (2008 est.)

Total fertility rate: 5.6 children born/woman (2008 est.)

HIV/AIDS—adult prevalence rate: NA

HIV/AIDS—people living with HIV/AIDS: NA

HIV/AIDS—deaths: NA

Nationality:

noun: Mahorais (singular and plural)

adjective: Mahoran

Ethnic groups: NA

Religions: Muslim 97%, Christian (mostly Roman Catholic) 3%

Languages: Mahorian (a Swahili dialect), French (official language) spoken by 35% of the population

Literacy: NA

GOVERNMENT

Country name:

conventional long form: Territorial Collectivity of Mayotte

conventional short form: Mayotte

Dependency status: departmental collectivity of France

Government type: NA

Capital: *name:* Mamoudzou

geographic coordinates: 12 46 S, 45 13 E

time difference: UTC+3 (8 hours ahead of Washington, DC during Standard Time)

Administrative divisions: none (territorial overseas collectivity of France)

Independence: none (territorial overseas collectivity of France)

National holiday: Bastille Day, 14 July (1789)

Constitution: 4 October 1958 (French Constitution)

Legal system: the laws of France, where applicable, apply

Suffrage: 18 years of age; universal

Executive branch:

chief of state: President Nicolas SARKOZY (since 16 May 2007); represented by Prefect Jean-Paul KIHL (since 17 January 2005)

head of government: President of the General Council Said Omar OILI (since 8 April 2004)

cabinet: NA

elections: French president elected by popular vote for a five-year term; prefect appointed by the French president on the advice of the French Ministry of the Interior; president of the General Council elected by the members of the General Council for a six-year term; next election to be held in 2010

Legislative branch: unicameral General Council or Conseil General (19 seats; members are elected by popular vote to serve three-year terms)

elections: last held 21 and 28 March 2004 (next to be held in 2007)

election results: percent of vote by party—MDM 23.3%, UMP 22.8%, PS 10.2%, MRC 8.9%, FRAP 6.5%, MPM 1.2%, other 27.1%; seats by party—MDM 6, UMP 9, MRC 2, MPM 1, diverse left 1

note: Mayotte elects one member of the French Senate; elections last held 24 September 2001 (next to be held in September 2007); results—percent of vote by party—NA; seats by party—NA; Mayotte also elects one member to the French National Assembly; elections last held 16 June 2002 (next to be held in 2007); results—percent of vote by party—UMP-RPR 55.1%, UDF 44.9%; seats by party—UMP-RPR 1

Judicial branch: Supreme Court or Tribunal Superieur d'Appel

Political parties and leaders: Democratic Front or FD [Youssouf MOUSSA]; Mahoran Popular Movement or MPM [Ahmed MADI]; Federation of Mahorans or UMP-RPR [Mansour KAMARDINE]; Force of the Rally and the Alliance for Democracy or FRAP; Movement for Department Status Mayotte or MDM [Mouhoutar SALIM]; Renewed Communist Party of Mayotte or MRC [Omar SIMBA]; Socialist Party or PS [Ibrahim ABUBACAR] (local branch of French Parti Socialiste); Union for French Democracy or UDF [Henri JEAN-BAPTISTE]

Political pressure groups and leaders: NA

International organization participation: InOC, UPU

Diplomatic representation in the US: none (territorial overseas collectivity of France)

Diplomatic representation from the US: none (territorial overseas collectivity of France)

Flag description: unofficial, local flag with the coat of arms of Mayotte centered on a white field, above which the name of the island appears in red capital letters; the main elements of the coat of arms, flanked on either side by a seahorse, appear above a scroll with the motto RA HACHIRI (We are Vigilant); the only official flag is the national flag of France

ECONOMY

Economy—overview: Economic activity is based primarily on the agricultural sector, including fishing and livestock raising. Mayotte is not self-sufficient and must import a large portion of its food requirements, mainly from France. The economy and future development of the island are heavily dependent on French financial assistance, an important supplement to GDP. Mayotte's remote location is an obstacle to the development of tourism.

GDP (purchasing power parity): $953.6 million (2005 est.)

GDP (official exchange rate): $NA

GDP—real growth rate: NA%

GDP—per capita (PPP): $4,900 (2005 est.)

GDP—composition by sector:

agriculture: NA%

industry: NA%

services: NA%

Labor force: 44,560 (2002)

Unemployment rate: 25.4% (2005)

Population below poverty line: NA%

Household income or consumption by percentage share:

lowest 10%: NA%

highest 10%: NA%

Inflation rate (consumer prices): 1.7% (2005)

Budget:

revenues: $420 million

expenditures: $394 million (2005)

Agriculture—products: vanilla, ylang-ylang (perfume essence), coffee, copra, fish, livestock

Industries: newly created lobster and shrimp industry, construction
Industrial production growth rate: NA%
Electricity—production: NA kWh
Electricity—production by source:
fossil fuel: 0%
hydro: 0%
nuclear: 0%
other: 0%
Electricity—consumption: 139.2 million kWh (2005)
Exports: $6.5 million f.o.b. (2005)
Exports—commodities: ylang-ylang (perfume essence), vanilla, copra, coconuts, coffee, cinnamon
Exports—partners: France 43%, Comoros 36%, Reunion 15% (2006)
Imports: $341 million f.o.b.; note—excludes petroleum imports (2005)
Imports—commodities: food, machinery and equipment, transportation equipment, metals, chemicals
Imports—partners: France 49%, Seychelles 8.8%, China 4.1% (2006)
Economic aid—recipient: $201.3 million; note—extensive French financial assistance (2005)
Debt—external: $NA
Market value of publicly traded shares: $NA
Currency (code): euro (EUR)
Currency code: EUR
Exchange rates: euros per US dollar—0.7345 (2007), 0.7964 (2006), 0.8041 (2005), 0.8054 (2004), 0.886 (2003)
Fiscal year: calendar year

COMMUNICATIONS

Telephones—main lines in use: 10,000 (2002)
Telephones—mobile cellular: 48,100 (2005)
Telephone system:
general assessment: small system administered by French Department of Posts and Telecommunications
domestic: NA
international: country code—262; microwave radio relay and HF radiotelephone communications to Comoros
Radio broadcast stations: AM 1, FM 5, shortwave 0 (2001)

Radios: NA
Television broadcast stations: 3 (2001)
Televisions: 3,500 (1994)
Internet country code: .yt
Internet hosts: 1 (2007)
Internet Service Providers (ISPs): NA
Internet users: NA

TRANSPORTATION

Airports: 1 (2007)
Airports—with paved runways:
total: 1
1,524 to 2,437 m: 1 (2007)
Roadways:
total: 100 km
Ports and terminals: Dzaoudzi

MILITARY

Military—note: defense is the responsibility of France; a small contingent of French forces is stationed on the island

TRANSNATIONAL ISSUES

Disputes—international: claimed by Comoros

MEXICO

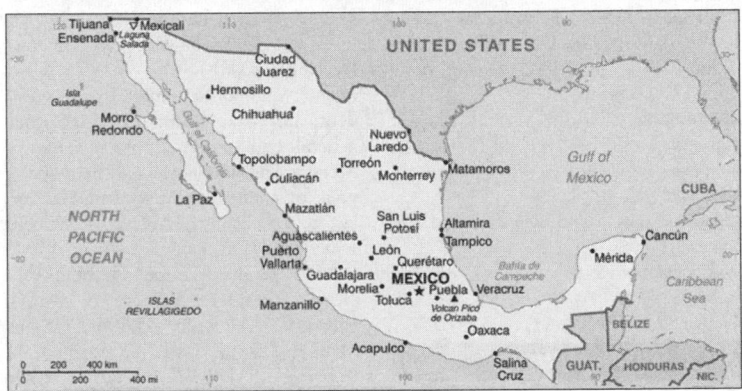

INTRODUCTION

Background: The site of advanced Amerindian civilizations, Mexico came under Spanish rule for three centuries before achieving independence early in the 19th century. A devaluation of the peso in late 1994 threw Mexico into economic turmoil, triggering the worst recession in over half a century. The nation continues to make an impressive recovery. Ongoing economic and social concerns include low real wages, underemployment for a large segment of the population, inequitable income distribution, and few advancement opportunities for the largely Amerindian population in the impoverished southern states. The elections held in 2000 marked the first time since the 1910 Mexican Revolution that an opposition candidate—Vicente FOX of the National Action Party (PAN)—defeated the party in government, the Institutional Revolutionary Party (PRI). He was succeeded in 2006 by another PAN candidate Felipe CALDERON.

GEOGRAPHY

Location: Middle America, bordering the Caribbean Sea and the Gulf of Mexico, between Belize and the US and bordering the North Pacific Ocean, between Guatemala and the US
Geographic coordinates: 23 00 N, 102 00 W
Map references: North America
Area: *total:* 1,972,550 sq km

land: 1,923,040 sq km
water: 49,510 sq km
Area—comparative: slightly less than three times the size of Texas
Land boundaries:
total: 4,353 km
border countries: Belize 250 km, Guatemala 962 km, US 3,141 km
Coastline: 9,330 km
Maritime claims:
territorial sea: 12 nm
contiguous zone: 24 nm
exclusive economic zone: 200 nm
continental shelf: 200 nm or to the edge of the continental margin
Climate: varies from tropical to desert
Terrain: high, rugged mountains; low coastal plains; high plateaus; desert
Elevation extremes:
lowest point: Laguna Salada -10 m
highest point: Volcan Pico de Orizaba 5,700 m
Natural resources: petroleum, silver, copper, gold, lead, zinc, natural gas, timber
Land use: *arable land:* 12.66%
permanent crops: 1.28%
other: 86.06% (2005)
Irrigated land: 63,200 sq km (2003)
Total renewable water resources: 457.2 cu km (2000)
Freshwater withdrawal (domestic/industrial/agricultural): *total:* 78.22 cu km/yr (17%/5%/77%)
per capita: 731 cu m/yr (2000)

Natural hazards: tsunamis along the Pacific coast, volcanoes and destructive earthquakes in the center and south, and hurricanes on the Pacific, Gulf of Mexico, and Caribbean coasts

Environment—current issues: scarcity of hazardous waste disposal facilities; rural to urban migration; natural fresh water resources scarce and polluted in north, inaccessible and poor quality in center and extreme southeast; raw sewage and industrial effluents polluting rivers in urban areas; deforestation; widespread erosion; desertification; deteriorating agricultural lands; serious air and water pollution in the national capital and urban centers along US-Mexico border; land subsidence in Valley of Mexico caused by groundwater depletion *note:* the government considers the lack of clean water and deforestation national security issues

Environment—international agreements: *party to:* Biodiversity, Climate Change, Climate Change-Kyoto Protocol, Desertification, Endangered Species, Hazardous Wastes, Law of the Sea, Marine Dumping, Marine Life Conservation, Ozone Layer Protection, Ship Pollution, Wetlands, Whaling *signed, but not ratified:* none of the selected agreements

Geography—note: strategic location on southern border of US; corn (maize), one of the world's major grain crops, is thought to have originated in Mexico

PEOPLE

Population: 109,955,400 (July 2008 est.)
Age structure:
0–14 years: 29.6% (male 16,619,995/female 15,936,154)
15–64 years: 64.3% (male 34,179,440/female 36,530,154)
65 years and over: 6.1% (male 3,023,185/female 3,666,472) (2008 est.)
Median age:
total: 26 years
male: 24.9 years
female: 27 years (2008 est.)
Population growth rate: 1.142% (2008 est.)
Birth rate: 20.04 births/1,000 population (2008 est.)
Death rate: 4.78 deaths/1,000 population (2008 est.)
Net migration rate: -3.84 migrant(s)/1,000 population (2008 est.)
Sex ratio:
at birth: 1.05 male(s)/female
under 15 years: 1.04 male(s)/female
15–64 years: 0.94 male(s)/female
65 years and over: 0.82 male(s)/female
total population: 0.96 male(s)/female (2008 est.)

Infant mortality rate:
total: 19.01 deaths/1,000 live births
male: 20.91 deaths/1,000 live births
female: 17.02 deaths/1,000 live births (2008 est.)
Life expectancy at birth:
total population: 75.84 years
male: 73.05 years
female: 78.78 years (2008 est.)
Total fertility rate: 2.37 children born/woman (2008 est.)
HIV/AIDS—adult prevalence rate: 0.3% (2003 est.)
HIV/AIDS—people living with HIV/AIDS: 160,000 (2003 est.)
HIV/AIDS—deaths: 5,000 (2003 est.)
Major infectious diseases:
degree of risk: intermediate
food or waterborne diseases: bacterial diarrhea, hepatitis A, and typhoid fever
vectorborne disease: dengue fever
water contact disease: leptospirosis (2008)
Nationality:
noun: Mexican(s)
adjective: Mexican
Ethnic groups: mestizo (Amerindian-Spanish) 60%, Amerindian or predominantly Amerindian 30%, white 9%, other 1%
Religions: Roman Catholic 76.5%, Protestant 6.3% (Pentecostal 1.4%, Jehovah's Witnesses 1.1%, other 3.8%), other 0.3%, unspecified 13.8%, none 3.1% (2000 census)
Languages: Spanish, various Mayan, Nahuatl, and other regional indigenous languages
Literacy:
definition: age 15 and over can read and write
total population: 91%
male: 92.4%
female: 89.6% (2004 est.)

GOVERNMENT

Country name:
conventional long form: United Mexican States
conventional short form: Mexico
local long form: Estados Unidos Mexicanos
local short form: Mexico
Government type: federal republic
Capital: *name:* Mexico (Distrito Federal)
geographic coordinates: 19 26 N, 99 08 W
time difference: UTC-6 (1 hour behind Washington, DC during Standard Time)
daylight saving time: +1hr, begins first Sunday in April; ends last Sunday in October
note: Mexico is divided into three time zones
Administrative divisions: 31 states (estados, singular—estado) and 1 federal district* (distrito federal); Aguascal-

ientes, Baja California, Baja California Sur, Campeche, Chiapas, Chihuahua, Coahuila de Zaragoza, Colima, Distrito Federal*, Durango, Guanajuato, Guerrero, Hidalgo, Jalisco, Mexico, Michoacan de Ocampo, Morelos, Nayarit, Nuevo Leon, Oaxaca, Puebla, Queretaro de Arteaga, Quintana Roo, San Luis Potosi, Sinaloa, Sonora, Tabasco, Tamaulipas, Tlaxcala, Veracruz-Llave, Yucatan, Zacatecas
Independence: 16 September 1810 (declared); 27 September 1821 (recognized by Spain)
National holiday: Independence Day, 16 September (1810)
Constitution: 5 February 1917
Legal system: mixture of US constitutional theory and civil law system; judicial review of legislative acts; accepts compulsory ICJ jurisdiction with reservations
Suffrage: 18 years of age; universal and compulsory (but not enforced)
Executive branch:
chief of state: President Felipe de Jesus CALDERON Hinojosa (since 1 December 2006); note—the president is both the chief of state and head of government
head of government: President Felipe de Jesus CALDERON Hinojosa (since 1 December 2006)
cabinet: Cabinet appointed by the president; note—appointment of attorney general requires consent of the Senate
elections: president elected by popular vote for a single six-year term; election last held on 2 July 2006 (next to be held 1 July 2012)
election results: Felipe CALDERON elected president; percent of vote—Felipe CALDERON 35.89%, Andres Manuel LOPEZ OBRADOR 35.31%, Roberto MADRAZO 22.26%, other 6.54%
Legislative branch: bicameral National Congress or Congreso de la Union consists of the Senate or Camara de Senadores (128 seats; 96 members are elected by popular vote to serve six-year terms, and 32 seats are allocated on the basis of each party's popular vote) and the Federal Chamber of Deputies or Camara Federal de Diputados (500 seats; 300 members are elected by popular vote; remaining 200 members are allocated on the basis of each party's popular vote; to serve three-year terms)
elections: Senate—last held 2 July 2006 for all of the seats (next to be held 1 July 2012); Chamber of Deputies—last held 2 July 2006 (next to be held 5 July 2009)
election results: Senate—percent of vote by party—NA; seats by party—PAN 52,

PRI 33, PRD 26, PVEM 6, CD 5, PT 5, independent 1; Chamber of Deputies—percent of vote by party—NA; seats by party—PAN 207, PRD 127, PRI 106, PVEM 17, CD 17, PT 11, other 15

Judicial branch: Supreme Court of Justice or Suprema Corte de Justicia de la Nacion (justices or ministros are appointed by the president with consent of the Senate)

Political parties and leaders: Convergence for Democracy or CD [Luis MALDONADO Venegas]; Institutional Revolutionary Party or PRI [Beatriz PAREDES]; Labor Party or PT [Alberto ANAYA Gutierrez]; Mexican Green Ecological Party or PVEM [Jorge Emilio GONZALEZ Martinez]; National Action Party (Partido Accion Nacional) or PAN [German MARTINEZ Cazares]; New Alliance Party (Partido Nueva Alianza) or PNA [Jorge Antonio KAHWAGI Macari]; Party of the Democratic Revolution (Partido de la Revolucion Democratica) or PRD [Leonel COTA Montano]; Social Democratic and Peasant Alternative Party (Partido Alternativa Socialdemocrata y Campesina) or Alternativa [Alberto BEGNE Guerra]

Political pressure groups and leaders: Broad Progressive Front or FAP; Businessmen's Coordinating Council or CCE; Confederation of Employers of the Mexican Republic or COPARMEX; Confederation of Industrial Chambers or CONCAMIN; Confederation of Mexican Workers or CTM; Confederation of National Chambers of Commerce or CONCANACO; Coordinator for Foreign Trade Business Organizations or COECE; Federation of Unions Providing Goods and Services or FESEBES; National Chamber of Transformation Industries or CANACINTRA; National Peasant Confederation or CNC; National Small Business Chamber or CANACOPE; National Syndicate of Education Workers or SNTE; National Union of Workers or UNT; Popular Assembly of the People of Oaxaca or APPO; Roman Catholic Church

International organization participation: APEC, BCIE, BIS, CAN (observer), Caricom (observer), CDB, CE (observer), CSN (observer), EBRD, FAO, G-3, G-15, G-24, IADB, IAEA, IBRD, ICAO, ICC, ICCt, ICRM, IDA, IFAD, IFC, IFRCS, IHO, ILO, IMF, IMO, IMSO, Interpol, IOC, IOM, IPU, ISO, ITSO, ITU, ITUC, LAES, LAIA, NAFTA, NAM (observer), NEA, OAS, OECD, OPANAL, OPCW, PCA, RG, UN, UNCTAD, UNESCO, UNHCR, UNIDO, Union Latina, UNITAR, UNWTO, UPU, WCL, WCO, WFTU, WHO, WIPO, WMO, WTO

Diplomatic representation in the US:
chief of mission: Ambassador Arturo SARUKHAN Casamitjana
chancery: 1911 Pennsylvania Avenue NW, Washington, DC 20006
telephone: [1] (202) 728-1600
FAX: [1] (202) 728-1698
consulate(s) general: Atlanta, Austin, Boston, Chicago, Dallas, Denver, El Paso, Houston, Laredo (Texas), Los Angeles, Miami, New Orleans, New York, Nogales (Arizona), Omaha, Orlando, Phoenix, Sacramento, San Antonio, San Diego, San Francisco, San Jose, San Juan (Puerto Rico)
consulate(s): Albuquerque, Brownsville (Texas), Calexico (California), Del Rio (Texas), Detroit, Douglas (Arizona), Eagle Pass (Texas), Fresno (California), Indianapolis (Indiana), Kansas City (Missouri), Laredo (Texas), Las Vegas, Little Rock (Arkansas), McAllen (Texas), New Orleans, Omaha, Orlando, Oxnard (California), Philadelphia, Portland (Oregon), Presidio (Texas), Raleigh, Saint Paul (Minnesota), Salt Lake City, San Bernardino, Santa Ana (California), Seattle, Tucson, Yuma (Arizona)

Diplomatic representation from the US:
chief of mission: Ambassador Antonio O. GARZA, Jr.
embassy: Paseo de la Reforma 305, Colonia Cuauhtemoc, 06500 Mexico, Distrito Federal
mailing address: P. O. Box 9000, Brownsville, TX 78520-9000
telephone: [52] (55) 5080-2000
FAX: [52] (55) 5511-9980
consulate(s) general: Ciudad Juarez, Guadalajara, Monterrey, Tijuana
consulate(s): Hermosillo, Matamoros, Merida, Nogales, Nuevo Laredo

Flag description: three equal vertical bands of green (hoist side), white, and red; the coat of arms (an eagle perched on a cactus with a snake in its beak) is centered in the white band

ECONOMY

Economy—overview: Mexico has a free market economy in the trillion dollar class. It contains a mixture of modern and outmoded industry and agriculture, increasingly dominated by the private sector. Recent administrations have expanded competition in seaports, railroads, telecommunications, electricity generation, natural gas distribution, and airports. Per capita income is one-fourth that of the US; income distribution remains highly unequal. Trade with the US and Canada has tripled since the implementation of NAFTA in 1994. Mexico has 12 free trade agreements with over 40 countries including, Guatemala, Honduras, El Salvador, the European Free Trade Area, and Japan, putting more than 90% of trade under free trade agreements. In 2007, during his first year in office, the Felipe CALDERON administration was able to garner support from the opposition to successfully pass a pension and a fiscal reform. The administration continues to face many economic challenges including the need to upgrade infrastructure, modernize labor laws, and allow private investment in the energy sector. CALDERON has stated that his top economic priorities remain reducing poverty and creating jobs.

GDP (purchasing power parity): $1.346 trillion (2007 est.)

GDP (official exchange rate): $893.4 billion (2007 est.)

GDP—real growth rate: 3.3% (2007 est.)

GDP—per capita (PPP): $12,800 (2007 est.)

GDP—composition by sector:
agriculture: 4%
industry: 26.6%
services: 69.5% (2007 est.)

Labor force: 44.71 million (2007 est.)

Labor force—by occupation: *agriculture:* 18%
industry: 24%
services: 58% (2003)

Unemployment rate: 3.7% plus underemployment of perhaps 25% (2007 est.)

Population below poverty line: 13.8% using food-based definition of poverty; asset based poverty amounted to more than 40% (2006)

Household income or consumption by percentage share:
lowest 10%: 1.2%
highest 10%: 37% (2006)

Distribution of family income—Gini index: 50.9 (2005)

Inflation rate (consumer prices): 4% (2007)

Investment (gross fixed): 20.8% of GDP (2007 est.)

Budget:
revenues: $209.2 billion
expenditures: $209.2 billion (2007 est.)

Public debt: 22.8% of GDP (2007 est.)

Agriculture—products: corn, wheat, soybeans, rice, beans, cotton, coffee, fruit, tomatoes; beef, poultry, dairy products; wood products

Industries: food and beverages, tobacco, chemicals, iron and steel, petroleum, mining, textiles, clothing, motor vehicles, consumer durables, tourism

Industrial production growth rate: 1.4% (2007 est.)

Electricity—production: 222.4 billion kWh (2005)

Electricity—production by source:
fossil fuel: 78.7%
hydro: 14.2%
nuclear: 4.2%
other: 2.9% (2001)

Electricity—consumption: 183.3 billion kWh (2005)

Electricity—exports: 1.597 billion kWh (2005)

Electricity—imports: 470.7 million kWh (2005)

Oil—production: 3.784 million bbl/day (2005 est.)

Oil—consumption: 2.078 million bbl/day (2005 est.)

Oil—exports: 2.268 million bbl/day (2004)

Oil—imports: 308,500 bbl/day (2004)

Oil—proved reserves: 14.7 billion bbl (2007 est.)

Natural gas—production: 41.37 billion cu m (2005 est.)

Natural gas—consumption: 47.5 billion cu m (2005 est.)

Natural gas—exports: 282.9 million cu m (2005 est.)

Natural gas—imports: 9.717 billion cu m (2005)

Natural gas—proved reserves: 434.1 billion cu m (1 January 2006 est.)

Current account balance: -$7.37 billion (2007 est.)

Exports: $271.9 billion f.o.b. (2007 est.)

Exports—commodities: manufactured goods, oil and oil products, silver, fruits, vegetables, coffee, cotton

Exports—partners: US 84.7%, Canada 2.1%, Spain 1.3% (2006)

Imports: $283 billion f.o.b. (2007 est.)

Imports—commodities: metalworking machines, steel mill products, agricultural machinery, electrical equipment, car parts for assembly, repair parts for motor vehicles, aircraft, and aircraft parts

Imports—partners: US 50.9%, China 9.5%, Japan 6%, South Korea 4.2% (2006)

Economic aid—recipient: $189.4 million (2005)

Reserves of foreign exchange and gold: $87.19 billion (31 December 2007 est.)

Debt—external: $179.7 billion (31 December 2007)

Stock of direct foreign investment—at home: $259.5 billion (2007 est.)

Stock of direct foreign investment—abroad: $36.23 billion (2007 est.)

Market value of publicly traded shares: $348.3 billion (2006)

Currency (code): Mexican peso (MXN)

Currency code: MXN

Exchange rates: Mexican pesos per US dollar—10.8 (2007), 10.899 (2006), 10.898 (2005), 11.286 (2004), 10.789 (2003)

Fiscal year: calendar year

COMMUNICATIONS

Telephones—main lines in use: 19.861 million (2006)

Telephones—mobile cellular: 57.016 million (2006)

Telephone system:
general assessment: adequate telephone service for business and government, but the population is poorly served; mobile subscribers far outnumber fixed-line subscribers; domestic satellite system with 120 earth stations; extensive microwave radio relay network; considerable use of fiber-optic cable and coaxial cable
domestic: low telephone density with about 18 fixed lines per 100 persons; privatized in December 1990; despite the opening to competition in January 1997, Telmex remains dominant; legal challenges to Telmex's alleged anti-competitive behavior in the mobile and fixed-line markets culminated in a World Trade Organization ruling in 2004 against Mexico prompting some strengthening of the powers granted Mexico's telecom regulator
international: country code—52; Columbus-2 fiber-optic submarine cable with access to the US, Virgin Islands, Canary Islands, Spain, and Italy; the Americas Region Caribbean Ring System (ARCOS-1) and the MAYA-1 submarine cable system together provide access to Central America, parts of South America and the Caribbean, and the US; satellite earth stations—120 (32 Intelsat, 2 Solidaridad (giving Mexico improved access to South America, Central America, and much of the US as well as enhancing domestic communications), 1 Panamsat, numerous Inmarsat mobile earth stations); linked to Central American Microwave System of trunk connections (2005)

Radio broadcast stations: AM 850, FM 545, shortwave 15 (2003)

Radios: 31 million (1997)

Television broadcast stations: 236 (plus repeaters) (1997)

Televisions: 25.6 million (1997)

Internet country code: .mx

Internet hosts: 7.629 million (2007)

Internet Service Providers (ISPs): 51 (2000)

Internet users: 22 million (2006)

TRANSPORTATION

Airports: 1,834 (2007)

Airports—with paved runways:
total: 231
over 3,047 m: 12
2,438 to 3,047 m: 29
1,524 to 2,437 m: 84
914 to 1,523 m: 77
under 914 m: 29 (2007)

Airports—with unpaved runways:
total: 1,603
over 3,047 m: 1
1,524 to 2,437 m: 63
914 to 1,523 m: 408
under 914 m: 1,131 (2007)

Heliports: 1 (2007)

Pipelines: gas 22,705 km; liquid petroleum gas 1,875 km; oil 8,688 km; oil/gas/water 228 km; refined products 6,520 km (2006)

Railways: *total:* 17,665 km
standard gauge: 17,665 km 1.435-m gauge (2006)

Roadways: *total:* 235,670 km
paved: 116,751 km (includes 6,144 km of expressways)
unpaved: 118,919 km (2004)

Waterways: 2,900 km (navigable rivers and coastal canals) (2007)

Merchant marine:
total: 60 ships (1000 GRT or over) 802,128 GRT/1,157,971 DWT
by type: bulk carrier 2, cargo 7, chemical tanker 6, liquefied gas 4, passenger/cargo 11, petroleum tanker 25, roll on/roll off 5
foreign-owned: 4 (Denmark 2, Norway 1, UAE 1)
registered in other countries: 14 (Brazil 1, Honduras 1, Liberia 1, Panama 4, Portugal 1, Spain 3, Venezuela 3) (2007)

Ports and terminals: Altamira, Coatzacoalcos, Manzanillo, Morro Redondo, Salina Cruz, Tampico, Veracruz

MILITARY

Military branches: Secretariat of National Defense (Secretaria de Defensa Nacional, Sedena): Army (Ejercito, includes Mexican Air Force (Fuerza Aerea Mexicana, FAM)); Secretariat of the Navy (Secretaria de Marina, Semar): Mexican Navy (Armada de Mexico, ARM, includes Naval Air Force (FAN) and naval infantry) (2008)

Military service age and obligation: 18 years of age for compulsory military service, conscript service obligation—12 months; 16 years of age with consent for voluntary enlistment; conscripts serve only in the Army; Navy and Air Force service is all voluntary; women are eligible for voluntary military service (2007)

Manpower available for military service: *males age 16–49:* 27,774,688
females age 16–49: 29,376,791 (2008 est.)

Manpower fit for military service:
males age 16–49: 22,188,284
females age 16–49: 24,884,614 (2008 est.)
Manpower reaching militarily significant age annually:
males age 16–49: 1,110,544
females age 16–49: 1,073,223 (2008 est.)
Military expenditures—percent of GDP: 0.5% (2006 est.)

TRANSNATIONAL ISSUES

Disputes—international: abundant rainfall in recent years along much of the Mexico-US border region has ameliorated periodically strained water-sharing arrangements; the US has intensified security measures to monitor and control legal and illegal personnel, transport, and commodities across its border with Mexico; Mexico must deal with thousands of impoverished Guatemalans and other Central Americans who cross the porous border looking for work in Mexico and the United States

Refugees and internally displaced persons: *IDPs:* 5,500–10,000 (government's quashing of Zapatista uprising in 1994 in eastern Chiapas Region) (2007)

Trafficking in persons: *current situation:* Mexico is a source, transit, and destination country for persons trafficked for sexual exploitation and labor; while the vast majority of victims are Central Americans trafficked along Mexico's southern border, other source regions include South America, the Caribbean, Eastern Europe, Africa, and Asia; women and children are trafficked from rural regions to urban centers and tourist areas for sexual exploitation, often through fraudulent offers of employment or through threats of physical violence; the Mexican trafficking problem is often conflated with alien smuggling, and frequently the same criminal networks are involved; pervasive corruption among state and local law enforcement often impedes investigations
tier rating: Tier 2 Watch List—Mexico remains on the Tier 2 Watch List for the third consecutive year based on future commitments to undertake additional efforts in prosecution, protection, and prevention of trafficking in persons, and the failure of the government to provide critical law enforcement data

Illicit drugs: major drug-producing nation; cultivation of opium poppy in 2005 amounted to 3,300 hectares yielding a potential production of 8 metric tons of pure heroin, or 17 metric tons of "black tar" heroin, the dominant form of Mexican heroin in the western United States; marijuana cultivation decreased 3% to 5,600 hectares in 2005—just two years after a decade-high cultivation peak in 2003—and yielded a potential production of 10,100 metric tons; government conducts the largest independent illicit-crop eradication program in the world; continues as the primary transshipment country for US-bound cocaine from South America, with an estimated 90% of annual cocaine movements towards the US stopping in Mexico; major drug syndicates control majority of drug trafficking throughout the country; producer and distributor of ecstasy; significant money-laundering center; major supplier of heroin and largest foreign supplier of marijuana and methamphetamine to the US market

MICRONESIA, FEDERATED STATES OF

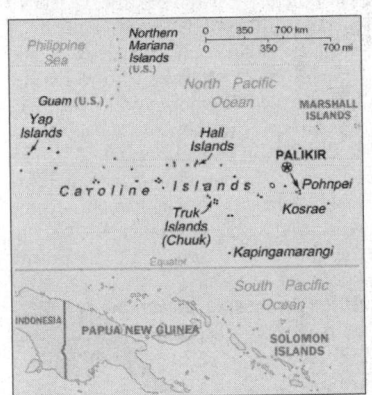

INTRODUCTION

Background: In 1979 the Federated States of Micronesia, a UN Trust Territory under US administration, adopted a constitution. In 1986 independence was attained under a Compact of Free Association with the US, which was amended and renewed in 2004. Present concerns include large-scale unemployment, overfishing, and overdependence on US aid.

GEOGRAPHY

Location: Oceania, island group in the North Pacific Ocean, about three-quarters of the way from Hawaii to Indonesia

Geographic coordinates: 6 55 N, 158 15 E
Map references: Oceania
Area:
total: 702 sq km
land: 702 sq km
water: 0 sq km (fresh water only)
note: includes Pohnpei (Ponape), Chuuk (Truk) Islands, Yap Islands, and Kosrae (Kosaie)
Area—comparative: four times the size of Washington, DC (land area only)
Land boundaries: 0 km
Coastline: 6,112 km
Maritime claims:
territorial sea: 12 nm
exclusive economic zone: 200 nm
Climate: tropical; heavy year-round rainfall, especially in the eastern islands; located on southern edge of the typhoon belt with occasionally severe damage
Terrain: islands vary geologically from high mountainous islands to low, coral atolls; volcanic outcroppings on Pohnpei, Kosrae, and Chuuk
Elevation extremes:
lowest point: Pacific Ocean 0 m
highest point: Dolohmwar (Totolom) 791 m
Natural resources: forests, marine products, deep-seabed minerals, phosphate
Land use:
arable land: 5.71%

permanent crops: 45.71%
other: 48.58% (2005)
Irrigated land: NA
Natural hazards: typhoons (June to December)
Environment—current issues: overfishing, climate change, pollution
Environment—international agreements: *party to:* Biodiversity, Climate Change, Climate Change-Kyoto Protocol, Desertification, Hazardous Wastes, Law of the Sea, Ozone Layer Protection
signed, but not ratified: none of the selected agreements
Geography—note: four major island groups totaling 607 islands

PEOPLE

Population: 107,665 (July 2008 est.)
Age structure:
0–14 years: 35.3% (male 19,344/female 18,687)
15–64 years: 61.8% (male 33,142/female 33,389)
65 years and over: 2.9% (male 1,320/female 1,783) (2008 est.)
Median age:
total: 21.6 years
male: 21.1 years
female: 22.1 years (2008 est.)
Population growth rate: -0.191% (2008 est.)

Birth rate: 23.66 births/1,000 population (2008 est.)

Death rate: 4.53 deaths/1,000 population (2008 est.)

Net migration rate: -21.04 migrant(s)/1,000 population (2008 est.)

Sex ratio: NA

Infant mortality rate:

total: 27.03 deaths/1,000 live births

male: 29.8 deaths/1,000 live births

female: 24.13 deaths/1,000 live births (2008 est.)

Life expectancy at birth:

total population: 70.65 years

male: 68.79 years

female: 72.61 years (2008 est.)

Total fertility rate: 2.98 children born/woman (2008 est.)

HIV/AIDS—adult prevalence rate: NA

HIV/AIDS—people living with HIV/AIDS: NA

HIV/AIDS—deaths: NA

Nationality:

noun: Micronesian(s)

adjective: Micronesian; Chuukese, Kosraen(s), Pohnpeian(s), Yapese

Ethnic groups: Chuukese 48.8%, Pohnpeian 24.2%, Kosraean 6.2%, Yapese 5.2%, Yap outer islands 4.5%, Asian 1.8%, Polynesian 1.5%, other 6.4%, unknown 1.4% (2000 census)

Religions: Roman Catholic 50%, Protestant 47%, other 3%

Languages: English (official and common language), Trukese, Pohnpeian, Yapese, Kosrean, Ulithian, Woleaian, Nukuoro, Kapingamarangi

Literacy: *definition:* age 15 and over can read and write

total population: 89%

male: 91%

female: 88% (1980 est.)

GOVERNMENT

Country name:

conventional long form: Federated States of Micronesia

conventional short form: none

local long form: Federated States of Micronesia

local short form: none

former: Trust Territory of the Pacific Islands, Ponape, Truk, and Yap Districts

abbreviation: FSM

Government type: constitutional government in free association with the US; the Compact of Free Association entered into force 3 November 1986 and the Amended Compact entered into force May 2004

Capital: *name:* Palikir

geographic coordinates: 6 55 N, 158 09 E

time difference: UTC+11 (16 hours ahead of Washington, DC during Standard Time)

Administrative divisions: 4 states; Chuuk (Truk), Kosrae (Kosaie), Pohnpei (Ponape), Yap

Independence: 3 November 1986 (from the US-administered UN trusteeship)

National holiday: Constitution Day, 10 May (1979)

Constitution: 10 May 1979

Legal system: based on adapted Trust Territory laws, acts of the legislature, municipal, common, and customary laws; has not accepted compulsory ICJ jurisdiction

Suffrage: 18 years of age; universal

Executive branch:

chief of state: President Emanuel MORI (since 11 May 2007); Vice President Alik L. ALIK (since 11 May 2007); note—the president is both the chief of state and head of government

head of government: President Emanuel MORI (since 11 May 2007); Vice President Alik L. ALIK (since 11 May 2007)

cabinet: Cabinet includes the vice president and the heads of the eight executive departments

elections: president and vice president elected by Congress from among the four senators at large for a four-year term (eligible for a second term); election last held 11 May 2007 (next to be held May 2011); note—a proposed constitutional amendment to establish popular elections for president and vice president failed

election results: Emanuel MORI elected president; percent of Congress vote—NA; Alik L. ALIK elected vice president; percent of Congress vote—NA

Legislative branch: unicameral Congress (14 seats; 4—one elected from each state to serve four-year terms and 10—elected from single-member districts delineated by population to serve two-year terms; members elected by popular vote)

elections: last held 6 March 2007 (next to be held in March 2009)

election results: percent of vote—NA%; seats—independents 14

Judicial branch: Supreme Court

Political parties and leaders: no formal parties

International organization participation: ACP, ADB, FAO, G-77, IBRD, ICAO, ICRM, IDA, IFC, IFRCS, IMF, IOC, ITSO, ITU, MIGA, OPCW, PIF, Sparteca, SPC, UN, UNCTAD, UNESCO, WHO, WMO

Diplomatic representation in the US:

chief of mission: Ambassador Jesse Bibiano MAREHALAU

chancery: 1725 N Street NW, Washington, DC 20036

telephone: [1] (202) 223-4383

FAX: [1] (202) 223-4391

consulate(s) general: Honolulu, Tamuning (Guam)

Diplomatic representation from the US:

chief of mission: Ambassador Miriam K. HUGHES

embassy: 101 Upper Pics Road, Kolonia

mailing address: P. O. Box 1286, Kolonia, Pohnpei, 96941

telephone: [691] 320-2187

FAX: [691] 320-2186

Flag description: light blue with four white five-pointed stars centered; the stars are arranged in a diamond pattern

ECONOMY

Economy—overview: Economic activity consists primarily of subsistence farming and fishing. The islands have few mineral deposits worth exploiting, except for high-grade phosphate. The potential for a tourist industry exists, but the remote location, a lack of adequate facilities, and limited air connections hinder development. Under the original terms of the Compact of Free Association, the US provided $1.3 billion in grant aid during the period 1986–2001; the level of aid has been subsequently reduced. The Amended Compact of Free Association with the US guarantees the Federated States of Micronesia (FSM) millions of dollars in annual aid through 2023, and establishes a Trust Fund into which the US and the FSM make annual contributions in order to provide annual payouts to the FSM in perpetuity after 2023. The country's medium-term economic outlook appears fragile due not only to the reduction in US assistance but also to the current slow growth of the private sector.

GDP (purchasing power parity): $277 million; note—supplemented by grant aid, averaging perhaps $100 million annually (2002 est.)

GDP (official exchange rate): $232 million (2005)

GDP—real growth rate: 0.3% (2005 est.)

GDP—per capita (PPP): $2,300 (2005 est.)

GDP—composition by sector:

agriculture: 28.9%

industry: 15.2%

services: 55.9% (2004 est.)

Labor force: 37,410 (2000)

Labor force—by occupation: *agriculture:* 0.9%

industry: 34.4%

services: 64.7%

note: two-thirds are government employees (FY05 est.)

Unemployment rate: 22% (2000 est.)

Population below poverty line: 26.7% (2000)
Household income or consumption by percentage share:
lowest 10%: NA%
highest 10%: NA%
Inflation rate (consumer prices): 2.2% (2005)
Budget:
revenues: $127.3 million ($69 million less grants)
expenditures: $144.2 million (FY05 est.)
Agriculture—products: black pepper, tropical fruits and vegetables, coconuts, bananas, cassava (tapioca), sakau (kava), betel nuts, sweet potatoes; pigs, chickens; fish
Industries: tourism, construction; fish processing, specialized aquaculture; craft items (from shell, wood, and pearls)
Industrial production growth rate: NA%
Electricity—production: 192 million kWh (2002)
Electricity—production by source: NA
Electricity—consumption: 178.6 million kWh (2002)
Electricity—exports: 0 kWh (2002)
Electricity—imports: 0 kWh (2002)
Current account balance: -$34.3 million (FY05 est.)
Exports: $14 million (f.o.b.) (2004 est.)
Exports—commodities: fish, garments, bananas, black pepper, sakau (kava), betel nut
Exports—partners: Japan, US, Guam (2006)
Imports: $132.7 million f.o.b. (2004)

Imports—commodities: food, manufactured goods, machinery and equipment, beverages
Imports—partners: US, Japan, Hong Kong (2006)
Economic aid—recipient: $106.4 million (2005)
Debt—external: $60.8 million (FY05 est.)
Market value of publicly traded shares: $NA
Currency (code): US dollar (USD)
Currency code: USD
Exchange rates: the US dollar is used
Fiscal year: 1 October—30 September

COMMUNICATIONS

Telephones—main lines in use: 12,400 (2005)
Telephones—mobile cellular: 14,100 (2005)
Telephone system:
general assessment: adequate system
domestic: islands interconnected by shortwave radiotelephone (used mostly for government purposes), satellite (Intelsat) ground stations, and some coaxial and fiber-optic cable; cellular service available on Kosrae, Pohnpei, and Yap
international: country code—691; satellite earth stations—5 Intelsat (Pacific Ocean) (2002)
Radio broadcast stations: AM 5, FM 1, shortwave 0 (2004)
Radios: 9,400 (1996)
Television broadcast stations: 3 (cable TV also available) (2004)
Televisions: 2,800 (1999)

Internet country code: .fm
Internet hosts: 632 (2007)
Internet Service Providers (ISPs): 1 (2000)
Internet users: 16,000 (2006)

TRANSPORTATION

Airports: 6 (2007)
Airports—with paved runways:
total: 6
1,524 to 2,437 m: 4
914 to 1,523 m: 2 (2007)
Roadways:
total: 240 km
paved: 42 km
unpaved: 198 km (1999)
Merchant marine:
total: 3 ships (1000 GRT or over) 3,560 GRT/2,060 DWT
by type: cargo 1, passenger/cargo 2 (2007)
Ports and terminals: Tomil Harbor

MILITARY

Military branches: no regular military forces
Manpower available for military service:
males age 16–49: 26,686 (2008 est.)
Manpower fit for military service:
males age 16–49: 21,748 (2008 est.)
Manpower reaching militarily significant age annually:
males age 16–49: 1,310 (2008 est.)
Military—note: defense is the responsibility of the US

TRANSNATIONAL ISSUES

Disputes—international: none
Illicit drugs: major consumer of cannabis

MOLDOVA

INTRODUCTION

Background: Formerly part of Romania, Moldova was incorporated into the Soviet Union at the close of World War II. Although independent from the USSR since 1991, Russian forces have remained on Moldovan territory east of the Dniester River supporting the Slavic majority population, mostly Ukrainians and Russians, who have proclaimed a "Transnistria" republic. One of the poorest nations in Europe, Moldova became the first former Soviet state to elect a Communist as its president in 2001.

GEOGRAPHY

Location: Eastern Europe, northeast of Romania
Geographic coordinates: 47 00 N, 29 00 E

Map references: Europe
Area:
total: 33,843 sq km
land: 33,371 sq km
water: 472 sq km
Area—comparative: slightly larger than Maryland
Land boundaries:
total: 1,389 km
border countries: Romania 450 km, Ukraine 939 km
Coastline: 0 km (landlocked)
Maritime claims: none (landlocked)
Climate: moderate winters, warm summers
Terrain: rolling steppe, gradual slope south to Black Sea
Elevation extremes:
lowest point: Dniester River 2 m
highest point: Dealul Balanesti 430 m

Natural resources: lignite, phosphorites, gypsum, arable land, limestone
Land use:
arable land: 54.52%
permanent crops: 8.81%
other: 36.67% (2005)
Irrigated land: 3,000 sq km (2003)
Total renewable water resources: 11.7 cu km (1997)
Freshwater withdrawal (domestic/industrial/agricultural): *total:* 2.31 cu km/yr (10%/58%/33%)
per capita: 549 cu m/yr (2000)
Natural hazards: landslides
Environment—current issues: heavy use of agricultural chemicals, including banned pesticides such as DDT, has contaminated soil and groundwater; extensive soil erosion from poor farming methods

Environment—international agreements: *party to:* Air Pollution, Air Pollution-Persistent Organic Pollutants, Biodiversity, Climate Change, Climate Change-Kyoto Protocol, Desertification, Endangered Species, Hazardous Wastes, Ozone Layer Protection, Ship Pollution, Wetlands

signed, but not ratified: none of the selected agreements

Geography—note: landlocked; well endowed with various sedimentary rocks and minerals including sand, gravel, gypsum, and limestone

PEOPLE

Population: 4,324,450 (July 2008 est.)

Age structure:
0–14 years: 16.3% (male 361,000/female 341,785)
15–64 years: 72.9% (male 1,528,080/female 1,622,620)
65 years and over: 10.9% (male 174,448/female 296,517) (2008 est.)

Median age: *total:* 34.3 years
male: 32.4 years
female: 36.4 years (2008 est.)

Population growth rate: -0.092% (2008 est.)

Birth rate: 11.01 births/1,000 population (2008 est.)

Death rate: 10.8 deaths/1,000 population (2008 est.)

Net migration rate: -1.13 migrant(s)/1,000 population (2008 est.)

Sex ratio:
at birth: 1.06 male(s)/female
under 15 years: 1.06 male(s)/female
15–64 years: 0.94 male(s)/female
65 years and over: 0.59 male(s)/female
total population: 0.91 male(s)/female (2008 est.)

Infant mortality rate:
total: 13.5 deaths/1,000 live births
male: 14.95 deaths/1,000 live births
female: 11.96 deaths/1,000 live births (2008 est.)

Life expectancy at birth:
total population: 70.5 years
male: 66.81 years
female: 74.41 years (2008 est.)

Total fertility rate: 1.26 children born/woman (2008 est.)

HIV/AIDS—adult prevalence rate: 0.2% (2001 est.)

HIV/AIDS—people living with HIV/AIDS: 5,500 (2001 est.)

HIV/AIDS—deaths: fewer than 300 (2001 est.)

Nationality: *noun:* Moldovan(s)
adjective: Moldovan

Ethnic groups: Moldovan/Romanian 78.2%, Ukrainian 8.4%, Russian 5.8%, Gagauz 4.4%, Bulgarian 1.9%, other 1.3% (2004 census)
note: internal disputes with ethnic Slavs in the Transnistrian region

Religions: Eastern Orthodox 98%, Jewish 1.5%, Baptist and other 0.5% (2000)

Languages: Moldovan (official, virtually the same as the Romanian language), Russian, Gagauz (a Turkish dialect)

Literacy: *definition:* age 15 and over can read and write
total population: 99.1%
male: 99.7%
female: 98.6% (2005 est.)

GOVERNMENT

Country name:
conventional long form: Republic of Moldova
conventional short form: Moldova
local long form: Republica Moldova
local short form: Moldova
former: Moldavian Soviet Socialist Republic, Moldovan Soviet Socialist Republic

Government type: republic

Capital: *name:* Chisinau (Kishinev)
note: pronounced kee-shee-now
geographic coordinates: 47 00 N, 28 51 E
time difference: UTC+2 (7 hours ahead of Washington, DC during Standard Time)
daylight saving time: +1hr, begins last Sunday in March; ends last Sunday in October

Administrative divisions: 32 raions (raioane, singular—raionul), 3 municipalities (municipiul), 1 autonomous territorial unit (unitatea teritoriala autonoma), and 1 territorial unit (unitatea teritoriala)
raions: Anenii Noi, Basarabeasca, Briceni, Cahul, Cantemir, Calarasi, Causeni, Cimislia, Criuleni, Donduseni, Drochia, Dubasari, Edinet, Falesti, Floresti, Glodeni, Hincesti, Ialoveni, Leova, Nisporeni, Ocnita, Orhei, Rezina, Riscani, Singerei, Soldanesti, Soroca, Stefan-Voda, Straseni, Taraclia, Telenesti, Ungheni
municipalities: Balti, Bender, Chisinau
autonomous territorial unit: Gagauzia
territorial unit: Stinga Nistrului

Independence: 27 August 1991 (from Soviet Union)

National holiday: Independence Day, 27 August (1991)

Constitution: new constitution adopted 29 July 1994, effective 27 August 1994; replaced old Soviet constitution of 1979

Legal system: based on civil law system; Constitutional Court reviews legality of legislative acts and governmental decisions of resolution; accepts many UN and Organization for Security and Cooperation in Europe (OSCE) documents; has not accepted compulsory ICJ jurisdiction

Suffrage: 18 years of age; universal

Executive branch:
chief of state: President Vladimir VORONIN (since 4 April 2001)
head of government: Prime Minister Zinaida GRECEANII (since 31 March 2008); First Deputy Prime Minister Igor DODON (since 31 March 2008)
cabinet: Cabinet selected by president, subject to approval of Parliament
elections: president elected by Parliament for a four-year term (eligible for a second term); election last held 4 April 2005 (next to be held in 2009); note—prime minister designated by the president upon consultation with Parliament; within 15 days from designation, the prime minister-designate must request a vote of confidence from the Parliament regarding his/her work program and entire cabinet; prime minister designated 21 March 2008; cabinet received a vote of confidence 31 March 2008
election results: Vladimir VORONIN reelected president; parliamentary votes—Vladimir VORONIN 75, Gheorghe DUCA 1; Zinaida GRECEANII designated prime minister; parliamentary votes of confidence—56 of 101

Legislative branch: unicameral Parliament or Parlamentul (101 seats;

parties and electoral blocs elected by popular vote to serve four-year terms)

elections: last held 6 March 2005 (next to be held in 2009)

election results: percent of vote by party—PCRM 46.1%, Democratic Moldova Bloc (AMN, PD, PSL) 28.4%, PPCD 9.1%, other parties 16.4%; seats by party—PCRM 56, Democratic Moldova Bloc (AMN, PD, PSL) 34, PPCD 11

Judicial branch: Supreme Court; Constitutional Court (the sole authority for constitutional judicature)

Political parties and leaders: Christian Democratic People's Party or PPCD [Iurie ROSCA]; Communist Party of the Republic of Moldova or PCRM [Vladimir VORONIN]; Democratic Party or PD [Dumitru DIACOV]; Liberal Democratic Party or PLDM [Vladmir FILAT]; National Liberal Party or PNL [Vitalia PAVLICENKO]; Our Moldova Alliance or AMN [Serafim URE-CHEAN]; Party for Social Democracy or PDSM [Dumitru BRAGHIS]; Social Liberal Party or PSL [Oleg SERE-BRIAN]

Political pressure groups and leaders: NA

International organization participation: ACCT, BSEC, CE, CEI, CIS, EAEC (observer), EAPC, EBRD, FAO, GCTU, GUAM, IAEA, IBRD, ICAO, ICCt (signatory), IDA, IFAD, IFC, IFRCS, ILO, IMF, IMO, Interpol, IOC, IOM, IPU, ISO (correspondent), ITU, MIGA, OIF, OPCW, OSCE, PFP, SECI, UN, UNCTAD, UNESCO, UNIDO, Union Latina, UNMIL, UNMIS, UNOCI, UNOMIG, UNWTO, UPU, WCO, WHO, WIPO, WMO, WTO

Diplomatic representation in the US:
chief of mission: Ambassador Nicolae CHIRTOACA
chancery: 2101 S Street NW, Washington, DC 20008
telephone: [1] (202) 667-1130
FAX: [1] (202) 667-1204

Diplomatic representation from the US:
chief of mission: Ambassador Michael D. KIRBY
embassy: 103 Mateevici Street, Chisinau MD-2009
mailing address: use embassy street address
telephone: [373] (22) 40-8300
FAX: [373] (22) 23-3044

Flag description: three equal vertical bands of blue (hoist side), yellow, and red; emblem in center of flag is of a Roman eagle of gold outlined in black with a red beak and talons carrying a yellow cross in its beak and a green olive branch in its right talons and a yellow scepter in its left talons; on its breast is a shield divided horizontally red over blue with a stylized ox head, star, rose, and crescent all in black-outlined yellow; same color scheme as Romania

ECONOMY

Economy—overview: Moldova remains one of the poorest countries in Europe despite recent progress from its small economic base. It enjoys a favorable climate and good farmland but has no major mineral deposits. As a result, the economy depends heavily on agriculture, featuring fruits, vegetables, wine, and tobacco. Moldova must import almost all of its energy supplies. Moldova's dependence on Russian energy was underscored at the end of 2005, when a Russian-owned electrical station in Moldova's separatist Transnistria region cut off power to Moldova and Russia's Gazprom cut off natural gas in disputes over pricing. Russia's decision to ban Moldovan wine and agricultural products, coupled with its decision to double the price Moldova paid for Russian natural gas, slowed GDP growth in 2006. However, in 2007 growth returned to the 6% level Moldova had achieved in 2000–05, boosted by Russia's partial removal of the bans, solid fixed capital investment, and strong domestic demand driven by remittances from abroad. Economic reforms have been slow because of corruption and strong political forces backing government controls. Nevertheless, the government's primary goal of EU integration has resulted in some market-oriented progress. The granting of EU trade preferences and increased exports to Russia will encourage higher growth rates in 2008, but the agreements are unlikely to serve as a panacea, given the extent to which export success depends on higher quality standards and other factors. The economy remains vulnerable to higher fuel prices, poor agricultural weather, and the skepticism of foreign investors. Also, the presence of an illegal separatist regime in Moldova's Transnistria region continues to be a drag on the Moldovan economy.

GDP (purchasing power parity): $9.821 billion (2007 est.)

GDP (official exchange rate): $4.227 billion (2007 est.)

GDP—real growth rate: 5% (2007 est.)

GDP—per capita (PPP): $2,900 (2007 est.)

GDP—composition by sector:
agriculture: 17.8%
industry: 21.7%
services: 60.5% (2007 est.)

Labor force: 1.333 million (2007 est.)

Labor force—by occupation: *agriculture:* 40.7%

industry: 12.1%
services: 47.2% (2005)

Unemployment rate: 2.1%; note—roughly 25% of working age Moldovans are employed abroad (2007 est.)

Population below poverty line: 29.5% (2005)

Household income or consumption by percentage share:
lowest 10%: 3.2%
highest 10%: 26.4% (2003)

Distribution of family income—Gini index: 33.2 (2003)

Inflation rate (consumer prices): 12.6% (2007 est.)

Investment (gross fixed): 33.3% of GDP (2007 est.)

Budget: *revenues:* $1.83 billion
expenditures: $1.841 billion (2007 est.)

Public debt: 23.3% of GDP (2007 est.)

Agriculture—products: vegetables, fruits, wine, grain, sugar beets, sunflower seed, tobacco; beef, milk

Industries: sugar, vegetable oil, food processing, agricultural machinery; foundry equipment, refrigerators and freezers, washing machines; hosiery, shoes, textiles

Industrial production growth rate: 1% (2007 est.)

Electricity—production: 3.881 billion kWh (2005)

Electricity—production by source:
fossil fuel: 90.6%
hydro: 9.4%
nuclear: 0%
other: 0% (2001)

Electricity—consumption: 5.551 billion kWh (2005)

Electricity—exports: 220 million kWh (2005)

Electricity—imports: 3.361 billion kWh (2005)

Oil—production: 0 bbl/day (2005 est.)

Oil—consumption: 14,500 bbl/day (2005 est.)

Oil—exports: 31.69 bbl/day (2004)

Oil—imports: 14,200 bbl/day (2004)

Oil—proved reserves: 0 bbl (1 January 2006 est.)

Natural gas—production: 0 cu m (2005 est.)

Natural gas—consumption: 2.35 billion cu m (2005 est.)

Natural gas—exports: 0 cu m (2005 est.)

Natural gas—imports: 2.35 billion cu m (2005)

Natural gas—proved reserves: 0 cu m (1 January 2006 est.)

Current account balance: -$410 million (2007 est.)

Exports: $1.361 billion f.o.b. (2007 est.)

Exports—commodities: foodstuffs, textiles, machinery

Exports—partners: Russia 23.7%, Italy 11.5%, Romania 10.1%, Ukraine 10%, Germany 8.6%, Belarus 6% (2006)
Imports: $3.677 billion f.o.b. (2007 est.)
Imports—commodities: mineral products and fuel, machinery and equipment, chemicals, textiles
Imports—partners: Russia 20.8%, Ukraine 16.9%, Romania 13.4%, Germany 8.7%, Italy 6.1%, Poland 4.4% (2006)
Economic aid—recipient: $191.8 million (2005)
Reserves of foreign exchange and gold: $1.334 billion (31 December 2007 est.)
Debt—external: $3.3 billion (31 December 2007)
Stock of direct foreign investment—at home: $NA
Stock of direct foreign investment—abroad: $NA
Market value of publicly traded shares: $573.9 million (2004)
Currency (code): Moldovan leu (MDL)
Currency code: MDL
Exchange rates: lei per US dollar— 12.177 (2007), 13.131 (2006), 12.6 (2005), 12.33 (2004), 13.945 (2003)
Fiscal year: calendar year

Telephones—main lines in use: 1.018 million (2006)
Telephones—mobile cellular: 1.358 million (2006)
Telephone system:
general assessment: inadequate, outmoded, poor service outside Chisinau; some modernization is under way
domestic: depending on location, new subscribers may face long wait for service; multiple private operators of GSM mobile-cellular telephone service are operating; GPRS system is being introduced; a CDMA mobile telephone network began operations in 2007
international: country code—373; service through Romania and Russia via landline; satellite earth stations—at least 3 (Intelsat, Eutelsat, and Intersputnik) (2006)
Radio broadcast stations: AM 2, FM 29, shortwave NA (2006)
Radios: 3.22 million (1997)
Television broadcast stations: 40 (2006)
Televisions: 1.26 million (1997)
Internet country code: .md
Internet hosts: 112,026 (2007)
Internet Service Providers (ISPs): 2 (1999)
Internet users: 727,700 (2006)

Airports: 10 (2007)
Airports—with paved runways:
total: 6
over 3,047 m: 1
2,438 to 3,047 m: 2
1,524 to 2,437 m: 2
under 914 m: 1 (2007)
Airports—with unpaved runways:
total: 4
1,524 to 2,437 m: 2
914 to 1,523 m: 1
under 914 m: 1 (2007)
Pipelines: gas 1,980 km (2007)
Railways:
total: 1,138 km
broad gauge: 1,124 km 1.520-m gauge
standard gauge: 14 km 1.435-m gauge (2006)
Roadways:
total: 12,733 km
paved: 10,976 km

unpaved: 1,757 km (2004)
Waterways: 424 km (on Dniester and Prut rivers) (2007)
Merchant marine:
total: 8 ships (1000 GRT or over) 15,668 GRT/17,585 DWT
by type: cargo 8
foreign-owned: 3 (Ukraine 3) (2007)

Military branches: National Army: Ground Forces, Rapid Reaction Forces, Air and Air Defense Forces (2008)
Military service age and obligation: 18 years of age for compulsory military service; 12-month service obligation (2006)
Manpower available for military service: *males age 16–49:* 1,161,924 *females age 16–49:* 1,187,771 (2008 est.)
Manpower fit for military service: *males age 16–49:* 877,070 *females age 16–49:* 994,091 (2008 est.)
Manpower reaching militarily significant age annually: *males age 16–49:* 33,053 *females age 16–49:* 31,712 (2008 est.)
Military expenditures—percent of GDP: 0.4% (2005 est.)

Disputes—international: Moldova and Ukraine operate joint customs posts to monitor the transit of people and commodities through Moldova's break-away Transnistria region, which remains under OSCE supervision
Illicit drugs: limited cultivation of opium poppy and cannabis, mostly for CIS consumption; transshipment point for illicit drugs from Southwest Asia via Central Asia to Russia, Western Europe, and possibly the US; widespread crime and underground economic activity

MONACO

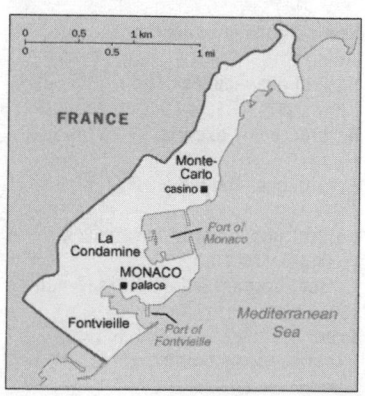

Background: The Genoese built a fortress on the site of present-day Monaco in 1215. The current ruling Grimaldi family secured control in the late 13th century, and a principality was established in 1338. Economic development was spurred in the late 19th century with a railroad linkup to France and the opening of a casino. Since then, the principality's mild climate, splendid scenery, and gambling facilities have made Monaco world famous as a tourist and recreation center.

Location: Western Europe, bordering the Mediterranean Sea on the southern coast of France, near the border with Italy
Geographic coordinates: 43 44 N, 7 24 E
Map references: Europe
Area:
total: 1.95 sq km
land: 1.95 sq km
water: 0 sq km
Area—comparative: about three times the size of The Mall in Washington, DC
Land boundaries: *total:* 4.4 km

border countries: France 4.4 km
Coastline: 4.1 km
Maritime claims:
territorial sea: 12 nm
exclusive economic zone: 12 nm
Climate: Mediterranean with mild, wet winters and hot, dry summers
Terrain: hilly, rugged, rocky
Elevation extremes:
lowest point: Mediterranean Sea 0 m
highest point: Mont Agel 140 m
Natural resources: none
Land use:
arable land: 0%
permanent crops: 0%
other: 100% (urban area) (2005)
Irrigated land: NA
Natural hazards: NA
Environment—current issues: NA
Environment—international agreements: *party to:* Air Pollution, Air Pollution-Sulfur 94, Air Pollution-Volatile Organic Compounds, Biodiversity, Climate Change, Climate Change-Kyoto Protocol, Desertification, Endangered Species, Hazardous Wastes, Law of the Sea, Marine Dumping, Ozone Layer Protection, Ship Pollution, Wetlands, Whaling
signed, but not ratified: none of the selected agreements
Geography—note: second-smallest independent state in the world (after Holy See); almost entirely urban

PEOPLE

Population: 32,796 (July 2008 est.)
Age structure:
0–14 years: 14.8% (male 2,488/female 2,369)
15–64 years: 62.4% (male 10,110/female 10,353)
65 years and over: 22.8% (male 3,048/female 4,428) (2008 est.)
Median age:
total: 45.5 years
male: 43.5 years
female: 47.5 years (2008 est.)
Population growth rate: 0.375% (2008 est.)
Birth rate: 9.09 births/1,000 population (2008 est.)
Death rate: 12.96 deaths/1,000 population (2008 est.)
Net migration rate: 7.62 migrant(s)/1,000 population (2008 est.)
Sex ratio:
at birth: 1.06 male(s)/female
under 15 years: 1.05 male(s)/female
15–64 years: 0.98 male(s)/female
65 years and over: 0.69 male(s)/female
total population: 0.91 male(s)/female (2008 est.)
Infant mortality rate:
total: 5.18 deaths/1,000 live births

male: 6 deaths/1,000 live births
female: 4.33 deaths/1,000 live births (2008 est.)
Life expectancy at birth:
total population: 79.96 years
male: 76.14 years
female: 83.97 years (2008 est.)
Total fertility rate: 1.75 children born/woman (2008 est.)
HIV/AIDS—adult prevalence rate: NA
HIV/AIDS—people living with HIV/AIDS: NA
HIV/AIDS—deaths: NA
Nationality:
noun: Monegasque(s) or Monacan(s)
adjective: Monegasque or Monacan
Ethnic groups: French 47%, Monegasque 16%, Italian 16%, other 21%
Religions: Roman Catholic 90%, other 10%
Languages: French (official), English, Italian, Monegasque
Literacy:
definition: age 15 and over can read and write
total population: 99%
male: 99%
female: 99% (2003 est.)

GOVERNMENT

Country name:
conventional long form: Principality of Monaco
conventional short form: Monaco
local long form: Principaute de Monaco
local short form: Monaco
Government type: constitutional monarchy
Capital: *name:* Monaco
geographic coordinates: 43 44 N, 7 25 E
time difference: UTC+1 (6 hours ahead of Washington, DC during Standard Time)
daylight saving time: +1hr, begins last Sunday in March; ends last Sunday in October
Administrative divisions: none; there are no first-order administrative divisions as defined by the US Government, but there are four quarters (quartiers, singular—quartier); Fontvieille, La Condamine, Monaco-Ville, Monte-Carlo
Independence: 1419 (beginning of rule by the House of Grimaldi)
National holiday: National Day (Saint Rainier's Day), 19 November (1857)
Constitution: 17 December 1962
Legal system: based on French law; has not accepted compulsory ICJ jurisdiction
Suffrage: 18 years of age; universal
Executive branch:
chief of state: Prince ALBERT II (since 6 April 2005)
head of government: Minister of State Jean-Paul PROUST (since 1 June 2005)

cabinet: Council of Government is under the authority of the monarch
elections: the monarchy is hereditary; minister of state appointed by the monarch from a list of three French national candidates presented by the French Government
Legislative branch: unicameral National Council or Conseil National (24 seats; 16 members elected by list majority system, 8 by proportional representation; to serve five-year terms)
elections: last held 3 February 2008 (next to be held February 2013)
election results: percent of vote by party— UPM 52.2%, REM 40.5%, Monaco Together 7.3%; seats by party—UPM 21, REM 3
Judicial branch: Supreme Court or Tribunal Supreme (judges appointed by the monarch on the basis of nominations by the National Council)
Political parties and leaders: Union for Monaco or UPM (including National Union for the Future of Monaco or UNAM); Rally and Issues for Monaco or REM; Monaco Together
Political pressure groups and leaders: NA
International organization participation: ACCT, CE, FAO, IAEA, ICAO, ICC, ICCt (signatory), ICRM, IFRCS, IHO, IMO, IMSO, Interpol, IOC, IPU, ITSO, ITU, OIF, OPCW, OSCE, UN, UNCTAD, UNESCO, UNIDO, Union Latina, UNWTO, UPU, WHO, WIPO, WMO
Diplomatic representation in the US:
chief of mission: Ambassador to the US and UN Gilles NOGHES
chancery: 565 Fifth Avenue, 3rd floor, New York, NY 10017
telephone: (212) 286-0500
FAX: (212) 286-1574
Diplomatic representation from the US: the US does not have an embassy in Monaco; the US Ambassador to France is accredited to Monaco; the US Consul General in Marseille (France), under the authority of the US ambassador to France, handles routine diplomatic and consular matters concerning Monaco
Flag description: two equal horizontal bands of red (top) and white; similar to the flag of Indonesia which is longer and the flag of Poland which is white (top) and red

ECONOMY

Economy—overview: Monaco, bordering France on the Mediterranean coast, is a popular resort, attracting tourists to its casino and pleasant climate. The principality also is a major banking center and has successfully sought to diversify into services and small, high-value-added, nonpolluting industries. The state has no income tax and low business taxes and thrives as a tax haven both for individuals who have established residence and for foreign companies that have set up businesses and offices. The state retains monopolies in a number of sectors, including tobacco, the telephone network, and the postal service. Living standards are high, roughly comparable to those in prosperous French metropolitan areas.

GDP (purchasing power parity): $976.3 million

note: Monaco does not publish national income figures; the estimates are extremely rough (2006 est.)

GDP (official exchange rate): $NA

GDP—real growth rate: 0.9% (2000 est.)

GDP—per capita (PPP): $30,000 (2006 est.)

GDP—composition by sector:

agriculture: 0%

industry: 4.9%

services: 95.1% (2005)

Labor force: 44,000

note: includes workers from all foreign countries (2005 est.)

Unemployment rate: 0% (2005)

Population below poverty line: NA%

Household income or consumption by percentage share:

lowest 10%: NA%

highest 10%: NA%

Inflation rate (consumer prices): 1.9% (2000)

Budget:

revenues: $863 million

expenditures: $920.6 million (2005 est.)

Agriculture—products: none

Industries: tourism, construction, small-scale industrial and consumer products

Industrial production growth rate: NA%

Electricity—consumption: NA kWh

Electricity—imports: NA kWh; note—electricity supplied by France

Exports: $716.3 million

note: full customs integration with France, which collects and rebates Monegasque trade duties; also participates in EU market system through customs union with France (2005)

Imports: $916.1 million

note: full customs integration with France, which collects and rebates Monegasque trade duties; also participates in EU market system through customs union with France (2005)

Economic aid—recipient: $NA

Debt—external: $18 billion (2000 est.)

Market value of publicly traded shares: $NA

Currency (code): euro (EUR)

Currency code: EUR

Exchange rates: euros per US dollar—0.7345 (2007), 0.7964 (2006), 0.8041 (2005), 0.8054 (2004), 0.886 (2003)

Fiscal year: calendar year

COMMUNICATIONS

Telephones—main lines in use: 34,000 (2005)

Telephones—mobile cellular: 17,200 (2005)

Telephone system:

general assessment: modern automatic telephone system

domestic: NA

international: country code—377; no satellite earth stations; connected by cable into the French communications system

Radio broadcast stations: AM 1, FM NA, shortwave 8 (1998)

Radios: 34,000 (1997)

Television broadcast stations: 5 (1998)

Televisions: 25,000 (1997)

Internet country code: .mc

Internet hosts: 14,520 (2007)

Internet Service Providers (ISPs): 2 (2000)

Internet users: 20,000 (2006)

TRANSPORTATION

Heliports: 1 (2007)

Roadways:

total: 50 km

paved: 50 km (1999)

Merchant marine:

registered in other countries: 64 (Bahamas 11, Barbados 1, Georgia 10, Isle of Man 3, Liberia 8, Malta 1, Marshall Islands 7, Norway 5, Panama 11, St Kitts and Nevis 1, St Vincent and The Grenadines 6, unknown 1) (2007)

Ports and terminals: Monaco

MILITARY

Military branches: no regular military forces; the Palace Guard performs ceremonial duties

Manpower available for military service:

males age 16–49: 6,687 (2008 est.)

Manpower fit for military service:

males age 16–49: 5,376 (2008 est.)

Manpower reaching militarily significant age annually:

males age 16–49: 191 (2008 est.)

Military—note: defense is the responsibility of France

TRANSNATIONAL ISSUES

Disputes—international: none

MONGOLIA

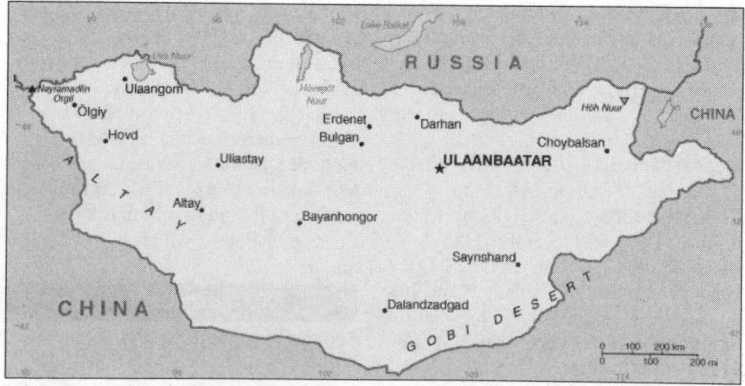

INTRODUCTION

Background: The Mongols gained fame in the 13th century when under Chinggis KHAN they conquered a huge Eurasian empire. After his death the empire was divided into several powerful Mongol states, but these broke apart in the 14th century. The Mongols eventually retired to their original steppe homelands and in the late 17th century came under Chinese rule. Mongolia won its independence in 1921 with Soviet backing. A Communist regime was installed in 1924. Following a peaceful democratic revolution, the ex-Communist Mongolian People's Revolutionary Party (MPRP) won elections in 1990 and 1992, but was defeated by the Democratic Union Coalition (DUC) in the 1996 parliamentary election. Since then, parliamentary elections returned the MPRP overwhelmingly to power in 2000, but 2004 elections reduced MPRP representation and, therefore, its authority.

GEOGRAPHY

Location: Northern Asia, between China and Russia
Geographic coordinates: 46 00 N, 105 00 E
Map references: Asia
Area:
total: 1,564,116 sq km
land: 1,554,731 sq km
water: 9,385 sq km
Area—comparative: slightly smaller than Alaska
Land boundaries:
total: 8,220 km
border countries: China 4,677 km, Russia 3,543 km
Coastline: 0 km (landlocked)
Maritime claims: none (landlocked)
Climate: desert; continental (large daily and seasonal temperature ranges)
Terrain: vast semidesert and desert plains, grassy steppe, mountains in west and southwest; Gobi Desert in south-central
Elevation extremes:
lowest point: Hoh Nuur 518 m
highest point: Nayramadlin Orgil (Huyten Orgil) 4,374 m
Natural resources: oil, coal, copper, molybdenum, tungsten, phosphates, tin, nickel, zinc, fluorspar, gold, silver, iron
Land use:
arable land: 0.76%
permanent crops: 0%
other: 99.24% (2005)
Irrigated land: 840 sq km (2003)
Total renewable water resources: 34.8 cu km (1999)
Freshwater withdrawal (domestic/industrial/agricultural): *total:* 0.44 cu km/yr (20%/27%/52%)
per capita: 166 cu m/yr (2000)
Natural hazards: dust storms, grassland and forest fires, drought, and "zud," which is harsh winter conditions
Environment—current issues: limited natural fresh water resources in some areas; the policies of former Communist regimes promoted rapid urbanization and industrial growth that had negative effects on the environment; the burning of soft coal in power plants and the lack of enforcement of environmental laws severely polluted the air in Ulaanbaatar; deforestation, overgrazing, and the converting of virgin land to agricultural production increased soil erosion from wind and rain; desertification and mining activities had a deleterious effect on the environment
Environment—international agreements: *party to:* Biodiversity, Climate Change, Climate Change-Kyoto Protocol, Desertification, Endangered Species, Environmental Modification, Hazardous Wastes, Law of the Sea, Ozone Layer Protection, Ship Pollution, Wetlands, Whaling
signed, but not ratified: none of the selected agreements
Geography—note: landlocked; strategic location between China and Russia

PEOPLE

Population: 2,996,081 (July 2008 est.)
Age structure:
0–14 years: 28.4% (male 433,835/female 416,549)
15–64 years: 67.7% (male 1,013,215/female 1,015,221)
65 years and over: 3.9% (male 51,093/female 66,168) (2008 est.)
Median age:
total: 24.9 years
male: 24.6 years
female: 25.3 years (2008 est.)
Population growth rate: 1.493% (2008 est.)
Birth rate: 21.09 births/1,000 population (2008 est.)
Death rate: 6.16 deaths/1,000 population (2008 est.)
Net migration rate: NA
Sex ratio:
at birth: 1.05 male(s)/female
under 15 years: 1.04 male(s)/female
15–64 years: 1 male(s)/female
65 years and over: 0.77 male(s)/female
total population: 1 male(s)/female (2008 est.)
Infant mortality rate:
total: 41.24 deaths/1,000 live births
male: 44.41 deaths/1,000 live births
female: 37.92 deaths/1,000 live births (2008 est.)
Life expectancy at birth:
total population: 67.32 years
male: 64.92 years
female: 69.84 years (2008 est.)
Total fertility rate: 2.24 children born/woman (2008 est.)
HIV/AIDS—adult prevalence rate: less than 0.1% (2003 est.)
HIV/AIDS—people living with HIV/AIDS: fewer than 500 (2003 est)
HIV/AIDS—deaths: fewer than 200 (2003 est.)
Nationality:
noun: Mongolian(s)
adjective: Mongolian
Ethnic groups: Mongol (mostly Khalkha) 94.9%, Turkic (mostly Kazakh) 5%, other (including Chinese and Russian) 0.1% (2000)
Religions: Buddhist Lamaist 50%, Shamanist and Christian 6%, Muslim 4%, none 40% (2004)
Languages: Khalkha Mongol 90%, Turkic, Russian (1999)

Literacy:
definition: age 15 and over can read and write
total population: 97.8%
male: 98%
female: 97.5% (2000 census)

GOVERNMENT

Country name:
conventional long form: none
conventional short form: Mongolia
local long form: none
local short form: Mongol Uls
former: Outer Mongolia
Government type: mixed parliamentary/presidential
Capital: *name:* Ulaanbaatar
geographic coordinates: 47 55 N, 106 55 E
time difference: UTC+8 (13 hours ahead of Washington, DC during Standard Time)
Administrative divisions: 21 provinces (aymguud, singular—aymag) and 1 municipality* (singular—hot); Arhangay, Bayanhongor, Bayan-Olgiy, Bulgan, Darhan-Uul, Dornod, Dornogovi, Dundgovi, Dzavhan, Govi-Altay, Govisumber, Hentiy, Hovd, Hovsgol, Omnogovi, Orhon, Ovorhangay, Selenge, Suhbaatar, Tov, Ulaanbaatar*, Uvs
Independence: 11 July 1921 (from China)
National holiday: Independence Day/Revolution Day, 11 July (1921)
Constitution: 12 February 1992
Legal system: blend of Soviet, German, and US systems that combine "continental" or "civil" code and case-precedent; constitution ambiguous on judicial review of legislative acts; has not accepted compulsory ICJ jurisdiction
Suffrage: 18 years of age; universal
Executive branch:
chief of state: President Nambaryn ENKHBAYAR (since 24 June 2005)
head of government: Prime Minister Sanjaa BAYAR (since 22 November 2007); Deputy Prime Minister Miegombyn ENKHBOLD (since 6 December 2007)
cabinet: Cabinet nominated by the prime minister in consultation with the president and confirmed by the State Great Hural (parliament)
elections: presidential candidates nominated by political parties represented in State Great Hural and elected by popular vote for a four-year term (eligible for a second term); election last held 22 May 2005 (next to be held in May 2009); following legislative elections, leader of majority party or majority coalition is usually elected prime minister by State Great Hural

election results: Nambaryn ENKHBAYAR elected president; percent of vote—Nambaryn ENKHBAYAR 53.44%, Mendsaikhanin ENKHSAIKHAN 20.05%, Bazarsadyn JARGALSAIKHAN 13.92%, Badarchyn ERDENEBAT 12.59%; Miegombyn ENKHBOLD elected prime minister by the State Great Hural 56 to 10

Legislative branch: unicameral State Great Hural 76 seats; members elected by popular vote to serve four-year terms *elections:* last held 27 June 2004 (next to be held on 29 June 2008)

election results: percent of vote by party—MPRP 48.8%, MDC 44.8%, independents 3.5%, Republican Party 1.5%, others 1.4%; seats by party—MPRP 36, MDC 34, others 4; note—2 seats disputed and unfilled; following June 2004 election MDC collapsed

Judicial branch: Supreme Court (serves as appeals court for people's and provincial courts but rarely overturns verdicts of lower courts; judges are nominated by the General Council of Courts and approved by the president)

Political parties and leaders: Citizens Will Party [Sanjaasurengiin OYUN] (also called Civil Will); Democratic Party or DP [Tsakhiagiyn ELBEGDORJ]; Motherland-Mongolian New Socialist Democratic Party or M-MNSDP [Badarchyn ERDENEBAT]; Mongolian People's Revolutionary Party or MPRP [Sanji BAYAR]; Mongolian Republican Party or MRP [Bazarsadyn JARGALSAIKHAN]; People's Party or PP [Lamjav GUNDALAI]

note: DP and Motherland Party formed Motherland-Democracy Coalition (MDC) in 2003 and with cooperation from Civil Will and Republican parties contested June 2004 elections as single party; coalition was dissolved in December 2004

Political pressure groups and leaders: NA

International organization participation: ADB, ARF, CP, EBRD, FAO, G-77, IAEA, IBRD, ICAO, ICCt, ICRM, IDA, IFAD, IFC, IFRCS, ILO, IMF, IMO, Interpol, IOC, IPU, ISO, ITSO, ITU, ITUC, MIGA, MINURSO, NAM, OPCW, OSCE (partner), SCO (observer), UN, UNCTAD, UNESCO, UNIDO, UNMIL, UNMIS, UNOMIG, UNWTO, UPU, WCO, WHO, WIPO, WMO, WTO

Diplomatic representation in the US: *chief of mission:* Ambassador Khasbazaryn BEKHBAT

chancery: 2833 M Street NW, Washington, DC 20007

telephone: [1] (202) 333-7117

FAX: [1] (202) 298-9227

Diplomatic representation from the US: *chief of mission:* Ambassador Mark C. MINTON

embassy: Big Ring Road, 11th Micro Region, Ulaanbaatar

mailing address: PSC 461, Box 300, FPO AP 96521-0002; P.O. Box 1021, Ulaanbaatar-13

telephone: [976] (11) 329-095

FAX: [976] (11) 320-776

Flag description: three equal, vertical bands of red (hoist side), blue, and red; centered on the hoist-side red band in yellow is the national emblem ("soyombo"—a columnar arrangement of abstract and geometric representation for fire, sun, moon, earth, water, and the yin-yang symbol)

ECONOMY

Economy—overview: Economic activity in Mongolia has traditionally been based on herding and agriculture. Mongolia has extensive mineral deposits. Copper, coal, gold, molybdenum, fluorspar, uranium, tin, and tungsten account for a large part of industrial production and foreign direct investment. Soviet assistance, at its height one-third of GDP, disappeared almost overnight in 1990 and 1991 at the time of the dismantlement of the USSR. The following decade saw Mongolia endure both deep recession because of political inaction and natural disasters, as well as economic growth because of reform-embracing, free-market economics and extensive privatization of the formerly state-run economy. Severe winters and summer droughts in 2000–02 resulted in massive livestock die-off and zero or negative GDP growth. This was compounded by falling prices for Mongolia's primary sector exports and widespread opposition to privatization. Growth was 10.6% in 2004, 5.5% in 2005, 7.5% in 2006, and 9.9% in 2007 largely because of high copper prices and new gold production. Mongolia is experiencing its highest inflation rate in over a decade as consumer prices in 2007 rose 15%, largely because of increased fuel and food costs. Mongolia's economy continues to be heavily influenced by its neighbors. For example, Mongolia purchases 95% of its petroleum products and a substantial amount of electric power from Russia, leaving it vulnerable to price increases. Trade with China represents more than half of Mongolia's total external trade—China receives about 70% of Mongolia's exports. Remittances from Mongolians working abroad both legally and illegally are sizable, and money laundering is a growing concern. Mongolia settled its $11 billion debt with Russia at the end of 2003 on favorable terms. Mongolia, which joined the World Trade Organization in 1997, seeks to expand its participation and integration into Asian regional economic and trade regimes.

GDP (purchasing power parity): $8.42 billion (2007 est.)

GDP (official exchange rate): $3.905 billion (2007 est.)

GDP—real growth rate: 9.9% (2007 est.)

GDP—per capita (PPP): $3,200 (2007 est.)

GDP—composition by sector: *agriculture:* 18.8%

industry: 40.4%

services: 40.8% (2006)

Labor force: 1.042 million (2006)

Labor force—by occupation: *agriculture:* 39.9%

industry: 11.7%

services: 49.4% (2006)

Unemployment rate: 3% (2007)

Population below poverty line: 36.1% (2004)

Household income or consumption by percentage share: *lowest 10%:* 3%

highest 10%: 24.6% (2002)

Distribution of family income—Gini index: 32.8 (2002)

Inflation rate (consumer prices): 9% (2007)

Budget: *revenues:* $1.58 billion

expenditures: $1.497 billion (2007)

Agriculture—products: wheat, barley, vegetables, forage crops; sheep, goats, cattle, camels, horses

Industries: construction and construction materials; mining (coal, copper, molybdenum, fluorspar, tin, tungsten, and gold); oil; food and beverages; processing of animal products, cashmere and natural fiber manufacturing

Industrial production growth rate: 3% (2006 est.)

Electricity—production: 3.43 billion kWh (2006)

Electricity—production by source: *fossil fuel:* 100%

hydro: 0%

nuclear: 0%

other: 0% (2001)

Electricity—consumption: 2.94 billion kWh (2006)

Electricity—exports: 15.95 million kWh (2006)

Electricity—imports: 125 million kWh (2006)

Oil—production: 0 bbl/day (2005 est.)

Oil—consumption: 12,000 bbl/day (2005 est.)

Oil—exports: 821.9 bbl/day (2005 est.)
Oil—imports: 12,280 bbl/day (2004 est.)
Oil—proved reserves: 0 bbl (1 January 2006 est.)
Natural gas—production: 0 cu m (2005 est.)
Natural gas—consumption: 0 cu m (2005)
Natural gas—exports: 0 cu m (2005 est.)
Natural gas—imports: 0 cu m (2005)
Natural gas—proved reserves: 0 cu m (1 January 2006 est.)
Current account balance: -$23 million (2007 est.)
Exports: $1.889 billion f.o.b. (2007)
Exports—commodities: copper, apparel, livestock, animal products, cashmere, wool, hides, fluorspar, other nonferrous metals
Exports—partners: China 71.8%, Canada 11.7%, US 7.3% (2006)
Imports: $2.117 billion c.i.f. (2007)
Imports—commodities: machinery and equipment, fuel, cars, food products, industrial consumer goods, chemicals, building materials, sugar, tea
Imports—partners: Russia 29.8%, China 29.5%, Japan 11.9% (2006)
Economic aid—recipient: $159.5 million (2006)
Debt—external: $1.438 billion (2007)
Stock of direct foreign investment—at home: $NA
Stock of direct foreign investment—abroad: $NA
Market value of publicly traded shares: $613.3 million (2007)
Currency (code): togrog/tugrik (MNT)
Currency code: MNT
Exchange rates: togrogs/tugriks per US dollar—1,170 (2007), 1,179.6 (2006), 1,205 (2005), 1,185.3 (2004), 1,146.5 (2003)
Fiscal year: calendar year

COMMUNICATIONS

Telephones—main lines in use: 158,900 (2006)
Telephones—mobile cellular: 775,300 (2006)

Telephone system:
general assessment: network is improving with international direct dialing available in many areas
domestic: very low fixed-line density; there are multiple mobile cellular service providers and subscribership is increasing rapidly; a fiber-optic network is also being installed that will improve broadband and communication services between major urban centers
international: country code—976; satellite earth stations—7
Radio broadcast stations: AM 7, FM 115 (includes 20 National radio broadcaster repeaters), shortwave 4 (2006)
Radios: 155,900 (1999)
Television broadcast stations: 456 (including provincial and low-power repeaters) (2006)
Televisions: 168,800 (1999)
Internet country code: .mn
Internet hosts: 298 (2007)
Internet Service Providers (ISPs): 5 (2001)
Internet users: 268,300 (2005)

TRANSPORTATION

Airports: 44 (2007)
Airports—with paved runways:
total: 13
over 3,047 m: 1
2,438 to 3,047 m: 10
1,524 to 2,437 m: 2 (2007)
Airports—with unpaved runways:
total: 31
over 3,047 m: 1
2,438 to 3,047 m: 5
1,524 to 2,437 m: 23
914 to 1,523 m: 1
under 914 m: 1 (2007)
Heliports: 1 (2007)
Railways:
total: 1,810 km
broad gauge: 1,810 km 1.524-m gauge (2006)
Roadways:
total: 49,250 km
paved: 1,724 km
unpaved: 47,526 km (2002)

Waterways: 580 km
note: only waterway in operation is Lake Hovsgol (135 km); Selenge River (270 km) and Orhon River (175 km) are navigable but carry little traffic; lakes and rivers freeze in winter, are open from May to September (2004)
Merchant marine:
total: 73 ships (1000 GRT or over) 448,252 GRT/668,689 DWT
by type: bulk carrier 12, cargo 52, chemical tanker 1, liquefied gas 1, passenger/cargo 1, petroleum tanker 1, roll on/roll off 5
foreign-owned: 62 (Bulgaria 2, China 3, Hong Kong 1, Japan 1, Lebanon 1, Malaysia 1, Russia 17, Singapore 12, Syria 1, Thailand 1, Ukraine 3, UAE 5, Vietnam 14) (2007)

MILITARY

Military branches: Mongolian Armed Forces: Mongolian Army, Mongolian Air Force; there is no navy (2008)
Military service age and obligation: 18–25 years of age for compulsory military service; conscript service obligation—12 months in land or air defense forces or police; a small portion of Mongolian land forces (2.5 percent) is comprised of contract soldiers; women cannot be deployed overseas for military operations (2006)
Manpower available for military service:
males age 16–49: 865,425
females age 16–49: 860,669 (2008 est.)
Manpower fit for military service:
males age 16–49: 696,652
females age 16–49: 731,480 (2008 est.)
Manpower reaching militarily significant age annually:
males age 16–49: 29,990
females age 16–49: 29,256 (2008 est.)
Military expenditures—percent of GDP: 1.4% (2006)

TRANSNATIONAL ISSUES

Disputes—international: none

MONTENEGRO

INTRODUCTION

Background: The use of the name Montenegro began in the 15th century when the Crnojevic dynasty began to rule the Serbian principality of Zeta; over subsequent centuries Montenegro was able to maintain its independence from the Ottoman Empire. From the 16th to 19th centuries, Montenegro became a theocracy ruled by a series of bishop princes; in 1852, it was transformed into a secular principality. After World War I, Montenegro was absorbed by the Kingdom of Serbs, Croats, and Slovenes, which became the Kingdom of Yugoslavia in 1929; at the conclusion of World War II, it became a constituent republic of the Socialist Federal Republic of Yugoslavia. When the latter dissolved in 1992, Montenegro federated with Serbia, first as the Federal Republic of Yugoslavia and, after 2003, in a looser union of Serbia and Montenegro. In May 2006, Montenegro invoked its right under the Constitutional Charter of Serbia and Montenegro to hold a referendum on independence from the state union. The vote for severing ties with Serbia exceeded 55%—the threshold set by the EU—allowing Montenegro to formally declare its independence on 3 June 2006.

441

GEOGRAPHY

Location: Southeastern Europe, between the Adriatic Sea and Serbia
Geographic coordinates: 42 30 N, 19 18 E
Map references: Europe
Area:
total: 14,026 sq km
land: 13,812 sq km
water: 214 sq km
Area—comparative: slightly smaller than Connecticut
Land boundaries:
total: 625 km
border countries: Albania 172 km, Bosnia and Herzegovina 225 km, Croatia 25 km, Kosovo 79 km, Serbia 124 km
Coastline: 293.5 km
Maritime claims:
territorial sea: 12 nm
continental shelf: defined by treaty
Climate: Mediterranean climate, hot dry summers and autumns and relatively cold winters with heavy snowfalls inland
Terrain: highly indented coastline with narrow coastal plain backed by rugged high limestone mountains and plateaus
Elevation extremes:
lowest point: Adriatic Sea 0 m
highest point: Bobotov Kuk 2,522 m
Natural resources: bauxite, hydroelectricity
Land use:
arable land: 13.7%
permanent crops: 1%
other: 85.3%
Irrigated land: NA
Natural hazards: destructive earthquakes
Environment—current issues: pollution of coastal waters from sewage outlets, especially in tourist-related areas such as Kotor
Environment—international agreements: *party to:* Climate Change, Climate Change-Kyoto Protocol, Hazardous Wastes, Law of the Sea, Marine Dumping, Ship Pollution

Geography—note: strategic location along the Adriatic coast

PEOPLE

Population: 678,177 (July 2008 est.)
Population growth rate: -0.925% (2008 est.)
Birth rate: 11.17 births/1,000 population (2008 est.)
Death rate: 8.51 deaths/1,000 population (2008 est.)
Major infectious diseases:
degree of risk: intermediate
food or waterborne diseases: bacterial diarrhea and hepatitis A
vectorborne disease: Crimean Congo hemorrhagic fever (2008)
Nationality: *noun:* Montenegrin(s)
adjective: Montenegrin
Ethnic groups: Montenegrin 43%, Serbian 32%, Bosniak 8%, Albanian 5%, other (Muslims, Croats, Roma (Gypsy)) 12%
Religions: Orthodox, Muslim, Roman Catholic
Languages: Montenegrin (official), Serbian, Bosnian, Albanian, Croatian

GOVERNMENT

Country name:
conventional long form: none
conventional short form: Montenegro
local long form: none
local short form: Crna Gora
former: People's Republic of Montenegro, Socialist Republic of Montenegro, Republic of Montenegro
Government type: republic
Capital: *name:* Podgorica
geographic coordinates: 42 26 N, 19 16 E
time difference: UTC+1 (6 hours ahead of Washington, DC during Standard Time)
daylight saving time: +1 hr, begins last Sunday in March; ends last Sunday in October
Administrative divisions: 21 municipalities (opstine, singular—opstina); Andrijevica, Bar, Berana, Bijelo Polje, Budva, Cetinje, Danilovgrad, Herceg Novi, Kolasin, Kotor, Mojkovac, Niksic, Plav, Pljevlja, Pluzine, Podgorica, Rozaje, Savnik, Tivat, Ulcinj, Zabljak
Independence: 3 June 2006 (from Serbia and Montenegro)
National holiday: National Day, 13 July (1878)
Constitution: 19 October 2007 (was approved by the Assembly)
Legal system: based on civil law system; has not accepted compulsory ICJ jurisdiction
Suffrage: 18 years of age; universal
Executive branch:
chief of state: President Filip VUJANOVIC (since 11 May 2003)

head of government: Prime Minister Milo DJUKANOVIC (since 29 February 2008)
cabinet: Ministries act as cabinet
elections: president elected by direct vote for five-year term (eligible for a second term); election last held 6 April 2008 (next to be held in 2013); prime minister proposed by president, accepted by Assembly
election results: Filip VUJANOVIC reelected president; Filip VUJANOVIC 51.89%, Andrija MANDIC 19.55%, Nebojsa MEDOJEVIC 16.64%, Srdan MILIC 11.92%
Legislative branch: unicameral Assembly (81 seats; members elected by direct vote for four-year terms; changed from 74 seats in 2006)
elections: last held 10 September 2006 (next to be held 2010)
election results: percent of vote by party—Coalition for European Montenegro 47.7%, Serbian List 14.4%, Coalition SNP-NS-DSS 13.8%, PZP 12.9%, Liberals and Bosniaks 3.7%, other (including Albanian minority parties) 7.5%; seats by party—Coalition for European Montenegro 41, Serbian List 12, Coalition SNP/NS/DSS 11, PZP 11, Liberals and Bosniaks 3, Albanian minority parties 3
Judicial branch: Constitutional Court (five judges with nine-year terms); Supreme Court (judges have life tenure)
Political parties and leaders: Albanian Alternative or AA [Vesel SINISHTAJ]; Coalition for European Montenegro or DPS-SDP (bloc) [Milo DJUKANOVIC] (includes Democratic Party of Socialists or DPS [Milo DJUKANOVIC] and Social Democratic Party of SDP [Ranko KRIVOKAPIC]); Coalition SNP-NS-DSS (bloc) (includes Socialist People's Party or SNP [Srdjan MILIC], People's Party of Montenegro or NS [Predrag POPOVIC], and Democratic Serbian Party of Montenegro or DSS [Ranko KADIC]); Democratic League-Party of Democratic Prosperity or SPP [Mehmet BARHDI]; Democratic Union of Albanians or DUA [Ferhat DINOSA]; Liberals and the Bosniak Party (bloc) [Miodrag ZIVKOVIC (includes Liberal Party of Montenegro or LP [Miodrag ZIVKOVIC] and Bosniak Party or BS [Rafet HUSOVIC]); Movement for Changes or PZP [Nebojsa MEDOJEVIC]; Serbian List (bloc) [Andrija MANDIC] (includes Party of Serb Radicals or SSR [Dusko SEKULIC], People's Socialist Party or NSS [Emilo LABUDOVIC], and Serbian People's Party of Montenegro or SNS [Andrija MANDIC])

International organization participation: CE, CEI, EAPC, EBRD, FAO, IBRD, ICAO, ICCt, ICRM, IDA, IFC, IFRCS, ILO, IMF, IMO, Interpol, IOC, IOM, IPU, ISO (correspondent), ITU, ITUC, MIGA, OPCW, OSCE, PCA, PFP, UN, UNESCO, UNIDO, UNWTO, UPU, WCO, WHO, WIPO, WMO, WTO (observer)

Diplomatic representation in the US:
chief of mission: Ambassador Miodrag VLAHOVIC
chancery: 1610 New Hampshire Avenue NW, Washington, DC, 20009
telephone: [1] (202) 234-6108
FAX: [1] (202) 234-6109
consulate(s) general: New York

Diplomatic representation from the US:
chief of mission: Ambassador Roderick W. MOORE
embassy: Ljubljanska bb, 81000 Podgorica, Montenegro
mailing address: use embassy street address
telephone: [382] 81 225 417
FAX: [382] 81 241 358

Flag description: a red field bordered by a narrow golden-yellow stripe with the Montenegrin coat of arms centered

ECONOMY

Economy—overview: Montenegro severed its economy from federal control and from Serbia during the MILOSEVIC era and maintained its own central bank, used the euro instead of the Yugoslav dinar as official currency, collected customs tariffs, and managed its own budget. The dissolution of the loose political union between Serbia and Montenegro in 2006 led to separate membership in several international financial institutions, such as the European Bank for Reconstruction and Development. On 18 January 2007, Montenegro joined the World Bank and IMF. Montenegro is pursuing its own membership in the World Trade Organization as well as negotiating a Stabilization and Association agreement with the European Union in anticipation of eventual membership. Severe unemployment remains a key political and economic problem for this entire region. Montenegro has privatized its large aluminum complex—the dominant industry—as well as most of its financial sector, and has begun to attract foreign direct investment in the tourism sector.

GDP (purchasing power parity): $5.918 billion (2007 est.)
GDP (official exchange rate): $2.974 billion (2007 est.)
GDP—real growth rate: 7.5% (2007 est.)
GDP—per capita (PPP): $3,800 (2005 est.)

GDP—composition by sector:
agriculture: NA%
industry: NA%
services: NA%
Labor force: 259,100 (2004)
Labor force—by occupation:
agriculture: 2%
industry: 30%
services: 68% (2004 est.)
Unemployment rate: 14.7% (2007 est.)
Population below poverty line: 7% (2007 est.)
Distribution of family income—Gini index: 30 (2003)
Inflation rate (consumer prices): 3.4% (2007)
Investment (gross fixed): 30.5% of GDP (2006 est.)
Budget: *revenues:* NA
expenditures: NA
Public debt: 38% of GDP (2006)
Agriculture—products: grains, tobacco, potatoes, citrus fruits, olives, grapes; sheepherding; commercial fishing negligible
Industries: steelmaking, aluminum, agricultural processing, consumer goods, tourism
Electricity—production: 2.864 billion kWh (2005 est.)
Electricity—consumption: 18.6 million kWh (2005)
Electricity—exports: 0 kWh (2005)
Electricity—imports: 0 kWh (2005)
Oil—production: 0 bbl/day (2004)
Oil—consumption: 450 bbl/day (2004)
Oil—proved reserves: 0 bbl (1 January 2006 est.)
Natural gas—consumption: NA cu m
Current account balance: $NA
Exports: $171.3 million (2003)
Exports—partners: Switzerland 83.9%, Italy 6.1%, Bosnia and Herzegovina 1.3% (2006)
Imports: $601.7 million (2003)
Imports—partners: Greece 10.2%, Italy 10.2%, Germany 9.6%, Bosnia and Herzegovina 9.2% (2006)
Economic aid—recipient: $NA
Reserves of foreign exchange and gold: $NA
Debt—external: $650 million (2006)
Market value of publicly traded shares: $NA
Currency (code): euro (EUR)
Exchange rates: euros per US dollar—0.7345 (2007), 0.7964 (2006), 0.8041 (2005), 0.8054 (2004), 0.886 (2003)
Fiscal year: calendar year

COMMUNICATIONS

Telephones—main lines in use: 353,300 (2006)
Telephones—mobile cellular: 821,800 (2006)

Telephone system:
general assessment: modern telecommunications system with access to European satellites
domestic: GSM wireless service, available through 2 providers with national coverage, is growing rapidly
international: country code—382; 2 international switches connect the national system
Radio broadcast stations: 31 (station types NA) (2004)
Television broadcast stations: 13 (2004)
Internet country code: .me
Internet users: 266,000 (2006)

TRANSPORTATION

Airports: 5 (2007)
Airports—with paved runways:
total: 3
2,438 to 3,047 m: 2
1,524 to 2,437 m: 1 (2007)
Airports—with unpaved runways:
total: 2
914 to 1,523 m: 1
under 914 m: 1 (2007)
Heliports: 1 (2007)
Railways:
total: 250 km
standard gauge: 250 km 1.435-m gauge (electrified 169 km) (2006)
Roadways:
total: 7,353 km
paved: 4,274 km
unpaved: 3,079 km (2005)
Merchant marine:
total: 4 ships (1000 GRT or over) 9,458 GRT/10,172 DWT
by type: cargo 4
registered in other countries: 3 (Bahamas 2, St Vincent and The Grenadines 1) (2007)
Ports and terminals: Bar

MILITARY

Military service age and obligation: compulsory national military service abolished August 2006
Military—note: Montenegrin plans call for the establishment of a fully professional armed forces

TRANSNATIONAL ISSUES

Disputes—international: none
Refugees and internally displaced persons: *refugees (country of origin):* 7,000 (Kosovo); note—mostly ethnic Serbs and Roma who fled Kosovo in 1999
IDPs: 16,192 (ethnic conflict in 1999 and riots in 2004) (2007)

443

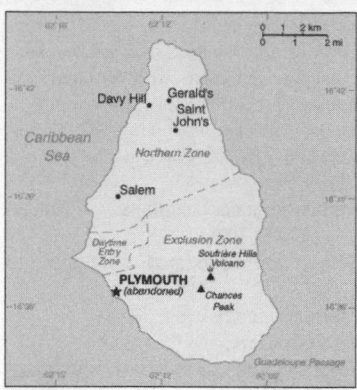

Background: English and Irish colonists from St. Kitts first settled on Montserrat in 1632; the first African slaves arrived three decades later. The British and French fought for possession of the island for most of the 18th century, but it finally was confirmed as a British possession in 1783. The island's sugar plantation economy was converted to small farm landholdings in the mid 19th century. Much of this island was devastated and two-thirds of the population fled abroad because of the eruption of the Soufriere Hills Volcano that began on 18 July 1995. Montserrat has endured volcanic activity since, with the last eruption occurring in July 2003.

GEOGRAPHY

Location: Caribbean, island in the Caribbean Sea, southeast of Puerto Rico
Geographic coordinates: 16 45 N, 62 12 W
Map references: Central America and the Caribbean
Area:
total: 102 sq km
land: 102 sq km
water: 0 sq km
Area—comparative: about 0.6 times the size of Washington, DC
Land boundaries: 0 km
Coastline: 40 km
Maritime claims:
territorial sea: 3 nm
exclusive fishing zone: 200 nm
Climate: tropical; little daily or seasonal temperature variation
Terrain: volcanic island, mostly mountainous, with small coastal lowland
Elevation extremes:
lowest point: Caribbean Sea 0 m
highest point: lava dome in English's Crater (in the Soufriere Hills volcanic

complex) estimated at over 930 m (2006)
Natural resources: NEGL
Land use:
arable land: 20%
permanent crops: 0%
other: 80% (2005)
Irrigated land: NA
Natural hazards: severe hurricanes (June to November); volcanic eruptions (Soufriere Hills volcano has erupted continuously since 1995)
Environment—current issues: land erosion occurs on slopes that have been cleared for cultivation
Geography—note: the island is entirely volcanic in origin and comprised of three major volcanic centers of differing ages

PEOPLE

Population: 9,638
note: an estimated 8,000 refugees left the island following the resumption of volcanic activity in July 1995; some have returned (July 2008 est.)
Age structure:
0–14 years: 23.5% (male 1,159/female 1,108)
15–64 years: 65.9% (male 3,027/female 3,323)
65 years and over: 10.6% (male 521/female 500) (2008 est.)
Median age:
total: 29.7 years
male: 29.3 years
female: 30.2 years (2008 est.)
Population growth rate: 1.038% (2008 est.)
Birth rate: 17.33 births/1,000 population (2008 est.)
Death rate: 6.95 deaths/1,000 population (2008 est.)
Net migration rate: NA
Sex ratio:
at birth: 1.05 male(s)/female
under 15 years: 1.05 male(s)/female
15–64 years: 0.91 male(s)/female
65 years and over: 1.04 male(s)/female
total population: 0.95 male(s)/female (2008 est.)
Infant mortality rate:
total: 6.86 deaths/1,000 live births
male: 7.95 deaths/1,000 live births
female: 5.71 deaths/1,000 live births (2008 est.)
Life expectancy at birth:
total population: 79.15 years
male: 76.93 years
female: 81.47 years (2008 est.)
Total fertility rate: 1.76 children born/woman (2008 est.)
HIV/AIDS—adult prevalence rate: NA

HIV/AIDS—people living with HIV/AIDS: NA
HIV/AIDS—deaths: NA
Nationality:
noun: Montserratian(s)
adjective: Montserratian
Ethnic groups: black, white
Religions: Anglican, Methodist, Roman Catholic, Pentecostal, Seventh-Day Adventist, other Christian denominations
Languages: English
Literacy:
definition: age 15 and over has ever attended school
total population: 97%
male: 97%
female: 97% (1970 est.)

GOVERNMENT

Country name:
conventional long form: none
conventional short form: Montserrat
Dependency status: overseas territory of the UK
Government type: NA
Capital: *name:* Plymouth
geographic coordinates: 16 42 N, 62 13 W
time difference: UTC-4 (1 hour ahead of Washington, DC during Standard Time)
note: Plymouth was abandoned in 1997 because of volcanic activity; interim government buildings have been built at Brades Estate in the Carr's Bay/Little Bay vicinity at the northwest end of Montserrat
Administrative divisions: 3 parishes; Saint Anthony, Saint Georges, Saint Peter
Independence: none (overseas territory of the UK)
National holiday: Birthday of Queen ELIZABETH II, second Saturday in June (1926)
Constitution: effective 19 December 1989
Legal system: English common law and statutory law
Suffrage: 18 years of age; universal
Executive branch:
chief of state: Queen ELIZABETH II (since 6 February 1952); represented by Governor Peter A. WATERWORTH (since 27 July 2007)
head of government: Chief Minister Lowell LEWIS (since 2 June 2006)
cabinet: Executive Council consists of the governor, the chief minister, three other ministers, the attorney general, and the finance secretary
elections: the monarch is hereditary; governor appointed by the monarch; fol-

lowing legislative elections, the leader of the majority party usually becomes chief minister

Legislative branch: unicameral Legislative Council (11 seats, 9 popularly elected; members serve five-year terms)
note: expanded in 2001 from 7 to 9 elected members with attorney general and financial secretary sitting as ex-officio members
elections: last held 31 May 2006 (next to be held by 2011)
election results: percent of vote by party—MCAP 36.1%, NPLM 29.4%, MDP 24.4%, independents 10.1%; seats by party—MCAP 4, NPLM 3, MDP 1, independents 1
note: in 2001, the Elections Commission instituted a single constituency/voter-at-large system whereby all eligible voters cast ballots for all nine seats of the Legislative Council

Judicial branch: Eastern Caribbean Supreme Court (based in Saint Lucia, one judge of the Supreme Court is a resident of the islands and presides over the High Court)

Political parties and leaders: Montserrat Democratic Party or MDP [Lowell LEWIS]; Movement for Change and Prosperity or MCAP [Roselyn CASSELL-SEALY]; New People's Liberation Movement or NPLM [John A. OSBORNE]

Political pressure groups and leaders: NA

International organization participation: Caricom, CDB, Interpol (subbureau), OECS, UPU

Diplomatic representation in the US: none (overseas territory of the UK)

Diplomatic representation from the US: none (overseas territory of the UK)

Flag description: blue, with the flag of the UK in the upper hoist-side quadrant and the Montserratian coat of arms centered in the outer half of the flag; the coat of arms features a woman standing beside a yellow harp with her arm around a black cross

ECONOMY

Economy—overview: Severe volcanic activity, which began in July 1995, has put a damper on this small, open economy. A catastrophic eruption in June 1997 closed the airports and seaports, causing further economic and social dislocation. Two-thirds of the 12,000 inhabitants fled the island. Some began to return in 1998, but lack of housing limited the number. The agriculture sector continued to be affected by the lack of suitable land for farming and the destruction of crops. Prospects for the economy depend largely on developments in relation to the volcanic activity and on public sector construction activity. The UK has launched a three-year $122.8 million aid program to help reconstruct the economy. Half of the island is expected to remain uninhabitable for another decade.

GDP (purchasing power parity): $29 million (2002 est.)
GDP (official exchange rate): $NA
GDP—real growth rate: -1% (2002 est.)
GDP—per capita (PPP): $3,400 (2002 est.)
GDP—composition by sector:
agriculture: 1.2%
industry: 23.1%
services: 75.7% (1999 est.)
Labor force: 4,521
note: lowered by flight of people from volcanic activity (2000 est.)
Unemployment rate: 6% (1998 est.)
Population below poverty line: NA%
Household income or consumption by percentage share:
lowest 10%: NA%
highest 10%: NA%
Inflation rate (consumer prices): 2.6% (2002 est.)
Budget:
revenues: $31.4 million
expenditures: $31.6 million (1997 est.)
Agriculture—products: cabbages, carrots, cucumbers, tomatoes, onions, peppers; livestock products
Industries: tourism, rum, textiles, electronic appliances
Industrial production growth rate: NA%
Electricity—production: 20 million kWh (2005)
Electricity—production by source:
fossil fuel: 100%
hydro: 0%
nuclear: 0%
other: 0% (2001)
Electricity—consumption: 18.6 million kWh (2005)
Electricity—exports: 0 kWh (2005)
Electricity—imports: 0 kWh (2005)
Oil—production: 0 bbl/day (2005 est.)
Oil—consumption: 480 bbl/day (2005 est.)
Oil—exports: 0 bbl/day (2004)
Oil—imports: 458 bbl/day (2004)
Oil—proved reserves: 0 bbl (1 January 2006 est.)
Natural gas—production: 0 cu m (2005 est.)
Natural gas—consumption: 0 cu m (2005 est.)
Natural gas—exports: 0 cu m (2005 est.)
Natural gas—imports: 0 cu m (2005)
Natural gas—proved reserves: 0 cu m

(1 January 2006 est.)
Exports: $700,000 (2001)
Exports—commodities: electronic components, plastic bags, apparel; hot peppers, limes, live plants; cattle
Exports—partners: US, Antigua and Barbuda (2006)
Imports: $17 million (2001)
Imports—commodities: machinery and transportation equipment, foodstuffs, manufactured goods, fuels, lubricants, and related materials
Imports—partners: US, UK, Trinidad and Tobago, Japan, Canada (2006)
Economic aid—recipient: Country Policy Plan (2001) is a three-year program for spending $122.8 million in British budgetary assistance (2002 est.)
Debt—external: $8.9 million (1997)
Currency (code): East Caribbean dollar (XCD)
Currency code: XCD
Exchange rates: East Caribbean dollars per US dollar—2.7 (2007), 2.7 (2006), 2.7 (2005), 2.7 (2004), 2.7 (2003)
note: fixed rate since 1976
Fiscal year: 1 April—31 March

COMMUNICATIONS

Telephones—main lines in use: NA
Telephones—mobile cellular: NA
Telephone system:
general assessment: modern and fully digitalized
domestic: NA
international: country code—1-664; landing point for the East Caribbean Fiber System (ECFS) optic submarine cable with links to 13 other islands in the eastern Caribbean extending from the British Virgin Islands to Trinidad
Radio broadcast stations: AM 1, FM 2, shortwave 0 (1998)
Radios: 7,000 (1997)
Television broadcast stations: 1 (1997)
Televisions: 3,000 (1997)
Internet country code: .ms
Internet hosts: 367 (2007)
Internet Service Providers (ISPs): 17 (2000)
Internet users: NA

TRANSPORTATION

Airports: 2 (2007)
Airports—with paved runways:
total: 2
under 914 m: 2 (2007)
Roadways:
total: 227 km
note: volcanic eruptions that began in 1995 destroyed most of the road system (2003)
Ports and terminals: Little Bay, Plymouth

445

MILITARY

Military branches: no regular military forces; Royal Montserrat Police Force (2008)
Manpower available for military service: *males age 16–49:* 2,528 (2008 est.)

Manpower fit for military service: *males age 16–49:* 2,097 (2008 est.)
Manpower reaching militarily significant age annually: *males age 16–49:* 67 (2008 est.)
Military—note: defense is the responsibility of the UK

TRANSNATIONAL ISSUES

Disputes—international: none
Illicit drugs: transshipment point for South American narcotics destined for the US and Europe

MOROCCO

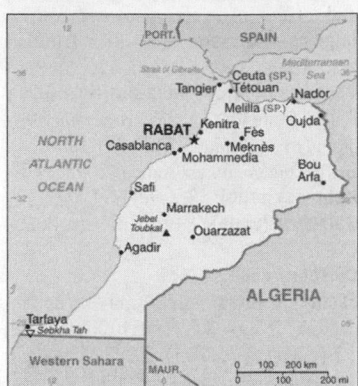

INTRODUCTION

Background: In 788, about a century after the Arab conquest of North Africa, successive Moorish dynasties began to rule in Morocco. In the 16th century, the Sa'adi monarchy, particularly under Ahmad AL-MANSUR (1578–1603), repelled foreign invaders and inaugurated a golden age. In 1860, Spain occupied northern Morocco and ushered in a half century of trade rivalry among European powers that saw Morocco's sovereignty steadily erode; in 1912, the French imposed a protectorate over the country. A protracted independence struggle with France ended successfully in 1956. The internationalized city of Tangier and most Spanish possessions were turned over to the new country that same year. Morocco virtually annexed Western Sahara during the late 1970s, but final resolution on the status of the territory remains unresolved. Gradual political reforms in the 1990s resulted in the establishment of a bicameral legislature, which first met in 1997. Improvements in human rights have occurred and there is a largely free press. Despite the continuing reforms, ultimate authority remains in the hands of the monarch.

GEOGRAPHY

Location: Northern Africa, bordering the North Atlantic Ocean and the Mediterranean Sea, between Algeria and Western Sahara

Geographic coordinates: 32 00 N, 5 00 W
Map references: Africa
Area:
total: 446,550 sq km
land: 446,300 sq km
water: 250 sq km
Area—comparative: slightly larger than California
Land boundaries:
total: 2,017.9 km
border countries: Algeria 1,559 km, Western Sahara 443 km, Spain (Ceuta) 6.3 km, Spain (Melilla) 9.6 km
Coastline: 1,835 km
Maritime claims:
territorial sea: 12 nm
contiguous zone: 24 nm
exclusive economic zone: 200 nm
continental shelf: 200 m depth or to the depth of exploitation
Climate: Mediterranean, becoming more extreme in the interior
Terrain: northern coast and interior are mountainous with large areas of bordering plateaus, intermontane valleys, and rich coastal plains
Elevation extremes:
lowest point: Sebkha Tah -55 m
highest point: Jebel Toubkal 4,165 m
Natural resources: phosphates, iron ore, manganese, lead, zinc, fish, salt
Land use:
arable land: 19%
permanent crops: 2%
other: 79% (2005)
Irrigated land: 14,450 sq km (2003)
Total renewable water resources: 29 cu km (2003)
Freshwater withdrawal (domestic/industrial/agricultural): *total:* 12.6 cu km/yr (10%/3%/87%)
per capita: 400 cu m/yr (2000)
Natural hazards: northern mountains geologically unstable and subject to earthquakes; periodic droughts
Environment—current issues: land degradation/desertification (soil erosion resulting from farming of marginal areas, overgrazing, destruction of vegetation); water supplies contaminated by raw sewage; siltation of reservoirs; oil pollution of coastal waters

Environment—international agreements: *party to:* Biodiversity, Climate Change, Climate Change-Kyoto Protocol, Desertification, Endangered Species, Hazardous Wastes, Law of the Sea, Marine Dumping, Ozone Layer Protection, Ship Pollution, Wetlands, Whaling
signed, but not ratified: Environmental Modification
Geography—note: strategic location along Strait of Gibraltar

PEOPLE

Population: 34,343,219 (July 2008 est.)
Age structure:
0–14 years: 30.5% (male 5,337,322/female 5,136,156)
15–64 years: 64.3% (male 11,015,409/female 11,069,038)
65 years and over: 5.2% (male 765,882/female 1,019,412) (2008 est.)
Median age:
total: 24.7 years
male: 24.1 years
female: 25.2 years (2008 est.)
Population growth rate: 1.505% (2008 est.)
Birth rate: 21.31 births/1,000 population (2008 est.)
Death rate: 5.49 deaths/1,000 population (2008 est.)
Net migration rate: -0.77 migrant(s)/1,000 population (2008 est.)
Sex ratio:
at birth: 1.05 male(s)/female
under 15 years: 1.04 male(s)/female
15–64 years: 1 male(s)/female
65 years and over: 0.75 male(s)/female
total population: 0.99 male(s)/female (2008 est.)
Infant mortality rate:
total: 38.22 deaths/1,000 live births
male: 41.74 deaths/1,000 live births
female: 34.53 deaths/1,000 live births (2008 est.)
Life expectancy at birth:
total population: 71.52 years
male: 69.16 years
female: 74 years (2008 est.)
Total fertility rate: 2.57 children born/woman (2008 est.)
HIV/AIDS—adult prevalence rate: 0.1% (2001 est.)

HIV/AIDS—people living with HIV/AIDS: 15,000 (2001 est.)

HIV/AIDS—deaths: NA

Nationality:

noun: Moroccan(s)

adjective: Moroccan

Ethnic groups: Arab-Berber 99.1%, other 0.7%, Jewish 0.2%

Religions: Muslim 98.7%, Christian 1.1%, Jewish 0.2%

Languages: Arabic (official), Berber dialects, French often the language of business, government, and diplomacy

Literacy:

definition: age 15 and over can read and write

total population: 52.3%

male: 65.7%

female: 39.6% (2004 census)

GOVERNMENT

Country name:

conventional long form: Kingdom of Morocco

conventional short form: Morocco

local long form: Al Mamlakah al Maghribiyah

local short form: Al Maghrib

Government type: constitutional monarchy

Capital: *name:* Rabat

geographic coordinates: 34 01 N, 6 49 W

time difference: UTC 0 (5 hours ahead of Washington, DC during Standard Time)

Administrative divisions: 15 regions; Grand Casablanca, Chaouia-Ouardigha, Doukkala-Abda, Fes-Boulemane, Gharb-Chrarda-Beni Hssen, Guelmim-Es Smara, Laayoune-Boujdour-Sakia El Hamra, Marrakech-Tensift-Al Haouz, Meknes-Tafilalet, Oriental, Rabat-Sale-Zemmour-Zaer, Souss-Massa-Draa, Tadla-Azilal, Tanger-Tetouan, Taza-Al Hoceima-Taounate

note: Morocco claims the territory of Western Sahara, the political status of which is considered undetermined by the US Government; portions of the regions Guelmim-Es Smara and Laayoune-Boujdour-Sakia El Hamra as claimed by Morocco lie within Western Sahara; Morocco claims another region, Oued Eddahab-Lagouira, which falls entirely within Western Sahara

Independence: 2 March 1956 (from France)

National holiday: Throne Day (accession of King MOHAMED VI to the throne), 30 July (1999)

Constitution: 10 March 1972; revised 4 September 1992, amended (to create bicameral legislature) September 1996

Legal system: based on Islamic law and French and Spanish civil law systems; judicial review of legislative acts in Constitutional Chamber of Supreme Court; has not accepted compulsory ICJ jurisdiction

Suffrage: 18 years of age; universal (as of January 2003)

Executive branch:

chief of state: King MOHAMED VI (since 30 July 1999)

head of government: Prime Minister Abbas EL FASSI (since 19 September 2007)

cabinet: Council of Ministers appointed by the monarch

elections: the monarch is hereditary; prime minister appointed by the monarch following legislative elections

Legislative branch: bicameral Parliament consists of a Chamber of Counselors (or upper house) (270 seats; members elected indirectly by local councils, professional organizations, and labor syndicates for nine-year terms; one-third of the members are elected every three years) and Chamber of Representatives (or lower house) (325 seats; 295 members elected by multi-seat constituencies and 30 from national lists of women; members elected by popular vote for five-year terms)

elections: Chamber of Counselors—last held 8 September 2006 (next to be held in 2009); Chamber of Representatives—last held 7 September 2007 (next to be held in 2012)

election results: Chamber of Counselors—percent of vote by party—NA; seats by party—PI 17, MP 14, RNI 13, USFP 11, UC 6, PND 4, PPS 4, Al Ahd 4, other 17; Chamber of Representatives—percent of vote by party—NA; seats by party—PI 52, PJD 46, MP 41, RNI 39, USFP 38, UC 27, PPS 17, FFD 9, MDS 9, Al Ahd 8, other 39

Judicial branch: Supreme Court (judges are appointed on the recommendation of the Supreme Council of the Judiciary, presided over by the monarch)

Political parties and leaders: Action Party or PA [Muhammad EL IDRISSI]; Alliance of Liberties or ADL [Ali BELHAJ]; Annahj Addimocrati or Annahj [Abdellah EL HARIF]; Avant Garde Social Democratic Party or PADS [Ahmed BENJELLOUN]; Citizen Forces or FC [Abderrahman LAHJOUJI]; Citizen's Initiatives for Development [Mohamed BENHAMOU]; Constitutional Union or UC [Mohamed ABIED]; Democratic and Independence Party or PDI [Abdelwahed MAACH]; Democratic and Social Movement or MDS [Mahmoud ARCHANE]; Democratic Forces Front or FFD; Democratic Socialist Party or PSD [Aissa OUARDIGHI]; Democratic Society Party or PSD [Zhor CHEKKAFI]; Democratic Union or UD [Bouazza IKKEN]; Environment and Development Party or PED [Ahmed EL ALAMI]; Front of Democratic Forces or FFD [Thami EL KHYARI]; Independence Party (Istiqlal) or PI [Abbas EL FASSI]; Justice and Development Party or PJD [Saad Eddine EL OTHMANI]; Labor Party [Abdelkrim BENATIK]; Moroccan Liberal Party or PML [Mohamed ZIANE]; National Democratic Party or PND [Abdallah KADIRI]; National Ittihadi Congress Party or CNI [Abdelmajid BOUZOUBAA]; National Rally of Independents or RNI [Mustapha EL MANSOURI]; National Union of Popular Forces or UNFP [Abdellah IBRAHIM]; Parti Al Ahd or Al Ahd [Najib EL OUAZZANI]; Party of Progress and Socialism or PPS [Ismail ALAOUI]; Party of Renewal and Equity or PRE [Chakir ACHABAR]; Party of the Unified Socialist Left or GSU [Mohamed Ben Said AIT IDDER]; Popular Movement or MP [Mohamed LAENSER]; Reform and Development Party or PRD [Abderrahmane EL KOUHEN]; Social Center Party or PSC [Lahcen MADIH]; Socialist Union of Popular Forces or USFP

Political pressure groups and leaders: Democratic Confederation of Labor or CDT [Noubir AMAOUI]; General Union of Moroccan Workers or UGTM [Abderrazzak AFILAL]; Moroccan Employers Association or CGEM [Hassan CHAMI]; National Labor Union of Morocco or UNMT [Abdelslam MAATI]; Union of Moroccan Workers or UMT [Mahjoub BENSEDDIK]

International organization participation: ABEDA, ACCT, AfDB, AFESD, AMF, AMU, EBRD, FAO, G-77, IAEA, IBRD, ICAO, ICC, ICCt (signatory), ICRM, IDA, IDB, IFAD, IFC, IFRCS, IHO, ILO, IMF, IMO, IMSO, Interpol, IOC, IOM, IPU, ISO, ITSO, ITU, ITUC, LAS, MIGA, MONUC, NAM, OAS (observer), OIC, OIF, OPCW, OSCE (partner), PCA, UN, UNCTAD, UNESCO, UNHCR, UNIDO, UNITAR, UNOCI, UNWTO, UPU, WCL, WCO, WHO, WIPO, WMO, WTO

Diplomatic representation in the US:

chief of mission: Ambassador Aziz MEKOUAR

chancery: 1601 21st Street NW, Washington, DC 20009

telephone: [1] (202) 462-7979

FAX: [1] (202) 265-0161

consulate(s) general: New York

Diplomatic representation from the US:
chief of mission: Ambassador Thomas T. RILEY
embassy: 2 Avenue de Mohamed El Fassi, Rabat
mailing address: PSC 74, Box 021, APO AE 09718
telephone: [212] (37) 76 22 65
FAX: [212] (37) 76 56 61
consulate(s) general: Casablanca
Flag description: red with a green pentacle (five-pointed, linear star) known as Sulayman's (Solomon's) seal in the center of the flag; red and green are traditional colors in Arab flags, although the use of red is more commonly associated with the Arab states of the Persian gulf; design dates to 1912

ECONOMY

Economy—overview: Moroccan economic policies brought macroeconomic stability to the country in the early 1990s but have not spurred growth sufficient to reduce unemployment—nearing 20% in urban areas—despite the Moroccan Government's ongoing efforts to diversify the economy. Morocco's GDP growth rate slowed to 2.1% in 2007 as a result of a draught that severely reduced agricultural output and necessitated wheat imports at rising world prices. Continued dependence on foreign energy and Morocco's inability to develop small and medium size enterprises also contributed to the slowdown. Moroccan authorities understand that reducing poverty and providing jobs are key to domestic security and development. In 2005, Morocco launched the National Initiative for Human Development (INDH), a $2 billion social development plan to address poverty and unemployment and to improve the living conditions of the country's urban slums. Moroccan authorities are implementing reform efforts to open the economy to international investors. Despite structural adjustment programs supported by the IMF, the World Bank, and the Paris Club, the dirham is only fully convertible for current account transactions. In 2000, Morocco entered an Association Agreement with the EU and, in 2006, entered a Free Trade Agreement (FTA) with the US. Long-term challenges include improving education and job prospects for Morocco's youth, and closing the income gap between the rich and the poor, which the government hopes to achieve by increasing tourist arrivals and boosting competitiveness in textiles.
GDP (purchasing power parity): $125.3 billion (2007 est.)

GDP (official exchange rate): $73.43 billion (2007 est.)
GDP—real growth rate: 2.2% (2007 est.)
GDP—per capita (PPP): $4,100 (2007 est.)
GDP—composition by sector:
agriculture: 14.5%
industry: 37.9%
services: 47.7% (2007 est.)
Labor force: 11.05 million (2007 est.)
Labor force—by occupation: *agriculture:* 40%
industry: 15%
services: 45% (2003 est.)
Unemployment rate: 2.1% (2007 est.)
Population below poverty line: 15% (2007 est.)
Household income or consumption by percentage share:
lowest 10%: 2.6%
highest 10%: 30.9% (1999)
Distribution of family income—Gini index: 40 (2005 est.)
Inflation rate (consumer prices): 2% (2007 est.)
Investment (gross fixed): 25.8% of GDP (2007 est.)
Budget:
revenues: $20.58 billion
expenditures: $21.71 billion (2007 est.)
Public debt: 72.4% of GDP (2007 est.)
Agriculture—products: barley, wheat, citrus, wine, vegetables, olives; livestock
Industries: phosphate rock mining and processing, food processing, leather goods, textiles, construction, tourism
Industrial production growth rate: 4% (2007 est.)
Electricity—production: 21.37 billion kWh (2005)
Electricity—production by source:
fossil fuel: 95.4%
hydro: 4.6%
nuclear: 0%
other: 0% (2001)
Electricity—consumption: 20.67 billion kWh (2005)
Electricity—exports: 0 kWh (2005)
Electricity—imports: 802 million kWh (2005)
Oil—production: 3,746 bbl/day (2005 est.)
Oil—consumption: 176,000 bbl/day (2005 est.)
Oil—exports: 21,890 bbl/day (2004 est.)
Oil—imports: 186,100 bbl/day (2004 est.)
Oil—proved reserves: 100 million bbl (2007 est.)
Natural gas—production: 47.95 million cu m (2005 est.)
Natural gas—consumption: 47.95 million cu m (2005 est.)
Natural gas—exports: 0 cu m (2005 est.)

Natural gas—imports: 0 cu m (2005)
Natural gas—proved reserves: 1.629 billion cu m (1 January 2006 est.)
Current account balance: -$71 million (2007 est.)
Exports: $12.75 billion f.o.b. (2007 est.)
Exports—commodities: clothing and textiles, electric components, inorganic chemicals, transistors, crude minerals, fertilizers (including phosphates), petroleum products, citrus fruits, vegetables, fish
Exports—partners: Spain 20.6%, France 20.5%, UK 4.8%, Italy 4.7%, India 4% (2006)
Imports: $27.14 billion f.o.b. (2007 est.)
Imports—commodities: crude petroleum, textile fabric, telecommunications equipment, wheat, gas and electricity, transistors, plastics
Imports—partners: France 17.5%, Spain 13.9%, Saudi Arabia 6.9%, China 6.9%, Italy 6.3%, Germany 6% (2006)
Economic aid—recipient: ODA, $651.8 million (2005)
Reserves of foreign exchange and gold: $24.29 billion (31 December 2007 est.)
Debt—external: $19.91 billion (31 December 2007 est.)
Stock of direct foreign investment—at home: $26.52 billion (2007 est.)
Stock of direct foreign investment—abroad: $567 million (2006 est.)
Market value of publicly traded shares: $49.6 billion (2006)
Currency (code): Moroccan dirham (MAD)
Currency code: MAD
Exchange rates: Moroccan dirhams per US dollar—8.3563 (2007), 8.7722 (2006), 8.865 (2005), 8.868 (2004), 9.574 (2003)
Fiscal year: calendar year

COMMUNICATIONS

Telephones—main lines in use: 1.266 million (2006)
Telephones—mobile cellular: 16.005 million (2006)
Telephone system:
general assessment: modern system with all important capabilities; however, density is low with only 4 fixed lines available for each 100 persons; mobile-cellular subscribership is approaching 50 per 100 persons
domestic: good system composed of open-wire lines, cables, and microwave radio relay links; Internet available but expensive; principal switching centers are Casablanca and Rabat; national network nearly 100% digital using fiber-optic links; improved rural service employs microwave radio relay

international: country code—212; landing point for the SEA-ME-WE-3 optical telecommunications submarine cable that provides connectivity to Asia, the Middle East, and Europe; satellite earth stations—2 Intelsat (Atlantic Ocean) and 1 Arabsat; microwave radio relay to Gibraltar, Spain, and Western Sahara; coaxial cable and microwave radio relay to Algeria; participant in Medarabtel; fiber-optic cable link from Agadir to Algeria and Tunisia

Radio broadcast stations: AM 27, FM 25, shortwave 6 (1998)

Radios: 6.64 million (1997)

Television broadcast stations: 35 (plus 66 repeaters) (1995)

Televisions: 3.1 million (1997)

Internet country code: .ma

Internet hosts: 137,187 (2007)

Internet Service Providers (ISPs): 8 (2000)

Internet users: 6.1 million (2006)

TRANSPORTATION

Airports: 60 (2007)

Airports—with paved runways: *total:* 27
over 3,047 m: 11
2,438 to 3,047 m: 6
1,524 to 2,437 m: 7
914 to 1,523 m: 1
under 914 m: 2 (2007)

Airports—with unpaved runways: *total:* 33
2,438 to 3,047 m: 2
1,524 to 2,437 m: 9
914 to 1,523 m: 11
under 914 m: 11 (2007)

Heliports: 1 (2007)

Pipelines: gas 720 km; oil 439 km (2007)

Railways:
total: 1,907 km
standard gauge: 1,907 km 1.435-m gauge (1,003 km electrified) (2006)

Roadways:
total: 57,493 km
paved: 32,716 km (includes 507 km of expressways)
unpaved: 24,777 km (2004)

Merchant marine:
total: 35 ships (1000 GRT or over) 344,445 GRT/252,341 DWT
by type: cargo 3, chemical tanker 6, container 8, passenger/cargo 12, petroleum tanker 1, refrigerated cargo 1, roll on/roll off 4
foreign-owned: 14 (France 13, Germany 1) (2007)

Ports and terminals: Agadir, Casablanca, Mohammedia, Safi

MILITARY

Military branches: Royal Armed Forces (Forces Armees Royales, FAR): Royal Moroccan Army (includes Air Defense), Navy (includes Marines), Royal Moroccan Air Force (Al Quwwat al Jawyiya al Malakiya Marakishiya; Force Aerienne Royale Marocaine) (2008)

Military service age and obligation: 18 years of age for compulsory and voluntary military service; conscript service obligation—18 months (2004)

Manpower available for military service:
males age 16–49: 9,152,580
females age 16–49: 9,080,830 (2008 est.)

Manpower fit for military service:
males age 16–49: 7,627,988

females age 16–49: 7,754,873 (2008 est.)

Manpower reaching militarily significant age annually:
males age 16–49: 355,479
females age 16–49: 343,016 (2008 est.)

Military expenditures—percent of GDP: 5% (2003 est.)

TRANSNATIONAL ISSUES

Disputes—international: claims and administers Western Sahara whose sovereignty remains unresolved—UN-administered cease-fire has remained in effect since September 1991, but attempts to hold a referendum have failed and parties thus far have rejected all brokered proposals; Morocco protests Spain's control over the coastal enclaves of Ceuta, Melilla, and Penon de Velez de la Gomera, the islands of Penon de Alhucemas and Islas Chafarinas, and surrounding waters; discussions have not progressed on a comprehensive maritime delimitation, setting limits on resource exploration and refugee interdiction, since Morocco's 2002 rejection of Spain's unilateral designation of a median line from the Canary Islands; Morocco serves as one of the primary launching areas of illegal migration into Spain from North Africa

Illicit drugs: one of the world's largest producers of illicit hashish; shipments of hashish mostly directed to Western Europe; transit point for cocaine from South America destined for Western Europe; significant consumer of cannabis

MOZAMBIQUE

INTRODUCTION

Background: Almost five centuries as a Portuguese colony came to a close with independence in 1975. Large-scale emigration by whites, economic dependence on South Africa, a severe drought, and a prolonged civil war hindered the country's development until the mid 1990's. The ruling Front for the Liberation of Mozambique (FRELIMO) party formally abandoned Marxism in 1989, and a new constitution the following year provided for multiparty elections and a free market economy. A UN-negotiated peace agreement between FRELIMO and rebel Mozambique National Resistance (RENAMO) forces ended the fighting in 1992. In December 2004, Mozambique underwent a delicate transition as

Joaquim CHISSANO stepped down after 18 years in office. His elected successor, Armando Emilio GUEBUZA, promised to continue the sound economic policies that have encouraged foreign investment. Mozambique has seen very strong economic growth since the end of the civil war largely due to post-conflict reconstruction.

GEOGRAPHY

Location: Southeastern Africa, bordering the Mozambique Channel, between South Africa and Tanzania

Geographic coordinates: 18 15 S, 35 00 E

Map references: Africa

Area:
total: 801,590 sq km
land: 784,090 sq km
water: 17,500 sq km

Area—comparative: slightly less than twice the size of California

Land boundaries:
total: 4,571 km
border countries: Malawi 1,569 km, South Africa 491 km, Swaziland 105 km, Tanzania 756 km, Zambia 419 km, Zimbabwe 1,231 km

Coastline: 2,470 km

Maritime claims:
territorial sea: 12 nm
exclusive economic zone: 200 nm

Climate: tropical to subtropical

Terrain: mostly coastal lowlands, uplands in center, high plateaus in northwest, mountains in west

Elevation extremes:
lowest point: Indian Ocean 0 m
highest point: Monte Binga 2,436 m

Natural resources: coal, titanium, natural gas, hydropower, tantalum, graphite

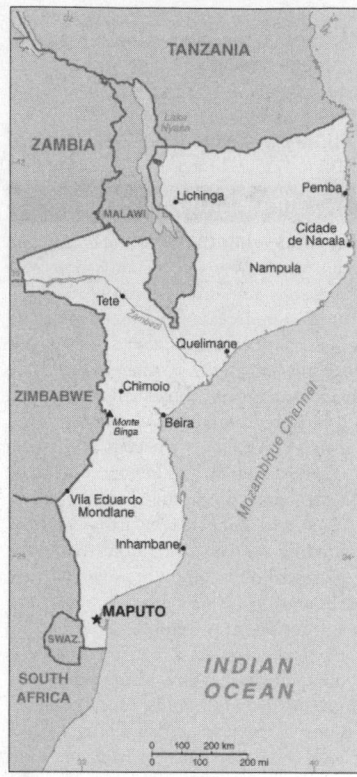

Land use: *arable land:* 5.43%
permanent crops: 0.29%
other: 94.28% (2005)
Irrigated land: 1,180 sq km (2003)
Total renewable water resources: 216 cu km (1992)
Freshwater withdrawal (domestic/industrial/agricultural): *total:* 0.63 cu km/yr (11%/2%/87%)
per capita: 32 cu m/yr (2000)
Natural hazards: severe droughts; devastating cyclones and floods in central and southern provinces
Environment—current issues: a long civil war and recurrent drought in the hinterlands have resulted in increased migration of the population to urban and coastal areas with adverse environmental consequences; desertification; pollution of surface and coastal waters; elephant poaching for ivory is a problem
Environment—international agreements: *party to:* Biodiversity, Climate Change, Climate Change-Kyoto Protocol, Desertification, Endangered Species, Hazardous Wastes, Law of the Sea, Ozone Layer Protection, Ship Pollution, Wetlands
signed, but not ratified: none of the selected agreements
Geography—note: the Zambezi flows through the north-central and most fertile part of the country

PEOPLE

Population: 21,284,701
note: estimates for this country explicitly take into account the effects of excess mortality due to AIDS; this can result in lower life expectancy, higher infant mortality, higher death rates, lower population growth rates, and changes in the distribution of population by age and sex than would otherwise be expected; the 1997 Mozambican census reported a population of 16,099,246 (July 2008 est.)
Age structure:
0–14 years: 44.5% (male 4,762,335/female 4,711,422)
15–64 years: 52.7% (male 5,472,184/female 5,736,154)
65 years and over: 2.8% (male 251,026/female 351,580) (2008 est.)
Median age:
total: 17.4 years
male: 17 years
female: 17.8 years (2008 est.)
Population growth rate: 1.792% (2008 est.)
Birth rate: 38.21 births/1,000 population (2008 est.)
Death rate: 20.29 deaths/1,000 population (2008 est.)
Net migration rate: NA
Sex ratio:
at birth: 1.02 male(s)/female
under 15 years: 1.01 male(s)/female
15–64 years: 0.95 male(s)/female
65 years and over: 0.71 male(s)/female
total population: 0.97 male(s)/female (2008 est.)
Infant mortality rate:
total: 107.84 deaths/1,000 live births
male: 110.67 deaths/1,000 live births
female: 104.97 deaths/1,000 live births (2008 est.)
Life expectancy at birth:
total population: 41.04 years
male: 41.62 years
female: 40.44 years (2008 est.)
Total fertility rate: 5.24 children born/woman (2008 est.)
HIV/AIDS—adult prevalence rate: 12.2% (2003 est.)
HIV/AIDS—people living with HIV/AIDS: 1.3 million (2003 est.)
HIV/AIDS—deaths: 110,000 (2003 est.)
Major infectious diseases:
degree of risk: very high
food or waterborne diseases: bacterial and protozoal diarrhea, hepatitis A, and typhoid fever
vectorborne diseases: malaria and plague
water contact disease: schistosomiasis (2008)
Nationality:
noun: Mozambican(s)
adjective: Mozambican

Ethnic groups: African 99.66% (Makhuwa, Tsonga, Lomwe, Sena, and others), Europeans 0.06%, Euro-Africans 0.2%, Indians 0.08%
Religions: Catholic 23.8%, Muslim 17.8%, Zionist Christian 17.5%, other 17.8%, none 23.1% (1997 census)
Languages: Emakhuwa 26.1%, Xichangana 11.3%, Portuguese 8.8% (official; spoken by 27% of population as a second language), Elomwe 7.6%, Cisena 6.8%, Echuwabo 5.8%, other Mozambican languages 32%, other foreign languages 0.3%, unspecified 1.3% (1997 census)
Literacy:
definition: age 15 and over can read and write
total population: 47.8%
male: 63.5%
female: 32.7% (2003 est.)

GOVERNMENT

Country name:
conventional long form: Republic of Mozambique
conventional short form: Mozambique
local long form: Republica de Mocambique
local short form: Mocambique
former: Portuguese East Africa
Government type: republic
Capital: *name:* Maputo
geographic coordinates: 25 57 S, 32 35 E
time difference: UTC+2 (7 hours ahead of Washington, DC during Standard Time)
Administrative divisions: 10 provinces (provincias, singular—provincia), 1 city (cidade)*; Cabo Delgado, Gaza, Inhambane, Manica, Maputo, Cidade de Maputo*, Nampula, Niassa, Sofala, Tete, Zambezia
Independence: 25 June 1975 (from Portugal)
National holiday: Independence Day, 25 June (1975)
Constitution: 30 November 1990
Legal system: based on Portuguese civil law system and customary law; has not accepted compulsory ICJ jurisdiction
Suffrage: 18 years of age; universal
Executive branch:
chief of state: President Armando GUEBUZA (since 2 February 2005)
head of government: Prime Minister Luisa DIOGO (since 17 February 2004)
cabinet: Cabinet
elections: president elected by popular vote for a five-year term (eligible for a second term); election last held 1–2 December 2004 (next to be held in December 2009); prime minister appointed by the president
election results: Armando GUEBUZA elected president; percent of vote—

Armando GUEBUZA 63.7%, Afonso DHLAKAMA 31.7%

Legislative branch: unicameral Assembly of the Republic or Assembleia da Republica (250 seats; members are directly elected by popular vote to serve five-year terms)

elections: last held 1–2 December 2004 (next to be held in December 2009)

election results: percent of vote by party—FRELIMO 62%, RENAMO 29.7%, other 8.3%; seats by party—FRELIMO 160, RENAMO 90

Judicial branch: Supreme Court (the court of final appeal; some of its professional judges are appointed by the president and some are elected by the Assembly); other courts include an Administrative Court, customs courts, maritime courts, courts marshal, labor courts

note: although the constitution provides for a separate Constitutional Court, one has never been established; in its absence the Supreme Court reviews constitutional cases

Political parties and leaders: Front for the Liberation of Mozambique (Frente de Liberatacao de Mocambique) or FRELIMO [Armando Emilio GUEBUZA]; Mozambique National Resistance-Electoral Union (Resistencia Nacional Mocambicana-Uniao Eleitoral) or RENAMO-UE [Afonso DHLAKAMA]

Political pressure groups and leaders: Institute for Peace and Democracy (Instituto para Paz e Democracia) or IPADE [Raul DOMINGOS, president]; Etica [Abdul CARIMO Issa, chairman]; Movement for Peace and Citizenship (Movimento para Paz e Cidadania); Mozambican League of Human Rights (Liga Mocambicana dos Direitos Humanos) or LDH [Alice MABOTE, president]; Human Rights and Development (Direitos Humanos e Desenvolvimento) or DHD [Artemisia FRANCO, secretary general]

International organization participation: ACP, AfDB, AU, C, CPLP, FAO, G-77, IAEA, IBRD, ICAO, ICCt (signatory), ICRM, IDA, IDB, IFAD, IFC, IFRCS, IHO, ILO, IMF, IMO, IMSO, Interpol, IOC, IOM (observer), IPU, ISO (correspondent), ITSO, ITU, ITUC, MIGA, NAM, OIC, OIF (observer), OPCW, SADC, UN, UNCTAD, UNESCO, UNHCR, UNIDO, Union Latina, UNMIS, UNWTO, UPU, WCO, WFTU, WHO, WIPO, WMO, WTO

Diplomatic representation in the US:

chief of mission: Ambassador Marcos Geraldo NAMASHULUA

chancery: 1525 New Hampshire Avenue, Washington, DC 20036

telephone: [1] (202) 293-7146

FAX: [1] (202) 835-0245

Diplomatic representation from the US:

chief of mission: Ambassador (vacant); Charge d'Affaires Todd C. CHAPMAN

embassy: Avenida Kenneth Kuanda 193, Maputo

mailing address: P. O. Box 783, Maputo

telephone: [258] (21) 492797

FAX: [258] (21) 490448

Flag description: three equal horizontal bands of green (top), black, and yellow with a red isosceles triangle based on the hoist side; the black band is edged in white; centered in the triangle is a yellow five-pointed star bearing a crossed rifle and hoe in black superimposed on an open white book

ECONOMY

Economy—overview: At independence in 1975, Mozambique was one of the world's poorest countries. Socialist mismanagement and a brutal civil war from 1977–92 exacerbated the situation. In 1987, the government embarked on a series of macroeconomic reforms designed to stabilize the economy. These steps, combined with donor assistance and with political stability since the multi-party elections in 1994, have led to dramatic improvements in the country's growth rate. Inflation was reduced to single digits during the late 1990s, and although it returned to double digits in 2000–06, in 2007 inflation had slowed to 8%, while GDP growth reached 7.5%. Fiscal reforms, including the introduction of a value-added tax and reform of the customs service, have improved the government's revenue collection abilities. In spite of these gains, Mozambique remains dependent upon foreign assistance for much of its annual budget, and the majority of the population remains below the poverty line. Subsistence agriculture continues to employ the vast majority of the country's work force. A substantial trade imbalance persists although the opening of the Mozal aluminum smelter, the country's largest foreign investment project to date, has increased export earnings. At the end of 2007, and after years of negotiations, the government took over Portugal's majority share of the Cahora Bassa Hydroelectricity (HCB) company, a dam that was not transferred to Mozambique at independence because of the ensuing civil war and unpaid debts. More power is needed for additional investment projects in titanium extraction and processing and garment manufacturing that could further close the import/export gap. Mozambique's once substantial foreign debt has been reduced through forgiveness and rescheduling under the IMF's Heavily Indebted Poor Countries (HIPC) and Enhanced HIPC initiatives, and is now at a manageable level. In July 2007 the Millennium Challenge Corporation (MCC) signed a Compact with Mozambique; the Mozambican government moved rapidly to ratify the Compact and propose a plan for funding.

GDP (purchasing power parity): $17.02 billion (2007 est.)

GDP (official exchange rate): $7.559 billion (2007 est.)

GDP—real growth rate: 7% (2007 est.)

GDP—per capita (PPP): $800 (2007 est.)

GDP—composition by sector:

agriculture: 23%

industry: 30.1%

services: 46.8% (2007 est.)

Labor force: 9.6 million (2007 est.)

Labor force—by occupation:

agriculture: 81%

industry: 6%

services: 13% (1997 est.)

Unemployment rate: 21% (1997 est.)

Population below poverty line: 70% (2001 est.)

Household income or consumption by percentage share:

lowest 10%: 2.1%

highest 10%: 39.4% (2002)

Distribution of family income—Gini index: 47.3 (2002)

Inflation rate (consumer prices): 7.9% (2007 est.)

Investment (gross fixed): 20.5% of GDP (2007 est.)

Budget: *revenues:* $2.325 billion

expenditures: $2.773 billion (2007 est.)

Public debt: 22.2% of GDP (2007 est.)

Agriculture—products: cotton, cashew nuts, sugarcane, tea, cassava (tapioca), corn, coconuts, sisal, citrus and tropical fruits, potatoes, sunflowers; beef, poultry

Industries: food, beverages, chemicals (fertilizer, soap, paints), aluminum, petroleum products, textiles, cement, glass, asbestos, tobacco

Industrial production growth rate: 10% (2007 est.)

Electricity—production: 13.17 billion kWh (2005)

Electricity—production by source:

fossil fuel: 2.9%

hydro: 97.1%

nuclear: 0%

other: 0% (2001)

Electricity—consumption: 9.127 billion kWh (2005)

Electricity—exports: 12 billion kWh (2005)

451

Electricity—imports: 9.588 billion kWh (2005)

Oil—production: 0 bbl/day (2005 est.)

Oil—consumption: 13,000 bbl/day (2005 est.)

Oil—exports: 0 bbl/day (2004)

Oil—imports: 13,320 bbl/day (2004)

Oil—proved reserves: 0 bbl (1 January 2006 est.)

Natural gas—production: 191.8 million cu m (2005 est.)

Natural gas—consumption: 191.8 million cu m (2005 est.)

Natural gas—exports: 0 cu m (2005 est.)

Natural gas—imports: 0 cu m (2005)

Natural gas—proved reserves: 122.2 billion cu m (1 January 2006 est.)

Current account balance: -$713 million (2007 est.)

Exports: $2.699 billion f.o.b. (2007 est.)

Exports—commodities: aluminum, prawns, cashews, cotton, sugar, citrus, timber; bulk electricity

Exports—partners: Netherlands 59.7%, South Africa 15.2%, Zimbabwe 3.2% (2006)

Imports: $2.997 billion f.o.b. (2007 est.)

Imports—commodities: machinery and equipment, vehicles, fuel, chemicals, metal products, foodstuffs, textiles

Imports—partners: South Africa 36.3%, Netherlands 15.6%, Portugal 3.3% (2006)

Economic aid—recipient: $1.286 billion (2005)

Reserves of foreign exchange and gold: $1.445 billion (31 December 2007 est.)

Debt—external: $4.168 billion (31 December 2007 est.)

Market value of publicly traded shares: $NA

Currency (code): metical (MZM)

Currency code: MZM

Exchange rates: meticais per US dollar—26.264 (2007), 25.4 (2006), 23,061 (2005), 22,581 (2004), 23,782 (2003)

note: in 2006 Mozambique revalued its currency, with 1000 old meticais equal to 1 new meticais

Fiscal year: calendar year

COMMUNICATIONS

Telephones—main lines in use: 67,000 (2006)

Telephones—mobile cellular: 2.339 million (2006)

Telephone system:

general assessment: fair system with an extremely low density of less than 1 fixed line per 100 persons

domestic: the telecommunications sector is shackled with a heavy state presence, lack of competition, and high operating costs and charges; stagnation in the fixed-line network contrasts with rapid growth in the mobile-cellular network; mobile-cellular coverage now includes all the main cities and key roads, including those from Maputo to the South African and Swaziland borders, the national highway through Gaza and Inhambane provinces, the Beira corridor, and from Nampula to Nacala

international: country code—258; satellite earth stations—5 Intelsat (2 Atlantic Ocean and 3 Indian Ocean)

Radio broadcast stations: AM 13, FM 17, shortwave 11 (2001)

Radios: 730,000 (1997)

Television broadcast stations: 1 (2000)

Televisions: 67,600 (2000)

Internet country code: .mz

Internet hosts: 15,231 (2007)

Internet Service Providers (ISPs): 11 (2002)

Internet users: 178,000 (2005)

TRANSPORTATION

Airports: 147 (2007)

Airports—with paved runways:

total: 22

over 3,047 m: 1

2,438 to 3,047 m: 3

1,524 to 2,437 m: 10

914 to 1,523 m: 3

under 914 m: 5 (2007)

Airports—with unpaved runways:

total: 125

2,438 to 3,047 m: 1

1,524 to 2,437 m: 9

914 to 1,523 m: 36

under 914 m: 79 (2007)

Pipelines: gas 964 km; refined products 278 km (2007)

Railways: *total:* 3,123 km

narrow gauge: 2,983 km 1.067-m gauge; 140 km 0.762-m gauge (2006)

Roadways: *total:* 30,400 km

paved: 5,685 km

unpaved: 24,715 km (1999)

Waterways: 460 km (Zambezi River navigable to Tete and along Cahora Bassa Lake) (2007)

Merchant marine:

total: 2 ships (1000 GRT or over) 2,964 GRT/5,324 DWT

by type: cargo 2

foreign-owned: 2 (Belgium 2) (2007)

Ports and terminals: Beira, Maputo, Nacala

MILITARY

Military branches: Mozambique Armed Defense Forces (FADM): Mozambique Army, Mozambique Navy (Marinha Mocambique, MM), Mozambique Air Force (Forca Aerea de Mocambique, FAM) (2006)

Military service age and obligation: 18–30 years of age for compulsory military service; 2-year service obligation (2006)

Manpower available for military service: *males age 16–49:* 4,545,975 (2008 est.)

Manpower fit for military service: *males age 16–49:* 2,287,526 (2008 est.)

Manpower reaching militarily significant age annually: *males age 16–49:* 257,261 (2008 est.)

Military expenditures—percent of GDP: 0.8% (2006)

TRANSNATIONAL ISSUES

Disputes—international: none

Illicit drugs: southern African transit point for South Asian hashish and heroin, and South American cocaine probably destined for the European and South African markets; producer of cannabis (for local consumption) and methaqualone (for export to South Africa); corruption and poor regulatory capability makes the banking system vulnerable to money laundering, but the lack of a well-developed financial infrastructure limits the country's utility as a money-laundering center

INTRODUCTION

Background: South Africa occupied the German colony of South-West Africa during World War I and administered it as a mandate until after World War II, when it annexed the territory. In 1966 the Marxist South-West Africa People's Organization (SWAPO) guerrilla group launched a war of independence for the area that was soon named Namibia, but it was not until 1988 that South Africa agreed to end its administration in accordance with a UN peace plan for the entire region. Namibia has been governed by SWAPO since the country won independence in 1990. Hifikepunye POHAMBA was elected president in November 2004 in a landslide victory replacing Sam NUJOMA who led the country during its first 14 years of self rule.

GEOGRAPHY

Location: Southern Africa, bordering the South Atlantic Ocean, between Angola and South Africa
Geographic coordinates: 22 00 S, 17 00 E
Map references: Africa
Area:
total: 825,418 sq km
land: 825,418 sq km
water: 0 sq km
Area—comparative: slightly more than half the size of Alaska
Land boundaries:
total: 3,936 km
border countries: Angola 1,376 km, Botswana 1,360 km, South Africa 967 km, Zambia 233 km
Coastline: 1,572 km
Maritime claims:
territorial sea: 12 nm
contiguous zone: 24 nm
exclusive economic zone: 200 nm
Climate: desert; hot, dry; rainfall sparse and erratic

Terrain: mostly high plateau; Namib Desert along coast; Kalahari Desert in east
Elevation extremes:
lowest point: Atlantic Ocean 0 m
highest point: Konigstein 2,606 m
Natural resources: diamonds, copper, uranium, gold, silver, lead, tin, lithium, cadmium, tungsten, zinc, salt, hydropower, fish
note: suspected deposits of oil, coal, and iron ore
Land use:
arable land: 0.99%
permanent crops: 0.01%
other: 99% (2005)
Irrigated land: 80 sq km (2003)
Total renewable water resources: 45.5 cu km (1991)
Freshwater withdrawal (domestic/industrial/agricultural): *total:* 0.3 cu km/yr (24%/5%/71%)
per capita: 148 cu m/yr (2000)
Natural hazards: prolonged periods of drought
Environment—current issues: limited natural fresh water resources; desertification; wildlife poaching; land degradation has led to few conservation areas
Environment—international agreements: *party to:* Antarctic-Marine Living Resources, Biodiversity, Climate Change, Climate Change-Kyoto Protocol, Desertification, Endangered Species, Hazardous Wastes, Law of the Sea, Ozone Layer Protection, Wetlands
signed, but not ratified: none of the selected agreements
Geography—note: first country in the world to incorporate the protection of the environment into its constitution; some 14% of the land is protected, including virtually the entire Namib Desert coastal strip

PEOPLE

Population: 2,088,669
note: estimates for this country explicitly take into account the effects of excess mortality due to AIDS; this can result in lower life expectancy, higher infant mortality, higher death rates, lower population growth rates, and changes in the distribution of population by age and sex than would otherwise be expected (July 2008 est.)
Age structure:
0–14 years: 36.7% (male 386,252/female 379,426)
15–64 years: 59.5% (male 627,752/female 615,241)
65 years and over: 3.8% (male 35,960/female 44,038) (2008 est.)

Median age:
total: 20.7 years
male: 20.6 years
female: 20.8 years (2008 est.)
Population growth rate: 0.947% (2008 est.)
Birth rate: 23.19 births/1,000 population (2008 est.)
Death rate: 14.07 deaths/1,000 population (2008 est.)
Net migration rate: 0.35 migrant(s)/1,000 population (2008 est.)
Sex ratio:
at birth: 1.03 male(s)/female
under 15 years: 1.02 male(s)/female
15–64 years: 1.02 male(s)/female
65 years and over: 0.82 male(s)/female
total population: 1.01 male(s)/female (2008 est.)
Infant mortality rate:
total: 45.64 deaths/1,000 live births
male: 49.24 deaths/1,000 live births
female: 41.93 deaths/1,000 live births (2008 est.)
Life expectancy at birth:
total population: 49.89 years
male: 50.39 years
female: 49.38 years (2008 est.)
Total fertility rate: 2.81 children born/woman (2008 est.)
HIV/AIDS—adult prevalence rate: 21.3% (2003 est.)
HIV/AIDS—people living with HIV/AIDS: 210,000 (2001 est.)
HIV/AIDS—deaths: 16,000 (2003 est.)
Major infectious diseases:
degree of risk: high
food or waterborne diseases: bacterial diarrhea, hepatitis A, and typhoid fever
vectorborne disease: malaria
water contact disease: schistosomiasis (2008)
Nationality: *noun:* Namibian(s)
adjective: Namibian
Ethnic groups: black 87.5%, white 6%, mixed 6.5%
note: about 50% of the population belong to the Ovambo tribe and 9% to the Kavangos tribe; other ethnic groups include Herero 7%, Damara 7%, Nama 5%, Caprivian 4%, Bushmen 3%, Baster 2%, Tswana 0.5%
Religions: Christian 80% to 90% (Lutheran 50% at least), indigenous beliefs 10% to 20%
Languages: English 7% (official), Afrikaans common language of most of the population and about 60% of the white population, German 32%, indigenous languages 1% (includes Oshivambo, Herero, Nama)
Literacy: *definition:* age 15 and over can read and write

total population: 85%
male: 86.8%
female: 83.5% (2001 census)

GOVERNMENT

Country name:
conventional long form: Republic of Namibia
conventional short form: Namibia
local long form: Republic of Namibia
local short form: Namibia
former: German Southwest Africa, South-West Africa
Government type: republic
Capital: *name:* Windhoek
geographic coordinates: 22 34 S, 17 05 E
time difference: UTC+1 (6 hours ahead of Washington, DC during Standard Time)
daylight saving time: +1hr, begins first Sunday in September; ends first Sunday in April
Administrative divisions: 13 regions; Caprivi, Erongo, Hardap, Karas, Khomas, Kunene, Ohangwena, Okavango, Omaheke, Omusati, Oshana, Oshikoto, Otjozondjupa
Independence: 21 March 1990 (from South African mandate)
National holiday: Independence Day, 21 March (1990)
Constitution: ratified 9 February 1990, effective 12 March 1990
Legal system: based on Roman-Dutch law and 1990 constitution; has not accepted compulsory ICJ jurisdiction
Suffrage: 18 years of age; universal
Executive branch:
chief of state: President Hifikepunye POHAMBA (since 21 March 2005)
head of government: Prime Minister Nahas ANGULA (since 21 March 2005)
cabinet: Cabinet appointed by the president from among the members of the National Assembly
elections: president elected by popular vote for a five-year term (eligible for a second term); election last held 15 November 2004 (next to be held in November 2009)
election results: Hifikepunye POHAMBA elected president; percent of vote—Hifikepunye POHAMBA 76.4%, Den ULENGA 7.3%, Katuutire KAURA 5.1%, Kuaima RIRUAKO 4.2%, Justus GAROEB 3.8%, other 3.2%
Legislative branch: bicameral legislature consists of the National Council (26 seats; two members are chosen from each regional council to serve six-year terms) and the National Assembly (72 seats; members are elected by popular vote to serve five-year terms)
elections: National Council—elections for regional councils to determine members of the National Council held 29–30 November 2004 (next to be held in November 2010); National Assembly—

last held 15–16 November 2004 (next to be held in November 2009)
election results: National Council—percent of vote by party—SWAPO 89.7%, UDF 4.7%, NUDO 2.8%, DTA 1.9%, other 0.9%; seats by party—SWAPO 24, UDF 1, DTA 1; National Assembly—percent of vote by party—SWAPO 76.1%, COD 7.3%, DTA 5.1%, NUDO 4.2%, UDF 3.6%, RP 1.9%, MAG 0.8%, other 1.0%; seats by party—SWAPO 55, COD 5, DTA 4, NUDO 3, UDF 3, RP 1, MAG 1
note: the National Council is primarily an advisory body
Judicial branch: Supreme Court (judges appointed by the president on the recommendation of the Judicial Service Commission)
Political parties and leaders: Congress of Democrats or COD [Ben ULENGA]; Democratic Turnhalle Alliance of Namibia or DTA [Katuutire KAURA]; Monitor Action Group or MAG [Jurie VILJOEN]; National Democratic Movement for Change or NamDMC; National Unity Democratic Organization or NUDO [Kuaima RIRUAKO]; Rally for Democracy and Progress or RDP [Hidipo HAMUTENYA]; Republican Party or RP [Henk MUDGE]; South West Africa National Union or SWANU [Rihupisa KANDANDO]; South West Africa People's Organization or SWAPO [Hifikepunye POHAMBA]; United Democratic Front or UDF [Justus GAROEB]
Political pressure groups and leaders: NA
International organization participation: ACP, AfDB, AU, C, FAO, G-77, IAEA, IBRD, ICAO, ICCt, ICRM, IDA, IFAD, IFC, IFRCS, ILO, IMF, IMO, Interpol, IOC, IOM (observer), IPU, ISO (correspondent), ITSO, ITU, MIGA, NAM, OPCW, SACU, SADC, UN, UNCTAD, UNESCO, UNHCR, UNIDO, UNMEE, UNMIL, UNMIS, UNOCI, UNWTO, UPU, WCL, WCO, WHO, WIPO, WMO, WTO
Diplomatic representation in the US:
chief of mission: Ambassador Patrick NANDAGO
chancery: 1605 New Hampshire Avenue NW, Washington, DC 20009
telephone: [1] (202) 986-0540
FAX: [1] (202) 986-0443
Diplomatic representation from the US:
chief of mission: Ambassador G. Dennise MATHIEU
embassy: 14 Lossen Street, Windhoek
mailing address: Private Bag 12029 Ausspannplatz, Windhoek
telephone: [264] (61) 295-8500
FAX: [264] (61) 295-8603
Flag description: a wide red stripe edged by narrow white stripes divides the flag diagonally from lower hoist corner to

upper fly corner; the upper hoist-side triangle is blue and charged with a yellow, 12-rayed sunburst; the lower fly-side triangle is green

ECONOMY

Economy—overview: The economy is heavily dependent on the extraction and processing of minerals for export. Mining accounts for 8% of GDP, but provides more than 50% of foreign exchange earnings. Rich alluvial diamond deposits make Namibia a primary source for gem-quality diamonds. Namibia is the fourth-largest exporter of nonfuel minerals in Africa, the world's fifth-largest producer of uranium, and the producer of large quantities of lead, zinc, tin, silver, and tungsten. The mining sector employs only about 3% of the population while about half of the population depends on subsistence agriculture for its livelihood. Namibia normally imports about 50% of its cereal requirements; in drought years food shortages are a major problem in rural areas. A high per capita GDP, relative to the region, hides one of the world's most unequal income distributions. The Namibian economy is closely linked to South Africa with the Namibian dollar pegged one-to-one to the South African rand. Increased payments from the Southern African Customs Union (SACU) put Namibia's budget into surplus in 2007 for the first time since independence, but SACU payments will decline after 2008 as part of a new revenue sharing formula. Increased fish production and mining of zinc, copper, uranium, and silver spurred growth in 2003–07, but growth in recent years was undercut by poor fish catches and high costs for metal inputs.
GDP (purchasing power parity): $10.72 billion (2007 est.)
GDP (official exchange rate): $7.4 billion (2007 est.)
GDP—real growth rate: 4.4% (2007 est.)
GDP—per capita (PPP): $5,200 (2007 est.)
GDP—composition by sector:
agriculture: 10.8%
industry: 33.4%
services: 55.8% (2007 est.)
Labor force: 660,000 (2007 est.)
Labor force—by occupation:
agriculture: 47%
industry: 20%
services: 33% (1999 est.)
Unemployment rate: 5.2% (2007 est.)
Population below poverty line: the UNDP's 2005 Human Development Report indicated that 34.9% of the population live on $1 per day and 55.8% live on $2 per day
Household income or consumption by percentage share: *lowest 10%:* 0.5%
highest 10%: 64.5% (2003)

Distribution of family income—Gini index: 70.7 (2003)

Inflation rate (consumer prices): 6.7% (2007 est.)

Investment (gross fixed): 27.7% of GDP (2007 est.)

Budget: *revenues:* $2.765 billion *expenditures:* $2.515 billion (2007 est.)

Public debt: 21.8% of GDP (2007 est.)

Agriculture—products: millet, sorghum, peanuts, grapes; livestock; fish

Industries: meatpacking, fish processing, dairy products; mining (diamonds, lead, zinc, tin, silver, tungsten, uranium, copper)

Industrial production growth rate: 4% (2007 est.)

Electricity—production: 1.688 billion kWh (2005)

Electricity—production by source: NA

Electricity—consumption: 2.863 billion kWh (2005)

Electricity—exports: 78 million kWh (2005)

Electricity—imports: 1.567 billion kWh; note—electricity supplied by South Africa (2005)

Oil—production: 0 bbl/day (2005 est.)

Oil—consumption: 18,400 bbl/day (2005 est.)

Oil—exports: 0 bbl/day (2004)

Oil—imports: 17,580 bbl/day (2004)

Oil—proved reserves: 0 bbl (1 January 2006 est.)

Natural gas—production: 0 cu m (2005 est.)

Natural gas—consumption: 0 cu m (2005 est.)

Natural gas—exports: 0 cu m (2005 est.)

Natural gas—imports: 0 cu m (2005)

Natural gas—proved reserves: 59.75 billion cu m (1 January 2006 est.)

Current account balance: $1.36 billion (2007 est.)

Exports: $2.916 billion f.o.b. (2007 est.)

Exports—commodities: diamonds, copper, gold, zinc, lead, uranium; cattle, processed fish, karakul skins

Exports—partners: South Africa 33.4%, US 4% (2006)

Imports: $2.977 billion f.o.b. (2007 est.)

Imports—commodities: foodstuffs; petroleum products and fuel, machinery and equipment, chemicals

Imports—partners: South Africa 85.2%, US (2006)

Economic aid—recipient: ODA, $123.4 million (2005 est.)

Reserves of foreign exchange and gold: $896 million (31 December 2007 est.)

Debt—external: $941.1 million (31 December 2007 est.)

Stock of direct foreign investment—at home: $NA

Stock of direct foreign investment—abroad: $NA

Market value of publicly traded shares: $541.8 million (2006)

Currency (code): Namibian dollar (NAD); South African rand (ZAR)

Currency code: NAD; ZAR

Exchange rates: Namibian dollars per US dollar—7.18 (2007), 6.7649 (2006), 6.3593 (2005), 6.4597 (2004), 7.5648 (2003)

Fiscal year: 1 April—31 March

COMMUNICATIONS

Telephones—main lines in use: 138,900 (2005)

Telephones—mobile cellular: 495,000 (2005)

Telephone system:

general assessment: good system with a combined fixed-line and mobile-cellular teledensity of about 30 per 100 persons

domestic: core fiber-optic network links most centers and connections are now digital; Namibia's first mobile-cellular network, launched in 1994, provides coverage to 86 percent of Namibia by area

international: country code—264; fiber-optic cable to South Africa, microwave radio relay link to Botswana, direct links to other neighboring countries; connected to the South African Far East (SAFE) submarine cable through South Africa; satellite earth stations—4 Intelsat

Radio broadcast stations: AM 2, FM 39, shortwave 4 (2001)

Radios: 232,000 (1997)

Television broadcast stations: 2 (2007)

Televisions: 60,000 (1997)

Internet country code: .na

Internet hosts: 3,717 (2007)

Internet Service Providers (ISPs): 2 (2000)

Internet users: 80,600 (2005)

TRANSPORTATION

Airports: 137 (2007)

Airports—with paved runways:

total: 21

over 3,047 m: 3

2,438 to 3,047 m: 2

1,524 to 2,437 m: 13

914 to 1,523 m: 3 (2007)

Airports—with unpaved runways:

total: 116

2,438 to 3,047 m: 2

1,524 to 2,437 m: 22

914 to 1,523 m: 72

under 914 m: 20 (2007)

Railways:

total: 2,382 km

narrow gauge: 2,382 km 1.067-m gauge (2006)

Roadways:

total: 42,237 km

paved: 5,406 km

unpaved: 36,831 km (2002)

Merchant marine:

total: 1 ship (1000 GRT or over) 2,265 GRT/3,605 DWT

by type: cargo 1 (2007)

Ports and terminals: Luderitz, Walvis Bay

MILITARY

Military branches: Namibian Defense Force: Army, Navy, Air Wing (2008)

Military service age and obligation: 18–25 years of age for voluntary military service; no conscription (2008)

Manpower available for military service:

males age 16–49: 527,948 (2008 est.)

Manpower fit for military service:

males age 16–49: 313,497 (2008 est.)

Military expenditures—percent of GDP: 3.7% (2006)

TRANSNATIONAL ISSUES

Disputes—international: concerns from international experts and local populations over the Okavango Delta ecology in Botswana and human displacement scuttled Namibian plans to construct a hydroelectric dam on Popa Falls along the Angola-Namibia border; managed dispute with South Africa over the location of the boundary in the Orange River; Namibia has supported, and in 2004 Zimbabwe dropped objections to, plans between Botswana and Zambia to build a bridge over the Zambezi River, thereby de facto recognizing a short, but not clearly delimited, Botswana-Zambia boundary in the river

Refugees and internally displaced persons: *refugees (country of origin):* 4,700 (Angola) (2007)

NAURU

INTRODUCTION

Background: The exact origins of the Nauruans are unclear, since their language does not resemble any other in the Pacific. The island was annexed by Germany in 1888 and its phosphate deposits began to be mined early in the 20th century by a German-British consortium. Nauru was occupied by Australian forces in World War I and subsequently became a League of Nations mandate. After the Second World War—and a brutal occupation by Japan—Nauru became a UN trust terri-

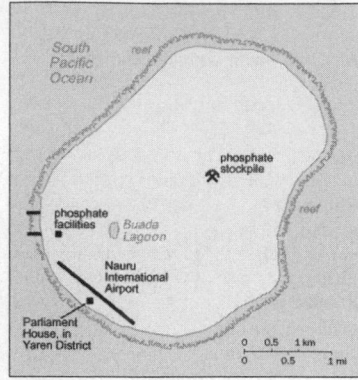

tory. It achieved its independence in 1968 and joined the UN in 1999 as the world's smallest independent republic.

GEOGRAPHY

Location: Oceania, island in the South Pacific Ocean, south of the Marshall Islands
Geographic coordinates: 0 32 S, 166 55 E
Map references: Oceania
Area:
total: 21 sq km
land: 21 sq km
water: 0 sq km
Area—comparative: about 0.1 times the size of Washington, DC
Land boundaries: 0 km
Coastline: 30 km
Maritime claims:
territorial sea: 12 nm
contiguous zone: 24 nm
exclusive economic zone: 200 nm
Climate: tropical with a monsoonal pattern; rainy season (November to February)
Terrain: sandy beach rises to fertile ring around raised coral reefs with phosphate plateau in center
Elevation extremes:
lowest point: Pacific Ocean 0 m
highest point: unnamed location along plateau rim 61 m
Natural resources: phosphates, fish
Land use:
arable land: 0%
permanent crops: 0%
other: 100% (2005)
Irrigated land: NA
Natural hazards: periodic droughts
Environment—current issues: limited natural fresh water resources, roof storage tanks collect rainwater, but mostly dependent on a single, aging desalination plant; intensive phosphate mining during the past 90 years—mainly by a UK, Australia, and NZ consortium—has left the central 90% of Nauru a wasteland and threatens limited remaining land resources

Environment—international agreements: *party to:* Biodiversity, Climate Change, Climate Change-Kyoto Protocol, Desertification, Hazardous Wastes, Law of the Sea, Marine Dumping, Ozone Layer Protection, Whaling
signed, but not ratified: none of the selected agreements
Geography—note: Nauru is one of the three great phosphate rock islands in the Pacific Ocean—the others are Banaba (Ocean Island) in Kiribati and Makatea in French Polynesia; only 53 km south of Equator

PEOPLE

Population: 13,770 (July 2008 est.)
Age structure:
0–14 years: 35.5% (male 2,492/female 2,393)
15–64 years: 62.5% (male 4,237/female 4,363)
65 years and over: 2.1% (male 148/female 137) (2008 est.)
Median age:
total: 21.3 years
male: 20.7 years
female: 21.9 years (2008 est.)
Population growth rate: 1.772% (2008 est.)
Birth rate: 24.26 births/1,000 population (2008 est.)
Death rate: 6.54 deaths/1,000 population (2008 est.)
Net migration rate: NA
Sex ratio:
at birth: 1.05 male(s)/female
under 15 years: 1.04 male(s)/female
15–64 years: 0.97 male(s)/female
65 years and over: 1.08 male(s)/female
total population: 1 male(s)/female (2008 est.)
Infant mortality rate:
total: 9.43 deaths/1,000 live births
male: 11.84 deaths/1,000 live births
female: 6.9 deaths/1,000 live births (2008 est.)
Life expectancy at birth:
total population: 63.81 years
male: 60.2 years
female: 67.6 years (2008 est.)
Total fertility rate: 2.94 children born/woman (2008 est.)
HIV/AIDS—adult prevalence rate: NA
HIV/AIDS—people living with HIV/AIDS: NA
HIV/AIDS—deaths: NA
Nationality:
noun: Nauruan(s)
adjective: Nauruan
Ethnic groups: Nauruan 58%, other Pacific Islander 26%, Chinese 8%, European 8%
Religions: Christian (two-thirds Protestant, one-third Roman Catholic)

Languages: Nauruan (official; a distinct Pacific Island language), English widely understood, spoken, and used for most government and commercial purposes
Literacy: NA

GOVERNMENT

Country name:
conventional long form: Republic of Nauru
conventional short form: Nauru
local long form: Republic of Nauru
local short form: Nauru
former: Pleasant Island
Government type: republic
Capital: no official capital; government offices in Yaren District
time difference: UTC+12 (17 hours ahead of Washington, DC during Standard Time)
Administrative divisions: 14 districts; Aiwo, Anabar, Anetan, Anibare, Baiti, Boe, Buada, Denigomodu, Ewa, Ijuw, Meneng, Nibok, Uaboe, Yaren
Independence: 31 January 1968 (from the Australia-, NZ-, and UK-administered UN trusteeship)
National holiday: Independence Day, 31 January (1968)
Constitution: 29 January 1968; amended 17 May 1968 (Constitution Day)
Legal system: acts of the Nauru Parliament and British common law; accepts compulsory ICJ jurisdiction with reservations
Suffrage: 20 years of age; universal and compulsory
Executive branch:
chief of state: President Marcus STEPHEN (since 19 December 2007); note—the president is both the chief of state and head of government
head of government: President Marcus STEPHEN (since 19 December 2007); note—President Ludwig SCOTTY defeated in a no confidence vote in parliament on 19 December 2007
cabinet: Cabinet appointed by the president from among the members of Parliament
elections: president elected by Parliament for a three-year term; election last held 19 December 2007 (next to be held in 2010)
election results: NA
Legislative branch: unicameral Parliament (18 seats; members elected by popular vote to serve three-year terms)
elections: last held 26 April 2008 (next to be held in 2011)
election results: percent of vote—NA; seats—independents 18; note—President Marcus STEPHEN called a snap election to break a parliamentary stalemate blocking legislative action
Judicial branch: Supreme Court

Political parties and leaders:
Democratic Party [Kennan ADEANG];
Nauru Party (informal); Nauru First
(Naoero Amo) Party; note—loose multi-party system
Political pressure groups and leaders:
NA
International organization participation: ACP, ADB, C, FAO, ICAO, ICCt,
Interpol, IOC, ITU, OPCW, PIF,
Sparteca, SPC, UN, UNCTAD,
UNESCO, UPU, WHO
Diplomatic representation in the US:
chief of mission: Ambassador Marlene I.
MOSES
chancery: 800 2nd Avenue, Suite 400 D,
New York, NY 10017
telephone: [1] (212) 937-0074
FAX: [1] (212) 937-0079
consulate(s): Agana (Guam)
Diplomatic representation from the US:
the US does not have an embassy in
Nauru; the US Ambassador to Fiji is
accredited to Nauru
Flag description: blue with a narrow,
horizontal, yellow stripe across the
center and a large white 12-pointed star
below the stripe on the hoist side; the
star indicates the country's location in
relation to the Equator (the yellow
stripe) and the 12 points symbolize the
12 original tribes of Nauru

ECONOMY

Economy—overview: Revenues of this
tiny island have traditionally come from
exports of phosphates, now significantly
depleted. An Australian company in
2005 entered into an agreement
intended to exploit remaining supplies.
Few other resources exist with most
necessities being imported, mainly from
Australia, its former occupier and later
major source of support. The rehabilitation of mined land and the replacement
of income from phosphates are serious
long-term problems. In anticipation of
the exhaustion of Nauru's phosphate
deposits, substantial amounts of phosphate income were invested in trust
funds to help cushion the transition and
provide for Nauru's economic future. As
a result of heavy spending from the trust
funds, the government faces virtual
bankruptcy. To cut costs the government
has frozen wages and reduced overstaffed
public service departments. In 2005, the
deterioration in housing, hospitals, and
other capital plant continued, and the
cost to Australia of keeping the government and economy afloat continued to
climb. Few comprehensive statistics on
the Nauru economy exist, with estimates
of Nauru's GDP varying widely.
GDP (purchasing power parity): $60
million (2005 est.)

GDP (official exchange rate): $NA
GDP—real growth rate: NA%
GDP—per capita (PPP): $5,000 (2005 est.)
GDP—composition by sector:
agriculture: NA%
industry: NA%
services: NA%
Labor force—by occupation: *note:*
employed in mining phosphates, public
administration, education, and transportation (1992)
Unemployment rate: 90% (2004 est.)
Population below poverty line: NA%
Household income or consumption by percentage share:
lowest 10%: NA%
highest 10%: NA%
Inflation rate (consumer prices): -3.6%
(1993)
Budget: *revenues:* $13.5 million
expenditures: $13.5 million (2005)
Agriculture—products: coconuts
Industries: phosphate mining, offshore
banking, coconut products
Industrial production growth rate:
NA%
Electricity—production: 30 million
kWh (2005)
Electricity—production by source:
fossil fuel: 100%
hydro: 0%
nuclear: 0%
other: 0% (2001)
Electricity—consumption: 27.9 million
kWh (2005)
Electricity—exports: 0 kWh (2005)
Electricity—imports: 0 kWh (2005)
Oil—production: 0 bbl/day (2005 est.)
Oil—consumption: 1,050 bbl/day (2005 est.)
Oil—exports: 0 bbl/day (2004)
Oil—imports: 1,023 bbl/day (2004)
Oil—proved reserves: 0 bbl (1 January 2006 est.)
Natural gas—production: 0 cu m (2005 est.)
Natural gas—consumption: 0 cu m (2005 est.)
Natural gas—exports: 0 cu m (2005 est.)
Natural gas—imports: 0 cu m (2005)
Natural gas—proved reserves: 0 cu m (1 January 2006 est.)
Exports: $64,000 f.o.b. (2005 est.)
Exports—commodities: phosphates
Exports—partners: South Africa 63.7%,
South Korea 7.6%, Canada 6.6% (2006)
Imports: $20 million c.i.f. (2004 est.)
Imports—commodities: food, fuel,
manufactures, building materials,
machinery
Imports—partners: South Korea 43.8%,
Australia 36.2%, US 5.9%, Germany
4.3% (2006)

Economic aid—recipient: $20 million
mostly from Australia (2005)
Debt—external: $33.3 million (2002)
Currency (code): Australian dollar
(AUD)
Currency code: AUD
Exchange rates: Australian dollars per
US dollar—1.2137 (2007), 1.3285
(2006), 1.3095 (2005), 1.3598 (2004),
1.5419 (2003)
Fiscal year: 1 July—30 June

COMMUNICATIONS

Telephones—main lines in use: 1,900
(2002)
Telephones—mobile cellular: 1,500
(2002)
Telephone system:
general assessment: adequate local and
international radiotelephone communication provided via Australian facilities
domestic: NA
international: country code—674; satellite earth station—1 Intelsat (Pacific
Ocean)
Radio broadcast stations: AM 1, FM 0,
shortwave 0 (1998)
Radios: 7,000 (1997)
Television broadcast stations: 1 (1997)
Televisions: 500 (1997)
Internet country code: .nr
Internet hosts: 53 (2007)
Internet Service Providers (ISPs): 1
(2000)
Internet users: 300 (2002)

TRANSPORTATION

Airports: 1 (2007)
Airports—with paved runways:
total: 1
1,524 to 2,437 m: 1 (2007)
Roadways:
total: 30 km
Ports and terminals: Nauru

MILITARY

Military branches: no regular military
forces; Nauru Police Force (2008)
Manpower available for military service:
males age 16–49: 3,470 (2008 est.)
Military expenditures—percent of GDP: NA
Military—note: Nauru maintains no
defense forces; under an informal agreement, defense is the responsibility of
Australia

TRANSNATIONAL ISSUES

Disputes—international: none

NAVASSA ISLAND

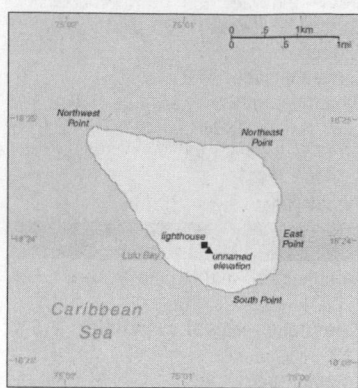

INTRODUCTION

Background: This uninhabited island was claimed by the US in 1857 for its guano. Mining took place between 1865 and 1898. The lighthouse, built in 1917, was shut down in 1996 and administration of Navassa Island transferred from the Coast Guard to the Department of the Interior. A 1998 scientific expedition to the island described it as a unique preserve of Caribbean biodiversity; the following year it became a National Wildlife Refuge and annual scientific expeditions have continued.

GEOGRAPHY

Location: Caribbean, island in the Caribbean Sea, 35 miles west of Tiburon Peninsula of Haiti
Geographic coordinates: 18 25 N, 75 02 W
Map references: Central America and the Caribbean

Area:
total: 5.4 sq km
land: 5.4 sq km
water: 0 sq km
Area—comparative: about nine times the size of The Mall in Washington, DC
Land boundaries: 0 km
Coastline: 8 km
Maritime claims:
territorial sea: 12 nm
exclusive economic zone: 200 nm
Climate: marine, tropical
Terrain: raised coral and limestone plateau, flat to undulating; ringed by vertical white cliffs (9 to 15 m high)
Elevation extremes:
lowest point: Caribbean Sea 0 m
highest point: unnamed location on southwest side 77 m
Natural resources: guano
Land use:
arable land: 0%
permanent crops: 0%
other: 100% (2005)
Natural hazards: hurricanes
Environment—current issues: NA
Geography—note: strategic location 160 km south of the US Naval Base at Guantanamo Bay, Cuba; mostly exposed rock with numerous solution holes but with enough grassland to support goat herds; dense stands of fig trees, scattered cactus

PEOPLE

Population: uninhabited
note: transient Haitian fishermen and others camp on the island

GOVERNMENT

Country name:
conventional long form: none
conventional short form: Navassa Island
Dependency status: unorganized, unincorporated territory of the US; administered by the Fish and Wildlife Service, US Department of the Interior, from the Caribbean Islands National Wildlife Refuge in Boqueron, Puerto Rico; in September 1996, the Coast Guard ceased operations and maintenance of Navassa Island Light, a 46-meter-tall lighthouse on the southern side of the island; there has also been a private claim advanced against the island
Legal system: the laws of the US, where applicable, apply
Diplomatic representation from the US: none (territory of the US)
Flag description: the flag of the US is used

ECONOMY

Economy—overview: Subsistence fishing and commercial trawling occur within refuge waters.

TRANSPORTATION

Ports and terminals: none; offshore anchorage only

MILITARY

Military—note: defense is the responsibility of the US

TRANSNATIONAL ISSUES

Disputes—international: claimed by Haiti, source of subsistence fishing

NEPAL

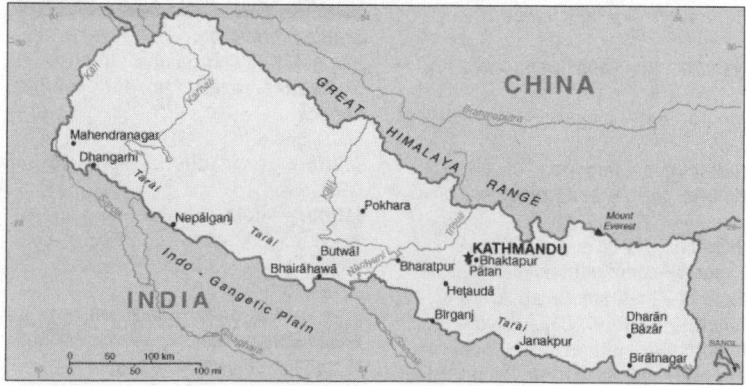

INTRODUCTION

Background: In 1951, the Nepalese monarch ended the century-old system of rule by hereditary premiers and instituted a cabinet system of government. Reforms in 1990 established a multiparty democracy within the framework of a constitutional monarchy. A Maoist insurgency, launched in 1996, gained traction and threatened to bring down the regime, especially after a negotiated cease-fire between the Maoists and government forces broke down in August 2003. In 2001, the crown prince massacred ten members of the royal family, including the king and queen, and then took his own life. In October 2002, the new king dismissed the prime minister and his cabinet for "incompetence" after they dissolved the parliament and were subsequently unable to hold elections because of the ongoing insurgency. While stopping short of reestablishing parliament, the king in June 2004 reinstated the most recently elected prime minister who formed a four-party coalition government. Citing dissatisfaction with the government's lack of progress in addressing the Maoist insurgency and corruption, the king in February 2005 dissolved the government, declared a state of emergency, imprisoned party leaders, and assumed power. The king's government subsequently released party leaders and officially ended the state of emergency in May 2005, but the monarch retained absolute power until April 2006. After nearly three weeks of mass protests organized by the seven-party opposition and the Maoists, the king allowed parliament to reconvene in April 2006. Following a November 2006 peace accord between the government and the Maoists, an interim constitution was promulgated and the Maoists were allowed to enter parliament in January 2007. The peace accord calls for the creation of a Constituent Assembly to draft a new constitution. The Constituent Assembly elections, twice delayed, were held 10 April 2008. The Assembly will meet for the first time on 28 May 2008.

GEOGRAPHY

Location: Southern Asia, between China and India
Geographic coordinates: 28 00 N, 84 00 E
Map references: Asia
Area: *total:* 147,181 sq km
land: 143,181 sq km
water: 4,000 sq km
Area—comparative: slightly larger than Arkansas
Land boundaries:
total: 2,926 km
border countries: China 1,236 km, India 1,690 km
Coastline: 0 km (landlocked)
Maritime claims: none (landlocked)
Climate: varies from cool summers and severe winters in north to subtropical summers and mild winters in south
Terrain: Tarai or flat river plain of the Ganges in south, central hill region, rugged Himalayas in north
Elevation extremes:
lowest point: Kanchan Kalan 70 m
highest point: Mount Everest 8,850 m
Natural resources: quartz, water, timber, hydropower, scenic beauty, small deposits of lignite, copper, cobalt, iron ore
Land use: *arable land:* 16.07%
permanent crops: 0.85%
other: 83.08% (2005)
Irrigated land: 11,700 sq km (2003)
Total renewable water resources: 210.2 cu km (1999)
Freshwater withdrawal (domestic/industrial/agricultural): *total:* 10.18 cu km/yr (3%/1%/96%)
per capita: 375 cu m/yr (2000)

Natural hazards: severe thunderstorms, flooding, landslides, drought, and famine depending on the timing, intensity, and duration of the summer monsoons
Environment—current issues: deforestation (overuse of wood for fuel and lack of alternatives); contaminated water (with human and animal wastes, agricultural runoff, and industrial effluents); wildlife conservation; vehicular emissions
Environment—international agreements: *party to:* Biodiversity, Climate Change, Climate Change-Kyoto Protocol, Desertification, Endangered Species, Hazardous Wastes, Law of the Sea, Ozone Layer Protection, Tropical Timber 83, Tropical Timber 94, Wetlands
signed, but not ratified: Marine Life Conservation
Geography—note: landlocked; strategic location between China and India; contains eight of world's 10 highest peaks, including Mount Everest and Kanchenjunga—the world's tallest and third tallest—on the borders with China and India respectively

PEOPLE

Population: 29,519,114 (July 2008 est.)
Age structure:
0–14 years: 38% (male 5,792,042/female 5,427,370)
15–64 years: 58.2% (male 8,832,488/female 8,345,724)
65 years and over: 3.8% (male 542,192/female 579,298) (2008 est.)
Median age:
total: 20.7 years
male: 20.5 years
female: 20.8 years (2008 est.)
Population growth rate: 2.095% (2008 est.)
Birth rate: 29.92 births/1,000 population (2008 est.)
Death rate: 8.97 deaths/1,000 population (2008 est.)
Net migration rate: NA
Sex ratio:
at birth: 1.05 male(s)/female
under 15 years: 1.07 male(s)/female
15–64 years: 1.06 male(s)/female
65 years and over: 0.94 male(s)/female
total population: 1.06 male(s)/female (2008 est.)
Infant mortality rate:
total: 62 deaths/1,000 live births
male: 60.18 deaths/1,000 live births
female: 63.91 deaths/1,000 live births (2008 est.)
Life expectancy at birth:
total population: 60.94 years
male: 61.12 years
female: 60.75 years (2008 est.)

Total fertility rate: 3.91 children born/woman (2008 est.)
HIV/AIDS—adult prevalence rate: 0.5% (2001 est.)
HIV/AIDS—people living with HIV/AIDS: 61,000 (2001 est.)
HIV/AIDS—deaths: 3,100 (2003 est.)
Major infectious diseases:
degree of risk: intermediate
food or waterborne diseases: bacterial diarrhea, hepatitis A, and typhoid fever
vectorborne disease: Japanese encephalitis and malaria (2008)
Nationality:
noun: Nepalese (singular and plural)
adjective: Nepalese
Ethnic groups: Chhettri 15.5%, Brahman-Hill 12.5%, Magar 7%, Tharu 6.6%, Tamang 5.5%, Newar 5.4%, Muslim 4.2%, Kami 3.9%, Yadav 3.9%, other 32.7%, unspecified 2.8% (2001 census)
Religions: Hindu 80.6%, Buddhist 10.7%, Muslim 4.2%, Kirant 3.6%, other 0.9% (2001 census)
note: only official Hindu state in the world
Languages: Nepali 47.8%, Maithali 12.1%, Bhojpuri 7.4%, Tharu (Dagaura/Rana) 5.8%, Tamang 5.1%, Newar 3.6%, Magar 3.3%, Awadhi 2.4%, other 10%, unspecified 2.5% (2001 census)
note: many in government and business also speak English (2001 est.)
Literacy: *definition:* age 15 and over can read and write
total population: 48.6%
male: 62.7%
female: 34.9% (2001 census)

GOVERNMENT

Country name:
conventional long and short form: Nepal
local long and short form: Nepal
Government type: constitutional monarchy
Capital: *name:* Kathmandu
geographic coordinates: 27 43 N, 85 19 E
time difference: UTC+5.75 (10.75 hours ahead of Washington, DC during Standard Time)
Administrative divisions: 14 zones (anchal, singular and plural); Bagmati, Bheri, Dhawalagiri, Gandaki, Janakpur, Karnali, Kosi, Lumbini, Mahakali, Mechi, Narayani, Rapti, Sagarmatha, Seti
Independence: 1768 (unified by Prithvi Narayan SHAH)
National holiday: NA; note—in 2006, Parliament abolished the birthday of King GYANENDRA (7 July) and Constitution Day (9 November) as national holidays

459

Constitution: 9 November 1990; note—a new interim constitution was promulgated in January 2007; the November 2006 peace agreement calls for the election of a Constituent Assembly to draft a new permanent constitution

Legal system: based on Hindu legal concepts and English common law; has not accepted compulsory ICJ jurisdiction

Suffrage: 18 years of age; universal

Executive branch:

chief of state: Prime Minister Girija Prasad KOIRALA (since 30 April 2006); note—the prime minister is both the chief of state and head of government

head of government: Prime Minister Girija Prasad KOIRALA (since 30 April 2006)

cabinet: Cabinet historically appointed by the monarch on the recommendation of the prime minister; note—the prime minister selected the Cabinet in May 2006 in consultation with the political parties

elections: following legislative elections, the leader of the majority party or leader of a majority coalition historically has been appointed prime minister by the monarch

Legislative branch: unicameral Constituent Assembly (601 seats; 240 seats decided by direct popular vote; 335 seats by proportional representation; 26 appointed by the Cabinet (Council of Ministers))

note: KOIRALA has called first sitting of the Assembly on 28 May 2008

elections: last held 10 April 2008 (next to be held NA)

election results: percent of vote by party—NA; seats by party—CPN-M 220, NC 110, CPN-UML 103, Madhesi Jana Adhikar Forum 52, Terai Madhesi Democratic Party/Nepal Sadbhawana Party 29, other smaller parties 61; note—26 seats to be filled by the new Cabinet

Judicial branch: Supreme Court or Sarbochha Adalat (chief justice is appointed by the monarch on recommendation of the Constitutional Council; the other judges are appointed by the monarch on the recommendation of the Judicial Council)

Political parties and leaders: Chure Bhawar Rastriya Ekata Party [Keshav Prasad MAINALI]; Communist Party of Nepal (Maoist) [Pushpa Kamal DAHAL, also known as PRACHANDA, chairman; Dr. Baburam BHATTARAI]; Communist Party of Nepal (ML) [C.P. MAINALI]; Communist Party of Nepal (Unified) [Raj Singh SHRIS]; Communist Party of Nepal (United) [Ganesh SHAH]; Communist Party of Nepal/United Marxist-Leninist or CPN/UML [Amrit Kumar BOHARA];

Dalit Janajati Party [Vishwendraman PASHWAN]; Janamorcha Nepal [Amik SHERCHAN]; Madhesi Jana Adhikar Forum [Upendra YADAV]; National Democratic Party or NDP [Pashupati Shumsher RANA] (also called Rastriya Prajatantra Party or RPP); Nepal Loktantrik Samajbadi Dal [Upendra GACHCHHADAR]; Nepal Pariwar Dal [Vinod DANGI]; Nepal Rastriya Party [Khushilal YADAV]; Nepal Sadbhawana Party (Anandi Devi) [Shyam Sundar GUPTA]; Nepal Workers and Peasants Party or NWPP [Narayan Man BIJUKCHHE]; Nepali Congress Party or NCP [Girija Prasad KOIRALA]; Nepali Janata Dal [Bharat Prasad MAHATO]; Rastriya Janamorcha [Chitra BAHADUR K.C.]; Rastriya Janamukti Party [Malwar Singh THAPA]; Rastriya Janashakti Party or RJP [Surya Bahadur THAPA] (split from RPP in March 2005); Rastriya Prajatantra Party Nepal [Kamal THAPA]; Sadbhavana Party (Mahato) [Rajendra MAHATO]; Samajbadi Prajatantrik Janata Party Nepal [Prem Bahadur SINGH]; Sanghiya Loktantrik Rastriya Manch [Kamal CHHARAHANG]; Terai Madhesi Democratic Party [Mahantha THAKUR]

Political pressure groups and leaders: several small armed Madhesi groups along the southern border with India; a variety of groups advocating regional autonomy for individual ethnic groups

International organization participation: ADB, BIMSTEC, CP, FAO, G-77, IBRD, ICAO, ICC, ICRM, IDA, IFAD, IFC, IFRCS, ILO, IMF, IMO, Interpol, IOC, IOM, IPU, ISO (correspondent), ITSO, ITU, ITUC, MIGA, MINUSTAH, MONUC, NAM, OPCW, SAARC, SACEP, UN, UNAMID, UNCTAD, UNESCO, UNIDO, UNIFIL, UNMEE, UNMIL, UNMIS, UNOCI, UNOMIG, UNTSO, UNWTO, UPU, WCL, WCO, WFTU, WHO, WIPO, WMO, WTO

Diplomatic representation in the US:

chief of mission: Ambassador Suresh Chandra CHALISE

chancery: 2131 Leroy Place NW, Washington, DC 20008

telephone: [1] (202) 667-4550

FAX: [1] (202) 667-5534

consulate(s) general:
New York:

Diplomatic representation from the US:

chief of mission: Ambassador Nancy J. POWELL

embassy: Maharajgunj, Kathmandu

mailing address: use embassy street address

telephone: [977] (1) 400-7200

FAX: [977] (1) 400-7272

Flag description: red with a blue border around the unique shape of two overlapping right triangles; the smaller, upper triangle bears a white stylized moon and the larger, lower triangle bears a white 12-pointed sun

ECONOMY

Economy—overview: Nepal is among the poorest and least developed countries in the world with almost one-third of its population living below the poverty line. Agriculture is the mainstay of the economy, providing a livelihood for three-fourths of the population and accounting for 38% of GDP. Industrial activity mainly involves the processing of agricultural produce including jute, sugarcane, tobacco, and grain. Security concerns relating to the Maoist conflict have led to a decrease in tourism, a key source of foreign exchange. Nepal has considerable scope for exploiting its potential in hydropower and tourism, areas of recent foreign investment interest. Prospects for foreign trade or investment in other sectors will remain poor, however, because of the small size of the economy, its technological backwardness, its remoteness, its landlocked geographic location, its civil strife, and its susceptibility to natural disaster.

GDP (purchasing power parity): $29.04 billion (2007 est.)

GDP (official exchange rate): $9.627 billion (2007 est.)

GDP—real growth rate: 2.5% (2007 est.)

GDP—per capita (PPP): $1,200 (2007 est.)

GDP—composition by sector:

agriculture: 38%

industry: 20%

services: 42% (FY05/06 est.)

Labor force: 11.11 million

note: severe lack of skilled labor (2006 est.)

Labor force—by occupation: *agriculture:* 76%

industry: 6%

services: 18% (2004 est.)

Unemployment rate: 42% (2004 est.)

Population below poverty line: 30.9% (2004)

Household income or consumption by percentage share:

lowest 10%: 2.6%

highest 10%: 40.6% (2004)

Distribution of family income—Gini index: 47.2 (2004)

Inflation rate (consumer prices): 6.4% (2007 est.)

Budget:

revenues: $1.153 billion

expenditures: $1.927 billion (FY06/07)

Agriculture—products: rice, corn, wheat, sugarcane, jute, root crops; milk, water buffalo meat

Industries: tourism, carpets, textiles; small rice, jute, sugar, and oilseed mills; cigarettes, cement and brick production

Industrial production growth rate: 2.2% (FY05/06)

Electricity—production: 2.511 billion kWh (2006)

Electricity—production by source:
fossil fuel: 8.5%
hydro: 91.5%
nuclear: 0%
other: 0% (2001)

Electricity—consumption: 1.96 billion kWh (2006)

Electricity—exports: 101 million kWh (2006)

Electricity—imports: 266 million kWh (2006)

Oil—production: 0 bbl/day (2005 est.)

Oil—consumption: 11,550 bbl/day (2006 est.)

Oil—exports: 0 bbl/day (2004)

Oil—imports: 11,530 bbl/day (2006 est.)

Oil—proved reserves: 0 bbl (1 January 2006 est.)

Natural gas—production: 0 cu m (2005 est.)

Natural gas—consumption: 0 cu m (2005 est.)

Natural gas—exports: 0 cu m (2005 est.)

Natural gas—imports: 0 cu m (2005)

Natural gas—proved reserves: 0 cu m (1 January 2006 est.)

Current account balance: $58 million (2007)

Exports: $830 million f.o.b.; note—does not include unrecorded border trade with India (2006)

Exports—commodities: carpets, clothing, leather goods, jute goods, grain

Exports—partners: India 67.9%, US 11.7%, Germany 4.7% (2006)

Imports: $2.398 billion f.o.b. (2006)

Imports—commodities: gold, machinery and equipment, petroleum products, fertilizer

Imports—partners: India 61.8%, China 3.8%, Indonesia 3.3% (2006)

Economic aid—recipient: $427.9 million (2005)

Debt—external: $3.07 billion (March 2006)

Stock of direct foreign investment—at home: $NA

Stock of direct foreign investment—abroad: $NA

Market value of publicly traded shares: $963.5 million (2005)

Currency (code): Nepalese rupee (NPR)

Currency code: NPR

Exchange rates: Nepalese rupees per US dollar—NA (2007), 72.446 (2006), 72.16 (2005), 73.674 (2004), 76.141 (2003)

Fiscal year: 16 July—15 July

COMMUNICATIONS

Telephones—main lines in use: 595,800 (2006)

Telephones—mobile cellular: 1.042 million (2006)

Telephone system:
general assessment: poor telephone and telegraph service; fair radiotelephone communication service and mobile-cellular telephone network
domestic: NA
international: country code—977; radiotelephone communications; microwave landline to India; satellite earth station—1 Intelsat (Indian Ocean)

Radio broadcast stations: AM 6, FM 5, shortwave 1 (2000)

Radios: 840,000 (1997)

Television broadcast stations: 1 (plus 9 repeaters) (1998)

Televisions: 130,000 (1997)

Internet country code: .np

Internet hosts: 18,733 (2007)

Internet Service Providers (ISPs): 6 (2000)

Internet users: 249,400 (2006)

TRANSPORTATION

Airports: 47 (2007)

Airports—with paved runways:
total: 10
over 3,047 m: 1
914 to 1,523 m: 8
under 914 m: 1 (2007)

Airports—with unpaved runways:
total: 37
1,524 to 2,437 m: 1
914 to 1,523 m: 6
under 914 m: 30 (2007)

Railways: *total:* 59 km

narrow gauge: 59 km 0.762-m gauge (2006)

Roadways:
total: 17,380 km
paved: 9,886 km
unpaved: 7,494 km (2004)

MILITARY

Military branches: Nepalese Army, Armed Police Force (2008)

Military service age and obligation: 18 years of age for voluntary military service; 15 years of age for military training; no conscription (2008)

Manpower available for military service:
males age 16–49: 7,322,965
females age 16–49: 6,859,064 (2008 est.)

Manpower fit for military service:
males age 16–49: 5,146,958
females age 16–49: 4,724,495 (2008 est.)

Manpower reaching militarily significant age annually:
males age 16–49: 335,747
females age 16–49: 312,297 (2008 est.)

Military expenditures—percent of GDP: 1.6% (2006)

TRANSNATIONAL ISSUES

Disputes—international: joint border commission continues to work on contested sections of boundary with India, including the 400 square kilometer dispute over the source of the Kalapani River; India has instituted a stricter border regime to restrict transit of Maoist insurgents and illegal cross-border activities; approximately 106,000 Bhutanese Lhotshampas (Hindus) have been confined in refugee camps in southeastern Nepal since 1990

Refugees and internally displaced persons: *refugees (country of origin):* 107,803 (Bhutan); 20,153 (Tibet/China)
IDPs: 50,000–70,000 (remaining from ten-year Maoist insurgency that officially ended in 2006; displacement spread across the country) (2007)

Illicit drugs: illicit producer of cannabis and hashish for the domestic and international drug markets; transit point for opiates from Southeast Asia to the West

NETHERLANDS

INTRODUCTION

Background: The Dutch United Provinces declared their independence from Spain in 1579; during the 17th century, they became a leading seafaring and commercial power, with settlements and colonies around the world. After a 20-year French occupation, a Kingdom of the Netherlands was formed in 1815. In 1830 Belgium seceded and formed a separate kingdom. The Netherlands remained neutral in World War I, but suffered invasion and occupation by Germany in World War II. A modern, industrialized nation, the Netherlands is also a large exporter of agricultural products. The country was a founding member of NATO and the EEC (now the EU), and participated in the introduction of the euro in 1999.

GEOGRAPHY

Location: Western Europe, bordering the North Sea, between Belgium and Germany

Geographic coordinates: 52 30 N, 5 45 E

Map references: Europe

Area:

total: 41,526 sq km

land: 33,883 sq km

water: 7,643 sq km

Area—comparative: slightly less than twice the size of New Jersey

Land boundaries:

total: 1,027 km

border countries: Belgium 450 km, Germany 577 km

Coastline: 451 km

Maritime claims:

territorial sea: 12 nm

contiguous zone: 24 nm

exclusive fishing zone: 200 nm

Climate: temperate; marine; cool summers and mild winters

Terrain: mostly coastal lowland and reclaimed land (polders); some hills in southeast

Elevation extremes:

lowest point: Zuidplaspolder -7 m

highest point: Vaalserberg 322 m

Natural resources: natural gas, petroleum, peat, limestone, salt, sand and gravel, arable land

Land use:

arable land: 21.96%

permanent crops: 0.77%

other: 77.27% (2005)

Irrigated land: 5,650 sq km (2003)

Total renewable water resources: 89.7 cu km (2005)

Freshwater withdrawal (domestic/industrial/agricultural): *total:* 8.86 cu km/yr (6%/60%/34%)

per capita: 544 cu m/yr (2001)

Natural hazards: flooding

Environment—current issues: water pollution in the form of heavy metals, organic compounds, and nutrients such as nitrates and phosphates; air pollution from vehicles and refining activities; acid rain

Environment—international agreements: *party to:* Air Pollution, Air Pollution-Nitrogen Oxides, Air Pollution-Persistent Organic Pollutants, Air Pollution-Sulfur 85, Air Pollution-Sulfur 94, Air Pollution-Volatile Organic Compounds, Antarctic-Environmental Protocol, Antarctic-Marine Living Resources, Antarctic Treaty, Biodiversity, Climate Change, Climate Change-Kyoto Protocol, Desertification, Endangered Species, Environmental Modification, Hazardous Wastes, Kyoto Protocol, Law of the Sea, Marine Dumping, Marine Life Conservation, Ozone Layer Protection, Ship Pollution, Tropical Timber 83, Tropical Timber 94, Wetlands, Whaling

Geography—note: located at mouths of three major European rivers (Rhine, Maas or Meuse, and Schelde)

PEOPLE

Population: 16,645,313 (July 2008 est.)

Age structure:

0–14 years: 17.6% (male 1,496,348/female 1,427,297)

15–64 years: 67.8% (male 5,705,003/female 5,583,787)

65 years and over: 14.6% (male 1,040,932/female 1,391,946) (2008 est.)

Median age: *total:* 40 years

male: 39.2 years

female: 40.9 years (2008 est.)

Population growth rate: 0.436% (2008 est.)

Birth rate: 10.53 births/1,000 population (2008 est.)

Death rate: 8.71 deaths/1,000 population (2008 est.)

Net migration rate: 2.55 migrant(s)/1,000 population (2008 est.)

Sex ratio:

at birth: 1.05 male(s)/female

under 15 years: 1.05 male(s)/female

15–64 years: 1.02 male(s)/female

65 years and over: 0.75 male(s)/female

total population: 0.98 male(s)/female (2008 est.)

Infant mortality rate:

total: 4.81 deaths/1,000 live births

male: 5.34 deaths/1,000 live births

female: 4.25 deaths/1,000 live births (2008 est.)

Life expectancy at birth:

total population: 79.25 years

male: 76.66 years

female: 81.98 years (2008 est.)

Total fertility rate: 1.66 children born/woman (2008 est.)

HIV/AIDS—adult prevalence rate: 0.2% (2001 est.)

HIV/AIDS—people living with HIV/AIDS: 19,000 (2001 est.)

HIV/AIDS—deaths: fewer than 100 (2003 est.)

Nationality:

noun: Dutchman(men), Dutchwoman (women)

adjective: Dutch

Ethnic groups: Dutch 83%, other 17% (of which 9% are non-Western origin mainly Turks, Moroccans, Antilleans, Surinamese, and Indonesians) (1999 est.)

Religions: Roman Catholic 31%, Dutch Reformed 13%, Calvinist 7%, Muslim 5.5%, other 2.5%, none 41% (2002)

Languages: Dutch (official), Frisian (official)

Literacy: *definition:* age 15 and over can read and write

total population: 99%

male: 99%

female: 99% (2003 est.)

GOVERNMENT

Country name:

conventional long form: Kingdom of the Netherlands

conventional short form: Netherlands

local long form: Koninkrijk der Nederlanden

local short form: Nederland

Government type: constitutional monarchy

Capital: *name:* Amsterdam

geographic coordinates: 52 23 N, 4 54 E

time difference: UTC+1 (6 hours ahead of Washington, DC during Standard Time)

daylight saving time: +1hr, begins last Sunday in March; ends last Sunday in October

note: The Hague (seat of government)

Administrative divisions: 12 provinces (provincies, singular—provincie); Drenthe, Flevoland, Friesland (Fryslan), Gelderland, Groningen, Limburg, Noord-Brabant (North Brabant), Noord-Holland (North Holland), Overijssel, Utrecht, Zeeland, Zuid-Holland (South Holland)

Dependent areas: Aruba, Netherlands Antilles

Independence: 23 January 1579 (the northern provinces of the Low Countries conclude the Union of Utrecht breaking with Spain; on 26 July 1581 they formally declared their independence with an Act of Abjuration; however, it was not until 30 January 1648 and the Peace of Westphalia that Spain recognized this independence)

National holiday: Queen's Day (Birthday of Queen-Mother JULIANA and accession to the throne of her oldest daughter BEATRIX), 30 April (1909 and 1980)

Constitution: adopted 1815; amended many times, most recently in 2002

Legal system: based on civil law system incorporating French penal theory; constitution does not permit judicial review of acts of the States General; accepts compulsory ICJ jurisdiction with reservations

Suffrage: 18 years of age; universal

Executive branch:

chief of state: Queen BEATRIX (since 30 April 1980); Heir Apparent WILLEM-ALEXANDER (born 27 April 1967), son of the monarch

head of government: Prime Minister Jan Peter BALKENENDE (since 22 July 2002); Deputy Prime Ministers Wouter BOS (since 22 February 2007) and Andre ROUVOET (since 22 February 2007)

cabinet: Council of Ministers appointed by the monarch

elections: the monarchy is hereditary; following Second Chamber elections, the leader of the majority party or leader of a majority coalition is usually appointed prime minister by the monarch; deputy prime ministers appointed by the monarch

note: there is also a Council of State composed of the monarch, heir apparent, and councilors that provides consultations to the cabinet on legislative and administrative policy

Legislative branch: bicameral States General or Staten Generaal consists of the First Chamber or Eerste Kamer (75 seats; members indirectly elected by the country's 12 provincial councils to serve four-year terms) and the Second Chamber or Tweede Kamer (150 seats; members directly elected by popular vote to serve four-year terms)

elections: First Chamber—last held 29 May 2007 (next to be held in May 2011); Second Chamber—last held 22 November 2006 (next to be held by early 2011)

election results: First Chamber—percent of vote by party—NA%; seats by party—CDA 21, PvdA 14, VVD 14, Socialist Party 11, Christian Union 4, Green Left Party 4, D66 2, other 5; Second Chamber—percent of vote by party—CDA 26.5%, PvdA 21.2%, Socialist Party 16.6%, VVD 14.6%, Party for Freedom 5.9%, Green Party 4.6%, Christian Union 4.0%, other 6.6%; seats by party—CDA 41, PvdA 33, Socialist Party 25, VVD 22, Party for Freedom 9, Green Party 7, Christian Union 6, other 7

Judicial branch: Supreme Court or Hoge Raad (justices are nominated for life by the monarch)

Political parties and leaders: Christian Democratic Appeal or CDA [Jan Peter BALKENENDE]; Christian Union Party [Andre ROUVOET]; Democrats 66 or D66 [Alexander PECHTOLD]; Green Left Party [Femke HALSEMA]; Labor Party or PvdA [Wouter BOS]; Party for Freedom or PVV [Geert WILDERS]; Party for the Animals or PvdD [Marianne THIEME]; People's Party for Freedom and Democracy (Liberal) or VVD [Mark RUTTE]; Reformed Political Party of SGP [Bas VAN DER VLIES]; Socialist Party [Jan MARIJNISSEN]; plus a few minor parties

Political pressure groups and leaders: Christian Trade Union Federation or CNV [Rene PAAS]; Confederation of Netherlands Industry and Employers or VNO-NCW [Bernard WIENTJES]; Federation for Small and Medium-sized businesses or MKB [Loek HERMANS]; Netherlands Trade Union Federation or FNV [Agnes JONGERIUS]; Social Economic Council or SER [Alexander RINNOOY Kan]; Trade Union Federation of Middle and High Personnel or MHP [Ad VERHOEVEN]

International organization participation: ADB (nonregional members), AfDB, Arctic Council (observer), Australia Group, Benelux, BIS, CBSS (observer), CE, CERN, EAPC, EBRD, EIB, EMU, ESA, EU, FAO, G-10, IADB, IAEA, IBRD, ICAO, ICC, ICCt, ICRM, IDA, IEA, IFAD, IFC, IFRCS, IHO, ILO, IMF, IMO, IMSO, Interpol, IOC, IOM, IPU, ISO, ITSO, ITU, ITUC, MIGA, NAM (guest), NATO, NEA, NSG, OAS (observer), OECD, OPCW, OSCE, Paris Club, PCA, Schengen Convention, SECI (observer), UN, UNAMID, UNCTAD, UNESCO, UNHCR, UNIDO, UNIFIL, UNMIS, UNRWA, UNTSO, UNWTO, UPU, WCL, WCO, WEU, WHO, WIPO, WMO, WTO, ZC

Diplomatic representation in the US:

chief of mission: Ambassador Christiaan Mark Johan KROENER

chancery: 4200 Linnean Avenue NW, Washington, DC 20008

telephone: [1] (202) 244-5300, [1] 877-388-2443

FAX: [1] (202) 362-3430

consulate(s) general: Chicago, Los Angeles, Miami, New York

Diplomatic representation from the US:

chief of mission: Ambassador (vacant); Charge d'Affaires Michael GALLAGHER

embassy: Lange Voorhout 102, 2514 EJ, The Hague

mailing address: PSC 71, Box 1000, APO AE 09715

telephone: [31] (70) 310-2209

FAX: [31] (70) 361-4688

consulate(s) general: Amsterdam

Flag description: three equal horizontal bands of red (top), white, and blue; similar to the flag of Luxembourg, which uses a lighter blue and is longer; one of the oldest flags in constant use, originating with WILLIAM I, Prince of Orange, in the latter half of the 16th century

ECONOMY

Economy—overview: The Netherlands has a prosperous and open economy, which depends heavily on foreign trade. The economy is noted for stable industrial relations, moderate unemployment and inflation, a sizable current account surplus, and an important role as a European transportation hub. Industrial activity is predominantly in food processing, chemicals, petroleum refining, and electrical machinery. A highly mechanized agricultural sector employs no more than 3% of the labor force but provides large surpluses for the food-processing industry and for exports. The Netherlands, along with 11 of its EU partners, began circulating the euro currency on 1 January 2002. The country continues to be one of the leading European nations for attracting foreign direct investment and is one of the five largest investors in the US. The economy experienced a slowdown in 2005 but in 2006 recovered to the fastest pace in six years on the back of increased exports and strong investment. The pace of job growth reached 10-year highs in 2007.

GDP (purchasing power parity): $639.5 billion (2007 est.)

GDP (official exchange rate): $768.7 billion (2007 est.)

GDP—real growth rate: 3.5% (2007 est.)

GDP—per capita (PPP): $38,500 (2007 est.)

GDP—composition by sector:

agriculture: 2.1%

industry: 24.1%

services: 73.7% (2007 est.)

Labor force: 7.5 million (2007 est.)

Labor force—by occupation: *agriculture:* 3%

industry: 21%

services: 76% (2005 est.)

Unemployment rate: 3.2% (2007 est.)

Population below poverty line: 10.5% (2005)

Household income or consumption by percentage share:

lowest 10%: 2.5%

highest 10%: 22.9% (1999)

Distribution of family income—Gini index: 30.9 (2005)
Inflation rate (consumer prices): 1.6% (2007 est.)
Investment (gross fixed): 20% of GDP (2007 est.)
Budget: *revenues:* $349.9 billion
expenditures: $350.4 billion (2007 est.)
Public debt: 46.2% of GDP (2007 est.)
Agriculture—products: grains, potatoes, sugar beets, fruits, vegetables; livestock
Industries: agroindustries, metal and engineering products, electrical machinery and equipment, chemicals, petroleum, construction, microelectronics, fishing
Industrial production growth rate: 3.1% (2007 est.)
Electricity—production: 94.34 billion kWh (2005)
Electricity—production by source:
fossil fuel: 89.9%
hydro: 0.1%
nuclear: 4.3%
other: 5.7% (2001)
Electricity—consumption: 108.2 billion kWh (2005)
Electricity—exports: 5.398 billion kWh (2005)
Electricity—imports: 23.69 billion kWh (2005)
Oil—production: 76,000 bbl/day (2006)
Oil—consumption: 1.011 million bbl/day (2006)
Oil—exports: 1.546 million bbl/day (2004)
Oil—imports: 2.465 million bbl/day (2004)
Oil—proved reserves: 106 million bbl (1 January 2006 est.)
Natural gas—production: 77.3 billion cu m (2006)
Natural gas—consumption: 47.8 billion cu m (2006)
Natural gas—exports: 50.21 billion cu m (2005 est.)
Natural gas—imports: 22.08 billion cu m (2005)
Natural gas—proved reserves: 1.684 trillion cu m (1 January 2006 est.)
Current account balance: $50.93 billion (2007 est.)
Exports: $457.2 billion f.o.b. (2007 est.)
Exports—commodities: machinery and equipment, chemicals, fuels; foodstuffs
Exports—partners: Germany 25.5%, Belgium 13.9%, UK 8.9%, France 8.6%, Italy 5.1%, US 4.5% (2006)
Imports: $404.7 billion f.o.b. (2007 est.)
Imports—commodities: machinery and transport equipment, chemicals, fuels, foodstuffs, clothing
Imports—partners: Germany 17.1%, Belgium 9.4%, China 9.4%, US 7.8%, UK 5.9%, Russia 5.1%, France 4.5% (2006)
Economic aid—donor: ODA, $5.452 billion (2006)

Reserves of foreign exchange and gold: $23.9 billion (2006 est.)
Debt—external: $2.277 trillion (30 June 2007)
Stock of direct foreign investment—at home: $535.1 billion (2007 est.)
Stock of direct foreign investment—abroad: $811.4 billion (2007 est.)
Market value of publicly traded shares: $924.4 billion (November 2007)
Currency (code): euro (EUR)
Currency code: EUR
Exchange rates: euros per US dollar— 0.7345 (2007), 0.7964 (2006), 0.8041 (2005), 0.8054 (2004), 0.886 (2003)
Fiscal year: calendar year

COMMUNICATIONS

Telephones—main lines in use: 7.6 million (2005)
Telephones—mobile cellular: 15.834 million (2005)
Telephone system:
general assessment: highly developed and well maintained
domestic: extensive fixed-line fiber-optic network; cellular telephone system is one of the largest in Europe with 5 major network operators utilizing the third generation of the Global System for Mobile Communications (GSM)
international: country code—31; submarine cables provide links to the US and Europe; satellite earth stations—5 (3 Intelsat—1 Indian Ocean and 2 Atlantic Ocean, 1 Eutelsat, and 1 Inmarsat (2004)
Radio broadcast stations: AM 4, FM 246, shortwave 3 (2004)
Radios: 15.3 million (1996)
Television broadcast stations: 21 (plus 26 repeaters) (1995)
Televisions: 8.1 million (1997)
Internet country code: .nl
Internet hosts: 11.17 million (2007)
Internet Service Providers (ISPs): 52 (2000)
Internet users: 14.544 million (2006)

TRANSPORTATION

Airports: 27 (2007)
Airports—with paved runways:
total: 20
over 3,047 m: 2
2,438 to 3,047 m: 9
1,524 to 2,437 m: 3
914 to 1,523 m: 4
under 914 m: 2 (2007)
Airports—with unpaved runways:
total: 7
914 to 1,523 m: 3
under 914 m: 4 (2007)
Heliports: 1 (2007)
Pipelines: condensate 81 km; gas 7,394 km; oil 578 km; refined products 716 km (2007)
Railways: *total:* 2,811 km
standard gauge: 2,811 km 1.435-m gauge (2,064 km electrified) (2006)

Roadways:
total: 134,000 km (includes 3,270 km of expressways) (2004)
Waterways: 6,183 km (navigable for ships of 50 tons) (2005)
Merchant marine:
total: 566 ships (1000 GRT or over) 5,210,664 GRT/5,217,874 DWT
by type: bulk carrier 9, cargo 346, carrier 19, chemical tanker 39, container 63, liquefied gas 13, passenger 14, passenger/cargo 16, petroleum tanker 12, refrigerated cargo 11, roll on/roll off 20, specialized tanker 4
foreign-owned: 172 (Belgium 2, Denmark 19, Finland 14, France 1, Germany 70, Ireland 9, South Korea 1, Norway 9, Sweden 27, UK 7, US 13)
registered in other countries: 220 (Antigua and Barbuda 19, Australia 2, Austria 2, Bahamas 24, Canada 1, Cyprus 23, Gibraltar 11, Isle of Man 1, Liberia 28, Luxembourg 1, Malta 3, Marshall Islands 5, Netherlands Antilles 53, Norway 1, Panama 14, Paraguay 1, Philippines 22, Portugal 1, St Vincent and The Grenadines 5, UK 2, US 1, unknown 1) (2007)
Ports and terminals: Amsterdam, IJmuiden, Rotterdam, Terneuzen, Vlissingen

MILITARY

Military branches: Royal Netherlands Army, Royal Netherlands Navy (includes Naval Air Service and Marine Corps), Royal Netherlands Air Force (Koninklijke Luchtmacht, KLu), Royal Military Police (2008)
Military service age and obligation: 20 years of age for an all-volunteer force (2004)
Manpower available for military service: *males age 16–49:* 3,950,825
females age 16–49: 3,850,800 (2008 est.)
Manpower fit for military service:
males age 16–49: 3,233,773
females age 16–49: 3,150,790 (2008 est.)
Manpower reaching militarily significant age annually:
males age 16–49: 105,735
females age 16–49: 100,747 (2008 est.)
Military expenditures—percent of GDP: 1.6% (2005 est.)

TRANSNATIONAL ISSUES

Disputes—international: none
Illicit drugs: major European producer of synthetic drugs, including ecstasy, and cannabis cultivator; important gateway for cocaine, heroin, and hashish entering Europe; major source of US-bound ecstasy; large financial sector vulnerable to money laundering; significant consumer of ecstasy

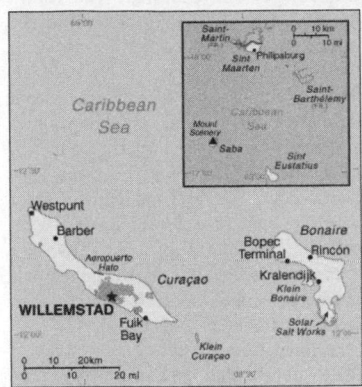

INTRODUCTION

Background: Once the center of the Caribbean slave trade, the island of Curacao was hard hit by the abolition of slavery in 1863. Its prosperity (and that of neighboring Aruba) was restored in the early 20th century with the construction of oil refineries to service the newly discovered Venezuelan oil fields. The island of Saint Martin is shared with France; its southern portion is named Sint Maarten and is part of the Netherlands Antilles; its northern portion, called Saint Martin, is an overseas collectivity of France.

GEOGRAPHY

Location: Caribbean, two island groups in the Caribbean Sea—composed of five islands, Curacao and Bonaire located off the coast of Venezuela, and Sint Maarten, Saba, and Sint Eustatius lie east of the US Virgin Islands

Geographic coordinates: 12 15 N, 68 45 W

Map references: Central America and the Caribbean

Area:
total: 960 sq km
land: 960 sq km
water: 0 sq km
note: includes Bonaire, Curacao, Saba, Sint Eustatius, and Sint Maarten (Dutch part of the island of Saint Martin)

Area—comparative: more than five times the size of Washington, DC

Land boundaries:
total: 15 km
border countries: Saint Martin 15 km

Coastline: 364 km

Maritime claims:
territorial sea: 12 nm
exclusive fishing zone: 12 nm

Climate: tropical; ameliorated by northeast trade winds

Terrain: generally hilly, volcanic interiors

Elevation extremes:
lowest point: Caribbean Sea 0 m
highest point: Mount Scenery 862 m

Natural resources: phosphates (Curacao only), salt (Bonaire only)

Land use:
arable land: 10%
permanent crops: 0%
other: 90% (2005)

Irrigated land: NA

Natural hazards: Sint Maarten, Saba, and Sint Eustatius are subject to hurricanes from July to October; Curacao and Bonaire are south of Caribbean hurricane belt and are rarely threatened

Environment—current issues: NA

Geography—note: the five islands of the Netherlands Antilles are divided geographically into the Leeward Islands (northern) group (Saba, Sint Eustatius, and Sint Maarten) and the Windward Islands (southern) group (Bonaire and Curacao); the island of Saint Martin is the smallest landmass in the world shared by two independent states, the French territory of Saint Martin and the Dutch territory of Sint Maarten

PEOPLE

Population: 225,369 (July 2008 est.)

Age structure:
0–14 years: 23.2% (male 26,749/female 25,467)
15–64 years: 67.5% (male 73,319/female 78,842)
65 years and over: 9.3% (male 8,541/female 12,451) (2008 est.)

Median age:
total: 33.4 years
male: 31.6 years
female: 35.2 years (2008 est.)

Population growth rate: 0.754% (2008 est.)

Birth rate: 14.37 births/1,000 population (2008 est.)

Death rate: 6.43 deaths/1,000 population (2008 est.)

Net migration rate: -0.39 migrant(s)/1,000 population (2008 est.)

Sex ratio:
at birth: 1.05 male(s)/female
under 15 years: 1.05 male(s)/female
15–64 years: 0.93 male(s)/female
65 years and over: 0.69 male(s)/female
total population: 0.93 male(s)/female (2008 est.)

Infant mortality rate:
total: 9.36 deaths/1,000 live births
male: 10.04 deaths/1,000 live births
female: 8.64 deaths/1,000 live births (2008 est.)

Life expectancy at birth:
total population: 76.45 years
male: 74.15 years
female: 78.87 years (2008 est.)

Total fertility rate: 1.98 children born/woman (2008 est.)

HIV/AIDS—adult prevalence rate: NA
HIV/AIDS—people living with HIV/AIDS: NA
HIV/AIDS—deaths: NA

Nationality:
noun: Dutch Antillean(s)
adjective: Dutch Antillean

Ethnic groups: mixed black 85%, other 15% (includes Carib Amerindian, white, East Asian)

Religions: Roman Catholic 72%, Pentecostal 4.9%, Protestant 3.5%, Seventh-Day Adventist 3.1%, Methodist 2.9%, Jehovah's Witnesses 1.7%, other Christian 4.2%, Jewish 1.3%, other or unspecified 1.2%, none 5.2% (2001 census)

Languages: Papiamento 65.4% (a Spanish-Portuguese-Dutch-English dialect), English 15.9% (widely spoken), Dutch 7.3% (official), Spanish 6.1%, Creole 1.6%, other 1.9%, unspecified 1.8% (2001 census)

Literacy:
definition: age 15 and over can read and write
total population: 96.7%
male: 96.7%
female: 96.8% (2003 est.)

GOVERNMENT

Country name:
conventional long form: none
conventional short form: Netherlands Antilles
local long form: none
local short form: Nederlandse Antillen
former: Curacao and Dependencies

Dependency status: an autonomous country within the Kingdom of the Netherlands; full autonomy in internal affairs granted in 1954; Dutch Government responsible for defense and foreign affairs

Government type: parliamentary

Capital: *name:* Willemstad (on Curacao)
geographic coordinates: 12 06 N, 68 56 W
time difference: UTC-4 (1 hour ahead of Washington, DC during Standard Time)

Administrative divisions: none (part of the Kingdom of the Netherlands)
note: each island has its own government

Independence: none (part of the Kingdom of the Netherlands)

National holiday: Queen's Day (Birthday of Queen-Mother JULIANA and accession to the throne of her oldest daughter BEATRIX), 30 April (1909 and 1980)

Constitution: 29 December 1954, Statute of the Realm of the Netherlands, as amended

Legal system: based on Dutch civil law system with some English common law influence

Suffrage: 18 years of age; universal

Executive branch:

chief of state: Queen BEATRIX of the Netherlands (since 30 April 1980); represented by Governor General Frits GOEDGEDRAG (since 1 July 2002)

head of government: Prime Minister Emily de JONGH-ELHAGE (since 26 March 2006)

cabinet: Council of Ministers elected by the Staten (legislature)

elections: the monarch is hereditary; governor general appointed by the monarch for a six-year term; following legislative elections, the leader of the majority party is usually elected prime minister by the Staten; election last held 27 January 2006 (next to be held by 2010)

note: government coalition—PAR, PNP, DP-St. M, UPB, WIPM Saba, DP-St. E

Legislative branch: unicameral States or Staten (22 seats, Curacao 14, Bonaire 3, St. Maarten 3, St. Eustatius 1, Saba 1; members are elected by popular vote to serve four-year terms)

elections: last held 27 January 2006 (next to be held in 2010)

election results: percent of vote by party— NA; seats by party—PAR 5, MAN 3, FOL 2, Forsa Korsou 2, National Alliance 2, PNP 2, UPB 2, DP-St. E 1, DP-St. M 1, PDB 1, WIPM 1

note: the government is a coalition of several parties

Judicial branch: Joint High Court of Justice (judges appointed by the monarch)

Political parties and leaders: *Bonaire:* Democratic Party of Bonaire or PDB [Jopi ABRAHAM]; Patriotic Union of Bonaire or UPB [Ramonsito BOOI]

Curacao: Ban Vota [Norbert GEORGE]; C-93 [Stanley BROWN]; Democratic Party of Curacao or DP [Errol HERNANDEZ]; E Mayoria [Aurelio PEDRO]; Forsa Korsou [Nelson NAVARRO]; Liste Ni'un Paso Atras [Nelson PIERRE]; Movemiento Patriotiko Korsou [Reginald LAK]; New Antilles Movement or MAN [Charles COOPER]; Partido Akshon Pa Prosperidat I Seguridat [Sonja BERKE-MEYER]; Partido Laboral Krusada Popular or PLKP [Errol COVA]; Party for the Restructured Antilles or PAR [Emily de JONGH-ELHAGE]; People's

National Party or PNP [Ersilia DE LAN-NOOY]; Pidjin [Jasmin PINEDO]; Pueblo Soberano [Herman WIELS]; Workers' Liberation Front or FOL [Anthony GODETT]

Saba: Saba Labor Party [Akilah LEVEN-STONE]; Windward Islands People's Movement or WIPM [Ray HASSELL]

Sint Eustatius: Democratic Party of Sint Eustatius or DP-St. E [Julian WOODLEY]; Progressive Labor Party [Clyde VAN PUTTEN]; St. Eustatius Alliance [Ingrid HOUTMAN-WHIT-FIELD]

Sint Maarten: Democratic Party of Sint Maarten or DP-St. M [Sarah WESCOTT-WILLIAMS]; Freedom Slate of National Democratic Party [Theophilus PRIEST]; National Alliance or NA [William MARLIN]; People's Progressive Alliance or PPA [Gracita ARRINDELL]; St. Maarten People's Party [Johan LEONARD]; United People's Labor Party [Bienvenido RICHARDSON]

note: political parties are indigenous to each island

Political pressure groups and leaders: Unions (AVBO) and Employers Association (VBC)

International organization participation: Caricom (observer), ILO, IMF, Interpol, IOC, UNESCO (associate), UNWTO (associate), UPU, WCL, WCO, WMO

Diplomatic representation in the US: none (represented by the Kingdom of the Netherlands); note—Mr. Jeffrey COR-RION, Minister Plenipotentiary for Aruba at the Embassy of the Kingdom of the Netherlands

Diplomatic representation from the US:

chief of mission: Consul General Robert E. SORENSON

consulate(s) general: J. B. Gorsiraweg #1, Willemstad, Curacao

mailing address: P. O. Box 158, Willemstad, Curacao

telephone: [599] (9) 4613066

FAX: [599] (9) 4616489

Flag description: white, with a horizontal blue stripe in the center superimposed on a vertical red band, also centered; five white, five-pointed stars are arranged in an oval pattern in the center of the blue band; the five stars represent the five main islands of Bonaire, Curacao, Saba, Sint Eustatius, and Sint Maarten

ECONOMY

Economy—overview: Tourism, petroleum refining, and offshore finance are the mainstays of this small economy, which is closely tied to the outside world.

Although GDP has declined or grown slightly in each of the past eight years, the islands enjoy a high per capita income and a well-developed infrastructure compared with other countries in the region. Most of the oil Netherlands Antilles imports for its refineries come from Venezuela. Almost all consumer and capital goods are imported, the US, Italy, and Mexico being the major suppliers. Poor soils and inadequate water supplies hamper the development of agriculture. Budgetary problems hamper reform of the health and pension systems of an aging population. The Netherlands provides financial aid to support the economy.

GDP (purchasing power parity): $2.8 billion (2004 est.)

GDP (official exchange rate): $NA

GDP—real growth rate: 1% (2004 est.)

GDP—per capita (PPP): $16,000 (2004 est.)

GDP—composition by sector:

agriculture: 1%

industry: 15%

services: 84% (2000 est.)

Labor force: 83,600 (2005)

Labor force—by occupation: *agriculture:* 1%

industry: 20%

services: 79% (2005 est.)

Unemployment rate: 17% (2002 est.)

Population below poverty line: NA%

Household income or consumption by percentage share:

lowest 10%: NA%

highest 10%: NA%

Inflation rate (consumer prices): 2.1% (2003 est.)

Budget: *revenues:* $757.9 million

expenditures: $949.5 million (2004)

Agriculture—products: aloes, sorghum, peanuts, vegetables, tropical fruit

Industries: tourism (Curacao, Sint Maarten, and Bonaire), petroleum refining (Curacao), petroleum transshipment facilities (Curacao and Bonaire), light manufacturing (Curacao)

Industrial production growth rate: NA%

Electricity—production: 1.175 billion kWh (2005)

Electricity—production by source:

fossil fuel: 100%

hydro: 0%

nuclear: 0%

other: 0% (2001)

Electricity—consumption: 891 million kWh (2005)

Electricity—exports: 0 kWh (2005)

Electricity—imports: 0 kWh (2005)

Oil—production: 0 bbl/day (2005 est.)

Oil—consumption: 68,000 bbl/day (2005 est.)

Oil—exports: 217,800 bbl/day (2004)
Oil—imports: 282,500 bbl/day (2004)
Oil—proved reserves: 0 bbl (1 January 2006 est.)
Natural gas—production: 0 cu m (2005 est.)
Natural gas—consumption: 0 cu m (2005 est.)
Natural gas—exports: 0 cu m (2005 est.)
Natural gas—imports: 0 cu m (2005)
Natural gas—proved reserves: 0 cu m (1 January 2006 est.)
Exports: $3.71 billion f.o.b. (2006)
Exports—commodities: petroleum products
Exports—partners: US 27.2%, Panama 11.4%, Mexico 9%, Germany 6.2%, Haiti 5.3%, Singapore 4.8%, The Bahamas 4.2% (2006)
Imports: $15.74 billion f.o.b. (2006)
Imports—commodities: crude petroleum, food, manufactures
Imports—partners: Venezuela 71.1%, US 10.4%, Italy 3.7% (2006)
Economic aid—recipient: $21.32 million (2004)
Debt—external: $2.68 billion (2004)
Stock of direct foreign investment—at home: $NA
Stock of direct foreign investment—abroad: $NA
Market value of publicly traded shares: $488.6 billion (2003)
Currency (code): Netherlands Antillean guilder (ANG)
Currency code: ANG
Exchange rates: Netherlands Antillean guilders per US dollar—NA (2007), 1.79 (2006), 1.79 (2005), 1.79 (2004), 1.79 (2003)
Fiscal year: calendar year

COMMUNICATIONS

Telephones—main lines in use: 81,000 (2001)
Telephones—mobile cellular: 200,000 (2004)
Telephone system:
general assessment: generally adequate facilities
domestic: extensive interisland microwave radio relay links
international: country code—599; the Americas Region Caribbean Ring System (ARCOS-1) and the Americas-2 submarine cable systems provide connectivity to Central America, parts of South America and the Caribbean, and the US; satellite earth stations—2 Intelsat (Atlantic Ocean)
Radio broadcast stations: AM 8, FM 19, shortwave 0 (2003)
Radios: 217,000 (1997)
Television broadcast stations: 3 (there is also a cable service that supplies programs received from various US satellite networks and 4 Venezuelan channels) (2003)
Televisions: 69,000 (1997)
Internet country code: .an
Internet hosts: 31,812 (2007)
Internet Service Providers (ISPs): 6
Internet users: 2,000 (2000)

TRANSPORTATION

Airports: 5 (2007)
Airports—with paved runways:
total: 5
over 3,047 m: 1
2,438 to 3,047 m: 1
1,524 to 2,437 m: 1
914 to 1,523 m: 1
under 914 m: 1 (2007)

Roadways:
total: 845
Merchant marine:
total: 138 ships (1000 GRT or over) 1,096,005 GRT/1,437,692 DWT
by type: barge carrier 2, bulk carrier 4, cargo 70, carrier 12, chemical tanker 3, container 10, liquefied gas 1, passenger 2, petroleum tanker 2, refrigerated cargo 25, roll on/roll off 4, specialized tanker 3
foreign-owned: 125 (Belgium 1, Cuba 1, Denmark 1, Germany 48, Netherlands 53, Norway 5, Sweden 3, Turkey 12, US 1) (2007)
Ports and terminals: Bopec Terminal, Willemstad

MILITARY

Military branches: no regular military forces; National Guard (2008)
Military service age and obligation: 16 years of age for National Guard recruitment; no conscription (2004)
Manpower available for military service:
males age 16–49: 55,365
females age 16–49: 57,060 (2008 est.)
Manpower fit for military service:
males age 16–49: 46,102
females age 16–49: 47,219 (2008 est.)
Manpower reaching militarily significant age annually:
males age 16–49: 1,855
females age 16–49: 1,760 (2008 est.)
Military—note: defense is the responsibility of the Kingdom of the Netherlands

TRANSNATIONAL ISSUES

Disputes—international: none
Illicit drugs: transshipment point for South American drugs bound for the US and Europe; money-laundering center

NEW CALEDONIA

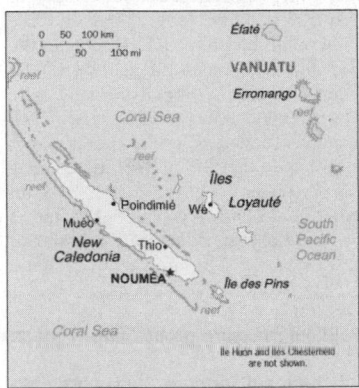

INTRODUCTION

Background: Settled by both Britain and France during the first half of the 19th century, the island was made a French possession in 1853. It served as a penal colony for four decades after 1864. Agitation for independence during the 1980s and early 1990s ended in the 1998 Noumea Accord, which over a period of 15 to 20 years will transfer an increasing amount of governing responsibility from France to New Caledonia. The agreement also commits France to conduct as many as three referenda between 2013 and 2018, to decide whether New Caledonia should assume full sovereignty and independence.

GEOGRAPHY

Location: Oceania, islands in the South Pacific Ocean, east of Australia

Geographic coordinates: 21 30 S, 165 30 E
Map references: Oceania
Area: *total:* 19,060 sq km
land: 18,575 sq km
water: 485 sq km
Area—comparative: slightly smaller than New Jersey
Land boundaries: 0 km
Coastline: 2,254 km
Maritime claims:
territorial sea: 12 nm
exclusive economic zone: 200 nm
Climate: tropical; modified by southeast trade winds; hot, humid
Terrain: coastal plains with interior mountains
Elevation extremes:
lowest point: Pacific Ocean 0 m

highest point: Mont Panie 1,628 m
Natural resources: nickel, chrome, iron, cobalt, manganese, silver, gold, lead, copper
Land use:
arable land: 0.32%
permanent crops: 0.22%
other: 99.46% (2005)
Irrigated land: 100 sq km (2003)
Natural hazards: cyclones, most frequent from November to March
Environment—current issues: erosion caused by mining exploitation and forest fires
Geography—note: consists of the main island of New Caledonia (one of the largest in the Pacific Ocean), the archipelago of Iles Loyaute, and numerous small, sparsely populated islands and atolls

PEOPLE

Population: 224,824 (July 2008 est.)
Age structure:
0–14 years: 27.3% (male 31,376/female 30,064)
15–64 years: 65.6% (male 74,064/female 73,369)
65 years and over: 7.1% (male 7,377/female 8,574) (2008 est.)
Median age:
total: 28.4 years
male: 28 years
female: 28.8 years (2008 est.)
Population growth rate: 1.175% (2008 est.)
Birth rate: 17.39 births/1,000 population (2008 est.)
Death rate: 5.64 deaths/1,000 population (2008 est.)
Net migration rate: NA
note: there has been steady emigration from Wallis and Futuna to New Caledonia (2008 est.)
Sex ratio:
at birth: 1.05 male(s)/female
under 15 years: 1.04 male(s)/female
15–64 years: 1.01 male(s)/female
65 years and over: 0.86 male(s)/female
total population: 1.01 male(s)/female (2008 est.)
Infant mortality rate:
total: 7.19 deaths/1,000 live births
male: 7.85 deaths/1,000 live births
female: 6.5 deaths/1,000 live births (2008 est.)
Life expectancy at birth:
total population: 74.75 years
male: 71.76 years
female: 77.88 years (2008 est.)
Total fertility rate: 2.21 children born/woman (2008 est.)
HIV/AIDS—adult prevalence rate: NA
HIV/AIDS—people living with HIV/AIDS: NA
HIV/AIDS—deaths: NA
Nationality: *noun:* New Caledonian(s)
adjective: New Caledonian

Ethnic groups: Melanesian 42.5%, European 37.1%, Wallisian 8.4%, Polynesian 3.8%, Indonesian 3.6%, Vietnamese 1.6%, other 3%
Religions: Roman Catholic 60%, Protestant 30%, other 10%
Languages: French (official), 33 Melanesian-Polynesian dialects
Literacy:
definition: age 15 and over can read and write
total population: 96.2%
male: 96.8%
female: 95.5% (1996 census)

GOVERNMENT

Country name:
conventional long form: Territory of New Caledonia and Dependencies
conventional short form: New Caledonia
local long form: Territoire des Nouvelle-Caledonie et Dependances
local short form: Nouvelle-Caledonie
Dependency status: territorial collectivity of France since 1998
Government type: NA
Capital: *name:* Noumea
geographic coordinates: 22 16 S, 166 27 E
time difference: UTC+11 (16 hours ahead of Washington, DC during Standard Time)
Administrative divisions: none (overseas territory of France); there are no first-order administrative divisions as defined by the US Government, but there are 3 provinces named Province des Iles, Province Nord, and Province Sud
Independence: none (overseas territory of France); note—a referendum on independence was held in 1998 but did not pass; a new referendum is scheduled for 2014
National holiday: Bastille Day, 14 July (1789)
Constitution: 4 October 1958 (French Constitution)
Legal system: based on French civil law; the 1988 Matignon Accords grant substantial autonomy to the islands
Suffrage: 18 years of age; universal
Executive branch:
chief of state: President Nicolas SARKOZY (since 16 May 2007); represented by High Commissioner Yves DASSONVILLE (since 9 November 2007)
head of government: President of the Government Harold MARTIN (since 7 August 2007)
cabinet: Cabinet consisting of 11 members elected from and by the Territorial Congress
elections: French president elected by popular vote for a five-year term; high commissioner appointed by the French president on the advice of the French Ministry of Interior; president of the gov-

ernment elected by the members of the Territorial Congress for a five-year term (no term limits); note—last election held 7 August 2007 when Harold MARTIN was elected following the resignation of Marie-Noelle THEMEREAU as president on 24 July 2007 (next to be held in 2012)
Legislative branch: unicameral Territorial Congress or Congres du territoire (54 seats; members belong to the three Provincial Assemblies or Assemblees Provinciales elected by popular vote to serve five-year terms)
elections: last held 9 May 2004 (next to be held in 2009)
election results: percent of vote by party—NA; seats by party—RPCR-UMP 16, AE 16, UNI-FLNKS 8, UC 7, FN 4, others 3
note: New Caledonia currently holds one seat in the French Senate; by 2010, New Caledonia will gain a second seat in the French Senate; elections last held 24 September 2001 (next to be held not later than September 2007); results—percent of vote by party—NA; seats by party—UMP 1; New Caledonia also elects two seats to the French National Assembly; elections last held 10 and 17 June 2007 (next to be held on June 2012); results—percent of vote by party—NA; seats by party—UMP 2
Judicial branch: Court of Appeal or Cour d'Appel; County Courts; Joint Commerce Tribunal Court; Children's Court
Political parties and leaders: Alliance pour la Caledonie or APLC [Didier LE ROUX]; Caledonian Union or UC; Federation des Comites de Coordination des Independantistes or FCCI [Francois BURCK]; Front National or FN [Guy GEORGE]; Front Uni de Liberation Kanak or FULK [Ernest UNE]; Kanak Socialist Front for National Liberation or FLNKS (includes PALIKA, UNI, UC, and UPM); Parti de Liberation Kanak or PALIKA [Paul NEAOUTYINE and Elie POIGOUNE]; Rally for Caledonia in the Republic (anti independence) or RPCR-UMP [Jacques LAFLEUR]; The Future Together or AE [Harold MARTIN]; Union Nationale pour l'Independance or UNI [Paul NEAOUTYINE]; note—may no longer exist, but Paul NEAOUTYINE has since become a president of Parti de Liberation Kanak or PALIKA; Union Progressiste Melanesienne or UPM [Victor TUTU-GORO]
Political pressure groups and leaders: NA
International organization participation: ITUC, PIF (associate member), SPC, UPU, WFTU, WMO
Diplomatic representation in the US: none (overseas territory of France)

Diplomatic representation from the US: none (overseas territory of France)
Flag description: the flag of France is used

ECONOMY

ECONOMY

Economy—overview: New Caledonia has about 25% of the world's known nickel resources. Only a small amount of the land is suitable for cultivation, and food accounts for about 20% of imports. In addition to nickel, substantial financial support from France—equal to more than 15% of GDP—and tourism are keys to the health of the economy. Substantial new investment in the nickel industry, combined with the recovery of global nickel prices, brightens the economic outlook for the next several years.
GDP (purchasing power parity): $3.158 billion (2003 est.)
GDP (official exchange rate): $3.3 billion (2003 est.)
GDP—real growth rate: NA%
GDP—per capita (PPP): $15,000 (2003 est.)
GDP—composition by sector:
agriculture: 15%
industry: 8.8%
services: 76.2% (2003)
Labor force: 78,990 (2004)
Labor force—by occupation: *agriculture:* 20%
industry: 20%
services: 60% (2002)
Unemployment rate: 17.1% (2004)
Population below poverty line: NA%
Household income or consumption by percentage share:
lowest 10%: NA%
highest 10%: NA%
Inflation rate (consumer prices): 1.4% (2000 est.)
Budget:
revenues: $996 million
expenditures: $1.072 billion (2001 est.)
Agriculture—products: vegetables; beef, deer, other livestock products; fish
Industries: nickel mining and smelting
Industrial production growth rate: -0.6% (1996)
Electricity—production: 1.508 billion kWh (2005)
Electricity—production by source:
fossil fuel: 76.3%
hydro: 23.7%
nuclear: 0%
other: 0% (2001)

Electricity—consumption: 1.403 billion kWh (2005)
Electricity—exports: 0 kWh (2005)
Electricity—imports: 0 kWh (2005)
Oil—production: 0 bbl/day (2005 est.)
Oil—consumption: 11,000 bbl/day (2005 est.)
Oil—exports: 605.7 bbl/day (2004)
Oil—imports: 11,980 bbl/day (2004)
Oil—proved reserves: 0 bbl (1 January 2006 est.)
Natural gas—production: 0 cu m (2005 est.)
Natural gas—consumption: 0 cu m (2005 est.)
Natural gas—exports: 0 cu m (2005 est.)
Natural gas—imports: 0 cu m (2005)
Natural gas—proved reserves: 0 cu m (1 January 2006 est.)
Exports: $1.341 billion f.o.b. (2006)
Exports—commodities: ferronickels, nickel ore, fish
Exports—partners: Japan 17.4%, France 15.9%, Taiwan 14.5%, China 10.8%, Spain 9.4%, Belgium 7.3%, Italy 6%, Australia 4.6% (2006)
Imports: $1.998 billion f.o.b. (2006)
Imports—commodities: machinery and equipment, fuels, chemicals, foodstuffs
Imports—partners: France 39.5%, Singapore 15.1%, Australia 11.3%, NZ 4.8% (2006)
Economic aid—recipient: $524.3 million annual subsidy from France (2004)
Debt—external: $79 million (1998 est.)
Market value of publicly traded shares: $NA
Currency (code): Comptoirs Francais du Pacifique franc (XPF)
Currency code: XPF
Exchange rates: Comptoirs Francais du Pacifique francs (XPF) per US dollar—87.59 (2007), 95.025 (2006), 95.89 (2005), 96.04 (2004), 105.66 (2003)
Fiscal year: calendar year

COMMUNICATIONS

Telephones—main lines in use: 55,300 (2005)
Telephones—mobile cellular: 134,300 (2005)
Telephone system:
general assessment: NA
domestic: a submarine cable network connection between New Caledonia and Australia, scheduled for completion in 2008, will improve high-speed connectivity and access to international networks

international: country code—687; satellite earth station—1 Intelsat (Pacific Ocean)
Radio broadcast stations: AM 1, FM 5, shortwave 0 (1998)
Radios: 107,000 (1997)
Television broadcast stations: 6 (plus 25 repeaters) (1997)
Televisions: 52,000 (1997)
Internet country code: .nc
Internet hosts: 14,252 (2007)
Internet Service Providers (ISPs): 1 (2000)
Internet users: 80,000 (2006)

TRANSPORTATION

Airports: 25 (2007)
Airports—with paved runways:
total: 12
over 3,047 m: 1
914 to 1,523 m: 9
under 914 m: 2 (2007)
Airports—with unpaved runways:
total: 13
914 to 1,523 m: 7
under 914 m: 6 (2007)
Heliports: 6 (2007)
Roadways:
total: 5,432 km (2000)
Merchant marine:
total: 2 ships (1000 GRT or over) 3,566 GRT/2,543 DWT
by type: cargo 1, passenger/cargo 1 (2007)
Ports and terminals: Noumea

MILITARY

Military branches: no regular indigenous military forces; French Armed Forces (includes Army, Navy, Air Force, Gendarmerie); Police Force
Manpower available for military service:
males age 16–49: 57,738 (2008 est.)
Manpower fit for military service:
males age 16–49: 47,342 (2008 est.)
Manpower reaching militarily significant age annually:
males age 16–49: 2,202 (2008 est.)
Military expenditures—percent of GDP: NA
Military—note: defense is the responsibility of France

TRANSNATIONAL ISSUES

Disputes—international: Matthew and Hunter Islands east of New Caledonia claimed by France and Vanuatu

NEW ZEALAND

INTRODUCTION

Background: The Polynesian Maori reached New Zealand in about A.D. 800. In 1840, their chieftains entered into a compact with Britain, the Treaty of Waitangi, in which they ceded sovereignty to Queen Victoria while retaining territorial rights. In that same year, the British began the first organized colonial settlement. A series of land wars between 1843 and 1872 ended with the defeat of the native peoples. The British colony of New Zealand became an independent dominion in 1907 and supported the UK

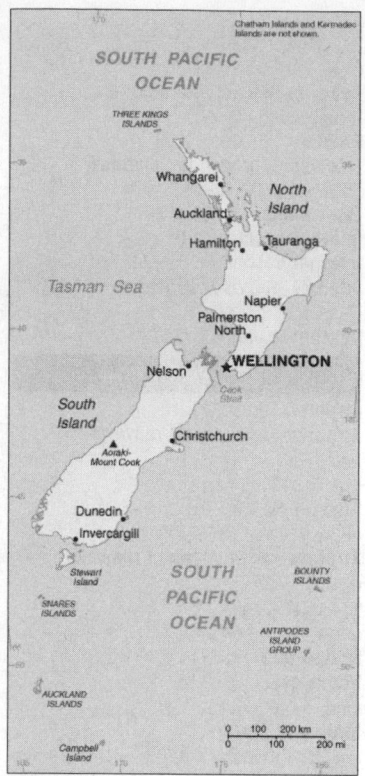

militarily in both World Wars. New Zealand's full participation in a number of defense alliances lapsed by the 1980s. In recent years, the government has sought to address longstanding Maori grievances.

GEOGRAPHY

Location: Oceania, islands in the South Pacific Ocean, southeast of Australia
Geographic coordinates: 41 00 S, 174 00 E
Map references: Oceania
Area:
total: 268,680 sq km
land: 268,021 sq km
water: NA
note: includes Antipodes Islands, Auckland Islands, Bounty Islands, Campbell Island, Chatham Islands, and Kermadec Islands
Area—comparative: about the size of Colorado
Land boundaries: 0 km
Coastline: 15,134 km
Maritime claims:
territorial sea: 12 nm
contiguous zone: 24 nm
exclusive economic zone: 200 nm
continental shelf: 200 nm or to the edge of the continental margin
Climate: temperate with sharp regional contrasts

Terrain: predominately mountainous with some large coastal plains
Elevation extremes:
lowest point: Pacific Ocean 0 m
highest point: Aoraki-Mount Cook 3,754 m
Natural resources: natural gas, iron ore, sand, coal, timber, hydropower, gold, limestone
Land use:
arable land: 5.54%
permanent crops: 6.92%
other: 87.54% (2005)
Irrigated land: 2,850 sq km (2003)
Total renewable water resources: 397 cu km (1995)
Freshwater withdrawal (domestic/industrial/agricultural): *total:* 2.11 cu km/yr (48%/9%/42%)
per capita: 524 cu m/yr (2000)
Natural hazards: earthquakes are common, though usually not severe; volcanic activity
Environment—current issues: deforestation; soil erosion; native flora and fauna hard-hit by invasive species
Environment—international agreements: *party to:* Antarctic-Environmental Protocol, Antarctic-Marine Living Resources, Antarctic Treaty, Biodiversity, Climate Change, Climate Change-Kyoto Protocol, Desertification, Endangered Species, Environmental Modification, Hazardous Wastes, Law of the Sea, Marine Dumping, Ozone Layer Protection, Ship Pollution, Tropical Timber 83, Tropical Timber 94, Wetlands, Whaling
signed, but not ratified: Antarctic Seals, Marine Life Conservation
Geography—note: about 80% of the population lives in cities; Wellington is the southernmost national capital in the world

PEOPLE

Population: 4,173,460 (July 2008 est.)
Age structure:
0–14 years: 20.9% (male 446,883/female 424,240)
15–64 years: 66.5% (male 1,390,669/female 1,385,686)
65 years and over: 12.6% (male 238,560/female 287,422) (2008 est.)
Median age:
total: 36.3 years
male: 35.6 years
female: 37.1 years (2008 est.)
Population growth rate: 0.971% (2008 est.)
Birth rate: 14.09 births/1,000 population (2008 est.)
Death rate: 7 deaths/1,000 population (2008 est.)
Net migration rate: 2.62 migrant(s)/1,000 population (2008 est.)

Sex ratio:
at birth: 1.05 male(s)/female
under 15 years: 1.05 male(s)/female
15–64 years: 1 male(s)/female
65 years and over: 0.83 male(s)/female
total population: 0.99 male(s)/female (2008 est.)
Infant mortality rate:
total: 4.99 deaths/1,000 live births
male: 5.62 deaths/1,000 live births
female: 4.33 deaths/1,000 live births (2008 est.)
Life expectancy at birth:
total population: 80.24 years
male: 78.33 years
female: 82.25 years (2008 est.)
Total fertility rate: 2.11 children born/woman (2008 est.)
HIV/AIDS—adult prevalence rate: 0.1% (2003 est.)
HIV/AIDS—people living with HIV/AIDS: 1,400 (2003 est.)
HIV/AIDS—deaths: fewer than 200 (2003 est.)
Nationality:
noun: New Zealander(s)
adjective: New Zealand
Ethnic groups: European 69.8%, Maori 7.9%, Asian 5.7%, Pacific islander 4.4%, other 0.5%, mixed 7.8%, unspecified 3.8% (2001 census)
Religions: Anglican 14.9%, Roman Catholic 12.4%, Presbyterian 10.9%, Methodist 2.9%, Pentecostal 1.7%, Baptist 1.3%, other Christian 9.4%, other 3.3%, unspecified 17.2%, none 26% (2001 census)
Languages: English (official), Maori (official), Sign Language (official)
Literacy:
definition: age 15 and over can read and write
total population: 99%
male: 99%
female: 99% (2003 est.)

GOVERNMENT

Country name:
conventional long form: none
conventional short form: New Zealand
abbreviation: NZ
Government type: parliamentary democracy
Capital: *name:* Wellington
geographic coordinates: 41 28 S, 174 51 E
time difference: UTC+12 (17 hours ahead of Washington, DC during Standard Time)
daylight saving time: +1hr, begins first Sunday in October; ends third Sunday in March
note: New Zealand is divided into two time zones, including Chatham Island
Administrative divisions: 16 regions and 1 territory*; Auckland, Bay of Plenty,

Canterbury, Chatham Islands*, Gisborne, Hawke's Bay, Manawatu-Wanganui, Marlborough, Nelson, Northland, Otago, Southland, Taranaki, Tasman, Waikato, Wellington, West Coast

Dependent areas: Cook Islands, Niue, Tokelau

Independence: 26 September 1907 (from UK)

National holiday: Waitangi Day (Treaty of Waitangi established British sovereignty over New Zealand), 6 February (1840); ANZAC Day (commemorated as the anniversary of the landing of troops of the Australian and New Zealand Army Corps during World War I at Gallipoli, Turkey), 25 April (1915)

Constitution: consists of a series of legal documents, including certain acts of the UK and New Zealand Parliaments, as well as The Constitution Act 1986, which is the principal formal charter; adopted 1 January 1987, effective 1 January 1987

Legal system: based on English law, with special land legislation and land courts for the Maori; accepts compulsory ICJ jurisdiction with reservations

Suffrage: 18 years of age; universal

Executive branch:

chief of state: Queen ELIZABETH II (since 6 February 1952); represented by Governor General Anand SATYANAND (since 23 August 2006)

head of government: Prime Minister Helen CLARK (since 10 December 1999); Deputy Prime Minister Michael CULLEN (since July 2002)

cabinet: Executive Council appointed by the governor general on the recommendation of the prime minister

elections: the monarch is hereditary; governor general appointed by the monarch; following legislative elections, the leader of the majority party or the leader of a majority coalition is usually appointed prime minister by the governor general; deputy prime minister appointed by the governor general

Legislative branch: unicameral House of Representatives—commonly called Parliament (120 seats; 69 members elected by popular vote in single-member constituencies including 7 Maori constituencies, and 51 proportional seats chosen from party lists; to serve three-year terms)

elections: last held 17 September 2005 (next to be held not later than 15 November 2008)

election results: percent of vote by party—NZLP 41.1%, NP 39.1%, NZFP 5.7%, Green Party 5.3%, Maori 2.1%, UF 2.7%, ACT New Zealand 1.5%, Progressive 1.2%, other 1.3%; seats by party—NZLP 50, NP 48, NZFP 7, Green Party 6, Maori 4, UF 3, ACT New Zealand 2, Progressive 1

note: results of 2005 election saw the total number of seats increase to 121 because the Maori Party won one more electorate seat than its entitlement under the party vote

Judicial branch: Supreme Court; Court of Appeal; High Court; note—judges appointed by the Governor-General

Political parties and leaders: ACT New Zealand [Rodney HIDE]; Green Party [Jeanette FITZSIMONS]; Maori Party [Whatarangi WINIATA]; National Party or NP [John KEY]; New Zealand First Party or NZFP [Winston PETERS]; New Zealand Labor Party or NZLP [Helen CLARK]; Progressive Party [James (Jim) ANDERTON]; United Future or UF [Peter DUNNE]

Political pressure groups and leaders: NA

International organization participation: ADB, ANZUS (US suspended security obligations to NZ on 11 August 1986), APEC, ARF, ASEAN (dialogue partner), Australia Group, BIS, C, CP, EAS, EBRD, FAO, IAEA, IBRD, ICAO, ICC, ICCt, ICRM, IEA, IFAD, IFC, IFRCS, IHO, ILO, IMF, IMO, IMSO, Interpol, IOC, IOM, IPU, ISO, ITSO, ITU, ITUC, NAM (guest), NSG, OECD, OPCW, PCA, PIF, Sparteca, SPC, UN, UNCTAD, UNESCO, UNHCR, UNIDO, UNMIS, UNMIT, UNTSO, UPU, WCO, WFTU, WHO, WIPO, WMO, WTO

Diplomatic representation in the US:

chief of mission: Ambassador Roy N. FERGUSON

chancery: 37 Observatory Circle NW, Washington, DC 20008

telephone: [1] (202) 328-4800

FAX: [1] (202) 667-5227

consulate(s) general: Los Angeles, New York

Diplomatic representation from the US:

chief of mission: Ambassador William P. McCORMICK

embassy: 29 Fitzherbert Terrace, Thorndon, Wellington

mailing address: P. O. Box 1190, Wellington; PSC 467, Box 1, APO AP 96531-1034

telephone: [64] (4) 462-6000

FAX: [64] (4) 499-0490

consulate(s) general: Auckland

Flag description: blue with the flag of the UK in the upper hoist-side quadrant with four red five-pointed stars edged in white centered in the outer half of the flag; the stars represent the Southern Cross constellation

Government—note: while not an official symbol, the Kiwi, a small native flightless bird, represents New Zealand

ECONOMY

Economy—overview: Over the past 20 years the government has transformed New Zealand from an agrarian economy dependent on concessionary British market access to a more industrialized, free market economy that can compete globally. This dynamic growth has boosted real incomes—but left behind many at the bottom of the ladder—and broadened and deepened the technological capabilities of the industrial sector. Per capita income has risen for eight consecutive years and reached $27,300 in 2007 in purchasing power parity terms. Consumer and government spending have driven growth in recent years, and exports picked up in 2006 after struggling for several years. Exports were equal to about 22% of GDP in 2007, down from 33% of GDP in 2001. Thus far the economy has been resilient, and the Labor Government promises that expenditures on health, education, and pensions will increase proportionately to output. Inflationary pressures have built in recent years and the central bank raised its key rate 13 times since January 2004 to finish 2007 at 8.25%. A large balance of payments deficit poses another challenge in managing the economy.

GDP (purchasing power parity): $111.7 billion (2007 est.)

GDP (official exchange rate): $128.1 billion (2007 est.)

GDP—real growth rate: 3% (2007 est.)

GDP—per capita (PPP): $26,400 (2007 est.)

GDP—composition by sector:

agriculture: 4.8%

industry: 26%

services: 69.3% (2007 est.)

Labor force: 2.233 million (2007 est.)

Labor force—by occupation: *agriculture:* 7%

industry: 19%

services: 74% (2006 est.)

Unemployment rate: 3.6% (2007 est.)

Population below poverty line: NA%

Household income or consumption by percentage share:

lowest 10%: %NA

highest 10%: %NA

Distribution of family income—Gini index: 36.2 (1997)

Inflation rate (consumer prices): 2.4% (2007 est.)

Investment (gross fixed): 23.1% of GDP (2007 est.)

Budget: *revenues:* $57.84 billion

expenditures: $53.7 billion (2007 est.)

Public debt: 20.7% of GDP (2007 est.)

Agriculture—products: dairy products, lamb and mutton; wheat, barley, potatoes, pulses, fruits, vegetables; wool, beef; fish

Industries: food processing, wood and paper products, textiles, machinery, transportation equipment, banking and insurance, tourism, mining

Industrial production growth rate: 0.3% (2007 est.)

Electricity—production: 42.06 billion kWh (2006 est.)

Electricity—production by source:
fossil fuel: 31.6%
hydro: 57.8%
nuclear: 0%
other: 10.7% (2001)

Electricity—consumption: 37.39 billion kWh (2006 est.)

Electricity—exports: 0 kWh (2005)

Electricity—imports: 0 kWh (2005)

Oil—production: 25,880 bbl/day (2006 est.)

Oil—consumption: 156,000 bbl/day (2006 est.)

Oil—exports: 15,720 bbl/day (2004)

Oil—imports: 140,900 bbl/day (2004)

Oil—proved reserves: 55.5 million bbl (1 January 2006 est.)

Natural gas—production: 3.9 billion cu m (2006 est.)

Natural gas—consumption: 3.7 billion cu m (2006 est.)

Natural gas—exports: 0 cu m (2005 est.)

Natural gas—imports: 0 cu m (2005)

Natural gas—proved reserves: 29.67 billion cu m (1 January 2006 est.)

Current account balance: -$10.38 billion (2007 est.)

Exports: $27.26 billion (2007 est.)

Exports—commodities: dairy products, meat, wood and wood products, fish, machinery

Exports—partners: Australia 20.5%, US 13.1%, Japan 10.3%, China 5.4%, UK 4.9% (2006)

Imports: $28.97 billion (2007 est.)

Imports—commodities: machinery and equipment, vehicles and aircraft, petroleum, electronics, textiles, plastics

Imports—partners: Australia 20.5%, China 12.3%, US 11.8%, Japan 9.2%, Germany 4.4%, Singapore 4.4% (2006)

Economic aid—donor: ODA, $259 million (2006)

Reserves of foreign exchange and gold: $17.25 billion (31 December 2007 est.)

Debt—external: $50.07 billion (31 December 2007 est.)

Stock of direct foreign investment—at home: $66.92 billion (2007 est.)

Stock of direct foreign investment—abroad: $NA

Market value of publicly traded shares: $40.62 billion (2005)

Currency (code): New Zealand dollar (NZD)

Currency code: NZD

Exchange rates: New Zealand dollars per US dollar—1.3811 (2007), 1.5408 (2006), 1.4203 (2005), 1.5087 (2004), 1.7221 (2003)

Fiscal year: 1 April—31 March
note: this is the fiscal year for tax purposes

COMMUNICATIONS

Telephones—main lines in use: 1.729 million (2005)

Telephones—mobile cellular: 3.53 million (2005)

Telephone system:
general assessment: excellent domestic and international systems
domestic: NA
international: country code—64; the Southern Cross submarine cable system provides links to Australia, Fiji, and the US; satellite earth stations—8 (1 Inmarsat—Pacific Ocean, 7 other)

Radio broadcast stations: AM 124, FM 290, shortwave 4 (1998)

Radios: 3.75 million (1997)

Television broadcast stations: 41 (plus about 700 repeaters) (1997)

Televisions: 1.926 million (1997)

Internet country code: .nz

Internet hosts: 1.433 million (2007)

Internet Service Providers (ISPs): 36 (2000)

Internet users: 3.2 million (2006)

TRANSPORTATION

Airports: 121 (2007)

Airports—with paved runways:
total: 41
over 3,047 m: 2
2,438 to 3,047 m: 1
1,524 to 2,437 m: 11
914 to 1,523 m: 26
under 914 m: 1 (2007)

Airports—with unpaved runways:
total: 80
1,524 to 2,437 m: 3

914 to 1,523 m: 31
under 914 m: 46 (2007)

Pipelines: condensate 331 km; gas 1,896 km; liquid petroleum gas 172 km; oil 288 km; refined products 260 km (2007)

Railways:
total: 4,128 km
narrow gauge: 4,128 km 1.067-m gauge (506 km electrified) (2006)

Roadways:
total: 92,931 km
paved: 59,783 km (includes 171 km of expressways)
unpaved: 33,148 km (2003)

Merchant marine:
total: 11 ships (1000 GRT or over) 108,667 GRT/89,458 DWT
by type: bulk carrier 3, cargo 1, passenger/cargo 4, petroleum tanker 1, roll on/roll off 2
foreign-owned: 1 (Germany 1)
registered in other countries: 8 (Antigua and Barbuda 2, Cook Islands 1, Dominica 3, France 1, UK 1) (2007)

Ports and terminals: Auckland, Lyttelton, Marsden Point, Tauranga, Wellington, Whangarei

MILITARY

Military branches: New Zealand Defense Force (NZDF): New Zealand Army, Royal New Zealand Navy, Royal New Zealand Air Force (2008)

Military service age and obligation: 17 years of age for voluntary military service; soldiers cannot be deployed until the age of 18; no conscription (2008)

Manpower available for military service:
males age 16–49: 1,009,298
females age 16–49: 997,134 (2008 est.)

Manpower fit for military service:
males age 16–49: 833,073
females age 16–49: 822,807 (2008 est.)

Manpower reaching militarily significant age annually:
males age 16–49: 31,834
females age 16–49: 30,243 (2008 est.)

Military expenditures—percent of GDP: 1% (2005 est.)

TRANSNATIONAL ISSUES

Disputes—international: asserts a territorial claim in Antarctica (Ross Dependency)

Illicit drugs: significant consumer of amphetamines

NICARAGUA

INTRODUCTION

Background: The Pacific coast of Nicaragua was settled as a Spanish colony from Panama in the early 16th century. Independence from Spain was declared in 1821 and the country became an independent republic in 1838. Britain occupied the Caribbean Coast in the first half of the 19th century, but gradually ceded control of the region in subsequent decades. Violent opposition to governmental manipulation and corruption spread to all classes by 1978 and resulted in a short-lived civil war that brought the Marxist Sandinista guerrillas to power in 1979. Nicaraguan aid to leftist rebels in El Salvador caused the US to

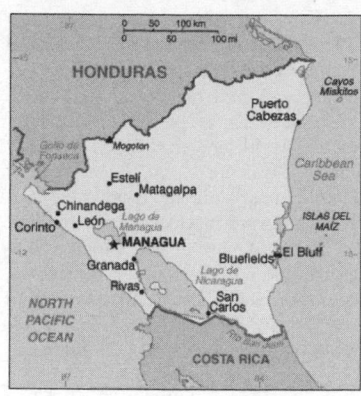

sponsor anti-Sandinista contra guerrillas through much of the 1980s. Free elections in 1990, 1996, and 2001, saw the Sandinistas defeated, but voting in 2006 announced the return of former Sandinista President Daniel ORTEGA Saavedra. Nicaragua's infrastructure and economy—hard hit by the earlier civil war and by Hurricane Mitch in 1998—are slowly being rebuilt.

GEOGRAPHY

Location: Central America, bordering both the Caribbean Sea and the North Pacific Ocean, between Costa Rica and Honduras
Geographic coordinates: 13 00 N, 85 00 W
Map references: Central America and the Caribbean
Area:
total: 129,494 sq km
land: 120,254 sq km
water: 9,240 sq km
Area—comparative: slightly smaller than the state of New York
Land boundaries:
total: 1,231 km
border countries: Costa Rica 309 km, Honduras 922 km
Coastline: 910 km
Maritime claims:
territorial sea: 12 nm
contiguous zone: 24 nm
continental shelf: natural prolongation
Climate: tropical in lowlands, cooler in highlands
Terrain: extensive Atlantic coastal plains rising to central interior mountains; narrow Pacific coastal plain interrupted by volcanoes
Elevation extremes:
lowest point: Pacific Ocean 0 m
highest point: Mogoton 2,438 m
Natural resources: gold, silver, copper, tungsten, lead, zinc, timber, fish
Land use:
arable land: 14.81%
permanent crops: 1.82%

other: 83.37% (2005)
Irrigated land: 610 sq km (2003)
Total renewable water resources: 196.7 cu km (2000)
Freshwater withdrawal (domestic/ industrial/agricultural): *total:* 1.3 cu km/yr (15%/2%/83%)
per capita: 237 cu m/yr (2000)
Natural hazards: destructive earthquakes, volcanoes, landslides; extremely susceptible to hurricanes
Environment—current issues: deforestation; soil erosion; water pollution
Environment—international agreements: *party to:* Biodiversity, Climate Change, Climate Change-Kyoto Protocol, Desertification, Endangered Species, Hazardous Wastes, Law of the Sea, Ozone Layer Protection, Ship Pollution, Wetlands, Whaling
signed, but not ratified: Environmental Modification
Geography—note: largest country in Central America; contains the largest freshwater body in Central America, Lago de Nicaragua

PEOPLE

Population: 5,785,846 (July 2008 est.)
Age structure:
0–14 years: 34.6% (male 1,019,281/female 981,903)
15–64 years: 62.1% (male 1,792,398/female 1,803,133)
65 years and over: 3.3% (male 82,840/female 106,291) (2008 est.)
Median age: *total:* 21.7 years
male: 21.3 years
female: 22.1 years (2008 est.)
Population growth rate: 1.825% (2008 est.)
Birth rate: 23.7 births/1,000 population (2008 est.)
Death rate: 4.33 deaths/1,000 population (2008 est.)
Net migration rate: -1.13 migrant(s)/ 1,000 population (2008 est.)
Sex ratio:
at birth: 1.05 male(s)/female
under 15 years: 1.04 male(s)/female
15–64 years: 0.99 male(s)/female
65 years and over: 0.78 male(s)/female
total population: 1 male(s)/female (2008 est.)
Infant mortality rate:
total: 25.91 deaths/1,000 live births
male: 29.06 deaths/1,000 live births
female: 22.6 deaths/1,000 live births (2008 est.)
Life expectancy at birth:
total population: 71.21 years
male: 69.08 years
female: 73.44 years (2008 est.)
Total fertility rate: 2.63 children born/woman (2008 est.)

HIV/AIDS—adult prevalence rate: 0.2% (2003 est.)
HIV/AIDS—people living with HIV/AIDS: 6,400 (2003 est.)
HIV/AIDS—deaths: fewer than 500 (2003 est.)
Major infectious diseases:
degree of risk: intermediate
food or waterborne diseases: bacterial diarrhea, hepatitis A, and typhoid fever
vectorborne disease: dengue fever and malaria
water contact disease: leptospirosis (2008)
Nationality:
noun: Nicaraguan(s)
adjective: Nicaraguan
Ethnic groups: mestizo (mixed Amerindian and white) 69%, white 17%, black 9%, Amerindian 5%
Religions: Roman Catholic 72.9%, Evangelical 15.1%, Moravian 1.5%, Episcopal 0.1%, other 1.9%, none 8.5% (1995 census)
Languages: Spanish 97.5% (official), Miskito 1.7%, other 0.8% (1995 census)
note: English and indigenous languages on Atlantic coast
Literacy:
definition: age 15 and over can read and write
total population: 67.5%
male: 67.2%
female: 67.8% (2003 est.)

GOVERNMENT

Country name:
conventional long form: Republic of Nicaragua
conventional short form: Nicaragua
local long form: Republica de Nicaragua
local short form: Nicaragua
Government type: republic
Capital: *name:* Managua
geographic coordinates: 12 09 N, 86 17 W
time difference: UTC-6 (1 hour behind Washington, DC during Standard Time)
Administrative divisions: 15 departments (departamentos, singular—departamento) and 2 autonomous regions* (regiones autonomistas, singular—region autonoma); Atlantico Norte*, Atlantico Sur*, Boaco, Carazo, Chinandega, Chontales, Esteli, Granada, Jinotega, Leon, Madriz, Managua, Masaya, Matagalpa, Nueva Segovia, Rio San Juan, Rivas
Independence: 15 September 1821 (from Spain)
National holiday: Independence Day, 15 September (1821)
Constitution: 9 January 1987; reforms in 1995, 2000, and 2005
Legal system: civil law system; Supreme Court may review administrative acts; accepts compulsory ICJ jurisdiction

Suffrage: 16 years of age; universal
Executive branch:
chief of state: President Daniel ORTEGA Saavedra (since 10 January 2007); Vice President Jaime MORALES Carazo (since 10 January 2007); note—the president is both chief of state and head of government
head of government: President Daniel ORTEGA Saavedra (since 10 January 2007); Vice President Jaime MORALES Carazo (since 10 January 2007)
cabinet: Council of Ministers appointed by the president
elections: president and vice president elected on the same ticket by popular vote for a five-year term (eligible for a second term so long as it is not consecutive); election last held 5 November 2006 (next to be held by November 2011)
election results: Daniel ORTEGA Saavedra elected president—38.07%, Eduardo MONTEALEGRE 29%, Jose RIZO 26.21%, Edmundo JARQUIN 6.44%
Legislative branch: unicameral National Assembly or Asamblea Nacional (92 seats; 90 members are elected by proportional representation and party lists to serve five-year terms; 1 seat for the previous president, 1 seat for the runner-up in previous presidential election)
elections: last held 5 November 2006 (next to be held by November 2011)
election results: percent of vote by party— NA; seats by party—FSLN 38, PLC 25, ALN 23 (22 plus one for presidential candidate Eduardo MONTEALEGRE, runner-up in the 2006 presidential election), MRS 5, APRE 1 (outgoing President Enrique BOLANOS)
Judicial branch: Supreme Court or Corte Suprema de Justicia (16 judges elected for five-year terms by the National Assembly)
Political parties and leaders: Conservative Party or PC [Azalia AVILES Salmeron]; Liberal Constitutionalist Party or PLC [Jorge CASTILLO Quant]; Nicaraguan Liberal Alliance or ALN [Eduardo MONTEALEGRE]; Sandinista National Liberation Front or FSLN [Daniel ORTEGA Saavedra]; Sandinista Renovation Movement or MRS [Enrique SAENZ Navarrete]
Political pressure groups and leaders: National Workers Front or FNT is a Sandinista umbrella group of eight labor unions including—Farm Workers Association or ATC, Health Workers Federation or FETASALUD, Heroes and Martyrs Confederation of Professional Associations or CONAPRO, National

Association of Educators of Nicaragua or ANDEN, National Union of Employees or UNE, National Union of Farmers and Ranchers or UNAG, Sandinista Workers Central or CST, and Union of Journalists of Nicaragua or UPN; Permanent Congress of Workers or CPT is an umbrella group of four non-Sandinista labor unions including—Autonomous Nicaraguan Workers Central or CTN-A, Confederation of Labor Unification or CUS, Independent General Confederation of Labor or CGT-I, and Labor Action and Unity Central or CAUS; Nicaraguan Workers' Central or CTN is an independent labor union; Superior Council of Private Enterprise or COSEP is a confederation of business groups
International organization participation: BCIE, CACM, FAO, G-77, IADB, IAEA, IBRD, ICAO, ICRM, IDA, IFAD, IFC, IFRCS, ILO, IMF, IMO, Interpol, IOC, IOM, IPU, ISO (correspondent), ITSO, ITU, ITUC, LAES, LAIA (observer), MIGA, NAM, OAS, OPANAL, OPCW, PCA, RG, UN, UNCTAD, UNESCO, UNHCR, UNIDO, Union Latina, UNWTO, UPU, WCL, WCO, WHO, WIPO, WMO, WTO
Diplomatic representation in the US:
chief of mission: Ambassador Arturo CRUZ Sequeira, Jr.
chancery: 1627 New Hampshire Avenue NW, Washington, DC 20009
telephone: [1] (202) 939-6570, [1] (202) 939-6573
FAX: [1] (202) 939-6545
consulate(s) general: Houston, Los Angeles, Miami, New York, San Francisco
Diplomatic representation from the US:
chief of mission: Ambassador Paul A. TRIVELLI
embassy: Kilometer 4.5 Carretera Sur, Managua
mailing address: P.O. Box 327
telephone: [505] 266-6010
FAX: [505] 266-3861
Flag description: three equal horizontal bands of blue (top), white, and blue with the national coat of arms centered in the white band; the coat of arms features a triangle encircled by the words REPUBLICA DE NICARAGUA on the top and AMERICA CENTRAL on the bottom; similar to the flag of El Salvador, which features a round emblem encircled by the words REPUBLICA DE EL SALVADOR EN LA AMERICA CENTRAL centered in the white band; also similar to the flag of Honduras, which has five blue stars arranged in an X pattern centered in the white band

Economy—overview: Nicaragua has widespread underemployment, one of the highest degrees of income inequality in the world, and the third lowest per capita income in the Western Hemisphere. While the country has progressed toward macroeconomic stability in the past few years, annual GDP growth has been far too low to meet the country's needs, forcing the country to rely on international economic assistance to meet fiscal and debt financing obligations. In early 2004, Nicaragua secured some $4.5 billion in foreign debt reduction under the Heavily Indebted Poor Countries (HIPC) initiative, and in October 2007, the IMF approved a new poverty reduction and growth facility (PRGF) program that should create fiscal space for social spending and investment. The continuity of a relationship with the IMF reinforces donor confidence, despite private sector concerns surrounding ORTEGA, which has dampened investment. The US-Central America Free Trade Agreement (CAFTA) has been in effect since April 2006 and has expanded export opportunities for many agricultural and manufactured goods. Energy shortages fueled by high oil prices, however, are a serious bottleneck to growth.
GDP (purchasing power parity): $15.84 billion (2007 est.)
GDP (official exchange rate): $5.723 billion (2007 est.)
GDP—real growth rate: 3.8% (2007 est.)
GDP—per capita (PPP): $2,600 (2007 est.)
GDP—composition by sector:
agriculture: 17.1%
industry: 25.9%
services: 57% (2007 est.)
Labor force: 2.262 million (2007 est.)
Labor force—by occupation: *agriculture:* 29%
industry: 19%
services: 52% (2006 est.)
Unemployment rate: 3.6% plus underemployment of 46.5% (2007 est.)
Population below poverty line: 48% (2005)
Household income or consumption by percentage share:
lowest 10%: 2.2%
highest 10%: 33.8% (2001)
Distribution of family income—Gini index: 43.1 (2001)
Inflation rate (consumer prices): 11.1% (2007 est.)
Investment (gross fixed): 28.9% of GDP (2007 est.)

Budget:
revenues: $1.115 billion
expenditures: $1.291 billion (2007 est.)
Public debt: 63% of GDP (2007 est.)
Agriculture—products: coffee, bananas, sugarcane, cotton, rice, corn, tobacco, sesame, soya, beans; beef, veal, pork, poultry, dairy products; shrimp, lobsters
Industries: food processing, chemicals, machinery and metal products, textiles, clothing, petroleum refining and distribution, beverages, footwear, wood
Industrial production growth rate: 3% (2007 est.)
Electricity—production: 2.778 billion kWh (2006)
Electricity—production by source:
fossil fuel: 83.9%
hydro: 7.7%
nuclear: 0%
other: 8.4% (2001)
Electricity—consumption: 2.929 billion kWh (2006)
Electricity—exports: 8 million kWh (2005)
Electricity—imports: 69.34 million kWh (2006)
Oil—production: 0 bbl/day (2005 est.)
Oil—consumption: 28,000 bbl/day (2005 est.)
Oil—exports: 1,397 bbl/day (2004)
Oil—imports: 15,560 bbl/day (2005 est.)
Oil—proved reserves: 0 bbl (1 January 2006 est.)
Natural gas—production: 0 cu m (2005 est.)
Natural gas—consumption: 0 cu m (2005 est.)
Natural gas—exports: 0 cu m (2005 est.)
Natural gas—imports: 0 cu m (2005)
Natural gas—proved reserves: 0 cu m (1 January 2006 est.)
Current account balance: -$989 million (2007 est.)
Exports: $2.235 billion f.o.b.; note—includes free trade zones (2007 est.)
Exports—commodities: coffee, beef, shrimp and lobster, tobacco, sugar, gold, peanuts
Exports—partners: US 65.2%, El Salvador 6.9%, Honduras 3.8% (2006)
Imports: $3.935 billion f.o.b. (2007 est.)
Imports—commodities: consumer goods, machinery and equipment, raw materials, petroleum products
Imports—partners: US 20.1%, Mexico 13.9%, Venezuela 9.4%, Costa Rica 6.9%, Guatemala 5.4%, China 4.3% (2006)
Economic aid—recipient: $471 million (2006 est.)

Reserves of foreign exchange and gold: $1.103 billion (31 December 2007 est.)
Debt—external: $3.335 billion (31 December 2007 est.)
Market value of publicly traded shares: $NA
Currency (code): gold cordoba (NIO)
Currency code: NIO
Exchange rates: gold cordobas per US dollar—18.457 (2007), 17.582 (2006), 16.733 (2005), 15.937 (2004), 15.105 (2003)
Fiscal year: calendar year

COMMUNICATIONS

Telephones—main lines in use: 247,900 (2006)
Telephones—mobile cellular: 1.83 million (2006)
Telephone system:
general assessment: system being upgraded by foreign investment; nearly all installed telecommunications capacity now uses digital technology, owing to investments since privatization of the formerly state-owned telecommunications company
domestic: since privatization, access to fixed-line and mobile-cellular services has improved but teledensity still lags behind other Central American countries; connected to Central American Microwave System
international: country code—505; the Americas Region Caribbean Ring System (ARCOS-1) fiber optic submarine cable provides connectivity to South and Central America, parts of the Caribbean, and the US; satellite earth stations—1 Intersputnik (Atlantic Ocean region) and 1 Intelsat (Atlantic Ocean)
Radio broadcast stations: AM 63, FM 32, shortwave 1 (1998)
Radios: 1.24 million (1997)
Television broadcast stations: 3 (plus 7 repeaters) (1997)
Televisions: 320,000 (1997)
Internet country code: .ni
Internet hosts: 27,941 (2007)
Internet Service Providers (ISPs): 3 (2000)
Internet users: 155,000 (2006)

TRANSPORTATION

Airports: 163 (2007)
Airports—with paved runways:
total: 11
2,438 to 3,047 m: 3
1,524 to 2,437 m: 2
914 to 1,523 m: 3

under 914 m: 3 (2007)
Airports—with unpaved runways:
total: 152
1,524 to 2,437 m: 1
914 to 1,523 m: 16
under 914 m: 135 (2007)
Pipelines: oil 54 km (2007)
Railways:
total: 6 km
narrow gauge: 6 km 1.067-m gauge (2006)
Roadways:
total: 19,036 km
paved: 2,299 km
unpaved: 16,737 km (2005)
Waterways: 2,220 km (including lakes Managua and Nicaragua) (2007)
Ports and terminals: Bluefields, Corinto, El Bluff

MILITARY

Military branches: National Army of Nicaragua (ENN; includes Navy, Air Force) (2008)
Military service age and obligation: 17 years of age for voluntary military service; tour of duty 18–36 months (2008)
Manpower available for military service:
males age 16–49: 1,513,312
females age 16–49: 1,507,999 (2008 est.)
Manpower fit for military service:
males age 16–49: 1,235,400
females age 16–49: 1,302,318 (2008 est.)
Manpower reaching militarily significant age annually:
males age 16–49: 72,689
females age 16–49: 70,452 (2008 est.)
Military expenditures—percent of GDP: 0.6% (2006)

TRANSNATIONAL ISSUES

Disputes—international: memorials and countermemorials were filed by the parties in Nicaragua's 1999 and 2001 proceedings against Honduras and Colombia at the ICJ over the maritime boundary and territorial claims in the western Caribbean Sea, final public hearings are scheduled for 2007; the 1992 ICJ ruling for El Salvador and Honduras advised a tripartite resolution to establish a maritime boundary in the Gulf of Fonseca, which considers Honduran access to the Pacific; legal dispute over navigational rights of San Juan River on border with Costa Rica
Illicit drugs: transshipment point for cocaine destined for the US and transshipment point for arms-for-drugs dealing

NIGER

INTRODUCTION

Background: Niger became independent from France in 1960 and experienced single-party and military rule until 1991, when Gen. Ali SAIBOU was forced by public pressure to allow multiparty elections, which resulted in a democratic government in 1993. Political infighting

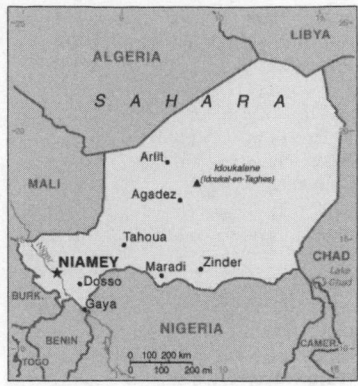

brought the government to a standstill and in 1996 led to a coup by Col. Ibrahim BARE. In 1999 BARE was killed in a coup by military officers who promptly restored democratic rule and held elections that brought Mamadou TANDJA to power in December of that year. TANDJA was reelected in 2004. Niger is one of the poorest countries in the world with minimal government services and insufficient funds to develop its resource base. The largely agrarian and subsistence-based economy is frequently disrupted by extended droughts common to the Sahel region of Africa. A predominately Tuareg ethnic group emerged in February 2007, the Nigerien Movement for Justice (MNJ), and attacked several military targets in Niger's northern region throughout 2007. Events have since evolved into a budding insurrection.

GEOGRAPHY

Location: Western Africa, southeast of Algeria

Geographic coordinates: 16 00 N, 8 00 E

Map references: Africa

Area:
total: 1.267 million sq km
land: 1,266,700 sq km
water: 300 sq km

Area—comparative: slightly less than twice the size of Texas

Land boundaries:
total: 5,697 km
border countries: Algeria 956 km, Benin 266 km, Burkina Faso 628 km, Chad 1,175 km, Libya 354 km, Mali 821 km, Nigeria 1,497 km

Coastline: 0 km (landlocked)

Maritime claims: none (landlocked)

Climate: desert; mostly hot, dry, dusty; tropical in extreme south

Terrain: predominately desert plains and sand dunes; flat to rolling plains in south; hills in north

Elevation extremes:
lowest point: Niger River 200 m

highest point: Mont Bagzane 2,022 m

Natural resources: uranium, coal, iron ore, tin, phosphates, gold, molybdenum, gypsum, salt, petroleum

Land use: *arable land:* 11.43%
permanent crops: 0.01%
other: 88.56% (2005)

Irrigated land: 730 sq km (2003)

Total renewable water resources: 33.7 cu km (2003)

Freshwater withdrawal (domestic/industrial/agricultural): *total:* 2.18 cu km/yr (4%/0%/95%)
per capita: 156 cu m/yr (2000)

Natural hazards: recurring droughts

Environment—current issues: overgrazing; soil erosion; deforestation; desertification; wildlife populations (such as elephant, hippopotamus, giraffe, and lion) threatened because of poaching and habitat destruction

Environment—international agreements: *party to:* Biodiversity, Climate Change, Climate Change-Kyoto Protocol, Desertification, Endangered Species, Environmental Modification, Hazardous Wastes, Ozone Layer Protection, Wetlands
signed, but not ratified: Law of the Sea

Geography—note: landlocked; one of the hottest countries in the world; northern four-fifths is desert, southern one-fifth is savanna, suitable for livestock and limited agriculture

PEOPLE

Population: 13,272,679 (July 2008 est.)

Age structure:
0–14 years: 47% (male 3,174,834/female 3,057,003)
15–64 years: 50.6% (male 3,450,393/female 3,267,496)
65 years and over: 2.4% (male 159,945/female 163,008) (2008 est.)

Median age: *total:* 16.4 years
male: 16.5 years
female: 16.4 years (2008 est.)

Population growth rate: 2.878% (2008 est.)

Birth rate: 49.62 births/1,000 population (2008 est.)

Death rate: 20.26 deaths/1,000 population (2008 est.)

Net migration rate: -0.57 migrant(s)/1,000 population (2008 est.)

Sex ratio:
at birth: 1.03 male(s)/female
under 15 years: 1.04 male(s)/female
15–64 years: 1.06 male(s)/female
65 years and over: 0.98 male(s)/female
total population: 1.05 male(s)/female (2008 est.)

Infant mortality rate:
total: 115.42 deaths/1,000 live births
male: 119.3 deaths/1,000 live births

female: 111.42 deaths/1,000 live births (2008 est.)

Life expectancy at birth:
total population: 44.28 years
male: 44.3 years
female: 44.26 years (2008 est.)

Total fertility rate: 7.29 children born/woman (2008 est.)

HIV/AIDS—adult prevalence rate: 1.2% (2003 est.)

HIV/AIDS—people living with HIV/AIDS: 70,000 (2003 est.)

HIV/AIDS—deaths: 4,800 (2003 est.)

Major infectious diseases:
degree of risk: very high
food or waterborne diseases: bacterial and protozoal diarrhea, hepatitis A, and typhoid fever
vectorborne disease: malaria
respiratory disease: meningococcal meningitis
note: highly pathogenic H5N1 avian influenza has been identified in this country; it poses a negligible risk with extremely rare cases possible among US citizens who have close contact with birds (2008)

Nationality:
noun: Nigerien(s)
adjective: Nigerien

Ethnic groups: Haoussa 55.4%, Djerma Sonrai 21%, Tuareg 9.3%, Peuhl 8.5%, Kanouri Manga 4.7%, other 1.2% (2001 census)

Religions: Muslim 80%, other (includes indigenous beliefs and Christian) 20%

Languages: French (official), Hausa, Djerma

Literacy:
definition: age 15 and over can read and write
total population: 28.7%
male: 42.9%
female: 15.1% (2005 est.)

GOVERNMENT

Country name:
conventional long form: Republic of Niger
conventional short form: Niger
local long form: Republique du Niger
local short form: Niger

Government type: republic

Capital: *name:* Niamey
geographic coordinates: 13 31 N, 2 07 E
time difference: UTC+1 (6 hours ahead of Washington, DC during Standard Time)

Administrative divisions: 8 regions (regions, singular—region) includes 1 capital district* (communite urbaine); Agadez, Diffa, Dosso, Maradi, Niamey*, Tahoua, Tillaberi, Zinder

Independence: 3 August 1960 (from France)

National holiday: Republic Day, 18 December (1958)

Constitution: new constitution adopted 18 July 1999

Legal system: based on French civil law system and customary law; has not accepted compulsory ICJ jurisdiction

Suffrage: 18 years of age; universal

Executive branch:

chief of state: President Mamadou TANDJA (since 22 December 1999)

head of government: Prime Minister Seyni OUMAROU (since 3 June 2007); appointed by the president and shares some executive responsibilities with the president

cabinet: 26-member Cabinet appointed by the president

elections: president elected by popular vote for a five-year term (eligible for a second term); second round of election last held 4 December 2004 (next to be held December 2009)

election results: Mamadou TANDJA reelected president; percent of vote— Mamadou TANDJA 65.5%, Mahamadou ISSOUFOU 34.5%

Legislative branch: unicameral National Assembly (113 seats; members elected by popular vote to serve five-year terms)

elections: last held 4 December 2004 (next to be held in December 2009)

election results: percent of vote by party— NA; seats by party—MNSD 47, PNDS 25, CDS 22, RSD 7, RDP 6, ANDP 5, PSDN 1

Judicial branch: State Court or Cour d'Etat; Court of Appeals or Cour d'Appel

Political parties and leaders: Democratic and Social Convention-Rahama or CDS-Rahama [Mahamane OUSMANE]; National Movement for a Developing Society-Nassara or MNSD-Nassara [Hama AMADOU]; Niger Social Democratic Party or PSDN; Nigerien Alliance for Democracy and Social Progress-Zaman Lahiya or ANDP-Zaman Lahiya [Moumouni DJER-MAKOYE]; Nigerien Party for Autonomy or PNA-Alouma'a [Sanousi JACKOU]; Nigerien Party for Democracy and Socialism or PNDS-Tarrayya [Issifou MAHAMADOU]; Nigerien Progressive Party or PPN-RDA [Abdoulaye DIORI]; Rally for Democracy and Progress or RDP-jama'a [Hamid ALGABID]; Social and Democratic Rally or RSD-Gaskiyya [Cheiffou AMADOU]

Political pressure groups and leaders: The Nigerien Movement for Justice or MNJ, a predominantly Tuareg rebel group demanding concessions including a greater share of the country's uranium revenues from the government

International organization participation: ACCT, ACP, AfDB, AU, ECOWAS, Entente, FAO, FZ, G-77, IAEA, IBRD, ICAO, ICCt, ICRM, IDA, IDB, IFAD, IFC, IFRCS, ILO, IMF, Interpol, IOC, IOM, IPU, ITSO, ITU, ITUC, NAM, OIC, OIF, OPCW, UN, UNCTAD, UNESCO, UNIDO, UNMIL, UNOCI, UNWTO, UPU, WADB (regional), WAEMU, WCL, WCO, WFTU, WHO, WIPO, WMO, WTO

Diplomatic representation in the US:

chief of mission: Ambassador Aminata Djibrilla Maiga TOURE

chancery: 2204 R Street NW, Washington, DC 20008

telephone: [1] (202) 483-4224 through 4227

FAX: [1] (202)483-3169

Diplomatic representation from the US:

chief of mission: Ambassador Bernadette M. ALLEN

embassy: Rue Des Ambassades, Niamey

mailing address: B. P. 11201, Niamey

telephone: [227] 20-73-31-69

FAX: [227] 20-73-55-60

Flag description: three equal horizontal bands of orange (top), white, and green with a small orange disk (representing the sun) centered in the white band; similar to the flag of India, which has a blue spoked wheel centered in the white band

ECONOMY

Economy—overview: Niger is one of the poorest countries in the world, ranking near last on the United Nations Development Fund index of human development. It is a landlocked, Sub-Saharan nation, whose economy centers on subsistence crops, livestock, and some of the world's largest uranium deposits. Drought cycles, desertification, and a 2.9% population growth rate, have undercut the economy. Niger shares a common currency, the CFA franc, and a common central bank, the Central Bank of West African States (BCEAO), with seven other members of the West African Monetary Union. In December 2000, Niger qualified for enhanced debt relief under the International Monetary Fund program for Highly Indebted Poor Countries (HIPC) and concluded an agreement with the Fund on a Poverty Reduction and Growth Facility (PRGF). Debt relief provided under the enhanced HIPC initiative significantly reduces Niger's annual debt service obligations, freeing funds for expenditures on basic health care, primary education, HIV/AIDS prevention, rural infrastructure, and other programs geared at poverty reduction. In December 2005, Niger received 100% multilateral debt relief from the IMF, which translates into the forgiveness of approximately US $86 million in debts to the IMF, excluding the remaining assistance under HIPC. Nearly half of the government's budget is derived from foreign donor resources. Future growth may be sustained by exploitation of oil, gold, coal, and other mineral resources. Uranium prices have increased sharply in the last few years. A drought and locust infestation in 2005 led to food shortages for as many as 2.5 million Nigeriens.

GDP (purchasing power parity): $8.902 billion (2007 est.)

GDP (official exchange rate): $4.174 billion (2007 est.)

GDP—real growth rate: 3.2% (2007 est.)

GDP—per capita (PPP): $700 (2007 est.)

GDP—composition by sector:

agriculture: 39%

industry: 17%

services: 44% (2001)

Labor force: 70,000 salaried workers, 60% of whom are employed in the public sector (1995)

Labor force—by occupation: *agriculture:* 90%

industry: 6%

services: 4% (1995)

Unemployment rate: NA%

Population below poverty line: 63% (1993 est.)

Household income or consumption by percentage share:

lowest 10%: 0.8%

highest 10%: 35.4% (1995)

Distribution of family income—Gini Index: 50.5 (1995)

Inflation rate (consumer prices): 0.1% (2007 est.)

Budget:

revenues: $320 million (includes $134 million from foreign sources)

expenditures: $320 million (2002 est.)

Agriculture—products: cowpeas, cotton, peanuts, millet, sorghum, cassava (tapioca), rice; cattle, sheep, goats, camels, donkeys, horses, poultry

Industries: uranium mining, cement, brick, soap, textiles, food processing, chemicals, slaughterhouses

Industrial production growth rate: 5.1% (2003 est.)

Electricity—production: 234.1 million kWh (2005)

Electricity—production by source:

fossil fuel: 100%

hydro: 0%

nuclear: 0%

other: 0% (2001)

Electricity—consumption: 437.7 million kWh (2005)

Electricity—exports: 0 kWh (2005)

Electricity—imports: 220 million kWh (2005)

Oil—production: 0 bbl/day (2005 est.)

Oil—consumption: 5,450 bbl/day (2005 est.)

Oil—exports: 0 bbl/day (2004)

Oil—imports: 5,412 bbl/day (2004)

Oil—proved reserves: NA bbl

Natural gas—production: 0 cu m (2005 est.)

Natural gas—consumption: 0 cu m (2005 est.)

Natural gas—exports: 0 cu m (2005 est.)

Natural gas—imports: 0 cu m (2005)

Natural gas—proved reserves: 0 cu m (1 January 2006 est.)

Current account balance: -$321 million (2007 est.)

Exports: $428 million f.o.b. (2006)

Exports—commodities: uranium ore, livestock, cowpeas, onions

Exports—partners: France 34.8%, US 26.6%, Nigeria 18.3%, Russia 11.3% (2006)

Imports: $800 million f.o.b. (2006)

Imports—commodities: foodstuffs, machinery, vehicles and parts, petroleum, cereals

Imports—partners: US 14%, France 12.1%, China 7.8%, Nigeria 7.7%, French Polynesia 7.7%, Cote d'Ivoire 4.9% (2006)

Economic aid—recipient: $515.4 million (2005)

Debt—external: $2.1 billion (2003 est.)

Market value of publicly traded shares: $NA

Currency (code): Communaute Financiere Africaine franc (XOF); note—responsible authority is the Central Bank of the West African States

Currency code: XOF

Exchange rates: Communaute Financiere Africaine francs (XOF) per US dollar—493.51 (2007), 522.59 (2006), 527.47 (2005), 528.29 (2004), 581.2 (2003)

note: since 1 January 1999, the XOF franc has been pegged to the euro at a rate of 655.957 XOF francs per euro

Fiscal year: calendar year

COMMUNICATIONS

Telephones—main lines in use: 24,000 (2005)

Telephones—mobile cellular: 323,900 (2005)

Telephone system:

general assessment: inadequate; small system of wire, radio telephone communications, and microwave radio relay links concentrated in the southwestern area of Niger

domestic: combined fixed-line and mobile-cellular teledensity is less than 3 per 100 persons; domestic satellite system with 3 earth stations and 1 planned

international: country code—227; satellite earth stations—2 Intelsat (1 Atlantic Ocean and 1 Indian Ocean)

Radio broadcast stations: AM 5, FM 6, shortwave 4 (2001)

Radios: 680,000 (1997)

Television broadcast stations: 5 (2007)

Televisions: 125,000 (1997)

Internet country code: .ne

Internet hosts: 200 (2007)

Internet Service Providers (ISPs): 1 (2002)

Internet users: 40,000 (2006)

TRANSPORTATION

Airports: 28 (2007)

Airports—with paved runways:

total: 9

2,438 to 3,047 m: 3

1,524 to 2,437 m: 5

under 914 m: 1 (2007)

Airports—with unpaved runways:

total: 19

1,524 to 2,437 m: 2

914 to 1,523 m: 14

under 914 m: 3 (2007)

Roadways:

total: 14,565 km

paved: 3,641 km

unpaved: 10,924 km (2004)

Waterways: 300 km (the Niger, the only major river, is navigable to Gaya between September and March) (2005)

MILITARY

Military branches: Nigerien Armed Forces (Forces Armees Nigeriennes, FAN): Army, Niger Air Force (2008)

Military service age and obligation: 18 years of age for compulsory military service; 2-year conscript service obligation (2006)

Manpower available for military service:

males age 16–49: 2,871,868

females age 16–49: 2,696,966 (2008 est.)

Manpower fit for military service:

males age 16–49: 1,665,108

females age 16–49: 1,548,965 (2008 est.)

Manpower reaching militarily significant age annually:

males age 16–49: 150,728

females age 16–49: 143,379 (2008 est.)

Military expenditures—percent of GDP: 1.3% (2006)

TRANSNATIONAL ISSUES

Disputes—international: Libya claims about 25,000 sq km in a currently dormant dispute in the Tommo region; much of Benin-Niger boundary, including tripoint with Nigeria, remains undemarcated; only Nigeria and Cameroon have heeded the Lake Chad Commission's admonition to ratify the delimitation treaty which also includes the Chad-Niger and Niger-Nigeria boundaries

NIGERIA

INTRODUCTION

Background: British influence and control over what would become Nigeria grew through the 19th century. A series of constitutions after World War II granted Nigeria greater autonomy; independence came in 1960. Following nearly 16 years of military rule, a new constitution was adopted in 1999, and a peaceful transition to civilian government was completed. The government continues to face the daunting task of reforming a petroleum-based economy, whose revenues have been squandered through corruption and mismanagement, and institutionalizing democracy. In addition, Nigeria continues to experience longstanding ethnic and religious tensions. Although both the 2003 and 2007 presidential elections were marred by significant irregularities and violence, Nigeria is currently experiencing its longest period of civilian rule since independence. The general elections of April 2007 marked the first civilian-to-civilian transfer of power in the country's history.

GEOGRAPHY

Location: Western Africa, bordering the

Gulf of Guinea, between Benin and Cameroon

Geographic coordinates: 10 00 N, 8 00 E

Map references: Africa

Area:

total: 923,768 sq km

land: 910,768 sq km

water: 13,000 sq km

Area—comparative: slightly more than twice the size of California

Land boundaries:

total: 4,047 km

border countries: Benin 773 km, Cameroon 1,690 km, Chad 87 km, Niger 1,497 km

Coastline: 853 km

Maritime claims:

territorial sea: 12 nm

exclusive economic zone: 200 nm

continental shelf: 200 m depth or to the depth of exploitation

Climate: varies; equatorial in south, tropical in center, arid in north

Terrain: southern lowlands merge into central hills and plateaus; mountains in southeast, plains in north

Elevation extremes:

lowest point: Atlantic Ocean 0 m

highest point: Chappal Waddi 2,419 m

Natural resources: natural gas, petroleum, tin, iron ore, coal, limestone, niobium, lead, zinc, arable land

Land use:

arable land: 33.02%

permanent crops: 3.14%

other: 63.84% (2005)

Irrigated land: 2,820 sq km (2003)

Total renewable water resources: 286.2 cu km (2003)

Freshwater withdrawal (domestic/industrial/agricultural): *total:* 8.01 cu km/yr (21%/10%/69%)

per capita: 61 cu m/yr (2000)

Natural hazards: periodic droughts; flooding

Environment—current issues: soil degradation; rapid deforestation; urban air and water pollution; desertification; oil pollution—water, air, and soil; has suffered serious damage from oil spills; loss of arable land; rapid urbanization

Environment—international agreements: *party to:* Biodiversity, Climate Change, Climate Change-Kyoto Protocol, Desertification, Endangered Species, Hazardous Wastes, Law of the Sea, Marine Dumping, Marine Life Conservation, Ozone Layer Protection, Ship Pollution, Wetlands

signed, but not ratified: none of the selected agreements

Geography—note: the Niger enters the country in the northwest and flows southward through tropical rain forests and swamps to its delta in the Gulf of Guinea

PEOPLE

Population: 138,283,240

note: estimates for this country explicitly take into account the effects of excess mortality due to AIDS; this can result in lower life expectancy, higher infant mortality, higher death rates, lower population growth rates, and changes in the distribution of population by age and sex than would otherwise be expected (July 2008 est.)

Age structure:

0–14 years: 42.2% (male 29,378,127/female 28,953,864)

15–64 years: 54.7% (male 38,466,129/female 37,172,355)

65 years and over: 3.1% (male 2,046,309/female 2,266,456) (2008 est.)

Median age:

total: 18.7 years

male: 18.8 years

female: 18.6 years (2008 est.)

Population growth rate: 2.382% (2008 est.)

Birth rate: 39.98 births/1,000 population (2008 est.)

Death rate: 16.41 deaths/1,000 population (2008 est.)

Net migration rate: 0.25 migrant(s)/1,000 population (2008 est.)

Sex ratio:

at birth: 1.03 male(s)/female

under 15 years: 1.01 male(s)/female

15–64 years: 1.03 male(s)/female

65 years and over: 0.9 male(s)/female

total population: 1.02 male(s)/female (2008 est.)

Infant mortality rate:

total: 93.93 deaths/1,000 live births

male: 100.87 deaths/1,000 live births

female: 86.79 deaths/1,000 live births (2008 est.)

Life expectancy at birth:

total population: 47.81 years

male: 47.15 years

female: 48.5 years (2008 est.)

Total fertility rate: 5.41 children born/woman (2008 est.)

HIV/AIDS—adult prevalence rate: 5.4% (2003 est.)

HIV/AIDS—people living with HIV/AIDS: 3.6 million (2003 est.)

HIV/AIDS—deaths: 310,000 (2003 est.)

Major infectious diseases:

degree of risk: very high

food or waterborne diseases: bacterial and protozoal diarrhea, hepatitis A, and typhoid fever

vectorborne disease: malaria and yellow fever

respiratory disease: meningococcal meningitis

aerosolized dust or soil contact disease: one of the most highly endemic areas for Lassa fever

water contact disease: leptospirosis and shistosomiasis

note: highly pathogenic H5N1 avian influenza has been identified in this country; it poses a negligible risk with extremely rare cases possible among US citizens who have close contact with birds (2008)

Nationality:

noun: Nigerian(s)

adjective: Nigerian

Ethnic groups: Nigeria, Africa's most populous country, is composed of more than 250 ethnic groups; the following are the most populous and politically influential: Hausa and Fulani 29%, Yoruba 21%, Igbo (Ibo) 18%, Ijaw 10%, Kanuri 4%, Ibibio 3.5%, Tiv 2.5%

Religions: Muslim 50%, Christian 40%, indigenous beliefs 10%

Languages: English (official), Hausa, Yoruba, Igbo (Ibo), Fulani

Literacy:

definition: age 15 and over can read and write

total population: 68%

male: 75.7%

female: 60.6% (2003 est.)

GOVERNMENT

Country name:

conventional long form: Federal Republic of Nigeria

conventional short form: Nigeria

Government type: federal republic

Capital: *name:* Abuja

geographic coordinates: 9 12 N, 7 11 E

time difference: UTC+1 (6 hours ahead of Washington, DC during Standard Time)

Administrative divisions: 36 states and 1 territory*; Abia, Adamawa, Akwa Ibom, Anambra, Bauchi, Bayelsa, Benue, Borno, Cross River, Delta, Ebonyi, Edo, Ekiti, Enugu, Federal Capital Territory*, Gombe, Imo, Jigawa, Kaduna, Kano, Katsina, Kebbi, Kogi, Kwara, Lagos, Nassarawa, Niger, Ogun, Ondo, Osun, Oyo, Plateau, Rivers, Sokoto, Taraba, Yobe, Zamfara

Independence: 1 October 1960 (from UK)

National holiday: Independence Day (National Day), 1 October (1960)

Constitution: new constitution adopted 5 May 1999; effective 29 May 1999

Legal system: based on English common law, Islamic law (in 12 northern states), and traditional law; accepts compulsory ICJ jurisdiction with reservations

Suffrage: 18 years of age; universal

Executive branch:

chief of state: President Umaru Musa

YAR'ADUA (since 29 May 2007); note—the president is both the chief of state and head of government

head of government: President Umaru Musa YAR'ADUA (since 29 May 2007)

cabinet: Federal Executive Council

elections: president is elected by popular vote for a four-year term (eligible for a second term); election last held 21 April 2007 (next to be held in April 2011)

election results: Umaru Musa YAR'ADUA elected president; percent of vote—Umaru Musa YAR'ADUA 69.8%, Muhammadu BUHARI 18.7%, Atiku ABUBAKAR 7.5%, Orji Uzor KALU 1.7%, other 2.3%

Legislative branch: bicameral National Assembly consists of the Senate (109 seats, 3 from each state plus 1 from Abuja; members elected by popular vote to serve four-year terms) and House of Representatives (360 seats; members elected by popular vote to serve four-year terms)

elections: Senate—last held 21 April 2007 (next to be held in April 2011); House of Representatives—last held 21 April 2007 (next to be held in April 2011)

election results: Senate—percent of vote by party—PDP 53.7%, ANPP 27.9%, AD 9.7%, other 8.7%; seats by party—PDP 76, ANPP 27, AD 6; House of Representatives—percent of vote by party—PDP 54.5%, ANPP 27.4%, AD 8.8%, UNPP 2.8%, NPD 1.9%, APGA 1.6%, PRP 0.8%; seats by party—PDP 76, ANPP 27, AD 6, UNPP 2, APGA 2, NPD 1, PRP 1, vacant 1

Judicial branch: Supreme Court (judges appointed by the President); Federal Court of Appeal (judges are appointed by the federal government on the advice of the Advisory Judicial Committee)

Political parties and leaders: Accord Party [Ikra Aliyu BILBIS]; Action Congress or AC [Hassan ZUMI]; Alliance for Democracy or AD [Mojisoluwa AKINFENWA]; All Nigeria Peoples' Party or ANPP [Edwin UME-EZEOKE]; All Progressives Grand Alliance or APGA [Victor C. UMEH]; Democratic People's Party or DPP [Jeremiah USENI]; Fresh Democratic Party [Chris OKOTIE]; Labor Party [Dan NWANYANWU]; Movement for the Restoration and Defense of Democracy or MRDD [Mohammed Gambo JIMETA]; National Democratic Party or NDP [Aliyu Habu FARI]; Peoples Democratic Party or PDP [vacant]; Peoples Progressive Alliance [Clement EBRI]; Peoples Redemption Party or PRP [Abdulkadir Balarabe MUSA]; Peoples Salvation Party or PSP [Lawal MAITURARE]; United Nigeria Peoples Party or UNPP [Mallam Selah JAMBO]

Political pressure groups and leaders: NA

International organization participation: ACP, AfDB, AU, C, ECOWAS, FAO, G-15, G-24, G-77, IAEA, IBRD, ICAO, ICC, ICCt, ICRM, IDA, IDB, IFAD, IFC, IFRCS, IHO, ILO, IMF, IMO, IMSO, Interpol, IOC, IOM, IPU, ISO, ITSO, ITU, ITUC, MIGA, MIN-URSO, NAM, OAS (observer), OIC, OPCW, OPEC, PCA, UN, UNAMID, UNCTAD, UNESCO, UNHCR, UNIDO, UNITAR, UNMEE, UNMIL, UNMIS, UNOCI, UNOMIG, UNWTO, UPU, WCO, WFTU, WHO, WIPO, WMO, WTO

Diplomatic representation in the US:
chief of mission: Ambassador Oluwole ROTIMI
chancery: 3519 International Court NW, Washington, DC 20008
telephone: [1] (202) 986-8400
FAX: [1] (202) 775-1385
consulate(s) general: Atlanta, New York

Diplomatic representation from the US:
chief of mission: Ambassador Robin SANDERS
embassy: 1075 Diplomatic Drive, Abuja
mailing address: P. O. Box 5760, Garki, Abuja
telephone: [234] (9) 461-4000
FAX: [234] (9) 461-4036/4273

Flag description: three equal vertical bands of green (hoist side), white, and green

ECONOMY

Economy—overview: Oil-rich Nigeria, long hobbled by political instability, corruption, inadequate infrastructure, and poor macroeconomic management, is undertaking some reforms under a new reform-minded administration. Nigeria's former military rulers failed to diversify the economy away from its overdependence on the capital-intensive oil sector, which provides 20% of GDP, 95% of foreign exchange earnings, and about 80% of budgetary revenues. The largely subsistence agricultural sector has failed to keep up with rapid population growth—Nigeria is Africa's most populous country—and the country, once a large net exporter of food, now must import food. Following the signing of an IMF stand-by agreement in August 2000, Nigeria received a debt-restructuring deal from the Paris Club and a $1 billion credit from the IMF, both contingent on economic reforms. Nigeria pulled out of its IMF program in April 2002, after failing to meet spending and exchange rate targets, making it ineligible for addi-tional debt forgiveness from the Paris Club. In the last year the government has begun showing the political will to implement the market-oriented reforms urged by the IMF, such as to modernize the banking system, to curb inflation by blocking excessive wage demands, and to resolve regional disputes over the distribution of earnings from the oil industry. In 2003, the government began deregulating fuel prices, announced the privatization of the country's four oil refineries, and instituted the National Economic Empowerment Development Strategy, a domestically designed and run program modeled on the IMF's Poverty Reduction and Growth Facility for fiscal and monetary management. In November 2005, Abuja won Paris Club approval for a debt-relief deal that eliminated $18 billion of debt in exchange for $12 billion in payments—a total package worth $30 billion of Nigeria's total $37 billion external debt. The deal requires Nigeria to be subject to stringent IMF reviews. GDP rose strongly in 2007, based largely on increased oil exports and high global crude prices. Newly-elected President YAR'ADUA has pledged to continue the economic reforms of his predecessor and the proposed budget for 2008 reflects the administrations emphasis on infrastructure improvements. Infrastructure is the main impediment to growth. The government is working toward developing stronger public-private partnerships for electricity and roads.

GDP (purchasing power parity): $292.7 billion (2007 est.)

GDP (official exchange rate): $166.8 billion (2007 est.)

GDP—real growth rate: 6.4% (2007 est.)

GDP—per capita (PPP): $2,000 (2007 est.)

GDP—composition by sector:
agriculture: 17.6%
industry: 52.7%
services: 29.7% (2007 est.)

Labor force: 50.13 million (2007 est.)

Labor force—by occupation: *agriculture:* 70%
industry: 10%
services: 20% (1999 est.)

Unemployment rate: 4.9% (2007 est.)

Population below poverty line: 70% (2007 est.)

Household income or consumption by percentage share:
lowest 10%: 1.9%
highest 10%: 33.2% (2003)

Distribution of family income—Gini index: 43.7 (2003)

Inflation rate (consumer prices): 5.5% (2007 est.)

Investment (gross fixed): 24.9% of GDP (2007 est.)

Budget:

revenues: $19.65 billion

expenditures: $21.68 billion (2007 est.)

Public debt: 14.5% of GDP (2007 est.)

Agriculture—products: cocoa, peanuts, palm oil, corn, rice, sorghum, millet, cassava (tapioca), yams, rubber; cattle, sheep, goats, pigs; timber; fish

Industries: crude oil, coal, tin, columbite; palm oil, peanuts, cotton, rubber, wood; hides and skins, textiles, cement and other construction materials, food products, footwear, chemicals, fertilizer, printing, ceramics, steel, small commercial ship construction and repair

Industrial production growth rate: 3.4% (2007 est.)

Electricity—production: 22.53 billion kWh (2005)

Electricity—production by source:

fossil fuel: 61.9%

hydro: 38.1%

nuclear: 0%

other: 0% (2001)

Electricity—consumption: 16.88 billion kWh (2005)

Electricity—exports: 0 kWh (2005)

Electricity—imports: 0 kWh (2005)

Oil—production: 2.44 million bbl/day (2006 est.)

Oil—consumption: 302,000 bbl/day (2006 est.)

Oil—exports: 2.141 million bbl/day (2006)

Oil—imports: 167,900 bbl/day (2004)

Oil—proved reserves: 37.25 billion bbl (2007 est.)

Natural gas—production: 21.48 billion cu m (2005 est.)

Natural gas—consumption: 9.936 billion cu m (2005 est.)

Natural gas—exports: 11.55 billion cu m (2005 est.)

Natural gas—imports: 0 cu m (2005)

Natural gas—proved reserves: 5.015 trillion cu m (1 January 2006 est.)

Current account balance: $1.205 billion (2007 est.)

Exports: $62.42 billion f.o.b. (2007 est.)

Exports—commodities: petroleum and petroleum products 95%, cocoa, rubber

Exports—partners: US 48.8%, Spain 8%, Brazil 7.3%, France 4.2% (2006)

Imports: $38.83 billion f.o.b. (2007 est.)

Imports—commodities: machinery, chemicals, transport equipment, manufactured goods, food and live animals

Imports—partners: China 10.7%, US 8.4%, Netherlands 6.2%, UK 5.8%, France 5.6%, Brazil 5.1%, Germany 4.5% (2006)

Economic aid—recipient: $6.437 billion (2005)

Reserves of foreign exchange and gold: $51.33 billion (31 December 2007 est.)

Debt—external: $8.031 billion (31 December 2007 est.)

Stock of direct foreign investment—at home: $33.64 billion (2007 est.)

Stock of direct foreign investment—abroad: $12.63 billion (2007 est.)

Market value of publicly traded shares: $32.82 billion (2006)

Currency (code): naira (NGN)

Currency code: NGN

Exchange rates: nairas per US dollar—127.46 (2007), 127.38 (2006), 132.59 (2005), 132.89 (2004), 129.22 (2003)

Fiscal year: calendar year

COMMUNICATIONS

Telephones—main lines in use: 1.688 million (2006)

Telephones—mobile cellular: 32.322 million (2006)

Telephone system:

general assessment: further expansion and modernization of the fixed-line telephone network is needed

domestic: the addition of a second fixed-line provider in 2002 resulted in faster growth of this service with fixed-line subscribership nearly tripling over the past five years; wireless telephony has grown rapidly, in part responding to the shortcomings of the fixed-line network; multiple service providers operate nationally; combined fixed-line and mobile-cellular teledensity reached 25 per 100 persons in 2006

international: country code—234; landing point for the SAT-3/WASC fiber-optic submarine cable that provides connectivity to Europe and Asia; satellite earth stations—3 Intelsat (2 Atlantic Ocean and 1 Indian Ocean)

Radio broadcast stations: AM 83, FM 36, shortwave 11 (2001)

Radios: 23.5 million (1997)

Television broadcast stations: 3 (the government controls 2 of the broadcasting stations and 15 repeater stations) (2001)

Televisions: 6.9 million (1997)

Internet country code: .ng

Internet hosts: 1,968 (2007)

Internet Service Providers (ISPs): 11 (2000)

Internet users: 8 million (2006)

TRANSPORTATION

Airports: 70 (2007)

Airports—with paved runways:

total: 36

over 3,047 m: 6

2,438 to 3,047 m: 12

1,524 to 2,437 m: 10

914 to 1,523 m: 6

under 914 m: 2 (2007)

Airports—with unpaved runways:

total: 34

1,524 to 2,437 m: 1

914 to 1,523 m: 14

under 914 m: 19 (2007)

Heliports: 2 (2007)

Pipelines: condensate 124 km; gas 3,071 km; liquid petroleum gas 156 km; oil 4,347 km; refined products 3,949 km (2007)

Railways:

total: 3,505 km

narrow gauge: 3,505 km 1.067-m gauge (2006)

Roadways:

total: 194,394 km

paved: 60,068 km

unpaved: 134,326 km (1999)

Waterways: 8,600 km (Niger and Benue rivers and smaller rivers and creeks) (2007)

Merchant marine:

total: 55 ships (1000 GRT or over) 284,400 GRT/483,316 DWT

by type: cargo 5, chemical tanker 8, combination ore/oil 1, liquefied gas 1, passenger/cargo 1, petroleum tanker 37, specialized tanker 2

foreign-owned: 3 (Norway 1, Singapore 1, Spain 1)

registered in other countries: 23 (Bahamas 2, Bermuda 11, Cambodia 2, Panama 6, Poland 1, Seychelles 1, unknown 2) (2007)

Ports and terminals: Bonny Inshore Terminal, Calabar, Lagos

MILITARY

Military branches: Nigerian Armed Forces: Army, Navy, Air Force (2008)

Military service age and obligation: 18 years of age for voluntary military service (2007)

Manpower available for military service:

males age 16–49: 31,929,204

females age 16–49: 30,638,979 (2008 est.)

Manpower fit for military service:

males age 16–49: 18,556,755

females age 16–49: 17,288,225 (2008 est.)

Manpower reaching militarily significant age annually:

males age 16–49: 1,533,974

females age 16–49: 1,509,619 (2008 est.)

Military expenditures—percent of GDP: 1.5% (2006)

TRANSNATIONAL ISSUES

Disputes—international: Joint Border Commission with Cameroon reviewed 2002 ICJ ruling on the entire boundary and bilaterally resolved differences, including June 2006 Greentree Agreement that immediately cedes sovereignty of the Bakassi Peninsula to Cameroon with a phase-out of Nigerian control within two years while resolving patriation issues; the ICJ ruled on an equidistance settlement of Cameroon-Equatorial Guinea-Nigeria maritime boundary in the Gulf of Guinea, but imprecisely defined coordinates in the ICJ decision and a sovereignty dispute between Equatorial Guinea and Cameroon over an island at the mouth of the Ntem River all contribute to the delay in implementation; only Nigeria and Cameroon have heeded the Lake Chad Commission's admonition to ratify the delimitation treaty which also includes the Chad-Niger and Niger-Nigeria boundaries

Refugees and internally displaced persons: *refugees (country of origin):* 5,778 (Liberia) *IDPs:* undetermined (communal violence between Christians and Muslims since President OBASANJO's election in 1999; displacement is mostly short-term) (2007)

Illicit drugs: a transit point for heroin and cocaine intended for European, East Asian, and North American markets; consumer of amphetamines; safe haven for Nigerian narcotraffickers operating worldwide; major money-laundering center; massive corruption and criminal activity; Nigeria has improved some anti-money-laundering controls, resulting in its removal from the Financial Action Task Force's (FATF's) Noncooperative Countries and Territories List in June 2006; Nigeria's anti-money-laundering regime continues to be monitored by FATF

NIUE

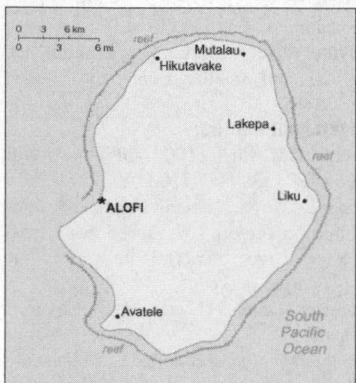

INTRODUCTION

Background: Niue's remoteness, as well as cultural and linguistic differences between its Polynesian inhabitants and those of the rest of the Cook Islands, have caused it to be separately administered. The population of the island continues to drop (from a peak of 5,200 in 1966 to an estimated 1,444 in 2008), with substantial emigration to New Zealand, 2,400 km to the southwest.

GEOGRAPHY

Location: Oceania, island in the South Pacific Ocean, east of Tonga
Geographic coordinates: 19 02 S, 169 52 W
Map references: Oceania
Area:
total: 260 sq km
land: 260 sq km
water: 0 sq km
Area—comparative: 1.5 times the size of Washington, DC
Land boundaries: 0 km
Coastline: 64 km
Maritime claims:
territorial sea: 12 nm
exclusive economic zone: 200 nm

Climate: tropical; modified by southeast trade winds
Terrain: steep limestone cliffs along coast, central plateau
Elevation extremes:
lowest point: Pacific Ocean 0 m
highest point: unnamed location near Mutalau settlement 68 m
Natural resources: fish, arable land
Land use:
arable land: 11.54%
permanent crops: 15.38%
other: 73.08% (2005)
Irrigated land: NA
Natural hazards: typhoons
Environment—current issues: increasing attention to conservationist practices to counter loss of soil fertility from traditional slash and burn agriculture
Environment—international agreements: *party to:* Biodiversity, Climate Change, Climate Change-Kyoto Protocol, Desertification, Law of the Sea *signed, but not ratified:* none of the selected agreements
Geography—note: one of world's largest coral islands

PEOPLE

Population: 1,444 (July 2008 est.)
Age structure:
0–14 years: NA
15–64 years: NA
65 years and over: NA
Population growth rate: -0.032% (2008 est.)
Birth rate: NA
Death rate: NA (2008 est.)
Net migration rate: NA
Sex ratio:
NA
Infant mortality rate:
total: NA
male: NA
female: NA (2008 est.)

Life expectancy at birth:
total population: NA
male: NA
female: NA (2008 est.)
Total fertility rate: NA (2008 est.)
HIV/AIDS—adult prevalence rate: NA
HIV/AIDS—people living with HIV/AIDS: NA
HIV/AIDS—deaths: NA
Nationality:
noun: Niuean(s)
adjective: Niuean
Ethnic groups: Niuen 78.2%, Pacific islander 10.2%, European 4.5%, mixed 3.9%, Asian 0.2%, unspecified 3% (2001 census)
Religions: Ekalesia Niue (Niuean Church—a Protestant church closely related to the London Missionary Society) 61.1%, Latter-Day Saints 8.8%, Roman Catholic 7.2%, Jehovah's Witnesses 2.4%, Seventh-Day Adventist 1.4%, other 8.4%, unspecified 8.7%, none 1.9% (2001 census)
Languages: Niuean, a Polynesian language closely related to Tongan and Samoan; English
Literacy:
definition: NA
total population: 95%
male: NA
female: NA

GOVERNMENT

Country name:
conventional long form: none
conventional short form: Niue
note: pronunciation falls between nyu-way and new-way, but not like new-wee
former: Savage Island
Dependency status: self-governing in free association with New Zealand since 1974; Niue fully responsible for internal affairs; New Zealand retains responsibility for external affairs and defense; however, these responsibilities confer no

rights of control and are only exercised at the request of the Government of Niue

Government type: self-governing parliamentary democracy

Capital: *name:* Alofi

geographic coordinates: 19 01 S, 169 55 W

time difference: UTC-11 (6 hours behind Washington, DC during Standard Time)

Administrative divisions: none; note—there are no first-order administrative divisions as defined by the US Government, but there are 14 villages at the second order

Independence: on 19 October 1974, Niue became a self-governing parliamentary government in free association with New Zealand

National holiday: Waitangi Day (Treaty of Waitangi established British sovereignty over New Zealand), 6 February (1840)

Constitution: 19 October 1974 (Niue Constitution Act)

Legal system: English common law; note—Niue is self-governing, with the power to make its own laws

Suffrage: 18 years of age; universal

Executive branch:

chief of state: Queen ELIZABETH II (since 6 February 1952); represented by Governor General of New Zealand Anand SATYANAND (since 23 August 2006); the UK and New Zealand are represented by New Zealand High Commissioner John BRYAN (since May 2000)

head of government: Premier Young VIVIAN (since 1 May 2002)

cabinet: Cabinet consists of the premier and three ministers

elections: the monarch is hereditary; premier elected by the Legislative Assembly for a three-year term; election last held 12 May 2005 (next to be held in May 2008)

election results: Young VIVIAN reelected premier; percent of Legislative Assembly vote—Young VIVIAN 85%, O'Love JACOBSEN 15%

Legislative branch: unicameral Legislative Assembly (20 seats; members elected by popular vote to serve three-year terms; six elected from a common roll and 14 are village representatives)

elections: last held 30 April 2005 (next to be held in April 2008)

election results: percent of vote by party—NA; seats by party—NA

Judicial branch: Supreme Court of New Zealand; High Court of Niue

Political parties and leaders: Alliance of Independents or AI; Niue People's Action Party or NPP [Young VIVIAN]

Political pressure groups and leaders: NA

International organization participation: ACP, FAO, IFAD, OPCW, PIF, Sparteca, SPC, UNESCO, UPU, WHO, WMO

Diplomatic representation in the US: none (self-governing territory in free association with New Zealand)

Diplomatic representation from the US: none (self-governing territory in free association with New Zealand)

Flag description: yellow with the flag of the UK in the upper hoist-side quadrant; the flag of the UK bears five yellow five-pointed stars—a large star on a blue disk in the center and a smaller star on each arm of the bold red cross

ECONOMY

Economy—overview: The economy suffers from the typical Pacific island problems of geographic isolation, few resources, and a small population. Government expenditures regularly exceed revenues, and the shortfall is made up by critically needed grants from New Zealand that are used to pay wages to public employees. Niue has cut government expenditures by reducing the public service by almost half. The agricultural sector consists mainly of subsistence gardening, although some cash crops are grown for export. Industry consists primarily of small factories to process passion fruit, lime oil, honey, and coconut cream. The sale of postage stamps to foreign collectors is an important source of revenue. The island in recent years has suffered a serious loss of population because of emigration to New Zealand. Efforts to increase GDP include the promotion of tourism and a financial services industry, although the International Banking Repeal Act of 2002 resulted in the termination of all offshore banking licenses. Economic aid from New Zealand in 2002 was US$2.6 million. Niue suffered a devastating typhoon in January 2004, which decimated nascent economic programs. While in the process of rebuilding, Niue has been dependent on foreign aid.

GDP (purchasing power parity): $7.6 million (2000 est.)

GDP (official exchange rate): $10.01 million (2003)

GDP—real growth rate: 6.2% (2003 est.)

GDP—per capita (PPP): $5,800 (2003 est.)

GDP—composition by sector:

agriculture: 23.5%

industry: 26.9%

services: 49.5% (2003)

Labor force: 663 (2001)

Labor force—by occupation: *note:* most work on family plantations; paid work

exists only in government service, small industry, and the Niue Development Board

Unemployment rate: 12% (2001)

Population below poverty line: NA%

Household income or consumption by percentage share:

lowest 10%: NA%

highest 10%: NA%

Inflation rate (consumer prices): 4% (2005)

Budget:

revenues: $15.07 million

expenditures: $16.33 million (FY0405)

Agriculture—products: coconuts, passion fruit, honey, limes, taro, yams, cassava (tapioca), sweet potatoes; pigs, poultry, beef cattle

Industries: tourism, handicrafts, food processing

Industrial production growth rate: NA%

Electricity—production: 3 million kWh (2005)

Electricity—production by source:

fossil fuel: 100%

hydro: 0%

nuclear: 0%

other: 0% (2001)

Electricity—consumption: 2.79 million kWh (2005)

Electricity—exports: 0 kWh (2005)

Electricity—imports: 0 kWh (2005)

Oil—production: 0 bbl/day (2005 est.)

Oil—consumption: 20 bbl/day (2005 est.)

Oil—exports: 0 bbl/day (2004)

Oil—imports: 20.38 bbl/day (2004)

Oil—proved reserves: 0 bbl (1 January 2006 est.)

Natural gas—production: 0 cu m (2005 est.)

Natural gas—consumption: 0 cu m (2005 est.)

Natural gas—exports: 0 cu m (2005 est.)

Natural gas—imports: 0 cu m (2005)

Natural gas—proved reserves: 0 cu m (1 January 2006 est.)

Exports: $201,400 (2004)

Exports—commodities: canned coconut cream, copra, honey, vanilla, passion fruit products, pawpaws, root crops, limes, footballs, stamps, handicrafts

Exports—partners: New Zealand mainly, Fiji, Cook Islands, Australia (2006)

Imports: $9.038 million (2004)

Imports—commodities: food, live animals, manufactured goods, machinery, fuels, lubricants, chemicals, drugs

Imports—partners: New Zealand mainly, Fiji, Japan, Samoa, Australia, US (2006)

Economic aid—recipient: $2.6 million from New Zealand (2002)
Debt—external: $418,000 (2002 est.)
Currency (code): New Zealand dollar (NZD)
Currency code: NZD
Exchange rates: New Zealand dollars per US dollar—1.3811 (2007), 1.5408 (2006), 1.4203 (2005), 1.5087 (2004), 1.7221 (2003)
Fiscal year: 1 April—31 March

COMMUNICATIONS

Telephones—main lines in use: 1,100 (2002 est.)
Telephones—mobile cellular: 400 (2002)

Telephone system:
domestic: single-line telephone system connects all villages on island
international: country code—683 (2001)
Radio broadcast stations: AM 1, FM 1, shortwave 0 (1998)
Radios: 1,000 (1997)
Television broadcast stations: 1 (1997)
Televisions: NA
Internet country code: .nu
Internet Service Providers (ISPs): 1 (2000)
Internet users: 900 (2002)

TRANSPORTATION

Airports: 1 (2007)
Airports—with paved runways:

total: 1
1,524 to 2,437 m: 1 (2007)
Roadways:
total: 234 km
paved: 86 km
unpaved: 148 km (2001)
Ports and terminals: none; offshore anchorage only

MILITARY

Military branches: no regular indigenous military forces; Police Force
Military—note: defense is the responsibility of New Zealand

TRANSNATIONAL ISSUES

Disputes—international: none

NORFOLK ISLAND

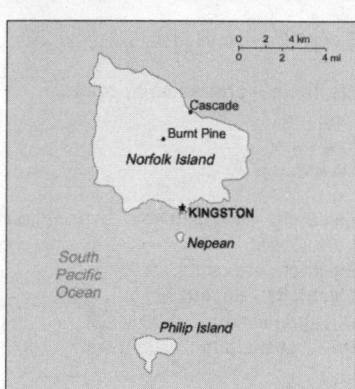

INTRODUCTION

Background: Two British attempts at establishing the island as a penal colony (1788–1814 and 1825–55) were ultimately abandoned. In 1856, the island was resettled by Pitcairn Islanders, descendants of the Bounty mutineers and their Tahitian companions.

GEOGRAPHY

Location: Oceania, island in the South Pacific Ocean, east of Australia
Geographic coordinates: 29 02 S, 167 57 E
Map references: Oceania
Area:
total: 34.6 sq km
land: 34.6 sq km
water: 0 sq km
Area—comparative: about 0.2 times the size of Washington, DC
Land boundaries: 0 km
Coastline: 32 km
Maritime claims:
territorial sea: 12 nm
exclusive fishing zone: 200 nm

Climate: subtropical; mild, little seasonal temperature variation
Terrain: volcanic formation with mostly rolling plains
Elevation extremes:
lowest point: Pacific Ocean 0 m
highest point: Mount Bates 319 m
Natural resources: fish
Land use:
arable land: 0%
permanent crops: 0%
other: 100% (2005)
Irrigated land: NA
Natural hazards: typhoons (especially May to July)
Environment—current issues: NA
Geography—note: most of the 32 km coastline consists of almost inaccessible cliffs, but the land slopes down to the sea in one small southern area on Sydney Bay, where the capital of Kingston is situated

PEOPLE

Population: 2,128 (July 2008 est.)
Age structure:
0–14 years: 20.2%
15–64 years: 63.9%
65 years and over: 15.9% (2007 est.)
Population growth rate: 0.006% (2008 est.)
Birth rate: NA
Death rate: NA (2008 est.)
Net migration rate: NA
Sex ratio:
NA
Infant mortality rate: *total:* NA
male: NA
female: NA (2008 est.)
Life expectancy at birth:
total population: NA
male: NA
female: NA (2008 est.)

Total fertility rate: NA (2008 est.)
HIV/AIDS—adult prevalence rate: NA
HIV/AIDS—people living with HIV/AIDS: NA
HIV/AIDS—deaths: NA
Nationality:
noun: Norfolk Islander(s)
adjective: Norfolk Islander(s)
Ethnic groups: descendants of the Bounty mutineers, Australian, New Zealander, Polynesian
Religions: Anglican 34.9%, Roman Catholic 11.7%, Uniting Church in Australia 11.2%, Seventh-Day Adventist 2.8%, Australian Christian 2.4%, Jehovah's Witness 0.9%, other 2.7%, unspecified 15.2%, none 18.1% (2001 census)
Languages: English (official), Norfolk—a mixture of 18th century English and ancient Tahitian
Literacy:
NA

GOVERNMENT

Country name:
conventional long form: Territory of Norfolk Island
conventional short form: Norfolk Island
Dependency status: self governing territory of Australia; administered from Canberra by the Australian Attorney-General's Department
Government type: NA
Capital: *name:* Kingston
geographic coordinates: 29 03 S, 167 58 E
time difference: UTC+11.5 (16.5 hours ahead of Washington, DC during Standard Time)
Administrative divisions: none (territory of Australia)
Independence: none (territory of Australia)

National holiday: Bounty Day (commemorates the arrival of Pitcairn Islanders), 8 June (1856)

Constitution: Norfolk Island Act of 1979, as amended in 2005

Legal system: based on the laws of Australia, local ordinances and acts; English common law applies in matters not covered by either Australian or Norfolk Island law

Suffrage: 18 years of age; universal

Executive branch:

chief of state: Queen ELIZABETH II (since 6 February 1952); represented by the Australian governor general

head of government: Acting Administrator Owen WALSH (since October 2007)

cabinet: Executive Council is made up of four of the nine members of the Legislative Assembly; the council devises government policy and acts as an advisor to the administrator

elections: the monarch is hereditary; administrator appointed by the governor general of Australia and represents the monarch and Australia

Legislative branch: unicameral Legislative Assembly (9 seats; members elected by electors who have nine equal votes each but only four votes can be given to any one candidate; to serve three-year terms)

elections: last held 21 March 2007 (next to be held by 28 March 2010)

election results: seats—independents 9 (note—no political parties)

Judicial branch: Supreme Court; Court of Petty Sessions

Political parties and leaders: none

Political pressure groups and leaders: none

International organization participation: UPU

Diplomatic representation in the US: none (territory of Australia)

Diplomatic representation from the US: none (territory of Australia)

Flag description: three vertical bands of green (hoist side), white, and green with a large green Norfolk Island pine tree centered in the slightly wider white band

ECONOMY

Economy—overview: Tourism, the primary economic activity, has steadily increased over the years and has brought a level of prosperity unusual among inhabitants of the Pacific islands. The agricultural sector has become self-sufficient in the production of beef, poultry, and eggs.

GDP (purchasing power parity): $NA

Labor force: NA

Labor force—by occupation:

agriculture: 10%

industry and services: 90%

Budget:

revenues: $4.6 million

expenditures: $4.8 million (FY99/00)

Agriculture—products: Norfolk Island pine seed, Kentia palm seed, cereals, vegetables, fruit; cattle, poultry

Industries: tourism, light industry, ready mixed concrete

Electricity—production: NA kWh

Electricity—production by source:

fossil fuel: 0%

hydro: 0%

nuclear: 0%

other: 0% (2002)

Electricity—consumption: NA kWh

Exports: $1.5 million f.o.b. (FY91/92)

Exports—commodities: postage stamps, seeds of the Norfolk Island pine and Kentia palm, small quantities of avocados

Exports—partners: Australia, other Pacific island countries, NZ, Asia, Europe (2006)

Imports: $17.9 million c.i.f. (FY91/92)

Imports—commodities: NA

Imports—partners: Australia, other Pacific island countries, NZ, Asia, Europe (2006)

Economic aid—recipient: $NA

Debt—external: $NA

Currency (code): Australian dollar (AUD)

Currency code: AUD

Exchange rates: Australian dollars per US dollar—1.2137 (2007), 1.3285 (2006), 1.3095 (2005), 1.3598 (2004), 1.5419 (2003)

Fiscal year: 1 July—30 June

COMMUNICATIONS

Telephones—main lines in use: 2,532; note—a mix of analog (2500) and digital (32) circuits (2004)

Telephones—mobile cellular: 0; note—proposed cellular service disallowed in August 2002 island referendum (2002)

Telephone system:

general assessment: adequate

domestic: free local calls

international: country code—672; undersea coaxial cable links with Australia and New Zealand; satellite earth station—1

Radio broadcast stations: AM 1, FM 3, shortwave 0 (2005)

Radios: 2,500 (1996)

Television broadcast stations: 1 (local programming station plus 2 repeaters that air Australian programs by satellite) (2005)

Televisions: 1,200 (1996)

Internet country code: .nf

Internet hosts: 96 (2007)

Internet Service Providers (ISPs): 2 (2000)

Internet users: 700 (2002 est.)

TRANSPORTATION

Airports: 1 (2007)

Airports—with paved runways:

total: 1

1,524 to 2,437 m: 1 (2007)

Roadways:

total: 80 km

paved: 53 km

unpaved: 27 km (2002)

Ports and terminals: none; loading jetties at Kingston and Cascade

MILITARY

Military—note: defense is the responsibility of Australia

TRANSNATIONAL ISSUES

Disputes—international: none

NORTHERN MARIANA ISLANDS

INTRODUCTION

Background: Under US administration as part of the UN Trust Territory of the Pacific, the people of the Northern Mariana Islands decided in the 1970s not to seek independence but instead to forge closer links with the US. Negotiations for territorial status began in 1972. A covenant to establish a commonwealth in political union with the US was approved in 1975, and came into force on 24 March 1976. A new government and constitution went into effect in 1978.

GEOGRAPHY

Location: Oceania, islands in the North Pacific Ocean, about three-quarters of the way from Hawaii to the Philippines

Geographic coordinates: 15 12 N, 145 45 E

Map references: Oceania

Area:

total: 477 sq km

land: 477 sq km

water: 0 sq km

note: includes 14 islands including Saipan, Rota, and Tinian

Area—comparative: 2.5 times the size of Washington, DC

Land boundaries: 0 km

Coastline: 1,482 km

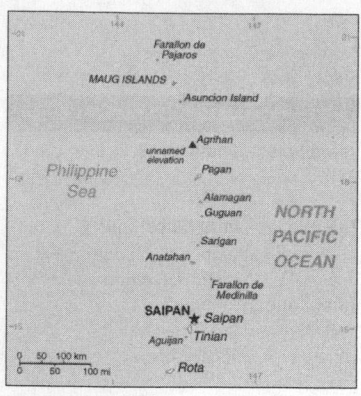

Maritime claims:

territorial sea: 12 nm

exclusive economic zone: 200 nm

Climate: tropical marine; moderated by northeast trade winds, little seasonal temperature variation; dry season December to June, rainy season July to October

Terrain: southern islands are limestone with level terraces and fringing coral reefs; northern islands are volcanic

Elevation extremes:

lowest point: Pacific Ocean 0 m

highest point: unnamed location on Agrihan 965 m

Natural resources: arable land, fish

Land use:

arable land: 13.04%

permanent crops: 4.35%

other: 82.61% (2005)

Irrigated land: NA

Natural hazards: active volcanoes on Pagan and Agrihan; typhoons (especially August to November)

Environment—current issues: contamination of groundwater on Saipan may contribute to disease; clean-up of landfill; protection of endangered species conflicts with development

Geography—note: strategic location in the North Pacific Ocean

PEOPLE

Population: 86,616 (July 2008 est.)

Age structure:

0–14 years: 18.4% (male 8,342/female 7,594)

15–64 years: 79.9% (male 27,996/female 41,245)

65 years and over: 1.7% (male 740/female 699) (2008 est.)

Median age:

total: 29.9 years

male: 32 years

female: 28.9 years (2008 est.)

Population growth rate: 2.377% (2008 est.)

Birth rate: 19.04 births/1,000 population (2008 est.)

Death rate: 2.31 deaths/1,000 population (2008 est.)

Net migration rate: 7.04 migrant(s)/1,000 population (2008 est.)

Sex ratio:

at birth: 1.06 male(s)/female

under 15 years: 1.1 male(s)/female

15–64 years: 0.68 male(s)/female

65 years and over: 1.06 male(s)/female

total population: 0.75 male(s)/female (2008 est.)

Infant mortality rate:

total: 6.72 deaths/1,000 live births

male: 6.68 deaths/1,000 live births

female: 6.76 deaths/1,000 live births (2008 est.)

Life expectancy at birth:

total population: 76.5 years

male: 73.89 years

female: 79.26 years (2008 est.)

Total fertility rate: 1.18 children born/woman (2008 est.)

HIV/AIDS—adult prevalence rate: NA

HIV/AIDS—people living with HIV/AIDS: NA

HIV/AIDS—deaths: NA

Nationality:

noun: NA (US citizens)

adjective: NA

Ethnic groups: Asian 56.3%, Pacific islander 36.3%, Caucasian 1.8%, other 0.8%, mixed 4.8% (2000 census)

Religions: Christian (Roman Catholic majority, although traditional beliefs and taboos may still be found)

Languages: Philippine languages 24.4%, Chinese 23.4%, Chamorro 22.4%, English 10.8%, other Pacific island languages 9.5%, other 9.6% (2000 census)

Literacy:

definition: age 15 and over can read and write

total population: 97%

male: 97%

female: 96% (1980 est.)

GOVERNMENT

Country name:

conventional long form: Commonwealth of the Northern Mariana Islands

conventional short form: Northern Mariana Islands

abbreviation: CNMI

former: Trust Territory of the Pacific Islands, Mariana Islands District

Dependency status: commonwealth in political union with the US; federal funds to the Commonwealth administered by the US Department of the Interior, Office of Insular Affairs

Government type: commonwealth; self-governing with locally elected governor, lieutenant governor, and legislature

Capital: *name:* Saipan

geographic coordinates: 15 12 N, 145 45 E

time difference: UTC+10 (15 hours ahead of Washington, DC during Standard Time)

Administrative divisions: none (commonwealth in political union with the US); there are no first-order administrative divisions as defined by the US Government, but there are four municipalities at the second order: Northern Islands, Rota, Saipan, Tinian

Independence: none (commonwealth in political union with the US)

National holiday: Commonwealth Day, 8 January (1978)

Constitution: Constitution of the Commonwealth of the Northern Mariana Islands effective 1 January 1978; Covenant Agreement fully effective 4 November 1986

Legal system: based on US system, except for customs, wages, immigration laws, and taxation

Suffrage: 18 years of age; universal; indigenous inhabitants are US citizens but do not vote in US presidential elections

Executive branch:

chief of state: President George W. BUSH of the US (since 20 January 2001); Vice President Richard B. CHENEY (since 20 January 2001)

head of government: Governor Benigno R. FITIAL (since 9 January 2006); Lieutenant Governor Timothy P. VILLAGOMEZ (since 9 January 2006)

cabinet: the cabinet consists of the heads of the 10 principal departments under the executive branch who are appointed by the governor with the advice and consent of the Senate; other members include Special Assistants to the governor and office heads appointed by and reporting directly to the governor

elections: under the US Constitution, residents of unincorporated territories, such as the Commonwealth of the Northern Mariana Islands, do not vote in elections for US president and vice president; however, they may vote in the Democratic and Republican presidential primary elections; governor and lieutenant governor elected on the same ticket by popular vote for four-year terms (eligible for a second term); election last held 5 November 2005 (next to be held in November 2009)

election results: Benigno R. FITIAL elected governor in a four-way race; percent of vote—Benigno R. FITIAL 28.07%, Heinz HOFSCHNEIDER 27.34%, Juan BABAUTA 26.6%, Froilan TENORIO 17.99%

Legislative branch: bicameral Legislature consists of the Senate (9 seats; members are elected by popular vote to serve four-year staggered terms) and the House of Representatives (20 seats; members are elected by popular vote to serve two-year terms)

elections: Senate—last held 3 November 2007 (next to be held in November 2009); House of Representatives—last held 3 November 2007 (next to be held in November 2009)

election results: Senate—percent of vote by party—NA; seats by party—Covenant Party 3, Republican Party 3, Democratic Party 1, independents 2; House of Representatives—percent of vote by party—NA; seats by party—Republican Party 12, Covenant Party 4, Democratic Party 1, independents 3

note: the Northern Mariana Islands does not have a nonvoting delegate in the US Congress; instead, it has an elected official or "resident representative" in Washington, DC; seats by party—Republican Party 1 (Pedro A. TENORIO)

Judicial branch: Commonwealth Supreme Court; Superior Court; Federal District Court

Political parties and leaders: Covenant Party [Benigno R. FITIAL]; Democratic Party [Dr. Carlos S. CAMACHO]; Republican Party [Juan S. REYES]

Political pressure groups and leaders: NA

International organization participation: Interpol (subbureau), SPC, UPU

Flag description: blue, with a white, five-pointed star superimposed on the gray silhouette of a latte stone (a traditional foundation stone used in building) in the center, surrounded by a wreath

ECONOMY

Economy—overview: The economy benefits substantially from financial assistance from the US. The rate of funding has declined as locally generated government revenues have grown. The key tourist industry employs about 50% of the work force and accounts for roughly one-fourth of GDP. Japanese tourists predominate. Annual tourist entries have exceeded one-half million in recent years, but financial difficulties in Japan have caused a temporary slowdown. The agricultural sector is made up

of cattle ranches and small farms producing coconuts, breadfruit, tomatoes, and melons. Garment production is by far the most important industry with the employment of 17,500 mostly Chinese workers and sizable shipments to the US under duty and quota exemptions.

GDP (purchasing power parity): $900 million

note: GDP estimate includes US subsidy (2000 est.)

GDP (official exchange rate): $633.4 million (2000)

GDP—real growth rate: NA%

GDP—per capita (PPP): $12,500 (2000 est.)

GDP—composition by sector:
agriculture: NA%
industry: NA%
services: NA%

Labor force: 44,470 total indigenous labor force; 2,699 unemployed; 28,717 foreign workers (2000)

Labor force—by occupation: *agriculture:* NA%
industry: NA%
services: NA%

Unemployment rate: 3.9% (2001)

Population below poverty line: NA%

Household income or consumption by percentage share:
lowest 10%: NA%
highest 10%: NA%

Inflation rate (consumer prices): -0.8% (2000)

Budget:
revenues: $193 million
expenditures: $223 million (FY01/02 est.)

Agriculture—products: coconuts, fruits, vegetables; cattle

Industries: tourism, construction, garments, handicrafts

Industrial production growth rate: NA%

Electricity—production: NA kWh
Electricity—consumption: NA kWh
Electricity—exports: 0 kWh (2007 est.)
Electricity—imports: 0 kWh (2007 est.)

Exports: $NA
Exports—commodities: garments
Exports—partners: US (2006)
Imports: $214.4 million (2001)
Imports—commodities: food, construction equipment and materials, petroleum products

Imports—partners: US, Japan (2006)
Economic aid—recipient: extensive funding from US
Debt—external: $NA
Currency (code): US dollar (USD)
Currency code: USD
Exchange rates: the US dollar is used
Fiscal year: 1 October—30 September

COMMUNICATIONS

Telephones—main lines in use: 21,000 (2000)
Telephones—mobile cellular: 20,500 (2004)
Telephone system:
general assessment: NA
domestic: NA
international: country code—1-670; satellite earth stations—2 Intelsat (Pacific Ocean)
Radio broadcast stations: AM 1, FM 6, shortwave 1 (2005)
Radios: NA
Television broadcast stations: 1 (on Saipan; in addition, 2 cable services on Saipan provide varied programming from satellite networks) (2006)
Televisions: NA
Internet country code: .mp
Internet hosts: 5 (2007)
Internet Service Providers (ISPs): 1 (2001)
Internet users: 10,000 (2003)

TRANSPORTATION

Airports: 5 (2007)
Airports—with paved runways:
total: 3
2,438 to 3,047 m: 2
1,524 to 2,437 m: 1 (2007)
Airports—with unpaved runways:
total: 2
2,438 to 3,047 m: 1
under 914 m: 1 (2007)
Heliports: 1 (2007)
Roadways:
total: 536 km (2004)
Ports and terminals: Saipan, Tinian

MILITARY

Military—note: defense is the responsibility of the US

TRANSNATIONAL ISSUES

Disputes—international: none

NORWAY

INTRODUCTION

Background: Two centuries of Viking raids into Europe tapered off following the adoption of Christianity by King Olav TRYGGVASON in 994.

Conversion of the Norwegian kingdom occurred over the next several decades. In 1397, Norway was absorbed into a union with Denmark that lasted more than four centuries. In 1814, Norwegians resisted the cession of their country to

Sweden and adopted a new constitution. Sweden then invaded Norway but agreed to let Norway keep its constitution in return for accepting the union under a Swedish king. Rising nationalism throughout the 19th century led to a

1905 referendum granting Norway independence. Although Norway remained neutral in World War I, it suffered heavy losses to its shipping. Norway proclaimed its neutrality at the outset of World War II, but was nonetheless occupied for five years by Nazi Germany (1940–45). In 1949, neutrality was abandoned and Norway became a member of NATO. Discovery of oil and gas in adjacent waters in the late 1960s boosted Norway's economic fortunes. The current focus is on containing spending on the extensive welfare system and planning for the time when petroleum reserves are depleted. In referenda held in 1972 and 1994, Norway rejected joining the EU.

GEOGRAPHY

Location: Northern Europe, bordering the North Sea and the North Atlantic Ocean, west of Sweden
Geographic coordinates: 62 00 N, 10 00 E
Map references: Europe
Area:
total: 323,802 sq km
land: 307,442 sq km
water: 16,360 sq km
Area—comparative: slightly larger than New Mexico
Land boundaries: total: 2,542 km

border countries: Finland 727 km, Sweden 1,619 km, Russia 196 km
Coastline: 25,148 km (includes mainland 2,650 km, as well as long fjords, numerous small islands, and minor indentations 22,498 km; length of island coastlines 58,133 km)
Maritime claims:
territorial sea: 12 nm
contiguous zone: 10 nm
exclusive economic zone: 200 nm
continental shelf: 200 nm
Climate: temperate along coast, modified by North Atlantic Current; colder interior with increased precipitation and colder summers; rainy year-round on west coast
Terrain: glaciated; mostly high plateaus and rugged mountains broken by fertile valleys; small, scattered plains; coastline deeply indented by fjords; arctic tundra in north
Elevation extremes:
lowest point: Norwegian Sea 0 m
highest point: Galdhopiggen 2,469 m
Natural resources: petroleum, natural gas, iron ore, copper, lead, zinc, titanium, pyrites, nickel, fish, timber, hydropower
Land use:
arable land: 2.7%
permanent crops: 0%
other: 97.3% (2005)
Irrigated land: 1,270 sq km (2003)
Total renewable water resources: 381.4 cu km (2005)
Freshwater withdrawal (domestic/industrial/agricultural): *total:* 2.4 cu km/yr (23%/67%/10%)
per capita: 519 cu m/yr (1996)
Natural hazards: rockslides, avalanches
Environment—current issues: water pollution; acid rain damaging forests and adversely affecting lakes, threatening fish stocks; air pollution from vehicle emissions
Environment—international agreements: *party to:* Air Pollution, Air Pollution-Nitrogen Oxides, Air Pollution-Persistent Organic Pollutants, Air Pollution-Sulfur 85, Air Pollution-Sulfur 94, Air Pollution-Volatile Organic Compounds, Antarctic-Environmental Protocol, Antarctic-Marine Living Resources, Antarctic Seals, Antarctic Treaty, Biodiversity, Climate Change, Climate Change-Kyoto Protocol, Desertification, Endangered Species, Environmental Modification, Hazardous Wastes, Law of the Sea, Marine Dumping, Ozone Layer Protection, Ship Pollution, Tropical Timber 83, Tropical Timber 94, Wetlands, Whaling
signed, but not ratified: none of the selected agreements

Geography—note: about two-thirds mountains; some 50,000 islands off its much indented coastline; strategic location adjacent to sea lanes and air routes in North Atlantic; one of most rugged and longest coastlines in the world

PEOPLE

Population: *4,644,457* (July 2008 est.)
Age structure:
0–14 years: 18.8% (male 446,146/female 426,166)
15–64 years: 66.2% (male 1,559,750/female 1,516,217)
65 years and over: 15% (male 297,175/female 399,003) (2008 est.)
Median age:
total: 39 years
male: 38.2 years
female: 39.9 years (2008 est.)
Population growth rate: 0.35% (2008 est.)
Birth rate: 11.12 births/1,000 population (2008 est.)
Death rate: 9.33 deaths/1,000 population (2008 est.)
Net migration rate: 1.71 migrant(s)/1,000 population (2008 est.)
Sex ratio:
at birth: 1.05 male(s)/female
under 15 years: 1.05 male(s)/female
15–64 years: 1.03 male(s)/female
65 years and over: 0.74 male(s)/female
total population: 0.98 male(s)/female (2008 est.)
Infant mortality rate:
total: 3.61 deaths/1,000 live births
male: 3.96 deaths/1,000 live births
female: 3.24 deaths/1,000 live births (2008 est.)
Life expectancy at birth:
total population: 79.81 years
male: 77.16 years
female: 82.6 years (2008 est.)
Total fertility rate: 1.78 children born/woman (2008 est.)
HIV/AIDS—adult prevalence rate: 0.1% (2001 est.)
HIV/AIDS—people living with HIV/AIDS: 2,100 (2001 est.)
HIV/AIDS—deaths: fewer than 100 (2003 est.)
Nationality:
noun: Norwegian(s)
adjective: Norwegian
Ethnic groups: Norwegian, Sami 20,000
Religions: Church of Norway 85.7%, Pentecostal 1%, Roman Catholic 1%, other Christian 2.4%, Muslim 1.8%, other 8.1% (2004)
Languages: Bokmal Norwegian (official), Nynorsk Norwegian (official), small Sami- and Finnish-speaking minorities; note—Sami is official in six municipalities

Literacy:
definition: age 15 and over can read and write
total population: 100%
male: 100%
female: 100%

GOVERNMENT

Country name:
conventional long form: Kingdom of Norway
conventional short form: Norway
local long form: Kongeriket Norge
local short form: Norge
Government type: constitutional monarchy
Capital: *name:* Oslo
geographic coordinates: 59 55 N, 10 45 E
time difference: UTC+1 (6 hours ahead of Washington, DC during Standard Time)
daylight saving time: +1hr, begins last Sunday in March; ends last Sunday in October
Administrative divisions: 19 counties (fylker, singular—fylke); Akershus, Aust-Agder, Buskerud, Finnmark, Hedmark, Hordaland, More og Romsdal, Nordland, Nord-Trondelag, Oppland, Oslo, Ostfold, Rogaland, Sogn og Fjordane, Sor-Trondelag, Telemark, Troms, Vest-Agder, Vestfold
Dependent areas: Bouvet Island, Jan Mayen, Svalbard
Independence: 7 June 1905 (Norway declared the union with Sweden dissolved); 26 October 1905 (Sweden agreed to the repeal of the union)
National holiday: Constitution Day, 17 May (1814)
Constitution: 17 May 1814; amended many times
Legal system: mixture of customary law, civil law system, and common law traditions; Supreme Court renders advisory opinions to legislature when asked; accepts compulsory ICJ jurisdiction with reservations
Suffrage: 18 years of age; universal
Executive branch:
chief of state: King HARALD V (since 17 January 1991); Heir Apparent Crown Prince HAAKON MAGNUS, son of the monarch (born 20 July 1973)
head of government: Prime Minister Jens STOLTENBERG (since 17 October 2005)
cabinet: State Council appointed by the monarch with the approval of parliament
elections: the monarch is hereditary; following parliamentary elections, the leader of the majority party or the leader of the majority coalition is usually appointed prime minister by the monarch with the approval of the parliament

Legislative branch: modified unicameral Parliament or Storting (169 seats; members are elected by popular vote by proportional representation to serve four-year terms); note—in 2009 the number of seats will change to 165
elections: last held 12 September 2005 (next to be held in September 2009)
election results: percent of vote by party—Labor Party 32.7%, Progress Party 22.1%, Conservative Party 14.1%, Socialist Left Party 8.8%, Christian People's Party 6.8%, Center Party 6.5%, Liberal Party 5.9%, other 3.1%; seats by party—Labor Party 61, Progress Party 38, Conservative Party 23, Socialist Left Party 15, Christian People's Party 11, Center Party 11, Liberal Party 10
note: for certain purposes, the parliament divides itself into two chambers and elects one-fourth of its membership in the Lagting and three-fourths of its membership in the Odelsting
Judicial branch: Supreme Court or Hoyesterett (justices appointed by the monarch)
Political parties and leaders: Center Party [Aslaug Marie HAGA]; Christian People's Party [Dagfinn HOYBRATEN]; Conservative Party [Erna SOLBERG]; Labor Party [Jens STOLTENBERG]; Liberal Party [Lars SPONHEIM]; Progress Party [Siv JENSEN]; Socialist Left Party [Kristin HALVORSEN]
Political pressure groups and leaders: NA
International organization participation: ADB (nonregional members), AfDB, Arctic Council, Australia Group, BIS, CBSS, CE, CERN, EAPC, EBRD, EFTA, ESA, FAO, IADB, IAEA, IBRD, ICAO, ICC, ICCt, ICRM, IDA, IEA, IFAD, IFC, IFRCS, IHO, ILO, IMF, IMO, IMSO, Interpol, IOC, IOM, IPU, ISO, ITSO, ITU, ITUC, MIGA, NAM (guest), NATO, NC, NEA, NIB, NSG, OAS (observer), OECD, OPCW, OSCE, Paris Club, PCA, Schengen Convention, UN, UNCTAD, UNESCO, UNHCR, UNIDO, UNMEE, UNMIS, UNRWA, UNTSO, UPU, WCO, WEU (associate), WHO, WIPO, WMO, WTO, ZC
Diplomatic representation in the US:
chief of mission: Ambassador Wegger C. STROMMEN
chancery: 2720 34th Street NW, Washington, DC 20008
telephone: [1] (202) 333-6000
FAX: [1] (202) 337-0870
consulate(s) general: Houston, Minneapolis, New York, San Francisco
Diplomatic representation from the US:
chief of mission: Ambassador Benson K. WHITNEY

embassy: Henrik Ibsens gate 48, 0244 Oslo; note—the embassy will move to Huseby in the near future
mailing address: PSC 69, Box 1000, APO AE 09707
telephone: [47] (22) 44 85 50
FAX: [47] (22) 44 33 63, 56 27 51
Flag description: red with a blue cross outlined in white that extends to the edges of the flag; the vertical part of the cross is shifted to the hoist side in the style of the Dannebrog (Danish flag)

ECONOMY

Economy—overview: The Norwegian economy is a prosperous bastion of welfare capitalism, featuring a combination of free market activity and government intervention. The government controls key areas, such as the vital petroleum sector, through large-scale state enterprises. The country is richly endowed with natural resources—petroleum, hydropower, fish, forests, and minerals—and is highly dependent on its oil production and international oil prices, with oil and gas accounting for one-third of exports. Only Saudi Arabia and Russia export more oil than Norway. Norway opted to stay out of the EU during a referendum in November 1994; nonetheless, as a member of the European Economic Area, it contributes sizably to the EU budget. The government has moved ahead with privatization. Although Norwegian oil production peaked in 2000, natural gas production is still rising. Norwegians realize that once their gas production peaks they will eventually face declining oil and gas revenues; accordingly, Norway has been saving its oil-and-gas-boosted budget surpluses in a Government Petroleum Fund, which is invested abroad and now is valued at more than $250 billion. After lackluster growth of less than 1% in 2002–03, GDP growth picked up to 3–5% in 2004–07, partly due to higher oil prices. Norway's economy remains buoyant. Domestic economic activity is, and will continue to be, the main driver of growth, supported by high consumer confidence and strong investment spending in the offshore oil and gas sector. Norway's record high budget surplus and upswing in the labor market in 2007 highlight the strength of its economic position going into 2008.
GDP (purchasing power parity): $247.4 billion (2007 est.)
GDP (official exchange rate): $391.5 billion (2007 est.)
GDP—real growth rate: 3.5% (2007 est.)
GDP—per capita (PPP): $53,000 (2007 est.)

GDP—composition by sector:
agriculture: 2.4%
industry: 42.9%
services: 54.7% (2007 est.)
Labor force: 2.507 million (2007 est.)
Labor force—by occupation:
agriculture: 4%
industry: 22%
services: 74% (1995)
Unemployment rate: 2.5% (2007 est.)
Population below poverty line: NA%
Household income or consumption by percentage share:
lowest 10%: 3.9%
highest 10%: 23.4% (2000)
Distribution of family income—Gini index: 28 (2005)
Inflation rate (consumer prices): 0.8% (2007 est.)
Investment (gross fixed): 20.7% of GDP (2007 est.)
Budget:
revenues: $224.2 billion
expenditures: $158 billion (2007 est.)
Public debt: 75.1% of GDP (2007 est.)
Agriculture—products: barley, wheat, potatoes; pork, beef, veal, milk; fish
Industries: petroleum and gas, food processing, shipbuilding, pulp and paper products, metals, chemicals, timber, mining, textiles, fishing
Industrial production growth rate: 1% (2007 est.)
Electricity—production: 135.8 billion kWh (2005)
Electricity—production by source:
fossil fuel: 0.4%
hydro: 99.3%
nuclear: 0%
other: 0.4% (2001)
Electricity—consumption: 113.9 billion kWh (2005)
Electricity—exports: 15.7 billion kWh (2005)
Electricity—imports: 3.652 billion kWh (2005)
Oil—production: 2.978 million bbl/day (2005 est.)
Oil—consumption: 228,400 bbl/day (2005 est.)
Oil—exports: 3.018 million bbl/day (2004)
Oil—imports: 91,930 bbl/day (2004)
Oil—proved reserves: 7.705 billion bbl (1 January 2006 est.)
Natural gas—production: 83.44 billion cu m (2005 est.)
Natural gas—consumption: 5.342 billion cu m (2005 est.)
Natural gas—exports: 78.1 billion cu m (2005 est.)
Natural gas—imports: 0 cu m (2005)
Natural gas—proved reserves: 2.288 trillion cu m (1 January 2006 est.)
Current account balance: $63.66 billion (2007 est.)

Exports: $139.4 billion f.o.b. (2007 est.)
Exports—commodities: petroleum and petroleum products, machinery and equipment, metals, chemicals, ships, fish
Exports—partners: UK 26.8%, Germany 12.3%, Netherlands 10.3%, France 8.2%, Sweden 6.4%, US 5.7% (2006)
Imports: $78.11 billion f.o.b. (2007 est.)
Imports—commodities: machinery and equipment, chemicals, metals, foodstuffs
Imports—partners: Sweden 15%, Germany 13.5%, Denmark 6.9%, UK 6.4%, China 5.7%, US 5.3%, Netherlands 4.1% (2006)
Economic aid—donor: ODA, $2.954 billion (2006)
Reserves of foreign exchange and gold: $60.84 billion (2006 est.)
Debt—external: $469.1 billion; note—Norway is a net external creditor (30 June 2007)
Stock of direct foreign investment—at home: $62.56 billion (2007 est.)
Stock of direct foreign investment—abroad: $124.7 billion (2007 est.)
Market value of publicly traded shares: $191 billion (2005)
Currency (code): Norwegian krone (NOK)
Currency code: NOK
Exchange rates: Norwegian kroner per US dollar—5.8396 (2007), 6.4117 (2006), 6.4425 (2005), 6.7408 (2004), 7.0802 (2003)
Fiscal year: calendar year

COMMUNICATIONS

Telephones—main lines in use: 2.055 million (2006)
Telephones—mobile cellular: 5.041 million (2006)
Telephone system:
general assessment: modern in all respects; one of the most advanced telecommunications networks in Europe
domestic: Norway has a domestic satellite system; moreover, the prevalence of rural areas encourages the wide use of cellular-mobile systems instead of fixed-wire systems
international: country code—47; 2 buried coaxial cable systems; submarine cables provide links to other Nordic countries and Europe; satellite earth stations—NA Eutelsat, NA Intelsat (Atlantic Ocean), and 1 Inmarsat (Atlantic and Indian Ocean regions); note—Norway shares the Inmarsat earth station with the other Nordic countries (Denmark, Finland, Iceland, and Sweden) (1999)
Radio broadcast stations: AM 5, FM at least 650, shortwave 1 (1998)
Radios: 4.03 million (1997)
Television broadcast stations: 360 (plus

2,729 repeaters) (1995)
Televisions: 2.03 million (1997)
Internet country code: .no
Internet hosts: 2.084 million (2007)
Internet Service Providers (ISPs): 13 (2000)
Internet users: 4.074 million (2006)

TRANSPORTATION

Airports: 98 (2007)
Airports—with paved runways:
total: 67
over 3,047 m: 1
2,438 to 3,047 m: 12
1,524 to 2,437 m: 12
914 to 1,523 m: 13
under 914 m: 29 (2007)
Airports—with unpaved runways:
total: 31
914 to 1,523 m: 6
under 914 m: 25 (2007)
Heliports: 1 (2007)
Pipelines: condensate 508 km; gas 6,529 km; oil 2,444 km; oil/gas/water 457 km (2007)
Railways:
total: 4,043 km
standard gauge: 4,043 km 1.435-m gauge (2,509 km electrified) (2006)
Roadways:
total: 92,513 km
paved: 71,832 km (includes 664 km of expressways)
unpaved: 20,681 km (2005)
Waterways: 1,577 km (2007)
Merchant marine:
total: 715 ships (1000 GRT or over) 16,511,659 GRT/22,299,832 DWT
by type: bulk carrier 49, cargo 151, carrier 1, chemical tanker 146, combination ore/oil 12, container 5, liquefied gas 72, passenger/cargo 122, petroleum tanker 79, refrigerated cargo 12, roll on/roll off 16, specialized tanker 1, vehicle carrier 49
foreign-owned: 174 (China 47, Cyprus 2, Denmark 26, Estonia 1, Finland 1, France 3, Germany 2, Greece 6, Hong Kong 5, Iceland 3, Italy 4, Japan 1, Lithuania 1, Monaco 5, Netherlands 1, Poland 3, Saudi Arabia 3, Singapore 1, Sweden 31, UAE 1, UK 9, US 18)
registered in other countries: 872 (Antigua and Barbuda 7, Australia 1, Bahamas 232, Barbados 35, Belize 3, Bermuda 5, Brazil 1, Canada 1, Cayman Islands 2, China 1, Comoros 1, Cook Islands 1, Cyprus 17, Denmark 1, Dominica 1, Estonia 2, Faroe Islands 4, Finland 1, France 17, Gibraltar 27, Hong Kong 30, Isle of Man 33, Liberia 42, Libya 1, Malta 71, Marshall Islands 62, Mexico 1, Netherlands 9, Netherlands Antilles 5, Nigeria 1, Panama 60, Philippines 2, Portugal 3, Singapore 125, Spain 6, St

Vincent and The Grenadines 19, Sweden 5, UK 33, US 4, unknown 2) (2007)

Ports and terminals: Bergen, Borg Havn, Haugesund, Maaloy, Mongstad, Narvik, Oslo, Sture

MILITARY

Military branches: Norwegian Army (Haeren), Royal Norwegian Navy (Kongelige Norske Sjoeforsvaret, RNoN; includes Coastal Rangers and Coast Guard (Kystvakt)), Royal Norwegian Air Force (Kongelige Norske Luftforsvaret, RNoAF), Home Guard (Heimevernet, HV) (2007)

Military service age and obligation: 18–44 years of age for male compulsory military service; 16 years of age in wartime; 17 years of age for male volunteers; 18 years of age for women; 12-month service obligation, in practice shortened to 8 to 9 months; although all males between ages of 18 and 44 are liable for service, in practice they are seldom called to duty after age 30; reserve obligation to age 35–60; 16 years of age for volunteers to the Home Guard, who serve 6-month duty tours (2006)

Manpower available for military service:
males age 16–49: 1,078,181
females age 16–49: 1,046,550 (2008 est.)

Manpower fit for military service:
males age 16–49: 888,101
females age 16–49: 862,159 (2008 est.)

Manpower reaching militarily significant age annually:
males age 16–49: 32,185
females age 16–49: 30,683 (2008 est.)

Military expenditures—percent of GDP: 1.9% (2005 est.)

TRANSNATIONAL ISSUES

Disputes—international: Norway asserts a territorial claim in Antarctica (Queen Maud Land and its continental shelf); despite dialogue, Russia and Norway continue to dispute their maritime limits in the Barents Sea and Russia's fishing rights beyond Svalbard's territorial limits within the Svalbard Treaty zone

OMAN

INTRODUCTION

Background: The inhabitants of the area of Oman have long prospered on Indian Ocean trade. In the late 18th century, a newly established sultanate in Muscat signed the first in a series of friendship treaties with Britain. Over time, Oman's dependence on British political and military advisors increased, but it never became a British colony. In 1970, QABOOS bin Said al-Said overthrew the restrictive rule of his father; he has ruled as sultan ever since. His extensive modernization program has opened the country to the outside world while preserving the longstanding close ties with the UK. Oman's moderate, independent foreign policy has sought to maintain good relations with all Middle Eastern countries.

GEOGRAPHY

Location: Middle East, bordering the Arabian Sea, Gulf of Oman, and Persian Gulf, between Yemen and UAE
Geographic coordinates: 21 00 N, 57 00 E
Map references: Middle East
Area: *total:* 212,460 sq km
land: 212,460 sq km
water: 0 sq km
Area—comparative: slightly smaller than Kansas
Land boundaries:
total: 1,374 km
border countries: Saudi Arabia 676 km, UAE 410 km, Yemen 288 km
Coastline: 2,092 km
Maritime claims:
territorial sea: 12 nm
contiguous zone: 24 nm
exclusive economic zone: 200 nm
Climate: dry desert; hot, humid along coast; hot, dry interior; strong southwest summer monsoon (May to September) in far south

Terrain: central desert plain, rugged mountains in north and south
Elevation extremes:
lowest point: Arabian Sea 0 m
highest point: Jabal Shams 2,980 m
Natural resources: petroleum, copper, asbestos, some marble, limestone, chromium, gypsum, natural gas
Land use:
arable land: 0.12%
permanent crops: 0.14%
other: 99.74% (2005)
Irrigated land: 720 sq km (2003)
Total renewable water resources: 1 cu km (1997)
Freshwater withdrawal (domestic/industrial/agricultural): *total:* 1.36 cu km/yr (7%/2%/90%)
per capita: 529 cu m/yr (2000)
Natural hazards: summer winds often raise large sandstorms and dust storms in interior; periodic droughts
Environment—current issues: rising soil salinity; beach pollution from oil spills; limited natural fresh water resources
Environment—international agreements: *party to:* Biodiversity, Climate Change, Climate Change-Kyoto Protocol, Desertification, Hazardous Wastes, Law of the Sea, Marine Dumping, Ozone Layer Protection, Ship Pollution, Whaling
signed, but not ratified: none of the selected agreements
Geography—note: strategic location on Musandam Peninsula adjacent to Strait of Hormuz, a vital transit point for world crude oil

PEOPLE

Population: 3,311,640
note: includes 577,293 non-nationals (July 2008 est.)
Age structure:
0–14 years: 42.7% (male 721,796/female 692,699)
15–64 years: 54.5% (male 1,053,040/female 752,962)
65 years and over: 2.8% (male 51,290/female 39,853) (2008 est.)
Median age:
total: 18.9 years
male: 21.3 years
female: 16.6 years (2008 est.)
Population growth rate: 3.19% (2008 est.)
Birth rate: 35.26 births/1,000 population (2008 est.)
Death rate: 3.68 deaths/1,000 population (2008 est.)
Net migration rate: 0.33 migrant(s)/1,000 population (2008 est.)

Sex ratio:
at birth: 1.05 male(s)/female
under 15 years: 1.04 male(s)/female
15–64 years: 1.4 male(s)/female
65 years and over: 1.29 male(s)/female
total population: 1.23 male(s)/female (2008 est.)
Infant mortality rate:
total: 17.45 deaths/1,000 live births
male: 19.95 deaths/1,000 live births
female: 14.83 deaths/1,000 live births (2008 est.)
Life expectancy at birth:
total population: 73.91 years
male: 71.64 years
female: 76.29 years (2008 est.)
Total fertility rate: 5.62 children born/woman (2008 est.)
HIV/AIDS—adult prevalence rate: 0.1% (2001 est.)
HIV/AIDS—people living with HIV/AIDS: 1,300 (2001 est.)
HIV/AIDS—deaths: fewer than 200 (2003 est.)
Nationality:
noun: Omani(s)
adjective: Omani
Ethnic groups: Arab, Baluchi, South Asian (Indian, Pakistani, Sri Lankan, Bangladeshi), African
Religions: Ibadhi Muslim 75%, other (includes Sunni Muslim, Shi'a Muslim, Hindu) 25%
Languages: Arabic (official), English, Baluchi, Urdu, Indian dialects
Literacy:
definition: NA
total population: 81.4%
male: 86.8%
female: 73.5% (2003 est.)

GOVERNMENT

Country name:
conventional long form: Sultanate of Oman
conventional short form: Oman
local long form: Saltanat Uman
local short form: Uman
former: Muscat and Oman
Government type: monarchy
Capital: *name:* Muscat
geographic coordinates: 23 37 N, 58 35 E
time difference: UTC+4 (9 hours ahead of Washington, DC during Standard Time)
Administrative divisions: 5 regions (manatiq, singular—mintaqat) and 4 governorates* (muhafazat, singular—muhafazat) Ad Dakhiliyah, Al Batinah, Al Buraymi*, Al Wusta, Ash Sharqiyah, Az Zahirah, Masqat (Muscat)*, Musandam*, Zufar (Dhofar)*
Independence: 1650 (expulsion of the Portuguese)

National holiday: Birthday of Sultan QABOOS, 18 November (1940)

Constitution: none; note—on 6 November 1996, Sultan QABOOS issued a royal decree promulgating a basic law considered by the government to be a constitution which, among other things, clarifies the royal succession, provides for a prime minister, bars ministers from holding interests in companies doing business with the government, establishes a bicameral legislature, and guarantees basic civil liberties for Omani citizens

Legal system: based on English common law and Islamic law; ultimate appeal to the monarch; has not accepted compulsory ICJ jurisdiction

Suffrage: 21 years of age; universal; note—members of the military and security forces are not allowed to vote

Executive branch:

chief of state: Sultan and Prime Minister QABOOS bin Said al-Said (sultan since 23 July 1970 and prime minister since 23 July 1972); note—the monarch is both the chief of state and head of government

head of government: Sultan and Prime Minister QABOOS bin Said al-Said (sultan since 23 July 1970 and prime minister since 23 July 1972)

cabinet: Cabinet appointed by the monarch

elections: the monarch is hereditary

Legislative branch: bicameral Majlis Oman consists of Majlis al-Dawla or upper chamber (70 seats; members appointed by the monarch; has advisory powers only) and Majlis al-Shura or lower chamber (84 seats; members elected by popular vote to serve four-year terms; body has some limited power to propose legislation, but otherwise has only advisory powers)

elections: last held 27 October 2007 (next to be held in 2011)

election results: new candidates won 46 seats and 38 members of the outgoing Majlis kept their positions; none of the 20 female candidates were elected

Judicial branch: Supreme Court

note: the nascent civil court system, administered by region, has judges who practice secular and Sharia law

Political parties and leaders: none

Political pressure groups and leaders: none

International organization participation: ABEDA, AFESD, AMF, FAO, G-77, GCC, IBRD, ICAO, ICCt (signatory), IDA, IDB, IFAD, IFC, IHO, ILO, IMF, IMO, IMSO, Interpol, IOC, ISO, ITSO, ITU, LAS, MIGA, NAM, OIC, OPCW, UN, UNCTAD, UNESCO, UNIDO, UNWTO, UPU, WCO, WFTU, WHO, WIPO, WMO, WTO

Diplomatic representation in the US:

chief of mission: Ambassador Hunaina bint Sultan bin Ahmad al-MUGHAIRI

chancery: 2535 Belmont Road, NW, Washington, DC 20008

telephone: [1] (202) 387-1980

FAX: [1] (202) 745-4933

Diplomatic representation from the US:

chief of mission: Ambassador Gary A. GRAPPO

embassy: Jameat A'Duwal Al Arabiya Street, Al Khuwair area, Muscat

mailing address: P. O. Box 202, P.C. 115, Madinat Sultan Qaboos, Muscat

telephone: [968] 24-643-400

FAX: [968] 24-699771

Flag description: three horizontal bands of white, red, and green of equal width with a broad, vertical, red band on the hoist side; the national emblem (a khanjar dagger in its sheath superimposed on two crossed swords in scabbards) in white is centered near the top of the vertical band

ECONOMY

Economy—overview: Oman is a middle-income economy that is heavily dependent on dwindling oil resources, but sustained high oil prices in recent years have helped build Oman's budget and trade surpluses and foreign reserves. Oman joined the World Trade Organization in November 2000 and continues to liberalize its markets. It ratified a free trade agreement with the US in September 2006, and, through the Gulf Cooperation Council, seeks similar agreements with the EU, China and Japan. As a result of its dwindling oil resources, Oman is actively pursuing a development plan that focuses on diversification, industrialization, and privatization, with the objective of reducing the oil sector's contribution to GDP to 9 percent by 2020. Muscat is attempting to "Omanize" the labor force by replacing foreign expatriate workers with local workers. Oman actively seeks private foreign investors, especially in the industrial, information technology, tourism, and higher education fields. Industrial development plans focus on gas resources, metal manufacturing, petrochemicals, and international transshipment ports.

GDP (purchasing power parity): $61.61 billion (2007 est.)

GDP (official exchange rate): $40.06 billion (2007 est.)

GDP—real growth rate: 6.4% (2007 est.)

GDP—per capita (PPP): $24,000 (2007 est.)

GDP—composition by sector:

agriculture: 2.2%

industry: 38.3%

services: 59.5% (2007 est.)

Labor force: 920,000 (2002 est.)

Labor force—by occupation: *agriculture:* NA%

industry: NA%

services: NA%

Unemployment rate: 15% (2004 est.)

Population below poverty line: NA%

Household income or consumption by percentage share:

lowest 10%: NA%

highest 10%: NA%

Inflation rate (consumer prices): 5.5% (2007 est.)

Investment (gross fixed): 20.3% of GDP (2007 est.)

Budget:

revenues: $13.99 billion

expenditures: $13.68 billion (2007 est.)

Public debt: 3.8% of GDP (2007 est.)

Agriculture—products: dates, limes, bananas, alfalfa, vegetables; camels, cattle; fish

Industries: crude oil production and refining, natural and liquefied natural gas (LNG) production; construction, cement, copper, steel, chemicals, optic fiber

Industrial production growth rate: 3.2% (2007 est.)

Electricity—production: 11.89 billion kWh (2005)

Electricity—production by source:

fossil fuel: 100%

hydro: 0%

nuclear: 0%

other: 0% (2001)

Electricity—consumption: 8.661 billion kWh (2005)

Electricity—exports: 0 kWh (2005)

Electricity—imports: 0 kWh (2005)

Oil—production: 740,000 bbl/day (2006 est.)

Oil—consumption: 66,000 bbl/day (2005 est.)

Oil—exports: 733,100 bbl/day (2004)

Oil—imports: 15,440 bbl/day (2004)

Oil—proved reserves: 4.85 billion bbl (2007 est.)

Natural gas—production: 18.98 billion cu m (2005 est.)

Natural gas—consumption: 8.795 billion cu m (2005 est.)

Natural gas—exports: 10.19 billion cu m (2005 est.)

Natural gas—imports: 0 cu m (2005)

Natural gas—proved reserves: 795.2 billion cu m (1 January 2006 est.)

Current account balance: $3.996 billion (2007 est.)

Exports: $22.89 billion f.o.b. (2007 est.)
Exports—commodities: petroleum, reexports, fish, metals, textiles
Exports—partners: China 23.6%, South Korea 17.9%, Japan 10.9%, Thailand 10.7%, South Africa 7.7%, UAE 6.3% (2006)
Imports: $11 billion f.o.b. (2007 est.)
Imports—commodities: machinery and transport equipment, manufactured goods, food, livestock, lubricants
Imports—partners: UAE 22.5%, Japan 16.5%, US 8.1%, Germany 5.4%, India 4.4% (2006)
Economic aid—recipient: $30.68 million (2005)
Reserves of foreign exchange and gold: $9.524 billion (31 December 2007 est.)
Debt—external: $5.297 billion (31 December 2007 est.)
Stock of direct foreign investment—at home: $NA
Stock of direct foreign investment—abroad: $NA
Market value of publicly traded shares: $16.16 billion (2006)
Currency (code): Omani rial (OMR)
Currency code: OMR
Exchange rates: Omani rials per US dollar—0.3845 (2007), 0.3845 (2006), 0.3845 (2005), 0.3845 (2004), 0.3845 (2003)
Fiscal year: calendar year

COMMUNICATIONS

Telephones—main lines in use: 278,300 (2006)
Telephones—mobile cellular: 1.818 million (2006)
Telephone system:
general assessment: modern system consisting of open-wire, microwave, and radiotelephone communication stations; limited coaxial cable
domestic: fixed-line and mobile-cellular subscribership both increasing; open-wire, microwave, radiotelephone communications, and a domestic satellite system with 8 earth stations
international: country code—968; the Fiber-Optic Link Around the Globe (FLAG) and the SEA-ME-WE-3 submarine cable provide connectivity to Asia, the Middle East, and Europe; satellite earth stations—2 Intelsat (Indian Ocean), 1 Arabsat
Radio broadcast stations: AM 3, FM 9, shortwave 2 (1999)
Radios: 1.4 million (1997)
Television broadcast stations: 13 (plus 25 repeaters) (1999)
Televisions: 1.6 million (1997)
Internet country code: .om
Internet hosts: 3,763 (2007)
Internet Service Providers (ISPs): 1 (2000)
Internet users: 319,200 (2006)

TRANSPORTATION

Airports: 137 (2007)
Airports—with paved runways:
total: 7
over 3,047 m: 4
2,438 to 3,047 m: 1
1,524 to 2,437 m: 1
914 to 1,523 m: 1 (2007)
Airports—with unpaved runways:
total: 130
over 3,047 m: 2
2,438 to 3,047 m: 8
1,524 to 2,437 m: 51
914 to 1,523 m: 35
under 914 m: 34 (2007)
Heliports: 2 (2007)
Pipelines: gas 4,126 km; oil 3,558 km (2007)
Roadways:
total: 34,965 km
paved: 9,673 km (includes 550 km of expressways)
unpaved: 25,292 km (2001)
Merchant marine:
total: 2 ships (1000 GRT or over) 12,155 GRT/7,244 DWT
by type: chemical tanker 1, passenger 1
registered in other countries: 1 (Panama 1) (2007)
Ports and terminals: Mina' Qabus, Salalah

MILITARY

Military branches: Sultan's Armed Forces (SAF): Royal Army of Oman, Royal Navy of Oman, Royal Air Force of Oman (2008)
Military service age and obligation: 18–30 years of age for voluntary military service; no conscription (2008)
Manpower available for military service:
males age 16–49: 802,455
females age 16–49: 626,841 (2008 est.)
Manpower fit for military service:
males age 16–49: 663,881
females age 16–49: 543,410 (2008 est.)
Manpower reaching militarily significant age annually:
males age 16–49: 34,238
females age 16–49: 33,139 (2008 est.)
Military expenditures—percent of GDP: 11.4% (2005 est.)

TRANSNATIONAL ISSUES

Disputes—international: boundary agreement reportedly signed and ratified with UAE in 2003 for entire border, including Oman's Musandam Peninsula and Al Madhah exclave, but details of the alignment have not been made public
Trafficking in persons: current situation: Oman is a destination country for men and women primarily from Bangladesh, India, Sri Lanka, and Pakistan who migrate willingly, some of whom become victims of trafficking when subjected to conditions of involuntary servitude as domestic workers and laborers, including non-payment of wages, restrictions on movement and withholding of passports, threats, and physical or sexual abuse; Oman may also be a destination country for women from Asia, Eastern Europe, and North Africa for commercial sexual exploitation
tier rating: Tier 3—Oman was downgraded to Tier 3 in the 2007 report because it did not report any law enforcement efforts to prosecute and punish trafficking offenses in 2006 and continues to lack victim protection services or a systematic procedure to identify victims of trafficking

PACIFIC OCEAN

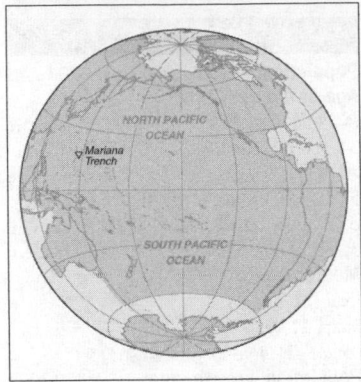

Background: The Pacific Ocean is the largest of the world's five oceans (followed by the Atlantic Ocean, Indian Ocean, Southern Ocean, and Arctic Ocean). Strategically important access waterways include the La Perouse, Tsugaru, Tsushima, Taiwan, Singapore, and Torres Straits. The decision by the International Hydrographic Organization in the spring of 2000 to delimit a fifth ocean, the Southern Ocean, removed the portion of the Pacific Ocean south of 60 degrees south.

Location: body of water between the Southern Ocean, Asia, Australia, and the Western Hemisphere

Geographic coordinates: 0 00 N, 160 00 W

Map references: Political Map of the World

Area:

total: 155.557 million sq km

note: includes Bali Sea, Bering Sea, Bering Strait, Coral Sea, East China Sea, Gulf of Alaska, Gulf of Tonkin, Philippine Sea, Sea of Japan, Sea of Okhotsk, South China Sea, Tasman Sea, and other tributary water bodies

Area—comparative: about 15 times the size of the US; covers about 28% of the global surface; larger than the total land area of the world

Coastline: 135,663 km

Climate: planetary air pressure systems and resultant wind patterns exhibit remarkable uniformity in the south and east; trade winds and westerly winds are well-developed patterns, modified by seasonal fluctuations; tropical cyclones (hurricanes) may form south of Mexico from June to October and affect Mexico and Central America; continental influences cause climatic uniformity to be much less pronounced in the eastern and western regions at the same latitude in the North Pacific Ocean; the western Pacific is monsoonal—a rainy season occurs during the summer months, when moisture-laden winds blow from the ocean over the land, and a dry season during the winter months, when dry winds blow from the Asian landmass back to the ocean; tropical cyclones (typhoons) may strike southeast and east Asia from May to December

Terrain: surface currents in the northern Pacific are dominated by a clockwise, warm-water gyre (broad circular system of currents) and in the southern Pacific by a counterclockwise, cool-water gyre; in the northern Pacific, sea ice forms in the Bering Sea and Sea of Okhotsk in winter; in the southern Pacific, sea ice from Antarctica reaches its northernmost extent in October; the ocean floor in the eastern Pacific is dominated by the East Pacific Rise, while the western Pacific is dissected by deep trenches, including the Mariana Trench, which is the world's deepest

Elevation extremes:

lowest point: Challenger Deep in the Mariana Trench -10,924 m

highest point: sea level 0 m

Natural resources: oil and gas fields, polymetallic nodules, sand and gravel aggregates, placer deposits, fish

Natural hazards: surrounded by a zone of violent volcanic and earthquake activity sometimes referred to as the "Pacific Ring of Fire"; subject to tropical cyclones (typhoons) in southeast and east Asia from May to December (most frequent from July to October); tropical cyclones (hurricanes) may form south of Mexico and strike Central America and Mexico from June to October (most common in August and September); cyclical El Nino/La Nina phenomenon occurs in the equatorial Pacific, influencing weather in the Western Hemisphere and the western Pacific; ships subject to superstructure icing in extreme north from October to May; persistent fog in the northern Pacific can be a maritime hazard from June to December

Environment—current issues: endangered marine species include the dugong, sea lion, sea otter, seals, turtles, and whales; oil pollution in Philippine Sea and South China Sea

Geography—note: the major chokepoints are the Bering Strait, Panama Canal, Luzon Strait, and the Singapore Strait; the Equator divides the Pacific Ocean into the North Pacific Ocean and the South Pacific Ocean; dotted with low coral islands and rugged volcanic islands in the southwestern Pacific Ocean

Economy—overview: The Pacific Ocean is a major contributor to the world economy and particularly to those nations its waters directly touch. It provides low-cost sea transportation between East and West, extensive fishing grounds, offshore oil and gas fields, minerals, and sand and gravel for the construction industry. In 1996, over 60% of the world's fish catch came from the Pacific Ocean. Exploitation of offshore oil and gas reserves is playing an ever-increasing role in the energy supplies of the US, Australia, NZ, China, and Peru. The high cost of recovering offshore oil and gas, combined with the wide swings in world prices for oil since 1985, has led to fluctuations in new drillings.

Ports and terminals: Bangkok (Thailand), Hong Kong (China), Kaohsiung (Taiwan), Los Angeles (US), Manila (Philippines), Pusan (South Korea), San Francisco (US), Seattle (US), Shanghai (China), Singapore, Sydney (Australia), Vladivostok (Russia), Wellington (NZ), Yokohama (Japan)

Transportation—note: Inside Passage offers protected waters from southeast Alaska to Puget Sound (Washington state)

Disputes—international: some maritime disputes (see littoral states)

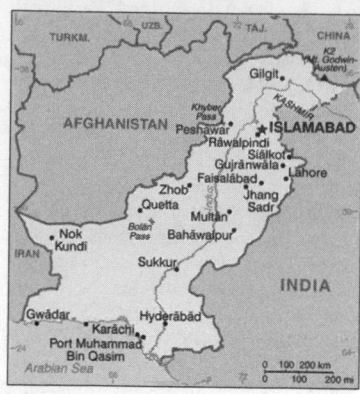

INTRODUCTION

Background: The Indus Valley civilization, one of the oldest in the world and dating back at least 5,000 years, spread over much of what is presently Pakistan. During the second millennium B.C., remnants of this culture fused with the migrating Indo-Aryan peoples. The area underwent successive centuries from the Persians, Greeks, Scythians, Arabs (who brought Islam), Afghans, and Turks. The Mughal Empire flourished in the 16th and 17th centuries; the British came to dominate the region in the 18th century. The separation in 1947 of British India into the Muslim state of Pakistan (with West and East sections) and largely Hindu India was never satisfactorily resolved, and India and Pakistan fought two wars—in 1947-48 and 1965—over the disputed Kashmir territory. A third war between these countries in 1971—in which India capitalized on Islamabad's marginalization of Bengalis in Pakistani politics—resulted in East Pakistan becoming the separate nation of Bangladesh. In response to Indian nuclear weapons testing, Pakistan conducted its own tests in 1998. The dispute over the state of Kashmir is ongoing, but discussions and confidence-building measures have led to decreased tensions since 2002.

GEOGRAPHY

Location: Southern Asia, bordering the Arabian Sea, between India on the east and Iran and Afghanistan on the west and China in the north
Geographic coordinates: 30 00 N, 70 00 E
Map references: Asia
Area: *total:* 803,940 sq km
land: 778,720 sq km
water: 25,220 sq km

Area—comparative: slightly less than twice the size of California
Land boundaries:
total: 6,774 km
border countries: Afghanistan 2,430 km, China 523 km, India 2,912 km, Iran 909 km
Coastline: 1,046 km
Maritime claims:
territorial sea: 12 nm
contiguous zone: 24 nm
exclusive economic zone: 200 nm
continental shelf: 200 nm or to the edge of the continental margin
Climate: mostly hot, dry desert; temperate in northwest; arctic in north
Terrain: flat Indus plain in east; mountains in north and northwest; Balochistan plateau in west
Elevation extremes:
lowest point: Indian Ocean 0 m
highest point: K2 (Mt. Godwin-Austen) 8,611 m
Natural resources: land, extensive natural gas reserves, limited petroleum, poor quality coal, iron ore, copper, salt, limestone
Land use: *arable land:* 24.44%
permanent crops: 0.84%
other: 74.72% (2005)
Irrigated land: 182,300 sq km (2003)
Total renewable water resources: 233.8 cu km (2003)
Freshwater withdrawal (domestic/industrial/agricultural): *total:* 169.39 cu km/yr (2%/2%/96%)
per capita: 1,072 cu m/yr (2000)
Natural hazards: frequent earthquakes, occasionally severe especially in north and west; flooding along the Indus after heavy rains (July and August)
Environment—current issues: water pollution from raw sewage, industrial wastes, and agricultural runoff; limited natural fresh water resources; most of the population does not have access to potable water; deforestation; soil erosion; desertification
Environment—international agreements: *party to:* Biodiversity, Climate Change, Climate Change-Kyoto Protocol, Desertification, Endangered Species, Environmental Modification, Hazardous Wastes, Law of the Sea, Marine Dumping, Ozone Layer Protection, Ship Pollution, Wetlands
signed, but not ratified: Marine Life Conservation
Geography—note: controls Khyber Pass and Bolan Pass, traditional invasion routes between Central Asia and the Indian Subcontinent

PEOPLE

Population: 167,762,040 (July 2008 est.)
Age structure:
0–14 years: 36.3% (male 31,316,803/female 29,567,622)
15–64 years: 59.4% (male 51,000,863/female 48,648,480)
65 years and over: 4.3% (male 3,409,246/female 3,819,026) (2008 est.)
Median age:
total: 21.2 years
male: 21 years
female: 21.4 years (2008 est.)
Population growth rate: 1.805% (2008 est.)
Birth rate: 26.93 births/1,000 population (2008 est.)
Death rate: 7.83 deaths/1,000 population (2008 est.)
Net migration rate: -1.05 migrant(s)/1,000 population (2008 est.)
Sex ratio:
at birth: 1.05 male(s)/female
under 15 years: 1.06 male(s)/female
15–64 years: 1.05 male(s)/female
65 years and over: 0.89 male(s)/female
total population: 1.05 male(s)/female (2008 est.)
Infant mortality rate:
total: 66.95 deaths/1,000 live births
male: 67.05 deaths/1,000 live births
female: 66.85 deaths/1,000 live births (2008 est.)
Life expectancy at birth:
total population: 64.13 years
male: 63.07 years
female: 65.24 years (2008 est.)
Total fertility rate: 3.58 children born/woman (2008 est.)
HIV/AIDS—adult prevalence rate: 0.1% (2001 est.)
HIV/AIDS—people living with HIV/AIDS: 74,000 (2001 est.)
HIV/AIDS—deaths: 4,900 (2003 est.)
Major infectious diseases:
degree of risk: high
food or waterborne diseases: bacterial diarrhea, hepatitis A and E, and typhoid fever
vectorborne diseases: dengue fever and malaria
animal contact disease: rabies
note: highly pathogenic H5N1 avian influenza has been identified in this country; it poses a negligible risk with extremely rare cases possible among US citizens who have close contact with birds (2008)
Nationality:
noun: Pakistani(s)
adjective: Pakistani

Ethnic groups: Punjabi, Sindhi, Pashtun (Pathan), Baloch, Muhajir (immigrants from India at the time of partition and their descendants)

Religions: Muslim 97% (Sunni 77%, Shi'a 20%), other (includes Christian and Hindu) 3%

Languages: Punjabi 48%, Sindhi 12%, Siraiki (a Punjabi variant) 10%, Pashtu 8%, Urdu (official) 8%, Balochi 3%, Hindko 2%, Brahui 1%, English (official; lingua franca of Pakistani elite and most government ministries), Burushaski and other 8%

Literacy: *definition:* age 15 and over can read and write
total population: 49.9%
male: 63%
female: 36% (2005 est.)

GOVERNMENT

Country name:
conventional long form: Islamic Republic of Pakistan
conventional short form: Pakistan
local long form: Jamhuryat Islami Pakistan
local short form: Pakistan
former: West Pakistan

Government type: federal republic

Capital: *name:* Islamabad
geographic coordinates: 33 42 N, 73 10 E
time difference: UTC+5 (10 hours ahead of Washington, DC during Standard Time)

Administrative divisions: 4 provinces, 1 territory*, and 1 capital territory**; Balochistan, Federally Administered Tribal Areas*, Islamabad Capital Territory**, North-West Frontier Province, Punjab, Sindh
note: the Pakistani-administered portion of the disputed Jammu and Kashmir region consists of two administrative entities: Azad Kashmir and Northern Areas

Independence: 14 August 1947 (from British India)

National holiday: Republic Day, 23 March (1956)

Constitution: 12 April 1973; suspended 5 July 1977, restored 30 December 1985; suspended 15 October 1999, restored in stages in 2002; amended 31 December 2003; suspended 3 November 2007; restored on 15 December 2007

Legal system: based on English common law with provisions to accommodate Pakistan's status as an Islamic state; accepts compulsory ICJ jurisdiction with reservations

Suffrage: 18 years of age; universal; joint electorates and reserved parliamentary seats for women and non-Muslims

Executive branch:
chief of state: President Pervez MUSHARRAF (since 20 June 2001)

note: following an October 1999 military coup, General Pervez MUSHARRAF suspended Pakistan's constitution and assumed the additional title of Chief Executive; in May 2000, Pakistan's Supreme Court validated the 1999 coup and granted MUSHARRAF executive and legislative authority for three years following the coup; in June 2001, MUSHARRAF named himself president, replacing Mohammad Rafiq TARAR; an April 2002 referendum extended MUSHARRAF's presidency by five years; on 6 October 2007, MUSHARRAF was reelected President of Pakistan, although the Supreme Court was reviewing a challenge to his eligibility to serve another term; MUSHARRAF declared emergency rule from 3 November to 15 December, during which time he replaced several Supreme Court Justices; the reconstituted court upheld his presidency on 22 November 2007
head of government: Syed Yousuf Raza GILANI (since 25 March 2008)
cabinet: Cabinet appointed by the President upon the advice of the prime minister
elections: the president is elected by secret ballot (1,170 votes total) through an Electoral College comprising the members of the Senate, National Assembly, and the provincial assemblies for a five-year term; election last held on 6 October 2007 (next to be held in October 2012); the prime minister is selected by the National Assembly; election last held on 24 March 2008
election results: MUSHARRAF reelected; MUSHARRAF 671 votes; Wajihuddin AHMED 8 votes; 6 votes invalid; GILANI elected prime minister GILANI 264 votes; Pervaiz ELAHI 42 votes; several abstentions

Legislative branch: bicameral Parliament or Majlis-e-Shoora consists of the Senate (100 seats; members indirectly elected by provincial assemblies and the territories' representatives in the National Assembly to serve six-year terms; one half are elected every three years) and the National Assembly (342 seats; 272 members elected by popular vote; 60 seats reserved for women; 10 seats reserved for non-Muslims; to serve five-year terms)
elections: Senate—last held in March 2006 (next to be held in March 2009); National Assembly—last held 18 February 2008 (next to be held in 2013)
election results: Senate results—percent of vote by party—NA; seats by party— PML 38, MMA 18, PPPP 10, MQM 6, PML-N 4, PKMAP 3, ANP 2, PPP-S 2,

BNP-A 1, BNP-M 1, JWP 1, NA 1, PML-F 1, independents 12; National Assembly results (as of 18 March 2008)—percent of votes by party—NA; seats by party—PPPP 121, PML-N 91, PML 54, MQM 25, ANP 13, MMA 6, PML-F 5, BNP-A 1, NPP 1, PPP-S 1, independents 18; note—by-elections for the remaining seats of the National Assembly were to be held in mid-April 2008

Judicial branch: Supreme Court (justices appointed by the president); Federal Islamic or Shari'a Court

Political parties and leaders: Awami National Party or ANP [Asfandyar Wali KHAN]; Balochistan National Party-Hayee Group or BNP-H [Dr. Hayee BALOCH]; Balochistan National Party-Awami or BNP-A [Moheem Khan BALOCH]; Balochistan National Party-Mengal or BNP-M [Sardar Ataullah MENGAL]; Jamhoori Watan Party or JWP; Jamiat Ahle Hadith or JAH [Sajid MIR]; Jamaat-i Islami or JI [Qazi Hussain AHMED]; Jamiat Ulema-i Islam Fazlur Rehman or JUI-F [Fazlur REHMAN]; Jamiat Ulema-i Islam Sami-ul HAQ or JUI-S [Sami ul-HAQ]; Jamiat Ulema-i Pakistan or JUP [Shah Faridul HAQ]; Muttahida Majlis-e Amal or MMA [Qazi Hussain AHMED]; Muttahida Qaumi Movement or MQM [Altaf HUSSAIN]; National Alliance or NA [Ghulam Mustapha JATOI] (merged with PML); National Peoples Party or NPP; Pakhtun Khwa Milli Awami Party or PKMAP [Mahmood Khan ACHAKZAI]; Pakistan Awami Tehrik or PAT [Tahir ul QADRI]; Pakistan Muslim League-Functional or PML-F [Pir PAGARO]; Pakistan Muslim League-Nawaz Sharif or PML-N [Nawaz SHARIF]; Pakistan Muslim League or PML [Chaudhry Shujaat HUSSAIN]; Pakistan Peoples Party-SHERPAO or PPP-S [Aftab Ahmed Khan SHERPAO]; Pakistan Peoples Party Parliamentarians or PPPP [Bilawal Bhutto ZARDARI, chairman; Asif Ali ZARDARI, co-chairman]; Pakistan Tehrik-e Insaaf or PTI [Imran KHAN]; Tehrik-i Islami [Allama Sajid NAQVI]
note: political alliances in Pakistan can shift frequently

Political pressure groups and leaders: military remains most important political force; ulema (clergy), landowners, industrialists, and small merchants also influential

International organization participation: ADB, ARF, C (reinstated 2004), CP, ECO, FAO, G-24, G-77, IAEA, IBRD, ICAO, ICC, ICRM, IDA, IDB, IFAD, IFC, IFRCS, IHO, ILO, IMF,

IMO, IMSO, Interpol, IOC, IOM, IPU, ISO, ITSO, ITU, ITUC, MIGA, MINURSO, MINUSTAH, MONUC, NAM, OAS (observer), OIC, OPCW, PCA, SAARC, SACEP, SCO (observer), UN, UNAMID, UNCTAD, UNESCO, UNHCR, UNIDO, UNMEE, UNMIL, UNMIS, UNMIT, UNOCI, UNOMIG, UNWTO, UPU, WCL, WCO, WFTU, WHO, WIPO, WMO, WTO

Diplomatic representation in the US:
chief of mission: Ambassador Mahmud Ali DURRANI
chancery: 3517 International Court, Washington, DC 20008
telephone: [1] (202) 243-6500
FAX: [1] (202) 686-1544
consulate(s) general: Boston, Chicago, Houston, Los Angeles, New York, Sunnyvale (California)

Diplomatic representation from the US:
chief of mission: Ambassador Anne W. PATTERSON
embassy: Diplomatic Enclave, Ramna 5, Islamabad
mailing address: P. O. Box 1048, Unit 62200, APO AE 09812-2200
telephone: [92] (51) 208-0000
FAX: [92] (51) 2276427
consulate(s) general: Karachi
consulate(s): Lahore, Peshawar

Flag description: green with a vertical white band (symbolizing the role of religious minorities) on the hoist side; a large white crescent and star are centered in the green field; the crescent, star, and color green are traditional symbols of Islam

ECONOMY

Economy—overview: Pakistan, an impoverished and underdeveloped country, has suffered from decades of internal political disputes, low levels of foreign investment, and a costly, ongoing confrontation with neighboring India. However, since 2001, IMF-approved reforms—most notably, privatization of the banking sector—bolstered by generous foreign assistance and renewed access to global markets, have generated macroeconomic recovery. Pakistan has experienced GDP growth in the 6–8% range in 2004-07, spurred by gains in the industrial and service sectors. Poverty levels have decreased by 10% since 2001, and Islamabad has steadily raised development spending in recent years, including a 52% real increase in the budget allocation for development in FY07. In 2007 the fiscal deficit—a result of chronically low tax collection and increased spending—exceeded Islamabad's target of 4% of GDP. Inflation remains the top concern among the public, jumping from 7.7% in 2007 to more than 11% during the first few months of 2008, primarily because of rising world commodity prices. The Pakistani rupee has depreciated since the proclamation of emergency rule in November 2007.

GDP (purchasing power parity): $410 billion (2007 est.)

GDP (official exchange rate): $143.8 billion (2007 est.)

GDP—real growth rate: 6.4% (2007 est.)

GDP—per capita (PPP): $2,600 (2007 est.)

GDP—composition by sector:
agriculture: 19.6%
industry: 26.8%
services: 53.7% (2007 est.)

Labor force: 49.18 million
note: extensive export of labor, mostly to the Middle East, and use of child labor (2007 est.)

Labor force—by occupation: *agriculture:* 42%
industry: 20%
services: 38% (2004 est.)

Unemployment rate: 7.5% plus substantial underemployment (2007 est.)

Population below poverty line: 24% (FY05/06 est.)

Household income or consumption by percentage share:
lowest 10%: 4%
highest 10%: 26.3% (2002)

Distribution of family income—Gini index: 30.6 (2002)

Inflation rate (consumer prices): 7.8% (2007 est.)

Investment (gross fixed): 21.4% of GDP (2007 est.)

Budget:
revenues: $21.95 billion
expenditures: $27.62 billion (2007 est.)

Public debt: 52.8% of GDP (2007 est.)

Agriculture—products: cotton, wheat, rice, sugarcane, fruits, vegetables; milk, beef, mutton, eggs

Industries: textiles and apparel, food processing, pharmaceuticals, construction materials, paper products, fertilizer, shrimp

Industrial production growth rate: 6.8% (2007 est.)

Electricity—production: 89.82 billion kWh (2005)

Electricity—production by source:
fossil fuel: 68.8%
hydro: 28.2%
nuclear: 3%
other: 0% (2001)

Electricity—consumption: 67.06 billion kWh (2005)

Electricity—exports: 0 kWh (2005)

Electricity—imports: 0 kWh (2005)

Oil—production: 68,220 bbl/day (2005 est.)

Oil—consumption: 345,000 bbl/day (2005 est.)

Oil—exports: 23,230 bbl/day (2004)

Oil—imports: 278,900 bbl/day (2004)

Oil—proved reserves: 376.8 million bbl (2007 est.)

Natural gas—production: 29.54 billion cu m (2005 est.)

Natural gas—consumption: 29.54 billion cu m (2005 est.)

Natural gas—exports: 0 cu m (2005 est.)

Natural gas—imports: 0 cu m (2005)

Natural gas—proved reserves: 764.6 billion cu m (1 January 2006 est.)

Current account balance: -$7.105 billion (2007 est.)

Exports: $16.31 billion f.o.b. (2007 est.)

Exports—commodities: textiles (garments, bed linen, cotton cloth, yarn), rice, leather goods, sports goods, chemicals, manufactures, carpets and rugs

Exports—partners: US 21%, UAE 9%, Afghanistan 7.7%, China 5.3%, UK 5.1% (2006)

Imports: $30.33 billion f.o.b. (2007 est.)

Imports—commodities: petroleum, petroleum products, machinery, plastics, transportation equipment, edible oils, paper and paperboard, iron and steel, tea

Imports—partners: China 13.8%, Saudi Arabia 10.5%, UAE 9.7%, US 6.5%, Japan 5.7%, Kuwait 4.7%, Germany 4.1% (2006)

Economic aid—recipient: $1.666 billion (2005)

Reserves of foreign exchange and gold: $15.69 billion (31 December 2007 est.)

Debt—external: $39.23 billion (31 December 2007 est.)

Stock of direct foreign investment—at home: $20.01 billion (2007 est.)

Stock of direct foreign investment—abroad: $952 million (2007 est.)

Market value of publicly traded shares: $45.52 billion (2006)

Currency (code): Pakistani rupee (PKR)

Currency code: PKR

Exchange rates: Pakistani rupees per US dollar—60.6295 (2007), 60.35 (2006), 59.515 (2005), 58.258 (2004), 57.752 (2003)

Fiscal year: 1 July—30 June

COMMUNICATIONS

Telephones—main lines in use: 5.24 million (2006)

Telephones—mobile cellular: 63.16 million (2007)

Telephone system:
general assessment: the telecommunications infrastructure is improving dramatically with foreign and domestic investments into fixed-line and mobile networks; mobile-cellular subscribership has skyrocketed, reaching some 63 million in mid-2007, up from only about 300,000 in 2000; fiber systems are being constructed throughout the country to aid in network growth; main line availability has risen only marginally over the

same period and there are still difficulties getting main line service to rural areas
domestic: microwave radio relay, coaxial cable, fiber-optic cable, cellular, and satellite networks
international: country code—92; landing point for the SEA-ME-WE-3 and SEA-ME-WE-4 submarine cable systems that provide links to Asia, the Middle East, and Europe; satellite earth stations—3 Intelsat (1 Atlantic Ocean and 2 Indian Ocean); 3 operational international gateway exchanges (1 at Karachi and 2 at Islamabad); microwave radio relay to neighboring countries (2006)

Radio broadcast stations: AM 31, FM 68, shortwave NA (2006)
Radios: 13.5 million (1997)
Television broadcast stations: 20 (5 state-run channels and 15 privately-owned satellite channels) (2006)
Televisions: 3.1 million (1997)
Internet country code: .pk
Internet hosts: 164,067 (2007)
Internet Service Providers (ISPs): 30 (2000)
Internet users: 12 million (2006)

TRANSPORTATION

Airports: 146 (2007)
Airports—with paved runways:
total: 92
over 3,047 m: 16
2,438 to 3,047 m: 19
1,524 to 2,437 m: 29
914 to 1,523 m: 18
under 914 m: 10 (2007)
Airports—with unpaved runways:
total: 54
2,438 to 3,047 m: 1
1,524 to 2,437 m: 16
914 to 1,523 m: 13
under 914 m: 24 (2007)
Heliports: 18 (2007)
Pipelines: gas 10,398 km; oil 2,076 km (2007)
Railways:
total: 8,163 km
broad gauge: 7,718 km 1.676-m gauge (293 km electrified)
narrow gauge: 445 km 1.000-m gauge (2006)
Roadways:
total: 258,340 km
paved: 167,146 km (includes 711 km of expressways)
unpaved: 91,194 km (2004)

Merchant marine:
total: 14 ships (1000 GRT or over) 325,254 GRT/536,876 DWT
by type: bulk carrier 1, cargo 10, petroleum tanker 3
registered in other countries: 12 (Comoros 2, Hong Kong 1, North Korea 1, Malta 2, Panama 5, St Vincent and The Grenadines 1) (2007)
Ports and terminals: Karachi, Port Muhammad Bin Qasim

MILITARY

Military branches: Army (includes National Guard), Navy (includes Marines and Maritime Security Agency), Pakistan Air Force (Pakistan Fiza'ya) (2008)
Military service age and obligation: 16 years of age for voluntary military service; soldiers cannot be deployed for combat until age of 18; the Pakistani Air Force and Pakistani Navy have inducted their first female pilots and sailors (2006)
Manpower available for military service:
males age 16–49: 42,633,765
females age 16–49: 40,114,017 (2008 est.)
Manpower fit for military service:
males age 16–49: 32,453,913
females age 16–49: 31,369,057 (2008 est.)
Manpower reaching militarily significant age annually:
males age 16–49: 1,976,444
females age 16–49: 1,856,505 (2008 est.)
Military expenditures—percent of GDP: 3% (2007 est.)

TRANSNATIONAL ISSUES

Disputes—international: various talks and confidence-building measures cautiously have begun to defuse tensions over Kashmir, particularly since the October 2005 earthquake in the region; Kashmir nevertheless remains the site of the world's largest and most militarized territorial dispute with portions under the de facto administration of China (Aksai Chin), India (Jammu and Kashmir), and Pakistan (Azad Kashmir and Northern Areas); UN Military Observer Group in India and Pakistan (UNMOGIP) has maintained a small group of peacekeepers since 1949; India does not recognize Pakistan's ceding historic Kashmir lands to China in 1964; India and Pakistan have maintained their 2004 cease fire in Kashmir and initiated discussions on defusing the armed stand-off in the Siachen glacier region; Pakistan protests India's fencing the highly militarized Line of Control and construction of the Baglihar Dam on the Chenab River in Jammu and Kashmir, which is part of the larger dispute on water sharing of the Indus River and its tributaries; to defuse tensions and prepare for discussions on a maritime boundary, India and Pakistan seek technical resolution of the disputed boundary in Sir Creek estuary at the mouth of the Rann of Kutch in the Arabian Sea; Pakistani maps continue to show the Junagadh claim in India's Gujarat State; by 2005, Pakistan, with UN assistance, repatriated 2.3 million Afghan refugees leaving slightly more than a million, many of whom remain at their own choosing; Pakistan has proposed and Afghanistan protests construction of a fence and laying of mines along portions of their porous border; Pakistan has sent troops into remote tribal areas to monitor and control the border with Afghanistan and to stem terrorist or other illegal activities
Refugees and internally displaced persons: *refugees (country of origin):* 1,043,984 (Afghanistan)
IDPs: undetermined (government strikes on Islamic militants in South Waziristan); 34,000 (October 2005 earthquake; most of those displaced returned to their home villages in the spring of 2006) (2007)
Illicit drugs: opium poppy cultivation estimated to be 800 hectares in 2005 yielding a potential production of 4 metric tons of pure heroin; federal and provincial authorities continue to conduct anti-poppy campaigns that force eradication—fines and arrests will take place if the ban on poppy cultivation is not observed; key transit point for Afghan drugs, including heroin, opium, morphine, and hashish, bound for Western markets, the Gulf States, and Africa; financial crimes related to drug trafficking, terrorism, corruption, and smuggling remain problems

PALAU

INTRODUCTION

Background: After three decades as part of the UN Trust Territory of the Pacific under US administration, this westernmost cluster of the Caroline Islands opted for independence in 1978 rather than join the Federated States of Micronesia. A Compact of Free Association with the US was approved in 1986, but not ratified until 1993. It entered into force the following year, when the islands gained independence.

GEOGRAPHY

Location: Oceania, group of islands in the North Pacific Ocean, southeast of the Philippines
Geographic coordinates: 7 30 N, 134 30 E

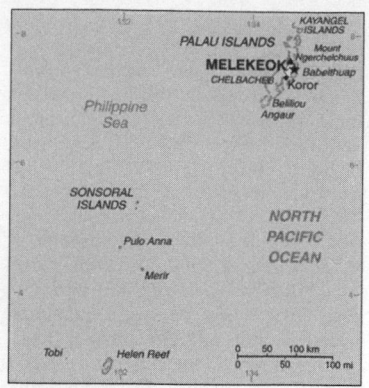

Map references: Oceania
Area: *total:* 458 sq km
land: 458 sq km
water: 0 sq km
Area—comparative: slightly more than 2.5 times the size of Washington, DC
Land boundaries: 0 km
Coastline: 1,519 km
Maritime claims:
territorial sea: 3 nm
exclusive fishing zone: 200 nm
Climate: tropical; hot and humid; wet season May to November
Terrain: varying geologically from the high, mountainous main island of Babelthuap to low, coral islands usually fringed by large barrier reefs
Elevation extremes:
lowest point: Pacific Ocean 0 m
highest point: Mount Ngerchelchuus 242 m
Natural resources: forests, minerals (especially gold), marine products, deep-seabed minerals
Land use: *arable land:* 8.7%
permanent crops: 4.35%
other: 86.95% (2005)
Irrigated land: NA
Natural hazards: typhoons (June to December)
Environment—current issues: inadequate facilities for disposal of solid waste; threats to the marine ecosystem from sand and coral dredging, illegal fishing practices, and overfishing
Environment—international agreements: *party to:* Biodiversity, Climate Change, Climate Change-Kyoto Protocol, Desertification, Law of the Sea, Ozone Layer Protection, Wetlands, Whaling
signed, but not ratified: none of the selected agreements
Geography—note: westernmost archipelago in the Caroline chain, consists of six island groups totaling more than 300 islands; includes World War II battleground of Beliliou (Peleliu) and world-famous rock islands

PEOPLE

Population: 21,093 (July 2008 est.)
Age structure:
0–14 years: 25.8% (male 2,797/female 2,637)
15–64 years: 69.4% (male 7,864/female 6,779)
65 years and over: 4.8% (male 482/female 534) (2008 est.)
Median age:
total: 32.3 years
male: 33.3 years
female: 31.3 years (2008 est.)
Population growth rate: 1.157% (2008 est.)
Birth rate: 17.4 births/1,000 population (2008 est.)
Death rate: 6.73 deaths/1,000 population (2008 est.)
Net migration rate: 0.9 migrant(s)/1,000 population (2008 est.)
Sex ratio:
at birth: 1.06 male(s)/female
under 15 years: 1.06 male(s)/female
15–64 years: 1.16 male(s)/female
65 years and over: 0.9 male(s)/female
total population: 1.12 male(s)/female (2008 est.)
Infant mortality rate:
total: 13.69 deaths/1,000 live births
male: 15.37 deaths/1,000 live births
female: 11.9 deaths/1,000 live births (2008 est.)
Life expectancy at birth:
total population: 71 years
male: 67.82 years
female: 74.36 years (2008 est.)
Total fertility rate: 2.45 children born/woman (2008 est.)
HIV/AIDS—adult prevalence rate: NA
HIV/AIDS—people living with HIV/AIDS: NA
HIV/AIDS—deaths: NA
Nationality:
noun: Palauan(s)
adjective: Palauan
Ethnic groups: Palauan (Micronesian with Malayan and Melanesian admixtures) 69.9%, Filipino 15.3%, Chinese 4.9%, other Asian 2.4%, white 1.9%, Carolinian 1.4%, other Micronesian 1.1%, other or unspecified 3.2% (2000 census)
Religions: Roman Catholic 41.6%, Protestant 23.3%, Modekngei 8.8% (indigenous to Palau), Seventh-Day Adventist 5.3%, Jehovah's Witness 0.9%, Latter-Day Saints 0.6%, other 3.1%, unspecified or none 16.4% (2000 census)
Languages: Palauan 64.7% official in all islands except Sonsoral (Sonsoralese and English are official), Tobi (Tobi and English are official), and Angaur (Angaur, Japanese, and English are official), Filipino 13.5%, English 9.4%, Chinese 5.7%, Carolinian 1.5%, Japanese 1.5%, other Asian 2.3%, other languages 1.5% (2000 census)
Literacy: *definition:* age 15 and over can read and write
total population: 92%
male: 93%
female: 90% (1980 est.)

GOVERNMENT

Country name:
conventional long form: Republic of Palau
conventional short form: Palau
local long form: Beluu er a Belau
local short form: Belau
former: Trust Territory of the Pacific Islands, Palau District
Government type: constitutional government in free association with the US; the Compact of Free Association entered into force 1 October 1994
Capital: *name:* Melekeok
geographic coordinates: 7 29 N, 134 38 E
time difference: UTC+9 (14 hours ahead of Washington, DC during Standard Time)
Administrative divisions: 16 states; Aimeliik, Airai, Angaur, Hatohobei, Kayangel, Koror, Melekeok, Ngaraard, Ngarchelong, Ngardmau, Ngatpang, Ngchesar, Ngeremlengui, Ngiwal, Peleliu, Sonsorol
Independence: 1 October 1994 (from the US-administered UN trusteeship)
National holiday: Constitution Day, 9 July (1979)
Constitution: 1 January 1981
Legal system: based on Trust Territory laws, acts of the legislature, municipal, common, and customary laws; has not accepted compulsory ICJ jurisdiction
Suffrage: 18 years of age; universal
Executive branch:
chief of state: President Tommy Esang REMENGESAU, Jr. (since 19 January 2001); Vice President Camsek CHIN (since 1 January 2005); note—the president is both the chief of state and head of government
head of government: President Tommy Esang REMENGESAU, Jr. (since 19 January 2001); Vice President Camsek CHIN (since 1 January 2005)
cabinet: NA
elections: president and vice president elected on separate tickets by popular vote for four-year terms (eligible for a second term); election last held 2 November 2004 (next to be held in November 2008)
election results: Tommy Esang REMENGESAU, Jr. reelected president; percent of vote—Tommy Esang

REMENGESAU, Jr. 64%, Polycarp BASILIUS 33%; Elias Camsek CHIN elected vice president; percent of vote—Elias Camsek CHIN 70%, Sandra PIERANTOZZI 29%

Legislative branch: bicameral National Congress or Olbiil Era Kelulau (OEK) consists of the Senate (9 seats; members elected by popular vote on a population basis to serve four-year terms) and the House of Delegates (16 seats; members elected by popular vote to serve four-year terms)

elections: Senate—last held 2 November 2004 (next to be held in November 2008); House of Delegates—last held 2 November 2004 (next to be held in November 2008)

election results: Senate—percent of vote—NA; seats—independents 9; House of Delegates—percent of vote—NA; seats—independents 16

Judicial branch: Supreme Court; Court of Common Pleas; Land Court

Political parties and leaders: none

Political pressure groups and leaders: NA

International organization participation: ACP, ADB, FAO, IAEA, IBRD, ICAO, ICRM, IDA, IFC, IFRCS, IMF, IOC, IPU, MIGA, OPCW, PIF, Sparteca, SPC, UN, UNCTAD, UNESCO, WHO

Diplomatic representation in the US:
chief of mission: Ambassador Hersey KYOTA
chancery: 1700 Pennsylvania Avenue NW, Suite 400, Washington, DC 20006
telephone: [1] (202) 452-6814
FAX: [1] (202) 452-6281
consulate(s) general: Honolulu
consulate(s): Tamuning (Guam)

Diplomatic representation from the US:
chief of mission: Charge d'Affaires Mark BEZNER
embassy: Koror (no street address)
mailing address: P. O. Box 6028, Republic of Palau 96940
telephone: [680] 488-2920, 2990
FAX: [680] 488-2911

Flag description: light blue with a large yellow disk (representing the moon) shifted slightly to the hoist side

ECONOMY

Economy—overview: The economy consists primarily of tourism, subsistence agriculture, and fishing. The government is the major employer of the work force relying heavily on financial assistance from the US. The Compact of Free Association with the US, entered into after the end of the UN trusteeship on 1 October 1994, provided Palau with up to $700 million in US aid for the following 15 years in return for furnishing military facilities. Business and tourist arrivals numbered 63,000 in 2003. The population enjoys a per capita income roughly 50% higher than that of the Philippines and much of Micronesia. Long-run prospects for the key tourist sector have been greatly bolstered by the expansion of air travel in the Pacific, the rising prosperity of leading East Asian countries, and the willingness of foreigners to finance infrastructure development.

GDP (purchasing power parity): $124.5 million
note: GDP estimates includes US subsidy (2004 est.)

GDP (official exchange rate): $145 million (2005)

GDP—real growth rate: 5.5% (2005 est.)

GDP—per capita (PPP): $7,600 (2005 est.)

GDP—composition by sector:
agriculture: 6.2%
industry: 12%
services: 81.8% (2003)

Labor force: 9,777 (2005)

Labor force—by occupation: *agriculture:* 20%
industry: NA%
services: NA%

Unemployment rate: 4.2% (2005 est.)

Population below poverty line: NA%

Household income or consumption by percentage share:
lowest 10%: NA%
highest 10%: NA%

Inflation rate (consumer prices): 2.7% (2005 est.)

Budget:
revenues: $72.07 million
expenditures: $72.43 million (FY04/05 est.)

Agriculture—products: coconuts, copra, cassava (tapioca), sweet potatoes; fish

Industries: tourism, craft items (from shell, wood, pearls), construction, garment making

Industrial production growth rate: NA%

Electricity—production by source: NA

Current account balance: $15.09 million (FY03/04)

Exports: $5.882 million f.o.b. (2004 est.)

Exports—commodities: shellfish, tuna, copra, garments

Exports—partners: US, Japan, Singapore (2006)

Imports: $107.3 million f.o.b. (2004 est.)

Imports—commodities: machinery and equipment, fuels, metals; foodstuffs

Imports—partners: US, Singapore, Japan, South Korea (2006)

Economic aid—recipient: $23.46 million (2005)

Debt—external: $0 (FY99/00)

Market value of publicly traded shares: $NA

Currency (code): US dollar (USD)

Currency code: USD

Exchange rates: the US dollar is used

Fiscal year: 1 October—30 September

COMMUNICATIONS

Telephones—main lines in use: 6,700 (2002)

Telephones—mobile cellular: 1,000 (2002)

Telephone system:
general assessment: NA
domestic: NA
international: country code—680; satellite earth station—1 Intelsat (Pacific Ocean)

Radio broadcast stations: AM 1, FM 4, shortwave 1 (2001)

Radios: 12,000 (1997)

Television broadcast stations: 1 (cable) (2005)

Televisions: 11,000 (1997)

Internet country code: .pw

Internet hosts: 1 (2007)

Internet Service Providers (ISPs): 1 (2002)

TRANSPORTATION

Airports: 3 (2007)

Airports—with paved runways:
total: 1
1,524 to 2,437 m: 1 (2007)

Airports—with unpaved runways:
total: 2
1,524 to 2,437 m: 2 (2007)

Roadways:
total: 60 km

Ports and terminals: Koror

MILITARY

Military branches: no regular military forces; Palau National Police (2008)

Manpower available for military service:
males age 16–49: 5,973 (2008 est.)

Manpower fit for military service:
males age 16–49: 4,397 (2008 est.)

Manpower reaching militarily significant age annually:
males age 16–49: 179 (2008 est.)

Military expenditures—percent of GDP: NA

Military—note: defense is the responsibility of the US; under a Compact of Free Association between Palau and the US, the US military is granted access to the islands for 50 years, but it has not stationed any military forces there (2008)

TRANSNATIONAL ISSUES

Disputes—international: maritime delineation negotiations continue with Philippines, Indonesia

501

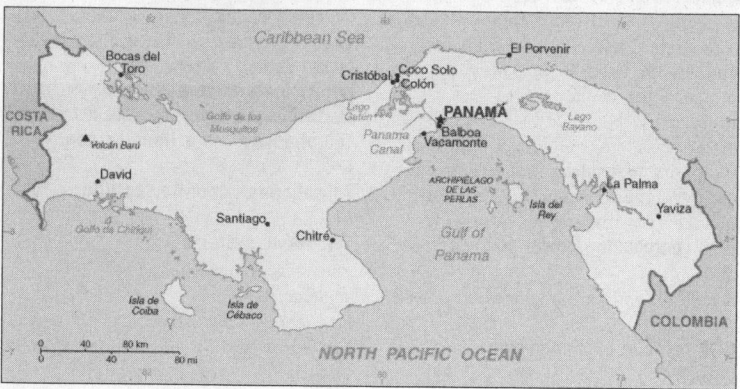

Background: Explored and settled by the Spanish in the 16th century, Panama broke with Spain in 1821 and joined a union of Colombia, Ecuador, and Venezuela—named the Republic of Gran Colombia. When the latter dissolved in 1830, Panama remained part of Colombia. With US backing, Panama seceded from Colombia in 1903 and promptly signed a treaty with the US allowing for the construction of a canal and US sovereignty over a strip of land on either side of the structure (the Panama Canal Zone). The Panama Canal was built by the US Army Corps of Engineers between 1904 and 1914. In 1977, an agreement was signed for the complete transfer of the Canal from the US to Panama by the end of the century. Certain portions of the Zone and increasing responsibility over the Canal were turned over in the subsequent decades. With US help, dictator Manuel NORIEGA was deposed in 1989. The entire Panama Canal, the area supporting the Canal, and remaining US military bases were transferred to Panama by the end of 1999. In October 2006, Panamanians approved an ambitious plan to expand the Canal. The project, which began in 2007 and could double the Canal's capacity, is expected to be completed in 2014–15.

GEOGRAPHY

Location: Central America, bordering both the Caribbean Sea and the North Pacific Ocean, between Colombia and Costa Rica
Geographic coordinates: 9 00 N, 80 00 W
Map references: Central America and the Caribbean
Area:
total: 78,200 sq km
land: 75,990 sq km
water: 2,210 sq km
Area—comparative: slightly smaller than South Carolina
Land boundaries:
total: 555 km
border countries: Colombia 225 km, Costa Rica 330 km
Coastline: 2,490 km
Maritime claims:
territorial sea: 12 nm
contiguous zone: 24 nm
exclusive economic zone: 200 nm or edge of continental margin
Climate: tropical maritime; hot, humid, cloudy; prolonged rainy season (May to January), short dry season (January to May)
Terrain: interior mostly steep, rugged mountains and dissected, upland plains; coastal areas largely plains and rolling hills
Elevation extremes:
lowest point: Pacific Ocean 0 m
highest point: Volcan Baru 3,475 m
Natural resources: copper, mahogany forests, shrimp, hydropower
Land use: *arable land:* 7.26%
permanent crops: 1.95%
other: 90.79% (2005)
Irrigated land: 430 sq km (2003)
Total renewable water resources: 148 cu km (2000)
Freshwater withdrawal (domestic/industrial/agricultural): *total:* 0.82 cu km/yr (67%/5%/28%)
per capita: 254 cu m/yr (2000)
Natural hazards: occasional severe storms and forest fires in the Darien area
Environment—current issues: water pollution from agricultural runoff threatens fishery resources; deforestation of tropical rain forest; land degradation and soil erosion threatens siltation of Panama Canal; air pollution in urban areas; mining threatens natural resources

Environment—international agreements: *party to:* Biodiversity, Climate Change, Climate Change-Kyoto Protocol, Desertification, Endangered Species, Hazardous Wastes, Law of the Sea, Marine Dumping, Ozone Layer Protection, Ship Pollution, Tropical Timber 83, Tropical Timber 94, Wetlands, Whaling
signed, but not ratified: Marine Life Conservation
Geography—note: strategic location on eastern end of isthmus forming land bridge connecting North and South America; controls Panama Canal that links North Atlantic Ocean via Caribbean Sea with North Pacific Ocean

PEOPLE

Population: 3,292,693 (July 2008 est.)
Age structure:
0–14 years: 29.8% (male 500,579/female 480,635)
15–64 years: 63.6% (male 1,061,446/female 1,033,675)
65 years and over: 6.6% (male 100,780/female 115,578) (2008 est.)
Median age:
total: 26.6 years
male: 26.3 years
female: 26.9 years (2008 est.)
Population growth rate: 1.528% (2008 est.)
Birth rate: 21.15 births/1,000 population (2008 est.)
Death rate: 5.52 deaths/1,000 population (2008 est.)
Net migration rate: -0.34 migrant(s)/1,000 population (2008 est.)
Sex ratio: *at birth:* 1.04 male(s)/female
under 15 years: 1.04 male(s)/female
15–64 years: 1.03 male(s)/female
65 years and over: 0.87 male(s)/female
total population: 1.02 male(s)/female (2008 est.)
Infant mortality rate:
total: 15.62 deaths/1,000 live births
male: 16.95 deaths/1,000 live births
female: 14.22 deaths/1,000 live births (2008 est.)
Life expectancy at birth:
total population: 75.17 years
male: 72.71 years
female: 77.73 years (2008 est.)
Total fertility rate: 2.65 children born/woman (2008 est.)
HIV/AIDS—adult prevalence rate: 0.9% (2003 est.)
HIV/AIDS—people living with HIV/AIDS: 16,000 (2003 est.)
HIV/AIDS—deaths: fewer than 500 (2003 est.)

Major infectious diseases:
degree of risk: intermediate
food or waterborne diseases: bacterial diarrhea and hepatitis A
vectorborne disease: dengue fever and malaria
water contact disease: leptospirosis (2008)
Nationality:
noun: Panamanian(s)
adjective: Panamanian
Ethnic groups: mestizo (mixed Amerindian and white) 70%, Amerindian and mixed (West Indian) 14%, white 10%, Amerindian 6%
Religions: Roman Catholic 85%, Protestant 15%
Languages: Spanish (official), English 14%; note—many Panamanians bilingual
Literacy:
definition: age 15 and over can read and write
total population: 91.9%
male: 92.5%
female: 91.2% (2000 census)

GOVERNMENT

Country name:
conventional long form: Republic of Panama
conventional short form: Panama
local long form: Republica de Panama
local short form: Panama
Government type: constitutional democracy
Capital: *name:* Panama
geographic coordinates: 8 58 N, 79 32 W
time difference: UTC-5 (same time as Washington, DC during Standard Time)
Administrative divisions: 11 provinces (provincias, singular—provincia) and 1 territory* (comarca); Bocas del Toro, Comarca Kuna Yala, Comarca Ngobe-Bugle, Chiriqui, Cocle, Colon, Darien, Herrera, Los Santos, Panama, San Blas*(Kuna Yala), and Veraguas
Independence: 3 November 1903 (from Colombia; became independent from Spain 28 November 1821)
National holiday: Independence Day, 3 November (1903)
Constitution: 11 October 1972; major reforms adopted 1978, 1983, 1994, and 2004
Legal system: based on civil law system; judicial review of legislative acts in the Supreme Court of Justice; accepts compulsory ICJ jurisdiction with reservations
Suffrage: 18 years of age; universal and compulsory
Executive branch:
chief of state: President Martin TORRIJOS Espino (since 1 September 2004); First Vice President Samuel LEWIS Navarro (since 1 September 2004);

Second Vice President Ruben AROSEMENA Valdes (since 1 September 2004); note—the president is both the chief of state and head of government
head of government: President Martin TORRIJOS Espino (since 1 September 2004); First Vice President Samuel LEWIS Navarro (since 1 September 2004); Second Vice President Ruben AROSEMENA Valdes (since 1 September 2004)
cabinet: Cabinet appointed by the president
elections: president and vice presidents elected on the same ticket by popular vote for five-year terms (not eligible for immediate reelection; president and vice presidents must sit out two additional terms (10 years) before becoming eligible for reelection); election last held 2 May 2004 (next to be held on 3 May 2009); note—beginning in 2009, Panama will have only one vice president
election results: Martin TORRIJOS Espino elected president; percent of vote Martin TORRIJOS Espino 47.5%, Guillermo ENDARA Galimany 30.6%, Jose Miguel ALEMAN 17%, Ricardo MARTINELLI 4.9%
note: government coalition—PRD (Democratic Revolutionary Party), PP (Popular Party)
Legislative branch: unicameral National Assembly or Asamblea Nacional (78 seats; members are elected by popular vote to serve five-year terms); note—in 2009, the number of seats will change to 71
elections: last held 2 May 2004 (next to be held 3 May 2009)
election results: percent of vote by party—NA; seats by party—PRD 41, PA 17, PS 9, MOLIRENA 4, CD 3, PLN 3, PP 1
note: legislators from outlying rural districts are chosen on a plurality basis while districts located in more populous towns and cities elect multiple legislators by means of a proportion-based formula
Judicial branch: Supreme Court of Justice or Corte Suprema de Justicia (nine judges appointed for 10-year terms); five superior courts; three courts of appeal
Political parties and leaders: Democratic Change or CD [Ricardo MARTINELLI]; Democratic Revolutionary Party or PRD [Hugo GUIRAUD]; Nationalist Republican Liberal Movement or MOLIRENA [Gisela CHUNG]; Panamenista Party or PA [Juan Carlos VARELA] (formerly the Arnulfista Party); Patriotic Union Party or PU (combination of the Liberal National Party or PLN and the Solidarity Party or PS)[Jose Raul MULINO and

Anibal GALINDO]; Popular Party or PP [Rene ORILLAC] (formerly Christian Democratic Party or PDC)
Political pressure groups and leaders: Chamber of Commerce; National Civic Crusade; National Council of Organized Workers or CONATO; National Council of Private Enterprise or CONEP; National Union of Construction and Similar Workers (SUNTRACS); Panamanian Association of Business Executives or APEDE; Panamanian Industrialists Society or SIP; Workers Confederation of the Republic of Panama or CTRP
International organization participation: BCIE, CAN (observer), CSN (observer), FAO, G-77, IADB, IAEA, IBRD, ICAO, ICC, ICCt, ICRM, IDA, IFAD, IFC, IFRCS, ILO, IMF, IMO, IMSO, Interpol, IOC, IOM, IPU, ISO, ITSO, ITU, ITUC, LAES, LAIA (observer), MIGA, NAM, OAS, OPANAL, OPCW, PCA, RG, UN, UN Security Council (temporary), UNCTAD, UNESCO, UNIDO, Union Latina, UNWTO, UPU, WCL, WCO, WFTU, WHO, WIPO, WMO, WTO
Diplomatic representation in the US:
chief of mission: Ambassador Federico HUMBERT Arias
chancery: 2862 McGill Terrace NW, Washington, DC 20008
telephone: [1] (202) 483-1407
FAX: [1] (202) 483-8416
consulate(s) general: Atlanta, Honolulu, Houston, Miami, New Orleans, New York, Philadelphia, San Diego, San Francisco, San Juan (Puerto Rico), Tampa
Diplomatic representation from the US:
chief of mission: Ambassador William A. EATON
embassy: Edificio 783, Avenida Demetrio Basilio Lakas Panama, Apartado Postal 0816-02561, Zona 5, Panama City 5
mailing address: American Embassy Panama, Unit 0945, APO AA 34002
telephone: [507] 207-7000
FAX: [507] 227-1964
Flag description: divided into four, equal rectangles; the top quadrants are white (hoist side) with a blue five-pointed star in the center and plain red; the bottom quadrants are plain blue (hoist side) and white with a red five-pointed star in the center

ECONOMY

Economy—overview: Panama's dollarized economy rests primarily on a well-developed services sector that accounts for three-fourths of GDP. Services include operating the Panama Canal, banking, the Colon Free Zone, insur-

503

ance, container ports, flagship registry, and tourism. Economic growth will be bolstered by the Panama Canal expansion project that began in 2007 and should be completed by 2014 at a cost of $5.3 billion (about 30% of current GDP). The expansion project will more than double the Canal's capacity, enabling it to accommodate ships that are now too large to transverse the transoceanic crossway and should help to reduce the high unemployment rate. The government has implemented tax reforms, as well as social security reforms, and backs regional trade agreements and development of tourism. Not a CAFTA signatory, Panama in December 2006 independently negotiated a free trade agreement with the US, which, when implemented, will help promote the country's economic growth.

GDP (purchasing power parity): $34.51 billion (2007 est.)

GDP (official exchange rate): $19.74 billion (2007 est.)

GDP—real growth rate: 11.2% (2007 est.)

GDP—per capita (PPP): $10,300 (2007 est.)

GDP—composition by sector:
agriculture: 6.6%
industry: 16.4%
services: 77% (2007 est.)

Labor force: 1.362 million
note: shortage of skilled labor, but an oversupply of unskilled labor (2007 est.)

Labor force—by occupation: *agriculture:* 15%
industry: 18%
services: 67% (2006)

Unemployment rate: 6.4% (2007 est.)

Population below poverty line: 37% (1999 est.)

Household income or consumption by percentage share:
lowest 10%: 0.7%
highest 10%: 43% (2003)

Distribution of family income—Gini index: 56.1 (2003)

Inflation rate (consumer prices): 4.2% (2007 est.)

Investment (gross fixed): 20.1% of GDP (2007 est.)

Budget:
revenues: $5.505 billion
expenditures: $4.822 billion (2007 est.)

Public debt: 52.8% of GDP (2007 est.)

Agriculture—products: bananas, rice, corn, coffee, sugarcane, vegetables; livestock; shrimp

Industries: construction, brewing, cement and other construction materials, sugar milling

Industrial production growth rate: 10.5% (2007 est.)

Electricity—production: 5.661 billion kWh (2005)

Electricity—production by source:
fossil fuel: 37%
hydro: 61.3%
nuclear: 0%
other: 1.7% (2001)

Electricity—consumption: 4.735 billion kWh (2005)

Electricity—exports: 51 million kWh (2005)

Electricity—imports: 55 million kWh (2005)

Oil—production: 0 bbl/day (2005 est.)

Oil—consumption: 93,000 bbl/day (2006 est.)

Oil—exports: 4,140 bbl/day (2004)

Oil—imports: 92,170 bbl/day (2004)

Oil—proved reserves: 0 bbl (1 January 2006 est.)

Natural gas—production: 0 cu m (2005 est.)

Natural gas—consumption: 0 cu m (2005 est.)

Natural gas—exports: 0 cu m (2005 est.)

Natural gas—imports: 0 cu m (2005)

Natural gas—proved reserves: 0 cu m (1 January 2006 est.)

Current account balance: -$1.579 billion (2007 est.)

Exports: $9.312 billion f.o.b.; note—includes the Colon Free Zone (2007 est.)

Exports—commodities: bananas, shrimp, sugar, coffee, clothing

Exports—partners: US 39.8%, Spain 8.1%, Netherlands 6.7%, Sweden 5.6%, Costa Rica 4.5% (2006)

Imports: $12.63 billion f.o.b.
note: includes the Colon Free Zone (2007 est.)

Imports—commodities: capital goods, foodstuffs, consumer goods, chemicals

Imports—partners: US 27%, Netherlands Antilles 10.1%, Costa Rica 5.1%, Japan 4.7% (2006)

Economic aid—recipient: $19.54 million (2005)

Reserves of foreign exchange and gold: $1.935 billion (31 December 2007 est.)

Debt—external: $10.45 billion (31 December 2007 est.)

Stock of direct foreign investment—at home: $NA

Stock of direct foreign investment—abroad: $NA

Market value of publicly traded shares: $5.074 billion (2005)

Currency (code): balboa (PAB); US dollar (USD)

Currency code: PAB; USD

Exchange rates: balboas per US dollar—1 (2007), 1 (2006), 1 (2005), 1 (2004), 1 (2003)

Fiscal year: calendar year

COMMUNICATIONS

Telephones—main lines in use: 432,900 (2006)

Telephones—mobile cellular: 1.694 million (2005)

Telephone system:
general assessment: domestic and international facilities well developed
domestic: combined fixed-line and mobile-cellular teledensity is approaching 70 per 100 persons
international: country code—507; landing point for the Americas Region Caribbean Ring System (ARCOS-1), the MAYA-1, and PAN-AM submarine cable systems that together provide links to the US, and parts of the Caribbean, Central America, and South America; satellite earth stations—2 Intelsat (Atlantic Ocean); connected to the Central American Microwave System

Radio broadcast stations: AM 101, FM 134, shortwave 0 (1998)

Radios: 815,000 (1997)

Television broadcast stations: 38 (including repeaters) (1998)

Televisions: 510,000 (1997)

Internet country code: .pa

Internet hosts: 7,078 (2007)

Internet Service Providers (ISPs): 6 (2000)

Internet users: 220,000 (2006)

TRANSPORTATION

Airports: 116 (2007)

Airports—with paved runways:
total: 54
over 3,047 m: 1
2,438 to 3,047 m: 1
1,524 to 2,437 m: 5
914 to 1,523 m: 18
under 914 m: 29 (2007)

Airports—with unpaved runways:
total: 62
1,524 to 2,437 m: 1
914 to 1,523 m: 11
under 914 m: 50 (2007)

Heliports: 2 (2007)

Railways:
total: 355 km
standard gauge: 77 km 1.435-m gauge
narrow gauge: 278 km 0.914-m gauge (2006)

Roadways:
total: 11,643 km
paved: 4,028 km
unpaved: 7,615 km (2000)

Waterways: 800 km (includes 82 km Panama Canal) (2007)

Merchant marine:
total: 5,764 ships (1000 GRT or over) 159,649,801 GRT/240,190,316 DWT
by type: barge carrier 2, bulk carrier

1,940, cargo 1,034, carrier 3, chemical tanker 507, combination ore/oil 6, container 710, liquefied gas 191, livestock carrier 7, passenger 46, passenger/cargo 72, petroleum tanker 522, refrigerated cargo 288, roll on/roll off 129, specialized tanker 22, vehicle carrier 285

foreign-owned: 4,949 (Albania 1, Argentina 8, Australia 4, Bahamas 2, Bangladesh 1, Belgium 11, Bulgaria 1, Canada 17, Chile 8, China 473, Colombia 4, Croatia 6, Cuba 11, Cyprus 15, Denmark 32, Dominican Republic 1, Ecuador 2, Egypt 13, Estonia 3, France 15, Gabon 1, Germany 38, Greece 505, Hong Kong 137, India 25, Indonesia 37, Iran 4, Ireland 1, Israel 2, Italy 10, Jamaica 1, Japan 2,151, Jordan 11, South Korea 316, Kuwait 1, Latvia 5, Lebanon 3, Lithuania 5, Malaysia 14, Maldives 1, Malta 2, Mexico 4, Monaco 11, Netherlands 14, Nigeria 6, Norway 60, Oman 1, Pakistan 5, Peru 15, Philippines 12, Poland 15, Portugal 9, Qatar 1, Romania 8, Russia 9, Saudi Arabia 14, Singapore 83, Spain 61, Sri Lanka 3, Sweden 9, Switzerland 26, Syria 24, Taiwan 306, Thailand 10, Turkey 53, Turks and Caicos Islands 1, Ukraine 8, UAE 108, UK 35, US 115, Venezuela 10, Vietnam 10, Yemen 5)

registered in other countries: 1 (Venezuela 1) (2007)

Ports and terminals: Balboa, Colon, Cristobal

MILITARY

Military branches: no regular military forces; Panamanian Public Forces or PPF includes the Panamanian National Police (PNP), National Maritime Service (NMS), and National Air Service (NAS) (2008)

Manpower available for military service:
males age 16–49: 851,044 (2008 est.)

Manpower fit for military service:
males age 16–49: 673,103 (2008 est.)

Manpower reaching militarily significant age annually:
males age 16–49: 30,348 (2008 est.)

Military expenditures—percent of GDP: 1% (2006)

Military—note: on 10 February 1990, the government of then President ENDARA abolished Panama's military and reformed the security apparatus by creating the Panamanian Public Forces; in October 1994, Panama's Legislative Assembly approved a constitutional amendment prohibiting the creation of a standing military force but allowing the temporary establishment of special police units to counter acts of "external aggression"

TRANSNATIONAL ISSUES

Disputes—international: organized illegal narcotics operations in Colombia operate within the remote border region with Panama

Illicit drugs: major cocaine transshipment point and primary money-laundering center for narcotics revenue; money-laundering activity is especially heavy in the Colon Free Zone; offshore financial center; negligible signs of coca cultivation; monitoring of financial transactions is improving; official corruption remains a major problem

PAPUA NEW GUINEA

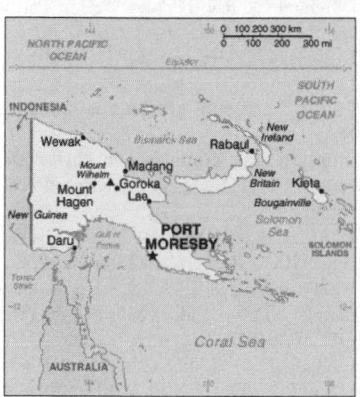

INTRODUCTION

Background: The eastern half of the island of New Guinea—second largest in the world—was divided between Germany (north) and the UK (south) in 1885. The latter area was transferred to Australia in 1902, which occupied the northern portion during World War I and continued to administer the combined areas until independence in 1975. A nine-year secessionist revolt on the island of Bougainville ended in 1997 after claiming some 20,000 lives.

GEOGRAPHY

Location: Oceania, group of islands including the eastern half of the island of New Guinea between the Coral Sea and the South Pacific Ocean, east of Indonesia

Geographic coordinates: 6 00 S, 147 00 E

Map references: Oceania

Area: *total:* 462,840 sq km
land: 452,860 sq km
water: 9,980 sq km

Area—comparative: slightly larger than California

Land boundaries:
total: 820 km
border countries: Indonesia 820 km

Coastline: 5,152 km

Maritime claims: measured from claimed archipelagic baselines
territorial sea: 12 nm
continental shelf: 200-m depth or to the depth of exploitation
exclusive fishing zone: 200 nm

Climate: tropical; northwest monsoon (December to March), southeast monsoon (May to October); slight seasonal temperature variation

Terrain: mostly mountains with coastal lowlands and rolling foothills

Elevation extremes:
lowest point: Pacific Ocean 0 m
highest point: Mount Wilhelm 4,509 m

Natural resources: gold, copper, silver, natural gas, timber, oil, fisheries

Land use:
arable land: 0.49%
permanent crops: 1.4%
other: 98.11% (2005)

Irrigated land: NA

Total renewable water resources: 801 cu km (1987)

Freshwater withdrawal (domestic/industrial/agricultural): *total:* 0.1 cu km/yr (56%/43%/1%)
per capita: 17 cu m/yr (1987)

Natural hazards: active volcanism; situated along the Pacific "Ring of Fire"; the country is subject to frequent and sometimes severe earthquakes; mud slides; tsunamis

Environment—current issues: rain forest subject to deforestation as a result of growing commercial demand for tropical timber; pollution from mining projects; severe drought

Environment—international agreements: *party to:* Antarctic Treaty, Biodiversity, Climate Change, Climate Change-Kyoto Protocol, Desertification, Endangered Species, Environmental Modification, Hazardous Wastes, Law of the Sea, Marine Dumping, Ozone Layer Protection, Ship Pollution, Tropical Timber 83, Tropical Timber 94, Wetlands
signed, but not ratified: none of the selected agreements

Geography—note: shares island of New Guinea with Indonesia; one of world's largest swamps along southwest coast

PEOPLE

Population: 5,931,769 (July 2008 est.)
Age structure:

0–14 years: 37.3% (male 1,124,174/female 1,086,478)
15–64 years: 58.7% (male 1,791,342/female 1,690,089)
65 years and over: 4% (male 111,023/female 128,663) (2008 est.)
Median age:
total: 21.5 years
male: 21.6 years
female: 21.4 years (2008 est.)
Population growth rate: 2.118% (2008 est.)
Birth rate: 28.14 births/1,000 population (2008 est.)
Death rate: 6.96 deaths/1,000 population (2008 est.)
Net migration rate: NA
Sex ratio:
at birth: 1.05 male(s)/female
under 15 years: 1.03 male(s)/female
15–64 years: 1.06 male(s)/female
65 years and over: 0.86 male(s)/female
total population: 1.04 male(s)/female (2008 est.)
Infant mortality rate:
total: 46.67 deaths/1,000 live births
male: 50.68 deaths/1,000 live births
female: 42.47 deaths/1,000 live births (2008 est.)
Life expectancy at birth:
total population: 66 years
male: 63.76 years
female: 68.35 years (2008 est.)
Total fertility rate: 3.71 children born/woman (2008 est.)
HIV/AIDS—adult prevalence rate: 0.6% (2003 est.)
HIV/AIDS—people living with HIV/AIDS: 60,000 (2005 est.)
HIV/AIDS—deaths: 600 (2003 est.)
Major infectious diseases:
degree of risk: very high
food or waterborne diseases: bacterial diarrhea, hepatitis A, and typhoid fever
vectorborne diseases: dengue fever and malaria (2008)
Nationality:
noun: Papua New Guinean(s)
adjective: Papua New Guinean
Ethnic groups: Melanesian, Papuan, Negrito, Micronesian, Polynesian
Religions: Roman Catholic 22%, Lutheran 16%, Presbyterian/Methodist/London Missionary Society 8%, Anglican 5%, Evangelical Alliance 4%, Seventh-Day Adventist 1%, other Protestant 10%, indigenous beliefs 34%

Languages: Melanesian Pidgin serves as the lingua franca, English spoken by 1%-2%, Motu spoken in Papua region
note: 820 indigenous languages spoken (over one-tenth of the world's total)
Literacy: *definition:* age 15 and over can read and write
total population: 57.3%
male: 63.4%
female: 50.9% (2000 census)

GOVERNMENT

Country name:
conventional long form: Independent State of Papua New Guinea
conventional short form: Papua New Guinea
local short form: Papuaniugini
former: Territory of Papua and New Guinea
abbreviation: PNG
Government type: constitutional parliamentary democracy
Capital: *name:* Port Moresby
geographic coordinates: 9 30 S, 147 10 E
time difference: UTC+10 (15 hours ahead of Washington, DC during Standard Time)
Administrative divisions: 20 provinces; Bougainville, Central, Chimbu, Eastern Highlands, East New Britain, East Sepik, Enga, Gulf, Madang, Manus, Milne Bay, Morobe, National Capital, New Ireland, Northern, Sandaun, Southern Highlands, Western, Western Highlands, West New Britain
Independence: 16 September 1975 (from the Australian-administered UN trusteeship)
National holiday: Independence Day, 16 September (1975)
Constitution: 16 September 1975
Legal system: based on English common law; has not accepted compulsory ICJ jurisdiction
Suffrage: 18 years of age; universal
Executive branch:
chief of state: Queen ELIZABETH II (since 6 February 1952); represented by governor general Sir Paulius MATANE (since 29 June 2004)
head of government: Prime Minister Sir Michael SOMARE (since 2 August 2002); Deputy Prime Minister Puka TEMU (since 29 August 2007)
cabinet: National Executive Council appointed by governor general on recommendation of prime minister
elections: monarch is hereditary; governor general nominated by parliament and appointed by chief of state; following legislative elections, leader of majority party or leader of majority coalition usually is appointed prime minister by governor general

Legislative branch: unicameral National Parliament (109 seats, 89 filled from open electorates and 20 from provinces and national capital district; members elected by popular vote to serve five-year terms); constitution allows up to 126 seats
elections: last held from 30 June to 10 July 2007; next to be held in June 2012
election results: percent of vote by party—NA; seats by party—National Alliance 27, PNGP 8, PAP 6, URP 6, PANGU 5, PDM 5, independents 19, others 33; note—election to 1 seat was nullified
note: 15 other parties won 4 or fewer seats; association with political parties is fluid
Judicial branch: Supreme Court (the chief justice is appointed by the governor general on the proposal of the National Executive Council after consultation with the minister responsible for justice; other judges are appointed by the Judicial and Legal Services Commission)
Political parties and leaders: National Alliance Party or NA [Michael SOMARE]; Papua and Niugini Union Party or PANGU PATI [Andrew KUMBAKOR]; Papua New Guinea Party or PNGP [Sir Mekere MORAUTA]; People's Democratic Movement or PDM [Michael OGIO]; People's Action Party or PAP [Gabriel KAPRIS]; United Resources Party or URP [William DUMA] (2007)
Political pressure groups and leaders: NA
International organization participation: ACP, ADB, APEC, ARF, ASEAN (observer), C, CP, FAO, G-77, IBRD, ICAO, ICRM, IDA, IFAD, IFC, IFRCS, IHO, ILO, IMF, IMO, Interpol, IOC, IOM (observer), IPU, ISO (correspondent), ITSO, ITU, MIGA, NAM, OPCW, PIF, Sparteca, SPC, UN, UNCTAD, UNESCO, UNIDO, UNWTO, UPU, WCO, WFTU, WHO, WIPO, WMO, WTO
Diplomatic representation in the US:
chief of mission: Ambassador Evan Jeremy PAKI
chancery: 1779 Massachusetts Avenue NW, Suite 805, Washington, DC 20036
telephone: [1] (202) 745-3680
FAX: [1] (202) 745-3679
Diplomatic representation from the US:
chief of mission: Ambassador Leslie W. Rowe
embassy: Douglas Street, Port Moresby, N.C.D.
mailing address: 4240 Port Moresby PI, US Department of State, Washington DC 20521-4240
telephone: [675] 321-1455
FAX: [675] 321-3423

Flag description: divided diagonally from upper hoist-side corner; the upper triangle is red with a soaring yellow bird of paradise centered; the lower triangle is black with five, white, five-pointed stars of the Southern Cross constellation centered

ECONOMY

Economy—overview: Papua New Guinea is richly endowed with natural resources, but exploitation has been hampered by rugged terrain and the high cost of developing infrastructure. Agriculture provides a subsistence livelihood for 85% of the population. Mineral deposits, including copper, gold, and oil, account for nearly two-thirds of export earnings. The government of Prime Minister SOMARE has expended much of its energy remaining in power. He was the first prime minister ever to serve a full five-year term. The government also brought stability to the national budget, largely through expenditure control; however, it relaxed spending constraints in 2006 and 2007 as elections approached. Numerous challenges still face the government including regaining investor confidence, restoring integrity to state institutions, promoting economic efficiency by privatizing moribund state institutions, and balancing relations with Australia, its former colonial ruler. Other socio-cultural challenges could upend the economy including a worsening HIV/AIDS epidemic and chronic law and order and land tenure issues. Australia will supply more than $300 million in aid in FY07/08, which accounts for nearly 20% of the national budget.

GDP (purchasing power parity): $11.94 billion (2007 est.)

GDP (official exchange rate): $6.001 billion (2007 est.)

GDP—real growth rate: 6.2% (2007 est.)

GDP—per capita (PPP): $2,000 (2007 est.)

GDP—composition by sector:
agriculture: 34%
industry: 37.3%
services: 28.7% (2007 est.)

Labor force: 3.557 million (2007 est.)

Labor force—by occupation: *agriculture:* 85%
industry: NA%
services: NA% (2005 est.)

Unemployment rate: 1.9% up to 80% in urban areas (2004)

Population below poverty line: 37% (2002 est.)

Household income or consumption by percentage share:
lowest 10%: 1.7%

highest 10%: 40.5% (1996)

Distribution of family income—Gini index: 50.9 (1996)

Inflation rate (consumer prices): 1.7% (2007 est.)

Investment (gross fixed): 20.2% of GDP (2007 est.)

Budget:
revenues: $2.347 billion
expenditures: $2.153 billion (2007 est.)

Public debt: 42% of GDP (2007 est.)

Agriculture—products: coffee, cocoa, copra, palm kernels, tea, sugar, rubber, sweet potatoes, fruit, vegetables, vanilla; shell fish, poultry, pork

Industries: copra crushing, palm oil processing, plywood production, wood chip production; mining of gold, silver, and copper; crude oil production, petroleum refining; construction, tourism

Industrial production growth rate: 6.4% (2007 est.)

Electricity—production: 3.698 billion kWh (2005)

Electricity—production by source:
fossil fuel: 54.1%
hydro: 45.9%
nuclear: 0%
other: 0% (2001)

Electricity—consumption: 3.439 billion kWh (2005)

Electricity—exports: 0 kWh (2005)

Electricity—imports: 0 kWh (2005)

Oil—production: 50,000 bbl/day (January 2006 est.)

Oil—consumption: 26,000 bbl/day (2005 est.)

Oil—exports: 44,580 bbl/day (2004)

Oil—imports: 24,020 bbl/day (2004)

Oil—proved reserves: 170 million bbl (2007 est.)

Natural gas—production: 95.91 million cu m (2005 est.)

Natural gas—consumption: 95.91 million cu m (2005 est.)

Natural gas—exports: 0 cu m (2005 est.)

Natural gas—imports: 0 cu m (2005)

Natural gas—proved reserves: 331.3 billion cu m (1 January 2006 est.)

Current account balance: $259 million (2007 est.)

Exports: $4.676 billion f.o.b. (2007 est.)

Exports—commodities: oil, gold, copper ore, logs, palm oil, coffee, cocoa, crayfish, prawns

Exports—partners: Australia 30.2%, Japan 8.2%, China 5.7% (2006)

Imports: $2.564 billion f.o.b. (2007 est.)

Imports—commodities: machinery and transport equipment, manufactured goods, food, fuels, chemicals

Imports—partners: Australia 52%, Singapore 12.6%, China 5.9%, Japan 4.3% (2006)

Economic aid—recipient: $266.1 million (2005)

Reserves of foreign exchange and gold: $2.108 billion (31 December 2007 est.)

Debt—external: $1.595 billion (31 December 2007 est.)

Stock of direct foreign investment—at home: $NA

Stock of direct foreign investment—abroad: $NA

Market value of publicly traded shares: $4.863 billion (2005)

Currency (code): kina (PGK)

Currency code: PGK

Exchange rates: kina per US dollar—3.03 (2007), 3.0643 (2006), 3.08 (2005), 3.2225 (2004), 3.5635 (2003)

Fiscal year: calendar year

COMMUNICATIONS

Telephones—main lines in use: 63,700 (2005)

Telephones—mobile cellular: 75,000 (2005)

Telephone system:
general assessment: services are minimal; facilities provide radiotelephone and telegraph, coastal radio, aeronautical radio, and international radio communication services
domestic: access to telephone services is not widely available; combined fixed-line and mobile-cellular teledensity is less than 3 per 100 persons
international: country code—675; submarine cables to Australia and Guam; satellite earth station—1 Intelsat (Pacific Ocean); international radio communication service

Radio broadcast stations: AM 8, FM 19, shortwave 28 (1998)

Radios: 410,000 (1997)

Television broadcast stations: 3 (all in the Port Moresby area; stations at Mt. Hagen, Goroka, Lae, and Rabaul are planned) (2004)

Televisions: 59,841 (1999)

Internet country code: .pg

Internet hosts: 2,436 (2007)

Internet Service Providers (ISPs): 3 (2000)

Internet users: 110,000 (2006)

TRANSPORTATION

Airports: 578 (2007)

Airports—with paved runways:
total: 21
2,438 to 3,047 m: 2
1,524 to 2,437 m: 14
914 to 1,523 m: 4
under 914 m: 1 (2007)

Airports—with unpaved runways:
total: 557
1,524 to 2,437 m: 10

914 to 1,523 m: 58
under 914 m: 489 (2007)
Heliports: 2 (2007)
Pipelines: oil 264 km (2007)
Roadways:
total: 19,600 km
paved: 686 km
unpaved: 18,914 km (1999)
Waterways: 11,000 km (2006)
Merchant marine:
total: 24 ships (1000 GRT or over)
56,157 GRT/72,821 DWT
by type: bulk carrier 3, cargo 20, petroleum tanker 1
foreign-owned: 6 (UK 6) (2007)
Ports and terminals: Kimbe, Lae,
Madang, Rabaul, Wewak

MILITARY

Military branches: Papua New Guinea
Defense Force (PNGDF; includes
Maritime Operations Element, Air
Operations Element) (2008)
Military service age and obligation: 16
years of age for voluntary military
service; no conscription (2008)
**Manpower available for military
service:** *males age 16–49:* 1,481,417
females age 16–49: 1,385,040 (2008 est.)
Manpower fit for military service:
males age 16–49: 1,080,466
females age 16–49: 1,092,000 (2008 est.)
**Military expenditures—percent of
GDP:** 1.4% (2005 est.)

TRANSNATIONAL ISSUES

Disputes—international: relies on assistance from Australia to keep out illegal
cross-border activities from primarily
Indonesia, including goods smuggling,
illegal narcotics trafficking, and squatters
and secessionists
**Refugees and internally displaced per-
sons:** *refugees (country of origin):* 10,177
(Indonesia) (2007)
Illicit drugs: major consumer of cannabis

PARACEL ISLANDS

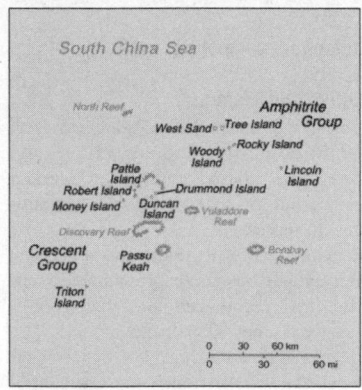

INTRODUCTION

Background: The Paracel Islands are surrounded by productive fishing grounds
and by potential oil and gas reserves. In
1932, French Indochina annexed the
islands and set up a weather station on
Pattle Island; maintenance was continued by its successor, Vietnam. China
has occupied the Paracel Islands since
1974, when its troops seized a South
Vietnamese garrison occupying the
western islands. China built a military
installation on Mischief Reef in 1999.
The islands are claimed by Taiwan and
Vietnam.

GEOGRAPHY

Location: Southeastern Asia, group of
small islands and reefs in the South
China Sea, about one-third of the way
from central Vietnam to the northern
Philippines
Geographic coordinates: 16 30 N, 112
00 E
Map references: Southeast Asia
Area: *total:* NA sq km
land: NA sq km
water: 0 sq km
Area—comparative: NA
Land boundaries: 0 km
Coastline: 518 km
Maritime claims: NA
Climate: tropical
Terrain: mostly low and flat
Elevation extremes:
lowest point: South China Sea 0 m
highest point: unnamed location on
Rocky Island 14 m
Natural resources: none
Land use:
arable land: 0%
permanent crops: 0%
other: 100% (2005)
Irrigated land: 0 sq km
Natural hazards: typhoons
Environment—current issues: NA
Geography—note: composed of 130

small coral islands and reefs divided into
the northeast Amphitrite Group and the
western Crescent Group

PEOPLE

Population: no indigenous inhabitants
note: there are scattered Chinese garrisons

GOVERNMENT

Country name:
conventional long form: none
conventional short form: Paracel Islands

ECONOMY

Economy—overview: China announced
plans in 1997 to open the islands for
tourism.

TRANSPORTATION

Airports: 1 (2007)
Airports—with paved runways: *total:* 1
1,524 to 2,437 m: 1 (2007)
Ports and terminals: small Chinese port
facilities on Woody Island and Duncan
Island being expanded

MILITARY

Military—note: occupied by China

TRANSNATIONAL ISSUES

Disputes—international: occupied by
China, also claimed by Taiwan and
Vietnam

PARAGUAY

INTRODUCTION

Background: In the disastrous War of the
Triple Alliance (1865–70)—between
Paraguay and Argentina, Brazil, and
Uruguay—Paraguay lost two-thirds of all
adult males and much of its territory. It

stagnated economically for the next half
century. In the Chaco War of 1932–35,
Paraguay won large, economically important areas from Bolivia. The 35-year military dictatorship of Alfredo
STROESSNER ended in 1989, and,
despite a marked increase in political

infighting in recent years, Paraguay has
held relatively free and regular presidential elections since then.

GEOGRAPHY

Location: Central South America,
northeast of Argentina

Geographic coordinates: 23 00 S, 58 00 W

Map references: South America

Area: *total:* 406,750 sq km
land: 397,300 sq km
water: 9,450 sq km

Area—comparative: slightly smaller than California

Land boundaries:
total: 3,995 km
border countries: Argentina 1,880 km, Bolivia 750 km, Brazil 1,365 km

Coastline: 0 km (landlocked)

Maritime claims: none (landlocked)

Climate: subtropical to temperate; substantial rainfall in the eastern portions, becoming semiarid in the far west

Terrain: grassy plains and wooded hills east of Rio Paraguay; Gran Chaco region west of Rio Paraguay mostly low, marshy plain near the river, and dry forest and thorny scrub elsewhere

Elevation extremes:
lowest point: junction of Rio Paraguay and Rio Parana 46 m
highest point: Cerro Pero (Cerro Tres Kandu) 842 m

Natural resources: hydropower, timber, iron ore, manganese, limestone

Land use:
arable land: 7.47%
permanent crops: 0.24%
other: 92.29% (2005)

Irrigated land: 670 sq km (2003)

Total renewable water resources: 336 cu km (2000)

Freshwater withdrawal (domestic/industrial/agricultural): *total:* 0.49 cu km/yr (20%/8%/71%)
per capita: 80 cu m/yr (2000)

Natural hazards: local flooding in southeast (early September to June); poorly drained plains may become boggy (early October to June)

Environment—current issues: deforestation; water pollution; inadequate means for waste disposal pose health risks for many urban residents; loss of wetlands

Environment—international agreements: *party to:* Biodiversity, Climate Change, Climate Change-Kyoto Protocol, Desertification, Endangered Species, Hazardous Wastes, Law of the Sea, Ozone Layer Protection, Wetlands *signed, but not ratified:* none of the selected agreements

Geography—note: landlocked; lies between Argentina, Bolivia, and Brazil; population concentrated in southern part of country

PEOPLE

Population: 6,831,306 (July 2008 est.)

Age structure:
0–14 years: 36.9% (male 1,283,311/female 1,240,769)
15–64 years: 57.9% (male 1,988,256/female 1,968,869)
65 years and over: 5.1% (male 161,811/female 188,290) (2008 est.)

Median age:
total: 21.7 years
male: 21.5 years
female: 22 years (2008 est.)

Population growth rate: 2.39% (2008 est.)

Birth rate: 28.47 births/1,000 population (2008 est.)

Death rate: 4.49 deaths/1,000 population (2008 est.)

Net migration rate: -0.07 migrant(s)/1,000 population (2008 est.)

Sex ratio:
at birth: 1.05 male(s)/female
under 15 years: 1.03 male(s)/female
15–64 years: 1.01 male(s)/female
65 years and over: 0.86 male(s)/female
total population: 1.01 male(s)/female (2008 est.)

Infant mortality rate:
total: 25.55 deaths/1,000 live births
male: 29.74 deaths/1,000 live births
female: 21.16 deaths/1,000 live births (2008 est.)

Life expectancy at birth:
total population: 75.56 years
male: 72.99 years
female: 78.26 years (2008 est.)

Total fertility rate: 3.8 children born/woman (2008 est.)

HIV/AIDS—adult prevalence rate: 0.5% (2003 est.)

HIV/AIDS—people living with HIV/AIDS: 15,000 (1999 est.)

HIV/AIDS—deaths: 600 (2003 est.)

Major infectious diseases:
degree of risk: intermediate
food or waterborne diseases: bacterial diarrhea, hepatitis A, and typhoid fever
vectorborne disease: dengue fever and malaria (2008)

Nationality: *noun:* Paraguayan(s)
adjective: Paraguayan

Ethnic groups: mestizo (mixed Spanish and Amerindian) 95%, other 5%

Religions: Roman Catholic 89.6%, Protestant 6.2%, other Christian 1.1%, other or unspecified 1.9%, none 1.1% (2002 census)

Languages: Spanish (official), Guarani (official)

Literacy:
definition: age 15 and over can read and write
total population: 94%
male: 94.9%
female: 93% (2003 est.)

GOVERNMENT

Country name:
conventional long form: Republic of Paraguay
conventional short form: Paraguay
local long form: Republica del Paraguay
local short form: Paraguay

Government type: constitutional republic

Capital: *name:* Asuncion
geographic coordinates: 25 16 S, 57 40 W
time difference: UTC-4 (1 hour ahead of Washington, DC during Standard Time)

Administrative divisions: 17 departments (departamentos, singular—departamento) and 1 capital city*; Alto Paraguay, Alto Parana, Amambay, Asuncion*, Boqueron, Caaguazu, Caazapa, Canindeyu, Central, Concepcion, Cordillera, Guaira, Itapua, Misiones, Neembucu, Paraguari, Presidente Hayes, San Pedro

Independence: 14 May 1811 (from Spain)

National holiday: Independence Day, 14 May 1811 (observed 15 May)

Constitution: promulgated 20 June 1992

Legal system: based on Argentine codes, Roman law, and French codes; judicial review of legislative acts in Supreme Court of Justice; accepts compulsory ICJ jurisdiction

Suffrage: 18 years of age; universal and compulsory up to age 75

Executive branch:
chief of state: President Nicanor DUARTE FRUTOS (since 15 August 2003); Vice President Francisco OVIEDO Britez (since 21 November 2007); note—the president is both the chief of state and head of government
head of government: President Nicanor DUARTE FRUTOS (since 15 August 2003); Vice President Francisco OVIEDO Britez (since 21 November 2007)
cabinet: Council of Ministers appointed by the president
elections: president and vice president elected on the same ticket by popular

509

vote for a single five-year term; election last held 20 April 2008 (next to be held April 2013)

election results: Fernando LUGO elected president; percent of vote—Fernando LUGO 40.8%, Blanca OVELAR 30.6%, Lino OVIEDO 21.9%, Pedro FADUL 2.4%, other 4.3%; note—LUGO will take office on 15 August 2008

Legislative branch: bicameral Congress or Congreso consists of the Chamber of Senators or Camara de Senadores (45 seats; members are elected by popular vote to serve five-year terms) and the Chamber of Deputies or Camara de Diputados (80 seats; members are elected by popular vote to serve five-year terms)

elections: Chamber of Senators—last held 20 April 2008 (next to be held in April 2013); Chamber of Deputies—last held 20 April 2008 (next to be held in April 2013)

election results: Chamber of Senators—percent of vote by party—NA; seats by party—ANR 15, PLRA 14, UNACE 9, PPQ 4, other 3; Chamber of Deputies—percent of vote by party—NA; seats by party—ANR 30, PLRA 27, UNACE 15, PPQ 3, APC 2, other 3

Judicial branch: Supreme Court of Justice or Corte Suprema de Justicia (nine judges appointed on the proposal of the Council of Magistrates or Consejo de la Magistratura)

Political parties and leaders: Alianza Patriotica por el Cambio (Patriotic Alliance for Change) or APC [Fernando LUGO]; Asociacion Nacional Republicana—Colorado Party or ANR [Jose Alberto ALDERETE]; Movimiento Union Nacional de Ciudadanos Eticos or UNACE [Enrique GONZALEZ Quintana]; Patria Querida (Beloved Fatherland Party) or PPQ [Pedro Nicolas Maraa FADUL Niella]; Partido del Movimiento al Socialismo or P-MAS; Partido Encuentro Nacional or PEN [Emilio CAMACHO Paredes]; Partido Liberal Radical Autentico or PLRA [Blas LLANO]; Partido Pais Solidario or PPS [Carlos Alberto FILIZZOLA Pallares]

Political pressure groups and leaders: Ahorristas Estafados or AE; National Coordinating Board of Campesino Organizations or MCNOC [Luis AGUAYO]; National Federation of Campesinos or FNC [Odilon ESPINOLA]; National Workers Central or CNT [Secretary General Juan TORRALES]; Paraguayan Workers Confederation or CPT; Roman Catholic Church; Unitary Workers Central or CUT [Jorge Guzman ALVARENGA Malgarejo]

International organization participation: CAN (associate), CSN, FAO, G-77, IADB, IAEA, IBRD, ICAO, ICCt, ICRM, IDA, IFAD, IFC, IFRCS, ILO, IMF, IMO, Interpol, IOC, IOM, IPU, ISO (correspondent), ITSO, ITU, ITUC, LAES, LAIA, Mercosur, MIGA, NAM (observer), OAS, OPANAL, OPCW, PCA, RG, UN, UNCTAD, UNESCO, UNFICYP, UNIDO, Union Latina, UNMEE, UNMIL, UNMIS, UNOCI, UNWTO, UPU, WCL, WCO, WHO, WIPO, WMO, WTO

Diplomatic representation in the US:
chief of mission: Ambassador James SPALDING Hellmers
chancery: 2400 Massachusetts Avenue NW, Washington, DC 20008
telephone: [1] (202) 483-6960 through 6962
FAX: [1] (202) 234-4508
consulate(s) general: Los Angeles, Miami, New York

Diplomatic representation from the US:
Ambassador James C. CASON
embassy: 1776 Avenida Mariscal Lopez, Casilla Postal 402, Asuncion
mailing address: Unit 4711, APO AA 34036-0001
telephone: [595] (21) 213-715
FAX: [595] (21) 213-728

Flag description: three equal, horizontal bands of red (top), white, and blue with an emblem centered in the white band; unusual flag in that the emblem is different on each side; the obverse (hoist side at the left) bears the national coat of arms (a yellow five-pointed star within a green wreath capped by the words REPUBLICA DEL PARAGUAY, all within two circles); the reverse (hoist side at the right) bears the seal of the treasury (a yellow lion below a red Cap of Liberty and the words Paz y Justicia (Peace and Justice) capped by the words REPUBLICA DEL PARAGUAY, all within two circles)

ECONOMY

Economy—overview: Landlocked Paraguay has a market economy marked by a large informal sector. This sector features both reexport of imported consumer goods to neighboring countries, as well as the activities of thousands of microenterprises and urban street vendors. Because of the importance of the informal sector, accurate economic measures are difficult to obtain. A large percentage of the population, especially in rural areas, derives its living from agricultural activity, often on a subsistence basis. On a per capita basis, real income has stagnated at 1980 levels. Most observers attribute Paraguay's poor economic performance to political uncertainty, corruption, limited progress on structural reform, and deficient infrastructure. The economy rebounded between 2003 and 2007, posting modest growth each year, as growing world demand for commodities combined with high prices and favorable weather to support Paraguay's commodity-based export expansion.

GDP (purchasing power parity): $27.08 billion (2007 est.)
GDP (official exchange rate): $10.87 billion (2007 est.)
GDP—real growth rate: 6.4% (2007 est.)
GDP—per capita (PPP): $4,500 (2007 est.)
GDP—composition by sector:
agriculture: 22.4%
industry: 17.6%
services: 60% (2007 est.)
Labor force: 2.787 million (2007 est.)
Labor force—by occupation: *agriculture:* 31%
industry: 17%
services: 52% (2007)
Unemployment rate: 5.6% (2007 est.)
Population below poverty line: 32% (2005 est.)
Household income or consumption by percentage share:
lowest 10%: 0.7%
highest 10%: 46.1% (2003)
Distribution of family income—Gini index: 56.8 (2008)
Inflation rate (consumer prices): 8.1% (2007)
Investment (gross fixed): 18.9% of GDP (2007 est.)
Budget:
revenues: $2.268 billion
expenditures: $2.469 billion (2007)
Public debt: 27.1% of GDP (2007)
Agriculture—products: cotton, sugarcane, soybeans, corn, wheat, tobacco, cassava (tapioca), fruits, vegetables; beef, pork, eggs, milk; timber
Industries: sugar, cement, textiles, beverages, wood products, steel, metallurgic, electric power
Industrial production growth rate: -1% (2007 est.)
Electricity—production: 70 billion kWh (2007)
Electricity—production by source:
fossil fuel: 0%
hydro: 99.9%
nuclear: 0%
other: 0.1% (2001)
Electricity—consumption: 6 billion kWh (2007)
Electricity—exports: 64 billion kWh (2007)
Electricity—imports: 0 kWh (2007)
Oil—production: 0 bbl/day (2007 est.)
Oil—consumption: 28,000 bbl/day (2007 est.)

Oil—exports: 0 bbl/day (2007)

Oil—imports: 25,940 bbl/day (2007)

Oil—proved reserves: 0 bbl (1 January 2006 est.)

Natural gas—production: 0 cu m (2007 est.)

Natural gas—consumption: 0 cu m (2007 est.)

Natural gas—exports: 0 cu m (2007 est.)

Natural gas—imports: 0 cu m (2007)

Natural gas—proved reserves: 0 cu m (1 January 2007 est.)

Current account balance: $162 million (2007 est.)

Exports: $6.898 billion f.o.b. (2007)

Exports—commodities: soybeans, feed, cotton, meat, edible oils, electricity, wood, leather

Exports—partners: Uruguay 22%, Brazil 17.2%, Russia 11.9%, Argentina 8.8%, Chile 6.9% (2006)

Imports: $7.012 billion f.o.b. (2007)

Imports—commodities: road vehicles, consumer goods, tobacco, petroleum products, electrical machinery, tractors, chemicals, vehicle parts

Imports—partners: China 27%, Brazil 20%, Argentina 13.6%, Japan 8.3%, US 6.4% (2006)

Economic aid—recipient: $51.09 million (2005)

Reserves of foreign exchange and gold: $2.463 billion (31 December 2007)

Debt—external: $3.605 billion (31 December 2007 est.)

Stock of direct foreign investment—at home: $2.057 million (2007)

Stock of direct foreign investment—abroad: $NA

Market value of publicly traded shares: $233.8 million (2005)

Currency (code): guarani (PYG)

Currency code: PYG

Exchange rates: guarani per US dollar—5,031 (2007), 5,672.8 (2006), 6,178 (2005), 5,974.6 (2004), 6,424.3 (2003)

Fiscal year: calendar year

COMMUNICATIONS

Telephones—main lines in use: 331,100 (2006)

Telephones—mobile cellular: 3.233 million (2006)

Telephone system:

general assessment: meager telephone service; principal switching center is in Asuncion

domestic: the fixed-line market is a state monopoly; deficiencies in provision of fixed-line service have resulted in a rapid expansion of mobile-cellular services fostered by competition among multiple providers

international: country code—595; satellite earth station—1 Intelsat (Atlantic Ocean)

Radio broadcast stations: AM 41, FM 121, shortwave 6 (3 inactive) (2006)

Radios: 925,000 (1997)

Television broadcast stations: 5 (2007)

Televisions: 990,000 (2001)

Internet country code: .py

Internet hosts: 12,497 (2007)

Internet Service Providers (ISPs): 4 (2000)

Internet users: 260,000 (2006)

TRANSPORTATION

Airports: 838 (2007)

Airports—with paved runways:

total: 13

over 3,047 m: 3

1,524 to 2,437 m: 5

914 to 1,523 m: 5 (2007)

Airports—with unpaved runways:

total: 825

1,524 to 2,437 m: 26

914 to 1,523 m: 267

under 914 m: 532 (2007)

Railways:

total: 36 km

standard gauge: 36 km 1.435-m gauge (2006)

Roadways:

total: 29,500 km

paved: 14,986 km

unpaved: 14,514 km (1999)

Waterways: 3,100 km (2007)

Merchant marine:

total: 22 ships (1000 GRT or over) 39,693 GRT/43,530 DWT

by type: cargo 16, container 1, livestock carrier 1, passenger 1, petroleum tanker 2, roll on/roll off 1

foreign-owned: 5 (Argentina 3, Netherlands 1, Switzerland 1) (2007)

Ports and terminals: Asuncion, Villeta, San Antonio, Encarnacion

MILITARY

Military branches: Army, National Navy (Armada Nacional, includes Naval Aviation, Marine Corps, General Naval Prefecture), Air Force (Fuerza Aerea Paraguay, FAP) (2008)

Military service age and obligation: 18 years of age for compulsory and voluntary military service; conscript service obligation—12 months for Army, 24 months for Navy (2006)

Manpower available for military service:

males age 16–49: 1,589,873

females age 16–49: 1,585,573 (2008 est.)

Manpower fit for military service:

males age 16–49: 1,327,730

females age 16–49: 1,356,989 (2008 est.)

Manpower reaching militarily significant age annually:

males age 16–49: 72,109

females age 16–49: 70,509 (2008 est.)

Military expenditures—percent of GDP: 1% (2006 est.)

TRANSNATIONAL ISSUES

Disputes—international: unruly region at convergence of Argentina-Brazil-Paraguay borders is locus of money laundering, smuggling, arms and illegal narcotics trafficking, and fundraising for extremist organizations

Illicit drugs: major illicit producer of cannabis, most or all of which is consumed in Brazil, Argentina, and Chile; transshipment country for Andean cocaine headed for Brazil, other Southern Cone markets, and Europe; corruption and some money-laundering activity, especially in the Tri-Border Area; weak anti-money-laundering laws and enforcement

PERU

INTRODUCTION

Background: Ancient Peru was the seat of several prominent Andean civilizations, most notably that of the Incas whose empire was captured by the Spanish conquistadors in 1533. Peruvian independence was declared in 1821, and remaining Spanish forces defeated in 1824. After a dozen years of military rule, Peru returned to democratic leadership in 1980, but experienced economic problems and the growth of a violent insurgency. President Alberto FUJIMORI's election in 1990 ushered in a decade that saw a dramatic turnaround in the economy and significant progress in curtailing guerrilla activity. Nevertheless, the president's increasing reliance on authoritarian measures and an economic slump in the late 1990s generated mounting dissatisfaction with his regime, which led to his ouster in 2000. A caretaker government oversaw new elections in the spring of 2001, which ushered in Alejandro TOLEDO as the new head of government—Peru's first democratically elected president of Native American ethnicity. The presidential election of 2006 saw the return of Alan GARCIA who, after a disappointing presidential term from 1985 to 1990, returned to the presidency with promises to improve social conditions and maintain fiscal responsibility.

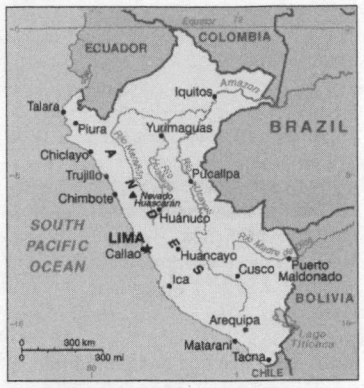

GEOGRAPHY

Location: Western South America, bordering the South Pacific Ocean, between Chile and Ecuador

Geographic coordinates: 10 00 S, 76 00 W

Map references: South America

Area: *total:* 1,285,220 sq km
land: 1.28 million sq km
water: 5,220 sq km

Area—comparative: slightly smaller than Alaska

Land boundaries:
total: 7,461 km
border countries: Bolivia 1,075 km, Brazil 2,995 km, Chile 171 km, Colombia 1,800 km, Ecuador 1,420 km

Coastline: 2,414 km

Maritime claims:
territorial sea: 200 nm
continental shelf: 200 nm

Climate: varies from tropical in east to dry desert in west; temperate to frigid in Andes

Terrain: western coastal plain (costa), high and rugged Andes in center (sierra), eastern lowland jungle of Amazon Basin (selva)

Elevation extremes:
lowest point: Pacific Ocean 0 m
highest point: Nevado Huascaran 6,768 m

Natural resources: copper, silver, gold, petroleum, timber, fish, iron ore, coal, phosphate, potash, hydropower, natural gas

Land use: *arable land:* 2.88%
permanent crops: 0.47%
other: 96.65% (2005)

Irrigated land: 12,000 sq km (2003)

Total renewable water resources: 1,913 cu km (2000)

Freshwater withdrawal (domestic/industrial/agricultural): *total:* 20.13 cu km/yr (8%/10%/82%)
per capita: 720 cu m/yr (2000)

Natural hazards: earthquakes, tsunamis, flooding, landslides, mild volcanic activity

Environment—current issues: deforestation (some the result of illegal logging); overgrazing of the slopes of the costa and sierra leading to soil erosion; desertification; air pollution in Lima; pollution of rivers and coastal waters from municipal and mining wastes

Environment—international agreements: *party to:* Antarctic-Environmental Protocol, Antarctic-Marine Living Resources, Antarctic Treaty, Biodiversity, Climate Change, Climate Change-Kyoto Protocol, Desertification, Endangered Species, Hazardous Wastes, Marine Dumping, Ozone Layer Protection, Ship Pollution, Tropical Timber 83, Tropical Timber 94, Wetlands, Whaling
signed, but not ratified: none of the selected agreements

Geography—note: shares control of Lago Titicaca, world's highest navigable lake, with Bolivia; a remote slope of Nevado Mismi, a 5,316 m peak, is the ultimate source of the Amazon River

PEOPLE

Population: 29,180,899 (July 2008 est.)

Age structure:
0–14 years: 29.7% (male 4,409,227/female 4,253,836)
15–64 years: 64.7% (male 9,501,597/female 9,381,139)
65 years and over: 5.6% (male 770,389/female 864,711) (2008 est.)

Median age:
total: 25.8 years
male: 25.5 years
female: 26.1 years (2008 est.)

Population growth rate: 1.264% (2008 est.)

Birth rate: 19.77 births/1,000 population (2008 est.)

Death rate: 6.16 deaths/1,000 population (2008 est.)

Net migration rate: -0.97 migrant(s)/1,000 population (2008 est.)

Sex ratio:
at birth: 1.05 male(s)/female
under 15 years: 1.04 male(s)/female
15–64 years: 1.01 male(s)/female
65 years and over: 0.89 male(s)/female
total population: 1.01 male(s)/female (2008 est.)

Infant mortality rate:
total: 29.53 deaths/1,000 live births
male: 32.02 deaths/1,000 live births
female: 26.93 deaths/1,000 live births (2008 est.)

Life expectancy at birth:
total population: 70.44 years
male: 68.61 years
female: 72.37 years (2008 est.)

Total fertility rate: 2.42 children born/woman (2008 est.)

HIV/AIDS—adult prevalence rate: 0.5% (2003 est.)

HIV/AIDS—people living with HIV/AIDS: 82,000 (2003 est.)

HIV/AIDS—deaths: 4,200 (2003 est.)

Major infectious diseases:
degree of risk: very high
food or waterborne diseases: bacterial, hepatitis A, and typhoid fever
vectorborne disease: dengue fever, malaria, Oroya fever, and yellow fever
water contact disease: leptospirosis (2008)

Nationality:
noun: Peruvian(s)
adjective: Peruvian

Ethnic groups: Amerindian 45%, mestizo (mixed Amerindian and white) 37%, white 15%, black, Japanese, Chinese, and other 3%

Religions: Roman Catholic 81%, Seventh Day Adventist 1.4%, other Christian 0.7%, other 0.6%, unspecified or none 16.3% (2003 est.)

Languages: Spanish (official), Quechua (official), Aymara, and a large number of minor Amazonian languages

Literacy:
definition: age 15 and over can read and write
total population: 87.7%
male: 93.5%
female: 82.1% (2004 est.)

GOVERNMENT

Country name:
conventional long form: Republic of Peru
conventional short form: Peru
local long form: Republica del Peru
local short form: Peru

Government type: constitutional republic

Capital: *name:* Lima
geographic coordinates: 12 03 S, 77 03 W
time difference: UTC-5 (same time as Washington, DC during Standard Time)

Administrative divisions: 25 regions (regiones, singular—region) and 1 province* (provincia); Amazonas, Ancash, Apurimac, Arequipa, Ayacucho, Cajamarca, Callao, Cusco, Huancavelica, Huanuco, Ica, Junin, La Libertad, Lambayeque, Lima, Lima*, Loreto, Madre de Dios, Moquegua, Pasco, Piura, Puno, San Martin, Tacna, Tumbes, Ucayali

Independence: 28 July 1821 (from Spain)

National holiday: Independence Day, 28 July (1821)

Constitution: 29 December 1993

Legal system: based on civil law system; accepts compulsory ICJ jurisdiction with reservations

Suffrage: 18 years of age; universal and compulsory until the age of 70; note—for the first time in recent elections, members of the military and national police were eligible to vote in the 2006 elections

Executive branch:

chief of state: President Alan GARCIA Perez (since 28 July 2006); First Vice President Luis GIAMPIETRI Rojas; Second Vice President Lourdes MENDOZA del Solar (since 28 July 2006); note—the president is both the chief of state and head of government

head of government: President Alan GARCIA Perez (since 28 July 2006); First Vice President Luis GIAMPIETRI Rojas; Second Vice President Lourdes MENDOZA del Solar (since 28 July 2006)

note: Prime Minister Jorge DEL CASTILLO Galvez (since 28 August 2006) does not exercise executive power; this power is in the hands of the president

cabinet: Council of Ministers appointed by the president

elections: president elected by popular vote for a five-year term (eligible for a nonconsecutive reelection); presidential and congressional elections held 9 April 2006 with runoff election held 4 June 2006; next to be held in April 2011

election results: Alan GARCIA elected president in runoff election; percent of vote—Alan GARCIA 52.5%, Ollanta HUMALA Tasso 47.5%

Legislative branch: unicameral Congress of the Republic of Peru or Congreso de la Republica del Peru (120 seats; members are elected by popular vote to serve five-year terms)

elections: last held 9 April 2006 (next to be held in April 2011)

election results: percent of vote by party—UPP 21.2%, PAP 20.6%, UN 15.3%, AF 13.1%, FC 7.1%, PP 4.1%, RN 4.0%, other 14.6%; seats by party—UPP 45, PAP 36, UN 17, AF 13, FC 5, PP 2, RN 2

Judicial branch: Supreme Court of Justice or Corte Suprema de Justicia (judges are appointed by the National Council of the Judiciary)

Political parties and leaders: Alliance For Progress (Alianza Para El Progreso) [Cesar ACUNA Peralta]; Alliance For The Future (Alianza Por El Futuro) or AF (a coalition of pro-FUJIMORI parties including Cambio 90, Nueva Mayoria, and Si Cumple); Centrist Front (Frente Del Centro) or FC (a coalition of Accion Popular, Somos Peru, and Coordinadora Nacional de Independientes); Independent Moralizing Front (Frente Independiente Moralizador) or FIM; National

Renovation Party (Partido Renovacion Nacional) [Rafael REY]; National Restoration (Restauracion Nacional) or RN [Humberto LAY Sun]; National Unity (Unidad Nacional) or UN (a coalition of Partido Popular Cristiano and Partido Solidaridad Nacional) [Lourdes FLORES Nano]; Peru Possible (Peru Posible) or PP [Alejandro TOLEDO Manrique]; Peruvian Aprista Party (Partido Aprista Peruano) or PAP [Alan GARCIA] (also referred to by its original name Alianza Popular Revolucionaria Americana or APRA); Peruvian Nationalist Party (Partido Nacionalista Peruano) or PNP [Ollanta HUMALA Tasso]; Union for Peru (Union por el Peru) or UPP [Aldo ESTRADA Choque]

Political pressure groups and leaders: leftist guerrilla groups include Shining Path [Abimael GUZMAN Reynoso (imprisoned), Gabriel MACARIO (top leader at-large)]; Tupac Amaru Revolutionary Movement or MRTA [Victor POLAY (imprisoned), Hugo AVALLENEDA Valdez (top leader at-large)]

International organization participation: APEC, CAN, CSN, FAO, G-15, G-24, G-77, IADB, IAEA, IBRD, ICAO, ICCt, ICRM, IDA, IFAD, IFC, IFRCS, IHO, ILO, IMF, IMO, IMSO, Interpol, IOC, IOM, IPU, ISO, ITSO, ITU, ITUC, LAES, LAIA, Mercosur (associate), MIGA, MINUSTAH, NAM, OAS, OPANAL, OPCW, PCA, RG, UN, UNCTAD, UNESCO, UNFICYP, UNIDO, Union Latina, UNMEE, UNMIL, UNMIS, UNOCI, UNWTO, UPU, WCL, WCO, WFTU, WHO, WIPO, WMO, WTO

Diplomatic representation in the US:

chief of mission: Ambassador Felipe ORTIZ de Zevallos

chancery: 1700 Massachusetts Avenue NW, Washington, DC 20036

telephone: [1] (202) 833-9860 through 9869

FAX: [1] (202) 659-8124

consulate(s) general: Atlanta, Boston, Chicago, Denver, Hartford, Houston, Los Angeles, Miami, New York, Paterson (New Jersey), San Francisco, Washington, DC

Diplomatic representation from the US:

chief of mission: Ambassador P. Michael MCKINLEY

embassy: Avenida La Encalada, Cuadra 17s/n, Surco, Lima 33

mailing address: P. O. Box 1995, Lima 1; American Embassy (Lima), APO AA 34031-5000

telephone: [51] (1) 434-3000

FAX: [51] (1) 618-2397

Flag description: three equal, vertical bands of red (hoist side), white, and red with the coat of arms centered in the white band; the coat of arms features a shield bearing a vicuna, cinchona tree (the source of quinine), and a yellow cornucopia spilling out gold coins, all framed by a green wreath

ECONOMY

Economy—overview: Peru's economy reflects its varied geography—an arid coastal region, the Andes further inland, and tropical lands bordering Colombia and Brazil. Abundant mineral resources are found in the mountainous areas, and Peru's coastal waters provide excellent fishing grounds. However, overdependence on minerals and metals subjects the economy to fluctuations in world prices, and a lack of infrastructure deters trade and investment. After several years of inconsistent economic performance, the Peruvian economy grew by more than 4% per year during the period 2002–06, with a stable exchange rate and low inflation. Growth jumped to 7.5% in 2007, driven by higher world prices for minerals and metals. Risk premiums on Peruvian bonds on secondary markets reached historically low levels in late 2004, reflecting investor optimism regarding the government's prudent fiscal policies and openness to trade and investment. Despite the strong macroeconomic performance, underemployment and poverty have stayed persistently high. Growth prospects depend on exports of minerals, textiles, and agricultural products, and by expectations for the Camisea natural gas megaproject and for other promising energy projects. Upon taking office, President GARCIA announced Sierra Exportadora, a program aimed at promoting economic growth in Peru's southern and central highlands.

GDP (purchasing power parity): $219 billion (2007 est.)

GDP (official exchange rate): $109.1 billion (2007 est.)

GDP—real growth rate: 9% (2007 est.)

GDP—per capita (PPP): $7,800 (2007 est.)

GDP—composition by sector:

agriculture: 8.4%

industry: 25.6%

services: 66% (2007 est.)

Labor force: 9.839 million (2007 est.)

Labor force—by occupation:

agriculture: 9%

industry: 18%

services: 73% (2001)

Unemployment rate: 6.9% in metropolitan Lima; widespread underemployment (2007 est.)

Population below poverty line: 44.5% (2006)
Household income or consumption by percentage share:
lowest 10%: 1.3%
highest 10%: 40.9% (2003)
Distribution of family income—Gini index: 52 (2003)
Inflation rate (consumer prices): 1.8% (2007 est.)
Investment (gross fixed): 23% of GDP (2007 est.)
Budget:
revenues: $32.48 billion
expenditures: $29.11 billion (2007 est.)
Public debt: 29.2% of GDP (2007 est.)
Agriculture—products: asparagus, coffee, cotton, sugarcane, rice, potatoes, corn, plantains, grapes, oranges, coca; poultry, beef, dairy products; fish, guinea pigs
Industries: mining and refining of minerals; steel, metal fabrication; petroleum extraction and refining, natural gas; fishing and fish processing, textiles, clothing, food processing
Industrial production growth rate: 9.3% (2007 est.)
Electricity—production: 24.97 billion kWh (2005 est.)
Electricity—production by source:
fossil fuel: 14.5%
hydro: 84.7%
nuclear: 0%
other: 0.8% (2001)
Electricity—consumption: 22.59 billion kWh (2005)
Electricity—exports: 0 kWh (2005)
Electricity—imports: 0 kWh (2005)
Oil—production: 110,700 bbl/day (2005 est.)
Oil—consumption: 166,000 bbl/day (2005 est.)
Oil—exports: 53,040 bbl/day (2004 est.)
Oil—imports: 121,500 bbl/day (2004)
Oil—proved reserves: 430.8 million bbl (2007 est.)
Natural gas—production: 1.515 billion cu m (2005 est.)
Natural gas—consumption: 1.515 billion cu m (2005 est.)
Natural gas—exports: 0 cu m (2005 est.)
Natural gas—imports: 0 cu m (2005)
Natural gas—proved reserves: 236.9 billion cu m (1 January 2006 est.)
Current account balance: $1.75 billion (2007 est.)
Exports: $27.96 billion f.o.b. (2007 est.)
Exports—commodities: copper, gold, zinc, crude petroleum and petroleum products, coffee, potatoes, asparagus, textiles, guinea pigs
Exports—partners: US 24.1%, China 9.6%, Switzerland 7.1%, Canada 6.8%, Chile 6%, Japan 5.2% (2006)

Imports: $19.6 billion f.o.b. (2007 est.)
Imports—commodities: petroleum and petroleum products, plastics, machinery, vehicles, iron and steel, wheat, paper
Imports—partners: US 16.5%, China 10.3%, Brazil 10.3%, Ecuador 7.2%, Colombia 6.1%, Chile 5.8%, Argentina 4.8%, Mexico 4% (2006)
Economic aid—recipient: $397.8 million (2005)
Reserves of foreign exchange and gold: $27.78 billion (31 December 2007 est.)
Debt—external: $31.07 billion (31 December 2007 est.)
Stock of direct foreign investment—at home: $24.72 billion (2007 est.)
Stock of direct foreign investment—abroad: $1.476 billion (2007 est.)
Market value of publicly traded shares: $59.66 billion (2006)
Currency (code): nuevo sol (PEN)
Currency code: PEN
Exchange rates: nuevo sol per US dollar—3.1731 (2007), 3.2742 (2006), 3.2958 (2005), 3.4132 (2004), 3.4785 (2003)
Fiscal year: calendar year

COMMUNICATIONS

Telephones—main lines in use: 2.332 million (2006)
Telephones—mobile cellular: 8.5 million (2006)
Telephone system:
general assessment: adequate for most requirements
domestic: fixed-line teledensity is only about 8 per 100 persons; mobile-cellular teledensity, spurred by competition among multiple providers, has increased to about 30 telephones per 100 persons; nationwide microwave radio relay system and a domestic satellite system with 12 earth stations
international: country code—51; the South America-1 (SAM-1) and Pan American (PAN-AM) submarine cable systems provide links to parts of Central and South America, the Caribbean, and US; satellite earth stations—2 Intelsat (Atlantic Ocean)
Radio broadcast stations: AM 472, FM 198, shortwave 189 (1999)
Radios: 6.65 million (1997)
Television broadcast stations: 13 (plus 112 repeaters) (1997)
Televisions: 3.06 million (1997)
Internet country code: .pe
Internet hosts: 270,193 (2007)
Internet Service Providers (ISPs): 10 (2000)
Internet users: 6.1 million (2006)

TRANSPORTATION

Airports: 237 (2007)

Airports—with paved runways:
total: 54
over 3,047 m: 6
2,438 to 3,047 m: 20
1,524 to 2,437 m: 14
914 to 1,523 m: 11
under 914 m: 3 (2007)
Airports—with unpaved runways:
total: 183
2,438 to 3,047 m: 2
1,524 to 2,437 m: 24
914 to 1,523 m: 40
under 914 m: 117 (2007)
Heliports: 1 (2007)
Pipelines: gas 1,181 km; gas/liquid petroleum gas 61 km; liquid natural gas 106 km; liquid petroleum gas 517 km; oil 1,749 km; refined products 13 km (2007)
Railways: *total:* 1,989 km
standard gauge: 1,726 km 1.435-m gauge
narrow gauge: 263 km 0.914-m gauge (2006)
Roadways: *total:* 78,829 km
paved: 11,351 km (includes 276 km of expressways)
unpaved: 67,478 km (2004)
Waterways: 8,808 km
note: 8,600 km of navigable tributaries of Amazon system and 208 km of Lago Titicaca (2007)
Merchant marine:
total: 6 ships (1000 GRT or over) 76,220 GRT/119,615 DWT
by type: cargo 3, petroleum tanker 3
foreign-owned: 1 (US 1)
registered in other countries: 16 (Belize 1, Panama 15) (2007)
Ports and terminals: Callao, Iquitos, Matarani, Paita, Pucallpa, Yurimaguas; note—Iquitos, Pucallpa, and Yurimaguas are on the upper reaches of the Amazon and its tributaries

MILITARY

Military branches: Peruvian Army (Ejercito Peruano), Peruvian Navy (Marina de Guerra del Peru, MGP (includes naval air, naval infantry, and coast guard)), Peruvian Air Force (Fuerza Aerea del Peru, FAP) (2008)
Military service age and obligation: 18–30 years of age for voluntary male and female military service; no conscription (2008)
Manpower available for military service:
males age 16–49: 7,653,898
females age 16–49: 7,531,329 (2008 est.)
Manpower fit for military service:
males age 16–49: 5,796,449
females age 16–49: 6,217,524 (2008 est.)
Manpower reaching militarily significant age annually:
males age 16–49: 306,260
females age 16–49: 296,819 (2008 est.)

Military expenditures—percent of GDP: 1.5% (2006)

Disputes—international: Chile and Ecuador rejected Peru's November 2005 unilateral legislation to shift the axis of their joint treaty-defined maritime boundaries along the parallels of latitude to equidistance lines which favor Peru; organized illegal narcotics operations in Colombia have penetrated Peru's shared border; Peru rejects Bolivia's claim to restore maritime access through a sovereign corridor through Chile along the Peruvian border

Refugees and internally displaced persons: *IDPs:* 60,000–150,000 (civil war from 1980–2000; most IDPs are indigenous peasants in Andean and Amazonian regions) (2007)

Illicit drugs: until 1996 the world's largest coca leaf producer, Peru is now the world's second largest producer of coca leaf, though it lags far behind Colombia; cultivation of coca in Peru rose 25% to 34,000 hectares in 2005; much of the cocaine base is shipped to neighboring Colombia for processing into cocaine, while finished cocaine is shipped out from Pacific ports to the international drug market; increasing amounts of base and finished cocaine, however, are being moved to Brazil and Bolivia for use in the Southern Cone or transshipped to Europe and Africa

PHILIPPINES

INTRODUCTION

Background: The Philippine Islands became a Spanish colony during the 16th century; they were ceded to the US in 1898 following the Spanish-American War. In 1935 the Philippines became a self-governing commonwealth. Manuel QUEZON was elected president and was tasked with preparing the country for independence after a 10-year transition. In 1942 the islands fell under Japanese occupation during World War II, and US forces and Filipinos fought together during 1944–45 to regain control. On 4 July 1946 the Republic of the Philippines attained its independence. The 20-year rule of Ferdinand MARCOS ended in 1986, when a "people power" movement in Manila ("EDSA 1") forced him into exile and installed Corazon AQUINO as president. Her presidency was hampered by several coup attempts, which prevented a return to full political stability and economic development. Fidel RAMOS was elected president in 1992 and his administration was marked by greater stability and progress on economic reforms. In 1992, the US closed its last military bases on the islands. Joseph ESTRADA was elected president in 1998, but was succeeded by his vice-president, Gloria MACAPAGAL-ARROYO, in January 2001 after ESTRADA's stormy impeachment trial on corruption charges broke down and another "people power" movement ("EDSA 2") demanded his resignation. MACAPAGAL-ARROYO was elected to a six-year term as president in May 2004. The Philippine Government faces threats from three terrorist groups on the US Government's Foreign Terrorist Organization list, but in 2006 and 2007 scored some major successes in capturing or killing key wanted terrorists. Decades of Muslim insurgency in the southern Philippines have led to a peace accord with one group and an ongoing cease-fire and peace talks with another.

GEOGRAPHY

Location: Southeastern Asia, archipelago between the Philippine Sea and the South China Sea, east of Vietnam

Geographic coordinates: 13 00 N, 122 00 E

Map references: Southeast Asia

Area:
total: 300,000 sq km
land: 298,170 sq km
water: 1,830 sq km

Area—comparative: slightly larger than Arizona

Land boundaries: 0 km

Coastline: 36,289 km

Maritime claims:
territorial sea: irregular polygon extending up to 100 nm from coastline as defined by 1898 treaty; since late 1970s has also claimed polygonal-shaped area in South China Sea up to 285 nm in breadth
exclusive economic zone: 200 nm
continental shelf: to depth of exploitation

Climate: tropical marine; northeast monsoon (November to April); southwest monsoon (May to October)

Terrain: mostly mountains with narrow to extensive coastal lowlands

Elevation extremes:
lowest point: Philippine Sea 0 m
highest point: Mount Apo 2,954 m

Natural resources: timber, petroleum, nickel, cobalt, silver, gold, salt, copper

Land use:
arable land: 19%
permanent crops: 16.67%
other: 64.33% (2005)

Irrigated land: 15,500 sq km (2003)

Total renewable water resources: 479 cu km (1999)

Freshwater withdrawal (domestic/industrial/agricultural): *total:* 28.52 cu km/yr (17%/9%/74%)
per capita: 343 cu m/yr (2000)

Natural hazards: astride typhoon belt, usually affected by 15 and struck by five to six cyclonic storms per year; landslides; active volcanoes; destructive earthquakes; tsunamis

Environment—current issues: uncontrolled deforestation especially in watershed areas; soil erosion; air and water pollution in major urban centers; coral reef degradation; increasing pollution of coastal mangrove swamps that are important fish breeding grounds

Environment—international agreements: *party to:* Biodiversity, Climate Change, Climate Change-Kyoto Protocol, Desertification, Endangered Species, Hazardous Wastes, Law of the

Sea, Marine Dumping, Ozone Layer Protection, Ship Pollution, Tropical Timber 83, Tropical Timber 94, Wetlands, Whaling
signed, but not ratified: Air Pollution-Persistent Organic Pollutants
Geography—note: the Philippine archipelago is made up of 7,107 islands; favorably located in relation to many of Southeast Asia's main water bodies: the South China Sea, Philippine Sea, Sulu Sea, Celebes Sea, and Luzon Strait

PEOPLE

Population: 92,681,453 (July 2008 est.)
Age structure:
0–14 years: 34.1% (male 16,121,508/female 15,487,841)
15–64 years: 61.7% (male 28,524,176/female 28,652,155)
65 years and over: 4.2% (male 1,690,006/female 2,205,767) (2008 est.)
Median age:
total: 23 years
male: 22.5 years
female: 23.5 years (2008 est.)
Population growth rate: 1.728% (2008 est.)
Birth rate: 24.07 births/1,000 population (2008 est.)
Death rate: 5.32 deaths/1,000 population (2008 est.)
Net migration rate: -1.47 migrant(s)/1,000 population (2008 est.)
Sex ratio:
at birth: 1.05 male(s)/female
under 15 years: 1.04 male(s)/female
15–64 years: 1 male(s)/female
65 years and over: 0.77 male(s)/female
total population: 1 male(s)/female (2008 est.)
Infant mortality rate:
total: 21.45 deaths/1,000 live births
male: 24.14 deaths/1,000 live births
female: 18.64 deaths/1,000 live births (2008 est.)
Life expectancy at birth:
total population: 70.8 years
male: 67.89 years
female: 73.85 years (2008 est.)
Total fertility rate: 3 children born/woman (2008 est.)
HIV/AIDS—adult prevalence rate: less than 0.1% (2003 est.)
HIV/AIDS—people living with HIV/AIDS: 9,000 (2003 est.)
HIV/AIDS—deaths: fewer than 500 (2003 est.)
Major infectious diseases:
degree of risk: high
food or waterborne diseases: bacterial diarrhea, hepatitis A, and typhoid fever
vectorborne diseases: dengue fever and malaria (2008)
Nationality: *noun:* Filipino(s)

adjective: Philippine
Ethnic groups: Tagalog 28.1%, Cebuano 13.1%, Ilocano 9%, Bisaya/Binisaya 7.6%, Hiligaynon Ilonggo 7.5%, Bikol 6%, Waray 3.4%, other 25.3% (2000 census)
Religions: Roman Catholic 80.9%, Muslim 5%, Evangelical 2.8%, Iglesia ni Kristo 2.3%, Aglipayan 2%, other Christian 4.5%, other 1.8%, unspecified 0.6%, none 0.1% (2000 census)
Languages: Filipino (official; based on Tagalog) and English (official); eight major dialects—Tagalog, Cebuano, Ilocano, Hiligaynon or Ilonggo, Bicol, Waray, Pampango, and Pangasinan
Literacy: *definition:* age 15 and over can read and write
total population: 92.6%
male: 92.5%
female: 92.7% (2000 census)

GOVERNMENT

Country name:
conventional long form: Republic of the Philippines
conventional short form: Philippines
local long form: Republika ng Pilipinas
local short form: Pilipinas
Government type: republic
Capital: *name:* Manila
geographic coordinates: 14 35 N, 121 00 E
time difference: UTC+8 (13 hours ahead of Washington, DC during Standard Time)
Administrative divisions: 81 provinces and 136 chartered cities
provinces: Abra, Agusan del Norte, Agusan del Sur, Aklan, Albay, Antique, Apayao, Aurora, Basilan, Bataan, Batanes, Batangas, Biliran, Benguet, Bohol, Bukidnon, Bulacan, Cagayan, Camarines Norte, Camarines Sur, Camiguin, Capiz, Catanduanes, Cavite, Cebu, Compostela, Davao del Norte, Davao del Sur, Davao Oriental, Dinagat Islands, Eastern Samar, Guimaras, Ifugao, Ilocos Norte, Ilocos Sur, Iloilo, Isabela, Kalinga, Laguna, Lanao del Norte, Lanao del Sur, La Union, Leyte, Maguindanao, Marinduque, Masbate, Mindoro Occidental, Mindoro Oriental, Misamis Occidental, Misamis Oriental, Mountain Province, Negros Occidental, Negros Oriental, North Cotabato, Northern Samar, Nueva Ecija, Nueva Vizcaya, Palawan, Pampanga, Pangasinan, Quezon, Quirino, Rizal, Romblon, Samar, Sarangani, Shariff Kabunsuan, Siquijor, Sorsogon, South Cotabato, Southern Leyte, Sultan Kudarat, Sulu, Surigao del Norte, Surigao del Sur, Tarlac, Tawi-Tawi, Zambales, Zamboanga del Norte, Zamboanga del Sur, Zamboanga Sibugay

chartered cities: Alaminos, Angeles, Antipolo, Bacolod, Bago, Baguio, Bais, Balanga, Batac, Batangas, Bayawan, Baybay, Bayugan, Bislig, Bogo, Borongan, Butuan, Cabadbaran, Cabanatuan, Cadiz, Cagayan de Oro, Calamba, Calapan, Calbayog, Candon, Canlaon, Carcar, Catbalogan, Cauayan, Cavite, Cebu, Cotabato, Dagupan, Danao, Dapitan, Davao, Digos, Dipolog, Dumaguete, Escalante, El Salvador, Gapan, General Santos, Gingoog, Guihulngan, Himamaylan, Iligan, Iloilo, Isabela, Iriga, Kabankalan, Kalookan, Kidapawan, Koronadal, La Carlota, Lamitan, Laoag, Lapu-Lapu, Las Pinas, Legazpi, Ligao, Lipa, Lucena, Maasin, Makati, Malabon, Malaybalay, Malolos, Mandaluyong, Mandaue, Manila, Marawi, Marikina, Masbate, Mati, Meycauayan, Muntinlupa, Munoz, Naga (Camarines Sur), Naga (Cebu), Navotas, Olongapo, Ormoc, Oroquieta, Ozamis, Pagadian, Palayan, Panabo, Paranaque, Pasay, Pasig, Passi, Puerto Princesa, Quezon, Roxas, Sagay, Samal, San Carlos (in Negros Occidental), San Carlos (in Pangasinan), San Fernando (in La Union), San Fernando (in Pampanga), San Jose, San Jose del Monte, San Juan, San Pablo, Santa Rosa, Santiago, Silay, Sipalay, Sorsogon, Surigao, Tabaco, Tabuk, Tacloban, Tacurong, Tagaytay, Tagbilaran, Taguig, Tagum, Talisay (in Cebu), Talisay (in Negros Occidental), Tanauan, Tandag, Tangub, Tanjay, Tarlac, Tayabas, Toledo, Tuguegarao, Trece Martires, Urdaneta, Valencia, Valenzuela, Victorias, Vigan, Zamboanga (2007)
Independence: 12 June 1898 (independence proclaimed from Spain); 4 July 1946 (from the US)
National holiday: Independence Day, 12 June (1898); note—12 June 1898 was date of declaration of independence from Spain; 4 July 1946 was date of independence from US
Constitution: 2 February 1987, effective 11 February 1987
Legal system: based on Spanish and Anglo-American law; accepts compulsory ICJ jurisdiction with reservations
Suffrage: 18 years of age; universal
Executive branch:
chief of state: President Gloria MACAPAGAL-ARROYO (since 20 January 2001); note—president is both chief of state and head of government
head of government: President Gloria MACAPAGAL-ARROYO (since 20 January 2001)
cabinet: Cabinet appointed by the president with consent of Commission of Appointments

elections: president and vice president (Manuel "Noli" DE CASTRO) elected on separate tickets by popular vote for a single six-year term; election last held on 10 May 2004 (next to be held in May 2010)

election results: Gloria MACAPAGAL-ARROYO elected president; percent of vote—Gloria MACAPAGAL-ARROYO 40%, Fernando POE 37%, three others 23%

Legislative branch: bicameral Congress or Kongreso consists of the Senate or Senado (24 seats—one-half elected every three years; members elected at large by popular vote to serve six-year terms) and the House of Representatives or Kapulungan Ng Mga Kinatawan (as a result of May 2007 election it has 239 seats including 218 members representing districts and 21 sectoral party-list members representing special minorities elected on the basis of 1 seat for every 2% of the total vote but limited to 3 seats; members elected by popular vote to serve three-year terms; note—the Constitution prohibits the House of Representatives from having more than 250 members)

elections: Senate—last held on 14 May 2007 (next to be held in May 2010); House of Representatives—elections last held on 14 May 2007 (next to be held in May 2010)

election results: Senate—percent of vote by party—NA; seats by party—Lakas-Kampi 4, LP 4, NPC 3, Nacionalista 2, independents 4, others 6; note—there are 23 rather than 24 sitting senators because one senator was elected mayor of Manila; House of Representatives—percent of vote by party—NA; seats by party—Lakas 86, Kampi 46, NPC 29, LP 21, Party-list 21, others 36

Judicial branch: Supreme Court (15 justices are appointed by the president on the recommendation of the Judicial and Bar Council and serve until 70 years of age); Court of Appeals; Sandigan-bayan (special court for hearing corruption cases of government officials)

Political parties and leaders: Genuine Opposition or GO (coalition of oppositon parties formed to contest the 2007 elections); Kabalikat Ng Malayang Pilipino or Kampi [Ronaldo PUNO]; Laban Ng Demokratikong Pilipino (Struggle of Filipino Democrats) or LDP [Edgardo ANGARA]; Lakas Ng Edsa (National Union of Christian Democrats) or Lakas [Jose DE VENECIA]; Liberal Party or LP [Manuel ROXAS]; Nacionalista [Manuel VILLAR]; National People's Coalition or NPC [Frisco SAN JUAN]; PDP-

Laban [Aquilino PIMENTEL]; People's Reform Party [Miriam Defensor SANTIAGO]; PROMDI [Emilio OSMENA]; Pwersa Ng Masang Pilipino (Party of the Philippine Masses) or PMP [Joseph ESTRADA]; Reporma [Renato DE VILLA]

Political pressure groups and leaders: AKBAYAN [Etta ROSALES, Mario AGUJA, and Risa HONTIVEROS-BARAQUIEL]; ALAGAD [Rodante MARROLITA]; ALIF [Acmad TOMAWIS]; An Waray [Horencio NOEL]; Anak Mindanao [Mujiv HATAMIN]; ANAKPAWIS [Crispin BELTRAN and Rafael MARIANO]; Association of Philippine Electric Cooperatives (APEC) [Sunny Rose MADAMBA, Ernesto PABLO, and Edgar VALDEZ]; AVE [Eulogio MAGSAYSAY]; Bayan Muna [Satur OCAMPO, Joel VIRADOR, and Teodoro CASINO, Jr.]; BUHAY [Rene VELARDE and Hans Christian SENERES]; BUTIL [Benjamin CRUZ]; CIBAC [Emmanuel Joel VILLANUEVA]; COOP-NATCO [Guillermo CUA]; GABRIELA [Liza MAZA]; Partido Ng Manggagawa [Renato MAGTUBO]; Veterans Federation of the Philippines [Ernesto GIDAYA]

International organization participation: ADB, APEC, APT, ARF, ASEAN, BIS, CP, EAS, FAO, G-24, G-77, IAEA, IBRD, ICAO, ICC, ICCt (signatory), ICRM, IDA, IFAD, IFC, IFRCS, IHO, ILO, IMF, IMO, IMSO, Interpol, IOC, IOM, IPU, ISO, ITSO, ITU, ITUC, MIGA, MINUSTAH, NAM, OAS (observer), OPCW, PIF (partner), UN, UNCTAD, UNESCO, UNHCR, UNIDO, Union Latina, UNMIL, UNMIS, UNMIT, UNOCI, UNWTO, UPU, WCL, WCO, WFTU, WHO, WIPO, WMO, WTO

Diplomatic representation in the US:
chief of mission: Ambassador Willy C. GAA
chancery: 1600 Massachusetts Avenue NW, Washington, DC 20036
telephone: [1] (202) 467-9300
FAX: [1] (202) 467-9417
consulate(s) general: Chicago, Honolulu, Los Angeles, New York, San Francisco, Saipan (Northern Mariana Islands), Tamuning (Guam)

Diplomatic representation from the US:
chief of mission: Ambassador Kristie A. KENNEY
embassy: 1201 Roxas Boulevard, Ermita 1000, Manila
mailing address: PSC 500, FPO AP 96515-1000
telephone: [63] (2) 301-2000

FAX: [63] (2) 522-4361

Flag description: two equal horizontal bands of blue (top; representing peace and justice) and red (representing courage); a white equilateral triangle based on the hoist side represents equality; the center of the triangle displays a yellow sun with eight primary rays, each representing one of the first eight provinces that sought independence from Spain; each corner of the triangle contains a small, yellow, five-pointed star representing the three major geographical divisions of the country: Luzon, Visayas, and Mindanao; the design of the flag dates to 1897; in wartime the flag is flown upside down with the red band at the top

ECONOMY

Economy—overview: The Philippine economy grew at its fastest pace in three decades with real GDP growth exceeding 7% in 2007. Higher government spending contributed to the growth, but a resilient service sector and large remittances from the millions of Filipinos who work abroad have played an increasingly important role. Economic growth has averaged 5% since President MACAPAGAL-ARROYO took office in 2001. Nevertheless, the Philippines will need still higher, sustained growth to make progress in alleviating poverty, given its high population growth and unequal distribution of income. MACAPAGAL-ARROYO averted a fiscal crisis by pushing for new revenue measures and, until recently, tightening expenditures. Declining fiscal deficits, tapering debt and debt service ratios, as well as recent efforts to increase spending on infrastructure and social services have heightened optimism over Philippine economic prospects. Although the general macroeconomic outlook has improved significantly, the Philippines continues to face important challenges and must maintain the reform momentum in order to catch up with regional competitors, improve employment opportunities, and alleviate poverty. Longer-term fiscal stability will require more sustainable revenue sources, rather than non-recurring revenues from privatization.

GDP (purchasing power parity): $299.6 billion (2007 est.)

GDP (official exchange rate): $144.1 billion (2007 est.)

GDP—real growth rate: 7.3% (2007 est.)

GDP—per capita (PPP): $3,400 (2007 est.)

GDP—composition by sector:
agriculture: 13.7%

industry: 31.4%
services: 54.8% (2007 est.)
Labor force: 36.22 million (2007 est.)
Labor force—by occupation: *agriculture:* 35%
industry: 15%
services: 50% (2007 est.)
Unemployment rate: 7.3% (2007 est.)
Population below poverty line: 30% (2003 est.)
Household income or consumption by percentage share:
lowest 10%: 2.4%
highest 10%: 31.2% (2006)
Distribution of family income—Gini index: 45.8 (2006)
Inflation rate (consumer prices): 2.8% (2007 est.)
Investment (gross fixed): 14.2% of GDP (2007 est.)
Budget:
revenues: $24.63 billion
expenditures: $24.9 billion (2007 est.)
Public debt: 55.8% of GDP (2007 est.)
Agriculture—products: sugarcane, coconuts, rice, corn, bananas, cassavas, pineapples, mangoes; pork, eggs, beef; fish
Industries: electronics assembly, garments, footwear, pharmaceuticals, chemicals, wood products, food processing, petroleum refining, fishing
Industrial production growth rate: 6.6% (2007 est.)
Electricity—production: 53.67 billion kWh (2005)
Electricity—production by source:
fossil fuel: 55.6%
hydro: 17.5%
nuclear: 0%
other: 26.9% (2001)
Electricity—consumption: 46.86 billion kWh (2005)
Electricity—exports: 0 kWh (2005)
Electricity—imports: 0 kWh (2005)
Oil—production: 24,310 bbl/day (2005 est.)
Oil—consumption: 340,000 bbl/day (2005 est.)
Oil—exports: 34,900 bbl/day (2004)
Oil—imports: 353,700 bbl/day (2004)
Oil—proved reserves: 152 million bbl (31 December 2006)
Natural gas—production: 2.781 billion cu m (2005 est.)
Natural gas—consumption: 2.781 billion cu m (2005 est.)
Natural gas—exports: 0 cu m (2005 est.)
Natural gas—imports: 0 cu m (2005)
Natural gas—proved reserves: 107.5 billion cu m (1 January 2006 est.)
Current account balance: $6.351 billion (2007 est.)
Exports: $49.32 billion f.o.b. (2007 est.)

Exports—commodities: semiconductors and electronic products, transport equipment, garments, copper products, petroleum products, coconut oil, fruits
Exports—partners: US 18.3%, Japan 16.5%, Netherlands 10.1%, China 9.8%, Hong Kong 7.8%, Singapore 7.3%, Malaysia 5.6%, Taiwan 4.3% (2006)
Imports: $57.56 billion f.o.b. (2007 est.)
Imports—commodities: electronic products, mineral fuels, machinery and transport equipment, iron and steel, textile fabrics, grains, chemicals, plastic
Imports—partners: US 16.3%, Japan 13.6%, Singapore 8.5%, Taiwan 8%, China 7.1%, South Korea 6.2%, Saudi Arabia 5.8%, Malaysia 4.1%, Thailand 4.1%, Hong Kong 4% (2006)
Economic aid—recipient: ODA, $451.4 million in commitments (2006)
Reserves of foreign exchange and gold: $33.71 billion (31 December 2007 est.)
Debt—external: $61.83 billion (31 December 2007 est.)
Stock of direct foreign investment—at home: $18.4 billion (2007 est.)
Stock of direct foreign investment—abroad: $5.28 billion (2007 est.)
Market value of publicly traded shares: $103.4 billion (2007)
Currency (code): Philippine peso (PHP)
Currency code: PHP
Exchange rates: Philippine pesos per US dollar—46.148 (2007), 51.246 (2006), 55.086 (2005), 56.04 (2004), 54.203 (2003)
Fiscal year: calendar year

COMMUNICATIONS

Telephones—main lines in use: 3.633 million (2006)
Telephones—mobile cellular: 42.869 million (2006)
Telephone system:
general assessment: good international radiotelephone and submarine cable services; domestic and interisland service adequate
domestic: domestic satellite system with 11 earth stations; cellular communications now dominate the industry; combined fixed-line and mobile-cellular telephone density exceeds 50 telephones per 100 persons with more than 10 mobile cellular subscribers for every fixed-line subscriber
international: country code—63; a series of submarine cables together provide connectivity to Asia, US, the Middle East, and Europe; multiple international gateways (2006)
Radio broadcast stations: AM 381, FM 628, shortwave 4 (each shortwave station operates on multiple frequencies in

the language of the target audience) (2007)
Radios: 11.5 million (1997)
Television broadcast stations: 250 (plus 1,501 CATV networks) (2007)
Televisions: 3.7 million (1997)
Internet country code: .ph
Internet hosts: 271,609 (2007)
Internet Service Providers (ISPs): 33 (2000)
Internet users: 4.615 million (2005)

TRANSPORTATION

Airports: 255 (2007)
Airports—with paved runways: *total:* 84
over 3,047 m: 4
2,438 to 3,047 m: 8
1,524 to 2,437 m: 26
914 to 1,523 m: 36
under 914 m: 10 (2007)
Airports—with unpaved runways:
total: 171
1,524 to 2,437 m: 4
914 to 1,523 m: 68
under 914 m: 99 (2007)
Heliports: 2 (2007)
Pipelines: gas 565 km; oil 135 km; refined products 105 km (2007)
Railways:
total: 897 km
narrow gauge: 897 km 1.067-m gauge (492 km are in operation) (2006)
Roadways:
total: 200,037 km
paved: 19,804 km
unpaved: 180,233 km (2003)
Waterways: 3,219 km (limited to vessels with draft less than 1.5 m) (2007)
Merchant marine:
total: 383 ships (1000 GRT or over) 4,542,681 GRT/6,164,312 DWT
by type: bulk carrier 75, cargo 120, chemical tanker 16, container 5, liquefied gas 5, livestock carrier 16, passenger 7, passenger/cargo 66, petroleum tanker 34, refrigerated cargo 14, roll on/roll off 13, vehicle carrier 12
foreign-owned: 135 (Bermuda 31, China 2, Greece 3, Hong Kong 2, Japan 69, Malaysia 2, Netherlands 22, Norway 2, Singapore 1, UAE 1)
registered in other countries: 34 (Australia 1, Bahamas 1, Belize 1, Comoros 1, Cyprus 1, Ecuador 1, Hong Kong 10, Indonesia 1, Panama 12, Singapore 4, St Vincent and The Grenadines 1) (2007)
Ports and terminals: Cagayan de Oro, Cebu, Davao, Liman, Manila, Nasipit Harbor

MILITARY

Military branches: Armed Forces of the Philippines (AFP): Army, Navy (includes Marine Corps and Coast Guard), Air Force (2008)

Military service age and obligation: 18–25 years of age (officers 21–29) for compulsory and voluntary military service; applicants must be single male or female Philippine citizens (2007)

Manpower available for military service:
males age 16–49: 23,547,252
females age 16–49: 23,177,487 (2008 est.)

Manpower fit for military service:
males age 16–49: 18,232,050
females age 16–49: 19,827,538 (2008 est.)

Manpower reaching militarily significant age annually:
males age 16–49: 1,003,836
females age 16–49: 968,845 (2008 est.)

Military expenditures—percent of GDP: 0.9% (2005 est.)

TRANSNATIONAL ISSUES

Disputes—international: Philippines claims sovereignty over certain of the Spratly Islands, known locally as the Kalayaan (Freedom) Islands, also claimed by China, Malaysia, Taiwan, and Vietnam; the 2002 "Declaration on the Conduct of Parties in the South China Sea," has eased tensions in the Spratly Islands but falls short of a legally binding "code of conduct" desired by several of the disputants; in March 2005, the national oil companies of China, the Philippines, and Vietnam signed a joint accord to conduct marine seismic activities in the Spratly Islands; Philippines retains a dormant claim to Malaysia's Sabah State in northern Borneo based on the Sultanate of Sulu's granting the Philippines Government power of attorney to pursue a sovereignty claim on his behalf; maritime delimitation negotiations continue with Palau

Refugees and internally displaced persons: *IDPs:* 300,000 (fighting between government troops and MILF and Abu Sayyaf groups) (2007)

Illicit drugs: domestic methamphetamine production has been a growing problem in recent years despite government crackdowns; major consumer of amphetamines; longstanding marijuana producer mainly in rural areas where Manila's control is limited

PITCAIRN ISLANDS

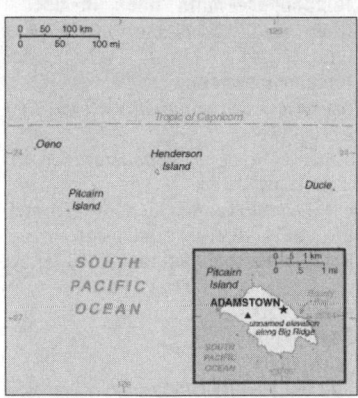

INTRODUCTION

Background: Pitcairn Island was discovered in 1767 by the British and settled in 1790 by the Bounty mutineers and their Tahitian companions. Pitcairn was the first Pacific island to become a British colony (in 1838) and today remains the last vestige of that empire in the South Pacific. Outmigration, primarily to New Zealand, has thinned the population from a peak of 233 in 1937 to less than 50 today.

GEOGRAPHY

Location: Oceania, islands in the South Pacific Ocean, about midway between Peru and New Zealand

Geographic coordinates: 25 04 S, 130 06 W

Map references: Oceania

Area: *total:* 47 sq km
land: 47 sq km
water: 0 sq km

Area—comparative: about 0.3 times the size of Washington, DC

Land boundaries: 0 km

Coastline: 51 km

Maritime claims:
territorial sea: 3 nm
exclusive economic zone: 200 nm

Climate: tropical; hot and humid; modified by southeast trade winds; rainy season (November to March)

Terrain: rugged volcanic formation; rocky coastline with cliffs

Elevation extremes:
lowest point: Pacific Ocean 0 m
highest point: Pawala Valley Ridge 347 m

Natural resources: miro trees (used for handicrafts), fish
note: manganese, iron, copper, gold, silver, and zinc have been discovered offshore

Land use: *arable land:* NA
permanent crops: NA
other: NA

Irrigated land: NA

Natural hazards: typhoons (especially November to March)

Environment—current issues: deforestation (only a small portion of the original forest remains because of burning and clearing for settlement)

Geography—note: Britain's most isolated dependency; only the larger island of Pitcairn is inhabited but it has no port or natural harbor; supplies must be transported by rowed longboat from larger ships stationed offshore

PEOPLE

Population: 48 (July 2008 est.)

Age structure:
0–14 years: NA
15–64 years: NA
65 years and over: NA

Population growth rate: 0% (2008 est.)

Birth rate: NA

Death rate: NA (2008 est.)

Net migration rate: NA

Sex ratio: NA

Infant mortality rate:
total: NA
male: NA
female: NA (2008 est.)

Life expectancy at birth:
total population: NA
male: NA
female: NA (2008 est.)

Total fertility rate: NA (2008 est.)

HIV/AIDS—adult prevalence rate: NA

HIV/AIDS—people living with HIV/AIDS: NA

HIV/AIDS—deaths: NA

Nationality:
noun: Pitcairn Islander(s)
adjective: Pitcairn Islander

Ethnic groups: descendants of the Bounty mutineers and their Tahitian wives

Religions: Seventh-Day Adventist 100%

Languages: English (official), Pitkern (mixture of an 18th century English dialect and a Tahitian dialect)

Literacy: NA

GOVERNMENT

Country name:
conventional long form: Pitcairn, Henderson, Ducie, and Oeno Islands
conventional short form: Pitcairn Islands

Dependency status: overseas territory of the UK

Government type: NA

Capital: *name:* Adamstown
geographic coordinates: 25 04 S, 130 05 W
time difference: UTC-9 (4 hours behind Washington, DC during Standard Time)

Administrative divisions: none (overseas territory of the UK)
Independence: none (overseas territory of the UK)
National holiday: Birthday of Queen ELIZABETH II, second Saturday in June (1926)
Constitution: 30 November 1838; reformed 1904 with additional reforms in 1940; further refined by the Local Government Ordinance of 1964
Legal system: local island by-laws
Suffrage: 18 years of age; universal with three years residency
Executive branch:
chief of state: Queen ELIZABETH II (since 6 February 1952); represented by UK High Commissioner to New Zealand and Governor (nonresident) of the Pitcairn Islands George FERGUSSON (since April 2006); Commissioner (nonresident) Leslie JAQUES (since September 2003) serves as liaison between the governor and the Island Council
head of government: Governor George FERGUSSON (since April 2006); Mayor and Chairman of the Island Council Mike WARREN (since 1 January 2008)
cabinet: NA
elections: the monarchy is hereditary; governor and commissioner appointed by the monarch; island mayor elected by popular vote for a three-year term; election last held December 2004 (next to be held in December 2007)
election results: Jay WARREN elected mayor and chairman of the Island Council
Legislative branch: unicameral Island Council (10 seats; 5 members elected by popular vote, 1 nominated by the 5 elected members, 2 appointed by the governor including 1 seat for the Island Secretary, the Island Mayor, and a commissioner liaising between the governor and council; elected members serve one-year terms)
elections: last held 24 December 2006 (next to be held in December 2007)
election results: percent of vote—NA; seats—all independents

Judicial branch: Magistrate's Court; Supreme Court; Court of Appeal; Judicial Officers are appointed by the Governor
Political parties and leaders: none
Political pressure groups and leaders: none
International organization participation: SPC, UPU
Diplomatic representation in the US: none (overseas territory of the UK)
Diplomatic representation from the US: none (overseas territory of the UK)
Flag description: blue with the flag of the UK in the upper hoist-side quadrant and the Pitcairn Islander coat of arms centered on the outer half of the flag; the coat of arms is yellow, green, and light blue with a shield featuring a yellow anchor

ECONOMY

Economy—overview: The inhabitants of this tiny isolated economy exist on fishing, subsistence farming, handicrafts, and postage stamps. The fertile soil of the valleys produces a wide variety of fruits and vegetables, including citrus, sugarcane, watermelons, bananas, yams, and beans. Bartering is an important part of the economy. The major sources of revenue are the sale of postage stamps to collectors and the sale of handicrafts to passing ships. In October 2004, more than one-quarter of Pitcairn's small labor force was arrested, putting the economy in a bind, since their services were required as lighter crew to load or unload passing ships.
GDP (purchasing power parity): $NA
Labor force: 15 able-bodied men (2004)
Labor force—by occupation: *note:* no business community in the usual sense; some public works; subsistence farming and fishing
Budget:
revenues: $746,000
expenditures: $1.028 million (FY04/05)
Agriculture—products: honey; wide variety of fruits and vegetables; goats, chickens, fish
Industries: postage stamps, handicrafts, beekeeping, honey

Electricity—production: NA kWh; note—electric power is provided by a small diesel-powered generator
Exports: $NA
Exports—commodities: fruits, vegetables, curios, stamps
Imports: $NA
Imports—commodities: fuel oil, machinery, building materials, flour, sugar, other foodstuffs
Economic aid—recipient: $3.465 million (2004)
Currency (code): New Zealand dollar (NZD)
Currency code: NZD
Exchange rates: New Zealand dollars per US dollar—1.3811 (2007), 1.5408 (2006), 1.4203 (2005), 1.5087 (2004), 1.7221 (2003)
Fiscal year: 1 April—31 March

COMMUNICATIONS

Telephones—main lines in use: 1 (there are 17 telephones on one party line); (2004)
Telephone system:
general assessment: satellite phone services
domestic: domestic communication via radio (CB)
international: country code—872; satellite earth station—1 (Inmarsat)
Radio broadcast stations: AM 1, FM 0, shortwave 0 (15 Ham radio operators (VP6)) (2004)
Radios: NA
Televisions: NA
Internet country code: .pn
Internet hosts: 9 (2007)
Internet Service Providers (ISPs): NA
Internet users: NA

TRANSPORTATION

Ports and terminals: Adamstown (on Bounty Bay)

MILITARY

Military—note: defense is the responsibility of the UK

TRANSNATIONAL ISSUES

Disputes—international: none

POLAND

INTRODUCTION

Background: Poland is an ancient nation that was conceived near the middle of the 10th century. Its golden age occurred in the 16th century. During the following century, the strengthening of the gentry and internal disorders weakened the nation. In a series of agreements between 1772 and 1795, Russia, Prussia, and Austria partitioned Poland amongst themselves. Poland regained its independence in 1918 only to be overrun by Germany and the Soviet Union in World War II. It became a Soviet satellite state following the war, but its government was comparatively tolerant and progressive. Labor turmoil in 1980 led to the formation of the independent trade union "Solidarity" that over time became a political force and by 1990 had swept parliamentary elections and the presidency. A "shock therapy" program during the early 1990s enabled the

country to transform its economy into one of the most robust in Central Europe, but Poland still faces the lingering challenges of high unemployment, underdeveloped and dilapidated infrastructure, and a poor rural underclass. Solidarity suffered a major defeat in the 2001 parliamentary elections when it failed to elect a single deputy to the lower house of Parliament, and the new leaders of the Solidarity Trade Union subsequently pledged to reduce the Trade Union's political role. Poland joined NATO in 1999 and the European Union in 2004. With its transformation to a democratic, market-oriented country largely completed, Poland is an increasingly active member of Euro-Atlantic organizations.

GEOGRAPHY

Location: Central Europe, east of Germany
Geographic coordinates: 52 00 N, 20 00 E
Map references: Europe
Area: *total:* 312,685 sq km
land: 304,465 sq km
water: 8,220 sq km
Area—comparative: slightly smaller than New Mexico
Land boundaries:
total: 3,056 km
border countries: Belarus 416 km, Czech Republic 790 km, Germany 467 km, Lithuania 103 km, Russia (Kaliningrad Oblast) 210 km, Slovakia 541 km, Ukraine 529 km
Coastline: 491 km
Maritime claims:
territorial sea: 12 nm
exclusive economic zone: defined by international treaties
Climate: temperate with cold, cloudy, moderately severe winters with frequent precipitation; mild summers with frequent showers and thundershowers
Terrain: mostly flat plain; mountains along southern border

Elevation extremes:
lowest point: near Raczki Elblaskie -2 m
highest point: Rysy 2,499 m
Natural resources: coal, sulfur, copper, natural gas, silver, lead, salt, amber, arable land
Land use:
arable land: 40.25%
permanent crops: 1%
other: 58.75% (2005)
Irrigated land: 1,000 sq km (2003)
Total renewable water resources: 63.1 cu km (2005)
Freshwater withdrawal (domestic/industrial/agricultural): *total:* 11.73 cu km/yr (13%/79%/8%)
per capita: 304 cu m/yr (2002)
Natural hazards: flooding
Environment—current issues: situation has improved since 1989 due to decline in heavy industry and increased environmental concern by post-Communist governments; air pollution nonetheless remains serious because of sulfur dioxide emissions from coal-fired power plants, and the resulting acid rain has caused forest damage; water pollution from industrial and municipal sources is also a problem, as is disposal of hazardous wastes; pollution levels should continue to decrease as industrial establishments bring their facilities up to EU code, but at substantial cost to business and the government
Environment—international agreements: *party to:* Air Pollution, Antarctic-Environmental Protocol, Antarctic-Marine Living Resources, Antarctic Seals, Antarctic Treaty, Biodiversity, Climate Change, Climate Change-Kyoto Protocol, Desertification, Endangered Species, Environmental Modification, Hazardous Wastes, Kyoto Protocol, Law of the Sea, Marine Dumping, Ozone Layer Protection, Ship Pollution, Wetlands
signed, but not ratified: Air Pollution-Nitrogen Oxides, Air Pollution-Persistent Organic Pollutants, Air Pollution-Sulfur 94
Geography—note: historically, an area of conflict because of flat terrain and the lack of natural barriers on the North European Plain

PEOPLE

Population: 38,500,696 (July 2008 est.)
Age structure:
0–14 years: 15.2% (male 3,013,109/female 2,849,977)
15–64 years: 71.4% (male 13,681,481/female 13,808,412)
65 years and over: 13.4% (male 1,964,477/female 3,183,240) (2008 est.)
Median age: *total:* 37.6 years

male: 35.8 years
female: 39.5 years (2008 est.)
Population growth rate: -0.045% (2008 est.)
Birth rate: 10.01 births/1,000 population (2008 est.)
Death rate: 9.99 deaths/1,000 population (2008 est.)
Net migration rate: -0.46 migrant(s)/1,000 population (2008 est.)
Sex ratio:
at birth: 1.06 male(s)/female
under 15 years: 1.06 male(s)/female
15–64 years: 0.99 male(s)/female
65 years and over: 0.62 male(s)/female
total population: 0.94 male(s)/female (2008 est.)
Infant mortality rate:
total: 6.93 deaths/1,000 live births
male: 7.66 deaths/1,000 live births
female: 6.17 deaths/1,000 live births (2008 est.)
Life expectancy at birth:
total population: 75.41 years
male: 71.42 years
female: 79.65 years (2008 est.)
Total fertility rate: 1.27 children born/woman (2008 est.)
HIV/AIDS—adult prevalence rate: 0.1%; note—no country specific models provided (2001 est.)
HIV/AIDS—people living with HIV/AIDS: 14,000 (2003 est.)
HIV/AIDS—deaths: 100 (2001 est.)
Major infectious diseases:
degree of risk: intermediate
food or waterborne diseases: bacterial diarrhea
vectorborne disease: tickborne encephalitis
note: highly pathogenic H5N1 avian influenza has been identified in this country; it poses a negligible risk with extremely rare cases possible among US citizens who have close contact with birds (2008)
Nationality:
noun: Pole(s)
adjective: Polish
Ethnic groups: Polish 96.7%, German 0.4%, Belarusian 0.1%, Ukrainian 0.1%, other and unspecified 2.7% (2002 census)
Religions: Roman Catholic 89.8% (about 75% practicing), Eastern Orthodox 1.3%, Protestant 0.3%, other 0.3%, unspecified 8.3% (2002)
Languages: Polish 97.8%, other and unspecified 2.2% (2002 census)
Literacy:
definition: age 15 and over can read and write
total population: 99.8%
male: 99.8%
female: 99.7% (2003 est.)

GOVERNMENT

Country name:
conventional long form: Republic of Poland
conventional short form: Poland
local long form: Rzeczpospolita Polska
local short form: Polska
Government type: republic
Capital: *name:* Warsaw
geographic coordinates: 52 15 N, 21 00 E
time difference: UTC+1 (6 hours ahead of Washington, DC during Standard Time)
daylight saving time: +1hr, begins last Sunday in March; ends last Sunday in October
Administrative divisions: 16 provinces (wojewodztwa, singular—wojewodztwo); Dolnoslaskie wojewodztwo, Kujawsko-Pomorskie wojewodztwo, Lodzkie wojewodztwo, Lubelskie wojewodztwo, Lubuskie wojewodztwo, Malopolskie wojewodztwo, Mazowieckie wojewodztwo, Opolskie wojewodztwo, Podkarpackie wojewodztwo, Podlaskie wojewodztwo, Pomorskie wojewodztwo, Slaskie wojewodztwo, Swietokrzyskie wojewodztwo, Warminsko-Mazurskie wojewodztwo, Wielkopolskie wojewodztwo, Zachodniopomorskie wojewodztwo
Independence: 11 November 1918 (republic proclaimed)
National holiday: Constitution Day, 3 May (1791)
Constitution: adopted by the National Assembly 2 April 1997; passed by national referendum 25 May 1997; effective 17 October 1997
Legal system: based on a mixture of Continental (Napoleonic) civil law and holdover Communist legal theory; changes being gradually introduced as part of broader democratization process; limited judicial review of legislative acts, but rulings of the Constitutional Tribunal are final; court decisions can be appealed to the European Court of Justice in Strasbourg; accepts compulsory ICJ jurisdiction with reservations
Suffrage: 18 years of age; universal
Executive branch:
chief of state: President Lech KACZYNSKI (since 23 December 2005)
head of government: Prime Minister Donald TUSK (since 16 November 2007); Deputy Prime Ministers Waldemar PAWLAK (since 16 November 2007) and Grzegorz SCHETYNA (since 16 November 2007)
cabinet: Council of Ministers responsible to the prime minister and the Sejm; the prime minister proposes, the president appoints, and the Sejm approves the Council of Ministers

elections: president elected by popular vote for a five-year term (eligible for a second term); election last held 9 and 23 October 2005 (next to be held in the fall 2010); prime minister and deputy prime ministers appointed by the president and confirmed by the Sejm
election results: Lech KACZYNSKI elected president; percent of popular vote—Lech KACZYNSKI 54%, Donald Tusk 46%
Legislative branch: bicameral National Assembly or Zgromadzenie Narodowe consists of the Senate or Senat (upper house) (100 seats; members are elected by a majority vote on a provincial basis to serve four-year terms), and the Sejm (lower house) (460 seats; members are elected under a complex system of proportional representation to serve four-year terms); the designation of National Assembly is only used on those rare occasions when the two houses meet jointly
elections: Senate—last held 21 October 2007 (next to be held by October 2011); Sejm elections last held 21 October 2007 (next to be held by October 2011)
election results: Senate—percent of vote by party—NA; seats by party—PO 60, PiS 39, independents 1; Sejm—percent of vote by party—PO 41.5%, PiS 32.1%, LiD 13.2%, PSL 8.9%, other 4.3%; seats by party—PO 209, PiS 166, LiD 53, PSL 31, German minorities 1; note—seats by party as of February 2008—PO 209, PiS 159, LiD 53, PSL 31, German minorities 1, nonaffiliated 7
note: one seat is assigned to ethnic minority parties in the Sejm only
Judicial branch: Supreme Court (judges are appointed by the president on the recommendation of the National Council of the Judiciary for an indefinite period); Constitutional Tribunal (judges are chosen by the Sejm for nine-year terms)
Political parties and leaders: Civic Platform or PO [Donald TUSK]; Democratic Left Alliance or SLD [Wojciech OLEJNICZAK]; Democratic Party or PD [Janusz ONYSZKIEWICZ]; German Minority of Lower Silesia or MNSO [Henryk KROLL]; Law and Justice or PiS [Jaroslaw KACZYNSKI]; League of Polish Families or LPR [Sylwester CHRUSZCZ]; Left and Democrats (LiD) (a coalition formed by the SLD, PD, SDPL, and UP) [Wojciech OLEJNICZAK]; Polish People's Party or PSL [Waldemar PAWLAK]; Samoobrona or SO [Andrzej LEPPER]; Social Democratic Party of Poland or SDPL [Marek BOROWSKI]; Union of Labor or UP [Andrzej SPYCHALSKI]
Political pressure groups and leaders: All Poland Trade Union Alliance or

OPZZ (trade union) [Jan GUZ]; Roman Catholic Church [Cardinal Stanislaw DZIWISZ, Archbishop Jozef MICHALIK]; Solidarity Trade Union [Janusz SNIADEK]
International organization participation: ACCT (observer), Arctic Council (observer), Australia Group, BIS, BSEC (observer), CBSS, CE, CEI, CERN, EAPC, EBRD, EIB, ESA (cooperating state), EU, FAO, IAEA, IBRD, ICAO, ICC, ICCt, ICRM, IDA, IFC, IFRCS, IHO, ILO, IMF, IMO, IMSO, Interpol, IOC, IOM, IPU, ISO, ITSO, ITU, ITUC, MIGA, MINURSO, NAM (guest), NATO, NSG, OAS (observer), OECD, OIF (observer), OPCW, OSCE, PCA, Schengen Convention, SECI (observer), UN, UNCTAD, UNDOF, UNESCO, UNHCR, UNIDO, UNIFIL, UNMEE, UNMIL, UNMIS, UNOCI, UNOMIG, UNWTO, UPU, WCL, WCO, WEU (associate), WFTU, WHO, WIPO, WMO, WTO, ZC
Diplomatic representation in the US:
chief of mission: Ambassador Robert KUPIECKI
chancery: 2640 16th Street NW, Washington, DC 20009
telephone: [1] (202) 234-3800 through 3802
FAX: [1] (202) 328-6271
consulate(s) general: Chicago, Los Angeles, New York
Diplomatic representation from the US:
chief of mission: Ambassador Victor ASHE
embassy: Aleje Ujazdowskie 29/31 00-540 Warsaw
mailing address: American Embassy Warsaw, US Department of State, Washington, DC 20521-5010 (pouch)
telephone: [48] (22) 504-2000
FAX: [48] (22) 504-2688
consulate(s) general: Krakow
Flag description: two equal horizontal bands of white (top) and red; similar to the flags of Indonesia and Monaco which are red (top) and white

ECONOMY

Economy—overview: Poland has pursued a policy of economic liberalization since 1990 and today stands out as a success story among transition economies. In 2007, GDP grew an estimated 6.5%, based on rising private consumption, a jump in corporate investment, and EU funds inflows. GDP per capita is still much below the EU average, but is similar to that of the three Baltic states. Since 2004, EU membership and access to EU structural funds have provided a major boost to the economy. Unemployment is falling rapidly, though at roughly 12.8% in 2007, it remains well above the EU average. Tightening labor markets, and rising global energy and food prices, pose a risk to consumer price stability. In December 2007 inflation reached 4.1% on a year-over-year basis, or higher than the upper limit of the National Bank of Poland's target range. Poland's economic performance could improve further if the country addresses some of the remaining deficiencies in its business environment. An inefficient commercial court system, a rigid labor code, bureaucratic red tape, and persistent low-level corruption keep the private sector from performing up to its full potential. Rising demands to fund health care, education, and the state pension system present a challenge to the Polish government's effort to hold the consolidated public sector budget deficit under 3.0% of GDP, a target which was achieved in 2007. The PO/PSL coalition government which came to power in November 2007 plans to further reduce the budget deficit with the aim of eventually adopting the euro. The new government has also announced its intention to enact business-friendly reforms, reduce public sector spending growth, lower taxes, and accelerate privatization. However, the government does not have the necessary two-thirds majority needed to override a presidential veto, and thus may have to water down initiatives in order to garner enough support to pass its pro-business policies.

GDP (purchasing power parity): $620.9 billion (2007 est.)

GDP (official exchange rate): $420.3 billion (2007 est.)

GDP—real growth rate: 6.5% (2007 est.)

GDP—per capita (PPP): $16,300 (2007 est.)

GDP—composition by sector:
agriculture: 4.1%
industry: 31.6%
services: 64.4% (2007 est.)

Labor force: 16.86 million (2007 est.)

Labor force—by occupation: *agriculture:* 16.1%
industry: 29%
services: 54.9% (2002)

Unemployment rate: 12.8% (2007 est.)

Population below poverty line: 17% (2003 est.)

Household income or consumption by percentage share:
lowest 10%: 3.1%
highest 10%: 27% (2002)

Distribution of family income—Gini index: 36 (2005)

Inflation rate (consumer prices): 2.5% (2007 est.)

Investment (gross fixed): 22.1% of GDP (2007 est.)

Budget: *revenues:* $85.25 billion
expenditures: $91.37 billion (2007 est.)

Public debt: 43.1% of GDP (2007 est.)

Agriculture—products: potatoes, fruits, vegetables, wheat; poultry, eggs, pork, dairy

Industries: machine building, iron and steel, coal mining, chemicals, shipbuilding, food processing, glass, beverages, textiles

Industrial production growth rate: 8.9% (2007 est.)

Electricity—production: 146.2 billion kWh (2005)

Electricity—production by source:
fossil fuel: 98.1%
hydro: 1.5%
nuclear: 0%
other: 0.4% (2001)

Electricity—consumption: 120.4 billion kWh (2005)

Electricity—exports: 16.19 billion kWh (2005)

Electricity—imports: 5.002 billion kWh (2005)

Oil—production: 32,800 bbl/day (2005 est.)

Oil—consumption: 462,700 bbl/day (2005 est.)

Oil—exports: 51,780 bbl/day (2004)

Oil—imports: 480,300 bbl/day (2004)

Oil—proved reserves: 96.38 million bbl (1 January 2006 est.)

Natural gas—production: 5.828 billion cu m (2005)

Natural gas—consumption: 15.58 billion cu m (2005 est.)

Natural gas—exports: 42.2 million cu m (2005 est.)

Natural gas—imports: 10.01 billion cu m (2005)

Natural gas—proved reserves: 158.1 billion cu m (1 January 2006 est.)

Current account balance: -$15.48 billion (2007 est.)

Exports: $144.6 billion f.o.b. (2007 est.)

Exports—commodities: machinery and transport equipment 37.8%, intermediate manufactured goods 23.7%, miscellaneous manufactured goods 17.1%, food and live animals 7.6% (2003)

Exports—partners: Germany 27.1%, Italy 6.5%, France 6.2%, UK 5.7%, Czech Republic 5.5%, Russia 4.3% (2006)

Imports: $160.2 billion f.o.b. (2007 est.)

Imports—commodities: machinery and transport equipment 38%, intermediate manufactured goods 21%, chemicals 14.8%, minerals, fuels, lubricants, and related materials 9.1% (2003)

Imports—partners: Germany 29%, Russia 9.5%, Italy 6.4%, Netherlands 5.7%, France 5.4% (2006)

Economic aid—recipient: $1.524 billion in available EU structural adjustment and cohesion funds (2004)

Reserves of foreign exchange and gold: $65.75 billion (31 December 2007 est.)

Debt—external: $169.5 billion (31 December 2007)

Stock of direct foreign investment—at home: $142.1 billion (2007 est.)

Stock of direct foreign investment—abroad: $19.64 billion (2007 est.)

Market value of publicly traded shares: $149.1 billion (2006)

Currency (code): zloty (PLN)

Currency code: PLN

Exchange rates: zlotych per US dollar—2.81 (2007), 3.1032 (2006), 3.2355 (2005), 3.6576 (2004), 3.8891 (2003)
note: zlotych is the plural form of zloty

Fiscal year: calendar year

COMMUNICATIONS

Telephones—main lines in use: 11.475 million (2006)

Telephones—mobile cellular: 36.746 million (2006)

Telephone system:
general assessment: modernization of the telecommunications network has accelerated with market based competition finalized in 2003; fixed-line service, dominated by the former state-owned company, is dwarfed by the growth in wireless telephony
domestic: mobile-cellular service available since 1993 and provided by three nation-wide networks with a fourth provider beginning operations in late 2006; cellular coverage is generally good with some gaps in the east; fixed-line service is growing slowly and still lags in rural areas
international: country code—48; international direct dialing with automated exchanges; satellite earth station—1 with access to Intelsat, Eutelsat, Inmarsat, and Intersputnik

Radio broadcast stations: AM 14, FM 777, shortwave 1 (1998)

Radios: 20.2 million (1997)

Television broadcast stations: 40 (2006)

Televisions: 13.05 million (1997)

Internet country code: .pl

Internet hosts: 5.681 million (2007)

Internet Service Providers (ISPs): 19 (2000)

Internet users: 11 million (2006)

TRANSPORTATION

Airports: 123 (2007)

Airports—with paved runways:
total: 83
over 3,047 m: 4

2,438 to 3,047 m: 30
1,524 to 2,437 m: 39
914 to 1,523 m: 7
under 914 m: 3 (2007)
Airports—with unpaved runways:
total: 40
2,438 to 3,047 m: 1
1,524 to 2,437 m: 4
914 to 1,523 m: 13
under 914 m: 22 (2007)
Heliports: 7 (2007)
Pipelines: gas 13,552 km; oil 1,384 km; refined products 777 km (2007)
Railways: *total:* 23,072 km
broad gauge: 629 km 1.524-m gauge
standard gauge: 22,443 km 1.435-m gauge (20,555 km operational; 11,910 km electrified) (2006)
Roadways: *total:* 423,997 km
paved: 295,356 km (includes 484 km of expressways)
unpaved: 128,641 km (2004)
Waterways: 3,997 km (navigable rivers and canals) (2006)
Merchant marine:
total: 11 ships (1000 GRT or over) 55,701 GRT/45,082 DWT
by type: cargo 6, chemical tanker 2, passenger/cargo 1, roll on/roll off 1, vehicle carrier 1

foreign-owned: 1 (Nigeria 1)
registered in other countries: 102 (Antigua and Barbuda 2, Bahamas 15, Cyprus 18, Liberia 14, Malta 25, Norway 3, Panama 15, Slovakia 2, St Vincent and The Grenadines 1, Vanuatu 7) (2007)
Ports and terminals: Gdansk, Gdynia, Swinoujscie, Szczecin

MILITARY

Military branches: Polish Armed Forces: Land Forces (includes Navy (Marynarka Wojenna, MW)), Polish Air Force (Sily Powietrzne Rzeczypospolitej Polskiej, SPRP) (2008)
Military service age and obligation: 17 years of age for male compulsory military service after January 1st of the year of 18th birthday; 17 years of age for voluntary military service; conscript service obligation shortened from 12 to 9 months in 2005; by 2008, plans call for at least 60% of military personnel to be volunteers; only soldiers who have completed their conscript service are allowed to volunteer for professional service; as of April 2004, women are only allowed to serve as officers and noncommissioned officers (2006)

Manpower available for military service:
males age 16–49: 9,741,508
females age 16–49: 9,514,843 (2008 est.)
Manpower fit for military service:
males age 16–49: 7,937,840
females age 16–49: 7,949,677 (2008 est.)
Manpower reaching militarily significant age annually:
males age 16–49: 257,605
females age 16–49: 245,832 (2008 est.)
Military expenditures—percent of GDP: 1.71% (2005 est.)

TRANSNATIONAL ISSUES

Disputes—international: as a member state that forms part of the EU's external border, Poland has implemented the strict Schengen border rules to restrict illegal immigration and trade along its eastern borders with Belarus and Ukraine
Illicit drugs: despite diligent counternarcotics measures and international information sharing on cross-border crimes, a major illicit producer of synthetic drugs for the international market; minor transshipment point for Southwest Asian heroin and Latin American cocaine to Western Europe

PORTUGAL

INTRODUCTION

Background: Following its heyday as a world power during the 15th and 16th centuries, Portugal lost much of its wealth and status with the destruction of Lisbon in a 1755 earthquake, occupation during the Napoleonic Wars, and the independence in 1822 of Brazil as a colony. A 1910 revolution deposed the monarchy; for most of the next six decades, repressive governments ran the country. In 1974, a left-wing military coup installed broad democratic reforms. The following year, Portugal granted independence to all of its African colonies. Portugal is a founding member of NATO and entered the EC (now the EU) in 1986.

GEOGRAPHY

Location: Southwestern Europe, bordering the North Atlantic Ocean, west of Spain
Geographic coordinates: 39 30 N, 8 00 W
Map references: Europe
Area:
total: 92,391 sq km
land: 91,951 sq km

water: 440 sq km
note: includes Azores and Madeira Islands
Area—comparative: slightly smaller than Indiana
Land boundaries:
total: 1,214 km
border countries: Spain 1,214 km
Coastline: 1,793 km
Maritime claims:
territorial sea: 12 nm
contiguous zone: 24 nm
exclusive economic zone: 200 nm
continental shelf: 200-m depth or to the depth of exploitation
Climate: maritime temperate; cool and rainy in north, warmer and drier in south
Terrain: mountainous north of the Tagus River, rolling plains in south
Elevation extremes:
lowest point: Atlantic Ocean 0 m
highest point: Ponta do Pico (Pico or Pico Alto) on Ilha do Pico in the Azores 2,351 m
Natural resources: fish, forests (cork), iron ore, copper, zinc, tin, tungsten, silver, gold, uranium, marble, clay, gypsum, salt, arable land, hydropower
Land use:
arable land: 17.29%

permanent crops: 7.84%

other: 74.87% (2005)

Irrigated land: 6,500 sq km (2003)

Total renewable water resources: 73.6 cu km (2005)

Freshwater withdrawal (domestic/industrial/agricultural): *total:* 11.09 cu km/yr (10%/12%/78%)

per capita: 1,056 cu m/yr (1998)

Natural hazards: Azores subject to severe earthquakes

Environment—current issues: soil erosion; air pollution caused by industrial and vehicle emissions; water pollution, especially in coastal areas

Environment—international agreements: *party to:* Air Pollution, Biodiversity, Climate Change, Climate Change-Kyoto Protocol, Desertification, Endangered Species, Hazardous Wastes, Law of the Sea, Marine Dumping, Marine Life Conservation, Ozone Layer Protection, Ship Pollution, Tropical Timber 83, Tropical Timber 94, Wetlands, Whaling

signed, but not ratified: Air Pollution-Persistent Organic Pollutants, Air Pollution-Volatile Organic Compounds, Environmental Modification

Geography—note: Azores and Madeira Islands occupy strategic locations along western sea approaches to Strait of Gibraltar

PEOPLE

Population: 10,676,910 (July 2008 est.)

Age structure:

0–14 years: 16.4% (male 912,995/female 835,715)

15–64 years: 66.2% (male 3,514,905/female 3,555,097)

65 years and over: 17.4% (male 764,443/female 1,093,755) (2008 est.)

Median age:

total: 39.1 years

male: 37 years

female: 41.3 years (2008 est.)

Population growth rate: 0.305% (2008 est.)

Birth rate: 10.45 births/1,000 population (2008 est.)

Death rate: 10.62 deaths/1,000 population (2008 est.)

Net migration rate: 3.23 migrant(s)/1,000 population (2008 est.)

Sex ratio:

at birth: 1.07 male(s)/female

under 15 years: 1.09 male(s)/female

15–64 years: 0.99 male(s)/female

65 years and over: 0.7 male(s)/female

total population: 0.95 male(s)/female (2008 est.)

Infant mortality rate:

total: 4.85 deaths/1,000 live births

male: 5.31 deaths/1,000 live births

female: 4.36 deaths/1,000 live births (2008 est.)

Life expectancy at birth:

total population: 78.04 years

male: 74.78 years

female: 81.53 years (2008 est.)

Total fertility rate: 1.49 children born/woman (2008 est.)

HIV/AIDS—adult prevalence rate: 0.4% (2001 est.)

HIV/AIDS—people living with HIV/AIDS: 22,000 (2001 est.)

HIV/AIDS—deaths: fewer than 1,000 (2003 est.)

Nationality:

noun: Portuguese (singular and plural)

adjective: Portuguese

Ethnic groups: homogeneous Mediterranean stock; citizens of black African descent who immigrated to mainland during decolonization number less than 100,000; since 1990 East Europeans have entered Portugal

Religions: Roman Catholic 84.5%, other Christian 2.2%, other 0.3%, unknown 9%, none 3.9% (2001 census)

Languages: Portuguese (official), Mirandese (official—but locally used)

Literacy: *definition:* age 15 and over can read and write

total population: 93.3%

male: 95.5%

female: 91.3% (2003 est.)

GOVERNMENT

Country name:

conventional long form: Portuguese Republic

conventional short form: Portugal

local long form: Republica Portuguesa

local short form: Portugal

Government type: republic; parliamentary democracy

Capital: *name:* Lisbon

geographic coordinates: 38 43 N, 9 08 W

time difference: UTC 0 (5 hours ahead of Washington, DC during Standard Time)

daylight saving time: +1hr, begins last Sunday in March; ends last Sunday in October

Administrative divisions: 18 districts (distritos, singular—distrito) and 2 autonomous regions* (regioes autonomas, singular—regiao autonoma); Aveiro, Acores (Azores)*, Beja, Braga, Braganca, Castelo Branco, Coimbra, Evora, Faro, Guarda, Leiria, Lisboa (Lisbon), Madeira*, Portalegre, Porto, Santarem, Setubal, Viana do Castelo, Vila Real, Viseu

Independence: 1143 (Kingdom of Portugal recognized); 5 October 1910 (republic proclaimed)

National holiday: Portugal Day (Day of Portugal), 10 June (1580); note—also called Camoes Day, the day that revered national poet Luis de Camoes (1524–80) died

Constitution: adopted 2 April 1976; note—subsequent revisions of the Constitution placed the military under strict civilian control, trimmed the powers of the president, and laid the groundwork for a stable, pluralistic liberal democracy; as well, they allowed for the privatization of nationalized firms and the government-owned communications media

Legal system: based on civil law system; the Constitutional Tribunal reviews the constitutionality of legislation; accepts compulsory ICJ jurisdiction with reservations

Suffrage: 18 years of age; universal

Executive branch:

chief of state: President Anibal CAVACO SILVA (since 9 March 2006)

head of government: Prime Minister Jose SOCRATES Carvalho Pinto de Sousa (since 12 March 2005)

cabinet: Council of Ministers appointed by the president on the recommendation of the prime minister

note: there is also a Council of State that acts as a consultative body to the president

elections: president elected by popular vote for a five-year term (eligible for a second term); election last held 22 January 2006 (next to be held in January 2011); following legislative elections, the leader of the majority party or leader of a majority coalition is usually appointed prime minister by the president

election results: Anibal CAVACO SILVA elected president; percent of vote— Anibal CAVACO SILVA 50.6%, Manuel ALEGRE 20.7%, Mario Alberto Nobre Lopes SOARES 14.3%, Jeronimo DE SOUSA 8.5%, Franciso LOUCA 5.3%

Legislative branch: unicameral Assembly of the Republic or Assembleia da Republica (230 seats; members are elected by popular vote to serve four-year terms)

elections: last held 20 February 2005 (next to be held in Fall 2009)

election results: percent of vote by party— PS 45.1%, PSD 28.7%, CDU 7.6%, CDS/PP 7.3%, BE 6.4%, other 4.9%; seats by party—PS 121, PSD 75, CDU 14, CDS/PP 12, BE 8

Judicial branch: Supreme Court or Supremo Tribunal de Justica (judges appointed for life by the Conselho Superior da Magistratura)

Political parties and leaders: Democratic and Social Center/Popular

Party or CDS/PP [Paulo PORTAS]; Green Ecologist Party (The Greens) or PEV [leadership commission elected by members]; Portuguese Communist Party or PCP [Jeronimo DE SOUSA]; Portuguese Socialist Party or PS [Jose SOCRATES Carvalho Pinto de Sousa]; Social Democratic Party or PSD [Luis Filipe MENEZES]; The Left Bloc or BE [Franciso Anacleto LOUCA]; Unitarian Democratic Coalition or CDU [Jeronimo DE SOUSA] (includes PCP and PEV)

Political pressure groups and leaders: NA

International organization participation: ABEDA, ADB (nonregional members), AfDB, Australia Group, BIS, CE, CERN, CPLP, EAPC, EBRD, EIB, EMU, ESA, EU, FAO, IADB, IAEA, IBRD, ICAO, ICC, ICCt, ICRM, IDA, IEA, IFAD, IFC, IFRCS, IHO, ILO, IMF, IMO, IMSO, Interpol, IOC, IOM, IPU, ISO, ITSO, ITU, ITUC, LAIA (observer), MIGA, NAM (guest), NATO, NEA, NSG, OAS (observer), OECD, OPCW, OSCE, PCA, Schengen Convention, SECI (observer), UN, UNCTAD, UNESCO, UNHCR, UNIDO, UNIFIL, Union Latina, UNMIT, UNWTO, UPU, WCL, WCO, WEU, WFTU, WHO, WIPO, WMO, WTO, ZC

Diplomatic representation in the US:
chief of mission: Ambassador Joao DE VALLERA
chancery: 2012 Massachusetts Avenue NW, Washington, DC 20036
telephone: [1] (202) 350-5400
FAX: [1] (202) 462-3726
consulate(s) general: Boston, New York, Newark (New Jersey), San Francisco
consulate(s): New Bedford (Massachusetts), Providence (Rhode Island)

Diplomatic representation from the US:
chief of mission: Ambassador Thomas F. STEPHENSON
embassy: Avenida das Forcas Armadas, 1600-081 Lisbon
mailing address: Apartado 43033, 1601-301 Lisboa; PSC 83, APO AE 09726
telephone: [351] (21) 727-3300
FAX: [351] (21) 726-9109
consulate(s): Ponta Delgada (Azores)

Flag description: two vertical bands of green (hoist side, two-fifths) and red (three-fifths) with the Portuguese coat of arms centered on the dividing line

ECONOMY

Economy—overview: Portugal has become a diversified and increasingly service-based economy since joining the European Community in 1986. Over the past two decades, successive govern-ments have privatized many state-controlled firms and liberalized key areas of the economy, including the financial and telecommunications sectors. The country qualified for the European Monetary Union (EMU) in 1998 and began circulating the euro on 1 January 2002 along with 11 other EU member economies. Economic growth had been above the EU average for much of the 1990s, but fell back in 2001–07. GDP per capita stands at roughly two-thirds of the EU-27 average. A poor educational system, in particular, has been an obstacle to greater productivity and growth. Portugal has been increasingly overshadowed by lower-cost producers in Central Europe and Asia as a target for foreign direct investment. The budget deficit surged to an all-time high of 6% of GDP in 2005, but the government reduced the deficit to 2.6% in 2007—a year ahead of Portugal's targeted schedule. Nonetheless, the government faces tough choices in its attempts to boost Portugal's economic competitiveness while keeping the budget deficit within the eurozone's 3%-of-GDP ceiling.

GDP (purchasing power parity): $230.5 billion (2007 est.)

GDP (official exchange rate): $223.3 billion (2007 est.)

GDP—real growth rate: 1.9% (2007 est.)

GDP—per capita (PPP): $21,700 (2007 est.)

GDP—composition by sector:
agriculture: 8.1%
industry: 25.4%
services: 66.5% (2007 est.)

Labor force: 5.62 million (2007 est.)

Labor force—by occupation:
agriculture: 10%
industry: 30%
services: 60% (2007 est.)

Unemployment rate: 7.7% (2007 est.)

Population below poverty line: 18% (2006)

Household income or consumption by percentage share: *lowest 10%:* 3.1%
highest 10%: 28.4% (1995 est.)

Distribution of family income—Gini index: 38.5 (2007)

Inflation rate (consumer prices): 2.4% (2007)

Investment (gross fixed): 21.7% of GDP (2007 est.)

Budget: *revenues:* $92.35 billion
expenditures: $98 billion (2007 est.)

Public debt: 63.6% of GDP (2007 est.)

Agriculture—products: grain, potatoes, tomatoes, olives, grapes; sheep, cattle, goats, swine, poultry, dairy products; fish

Industries: textiles, clothing, footwear, wood and cork, paper, chemicals, auto-parts manufacturing, base metals, diary products, wine and other foods, porcelain and ceramics, glassware, technology, telecommunications; ship construction and refurbishment; tourism

Industrial production growth rate: 2.5% (2007 est.)

Electricity—production: 49.04 billion kWh (2006)

Electricity—production by source:
fossil fuel: 64.5%
hydro: 31.3%
nuclear: 0%
other: 4.1% (2001)

Electricity—consumption: 48.55 billion kWh (2006)

Electricity—exports: 3.138 billion kWh (2006)

Electricity—imports: 8.624 billion kWh (2006)

Oil—production: 9,500 bbl/day (2006 est.)

Oil—consumption: 305,800 bbl/day (2006 est.)

Oil—exports: 43,070 bbl/day (2004)

Oil—imports: 361,300 bbl/day (2004)

Oil—proved reserves: NA bbl

Natural gas—production: 0 cu m (2007 est.)

Natural gas—consumption: 3.86 billion cu m (2006)

Natural gas—exports: 0 cu m (2007 est.)

Natural gas—imports: 4.082 billion cu m (2006)

Natural gas—proved reserves: 0 cu m (1 January 2006 est.)

Current account balance: -$20.89 billion (2007 est.)

Exports: $51.5 billion f.o.b. (2007 est.)

Exports—commodities: agricultural products, food products, oil products, chemical products, plastics and rubber, skins and leather, wood and cork, wood pulp and paper, textile materials, clothing, footwear, minerals and mineral products, base metals, machinery and tools, vehicles and other transport material, and optical and precision

Exports—partners: Spain 26.5%, Germany 12.9%, France 12%, UK 6.7%, US 6.1% (2006)

Imports: $75.3 billion f.o.b. (2007 est.)

Imports—commodities: agricultural products, food products, oil products, chemical products, plastics and rubber, skins and leather, wood and cork, wood pulp and paper, textile materials, clothing, footwear, minerals and mineral products, base metals, machinery and tools, vehicles and other transport material, and optical and precision instruments, computer accessories and parts, semi-conductors and related devices, household goods, passenger cars new and used, and wine products

Imports—partners: Spain 29%, Germany 13.1%, France 8.1%, Italy 5.6%, Netherlands 4.4% (2006)

Economic aid—donor: ODA, $396 million (2006)

Reserves of foreign exchange and gold: $11.55 billion (31 December 2007 est.)

Debt—external: $389.5 billion (31 December 2007)

Stock of direct foreign investment—at home: $89.2 billion (2007 est.)

Stock of direct foreign investment—abroad: $54.85 billion (2007 est.)

Market value of publicly traded shares: $66.98 billion (2005)

Currency (code): euro (EUR)

Currency code: EUR

Exchange rates: euros per US dollar— 0.7345 (2007), 0.7964 (2006), 0.8041 (2005), 0.8054 (2004), 0.886 (2003)

Fiscal year: calendar year

COMMUNICATIONS

Telephones—main lines in use: 4.231 million (2006)

Telephones—mobile cellular: 12.226 million (2006)

Telephone system:
general assessment: Portugal's telephone system has achieved a state-of-the-art network with broadband, high-speed capabilities
domestic: integrated network of coaxial cables, open-wire, microwave radio relay, and domestic satellite earth stations
international: country code—351; a combination of submarine cables provide connectivity to Europe, North and East Africa, South Africa, the Middle East, Asia, and the US; satellite earth stations—3 Intelsat (2 Atlantic Ocean and 1 Indian Ocean), NA Eutelsat; tropospheric scatter to Azores (1998)

Radio broadcast stations: AM 47, FM 172 (many are repeaters), shortwave 2 (1998)

Radios: 3.02 million (1997)

Television broadcast stations: 62 (plus 166 repeaters; includes Azores and Madeira Islands) (1995)

Televisions: 3.31 million (1997)

Internet country code: .pt

Internet hosts: 836,616 (2007)

Internet Service Providers (ISPs): 16 (2000)

Internet users: 3.213 million (2006)

TRANSPORTATION

Airports: 66 (2007)

Airports—with paved runways:
total: 44
over 3,047 m: 5
2,438 to 3,047 m: 9
1,524 to 2,437 m: 5
914 to 1,523 m: 13
under 914 m: 12 (2007)

Airports—with unpaved runways:
total: 22
914 to 1,523 m: 1
under 914 m: 21 (2007)

Pipelines: gas 1,098 km; oil 11 km; refined products 188 km (2007)

Railways:
total: 2,786 km
broad gauge: 2,603 km 1.668-m gauge (1,351 km electrified)
narrow gauge: 183 km 1.000-m gauge (2006)

Roadways:
total: 78,470 km
paved: 67,484 km (includes 2,002 km of expressways)
unpaved: 10,986 km (2004)

Waterways: 210 km (on Douro River from Porto) (2006)

Merchant marine:
total: 117 ships (1000 GRT or over) 1,022,783 GRT/1,287,951 DWT
by type: bulk carrier 10, cargo 37, carrier 1, chemical tanker 16, container 6, liquefied gas 9, passenger 10, passenger/cargo 10, petroleum tanker 6, roll on/roll off 1, specialized tanker 1, vehicle carrier 10
foreign-owned: 80 (Belgium 9, Denmark 3, Germany 22, Greece 4, Italy 11, Japan 10, Malta 1, Mexico 1, Netherlands 1, Norway 3, Spain 10, Sweden 2, Switzerland 2, US 1)
registered in other countries: 15 (Cyprus 1, Hong Kong 1, Malta 3, Panama 9, St Vincent and The Grenadines 1) (2007)

Ports and terminals: Leixoes, Lisbon, Setubal, Sines

MILITARY

Military branches: Portuguese Army (Exercito Portugues), Portuguese Navy (Marinha Portuguesa; includes Marine Corps), Portuguese Air Force (Forca Aerea Portuguesa, FAP) (2008)

Military service age and obligation: 18 years of age for voluntary military service; compulsory military service ended in 2004; women serve in the armed forces, on naval ships since 1993, but are prohibited from serving in some combatant specialties (2005)

Manpower available for military service:
males age 16–49: 2,573,913
females age 16–49: 2,498,262 (2008 est.)

Manpower fit for military service:
males age 16–49: 2,099,647
females age 16–49: 2,060,559 (2008 est.)

Manpower reaching militarily significant age annually:
males age 16–49: 64,910
females age 16–49: 58,599 (2008 est.)

Military expenditures—percent of GDP: 2.3% (2005 est.)

TRANSNATIONAL ISSUES

Disputes—international: Portugal does not recognize Spanish sovereignty over the territory of Olivenza based on a difference of interpretation of the 1815 Congress of Vienna and the 1801 Treaty of Badajoz

Illicit drugs: seizing record amounts of Latin American cocaine destined for Europe; a European gateway for Southwest Asian heroin; transshipment point for hashish from North Africa to Europe; consumer of Southwest Asian heroin

PUERTO RICO

INTRODUCTION

Background: Populated for centuries by aboriginal peoples, the island was claimed by the Spanish Crown in 1493 following COLUMBUS' second voyage to the Americas. In 1898, after 400 years of colonial rule that saw the indigenous population nearly exterminated and African slave labor introduced, Puerto Rico was ceded to the US as a result of the Spanish-American War. Puerto Ricans were granted US citizenship in 1917. Popularly-elected governors have served since 1948. In 1952, a constitution was enacted providing for internal self government. In plebiscites held in 1967, 1993, and 1998, voters chose not to alter the existing political status.

GEOGRAPHY

Location: Caribbean, island between the Caribbean Sea and the North Atlantic Ocean, east of the Dominican Republic

Geographic coordinates: 18 15 N, 66 30 W

Map references: Central America and the Caribbean

Area:
total: 13,790 sq km
land: 8,870 sq km
water: 4,921 sq km

Area—comparative: slightly less than three times the size of Rhode Island

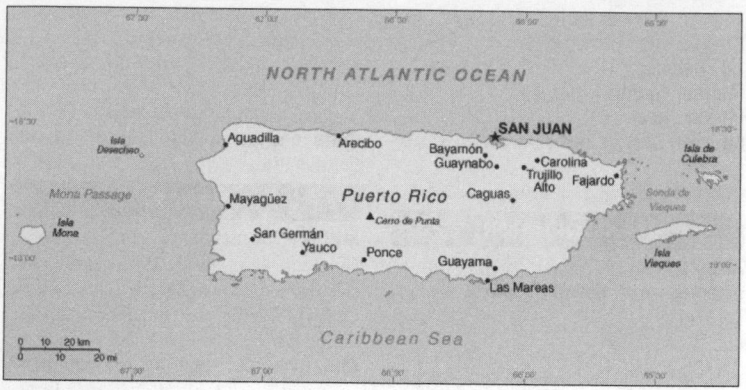

Land boundaries: 0 km
Coastline: 501 km
Maritime claims:
territorial sea: 12 nm
exclusive economic zone: 200 nm
Climate: tropical marine, mild; little seasonal temperature variation
Terrain: mostly mountains with coastal plain belt in north; mountains precipitous to sea on west coast; sandy beaches along most coastal areas
Elevation extremes:
lowest point: Caribbean Sea 0 m
highest point: Cerro de Punta 1,339 m
Natural resources: some copper and nickel; potential for onshore and offshore oil
Land use: *arable land:* 3.69%
permanent crops: 5.59%
other: 90.72% (2005)
Irrigated land: 400 sq km (2003)
Natural hazards: periodic droughts; hurricanes
Environment—current issues: erosion; occasional drought causing water shortages
Geography—note: important location along the Mona Passage—a key shipping lane to the Panama Canal; San Juan is one of the biggest and best natural harbors in the Caribbean; many small rivers and high central mountains ensure land is well watered; south coast relatively dry; fertile coastal plain belt in north

PEOPLE

Population: 3,958,128 (July 2008 est.)
Age structure:
0–14 years: 20.5% (male 415,141/female 396,782)
15–64 years: 66% (male 1,254,416/female 1,358,229)
65 years and over: 13.5% (male 229,727/female 303,833) (2008 est.)
Median age: *total:* 35.6 years
male: 33.8 years
female: 37.3 years (2008 est.)
Population growth rate: 0.369% (2008 est.)

Birth rate: 12.61 births/1,000 population (2008 est.)
Death rate: 7.88 deaths/1,000 population (2008 est.)
Net migration rate: -1.03 migrant(s)/1,000 population (2008 est.)
Sex ratio:
at birth: 1.05 male(s)/female
under 15 years: 1.05 male(s)/female
15–64 years: 0.92 male(s)/female
65 years and over: 0.76 male(s)/female
total population: 0.92 male(s)/female (2008 est.)
Infant mortality rate:
total: 8.65 deaths/1,000 live births
male: 9.15 deaths/1,000 live births
female: 8.13 deaths/1,000 live births (2008 est.)
Life expectancy at birth:
total population: 78.58 years
male: 74.64 years
female: 82.73 years (2008 est.)
Total fertility rate: 1.76 children born/woman (2008 est.)
HIV/AIDS—adult prevalence rate: NA
HIV/AIDS—people living with HIV/AIDS: 7,397 (1997)
HIV/AIDS—deaths: NA
Nationality:
noun: Puerto Rican(s) (US citizens)
adjective: Puerto Rican
Ethnic groups: white (mostly Spanish origin) 80.5%, black 8%, Amerindian 0.4%, Asian 0.2%, mixed and other 10.9%
Religions: Roman Catholic 85%, Protestant and other 15%
Languages: Spanish, English
Literacy: *definition:* age 15 and over can read and write
total population: 94.1%
male: 93.9%
female: 94.4% (2002 est.)

GOVERNMENT

Country name:
conventional long form: Commonwealth of Puerto Rico
conventional short form: Puerto Rico

Dependency status: unincorporated, organized territory of the US with commonwealth status; policy relations between Puerto Rico and the US conducted under the jurisdiction of the Office of the President
Government type: commonwealth
Capital: *name:* San Juan
geographic coordinates: 18 28 N, 66 07 W
time difference: UTC-4 (1 hour ahead of Washington, DC during Standard Time)
Administrative divisions: none (territory of the US with commonwealth status); there are no first-order administrative divisions as defined by the US Government, but there are 78 municipalities (municipios, singular—municipio) at the second order; Adjuntas, Aguada, Aguadilla, Aguas Buenas, Aibonito, Anasco, Arecibo, Arroyo, Barceloneta, Barranquitas, Bayamon, Cabo Rojo, Caguas, Camuy, Canovanas, Carolina, Catano, Cayey, Ceiba, Ciales, Cidra, Coamo, Comerio, Corozal, Culebra, Dorado, Fajardo, Florida, Guanica, Guayama, Guayanilla, Guaynabo, Gurabo, Hatillo, Hormigueros, Humacao, Isabela, Jayuya, Juana Diaz, Juncos, Lajas, Lares, Las Marias, Las Piedras, Loiza, Luquillo, Manati, Maricao, Maunabo, Mayaguez, Moca, Morovis, Naguabo, Naranjito, Orocovis, Patillas, Penuelas, Ponce, Quebradillas, Rincon, Rio Grande, Sabana Grande, Salinas, San German, San Juan, San Lorenzo, San Sebastian, Santa Isabel, Toa Alta, Toa Baja, Trujillo Alto, Utuado, Vega Alta, Vega Baja, Vieques, Villalba, Yabucoa, Yauco
Independence: none (territory of the US with commonwealth status)
National holiday: US Independence Day, 4 July (1776); Puerto Rico Constitution Day, 25 July (1952)
Constitution: ratified 3 March 1952; approved by US Congress 3 July 1952; effective 25 July 1952
Legal system: based on Spanish civil code and within the US Federal system of justice
Suffrage: 18 years of age; universal; island residents are US citizens but do not vote in US presidential elections
Executive branch:
chief of state: President George W. BUSH of the US (since 20 January 2001); Vice President Richard B. CHENEY (since 20 January 2001)
head of government: Governor Anibal ACEVEDO-VILA (since 2 January 2005)
cabinet: Cabinet appointed by the governor with the consent of the legislature
elections: under the US Constitution, residents of unincorporated territories, such

as Puerto Rico, do not vote in elections for US president and vice president; however, they may vote in Democratic and Republican presidential primary elections; governor elected by popular vote for a four-year term (no term limits); election last held 2 November 2004 (next to be held in November 2008)

election results: Anibal ACEVEDO-VILA elected governor; percent of vote—48.4%

Legislative branch: bicameral Legislative Assembly consists of the Senate (at least 27 seats—currently 29; members are directly elected by popular vote to serve four-year terms) and the House of Representatives (51 seats; members are elected by popular vote to serve four-year terms)

elections: Senate—last held 2 November 2004 (next to be held November 2008); House of Representatives—last held 2 November 2004 (next to be held in November 2008)

election results: Senate—percent of vote by party—PNP 43.4%, PPD 40.3%, PIP 9.4%; seats by party—PNP 17, PPD 9, PIP 1; House of Representatives—percent of vote by party—PNP 46.3%, PPD 43.1%, PIP 9.7%; seats by party—PNP 32, PPD 18, PIP 1

note: Puerto Rico elects, by popular vote, a resident commissioner to serve a four-year term as a nonvoting representative in the US House of Representatives; aside from not voting on the House floor, he enjoys all the rights of a member of Congress; elections last held 2 November 2004 (next to be held in November 2008); results—percent of vote by party—PNP 48.6%, other 51.4%; seats by party—PNP 1

Judicial branch: Supreme Court; Appellate Court; Court of First Instance composed of two sections: a Superior Court and a Municipal Court (justices for all these courts appointed by the governor with the consent of the Senate)

Political parties and leaders: National Democratic Party [Roberto PRATS]; National Republican Party of Puerto Rico [Dr. Tiody FERRE]; New Progressive Party or PNP [Pedro ROSSELLO] (pro-US statehood); Popular Democratic Party or PPD [Anibal ACEVEDO-VILA] (pro-commonwealth); Puerto Rican Independence Party or PIP [Ruben BERRIOS Martinez] (pro independence)

Political pressure groups and leaders: Boricua Popular Army or EPB (a revolutionary group also known as Los Macheteros); note—the following rad-

ical groups are considered dormant by Federal law enforcement: Armed Forces for National Liberation or FALN, Armed Forces of Popular Resistance, Volunteers of the Puerto Rican Revolution

International organization participation: Caricom (observer), Interpol (subbureau), IOC, ITUC, UNWTO (associate), UPU, WCL, WFTU

Diplomatic representation in the US: none (territory of the US with commonwealth status)

Diplomatic representation from the US: none (territory of the US with commonwealth status)

Flag description: five equal horizontal bands of red (top and bottom) alternating with white; a blue isosceles triangle based on the hoist side bears a large, white, five-pointed star in the center; design initially influenced by the US flag, but similar to the Cuban flag, with the colors of the bands and triangle reversed

ECONOMY

Economy—overview: Puerto Rico has one of the most dynamic economies in the Caribbean region. A diverse industrial sector has far surpassed agriculture as the primary locus of economic activity and income. Encouraged by duty-free access to the US and by tax incentives, US firms have invested heavily in Puerto Rico since the 1950s. US minimum wage laws apply. Sugar production has lost out to dairy production and other livestock products as the main source of income in the agricultural sector. Tourism has traditionally been an important source of income, with estimated arrivals of nearly 5 million tourists in 2004. Growth fell off in 2001–03, largely due to the slowdown in the US economy, recovered in 2004–05, but declined again in 2006–07.

GDP (purchasing power parity): $77.41 billion (2007 est.)

GDP (official exchange rate): $NA (2007 est.)

GDP—real growth rate: -1.2% (2007 est.)

GDP—per capita (PPP): $19,600 (2007 est.)

GDP—composition by sector:
agriculture: 1%
industry: 45%
services: 54% (2002 est.)

Labor force: 1.3 million (2000)

Labor force—by occupation: *agriculture:* 3%
industry: 20%
services: 77% (2000 est.)

Unemployment rate: 12% (2002)

Population below poverty line: NA%

Household income or consumption by percentage share:
lowest 10%: NA%
highest 10%: NA%

Inflation rate (consumer prices): 6.5% (2003 est.)

Budget:
revenues: $6.7 billion
expenditures: $9.6 billion (FY99/00)

Agriculture—products: sugarcane, coffee, pineapples, plantains, bananas; livestock products, chickens

Industries: pharmaceuticals, electronics, apparel, food products, tourism

Industrial production growth rate: NA%

Electricity—production: 24.96 billion kWh (2005)

Electricity—production by source:
fossil fuel: 99.2%
hydro: 0.8%
nuclear: 0%
other: 0% (2001)

Electricity—consumption: 23.21 billion kWh (2005)

Electricity—exports: 0 kWh (2005)

Electricity—imports: 0 kWh (2005)

Oil—production: 1,354 bbl/day (2005 est.)

Oil—consumption: 230,000 bbl/day (2005 est.)

Oil—exports: 10,580 bbl/day (2004)

Oil—imports: 230,100 bbl/day (2004)

Oil—proved reserves: 0 bbl (1 January 2006 est.)

Natural gas—production: 0 cu m (2005 est.)

Natural gas—consumption: 642.6 million cu m (2005 est.)

Natural gas—exports: 0 cu m (2005 est.)

Natural gas—imports: 642.6 million cu m (2005)

Natural gas—proved reserves: 0 cu m (1 January 2006 est.)

Exports: $46.9 billion f.o.b. (2001)

Exports—commodities: chemicals, electronics, apparel, canned tuna, rum, beverage concentrates, medical equipment

Exports—partners: US 90.3%, UK 1.6%, Netherlands 1.4%, Dominican Republic 1.4% (2006)

Imports: $29.1 billion c.i.f. (2001)

Imports—commodities: chemicals, machinery and equipment, clothing, food, fish, petroleum products

Imports—partners: US 55.0%, Ireland 23.7%, Japan 5.4% (2006)

Economic aid—recipient: $NA

Debt—external: $NA

Market value of publicly traded shares: $NA

Currency (code): US dollar (USD)

Currency code: USD

Exchange rates: the US dollar is used
Fiscal year: 1 July—30 June

COMMUNICATIONS

Telephones—main lines in use: 1.038 million (2005)
Telephones—mobile cellular: 3.354 million (2005)
Telephone system:
general assessment: modern system integrated with that of the US by high-capacity submarine cable and Intelsat with high-speed data capability
domestic: digital telephone system; cellular telephone service
international: country code—1-787, 939; submarine cables provide connectivity to the US, Caribbean, Central and South America; satellite earth station—1 Intelsat
Radio broadcast stations: AM 74, FM 53, shortwave 0 (2005)
Radios: 2.7 million (1997)
Television broadcast stations: 32 (2006)
Televisions: 1.021 million (1997)

Internet country code: .pr
Internet hosts: 413 (2007)
Internet Service Providers (ISPs): 76 (2000)
Internet users: 915,600 (2005)

TRANSPORTATION

Airports: 29 (2007)
Airports—with paved runways:
total: 17
over 3,047 m: 3
1,524 to 2,437 m: 2
914 to 1,523 m: 7
under 914 m: 5 (2007)
Airports—with unpaved runways:
total: 12
1,524 to 2,437 m: 1
914 to 1,523 m: 1
under 914 m: 10 (2007)
Railways: *total:* 96 km
narrow gauge: 96 km 1.000-m gauge (2006)
Roadways: *total:* 25,735 km
paved: 24,353 km (includes 427 km of expressways)
unpaved: 1,382 km (2005)

Merchant marine:
total: 3 ships (1000 GRT or over) 77,177 GRT/50,138 DWT
by type: roll on/roll off 3
foreign-owned: 3 (US 3)
registered in other countries: 1 (St Vincent and The Grenadines 1) (2007)
Ports and terminals: Guayanilla, Mayaguez, San Juan

MILITARY

Military branches: no regular indigenous military forces; paramilitary National Guard, Police Force
Military—note: defense is the responsibility of the US

TRANSNATIONAL ISSUES

Disputes—international: increasing numbers of illegal migrants from the Dominican Republic cross the Mona Passage to Puerto Rico each year looking for work

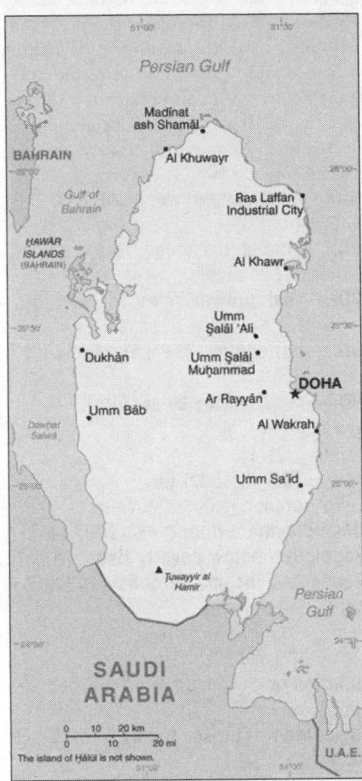

INTRODUCTION

Background: Ruled by the al-Thani family since the mid-1800s, Qatar transformed itself from a poor British protectorate noted mainly for pearling into an independent state with significant oil and natural gas revenues. During the late 1980s and early 1990s, the Qatari economy was crippled by a continuous siphoning off of petroleum revenues by the Amir, who had ruled the country since 1972. His son, the current Amir HAMAD bin Khalifa al-Thani, overthrew him in a bloodless coup in 1995. In 2001, Qatar resolved its longstanding border disputes with both Bahrain and Saudi Arabia. Oil and natural gas revenues enable Qatar to have one of the highest per capita incomes in the world.

GEOGRAPHY

Location: Middle East, peninsula bordering the Persian Gulf and Saudi Arabia

Geographic coordinates: 25 30 N, 51 15 E

Map references: Middle East

Area:
total: 11,437 sq km
land: 11,437 sq km
water: 0 sq km

Area—comparative: slightly smaller than Connecticut

Land boundaries:
total: 60 km
border countries: Saudi Arabia 60 km

Coastline: 563 km

Maritime claims:
territorial sea: 12 nm
contiguous zone: 24 nm
exclusive economic zone: as determined by bilateral agreements or the median line

Climate: arid; mild, pleasant winters; very hot, humid summers

Terrain: mostly flat and barren desert covered with loose sand and gravel

Elevation extremes:
lowest point: Persian Gulf 0 m
highest point: Qurayn Abu al Bawl 103 m

Natural resources: petroleum, natural gas, fish

Land use:
arable land: 1.64%
permanent crops: 0.27%
other: 98.09% (2005)

Irrigated land: 130 sq km (2002)

Total renewable water resources: 0.1 cu km (1997)

Freshwater withdrawal (domestic/industrial/agricultural): *total:* 0.29 cu km/yr (24%/3%/72%)
per capita: 358 cu m/yr (2000)

Natural hazards: haze, dust storms, sandstorms common

Environment—current issues: limited natural fresh water resources are increasing dependence on large-scale desalination facilities

Environment—international agreements: *party to:* Biodiversity, Climate Change, Climate Change-Kyoto Protocol, Desertification, Endangered Species, Hazardous Wastes, Law of the Sea, Ozone Layer Protection, Ship Pollution
signed, but not ratified: none of the selected agreements

Geography—note: strategic location in central Persian Gulf near major petroleum deposits

PEOPLE

Population: 928,635 (July 2008 est.)

Age structure:
0–14 years: 22.8% (male 108,063/female 103,887)
15–64 years: 72.9% (male 463,942/female 213,137)
65 years and over: 4.3% (male 29,515/female 10,091) (2008 est.)

Median age: *total:* 32.1 years
male: 37.5 years
female: 23.4 years (2008 est.)

Population growth rate: 2.279% (2008 est.)

Birth rate: 15.56 births/1,000 population (2008 est.)

Death rate: 4.94 deaths/1,000 population (2008 est.)

Net migration rate: 12.18 migrant(s)/1,000 population (2008 est.)

Sex ratio:
at birth: 1.05 male(s)/female
under 15 years: 1.04 male(s)/female
15–64 years: 2.18 male(s)/female
65 years and over: 2.92 male(s)/female
total population: 1.84 male(s)/female (2008 est.)

Infant mortality rate:
total: 16.88 deaths/1,000 live births
male: 19.93 deaths/1,000 live births
female: 13.68 deaths/1,000 live births (2008 est.)

Life expectancy at birth:
total population: 74.38 years
male: 71.82 years
female: 77.07 years (2008 est.)

Total fertility rate: 2.7 children born/woman (2008 est.)

HIV/AIDS—adult prevalence rate: 0.09% (2001 est.)

HIV/AIDS—people living with HIV/AIDS: NA

HIV/AIDS—deaths: NA

Nationality:
noun: Qatari(s)
adjective: Qatari

Ethnic groups: Arab 40%, Indian 18%, Pakistani 18%, Iranian 10%, other 14%

Religions: Muslim 77.5%, Christian 8.5%, other 14% (2004 census)

Languages: Arabic (official), English commonly used as a second language

Literacy: *definition:* age 15 and over can read and write
total population: 89%
male: 89.1%
female: 88.6% (2004 census)

GOVERNMENT

Country name:
conventional long form: State of Qatar
conventional short form: Qatar
local long form: Dawlat Qatar
local short form: Qatar
note: closest approximation of the native pronunciation falls between cutter and gutter, but not like guitar

Government type: emirate

Capital: *name:* Doha
geographic coordinates: 25 17 N, 51 32 E
time difference: UTC+3 (8 hours ahead of Washington, DC during Standard Time)

Administrative divisions: 10 municipalities (baladiyat, singular—baladiyah);

Ad Dawhah, Al Ghuwayriyah, Al Jumayliyah, Al Khawr, Al Wakrah, Ar Rayyan, Jarayan al Batinah, Madinat ash Shamal, Umm Sa'id, Umm Salal
Independence: 3 September 1971 (from UK)
National holiday: Independence Day, 3 September (1971); also observed is National Day, 18 December
Constitution: ratified by public referendum on 29 April 2003, endorsed by the Amir on 8 June 2004, effective on 9 June 2005
Legal system: based on Islamic and civil law codes; discretionary system of law controlled by the Amir, although civil codes are being implemented; Islamic law dominates family and personal matters; has not accepted compulsory ICJ jurisdiction
Suffrage: 18 years of age; universal
Executive branch:
chief of state: Amir HAMAD bin Khalifa al-Thani (since 27 June 1995 when, as heir apparent, he ousted his father, Amir KHALIFA bin Hamad al-Thani, in a bloodless coup); Heir Apparent TAMIM bin Hamad bin Khalifa al-Thani, fourth son of the monarch (selected Heir Apparent by the monarch on 5 August 2003); note—Amir HAMAD also holds the positions of Minister of Defense and Commander-in-Chief of the Armed Forces
head of government: Prime Minister HAMAD bin Jasim bin Jabir al-Thani (since 3 April 2007); Deputy Prime Minister Abdallah bin Hamad al-ATIYAH (since 3 April 2007)
cabinet: Council of Ministers appointed by the monarch
elections: the monarch is hereditary
note: in April 2007, Qatar held nationwide elections for a 29-member Central Municipal Council (CMC), which has limited consultative powers aimed at improving the provision of municipal services; the first election for the CMC was held in March 1999
Legislative branch: unicameral Advisory Council or Majlis al-Shura (35 seats; members appointed)
note: no legislative elections have been held since 1970 when there were partial elections to the body; Council members have had their terms extended every year since the new constitution came into force on 9 June 2005; the constitution provides for a new 45-member Advisory Council or Majlis al-Shura; the public would elect two-thirds of the Majlis al-Shura; the Amir would appoint the remaining members; preparations are underway to conduct elections to the Majlis al-Shura

Judicial branch: Courts of First Instance, Appeal, and Cassation; an Administrative Court and a Constitutional Court were established in 2007; note—all judges are appointed by Amiri Decree based on the recommendation of the Supreme Judiciary Council for renewable three-year terms
Political parties and leaders: none
Political pressure groups and leaders: none
International organization participation: ABEDA, ACC, AFESD, AMF, FAO, G-77, GCC, IAEA, IBRD, ICAO, ICC, ICRM, IDA, IDB, IFAD, IFRCS, IHO, ILO, IMF, IMO, IMSO, Interpol, IOC, IPU, ISO, ITSO, ITU, LAS, MIGA, NAM, OAPEC, OAS (observer), OIC, OPCW, OPEC, PCA, UN, UN Security Council (temporary), UNCTAD, UNESCO, UNIDO, UNIFIL, UNWTO, UPU, WCO, WHO, WIPO, WMO, WTO
Diplomatic representation in the US:
chief of mission: Ambassador Ali Fahad al-Shahwany al-HAJRI
chancery: 2555 M Street NW, Washington, DC 20037
telephone: [1] (202) 274-1600 and 274-1603
FAX: [1] (202) 237-0061
consulate(s) general: Houston
Diplomatic representation from the US:
chief of mission: Ambassador (vacant); charge d'Affaires Michael A. RATNEY
embassy: Al-Luqta District, 22 February Road, Doha
mailing address: P. O. Box 2399, Doha
telephone: [974] 488 4298
FAX: [974] 488 4176
Flag description: maroon with a broad white serrated band (nine white points) on the hoist side

ECONOMY

Economy—overview: Qatar is in the midst of an economic boom supported by its expanding production of natural gas and oil. Economic policy is focused on development of Qatar's nonassociated natural gas reserves and increasing private and foreign investment in non-energy sectors. Oil and gas account for more than 60% of GDP, roughly 85% of export earnings, and 70% of government revenues. Oil and gas have made Qatar one of the world's faster growing and higher per-capita income countries—equal to the EU in 2007 per-capita income. Sustained high oil prices and increased natural gas exports in recent years have helped build Qatar's budget and trade surpluses and foreign reserves. Proved oil reserves of more than 15 billion barrels should ensure continued

output at current levels for 22 years. Qatar's proved reserves of natural gas are roughly 25 trillion cubic meters, about 15% of the world total and third largest in the world. Qatar has permitted substantial foreign investment in the development of its gas fields during the last decade and became the world's top liquefied natural gas (LNG) exporter in 2007.
GDP (purchasing power parity): $57.69 billion (2007)
GDP (official exchange rate): $67.76 billion (2007 est.)
GDP—real growth rate: 14.2% (2007 est.)
GDP—per capita (PPP): $80,900 (2007 est.)
GDP—composition by sector:
agriculture: 0.1%
industry: 71.2%
services: 28.7% (2007 est.)
Labor force: 638,000 (2007 est.)
Unemployment rate: 0.7% (2007 est.)
Population below poverty line: NA%
Household income or consumption by percentage share:
lowest 10%: NA%
highest 10%: NA%
Inflation rate (consumer prices): 13.8% (2007 est.)
Investment (gross fixed): 46.1% of GDP (2007 est.)
Budget:
revenues: $27.22 billion
expenditures: $22.55 billion (2007 est.)
Public debt: 11.7% of GDP (2007 est.)
Agriculture—products: fruits, vegetables; poultry, dairy products, beef; fish
Industries: crude oil production and refining, ammonia, fertilizers, petrochemicals, steel reinforcing bars, cement, commercial ship repair
Industrial production growth rate: 8% (2007 est.)
Electricity—production: 13.54 billion kWh (2005)
Electricity—production by source:
fossil fuel: 100%
hydro: 0%
nuclear: 0%
other: 0% (2001)
Electricity—consumption: 12.52 billion kWh (2005)
Electricity—exports: 0 kWh (2005)
Electricity—imports: 0 kWh (2005)
Oil—production: 1.111 million bbl/day (2005 est.)
Oil—consumption: 95,000 bbl/day (2005 est.)
Oil—exports: 960,600 bbl/day (2004)
Oil—imports: 0 bbl/day (2004)
Oil—proved reserves: 15.2 billion bbl (2007 est.)
Natural gas—production: 43.93 billion cu m (2005 est.)

Natural gas—consumption: 17.93 billion cu m (2005 est.)

Natural gas—exports: 25.99 billion cu m (2005 est.)

Natural gas—imports: 0 cu m (2005)

Natural gas—proved reserves: 25.79 trillion cu m (1 January 2007 est.)

Current account balance: $23.44 billion (2007 est.)

Exports: $37.95 billion f.o.b. (2007 est.)

Exports—commodities: liquefied natural gas (LNG), petroleum products, fertilizers, steel

Exports—partners: Japan 41%, South Korea 16.7%, Singapore 6.6%, Thailand 4.2% (2006)

Imports: $19.86 billion f.o.b. (2007 est.)

Imports—commodities: machinery and transport equipment, food, chemicals

Imports—partners: France 13.3%, Japan 10.2%, US 9.3%, Italy 8.9%, Germany 7.8%, UK 6.2%, Saudi Arabia 5.8%, South Korea 4.7% (2006)

Economic aid—recipient: $2.18 million (2004)

Reserves of foreign exchange and gold: $9.788 billion (31 December 2007 est.)

Debt—external: $33.09 billion (31 December 2007 est.)

Stock of direct foreign investment—at home: $11.18 billion (2007 est.)

Stock of direct foreign investment—abroad: $5.625 billion (2007 est.)

Market value of publicly traded shares: $61.56 billion (2006)

Currency (code): Qatari rial (QAR)

Currency code: QAR

Exchange rates: Qatari rials per US dollar—3.64 (2007), 3.64 (2006), 3.64 (2005), 3.64 (2004), 3.64 (2003)

Fiscal year: 1 April—31 March

COMMUNICATIONS

Telephones—main lines in use: 228,300 (2006)

Telephones—mobile cellular: 919,800 (2006)

Telephone system:
general assessment: modern system centered in Doha

domestic: combined fixed-line and mobile-cellular telephone density is roughly 130 telephones per 100 persons
international: country code—974; landing point for the Fiber-Optic Link Around the Globe (FLAG) submarine cable network that provides links to Asia, Middle East, Europe, and the US; tropospheric scatter to Bahrain; microwave radio relay to Saudi Arabia and the UAE; satellite earth stations—2 Intelsat (1 Atlantic Ocean and 1 Indian Ocean) and 1 Arabsat

Radio broadcast stations: AM 6, FM 5, shortwave 1 (1998)

Radios: 256,000 (1997)

Television broadcast stations: 1 (plus 3 repeaters) (2001)

Televisions: 230,000 (1997)

Internet country code: .qa

Internet hosts: 19 (2007)

Internet Service Providers (ISPs): 1 (2000)

Internet users: 289,900 (2006)

TRANSPORTATION

Airports: 5 (2007)

Airports—with paved runways:
total: 3
over 3,047 m: 2
1,524 to 2,437 m: 1 (2007)

Airports—with unpaved runways:
total: 2
914 to 1,523 m: 1
under 914 m: 1 (2007)

Heliports: 1 (2007)

Pipelines: condensate 322 km; condensate/gas 209 km; gas 1,970 km; liquid petroleum gas 87 km; oil 741 km (2007)

Roadways: *total:* 1,230 km
paved: 1,107 km
unpaved: 123 km (1999)

Merchant marine:
total: 20 ships (1000 GRT or over) 574,969 GRT/856,057 DWT
by type: bulk carrier 1, cargo 2, chemical tanker 2, container 8, liquefied gas 2, petroleum tanker 4, roll on/roll off 1
foreign-owned: 7 (Kuwait 7)
registered in other countries: 3 (Liberia 2, Panama 1) (2007)

Ports and terminals: Doha, Ra's Laffan

MILITARY

Military branches: Qatari Amiri Land Force (QALF), Qatari Amiri Navy (QAN), Qatari Amiri Air Force (QAAF) (2007)

Military service age and obligation: 18 years of age for voluntary military service; no conscription (2008)

Manpower available for military service:
males age 16–49: 320,383
females age 16–49: 167,475 (2008 est.)

Manpower fit for military service:
males age 16–49: 258,159
females age 16–49: 143,999 (2008 est.)

Manpower reaching militarily significant age annually:
males age 16–49: 7,539
females age 16–49: 7,022 (2008 est.)

Military expenditures—percent of GDP: 10% (2005 est.)

TRANSNATIONAL ISSUES

Disputes—international: none

Trafficking in persons: *current situation:* Qatar is a destination country for men and women from South and Southeast Asia who migrate willingly, but are subsequently trafficked into involuntary servitude as domestic workers and laborers, and, to a lesser extent, commercial sexual exploitation; the most common offense was forcing workers to accept worse contract terms than those under which they were recruited; other conditions include bonded labor, withholding of pay, restrictions on movement, arbitrary detention, and physical, mental, and sexual abuse
tier rating: Tier 3—Qatar's rating was downgraded to Tier 3 in the 2007 report for continuing to detain and deport victims rather than providing them protection; the government also failed to increase prosecutions for trafficking in a meaningful way in 2006; workers complaining of working conditions or non-payment of wages were sometimes prosecuted under false charges in retaliation

INTRODUCTION

Background: The principalities of Wallachia and Moldavia—for centuries under the suzerainty of the Turkish Ottoman Empire—secured their autonomy in 1856; they united in 1859 and a few years later adopted the new name of Romania. The country gained recognition of its independence in 1878. It joined the Allied Powers in World War I and acquired new territories—most notably Transylvania—following the conflict. In 1940, Romania allied with the Axis powers and participated in the 1941 German invasion of the USSR. Three years later, overrun by the Soviets, Romania signed an armistice. The postwar Soviet occupation led to the formation of a Communist "people's republic" in 1947 and the abdication of the king. The decades-long rule of dictator Nicolae CEAUSESCU, who took power in 1965, and his Securitate police state became increasingly oppressive and draconian through the 1980s. CEAUSESCU was overthrown and executed in late 1989. Former Communists dominated the government until 1996 when they were swept from power. Romania joined NATO in 2004 and the EU in 2007.

GEOGRAPHY

Location: Southeastern Europe, bordering the Black Sea, between Bulgaria and Ukraine

Geographic coordinates: 46 00 N, 25 00 E

Map references: Europe

Area:
total: 237,500 sq km
land: 230,340 sq km
water: 7,160 sq km

Area—comparative: slightly smaller than Oregon

Land boundaries: *total:* 2,508 km

border countries: Bulgaria 608 km, Hungary 443 km, Moldova 450 km, Serbia 476 km, Ukraine (north) 362 km, Ukraine (east) 169 km

Coastline: 225 km

Maritime claims:
territorial sea: 12 nm
contiguous zone: 24 nm
exclusive economic zone: 200 nm
continental shelf: 200-m depth or to the depth of exploitation

Climate: temperate; cold, cloudy winters with frequent snow and fog; sunny summers with frequent showers and thunderstorms

Terrain: central Transylvanian Basin is separated from the Plain of Moldavia on the east by the Carpathian Mountains and separated from the Walachian Plain on the south by the Transylvanian Alps

Elevation extremes:
lowest point: Black Sea 0 m
highest point: Moldoveanu 2,544 m

Natural resources: petroleum (reserves declining), timber, natural gas, coal, iron ore, salt, arable land, hydropower

Land use: *arable land:* 39.49%
permanent crops: 1.92%
other: 58.59% (2005)

Irrigated land: 30,770 sq km (2003)

Total renewable water resources: 42.3 cu km (2003)

Freshwater withdrawal (domestic/industrial/agricultural): *total:* 6.5 cu km/yr (9%/34%/57%)
per capita: 299 cu m/yr (2003)

Natural hazards: earthquakes, most severe in south and southwest; geologic structure and climate promote landslides

Environment—current issues: soil erosion and degradation; water pollution; air pollution in south from industrial effluents; contamination of Danube delta wetlands

Environment—international agreements: *party to:* Air Pollution, Air Pollution-Persistent Organic Pollutants, Antarctic-Environmental Protocol, Antarctic Treaty, Biodiversity, Climate Change, Climate Change-Kyoto Protocol, Desertification, Endangered Species, Environmental Modification, Hazardous Wastes, Law of the Sea, Ozone Layer Protection, Ship Pollution, Wetlands
signed, but not ratified: none of the selected agreements

Geography—note: controls most easily traversable land route between the Balkans, Moldova, and Ukraine

PEOPLE

Population: 22,246,862 (July 2008 est.)

Age structure:
0–14 years: 15.6% (male 1,778,864/female 1,687,659)
15–64 years: 69.7% (male 7,718,125/female 7,791,102)
65 years and over: 14.7% (male 1,337,915/female 1,933,197) (2008 est.)

Median age: *total:* 37.3 years
male: 35.9 years
female: 38.7 years (2008 est.)

Population growth rate: -0.136% (2008 est.)

Birth rate: 10.61 births/1,000 population (2008 est.)

Death rate: 11.84 deaths/1,000 population (2008 est.)

Net migration rate: -0.13 migrant(s)/1,000 population (2008 est.)

Sex ratio:
at birth: 1.06 male(s)/female
under 15 years: 1.05 male(s)/female
15–64 years: 0.99 male(s)/female
65 years and over: 0.69 male(s)/female
total population: 0.95 male(s)/female (2008 est.)

Infant mortality rate:
total: 23.73 deaths/1,000 live births
male: 26.81 deaths/1,000 live births
female: 20.46 deaths/1,000 live births (2008 est.)

Life expectancy at birth:
total population: 72.18 years
male: 68.69 years
female: 75.89 years (2008 est.)

Total fertility rate: 1.38 children born/woman (2008 est.)

HIV/AIDS—adult prevalence rate: less than 0.1% (2001 est.)

HIV/AIDS—people living with HIV/AIDS: 6,500 (2001 est.)

HIV/AIDS—deaths: 350 (2001 est.)

Nationality:
noun: Romanian(s)
adjective: Romanian

Ethnic groups: Romanian 89.5%, Hungarian 6.6%, Roma 2.5%, Ukrainian 0.3%, German 0.3%, Russian 0.2%, Turkish 0.2%, other 0.4% (2002 census)

Religions: Eastern Orthodox (including all sub-denominations) 86.8%, Protestant (various denominations including Reformate and Pentecostal) 7.5%, Roman Catholic 4.7%, other (mostly Muslim) and unspecified 0.9%, none 0.1% (2002 census)

Languages: Romanian 91% (official), Hungarian 6.7%, Romany (Gypsy) 1.1%, other 1.2%

Literacy: *definition:* age 15 and over can read and write
total population: 97.3%
male: 98.4%
female: 96.3% (2002 census)

GOVERNMENT

Country name:
conventional long form: none
conventional short form: Romania
local long form: none
local short form: Romania
Government type: republic
Capital: *name:* Bucharest
geographic coordinates: 44 26 N, 26 06 E
time difference: UTC+2 (7 hours ahead of Washington, DC during Standard Time)
daylight saving time: +1hr, begins last Sunday in March; ends last Sunday in October
Administrative divisions: 41 counties (judete, singular—judet) and 1 municipality* (municipiu; Alba, Arad, Arges, Bacau, Bihor, Bistrita-Nasaud, Botosani, Braila, Brasov, Bucuresti (Bucharest)*, Buzau, Calarasi, Caras-Severin, Cluj, Constanta, Covasna, Dimbovita, Dolj, Galati, Gorj, Giurgiu, Harghita, Hunedoara, Ialomita, Iasi, Ilfov, Maramures, Mehedinti, Mures, Neamt, Olt, Prahova, Salaj, Satu Mare, Sibiu, Suceava, Teleorman, Timis, Tulcea, Vaslui, Vilcea, Vrancea
Independence: 9 May 1877 (independence proclaimed from the Ottoman Empire; independence recognized 13 July 1878 by the Treaty of Berlin); 26 March 1881 (kingdom proclaimed); 30 December 1947 (republic proclaimed)
National holiday: Unification Day (of Romania and Transylvania), 1 December (1918)
Constitution: 8 December 1991; revision effective 29 October 2003
Legal system: based on civil law system; has not accepted compulsory ICJ jurisdiction
Suffrage: 18 years of age; universal
Executive branch:
chief of state: President Traian BASESCU (since 20 December 2004); note—President Traian BASESCU was suspended by vote of parliament on 19 April 2007, but resumed his duties on 23 May 2007 after a popular referendum confirmed that his impeachment should not stand
head of government: Prime Minister Calin Popescu-TARICEANU (since 29 December 2004)
cabinet: Council of Ministers appointed by the prime minister
elections: president elected by popular vote for a five-year term (eligible for a second term); election last held 28 November 2004 with runoff between the top two candidates held 12 December 2004 (next to be held in November-December 2009); prime minister appointed by the president with the consent of the Parliament

election results: percent of vote—Traian BASESCU 51.23%, Adrian NASTASE 48.77%
Legislative branch: bicameral Parliament or Parlament consists of the Senate or Senat (137 seats; members are elected by popular vote on a proportional representation basis to serve four-year terms) and the Chamber of Deputies or Camera Deputatilor (332 seats; members are elected by popular vote on a proportional representation basis to serve four-year terms)
elections: Senate—last held 28 November 2004 (next expected to be held in November 2008); Chamber of Deputies—last held 28 November 2004 (next expected to be held November 2008)
election results: Senate—percent of vote by alliance/party—PSD-PUR 37.1%, PNL-PD 31.8%, PRM 13.6%, UDMR 6.2%, other 11.3%; seats by party—PSD 44, PNL 30, PD 20, PRM 20, PC 11, UDMR 10, independents 2; seats by party as of February 2008—PSD 44, PDL 27, PNL 24, PRM 16, PC 10, UDMR 10, independents 6; Chamber of Deputies—percent of vote by alliance/party—PSD-PUR 36.8%, PNL-PD 31.5%, PRM 13%, UDMR 6.2%, other 12.5%; seats by party—PSD 111, PNL 66, PD 45, PRM 34, UDMR 22, PC 20, ex-PRM (Ciontu Group) 12, PIN (GUSA Group) 3, independent 1, ethnic minorities 18; seats by party as of February 2008—PSD 104, PDL 73, PNL 56, PRM 25, UDMR 22, PC 16, independents 18, ethnic minorities 18
Judicial branch: Supreme Court of Justice (comprised of 11 judges appointed for three-year terms by the president in consultation with the Superior Council of Magistrates, which is comprised of the minister of justice, the prosecutor general, two civil society representatives appointed by the Senate, and 14 judges and prosecutors elected by their peers); a separate body, the Constitutional Court, validates elections and makes decisions regarding the constitutionality of laws, treaties, ordinances, and internal rules of the Parliament; it is comprised of nine members serving nine-year terms, with three members each appointed by the president, the Senate, and the Chamber of Deputies
Political parties and leaders: Conservative Party or PC [Daniela POPA] (formerly Humanist Party or PUR); Democratic Liberal Party or PDL [Emil BOC]; Democratic Union of Hungarians in Romania or UDMR [Bela MARKO]; National Liberal Party or

PNL [Calin Popescu-TARICEANU]; Romania Mare Party (Greater Romania Party) or PRM [Corneliu Vadim TUDOR]; Social Democratic Party or PSD [Mircea Dan GEOANA] (formerly Party of Social Democracy in Romania or PDSR)
Political pressure groups and leaders: various human rights and professional associations
International organization participation: ACCT, Australia Group, BIS, BSEC, CE, CEI, EAPC, EBRD, EIB, ESA (cooperating state), EU (new member), FAO, G-9, G-77, IAEA, IBRD, ICAO, ICC, ICCt, ICRM, IDA, IFAD, IFC, IFRCS, IHO, ILO, IMF, IMO, IMSO, Interpol, IOC, IOM, IPU, ISO, ITSO, ITU, ITUC, LAIA (observer), MIGA, NAM (guest), NATO, NSG, OAS (observer), OIF, OPCW, OSCE, PCA, SECI, UN, UNCTAD, UNESCO, UNHCR, UNIDO, Union Latina, UNMEE, UNMIL, UNMIS, UNOCI, UNOMIG, UNWTO, UPU, WCL, WCO, WEU (associate partner), WFTU, WHO, WIPO, WMO, WTO, ZC
Diplomatic representation in the US:
chief of mission: Ambassador Adrian Cosmin VIERITA
chancery: 1607 23rd Street NW, Washington, DC 20008
telephone: [1] (202) 332-4846, 4848, 4851, 4852
FAX: [1] (202) 232-4748
consulate(s) general: Chicago, Los Angeles, New York
Diplomatic representation from the US:
chief of mission: Ambassador Nicholas F. TAUBMAN
embassy: Strada Tudor Arghezi 7-9, Bucharest
mailing address: pouch: American Embassy Bucharest, US Department of State, 5260 Bucharest Place, Washington, DC 20521-5260 (pouch)
telephone: [40] (21) 200-3300
FAX: [40] (21) 200-3442
Flag description: three equal vertical bands of blue (hoist side), yellow, and red; the national coat of arms that used to be centered in the yellow band has been removed; now similar to the flag of Chad, also resembles the flags of Andorra and Moldova

ECONOMY

Economy—overview: Romania, which joined the European Union on 1 January 2007, began the transition from Communism in 1989 with a largely obsolete industrial base and a pattern of output unsuited to the country's needs. The country emerged in 2000 from a punishing three-year recession thanks to strong demand in EU export markets. Domestic consumption and investment have fueled strong GDP growth in recent years, but have led to large current account imbalances. Romania's macroeconomic gains have only recently started to spur creation of a middle class and address Romania's widespread poverty. Corruption and red tape continue to handicap its business environment. Inflation rose in 2007 for the first time in eight years, driven in part by the depreciation of the currency, rising energy costs, a nation-wide drought affecting food prices, and a relaxation of fiscal discipline. Romania hopes to adopt the euro by 2014.

GDP (purchasing power parity): $245.5 billion (2007 est.)

GDP (official exchange rate): $166 billion (2007 est.)

GDP—real growth rate: 6% (2007 est.)

GDP—per capita (PPP): $11,400 (2007 est.)

GDP—composition by sector:
agriculture: 7.9%
industry: 35.6%
services: 56.5% (2007 est.)

Labor force: 9.35 million (2007 est.)

Labor force—by occupation:
agriculture: 29.7%
industry: 23.2%
services: 47.1% (2006)

Unemployment rate: 4.1% (2007 est.)

Population below poverty line: 25% (2005 est.)

Household income or consumption by percentage share:
lowest 10%: 1.2%
highest 10%: 20.8% (2006)

Distribution of family income—Gini index: 31 (2005)

Inflation rate (consumer prices): 4.8% (2007 est.)

Investment (gross fixed): 28% of GDP (2007 est.)

Budget: *revenues:* $56.29 billion
expenditures: $60.41 billion (2007)

Public debt: 18.2% of GDP (2007 est.)

Agriculture—products: wheat, corn, barley, sugar beets, sunflower seed, potatoes, grapes; eggs, sheep

Industries: electric machinery and equipment, textiles and footwear, light machinery and auto assembly, mining, timber, construction materials, metallurgy, chemicals, food processing, petroleum refining

Industrial production growth rate: 10.6% (2007 est.)

Electricity—production: 60.52 billion kWh (2007)

Electricity—production by source:
fossil fuel: 62.5%

hydro: 27.6%
nuclear: 9.9%
other: 0% (2001)

Electricity—consumption: 58.49 billion kWh (2007)

Electricity—exports: 3.33 billion kWh (2007)

Electricity—imports: 1.29 billion kWh (2007)

Oil—production: 122,700 bbl/day (2005 est.)

Oil—consumption: 236,000 bbl/day (2005 est.)

Oil—exports: 92,510 bbl/day (2004)

Oil—imports: 181,100 bbl/day (2004)

Oil—proved reserves: 955.6 million bbl (1 January 2006 est.)

Natural gas—production: 12.24 billion cu m (2007)

Natural gas—consumption: 17.09 billion cu m (2007)

Natural gas—exports: 0 cu m (2007)

Natural gas—imports: 4.851 billion cu m (2007)

Natural gas—proved reserves: 96.41 billion cu m (1 January 2006 est.)

Current account balance: -$23.13 billion (2007)

Exports: $40.25 billion f.o.b. (2007)

Exports—commodities: machinery and equipment, textiles and footwear, metals and metal products, machinery and equipment, minerals and fuels, chemicals, agricultural products

Exports—partners: Italy 17.9%, Germany 15.7%, Turkey 7.7%, France 7.5%, Hungary 4.9%, UK 4.7% (2006)

Imports: $64.33 billion f.o.b. (2007)

Imports—commodities: machinery and equipment, fuels and minerals, chemicals, textile and products, metals, agricultural products

Imports—partners: Germany 15.2%, Italy 14.5%, Russia 7.8%, France 6.5%, Turkey 4.9%, China 4.3% (2006)

Economic aid—recipient: $914.3 million (2004)

Reserves of foreign exchange and gold: $37.24 billion (31 December 2007)

Debt—external: $72.36 billion (31 December 2007)

Stock of direct foreign investment—at home: $50.39 billion (2007 est.)

Stock of direct foreign investment—abroad: $288 million (2007 est.)

Market value of publicly traded shares: $45.42 billion (2007)

Currency (code): "new" leu (RON) was introduced in 2005; "old" leu (ROL) was phased out in 2006; note—because of currency revaluation, 10,000 ROL = 1 RON

Currency code: ROL

Exchange rates: lei per US dollar—2.43 (2007), 2.809 (2006), 3 (2005), 3 (2004), 3 (2003)

Fiscal year: calendar year

COMMUNICATIONS

Telephones—main lines in use: 4.231 million (2006)

Telephones—mobile cellular: 17.4 million (2006)

Telephone system:
general assessment: rapidly improving domestic and international service, especially in wireless telephony
domestic: more than 90 percent of telephone network is automatic; liberalization in 2003 is transforming telecommunications; fixed-line teledensity is roughly 20 telephones per 100 persons; mobile-cellular teledensity is approaching 80 telephones per 100 persons
international: country code—40; the Black Sea Fiber Optic System provides connectivity to Bulgaria and Turkey; satellite earth stations—10; digital, international, direct-dial exchanges operate in Bucharest (2005)

Radio broadcast stations: 698 (frequency type NA) (2006)

Radios: 7.2 million (1997)

Television broadcast stations: 623 (plus 200 repeaters) (2006)

Televisions: 5.25 million (1997)

Internet country code: .ro

Internet hosts: 1.406 million (2007)

Internet Service Providers (ISPs): 38 (2000)

Internet users: 5.063 million (2006)

TRANSPORTATION

Airports: 61 (2007)

Airports—with paved runways:
total: 25
over 3,047 m: 3
2,438 to 3,047 m: 9
1,524 to 2,437 m: 12
914 to 1,523 m: 1 (2007)

Airports—with unpaved runways:
total: 36
1,524 to 2,437 m: 2
914 to 1,523 m: 12
under 914 m: 22 (2007)

Heliports: 2 (2007)

Pipelines: gas 3,674 km; oil 2,424 km (2007)

Railways:
total: 11,385 km
broad gauge: 60 km 1.524-m gauge
standard gauge: 10,898 km 1.435-m gauge (3,888 km electrified)
narrow gauge: 427 km 0.760-m gauge (2006)

Roadways:
total: 198,817 km
paved: 60,043 km (includes 228 km of expressways)
unpaved: 138,774 km (2004)

Waterways: 1,731 km

note: includes 1,075 km on Danube River, 524 km on secondary branches, and 132 km on canals (2006)

Merchant marine:

total: 19 ships (1000 GRT or over) 146,307 GRT/165,548 DWT

by type: cargo 13, passenger 1, passenger/cargo 2, petroleum tanker 2, roll on/roll off 1

registered in other countries: 50 (Cambodia 1, Georgia 15, North Korea 6, Malta 10, Marshall Islands 1, Panama 8, Sierra Leone 2, St Kitts and Nevis 1, St Vincent and The Grenadines 1, Syria 4, Tuvalu 1, unknown 4) (2007)

Ports and terminals: Braila, Constanta, Galati, Tulcea

MILITARY

Military branches: Land Forces, Naval Forces, Romanian Air Force (Fortele Aeriene Romane, FAR), Special Operations (2008)

Military service age and obligation: 18 years of age for voluntary military service; conscription officially ended October 2006; all military inductees (including women) contract for an initial 5-year term of service; subsequent voluntary service contracts are for successive 3-year terms until the age of 36 (2006)

Manpower available for military service:

males age 16–49: 5,682,299

females age 16–49: 5,557,098 (2008 est.)

Manpower fit for military service:

males age 16–49: 4,572,017

females age 16–49: 4,644,474 (2008 est.)

Manpower reaching militarily significant age annually:

males age 16–49: 127,706

females age 16–49: 121,852 (2008 est.)

Military expenditures—percent of GDP: 1.9% (2007 est.)

TRANSNATIONAL ISSUES

Disputes—international: the ICJ gave Ukraine until December 2006 to reply, and Romania until June 2007 to issue a rejoinder, in their dispute submitted in 2004 over Ukrainian-administered Zmiyinyy/Serpilor (Snake) Island and Black Sea maritime boundary delimitation; Romania also opposes Ukraine's reopening of a navigation canal from the Danube border through Ukraine to the Black Sea

Illicit drugs: major transshipment point for Southwest Asian heroin transiting the Balkan route and small amounts of Latin American cocaine bound for Western Europe; although not a significant financial center, role as a narcotics conduit leaves it vulnerable to laundering, which occurs via the banking system, currency exchange houses, and casinos

RUSSIA

INTRODUCTION

Background: Founded in the 12th century, the Principality of Muscovy, was able to emerge from over 200 years of Mongol domination (13th-15th centuries) and to gradually conquer and absorb surrounding principalities. In the early 17th century, a new Romanov Dynasty continued this policy of expansion across Siberia to the Pacific. Under PETER I (ruled 1682–1725), hegemony was extended to the Baltic Sea and the country was renamed the Russian Empire. During the 19th century, more territorial acquisitions were made in Europe and Asia. Defeat in the Russo-Japanese War of 1904–05 contributed to the Revolution of 1905, which resulted in the formation of a parliament and other reforms. Repeated devastating defeats of the Russian army in World War I led to widespread rioting in the major cities of the Russian Empire and to the overthrow in 1917 of the imperial household. The Communists under Vladimir LENIN seized power soon after and formed the USSR. The brutal rule of Iosif STALIN (1928–53) strengthened Communist rule and Russian dominance of the Soviet Union at a cost of tens of millions of lives. The Soviet economy and society stagnated in the following decades until General Secretary Mikhail GORBACHEV (1985–91) introduced glasnost (openness) and perestroika (restructuring) in an attempt to modernize Communism, but his initiatives inadvertently released forces that by December 1991 splintered the USSR into Russia and 14 other independent republics. Since then, Russia has struggled in its efforts to build a democratic political system and market economy to replace the social, political, and economic controls of the Communist period. In tandem with its prudent management of Russia's windfall energy wealth, which has helped the country rebound from the economic collapse of the 1990s, the Kremlin in recent years has overseen a recentralization of power that has undermined democratic institutions. Russia has severely disabled the Chechen rebel movement, although violence still occurs throughout the North Caucasus.

GEOGRAPHY

Location: Northern Asia (the area west of the Urals is considered part of Europe), bordering the Arctic Ocean, between Europe and the North Pacific Ocean

Geographic coordinates: 60 00 N, 100 00 E

Map references: Asia

Area:

total: 17,075,200 sq km

land: 16,995,800 sq km

water: 79,400 sq km

Area—comparative: approximately 1.8 times the size of the US

Land boundaries:

total: 20,096.5 km

border countries: Azerbaijan 284 km, Belarus 959 km, China (southeast) 3,605 km, China (south) 40 km, Estonia 294 km, Finland 1,340 km, Georgia 723 km,

Kazakhstan 6,846 km, North Korea 19 km, Latvia 217 km, Lithuania (Kaliningrad Oblast) 280.5 km, Mongolia 3,485 km, Norway 196 km, Poland (Kaliningrad Oblast) 232 km, Ukraine 1,576 km

Coastline: 37,653 km

Maritime claims:

territorial sea: 12 nm

contiguous zone: 24 nm

exclusive economic zone: 200 nm

continental shelf: 200-m depth or to the depth of exploitation

Climate: ranges from steppes in the south through humid continental in much of European Russia; subarctic in Siberia to tundra climate in the polar north; winters vary from cool along Black Sea coast to frigid in Siberia; summers vary from warm in the steppes to cool along Arctic coast

Terrain: broad plain with low hills west of Urals; vast coniferous forest and tundra in Siberia; uplands and mountains along southern border regions

Elevation extremes:

lowest point: Caspian Sea -28 m

highest point: Gora El'brus 5,633 m

Natural resources: wide natural resource base including major deposits of oil, natural gas, coal, and many strategic minerals, timber

note: formidable obstacles of climate, terrain, and distance hinder exploitation of natural resources

Land use:

arable land: 7.17%

permanent crops: 0.11%

other: 92.72% (2005)

Irrigated land: 46,000 sq km (2003)

Total renewable water resources: 4,498 cu km (1997)

Freshwater withdrawal (domestic/industrial/agricultural): *total:* 76.68 cu km/yr (19%/63%/18%)

per capita: 535 cu m/yr (2000)

Natural hazards: permafrost over much of Siberia is a major impediment to development; volcanic activity in the Kuril Islands; volcanoes and earthquakes on the Kamchatka Peninsula; spring floods and summer/autumn forest fires throughout Siberia and parts of European Russia

Environment—current issues: air pollution from heavy industry, emissions of coal-fired electric plants, and transportation in major cities; industrial, municipal, and agricultural pollution of inland waterways and seacoasts; deforestation; soil erosion; soil contamination from improper application of agricultural chemicals; scattered areas of sometimes intense radioactive contamination; groundwater contamination from toxic waste; urban solid waste management; abandoned stocks of obsolete pesticides

Environment—international agreements: *party to:* Air Pollution, Air Pollution-Nitrogen Oxides, Air Pollution-Sulfur 85, Antarctic-Environmental Protocol, Antarctic-Marine Living Resources, Antarctic Seals, Antarctic Treaty, Biodiversity, Climate Change, Climate Change-Kyoto Protocol, Endangered Species, Environmental Modification, Hazardous Wastes, Law of the Sea, Marine Dumping, Ozone Layer Protection, Ship Pollution, Tropical Timber 83, Wetlands, Whaling

signed, but not ratified: Air Pollution-Sulfur 94

Geography—note: largest country in the world in terms of area but unfavorably located in relation to major sea lanes of the world; despite its size, much of the country lacks proper soils and climates (either too cold or too dry) for agriculture; Mount El'brus is Europe's tallest peak

PEOPLE

Population: 140,702,094 (July 2008 est.)

Age structure:

0–14 years: 14.6% (male 10,577,858/female 10,033,254)

15–64 years: 71.2% (male 48,187,807/female 52,045,102)

65 years and over: 14.1% (male 6,162,400/female 13,695,673) (2008 est.)

Median age:

total: 38.3 years

male: 35.1 years

female: 41.4 years (2008 est.)

Population growth rate: -0.474% (2008 est.)

Birth rate: 11.03 births/1,000 population (2008 est.)

Death rate: 16.06 deaths/1,000 population (2008 est.)

Net migration rate: 0.28 migrant(s)/1,000 population (2008 est.)

Sex ratio:

at birth: 1.06 male(s)/female

under 15 years: 1.05 male(s)/female

15–64 years: 0.93 male(s)/female

65 years and over: 0.45 male(s)/female

total population: 0.86 male(s)/female (2008 est.)

Infant mortality rate:

total: 10.81 deaths/1,000 live births

male: 12.34 deaths/1,000 live births

female: 9.18 deaths/1,000 live births (2008 est.)

Life expectancy at birth:

total population: 65.94 years

male: 59.19 years

female: 73.1 years (2008 est.)

Total fertility rate: 1.4 children born/woman (2008 est.)

HIV/AIDS—adult prevalence rate: 1.1% (2001 est.)

HIV/AIDS—people living with HIV/AIDS: 860,000 (2001 est.)

HIV/AIDS—deaths: 9,000 (2001 est.)

Major infectious diseases:

degree of risk: intermediate

food or waterborne diseases: bacterial diarrhea and hepatitis A

vectorborne disease: Crimean Congo hemorrhagic fever and tickborne encephalitis

note: highly pathogenic H5N1 avian influenza has been identified in this country; it poses a negligible risk with extremely rare cases possible among US citizens who have close contact with birds (2008)

Nationality:

noun: Russian(s)

adjective: Russian

Ethnic groups: Russian 79.8%, Tatar 3.8%, Ukrainian 2%, Bashkir 1.2%, Chuvash 1.1%, other or unspecified 12.1% (2002 census)

Religions: Russian Orthodox 15–20%, Muslim 10–15%, other Christian 2% (2006 est.)

note: estimates are of practicing worshipers; Russia has large populations of non-practicing believers and non-believers, a legacy of over seven decades of Soviet rule

Languages: Russian, many minority languages

Literacy:

definition: age 15 and over can read and write

total population: 99.4%

male: 99.7%

female: 99.2% (2002 census)

GOVERNMENT

Country name:

conventional long form: Russian Federation

conventional short form: Russia

local long form: Rossiyskaya Federatsiya

local short form: Rossiya

former: Russian Empire, Russian Soviet Federative Socialist Republic

Government type: federation

Capital: *name:* Moscow

geographic coordinates: 55 45 N, 37 35 E

time difference: UTC+3 (8 hours ahead of Washington, DC during Standard Time)

daylight saving time: +1hr, begins last Sunday in March; ends last Sunday in October

note: Russia is divided into 11 time zones

Administrative divisions: 46 oblasts (oblastey, singular—oblast), 21 republics (respublik, singular—respublika), 4 autonomous okrugs (avtonomnykh okrugov, singular—avtonomnyy okrug), 9 krays (krayev, singular—kray), 2 federal cities (goroda, singular—gorod), and 1 autonomous oblast (avtonomnaya oblast')

oblasts: Amur (Blagoveshchensk), Arkhangel'sk, Astrakhan', Belgorod, Bryansk, Chelyabinsk, Irkutsk, Ivanovo, Kaliningrad, Kaluga, Kemerovo, Kirov, Kostroma, Kurgan, Kursk, Leningrad, Lipetsk, Magadan, Moscow, Murmansk, Nizhniy Novgorod, Novgorod, Novosibirsk, Omsk, Orenburg, Orel, Penza, Pskov, Rostov, Ryazan', Sakhalin (Yuzhno-Sakhalinsk), Samara, Saratov, Smolensk, Sverdlovsk (Yekaterinburg), Tambov, Tomsk, Tula, Tver', Tyumen', Ul'yanovsk, Vladimir, Volgograd, Vologda, Voronezh, Yaroslavl'

republics: Adygeya (Maykop), Altay (Gorno-Altaysk), Bashkortostan (Ufa), Buryatiya (Ulan-Ude), Chechnya (Groznyy), Chuvashiya (Cheboksary), Dagestan (Makhachkala), Ingushetiya (Magas), Kabardino-Balkariya (Nal'chik), Kalmykiya (Elista), Karachayevo-Cherkesiya (Cherkessk), Kareliya (Petrozavodsk), Khakasiya (Abakan), Komi (Syktyvkar), Mariy-El (Yoshkar-Ola), Mordoviya (Saransk), North Ossetia (Vladikavkaz), Sakha [Yakutiya] (Yakutsk), Tatarstan (Kazan'), Tyva (Kyzyl), Udmurtiya (Izhevsk)

autonomous okrugs: Chukotka (Anadyr'), Khanty-Mansi (Khanty-Mansiysk), Nenets (Nar'yan-Mar), Yamalo-Nenets (Salekhard)

krays: Altay (Barnaul), Kamchatka (Petropavlovsk-Kamchatskiy), Khabarovsk, Krasnodar, Krasnoyarsk, Perm', Primorsk (Vladivostok), Stavropol', Zabaykal'skiy (Chita)

federal cities: Moscow (Moskva), Saint Petersburg (Sankt-Peterburg)

autonomous oblast: Yevrey [Jewish] (Birobidzhan)

note: administrative divisions have the same names as their administrative centers (exceptions have the administrative center name following in parentheses)

Independence: 24 August 1991 (from Soviet Union)

National holiday: Russia Day, 12 June (1990)

Constitution: adopted 12 December 1993

Legal system: based on civil law system; judicial review of legislative acts; has not accepted compulsory ICJ jurisdiction

Suffrage: 18 years of age; universal

Executive branch:

chief of state: President Dmitriy Anatolyevich MEDVEDEV (since 7 May 2008)

head of government: Premier Vladimir Vladimirovich PUTIN (since 8 May 2008); First Deputy Premiers Igor Ivanovich SHUVALOV and Viktor Alekseyevich ZUBKOV (since 12 May 2008); Deputy Premiers Sergey Borisovich IVANOV (since 12 May 2008), Aleksey Leonidovich KUDRIN (since 24 September 2007), Igor Ivanovich SECHIN (since 12 May 2008), Sergey Semenovich SOBYANIN (since 12 May 2008), and Aleksandr Dmitriyevich ZHUKOV (since 9 March 2004)

cabinet: Ministries of the Government or "Government" composed of the premier and his deputies, ministers, and selected other individuals; all are appointed by the president

note: there is also a Presidential Administration (PA) that provides staff and policy support to the president, drafts presidential decrees, and coordinates policy among government agencies; a Security Council also reports directly to the president

elections: president elected by popular vote for a four-year term (eligible for a second term); election last held 2 March 2008 (next to be held in March 2012); note—no vice president; if the president dies in office, cannot exercise his powers because of ill health, is impeached, or resigns, the premier serves as acting president until a new presidential election is held, which must be within three months; premier appointed by the president with the approval of the Duma

election results: Dmitriy MEDVEDEV elected president; percent of vote—Dmitry MEDVEDEV 70.2%, Gennady ZYUGANOV 17.7%, Vladimir ZHIRINOVSKY 9.4%

Legislative branch: bicameral Federal Assembly or Federalnoye Sobraniye consists of the Federation Council or Sovet Federatsii (168 seats; as of July 2000, members appointed by the top executive and legislative officials in each of the 84 federal administrative units—oblasts, krays, republics, autonomous okrugs and oblasts, and the federal cities of Moscow and Saint Petersburg; to serve four-year terms) and the State Duma or Gosudarstvennaya Duma (450 seats; as of 2007, all members elected by proportional representation from party lists winning at least 7% of the vote; members elected by popular vote to serve four-year terms)

elections: State Duma—last held 2 December 2007 (next to be held in December 2011)

election results: State Duma—United Russia 64.3%, CPRF 11.5%, LDPR 8.1%, JR 7.7%, other 8.4%; total seats by party—United Russia 315, CPRF 57, LDPR 40, JR 38

Judicial branch: Constitutional Court; Supreme Court; Supreme Arbitration Court; judges for all courts are appointed for life by the Federation Council on the recommendation of the president

Political parties and leaders: Agrarian Party [Vladimir PLOTNIKOV]; A Just Russia or JR [Sergey MIRONOV] (formed from the merger of three small political parties: Rodina (Motherland), Pensioners Party, and Party of Life); Civic Force [Mikhail BARSHCHEVSKIY]; Communist Party of the Russian Federation or CPRF [Gennadiy Andreyevich ZYUGANOV]; Democratic Party [Andrey BOGDANOV]; Green Party [Anatoliy PANFILOV]; Liberal Democratic Party of Russia or LDPR [Vladimir Volfovich ZHIRINOVSKIY]; Party of Russia's Rebirth [Gennadiy SELEZNEV]; Patriots of Russia [Gennadiy SEMIGIN]; Peace and Unity Party [Sazhi UMALATOVA]; People's Union [Sergey BABURIN]; Social Justice Party [Arkadiy GAYDAMAK]; Union of Right Forces or SPS [Nikita BELYKH]; United Russia or UR [Boris Vyacheslavovich GRYZLOV]; Yabloko Party [Grigoriy Alekseyevich YAVLINSKIY]

Political pressure groups and leaders: NA

International organization participation: APEC, Arctic Council, ARF, ASEAN (dialogue partner), BIS, BSEC, CBSS, CE, CERN (observer), CIS, CSTO, EAEC, EAPC, EBRD, G-8, GCTU, IAEA, IBRD, ICAO, ICC, ICCt (signatory), ICRM, IDA, IFC, IFRCS, IHO, ILO, IMF, IMO, IMSO, Interpol, IOC, IOM (observer), IPU, ISO, ITSO, ITU, ITUC, LAIA (observer), MIGA, MINURSO, NAM (guest), NSG, OAS (observer), OIC (observer), OPCW, OSCE, Paris Club, PCA, PFP, SCO, UN, UN Security Council, UNCTAD, UNESCO, UNHCR, UNIDO, UNITAR, UNMEE, UNMIL, UNMIS, UNOCI, UNOMIG, UNTSO, UNWTO, UPU, WCO, WFTU, WHO, WIPO, WMO, WTO (observer), ZC

Diplomatic representation in the US:

chief of mission: Ambassador Yuriy Viktorovich USHAKOV

chancery: 2650 Wisconsin Avenue NW, Washington, DC 20007

telephone: [1] (202) 298-5700, 5701, 5704, 5708

FAX: [1] (202) 298-5735

consulate(s) general: Houston, New York, San Francisco, Seattle

Diplomatic representation from the US:
chief of mission: Ambassador William J. BURNS
embassy: Bolshoy Deviatinskiy Pereulok No. 8, 121099 Moscow
mailing address: PSC-77, APO AE 09721
telephone: [7] (495) 728-5000
FAX: [7] (495) 728-5090
consulate(s) general: Saint Petersburg, Vladivostok, Yekaterinburg
Flag description: three equal horizontal bands of white (top), blue, and red

ECONOMY

Economy—overview: Russia ended 2007 with its ninth straight year of growth, averaging 7% annually since the financial crisis of 1998. Although high oil prices and a relatively cheap ruble initially drove this growth, since 2003 consumer demand and, more recently, investment have played a significant role. Over the last six years, fixed capital investments have averaged real gains greater than 10% per year and personal incomes have achieved real gains more than 12% per year. During this time, poverty has declined steadily and the middle class has continued to expand. Russia has also improved its international financial position since the 1998 financial crisis. The federal budget has run surpluses since 2001 and ended 2007 with a surplus of about 3% of GDP. Over the past several years, Russia has used its stabilization fund based on oil taxes to prepay all Soviet-era sovereign debt to Paris Club creditors and the IMF. Foreign debt is approximately one-third of GDP. The state component of foreign debt has declined, but commercial debt to foreigners has risen strongly. Oil export earnings have allowed Russia to increase its foreign reserves from $12 billion in 1999 to some $470 billion at yearend 2007, the third largest reserves in the world. During PUTIN's first administration, a number of important reforms were implemented in the areas of tax, banking, labor, and land codes. These achievements have raised business and investor confidence in Russia's economic prospects, with foreign direct investment rising from $14.6 billion in 2005 to approximately $45 billion in 2007. In 2007, Russia's GDP grew 8.1%, led by non-tradable services and goods for the domestic market, as opposed to oil or mineral extraction and exports. Rising inflation returned in the second half of 2007, driven largely by unsterilized capital inflows and by rising food costs, and approached 12% by year-end. In 2006, Russia signed a bilateral market access agreement with the US as a prelude to possible WTO entry, and its companies are involved in global merger and acquisition activity in the oil and gas, metals, and telecom sectors. Despite Russia's recent success, serious problems persist. Oil, natural gas, metals, and timber account for more than 80% of exports and 30% of government revenues, leaving the country vulnerable to swings in world commodity prices. Russia's manufacturing base is dilapidated and must be replaced or modernized if the country is to achieve broad-based economic growth. The banking system, while increasing consumer lending and growing at a high rate, is still small relative to the banking sectors of Russia's emerging market peers. Political uncertainties associated with this year's power transition, corruption, and lack of trust in institutions continue to dampen domestic and foreign investor sentiment. President PUTIN has granted more influence to forces within his government that desire to reassert state control over the economy. Russia has made little progress in building the rule of law, the bedrock of a modern market economy. The government has promised additional legislative amendments to make its intellectual property protection WTO-consistent, but enforcement remains problematic.

GDP (purchasing power parity): $2.088 trillion (2007 est.)
GDP (official exchange rate): $1.286 trillion (2007 est.)
GDP—real growth rate: 8.1% (2007 est.)
GDP—per capita (PPP): $14,700 (2007 est.)
GDP—composition by sector:
agriculture: 4.7%
industry: 39.1%
services: 56.2% (2007 est.)
Labor force: 74.1 million (2007 est.)
Labor force—by occupation: *agriculture:* 10.8%
industry: 28.8%
services: 60.5% (November 2007 est.)
Unemployment rate: 6.2% (2007 est.)
Population below poverty line: 15.8% (November 2007)
Household income or consumption by percentage share:
lowest 10%: 1.9%
highest 10%: 30.4% (September 2007)
Distribution of family income—Gini index: 41.3 (September 2007)
Inflation rate (consumer prices): 11.9% annual average
note: 12% at year-end (2007 est.)
Investment (gross fixed): 21% of GDP (2007 est.)
Budget: *revenues:* $299 billion
expenditures: $262 billion (2007 est.)

Public debt: 5.9% of GDP (2007 est.)
Agriculture—products: grain, sugar beets, sunflower seed, vegetables, fruits; beef, milk
Industries: complete range of mining and extractive industries producing coal, oil, gas, chemicals, and metals; all forms of machine building from rolling mills to high-performance aircraft and space vehicles; defense industries including radar, missile production, and advanced electronic components, shipbuilding; road and rail transportation equipment; communications equipment; agricultural machinery, tractors, and construction equipment; electric power generating and transmitting equipment; medical and scientific instruments; consumer durables, textiles, foodstuffs, handicrafts
Industrial production growth rate: 7.4% (2007 est.)
Electricity—production: 1 trillion kWh (2007 est.)
Electricity—production by source:
fossil fuel: 66.3%
hydro: 17.2%
nuclear: 16.4%
other: 0.1% (2003)
Electricity—consumption: 985.2 billion kWh (2007 est.)
Electricity—exports: 18 billion kWh (2007)
Electricity—imports: 2.9 billion kWh (2007 est.)
Oil—production: 9.87 million bbl/day (2007)
Oil—consumption: 2.916 million bbl/day (2006)
Oil—exports: 5.08 million bbl/day (2007)
Oil—imports: 100,000 bbl/day (2005)
Oil—proved reserves: 60 billion bbl (1 January 2006 est.)
Natural gas—production: 656.2 billion cu m (2007 est.)
Natural gas—consumption: 610 billion cu m (2007 est.)
Natural gas—exports: 182 billion cu m (2007 est.)
Natural gas—imports: 37.5 billion cu m (2005)
Natural gas—proved reserves: 47.57 trillion cu m (1 January 2006)
Current account balance: $76.6 billion (2007 est.)
Exports: $365 billion (2007 est.)
Exports—commodities: petroleum and petroleum products, natural gas, wood and wood products, metals, chemicals, and a wide variety of civilian and military manufactures
Exports—partners: Netherlands 12.3%, Italy 8.6%, Germany 8.4%, China 5.4%, Ukraine 5.1%, Turkey 4.9%, Switzerland 4.1% (2006)

Imports: $260.4 billion (2007 est.)

Imports—commodities: machinery and equipment, consumer goods, medicines, meat, sugar, semifinished metal products

Imports—partners: Germany 13.9%, China 9.7%, Ukraine 7%, Japan 5.9%, South Korea 5.1%, US 4.8%, France 4.4%, Italy 4.3% (2006)

Economic aid—recipient: $982.7 million in FY06 from US, including $847 million in non-proliferation subsidies

Reserves of foreign exchange and gold: $476.4 billion (31 December 2007 est.)

Debt—external: $343.1 billion (31 December 2007)

Stock of direct foreign investment—at home: $271.6 billion (2006)

Stock of direct foreign investment—abroad: $209.6 billion (2006)

Market value of publicly traded shares: $1.322 trillion (2006)

Currency (code): Russian ruble (RUB)

Currency code: RUR

Exchange rates: Russian rubles per US dollar—25.659 (2007), 27.19 (2006), 28.284 (2005), 28.814 (2004), 30.692 (2003)

Fiscal year: calendar year

COMMUNICATIONS

Telephones—main lines in use: 40.1 million (2005)

Telephones—mobile cellular: 150 million (2006)

Telephone system:

general assessment: the telephone system is experiencing significant changes; there are more than 1,000 companies licensed to offer communication services; access to digital lines has improved, particularly in urban centers; Internet and e-mail services are improving; Russia has made progress toward building the telecommunications infrastructure necessary for a market economy; the estimated number of mobile subscribers jumped from fewer than 1 million in 1998 to 150 million in 2006; a large demand for main line service remains unsatisfied, but fixed-line operators continue to grow their services

domestic: cross-country digital trunk lines run from Saint Petersburg to Khabarovsk, and from Moscow to Novorossiysk; the telephone systems in 60 regional capitals have modern digital infrastructures; cellular services, both analog and digital, are available in many areas; in rural areas, the telephone services are still outdated, inadequate, and low density

international: country code—7; Russia is connected internationally by undersea fiber optic cables; digital switches in sev-

eral cities provide more than 50,000 lines for international calls; satellite earth stations provide access to Intelsat, Intersputnik, Eutelsat, Inmarsat, and Orbita systems

Radio broadcast stations: AM 323, FM 1,500 est., shortwave 62 (2004)

Radios: 61.5 million (1997)

Television broadcast stations: 7,306 (1998)

Televisions: 60.5 million (1997)

Internet country code: .ru; note—Russia also has responsibility for a legacy domain ".su" that was allocated to the Soviet Union and is being phased out

Internet hosts: 2.844 million (2007)

Internet Service Providers (ISPs): 300 (June 2000)

Internet users: 25.689 million (2006)

TRANSPORTATION

Airports: 1,260 (2007)

Airports—with paved runways:
total: 601
over 3,047 m: 51
2,438 to 3,047 m: 197
1,524 to 2,437 m: 129
914 to 1,523 m: 102
under 914 m: 122 (2007)

Airports—with unpaved runways:
total: 659
over 3,047 m: 4
2,438 to 3,047 m: 13
1,524 to 2,437 m: 69
914 to 1,523 m: 89
under 914 m: 484 (2007)

Heliports: 47 (2007)

Pipelines: condensate 122 km; gas 158,699 km; oil 72,347 km; refined products 13,658 km (2007)

Railways:
total: 87,157 km
broad gauge: 86,200 km 1.520-m gauge (40,300 km electrified)
narrow gauge: 957 km 1.067-m gauge (on Sakhalin Island)
note: an additional 30,000 km of non-common carrier lines serve industries (2006)

Roadways:
total: 871,000 km
paved: 738,000 km (includes 29,000 km of expressways)
unpaved: 133,000 km
note: includes public and departmental roads (2004)

Waterways: 102,000 km (including 33,000 km with guaranteed depth)
note: 72,000 km system in European Russia links Baltic Sea, White Sea, Caspian Sea, Sea of Azov, and Black Sea (2006)

Merchant marine:
total: 1,130 ships (1000 GRT or over) 4,712,349 GRT/5,747,083 DWT

by type: bulk carrier 28, cargo 718, carrier 2, chemical tanker 27, combination ore/oil 35, container 10, passenger 15, passenger/cargo 8, petroleum tanker 215, refrigerated cargo 51, roll on/roll off 14, specialized tanker 7

foreign-owned: 101 (Belgium 6, Cyprus 2, Germany 2, Greece 1, South Korea 1, Latvia 2, Switzerland 6, Turkey 70, Ukraine 10, US 1)

registered in other countries: 469 (Antigua and Barbuda 5, Bahamas 5, Belize 39, Bulgaria 1, Cambodia 112, Comoros 9, Cyprus 50, Dominica 2, Georgia 17, North Korea 1, Liberia 87, Malta 66, Marshall Islands 4, Mongolia 17, Panama 9, Sierra Leone 5, St Kitts and Nevis 14, St Vincent and The Grenadines 19, Thailand 1, Tuvalu 4, Vanuatu 1, Venezuela 1, unknown 21) (2007)

Ports and terminals: Azov, Kaliningrad, Kavkaz, Nakhodka, Novorossiysk, Primorsk, Saint Petersburg, Vostochnyy

MILITARY

Military branches: Ground Forces (SV), Navy (VMF), Air Forces (Voyenno-Vozdushniye Sily, VVS); Airborne Troops (VDV), Strategic Rocket Troops (RVSN), and Space Troops (KV) are independent "combat arms," not subordinate to any of the three branches; Russian Ground Forces include the following combat arms: motorized-rifle troops, tank troops, missile and artillery troops, air defense of ground troops (2007)

Military service age and obligation: 18–27 years of age for compulsory or voluntary military service; males are registered for the draft at 17 years of age; service obligation—1 year; reserve obligation to age 50; foreign citizens and dual-nationality Russians are precluded from contract military service (2008)

Manpower available for military service:
males age 16–49: 36,219,908
females age 16–49: 37,019,853 (2008 est.)

Manpower fit for military service:
males age 16–49: 21,488,878
females age 16–49: 28,760,976 (2008 est.)

Manpower reaching militarily significant age annually:
males age 16–49: 821,103
females age 16–49: 781,570 (2008 est.)

Military expenditures—percent of GDP: 3.9% (2005)

TRANSNATIONAL ISSUES

Disputes—international: China and Russia have demarcated the once disputed islands at the Amur and Ussuri confluence and in the Argun River in accordance with the 2004 Agreement, ending their centuries-long border disputes; the sovereignty dispute over the islands of Etorofu, Kunashiri, Shikotan, and the Habomai group, known in Japan as the "Northern Territories" and in Russia as the "Southern Kurils," occupied by the Soviet Union in 1945, now administered by Russia, and claimed by Japan, remains the primary sticking point to signing a peace treaty formally ending World War II hostilities; Russia and Georgia agree on delimiting all but small, strategic segments of the land boundary and the maritime boundary; OSCE observers monitor volatile areas such as the Pankisi Gorge in the Akhmeti region and the Kodori Gorge in Abkhazia; Azerbaijan, Kazakhstan, and Russia signed equidistance boundaries in the Caspian seabed but the littoral states have no consensus on dividing the water column; Russia and Norway dispute their maritime limits in the Barents Sea and Russia's fishing rights beyond Svalbard's territorial limits within the Svalbard Treaty zone; various groups in Finland advocate restoration of Karelia (Kareliya) and other areas ceded to the Soviet Union following the Second World War but the Finnish Government asserts no territorial demands; in May 2005, Russia recalled its signatures to the 1996 border agreements with Estonia (1996) and Latvia (1997), when the two Baltic states announced issuance of unilateral declarations referencing Soviet occupation and ensuing territorial losses;

Russia demands better treatment of ethnic Russians in Estonia and Latvia; Estonian citizen groups continue to press for realignment of the boundary based on the 1920 Tartu Peace Treaty that would bring the now divided ethnic Setu people and parts of the Narva region within Estonia; Lithuania and Russia committed to demarcating their boundary in 2006 in accordance with the land and maritime treaty ratified by Russia in May 2003 and by Lithuania in 1999; Lithuania operates a simplified transit regime for Russian nationals traveling from the Kaliningrad coastal exclave into Russia, while still conforming, as an EU member state with an EU external border, where strict Schengen border rules apply; preparations for the demarcation delimitation of land boundary with Ukraine have commenced; the dispute over the boundary between Russia and Ukraine through the Kerch Strait and Sea of Azov remains unresolved despite a December 2003 framework agreement and on-going expert-level discussions; Kazakhstan and Russia boundary delimitation was ratified on November 2005 and field demarcation should commence in 2007; Russian Duma has not yet ratified 1990 Bering Sea Maritime Boundary Agreement with the US

Refugees and internally displaced persons: *IDPs:* 18,000–160,000 (displacement from Chechnya and North Ossetia) (2007)

Trafficking in persons: *current situation:* Russia is a source, transit, and destination country for men, women, and chil-

dren trafficked for various purposes; it remains a significant source of women trafficked to over 50 countries for commercial sexual exploitation; Russia is also a transit and destination country for men and women trafficked from Central Asia, Eastern Europe, and North Korea to Central and Western Europe and the Middle East for purposes of forced labor and sexual exploitation; internal trafficking remains a problem in Russia with women trafficked from rural areas to urban centers for commercial sexual exploitation, and men trafficked internally and from Central Asia for forced labor in the construction and agricultural industries; debt bondage is common among trafficking victims, and child sex tourism remains a concern

tier rating: Tier 2 Watch List—Russia is placed on the Tier 2 Watch List for a third consecutive year for its continued failure to show evidence of increasing efforts to combat trafficking, particularly in the area of victim protection and assistance

Illicit drugs: limited cultivation of illicit cannabis and opium poppy and producer of methamphetamine, mostly for domestic consumption; government has active illicit crop eradication program; used as transshipment point for Asian opiates, cannabis, and Latin American cocaine bound for growing domestic markets, to a lesser extent Western and Central Europe, and occasionally to the US; major source of heroin precursor chemicals; corruption and organized crime are key concerns; major consumer of opiates

RWANDA

INTRODUCTION

Background: In 1959, three years before independence from Belgium, the majority ethnic group, the Hutus, over-

threw the ruling Tutsi king. Over the next several years, thousands of Tutsis were killed, and some 150,000 driven into exile in neighboring countries. The children of these exiles later formed a rebel group, the Rwandan Patriotic Front (RPF), and began a civil war in 1990. The war, along with several political and economic upheavals, exacerbated ethnic tensions, culminating in April 1994 in the genocide of roughly 800,000 Tutsis and moderate Hutus. The Tutsi rebels defeated the Hutu regime and ended the killing in July 1994, but approximately 2 million Hutu refugees—many fearing Tutsi retribution—fled to neighboring Burundi, Tanzania, Uganda, and Zaire. Since then, most of the refugees have returned to Rwanda, but several thousand remained in the neighboring Democratic Republic of the Congo

(DRC; the former Zaire) and formed an extremist insurgency bent on retaking Rwanda, much as the RPF tried in 1990. Despite substantial international assistance and political reforms—including Rwanda's first local elections in March 1999 and its first post-genocide presidential and legislative elections in August and September 2003—the country continues to struggle to boost investment and agricultural output, and ethnic reconciliation is complicated by the real and perceived Tutsi political dominance. Kigali's increasing centralization and intolerance of dissent, the nagging Hutu extremist insurgency across the border, and Rwandan involvement in two wars in recent years in the neighboring DRC continue to hinder Rwanda's efforts to escape its bloody legacy.

GEOGRAPHY

Location: Central Africa, east of Democratic Republic of the Congo
Geographic coordinates: 2 00 S, 30 00 E
Map references: Africa
Area:
total: 26,338 sq km
land: 24,948 sq km
water: 1,390 sq km
Area—comparative: slightly smaller than Maryland
Land boundaries:
total: 893 km
border countries: Burundi 290 km, Democratic Republic of the Congo 217 km, Tanzania 217 km, Uganda 169 km
Coastline: 0 km (landlocked)
Maritime claims: none (landlocked)
Climate: temperate; two rainy seasons (February to April, November to January); mild in mountains with frost and snow possible
Terrain: mostly grassy uplands and hills; relief is mountainous with altitude declining from west to east
Elevation extremes:
lowest point: Rusizi River 950 m
highest point: Volcan Karisimbi 4,519 m
Natural resources: gold, cassiterite (tin ore), wolframite (tungsten ore), methane, hydropower, arable land
Land use:
arable land: 45.56%
permanent crops: 10.25%
other: 44.19% (2005)
Irrigated land: 90 sq km (2003)
Total renewable water resources: 5.2 cu km (2003)
Freshwater withdrawal (domestic/industrial/agricultural): *total:* 0.15 cu km/yr (24%/8%/68%)
per capita: 17 cu m/yr (2000)
Natural hazards: periodic droughts; the volcanic Virunga mountains are in the northwest along the border with Democratic Republic of the Congo
Environment—current issues: deforestation results from uncontrolled cutting of trees for fuel; overgrazing; soil exhaustion; soil erosion; widespread poaching
Environment—international agreements: *party to:* Biodiversity, Climate Change, Climate Change-Kyoto Protocol, Desertification, Endangered Species, Hazardous Wastes, Ozone Layer Protection, Wetlands
signed, but not ratified: Law of the Sea
Geography—note: landlocked; most of the country is savanna grassland with the population predominantly rural

PEOPLE

Population: 10,186,063

note: estimates for this country explicitly take into account the effects of excess mortality due to AIDS; this can result in lower life expectancy, higher infant mortality, higher death rates, lower population growth rates, and changes in the distribution of population by age and sex than would otherwise be expected (July 2008 est.)
Age structure:
0–14 years: 41.9% (male 2,143,479/female 2,124,588)
15–64 years: 55.7% (male 2,826,557/female 2,842,020)
65 years and over: 2.4% (male 99,721/female 149,698) (2008 est.)
Median age:
total: 18.7 years
male: 18.5 years
female: 18.9 years (2008 est.)
Population growth rate: 2.779% (2008 est.)
Birth rate: 39.97 births/1,000 population (2008 est.)
Death rate: 14.46 deaths/1,000 population (2008 est.)
Net migration rate: 2.29 migrant(s)/1,000 population (2008 est.)
Sex ratio:
at birth: 1.03 male(s)/female
under 15 years: 1.01 male(s)/female
15–64 years: 0.99 male(s)/female
65 years and over: 0.67 male(s)/female
total population: 0.99 male(s)/female (2008 est.)
Infant mortality rate:
total: 83.42 deaths/1,000 live births
male: 88.53 deaths/1,000 live births
female: 78.16 deaths/1,000 live births (2008 est.)
Life expectancy at birth:
total population: 49.76 years
male: 48.56 years
female: 51 years (2008 est.)
Total fertility rate: 5.31 children born/woman (2008 est.)
HIV/AIDS—adult prevalence rate: 5.1% (2003 est.)
HIV/AIDS—people living with HIV/AIDS: 250,000 (2003 est.)
HIV/AIDS—deaths: 22,000 (2003 est.)
Major infectious diseases:
degree of risk: very high
food or waterborne diseases: bacterial diarrhea, hepatitis A, and typhoid fever
vectorborne disease: malaria (2008)
Nationality:
noun: Rwandan(s)
adjective: Rwandan
Ethnic groups: Hutu (Bantu) 84%, Tutsi (Hamitic) 15%, Twa (Pygmy) 1%
Religions: Roman Catholic 56.5%, Protestant 26%, Adventist 11.1%, Muslim 4.6%, indigenous beliefs 0.1%, none 1.7% (2001)

Languages: Kinyarwanda (official) universal Bantu vernacular, French (official), English (official), Kiswahili (Swahili) used in commercial centers
Literacy: *definition:* age 15 and over can read and write
total population: 70.4%
male: 76.3%
female: 64.7% (2003 est.)
People—note: Rwanda is the most densely populated country in Africa

GOVERNMENT

Country name:
conventional long form: Republic of Rwanda
conventional short form: Rwanda
local long form: Republika y'u Rwanda
local short form: Rwanda
former: Ruanda, German East Africa
Government type: republic; presidential, multiparty system
Capital: *name:* Kigali
geographic coordinates: 1 57 S, 30 04 E
time difference: UTC+2 (7 hours ahead of Washington, DC during Standard Time)
Administrative divisions: 4 provinces (in French—provinces, singular—province; in Kinyarwanda—intara for singular and plural) and 1 city* (in French—ville; in Kinyarwanda—umujyi); Est (Eastern), Kigali*, Nord (Northern), Ouest (Western), Sud (Southern)
Independence: 1 July 1962 (from Belgium-administered UN trusteeship)
National holiday: Independence Day, 1 July (1962)
Constitution: new constitution passed by referendum 26 May 2003
Legal system: based on German and Belgian civil law systems and customary law; judicial review of legislative acts in the Supreme Court; has not accepted compulsory ICJ jurisdiction
Suffrage: 18 years of age; universal
Executive branch:
chief of state: President Paul KAGAME (since 22 April 2000)
head of government: Prime Minister Bernard MAKUZA (since 8 March 2000)
cabinet: Council of Ministers appointed by the president
elections: President elected by popular vote for a seven-year term (eligible for a second term); elections last held 25 August 2003 (next to be held in 2010)
election results: Paul KAGAME elected president in first direct popular vote; Paul KAGAME 95.05%, Faustin TWAGIRAMUNGU 3.62%, Jean-Nepomuscene NAYINZIRA 1.33%
Legislative branch: bicameral Parliament consists of Senate (26 seats;

12 members elected by local councils, 8 appointed by the president, 4 by the Political Organizations Forum, 2 represent institutions of higher learning; to serve eight-year terms) and Chamber of Deputies (80 seats; 53 members elected by popular vote, 24 women elected by local bodies, 3 selected by youth and disability organizations; to serve five-year terms)

elections: Senate—members appointed as part of the transitional government (next to be held in 2011); Chamber of Deputies—last held 29 September 2003 (next to be held in 2008)

election results: seats by party under the 2003 Constitution—RPF 40, PSD 7, PL 6, additional 27 members indirectly elected

Judicial branch: Supreme Court; High Courts of the Republic; Provincial Courts; District Courts; mediation committees

Political parties and leaders: Centrist Democratic Party or PDC [Alfred MUKEZAMFURA]; Democratic Popular Union of Rwanda or UDPR [Adrien RANGIRA]; Democratic Republican Movement or MDR [Celestin KABANDA] (officially banned); Islamic Democratic Party or PDI [Andre BUMAYA]; Liberal Party or PL [Protais MITALI]; Party for Democratic Renewal (officially banned); Rwandan Patriotic Front or RPF [Paul KAGAME]; Social Democratic Party or PSD [Vincent BIRUTA]

Political pressure groups and leaders: IBUKA—association of genocide survivors

International organization participation: ACCT, ACP, AfDB, AU, CEPGL, COMESA, EAC, FAO, G-77, IBRD, ICAO, ICRM, IDA, IFAD, IFC, IFRCS, ILO, IMF, Interpol, IOC, IOM, IPU, ISO (correspondent), ITSO, ITU, ITUC, MIGA, NAM, OIF, OPCW, UN, UNAMID, UNCTAD, UNESCO, UNIDO, UNMIS, UNWTO, UPU, WCL, WCO, WHO, WIPO, WMO, WTO

Diplomatic representation in the US:
chief of mission: Ambassador James KOMONYO
chancery: 1714 New Hampshire Avenue NW, Washington, DC 20009
telephone: [1] (202) 232-2882
FAX: [1] (202) 232-4544

Diplomatic representation from the US:
chief of mission: Ambassador Michael ARIETTI
embassy: 337 Boulevard de la Revolution, Kigali
mailing address: B. P. 28, Kigali
telephone: [250] 50 56 01 through 03

FAX: [250] 57 2128

Flag description: three horizontal bands of sky blue (top, double width), yellow, and green, with a golden sun with 24 rays near the fly end of the blue band

ECONOMY

Economy—overview: Rwanda is a poor rural country with about 90% of the population engaged in (mainly subsistence) agriculture. It is the most densely populated country in Africa and is landlocked with few natural resources and minimal industry. Primary foreign exchange earners are coffee and tea. The 1994 genocide decimated Rwanda's fragile economic base, severely impoverished the population, particularly women, and eroded the country's ability to attract private and external investment. However, Rwanda has made substantial progress in stabilizing and rehabilitating its economy to pre-1994 levels, although poverty levels are higher now. GDP has rebounded and inflation has been curbed. Despite Rwanda's fertile ecosystem, food production often does not keep pace with population growth, requiring food imports. Rwanda continues to receive substantial aid money and obtained IMF-World Bank Heavily Indebted Poor Country (HIPC) initiative debt relief in 2005–06. Rwanda also received Millennium Challenge Account Threshold status in 2006. The government has embraced an expansionary fiscal policy to reduce poverty by improving education, infrastructure, and foreign and domestic investment and pursuing market-oriented reforms, although energy shortages, instability in neighboring states, and lack of adequate transportation linkages to other countries continue to handicap growth.

GDP (purchasing power parity): $8.44 billion (2007 est.)

GDP (official exchange rate): $3.32 billion (2007 est.)

GDP—real growth rate: 6% (2007 est.)

GDP—per capita (PPP): $900 (2007 est.)

GDP—composition by sector:
agriculture: 36.9%
industry: 21.7%
services: 41.4% (2007 est.)

Labor force: 4.6 million (2000)

Labor force—by occupation: *agriculture:* 90%
industry and services: 10% (2000)

Unemployment rate: NA%

Population below poverty line: 60% (2001 est.)

Household income or consumption by percentage share:
lowest 10%: 2.1%

highest 10%: 38.2% (2000)

Distribution of family income—Gini index: 46.8 (2000)

Inflation rate (consumer prices): 9.4% (2007 est.)

Investment (gross fixed): 22.1% of GDP (2007 est.)

Budget:
revenues: $797 million
expenditures: $873.1 million; including capital expenditures of $NA (2007 est.)

Agriculture—products: coffee, tea, pyrethrum (insecticide made from chrysanthemums), bananas, beans, sorghum, potatoes; livestock

Industries: cement, agricultural products, small-scale beverages, soap, furniture, shoes, plastic goods, textiles, cigarettes

Industrial production growth rate: 13.1% (2007 est.)

Electricity—production: 95 million kWh (2005)

Electricity—production by source:
fossil fuel: 2.3%
hydro: 97.7%
nuclear: 0%
other: 0% (2001)

Electricity—consumption: 198.4 million kWh (2005)

Electricity—exports: 10 million kWh (2005 est.)

Electricity—imports: 120 million kWh (2005)

Oil—production: 0 bbl/day (2005 est.)

Oil—consumption: 5,300 bbl/day (2005 est.)

Oil—exports: 0 bbl/day (2004)

Oil—imports: 5,165 bbl/day (2004)

Oil—proved reserves: 0 bbl (1 January 2006 est.)

Natural gas—production: 0 cu m (2005 est.)

Natural gas—consumption: 0 cu m (2005 est.)

Natural gas—exports: 0 cu m (2005 est.)

Natural gas—imports: 0 cu m (2005)

Natural gas—proved reserves: 54.32 billion cu m (1 January 2006 est.)

Current account balance: -$161 million (2007 est.)

Exports: $167 million f.o.b. (2007 est.)

Exports—commodities: coffee, tea, hides, tin ore

Exports—partners: China 10.3%, Germany 9.7%, US 4.3% (2006)

Imports: $585 million f.o.b. (2007 est.)

Imports—commodities: foodstuffs, machinery and equipment, steel, petroleum products, cement and construction material

Imports—partners: Kenya 19.6%, Germany 7.8%, Uganda 6.8%, Belgium 5.1% (2006)

Economic aid—recipient: $576 million (2005)

Reserves of foreign exchange and gold: $517 million (31 December 2007 est.)

Debt—external: $1.4 billion (2004 est.)

Market value of publicly traded shares: $NA

Currency (code): Rwandan franc (RWF)

Currency code: RWF

Exchange rates: Rwandan francs per US dollar—585 (2007), 560 (2006), 610 (2005), 574.62 (2004), 537.66 (2003)

Fiscal year: calendar year

COMMUNICATIONS

Telephones—main lines in use: 22,000 (2005)

Telephones—mobile cellular: 290,000 (2005)

Telephone system:

general assessment: small, inadequate telephone system primarily serves business and government

domestic: the capital, Kigali, is connected to the centers of the provinces by microwave radio relay and, recently, by cellular telephone service; much of the network depends on wire and HF radiotelephone; combined fixed-line and mobile-cellular telephone density is only about 4 telephones per 100 persons

international: country code—250; international connections employ microwave radio relay to neighboring countries and satellite communications to more distant countries; satellite earth stations—1 Intelsat (Indian Ocean) in Kigali (includes telex and telefax service)

Radio broadcast stations: AM 0, FM 8 (two main FM programs are broadcast through a system of repeaters, three international FM programs include the BBC, VOA, and Deutchewelle), shortwave 1 (2005)

Radios: 601,000 (1997)

Television broadcast stations: 2 (2004)

Televisions: NA; probably less than 1,000 (1997)

Internet country code: .rw

Internet hosts: 1,592 (2007)

Internet Service Providers (ISPs): 2 (2002)

Internet users: 65,000 (2006)

TRANSPORTATION

Airports: 9 (2007)

Airports—with paved runways: *total:* 4
over 3,047 m: 1
914 to 1,523 m: 2
under 914 m: 1 (2007)

Airports—with unpaved runways:
total: 5
914 to 1,523 m: 2
under 914 m: 3 (2007)

Roadways:
total: 14,008 km
paved: 2,662 km
unpaved: 11,346 km (2004)

Waterways: Lac Kivu navigable by shallow-draft barges and native craft (2006)

Ports and terminals: Cyangugu, Gisenyi, Kibuye

MILITARY

Military branches: Rwandan Defense Forces: Army, Air Force

Military service age and obligation: 18 years of age for voluntary military service; no conscription (2008)

Manpower available for military service:
males age 16–49: 2,430,469
females age 16–49: 2,392,933 (2008 est.)

Manpower fit for military service:
males age 16–49: 1,404,066
females age 16–49: 1,403,700 (2008 est.)

Military expenditures—percent of GDP: 2.9% (2006 est.)

TRANSNATIONAL ISSUES

Disputes—-international: fighting among ethnic groups—loosely associated political rebels, armed gangs, and various government forces in Great Lakes region transcending the boundaries of Burundi, Democratic Republic of the Congo, Rwanda, and Uganda—abated substantially from a decade ago due largely to UN peacekeeping, international mediation, and efforts by local governments to create civil societies; nonetheless, 57,000 Rwandan refugees still reside in 21 African states, including Zambia, Gabon, and 20,000 who fled to Burundi in 2005 and 2006 to escape drought and recriminations from traditional courts investigating the 1994 massacres; the 2005 DROC and Rwanda border verification mechanism to stem rebel actions on both sides of the border remains in place

Refugees and internally displaced persons: *refugees (country of origin):* 46,272 (Democratic Republic of the Congo); 4,400 (Burundi) (2007)

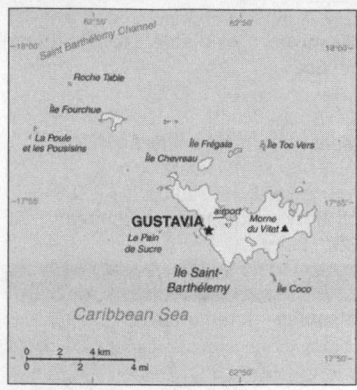

INTRODUCTION

Background: Discovered in 1493 by Christopher COLUMBUS who named it for his brother Bartolomeo, St. Barthelemy was first settled by the French in 1648. In 1784, the French sold the island to Sweden, who renamed the largest town Gustavia, after the Swedish King GUSTAV III, and made it a free port; the island prospered as a trade and supply center during the colonial wars of the 18th century. France repurchased the island in 1878 and placed it under the administration of Guadeloupe. St. Barthelemy retained its free port status along with various Swedish appelations such as Swedish street and town names, and the three-crown symbol on the coat of arms. In 2003, the populace of the island voted to secede from Guadeloupe and in 2007, the island became a French overseas collectivity.

GEOGRAPHY

Location: located approximately 125 miles northwest of Guadeloupe
Geographic coordinates: 17 90 N, 62 85 W
Map references: Central America and the Caribbean
Area: 21 sq km
Area—comparative: less than an eighth of the size of Washington, DC
Land boundaries: 0 km
Climate: tropical, with practically no variation in temperature; has two seasons (dry and humid)
Terrain: hilly, almost completely surrounded by shallow-water reefs, with 20 beaches
Elevation extremes:
lowest point: Caribbean Ocean 0 m
highest point: Morne du Vitet 286 m
Natural resources: has few natural resouces, its beaches being the most important

Environment—current issues: with no natural rivers or streams, fresh water is in short supply, especially in summer, and provided by desalinization of sea water, collection of rain water, or imported via water tanker

PEOPLE

Population: 7,492 (July 2008 est.)
Ethnic groups: white, Creole (mulatto), black, Guadeloupe Mestizo (French-East Asia)
Religions: Roman Catholic, Protestant, Jehovah's Witness
Languages: French (primary), English

GOVERNMENT

Country name:
conventional long form: Overseas Collectivity of Saint Barthelemy
conventional short form: Saint Barthelemy
local long form: Collectivite d'outre mer de Saint-Barthelemy
local short form: Saint-Barthelemy
Dependency status: overseas collectivity of France
Capital: *name:* Gustavia
geographic coordinates: 17 53 N, 62 51 W
time difference: UTC-4 (1 hour ahead of Washington, DC, during Standard Time)
Independence: none (overseas collectivity of France)
National holiday: Bastille Day, 14 July (1789); note—local holiday is St. Barthelemy Day, 24 August
Constitution: 4 October 1958 (French Constitution)
Legal system: the laws of France, where applicable, apply
Suffrage: 18 years of age, universal
Executive branch:
chief of state: President Nicolas SARKOZY (since 16 May 2007), represented by Prefect Dominique LACROIX (since 21 March 2007)
head of government: President of the Territorial Council Bruno MAGRAS (since 16 July 2007)
cabinet: Executive Council; note—there is also an advisory, economic, social, and cultural council
elections: French president elected by popular vote for a five-year term; prefect appointed by the French president on the advice of the French Ministry of Interior; president of the Territorial Council is elected by the members of the Council for a five-year term
election results: Bruno MAGRAS unanimously elected president by the Territorial Council on 16 July 2007
Legislative branch: unicameral Territorial Council (19 seats; members

are elected by popular vote to serve five-year terms)
elections: last held 1 and 8 July 2007 (next to be held July 2012)
election results: percent of vote by party—SBA 72.2%, Action-Equilibre-Transparence 9.9%, Ensemble pour Saint-Barthelemy 7.9%, Tous Unis pour Saint-Barthelemy 9.9%; seats by party—SBA 16, Action-Equilibre-Transparence 1, Ensemble pour Saint-Barthelemy 1, Tous Unis pour Saint-Barthelemy 1
Political parties and leaders: Action-Equilibre-Transparence [Maxime DESOUCHES]; Ensemble pour Saint-Barthelemy [Benoit CHAUVIN]; Saint-Barth d'Abord! or SBA [Bruno MAGRAS]; Tous Unis pour Saint-Barthelemy [Karine MIOT-RICHARD]
International organization participation: UPU
Diplomatic representation in the US: none (overseas collectivity of France)
Diplomatic representation from the US: none (overseas collectivity of France)
Flag description: the flag of France is used

ECONOMY

Economy—overview: The economy of Saint Barthelemy is based upon high-end tourism and duty-free luxury commerce, serving visitors primarily from North America. The luxury hotels and villas host 70,000 visitors each year with another 130,000 arriving by boat. The relative isolation and high cost of living inhibits mass tourism. The construction and public sectors also enjoy significant investment in support of tourism. With limited fresh water resources, all food must be imported, as must all energy resources and most manufactured goods. Employment is strong and attracts labor from Brazil and Portugal.
Currency (code): euro (EUR); note—US dollar (USD) widely used
Exchange rates: euros per US dollar—0.7345 (2007), 0.7964 (2006), 0.8041 (2005), 0.8054 (2004), 0.886 (2003)

COMMUNICATIONS

Telephone system:
general assessment: fully integrated access
domestic: direct dial capability with both fixed and wireless systems
international: country code—590; undersea fiber-optic cable provides voice and data connectivity to Puerto Rico and Guadeloupe
Internet country code: .bl; note—.gp, the ccTLD for Guadeloupe, and .fr, the ccTLD for France, might also be encountered

under 914 m: 1

TRANSPORTATION

Airports: 1
Airports—with paved runways:
total: 1

Transportation—note: nearest airport for international flights is Princess Juliana International Airport (SXM) located in Sint Maarten (Netherlands Antilles)

MILITARY

Military—note: defense is the responsibility of France

SAINT HELENA

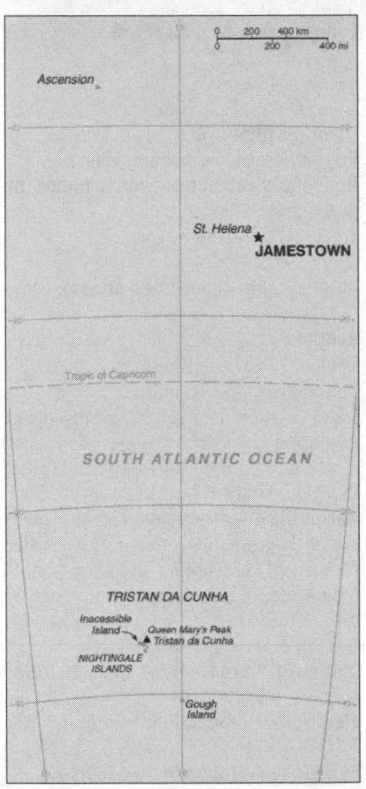

INTRODUCTION

Background: Saint Helena is a British Overseas Territory consisting of Saint Helena and Ascension Islands, and the island group of Tristan da Cunha.
Saint Helena: Uninhabited when first discovered by the Portuguese in 1502, Saint Helena was garrisoned by the British during the 17th century. It acquired fame as the place of Napoleon BONAPARTE's exile, from 1815 until his death in 1821, but its importance as a port of call declined after the opening of the Suez Canal in 1869. During the Anglo-Boer War in South Africa, several thousand Boer prisoners were confined on the island between 1900 and 1903.
Ascension Island: This barren and uninhabited island was discovered and named by the Portuguese in 1503. The British garrisoned the island in 1815 to prevent a rescue of Napoleon from Saint Helena and it served as a provisioning station for

the Royal Navy's West Africa Squadron on anti-slavery patrol. The island remained under Admiralty control until 1922, when it became a dependency of Saint Helena. During World War II, the UK permitted the US to construct an airfield on Ascension in support of trans-Atlantic flights to Africa and anti-submarine operations in the South Atlantic. In the 1960s the island became an important space tracking station for the US. In 1982, Ascension was an essential staging area for British forces during the Falklands War, and it remains a critical refueling point in the air-bridge from the UK to the South Atlantic.
Tristan da Cunha: The island group consists of the islands of Tristan da Cunha, Nightingale, Inaccessible, and Gough. Tristan da Cunha is named after its Portuguese discoverer (1506); it was garrisoned by the British in 1816 to prevent any attempt to rescue Napoleon from Saint Helena. Gough and Inaccessible Islands have been designated World Heritage Sites. South Africa leases a site for a meteorological station on Gough Island.

GEOGRAPHY

Location: islands in the South Atlantic Ocean, about midway between South America and Africa; Ascension Island lies 700 nm northwest of Saint Helena; Tristan da Cunha lies 2300 nm southwest of Saint Helena
Geographic coordinates: *Saint Helena:* 15 57 S, 5 42 W
Ascension Island: 7 57 S, 14 22 W
Tristan da Cunha island group: 37 15 S, 12 30 W
Map references: Africa
Area: *total:* 413 sq km
land: Saint Helena Island 122 sq km; Ascension Island 90 sq km; Tristan da Cunha island group 201 sq km
water: 0 sq km
Area—comparative: slightly more than twice the size of Washington, DC
Land boundaries: 0 km
Coastline: *Saint Helena:* 60 km
Ascension Island: NA
Tristan da Cunha: 40 km
Maritime claims:
territorial sea: 12 nm

exclusive fishing zone: 200 nm
Climate: *Saint Helena:* tropical marine; mild, tempered by trade winds
Ascension Island: tropical marine; mild, semi-arid
Tristan da Cunha: temperate marine; mild, tempered by trade winds (tends to be cooler than Saint Helena)
Terrain: the islands of this group result from volcanic activity associated with the Atlantic Mid-Ocean Ridge
Saint Helena: rugged, volcanic; small scattered plateaus and plains
Ascension: surface covered by lava flows and cinder cones of 44 dormant volcanoes; ground rises to the east
Tristan da Cunha: sheer cliffs line the coastline of the nearly circular island; the flanks of the central volcanic peak are deeply dissected; narrow coastal plain lies between The Peak and the coastal cliffs
Elevation extremes:
lowest point: Atlantic Ocean 0 m
highest point: Queen Mary's Peak on Tristan da Cunha 2,062 m; Green Mountain on Ascension Island 859 m; Mount Actaeon on Saint Helena Island 818 m
Natural resources: fish, lobster
Land use: *arable land:* 12.9%
permanent crops: 0%
other: 87.1% (2005)
Irrigated land: NA
Natural hazards: active volcanism on Tristan da Cunha, last eruption in 1961
Environment—current issues: NA
Geography—note: Saint Helena harbors at least 40 species of plants unknown anywhere else in the world; Ascension is a breeding ground for sea turtles and sooty terns; Queen Mary's Peak on Tristan da Cunha is the highest island mountain in the South Atlantic and a prominent landmark on the sea lanes around southern Africa

PEOPLE

Population: 7,601
note: only Saint Helena, Ascension, and Tristan da Cunha islands are inhabited (July 2008 est.)
Age structure:
0–14 years: 18.5% (male 716/female 690)

15–64 years: 70.7% (male 2,754/female 2,618)

65 years and over: 10.8% (male 381/female 442) (2008 est.)

Median age:
total: 37.1 years
male: 37.2 years
female: 37 years (2008 est.)

Population growth rate: 0.487% (2008 est.)

Birth rate: 11.45 births/1,000 population (2008 est.)

Death rate: 6.58 deaths/1,000 population (2008 est.)

Net migration rate: NA

Sex ratio:
at birth: 1.05 male(s)/female
under 15 years: 1.04 male(s)/female
15–64 years: 1.05 male(s)/female
65 years and over: 0.86 male(s)/female
total population: 1.03 male(s)/female (2008 est.)

Infant mortality rate:
total: 18.31 deaths/1,000 live births
male: 21.47 deaths/1,000 live births
female: 14.98 deaths/1,000 live births (2008 est.)

Life expectancy at birth:
total population: 78.27 years
male: 75.36 years
female: 81.33 years (2008 est.)

Total fertility rate: 1.56 children born/woman (2008 est.)

HIV/AIDS—adult prevalence rate: NA
HIV/AIDS—people living with HIV/AIDS: NA
HIV/AIDS—deaths: NA

Nationality:
noun: Saint Helenian(s)
adjective: Saint Helenian
note: referred to locally as "Saints"

Ethnic groups: African descent 50%, white 25%, Chinese 25%

Religions: Anglican (majority), Baptist, Seventh-Day Adventist, Roman Catholic

Languages: English

Literacy:
definition: age 20 and over can read and write
total population: 97%
male: 97%
female: 98% (1987 est.)

GOVERNMENT

Country name:
conventional long form: none
conventional short form: Saint Helena

Dependency status: overseas territory of the UK

Government type: NA

Capital: *name:* Jamestown
geographic coordinates: 15 56 S, 5 44 W
time difference: UTC 0 (5 hours ahead of Washington, DC during Standard Time)

Administrative divisions: 1 administrative area and 2 dependencies*; Ascension*, Saint Helena, Tristan da Cunha*

Independence: none (overseas territory of the UK)

National holiday: Birthday of Queen ELIZABETH II, second Saturday in June (1926)

Constitution: 1 January 1989

Legal system: English common law and statutes, supplemented by local statutes

Suffrage: NA years of age

Executive branch:
chief of state: Queen ELIZABETH II (since 6 February 1952)
head of government: Governor and Commander in Chief Andrew GURR (since 11 November 2007)
cabinet: Executive Council consists of the governor, three ex-officio officers, and five elected members of the Legislative Council
elections: the monarch is hereditary; governor is appointed by the monarch

Legislative branch: unicameral Legislative Council (16 seats, including the speaker, three ex officio and 12 elected members; members are elected by popular vote to serve four-year terms)
elections: last held 31 August 2005 (next to be held in 2009)
election results: percent of vote—NA; seats—independents 12

Judicial branch: Magistrate's Court; Supreme Court; Court of Appeal

Political parties and leaders: none

Political pressure groups and leaders: none

International organization participation: UPU

Diplomatic representation in the US: none (overseas territory of the UK)

Diplomatic representation from the US: none (overseas territory of the UK)

Flag description: blue with the flag of the UK in the upper hoist-side quadrant and the Saint Helenian shield centered on the outer half of the flag; the shield features a rocky coastline and three-masted sailing ship

ECONOMY

Economy—overview: The economy depends largely on financial assistance from the UK, which will amount to about $27 million in FY06/07 or almost 70% of annual budgetary revenues. The local population earns income from fishing, raising livestock, and sales of handicrafts. Because there are few jobs, 25% of the work force has left to seek employment on Ascension Island, on the Falklands, and in the UK.

GDP (purchasing power parity): $18 million (1998 est.)

GDP (official exchange rate): $NA

GDP—real growth rate: NA%

GDP—per capita (PPP): $2,500 (1998 est.)

GDP—composition by sector:
agriculture: NA%
industry: NA%
services: NA%

Labor force: 2,486
note: 1,200 work offshore (1998 est.)

Labor force—by occupation: *agriculture:* 6%
industry: 48%
services: 46% (1987 est.)

Unemployment rate: 14% (1998 est.)

Population below poverty line: NA%

Household income or consumption by percentage share:
lowest 10%: NA%
highest 10%: NA%

Inflation rate (consumer prices): 3.2% (1997 est.)

Budget:
revenues: $13.09 million
expenditures: $32.16 million
note: revenue data reflect locally raised revenues only; the budget deficit is resolved by grant aid from the United Kingdom (FY06/07 est.)

Agriculture—products: coffee, corn, potatoes, vegetables; timber; fish, lobster (on Tristan da Cunha); livestock

Industries: construction, crafts (furniture, lacework, fancy woodwork), fishing, philatelic sales

Industrial production growth rate: NA%

Electricity—production: 8 million kWh (2005)

Electricity—production by source:
fossil fuel: 100%
hydro: 0%
nuclear: 0%
other: 0% (2001)

Electricity—consumption: 7.44 million kWh (2005)

Electricity—exports: 0 kWh (2005)

Electricity—imports: 0 kWh (2005)

Oil—production: 0 bbl/day (2005 est.)

Oil—consumption: 70 bbl/day (2005 est.)

Oil—exports: 0 bbl/day (2004)

Oil—imports: 64.07 bbl/day (2004)

Oil—proved reserves: 0 bbl (1 January 2006 est.)

Natural gas—production: 0 cu m (2005 est.)

Natural gas—consumption: 0 cu m (2005 est.)

Natural gas—exports: 0 cu m (2005 est.)

Natural gas—imports: 0 cu m (2005)

Natural gas—proved reserves: 0 cu m (1 January 2006 est.)

Exports: $19 million f.o.b. (2004 est.)

Exports—commodities: fish (frozen, canned, and salt-dried skipjack, tuna), coffee, handicrafts

Exports—partners: Tanzania 37.7%, US 17.4%, Japan 15.2%, UK 8.4%, Nigeria 4.8%, Spain 4.5% (2006)

Imports: $45 million c.i.f. (2004 est.)

Imports—commodities: food, beverages, tobacco, fuel oils, animal feed, building materials, motor vehicles and parts, machinery and parts

Imports—partners: UK 53.5%, South Africa 14.3%, Spain 10.3%, Tanzania 8.5%, US 4.6% (2006)

Economic aid—recipient: $29.56 million obtained in a grant from the United Kingdom (FY06/07)

Debt—external: $NA

Currency (code): Saint Helenian pound (SHP)

Currency code: SHP

Exchange rates: Saint Helenian pounds per US dollar—0.4993 (2007), 0.5434 (2006), 0.5493 (2005), 0.5462 (2004), 0.6125 (2003)

note: the Saint Helenian pound is on par with the British pound

Fiscal year: 1 April—31 March

COMMUNICATIONS

Telephones—main lines in use: 2,200 (2002)

Telephone system:

general assessment: can communicate worldwide

domestic: automatic digital network

international: country code (Saint Helena)—290, (Ascension Island)—247; international direct dialing; satellite voice and data communications; satellite earth stations—5 (Ascension Island—4, Saint Helena—1)

Radio broadcast stations: *Saint Helena:* AM 1, FM 1, shortwave 0

Ascension: AM 1, FM 1, shortwave 1 (2005)

Radios: 3,000 (1997)

Television broadcast stations: 0 (3 television channels are received via satellite and distributed by UHF) (2005)

Televisions: 2,000 (1997)

Internet country code: .sh; note—Ascension Island assigned .ac

Internet hosts: 283 (2007)

Internet Service Providers (ISPs): 1 (2000)

Internet users: 1,000; note—includes Ascension Island (2003)

Communications—note: South Africa maintains a meteorological station on Gough Island

TRANSPORTATION

Airports: 1 (2007)

Airports—with paved runways:

total: 1

over 3,047 m: 1 (2007)

Roadways:

total: 198 km (Saint Helena 138 km, Ascension 40 km, Tristan da Cunha 20 km)

paved: 168 km (Saint Helena 118 km, Ascension 40 km, Tristan da Cunha 10 km)

unpaved: 30 km (Saint Helena 20 km, Ascension 0 km, Tristan da Cunha 10 km) (2002)

Ports and terminals: *Saint Helena:* Jamestown

Ascension Island: Georgetown

Tristan da Cunha: Calshot Harbor

Transportation—note: there is no air connection to Saint Helena or Tristan da Cunha; an international airport for Saint Helena is in development for 2010

MILITARY

Military—note: defense is the responsibility of the UK

TRANSNATIONAL ISSUES

Disputes—international: none

SAINT KITTS AND NEVIS

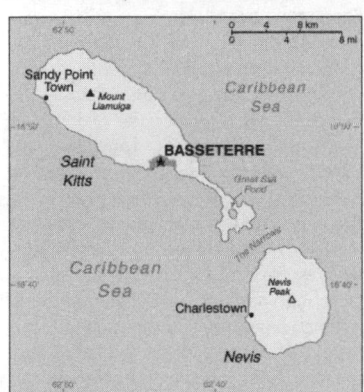

INTRODUCTION

Background: First settled by the British in 1623, the islands became an associated state with full internal autonomy in 1967. The island of Anguilla rebelled and was allowed to secede in 1971. Saint Kitts and Nevis achieved independence in 1983. In 1998, a vote in Nevis on a referendum to separate from Saint Kitts fell short of the two-thirds majority needed. Nevis continues in its efforts to try and separate from Saint Kitts.

GEOGRAPHY

Location: Caribbean, islands in the Caribbean Sea, about one-third of the way from Puerto Rico to Trinidad and Tobago

Geographic coordinates: 17 20 N, 62 45 W

Map references: Central America and the Caribbean

Area:

total: 261 sq km (Saint Kitts 168 sq km; Nevis 93 sq km)

land: 261 sq km

water: 0 sq km

Area—comparative: 1.5 times the size of Washington, DC

Land boundaries: 0 km

Coastline: 135 km

Maritime claims:

territorial sea: 12 nm

contiguous zone: 24 nm

exclusive economic zone: 200 nm

continental shelf: 200 nm or to the edge of the continental margin

Climate: tropical, tempered by constant sea breezes; little seasonal temperature variation; rainy season (May to November)

Terrain: volcanic with mountainous interiors

Elevation extremes:

lowest point: Caribbean Sea 0 m

highest point: Mount Liamuiga 1,156 m

Natural resources: arable land

Land use:

arable land: 19.44%

permanent crops: 2.78%

other: 77.78% (2005)

Irrigated land: NA

Total renewable water resources: 0.02 cu km (2000)

Natural hazards: hurricanes (July to October)

Environment—current issues: NA

Environment—international agreements: *party to:* Biodiversity, Climate Change, Climate Change-Kyoto Protocol, Desertification, Endangered Species, Hazardous Wastes, Law of the Sea, Marine Dumping, Ozone Layer Protection, Ship Pollution, Whaling

signed, but not ratified: none of the selected agreements

Geography—note: with coastlines in the shape of a baseball bat and ball, the two volcanic islands are separated by a 3-km-wide channel called The Narrows; on the southern tip of long, baseball bat-shaped Saint Kitts lies the Great Salt

Pond; Nevis Peak sits in the center of its almost circular namesake island and its ball shape complements that of its sister island

PEOPLE

Population: 39,619 (July 2008 est.)
Age structure:
0–14 years: 26.8% (male 5,441/female 5,184)
15–64 years: 65.4% (male 12,966/female 12,937)
65 years and over: 7.8% (male 1,286/female 1,805) (2008 est.)
Median age: *total:* 28.3 years
male: 27.6 years
female: 29 years (2008 est.)
Population growth rate: 0.745% (2008 est.)
Birth rate: 17.77 births/1,000 population (2008 est.)
Death rate: 8 deaths/1,000 population (2008 est.)
Net migration rate: -2.32 migrant(s)/1,000 population (2008 est.)
Sex ratio:
at birth: 1.06 male(s)/female
under 15 years: 1.05 male(s)/female
15–64 years: 1 male(s)/female
65 years and over: 0.71 male(s)/female
total population: 0.99 male(s)/female (2008 est.)
Infant mortality rate:
total: 13.36 deaths/1,000 live births
male: 15.04 deaths/1,000 live births
female: 11.57 deaths/1,000 live births (2008 est.)
Life expectancy at birth:
total population: 72.93 years
male: 70.06 years
female: 75.96 years (2008 est.)
Total fertility rate: 2.28 children born/woman (2008 est.)
HIV/AIDS—adult prevalence rate: NA
HIV/AIDS—people living with HIV/AIDS: NA
HIV/AIDS—deaths: NA
Nationality:
noun: Kittitian(s), Nevisian(s)
adjective: Kittitian, Nevisian
Ethnic groups: predominantly black; some British, Portuguese, and Lebanese
Religions: Anglican, other Protestant, Roman Catholic
Languages: English
Literacy: *definition:* age 15 and over has ever attended school
total population: 97.8%
male: NA%
female: NA% (2003 est.)

GOVERNMENT

Country name:
conventional long form: Federation of Saint Kitts and Nevis

conventional short form: Saint Kitts and Nevis
former: Federation of Saint Christopher and Nevis
Government type: parliamentary democracy
Capital: *name:* Basseterre
geographic coordinates: 17 18 N, 62 43 W
time difference: UTC-4 (1 hour ahead of Washington, DC during Standard Time)
Administrative divisions: 14 parishes; Christ Church Nichola Town, Saint Anne Sandy Point, Saint George Basseterre, Saint George Gingerland, Saint James Windward, Saint John Capesterre, Saint John Figtree, Saint Mary Cayon, Saint Paul Capesterre, Saint Paul Charlestown, Saint Peter Basseterre, Saint Thomas Lowland, Saint Thomas Middle Island, Trinity Palmetto Point
Independence: 19 September 1983 (from UK)
National holiday: Independence Day, 19 September (1983)
Constitution: 19 September 1983
Legal system: based on English common law; has not accepted compulsory ICJ jurisdiction
Suffrage: 18 years of age; universal
Executive branch:
chief of state: Queen ELIZABETH II (since 6 February 1952); represented by Governor General Cuthbert Montraville SEBASTIAN (since 1 January 1996)
head of government: Prime Minister Dr. Denzil DOUGLAS (since 6 July 1995); Deputy Prime Minister Sam CONDOR (since 6 July 1995)
cabinet: Cabinet appointed by the governor general in consultation with the prime minister
elections: the monarch is hereditary; the governor general is appointed by the monarch; following legislative elections, the leader of the majority party or leader of a majority coalition is usually appointed prime minister by the governor general; deputy prime minister appointed by the governor general
Legislative branch: unicameral National Assembly (14 seats, 3 appointed and 11 popularly elected from single-member constituencies; members serve five-year terms)
elections: last held 25 October 2004 (next to be held by 2009)
election results: percent of vote by party—NA; seats by party—SKNLP 7, CCM 2, NRP 1, PAM 1
Judicial branch: Eastern Caribbean Supreme Court (based on Saint Lucia; one judge of the Supreme Court resides in Saint Kitts and Nevis)
Political parties and leaders: Concerned Citizens Movement or CCM [Vance AMORY]; Nevis Reformation

Party or NRP [Joseph PARRY]; People's Action Movement or PAM [Lindsay GRANT]; Saint Kitts and Nevis Labor Party or SKNLP [Dr. Denzil DOUGLAS]
Political pressure groups and leaders: NA
International organization participation: ACP, C, Caricom, CDB, FAO, G-77, IBRD, ICAO, ICCt, ICRM, IDA, IFAD, IFC, IFRCS, ILO, IMF, IMO, Interpol, IOC, ITU, MIGA, NAM, OAS, OECS, OPANAL, OPCW, UN, UNCTAD, UNESCO, UNIDO, UPU, WHO, WIPO, WTO
Diplomatic representation in the US:
chief of mission: Ambassador Dr. Izben Cordinal WILLIAMS
chancery: 3216 New Mexico Avenue NW, Washington, DC 20016
telephone: [1] (202) 686-2636
FAX: [1] (202) 686-5740
consulate(s) general: New York
Diplomatic representation from the US: the US does not have an embassy in Saint Kitts and Nevis; the US Ambassador to Barbados is accredited to Saint Kitts and Nevis
Flag description: divided diagonally from the lower hoist side by a broad black band bearing two white, five-pointed stars; the black band is edged in yellow; the upper triangle is green, the lower triangle is red

ECONOMY

Economy—overview: Sugar was the traditional mainstay of the Saint Kitts economy until the 1970s. Following the 2005 harvest, the government closed the sugar industry after decades of losses of 3–4% of GDP annually. To compensate for employment losses, the government has embarked on a program to diversify the agricultural sector and to stimulate other sectors of the economy. Activities such as tourism, export-oriented manufacturing, and offshore banking have assumed larger roles in the economy and have contributed to the recent robust growth. Tourism revenues are now the chief source of the islands' foreign exchange; about 341,800 tourists visited Nevis in 2005. The current government is constrained by a high debt burden, public debt reached 190% of GDP by the end of 2005, largely attributable to public enterprise losses.
GDP (purchasing power parity): $721 million (2007 est.)
GDP (official exchange rate): $527 million (2007 est.)
GDP—real growth rate: 3.3% (2007 est.)
GDP—per capita (PPP): $13,900 (2007 est.)

GDP—composition by sector:
agriculture: 3.5%
industry: 25.8%
services: 70.7% (2001)
Labor force: 18,170 (June 1995)
Unemployment rate: 4.5% (1997)
Population below poverty line: NA%
Household income or consumption by percentage share:
lowest 10%: NA%
highest 10%: NA%
Inflation rate (consumer prices): 4.5% (2007 est.)
Budget:
revenues: $89.7 million
expenditures: $128.2 million (2003 est.)
Agriculture—products: sugarcane, rice, yams, vegetables, bananas; fish
Industries: tourism, cotton, salt, copra, clothing, footwear, beverages
Industrial production growth rate: NA%
Electricity—production: 125 million kWh (2005)
Electricity—production by source:
fossil fuel: 100%
hydro: 0%
nuclear: 0%
other: 0% (2001)
Electricity—consumption: 116.3 million kWh (2005)
Electricity—exports: 0 kWh (2005)
Electricity—imports: 0 kWh (2005)
Oil—production: 0 bbl/day (2005 est.)
Oil—consumption: 900 bbl/day (2005 est.)
Oil—exports: 0 bbl/day (2004)
Oil—imports: 871.6 bbl/day (2004)
Oil—proved reserves: 0 bbl (1 January 2006 est.)
Natural gas—production: 0 cu m (2005 est.)
Natural gas—consumption: 0 cu m (2005 est.)
Natural gas—exports: 0 cu m (2005 est.)
Natural gas—imports: 0 cu m (2005)
Natural gas—proved reserves: 0 cu m (1 January 2006 est.)
Current account balance: -$163 million (2007 est.)
Exports: $84 million (2006)
Exports—commodities: machinery, food, electronics, beverages, tobacco

Exports—partners: US 62%, Canada 9.4%, Netherlands 6.6%, Azerbaijan 5% (2006)
Imports: $383 million (2006)
Imports—commodities: machinery, manufactures, food, fuels
Imports—partners: US 48.9%, Trinidad and Tobago 13.1%, Spain 4.6%, UK 4.5% (2006)
Economic aid—recipient: $3.52 million (2005)
Debt—external: $314 million (2004)
Currency (code): East Caribbean dollar (XCD)
Currency code: XCD
Exchange rates: East Caribbean dollars per US dollar—2.7 (2007), 2.7 (2006), 2.7 (2005), 2.7 (2004), 2.7 (2003)
Fiscal year: calendar year

COMMUNICATIONS

Telephones—main lines in use: 25,000 (2004)
Telephones—mobile cellular: 10,000 (2004)
Telephone system:
general assessment: good interisland and international connections
domestic: interisland links via Eastern Caribbean Fiber Optic cable; construction of enhanced wireless infrastructure launched in November 2004
international: country code—1-869; connected internationally by the East Caribbean Fiber Optic System (ECFS) and Southern Caribbean fiber optic system (SCF) submarine cables
Radio broadcast stations: AM 3, FM 3, shortwave 0 (2003)
Radios: 28,000 (1997)
Television broadcast stations: 1 (plus 3 repeaters) (2003)
Televisions: 10,000 (1997)
Internet country code: .kn
Internet hosts: 45 (2007)
Internet Service Providers (ISPs): 16 (2000)
Internet users: 10,000 (2002)

TRANSPORTATION

Airports: 2 (2007)
Airports—with paved runways:
total: 2
1,524 to 2,437 m: 1
914 to 1,523 m: 1 (2007)

Railways:
total: 50 km
narrow gauge: 50 km 0.762-m gauge on Saint Kitts for tourists (2006)
Roadways:
total: 320 km
Merchant marine:
total: 104 ships (1000 GRT or over) 465,056 GRT/663,511 DWT
by type: bulk carrier 3, cargo 66, chemical tanker 8, container 1, passenger/cargo 2, petroleum tanker 15, refrigerated cargo 5, roll on/roll off 2, specialized tanker 2
foreign-owned: 76 (Belgium 1, Egypt 2, Estonia 1, Greece 2, India 1, Iran 1, Latvia 4, Monaco 1, Romania 1, Russia 14, Spain 1, Syria 5, Tanzania 1, Turkey 13, Ukraine 5, UAE 22, Yemen 1) (2007)
Ports and terminals: Basseterre

MILITARY

Military branches: Saint Kitts and Nevis Defense Force (includes Coast Guard), Royal Saint Kitts and Nevis Police Force
Military service age and obligation: 18 years of age for voluntary military service; no conscription (2008)
Manpower available for military service:
males age 16–49: 10,095
females age 16–49: 10,081 (2008 est.)
Manpower fit for military service:
males age 16–49: 8,064
females age 16–49: 8,464 (2008 est.)
Manpower reaching militarily significant age annually:
males age 16–49: 366
females age 16–49: 354 (2008 est.)
Military expenditures—percent of GDP: NA

TRANSNATIONAL ISSUES

Disputes—international: joins other Caribbean states to counter Venezuela's claim that Aves Island sustains human habitation, a criterion under UNCLOS, which permits Venezuela to extend its EEZ/continental shelf over a large portion of the eastern Caribbean Sea
Illicit drugs: transshipment point for South American drugs destined for the US and Europe; some money-laundering activity

SAINT LUCIA

INTRODUCTION

Background: The island, with its fine natural harbor at Castries, was contested between England and France throughout the 17th and early 18th centuries (changing possession 14 times); it was finally ceded to the UK in 1814. Even after the abolition of slavery on its plantations in 1834, Saint Lucia remained an agricultural island, dedicated to producing tropical commodity crops. Self-government was granted in 1967 and independence in 1979.

GEOGRAPHY

Location: Caribbean, island between the Caribbean Sea and North

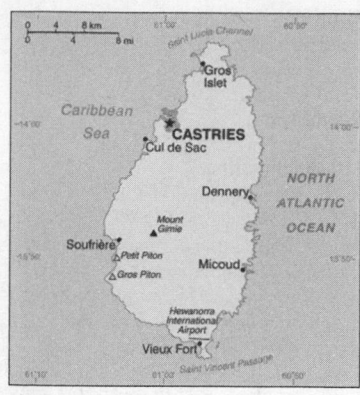

Atlantic Ocean, north of Trinidad and Tobago

Geographic coordinates: 13 53 N, 60 58 W

Map references: Central America and the Caribbean

Area:
total: 616 sq km
land: 606 sq km
water: 10 sq km

Area—comparative: 3.5 times the size of Washington, DC

Land boundaries: 0 km

Coastline: 158 km

Maritime claims:
territorial sea: 12 nm
contiguous zone: 24 nm
exclusive economic zone: 200 nm
continental shelf: 200 nm or to the edge of the continental margin

Climate: tropical, moderated by northeast trade winds; dry season January to April, rainy season May to August

Terrain: volcanic and mountainous with some broad, fertile valleys

Elevation extremes:
lowest point: Caribbean Sea 0 m
highest point: Mount Gimie 950 m

Natural resources: forests, sandy beaches, minerals (pumice), mineral springs, geothermal potential

Land use:
arable land: 6.45%
permanent crops: 22.58%
other: 70.97% (2005)

Irrigated land: 30 sq km (2003)

Freshwater withdrawal (domestic/industrial/agricultural):
total: 0.01
per capita: 81 cu m/yr (1997)

Natural hazards: hurricanes and volcanic activity

Environment—current issues: deforestation; soil erosion, particularly in the northern region

Environment—international agreements: *party to:* Biodiversity, Climate Change, Climate Change-Kyoto Protocol, Desertification, Endangered

Species, Environmental Modification, Hazardous Wastes, Law of the Sea, Marine Dumping, Ozone Layer Protection, Ship Pollution, Wetlands, Whaling
signed, but not ratified: none of the selected agreements

Geography—note: the twin Pitons (Gros Piton and Petit Piton), striking cone-shaped peaks south of Soufriere, are one of the scenic natural highlights of the Caribbean

PEOPLE

Population: 172,884 (July 2008 est.)

Age structure:
0–14 years: 28.9% (male 25,786/female 24,169)
15–64 years: 66% (male 56,346/female 57,725)
65 years and over: 5.1% (male 3,212/female 5,646) (2008 est.)

Median age:
total: 26 years
male: 25.2 years
female: 26.9 years (2008 est.)

Population growth rate: 1.305% (2008 est.)

Birth rate: 18.89 births/1,000 population (2008 est.)

Death rate: 4.99 deaths/1,000 population (2008 est.)

Net migration rate: -0.84 migrant(s)/1,000 population (2008 est.)

Sex ratio:
at birth: 1.07 male(s)/female
under 15 years: 1.07 male(s)/female
15–64 years: 0.98 male(s)/female
65 years and over: 0.57 male(s)/female
total population: 0.97 male(s)/female (2008 est.)

Infant mortality rate:
total: 12.46 deaths/1,000 live births
male: 13.56 deaths/1,000 live births
female: 11.27 deaths/1,000 live births (2008 est.)

Life expectancy at birth:
total population: 74.32 years
male: 70.77 years
female: 78.12 years (2008 est.)

Total fertility rate: 2.11 children born/woman (2008 est.)

HIV/AIDS—adult prevalence rate: NA

HIV/AIDS—people living with HIV/AIDS: NA

HIV/AIDS—deaths: NA

Nationality:
noun: Saint Lucian(s)
adjective: Saint Lucian

Ethnic groups: black 82.5%, mixed 11.9%, East Indian 2.4%, other or unspecified 3.1% (2001 census)

Religions: Roman Catholic 67.5%, Seventh Day Adventist 8.5%, Pentecostal 5.7%, Rastafarian 2.1%,

Anglican 2%, Evangelical 2%, other Christian 5.1%, other 1.1%, unspecified 1.5%, none 4.5% (2001 census)

Languages: English (official), French patois

Literacy:
definition: age 15 and over has ever attended school
total population: 90.1%
male: 89.5%
female: 90.6% (2001 est.)

GOVERNMENT

Country name:
conventional long form: none
conventional short form: Saint Lucia

Government type: parliamentary democracy

Capital: *name:* Castries
geographic coordinates: 14 01 N, 61 00 W
time difference: UTC-4 (1 hour ahead of Washington, DC during Standard Time)

Administrative divisions: 11 quarters; Anse-la-Raye, Castries, Choiseul, Dauphin, Dennery, Gros-Islet, Laborie, Micoud, Praslin, Soufriere, Vieux-Fort

Independence: 22 February 1979 (from UK)

National holiday: Independence Day, 22 February (1979)

Constitution: 22 February 1979

Legal system: based on English common law; has not accepted compulsory ICJ jurisdiction

Suffrage: 18 years of age; universal

Executive branch:
chief of state: Queen ELIZABETH II (since 6 February 1952); represented by Governor General Dame Pearlette LOUISY (since September 1997)
head of government: Prime Minister Stephenson KING (since 9 September 2007); note—Sir John COMPTON died in office Friday, 7 September 2007
cabinet: Cabinet appointed by the governor general on the advice of the prime minister
elections: the monarch is hereditary; the governor general is appointed by the monarch; following legislative elections, the leader of the majority party or the leader of a majority coalition is usually appointed prime minister by the governor general; deputy prime minister appointed by the governor general

Legislative branch: bicameral Parliament consists of the Senate (11 seats; six members appointed on the advice of the prime minister, three on the advice of the leader of the opposition, and two after consultation with religious, economic, and social groups) and the House of Assembly (17 seats; members are elected by popular vote to serve five-year terms) *elections*: House of Assembly—last held 11 December 2006 (next to be held in December 2011)

election results: House of Assembly—percent of vote by party—UWP 50%, SLP 46.9%, other 3.1%; seats by party— UWP 11, SLP 6

Judicial branch: Eastern Caribbean Supreme Court (jurisdiction extends to Anguilla, Antigua and Barbuda, the British Virgin Islands, Dominica, Grenada, Montserrat, Saint Kitts and Nevis, Saint Lucia, and Saint Vincent and the Grenadines)

Political parties and leaders: National Alliance or NA [George ODLUM]; Saint Lucia Freedom Party or SFP [Martinus FRANCOIS]; Saint Lucia Labor Party or SLP [Kenneth ANTHONY]; Sou Tout Apwe Fete Fini or STAFF [Christopher HUNTE]; United Workers Party or UWP [Stephenson KING]

Political pressure groups and leaders: NA

International organization participation: ACCT, ACP, C, Caricom, CDB, FAO, G-77, IBRD, ICAO, ICCt (signatory), ICRM, IDA, IFAD, IFC, IFRCS, ILO, IMF, IMO, Interpol, IOC, ISO, ITU, ITUC, MIGA, NAM, OAS, OECS, OIF, OPANAL, OPCW, UN, UNCTAD, UNESCO, UNIDO, UPU, WCL, WCO, WFTU, WHO, WIPO, WMO, WTO

Diplomatic representation in the US:

chief of mission: Ambassador Sonia Merlyn JOHNNY

chancery: 3216 New Mexico Avenue NW, Washington, DC 20016

telephone: [1] (202) 364-6792 through 6795

FAX: [1] (202) 364-6723

consulate(s) general: Miami, New York

Diplomatic representation from the US: the US does not have an embassy in Saint Lucia; the US Ambassador to Barbados is accredited to Saint Lucia

Flag description: blue, with a gold isosceles triangle below a black arrowhead; the upper edges of the arrowhead have a white border

ECONOMY

Economy—overview: The island nation has been able to attract foreign business and investment, especially in its offshore banking and tourism industries, with a surge in foreign direct investment in 2006, attributed to the construction of several tourism projects. Tourism is the main source of foreign exchange, with almost 900,000 arrivals in 2007. The manufacturing sector is the most diverse in the Eastern Caribbean area, and the government is trying to revitalize the banana industry. Saint Lucia is vulnerable to a variety of external shocks including declines in European Union banana preferences, volatile tourism receipts, natural disasters, and dependence on foreign oil. High debt servicing obligations constrain the KING administration's ability to respond to adverse external shocks. Economic fundamentals remain solid, even though unemployment needs to be reduced.

GDP (purchasing power parity): $1.794 billion (2007 est.)

GDP (official exchange rate): $958 million (2007 est.)

GDP—real growth rate: 3.2% (2007 est.)

GDP—per capita (PPP): $10,700 (2007 est.)

GDP—composition by sector:

agriculture: 5%

industry: 15%

services: 80% (2005 est.)

Labor force: 43,800 (2001 est.)

Labor force—by occupation: *agriculture*: 21.7%

industry: 24.7%

services: 53.6% (2002 est.)

Unemployment rate: 20% (2003 est.)

Population below poverty line: NA%

Household income or consumption by percentage share:

lowest 10%: NA%

highest 10%: NA%

Inflation rate (consumer prices): 1.9% (2007 est.)

Budget:

revenues: $141.2 million

expenditures: $146.7 million (2000 est.)

Agriculture—products: bananas, coconuts, vegetables, citrus, root crops, cocoa

Industries: clothing, assembly of electronic components, beverages, corrugated cardboard boxes, tourism; lime processing, coconut processing

Industrial production growth rate: -8.9% (1997 est.)

Electricity—production: 304.2 million kWh (2005)

Electricity—production by source:

fossil fuel: 100%

hydro: 0%

nuclear: 0%

other: 0% (2001)

Electricity—consumption: 282.9 million kWh (2005)

Electricity—exports: 0 kWh (2005)

Electricity—imports: 0 kWh (2005)

Oil—production: 0 bbl/day (2005 est.)

Oil—consumption: 2,700 bbl/day (2005 est.)

Oil—exports: 0 bbl/day (2004)

Oil—imports: 2,678 bbl/day (2004)

Oil—proved reserves: 0 bbl (1 January 2006 est.)

Natural gas—production: 0 cu m (2005 est.)

Natural gas—consumption: 0 cu m (2005 est.)

Natural gas—exports: 0 cu m (2005 est.)

Natural gas—imports: 0 cu m (2005)

Natural gas—proved reserves: 0 cu m (1 January 2006 est.)

Current account balance: -$199 million (2007 est.)

Exports: $288 million (2006)

Exports—commodities: bananas 41%, clothing, cocoa, vegetables, fruits, coconut oil

Exports—partners: France 69.7%, US 10.2%, UK 8.8% (2006)

Imports: $791 million (2006)

Imports—commodities: food 23%, manufactured goods 21%, machinery and transportation equipment 19%, chemicals, fuels

Imports—partners: US 21.8%, Trinidad and Tobago 15.4%, France 12.2%, Italy 9.3%, Venezuela 7.4%, UK 7.1%, Netherlands 6% (2006)

Economic aid—recipient: $11.06 million (2005)

Debt—external: $257 million (2004)

Currency (code): East Caribbean dollar (XCD)

Currency code: XCD

Exchange rates: East Caribbean dollars per US dollar—2.7 (2007), 2.7 (2006), 2.7 (2005), 2.7 (2004), 2.7 (2003)

Fiscal year: 1 April—31 March

COMMUNICATIONS

Telephones—main lines in use: 51,100 (2002)

Telephones—mobile cellular: 105,700 (2005)

Telephone system:

general assessment: adequate system

domestic: system is automatically switched

international: country code—1-758; the East Caribbean Fiber Optic System (ECFS) and Southern Caribbean fiber optic system (SCF) submarine cables, along with Intelsat from Martinique, carry calls internationally; direct microwave radio relay link with Martinique and Saint Vincent and the Grenadines; tropospheric scatter to Barbados

Radio broadcast stations: AM 2, FM 7, shortwave 0 (2003)

Radios: 111,000 (1997)

Television broadcast stations: 2 (1 commercial broadcast station and 1 community antenna television or CATV channel) (2003)

553

Televisions: 32,000 (1997)
Internet country code: .lc
Internet hosts: 15 (2007)
Internet Service Providers (ISPs): 15 (2000)
Internet users: 55,000 (2004)

TRANSPORTATION

Airports: 2 (2007)
Airports—with paved runways:
total: 2
2,438 to 3,047 m: 1
1,524 to 2,437 m: 1 (2007)
Roadways:
total: 910 km
paved: 48 km

unpaved: 862 km (2000)
Ports and terminals: Castries, Cul-de-Sac, Vieux-Fort

MILITARY

Military branches: no regular military forces; Royal Saint Lucia Police Force (includes Special Service Unit, Coast Guard) (2007)
Manpower available for military service:
males age 16–49: 48,358 (2008 est.)
Manpower fit for military service:
males age 16–49: 38,660 (2008 est.)
Manpower reaching militarily significant age annually:

males age 16–49: 1,706 (2008 est.)
Military expenditures—percent of GDP: NA

TRANSNATIONAL ISSUES

Disputes—international: joins other Caribbean states to counter Venezuela's claim that Aves Island sustains human habitation, a criterion under UNCLOS, which permits Venezuela to extend its EEZ/continental shelf over a large portion of the eastern Caribbean Sea
Illicit drugs: transit point for South American drugs destined for the US and Europe

SAINT MARTIN

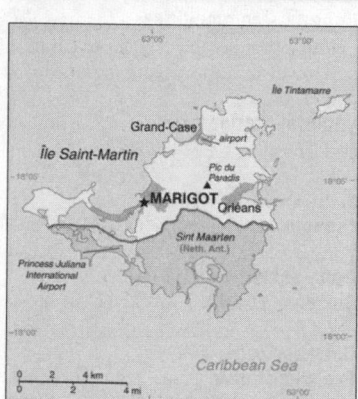

INTRODUCTION

Background: Although sighted by Christopher COLUMBUS in 1493 and claimed for Spain, it was the Dutch who occupied the island in 1631 and set about exploiting its salt deposits. The Spanish retook the island in 1633, but continued to be harassed by the Dutch. The Spanish finally relinquished St. Martin to the French and Dutch, who divided it amongst themselves in 1648. The cultivation of sugar cane introduced slavery to the island in the late 18th century; the practice was not abolished until 1848. The island became a free port in 1939; the tourism industry was dramatically expanded during the 1970s and 1980s. In 2003, the populace of St. Martin voted to secede from Guadeloupe and in 2007, the northern portion of the island became a French overseas collectivity.

GEOGRAPHY

Location: island 300 km southeast of Puerto Rico
Geographic coordinates: 18 05 N, 63 57 W

Map references: Central America and the Caribbean
Area: *total:* 54.4 sq km
land: 54.4 sq km
water: NEGL
Area—comparative: more than one-third the size of Washington, DC
Land boundaries: *total:* 15 km
border countries: Netherlands Antilles (Sint Maarten) 15 km
Coastline: 58.9 km (for entire island)
Climate: temperature averages 80–85 degrees all year long; low humidity, gentle trade winds, brief, intense rain showers; July-Novemeber is the hurricane season
Elevation extremes:
lowest point: Caribbean Ocean 0 m
highest point: Pic du Paradis 424 m
Natural resources: salt
Environment—current issues: fresh water supply is dependent on desalinization of sea water
Geography—note: the island of Saint Martin is the smallest landmass in the world shared by two independent states, the French territory of Saint Martin and the Dutch territory of Sint Maarten

PEOPLE

Population: 29,376 (July 2008 est.)
Ethnic groups: creole (mulatto), black, Guadeloupe Mestizo (French-East Asia), white, East Indian
Religions: Roman Catholic, Jehovah's Witness, Protestant, Hindu
Languages: French (official language), English, Dutch, French Patois, Spanish, Papiamento (dialect of Netherlands Antilles)

GOVERNMENT

Country name: *conventional long form:* Overseas Collectivity of Saint Martin

conventional short form: Saint Martin
local long form: Collectivity d'outre mer de Saint-Martin
local short form: Saint-Martin
Dependency status: overseas collectivity of France
Capital: *name:* Marigot
geographical coordinates: 18 04 N, 63 05 W
time difference: UTC-4 (1 hour behind Washington, DC, during Standard Time)
daylight savings: +1 hour
Independence: none (overseas collectivity of France)
National holiday: Bastille Day, 14 July (1789); note—local holiday is Schoalcher Day (Slavery Abolition Day) 12 July (1848)
Constitution: 4 October 1958 (French Constitution)
Legal system: the laws of France, where applicable, apply
Suffrage: 18 years of age, universal
Executive branch:
chief of state: President Nicolas SARKOZY (since 16 May 2007), represented by Prefect Dominique LACROIX (since 21 March 2007)
head of government: President of the Territorial Council Louis-Constant FLEMING (since 16 July 2007)
cabinet: Executive Council; note—there is also an advisory economic, social, and cultural council
election: French president elected by popular vote to a five-year term; prefect appointed by the French president on the advice of the French Ministry of Interior; president of the Territorial Council is elected by the members of the Council for a five-year term
election results: Louis-Constant FLEMING unanimously elected president by the Territorial Council on 16 July 2007

Legislative branch: unicameral Territorial Council (23 seats; members are elected by popular vote to serve five-year terms)

elections: last held 1 and 8 July 2007 (next to be held July 2012)

election results: percent of seats by party—UPP 49%, RRR 42.2%, Reussir Saint-Martin 8.9%; seats by party—UPP 16, RRR 6, Reussir Saint-Martin 1

Political parties and leaders: Union Pour le Progres or UPP [Louis-Constant FLEMING]; Rassemblement Responsabilite Reussite or RRR [Alain RICHARDSON]; Reussir Saint-Martin [Jean-Luc HAMLET]

International organization participation: UPU

Diplomatic representation in the US: none (overseas collectivity of France)

Diplomatic representation from the US: none (overseas collectivity of France)

Flag description: the flag of France is used

ECONOMY

Economy—overview: The economy of Saint Martin centers around tourism with 85% of the labor force engaged in this sector. Over one million visitors come to the island each year with most arriving through the Princess Juliana International Airport in Sint Maarten. No significant agriculture and limited local fishing means that almost all food must be imported. Energy resources and manufactured goods are also imported, primarily from Mexico and the United States. Saint Martin is reported to have the highest per capita income in the Caribbean.

GDP—composition by sector:

agriculture: 1%

industry: 15%

services: 84% (2000)

Labor force—by occupation: 85% directly or indirectly employed in tourist industry

Industries: tourism, light industry and manufacturing, heavy industry

Imports—commodities: crude petroleum, food, manufactured items

Imports—partners: US, Mexico (2006)

Currency (code): euro (EUR); note—US dollar (USD) widely used

Exchange rates: euros per US dollar—0.7345 (2007), 0.7964 (2006), 0.8041 (2005), 0.8054 (2004), 0.886 (2003)

COMMUNICATIONS

Telephone system:

general assessment: fully integrated access

domestic: direct dial capability with both fixed and wireless systems

international: country code—590; undersea fiber-optic cable provides voice and data connectivity to Puerto Rico and Guadaloupe

Radio broadcast stations: FM 3 (2007)

Internet country code: .mf; note—.gp, the ccTLD for Guadeloupe, and .fr, the ccTLD for France, might also be encountered

TRANSPORTATION

Airports: 1

Airports—with paved runways:

total: 1

914 to 1,523 m: 1

Transportation—note: nearest airport for international flights is Princess Juliana International Airport (SXM) located in Sint Maarten

MILITARY

Military—note: defense is the responsibility of France

SAINT PIERRE AND MIQUELON

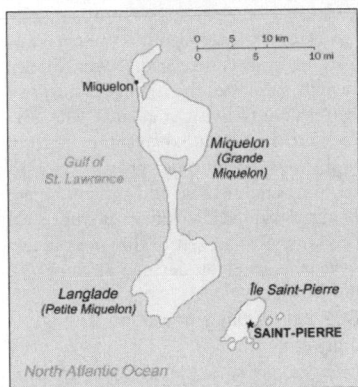

INTRODUCTION

Background: First settled by the French in the early 17th century, the islands represent the sole remaining vestige of France's once vast North American possessions.

GEOGRAPHY

Location: Northern North America, islands in the North Atlantic Ocean, south of Newfoundland (Canada)

Geographic coordinates: 46 50 N, 56 20 W

Map references: North America

Area:

total: 242 sq km

land: 242 sq km

water: 0 sq km

note: includes eight small islands in the Saint Pierre and the Miquelon groups

Area—comparative: 1.5 times the size of Washington, DC

Land boundaries: 0 km

Coastline: 120 km

Maritime claims:

territorial sea: 12 nm

exclusive economic zone: 200 nm

Climate: cold and wet, with much mist and fog; spring and autumn are windy

Terrain: mostly barren rock

Elevation extremes:

lowest point: Atlantic Ocean 0 m

highest point: Morne de la Grande Montagne 240 m

Natural resources: fish, deepwater ports

Land use:

arable land: 12.5%

permanent crops: 0%

other: 87.5% (2005)

Irrigated land: NA

Natural hazards: persistent fog throughout the year can be a maritime hazard

Environment—current issues: recent test drilling for oil in waters around Saint Pierre and Miquelon may bring future development that would impact the environment

Geography—note: vegetation scanty

PEOPLE

Population: 7,044 (July 2008 est.)

Age structure:

0–14 years: 22.4% (male 806/female 772)

15–64 years: 66.3% (male 2,370/female 2,301)

65 years and over: 11.3% (male 366/female 429) (2008 est.)

Median age:

total: 34.9 years

male: 34.3 years

female: 35.3 years (2008 est.)

Population growth rate: 0.114% (2008 est.)

Birth rate: 12.92 births/1,000 population (2008 est.)

Death rate: 6.81 deaths/1,000 population (2008 est.)

Net migration rate: -4.97 migrant(s)/1,000 population (2008 est.)

Sex ratio:

at birth: 1.07 male(s)/female

under 15 years: 1.04 male(s)/female

15–64 years: 1.03 male(s)/female

65 years and over: 0.85 male(s)/female
total population: 1.01 male(s)/female (2008 est.)
Infant mortality rate:
total: 7.04 deaths/1,000 live births
male: 8.06 deaths/1,000 live births
female: 5.96 deaths/1,000 live births (2008 est.)
Life expectancy at birth:
total population: 78.91 years
male: 76.55 years
female: 81.4 years (2008 est.)
Total fertility rate: 1.98 children born/woman (2008 est.)
HIV/AIDS—adult prevalence rate: NA
HIV/AIDS—people living with HIV/AIDS: NA
HIV/AIDS—deaths: NA
Nationality:
noun: Frenchman(men), Frenchwoman(women)
adjective: French
Ethnic groups: Basques and Bretons (French fishermen)
Religions: Roman Catholic 99%, other 1%
Languages: French (official)
Literacy: *definition:* age 15 and over can read and write
total population: 99%
male: 99%
female: 99% (1982 est.)

GOVERNMENT

Country name:
conventional long form: Territorial Collectivity of Saint Pierre and Miquelon
conventional short form: Saint Pierre and Miquelon
local long form: Departement de Saint-Pierre et Miquelon
local short form: Saint-Pierre et Miquelon
Dependency status: self-governing territorial overseas collectivity of France
Government type: NA
Capital: *name:* Saint-Pierre
geographic coordinates: 46 46 N, 56 11 W
time difference: UTC-3 (2 hours ahead of Washington, DC during Standard Time)
daylight saving time: +1hr, begins second Sunday in March; ends first Sunday in November
Administrative divisions: none (territorial overseas collectivity of France); note—there are no first-order administrative divisions as defined by the US Government, but there are two communes—Saint Pierre, Miquelon at the second order
Independence: none (territorial collectivity of France; has been under French control since 1763)
National holiday: Bastille Day, 14 July (1789)

Constitution: 4 October 1958 (French Constitution)
Legal system: the laws of France, where applicable, apply
Suffrage: 18 years of age; universal
Executive branch:
chief of state: President Nicolas SARKOZY (since 16 May 2007); represented by Prefect Yves FAUQUEUR (since 28 August 2006)
head of government: President of the Territorial Council Stephane ARTANO (since 21 February 2007)
cabinet: NA
elections: French president elected by popular vote for a five-year term; election last held 6 May 2007 (next to be held in 2012); prefect appointed by the French president on the advice of the French Ministry of Interior; president of the Territorial Council is elected by the members of the council
Legislative branch: unicameral Territorial Council or Conseil Territorial (19 seats, 15 from Saint Pierre and four from Miquelon; members are elected by popular vote to serve six-year terms)
elections: elections last held 19 and 26 in March 2006 (next to be held in March 2012)
election results: percent of vote by party—NA; seats by party—AD 16, Cap sur l'Avenir 2, SPM 2000/AM 1
note: Saint Pierre and Miquelon elect one seat to the French Senate; elections last held 26 September 2004 (next to be held in September 2013); results—percent of vote by party—NA; seats by party—UMP 1; Saint Pierre and Miquelon also elects one seat to the French National Assembly; elections last held, first round—10 June 2007, second round—17 June 2007 (next to be held in 2012); results—percent of vote by party—NA; seats by party—Left Radical Party 1
Judicial branch: Superior Tribunal of Appeals or Tribunal Superieur d'Appel
Political parties and leaders: Archipelago Tomorrow or AD affiliated with UDF/RPR list; Cap sur l'Avenir affiliated with PRG; Left Radical Party or PRG; Rassemblement pour la Republique or RPR (now UMP); Saint Pierre and Miquelon 2000/Avenir Miquelon or SPM 2000/AM; Socialist Party or PS; Union pour la Democratie Francaise or UDF
Political pressure groups and leaders: NA
International organization participation: UPU, WFTU
Diplomatic representation in the US: none (territorial overseas collectivity of France)

Diplomatic representation from the US: none (territorial overseas collectivity of France)
Flag description: a yellow sailing ship facing the hoist side rides on a dark blue background with yellow wavy lines under the ship; on the hoist side, a vertical band is divided into three parts: the top part (called ikkurina) is red with a green diagonal cross extending to the corners overlaid by a white cross dividing the rectangle into four sections; the middle part has a white background with an ermine pattern; the third part has a red background with two stylized yellow lions outlined in black, one above the other; these three heraldic arms represent settlement by colonists from the Basque Country (top), Brittany, and Normandy; the flag of France is used for official occasions

ECONOMY

Economy—overview: The inhabitants have traditionally earned their livelihood by fishing and by servicing fishing fleets operating off the coast of Newfoundland. The economy has been declining, however, because of disputes with Canada over fishing quotas and a steady decline in the number of ships stopping at Saint Pierre. In 1992, an arbitration panel awarded the islands an exclusive economic zone of 12,348 sq km to settle a longstanding territorial dispute with Canada, although it represents only 25% of what France had sought. France heavily subsidizes the islands to the great betterment of living standards. The government hopes an expansion of tourism will boost economic prospects. Fish farming, crab fishing, and agriculture are being developed to diversify the local economy. Recent test drilling for oil may pave the way for development of the energy sector.
GDP (purchasing power parity): $48.3 million
note: supplemented by annual payments from France of about $60 million (2003 est.)
GDP (official exchange rate): $NA
GDP—real growth rate: NA%
GDP—per capita (PPP): $7,000 (2001 est.)
GDP—composition by sector:
agriculture: NA%
industry: NA%
services: NA%
Labor force: 3,450 (2005)
Labor force—by occupation: *agriculture:* 18%
industry: 41%
services: 41% (1996 est.)
Unemployment rate: 10.3% (1999)

Population below poverty line: NA%

Household income or consumption by percentage share:

lowest 10%: NA%

highest 10%: NA%

Inflation rate (consumer prices): 8.1% (2005)

Budget: *revenues:* $70 million

expenditures: $60 million (1996 est.)

Agriculture—products: vegetables; poultry, cattle, sheep, pigs; fish

Industries: fish processing and supply base for fishing fleets; tourism

Industrial production growth rate: NA%

Electricity—production: 50 million kWh (2005)

Electricity—production by source:

fossil fuel: 100%

hydro: 0%

nuclear: 0%

other: 0% (2001)

Electricity—consumption: 46.5 million kWh (2005)

Electricity—exports: 0 kWh (2005)

Electricity—imports: 0 kWh (2005)

Oil—production: 0 bbl/day (2005 est.)

Oil—consumption: 550 bbl/day (2005 est.)

Oil—exports: 0 bbl/day (2004)

Oil—imports: 541.6 bbl/day (2004)

Oil—proved reserves: 0 bbl (1 January 2006 est.)

Natural gas—production: 0 cu m (2005 est.)

Natural gas—consumption: 0 cu m (2005 est.)

Natural gas—exports: 0 cu m (2005 est.)

Natural gas—imports: 0 cu m (2005)

Natural gas—proved reserves: 0 cu m (1 January 2006 est.)

Exports: $5.5 million f.o.b. (2005 est.)

Exports—commodities: fish and fish products, soybeans, animal feed, mollusks and crustaceans, fox and mink pelts

Exports—partners: Spain 33.6%, Belgium 21.8%, India 18.3%, France 9.4%, US 7.5% (2006)

Imports: $68.2 million f.o.b. (2005 est.)

Imports—commodities: meat, clothing, fuel, electrical equipment, machinery, building materials

Imports—partners: France 51.3%, Canada 31.8%, Belgium 4.1% (2006)

Economic aid—recipient: approximately $60 million in annual grants from France

Debt—external: $NA

Currency (code): euro (EUR)

Currency code: EUR

Exchange rates: euros per US dollar— 0.7345 (2007), 0.7964 (2006), 0.8041 (2005), 0.8054 (2004), 0.886 (2003)

Fiscal year: calendar year

COMMUNICATIONS

Telephones—main lines in use: 4,800 (2002)

Telephone system:

general assessment: adequate

domestic: NA

international: country code—508; radiotelephone communication with most countries in the world; satellite earth station—1 in French domestic satellite system

Radio broadcast stations: AM 1, FM 4, shortwave 0 (1998)

Radios: 4,000 (1997)

Television broadcast stations: 0 (2 repeaters rebroadcast programs from France, Canada, and the US) (1997)

Televisions: 4,000 (1997)

Internet country code: .pm

Internet hosts: 0 (2007)

Internet Service Providers (ISPs): 1 (2000)

Internet users: NA

TRANSPORTATION

Airports: 2 (2007)

Airports—with paved runways:

total: 2

1,524 to 2,437 m: 1

914 to 1,523 m: 1 (2007)

Ports and terminals: Saint-Pierre

MILITARY

Military—note: defense is the responsibility of France

TRANSNATIONAL ISSUES

Disputes—international: none

SAINT VINCENT AND THE GRENADINES

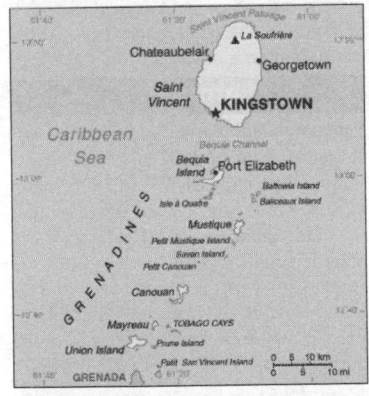

INTRODUCTION

Background: Resistance by native Caribs prevented colonization on St. Vincent until 1719. Disputed between France and the United Kingdom for most of the 18th century, the island was ceded to the latter in 1783. Between 1960 and 1962, Saint Vincent and the Grenadines was a separate administrative unit of the Federation of the West Indies. Autonomy was granted in 1969 and independence in 1979.

GEOGRAPHY

Location: Caribbean, islands between the Caribbean Sea and North Atlantic Ocean, north of Trinidad and Tobago

Geographic coordinates: 13 15 N, 61 12 W

Map references: Central America and the Caribbean

Area:

total: 389 sq km (Saint Vincent 344 sq km)

land: 389 sq km

water: 0 sq km

Area—comparative: twice the size of Washington, DC

Land boundaries: 0 km

Coastline: 84 km

Maritime claims:

territorial sea: 12 nm

contiguous zone: 24 nm

exclusive economic zone: 200 nm

continental shelf: 200 nm

Climate: tropical; little seasonal temperature variation; rainy season (May to November)

Terrain: volcanic, mountainous

Elevation extremes:

lowest point: Caribbean Sea 0 m

highest point: La Soufriere 1,234 m

Natural resources: hydropower, cropland

Land use:

arable land: 17.95%

permanent crops: 17.95%

other: 64.1% (2005)

Irrigated land: 10 sq km (2003)

Freshwater withdrawal (domestic/industrial/agricultural):

total: 0.01

per capita: 83 cu m/yr (1995)

Natural hazards: hurricanes; Soufriere volcano on the island of Saint Vincent is a constant threat

Environment—current issues: pollution of coastal waters and shorelines from discharges by pleasure yachts and other effluents; in some areas, pollution is severe enough to make swimming prohibitive

Environment—international agreements: *party to:* Biodiversity, Climate Change, Climate Change-Kyoto Protocol, Desertification, Endangered Species, Environmental Modification, Hazardous Wastes, Law of the Sea, Marine Dumping, Ozone Layer Protection, Ship Pollution, Whaling *signed, but not ratified:* none of the selected agreements

Geography—note: the administration of the islands of the Grenadines group is divided between Saint Vincent and the Grenadines and Grenada; Saint Vincent and the Grenadines is comprised of 32 islands and cays

PEOPLE

Population: 118,432 (July 2008 est.)
Age structure:
0–14 years: 25.1% (male 15,161/female 14,600)
15–64 years: 68.4% (male 41,855/female 39,105)
65 years and over: 6.5% (male 3,402/female 4,309) (2008 est.)
Median age:
total: 28 years
male: 27.8 years
female: 28.1 years (2008 est.)
Population growth rate: 0.231% (2008 est.)
Birth rate: 15.82 births/1,000 population (2008 est.)
Death rate: 5.96 deaths/1,000 population (2008 est.)
Net migration rate: -7.56 migrant(s)/1,000 population (2008 est.)
Sex ratio: *at birth:* 1.03 male(s)/female
under 15 years: 1.04 male(s)/female
15–64 years: 1.07 male(s)/female
65 years and over: 0.79 male(s)/female
total population: 1.04 male(s)/female (2008 est.)
Infant mortality rate:
total: 13.62 deaths/1,000 live births
male: 14.83 deaths/1,000 live births
female: 12.36 deaths/1,000 live births (2008 est.)
Life expectancy at birth:
total population: 74.34 years
male: 72.42 years
female: 76.31 years (2008 est.)
Total fertility rate: 1.79 children born/woman (2008 est.)
HIV/AIDS—adult prevalence rate: NA
HIV/AIDS—people living with HIV/AIDS: NA
HIV/AIDS—deaths: NA
Nationality: *noun:* Saint Vincentian(s) or Vincentian(s)

adjective: Saint Vincentian or Vincentian
Ethnic groups: black 66%, mixed 19%, East Indian 6%, Carib Amerindian 2%, other 7%
Religions: Anglican 47%, Methodist 28%, Roman Catholic 13%, other (includes Hindu, Seventh-Day Adventist, other Protestant) 12%
Languages: English, French patois
Literacy: *definition:* age 15 and over has ever attended school
total population: 96%
male: 96%
female: 96% (1970 est.)

GOVERNMENT

Country name:
conventional long form: none
conventional short form: Saint Vincent and the Grenadines
Government type: parliamentary democracy
Capital: *name:* Kingstown
geographic coordinates: 13 09 N, 61 14 W
time difference: UTC-4 (1 hour ahead of Washington, DC during Standard Time)
Administrative divisions: 6 parishes; Charlotte, Grenadines, Saint Andrew, Saint David, Saint George, Saint Patrick
Independence: 27 October 1979 (from UK)
National holiday: Independence Day, 27 October (1979)
Constitution: 27 October 1979
Legal system: based on English common law; has not accepted compulsory ICJ jurisdiction
Suffrage: 18 years of age; universal
Executive branch:
chief of state: Queen ELIZABETH II (since 6 February 1952); represented by Governor General Sir Fredrick Nathaniel BALLANTYNE (since 2 September 2002)
head of government: Prime Minister Ralph E. GONSALVES (since 29 March 2001)
cabinet: Cabinet appointed by the governor general on the advice of the prime minister
elections: the monarch is hereditary; the governor general is appointed by the monarch; following legislative elections, the leader of the majority party is usually appointed prime minister by the governor general; deputy prime minister appointed by the governor general on the advice of the prime minister
Legislative branch: unicameral House of Assembly (21 seats, 15 elected representatives and six appointed senators; representatives are elected by popular vote to serve five-year terms)
elections: last held 7 December 2005 (next to be held in 2010)
election results: percent of vote by party—

ULP 55.3%, NDP 44.7%; seats by party—ULP 12, NDP 3
Judicial branch: Eastern Caribbean Supreme Court (based on Saint Lucia; one judge of the Supreme Court resides in Saint Vincent and the Grenadines)
Political parties and leaders: New Democratic Party or NDP [Arnhim EUSTACE]; Unity Labor Party or ULP [Ralph GONSALVES] (formed by the coalition of Saint Vincent Labor Party or SVLP and the Movement for National Unity or MNU)
Political pressure groups and leaders: NA
International organization participation: ACP, C, Caricom, CDB, FAO, G-77, IBRD, ICAO, ICCt, ICRM, IDA, IFAD, IFRCS, ILO, IMF, IMO, Interpol, IOC, ISO (subscriber), ITU, ITUC, MIGA, NAM, OAS, OECS, OPANAL, OPCW, UN, UNCTAD, UNESCO, UNIDO, UPU, WCL, WFTU, WHO, WIPO, WTO
Diplomatic representation in the US:
chief of mission: Ambassador Ellsworth I. A. JOHN
chancery: 3216 New Mexico Avenue NW, Washington, DC 20016
telephone: [1] (202) 364-6730
FAX: [1] (202) 364-6736
consulate(s) general: New York
Diplomatic representation from the US: the US does not have an embassy in Saint Vincent and the Grenadines; the US Ambassador to Barbados is accredited to Saint Vincent and the Grenadines
Flag description: three vertical bands of blue (hoist side), gold (double width), and green; the gold band bears three green diamonds arranged in a V pattern

ECONOMY

Economy—overview: Economic growth slowed slightly in 2007 after reaching a 10 year high of nearly 7% in 2006, but is expected to remain robust, hinging upon seasonal variations in the agricultural and tourism sectors and a recent increase in construction activity. This lower-middle-income country is vulnerable to natural disasters—tropical storms wiped out substantial portions of crops in 1994, 1995, and 2002. In 2007, the islands had more than 200,000 tourist arrivals, mostly to the Grenadines. Saint Vincent is home to a small offshore banking sector and has moved to adopt international regulatory standards. The government's ability to invest in social programs and respond to external shocks is constrained by its high debt burden—25 percent of current revenues are directed towards debt servicing.

GDP (purchasing power parity): $1.042 billion (2007 est.)

GDP (official exchange rate): $559 million (2007 est.)

GDP—real growth rate: 6.6% (2007 est.)

GDP—per capita (PPP): $9,800 (2007 est.)

GDP—composition by sector:
agriculture: 10%
industry: 26%
services: 64% (2001 est.)

Labor force: 41,680 (1991 est.)

Labor force—by occupation: *agriculture:* 26%
industry: 17%
services: 57% (1980 est.)

Unemployment rate: 15% (2001 est.)

Population below poverty line: NA%

Household income or consumption by percentage share:
lowest 10%: NA%
highest 10%: NA%

Inflation rate (consumer prices): 6.1% (2007 est.)

Budget:
revenues: $94.6 million
expenditures: $85.8 million (2000 est.)

Agriculture—products: bananas, coconuts, sweet potatoes, spices; small numbers of cattle, sheep, pigs, goats; fish

Industries: food processing, cement, furniture, clothing, starch

Industrial production growth rate: -0.9% (1997 est.)

Electricity—production: 115 million kWh (2005)

Electricity—production by source:
fossil fuel: 69.3%
hydro: 30.7%
nuclear: 0%
other: 0% (2001)

Electricity—consumption: 107 million kWh (2005)

Electricity—exports: 0 kWh (2005)

Electricity—imports: 0 kWh (2005)

Oil—production: 0 bbl/day (2005 est.)

Oil—consumption: 1,500 bbl/day (2005 est.)

Oil—exports: 0 bbl/day (2004)

Oil—imports: 1,468 bbl/day (2004)

Oil—proved reserves: 0 bbl (1 January 2006 est.)

Natural gas—production: 0 cu m (2005 est.)

Natural gas—consumption: 0 cu m (2005 est.)

Natural gas—exports: 0 cu m (2005 est.)

Natural gas—imports: 0 cu m (2005)

Natural gas—proved reserves: 0 cu m (1 January 2006 est.)

Current account balance: -$149 million (2007 est.)

Exports: $193 million (2006)

Exports—commodities: bananas, eddoes and dasheen (taro), arrowroot starch; tennis racquets

Exports—partners: France 26.2%, Greece 21.3%, Italy 18.9%, Russia 7.2%, UK 6.8% (2006)

Imports: $578 million (2006)

Imports—commodities: foodstuffs, machinery and equipment, chemicals and fertilizers, minerals and fuels

Imports—partners: Singapore 17.3%, Trinidad and Tobago 12.1%, US 11.1%, Italy 10.9%, Spain 9.5%, Turkey 4.6%, Germany 4.4% (2006)

Economic aid—recipient: $4.89 million (1995); note—EU $34.5 million (2005)

Debt—external: $223 million (2004)

Currency (code): East Caribbean dollar (XCD)

Currency code: XCD

Exchange rates: East Caribbean dollars per US dollar—2.7 (2007), 2.7 (2006), 2.7 (2005), 2.7 (2004), 2.7 (2003)

Fiscal year: calendar year

COMMUNICATIONS

Telephones—main lines in use: 22,600 (2006)

Telephones—mobile cellular: 87,600 (2006)

Telephone system:
general assessment: adequate system
domestic: islandwide, fully automatic telephone system; VHF/UHF radiotelephone from Saint Vincent to the other islands of the Grenadines; mobile-cellular teledensity about 75 telephones per 100 persons
international: country code—1-784; the East Caribbean Fiber Optic System (ECFS) and Southern Caribbean fiber optic system (SCF) submarine cables carry international calls; connectivity also provided by VHF/UHF radiotelephone from Saint Vincent to Barbados; SHF radiotelephone to Grenada and Saint Lucia; access to Intelsat earth station in Martinique through Saint Lucia

Radio broadcast stations: AM 1, FM 6, shortwave 0 (2004)

Radios: 77,000 (1997)

Television broadcast stations: 1 (plus 3 repeaters) (2004)

Televisions: 18,000 (1997)

Internet country code: .vc

Internet hosts: 97 (2007)

Internet Service Providers (ISPs): 15 (2000)

Internet users: 10,000 (2005)

TRANSPORTATION

Airports: 6 (2007)

Airports—with paved runways:
total: 5
914 to 1,523 m: 4

under 914 m: 1 (2007)

Airports—with unpaved runways:
total: 1
under 914 m: 1 (2007)

Roadways:
total: 829 km
paved: 580 km
unpaved: 249 km (2003)

Merchant marine:
total: 582 ships (1000 GRT or over) 5,598,917 GRT/8,255,014 DWT
by type: bulk carrier 92, cargo 353, carrier 19, chemical tanker 4, container 17, liquefied gas 6, livestock carrier 1, passenger 5, passenger/cargo 11, petroleum tanker 19, refrigerated cargo 31, roll on/roll off 21, specialized tanker 3
foreign-owned: 536 (Austria 2, Bangladesh 1, Barbados 1, Belgium 9, Bulgaria 13, Canada 6, China 106, Croatia 7, Cyprus 3, Czech Republic 1, Denmark 16, Egypt 4, Estonia 20, France 7, Germany 3, Greece 81, Guyana 2, Hong Kong 7, Iceland 15, India 5, Iran 1, Israel 4, Italy 19, Kenya 2, Latvia 20, Lebanon 7, Lithuania 7, Malta 1, Monaco 6, Montenegro 1, Netherlands 5, Norway 19, Pakistan 1, Philippines 1, Poland 1, Portugal 1, Puerto Rico 1, Romania 1, Russia 19, Singapore 6, Slovenia 5, Sweden 2, Switzerland 12, Syria 11, Turkey 20, Ukraine 12, UAE 12, UK 9, US 21) (2007)

Ports and terminals: Kingstown

MILITARY

Military branches: no regular military forces; Royal Saint Vincent and the Grenadines Police Force, Coast Guard (2007)

Manpower available for military service:
males age 16–49: 34,373 (2008 est.)

Manpower fit for military service:
males age 16–49: 28,518 (2008 est.)

Manpower reaching militarily significant age annually:
males age 16–49: 1,224 (2008 est.)

Military expenditures—percent of GDP: NA

TRANSNATIONAL ISSUES

Disputes—international: joins other Caribbean states to counter Venezuela's claim that Aves Island sustains human habitation, a criterion under UNCLOS, which permits Venezuela to extend its EEZ/continental shelf over a large portion of the eastern Caribbean Sea

Illicit drugs: transshipment point for South American drugs destined for the US and Europe; small-scale cannabis cultivation

559

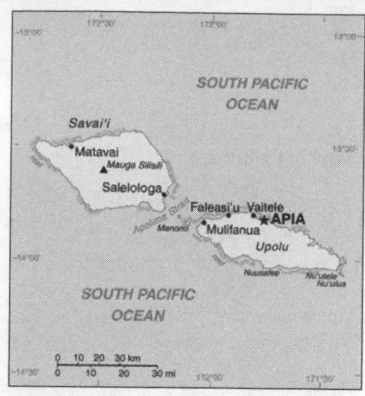

INTRODUCTION

Background: New Zealand occupied the German protectorate of Western Samoa at the outbreak of World War I in 1914. It continued to administer the islands as a mandate and then as a trust territory until 1962, when the islands became the first Polynesian nation to reestablish independence in the 20th century. The country dropped the "Western" from its name in 1997.

GEOGRAPHY

Location: Oceania, group of islands in the South Pacific Ocean, about half way between Hawaii and New Zealand
Geographic coordinates: 13 35 S, 172 20 W
Map references: Oceania
Area:
total: 2,944 sq km
land: 2,934 sq km
water: 10 sq km
Area—comparative: slightly smaller than Rhode Island
Land boundaries: 0 km
Coastline: 403 km
Maritime claims:
territorial sea: 12 nm
contiguous zone: 24 nm
exclusive economic zone: 200 nm
Climate: tropical; rainy season (November to April), dry season (May to October)
Terrain: two main islands (Savaii, Upolu) and several smaller islands and uninhabited islets; narrow coastal plain with volcanic, rocky, rugged mountains in interior
Elevation extremes:
lowest point: Pacific Ocean 0 m
highest point: Mauga Silisili (Savaii) 1,857 m
Natural resources: hardwood forests, fish, hydropower

Land use:
arable land: 21.13%
permanent crops: 24.3%
other: 54.57% (2005)
Irrigated land: NA
Natural hazards: occasional typhoons; active volcanism
Environment—current issues: soil erosion, deforestation, invasive species, overfishing
Environment—international agreements: *party to:* Biodiversity, Climate Change, Climate Change-Kyoto Protocol, Desertification, Hazardous Wastes, Law of the Sea, Ozone Layer Protection, Ship Pollution, Wetlands
signed, but not ratified: none of the selected agreements
Geography—note: occupies an almost central position within Polynesia

PEOPLE

Population: 217,083
note: prior estimates used official net migration data by sex, but a highly unusual pattern for 1993 lead to a significant imbalance in the sex ratios (more men and fewer women) and a seeming reduction in the female population; the revised total was calculated using a 1993 number that was an average of the 1992 and 1994 migration figures (July 2008 est.)
Age structure:
0–14 years: 37.9% (male 41,834/female 40,343)
15–64 years: 56.5% (male 64,402/female 58,257)
65 years and over: 5.6% (male 5,481/female 6,766) (2008 est.)
Median age:
total: 20.6 years
male: 20.8 years
female: 20.4 years (2008 est.)
Population growth rate: 1.322% (2008 est.)
Birth rate: 28.2 births/1,000 population (2008 est.)
Death rate: 5.84 deaths/1,000 population (2008 est.)
Net migration rate: -9.14 migrant(s)/1,000 population (2008 est.)
Sex ratio:
at birth: 1.05 male(s)/female
under 15 years: 1.04 male(s)/female
15–64 years: 1.11 male(s)/female
65 years and over: 0.81 male(s)/female
total population: 1.06 male(s)/female (2008 est.)
Infant mortality rate:
total: 25.04 deaths/1,000 live births
male: 29.56 deaths/1,000 live births

female: 20.29 deaths/1,000 live births (2008 est.)
Life expectancy at birth:
total population: 71.58 years
male: 68.76 years
female: 74.55 years (2008 est.)
Total fertility rate: 4.18 children born/woman (2008 est.)
HIV/AIDS—adult prevalence rate: NA
HIV/AIDS—people living with HIV/AIDS: NA
HIV/AIDS—deaths: NA
Nationality:
noun: Samoan(s)
adjective: Samoan
Ethnic groups: Samoan 92.6%, Euronesians 7% (persons of European and Polynesian blood), Europeans 0.4%
Religions: Congregationalist 34.8%, Roman Catholic 19.6%, Methodist 15%, Latter-Day Saints 12.7%, Assembly of God 6.6%, Seventh-Day Adventist 3.5%, Worship Centre 1.3%, other Christian 4.5%, other 1.9%, unspecified 0.1% (2001 census)
Languages: Samoan (Polynesian), English
Literacy:
definition: age 15 and over can read and write
total population: 99.7%
male: 99.6%
female: 99.7% (2003 est.)

GOVERNMENT

Country name:
conventional long form: Independent State of Samoa
conventional short form: Samoa
local long form: Malo Sa'oloto Tuto'atasi o Samoa
local short form: Samoa
former: Western Samoa
Government type: parliamentary democracy
Capital: *name:* Apia
geographic coordinates: 13 50 S, 171 44 W
time difference: UTC-11 (6 hours behind Washington, DC during Standard Time)
Administrative divisions: 11 districts; A'ana, Aiga-i-le-Tai, Atua, Fa'asaleleaga, Gaga'emauga, Gagaifomauga, Palauli, Satupa'itea, Tuamasaga, Va'a-o-Fonoti, Vaisigano
Independence: 1 January 1962 (from New Zealand-administered UN trusteeship)
National holiday: Independence Day Celebration, 1 June (1962); note—1 January 1962 is the date of independence from the New Zealand-administered UN trusteeship; it is observed in June

Constitution: 1 January 1962
Legal system: based on English common law and local customs; judicial review of legislative acts with respect to fundamental rights of the citizen; has not accepted compulsory ICJ jurisdiction
Suffrage: 21 years of age; universal
Executive branch:
chief of state: TUIATUA Tupua Tamasese Efi (since 20 June 2007)
head of government: Prime Minister Sailele Malielegaoi TUILA'EPA (since 1998); Deputy Prime Minister MISA Telefoni (since 2001)
cabinet: Cabinet consists of 12 members appointed by the chief of state on the prime minister's advice
elections: chief of state is elected by the Legislative Assembly to serve a five-year term (no term limits); election last held 15 June 2007 (next to be held in 2012); following legislative elections, the leader of the majority party is usually appointed prime minister by the chief of state with the approval of the Legislative Assembly
election results: TUIATUA Tupua Tamasese Efi unanimously elected by the Legislative Assembly
Legislative branch: unicameral Legislative Assembly or Fono (49 seats, 47 elected by voters affiliated with traditional village-based electoral districts, 2 elected by independent, mostly non-Samoan or part-Samoan, voters who cannot, (or choose not to) establish a village affiliation; only chiefs (matai) may stand for election to the Fono from the 47 village-based electorates; members serve five-year terms)
elections: election last held 31 March 2006 (next election to be held not later than March 2011)
election results: percent of vote by party—NA; seats by party—HRPP 35, SDUP 10, independents 4
Judicial branch: Court of Appeal; Supreme Court; District Court; Land and Titles Court
Political parties and leaders: Human Rights Protection Party or HRPP [Sailele Malielegaoi TUILA'EPA]; Samoa Christian Party or TCP [Tuala Tiresa MALIETOA]; Samoa Democratic United Party or SDUP [LE MAMEA Ropati]; Samoa Party or SP [Su'a Rimoni Ah CHONG]; Samoa Progressive Political Party or SPPP [Toeolesulusulu SIUEVA]
Political pressure groups and leaders: NA
International organization participation: ACP, ADB, C, FAO, G-77, IBRD, ICAO, ICCt, ICRM, IDA, IFAD, IFC, IFRCS, ILO, IMF, IMO, IOC, IPU, ITU, ITUC, MIGA, OPCW, PIF, Sparteca, SPC, UN, UNCTAD, UNESCO, UPU, WCO, WHO, WIPO, WMO, WTO (observer)
Diplomatic representation in the US:
chief of mission: Ambassador Aliioaiga Feturi ELISAIA
chancery: 800 Second Avenue, Suite 400D, New York, NY 10017
telephone: [1] (212) 599-6196, 6197
FAX: [1] (212) 599-0797
Diplomatic representation from the US:
chief of mission: none; US Ambassador to New Zealand is accredited to Samoa
embassy: Accident Compensation Board (ACB) Building, 5th Floor, Beach Road, Apia
mailing address: P. O. Box 3430, Apia, 0815
telephone: [685] 21436/21452/21631/22696
FAX: [685] 22030
Flag description: red with a blue rectangle in the upper hoist-side quadrant bearing five white five-pointed stars representing the Southern Cross constellation

ECONOMY

Economy—overview: The economy of Samoa has traditionally been dependent on development aid, family remittances from overseas, agriculture, and fishing. The country is vulnerable to devastating storms. Agriculture employs two-thirds of the labor force and furnishes 90% of exports, featuring coconut cream, coconut oil, and copra. The fish catch declined during the El Nino of 2002–03 but returned to normal by mid-2005. The manufacturing sector mainly processes agricultural products. One factory in the Foreign Trade Zone employs 3,000 people to make automobile electrical harnesses for an assembly plant in Australia. Tourism is an expanding sector, accounting for 25% of GDP; 116,000 tourists visited the islands in 2006. The Samoan Government has called for deregulation of the financial sector, encouragement of investment, and continued fiscal discipline, while at the same time protecting the environment. Observers point to the flexibility of the labor market as a basic strength for future economic advances. Foreign reserves are in a relatively healthy state, the external debt is stable, and inflation is low.
GDP (purchasing power parity): $1.029 billion (2007 est.)
GDP (official exchange rate): $397 million (2007 est.)
GDP—real growth rate: 6% (2007 est.)
GDP—per capita (PPP): $5,400 (2007 est.)
GDP—composition by sector:
agriculture: 11.4%
industry: 58.4%
services: 30.2% (2004 est.)
Labor force: 90,000 (2000 est.)
Labor force—by occupation: *agriculture:* NA%
industry: NA%
services: NA%
Unemployment rate: NA%
Population below poverty line: NA%
Household income or consumption by percentage share:
lowest 10%: NA%
highest 10%: NA%
Inflation rate (consumer prices): 6% (2007 est.)
Budget: *revenues:* $171.3 million
expenditures: $78.1 million (FY04/05 est.)
Agriculture—products: coconuts, bananas, taro, yams, coffee, cocoa
Industries: food processing, building materials, auto parts
Industrial production growth rate: 2.8% (2000)
Electricity—production: 105 million kWh (2005)
Electricity—production by source:
fossil fuel: 58%
hydro: 42%
nuclear: 0%
other: 0% (2001)
Electricity—consumption: 97.65 million kWh (2005)
Electricity—exports: 0 kWh (2005)
Electricity—imports: 0 kWh (2005)
Oil—production: 0 bbl/day (2005 est.)
Oil—consumption: 1,100 bbl/day (2005 est.)
Oil—exports: 0 bbl/day (2004)
Oil—imports: 1,060 bbl/day (2004)
Oil—proved reserves: 0 bbl (1 January 2006 est.)
Natural gas—production: 0 cu m (2005 est.)
Natural gas—consumption: 0 cu m (2005 est.)
Natural gas—exports: 0 cu m (2005 est.)
Natural gas—imports: 0 cu m (2005)
Natural gas—proved reserves: 0 cu m (1 January 2006 est.)
Current account balance: -$24 million (FY03/04)
Exports: $131 million f.o.b. (2006)
Exports—commodities: fish, coconut oil and cream, copra, taro, automotive parts, garments, beer
Exports—partners: Australia 44.1%, American Samoa 29.9%, Taiwan 11.3% (2006)
Imports: $324 million f.o.b. (2006)
Imports—commodities: machinery and equipment, industrial supplies, foodstuffs

Imports—partners: NZ 21.5%, Fiji 14.8%, Singapore 13.2%, Australia 8.6%, Japan 8.6%, US 6.2%, Indonesia 5%, China 4.4% (2006)

Economic aid—recipient: $43.95 million (2005)

Reserves of foreign exchange and gold: $70.15 million (FY03/04)

Debt—external: $177 million (2004)

Market value of publicly traded shares: $NA

Currency (code): tala (SAT)

Currency code: SAT (former WST code is still in wide use)

Exchange rates: tala per US dollar— NA (2007), 2.7594 (2006), 2.7103 (2005), 2.7807 (2004), 2.9732 (2003)

Fiscal year: June 1—May 31

COMMUNICATIONS

Telephones—main lines in use: 19,500 (2005)

Telephones—mobile cellular: 24,000 (2005)

Telephone system:
general assessment: adequate
domestic: NA
international: country code—685; satellite earth station—1 Intelsat (Pacific Ocean)

Radio broadcast stations: AM 2, FM 5, shortwave 0 (2004)

Radios: 174,849 (1997)

Television broadcast stations: 2 (2002)

Televisions: 8,634 (1999)

Internet country code: .ws

Internet hosts: 10,156 (2007)

Internet Service Providers (ISPs): 2 (2000)

Internet users: 8,000 (2006)

TRANSPORTATION

Airports: 4 (2007)

Airports—with paved runways:
total: 3
2,438 to 3,047 m: 1
under 914 m: 2 (2007)

Airports—with unpaved runways:
total: 1
under 914 m: 1 (2007)

Roadways:
total: 2,337 km
paved: 332 km
unpaved: 2,005 km (2004)

Merchant marine:
total: 1 ship (1000 GRT or over) 7,091 GRT/8,127 DWT
by type: cargo 1
foreign-owned: 1 (Cyprus 1) (2007)

Ports and terminals: Apia

MILITARY

Military branches: no regular military forces; Samoa Police Force (2008)

Manpower available for military service:
males age 16–49: 53,417 (2008 est.)

Manpower fit for military service:
males age 16–49: 42,359 (2008 est.)

Manpower reaching militarily significant age annually:
males age 16–49: 2,571 (2008 est.)

Military expenditures—percent of GDP: NA

Military—note: Samoa has no formal defense structure or regular armed forces; informal defense ties exist with NZ, which is required to consider any Samoan request for assistance under the 1962 Treaty of Friendship

TRANSNATIONAL ISSUES

Disputes—international: none

SAN MARINO

INTRODUCTION

Background: The third smallest state in Europe (after the Holy See and Monaco), San Marino also claims to be the world's oldest republic. According to tradition, it was founded by a Christian stonemason named Marinus in A.D. 301. San Marino's foreign policy is aligned with that of Italy; social and political trends in the republic also track closely with those of its larger neighbor.

GEOGRAPHY

Location: Southern Europe, an enclave in central Italy

Geographic coordinates: 43 46 N, 12 25 E

Map references: Europe

Area:
total: 61.2 sq km
land: 61.2 sq km
water: 0 sq km

Area—comparative: about one third times the size of Washington, DC

Land boundaries:
total: 39 km
border countries: Italy 39 km

Coastline: 0 km (landlocked)

Maritime claims: none (landlocked)

Climate: Mediterranean; mild to cool winters; warm, sunny summers

Terrain: rugged mountains

Elevation extremes:
lowest point: Torrente Ausa 55 m
highest point: Monte Titano 755 m

Natural resources: building stone

Land use:
arable land: 16.67%
permanent crops: 0%
other: 83.33% (2005)

Irrigated land: NA

Natural hazards: NA

Environment—current issues: NA

Environment—international agreements: *party to:* Biodiversity, Climate Change, Desertification, Whaling
signed, but not ratified: Air Pollution

Geography—note: landlocked; smallest independent state in Europe after the Holy See and Monaco; dominated by the Apennines

PEOPLE

Population: 29,973 (July 2008 est.)

Age structure:
0–14 years: 16.8% (male 2,608/female 2,430)
15–64 years: 66% (male 9,464/female 10,304)
65 years and over: 17.2% (male 2,229/female 2,938) (2008 est.)

Median age:
total: 41.2 years
male: 40.9 years
female: 41.6 years (2008 est.)

Population growth rate: 1.181% (2008 est.)

Birth rate: 9.74 births/1,000 population (2008 est.)

Death rate: 8.37 deaths/1,000 population (2008 est.)

Net migration rate: 10.44 migrant(s)/1,000 population (2008 est.)

Sex ratio: *at birth:* 1.09 male(s)/female
under 15 years: 1.07 male(s)/female
15–64 years: 0.92 male(s)/female
65 years and over: 0.76 male(s)/female
total population: 0.91 male(s)/female (2008 est.)

Infant mortality rate:
total: 5.44 deaths/1,000 live births
male: 5.86 deaths/1,000 live births
female: 4.98 deaths/1,000 live births
(2008 est.)

Life expectancy at birth:
total population: 81.88 years
male: 78.43 years
female: 85.64 years (2008 est.)

Total fertility rate: 1.35 children born/woman (2008 est.)

HIV/AIDS—adult prevalence rate: NA

HIV/AIDS—people living with HIV/AIDS: NA

HIV/AIDS—deaths: NA

Nationality:
noun: Sammarinese (singular and plural)
adjective: Sammarinese

Ethnic groups: Sammarinese, Italian

Religions: Roman Catholic

Languages: Italian

Literacy:
definition: age 10 and over can read and write
total population: 96%
male: 97%
female: 95%

GOVERNMENT

Country name:
conventional long form: Republic of San Marino
conventional short form: San Marino
local long form: Repubblica di San Marino
local short form: San Marino

Government type: republic

Capital: *name:* San Marino
geographic coordinates: 43 56 N, 12 25 E
time difference: UTC+1 (6 hours ahead of Washington, DC during Standard Time)
daylight saving time: +1hr, begins last Sunday in March; ends last Sunday in October

Administrative divisions: 9 municipalities (castelli, singular—castello); Acquaviva, Borgo Maggiore, Chiesanuova, Domagnano, Faetano, Fiorentino, Montegiardino, San Marino Citta, Serravalle

Independence: 3 September AD 301

National holiday: Founding of the Republic, 3 September (AD 301)

Constitution: 8 October 1600; electoral law of 1926 serves some of the functions of a constitution

Legal system: based on civil law system with Italian law influences; has not accepted compulsory ICJ jurisdiction

Suffrage: 18 years of age; universal

Executive branch:
chief of state: Co-chiefs of State Captain Regent Federico Pedini AMATI and Captain Regent Rosa ZAFFERANI (for the period 1 April-30 September 2008)

head of government: Secretary of State for Foreign and Political Affairs Fiorenzo STOLFI (since 27 July 2006)
cabinet: Congress of State elected by the Great and General Council for a five-year term
elections: co-chiefs of state (captains regent) elected by the Great and General Council for a six-month term; election last held in September 2007 (next to be held in March 2008); secretary of state for foreign and political affairs elected by the Great and General Council for a five-year term; election last held 27 July 2006 (next to be held by 2011)
election results: Mirko TOMASSONI and Alberto SELVA elected captains regent; percent of legislative vote—NA; Fiorenzo STOLFI elected secretary of state for foreign and political affairs; percent of legislative vote—NA
note: the popularly elected parliament (Grand and General Council) selects two of its members to serve as the Captains Regent (co-chiefs of state) for a six-month period; they preside over meetings of the Grand and General Council and its cabinet (Congress of State), which has 10 other members, all selected by the Grand and General Council; assisting the captains regent are 10 secretaries of state; the secretary of state for Foreign Affairs has assumed some prime ministerial roles

Legislative branch: unicameral Grand and General Council or Consiglio Grande e Generale (60 seats; members are elected by popular vote to serve five-year terms)
elections: last held 4 June 2006 (next to be held by June 2011)
election results: percent of vote by party—PDCS 32.9%, Party of Socialists and Democrats 31.9%, AP 11.9%, United Left 8.7%, New Socialist Party 5.4%, other parties 9.2%; seats by party—PDCS 21, Party of Socialists and Democrats 20, AP 7, United Left 5, New Socialist Party 3, others 4; note—following a government reshuffle on 28 NOvember 2007, a splinter party of the PDCS joined the center-left coalition formed by the Party of Socialists and Democrats, the APDS, and the United Left strengthening the government's parliamentary majority to 36 seats out of 60

Judicial branch: Council of Twelve or Consiglio dei XII

Political parties and leaders: Communist Refoundation or RC [Ivan FOSHI]; Ideas in Movement or IM [Alessandro ROSSI]; National Alliance or AN [Glauco SANSOVINI]; New Socialist Party [Augusto CASALI]; Party of Socialists and Democrats [Claudio

FELICI]; San Marino Christian Democratic Party or PDCS [Pier Marino MENICUCCI]; San Marino Popular Alliance of Democrats or AP [Roberto GIORGETTI]; San Marino Socialist Party or PSS [Alberto CECCHETTI]; Socialists for Reform or SR [Renzo GIARDI]; United Left

Political pressure groups and leaders: NA

International organization participation: CE, FAO, IBRD, ICAO, ICCt, ICRM, IFRCS, ILO, IMF, IMO, Interpol, IOC, IOM (observer), IPU, ITU, ITUC, OPCW, OSCE, UN, UNCTAD, UNESCO, Union Latina, UNWTO, UPU, WHO, WIPO

Diplomatic representation in the US: San Marino does not have an embassy in the US
honorary consulate(s) general: New York, Washington, DC
honorary consulate(s): Detroit, Honolulu

Diplomatic representation from the US: the US does not have an embassy in San Marino; the ambassador to Italy is accredited to San Marino

Flag description: two equal horizontal bands of white (top) and light blue with the national coat of arms superimposed in the center; the coat of arms has a shield (featuring three towers on three peaks) flanked by a wreath, below a crown and above a scroll bearing the word LIBERTAS (Liberty)

ECONOMY

Economy—overview: The tourist sector contributes over 50% of GDP. In 2006 more than 2.1 million tourists visited San Marino. The key industries are banking, clothing and apparel, electronics, and ceramics. Main agricultural products are wine and cheeses. The per capita level of output and standard of living are comparable to those of the most prosperous regions of Italy, which supplies much of its food.

GDP (purchasing power parity): $850 million (2004 est.)

GDP (official exchange rate): $1.048 billion (2004)

GDP—real growth rate: 4.6% (2004 est.)

GDP—per capita (PPP): $34,100 (2004 est.)

GDP—composition by sector:
agriculture: NA%
industry: NA%
services: NA%

Labor force: 20,470 (2004)

Labor force—by occupation: *agriculture:* 0.2%
industry: 40.1%
services: 59.7% (2006 est.)

Unemployment rate: 3.8% (2004)
Population below poverty line: NA%
Household income or consumption by percentage share:
lowest 10%: NA%
highest 10%: NA%
Inflation rate (consumer prices): -1.5% (2006)
Budget: *revenues:* $709.6 million
expenditures: $672.3 million (2004)
Agriculture—products: wheat, grapes, corn, olives; cattle, pigs, horses, beef, cheese, hides
Industries: tourism, banking, textiles, electronics, ceramics, cement, wine
Industrial production growth rate: 5.6% (2005 est.)
Exports: $1.291 billion (2004)
Exports—commodities: building stone, lime, wood, chestnuts, wheat, wine, baked goods, hides, ceramics
Imports: $2.035 billion (2004)
Imports—commodities: wide variety of consumer manufactures, food
Economic aid—recipient: $NA
Debt—external: $NA
Market value of publicly traded shares: $NA
Currency (code): euro (EUR)
Currency code: EUR
Exchange rates: euros per US dollar— 0.7345 (2007), 0.7964 (2006), 0.8041 (2005), 0.8054 (2004), 0.886 (2003)
Fiscal year: calendar year

COMMUNICATIONS

Telephones—main lines in use: 21,000 (2006)
Telephones—mobile cellular: 17,390 (2006)
Telephone system:
general assessment: adequate connections
domestic: automatic telephone system completely integrated into Italian system; combined fixed-line and mobile-cellular teledensity exceeds 130 telephones per 100 persons
international: country code—378; connected to Italian international network
Radio broadcast stations: AM 0, FM 3, shortwave 0 (1998)
Radios: 16,000 (1997)
Television broadcast stations: 1 (San Marino residents also receive broadcasts from Italy) (1997)
Televisions: 9,000 (1997)
Internet country code: .sm
Internet hosts: 3,344 (2007)
Internet Service Providers (ISPs): 2 (2000)
Internet users: 15,400 (2006)

TRANSPORTATION

Roadways: *total:* 104 km
paved: 104 km (2003)

MILITARY

Military branches: no regular military forces; Voluntary Military Force (Corpi Militari Voluntar) performs ceremonial duties and limited police functions (2006)
Military service age and obligation: 16–55 for voluntary service in Voluntary Military Force (2006)
Manpower available for military service:
males age 16–49: 6,613 (2008 est.)
Manpower fit for military service:
males age 16–49: 5,345 (2008 est.)
Manpower reaching militarily significant age annually:
males age 16–49: 156 (2008 est.)
Military expenditures—percent of GDP: NA
Military—note: defense is the responsibility of Italy

TRANSNATIONAL ISSUES

Disputes—international: none

SAO TOME AND PRINCIPE

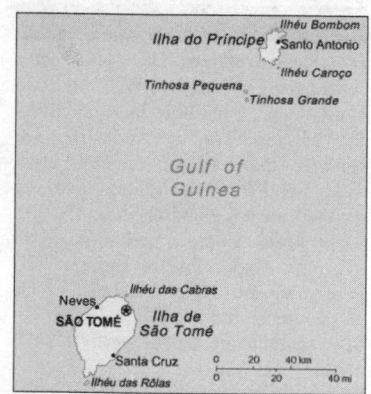

INTRODUCTION

Background: Discovered and claimed by Portugal in the late 15th century, the islands' sugar-based economy gave way to coffee and cocoa in the 19th century—all grown with plantation slave labor, a form of which lingered into the 20th century. While independence was achieved in 1975, democratic reforms were not instituted until the late 1980s. The country held its first free elections in 1991, but frequent internal wrangling between the various political parties precipitated repeated changes in leadership and two failed coup attempts in 1995 and 2003. The recent discovery of oil in the Gulf of Guinea promises to attract increased attention to the small island nation.

GEOGRAPHY

Location: Western Africa, islands in the Gulf of Guinea, straddling the Equator, west of Gabon
Geographic coordinates: 1 00 N, 7 00 E
Map references: Africa
Area:
total: 1,001 sq km
land: 1,001 sq km
water: 0 sq km
Area—comparative: more than five times the size of Washington, DC
Land boundaries: 0 km
Coastline: 209 km
Maritime claims: measured from claimed archipelagic baselines
territorial sea: 12 nm
exclusive economic zone: 200 nm
Climate: tropical; hot, humid; one rainy season (October to May)
Terrain: volcanic, mountainous

Elevation extremes:
lowest point: Atlantic Ocean 0 m
highest point: Pico de Sao Tome 2,024 m
Natural resources: fish, hydropower
Land use: *arable land:* 8.33%
permanent crops: 48.96%
other: 42.71% (2005)
Irrigated land: 100 sq km (2003)
Natural hazards: NA
Environment—current issues: deforestation; soil erosion and exhaustion
Environment—international agreements: *party to:* Biodiversity, Climate Change, Desertification, Endangered Species, Environmental Modification, Law of the Sea, Ozone Layer Protection, Ship Pollution, Wetlands
signed, but not ratified: none of the selected agreements
Geography—note: the smallest country in Africa; the two main islands form part of a chain of extinct volcanoes and both are mountainous

PEOPLE

Population: 206,178 (July 2008 est.)
Age structure:
0–14 years: 47.1% (male 49,196/female 47,941)

15–64 years: 49.3% (male 49,326/female 52,324)

65 years and over: 3.6% (male 3,350/female 4,041) (2008 est.)

Median age:

total: 16.3 years

male: 15.8 years

female: 16.9 years (2008 est.)

Population growth rate: 3.116% (2008 est.)

Birth rate: 39.12 births/1,000 population (2008 est.)

Death rate: 5.98 deaths/1,000 population (2008 est.)

Net migration rate: -1.97 migrant(s)/ 1,000 population (2008 est.)

Sex ratio:

at birth: 1.03 male(s)/female

under 15 years: 1.03 male(s)/female

15–64 years: 0.94 male(s)/female

65 years and over: 0.83 male(s)/female

total population: 0.98 male(s)/female (2008 est.)

Infant mortality rate:

total: 38.36 deaths/1,000 live births

male: 40.11 deaths/1,000 live births

female: 36.55 deaths/1,000 live births (2008 est.)

Life expectancy at birth:

total population: 68 years

male: 66.35 years

female: 69.69 years (2008 est.)

Total fertility rate: 5.43 children born/woman (2008 est.)

HIV/AIDS—adult prevalence rate: NA

HIV/AIDS—people living with HIV/AIDS: NA

HIV/AIDS—deaths: NA

Major infectious diseases:

degree of risk: high

food or waterborne diseases: bacterial diarrhea, hepatitis A, and typhoid fever

vectorborne disease: malaria (2008)

Nationality:

noun: Sao Tomean(s)

adjective: Sao Tomean

Ethnic groups: mestico, angolares (descendants of Angolan slaves), forros (descendants of freed slaves), servicais (contract laborers from Angola, Mozambique, and Cape Verde), tongas (children of servicais born on the islands), Europeans (primarily Portuguese)

Religions: Catholic 70.3%, Evangelical 3.4%, New Apostolic 2%, Adventist 1.8%, other 3.1%, none 19.4% (2001 census)

Languages: Portuguese (official)

Literacy:

definition: age 15 and over can read and write

total population: 84.9%

male: 92.2%

female: 77.9% (2001 census)

GOVERNMENT

Country name:

conventional long form: Democratic Republic of Sao Tome and Principe

conventional short form: Sao Tome and Principe

local long form: Republica Democratica de Sao Tome e Principe

local short form: Sao Tome e Principe

Government type: republic

Capital: *name:* Sao Tome

geographic coordinates: 0 12 N, 6 39 E

time difference: UTC 0 (5 hours ahead of Washington, DC during Standard Time)

Administrative divisions: 2 provinces; Principe, Sao Tome

note: Principe has had self government since 29 April 1995

Independence: 12 July 1975 (from Portugal)

National holiday: Independence Day, 12 July (1975)

Constitution: approved March 1990, effective 10 September 1990

Legal system: based on Portuguese legal system and customary law; has not accepted compulsory ICJ jurisdiction

Suffrage: 18 years of age; universal

Executive branch:

chief of state: President Fradique DE MENEZES (since 3 September 2001)

head of government: Prime Minister Patrice TROVOADA (since 14 February 2008)

cabinet: Council of Ministers appointed by the president on the proposal of the prime minister

elections: president elected by popular vote for a five-year term (eligible for a second term); election last held 30 July 2006 (next to be held July 2011); prime minister chosen by the National Assembly and approved by the president

election results: Fradique DE MENEZES elected president; percent of vote— Fradique DE MENEZES 60%, Patrice TROVOADA 38.5%

Legislative branch: unicameral National Assembly or Assembleia Nacional (55 seats; members are elected by popular vote to serve four-year terms)

elections: last held on 26 March 2006 (next to be held in March 2010)

election results: percent of vote by party— MDFM-PCD 37.2%, MLSTP 28.9%, ADI 20.0%, NR 4.7%, others 9.2%; seats by party—MDFM-PCD 23, MLSTP 19, ADI 12, NR 1

Judicial branch: Supreme Court (judges are appointed by the National Assembly)

Political parties and leaders: Force for Change Democratic Movement or MDFM [Tome Soares da VERA CRUZ]; Independent Democratic Action or ADI [[Patrice TROVOADA]; Movement for the Liberation of Sao Tome and Principe-Social Democratic Party or MLSTP-PSD [Rafael BRANCO]; New Way Movement or NR; Party for Democratic Convergence or PCD [Delfim NEVES]; Ue-Kedadji coalition; other small parties

Political pressure groups and leaders: NA

International organization participation: ACCT, ACP, AfDB, AU, CPLP, FAO, G-77, IBRD, ICAO, ICCt (signatory), ICRM, IDA, IFAD, IFRCS, ILO, IMF, IMO, Interpol, IOC, IOM (observer), IPU, ITU, ITUC, NAM, OIF, OPCW, UN, UNCTAD, UNESCO, UNIDO, Union Latina, UNWTO, UPU, WCL, WHO, WIPO, WMO, WTO (observer)

Diplomatic representation in the US:

chief of mission: First Secretary Domingos Augusto FERREIRA

chancery: 400 Park Avenue, 7th Floor, New York, NY 10022

telephone: [1] (212) 317-0580

FAX: [1] (212) 935-7348

consulate(s): Atlanta

Diplomatic representation from the US: the US does not have an embassy in Sao Tome and Principe; the Ambassador to Gabon is accredited to Sao Tome and Principe on a nonresident basis and makes periodic visits to the islands

Flag description: three horizontal bands of green (top), yellow (double width), and green with two black five-pointed stars placed side by side in the center of the yellow band and a red isosceles triangle based on the hoist side; uses the popular pan-African colors of Ethiopia

ECONOMY

Economy—overview: This small, poor island economy has become increasingly dependent on cocoa since independence in 1975. Cocoa production has substantially declined in recent years because of drought and mismanagement. Sao Tome has to import all fuels, most manufactured goods, consumer goods, and a substantial amount of food. Over the years, it has had difficulty servicing its external debt and has relied heavily on concessional aid and debt rescheduling. Sao Tome benefited from $200 million in debt relief in December 2000 under the Highly Indebted Poor Countries (HIPC) program, which helped bring down the country's $300 million debt burden. In August 2005, Sao Tome signed on to a new 3-year IMF Poverty Reduction and Growth Facility (PRGF) program worth $4.3 million. Considerable potential exists for development of a tourist industry, and the government has taken steps to expand facilities in recent years. The government also has attempted to reduce price controls and subsidies. Sao Tome is optimistic about the development of petroleum resources in its territorial waters in the oil-rich Gulf of Guinea, which are being jointly developed in a 60-40 split with Nigeria. The first production licenses were sold in 2004, though a dispute over licensing with Nigeria delayed Sao Tome's receipt of more than $20 million in signing bonuses for almost a year. Real GDP growth exceeded 6% in 2007, as a result of increases in public expenditures and oil-related capital investment.

GDP (purchasing power parity): $256 million (2007 est.)

GDP (official exchange rate): $144 million (2007 est.)

GDP—real growth rate: 6% (2007 est.)

GDP—per capita (PPP): $1,600 (2007 est.)

GDP—composition by sector:
agriculture: 14.9%
industry: 14%
services: 71% (2007 est.)

Labor force: 35,050 (1991)

Labor force—by occupation: *note:* population mainly engaged in subsistence agriculture and fishing; shortages of skilled workers

Unemployment rate: NA%

Population below poverty line: 54% (2004 est.)

Household income or consumption by percentage share:
lowest 10%: NA%
highest 10%: NA%

Inflation rate (consumer prices): 19.9% (2007 est.)

Investment (gross fixed): 43.5% of GDP (2007 est.)

Budget:
revenues: $74.59 million
expenditures: $57.25 million (2007 est.)

Agriculture—products: cocoa, coconuts, palm kernels, copra, cinnamon, pepper, coffee, bananas, papayas, beans; poultry; fish

Industries: light construction, textiles, soap, beer, fish processing, timber

Industrial production growth rate: 7% (2007 est.)

Electricity—production: 18 million kWh (2005)

Electricity—production by source:
fossil fuel: 41.2%
hydro: 58.8%
nuclear: 0%
other: 0% (2001)

Electricity—consumption: 16.74 million kWh (2005)

Electricity—exports: 0 kWh (2005)

Electricity—imports: 0 kWh (2005)

Oil—production: 0 bbl/day (2005 est.)

Oil—consumption: 650 bbl/day (2005 est.)

Oil—exports: 0 bbl/day (2004)

Oil—imports: 634.4 bbl/day (2004)

Oil—proved reserves: 0 bbl (1 January 2006 est.)

Natural gas—production: 0 cu m (2005 est.)

Natural gas—consumption: 0 cu m (2005 est.)

Natural gas—exports: 0 cu m (2005 est.)

Natural gas—imports: 0 cu m (2005)

Natural gas—proved reserves: 0 cu m (1 January 2006 est.)

Current account balance: -$51 million (2007 est.)

Exports: $4.384 million f.o.b. (2007 est.)

Exports—commodities: cocoa 80%, copra, coffee, palm oil

Exports—partners: Netherlands 47.9%, Belgium 19%, Portugal 9.3% (2006)

Imports: $71.2 million f.o.b. (2007 est.)

Imports—commodities: machinery and electrical equipment, food products, petroleum products

Imports—partners: Portugal 48.8%, France 19.7%, Belgium 5.1%, US 5.1% (2006)

Economic aid—recipient: $31.9 million in December 2000 under the Heavily Indebted Poor Country Initiative (HIPC) program (2005)

Reserves of foreign exchange and gold: $34.6 million (31 December 2007 est.)

Debt—external: $318 million (2002)

Market value of publicly traded shares: $NA

Currency (code): dobra (STD)

Currency code: STD

Exchange rates: dobras per US dollar—13,700 (2007), 12,050 (2006), 9,900.4 (2005), 9,902.3 (2004), 9,347.6 (2003)

Fiscal year: calendar year

COMMUNICATIONS

Telephones—main lines in use: 7,100 (2005)

Telephones—mobile cellular: 12,000 (2005)

Telephone system:
general assessment: adequate facilities
domestic: minimal system
international: country code—239; satellite earth station—1 Intelsat (Atlantic Ocean)

Radio broadcast stations: AM 1, FM 5, shortwave 1 (2001)

Radios: 38,000 (1997)

Television broadcast stations: 2 (2001)

Televisions: 23,000 (1997)

Internet country code: .st

Internet hosts: 996 (2007)

Internet Service Providers (ISPs): 1 (2002)

Internet users: 23,000 (2005)

TRANSPORTATION

Airports: 2 (2007)

Airports—with paved runways:
total: 2
1,524 to 2,437 m: 1
914 to 1,523 m: 1 (2007)

Roadways:
total: 320 km
paved: 218 km
unpaved: 102 km (1999)

Merchant marine:
total: 7 ships (1000 GRT or over) 20,455 GRT/27,871 DWT
by type: bulk carrier 1, cargo 6
foreign-owned: 2 (Egypt 1, Greece 1) (2007)

Ports and terminals: Sao Tome

MILITARY

Military branches: Armed Forces of Sao Tome and Principe (FASTP): Army, Coast Guard of Sao Tome e Principe (Guarda Costeira de Sao Tome e Principe, GCSTP), Presidential Guard (2007)

Military service age and obligation: 18 years of age (est.) (2004)

Manpower available for military service:
males age 16–49: 42,340
females age 16–49: 43,781 (2008 est.)

Manpower fit for military service:
males age 16–49: 33,735
females age 16–49: 36,779 (2008 est.)

Military expenditures—percent of GDP: 0.8% (2006)

Military—note: Sao Tome and Principe's army is a tiny force with almost no resources at its disposal and would be wholly ineffective operating unilaterally; infantry equipment is considered simple to operate and maintain but may require refurbishment or replacement after 25 years in tropical climates; poor pay, working conditions, and alleged nepotism in the promotion of officers have been problems in the past, as reflected in the 1995 and 2003 coups; these issues are being addressed with foreign assistance aimed at improving the army and its focus on realistic security concerns; command is exercised from the president, through the Minister of Defense, to the Chief of the Armed Forces staff (2005)

TRANSNATIONAL ISSUES

Disputes—international: none

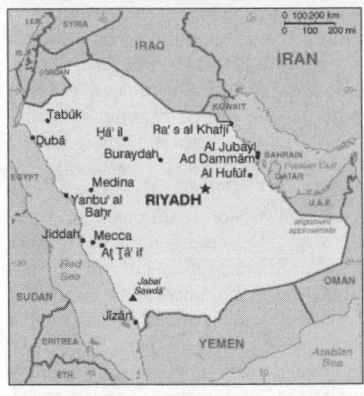

INTRODUCTION

Background: Saudi Arabia is the birthplace of Islam and home to Islam's two holiest shrines in Mecca and Medina. The king's official title is the Custodian of the Two Holy Mosques. The modern Saudi state was founded in 1932 by ABD AL-AZIZ bin Abd al-Rahman AL SAUD (Ibn Saud) after a 30-year campaign to unify most of the Arabian Peninsula. A male descendent of Ibn Saud, his son ABDALLAH bin Abd al-Aziz, rules the country today as required by the country's 1992 Basic Law. Following Iraq's invasion of Kuwait in 1990, Saudi Arabia accepted the Kuwaiti royal family and 400,000 refugees while allowing Western and Arab troops to deploy on its soil for the liberation of Kuwait the following year. The continuing presence of foreign troops on Saudi soil after the liberation of Kuwait became a source of tension between the royal family and the public until all operational US troops left the country in 2003. Major terrorist attacks in May and November 2003 spurred a strong ongoing campaign against domestic terrorism and extremism. King ABDALLAH has continued the cautious reform program begun when he was crown prince. To promote increased political participation, the government held elections nationwide from February through April 2005 for half the members of 179 municipal councils. In December 2005, King ABDALLAH completed the process by appointing the remaining members of the advisory municipal councils. The country remains a leading producer of oil and natural gas and holds more than 20% of the world's proven oil reserves. The government continues to pursue economic reform and diversification, particularly since Saudi Arabia's accession to the WTO in December 2005, and promotes foreign investment in the kingdom. A burgeoning population, aquifer depletion, and an economy largely dependent on petroleum output and prices are all ongoing governmental concerns.

GEOGRAPHY

Location: Middle East, bordering the Persian Gulf and the Red Sea, north of Yemen

Geographic coordinates: 25 00 N, 45 00 E

Map references: Middle East

Area: *total:* 2,149,690 sq km
land: 2,149,690 sq km
water: 0 sq km

Area—comparative: slightly more than one-fifth the size of the US

Land boundaries:
total: 4,431 km
border countries: Iraq 814 km, Jordan 744 km, Kuwait 222 km, Oman 676 km, Qatar 60 km, UAE 457 km, Yemen 1,458 km

Coastline: 2,640 km

Maritime claims:
territorial sea: 12 nm
contiguous zone: 18 nm
continental shelf: not specified

Climate: harsh, dry desert with great temperature extremes

Terrain: mostly uninhabited, sandy desert

Elevation extremes:
lowest point: Persian Gulf 0 m
highest point: Jabal Sawda' 3,133 m

Natural resources: petroleum, natural gas, iron ore, gold, copper

Land use:
arable land: 1.67%
permanent crops: 0.09%
other: 98.24% (2005)

Irrigated land: 16,200 sq km (2003)

Total renewable water resources: 2.4 cu km (1997)

Freshwater withdrawal (domestic/industrial/agricultural): *total:* 17.32 cu km/yr (10%/1%/89%)
per capita: 705 cu m/yr (2000)

Natural hazards: frequent sand and dust storms

Environment—current issues: desertification; depletion of underground water resources; the lack of perennial rivers or permanent water bodies has prompted the development of extensive seawater desalination facilities; coastal pollution from oil spills

Environment—international agreements: *party to:* Biodiversity, Climate Change, Climate Change-Kyoto Protocol, Desertification, Endangered Species, Hazardous Wastes, Law of the Sea, Marine Dumping, Ozone Layer Protection, Ship Pollution
signed, but not ratified: none of the selected agreements

Geography—note: extensive coastlines on Persian Gulf and Red Sea provide great leverage on shipping (especially crude oil) through Persian Gulf and Suez Canal

PEOPLE

Population: 28,161,417
note: includes 5,576,076 non-nationals (July 2008 est.)

Age structure:
0–14 years: 38.1% (male 5,469,641/female 5,258,508)
15–64 years: 59.5% (male 9,467,325/female 7,284,077)
65 years and over: 2.4% (male 355,173/female 326,693) (2008 est.)

Median age:
total: 21.5 years
male: 22.9 years
female: 19.7 years (2008 est.)

Population growth rate: 1.945% (2008 est.)

Birth rate: 28.83 births/1,000 population (2008 est.)

Death rate: 2.52 deaths/1,000 population (2008 est.)

Net migration rate: -6.86 migrant(s)/1,000 population (2008 est.)

Sex ratio:
at birth: 1.05 male(s)/female
under 15 years: 1.04 male(s)/female
15–64 years: 1.3 male(s)/female
65 years and over: 1.09 male(s)/female
total population: 1.19 male(s)/female (2008 est.)

Infant mortality rate:
total: 12.01 deaths/1,000 live births
male: 13.79 deaths/1,000 live births
female: 10.15 deaths/1,000 live births (2008 est.)

Life expectancy at birth:
total population: 76.09 years
male: 74.04 years
female: 78.25 years (2008 est.)

Total fertility rate: 3.89 children born/woman (2008 est.)

HIV/AIDS—adult prevalence rate: 0.01% (2001 est.)

HIV/AIDS—people living with HIV/AIDS: NA

HIV/AIDS—deaths: NA

Nationality:
noun: Saudi(s)
adjective: Saudi or Saudi Arabian

Ethnic groups: Arab 90%, Afro-Asian 10%
Religions: Muslim 100%
Languages: Arabic
Literacy:
definition: age 15 and over can read and write
total population: 78.8%
male: 84.7%
female: 70.8% (2003 est.)

GOVERNMENT

Country name:
conventional long form: Kingdom of Saudi Arabia
conventional short form: Saudi Arabia
local long form: Al Mamlakah al Arabiyah as Suudiyah
local short form: Al Arabiyah as Suudiyah
Government type: monarchy
Capital: *name:* Riyadh
geographic coordinates: 24 38 N, 46 43 E
time difference: UTC+3 (8 hours ahead of Washington, DC during Standard Time)
Administrative divisions: 13 provinces (mintaqat, singular—mintaqah); Al Bahah, Al Hudud ash Shamaliyah, Al Jawf, Al Madinah, Al Qasim, Ar Riyad, Ash Sharqiyah (Eastern Province), 'Asir, Ha'il, Jizan, Makkah, Najran, Tabuk
Independence: 23 September 1932 (unification of the kingdom)
National holiday: Unification of the Kingdom, 23 September (1932)
Constitution: governed according to Islamic law; the Basic Law that articulates the government's rights and responsibilities was promulgated by royal decree in 1992
Legal system: based on Shari'a law, several secular codes have been introduced; commercial disputes handled by special committees; has not accepted compulsory ICJ jurisdiction
Suffrage: 21 years of age; male
Executive branch:
chief of state: King and Prime Minister ABDALLAH bin Abd al-Aziz Al Saud (since 1 August 2005); Heir Apparent Crown Prince SULTAN bin Abd al-Aziz Al Saud (half brother of the monarch, born 5 January 1928); note—the monarch is both the chief of state and head of government
head of government: King and Prime Minister ABDALLAH bin Abd al-Aziz Al Saud (since 1 August 2005)
cabinet: Council of Ministers is appointed by the monarch every four years and includes many royal family members
elections: none; the monarchy is hereditary; note—a new Allegiance Commission created by royal decree in October 2006 established a committee of

Saudi princes that will play a role in selecting future Saudi kings, but the new system will not take effect until after Crown Prince Sultan becomes king
Legislative branch: Consultative Council or Majlis al-Shura (150 members and a chairman appointed by the monarch for four-year terms); note—though the Council of Ministers announced in October 2003 its intent to introduce elections for half of the members of local and provincial assemblies and a third of the members of the national Consultative Council or Majlis al-Shura, incrementally over a period of four to five years, to date no such elections have been held or announced
Judicial branch: Supreme Council of Justice
Political parties and leaders: none
Political pressure groups and leaders: none
International organization participation: ABEDA, AfDB, AFESD, AMF, BIS, FAO, G-77, GCC, IAEA, IBRD, ICAO, ICC, ICRM, IDB, IFAD, IFC, IFRCS, IHO, ILO, IMF, IMO, IMSO, Interpol, IOC, IPU, ISO, ITSO, ITU, LAS, MIGA, NAM, OAPEC, OAS (observer), OIC, OPCW, OPEC, PCA, UN, UNCTAD, UNESCO, UNIDO, UNRWA, UNWTO, UPU, WCO, WFTU, WHO, WIPO, WMO, WTO
Diplomatic representation in the US:
chief of mission: Ambassador Adil al-Ahmad al-JUBAYR
chancery: 601 New Hampshire Avenue NW, Washington, DC 20037
telephone: [1] (202) 342-3800
FAX: [1] (202) 944-3113
consulate(s) general: Houston, Los Angeles, New York
Diplomatic representation from the US:
chief of mission: Ambassador Ford FRAKER
embassy: Collector Road M, Diplomatic Quarter, Riyadh
mailing address: American Embassy, Unit 61307, APO AE 09803-1307; International Mail: P. O. Box 94309, Riyadh 11693
telephone: [966] (1) 488-3800
FAX: [966] (1) 488-7360
consulate(s) general: Dhahran, Jiddah (Jeddah)
Flag description: green, a traditional color in Islamic flags, with the Shahada or Muslim creed in large white Arabic script (translated as "There is no god but God; Muhammad is the Messenger of God") above a white horizontal saber (the tip points to the hoist side); design dates to the early twentieth century and is closely associated with the Al Saud family which established the kingdom in 1932

ECONOMY

Economy—overview: Saudi Arabia has an oil-based economy with strong government controls over major economic activities. It possesses more than 20% of the world's proven petroleum reserves, ranks as the largest exporter of petroleum, and plays a leading role in OPEC. The petroleum sector accounts for roughly 75% of budget revenues, 45% of GDP, and 90% of export earnings. About 40% of GDP comes from the private sector. Roughly 5.5 million foreign workers play an important role in the Saudi economy, particularly in the oil and service sectors. High oil prices have boosted growth, government revenues, and Saudi ownership of foreign assets, while enabling Riyadh to pay down domestic debt. The government is encouraging private sector growth—especially in power generation, telecommunications, natural gas exploration, and petrochemicals—to lessen the kingdom's dependence on oil exports and to increase employment opportunities for the swelling Saudi population, nearly 40% of which are youths under 15 years old. Unemployment is high, and the large youth population generally lacks the education and technical skills the private sector needs. Riyadh has substantially boosted spending on job training and education, infrastructure development, and government salaries. As part of its effort to attract foreign investment and diversify the economy, Saudi Arabia acceded to the WTO in December 2005 after many years of negotiations. The government has announced plans to establish six "economic cities" in different regions of the country to promote development and diversification.
GDP (purchasing power parity): $564.6 billion (2007 est.)
GDP (official exchange rate): $376 billion (2007 est.)
GDP—real growth rate: 4.1% (2007 est.)
GDP—per capita (PPP): $23,200 (2007 est.)
GDP—composition by sector:
agriculture: 3%
industry: 65.9%
services: 31.1% (2007 est.)
Labor force: 6.563 million
note: about one-third of the population in the 15–64 age group is non-national (2007 est.)
Labor force—by occupation: *agriculture:* 12%
industry: 25%
services: 63% (1999 est.)

Unemployment rate: 13% among Saudi males only (local bank estimate; some estimates range as high as 25%) (2004 est.)

Population below poverty line: NA%

Household income or consumption by percentage share:
lowest 10%: NA%
highest 10%: NA%

Inflation rate (consumer prices): 4.1% (2007 est.)

Investment (gross fixed): 18.8% of GDP (2007 est.)

Budget: *revenues:* $165.4 billion
expenditures: $118.3 billion (2007 est.)

Public debt: 23.3% of GDP (2007 est.)

Agriculture—products: wheat, barley, tomatoes, melons, dates, citrus; mutton, chickens, eggs, milk

Industries: crude oil production, petroleum refining, basic petrochemicals, ammonia, industrial gases, sodium hydroxide (caustic soda), cement, fertilizer, plastics, metals, commercial ship repair, commercial aircraft repair, construction

Industrial production growth rate: -1.1% (2007 est.)

Electricity—production: 165.6 billion kWh (2005)

Electricity—production by source:
fossil fuel: 100%
hydro: 0%
nuclear: 0%
other: 0% (2001)

Electricity—consumption: 146.9 billion kWh (2005)

Electricity—exports: 0 kWh (2005)
Electricity—imports: 0 kWh (2005)

Oil—production: 11 million bbl/day (2007 est.)

Oil—consumption: 2 million bbl/day (2005)

Oil—exports: 8.9 million bbl/day (2007 est.)

Oil—imports: 0 bbl/day (2004)

Oil—proved reserves: 264.3 billion bbl (2007 est.)

Natural gas—production: 68.32 billion cu m (2005 est.)

Natural gas—consumption: 68.32 billion cu m (2005 est.)

Natural gas—exports: 0 cu m (2005 est.)

Natural gas—imports: 0 cu m (2005)

Natural gas—proved reserves: 6.568 trillion cu m (1 January 2006 est.)

Current account balance: $100.8 billion (2007 est.)

Exports: $230 billion f.o.b. (2007 est.)

Exports—commodities: petroleum and petroleum products 90%

Exports—partners: Japan 17.7%, US 15.9%, South Korea 9.1%, China 7.2%, Taiwan 4.7%, Singapore 4.5% (2006)

Imports: $81.17 billion f.o.b. (2007 est.)

Imports—commodities: machinery and equipment, foodstuffs, chemicals, motor vehicles, textiles

Imports—partners: US 12.3%, Germany 8.6%, China 7.9%, Japan 7.3%, UK 4.9%, Italy 4.8%, South Korea 4.1% (2006)

Economic aid—donor: since 2002, Saudi Arabia has provided more than $480 million in budgetary support to the Palestinian Authority, supported Palestinian refugees through contributions to the UN Relief and Works Agency (UNRWA), provided more than $250 million to Arab League funds for the Palestinians, and pledged $500 million in assistance over the next three years at the Donors Conference in Dec 2007; pledged $230 million to development in Afghanistan; pledged $1 billion in export guarantees and soft loans to Iraq; pledged $133 million in direct grant aid, $187 million in concessional loans, and $153 million in export credits for Pakistan earthquake relief; pledged a total of $1.59 billion to Lebanon in assistance and deposits to the Central Bank of Lebanon in 2006 and pledged an additional $1.1 billion in early 2007

Economic aid—recipient: $26.29 million (2005)

Reserves of foreign exchange and gold: $34.01 billion (31 December 2007 est.)

Debt—external: $52.08 billion (31 December 2007 est.)

Stock of direct foreign investment—at home: $NA

Stock of direct foreign investment—abroad: $NA

Market value of publicly traded shares: $326.9 billion (2006)

Currency (code): Saudi riyal (SAR)
Currency code: SAR

Exchange rates: Saudi riyals per US dollar—3.745 (2007), 3.745 (2006), 3.747 (2005), 3.75 (2004), 3.75 (2003)

Fiscal year: calendar year

COMMUNICATIONS

Telephones—main lines in use: 4.5 million (2006)

Telephones—mobile cellular: 19.663 million (2006)

Telephone system:
general assessment: modern system
domestic: extensive microwave radio relay, coaxial cable, and fiber-optic cable systems; mobile-cellular subscribership has been increasing rapidly
international: country code—966; landing point for the international submarine cable Fiber-Optic Link Around the Globe (FLAG) and for both the SEA-ME-WE-3 and SEA-ME-WE-4 submarine cable networks providing connectivity to Asia, Middle East, Europe, and US; microwave radio relay to Bahrain, Jordan, Kuwait, Qatar, UAE, Yemen, and Sudan; coaxial cable to Kuwait and Jordan; satellite earth stations—5 Intelsat (3 Atlantic Ocean and 2 Indian Ocean), 1 Arabsat, and 1 Inmarsat (Indian Ocean region)

Radio broadcast stations: AM 43, FM 31, shortwave 2 (1998)

Radios: 6.25 million (1997)

Television broadcast stations: 117 (1997)

Televisions: 5.1 million (1997)

Internet country code: .sa

Internet hosts: 18,369 (2007)

Internet Service Providers (ISPs): 22 (2003)

Internet users: 4.7 million (2006)

TRANSPORTATION

Airports: 213 (2007)

Airports—with paved runways:
total: 77
over 3,047 m: 32
2,438 to 3,047 m: 15
1,524 to 2,437 m: 26
914 to 1,523 m: 2
under 914 m: 2 (2007)

Airports—with unpaved runways:
total: 136
over 3,047 m: 1
2,438 to 3,047 m: 8
1,524 to 2,437 m: 73
914 to 1,523 m: 39
under 914 m: 15 (2007)

Heliports: 8 (2007)

Pipelines: condensate 212 km; gas 1,880 km; liquid petroleum gas 1,183 km; oil 4,521 km; refined products 1,148 km (2007)

Railways: *total:* 1,392 km
standard gauge: 1,392 km 1.435-m gauge (with branch lines and sidings) (2006)

Roadways:
total: 152,044 km
paved: 45,461 km
unpaved: 106,583 km (2000)

Merchant marine:
total: 59 ships (1000 GRT or over) 847,094 GRT/1,059,026 DWT
by type: cargo 5, chemical tanker 15, container 4, passenger/cargo 8, petroleum tanker 16, refrigerated cargo 3, roll on/roll off 8
foreign-owned: 10 (Egypt 1, Greece 2, Kuwait 6, UAE 1)
registered in other countries: 63 (Bahamas 15, Comoros 1, Dominica 1, France 1, Liberia 24, Marshall Islands 4, Norway 3, Panama 14) (2007)

Ports and terminals: Ad Dammam, Al Jubayl, Jiddah, Yanbu' al Sinaiyah

MILITARY

Military branches: Land Forces (Army), Navy, Air Force, Air Defense Force, National Guard, Ministry of Interior Forces (paramilitary)

Military service age and obligation: 18 years of age (est.); no conscription (2004)

Manpower available for military service:
males age 16–49: 8,547,441
females age 16–49: 6,381,098 (2008 est.)

Manpower fit for military service:
males age 16–49: 7,398,417
females age 16–49: 5,525,357 (2008 est.)

Manpower reaching militarily significant age annually:
males age 16–49: 272,046
females age 16–49: 261,991 (2008 est.)

Military expenditures—percent of GDP: 10% (2005 est.)

TRANSNATIONAL ISSUES

Disputes—international: Saudi Arabia has reinforced its concrete-filled security barrier along sections of the now fully demarcated border with Yemen to stem illegal cross-border activities; Kuwait and Saudi Arabia continue discussions on a maritime boundary with Iran

Refugees and internally displaced persons: *refugees (country of origin):* 240,015 (Palestinian Territories) (2007)

Trafficking in persons: *current situation:* Saudi Arabia is a destination country for workers from South and Southeast Asia who are subjected to conditions that constitute involuntary servitude including being subjected to physical and sexual abuse, non-payment of wages, confinement, and withholding of passports as a restriction on their movement; domestic workers are particularly vulnerable because some are confined to the house in which they work unable to seek help; Saudi Arabia is also a destination country for Nigerian, Yemeni, Pakistani, Afghan, Somali, Malian, and Sudanese children trafficked for forced begging and involuntary servitude as street vendors; some Nigerian women were reportedly trafficked into Saudi Arabia for commercial sexual exploitation

tier rating: Tier 3—Saudi Arabia does not fully comply with the minimum standards for the elimination of trafficking and is not making significant efforts to do so

Illicit drugs: death penalty for traffickers; improving anti-money-laundering legislation and enforcement

SENEGAL

INTRODUCTION

Background: The French colonies of Senegal and the French Sudan were merged in 1959 and granted their independence as the Mali Federation in 1960. The union broke up after only a few months. Senegal joined with The Gambia to form the nominal confederation of Senegambia in 1982, but the envisaged integration of the two countries was never carried out, and the union was dissolved in 1989. The Movement of Democratic Forces in the Casamance (MFDC) has led a low-level separatist insurgency in southern Senegal since the 1980s, and several peace deals have failed to resolve the conflict. Nevertheless, Senegal remains one of the most stable democracies in Africa. Senegal was ruled by a Socialist Party for 40 years until current President Abdoulaye WADE was elected in 2000.

He was reelected in February 2007, but complaints of fraud led opposition parties to boycott June 2007 legislative polls. Senegal has a long history of participating in international peacekeeping.

GEOGRAPHY

Location: Western Africa, bordering the North Atlantic Ocean, between Guinea-Bissau and Mauritania

Geographic coordinates: 14 00 N, 14 00 W

Map references: Africa

Area: *total:* 196,190 sq km
land: 192,000 sq km
water: 4,190 sq km

Area—comparative: slightly smaller than South Dakota

Land boundaries:
total: 2,640 km
border countries: The Gambia 740 km, Guinea 330 km, Guinea-Bissau 338 km, Mali 419 km, Mauritania 813 km

Coastline: 531 km

Maritime claims:
territorial sea: 12 nm
contiguous zone: 24 nm
exclusive economic zone: 200 nm
continental shelf: 200 nm or to the edge of the continental margin

Climate: tropical; hot, humid; rainy season (May to November) has strong southeast winds; dry season (December to April) dominated by hot, dry, harmattan wind

Terrain: generally low, rolling, plains rising to foothills in southeast

Elevation extremes:
lowest point: Atlantic Ocean 0 m

highest point: unnamed feature near Nepen Diakha 581 m

Natural resources: fish, phosphates, iron ore

Land use: *arable land:* 12.51%
permanent crops: 0.24%
other: 87.25% (2005)

Irrigated land: 1,200 sq km (2003)

Total renewable water resources: 39.4 cu km (1987)

Freshwater withdrawal (domestic/industrial/agricultural): *total:* 2.22 cu km/yr (4%/3%/93%)
per capita: 190 cu m/yr (2002)

Natural hazards: lowlands seasonally flooded; periodic droughts

Environment—current issues: wildlife populations threatened by poaching; deforestation; overgrazing; soil erosion; desertification; overfishing

Environment—international agreements: *party to:* Biodiversity, Climate Change, Climate Change-Kyoto Protocol, Desertification, Endangered Species, Hazardous Wastes, Law of the Sea, Marine Life Conservation, Ozone Layer Protection, Ship Pollution, Wetlands, Whaling

Geography—note: westernmost country on the African continent; The Gambia is almost an enclave within Senegal

PEOPLE

Population: 12,853,259 (July 2008 est.)

Age structure:
0–14 years: 41.9% (male 2,717,257/female 2,668,602)
15–64 years: 55.1% (male 3,524,683/female 3,552,643)

65 years and over: 3% (male 183,188/female 206,886) (2008 est.)

Median age:

total: 18.8 years

male: 18.6 years

female: 19 years (2008 est.)

Population growth rate: 2.58% (2008 est.)

Birth rate: 36.52 births/1,000 population (2008 est.)

Death rate: 10.72 deaths/1,000 population (2008 est.)

Net migration rate: 0 migrant(s)/1,000 population (2008 est.)

Sex ratio:

at birth: 1.03 male(s)/female

under 15 years: 1.02 male(s)/female

15–64 years: 0.99 male(s)/female

65 years and over: 0.89 male(s)/female

total population: 1 male(s)/female (2008 est.)

Infant mortality rate:

total: 58.93 deaths/1,000 live births

male: 62.79 deaths/1,000 live births

female: 54.96 deaths/1,000 live births (2008 est.)

Life expectancy at birth:

total population: 57.08 years

male: 55.7 years

female: 58.5 years (2008 est.)

Total fertility rate: 4.86 children born/woman (2008 est.)

HIV/AIDS—adult prevalence rate: 0.8% (2003 est.)

HIV/AIDS—people living with HIV/AIDS: 44,000 (2003 est.)

HIV/AIDS—deaths: 3,500 (2003 est.)

Major infectious diseases:

degree of risk: very high

food or waterborne diseases: bacterial and protozoal diarrhea, hepatitis A, and typhoid fever

vectorborne diseases: Crimean-Congo hemorrhagic fever, dengue fever, malaria, Rift Valley fever, and yellow fever

water contact disease: schistosomiasis

respiratory disease: meningococcal meningitis (2008)

Nationality:

noun: Senegalese (singular and plural)

adjective: Senegalese

Ethnic groups: Wolof 43.3%, Pular 23.8%, Serer 14.7%, Jola 3.7%, Mandinka 3%, Soninke 1.1%, European and Lebanese 1%, other 9.4%

Religions: Muslim 94%, Christian 5% (mostly Roman Catholic), indigenous beliefs 1%

Languages: French (official), Wolof, Pulaar, Jola, Mandinka

Literacy: *definition:* age 15 and over can read and write

total population: 39.3%

male: 51.1%

female: 29.2% (2002 est.)

GOVERNMENT

Country name: *conventional long form:* Republic of Senegal

conventional short form: Senegal

local long form: Republique du Senegal

local short form: Senegal

former: Senegambia (along with The Gambia), Mali Federation

Government type: republic

Capital: *name:* Dakar

geographic coordinates: 14 40 N, 17 26 W

time difference: UTC 0 (5 hours ahead of Washington, DC during Standard Time)

Administrative divisions: 11 regions (regions, singular—region); Dakar, Diourbel, Fatick, Kaolack, Kolda, Louga, Matam, Saint-Louis, Tambacounda, Thies, Ziguinchor

Independence: 4 April 1960 (from France); note—complete independence achieved upon dissolution of federation with Mali on 20 August 1960

National holiday: Independence Day, 4 April (1960)

Constitution: adopted 7 January 2001

Legal system: based on French civil law system; judicial review of legislative acts in Constitutional Court; the Council of State audits the government's accounting office; accepts compulsory ICJ jurisdiction with reservations

Suffrage: 18 years of age; universal

Executive branch:

chief of state: President Abdoulaye WADE (since 1 April 2000)

head of government: Prime Minister Cheikh Hadjibou SOUMARE (since 19 June 2007)

cabinet: Council of Ministers appointed by the prime minister in consultation with the president

elections: president elected by popular vote for a five-year term (eligible for a second term) under new constitution; election last held on 25 February 2007 (next to be held in 2012); prime minister appointed by the president

election results: Abdoulaye WADE reelected president in the first round of voting; percent of vote—Abdoulaye WADE 55.9%, Idrissa SECK 14.9%, Ousmane Tanor DIENG 13.6%, Moustapha NIASSE 5.9%, other 9.7%

Legislative branch: bicameral Parliament consisting of the National Assembly or Assemblee Nationale (150 seats; 90 members elected by direct popular vote with the remaining members elected by proportional representation from party lists to serve five-year terms) and the Senate reinstituted in 2007 (100 seats; 35 indirectly elected with the remaining 65 members to be appointed by the president)

elections: National Assembly—last held on 3 June 2007 (next to be held 2012); note—the National Assembly in December 2005 voted to postpone legislative elections originally scheduled for 2006; legislative elections were first rescheduled to coincide with the 25 February 2007 presidential elections and later rescheduled for 3 June 2007; the June election was boycotted by 12 opposition parties, including the former ruling Socialist Party, that resulted in a record-low, 35-percent voter turnout; Senate—last held 19 August 2007 (next to be held—NA)

election results: National Assembly results—percent of vote by party—NA; seats by party—SOPI Coalition 131, other 19; Senate results—percent of vote by party—NA; seats by party—PDS 34, AJ/PADS 1, 65 appointed by the president

Judicial branch: Constitutional Court; Council of State; Court of Final Appeals or Cour de Cassation; Court of Appeals

Political parties and leaders: African Party of Independence [Majhemout DIOP]; And-Jef/African Party for Democracy and Socialism or AJ/PADS [Landing SAVANE]; Alliance of Forces of Progress or AFP [Moustapha NIASSE]; Democratic League-Labor Party Movement or LD-MPT [Dr. Abdoulaye BATHILY]; Front for Socialism and Democracy/Benno Jubel or FSD/BJ [Cheikh Abdoulaye Bamba DIEYE]; Gainde Centrist Bloc or BGC [Jean-Paul DIAS]; Independence and Labor Party or PIT [Amath DANSOKHO]; Jef-Jel [Talla SYLLA]; National Democratic Rally or RND [Madior DIOUF]; People's Labor Party or PTP [Elhadji DIOUF]; Reform Party or PR [Abdourahim AGNE]; Senegalese Democratic Party or PDS [Abdoulaye WADE]; Socialist Party or PS [Ousmane Tanor DIENG]; SOPI Coalition [Abdoulaye WADE] (a coalition led by the PDS); Union for Democratic Renewal or URD [Djibo Leyti KA]

Political pressure groups and leaders: labor; Sufi brotherhoods, including the Mourides and Tidjanes; students; teachers

International organization participation: ACCT, ACP, AfDB, AU, ECOWAS, FAO, FZ, G-15, G-77, IAEA, IBRD, ICAO, ICC, ICCt, ICRM, IDA, IDB, IFAD, IFC, IFRCS, ILO, IMF, IMO, IMSO, Interpol, IOC, IOM, IPU, ISO (correspondent), ITSO, ITU, ITUC, MIGA, MINURCAT, MONUC, NAM, OIC, OIF, OPCW, PCA, UN, UNAMID, UNCTAD, UNESCO, UNIDO, Union Latina, UNMIL,

UNOCI, UNWTO, UPU, WADB (regional), WAEMU, WCL, WCO, WFTU, WHO, WIPO, WMO, WTO

Diplomatic representation in the US:
chief of mission: Ambassador Amadou Lamine BA
chancery: 2112 Wyoming Avenue NW, Washington, DC 20008
telephone: [1] (202) 234-0540
FAX: [1] (202) 332-6315
consulate(s) general: Houston, New York

Diplomatic representation from the US:
chief of mission: Ambassador (vacant); Charge d'Affaires Jay Thomas Smith
embassy: Avenue Jean XXIII at the corner of Rue Kleber, Dakar
mailing address: B. P. 49, Dakar
telephone: [221] 33-823-4296
FAX: [221] 33-822-2991

Flag description: three equal vertical bands of green (hoist side), yellow, and red with a small green five-pointed star centered in the yellow band; uses the popular pan-African colors of Ethiopia

ECONOMY

Economy—overview: In January 1994, Senegal undertook a bold and ambitious economic reform program with the support of the international donor community. This reform began with a 50% devaluation of Senegal's currency, the CFA franc, which was linked at a fixed rate to the French franc. Government price controls and subsidies have been steadily dismantled. After seeing its economy contract by 2.1% in 1993, Senegal made an important turnaround, thanks to the reform program, with real growth in GDP averaging over 5% annually during 1995–2007. Annual inflation had been pushed down to the low single digits. As a member of the West African Economic and Monetary Union (WAEMU), Senegal is working toward greater regional integration with a unified external tariff and a more stable monetary policy. High unemployment, however, continues to prompt illegal migrants to flee Senegal in search of better job opportunities in Europe. Senegal was also beset by an energy crisis that caused widespread blackouts in 2006 and 2007. The phosphate industry has struggled for two years to secure capital, and reduced output has directly impacted GDP. In 2007, Senegal signed agreements for major new mining concessions for iron, zircon, and gold with foreign companies. Firms from Dubai have agreed to manage and modernize Dakar's maritime port, and create a new special economic zone. Senegal still relies heavily upon outside donor assistance. Under the IMF's Highly Indebted Poor Countries (HIPC) debt relief program, Senegal has benefited from eradication of two-thirds of its bilateral, multilateral, and private-sector debt. In 2007, Senegal and the IMF agreed to a new, non-disbursing, Policy Support Initiative program.

GDP (purchasing power parity): $20.6 billion (2007 est.)
GDP (official exchange rate): $11.12 billion (2007)
GDP—real growth rate: 5% (2007 est.)
GDP—per capita (PPP): $1,700 (2007 est.)
GDP—composition by sector:
agriculture: 16.7%
industry: 18.9%
services: 64.4% (2007 est.)
Labor force: 4.85 million (2007 est.)
Labor force—by occupation: *agriculture:* 77.5%
industry and services: 22.5% (2007 est.)
Unemployment rate: 48% (2007 est.)
Population below poverty line: 54% (2001 est.)
Household income or consumption by percentage share:
lowest 10%: 2.7%
highest 10%: 33.4% (2001)
Distribution of family income—Gini index: 41.3 (2001)
Inflation rate (consumer prices): 5.9% (2007)
Investment (gross fixed): 12.2% of GDP (2007 est.)
Budget:
revenues: $2.271 billion
expenditures: $2.815 billion (2007 est.)
Public debt: 22.6% of GDP (2007)
Agriculture—products: peanuts, millet, corn, sorghum, rice, cotton, tomatoes, green vegetables; cattle, poultry, pigs; fish
Industries: agricultural and fish processing, phosphate mining, fertilizer production, petroleum refining; iron ore, zircon, and gold mining, construction materials, ship construction and repair
Industrial production growth rate: 2.7% (2007 est.)
Electricity—production: 2.159 billion kWh (2006)
Electricity—production by source:
fossil fuel: 100%
hydro: 0%
nuclear: 0%
other: 0% (2001)
Electricity—consumption: 1.859 billion kWh (2006)
Electricity—exports: 0 kWh (2005)
Electricity—imports: 0 kWh (2005)
Oil—production: 0 bbl/day (2005 est.)
Oil—consumption: 35,000 bbl/day (2005 est.)
Oil—exports: 3,889 bbl/day (2004)

Oil—imports: 37,180 bbl/day (2004)
Oil—proved reserves: 0 bbl (1 January 2006 est.)
Natural gas—production: 2 billion cu m (2006 est.)
Natural gas—consumption: 2 billion cu m (2006 est.)
Natural gas—exports: NA cu m
Natural gas—imports: NA cu m
Natural gas—proved reserves: NA cu m
Current account balance: -$906 million (2007 est.)
Exports: $1.604 billion f.o.b. (2007 est.)
Exports—commodities: fish, groundnuts (peanuts), petroleum products, phosphates, cotton
Exports—partners: Mali 19.2%, France 8.3%, India 5.8%, The Gambia 5.3%, Spain 5.1%, Italy 4.9% (2006)
Imports: $3.27 billion f.o.b. (2007 est.)
Imports—commodities: food and beverages, capital goods, fuels
Imports—partners: France 25.1%, UK 5.2%, Thailand 4.8%, China 4.5%, Spain 4% (2006)
Economic aid—recipient: $477 million (2007 est.)
Reserves of foreign exchange and gold: $1.66 billion (31 December 2007 est.)
Debt—external: $2.13 billion (31 December 2007)
Market value of publicly traded shares: $NA
Currency (code): Communaute Financiere Africaine franc (XOF); note—responsible authority is the Central Bank of the West African States
Currency code: XOF
Exchange rates: Communaute Financiere Africaine francs (XOF) per US dollar—481.83 (2007), 522.89 (2006), 527.47 (2005), 528.29 (2004), 581.2 (2003)
note: since 1 January 1999, the XOF franc has been pegged to the euro at a rate of 655.957 XOF francs per euro
Fiscal year: calendar year

COMMUNICATIONS

Telephones—main lines in use: 282,600 (2006)
Telephones—mobile cellular: 2.983 million (2006)
Telephone system:
general assessment: good system
domestic: above-average urban system; more than half of all fixed-line connections are in Dakar with expansion of fixed-line services in rural areas needed; mobile-cellular service is expanding rapidly; microwave radio relay, coaxial cable and fiber-optic cable in trunk system
international: country code—221; the

SAT-3/WASC fiber optic cable provides connectivity to Europe and Asia while Atlantis-2 provides connectivity to South America; satellite earth station—1 Intelsat (Atlantic Ocean)

Radio broadcast stations: AM 8, FM 20, shortwave 1 (2001)

Radios: 1.24 million (1997)

Television broadcast stations: 4 (2007)

Televisions: 361,000 (1997)

Internet country code: .sn

Internet hosts: 199 (2007)

Internet Service Providers (ISPs): 1 (2002)

Internet users: 650,000 (2006)

TRANSPORTATION

Airports: 20 (2007)

Airports—with paved runways:
total: 9
over 3,047 m: 1
1,524 to 2,437 m: 7
914 to 1,523 m: 1 (2007)

Airports—with unpaved runways:
total: 11
1,524 to 2,437 m: 6
914 to 1,523 m: 4
under 914 m: 1 (2007)

Pipelines: gas 43 km (2007)

Railways: *total:* 906 km

narrow gauge: 906 km 1.000 meter gauge (2006)

Roadways:
total: 13,576 km
paved: 3,972 km (includes 7 km of expressways)
unpaved: 9,604 km (2003)

Waterways: 1,000 km (primarily on Senegal, Saloum, and Casamance rivers) (2005)

Ports and terminals: Dakar

MILITARY

Military branches: Army, Senegalese Navy (Marine Senegalaise), Senegalese Air Force (Armee de l'Air du Senegal) (2008)

Military service age and obligation: 18 years of age for compulsory and voluntary military service; conscript service obligation—2 years (2004)

Manpower available for military service:
males age 16–49: 2,943,619
females age 16–49: 2,955,179 (2008 est.)

Manpower fit for military service:
males age 16–49: 1,866,602
females age 16–49: 1,947,076 (2008 est.)

Manpower reaching militarily significant age annually:

males age 16–49: 141,832
females age 16–49: 139,541 (2008 est.)

Military expenditures—percent of GDP: 1.4% (2005 est.)

TRANSNATIONAL ISSUES

Disputes—international: The Gambia and Guinea-Bissau attempt to stem separatist violence, cross border raids, and arms smuggling into their countries from Senegal's Casamance region, and in 2006, respectively accepted 6,000 and 10,000 Casamance residents fleeing the conflict; 2,500 Guinea-Bissau residents fled into Senegal in 2006 to escape armed confrontations along the border

Refugees and internally displaced persons: *refugees (country of origin):* 19,630 (Mauritania)
IDPs: 22,400 (approximately 65% of the IDP population returned in 2005, but new displacement is occurring due to clashes between government troops and separatists in Casamance region) (2007)

Illicit drugs: transshipment point for Southwest and Southeast Asian heroin and South American cocaine moving to Europe and North America; illicit cultivator of cannabis

SERBIA

INTRODUCTION

Background: The Kingdom of Serbs, Croats, and Slovenes was formed in 1918; its name was changed to Yugoslavia in 1929. Various paramilitary bands resisted Nazi Germany's occupation and division of Yugoslavia from 1941 to 1945, but fought each other and ethnic opponents as much as the invaders. The military and political movement headed by Josip TITO (Partisans) took full control of Yugoslavia when German and Croatian

separatist forces were defeated in 1945. Although Communist, TITO's new government and his successors (he died in 1980) managed to steer their own path between the Warsaw Pact nations and the West for the next four and a half decades. In 1989, Slobodan MILO-SEVIC became president of the Serbian Republic and his ultranationalist calls for Serbian domination led to the violent breakup of Yugoslavia along ethnic lines. In 1991, Croatia, Slovenia, and Macedonia declared independence, followed by Bosnia in 1992. The remaining republics of Serbia and Montenegro declared a new Federal Republic of Yugoslavia (FRY) in April 1992 and under MILOSEVIC's leadership, Serbia led various military campaigns to unite ethnic Serbs in neighboring republics into a "Greater Serbia." These actions led to Yugoslavia being ousted from the UN in 1992, but Serbia continued its—ultimately unsuccessful—campaign until signing the Dayton Peace Accords in 1995. MILOSEVIC kept tight control over Serbia and eventually became president of the FRY in 1997. In 1998, an ethnic Albanian insurgency in the formerly autonomous Serbian province of

Kosovo provoked a Serbian counterinsurgency campaign that resulted in massacres and massive expulsions of ethnic Albanians living in Kosovo. The MILO-SEVIC government's rejection of a proposed international settlement led to NATO's bombing of Serbia in the spring of 1999 and to the eventual withdrawal of Serbian military and police forces from Kosovo in June 1999. UNSC Resolution 1244 in June 1999 authorized the stationing of a NATO-led force (KFOR) in Kosovo to provide a safe and secure environment for the region's ethnic communities, created a UN interim Administration Mission in Kosovo (UNMIK) to foster self-governing institutions, and reserved the issue of Kosovo's final status for an unspecified date in the future. In 2001, UNMIK promulgated a constitutional framework that allowed Kosovo to establish institutions of self-government and led to Kosovo's first parliamentary election. FRY elections in September 2000 led to the ouster of MILOSEVIC and installed Vojislav KOSTUNICA as president. A broad coalition of democratic reformist parties known as DOS (the Democratic Opposition of Serbia) was subsequently elected to parliament in December 2000 and took control of the government. DOS arrested MILOSEVIC in 2001 and allowed for him to be tried in The Hague for crimes against humanity. (MILOSEVIC died in March 2006 before the completion of his trial.) In 2001, the country's suspension from the UN was lifted. In 2003, the FRY became Serbia and Montenegro, a loose federation of the two republics with a federal level parliament. Widespread violence predominantly targeting ethnic Serbs in Kosovo in March 2004 caused the international community to open negotiations on the future status of Kosovo in January 2006. In May 2006, Montenegro invoked its right to secede from the federation and—following a successful referendum—it declared itself an independent nation on 3 June 2006. Two days later, Serbia declared that it was the successor state to the union of Serbia and Montenegro. A new Serbian constitution was approved in October 2006 and adopted the following month. After 15 months of inconclusive negotiations mediated by the UN and four months of further inconclusive negotiations mediated by the US, EU, and Russia, on 17 February 2008, the UNMIK-administered province of Kosovo declared itself independent of Serbia.

GEOGRAPHY

Location: Southeastern Europe, between Macedonia and Hungary
Geographic coordinates: 44 00 N, 21 00 E
Map references: Europe
Area: *total:* 77,474 sq km
land: 77,474 sq km
water: 0 sq km
Area—comparative: slightly smaller than South Carolina
Land boundaries:
total: 2,026 km
border countries: Bosnia and Herzegovina 302 km, Bulgaria 318 km, Croatia 241 km, Hungary 151 km, Kosovo 352 km, Macedonia 62 km, Montenegro 124 km, Romania 476 km
Coastline: 0 km (landlocked)
Maritime claims: none (landlocked)
Climate: in the north, continental climate (cold winters and hot, humid summers with well distributed rainfall); in other parts, continental and Mediterranean climate (relatively cold winters with heavy snowfall and hot, dry summers and autumns)
Terrain: extremely varied; to the north, rich fertile plains; to the east, limestone ranges and basins; to the southeast, ancient mountains and hills

Elevation extremes:
lowest point: NA
highest point: Midzor 2,169 m
Natural resources: oil, gas, coal, iron ore, copper, zinc, antimony, chromite, gold, silver, magnesium, pyrite, limestone, marble, salt, arable land
Land use: *arable land:* NA
permanent crops: NA
other: NA
Irrigated land: NA
Total renewable water resources: 208.5 cu km (note—includes Kosovo) (2003)
Natural hazards: destructive earthquakes
Environment—current issues: air pollution around Belgrade and other industrial cities; water pollution from industrial wastes dumped into the Sava which flows into the Danube
Environment—international agreements: *party to:* Air Pollution, Biodiversity, Climate Change, Climate Change-Kyoto Protocol, Endangered Species, Hazardous Wastes, Law of the Sea, Marine Dumping, Marine Life Conservation, Ozone Layer Protection, Ship Pollution, Wetlands
signed, but not ratified: none of the selected agreements
Geography—note: controls one of the major land routes from Western Europe to Turkey and the Near East

PEOPLE

Population: 10,159,046
note: all population data includes Kosovo (July 2008 est.)
Median age: *total:* 37.5 years
male: 36.1 years
female: 39 years (2008 est.)
Life expectancy at birth:
total population: 75.29 years
male: 72.7 years
female: 78.09 years (2008 est.)
Total fertility rate: 1.69 children born/woman (2008 est.)
Major infectious diseases:
degree of risk: intermediate
food or waterborne diseases: bacterial diarrhea and hepatitis A
vectorborne disease: Crimean Congo hemorrhagic fever
note: highly pathogenic H5N1 avian influenza has been identified in this country; it poses a negligible risk with extremely rare cases possible among US citizens who have close contact with birds (2008)
Nationality: *noun:* Serb(s)
adjective: Serbian
Ethnic groups: Serb 82.9%, Hungarian 3.9%, Romany (Gypsy) 1.4%, Yugoslavs 1.1%, Bosniaks 1.8%, Montenegrin 0.9%, other 8% (2002 census)

Religions: Serbian Orthodox 85%, Catholic 5.5%, Protestant 1.1%, Muslim 3.2%, unspecified 2.6%, other, unknown, or atheist 2.6% (2002 census)
Languages: Serbian 88.3% (official), Hungarian 3.8%, Bosniak 1.8%, Romany (Gypsy) 1.1%, other 4.1%, unknown 0.9% (2002 census)
note: Romanian, Hungarian, Slovak, Ukrainian, and Croatian all official in Vojvodina
Literacy: *definition:* age 15 and over can read and write
total population: 96.4%
male: 98.9%
female: 94.1% (2003 census)
note: includes Montenegro

GOVERNMENT

Country name:
conventional long form: Republic of Serbia
conventional short form: Serbia
local long form: Republika Srbija
local short form: Srbija
former: People's Republic of Serbia, Socialist Republic of Serbia
Government type: republic
Capital: *name:* Belgrade (Beograd)
geographic coordinates: 44 50 N, 20 30 E
time difference: UTC+1 (6 hours ahead of Washington, DC during Standard Time)
daylight saving time: +1hr, begins last Sunday in March; ends last Sunday in October
Administrative divisions: 161 municipalities (opcstine, singular—opcstina)
Serbia Proper: Beograd: Barajevo, Cukavica, Grocka, Lazarevac, Mladnovac, Novi Beograd, Obrenovac, Palilula, Rakovica, Savski Venac, Sopot, Stari Grad, Surcin, Vozdovac, Vracar, Zemun, Zrezdara; Borski Okrug: Bor, Kladovo, Majdanpek, Negotin; Branicevski Okrug: Golubac, Kucevo, Malo Crnice, Petrovac, Pozarevac, Veliko Gradiste, Zabari, Zagubica; Jablanicki Okrug: Bojnik, Crna Trava, Lebane, Leskovac, Medvedja, Vlasotince; Kolubarski Okrug: Lajkovac, Ljig, Mionica, Osecina, Ub, Valjevo; Macvanski Okrug: Bogatic, Koceljeva, Krupanj, Ljubovija, Loznica, Mali Zvornik, Sabac, Vladimirci; Moravicki Okrug: Cacak, Gornkji Milanovac, Ivanjica, Lucani; Nisavski Okrug: Aleksinac, Doljevac, Gadzin Han, Merosina, Nis, Razanj, Svrljig; Pcinjski Okrug: Bosilegrad, Bujanovac, Presevo, Surdulica, Trgoviste, Vladicin Han, Vranje; Pirotski Okrug: Babusnica, Bela Palanka, Dimitrovgrad, Pirot; Podunavski Okrug: Smederevo, Smederevskia Palanka, Velika Plana; Pomoravski Okrug: Cuprija, Despotovac, Jagodina, Paracin, Rckovac, Svilajnac;

Rasinski Okrug: Aleksandrovac, Brus, Cicevac, Krusevac, Trstenik, Varvarin; Raski Okrug: Kraljevo, Novi Pazar, Raska, Tutin, Vrnjacka Banja; Sumadijski Okrug: Arandjelovac, Batocina, Knic, Kragujevac, Lapovo, Raca, Topola; Toplicki Okrug: Blace, Kursumlija, Prokuplje, Zitoradja; Zajecarski Okrug: Boljevac, Knjazevac, Sokobanja, Zalecar; Zlatiborski Okrug: Arilje, Bajina Basta, Cajetina, Kosjeric, Nova Varos, Pozega, Priboj, Prijepolje, Sjenica, Uzice

Vojvodina Autonomous Province: Juzno-Backi Okrug: Backi Petrovac, Beocin, Novi Sad, Sremski Karlovci, Temerin, Titel, Zabalj; Juzno Banatski Okrug: Alibunar, Bela Crkva, Kovacica, Kovin, Opovo, Pancevo, Plandiste, Vrsac; Severno-Backi Okrug: Backa Topola, Mali Idjos, Subotica; Severno-Banatski Okrug: Ada, Coka, Kanjiza, Kikinda, Novi Knezevac, Senta; Srednje-Banatski Okrug: Nova Crnja, Novi Becej, Secanj, Zitiste, Zrenjanin; Sremski Okrug: Indjija, Irig, Pecinci, Ruma, Sid, Sremska Mitrovica, Stara Pazova; Zapadno-Backi Okrug: Apatin, Kula, Odzaci, Sombor

Independence: 5 June 2006 (from Serbia and Montenegro)

National holiday: National Day, 15 February

Constitution: adopted 8 November 2006; effective 10 November 2006

Legal system: based on civil law system; has not accepted compulsory ICJ jurisdiction

Suffrage: 18 years of age; universal

Executive branch:

chief of state: President Boris TADIC (since 11 July 2004)

head of government: Prime Minister Vojislav KOSTUNICA (since 3 March 2004); resigned 8 March 2008

cabinet: Federal Ministries act as cabinet

elections: president elected by direct vote for a five-year term (eligible for a second term); election last held 3 February 2008 (next to be held in 2013); prime minister elected by the Assembly

election results: Boris TADIC elected president in the second round of voting; Boris TADIC received 51.2% of the vote and Tomislav NIKOLIC 48.8%

Legislative branch: unicameral National Assembly (250 seats; deputies elected by direct vote to serve four-year terms)

elections: last held on 11 May 2008 (next to be held in May 2012)

election results: percent of vote by party— For a European Serbia 38.7%, SRS 29.1%, DSS-NS 11.3%, coalition led by the SPS 7.9%, LPD 5.2%, other 7.8%; seats by party—For a European Serbia

103, SRS 77, DSS-NS 30, coalition led by the SPS 20, LDP 13, other 7

Judicial branch: Constitutional Court, Supreme Court (to become court of cassation under new constitution), appellate courts, district courts, municipal courts

Political parties and leaders: Coalition of Albanians of the Presevo Valley or KAPD [Riza HALIMI]; Coalition for Sandzak or KZS [Sulejman UGLJANIN]; Democratic Party of Albanians or PDSh [Ragmi MUSTAFA]; Democratic Party of Serbia or DSS [Vojislav KOSTUNICA]; Democratic Party or DS [Boris TADIC]; Democratic Union of the Valley or BDL [Skender DESTANI]; For a European Serbia [Boris TADIC]; Force of Serbia Movement or PSS [Bogoljub KARIC]; G17 Plus [Mladjan DINKIC]; League of Vojvodina Hungarians or SVM [Istvan PASTOR]; Liberal Democratic Party or LDP [Cedomir JOVANOVIC]; Movement for Democratic Progress of LPD [Jonuz MUSLIU]; New Serbia or NS [Velimir ILIC]; Party of Democratic Action or PVD [Riza HALIMI]; Roma Party or RP [Srdjan SAJN]; Serbian Radical Party or SRS [Vojislav SESELJ (currently on trial at The Hague), but Tomislav NIKOLIC is acting leader]; Socialist Party of Serbia or SPS [Ivica DACIC]; Union of Roma of Serbia or URS [Rajko DJURIC]

International organization participation: ABEDA, BIS, CE, CEI, EAPC, EBRD, FAO, G-9, IAEA, IBRD, ICAO, ICC, ICCt, ICRM, IDA, IFAD (suspended), IFC, IFRCS, IHO, ILO, IMF, IMO, IMSO, Interpol, IOC, IOM, IPU, ISO, ITSO, ITU, ITUC, MIGA, MONUC, NAM (observer), OAS (observer), OIF (observer), OPCW, OSCE, PCA, PFP, SECI, UN, UNCTAD, UNESCO, UNHCR, UNIDO, UNMIL, UNOCI, UNWTO, UPU, WCL, WCO, WHO, WIPO, WMO, WTO (observer)

Diplomatic representation in the US:

chief of mission: Ambassador Ivan VUJACIC

chancery: 2134 Kalorama Road NW, Washington, DC 20008

telephone: [1] (202) 332-0333

FAX: [1] (202) 332-3933

consulate(s) general: Chicago, New York

Diplomatic representation from the US:

chief of mission: Ambassador Cameron MUNTER

embassy: Kneza Milosa 50, 11000 Belgrade

mailing address: 5070 Belgrade Place, Washington, DC 20521-5070

telephone: [381] (11) 361-9344

FAX: [381] (11) 361-8230

Flag description: three equal horizontal stripes of red (top), blue, and white; charged with the coat of arms of Serbia shifted slightly to the hoist side

ECONOMY

Economy—overview: MILOSEVIC-era mismanagement of the economy, an extended period of economic sanctions, and the damage to Yugoslavia's infrastructure and industry during the NATO airstrikes in 1999 left the economy only half the size it was in 1990. After the ousting of former Federal Yugoslav President MILOSEVIC in September 2000, the Democratic Opposition of Serbia (DOS) coalition government implemented stabilization measures and embarked on a market reform program. After renewing its membership in the IMF in December 2000, a down-sized Yugoslavia continued to reintegrate into the international community by rejoining the World Bank (IBRD) and the European Bank for Reconstruction and Development (EBRD). A World Bank-European Commission sponsored Donors' Conference held in June 2001 raised $1.3 billion for economic restructuring. In November 2001, the Paris Club agreed to reschedule the country's $4.5 billion public debt and wrote off 66% of the debt. In July 2004, the London Club of private creditors forgave $1.7 billion of debt just over half the total owed. Belgrade has made only minimal progress in restructuring and privatizing its holdings in major sectors of the economy, including energy and telecommunications. It has made halting progress towards EU membership and is currently pursuing a Stabilization and Association Agreement with Brussels. Serbia is also pursuing membership in the World Trade Organization. Unemployment remains an ongoing political and economic problem.

GDP (purchasing power parity): $77.28 billion (2007 est.)

GDP (official exchange rate): $41.68 billion (2007 est.)

GDP—real growth rate: 7.3% (2007 est.)

GDP—per capita (PPP): $10,400 (2007 est.)

GDP—composition by sector:

agriculture: 12.3%

industry: 24.2%

services: 63.5% (2007 est.)

Labor force: 2.961 million (2002 est.)

Labor force—by occupation: *agriculture:* 30%

industry: 46%

services: 24% (2002)

Unemployment rate: 18.8% (2007 est.)

Population below poverty line: 6.5% (2007 est.)

Distribution of family income—Gini index: 30 (2003)

Inflation rate (consumer prices): 6.8% (2007)

Investment (gross fixed): 20.1% of GDP (2007 est.)

Budget:
revenues: $9.6 billion
expenditures: $9.8 billion (2007 est.)

Public debt: 37% of GDP (2007 est.)

Agriculture—products: wheat, maize, sugar beets, sunflower, raspberries, beef, pork, milk

Industries: sugar, agricultural machinery, electrical and communication equipment, paper and pulp, lead, transportation equipment

Industrial production growth rate: 1.8% (2007 est.)

Electricity—production: 33.87 billion kWh (2004)

Electricity—consumption: NA kWh

Electricity—exports: 12.05 billion kWh (2004)

Electricity—imports: 11.23 billion kWh (2004)

Oil—production: 14,660 bbl/day (2003)

Oil—consumption: 85,000 bbl/day (2003 est.)

Oil—exports: NA bbl/day

Oil—imports: NA bbl/day

Oil—proved reserves: 77.5 million bbl (1 January 2006 est.)

Natural gas—production: 650 million cu m (2005 est.)

Natural gas—consumption: 2.55 billion cu m (2005 est.)

Natural gas—exports: 0 cu m (2005 est.)

Natural gas—imports: 2.1 billion cu m (2004 est.)

Natural gas—proved reserves: 46.17 billion cu m (1 January 2006)

Current account balance: -$6.889 billion (2007 est.)

Exports: $8.824 billion (2007 est.)

Exports—commodities: manufactured goods, food and live animals, machinery and transport equipment

Imports: $18.35 billion (2007 est.)

Economic aid—recipient: $2 billion pledged in 2001 to Serbia and Montenegro (disbursements to follow over several years; some aid pledged by EU and US has been placed on hold because of lack of cooperation by Serbia

in handing over General Ratko MLADIC to the criminal court in The Hague)

Reserves of foreign exchange and gold: $14.22 billion (2007 est.)

Debt—external: $26.24 billion (includes debt for Montenegro and Kosovo) (2007 est.)

Stock of direct foreign investment—at home: $11.95 billion (2006 est.)

Stock of direct foreign investment—abroad: $NA

Market value of publicly traded shares: $5.409 billion (2005)

Currency (code): Serbian dinar (RSD)

Exchange rates: Serbian dinars per US dollar—54.5 (2007), 59.98 (2006)

COMMUNICATIONS

Telephones—main lines in use: 2.719 million (2006)

Telephones—mobile cellular: 6.644 million (2006)

Telephone system:
general assessment: modernization of the telecommunications network has been slow as a result of damage stemming from the 1999 war and transition to a competitive market-based system; network was only 65% digitalized in 2005
domestic: teledensity remains below the average for neighboring states; GSM wireless service, available through multiple providers with national coverage, is growing very rapidly; best telecommunications service limited to urban centers
international: country code—381

Radio broadcast stations: 153 (station types NA) (2001)

Internet country code: .rs

Internet hosts: NA

Internet users: 1.4 million (2006)

TRANSPORTATION

Airports: 39 (2007)

Airports—with paved runways:
total: 16
over 3,047 m: 2
2,438 to 3,047 m: 4
1,524 to 2,437 m: 4
914 to 1,523 m: 2
under 914 m: 4 (2007)

Airports—with unpaved runways:
total: 23
1,524 to 2,437 m: 2
914 to 1,523 m: 9
under 914 m: 12 (2007)

Heliports: 2 (2007)

Pipelines: gas 1,921 km; oil 393 km (2007)

Railways:
total: 3,379 km
standard gauge: 3,379 km 1.435-m gauge (electrified 1,254 km) (2006)

Roadways:
total: 37,841 km
paved: 32,100 km
unpaved: 5,741 km (2006)

Waterways: 587 km (primarily on Danube and Sava rivers) (2005)

MILITARY

Military branches: Serbian Armed Forces (Vojska Srbije, VS): Land Forces Command (includes Serbian naval force, consisting of a river flotilla on the Danube), Joint Operations Command, Air and Air Defense Forces Command (2007)

Military service age and obligation: 19–35 years of age for compulsory military service; under a state of war or impending war, conscription can begin at age 16; conscription is to be abolished in 2010; 9-month service obligation, with a reserve obligation to age 60 for men and 50 for women (2007)

TRANSNATIONAL ISSUES

Disputes—international: Serbia with several other states protest the U.S. and other states' recognition of Kosovo's declaring itself as a sovereign and independent state in February 2008; ethnic Serbian municipalities along Kosovo's northern border challenge final status of Kosovo-Serbia boundary; several thousand NATO-led KFOR peacekeepers under UNMIK authority continue to keep the peace within Kosovo between the ethnic Albanian majority and the Serb minority in Kosovo; Serbia delimited about half of the boundary with Bosnia and Herzegovina, but sections along the Drina River remain in dispute

Refugees and internally displaced persons: *refugees (country of origin):* 71,111 (Croatia); 27,414 (Bosnia and Herzegovina); 206,000 (Kosovo), note—mostly ethnic Serbs and Roma who fled Kosovo in 1999 (2007)

Illicit drugs: transshipment point for Southwest Asian heroin moving to Western Europe on the Balkan route; economy vulnerable to money laundering

SEYCHELLES

INTRODUCTION

Background: A lengthy struggle between France and Great Britain for the islands ended in 1814, when they were ceded to the latter. Independence came in 1976.

Socialist rule was brought to a close with a new constitution and free elections in 1993. President France-Albert RENE, who had served since 1977, was re-elected in 2001, but stepped down in 2004. Vice President James MICHEL

took over the presidency and in July 2006 was elected to a new five-year term.

GEOGRAPHY

Location: archipelago in the Indian Ocean, northeast of Madagascar

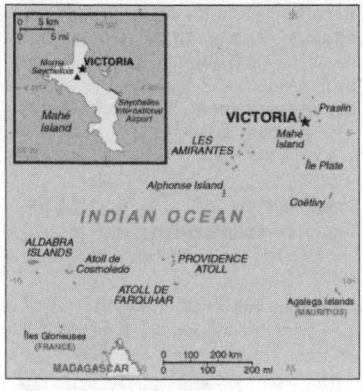

Geographic coordinates: 4 35 S, 55 40 E
Map references: Africa
Area:
total: 455 sq km
land: 455 sq km
water: 0 sq km
Area—comparative: 2.5 times the size of Washington, DC
Land boundaries: 0 km
Coastline: 491 km
Maritime claims:
territorial sea: 12 nm
contiguous zone: 24 nm
exclusive economic zone: 200 nm
continental shelf: 200 nm or to the edge of the continental margin
Climate: tropical marine; humid; cooler season during southeast monsoon (late May to September); warmer season during northwest monsoon (March to May)
Terrain: Mahe Group is granitic, narrow coastal strip, rocky, hilly; others are coral, flat, elevated reefs
Elevation extremes:
lowest point: Indian Ocean 0 m
highest point: Morne Seychellois 905 m
Natural resources: fish, copra, cinnamon trees
Land use:
arable land: 2.17%
permanent crops: 13.04%
other: 84.79% (2005)
Irrigated land: NA
Natural hazards: lies outside the cyclone belt, so severe storms are rare; short droughts possible
Environment—current issues: water supply depends on catchments to collect rainwater
Environment—international agreements: *party to:* Biodiversity, Climate Change, Climate Change-Kyoto Protocol, Desertification, Endangered Species, Hazardous Wastes, Law of the Sea, Marine Dumping, Ozone Layer Protection, Ship Pollution, Wetlands
signed, but not ratified: none of the selected agreements
Geography—note: 41 granitic and about 75 coralline islands

PEOPLE

Population: 82,247 (July 2008 est.)
Age structure:
0–14 years: 24.9% (male 10,337/female 10,108)
15–64 years: 69.1% (male 27,752/female 29,048)
65 years and over: 6.1% (male 1,575/female 3,427) (2008 est.)
Median age:
total: 28.7 years
male: 27.6 years
female: 29.8 years (2008 est.)
Population growth rate: 0.428% (2008 est.)
Birth rate: 15.6 births/1,000 population (2008 est.)
Death rate: 6.21 deaths/1,000 population (2008 est.)
Net migration rate: -5.11 migrant(s)/1,000 population (2008 est.)
Sex ratio:
at birth: 1.03 male(s)/female
under 15 years: 1.02 male(s)/female
15–64 years: 0.96 male(s)/female
65 years and over: 0.46 male(s)/female
total population: 0.93 male(s)/female (2008 est.)
Infant mortality rate:
total: 14.36 deaths/1,000 live births
male: 18.18 deaths/1,000 live births
female: 10.42 deaths/1,000 live births (2008 est.)
Life expectancy at birth:
total population: 72.6 years
male: 67.27 years
female: 78.1 years (2008 est.)
Total fertility rate: 1.73 children born/woman (2008 est.)
HIV/AIDS—adult prevalence rate: NA
HIV/AIDS—people living with HIV/AIDS: NA
HIV/AIDS—deaths: NA
Nationality:
noun: Seychellois (singular and plural)
adjective: Seychellois
Ethnic groups: mixed French, African, Indian, Chinese, and Arab
Religions: Roman Catholic 82.3%, Anglican 6.4%, Seventh Day Adventist 1.1%, other Christian 3.4%, Hindu 2.1%, Muslim 1.1%, other non-Christian 1.5%, unspecified 1.5%, none 0.6% (2002 census)
Languages: Creole 91.8%, English 4.9% (official), other 3.1%, unspecified 0.2% (2002 census)
Literacy: *definition:* age 15 and over can read and write
total population: 91.8%
male: 91.4%
female: 92.3% (2002 census)

GOVERNMENT

Country name:
conventional long form: Republic of Seychelles
conventional short form: Seychelles
local long form: Republic of Seychelles
local short form: Seychelles
Government type: republic
Capital: *name:* Victoria
geographic coordinates: 4 38 S, 55 27 E
time difference: UTC+4 (9 hours ahead of Washington, DC during Standard Time)
Administrative divisions: 23 administrative districts; Anse aux Pins, Anse Boileau, Anse Etoile, Anse Louis, Anse Royale, Baie Lazare, Baie Sainte Anne, Beau Vallon, Bel Air, Bel Ombre, Cascade, Glacis, Grand' Anse (on Mahe), Grand' Anse (on Praslin), La Digue, La Riviere Anglaise, Mont Buxton, Mont Fleuri, Plaisance, Pointe La Rue, Port Glaud, Saint Louis, Takamaka
Independence: 29 June 1976 (from UK)
National holiday: Constitution Day (National Day), 18 June (1993)
Constitution: 18 June 1993
Legal system: based on English common law, French civil law, and customary law; has not accepted compulsory ICJ jurisdiction
Suffrage: 17 years of age; universal
Executive branch:
chief of state: President James Alix MICHEL (since 14 April 2004); note—the president is both the chief of state and head of government
head of government: President James MICHEL (since 14 April 2004)
cabinet: Council of Ministers appointed by the president
elections: president elected by popular vote for a five-year term (eligible for two more terms); election last held 28–30 July 2006 (next to be held in 2011)
election results: President James MICHEL elected president; percent of vote—James MICHEL 53.73%, Wavel RAMKALAWAN 45.71%, Philippe BOULLE 0.56%; note—this was the first election in which President James MICHEL participated; he was originally sworn in as president after former president France Albert RENE stepped down in April 2004
Legislative branch: unicameral National Assembly or Assemblee Nationale (34 seats; 25 members elected by popular vote, 9 allocated on a proportional basis to parties winning at least 10% of the vote; to serve five-year terms)
elections: last held 10–12 May 2007 (next to be held in 2012)
election results: percent of vote by party—SPPF 56.2%, SNP 43.8%; seats by party—SPPF 23, SNP 11
Judicial branch: Court of Appeal; Supreme Court; judges for both courts are appointed by the president
Political parties and leaders: Democratic Party or DP [James MANCHAM, Paul CHOW]; Seychelles

National Party or SNP [Wavel RAMKALAWAN] (formerly the United Opposition or UO); Seychelles People's Progressive Front or SPPF [France Albert RENE, James MICHEL] (the governing party)

Political pressure groups and leaders: Roman Catholic Church; trade unions

International organization participation: ACCT, ACP, AfDB, AU, C, COMESA, FAO, G-77, IAEA, IBRD, ICAO, ICCt (signatory), ICRM, IFAD, IFC, IFRCS, ILO, IMF, IMO, InOC, Interpol, IOC, ISO (correspondent), ITU, ITUC, MIGA, NAM, OIF, OPCW, UN, UNCTAD, UNESCO, UNIDO, UNWTO, UPU, WCO, WHO, WIPO, WMO, WTO (observer)

Diplomatic representation in the US:
chief of mission: Ambassador Jean Ronald JUMEAU
chancery: 800 Second Avenue, Suite 400C, New York, NY 10017
telephone: [1] (212) 972-1785
FAX: [1] (212) 972-1786

Diplomatic representation from the US: the US does not have an embassy in Seychelles; the ambassador to Mauritius is accredited to Seychelles

Flag description: five oblique bands of blue (hoist side), yellow, red, white, and green (bottom) radiating from the bottom of the hoist side

ECONOMY

Economy—overview: Since independence in 1976, per capita output in this Indian Ocean archipelago has expanded to roughly seven times the pre-independence, near-subsistence level, moving the island into the upper-middle income group of countries. Growth has been led by the tourist sector, which employs about 30% of the labor force and provides more than 70% of hard currency earnings, and by tuna fishing. In recent years, the government has encouraged foreign investment to upgrade hotels and other services. At the same time, the government has moved to reduce the dependence on tourism by promoting the development of farming, fishing, and small-scale manufacturing. Sharp drops illustrated the vulnerability of the tourist sector in 1991–92 due largely to the Gulf War and once again following the 11 September 2001 terrorist attacks on the US. Economic growth slowed in 1998–2002 and fell in 2003–04, due to sluggish tourist and tuna sectors, but resumed in 2005–07. Real GDP grew by 5.8% in 2007, driven by tourism and a boom in tourism-related construction. The Seychelles rupee was allowed to depreciate in 2006 after being overvalued for years and fell by 10% in the first 9 months of 2007.

GDP (purchasing power parity): $1.378 billion (2007 est.)

GDP (official exchange rate): $710 million (2007 est.)

GDP—real growth rate: 5.3% (2007 est.)

GDP—per capita (PPP): $16,600 (2007 est.)

GDP—composition by sector:
agriculture: 2.4%
industry: 25.7%
services: 71.9% (2007 est.)

Labor force: 39,560 (2006)

Labor force—by occupation: *agriculture:* 3%
industry: 23%
services: 74% (2006)

Unemployment rate: 2% (2006 est.)

Population below poverty line: NA%

Household income or consumption by percentage share:
lowest 10%: NA%
highest 10%: NA%

Inflation rate (consumer prices): 5.7% (2007 est.)

Investment (gross fixed): 14% of GDP (2007 est.)

Budget: *revenues:* $372.4 million
expenditures: $355.2 million (2007 est.)

Public debt: 144.3% of GDP (2007 est.)

Agriculture—products: coconuts, cinnamon, vanilla, sweet potatoes, cassava (tapioca), bananas; poultry; tuna

Industries: fishing, tourism, processing of coconuts and vanilla, coir (coconut fiber) rope, boat building, printing, furniture; beverages

Industrial production growth rate: 1% (2007 est.)

Electricity—production: 252 million kWh (2006)

Electricity—production by source:
fossil fuel: 100%
hydro: 0%
nuclear: 0%
other: 0% (2001)

Electricity—consumption: 216.6 million kWh (2006)

Electricity—exports: 0 kWh (2005)

Electricity—imports: 0 kWh (2005)

Oil—production: 0 bbl/day (2006)

Oil—consumption: 6,453 bbl/day (2006)

Oil—exports: 0 bbl/day (2006)

Oil—imports: 6,453 bbl/day (2006)

Oil—proved reserves: 0 bbl (1 January 2006)

Natural gas—production: 0 cu m (2005 est.)

Natural gas—consumption: 0 cu m (2005 est.)

Natural gas—exports: 0 cu m (2005 est.)

Natural gas—imports: 0 cu m (2005)

Natural gas—proved reserves: 0 cu m (1 January 2006 est.)

Current account balance: -$275 million (2007 est.)

Exports: $400 million f.o.b. (2007 est.)

Exports—commodities: canned tuna, frozen fish, cinnamon bark, copra, petroleum products (reexports)

Exports—partners: UK 25.5%, France 17.5%, Italy 11.9%, Mauritius 8.5%, Japan 8.3%, Spain 8.2%, Netherlands 4.3% (2006)

Imports: $720 million f.o.b. (2007 est.)

Imports—commodities: machinery and equipment, foodstuffs, petroleum products, chemicals

Imports—partners: Saudi Arabia 17.7%, South Africa 9.7%, Spain 8.1%, France 7.8%, Singapore 7.2%, Italy 4.8%, UK 4% (2006)

Economic aid—recipient: $18.81 million (2005)

Reserves of foreign exchange and gold: $41.61 million (31 December 2007 est.)

Debt—external: $1.059 billion (31 December 2007 est.)

Market value of publicly traded shares: $NA

Currency (code): Seychelles rupee (SCR)

Currency code: SCR

Exchange rates: Seychelles rupees per US dollar—6.5 (2007), 5.5 (2006), 5.5 (2005), 5.5 (2004), 5.4007 (2003)

Fiscal year: calendar year

COMMUNICATIONS

Telephones—main lines in use: 20,700 (2006)

Telephones—mobile cellular: 70,300 (2006)

Telephone system:
general assessment: effective system
domestic: combined fixed-line and mobile-cellular teledensity is roughly 110 telephones per 100 persons; radiotelephone communications between islands in the archipelago
international: country code—248; direct radiotelephone communications with adjacent island countries and African coastal countries; satellite earth station—1 Intelsat (Indian Ocean)

Radio broadcast stations: AM 1, FM 1, shortwave 2 (2001)

Radios: 42,000 (1997)

Television broadcast stations: 2 (plus 9 repeaters) (1997)

Televisions: 11,000 (1997)

Internet country code: .sc

Internet hosts: 187 (2007)

Internet Service Providers (ISPs): 1 (2000)

Internet users: 29,000 (2006)

TRANSPORTATION

Airports: 15 (2007)

Airports—with paved runways:
total: 9
2,438 to 3,047 m: 1

914 to 1,523 m: 6
under 914 m: 2 (2007)
Airports—with unpaved runways:
total: 6
914 to 1,523 m: 2
under 914 m: 4 (2007)
Roadways:
total: 458 km
paved: 440 km
unpaved: 18 km (2003)
Merchant marine:
total: 6 ships (1000 GRT or over)
108,348 GRT/165,593 DWT
by type: cargo 1, carrier 1, chemical
tanker 4

foreign-owned: 3 (Hong Kong 1, Nigeria
1, South Africa 1) (2007)
Ports and terminals: Victoria

MILITARY

Military branches: Seychelles Defense
Force: Army, Coast Guard (includes
Naval Wing, Air Wing), National Guard
(2005)
Military service age and obligation: 18
years of age for voluntary military service
(younger with parental consent); no
conscription (2008)
**Manpower available for military
service:** *males age 16–49:* 23,598

females age 16–49: 24,424 (2008 est.)
Manpower fit for military service:
males age 16–49: 17,942
females age 16–49: 20,436 (2008 est.)
**Military expenditures—percent of
GDP:** 2% (2006 est.)

TRANSNATIONAL ISSUES

Disputes—international: together with
Mauritius, Seychelles claims the Chagos
Archipelago (UK-administered British
Indian Ocean Territory)

SIERRA LEONE

INTRODUCTION

Background: Democracy is slowly being
reestablished after the civil war from
1991 to 2002 that resulted in tens of
thousands of deaths and the displacement of more than 2 million people
(about one-third of the population). The
military, which took over full responsibility for security following the departure
of UN peacekeepers at the end of 2005,
is increasingly developing as a guarantor
of the country's stability. The armed
forces remained on the sideline during
the 2007 presidential election, but still
look to the UN Integrated Office in
Sierra Leone (UNIOSIL)—a civilian
UN mission—to support efforts to consolidate peace. The new government's
priorities include furthering development, creating jobs, and stamping out
endemic corruption.

GEOGRAPHY

Location: Western Africa, bordering the
North Atlantic Ocean, between Guinea
and Liberia
Geographic coordinates: 8 30 N, 11 30
W

Map references: Africa
Area: *total:* 71,740 sq km
land: 71,620 sq km
water: 120 sq km
Area—comparative: slightly smaller
than South Carolina
Land boundaries:
total: 958 km
border countries: Guinea 652 km, Liberia
306 km
Coastline: 402 km
Maritime claims:
territorial sea: 12 nm
contiguous zone: 24 nm
exclusive economic zone: 200 nm
continental shelf: 200 nm
Climate: tropical; hot, humid; summer
rainy season (May to December); winter
dry season (December to April)
Terrain: coastal belt of mangrove
swamps, wooded hill country, upland
plateau, mountains in east
Elevation extremes:
lowest point: Atlantic Ocean 0 m
highest point: Loma Mansa (Bintimani)
1,948 m
Natural resources: diamonds, titanium
ore, bauxite, iron ore, gold, chromite
Land use:
arable land: 7.95%
permanent crops: 1.05%
other: 91% (2005)
Irrigated land: 300 sq km (2003)
Total renewable water resources: 160
cu km (1987)
**Freshwater withdrawal (domestic/
industrial/agricultural):** *total:* 0.38 cu
km/yr (5%/3%/92%)
per capita: 69 cu m/yr (2000)
Natural hazards: dry, sand-laden harmattan winds blow from the Sahara
(December to February); sandstorms,
dust storms
Environment—current issues: rapid
population growth pressuring the envi-

ronment; overharvesting of timber,
expansion of cattle grazing, and slash-
and-burn agriculture have resulted in
deforestation and soil exhaustion; civil
war depleted natural resources; overfishing
**Environment—international agree-
ments:** *party to:* Biodiversity, Climate
Change, Climate Change-Kyoto
Protocol, Desertification, Endangered
Species, Law of the Sea, Marine Life
Conservation, Ozone Layer Protection,
Ship Pollution, Wetlands
signed, but not ratified: Environmental
Modification
Geography—note: rainfall along the
coast can reach 495 cm (195 inches) a
year, making it one of the wettest places
along coastal, western Africa

PEOPLE

Population: 6,294,774 (July 2008 est.)
Age structure:
0–14 years: 44.6% (male
1,377,981/female 1,429,993)
15–64 years: 52.2% (male
1,573,990/female 1,708,840)
65 years and over: 3.2% (male
94,359/female 109,611) (2008 est.)
Median age:
total: 17.5 years
male: 17.2 years
female: 17.8 years (2008 est.)
Population growth rate: 2.282% (2008
est.)
Birth rate: 45.08 births/1,000 population
(2008 est.)
Death rate: 22.26 deaths/1,000 population (2008 est.)
Net migration rate: 1 migrant(s)/1,000
population
note: refugees currently in surrounding
countries are slowly returning (2005 est.)
Sex ratio:
at birth: 1.03 male(s)/female

579

under 15 years: 0.96 male(s)/female
15–64 years: 0.92 male(s)/female
65 years and over: 0.86 male(s)/female
total population: 0.94 male(s)/female
(2008 est.)
Infant mortality rate:
total: 156.48 deaths/1,000 live births
male: 173.59 deaths/1,000 live births
female: 138.85 deaths/1,000 live births
(2008 est.)
Life expectancy at birth:
total population: 40.93 years
male: 38.64 years
female: 43.28 years (2008 est.)
Total fertility rate: 5.95 children
born/woman (2008 est.)
HIV/AIDS—adult prevalence rate: 7%
(2001 est.)
**HIV/AIDS—people living with
HIV/AIDS:** 170,000 (2001 est.)
HIV/AIDS—deaths: 11,000 (2001 est.)
Major infectious diseases:
degree of risk: very high
food or waterborne diseases: bacterial and
protozoal diarrhea, hepatitis A, and
typhoid fever
vectorborne diseases: malaria and yellow
fever
water contact disease: schistosomiasis
aerosolized dust or soil contact disease:
Lassa fever (2008)
Nationality: *noun:* Sierra Leonean(s)
adjective: Sierra Leonean
Ethnic groups: 20 African ethnic groups
90% (Temne 30%, Mende 30%, other
30%), Creole (Krio) 10% (descendants
of freed Jamaican slaves who were settled
in the Freetown area in the late-18th
century), refugees from Liberia's recent
civil war, small numbers of Europeans,
Lebanese, Pakistanis, and Indians
Religions: Muslim 60%, Christian 10%,
indigenous beliefs 30%
Languages: English (official, regular use
limited to literate minority), Mende
(principal vernacular in the south),
Temne (principal vernacular in the
north), Krio (English-based Creole,
spoken by the descendants of freed
Jamaican slaves who were settled in the
Freetown area, a lingua franca and a first
language for 10% of the population but
understood by 95%)
Literacy: *definition:* age 15 and over can
read and write English, Mende, Temne,
or Arabic
total population: 35.1%
male: 46.9%
female: 24.4% (2004 est.)

GOVERNMENT

Country name:
conventional long form: Republic of Sierra
Leone
conventional short form: Sierra Leone

local long form: Republic of Sierra Leone
local short form: Sierra Leone
Government type: constitutional democracy
Capital: *name:* Freetown
geographic coordinates: 8 30 N, 13 15 W
time difference: UTC 0 (5 hours ahead of
Washington, DC during Standard Time)
Administrative divisions: 3 provinces
and 1 area*; Eastern, Northern,
Southern, Western*
Independence: 27 April 1961 (from
UK)
National holiday: Independence Day, 27
April (1961)
Constitution: 1 October 1991; subsequently amended several times
Legal system: based on English law and
customary laws indigenous to local
tribes; has not accepted compulsory ICJ
jurisdiction
Suffrage: 18 years of age; universal
Executive branch:
chief of state: President Ernest Bai
KOROMA (since 17 September 2007);
note—the president is both the chief of
state and head of government
head of government: President Ernest Bai
KOROMA (since 17 September 2007)
cabinet: Ministers of State appointed by
the president with the approval of the
House of Representatives; the cabinet is
responsible to the president
elections: president elected by popular
vote for a five-year term (eligible for a
second term); election last held 11
August 2007 and 8 September 2007
(next to be held in 2012)
election results: second round results; percent of vote—Ernest Bai KOROMA
54.6%, Solomon BEREWA 45.4%
Legislative branch: unicameral
Parliament (124 seats; 112 members
elected by popular vote, 12 filled by paramount chiefs elected in separate elections; to serve five-year terms)
elections: last held on 11 August 2007
(next to be held in 2012)
election results: percent of vote by party—
NA; seats by party—APC 59, SLPP 43,
PMDC 10
Judicial branch: Supreme Court;
Appeals Court; High Court
Political parties and leaders: All
People's Congress or APC [Ernest Bai
KOROMA]; Peace and Liberation Party
or PLP [Darlington MORRISON];
People's Movement for Democratic
Change or PMDC [Charles MARGAI];
Sierra Leone People's Party or SLPP
[Solomon BEREWA]; numerous others
Political pressure groups and leaders:
trade unions and student unions
International organization participation: ACP, AfDB, AU, C, ECOWAS,

FAO, G-77, IAEA, IBRD, ICAO, ICCt,
ICRM, IDA, IDB, IFAD, IFC, IFRCS,
ILO, IMF, IMO, Interpol, IOC, IOM,
ITU, ITUC, MIGA, NAM, OIC,
OPCW, UN, UNCTAD, UNESCO,
UNIDO, UNMIT, UNWTO, UPU,
WCL, WCO, WFTU, WHO, WIPO,
WMO, WTO
Diplomatic representation in the US:
chief of mission: Ambassador Bockari
Kortu STEVENS
chancery: 1701 19th Street NW,
Washington, DC 20009
telephone: [1] (202) 939-9261 through
9263
FAX: [1] (202) 483-1793
Diplomatic representation from the US:
chief of mission: Ambassador Thomas N.
HULL
embassy: Corner of Walpole and Siaka
Stevens Streets, Freetown
mailing address: use embassy street address
telephone: [232] (22) 515 000 or [232]
(76) 515 000
FAX: [232] (22) 225471
Flag description: three equal horizontal
bands of light green (top), white, and
light blue

ECONOMY

Economy—overview: Sierra Leone is an
extremely poor nation with tremendous
inequality in income distribution. While
it possesses substantial mineral, agricultural, and fishery resources, its physical
and social infrastructure is not well developed, and serious social disorders continue to hamper economic development.
Nearly half of the working-age population engages in subsistence agriculture.
Manufacturing consists mainly of the
processing of raw materials and of light
manufacturing for the domestic market.
Alluvial diamond mining remains the
major source of hard currency earnings
accounting for nearly half of Sierra
Leone's exports. The fate of the economy
depends upon the maintenance of
domestic peace and the continued receipt
of substantial aid from abroad, which is
essential to offset the severe trade imbalance and supplement government revenues. The IMF has completed a Poverty
Reduction and Growth Facility program
that helped stabilize economic growth
and reduce inflation. A recent increase in
political stability has led to a revival of
economic activity such as the rehabilitation of bauxite and rutile mining.
GDP (purchasing power parity): $3.971
billion (2007 est.)
GDP (official exchange rate): $1.664
billion (2007 est.)
GDP—real growth rate: 6.8% (2007 est.)
GDP—per capita (PPP): $700 (2007 est.)

GDP—composition by sector:
agriculture: 49%
industry: 31%
services: 21% (2001 est.)
Labor force: 1.369 million (1981 est.)
Labor force—by occupation:
agriculture: NA%
industry: NA%
services: NA%
Unemployment rate: NA%
Population below poverty line: 70.2% (2004)
Household income or consumption by percentage share:
lowest 10%: 0.5%
highest 10%: 43.6% (1989)
Distribution of family income—Gini index: 62.9 (1989)
Inflation rate (consumer prices): 11.7% (2007 est.)
Budget: *revenues:* $96 million
expenditures: $351 million (2000 est.)
Agriculture—products: rice, coffee, cocoa, palm kernels, palm oil, peanuts; poultry, cattle, sheep, pigs; fish
Industries: diamond mining; small-scale manufacturing (beverages, textiles, cigarettes, footwear); petroleum refining, small commercial ship repair
Industrial production growth rate: NA%
Electricity—production: 245 million kWh (2005)
Electricity—production by source:
fossil fuel: 100%
hydro: 0%
nuclear: 0%
other: 0% (2001)
Electricity—consumption: 227.9 million kWh (2005)
Electricity—exports: 0 kWh (2005)
Electricity—imports: 0 kWh (2005)
Oil—production: 0.7008 bbl/day (2005 est.)
Oil—consumption: 8,000 bbl/day (2005 est.)
Oil—exports: 431.1 bbl/day (2004)
Oil—imports: 8,864 bbl/day (2004)
Oil—proved reserves: 0 bbl (1 January 2006 est.)
Natural gas—production: 0 cu m (2005 est.)
Natural gas—consumption: 0 cu m (2005 est.)
Natural gas—exports: 0 cu m (2005 est.)
Natural gas—imports: 0 cu m (2005)
Natural gas—proved reserves: 0 cu m (1 January 2006 est.)
Current account balance: -$63 million (2007 est.)

Exports: $216 million f.o.b. (2006)
Exports—commodities: diamonds, rutile, cocoa, coffee, fish
Exports—partners: Belgium 52.1%, US 19.1%, Netherlands 6.8% (2006)
Imports: $560 million f.o.b. (2006)
Imports—commodities: foodstuffs, machinery and equipment, fuels and lubricants, chemicals
Imports—partners: Cote d'Ivoire 9.3%, US 7.7%, China 7.7%, Brazil 6.9%, UK 6.7%, Netherlands 5.5%, South Africa 4.5%, India 4.4%, France 4.2% (2006)
Economic aid—recipient: $343.4 million (2005 est.)
Debt—external: $1.61 billion (2003 est.)
Market value of publicly traded shares: $NA
Currency (code): leone (SLL)
Currency code: SLL
Exchange rates: leones per US dollar—NA (2007), 2,961.7 (2006), 2,889.6 (2005), 2,701.3 (2004), 2,347.9 (2003)
Fiscal year: calendar year

COMMUNICATIONS

Telephones—main lines in use: 24,000 (2002)
Telephones—mobile cellular: 113,200 (2003)
Telephone system:
general assessment: marginal telephone service
domestic: the national microwave radio relay trunk system connects Freetown to Bo and Kenema
international: country code—232; satellite earth station—1 Intelsat (Atlantic Ocean) (2000)
Radio broadcast stations: AM 1, FM 9, shortwave 1 (2001)
Radios: 1.12 million (1997)
Television broadcast stations: 2 (1999)
Televisions: 53,000 (1997)
Internet country code: .sl
Internet hosts: 46 (2007)
Internet Service Providers (ISPs): 1 (2001)
Internet users: 10,000 (2005)

TRANSPORTATION

Airports: 10 (2007)
Airports—with paved runways:
total: 1
over 3,047 m: 1 (2007)
Airports—with unpaved runways:
total: 9
914 to 1,523 m: 7
under 914 m: 2 (2007)

Heliports: 2 (2007)
Roadways:
total: 11,300 km
paved: 904 km
unpaved: 10,396 km (2002)
Waterways: 800 km (600 km year round) (2005)
Merchant marine:
total: 113 ships (1000 GRT or over) 314,549 GRT/419,409 DWT
by type: bulk carrier 1, cargo 85, chemical tanker 4, combination ore/oil 1, container 4, liquefied gas 1, livestock carrier 1, passenger 1, passenger/cargo 4, petroleum tanker 7, roll on/roll off 4
foreign-owned: 47 (Belgium 1, China 8, Greece 1, Romania 2, Russia 5, Syria 8, Turkey 7, Ukraine 8, UAE 7) (2007)
Ports and terminals: Freetown, Pepel, Sherbro Islands

MILITARY

Military branches: Republic of Sierra Leone Armed Forces (RSLAF): Army (includes Navy (Maritime Wing), Air Wing) (2008)
Military service age and obligation: 17 years 6 months of age for voluntary military service (younger with parental consent); no conscription (2008)
Manpower available for military service:
males age 16–49: 1,315,561 (2008 est.)
Manpower fit for military service:
males age 16–49: 671,418 (2008 est.)
Military expenditures—percent of GDP: 2.3% (2006)

TRANSNATIONAL ISSUES

Disputes—international: as domestic fighting among disparate ethnic groups, rebel groups, warlords, and youth gangs in Cote d'Ivoire, Guinea, Liberia, and Sierra Leone gradually abate, the number of refugees in border areas has begun to slowly dwindle; UN Mission in Sierra Leone (UNAMSIL) has maintained over 4,000 peacekeepers in Sierra Leone since 1999; Sierra Leone considers excessive Guinea's definition of the flood plain limits to define the left bank boundary of the Makona and Moa rivers and protests Guinea's continued occupation of these lands including the hamlet of Yenga occupied since 1998
Refugees and internally displaced persons: *refugees (country of origin):* 27,311 (Liberia) (2007)

SINGAPORE

INTRODUCTION

Background: Singapore was founded as a British trading colony in 1819. It joined the Malaysian Federation in 1963 but separated two years later and became independent. Singapore subsequently became one of the world's most prosperous countries with strong international trading links (its port is one of the world's busiest in terms of tonnage handled) and with per capita GDP equal to

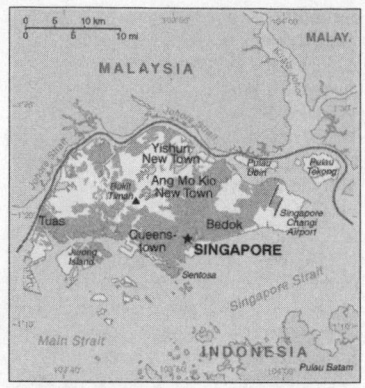

that of the leading nations of Western Europe.

GEOGRAPHY

Location: Southeastern Asia, islands between Malaysia and Indonesia
Geographic coordinates: 1 22 N, 103 48 E
Map references: Southeast Asia
Area: *total:* 692.7 sq km
land: 682.7 sq km
water: 10 sq km
Area—comparative: slightly more than 3.5 times the size of Washington, DC
Land boundaries: 0 km
Coastline: 193 km
Maritime claims:
territorial sea: 3 nm
exclusive fishing zone: within and beyond territorial sea, as defined in treaties and practice
Climate: tropical; hot, humid, rainy; two distinct monsoon seasons—Northeastern monsoon (December to March) and Southwestern monsoon (June to September); inter-monsoon—frequent afternoon and early evening thunderstorms
Terrain: lowland; gently undulating central plateau contains water catchment area and nature preserve
Elevation extremes:
lowest point: Singapore Strait 0 m
highest point: Bukit Timah 166 m
Natural resources: fish, deepwater ports
Land use: *arable land:* 1.47%
permanent crops: 1.47%
other: 97.06% (2005)
Irrigated land: NA
Total renewable water resources: 0.6 cu km (1975)
Freshwater withdrawal (domestic/industrial/agricultural): *total:* 0.19 cu km/yr (45%/51%/4%)
per capita: 44 cu m/yr (1975)
Natural hazards: NA
Environment—current issues: industrial pollution; limited natural fresh water resources; limited land availability

presents waste disposal problems; seasonal smoke/haze resulting from forest fires in Indonesia
Environment—international agreements: *party to:* Biodiversity, Climate Change, Climate Change-Kyoto Protocol, Desertification, Endangered Species, Hazardous Wastes, Law of the Sea, Ozone Layer Protection, Ship Pollution
signed, but not ratified: none of the selected agreements
Geography—note: focal point for Southeast Asian sea routes

PEOPLE

Population: 4,608,167 (July 2008 est.)
Age structure:
0–14 years: 14.8% (male 353,333/female 329,005)
15–64 years: 76.5% (male 1,717,357/female 1,809,462)
65 years and over: 8.7% (male 177,378/female 221,632) (2008 est.)
Median age:
total: 38.4 years
male: 38 years
female: 38.8 years (2008 est.)
Population growth rate: 1.135% (2008 est.)
Birth rate: 8.99 births/1,000 population (2008 est.)
Death rate: 4.53 deaths/1,000 population (2008 est.)
Net migration rate: 6.88 migrant(s)/1,000 population (2008 est.)
Sex ratio:
at birth: 1.08 male(s)/female
under 15 years: 1.07 male(s)/female
15–64 years: 0.95 male(s)/female
65 years and over: 0.8 male(s)/female
total population: 0.95 male(s)/female (2008 est.)
Infant mortality rate:
total: 2.3 deaths/1,000 live births
male: 2.51 deaths/1,000 live births
female: 2.08 deaths/1,000 live births (2008 est.)
Life expectancy at birth:
total population: 81.89 years
male: 79.29 years
female: 84.68 years (2008 est.)
Total fertility rate: 1.08 children born/woman (2008 est.)
HIV/AIDS—adult prevalence rate: 0.2% (2003 est.)
HIV/AIDS—people living with HIV/AIDS: 4,100 (2003 est.)
HIV/AIDS—deaths: fewer than 200 (2003 est.)
Nationality: *noun:* Singaporean(s)
adjective: Singapore
Ethnic groups: Chinese 76.8%, Malay 13.9%, Indian 7.9%, other 1.4% (2000 census)

Religions: Buddhist 42.5%, Muslim 14.9%, Taoist 8.5%, Hindu 4%, Catholic 4.8%, other Christian 9.8%, other 0.7%, none 14.8% (2000 census)
Languages: Mandarin 35%, English 23%, Malay 14.1%, Hokkien 11.4%, Cantonese 5.7%, Teochew 4.9%, Tamil 3.2%, other Chinese dialects 1.8%, other 0.9% (2000 census)
Literacy:
definition: age 15 and over can read and write
total population: 92.5%
male: 96.6%
female: 88.6% (2000 census)

GOVERNMENT

Country name:
conventional long form: Republic of Singapore
conventional short form: Singapore
local long form: Republic of Singapore
local short form: Singapore
Government type: parliamentary republic
Capital: *name:* Singapore
geographic coordinates: 1 17 N, 103 51 E
time difference: UTC+8 (13 hours ahead of Washington, DC during Standard Time)
Administrative divisions: none
Independence: 9 August 1965 (from Malaysian Federation)
National holiday: National Day, 9 August (1965)
Constitution: 3 June 1959; amended 1965 (based on preindependence State of Singapore Constitution)
Legal system: based on English common law; has not accepted compulsory ICJ jurisdiction
Suffrage: 21 years of age; universal and compulsory
Executive branch:
chief of state: President S R NATHAN (since 1 September 1999)
note: uses S R NATHAN but his full name and the one used in formal communications is Sellapan RAMANATHAN
head of government: Prime Minister LEE Hsien Loong (since 12 August 2004); Senior Minister GOH Chok Tong (since 12 August 2004); Minister Mentor LEE Kuan Yew (since 12 August 2004); Deputy Prime Ministers Shunmugam JAYAKUMAR (since 12 August 2004) and WONG Kan Seng (since 1 September 2005)
cabinet: appointed by president, responsible to parliament
elections: president elected by popular vote for six-year term; appointed on 17 August 2005 (next election to be held by August 2011); following legislative elec-

tions, leader of majority party or leader of majority coalition is usually appointed prime minister by president; deputy prime ministers appointed by president
election results: Sellapan Rama (S R) NATHAN appointed president in August 2005 after Presidential Elections Committee disqualified three other would-be candidates; scheduled election not held
Legislative branch: unicameral Parliament (84 seats; members elected by popular vote to serve five-year terms); note—in addition, there are up to nine nominated members; up to three losing opposition candidates who came closest to winning seats may be appointed as "nonconstituency" members
elections: last held on 6 May 2006 (next to be held by 2011)
election results: percent of vote by party—PAP 66.6%, WP 16.3%, SDA 13%, SDP 4.1%; seats by party—PAP 82, WP 1, SDA 1
Judicial branch: Supreme Court (chief justice is appointed by the president with the advice of the prime minister, other judges are appointed by the president with the advice of the chief justice); Court of Appeals
Political parties and leaders: People's Action Party or PAP [LEE Hsien Loong]; Singapore Democratic Alliance or SDA [CHIAM See Tong]; Singapore Democratic Party or SDP [CHEE Soon Juan]; Workers' Party or WP [Sylvia LIM Swee Lian]
note: SDA includes Singapore Justice Party or SJP, Singapore National Malay Organization or PKMS, Singapore People's Party or SPP
Political pressure groups and leaders: NA
International organization participation: ADB, APEC, APT, ARF, ASEAN, BIS, C, CP, EAS, G-77, IAEA, IBRD, ICAO, ICC, ICRM, IDA, IFC, IFRCS, IHO, ILO, IMF, IMO, IMSO, Interpol, IOC, IPU, ISO, ITSO, ITU, ITUC, MIGA, NAM, OPCW, PCA, UN, UNCTAD, UNESCO, UNMIT, UPU, WCL, WCO, WHO, WIPO, WMO, WTO
Diplomatic representation in the US:
chief of mission: Ambassador CHAN Heng Chee
chancery: 3501 International Place NW, Washington, DC 20008
telephone: [1] (202) 537-3100
FAX: [1] (202) 537-0876
consulate(s) general: San Francisco
consulate(s): New York
Diplomatic representation from the US:
chief of mission: Ambassador Patricia L. HERBOLD

embassy: 27 Napier Road, Singapore 258508
mailing address: FPO AP 96507-0001
telephone: [65] 6476-9100
FAX: [65] 6476-9340
Flag description: two equal horizontal bands of red (top) and white; near the hoist side of the red band, there is a vertical, white crescent (closed portion is toward the hoist side) partially enclosing five white five-pointed stars arranged in a circle

ECONOMY

Economy—overview: Singapore has a highly developed and successful free-market economy. It enjoys a remarkably open and corruption-free environment, stable prices, and a per capita GDP equal to that of the four largest West European countries. The economy depends heavily on exports, particularly in consumer electronics and information technology products. It was hard hit from 2001–03 by the global recession, by the slump in the technology sector, and by an outbreak of Severe Acute Respiratory Syndrome (SARS) in 2003, which curbed tourism and consumer spending. Fiscal stimulus, low interest rates, a surge in exports, and internal flexibility led to vigorous growth in 2004–07 with real GDP growth averaging 7% annually. The government hopes to establish a new growth path that will be less vulnerable to the global demand cycle for information technology products—it has attracted major investments in pharmaceuticals and medical technology production—and will continue efforts to establish Singapore as Southeast Asia's financial and high-tech hub.
GDP (purchasing power parity): $228.1 billion (2007 est.)
GDP (official exchange rate): $161.3 billion (2007 est.)
GDP—real growth rate: 7.7% (2007 est.)
GDP—per capita (PPP): $49,700 (2007 est.)
GDP—composition by sector:
agriculture: 0%
industry: 31.2%
services: 68.8% (2007 est.)
Labor force: 2.751 million (2007 est.)
Labor force—by occupation: manufacturing 21%, construction 5%, transportation and communication 7%, financial, business, and other services 42%, other 25% (2006)
Unemployment rate: 2.1% (2007 est.)
Population below poverty line: NA%
Household income or consumption by percentage share:
lowest 10%: 1.9%

highest 10%: 32.8% (1998)
Distribution of family income—Gini index: 52.2 (2005)
Inflation rate (consumer prices): 2.1% (2007)
Investment (gross fixed): 24.9% of GDP (2007 est.)
Budget:
revenues: $27 billion
expenditures: $21.5 billion (2007 est.)
Public debt: 101.2% of GDP (2007 est.)
Agriculture—products: rubber, copra, fruit, orchids, vegetables; poultry, eggs; fish, ornamental fish
Industries: electronics, chemicals, financial services, oil drilling equipment, petroleum refining, rubber processing and rubber products, processed food and beverages, ship repair, offshore platform construction, life sciences, entrepot trade
Industrial production growth rate: 7.4% (2007 est.)
Electricity—production: 39.44 billion kWh (2006)
Electricity—production by source:
fossil fuel: 100%
hydro: 0%
nuclear: 0%
other: 0% (2001)
Electricity—consumption: 35.92 billion kWh (2006)
Electricity—exports: 0 kWh (2006)
Electricity—imports: 0 kWh (2006)
Oil—production: 9,836 bbl/day (2005 est.)
Oil—consumption: 802,000 bbl/day (2005 est.)
Oil—exports: 1.073 million bbl/day (2004)
Oil—imports: 1.83 million bbl/day (2004)
Oil—proved reserves: 0 bbl (1 January 2006 est.)
Natural gas—production: 0 cu m (2006 est.)
Natural gas—consumption: 6.8 billion cu m (2006 est.)
Natural gas—exports: 0 cu m (2006 est.)
Natural gas—imports: 6.339 billion cu m
note: from Indonesia and Malaysia (2005)
Natural gas—proved reserves: 0 cu m (1 January 2006 est.)
Current account balance: $39.16 billion (2007 est.)
Exports: $450.6 billion f.o.b. (2007 est.)
Exports—commodities: machinery and equipment (including electronics), consumer goods, chemicals, mineral fuels
Exports—partners: Malaysia 13.1%, US 10.2%, Hong Kong 10.1%, China 9.7%, Indonesia 9.2%, Japan 5.5%, Thailand 4.2% (2006)

583

Imports: $396 billion (2007 est.)

Imports—commodities: machinery and equipment, mineral fuels, chemicals, foodstuffs

Imports—partners: Malaysia 13%, US 12.7%, China 11.4%, Japan 8.3%, Taiwan 6.4%, Indonesia 6.2%, South Korea 4.4% (2006)

Economic aid—recipient: $0 (2007)

Reserves of foreign exchange and gold: $163 billion (31 December 2007 est.)

Debt—external: $25.59 billion (31 December 2007 est.)

Stock of direct foreign investment—at home: $214.5 billion (2007 est.)

Stock of direct foreign investment—abroad: $111.2 billion (2005)

Market value of publicly traded shares: $382.4 billion (2007)

Currency (code): Singapore dollar (SGD)

Currency code: SGD

Exchange rates: Singapore dollars per US dollar—1.507 (2007), 1.5889 (2006), 1.6644 (2005), 1.6902 (2004), 1.7422 (2003)

Fiscal year: 1 April—31 March

COMMUNICATIONS

Telephones—main lines in use: 1.854 million (2006)

Telephones—mobile cellular: 4.789 million (2006)

Telephone system:
general assessment: excellent service
domestic: excellent domestic facilities; launched 3G wireless service in February 2005; combined fixed-line and mobile-cellular teledensity is about 150 telephones per 100 persons
international: country code—65; numerous submarine cables provide links throughout Asia, Australia, the Middle East, Europe, and US; satellite earth stations -4; supplemented by VSAT coverage (2003)

Radio broadcast stations: AM 0, FM 17, shortwave 2 (2003)

Radios: 2.6 million (2000)

Television broadcast stations: 1 (broadcasting on six channels); additional reception of numerous UHF and VHF signals originating in Malaysia and Indonesia (2006)

Televisions: 1.33 million (1997)

Internet country code: .sg

Internet hosts: 954,475 (2007)

Internet Service Providers (ISPs): 9 (2000)

Internet users: 1.717 million (2006)

TRANSPORTATION

Airports: 8 (2007)

Airports—with paved runways:
total: 8
over 3,047 m: 2
2,438 to 3,047 m: 1
1,524 to 2,437 m: 4
914 to 1,523 m: 1 (2007)

Pipelines: gas 139 km; refined products 8 km (2007)

Roadways:
total: 3,234 km
paved: 3,234 km (includes 150 km of expressways) (2005)

Merchant marine:
total: 1,131 ships (1000 GRT or over) 33,237,005 GRT/52,487,127 DWT
by type: bulk carrier 167, cargo 85, carrier 1, chemical tanker 156, container 231, liquefied gas 72, livestock carrier 2, petroleum tanker 355, refrigerated cargo 6, roll on/roll off 3, specialized tanker 7, vehicle carrier 46
foreign-owned: 652 (Australia 6, Bangladesh 2, Belgium 8, China 19, Denmark 68, France 1, Germany 18, Greece 14, Hong Kong 37, India 9, Indonesia 56, Italy 4, Japan 108, South Korea 7, Malaysia 28, Norway 125, Philippines 4, Slovenia 1, Sweden 17, Switzerland 2, Taiwan 60, Thailand 20, UAE 8, UK 13, US 17)
registered in other countries: 293 (Bahamas 9, Belize 3, Bermuda 1, Bolivia 1, Cambodia 2, Cayman Islands 10, Cyprus 1, Dominica 8, France 2, Honduras 10, Hong Kong 11, Indonesia 26, Isle of Man 2, Kiribati 1, Liberia 42, Malaysia 22, Marshall Islands 12, Mongolia 12, Nigeria 1, Norway 1, Panama 83, Philippines 1, St Vincent and The Grenadines 6, Thailand 2, Tuvalu 13, US 11, unknown 4) (2007)

Ports and terminals: Singapore

MILITARY

Military branches: Singapore Armed Forces: Army, Navy, Air Force (includes Air Defense) (2008)

Military service age and obligation: 18 years of age for male compulsory military service; 16 years of age for volunteers; 2-year conscript service obligation, with a reserve obligation to age 40 (enlisted) or age 50 (officers) (2008)

Manpower available for military service:
males age 16–49: 1,277,862 (2008 est.)

Manpower fit for military service:
males age 16–49: 1,038,603 (2008 est.)

Military expenditures—percent of GDP: 4.9% (2005 est.)

TRANSNATIONAL ISSUES

Disputes—international: disputes persist with Malaysia over deliveries of fresh water to Singapore, Singapore's extensive land reclamation works, bridge construction, and maritime boundaries in the Johor and Singapore Straits; in November 2007, the ICJ will hold public hearings as a consequence of the Memorials and Countermemorials filed by the parties in 2003 and 2005 over sovereignty of Pedra Branca Island/Pulau Batu Puteh, Middle Rocks and South Ledge; Indonesia and Singapore continue to work on finalization of their 1973 maritime boundary agreement by defining unresolved areas north of Indonesia's Batam Island; piracy remains a problem in the Malacca Strait

Illicit drugs: drug abuse limited because of aggressive law enforcement efforts; as a transportation and financial services hub, Singapore is vulnerable, despite strict laws and enforcement, as a venue for money laundering

SLOVAKIA

INTRODUCTION

Background: The dissolution of the Austro-Hungarian Empire at the close of World War I allowed the Slovaks to join the closely related Czechs to form Czechoslovakia. Following the chaos of World War II, Czechoslovakia became a Communist nation within Soviet-dominated Eastern Europe. Soviet influence collapsed in 1989 and Czechoslovakia once more became free. The Slovaks and the Czechs agreed to separate peacefully on 1 January 1993. Slovakia joined both NATO and the EU in the spring of 2004.

GEOGRAPHY

Location: Central Europe, south of Poland

Geographic coordinates: 48 40 N, 19 30 E

Map references: Europe

Area: *total:* 48,845 sq km
land: 48,800 sq km
water: 45 sq km

Area—comparative: about twice the size of New Hampshire

Land boundaries: *total:* 1,524 km
border countries: Austria 91 km, Czech Republic 215 km, Hungary 677 km, Poland 444 km, Ukraine 97 km

Coastline: 0 km (landlocked)
Maritime claims: none (landlocked)
Climate: temperate; cool summers; cold, cloudy, humid winters
Terrain: rugged mountains in the central and northern part and lowlands in the south
Elevation extremes:
lowest point: Bodrok River 94 m
highest point: Gerlachovsky Stit 2,655 m
Natural resources: brown coal and lignite; small amounts of iron ore, copper and manganese ore; salt; arable land
Land use:
arable land: 29.23%
permanent crops: 2.67%
other: 68.1% (2005)
Irrigated land: 1,830 sq km (2003)
Total renewable water resources: 50.1 cu km (2003)
Freshwater withdrawal (domestic/industrial/agricultural): *total:* 1.04
per capita: 193 cu m/yr (2003)
Natural hazards: NA
Environment—current issues: air pollution from metallurgical plants presents human health risks; acid rain damaging forests
Environment—international agreements: *party to:* Air Pollution, Air Pollution-Nitrogen Oxides, Air Pollution-Persistent Organic Pollutants, Air Pollution-Sulfur 85, Air Pollution-Sulfur 94, Air Pollution-Volatile Organic Compounds, Antarctic Treaty, Biodiversity, Climate Change, Climate Change-Kyoto Protocol, Desertification, Endangered Species, Environmental Modification, Hazardous Wastes, Law of the Sea, Ozone Layer Protection, Ship Pollution, Wetlands, Whaling
signed, but not ratified: none of the selected agreements
Geography—note: landlocked; most of the country is rugged and mountainous; the Tatra Mountains in the north are interspersed with many scenic lakes and valleys

PEOPLE

Population: 5,455,407 (July 2008 est.)
Age structure:
0–14 years: 16.1% (male 448,083/female 427,643)
15–64 years: 71.7% (male 1,947,112/female 1,961,788)
65 years and over: 12.3% (male 250,787/female 419,994) (2008 est.)
Median age:
total: 36.5 years
male: 34.8 years
female: 38.2 years (2008 est.)
Population growth rate: 0.143% (2008 est.)
Birth rate: 10.64 births/1,000 population (2008 est.)
Death rate: 9.5 deaths/1,000 population (2008 est.)
Net migration rate: 0.3 migrant(s)/1,000 population (2008 est.)
Sex ratio:
at birth: 1.05 male(s)/female
under 15 years: 1.05 male(s)/female
15–64 years: 0.99 male(s)/female
65 years and over: 0.6 male(s)/female
total population: 0.94 male(s)/female (2008 est.)
Infant mortality rate:
total: 6.98 deaths/1,000 live births
male: 8.15 deaths/1,000 live births
female: 5.75 deaths/1,000 live births (2008 est.)
Life expectancy at birth:
total population: 75.17 years
male: 71.23 years
female: 79.32 years (2008 est.)
Total fertility rate: 1.34 children born/woman (2008 est.)
HIV/AIDS—adult prevalence rate: less than 0.1% (2001 est.)
HIV/AIDS—people living with HIV/AIDS: fewer than 200 (2003 est.)
HIV/AIDS—deaths: fewer than 100 (2001 est.)
Nationality:
noun: Slovak(s)
adjective: Slovak

Ethnic groups: Slovak 85.8%, Hungarian 9.7%, Roma 1.7%, Ruthenian/Ukrainian 1%, other and unspecified 1.8% (2001 census)
Religions: Roman Catholic 68.9%, Protestant 10.8%, Greek Catholic 4.1%, other or unspecified 3.2%, none 13% (2001 census)
Languages: Slovak (official) 83.9%, Hungarian 10.7%, Roma 1.8%, Ukrainian 1%, other or unspecified 2.6% (2001 census)
Literacy:
definition: age 15 and over can read and write
total population: 99.6%
male: 99.7%
female: 99.6% (2001 est.)

GOVERNMENT

Country name:
conventional long form: Slovak Republic
conventional short form: Slovakia
local long form: Slovenska Republika
local short form: Slovensko
Government type: parliamentary democracy
Capital: *name:* Bratislava
geographic coordinates: 48 09 N, 17 07 E
time difference: UTC+1 (6 hours ahead of Washington, DC during Standard Time)
daylight saving time: +1hr, begins last Sunday in March; ends last Sunday in October
Administrative divisions: 8 regions (kraje, singular—kraj); Banskobystricky kraj, Bratislavsky kraj, Kosicky kraj, Nitriansky kraj, Presovsky kraj, Trenciansky kraj, Trnavsky kraj, Zilinsky kraj
Independence: 1 January 1993 (Czechoslovakia split into the Czech Republic and Slovakia)
National holiday: Constitution Day, 1 September (1992)
Constitution: ratified 1 September 1992, effective 1 January 1993; changed in September 1998 to allow direct election of the president; amended February 2001 to allow Slovakia to apply for NATO and EU membership
Legal system: civil law system based on Austro-Hungarian codes; accepts compulsory ICJ jurisdiction with reservations; legal code modified to comply with the obligations of Organization on Security and Cooperation in Europe (OSCE) and to expunge Marxist-Leninist legal theory
Suffrage: 18 years of age; universal
Executive branch:
chief of state: President Ivan GASPAROVIC (since 15 June 2004)
head of government: Prime Minister Robert FICO (since 4 July 2006); Deputy

Prime Ministers Dusan CAPLOVIC, Robert KALINAK, Stefan HARABIN, Jan MIKOLAJ (since 4 July 2006)
cabinet: Cabinet appointed by the president on the recommendation of the prime minister
elections: president elected by popular vote for a five-year term (eligible for a second term); election last held 3 April and 17 April 2004 (next to be held in April 2009); following National Council elections, the leader of the majority party or the leader of a majority coalition is usually appointed prime minister by the president
election results: Ivan GASPAROVIC elected president in runoff; percent of vote—Ivan GASPAROVIC 59.9%, Vladimir MECIAR 40.1%
Legislative branch: unicameral National Council of the Slovak Republic or Narodna Rada Slovenskej Republiky (150 seats; members are elected on the basis of proportional representation to serve four-year terms)
elections: last held 17 June 2006 (next to be held in 2010)
election results: percent of vote by party—Smer 29.1%, SDKU 18.4%, SMK 11.7%, SNS 11.7%, LS-HZDS 8.8%, KDH 8.3%, other 12%; seats by party—Smer 50, SDKU 31, SMK 20, SNS 19, LS-HZDS 16, KDH 14
Judicial branch: Supreme Court (judges are elected by the National Council); Constitutional Court (judges appointed by president from group of nominees approved by the National Council); Special Court (judges elected by a council of judges and appointed by president)
Political parties and leaders: Parties in the Parliament: Christian Democratic Movement or KDH [Pavol HRUSOVSKY]; Direction-Social Democracy or Smer-SD [Robert FICO]; Party of the Hungarian Coalition or SMK [Pal CSAKY]; People's Party-Movement for a Democratic Slovakia or LS-HZDS [Vladimir MECIAR]; Slovak Democratic and Christian Union or SDKU-DS [Mikulas DZURINDA]; Slovak National Party or SNS [Jan SLOTA]; Parties outside the Parliament: Agrarian Party of the Provinces or ASV [Jozef VASKEBA]; Civic Conservative Party or OKS [Peter TATAR]; Free Forum [Zuzana MARTINAKOVA]; Hope or NADEJ [Alexandra NOVOTNA]; Left-wing Bloc or LB [Jozef KALMAN]; Mission 21—New Christian Democracy or MISIA 21 [Ivan SIMKO]; Movement for Democracy or HZD [Jozef GRAPA]; New Citizens Alliance or ANO [Pavol RUSKO]; Party

of the Democratic Left or SDL [Ladislav KOZMON]; Prosperita Slovenska or PS [Frantisek A. ZVRSKOVEC]; Slovak Communist Party or KSS [Vladimir DADO]; Slovak National Coalition or SLNKO [Vitazoslav MORIC]; Slovak People's Party or SLS [Jozef SASIK]; Union of the Workers of Slovakia or ZRS [Jan LUPTAK]
Political pressure groups and leaders: Federation of Employers' Associations of the Slovak Republic; Association of Towns and Villages or ZMOS; Confederation of Trade Unions or KOZ; National Union of Employers or RUZ; Slovak Chamber of Commerce and Industry or SOPK; Entrepreneurs Association of Slovakia or ZPS; The Business Alliance of Slovakia or PAS
International organization participation: ACCT (observer), Australia Group, BIS, BSEC (observer), CE, CEI, CERN, EAPC, EBRD, EIB, EU, FAO, IAEA, IBRD, ICAO, ICC, ICCt, ICRM, IDA, IEA, IFC, IFRCS, ILO, IMF, IMO, IMSO, Interpol, IOC, IOM, IPU, ISO, ITU, ITUC, MIGA, NAM (guest), NATO, NEA, NSG, OAS (observer), OECD, OIF (observer), OPCW, OSCE, PCA, Schengen Convention, SECI (observer), UN, UNCTAD, UNDOF, UNESCO, UNFICYP, UNIDO, UNTSO, UNWTO, UPU, WCL, WCO, WEU (associate partner), WFTU, WHO, WIPO, WMO, WTO, ZC
Diplomatic representation in the US:
chief of mission: Ambassador Rastislav KACER
chancery: 3523 International Court NW, Washington, DC 20008
telephone: [1] (202) 237-1054
FAX: [1] (202) 237-6438
consulate(s) general: Los Angeles, New York
Diplomatic representation from the US:
chief of mission: Ambassador Vincent OBSITNIK
embassy: Hviezdoslavovo Namestie 4, 81102 Bratislava
mailing address: P.O. Box 309, 814 99 Bratislava
telephone: [421] (2) 5443-3338
FAX: [421] (2) 5441-8861
Flag description: three equal horizontal bands of white (top), blue, and red superimposed with the coat of arms of Slovakia (consisting of a red shield bordered in white and bearing a white Cross of Lorraine surmounting three blue hills); the coat of arms is centered vertically and offset slightly to the hoist side

ECONOMY

Economy—overview: Slovakia has mas-

tered much of the difficult transition from a centrally planned economy to a modern market economy. The DZURINDA government made excellent progress during 2001–04 in macroeconomic stabilization and structural reform. Major privatizations are nearly complete, the banking sector is almost completely in foreign hands, and the government has helped facilitate a foreign investment boom with business friendly policies such as labor market liberalization and a 19% flat tax. Foreign investment in the automotive sector has been strong. Slovakia's economic growth exceeded expectations in 2001–07 despite the general European slowdown. Unemployment, at an unacceptable 18% in 2003–04, dropped to 8.6% in 2007 but remains the economy's Achilles heel. Slovakia joined the EU on 1 May 2004 and will be the second of the new EU member states to adopt the euro in 2009 if it continues to meet euro adoption criteria in 2008. Despite its 2006 pre-election promises to loosen fiscal policy and reverse the previous DZURINDA government's pro-market reforms, FICO's cabinet has thus far been careful to keep a lid on spending in order to meet euro adoption criteria. The FICO government is pursuing a state-interventionist economic policy, however, and has pushed to regulate energy and food prices.
GDP (purchasing power parity): $109.6 billion (2007 est.)
GDP (official exchange rate): $74.99 billion (2007 est.)
GDP—real growth rate: 10.4% (2007 est.)
GDP—per capita (PPP): $20,300 (2007 est.)
GDP—composition by sector:
agriculture: 2.6%
industry: 33.5%
services: 63.9% (2007 est.)
Labor force: 2.661 million (2007 est.)
Labor force—by occupation: agriculture 5.8%, industry 29.3%, construction 9%, services 55.9% (2003)
Unemployment rate: 8.4% (2007 est.)
Population below poverty line: 21% (2002)
Household income or consumption by percentage share:
lowest 10%: 3.1%
highest 10%: 20.9% (1996)
Distribution of family income—Gini index: 26 (2005)
Inflation rate (consumer prices): 2.8% (2007 est.)
Investment (gross fixed): 25.7% of GDP (2007 est.)
Budget:
revenues: $34.34 billion

expenditures: $35.99 billion (2007 est.)
Public debt: 35.9% of GDP (2007 est.)
Agriculture—products: grains, potatoes, sugar beets, hops, fruit; pigs, cattle, poultry; forest products
Industries: metal and metal products; food and beverages; electricity, gas, coke, oil, nuclear fuel; chemicals and man-made fibers; machinery; paper and printing; earthenware and ceramics; transport vehicles; textiles; electrical and optical apparatus; rubber products
Industrial production growth rate: 17.2% (2007 est.)
Electricity—production: 29.89 billion kWh (2005)
Electricity—production by source:
fossil fuel: 30.3%
hydro: 16%
nuclear: 53.6%
other: 0% (2001)
Electricity—consumption: 24.93 billion kWh (2005)
Electricity—exports: 11.27 billion kWh (2005)
Electricity—imports: 8.005 billion kWh (2005)
Oil—production: 12,840 bbl/day (2005 est.)
Oil—consumption: 79,350 bbl/day (2005 est.)
Oil—exports: 77,660 bbl/day (2004)
Oil—imports: 138,200 bbl/day (2004)
Oil—proved reserves: 9 million bbl (1 January 2006 est.)
Natural gas—production: 141.9 million cu m (2005 est.)
Natural gas—consumption: 6.231 billion cu m (2005 est.)
Natural gas—exports: 354.9 million cu m (2005 est.)
Natural gas—imports: 6.396 billion cu m (2005)
Natural gas—proved reserves: 14.39 billion cu m (1 January 2006 est.)
Current account balance: -$3.998 billion (2007 est.)
Exports: $57.53 billion f.o.b. (2007 est.)
Exports—commodities: vehicles 25.9%, machinery and electrical equipment 21.3%, base metals 14.6%, chemicals and minerals 10.1%, plastics 5.4% (2004)
Exports—partners: Germany 23.4%, Czech Republic 13.7%, Italy 6.5%, Poland 6.2%, Hungary 6.1%, Austria 6%, France 4.3%, Netherlands 4.2% (2006)
Imports: $58.4 billion f.o.b. (2007 est.)
Imports—commodities: machinery and transport equipment 41.1%, intermediate manufactured goods 19.3%, fuels 12.3%, chemicals 9.8%, miscellaneous manufactured goods 10.2% (2003)
Imports—partners: Germany 23%, Czech Republic 18%, Russia 11.2%, Hungary 6.1%, Austria 5.5%, Poland 4.9%, Italy 4.4% (2006)

Economic aid—recipient: $235 million in available EU structural adjustment and cohesion funds (2004)
Reserves of foreign exchange and gold: $18.98 billion (31 December 2007 est.)
Debt—external: $36.63 billion (31 December 2007)
Stock of direct foreign investment—at home: $24.37 billion (2007 est.)
Stock of direct foreign investment—abroad: $1.328 billion (2007 est.)
Market value of publicly traded shares: $5.574 billion (2006)
Currency (code): Slovak koruna (SKK)
Currency code: SKK
Exchange rates: koruny per US dollar—24.919 (2007), 29.611 (2006), 31.018 (2005), 32.257 (2004), 36.773 (2003)
Fiscal year: calendar year

COMMUNICATIONS

Telephones—main lines in use: 1.167 million (2006)
Telephones—mobile cellular: 4.893 million (2006)
Telephone system:
general assessment: Slovakia has a modern telecommunications system that has expanded dramatically in recent years with the growth in cellular services
domestic: analog system is now receiving digital equipment and is being enlarged with fiber-optic cable, especially in the larger cities; 3 companies provide nationwide cellular services
international: country code—421; 3 international exchanges (1 in Bratislava and 2 in Banska Bystrica) are available; Slovakia is participating in several international telecommunications projects that will increase the availability of external services
Radio broadcast stations: AM 15, FM 78, shortwave 2 (1998)
Radios: 3.12 million (1997)
Television broadcast stations: 80 (national broadcasting 6, regional 7, local 67) (2004)
Televisions: 2.62 million (1997)
Internet country code: .sk
Internet hosts: 821,816 (2007)
Internet Service Providers (ISPs): 6 (2000)
Internet users: 2.256 million (2006)

TRANSPORTATION

Airports: 35 (2007)
Airports—with paved runways:
total: 20
over 3,047 m: 2
2,438 to 3,047 m: 2
1,524 to 2,437 m: 3
914 to 1,523 m: 3
under 914 m: 10 (2007)
Airports—with unpaved runways:
total: 15
914 to 1,523 m: 8

under 914 m: 7 (2007)
Heliports: 1 (2007)
Pipelines: gas 6,769 km; oil 416 km (2007)
Railways:
total: 3,662 km
broad gauge: 100 km 1.520-m gauge
standard gauge: 3,512 km 1.435-m gauge (1,588 km electrified)
narrow gauge: 50 km (1.000-m or 0.750-m gauge) (2006)
Roadways:
total: 42,993 km
paved: 37,533 km (includes 316 km of expressways)
unpaved: 5,460 km (2004)
Waterways: 172 km (on Danube River) (2005)
Merchant marine:
total: 54 ships (1000 GRT or over) 260,766 GRT/361,651 DWT
by type: bulk carrier 6, cargo 45, refrigerated cargo 3
foreign-owned: 46 (Bulgaria 7, Estonia 2, Greece 4, Israel 6, Italy 1, Poland 2, Syria 2, Turkey 11, Ukraine 10, UK 1) (2007)
Ports and terminals: Bratislava, Komarno

MILITARY

Military branches: Armed Forces of the Slovak Republic (Ozbrojene Sily Slovenskej Republiky): Land Forces (Pozemne Sily), Air Forces (Vzdusne Sily) (2005)
Military service age and obligation: 17–30 years of age for voluntary military service; conscription abolished in 2006; women are eligible to serve (2007)
Manpower available for military service:
males age 16–49: 1,420,966
females age 16–49: 1,386,259 (2008 est.)
Manpower fit for military service:
males age 16–49: 1,166,833
females age 16–49: 1,156,874 (2008 est.)
Manpower reaching militarily significant age annually:
males age 16–49: 38,183
females age 16–49: 36,388 (2008 est.)
Military expenditures—percent of GDP: 1.87% (2005 est.)

TRANSNATIONAL ISSUES

Disputes—international: bilateral government, legal, technical and economic working group negotiations continued in 2006 between Slovakia and Hungary over Hungary's completion of its portion of the Gabcikovo-Nagymaros hydroelectric dam project along the Danube; as a member state that forms part of the EU's external border, Slovakia has implemented the strict Schengen border rules
Illicit drugs: transshipment point for Southwest Asian heroin bound for Western Europe; producer of synthetic drugs for regional market; consumer of ecstasy

INTRODUCTION

Background: The Slovene lands were part of the Austro-Hungarian Empire until the latter's dissolution at the end of World War I. In 1918, the Slovenes joined the Serbs and Croats in forming a new multinational state, which was named Yugoslavia in 1929. After World War II, Slovenia became a republic of the renewed Yugoslavia, which though Communist, distanced itself from Moscow's rule. Dissatisfied with the exercise of power by the majority Serbs, the Slovenes succeeded in establishing their independence in 1991 after a short 10-day war. Historical ties to Western Europe, a strong economy, and a stable democracy have assisted in Slovenia's transformation to a modern state. Slovenia acceded to both NATO and the EU in the spring of 2004.

GEOGRAPHY

Location: Central Europe, eastern Alps bordering the Adriatic Sea, between Austria and Croatia
Geographic coordinates: 46 07 N, 14 49 E
Map references: Europe
Area:
total: 20,273 sq km
land: 20,151 sq km
water: 122 sq km
Area—comparative: slightly smaller than New Jersey
Land boundaries:
total: 1,382 km
border countries: Austria 330 km, Croatia 670 km, Hungary 102 km, Italy 280 km
Coastline: 46.6 km
Maritime claims: *territorial sea:* 12 nm
Climate: Mediterranean climate on the coast, continental climate with mild to hot summers and cold winters in the plateaus and valleys to the east

Terrain: a short coastal strip on the Adriatic, an alpine mountain region adjacent to Italy and Austria, mixed mountains and valleys with numerous rivers to the east
Elevation extremes:
lowest point: Adriatic Sea 0 m
highest point: Triglav 2,864 m
Natural resources: lignite coal, lead, zinc, building stone, hydropower, forests
Land use:
arable land: 8.53%
permanent crops: 1.43%
other: 90.04% (2005)
Irrigated land: 30 sq km (2003)
Total renewable water resources: 32.1 cu km (2005)
Freshwater withdrawal (domestic/industrial/agricultural): *total:* 0.9
per capita: 457 cu m/yr (2002)
Natural hazards: flooding and earthquakes
Environment—current issues: Sava River polluted with domestic and industrial waste; pollution of coastal waters with heavy metals and toxic chemicals; forest damage near Koper from air pollution (originating at metallurgical and chemical plants) and resulting acid rain
Environment—international agreements: *party to:* Air Pollution, Air Pollution-Nitrogen Oxides, Air Pollution-Sulfur 94, Biodiversity, Climate Change, Climate Change-Kyoto Protocol, Desertification, Endangered Species, Hazardous Wastes, Law of the Sea, Marine Dumping, Ozone Layer Protection, Ship Pollution, Wetlands, Whaling
signed, but not ratified: none of the selected agreements
Geography—note: despite its small size, this eastern Alpine country controls some of Europe's major transit routes

PEOPLE

Population: 2,007,711 (July 2008 est.)
Age structure:
0–14 years: 13.6% (male 140,686/female 132,778)
15–64 years: 70.1% (male 709,689/female 697,862)
65 years and over: 16.3% (male 127,313/female 199,383) (2008 est.)
Median age:
total: 41.4 years
male: 39.8 years
female: 42.9 years (2008 est.)
Population growth rate: -0.088% (2008 est.)
Birth rate: 8.99 births/1,000 population (2008 est.)

Death rate: 10.51 deaths/1,000 population (2008 est.)
Net migration rate: 0.64 migrant(s)/1,000 population (2008 est.)
Sex ratio:
at birth: 1.07 male(s)/female
under 15 years: 1.06 male(s)/female
15–64 years: 1.02 male(s)/female
65 years and over: 0.64 male(s)/female
total population: 0.95 male(s)/female (2008 est.)
Infant mortality rate:
total: 4.3 deaths/1,000 live births
male: 4.87 deaths/1,000 live births
female: 3.69 deaths/1,000 live births (2008 est.)
Life expectancy at birth:
total population: 76.73 years
male: 73.04 years
female: 80.66 years (2008 est.)
Total fertility rate: 1.27 children born/woman (2008 est.)
HIV/AIDS—adult prevalence rate: less than 0.1% (2001 est.)
HIV/AIDS—people living with HIV/AIDS: 280 (2001 est.)
HIV/AIDS—deaths: fewer than 100 (2003 est.)
Nationality:
noun: Slovene(s)
adjective: Slovenian
Ethnic groups: Slovene 83.1%, Serb 2%, Croat 1.8%, Bosniak 1.1%, other or unspecified 12% (2002 census)
Religions: Catholic 57.8%, Muslim 2.4%, Orthodox 2.3%, other Christian 0.9%, unaffiliated 3.5%, other or unspecified 23%, none 10.1% (2002 census)
Languages: Slovenian 91.1%, Serbo-Croatian 4.5%, other or unspecified 4.4% (2002 census)
Literacy:
definition: NA
total population: 99.7%
male: 99.7%
female: 99.6%

GOVERNMENT

Country name:
conventional long form: Republic of Slovenia
conventional short form: Slovenia
local long form: Republika Slovenija
local short form: Slovenija
former: People's Republic of Slovenia, Socialist Republic of Slovenia
Government type: parliamentary republic
Capital: *name:* Ljubljana
geographic coordinates: 46 03 N, 14 31 E
time difference: UTC+1 (6 hours ahead of Washington, DC during Standard Time)

daylight saving time: +1hr, begins last Sunday in March; ends last Sunday in October

Administrative divisions: 182 municipalities (obcine, singular—obcina) and 11 urban municipalities* (mestne obcine, singular—mestna obcina) Ajdovscina, Beltinci, Benedikt, Bistrica ob Sotli, Bled, Bloke, Bohinj, Borovnica, Bovec, Braslovce, Brda, Brezice, Brezovica, Cankova, Celje*, Cerklje na Gorenjskem, Cerknica, Cerkno, Cerkvenjak, Crensovci, Crna na Koroskem, Crnomelj, Destrnik, Divaca, Dobje, Dobrepolje, Dobrna, Dobrova-Horjul-Polhov Gradec, Dobrovnik-Dobronak, Dolenjske Toplice, Dol pri Ljubljani, Domzale, Dornava, Dravograd, Duplek, Gorenja Vas-Poljane, Gorisnica, Gornja Radgona, Gornji Grad, Gornji Petrovci, Grad, Grosuplje, Hajdina, Hoce-Slivnica, Hodos-Hodos, Horjul, Hrastnik, Hrpelje-Kozina, Idrija, Ig, Ilirska Bistrica, Ivancna Gorica, Izola-Isola, Jesenice, Jezersko, Jursinci, Kamnik, Kanal, Kidricevo, Kobarid, Kobilje, Kocevje, Komen, Komenda, Koper-Capodistria*, Kostel, Kozje, Kranj*, Kranjska Gora, Krizevci, Krsko, Kungota, Kuzma, Lasko, Lenart, Lendava-Lendva, Litija, Ljubljana*, Ljubno, Ljutomer, Logatec, Loska Dolina, Loski Potok, Lovrenc na Pohorju, Luce, Lukovica, Majsperk, Maribor*, Markovci, Medvode, Menges, Metlika, Mezica, Miklavz na Dravskem Polju, Miren-Kostanjevica, Mirna Pec, Mislinja, Moravce, Moravske Toplice, Mozirje, Murska Sobota*, Muta, Naklo, Nazarje, Nova Gorica*, Novo Mesto*, Odranci, Oplotnica, Ormoz, Osilnica, Pesnica, Piran-Pirano, Pivka, Podcetrtek, Podlehnik, Podvelka, Polzela, Postojna, Prebold, Preddvor, Prevalje, Ptuj*, Puconci, Race-Fram, Radece, Radenci, Radlje ob Dravi, Radovljica, Ravne na Koroskem, Razkrizje, Ribnica, Ribnica na Pohorju, Rogasovci, Rogaska Slatina, Rogatec, Ruse, Salovci, Selnica ob Dravi, Semic, Sempeter-Vrtojba, Sencur, Sentilj, Sentjernej, Sentjur pri Celju, Sevnica, Sezana, Skocjan, Skofja Loka, Skofljica, Slovenj Gradec*, Slovenska Bistrica, Slovenske Konjice, Smarje pri Jelsah, Smartno ob Paki, Smartno pri Litiji, Sodrazica, Solcava, Sostanj, Starse, Store, Sveta Ana, Sveti Andraz v Slovenskih Goricah, Sveti Jurij, Tabor, Tisina, Tolmin, Trbovlje, Trebnje, Trnovska Vas, Trzic, Trzin, Turnisce, Velenje*, Velika Polana, Velike Lasce, Verzej, Videm, Vipava, Vitanje, Vodice, Vojnik, Vransko, Vrhnika, Vuzenica,

Zagorje ob Savi, Zalec, Zavrc, Zelezniki, Zetale, Ziri, Zirovnica, Zuzemberk, Zrece
note: the Government of Slovenia has reported 210 municipalities

Independence: 25 June 1991 (from Yugoslavia)

National holiday: Independence Day/Statehood Day, 25 June (1991)

Constitution: adopted 23 December 1991

Legal system: based on civil law system; has not accepted compulsory ICJ jurisdiction

Suffrage: 18 years of age; universal (16 years of age, if employed)

Executive branch:
chief of state: President Danilo TURK (since 22 December 2007)
head of government: Prime Minister Janez JANSA (since 9 November 2004)
cabinet: Council of Ministers nominated by the prime minister and elected by the National Assembly
elections: president elected by popular vote for a five-year term (eligible for a second term); election last held 21 October and 11 November 2007 (next to be held in the fall of 2012); following National Assembly elections, the leader of the majority party or the leader of a majority coalition is usually nominated to become prime minister by the president and elected by the National Assembly; election last held on 9 November 2004 (next National Assembly elections to be held in October 2008)
election results: Danilo TURK elected president; percent of vote—Danilo TURK 68.2%, Alojze PETERLE 31.8%; Janez JANSA elected prime minister by National Assembly vote—57 to 27 in 2004

Legislative branch: bicameral Parliament consists of a National Assembly or Drzavni Zbor (90 seats; 40 members are directly elected and 50 are elected on a proportional basis; note—the number of directly elected and proportionally elected seats varies with each election; the constitution mandates 1 seat each for Slovenia's Hungarian and Italian minorities; members are elected by popular vote to serve four-year terms) and the National Council or Drzavni Svet (40 seats; members indirectly elected by an electoral college to serve five-year terms; note—this is primarily an advisory body with limited legislative powers; it may propose laws, ask to review any National Assembly decision, and call national referenda)
elections: National Assembly—last held 3 October 2004 (next to be held 8 October 2008)

election results: percent of vote by party—SDS 29.1%, LDS 22.8%, ZLSD 10.2%, NSi 9%, SLS 6.8%, SNS 6.3%, DeSUS 4.1%, other 11.7%; seats by party—SDS 29, LDS 23, ZLSD 10, NSi 9, SLS 7, SNS 6, DeSUS 4, Hungarian minority 1, Italian minority 1

Judicial branch: Supreme Court (judges are elected by the National Assembly on the recommendation of the Judicial Council); Constitutional Court (judges elected for nine-year terms by the National Assembly and nominated by the president)

Political parties and leaders: Liberal Democracy of Slovenia or LDS [Katarina KRESAL]; New Slovenia or NSi [Andrej BAJUK]; Slovenian Democratic Party or SDS [Janez JANSA]; Democratic Party of Pensioners of Slovenia or DeSUS [Karl ERJAVEC]; Slovene National Party or SNS [Zmago JELINCIC]; Slovene People's Party or SLS [Bojan SROT]; Slovene Youth Party or SMS [Darko KRANJC]; Social Democrats or SD [Borut PAHOR] (formerly ZLSD); ZARES [Gregor Golobic]

Political pressure groups and leaders: NA

International organization participation: ACCT (observer), Australia Group, BIS, CE, CEI, EAPC, EBRD, EIB, EMU, EU, FAO, IADB, IAEA, IBRD, ICAO, ICC, ICCt, ICRM, IDA, IFC, IFRCS, IHO, ILO, IMF, IMO, Interpol, IOC, IOM, IPU, ISO, ITU, MIGA, NAM (guest), NATO, NSG, OAS (observer), OIF (observer), OPCW, OSCE, PCA, Schengen Convention, SECI, UN, UNCTAD, UNESCO, UNIDO, UNTSO, UNWTO, UPU, WCO, WEU (associate partner), WHO, WIPO, WMO, WTO, ZC

Diplomatic representation in the US:
chief of mission: Ambassador Samuel ZBOGAR
chancery: 2410 California Street N.W., Washington, DC 20008
telephone: [1] (202) 386-6601
FAX: [1] (202) 386-6633
consulate(s) general: Cleveland, New York

Diplomatic representation from the US:
chief of mission: Ambassador (vacant); Charge d'Affaires Maryruth COLEMAN
embassy: Presernova 31, 1000 Ljubljana
mailing address: American Embassy Ljubljana, US Department of State, 7140 Ljubljana Place, Washington, DC 20521-7140
telephone: [386] (1) 200-5500
FAX: [386] (1) 200-5555

Flag description: three equal horizontal bands of white (top), blue, and red, with the Slovenian seal (a shield with the image of Triglav, Slovenia's highest peak, in white against a blue background at the center; beneath it are two wavy blue lines depicting seas and rivers, and above it are three six-pointed stars arranged in an inverted triangle, which are taken from the coat of arms of the Counts of Celje, the great Slovene dynastic house of the late 14th and early 15th centuries); the seal is in the upper hoist side of the flag centered on the white and blue bands

ECONOMY

Economy—overview: Slovenia, which on 1 January 2007 became the first 2004 European Union entrant to adopt the euro, is a model of economic success and stability for the region. With the highest per capita GDP in Central Europe, Slovenia has excellent infrastructure, a well-educated work force, and a strategic location between the Balkans and Western Europe. Privatization has lagged since 2002, and the economy has one of highest levels of state control in the EU. Structural reforms to improve the business environment have allowed for somewhat greater foreign participation in Slovenia's economy and have helped to lower unemployment. In March 2004, Slovenia became the first transition country to graduate from borrower status to donor partner at the World Bank. In December 2007, Slovenia was invited to begin the accession process for joining the OECD. Despite its economic success, foreign direct investment (FDI) in Slovenia has lagged behind the region average, and taxes remain relatively high. Furthermore, the labor market is often seen as inflexible, and legacy industries are losing sales to more competitive firms in China, India, and elsewhere.

GDP (purchasing power parity): $54.67 billion (2007 est.)

GDP (official exchange rate): $46.08 billion (2007 est.)

GDP—real growth rate: 6.1% (2007 est.)

GDP—per capita (PPP): $27,200 (2007 est.)

GDP—composition by sector:
agriculture: 2%
industry: 34.4%
services: 63.5% (2007 est.)

Labor force: 925,000 (2007 est.)

Labor force—by occupation: *agriculture:* 2.5%
industry: 36%
services: 61.5% (2007)

Unemployment rate: 4.8% (2007 est.)

Population below poverty line: 12.9% (2004)

Household income or consumption by percentage share:
lowest 10%: 3.6%
highest 10%: 21.4% (1998)

Distribution of family income—Gini index: 24 (2005)

Inflation rate (consumer prices): 3.6% (2007 est.)

Investment (gross fixed): 28.7% of GDP (2007 est.)

Budget:
revenues: $19.17 billion
expenditures: $19.04 billion (2007 est.)

Public debt: 24.1% of GDP (2007 est.)

Agriculture—products: potatoes, hops, wheat, sugar beets, corn, grapes; cattle, sheep, poultry

Industries: ferrous metallurgy and aluminum products, lead and zinc smelting; electronics (including military electronics), trucks, automobiles, electric power equipment, wood products, textiles, chemicals, machine tools

Industrial production growth rate: 10.1% (2007 est.)

Electricity—production: 14.9 billion kWh (2006)

Electricity—production by source:
fossil fuel: 35.2%
hydro: 27.3%
nuclear: 36.8%
other: 0.7% (2001)

Electricity—consumption: 13.71 billion kWh (2006)

Electricity—exports: 4.8 billion kWh (2006)

Electricity—imports: 4.07 billion kWh (2006)

Oil—production: 5 bbl/day (2005 est.)

Oil—consumption: 54,000 bbl/day (2005 est.)

Oil—exports: 2,276 bbl/day (2004)

Oil—imports: 55,880 bbl/day (2004)

Oil—proved reserves: 0 bbl (1 January 2006 est.)

Natural gas—production: 4.795 million cu m (2005 est.)

Natural gas—consumption: 1.078 billion cu m (2005 est.)

Natural gas—exports: 0 cu m (2005 est.)

Natural gas—imports: 1.073 billion cu m (2005)

Natural gas—proved reserves: 0 cu m (1 January 2006 est.)

Current account balance: -$2.222 billion (2007 est.)

Exports: $27.01 billion f.o.b. (2007 est.)

Exports—commodities: manufactured goods, machinery and transport equipment, chemicals, food

Exports—partners: Germany 19.2%, Italy 12.5%, Croatia 8.8%, Austria 8.5%, France 6.3%, Russia 4.2% (2006)

Imports: $29.28 billion f.o.b. (2007 est.)

Imports—commodities: machinery and transport equipment, manufactured goods, chemicals, fuels and lubricants, food

Imports—partners: Germany 19.5%, Italy 17.8%, Austria 11.7%, France 5.9%, Croatia 4.6% (2006)

Economic aid—recipient: ODA, $484 million (2004–06)

Reserves of foreign exchange and gold: $5.643 billion (30 September 2007 est.)

Debt—external: $40.42 billion (30 June 2007)

Stock of direct foreign investment—at home: $8.19 billion (2007 est.)

Stock of direct foreign investment—abroad: $4.903 billion (2007 est.)

Market value of publicly traded shares: $15.18 billion (2006)

Currency (code): euro (EUR)
note: on 1 January 2007, the euro became Slovenia's currency; both the tolar and the euro were in circulation from 1 January until 15 January 2007

Currency code: SIT

Exchange rates: euros per US dollar— 0.7345 (2007), tolars per US dollar— 190.85 (2006), 192.71 (2005), 192.38 (2004), 207.11 (2003)
note: Slovenia adopted the euro as its currency on 1 January 2007

Fiscal year: calendar year

COMMUNICATIONS

Telephones—main lines in use: 837,500 (2006)

Telephones—mobile cellular: 1.82 million (2006)

Telephone system:
general assessment: well-developed telecommunications infrastructure
domestic: combined fixed-line and mobile-cellular teledensity exceeds 130 telephones per 100 persons
international: country code—386

Radio broadcast stations: AM 10, FM 230, shortwave 0 (2006)

Radios: 805,000 (1997)

Television broadcast stations: 31 (2006)

Televisions: 710,000 (1997)

Internet country code: .si

Internet hosts: 134,266 (2007)

Internet Service Providers (ISPs): 11 (2000)

Internet users: 1.251 million (2006)

TRANSPORTATION

Airports: 14 (2007)

Airports—with paved runways:
total: 6
over 3,047 m: 1
2,438 to 3,047 m: 1
1,524 to 2,437 m: 1

914 to 1,523 m: 2
under 914 m: 1 (2007)
Airports—with unpaved runways:
total: 8
1,524 to 2,437 m: 2
914 to 1,523 m: 2
under 914 m: 4 (2007)
Pipelines: gas 840 km; oil 11 km (2007)
Railways:
total: 1,229 km
standard gauge: 1,229 km 1.435-m gauge
(504 km electrified) (2006)
Roadways:
total: 38,451 km
paved: 38,451 km (includes 483 km of
expressways) (2004)
Merchant marine:
registered in other countries: 26 (Antigua
and Barbuda 6, Bahamas 1, Cyprus 4,
Georgia 2, Liberia 1, Malta 3, Marshall
Islands 3, Singapore 1, St Vincent and
The Grenadines 5) (2007)

Ports and terminals: Koper

MILITARY

Military branches: Slovenian Army
(includes air and naval forces)
Military service age and obligation: 17
years of age for voluntary military
service; conscription abolished in 2003
(2007)
**Manpower available for military
service:**
males age 16–49: 494,496
females age 16–49: 481,180 (2008 est.)
Manpower fit for military service:
males age 16–49: 406,951
females age 16–49: 395,444 (2008 est.)
**Manpower reaching militarily signifi-
cant age annually:**
males age 16–49: 10,516
females age 16–49: 9,934 (2008 est.)
**Military expenditures—percent of
GDP:** 1.7% (2005 est.)

TRANSNATIONAL ISSUES

Disputes—international: the Croatia-
Slovenia land and maritime boundary
agreement, which would have ceded
most of Piran Bay and maritime access to
Slovenia and several villages to Croatia,
remains unratified and in dispute;
Slovenia also protests Croatia's 2003
claim to an exclusive economic zone in
the Adriatic; as a member state that
forms part of the EU's external border,
Slovenia has implemented the strict
Schengen border rules to curb illegal
migration and commerce through southeastern Europe while encouraging close
cross-border ties with Croatia
Illicit drugs: minor transit point for
cocaine and Southwest Asian heroin
bound for Western Europe, and for precursor chemicals

SOLOMON ISLANDS

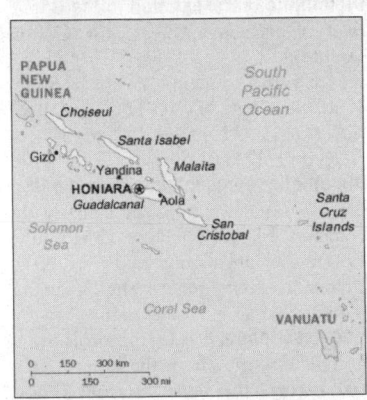

INTRODUCTION

Background: The UK established a protectorate over the Solomon Islands in
the 1890s. Some of the bitterest fighting
of World War II occurred on this archipelago. Self-government was achieved in
1976 and independence two years later.
Ethnic violence, government malfea-
sance, and endemic crime have undermined stability and civil society. In June
2003, then Prime Minister Sir Allan
KEMAKEZA sought the assistance of
Australia in reestablishing law and order;
the following month, an Australian-led
multinational force arrived to restore
peace and disarm ethnic militias. The
Regional Assistance Mission to the
Solomon Islands (RAMSI) has generally
been effective in restoring law and order
and rebuilding government institutions.

GEOGRAPHY

Location: Oceania, group of islands in

the South Pacific Ocean, east of Papua
New Guinea
Geographic coordinates: 8 00 S, 159 00
E
Map references: Oceania
Area:
total: 28,450 sq km
land: 27,540 sq km
water: 910 sq km
Area—comparative: slightly smaller
than Maryland
Land boundaries: 0 km
Coastline: 5,313 km
Maritime claims: measured from
claimed archipelagic baselines
territorial sea: 12 nm
exclusive economic zone: 200 nm
continental shelf: 200 nm
Climate: tropical monsoon; few
extremes of temperature and weather
Terrain: mostly rugged mountains with
some low coral atolls
Elevation extremes:
lowest point: Pacific Ocean 0 m
highest point: Mount Makarakomburu
2,447 m
Natural resources: fish, forests, gold,
bauxite, phosphates, lead, zinc, nickel
Land use:
arable land: 0.62%
permanent crops: 2.04%
other: 97.34% (2005)
Irrigated land: NA
Total renewable water resources: 11.7
cu km (1987)
Natural hazards: typhoons, but rarely
destructive; geologically active region
with frequent earthquakes, tremors, and
volcanic activity; tsunamis

Environment—current issues: defor-
estation; soil erosion; many of the sur-
rounding coral reefs are dead or dying
**Environment—international agree-
ments:** *party to:* Biodiversity, Climate
Change, Climate Change-Kyoto
Protocol, Desertification, Environmental
Modification, Law of the Sea, Marine
Dumping, Marine Life Conservation,
Ozone Layer Protection, Whaling
signed, but not ratified: none of the
selected agreements
Geography—note: strategic location on
sea routes between the South Pacific
Ocean, the Solomon Sea, and the Coral
Sea; on 2 April 2007 an undersea earth-
quake measuring 8.1 on the Richter scale
occurred 345 km WNW of the capital
Honiara, the resulting tsunami devas-
tated coastal areas of Western and
Choiseul provinces with dozens of deaths
and thousands dislocated; the provincial
capital of Gizo was especially hard hit

PEOPLE

Population: 581,318 (July 2008 est.)
Age structure:
0–14 years: 40.1% (male 118,856/female
114,173)
15–64 years: 56.5% (male
166,004/female 162,317)
65 years and over: 3.4% (male
9,487/female 10,481) (2008 est.)
Median age:
total: 19.4 years
male: 19.3 years
female: 19.6 years (2008 est.)
Population growth rate: 2.467% (2008
est.)

Birth rate: 28.48 births/1,000 population (2008 est.)

Death rate: 3.81 deaths/1,000 population (2008 est.)

Net migration rate: NA

Sex ratio:

at birth: 1.05 male(s)/female

under 15 years: 1.04 male(s)/female

15–64 years: 1.02 male(s)/female

65 years and over: 0.91 male(s)/female

total population: 1.03 male(s)/female (2008 est.)

Infant mortality rate:

total: 19.67 deaths/1,000 live births

male: 22.36 deaths/1,000 live births

female: 16.84 deaths/1,000 live births (2008 est.)

Life expectancy at birth:

total population: 73.44 years

male: 70.9 years

female: 76.1 years (2008 est.)

Total fertility rate: 3.65 children born/woman (2008 est.)

HIV/AIDS—adult prevalence rate: NA

HIV/AIDS—people living with HIV/AIDS: NA

HIV/AIDS—deaths: NA

Nationality:

noun: Solomon Islander(s)

adjective: Solomon Islander

Ethnic groups: Melanesian 94.5%, Polynesian 3%, Micronesian 1.2%, other 1.1%, unspecified 0.2% (1999 census)

Religions: Church of Melanesia 32.8%, Roman Catholic 19%, South Seas Evangelical 17%, Seventh-Day Adventist 11.2%, United Church 10.3%, Christian Fellowship Church 2.4%, other Christian 4.4%, other 2.4%, unspecified 0.3%, none 0.2% (1999 census)

Languages: Melanesian pidgin in much of the country is lingua franca; English (official; but spoken by only 1%-2% of the population); 120 indigenous languages

Literacy:

NA

GOVERNMENT

Country name:

conventional long form: none

conventional short form: Solomon Islands

local long form: none

local short form: Solomon Islands

former: British Solomon Islands

Government type: parliamentary democracy

Capital: *name:* Honiara

geographic coordinates: 9 26 S, 159 57 E

time difference: UTC+11 (16 hours ahead of Washington, DC during Standard Time)

Administrative divisions: 9 provinces and 1 capital territory*; Central, Choiseul, Guadalcanal, Honiara*, Isabel, Makira, Malaita, Rennell and Bellona, Temotu, Western

Independence: 7 July 1978 (from UK)

National holiday: Independence Day, 7 July (1978)

Constitution: 7 July 1978

Legal system: English common law, which is widely disregarded; has not accepted compulsory ICJ jurisdiction

Suffrage: 21 years of age; universal

Executive branch:

chief of state: Queen ELIZABETH II (since 6 February 1952); represented by Governor General Nathaniel WAENA (since 7 July 2004)

head of government: Prime Minister Derek SIKUA (since 20 December 2007); note—Prime Minister Manasseh SOGAVARE defeated in a no confidence vote in parliament on 13 December 2007; SIKUA elected on 20 December 2007

cabinet: Cabinet consists of 20 members appointed by the governor general on the advice of the prime minister from among the members of Parliament

elections: the monarch is hereditary; governor general appointed by the monarch on the advice of Parliament for up to five years (eligible for a second term); following legislative elections, the leader of the majority party or the leader of a majority coalition is usually elected prime minister by Parliament; deputy prime minister appointed by the governor general on the advice of the prime minister from among the members of Parliament

Legislative branch: unicameral National Parliament (50 seats; members elected from single-member constituencies by popular vote to serve four-year terms)

elections: last held on 5 April 2006 (next to be held in 2010)

election results: percent of vote by party—National Party 6.9%, PAP 6.3%, SIPRA 6.3%, Liberal 5%, Democratic 4.9%, SOCRED 4.3%, LAFARI 2.8%, independents 60.3%; seats by party—National Party 4, SIPRA 4, Democratic 3, PAP 3, LAFARI 2, Liberal 2, SOCRED 2, independents 30

Judicial branch: Court of Appeal

Political parties and leaders: Association of Independent Members or AIM [Thomas CHAN]; Christian Alliance Solomon Islands or CASI [Edward RONIA]; LAFARI Party [John GARO]; National Party [Francis HILLY]; People's Alliance Party or PAP [Sir Allan KEMAKEZA]; Social Credit Party or SOCRED [Manasseh Damukana SOGAVARE]; Solomon First Party [David QUAN]; Solomon Islands Democratic Party [Gabriel SURI]; Solomon Islands Labor Party or SILP [Joses TUHANUKU]; Solomon Islands Liberal Party [Bartholomew ULU-FA'ALU]; Solomon Islands Party for Rural Advancement or SIPRA [Job D. TAUSINGA]; United Party [Sir Peter KENILOREA]

note: in general, Solomon Islands politics is characterized by fluid coalitions

Political pressure groups and leaders: Isatabu Freedom Movement (IFM); Malaita Eagle Force (MEF); note—these rival armed ethnic factions crippled the Solomon Islands in a wave of violence from 1999 to 2003

International organization participation: ACP, ADB, C, ESCAP, FAO, G-77, IBRD, ICAO, ICCt (signatory), ICRM, IDA, IFAD, IFC, IFRCS, ILO, IMF, IMO, IOC, ITU, MIGA, OPCW, PIF, Sparteca, SPC, UN, UNCTAD, UNESCO, UPU, WFTU, WHO, WMO, WTO

Diplomatic representation in the US:

chief of mission: Ambassador Collin David BECK

chancery: 800 Second Avenue, Suite 400L, New York, NY 10017

telephone: [1] (212) 599-6192, 6193

FAX: [1] (212) 661-8925

Diplomatic representation from the US: the US does not have an embassy in Solomon Islands (embassy closed July 1993); the ambassador to Papua New Guinea is accredited to the Solomon Islands

Flag description: divided diagonally by a thin yellow stripe from the lower hoist-side corner; the upper triangle (hoist side) is blue with five white five-pointed stars arranged in an X pattern; the lower triangle is green

Government—note: by the end of 2007, the Regional Assistance Mission to the Solomon Islands (RAMSI)—originally made up of police and troops from Australia, NZ, Fiji, Papua New Guinea, and Tonga—had been scaled back to 303 police officers, 197 civilian technical advisers, and 72 military advisers from 15 countries across the region

ECONOMY

Economy—overview: The bulk of the population depends on agriculture, fishing, and forestry for at least part of its livelihood. Most manufactured goods and petroleum products must be imported. The islands are rich in undeveloped mineral resources such as lead, zinc, nickel, and gold. Prior to the arrival of the Regional Assistance Mission to the Solomon Islands (RAMSI), severe ethnic violence, the closing of key businesses, and an empty government treasury culminated in economic collapse. RAMSI's efforts to restore law and order and economic stability have led to modest growth as the economy rebuilds.

GDP (purchasing power parity): $948 million (2007 est.)

GDP (official exchange rate): $358 million (2007 est.)

GDP—real growth rate: 5.4% (2007 est.)

GDP—per capita (PPP): $1,900 (2007 est.)

GDP—composition by sector:
agriculture: 42%
industry: 11%
services: 47% (2000 est.)

Labor force: 249,200 (1999)

Labor force—by occupation: agriculture: 75%
industry: 5%
services: 20% (2000 est.)

Unemployment rate: NA%

Population below poverty line: NA%

Household income or consumption by percentage share:
lowest 10%: NA%
highest 10%: NA%

Inflation rate (consumer prices): 6.3% (2007 est.)

Budget:
revenues: $49.7 million
expenditures: $75.1 million (2003)

Agriculture—products: cocoa beans, coconuts, palm kernels, rice, potatoes, vegetables, fruit; timber; cattle, pigs; fish

Industries: fish (tuna), mining, timber

Industrial production growth rate: NA%

Electricity—production: 60 million kWh (2005)

Electricity—production by source:
fossil fuel: 100%
hydro: 0%
nuclear: 0%
other: 0% (2001)

Electricity—consumption: 55.8 million kWh (2005)

Electricity—exports: 0 kWh (2005)

Electricity—imports: 0 kWh (2005)

Oil—production: 0 bbl/day (2005 est.)

Oil—consumption: 1,300 bbl/day (2005 est.)

Oil—exports: 0 bbl/day (2004)

Oil—imports: 1,296 bbl/day (2004)

Oil—proved reserves: 0 bbl (1 January 2006 est.)

Natural gas—production: 0 cu m (2005 est.)

Natural gas—consumption: 0 cu m (2005 est.)

Natural gas—exports: 0 cu m (2005 est.)

Natural gas—imports: 0 cu m (2005)

Natural gas—proved reserves: 0 cu m (1 January 2006 est.)

Current account balance: -$143 million (2007 est.)

Exports: $237 million f.o.b. (2006)

Exports—commodities: timber, fish, copra, palm oil, cocoa

Exports—partners: China 48%, South Korea 9.5%, Japan 8.9%, Thailand 4.7%, Italy 4.4%, Philippines 4.2% (2006)

Imports: $256 million f.o.b. (2006)

Imports—commodities: food, plant and equipment, manufactured goods, fuels, chemicals

Imports—partners: Australia 25.5%, Singapore 23.5%, Japan 7.8%, NZ 5.1%, Fiji 4.2%, Papua New Guinea 4.1% (2006)

Economic aid—recipient: $198.2 million annually, mainly from Australia (2005 est.)

Debt—external: $166 million (2004)

Currency (code): Solomon Islands dollar (SBD)

Currency code: SBD

Exchange rates: Solomon Islands dollars per US dollar—NA (2007), 7.3447 (2006), 7.5299 (2005), 7.4847 (2004), 7.5059 (2003)

Fiscal year: calendar year

COMMUNICATIONS

Telephones—main lines in use: 7,400 (2005)

Telephones—mobile cellular: 6,000 (2005)

Telephone system:
general assessment: NA
domestic: NA
international: country code—677; satellite earth station—1 Intelsat (Pacific Ocean)

Radio broadcast stations: AM 1, FM 1, shortwave 1 (2004)

Radios: 57,000 (1997)

Televisions: 3,000 (1997)

Internet country code: .sb

Internet hosts: 3,414 (2007)

Internet Service Providers (ISPs): 1 (2000)

Internet users: 8,000 (2006)

TRANSPORTATION

Airports: 35 (2007)

Airports—with paved runways:
total: 2
1,524 to 2,437 m: 1
914 to 1,523 m: 1 (2007)

Airports—with unpaved runways:
total: 33
1,524 to 2,437 m: 1
914 to 1,523 m: 9
under 914 m: 23 (2007)

Heliports: 3 (2007)

Roadways:
total: 1,360 km
paved: 34 km
unpaved: 1,326 km (1999)

Ports and terminals: Honiara, Malloco Bay, Viru Harbor

MILITARY

Military branches: no regular military forces; Solomon Islands Police Force (2008)

Manpower available for military service:
males age 16–49: 141,051 (2008 est.)

Manpower fit for military service:
males age 16–49: 116,891 (2008 est.)

Manpower reaching militarily significant age annually:
males age 16–49: 6,924 (2008 est.)

Military expenditures—percent of GDP: 3% (2006)

TRANSNATIONAL ISSUES

Disputes—international: since 2003, the Regional Assistance Mission to the Solomon Islands (RAMSI), consisting of police, military, and civilian advisors drawn from 15 countries, has assisted in reestablishing and maintaining civil and political order while reinforcing regional stability and security

Refugees and internally displaced persons: IDPs: 5,400 (displaced by tsunami on 2 April 2007) (2007)

SOMALIA

INTRODUCTION

Background: Britain withdrew from British Somaliland in 1960 to allow its protectorate to join with Italian Somaliland and form the new nation of Somalia. In 1969, a coup headed by Mohamed SIAD Barre ushered in an authoritarian socialist rule that managed to impose a degree of stability in the country for a couple of decades. After the regime's collapse early in 1991, Somalia descended into turmoil, factional fighting, and anarchy. In May 1991, northern clans declared an independent Republic of Somaliland that now includes the administrative regions of Awdal, Woqooyi Galbeed, Togdheer,

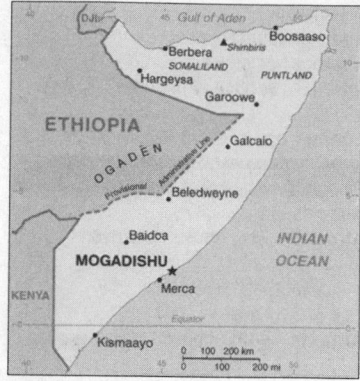

Sanaag, and Sool. Although not recognized by any government, this entity has maintained a stable existence and continues efforts to establish a constitutional democracy, including holding municipal, parliamentary, and presidential elections. The regions of Bari, Nugaal, and northern Mudug comprise a neighboring self-declared autonomous state of Puntland, which has been self-governing since 1998 but does not aim at independence; it has also made strides toward reconstructing a legitimate, representative government but has suffered some civil strife. Puntland disputes its border with Somaliland as it also claims portions of eastern Sool and Sanaag. Beginning in 1993, a two-year UN humanitarian effort (primarily in the south) was able to alleviate famine conditions, but when the UN withdrew in 1995, having suffered significant casualties, order still had not been restored. A two-year peace process, led by the Government of Kenya under the auspices of the Intergovernmental Authority on Development (IGAD), concluded in October 2004 with the election of Abdullahi YUSUF Ahmed as President of the Transitional Federal Government (TFG) of Somalia and the formation of an interim government, known as the Somalia Transitional Federal Institutions (TFIs). The Somalia TFIs include a 275-member parliamentary body, known as the Transitional Federal Assembly (TFA), a transitional Prime Minister, Nur "Adde" Hassan HUSSEIN, and a 90-member cabinet. The TFIs are based on the Transitional Federal Charter, which outlines a five-year mandate leading to the establishment of a new Somali constitution and a transition to a representative government following national elections. While its institutions remain weak, the TFG continues to reach out to Somali stakeholders and work with international donors to help build the gover-

nance capacity of the TFIs and work towards national elections in 2009. In June 2006, a loose coalition of clerics, business leaders, and Islamic court militias known as the Council of Islamic Courts (CIC) defeated powerful Mogadishu warlords and took control of the capital. The Courts continued to expand militarily throughout much of southern Somalia and threatened to overthrow the TFG in Baidoa. Ethiopian and TFG forces, concerned over links between some CIC factions and the al-Qaida East Africa network and the al-Qaida operatives responsible for the bombings of the US embassies in Tanzania and Kenya in 1998, intervened in late December 2006, resulting in the collapse of the CIC as an organization. However, the TFG continues to face violent resistance from extremist elements, such as the al-Shabaab militia previously affiliated with the now-defunct CIC.

GEOGRAPHY

Location: Eastern Africa, bordering the Gulf of Aden and the Indian Ocean, east of Ethiopia
Geographic coordinates: 10 00 N, 49 00 E
Map references: Africa
Area:
total: 637,657 sq km
land: 627,337 sq km
water: 10,320 sq km
Area—comparative: slightly smaller than Texas
Land boundaries:
total: 2,340 km
border countries: Djibouti 58 km, Ethiopia 1,600 km, Kenya 682 km
Coastline: 3,025 km
Maritime claims: *territorial sea:* 200 nm
Climate: principally desert; northeast monsoon (December to February), moderate temperatures in north and hot in south; southwest monsoon (May to October), torrid in the north and hot in the south, irregular rainfall, hot and humid periods (tangambili) between monsoons
Terrain: mostly flat to undulating plateau rising to hills in north
Elevation extremes:
lowest point: Indian Ocean 0 m
highest point: Shimbiris 2,416 m
Natural resources: uranium and largely unexploited reserves of iron ore, tin, gypsum, bauxite, copper, salt, natural gas, likely oil reserves
Land use:
arable land: 1.64%
permanent crops: 0.04%
other: 98.32% (2005)
Irrigated land: 2,000 sq km (2003)

Total renewable water resources: 15.7 cu km (1997)
Freshwater withdrawal (domestic/industrial/agricultural): *total:* 3.29 cu km/yr (0%/0%/100%)
per capita: 400 cu m/yr (2000)
Natural hazards: recurring droughts; frequent dust storms over eastern plains in summer; floods during rainy season
Environment—current issues: famine; use of contaminated water contributes to human health problems; deforestation; overgrazing; soil erosion; desertification
Environment—international agreements: *party to:* Biodiversity, Climate Change, Endangered Species, Law of the Sea, Ozone Layer Protection
Geography—note: strategic location on Horn of Africa along southern approaches to Bab el Mandeb and route through Red Sea and Suez Canal

PEOPLE

Population: 9,558,666
note: this estimate was derived from an official census taken in 1975 by the Somali Government; population counting in Somalia is complicated by the large number of nomads and by refugee movements in response to famine and clan warfare (July 2008 est.)
Age structure:
0–14 years: 44.7% (male 2,143,758/female 2,132,869)
15–64 years: 52.8% (male 2,525,562/female 2,516,879)
65 years and over: 2.5% (male 100,655/female 138,943) (2008 est.)
Median age:
total: 17.5 years
male: 17.4 years
female: 17.6 years (2008 est.)
Population growth rate: 2.824% (2008 est.)
Birth rate: 44.12 births/1,000 population (2008 est.)
Death rate: 15.89 deaths/1,000 population (2008 est.)
Net migration rate: 5 migrant(s)/1,000 population (2005 est.)
Sex ratio:
at birth: 1.03 male(s)/female
under 15 years: 1.01 male(s)/female
15–64 years: 1 male(s)/female
65 years and over: 0.72 male(s)/female
total population: 1 male(s)/female (2008 est.)
Infant mortality rate:
total: 110.97 deaths/1,000 live births
male: 120.17 deaths/1,000 live births
female: 101.5 deaths/1,000 live births (2008 est.)
Life expectancy at birth:
total population: 49.25 years
male: 47.43 years

female: 51.12 years (2008 est.)
Total fertility rate: 6.6 children born/woman (2008 est.)
HIV/AIDS—adult prevalence rate: 1% (2001 est.)
HIV/AIDS—people living with HIV/AIDS: 43,000 (2001 est.)
HIV/AIDS—deaths: NA
Major infectious diseases:
degree of risk: high
food or waterborne diseases: bacterial and protozoal diarrhea, hepatitis A and E, and typhoid fever
vectorborne diseases: dengue fever, malaria, and Rift Valley fever
water contact disease: schistosomiasis
animal contact disease: rabies (2008)
Nationality:
noun: Somali(s)
adjective: Somali
Ethnic groups: Somali 85%, Bantu and other non-Somali 15% (including Arabs 30,000)
Religions: Sunni Muslim
Languages: Somali (official), Arabic, Italian, English
Literacy:
definition: age 15 and over can read and write
total population: 37.8%
male: 49.7%
female: 25.8% (2001 est.)

GOVERNMENT

Country name:
conventional long form: none
conventional short form: Somalia
local long form: Jamhuuriyada Demuqraadiga Soomaaliyeed
local short form: Soomaaliya
former: Somali Republic, Somali Democratic Republic
Government type: no permanent national government; transitional, parliamentary federal government
Capital: *name:* Mogadishu
geographic coordinates: 2 04 N, 45 22 E
time difference: UTC+3 (8 hours ahead of Washington, DC during Standard Time)
Administrative divisions: 18 regions (plural—NA, singular—gobolka); Awdal, Bakool, Banaadir, Bari, Bay, Galguduud, Gedo, Hiiraan, Jubbada Dhexe, Jubbada Hoose, Mudug, Nugaal, Sanaag, Shabeellaha Dhexe, Shabeellaha Hoose, Sool, Togdheer, Woqooyi Galbeed
Independence: 1 July 1960 (from a merger of British Somaliland, which became independent from the UK on 26 June 1960, and Italian Somaliland, which became independent from the Italian-administered UN trusteeship on 1 July 1960, to form the Somali Republic)

National holiday: Foundation of the Somali Republic, 1 July (1960); note—26 June (1960) in Somaliland
Constitution: 25 August 1979, presidential approval 23 September 1979
note: the formation of transitional governing institutions, known as the Transitional Federal Government, is currently ongoing
Legal system: no national system; a mixture of English common law, Italian law, Islamic Shari'a, and Somali customary law; accepts compulsory ICJ jurisdiction with reservations
Suffrage: 18 years of age; universal
Executive branch:
chief of state: Transitional Federal President Abdullahi YUSUF Ahmed (since 14 October 2004); note—a transitional governing entity with a five-year mandate, known as the Transitional Federal Institutions (TFIs), was established in October 2004; the TFIs relocated to Somalia in June 2004
head of government: Prime Minister Nur "Adde" HASSAN Hussein (since 24 November 2007)
cabinet: Cabinet appointed by the prime minister and approved by the Transitional Federal Assembly
election results: Abdullahi YUSUF Ahmed, the former leader of the semi-autonomous Puntland region of Somalia, was elected president by the Transitional Federal Assembly
Legislative branch: unicameral National Assembly
note: unicameral Transitional Federal Assembly (TFA) (275 seats; 244 members appointed by the four major clans (61 for each clan), 31 seats allocated to smaller clans and subclans)
Judicial branch: following the breakdown of the central government, most regions have reverted to local forms of conflict resolution, either secular, traditional Somali customary law, or Shari'a (Islamic) law with a provision for appeal of all sentences
Political parties and leaders: none
Political pressure groups and leaders: numerous clan and sub-clan factions exist both in support and in opposition to the transitional government
International organization participation: ACP, AfDB, AFESD, AMF, AU, CAEU, FAO, G-77, IBRD, ICAO, ICRM, IDA, IDB, IFAD, IFC, IFRCS, IGAD, ILO, IMF, IMO, Interpol, IOC, IOM (observer), IPU, ITSO, ITU, LAS, NAM, OIC, UN, UNCTAD, UNESCO, UNHCR, UNIDO, UPU, WFTU, WHO, WIPO, WMO
Diplomatic representation in the US: Somalia does not have an embassy in the

US (ceased operations on 8 May 1991); note—the TFG is represented in the United States through its Permanent Mission to the United Nations
Diplomatic representation from the US: the US does not have an embassy in Somalia; US interests are represented by the US Embassy in Nairobi, Kenya at United Nations Avenue, Nairobi; mailing address: Unit 64100, Nairobi; APO AE 09831; telephone: [254] (20) 363-6000; FAX [254] (20) 363-6157
Flag description: light blue with a large white five-pointed star in the center; blue field influenced by the flag of the UN
Government—note: although an interim government was created in 2004, other regional and local governing bodies continue to exist and control various regions of the country, including the self-declared Republic of Somaliland in northwestern Somalia and the semi-autonomous State of Puntland in north-eastern Somalia

ECONOMY

Economy—overview: Despite the lack of effective national governance, Somalia has maintained a healthy informal economy, largely based on livestock, remittance/money transfer companies, and telecommunications. Agriculture is the most important sector, with livestock normally accounting for about 40% of GDP and about 65% of export earnings. Nomads and semi-pastoralists, who are dependent upon livestock for their livelihood, make up a large portion of the population. Livestock, hides, fish, charcoal, and bananas are Somalia's principal exports, while sugar, sorghum, corn, qat, and machined goods are the principal imports. Somalia's small industrial sector, based on the processing of agricultural products, has largely been looted and sold as scrap metal. Somalia's service sector also has grown. Telecommunication firms provide wireless services in most major cities and offer the lowest international call rates on the continent. In the absence of a formal banking sector, money exchange services have sprouted throughout the country, handling between $500 million and $1 billion in remittances annually. Mogadishu's main market offers a variety of goods from food to the newest electronic gadgets. Hotels continue to operate and are supported with private-security militias. Somalia's arrears to the IMF continued to grow in 2006–07. Statistics on Somalia's GDP, growth, per capita income, and inflation should be viewed skeptically. In late December 2004, a major tsunami caused an estimated 150 deaths and resulted in destruction of property in coastal areas.

GDP (purchasing power parity): $5.575 billion (2007 est.)

GDP (official exchange rate): $2.483 billion (2007 est.)

GDP—real growth rate: 2.6% (2007 est.)

GDP—per capita (PPP): $600 (2007 est.)

GDP—composition by sector:
agriculture: 65%
industry: 10%
services: 25% (2000 est.)

Labor force: 3.7 million (few skilled laborers) (1975)

Labor force—by occupation: *agriculture:* 71%
industry and services: 29% (1975)

Unemployment rate: NA%

Population below poverty line: NA%

Household income or consumption by percentage share:
lowest 10%: NA%
highest 10%: NA%

Inflation rate (consumer prices): NA%; note—businesses print their own money, so inflation rates cannot be easily determined

Budget:
revenues: $NA
expenditures: $NA

Agriculture—products: bananas, sorghum, corn, coconuts, rice, sugarcane, mangoes, sesame seeds, beans; cattle, sheep, goats; fish

Industries: a few light industries, including sugar refining, textiles, wireless communication

Industrial production growth rate: NA%

Electricity—production: 270 million kWh (2005)

Electricity—production by source:
fossil fuel: 100%
hydro: 0%
nuclear: 0%
other: 0% (2001)

Electricity—consumption: 251.1 million kWh (2005)

Electricity—exports: 0 kWh (2005)

Electricity—imports: 0 kWh (2005)

Oil—production: 0 bbl/day (2005 est.)

Oil—consumption: 5,000 bbl/day (2005 est.)

Oil—exports: 0 bbl/day (2004)

Oil—imports: 4,800 bbl/day (2004)

Oil—proved reserves: 0 bbl (1 January 2006 est.)

Natural gas—production: 0 cu m (2005 est.)

Natural gas—consumption: 0 cu m (2005 est.)

Natural gas—exports: 0 cu m (2005 est.)

Natural gas—imports: 0 cu m (2005)

Natural gas—proved reserves: 5.432 billion cu m (1 January 2006 est.)

Exports: $300 million f.o.b. (2006)

Exports—commodities: livestock, bananas, hides, fish, charcoal, scrap metal

Exports—partners: UAE 49.7%, Yemen 21.4%, Oman 5.9% (2006)

Imports: $798 million f.o.b. (2006)

Imports—commodities: manufactures, petroleum products, foodstuffs, construction materials, qat

Imports—partners: Djibouti 30.8%, Brazil 8.5%, India 8.2%, Kenya 8.1%, Oman 5.5%, UAE 5.2%, Yemen 5% (2006)

Economic aid—recipient: $236.4 million (2005 est.)

Debt—external: $3 billion (2001 est.)

Currency (code): Somali shilling (SOS)

Currency code: SOS

Exchange rates: Somali shillings per US dollar—NA (2007), 1,438.3 (2006) official rate; the unofficial black market rate was about 23,000 shillings per dollar as of February 2007
note: the Republic of Somaliland, a self-declared independent country not recognized by any foreign government, issues its own currency, the Somaliland shilling

Fiscal year: NA

COMMUNICATIONS

Telephones—main lines in use: 100,000 (2005)

Telephones—mobile cellular: 500,000 (2005)

Telephone system: *ngeneral assessment:* the public telecommunications system was almost completely destroyed or dismantled during the civil war; private wireless companies offer service in most major cities and charge the lowest international rates on the continent
domestic: local cellular telephone systems have been established in Mogadishu and in several other population centers
international: country code—252; international connections are available from Mogadishu by satellite (2001)

Radio broadcast stations: AM 0, FM 11 (also 1 station each in Puntland and Somaliland), shortwave 1 (in Mogadishu) (2001)

Radios: 470,000 (1997)

Television broadcast stations: 4 (2 in Mogadishu and 2 in Hargeisa) (2001)

Televisions: 135,000 (1997)

Internet country code: .so

Internet hosts: 0 (2007)

Internet Service Providers (ISPs): 3 (one each in Boosaaso, Hargeisa, and Mogadishu) (2000)

Internet users: 94,000 (2006)

TRANSPORTATION

Airports: 67 (2007)

Airports—with paved runways:
total: 7
over 3,047 m: 4
2,438 to 3,047 m: 2
1,524 to 2,437 m: 1 (2007)

Airports—with unpaved runways:
total: 60
over 3,047 m: 1
2,438 to 3,047 m: 3
1,524 to 2,437 m: 20
914 to 1,523 m: 29
under 914 m: 7 (2007)

Roadways:
total: 22,100 km
paved: 2,608 km
unpaved: 19,492 km (1999)

Merchant marine:
total: 1 ship (1000 GRT or over) 2,659 GRT/2,540 DWT
by type: cargo 1
foreign-owned: 1 (UAE 1) (2007)

Ports and terminals: Berbera, Kismaayo

MILITARY

Military branches: no national-level armed forces (2008)

Manpower available for military service:
males age 16–49: 2,181,050
females age 16–49: 2,125,558 (2008 est.)

Manpower fit for military service:
males age 16–49: 1,274,783
females age 16–49: 1,317,991 (2008 est.)

Military expenditures—percent of GDP: 0.9% (2005 est.)

TRANSNATIONAL ISSUES

Disputes—international: Ethiopian forces invaded southern Somalia and routed Islamist Courts from Mogadishu in January 2007; "Somaliland" secessionists provide port facilities in Berbera to landlocked Ethiopia and have established commercial ties with other regional states; "Puntland" and "Somaliland" "governments" seek international support in their secessionist aspirations and overlapping border claims; the undemarcated former British administrative line has little meaning as a political separation to rival clans within Ethiopia's Ogaden and southern Somalia's Oromo region; Kenya works hard to prevent the clan and militia fighting in Somalia from spreading south across the border, which has long been open to nomadic pastoralists

Refugees and internally displaced persons: *IDPs:* 1.1 million (civil war since 1988, clan-based competition for resources) (2007)

Background: Dutch traders landed at the southern tip of modern day South Africa in 1652 and established a stopover point on the spice route between the Netherlands and the East, founding the city of Cape Town. After the British seized the Cape of Good Hope area in 1806, many of the Dutch settlers (the Boers) trekked north to found their own republics. The discovery of diamonds (1867) and gold (1886) spurred wealth and immigration and intensified the subjugation of the native inhabitants. The Boers resisted British encroachments but were defeated in the Boer War (1899–1902); however, the British and the Afrikaners, as the Boers became known, ruled together under the Union of South Africa. In 1948, the National Party was voted into power and instituted a policy of apartheid—the separate development of the races. The first multi-racial elections in 1994 brought an end to apartheid and ushered in black majority rule.

GEOGRAPHY

Location: Southern Africa, at the southern tip of the continent of Africa
Geographic coordinates: 29 00 S, 24 00 E
Map references: Africa
Area:
total: 1,219,912 sq km
land: 1,219,912 sq km
water: 0 sq km
note: includes Prince Edward Islands (Marion Island and Prince Edward Island)
Area—comparative: slightly less than twice the size of Texas
Land boundaries:
total: 4,862 km
border countries: Botswana 1,840 km,

Lesotho 909 km, Mozambique 491 km, Namibia 967 km, Swaziland 430 km, Zimbabwe 225 km
Coastline: 2,798 km
Maritime claims:
territorial sea: 12 nm
contiguous zone: 24 nm
exclusive economic zone: 200 nm
continental shelf: 200 nm or to edge of the continental margin
Climate: mostly semiarid; subtropical along east coast; sunny days, cool nights
Terrain: vast interior plateau rimmed by rugged hills and narrow coastal plain
Elevation extremes:
lowest point: Atlantic Ocean 0 m
highest point: Njesuthi 3,408 m
Natural resources: gold, chromium, antimony, coal, iron ore, manganese, nickel, phosphates, tin, uranium, gem diamonds, platinum, copper, vanadium, salt, natural gas
Land use:
arable land: 12.1%
permanent crops: 0.79%
other: 87.11% (2005)
Irrigated land: 14,980 sq km (2003)
Total renewable water resources: 50 cu km (1990)
Freshwater withdrawal (domestic/ industrial/agricultural): *total:* 12.5 cu km/yr (31%/6%/63%)
per capita: 264 cu m/yr (2000)
Natural hazards: prolonged droughts
Environment—current issues: lack of important arterial rivers or lakes requires extensive water conservation and control measures; growth in water usage outpacing supply; pollution of rivers from agricultural runoff and urban discharge; air pollution resulting in acid rain; soil erosion; desertification
Environment—international agreements: *party to:* Antarctic-Environmental Protocol, Antarctic-Marine Living Resources, Antarctic Seals, Antarctic Treaty, Biodiversity, Climate Change, Climate Change-Kyoto Protocol, Desertification, Endangered Species, Hazardous Wastes, Law of the Sea, Marine Dumping, Marine Life Conservation, Ozone Layer Protection, Ship Pollution, Wetlands, Whaling
signed, but not ratified: none of the selected agreements
Geography—note: South Africa completely surrounds Lesotho and almost completely surrounds Swaziland

PEOPLE

Population: 43,786,115

note: estimates for this country explicitly take into account the effects of excess mortality due to AIDS; this can result in lower life expectancy, higher infant mortality, higher death rates, lower population growth rates, and changes in the distribution of population by age and sex than would otherwise be expected (July 2008 est.)
Age structure:
0–14 years: 28.6% (male 6,295,422/female 6,219,283)
15–64 years: 65.9% (male 14,114,838/female 14,737,791)
65 years and over: 5.5% (male 927,932/female 1,490,849) (2008 est.)
Median age:
total: 24.5 years
male: 23.8 years
female: 25.3 years (2008 est.)
Population growth rate: -0.501% (2008 est.)
Birth rate: 17.71 births/1,000 population (2008 est.)
Death rate: 22.7 deaths/1,000 population (2008 est.)
Net migration rate: -0.02 migrant(s)/ 1,000 population
note: there is an increasing flow of Zimbabweans into South Africa and Botswana in search of better economic opportunities (2008 est.)
Sex ratio: *at birth:* 1.02 male(s)/female
under 15 years: 1.01 male(s)/female
15–64 years: 0.96 male(s)/female
65 years and over: 0.62 male(s)/female
total population: 0.95 male(s)/female (2008 est.)
Infant mortality rate:
total: 58.26 deaths/1,000 live births
male: 61.64 deaths/1,000 live births
female: 54.81 deaths/1,000 live births (2008 est.)
Life expectancy at birth:
total population: 42.37 years
male: 43.3 years
female: 41.42 years (2008 est.)
Total fertility rate: 2.11 children born/woman (2008 est.)
HIV/AIDS—adult prevalence rate: 21.5% (2003 est.)
HIV/AIDS—people living with HIV/AIDS: 5.3 million (2003 est.)
HIV/AIDS—deaths: 370,000 (2003 est.)
Major infectious diseases:
degree of risk: intermediate
food or waterborne diseases: bacterial diarrhea, hepatitis A, and typhoid fever
vectorborne disease: Crimean Congo hemorrhagic fever and malaria
water contact disease: schistosomiasis (2008)

Nationality:

noun: South African(s)

adjective: South African

Ethnic groups: black African 79%, white 9.6%, colored 8.9%, Indian/Asian 2.5% (2001 census)

Religions: Zion Christian 11.1%, Pentecostal/Charismatic 8.2%, Catholic 7.1%, Methodist 6.8%, Dutch Reformed 6.7%, Anglican 3.8%, Muslim 1.5%, other Christian 36%, other 2.3%, unspecified 1.4%, none 15.1% (2001 census)

Languages: IsiZulu 23.8%, IsiXhosa 17.6%, Afrikaans 13.3%, Sepedi 9.4%, English 8.2%, Setswana 8.2%, Sesotho 7.9%, Xitsonga 4.4%, other 7.2% (2001 census)

Literacy:

definition: age 15 and over can read and write

total population: 86.4%

male: 87%

female: 85.7% (2003 est.)

GOVERNMENT

Country name:

conventional long form: Republic of South Africa

conventional short form: South Africa

former: Union of South Africa

abbreviation: RSA

Government type: republic

Capital: *name:* Pretoria (administrative capital)

geographic coordinates: 25 42 S, 28 13 E

time difference: UTC+2 (7 hours ahead of Washington, DC during Standard Time)

note: Cape Town (legislative capital); Bloemfontein (judicial capital)

Administrative divisions: 9 provinces; Eastern Cape, Free State, Gauteng, KwaZulu-Natal, Limpopo, Mpumalanga, Northern Cape, North-West, Western Cape

Independence: 31 May 1910 (Union of South Africa formed from four British colonies: Cape Colony, Natal, Transvaal, and Orange Free State); 31 May 1961 (republic declared) 27 April 1994 (majority rule)

National holiday: Freedom Day, 27 April (1994)

Constitution: 10 December 1996; this new constitution was certified by the Constitutional Court on 4 December 1996, was signed by then President MANDELA on 10 December 1996, and entered into effect on 4 February 1997

Legal system: based on Roman-Dutch law and English common law; has not accepted compulsory ICJ jurisdiction

Suffrage: 18 years of age; universal

Executive branch:

chief of state: President Thabo MBEKI (since 16 June 1999); Executive Deputy President Phumzile MLAMBO-NGCUKA (since 23 June 2005); note—the president is both the chief of state and head of government

head of government: President Thabo MBEKI (since 16 June 1999); Executive Deputy President Phumzile MLAMBO-NGCUKA (since 23 June 2005)

cabinet: Cabinet appointed by the president

elections: president elected by the National Assembly for a five-year term (eligible for a second term); election last held on 24 April 2004 (next to be held in April 2009)

election results: Thabo MBEKI elected president; percent of National Assembly vote—100% (by acclamation)

Legislative branch: bicameral Parliament consisting of the National Assembly (400 seats; members are elected by popular vote under a system of proportional representation to serve five-year terms) and the National Council of Provinces (90 seats, 10 members elected by each of the nine provincial legislatures for five-year terms; has special powers to protect regional interests, including the safeguarding of cultural and linguistic traditions among ethnic minorities); note—following the implementation of the new constitution on 4 February 1997, the former Senate was disbanded and replaced by the National Council of Provinces with essentially no change in membership and party affiliations, although the new institution's responsibilities have been changed somewhat by the new constitution

elections: National Assembly and National Council of Provinces—last held on 14 April 2004 (next to be held in 2009)

election results: National Assembly—percent of vote by party—ANC 69.7%, DA 12.4%, IFP 7%, UDM 2.3%, NNP 1.7%, ACDP 1.6%, other 5.3%; seats by party—ANC 279, DA 50, IFP 28, UDM 9, NNP 7, ACDP 6, other 21; National Council of Provinces—percent of vote by party—NA; seats by party—NA

Judicial branch: Constitutional Court; Supreme Court of Appeals; High Courts; Magistrate Courts

Political parties and leaders: African Christian Democratic Party or ACDP [Kenneth MESHOE]; African National Congress or ANC [Jacob ZUMA]; Democratic Alliance or DA [Helen ZILLE]; Freedom Front Plus or FF+ [Pieter MULDER]; Inkatha Freedom Party or IFP [Mangosuthu BUTHELEZI]; New National Party or NNP; Pan-Africanist Congress or PAC [Motsoko PHEKO]; United Democratic Movement or UDM [Bantu HOLOMISA]

Political pressure groups and leaders: Congress of South African Trade Unions or COSATU [Zwelinzima VAVI, general secretary]; South African Communist Party or SACP [Blade NZIMANDE, general secretary]; South African National Civics Organization or SANCO [Mlungisi HLONGWANE, national president]; note—COSATU and SACP are in a formal alliance with the ANC

International organization participation: ACP, AfDB, AU, BIS, C, FAO, G-24, G-77, IAEA, IBRD, ICAO, ICC, ICCt, ICRM, IDA, IFAD, IFC, IFRCS, IHO, ILO, IMF, IMO, IMSO, Interpol, IOC, IOM, IPU, ISO, ITSO, ITU, ITUC, MIGA, MONUC, NAM, NSG, OPCW, PCA, SACU, SADC, UN, UN Security Council (temporary), UNAMID, UNCTAD, UNESCO, UNHCR, UNIDO, UNITAR, UNMEE, UNWTO, UPU, WCL, WCO, WFTU, WHO, WIPO, WMO, WTO, ZC

Diplomatic representation in the US:

chief of mission: Ambassador Welile Augustine NHLAPO

chancery: 3051 Massachusetts Avenue NW, Washington, DC 20008

telephone: [1] (202) 232-4400

FAX: [1] (202) 265-1607

consulate(s) general: Chicago, Los Angeles, New York

Diplomatic representation from the US:

chief of mission: Ambassador Eric BOST

embassy: 877 Pretorius Street, Pretoria

mailing address: P. O. Box 9536, Pretoria 0001

telephone: [27] (12) 342-1048

FAX: [27] (12) 342-2244

consulate(s) general: Cape Town, Durban, Johannesburg

Flag description: two equal width horizontal bands of red (top) and blue separated by a central green band that splits into a horizontal Y, the arms of which end at the corners of the hoist side; the Y embraces a black isosceles triangle from which the arms are separated by narrow yellow bands; the red and blue bands are separated from the green band and its arms by narrow white stripes

ECONOMY

Economy—overview: South Africa is a middle-income, emerging market with an abundant supply of natural resources; well-developed financial, legal, communications, energy, and transport sectors; a stock exchange that is 17th largest in the world; and modern infrastructure supporting an efficient distribution of goods to major urban centers throughout the region. Growth has been robust since 2004, as South Africa has reaped the benefits of macroeconomic stability and a global commodities boom. However, unemployment remains high and outdated infrastructure has constrained growth. At the end of 2007, South Africa began to experience an electricity crisis because state power supplier Eskom suffered supply problems with aged plants, necessitating "load-shedding" cuts to residents and businesses in the major cities. Daunting economic problems remain from the apartheid era—especially poverty, lack of economic empowerment among the disadvantaged groups, and a shortage of public transportation. South African economic policy is fiscally conservative but pragmatic, focusing on controlling inflation, maintaining a budget surplus, and using state-owned enterprises to deliver basic services to low-income areas as a means to increase job growth and household income.

GDP (purchasing power parity): $467.1 billion (2007 est.)

GDP (official exchange rate): $282.6 billion (2007 est.)

GDP—real growth rate: 5.1% (2007 est.)

GDP—per capita (PPP): $9,800 (2007 est.)

GDP—composition by sector:
agriculture: 3.2%
industry: 31.3%
services: 65.5% (2007 est.)

Labor force: 20.49 million economically active (2007 est.)

Labor force—by occupation: *agriculture:* 9%
industry: 26%
services: 65% (2007 est.)

Unemployment rate: 24.3% (2007 est.)

Population below poverty line: 50% (2000 est.)

Household income or consumption by percentage share:
lowest 10%: 1.4%
highest 10%: 44.7% (2000)

Distribution of family income—Gini index: 65 (2005)

Inflation rate (consumer prices): 7.1% (2007 est.)

Investment (gross fixed): 20.6% of GDP (2007 est.)

Budget:
revenues: $68.2 billion
expenditures: $66.7 billion (2007 est.)

Public debt: 31.3% of GDP (2007 est.)

Agriculture—products: corn, wheat, sugarcane, fruits, vegetables; beef, poultry, mutton, wool, dairy products

Industries: mining (world's largest pro-ducer of platinum, gold, chromium), automobile assembly, metalworking, machinery, textiles, iron and steel, chemicals, fertilizer, foodstuffs, commercial ship repair

Industrial production growth rate: 4.4% (2007 est.)

Electricity—production: 264 billion kWh (2007)

Electricity—production by source:
fossil fuel: 93.5%
hydro: 1.1%
nuclear: 5.5%
other: 0% (2001)

Electricity—consumption: 241.4 billion kWh (2007)

Electricity—exports: 13.42 billion kWh (2005)

Electricity—imports: 11.32 billion kWh (2007)

Oil—production: 200,000 bbl/day (2006 est.)

Oil—consumption: 519,000 bbl/day (2006 est.)

Oil—exports: 217,700 bbl/day (2004)

Oil—imports: 319,000 bbl/day (2006 est.)

Oil—proved reserves: 15 million bbl (1 January 2007 est.)

Natural gas—production: 2.11 billion cu m (2005 est.)

Natural gas—consumption: 2.11 billion cu m (2005 est.)

Natural gas—exports: 0 cu m (2005 est.)

Natural gas—imports: 0 cu m (2005)

Natural gas—proved reserves: 27.16 million cu m (1 January 2006 est.)

Current account balance: -$20.56 billion (2007 est.)

Exports: $76.27 billion f.o.b. (2007 est.)

Exports—commodities: gold, diamonds, platinum, other metals and minerals, machinery and equipment

Exports—partners: Japan 12.1%, US 11.8%, UK 9%, Germany 7.6%, Netherlands 5.3%, China 4% (2006)

Imports: $82.12 billion f.o.b. (2007 est.)

Imports—commodities: machinery and equipment, chemicals, petroleum products, scientific instruments, foodstuffs

Imports—partners: Germany 12.6%, China 10%, US 7.6%, Japan 6.6%, Saudi Arabia 5.3%, UK 5% (2006)

Economic aid—recipient: $700 million (2005)

Reserves of foreign exchange and gold: $32.98 billion (31 December 2007)

Debt—external: $39.71 billion (31 December 2007)

Stock of direct foreign investment—at home: $93.47 billion (2007 est.)

Stock of direct foreign investment—abroad: $53.93 billion (2007 est.)

Market value of publicly traded shares: $842 billion (January 2008)

Currency (code): rand (ZAR)

Currency code: ZAR

Exchange rates: rand per US dollar—7.05 (2007), 6.7649 (2006), 6.3593 (2005), 6.4597 (2004), 7.5648 (2003)

Fiscal year: 1 April—31 March

COMMUNICATIONS

Telephones—main lines in use: 4.729 million (2005)

Telephones—mobile cellular: 39.66 million (2006)

Telephone system:
general assessment: the system is the best developed and most modern in Africa
domestic: combined fixed-line and mobile-cellular teledensity roughly 100 telephones per 100 persons; consists of carrier-equipped open-wire lines, coaxial cables, microwave radio relay links, fiber-optic cable, radiotelephone communication stations, and wireless local loops; key centers are Bloemfontein, Cape Town, Durban, Johannesburg, Port Elizabeth, and Pretoria
international: country code—27; the SAT-3/WASC and SAFE fiber optic cable systems connect in South Africa providing connectivity to Europe and Asia; satellite earth stations—3 Intelsat (1 Indian Ocean and 2 Atlantic Ocean)

Radio broadcast stations: AM 14, FM 347 (plus 243 repeaters), shortwave 1 (1998)

Radios: 17 million (2001)

Television broadcast stations: 556 (plus 144 network repeaters) (1997)

Televisions: 6 million (2000)

Internet country code: .za

Internet hosts: 1.088 million (2007)

Internet Service Providers (ISPs): 150 (2001)

Internet users: 5.1 million (2005)

TRANSPORTATION

Airports: 728 (2007)

Airports—with paved runways:
total: 146
over 3,047 m: 10
2,438 to 3,047 m: 5
1,524 to 2,437 m: 51
914 to 1,523 m: 67
under 914 m: 13 (2007)

Airports—with unpaved runways:
total: 582
1,524 to 2,437 m: 34
914 to 1,523 m: 300
under 914 m: 248 (2007)

Heliports: 1 (2007)

Pipelines: condensate 100 km; gas 1,177 km; oil 992 km; refined products 1,379 km (2007)

Railways:
total: 20,872 km

narrow gauge: 20,436 km 1.065-m gauge (8,931 km electrified); 436 km 0.610-m gauge (2006)

Roadways:
total: 362,099 km
paved: 73,506 km (includes 239 km of expressways)
unpaved: 288,593 km (2002)

Merchant marine:
total: 2 ships (1000 GRT or over) 28,722 GRT/32,226 DWT
by type: container 1, petroleum tanker 1
foreign-owned: 1 (Denmark 1)
registered in other countries: 7 (Bahamas 1, Seychelles 1, UK 4, unknown 1) (2007)

Ports and terminals: Cape Town, Durban, Port Elizabeth, Richards Bay, Saldanha Bay

MILITARY

Military branches: South African National Defense Force (SANDF): South African Army, South African Navy (SAN), South African Air Force (SAAF), Joint Operations Command, Military Intelligence, Military Health Service (2008)

Military service age and obligation: 18 years of age for voluntary military service; women have a long history of military service in noncombat roles dating back to World War I (2004)

Manpower available for military service:
males age 16–49: 11,622,507
females age 16–49: 11,501,537 (2008 est.)

Manpower fit for military service:
males age 16–49: 6,042,498
females age 16–49: 5,471,103 (2008 est.)

Manpower reaching militarily significant age annually:
males age 16–49: 496,337
females age 16–49: 492,024 (2008 est.)

Military expenditures—percent of GDP: 1.7% (2006)

Military—note: with the end of apartheid and the establishment of majority rule, former military, black homelands forces, and ex-opposition forces were integrated into the South African National Defense Force (SANDF); as of 2003 the integration process was considered complete

TRANSNATIONAL ISSUES

Disputes—international: South Africa has placed military along the border to apprehend the thousands of Zimbabweans fleeing economic dysfunction and political persecution; as of January 2007, South Africa also supports large numbers of refugees and asylum seekers from the Democratic Republic of the Congo (33,000), Somalia (20,000), Burundi (6,500), and other states in Africa (26,000); managed dispute with Namibia over the location of the boundary in the Orange River; in 2006, Swazi king advocates resort to ICJ to claim parts of Mpumalanga and KwaZulu-Natal from South Africa

Refugees and internally displaced persons: *refugees (country of origin):* 10,772 (Democratic Republic of Congo); 7,818 (Somalia); 5,759 (Angola) (2007)

Trafficking in persons: *current situation:* South Africa is a source, transit, and destination country for men, women, and children trafficked for forced labor and sexual exploitation; women and girls are trafficked internally—and occasionally to European and Asian countries—for sexual exploitation; women from other African countries are trafficked to South Africa and, less frequently, onward to Europe for sexual exploitation; men and boys are trafficked from neighboring countries for forced agricultural labor; Asian and Eastern European women are trafficked to South Africa for debt-bonded sexual exploitation
tier rating: Tier 2 Watch List—South Africa is placed on the Tier 2 Watch List for its failure to show increasing efforts to address trafficking in 2005

Illicit drugs: transshipment center for heroin, hashish, and cocaine, as well as a major cultivator of marijuana in its own right; cocaine and heroin consumption on the rise; world's largest market for illicit methaqualone, usually imported illegally from India through various east African countries, but increasingly producing its own synthetic drugs for domestic consumption; attractive venue for money launderers given the increasing level of organized criminal and narcotics activity in the region and the size of the South African economy

SOUTH GEORGIA AND THE SOUTH SANDWICH ISLANDS

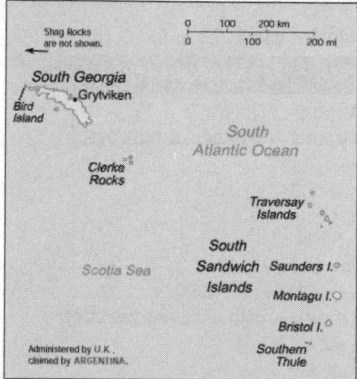

INTRODUCTION

Background: The islands, which have large bird and seal populations, lie approximately 1,000 km east of the Falkland Islands and have been under British administration since 1908— except for a brief period in 1982 when Argentina occupied them. Grytviken, on South Georgia, was a 19th and early 20th century whaling station. Famed explorer Ernest SHACKLETON stopped there in 1914 en route to his ill-fated attempt to cross Antarctica on foot. He returned some 20 months later with a few companions in a small boat and arranged a successful rescue for the rest of his crew, stranded off the Antarctic Peninsula. He died in 1922 on a subsequent expedition and is buried in Grytviken. Today, the station houses scientists from the British Antarctic Survey. Recognizing the importance of preserving the marine stocks in adjacent waters, the UK, in 1993, extended the exclusive fishing zone from 12 nm to 200 nm around each island.

GEOGRAPHY

Location: Southern South America, islands in the South Atlantic Ocean, east of the tip of South America

Geographic coordinates: 54 30 S, 37 00 W

Map references: Antarctic Region

Area:
total: 3,903 sq km
land: 3,903 sq km
water: 0 sq km
note: includes Shag Rocks, Black Rock, Clerke Rocks, South Georgia Island, Bird Island, and the South Sandwich Islands, which consist of 11 islands

Area—comparative: slightly larger than Rhode Island

Land boundaries: 0 km

Coastline: NA km

Maritime claims:
territorial sea: 12 nm
exclusive fishing zone: 200 nm

Climate: variable, with mostly westerly winds throughout the year interspersed with periods of calm; nearly all precipitation falls as snow

Terrain: most of the islands, rising steeply from the sea, are rugged and mountainous; South Georgia is largely barren and has steep, glacier-covered mountains; the South Sandwich Islands are of volcanic origin with some active volcanoes

Elevation extremes:

lowest point: Atlantic Ocean 0 m

highest point: Mount Paget (South Georgia) 2,934 m

Natural resources: fish

Land use:

arable land: 0%

permanent crops: 0%

other: 100% (largely covered by permanent ice and snow with some sparse vegetation consisting of grass, moss, and lichen) (2005)

Irrigated land: 0 sq km

Natural hazards: the South Sandwich Islands have prevailing weather conditions that generally make them difficult to approach by ship; they are also subject to active volcanism

Environment—current issues: NA

Geography—note: the north coast of South Georgia has several large bays, which provide good anchorage; reindeer, introduced early in the 20th century, live on South Georgia

PEOPLE

Population: no indigenous inhabitants

note: the small military garrison on South Georgia withdrew in March 2001 replaced by a permanent group of scientists of the British Antarctic Survey, which also has a biological station on Bird Island; the South Sandwich Islands are uninhabited

GOVERNMENT

Country name:

conventional long form: South Georgia and the South Sandwich Islands

conventional short form: none

abbreviation: SGSSI

Dependency status: overseas territory of the UK, also claimed by Argentina; administered from the Falkland Islands by a commissioner, who is concurrently governor of the Falkland Islands, representing Queen ELIZABETH II

Legal system: the laws of the UK, where applicable, apply; the senior magistrate from the Falkland Islands presides over the Magistrates Court

Diplomatic representation in the US: none (overseas territory of the UK, also claimed by Argentina)

Diplomatic representation from the US: none (overseas territory of the UK, also claimed by Argentina)

Flag description: blue, with the flag of the UK in the upper hoist-side quadrant and the South Georgia and the South Sandwich Islands coat of arms centered on the outer half of the flag; the coat of arms features a shield with a golden lion centered; the shield is supported by a fur seal on the left and a penguin on the right; a reindeer appears above the shield, and below it on a scroll is the motto LEO TERRAM PROPRIAM PROTEGAT (Let the Lion Protect its Own Land)

ECONOMY

Economy—overview: Some fishing takes place in adjacent waters. There is a potential source of income from harvesting finfish and krill. The islands receive income from postage stamps produced in the UK, sale of fishing licenses, and harbor and landing fees from tourist vessels. Tourism from specialized cruise ships is increasing rapidly.

COMMUNICATIONS

Telephone system:

general assessment: NA

domestic: NA

international: coastal radiotelephone station at Grytviken

Radio broadcast stations: 0 (2003)

Television broadcast stations: 0 (2003)

Internet country code: .gs

Internet hosts: 193 (2007)

TRANSPORTATION

Ports and terminals: Grytviken

MILITARY

Military—note: defense is the responsibility of the UK

TRANSNATIONAL ISSUES

Disputes—international: Argentina, which claims the islands in its constitution and briefly occupied them by force in 1982, agreed in 1995 to no longer seek settlement by force

SOUTHERN OCEAN

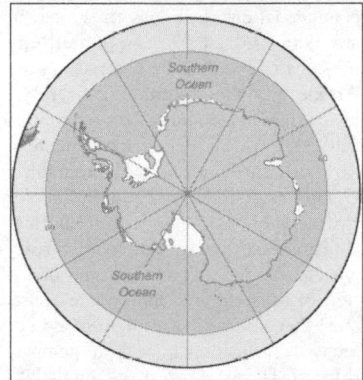

INTRODUCTION

Background: A large body of recent oceanographic research has shown that the Antarctic Circumpolar Current (ACC), an ocean current that flows from west to east around Antarctica, plays a crucial role in global ocean circulation. The region where the cold waters of the ACC meet and mingle with the warmer waters of the north defines a distinct border—the Antarctic Convergence—which fluctuates with the seasons, but which encompasses a discrete body of water and a unique ecologic region. The Convergence concentrates nutrients, which promotes marine plant life, and which in turn allows for a greater abundance of animal life. In the spring of 2000, the International Hydrographic Organization decided to delimit the waters within the Convergence as a fifth world ocean—the Southern Ocean—by combining the southern portions of the Atlantic Ocean, Indian Ocean, and Pacific Ocean. The Southern Ocean extends from the coast of Antarctica north to 60 degrees south latitude, which coincides with the Antarctic Treaty Limit and which approximates the extent of the Antarctic Convergence. As such, the Southern Ocean is now the fourth largest of the world's five oceans (after the Pacific Ocean, Atlantic Ocean, and Indian Ocean, but larger than the Arctic Ocean). It should be noted that inclusion of the Southern Ocean does not imply recognition of this feature as one of the world's primary oceans by the US Government.

GEOGRAPHY

Location: body of water between 60 degrees south latitude and Antarctica

Geographic coordinates: 60 00 S, 90 00 E (nominally), but the Southern Ocean has the unique distinction of being a large circumpolar body of water totally encircling the continent of Antarctica; this ring of water lies between 60 degrees south latitude and the coast of Antarctica and encompasses 360 degrees of longitude

Map references: Antarctic Region

Area:

total: 20.327 million sq km

note: includes Amundsen Sea, Bellingshausen Sea, part of the Drake Passage, Ross Sea, a small part of the Scotia Sea, Weddell Sea, and other tributary water bodies

Area—comparative: slightly more than twice the size of the US

Coastline: 17,968 km

Climate: sea temperatures vary from about 10 degrees Celsius to -2 degrees Celsius; cyclonic storms travel eastward around the continent and frequently are intense because of the temperature contrast between ice and open ocean; the ocean area from about latitude 40 south to the Antarctic Circle has the strongest average winds found anywhere on Earth; in winter the ocean freezes outward to 65 degrees south latitude in the Pacific sector and 55 degrees south latitude in the Atlantic sector, lowering surface temperatures well below 0 degrees Celsius; at some coastal points intense persistent drainage winds from the interior keep the shoreline ice-free throughout the winter

Terrain: the Southern Ocean is deep, 4,000 to 5,000 m over most of its extent with only limited areas of shallow water; the Antarctic continental shelf is generally narrow and unusually deep, its edge lying at depths of 400 to 800 m (the global mean is 133 m); the Antarctic icepack grows from an average minimum of 2.6 million sq km in March to about 18.8 million sq km in September, better than a sixfold increase in area; the Antarctic Circumpolar Current (21,000 km in length) moves perpetually eastward; it is the world's largest ocean current, transporting 130 million cubic meters of water per second—100 times the flow of all the world's rivers

Elevation extremes:

lowest point: -7,235 m at the southern end of the South Sandwich Trench

highest point: sea level 0 m

Natural resources: probable large and possible giant oil and gas fields on the continental margin; manganese nodules, possible placer deposits, sand and gravel, fresh water as icebergs; squid, whales, and seals—none exploited; krill, fish

Natural hazards: huge icebergs with drafts up to several hundred meters; smaller bergs and iceberg fragments; sea ice (generally 0.5 to 1 m thick) with sometimes dynamic short-term variations and with large annual and interan-nual variations; deep continental shelf floored by glacial deposits varying widely over short distances; high winds and large waves much of the year; ship icing, especially May-October; most of region is remote from sources of search and rescue

Environment—current issues: increased solar ultraviolet radiation resulting from the Antarctic ozone hole in recent years, reducing marine primary productivity (phytoplankton) by as much as 15% and damaging the DNA of some fish; illegal, unreported, and unregulated fishing in recent years, especially the landing of an estimated five to six times more Patagonian toothfish than the regulated fishery, which is likely to affect the sustainability of the stock; large amount of incidental mortality of seabirds resulting from long-line fishing for toothfish

note: the now-protected fur seal population is making a strong comeback after severe overexploitation in the 18th and 19th centuries

Environment—international agreements: the Southern Ocean is subject to all international agreements regarding the world's oceans; in addition, it is subject to these agreements specific to the Antarctic region: International Whaling Commission (prohibits commercial whaling south of 40 degrees south [south of 60 degrees south between 50 degrees and 130 degrees west]); Convention on the Conservation of Antarctic Seals (limits sealing); Convention on the Conservation of Antarctic Marine Living Resources (regulates fishing)

note: many nations (including the US) prohibit mineral resource exploration and exploitation south of the fluctuating Polar Front (Antarctic Convergence), which is in the middle of the Antarctic Circumpolar Current and serves as the dividing line between the cold polar surface waters to the south and the warmer waters to the north

Geography—note: the major chokepoint is the Drake Passage between South America and Antarctica; the Polar Front (Antarctic Convergence) is the best natural definition of the northern extent of the Southern Ocean; it is a distinct region at the middle of the Antarctic Circumpolar Current that separates the cold polar surface waters to the south from the warmer waters to the north; the Front and the Current extend entirely around Antarctica, reaching south of 60 degrees south near New Zealand and near 48 degrees south in the far South Atlantic coinciding with the path of the maximum westerly winds

ECONOMY

Economy—overview: Fisheries in 2005–06 landed 128,081 metric tons, of which 83% (106,591 tons) was krill (Euphausia superba) and 9.7% (12,364 tons) Patagonian toothfish (Dissostichus eleginoides), compared to 147,506 tons in 2004–05 of which 86% (127,035 tons) was krill and 8% (11,821 tons) Patagonian toothfish (estimated fishing from the area covered by the Convention of the Conservation of Antarctic Marine Living Resources (CCAMLR), which extends slightly beyond the Southern Ocean area). International agreements were adopted in late 1999 to reduce illegal, unreported, and unregulated fishing, which in the 2000–01 season landed, by one estimate, 8,376 metric tons of Patagonian and Antarctic toothfish. In the 2006–07 Antarctic summer, 35,552 tourists visited the Southern Ocean, compared to 29,799 in 2005–2006 (estimates provided to the Antarctic Treaty by the International Association of Antarctica Tour Operators (IAATO), and does not include passengers on overflights and those flying directly in and out of Antarctica).

TRANSPORTATION

Ports and terminals: McMurdo, Palmer, and offshore anchorages in Antarctica

note: few ports or harbors exist on southern side of Southern Ocean; ice conditions limit use of most to short periods in midsummer; even then some cannot be entered without icebreaker escort; most Antarctic ports are operated by government research stations and, except in an emergency, are not open to commercial or private vessels; vessels in any port south of 60 degrees south are subject to inspection by observers under Article 7 of the Antarctic Treaty; The Hydrographic Committee on Antarctica (HCA), a special hydrographic commission of International Hydrographic Organization (IHO), is responsible for hydrographic surveying and nautical charting matters in Antarctic Treaty area; it coordinates and facilitates provision of accurate and appropriate charts and other aids to navigation in support of safety of navigation in region; membership of HCA is open to any IHO Member State whose government has acceded to the Antarctic Treaty and which contributes resources and/or data to IHO Chart coverage of the area; members of HCA are Argentina, Australia, Brazil, Chile, China, Ecuador, France, Germany, Greece, India, Italy,

NZ, Norway, Russia, South Africa, Spain, UK, and US (2007)
Transportation—note: Drake Passage offers alternative to transit through the Panama Canal

TRANSNATIONAL ISSUES

Disputes—international: Antarctic Treaty defers claims (see Antarctica entry), but Argentina, Australia, Chile, France, NZ, Norway, and UK assert claims (some overlapping), including the continental shelf in the Southern Ocean; several states have expressed an interest in extending those continental shelf claims under the United Nations Convention on the Law of the Sea (UNCLOS) to include undersea ridges; the US and most other states do not recognize the land or maritime claims of other states and have made no claims themselves (the US and Russia have reserved the right to do so); no formal claims exist in the waters in the sector between 90 degrees west and 150 degrees west

SPAIN

INTRODUCTION

Background: Spain's powerful world empire of the 16th and 17th centuries ultimately yielded command of the seas to England. Subsequent failure to embrace the mercantile and industrial revolutions caused the country to fall behind Britain, France, and Germany in economic and political power. Spain remained neutral in World Wars I and II but suffered through a devastating civil war (1936–39). A peaceful transition to democracy following the death of dictator Francisco FRANCO in 1975, and rapid economic modernization (Spain joined the EU in 1986) have given Spain one of the most dynamic economies in Europe and made it a global champion of freedom. Continuing challenges include Basque Fatherland and Liberty (ETA) terrorism, illegal immigration, and slowing economic growth.

GEOGRAPHY

Location: Southwestern Europe, bordering the Bay of Biscay, Mediterranean Sea, North Atlantic Ocean, and Pyrenees Mountains, southwest of France
Geographic coordinates: 40 00 N, 4 00 W
Map references: Europe
Area:
total: 504,782 sq km
land: 499,542 sq km
water: 5,240 sq km
note: there are two autonomous cities—Ceuta and Melilla—and 17 autonomous communities including Balearic Islands and Canary Islands, and three small Spanish possessions off the coast of Morocco—Islas Chafarinas, Penon de Alhucemas, and Penon de Velez de la Gomera
Area—comparative: slightly more than twice the size of Oregon
Land boundaries:
total: 1,917.8 km
border countries: Andorra 63.7 km, France 623 km, Gibraltar 1.2 km, Portugal 1,214 km, Morocco (Ceuta) 6.3 km, Morocco (Melilla) 9.6 km
Coastline: 4,964 km
Maritime claims:
territorial sea: 12 nm
contiguous zone: 24 nm
exclusive economic zone: 200 nm (applies only to the Atlantic Ocean)
Climate: temperate; clear, hot summers in interior, more moderate and cloudy along coast; cloudy, cold winters in interior, partly cloudy and cool along coast
Terrain: large, flat to dissected plateau surrounded by rugged hills; Pyrenees in north
Elevation extremes:
lowest point: Atlantic Ocean 0 m
highest point: Pico de Teide (Tenerife) on Canary Islands 3,718 m
Natural resources: coal, lignite, iron ore, copper, lead, zinc, uranium, tungsten, mercury, pyrites, magnesite, fluorspar, gypsum, sepiolite, kaolin, potash, hydropower, arable land
Land use:
arable land: 27.18%
permanent crops: 9.85%
other: 62.97% (2005)
Irrigated land: 37,800 sq km (2003)
Total renewable water resources: 111.1 cu km (2005)
Freshwater withdrawal (domestic/industrial/agricultural): *total:* 37.22 cu km/yr (13%/19%/68%)
per capita: 864 cu m/yr (2002)

Natural hazards: periodic droughts
Environment—current issues: pollution of the Mediterranean Sea from raw sewage and effluents from the offshore production of oil and gas; water quality and quantity nationwide; air pollution; deforestation; desertification
Environment—international agreements: *party to:* Air Pollution, Air Pollution-Nitrogen Oxides, Air Pollution-Sulfur 94, Air Pollution-Volatile Organic Compounds, Antarctic-Environmental Protocol, Antarctic-Marine Living Resources, Antarctic Treaty, Biodiversity, Climate Change, Climate Change-Kyoto Protocol, Desertification, Endangered Species, Environmental Modification, Hazardous Wastes, Law of the Sea, Marine Dumping, Marine Life Conservation, Ozone Layer Protection, Ship Pollution, Tropical Timber 83, Tropical Timber 94, Wetlands, Whaling *signed, but not ratified:* Air Pollution-Persistent Organic Pollutants
Geography—note: strategic location along approaches to Strait of Gibraltar

PEOPLE

Population: 40,491,051 (July 2008 est.)
Age structure:
0–14 years: 14.4% (male 3,011,815/female 2,832,788)
15–64 years: 67.6% (male 13,741,493/female 13,641,914)
65 years and over: 17.9% (male 3,031,597/female 4,231,444) (2008 est.)
Median age:
total: 40.7 years
male: 39.3 years
female: 42.1 years (2008 est.)
Population growth rate: 0.096% (2008 est.)
Birth rate: 9.87 births/1,000 population (2008 est.)
Death rate: 9.9 deaths/1,000 population (2008 est.)
Net migration rate: 0.99 migrant(s)/1,000 population (2008 est.)
Sex ratio:
at birth: 1.07 male(s)/female

under 15 years: 1.06 male(s)/female
15-64 years: 1.01 male(s)/female
65 years and over: 0.72 male(s)/female
total population: 0.96 male(s)/female
(2008 est.)

Infant mortality rate:
total: 4.26 deaths/1,000 live births
male: 4.65 deaths/1,000 live births
female: 3.85 deaths/1,000 live births
(2008 est.)

Life expectancy at birth:
total population: 79.92 years
male: 76.6 years
female: 83.45 years (2008 est.)

Total fertility rate: 1.3 children born/woman (2008 est.)

HIV/AIDS—adult prevalence rate: 0.7% (2001 est.)

HIV/AIDS—people living with HIV/AIDS: 140,000 (2001 est.)

HIV/AIDS—deaths: fewer than 1,000 (2003 est.)

Nationality:
noun: Spaniard(s)
adjective: Spanish

Ethnic groups: composite of Mediterranean and Nordic types

Religions: Roman Catholic 94%, other 6%

Languages: Castilian Spanish (official) 74%, Catalan 17%, Galician 7%, Basque 2%, are official regionally

Literacy:
definition: age 15 and over can read and write
total population: 97.9%
male: 98.7%
female: 97.2% (2003 est.)

GOVERNMENT

Country name:
conventional long form: Kingdom of Spain
conventional short form: Spain
local long form: Reino de Espana
local short form: Espana

Government type: parliamentary monarchy

Capital: *name:* Madrid
geographic coordinates: 40 24 N, 3 41 W
time difference: UTC+1 (6 hours ahead of Washington, DC during Standard Time)
daylight saving time: +1hr, begins last Sunday in March; ends last Sunday in October
note: Spain is divided into two time zones including the Canary Islands

Administrative divisions: 17 autonomous communities (comunidades autonomas, singular—comunidad autonoma)and 2 autonomous cities* (ciudades autonomas, singular—ciudad autonoma); Andalucia, Aragon, Asturias, Baleares (Balearic Islands), Ceuta*, Canarias (Canary Islands), Cantabria, Castilla-La Mancha, Castilla

y Leon, Cataluna, Comunidad Valenciana, Extremadura, Galicia, La Rioja, Madrid, Melilla*, Murcia, Navarra, Pais Vasco (Basque Country)
note: the autonomous cities of Ceuta and Melilla plus three small islands of Islas Chafarinas, Penon de Alhucemas, and Penon de Velez de la Gomera, administered directly by the Spanish central government, are all along the coast of Morocco and are collectively referred to as Places of Sovereignty (Plazas de Soberania)

Independence: the Iberian peninsula was characterized by a variety of independent kingdoms prior to the Muslim occupation that began in the early 8th century A.D. and lasted nearly seven centuries; the small Christian redoubts of the north began the reconquest almost immediately, culminating in the seizure of Granada in 1492; this event completed the unification of several kingdoms and is traditionally considered the forging of present-day Spain

National holiday: National Day, 12 October (1492); year when Columbus first set foot in the Americas

Constitution: approved by legislature 31 October 1978; passed by referendum 6 December 1978, effective 29 December 1978

Legal system: civil law system, with regional applications; accepts compulsory ICJ jurisdiction with reservations

Suffrage: 18 years of age; universal

Executive branch:
chief of state: King JUAN CARLOS I (since 22 November 1975); Heir Apparent Prince FELIPE, son of the monarch, born 30 January 1968
head of government: President of the Government (Prime Minister equivalent) Jose Luis RODRIGUEZ ZAPATERO (since 17 April 2004); First Vice President (and Minister of the Presidency) Maria Teresa FERNANDEZ DE LA VEGA (since 18 April 2004) and Second Vice President (and Minister of Economy and Finance) Pedro SOLBES (since 18 April 2004)
cabinet: Council of Ministers designated by the president
note: there is also a Council of State that is the supreme consultative organ of the government, but its recommendations are non-binding
elections: the monarchy is hereditary; following legislative elections, the leader of the majority party or the leader of the majority coalition is usually proposed president by the monarch and elected by the National Assembly; election last held on 9 and 11 April 2008 (next to be held in March 2012); vice presidents

appointed by the monarch on the proposal of the president
election results: Jose Luis RODRIGUEZ ZAPATERO reelected President of the Government; percent of National Assembly vote—46.94%

Legislative branch: bicameral; General Courts or National Assembly or Las Cortes Generales consists of the Senate or Senado (264 seats as of 2008; 208 members directly elected by popular vote and the other 56—as of 2008—appointed by the regional legislatures; to serve four-year terms) and the Congress of Deputies or Congreso de los Diputados (350 seats; each of the 50 electoral provinces fills a minimum of two seats and the North African enclaves of Ceuta and Melilla fill one seat each with members serving a four-year term; the other 248 members are determined by proportional representation based on popular vote on block lists who serve four-year terms)
elections: Senate—last held on 9 March 2008 (next to be held in March 2012); Congress of Deputies—last held on 9 March 2008 (next to be held in March 2012)
election results: Senate—percent of vote by party—NA; seats by party—PP 101, PSOE 88, Entesa Catalona de Progress 12, CiU 4, PNV 2, CC 1, members appointed by regional legislatures 56; Congress of Deputies—percent of vote by party—PSOE 43.6%, PP 40.1%, CiU 3.1%, PNV 1.2%, ERC 1.2%, other 10.8%; seats by party—PSOE 169, PP 154, CiU 10, PNV 6, ERC 3, other 8

Judicial branch: Supreme Court or Tribunal Supremo

Political parties and leaders: Aragonese Party or CHA [Bizen FUSTER]; Basque Nationalist Party or PNV [Inigo URKULLU]; Basque Solidarity or EA [Begona ERRAZTI]; Canarian Coalition or CC [Jose Torres STINGA] (a coalition of five parties); Convergence and Union or CiU [Artur MAS i Gavarro] (a coalition of the Democratic Convergence of Catalonia or CDC [Artur MAS i Gavarro] and the Democratic Union of Catalonia or UDC [Josep Antoni DURAN i LLEIDA]); Entesa Catalonia de Progress (a Senate coalition grouping four Catalan parties—PSC, ERC, ICV, EUA); Galician Nationalist Bloc or BNG [Anxo Manuel QUINTANA Gonzalez]; Initiative for Catalonia Greens or ICV [Joan SAURA i Laporta]; Navarra yes or Na Bai [Uxue BARKOS Berruezo] (a coalition of four Navarran parties); Popular Party or PP [Mariano RAJOY Brey]; Republican Left of Catalonia or ERC [Josep-Lluis

CAROD-ROVIRA]; Spanish Socialist Workers Party or PSOE [Jose Luis RODRIGUEZ ZAPATERO]; United Left or IU [Gaspar LLAMAZARES Trigo] (a coalition of parties including the PCE and other small parties)

Political pressure groups and leaders: Association for Victims of Terrorism or AVT (grassroots organization devoted primarily to opposing ETA terrorist attacks and supporting its victims); Basta Ya (Spanish for "Enough is Enough"; grassroots organization devoted primarily to opposing ETA terrorist attacks and supporting its victims); business and landowning interests; Catholic Church; free labor unions (authorized in April 1977); Nunca Mais (Galician for "Never Again"; formed in response to the oil Tanker Prestige oil spill); Socialist General Union of Workers or UGT and the smaller independent Workers Syndical Union or USO; university students; Trade Union Confederation of Workers' Commissions or CC.OO.

International organization participation: ADB (nonregional members), AfDB, Australia Group, BCIE, BIS, CE, CERN, EAPC, EBRD, EIB, EMU, ESA, EU, FAO, IADB, IAEA, IBRD, ICAO, ICC, ICCt, ICRM, IDA, IEA, IFAD, IFC, IFRCS, IHO, ILO, IMF, IMO, IMSO, Interpol, IOC, IOM, IPU, ISO, ITSO, ITU, ITUC, LAIA (observer), MIGA, NAM (guest), NATO, NEA, NSG, OAS (observer), OECD, OPCW, OSCE, Paris Club, PCA, Schengen Convention, SECI (observer), UN, UNCTAD, UNESCO, UNHCR, UNIDO, UNIFIL, Union Latina, UNMEE, UNRWA, UNWTO, UPU, WCL, WCO, WEU, WHO, WIPO, WMO, WTO, ZC

Diplomatic representation in the US:
chief of mission: Ambassador Carlos WESTENDORP
chancery: 2375 Pennsylvania Avenue NW, Washington, DC 20037
telephone: [1] (202) 452-0100, 728-2340
FAX: [1] (202) 833-5670
consulate(s) general: Boston, Chicago, Houston, Los Angeles, Miami, New Orleans, New York, San Francisco, San Juan (Puerto Rico)

Diplomatic representation from the US:
chief of mission: Ambassador Eduardo AGUIRRE, Jr.
embassy: Serrano 75, 28006 Madrid
mailing address: PSC 61, APO AE 09642
telephone: [34] (91) 587-2200
FAX: [34] (91) 587-2303
consulate(s) general: Barcelona

Flag description: three horizontal bands of red (top), yellow (double width), and red with the national coat of arms on the hoist side of the yellow band; the coat of arms includes the royal seal framed by the Pillars of Hercules, which are the two promontories (Gibraltar and Ceuta) on either side of the eastern end of the Strait of Gibraltar

ECONOMY

Economy—overview: The Spanish economy boomed from 1986 to 1990 averaging 5% annual growth. After a European-wide recession in the early 1990s, the Spanish economy resumed moderate growth starting in 1994. Spain's mixed capitalist economy supports a GDP that on a per capita basis is equal to that of the leading West European economies. The center-right government of former President Jose Maria AZNAR successfully worked to gain admission to the first group of countries launching the European single currency (the euro) on 1 January 1999. The AZNAR administration continued to advocate liberalization, privatization, and deregulation of the economy and introduced some tax reforms to that end. Unemployment fell steadily under the AZNAR administration but remains high at 7.6%. Growth averaging more than 3% annually during 2003–07 was satisfactory given the background of a faltering European economy. The Socialist president, RODRIGUEZ ZAPATERO, has made mixed progress in carrying out key structural reforms, which need to be accelerated and deepened to sustain Spain's economic growth. Despite the economy's relative solid footing significant downside risks remain including Spain's continued loss of competitiveness, the potential for a housing market collapse, the country's changing demographic profile, and a decline in EU structural funds.

GDP (purchasing power parity): $1.352 trillion (2007 est.)
GDP (official exchange rate): $1.439 trillion (2007 est.)
GDP—real growth rate: 3.8% (2007 est.)
GDP—per capita (PPP): $30,100 (2007 est.)
GDP—composition by sector:
agriculture: 3.5%
industry: 29.8%
services: 66.6% (2007 est.)
Labor force: 22.19 million (2007 est.)
Labor force—by occupation: *agriculture:* 5.3%
industry: 30.1%
services: 64.6% (2004 est.)
Unemployment rate: 8.3% (2007 est.)
Population below poverty line: 19.8% (2005)

Household income or consumption by percentage share:
lowest 10%: 2.6%
highest 10%: 26.6% (2000)
Distribution of family income—Gini index: 32 (2005)
Inflation rate (consumer prices): 2.8% (2007 est.)
Investment (gross fixed): 31.1% of GDP (2007 est.)
Budget:
revenues: $589.2 billion
expenditures: $556.5 billion (2007 est.)
Public debt: 35.2% of GDP (2007 est.)
Agriculture—products: grain, vegetables, olives, wine grapes, sugar beets, citrus; beef, pork, poultry, dairy products; fish
Industries: textiles and apparel (including footwear), food and beverages, metals and metal manufactures, chemicals, shipbuilding, automobiles, machine tools, tourism, clay and refractory products, footwear, pharmaceuticals, medical equipment
Industrial production growth rate: 3% (2007 est.)
Electricity—production: 270.3 billion kWh (2005)
Electricity—production by source:
fossil fuel: 50.4%
hydro: 18.2%
nuclear: 27.2%
other: 4.1% (2001)
Electricity—consumption: 243 billion kWh (2005)
Electricity—exports: 11.56 billion kWh (2005)
Electricity—imports: 10.21 billion kWh (2005)
Oil—production: 29,350 bbl/day (2005 est.)
Oil—consumption: 1.6 million bbl/day (2005 est.)
Oil—exports: 175,200 bbl/day (2004)
Oil—imports: 1.714 million bbl/day (2004)
Oil—proved reserves: 157.6 million bbl (1 January 2006 est.)
Natural gas—production: 151.5 million cu m (2005 est.)
Natural gas—consumption: 30.58 billion cu m (2005 est.)
Natural gas—exports: 0 cu m (2005 est.)
Natural gas—imports: 31.76 billion cu m (2005)
Natural gas—proved reserves: 2.444 billion cu m (1 January 2006 est.)
Current account balance: -$145.6 billion (2007 est.)
Exports: $252.4 billion f.o.b. (2007 est.)
Exports—commodities: machinery, motor vehicles; foodstuffs, pharmaceuticals, medicines, other consumer goods

Exports—partners: France 18.8%, Germany 11%, Portugal 9%, Italy 8.6%, UK 8%, US 4.4% (2006)
Imports: $373.6 billion f.o.b. (2007 est.)
Imports—commodities: machinery and equipment, fuels, chemicals, semifinished goods, foodstuffs, consumer goods, measuring and medical control instruments
Imports—partners: Germany 14.8%, France 13.4%, Italy 8.3%, UK 5.2%, Netherlands 4.9%, China 4.6% (2006)
Economic aid—donor: ODA, $3.814 billion (2006)
Reserves of foreign exchange and gold: $19.05 billion (31 December 2007 est.)
Debt—external: $2.047 trillion (30 June 2007 est.)
Stock of direct foreign investment—at home: $487.8 billion (2007 est.)
Stock of direct foreign investment—abroad: $613.9 billion (2007 est.)
Market value of publicly traded shares: $960 billion (2005)
Currency (code): euro (EUR)
Currency code: EUR
Exchange rates: euros per US dollar—0.7345 (2007), 0.7964 (2006), 0.8041 (2005), 0.8054 (2004), 0.886 (2003)
Fiscal year: calendar year

COMMUNICATIONS

Telephones—main lines in use: 18.385 million (2006)
Telephones—mobile cellular: 46.152 million (2006)
Telephone system:
general assessment: well developed, modern facilities; fixed-line teledensity is 45 per 100 persons
domestic: combined fixed-line and mobile-cellular teledensity is 160 telephones per 100 persons
international: country code—34; submarine cables provide connectivity to Europe, Middle East, Asia, and US; satellite earth stations—2 Intelsat (1 Atlantic Ocean and 1 Indian Ocean), NA Eutelsat; tropospheric scatter to adjacent countries
Radio broadcast stations: AM 208, FM 715, shortwave 1 (1998)
Radios: 13.1 million (1997)
Television broadcast stations: 224 (plus 2,105 repeaters; includes 11 television broadcast stations and 88 repeaters in the Canary Islands) (1995)
Televisions: 16.2 million (1997)

Internet country code: .es
Internet hosts: 2.552 million (2007)
Internet Service Providers (ISPs): 56 (2000)
Internet users: 18.578 million (2006)

TRANSPORTATION

Airports: 154 (2007)
Airports—with paved runways:
total: 96
over 3,047 m: 18
2,438 to 3,047 m: 11
1,524 to 2,437 m: 18
914 to 1,523 m: 25
under 914 m: 24 (2007)
Airports—with unpaved runways:
total: 58
1,524 to 2,437 m: 2
914 to 1,523 m: 14
under 914 m: 42 (2007)
Heliports: 8 (2007)
Pipelines: gas 7,858 km; oil 622 km; refined products 3,445 km (2007)
Railways:
total: 14,974 km
broad gauge: 11,919 km 1.668-m gauge (6,950 km electrified)
standard gauge: 1,099 km 1.435-m gauge (1,054 km electrified)
narrow gauge: 1,928 km 1.000-m gauge (815 km electrified); 28 km 0.914-m gauge (28 km electrified) (2006)
Roadways:
total: 666,292 km
paved: 659,629 km (includes 12,009 km of expressways)
unpaved: 6,663 km (2003)
Waterways: 1,000 km (2003)
Merchant marine:
total: 167 ships (1000 GRT or over) 2,365,450 GRT/2,282,245 DWT
by type: bulk carrier 9, cargo 13, chemical tanker 13, container 25, liquefied gas 10, passenger 1, passenger/cargo 52, petroleum tanker 15, refrigerated cargo 5, roll on/roll off 17, specialized tanker 2, vehicle carrier 5
foreign-owned: 33 (Cuba 1, Denmark 2, Germany 9, Italy 1, Mexico 3, Norway 6, US 9, Uruguay 2)
registered in other countries: 106 (Angola 1, Bahamas 11, Belize 2, Brazil 4, Cape Verde 1, Cuba 1, Cyprus 7, Ireland 1, Malta 1, Marshall Islands 3, Nigeria 1, Panama 61, Portugal 10, St Kitts and Nevis 1, Venezuela 1) (2007)
Ports and terminals: Algeciras, Barcelona, Bilbao, Cartagena, Huelva, Tarragona, Valencia

MILITARY

Military branches: Spanish Armed Forces: Army (Ejercito de Tierra), Spanish Navy (Armada Espanola, AE; includes Marine Corps), Spanish Air Force (Ejercito del Aire Espanola, EdA) (2007)
Military service age and obligation: 20 years of age (2004)
Manpower available for military service:
males age 16–49: 10,033,069
females age 16–49: 9,764,937 (2008 est.)
Manpower fit for military service:
males age 16–49: 8,228,426
females age 16–49: 7,990,678 (2008 est.)
Manpower reaching militarily significant age annually:
males age 16–49: 203,650
females age 16–49: 191,352 (2008 est.)
Military expenditures—percent of GDP: 1.2% (2005 est.)

TRANSNATIONAL ISSUES

Disputes—international: in 2002, Gibraltar residents voted overwhelmingly by referendum to remain a British colony and against a "total shared sovereignty" arrangement while demanding participation in talks between the UK and Spain; Spain disapproves of UK plans to grant Gibraltar greater autonomy; Morocco protests Spain's control over the coastal enclaves of Ceuta, Melilla, and the islands of Penon de Velez de la Gomera, Penon de Alhucemas and Islas Chafarinas, and surrounding waters; Morocco serves as the primary launching site of illegal migration into Spain from North Africa; Portugal does not recognize Spanish sovereignty over the territory of Olivenza based on a difference of interpretation of the 1815 Congress of Vienna and the 1801 Treaty of Badajoz
Illicit drugs: despite rigorous law enforcement efforts, North African, Latin American, Galician, and other European traffickers take advantage of Spain's long coastline to land large shipments of cocaine and hashish for distribution to the European market; consumer for Latin American cocaine and North African hashish; destination and minor transshipment point for Southwest Asian heroin; money-laundering site for Colombian narcotics trafficking organizations and organized crime

SPRATLY ISLANDS

INTRODUCTION

Background: The Spratly Islands consist of more than 100 small islands or reefs. They are surrounded by rich fishing grounds and potentially by gas and oil deposits. They are claimed in their entirety by China, Taiwan, and Vietnam, while portions are claimed by Malaysia and the Philippines. About 45 islands are occupied by relatively small numbers of military forces from China, Malaysia, the Philippines, Taiwan, and Vietnam. Brunei has established a fishing zone that overlaps a southern reef but has not made any formal claim.

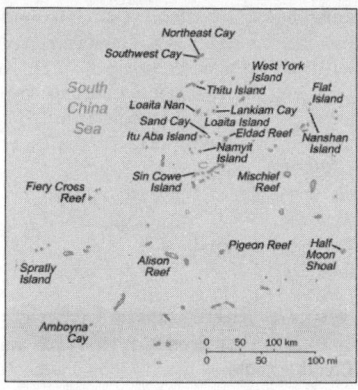

lowest point: South China Sea 0 m
highest point: unnamed location on Southwest Cay 4 m
Natural resources: fish, guano, undetermined oil and natural gas potential
Land use:
arable land: 0%
permanent crops: 0%
other: 100% (2005)
Irrigated land: 0 sq km
Natural hazards: typhoons; numerous reefs and shoals pose a serious maritime hazard
Environment—current issues: NA
Geography—note: strategically located near several primary shipping lanes in the central South China Sea; includes numerous small islands, atolls, shoals, and coral reefs

GEOGRAPHY

Location: Southeastern Asia, group of reefs and islands in the South China Sea, about two-thirds of the way from southern Vietnam to the southern Philippines
Geographic coordinates: 8 38 N, 111 55 E
Map references: Southeast Asia
Area: *total:* less than 5 sq km
land: less than 5 sq km
water: 0 sq km
note: includes 100 or so islets, coral reefs, and sea mounts scattered over an area of nearly 410,000 sq km of the central South China Sea
Area—comparative: NA
Land boundaries: 0 km
Coastline: 926 km
Maritime claims: NA
Climate: tropical
Terrain: flat
Elevation extremes:

PEOPLE

Population: no indigenous inhabitants
note: there are scattered garrisons occupied by personnel of several claimant states

GOVERNMENT

Country name:
conventional long form: none
conventional short form: Spratly Islands

ECONOMY

Economy—overview: Economic activity is limited to commercial fishing. The proximity to nearby oil- and gas-producing sedimentary basins suggests the potential for oil and gas deposits, but the region is largely unexplored. There are no reliable estimates of potential reserves. Commercial exploitation has yet to be developed.

TRANSPORTATION

Airports: 3 (2007)
Airports—with paved runways:
total: 2
914 to 1,523 m: 1
under 914 m: 1 (2007)
Airports—with unpaved runways:
total: 1
914 to 1,523 m: 1 (2007)
Heliports: 3 (2007)
Ports and terminals: none; offshore anchorage only

MILITARY

Military—note: Spratly Islands consist of more than 100 small islands or reefs of which about 45 are claimed and occupied by China, Malaysia, the Philippines, Taiwan, and Vietnam

TRANSNATIONAL ISSUES

Disputes—international: all of the Spratly Islands are claimed by China, Taiwan, and Vietnam; parts of them are claimed by Malaysia and the Philippines; in 1984, Brunei established an exclusive fishing zone that encompasses Louisa Reef in the southern Spratly Islands but has not publicly claimed the reef; claimants in November 2002 signed the "Declaration on the Conduct of Parties in the South China Sea," which has eased tensions but falls short of a legally binding "code of conduct"; in March 2005, the national oil companies of China, the Philippines, and Vietnam signed a joint accord to conduct marine seismic activities in the Spratly Islands

SRI LANKA

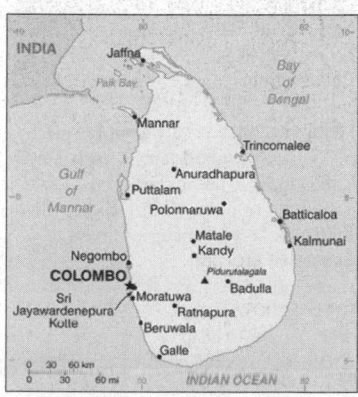

INTRODUCTION

Background: The first Sinhalese arrived in Sri Lanka late in the 6th century B.C.

probably from northern India. Buddhism was introduced in about the mid-third century B.C., and a great civilization developed at the cities of Anuradhapura (kingdom from circa 200 B.C. to circa A.D. 1000) and Polonnaruwa (from about 1070 to 1200). In the 14th century, a south Indian dynasty established a Tamil kingdom in northern Sri Lanka. The coastal areas of the island were controlled by the Portuguese in the 16th century and by the Dutch in the 17th century. The island was ceded to the British in 1796, became a crown colony in 1802, and was united under British rule by 1815. As Ceylon, it became independent in 1948; its name was changed to Sri Lanka in 1972. Tensions between the Sinhalese majority and Tamil separatists erupted into war in 1983. Tens of thousands have died in the ethnic conflict that continues to fester. After two decades of fighting, the government and Liberation Tigers of Tamil Eelam (LTTE) formalized a cease-fire in February 2002 with Norway brokering peace negotiations. Violence between the LTTE and government forces intensified in 2006 and the government regained control of the Eastern Province in 2007. In January 2008, the government officially withdrew from the ceasefire, and has begun engaging the LTTE in the northern portion of the country.

GEOGRAPHY

Location: Southern Asia, island in the Indian Ocean, south of India
Geographic coordinates: 7 00 N, 81 00 E

Map references: Asia
Area:
total: 65,610 sq km
land: 64,740 sq km
water: 870 sq km
Area—comparative: slightly larger than West Virginia
Land boundaries: 0 km
Coastline: 1,340 km
Maritime claims:
territorial sea: 12 nm
contiguous zone: 24 nm
exclusive economic zone: 200 nm
continental shelf: 200 nm or to the edge of the continental margin
Climate: tropical monsoon; northeast monsoon (December to March); southwest monsoon (June to October)
Terrain: mostly low, flat to rolling plain; mountains in south-central interior
Elevation extremes:
lowest point: Indian Ocean 0 m
highest point: Pidurutalagala 2,524 m
Natural resources: limestone, graphite, mineral sands, gems, phosphates, clay, hydropower
Land use: *arable land:* 13.96%
permanent crops: 15.24%
other: 70.8% (2005)
Irrigated land: 7,430 sq km (2003)
Total renewable water resources: 50 cu km (1999)
Freshwater withdrawal (domestic/industrial/agricultural): *total:* 12.61 cu km/yr (2%/2%/95%)
per capita: 608 cu m/yr (2000)
Natural hazards: occasional cyclones and tornadoes
Environment—current issues: deforestation; soil erosion; wildlife populations threatened by poaching and urbanization; coastal degradation from mining activities and increased pollution; freshwater resources being polluted by industrial wastes and sewage runoff; waste disposal; air pollution in Colombo
Environment—international agreements: *party to:* Biodiversity, Climate Change, Climate Change-Kyoto Protocol, Desertification, Endangered Species, Environmental Modification, Hazardous Wastes, Law of the Sea, Ozone Layer Protection, Ship Pollution, Wetlands
signed, but not ratified: Marine Life Conservation
Geography—note: strategic location near major Indian Ocean sea lanes

PEOPLE

Population: 21,128,773
note: since the outbreak of hostilities between the government and armed Tamil separatists in the mid-1980s, several hundred thousand Tamil civilians

have fled the island and more than 200,000 Tamils have sought refuge in the West (July 2008 est.)
Age structure:
0–14 years: 24.1% (male 2,596,463/female 2,495,136)
15–64 years: 68% (male 7,019,446/female 7,340,809)
65 years and over: 7.9% (male 783,823/female 893,096) (2008 est.)
Median age:
total: 30.4 years
male: 29.5 years
female: 31.4 years (2008 est.)
Population growth rate: 0.943% (2008 est.)
Birth rate: 16.63 births/1,000 population (2008 est.)
Death rate: 6.07 deaths/1,000 population (2008 est.)
Net migration rate: -1.12 migrant(s)/1,000 population (2008 est.)
Sex ratio:
at birth: 1.04 male(s)/female
under 15 years: 1.04 male(s)/female
15–64 years: 0.96 male(s)/female
65 years and over: 0.88 male(s)/female
total population: 0.97 male(s)/female (2008 est.)
Infant mortality rate:
total: 19.01 deaths/1,000 live births
male: 20.76 deaths/1,000 live births
female: 17.17 deaths/1,000 live births (2008 est.)
Life expectancy at birth:
total population: 74.97 years
male: 72.95 years
female: 77.08 years (2008 est.)
Total fertility rate: 2.02 children born/woman (2008 est.)
HIV/AIDS—adult prevalence rate: less than 0.1% (2001 est.)
HIV/AIDS—people living with HIV/AIDS: 3,500 (2001 est.)
HIV/AIDS—deaths: fewer than 200 (2003 est.)
Major infectious diseases:
degree of risk: high
food or waterborne diseases: bacterial diarrhea and hepatitis A
vectorborne disease: dengue fever and malaria
water contact disease: leptospirosis (2008)
Nationality:
noun: Sri Lankan(s)
adjective: Sri Lankan
Ethnic groups: Sinhalese 73.8%, Sri Lankan Moors 7.2%, Indian Tamil 4.6%, Sri Lankan Tamil 3.9%, other 0.5%, unspecified 10% (2001 census provisional data)
Religions: Buddhist 69.1%, Muslim 7.6%, Hindu 7.1%, Christian 6.2%, unspecified 10% (2001 census provisional data)

Languages: Sinhala (official and national language) 74%, Tamil (national language) 18%, other 8%
note: English is commonly used in government and is spoken competently by about 10% of the population
Literacy:
definition: age 15 and over can read and write
total population: 90.7%
male: 92.3%
female: 89.1% (2001 census)

GOVERNMENT

Country name:
conventional long form: Democratic Socialist Republic of Sri Lanka
conventional short form: Sri Lanka
local long form: Shri Lamka Prajatantrika Samajaya di Janarajaya/Ilankai Jananayaka Choshalichak Kutiyarachu
local short form: Shri Lamka/Ilankai
former: Serendib, Ceylon
Government type: republic
Capital: *name:* Colombo
geographic coordinates: 6 56 N, 79 51 E
time difference: UTC+5.5 (10.5 hours ahead of Washington, DC during Standard Time)
note: Sri Jayewardenepura Kotte (legislative capital)
Administrative divisions: 8 provinces; Central, North Central, North Eastern, North Western, Sabaragamuwa, Southern, Uva, Western
note: in October 2006, the Sri Lankan Supreme Court ruled voided a presidential directive merging the North and Eastern Provinces; many have defended the merger as a prerequisite to a negotiated settlement to the ethnic conflict; a parliamentary decision on the issue is pending
Independence: 4 February 1948 (from UK)
National holiday: Independence Day, 4 February (1948)
Constitution: adopted 16 August 1978, certified 31 August 1978
Legal system: a highly complex mixture of English common law, Roman-Dutch, Kandyan, and Jaffna Tamil law; has not accepted compulsory ICJ jurisdiction
Suffrage: 18 years of age; universal
Executive branch:
chief of state: President Mahinda RAJAPAKSA (since 19 November 2005); note—the president is both the chief of state and head of government; Ratnasiri WICKREMANAYAKE (since 21 November 2005) holds the largely ceremonial title of prime minister
head of government: President Mahinda RAJAPAKSA (since 19 November 2005)

cabinet: Cabinet appointed by the president in consultation with the prime minister

elections: president elected by popular vote for a six-year term (eligible for a second term); election last held on 17 November 2005 (next to be held in 2011)

election results: Mahinda RAJAPAKSA elected president; percent of vote— Mahinda RAJAPAKSA 50.3%, Ranil WICKREMESINGHE 48.4%, other 1.3%

Legislative branch: unicameral Parliament (225 seats; members elected by popular vote on the basis of an open-list, proportional representation system by electoral district to serve six-year terms)

elections: last held on 2 April 2004 (next to be held by 2010)

election results: percent of vote by party or electoral alliance—SLFP and JVP (no longer in formal UPFA alliance) 45.6%, UNP 37.8%, TNA 6.8%, JHU 6%, SLMC 2%, UPF 0.5%, EPDP 0.3%, other 1%; seats by party—UNP 68, SLFP 57, JVP 39, TNA 22, CWC 8, JHU 7, SLMC 6, SLMC dissidents 4, Communist Party 2, JHU dissidents 2, LSSP 2, MEP 2, NUA 2, UPF 2, EPDP 1, UNP dissident 1

Judicial branch: Supreme Court; Court of Appeals; judges for both courts are appointed by the president

Political parties and leaders: All Ceylon Tamil Congress or ACTC [G.PONNAMBALAM]; Ceylon Workers Congress or CWC [Arumugam THONDAMAN]; Communist Party or CP [D. GUNASEKERA]; Eelam People's Democratic Party or EPDP [Douglas DEVANANDA]; Eelam People's Revolutionary Liberation Front or EPRLF [Suresh PREMACHAN-DRAN]; Janatha Vimukthi Peramuna or JVP [Somawansa AMARASINGHE]; Lanka Sama Samaja Party or LSSP [Tissa VITHARANA]; Mahajana Eksath Peramuna (People's United Front) or MEP [D. GUNAWARDENE]; National Heritage Party or JHU [Ellawala METHANANDA]; National Unity Alliance or NUA [Ferial ASHRAFF]; People's Liberation Organization of Tamil Eelam or PLOTE [D. SID-HARTHAN]; Sri Lanka Freedom Party or SLFP [Mahinda RAJAPAKSA]; Sri Lanka Muslim Congress or SLMC [Rauff HAKEEM]; Tamil Eelam Liberation Organization or TELO [Selvam ADAIKALANATHAN]; Tamil National Alliance or TNA [R. SAM-PANTHAN]; Tamil United Liberation Front or TULF [V. ANANDASAN-GAREE]; United National Party or UNP [Ranil WICKREMASINGHE]; Up-country People's Front or UPF [P. CHANDRASEKARAN]

Political pressure groups and leaders: Buddhist clergy; labor unions; Liberation Tigers of Tamil Eelam or LTTE [Velupillai PRAB-HAKARAN](insurgent group fighting for a separate state); radical chauvinist Sinhalese groups such as the National Movement Against Terrorism; Sinhalese Buddhist lay groups; Tamil Makkal Viduthalai Pulikal (TMVP) or Karuna Faction [Vinayagamurthi MURALITHARAN] (paramilitary breakaway from LTTE and fighting LTTE)

International organization participation: ADB, BIMSTEC, C, CP, FAO, G-15, G-24, G-77, IAEA, IBRD, ICAO, ICC, ICRM, IDA, IFAD, IFC, IFRCS, IHO, ILO, IMF, IMO, IMSO, Interpol, IOC, IOM, IPU, ISO, ITSO, ITU, ITUC, MIGA, MINURSO, MINUSTAH, NAM, OAS (observer), OPCW, PCA, SAARC, SACEP, UN, UNCTAD, UNESCO, UNIDO, UNMEE, UNMIS, UNWTO, UPU, WCL, WCO, WFTU, WHO, WIPO, WMO, WTO

Diplomatic representation in the US:
chief of mission: Ambassador Bernard GOONETILLEKE
chancery: 2148 Wyoming Avenue NW, Washington, DC 20008
telephone: [1] (202) 483-4025 (through 4028)
FAX: [1] (202) 232-7181
consulate(s) general: Los Angeles
consulate(s): New York

Diplomatic representation from the US:
chief of mission: Ambassador Robert O. BLAKE, Jr.
embassy: 210 Galle Road, Colombo 3
mailing address: P. O. Box 106, Colombo
telephone: [94] (11) 249-8500
FAX: [94] (11) 243-7345

Flag description: yellow with two panels; the smaller hoist-side panel has two equal vertical bands of green (hoist side) and orange; the other panel is a large dark red rectangle with a yellow lion holding a sword, and there is a yellow bo leaf in each corner; the yellow field appears as a border around the entire flag and extends between the two panels

ECONOMY

Economy—overview: In 1977, Colombo abandoned statist economic policies and its import substitution trade policy for more market-oriented policies, export-oriented trade, and encouragement of foreign investment. Recent changes in government, however, have brought some policy reversals. Currently, the ruling Sri Lanka Freedom Party has a more statist economic approach, which seeks to reduce poverty by steering investment to disadvantaged areas, developing small and medium enterprises, promoting agriculture, and expanding the already enormous civil service. The government has halted privatizations. Although suffering a brutal civil war that began in 1983, Sri Lanka saw GDP growth average 4.5% in the last 10 years with the exception of a recession in 2001. In late December 2004, a major tsunami took about 31,000 lives, left more than 6,300 missing and 443,000 displaced, and destroyed an estimated $1.5 billion worth of property. Government spending and reconstruction drove growth to more than 7% in 2006 but reduced agriculture output probably slowed growth to about 6 percent in 2007. Government spending and loose monetary policy drove inflation to nearly 16% in 2007. Sri Lanka's most dynamic sectors now are food processing, textiles and apparel, food and beverages, port construction, telecommunications, and insurance and banking. In 2006, plantation crops made up only about 15% of exports (compared with more than 90% in 1970), while textiles and garments accounted for more than 60%. About 800,000 Sri Lankans work abroad, 90% of them in the Middle East. They send home more than $1 billion a year. The struggle by the Tamil Tigers of the north and east for an independent homeland continues to cast a shadow over the economy.

GDP (purchasing power parity): $81.29 billion (2007 est.)

GDP (official exchange rate): $30.01 billion (2007 est.)

GDP—real growth rate: 6.3% (2007 est.)

GDP—per capita (PPP): $4,100 (2007 est.)

GDP—composition by sector:
agriculture: 16.5%
industry: 26.9%
services: 56.5% (2007 est.)

Labor force: 7.67 million (2007 est.)

Labor force—by occupation: *agriculture:* 34.3%
industry: 25.3%
services: 40.4% (30 June 2006 est.)

Unemployment rate: 5.7% (2007 est.)

Population below poverty line: 22% (2002 est.)

Household income or consumption by percentage share:
lowest 10%: 1.1%

highest 10%: 39.7% (FY03/04)
Distribution of family income—Gini index: 50 (FY03/04)
Inflation rate (consumer prices): 19.7% (2007 est.)
Investment (gross fixed): 28.9% of GDP (2007 est.)
Budget:
revenues: $5.64 billion
expenditures: $7.77 billion (2007 est.)
Public debt: 83.9% of GDP (2007 est.)
Agriculture—products: rice, sugarcane, grains, pulses, oilseed, spices, tea, rubber, coconuts; milk, eggs, hides, beef; fish
Industries: processing of rubber, tea, coconuts, tobacco and other agricultural commodities; telecommunications, insurance, banking; clothing, textiles; cement, petroleum refining
Industrial production growth rate: 7.4% (2007 est.)
Electricity—production: 8.411 billion kWh (2005)
Electricity—production by source:
fossil fuel: 51.7%
hydro: 48.3%
nuclear: 0%
other: 0% (2001)
Electricity—consumption: 7.072 billion kWh (2005)
Electricity—exports: 0 kWh (2005)
Electricity—imports: 0 kWh (2005)
Oil—production: 0 bbl/day (2005 est.)
Oil—consumption: 84,000 bbl/day (2005 est.)
Oil—exports: 691.5 bbl/day (2004)
Oil—imports: 82,390 bbl/day (2004)
Oil—proved reserves: 0 bbl (1 January 2006 est.)
Natural gas—production: 0 cu m (2005 est.)
Natural gas—consumption: 0 cu m (2005 est.)
Natural gas—exports: 0 cu m (2005 est.)
Natural gas—proved reserves: 0 cu m (1 January 2006 est.)
Current account balance: -$1.369 billion (2007 est.)
Exports: $8.139 billion f.o.b. (2007 est.)
Exports—commodities: textiles and apparel, tea and spices; diamonds, emeralds, rubies; coconut products, rubber manufactures, fish
Exports—partners: US 27.7%, UK 11.3%, India 9.3%, Belgium 4.8%, Germany 4% (2006)
Imports: $10.61 billion f.o.b. (2007 est.)
Imports—commodities: textile fabrics, mineral products, petroleum, foodstuffs, machinery and transportation equipment

Imports—partners: India 19.6%, China 10.5%, Singapore 8.8%, Iran 5.7%, Malaysia 5.1%, Hong Kong 4.2%, Japan 4.1% (2006)
Economic aid—recipient: $1.189 billion (2005)
Reserves of foreign exchange and gold: $3.417 billion (31 December 2007 est.)
Debt—external: $12.19 billion (31 December 2007 est.)
Stock of direct foreign investment—at home: $NA
Stock of direct foreign investment—abroad: $NA
Market value of publicly traded shares: $7.769 billion (2006)
Currency (code): Sri Lankan rupee (LKR)
Currency code: LKR
Exchange rates: Sri Lankan rupees per US dollar—110.78 (2007), 103.99 (2006), 100.498 (2005), 101.194 (2004), 96.521 (2003)
Fiscal year: calendar year

COMMUNICATIONS

Telephones—main lines in use: 2.742 million (2007)
Telephones—mobile cellular: 7.983 million (2007)
Telephone system:
general assessment: telephone services have improved significantly and are available in most parts of the country
domestic: national trunk network consists mostly of digital microwave radio relay; fiber-optic links now in use in Colombo area and 2 fixed wireless local loops have been installed; competition is strong in mobile cellular systems and mobile cellular subscribership is increasing; combined fixed-line and mobile-cellular teledensity is about 50 per 100 persons
international: country code—94; the SEA-ME-WE-3 and SEA-ME-WE-4 submarine cables provide connectivity to Asia, Australia, Middle East, Europe, US; satellite earth stations—2 Intelsat (Indian Ocean)
Radio broadcast stations: AM 15, FM 52, shortwave 4 (2007)
Radios: 3.85 million (1997)
Television broadcast stations: 14 (2006)
Televisions: 1.53 million (1997)
Internet country code: .lk
Internet hosts: 6,198 (2007)
Internet Service Providers (ISPs): 5 (2000)
Internet users: 428,000 (2006)

TRANSPORTATION

Airports: 18 (2007)
Airports—with paved runways:
total: 14
over 3,047 m: 1
1,524 to 2,437 m: 6
914 to 1,523 m: 7 (2007)
Airports—with unpaved runways:
total: 4
914 to 1,523 m: 1
under 914 m: 3 (2007)
Railways:
total: 1,449 km
broad gauge: 1,449 km 1.676-m gauge (2006)
Roadways:
total: 97,287 km
paved: 78,802 km
unpaved: 18,485 km (2003)
Waterways: 160 km (primarily on rivers in southwest) (2006)
Merchant marine:
total: 24 ships (1000 GRT or over) 162,280 GRT/227,478 DWT
by type: bulk carrier 2, cargo 18, container 2, petroleum tanker 2
foreign-owned: 6 (Germany 6)
registered in other countries: 3 (Panama 3) (2007)
Ports and terminals: Colombo

MILITARY

Military branches: Sri Lanka Army, Sri Lanka Navy, Sri Lanka Air Force (2008)
Military service age and obligation: 18 years of age for voluntary military service (2007)
Manpower available for military service:
males age 16–49: 5,458,720
females age 16–49: 5,594,006 (2008 est.)
Manpower fit for military service:
males age 16–49: 4,477,437
females age 16–49: 4,683,716 (2008 est.)
Manpower reaching militarily significant age annually:
males age 16–49: 174,065
females age 16–49: 168,593 (2008 est.)
Military expenditures—percent of GDP: 2.6% (2006)

TRANSNATIONAL ISSUES

Disputes—international: none
Refugees and internally displaced persons: *IDPs:* 460,000 (both Tamils and non-Tamils displaced due to long-term civil war between the government and the separatist Liberation Tigers of Tamil Eelam (LTTE)) (2007)

SUDAN

INTRODUCTION

Background: Military regimes favoring Islamic-oriented governments have dominated national politics since independence from the UK in 1956. Sudan was embroiled in two prolonged civil wars during most of the remainder of the 20th century. These conflicts were

rooted in northern economic, political, and social domination of largely non-Muslim, non-Arab southern Sudanese. The first civil war ended in 1972 but broke out again in 1983. The second war and famine-related effects resulted in more than four million people displaced and, according to rebel estimates, more than two million deaths over a period of two decades. Peace talks gained momentum in 2002–04 with the signing of several accords. The final North/South Comprehensive Peace Agreement (CPA), signed in January 2005, granted the southern rebels autonomy for six years. After which, a referendum for independence is scheduled to be held. A separate conflict, which broke out in the western region of Darfur in 2003, has displaced nearly two million people and caused an estimated 200,000 to 400,000 deaths. The UN took command of the Darfur peacekeeping operation from the African Union on 31 December 2007. As of early 2008, peacekeeping troops were struggling to stabilize the situation, which has become increasingly regional in scope, and has brought instability to eastern Chad, and Sudanese incursions into the Central African Republic. Sudan also has faced large refugee influxes from neighboring countries, primarily Ethiopia and Chad. Armed conflict, poor transport infrastructure, and lack of government support have chronically obstructed the provision of humanitarian assistance to affected populations.

GEOGRAPHY

Location: Northern Africa, bordering the Red Sea, between Egypt and Eritrea
Geographic coordinates: 15 00 N, 30 00 E
Map references: Africa
Area:
total: 2,505,810 sq km
land: 2.376 million sq km
water: 129,810 sq km

Area—comparative: slightly more than one-quarter the size of the US
Land boundaries:
total: 7,687 km
border countries: Central African Republic 1,165 km, Chad 1,360 km, Democratic Republic of the Congo 628 km, Egypt 1,273 km, Eritrea 605 km, Ethiopia 1,606 km, Kenya 232 km, Libya 383 km, Uganda 435 km
Coastline: 853 km
Maritime claims:
territorial sea: 12 nm
contiguous zone: 18 nm
continental shelf: 200 m depth or to the depth of exploitation
Climate: tropical in south; arid desert in north; rainy season varies by region (April to November)
Terrain: generally flat, featureless plain; mountains in far south, northeast and west; desert dominates the north
Elevation extremes:
lowest point: Red Sea 0 m
highest point: Kinyeti 3,187 m
Natural resources: petroleum; small reserves of iron ore, copper, chromium ore, zinc, tungsten, mica, silver, gold, hydropower
Land use:
arable land: 6.78%
permanent crops: 0.17%
other: 93.05% (2005)
Irrigated land: 18,630 sq km (2003)
Total renewable water resources: 154 cu km (1997)
Freshwater withdrawal (domestic/industrial/agricultural): *total:* 37.32 cu km/yr (3%/1%/97%)
per capita: 1,030 cu m/yr (2000)
Natural hazards: dust storms and periodic persistent droughts
Environment—current issues: inadequate supplies of potable water; wildlife populations threatened by excessive hunting; soil erosion; desertification; periodic drought
Environment—international agreements: *party to:* Biodiversity, Climate Change, Climate Change-Kyoto Protocol, Desertification, Endangered Species, Hazardous Wastes, Law of the Sea, Ozone Layer Protection, Wetlands
signed, but not ratified: none of the selected agreements
Geography—note: largest country in Africa; dominated by the Nile and its tributaries

PEOPLE

Population: 40,218,455 (July 2008 est.)
Age structure:
0–14 years: 41.1% (male 8,451,576/female 8,093,609)
15–64 years: 56.4% (male 11,407,233/

female 11,275,685)
65 years and over: 2.5% (male 518,822/female 471,530) (2008 est.)
Median age:
total: 18.9 years
male: 18.7 years
female: 19.1 years (2008 est.)
Population growth rate: 2.134% (2008 est.)
Birth rate: 34.31 births/1,000 population (2008 est.)
Death rate: 13.64 deaths/1,000 population (2008 est.)
Net migration rate: 0.67 migrant(s)/1,000 population (2008 est.)
Sex ratio:
at birth: 1.05 male(s)/female
under 15 years: 1.04 male(s)/female
15–64 years: 1.01 male(s)/female
65 years and over: 1.1 male(s)/female
total population: 1.03 male(s)/female (2008 est.)
Infant mortality rate:
total: 86.98 deaths/1,000 live births
male: 87.09 deaths/1,000 live births
female: 86.86 deaths/1,000 live births (2008 est.)
Life expectancy at birth:
total population: 50.28 years
male: 49.38 years
female: 51.23 years (2008 est.)
Total fertility rate: 4.58 children born/woman (2008 est.)
HIV/AIDS—adult prevalence rate: 2.3% (2001 est.)
HIV/AIDS—people living with HIV/AIDS: 400,000 (2001 est.)
HIV/AIDS—deaths: 23,000 (2003 est.)
Major infectious diseases:
degree of risk: very high
food or waterborne diseases: bacterial and protozoal diarrhea, hepatitis A, and typhoid fever
vectorborne diseases: malaria, dengue fever, African trypanosomiasis (sleeping sickness)
water contact disease: schistosomiasis
respiratory disease: meningococcal meningitis
note: highly pathogenic H5N1 avian influenza has been identified in this country; it poses a negligible risk with extremely rare cases possible among US citizens who have close contact with birds (2008)
Nationality:
noun: Sudanese (singular and plural)
adjective: Sudanese
Ethnic groups: black 52%, Arab 39%, Beja 6%, foreigners 2%, other 1%
Religions: Sunni Muslim 70% (in north), Christian 5% (mostly in south and Khartoum), indigenous beliefs 25%
Languages: Arabic (official), Nubian, Ta Bedawie, diverse dialects of Nilotic,

Nilo-Hamitic, Sudanic languages, English

note: program of "Arabization" in process

Literacy:

definition: age 15 and over can read and write

total population: 61.1%

male: 71.8%

female: 50.5% (2003 est.)

GOVERNMENT

Country name:

conventional long form: Republic of the Sudan

conventional short form: Sudan

local long form: Jumhuriyat as-Sudan

local short form: As-Sudan

former: Anglo-Egyptian Sudan

Government type: Government of National Unity (GNU)—the National Congress Party (NCP) and Sudan People's Liberation Movement (SPLM) formed a power-sharing government under the 2005 Comprehensive Peace Agreement (CPA); the NCP, which came to power by military coup in 1989, is the majority partner; the agreement stipulates national elections in 2009

Capital: *name:* Khartoum

geographic coordinates: 15 36 N, 32 32 E

time difference: UTC+3 (8 hours ahead of Washington, DC during Standard Time)

Administrative divisions: 25 states (wilayat, singular—wilayah); A'ali an Nil (Upper Nile), Al Bahr al Ahmar (Red Sea), Al Buhayrat (Lakes), Al Jazirah (El Gezira), Al Khartum (Khartoum), Al Qadarif (Gedaref), Al Wahdah (Unity), An Nil al Abyad (White Nile), An Nil al Azraq (Blue Nile), Ash Shamaliyah (Northern), Bahr al Jabal (Bahr al Jabal), Gharb al Istiwa'iyah (Western Equatoria), Gharb Bahr al Ghazal (Western Bahr al Ghazal), Gharb Darfur (Western Darfur), Janub Darfur (Southern Darfur), Janub Kurdufan (Southern Kordofan), Junqali (Jonglei), Kassala (Kassala), Nahr an Nil (Nile), Shamal Bahr al Ghazal (Northern Bahr al Ghazal), Shamal Darfur (Northern Darfur), Shamal Kurdufan (Northern Kordofan), Sharq al Istiwa'iyah (Eastern Equatoria), Sinnar (Sinnar), Warab (Warab)

Independence: 1 January 1956 (from Egypt and UK)

National holiday: Independence Day, 1 January (1956)

Constitution: constitution implemented on 30 June 1998, partially suspended 12 December 1999 by President BASHIR; under the CPA, Interim National Constitution ratified 5 July 2005; Constitution of Southern Sudan signed December 2005

Legal system: based on English common law and Islamic law; as of 20 January 1991, the now defunct Revolutionary Command Council imposed Islamic law in the northern states; Islamic law applies to all residents of the northern states regardless of their religion; however, the CPA establishes some protections for non-Muslims in Khartoum; some separate religious courts; accepts compulsory ICJ jurisdiction with reservations; the southern legal system is still developing under the CPA following the civil war; Islamic law will not apply to the southern states

Suffrage: 17 years of age; universal

Executive branch:

chief of state: President Umar Hassan Ahmad al-BASHIR (since 16 October 1993); First Vice President Salva KIIR (since 4 August 2005), Vice President Ali Osman TAHA (since 20 September 2005); note—the president is both the chief of state and head of government

head of government: President Umar Hassan Ahmad al-BASHIR (since 16 October 1993); First Vice President Salva KIIR (since 4 August 2005), Vice President Ali Osman TAHA (since 20 September 2005)

cabinet: Council of Ministers appointed by the president; note—the National Congress Party or NCP (formerly the National Islamic Front or NIF) dominates al-BASHIR's cabinet

elections: election last held 13–23 December 2000; next to be held no later than July 2009 under terms of the 2005 Comprehensive Peace Agreement

election results: Umar Hassan Ahmad al-BASHIR reelected president; percent of vote—Umar Hassan Ahmad al-BASHIR 86.5%, Ja'afar Muhammed NUMAYRI 9.6%, three other candidates received a combined vote of 3.9%; election widely viewed as rigged; all popular opposition parties boycotted elections because of a lack of guarantees for a free and fair election

note: al-BASHIR assumed power as chairman of Sudan's Revolutionary Command Council for National Salvation (RCC) in June 1989 and served concurrently as chief of state, chairman of the RCC, prime minister, and minister of defense until mid-October 1993 when he was appointed president by the RCC; he was elected president by popular vote for the first time in March 1996

Legislative branch: bicameral National Legislature consists of a Council of States (50 seats; members indirectly elected by state legislatures to serve six-year terms) and a National Assembly (450 seats;

members presently appointed, but in the future 75% of members to be directly elected and 25% elected in special or indirect elections; to serve six-year terms)

elections: last held 13–22 December 2000 (next to be held 2009)

election results: NCP 355, others 5; note—replaced by appointments under the 2005 Comprehensive Peace Agreement

Judicial branch: Constitutional Court of nine justices; National Supreme Court; National Courts of Appeal; other national courts; National Judicial Service Commission will undertake overall management of the National Judiciary

Political parties and leaders: National Congress Party or NCP [Umar Hassan al-BASHIR]; Sudan People's Liberation Movement or SPLM [Salva Mayardit KIIR]; and elements of the National Democratic Alliance or NDA including factions of the Democratic Union Party [Muhammad Uthman al-MIRGHANI] and Umma Party [SADIQ Siddiq al-Mahdi]; note—all political parties listed above in the Government of National Unity

Political pressure groups and leaders: Umma Party [Sadiq al-MAHDI]; Popular Congress Party or PCP [Hassan al-TURABI]

International organization participation: ABEDA, ACP, AfDB, AFESD, AMF, AU, CAEU, COMESA, FAO, G-77, IAEA, IBRD, ICAO, ICCt (signatory), ICRM, IDA, IDB, IFAD, IFC, IFRCS, IGAD, ILO, IMF, IMO, Interpol, IOC, IOM, IPU, ISO, ITSO, ITU, LAS, MIGA, NAM, OIC, OPCW, PCA, UN, UNCTAD, UNESCO, UNHCR, UNIDO, UNWTO, UPU, WCO, WFTU, WHO, WIPO, WMO, WTO (observer)

Diplomatic representation in the US:

chief of mission: Ambassador (vacant); Charge d'Affaires John UKEC Lueth

chancery: 2210 Massachusetts Avenue NW, Washington, DC 20008

telephone: [1] (202) 338-8565

FAX: [1] (202) 667-2406

Diplomatic representation from the US:

chief of mission: Ambassador (vacant); Charge d'Affaires Alberto M. FERNANDEZ

embassy: Sharia Ali Abdul Latif Avenue, Khartoum

mailing address: P. O. Box 699, Khartoum; APO AE 09829

telephone: [249] (183) 774701/2/3

FAX: [249] (183) 774137

note: US Consul in Cairo provides backup service for Khartoum

Flag description: three equal horizontal bands of red (top), white, and black with a green isosceles triangle based on the hoist side

ECONOMY

Economy—overview: Sudan's economy is booming on the back of increases in oil production, high oil prices, and large inflows of foreign direct investment. GDP growth registered more than 10% per year in 2006 and 2007. From 1997 to date, Sudan has been working with the IMF to implement macroeconomic reforms, including a managed float of the exchange rate. Sudan began exporting crude oil in the last quarter of 1999. Agricultural production remains important, because it employs 80% of the work force and contributes a third of GDP. The Darfur conflict, the aftermath of two decades of civil war in the south, the lack of basic infrastructure in large areas, and a reliance by much of the population on subsistence agriculture ensure much of the population will remain at or below the poverty line for years despite rapid rises in average per capita income. In January 2007, the government introduced a new currency, the Sudanese Pound, at an initial exchange rate of $1.00 equals 2 Sudanese Pounds.

GDP (purchasing power parity): $80.71 billion (2007 est.)

GDP (official exchange rate): $49.71 billion (2007 est.)

GDP—real growth rate: 10.5% (2007 est.)

GDP—per capita (PPP): $2,200 (2007 est.)

GDP—composition by sector:
agriculture: 31.8%
industry: 34.2%
services: 33.9% (2007 est.)

Labor force: 7.415 million (1996 est.)

Labor force—by occupation: *agriculture:* 80%
industry: 7%
services: 13% (1998 est.)

Unemployment rate: 18.7% (2002 est.)

Population below poverty line: 40% (2004 est.)

Household income or consumption by percentage share:
lowest 10%: NA%
highest 10%: NA%

Inflation rate (consumer prices): 8% (2007 est.)

Investment (gross fixed): 18.7% of GDP (2007 est.)

Budget: *revenues:* $8.665 billion
expenditures: $10.89 billion (2007 est.)

Public debt: 98.9% of GDP (2007 est.)

Agriculture—products: cotton, groundnuts (peanuts), sorghum, millet, wheat, gum arabic, sugarcane, cassava (tapioca), mangos, papaya, bananas, sweet potatoes, sesame; sheep, livestock

Industries: oil, cotton ginning, textiles, cement, edible oils, sugar, soap distilling, shoes, petroleum refining, pharmaceuticals, armaments, automobile/light truck assembly

Industrial production growth rate: 30% (2007 est.)

Electricity—production: 3.944 billion kWh (2005)

Electricity—production by source:
fossil fuel: 52.1%
hydro: 47.9%
nuclear: 0%
other: 0% (2001)

Electricity—consumption: 3.298 billion kWh (2005)

Electricity—exports: 0 kWh (2005)

Electricity—imports: 0 kWh (2005)

Oil—production: 397,000 bbl/day (2006 est.)

Oil—consumption: 79,760 bbl/day (2006 est.)

Oil—exports: 279,100 bbl/day (2004)

Oil—imports: 7,945 bbl/day (2004)

Oil—proved reserves: 6.49 billion bbl (2007 est.)

Natural gas—production: 0 cu m (2006 est.)

Natural gas—consumption: 0 cu m (2006 est.)

Natural gas—exports: 0 cu m (2006 est.)

Natural gas—imports: 0 cu m (2006 est.)

Natural gas—proved reserves: 86 billion cu m (1 January 2006 est.)

Current account balance: -$5.432 billion (2007 est.)

Exports: $8.878 billion f.o.b. (2007 est.)

Exports—commodities: oil and petroleum products; cotton, sesame, livestock, groundnuts, gum arabic, sugar

Exports—partners: Japan 48%, China 31%, South Korea 3.8% (2006)

Imports: $7.722 billion f.o.b. (2007 est.)

Imports—commodities: foodstuffs, manufactured goods, refinery and transport equipment, medicines and chemicals, textiles, wheat

Imports—partners: China 20.8%, Saudi Arabia 8%, India 7.4%, Japan 6.6%, UAE 5.5%, Egypt 5.3%, France 4.6%, South Korea 4.2% (2006)

Economic aid—recipient: $1.829 billion (2005)

Reserves of foreign exchange and gold: $1.378 billion (31 December 2007 est.)

Debt—external: $29.51 billion (31 December 2007 est.)

Market value of publicly traded shares: $NA

Currency (code): Sudanese pounds (SDG)

Currency code: SDD

Exchange rates: Sudanese pounds per US dollar—2.06 (2007), 2.172 (2006), 2.4361 (2005), 2.5791 (2004), 2.6098 (2003)

note: in October 2007 Sudan redenominated its currency by transforming 100 units of Sudanese dinar into one unit of Sudanese pound

Fiscal year: calendar year

COMMUNICATIONS

Telephones—main lines in use: 636,900 (2006)

Telephones—mobile cellular: 4.683 million (2006)

Telephone system:
general assessment: large, well-equipped system by regional standards and being upgraded; cellular communications started in 1996 and have expanded substantially
domestic: consists of microwave radio relay, cable, radiotelephone communications, tropospheric scatter, and a domestic satellite system with 14 earth stations
international: country code—249; linked to international submarine cable Fiber-Optic Link Around the Globe (FLAG); satellite earth stations—1 Intelsat (Atlantic Ocean), 1 Arabsat (2000)

Radio broadcast stations: AM 12, FM 1, shortwave 1 (1998)

Radios: 7.55 million (1997)

Television broadcast stations: 3 (1997)

Televisions: 2.38 million (1997)

Internet country code: .sd

Internet hosts: 21 (2007)

Internet Service Providers (ISPs): 2 (2002)

Internet users: 3.5 million (2006)

TRANSPORTATION

Airports: 101 (2007)

Airports—with paved runways:
total: 16
over 3,047 m: 2
2,438 to 3,047 m: 9
1,524 to 2,437 m: 4
under 914 m: 1 (2007)

Airports—with unpaved runways:
total: 85
over 3,047 m: 1
1,524 to 2,437 m: 20
914 to 1,523 m: 37
under 914 m: 27 (2007)

Heliports: 4 (2007)

Pipelines: gas 156 km; oil 4,070 km; refined products 1,613 km (2007)

Railways:
total: 5,978 km
narrow gauge: 4,578 km 1.067-m gauge;

1,400 km 0.600-m gauge for cotton plantations (2006)

Roadways:
total: 11,900 km
paved: 4,320 km
unpaved: 7,580 km (1999)

Waterways: 4,068 km (1,723 km open year round on White and Blue Nile rivers) (2006)

Merchant marine:
total: 3 ships (1000 GRT or over) 21,311 GRT/26,179 DWT
by type: cargo 2, livestock carrier 1 (2007)

Ports and terminals: Port Sudan

MILITARY

Military branches: Sudanese People's Armed Forces (SPAF): Land Forces, Navy, Air Force, Popular Defense Forces; Sudan People's Liberation Army (SPLA): Land Forces (2008)

Military service age and obligation: 18–30 years of age for compulsory military service; 2-year service obligation (2006)

Manpower available for military service:
males age 16–49: 9,639,923
females age 16–49: 9,321,106 (2008 est.)

Manpower fit for military service:
males age 16–49: 5,586,468
females age 16–49: 5,678,427 (2008 est.)

Manpower reaching militarily significant age annually:
males age 16–49: 488,679
females age 16–49: 469,547 (2008 est.)

Military expenditures—percent of GDP: 3% (2005 est.)

TRANSNATIONAL ISSUES

Disputes—international: the effects of Sudan's almost constant ethnic and rebel militia fighting since the mid-20th century have penetrated all of the neighboring states; as of 2006, Chad, Ethiopia, Kenya, Central African Republic, Democratic Republic of the Congo, and Uganda provided shelter for over half a million Sudanese refugees, which includes 240,000 Darfur residents driven from their homes by Janjawid armed militia and the Sudanese military forces; Sudan, in turn, hosted about 116,000 Eritreans, 20,000 Chadians, and smaller numbers of Ethiopians, Ugandans, Central Africans, and Congolese as refugees; in February 2006, Sudan and DROC signed an agreement to repatriate 13,300 Sudanese and 6,800 Congolese; Sudan accuses Eritrea of supporting Sudanese rebel groups; efforts to demarcate the porous boundary with Ethiopia proceed slowly due to civil and ethnic fighting in eastern Sudan; the boundary that separates Kenya and Sudan's sovereignty is unclear in the "Ilemi Triangle," which Kenya has administered since colonial times; while Sudan claims to administer the Hala'ib Triangle north of the 1899 Treaty boundary along the 22nd Parallel; both states withdrew their military presence in the 1990s, and Egypt has invested in and effectively administers the area; periodic violent skirmishes with Sudanese residents over water and grazing rights persist among related pastoral populations along the border with the Central African Republic

Refugees and internally displaced persons: *refugees (country of origin):* 157,220 (Eritrea); 25,023 (Chad); 11,009 (Ethiopia); 7,895 (Uganda); 5,023 (Central African Republic)
IDPs: 5.3–6.2 million (civil war 1983–2005; ongoing conflict in Darfur region) (2007)

Trafficking in persons: *current situation:* Sudan is a source country for men, women, and children trafficked for the purposes of forced labor and sexual exploitation; Sudan may also be a transit and destination country for Ethiopian women trafficked for domestic servitude; boys are trafficked to the Middle East, particularly Qatar and the United Arab Emirates, for use as camel jockeys; small numbers of girls are reportedly trafficked within Sudan for domestic servitude as well as for commercial sexual exploitation in small brothels in internally displaced persons (IDP) camps; the terrorist rebel organization "Lord's Resistance Army" (LRA) continues to abduct and forcibly conscript small numbers of children in Southern Sudan for use as cooks, porters, and combatants in its ongoing war against Uganda; some of these children are then trafficked across borders into Uganda or possibly the Democratic Republic of the Congo; children are utilized by rebel groups and the Sudanese Armed Forces and associated militias in the ongoing conflict in Darfur; during the decades of civil war, thousands of Dinka women and children were enslaved by members of Baggara tribes and subjected to various forms of forced labor without remuneration as well as physical and sexual abuse; with the cessation of the North-South conflict and the ongoing peace process, there were no known new abductions of Dinka by Baggara tribes during 2005; however, inter-tribal abductions of a different nature continue in Southern Sudan and warrant further investigation
tier rating: Tier 3—Sudan does not fully comply with the minimum standards for the elimination of trafficking and is not making significant efforts to do so

SURINAME

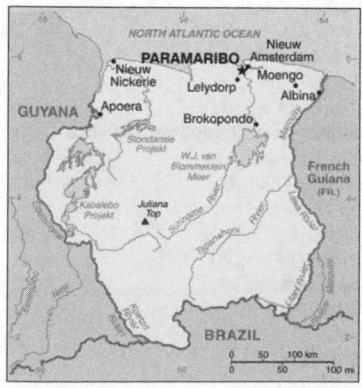

INTRODUCTION

Background: First explored by the Spaniards in the 16th century and then settled by the English in the mid-17th century, Suriname became a Dutch colony in 1667. With the abolition of slavery in 1863, workers were brought in from India and Java. Independence from the Netherlands was granted in 1975. Five years later the civilian government was replaced by a military regime that soon declared a socialist republic. It continued to exert control through a succession of nominally civilian administrations until 1987, when international pressure finally forced a democratic election. In 1990, the military overthrew the civilian leadership, but a democratically elected government—a four-party New Front coalition—returned to power in 1991 and has ruled since, expanding to eight parties in 2005.

GEOGRAPHY

Location: Northern South America, bordering the North Atlantic Ocean, between French Guiana and Guyana
Geographic coordinates: 4 00 N, 56 00 W
Map references: South America
Area: *total:* 163,270 sq km

land: 161,470 sq km
water: 1,800 sq km
Area—comparative: slightly larger than Georgia
Land boundaries:
total: 1,703 km
border countries: Brazil 593 km, French Guiana 510 km, Guyana 600 km
Coastline: 386 km
Maritime claims:
territorial sea: 12 nm
exclusive economic zone: 200 nm
Climate: tropical; moderated by trade winds
Terrain: mostly rolling hills; narrow coastal plain with swamps
Elevation extremes:
lowest point: unnamed location in the coastal plain -2 m
highest point: Juliana Top 1,230 m
Natural resources: timber, hydropower, fish, kaolin, shrimp, bauxite, gold, and small amounts of nickel, copper, platinum, iron ore
Land use:
arable land: 0.36%
permanent crops: 0.06%
other: 99.58% (2005)
Irrigated land: 510 sq km (2003)
Total renewable water resources: 122 cu km (2003)
Freshwater withdrawal (domestic/industrial/agricultural): *total:* 0.67 cu km/yr (4%/3%/93%)
per capita: 1,489 cu m/yr (2000)
Natural hazards: NA
Environment—current issues: deforestation as timber is cut for export; pollution of inland waterways by small-scale mining activities
Environment—international agreements: *party to:* Biodiversity, Climate Change, Climate Change-Kyoto Protocol, Desertification, Endangered Species, Law of the Sea, Marine Dumping, Ozone Layer Protection, Ship Pollution, Tropical Timber 94, Wetlands, Whaling
signed, but not ratified: none of the selected agreements
Geography—note: smallest independent country on South American continent; mostly tropical rain forest; great diversity of flora and fauna that, for the most part, is increasingly threatened by new development; relatively small population, mostly along the coast

PEOPLE

Population: 475,996 (July 2008 est.)
Age structure:
0–14 years: 27.5% (male 66,695/female 64,356)
15–64 years: 66.2% (male 156,961/female 158,234)

65 years and over: 6.3% (male 12,868/female 16,882) (2008 est.)
Median age:
total: 27.5 years
male: 27.1 years
female: 27.9 years (2008 est.)
Population growth rate: 1.099% (2008 est.)
Birth rate: 17.02 births/1,000 population (2008 est.)
Death rate: 5.51 deaths/1,000 population (2008 est.)
Net migration rate: -0.52 migrant(s)/1,000 population (2008 est.)
Sex ratio:
at birth: 1.07 male(s)/female
under 15 years: 1.04 male(s)/female
15–64 years: 0.99 male(s)/female
65 years and over: 0.76 male(s)/female
total population: 0.99 male(s)/female (2008 est.)
Infant mortality rate:
total: 19.45 deaths/1,000 live births
male: 22.96 deaths/1,000 live births
female: 15.71 deaths/1,000 live births (2008 est.)
Life expectancy at birth:
total population: 73.48 years
male: 70.76 years
female: 76.39 years (2008 est.)
Total fertility rate: 2.01 children born/woman (2008 est.)
HIV/AIDS—adult prevalence rate: 1.7% (2001 est.)
HIV/AIDS—people living with HIV/AIDS: 5,200 (2001 est.)
HIV/AIDS—deaths: fewer than 500 (2003 est.)
Major infectious diseases:
degree of risk: high
food or waterborne diseases: bacterial and protozoal diarrhea, hepatitis A, and typhoid fever
vectorborne disease: dengue fever, Mayaro virus, and malaria
water contact disease: leptospirosis (2008)
Nationality:
noun: Surinamer(s)
adjective: Surinamese
Ethnic groups: Hindustani (also known locally as "East Indians"; their ancestors emigrated from northern India in the latter part of the 19th century) 37%, Creole (mixed white and black) 31%, Javanese 15%, "Maroons" (their African ancestors were brought to the country in the 17th and 18th centuries as slaves and escaped to the interior) 10%, Amerindian 2%, Chinese 2%, white 1%, other 2%
Religions: Hindu 27.4%, Protestant 25.2% (predominantly Moravian), Roman Catholic 22.8%, Muslim 19.6%, indigenous beliefs 5%
Languages: Dutch (official), English

(widely spoken), Sranang Tongo (Surinamese, sometimes called Taki-Taki, is native language of Creoles and much of the younger population and is lingua franca among others), Caribbean Hindustani (a dialect of Hindi), Javanese
Literacy:
definition: age 15 and over can read and write
total population: 89.6%
male: 92%
female: 87.2% (2004 census)

GOVERNMENT

Country name:
conventional long form: Republic of Suriname
conventional short form: Suriname
local long form: Republiek Suriname
local short form: Suriname
former: Netherlands Guiana, Dutch Guiana
Government type: constitutional democracy
Capital: *name:* Paramaribo
geographic coordinates: 5 50 N, 55 10 W
time difference: UTC-3 (2 hours ahead of Washington, DC during Standard Time)
Administrative divisions: 10 districts (distrikten, singular—distrikt); Brokopondo, Commewijne, Coronie, Marowijne, Nickerie, Para, Paramaribo, Saramacca, Sipaliwini, Wanica
Independence: 25 November 1975 (from the Netherlands)
National holiday: Independence Day, 25 November (1975)
Constitution: ratified 30 September 1987; effective 30 October 1987
Legal system: based on Dutch legal system incorporating French penal theory; accepts compulsory ICJ jurisdiction with reservations
Suffrage: 18 years of age; universal
Executive branch:
chief of state: President Runaldo Ronald VENETIAAN (since 12 August 2000); Vice President Ramdien SARDJOE (since 3 August 2005); note—the president is both the chief of state and head of government
head of government: President Runaldo Ronald VENETIAAN (since 12 August 2000); Vice President Ram SARDJOE (since 3 August 2005)
cabinet: Cabinet of Ministers appointed by the president
elections: president and vice president elected by the National Assembly or, if no presidential or vice presidential candidate receives a two-thirds constitutional majority in the National Assembly after two votes, by a simple majority in the larger United People's Assembly (893 representatives from the

615

national, local, and regional councils), for five-year terms (no term limits); election last held on 25 May 2005 (next to be held in 2010)

election results: Runaldo Ronald VENETIAAN reelected president; percent of vote—Runaldo Ronald VENETIAAN 62.9%, Rabin PARMESSAR 35.4%, other 1.7%; note—after two votes in the parliament failed to secure a two-thirds majority for a candidate, the vote then went to a special session of the United People's Assembly on 3 August 2005

Legislative branch: unicameral National Assembly or Nationale Assemblee (51 seats; members are elected by popular vote to serve five-year terms)

elections: last held on 25 May 2005 (next to be held in 2010)

election results: percent of vote by party—NF 39.7%, NDP 22.2%, VVV 13.8%, A-Com 7.2%, A-1 5.9%, other 11.2%; seats by party—NF 23, NDP 15, VVV 5, A-Com 5, A-1 3

Judicial branch: Cantonal Courts and a Court of Justice as an appellate court (justices are nominated for life)

Political parties and leaders: Alternative-1 or A-1 (a coalition of Amazone Party of Suriname or APS [Kenneth VAN GENDEREN], Democrats of the 21st Century or D-21 [Soewarto MOESTADJA], Nieuw Suriname or NS [Radjen Nanan PANDAY], Political Wing of the FAL or PVF [Jiwan SITAL], Trefpunt 2000 or T-2000 [Arti JESSURUN]); General Interior Development Party or ABOP [Ronnie BRUNSWIJK]; National Democratic Party or NDP [Desire BOUTERSE]; New Front for Democracy and Development or NF (a coalition that includes A-Combination or A-Com, Democratic Alternative 1991 or DA-91, an independent, business-oriented party [Winston JESSURUN], National Party Suriname or NPS [Ronald VENETIAAN], United Reform Party or VHP [Ramdien SARDJOE], Pertjaja Luhur or PL [Salam Paul SOMOHARDJO], Surinamese Labor Party or SPA [Siegfried GILDS]); Party for Democracy and Development in Unity or DOE [Marten SCHALKWIJK]; People's Alliance for Progress or VVV (a coalition of Democratic National Platform 2000 or DNP-2000 [Jules WIJDENBOSCH], Grassroots Party for Renewal and Democracy or BVD [Tjan GOBARDHAN], Party for National Unity and Solidarity of the Highest Order or KTPI [Willy SOEMITA], Party for Progression, Justice, and Perseverance or PPRS [Renee KAIMAN], Pendawalima

or PL [Raymond SAPOEN]); Progressive Laborers and Farmers Union or PALU [Jim HOK]; Progressive Political Party or PPP [Surinder MUNGRA]; Seeka [Paul ABENA]; Union of Progressive Surinamers or UPS [Sheoradj PANDAY]

Political pressure groups and leaders: Association of Indigenous Village Chiefs [Ricardo PANE]; Association of Saramaccan Authorities or Maroon [Head Captain WASE]; Women's Parliament Forum or PVF [Iris GILLIAD]

International organization participation: ACP, Caricom, CSN, FAO, G-77, IADB, IBRD, ICAO, ICRM, IDA, IDB, IFAD, IFRCS, IHO (suspended), ILO, IMF, IMO, Interpol, IOC, IPU, ISO (subscriber), ITU, ITUC, LAES, MIGA, NAM, OAS, OIC, OPANAL, OPCW, PCA, UN, UNCTAD, UNESCO, UNIDO, UPU, WCL, WHO, WIPO, WMO, WTO

Diplomatic representation in the US:
chief of mission: Ambassador Jacques Ruben Constantijn KROSS
chancery: Suite 460, 4301 Connecticut Avenue NW, Washington, DC 20008
telephone: [1] (202) 244-7488
FAX: [1] (202) 244-5878
consulate(s) general: Miami

Diplomatic representation from the US:
chief of mission: Ambassador Lisa Bobbie SCHREIBER HUGHES
embassy: Dr. Sophie Redmondstraat 129, Paramaribo
mailing address: US Department of State, 3390 Paramaribo Place, Washington, DC, 20521-3390
telephone: [597] 472-900
FAX: [597] 425-690

Flag description: five horizontal bands of green (top, double width), white, red (quadruple width), white, and green (double width); there is a large, yellow, five-pointed star centered in the red band

ECONOMY

Economy—overview: The economy is dominated by the mining industry, with exports of alumina, gold, and oil accounting for about 85% of exports and 25% of government revenues, making the economy highly vulnerable to mineral price volatility. The short-term economic outlook depends on the government's ability to control inflation and on the development of projects in the bauxite and gold mining sectors. Suriname has received aid for these projects from Netherlands, Belgium, and the European Development Fund. Suriname's economic prospects for the medium term will depend on continued

commitment to responsible monetary and fiscal policies and to the introduction of structural reforms to liberalize markets and promote competition. In 2000, the government of Ronald VENETIAAN, returned to office and inherited an economy with inflation of over 100% and a growing fiscal deficit. He quickly implemented an austerity program, raised taxes, attempted to control spending, and tamed inflation. These economic policies are likely to remain in effect during VENETIAAN's third term. Prospects for local onshore oil production are good as a drilling program is underway. Offshore oil drilling was given a boost in 2004 when the State Oil Company (Staatsolie) signed exploration agreements with Repsol, Maersk, and Occidental. Bidding on these new offshore blocks was completed in July 2006.

GDP (purchasing power parity): $4.073 billion (2007 est.)

GDP (official exchange rate): $2.404 billion (2007 est.)

GDP—real growth rate: 5.5% (2007 est.)

GDP—per capita (PPP): $7,800 (2007 est.)

GDP—composition by sector:
agriculture: 10.8%
industry: 24.4%
services: 64.8% (2005 est.)

Labor force: 156,700 (2004)

Labor force—by occupation: *agriculture:* 8%
industry: 14%
services: 78% (2004)

Unemployment rate: 9.5% (2004)

Population below poverty line: 70% (2002 est.)

Household income or consumption by percentage share:
lowest 10%: NA%
highest 10%: NA%

Inflation rate (consumer prices): 6.4% (2007 est.)

Budget:
revenues: $392.6 million
expenditures: $425.9 million (2004)

Agriculture—products: paddy rice, bananas, palm kernels, coconuts, plantains, peanuts; beef, chickens; shrimp; forest products

Industries: bauxite and gold mining, alumina production; oil, lumbering, food processing, fishing

Industrial production growth rate: 6.5% (1994 est.)

Electricity—production: 1.53 billion kWh (2005)

Electricity—production by source:
fossil fuel: 25.2%
hydro: 74.8%

nuclear: 0%
other: 0% (2001)
Electricity—consumption: 1.423 billion kWh (2005)
Electricity—exports: 0 kWh (2005)
Electricity—imports: 0 kWh (2005)
Oil—production: 9,461 bbl/day (2005 est.)
Oil—consumption: 12,000 bbl/day (2005 est.)
Oil—exports: 3,151 bbl/day (2004)
Oil—imports: 6,032 bbl/day (2004)
Oil—proved reserves: 111 million bbl (1 January 2006 est.)
Natural gas—production: 0 cu m (2005 est.)
Natural gas—consumption: 0 cu m (2005 est.)
Natural gas—exports: 0 cu m (2005 est.)
Natural gas—imports: 0 cu m (2005)
Natural gas—proved reserves: 0 cu m (1 January 2006 est.)
Current account balance: $24 million (2007 est.)
Exports: $1.391 billion f.o.b. (2006 est.)
Exports—commodities: alumina, gold, crude oil, lumber, shrimp and fish, rice, bananas
Exports—partners: Norway 23%, Canada 15.5%, US 12.6%, Belgium 10.1%, France 8.5%, UAE 6.9%, Iceland 4.2% (2006)
Imports: $1.297 billion f.o.b. (2006 est.)
Imports—commodities: capital equipment, petroleum, foodstuffs, cotton, consumer goods
Imports—partners: US 29.4%, Netherlands 18.9%, Trinidad and Tobago 14.9%, Japan 5.1%, China 4.9% (2006)
Economic aid—recipient: $43.97 million (2005)
Reserves of foreign exchange and gold: $263.3 million (2006)
Debt—external: $504.3 million (2005 est.)
Market value of publicly traded shares: $NA

Currency (code): Surinam dollar (SRD)
Currency code: SRG
Exchange rates: Surinamese dollars per US dollar—2.745 (2007), 2.745 (2006), 2.7317 (2005), 2.7336 (2004), 2.6013 (2003)
note: in January 2004, the government replaced the guilder with the Surinamese dollar, tied to a US dollar-dominated currency basket
Fiscal year: calendar year

COMMUNICATIONS

Telephones—main lines in use: 81,500 (2006)
Telephones—mobile cellular: 320,000 (2006)
Telephone system:
general assessment: international facilities are good
domestic: combined fixed-line and mobile-cellular teledensity about 90 telephones per 100 persons; microwave radio relay network
international: country code—597; satellite earth stations—2 Intelsat (Atlantic Ocean)
Radio broadcast stations: AM 4, FM 13, shortwave 1 (1998)
Radios: 300,000 (1997)
Television broadcast stations: 3 (plus 7 repeaters) (2000)
Televisions: 63,000 (1997)
Internet country code: .sr
Internet hosts: 28 (2007)
Internet Service Providers (ISPs): 2 (2000)
Internet users: 32,000 (2005)

TRANSPORTATION

Airports: 50 (2007)
Airports—with paved runways:
total: 5
over 3,047 m: 1
under 914 m: 4 (2007)
Airports—with unpaved runways:
total: 45
914 to 1,523 m: 5
under 914 m: 40 (2007)

Pipelines: oil 50 km (2007)
Roadways:
total: 4,304 km
paved: 1,130 km
unpaved: 3,174 km (2003)
Waterways: 1,200 km (most navigable by ships with drafts up to 7 m) (2005)
Merchant marine:
total: 1 ship (1000 GRT or over) 1,078 GRT/1,214 DWT
by type: cargo 1 (2007)
Ports and terminals: Paramaribo, Wageningen

MILITARY

Military branches: National Army (Nationaal Leger, NL; includes Naval Wing, Air Wing) (2007)
Military service age and obligation: 18 years of age (est.); no conscription
Manpower available for military service:
males age 16–49: 130,534
females age 16–49: 130,243 (2008 est.)
Manpower fit for military service:
males age 16–49: 105,770
females age 16–49: 109,666 (2008 est.)
Military expenditures—percent of GDP: 0.6% (2006 est.)

TRANSNATIONAL ISSUES

Disputes—international: area claimed by French Guiana between Riviere Litani and Riviere Marouini (both headwaters of the Lawa); Suriname claims a triangle of land between the New and Kutari/Koetari rivers in a historic dispute over the headwaters of the Courantyne; Guyana seeks United Nations Convention on the Law of the Sea (UNCLOS) arbitration to resolve the long-standing dispute with Suriname over the axis of the territorial sea boundary in potentially oil-rich waters
Illicit drugs: growing transshipment point for South American drugs destined for Europe via the Netherlands and Brazil; transshipment point for arms-for-drugs dealing

SVALBARD

INTRODUCTION

Background: First discovered by the Norwegians in the 12th century, the islands served as an international whaling base during the 17th and 18th centuries. Norway's sovereignty was recognized in 1920; five years later it officially took over the territory.

GEOGRAPHY

Location: Northern Europe, islands between the Arctic Ocean, Barents Sea, Greenland Sea, and Norwegian Sea, north of Norway
Geographic coordinates: 78 00 N, 20 00 E
Map references: Arctic Region
Area:
total: 61,020 sq km
land: 61,020 sq km
water: 0 sq km
note: includes Spitsbergen and Bjornoya (Bear Island)

Area—comparative: slightly smaller than West Virginia
Land boundaries: 0 km
Coastline: 3,587 km
Maritime claims:
territorial sea: 4 nm
exclusive fishing zone: 200 nm unilaterally claimed by Norway but not recognized by Russia
Climate: arctic, tempered by warm North Atlantic Current; cool summers, cold winters; North Atlantic Current

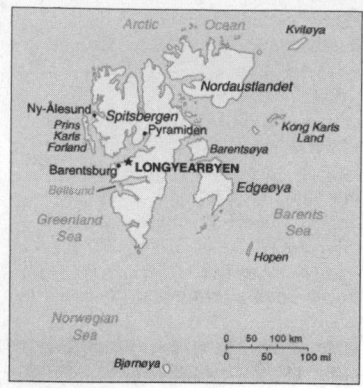

flows along west and north coasts of Spitsbergen, keeping water open and navigable most of the year

Terrain: wild, rugged mountains; much of high land ice covered; west coast clear of ice about one-half of the year; fjords along west and north coasts

Elevation extremes:
lowest point: Arctic Ocean 0 m
highest point: Newtontoppen 1,717 m

Natural resources: coal, iron ore, copper, zinc, phosphate, wildlife, fish

Land use: *arable land:* 0%
permanent crops: 0%
other: 100% (no trees; the only bushes are crowberry and cloudberry) (2005)

Irrigated land: NA

Natural hazards: ice floes often block the entrance to Bellsund (a transit point for coal export) on the west coast and occasionally make parts of the northeastern coast inaccessible to maritime traffic

Environment—current issues: NA

Geography—note: northernmost part of the Kingdom of Norway; consists of nine main islands; glaciers and snowfields cover 60% of the total area; Spitsbergen Island is the site of the Svalbard Global Seed Vault, a seed repository established by the Global Crop Diversity Trust and the Norwegian Government

PEOPLE

Population: 2,165 (July 2008 est.)
Age structure: *0–14 years:* NA
15–64 years: NA
65 years and over: NA
Population growth rate: -0.023% (2008 est.)
Birth rate: NA
Death rate: NA (2008 est.)
Net migration rate: NA
Sex ratio: NA
Infant mortality rate:
total: NA
male: NA
female: NA (2008 est.)
Life expectancy at birth:
total population: NA
male: NA
female: NA (2008 est.)

Total fertility rate: NA (2008 est.)
HIV/AIDS—adult prevalence rate: 0% (2001)
HIV/AIDS—people living with HIV/AIDS: 0 (2001)
HIV/AIDS—deaths: 0 (2001)
Ethnic groups: Norwegian 55.4%, Russian and Ukrainian 44.3%, other 0.3% (1998)
Languages: Norwegian, Russian
Literacy: NA

GOVERNMENT

Country name:
conventional long form: none
conventional short form: Svalbard (sometimes referred to as Spitzbergen)
Dependency status: territory of Norway; administered by the Polar Department of the Ministry of Justice, through a governor (sysselmann) residing in Longyearbyen, Spitsbergen; by treaty (9 February 1920) sovereignty was awarded to Norway
Government type: NA
Capital: *name:* Longyearbyen
geographic coordinates: 78 13 N, 15 33 E
time difference: UTC+1 (6 hours ahead of Washington, DC during Standard Time)
daylight saving time: +1hr, begins last Sunday in March; ends last Sunday in October
Independence: none (territory of Norway)
Legal system: the laws of Norway, where applicable, apply
Executive branch:
chief of state: King HARALD V of Norway (since 17 January 1991)
head of government: Governor Per SEFLAND (since 1 October 2005); Assistant Governor Rune Baard HANSEN (since 2003)
elections: none; the monarch is hereditary; governor and assistant governor responsible to the Polar Department of the Ministry of Justice
International organization participation: none
Flag description: the flag of Norway is used

ECONOMY

Economy—overview: Coal mining is the major economic activity on Svalbard. The treaty of 9 February 1920 gave the 41 signatories equal rights to exploit mineral deposits, subject to Norwegian regulation. Although US, UK, Dutch, and Swedish coal companies have mined in the past, the only companies still mining are Norwegian and Russian. The settlements on Svalbard are essentially company towns. The Norwegian state-owned coal company employs nearly 60% of the Norwegian population on the island, runs many of the local services, and provides most of the local infrastructure. There is also some hunting of seal, reindeer, and fox.

GDP (purchasing power parity): $NA
GDP—real growth rate: NA%
Labor force: NA
Budget:
revenues: $25.07 million
expenditures: $NA (2004 est.)
Electricity—production by source:
fossil fuel: 57.9984%
hydro: 42.0016%
nuclear: 0%
other: 0%
Exports: $197.6 million (2004)
Imports: $NA
Economic aid—recipient: $8.2 million from Norway (1998)
Currency (code): Norwegian krone (NOK)
Currency code: NOK
Exchange rates: Norwegian kroner per US dollar—5.8396 (2007), 6.4117 (2006), 6.4425 (2005), 6.7408 (2004), 7.0802 (2003)

COMMUNICATIONS

Telephones—main lines in use: NA
Telephone system:
general assessment: probably adequate
domestic: local telephone service
international: country code—47-790; satellite earth station—1 of unknown type (for communication with Norwegian mainland only)
Radio broadcast stations: AM 1, FM 1 (plus 2 repeaters), shortwave 0 (1998)
Radios: NA
Television broadcast stations: NA
Televisions: NA
Internet country code: .sj
Internet Service Providers (ISPs): 13 (Svalbard and Jan Mayen) (2000)
Internet users: NA

TRANSPORTATION

Airports: 4 (2007)
Airports—with paved runways:
total: 1
1,524 to 2,437 m: 1 (2007)
Airports—with unpaved runways:
total: 3
under 914 m: 3 (2007)
Heliports: 1 (2007)
Ports and terminals: Barentsburg, Longyearbyen, Ny-Alesund, Pyramiden

MILITARY

Military branches: no regular military forces
Military—note: Svalbard is a territory of Norway, demilitarized by treaty on 9 February 1920

TRANSNATIONAL ISSUES

Disputes—international: despite recent discussions, Russia and Norway dispute their maritime limits in the Barents Sea and Russia's fishing rights beyond Svalbard's territorial limits within the Svalbard Treaty zone

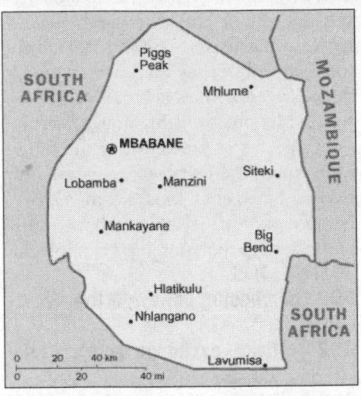

highest point: Emlembe 1,862 m
Natural resources: asbestos, coal, clay, cassiterite, hydropower, forests, small gold and diamond deposits, quarry stone, and talc
Land use:
arable land: 10.25%
permanent crops: 0.81%
other: 88.94% (2005)
Irrigated land: 500 sq km (2003)
Total renewable water resources: 4.5 cu km (1987)
Freshwater withdrawal (domestic/industrial/agricultural): *total:* 1.04 cu km/yr (2%/1%/97%)
per capita: 1,010 cu m/yr (2000)
Natural hazards: drought
Environment—current issues: limited supplies of potable water; wildlife populations being depleted because of excessive hunting; overgrazing; soil degradation; soil erosion
Environment—international agreements: *party to:* Biodiversity, Climate Change, Climate Change-Kyoto Protocol, Desertification, Endangered Species, Hazardous Wastes, Ozone Layer Protection
signed, but not ratified: Law of the Sea
Geography—note: landlocked; almost completely surrounded by South Africa

INTRODUCTION

Background: Autonomy for the Swazis of southern Africa was guaranteed by the British in the late 19th century; independence was granted in 1968. Student and labor unrest during the 1990s pressured King MSWATI III, the world's last absolute monarch, to grudgingly allow political reform and greater democracy, although he has backslid on these promises in recent years. A constitution came into effect in 2006, but political parties remain banned. The African United Democratic Party tried unsuccessfully to register as an official political party in mid 2006. Talks over the constitution broke down between the government and progressive groups in 2007. Swaziland recently surpassed Botswana as the country with the world's highest known HIV/AIDS prevalence rate.

GEOGRAPHY

Location: Southern Africa, between Mozambique and South Africa
Geographic coordinates: 26 30 S, 31 30 E
Map references: Africa
Area: *total:* 17,363 sq km
land: 17,203 sq km
water: 160 sq km
Area—comparative: slightly smaller than New Jersey
Land boundaries:
total: 535 km
border countries: Mozambique 105 km, South Africa 430 km
Coastline: 0 km (landlocked)
Maritime claims: none (landlocked)
Climate: varies from tropical to near temperate
Terrain: mostly mountains and hills; some moderately sloping plains
Elevation extremes:
lowest point: Great Usutu River 21 m

PEOPLE

Population: 1,128,814
note: estimates for this country explicitly take into account the effects of excess mortality due to AIDS; this can result in lower life expectancy, higher infant mortality, higher death rates, lower population growth rates, and changes in the distribution of population by age and sex than would otherwise be expected (July 2008 est.)
Age structure:
0–14 years: 39.9% (male 226,947/female 222,922)
15–64 years: 56.5% (male 306,560/female 331,406)
65 years and over: 3.6% (male 15,594/female 25,385) (2008 est.)
Median age:
total: 18.7 years
male: 18 years
female: 19.4 years (2008 est.)
Population growth rate: -0.41% (2008 est.)
Birth rate: 26.6 births/1,000 population (2008 est.)
Death rate: 30.7 deaths/1,000 population (2008 est.)
Net migration rate: NA
Sex ratio: *at birth:* 1.03 male(s)/female

under 15 years: 1.02 male(s)/female
15–64 years: 0.93 male(s)/female
65 years and over: 0.61 male(s)/female
total population: 0.95 male(s)/female (2008 est.)
Infant mortality rate:
total: 69.59 deaths/1,000 live births
male: 72.87 deaths/1,000 live births
female: 66.2 deaths/1,000 live births (2008 est.)
Life expectancy at birth:
total population: 31.99 years
male: 31.69 years
female: 32.3 years (2008 est.)
Total fertility rate: 3.34 children born/woman (2008 est.)
HIV/AIDS—adult prevalence rate: 38.8% (2003 est.)
HIV/AIDS—people living with HIV/AIDS: 220,000 (2003 est.)
HIV/AIDS—deaths: 17,000 (2003 est.)
Major infectious diseases:
degree of risk: intermediate
food or waterborne diseases: bacterial diarrhea, hepatitis A, and typhoid fever
vectorborne disease: malaria
water contact disease: schistosomiasis (2008)
Nationality:
noun: Swazi(s)
adjective: Swazi
Ethnic groups: African 97%, European 3%
Religions: Zionist 40% (a blend of Christianity and indigenous ancestral worship), Roman Catholic 20%, Muslim 10%, other (includes Anglican, Bahai, Methodist, Mormon, Jewish) 30%
Languages: English (official, government business conducted in English), siSwati (official)
Literacy:
definition: age 15 and over can read and write
total population: 81.6%
male: 82.6%
female: 80.8% (2003 est.)

GOVERNMENT

Country name:
conventional long form: Kingdom of Swaziland
conventional short form: Swaziland
local long form: Umbuso weSwatini
local short form: eSwatini
Government type: monarchy
Capital: *name:* Mbabane
geographic coordinates: 26 18 S, 31 06 E
time difference: UTC+2 (7 hours ahead of Washington, DC during Standard Time)
note: Lobamba (royal and legislative capital)

Administrative divisions: 4 districts; Hhohho, Lubombo, Manzini, Shiselweni
Independence: 6 September 1968 (from UK)
National holiday: Independence Day, 6 September (1968)
Constitution: signed by the King in July 2005 went into effect on 8 February 2006
Legal system: based on South African Roman-Dutch law in statutory courts and Swazi traditional law and custom in traditional courts; accepts compulsory ICJ jurisdiction with reservations
Suffrage: 18 years of age
Executive branch:
chief of state: King MSWATI III (since 25 April 1986)
head of government: Prime Minister Absolom Themba DLAMINI (since 14 November 2003)
cabinet: Cabinet recommended by the prime minister and confirmed by the monarch
elections: the monarch is hereditary; prime minister appointed by the monarch from among the elected members of the House of Assembly
Legislative branch: bicameral Parliament or Libandla consists of the Senate (30 seats; 10 members appointed by the House of Assembly and 20 appointed by the monarch; to serve five-year terms) and the House of Assembly (65 seats; 10 members appointed by the monarch and 55 elected by popular vote; to serve five-year terms)
elections: House of Assembly—last held 18 October 2003 (next to be held in October 2008)
election results: House of Assembly—balloting is done on a nonparty basis; candidates for election are nominated by the local council of each constituency and for each constituency the three candidates with the most votes in the first round of voting are narrowed to a single winner by a second round
Judicial branch: High Court; Supreme Court; judges for both courts are appointed by the monarch
Political parties and leaders: the status of political parties, previously banned, is unclear under the new (2006) Constitution and currently being debated—the following are considered political associations; African United Democratic Party or AUDP [Stanley MAUNDZISA, president]; Imbokodvo National Movement or INM; Ngwane National Liberatory Congress or NNLC [Obed DLAMINI, president]; People's United Democratic Movement or PUDEMO [Mario MASUKU, president]
Political pressure groups and leaders: NA

International organization participation: ACP, AfDB, AU, C, COMESA, FAO, G-77, IBRD, ICAO, ICRM, IDA, IFAD, IFC, IFRCS, ILO, IMF, Interpol, IOC, ISO (correspondent), ITSO, ITU, ITUC, MIGA, NAM, OPCW, PCA, SACU, SADC, UN, UNCTAD, UNESCO, UNIDO, UNWTO, UPU, WCO, WHO, WIPO, WMO, WTO
Diplomatic representation in the US:
chief of mission: Ambassador Ephraim Mandla HLOPHE
chancery: 1712 New Hampshire Avenue, NW, Washington, DC 20009
telephone: [1] (202) 234-5002
FAX: [1] (202) 234-8254
Diplomatic representation from the US:
chief of mission: Ambassador Maurice S. PARKER
embassy: Central Bank Building, Mahlokahla Street, Mbabane
mailing address: P. O. Box 199, Mbabane
telephone: [268] 404-6441 through 404-6445
FAX: [268] 404-5959
Flag description: three horizontal bands of blue (top), red (triple width), and blue; the red band is edged in yellow; centered in the red band is a large black and white shield covering two spears and a staff decorated with feather tassels, all placed horizontally

ECONOMY

Economy—overview: In this small, landlocked economy, subsistence agriculture occupies approximately 70% of the population. The manufacturing sector has diversified since the mid-1980s. Sugar and wood pulp remain important foreign exchange earners. In 2007, the sugar industry increased efficiency and diversification efforts, in response to a 17% decline in EU sugar prices. Mining has declined in importance in recent years with only coal and quarry stone mines remaining active. Surrounded by South Africa, except for a short border with Mozambique, Swaziland is heavily dependent on South Africa from which it receives more than nine-tenths of its imports and to which it sends 60% of its exports. Swaziland's currency is pegged to the South African rand, subsuming Swaziland's monetary policy to South Africa. Customs duties from the Southern African Customs Union, which may equal as much as 70% of government revenue this year, and worker remittances from South Africa substantially supplement domestically earned income. Swaziland is not poor enough to merit an IMF program; however, the country is struggling to reduce the size of the civil service and control costs at

public enterprises. The government is trying to improve the atmosphere for foreign investment. With an estimated 40% unemployment rate, Swaziland's need to increase the number and size of small and medium enterprises and attract foreign direct investment is acute. Overgrazing, soil depletion, drought, and sometimes floods persist as problems for the future. More than one-fourth of the population needed emergency food aid in 2006–07 because of drought, and nearly two-fifths of the adult population has been infected by HIV/AIDS.
GDP (purchasing power parity): $5.626 billion (2007 est.)
GDP (official exchange rate): $2.936 billion (2007 est.)
GDP—real growth rate: 2.4% (2007 est.)
GDP—per capita (PPP): $4,800 (2007 est.)
GDP—composition by sector:
agriculture: 11.8%
industry: 45.7%
services: 42.5% (2007 est.)
Labor force: 300,000 (2006)
Labor force—by occupation: *agriculture:* NA%
industry: NA%
services: NA%
Unemployment rate: 40% (2006 est.)
Population below poverty line: 69% (2006)
Household income or consumption by percentage share:
lowest 10%: 1.6%
highest 10%: 40.7% (2001)
Distribution of family income—Gini index: 50.4 (2001)
Inflation rate (consumer prices): 8.2% (2007 est.)
Investment (gross fixed): 18.7% of GDP (2007 est.)
Budget:
revenues: $1.13 billion
expenditures: $1.143 billion (2007 est.)
Agriculture—products: sugarcane, cotton, corn, tobacco, rice, citrus, pineapples, sorghum, peanuts; cattle, goats, sheep
Industries: coal, wood pulp, sugar, soft drink concentrates, textiles and apparel
Industrial production growth rate: 1.1% (2007 est.)
Electricity—production: 460 million kWh (2007)
Electricity—production by source:
fossil fuel: 58%
hydro: 42%
nuclear: 0%
other: 0% (2001)
Electricity—consumption: 1.2 billion kWh (2007)
Electricity—exports: 0 kWh (2007)

Electricity—imports: 872 million kWh; note—electricity supplied by South Africa (2007)
Oil—production: 0 bbl/day (2005 est.)
Oil—consumption: 3,500 bbl/day (2005 est.)
Oil—exports: 0 bbl/day (2004)
Oil—imports: 3,530 bbl/day (2004)
Oil—proved reserves: 0 bbl (1 January 2006 est.)
Natural gas—production: 0 cu m (2005 est.)
Natural gas—consumption: 0 cu m (2005 est.)
Natural gas—exports: 0 cu m (2005 est.)
Natural gas—imports: 0 cu m (2005)
Natural gas—proved reserves: 0 cu m (1 January 2006 est.)
Current account balance: $36 million (2007 est.)
Exports: $1.926 billion f.o.b. (2007 est.)
Exports—commodities: soft drink concentrates, sugar, wood pulp, cotton yarn, refrigerators, citrus and canned fruit
Exports—partners: South Africa 59.7%, EU 8.8%, US 8.8%, Mozambique 6.2% (2006)
Imports: $1.929 billion f.o.b. (2007 est.)
Imports—commodities: motor vehicles, machinery, transport equipment, foodstuffs, petroleum products, chemicals
Imports—partners: South Africa 95.6%, EU 0.9%, Japan 0.9% (2006)
Economic aid—recipient: $46.03 million (2005)
Reserves of foreign exchange and gold: $762.7 million (31 December 2007 est.)
Debt—external: $524 million (31 December 2007 est.)

Stock of direct foreign investment—at home: $NA
Stock of direct foreign investment—abroad: $NA
Market value of publicly traded shares: $196.8 million (2005)
Currency (code): lilangeni (SZL)
Currency code: SZL
Exchange rates: lilangeni per US dollar—7.4 (2007), 6.85 (2006), 6.3593 (2005), 6.4597 (2004), 7.5648 (2003)
Fiscal year: 1 April—31 March

COMMUNICATIONS

Telephones—main lines in use: 44,000 (2006)
Telephones—mobile cellular: 250,000 (2006)
Telephone system:
general assessment: a somewhat modern but not an advanced system
domestic: mobile-cellular subscribership is increasing; combined fixed-line and mobile cellular teledensity about 25 telephones per 100 persons; telephone system consists of carrier-equipped, open-wire lines and low-capacity, microwave radio relay
international: country code—268; satellite earth station—1 Intelsat (Atlantic Ocean)
Radio broadcast stations: AM 3, FM 2 (plus 4 repeaters), shortwave 3 (2004)
Radios: 170,000 (1999)
Television broadcast stations: 12 (includes 7 relay stations) (2004)
Televisions: 23,000 (2000)
Internet country code: .sz
Internet hosts: 2,672 (2007)
Internet Service Providers (ISPs): 5 (2002)

Internet users: 41,600 (2005)

TRANSPORTATION

Airports: 18 (2007)
Airports—with paved runways:
total: 1
2,438 to 3,047 m: 1 (2007)
Airports—with unpaved runways:
total: 17
914 to 1,523 m: 7
under 914 m: 10 (2007)
Railways:
total: 301 km
narrow gauge: 301 km 1.067-m gauge (2006)
Roadways:
total: 3,594 km
paved: 1,078 km
unpaved: 2,516 km (2002)

MILITARY

Military branches: Umbutfo Swaziland Defense Force (USDF): Ground Force (includes air wing) (2008)
Military service age and obligation: 18–30 years of age for male and female voluntary military service; no conscription (2008)
Manpower available for military service:
males age 16–49: 266,311 (2008 est.)
Manpower fit for military service:
males age 16–49: 122,260 (2008 est.)
Military expenditures—percent of GDP: 4.7% (2006)

TRANSNATIONAL ISSUES

Disputes—international: in 2006, Swazi king advocates resort to ICJ to claim parts of Mpumalanga and KwaZulu-Natal from South Africa

SWEDEN

INTRODUCTION

Background: A military power during the 17th century, Sweden has not participated in any war in almost two centuries. An armed neutrality was preserved in both World Wars. Sweden's long-successful economic formula of a capitalist system interlarded with substantial welfare elements was challenged in the 1990s by high unemployment and in 2000–02 by the global economic downturn, but fiscal discipline over the past several years has allowed the country to weather economic vagaries. Sweden joined the EU in 1995, but the public rejected the introduction of the euro in a 2003 referendum.

GEOGRAPHY

Location: Northern Europe, bordering the Baltic Sea, Gulf of Bothnia, Kattegat, and Skagerrak, between Finland and Norway
Geographic coordinates: 62 00 N, 15 00 E
Map references: Europe
Area:
total: 449,964 sq km
land: 410,934 sq km
water: 39,030 sq km
Area—comparative: slightly larger than California
Land boundaries:
total: 2,233 km
border countries: Finland 614 km, Norway 1,619 km
Coastline: 3,218 km
Maritime claims:
territorial sea: 12 nm (adjustments made to return a portion of straits to high seas)

exclusive economic zone: agreed boundaries or midlines
continental shelf: 200-m depth or to the depth of exploitation
Climate: temperate in south with cold, cloudy winters and cool, partly cloudy summers; subarctic in north
Terrain: mostly flat or gently rolling lowlands; mountains in west
Elevation extremes:
lowest point: reclaimed bay of Lake Hammarsjon, near Kristianstad -2.41 m
highest point: Kebnekaise 2,111 m
Natural resources: iron ore, copper, lead, zinc, gold, silver, tungsten, uranium, arsenic, feldspar, timber, hydropower
Land use: *arable land:* 5.93%
permanent crops: 0.01%
other: 94.06% (2005)

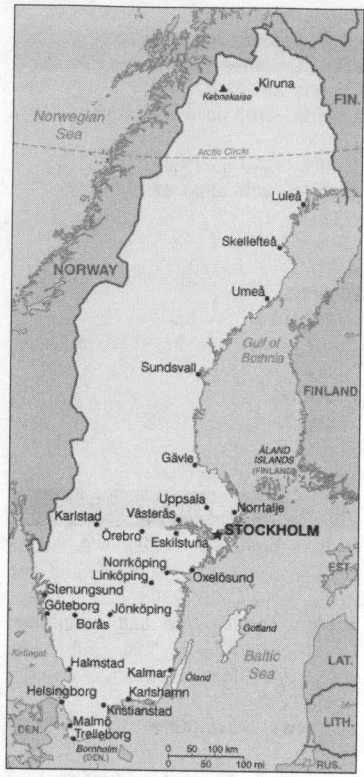

Irrigated land: 1,150 sq km (2003)

Total renewable water resources: 179 cu km (2005)

Freshwater withdrawal (domestic/industrial/agricultural): *total:* 2.68 cu km/yr (37%/54%/9%)

per capita: 296 cu m/yr (2002)

Natural hazards: ice floes in the surrounding waters, especially in the Gulf of Bothnia, can interfere with maritime traffic

Environment—current issues: acid rain damage to soils and lakes; pollution of the North Sea and the Baltic Sea

Environment—international agreements: *party to:* Air Pollution, Air Pollution-Nitrogen Oxides, Air Pollution-Persistent Organic Pollutants, Air Pollution-Sulfur 85, Air Pollution-Sulfur 94, Air Pollution-Volatile Organic Compounds, Antarctic-Environmental Protocol, Antarctic-Marine Living Resources, Antarctic Treaty, Biodiversity, Climate Change, Climate Change-Kyoto Protocol, Desertification, Endangered Species, Environmental Modification, Hazardous Wastes, Law of the Sea, Marine Dumping, Ozone Layer Protection, Ship Pollution, Tropical Timber 83, Tropical Timber 94, Wetlands, Whaling

signed, but not ratified: none of the selected agreements

Geography—note: strategic location along Danish Straits linking Baltic and North Seas

PEOPLE

Population: 9,045,389 (July 2008 est.)

Age structure:

0–14 years: 16% (male 745,110/female 703,857)

15–64 years: 65.6% (male 3,008,148/female 2,928,930)

65 years and over: 18.3% (male 729,500/female 929,844) (2008 est.)

Median age:

total: 41.3 years

male: 40.2 years

female: 42.4 years (2008 est.)

Population growth rate: 0.157% (2008 est.)

Birth rate: 10.15 births/1,000 population (2008 est.)

Death rate: 10.24 deaths/1,000 population (2008 est.)

Net migration rate: 1.66 migrant(s)/1,000 population (2008 est.)

Sex ratio:

at birth: 1.06 male(s)/female

under 15 years: 1.06 male(s)/female

15–64 years: 1.03 male(s)/female

65 years and over: 0.78 male(s)/female

total population: 0.98 male(s)/female (2008 est.)

Infant mortality rate:

total: 2.75 deaths/1,000 live births

male: 2.91 deaths/1,000 live births

female: 2.58 deaths/1,000 live births (2008 est.)

Life expectancy at birth:

total population: 80.74 years

male: 78.49 years

female: 83.13 years (2008 est.)

Total fertility rate: 1.67 children born/woman (2008 est.)

HIV/AIDS—adult prevalence rate: 0.1% (2001 est.)

HIV/AIDS—people living with HIV/AIDS: 3,600 (2001 est.)

HIV/AIDS—deaths: fewer than 100 (2003 est.)

Nationality:

noun: Swede(s)

adjective: Swedish

Ethnic groups: indigenous **Population:** Swedes with Finnish and Sami minorities; foreign-born or first-generation immigrants: Finns, Yugoslavs, Danes, Norwegians, Greeks, Turks

Religions: Lutheran 87%, other (includes Roman Catholic, Orthodox, Baptist, Muslim, Jewish, and Buddhist) 13%

Languages: Swedish, small Sami- and Finnish-speaking minorities

Literacy: *definition:* age 15 and over can read and write

total population: 99%

male: 99%

female: 99% (2003 est.)

GOVERNMENT

Country name:

conventional long form: Kingdom of Sweden

conventional short form: Sweden

local long form: Konungariket Sverige

local short form: Sverige

Government type: constitutional monarchy

Capital: *name:* Stockholm

geographic coordinates: 59 20 N, 18 03 E

time difference: UTC+1 (6 hours ahead of Washington, DC during Standard Time)

daylight saving time: +1hr, begins last Sunday in March; ends last Sunday in October

Administrative divisions: 21 counties (lan, singular and plural); Blekinge, Dalarnas, Gavleborgs, Gotlands, Hallands, Jamtlands, Jonkopings, Kalmar, Kronobergs, Norrbottens, Orebro, Ostergotlands, Skane, Sodermanlands, Stockholm, Uppsala, Varmlands, Vasterbottens, Vasternorrlands, Vastmanlands, Vastra Gotalands

Independence: 6 June 1523 (Gustav VASA elected king)

National holiday: Swedish Flag Day, 6 June (1916); National Day, 6 June (1983)

Constitution: 1 January 1975

Legal system: civil law system influenced by customary law; accepts compulsory ICJ jurisdiction with reservations

Suffrage: 18 years of age; universal

Executive branch:

chief of state: King CARL XVI GUSTAF (since 19 September 1973); Heir Apparent Princess VICTORIA Ingrid Alice Desiree, daughter of the monarch (born 14 July 1977)

head of government: Prime Minister Fredrik REINFELDT (since 5 October 2006)

cabinet: Cabinet appointed by the prime minister

elections: the monarchy is hereditary; following legislative elections, the prime minister is elected by the parliament; election last held on 17 September 2006 (next to be held in September 2010)

election results: Fredrik REINFELDT elected prime minister with 175 out of 349 votes

Legislative branch: unicameral Parliament or Riksdag (349 seats; members are elected by popular vote on a proportional representation basis to serve four-year terms)

elections: last held on 17 September 2006

(next to be held in September 2010)

election results: percent of vote by party—Social Democrats 37.2%, Moderates 27.8%, Center Party 8.3%, Liberal People's Party 8.0%, Christian Democrats 6.9%, Left Party 6.3%, Greens 5.4%; seats by party—Social Democrats 130, Moderates 97, Center Party 29, Liberal People's Party 28, Christian Democrats 24, Left Party 22, Greens 19

Judicial branch: Supreme Court or Hogsta Domstolen (judges are appointed by the prime minister and the cabinet)

Political parties and leaders: Center Party [Maud OLOFSSON]; Christian Democratic Party [Goran HAGGLUND]; Environment Party the Greens [no formal leader but party spokespersons are Maria WETTERSTRAND and Peter ERIKSSON]; Left Party or V (formerly Communist) [Lars OHLY]; Liberal People's Party [Jan BJORKLUND]; Moderate Party (conservative) [Fredrik REINFELDT]; Social Democratic Party [Mona SAHLIN]

Political pressure groups and leaders: NA

International organization participation: ADB (nonregional members), AfDB, Arctic Council, Australia Group, BIS, CBSS, CE, CERN, EAPC, EBRD, EIB, ESA, EU, FAO, G-9, G-10, IADB, IAEA, IBRD, ICAO, ICC, ICCt, ICRM, IDA, IEA, IFAD, IFC, IFRCS, IHO, ILO, IMF, IMO, IMSO, Interpol, IOC, IOM, IPU, ISO, ITSO, ITU, ITUC, MIGA, MINURCAT, NAM (guest), NC, NEA, NIB, NSG, OAS (observer), OECD, OPCW, OSCE, Paris Club, PCA, PFP, Schengen Convention, UN, UNCTAD, UNESCO, UNHCR, UNIDO, UNMEE, UNMIL, UNMIS, UNMOGIP, UNOMIG, UNRWA, UNTSO, UPU, WCO, WEU (observer), WFTU, WHO, WIPO, WMO, WTO, ZC

Diplomatic representation in the US:
chief of mission: Ambassador Jonas HAFSTROM
chancery: 2900 K Street NW, Washington, DC 20007
telephone: [1] (202) 467-2600
FAX: [1] (202) 467-2699
consulate(s) general: Chicago, Los Angeles, New York

Diplomatic representation from the US:
chief of mission: Ambassador Michael M. WOOD
embassy: Dag Hammarskjolds Vag 31, SE-11589 Stockholm
mailing address: American Embassy Stockholm, US Department of State, 5750 Stockholm Place, Washington, DC 20521-5750

telephone: [46] (08) 783 53 00
FAX: [46] (08) 661 19 64
Flag description: blue with a golden yellow cross extending to the edges of the flag; the vertical part of the cross is shifted to the hoist side in the style of the Dannebrog (Danish flag)

ECONOMY

Economy—overview: Aided by peace and neutrality for the whole of the 20th century, Sweden has achieved an enviable standard of living under a mixed system of high-tech capitalism and extensive welfare benefits. It has a modern distribution system, excellent internal and external communications, and a skilled labor force. Timber, hydropower, and iron ore constitute the resource base of an economy heavily oriented toward foreign trade. Privately owned firms account for about 90% of industrial output, of which the engineering sector accounts for 50% of output and exports. Agriculture accounts for only 1% of GDP and 2% of employment. Sweden is in the midst of a sustained economic upswing, boosted by increased domestic demand and strong exports. This and robust finances have offered the center-right government considerable scope to implement its reform program aimed at increasing employment, reducing welfare dependence, and streamlining the state's role in the economy. The govenment plans to sell $31 billion in state assets during the next three years to further stimulate growth and raise revenue to pay down the federal debt. In September 2003, Swedish voters turned down entry into the euro system concerned about the impact on the economy and sovereignty.

GDP (purchasing power parity): $334.6 billion (2007 est.)
GDP (official exchange rate): $455.3 billion (2007 est.)
GDP—real growth rate: 2.6% (2007 est.)
GDP—per capita (PPP): $36,500 (2007 est.)
GDP—composition by sector:
agriculture: 1.4%
industry: 28.9%
services: 69.8% (2007 est.)
Labor force: 4.839 million (2007 est.)
Labor force—by occupation: *agriculture:* 2%
industry: 24%
services: 74% (2000 est.)
Unemployment rate: 6.1% (2007 est.)
Population below poverty line: NA%
Household income or consumption by percentage share:
lowest 10%: 3.6%

highest 10%: 22.2% (2000)
Distribution of family income—Gini index: 23 (2005)
Inflation rate (consumer prices): 1.7% (2007 est.)
Investment (gross fixed): 18.9% of GDP (2007 est.)
Budget:
revenues: $253.4 billion
expenditures: $240.5 billion (2007 est.)
Public debt: 41.2% of GDP (2007 est.)
Agriculture—products: barley, wheat, sugar beets; meat, milk
Industries: iron and steel, precision equipment (bearings, radio and telephone parts, armaments), wood pulp and paper products, processed foods, motor vehicles
Industrial production growth rate: 4% (2007 est.)
Electricity—production: 153.2 billion kWh (2005)
Electricity—production by source:
fossil fuel: 4%
hydro: 50.8%
nuclear: 43%
other: 2.3% (2001)
Electricity—consumption: 134.1 billion kWh (2005)
Electricity—exports: 21.97 billion kWh (2005)
Electricity—imports: 14.58 billion kWh (2005)
Oil—production: 2,350 bbl/day (2005 est.)
Oil—consumption: 363,200 bbl/day (2005 est.)
Oil—exports: 231,100 bbl/day (2004)
Oil—imports: 580,600 bbl/day (2004)
Oil—proved reserves: 0 bbl (1 January 2006 est.)
Natural gas—production: 0 cu m (2005 est.)
Natural gas—consumption: 893.9 million cu m (2005 est.)
Natural gas—exports: 0 cu m (2005 est.)
Natural gas—imports: 893.9 million cu m (2005)
Natural gas—proved reserves: 0 cu m (1 January 2006 est.)
Current account balance: $37.99 billion (2007 est.)
Exports: $170.2 billion f.o.b. (2007 est.)
Exports—commodities: machinery 35%, motor vehicles, paper products, pulp and wood, iron and steel products, chemicals
Exports—partners: Germany 9.8%, US 9.3%, Norway 9.1%, UK 7.1%, Denmark 6.9%, Finland 6%, France 4.9%, Netherlands 4.7%, Belgium 4.5% (2006)
Imports: $150.6 billion f.o.b. (2007 est.)
Imports—commodities: machinery, petroleum and petroleum products,

chemicals, motor vehicles, iron and steel; foodstuffs, clothing

Imports—partners: Germany 17.3%, Denmark 9.1%, Norway 8.2%, UK 6%, Netherlands 5.8%, Finland 5.7%, France 4.6%, Belgium 4.1% (2006)

Economic aid—donor: ODA, $3.955 billion (2006)

Reserves of foreign exchange and gold: $28.02 billion (2006 est.)

Debt—external: $598.2 billion (30 June 2006)

Stock of direct foreign investment—at home: $210.8 billion (2007 est.)

Stock of direct foreign investment—abroad: $257.4 billion (2007 est.)

Market value of publicly traded shares: $403.9 billion (2005)

Currency (code): Swedish krona (SEK)

Currency code: SEK

Exchange rates: Swedish kronor per US dollar—6.7629 (2007), 7.3731 (2006), 7.4731 (2005), 7.3489 (2004), 8.0863 (2003)

Fiscal year: calendar year

COMMUNICATIONS

Telephones—main lines in use: 6.379 million (2005)

Telephones—mobile cellular: 9.087 million (2005)

Telephone system:
general assessment: highly developed telecommunications infrastructure; ranked among leading countries for fixed-line, mobile-cellular, Internet and broadband penetration
domestic: coaxial and multiconductor cables carry most of the voice traffic; parallel microwave radio relay systems carry some additional telephone channels
international: country code—46; submarine cables provide links to other Nordic countries and Europe; satellite earth stations—1 Intelsat (Atlantic Ocean), 1 Eutelsat, and 1 Inmarsat (Atlantic and Indian Ocean regions); note—Sweden shares the Inmarsat earth station with the other Nordic countries (Denmark, Finland, Iceland, and Norway)

Radio broadcast stations: AM 1, FM 265, shortwave 1 (1998)

Radios: 8.25 million (1997)

Television broadcast stations: 169 (plus 1,299 repeaters) (1995)

Televisions: 4.6 million (1997)

Internet country code: .se

Internet hosts: 3.318 million (2007)

Internet Service Providers (ISPs): 29 (2000)

Internet users: 6.981 million (2006)

TRANSPORTATION

Airports: 250 (2007)

Airports—with paved runways:
total: 152
over 3,047 m: 3
2,438 to 3,047 m: 12
1,524 to 2,437 m: 75
914 to 1,523 m: 24
under 914 m: 38 (2007)

Airports—with unpaved runways:
total: 98
914 to 1,523 m: 6
under 914 m: 92 (2007)

Heliports: 2 (2007)

Pipelines: gas 798 km (2007)

Railways:
total: 11,528 km
standard gauge: 11,528 km 1.435-m gauge (7,527 km electrified) (2006)

Roadways:
total: 424,947 km
paved: 129,651 km (includes 1,591 km of expressways)
unpaved: 295,296 km (2004)

Waterways: 2,052 km (2005)

Merchant marine:
total: 194 ships (1000 GRT or over) 3,883,695 GRT/2,451,123 DWT
by type: bulk carrier 7, cargo 23, carrier 1, chemical tanker 49, passenger 2, passenger/cargo 37, petroleum tanker 15, roll on/roll off 35, specialized tanker 3, vehicle carrier 22
foreign-owned: 34 (Denmark 4, Finland

10, Germany 4, Italy 7, Japan 1, Norway 5, UK 2, US 1)
registered in other countries: 198 (Antigua and Barbuda 1, Bahamas 5, Barbados 5, Bermuda 15, Cayman Islands 1, Cook Islands 9, Cyprus 2, Denmark 4, Finland 2, France 10, Gibraltar 10, Isle of Man 3, Italy 1, South Korea 2, Liberia 11, Malta 1, Marshall Islands 1, Netherlands 27, Netherlands Antilles 3, Norway 31, Panama 9, Portugal 2, Singapore 17, St Vincent and The Grenadines 2, UK 19, US 5) (2007)

Ports and terminals: Brofjorden, Goteborg, Helsingborg, Lulea, Malmo, Stenungsund, Stockholm, Trelleborg, Visby

MILITARY

Military branches: Swedish Armed Forces (Forsvarsmakten): Army (Armen), Royal Swedish Navy (Marinen), Swedish Air Force (Svenska Flygvapnet) (2007)

Military service age and obligation: 19 years of age for compulsory military service; conscript service obligation: 7–15 months (Navy), 8–12 months (Air Force); after completing initial service, soldiers have a reserve commitment until age 47 (2006)

Manpower available for military service:
males age 16–49: 2,052,890
females age 16–49: 1,980,550 (2008 est.)

Manpower fit for military service:
males age 16–49: 1,699,115
females age 16–49: 1,637,868 (2008 est.)

Manpower reaching militarily significant age annually:
males age 16–49: 64,605
females age 16–49: 61,110 (2008 est.)

Military expenditures—percent of GDP: 1.5% (2005 est.)

TRANSNATIONAL ISSUES

Disputes—international: none

SWITZERLAND

INTRODUCTION

Background: The Swiss Confederation was founded in 1291 as a defensive alliance among three cantons. In succeeding years, other localities joined the original three. The Swiss Confederation secured its independence from the Holy Roman Empire in 1499. A constitution of 1848, subsequently modified in 1874, replaced the confederation with a centralized federal government. Switzerland's sovereignty and neutrality have long been honored by the major European powers, and the country was not involved in either of the two World Wars. The political and economic integration of Europe over the past half century, as well as Switzerland's role in many UN and international organizations, has strengthened Switzerland's ties with its neighbors. However, the country did not officially become a UN member until 2002. Switzerland remains active in many UN and international organizations but retains a strong commitment to neutrality.

GEOGRAPHY

Location: Central Europe, east of France, north of Italy

Geographic coordinates: 47 00 N, 8 00 E

Map references: Europe

Area:
total: 41,290 sq km
land: 39,770 sq km
water: 1,520 sq km

Area—comparative: slightly less than twice the size of New Jersey

Land boundaries:

total: 1,852 km
border countries: Austria 164 km, France 573 km, Italy 740 km, Liechtenstein 41 km, Germany 334 km

Coastline: 0 km (landlocked)
Maritime claims: none (landlocked)
Climate: temperate, but varies with altitude; cold, cloudy, rainy/snowy winters; cool to warm, cloudy, humid summers with occasional showers
Terrain: mostly mountains (Alps in south, Jura in northwest) with a central plateau of rolling hills, plains, and large lakes

Elevation extremes:

lowest point: Lake Maggiore 195 m
highest point: Dufourspitze 4,634 m
Natural resources: hydropower potential, timber, salt

Land use:

arable land: 9.91%
permanent crops: 0.58%
other: 89.51% (2005)
Irrigated land: 250 sq km (2003)
Total renewable water resources: 53.3 cu km (2005)
Freshwater withdrawal (domestic/industrial/agricultural): *total:* 2.52 cu km/yr (24%/74%/2%)
per capita: 348 cu m/yr (2002)
Natural hazards: avalanches, landslides, flash floods
Environment—current issues: air pollution from vehicle emissions and open-air burning; acid rain; water pollution from increased use of agricultural fertilizers; loss of biodiversity
Environment—international agreements: *party to:* Air Pollution, Air Pollution-Nitrogen Oxides, Air Pollution-Persistent Organic Pollutants, Air Pollution-Sulfur 85, Air Pollution-Sulfur 94, Air Pollution-Volatile Organic Compounds, Antarctic Treaty, Biodiversity, Climate Change, Climate Change-Kyoto Protocol, Desertification, Endangered Species, Environmental Modification, Hazardous Wastes, Marine Dumping, Marine Life Conservation, Ozone Layer Protection, Ship Pollution, Tropical Timber 83, Tropical Timber 94, Wetlands, Whaling
signed, but not ratified: Law of the Sea
Geography—note: landlocked; crossroads of northern and southern Europe; along with southeastern France, northern Italy, and southwestern Austria, has the highest elevations in the Alps

PEOPLE

Population: 7,581,520 (July 2008 est.)
Age structure:
0–14 years: 15.8% (male 623,213/female 577,430)
15–64 years: 68.2% (male 2,605,044/female 2,562,354)
65 years and over: 16% (male 501,699/female 711,780) (2008 est.)
Median age:
total: 40.7 years
male: 39.6 years
female: 41.7 years (2008 est.)
Population growth rate: 0.329% (2008 est.)
Birth rate: 9.62 births/1,000 population (2008 est.)
Death rate: 8.54 deaths/1,000 population (2008 est.)
Net migration rate: 2.21 migrant(s)/1,000 population (2008 est.)
Sex ratio:
at birth: 1.05 male(s)/female
under 15 years: 1.08 male(s)/female
15–64 years: 1.02 male(s)/female
65 years and over: 0.7 male(s)/female
total population: 0.97 male(s)/female (2008 est.)
Infant mortality rate:
total: 4.23 deaths/1,000 live births
male: 4.71 deaths/1,000 live births
female: 3.73 deaths/1,000 live births (2008 est.)
Life expectancy at birth:
total population: 80.74 years
male: 77.91 years
female: 83.71 years (2008 est.)

Total fertility rate: 1.44 children born/woman (2008 est.)
HIV/AIDS—adult prevalence rate: 0.4% (2001 est.)
HIV/AIDS—people living with HIV/AIDS: 13,000 (2001 est.)
HIV/AIDS—deaths: fewer than 100 (2003 est.)
Nationality:
noun: Swiss (singular and plural)
adjective: Swiss
Ethnic groups: German 65%, French 18%, Italian 10%, Romansch 1%, other 6%
Religions: Roman Catholic 41.8%, Protestant 35.3%, Muslim 4.3%, Orthodox 1.8%, other Christian 0.4%, other 1%, unspecified 4.3%, none 11.1% (2000 census)
Languages: German (official) 63.7%, French (official) 20.4%, Italian (official) 6.5%, Serbo-Croatian 1.5%, Albanian 1.3%, Portuguese 1.2%, Spanish 1.1%, English 1%, Romansch (official) 0.5%, other 2.8% (2000 census)
note: German, French, Italian, and Romansch are all national and official languages
Literacy:
definition: age 15 and over can read and write
total population: 99%
male: 99%
female: 99% (2003 est.)

GOVERNMENT

Country name:
conventional long form: Swiss Confederation
conventional short form: Switzerland
local long form: Schweizerische Eidgenossenschaft (German); Confederation Suisse (French); Confederazione Svizzera (Italian); Confederaziun Svizra (Romansh)
local short form: Schweiz (German); Suisse (French); Svizzera (Italian); Svizra (Romansh)
Government type: formally a confederation but similar in structure to a federal republic
Capital: *name:* Bern
geographic coordinates: 46 57 N, 7 26 E
time difference: UTC+1 (6 hours ahead of Washington, DC during Standard Time)
daylight saving time: +1hr, begins last Sunday in March; ends last Sunday in October
Administrative divisions: 26 cantons (cantons, singular—canton in French; cantoni, singular—cantone in Italian; Kantone, singular—Kanton in German); Aargau, Appenzell Ausser-Rhoden, Appenzell Inner-Rhoden, Basel-Landschaft, Basel-Stadt, Bern, Fribourg,

Geneve, Glarus, Graubunden, Jura, Luzern, Neuchatel, Nidwalden, Obwalden, Sankt Gallen, Schaffhausen, Schwyz, Solothurn, Thurgau, Ticino, Uri, Valais, Vaud, Zug, Zurich

Independence: 1 August 1291 (founding of the Swiss Confederation)

National holiday: Founding of the Swiss Confederation, 1 August (1291)

Constitution: revision of Constitution of 1874 approved by the Federal Parliament 18 December 1998, adopted by referendum 18 April 1999, officially entered into force 1 January 2000

Legal system: civil law system influenced by customary law; judicial review of legislative acts, except with respect to federal decrees of general obligatory character; accepts compulsory ICJ jurisdiction with reservations

Suffrage: 18 years of age; universal

Executive branch:

chief of state: President Pascal COUCHEPIN (since 1 January 2008); Vice President Hans-Rudolf MERZ (since 1 January 2008); note—the president is both the chief of state and head of government representing the Federal Council; the Federal Council is the formal chief of state and head of government whose council members, rotating in one-year terms as federal president, represent the Council

head of government: President Pascal COUCHEPIN (since 1 January 2008); Vice President Hans-Rudolf MERZ (since 1 January 2008)

cabinet: Federal Council or Bundesrat (in German), Conseil Federal (in French), Consiglio Federale (in Italian) elected by the Federal Assembly usually from among its members for a four-year term

elections: president and vice president elected by the Federal Assembly from among the members of the Federal Council for a one-year term (they may not serve consecutive terms); election last held on 12 December 2007 (next to be held in December 2008)

election results: Pascal COUCHEPIN elected president; percent of Federal Assembly vote—80.0%; Hans-Rudolf MERZ elected vice president; percent of Federal Assembly vote—86.5%

Legislative branch: bicameral Federal Assembly or Bundesversammlung (in German), Assemblee Federale (in French), Assemblea Federale (in Italian) consists of the Council of States or Standerat (in German), Conseil des Etats (in French), Consiglio degli Stati (in Italian) (46 seats; membership consists of 2 representatives from each canton and 1 from each half canton; to serve four-year terms) and the National Council or Nationalrat (in German), Conseil National (in French), Consiglio Nazionale (in Italian) (200 seats; mem-

bers are elected by popular vote on the basis of proportional representation to serve four-year terms)

elections: Council of States—last held in most cantons on 19 October 2003 (each canton determines when the next election will be held); National Council—last held on 21 October 2007 (next to be held in October 2011)

election results: Council of States—percent of vote by party—NA; seats by party—CVP 15, FDP 14, SVP 8, SPS 6, other 3; National Council—percent of vote by party—SVP 29%, SPS 19.5%, FDP 15.6%, CVP 14.6%, Greens 9.6%, other 11.7%; seats by party—SVP 62, SPS 43, FDP 31, CVP 31, Green Party 20, other small parties 13; note—seating for the Council of States as of December 2007 is CVP 16, FDP 12, SVP 7, SPS 9, other 2

Judicial branch: Federal Supreme Court (judges elected for six-year terms by the Federal Assembly)

Political parties and leaders: Green Party (Gruene Partei der Schweiz or Gruene, Parti Ecologiste Suisse or Les Verts, Partito Ecologista Svizzero or I Verdi, Partida Ecologica Svizra or La Verda) [Ruth GENNER]; Christian Democratic People's Party (Christlichdemokratische Volkspartei der Schweiz or CVP, Parti Democrate-Chretien Suisse or PDC, Partito Democratico-Cristiano Popolare Svizzero or PDC, Partida Cristiandemocratica dalla Svizra or PCD) [Christophe DARBELLAY]; Radical Free Democratic Party (Freisinnig-Demokratische Partei der Schweiz or FDP, Parti Radical-Democratique Suisse or PRD, Partitio Liberal-Radicale Svizzero or PLR) [Fulvio PELLI]; Social Democratic Party (Sozialdemokratische Partei der Schweiz or SPS, Parti Socialist Suisse or PSS, Partito Socialista Svizzero or PSS, Partida Socialdemocratica de la Svizra or PSS) [Hans-Juerg FEHR]; Swiss People's Party (Schweizerische Volkspartei or SVP, Union Democratique du Centre or UDC, Unione Democratica de Centro or UDC, Uniun Democratica dal Center or UDC) [Ueli MAURER]; and other minor parties

Political pressure groups and leaders: NA

International organization participation: ADB (nonregional members), AfDB, Australia Group, BIS, CE, CERN, EAPC, EBRD, EFTA, ESA, FAO, G-10, IADB, IAEA, IBRD, ICAO, ICC, ICCt, ICRM, IDA, IEA, IFAD, IFC, IFRCS, ILO, IMF, IMO, IMSO, Interpol, IOC, IOM, IPU, ISO, ITSO, ITU, ITUC, LAIA (observer), MIGA, NAM (guest), NEA, NSG, OAS (observer), OECD, OIF, OPCW, OSCE,

Paris Club, PCA, PFP, UN, UNCTAD, UNESCO, UNHCR, UNIDO, UNITAR, UNOMIG, UNRWA, UNTSO, UNWTO, UPU, WCL, WCO, WHO, WIPO, WMO, WTO, ZC

Diplomatic representation in the US:

chief of mission: Ambassador Urs ZISWILER

chancery: 2900 Cathedral Avenue NW, Washington, DC 20008

telephone: [1] (202) 745-7900

FAX: [1] (202) 387-2564

consulate(s) general: Atlanta, Chicago, Los Angeles, New York, San Francisco

Diplomatic representation from the US:

chief of mission: Ambassador Peter R. CONEWAY

embassy: Jubilaeumsstrasse 93, CH-3001 Bern

mailing address: use embassy street address

telephone: [41] (031) 357 70 11

FAX: [41] (031) 357 73 44

Flag description: red square with a bold, equilateral white cross in the center that does not extend to the edges of the flag

ECONOMY

Economy—overview: Switzerland is a peaceful, prosperous, and stable modern market economy with low unemployment, a highly skilled labor force, and a per capita GDP larger than that of the big Western European economies. The Swiss in recent years have brought their economic practices largely into conformity with the EU's to enhance their international competitiveness. Switzerland remains a safehaven for investors, because it has maintained a degree of bank secrecy and has kept up the franc's long-term external value. Reflecting the anemic economic conditions of Europe, GDP growth stagnated during the 2001–03 period, improved during 2004–05, and jumped to 2.9% in 2006, and 2.6% in 2007. Unemployment has remained at less than half the EU average.

GDP (purchasing power parity): $300.2 billion (2007 est.)

GDP (official exchange rate): $423.9 billion (2007 est.)

GDP—real growth rate: 3.1% (2007 est.)

GDP—per capita (PPP): $41,100 (2007 est.)

GDP—composition by sector:

agriculture: 1.5%

industry: 34%

services: 64.5% (2003 est.)

Labor force: 3.95 million (2007 est.)

Labor force—by occupation: *agriculture:* 4.6%

industry: 26.3%

services: 69.1% (1998)

Unemployment rate: 2.5% (2007 est.)

Population below poverty line: NA%

Household income or consumption by percentage share: *lowest 10%:* 2.9%

highest 10%: 25.9% (2000)

Distribution of family income—Gini index: 33.7 (2000)
Inflation rate (consumer prices): 0.9% (2007 est.)
Investment (gross fixed): 21.5% of GDP (2007 est.)
Budget:
revenues: $149.5 billion
expenditures: $143.1 billion (2007 est.)
Public debt: 45.3% of GDP (2007 est.)
Agriculture—products: grains, fruits, vegetables; meat, eggs
Industries: machinery, chemicals, watches, textiles, precision instruments, tourism, banking, and insurance
Industrial production growth rate: 6.5% (2006 est.)
Electricity—production: 56.1 billion kWh (2005)
Electricity—production by source:
fossil fuel: 1.3%
hydro: 59.5%
nuclear: 37.1%
other: 2% (2001)
Electricity—consumption: 58.26 billion kWh (2005)
Electricity—exports: 32 billion kWh (2005)
Electricity—imports: 38.35 billion kWh (2005)
Oil—production: 3,202 bbl/day (2005 est.)
Oil—consumption: 275,000 bbl/day (2005 est.)
Oil—exports: 11,360 bbl/day (2004)
Oil—imports: 267,000 bbl/day (2004)
Oil—proved reserves: NA
Natural gas—production: 0 cu m (2005 est.)
Natural gas—consumption: 3.26 billion cu m (2005 est.)
Natural gas—exports: 0 cu m (2005 est.)
Natural gas—imports: 3.26 billion cu m (2005)
Natural gas—proved reserves: 0 cu m (1 January 2006 est.)
Current account balance: $72.84 billion (2007 est.)
Exports: $202.8 billion f.o.b. (2007 est.)
Exports—commodities: machinery, chemicals, metals, watches, agricultural products
Exports—partners: Germany 19.7%, US 11.1%, Italy 8.8%, France 8.6%, UK 4.8% (2006)
Imports: $184.9 billion f.o.b. (2007 est.)
Imports—commodities: machinery, chemicals, vehicles, metals; agricultural products, textiles
Imports—partners: Germany 31.7%, Italy 10.6%, France 10%, US 6.2%, Netherlands 4.7%, Austria 4.3% (2006)
Economic aid—donor: ODA, $1.646 billion (2006)
Reserves of foreign exchange and gold: $64.5 billion (2006 est.)
Debt—external: $1.34 trillion (30 June 2007)

Stock of direct foreign investment—at home: $265.7 billion (2007 est.)
Stock of direct foreign investment—abroad: $591.5 billion (2007 est.)
Market value of publicly traded shares: $938.6 billion (2005)
Currency (code): Swiss franc (CHF)
Currency code: CHF
Exchange rates: Swiss francs per US dollar—1.1973 (2007), 1.2539 (2006), 1.2452 (2005), 1.2435 (2004), 1.3467 (2003)
Fiscal year: calendar year

COMMUNICATIONS

Telephones—main lines in use: 5.04 million (2006)
Telephones—mobile cellular: 7.418 million (2006)
Telephone system:
general assessment: highly developed telecommunications infrastructure with excellent domestic and international services
domestic: ranked among leading countries for fixed-line teledensity and infrastructure; mobile-cellular subscribership roughly 100 per 100 persons; extensive cable and microwave radio relay networks
international: country code—41; satellite earth stations—2 Intelsat (Atlantic Ocean and Indian Ocean)
Radio broadcast stations: AM 4, FM 113 (plus many low-power stations), shortwave 2 (1998)
Radios: 7.1 million (1997)
Television broadcast stations: 115 (plus 1,919 repeaters) (1995)
Televisions: 3.31 million (1997)
Internet country code: .ch
Internet hosts: 1 405 million (2007)
Internet Service Providers (ISPs): 44 (Switzerland and Liechtenstein) (2000)
Internet users: 4.36 million (2006)

TRANSPORTATION

Airports: 65 (2007)
Airports—with paved runways:
total: 42
over 3,047 m: 3
2,438 to 3,047 m: 4
1,524 to 2,437 m: 12
914 to 1,523 m: 7
under 914 m: 16 (2007)
Airports—with unpaved runways:
total: 23
under 914 m: 23 (2007)
Heliports: 2 (2007)
Pipelines: gas 1,781 km; oil 94 km; refined products 7 km (2007)
Railways:
total: 4,839 km
standard gauge: 3,561 km 1.435-m gauge (3,195 km electrified)
narrow gauge: 1,268 km 1.000-m gauge (1,274 km electrified); 10 km 0.800-m gauge (10 km electrified) (2006)

Roadways:
total: 71,297 km
paved: 71,297 km (includes 1,728 of expressways) (2004)
Waterways: 65 km (Rhine River between Basel-Rheinfelden and Schaffhausen-Bodensee) (2003)
Merchant marine:
total: 32 ships (1000 GRT or over) 577,765 GRT/918,974 DWT
by type: bulk carrier 13, cargo 8, chemical tanker 4, container 6, specialized tanker 1
registered in other countries: 121 (Antigua and Barbuda 5, Bahamas 2, Cyprus 3, France 3, Indonesia 3, Italy 5, Liberia 11, Malta 22, Marshall Islands 14, Panama 26, Paraguay 1, Portugal 2, Russia 6, Singapore 2, St Vincent and The Grenadines 12, Tonga 1, UK 1, Vanuatu 2) (2007)
Ports and terminals: Basel

MILITARY

Military branches: Swiss Armed Forces: Land Forces, Swiss Air Force (Schweizer Luftwaffe) (2007)
Military service age and obligation: 19 years of age for male compulsory military service; 18 years of age for voluntary male and female military service; the Swiss Constitution states that "every Swiss male is obliged to do military service"; every Swiss male has to serve at least 260 days in the armed forces; conscripts receive 18 weeks of mandatory training, followed by seven 3-week intermittent recalls for training during the next 10 years (2008)
Manpower available for military service:
males age 16–49: 1,852,580
females age 16–49: 1,807,667 (2008 est.)
Manpower fit for military service:
males age 16–49: 1,513,984
females age 16–49: 1,478,761 (2008 est.)
Manpower reaching militarily significant age annually:
males age 16–49: 49,205
females age 16–49: 45,220 (2008 est.)
Military expenditures—percent of GDP: 1% (2005 est.)

TRANSNATIONAL ISSUES

Disputes—international: none
Illicit drugs: a major international financial center vulnerable to the layering and integration stages of money laundering; despite significant legislation and reporting requirements, secrecy rules persist and nonresidents are permitted to conduct business through offshore entities and various intermediaries; transit country for and consumer of South American cocaine, Southwest Asian heroin, and Western European synthetics; domestic cannabis cultivation and limited ecstasy production

627

SYRIA

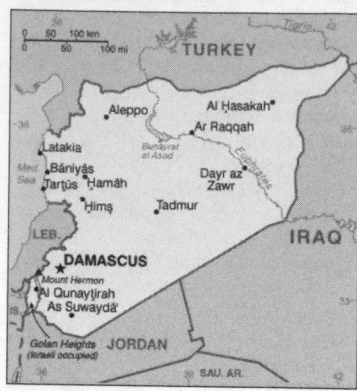

INTRODUCTION

Background: Following the breakup of the Ottoman Empire during World War I, France administered Syria until its independence in 1946. The country lacked political stability, however, and experienced a series of military coups during its first decades. Syria united with Egypt in February 1958 to form the United Arab Republic. In September 1961, the two entities separated, and the Syrian Arab Republic was reestablished. In November 1970, Hafiz al-ASAD, a member of the Socialist Ba'th Party and the minority Alawite sect, seized power in a bloodless coup and brought political stability to the country. In the 1967 Arab-Israeli War, Syria lost the Golan Heights to Israel. During the 1990s, Syria and Israel held occasional peace talks over its return. Following the death of President al-ASAD, his son, Bashar al-ASAD, was approved as president by popular referendum in July 2000. Syrian troops—stationed in Lebanon since 1976 in an ostensible peacekeeping role—were withdrawn in April 2005. During the July-August 2006 conflict between Israel and Hizballah, Syria placed its military forces on alert but did not intervene directly on behalf of its ally Hizballah.

GEOGRAPHY

Location: Middle East, bordering the Mediterranean Sea, between Lebanon and Turkey
Geographic coordinates: 35 00 N, 38 00 E
Map references: Middle East
Area: *total:* 185,180 sq km
land: 184,050 sq km
water: 1,130 sq km
note: includes 1,295 sq km of Israeli-occupied territory

Area—comparative: slightly larger than North Dakota
Land boundaries:
total: 2,253 km
border countries: Iraq 605 km, Israel 76 km, Jordan 375 km, Lebanon 375 km, Turkey 822 km
Coastline: 193 km
Maritime claims:
territorial sea: 12 nm
contiguous zone: 24 nm
Climate: mostly desert; hot, dry, sunny summers (June to August) and mild, rainy winters (December to February) along coast; cold weather with snow or sleet periodically in Damascus
Terrain: primarily semiarid and desert plateau; narrow coastal plain; mountains in west
Elevation extremes:
lowest point: unnamed location near Lake Tiberias -200 m
highest point: Mount Hermon 2,814 m
Natural resources: petroleum, phosphates, chrome and manganese ores, asphalt, iron ore, rock salt, marble, gypsum, hydropower
Land use:
arable land: 24.8%
permanent crops: 4.47%
other: 70.73% (2005)
Irrigated land: 13,330 sq km (2003)
Total renewable water resources: 46.1 cu km (1997)
Freshwater withdrawal (domestic/industrial/agricultural): *total:* 19.95 cu km/yr (3%/2%/95%)
per capita: 1,048 cu m/yr (2000)
Natural hazards: dust storms, sandstorms
Environment—current issues: deforestation; overgrazing; soil erosion; desertification; water pollution from raw sewage and petroleum refining wastes; inadequate potable water
Environment—international agreements: *party to:* Biodiversity, Climate Change, Climate Change-Kyoto Protocol, Desertification, Endangered Species, Hazardous Wastes, Ozone Layer Protection, Ship Pollution, Wetlands
signed, but not ratified: Environmental Modification
Geography—note: there are 42 Israeli settlements and civilian land use sites in the Israeli-occupied Golan Heights (August 2005 est.)

PEOPLE

Population: 19,747,586
note: in addition, about 40,000 people live in the Israeli-occupied Golan Heights—20,000 Arabs (18,000 Druze and 2,000 Alawites) and about 20,000 Israeli settlers (July 2008 est.)
Age structure:
0–14 years: 36.2% (male 3,679,473/female 3,467,096)
15–64 years: 60.5% (male 6,119,459/female 5,822,376)
65 years and over: 3.3% (male 310,838/female 348,344) (2008 est.)
Median age:
total: 21.4 years
male: 21.3 years
female: 21.5 years (2008 est.)
Population growth rate: 2.189% (2008 est.)
Birth rate: 26.57 births/1,000 population (2008 est.)
Death rate: 4.68 deaths/1,000 population (2008 est.)
Net migration rate: NA
Sex ratio:
at birth: 1.06 male(s)/female
under 15 years: 1.06 male(s)/female
15–64 years: 1.05 male(s)/female
65 years and over: 0.89 male(s)/female
total population: 1.05 male(s)/female (2008 est.)
Infant mortality rate:
total: 26.78 deaths/1,000 live births
male: 27.04 deaths/1,000 live births
female: 26.52 deaths/1,000 live births (2008 est.)
Life expectancy at birth:
total population: 70.9 years
male: 69.53 years
female: 72.35 years (2008 est.)
Total fertility rate: 3.21 children born/woman (2008 est.)
HIV/AIDS—adult prevalence rate: less than 0.1% (2001 est.)
HIV/AIDS—people living with HIV/AIDS: fewer than 500 (2003 est.)
HIV/AIDS—deaths: fewer than 200 (2003 est.)
Nationality:
noun: Syrian(s)
adjective: Syrian
Ethnic groups: Arab 90.3%, Kurds, Armenians, and other 9.7%
Religions: Sunni Muslim 74%, other Muslim (includes Alawite, Druze) 16%, Christian (various denominations) 10%, Jewish (tiny communities in Damascus, Al Qamishli, and Aleppo)
Languages: Arabic (official); Kurdish, Armenian, Aramaic, Circassian widely understood; French, English somewhat understood
Literacy:
definition: age 15 and over can read and write

total population: 79.6%
male: 86%
female: 73.6% (2004 census)

GOVERNMENT

Country name:
conventional long form: Syrian Arab Republic
conventional short form: Syria
local long form: Al Jumhuriyah al Arabiyah as Suriyah
local short form: Suriyah
former: United Arab Republic (with Egypt)
Government type: republic under an authoritarian military-dominated regime
Capital: *name:* Damascus
geographic coordinates: 33 30 N, 36 18 E
time difference: UTC+2 (7 hours ahead of Washington, DC during Standard Time)
daylight saving time: +1hr, begins 1 April; ends 30 September
Administrative divisions: 14 provinces (muhafazat, singular—muhafazah); Al Hasakah, Al Ladhiqiyah, Al Qunaytirah, Ar Raqqah, As Suwayda', Dar'a, Dayr az Zawr, Dimashq, Halab, Hamah, Hims, Idlib, Rif Dimashq, Tartus
Independence: 17 April 1946 (from League of Nations mandate under French administration)
National holiday: Independence Day, 17 April (1946)
Constitution: 13 March 1973
Legal system: based on a combination of French and Ottoman civil law; Islamic law is used in the family court system; has not accepted compulsory ICJ jurisdiction
Suffrage: 18 years of age; universal
Executive branch:
chief of state: President Bashar al-ASAD (since 17 July 2000); Vice President Farouk al-SHARA (since 11 February 2006) oversees foreign policy; Vice President Najah al-ATTAR (since 23 March 2006) oversees cultural policy
head of government: Prime Minister Muhammad Naji al-UTRI (since 10 September 2003); Deputy Prime Minister for Economic Affairs Abdallah al-DARDARI (since 14 June 2005)
cabinet: Council of Ministers appointed by the president
elections: president approved by popular referendum for a second seven-year term (no term limits); referendum last held on 27 May 2007 (next to be held in May 2014); the president appoints the vice presidents, prime minister, and deputy prime ministers
election results: Bashar al-ASAD approved as president; percent of vote—Bashar al-ASAD 97.6%
Legislative branch: unicameral People's Council or Majlis al-Shaab (250 seats;

members elected by popular vote to serve four-year terms)
elections: last held on 22–23 April 2007 (next to be held in 2011)
election results: percent of vote by party—NA; seats by party—NPF 172, independents 78
Judicial branch: Supreme Judicial Council (appoints and dismisses judges; headed by the president); national level—Supreme Constitutional Court (adjudicates electoral disputes and rules on constitutionality of laws and decrees; justices appointed for four-year terms by the President); Court of Cassation; Appeals Courts (Appeals Courts represent an intermediate level between the Court of Cassation and local level courts); local level—Magistrate Courts; Courts of First Instance; Juvenile Courts; Customs Courts; specialized courts—Economic Security Courts (hear cases related to economic crimes); Supreme State Security Court (hear cases related to national security); Personal Status Courts (religious; hear cases related to marriage and divorce)
Political parties and leaders: *legal parties:* National Progressive Front or NPF [President Bashar al-ASAD, Dr. Suleiman QADDAH] (includes Arab Socialist Renaissance (Ba'th) Party [President Bashar al-ASAD]; Socialist Unionist Democratic Party [Fadlallah Nasr Al-DIN]; Syrian Arab Socialist Union or ASU [Safwan QUDSI]; Syrian Communist Party (two branches) [Wissal Farha BAKDASH, Yusuf Rashid FAYSAL]; Syrian Social Nationalist Party [Ali QANSU]; Unionist Socialist Party [Fayez ISMAIL])
opposition parties not legally recognized:: Arab Democratic Socialist Union Party [Hasan Abdul AZIM]; Arab Socialist Movement; Democratic Ba'th Party [Ibrahim MAHKOS]; People's Democratic Party [Riad al TURK]; Revolutionary Workers' Party [Abdul Hafeez al HAFEZ]
Kurdish parties (considered illegal): Kurdish Democratic Front [Abdul Hamid DARWISH] (includes four parties); Kurdish Coordination [Abdul Hakim BASHAR] (includes Azadi Party [Kheirudin MURAD], Future Party [Masha'l TAMMO], Yekity Party [Hasam SALE])
other parties: Nahda Party [Abdul Aziz al MISLET]; Syrian Democratic Party [Mustafa QALAAJI]
Political pressure groups and leaders: Damascus Declaration National Council [Riyad SEIF, secretary general] (a broad alliance of opposition groups and individuals including: Committee for

Revival of Civil Society [Michel KILO, Riyad SEIF]; Communist Action Party [Fateh JAMOUS]; Kurdish Democratic Alliance; Kurdish Democratic Front; Liberal Nationalists' Movement; National Democratic Front; National Democratic Rally; and Syrian Human Rights Society or HRAS [Fawed FAWUZ]); National Salvation Front (alliance between former Vice President Abd al-Halim KHADDAM, the SMB, and other small opposition groups); Syrian Muslim Brotherhood or SMB [Sadr al-Din al-BAYANUNI] (operates in exile in London; endorsed the Damascus Declaration but is not an official member)
International organization participation: ABEDA, AFESD, AMF, CAEU, FAO, G-24, G-77, IAEA, IBRD, ICAO, ICC, ICCt (signatory), ICRM, IDA, IDB, IFAD, IFC, IFRCS, IHO, ILO, IMF, IMO, Interpol, IOC, IPU, ISO, ITSO, ITU, LAS, MIGA, NAM, OAPEC, OIC, UN, UNCTAD, UNESCO, UNIDO, UNRWA, UNWTO, UPU, WCO, WFTU, WHO, WIPO, WMO
Diplomatic representation in the US:
chief of mission: Ambassador Imad MUSTAFA
chancery: 2215 Wyoming Avenue NW, Washington, DC 20008
telephone: [1] (202) 232-6313
FAX: [1] (202) 234-9548
Diplomatic representation from the US:
chief of mission: Ambassador (vacant); Charge d'Affaires Michael CORBIN
embassy: Abou Roumaneh, Al-Mansour Street, No. 2, Damascus
mailing address: P. O. Box 29, Damascus
telephone: [963] (11) 3391-4444
FAX: [963] (11) 3391-3999
Flag description: three equal horizontal bands of red (top), white, and black, colors associated with the Arab Liberation flag; two small, green, five-pointed stars in a horizontal line centered in the white band; former flag of the United Arab Republic where the two stars represented the constituent states of Syria and Egypt; similar to the flag of Yemen, which has a plain white band, Iraq, which has an Arabic inscription centered in the white band, and that of Egypt, which has a gold Eagle of Saladin centered in the white band; the current design dates to 1980

ECONOMY

Economy—overview: The Syrian economy grew by an estimated 3.3% in real terms in 2007 led by the petroleum and agricultural sectors, which together account for about one-half of GDP. Higher crude oil prices countered declining oil production and led to higher budgetary and export receipts. Damascus has implemented modest economic reforms in the past few years, including cutting lending interest rates, opening private banks, consolidating all of the multiple exchange rates, raising prices on some subsidized items, most notably gasoline and cement, and establishing the Damascus Stock Exchange—which is set to begin operations in 2009. In October 2007, for example, Damascus raised the price of subsidized gasoline by 20%, and may institute a rationing system in 2008. In addition, President ASAD signed legislative decrees to encourage corporate ownership reform, and to allow the Central Bank to issue Treasury bills and bonds for government debt. Nevertheless, the economy remains highly controlled by the government. Long-run economic constraints include declining oil production, high unemployment and inflation, rising budget deficits, and increasing pressure on water supplies caused by heavy use in agriculture, rapid population growth, industrial expansion, and water pollution.

GDP (purchasing power parity): $87.09 billion (2007 est.)

GDP (official exchange rate): $37.76 billion (2007 est.)

GDP—real growth rate: 3.9% (2007 est.)

GDP—per capita (PPP): $4,500 (2007 est.)

GDP—composition by sector:
agriculture: 23.6%
industry: 27.5%
services: 48.9% (2007 est.)

Labor force: 5.462 million (2007 est.)

Labor force—by occupation: agriculture: 19.2%
industry: 14.5%
services: 66.3% (2006 est.)

Unemployment rate: 9% (2007 est.)

Population below poverty line: 11.9% (2006 est.)

Household income or consumption by percentage share:
lowest 10%: NA%
highest 10%: NA%

Inflation rate (consumer prices): 7% (2007 est.)

Investment (gross fixed): 21.3% of GDP (2007 est.)

Budget:
revenues: $8.848 billion
expenditures: $11.21 billion (2007 est.)

Public debt: 37.8% of GDP (2007 est.)

Agriculture—products: wheat, barley, cotton, lentils, chickpeas, olives, sugar beets; beef, mutton, eggs, poultry, milk

Industries: petroleum, textiles, food processing, beverages, tobacco, phosphate

rock mining, cement, oil seeds crushing, car assembly

Industrial production growth rate: 2.5% (2007 est.)

Electricity—production: 34.94 billion kWh (2007 est.)

Electricity—production by source:
fossil fuel: 57.6%
hydro: 42.4%
nuclear: 0%
other: 0% (2001)

Electricity—consumption: 34 billion kWh (2007 est.)

Electricity—exports: 986 million kWh (2006)

Electricity—imports: 0 kWh (2007 est.)

Oil—production: 380,000 bbl/day (2007 est.)

Oil—consumption: 229,000 bbl/day (2007 est.)

Oil—exports: 150,000 bbl/day (2007 est.)

Oil—imports: 160,000 bbl/day (2007 est.)

Oil—proved reserves: 2.5 billion bbl (2007 est.)

Natural gas—production: 7.8 billion cu m (2007 est.)

Natural gas—consumption: 4.4 billion cu m (2007 est.)

Natural gas—exports: NA cu m

Natural gas—imports: 0 cu m (2007 est.)

Natural gas—proved reserves: 240 billion cu m (1 January 2007 est.)

Current account balance: -$2.181 billion (2007 est.)

Exports: $10.58 billion f.o.b. (2007 est.)

Exports—commodities: crude oil, minerals, petroleum products, fruits and vegetables, cotton fiber, textiles, clothing, meat and live animals, wheat

Exports—partners: Iraq 27.4%, Germany 12.1%, Lebanon 9.5%, Italy 6.6%, Egypt 5.3%, Saudi Arabia 4.8% (2006)

Imports: $12.38 billion f.o.b. (2007 est.)

Imports—commodities: machinery and transport equipment, electric power machinery, food and livestock, metal and metal products, chemicals and chemical products, plastics, yarn, paper

Imports—partners: Saudi Arabia 12.3%, China 7.9%, Egypt 6.2%, UAE 6%, Italy 4.8%, Ukraine 4.8%, Germany 4.8%, Iran 4.5% (2006)

Economic aid—recipient: $213 million (2008 est.)

Reserves of foreign exchange and gold: $6.039 billion (31 December 2007 est.)

Debt—external: $6.465 billion (31 December 2007 est.)

Market value of publicly traded shares: $NA

Currency (code): Syrian pound (SYP)

Currency code: SYP

Exchange rates: Syrian pounds per US dollar—50.0085 (2007), 51.689 (2006), 50 (2005), 48.5 (2004), 52.8 (2003)
note: data for 2004–06 are the public sector rate; data for 2002–03 are the parallel market rate in 'Amman and Beirut; the official rate for repaying loans was 11.25 Syrian pounds per US dollars during 2004–06

Fiscal year: calendar year

COMMUNICATIONS

Telephones—main lines in use: 3.243 million (2006)

Telephones—mobile cellular: 4.675 million (2006)

Telephone system:
general assessment: fair system currently undergoing significant improvement and digital upgrades, including fiber-optic technology
domestic: the number of fixed-line connections has increased markedly since 2000; mobile-cellular service growing rapidly and teledensity has reached 25 wireless telephones per 100 persons; coaxial cable and microwave radio relay network
international: country code—963; submarine cable connection to Cyprus; satellite earth stations—1 Intelsat (Indian Ocean) and 1 Intersputnik (Atlantic Ocean region); coaxial cable and microwave radio relay to Iraq, Jordan, Lebanon, and Turkey; participant in Medarabtel

Radio broadcast stations: AM 14, FM 2, shortwave 1 (1998)

Radios: 4.15 million (1997)

Television broadcast stations: 44 (plus 17 repeaters) (1995)

Televisions: 1.05 million (1997)

Internet country code: .sy

Internet hosts: 119 (2007)

Internet Service Providers (ISPs): 1 (2000)

Internet users: 1.5 million (2006)

TRANSPORTATION

Airports: 90 (2007)

Airports—with paved runways: total: 26
over 3,047 m: 6
2,438 to 3,047 m: 15
914 to 1,523 m: 3
under 914 m: 2 (2007)

Airports—with unpaved runways:
total: 64
1,524 to 2,437 m: 1
914 to 1,523 m: 11
under 914 m: 52 (2007)

Heliports: 7 (2007)

Pipelines: gas 2,794 km; oil 2,000 km (2007)

Railways:

total: 2,711 km

standard gauge: 2,460 km 1.435-m gauge

narrow gauge: 251 km 1.050-m gauge (2006)

Roadways:

total: 94,890 km

paved: 19,073 km

unpaved: 75,817 km (2004)

Waterways: 900 km (not economically significant) (2005)

Merchant marine:

total: 96 ships (1000 GRT or over) 353,351 GRT/512,597 DWT

by type: bulk carrier 7, cargo 82, container 1, livestock carrier 4, petroleum tanker 1, roll on/roll off 1

foreign-owned: 10 (Jordan 2, Lebanon 4, Romania 4)

registered in other countries: 164 (Bolivia 1, Cambodia 32, Comoros 8, Cyprus 2, Dominica 2, Georgia 54, Hong Kong 1, North Korea 7, Lebanon 1, Libya 1, Malta 4, Mongolia 1, Panama 24, Sierra Leone 8, Slovakia 2, St Kitts and Nevis 5, St Vincent and The Grenadines 11, unknown 2) (2007)

Ports and terminals: Latakia, Tartus

MILITARY

Military branches: Syrian Armed Forces: Syrian Arab Army (includes Syrian Arab Navy), Syrian Arab Air and Air Defense Force (includes Air Defense Command) (2008)

Military service age and obligation: 18 years of age for compulsory military service; conscript service obligation—30 months (18 months in the Syrian Arab Navy); women are not conscripted but may volunteer to serve (2004)

Manpower available for military service:

males age 16–49: 5,251,875

females age 16–49: 4,966,367 (2008 est.)

Manpower fit for military service:

males age 16–49: 4,242,401

females age 16–49: 4,218,648 (2008 est.)

Manpower reaching militarily significant age annually:

males age 16–49: 215,734

females age 16–49: 203,106 (2008 est.)

Military expenditures—percent of GDP: 5.9% (2005 est.)

TRANSNATIONAL ISSUES

Disputes—international: Golan Heights is Israeli-occupied with the almost 1,000-strong UN Disengagement Observer Force (UNDOF) patrolling a buffer zone since 1964; lacking a treaty or other documentation describing the boundary, portions of the Lebanon-Syria boundary are unclear with several sections in dispute; since 2000, Lebanon has claimed Shaba'a farms in the Golan Heights; 2004 Agreement and pending demarcation settles border dispute with Jordan; approximately two million Iraqis have fled the conflict in Iraq with the majority taking refuge in Syria and Jordan

Refugees and internally displaced persons: *refugees (country of origin):* 1–1.4 million (Iraq); 522,100 (Palestinian Refugees (UNRWA))

IDPs: 305,000 (most displaced from Golan Heights during 1967 Arab-Israeli War) (2007)

Trafficking in persons: *current situation:* Syria is a destination country for women from South and Southeast Asia and Africa for domestic servitude and from Eastern Europe and Iraq for sexual exploitation; women are recruited for work in Syria as domestic servants, but some face conditions of exploitation and involuntary servitude including long hours, non-payment of wages, withholding of passports and other restrictions on movement, and physical and sexual abuse; Eastern European women recruited for work in Syria as cabaret dancers are not permitted to leave their work premises without permission and have their passports withheld; some displaced Iraqi women and children are reportedly forced into sexual exploitation

tier rating: Tier 3—Syria does not fully comply with the minimum standards for the elimination of trafficking and is not making significant efforts to do so

Illicit drugs: a transit point for opiates, hashish, and cocaine bound for regional and Western markets; weak anti-money-laundering controls and bank privatization may leave it vulnerable to money laundering

TAJIKISTAN

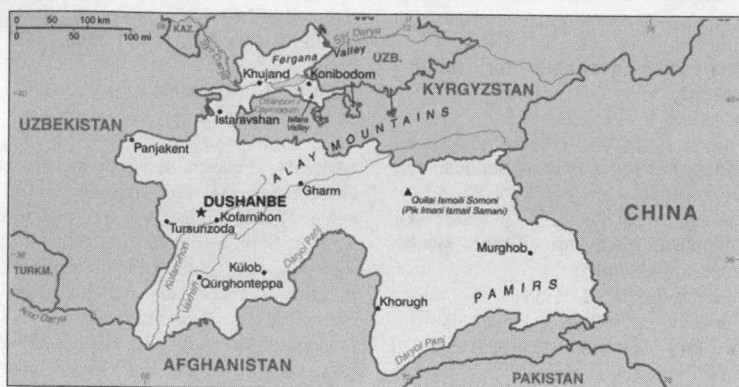

INTRODUCTION

Background: The Tajik people came under Russian rule in the 1860s and 1870s, but Russia's hold on Central Asia weakened following the Revolution of 1917. Bolshevik control of the area was fiercely contested and not fully reestablished until 1925. Much of present-day Sughd province was transferred from the Uzbekistan SSR to newly formed Tajikistan SSR in 1929. Ethnic Uzbeks form a substantial minority in Sughd province. Tajikistan became independent in 1991 following the breakup of the Soviet Union, and it is now in the process of strengthening its democracy and transitioning to a free market economy after its 1992–97 civil war. There have been no major security incidents in recent years, although the country remains the poorest in the former Soviet sphere. Attention by the international community in the wake of the war in Afghanistan has brought increased economic development and security assistance, which could create jobs and increase stability in the long term. Tajikistan is in the early stages of seeking World Trade Organization membership and has joined NATO's Partnership for Peace.

GEOGRAPHY

Location: Central Asia, west of China
Geographic coordinates: 39 00 N, 71 00 E
Map references: Asia
Area:
total: 143,100 sq km
land: 142,700 sq km
water: 400 sq km
Area—comparative: slightly smaller than Wisconsin
Land boundaries:
total: 3,651 km
border countries: Afghanistan 1,206 km,

China 414 km, Kyrgyzstan 870 km, Uzbekistan 1,161 km
Coastline: 0 km (landlocked)
Maritime claims: none (landlocked)
Climate: midlatitude continental, hot summers, mild winters; semiarid to polar in Pamir Mountains
Terrain: Pamir and Alay Mountains dominate landscape; western Fergana Valley in north, Kofarnihon and Vakhsh Valleys in southwest
Elevation extremes:
lowest point: Syr Darya (Sirdaryo) 300 m
highest point: Qullai Ismoili Somoni 7,495 m
Natural resources: hydropower, some petroleum, uranium, mercury, brown coal, lead, zinc, antimony, tungsten, silver, gold
Land use:
arable land: 6.52%
permanent crops: 0.89%
other: 92.59% (2005)
Irrigated land: 7,220 sq km (2003)
Total renewable water resources: 99.7 cu km (1997)
Freshwater withdrawal (domestic/industrial/agricultural): *total:* 11.96 cu km/yr (4%/5%/92%)
per capita: 1,837 cu m/yr (2000)
Natural hazards: earthquakes and floods
Environment—current issues: inadequate sanitation facilities; increasing levels of soil salinity; industrial pollution; excessive pesticides
Environment—international agreements: *party to:* Biodiversity, Climate Change, Desertification, Environmental Modification, Ozone Layer Protection, Wetlands
signed, but not ratified: none of the selected agreements
Geography—note: landlocked; mountainous region dominated by the Trans-Alay Range in the north and the Pamirs in the southeast; highest point, Qullai

Ismoili Somoni (formerly Communism Peak), was the tallest mountain in the former USSR

PEOPLE

Population: 7,211,884 (July 2008 est.)
Age structure:
0–14 years: 34.6% (male 1,270,289/female 1,226,954)
15–64 years: 61.7% (male 2,203,720/female 2,244,660)
65 years and over: 3.7% (male 113,156/female 153,105) (2008 est.)
Median age:
total: 21.6 years
male: 21.2 years
female: 22.1 years (2008 est.)
Population growth rate: 1.893% (2008 est.)
Birth rate: 27.18 births/1,000 population (2008 est.)
Death rate: 6.94 deaths/1,000 population (2008 est.)
Net migration rate: -1.31 migrant(s)/1,000 population (2008 est.)
Sex ratio:
at birth: 1.05 male(s)/female
under 15 years: 1.04 male(s)/female
15–64 years: 0.98 male(s)/female
65 years and over: 0.74 male(s)/female
total population: 0.99 male(s)/female (2008 est.)
Infant mortality rate:
total: 42.31 deaths/1,000 live births
male: 47.3 deaths/1,000 live births
female: 37.08 deaths/1,000 live births (2008 est.)
Life expectancy at birth:
total population: 64.97 years
male: 61.95 years
female: 68.15 years (2008 est.)
Total fertility rate: 3.04 children born/woman (2008 est.)
HIV/AIDS—adult prevalence rate: less than 0.1% (2001 est.)
HIV/AIDS—people living with HIV/AIDS: fewer than 200 (2003 est.)
HIV/AIDS—deaths: fewer than 100 (2001 est.)
Major infectious diseases:
degree of risk: high
food or waterborne diseases: bacterial diarrhea, hepatitis A, and typhoid fever
vectorborne disease: malaria (2008)
Nationality: *noun:* Tajikistani(s)
adjective: Tajikistani
Ethnic groups: Tajik 79.9%, Uzbek 15.3%, Russian 1.1%, Kyrgyz 1.1%, other 2.6% (2000 census)
Religions: Sunni Muslim 85%, Shi'a Muslim 5%, other 10% (2003 est.)
Languages: Tajik (official), Russian widely used in government and business

Literacy:
definition: age 15 and over can read and write
total population: 99.5%
male: 99.7%
female: 99.2% (2000 census)

Country name:
conventional long form: Republic of Tajikistan
conventional short form: Tajikistan
local long form: Jumhurii Tojikiston
local short form: Tojikiston
former: Tajik Soviet Socialist Republic

Government type: republic

Capital: *name:* Dushanbe
geographic coordinates: 38 35 N, 68 48 E
time difference: UTC+5 (10 hours ahead of Washington, DC during Standard Time)

Administrative divisions: 2 provinces
(viloyatho, singular—viloyat) and 1 autonomous province* (viloyati mukhtor); Viloyati Khatlon (Qurghonteppa), Viloyati Mukhtori Kuhistoni Badakhshon* [Gorno-Badakhshan] (Khorugh), Viloyati Sughd (Khujand)
note: the administrative center name follows in parentheses

Independence: 9 September 1991 (from Soviet Union)

National holiday: Independence Day (or National Day), 9 September (1991)

Constitution: 6 November 1994

Legal system: based on civil law system; no judicial review of legislative acts; has not accepted compulsory ICJ jurisdiction

Suffrage: 18 years of age; universal

Executive branch:
chief of state: President Emomali RAHMON (since 6 November 1994; head of state and Supreme Assembly chairman since 19 November 1992)
head of government: Prime Minister Oqil OQILOV (since 20 January 1999)
cabinet: Council of Ministers appointed by the president, approved by the Supreme Assembly
elections: president elected by popular vote for a seven-year term (eligible for a second term); election last held 6 November 2006 (next to be held in November 2013); prime minister appointed by the president
election results: Emomali RAHMON reelected president; percent of vote— Emomali RAHMON 79.3%, Olimzon BOBOYEV 6.2%, other 14.5%

Legislative branch: bicameral Supreme Assembly or Majlisi Oli consists of the National Assembly (upper chamber) or Majlisi Milliy (34 seats; 25 members selected by local deputies, 8 appointed by the president; 1 seat reserved for the former president; to serve five-year terms) and the Assembly of Representatives (lower chamber) or Majlisi Namoyandagon (63 seats; members are elected by popular vote to serve five-year terms)
elections: National Assembly—last held 25 March 2005 (next to be held in February 2010); Assembly of Representatives 27 February and 13 March 2005 (next to be held in February 2010)
election results: National Assembly—percent of vote by party—NA; seats by party—PDPT 29, CPT 2, independents 3; Assembly of Representatives—percent of vote by party—PDPT 74.9%, CPT 13.6%, Islamic Revival Party 8.9%, other 2.5%; seats by party—PDPT 51, CPT 5, Islamic Revival Party 2, independents 5

Judicial branch: Supreme Court (judges are appointed by the president)

Political parties and leaders: Agrarian Party of Tajikistan or APT [Amir KARAKULOV]; Democratic Party or DPT [Mahmadruzi ISKANDAROV (imprisoned October 2005); Rahmatullo VALIYEV, deputy]; Islamic Revival Party [Muhiddin KABIRI]; Party of Economic Reform or PER [Olimzon BOBOYEV]; People's Democratic Party of Tajikistan or PDPT [Emomali RAHMON]; Social Democratic Party or SDPT [Rahmatullo ZOYIROV]; Socialist Party or SPT [Mirhuseyn NARZIYEV]; Tajik Communist Party or CPT [Shodi SHABDOLOV]

Political pressure groups and leaders: splinter parties recognized by the government but not by the base of the party: Socialist Party or SPT [Abdualim GHAFFOROV; splintered from Narziyev's SPT]; Democratic Party or DPT [Masud SOBIROV; splintered from Iskanderov's DPT]; unregistered political parties: Agrarian Party [Hikmatullo NASREDDINOV]; Progressive Party [Sulton QUVVATOV]; Unity Party [Hikmatullo SAIDOV]

International organization participation: ADB, CIS, CSTO, EAEC, EAPC, EBRD, ECO, FAO, GCTU, IAEA, IBRD, ICAO, ICCt, ICRM, IDA, IDB, IFAD, IFC, IFRCS, ILO, IMF, Interpol, IOC, IOM, IPU, ISO (correspondent), ITSO, ITU, MIGA, OIC, OPCW, OSCE, PFP, SCO, UN, UNCTAD, UNESCO, UNIDO, UNWTO, UPU, WCO, WFTU, WHO, WIPO, WMO, WTO (observer)

Diplomatic representation in the US:
chief of mission: Ambassador Abdujabbor SHIRINOV
chancery: 1005 New Hampshire Avenue NW, Washington, DC 20037
telephone: [1] (202) 223-6090
FAX: [1] (202) 223-6091

Diplomatic representation from the US:
chief of mission: Ambassador Tracey Ann JACOBSON
embassy: 109-A Ismoili Somoni Avenue, Dushanbe 734019
mailing address: 7090 Dushanbe Place, Dulles, VA 20189
telephone: [992] (37) 229-20-00
FAX: [992] (37) 229-20-50

Flag description: three horizontal stripes of red (top), a wider stripe of white, and green; a gold crown surmounted by seven gold, five-pointed stars is located in the center of the white stripe

Economy—overview: Tajikistan has one of the lowest per capita GDPs among the 15 former Soviet republics. Only 7% of the land area is arable. Cotton is the most important crop, but this sector is burdened with debt and an obsolete infrastructure. Mineral resources include silver, gold, uranium, and tungsten. Industry consists only of a large aluminum plant, hydropower facilities, and small obsolete factories mostly in light industry and food processing. The civil war (1992–97) severely damaged the already weak economic infrastructure and caused a sharp decline in industrial and agricultural production. While Tajikistan has experienced steady economic growth since 1997, nearly two-thirds of the population continues to live in abject poverty. Economic growth reached 10.6% in 2004, but dropped to 8% in 2005, 7% in 2006, and 7.8% in 2007. Tajikistan's economic situation remains fragile due to uneven implementation of structural reforms, corruption, weak governance, widespread unemployment, seasonal power shortages, and the external debt burden. Continued privatization of medium and large state-owned enterprises could increase productivity. A debt restructuring agreement was reached with Russia in December 2002 including a $250 million write-off of Tajikistan's $300 million debt. Tajikistan ranks third in the world in terms of water resources per head, but suffers winter power shortages due to poor management of water levels in rivers and reservoirs. Completion of the Sangtuda I hydropower dam—built with Russian investment—and the Sangtuda II and Rogun dams will add substantially to electricity output. If finished according to Tajik plans, Rogun will be the world's tallest dam. Tajikistan has also received substantial infrastructure development loans from the Chinese government to improve roads and an electricity transmission network. To help increase north-south trade, the US funded a $36 million bridge which opened in August 2007 and links Tajikistan and Afghanistan.

GDP (purchasing power parity): $11.82 billion (2007 est.)

GDP (official exchange rate): $3.712 billion (2007 est.)

GDP—real growth rate: 7.8% (2007 est.)

GDP—per capita (PPP): $1,800 (2007 est.)

GDP—composition by sector:
agriculture: 23.4%
industry: 30.4%
services: 46.1% (2007 est.)

Labor force: 2.1 million (2007)

Labor force—by occupation: *agriculture:* 67.2%
industry: 7.5%
services: 25.3% (2000 est.)

Unemployment rate: 2.4% official rate; actual unemployment is higher (2007 est.)

Population below poverty line: 60% (2007 est.)

Household income or consumption by percentage share:
lowest 10%: 3.3%
highest 10%: 25.6% (2007 est.)

Distribution of family income—Gini index: 32.6 (2003)

Inflation rate (consumer prices): 13.2% (2007 est.)

Investment (gross fixed): 13.4% of GDP (2007 est.)

Budget: *revenues:* $700 million
expenditures: $673 million (2007 est.)

Agriculture—products: cotton, grain, fruits, grapes, vegetables; cattle, sheep, goats

Industries: aluminum, zinc, lead; chemicals and fertilizers, cement, vegetable oil, metal-cutting machine tools, refrigerators and freezers

Industrial production growth rate: 9% (2007 est.)

Electricity—production: 17.4 billion kWh (2007)

Electricity—production by source:
fossil fuel: 1.9%
hydro: 98.1%
nuclear: 0%
other: 0% (2001)

Electricity—consumption: 17.9 billion kWh (2007)

Electricity—exports: 4.259 billion kWh (2007)

Electricity—imports: 4.36 billion kWh (2007 est.)

Oil—production: 282.1 bbl/day (2005 est.)

Oil—consumption: 8,000 bbl/day (2007 est.)

Oil—exports: 0 bbl/day (2007)

Oil—imports: 7,600 bbl/day (2007)

Oil—proved reserves: 12 million bbl (1 January 2006 est.)

Natural gas—production: 39.32 million cu m (2005 est.)

Natural gas—consumption: 689 million cu m (2007 est.)

Natural gas—exports: 0 cu m (2005 est.)

Natural gas—imports: 650 million cu m (2007 est.)

Natural gas—proved reserves: 5.432 billion cu m (1 January 2006 est.)

Current account balance: -$351 million (2007 est.)

Exports: $1.625 billion f.o.b. (2007 est.)

Exports—commodities: aluminum, electricity, cotton, fruits, vegetable oil, textiles

Exports—partners: Netherlands 40.7%, Turkey 31.7%, Iran 5.4%, Uzbekistan 4.8%, Russia 4.7% (2006)

Imports: $2.736 billion f.o.b. (2007 est.)

Imports—commodities: electricity, petroleum products, aluminum oxide, machinery and equipment, foodstuffs

Imports—partners: Russia 24.6%, Kazakhstan 10.8%, Uzbekistan 10.2%, China 8.6%, Azerbaijan 8% (2006)

Economic aid—recipient: $241.4 million from US (2005)

Reserves of foreign exchange and gold: $235 million (31 December 2007 est.)

Debt—external: $1.56 billion (31 December 2007 est.)

Market value of publicly traded shares: $NA

Currency (code): somoni (TJS)

Currency code: TJS

Exchange rates: Tajikistani somoni per US dollar—3.4418 (2007), 3.3 (2006), 3.1166 (2005), 2.9705 (2004), 3.0614 (2003)

Fiscal year: calendar year

COMMUNICATIONS

Telephones—main lines in use: 280,200 (2005)

Telephones—mobile cellular: 265,000 (2005)

Telephone system:
general assessment: poorly developed and not well maintained; many towns are not linked to the national network
domestic: the domestic telecommunications network has historically been under funded and poorly maintained; main line availability has not changed significantly since 1998; cellular telephone use is growing but geographic coverage remains limited
international: country code—992; linked by cable and microwave radio relay to other CIS republics and by leased connections to the Moscow international gateway switch; Dushanbe linked by Intelsat to international gateway switch in Ankara (Turkey); satellite earth stations—3 (2 Intelsat and 1 Orbita) (2006)

Radio broadcast stations: AM 8, FM 10, shortwave 2 (2002)

Radios: 1.291 million (1991)

Television broadcast stations: 6 (2006)

Televisions: 820,000 (1997)

Internet country code: .tj

Internet hosts: 2,050 (2007)

Internet Service Providers (ISPs): 4 (2002)

Internet users: 19,500 (2005)

TRANSPORTATION

Airports: 26 (2007)

Airports—with paved runways: *total:* 18
over 3,047 m: 2
2,438 to 3,047 m: 4
1,524 to 2,437 m: 6
914 to 1,523 m: 3
under 914 m: 3 (2007)

Airports—with unpaved runways:
total: 8
under 914 m: 8 (2007)

Pipelines: gas 549 km; oil 38 km (2007)

Railways: *total:* 482 km
broad gauge: 482 km 1.520-m gauge (2006)

Roadways: *total:* 27,767 km (2000)

Waterways: 200 km (along Vakhsh River) (2006)

MILITARY

Military branches: Ground Forces, Air and Air Defense Forces, Mobile Force (2008)

Military service age and obligation: 18 years of age for compulsory military service; 2-year conscript service obligation (2006)

Manpower available for military service: *males age 16–49:* 1,897,356
females age 16–49: 1,911,594 (2008 est.)

Manpower fit for military service:
males age 16–49: 1,391,287
females age 16–49: 1,561,826 (2008 est.)

Manpower reaching militarily significant age annually:
males age 16–49: 84,137
females age 16–49: 81,777 (2008 est.)

Military expenditures—percent of GDP: 3.9% (2005 est.)

TRANSNATIONAL ISSUES

Disputes—international: in 2006, China and Tajikistan pledged to commence demarcation of the revised boundary agreed to in the delimitation of 2002; talks continue with Uzbekistan to delimit border and remove minefields; disputes in Isfara Valley delay delimitation with Kyrgyzstan

Illicit drugs: major transit country for Afghan narcotics bound for Russian and, to a lesser extent, Western European markets; limited illicit cultivation of opium poppy for domestic consumption; Tajikistan seizes roughly 80% of all drugs captured in Central Asia and stands third worldwide in seizures of opiates (heroin and raw opium); significant consumer of opiates

TANZANIA

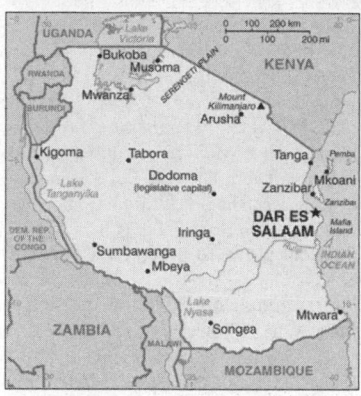

INTRODUCTION

Background: Shortly after achieving independence from Britain in the early 1960s, Tanganyika and Zanzibar merged to form the nation of Tanzania in 1964. One-party rule came to an end in 1995 with the first democratic elections held in the country since the 1970s. Zanzibar's semi-autonomous status and popular opposition have led to two contentious elections since 1995, which the ruling party won despite international observers' claims of voting irregularities.

GEOGRAPHY

Location: Eastern Africa, bordering the Indian Ocean, between Kenya and Mozambique

Geographic coordinates: 6 00 S, 35 00 E

Map references: Africa

Area:

total: 945,087 sq km

land: 886,037 sq km

water: 59,050 sq km

note: includes the islands of Mafia, Pemba, and Zanzibar

Area—comparative: slightly larger than twice the size of California

Land boundaries:

total: 3,861 km

border countries: Burundi 451 km, Democratic Republic of the Congo 459 km, Kenya 769 km, Malawi 475 km, Mozambique 756 km, Rwanda 217 km, Uganda 396 km, Zambia 338 km

Coastline: 1,424 km

Maritime claims:

territorial sea: 12 nm

exclusive economic zone: 200 nm

Climate: varies from tropical along coast to temperate in highlands

Terrain: plains along coast; central plateau; highlands in north, south

Elevation extremes:

lowest point: Indian Ocean 0 m

highest point: Kilimanjaro 5,895 m

Natural resources: hydropower, tin, phosphates, iron ore, coal, diamonds, gemstones, gold, natural gas, nickel

Land use:

arable land: 4.23%

permanent crops: 1.16%

other: 94.61% (2005)

Irrigated land: 1,840 sq km (2003)

Total renewable water resources: 91 cu km (2001)

Freshwater withdrawal (domestic/industrial/agricultural): *total:* 5.18 cu km/yr (10%/0%/89%)

per capita: 135 cu m/yr (2000)

Natural hazards: flooding on the central plateau during the rainy season; drought

Environment—current issues: soil degradation; deforestation; desertification; destruction of coral reefs threatens marine habitats; recent droughts affected marginal agriculture; wildlife threatened by illegal hunting and trade, especially for ivory

Environment—international agreements: *party to:* Biodiversity, Climate Change, Climate Change-Kyoto Protocol, Desertification, Endangered Species, Hazardous Wastes, Law of the Sea, Ozone Layer Protection, Wetlands

signed, but not ratified: none of the selected agreements

Geography—note: Kilimanjaro is highest point in Africa; bordered by three of the largest lakes on the continent: Lake Victoria (the world's second-largest freshwater lake) in the north, Lake Tanganyika (the world's second deepest) in the west, and Lake Nyasa in the southwest

PEOPLE

Population: 40,213,162

note: estimates for this country explicitly take into account the effects of excess mortality due to AIDS; this can result in lower life expectancy, higher infant mortality, higher death rates, lower population growth rates, and changes in the distribution of population by age and sex than would otherwise be expected (July 2008 est.)

Age structure:

0–14 years: 43.5% (male 8,763,471/female 8,719,198)

15–64 years: 53.7% (male 10,638,666/female 10,947,190)

65 years and over: 2.8% (male 502,368/female 642,269) (2008 est.)

Median age:

total: 17.8 years

male: 17.6 years

female: 18.1 years (2008 est.)

Population growth rate: 2.072% (2008 est.)

Birth rate: 35.12 births/1,000 population (2008 est.)

Death rate: 12.92 deaths/1,000 population (2008 est.)

Net migration rate: -1.48 migrant(s)/1,000 population (2008 est.)

Sex ratio:

at birth: 1.03 male(s)/female

under 15 years: 1.01 male(s)/female

15–64 years: 0.97 male(s)/female

65 years and over: 0.78 male(s)/female

total population: 0.98 male(s)/female (2008 est.)

Infant mortality rate:

total: 70.46 deaths/1,000 live births

male: 77.51 deaths/1,000 live births

female: 63.19 deaths/1,000 live births (2008 est.)

Life expectancy at birth:

total population: 51.45 years

male: 50.06 years

female: 52.88 years (2008 est.)

Total fertility rate: 4.62 children born/woman (2008 est.)

HIV/AIDS—adult prevalence rate: 8.8% (2003 est.)

HIV/AIDS—people living with HIV/AIDS: 1.6 million (2003 est.)

HIV/AIDS—deaths: 160,000 (2003 est.)

Major infectious diseases:

degree of risk: very high

food or waterborne diseases: bacterial diarrhea, hepatitis A, and typhoid fever

vectorborne diseases: malaria and plague

water contact disease: schistosomiasis (2008)

Nationality:

noun: Tanzanian(s)

adjective: Tanzanian

Ethnic groups: mainland—African 99% (of which 95% are Bantu consisting of more than 130 tribes), other 1% (consisting of Asian, European, and Arab); Zanzibar—Arab, African, mixed Arab and African

Religions: mainland—Christian 30%, Muslim 35%, indigenous beliefs 35%; Zanzibar—more than 99% Muslim

Languages: Kiswahili or Swahili (official), Kiunguja (name for Swahili in Zanzibar), English (official, primary language of commerce, administration, and higher education), Arabic (widely spoken in Zanzibar), many local languages

note: Kiswahili (Swahili) is the mother tongue of the Bantu people living in Zanzibar and nearby coastal Tanzania; although Kiswahili is Bantu in structure and origin, its vocabulary draws on a variety of sources including Arabic and English; it has become the lingua franca of central and eastern Africa; the first language of most people is one of the local languages

Literacy:
definition: age 15 and over can read and write Kiswahili (Swahili), English, or Arabic
total population: 69.4%
male: 77.5%
female: 62.2% (2002 census)

GOVERNMENT

Country name:
conventional long form: United Republic of Tanzania
conventional short form: Tanzania
local long form: Jamhuri ya Muungano wa Tanzania
local short form: Tanzania
former: United Republic of Tanganyika and Zanzibar

Government type: republic

Capital: *name:* Dar es Salaam
geographic coordinates: 6 48 S, 39 17 E
time difference: UTC+3 (8 hours ahead of Washington, DC during Standard Time)
note: legislative offices have been transferred to Dodoma, which is planned as the new national capital; the National Assembly now meets there on a regular basis

Administrative divisions: 26 regions; Arusha, Dar es Salaam, Dodoma, Iringa, Kagera, Kigoma, Kilimanjaro, Lindi, Manyara, Mara, Mbeya, Morogoro, Mtwara, Mwanza, Pemba North, Pemba South, Pwani, Rukwa, Ruvuma, Shinyanga, Singida, Tabora, Tanga, Zanzibar Central/South, Zanzibar North, Zanzibar Urban/West

Independence: 26 April 1964; Tanganyika became independent 9 December 1961 (from UK-administered UN trusteeship); Zanzibar became independent 19 December 1963 (from UK); Tanganyika united with Zanzibar 26 April 1964 to form the United Republic of Tanganyika and Zanzibar; renamed United Republic of Tanzania 29 October 1964

National holiday: Union Day (Tanganyika and Zanzibar), 26 April (1964)

Constitution: 25 April 1977; major revisions October 1984

Legal system: based on English common law; judicial review of legislative acts limited to matters of interpretation; has not accepted compulsory ICJ jurisdiction

Suffrage: 18 years of age; universal

Executive branch:
chief of state: President Jakaya KIKWETE

(since 21 December 2005); Vice President Dr. Ali Mohammed SHEIN (since 5 July 2001); note—the president is both chief of state and head of government
head of government: President Jakaya KIKWETE (since 21 December 2005); Vice President Dr. Ali Mohammed SHEIN (since 5 July 2001)
note: Zanzibar elects a president who is head of government for matters internal to Zanzibar; Amani Abeid KARUME was reelected to that office on 30 October 2005
cabinet: Cabinet appointed by the president from among the members of the National Assembly
elections: president and vice president elected on the same ballot by popular vote for five-year terms (eligible for a second term); election last held 14 December 2005 (next to be held in December 2010); prime minister appointed by the president
election results: Jakaya KIKWETE elected president; percent of vote—Jakaya KIKWETE 80.3%, Ibrahim LIPUMBA 11.7%, Freeman MBOWE 5.9%

Legislative branch: unicameral National Assembly or Bunge (274 seats; 232 members elected by popular vote, 37 allocated to women nominated by the president, 5 to members of the Zanzibar House of Representatives; to serve five-year terms); note—in addition to enacting laws that apply to the entire United Republic of Tanzania, the Assembly enacts laws that apply only to the mainland; Zanzibar has its own House of Representatives to make laws especially for Zanzibar (the Zanzibar House of Representatives has 50 seats elected by universal suffrage to serve five-year terms)
elections: last held 14 December 2005 (next to be held in December 2010)
election results: National Assembly—percent of vote by party—NA; seats by party—CCM 206, CUF 19, CHADEMA 5, other 2, women appointed by the president 37, Zanzibar representatives 5 Zanzibar House of Representatives—percent of vote by party—NA; seats by party—CCM 30, CUF 19; 1 seat was nullified with a rerun to take place soon

Judicial branch: Permanent Commission of Enquiry (official ombudsman); Court of Appeal (consists of a chief justice and four judges); High Court (consists of a Jaji Kiongozi and 29 judges appointed by the president; holds regular sessions in all regions); District Courts; Primary Courts (limited jurisdiction and appeals can be made to the higher courts)

Political parties and leaders: Chama Cha Demokrasia na Maendeleo (Party of Democracy and Development) or CHADEMA [Bob MAKANI]; Chama Cha Mapinduzi or CCM (Revolutionary Party) [Jakaya Mrisho KIKWETE]; Civic United Front or CUF [Ibrahim LIPUMBA]; Democratic Party [Christopher MTIKLA] (unregistered); Tanzania Labor Party or TLP [Augustine Lyatonga MREME]; United Democratic Party or UDP [John CHEYO]

Political pressure groups and leaders: NA

International organization participation: ACP, AfDB, AU, C, EAC, EADB, FAO, G-6, G-77, IAEA, IBRD, ICAO, ICC, ICCt, ICRM, IDA, IFAD, IFC, IFRCS, ILO, IMF, IMO, IMSO, Interpol, IOC, IOM, IPU, ISO, ITSO, ITU, ITUC, MIGA, NAM, OPCW, SADC, UN, UN Security Council (temporary), UNCTAD, UNESCO, UNHCR, UNIDO, UNIFIL, UNMEE, UNMIS, UNOCI, UNWTO, UPU, WCO, WFTU, WHO, WIPO, WMO, WTO

Diplomatic representation in the US:
chief of mission: Ambassador Ombeni Yohana SEFUE
chancery: 2139 R Street NW, Washington, DC 20008
telephone: [1] (202) 939-6125
FAX: [1] (202) 797-7408

Diplomatic representation from the US:
chief of mission: Ambassador Mark GREEN
embassy: 140 Msese Road, Kinondoni District, Dar es Salaam
mailing address: P. O. Box 9123, Dar es Salaam
telephone: [255] (22) 2666-010 through 2666-015
FAX: [255] (22) 2666-701, 2668-501

Flag description: divided diagonally by a yellow-edged black band from the lower hoist-side corner; the upper triangle (hoist side) is green and the lower triangle is blue

ECONOMY

Economy—overview: Tanzania is one of the poorest countries in the world. The economy depends heavily on agriculture, which accounts for more than 40% of GDP, provides 85% of exports, and employs 80% of the work force. Topography and climatic conditions, however, limit cultivated crops to only 4% of the land area. Industry traditionally featured the processing of agricultural products and light consumer goods. The World Bank, the IMF, and bilateral donors have provided funds to rehabilitate Tanzania's out-of-date economic infrastructure and to alleviate poverty.

Long-term growth through 2005 featured a pickup in industrial production and a substantial increase in output of minerals led by gold. Recent banking reforms have helped increase private-sector growth and investment. Continued donor assistance and solid macroeconomic policies supported real GDP growth of nearly 7% in 2007.

GDP (purchasing power parity): $48.94 billion (2007 est.)

GDP (official exchange rate): $16.18 billion (2007 est.)

GDP—real growth rate: 7.3% (2007 est.)

GDP—per capita (PPP): $1,300 (2007 est.)

GDP—composition by sector:
agriculture: 42.8%
industry: 18.4%
services: 38.7% (2007 est.)

Labor force: 20.04 million (2007 est.)

Labor force—by occupation: *agriculture:* 80%
industry and services: 20% (2002 est.)

Unemployment rate: NA%

Population below poverty line: 36% (2002 est.)

Household income or consumption by percentage share:
lowest 10%: 2.9%
highest 10%: 26.9% (2000)

Distribution of family income—Gini index: 34.6 (2000)

Inflation rate (consumer prices): 7% (2007 est.)

Investment (gross fixed): 21.3% of GDP (2007 est.)

Budget:
revenues: $3.148 billion
expenditures: $3.577 billion (2007 est.)

Public debt: 19.7% of GDP (2007 est.)

Agriculture—products: coffee, sisal, tea, cotton, pyrethrum (insecticide made from chrysanthemums), cashew nuts, tobacco, cloves, corn, wheat, cassava (tapioca), bananas, fruits, vegetables; cattle, sheep, goats

Industries: agricultural processing (sugar, beer, cigarettes, sisal twine); diamond, gold, and iron mining, salt, soda ash; cement, oil refining, shoes, apparel, wood products, fertilizer

Industrial production growth rate: 8.2% (2007 est.)

Electricity—production: 1.88 billion kWh (2005)

Electricity—production by source:
fossil fuel: 18.9%
hydro: 81.1%
nuclear: 0%
other: 0% (2001)

Electricity—consumption: 1.199 billion kWh (2005)

Electricity—exports: 0 kWh (2005 est.)

Electricity—imports: 136 million kWh (2005)

Oil—production: 0 bbl/day (2005 est.)

Oil—consumption: 25,000 bbl/day (2005 est.)

Oil—exports: 0 bbl/day (2004)

Oil—imports: 24,800 bbl/day (2004)

Oil—proved reserves: 0 bbl (1 January 2006 est.)

Natural gas—production: 0 cu m (2005 est.)

Natural gas—consumption: 0 cu m (2005 est.)

Natural gas—exports: 0 cu m (2005 est.)

Natural gas—imports: 0 cu m (2005)

Natural gas—proved reserves: 21.73 billion cu m (1 January 2006 est.)

Current account balance: -$1.495 billion (2007 est.)

Exports: $2.134 billion f.o.b. (2007 est.)

Exports—commodities: gold, coffee, cashew nuts, manufactures, cotton

Exports—partners: China 8.8%, India 8.8%, Netherlands 6.2%, Japan 5.3%, UAE 4.2%, Germany 4.2% (2006)

Imports: $4.835 billion f.o.b. (2007 est.)

Imports—commodities: consumer goods, machinery and transportation equipment, industrial raw materials, crude oil

Imports—partners: South Africa 9.8%, China 9.4%, Kenya 7.8%, India 6.8%, UAE 5.9%, Zambia 5.7% (2006)

Economic aid—recipient: $1.505 billion (2005)

Reserves of foreign exchange and gold: $2.908 billion (31 December 2007 est.)

Debt—external: $4.379 billion (31 December 2007 est.)

Stock of direct foreign investment—at home: $NA

Stock of direct foreign investment—abroad: $NA

Market value of publicly traded shares: $587.9 million (2005)

Currency (code): Tanzanian shilling (TZS)

Currency code: TZS

Exchange rates: Tanzanian shillings per US dollar—1,255 (2007), 1,251.9 (2006), 1,128.93 (2005), 1,089.33 (2004), 1,038.42 (2003)

Fiscal year: 1 July—30 June

COMMUNICATIONS

Telephones—main lines in use: 169,135 (2007)

Telephones—mobile cellular: 6.72 million (2007)

Telephone system:
general assessment: telecommunications services are inadequate; system operating below capacity and being modernized for better service; small aperture terminal (VSAT) system under construction
domestic: fixed-line telephone network inadequate with less than 1 connection per 100 persons; mobile-cellular service, aided by multiple providers, is increasing; trunk service provided by open-wire, microwave radio relay, tropospheric scatter, and fiber-optic cable; some links being made digital
international: country code—255; satellite earth stations—2 Intelsat (1 Indian Ocean, 1 Atlantic Ocean)

Radio broadcast stations: AM 12, FM 11, shortwave 2 (1998)

Radios: 8.8 million (1997)

Television broadcast stations: 3 (1999)

Televisions: 103,000 (1997)

Internet country code: .tz

Internet hosts: 20,757 (2007)

Internet Service Providers (ISPs): 6 (2000)

Internet users: 384,300 (2005)

TRANSPORTATION

Airports: 124 (2007)

Airports—with paved runways:
total: 10
over 3,047 m: 2
2,438 to 3,047 m: 2
1,524 to 2,437 m: 5
914 to 1,523 m: 1 (2007)

Airports—with unpaved runways:
total: 114
1,524 to 2,437 m: 17
914 to 1,523 m: 63
under 914 m: 34 (2007)

Pipelines: gas 287 km; oil 891 km (2007)

Railways:
total: 3,690 km
narrow gauge: 969 km 1.067-m gauge; 2,721 km 1.000-m gauge (2006)

Roadways:
total: 78,891 km
paved: 6,808 km
unpaved: 72,083 km (2003)

Waterways: Lake Tanganyika, Lake Victoria, and Lake Nyasa principal avenues of commerce with neighboring countries; rivers not navigable (2005)

Merchant marine:
total: 9 ships (1000 GRT or over) 24,801 GRT/31,507 DWT
by type: cargo 1, passenger/cargo 4, petroleum tanker 4
registered in other countries: 2 (Honduras 1, St Kitts and Nevis 1) (2007)

Ports and terminals: Dar es Salaam

MILITARY

Military branches: Tanzanian People's Defense Force (Jeshi la Wananchi la Tanzania, JWTZ): Army, Naval Wing (includes Coast Guard), Air Defense

Command (includes Air Wing), National Service (2007)
Military service age and obligation: 18 years of age for voluntary military service (2007)
Manpower available for military service:
males age 16–49: 9,108,177 (2008 est.)
Manpower fit for military service:
males age 16–49: 5,278,833 (2008 est.)
Military expenditures—percent of GDP: 0.2% (2005 est.)

Disputes—international: Tanzania still hosts more than a half-million refugees, more than any other African country, mainly from Burundi and the Democratic Republic of the Congo, despite the international community's efforts at repatriation; disputes with Malawi over the boundary in Lake Nyasa (Lake Malawi) and the meandering Songwe River remain dormant

Refugees and internally displaced persons: *refugees (country of origin):* 352,640 (Burundi); 127,973 (Democratic Republic of the Congo) (2007)
Illicit drugs: growing role in transshipment of Southwest and Southeast Asian heroin and South American cocaine destined for South African, European, and US markets and of South Asian methaqualone bound for southern Africa; money laundering remains a problem

THAILAND

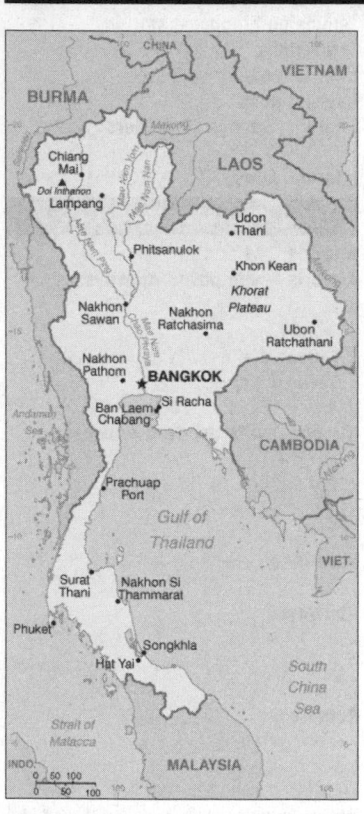

INTRODUCTION

Background: A unified Thai kingdom was established in the mid-14th century. Known as Siam until 1939, Thailand is the only Southeast Asian country never to have been taken over by a European power. A bloodless revolution in 1932 led to a constitutional monarchy. In alliance with Japan during World War II, Thailand became a US ally following the conflict. Thailand is currently facing separatist violence in its southern ethnic Malay-Muslim provinces.

GEOGRAPHY

Location: Southeastern Asia, bordering the Andaman Sea and the Gulf of Thailand, southeast of Burma
Geographic coordinates: 15 00 N, 100 00 E
Map references: Southeast Asia
Area:
total: 514,000 sq km
land: 511,770 sq km
water: 2,230 sq km
Area—comparative: slightly more than twice the size of Wyoming
Land boundaries:
total: 4,863 km
border countries: Burma 1,800 km, Cambodia 803 km, Laos 1,754 km, Malaysia 506 km
Coastline: 3,219 km
Maritime claims:
territorial sea: 12 nm
exclusive economic zone: 200 nm
continental shelf: 200-m depth or to the depth of exploitation
Climate: tropical; rainy, warm, cloudy southwest monsoon (mid-May to September); dry, cool northeast monsoon (November to mid-March); southern isthmus always hot and humid
Terrain: central plain; Khorat Plateau in the east; mountains elsewhere
Elevation extremes:
lowest point: Gulf of Thailand 0 m
highest point: Doi Inthanon 2,576 m
Natural resources: tin, rubber, natural gas, tungsten, tantalum, timber, lead, fish, gypsum, lignite, fluorite, arable land
Land use: *arable land:* 27.54%
permanent crops: 6.93%
other: 65.53% (2005)
Irrigated land: 49,860 sq km (2003)
Total renewable water resources: 409.9 cu km (1999)
Freshwater withdrawal (domestic/industrial/agricultural): *total:* 82.75 cu km/yr (2%/2%/95%)
per capita: 1,288 cu m/yr (2000)
Natural hazards: land subsidence in Bangkok area resulting from the depletion of the water table; droughts

Environment—current issues: air pollution from vehicle emissions; water pollution from organic and factory wastes; deforestation; soil erosion; wildlife populations threatened by illegal hunting
Environment—international agreements: *party to:* Biodiversity, Climate Change, Climate Change-Kyoto Protocol, Desertification, Endangered Species, Hazardous Wastes, Marine Life Conservation, Ozone Layer Protection, Tropical Timber 83, Tropical Timber 94, Wetlands
signed, but not ratified: Law of the Sea
Geography—note: controls only land route from Asia to Malaysia and Singapore

PEOPLE

Population: 65,493,298
note: estimates for this country explicitly take into account the effects of excess mortality due to AIDS; this can result in lower life expectancy, higher infant mortality, higher death rates, lower population growth rates, and changes in the distribution of population by age and sex than would otherwise be expected (July 2008 est.)
Age structure:
0–14 years: 21.2% (male 7,104,776/female 6,781,453)
15–64 years: 70.3% (male 22,763,274/female 23,304,793)
65 years and over: 8.5% (male 2,516,721/female 3,022,281) (2008 est.)
Median age:
total: 32.8 years
male: 32 years
female: 33.7 years (2008 est.)
Population growth rate: 0.64% (2008 est.)
Birth rate: 13.57 births/1,000 population (2008 est.)
Death rate: 7.17 deaths/1,000 population (2008 est.)
Net migration rate: NA
Sex ratio: *at birth:* 1.05 male(s)/female

under 15 years: 1.05 male(s)/female
15–64 years: 0.98 male(s)/female
65 years and over: 0.83 male(s)/female
total population: 0.98 male(s)/female
(2008 est.)

Infant mortality rate:
total: 18.23 deaths/1,000 live births
male: 19.5 deaths/1,000 live births
female: 16.89 deaths/1,000 live births
(2008 est.)

Life expectancy at birth:
total population: 72.83 years
male: 70.51 years
female: 75.27 years (2008 est.)

Total fertility rate: 1.64 children born/woman (2008 est.)

HIV/AIDS—adult prevalence rate: 1.5% (2003 est.)

HIV/AIDS—people living with HIV/AIDS: 570,000 (2003 est.)

HIV/AIDS—deaths: 58,000 (2003 est.)

Major infectious diseases:
degree of risk: high
food or waterborne diseases: bacterial diarrhea and hepatitis A
vectorborne diseases: dengue fever, Japanese encephalitis, and malaria
animal contact disease: rabies
water contact disease: leptospirosis
note: highly pathogenic H5N1 avian influenza has been identified in this country; it poses a negligible risk with extremely rare cases possible among US citizens who have close contact with birds (2008)

Nationality:
noun: Thai (singular and plural)
adjective: Thai

Ethnic groups: Thai 75%, Chinese 14%, other 11%

Religions: Buddhist 94.6%, Muslim 4.6%, Christian 0.7%, other 0.1% (2000 census)

Languages: Thai, English (secondary language of the elite), ethnic and regional dialects

Literacy: *definition:* age 15 and over can read and write
total population: 92.6%
male: 94.9%
female: 90.5% (2000 census)

GOVERNMENT

Country name:
conventional long form: Kingdom of Thailand
conventional short form: Thailand
local long form: Ratcha Anachak Thai
local short form: Prathet Thai
former: Siam

Government type: constitutional monarchy

Capital: *name:* Bangkok
geographic coordinates: 13 45 N, 100 31 E
time difference: UTC+7 (12 hours ahead

of Washington, DC during Standard Time)

Administrative divisions: 76 provinces (changwat, singular and plural); Amnat Charoen, Ang Thong, Buriram, Chachoengsao, Chai Nat, Chaiyaphum, Chanthaburi, Chiang Mai, Chiang Rai, Chon Buri, Chumphon, Kalasin, Kamphaeng Phet, Kanchanaburi, Khon Kaen, Krabi, Krung Thep Mahanakhon (Bangkok), Lampang, Lamphun, Loei, Lop Buri, Mae Hong Son, Maha Sarakham, Mukdahan, Nakhon Nayok, Nakhon Pathom, Nakhon Phanom, Nakhon Ratchasima, Nakhon Sawan, Nakhon Si Thammarat, Nan, Narathiwat, Nong Bua Lamphu, Nong Khai, Nonthaburi, Pathum Thani, Pattani, Phangnga, Phatthalung, Phayao, Phetchabun, Phetchaburi, Phichit, Phitsanulok, Phra Nakhon Si Ayutthaya, Phrae, Phuket, Prachin Buri, Prachuap Khiri Khan, Ranong, Ratchaburi, Rayong, Roi Et, Sa Kaeo, Sakon Nakhon, Samut Prakan, Samut Sakhon, Samut Songkhram, Sara Buri, Satun, Sing Buri, Sisaket, Songkhla, Sukhothai, Suphan Buri, Surat Thani, Surin, Tak, Trang, Trat, Ubon Ratchathani, Udon Thani, Uthai Thani, Uttaradit, Yala, Yasothon

Independence: 1238 (traditional founding date; never colonized)

National holiday: Birthday of King PHUMIPHON (BHUMIBOL), 5 December (1927)

Constitution: constitution signed by King PHUMIPHON (BHUMIBOL) on 24 August 2007

Legal system: based on civil law system, with influences of common law; has not accepted compulsory ICJ jurisdiction

Suffrage: 18 years of age; universal and compulsory

Executive branch:
chief of state: King PHUMIPHON Adunyadet (BHUMIBOL Adulyadej) (since 9 June 1946)
head of government: Prime Minister SAMAK Sundavavej (since 29 January 2008); Deputy Prime Minister MINGKWAN Saengsuwan (MINGKWAN Sangsuwan) (since 7 February 2008); Deputy Prime Minister SAHAS Banditkun (SAHAS Banditkul) (since 7 February 2008); Deputy Prime Minister SANAN Kachornprasat (ANA Kachornparsart) (since 7 February 2008); Deputy Prime Minister SOMCHAI Wongsawat (since 7 February 2008); Deputy Prime Minister SURAPONG Suebwonglee (since 7 February 2008); Deputy Prime Minister SUWIT Khunkitti (since 7 February 2008)

cabinet: Council of Ministers
note: there is also a Privy Council
elections: monarch is hereditary; according to 2007 constitution, prime minister is designated from among members of House of Representatives; following national elections for House of Representatives, leader of party that could organize a majority coalition usually was appointed prime minister by king; prime minister is limited to two 4-year terms

Legislative branch: bicameral National Assembly or Rathasapha consisted of the Senate or Wuthisapha (150 seats; 76 members elected by popular vote representing 76 provinces, 74 appointed by judges and independent government bodies; all serve six-year terms) and the House of Representatives or Sapha Phuthaen Ratsadon (480 seats; 400 members elected from 157 multi-seat constituencies and 80 elected on proportional party-list basis of 10 per eight zones or groupings of provinces; all serve four-year terms)
elections: Senate—last held on 2 March 2008 (next to be held in March 2014); House of Representatives—last election held on 23 December 2007 (next to be held in December 2011)
election results: Senate—percent of vote by party—NA; seats by party—NA; House of Representatives—percent of vote by party—NA; seats by party—PPP 233, DP 164, TNP 34, Motherland 24, Middle Way 11, Unity 9, Royalist People's 5
note: 74 senators were appointed on 19 February 2008 by a seven-member committee headed by the chief of the Constitutional Court; 76 senators were elected on 2 March 2008; elections to the Senate are non-partisan; registered political party members are disqualified from being senators

Judicial branch: Supreme Court or Sandika (judges appointed by the monarch)

Political parties and leaders: Democrat Party or DP (Prachathipat Party) [ABHISIT Wetchachiwa] (ABHISIT Vejjajiva); Matchima Thippatai (Middle Way Party) [ANONGWAN Therpsuthin]; Motherland Party (Peua Pandin Party) [SUWIT Khunkitti]; People Power Party (Palang Prachachon Party) or PPP [SAMAK Sunthorawet] (SAMAK Sundaravej); Royalist People's Party (Pracharaj) [SANOH Thienthong]; Ruam Jai Thai Party (Thai Unity Party) [CHETTA Thanacharo] (CHETTHA Thanajaro); Thai Nation Party or TNP (Chat Thai Party) [BARNHARN SILPA-ARCHA]

Political pressure groups and leaders: NA

International organization participation: ADB, APEC, APT, ARF, ASEAN, BIMSTEC, BIS, CP, EAS, FAO, G-77, IAEA, IBRD, ICAO, ICC, ICCt (signatory), ICRM, IDA, IFAD, IFC, IFRCS, IHO, ILO, IMF, IMO, IMSO, Interpol, IOC, IOM, IPU, ISO, ITSO, ITU, ITUC, MIGA, NAM, OAS (observer), OIC (observer), OPCW, OSCE (partner), PCA, UN, UNAMID, UNCTAD, UNESCO, UNHCR, UNIDO, UNMIS, UNWTO, UPU, WCL, WCO, WFTU, WHO, WIPO, WMO, WTO

Diplomatic representation in the US:
chief of mission: Ambassador KRIT Kanchanakunchon (KRIT Garnjana-Goonchorn)
chancery: 1024 Wisconsin Avenue NW, Suite 401, Washington, DC 20007
telephone: [1] (202) 944-3600
FAX: [1] (202) 944-3611
consulate(s) general: Chicago, Los Angeles, New York

Diplomatic representation from the US:
chief of mission: Ambassador Eric G. JOHN
embassy: 120-122 Wireless Road, Bangkok 10330
mailing address: APO AP 96546
telephone: [66] (2) 205-4000
FAX: [66] (2) 254-2990, 205-4131
consulate(s) general: Chiang Mai

Flag description: five horizontal bands of red (top), white, blue (double width), white, and red

ECONOMY

Economy—overview: With a well-developed infrastructure, a free-enterprise economy, and generally pro-investment policies, Thailand appears to have fully recovered from the 1997–98 Asian Financial Crisis. The country was one of East Asia's best performers from 2002–04. Boosted by strong export growth, the Thai economy grew 4.5% in 2007. Bangkok has pursued preferential trade agreements with a variety of partners in an effort to boost exports and to maintain high growth. By 2007, the tourism sector had largely recovered from the major 2004 tsunami. Following the military coup in September 2006, investment and consumer confidence stagnated due to the uncertain political climate that lasted through the December 2007 elections. Foreign investor sentiment was further tempered by a 30% reserve requirement on capital inflows instituted in December 2006, and discussion of amending Thailand's rules governing foreign-owned businesses. Economic growth in 2007 was due almost entirely to robust export performance—despite the pressure of an appreciating currency. Exports have performed at record levels, rising nearly 17% in 2006 and 12% in 2007. Export-oriented manufacturing—in particular automobile production—and farm output are driving these gains.

GDP (purchasing power parity): $519.4 billion (2007 est.)
GDP (official exchange rate): $245.7 billion (2007 est.)
GDP—real growth rate: 4.8% (2007 est.)
GDP—per capita (PPP): $7,900 (2007 est.)
GDP—composition by sector:
agriculture: 11.4%
industry: 43.9%
services: 44.7% (2007 est.)
Labor force: 36.9 million (2007 est.)
Labor force—by occupation: *agriculture:* 49%
industry: 14%
services: 37% (2000 est.)
Unemployment rate: 1.4% (2007 est.)
Population below poverty line: 10% (2004 est.)
Household income or consumption by percentage share:
lowest 10%: 2.7%
highest 10%: 33.4% (2002)
Distribution of family income—Gini index: 42 (2002)
Inflation rate (consumer prices): 2.2% (2007 est.)
Investment (gross fixed): 26.8% of GDP (2007 est.)
Budget: *revenues:* $44.09 billion
expenditures: $49.84 billion (2007 est.)
Public debt: 37.9% of GDP (2007 est.)
Agriculture—products: rice, cassava (tapioca), rubber, corn, sugarcane, coconuts, soybeans
Industries: tourism, textiles and garments, agricultural processing, beverages, tobacco, cement, light manufacturing such as jewelry and electric appliances, computers and parts, integrated circuits, furniture, plastics, automobiles and automotive parts; world's second-largest tungsten producer and third-largest tin producer
Industrial production growth rate: 5.4% (2007 est.)
Electricity—production: 124.6 billion kWh (2005)
Electricity—production by source:
fossil fuel: 91.3%
hydro: 6.4%
nuclear: 0%
other: 2.4% (2001)
Electricity—consumption: 117.7 billion kWh (2005)

Electricity—exports: 642 million kWh (2005)
Electricity—imports: 4.419 billion kWh (2005)
Oil—production: 310,900 bbl/day (2005 est.)
Oil—consumption: 929,000 bbl/day (2005 est.)
Oil—exports: 225,700 bbl/day (2004)
Oil—imports: 893,400 bbl/day (2004)
Oil—proved reserves: 291 million bbl (1 January 2006 est.)
Natural gas—production: 22.73 billion cu m (2005 est.)
Natural gas—consumption: 31.23 billion cu m (2005 est.)
Natural gas—exports: 0 cu m (2005 est.)
Natural gas—imports: 8.497 billion cu m (2005)
Natural gas—proved reserves: 400.7 billion cu m (1 January 2006 est.)
Current account balance: $14.92 billion (2007 est.)
Exports: $151 billion f.o.b. (2007 est.)
Exports—commodities: textiles and footwear, fishery products, rice, rubber, jewelry, automobiles, computers and electrical appliances
Exports—partners: US 15%, Japan 12.6%, China 9%, Singapore 6.4%, Hong Kong 5.5%, Malaysia 5.1% (2006)
Imports: $125 billion f.o.b. (2007 est.)
Imports—commodities: capital goods, intermediate goods and raw materials, consumer goods, fuels
Imports—partners: Japan 19.9%, China 10.6%, US 7.5%, Malaysia 6.6%, UAE 5.5%, Singapore 4.4% (2006)
Economic aid—recipient: $171.1 million (2005)
Reserves of foreign exchange and gold: $87.46 billion (31 December 2007 est.)
Debt—external: $58.5 billion (31 December 2007)
Stock of direct foreign investment—at home: $80.84 billion (2007 est.)
Stock of direct foreign investment—abroad: $7.017 billion (2007 est.)
Market value of publicly traded shares: $139.6 billion (2006)
Currency (code): baht (THB)
Currency code: THB
Exchange rates: baht per US dollar—33.599 (2007), 37.882 (2006), 40.22 (2005), 40.222 (2004), 41.485 (2003)
Fiscal year: 1 October—30 September

COMMUNICATIONS

Telephones—main lines in use: 7.073 million (2006)
Telephones—mobile cellular: 40.816 million (2006)
Telephone system:

general assessment: high quality system, especially in urban areas like Bangkok
domestic: fixed line system provided by both a government owned and commercial provider; wireless service expanding rapidly and outpacing fixed lines
international: country code—66; connected to major submarine cable systems providing links throughout Asia, Australia, Middle East, Europe, and US; satellite earth stations—2 Intelsat (1 Indian Ocean, 1 Pacific Ocean)

Radio broadcast stations: AM 238, FM 351, shortwave 6 (2007)
Radios: 13.96 million (1997)
Television broadcast stations: 111 (2006)
Televisions: 15.19 million (1997)
Internet country code: .th
Internet hosts: 973,941 (2007)
Internet Service Providers (ISPs): 15 (2000)
Internet users: 8.466 million (2006)

TRANSPORTATION

Airports: 106 (2007)
Airports—with paved runways:
total: 65
over 3,047 m: 8
2,438 to 3,047 m: 11
1,524 to 2,437 m: 23
914 to 1,523 m: 17
under 914 m: 6 (2007)
Airports—with unpaved runways:
total: 41
1,524 to 2,437 m: 1
914 to 1,523 m: 12
under 914 m: 28 (2007)
Heliports: 3 (2007)
Pipelines: gas 4,381 km; refined products 320 km (2007)
Railways:
total: 4,071 km
narrow gauge: 4,071 km 1.000-m gauge (2006)
Roadways:
total: 57,403 km
paved: 56,542 km
unpaved: 861 km (2000)

Waterways: 4,000 km
note: 3,701 km navigable by boats with drafts up to 0.9 m (2005)
Merchant marine:
total: 405 ships (1000 GRT or over) 2,640,857 GRT/4,043,938 DWT
by type: bulk carrier 53, cargo 140, chemical tanker 16, container 21, liquefied gas 30, passenger/cargo 9, petroleum tanker 101, refrigerated cargo 32, specialized tanker 2, vehicle carrier 1
foreign-owned: 15 (China 1, Japan 4, Malaysia 3, Russia 1, Singapore 2, Taiwan 1, UK 3)
registered in other countries: 34 (Bahamas 1, Indonesia 1, Mongolia 1, Panama 10, Singapore 20, Tuvalu 1) (2007)
Ports and terminals: Bangkok, Laem Chabang, Prachuap Port, Si Racha

MILITARY

Military branches: Royal Thai Army (RTA), Royal Thai Navy (RTN, includes Royal Thai Marine Corps), Royal Thai Air Force (Knogtap Agard Thai, RTAF) (2008)
Military service age and obligation: 21 years of age for compulsory military service; 18 years of age for voluntary military service; males are registered at 18 years of age; 2-year conscript service obligation (2006)
Manpower available for military service:
males age 16–49: 17,553,410
females age 16–49: 17,751,268 (2008 est.)
Manpower fit for military service:
males age 16–49: 12,968,674
females age 16–49: 14,058,779 (2008 est.)
Manpower reaching militarily significant age annually:
males age 16–49: 531,315
females age 16–49: 511,288 (2008 est.)
Military expenditures—percent of GDP: 1.8% (2005 est.)

TRANSNATIONAL ISSUES

Disputes—international: separatist violence in Thailand's predominantly Muslim southern provinces prompt border closures and controls with Malaysia to stem terrorist activities; Southeast Asian states have enhanced border surveillance to check the spread of avian flu; talks continue on completion of demarcation with Laos but disputes remain over several islands in the Mekong River; despite continuing border committee talks, Thailand must deal with Karen and other ethnic rebels, refugees, and illegal cross-border activities, and as of 2006, over 116,000 Karen, Hmong, and other refugees and asylum seekers from Burma; Cambodia and Thailand dispute sections of historic boundary with missing boundary markers; Cambodia claims Thai encroachments into Cambodian territory and obstructing access to Preah Vihear temple ruins awarded to Cambodia by ICJ decision in 1962; Thailand is studying the feasibility of jointly constructing the Hatgyi Dam on the Salween river near the border with Burma; in 2004, international environmentalist pressure prompted China to halt construction of 13 dams on the Salween River that flows through China, Burma, and Thailand
Refugees and internally displaced persons: *refugees (country of origin):* 132,241 (Burma) (2007)
Illicit drugs: a minor producer of opium, heroin, and marijuana; transit point for illicit heroin en route to the international drug market from Burma and Laos; eradication efforts have reduced the area of cannabis cultivation and shifted some production to neighboring countries; opium poppy cultivation has been reduced by eradication efforts; also a drug money-laundering center; minor role in methamphetamine production for regional consumption; major consumer of methamphetamine since the 1990s despite a series of government crackdowns

TIMOR-LESTE

INTRODUCTION

Background: The Portuguese began to trade with the island of Timor in the early 16th century and colonized it in mid-century. Skirmishing with the Dutch in the region eventually resulted in an 1859 treaty in which Portugal ceded the western portion of the island. Imperial Japan occupied Portuguese Timor from 1942 to 1945, but Portugal resumed colonial authority after the Japanese defeat in World War II. East Timor declared itself independent from Portugal on 28 November 1975 and was invaded and occupied by Indonesian forces nine days later. It was incorporated into Indonesia in July 1976 as the province of Timor Timur (East Timor). An unsuccessful campaign of pacification followed over the next two decades, during which an estimated 100,000 to 250,000 individuals lost their lives. On 30 August 1999, in a UN-supervised popular referendum, an overwhelming majority of the people of Timor-Leste voted for independence from Indonesia. Between the referendum and the arrival of a multinational peacekeeping force in late September 1999, anti-independence Timorese militias—organized and supported by the Indonesian military—commenced a large-scale, scorched-earth

campaign of retribution. The militias killed approximately 1,400 Timorese and forcibly pushed 300,000 people into western Timor as refugees. The majority of the country's infrastructure, including homes, irrigation systems, water supply systems, and schools, and nearly 100% of the country's electrical grid were destroyed. On 20 September 1999 the Australian-led peacekeeping troops of the International Force for East Timor (INTERFET) deployed to the country and brought the violence to an end. On 20 May 2002, Timor-Leste was internationally recognized as an independent state. In late April 2006, internal tensions threatened the new nation's security when a military strike led to violence and a near breakdown of law and order in Dili. At the request of the Government of Timor-Leste, an Australian-led International Stabilization Force (ISF) deployed to Timor-Leste in late May. In August, the UN Security Council established the UN Integrated Mission in Timor-Leste (UNMIT), which included an authorized police presence of over 1,600 personnel. In subsequent months, many of the ISF soldiers were replaced by UN police officers; approximately 80 ISF officers remained as of January 2008. From April to June 2007, the Government of Timor-Leste held presidential and parliamentary elections in a largely peaceful atmosphere with the support and assistance of UNMIT and international donors.

GEOGRAPHY

Location: Southeastern Asia, northwest of Australia in the Lesser Sunda Islands at the eastern end of the Indonesian archipelago; note—Timor-Leste includes the eastern half of the island of Timor, the Oecussi (Ambeno) region on the northwest portion of the island of Timor, and the islands of Pulau Atauro and Pulau Jaco

Geographic coordinates: 8 50 S, 125 55 E

Map references: Southeast Asia
Area:
total: 15,007 sq km
land: NA sq km
water: NA sq km
Area—comparative: slightly larger than Connecticut
Land boundaries:
total: 228 km
border countries: Indonesia 228 km
Coastline: 706 km
Maritime claims:
territorial sea: 12 nm
contiguous zone: 24 nm
exclusive fishing zone: 200 nm
Climate: tropical; hot, humid; distinct rainy and dry seasons
Terrain: mountainous
Elevation extremes:
lowest point: Timor Sea, Savu Sea, and Banda Sea 0 m
highest point: Foho Tatamailau 2,963 m
Natural resources: gold, petroleum, natural gas, manganese, marble
Land use:
arable land: 8.2%
permanent crops: 4.57%
other: 87.23% (2005)
Irrigated land: 1,065 sq km (2003)
Natural hazards: floods and landslides are common; earthquakes, tsunamis, tropical cyclones
Environment—current issues: widespread use of slash and burn agriculture has led to deforestation and soil erosion
Environment—international agreements: *party to:* Climate Change, Desertification
Geography—note: Timor comes from the Malay word for "East"; the island of Timor is part of the Malay Archipelago and is the largest and easternmost of the Lesser Sunda Islands

PEOPLE

Population: 1,108,777
note: other estimates range as low as 800,000 (July 2008 est.)
Age structure:
0–14 years: 35.1% (male 197,975/female 191,716)
15–64 years: 61.6% (male 347,573/female 334,908)
65 years and over: 3.3% (male 17,578/female 19,027) (2008 est.)
Median age:
total: 21.5 years
male: 21.5 years
female: 21.5 years (2008 est.)
Population growth rate: 2.05% (2008 est.)
Birth rate: 26.52 births/1,000 population (2008 est.)
Death rate: 6.02 deaths/1,000 population (2008 est.)

Net migration rate: NA
Sex ratio:
at birth: 1.05 male(s)/female
under 15 years: 1.03 male(s)/female
15–64 years: 1.04 male(s)/female
65 years and over: 0.92 male(s)/female
total population: 1.03 male(s)/female (2008 est.)
Infant mortality rate:
total: 41.98 deaths/1,000 live births
male: 48.16 deaths/1,000 live births
female: 35.49 deaths/1,000 live births (2008 est.)
Life expectancy at birth:
total population: 66.94 years
male: 64.6 years
female: 69.39 years (2008 est.)
Total fertility rate: 3.36 children born/woman (2008 est.)
HIV/AIDS—adult prevalence rate: NA
HIV/AIDS—people living with HIV/AIDS: NA
HIV/AIDS—deaths: NA
Major infectious diseases:
degree of risk: high
food or waterborne diseases: bacterial and protozoal diarrhea, hepatitis A, and typhoid fever
vectorborne diseases: chikungunya, dengue fever and malaria (2008)
Nationality:
noun: Timorese
adjective: Timorese
Ethnic groups: Austronesian (Malayo-Polynesian), Papuan, small Chinese minority
Religions: Roman Catholic 98%, Muslim 1%, Protestant 1% (2005)
Languages: Tetum (official), Portuguese (official), Indonesian, English
note: there are about 16 indigenous languages; Tetum, Galole, Mambae, and Kemak are spoken by significant numbers of people
Literacy:
definition: age 15 and over can read and write
total population: 58.6%
male: NA
female: NA (2002)

GOVERNMENT

Country name:
conventional long form: Democratic Republic of Timor-Leste
conventional short form: Timor-Leste
local long form: Republika Demokratika Timor Lorosa'e [Tetum]; Republica Democratica de Timor-Leste [Portuguese]
local short form: Timor Lorosa'e [Tetum]; Timor-Leste [Portuguese]
former: East Timor, Portuguese Timor
Government type: republic
Capital: *name:* Dili

geographic coordinates: 8 35 S, 125 36 E

time difference: UTC+9 (14 hours ahead of Washington, DC during Standard Time)

Administrative divisions: 13 administrative districts; Aileu, Ainaro, Baucau, Bobonaro (Maliana), Cova-Lima (Suai), Dili, Ermera, Lautem (Los Palos), Liquica, Manatuto, Manufahi (Same), Oecussi (Ambeno), Viqueque

Independence: 28 November 1975 (independence proclaimed from Portugal); note—20 May 2002 is the official date of international recognition of Timor-Leste's independence from Indonesia

National holiday: Independence Day, 28 November (1975)

Constitution: 22 March 2002 (based on the Portuguese model)

Legal system: UN-drafted legal system based on Indonesian law remains in place but is to be replaced by civil and penal codes based on Portuguese law; these have passed but have not been promulgated; has not accepted compulsory ICJ jurisdiction

Suffrage: 17 years of age; universal

Executive branch:

chief of state: President Jose RAMOS-HORTA (since 20 May 2007); note—the president plays a largely symbolic role but is able to veto legislation, dissolve parliament, and call national elections

head of government: Prime Minister Kay Rala Xanana GUSMAO (since 8 August 2007), note—he formerly used the name Jose Alexandre GUSMAO; Deputy Prime Minister Jose Luis GUTERRES (since 8 August 2007)

cabinet: Council of Ministers

elections: president elected by popular vote for a five-year term (eligible for a second term); election last held on 9 April 2007 with run-off on 8 May 2007 (next to be held in May 2012); following elections, president appoints leader of majority party or majority coalition as prime minister

election results: Jose RAMOS-HORTA elected president; percent of vote—Jose RAMOS-HORTA 69.2%, Francisco GUTTERES 30.8%

Legislative branch: unicameral National Parliament (number of seats can vary from 52 to 65; members elected by popular vote to serve five-year terms)

elections: last held on 30 June 2007 (next elections due by June 2012)

election results: percent of vote by party—FRETILIN 29%, CNRT 24.1%, ASDT-PSD 15.8%, PD 11.3%, PUN 4.5%, KOTA-PPT (Democratic Alliance) 3.2%, UNTERDIM 3.2%, others 8.9%;

seats by party—FRETILIN 21, CNRT 18, ASDT-PSD 11, PD 8, PUN 3, KOTA-PPT 2, UNDERTIM 2

Judicial branch: Supreme Court of Justice—constitution calls for one judge to be appointed by National Parliament and rest appointed by Superior Council for Judiciary; note—until Supreme Court is established, Court of Appeals is highest court

Political parties and leaders: Democratic Party or PD [Fernando de ARAUJO]; National Congress for Timorese Reconstruction or CNRT [Xanana GUSMAO]; National Democratic Union of Timorese Resistance or UNDERTIM [Cornelio DA Conceicao GAMA]; National Unity Party or PUN [Fernanda BORGES]; People's Party of Timor or PPT [Jacob XAVIER]; Revolutionary Front of Independent Timor-Leste or FRETILIN [Mari ALKATIRI]; Social Democratic Association of Timor or ASDT [Francisco Xavier do AMARAL]; Social Democratic Party or PSD [Mario CARRASCALAO]; Sons of the Mountain Warriors or KOTA [Manuel TILMAN] (also known as Association of Timorese Heroes)

International organization participation: ACP, ADB, ARF, CPLP, FAO, G-77, IBRD, ICAO, ICCt, IDA, IFAD, IFC, IFRCS, ILO, IMF, IMO, Interpol, IOC, MIGA, NAM, OPCW, PIF (observer), UN, UNCTAD, UNESCO, UNIDO, Union Latina, UNWTO, UPU, WCO, WHO

Diplomatic representation in the US:

chief of mission: Ambassador (vacant); Charge d'Affaires Constancio PINTO

chancery: 4201 Connecticut Avenue NW, Washington, DC 20008

telephone: [1] (202) 966-3202

FAX: [1] (202) 966-3205

consulate(s) general: New York

Diplomatic representation from the US:

chief of mission: Ambassador Hans G. KLEMM

embassy: Avenida de Portugal, Praia dos Conqueiros, Dili

mailing address: US Department of State, 8250 Dili Place, Washington, DC 20521-8250

telephone: (670) 332-4684

FAX: (670) 331-3206

Flag description: red, with a black isosceles triangle (based on the hoist side) superimposed on a slightly longer yellow arrowhead that extends to the center of the flag; a white star is in the center of the black triangle

ECONOMY

Economy—overview: In late 1999, about 70% of the economic infrastructure of Timor-Leste was laid waste by Indonesian troops and anti-independence militias. Three hundred thousand people fled westward. Over the next three years a massive international program, manned by 5,000 peacekeepers (8,000 at peak) and 1,300 police officers, led to substantial reconstruction in both urban and rural areas. By the end of 2005, refugees had returned or had settled in Indonesia. The country continues to face great challenges in rebuilding its infrastructure, strengthening the civil administration, and generating jobs for young people entering the work force. The development of oil and gas resources in offshore waters has begun to supplement government revenues ahead of schedule and above expectations—the result of high petroleum prices. The technology-intensive industry, however, has done little to create jobs for the unemployed because there are no production facilities in Timor. Gas is piped to Australia. In June 2005 the National Parliament unanimously approved the creation of a Petroleum Fund to serve as a repository for all petroleum revenues and preserve the value of Timor-Leste's petroleum wealth for future generations. The Fund held assets of US$1.8 billion as of September 2007. The mid-2006 outbreak of violence and civil unrest disrupted both private and public sector economic activity and created 100,000 internally displaced persons—about 10 percent of the population. While real non-oil GDP growth in 2006 was negative, the economy probably rebounded in 2007. The underlying economic policy challenge the country faces remains how best to use oil-and-gas wealth to lift the non-oil economy onto a higher growth path and reduce poverty. In late 2007, the new government announced plans aimed at increasing spending, reducing poverty, and improving the country's infrastructure, but it continues to face capacity constraints. In the short term, the government must also address continuing problems related to the crisis of 2006, especially the displaced Timorese.

GDP (purchasing power parity): $2.608 billion (2007 est.)

GDP (official exchange rate): $459 million (2007 est.)

GDP—real growth rate: 19.8% (2007 est.)

GDP—per capita (PPP): $2,500 (2007 est.)

GDP—composition by sector:

agriculture: 32.2%

industry: 12.8%

services: 55% (2005)

Labor force: NA

Labor force—by occupation:

agriculture: NA%

industry: NA%

services: NA%

Unemployment rate: 50% estimated; note—unemployment in urban areas reached 20%; data do not include underemployed (2001 est.)

Population below poverty line: 42% (2003 est.)

Household income or consumption by percentage share:

lowest 10%: NA%

highest 10%: NA%

Distribution of family income—Gini index: 38 (2002 est.)

Inflation rate (consumer prices): 7.8% (2007 est.)

Budget:

revenues: $733 million

expenditures: $309 million

note: the government passed a transitional budget to cover the latter half of 2007 and has moved the fiscal cycle to a calendar year, starting with the budget they passed for 2008 (FY06/07 est.)

Agriculture—products: coffee, rice, corn, cassava, sweet potatoes, soybeans, cabbage, mangoes, bananas, vanilla

Industries: printing, soap manufacturing, handicrafts, woven cloth

Industrial production growth rate: 8.5% (2004 est.)

Electricity—production: NA kWh

Electricity—production by source:

fossil fuel: 100%

hydro: 0%

nuclear: 0%

other: 0% (2001)

Electricity—consumption: NA kWh

Electricity—exports: 0 kWh (2005)

Electricity—imports: 0 kWh

Oil—production: 94,420 bbl/day (2005)

Oil—proved reserves: NA

Natural gas—production: 0 cu m (2005)

Natural gas—consumption: 0 cu m (2005)

Natural gas—exports: 0 cu m (2005 est.)

Natural gas—imports: 0 cu m (2005)

Natural gas—proved reserves: 200 billion cu m (1 January 2006 est.)

Current account balance: $1.161 billion (2007 est.)

Exports: $10 million; note—excludes oil (2005 est.)

Exports—commodities: coffee, sandalwood, marble; note—potential for oil and vanilla exports

Exports—partners: US, Germany, Portugal, Australia, Indonesia (2006)

Imports: $202 million (2004 est.)

Imports—commodities: food, gasoline, kerosene, machinery

Economic aid—recipient: $184.7 million (2005 est.)

Market value of publicly traded shares: $NA

Currency (code): US dollar (USD)

Currency code: USD

Exchange rates: the US dollar is used

Fiscal year: calendar year

COMMUNICATIONS

Telephones—main lines in use: 2,500 (2006)

Telephones—mobile cellular: 49,100 (2006)

Telephone system:

general assessment: rudimentary service limited to urban areas

domestic: system suffered significant damage during the violence associated with independence; extremely limited fixed-line services; mobile-cellular services and coverage limited primarily to urban areas

international: country code—670; international service is available in major urban centers

Radio broadcast stations: at least 21 (Timor-Leste has one national public broadcaster and 20 community and church radio stations—frequency type NA)

Radios: NA

Television broadcast stations: 1 (Timor-Leste has one national public broadcaster)

Televisions: NA

Internet country code: .tl

Internet hosts: 94 (2007)

Internet Service Providers (ISPs): NA

Internet users: 1,000 (2004)

TRANSPORTATION

Airports: 8 (2007)

Airports—with paved runways:

total: 3

2,438 to 3,047 m: 1

1,524 to 2,437 m: 1

under 914 m: 1 (2007)

Airports—with unpaved runways:

total: 5

914 to 1,523 m: 3

under 914 m: 2 (2007)

Heliports: 9 (2007)

Roadways:

total: 6,040 km

paved: 2,600 km

unpaved: 3,440 km (2005)

Merchant marine:

by type: passenger/cargo 1 (2007)

Ports and terminals: Dili

MILITARY

Military branches: Timor-Leste Defense Force (Forcas de Defesa de Timor-L'este, Falintil (FDTL)): Army, Navy (Armada) (2008)

Military service age and obligation: 18 years of age for voluntary military service; no conscription (2008)

Manpower available for military service:

males age 16–49: 284,903

females age 16–49: 272,212 (2008 est.)

Manpower fit for military service:

males age 16–49: 224,096

females age 16–49: 231,901 (2008 est.)

Manpower reaching militarily significant age annually:

males age 16–49: 13,045

females age 16–49: 12,670 (2008 est.)

Military expenditures—percent of GDP: NA

TRANSNATIONAL ISSUES

Disputes—international: Timor-Leste-Indonesia Boundary Committee has resolved all but a small portion of the land boundary, but discussions on maritime boundaries are stalemated over sovereignty of the uninhabited coral island of Pulau Batek/Fatu Sinai in the north and alignment with Australian claims in the south; many refugees who left Timor-Leste in 2003 still reside in Indonesia and refuse repatriation; Australia and Timor-Leste agreed in 2005 to defer the disputed portion of the boundary for 50 years and to split hydrocarbon revenues evenly outside the Joint Petroleum Development Area covered by the 2002 Timor Sea Treaty

Refugees and internally displaced persons: *IDPs:* 100,000 (2007)

Illicit drugs: NA

TOGO

INTRODUCTION

Background: French Togoland became Togo in 1960. Gen. Gnassingbe EYADEMA, installed as military ruler in 1967, ruled Togo with a heavy hand for almost four decades. Despite the facade of multiparty elections instituted in the early 1990s, the government was largely dominated by President EYADEMA, whose Rally of the Togolese People (RPT) party has maintained power almost continually since 1967 and maintains a majority of seats in today's legislature. Upon EYADEMA's death in

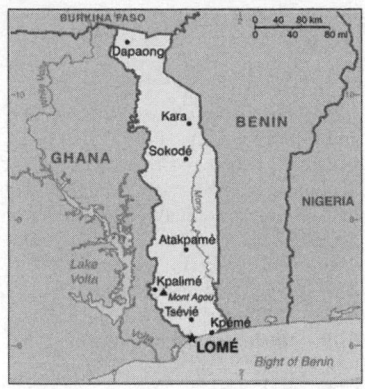

February 2005, the military installed the president's son, Faure GNASSINGBE, and then engineered his formal election two months later. Democratic gains since then allowed Togo to hold its first relatively free and fair legislative elections in October 2007. After years of political unrest and fire from international organizations for human rights abuses, Togo is finally being re-welcomed into the international community.

GEOGRAPHY

Location: Western Africa, bordering the Bight of Benin, between Benin and Ghana
Geographic coordinates: 8 00 N, 1 10 E
Map references: Africa
Area: *total:* 56,785 sq km
land: 54,385 sq km
water: 2,400 sq km
Area—comparative: slightly smaller than West Virginia
Land boundaries:
total: 1,647 km
border countries: Benin 644 km, Burkina Faso 126 km, Ghana 877 km
Coastline: 56 km
Maritime claims:
territorial sea: 30 nm
exclusive economic zone: 200 nm
Climate: tropical; hot, humid in south; semiarid in north
Terrain: gently rolling savanna in north; central hills; southern plateau; low coastal plain with extensive lagoons and marshes
Elevation extremes:
lowest point: Atlantic Ocean 0 m
highest point: Mont Agou 986 m
Natural resources: phosphates, limestone, marble, arable land
Land use:
arable land: 44.2%
permanent crops: 2.11%
other: 53.69% (2005)
Irrigated land: 70 sq km (2003)
Total renewable water resources: 14.7 cu km (2001)

Freshwater withdrawal (domestic/industrial/agricultural): *total:* 0.17 cu km/yr (53%/2%/45%)
per capita: 28 cu m/yr (2000)
Natural hazards: hot, dry harmattan wind can reduce visibility in north during winter; periodic droughts
Environment—current issues: deforestation attributable to slash-and-burn agriculture and the use of wood for fuel; water pollution presents health hazards and hinders the fishing industry; air pollution increasing in urban areas
Environment—international agreements: *party to:* Biodiversity, Climate Change, Climate Change-Kyoto Protocol, Desertification, Endangered Species, Law of the Sea, Ozone Layer Protection, Ship Pollution, Tropical Timber 83, Tropical Timber 94, Wetlands, Whaling
signed, but not ratified: none of the selected agreements
Geography—note: the country's length allows it to stretch through six distinct geographic regions; climate varies from tropical to savanna

PEOPLE

Population: 5,858,673
note: estimates for this country explicitly take into account the effects of excess mortality due to AIDS; this can result in lower life expectancy, higher infant mortality, higher death rates, lower population growth rates, and changes in the distribution of population by age and sex than would otherwise be expected (July 2008 est.)
Age structure:
0–14 years: 41.7% (male 1,226,320/female 1,218,182)
15–64 years: 55.6% (male 1,588,354/female 1,666,274)
65 years and over: 2.7% (male 63,508/female 96,035) (2008 est.)
Median age:
total: 18.6 years
male: 18.2 years
female: 19 years (2008 est.)
Population growth rate: 2.717% (2008 est.)
Birth rate: 36.66 births/1,000 population (2008 est.)
Death rate: 9.48 deaths/1,000 population (2008 est.)
Net migration rate: NA
Sex ratio:
at birth: 1.03 male(s)/female
under 15 years: 1.01 male(s)/female
15–64 years: 0.95 male(s)/female
65 years and over: 0.66 male(s)/female
total population: 0.97 male(s)/female (2008 est.)
Infant mortality rate:
total: 57.66 deaths/1,000 live births

male: 65.01 deaths/1,000 live births
female: 50.09 deaths/1,000 live births (2008 est.)
Life expectancy at birth:
total population: 58.28 years
male: 56.2 years
female: 60.43 years (2008 est.)
Total fertility rate: 4.85 children born/woman (2008 est.)
HIV/AIDS—adult prevalence rate: 4.1% (2003 est.)
HIV/AIDS—people living with HIV/AIDS: 110,000 (2003 est.)
HIV/AIDS—deaths: 10,000 (2003 est.)
Major infectious diseases:
degree of risk: very high
food or waterborne diseases: bacterial and protozoal diarrhea, hepatitis A, and typhoid fever
vectorborne diseases: malaria and yellow fever
water contact disease: schistosomiasis
respiratory disease: meningococcal meningitis
note: highly pathogenic H5N1 avian influenza has been identified in this country; it poses a negligible risk with extremely rare cases possible among US citizens who have close contact with birds (2008)
Nationality:
noun: Togolese (singular and plural)
adjective: Togolese
Ethnic groups: African (37 tribes; largest and most important are Ewe, Mina, and Kabre) 99%, European and Syrian-Lebanese less than 1%
Religions: Christian 29%, Muslim 20%, indigenous beliefs 51%
Languages: French (official and the language of commerce), Ewe and Mina (the two major African languages in the south), Kabye (sometimes spelled Kabiye) and Dagomba (the two major African languages in the north)
Literacy:
definition: age 15 and over can read and write
total population: 60.9%
male: 75.4%
female: 46.9% (2003 est.)

GOVERNMENT

Country name:
conventional long form: Togolese Republic
conventional short form: Togo
local long form: Republique togolaise
local short form: none
former: French Togoland
Government type: republic under transition to multiparty democratic rule
Capital: *name:* Lome
geographic coordinates: 6 08 N, 1 13 E
time difference: UTC 0 (5 hours ahead of Washington, DC during Standard Time)

Administrative divisions: 5 regions (regions, singular—region); Centrale, Kara, Maritime, Plateaux, Savanes

Independence: 27 April 1960 (from French-administered UN trusteeship)

National holiday: Independence Day, 27 April (1960)

Constitution: multiparty draft constitution approved by High Council of the Republic 1 July 1992, adopted by public referendum 27 September 1992

Legal system: French-based court system; accepts compulsory ICJ jurisdiction, with reservations

Suffrage: NA years of age; universal (adult)

Executive branch:

chief of state: President Faure GNASSINGBE (since 4 May 2005); note— Gnassingbe EYADEMA died on 5 February 2005 and was succeeded by his son, Faure GNASSINGBE, with the support of the military following international condemnation for the unconstitutional move he then stepped aside pending elections, and Abass BONFOH served as interim president; Faure GNASSINGBE later won popular elections in April 2005

head of government: Prime Minister Komlan MALLY (since 3 December 2007)

cabinet: Council of Ministers appointed by the president and the prime minister

elections: president elected by popular vote for a five-year term (no term limits); election last held 24 April 2005 (next to be held by 2010); prime minister appointed by the president

election results: Faure GNASSINGBE elected president; percent of vote— Faure GNASSINGBE 60.2%, Emmanuel Akitani BOB 38.3%, Nicolas LAWSON 1%, Harry OLYMPIO 0.5%

Legislative branch: unicameral National Assembly (81 seats; members are elected by popular vote to serve five-year terms)

elections: last held on 14 October 2007 (next to be held in 2012)

election results: percent of vote by party— RPT 39.4%, UFC 37.0%, CAR 8.2%, independents 2.5%, other 12.9%; seats by party—RPT 50, UFC 27, CAR 4

Judicial branch: Court of Appeal or Cour d'Appel; Supreme Court or Cour Supreme

Political parties and leaders: Action Committee for Renewal or CAR [Yawovi AGBOYIBO]; Democratic Convention of African Peoples or CDPA; Democratic Party for Renewal or PDR; Juvento [Monsilia DJATO]; Movement of the Believers of Peace and Equality or MOCEP; Pan-African Patriotic Convergence or CPP; Rally for the Support for Development and Democracy or RSDD [Harry OLYMPIO]; Rally of the Togolese People or RPT [Faure GNASSINGBE]; Socialist Pact for Renewal or PSR; Union for Democracy and Social Progress or UDPS [Gagou KOKOU]; Union of Forces for a Change or UFC [Gilchrist OLYMPIO]

Political pressure groups and leaders: NA

International organization participation: ABEDA, ACCT, ACP, AfDB, AU, ECOWAS, Entente, FAO, FZ, G-77, IBRD, ICAO, ICC, ICRM, IDA, IDB, IFAD, IFC, IFRCS, ILO, IMF, IMO, Interpol, IOC, IOM, IPU, ISO (correspondent), ITSO, ITU, ITUC, MIGA, NAM, OIC, OIF, OPCW, PCA, UN, UNCTAD, UNESCO, UNIDO, UNMIL, UNOCI, UNWTO, UPU, WADB (regional), WAEMU, WCL, WCO, WFTU, WHO, WIPO, WMO, WTO

Diplomatic representation in the US:

chief of mission: Ambassador (vacant); Charge d'Affaires Lorempo LANDJERGUE

chancery: 2208 Massachusetts Avenue NW, Washington, DC 20008

telephone: [1] (202) 234-4212

FAX: [1] (202) 232-3190

Diplomatic representation from the US:

chief of mission: Ambassador David B. DUNN

embassy: 4332 Blvd. Gnassingbe Eyadema, Cite OUA, Lome

mailing address: B. P. 852, Lome

telephone: [228] 261-5470

FAX: [228] 261-5501

Flag description: five equal horizontal bands of green (top and bottom) alternating with yellow; a white five-pointed star on a red square is in the upper hoist-side corner; uses the popular pan-African colors of Ethiopia

ECONOMY

Economy—overview: This small, sub-Saharan economy is heavily dependent on both commercial and subsistence agriculture, which provides employment for 65% of the labor force. Some basic foodstuffs must still be imported. Cocoa, coffee, and cotton generate about 40% of export earnings with cotton being the most important cash crop. Togo is the world's fourth-largest producer of phosphate. The government's decade-long effort, supported by the World Bank and the IMF, to implement economic reform measures, encourage foreign investment, and bring revenues in line with expenditures has moved slowly. Progress depends on follow through on privatization, increased openness in government financial operations, progress toward legislative elections, and continued support from foreign donors. Togo is working with donors to write a Poverty Reduction and Growth Facility (PRGF) that could eventually lead to a debt reduction plan. Economic growth remains marginal due to declining cotton production, underinvestment in phosphate mining, and strained relations with donors.

GDP (purchasing power parity): $5.208 billion (2007 est.)

GDP (official exchange rate): $2.497 billion (2007 est.)

GDP—real growth rate: 2.1% (2007 est.)

GDP—per capita (PPP): $800 (2007 est.)

GDP—composition by sector:
agriculture: 40%
industry: 25%
services: 35% (2003 est.)

Labor force: 1.302 million (1998)

Labor force—by occupation: *agriculture:* 65%
industry: 5%
services: 30% (1998 est.)

Unemployment rate: NA%

Population below poverty line: 32% (1989 est.)

Household income or consumption by percentage share:
lowest 10%: NA%
highest 10%: NA%

Inflation rate (consumer prices): 1% (2007 est.)

Investment (gross fixed): 24.2% of GDP (2007 est.)

Budget:
revenues: $481.5 million
expenditures: $427.7 million (2007 est.)

Agriculture—products: coffee, cocoa, cotton, yams, cassava (tapioca), corn, beans, rice, millet, sorghum; livestock; fish

Industries: phosphate mining, agricultural processing, cement, handicrafts, textiles, beverages

Industrial production growth rate: 3% (2007 est.)

Electricity—production: 176 million kWh (2005)

Electricity—production by source:
fossil fuel: 98.7%
hydro: 1.3%
nuclear: 0%
other: 0% (2001)

Electricity—consumption: 576 million kWh (2005)

Electricity—exports: 0 kWh (2005)

Electricity—imports: 486 million kWh; note—electricity supplied by Ghana (2005)

Oil—production: 0 bbl/day (2005 est.)

Oil—consumption: 16,000 bbl/day (2005 est.)
Oil—exports: 0 bbl/day (2004)
Oil—imports: 15,130 bbl/day (2004)
Oil—proved reserves: 0 bbl (1 January 2006 est.)
Natural gas—production: 0 cu m (2005 est.)
Natural gas—consumption: 0 cu m (2005 est.)
Natural gas—exports: 0 cu m (2005 est.)
Natural gas—imports: 0 cu m (2005)
Natural gas—proved reserves: 0 cu m (1 January 2006 est.)
Current account balance: -$160 million (2007 est.)
Exports: $690 million f.o.b. (2007 est.)
Exports—commodities: reexports, cotton, phosphates, coffee, cocoa
Exports—partners: Ghana 16.7%, Burkina Faso 14.4%, Benin 9.1%, Belgium 6.1%, Mali 5.8%, Germany 5.4%, India 4.6%, Netherlands 4.6% (2006)
Imports: $1.208 billion f.o.b. (2007 est.)
Imports—commodities: machinery and equipment, foodstuffs, petroleum products
Imports—partners: China 29.8%, UK 10.9%, France 8.9%, Netherlands 6%, Belgium 5.8%, US 4.6%, Estonia 4.2% (2006)
Economic aid—recipient: ODA, $86.71 million (2005 est.)
Reserves of foreign exchange and gold: $356 million (31 December 2007 est.)
Debt—external: $2 billion (2005)
Market value of publicly traded shares: $NA
Currency (code): Communaute Financiere Africaine franc (XOF); note—responsible authority is the Central Bank of the West African States
Currency code: XOF
Exchange rates: Communaute Financiere Africaine francs (XOF) per US dollar—482.71 (2007), 522.59

(2006), 527.47 (2005), 528.29 (2004), 581.2 (2003)
note: since 1 January 1999, the XOF franc has been pegged to the euro at a rate of 655.957 XOF francs per euro
Fiscal year: calendar year

COMMUNICATIONS

Telephones—main lines in use: 82,100 (2006)
Telephones—mobile cellular: 708,000 (2006)
Telephone system:
general assessment: fair system based on a network of microwave radio relay routes supplemented by open-wire lines and a mobile-cellular system
domestic: microwave radio relay and open-wire lines for conventional system; combined fixed-line and mobile-cellular teledensity roughly 15 telephones per 100 persons
international: country code—228; satellite earth stations—1 Intelsat (Atlantic Ocean), 1 Symphonie
Radio broadcast stations: AM 2, FM 9, shortwave 4 (1998)
Radios: 940,000 (1997)
Television broadcast stations: 3 (plus 2 repeaters) (1997)
Televisions: 73,000 (1997)
Internet country code: .tg
Internet hosts: 702 (2007)
Internet Service Providers (ISPs): 3 (2001)
Internet users: 320,000 (2006)

TRANSPORTATION

Airports: 9 (2007)
Airports—with paved runways:
total: 2
2,438 to 3,047 m: 2 (2007)
Airports—with unpaved runways:
total: 7
914 to 1,523 m: 4
under 914 m: 3 (2007)
Railways:
total: 568 km
narrow gauge: 568 km 1.000-m gauge (2006)

Roadways:
total: 7,520 km
paved: 2,376 km
unpaved: 5,144 km (1999)
Waterways: 50 km (seasonally on Mono River depending on rainfall) (2005)
Merchant marine:
total: 2 ships (1000 GRT or over) 3,918 GRT/3,852 DWT
by type: cargo 1, refrigerated cargo 1 (2007)
Ports and terminals: Kpeme, Lome

MILITARY

Military branches: Togolese Armed Forces: Ground Forces, Togolese Navy (Marine du Togo), Togolese Air Force (Force Aerienne Togolaise, FAT), National Gendarmerie (2008)
Military service age and obligation: 18 years of age for selective compulsory and voluntary military service; 2-year service obligation (2006)
Manpower available for military service:
males age 16–49: 1,365,505
females age 16–49: 1,374,993 (2008 est.)
Manpower fit for military service:
males age 16–49: 897,195
females age 16–49: 913,327 (2008 est.)
Military expenditures—percent of GDP: 1.6% (2005 est.)

TRANSNATIONAL ISSUES

Disputes—international: in 2001, Benin claimed Togo moved boundary monuments—joint commission continues to resurvey the boundary; in 2006 14,000 Togolese refugees remain in Benin and Ghana out of the 40,000 who fled there in 2005
Refugees and internally displaced persons: *refugees (country of origin):* 5,000 (Ghana)
IDPs: 1,500 (2007)
Illicit drugs: transit hub for Nigerian heroin and cocaine traffickers; money laundering not a significant problem

TOKELAU

INTRODUCTION

Background: Originally settled by Polynesian emigrants from surrounding island groups, the Tokelau Islands were made a British protectorate in 1889. They were transferred to New Zealand administration in 1925.

GEOGRAPHY

Location: Oceania, group of three atolls in the South Pacific Ocean, about one-

half of the way from Hawaii to New Zealand
Geographic coordinates: 9 00 S, 172 00 W
Map references: Oceania
Area:
total: 10 sq km
land: 10 sq km
water: 0 sq km
Area—comparative: about 17 times the size of The Mall in Washington, DC
Land boundaries: 0 km

Coastline: 101 km
Maritime claims:
territorial sea: 12 nm
exclusive economic zone: 200 nm
Climate: tropical; moderated by trade winds (April to November)
Terrain: low-lying coral atolls enclosing large lagoons
Elevation extremes:
lowest point: Pacific Ocean 0 m
highest point: unnamed location 5 m
Natural resources: NEGL

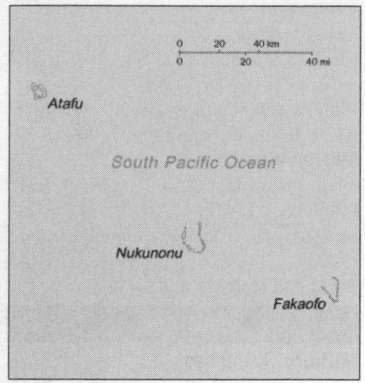

Land use:
arable land: 0% (soil is thin and infertile)
permanent crops: 0%
other: 100% (2005)
Irrigated land: NA
Natural hazards: lies in Pacific typhoon belt
Environment—current issues: limited natural resources and overcrowding are contributing to emigration to New Zealand
Geography—note: consists of three atolls (Atafu, Fakaofo, Nukunonu), each with a lagoon surrounded by a number of reef-bound islets of varying length and rising to over 3 m above sea level

PEOPLE

Population: 1,433 (July 2008 est.)
Age structure:
0–14 years: 42%
15–64 years: 53%
65 years and over: 5%
Population growth rate: -0.011% (2008 est.)
Birth rate: NA
Death rate: NA (2008 est.)
Net migration rate: NA
Sex ratio: NA
Infant mortality rate:
total: NA
male: NA
female: NA (2008 est.)
Life expectancy at birth:
total population: NA
male: NA
female: NA (2008 est.)
Total fertility rate: NA (2008 est.)
HIV/AIDS—adult prevalence rate: NA
HIV/AIDS—people living with HIV/AIDS: NA
HIV/AIDS—deaths: NA
Nationality:
noun: Tokelauan(s)
adjective: Tokelauan
Ethnic groups: Polynesian
Religions: Congregational Christian Church 70%, Roman Catholic 28%, other 2%

note: on Atafu, all Congregational Christian Church of Samoa; on Nukunonu, all Roman Catholic; on Fakaofo, both denominations, with the Congregational Christian Church predominant
Languages: Tokelauan (a Polynesian language), English
Literacy: NA

GOVERNMENT

Country name:
conventional long form: none
conventional short form: Tokelau
Dependency status: self-administering territory of New Zealand; note—Tokelau and New Zealand have agreed to a draft constitution as Tokelau moves toward free association with New Zealand; a UN sponsored referendum on self governance in October 2007 did not produce the two-thirds majority vote necessary for changing the political status
Government type: NA
Capital: none; each atoll has its own administrative center
time difference: UTC-11 (6 hours behind Washington, DC during Standard Time)
Administrative divisions: none (territory of New Zealand)
Independence: none (territory of New Zealand)
National holiday: Waitangi Day (Treaty of Waitangi established British sovereignty over New Zealand), 6 February (1840)
Constitution: administered under the Tokelau Islands Act of 1948; amended in 1970
Legal system: New Zealand and local statutes
Suffrage: 21 years of age; universal
Executive branch:
chief of state: Queen ELIZABETH II (since 6 February 1952); represented by Governor General of New Zealand Anand SATYANAND (since 23 August 2006); New Zealand is represented by Administrator David PAYTON (since 17 October 2006)
head of government: Pio TUIA (since 23 February 2008); note—position rotates annually among the three Faipule (village leaders)
cabinet: the Council for the Ongoing Government of Tokelau, consisting of three Faipule (village leaders) and three Pulenuku (village mayors), functions as a cabinet
elections: the monarch is hereditary; administrator appointed by the Minister of Foreign Affairs and Trade in New Zealand; the head of government is chosen from the Council of Faipule and serves a one-year term

Legislative branch: unicameral General Fono (20 seats; based upon proportional representation from the three islands elected by popular vote to serve three-year terms; Atafu has seven seats, Fakaofo has seven seats, Nukunonu has six seats); note—the Tokelau Amendment Act of 1996 confers limited legislative power on the General Fono
elections: last held 17–19 January 2008 (next to be held in 2011)
election results: independents 20
Judicial branch: Supreme Court in New Zealand exercises civil and criminal jurisdiction in Tokelau
Political parties and leaders: none
Political pressure groups and leaders: none
International organization participation: PIF (observer), SPC, UNESCO (associate), UPU
Diplomatic representation in the US: none (territory of New Zealand)
Diplomatic representation from the US: none (territory of New Zealand)
Flag description: the flag of New Zealand is used

ECONOMY

Economy—overview: Tokelau's small size (three villages), isolation, and lack of resources greatly restrain economic development and confine agriculture to the subsistence level. The people rely heavily on aid from New Zealand—about $4 million annually—to maintain public services with annual aid being substantially greater than GDP. The principal sources of revenue come from sales of copra, postage stamps, souvenir coins, and handicrafts. Money is also remitted to families from relatives in New Zealand.
GDP (purchasing power parity): $1.5 million (1993 est.)
GDP (official exchange rate): $NA
GDP—real growth rate: NA%
GDP—per capita (PPP): $1,000 (1993 est.)
GDP—composition by sector:
agriculture: NA%
industry: NA%
services: NA%
Labor force: 440 (2001)
Unemployment rate: NA%
Population below poverty line: NA%
Inflation rate (consumer prices): NA%
Budget:
revenues: $430,800
expenditures: $2.8 million (1987 est.)
Agriculture—products: coconuts, copra, breadfruit, papayas, bananas; pigs, poultry, goats; fish
Industries: small-scale enterprises for copra production, woodworking, plaited craft goods; stamps, coins; fishing

Electricity—production: NA kWh
Electricity—production by source:
fossil fuel: 100%
hydro: 0%
nuclear: 0%
other: 0% (2001)
Electricity—consumption: NA kWh
Exports: $0 (2002)
Exports—commodities: stamps, copra, handicrafts
Exports—partners: New Zealand (2006)
Imports: $969,200 c.i.f. (2002)
Imports—commodities: foodstuffs, building materials, fuel
Imports—partners: New Zealand (2006)
Currency (code): New Zealand dollar (NZD)
Currency code: NZD
Exchange rates: New Zealand dollars per US dollar—1.3811 (2007), 1.5408 (2006), 1.4203 (2005), 1.5087 (2004), 1.7221 (2003)
Fiscal year: 1 April—31 March

COMMUNICATIONS

Telephones—main lines in use: 300 (2002)
Telephone system:
general assessment: modern satellite-based communications system
domestic: radiotelephone service between islands
international: country code—690; radiotelephone service to Samoa; government-regulated telephone service (TeleTok); satellite earth stations—3
Radio broadcast stations: AM NA, FM NA, shortwave NA (one radio station provides service to all islands) (2002)
Radios: 1,000 (1997)
Internet country code: .tk
Internet hosts: 249 (2007)
Internet Service Providers (ISPs): 1 (2000)
Internet users: NA

TRANSPORTATION

Ports and terminals: none; offshore anchorage only

MILITARY

Military—note: defense is the responsibility of New Zealand

TRANSNATIONAL ISSUES

Disputes—international: Tokelau included American Samoa's Swains Island (Olohega) in its 2006 draft constitution

TONGA

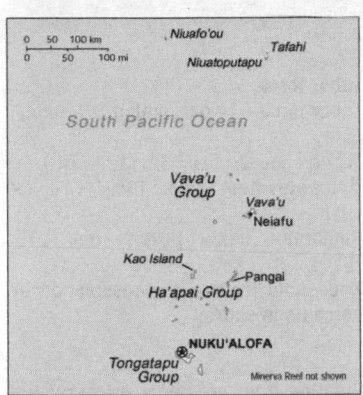

INTRODUCTION

Background: Tonga—unique among Pacific nations—never completely lost its indigenous governance. The archipelagos of "The Friendly Islands" were united into a Polynesian kingdom in 1845. Tonga became a constitutional monarchy in 1875 and a British protectorate in 1900; it withdrew from the protectorate and joined the Commonwealth of Nations in 1970. Tonga remains the only monarchy in the Pacific.

GEOGRAPHY

Location: Oceania, archipelago in the South Pacific Ocean, about two-thirds of the way from Hawaii to New Zealand
Geographic coordinates: 20 00 S, 175 00 W
Map references: Oceania
Area: *total:* 748 sq km
land: 718 sq km
water: 30 sq km

Area—comparative: four times the size of Washington, DC
Land boundaries: 0 km
Coastline: 419 km
Maritime claims:
territorial sea: 12 nm
exclusive economic zone: 200 nm
continental shelf: 200-m depth or to the depth of exploitation
Climate: tropical; modified by trade winds; warm season (December to May), cool season (May to December)
Terrain: most islands have limestone base formed from uplifted coral formation; others have limestone overlying volcanic base
Elevation extremes:
lowest point: Pacific Ocean 0 m
highest point: unnamed location on Kao Island 1,033 m
Natural resources: fish, fertile soil
Land use:
arable land: 20%
permanent crops: 14.67%
other: 65.33% (2005)
Irrigated land: NA
Natural hazards: cyclones (October to April); earthquakes and volcanic activity on Fonuafo'ou
Environment—current issues: deforestation results as more and more land is being cleared for agriculture and settlement; some damage to coral reefs from starfish and indiscriminate coral and shell collectors; overhunting threatens native sea turtle populations
Environment—international agreements: *party to:* Biodiversity, Climate Change, Climate Change-Kyoto Protocol, Desertification, Law of the Sea, Marine Dumping, Marine Life Conservation, Ozone Layer Protection, Ship Pollution
signed, but not ratified: none of the selected agreements
Geography—note: archipelago of 169 islands (36 inhabited)

PEOPLE

Population: 119,009 (July 2008 est.)
Age structure:
0–14 years: 33.7% (male 20,484/female 19,633)
15–64 years: 62% (male 36,699/female 37,108)
65 years and over: 4.3% (male 2,135/female 2,950) (2008 est.)
Median age:
total: 21.8 years
male: 21.3 years
female: 22.3 years (2008 est.)
Population growth rate: 1.669% (2008 est.)
Birth rate: 21.81 births/1,000 population (2008 est.)
Death rate: 5.12 deaths/1,000 population (2008 est.)
Net migration rate: NA
Sex ratio:
at birth: 1.05 male(s)/female
under 15 years: 1.04 male(s)/female
15–64 years: 0.99 male(s)/female
65 years and over: 0.72 male(s)/female
total population: 0.99 male(s)/female (2008 est.)
Infant mortality rate:
total: 11.88 deaths/1,000 live births
male: 13.07 deaths/1,000 live births
female: 10.63 deaths/1,000 live births (2008 est.)

Life expectancy at birth:
total population: 70.44 years
male: 67.9 years
female: 73.1 years (2008 est.)
Total fertility rate: 2.5 children born/woman (2008 est.)
HIV/AIDS—adult prevalence rate: NA
HIV/AIDS—people living with HIV/AIDS: NA
HIV/AIDS—deaths: NA
Nationality:
noun: Tongan(s)
adjective: Tongan
Ethnic groups: Polynesian, Europeans
Religions: Christian (Free Wesleyan Church claims over 30,000 adherents)
Languages: Tongan, English
Literacy:
definition: can read and write Tongan and/or English
total population: 98.9%
male: 98.8%
female: 99% (1999 est.)

GOVERNMENT

Country name:
conventional long form: Kingdom of Tonga
conventional short form: Tonga
local long form: Pule'anga Tonga
local short form: Tonga
former: Friendly Islands
Government type: constitutional monarchy
Capital: *name:* Nuku'alofa
geographic coordinates: 21 08 S, 175 12 W
time difference: UTC+13 (18 hours ahead of Washington, DC during Standard Time)
Administrative divisions: 3 island groups; Ha'apai, Tongatapu, Vava'u
Independence: 4 June 1970 (from UK protectorate)
National holiday: Emancipation Day, 4 June (1970)
Constitution: 4 November 1875; revised 1 January 1967
Legal system: based on English common law; has not accepted compulsory ICJ jurisdiction
Suffrage: 21 years of age; universal
Executive branch:
chief of state: King George TUPOU V (since 11 September 2006)
head of government: Prime Minister Dr. Feleti SEVELE (since 11 February 2006); Deputy Prime Minister Dr. Viliami TANGI (since 16 May 2006)
cabinet: Cabinet consists of 14 members, 10 appointed by the monarch for life; four appointed from among the elected members of the Legislative Assembly, including two each from the nobles' and peoples' representatives serving three-year terms

note: there is also a Privy Council that consists of the monarch, the cabinet, and two governors
elections: the monarch is hereditary; prime minister and deputy prime minister appointed by the monarch
Legislative branch: unicameral Legislative Assembly or Fale Alea (32 seats—14 reserved for cabinet ministers sitting ex officio, nine for nobles selected by the country's 33 nobles, and nine elected by popular vote; members serve three-year terms)
elections: last held on 21 March 2005 (next to be held in 2008)
election results: Peoples Representatives: percent of vote—HRDMT 70%, other 30%; seats—HRDMT 7, independents 2
Judicial branch: Supreme Court (judges are appointed by the monarch); Court of Appeal (Chief Justice and high court justices from overseas chosen and approved by Privy Council)
Political parties and leaders: People's Democratic Party [Tesina FUKO]
Political pressure groups and leaders: Human Rights and Democracy Movement Tonga or HRDMT [Rev. Simote VEA, chairman]; Public Servant's Association [Finau TUTONE]
International organization participation: ACP, ADB, C, FAO, G-77, IBRD, ICAO, ICRM, IDA, IFAD, IFC, IFRCS, IHO, IMF, IMO, Interpol, IOC, ITU, ITUC, OPCW, PIF, Sparteca, SPC, UN, UNCTAD, UNESCO, UNIDO, UPU, WCO, WHO, WIPO, WMO, WTO
Diplomatic representation in the US:
chief of mission: Ambassador Fekitamoeloa 'UTOIKAMANU
chancery: 250 East 51st Street, New York, NY 10022
telephone: [1] (917) 369-1025
FAX: [1] (917) 369-1024
consulate(s) general: San Francisco
Diplomatic representation from the US: the US does not have an embassy in Tonga; the ambassador to Fiji is accredited to Tonga
Flag description: red with a bold red cross on a white rectangle in the upper hoist-side corner

ECONOMY

Economy—overview: Tonga has a small, open, South Pacific island economy. It has a narrow export base in agricultural goods. Squash, vanilla beans, and yams are the main crops, and agricultural exports, including fish, make up two-thirds of total exports. The country must import a high proportion of its food, mainly from New Zealand. The country remains dependent on external aid and remittances from Tongan communities

overseas to offset its trade deficit. Tourism is the second-largest source of hard currency earnings following remittances. The government is emphasizing the development of the private sector, especially the encouragement of investment, and is committing increased funds for health and education. Tonga has a reasonably sound basic infrastructure and well-developed social services. High unemployment among the young, a continuing upturn in inflation, pressures for democratic reform, and rising civil service expenditures are major issues facing the government.
GDP (purchasing power parity): $526 million (2007 est.)
GDP (official exchange rate): $219 million (2007 est.)
GDP—real growth rate: -3.5% (2007 est.)
GDP—per capita (PPP): $5,100 (2007 est.)
GDP—composition by sector:
agriculture: 25%
industry: 17%
services: 57% (FY05/06 est.)
Labor force: 33,910 (2003)
Labor force—by occupation: *agriculture:* 65%
industry and services: 35% (1997 est.)
Unemployment rate: 13% (FY03/04 est.)
Population below poverty line: 24% (FY03/04)
Household income or consumption by percentage share:
lowest 10%: NA%
highest 10%: NA%
Inflation rate (consumer prices): 5.9% (2007 est.)
Budget:
revenues: $80.48 million
expenditures: $109.8 million (FY07/08)
Agriculture—products: squash, coconuts, copra, bananas, vanilla beans, cocoa, coffee, ginger, black pepper; fish
Industries: tourism, construction, fishing
Industrial production growth rate: 1% (2003 est.)
Electricity—production: 54 million kWh (2006)
Electricity—production by source:
fossil fuel: 100%
hydro: 0%
nuclear: 0%
other: 0% (2001)
Electricity—consumption: 47 million kWh (2006)
Electricity—exports: 0 kWh (2005)
Electricity—imports: 0 kWh (2005)
Oil—production: 0 bbl/day (2007 est.)
Oil—consumption: 880 bbl/day (2005 est.)
Oil—exports: 0 bbl/day (2007 est.)

Oil—imports: 842.3 bbl/day (2004)
Oil—proved reserves: 0 bbl (1 January 2007 est.)
Natural gas—production: 0 cu m (2007 est.)
Natural gas—consumption: 0 cu m (2007 est.)
Natural gas—exports: 0 cu m (2007 est.)
Natural gas—imports: 0 cu m (2007)
Natural gas—proved reserves: 0 cu m (1 January 2007 est.)
Current account balance: -$23 million (FY06/07)
Exports: $22 million f.o.b. (2006)
Exports—commodities: squash, fish, vanilla beans, root crops
Exports—partners: US 39.7%, Japan 27.8%, NZ 8.2%, South Korea 7.6% (2006)
Imports: $139 million f.o.b. (2006)
Imports—commodities: foodstuffs, machinery and transport equipment, fuels, chemicals
Imports—partners: Fiji 30.3%, NZ 27.7%, US 8.2%, Australia 7.5%, France 5.7%, UK 4.7% (2006)
Economic aid—recipient: $31.75 million (2005)
Reserves of foreign exchange and gold: $40.83 million (yearend)
Debt—external: $80.7 million (2004)
Market value of publicly traded shares: $NA
Currency (code): pa'anga (TOP)
Currency code: TOP
Exchange rates: pa'anga per US dollar—NA (2007), 2.0277 (2006), 1.96 (2005), 1.9716 (2004), 2.142 (2003)
Fiscal year: 1 July—30 June

COMMUNICATIONS

Telephones—main lines in use: 13,700 (2005)
Telephones—mobile cellular: 29,900 (2005)
Telephone system:
general assessment: competition between Tonga Telecommunications Corporation (TCC) and Shoreline Communications Tonga (SCT) is accelerating expansion of telecommunications; SCT recently granted authority to develop high-speed digital service for telephone, Internet, and television
domestic: combined fixed-line and mobile-cellular teledensity roughly 40 telephones per 100 persons; fully automatic switched network
international: country code—676; satellite earth station—1 Intelsat (Pacific Ocean) (2004)
Radio broadcast stations: AM 1, FM 4, shortwave 1 (2001)
Radios: 61,000 (1997)
Television broadcast stations: 3 (2004)
Televisions: 2,000 (1997)
Internet country code: .to
Internet hosts: 18,653 (2007)
Internet Service Providers (ISPs): 2 (2000)
Internet users: 3,100 (2006)

TRANSPORTATION

Airports: 6 (2007)
Airports—with paved runways:
total: 1
2,438 to 3,047 m: 1 (2007)
Airports—with unpaved runways:
total: 5
1,524 to 2,437 m: 1

914 to 1,523 m: 3
under 914 m: 1 (2007)
Roadways:
total: 680 km
paved: 184 km
unpaved: 496 km (1999)
Merchant marine:
total: 14 ships (1000 GRT or over) 58,756 GRT/67,889 DWT
by type: bulk carrier 1, cargo 9, liquefied gas 1, livestock carrier 1, passenger/cargo 1, refrigerated cargo 1
foreign-owned: 3 (Australia 1, Switzerland 1, UK 1) (2007)
Ports and terminals: Nuku'alofa

MILITARY

Military branches: Tonga Defense Services (TDS): Land Force (Royal Guard), Naval Force (includes Royal Marines, Air Wing) (2008)
Military service age and obligation: 18 years of age (est.); no conscription (2008)
Manpower available for military service:
males age 16–49: 32,053
females age 16–49: 30,981 (2008 est.)
Manpower fit for military service:
males age 16–49: 25,520
females age 16–49: 26,893 (2008 est.)
Manpower reaching militarily significant age annually:
males age 16–49: 1,464
females age 16–49: 1,412 (2008 est.)
Military expenditures—percent of GDP: 0.9% (2006 est.)

TRANSNATIONAL ISSUES

Disputes—international: none

TRINIDAD AND TOBAGO

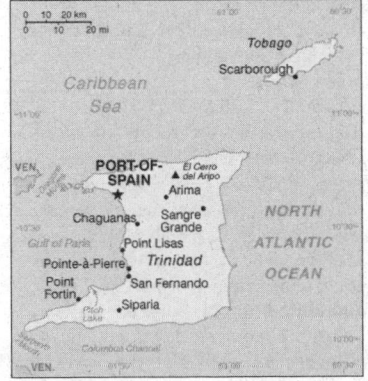

INTRODUCTION

Background: First colonized by the Spanish, the islands came under British control in the early 19th century. The islands' sugar industry was hurt by the emancipation of the slaves in 1834. Manpower was replaced with the importation of contract laborers from India between 1845 and 1917, which boosted sugar production as well as the cocoa industry. The discovery of oil on Trinidad in 1910 added another important export. Independence was attained in 1962. The country is one of the most prosperous in the Caribbean thanks largely to petroleum and natural gas production and processing. Tourism, mostly in Tobago, is targeted for expansion and is growing. The government is coping with a rise in violent crime.

GEOGRAPHY

Location: Caribbean, islands between the Caribbean Sea and the North Atlantic Ocean, northeast of Venezuela
Geographic coordinates: 11 00 N, 61 00 W
Map references: Central America and the Caribbean
Area:
total: 5,128 sq km
land: 5,128 sq km
water: 0 sq km
Area—comparative: slightly smaller than Delaware
Land boundaries: 0 km
Coastline: 362 km
Maritime claims: measured from claimed archipelagic baselines
territorial sea: 12 nm
contiguous zone: 24 nm
exclusive economic zone: 200 nm

continental shelf: 200 nm or to the outer edge of the continental margin
Climate: tropical; rainy season (June to December)
Terrain: mostly plains with some hills and low mountains
Elevation extremes:
lowest point: Caribbean Sea 0 m
highest point: El Cerro del Aripo 940 m
Natural resources: petroleum, natural gas, asphalt
Land use:
arable land: 14.62%
permanent crops: 9.16%
other: 76.22% (2005)
Irrigated land: 40 sq km (2003)
Total renewable water resources: 3.8 cu km (2000)
Freshwater withdrawal (domestic/industrial/agricultural): *total:* 0.31 cu km/yr (68%/26%/6%)
per capita: 237 cu m/yr (2000)
Natural hazards: outside usual path of hurricanes and other tropical storms
Environment—current issues: water pollution from agricultural chemicals, industrial wastes, and raw sewage; oil pollution of beaches; deforestation; soil erosion
Environment—international agreements: *party to:* Biodiversity, Climate Change, Climate Change-Kyoto Protocol, Desertification, Endangered Species, Hazardous Wastes, Law of the Sea, Marine Dumping, Marine Life Conservation, Ozone Layer Protection, Ship Pollution, Tropical Timber 83, Tropical Timber 94, Wetlands
signed, but not ratified: none of the selected agreements
Geography—note: Pitch Lake, on Trinidad's southwestern coast, is the world's largest natural reservoir of asphalt

PEOPLE

Population: 1,047,366 (July 2008 est.)
Age structure:
0–14 years: 19% (male 102,352/female 96,487)
15–64 years: 71.8% (male 396,352/female 356,080)
65 years and over: 9.2% (male 42,998/female 53,097) (2008 est.)
Median age:
total: 32.3 years
male: 31.9 years
female: 32.8 years (2008 est.)
Population growth rate: -0.891% (2008 est.)
Birth rate: 13.22 births/1,000 population (2008 est.)
Death rate: 10.93 deaths/1,000 population (2008 est.)
Net migration rate: -11.2 migrant(s)/1,000 population (2008 est.)

Sex ratio:
at birth: 1.04 male(s)/female
under 15 years: 1.06 male(s)/female
15–64 years: 1.11 male(s)/female
65 years and over: 0.81 male(s)/female
total population: 1.07 male(s)/female (2008 est.)
Infant mortality rate:
total: 23.59 deaths/1,000 live births
male: 25.34 deaths/1,000 live births
female: 21.76 deaths/1,000 live births (2008 est.)
Life expectancy at birth:
total population: 67 years
male: 66.07 years
female: 67.98 years (2008 est.)
Total fertility rate: 1.73 children born/woman (2008 est.)
HIV/AIDS—adult prevalence rate: 3.2% (2003 est.)
HIV/AIDS—people living with HIV/AIDS: 29,000 (2003 est.)
HIV/AIDS—deaths: 1,900 (2003 est.)
Nationality:
noun: Trinidadian(s), Tobagonian(s)
adjective: Trinidadian, Tobagonian
Ethnic groups: Indian (South Asian) 40%, African 37.5%, mixed 20.5%, other 1.2%, unspecified 0.8% (2000 census)
Religions: Roman Catholic 26%, Hindu 22.5%, Anglican 7.8%, Baptist 7.2%, Pentecostal 6.8%, Muslim 5.8%, Seventh Day Adventist 4%, other Christian 5.8%, other 10.8%, unspecified 1.4%, none 1.9% (2000 census)
Languages: English (official), Caribbean Hindustani (a dialect of Hindi), French, Spanish, Chinese
Literacy:
definition: age 15 and over can read and write
total population: 98.6%
male: 99.1%
female: 98% (2003 est.)

GOVERNMENT

Country name:
conventional long form: Republic of Trinidad and Tobago
conventional short form: Trinidad and Tobago
Government type: parliamentary democracy
Capital: *name:* Port-of-Spain
geographic coordinates: 10 39 N, 61 31 W
time difference: UTC-4 (1 hour ahead of Washington, DC during Standard Time)
Administrative divisions: 9 regional corporations, 2 city corporations, 3 borough corporations, 1 ward
regional corporations: Couva/Tabaquite/Talparo, Diego Martin, Mayaro/Rio Claro, Penal/Debe, Princes Town, Sangre Grande, San Juan/Laventille,

Siparia, Tunapuna/Piarco
city corporations: Port-of-Spain, San Fernando
borough corporations: Arima, Chaguanas, Point Fortin
ward: Tobago
Independence: 31 August 1962 (from UK)
National holiday: Independence Day, 31 August (1962)
Constitution: 1 August 1976
Legal system: based on English common law; judicial review of legislative acts in the Supreme Court; has not accepted compulsory ICJ jurisdiction
Suffrage: 18 years of age; universal
Executive branch:
chief of state: President George Maxwell RICHARDS (since 17 March 2003)
head of government: Prime Minister Patrick MANNING (since 24 December 2001)
cabinet: Cabinet appointed from among the members of Parliament
elections: president elected by an electoral college, which consists of the members of the Senate and House of Representatives, for a five-year term (eligible for a second term); election last held on 11 February 2008 (next to be held by February 2013); the president usually appoints as prime minister the leader of the majority party in the House of Representatives
election results: George Maxwell RICHARDS reelected president; percent of electoral college vote—NA
Legislative branch: bicameral Parliament consists of the Senate (31 seats; 16 members appointed by the ruling party, nine by the President, six by the opposition party to serve a maximum term of five years) and the House of Representatives (41 seats; members are elected by popular vote to serve five-year terms)
elections: House of Representatives—last held on 5 November 2007 (next to be held in 2012)
election results: House of Representatives—percent of vote—PNM 46%, UNC 29.7%; seats by party—PNM 26, UNC 15
note: Tobago has a unicameral House of Assembly with 12 members serving four-year terms; last election held in January 2005; seats by party—PNM 11, DAC 1
Judicial branch: Supreme Court of Judicature (comprised of the High Court of Justice and the Court of Appeals; the chief justice is appointed by the president after consultation with the prime minister and the leader of the opposition; other justices are appointed by the president on the advice of the Judicial

and Legal Service Commission); High Court of Justice; Caribbean Court of Appeals member; Court of Appeals; the highest court of appeal is the Privy Council in London

Political parties and leaders: Congress of the People [Winston DOOKERAN]; Democratic Action Congress or DAC [Hochoy CHARLES] (only active in Tobago); Democratic National Alliance or DNA [Gerald YETMING] (coalition of NAR, DDPT, MND); Movement for National Development or MND [Garvin NICHOLAS]; National Alliance for Reconstruction or NAR [Dr. Carson CHARLES]; People's National Movement or PNM [Patrick MAN-NING]; United National Congress or UNC [Basdeo PANDAY]

Political pressure groups and leaders: Jamaat-al Muslimeen [Yasin BAKR]

International organization participation: ACP, C, Caricom, CDB, FAO, G-24, G-77, IADB, IBRD, ICAO, ICCt, ICRM, IDA, IFAD, IFC, IFRCS, IHO, ILO, IMF, IMO, Interpol, IOC, ISO, ITSO, ITU, ITUC, LAES, MIGA, NAM, OAS, OPANAL, OPCW, UN, UNCTAD, UNESCO, UNIDO, UPU, WCL, WCO, WFTU, WHO, WIPO, WMO, WTO

Diplomatic representation in the US:
chief of mission: Ambassador Marina Annette VALERE
chancery: 1708 Massachusetts Avenue NW, Washington, DC 20036
telephone: [1] (202) 467-6490
FAX: [1] (202) 785-3130
consulate(s) general: Miami, New York

Diplomatic representation from the US:
chief of mission: Ambassador Roy L. AUSTIN
embassy: 15 Queen's Park West, Port-of-Spain
mailing address: P. O. Box 752, Port-of-Spain
telephone: [1] (868) 622-6371 through 6376
FAX: [1] (868) 628-5462

Flag description: red with a white-edged black diagonal band from the upper hoist side to the lower fly side

ECONOMY

Economy—overview: Trinidad and Tobago has earned a reputation as an excellent investment site for international businesses and has one of the highest growth rates and per capita incomes in Latin America. Recent growth has been fueled by investments in liquefied natural gas (LNG), petrochemicals, and steel. Additional petrochemical, aluminum, and plastics projects are in various stages of planning. Trinidad and Tobago is the leading Caribbean producer of oil and gas, and its economy is heavily dependent upon these resources but it also supplies manufactured goods, notably food and beverages, as well as cement to the Caribbean region. Oil and gas account for about 40% of GDP and 80% of exports, but only 5% of employment. The country is also a regional financial center, and tourism is a growing sector, although it is not proportionately as important as in many other Caribbean islands. The economy benefits from a growing trade surplus. Economic growth reached 12.6% in 2006 and 5.5% in 2007 as prices for oil, petrochemicals, and LNG remained high, and as foreign direct investment continued to grow to support expanded capacity in the energy sector.

GDP (purchasing power parity): $23.79 billion (2007 est.)

GDP (official exchange rate): $20.7 billion (2007 est.)

GDP—real growth rate: 5.5% (2007 est.)

GDP—per capita (PPP): $18,300 (2007 est.)

GDP—composition by sector:
agriculture: 0.6%
industry: 61.9%
services: 37.5% (2007 est.)

Labor force: 615,000 (2007 est.)

Labor force—by occupation: agriculture 4%, manufacturing, mining, and quarrying 12.9%, construction and utilities 17.5%, services 65.6% (2006 est.)

Unemployment rate: 6.5% (2007 est.)

Population below poverty line: 17% (2007 est.)

Household income or consumption by percentage share:
lowest 10%: NA%
highest 10%: NA%

Inflation rate (consumer prices): 7.9% (2007 est.)

Investment (gross fixed): 23.5% of GDP (2007 est.)

Budget:
revenues: $6.415 billion
expenditures: $6.214 billion (2007 est.)

Public debt: 26.6% of GDP (2007 est.)

Agriculture—products: cocoa, rice, citrus, coffee, vegetables; poultry

Industries: petroleum, chemicals, tourism, food processing, cement, beverage, cotton textiles

Industrial production growth rate: 8% (2007 est.)

Electricity—production: 7.704 billion kWh (2007)

Electricity—production by source:
fossil fuel: 99.8%
hydro: 0%
nuclear: 0%
other: 0.2% (2001)

Electricity—consumption: 7.083 billion kWh (2007)

Electricity—exports: 0 kWh (2005)

Electricity—imports: 0 kWh (2005)

Oil—production: 120,000 bbl/day (2007 est.)

Oil—consumption: 24,770 bbl/day (2007 est.)

Oil—exports: 202,100 bbl/day (2004)

Oil—imports: 91,780 bbl/day (2004)

Oil—proved reserves: 605.8 million bbl (1 January 2007 est.)

Natural gas—production: 39.92 billion cu m (2007 est.)

Natural gas—consumption: 37.29 billion cu m (2007 est.)

Natural gas—exports: 21.03 billion cu m (2007 est.)

Natural gas—imports: 0 cu m (2007)

Natural gas—proved reserves: 702.8 billion cu m (1 January 2007 est.)

Current account balance: $4.171 billion (2007 est.)

Exports: $12.02 billion f.o.b. (2007 est.)

Exports—commodities: petroleum and petroleum products, liquefied natural gas (LNG), methanol, ammonia, urea, steel products, beverages, cereal and cereal products, sugar, cocoa, coffee, citrus fruit, vegetables, flowers

Exports—partners: US 59.8%, Spain 5.3%, Jamaica 5.2% (2006)

Imports: $7.485 billion f.o.b. (2007 est.)

Imports—commodities: mineral fuels, lubricants, machinery, transportation equipment, manufactured goods, food, live animals, grain

Imports—partners: US 30.6%, Brazil 12%, Venezuela 6.8%, Gabon 4.8%, Colombia 4.6% (2006)

Economic aid—recipient: $200,000 (2007 est.)

Reserves of foreign exchange and gold: $6.761 billion (31 December 2007 est.)

Debt—external: $2.826 billion (31 December 2007 est.)

Stock of direct foreign investment—at home: $12.44 billion (2007)

Stock of direct foreign investment—abroad: $1.419 billion (2007)

Market value of publicly traded shares: $15.57 billion (2006)

Currency (code): Trinidad and Tobago dollar (TTD)

Currency code: TTD

Exchange rates: Trinidad and Tobago dollars per US dollar—6.3275 (2007), 6.3107 (2006), 6.2842 (2005), 6.299 (2004), 6.2951 (2003)

Fiscal year: 1 October—30 September

COMMUNICATIONS

Telephones—main lines in use: 325,500 (2006)

Telephones—mobile cellular: 1.655 million (2006)

Telephone system:
general assessment: excellent international service; good local service
domestic: mobile-cellular teledensity exceeds 150 telephones per 100 persons
international: country code—1-868; submarine cable systems provide connectivity to US and parts of the Caribbean and South America; satellite earth station—1 Intelsat (Atlantic Ocean); tropospheric scatter to Barbados and Guyana

Radio broadcast stations: AM 4, FM 18, shortwave 0 (2001)

Radios: 680,000 (1997)

Television broadcast stations: 6 (2005)

Televisions: 425,000 (1997)

Internet country code: .tt

Internet hosts: 24,681 (2007)

Internet Service Providers (ISPs): 17 (2000)

Internet users: 163,000 (2005)

TRANSPORTATION

Airports: 6 (2007)

Airports—with paved runways:
total: 3
over 3,047 m: 1
2,438 to 3,047 m: 1
1,524 to 2,437 m: 1 (2007)

Airports—with unpaved runways:
total: 3
914 to 1,523 m: 1
under 914 m: 2 (2007)

Pipelines: condensate 245 km; gas 1,320 km; oil 563 km (2007)

Roadways:
total: 8,320 km
paved: 4,252 km
unpaved: 4,068 km (1999)

Merchant marine:
total: 9 ships (1000 GRT or over) 27,599 GRT/8,081 DWT
by type: passenger 2, passenger/cargo 5, petroleum tanker 2
foreign-owned: 1 (US 1)
registered in other countries: 1 (Bahamas 1, unknown 1) (2007)

Ports and terminals: Point Fortin, Point Lisas, Port-of-Spain

MILITARY

Military branches: Trinidad and Tobago Defense Force: Ground Force, Coast Guard (includes air wing) (2004)

Military service age and obligation: 18 years of age for voluntary military service (16 years of age with parental consent); no conscription (2008)

Manpower available for military service:
males age 16–49: 301,561

females age 16–49: 264,225 (2008 est.)

Manpower fit for military service:
males age 16–49: 215,310
females age 16–49: 180,526 (2008 est.)

Military expenditures—percent of GDP: 0.3% (2006)

TRANSNATIONAL ISSUES

Disputes—international: in April 2006, the Permanent Court of Arbitration issued a decision that delimited a maritime boundary with Trinidad and Tobago and compelled Barbados to enter a fishing agreement that limited Barbadian fishermen's catches of flying fish in Trinidad and Tobago's exclusive economic zone; in 2005, Barbados and Trinidad and Tobago agreed to compulsory international arbitration under UNCLOS challenging whether the northern limit of Trinidad and Tobago's and Venezuela's maritime boundary extends into Barbadian waters; Guyana has also expressed its intention to include itself in the arbitration as the Trinidad and Tobago-Venezuela maritime boundary may extend into its waters as well

Illicit drugs: transshipment point for South American drugs destined for the US and Europe; producer of cannabis

TUNISIA

INTRODUCTION

Background: Rivalry between French and Italian interests in Tunisia culminated in a French invasion in 1881 and the creation of a protectorate. Agitation for independence in the decades following World War I was finally successful in getting the French to recognize Tunisia as an independent state in 1956. The country's first president, Habib BOURGUIBA, established a strict one-party state. He dominated the country for 31 years, repressing Islamic fundamentalism and establishing rights for women unmatched by any other Arab nation. In November 1987, BOURGUIBA was removed from office and replaced by Zine el Abidine BEN ALI in a bloodless coup. BEN ALI is currently serving his fourth consecutive five-year term as president; the next elections are scheduled for October 2009. Tunisia has long taken a moderate, non-aligned stance in its foreign relations. Domestically, it has sought to defuse rising pressure for a more open political society.

GEOGRAPHY

Location: Northern Africa, bordering the Mediterranean Sea, between Algeria and Libya

Geographic coordinates: 34 00 N, 9 00 E

Map references: Africa

Area:
total: 163,610 sq km
land: 155,360 sq km
water: 8,250 sq km

Area—comparative: slightly larger than Georgia

Land boundaries:
total: 1,424 km
border countries: Algeria 965 km, Libya 459 km

Coastline: 1,148 km

Maritime claims:
territorial sea: 12 nm
contiguous zone: 24 nm
exclusive economic zone: 12 nm

Climate: temperate in north with mild, rainy winters and hot, dry summers; desert in south

Terrain: mountains in north; hot, dry central plain; semiarid south merges into the Sahara

Elevation extremes:
lowest point: Shatt al Gharsah -17 m
highest point: Jebel ech Chambi 1,544 m

Natural resources: petroleum, phosphates, iron ore, lead, zinc, salt

Land use: *arable land:* 17.05%
permanent crops: 13.08%
other: 69.87% (2005)

Irrigated land: 3,940 sq km (2003)

Total renewable water resources: 4.6 cu km (2003)

Freshwater withdrawal (domestic/industrial/agricultural): *total:* 2.64 cu km/yr (14%/4%/82%)
per capita: 261 cu m/yr (2000)

Natural hazards: NA

Environment—current issues: toxic and hazardous waste disposal is ineffective and poses health risks; water pollution from raw sewage; limited natural fresh water resources; deforestation; overgrazing; soil erosion; desertification

Environment—international agreements: *party to:* Biodiversity, Climate Change, Climate Change-Kyoto Protocol, Desertification, Endangered Species, Environmental Modification, Hazardous Wastes, Law of the Sea, Marine Dumping, Ozone Layer Protection, Ship Pollution, Wetlands
signed, but not ratified: Marine Life Conservation

Geography—note: strategic location in central Mediterranean; Malta and Tunisia are discussing the commercial exploitation of the continental shelf between their countries, particularly for oil exploration

PEOPLE

Population: 10,383,577 (July 2008 est.)
Age structure:
0–14 years: 23.2% (male 1,246,105/female 1,167,379)
15–64 years: 69.7% (male 3,638,062/female 3,595,254)
65 years and over: 7.1% (male 345,590/female 391,187) (2008 est.)
Median age:
total: 28.8 years
male: 28.2 years
female: 29.3 years (2008 est.)
Population growth rate: 0.989% (2008 est.)
Birth rate: 15.5 births/1,000 population (2008 est.)
Death rate: 5.17 deaths/1,000 population (2008 est.)
Net migration rate: -0.44 migrant(s)/1,000 population (2008 est.)
Sex ratio:
at birth: 1.07 male(s)/female
under 15 years: 1.07 male(s)/female
15–64 years: 1.01 male(s)/female
65 years and over: 0.88 male(s)/female

total population: 1.01 male(s)/female (2008 est.)
Infant mortality rate:
total: 23.43 deaths/1,000 live births
male: 25.7 deaths/1,000 live births
female: 20.98 deaths/1,000 live births (2008 est.)
Life expectancy at birth:
total population: 75.56 years
male: 73.79 years
female: 77.46 years (2008 est.)
Total fertility rate: 1.73 children born/woman (2008 est.)
HIV/AIDS—adult prevalence rate: less than 0.1% (2005 est.)
HIV/AIDS—people living with HIV/AIDS: 1,000 (2003 est.)
HIV/AIDS—deaths: fewer than 200 (2003 est.)
Nationality:
noun: Tunisian(s)
adjective: Tunisian
Ethnic groups: Arab 98%, European 1%, Jewish and other 1%
Religions: Muslim 98%, Christian 1%, Jewish and other 1%
Languages: Arabic (official and one of the languages of commerce), French (commerce)
Literacy:
definition: age 15 and over can read and write
total population: 74.3%
male: 83.4%
female: 65.3% (2004 census)

GOVERNMENT

Country name:
conventional long form: Tunisian Republic
conventional short form: Tunisia
local long form: Al Jumhuriyah at Tunisiyah
local short form: Tunis
Government type: republic
Capital: name: Tunis
geographic coordinates: 36 48 N, 10 11 E
time difference: UTC+1 (6 hours ahead of Washington, DC during Standard Time)
daylight saving time: +1hr, begins last Sunday in March; ends last Sunday in October
Administrative divisions: 24 governorates; Ariana (Aryanah), Beja (Bajah), Ben Arous (Bin 'Arus), Bizerte (Banzart), Gabes (Qabis), Gafsa (Qafsah), Jendouba (Jundubah), Kairouan (Al Qayrawan), Kasserine (Al Qasrayn), Kebili (Qibili), Kef (Al Kaf), Mahdia (Al Mahdiyah), Manouba (Manubah), Medenine (Madanin), Monastir (Al Munastir), Nabeul (Nabul), Sfax (Safaqis), Sidi Bou Zid (Sidi Bu Zayd), Siliana (Silyanah), Sousse (Susah), Tataouine (Tatawin), Tozeur (Tawzar), Tunis, Zaghouan (Zaghwan)

Independence: 20 March 1956 (from France)
National holiday: Independence Day, 20 March (1956); also the anniversary of BEN ALI's assumption of the presidency, 7 November (1987)
Constitution: 1 June 1959; amended 1988, 2002
Legal system: based on French civil law system and Islamic law; some judicial review of legislative acts in the Supreme Court in joint session; has not accepted compulsory ICJ jurisdiction
Suffrage: 18 years of age; universal except for active government security forces (including the police and the military), people with mental disabilities, people who have served more than three months in prison (criminal cases only), and people given a suspended sentence of more than six months
Executive branch:
chief of state: President Zine el Abidine BEN ALI (since 7 November 1987)
head of government: Prime Minister Mohamed GHANNOUCHI (since 17 November 1999)
cabinet: Council of Ministers appointed by the president
elections: president elected by popular vote for a five-year term (no term limits); election last held on 24 October 2004 (next to be held in October 2009); prime minister appointed by the president
election results: President Zine El Abidine BEN ALI reelected for a fourth term; percent of vote—Zine El Abidine BEN ALI 94.5%, Mohamed BOUCHIHA 3.8%, Mohamed Ali HALOUANI 1%
Legislative branch: bicameral system consists of the Chamber of Deputies or Majlis al-Nuwaab (189 seats; members elected by popular vote to serve five-year terms) and the Chamber of Advisors (126 seats; 85 members elected by municipal counselors, deputies, mayors, and professional associations and trade unions; 41 members are presidential appointees; members serve six-year terms)
elections: Chamber of Deputies—last held on 24 October 2004 (next to be held in October 2009); Chamber of Advisors—last held on 3 July 2005 (next to be held in July 2011)
election results: Chamber of Deputies—percent of vote by party—NA; seats by party—RCD 152, MDS 14, PUP 11, UDU 7, Al-Tajdid 3, PSL 2; Chamber of Advisors—percent of vote by party—NA; seats by party—RCD 71 (14 trade union seats vacant due to boycott))
Judicial branch: Court of Cassation or Cour de Cassation
Political parties and leaders: Al-Tajdid Movement [Ahmed IBRAHIM]; Constitutional Democratic Rally Party

(Rassemblement Constitutionnel Democratique) or RCD (official ruling party) [President Zine El Abidine BEN ALI]; Democratic Forum for Labor and Liberties or FDTL [Mustapha Ben JAFAAR]; Green Party for Progress or PVP [Mongi KHAMASSI]; Liberal Social Party or PSL [Mondher THABET]; Movement of Socialist Democrats or MDS [Ismail BOULAHYA]; Popular Unity Party or PUP [Mohamed BOUCHIHA]; Progressive Democratic Party [Maya JERIBI]; Unionist Democratic Union or UDU [Ahmed INOUBLI]

Political pressure groups and leaders: 18 October Group [collective leadership]; Tunisian League for Human Rights or LTDH [Mokhtar TRIFI]; note—the Islamist Party, Al Nahda (Renaissance), is outlawed

International organization participation: ABEDA, ACCT, AfDB, AFESD, AMF, AMU, AU, BSEC (observer), FAO, G-77, IAEA, IBRD, ICAO, ICC, ICRM, IDA, IDB, IFAD, IFC, IFRCS, IHO, ILO, IMF, IMO, IMSO, Interpol, IOC, IOM, IPU, ISO, ITSO, ITU, ITUC, LAS, MIGA, MONUC, NAM, OAPEC (suspended), OAS (observer), OIC, OIF, OPCW, OSCE (partner), UN, UNCTAD, UNESCO, UNHCR, UNIDO, UNMEE, UNWTO, UPU, WCO, WFTU, WHO, WIPO, WMO, WTO

Diplomatic representation in the US:
chief of mission: Ambassador Nejib HACHANA
chancery: 1515 Massachusetts Avenue NW, Washington, DC 20005
telephone: [1] (202) 862-1850
FAX: [1] (202) 862-1858

Diplomatic representation from the US:
chief of mission: Ambassador Robert F. GODEC
embassy: Zone Nord-Est des Berges du Lac Nord de Tunis 1053
mailing address: use embassy street address
telephone: [216] 71 107-000
FAX: [216] 71 107-090

Flag description: red with a white disk in the center bearing a red crescent nearly encircling a red five-pointed star; the crescent and star are traditional symbols of Islam

ECONOMY

Economy—overview: Tunisia has a diverse economy, with important agricultural, mining, tourism, and manufacturing sectors. Governmental control of economic affairs while still heavy has gradually lessened over the past decade with increasing privatization, simplification of the tax structure, and a prudent

approach to debt. Progressive social policies also have helped raise living conditions in Tunisia relative to the region. Real growth, which averaged almost 5% over the past decade, reached 6.3% in 2007 because of development in non-textile manufacturing, a recovery in agricultural production, and strong growth in the services sector. However, Tunisia will need to reach even higher growth levels to create sufficient employment opportunities for an already large number of unemployed as well as the growing population of university graduates. Broader privatization, further liberalization of the investment code to increase foreign investment, improvements in government efficiency, and reduction of the trade deficit are among the challenges ahead.

GDP (purchasing power parity): $77 billion (2007 est.)

GDP (official exchange rate): $35.01 billion (2007 est.)

GDP—real growth rate: 6.3% (2007 est.)

GDP—per capita (PPP): $7,500 (2007 est.)

GDP—composition by sector:
agriculture: 11.6%
industry: 25.7%
services: 62.8% (2007 est.)

Labor force: 3.593 million (2007 est.)

Labor force—by occupation: *agriculture:* 55%
industry: 23%
services: 22% (1995 est.)

Unemployment rate: 14.1% (2007 est.)

Population below poverty line: 7.4% (2005 est.)

Household income or consumption by percentage share:
lowest 10%: 2.3%
highest 10%: 31.5% (2000)

Distribution of family income—Gini index: 40 (2005 est.)

Inflation rate (consumer prices): 3.1% (2007 est.)

Investment (gross fixed): 23.6% of GDP (2007 est.)

Budget:
revenues: $8.466 billion
expenditures: $9.475 billion (2007 est.)

Public debt: 55.4% of GDP (2007 est.)

Agriculture—products: olives, olive oil, grain, tomatoes, citrus fruit, sugar beets, dates, almonds; beef, dairy products

Industries: petroleum, mining (particularly phosphate and iron ore), tourism, textiles, footwear, agribusiness, beverages

Industrial production growth rate: 7.2% (2007 est.)

Electricity—production: 12.85 billion kWh (2005)

Electricity—production by source:

fossil fuel: 99.5%
hydro: 0.5%
nuclear: 0%
other: 0% (2001)

Electricity—consumption: 11.17 billion kWh (2005)

Electricity—exports: 0 kWh (2005)

Electricity—imports: 0 kWh (2005)

Oil—production: 76,900 bbl/day (2005 est.)

Oil—consumption: 90,000 bbl/day (2005 est.)

Oil—exports: 75,060 bbl/day (2004)

Oil—imports: 85,680 bbl/day (2004)

Oil—proved reserves: 1.7 billion bbl (2007 est.)

Natural gas—production: 2.398 billion cu m (2005 est.)

Natural gas—consumption: 4.124 billion cu m (2005 est.)

Natural gas—exports: 0 cu m (2005 est.)

Natural gas—imports: 1.726 billion cu m (2005)

Natural gas—proved reserves: 74.68 billion cu m (1 January 2006 est.)

Current account balance: -$860 million (2007 est.)

Exports: $15.15 billion f.o.b. (2007 est.)

Exports—commodities: clothing, semi-finished goods and textiles, agricultural products, mechanical goods, phosphates and chemicals, hydrocarbons, electrical equipment

Exports—partners: France 29%, Italy 20.3%, Germany 8.5%, Spain 6.1%, Libya 4.9%, US 4% (2006)

Imports: $18.03 billion f.o.b. (2007 est.)

Imports—commodities: textiles, machinery and equipment, hydrocarbons, chemicals, foodstuffs

Imports—partners: France 25.1%, Italy 21.8%, Germany 9.5%, Spain 4.9% (2006)

Economic aid—recipient: $376.5 million (2005)

Reserves of foreign exchange and gold: $7.854 billion (31 December 2007 est.)

Debt—external: $19.27 billion (December 2007)

Stock of direct foreign investment—at home: $21.78 billion (2007 est.)

Stock of direct foreign investment—abroad: $88 million (2007 est.)

Market value of publicly traded shares: $4.446 billion (2006)

Currency (code): Tunisian dinar (TND)

Currency code: TND

Exchange rates: Tunisian dinars per US dollar—1.2776 (2007), 1.331 (2006), 1.2974 (2005), 1.2455 (2004), 1.2885 (2003)

Fiscal year: calendar year

COMMUNICATIONS

Telephones—main lines in use: 1.268 million (2006)
Telephones—mobile cellular: 7.339 million (2006)
Telephone system:
general assessment: above the African average and continuing to be upgraded; key centers are Sfax, Sousse, Bizerte, and Tunis; Internet access available
domestic: in an effort jumpstart expansion of the fixed-line network, the government has awarded a concession to build and operate a VSAT network with international connectivity; competition between the two mobile-cellular service providers has resulted in lower activation and usage charges and a strong surge in subscribership; overall fixed-line and mobile-cellular teledensity is about 85 telephones per 100 persons
international: country code—216; a landing point for the SEA-ME-WE-4 submarine cable system that provides links to Europe, Middle East, and Asia; satellite earth stations—1 Intelsat (Atlantic Ocean) and 1 Arabsat; coaxial cable and microwave radio relay to Algeria and Libya; participant in Medarabtel; 2 international gateway digital switches
Radio broadcast stations: AM 7, FM 38, shortwave 2 (2007)
Radios: 2.06 million (1997)
Television broadcast stations: 26 (plus 76 repeaters) (1995)
Televisions: 920,000 (1997)
Internet country code: .tn
Internet hosts: 1,163 (2007)
Internet Service Providers (ISPs): 1 (2000)
Internet users: 1.295 million (2006)

TRANSPORTATION

Airports: 30 (2007)
Airports—with paved runways:
total: 14
over 3,047 m: 3
2,438 to 3,047 m: 6
1,524 to 2,437 m: 2
914 to 1,523 m: 3 (2007)
Airports—with unpaved runways:
total: 16
1,524 to 2,437 m: 2
914 to 1,523 m: 7
under 914 m: 7 (2007)
Pipelines: gas 2,665 km; oil 1,235 km; refined products 353 km (2007)
Railways:
total: 2,153 km
standard gauge: 471 km 1.435-m gauge
narrow gauge: 1,674 km 1.000-m gauge (65 km electrified)
dual gauge: 8 km 1.435 m and 1.000-m gauges (three rails) (2006)
Roadways:
total: 19,232 km
paved: 12,655 km (includes 262 km of expressways)
unpaved: 6,577 km (2004)

Merchant marine:
total: 8 ships (1000 GRT or over) 130,475 GRT/91,013 DWT
by type: bulk carrier 1, cargo 1, chemical tanker 2, passenger/cargo 4
foreign-owned: 1 (Libya 1) (2007)
Ports and terminals: Bizerte, Gabes, La Goulette, Rades, Sfax, Skhira

MILITARY

Military branches: Army, Navy, Republic of Tunisia Air Force (Al-Quwwat al-Jawwiya al-Jamahiriyah At'tunisia) (2008)
Military service age and obligation: 20 years of age for compulsory military service; conscript service obligation—12 months; 18 years of age for voluntary military service (2007)
Manpower available for military service:
males age 16–49: 2,992,249
females age 16–49: 2,912,819 (2008 est.)
Manpower fit for military service:
males age 16–49: 2,539,962
females age 16–49: 2,465,295 (2008 est.)
Manpower reaching militarily significant age annually:
males age 16–49: 101,794
females age 16–49: 95,198 (2008 est.)
Military expenditures—percent of GDP: 1.4% (2006)

TRANSNATIONAL ISSUES

Disputes—international: none

TURKEY

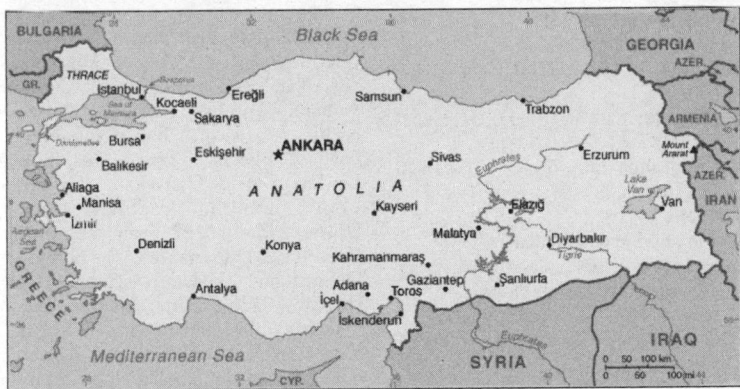

INTRODUCTION

Background: Modern Turkey was founded in 1923 from the Anatolian remnants of the defeated Ottoman Empire by national hero Mustafa KEMAL, who was later honored with the title Ataturk or "Father of the Turks." Under his authoritarian leadership, the country adopted wide-ranging social, legal, and political reforms. After a period of one-party rule, an experiment with multi party politics led to the 1950 election victory of the opposition Democratic Party and the peaceful transfer of power. Since then, Turkish political parties have multiplied, but democracy has been fractured by periods of instability and intermittent military coups (1960, 1971, 1980), which in each case eventually resulted in a return of political power to civilians. In 1997, the military again helped engineer the ouster—popularly dubbed a "postmodern coup"—of the then Islamic-oriented government. Turkey intervened militarily on Cyprus in 1974 to prevent a Greek takeover of the island and has since acted as patron state to the "Turkish Republic of Northern Cyprus," which only Turkey recognizes. A separatist insurgency begun in 1984 by the Kurdistan Workers' Party (PKK)—now known as the People's Congress of Kurdistan or Kongra-Gel (KGK)—has dominated the Turkish military's attention and claimed more than 30,000 lives. After the capture of the group's leader in 1999, the insurgents largely withdrew from Turkey mainly to northern Iraq. In 2004, KGK announced an end to its ceasefire and attacks attributed to the KGK increased. Turkey joined the UN in

1945 and in 1952 it became a member of NATO. In 1964, Turkey became an associate member of the European Community; over the past decade, it has undertaken many reforms to strengthen its democracy and economy enabling it to begin accession membership talks with the European Union.

GEOGRAPHY

Location: Southeastern Europe and Southwestern Asia (that portion of Turkey west of the Bosporus is geographically part of Europe), bordering the Black Sea, between Bulgaria and Georgia, and bordering the Aegean Sea and the Mediterranean Sea, between Greece and Syria

Geographic coordinates: 39 00 N, 35 00 E

Map references: Middle East

Area:
total: 780,580 sq km
land: 770,760 sq km
water: 9,820 sq km

Area—comparative: slightly larger than Texas

Land boundaries:
total: 2,648 km
border countries: Armenia 268 km, Azerbaijan 9 km, Bulgaria 240 km, Georgia 252 km, Greece 206 km, Iran 499 km, Iraq 352 km, Syria 822 km

Coastline: 7,200 km

Maritime claims:
territorial sea: 6 nm in the Aegean Sea; 12 nm in Black Sea and in Mediterranean Sea
exclusive economic zone: in Black Sea only: to the maritime boundary agreed upon with the former USSR

Climate: temperate; hot, dry summers with mild, wet winters; harsher in interior

Terrain: high central plateau (Anatolia); narrow coastal plain; several mountain ranges

Elevation extremes:
lowest point: Mediterranean Sea 0 m
highest point: Mount Ararat 5,166 m

Natural resources: coal, iron ore, copper, chromium, antimony, mercury, gold, barite, borate, celestite (strontium), emery, feldspar, limestone, magnesite, marble, perlite, pumice, pyrites (sulfur), clay, arable land, hydropower

Land use:
arable land: 29.81%
permanent crops: 3.39%
other: 66.8% (2005)

Irrigated land: 52,150 sq km (2003)

Total renewable water resources: 234 cu km (2003)

Freshwater withdrawal (domestic/industrial/agricultural): *total:* 39.78 cu km/yr (15%/11%/74%)

per capita: 544 cu m/yr (2001)

Natural hazards: severe earthquakes, especially in northern Turkey, along an arc extending from the Sea of Marmara to Lake Van

Environment—current issues: water pollution from dumping of chemicals and detergents; air pollution, particularly in urban areas; deforestation; concern for oil spills from increasing Bosporus ship traffic

Environment—international agreements: *party to:* Air Pollution, Antarctic Treaty, Biodiversity, Climate Change, Desertification, Endangered Species, Hazardous Wastes, Ozone Layer Protection, Ship Pollution, Wetlands
signed, but not ratified: Environmental Modification

Geography—note: strategic location controlling the Turkish Straits (Bosporus, Sea of Marmara, Dardanelles) that link Black and Aegean Seas; Mount Ararat, the legendary landing place of Noah's ark, is in the far eastern portion of the country

PEOPLE

Population: 71,892,807 (July 2008 est.)

Age structure:
0–14 years: 24.4% (male 8,937,515/female 8,608,375)
15–64 years: 68.6% (male 25,030,793/female 24,253,312)
65 years and over: 7% (male 2,307,236/female 2,755,576) (2008 est.)

Median age:
total: 29 years
male: 28.8 years
female: 29.2 years (2008 est.)

Population growth rate: 1.013% (2008 est.)

Birth rate: 16.15 births/1,000 population (2008 est.)

Death rate: 6.02 deaths/1,000 population (2008 est.)

Net migration rate: 0 migrant(s)/1,000 population (2008 est.)

Sex ratio:
at birth: 1.05 male(s)/female
under 15 years: 1.04 male(s)/female
15–64 years: 1.03 male(s)/female
65 years and over: 0.84 male(s)/female
total population: 1.02 male(s)/female (2008 est.)

Infant mortality rate:
total: 36.98 deaths/1,000 live births
male: 40.44 deaths/1,000 live births
female: 33.34 deaths/1,000 live births (2008 est.)

Life expectancy at birth:
total population: 73.14 years
male: 70.67 years
female: 75.73 years (2008 est.)

Total fertility rate: 1.87 children born/woman (2008 est.)

HIV/AIDS—adult prevalence rate: less than 0.1%; note—no country specific models provided (2001 est.)

HIV/AIDS—people living with HIV/AIDS: NA

HIV/AIDS—deaths: NA

Nationality:
noun: Turk(s)
adjective: Turkish

Ethnic groups: Turkish 80%, Kurdish 20% (estimated)

Religions: Muslim 99.8% (mostly Sunni), other 0.2% (mostly Christians and Jews)

Languages: Turkish (official), Kurdish, Dimli (or Zaza), Azeri, Kabardian
note: there is also a substantial Gagauz population in the European part of Turkey

Literacy:
definition: age 15 and over can read and write
total population: 87.4%
male: 95.3%
female: 79.6% (2004 est.)

GOVERNMENT

Country name:
conventional long form: Republic of Turkey
conventional short form: Turkey
local long form: Turkiye Cumhuriyeti
local short form: Turkiye

Government type: republican parliamentary democracy

Capital: *name:* Ankara
geographic coordinates: 39 56 N, 32 52 E
time difference: UTC+2 (7 hours ahead of Washington, DC during Standard Time)
daylight saving time: +1hr, begins last Sunday in March; ends last Sunday in October

Administrative divisions: 81 provinces (iller, singular—ili); Adana, Adiyaman, Afyonkarahisar, Agri, Aksaray, Amasya, Ankara, Antalya, Ardahan, Artvin, Aydin, Balikesir, Bartin, Batman, Bayburt, Bilecik, Bingol, Bitlis, Bolu, Burdur, Bursa, Canakkale, Cankiri, Corum, Denizli, Diyarbakir, Duzce, Edirne, Elazig, Erzincan, Erzurum, Eskisehir, Gaziantep, Giresun, Gumushane, Hakkari, Hatay, Icel (Mersin), Igdir, Isparta, Istanbul, Izmir (Smyrna), Kahramanmaras, Karabuk, Karaman, Kars, Kastamonu, Kayseri, Kilis, Kirikkale, Kirklareli, Kirsehir, Kocaeli, Konya, Kutahya, Malatya, Manisa, Mardin, Mugla, Mus, Nevsehir, Nigde, Ordu, Osmaniye, Rize, Sakarya, Samsun, Sanliurfa, Siirt, Sinop, Sirnak, Sivas, Tekirdag, Tokat, Trabzon (Trebizond), Tunceli, Usak, Van, Yalova, Yozgat, Zonguldak

Independence: 29 October 1923 (successor state to the Ottoman Empire)

National holiday: Republic Day, 29 October (1923)

Constitution: 7 November 1982

Legal system: civil law system derived from various European continental legal systems; note—member of the European Court of Human Rights (ECHR), although Turkey claims limited derogations on the ratified European Convention on Human Rights; has not accepted compulsory ICJ jurisdiction

Suffrage: 18 years of age; universal

Executive branch:

chief of state: President Abdullah GUL (since 28 August 2007)

head of government: Prime Minister Recep Tayyip ERDOGAN (since 14 March 2003); Deputy Prime Minister Cemil CICEK (since 29 August 2007); Deputy Prime Minister Hayati YAZICI (since 29 August 2007); Deputy Prime Minister Nazim EKREN (since 29 August 2007)

cabinet: Council of Ministers appointed by the president on the nomination of the prime minister

elections: president elected by the National Assembly for one seven-year terms; prime minister appointed by the president from among members of parliament

election results: Abdullah GUL received 339 votes in the third round of voting on 28 August 2007, after failing to garner the two thirds vote required by law in the first two rounds

note: president-elect must have a two-thirds majority of the National Assembly on the first two ballots and a simple majority on the third ballot

Legislative branch: unicameral Grand National Assembly of Turkey or Turkiye Buyuk Millet Meclisi (550 seats; members are elected by popular vote to serve five-year terms)

elections: last held on 22 July 2007 (next to be held on November 2012)

election results: percent of vote by party— AKP 46.7%, CHP 20.8%, MHP 14.3%, independents 5.2%, and other 13.0%; seats by party—AKP 341, CHP 112, MHP 71, independents 26; note—seats by party as of 17 December 2007—AKP 340, CHP 87, MHP 70, DTP 20, DSP 13, independents 6, other 12, vacant 2 (DTP entered parliament as independents; DSP entered parliament on CHP's party list); only parties surpassing the 10% threshold are entitled to parliamentary seats

Judicial branch: Constitutional Court; High Court of Appeals (Yargitay); Council of State (Danistay); Court of Accounts (Sayistay); Military High Court of Appeals; Military High Administrative Court

Political parties and leaders: Anavatan Partisi (Motherland Party) or Anavatan [Erkan MUMCU]; Democratic Left Party or DSP [Mehmet Zeki SEZER]; Democratic Society Party or DTP [Nurettin DEMIRTAS]; Felicity Party or SP [Recai KUTAN] (sometimes translated as Contentment Party); Justice and Development Party or AKP [Recep Tayyip ERDOGAN]; Nationalist Action Party or MHP [Devlet BAHCELI] (sometimes translated as Nationalist Movement Party); People's Rise Party (Halkin Yukselisi Partisi) or HYP [Yasar Nuri OZTURK]; Republican People's Party or CHP [Deniz BAYKAL]; Social Democratic People's Party or SHP [Murat KARAYALCIN]; True Path Party or DYP [Mehmet AGAR] (sometimes translated as Correct Way Party); Young Party or GP [Cem Cengiz UZAN]

note: the parties listed above are some of the more significant of the 49 parties that Turkey had on 1 December 2004

Political pressure groups and leaders: Confederation of Public Sector Unions or KESK [Ismail Hakki TOMBUL]; Confederation of Revolutionary Workers Unions or DISK [Suleyman CELEBI]; Independent Industrialists' and Businessmen's Association or MUSIAD [Omer BOLAT]; Moral Rights Workers Union or Hak-Is [Salim USLU]; Turkish Confederation of Employers' Unions or TISK [Tugurl KUDATGOBILIK]; Turkish Confederation of Labor or Turk-Is [Salih KILIC]; Turkish Confederation of Tradesmen and Craftsmen or TESK [Dervis GUNDAY]; Turkish Industrialists' and Businessmen's Association or TUSIAD [Omer SABANCI]; Turkish Union of Chambers of Commerce and Commodity Exchanges or TOBB [M. Rifat HISARCIKLIOGLU]

International organization participation: ADB (nonregional members), Australia Group, BIS, BSEC, CE, CERN (observer), EAPC, EBRD, ECO, EU (applicant), FAO, IAEA, IBRD, ICAO, ICC, ICRM, IDA, IDB, IEA, IFAD, IFC, IFRCS, IHO, ILO, IMF, IMO, IMSO, Interpol, IOC, IOM, IPU, ISO, ITSO, ITU, ITUC, MIGA, NATO, NEA, NSG, OAS (observer), OECD, OIC, OPCW, OSCE, PCA, SECI, UN, UNCTAD, UNESCO, UNHCR, UNIDO, UNIFIL, UNMIS, UNOCI, UNOMIG, UNRWA, UNWTO, UPU, WCO, WEU (associate), WFTU, WHO, WIPO, WMO, WTO, ZC

Diplomatic representation in the US:

chief of mission: Ambassador Nabi SENSOY

chancery: 2525 Massachusetts Avenue NW, Washington, DC 20008

telephone: [1] (202) 612-6700

FAX: [1] (202) 612-6744

consulate(s) general: Chicago, Houston, Los Angeles, New York

Diplomatic representation from the US:

chief of mission: Ambassador Ross WILSON

embassy: 110 Ataturk Boulevard, Kavaklidere, 06100 Ankara

mailing address: PSC 93, Box 5000, APO AE 09823

telephone: [90] (312) 455-5555

FAX: [90] (312) 467-0019

consulate(s) general: Istanbul

consulate(s): Adana; note—there is a Consular Agent in Izmir

Flag description: red with a vertical white crescent (the closed portion is toward the hoist side) and white five-pointed star centered just outside the crescent opening

ECONOMY

Economy—overview: Turkey's dynamic economy is a complex mix of modern industry and commerce along with a traditional agriculture sector that still accounts for more than 35% of employment. It has a strong and rapidly growing private sector, yet the state still plays a major role in basic industry, banking, transport, and communication. The largest industrial sector is textiles and clothing, which accounts for one-third of industrial employment; it faces stiff competition in international markets with the end of the global quota system. However, other sectors, notably the automotive and electronics industries, are rising in importance within Turkey's export mix. Real GNP growth has exceeded 6% in many years, but this strong expansion has been interrupted by sharp declines in output in 1994, 1999, and 2001. The economy is turning around with the implementation of economic reforms, and 2004 GDP growth reached 9%, followed by roughly 5% annual growth from 2005–07. Inflation fell to 7.7% in 2005—a 30-year low— but climbed back to 8.5% in 2007. Despite the strong economic gains from 2002–07, which were largely due to renewed investor interest in emerging markets, IMF backing, and tighter fiscal policy, the economy is still burdened by a high current account deficit and high external debt. Further economic and judicial reforms and prospective EU membership are expected to boost foreign direct investment. The stock value of FDI currently stands at about $85 billion. Privatization sales are currently approaching $21 billion. Oil began to flow through the Baku-Tblisi-Ceyhan pipeline in May 2006, marking a major milestone that will bring up to 1 million barrels per day from the Caspian to market. In 2007, Turkish financial markets weathered significant domestic political turmoil, including turbulence sparked by controversy over the selection of former Foreign Minister Abdullah GUL as Turkey's 11th president. Economic fundamentals are sound, marked by strong economic growth and foreign direct investment. Turkey's high current account deficit leaves the economy vulnerable to destabilizing shifts in investor confidence, however.

GDP (purchasing power parity): $888 billion (2007 est.)

GDP (official exchange rate): $663.4 billion (2007 est.)

GDP—real growth rate: 5% (2007 est.)

GDP—per capita (PPP): $12,900 (2007 est.)

GDP—composition by sector:
agriculture: 8.9%
industry: 28.3%
services: 62.8% (2007 est.)

Labor force: 23.53 million
note: about 1.2 million Turks work abroad (2007 est.)

Labor force—by occupation: *agriculture:* 35.9%
industry: 22.8%
services: 41.2% (3rd quarter)

Unemployment rate: 9.9% plus underemployment of 4% (2007 est.)

Population below poverty line: 20% (2002)

Household income or consumption by percentage share:
lowest 10%: 2%
highest 10%: 34.1% (2003)

Distribution of family income—Gini index: 43.6 (2003)

Inflation rate (consumer prices): 8.8% (2007 est.)

Investment (gross fixed): 21.5% of GDP (2007 est.)

Budget:
revenues: $145.5 billion
expenditures: $156.1 billion (2007 est.)

Public debt: 38.9% of GDP (2007 est.)

Agriculture—products: tobacco, cotton, grain, olives, sugar beets, pulse, citrus; livestock

Industries: textiles, food processing, autos, electronics, mining (coal, chromite, copper, boron), steel, petroleum, construction, lumber, paper

Industrial production growth rate: 5.4% (2007 est.)

Electricity—production: 154.2 billion kWh (2005)

Electricity—production by source:
fossil fuel: 79.3%
hydro: 20.4%
nuclear: 0%
other: 0.3% (2001)

Electricity—consumption: 129 billion kWh (2005)

Electricity—exports: 1.798 billion kWh (2005)

Electricity—imports: 636 million kWh (2005)

Oil—production: 45,460 bbl/day (2005 est.)

Oil—consumption: 660,800 bbl/day (2005 est.)

Oil—exports: 112,600 bbl/day (2004)

Oil—imports: 724,400 bbl/day (2004)

Oil—proved reserves: 300 million bbl (1 January 2006 est.)

Natural gas—production: 860.3 million cu m (2005 est.)

Natural gas—consumption: 26.25 billion cu m (2005 est.)

Natural gas—exports: 0 cu m (2005 est.)

Natural gas—imports: 25.48 billion cu m (2005)

Natural gas—proved reserves: 8.147 billion cu m (1 January 2006 est.)

Current account balance: -$38.03 billion (2007 est.)

Exports: $115.3 billion f.o.b. (2007 est.)

Exports—commodities: apparel, foodstuffs, textiles, metal manufactures, transport equipment

Exports—partners: Germany 11.3%, UK 8%, Italy 7.9%, US 6%, France 5.4%, Spain 4.4% (2006)

Imports: $162.1 billion f.o.b. (2007 est.)

Imports—commodities: machinery, chemicals, semi-finished goods, fuels, transport equipment

Imports—partners: Russia 12.8%, Germany 10.6%, China 6.9%, Italy 6.2%, France 5.2%, US 4.5%, Iran 4% (2006)

Economic aid—recipient: ODA, $464 million (2005)

Reserves of foreign exchange and gold: $76.51 billion (31 December 2007 est.)

Debt—external: $247.2 billion (31 December 2007)

Stock of direct foreign investment—at home: $106.3 billion (2007 est.)

Stock of direct foreign investment—abroad: $11.36 billion (2007 est.)

Market value of publicly traded shares: $162.4 billion (2006)

Currency (code): Turkish lira (TRY); old Turkish lira (TRL) before 1 January 2005

Currency code: TRL, YTL

Exchange rates: Turkish liras per US dollar—1.319 (2007), 1.4286 (2006), 1.3436 (2005), 1.4255 (2004), 1.5009 (2003)

note: on 1 January 2005 the old Turkish lira (TRL) was converted to new Turkish lira (TRY) at a rate of 1,000,000 old to 1 new Turkish lira

Fiscal year: calendar year

COMMUNICATIONS

Telephones—main lines in use: 18.978 million (2005)

Telephones—mobile cellular: 52.663 million (2006)

Telephone system:
general assessment: undergoing rapid modernization and expansion especially with cellular telephones
domestic: additional digital exchanges are permitting a rapid increase in subscribers; the construction of a network of technologically advanced intercity trunk lines, using both fiber-optic cable and digital microwave radio relay, is facilitating communication between urban centers; remote areas are reached by a domestic satellite system; the number of subscribers to mobile-cellular telephone service is growing rapidly
international: country code—90; international service is provided by the SEA-ME-WE-3 submarine cable and by submarine fiber-optic cables in the Mediterranean and Black Seas that link Turkey with Italy, Greece, Israel, Bulgaria, Romania, and Russia; satellite earth stations—12 Intelsat; mobile satellite terminals—328 in the Inmarsat and Eutelsat systems (2002)

Radio broadcast stations: AM 16, FM 107, shortwave 6 (2001)

Radios: 11.3 million (1997)

Television broadcast stations: 635 (plus 2,934 repeaters) (1995)

Televisions: 20.9 million (1997)

Internet country code: .tr

Internet hosts: 217,887 (2007)

Internet Service Providers (ISPs): 50 (2001)

Internet users: 12.284 million (2006)

TRANSPORTATION

Airports: 117 (2007)

Airports—with paved runways:
total: 90
over 3,047 m: 15
2,438 to 3,047 m: 33
1,524 to 2,437 m: 19
914 to 1,523 m: 19
under 914 m: 4 (2007)

Airports—with unpaved runways:
total: 27
over 3,047 m: 1
1,524 to 2,437 m: 2
914 to 1,523 m: 7
under 914 m: 17 (2007)

Heliports: 18 (2007)

Pipelines: gas 7,511 km; oil 3,636 km (2007)

Railways:
total: 8,697 km
standard gauge: 8,697 km 1.435-m gauge (1,920 km electrified) (2006)

Roadways:
total: 426,906 km
paved: 177,550 km (includes 1,892 km of expressways)
unpaved: 249,356 km (2004)
Waterways: 1,200 km (2005)
Merchant marine:
total: 565 ships (1000 GRT or over) 4,663,353 GRT/7,039,492 DWT
by type: bulk carrier 96, cargo 262, chemical tanker 58, combination ore/oil 1, container 30, liquefied gas 7, passenger 4, passenger/cargo 48, petroleum tanker 32, refrigerated cargo 1, roll on/roll off 25, specialized tanker 1
foreign-owned: 8 (China 1, Cyprus 2, Germany 1, Italy 3, UAE 1)
registered in other countries: 470 (Albania 1, Antigua and Barbuda 7, Bahamas 5, Belize 11, Cambodia 20, Comoros 8, Cyprus 1, Dominica 9, Georgia 23, Isle of Man 2, Italy 1, Kiribati 1, North Korea 1, Liberia 7, Malta 143, Marshall Islands 41, Netherlands Antilles 12, Panama 53, Russia 70, Sierra Leone 7, Slovakia 11, St Kitts and Nevis 13, St Vincent and The Grenadines 20, Tuvalu 1, UK 2, unknown 3) (2007)
Ports and terminals: Aliaga, Diliskelesi, Izmir, Kocaeli (Izmit), Mercin Limani, Nemrut Limani

MILITARY

Military branches: Turkish Armed Forces (TSK): Turkish Land Forces (Turk Kara Kuvvetleri, TKK), Turkish Naval Forces (Turk Deniz Kuvvetleri, TDK; includes naval air and naval infantry), Turkish Air Force (Turk Hava Kuvvetleri, THK) (2008)
Military service age and obligation: 20 years of age (2004)
Manpower available for military service:
males age 16–49: 20,213,205
females age 16–49: 19,432,688 (2008 est.)

Manpower fit for military service:
males age 16–49: 17,011,635
females age 16–49: 16,433,364 (2008 est.)
Manpower reaching militarily significant age annually:
males age 16–49: 660,452
females age 16–49: 638,527 (2008 est.)
Military expenditures—percent of GDP: 5.3% (2005 est.)
Military—note: a "National Security Policy Document" adopted in October 2005 increases the Turkish Armed Forces (TSK) role in internal security, augmenting the General Directorate of Security and Gendarmerie General Command (Jandarma); the TSK leadership continues to play a key role in politics and considers itself guardian of Turkey's secular state; in April 2007, it warned the ruling party about any pro-Islamic appointments; despite on-going negotiations on EU accession since October 2005, progress has been limited in establishing required civilian supremacy over the military; primary domestic threats are listed as fundamentalism (with the definition in some dispute with the civilian government), separatism (the Kurdish problem), and the extreme left wing; Ankara strongly opposed establishment of an autonomous Kurdish region; an overhaul of the Turkish Land Forces Command (TLFC) taking place under the "Force 2014" program is to produce 20–30% smaller, more highly trained forces characterized by greater mobility and firepower and capable of joint and combined operations; the TLFC has taken on increasing international peacekeeping responsibilities, and took charge of a NATO International Security Assistance Force (ISAF) command in Afghanistan in April 2007; the Turkish Navy is a regional naval power that wants to develop the capability to project power

beyond Turkey's coastal waters; the Navy is heavily involved in NATO, multinational, and UN operations; its roles include control of territorial waters and security for sea lines of communications; the Turkish Air Force adopted an "Aerospace and Missile Defense Concept" in 2002 and has initiated project work on an integrated missile defense system; Air Force priorities include attaining a modern deployable, survivable, and sustainable force structure, and establishing a sustainable command and control system (2008)

TRANSNATIONAL ISSUES

Disputes—international: complex maritime, air, and territorial disputes with Greece in the Aegean Sea; status of north Cyprus question remains; Syria and Iraq protest Turkish hydrological projects to control upper Euphrates waters; Turkey has expressed concern over the status of Kurds in Iraq; border with Armenia remains closed over Nagorno-Karabakh
Refugees and internally displaced persons: *IDPs:* 1–1.2 million (fighting 1984–99 between Kurdish PKK and Turkish military; most IDPs in southeastern provinces) (2007)
Illicit drugs: key transit route for Southwest Asian heroin to Western Europe and, to a lesser extent, the US—via air, land, and sea routes; major Turkish and other international trafficking organizations operate out of Istanbul; laboratories to convert imported morphine base into heroin exist in remote regions of Turkey and near Istanbul; government maintains strict controls over areas of legal opium poppy cultivation and over output of poppy straw concentrate; lax enforcement of money-laundering controls

TURKMENISTAN

INTRODUCTION

Background: Eastern Turkmenistan for centuries formed part of the Persian province of Khurasan; in medieval times Merv (today known as Mary) was one of the great cities of the Islamic world and an important stop on the Silk Road. Annexed by Russia between 1865 and 1885, Turkmenistan became a Soviet republic in 1924. It achieved independence upon the dissolution of the USSR in 1991. Extensive hydrocarbon/natural

gas reserves could prove a boon to this underdeveloped country if extraction and delivery projects were to be expanded. The Turkmenistan Government is actively seeking to develop alternative petroleum transportation routes to break Russia's pipeline monopoly. President for Life Saparmurat NYYAZOW died in December 2006, and Turkmenistan held its first multi-candidate presidential electoral process in February 2007. Gurbanguly BERDIMUHAMEDOW, a

vice premier under NYYAZOW, emerged as the country's new president.

GEOGRAPHY

Location: Central Asia, bordering the Caspian Sea, between Iran and Kazakhstan
Geographic coordinates: 40 00 N, 60 00 E
Map references: Asia
Area: *total:* 488,100 sq km
land: 488,100 sq km
water: NEGL

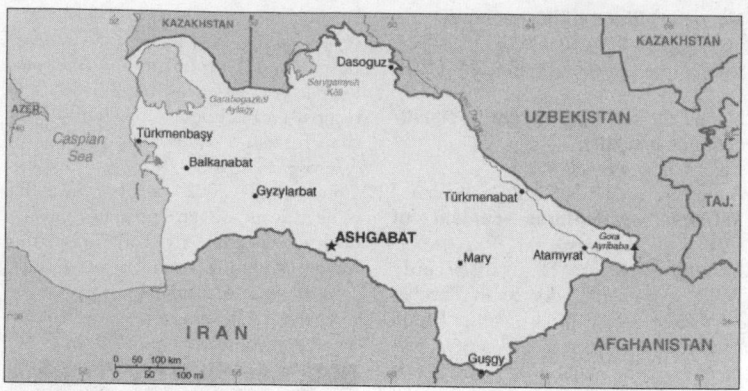

Area—comparative: slightly larger than California

Land boundaries:

total: 3,736 km

border countries: Afghanistan 744 km, Iran 992 km, Kazakhstan 379 km, Uzbekistan 1,621 km

Coastline: 0 km; note—Turkmenistan borders the Caspian Sea (1,768 km)

Maritime claims: none (landlocked)

Climate: subtropical desert

Terrain: flat-to-rolling sandy desert with dunes rising to mountains in the south; low mountains along border with Iran; borders Caspian Sea in west

Elevation extremes:

lowest point: Vpadina Akchanaya -81 m; note—Sarygamysh Koli is a lake in northern Turkmenistan with a water level that fluctuates above and below the elevation of Vpadina Akchanaya (the lake has dropped as low as -110 m)

highest point: Gora Ayribaba 3,139 m

Natural resources: petroleum, natural gas, sulfur, salt

Land use:

arable land: 4.51%

permanent crops: 0.14%

other: 95.35% (2005)

Irrigated land: 18,000 sq km (2003)

Total renewable water resources: 60.9 cu km (1997)

Freshwater withdrawal (domestic/industrial/agricultural): *total:* 24.65 cu km/yr (2%/1%/98%)

per capita: 5,104 cu m/yr (2000)

Natural hazards: NA

Environment—current issues: contamination of soil and groundwater with agricultural chemicals, pesticides; salination, water logging of soil due to poor irrigation methods; Caspian Sea pollution; diversion of a large share of the flow of the Amu Darya into irrigation contributes to that river's inability to replenish the Aral Sea; desertification

Environment—international agreements: *party to:* Biodiversity, Climate Change, Climate Change-Kyoto Protocol, Desertification, Hazardous Wastes, Ozone Layer Protection

signed, but not ratified: none of the selected agreements

Geography—note: landlocked; the western and central low-lying desolate portions of the country make up the great Garagum (Kara-Kum) desert, which occupies over 80% of the country; eastern part is plateau

PEOPLE

Population: 5,179,571 (July 2008 est.)

Age structure:

0–14 years: 34.2% (male 902,811/female 868,428)

15–64 years: 61.5% (male 1,577,187/female 1,607,353)

65 years and over: 4.3% (male 97,480/female 126,312) (2008 est.)

Median age:

total: 22.6 years

male: 22 years

female: 23.1 years (2008 est.)

Population growth rate: 1.596% (2008 est.)

Birth rate: 25.07 births/1,000 population (2008 est.)

Death rate: 6.11 deaths/1,000 population (2008 est.)

Net migration rate: -3 migrant(s)/1,000 population (2008 est.)

Sex ratio:

at birth: 1.05 male(s)/female

under 15 years: 1.04 male(s)/female

15–64 years: 0.98 male(s)/female

65 years and over: 0.77 male(s)/female

total population: 0.99 male(s)/female (2008 est.)

Infant mortality rate:

total: 51.81 deaths/1,000 live births

male: 56.01 deaths/1,000 live births

female: 47.4 deaths/1,000 live births (2008 est.)

Life expectancy at birth:

total population: 68.6 years

male: 65.53 years

female: 71.82 years (2008 est.)

Total fertility rate: 3.07 children born/woman (2008 est.)

HIV/AIDS—adult prevalence rate: less than 0.1% (2004 est.)

HIV/AIDS—people living with HIV/AIDS: fewer than 200 (2003 est.)

HIV/AIDS—deaths: fewer than 100 (2004 est.)

Nationality:

noun: Turkmen(s)

adjective: Turkmenistani

Ethnic groups: Turkmen 85%, Uzbek 5%, Russian 4%, other 6% (2003)

Religions: Muslim 89%, Eastern Orthodox 9%, unknown 2%

Languages: Turkmen 72%, Russian 12%, Uzbek 9%, other 7%

Literacy: *definition:* age 15 and over can read and write

total population: 98.8%

male: 99.3%

female: 98.3% (1999 est.)

GOVERNMENT

Country name:

conventional long form: none

conventional short form: Turkmenistan

local long form: none

local short form: Turkmenistan

former: Turkmen Soviet Socialist Republic

Government type: republic; authoritarian presidential rule, with little power outside the executive branch

Capital: *name:* Ashgabat (Ashkhabad)

geographic coordinates: 37 57 N, 58 23 E

time difference: UTC+5 (10 hours ahead of Washington, DC during Standard Time)

Administrative divisions: 5 provinces (welayatlar, singular—welayat) and 1 independent city*: Ahal Welayaty (Anew), Ashgabat*, Balkan Welayaty (Balkanabat), Dashoguz Welayaty, Lebap Welayaty (Turkmenabat), Mary Welayaty

note: administrative divisions have the same names as their administrative centers (exceptions have the administrative center name following in parentheses)

Independence: 27 October 1991 (from Soviet Union)

National holiday: Independence Day, 27 October (1991)

Constitution: adopted 18 May 1992

Legal system: based on civil law system and Islamic law; has not accepted compulsory ICJ jurisdiction

Suffrage: 18 years of age; universal

Executive branch:

chief of state: President Gurbanguly BERDIMUHAMEDOW (since 14 February 2007); note—the president is both the chief of state and head of government

head of government: President Gurbanguly BERDIMUHAMEDOW (since 14 February 2007) *cabinet:* Cabinet of Ministers appointed by the president

elections: president elected by popular vote for a five-year term; election last held on 11 February 2007 (next to be held in 2012)

election results: Gurbanguly BERDIMUHAMEDOW elected president; percent of vote—Gurbanguly BERDIMUHAMEDOW 89.2%, Amanyaz ATAJYKOW 3.2%, other candidates 7.6%

Legislative branch: two parliamentary bodies, a People's Council or Halk Maslahaty (supreme legislative body of about 2,500 delegates, some elected by popular vote and some appointed; meets at least yearly) and a National Assembly or Mejlis (50 seats; members are elected by popular vote to serve five-year terms)

elections: People's Council—last held in April 2003 (next to be held in December 2008); National Assembly—last held 19 December 2004 (next to be held in December 2008)

election results: People's Council—percent of vote by party—DPT 100%; seats by party—DPT 2,507; National Assembly—percent of vote by party—DPT 100%; seats by party—DPT 50; note—all elected officials are members of the Democratic Party of Turkmenistan and are preapproved by the president

note: in late 2003, a law was adopted reducing the powers of the National Assembly and making the People's Council the supreme legislative organ; the People's Council can now legally dissolve the National Assembly, and the president is now able to participate in the National Assembly as its supreme leader; the National Assembly can no longer adopt or amend the constitution or announce referendums or its elections; since the president is both the chairman of the People's Council and the supreme leader of the National Assembly, the 2003 law has the effect of making him the sole authority of both the executive and legislative branches of government

Judicial branch: Supreme Court (judges are appointed by the president)

Political parties and leaders: Democratic Party of Turkmenistan or DPT [Gurbanguly BERDIMUHAMEDOW]

note: formal opposition parties are outlawed; unofficial, small opposition movements exist underground or in foreign countries; the two most prominent opposition groups-in-exile have been National Democratic Movement of Turkmenistan (NDMT) and the United Democratic Party of Turkmenistan (UDPT); NDMT was led by former Foreign Minister Boris SHIKHMURADOV until his arrest and imprisonment in the wake of the 25 November 2002 attack on President NYYAZOW's motorcade

Political pressure groups and leaders: NA

International organization participation: ABEDA, ADB, CIS, EAPC, EBRD, ECO, FAO, G-77, IBRD, ICAO, ICRM, IDA, IDB, IFC, IFRCS, ILO, IMF, IMO, Interpol, IOC, IOM (observer), ISO (correspondent), ITU, MIGA, NAM, OIC, OPCW, OSCE, PFP, UN, UNCTAD, UNESCO, UNIDO, UNWTO, UPU, WCO, WFTU, WHO, WIPO, WMO

Diplomatic representation in the US:
chief of mission: Ambassador Meret Bairamovich ORAZOW
chancery: 2207 Massachusetts Avenue NW, Washington, DC 20008
telephone: [1] (202) 588-1500
FAX: [1] (202) 588-0697

Diplomatic representation from the US:
chief of mission: Ambassador (vacant); Charge d'Affaires Richard E. HOAGLAND
embassy: No. 9 1984 Street (formerly Pushkin Street), Ashgabat, Turkmenistan 744000
mailing address: 7070 Ashgabat Place, Washington, DC 20521-7070
telephone: [993] (12) 35-00-45
FAX: [993] (12) 39-26-14

Flag description: green field with a vertical red stripe near the hoist side, containing five tribal guls (designs used in producing carpets) stacked above two crossed olive branches similar to the olive branches on the UN flag; a white crescent moon representing Islam with five white stars representing the regions or welayats of Turkmenistan appear in the upper corner of the field just to the fly side of the red stripe

ECONOMY

Economy—overview: Turkmenistan is a largely desert country with intensive agriculture in irrigated oases and large gas and oil resources. One-half of its irrigated land is planted in cotton; formerly it was the world's 10th-largest producer. Poor harvests in recent years have led to an almost 50% decline in cotton exports. With an authoritarian ex-Communist regime in power and a tribally based social structure, Turkmenistan has taken a cautious approach to economic reform, hoping to use gas and cotton sales to sustain its inefficient economy. Privatization goals remain limited. From 1998–2005, Turkmenistan suffered from the continued lack of adequate export routes for natural gas and from obligations on extensive short-term external debt. At the same time, however, total exports rose by an average of roughly 15% per year from 2003–07, largely because of higher international oil and gas prices. Overall prospects in the near future are discouraging because of widespread internal poverty, a poor educational system, government misuse of oil and gas revenues, and Ashgabat's reluctance to adopt market-oriented reforms. In the past, Turkmenistan's economic statistics were state secrets. The new government has established a State Agency for Statistics, but GDP numbers and other figures are subject to wide margins of error. In particular, the rate of GDP growth is uncertain. Since his election, President BERDIMUHAMEDOW has sought to improve the health and education systems, ordered unification of the country's dual currency exchange rate, begun decreasing state subsidies for gasoline, signed an agreement to build a gas line to China, and created a special tourism zone on the Caspian Sea. All of these moves hint that the new post-NYYAZOW government will work to create a friendlier foreign investment environment.

GDP (purchasing power parity): $26.73 billion (2007 est.)

GDP (official exchange rate): $26.91 billion (2007 est.)

GDP—real growth rate: 11.6% (IMF estimate)
note: official government statistics are widely regarded as unreliable (2007 est.)

GDP—per capita (PPP): $5,200 (2007 est.)

GDP—composition by sector:
agriculture: 11.5%
industry: 40.8%
services: 47.7% (2007 est.)

Labor force: 2.089 million (2004 est.)

Labor force—by occupation: *agriculture:* 48.2%
industry: 14%
services: 37.8% (2004 est.)

Unemployment rate: 60% (2004 est.)

Population below poverty line: 30% (2004 est.)

Household income or consumption by percentage share:
lowest 10%: 2.6%
highest 10%: 31.7% (1998)

Distribution of family income—Gini index: 40.8 (1998)

Inflation rate (consumer prices): 6.4% (2007 est.)

Investment (gross fixed): 31.6% of GDP (2007 est.)

Budget:
revenues: $1.664 billion
expenditures: $1.624 billion (2007 est.)

Agriculture—products: cotton, grain; livestock

Industries: natural gas, oil, petroleum products, textiles, food processing

Industrial production growth rate: 10.3% (2007 est.)

Electricity—production: 12.05 billion kWh (2005 est.)

Electricity—production by source:
fossil fuel: 99.9%
hydro: 0.1%
nuclear: 0%
other: 0% (2001)
Electricity—consumption: 7.602 billion kWh (2005 est.)
Electricity—exports: 2.918 billion kWh (2005)
Electricity—imports: 0 kWh (2005)
Oil—production: 196,000 bbl/day (2007 est.)
Oil—consumption: 156,000 bbl/day (2007 est.)
Oil—exports: 40,000 bbl/day (2007 est.)
Oil—imports: 0 bbl/day (2007)
Oil—proved reserves: 500 million bbl (1 January 2007 est.)
Natural gas—production: 72.3 billion cu m (2007 est.)
Natural gas—consumption: 14.4 billion cu m (2007 est.)
Natural gas—exports: 58 billion cu m (2007 est.)
Natural gas—imports: 0 cu m (2007)
Natural gas—proved reserves: 2.86 trillion cu m (1 January 2007 est.)
Current account balance: $4.525 billion (2007 est.)
Exports: $7.567 billion f.o.b. (2007 est.)
Exports—commodities: gas, crude oil, petrochemicals, textiles, cotton fiber
Exports—partners: Ukraine 47.5%, Iran 16.4%, Azerbaijan 5.3% (2006)
Imports: $4.2 billion f.o.b. (2007 est.)
Imports—commodities: machinery and equipment, chemicals, foodstuffs
Imports—partners: UAE 15.5%, Turkey 11.1%, Ukraine 9.1%, Russia 9%, Germany 7.7%, Iran 7.6%, China 6.4%, US 4.5% (2006)
Economic aid—recipient: $28.25 million from the US (2005)
Reserves of foreign exchange and gold: $5.173 billion (31 December 2007 est.)
Debt—external: $1.4 billion to $5 billion (2004 est.)
Market value of publicly traded shares: $NA
Currency (code): Turkmen manat (TMM)
Currency code: TMM
Exchange rates: Turkmen manat per US$—6,250 (2007) official rate
note: the commercial rate was 19,800 Turkmen manat per US$ (2007)

Fiscal year: calendar year

COMMUNICATIONS

Telephones—main lines in use: 495,000 (2006)
Telephones—mobile cellular: 216,900 (2006)
Telephone system:
general assessment: poorly developed
domestic: Turkmentelekom, in cooperation with foreign investors, is planning to upgrade the country's telephone exchanges and install a new digital switching system; mobile-cellular useage remains limited
international: country code—993; linked by cable and microwave radio relay to other CIS republics and to other countries by leased connections to the Moscow international gateway switch; a new telephone link from Ashgabat to Iran has been established; a new exchange in Ashgabat switches international traffic through Turkey via Intelsat; satellite earth stations—1 Orbita and 1 Intelsat (2006)
Radio broadcast stations: AM 16, FM 8, shortwave 2 (1998)
Radios: 1.225 million (1997)
Television broadcast stations: 4 (government-owned and programmed) (2004)
Televisions: 820,000 (1997)
Internet country code: .tm
Internet hosts: 97 (2007)
Internet Service Providers (ISPs): 1
Internet users: 64,800 (2006)

TRANSPORTATION

Airports: 28 (2007)
Airports—with paved runways:
total: 22
over 3,047 m: 1
2,438 to 3,047 m: 11
1,524 to 2,437 m: 8
914 to 1,523 m: 2 (2007)
Airports—with unpaved runways:
total: 6
1,524 to 2,437 m: 2
under 914 m: 4 (2007)
Heliports: 1 (2007)
Pipelines: gas 6,441 km; oil 1,361 km (2007)
Railways:
total: 2,440 km
broad gauge: 2,440 km 1.520-m gauge (2006)

Roadways:
total: 24,000 km
paved: 19,488 km
unpaved: 4,512 km (1999)
Waterways: 1,300 km (Amu Darya and Kara Kum canal important inland waterways) (2006)
Merchant marine:
total: 8 ships (1000 GRT or over) 22,870 GRT/25,801 DWT
by type: cargo 4, combination ore/oil 1, petroleum tanker 2, refrigerated cargo 1 (2007)
Ports and terminals: Turkmenbasy

MILITARY

Military branches: Ground Forces, Navy, Air and Air Defense Forces (2007)
Military service age and obligation: 18–30 years of age for compulsory military service; 2-year conscript service obligation (2006)
Manpower available for military service:
males age 16–49: 1,316,698
females age 16–49: 1,331,005 (2008 est.)
Manpower fit for military service:
males age 16–49: 1,064,965
females age 16–49: 1,136,553 (2008 est.)
Manpower reaching militarily significant age annually:
males age 16–49: 57,615
females age 16–49: 55,426 (2008 est.)
Military expenditures—percent of GDP: 3.4% (2005 est.)

TRANSNATIONAL ISSUES

Disputes—international: cotton monoculture in Uzbekistan and Turkmenistan creates water-sharing difficulties for Amu Darya river states; field demarcation of the boundaries with Kazakhstan commenced in 2005, but Caspian seabed delimitation remains stalled with Azerbaijan, Iran, and Kazakhstan due to Turkmenistan's indecision over how to allocate the sea's waters and seabed
Refugees and internally displaced persons: *refugees (country of origin):* 11,173 (Tajikistan); less than 1,000 (Afghanistan) (2007)
Illicit drugs: transit country for Afghan narcotics bound for Russian and Western European markets; transit point for heroin precursor chemicals bound for Afghanistan

TURKS AND CAICOS ISLANDS

INTRODUCTION

Background: The islands were part of the UK's Jamaican colony until 1962, when they assumed the status of a separate crown colony upon Jamaica's independence. The governor of The Bahamas oversaw affairs from 1965 to 1973. With Bahamian independence, the islands received a separate governor in 1973. Although independence was agreed upon for 1982, the policy was reversed and the islands remain a British overseas territory.

GEOGRAPHY

Location: Caribbean, two island groups in the North Atlantic Ocean, southeast of The Bahamas, north of Haiti
Geographic coordinates: 21 45 N, 71 35 W

Map references: Central America and the Caribbean

Area:
total: 430 sq km
land: 430 sq km
water: 0 sq km

Area—comparative: 2.5 times the size of Washington, DC

Land boundaries: 0 km

Coastline: 389 km

Maritime claims:
territorial sea: 12 nm
exclusive fishing zone: 200 nm

Climate: tropical; marine; moderated by trade winds; sunny and relatively dry

Terrain: low, flat limestone; extensive marshes and mangrove swamps

Elevation extremes:
lowest point: Caribbean Sea 0 m
highest point: Blue Hills 49 m

Natural resources: spiny lobster, conch

Land use:
arable land: 2.33%
permanent crops: 0%
other: 97.67% (2005)

Irrigated land: NA

Natural hazards: frequent hurricanes

Environment—current issues: limited natural fresh water resources, private cisterns collect rainwater

Geography—note: about 40 islands (eight inhabited)

PEOPLE

Population: 22,352 (July 2008 est.)

Age structure:
0–14 years: 30.7% (male 3,497/female 3,374)
15–64 years: 65.2% (male 7,640/female 6,929)
65 years and over: 4.1% (male 435/female 477) (2008 cst.)

Median age:
total: 27.8 years
male: 28.5 years
female: 27 years (2008 est.)

Population growth rate: 2.644% (2008 est.)

Birth rate: 21.12 births/1,000 population (2008 est.)

Death rate: 4.16 deaths/1,000 population (2008 est.)

Net migration rate: 9.48 migrant(s)/1,000 population (2008 est.)

Sex ratio:
at birth: 1.05 male(s)/female
under 15 years: 1.04 male(s)/female
15–64 years: 1.1 male(s)/female
65 years and over: 0.91 male(s)/female
total population: 1.07 male(s)/female (2008 est.)

Infant mortality rate:
total: 14.35 deaths/1,000 live births
male: 16.56 deaths/1,000 live births
female: 12.04 deaths/1,000 live births (2008 est.)

Life expectancy at birth:
total population: 75.19 years
male: 72.91 years
female: 77.59 years (2008 est.)

Total fertility rate: 2.98 children born/woman (2008 est.)

HIV/AIDS—adult prevalence rate: NA

HIV/AIDS—people living with HIV/AIDS: NA

HIV/AIDS—deaths: NA

Nationality:
noun: none
adjective: none

Ethnic groups: black 90%, mixed, European, or North American 10%

Religions: Baptist 40%, Anglican 18%, Methodist 16%, Church of God 12%, other 14% (1990)

Languages: English (official)

Literacy:
definition: age 15 and over has ever attended school
total population: 98%
male: 99%
female: 98% (1970 est.)

People—note: destination and transit point for illegal Haitian immigrants bound for the Turks and Caicos Islands, The Bahamas, and the US

GOVERNMENT

Country name:
conventional long form: none
conventional short form: Turks and Caicos Islands
abbreviation: TCI

Dependency status: overseas territory of the UK

Government type: NA

Capital: *name:* Grand Turk (Cockburn Town)
geographic coordinates: 21 28 N, 71 08 W
time difference: UTC-5 (same time as Washington, DC during Standard Time)
daylight saving time: +1hr, begins first Sunday in April; ends last Sunday in October

Administrative divisions: none (overseas territory of the UK)

Independence: none (overseas territory of the UK)

National holiday: Constitution Day, 30 August (1976)

Constitution: Turks and Caicos Islands Constitution Order 2006 (effective 9 August 2006)

Legal system: based on laws of England and Wales, with a few adopted from Jamaica and The Bahamas

Suffrage: 18 years of age; universal

Executive branch:
chief of state: Queen ELIZABETH II (since 6 February 1952); represented by Governor Richard TAUWHARE (since 11 July 2005)
head of government: Premier Michael Eugene MISICK (since 15 August 2003); note—the office of premier was created in the 2006 constitution
cabinet: Cabinet consists of the governor, the premier, six ministers appointed by the governor from among the members of the House of Assembly, and the attorney general
elections: the monarch is hereditary; governor appointed by the monarch; following legislative elections, the leader of the majority party is appointed premier by the governor

Legislative branch: unicameral House of Assembly (21 seats of which 15 are popularly elected; members serve four-year terms)
elections: last held 9 February 2007 (next to be held in 2011)
election results: percent of vote by party—PNP 60%, PDM 40%; seats by party—PNP 13, PDM 2

Judicial branch: Supreme Court; Court of Appeal

Political parties and leaders: People's Democratic Movement or PDM [Floyd SEYMOUR]; Progressive National Party or PNP [Michael Eugene MISICK]

Political pressure groups and leaders: NA

International organization participation: Caricom (associate), CDB, Interpol (subbureau), UPU

Diplomatic representation in the US: none (overseas territory of the UK)

Diplomatic representation from the US: none (overseas territory of the UK)

Flag description: blue, with the flag of the UK in the upper hoist-side quadrant and the colonial shield centered on the outer half of the flag; the shield is yellow and contains a conch shell, lobster, and cactus

ECONOMY

Economy—overview: The Turks and Caicos economy is based on tourism, offshore financial services, and fishing.

Most capital goods and food for domestic consumption are imported. The US is the leading source of tourists, accounting for more than three-quarters of the 175,000 visitors that arrived in 2004. Major sources of government revenue also include fees from offshore financial activities and customs receipts.

GDP (purchasing power parity): $216 million (2002 est.)

GDP (official exchange rate): $NA

GDP—real growth rate: 4.9% (2000 est.)

GDP—per capita (PPP): $11,500 (2002 est.)

GDP—composition by sector:
agriculture: NA%
industry: NA%
services: NA%

Labor force: 4,848 (1990 est.)

Labor force—by occupation: *note:* about 33% in government and 20% in agriculture and fishing; significant numbers in tourism, financial, and other services

Unemployment rate: 10% (1997 est.)

Population below poverty line: NA%

Household income or consumption by percentage share:
lowest 10%: NA%
highest 10%: NA%

Inflation rate (consumer prices): 4% (1995)

Budget:
revenues: $47 million
expenditures: $33.6 million (1997–98 est.)

Agriculture—products: corn, beans, cassava (tapioca), citrus fruits; fish

Industries: tourism, offshore financial services

Industrial production growth rate: NA%

Electricity—production: 11.57 million kWh (2005)

Electricity—production by source:
fossil fuel: 100%
hydro: 0%
nuclear: 0%
other: 0% (2001)

Electricity—consumption: 10.76 million kWh (2005)

Electricity—exports: 0 kWh (2005 est.)

Electricity—imports: 0 kWh (2005)

Oil—production: 0 bbl/day (2005 est.)

Oil—consumption: 80 bbl/day (2005 est.)

Oil—exports: 0 bbl/day (2004)

Oil—imports: 83.55 bbl/day (2004)

Oil—proved reserves: 0 bbl (1 January 2006 est.)

Natural gas—production: 0 cu m (2005 est.)

Natural gas—consumption: 0 cu m (2005 est.)

Natural gas—exports: 0 cu m (2005 est.)

Natural gas—imports: 0 cu m (2005)

Natural gas—proved reserves: 0 cu m (1 January 2006 est.)

Exports: $169.2 million (2000)

Exports—commodities: lobster, dried and fresh conch, conch shells

Exports—partners: US, UK (2006)

Imports: $175.6 million (2000)

Imports—commodities: food and beverages, tobacco, clothing, manufactures, construction materials

Imports—partners: US, UK (2006)

Economic aid—recipient: $4.1 million (1997)

Debt—external: $NA

Currency (code): US dollar (USD)

Currency code: USD

Exchange rates: the US dollar is used

Fiscal year: calendar year

COMMUNICATIONS

Telephones—main lines in use: 5,700 (2002)

Telephones—mobile cellular: 1,700 (1999)

Telephone system:
general assessment: fully digital system with international direct dialing
domestic: full range of services available; GSM wireless service available
international: country code—1-649; the Americas Region Caribbean Ring System (ARCOS-1) fiber optic telecommunications submarine cable provides connectivity to South and Central America, parts of the Caribbean, and the US; satellite earth station—1 Intelsat (Atlantic Ocean)

Radio broadcast stations: AM 2, FM 7, shortwave 0 (2003)

Radios: 8,000 (1997)

Television broadcast stations: 0 (broadcasts received from The Bahamas; 2 cable television networks) (2003)

Televisions: NA

Internet country code: .tc

Internet hosts: 2,310 (2007)

Internet Service Providers (ISPs): 14 (2000)

Internet users: NA

TRANSPORTATION

Airports: 8 (2007)

Airports—with paved runways:
total: 6
1,524 to 2,437 m: 3
914 to 1,523 m: 1
under 914 m: 2 (2007)

Airports—with unpaved runways:
total: 2
under 914 m: 2 (2007)

Roadways:
total: 121 km
paved: 24 km
unpaved: 97 km (2003)

Merchant marine:
registered in other countries: 1 (Panama 1) (2007)

Ports and terminals: Grand Turk, Providenciales

MILITARY

Military—note: defense is the responsibility of the UK

TRANSNATIONAL ISSUES

Disputes—international: have received Haitians fleeing economic and civil disorder

Illicit drugs: transshipment point for South American narcotics destined for the US and Europe

TUVALU

INTRODUCTION

Background: In 1974, ethnic differences within the British colony of the Gilbert and Ellice Islands caused the Polynesians of the Ellice Islands to vote for separation from the Micronesians of the Gilbert Islands. The following year, the Ellice Islands became the separate British colony of Tuvalu. Independence was granted in 1978. In 2000, Tuvalu negotiated a contract leasing its Internet domain name ".tv" for $50 million in royalties over a 12-year period.

GEOGRAPHY

Location: Oceania, island group consisting of nine coral atolls in the South Pacific Ocean, about one-half of the way from Hawaii to Australia

Geographic coordinates: 8 00 S, 178 00 E

Map references: Oceania

Area: *total:* 26 sq km
land: 26 sq km
water: 0 sq km

Area—comparative: 0.1 times the size of Washington, DC

Land boundaries: 0 km

Coastline: 24 km

Maritime claims:
territorial sea: 12 nm
contiguous zone: 24 nm
exclusive economic zone: 200 nm

Climate: tropical; moderated by easterly trade winds (March to November); west-

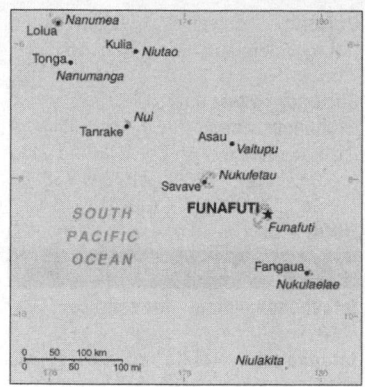

erly gales and heavy rain (November to March)

Terrain: very low-lying and narrow coral atolls

Elevation extremes:
lowest point: Pacific Ocean 0 m
highest point: unnamed location 5 m

Natural resources: fish

Land use:
arable land: 0%
permanent crops: 66.67%
other: 33.33% (2005)

Irrigated land: NA

Natural hazards: severe tropical storms are usually rare, but, in 1997, there were three cyclones; low level of islands make them sensitive to changes in sea level

Environment—current issues: since there are no streams or rivers and groundwater is not potable, most water needs must be met by catchment systems with storage facilities (the Japanese Government has built one desalination plant and plans to build one other); beachhead erosion because of the use of sand for building materials; excessive clearance of forest undergrowth for use as fuel; damage to coral reefs from the spread of the Crown of Thorns starfish; Tuvalu is concerned about global increases in greenhouse gas emissions and their effect on rising sea levels, which threaten the country's underground water table; in 2000, the government appealed to Australia and New Zealand to take in Tuvaluans if rising sea levels should make evacuation necessary

Environment—international agreements: *party to:* Biodiversity, Climate Change, Climate Change-Kyoto Protocol, Desertification, Law of the Sea, Ozone Layer Protection, Ship Pollution, Whaling
signed, but not ratified: none of the selected agreements

Geography—note: one of the smallest and most remote countries on Earth; six of the nine coral atolls—Nanumea, Nui, Vaitupu, Nukufetau, Funafuti, and Nukulaelae—have lagoons open to the ocean; Nanumaya and Niutao have land-locked lagoons; Niulakita does not have a lagoon

PEOPLE

Population: 12,177 (July 2008 est.)

Age structure:
0–14 years: 29.4% (male 1,826/female 1,754)
15–64 years: 65.4% (male 3,891/female 4,073)
65 years and over: 5.2% (male 236/female 397) (2008 est.)

Median age:
total: 25.2 years
male: 24.2 years
female: 26.4 years (2008 est.)

Population growth rate: 1.577% (2008 est.)

Birth rate: 22.75 births/1,000 population (2008 est.)

Death rate: 6.98 deaths/1,000 population (2008 est.)

Net migration rate: NA

Sex ratio:
at birth: 1.05 male(s)/female
under 15 years: 1.04 male(s)/female
15–64 years: 0.96 male(s)/female
65 years and over: 0.59 male(s)/female
total population: 0.96 male(s)/female (2008 est.)

Infant mortality rate:
total: 18.97 deaths/1,000 live births
male: 21.56 deaths/1,000 live births
female: 16.25 deaths/1,000 live births (2008 est.)

Life expectancy at birth:
total population: 68.97 years
male: 66.7 years
female: 71.36 years (2008 est.)

Total fertility rate: 2.94 children born/woman (2008 est.)

HIV/AIDS—adult prevalence rate: NA

HIV/AIDS—people living with HIV/AIDS: NA

HIV/AIDS—deaths: NA

Nationality:
noun: Tuvaluan(s)
adjective: Tuvaluan

Ethnic groups: Polynesian 96%, Micronesian 4%

Religions: Church of Tuvalu (Congregationalist) 97%, Seventh-Day Adventist 1.4%, Baha'i 1%, other 0.6%

Languages: Tuvaluan, English, Samoan, Kiribati (on the island of Nui)

Literacy:
NA

GOVERNMENT

Country name:
conventional long form: none
conventional short form: Tuvalu
local long form: none

local short form: Tuvalu
former: Ellice Islands
note: "Tuvalu" means "group of eight," referring to the country's eight traditionally inhabited islands

Government type: constitutional monarchy with a parliamentary democracy

Capital: *name:* Funafuti
geographic coordinates: 8 30 S, 179 12 E
time difference: UTC+12 (17 hours ahead of Washington, DC during Standard Time)
note: administrative offices are located in Vaiaku Village on Fongafale Islet

Administrative divisions: none

Independence: 1 October 1978 (from UK)

National holiday: Independence Day, 1 October (1978)

Constitution: 1 October 1978

Legal system: NA

Suffrage: 18 years of age; universal

Executive branch:
chief of state: Queen ELIZABETH II (since 6 February 1952); represented by Governor General Filoimea TELITO (since 15 April 2005)
head of government: Prime Minister Apisai IELEMIA (since 14 August 2006)
cabinet: Cabinet appointed by the governor general on the recommendation of the prime minister
elections: the monarch is hereditary; governor general appointed by the monarch on the recommendation of the prime minister; prime minister and deputy prime minister elected by and from the members of Parliament; election last held 14 August 2006 (next to be held following parliamentary elections in 2010)
election results: Apisai IELEMIA elected Prime Minister in a Parliamentary election on 14 August 2006

Legislative branch: unicameral Parliament or Fale I Fono, also called House of Assembly (15 seats; members elected by popular vote to serve four-year terms)
elections: last held 3 August 2006 (next to be held in 2010)
election results: percent of vote—NA; seats—independents 15

Judicial branch: High Court (a chief justice visits twice a year to preside over its sessions; its rulings can be appealed to the Court of Appeal in Fiji); eight Island Courts (with limited jurisdiction)

Political parties and leaders: there are no political parties but members of Parliament usually align themselves in informal groupings

Political pressure groups and leaders: none

International organization participation: ACP, ADB, C, FAO, IFRCS (observer), IMO, IOC, ITU, OPCW, PIF, Sparteca, SPC, UN, UNCTAD, UNESCO, UPU, WHO

Diplomatic representation in the US: Tuvalu does not have an embassy in the US—the country's only diplomatic post is in Fiji—Tuvalu does, however, have a UN office located at 800 2nd Avenue, Suite 400D, New York, NY 10017, telephone: [1] (212) 490-0534

Diplomatic representation from the US: the US does not have an embassy in Tuvalu; the US ambassador to Fiji is accredited to Tuvalu

Flag description: light blue with the flag of the UK in the upper hoist-side quadrant; the outer half of the flag represents a map of the country with nine yellow five-pointed stars symbolizing the nine islands

ECONOMY

Economy—overview: Tuvalu consists of a densely populated, scattered group of nine coral atolls with poor soil. The country has no known mineral resources and few exports. Subsistence farming and fishing are the primary economic activities. Fewer than 1,000 tourists, on average, visit Tuvalu annually. Job opportunities are scarce and public sector workers make up the majority of those employed. About 15% of the adult male population work as seamen on merchant ships abroad and remittances are a vital source of income, contributing around $4 million in 2006. Substantial income is received annually from the Tuvalu Trust Fund (TTF), an international trust fund established in 1987 by Australia, NZ, and the UK and supported also by Japan and South Korea. Thanks to wise investments and conservative withdrawals, this fund grew from an initial $17 million to an estimated value of $77 million in 2006. The TFF contributed nearly $9 million towards the government budget in 2006 and is an important cushion for meeting shortfalls in the government's budget. The US Government is also a major revenue source for Tuvalu because of payments from a 1988 treaty on fisheries. In an effort to ensure financial stability and sustainability, the government is pursuing public sector reforms, including privatization of some government functions and personnel cuts. Tuvalu also derives royalties from the lease of its ".tv" Internet domain name, with revenue of more than $2 million in 2006. A minor source of government revenue comes from the sale of stamps and coins. With merchandise exports only a fraction of merchandise imports, continued reliance must be placed on fishing and telecommunications license fees, remittances from overseas workers, official transfers, and income from overseas investments. Growing income disparities and the vulnerability of the country to climatic change are among leading concerns for the nation.

GDP (purchasing power parity): $14.94 million (2002 est.)

GDP (official exchange rate): $14.94 million (2002)

GDP—real growth rate: 3% (2006 est.)

GDP—per capita (PPP): $1,600 (2002 est.)

GDP—composition by sector:
agriculture: 16.6%
industry: 27.2%
services: 56.2% (2002)

Labor force: 3,615 (2004 est.)

Labor force—by occupation: *note:* people make a living mainly through exploitation of the sea, reefs, and atolls and from wages sent home by those abroad (mostly workers in the phosphate industry and sailors)

Unemployment rate: NA%

Population below poverty line: NA%

Household income or consumption by percentage share:
lowest 10%: NA%
highest 10%: NA%

Inflation rate (consumer prices): 3.8% (2006 est.)

Budget:
revenues: $21.54 million
expenditures: $23.05 million (2006)

Agriculture—products: coconuts; fish

Industries: fishing, tourism, copra

Industrial production growth rate: NA%

Electricity—production by source:
fossil fuel: NA
hydro: NA
nuclear: NA
other: NA

Current account balance: -$11.68 million (2003)

Exports: $1 million f.o.b. (2004 est.)

Exports—commodities: copra, fish

Exports—partners: Germany 60.5%, Italy 20.1%, Fiji 6.9% (2006)

Imports: $12.91 million c.i.f. (2005)

Imports—commodities: food, animals, mineral fuels, machinery, manufactured goods

Imports—partners: Fiji 46.1%, Japan 18.9%, China 18.2%, Australia 7.7%, NZ 4.1% (2006)

Economic aid—recipient: $10.49 million
note: includes distributions from the Tuvalu Trust Fund (2006)

Debt—external: $NA

Currency (code): Australian dollar (AUD); note—there is also a Tuvaluan dollar

Currency code: AUD

Exchange rates: Tuvaluan dollars or Australian dollars per US dollar—1.2137 (2007), 1.3285 (2006), 1.3095 (2005), 1.3598 (2004), 1.5419 (2003)

Fiscal year: calendar year

COMMUNICATIONS

Telephones—main lines in use: 900 (2005)

Telephones—mobile cellular: 1,300 (2005)

Telephone system:
general assessment: serves particular needs for internal communications
domestic: radiotelephone communications between islands
international: country code—688; international calls can be made by satellite

Radio broadcast stations: AM 1, FM 1, shortwave 0 (2004)

Radios: 4,000 (1997)

Television broadcast stations: 0 (2004)

Televisions: 800

Internet country code: .tv

Internet hosts: 30,200 (2007)

Internet Service Providers (ISPs): 1 (2000)

Internet users: 1,300 (2002)

TRANSPORTATION

Airports: 1 (2007)

Airports—with unpaved runways:
total: 1
1,524 to 2,437 m: 1 (2007)

Roadways:
total: 8 km
paved: 8 km (2002)

Merchant marine:
total: 74 ships (1000 GRT or over) 568,759 GRT/928,697 DWT
by type: bulk carrier 4, cargo 45, chemical tanker 5, container 2, passenger 2, passenger/cargo 1, petroleum tanker 13, refrigerated cargo 1, specialized tanker 1
foreign-owned: 61 (China 25, Hong Kong 10, Kenya 1, Maldives 1, Romania 1, Russia 4, Singapore 13, Thailand 1, Turkey 1, US 1, Vietnam 3) (2007)

Ports and terminals: Funafuti

MILITARY

Military branches: no regular military forces; Tuvalu Police Force (2008)

Military expenditures—percent of GDP: NA

TRANSNATIONAL ISSUES

Disputes—international: none

UGANDA

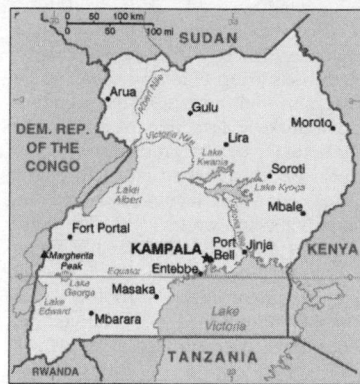

INTRODUCTION

Background: The colonial boundaries created by Britain to delimit Uganda grouped together a wide range of ethnic groups with different political systems and cultures. These differences prevented the establishment of a working political community after independence was achieved in 1962. The dictatorial regime of Idi AMIN (1971–79) was responsible for the deaths of some 300,000 opponents; guerrilla war and human rights abuses under Milton OBOTE (1980–85) claimed at least another 100,000 lives. The rule of Yoweri MUSEVENI since 1986 has brought relative stability and economic growth to Uganda. During the 1990s, the government promulgated non-party presidential and legislative elections.

GEOGRAPHY

Location: Eastern Africa, west of Kenya
Geographic coordinates: 1 00 N, 32 00 E
Map references: Africa
Area: *total:* 236,040 sq km
land: 199,710 sq km
water: 36,330 sq km
Area—comparative: slightly smaller than Oregon
Land boundaries:
total: 2,698 km
border countries: Democratic Republic of the Congo 765 km, Kenya 933 km, Rwanda 169 km, Sudan 435 km, Tanzania 396 km
Coastline: 0 km (landlocked)
Maritime claims: none (landlocked)
Climate: tropical; generally rainy with two dry seasons (December to February, June to August); semiarid in northeast
Terrain: mostly plateau with rim of mountains
Elevation extremes:
lowest point: Lake Albert 621 m
highest point: Margherita Peak on Mount Stanley 5,110 m

Natural resources: copper, cobalt, hydropower, limestone, salt, arable land
Land use:
arable land: 21.57%
permanent crops: 8.92%
other: 69.51% (2005)
Irrigated land: 90 sq km (2003)
Total renewable water resources: 66 cu km (1970)
Freshwater withdrawal (domestic/industrial/agricultural): *total:* 0.3 cu km/yr (43%/17%/40%)
per capita: 10 cu m/yr (2002)
Natural hazards: NA
Environment—current issues: draining of wetlands for agricultural use; deforestation; overgrazing; soil erosion; water hyacinth infestation in Lake Victoria; widespread poaching
Environment—international agreements: *party to:* Biodiversity, Climate Change, Climate Change-Kyoto Protocol, Desertification, Endangered Species, Hazardous Wastes, Law of the Sea, Marine Life Conservation, Ozone Layer Protection, Wetlands
signed, but not ratified: Environmental Modification
Geography—note: landlocked; fertile, well-watered country with many lakes and rivers

PEOPLE

Population: 31,367,972
note: estimates for this country explicitly take into account the effects of excess mortality due to AIDS; this can result in lower life expectancy, higher infant mortality, higher death rates, lower population growth rates, and changes in the distribution of population by age and sex than would otherwise be expected (July 2008 est.)
Age structure:
0–14 years: 50% (male 7,903,935/female 7,789,792)
15–64 years: 47.8% (male 7,528,073/female 7,469,938)
65 years and over: 2.2% (male 284,122/female 392,112) (2008 est.)
Median age:
total: 15 years
male: 14.9 years
female: 15.1 years (2008 est.)
Population growth rate: 3.603% (2008 est.)
Birth rate: 48.15 births/1,000 population (2008 est.)
Death rate: 12.32 deaths/1,000 population (2008 est.)
Net migration rate: 0.21 migrant(s)/1,000 population (2008 est.)
Sex ratio:
at birth: 1.03 male(s)/female
under 15 years: 1.01 male(s)/female
15–64 years: 1.01 male(s)/female

65 years and over: 0.72 male(s)/female
total population: 1 male(s)/female (2008 est.)
Infant mortality rate:
total: 65.99 deaths/1,000 live births
male: 69.65 deaths/1,000 live births
female: 62.21 deaths/1,000 live births (2008 est.)
Life expectancy at birth:
total population: 52.34 years
male: 51.31 years
female: 53.4 years (2008 est.)
Total fertility rate: 6.81 children born/woman (2008 est.)
HIV/AIDS—adult prevalence rate: 4.1% (2003 est.)
HIV/AIDS—people living with HIV/AIDS: 530,000 (2001 est.)
HIV/AIDS—deaths: 78,000 (2003 est.)
Major infectious diseases:
degree of risk: very high
food or waterborne diseases: bacterial diarrhea, hepatitis A, and typhoid fever
vectorborne diseases: chikungunya, malaria, plague, and African trypanosomiasis (sleeping sickness)
water contact disease: schistosomiasis (2008)
Nationality:
noun: Ugandan(s)
adjective: Ugandan
Ethnic groups: Baganda 16.9%, Banyakole 9.5%, Basoga 8.4%, Bakiga 6.9%, Iteso 6.4%, Langi 6.1%, Acholi 4.7%, Bagisu 4.6%, Lugbara 4.2%, Bunyoro 2.7%, other 29.6% (2002 census)
Religions: Roman Catholic 41.9%, Protestant 42% (Anglican 35.9%, Pentecostal 4.6%, Seventh Day Adventist 1.5%), Muslim 12.1%, other 3.1%, none 0.9% (2002 census)
Languages: English (official national language, taught in grade schools, used in courts of law and by most newspapers and some radio broadcasts), Ganda or Luganda (most widely used of the Niger-Congo languages, preferred for native language publications in the capital and may be taught in school), other Niger-Congo languages, Nilo-Saharan languages, Swahili, Arabic
Literacy:
definition: age 15 and over can read and write
total population: 66.8%
male: 76.8%
female: 57.7% (2002 census)

GOVERNMENT

Country name:
conventional long form: Republic of Uganda
conventional short form: Uganda
Government type: republic
Capital: *name:* Kampala

geographic coordinates: 0 19 N, 32 25 E

time difference: UTC+3 (8 hours ahead of Washington, DC during Standard Time)

Administrative divisions: 56 districts; Adjumani, Apac, Arua, Bugiri, Bundibugyo, Bushenyi, Busia, Gulu, Hoima, Iganga, Jinja, Kabale, Kabarole, Kaberamaido, Kalangala, Kampala, Kamuli, Kamwenge, Kanungu, Kapchorwa, Kasese, Katakwi, Kayunga, Kibale, Kiboga, Kisoro, Kitgum, Kotido, Kumi, Kyenjojo, Lira, Luwero, Masaka, Masindi, Mayuge, Mbale, Mbarara, Moroto, Moyo, Mpigi, Mubende, Mukono, Nakapiripirit, Nakasongola, Nebbi, Ntungamo, Pader, Pallisa, Rakai, Rukungiri, Sembabule, Sironko, Soroti, Tororo, Wakiso, Yumbe

note: as of a July 2005, 13 new districts were reportedly added bringing the total up to 69; the new districts are Amolatar, Amuria, Budaka, Butaleja, Ibanda, Kaabong, Kabingo, Kaliro, Kiruhura, Koboko, Manafwa, Mityana, Nakaseke; a total of ten more districts are in the process of being added

Independence: 9 October 1962 (from UK)

National holiday: Independence Day, 9 October (1962)

Constitution: 8 October 1995; in 2005 the constitution was amended removing presidential term limits and legalizing a multiparty political system

Legal system: in 1995, the government restored the legal system to one based on English common law and customary law; accepts compulsory ICJ jurisdiction, with reservations

Suffrage: 18 years of age; universal

Executive branch:

chief of state: President Lt. Gen. Yoweri Kaguta MUSEVENI (since seizing power 26 January 1986); note—the president is both chief of state and head of government

head of government: President Lt. Gen. Yoweri Kaguta MUSEVENI (since seizing power 26 January 1986); Prime Minister Apolo NSIBAMBI (since 5 April 1999); note—the prime minister assists the president in the supervision of the cabinet

cabinet: Cabinet appointed by the president from among elected legislators

elections: president reelected by popular vote for a five-year term; election last held 23 February 2006 (next to be held in 2011)

election results: Lt. Gen. Yoweri Kaguta MUSEVENI elected president; percent of vote—Lt. Gen. Yoweri Kaguta MUSEVENI 59.3%, Kizza BESIGYE 37.4%, other 3.3%

Legislative branch: unicameral National Assembly (332 seats; 215 members elected by popular vote, 104 nominated by legally established special interest groups [women 79, army 10, disabled 5, youth 5, labor 5], 13 ex officio members; to serve five-year terms)

elections: last held 23 February 2006 (next to be held in 2011)

election results: percent of vote by party—NA; seats by party—NRM 191, FDC 37, UPC 9, DP 8, CP 1, JEEMA 1, independents 36, other 49

Judicial branch: Court of Appeal (judges are appointed by the president and approved by the legislature); High Court (judges are appointed by the president)

Political parties and leaders: Conservative Party or CP [Ken LUKYA-MUZI]; Democratic Party or DP [Kizito SSEBAANA]; Forum for Democratic Change or FDC [Kizza BESIGYE]; Justice Forum or JEEMA [Muhammad Kibirige MAYANJA]; National Resistance Movement or NRM [Yoweri MUSEVENI]; Peoples Progressive Party or PPP [Bidandi SSALI]; Ugandan People's Congress or UPC [Miria OBOTE]

note: a national referendum in July 2005 opened the way for Uganda's transition to a multi-party political system

Political pressure groups and leaders:

International organization participation: ACP, AfDB, AU, C, COMESA, EAC, EADB, FAO, G-77, IAEA, IBRD, ICAO, ICCt, ICRM, IDA, IDB, IFAD, IFC, IFRCS, IGAD, ILO, IMF, Interpol, IOC, IOM, IPU, ISO (correspondent), ITSO, ITU, ITUC, MIGA, NAM, OIC, OPCW, PCA, UN, UNCTAD, UNESCO, UNHCR, UNIDO, UNMIS, UNOCI, UNWTO, UPU, WCO, WFTU, WHO, WIPO, WMO, WTO

Diplomatic representation in the US:

chief of mission: Ambassador Perezi Karukubiro KAMUNANWIRE

chancery: 5911 16th Street NW, Washington, DC 20011

telephone: [1] (202) 726-7100 through 7102, 0416

FAX: [1] (202) 726-1727

Diplomatic representation from the US:

chief of mission: Ambassador Steven BROWNING

embassy: 1577 Ggaba Road, Kampala

mailing address: P. O. Box 7007, Kampala

telephone: [256] (41) 234-142

FAX: [256] (41) 258-451

Flag description: six equal horizontal bands of black (top), yellow, red, black, yellow, and red; a white disk is superimposed at the center and depicts a red-crested crane (the national symbol) facing the hoist side

ECONOMY

Economy—overview: Uganda has substantial natural resources, including fertile soils, regular rainfall, and sizable mineral deposits of copper, cobalt, gold, and other minerals. Agriculture is the most important sector of the economy, employing over 80% of the work force. Coffee accounts for the bulk of export revenues. Since 1986, the government—with the support of foreign countries and international agencies—has acted to rehabilitate and stabilize the economy by undertaking currency reform, raising producer prices on export crops, increasing prices of petroleum products, and improving civil service wages. The policy changes are especially aimed at dampening inflation and boosting production and export earnings. During 1990–2001, the economy turned in a solid performance based on continued investment in the rehabilitation of infrastructure, improved incentives for production and exports, reduced inflation, gradually improved domestic security, and the return of exiled Indian-Ugandan entrepreneurs. Growth continues to be solid, despite variability in the price of coffee, Uganda's principal export, and a consistent upturn in Uganda's export markets. In 2000, Uganda qualified for enhanced Highly Indebted Poor Countries (HIPC) debt relief worth $1.3 billion and Paris Club debt relief worth $145 million. These amounts combined with the original HIPC debt relief added up to about $2 billion.

GDP (purchasing power parity): $29.04 billion (2007 est.)

GDP (official exchange rate): $11.23 billion (2007 est.)

GDP—real growth rate: 6.5% (2007 est.)

GDP—per capita (PPP): $900 (2007 est.)

GDP—composition by sector:

agriculture: 30.2%

industry: 24.7%

services: 45.1% (2007 est.)

Labor force: 14.02 million (2007 est.)

Labor force—by occupation: *agriculture:* 82%

industry: 5%

services: 13% (1999 est.)

Unemployment rate: NA%

Population below poverty line: 35% (2001 est.)

Household income or consumption by percentage share:

lowest 10%: 2.3%

highest 10%: 37.7% (2002)

Distribution of family income—Gini index: 45.7 (2002)

Inflation rate (consumer prices): 6.8% (2007 est.)

Investment (gross fixed): 25.4% of GDP (2007 est.)

Budget:

revenues: $2.248 billion

expenditures: $2.506 billion; including capital expenditures of $NA (2007 est.)

Public debt: 20.6% of GDP (2007 est.)

Agriculture—products: coffee, tea, cotton, tobacco, cassava (tapioca), potatoes, corn, millet, pulses, cut flowers; beef, goat meat, milk, poultry

Industries: sugar, brewing, tobacco, cotton textiles; cement, steel production

Industrial production growth rate: 5.8% (2007 est.)

Electricity—production: 1.983 billion kWh (2005)

Electricity—production by source:
fossil fuel: 0.9%
hydro: 99.1%
nuclear: 0%
other: 0% (2001)

Electricity—consumption: 1.674 billion kWh (2005)

Electricity—exports: 170 million kWh (2005)

Electricity—imports: 0 kWh (2005)

Oil—production: 0 bbl/day (2005 est.)

Oil—consumption: 11,000 bbl/day (2005 est.)

Oil—exports: 0 bbl/day (2004)

Oil—imports: 10,870 bbl/day (2004)

Oil—proved reserves: 0 bbl (1 January 2006 est.)

Natural gas—production: 0 cu m (2005 est.)

Natural gas—consumption: 0 cu m (2005 est.)

Natural gas—exports: 0 cu m (2005 est.)

Natural gas—imports: 0 cu m (2005)

Natural gas—proved reserves: 0 cu m (1 January 2006 est.)

Current account balance: -$224 million (2007 est.)

Exports: $1.626 billion f.o.b. (2007 est.)

Exports—commodities: coffee, fish and fish products, tea, cotton, flowers, horticultural products; gold

Exports—partners: Belgium 9.8%, Netherlands 9.3%, France 7.8%, Germany 7.6%, Rwanda 5.6%, Sudan 5.5% (2006)

Imports: $2.803 billion f.o.b. (2007 est.)

Imports—commodities: capital equipment, vehicles, petroleum, medical supplies; cereals

Imports—partners: Kenya 34.1%, UAE 8.5%, China 7.1%, India 5.6%, South Africa 5.4%, Japan 4.2% (2006)

Economic aid—recipient: $1.198 billion (2005)

Reserves of foreign exchange and gold: $2.1 billion (31 December 2007 est.)

Debt—external: $1.498 billion (31 December 2007 est.)

Stock of direct foreign investment—at home: $NA

Stock of direct foreign investment—abroad: $NA

Market value of publicly traded shares: $103.4 million (2005)

Currency (code): Ugandan shilling (UGX)

Currency code: UGX

Exchange rates: Ugandan shillings per US dollar—1,685.8 (2007), 1,834.9 (2006), 1,780.7 (2005), 1,810.3 (2004), 1,963.7 (2003)

Fiscal year: 1 July—30 June

COMMUNICATIONS

Telephones—main lines in use: 108,100 (2006)

Telephones—mobile cellular: 2.009 million (2006)

Telephone system:
general assessment: seriously inadequate; mobile cellular service is increasing rapidly, but the number of main lines is still deficient; e-mail and Internet services are available
domestic: intercity traffic by wire, microwave radio relay, and radiotelephone communication stations, fixed and mobile-cellular systems for short-range traffic
international: country code—256; satellite earth stations—1 Intelsat (Atlantic Ocean) and 1 Inmarsat; analog links to Kenya and Tanzania

Radio broadcast stations: AM 7, FM 33, shortwave 2 (2001)

Radios: 5 million (2001)

Television broadcast stations: 8 (plus 1 repeater) (2001)

Televisions: 500,000 (2001)

Internet country code: .ug

Internet hosts: 546 (2007)

Internet Service Providers (ISPs): 2 (2000)

Internet users: 750,000 (2006)

TRANSPORTATION

Airports: 32 (2007)

Airports—with paved runways:
total: 5
over 3,047 m: 3
1,524 to 2,437 m: 1
914 to 1,523 m: 1 (2007)

Airports—with unpaved runways:
total: 27
2,438 to 3,047 m: 1
1,524 to 2,437 m: 6
914 to 1,523 m: 11
under 914 m: 9 (2007)

Railways:
total: 1,244 km

narrow gauge: 1,244 km 1.000-m gauge (2006)

Roadways:
total: 70,746 km
paved: 16,272 km
unpaved: 54,474 km (2003)

Waterways: on Lake Victoria, 200 km on Lake Albert, Lake Kyoga, and parts of Albert Nile (2005)

Ports and terminals: Entebbe, Jinja, Port Bell

MILITARY

Military branches: Uganda Peoples Defense Force (UPDF): Army (includes Marine Unit), Air Force (2007)

Military service age and obligation: 18–26 years of age for compulsory and voluntary military duty; 18–30 years of age for professionals; 9-year service obligation; the government has stated that recruitment below 18 years of age could occur with proper consent and that "no person under the apparent age of 13 years shall be enrolled in the armed forces" (2007)

Manpower available for military service:
males age 16–49: 6,532,894
females age 16–49: 6,352,416 (2008 est.)

Manpower fit for military service:
males age 16–49: 3,856,365
females age 16–49: 3,769,120 (2008 est.)

Military expenditures—percent of GDP: 2.2% (2006)

TRANSNATIONAL ISSUES

Disputes—international: Uganda is subject to armed fighting among hostile ethnic groups, rebels, armed gangs, militias, and various government forces that extend across its borders; Uganda hosts 209,860 Sudanese, 27,560 Congolese, and 19,710 Rwandan refugees, while Ugandan refugees as well as members of the Lord's Resistance Army (LRA) seek shelter in southern Sudan and the Democratic Republic of the Congo's Garamba National Park; LRA forces have also attacked Kenyan villages across the border

Refugees and internally displaced persons: *refugees (country of origin):* 215,700 (Sudan); 28,880 (Democratic Republic of Congo); 24,900 (Rwanda)
IDPs: 1.27 million (350,000 IDPs returned in 2006 following ongoing peace talks between the Lord's Resistance Army (LRA) and the Government of Uganda) (2007)

UKRAINE

INTRODUCTION

Background: Ukraine was the center of the first eastern Slavic state, Kyivan Rus, which during the 10th and 11th centuries was the largest and most powerful state in Europe. Weakened by internecine quarrels and Mongol invasions, Kyivan Rus was incorporated into

the Grand Duchy of Lithuania and eventually into the Polish-Lithuanian Commonwealth. The cultural and religious legacy of Kyivan Rus laid the foundation for Ukrainian nationalism through subsequent centuries. A new Ukrainian state, the Cossack Hetmanate, was established during the mid-17th century after an uprising against the Poles. Despite continuous Muscovite pressure, the Hetmanate managed to remain autonomous for well over 100 years. During the latter part of the 18th century, most Ukrainian ethnographic territory was absorbed by the Russian Empire. Following the collapse of czarist Russia in 1917, Ukraine was able to bring about a short-lived period of independence (1917–20), but was reconquered and forced to endure a brutal Soviet rule that engineered two artificial famines (1921–22 and 1932–33) in which over 8 million died. In World War II, German and Soviet armies were responsible for some 7 to 8 million more deaths. Although final independence for Ukraine was achieved in 1991 with the dissolution of the USSR, democracy remained elusive as the legacy of state control and endemic corruption stalled efforts at economic reform, privatization, and civil liberties. A peaceful mass protest "Orange Revolution" in the closing months of 2004 forced the authorities to overturn a rigged presidential election and to allow a new internationally monitored vote that swept into power a reformist slate under Viktor YUSHCHENKO. Subsequent internal squabbles in the YUSHCHENKO camp allowed his rival Viktor YANUKOVYCH to stage a comeback in parliamentary elections and become prime minister in August of 2006. An early legislative election, brought on by a political crisis in the spring of 2007, saw Yuliya TYMOSHENKO, as head of an "Orange" coalition, installed as a new prime minister in December 2007.

GEOGRAPHY

Location: Eastern Europe, bordering the Black Sea, between Poland, Romania, and Moldova in the west and Russia in the east

Geographic coordinates: 49 00 N, 32 00 E

Map references: Asia, Europe

Area:
total: 603,700 sq km
land: 603,700 sq km
water: 0 sq km

Area—comparative: slightly smaller than Texas

Land boundaries:
total: 4,663 km
border countries: Belarus 891 km, Hungary 103 km, Moldova 939 km, Poland 526 km, Romania (south) 169 km, Romania (west) 362 km, Russia 1,576 km, Slovakia 97 km

Coastline: 2,782 km

Maritime claims:
territorial sea: 12 nm
exclusive economic zone: 200 nm
continental shelf: 200-m or to the depth of exploitation

Climate: temperate continental; Mediterranean only on the southern Crimean coast; precipitation disproportionately distributed, highest in west and north, lesser in east and southeast; winters vary from cool along the Black Sea to cold farther inland; summers are warm across the greater part of the country, hot in the south

Terrain: most of Ukraine consists of fertile plains (steppes) and plateaus, mountains being found only in the west (the Carpathians), and in the Crimean Peninsula in the extreme south

Elevation extremes:
lowest point: Black Sea 0 m
highest point: Hora Hoverla 2,061 m

Natural resources: iron ore, coal, manganese, natural gas, oil, salt, sulfur, graphite, titanium, magnesium, kaolin, nickel, mercury, timber, arable land

Land use:
arable land: 53.8%

permanent crops: 1.5%
other: 44.7% (2005)

Irrigated land: 22,080 sq km (2003)

Total renewable water resources: 139.5 cu km (1997)

Freshwater withdrawal (domestic/industrial/agricultural): *total:* 37.53 cu km/yr (12%/35%/52%)
per capita: 807 cu m/yr (2000)

Natural hazards: NA

Environment—current issues: inadequate supplies of potable water; air and water pollution; deforestation; radiation contamination in the northeast from 1986 accident at Chornobyl' Nuclear Power Plant

Environment—international agreements: *party to:* Air Pollution, Air Pollution-Nitrogen Oxides, Air Pollution-Sulfur 85, Antarctic-Environmental Protocol, Antarctic-Marine Living Resources, Antarctic Treaty, Biodiversity, Climate Change, Climate Change-Kyoto Protocol, Endangered Species, Environmental Modification, Hazardous Wastes, Law of the Sea, Marine Dumping, Ozone Layer Protection, Ship Pollution, Wetlands
signed, but not ratified: Air Pollution-Persistent Organic Pollutants, Air Pollution-Sulfur 94, Air Pollution-Volatile Organic Compounds

Geography—note: strategic position at the crossroads between Europe and Asia; second-largest country in Europe

PEOPLE

Population: 45,994,287 (July 2008 est.)

Age structure:
0–14 years: 13.9% (male 3,277,905/female 3,106,012)
15–64 years: 70% (male 15,443,818/female 16,767,931)
65 years and over: 16.1% (male 2,489,235/female 4,909,386) (2008 est.)

Median age: *total:* 39.4 years
male: 36.1 years
female: 42.5 years (2008 est.)

Population growth rate: -0.651% (2008 est.)

Birth rate: 9.55 births/1,000 population (2008 est.)

Death rate: 15.93 deaths/1,000 population (2008 est.)

Net migration rate: -0.12 migrant(s)/1,000 population (2008 est.)

Sex ratio:
at birth: 1.06 male(s)/female
under 15 years: 1.06 male(s)/female
15–64 years: 0.92 male(s)/female
65 years and over: 0.51 male(s)/female
total population: 0.86 male(s)/female (2008 est.)

Infant mortality rate:
total: 9.23 deaths/1,000 live births
male: 11.48 deaths/1,000 live births
female: 6.85 deaths/1,000 live births (2008 est.)

Life expectancy at birth:
total population: 68.06 years
male: 62.24 years
female: 74.24 years (2008 est.)
Total fertility rate: 1.25 children born/woman (2008 est.)
HIV/AIDS—adult prevalence rate: 1.4% (2003 est.)
HIV/AIDS—people living with HIV/AIDS: 360,000 (2001 est.)
HIV/AIDS—deaths: 20,000 (2003 est.)
Nationality:
noun: Ukrainian(s)
adjective: Ukrainian
Ethnic groups: Ukrainian 77.8%, Russian 17.3%, Belarusian 0.6%, Moldovan 0.5%, Crimean Tatar 0.5%, Bulgarian 0.4%, Hungarian 0.3%, Romanian 0.3%, Polish 0.3%, Jewish 0.2%, other 1.8% (2001 census)
Religions: Ukrainian Orthodox—Kyiv Patriarchate 50.4%, Ukrainian Orthodox—Moscow Patriarchate 26.1%, Ukrainian Greek Catholic 8%, Ukrainian Autocephalous Orthodox 7.2%, Roman Catholic 2.2%, Protestant 2.2%, Jewish 0.6%, other 3.2% (2006 est.)
Languages: Ukrainian (official) 67%, Russian 24%, other 9% (includes small Romanian-, Polish-, and Hungarian-speaking minorities)
Literacy: *definition:* age 15 and over can read and write
total population: 99.4%
male: 99.7%
female: 99.2% (2001 census)

GOVERNMENT

Country name:
conventional long form: none
conventional short form: Ukraine
local long form: none
local short form: Ukrayina
former: Ukrainian National Republic, Ukrainian State, Ukrainian Soviet Socialist Republic
Government type: republic
Capital: *name:* Kyiv (Kiev)
geographic coordinates: 50 26 N, 30 31 E
time difference: UTC+2 (7 hours ahead of Washington, DC during Standard Time)
daylight saving time: +1hr, begins last Sunday in March; ends last Sunday in October
Administrative divisions: 24 provinces (oblasti, singular—oblast'), 1 autonomous republic* (avtonomna respublika), and 2 municipalities (mista, singular—misto) with oblast status**; Cherkasy, Chernihiv, Chernivtsi, Crimea or Avtonomna Respublika Krym* (Simferopol'), Dnipropetrovs'k, Donets'k, Ivano-Frankivs'k, Kharkiv, Kherson, Khmel'nyts'kyy, Kirovohrad, Kyiv**, Kyiv, Luhans'k, L'viv, Mykolayiv, Odesa, Poltava, Rivne, Sevastopol'**, Sumy, Ternopil',

Vinnytsya, Volyn' (Luts'k), Zakarpattya (Uzhhorod), Zaporizhzhya, Zhytomyr
note: administrative divisions have the same names as their administrative centers (exceptions have the administrative center name following in parentheses)
Independence: 24 August 1991 (from Soviet Union)
National holiday: Independence Day, 24 August (1991); note—22 January 1918, the day Ukraine first declared its independence (from Soviet Russia) and the day the short-lived Western and Central Ukrainian republics united (1919), is now celebrated as Unity Day
Constitution: adopted 28 June 1996
Legal system: based on civil law system; judicial review of legislative acts; has not accepted compulsory ICJ jurisdiction
Suffrage: 18 years of age; universal
Executive branch:
chief of state: President Viktor A. YUSHCHENKO (since 23 January 2005)
head of government: Prime Minister Yuliya TYMOSHENKO (since 18 December 2007); First Deputy Prime Minister Oleksandr TURCHYNOV (since 18 December 2007); Deputy Prime Ministers Hryhoriy NEMYRYA and Ivan VASYUNYK (since 18 December 2007)
cabinet: Cabinet of Ministers selected by the prime minister; the only exceptions are the foreign and defense ministers, who are chosen by the president
note: there is also a National Security and Defense Council or NSDC originally created in 1992 as the National Security Council; the NSDC staff is tasked with developing national security policy on domestic and international matters and advising the president; a Presidential Secretariat helps draft presidential edicts and provides policy support to the president
elections: president elected by popular vote for a five-year term (eligible for a second term); note—a special repeat runoff presidential election between Viktor YUSHCHENKO and Viktor YANUKOVYCH took place on 26 December 2004 after the earlier 21 November 2004 contest—won by YANUKOVYCH—was invalidated by the Ukrainian Supreme Court because of widespread and significant violations; under constitutional reforms that went into effect 1 January 2006, the majority in parliament takes the lead in naming the prime minister
election results: Viktor YUSHCHENKO elected president; percent of vote—Viktor YUSHCHENKO 52%, Viktor YANUKOVYCH 44.2%
Legislative branch: unicameral Supreme Council or Verkhovna Rada (450 seats; members allocated on a proportional

basis to those parties that gain 3% or more of the national electoral vote; to serve five-year terms)
elections: last held 30 September 2007 (next to be held in 2012)
election results: percent of vote by party/bloc—Party of Regions 34.4%, Yuliya Tymoshenko Bloc 30.7%, Our Ukraine-People's Self Defense 14.2%, CPU 5.4%, Lytvyn bloc 4%, other parties 11.3%; seats by party/bloc—Party of Regions 175, Yuliya Tymoshenko Bloc 156, Our Ukraine-People's Self Defense 72, CPU 27, Lytvyn bloc 20
Judicial branch: Supreme Court; Constitutional Court
Political parties and leaders: Christian Democratic Union [Volodymyr STRETOVYCH]; Communist Party of Ukraine or CPU [Petro SYMONENKO]; European Party of Ukraine [Mykola KATERYNCHUK]; Fatherland Party (Batkivshchyna) [Yuliya TYMOSHENKO]; Forward Ukraine! [Viktor MUSIYAKA]; Labor Party of Ukraine [Mykola SYROTA]; People's Union Our Ukraine [Vyacheslav KYRYLENKO]; Party of Industrialists and Entrepreneurs [Anatoliy KINAKH]; Party of the Defenders of the Fatherland [Yuriy Karmazin]; People's Movement of Ukraine (Rukh) [Borys TARASYUK]; People's Party [Volodymyr LYTVYN]; PORA! (It's Time!) party [Vladyslav KASKIV]; Progressive Socialist Party [Natalya VITRENKO]; Reforms and Order Party [Viktor PYNZENYK]; Party of Regions [Viktor YANUKOVYCH]; Republican Party [Yuriy BOYKO]; Sobor [Anatoliy MATVIYENKO]; Social Democratic Party [Yevhen KORNICHUK]; Social Democratic Party (United) or SDPU(o) [Yuriy ZAHORODNIY]; Socialist Party of Ukraine or SPU [Oleksandr MOROZ]; Ukrainian People's Party [Yuriy KOSTENKO]; Viche [Inna BOHOSLOVSKA]
Political pressure groups and leaders: Committee of Voters of Ukraine [Ihor POPOV]; Peoples' Self-Defense [Yuriy LUTSENKO]
International organization participation: Australia Group, BSEC, CBSS (observer), CE, CEI, CIS, EAEC (observer), EAPC, EBRD, FAO, GCTU, GUAM, IAEA, IBRD, ICAO, ICC, ICCt (signatory), ICRM, IDA, IFC, IFRCS, IHO, ILO, IMF, IMO, IMSO, Interpol, IOC, IOM, IPU, ISO, ITU, ITUC, LAIA (observer), MIGA, NAM (observer), NSG, OAS (observer), OIF (observer), OPCW, OSCE, PCA, PFP, SECI (observer), UN, UNCTAD, UNESCO, UNIDO, UNMEE, UNMIL, UNMIS, UNOMIG, UNWTO, UPU, WCL, WCO, WFTU, WHO, WIPO, WMO, WTO (observer), ZC

Diplomatic representation in the US:
chief of mission: Ambassador Oleh V. SHAMSHUR
chancery: 3350 M Street NW, Washington, DC 20007
telephone: [1] (202) 333-0606
FAX: [1] (202) 333-0817
consulate(s) general: Chicago, New York, San Francisco

Diplomatic representation from the US:
chief of mission: Ambassador William B. TAYLOR Jr.
embassy: 10 Yurii Kotsiubynsky Street, 04053 Kyiv
mailing address: 5850 Kiev Place, Washington, DC 20521-5850
telephone: [380] (44) 490-4000
FAX: [380] (44) 490-4085
Flag description: two equal horizontal bands of azure (top) and golden yellow represent grain fields under a blue sky

ECONOMY

Economy—overview: After Russia, the Ukrainian republic was far and away the most important economic component of the former Soviet Union, producing about four times the output of the next-ranking republic. Its fertile black soil generated more than one-fourth of Soviet agricultural output, and its farms provided substantial quantities of meat, milk, grain, and vegetables to other republics. Likewise, its diversified heavy industry supplied the unique equipment (for example, large diameter pipes) and raw materials to industrial and mining sites (vertical drilling apparatus) in other regions of the former USSR. Shortly after independence was ratified in December 1991, the Ukrainian Government liberalized most prices and erected a legal framework for privatization, but widespread resistance to reform within the government and the legislature soon stalled reform efforts and led to some backtracking. Output by 1999 had fallen to less than 40% of the 1991 level. Ukraine's dependence on Russia for energy supplies and the lack of significant structural reform have made the Ukrainian economy vulnerable to external shocks. Ukraine depends on imports to meet about three-fourths of its annual oil and natural gas requirements. A dispute with Russia over pricing in late 2005 and early 2006 led to a temporary gas cut-off; Ukraine concluded a deal with Russia in January 2006 that almost doubled the price Ukraine pays for Russian gas. Outside institutions—particularly the IMF—have encouraged Ukraine to quicken the pace and scope of reforms. Ukrainian Government officials eliminated most tax and customs privileges in a March 2005 budget law, bringing more economic activity out of Ukraine's large shadow economy, but more improvements are needed, including fighting corruption, developing capital markets, and improving the legislative framework. Ukraine's economy remains buoyant despite political turmoil between the Prime Minister and President. Real GDP growth reached about 7% in 2006–07, fueled by high global prices for steel—Ukraine's top export—and by strong domestic consumption, spurred by rising pensions and wages. Although the economy is likely to expand in 2008, long-term growth could be threatened by the government's plans to reinstate tax, trade, and customs privileges and to maintain restrictive grain export quotas.

GDP (purchasing power parity): $320.1 billion (2007 est.)
GDP (official exchange rate): $140.5 billion (2007 est.)
GDP—real growth rate: 7.3% (2007 est.)
GDP—per capita (PPP): $6,900 (2007 est.)
GDP—composition by sector:
agriculture: 9.1%
industry: 32.3%
services: 58.7% (2007 est.)
Labor force: 21.58 million (2007 est.)
Labor force—by occupation: *agriculture:* 25%
industry: 20%
services: 55% (1996)
Unemployment rate: 2.3% officially registered; large number of unregistered or underemployed workers; the International Labor Organization calculates that Ukraine's real unemployment level is nearly 7% (2007 est.)
Population below poverty line: 37.7% (2003)
Household income or consumption by percentage share:
lowest 10%: 3.4%
highest 10%: 25.7% (2006)
Distribution of family income—Gini index: 31 (2006)
Inflation rate (consumer prices): 12.8% (2007 est.)
Investment (gross fixed): 26.3% of GDP (2007 est.)
Budget:
revenues: $43.54 billion
expenditures: $45.06 billion; note—this is the planned, consolidated budget (2007 est.)
Public debt: 11.7% of GDP (2007 est.)
Agriculture—products: grain, sugar beets, sunflower seeds, vegetables; beef, milk
Industries: coal, electric power, ferrous and nonferrous metals, machinery and transport equipment, chemicals, food processing (especially sugar)
Industrial production growth rate: 6% (2007 est.)
Electricity—production: 192.1 billion kWh (2006)

Electricity—production by source:
fossil fuel: 48.6%
hydro: 7.9%
nuclear: 43.5%
other: 0% (2001)
Electricity—consumption: 181.9 billion kWh (2006)
Electricity—exports: 10.07 billion kWh (2005)
Electricity—imports: 20 billion kWh (2006)
Oil—production: 90,400 bbl/day (2006)
Oil—consumption: 284,600 bbl/day (2006)
Oil—exports: 214,600 bbl/day (2004)
Oil—imports: 469,600 bbl/day (2004)
Oil—proved reserves: 395 million bbl (1 January 2006 est.)
Natural gas—production: 20.85 billion cu m (2006 est.)
Natural gas—consumption: 73.94 billion cu m (2006 est.)
Natural gas—exports: 4 billion cu m (2006 est.)
Natural gas—imports: 57.09 billion cu m (2006 est.)
Natural gas—proved reserves: 1.075 trillion cu m (1 January 2006 est.)
Current account balance: -$5.918 billion (2007 est.)
Exports: $49.84 billion (2007 est.)
Exports—commodities: ferrous and nonferrous metals, fuel and petroleum products, chemicals, machinery and transport equipment, food products
Exports—partners: Russia 21.4%, Turkey 7.1%, Italy 6.3%, US 4.1% (2006)
Imports: $60.41 billion (2007 est.)
Imports—commodities: energy, machinery and equipment, chemicals
Imports—partners: Russia 28.3%, Germany 11.7%, Poland 7.6%, China 7%, Turkmenistan 5.7% (2006)
Economic aid—recipient: $409.6 million (1995); IMF Extended Funds Facility $2.2 billion (2005)
Reserves of foreign exchange and gold: $32.48 billion (31 December 2007 est.)
Debt—external: $69.19 billion (31 December 2007)
Stock of direct foreign investment—at home: $31.08 billion (2007 est.)
Stock of direct foreign investment—abroad: $895 million (2007 est.)
Market value of publicly traded shares: $42.87 billion (2006)
Currency (code): hryvnia (UAH)
Currency code: UAH
Exchange rates: hryvnia per US dollar—5.05 (2007), 5.05 (2006), 5.1247 (2005), 5.3192 (2004), 5.3327 (2003)
Fiscal year: calendar year

COMMUNICATIONS

Telephones—main lines in use: 12.341 million (2006)

Telephones—mobile cellular: 49.076 million (2006)

Telephone system:

general assessment: Ukraine's telecommunication development plan emphasizes improving domestic trunk lines, international connections, and the mobile-cellular system

domestic: at independence in December 1991, Ukraine inherited a telephone system that was antiquated, inefficient, and in disrepair; more than 3.5 million applications for telephones could not be satisfied; telephone density is rising and the domestic trunk system is being improved; about one-third of Ukraine's networks are digital and a majority of regional centers now have digital switching stations; improvements in local networks and local exchanges continue to lag; the mobile-cellular telephone system is expanding rapidly

international: country code—380; 2 new domestic trunk lines are a part of the fiber-optic Trans-Asia-Europe (TAE) system and 3 Ukrainian links have been installed in the fiber-optic Trans-European Lines (TEL) project that connects 18 countries; additional international service is provided by the Italy-Turkey-Ukraine-Russia (ITUR) fiber-optic submarine cable and by an unknown number of earth stations in the Intelsat, Inmarsat, and Intersputnik satellite systems

Radio broadcast stations: 524 (station types NA) (2006)

Radios: 45.05 million (1997)

Television broadcast stations: 647 (2006)

Televisions: 18.05 million (1997)

Internet country code: .ua

Internet hosts: 234,349 (2007)

Internet Service Providers (ISPs): 260 (2001)

Internet users: 5.545 million (2006)

TRANSPORTATION

Airports: 437 (2007)

Airports—with paved runways:

total: 193
over 3,047 m: 13
2,438 to 3,047 m: 53
1,524 to 2,437 m: 27
914 to 1,523 m: 5

under 914 m: 95 (2007)

Airports—with unpaved runways:

total: 244
2,438 to 3,047 m: 3
1,524 to 2,437 m: 11
914 to 1,523 m: 13
under 914 m: 217 (2007)

Heliports: 10 (2007)

Pipelines: gas 33,721 km; oil 4,514 km; refined products 4,211 km (2007)

Railways:

total: 22,473 km
broad gauge: 22,473 km 1.524-m gauge (9,250 km electrified) (2006)

Roadways:

total: 169,477 km
paved: 164,732 km (includes 15 km of expressways)
unpaved: 4,745 km (2004)

Waterways: 2,253 km (most on Dnieper River) (2006)

Merchant marine:

total: 193 ships (1000 GRT or over) 763,293 GRT/899,859 DWT
by type: bulk carrier 6, cargo 145, container 3, passenger 6, passenger/cargo 4, petroleum tanker 9, refrigerated cargo 11, roll on/roll off 7, specialized tanker 2
registered in other countries: 194 (Belize 10, Cambodia 27, Comoros 13, Cyprus 6, Dominica 3, Georgia 24, Liberia 24, Malta 28, Moldova 3, Mongolia 3, Panama 8, Russia 10, Sierra Leone 8, Slovakia 10, St Kitts and Nevis 5, St Vincent and The Grenadines 12, unknown 3) (2007)

Ports and terminals: Feodosiya, Kerch, Kherson, Mariupol', Mykolayiv, Odesa, Yuzhnyy

MILITARY

Military branches: Ground Forces, Naval Forces, Air Forces (Viyskovo-Povitryani Syly), Air Defense Forces (2002)

Military service age and obligation: 18–25 years of age for compulsory and voluntary military service; conscript service obligation—18 months for Army and Air Force, 24 months for Navy (2004)

Manpower available for military service: *males age 16–49:* 11,457,562
females age 16–49: 11,767,357 (2008 est.)

Manpower fit for military service:
males age 16–49: 7,141,814
females age 16–49: 9,428,876 (2008 est.)

Manpower reaching militarily significant age annually:
males age 16–49: 288,605
females age 16–49: 276,324 (2008 est.)

Military expenditures—percent of GDP: 1.4% (2005 est.)

TRANSNATIONAL ISSUES

Disputes—international: 1997 boundary delimitation treaty with Belarus remains un-ratified due to unresolved financial claims, stalling demarcation and reducing border security; delimitation of land boundary with Russia is complete with preparations for demarcation underway; the dispute over the boundary between Russia and Ukraine through the Kerch Strait and Sea of Azov remains unresolved despite a December 2003 framework agreement and ongoing expert-level discussions; Moldova and Ukraine operate joint customs posts to monitor transit of people and commodities through Moldova's break-away Transnistria Region, which remains under OSCE supervision; the ICJ gave Ukraine until December 2006 to reply, and Romania until June 2007 to rejoin, in their dispute submitted in 2004 over Ukrainian-administered Zmiyinyy/Serpilor (Snake) Island and Black Sea maritime boundary; Romania opposes Ukraine's reopening of a navigation canal from the Danube border through Ukraine to the Black Sea

Illicit drugs: limited cultivation of cannabis and opium poppy, mostly for CIS consumption; some synthetic drug production for export to the West; limited government eradication program; used as transshipment point for opiates and other illicit drugs from Africa, Latin America, and Turkey to Europe and Russia; Ukraine has improved anti-money-laundering controls, resulting in its removal from the Financial Action Task Force's (FATF's) Noncooperative Countries and Territories List in February 2004; Ukraine's anti-money-laundering regime continues to be monitored by FATF

UNITED ARAB EMIRATES

INTRODUCTION

Background: The Trucial States of the Persian Gulf coast granted the UK control of their defense and foreign affairs in 19th century treaties. In 1971, six of these states—Abu Zaby, 'Ajman, Al Fujayrah, Ash Shariqah, Dubayy, and Umm al Qaywayn—merged to form the

United Arab Emirates (UAE). They were joined in 1972 by Ra's al Khaymah. The UAE's per capita GDP is on par with those of leading West European nations. Its generosity with oil revenues and its moderate foreign policy stance have allowed the UAE to play a vital role in the affairs of the region.

GEOGRAPHY

Location: Middle East, bordering the Gulf of Oman and the Persian Gulf, between Oman and Saudi Arabia

Geographic coordinates: 24 00 N, 54 00 E

Map references: Middle East

Area: *total:* 83,600 sq km

675

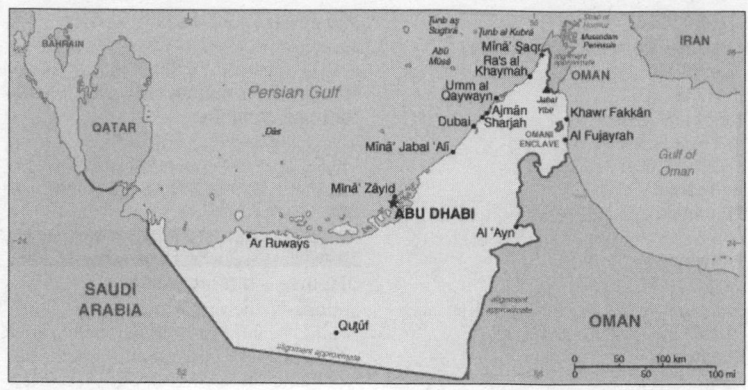

land: 83,600 sq km

water: 0 sq km

Area—comparative: slightly smaller than Maine

Land boundaries: *total:* 867 km

border countries: Oman 410 km, Saudi Arabia 457 km

Coastline: 1,318 km

Maritime claims:

territorial sea: 12 nm

contiguous zone: 24 nm

exclusive economic zone: 200 nm

continental shelf: 200 nm or to the edge of the continental margin

Climate: desert; cooler in eastern mountains

Terrain: flat, barren coastal plain merging into rolling sand dunes of vast desert wasteland; mountains in east

Elevation extremes:

lowest point: Persian Gulf 0 m

highest point: Jabal Yibir 1,527 m

Natural resources: petroleum, natural gas

Land use: *arable land:* 0.77%

permanent crops: 2.27%

other: 96.96% (2005)

Irrigated land: 760 sq km (2003)

Total renewable water resources: 0.2 cu km (1997)

Freshwater withdrawal (domestic/industrial/agricultural): *total:* 2.3 cu km/yr (23%/9%/68%)

per capita: 511 cu m/yr (2000)

Natural hazards: frequent sand and dust storms

Environment—current issues: lack of natural freshwater resources compensated by desalination plants; desertification; beach pollution from oil spills

Environment—international agreements: *party to:* Biodiversity, Climate Change, Climate Change-Kyoto Protocol, Desertification, Endangered Species, Hazardous Wastes, Marine Dumping, Ozone Layer Protection

signed, but not ratified: Law of the Sea

Geography—note: strategic location along southern approaches to Strait of Hormuz, a vital transit point for world crude oil

PEOPLE

Population: 4,621,399

note: estimate is based on the results of the 2005 census that included a significantly higher estimate of net inmigration of non-citizens than previous estimates (July 2008 est.)

Age structure:

0–14 years: 20.5% (male 484,102/female 462,405)

15–64 years: 78.6% (male 2,663,702/female 970,672)

65 years and over: 0.9% (male 26,244/female 14,274)

note: 73.9% of the population in the 15–64 age group is non-national (2008 est.)

Median age: *total:* 30.1 years

male: 32 years

female: 24.6 years (2008 est.)

Population growth rate: 3.833% (2008 est.)

Birth rate: 16.06 births/1,000 population (2008 est.)

Death rate: 2.13 deaths/1,000 population (2008 est.)

Net migration rate: 24.41 migrant(s)/1,000 population (2008 est.)

Sex ratio:

at birth: 1.05 male(s)/female

under 15 years: 1.05 male(s)/female

15–64 years: 2.74 male(s)/female

65 years and over: 1.84 male(s)/female

total population: 2.19 male(s)/female (2008 est.)

Infant mortality rate:

total: 13.11 deaths/1,000 live births

male: 15.32 deaths/1,000 live births

female: 10.8 deaths/1,000 live births (2008 est.)

Life expectancy at birth:

total population: 75.89 years

male: 73.35 years

female: 78.56 years (2008 est.)

Total fertility rate: 2.43 children born/woman (2008 est.)

HIV/AIDS—adult prevalence rate: 0.18% (2001 est.)

HIV/AIDS—people living with HIV/AIDS: NA

HIV/AIDS—deaths: NA

Nationality:

noun: Emirati(s)

adjective: Emirati

Ethnic groups: Emirati 19%, other Arab and Iranian 23%, South Asian 50%, other expatriates (includes Westerners and East Asians) 8% (1982)

note: less than 20% are UAE citizens (1982)

Religions: Muslim 96% (Shi'a 16%), other (includes Christian, Hindu) 4%

Languages: Arabic (official), Persian, English, Hindi, Urdu

Literacy:

definition: age 15 and over can read and write

total population: 77.9%

male: 76.1%

female: 81.7% (2003 est.)

GOVERNMENT

Country name:

conventional long form: United Arab Emirates

conventional short form: none

local long form: Al Imarat al Arabiyah al Muttahidah

local short form: none

former: Trucial Oman, Trucial States

abbreviation: UAE

Government type: federation with specified powers delegated to the UAE federal government and other powers reserved to member emirates

Capital: *name:* Abu Dhabi

geographic coordinates: 24 28 N, 54 22 E

time difference: UTC+4 (9 hours ahead of Washington, DC during Standard Time)

Administrative divisions: 7 emirates (imarat, singular—imarah); Abu Zaby (Abu Dhabi), 'Ajman, Al Fujayrah, Ash Shariqah (Sharjah), Dubayy (Dubai), Ra's al Khaymah, Umm al Qaywayn (Quwayn)

Independence: 2 December 1971 (from UK)

National holiday: Independence Day, 2 December (1971)

Constitution: 2 December 1971; made permanent in 1996

Legal system: based on a dual system of Shari'a and civil courts; has not accepted compulsory ICJ jurisdiction

Suffrage: none

Executive branch:

chief of state: President KHALIFA bin Zayid al-Nuhayyan (since 3 November 2004), ruler of Abu Zaby (Abu Dhabi) (since 4 November 2004); Vice President and Prime Minister MUHAMMAD bin Rashid al-Maktum (since 5 January 2006)

head of government: Prime Minister and Vice President MUHAMMAD bin Rashid al-Maktum (since 5 January 2006); Deputy Prime Ministers SULTAN bin Zayid al-Nuhayyan (since

20 November 1990) and HAMDAN bin Zayid al-Nuhayyan (since 20 October 2003)

cabinet: Council of Ministers appointed by the president

note: there is also a Federal Supreme Council (FSC) composed of the seven emirate rulers; the FSC is the highest constitutional authority in the UAE; establishes general policies and sanctions federal legislation; meets four times a year; Abu Zaby (Abu Dhabi) and Dubayy (Dubai) rulers have effective veto power

elections: president and vice president elected by the FSC for five-year terms (no term limits); election last held 3 November 2004 upon the death of the UAE's Founding Father and first President ZAYID bin Sultan Al Nuhayyan (next to be held in 2009); prime minister and deputy prime minister appointed by the president

election results: KHALIFA bin Zayid al-Nuhayyan elected president by a unanimous vote of the FSC; MUHAMMAD bin Rashid al-Maktum unanimously affirmed vice president after the 2006 death of his brother Sheikh Maktum bin Rashid al-Maktum

Legislative branch: unicameral Federal National Council (FNC) or Majlis al-Ittihad al-Watani (40 seats; 20 members appointed by the rulers of the constituent states, 20 members elected to serve two-year terms)

elections: elections for one half of the FNC (the other half remains appointed) held in the UAE on 18–20 December 2006; the new electoral college—a body of 6,689 Emiratis (including 1,189 women) appointed by the rulers of the seven emirates—were the only eligible voters and candidates; 456 candidates including 65 women ran for 20 contested FNC seats; one female from the Emirate of Abu Dhabi won a seat

note: reviews legislation but cannot change or veto

Judicial branch: Union Supreme Court (judges are appointed by the president)

Political parties and leaders: none

Political pressure groups and leaders: NA

International organization participation: ABEDA, AFESD, AMF, FAO, G-77, GCC, IAEA, IBRD, ICAO, ICC, ICCt (signatory), ICRM, IDA, IDB, IFAD, IFC, IFRCS, IHO, ILO, IMF, IMO, IMSO, Interpol, IOC, IPU, ISO, ITSO, ITU, LAS, MIGA, NAM, OAPEC, OIC, OPCW, OPEC, UN, UNCTAD, UNESCO, UNIDO, UPU, WCO, WHO, WIPO, WMO, WTO

Diplomatic representation in the US:

chief of mission: Ambassador Saqr Ghobash Said GHOBASH

chancery: 3522 International Court NW,

Suite 400, Washington, DC 20008

telephone: [1] (202) 243-2400

FAX: [1] (202) 243-2432

consulate(s): New York, Houston

Diplomatic representation from the US:

chief of mission: Ambassador (vacant); Charge d'Affaires Martin R. QUINN

embassy: Embassies District, Plot 38 Sector W59-02, Street No. 4, Abu Dhabi

mailing address: P. O. Box 4009, Abu Dhabi

telephone: [971] (2) 414-2200

FAX: [971] (2) 414-2603

consulate(s) general: Dubai

Flag description: three equal horizontal bands of green (top), white, and black with a wider vertical red band on the hoist side

ECONOMY

Economy—overview: The UAE has an open economy with a high per capita income and a sizable annual trade surplus. Despite largely successful efforts at economic diversification, nearly 40% of GDP is still directly based on oil and gas output. Since the discovery of oil in the UAE more than 30 years ago, the UAE has undergone a profound transformation from an impoverished region of small desert principalities to a modern state with a high standard of living. The government has increased spending on job creation and infrastructure expansion and is opening up utilities to greater private sector involvement. In April 2004, the UAE signed a Trade and Investment Framework Agreement with Washington and in November 2004 agreed to undertake negotiations toward a Free Trade Agreement with the US. The country's Free Trade Zones—offering 100% foreign ownership and zero taxes—are helping to attract foreign investors. Higher oil revenue, strong liquidity, housing shortages, and cheap credit in 2005–07 led to a surge in asset prices (shares and real estate) and consumer inflation. Rising prices are increasing the operating costs for businesses in the UAE and adversely impacting government employees and others on fixed incomes. Dependence on oil and a large expatriate workforce are significant long-term challenges. The UAE's strategic plan for the next few years focuses on diversification and creating more opportunities for nationals through improved education and increased private sector employment.

GDP (purchasing power parity): $167.3 billion (2007 est.)

GDP (official exchange rate): $192.6 billion (2007 est.)

GDP—real growth rate: 7.4% (2007 est.)

GDP—per capita (PPP): $37,300 (2007 est.)

GDP—composition by sector:

agriculture: 1.8%

industry: 59.3%

services: 38.9% (2007 est.)

Labor force: 3.065 million (2007 est.)

Labor force—by occupation: *agriculture:* 7%

industry: 15%

services: 78% (2000 est.)

Unemployment rate: 2.4% (2001)

Population below poverty line: 19.5% (2003)

Household income or consumption by percentage share:

lowest 10%: NA%

highest 10%: NA%

Inflation rate (consumer prices): 11% (2007 est.)

Investment (gross fixed): 21.8% of GDP (2007 est.)

Budget: *revenues:* $58.88 billion

expenditures: $38.06 billion (2007 est.)

Public debt: 22.9% of GDP (2007 est.)

Agriculture—products: dates, vegetables, watermelons; poultry, eggs, dairy products; fish

Industries: petroleum and petrochemicals; fishing, aluminum, cement, fertilizers, commercial ship repair, construction materials, some boat building, handicrafts, textiles

Industrial production growth rate: 4.3% (2007 est.)

Electricity—production: 57.06 billion kWh (2005)

Electricity—production by source:

fossil fuel: 100%

hydro: 0%

nuclear: 0%

other: 0% (2001)

Electricity—consumption: 52.62 billion kWh (2005)

Electricity—exports: 0 kWh (2005)

Electricity—imports: 0 kWh (2005)

Oil—production: 2.54 million bbl/day (2006 est.)

Oil—consumption: 372,000 bbl/day (2005 est.)

Oil—exports: 2.54 million bbl/day (2004 est.)

Oil—imports: 137,200 bbl/day (2004)

Oil—proved reserves: 97.8 billion bbl (2007 est.)

Natural gas—production: 45.07 billion cu m (2005 est.)

Natural gas—consumption: 39.56 billion cu m (2005 est.)

Natural gas—exports: 6.848 billion cu m (2005 est.)

Natural gas—imports: 1.343 billion cu m (2005)

Natural gas—proved reserves: 5.823 trillion cu m (1 January 2006 est.)

Current account balance: $41.67 billion (2007 est.)

Exports: $156.6 billion f.o.b. (2007 est.)

Exports—commodities: crude oil 45%, natural gas, reexports, dried fish, dates

Exports—partners: Japan 25.8%, South Korea 9.6%, Thailand 5.9%, India 4.5% (2006)

Imports: $101.6 billion f.o.b. (2007 est.)

Imports—commodities: machinery and transport equipment, chemicals, food

Imports—partners: US 11.5%, China 11%, India 9.8%, Germany 6.2%, Japan 5.8%, UK 5.5%, France 4.1%, Italy 4% (2006)

Economic aid—donor: since its founding in 1971, the Abu Dhabi Fund for Development has given about $5.2 billion in aid to 56 countries (2004)

Economic aid—recipient: $5.36 million (2004)

Reserves of foreign exchange and gold: $76.62 billion (31 December 2007 est.)

Debt—external: $57.52 billion (31 December 2007 est.)

Stock of direct foreign investment—at home: $44.37 billion (2007 est.)

Stock of direct foreign investment—abroad: $14.14 billion (2007 est.)

Market value of publicly traded shares: $138.5 billion (2006)

Currency (code): Emirati dirham (AED)

Currency code: AED

Exchange rates: Emirati dirhams per US dollar—3.673 (2007), 3.673 (2006), 3.6725 (2005), 3.6725 (2004), 3.6725 (2003)

note: officially pegged to the US dollar since February 2002

Fiscal year: calendar year

COMMUNICATIONS

Telephones—main lines in use: 1.31 million (2006)

Telephones—mobile cellular: 5.519 million (2006)

Telephone system:
general assessment: modern fiber-optic integrated services; digital network with rapidly growing use of mobile-cellular telephones; key centers are Abu Dhabi and Dubai
domestic: microwave radio relay, fiber optic and coaxial cable
international: country code—971; linked to the international submarine cable FLAG (Fiber-Optic Link Around the Globe); landing point for both the SEA-ME-WE-3 AND SEA-ME-WE-4 submarine cable networks; satellite earth stations—3 Intelsat (1 Atlantic Ocean and 2 Indian Ocean) and 1 Arabsat; tropospheric scatter to Bahrain; microwave radio relay to Saudi Arabia

Radio broadcast stations: AM 13, FM 8, shortwave 2 (2004)

Radios: 820,000 (1997)

Television broadcast stations: 15 (2004)

Televisions: 310,000 (1997)

Internet country code: .ae

Internet hosts: 6,001 (2007)

Internet Service Providers (ISPs): 1 (2000)

Internet users: 1.709 million (2006)

TRANSPORTATION

Airports: 39 (2007)

Airports—with paved runways:
total: 22
over 3,047 m: 10
2,438 to 3,047 m: 3
1,524 to 2,437 m: 2
914 to 1,523 m: 4
under 914 m: 3 (2007)

Airports—with unpaved runways:
total: 17
over 3,047 m: 2
2,438 to 3,047 m: 1
1,524 to 2,437 m: 4
914 to 1,523 m: 5
under 914 m: 5 (2007)

Heliports: 5 (2007)

Pipelines: condensate 520 km; gas 2,908 km; liquid petroleum gas 300 km; oil 2,950 km; oil/gas/water 5 km; refined products 156 km (2007)

Roadways: *total:* 1,088 km
paved: 1,088 km (includes 253 km of expressways) (1999)

Merchant marine:
total: 60 ships (1000 GRT or over) 617,519 GRT/858,519 DWT
by type: bulk carrier 6, cargo 10, chemical tanker 5, container 6, liquefied gas 1, passenger/cargo 1, petroleum tanker 25, roll on/roll off 5, specialized tanker 1
foreign-owned: 11 (Greece 3, Kuwait 8)
registered in other countries: 281 (Bahamas 20, Belize 4, Cambodia 2, Comoros 5, Cyprus 10, Georgia 1, Gibraltar 2, Hong Kong 1, India 2, Iran 1, Jordan 15, North Korea 4, Liberia 22, Malta 10, Marshall Islands 14, Mexico 1, Mongolia 5, Norway 1, Panama 108, Philippines 1, Saudi Arabia 1, Sierra Leone 7, Singapore 8, Somalia 1, St Kitts and Nevis 22, St Vincent and The Grenadines 12, Turkey 1, unknown 5) (2007)

Ports and terminals: Mina' Zayid (Abu Dhabi), Al Fujayrah, Mina' Jabal 'Ali (Dubai), Mina' Rashid (Dubai), Mina' Saqr (Ra's al Khaymah), Khawr Fakkan (Sharjah)

MILITARY

Military branches: United Arab Emirates Armed Forces: Army, Navy (includes Marines), Air Force and Air Defense, National Coast Guard (2008)

Military service age and obligation: 18 years of age (est.) for voluntary military service; 18 years of age for officers and women; no conscription (2008)

Manpower available for military service:
males age 16–49: 2,405,884 (includes non-nationals)
females age 16–49: 884,853 (2008 est.)

Manpower fit for military service:
males age 16–49: 2,004,558
females age 16–49: 760,637 (2008 est.)

Manpower reaching militarily significant age annually:
males age 16–49: 25,856
females age 16–49: 23,085 (2008 est.)

Military expenditures—percent of GDP: 3.1% (2005 est.)

TRANSNATIONAL ISSUES

Disputes—international: boundary agreement was signed and ratified with Oman in 2003 for entire border, including Oman's Musandam Peninsula and Al Madhah enclaves, but contents of the agreement and detailed maps showing the alignment have not been published; Iran and UAE dispute Tunb Islands and Abu Musa Island, which Iran occupies

Trafficking in persons: *current situation:* the United Arab Emirates is a destination country for men, women, and children trafficked from South and East Asia, Eastern Europe, Africa, and the Middle East for involuntary servitude and for sexual exploitation; an estimated 10,000 women from sub-Saharan Africa, Eastern Europe, South and East Asia, Iraq, Iran, and Morocco may be victims of sex trafficking in the UAE; women also migrate from Africa, and South and Southeast Asia to work as domestic servants, but may have their passports confiscated, be denied permission to leave the place of employment in the home, or face sexual or physical abuse by their employers; men from South Asia come to the UAE to work in the construction industry, but may be subjected to conditions of involuntary servitude as they are coerced to pay off recruitment and travel costs, sometimes having their wages denied for months at a time; victims of child camel jockey trafficking may still remain in the UAE, despite a July 2005 law banning the practice; while all identified victims were repatriated at the government's expense to their home countries, questions persist as to the effectiveness of the ban and the true number of victims
tier rating: Tier 2 Watch List—UAE is placed on the Tier 2 Watch List for its failure to show increased efforts to combat trafficking in 2005, particularly in its efforts to address the large-scale trafficking of foreign girls and women for commercial sexual exploitation

Illicit drugs: the UAE is a drug transshipment point for traffickers given its proximity to Southwest Asian drug-producing countries; the UAE's position as a major financial center makes it vulnerable to money laundering; anti-money-laundering controls improving, but informal banking remains unregulated

The island of Rockall is not shown.

Background: As the dominant industrial and maritime power of the 19th century, the United Kingdom of Great Britain and Ireland played a leading role in developing parliamentary democracy and in advancing literature and science. At its zenith, the British Empire stretched over one-fourth of the earth's surface. The first half of the 20th century saw the UK's strength seriously depleted in two World Wars and the Irish republic withdraw from the union. The second half witnessed the dismantling of the Empire and the UK rebuilding itself into a modern and prosperous European nation. As one of five permanent members of the UN Security Council, a founding member of NATO, and of the Commonwealth, the UK pursues a global approach to foreign policy; it currently is weighing the degree of its integration with continental Europe. A member of the EU, it chose to remain outside the Economic and Monetary Union for the time being. Constitutional reform is also a significant issue in the UK. The Scottish Parliament, the National Assembly for Wales, and the Northern Ireland Assembly were established in 1999, but the latter was suspended until May 2007 due to wrangling over the peace process.

GEOGRAPHY

Location: Western Europe, islands including the northern one-sixth of the island of Ireland between the North Atlantic Ocean and the North Sea, northwest of France

Geographic coordinates: 54 00 N, 2 00 W

Map references: Europe

Area:
total: 244,820 sq km
land: 241,590 sq km
water: 3,230 sq km
note: includes Rockall and Shetland Islands

Area—comparative: slightly smaller than Oregon

Land boundaries:
total: 360 km
border countries: Ireland 360 km

Coastline: 12,429 km

Maritime claims:
territorial sea: 12 nm
exclusive fishing zone: 200 nm
continental shelf: as defined in continental shelf orders or in accordance with agreed upon boundaries

Climate: temperate; moderated by prevailing southwest winds over the North Atlantic Current; more than one-half of the days are overcast

Terrain: mostly rugged hills and low mountains; level to rolling plains in east and southeast

Elevation extremes:
lowest point: The Fens -4 m
highest point: Ben Nevis 1,343 m

Natural resources: coal, petroleum, natural gas, iron ore, lead, zinc, gold, tin, limestone, salt, clay, chalk, gypsum, potash, silica sand, slate, arable land

Land use:
arable land: 23.23%
permanent crops: 0.2%
other: 76.57% (2005)

Irrigated land: 1,700 sq km (2003)

Total renewable water resources: 160.6 cu km (2005)

Freshwater withdrawal (domestic/industrial/agricultural): *total:* 11.75 cu km/yr (22%/75%/3%)
per capita: 197 cu m/yr (1994)

Natural hazards: winter windstorms; floods

Environment—current issues: continues to reduce greenhouse gas emissions (has met Kyoto Protocol target of a 12.5% reduction from 1990 levels and intends to meet the legally binding target and move toward a domestic goal of a 20% cut in emissions by 2010); by 2005 the government reduced the amount of industrial and commercial waste disposed of in landfill sites to 85% of 1998 levels and recycled or composted at least 25% of household waste, increasing to 33% by 2015

Environment—international agreements: *party to:* Air Pollution, Air Pollution-Nitrogen Oxides, Air Pollution-Persistent Organic Pollutants, Air Pollution-Sulfur 94, Air Pollution-Volatile Organic Compounds, Antarctic-Environmental Protocol, Antarctic-Marine Living Resources, Antarctic Seals, Antarctic Treaty, Biodiversity, Climate Change, Climate Change-Kyoto Protocol, Desertification, Endangered Species, Environmental Modification, Hazardous Wastes, Law of the Sea, Marine Dumping, Marine Life Conservation, Ozone Layer Protection, Ship Pollution, Tropical Timber 83, Tropical Timber 94, Wetlands, Whaling *signed, but not ratified:* none of the selected agreements

Geography—note: lies near vital North Atlantic sea lanes; only 35 km from France and linked by tunnel under the English Channel; because of heavily indented coastline, no location is more than 125 km from tidal waters

PEOPLE

Population: 60,943,912 (July 2008 est.)

Age structure:
0–14 years: 16.9% (male 5,287,590/female 5,036,881)
15–64 years: 67.1% (male 20,698,645/female 20,185,040)
65 years and over: 16% (male 4,186,561/female 5,549,195) (2008 est.)

Median age: *total:* 39.9 years
male: 38.8 years
female: 41 years (2008 est.)

Population growth rate: 0.276% (2008 est.)

Birth rate: 10.65 births/1,000 population (2008 est.)

Death rate: 10.05 deaths/1,000 population (2008 est.)

Net migration rate: 2.17 migrant(s)/1,000 population (2008 est.)

Sex ratio:
at birth: 1.05 male(s)/female
under 15 years: 1.05 male(s)/female
15–64 years: 1.03 male(s)/female
65 years and over: 0.75 male(s)/female
total population: 0.98 male(s)/female (2008 est.)

Infant mortality rate:
total: 4.93 deaths/1,000 live births
male: 5.49 deaths/1,000 live births
female: 4.34 deaths/1,000 live births (2008 est.)

Life expectancy at birth:
total population: 78.85 years

male: 76.37 years
female: 81.46 years (2008 est.)
Total fertility rate: 1.66 children born/woman (2008 est.)
HIV/AIDS—adult prevalence rate: 0.2% (2001 est.)
HIV/AIDS—people living with HIV/AIDS: 51,000 (2001 est.)
HIV/AIDS—deaths: fewer than 500 (2003 est.)
Nationality: *noun:* Briton(s), British (collective plural)
adjective: British
Ethnic groups: white (of which English 83.6%, Scottish 8.6%, Welsh 4.9%, Northern Irish 2.9%) 92.1%, black 2%, Indian 1.8%, Pakistani 1.3%, mixed 1.2%, other 1.6% (2001 census)
Religions: Christian (Anglican, Roman Catholic, Presbyterian, Methodist) 71.6%, Muslim 2.7%, Hindu 1%, other 1.6%, unspecified or none 23.1% (2001 census)
Languages: English, Welsh (about 26% of the population of Wales), Scottish form of Gaelic (about 60,000 in Scotland)
Literacy:
definition: age 15 and over has completed five or more years of schooling
total population: 99%
male: 99%
female: 99% (2003 est.)

GOVERNMENT

Country name:
conventional long form: United Kingdom of Great Britain and Northern Ireland; note—Great Britain includes England, Scotland, and Wales
conventional short form: United Kingdom
abbreviation: UK
Government type: constitutional monarchy
Capital: *name:* London
geographic coordinates: 51 30 N, 0 10 W
time difference: UTC 0 (5 hours ahead of Washington, DC during Standard Time)
daylight saving time: +1hr, begins last Sunday in March; ends last Sunday in October
Administrative divisions: *England:* 34 two-tier counties, 32 London boroughs and 1 City of London or Greater London, 36 metropolitan counties, 46 unitary authorities
two-tier counties: Bedfordshire, Buckinghamshire, Cambridgeshire, Cheshire, Cornwall and Isles of Scilly, Cumbria, Derbyshire, Devon, Dorset, Durham, East Sussex, Essex, Gloucestershire, Hampshire, Hertfordshire, Kent, Lancashire, Leicestershire, Lincolnshire, Norfolk, North Yorkshire, Northamptonshire, Northumberland, Nottinghamshire, Oxfordshire, Shropshire, Somerset, Staffordshire, Suffolk, Surrey,

Warwickshire, West Sussex, Wiltshire, Worcestershire
London boroughs and City of London or Greater London: Barking and Dagenham, Barnet, Bexley, Brent, Bromley, Camden, Croydon, Ealing, Enfield, Greenwich, Hackney, Hammersmith and Fulham, Haringey, Harrow, Havering, Hillingdon, Hounslow, Islington, Kensington and Chelsea, Kingston upon Thames, Lambeth, Lewisham, City of London, Merton, Newham, Redbridge, Richmond upon Thames, Southwark, Sutton, Tower Hamlets, Waltham Forest, Wandsworth, Westminster
metropolitan counties: Barnsley, Birmingham, Bolton, Bradford, Bury, Calderdale, Coventry, Doncaster, Dudley, Gateshead, Kirklees, Knowsley, Leeds, Liverpool, Manchester, Newcastle upon Tyne, North Tyneside, Oldham, Rochdale, Rotherham, Salford, Sandwell, Sefton, Sheffield, Solihull, South Tyneside, St. Helens, Stockport, Sunderland, Tameside, Trafford, Wakefield, Walsall, Wigan, Wirral, Wolverhampton
unitary authorities: Bath and North East Somerset, Blackburn with Darwen, Blackpool, Bournemouth, Bracknell Forest, Brighton and Hove, City of Bristol, Darlington, Derby, East Riding of Yorkshire, Halton, Hartlepool, County of Herefordshire, Isle of Wight, City of Kingston upon Hull, Leicester, Luton, Medway, Middlesbrough, Milton Keynes, North East Lincolnshire, North Lincolnshire, North Somerset, Nottingham, Peterborough, Plymouth, Poole, Portsmouth, Reading, Redcar and Cleveland, Rutland, Slough, South Gloucestershire, Southampton, Southend-on-Sea, Stockton-on-Tees, Stoke-on-Trent, Swindon, Telford and Wrekin, Thurrock, Torbay, Warrington, West Berkshire, Windsor and Maidenhead, Wokingham, York
Northern Ireland: 26 district council areas
district council areas: Antrim, Ards, Armagh, Ballymena, Ballymoney, Banbridge, Belfast, Carrickfergus, Castlereagh, Coleraine, Cookstown, Craigavon, Derry, Down, Dungannon, Fermanagh, Larne, Limavady, Lisburn, Magherafelt, Moyle, Newry and Mourne, Newtownabbey, North Down, Omagh, Strabane
Scotland: 32 unitary authorities
unitary authorities: Aberdeen City, Aberdeenshire, Angus, Argyll and Bute, Clackmannanshire, Dumfries and Galloway, Dundee City, East Ayrshire, East Dunbartonshire, East Lothian, East Renfrewshire, City of Edinburgh, Eilean Siar (Western Isles), Falkirk, Fife, Glasgow City, Highland, Inverclyde, Midlothian, Moray, North Ayrshire, North Lanarkshire, Orkney Islands,

Perth and Kinross, Renfrewshire, Shetland Islands, South Ayrshire, South Lanarkshire, Stirling, The Scottish Borders, West Dunbartonshire, West Lothian
Wales: 22 unitary authorities
unitary authorities: Blaenau Gwent; Bridgend; Caerphilly; Cardiff; Carmarthenshire; Ceredigion; Conwy; Denbighshire; Flintshire; Gwynedd; Isle of Anglesey; Merthyr Tydfil; Monmouthshire; Neath Port Talbot; Newport; Pembrokeshire; Powys; Rhondda, Cynon, Taff; Swansea; The Vale of Glamorgan; Torfaen; Wrexham
Dependent areas: Anguilla, Bermuda, British Indian Ocean Territory, British Virgin Islands, Cayman Islands, Falkland Islands, Gibraltar, Montserrat, Pitcairn Islands, Saint Helena, South Georgia and the South Sandwich Islands, Turks and Caicos Islands
Independence: England has existed as a unified entity since the 10th century; the union between England and Wales, begun in 1284 with the Statute of Rhuddlan, was not formalized until 1536 with an Act of Union; in another Act of Union in 1707, England and Scotland agreed to permanently join as Great Britain; the legislative union of Great Britain and Ireland was implemented in 1801, with the adoption of the name the United Kingdom of Great Britain and Ireland; the Anglo-Irish treaty of 1921 formalized a partition of Ireland; six northern Irish counties remained part of the United Kingdom as Northern Ireland and the current name of the country, the United Kingdom of Great Britain and Northern Ireland, was adopted in 1927
National holiday: the UK does not celebrate one particular national holiday
Constitution: unwritten; partly statutes, partly common law and practice
Legal system: based on common law tradition with early Roman and modern continental influences; has nonbinding judicial review of Acts of Parliament under the Human Rights Act of 1998; accepts compulsory ICJ jurisdiction, with reservations
Suffrage: 18 years of age; universal
Executive branch:
chief of state: Queen ELIZABETH II (since 6 February 1952); Heir Apparent Prince CHARLES (son of the queen, born 14 November 1948)
head of government: Prime Minister Gordon BROWN (since 27 June 2007)
cabinet: Cabinet of Ministers appointed by the prime minister
elections: the monarchy is hereditary; following legislative elections, the leader of the majority party or the leader of the majority coalition is usually the prime minister

Legislative branch: bicameral Parliament consists of House of Lords (618 seats; consisting of approximately 500 life peers, 92 hereditary peers, and 26 clergy) and House of Commons (646 seats since 2005 elections; members are elected by popular vote to serve five-year terms unless the House is dissolved earlier)

elections: House of Lords—no elections (note—in 1999, as provided by the House of Lords Act, elections were held in the House of Lords to determine the 92 hereditary peers who would remain there; elections are held only as vacancies in the hereditary peerage arise); House of Commons—last held 5 May 2005 (next to be held by May 2010)

election results: House of Commons—percent of vote by party—Labor 35.2%, Conservative 32.3%, Liberal Democrats 22%, other 10.5%; seats by party—Labor 355, Conservative 198, Liberal Democrat 62, other 31; seats by party in the House of Commons as of 4 June 2008—Labor 351, Conservative 192, Liberal Democrat 63, Scottish National Party/Plaid Cymru 9, Democratic Unionist 9, Sinn Fein 5, other 17

note: in 1998 elections were held for a Northern Ireland Assembly (because of unresolved disputes among existing parties, the transfer of power from London to Northern Ireland came only at the end of 1999 and has been suspended four times, the latest occurring in October 2002 and lasting until 8 May 2007); in 1999, the UK held the first elections for a Scottish Parliament and a Welsh Assembly, the most recent of which were held in May 2007

Judicial branch: House of Lords (highest court of appeal; several Lords of Appeal in Ordinary are appointed by the monarch for life); Supreme Courts of England, Wales, and Northern Ireland (comprising the Courts of Appeal, the High Courts of Justice, and the Crown Courts); Scotland's Court of Session and Court of the Justiciary

Political parties and leaders: Conservative [David CAMERON]; Democratic Unionist Party (Northern Ireland) [Peter ROBINSON]; Labor Party [Gordon BROWN]; Liberal Democrats [Nick CLEGG]; Party of Wales (Plaid Cymru) [Ieuan Wyn JONES]; Scottish National Party or SNP [Alex SALMOND]; Sinn Fein (Northern Ireland) [Gerry ADAMS]; Social Democratic and Labor Party or SDLP (Northern Ireland) [Mark DURKAN]; Ulster Unionist Party (Northern Ireland) [Sir Reg EMPEY]

Political pressure groups and leaders: Campaign for Nuclear Disarmament; Confederation of British Industry; National Farmers' Union; Trades Union Congress

International organization participation: ADB (nonregional members), AfDB, Arctic Council (observer), Australia Group, BIS, C, CBSS (observer), CDB, CE, CERN, EAPC, EBRD, EIB, ESA, EU, FAO, G-5, G-7, G-8, G-10, IADB, IAEA, IBRD, ICAO, ICC, ICCt, ICRM, IDA, IEA, IFAD, IFC, IFRCS, IHO, ILO, IMF, IMO, IMSO, Interpol, IOC, IOM, IPU, ISO, ITSO, ITU, MIGA, NATO, NEA, NSG, OAS (observer), OECD, OPCW, OSCE, Paris Club, PCA, PIF (partner), SECI (observer), UN, UN Security Council, UNCTAD, UNESCO, UNFICYP, UNHCR, UNIDO, UNMIL, UNMIS, UNOMIG, UNRWA, UNWTO, UPU, WCO, WEU, WHO, WIPO, WMO, WTO, ZC

Diplomatic representation in the US:
chief of mission: Ambassador Sir Nigel E. SHEINWALD
chancery: 3100 Massachusetts Avenue NW, Washington, DC 20008
telephone: [1] (202) 588-6500
FAX: [1] (202) 588-7870
consulate(s) general: Atlanta, Boston, Chicago, Houston, Los Angeles, Miami, New York, San Francisco
consulate(s): Denver, Orlando

Diplomatic representation from the US:
chief of mission: Ambassador Robert Holmes TUTTLE
embassy: 24 Grosvenor Square, London, W1A 1AE
mailing address: PSC 801, Box 40, FPO AE 09498-4040
telephone: [44] (0) 20 7499-9000
FAX: [44] (0) 20 7629-9124
consulate(s) general: Belfast, Edinburgh

Flag description: blue field with the red cross of Saint George (patron saint of England) edged in white superimposed on the diagonal red cross of Saint Patrick (patron saint of Ireland), which is superimposed on the diagonal white cross of Saint Andrew (patron saint of Scotland); properly known as the Union Flag, but commonly called the Union Jack; the design and colors (especially the Blue Ensign) have been the basis for a number of other flags including other Commonwealth countries and their constituent states or provinces, and British overseas territories

ECONOMY

Economy—overview: The UK, a leading trading power and financial center, is one of the quintet of trillion dollar economies of Western Europe. Over the past two decades, the government has greatly reduced public ownership and contained the growth of social welfare programs. Agriculture is intensive, highly mechanized, and efficient by European standards, producing about 60% of food needs with less than 2% of the labor force. The UK has large coal, natural gas, and oil reserves; primary energy production accounts for 10% of GDP, one of the highest shares of any industrial nation. Services, particularly banking, insurance, and business services, account by far for the largest proportion of GDP while industry continues to decline in importance. Since emerging from recession in 1992, Britain's economy has enjoyed the longest period of expansion on record; growth has remained in the 2–3% range since 2004, outpacing most of Europe. The economy's strength has complicated the Labor government's efforts to make a case for Britain to join the European Economic and Monetary Union (EMU). Critics point out that the economy is doing well outside of EMU, and public opinion polls show a majority of Britons are opposed to the euro. The BROWN government has been speeding up the improvement of education, health services, and affordable housing at a cost in higher taxes and a widening public deficit.

GDP (purchasing power parity): $2.137 trillion (2007 est.)

GDP (official exchange rate): $2.773 trillion (2007 est.)

GDP—real growth rate: 3.1% (2007 est.)

GDP—per capita (PPP): $35,100 (2007 est.)

GDP—composition by sector:
agriculture: 0.9%
industry: 23.4%
services: 75.7% (2007 est.)

Labor force: 30.87 million (2007 est.)

Labor force—by occupation: *agriculture:* 1.4%
industry: 18.2%
services: 80.4% (2006 est.)

Unemployment rate: 5.4% (2007 est.)

Population below poverty line: 14% (2006 est.)

Household income or consumption by percentage share:
lowest 10%: 2.1%
highest 10%: 28.5% (1999)

Distribution of family income—Gini index: 34 (2005)

Inflation rate (consumer prices): 2.3% (2007 est.)

Investment (gross fixed): 18.5% of GDP (2007 est.)

Budget:
revenues: $1.155 trillion
expenditures: $1.236 trillion (2007 est.)

Public debt: 43% of GDP (2007 est.)

Agriculture—products: cereals, oilseed, potatoes, vegetables; cattle, sheep, poultry; fish

Industries: machine tools, electric power equipment, automation equipment, railroad equipment, shipbuilding, aircraft, motor vehicles and parts, electronics and communications equipment, metals, chemicals, coal, petroleum, paper and paper products, food processing, textiles, clothing, other consumer goods

Industrial production growth rate: 0.5% (2007 est.)

Electricity—production: 372.6 billion kWh (2005)

Electricity—production by source:
fossil fuel: 73.8%
hydro: 0.9%
nuclear: 23.7%
other: 1.6% (2001)

Electricity—consumption: 348.7 billion kWh (2005)

Electricity—exports: 2.839 billion kWh (2005)

Electricity—imports: 11.16 billion kWh (2005)

Oil—production: 1.861 million bbl/day (2005 est.)

Oil—consumption: 1.82 million bbl/day (2005 est.)

Oil—exports: 1.956 million bbl/day (2004)

Oil—imports: 1.654 million bbl/day (2004)

Oil—proved reserves: 4.029 billion bbl (1 January 2006 est.)

Natural gas—production: 84.16 billion cu m (2005 est.)

Natural gas—consumption: 91.16 billion cu m (2005 est.)

Natural gas—exports: 8.843 billion cu m (2005 est.)

Natural gas—imports: 15.84 billion cu m (2005)

Natural gas—proved reserves: 509.2 billion cu m (1 January 2006 est.)

Current account balance: -$136.2 billion (2007 est.)

Exports: $441.4 billion f.o.b. (2007 est.)

Exports—commodities: manufactured goods, fuels, chemicals; food, beverages, tobacco

Exports—partners: US 13.9%, Germany 10.9%, France 10.4%, Ireland 7.1%, Netherlands 6.3%, Belgium 5.2%, Spain 4.5% (2006)

Imports: $616.8 billion f.o.b. (2007 est.)

Imports—commodities: manufactured goods, machinery, fuels; foodstuffs

Imports—partners: Germany 12.8%, US 8.9%, France 6.9%, Netherlands 6.6%, China 5.3%, Norway 4.9%, Belgium 4.5% (2006)

Economic aid—donor: ODA, $12.46 billion (2006)

Reserves of foreign exchange and gold: $57.3 billion (31 December 2007 est.)

Debt—external: $10.45 trillion (30 June 2007)

Stock of direct foreign investment—at home: $1.324 trillion (2007 est.)

Stock of direct foreign investment—abroad: $1.741 trillion (2007 est.)

Market value of publicly traded shares: $3.058 trillion (2005)

Currency (code): British pound (GBP)

Currency code: GBP

Exchange rates: British pounds per US dollar—0.4993 (2007), 0.5418 (2006), 0.5493 (2005), 0.5462 (2004), 0.6125 (2003)

Fiscal year: 6 April—5 April

COMMUNICATIONS

Telephones—main lines in use: 33.602 million (2006)

Telephones—mobile cellular: 69.657 million (2006)

Telephone system:
general assessment: technologically advanced domestic and international system
domestic: equal mix of buried cables, microwave radio relay, and fiber-optic systems
international: country code—44; numerous submarine cables provide links throughout Europe, Asia, Australia, the Middle East, and US; satellite earth stations—10 Intelsat (7 Atlantic Ocean and 3 Indian Ocean), 1 Inmarsat (Atlantic Ocean region), and 1 Eutelsat; at least 8 large international switching centers

Radio broadcast stations: AM 219, FM 431, shortwave 3 (1998)

Radios: 84.5 million (1997)

Television broadcast stations: 228 (plus 3,523 repeaters) (1995)

Televisions: 30.5 million (1997)

Internet country code: .uk

Internet hosts: 5.118 million (2007)

Internet Service Providers (ISPs): more than 400 (2000)

Internet users: 33.534 million (2006)

TRANSPORTATION

Airports: 449 (2007)

Airports—with paved runways:
total: 310
over 3,047 m: 8
2,438 to 3,047 m: 33
1,524 to 2,437 m: 131
914 to 1,523 m: 79
under 914 m: 59 (2007)

Airports—with unpaved runways:
total: 139
2,438 to 3,047 m: 1
1,524 to 2,437 m: 2
914 to 1,523 m: 23
under 914 m: 113 (2007)

Heliports: 11 (2007)

Pipelines: condensate 567 km; condensate/gas 22 km; gas 18,980 km; liquid petroleum gas 59 km; oil 4,930 km; oil/gas/water 165 km; refined products 4,444 km (2007)

Railways: *total:* 16,567 km
broad gauge: 303 km 1.600-m gauge (in Northern Ireland)
standard gauge: 16,264 km 1.435-m gauge (5,361 km electrified) (2006)

Roadways:
total: 388,008 km
paved: 388,008 km (includes 3,520 km of expressways) (2005)

Waterways: 3,200 km (620 km used for commerce) (2003)

Merchant marine:
total: 474 ships (1000 GRT or over) 11,723,618 GRT/12,315,588 DWT
by type: bulk carrier 26, cargo 60, carrier 4, chemical tanker 56, container 156, liquefied gas 18, passenger 10, passenger/cargo 62, petroleum tanker 27, refrigerated cargo 17, roll on/roll off 24, vehicle carrier 14
foreign-owned: 242 (Australia 1, Cyprus 1, Denmark 61, Finland 1, France 9, Germany 71, Greece 6, Hong Kong 2, Ireland 1, Italy 4, Japan 1, Netherlands 2, NZ 1, Norway 33, South Africa 4, Sweden 19, Switzerland 1, Taiwan 11, Turkey 2, US 11)
registered in other countries: 412 (Algeria 12, Antigua and Barbuda 4, Argentina 4, Australia 2, Bahamas 68, Barbados 3, Bermuda 20, Brunei 8, Cape Verde 1, Cayman Islands 9, Cyprus 21, Faroe Islands 1, Gibraltar 3, Greece 15, Hong Kong 32, India 1, Indonesia 3, Italy 7, South Korea 1, Liberia 74, Luxembourg 7, Malta 12, Marshall Islands 17, Netherlands 7, Norway 9, Panama 35, Papua New Guinea 6, Singapore 13, Slovakia 1, St Vincent and The Grenadines 9, Sweden 2, Thailand 3, Tonga 1, US 1, unknown 1) (2007)

Ports and terminals: Dover, Felixstowe, Immingham, Liverpool, London, Southampton, Teesport (England), Forth Ports, Hound Point (Scotland), Milford Haven (Wales)

MILITARY

Military branches: Army, Royal Navy (includes Royal Marines), Royal Air Force

Military service age and obligation: 16–33 years of age (officers 17–28) for voluntary military service (with parental consent under 18); women serve in military services, but are excluded from ground combat positions and some naval postings; must be citizen of the UK, Commonwealth, or Republic of Ireland; reservists serve a minimum of 3 years, to age 45 or 55; 16 years of age for voluntary military service by Nepalese citizens in the Brigade of the Gurkhas; 16–34 years of age for voluntary military service by Papua New Guinean citizens (2008)

Manpower available for military service: *males age 16–49:* 14,729,500
females age 16–49: 14,125,600 (2008 est.)

Manpower fit for military service:
males age 16–49: 12,121,602
females age 16–49: 11,616,582 (2008 est.)

Military expenditures—percent of GDP: 2.4% (2005 est.)

TRANSNATIONAL ISSUES

Disputes—international: in 2002, Gibraltar residents voted overwhelmingly by referendum to reject any "shared sovereignty" arrangement between the UK and Spain; the Government of Gibraltar insists on equal participation in talks between the two countries; Spain disapproves of UK plans to grant Gibraltar greater autonomy; Mauritius and Seychelles claim the Chagos Archipelago (British Indian Ocean Territory), and its former inhabitants since their eviction in 1965; most Chagossians reside in Mauritius, and in 2001 were granted UK citizenship, where some have since resettled; in May 2006, the High Court of London reversed the UK Government's 2004 orders of council that banned habitation on the islands; UK rejects sovereignty talks requested by Argentina, which still claims the Falkland Islands (Islas Malvinas) and South Georgia and the South Sandwich Islands; territorial claim in Antarctica (British Antarctic Territory) overlaps Argentine claim and partially overlaps Chilean claim; Iceland, the UK, and Ireland dispute Denmark's claim that the Faroe Islands' continental shelf extends beyond 200 nm

Illicit drugs: producer of limited amounts of synthetic drugs and synthetic precursor chemicals; major consumer of Southwest Asian heroin, Latin American cocaine, and synthetic drugs; money-laundering center

UNITED STATES

INTRODUCTION

Background: Britain's American colonies broke with the mother country in 1776 and were recognized as the new nation of the United States of America following the Treaty of Paris in 1783. During the 19th and 20th centuries, 37 new states were added to the original 13 as the nation expanded across the North American continent and acquired a number of overseas possessions. The two most traumatic experiences in the nation's history were the Civil War (1861–65) and the Great Depression of the 1930s. Buoyed by victories in World Wars I and II and the end of the Cold War in 1991, the US remains the world's most powerful nation state. The economy is marked by steady growth, low unemployment and inflation, and rapid advances in technology.

GEOGRAPHY

Location: North America, bordering both the North Atlantic Ocean and the North Pacific Ocean, between Canada and Mexico

Geographic coordinates: 38 00 N, 97 00 W

Map references: North America

Area: *total:* 9,826,630 sq km

land: 9,161,923 sq km

water: 664,707 sq km

note: includes only the 50 states and District of Columbia

Area—comparative: about half the size of Russia; about three-tenths the size of Africa; about half the size of South America (or slightly larger than Brazil); slightly larger than China; more than twice the size of the European Union

Land boundaries:

total: 12,034 km

border countries: Canada 8,893 km (including 2,477 km with Alaska), Mexico 3,141 km

note: US Naval Base at Guantanamo Bay, Cuba is leased by the US and is part of Cuba; the base boundary is 28 km

Coastline: 19,924 km

Maritime claims:

territorial sea: 12 nm

contiguous zone: 24 nm

exclusive economic zone: 200 nm

continental shelf: not specified

Climate: mostly temperate, but tropical in Hawaii and Florida, arctic in Alaska, semiarid in the great plains west of the Mississippi River, and arid in the Great Basin of the southwest; low winter temperatures in the northwest are ameliorated occasionally in January and February by warm chinook winds from the eastern slopes of the Rocky Mountains

Terrain: vast central plain, mountains in west, hills and low mountains in east; rugged mountains and broad river valleys in Alaska; rugged, volcanic topography in Hawaii

Elevation extremes:

lowest point: Death Valley -86 m

highest point: Mount McKinley 6,198 m

Natural resources: coal, copper, lead, molybdenum, phosphates, uranium, bauxite, gold, iron, mercury, nickel, potash, silver, tungsten, zinc, petroleum, natural gas, timber

Land use:

arable land: 18.01%

permanent crops: 0.21%

other: 81.78% (2005)

Irrigated land: 223,850 sq km (2003)

Total renewable water resources: 3,069 cu km (1985)

Freshwater withdrawal (domestic/industrial/agricultural): *total:* 477 cu km/yr (13%/46%/41%)

per capita: 1,600 cu m/yr (2000)

Natural hazards: tsunamis, volcanoes, and earthquake activity around Pacific Basin; hurricanes along the Atlantic and Gulf of Mexico coasts; tornadoes in the midwest and southeast; mud slides in California; forest fires in the west; flooding; permafrost in northern Alaska, a major impediment to development

Environment—current issues: air pollution resulting in acid rain in both the US and Canada; the US is the largest single emitter of carbon dioxide from the burning of fossil fuels; water pollution from runoff of pesticides and fertilizers; limited natural fresh water resources in much of the western part of the country require careful management; desertification

Environment—international agreements: *party to:* Air Pollution, Air Pollution-Nitrogen Oxides, Antarctic-Environmental Protocol, Antarctic-Marine Living Resources, Antarctic

Seals, Antarctic Treaty, Climate Change, Desertification, Endangered Species, Environmental Modification, Marine Dumping, Marine Life Conservation, Ozone Layer Protection, Ship Pollution, Tropical Timber 83, Tropical Timber 94, Wetlands, Whaling *signed, but not ratified:* Air Pollution-Persistent Organic Pollutants, Air Pollution-Volatile Organic Compounds, Biodiversity, Climate Change-Kyoto Protocol, Hazardous Wastes

Geography—note: world's third-largest country by size (after Russia and Canada) and by population (after China and India); Mt. McKinley is highest point in North America and Death Valley the lowest point on the continent

PEOPLE

Population: 303,824,646 (July 2008 est.)

Age structure:

0–14 years: 20.1% (male 31,257,108/female 29,889,645)

15–64 years: 67.1% (male 101,825,901/female 102,161,823)

65 years and over: 12.7% (male 16,263,255/female 22,426,914) (2008 est.)

Median age: *total:* 36.7 years

male: 35.4 years

female: 38.1 years (2008 est.)

Population growth rate: 0.883% (2008 est.)

Birth rate: 14.18 births/1,000 population (2008 est.)

Death rate: 8.27 deaths/1,000 population (2008 est.)

Net migration rate: 2.92 migrant(s)/1,000 population (2008 est.)

Sex ratio:

at birth: 1.05 male(s)/female

under 15 years: 1.05 male(s)/female

15–64 years: 1 male(s)/female

65 years and over: 0.73 male(s)/female

total population: 0.97 male(s)/female (2008 est.)

Infant mortality rate:

total: 6.3 deaths/1,000 live births

male: 6.95 deaths/1,000 live births

female: 5.62 deaths/1,000 live births (2008 est.)

Life expectancy at birth:

total population: 78.14 years

male: 75.29 years

female: 81.13 years (2008 est.)

Total fertility rate: 2.1 children born/woman (2008 est.)

HIV/AIDS—adult prevalence rate: 0.6% (2003 est.)

HIV/AIDS—people living with HIV/AIDS: 950,000 (2003 est.)

HIV/AIDS—deaths: 17,011 (2005 est.)

Nationality:

noun: American(s)

adjective: American

Ethnic groups: white 81.7%, black 12.9%, Asian 4.2%, Amerindian and Alaska native 1%, native Hawaiian and other Pacific islander 0.2% (2003 est.)

note: a separate listing for Hispanic is not included because the US Census Bureau considers Hispanic to mean a person of Latin American descent (including persons of Cuban, Mexican, or Puerto Rican origin) living in the US who may be of any race or ethnic group (white, black, Asian, etc.)

Religions: Protestant 51.3%, Roman Catholic 23.9%, Mormon 1.7%, other Christian 1.6%, Jewish 1.7%, Buddhist 0.7%, Muslim 0.6%, other or unspecified 2.5%, unaffiliated 12.1%, none 4% (2007 est.)

Languages: English 82.1%, Spanish 10.7%, other Indo-European 3.8%, Asian and Pacific island 2.7%, other 0.7% (2000 census)

note: Hawaiian is an official language in the state of Hawaii

Literacy: *definition:* age 15 and over can read and write

total population: 99%

male: 99%

female: 99% (2003 est.)

GOVERNMENT

Country name:

conventional long form: United States of America

conventional short form: United States

abbreviation: US or USA

Government type: Constitution-based federal republic; strong democratic tradition

Capital: *name:* Washington, DC

geographic coordinates: 38 53 N, 77 02 W

time difference: UTC-5 (during Standard Time)

daylight saving time: +1hr, begins second Sunday in March; ends first Sunday in November

note: the US is divided into six time zones

Administrative divisions: 50 states and 1 district*; Alabama, Alaska, Arizona, Arkansas, California, Colorado, Connecticut, Delaware, District of Columbia*, Florida, Georgia, Hawaii, Idaho, Illinois, Indiana, Iowa, Kansas, Kentucky, Louisiana, Maine, Maryland, Massachusetts, Michigan, Minnesota, Mississippi, Missouri, Montana, Nebraska, Nevada, New Hampshire, New Jersey, New Mexico, New York, North Carolina, North Dakota, Ohio, Oklahoma, Oregon, Pennsylvania, Rhode Island, South Carolina, South Dakota, Tennessee, Texas, Utah, Vermont, Virginia, Washington, West Virginia, Wisconsin, Wyoming

Dependent areas: American Samoa, Baker Island, Guam, Howland Island, Jarvis Island, Johnston Atoll, Kingman Reef, Midway Islands, Navassa Island, Northern Mariana Islands, Palmyra Atoll, Puerto Rico, Virgin Islands, Wake Island

note: from 18 July 1947 until 1 October 1994, the US administered the Trust Territory of the Pacific Islands; it entered into a political relationship with all four political units: the Northern Mariana Islands is a commonwealth in political union with the US (effective 3 November 1986); the Republic of the Marshall Islands signed a Compact of Free Association with the US (effective 21 October 1986); the Federated States of Micronesia signed a Compact of Free Association with the US (effective 3 November 1986); Palau concluded a Compact of Free Association with the US (effective 1 October 1994)

Independence: 4 July 1776 (from Great Britain)

National holiday: Independence Day, 4 July (1776)

Constitution: 17 September 1787, effective 4 March 1789

Legal system: federal court system based on English common law; each state has its own unique legal system, of which all but one (Louisiana, which is still influenced by the Napoleonic Code) is based on English common law; judicial review of legislative acts; has not accepted compulsory ICJ jurisdiction

Suffrage: 18 years of age; universal

Executive branch:

chief of state: President George W. BUSH (since 20 January 2001); Vice President Richard B. CHENEY (since 20 January 2001); note—the president is both the chief of state and head of government

head of government: President George W. BUSH (since 20 January 2001); Vice President Richard B. CHENEY (since 20 January 2001)

cabinet: Cabinet appointed by the president with Senate approval

elections: president and vice president elected on the same ticket by a college of representatives who are elected directly from each state; president and vice president serve four-year terms (eligible for a second term); election last held 2 November 2004 (next to be held on 4 November 2008)

election results: George W. BUSH reelected president; percent of popular vote—George W. BUSH 50.9%, John KERRY 48.1%, other 1.0%

Legislative branch: bicameral Congress consists of the Senate (100 seats, 2 members are elected from each state by popular vote to serve six-year terms; one-third are elected every two years) and the House of Representatives (435 seats; members are directly elected by popular vote to serve two-year terms)

elections: Senate—last held 7 November 2006 (next to be held November 2008); House of Representatives—last held 7

November 2006 (next to be held November 2008)

election results: Senate—percent of vote by party—NA; seats by party—Democratic Party 49, Republican Party 49, independent 2; House of Representatives—percent of vote by party—NA; seats by party—Democratic Party 233, Republican Party 202

Judicial branch: Supreme Court (nine justices; nominated by the president and confirmed with the advice and consent of the Senate; appointed to serve for life); United States Courts of Appeal; United States District Courts; State and County Courts

Political parties and leaders: Democratic Party [Howard DEAN]; Green Party; Libertarian Party [William (Bill) REDPATH]; Republican Party [Robert M. (Mike) DUNCAN]

Political pressure groups and leaders: NA

International organization participation: ADB (nonregional members), AfDB, ANZUS, APEC, Arctic Council, ARF, ASEAN (dialogue partner), Australia Group, BIS, BSEC (observer), CBSS (observer), CE (observer), CERN (observer), CP, EAPC, EBRD, FAO, G-5, G-7, G-8, G-10, IADB, IAEA, IBRD, ICAO, ICC, ICCt (signatory), ICRM, IDA, IEA, IFAD, IFC, IFRCS, IHO, ILO, IMF, IMO, IMSO, Interpol, IOC, IOM, ISO, ITSO, ITU, ITUC, MIGA, MINUSTAH, NAFTA, NATO, NEA, NSG, OAS, OECD, OPCW, OSCE, Paris Club, PCA, PIF (partner), SAARC (observer), SECI (observer), SPC, UN, UN Security Council, UNCTAD, UNESCO, UNHCR, UNITAR, UNMEE, UNMIL, UNOMIG, UNRWA, UNTSO, UPU, WCL, WCO, WHO, WIPO, WMO, WTO, ZC

Flag description: 13 equal horizontal stripes of red (top and bottom) alternating with white; there is a blue rectangle in the upper hoist-side corner bearing 50 small, white, five-pointed stars arranged in nine offset horizontal rows of six stars (top and bottom) alternating with rows of five stars; the 50 stars represent the 50 states, the 13 stripes represent the 13 original colonies; known as Old Glory; the design and colors have been the basis for a number of other flags, including Chile, Liberia, Malaysia, and Puerto Rico

ECONOMY

Economy—overview: The US has the largest and most technologically powerful economy in the world, with a per capita GDP of $46,000. In this market-oriented economy, private individuals and business firms make most of the decisions, and the federal and state governments buy needed goods and services predominantly in the private marketplace. US business firms enjoy greater flexibility than their counterparts in Western Europe and Japan in decisions to expand capital plant, to lay off surplus workers, and to develop new products. At the same time, they face higher barriers to enter their rivals' home markets than foreign firms face entering US markets. US firms are at or near the forefront in technological advances, especially in computers and in medical, aerospace, and military equipment; their advantage has narrowed since the end of World War II. The onrush of technology largely explains the gradual development of a "two-tier labor market" in which those at the bottom lack the education and the professional/technical skills of those at the top and, more and more, fail to get comparable pay raises, health insurance coverage, and other benefits. Since 1975, practically all the gains in household income have gone to the top 20% of households. The response to the terrorist attacks of 11 September 2001 showed the remarkable resilience of the economy. The war in March-April 2003 between a US-led coalition and Iraq, and the subsequent occupation of Iraq, required major shifts in national resources to the military. The rise in GDP in 2004–07 was undergirded by substantial gains in labor productivity. Hurricane Katrina caused extensive damage in the Gulf Coast region in August 2005, but had a small impact on overall GDP growth for the year. Soaring oil prices in 2005–2007 threatened inflation and unemployment, yet the economy continued to grow through year-end 2007. Imported oil accounts for about two-thirds of US consumption. Long-term problems include inadequate investment in economic infrastructure, rapidly rising medical and pension costs of an aging population, sizable trade and budget deficits, and stagnation of family income in the lower economic groups. The merchandise trade deficit reached a record $847 billion in 2007. Together, these problems caused a marked reduction in the value and status of the dollar worldwide in 2007.

GDP (purchasing power parity): $13.84 trillion (2007 est.)

GDP (official exchange rate): $13.84 trillion (2007 est.)

GDP—real growth rate: 2.2% (2007 est.)

GDP—per capita (PPP): $45,800 (2007 est.)

GDP—composition by sector:
agriculture: 0.9%
industry: 20.5%
services: 78.5% (2007 est.)

Labor force: 153.1 million (includes unemployed) (2007 est.)

Labor force—by occupation: farming, forestry, and fishing 0.6%, manufacturing, extraction, transportation, and crafts 22.6%, managerial, professional, and technical 35.5%, sales and office 24.8%, other services 16.5%
note: figures exclude the unemployed (2007)

Unemployment rate: 4.6% (2007 est.)

Population below poverty line: 12% (2004 est.)

Household income or consumption by percentage share:
lowest 10%: 2%
highest 10%: 30% (2007 est.)

Distribution of family income—Gini index: 45 (2007)

Inflation rate (consumer prices): 2.9% (2007 est.)

Investment (gross fixed): 18.8% of GDP (2007 est.)

Budget:
revenues: $2.568 trillion
expenditures: $2.73 trillion (2007 est.)

Public debt: 60.8% of GDP (2007 est.)

Agriculture—products: wheat, corn, other grains, fruits, vegetables, cotton; beef, pork, poultry, dairy products; fish; forest products

Industries: leading industrial power in the world, highly diversified and technologically advanced; petroleum, steel, motor vehicles, aerospace, telecommunications, chemicals, electronics, food processing, consumer goods, lumber, mining

Industrial production growth rate: 0.5% (2007 est.)

Electricity—production: 4.062 trillion kWh (2005)

Electricity—production by source:
fossil fuel: 71.4%
hydro: 5.6%
nuclear: 20.7%
other: 2.3% (2001)

Electricity—consumption: 3.816 trillion kWh (2005)

Electricity—exports: 19.8 billion kWh (2005)

Electricity—imports: 44.53 billion kWh (2005)

Oil—production: 8.322 million bbl/day (2005 est.)

Oil—consumption: 20.8 million bbl/day (2005 est.)

Oil—exports: 1.048 million bbl/day (2004)

Oil—imports: 13.15 million bbl/day (2004)

Oil—proved reserves: 21.76 billion bbl (1 January 2006 est.)

Natural gas—production: 490.8 billion cu m (2005 est.)

Natural gas—consumption: 604 billion cu m (2005 est.)

Natural gas—exports: 19.8 billion cu m (2005 est.)

Natural gas—imports: 117.9 billion cu m (2005)

Natural gas—proved reserves: 5.551 trillion cu m (1 January 2006 est.)
Current account balance: -$738.6 billion (2007 est.)
Exports: $1.149 trillion f.o.b. (2007 est.)
Exports—commodities: agricultural products (soybeans, fruit, corn) 9.2%, industrial supplies (organic chemicals) 26.8%, capital goods (transistors, aircraft, motor vehicle parts, computers, telecommunications equipment) 49.0%, consumer goods (automobiles, medicines) 15.0% (2003)
Exports—partners: Canada 22.2%, Mexico 12.9%, Japan 5.8%, China 5.3%, UK 4.4% (2006)
Imports: $1.965 trillion f.o.b. (2007 est.)
Imports—commodities: agricultural products 4.9%, industrial supplies 32.9% (crude oil 8.2%), capital goods 30.4% (computers, telecommunications equipment, motor vehicle parts, office machines, electric power machinery), consumer goods 31.8% (automobiles, clothing, medicines, furniture, toys) (2003)
Imports—partners: Canada 16%, China 15.9%, Mexico 10.4%, Japan 7.9%, Germany 4.8% (2006)
Economic aid—donor: ODA, $23.53 billion (2006)
Reserves of foreign exchange and gold: $70.57 billion (31 December 2007 est.)
Debt—external: $12.25 trillion (30 June 2007)
Stock of direct foreign investment—at home: $1.967 trillion (2007 est.)
Stock of direct foreign investment—abroad: $2.627 trillion (2007 est.)
Market value of publicly traded shares: $17 trillion (2005)
Currency (code): US dollar (USD)
Currency code: USD
Exchange rates: *British pounds per US dollar:* 0.4993 (2007), 0.5418 (2006), 0.5493 (2005), 0.5462 (2004), 0.6125 (2003)
Canadian dollars per US dollar: 1.0724 (2007), 1.1334 (2006), 1.2118 (2005), 1.3010 (2004), 1.4011 (2003)
Japanese yen per US dollar: 117.99 (2007), 116.18 (2006) 110.22 (2005), 108.19 (2004), 115.93 (2003)
euros per US dollar: 0.7345 (2007), 0.7964 (2006), 0.8041 (2005), 0.8054 (2004), 0.8860 (2003)
Chinese yuan per US dollar: 7.61 (2007), 7.97 (2006), 8.1943 (2005), 8.2768 (2004), 8.2770 (2003)
Fiscal year: 1 October—30 September

COMMUNICATIONS

Telephones—main lines in use: 172 million (2006)
Telephones—mobile cellular: 233 million (2006)
Telephone system:

general assessment: a large, technologically advanced, multipurpose communications system
domestic: a large system of fiber-optic cable, microwave radio relay, coaxial cable, and domestic satellites carries every form of telephone traffic; a rapidly growing cellular system carries mobile telephone traffic throughout the country
international: country code—1; multiple ocean cable systems provide international connectivity; satellite earth stations—61 Intelsat (45 Atlantic Ocean and 16 Pacific Ocean), 5 Intersputnik (Atlantic Ocean region), and 4 Inmarsat (Pacific and Atlantic Ocean regions) (2000)
Radio broadcast stations: AM 4,789, FM 8,961, shortwave 19 (2006)
Radios: 575 million (1997)
Television broadcast stations: 2,218 (2006)
Televisions: 219 million (1997)
Internet country code: .us
Internet hosts: 3.95 million (2007)
Internet Service Providers (ISPs): 7,000 (2002 est.)
Internet users: 208 million (2006)

TRANSPORTATION

Airports: 14,947 (2007)
Airports—with paved runways:
total: 5,143
over 3,047 m: 191
2,438 to 3,047 m: 224
1,524 to 2,437 m: 1,452
914 to 1,523 m: 2,323
under 914 m: 953 (2007)
Airports—with unpaved runways:
total: 9,804
2,438 to 3,047 m: 7
1,524 to 2,437 m: 153
914 to 1,523 m: 1,732
under 914 m: 7,912 (2007)
Heliports: 146 (2007)
Pipelines: petroleum products 244,620 km; natural gas 548,665 km (2006)
Railways:
total: 226,612 km
standard gauge: 226,612 km 1.435-m gauge (2005)
Roadways:
total: 6,430,366 km
paved: 4,165,110 km (includes 75,009 km of expressways)
unpaved: 2,265,256 km (2005)
Waterways: 41,009 km (19,312 km used for commerce)
note: Saint Lawrence Seaway of 3,769 km, including the Saint Lawrence River of 3,058 km, shared with Canada (2007)
Merchant marine:
total: 446 ships (1000 GRT or over) 10,308,428 GRT/12,616,742 DWT
by type: barge carrier 6, bulk carrier 64, cargo 82, carrier 2, chemical tanker 20, container 82, passenger 20, passenger/cargo 60, petroleum tanker 59,

refrigerated cargo 4, roll on/roll off 26, specialized tanker 1, vehicle carrier 20
foreign-owned: 67 (Australia 2, Canada 4, Denmark 29, Germany 6, Malaysia 4, Netherlands 1, Norway 4, Singapore 11, Sweden 5, UK 1)
registered in other countries: 785 (Antigua and Barbuda 8, Australia 5, Bahamas 162, Belize 3, Bermuda 23, Cambodia 6, Canada 3, Cayman Islands 41, Comoros 2, Cyprus 8, Ecuador 1, Greece 10, Honduras 1, Hong Kong 22, Ireland 2, Isle of Man 4, Italy 16, Liberia 103, South Korea 7, Luxembourg 3, Malta 11, Marshall Islands 129, Netherlands 13, Netherlands Antilles 1, Norway 18, Panama 115, Peru 1, Portugal 1, Puerto Rico 3, Russia 1, Singapore 17, Spain 9, St Vincent and The Grenadines 21, Sweden 1, Trinidad and Tobago 1, Tuvalu 1, UK 11, Vanuatu 1, unknown 4) (2007)
Ports and terminals: Corpus Christi, Duluth, Hampton Roads, Houston, Long Beach, Los Angeles, New York, Philadelphia, Tampa, Texas City

MILITARY

Military branches: US Army, US Navy (includes Marine Corps), US Air Force, US Coast Guard; note—Coast Guard administered in peacetime by the Department of Homeland Security, but in wartime reports to the Department of the Navy (2008)
Military service age and obligation: 18 years of age; 17 years of age with written parental consent (2006)
Manpower available for military service:
males age 16–49: 72,715,332
females age 16–49: 71,638,785 (2008 est.)
Manpower fit for military service:
males age 16–49: 59,413,358
females age 16–49: 59,187,183 (2008 est.)
Manpower reaching militarily significant age annually:
males age 16–49: 2,186,440
females age 16–49: 2,079,688 (2008 est.)
Military expenditures—percent of GDP: 4.06% (2005 est.)

TRANSNATIONAL ISSUES

Disputes—international: the U.S. has intensified domestic security measures and is collaborating closely with its neighbors, Canada and Mexico, to monitor and control legal and illegal personnel, transport, and commodities across the international borders; abundant rainfall in recent years along much of the Mexico-US border region has ameliorated periodically strained water-sharing arrangements; 1990 Maritime Boundary Agreement in the Bering Sea still awaits Russian Duma ratification;

managed maritime boundary disputes with Canada at Dixon Entrance, Beaufort Sea, Strait of Juan de Fuca, and around the disputed Machias Seal Island and North Rock; The Bahamas and US have not been able to agree on a maritime boundary; US Naval Base at Guantanamo Bay is leased from Cuba and only mutual agreement or US abandonment of the area can terminate the lease; Haiti claims US-administered Navassa Island; US has made no territo-rial claim in Antarctica (but has reserved the right to do so) and does not recognize the claims of any other states; Marshall Islands claims Wake Island; Tokelau included American Samoa's Swains Island among the islands listed in its 2006 draft constitution

Refugees and internally displaced persons: *refugees (country of origin)*: the US admitted 62,643 refugees during FY04/05 including; 10,586 (Somalia); 8,549 (Laos); 6,666 (Russia); 6,479 (Cuba); 3,100 (Haiti); 2,136 (Iran) (2006)

Illicit drugs: world's largest consumer of cocaine, shipped from Colombia through Mexico and the Caribbean; consumer of ecstasy and of Mexican heroin, marijuana and methamphetamine; minor consumer of high-quality Southeast Asian heroin; illicit producer of cannabis, marijuana, depressants, stimulants, hallucinogens, and methamphetamine; money-laundering center

UNITED STATES PACIFIC ISLAND WILDLIFE REFUGES

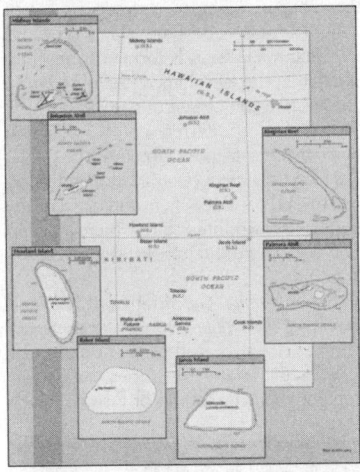

INTRODUCTION

Background: All of the following US Pacific island territories except Midway Atoll constitute the Pacific Remote Islands National Wildlife Refuge Complex and as such are managed by the Fish and Wildlife Service of the US Department of the Interior. Midway Atoll NWR has been included in a Refuge Complex with the Hawaiian Islands NWR and also designated as part of Papahanaumokuakea Marine National Monument. These remote refuges are the most widespread collection of marine- and terrestrial-life protected areas on the planet under a single country's jurisdiction. They protect many endemic species including corals, fish, shellfish, marine mammals, seabirds, water birds, land birds, insects, and vegetation not found elsewhere.

Baker Island: The US took possession of the island in 1857, and its guano deposits were mined by US and British companies during the second half of the 19th century. In 1935, a short-lived attempt at colonization began on this island but was disrupted by World War II and thereafter abandoned. The island was established as a National Wildlife Refuge in 1974.

Howland Island: Discovered by the US early in the 19th century, the uninhabited atoll was officially claimed by the US in 1857. Both US and British companies mined for guano deposits until about 1890. In 1935, a short-lived attempt at colonization began on this island, similar to the effort on nearby Baker Island, but was disrupted by World War II and thereafter abandoned. The famed American aviatrix Amelia EARHART disappeared while seeking out Howland Island as a refueling stop during her 1937 round-the-world flight; Earhart Light, a day beacon near the middle of the west coast, was named in her memory. The island was established as a National Wildlife Refuge in 1974.

Jarvis Island: First discovered by the British in 1821, the uninhabited island was annexed by the US in 1858, but abandoned in 1879 after tons of guano had been removed. The UK annexed the island in 1889, but never carried out plans for further exploitation. The US occupied and reclaimed the island in 1935 until it was abandoned in 1942 during World War II. The island was established as a National Wildlife Refuge in 1974.

Johnston Atoll: Both the US and the Kingdom of Hawaii annexed Johnston Atoll in 1858, but it was the US that mined the guano deposits until the late 1880s. Johnston and Sand Islands were designated wildlife refuges in 1926. The US Navy took over the atoll in 1934, and subsequently the US Air Force assumed control in 1948. The site was used for high-altitude nuclear tests in the 1950s and 1960s, and until late in 2000 the atoll was maintained as a storage and disposal site for chemical weapons. Munitions destruction, cleanup, and closure of the facility was completed by May 2005. The Fish and Wildlife Service and the US Air Force are currently discussing future management options; in the interim, Johnston Atoll and the three-mile Naval Defensive Sea around it remain under the jurisdiction and administrative control of the US Air Force.

Kingman Reef: The US annexed the reef in 1922. Its sheltered lagoon served as a way station for flying boats on Hawaii-to-American Samoa flights during the late 1930s. There are no terrestrial plants on the reef, which is frequently awash, but it does support abundant and diverse marine fauna and flora. In 2001, the waters surrounding the reef out to 12 nm were designated a US National Wildlife Refuge.

Midway Islands: The US took formal possession of the islands in 1867. The laying of the trans-Pacific cable, which passed through the islands, brought the first residents in 1903. Between 1935 and 1947, Midway was used as a refueling stop for trans-Pacific flights. The US naval victory over a Japanese fleet off Midway in 1942 was one of the turning points of World War II. The islands continued to serve as a naval station until closed in 1993. Today the islands are a National Wildlife Refuge and are the site of the world's largest Laysan albatross colony.

Palmyra Atoll: The Kingdom of Hawaii claimed the atoll in 1862, and the US included it among the Hawaiian Islands when it annexed the archipelago in 1898. The Hawaii Statehood Act of 1959 did not include Palmyra Atoll, which is now partly privately owned by the Nature Conservancy with the rest owned by the Federal government and managed by the US Fish and Wildlife Service. These organizations are managing the atoll as a wildlife refuge. The lagoons and surrounding waters within the 12 nm US territorial seas were transferred to the US Fish and Wildlife Service and designated as a National Wildlife Refuge in January 2001.

GEOGRAPHY

Location: Oceania

Baker Island: atoll in the North Pacific Ocean 1,830 nm (3,389 km) southwest of Honolulu, about half way between Hawaii and Australia

Howland Island: island in the North Pacific Ocean 1,815 nm (3,361 km) southwest of Honolulu, about half way between Hawaii and Australia

Jarvis Island: island in the South Pacific Ocean 1,305 nm (2,417 km) south of Honolulu, about half way between Hawaii and Cook Islands

Johnston Atoll: atoll in the North Pacific Ocean 717 nm (1,328 km) southwest of Honolulu, about one-third of the way from Hawaii to the Marshall Islands

Kingman Reef: reef in the North Pacific Ocean 930 nm (1,722 km) south of Honolulu, about half way between Hawaii and American Samoa

Midway Islands: atoll in the North Pacific Ocean 1,260 nm (2,334 km) northwest of Honolulu near the end of the Hawaiian Archipelago, about one-third of the way from Honolulu to Tokyo

Palmyra Atoll: atoll in the North Pacific Ocean 960 nm (1,778 km) south of Honolulu, about half way between Hawaii and American Samoa

Geographic coordinates: *Baker Island:* 0 13 N, 176 28 W

Howland Island: 0 48 N, 176 38 W

Jarvis Island: 0 23 S, 160 01 W

Johnston Atoll: 16 45 N, 169 31 W

Kingman Reef: 6 23 N, 162 25 W

Midway Islands: 28 12 N, 177 22 W

Palmyra Atoll: 5 53 N, 162 05 W

Map references: Oceania

Area:

total—6,959.41 sq km; emergent land—22.41 sq km; submerged—6,937 sq km

Baker Island: total—129.1 sq km; emergent land—2.1 sq km; submerged—127 sq km

Howland Island: total—138.6 sq km; emergent land—2.6 sq km; submerged—136 sq km

Jarvis Island: total—152 sq km; emergent land—5 sq km; submerged—147 sq km

Johnston Atoll: total—276.6 sq km; emergent land—2.6 sq km; submerged—274 sq km

Kingman Reef: total—1,958.01 sq km; emergent land—0.01 sq km; submerged—1,958 sq km

Midway Islands: total—2,355.2 sq km; emergent land—6.2 sq km; submerged—2,349 sq km

Palmyra Atoll: total—1,949.9 sq km; emergent land—3.9 sq km; submerged—1,946 sq km

Area—comparative: *Baker Island:* about two and a half times the size of The Mall in Washington, DC

Howland Island: about three times the size of The Mall in Washington, DC

Jarvis Island: about eight times the size of The Mall in Washington, DC

Johnston Atoll: about four and a half times the size of The Mall in Washington, DC

Kingman Reef: a little more than one and a half times the size of The Mall in Washington, DC

Midway Islands: about nine times the size of The Mall in Washington, DC

Palmyra Atoll: about 20 times the size of The Mall in Washington, DC

Land boundaries: none

Coastline: *Baker Island:* 4.8 km

Howland Island: 6.4 km

Jarvis Island: 8 km

Johnston Atoll: 34 km

Kingman Reef: 3 km

Midway Islands: 15 km

Palmyra Atoll: 14.5 km

Maritime claims:

territorial sea: 12 nm

exclusive economic zone: 200 nm

Climate: *Baker, Howland, and Jarvis Islands:* equatorial; scant rainfall, constant wind, burning sun

Johnston Atoll and Kingman Reef: tropical, but generally dry; consistent northeast trade winds with little seasonal temperature variation

Midway Islands: subtropical with cool, moist winters (December to February) and warm, dry summers (May to October); moderated by prevailing easterly winds; most of the 1,067 mm (42 in) of annual rainfall occurs during the winter

Palmyra Atoll: equatorial, hot; located within the low pressure area of the Intertropical Convergence Zone (ITCZ) where the northeast and southeast trade winds meet, it is extremely wet with between 4,000–5,000 mm (160–200 in) of rainfall each year

Terrain: low and nearly level sandy coral islands with narrow fringing reefs that have developed at the top of submerged volcanic mountains, which in most cases rise steeply from the ocean floor

Elevation extremes:

lowest point: Pacific Ocean 0 m

highest point: Baker Island, unnamed location—8 m; Howland Island, unnamed location—3 m; Jarvis Island, unnamed location—7 m; Johnston Atoll, Sand Island—10 m; Kingman Reef, unnamed location—less than 2 m; Midway Islands, unnamed location—13 m; Palmyra Atoll, unnamed location—3 m

Natural resources: terrestrial and aquatic wildlife

Land use:

arable land: 0%

permanent crops: 0%

other: 100% (2008)

Natural hazards: *Baker, Howland, and Jarvis Islands:* the narrow fringing reef surrounding the island can be a maritime hazard

Kingman Reef: wet or awash most of the time, maximum elevation of less than 2 m makes Kingman Reef a maritime hazard

Midway Islands, Johnston, and Palmyra Atolls: NA

Environment—current issues: *Baker, Howland, and Jarvis Islands, and Johnston Atoll:* no natural fresh water resources

Kingman Reef: none

Midway Islands and Palmyra Atoll: NA

Geography—note: *Baker, Howland, and Jarvis Islands:* scattered vegetation consisting of grasses, prostrate vines, and low growing shrubs; primarily a nesting, roosting, and foraging habitat for seabirds, shorebirds, and marine wildlife; closed to the public

Johnston Atoll: Johnston Island and Sand Island are natural islands, which have been expanded by coral dredging; North Island (Akau) and East Island (Hikina) are manmade islands formed from coral dredging; the egg-shaped reef is 34 km in circumference; closed to the public

Kingman Reef: barren coral atoll with deep interior lagoon; closed to the public

Midway Islands: a coral atoll managed as a national wildlife refuge and open to the public for wildlife-related recreation in the form of wildlife observation and photography

Palmyra Atoll: the high rainfall and resulting lush vegetation make the environment of this atoll unique among the US Pacific Island territories; supports a large undisturbed stand of Pisonia beach forest

PEOPLE

Population: no indigenous inhabitants

note: public entry is by special-use permit from US Fish and Wildlife Service only and generally restricted to scientists and educators; visited annually by US Fish and Wildlife Service

Johnston Atoll: in previous years, an average of 1,100 US military and civilian contractor personnel were present; as of May 2005 all US government personnel had left the island

Midway Islands: approximately 40 people make up the staff of US Fish and Wildlife Service and their services contractor living at the atoll

Palmyra Atoll: four to 20 Nature Conservancy, US Fish and Wildlife staff, and researchers

GOVERNMENT

Country name:

conventional long form: none

conventional short form: Baker Island; Howland Island; Jarvis Island; Johnston Atoll; Kingman Reef; Midway Islands; Palmyra Atoll

Dependency status: unincorporated territories of the US; administered from Washington, DC by the Fish and Wildlife Service of the US Department of the Interior as part of the National Wildlife Refuge system

note on Palmyra Atoll: incorporated Territory of the US; partly privately owned and partly federally owned; administered from Washington, DC by the Fish and Wildlife Service of the US Department of the Interior; the Office of Insular Affairs of the US Department of the Interior continues to administer nine excluded areas comprising certain tidal and submerged lands within the 12 nm territorial sea or within the lagoon
Legal system: the laws of the US, where applicable, apply
Diplomatic representation from the US: none (territories of the US)
Flag description: the flag of the US is used

ECONOMY

Economy—overview: no economic activity

TRANSPORTATION

Airports: *Baker Island:* one abandoned World War II runway of 1,665 m covered with vegetation and unusable
Howland Island: airstrip constructed in 1937 for scheduled refueling stop on the round-the-world flight of Amelia EARHART and Fred NOONAN; the aviators left Lae, New Guinea, for Howland Island but were never seen again; the airstrip is no longer serviceable
Johnston Atoll: one closed and not maintained
Kingman Reef: lagoon was used as a halfway station between Hawaii and American Samoa by Pan American Airways for flying boats in 1937 and 1938
Midway Islands: 3—one operational (2,409 m paved); no fuel for sale except emergencies
Palmyra Atoll: 1—1,846 m unpaved runway; privately owned (2008)
Ports and terminals: *Baker, Howland, and Jarvis Islands, and Kingman Reef:* none; offshore anchorage only
Johnston Atoll: Johnston Island
Midway Islands: Sand Island
Palmyra Atoll: West Lagoon

MILITARY

Military—note: defense is the responsibility of the US

TRANSNATIONAL ISSUES

Disputes—international: none

URUGUAY

INTRODUCTION

Background: Montevideo, founded by the Spanish in 1726 as a military stronghold, soon took advantage of its natural harbor to become an important commercial center. Claimed by Argentina but annexed by Brazil in 1821, Uruguay declared its independence four years later and secured its freedom in 1828 after a three-year struggle. The administrations of President Jose BATLLE in the early 20th century established widespread political, social, and economic reforms that established a statist tradition. A violent Marxist urban guerrilla movement named the Tupamaros, launched in the late 1960s, led Uruguay's president to cede control of the government to the military in 1973. By yearend, the rebels had been crushed, but the military continued to expand its hold over the government. Civilian rule was not restored until 1985. In 2004, the left-of-center Frente Amplio Coalition won national elections that effectively ended 170 years of political control previously held by the Colorado and Blanco parties. Uruguay's political and labor conditions are among the freest on the continent.

GEOGRAPHY

Location: Southern South America, bordering the South Atlantic Ocean, between Argentina and Brazil
Geographic coordinates: 33 00 S, 56 00 W
Map references: South America
Area:
total: 176,220 sq km
land: 173,620 sq km
water: 2,600 sq km
Area—comparative: slightly smaller than the state of Washington
Land boundaries:
total: 1,648 km
border countries: Argentina 580 km, Brazil 1,068 km
Coastline: 660 km
Maritime claims:
territorial sea: 12 nm
contiguous zone: 24 nm
exclusive economic zone: 200 nm
continental shelf: 200 nm or edge of continental margin
Climate: warm temperate; freezing temperatures almost unknown
Terrain: mostly rolling plains and low hills; fertile coastal lowland
Elevation extremes:
lowest point: Atlantic Ocean 0 m
highest point: Cerro Catedral 514 m
Natural resources: arable land, hydropower, minor minerals, fisheries
Land use:
arable land: 7.77%
permanent crops: 0.24%
other: 91.99% (2005)

Irrigated land: 2,100 sq km (2003)
Total renewable water resources: 139 cu km (2000)
Freshwater withdrawal (domestic/industrial/agricultural): *total:* 3.15 cu km/yr (2%/1%/96%)
per capita: 910 cu m/yr (2000)
Natural hazards: seasonally high winds (the pampero is a chilly and occasional violent wind that blows north from the Argentine pampas), droughts, floods; because of the absence of mountains, which act as weather barriers, all locations are particularly vulnerable to rapid changes from weather fronts
Environment—current issues: water pollution from meat packing/tannery industry; inadequate solid/hazardous waste disposal
Environment—international agreements: *party to:* Antarctic-Environmental Protocol, Antarctic-Marine Living Resources, Antarctic Treaty, Biodiversity, Climate Change, Climate Change-Kyoto Protocol, Desertification, Endangered Species, Environmental Modification, Hazardous Wastes, Law of the Sea, Ozone Layer Protection, Ship Pollution, Tropical Timber 94, Wetlands
signed, but not ratified: Marine Dumping, Marine Life Conservation
Geography—note: second-smallest South American country (after Suriname); most of the low-lying landscape (three-quarters of the country) is grassland, ideal for cattle and sheep raising

PEOPLE

Population: 3,477,778 (July 2008 est.)
Age structure: *0–14 years:* 22.7% (male 401,209/female 388,315)

15–64 years: 64% (male 1,105,891/female 1,120,858)

65 years and over: 13.3% (male 185,704/female 275,801) (2008 est.)

Median age:

total: 33.2 years

male: 31.8 years

female: 34.6 years (2008 est.)

Population growth rate: 0.486% (2008 est.)

Birth rate: 14.17 births/1,000 population (2008 est.)

Death rate: 9.12 deaths/1,000 population (2008 est.)

Net migration rate: -0.18 migrant(s)/1,000 population (2008 est.)

Sex ratio:

at birth: 1.04 male(s)/female

under 15 years: 1.03 male(s)/female

15–64 years: 0.99 male(s)/female

65 years and over: 0.67 male(s)/female

total population: 0.95 male(s)/female (2008 est.)

Infant mortality rate:

total: 11.66 deaths/1,000 live births

male: 13.1 deaths/1,000 live births

female: 10.17 deaths/1,000 live births (2008 est.)

Life expectancy at birth:

total population: 76.14 years

male: 72.89 years

female: 79.51 years (2008 est.)

Total fertility rate: 1.94 children born/woman (2008 est.)

HIV/AIDS—adult prevalence rate: 0.3% (2001 est.)

HIV/AIDS—people living with HIV/AIDS: 6,000 (2001 est.)

HIV/AIDS—deaths: fewer than 500 (2003 est.)

Nationality:

noun: Uruguayan(s)

adjective: Uruguayan

Ethnic groups: white 88%, mestizo 8%, black 4%, Amerindian (practically non-existent)

Religions: Roman Catholic 66% (less than half of the adult population attends church regularly), Protestant 2%, Jewish 1%, nonprofessing or other 31%

Languages: Spanish, Portunol, or Brazilero (Portuguese-Spanish mix on the Brazilian frontier)

Literacy:

definition: age 15 and over can read and write

total population: 98%

male: 97.6%

female: 98.4% (2003 est.)

GOVERNMENT

Country name:

conventional long form: Oriental Republic of Uruguay

conventional short form: Uruguay

local long form: Republica Oriental del Uruguay

local short form: Uruguay

former: Banda Oriental, Cisplatine Province

Government type: constitutional republic

Capital: *name:* Montevideo

geographic coordinates: 34 53 S, 56 11 W

time difference: UTC-3 (2 hours ahead of Washington, DC during Standard Time)

daylight saving time: +1hr, begins second Sunday in October; ends second Sunday in March

Administrative divisions: 19 departments (departamentos, singular—departamento); Artigas, Canelones, Cerro Largo, Colonia, Durazno, Flores, Florida, Lavalleja, Maldonado, Montevideo, Paysandu, Rio Negro, Rivera, Rocha, Salto, San Jose, Soriano, Tacuarembo, Treinta y Tres

Independence: 25 August 1825 (from Brazil)

National holiday: Independence Day, 25 August (1825)

Constitution: 27 November 1966, effective 15 February 1967; suspended 27 June 1973, new constitution rejected by referendum 30 November 1980; two constitutional reforms approved by plebiscite 26 November 1989 and 7 January 1997

Legal system: based on Spanish civil law system; accepts compulsory ICJ jurisdiction

Suffrage: 18 years of age; universal and compulsory

Executive branch:

chief of state: President Tabare VAZQUEZ Rosas (since 1 March 2005); Vice President Rodolfo NIN NOVOA (since 1 March 2005); note—the president is both the chief of state and head of government

head of government: President Tabare VAZQUEZ Rosas (since 1 March 2005); Vice President Rodolfo NIN NOVOA (since 1 March 2005)

cabinet: Council of Ministers appointed by the president with parliamentary approval

elections: president and vice president elected on the same ticket by popular vote for five-year terms (may not serve consecutive terms); election last held 31 October 2004 (next to be held in October 2009)

election results: Tabare VAZQUEZ elected president; percent of vote—Tabare VAZQUEZ 50.5%, Jorge LARRANAGA 35.1%, Guillermo STIRLING 10.3%; other 4.1%

Legislative branch: bicameral General Assembly or Asamblea General consists of Chamber of Senators or Camara de Senadores (30 seats; members are elected by popular vote to serve five-year terms; vice president has one vote in the Senate) and Chamber of Representatives or Camara de Representantes (99 seats; members are elected by popular vote to serve five-year terms)

elections: Chamber of Senators—last held 31 October 2004 (next to be held October 2009); Chamber of Representatives—last held 31 October 2004 (next to be held October 2009)

election results: Chamber of Senators—percent of vote by party—NA; seats by party—EP-FA 16, Blanco 11, Colorado Party 3; Chamber of Representatives—percent of vote by party—NA; seats by party—EP-FA 52, Blanco 36, Colorado Party 10, Independent Party 1

Judicial branch: Supreme Court (judges are nominated by the president and elected for 10-year terms by the General Assembly)

Political parties and leaders: Broad Front (Frente Amplio)—formerly known as the Progressive Encounter/Broad Front Coalition or EP-FA [Jorge BROVETTO] (a broad governing coalition that includes Movement of the Popular Participation or MPP [Jose MUJICA], New Space Party (Nuevo Espacio) [Rafael MICHELINI], Progressive Alliance (Alianza Progresista) [Rodolfo NIN NOVOA], Socialist Party [Eduardo FERNANDEZ], the Communist Party [Marina ARISMENDI], Uruguayan Assembly (Asamblea Uruguay) [Danilo ASTORI], and Vertiente Artiguista [Mariano ARANA]); Colorado Party (Foro Batllista) [Julio Maria SANGUINETTI]; National Party or Blanco [Luis Alberto LACALLE and Jorge LARRANAGA]

Political pressure groups and leaders: Architect's Society of Uruguay (professional organization); Catholic Church; Chamber of Uruguayan Industries (manufacturer's association); Chemist and Pharmaceutical Association (professional organization); Rural Association of Uruguay (rancher's association); students; umbrella labor organization PIT/CNT (powerful federation of Uruguayan Unions); Uruguayan Construction League; the Uruguayan Network of Political Women

International organization participation: CAN (associate), CSN, FAO, G-77, IADB, IAEA, IBRD, ICAO, ICC, ICCt, ICRM, IDA, IFAD, IFC, IFRCS, IHO, ILO, IMF, IMO, Interpol, IOC, IOM, IPU, ISO, ITSO, ITU, LAES, LAIA, Mercosur, MIGA, MINURSO, MINUSTAH, MONUC, NAM (observer), OAS, OPANAL, OPCW, PCA, RG, UN, UNCTAD, UNESCO, UNFICYP, UNIDO, Union Latina, UNMEE, UNMOGIP, UNOCI, UNOMIG, UNWTO, UPU, WCL, WCO, WFTU, WHO, WIPO, WMO, WTO

Diplomatic representation in the US:

chief of mission: Ambassador Carlos Alberto GIANELLI Derois *chancery:* 1913 I Street NW, Washington, DC 20006

telephone: [1] (202) 331-1313 through 1316

FAX: [1] (202) 331-8142

consulate(s) general: Chicago, Los Angeles, Miami, New York, Washington, DC

consulate(s): San Juan (Puerto Rico)

Diplomatic representation from the US:
chief of mission: Ambassador Frank E. BAXTER

embassy: Lauro Muller 1776, Montevideo 11200

mailing address: APO AA 34035

telephone: [598] (2) 418-7777

FAX: [598] (2) 418-8611

Flag description: nine equal horizontal stripes of white (top and bottom) alternating with blue; a white square in the upper hoist-side corner with a yellow sun bearing a human face known as the Sun of May with 16 rays that alternate between triangular and wavy

ECONOMY

Economy—overview: Uruguay's economy is characterized by an export-oriented agricultural sector, a well-educated work force, and high levels of social spending. After averaging growth of 5% annually during 1996–98, in 1999–2002 the economy suffered a major downturn, stemming largely from the spillover effects of the economic problems of its large neighbors, Argentina and Brazil. For instance, in 2001–02 Argentina made massive withdrawals of dollars deposited in Uruguayan banks, which led to a plunge in the Uruguayan peso and a massive rise in unemployment. Total GDP in these four years dropped by nearly 20%, with 2002 the worst year due to the banking crisis. The unemployment rate rose to nearly 20% in 2002, inflation surged, and the burden of external debt doubled. Cooperation with the IMF helped stem the damage. Uruguay in 2007 improved its debt profile by paying off $1.1 billion in IMF debt, and continues to follow the orthodox economic plan set by the Fund in 2005. The construction of a pulp mill in Fray Bentos, which represents the largest foreign direct investment in Uruguay's history at $1.2 billion, came online in November 2007 and is expected to add 1.6% to GDP and boost already rising exports. The economy has grown strongly since 2004 as a result of high commodity prices for Uruguayan exports, a strong peso, growth in the region, and low international interest rates.

GDP (purchasing power parity): $37.19 billion (2007 est.)

GDP (official exchange rate): $22.95 billion (2007 est.)

GDP—real growth rate: 7% (2007 est.)

GDP—per capita (PPP): $11,600 (2007 est.)

GDP—composition by sector:
agriculture: 10.1%
industry: 32%
services: 57.9% (2007 est.)

Labor force: 1.587 million (2007 est.)

Labor force—by occupation: *agriculture:* 9%
industry: 15%
services: 76% (2007 est.)

Unemployment rate: 9.2% (2007 est.)

Population below poverty line: 27.4% of households (2006)

Household income or consumption by percentage share:
lowest 10%: 1.9%
highest 10%: 34% (2003)

Distribution of family income—Gini index: 45.2 (2006)

Inflation rate (consumer prices): 8.1% (2007)

Investment (gross fixed): 13.9% of GDP (2007 est.)

Budget: *revenues:* $6.6 billion
expenditures: $6.3 billion (2007 est.)

Public debt: 67% of GDP (2007)

Agriculture—products: rice, wheat, soybeans, barley; livestock, beef; fish; forestry

Industries: food processing, electrical machinery, transportation equipment, petroleum products, textiles, chemicals, beverages

Industrial production growth rate: 7.9% (2007 est.)

Electricity—production: 9.2 billion kWh (2007)

Electricity—production by source:
fossil fuel: 0.7%
hydro: 99.1%
nuclear: 0%
other: 0.3% (2001)

Electricity—consumption: 7.03 billion kWh (2007)

Electricity—exports: 1 billion kWh (2007)

Electricity—imports: 780 million kWh (2007)

Oil—production: 27,830 bbl/day (2007 est.)

Oil—consumption: 33,400 bbl/day (2007 est.)

Oil—exports: 4,410 bbl/day (2007)

Oil—imports: 43,670 bbl/day (2007)

Oil—proved reserves: NA

Natural gas—production: 0 cu m (2007)

Natural gas—consumption: 102.8 million cu m (2007 est.)

Natural gas—exports: 0 cu m (2007 est.)

Natural gas—imports: 116.9 million cu m (2007)

Natural gas—proved reserves: 0 cu m (1 January 2006 est.)

Current account balance: -$184 million (2007 est.)

Exports: $5.063 billion f.o.b. (2007 est.)

Exports—commodities: meat, rice, leather products, wool, fish, dairy products

Exports—partners: Brazil 15%, US 12.1%, Argentina 6.8%, Mexico 6.4%, China 6%, Germany 5%, Russia 4.9% (2006)

Imports: $5.554 billion f.o.b. (2007 est.)

Imports—commodities: crude petroleum and petroleum products, machinery, chemicals, road vehicles, paper, plastics

Imports—partners: Argentina 20.4%, Brazil 17.1%, US 8.2%, Paraguay 7.2%, China 6.9%, Venezuela 4.8%, Nigeria 4.4% (2006)

Economic aid—recipient: $14.62 million (2005)

Reserves of foreign exchange and gold: $4.112 billion (December 2007 est.)

Debt—external: $11.42 billion (31 December 2007)

Stock of direct foreign investment—at home: $4.19 billion (2007)

Stock of direct foreign investment—abroad: $156 million (2007)

Market value of publicly traded shares: $224 million (2007)

Currency (code): Uruguayan peso (UYU)

Currency code: UYU

Exchange rates: Uruguayan pesos per US dollar—23.947 (2007), 24.048 (2006), 24.479 (2005), 28.704 (2004), 28.209 (2003)

Fiscal year: calendar year

COMMUNICATIONS

Telephones—main lines in use: 987,000 (2006)

Telephones—mobile cellular: 2.333 million (2006)

Telephone system:
general assessment: fully digitalized
domestic: most modern facilities concentrated in Montevideo; new nationwide microwave radio relay network; overall fixed-line and mobile-cellular teledensity is approaching 100 telephones per 100 persons
international: country code—598; the UNISOR submarine cable system provides direct connectivity to Brazil and Argentina; satellite earth stations—2 Intelsat (Atlantic Ocean) (2002)

Radio broadcast stations: AM 93, FM 191, shortwave 7 (2005)

Radios: 1.97 million (1997)

Television broadcast stations: 62 (2005)

Televisions: 782,000 (1997)

Internet country code: .uy

Internet hosts: 279,114 (2007)

Internet Service Providers (ISPs): 14 (2001)

Internet users: 756,000 (2006)

TRANSPORTATION

Airports: 60 (2007)
Airports—with paved runways:
total: 9
over 3,047 m: 1
1,524 to 2,437 m: 4
914 to 1,523 m: 2
under 914 m: 2 (2007)
Airports—with unpaved runways:
total: 51
1,524 to 2,437 m: 3
914 to 1,523 m: 19
under 914 m: 29 (2007)
Pipelines: gas 257 km; oil 160 km (2007)
Railways:
total: 2,073 km
standard gauge: 2,073 km 1.435-m gauge
note: 461 km have been taken out of service and 460 km are in partial use (2006)
Roadways:
total: 77,732 km
paved: 7,743 km

unpaved: 69,989 km (2004)
Waterways: 1,600 km (2005)
Merchant marine:
total: 14 ships (1000 GRT or over) 36,041 GRT/22,274 DWT
by type: cargo 2, chemical tanker 2, passenger/cargo 7, petroleum tanker 2, roll on/roll off 1
foreign-owned: 4 (Argentina 3, Greece 1)
registered in other countries: 7 (Argentina 1, Bahamas 1, Liberia 3, Spain 2) (2007)
Ports and terminals: Montevideo

MILITARY

Military branches: Uruguayan Armed Forces: Army, Navy (includes naval air arm, Marines, Maritime Prefecture in wartime), Air Force (Fuerza Aerea Uruguaya, FAU) (2008)
Military service age and obligation: 18 years of age for voluntary and compulsory military service; enlistment is voluntary in peacetime, but the government has the authority to conscript in emergencies (2007)

Manpower available for military service:
males age 16–49: 837,252
females age 16–49: 824,096 (2008 est.)
Manpower fit for military service:
males age 16–49: 703,955
females age 16–49: 690,296 (2008 est.)
Military expenditures—percent of GDP: 1.6% (2006)

TRANSNATIONAL ISSUES

Disputes—international: in Jan 2007, ICJ provisionally ruled Uruguay may begin construction of two paper mills on the Uruguay River, which forms the border with Argentina, while the court examines further whether Argentina has the legal right to stop such construction with potential environmental implications to both countries; uncontested dispute with Brazil over certain islands in the Quarai/Cuareim and Invernada streams and the resulting tripoint with Argentina

UZBEKISTAN

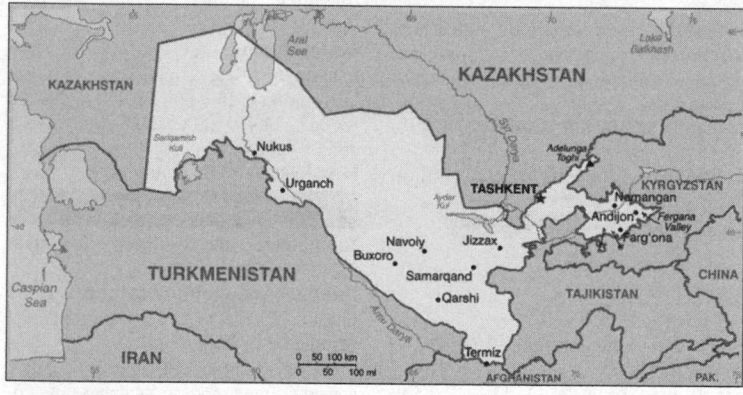

INTRODUCTION

Background: Russia conquered Uzbekistan in the late 19th century. Stiff resistance to the Red Army after World War I was eventually suppressed and a socialist republic set up in 1924. During the Soviet era, intensive production of "white gold" (cotton) and grain led to overuse of agrochemicals and the depletion of water supplies, which have left the land poisoned and the Aral Sea and certain rivers half dry. Independent since 1991, the country seeks to gradually lessen its dependence on agriculture while developing its mineral and petroleum reserves. Current concerns include terrorism by Islamic militants, economic stagnation, and the curtailment of human rights and democratization.

GEOGRAPHY

Location: Central Asia, north of Afghanistan
Geographic coordinates: 41 00 N, 64 00 E
Map references: Asia
Area: *total:* 447,400 sq km
land: 425,400 sq km
water: 22,000 sq km
Area—comparative: slightly larger than California
Land boundaries:
total: 6,221 km
border countries: Afghanistan 137 km, Kazakhstan 2,203 km, Kyrgyzstan 1,099 km, Tajikistan 1,161 km, Turkmenistan 1,621 km
Coastline: 0 km (doubly landlocked); note—Uzbekistan includes the southern portion of the Aral Sea with a 420 km

shoreline
Maritime claims: none (doubly landlocked)
Climate: mostly midlatitude desert, long, hot summers, mild winters; semiarid grassland in east
Terrain: mostly flat-to-rolling sandy desert with dunes; broad, flat intensely irrigated river valleys along course of Amu Darya, Syr Darya (Sirdaryo), and Zarafshon; Fergana Valley in east surrounded by mountainous Tajikistan and Kyrgyzstan; shrinking Aral Sea in west
Elevation extremes:
lowest point: Sariqarnish Kuli -12 m
highest point: Adelunga Toghi 4,301 m
Natural resources: natural gas, petroleum, coal, gold, uranium, silver, copper, lead and zinc, tungsten, molybdenum
Land use: *arable land:* 10.51%
permanent crops: 0.76%
other: 88.73% (2005)
Irrigated land: 42,810 sq km (2003)
Total renewable water resources: 72.2 cu km (2003)
Freshwater withdrawal (domestic/industrial/agricultural): *total:* 58.34 cu km/yr (5%/2%/93%)
per capita: 2,194 cu m/yr (2000)
Natural hazards: NA
Environment—current issues: shrinkage of the Aral Sea is resulting in growing concentrations of chemical pesticides and natural salts; these substances are then blown from the increasingly exposed lake bed and contribute to desertification; water pollution from industrial wastes and the heavy use of

fertilizers and pesticides is the cause of many human health disorders; increasing soil salination; soil contamination from buried nuclear processing and agricultural chemicals, including DDT

Environment—International agreements: *party to:* Biodiversity, Climate Change, Climate Change-Kyoto Protocol, Desertification, Endangered Species, Environmental Modification, Hazardous Wastes, Ozone Layer Protection, Wetlands

signed, but not ratified: none of the selected agreements

Geography—note: along with Liechtenstein, one of the only two doubly landlocked countries in the world

PEOPLE

Population: 28,268,440 (July 2008 est.)

Age structure:

0–14 years: 32.1% (male 4,618,815/female 4,445,571)

15–64 years: 63.3% (male 8,851,738/female 9,042,647)

65 years and over: 4.6% (male 538,699/female 770,970) (2008 est.)

Median age:

total: 23.2 years

male: 22.6 years

female: 23.8 years (2008 est.)

Population growth rate: 1.753% (2008 est.)

Birth rate: 26.45 births/1,000 population (2008 est.)

Death rate: 7.62 deaths/1,000 population (2008 est.)

Net migration rate: -1.3 migrant(s)/1,000 population (2008 est.)

Sex ratio:

at birth: 1.05 male(s)/female

under 15 years: 1.04 male(s)/female

15–64 years: 0.98 male(s)/female

65 years and over: 0.7 male(s)/female

total population: 0.98 male(s)/female (2008 est.)

Infant mortality rate:

total: 67.78 deaths/1,000 live births

male: 72.85 deaths/1,000 live births

female: 62.46 deaths/1,000 live births (2008 est.)

Life expectancy at birth:

total population: 65.38 years

male: 61.95 years

female: 68.99 years (2008 est.)

Total fertility rate: 2.86 children born/woman (2008 est.)

HIV/AIDS—adult prevalence rate: less than 0.1% (2001 est.)

HIV/AIDS—people living with HIV/AIDS: 11,000 (2003 est.)

HIV/AIDS—deaths: fewer than 500 (2003 est.)

Nationality:

noun: Uzbekistani

adjective: Uzbekistani

Ethnic groups: Uzbek 80%, Russian 5.5%, Tajik 5%, Kazakh 3%, Karakalpak 2.5%, Tatar 1.5%, other 2.5% (1996 est.)

Religions: Muslim 88% (mostly Sunnis), Eastern Orthodox 9%, other 3%

Languages: Uzbek 74.3%, Russian 14.2%, Tajik 4.4%, other 7.1%

Literacy:

definition: age 15 and over can read and write

total population: 99.3%

male: 99.6%

female: 99% (2003 est.)

GOVERNMENT

Country name:

conventional long form: Republic of Uzbekistan

conventional short form: Uzbekistan

local long form: Ozbekiston Respublikasi

local short form: Ozbekiston

former: Uzbek Soviet Socialist Republic

Government type: republic; authoritarian presidential rule, with little power outside the executive branch

Capital: *name:* Tashkent (Toshkent)

geographic coordinates: 41 20 N, 69 18 E

time difference: UTC+5 (10 hours ahead of Washington, DC during Standard Time)

Administrative divisions: 12 provinces (viloyatlar, singular—viloyat), 1 autonomous republic* (respublika), and 1 city** (shahar); Andijon Viloyati, Buxoro Viloyati, Farg'ona Viloyati, Jizzax Viloyati, Namangan Viloyati, Navoiy Viloyati, Qashqadaryo Viloyati (Qarshi), Qoraqalpog'iston Respublikasi* (Nukus), Samarqand Viloyati, Sirdaryo Viloyati (Guliston), Surxondaryo Viloyati (Termiz), Toshkent Shahri**, Toshkent Viloyati, Xorazm Viloyati (Urganch)

note: administrative divisions have the same names as their administrative centers (exceptions have the administrative center name following in parentheses)

Independence: 1 September 1991 (from Soviet Union)

National holiday: Independence Day, 1 September (1991)

Constitution: adopted 8 December 1992

Legal system: based on civil law system; has not accepted compulsory ICJ jurisdiction

Suffrage: 18 years of age; universal

Executive branch:

chief of state: President Islom KARIMOV (since 24 March 1990, when he was elected president by the then Supreme Soviet)

head of government: Prime Minister Shavkat MIRZIYOYEV (since 11 December 2003); First Deputy Prime Minister Rustam AZIMOV (since 2 January 2008)

cabinet: Cabinet of Ministers appointed by the president with approval of the Supreme Assembly

elections: president elected by popular vote for a seven-year term (eligible for a second term; previously was a five-year term, extended by constitutional amendment in 2002); election last held 23 December 2007 (next to be held in 2014); prime minister, ministers, and deputy ministers appointed by the president

election results: Islom KARIMOV reelected president; percent of vote— Islom KARIMOV 88.1%, Aslidden RUSTAMOV 3.2%, Dilorom TASH-MUKHAMEDOVA 2.9%, Akmal SAIDOV 2.6%

Legislative branch: bicameral Supreme Assembly or Oliy Majlis consists of an upper house or Senate (100 seats; 84 members are elected by regional governing councils and 16 appointed by the president; to serve five-year terms) and a lower house or Legislative Chamber (120 seats; members elected by popular vote to serve five-year terms)

elections: last held 26 December 2004 and 9 January 2005 (next to be held December 2009)

election results: Senate—percent of vote by party—NA; seats by party—NA; Legislative Chamber—percent of vote by party—NA; seats by party—LDPU 41, NDP 32, Fidokorlar 17, MTP 11, Adolat 9, unaffiliated 10

note: all parties in the Supreme Assembly support President KARIMOV

Judicial branch: Supreme Court (judges are nominated by the president and confirmed by the Supreme Assembly)

Political parties and leaders: Adolat (Justice) Social Democratic Party [Dilorom TASHMUHAMMEDOVA]; Democratic National Rebirth Party (Milly Tiklanish) or MTP [Hurshid DOSMUHAMMEDOV]; Fidokorlar National Democratic Party (Self-Sacrificers) [Ahtam TURSUNOV]; Liberal Democratic Party of Uzbekistan or LDPU [Adham SHADMANOV; People's Democratic Party or NDP (formerly Communist Party) [Asliddin RUSTAMOV]

Political pressure groups and leaders: Agrarian and Entrepreneurs' Party [Marat ZAHIDOV]; Birlik (Unity) Movement [Abdurakhim POLAT, chairman]; Committee for the Protection of Human Rights [Marat ZAHIDOV]; Erk (Freedom) Democratic Party [Muhammad SOLIH, chairman] was banned 9 December 1992; Ezgulik Human Rights Society [Vasila INOYA-TOVA]; Free Farmers' Party or Ozod Dehqonlar [Nigora KHIDOYATOVA]; Human Rights Society of Uzbekistan [Talib YAKUBOV, chairman]; Independent Human Rights Organization of Uzbekistan [Mikhail ARDZINOV, chairman]; Mazlum; Sunshine Coalition [Sanjar UMAROV,

THE CIA WORLD FACTBOOK

chairman]

International organization participation: ADB, CIS, CSTO, EAEC, EAPC, EBRD, ECO, FAO, GCTU, IAEA, IBRD, ICAO, ICCt (signatory), ICRM, IDA, IDB, IFC, IFRCS, ILO, IMF, Interpol, IOC, ISO, ITSO, ITU, MIGA, NAM, OIC, OPCW, OSCE, PFP, SCO, UN, UNCTAD, UNESCO, UNIDO, UNWTO, UPU, WCO, WFTU, WHO, WIPO, WMO, WTO (observer)

Diplomatic representation in the US:
chief of mission: Ambassador Abdulaziz KAMILOV
chancery: 1746 Massachusetts Avenue NW, Washington, DC 20036
telephone: [1] (202) 887-5300
FAX: [1] (202) 293-6804
consulate(s) general: New York

Diplomatic representation from the US:
chief of mission: Ambassador Richard B. NORLAND
embassy: 3 Moyqo'rq'on, 5th Block, Yunusobod District, Tashkent 100093
mailing address: use embassy street address
telephone: [998] (71) 120-5450
FAX: [998] (71) 120-6335

Flag description: three equal horizontal bands of blue (top), white, and green separated by red fimbriations with a white crescent moon and 12 white stars in the upper hoist-side quadrant

ECONOMY

Economy—overview: Uzbekistan is a dry, landlocked country of which 11% consists of intensely cultivated, irrigated river valleys. More than 60% of its population lives in densely populated rural communities. Uzbekistan is now the world's second-largest cotton exporter and fifth largest producer; it relies heavily on cotton production as the major source of export earnings. Other major export earners include gold, natural gas, and oil. Following independence in September 1991, the government sought to prop up its Soviet-style command economy with subsidies and tight controls on production and prices. While aware of the need to improve the investment climate, the government still sponsors measures that often increase, not decrease, its control over business decisions. A sharp increase in the inequality of income distribution has hurt the lower ranks of society since independence. In 2003, the government accepted Article VIII obligations under the IMF, providing for full currency convertibility. However, strict currency controls and tightening of borders have lessened the effects of convertibility and have also led to some shortages that have further stifled economic activity. The Central Bank often delays or restricts convertibility, especially for consumer goods. Potential investment by Russia

and China in Uzbekistan's gas and oil industry may boost growth prospects. In November 2005, Russian President Vladimir PUTIN and Uzbekistan President KARIMOV signed an "alliance," which included provisions for economic and business cooperation. Russian businesses have shown increased interest in Uzbekistan, especially in mining, telecom, and oil and gas. In 2006, Uzbekistan took steps to rejoin the Collective Security Treaty Organization (CSTO) and the Eurasian Economic Community (EurASEC), both organizations dominated by Russia. Uzbek authorities have accused US and other foreign companies operating in Uzbekistan of violating Uzbek tax laws and have frozen their assets.

GDP (purchasing power parity): $64.15 billion (2007 est.)

GDP (official exchange rate): $22.31 billion (2007 est.)

GDP—real growth rate: 9.5% (2007 est.)

GDP—per capita (PPP): $2,300 (2007 est.)

GDP—composition by sector:
agriculture: 29.4%
industry: 33.1%
services: 37.4% (2007 est.)

Labor force: 14.6 million (2007 est.)

Labor force—by occupation: *agriculture:* 44%
industry: 20%
services: 36% (1995)

Unemployment rate: 0.8% officially by the Ministry of Labor, plus another 20% underemployed (2007 est.)

Population below poverty line: 33% (2004 est.)

Household income or consumption by percentage share:
lowest 10%: 2.8%
highest 10%: 29.6% (2003)

Distribution of family income—Gini index: 36.8 (2003)

Inflation rate (consumer prices): 12.3% officially, but 38% based on analysis of consumer prices (2007 est.)

Budget: *revenues:* $6.478 billion
expenditures: $6.501 billion (2007 est.)

Public debt: 18.8% of GDP (2007 est.)

Agriculture—products: cotton, vegetables, fruits, grain; livestock

Industries: textiles, food processing, machine building, metallurgy, gold, petroleum, natural gas, chemicals

Industrial production growth rate: 12.1% (2007 est.)

Electricity—production: 49 billion kWh (2006 est.)

Electricity—production by source:
fossil fuel: 88.2%
hydro: 11.8%
nuclear: 0%
other: 0% (2001)

Electricity—consumption: 47 billion kWh (2006 est.)

Electricity—exports: 6.8 billion kWh (2006)

Electricity—imports: 10.5 billion kWh (2006 est.)

Oil—production: 124,900 bbl/day (2005)

Oil—consumption: 155,000 bbl/day (2005)

Oil—exports: 6,941 bbl/day (2004)

Oil—imports: 11,230 bbl/day (2004)

Oil—proved reserves: 594 million bbl (1 January 2006 est.)

Natural gas—production: 62.5 billion cu m (2006 est.)

Natural gas—consumption: 48.4 billion cu m (2006 est.)

Natural gas—exports: 12.5 billion cu m (2006 est.)

Natural gas—imports: 0 cu m (2005)

Natural gas—proved reserves: 1.798 trillion cu m (1 January 2006 est.)

Current account balance: $5.298 billion (2007 est.)

Exports: $8.05 billion f.o.b. (2007 est.)

Exports—commodities: cotton, gold, energy products, mineral fertilizers, ferrous and non-ferrous metals, textiles, food products, machinery, automobiles

Exports—partners: Russia 23.8%, Poland 11.7%, China 10.4%, Turkey 7.7%, Kazakhstan 5.9%, Ukraine 4.7%, Bangladesh 4.3% (2006)

Imports: $4.48 billion f.o.b. (2007 est.)

Imports—commodities: machinery and equipment, foodstuffs, chemicals, ferrous and non-ferrous metals

Imports—partners: Russia 27.8%, South Korea 15.2%, China 10.4%, Kazakhstan 7.3%, Germany 7.1%, Ukraine 4.8%, Turkey 4.5% (2006)

Economic aid—recipient: $172.3 million from the US (2005)

Reserves of foreign exchange and gold: $6.75 billion (31 December 2007 est.)

Debt—external: $3.927 billion (31 December 2007 est.)

Stock of direct foreign investment—at home: $NA

Stock of direct foreign investment—abroad: $NA

Market value of publicly traded shares: $36.89 million (2005)

Currency (code): soum (UZS)

Currency code: UZS

Exchange rates: Uzbekistani soum per US dollar—1,263.8 (2007), 1,219.8 (2006), 1,020 (2005), 971.265 (2004), 771.029 (2003)

Fiscal year: calendar year

COMMUNICATIONS

Telephones—main lines in use: 1.793 million (2005)

Telephones—mobile cellular: 5.8 million (2007)

Telephone system:
general assessment: antiquated and inade-

quate; in serious need of modernization *domestic:* the main line telecommunications system is dilapidated and telephone density is low; the state-owned telecommunications company, Uzbektelecom, is working on improving main line services; mobile services are growing swiftly, with the subscriber base more than doubling in 2007 to 5.8 million

international: country code—998; linked by landline or microwave radio relay with CIS member states and to other countries by leased connection via the Moscow international gateway switch; after the completion of the Uzbek link to the Trans-Asia-Europe (TAE) fiber-optic cable, Uzbekistan will be independent of Russian facilities for international communications (2006)

Radio broadcast stations: AM 4, FM 6, shortwave 3 (2006)

Radios: 10.8 million (1997)

Television broadcast stations: 28 (includes 1 cable rebroadcaster in Tashkent and approximately 20 stations in regional capitals) (2006)

Televisions: 6.4 million (1997)

Internet country code: .uz

Internet hosts: 11,832 (2007)

Internet Service Providers (ISPs): 42 (2000)

Internet users: 1.7 million (2006)

TRANSPORTATION

Airports: 54 (2007)

Airports—with paved runways:
total: 33
over 3,047 m: 6
2,438 to 3,047 m: 13
1,524 to 2,437 m: 5
914 to 1,523 m: 5
under 914 m: 4 (2007)

Airports—with unpaved runways:
total: 21

2,438 to 3,047 m: 2
under 914 m: 19 (2007)

Pipelines: gas 9,725 km; oil 868 km (2007)

Railways:
total: 3,950 km
broad gauge: 3,950 km 1.520-m gauge (620 km electrified) (2006)

Roadways: *total:* 81,600 km
paved: 71,237 km
unpaved: 10,363 km (1999)

Waterways: 1,100 km (2006)

Ports and terminals: Termiz (Amu Darya)

MILITARY

Military branches: Army, Air and Air Defense Forces, National Guard

Military service age and obligation: 18 years of age for compulsory military service; conscript service obligation—12 months (2004)

Manpower available for military service:
males age 16–49: 7,480,484
females age 16–49: 7,542,017 (2008 est.)

Manpower fit for military service:
males age 16–49: 5,684,540
females age 16–49: 6,432,976 (2008 est.)

Manpower reaching militarily significant age annually:
males age 16–49: 311,668
females age 16–49: 301,994 (2008 est.)

Military expenditures—percent of GDP: 2% (2005 est.)

TRANSNATIONAL ISSUES

Disputes—international: prolonged drought and cotton monoculture in Uzbekistan and Turkmenistan creates water-sharing difficulties for Amu Darya river states; field demarcation of the boundaries with Kazakhstan commenced in 2004; border delimitation of 130 km

of border with Kyrgyzstan is hampered by serious disputes around enclaves and other areas

Refugees and internally displaced persons: *refugees (country of origin):* 39,202 (Tajikistan); 1,060 (Afghanistan)
IDPs: 3,400 (forced population transfers by government from villages near Tajikistan border) (2007)

Trafficking in persons: *current situation:* Uzbekistan is a source and, to a lesser extent, a transit country for women trafficked to Asia and the Middle East for the purpose of sexual exploitation; women from other Central Asian countries and China are trafficked through Uzbekistan; men are trafficked for purposes of forced labor in the construction and agricultural industries to Ukraine, Russia, Kazakhstan, and Kyrgyzstan; men and women are also trafficked within the country

tier rating: Tier 3—Uzbekistan is placed on Tier 3 because it failed to fulfill commitments by the country to take additional steps during 2005, including the adoption of comprehensive anti-trafficking legislation, criminal code amendments to raise trafficking penalties, support to the country's first trafficking shelter, and approval of a national action plan

Illicit drugs: transit country for Afghan narcotics bound for Russian and, to a lesser extent, Western European markets; limited illicit cultivation of cannabis and small amounts of opium poppy for domestic consumption; poppy cultivation almost wiped out by government crop eradication program; transit point for heroin precursor chemicals bound for Afghanistan

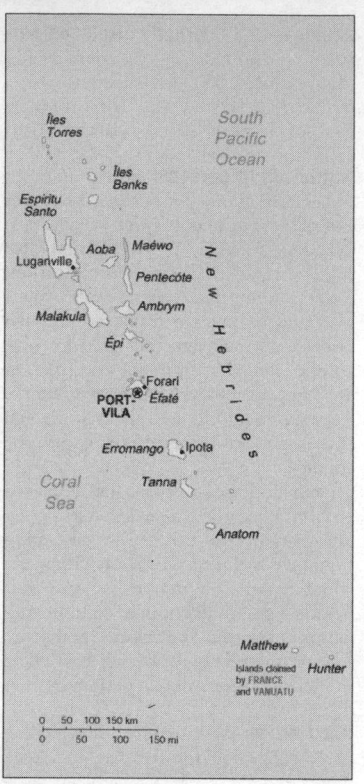

INTRODUCTION

Background: Multiple waves of colonizers, each speaking a distinct language, migrated to the New Hebrides in the millennia preceding European exploration in the 18th century. This settlement pattern accounts for the complex linguistic diversity found on the archipelago to this day. The British and French, who settled the New Hebrides in the 19th century, agreed in 1906 to an Anglo-French Condominium, which administered the islands until independence in 1980, when the new name of Vanuatu was adopted.

GEOGRAPHY

Location: Oceania, group of islands in the South Pacific Ocean, about three-quarters of the way from Hawaii to Australia

Geographic coordinates: 16 00 S, 167 00 E

Map references: Oceania

Area:
total: 12,200 sq km
land: 12,200 sq km
water: 0 sq km
note: includes more than 80 islands, about 65 of which are inhabited

Area—comparative: slightly larger than Connecticut

Land boundaries: 0 km

Coastline: 2,528 km

Maritime claims: measured from claimed archipelagic baselines
territorial sea: 12 nm
contiguous zone: 24 nm
exclusive economic zone: 200 nm
continental shelf: 200 nm or to the edge of the continental margin

Climate: tropical; moderated by southeast trade winds from May to October; moderate rainfall from November to April; may be affected by cyclones from December to April

Terrain: mostly mountainous islands of volcanic origin; narrow coastal plains

Elevation extremes:
lowest point: Pacific Ocean 0 m
highest point: Tabwemasana 1,877 m

Natural resources: manganese, hardwood forests, fish

Land use:
arable land: 1.64%
permanent crops: 6.97%
other: 91.39% (2005)

Irrigated land: NA

Natural hazards: tropical cyclones or typhoons (January to April); volcanic eruption on Aoba (Ambae) island began 27 November 2005, volcanism also causes minor earthquakes; tsunamis

Environment—current issues: most of the population does not have access to a reliable supply of potable water; deforestation

Environment—international agreements: *party to:* Antarctic-Marine Living Resources, Biodiversity, Climate Change, Climate Change-Kyoto Protocol, Desertification, Endangered Species, Law of the Sea, Marine Dumping, Ozone Layer Protection, Ship Pollution, Tropical Timber 94
signed, but not ratified: none of the selected agreements

Geography—note: a Y-shaped chain of four main islands and 80 smaller islands; several of the islands have active volcanoes

PEOPLE

Population: 215,446 (July 2008 est.)

Age structure:
0–14 years: 31.3% (male 34,441/female 33,000)
15–64 years: 64.8% (male 71,159/female 68,435)
65 years and over: 3.9% (male 4,352/female 4,059) (2008 est.)

Median age:
total: 23.8 years

male: 23.8 years
female: 23.8 years (2008 est.)

Population growth rate: 1.434% (2008 est.)

Birth rate: 21.95 births/1,000 population (2008 est.)

Death rate: 7.61 deaths/1,000 population (2008 est.)

Net migration rate: NA

Sex ratio:
at birth: 1.05 male(s)/female
under 15 years: 1.04 male(s)/female
15–64 years: 1.04 male(s)/female
65 years and over: 1.07 male(s)/female
total population: 1.04 male(s)/female (2008 est.)

Infant mortality rate:
total: 50.77 deaths/1,000 live births
male: 53.32 deaths/1,000 live births
female: 48.09 deaths/1,000 live births (2008 est.)

Life expectancy at birth:
total population: 63.61 years
male: 62.04 years
female: 65.27 years (2008 est.)

Total fertility rate: 2.57 children born/woman (2008 est.)

HIV/AIDS—adult prevalence rate: NA

HIV/AIDS—people living with HIV/AIDS: NA

HIV/AIDS—deaths: NA

Nationality:
noun: Ni-Vanuatu (singular and plural)
adjective: Ni-Vanuatu

Ethnic groups: Ni-Vanuatu 98.5%, other 1.5% (1999 Census)

Religions: Presbyterian 31.4%, Anglican 13.4%, Roman Catholic 13.1%, Seventh-Day Adventist 10.8%, other Christian 13.8%, indigenous beliefs 5.6% (including Jon Frum cargo cult), other 9.6%, none 1%, unspecified 1.3% (1999 Census)

Languages: local languages (more than 100) 72.6%, pidgin (known as Bislama or Bichelama) 23.1%, English 1.9%, French 1.4%, other 0.3%, unspecified 0.7% (1999 Census)

Literacy:
definition: age 15 and over can read and write
total population: 74%
male: NA
female: NA (1999 census)

GOVERNMENT

Country name:
conventional long form: Republic of Vanuatu
conventional short form: Vanuatu
local long form: Ripablik blong Vanuatu
local short form: Vanuatu
former: New Hebrides

Government type: parliamentary republic

Capital: *name:* Port-Vila (on Efate)
geographic coordinates: 17 44 S, 168 19 E
time difference: UTC+11 (16 hours ahead of Washington, DC during Standard Time)

Administrative divisions: 6 provinces; Malampa, Penama, Sanma, Shefa, Tafea, Torba

Independence: 30 July 1980 (from France and UK)

National holiday: Independence Day, 30 July (1980)

Constitution: 30 July 1980

Legal system: unified system being created from former dual French and British systems; has not accepted compulsory ICJ jurisdiction

Suffrage: 18 years of age; universal

Executive branch:
chief of state: President Kalkot Matas KELEKELE (since 16 August 2004)
head of government: Prime Minister Ham LINI (since 11 December 2004); Deputy Prime Minister Sato KILMAN (since 11 December 2004)
cabinet: Council of Ministers appointed by the prime minister, responsible to Parliament
elections: president elected for a five-year term by an electoral college consisting of Parliament and the presidents of the regional councils; election for president last held 16 August 2004 (next to be held in 2009); following legislative elections, the leader of the majority party or majority coalition is usually elected prime minister by Parliament from among its members; election for prime minister last held 29 July 2004 (next to be held following general elections in 2008)
election results: Kalkot Matas KELEKELE elected president, with 49 votes out of 56, after several ballots on 16 August 2004

Legislative branch: unicameral Parliament (52 seats; members elected by popular vote to serve four-year terms)
elections: last held 6 July 2004 (next to be held 2008)
election results: percent of vote by party—NA; seats by party—NUP 10, UMP 8, VP 8, VRP 4, MPP 3, VGP 3, other and independent 16; note—political party associations are fluid
note: the National Council of Chiefs advises on matters of culture and language

Judicial branch: Supreme Court (chief justice is appointed by the president after consultation with the prime minister and the leader of the opposition, three other justices are appointed by the president on the advice of the Judicial Service Commission)

Political parties and leaders: Jon Frum Movement [Song KEASPAI]; Melanesian Progressive Party or MPP [Barak SOPE]; National United Party or NUP [Hem LINI]; Union of Moderate Parties or UMP [Serge VOHOR]; Vanua'aku Pati (Our Land Party) or VP [Edward NATAPEI]; Vanuatu Greens Party or VGP [Moana CARCASSES]; Vanuatu Republican Party or VRP [Maxime Carlot KORMAN]

Political pressure groups and leaders: NA

International organization participation: ACCT, ACP, ADB, C, FAO, G-77, IBRD, ICAO, ICRM, IDA, IFC, IFRCS, ILO, IMF, IMO, IOC, ITU, ITUC, MIGA, NAM, OIF, OPCW, PIF, Sparteca, SPC, UN, UNCTAD, UNESCO, UNIDO, UPU, WFTU, WHO, WMO, WTO (observer)

Diplomatic representation in the US: Vanuatu does not have an embassy in the US; it does, however, have a Permanent Mission to the UN

Diplomatic representation from the US: the US does not have an embassy in Vanuatu; the ambassador to Papua New Guinea is accredited to Vanuatu

Flag description: two equal horizontal bands of red (top) and green with a black isosceles triangle (based on the hoist side) all separated by a black-edged yellow stripe in the shape of a horizontal Y (the two points of the Y face the hoist side and enclose the triangle); centered in the triangle is a boar's tusk encircling two crossed namele leaves, all in yellow

ECONOMY

Economy—overview: This South Pacific island economy is based primarily on small-scale agriculture, which provides a living for 65% of the population. Fishing, offshore financial services, and tourism, with more than 60,000 visitors in 2005, are other mainstays of the economy. Mineral deposits are negligible; the country has no known petroleum deposits. A small light industry sector caters to the local market. Tax revenues come mainly from import duties. Economic development is hindered by dependence on relatively few commodity exports, vulnerability to natural disasters, and long distances from main markets and between constituent islands. In response to foreign concerns, the government has promised to tighten regulation of its offshore financial center. In mid-2002 the government stepped up efforts to boost tourism through improved air connections, resort development, and cruise ship facilities.

Agriculture, especially livestock farming, is a second target for growth. Australia and New Zealand are the main suppliers of tourists and foreign aid.

GDP (purchasing power parity): $897 million (2007 est.)

GDP (official exchange rate): $455 million (2007 est.)

GDP—real growth rate: 5% (2007 est.)

GDP—per capita (PPP): $3,900 (2007 est.)

GDP—composition by sector:
agriculture: 26%
industry: 12%
services: 62% (2000 est.)

Labor force: 76,410 (1999)

Labor force—by occupation: *agriculture:* 65%
industry: 5%
services: 30% (2000 est.)

Unemployment rate: 1.7% (1999)

Population below poverty line: NA%

Household income or consumption by percentage share:
lowest 10%: NA%
highest 10%: NA%

Inflation rate (consumer prices): 3.9% (2007 est.)

Budget:
revenues: $78.7 million
expenditures: $72.23 million (2005)

Agriculture—products: copra, coconuts, cocoa, coffee, taro, yams, fruits, vegetables; beef; fish

Industries: food and fish freezing, wood processing, meat canning

Industrial production growth rate: 1% (1997 est.)

Electricity—production: 41 million kWh (2005)

Electricity—production by source:
fossil fuel: 100%
hydro: 0%
nuclear: 0%
other: 0% (2001)

Electricity—consumption: 38.13 million kWh (2005)

Electricity—exports: 0 kWh (2005 est.)

Electricity—imports: 0 kWh (2005)

Oil—production: 0 bbl/day (2005 est.)

Oil—consumption: 640 bbl/day (2005 est.)

Oil—exports: 0 bbl/day (2004)

Oil—imports: 628.5 bbl/day (2004)

Oil—proved reserves: 0 bbl (1 January 2006 est.)

Natural gas—production: 0 cu m (2005 est.)

Natural gas—consumption: 0 cu m (2005 est.)

Natural gas—exports: 0 cu m (2005 est.)

Natural gas—imports: 0 cu m (2005)

Natural gas—proved reserves: 0 cu m (1 January 2006 est.)

Current account balance: -$60 million (2003)

Exports: $40 million f.o.b. (2006)

Exports—commodities: copra, beef, cocoa, timber, kava, coffee

Exports—partners: Thailand 59.6%, India 16.8%, Japan 11.5% (2006)

Imports: $156 million c.i.f. (2006)

Imports—commodities: machinery and equipment, foodstuffs, fuels

Imports—partners: Australia 20.6%, Japan 19.7%, Singapore 12.1%, NZ 8.8%, Fiji 7.7%, China 7.4%, New Caledonia 4.3% (2006)

Economic aid—recipient: $39.48 million (2005)

Reserves of foreign exchange and gold: $40.54 million (2003)

Debt—external: $81.2 million (2004)

Market value of publicly traded shares: $NA

Currency (code): vatu (VUV)

Currency code: VUV

Exchange rates: vatu per US dollar—NA (2007), 111.93 (2006), NA (2005), 111.79 (2004), 122.19 (2003)

Fiscal year: calendar year

COMMUNICATIONS

Telephones—main lines in use: 7,000 (2005)

Telephones—mobile cellular: 12,700 (2005)

Telephone system:
general assessment: NA
domestic: NA
international: country code—678; satellite earth station—1 Intelsat (Pacific Ocean)

Radio broadcast stations: AM 2, FM 4, shortwave 1 (2001)

Radios: 67,000 (1997)

Television broadcast stations: 1 (2004)

Televisions: 2,300 (1999)

Internet country code: .vu

Internet hosts: 1,010 (2007)

Internet Service Providers (ISPs): 1 (2000)

Internet users: 7,500 (2004)

TRANSPORTATION

Airports: 31 (2007)

Airports—with paved runways:
total: 3
2,438 to 3,047 m: 1
1,524 to 2,437 m: 1
914 to 1,523 m: 1 (2007)

Airports—with unpaved runways:
total: 28
914 to 1,523 m: 6
under 914 m: 22 (2007)

Roadways:
total: 1,070 km
paved: 256 km

unpaved: 814 km (1999)

Merchant marine:
total: 51 ships (1000 GRT or over) 1,346,001 GRT/1,901,055 DWT
by type: bulk carrier 30, cargo 8, container 1, liquefied gas 2, petroleum tanker 1, refrigerated cargo 3, roll on/roll off 1, vehicle carrier 5
foreign-owned: 51 (Australia 2, Belgium 4, Canada 5, Estonia 1, Japan 28, Poland 7, Russia 1, Switzerland 2, US 1) (2007)

Ports and terminals: Forari, Port-Vila, Santo (Espiritu Santo)

MILITARY

Military branches: no regular military forces; Vanuatu Police Force (VPF), Vanuatu Mobile Force (VMF; includes Police Maritime Wing (PMW)) (2008)

Manpower available for military service:
males age 16–49: 58,900 (2008 est.)

Manpower fit for military service:
males age 16–49: 40,577 (2008 est.)

Military expenditures—percent of GDP: NA

TRANSNATIONAL ISSUES

Disputes—international: Matthew and Hunter Islands east of New Caledonia claimed by Vanuatu and France

VENEZUELA

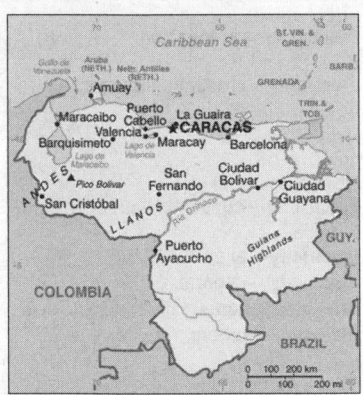

INTRODUCTION

Background: Venezuela was one of three countries that emerged from the collapse of Gran Colombia in 1830 (the others being Ecuador and New Granada, which became Colombia). For most of the first half of the 20th century, Venezuela was ruled by generally benevolent military strongmen, who promoted the oil industry and allowed for some social reforms. Democratically elected governments have held sway since 1959. Hugo

CHAVEZ, president since 1999, seeks to implement his "21st Century Socialism," which purports to alleviate social ills while at the same time attacking globalization and undermining regional stability. Current concerns include: a weakening of democratic institutions, political polarization, a politicized military, drug-related violence along the Colombian border, increasing internal drug consumption, overdependence on the petroleum industry with its price fluctuations, and irresponsible mining operations that are endangering the rain forest and indigenous peoples.

GEOGRAPHY

Location: Northern South America, bordering the Caribbean Sea and the North Atlantic Ocean, between Colombia and Guyana

Geographic coordinates: 8 00 N, 66 00 W

Map references: South America

Area: *total:* 912,050 sq km
land: 882,050 sq km
water: 30,000 sq km

Area—comparative: slightly more than twice the size of California

Land boundaries:
total: 4,993 km
border countries: Brazil 2,200 km, Colombia 2,050 km, Guyana 743 km

Coastline: 2,800 km

Maritime claims:
territorial sea: 12 nm
contiguous zone: 15 nm
exclusive economic zone: 200 nm
continental shelf: 200-m depth or to the depth of exploitation

Climate: tropical; hot, humid; more moderate in highlands

Terrain: Andes Mountains and Maracaibo Lowlands in northwest; central plains (llanos); Guiana Highlands in southeast

Elevation extremes:
lowest point: Caribbean Sea 0 m
highest point: Pico Bolivar (La Columna) 5,007 m

Natural resources: petroleum, natural gas, iron ore, gold, bauxite, other minerals, hydropower, diamonds

Land use:
arable land: 2.85%
permanent crops: 0.88%
other: 96.27% (2005)

Irrigated land: 5,750 sq km (2003)

Total renewable water resources: 1,233.2 cu km (2000)

Freshwater withdrawal (domestic/industrial/agricultural): *total:* 8.37 cu km/yr (6%/7%/47%)

per capita: 313 cu m/yr (2000)

Natural hazards: subject to floods, rockslides, mudslides; periodic droughts

Environment—current issues: sewage pollution of Lago de Valencia; oil and urban pollution of Lago de Maracaibo; deforestation; soil degradation; urban and industrial pollution, especially along the Caribbean coast; threat to the rainforest ecosystem from irresponsible mining operations

Environment—international agreements: *party to:* Antarctic Treaty, Biodiversity, Climate Change, Climate Change-Kyoto Protocol, Desertification, Endangered Species, Hazardous Wastes, Marine Life Conservation, Ozone Layer Protection, Ship Pollution, Tropical Timber 83, Tropical Timber 94, Wetlands

signed but not ratified:: none of the selected agreements

Geography—note: on major sea and air routes linking North and South America; Angel Falls in the Guiana Highlands is the world's highest waterfall

PEOPLE

Population: 26,414,815 (July 2008 est.)

Age structure:

0–14 years: 31% (male 4,162,862/female 4,034,044)

15–64 years: 63.8% (male 8,299,266/female 8,562,290)

65 years and over: 5.1% (male 602,725/female 753,628) (2008 est.)

Median age:

total: 25.2 years

male: 24.6 years

female: 25.8 years (2008 est.)

Population growth rate: 1.498% (2008 est.)

Birth rate: 20.92 births/1,000 population (2008 est.)

Death rate: 5.1 deaths/1,000 population (2008 est.)

Net migration rate: -0.84 migrant(s)/1,000 population (2008 est.)

Sex ratio:

at birth: 1.05 male(s)/female

under 15 years: 1.03 male(s)/female

15–64 years: 0.97 male(s)/female

65 years and over: 0.8 male(s)/female

total population: 0.98 male(s)/female (2008 est.)

Infant mortality rate:

total: 22.02 deaths/1,000 live births

male: 25.61 deaths/1,000 live births

female: 18.26 deaths/1,000 live births (2008 est.)

Life expectancy at birth:

total population: 73.45 years

male: 70.4 years

female: 76.65 years (2008 est.)

Total fertility rate: 2.52 children born/woman (2008 est.)

HIV/AIDS—adult prevalence rate: 0.7%; note—no country specific models provided (2001 est.)

HIV/AIDS—people living with HIV/AIDS: 110,000 (1999 est.)

HIV/AIDS—deaths: 4,100 (2003 est.)

Major infectious diseases:

degree of risk: high

food or waterborne diseases: bacterial diarrhea and hepatitis A

vectorborne disease: dengue fever, malaria, and Venezuelan equine encephalitis (2008)

Nationality:

noun: Venezuelan(s)

adjective: Venezuelan

Ethnic groups: Spanish, Italian, Portuguese, Arab, German, African, indigenous people

Religions: nominally Roman Catholic 96%, Protestant 2%, other 2%

Languages: Spanish (official), numerous indigenous dialects

Literacy:

definition: age 15 and over can read and write

total population: 93%

male: 93.3%

female: 92.7% (2001 census)

GOVERNMENT

Country name:

conventional long form: Bolivarian Republic of Venezuela

conventional short form: Venezuela

local long form: Republica Bolivariana de Venezuela

local short form: Venezuela

Government type: federal republic

Capital: *name:* Caracas

geographic coordinates: 10 30 N, 66 56 W

time difference: UTC-4.5 (half an hour ahead of Washington, DC during Standard Time)

Administrative divisions: 23 states (estados, singular—estado), 1 capital district* (distrito capital), and 1 federal dependency** (dependencia federal); Amazonas, Anzoategui, Apure, Aragua, Barinas, Bolivar, Carabobo, Cojedes, Delta Amacuro, Dependencias Federales**, Distrito Federal*, Falcon, Guarico, Lara, Merida, Miranda, Monagas, Nueva Esparta, Portuguesa, Sucre, Tachira, Trujillo, Vargas, Yaracuy, Zulia

note: the federal dependency consists of 11 federally controlled island groups with a total of 72 individual islands

Independence: 5 July 1811 (from Spain)

National holiday: Independence Day, 5 July (1811)

Constitution: 30 December 1999

Legal system: open, adversarial court system; has not accepted compulsory ICJ jurisdiction

Suffrage: 18 years of age; universal

Executive branch:

chief of state: President Hugo CHAVEZ Frias (since 3 February 1999); Executive Vice President Ramon Alonzo CARRIZALEZ Rengifo (since 4 January 2008); note—the president is both the chief of state and head of government

head of government: President Hugo CHAVEZ Frias (since 3 February 1999); Executive Vice President Ramon Alonzo CARRIZALEZ Rengifo (since 4 January 2008)

cabinet: Council of Ministers appointed by the president

elections: president elected by popular vote for a six-year term (eligible for a second term); election last held 3 December 2006 (next to be held in December 2012)

note: in 1999, a National Constituent Assembly drafted a new constitution that increased the presidential term to six years; an election was subsequently held on 30 July 2000 under the terms of this constitution

election results: Hugo CHAVEZ Frias reelected president; percent of vote—Hugo CHAVEZ Frias 62.9%, Manuel ROSALES 36.9%

Legislative branch: unicameral National Assembly or Asamblea Nacional (167 seats; members elected by popular vote to serve five-year terms; three seats reserved for the indigenous peoples of Venezuela)

elections: last held 4 December 2005 (next to be held in 2010)

election results: percent of vote by party—NA; seats by party—pro-government 167 (MVR 114, PODEMOS 15, PPT 11, indigenous 2, other 25), opposition 0; total seats by party as of 1 January 2008—pro-government 152 (PSUV 114, PPT 11, indigenous 2, other 25), PODEMOS 15

Judicial branch: Supreme Tribunal of Justice or Tribuna Suprema de Justicia (magistrates are elected by the National Assembly for a single 12-year term)

Political parties and leaders: A New Time or UNT [Manuel ROSALES]; Christian Democrats or COPEI [Cesar PEREZ Vivas]; Communist Party of Venezuela or PCV [Jeronimo CARRERA]; Democratic Action or AD [Henry RAMOS Allup]; Fatherland for All or PPT [Jose ALBORNOZ]; Justice

First [Julio BORGES]; Movement Toward Socialism or MAS [Hector MUJICA]; United Socialist Party of Venezuela or PSUV [Hugo CHAVEZ]; Venezuela Project or PV [Henrique SALAS Romer]; We Can or PODEMOS [Ismael GARCIA]

Political pressure groups and leaders: FEDECAMARAS, a conservative business group; VECINOS groups; Venezuelan Confederation of Workers or CTV (labor organization dominated by the Democratic Action)

International organization participation: CAN, Caricom (observer), CDB, CSN, FAO, G-3, G-15, G-24, G-77, IADB, IAEA, IBRD, ICAO, ICC, ICCt, ICRM, IDA, IFAD, IFC, IFRCS, IHO, ILO, IMF, IMO, Interpol, IOC, IOM, IPU, ISO, ITSO, ITU, ITUC, LAES, LAIA, LAS (observer), Mercosur (associate), MIGA, NAM, OAS, OPANAL, OPCW, OPEC, PCA, RG, UN, UNCTAD, UNESCO, UNHCR, UNIDO, Union Latina, UNWTO, UPU, WCL, WCO, WFTU, WHO, WIPO, WMO, WTO

Diplomatic representation in the US:
chief of mission: Ambassador Bernardo ALVAREZ Herrera
chancery: 1099 30th Street NW, Washington, DC 20007
telephone: [1] (202) 342-2214
FAX: [1] (202) 342-6820
consulate(s) general: Boston, Chicago, Houston, Miami, New Orleans, New York, San Francisco, San Juan (Puerto Rico)

Diplomatic representation from the US:
chief of mission: Ambassador Patrick DUDDY
embassy: Calle F con Calle Suapure, Urbanizacion Colinas de Valle Arriba, Caracas 1080
mailing address: P. O. Box 62291, Caracas 1060-A; APO AA 34037
telephone: [58] (212) 975-9234, 975-6411
FAX: [58] (212) 975-8991

Flag description: three equal horizontal bands of yellow (top), blue, and red with the coat of arms on the hoist side of the yellow band and an arc of eight white five-pointed stars centered in the blue band

ECONOMY

Economy—overview: Venezuela remains highly dependent on oil revenues, which account for roughly 90% of export earnings, more than 50% of the federal budget revenues, and around 30% of GDP. A nationwide strike between December 2002 and February 2003 had far-reaching economic consequences—real GDP declined by around 9% in 2002

and 8% in 2003—but economic output since then has recovered strongly. Fueled by high oil prices, record government spending helped to boost GDP in 2006 by about 9% and in 2007 by about 8%. This spending, combined with recent minimum wage hikes and improved access to domestic credit, has created a consumption boom but has come at the cost of higher inflation-roughly 20 percent in 2007. Imports also have jumped significantly. Embolden by his December 2006 reelection, President Hugo CHAVEZ in 2007 nationalized firms in the petroleum, communications, and electricity sectors, which reduced foreign influence in the economy. Although voters in December 2007 rejected CHAVEZ's proposed constitutional changes, CHAVEZ still has significant control of the economy and has indicated he intends to continue to consolidate and centralize authority over the economy by implementing "21st Century Socialism."

GDP (purchasing power parity): $334.6 billion (2007 est.)
GDP (official exchange rate): $236.4 billion (2007 est.)
GDP—real growth rate: 8.4% (2007 est.)
GDP—per capita (PPP): $12,200 (2007 est.)
GDP—composition by sector:
agriculture: 3.8%
industry: 38.4%
services: 57.8% (2007 est.)
Labor force: 12.37 million (2007 est.)
Labor force—by occupation: *agriculture:* 13%
industry: 23%
services: 64% (1997 est.)
Unemployment rate: 8.5% (2007 est.)
Population below poverty line: 37.9% (end 2005 est.)
Household income or consumption by percentage share:
lowest 10%: 0.7%
highest 10%: 35.2% (2003)
Distribution of family income—Gini index: 48.2 (2003)
Inflation rate (consumer prices): 18.7% (2007)
Investment (gross fixed): 23.7% of GDP (2007 est.)
Budget:
revenues: $65.83 billion
expenditures: $58.9 billion (2007 est.)
Public debt: 19.3% of GDP (2007 est.)
Agriculture—products: corn, sorghum, sugarcane, rice, bananas, vegetables, coffee; beef, pork, milk, eggs; fish
Industries: petroleum, construction materials, food processing, textiles; iron ore mining, steel, aluminum; motor

vehicle assembly
Industrial production growth rate: 3.9% (2007 est.)
Electricity—production: 99.2 billion kWh (2005)
Electricity—production by source:
fossil fuel: 31.7%
hydro: 68.3%
nuclear: 0%
other: 0% (2001)
Electricity—consumption: 73.36 billion kWh (2005)
Electricity—exports: 0 kWh (2005)
Electricity—imports: 0 kWh (2005)
Oil—production: 2.802 million bbl/day (2006 est.)
Oil—consumption: 599,000 bbl/day (2006 est.)
Oil—exports: 2.203 million bbl/day (2006 est.)
Oil—imports: 0 bbl/day (2006 est.)
Oil—proved reserves: 79.14 billion bbl (2007 est.)
Natural gas—production: 27.53 billion cu m (2005 est.)
Natural gas—consumption: 27.53 billion cu m (2005 est.)
Natural gas—exports: 0 cu m (2005 est.)
Natural gas—imports: 0 cu m (2005)
Natural gas—proved reserves: 4.112 trillion cu m (1 January 2006 est.)
Current account balance: $23.23 billion (2007 est.)
Exports: $69.17 billion f.o.b. (2007 est.)
Exports—commodities: petroleum, bauxite and aluminum, steel, chemicals, agricultural products, basic manufactures
Exports—partners: US 46.2%, Netherlands Antilles 13.5%, China 3.2% (2006)
Imports: $45.46 billion f.o.b. (2007 est.)
Imports—commodities: raw materials, machinery and equipment, transport equipment, construction materials
Imports—partners: US 30.6%, Colombia 10.2%, Brazil 10.1%, Mexico 5.9%, China 4.9%, Panama 4.8% (2006)
Economic aid—recipient: $48.66 million (2005)
Reserves of foreign exchange and gold: $33.48 billion (31 December 2007 est.)
Debt—external: $41.52 billion (31 December 2007 est.)
Stock of direct foreign investment—at home: $46.04 million (2007 est.)
Stock of direct foreign investment—abroad: $13.8 million (2007 est.)
Market value of publicly traded shares: $8.251 billion (2006)
Currency (code): bolivar (VEB)
Currency code: VEB
Exchange rates: bolivares per US dollar—2,147 (2007), 2,147 (2006),

2,089.8 (2005), 1,891.3 (2004), 1,607 (2003)

Fiscal year: calendar year

Telephones—main lines in use: 4.217 million (2006)

Telephones—mobile cellular: 18.79 million (2006)

Telephone system:

general assessment: modern and expanding

domestic: domestic satellite system with 3 earth stations; recent substantial improvement in telephone service in rural areas; substantial increase in digitalization of exchanges and trunk lines; installation of a national interurban fiber-optic network capable of digital multimedia services; fixed-line teledensity is 16 per 100 persons; mobile-cellular subscribership jumped 50 percent in 2006

international: country code—58; submarine cable systems provide connectivity to the Caribbean, Central and South America, and US; satellite earth stations—1 Intelsat (Atlantic Ocean) and 1 PanAmSat; participating with Colombia, Ecuador, Peru, and Bolivia in the construction of an international fiber-optic network

Radio broadcast stations: AM 201, FM NA (20 in Caracas), shortwave 11 (1998)

Radios: 10.75 million (1997)

Television broadcast stations: 66 (plus 45 repeaters) (1997)

Televisions: 4.1 million (1997)

Internet country code: .ve

Internet hosts: 126,500 (2007)

Internet Service Providers (ISPs): 16 (2000)

Internet users: 4.14 million (2006)

TRANSPORTATION

Airports: 390 (2007)

Airports—with paved runways:

total: 128

over 3,047 m: 5

2,438 to 3,047 m: 10

1,524 to 2,437 m: 34

914 to 1,523 m: 61

under 914 m: 18 (2007)

Airports—with unpaved runways:

total: 262

2,438 to 3,047 m: 1

1,524 to 2,437 m: 15

914 to 1,523 m: 97

under 914 m: 149 (2007)

Heliports: 2 (2007)

Pipelines: extra heavy crude oil 992 km; gas 5,400 km; oil 7,607 km; refined products 1,650 km; unknown (oil/water) 141 km (2007)

Railways:

total: 682 km

standard gauge: 682 km 1.435-m gauge (2006)

Roadways:

total: 96,155 km

paved: 32,308 km

unpaved: 63,847 km (1999)

Waterways: 7,100 km

note: Orinoco River (400 km) and Lake de Maracaibo navigable by oceangoing vessels (2005)

Merchant marine:

total: 59 ships (1000 GRT or over) 808,721 GRT/1,285,783 DWT

by type: bulk carrier 7, cargo 14, chemical tanker 3, container 1, liquefied gas 6, passenger/cargo 10, petroleum tanker 17, refrigerated cargo 1

foreign-owned: 12 (Denmark 3, Greece 3, Mexico 3, Panama 1, Russia 1, Spain 1)

registered in other countries: 11 (Bahamas 1, Panama 10) (2007)

Ports and terminals: La Guaira, Maracaibo, Puerto Cabello, Punta Cardon

MILITARY

Military branches: National Armed Forces (Fuerza Armada Nacionale, FAN): Ground Forces or Army (Fuerzas Terrestres or Ejercito), Naval Forces (Fuerzas Navales or Armada; includes Marines, Coast Guard), Air Force (Fuerzas Aereas or Aviacion), Armed Forces of Cooperation or National Guard (Fuerzas Armadas de Cooperacion or Guardia Nacional)

Military service age and obligation: 18 years of age for compulsory and voluntary military service; conscript service obligation—30 months; all citizens of military service age (between 18 and 50 years old) are obligated to register for military service (2007)

Manpower available for military service:

males age 16–49: 6,647,124

females age 16–49: 6,801,133 (2008 est.)

Manpower fit for military service:

males age 16–49: 5,280,974

females age 16–49: 5,768,814 (2008 est.)

Manpower reaching militarily significant age annually:

males age 16–49: 275,323

females age 16–49: 274,106 (2008 est.)

Military expenditures—percent of GDP: 1.2% (2005 est.)

TRANSNATIONAL ISSUES

Disputes—international: claims all of the area west of the Essequibo River in Guyana, preventing any discussion of a maritime boundary; Guyana has expressed its intention to join Barbados in asserting claims before the United Nations Convention on the Law of the Sea (UNCLOS) that Trinidad and Tobago's maritime boundary with Venezuela extends into their waters; dispute with Colombia over maritime boundary and Venezuelan-administered Los Monjes islands near the Gulf of Venezuela; Colombian-organized illegal narcotics and paramilitary activities penetrate Venezuela's shared border region; in 2006, an estimated 139,000 Colombians sought protection in 150 communities along the border in Venezuela; US, France, and the Netherlands recognize Venezuela's granting full effect to Aves Island, thereby claiming a Venezuelan EEZ/continental shelf extending over a large portion of the eastern Caribbean Sea; Dominica, Saint Kitts and Nevis, Saint Lucia, and Saint Vincent and the Grenadines protest Venezuela's full effect claim

Trafficking in persons: *current situation:* Venezuela is a source, transit, and destination country for women and children trafficked for the purposes of sexual exploitation and forced labor; women and children from Colombia, China, Peru, Ecuador, and the Dominican Republic are trafficked to and through Venezuela and subjected to commercial sexual exploitation or forced labor; Venezuelans are trafficked internally and to Western Europe, particularly Spain and the Netherlands, and to countries in the Caribbean region for commercial sexual exploitation; Venezuela is a transit country for illegal migrants from other countries in the region and for Asian nationals, some of whom are believed to be trafficking victims

tier rating: Tier 3—Venezuela does not fully comply with the minimum standards for the elimination of trafficking and is not making significant efforts to do so

Illicit drugs: small-scale illicit producer of opium and coca for the processing of opiates and coca derivatives; however, large quantities of cocaine, heroin, and marijuana transit the country from Colombia bound for US and Europe; significant narcotics-related money-laundering activity, especially along the border with Colombia and on Margarita Island; active eradication program primarily targeting opium; increasing signs of drug-related activities by Colombian insurgents on border

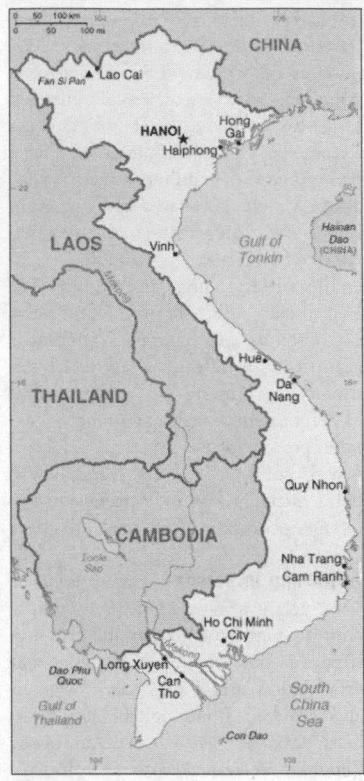

CHINA
Fan Si Pan Lao Cai
Hong Gai
HANOI
Haiphong
LAOS Vinh Gulf of Tonkin Hainan Dao (CHINA)
Hue
Da Nang
THAILAND
Quy Nhon
CAMBODIA
Nha Trang
Cam Ranh
Ho Chi Minh City
Dao Phu Quoc Long Xuyen
Can Tho South China Sea
Gulf of Thailand
Con Dao

INTRODUCTION

Background: The conquest of Vietnam by France began in 1858 and was completed by 1884. It became part of French Indochina in 1887. Vietnam declared independence after World War II, but France continued to rule until its 1954 defeat by Communist forces under Ho Chi MINH. Under the Geneva Accords of 1954, Vietnam was divided into the Communist North and anti-Communist South. US economic and military aid to South Vietnam grew through the 1960s in an attempt to bolster the government, but US armed forces were withdrawn following a cease-fire agreement in 1973. Two years later, North Vietnamese forces overran the South reuniting the country under Communist rule. Despite the return of peace, for over a decade the country experienced little economic growth because of conservative leadership policies. However, since the enactment of Vietnam's "doi moi" (renovation) policy in 1986, Vietnamese authorities have committed to increased economic liberalization and enacted structural reforms needed to modernize the economy and to produce more com-

petitive, export-driven industries. The country continues to experience protests from various groups—such as the Protestant Montagnard ethnic minority population of the Central Highlands and the Hoa Hao Buddhists in southern Vietnam over religious persecution. Montagnard grievances also include the loss of land to Vietnamese settlers.

GEOGRAPHY

Location: Southeastern Asia, bordering the Gulf of Thailand, Gulf of Tonkin, and South China Sea, alongside China, Laos, and Cambodia
Geographic coordinates: 16 00 N, 106 00 E
Map references: Southeast Asia
Area:
total: 329,560 sq km
land: 325,360 sq km
water: 4,200 sq km
Area—comparative: slightly larger than New Mexico
Land boundaries:
total: 4,639 km
border countries: Cambodia 1,228 km, China 1,281 km, Laos 2,130 km
Coastline: 3,444 km (excludes islands)
Maritime claims:
territorial sea: 12 nm
contiguous zone: 24 nm
exclusive economic zone: 200 nm
continental shelf: 200 nm or to the edge of the continental margin
Climate: tropical in south; monsoonal in north with hot, rainy season (May to September) and warm, dry season (October to March)
Terrain: low, flat delta in south and north; central highlands; hilly, mountainous in far north and northwest
Elevation extremes:
lowest point: South China Sea 0 m
highest point: Fan Si Pan 3,144 m
Natural resources: phosphates, coal, manganese, bauxite, chromate, offshore oil and gas deposits, forests, hydropower
Land use:
arable land: 20.14%
permanent crops: 6.93%
other: 72.93% (2005)
Irrigated land: 30,000 sq km (2003)
Total renewable water resources: 891.2 cu km (1999)
Freshwater withdrawal (domestic/industrial/agricultural): *total:* 71.39 cu km/yr (8%/24%/68%)
per capita: 847 cu m/yr (2000)
Natural hazards: occasional typhoons (May to January) with extensive flooding, especially in the Mekong River

delta
Environment—current issues: logging and slash-and-burn agricultural practices contribute to deforestation and soil degradation; water pollution and overfishing threaten marine life populations; groundwater contamination limits potable water supply; growing urban industrialization and population migration are rapidly degrading environment in Hanoi and Ho Chi Minh City
Environment—international agreements: *party to:* Biodiversity, Climate Change, Climate Change-Kyoto Protocol, Desertification, Endangered Species, Environmental Modification, Hazardous Wastes, Law of the Sea, Ozone Layer Protection, Ship Pollution, Wetlands
signed, but not ratified: none of the selected agreements
Geography—note: extending 1,650 km north to south, the country is only 50 km across at its narrowest point

PEOPLE

Population: 86,116,559 (July 2008 est.)
Age structure:
0–14 years: 25.6% (male 11,418,642/female 10,598,184)
15–64 years: 68.6% (male 29,341,216/female 29,777,696)
65 years and over: 5.8% (male 1,925,609/female 3,055,212) (2008 est.)
Median age:
total: 26.9 years
male: 25.8 years
female: 28 years (2008 est.)
Population growth rate: 0.99% (2008 est.)
Birth rate: 16.47 births/1,000 population (2008 est.)
Death rate: 6.18 deaths/1,000 population (2008 est.)
Net migration rate: -0.39 migrant(s)/1,000 population (2008 est.)
Sex ratio:
at birth: 1.07 male(s)/female
under 15 years: 1.08 male(s)/female
15–64 years: 0.99 male(s)/female
65 years and over: 0.63 male(s)/female
total population: 0.98 male(s)/female (2008 est.)
Infant mortality rate:
total: 23.61 deaths/1,000 live births
male: 24.01 deaths/1,000 live births
female: 23.19 deaths/1,000 live births (2008 est.)
Life expectancy at birth:
total population: 71.33 years
male: 68.52 years
female: 74.33 years (2008 est.)

Total fertility rate: 1.86 children born/woman (2008 est.)
HIV/AIDS—adult prevalence rate: 0.4% (2003 est.)
HIV/AIDS—people living with HIV/AIDS: 220,000 (2003 est.)
HIV/AIDS—deaths: 9,000 (2003 est.)
Major infectious diseases:
degree of risk: high
food or waterborne diseases: bacterial diarrhea, hepatitis A, and typhoid fever
vectorborne diseases: dengue fever, malaria, Japanese encephalitis, and plague
water contact disease: leptospirosis
note: highly pathogenic H5N1 avian influenza has been identified in this country; it poses a negligible risk with extremely rare cases possible among US citizens who have close contact with birds (2008)
Nationality:
noun: Vietnamese (singular and plural)
adjective: Vietnamese
Ethnic groups: Kinh (Viet) 86.2%, Tay 1.9%, Thai 1.7%, Muong 1.5%, Khome 1.4%, Hoa 1.1%, Nun 1.1%, Hmong 1%, others 4.1% (1999 census)
Religions: Buddhist 9.3%, Catholic 6.7%, Hoa Hao 1.5%, Cao Dai 1.1%, Protestant 0.5%, Muslim 0.1%, none 80.8% (1999 census)
Languages: Vietnamese (official), English (increasingly favored as a second language), some French, Chinese, and Khmer; mountain area languages (Mon-Khmer and Malayo-Polynesian)
Literacy: *definition:* age 15 and over can read and write
total population: 90.3%
male: 93.9%
female: 86.9% (2002 est.)

GOVERNMENT

Country name:
conventional long form: Socialist Republic of Vietnam
conventional short form: Vietnam
local long form: Cong Hoa Xa Hoi Chu Nghia Viet Nam
local short form: Viet Nam
abbreviation: SRV
Government type: Communist state
Capital: *name:* Hanoi
geographic coordinates: 21 02 N, 105 51 E
time difference: UTC+7 (12 hours ahead of Washington, DC during Standard Time)
Administrative divisions: 59 provinces (tinh, singular and plural) and 5 municipalities (thanh pho, singular and plural)
provinces: An Giang, Bac Giang, Bac Kan, Bac Lieu, Bac Ninh, Ba Ria-Vung Tau, Ben Tre, Binh Dinh, Binh Duong, Binh Phuoc, Binh Thuan, Ca Mau, Cao

Bang, Dac Lak, Dac Nong, Dien Bien, Dong Nai, Dong Thap, Gia Lai, Ha Giang, Ha Nam, Ha Tay, Ha Tinh, Hai Duong, Hau Giang, Hoa Binh, Hung Yen, Khanh Hoa, Kien Giang, Kon Tum, Lai Chau, Lam Dong, Lang Son, Lao Cai, Long An, Nam Dinh, Nghe An, Ninh Binh, Ninh Thuan, Phu Tho, Phu Yen, Quang Binh, Quang Nam, Quang Ngai, Quang Ninh, Quang Tri, Soc Trang, Son La, Tay Ninh, Thai Binh, Thai Nguyen, Thanh Hoa, Thua Thien-Hue, Tien Giang, Tra Vinh, Tuyen Quang, Vinh Long, Vinh Phuc, Yen Bai
municipalities: Can Tho, Da Nang, Hai Phong, Ha Noi, Ho Chi Minh
Independence: 2 September 1945 (from France)
National holiday: Independence Day, 2 September (1945)
Constitution: 15 April 1992
Legal system: based on communist legal theory and French civil law system; has not accepted compulsory ICJ jurisdiction
Suffrage: 18 years of age; universal
Executive branch:
chief of state: President Nguyen Minh TRIET (since 27 June 2006); Vice President Nguyen Thi DOAN (since 25 July 2007)
head of government: Prime Minister Nguyen Tan DUNG (since 27 June 2006); Permanent Deputy Prime Minister Nguyen Sinh HUNG (since 28 June 2006), Deputy Prime Minister Hoang Trung HAI (since 2 August 2007), Deputy Prime Minister Nguyen Thien NHAN (since 2 August 2007), Deputy Prime Minister Pham Gia KHIEM (since 28 June 2006), and Deputy Prime Minister Truong Vinh TRONG (since 28 June 2006)
cabinet: Cabinet appointed by president based on proposal of prime minister and confirmed by National Assembly
elections: president elected by the National Assembly from among its members for five-year term; last held 27 June 2006 (next to be held in 2011); prime minister appointed by the president from among the members of the National Assembly; deputy prime ministers appointed by the prime minister; appointment of prime minister and deputy prime ministers confirmed by National Assembly
election results: Nguyen Minh TRIET elected president; percent of National Assembly vote—94%; Nguyen Tan DUNG elected prime minister; percent of National Assembly vote—92%
Legislative branch: unicameral National Assembly or Quoc Hoi (500 seats; members elected by popular vote to serve five-year terms)

elections: last held 20 May 2007 (next to be held in May 2012)
election results: percent of vote by party—NA; seats by party—CPV 450, non-party CPV-approved 42, self-nominated 1; note—493 candidates were elected; CPV and non-party CPV-approved delegates were members of the Vietnamese Fatherland Front
Judicial branch: Supreme People's Court (chief justice is elected for a five-year term by the National Assembly on the recommendation of the president)
Political parties and leaders: Communist Party of Vietnam or CPV [Nong Duc MANH]; other parties proscribed
Political pressure groups and leaders: groups advocate democracy but are not recognized by government—8406 Bloc; Democratic Party of Vietnam or DPV; People's Democratic Party Vietnam or PDP-VN; Alliance for Democracy (2006)
International organization participation: ACCT (observer), ADB, APEC, APT, ARF, ASEAN, CP, EAS, FAO, G-77, IAEA, IBRD, ICAO, ICRM, IDA, IFAD, IFC, IFRCS, ILO, IMF, IMO, IMSO, Interpol, IOC, IOM, IPU, ISO, ITSO, ITU, MIGA, NAM, OIF, OPCW, UN, UN Security Council (temporary), UNCTAD, UNESCO, UNIDO, UNWTO, UPU, WCL, WCO, WFTU, WHO, WIPO, WMO, WTO
Diplomatic representation in the US:
chief of mission: Ambassador (Appointed) Le Cong PHUNG
chancery: 1233 20th Street NW, Suite 400, Washington, DC 20036
telephone: [1] (202) 861-0737
FAX: [1] (202) 861-0917
consulate(s) general: San Francisco
Diplomatic representation from the US:
chief of mission: Ambassador Michael W. MICHALAK
embassy: 7 Lang Ha Street, Ba Dinh District, Hanoi
mailing address: PSC 461, Box 400, FPO AP 96521-0002
telephone: [84] (4) 850-5000
FAX: [84] (4) 850-5010
consulate(s) general: Ho Chi Minh City
Flag description: red field with a large yellow five-pointed star in the center

ECONOMY

Economy—overview: Vietnam is a densely-populated developing country that in the last 30 years has had to recover from the ravages of war, the loss of financial support from the old Soviet Bloc, and the rigidities of a centrally-planned economy. Economic stagnation marked the period after reunification from 1975 to 1985. In 1986, the Sixth Party Congress approved a broad economic reform package that introduced market reforms and set the groundwork for Vietnam's improved investment climate. Substantial progress was achieved from 1986 to 1997 in moving forward from an extremely low level of development and significantly reducing poverty. The 1997 Asian financial crisis highlighted the problems in the Vietnamese economy and temporarily allowed opponents of reform to slow progress toward a market-oriented economy. GDP growth averaged 6.8% per year from 1997 to 2004 even against the background of the Asian financial crisis and a global recession. Since 2001, Vietnamese authorities have reaffirmed their commitment to economic liberalization and international integration. They have moved to implement the structural reforms needed to modernize the economy and to produce more competitive, export-driven industries. The economy grew 8.5% in 2007. Vietnam's membership in the ASEAN Free Trade Area (AFTA) and entry into force of the US-Vietnam Bilateral Trade Agreement in December 2001 have led to even more rapid changes in Vietnam's trade and economic regime. Vietnam's exports to the US increased 900% from 2001 to 2007. Vietnam joined the WTO in January 2007, following over a decade long negotiation process. WTO membership has provided Vietnam an anchor to the global market and reinforced the domestic economic reform process. Among other benefits, accession allows Vietnam to take advantage of the phase-out of the Agreement on Textiles and Clothing, which eliminated quotas on textiles and clothing for WTO partners on 1 January 2005. Agriculture's share of economic output has continued to shrink, from about 25% in 2000 to less than 20% in 2007. Deep poverty, defined as a percent of the population living under $1 per day, has declined significantly and is now smaller than that of China, India, and the Philippines. Vietnam is working to create jobs to meet the challenge of a labor force that is growing by more than one-and-a-half million people every year. In an effort to stem high inflation which took off in 2007, early in 2008 Vietnamese authorities began to raise benchmark interest rates and reserve requirements. Hanoi is targeting an economic growth rate of 7.5–8% during the next four years.

GDP (purchasing power parity): $221.4 billion (2007 est.)
GDP (official exchange rate): $70.02 billion (2007 est.)
GDP—real growth rate: 8.5% (2007 est.)

GDP—per capita (PPP): $2,600 (2007 est.)
GDP—composition by sector:
agriculture: 19.5%
industry: 42.3%
services: 38.2% (2007 est.)
Labor force: 46.42 million (2007 est.)
Labor force—by occupation: *agriculture:* 55.6%
industry: 18.9%
services: 25.5% (July 2005)
Unemployment rate: 5.3% (2007 est.)
Population below poverty line: 14.8% (2007 est.)
Household income or consumption by percentage share:
lowest 10%: 2.9%
highest 10%: 28.9% (2004)
Distribution of family income—Gini index: 37 (2004)
Inflation rate (consumer prices): 8.3% (2007 est.)
Investment (gross fixed): 40% of GDP (2007 est.)
Budget: *revenues:* $18.26 billion
expenditures: $19.63 billion (2007 est.)
Public debt: 42.8% of GDP (2007 est.)
Agriculture—products: paddy rice, coffee, rubber, cotton, tea, pepper, soybeans, cashews, sugar cane, peanuts, bananas; poultry; fish, seafood
Industries: food processing, garments, shoes, machine-building; mining, coal, steel; cement, chemical fertilizer, glass, tires, oil, paper
Industrial production growth rate: 17.1% (2007 est.)
Electricity—production: 59.01 billion kWh (2007)
Electricity—production by source:
fossil fuel: 43.7%
hydro: 56.3%
nuclear: 0%
other: 0% (2001)
Electricity—consumption: 51.35 billion kWh (2007)
Electricity—exports: 0 kWh (2007)
Electricity—imports: 0 kWh (2007)
Oil—production: 319,500 bbl/day (2007)
Oil—consumption: 271,100 bbl/day (2007 est.)
Oil—exports: 315,700 bbl/day (2007)
Oil—imports: 271,100 bbl/day (2007)
Oil—proved reserves: 3.3 billion bbl (2007 est.)
Natural gas—production: 6.86 billion cu m (2007 est.)
Natural gas—consumption: 6.86 billion cu m (2007 est.)
Natural gas—exports: 0 cu m (2007 est.)
Natural gas—imports: 0 cu m (2007)
Natural gas—proved reserves: 184.7 billion cu m (1 January 2006 est.)

Current account balance: -$6.722 billion (2007 est.)
Exports: $48.07 billion f.o.b. (2007 est.)
Exports—commodities: crude oil, marine products, rice, coffee, rubber, tea, garments, shoes
Exports—partners: US 21.2%, Japan 12.3%, Australia 9.4%, China 5.7%, Germany 4.5% (2006)
Imports: $52.28 billion f.o.b. (2007 est.)
Imports—commodities: machinery and equipment, petroleum products, fertilizer, steel products, raw cotton, grain, cement, motorcycles
Imports—partners: China 17.6%, Singapore 12.9%, Taiwan 11.5%, Japan 9.8%, South Korea 8.4%, Thailand 7.3%, Malaysia 4.2% (2006)
Economic aid—recipient: $5.4 billion in credits and grants pledged by the 2007 Consultative Group meeting in Hanoi (2007)
Reserves of foreign exchange and gold: $19.74 billion (31 December 2007 est.)
Debt—external: $21.69 billion (31 December 2007 est.)
Stock of direct foreign investment—at home: $33.74 billion (2007 est.)
Stock of direct foreign investment—abroad: $NA
Market value of publicly traded shares: $NA
Currency (code): dong (VND)
Currency code: VND
Exchange rates: dong per US dollar—16,119 (2007), 15,983 (2006), 15,746 (2005), NA (2004), 15,510 (2003)
Fiscal year: calendar year

COMMUNICATIONS

Telephones—main lines in use: 10.8 million (2007)
Telephones—mobile cellular: 33.2 million (2007)
Telephone system:
general assessment: Vietnam is putting considerable effort into modernization and expansion of its telecommunication system, but its performance continues to lag behind that of its more modern neighbors
domestic: all provincial exchanges are digitalized and connected to Hanoi, Da Nang, and Ho Chi Minh City by fiber-optic cable or microwave radio relay networks; main lines have been substantially increased, and the use of mobile telephones is growing rapidly
international: country code—84; a landing point for the SEA-ME-WE-3, the C2C, and Thailand-Vietnam-Hong Kong submarine cable systems; the Asia-America Gateway submarine cable system, scheduled for completion by the

end of 2008, will provide new access links to Asia and the US; satellite earth stations—2 Intersputnik (Indian Ocean region)

Radio broadcast stations: AM 65, FM 7, shortwave 29 (1999)

Radios: 8.2 million (1997)

Television broadcast stations: 67 (includes 61 relay, provincial, and city TV stations) (2006)

Televisions: 3.57 million (1997)

Internet country code: .vn

Internet hosts: 106,772 (2007)

Internet Service Providers (ISPs): 5 (2000)

Internet users: 17.87 million (2007)

TRANSPORTATION

Airports: 44 (2007)

Airports—with paved runways: *total:* 37

over 3,047 m: 9

2,438 to 3,047 m: 5

1,524 to 2,437 m: 13

914 to 1,523 m: 10 (2007)

Airports—with unpaved runways:

total: 7

1,524 to 2,437 m: 1

914 to 1,523 m: 3

under 914 m: 3 (2007)

Heliports: 1 (2007)

Pipelines: condensate/gas 432 km; gas 510 km; oil 49 km; refined products 206 km (2007)

Railways: *total:* 2,600 km

standard gauge: 178 km 1.435-m gauge

narrow gauge: 2,169 km 1.000-m gauge

dual gauge: 253 km three-rail track combining 1.435 m and 1.000-m gauges (2006)

Roadways:

total: 222,179 km

paved: 42,167 km

unpaved: 180,012 km (2004)

Waterways: 17,702 km (5,000 km navigable by vessels up to 1.8 m draft) (2005)

Merchant marine:

total: 314 ships (1000 GRT or over) 1,739,927 GRT/2,681,003 DWT

by type: barge carrier 1, bulk carrier 26, cargo 238, chemical tanker 7, container 6, liquefied gas 6, petroleum tanker 26, refrigerated cargo 2, roll on/roll off 1, specialized tanker 1

registered in other countries: 33 (Antigua and Barbuda 1, Honduras 1, South Korea 1, Liberia 3, Mongolia 14, Panama 10, Tuvalu 3, unknown 2) (2007)

Ports and terminals: Da Nang, Hai Phong, Ho Chi Minh City

MILITARY

Military branches: People's Armed Forces: People's Army of Vietnam (PAVN) (includes People's Navy Command (with naval infantry, coast guard), Air and Air Defense Force (Kon Quan Nhan Dan), Border Defense Command), People's Public Security Forces, Militia Force, Self-Defense Forces (2005)

Military service age and obligation: 18 years of age (male) for compulsory military service; females may volunteer for active duty military service; conscript service obligation—2 years (3 to 4 years in the navy); 18–45 years of age (male) or 18–40 years of age (female) for Militia Force or Self Defense Forces (2006)

Manpower available for military service: *males age 16–49:* 24,586,328

females age 16–49: 24,335,132 (2008 est.)

Manpower fit for military service:

males age 16–49: 18,849,274

females age 16–49: 20,575,884 (2008 est.)

Manpower reaching militarily significant age annually:

males age 16–49: 903,734

females age 16–49: 845,306 (2008 est.)

Military expenditures—percent of GDP: 2.5% (2005 est.)

TRANSNATIONAL ISSUES

Disputes—international: southeast Asian states have enhanced border surveillance to check the spread of avian flu; Cambodia and Laos protest Vietnamese squatters and armed encroachments along border; an estimated 300,000 Vietnamese refugees reside in China; establishment of a maritime boundary with Cambodia is hampered by unresolved dispute over the sovereignty of offshore islands; demarcation of the China-Vietnam boundary proceeds slowly and although the maritime boundary delimitation and fisheries agreements were ratified in June 2004, implementation has been delayed; China occupies the Paracel Islands also claimed by Vietnam and Taiwan; involved in complex dispute with China, Malaysia, Philippines, Taiwan, and possibly Brunei over the Spratly Islands; the 2002 "Declaration on the Conduct of Parties in the South China Sea" has eased tensions but falls short of a legally binding "code of conduct" desired by several of the disputants; Vietnam continues to expand construction of facilities in the Spratly Islands; in March 2005, the national oil companies of China, the Philippines, and Vietnam signed a joint accord to conduct marine seismic activities in the Spratly Islands

Illicit drugs: minor producer of opium poppy; probable minor transit point for Southeast Asian heroin; government continues to face domestic opium/heroin/methamphetamine addiction problems despite longstanding crackdowns

VIRGIN ISLANDS

INTRODUCTION

Background: During the 17th century, the archipelago was divided into two territorial units, one English and the other Danish. Sugarcane, produced by slave labor, drove the islands' economy during the 18th and early 19th centuries. In 1917, the US purchased the Danish portion, which had been in economic decline since the abolition of slavery in 1848.

GEOGRAPHY

Location: Caribbean, islands between the Caribbean Sea and the North Atlantic Ocean, east of Puerto Rico

Geographic coordinates: 18 20 N, 64 50 W

Map references: Central America and the Caribbean

Area: *total:* 1,910 sq km

land: 346 sq km

water: 1,564 sq km

Area—comparative: twice the size of Washington, DC

Land boundaries: 0 km

Coastline: 188 km

Maritime claims: *territorial sea:* 12 nm

exclusive economic zone: 200 nm

Climate: subtropical, tempered by easterly trade winds, relatively low humidity, little seasonal temperature variation; rainy season September to November

Terrain: mostly hilly to rugged and mountainous with little level land

Elevation extremes:

lowest point: Caribbean Sea 0 m

highest point: Crown Mountain 475 m

Natural resources: sun, sand, sea, surf

Land use: *arable land:* 5.71%

permanent crops: 2.86%

other: 91.43% (2005)

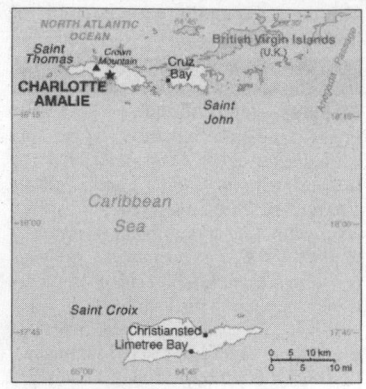

Irrigated land: NA

Natural hazards: several hurricanes in recent years; frequent and severe droughts and floods; occasional earthquakes

Environment—current issues: lack of natural freshwater resources

Geography—note: important location along the Anegada Passage—a key shipping lane for the Panama Canal; Saint Thomas has one of the best natural deepwater harbors in the Caribbean

PEOPLE

Population: 108,210 (July 2008 est.)

Age structure:

0–14 years: 21.2% (male 11,546/female 11,354)

15–64 years: 66.4% (male 34,096/female 37,760)

65 years and over: 12.4% (male 5,961/female 7,493) (2008 est.)

Median age:

total: 38.3 years

male: 37.2 years

female: 39.1 years (2008 est.)

Population growth rate: -0.27% (2008 est.)

Birth rate: 13.43 births/1,000 population (2008 est.)

Death rate: 6.79 deaths/1,000 population (2008 est.)

Net migration rate: -9.33 migrant(s)/1,000 population (2008 est.)

Sex ratio:

at birth: 1.06 male(s)/female

under 15 years: 1.02 male(s)/female

15–64 years: 0.9 male(s)/female

65 years and over: 0.8 male(s)/female

total population: 0.91 male(s)/female (2008 est.)

Infant mortality rate:

total: 7.53 deaths/1,000 live births

male: 8.57 deaths/1,000 live births

female: 6.43 deaths/1,000 live births (2008 est.)

Life expectancy at birth:

total population: 79.34 years

male: 75.56 years

female: 83.35 years (2008 est.)

Total fertility rate: 2.14 children born/woman (2008 est.)

HIV/AIDS—adult prevalence rate: NA

HIV/AIDS—people living with HIV/AIDS: NA

HIV/AIDS—deaths: NA

Nationality:

noun: Virgin Islander(s) (US citizens)

adjective: Virgin Islander

Ethnic groups: black 76.2%, white 13.1%, Asian 1.1%, other 6.1%, mixed 3.5% (2000 census)

Religions: Baptist 42%, Roman Catholic 34%, Episcopalian 17%, other 7%

Languages: English 74.7%, Spanish or Spanish Creole 16.8%, French or French Creole 6.6%, other 1.9% (2000 census)

Literacy:

definition: age 15 and over can read and write

total population: 90–95% est.

male: NA%

female: NA% (2005 est.)

GOVERNMENT

Country name:

conventional long form: United States Virgin Islands

conventional short form: Virgin Islands

former: Danish West Indies

abbreviation: USVI

Dependency status: organized, unincorporated territory of the US with policy relations between the Virgin Islands and the US under the jurisdiction of the Office of Insular Affairs, US Department of the Interior

Government type: NA

Capital: *name:* Charlotte Amalie

geographic coordinates: 18 21 N, 64 56 W

time difference: UTC-4 (1 hour ahead of Washington, DC during Standard Time)

Administrative divisions: none (territory of the US); there are no first-order administrative divisions as defined by the US Government, but there are three islands at the second order; Saint Croix, Saint John, Saint Thomas

Independence: none (territory of the US)

National holiday: Transfer Day (from Denmark to the US), 27 March (1917)

Constitution: Revised Organic Act of 22 July 1954

Legal system: based on US laws

Suffrage: 18 years of age; universal; island residents are US citizens but do not vote in US presidential elections

Executive branch:

chief of state: President George W. BUSH of the US (since 20 January 2001); Vice President Richard B. CHENEY (since 20 January 2001)

head of government: Governor John DeJONGH (since 1 January 2007)

cabinet: NA

elections: under the US Constitution, residents of unincorporated territories, such as the Virgin Islands, do not vote in elections for US president and vice president; however, they may vote in the Democratic and Republican presidential primary elections; governor and lieutenant governor elected on the same ticket by popular vote for four-year terms (eligible for a second term); election last held 7 and 21 November 2006 (next to be held November 2010)

election results: John DeJONGH elected governor; percent of vote—John DeJONGH 57.3%, Kenneth MAPP 42.7%

Legislative branch: unicameral Senate (15 seats; members are elected by popular vote to serve two-year terms)

elections: last held 7 November 2006 (next to be held November 2008)

election results: percent of vote by party—NA; seats by party—Democratic Party 8, ICM 4, independent 3

note: the Virgin Islands elects one nonvoting representative to the US House of Representatives; election last held 7 November 2006 (next to be held November 2008)

Judicial branch: US District Court of the Virgin Islands (under Third Circuit jurisdiction); Superior Court of the Virgin Islands (judges appointed by the governor for 10-year terms)

Political parties and leaders: Democratic Party [Arturo WATLINGTON]; Independent Citizens' Movement or ICM [Usie RICHARDS]; Republican Party [Gary SPRAUVE]

Political pressure groups and leaders: NA

International organization participation: IOC, UPU

Diplomatic representation in the US: none (territory of the US)

Diplomatic representation from the US: none (territory of the US)

Flag description: white field with a modified US coat of arms in the center between the large blue initials V and I; the coat of arms shows a yellow eagle holding an olive branch in one talon and three arrows in the other with a superimposed shield of vertical red and white stripes below a blue panel

ECONOMY

Economy—overview: Tourism is the primary economic activity, accounting for 80% of GDP and employment. The islands hosted 2.6 million visitors in

2005. The manufacturing sector consists of petroleum refining, textiles, electronics, pharmaceuticals, and watch assembly. One of the world's largest petroleum refineries is at Saint Croix. The agricultural sector is small, with most food being imported. International business and financial services are small but growing components of the economy. The islands are vulnerable to substantial damage from storms. The government is working to improve fiscal discipline, to support construction projects in the private sector, to expand tourist facilities, to reduce crime, and to protect the environment.

GDP (purchasing power parity): $1.577 billion (2004 est.)

GDP (official exchange rate): $NA

GDP—real growth rate: 2% (2002 est.)

GDP—per capita (PPP): $14,500 (2004 est.)

GDP—composition by sector:
agriculture: 1%
industry: 19%
services: 80% (2003 est.)

Labor force: 43,980 (2004 est.)

Labor force—by occupation: *agriculture:* 1%
industry: 19%
services: 80% (2003 est.)

Unemployment rate: 6.2% (2004)

Population below poverty line: 28.9% (2002)

Household income or consumption by percentage share:
lowest 10%: NA%
highest 10%: NA%

Inflation rate (consumer prices): 2.2% (2003)

Budget:
revenues: $NA
expenditures: $NA

Agriculture—products: fruit, vegetables, sorghum; Senepol cattle

Industries: tourism, petroleum refining, watch assembly, rum distilling, construction, pharmaceuticals, textiles, electronics

Industrial production growth rate: NA%

Electricity—production: 996.1 million kWh (2005)

Electricity—production by source:
fossil fuel: 100%
hydro: 0%
nuclear: 0%
other: 0% (2001)

Electricity—consumption: 926.4 million kWh (2005)

Electricity—exports: 0 kWh (2005)

Electricity—imports: 0 kWh (2005)

Oil—production: 17,620 bbl/day (2005 est.)

Oil—consumption: 98,000 bbl/day (2005 est.)

Oil—exports: 397,400 bbl/day (2004)

Oil—imports: 493,000 bbl/day (2004)

Oil—proved reserves: NA

Natural gas—production: 0 cu m (2005 est.)

Natural gas—consumption: 0 cu m (2005 est.)

Natural gas—exports: 0 cu m (2005 est.)

Natural gas—imports: 0 cu m (2005)

Natural gas—proved reserves: 0 cu m (1 January 2006 est.)

Exports: $4.234 billion (2001)

Exports—commodities: refined petroleum products

Exports—partners: US, Puerto Rico (2006)

Imports: $4.609 billion (2001)

Imports—commodities: crude oil, foodstuffs, consumer goods, building materials

Imports—partners: US, Puerto Rico (2006)

Economic aid—recipient: $NA

Debt—external: $NA

Currency (code): US dollar (USD)

Currency code: USD

Exchange rates: the US dollar is used

Fiscal year: 1 October—30 September

COMMUNICATIONS

Telephones—main lines in use: 71,700 (2005)

Telephones—mobile cellular: 80,300 (2005)

Telephone system:
general assessment: modern system with total digital switching, uses fiber-optic cable and microwave radio relay
domestic: full range of services available
international: country code—1-340; submarine cable connections to US, the Caribbean, Central and South America; satellite earth stations—NA

Radio broadcast stations: AM 6, FM 16, shortwave 0 (2005)

Radios: 107,000 (1997)

Television broadcast stations: 5 (2006)

Televisions: 68,000 (1997)

Internet country code: .vi

Internet hosts: 4,116 (2007)

Internet Service Providers (ISPs): 50 (2000)

Internet users: 30,000 (2005)

TRANSPORTATION

Airports: 2 (2007)

Airports—with paved runways:
total: 2
over 3,047 m: 1
1,524 to 2,437 m: 1 (2007)

Roadways:
total: 1,257 km (2004)

Ports and terminals: Charlotte Amalie, Limetree Bay

MILITARY

Military—note: defense is the responsibility of the US

TRANSNATIONAL ISSUES

Disputes—international: none

WAKE ISLAND

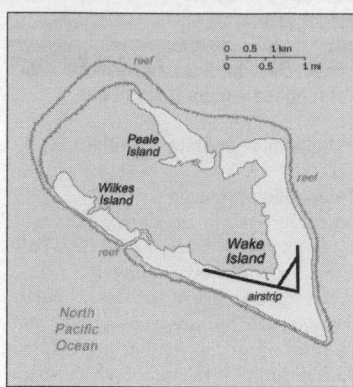

INTRODUCTION

Background: The US annexed Wake Island in 1899 for a cable station. An important air and naval base was constructed in 1940–41. In December 1941, the island was captured by the Japanese and held until the end of World War II. In subsequent years, Wake was developed as a stopover and refueling site for military and commercial aircraft transiting the Pacific. Since 1974, the island's airstrip has been used by the US military, as well as for emergency landings. All operations on the island were suspended and all personnel evacuated in August 2006 with the approach of super typhoon IOKE (category 5), which struck the island with sustained winds of 250 kph and a 6 m storm surge inflicting major damage. A US Air Force assessment and repair team returned to the island in September and restored limited function to the airfield and facilities. The future status of activities on the island will be determined upon completion of the survey and assessment.

GEOGRAPHY

Location: Oceania, atoll in the North Pacific Ocean, about two-thirds of the way from Hawaii to the Northern Mariana Islands
Geographic coordinates: 19 17 N, 166

39 E
Map references: Oceania
Area: *total:* 6.5 sq km
land: 6.5 sq km
water: 0 sq km
Area—comparative: about 11 times the size of The Mall in Washington, DC
Land boundaries: 0 km
Coastline: 19.3 km
Maritime claims:
territorial sea: 12 nm
exclusive economic zone: 200 nm
Climate: tropical
Terrain: atoll of three low coral islands, Peale, Wake, and Wilkes, built up on an underwater volcano; central lagoon is former crater, islands are part of the rim
Elevation extremes:
lowest point: Pacific Ocean 0 m
highest point: unnamed location 6 m
Natural resources: none
Land use:
arable land: 0%
permanent crops: 0%
other: 100% (2005)
Irrigated land: 0 sq km
Natural hazards: occasional typhoons
Environment—current issues: NA
Geography—note: strategic location in the North Pacific Ocean; emergency landing location for transpacific flights

PEOPLE

Population: no indigenous inhabitants
note: since super typhoon IOKE, a small military contingent along with 75 contractor personnel have returned to the island to conduct clean-up and restore basic operations on the island (July 2008 est.)

GOVERNMENT

Country name:
conventional long form: none
conventional short form: Wake Island
Dependency status: unorganized, unincorporated territory of the US; administered from Washington, DC, by the Department of the Interior; activities in the atoll are currently conducted by the US Air Force

Legal system: the laws of the US, where applicable, apply
Flag description: the flag of the US is used

ECONOMY

Economy—overview: Economic activity is limited to providing services to military personnel and contractors located on the island. All food and manufactured goods must be imported.
Electricity—production: NA kWh

COMMUNICATIONS

Telephone system:
general assessment: satellite communications; 2 DSN circuits off the Overseas Telephone System (OTS)
domestic: NA
international: NA
Radio broadcast stations: AM 0, FM 0, shortwave 0 (Armed Forces Radio/Television Service (AFRTS) radio service provided by satellite (2005)
Television broadcast stations: 0 (2005)

TRANSPORTATION

Airports: 1 (2007)
Airports—with paved runways:
total: 1
2,438 to 3,047 m: 1 (2007)
Ports and terminals: none; two offshore anchorages for large ships
Transportation—note: there are no commercial or civilian flights to and from Wake Island, except in direct support of island missions; emergency landing is available

MILITARY

Military—note: defense is the responsibility of the US; the US Air Force is responsible for overall administration and operation of the island; the launch support facility is administered by the US Missile Defense Agency (MDA)

TRANSNATIONAL ISSUES

Disputes—international: claimed by Marshall Islands

WALLIS AND FUTUNA

INTRODUCTION

Background: The Futuna island group was discovered by the Dutch in 1616 and Wallis by the British in 1767, but it was the French who declared a protectorate over the islands in 1842. In 1959, the inhabitants of the islands voted to become a French overseas territory.

GEOGRAPHY

Location: Oceania, islands in the South Pacific Ocean, about two-thirds of the way from Hawaii to New Zealand
Geographic coordinates: 13 18 S, 176 12 W
Map references: Oceania
Area: *total:* 274 sq km

land: 274 sq km
water: 0 sq km
note: includes Ile Uvea (Wallis Island), Ile Futuna (Futuna Island), Ile Alofi, and 20 islets
Area—comparative: 1.5 times the size of Washington, DC
Land boundaries: 0 km
Coastline: 129 km

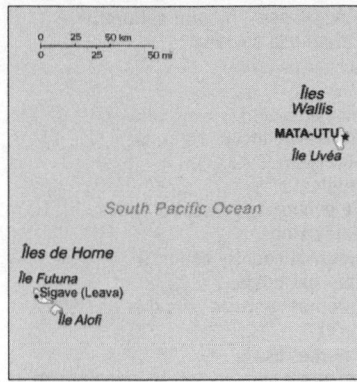

Maritime claims:
territorial sea: 12 nm
exclusive economic zone: 200 nm
Climate: tropical; hot, rainy season (November to April); cool, dry season (May to October); rains 2,500–3,000 mm per year (80% humidity); average temperature 26.6 degrees C
Terrain: volcanic origin; low hills
Elevation extremes:
lowest point: Pacific Ocean 0 m
highest point: Mont Singavi 765 m
Natural resources: NEGL
Land use:
arable land: 7.14%
permanent crops: 35.71%
other: 57.15% (2005)
Irrigated land: NA
Natural hazards: NA
Environment—current issues: deforestation (only small portions of the original forests remain) largely as a result of the continued use of wood as the main fuel source; as a consequence of cutting down the forests, the mountainous terrain of Futuna is particularly prone to erosion; there are no permanent settlements on Alofi because of the lack of natural fresh water resources
Geography—note: both island groups have fringing reefs

PEOPLE

Population: 16,448 (July 2008 est.)
Age structure: *0–14 years:* NA
15–64 years: NA
65 years and over: NA
Population growth rate: NA
Birth rate: NA
Death rate: NA (2008 est.)
Net migration rate: NA
note: there has been steady emigration from Wallis and Futuna to New Caledonia (2008 est.)
Infant mortality rate: *total:* NA
male: NA
female: NA (2008 est.)
Life expectancy at birth:
total population: NA
male: NA
female: NA (2008 est.)

Total fertility rate: NA (2008 est.)
HIV/AIDS—adult prevalence rate: NA
HIV/AIDS—people living with HIV/AIDS: NA
HIV/AIDS—deaths: NA
Nationality:
noun: Wallisian(s), Futunan(s), or Wallis and Futuna Islanders
adjective: Wallisian, Futunan, or Wallis and Futuna Islander
Ethnic groups: Polynesian
Religions: Roman Catholic 99%, other 1%
Languages: Wallisian 58.9% (indigenous Polynesian language), Futunian 30.1%, French 10.8%, other 0.2% (2003 census)
Literacy: *definition:* age 15 and over can read and write
total population: 50%
male: 50%
female: 50% (1969 est.)

GOVERNMENT

Country name:
conventional long form: Territory of the Wallis and Futuna Islands
conventional short form: Wallis and Futuna
local long form: Territoire des Iles Wallis et Futuna
local short form: Wallis et Futuna
Dependency status: overseas territory of France
Government type: NA
Capital: *name:* Mata-Utu (on Ile Uvea)
geographic coordinates: 13 57 S, 171 56 W
time difference: UTC+12 (17 hours ahead of Washington, DC during Standard Time)
Administrative divisions: none (overseas territory of France); there are no first-order administrative divisions as defined by the US Government, but there are three kingdoms at the second order named Alo, Sigave, Wallis
Independence: none (overseas territory of France)
National holiday: Bastille Day, 14 July (1789)
Constitution: 4 October 1958 (French Constitution)
Legal system: the laws of France, where applicable, apply
Suffrage: 18 years of age; universal
Executive branch:
chief of state: President Nicolas SARKOZY (since 16 May 2007); represented by High Administrator Richard DIDIER (since 19 July 2006)
head of government: President of the Territorial Assembly Patalione KANIMOA (since January 2001)
cabinet: Council of the Territory consists of three kings and three members appointed by the high administrator on the advice of the Territorial Assembly
note: there are three traditional kings with limited powers

elections: French president elected by popular vote for a five-year term; high administrator appointed by the French president on the advice of the French Ministry of the Interior; the presidents of the Territorial Government and the Territorial Assembly are elected by the members of the assembly
Legislative branch: unicameral Territorial Assembly or Assemblee Territoriale (20 seats; members are elected by popular vote to serve five-year terms)
elections: last held 11 March 2002 (next to be held 22 April 2007)
election results: percent of vote by party—NA; seats by party—RPR and affiliates 13, Socialists and affiliates 7
note: Wallis and Futuna elects one senator to the French Senate and one deputy to the French National Assembly; French Senate—elections last held 26 September 2004 (next to be held by September 2010); results—percent of vote by party—NA; seats—RPR (now UMP) 1; French National Assembly—elections last held 17 June 2007 (next to be held by 2012); results—percent of vote by party—NA; seats—PS 1
Judicial branch: justice generally administered under French law by the high administrator, but the three traditional kings administer customary law and there is a magistrate in Mata-Utu; a court of appeal is located in Noumea, New Caledonia
Political parties and leaders: Lua Kae Tahi (Giscardians); Mouvement des Radicaux de Gauche or MRG; Rally for the Republic or RPR (UMP) [Clovis LOGOLOGOFOLAU]; Socialist Party or PS; Taumu'a Lelei [Soane Muni UHILA]; Union Populaire Locale or UPL [Falakiko GATA]; Union Pour la Democratie Francaise or UDF
Political pressure groups and leaders: NA
International organization participation: SPC, UPU
Diplomatic representation in the US: none (overseas territory of France)
Diplomatic representation from the US: none (overseas territory of France)
Flag description: unofficial, local flag has a red field with four white isosceles triangles in the middle, representing the three native kings of the islands and the French administrator; the apexes of the triangles are oriented inward and at right angles to each other; the flag of France, outlined in white on two sides, is in the upper hoist quadrant; the flag of France is the only official flag

ECONOMY

Economy—overview: The economy is limited to traditional subsistence agriculture, with about 80% of labor force earnings from agriculture (coconuts and vegetables), livestock (mostly pigs), and fishing. About 4% of the population is employed in government. Revenues come from French Government subsidies, licensing of fishing rights to Japan and South Korea, import taxes, and remittances from expatriate workers in New Caledonia.

GDP (purchasing power parity): $60 million (2004 est.)

GDP (official exchange rate): $NA

GDP—real growth rate: NA%

GDP—per capita (PPP): $3,800 (2004 est.)

GDP—composition by sector:
agriculture: NA%
industry: NA%
services: NA%

Labor force: 3,104 (2003)

Labor force—by occupation: *agriculture:* 80%
industry: 4%
services: 16% (2001 est.)

Unemployment rate: 15.2% (2003)

Population below poverty line: NA%

Household income or consumption by percentage share:
lowest 10%: NA%
highest 10%: NA%

Inflation rate (consumer prices): 2.8% (2005)

Budget: *revenues:* $29,730
expenditures: $31,330 (2004)

Public debt: 5.6% of GDP (2004 est.)

Agriculture—products: breadfruit, yams, taro, bananas; pigs, goats; fish

Industries: copra, handicrafts, fishing, lumber

Industrial production growth rate: NA%

Electricity—production: NA kWh

Electricity—production by source:
fossil fuel: 0%
hydro: 0%
nuclear: 0%
other: 0%

Electricity—consumption: NA kWh

Electricity—exports: 0 kWh (2002)

Electricity—imports: 0 kWh (2002)

Exports: $47,450 f.o.b. (2004)

Exports—commodities: copra, chemicals, construction materials

Exports—partners: Italy 40%, Croatia 15%, US 14%, Denmark 13% (2006)

Imports: $61.17 million f.o.b. (2004)

Imports—commodities: chemicals, machinery, passenger ships, consumer goods

Imports—partners: France 97%, Australia 2%, NZ 1% (2006)

Economic aid—recipient: assistance from France, $NA

Debt—external: $3.67 million (2004)

Currency (code): Comptoirs Francais du Pacifique franc (XPF)

Currency code: XPF

Exchange rates: Comptoirs Francais du Pacifique francs (XPF) per US dollar— NA (2007), 95.03 (2006), 95.89 (2005), 96.04 (2004), 105.66 (2003)

Fiscal year: calendar year

COMMUNICATIONS

Telephones—main lines in use: 1,900 (2002)

Telephones—mobile cellular: NA

Telephone system:
general assessment: NA
domestic: NA
international: country code—681

Radio broadcast stations: AM 1, FM 0, shortwave 0 (2000)

Radios: NA

Television broadcast stations: 2 (2000)

Televisions: NA

Internet country code: .wf

Internet hosts: 1 (2007)

Internet Service Providers (ISPs): 1 (2000)

Internet users: 900 (2002)

TRANSPORTATION

Airports: 2 (2007)

Airports—with paved runways:
total: 1
1,524 to 2,437 m: 1 (2007)

Airports—with unpaved runways:
total: 1
914 to 1,523 m: 1 (2007)

Merchant marine:
total: 8 ships (1000 GRT or over) 92,346 GRT/98,307 DWT
by type: chemical tanker 2, passenger 6
foreign-owned: 8 (France 6, French Polynesia 2) (2007)

Ports and terminals: Leava, Mata-Utu

MILITARY

Military—note: defense is the responsibility of France

TRANSNATIONAL ISSUES

Disputes—international: none

WEST BANK

INTRODUCTION

Background: The September 1993 Israel-PLO Declaration of Principles on Interim Self-Government Arrangements provided for a transitional period of Palestinian self-rule in the West Bank and Gaza Strip. Under a series of agreements signed between May 1994 and September 1999, Israel transferred to the Palestinian Authority (PA) security and civilian responsibility for Palestinian-populated areas of the West Bank and Gaza. Negotiations to determine the permanent status of the West Bank and Gaza stalled following the outbreak of an intifada in September 2000, as Israeli forces reoccupied most Palestinian-controlled areas. In April 2003, the Quartet (US, EU, UN, and Russia) presented a roadmap to a final settlement of the conflict by 2005 based on reciprocal steps by the two parties leading to two states, Israel and a democratic Palestine. The proposed date for a permanent status agreement was postponed indefinitely due to violence and accusations that both sides had not followed through on their commitments. Following Palestinian leader Yasir ARAFAT's death in late 2004, Mahmud ABBAS was elected PA president in January 2005. A month later, Israel and the PA agreed to the Sharm el-Sheikh Commitments in an effort to move the peace process forward. In September 2005, Israel unilaterally withdrew all its settlers and soldiers and dismantled its military facilities in the Gaza Strip and withdrew settlers and redeployed soldiers from four small northern West Bank settlements. Nonetheless, Israel controls maritime, airspace, and most access to the Gaza Strip. A November 2005 PA-Israeli agreement authorized the reopening of the Rafah border crossing between the Gaza Strip and Egypt under joint PA and Egyptian control. In January 2006, the Islamic Resistance Movement, HAMAS, won control of the Palestinian Legislative Council (PLC). The international community refused to accept the HAMAS-led government because it did not recognize Israel, would not renounce violence, and refused to honor previous peace agreements between Israel and the PA. HAMAS took control of the PA government in March 2006, but President ABBAS had little success negotiating with HAMAS to present a political platform acceptable to the international community so as to lift economic sanctions on Palestinians. The PLC was unable to convene throughout most of 2006 as a result of Israel's detention of many HAMAS PLC members and Israeli-imposed travel restrictions on other PLC members. Violent clashes took place between Fatah and HAMAS supporters in the Gaza Strip in 2006 and early 2007, resulting in numerous Palestinian deaths and injuries. ABBAS and HAMAS Political Bureau Chief MISHAL in February 2007 signed the

West Bank is Israeli-occupied with current status subject to the Israeli-Palestinian Interim Agreement; permanent status to be determined through further negotiation.

Mecca Agreement in Saudi Arabia that resulted in the formation of a Palestinian National Unity Government (NUG) headed by HAMAS member Ismail HANIYA. However, fighting continued in the Gaza Strip, and in June, HAMAS militants succeeded in a violent takeover of all military and governmental institutions in the Gaza Strip. ABBAS dismissed the NUG and through a series of presidential decrees formed a PA government in the West Bank led by independent Salam FAYYAD. HAMAS rejected the NUG's dismissal and has called for resuming talks with Fatah, but ABBAS has ruled out negotiations until HAMAS agrees to a return of PA control over the Gaza Strip and recognizes the FAYYAD-led government. FAYYAD and his PA government initiated a series of security and economic reforms to improve conditions in the West Bank. ABBAS participated in talks with Israel's Prime Minister OLMERT and secured the release of some Palestinian prisoners and previously withheld customs revenue. During a November 2007 international meeting in Annapolis Maryland, ABBAS and OLMERT agreed to resume peace negotiations with the goal of reaching a final peace settlement by the end of 2008.

GEOGRAPHY

Location: Middle East, west of Jordan

Geographic coordinates: 32 00 N, 35 15 E

Map references: Middle East

Area:
total: 5,860 sq km
land: 5,640 sq km
water: 220 sq km
note: includes West Bank, Latrun Salient, and the northwest quarter of the Dead Sea, but excludes Mt. Scopus; East Jerusalem and Jerusalem No Man's Land are also included only as a means of depicting the entire area occupied by Israel in 1967

Area—comparative: slightly smaller than Delaware

Land boundaries:
total: 404 km
border countries: Israel 307 km, Jordan 97 km

Coastline: 0 km (landlocked)

Maritime claims: none (landlocked)

Climate: temperate; temperature and precipitation vary with altitude, warm to hot summers, cool to mild winters

Terrain: mostly rugged dissected upland, some vegetation in west, but barren in east

Elevation extremes:
lowest point: Dead Sea -408 m
highest point: Tall Asur 1,022 m

Natural resources: arable land

Land use:
arable land: 16.9%
permanent crops: 18.97%
other: 64.13% (2001)

Irrigated land: 150 sq km; note—includes Gaza Strip (2003)

Natural hazards: droughts

Environment—current issues: adequacy of fresh water supply; sewage treatment

Geography—note: landlocked; highlands are main recharge area for Israel's coastal aquifers; there are 242 West Bank settlements and 29 East Jerusalem settlements in addition to at least 20 occupied outposts (August 2005 est.)

PEOPLE

Population: 2,611,904
note: in addition, there are about 187,000 Israeli settlers in the West Bank and fewer than 177,000 in East Jerusalem (July 2008 est.)

Age structure:
0–14 years: 41.9% (male 560,604/female 533,590)
15–64 years: 54.8% (male 732,930/female 697,411)
65 years and over: 3.3% (male 37,015/female 50,354) (2008 est.)

Median age: *total:* 18.7 years
male: 18.6 years
female: 18.8 years (2008 est.)

Population growth rate: 2.919% (2008 est.)

Birth rate: 30.35 births/1,000 population (2008 est.)

Death rate: 3.79 deaths/1,000 population (2008 est.)

Net migration rate: 2.63 migrant(s)/1,000 population (2008 est.)

Sex ratio:
at birth: 1.06 male(s)/female
under 15 years: 1.05 male(s)/female
15–64 years: 1.05 male(s)/female
65 years and over: 0.74 male(s)/female
total population: 1.04 male(s)/female (2008 est.)

Infant mortality rate:
total: 18.21 deaths/1,000 live births
male: 20.09 deaths/1,000 live births
female: 16.21 deaths/1,000 live births (2008 est.)

Life expectancy at birth:
total population: 73.65 years
male: 71.85 years
female: 75.56 years (2008 est.)

Total fertility rate: 4.06 children born/woman (2008 est.)

HIV/AIDS—adult prevalence rate: NA

HIV/AIDS—people living with HIV/AIDS: NA

HIV/AIDS—deaths: NA

Nationality:
noun: NA
adjective: NA

Ethnic groups: Palestinian Arab and other 83%, Jewish 17%

Religions: Muslim 75% (predominantly Sunni), Jewish 17%, Christian and other 8%

Languages: Arabic, Hebrew (spoken by Israeli settlers and many Palestinians), English (widely understood)

Literacy:
definition: age 15 and over can read and write
total population: 92.4%
male: 96.7%
female: 88% (2004 est.)

GOVERNMENT

Country name:
conventional long form: none
conventional short form: West Bank

ECONOMY

Economy—overview: The West Bank—the larger of the two areas comprising the Palestinian Authority (PA)—has experienced a general decline in economic conditions since the second intifada began in September 2000. The downturn has been largely a result of Israeli closure policies—the imposition of closures and access restrictions in response to security concerns in Israel—which disrupted labor and trading relationships. In 2001, and even more severely in 2002, Israeli military measures in PA areas resulted in the destruction of capital, the disruption of administrative structures, and widespread business closures. International aid of at least $1.14 billion to the West Bank and Gaza Strip in 2004 prevented the complete collapse of the economy and allowed some reforms in the government's financial operations. In 2005, high unemployment and limited trade opportunities—due to continued closures both within the West Bank and externally—stymied growth. Israel's and the international community's financial embargo of the PA when HAMAS ran the PA during March 2006—June 2007 has interrupted the provision of PA social services and the payment of PA salaries. Since June the Fayyad government in the West Bank has restarted salary payments and the provision of services but would be unable to operate absent high levels of international assistance.

GDP (purchasing power parity): $5.034 billion (includes Gaza Strip) (2006 est.)
GDP (official exchange rate): $5.328 billion (includes Gaza Strip) (2006 est.)
GDP—real growth rate: -8% (includes Gaza Strip) (2006 est.)
GDP—per capita (PPP): $1,100 (includes Gaza Strip) (2006 est.)
GDP—composition by sector:
agriculture: 8%
industry: 13%
services: 79% (includes Gaza Strip) (2006 est.)
Labor force: 605,000 (2006)
Labor force—by occupation: *agriculture:* 18%
industry: 15%
services: 67% (2006)
Unemployment rate: 18.6% (2006)
Population below poverty line: 46% (2007 est.)
Household income or consumption by percentage share:
lowest 10%: NA%
highest 10%: NA%
Inflation rate (consumer prices): 3.6% (includes Gaza Strip) (2006)
Budget: *revenues:* $1.149 billion
expenditures: $2.31 billion
note: includes Gaza Strip (2006)
Agriculture—products: olives, citrus, vegetables; beef, dairy products
Industries: generally small family businesses that produce cement, textiles, soap, olive-wood carvings, and mother-of-pearl souvenirs; the Israelis have established some small-scale, modern industries in the settlements and industrial centers

Industrial production growth rate: 2.4% (includes Gaza Strip) (2005)
Electricity—production: NA kWh; note—most electricity imported from Israel; East Jerusalem Electric Company buys and distributes electricity to Palestinians in East Jerusalem and its concession in the West Bank; the Israel Electric Company directly supplies electricity to most Jewish residents and military facilities; some Palestinian municipalities, such as Nablus and Janin, generate their own electricity from small power plants
Electricity—production by source:
fossil fuel: 100%
hydro: 0%
nuclear: 0%
other: 0% (2001)
Electricity—consumption: NA kWh
Electricity—imports: NA kWh
Exports: $301 million f.o.b.; (includes Gaza Strip) (2005)
Exports—commodities: olives, fruit, vegetables, limestone
Exports—partners: Israel, Jordan, Gaza Strip (2006)
Imports: $2.44 billion c.i.f.; (includes Gaza Strip) (2005)
Imports—commodities: food, consumer goods, construction materials
Imports—partners: Israel, Jordan, Gaza Strip (2006)
Economic aid—recipient: $1.4 billion; (includes Gaza Strip) (2006 est.)
Debt—external: $NA
Market value of publicly traded shares: $4.461 billion (2005)
Currency (code): new Israeli shekel (ILS); Jordanian dinar (JOD)
Currency code: ILS; JOD
Exchange rates: new Israeli shekels per US dollar—4.14 (2007), 4.4565 (2006), 4.4877 (2005), 4.482 (2004), 4.5541 (2003)
Fiscal year: calendar year

COMMUNICATIONS

Telephones—main lines in use: 349,000 (includes Gaza Strip) (2005)
Telephones—mobile cellular: 1.095 million (includes Gaza Strip) (2005)
Telephone system:
general assessment: NA
domestic: Israeli company BEZEK and the Palestinian company PALTEL are responsible for fixed line services; the

Palestinian JAWAL company provides cellular services
international: country code—970 (2004)
Radio broadcast stations: AM 0, FM 25, shortwave 0 (2008)
Radios: NA; note—most Palestinian households have radios (1999)
Television broadcast stations: 30 (2008)
Televisions: NA; note—many Palestinian households have televisions (1999)
Internet country code: .ps; note—same as Gaza Strip
Internet Service Providers (ISPs): 8 (1999)
Internet users: 243,000 (includes Gaza Strip) (2005)

TRANSPORTATION

Airports: 3 (2007)
Airports—with paved runways:
total: 3
2,438 to 3,047 m: 1
1,524 to 2,437 m: 1
under 914 m: 1 (2007)
Roadways: *total:* 4,996 km
paved: 4,996 km
note: includes Gaza Strip (2004)

MILITARY

Military expenditures—percent of GDP: NA

TRANSNATIONAL ISSUES

Disputes—international: West Bank and Gaza Strip are Israeli-occupied with current status subject to the Israeli-Palestinian Interim Agreement—permanent status to be determined through further negotiation; Israel continues construction of a "seam line" separation barrier along parts of the Green Line and within the West Bank; Israel withdrew from four settlements in the northern West Bank in August 2005; since 1948, about 350 peacekeepers from the UN Truce Supervision Organization (UNTSO), headquartered in Jerusalem, monitor ceasefires, supervise armistice agreements, prevent isolated incidents from escalating, and assist other UN personnel in the region
Refugees and internally displaced persons: *refugees (country of origin):* 722,000 (Palestinian Refugees (UNRWA)) (2007)

WESTERN SAHARA

INTRODUCTION

Background: Morocco virtually annexed the northern two-thirds of Western Sahara (formerly Spanish Sahara) in 1976, and the rest of the territory in

1979, following Mauritania's withdrawal. A guerrilla war with the Polisario Front contesting Rabat's sovereignty ended in a 1991 UN-brokered cease-fire; a UN-organized referendum on final status has been repeatedly postponed. In April

2007, Morocco presented an autonomy plan for the territory to the UN, which the U.S. considers serious and credible. The Polisario also presented a plan to the UN in 2007. Since August 2007, representatives from the Government of

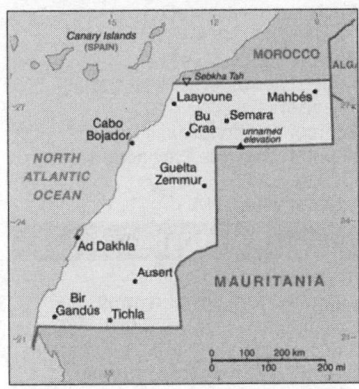

Morocco and the Polisario Front have met three times to negotiate the status of Western Sahara, with a fourth round of negotiations planned for March 2008.

GEOGRAPHY

Location: Northern Africa, bordering the North Atlantic Ocean, between Mauritania and Morocco
Geographic coordinates: 24 30 N, 13 00 W
Map references: Africa
Area:
total: 266,000 sq km
land: 266,000 sq km
water: 0 sq km
Area—comparative: about the size of Colorado
Land boundaries:
total: 2,046 km
border countries: Algeria 42 km, Mauritania 1,561 km, Morocco 443 km
Coastline: 1,110 km
Maritime claims: contingent upon resolution of sovereignty issue
Climate: hot, dry desert; rain is rare; cold offshore air currents produce fog and heavy dew
Terrain: mostly low, flat desert with large areas of rocky or sandy surfaces rising to small mountains in south and northeast
Elevation extremes:
lowest point: Sebjet Tah -55 m
highest point: unnamed elevation 805 m
Natural resources: phosphates, iron ore
Land use: *arable land:* 0.02%
permanent crops: 0%
other: 99.98% (2005)
Irrigated land: NA
Natural hazards: hot, dry, dust/sand-laden sirocco wind can occur during winter and spring; widespread harmattan haze exists 60% of time, often severely restricting visibility
Environment—current issues: sparse water and lack of arable land
Environment—international agreements: *party to:* none of the selected agreements
signed, but not ratified: none of the selected agreements

Geography—note: the waters off the coast are particularly rich fishing areas

PEOPLE

Population: 393,831
note: estimate is based on projections by age, sex, fertility, mortality, and migration; fertility and mortality are based on data from neighboring countries (July 2008 est.)
Age structure:
0–14 years: 45.1% (male 90,306/female 87,498)
15–64 years: 52.6% (male 101,730/female 105,313)
65 years and over: 2.3% (male 3,786/female 5,198) (2008 est.)
Population growth rate: 2.868% NA (2008 est.)
Birth rate: 39.95 births/1,000 population (2008 est.)
Death rate: 11.74 deaths/1,000 population (2008 est.)
Sex ratio:
NA
Infant mortality rate:
total: 71.13 deaths/1,000 live births
male: 71.22 deaths/1,000 live births
female: 71.04 deaths/1,000 live births (2008 est.)
Life expectancy at birth:
total population: 53.92 years NA
male: 51.64 years NA
female: 56.31 years NA (2008 est.)
Total fertility rate: NA 5.69 children born/woman (2008 est.)
HIV/AIDS—adult prevalence rate: NA
HIV/AIDS—people living with HIV/AIDS: NA
HIV/AIDS—deaths: NA
Nationality:
noun: Sahrawi(s), Sahraoui(s)
adjective: Sahrawi, Sahrawian, Sahraouian
Ethnic groups: Arab, Berber
Religions: Muslim
Languages: Hassaniya Arabic, Moroccan Arabic
Literacy:
NA

GOVERNMENT

Country name:
conventional long form: none
conventional short form: Western Sahara
former: Spanish Sahara
Government type: legal status of territory and issue of sovereignty unresolved; territory contested by Morocco and Polisario Front (Popular Front for the Liberation of the Saguia el Hamra and Rio de Oro), which in February 1976 formally proclaimed a government in exile of the Sahrawi Arab Democratic Republic (SADR), led by President Mohamed ABDELAZIZ; territory partitioned between Morocco and Mauritania in April 1976, with Morocco acquiring northern two-thirds; Mauritania, under pressure from Polisario guerrillas, abandoned all claims to its portion in August 1979; Morocco moved to occupy that sector shortly thereafter and has since asserted administrative control; the Polisario's government-in-exile was seated as an Organization of African Unity (OAU) member in 1984; guerrilla activities continued sporadically until a UN-monitored cease-fire was implemented on 6 September 1991 (Security Council Resolution 690) by the United Nations Mission for the Referendum in Western Sahara or MINURSO
Capital: none
time difference: UTC 0 (5 hours ahead of Washington, DC during Standard Time)
Administrative divisions: none (under de facto control of Morocco)
Suffrage: none; a UN-sponsored voter identification campaign not yet completed
Executive branch:
none
Political pressure groups and leaders:
none
International organization participation: none
Diplomatic representation in the US:
none
Diplomatic representation from the US:
none

ECONOMY

Economy—overview: Western Sahara depends on pastoral nomadism, fishing, and phosphate mining as the principal sources of income for the population. The territory lacks sufficient rainfall for sustainable agricultural production, and most of the food for the urban population must be imported. Incomes in Western Sahara are substantially below the Moroccan level. The Moroccan Government controls all trade and other economic activities in Western Sahara. Morocco and the EU signed a four-year agreement in July 2006 allowing European vessels to fish off the coast of Morocco, including the disputed waters off the coast of Western Sahara. Moroccan energy interests in 2001 signed contracts to explore for oil off the coast of Western Sahara, which has angered the Polisario. However, in 2006 the Polisario awarded similar exploration licenses in the disputed territory, which would come into force if Morocco and the Polisario resolve their dispute over Western Sahara.
GDP (purchasing power parity): $NA
GDP (official exchange rate): $NA
GDP—real growth rate: NA%
GDP—per capita (PPP): $NA
GDP—composition by sector:
agriculture: NA%
industry: NA%

services: 40%
Labor force: 12,000 (2005 est.)
Labor force—by occupation: *agriculture:* 50%
industry and services: 50% (2005 est.)
Unemployment rate: NA%
Population below poverty line: NA%
Household income or consumption by percentage share:
lowest 10%: NA%
highest 10%: NA%
Inflation rate (consumer prices): NA%
Budget:
revenues: $NA
expenditures: $NA
Agriculture—products: fruits and vegetables (grown in the few oases); camels, sheep, goats (kept by nomads); fish
Industries: phosphate mining, handicrafts
Industrial production growth rate: NA%
Electricity—production: 85 million kWh (2005)
Electricity—production by source:
fossil fuel: 100%
hydro: 0%
nuclear: 0%
other: 0% (2001)
Electricity—consumption: 79.05 million kWh (2005)
Electricity—exports: 0 kWh (2005)
Electricity—imports: 0 kWh (2005)
Oil—production: 0 bbl/day (2005 est.)
Oil—consumption: 1,750 bbl/day (2005 est.)
Oil—exports: 0 bbl/day (2004)
Oil—imports: 1,698 bbl/day (2004)
Oil—proved reserves: 0 bbl (1 January 2006 est.)
Natural gas—production: 0 cu m (2005 est.)
Natural gas—consumption: 0 cu m (2005 est.)

Natural gas—exports: 0 cu m (2005 est.)
Natural gas—imports: 0 cu m (2005)
Natural gas—proved reserves: 0 cu m (1 January 2006 est.)
Exports: $NA
Exports—commodities: phosphates 62%
Exports—partners: Morocco claims and administers Western Sahara, so trade partners are included in overall Moroccan accounts (2006)
Imports: $NA
Imports—commodities: fuel for fishing fleet, foodstuffs
Imports—partners: Morocco claims and administers Western Sahara, so trade partners are included in overall Moroccan accounts (2006)
Economic aid—recipient: $NA
Debt—external: $NA
Currency (code): Moroccan dirham (MAD)
Currency code: MAD
Exchange rates: Moroccan dirhams per US dollar—8.3563 (2007), 8.7722 (2006), 8.865 (2005), 8.868 (2004), 9.5744 (2003)
Fiscal year: calendar year

COMMUNICATIONS

Telephones—main lines in use: about 2,000 (1999 est.)
Telephones—mobile cellular: 0 (1999)
Telephone system:
general assessment: sparse and limited system
domestic: NA
international: country code—212; tied into Morocco's system by microwave radio relay, tropospheric scatter, and satellite; satellite earth stations—2 Intelsat (Atlantic Ocean) linked to Rabat, Morocco

Radio broadcast stations: AM 2, FM 0, shortwave 0 (1998)
Radios: 56,000 (1997)
Television broadcast stations: NA
Televisions: 6,000 (1997)
Internet country code: .eh
Internet Service Providers (ISPs): 1 (2000)
Internet users: NA

TRANSPORTATION

Airports: 9 (2007)
Airports—with paved runways:
total: 3
2,438 to 3,047 m: 3 (2007)
Airports—with unpaved runways:
total: 6
1,524 to 2,437 m: 1
914 to 1,523 m: 3
under 914 m: 2 (2007)
Ports and terminals: Ad Dakhla, Cabo Bojador, Laayoune (El Aaiun)

TRANSNATIONAL ISSUES

Disputes—international: Morocco claims and administers Western Sahara, whose sovereignty remains unresolved; UN-administered cease-fire has remained in effect since September 1991, administered by the UN Mission for the Referendum in Western Sahara (MINURSO), but attempts to hold a referendum have failed and parties thus far have rejected all brokered proposals; several states have extended diplomatic relations to the "Sahrawi Arab Democratic Republic" represented by the Polisario Front in exile in Algeria, while others recognize Moroccan sovereignty over Western Sahara; most of the approximately 102,000 Sahrawi refugees are sheltered in camps in Tindouf, Algeria

WORLD

INTRODUCTION

Background: Globally, the 20th century was marked by: (a) two devastating world wars; (b) the Great Depression of the 1930s; (c) the end of vast colonial empires; (d) rapid advances in science and technology, from the first airplane flight at Kitty Hawk, North Carolina (US) to the landing on the moon; (e) the Cold War between the Western alliance and the Warsaw Pact nations; (f) a sharp rise in living standards in North America, Europe, and Japan; (g) increased concerns about the environment, including loss of forests, shortages of energy and water, the decline in biological diversity, and air pollution; (h) the onset of the AIDS epidemic; and (i)

the ultimate emergence of the US as the only world superpower. The planet's population continues to explode: from 1 billion in 1820, to 2 billion in 1930, 3 billion in 1960, 4 billion in 1974, 5 billion in 1988, and 6 billion in 2000. For the 21st century, the continued exponential growth in science and technology raises both hopes (e.g., advances in medicine) and fears (e.g., development of even more lethal weapons of war).

GEOGRAPHY

Geographic overview: The surface of the earth is approximately 70.9% water and 29.1% land. The former portion is divided into large water bodies termed oceans. The World Factbook recognizes and describes five oceans, which are in

decreasing order of size: the Pacific Ocean, Atlantic Ocean, Indian Ocean, Southern Ocean, and Arctic Ocean. The land portion is generally divided into several, large, discrete landmasses termed continents. Depending on the convention used, the number of continents can vary from five to seven. The most common classification recognizes seven, which are (from largest to smallest): Asia, Africa, North America, South America, Antarctica, Europe, and Australia. Asia and Europe are sometimes lumped together into a Eurasian continent resulting in six continents. Alternatively, North and South America are sometimes grouped as simply the Americas, resulting in a continent total of six (or five, if the Eurasia designation

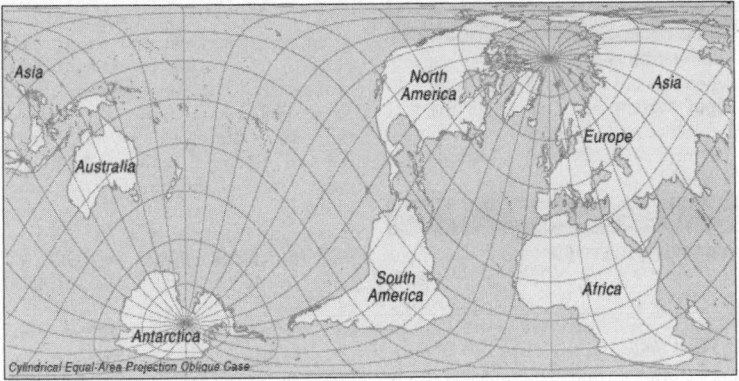

Cylindrical Equal-Area Projection Oblique Case

is used). North America is commonly understood to include the island of Greenland, the isles of the Caribbean, and to extend south all the way to the Isthmus of Panama. The easternmost extent of Europe is generally defined as being the Ural Mountains and the Ural River; on the southeast the Caspian Sea; and on the south the Caucasus Mountains, the Black Sea, and the Mediterranean. Africa's northeast extremity is frequently delimited at the Isthmus of Suez, but for geopolitical purposes, the Egyptian Sinai Peninsula is often included as part Africa. Asia usually incorporates all the islands of the Philippines, Malaysia, and Indonesia. The islands of the Pacific are often lumped with Australia into a "land mass" termed Oceania or Australasia. Although the above groupings are the most common, different continental dispositions are recognized or taught in certain parts of the world, with some arrangements more heavily based on cultural spheres rather than physical geographic considerations.

Map references: Physical Map of the World, Political Map of the World, Standard Time Zones of the World

Area:
total: 510.072 million sq km
land: 148.94 million sq km
water: 361.132 million sq km
note: 70.9% of the world's surface is water, 29.1% is land

Area—comparative: land area about 16 times the size of the US

Land boundaries: the land boundaries in the world total 251,060 km (not counting shared boundaries twice); two nations, China and Russia, each border 14 other countries

note: 45 nations and other areas are landlocked, these include: Afghanistan, Andorra, Armenia, Austria, Azerbaijan, Belarus, Bhutan, Bolivia, Botswana, Burkina Faso, Burundi, Central African Republic, Chad, Czech Republic, Ethiopia, Holy See (Vatican City), Hungary, Kazakhstan, Kosovo, Kyrgyzstan, Laos, Lesotho, Liechtenstein, Luxembourg, Macedonia, Malawi, Mali, Moldova, Mongolia, Nepal, Niger, Paraguay, Rwanda, San Marino, Serbia, Slovakia, Swaziland, Switzerland, Tajikistan, Turkmenistan, Uganda, Uzbekistan, West Bank, Zambia, Zimbabwe; two of these, Liechtenstein and Uzbekistan, are doubly landlocked

Coastline: 356,000 km

note: 94 nations and other entities are islands that border no other countries, they include: American Samoa, Anguilla, Antigua and Barbuda, Aruba, Ashmore and Cartier Islands, The Bahamas, Bahrain, Baker Island, Barbados, Bermuda, Bouvet Island, British Indian Ocean Territory, British Virgin Islands, Cape Verde, Cayman Islands, Christmas Island, Clipperton Island, Cocos (Keeling) Islands, Comoros, Cook Islands, Coral Sea Islands, Cuba, Cyprus, Dominica, Falkland Islands (Islas Malvinas), Faroe Islands, Fiji, French Polynesia, French Southern and Antarctic Lands, Greenland, Grenada, Guam, Guernsey, Heard Island and McDonald Islands, Howland Island, Iceland, Isle of Man, Jamaica, Jan Mayen, Japan, Jarvis Island, Jersey, Johnston Atoll, Kingman Reef, Kiribati, Madagascar, Maldives, Malta, Marshall Islands, Martinique, Mauritius, Mayotte, Federated States of Micronesia, Midway Islands, Montserrat, Nauru, Navassa Island, New Caledonia, New Zealand, Niue, Norfolk Island, Northern Mariana Islands, Palau, Palmyra Atoll, Paracel Islands, Philippines, Pitcairn Islands, Puerto Rico, Reunion, Saint Barthelemy, Saint Helena, Saint Kitts and Nevis, Saint Lucia, Saint Pierre and Miquelon, Saint Vincent and the Grenadines, Samoa, Sao Tome and Principe, Seychelles, Singapore, Solomon Islands, South Georgia and the South Sandwich Islands, Spratly Islands, Sri Lanka, Svalbard, Tokelau, Tonga, Trinidad and Tobago, Turks and Caicos Islands, Tuvalu, Vanuatu, Virgin Islands, Wake Island, Wallis and Futuna, Taiwan

Maritime claims: a variety of situations exist, but in general, most countries make the following claims measured from the mean low-tide baseline as described in the 1982 UN Convention on the Law of the Sea: territorial sea—12 nm, contiguous zone—24 nm, and exclusive economic zone—200 nm; additional zones provide for exploitation of continental shelf resources and an exclusive fishing zone; boundary situations with neighboring states prevent many countries from extending their fishing or economic zones to a full 200 nm

Climate: a wide equatorial band of hot and humid tropical climates—bordered north and south by subtropical temperate zones—that separate two large areas of cold and dry polar climates

Terrain: the greatest ocean depth is the Mariana Trench at 10,924 m in the Pacific Ocean

Elevation extremes:
lowest point: Bentley Subglacial Trench -2,540 m
note: in the oceanic realm, Challenger Deep in the Mariana Trench is the lowest point, lying -10,924 m below the surface of the Pacific Ocean
highest point: Mount Everest 8,850 m

Natural resources: the rapid depletion of nonrenewable mineral resources, the depletion of forest areas and wetlands, the extinction of animal and plant species, and the deterioration in air and water quality (especially in Eastern Europe, the former USSR, and China) pose serious long-term problems that governments and peoples are only beginning to address

Land use: arable land: 13.31%
permanent crops: 4.71%
other: 81.98% (2005)

Irrigated land: 2,770,980 sq km (2003)

Natural hazards: large areas subject to severe weather (tropical cyclones), natural disasters (earthquakes, landslides, tsunamis, volcanic eruptions)

Environment—current issues: large areas subject to overpopulation, industrial disasters, pollution (air, water, acid rain, toxic substances), loss of vegetation (overgrazing, deforestation, desertification), loss of wildlife, soil degradation, soil depletion, erosion; global warming becoming a greater concern

Geography—note: the world is now thought to be about 4.55 billion years old, just about one-third of the 13.7-billion-year age estimated for the universe

PEOPLE

Population: 6,677,563,921 (July 2008 est.)

Age structure:
0-14 years: 27% (male 933,716,943/ female 877,734,429)

15–64 years: 65% (male 2,205,342,972/female 2,153,959,605)
65 years and over: 8% (male 222,346,221/female 284,463,751) (2008 est.)

Median age: *total:*
male: 27.4 years
female: 28.7 years (2008 est.)

Population growth rate: 1.159% (2008 est.)

Birth rate: 19.97 births/1,000 population (2008 est.)

Death rate: 8.32 deaths/1,000 population (2008 est.)

Sex ratio: *at birth:* 1.07 male(s)/female
under 15 years: 1.06 male(s)/female
15–64 years: 1.02 male(s)/female
65 years and over: 0.78 male(s)/female
total population: 1.01 male(s)/female (2008 est.)

Infant mortality rate:
total: 42.64 deaths/1,000 live births
male: 45.42 deaths/1,000 live births
female: 39.67 deaths/1,000 live births (2008 est.)

Life expectancy at birth:
total population: 66.12 years
male: 64.18 years
female: 68.2 years (2008 est.)

Total fertility rate: 2.58 children born/woman (2008 est.)

HIV/AIDS—adult prevalence rate: NA
HIV/AIDS—people living with HIV/AIDS: NA
HIV/AIDS—deaths: NA

Religions: Christians 33.32% (of which Roman Catholics 16.99%, Protestants 5.78%, Orthodox 3.53%, Anglicans 1.25%), Muslims 21.01%, Hindus 13.26%, Buddhists 5.84%, Sikhs 0.35%, Jews 0.23%, Baha'is 0.12%, other religions 11.78%, non-religious 11.77%, atheists 2.32% (2007 est.)

Languages: Mandarin Chinese 13.22%, Spanish 4.88%, English 4.68%, Arabic 3.12%, Hindi 2.74%, Portuguese 2.69%, Bengali 2.59%, Russian 2.2%, Japanese 1.85%, Standard German 1.44%, Wu Chinese 1.17% (2005 est.)
note: percents are for "first language" speakers only

Literacy: *definition:* age 15 and over can read and write
total population: 82%
male: 87%
female: 77%
note: over two-thirds of the world's 785 million illiterate adults are found in only eight countries (India, China, Bangladesh, Pakistan, Nigeria, Ethiopia, Indonesia, and Egypt); of all the illiterate adults in the world, two-thirds are women; extremely low literacy rates are concentrated in three regions, South and West Asia, Sub-Saharan Africa, and the Arab states, where around one-third of the men and half of all women are illiterate (2005 est.)

GOVERNMENT

Administrative divisions: 266 nations, dependent areas, and other entities
Legal system: all members of the UN are parties to the statute that established the International Court of Justice (ICJ) or World Court

ECONOMY

Economy—overview: Global output rose by 5.2% in 2007, led by China (11.4%), India (9.2%), and Russia (8.1%). The 14 other successor nations of the USSR and the other old Warsaw Pact nations again experienced widely divergent growth rates; the three Baltic nations continued as strong performers, in the 8%-10% range of growth. From 2006 to 2007 growth rates slowed in all the major industrial countries except for the United Kingdom (3.0%). Analysts attribute the slowdown to uncertainties in the financial markets and lowered consumer confidence. Worldwide, nations varied widely in their growth results. Externally, the nation-state, as a bedrock economic-political institution, is steadily losing control over international flows of people, goods, funds, and technology. Internally, the central government often finds its control over resources slipping as separatist regional movements—typically based on ethnicity—gain momentum, e.g., in many of the successor states of the former Soviet Union, in the former Yugoslavia, in India, in Iraq, in Indonesia, and in Canada. Externally, the central government is losing decisionmaking powers to international bodies, notably the EU. In Western Europe, governments face the difficult political problem of channeling resources away from welfare programs in order to increase investment and strengthen incentives to seek employment. The addition of 80 million people each year to an already overcrowded globe is exacerbating the problems of pollution, desertification, underemployment, epidemics, and famine. Because of their own internal problems and priorities, the industrialized countries devote insufficient resources to deal effectively with the poorer areas of the world, which, at least from an economic point of view, are becoming further marginalized. The introduction of the euro as the common currency of much of Western Europe in January 1999, while paving the way for an integrated economic powerhouse, poses economic risks because of varying levels of income and cultural and political differences among the participating nations. The terrorist attacks on the US on 11 September 2001 accentuated a growing risk to global prosperity, illustrated, for example, by the reallocation of resources away from investment to anti-terrorist programs. The opening of war in March 2003 between a US-led coalition and Iraq added new uncertainties to global economic prospects. After the initial coalition victory, the complex political difficulties and the high economic cost of establishing domestic order in Iraq became major global problems that continued through 2007.

GDP (purchasing power parity): GWP (gross world product): $65.61 trillion (2007 est.)

GDP (official exchange rate): GWP (gross world product): $54.62 trillion (2007 est.)

GDP—real growth rate: 5.2% (2007 est.)

GDP—per capita (PPP): $10,000 (2007 est.)

GDP—composition by sector:
agriculture: 4%
industry: 32%
services: 64% (2007 est.)

Labor force: 3.131 billion (2007 est.)

Labor force—by occupation: *agriculture:* 40.2%
industry: 20.5%
services: 39.3% (2007 est.)

Unemployment rate: 30% combined unemployment and underemployment in many non-industrialized countries; developed countries typically 4%-12% unemployment (2007 est.)

Household income or consumption by percentage share:
lowest 10%: 2.5%
highest 10%: 29.8% (2002 est.)

Inflation rate (consumer prices): developed countries 1% to 4% typically; developing countries 5% to 20% typically; national inflation rates vary widely in individual cases, from declining prices in Japan to hyperinflation in one Third World country (Zimbabwe); inflation rates have declined for most countries for the last several years, held in check by increasing international competition from several low wage countries (2005 est.)

Investment (gross fixed): 22.7% of GDP (2007 est.)

Industries: dominated by the onrush of technology, especially in computers, robotics, telecommunications, and medicines and medical equipment; most of these advances take place in OECD nations; only a small portion of non-OECD countries have succeeded in rapidly adjusting to these technological forces; the accelerated development of new industrial (and agricultural) technology is complicating already grim environmental problems

Industrial production growth rate: 5% (2007 est.)

Electricity—production: 18.58 trillion kWh (2005 est.)

Electricity—production by source:
fossil fuel: NA
hydro: NA
nuclear: NA
other: NA
Electricity—consumption: 16.83 trillion kWh (2005 est.)
Electricity—exports: 634.8 billion kWh (2005)
Electricity—imports: 620.5 billion kWh (2005)
Oil—production: 78.9 million bbl/day (2005 est.)
Oil—consumption: 80.29 million bbl/day (2005 est.)
Oil—exports: 63.76 million bbl/day (2004)
Oil—imports: 63.18 million bbl/day (2004)
Oil—proved reserves: 1.331 trillion bbl (1 January 2006 est.)
Natural gas—production: 2.854 trillion cu m (2005 est.)
Natural gas—consumption: 3 trillion cu m (2005 est.)
Natural gas—exports: 808 billion cu m (2005 est.)
Natural gas—imports: 786.5 billion cu m (2005)
Natural gas—proved reserves: 172 trillion cu m (1 January 2006 est.)
Exports: $14.01 trillion f.o.b. (2006 est.)
Exports—commodities: the whole range of industrial and agricultural goods and services
top ten—share of world trade: electrical machinery, including computers 14.8%; mineral fuels, including oil, coal, gas, and refined products 14.4%; nuclear reactors, boilers, and parts 14.2%; cars, trucks, and buses 8.9%; scientific and precision instruments 3.5%; plastics 3.4%; iron and steel 2.7%; organic chemicals 2.6%; pharmaceutical products 2.6%; diamonds, pearls, and precious stones 1.9% (2006 est.)
Exports—partners: US 15%, Germany 7.4%, China 5.9%, France 4.6%, UK 4.5%, Japan 4.4% (2006)
Imports: $13.91 trillion f.o.b. (2006 est.)
Imports—commodities: the whole range of industrial and agricultural goods and services
top ten—share of world trade: see listing for exports
Imports—partners: China 9.8%, Germany 8.8%, US 8.5%, Japan 5.6%, France 4% (2006)
Economic aid—recipient: ODA, $106.4 billion (2005)
Debt—external: $53.97 trillion
note: this figure is the sum total of all countries' external debt, both public and private (2004 est.)
Stock of direct foreign investment—at home: World total DFI $14 trillion
top ten recipients of DFI: US $1.966 trillion; UK $1.324 trillion; France $872.4

billion; Germany $811.0 billion; HK $780.4 billion; China $758.9 billion; Belgium $703.9 billion; Netherlands $535.1 billion; Canada $527.4 billion; Spain $487.8 billion (year-end 2007 est.)
Stock of direct foreign investment—abroad: World total DFI $14 trillion
top ten sources of DFI: US $2.627 trillion; UK $1.741 trillion; France $1.211 trillion; Germany $1.123 trillion; Netherlands $811.4 billion; HK $716.2 billion; Spain $613.9 billion; Switzerland $591.5 billion; Belgium $537.6 billion; Japan $527.8 billion (year-end 2007 est.)
Market value of publicly traded shares: $43.64 trillion (2005 est.)

COMMUNICATIONS

Telephones—main lines in use: 1,263,367,600 (2005)
Telephones—mobile cellular: 2,168,433,600 (2005)
Telephone system:
general assessment: NA
domestic: NA
international: NA
Radio broadcast stations: AM NA, FM NA, shortwave NA
Radios: NA
Television broadcast stations: NA
Televisions: NA
Internet Service Providers (ISPs): 10,350 (2000 est.)
Internet users: 1,018,057,389 (2005)

TRANSPORTATION

Airports: total airports—49,024
top ten by passengers: Atlanta—84,846,639; Chicago—77,028,134; London—67,530,197; Tokyo—65,810,672; Los Angeles—61,041,066; Dallas/Fort Worth—60,226,138; Paris—56,849,567; Frankfurt—52,810,683; Beijing—48,654,770; Denver—47,325,016
top ten by cargo (metric tons): Memphis—3,692,081; Hong Kong—3,609,780; Anchorage—2,691,395; Seoul—2,336,572; Tokyo—2,280,830; Shanghai—2,168,122; Paris—2,130,724; Frankfurt—2,127,646; Louisville (US)—1,983,032; Singapore—1,931,881 (2006)
Heliports: 1,359 (2007)
Railways: *total:* 1,370,782 km (2006)
Roadways:
total: 32,345,165 km (2002)
Waterways: 671,886 km (2004)
Ports and terminals: *top ten container ports (TEUs):* Singapore—24,792,400; Hong Kong—23,539,000; Shanghai 21,710,000; Shenzhen (China)—18,468,890; Busan (South Korea)—12,030,000; Kaohsiung (Taiwan)—9,774,670;—Rotterdam—9,603,000; Dubai (UAE)—8,923,465; Hamburg—8,861,545; Los Angeles—8,469,853 (2006)

MILITARY

Military expenditures—percent of GDP: roughly 2% of gross world product (2005 est.)

TRANSNATIONAL ISSUES

Disputes—international: stretching over 250,000 km, the world's 322 international land boundaries separate 194 independent states and 70 dependencies, areas of special sovereignty, and other miscellaneous entities; ethnicity, culture, race, religion, and language have divided states into separate political entities as much as history, physical terrain, political fiat, or conquest, resulting in sometimes arbitrary and imposed boundaries; most maritime states have claimed limits that include territorial seas and exclusive economic zones; overlapping limits due to adjacent or opposite coasts create the potential for 430 bilateral maritime boundaries of which 209 have agreements that include contiguous and non-contiguous segments; boundary, borderland/resource, and territorial disputes vary in intensity from managed or dormant to violent or militarized; undemarcated, indefinite, porous, and unmanaged boundaries tend to encourage illegal cross-border activities, uncontrolled migration, and confrontation; territorial disputes may evolve from historical and/or cultural claims, or they may be brought on by resource competition; ethnic and cultural clashes continue to be responsible for much of the territorial fragmentation and internal displacement of the estimated 6.6 million people and cross-border displacements of 8.6 million refugees around the world as of early 2006; just over one million refugees were repatriated in the same period; other sources of contention include access to water and mineral (especially hydrocarbon) resources, fisheries, and arable land; armed conflict prevails not so much between the uniformed armed forces of independent states as between stateless armed entities that detract from the sustenance and welfare of local populations, leaving the community of nations to cope with resultant refugees, hunger, disease, impoverishment, and environmental degradation
Refugees and internally displaced persons: the United Nations High Commissioner for Refugees (UNHCR) estimated that in December 2006 there was a global population of 8.8 million registered refugees and as many as 24.5 million IDPs in more than 50 countries; the actual global population of refugees is probably closer to 10 million given the estimated 1.5 million Iraqi refugees displaced throughout the Middle East (2007)

Trafficking in persons: *current situation:* about 600,000 to 800,000 people, mostly women and children, are trafficked annually across national borders, not including millions trafficked within their own countries; at least 80% of the victims are female; 75% of all victims are trafficked into commercial sexual exploitation; roughly two-thirds of the global victims are trafficked intra-regionally within East Asia and the Pacific (260,000 to 280,000 people) and Europe and Eurasia (170,000 to 210,000 people) *Tier 2 Watch List:* Argentina, Armenia, Belarus, Burundi, Cambodia, Central African Republic, Chad, China, Cyprus, Djibouti, Dominican Republic, Egypt, Fiji, The Gambia, Guatemala, Guyana, Honduras, India, Kazakhstan, Kenya, Libya, Macau, Mauritania, Mexico, Moldova, Mozambique, Papua New Guinea, Russia, South Africa, Sri Lanka, Ukraine, United Arab Emirates *Tier 3:* Algeria, Bahrain, Burma, Cuba, Equatorial Guinea, Iran, Kuwait, Malaysia, North Korea, Oman, Qatar, Saudi Arabia, Sudan, Syria, Uzbekistan, Venezuela

Illicit drugs: *cocaine:* worldwide coca leaf cultivation in 2005 amounted to 208,500 hectares; Colombia produced slightly more than two-thirds of the worldwide crop, followed by Peru and Bolivia; potential pure cocaine production rose to 900 from 645 metric tons in 2005—partially due to improved methodologies used to calculate levels of production; Colombia conducts aggressive coca eradication campaign, but both Peruvian and Bolivian Governments are hesitant to eradicate coca in key growing areas; 551 metric tons of export-quality cocaine (85% pure) is documented to have been seized or destroyed in 2005; US consumption of export quality cocaine is estimated to have been in excess of 380 metric tons *opiates:* worldwide illicit opium poppy cultivation reached 208,500 hectares in 2005; potential opium production of 4,990 metric tons was only a 9% decrease over 2004's highest total recorded since estimates began in mid-1980s; Afghanistan is world's primary opium producer, accounting for 90% of the global supply; Southeast Asia—responsible for 9% of global opium—saw marginal increases in production; Latin America produced 1% of global opium, but most was refined into heroin destined for the US market; if all potential opium was processed into pure heroin, the potential global production would be 577 metric tons of heroin in 2005

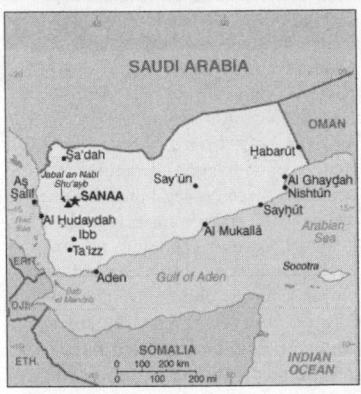

INTRODUCTION

Background: North Yemen became independent of the Ottoman Empire in 1918. The British, who had set up a protectorate area around the southern port of Aden in the 19th century, withdrew in 1967 from what became South Yemen. Three years later, the southern government adopted a Marxist orientation. The massive exodus of hundreds of thousands of Yemenis from the south to the north contributed to two decades of hostility between the states. The two countries were formally unified as the Republic of Yemen in 1990. A southern secessionist movement in 1994 was quickly subdued. In 2000, Saudi Arabia and Yemen agreed to a delimitation of their border.

GEOGRAPHY

Location: Middle East, bordering the Arabian Sea, Gulf of Aden, and Red Sea, between Oman and Saudi Arabia
Geographic coordinates: 15 00 N, 48 00 E
Map references: Middle East
Area:
total: 527,970 sq km
land: 527,970 sq km
water: 0 sq km
note: includes Perim, Socotra, the former Yemen Arab Republic (YAR or North Yemen), and the former People's Democratic Republic of Yemen (PDRY or South Yemen)
Area—comparative: slightly larger than twice the size of Wyoming
Land boundaries:
total: 1,746 km
border countries: Oman 288 km, Saudi Arabia 1,458 km
Coastline: 1,906 km
Maritime claims:
territorial sea: 12 nm
contiguous zone: 24 nm

exclusive economic zone: 200 nm
continental shelf: 200 nm or to the edge of the continental margin
Climate: mostly desert; hot and humid along west coast; temperate in western mountains affected by seasonal monsoon; extraordinarily hot, dry, harsh desert in east
Terrain: narrow coastal plain backed by flat-topped hills and rugged mountains; dissected upland desert plains in center slope into the desert interior of the Arabian Peninsula
Elevation extremes:
lowest point: Arabian Sea 0 m
highest point: Jabal an Nabi Shu'ayb 3,760 m
Natural resources: petroleum, fish, rock salt, marble; small deposits of coal, gold, lead, nickel, and copper; fertile soil in west
Land use:
arable land: 2.91%
permanent crops: 0.25%
other: 96.84% (2005)
Irrigated land: 5,500 sq km (2003)
Total renewable water resources: 4.1 cu km (1997)
Freshwater withdrawal (domestic/industrial/agricultural): *total:* 6.63 cu km/yr (4%/1%/95%)
per capita: 316 cu m/yr (2000)
Natural hazards: sandstorms and dust storms in summer
Environment—current issues: limited natural fresh water resources; inadequate supplies of potable water; overgrazing; soil erosion; desertification
Environment—international agreements: *party to:* Biodiversity, Climate Change, Climate Change-Kyoto Protocol, Desertification, Endangered Species, Environmental Modification, Hazardous Wastes, Law of the Sea, Ozone Layer Protection
signed, but not ratified: none of the selected agreements
Geography—note: strategic location on Bab el Mandeb, the strait linking the Red Sea and the Gulf of Aden, one of world's most active shipping lanes

PEOPLE

Population: 23,013,376 (July 2008 est.)
Age structure:
0–14 years: 46.2% (male 5,415,385/female 5,218,237)
15–64 years: 51.2% (male 5,996,202/female 5,795,779)
65 years and over: 2.6% (male 284,195/female 303,578) (2008 est.)
Median age: *total:* 16.7 years

male: 16.7 years
female: 16.8 years (2008 est.)
Population growth rate: 3.46% (2008 est.)
Birth rate: 42.42 births/1,000 population (2008 est.)
Death rate: 7.83 deaths/1,000 population (2008 est.)
Net migration rate: NA
Sex ratio:
at birth: 1.05 male(s)/female
under 15 years: 1.04 male(s)/female
15–64 years: 1.03 male(s)/female
65 years and over: 0.94 male(s)/female
total population: 1.03 male(s)/female (2008 est.)
Infant mortality rate:
total: 56.27 deaths/1,000 live births
male: 60.78 deaths/1,000 live births
female: 51.54 deaths/1,000 live births (2008 est.)
Life expectancy at birth:
total population: 62.9 years
male: 60.96 years
female: 64.94 years (2008 est.)
Total fertility rate: 6.41 children born/woman (2008 est.)
HIV/AIDS—adult prevalence rate: 0.1% (2001 est.)
HIV/AIDS—people living with HIV/AIDS: 12,000 (2001 est.)
HIV/AIDS—deaths: NA
Major infectious diseases:
degree of risk: high
food or waterborne diseases: bacterial and protozoal diarrhea, hepatitis A, and typhoid fever
vectorborne diseases: dengue fever and malaria
water contact disease: schistosomiasis (2008)
Nationality:
noun: Yemeni(s)
adjective: Yemeni
Ethnic groups: predominantly Arab; but also Afro-Arab, South Asians, Europeans
Religions: Muslim including Shaf'i (Sunni) and Zaydi (Shi'a), small numbers of Jewish, Christian, and Hindu
Languages: Arabic
Literacy:
definition: age 15 and over can read and write
total population: 50.2%
male: 70.5%
female: 30% (2003 est.)

GOVERNMENT

Country name:
conventional long form: Republic of Yemen

conventional short form: Yemen
local long form: Al Jumhuriyah al Yamaniyah
local short form: Al Yaman
former: Yemen Arab Republic [Yemen (Sanaa) or North Yemen] and People's Democratic Republic of Yemen [Yemen (Aden) or South Yemen]
Government type: republic
Capital: *name:* Sanaa
geographic coordinates: 15 21 N, 44 12 E
time difference: UTC+3 (8 hours ahead of Washington, DC during Standard Time)
Administrative divisions: 19 governorates (muhafazat, singular—muhafazah); Abyan, 'Adan, Ad Dali', Al Bayda', Al Hudaydah, Al Jawf, Al Mahrah, Al Mahwit, 'Amran, Dhamar, Hadramawt, Hajjah, Ibb, Lahij, Ma'rib, Sa'dah, San'a', Shabwah, Ta'izz
note: for electoral and administrative purposes, the capital city of Sanaa is treated as an additional governorate
Independence: 22 May 1990 (Republic of Yemen was established with the merger of the Yemen Arab Republic [Yemen (Sanaa) or North Yemen] and the Marxist-dominated People's Democratic Republic of Yemen [Yemen (Aden) or South Yemen]); note—previously North Yemen became independent in November 1918 (from the Ottoman Empire) and became a republic with the overthrow of the theocratic Imamate in 1962; South Yemen became independent on 30 November 1967 (from the UK)
National holiday: Unification Day, 22 May (1990)
Constitution: 16 May 1991; amended 29 September 1994 and February 2001
Legal system: based on Islamic law, Turkish law, English common law, and local tribal customary law; has not accepted compulsory ICJ jurisdiction
Suffrage: 18 years of age; universal
Executive branch:
chief of state: President Ali Abdallah SALIH (since 22 May 1990, the former president of North Yemen, assumed office upon the merger of North and South Yemen); Vice President Maj. Gen. Abd al-Rab Mansur al-HADI (since 3 October 1994)
head of government: Prime Minister Ali Muhammad MUJAWWAR (since 31 March 2007)
cabinet: Council of Ministers appointed by the president on the advice of the prime minister
elections: president elected by popular vote for a seven-year term; election last held 20 September 2006 (next to be held in September 2013); vice president appointed by the president; prime minister and deputy prime ministers

appointed by the president
election results: Ali Abdallah SALIH elected president; percent of vote—Ali Abdallah SALIH 77.2%, Faysal BIN SHAMLAN 21.8%
Legislative branch: a bicameral legislature consisting of a Shura Council (111 seats; members appointed by the president) and a House of Representatives (301 seats; members elected by popular vote to serve six-year terms)
elections: last held on 27 April 2003 (next to be held in April 2009)
election results: percent of vote by party—NA; seats by party—GPC 228, Islah 47, YSP 7, Nasserite Unionist Party 3, National Arab Socialist Ba'th Party 2, independents 14
Judicial branch: Supreme Court
Political parties and leaders: General People's Congress or GPC [Abdul-Kader BAJAMMAL]; Islamic Reform Grouping or Islah [Mohammed Abdullah AL-YADOUMI (acting)]; Nasserite Unionist Party [Abdal Malik al-MAKHLAFI]; National Arab Socialist Ba'th Party [Dr. Qasim SALAM]; Yemeni Socialist Party or YSP [Ali Salih MUQBIL]; note—there are at least seven more active political parties
Political pressure groups and leaders: NA
International organization participation: AFESD, AMF, CAEU, FAO, G-77, IAEA, IBRD, ICAO, ICCt (signatory), ICRM, IDA, IDB, IFAD, IFC, IFRCS, ILO, IMF, IMO, Interpol, IOC, IOM, IPU, ISO (correspondent), ITSO, ITU, ITUC, LAS, MIGA, MINURSO, NAM, OAS (observer), OIC, OPCW, UN, UNCTAD, UNESCO, UNHCR, UNIDO, UNMIS, UNOCI, UNOMIG, UNWTO, UPU, WCO, WFTU, WHO, WIPO, WMO, WTO (observer)
Diplomatic representation in the US:
chief of mission: Ambassador Abd al-Wahab Abdallah al-HAJRI
chancery: 2319 Wyoming Avenue NW, Washington, DC 20008
telephone: [1] (202) 965-4760
FAX: [1] (202) 337-2017
Diplomatic representation from the US:
chief of mission: Ambassador Stephen A. SECHE
embassy: Sa'awan Street, Sanaa
mailing address: P. O. Box 22347, Sanaa
telephone: [967] (1) 755-2000 ext. 2153 or 2266
FAX: [967] (1) 303-182
Flag description: three equal horizontal bands of red (top), white, and black; similar to the flag of Syria, which has two green stars in the white band, and of Iraq, which has an Arabic inscription centered in the white band; also similar

to the flag of Egypt, which has a heraldic eagle centered in the white band

ECONOMY

Economy—overview: Yemen, one of the poorest countries in the Arab world, reported average annual growth in the range of 3–4% from 2000 through 2007. Its economic fortunes depend mostly on declining oil resources, but the country is trying to diversify its earnings. In 2006 Yemen began an economic reform program designed to bolster non-oil sectors of the economy and foreign investment. As a result of the program, international donors pledged about $5 billion for development projects. In addition, Yemen has made some progress on reforms over the last year that will likely encourage foreign investment. Oil revenues probably increased in 2007 as a result of higher prices.
GDP (purchasing power parity): $52.05 billion (2007 est.)
GDP (official exchange rate): $21.66 billion (2007 est.)
GDP—real growth rate: 3.1% (2007 est.)
GDP—per capita (PPP): $2,300 (2007 est.)
GDP—composition by sector:
agriculture: 12.3%
industry: 41.5%
services: 46.2% (2007 est.)
Labor force: 6.316 million (2007 est.)
Labor force—by occupation: *note:* most people are employed in agriculture and herding; services, construction, industry, and commerce account for less than one-fourth of the labor force
Unemployment rate: 35% (2003 est.)
Population below poverty line: 45.2% (2003)
Household income or consumption by percentage share:
lowest 10%: 3%
highest 10%: 25.9% (2003)
Distribution of family income—Gini index: 33.4 (1998)
Inflation rate (consumer prices): 12.5% (2007 est.)
Investment (gross fixed): 25.2% of GDP (2007 est.)
Budget:
revenues: $7.407 billion
expenditures: $8.177 billion (2007 est.)
Public debt: 33.7% of GDP (2007 est.)
Agriculture—products: grain, fruits, vegetables, pulses, qat, coffee, cotton; dairy products, livestock (sheep, goats, cattle, camels), poultry; fish
Industries: crude oil production and petroleum refining; small-scale production of cotton textiles and leather goods; food processing; handicrafts; small alu-

minum products factory; cement; commercial ship repair

Industrial production growth rate: 3.2% (2007 est.)

Electricity—production: 4.456 billion kWh (2005 est.)

Electricity—production by source:
fossil fuel: 100%
hydro: 0%
nuclear: 0%
other: 0% (2001)

Electricity—consumption: 3.381 billion kWh (2005 est.)

Electricity—exports: 0 kWh (2005)

Electricity—imports: 0 kWh (2005)

Oil—production: 402,000 bbl/day (2005 est.)

Oil—consumption: 128,000 bbl/day (2005 est.)

Oil—exports: 320,600 bbl/day (2004)

Oil—imports: 58,100 bbl/day (2004)

Oil—proved reserves: 3.58 billion bbl (2007 est.)

Natural gas—production: 0 cu m (2005 est.)

Natural gas—consumption: 0 cu m (2005 est.)

Natural gas—exports: 0 cu m (2005 est.)

Natural gas—imports: 0 cu m (2005)

Natural gas—proved reserves: 459 billion cu m (1 January 2006 est.)

Current account balance: -$924 million (2007 est.)

Exports: $7.102 billion f.o.b. (2007 est.)

Exports—commodities: crude oil, coffee, dried and salted fish

Exports—partners: China 31.5%, India 17.5%, Thailand 16.7%, South Korea 7%, US 6.8%, UAE 4.1% (2006)

Imports: $6.735 billion f.o.b. (2007 est.)

Imports—commodities: food and live animals, machinery and equipment, chemicals

Imports—partners: UAE 16.4%, China 12.7%, Saudi Arabia 7.7%, Kuwait 5.8%, Brazil 4.5%, Malaysia 4.1%, US 4% (2006)

Economic aid—recipient: $2.3 billion (2003–07 disbursements)

Reserves of foreign exchange and gold: $7.76 billion (31 December 2007 est.)

Debt—external: $6.044 billion (31 December 2007 est.)

Market value of publicly traded shares: $NA

Currency (code): Yemeni rial (YER)

Currency code: YER

Exchange rates: Yemeni rials per US dollar—199.14 (2007), 197.18 (2006), 192.67 (2005), 184.78 (2004), 183.45 (2003)

Fiscal year: calendar year

COMMUNICATIONS

Telephones—main lines in use: 968,400 (2006)

Telephones—mobile cellular: 2 million (2006)

Telephone system:
general assessment: since unification in 1990, efforts have been made to create a national telecommunications network
domestic: the national network consists of microwave radio relay, cable, tropospheric scatter, and GSM mobile-cellular telephone systems; fixed-line and mobile-cellular teledensity remains low by regional standards
international: country code—967; landing point for the international submarine cable Fiber-Optic Link Around the Globe (FLAG); satellite earth stations—3 Intelsat (2 Indian Ocean and 1 Atlantic Ocean), 1 Intersputnik (Atlantic Ocean region), and 2 Arabsat; microwave radio relay to Saudi Arabia and Djibouti

Radio broadcast stations: AM 6, FM 1, shortwave 2 (1998)

Radios: 1.05 million (1997)

Television broadcast stations: 3 (including one Egypt-based station that broadcasts in Yemen); plus several repeaters (2007)

Televisions: 470,000 (1997)

Internet country code: .ye

Internet hosts: 5 (2007)

Internet Service Providers (ISPs): 1 (2000)

Internet users: 270,000 (2006)

TRANSPORTATION

Airports: 50 (2007)

Airports—with paved runways:
total: 17
over 3,047 m: 4
2,438 to 3,047 m: 8
1,524 to 2,437 m: 3
914 to 1,523 m: 1
under 914 m: 1 (2007)

Airports—with unpaved runways:
total: 33

over 3,047 m: 3
2,438 to 3,047 m: 8
1,524 to 2,437 m: 5
914 to 1,523 m: 13
under 914 m: 4 (2007)

Pipelines: gas 71 km; liquid petroleum gas 22 km; oil 1,309 km (2007)

Roadways:
total: 71,300 km
paved: 6,200 km
unpaved: 65,100 km (2005)

Merchant marine:
total: 4 ships (1000 GRT or over) 15,474 GRT/18,072 DWT
by type: cargo 1, chemical tanker 1, petroleum tanker 1, roll on/roll off 1
registered in other countries: 12 (Bolivia 1, Cambodia 3, North Korea 2, Panama 5, St Kitts and Nevis 1) (2007)

Ports and terminals: Aden, Hudaydah, Mukalla

MILITARY

Military branches: Army (includes Republican Guard), Navy (includes Marines), Yemen Air Force (Al Quwwat al Jawwiya al Jamahiriya al Yemeniya; includes Air Defense Force) (2008)

Military service age and obligation: voluntary military service program authorized in 2001; 2-year service obligation (2006)

Manpower available for military service:
males age 16–49: 5,080,038
females age 16–49: 4,852,555 (2008 est.)

Manpower fit for military service:
males age 16–49: 3,585,947
females age 16–49: 3,619,195 (2008 est.)

Manpower reaching militarily significant age annually:
males age 16–49: 268,468
females age 16–49: 258,196 (2008 est.)

Military expenditures—percent of GDP: 6.6% (2006)

Military—note: a Coast Guard was established in 2002

TRANSNATIONAL ISSUES

Disputes—international: Saudi Arabia has reinforced its concrete-filled security barrier along sections of the fully demarcated border with Yemen to stem illegal cross-border activities

Refugees and internally displaced persons: *refugees (country of origin):* 91,587 (Somalia) (2007)

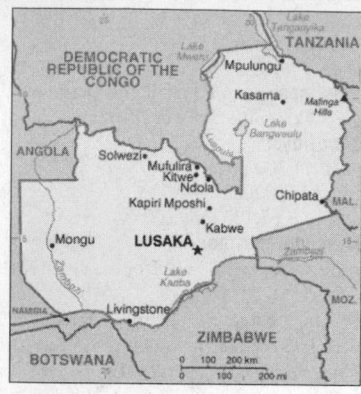

INTRODUCTION

Background: The territory of Northern Rhodesia was administered by the [British] South Africa Company from 1891 until it was taken over by the UK in 1923. During the 1920s and 1930s, advances in mining spurred development and immigration. The name was changed to Zambia upon independence in 1964. In the 1980s and 1990s, declining copper prices and a prolonged drought hurt the economy. Elections in 1991 brought an end to one-party rule, but the subsequent vote in 1996 saw blatant harassment of opposition parties. The election in 2001 was marked by administrative problems with three parties filing a legal petition challenging the election of ruling party candidate Levy MWANAWASA. The new president launched an anticorruption investigation in 2002 to probe high-level corruption during the previous administration. In 2006–07, this task force successfully prosecuted four cases, including a landmark civil case in the UK in which former President CHILUBA and numerous others were found liable for USD 41 million. MWANAWASA was reelected in 2006 in an election that was deemed free and fair.

GEOGRAPHY

Location: Southern Africa, east of Angola

Geographic coordinates: 15 00 S, 30 00 E

Map references: Africa

Area: *total:* 752,614 sq km
land: 740,724 sq km
water: 11,890 sq km

Area—comparative: slightly larger than Texas

Land boundaries:
total: 5,664 km
border countries: Angola 1,110 km, Democratic Republic of the Congo 1,930 km, Malawi 837 km, Mozambique 419 km, Namibia 233 km, Tanzania 338 km, Zimbabwe 797 km

Coastline: 0 km (landlocked)

Maritime claims: none (landlocked)

Climate: tropical; modified by altitude; rainy season (October to April)

Terrain: mostly high plateau with some hills and mountains

Elevation extremes:
lowest point: Zambezi river 329 m
highest point: unnamed location in Mafinga Hills 2,301 m

Natural resources: copper, cobalt, zinc, lead, coal, emeralds, gold, silver, uranium, hydropower

Land use: *arable land:* 6.99%
permanent crops: 0.04%
other: 92.97% (2005)

Irrigated land: 1,560 sq km (2003)

Total renewable water resources: 105.2 cu km (2001)

Freshwater withdrawal (domestic/industrial/agricultural): *total:* 1.74 cu km/yr (17%/7%/76%)
per capita: 149 cu m/yr (2000)

Natural hazards: periodic drought, tropical storms (November to April)

Environment—current issues: air pollution and resulting acid rain in the mineral extraction and refining region; chemical runoff into watersheds; poaching seriously threatens rhinoceros, elephant, antelope, and large cat populations; deforestation; soil erosion; desertification; lack of adequate water treatment presents human health risks

Environment—international agreements: *party to:* Biodiversity, Climate Change, Climate Change-Kyoto Protocol, Desertification, Endangered Species, Hazardous Wastes, Law of the Sea, Ozone Layer Protection, Wetlands
signed, but not ratified: none of the selected agreements

Geography—note: landlocked; the Zambezi forms a natural riverine boundary with Zimbabwe

PEOPLE

Population: 11,669,534
note: estimates for this country explicitly take into account the effects of excess mortality due to AIDS; this can result in lower life expectancy, higher infant mortality, higher death rates, lower population growth rates, and changes in the distribution of population by age and sex than would otherwise be expected (July 2008 est.)

Age structure:
0–14 years: 45.4% (male 2,659,572/female 2,634,379)
15–64 years: 52.3% (male 3,045,536/female 3,053,465)
65 years and over: 2.4% (male 115,662/female 160,920) (2008 est.)

Median age:
total: 16.9 years
male: 16.8 years
female: 17.1 years (2008 est.)

Population growth rate: 1.654% (2008 est.)

Birth rate: 40.52 births/1,000 population (2008 est.)

Death rate: 21.35 deaths/1,000 population (2008 est.)

Net migration rate: -2.63 migrant(s)/1,000 population (2008 est.)

Sex ratio:
at birth: 1.03 male(s)/female
under 15 years: 1.01 male(s)/female
15–64 years: 1 male(s)/female
65 years and over: 0.72 male(s)/female
total population: 1 male(s)/female (2008 est.)

Infant mortality rate:
total: 100.96 deaths/1,000 live births
male: 105.73 deaths/1,000 live births
female: 96.04 deaths/1,000 live births (2008 est.)

Life expectancy at birth:
total population: 38.59 years
male: 38.49 years
female: 38.7 years (2008 est.)

Total fertility rate: 5.23 children born/woman (2008 est.)

HIV/AIDS—adult prevalence rate: 16.5% (2003 est.)

HIV/AIDS—people living with HIV/AIDS: 920,000 (2003 est.)

HIV/AIDS—deaths: 89,000 (2003 est.)

Major infectious diseases:
degree of risk: very high
food or waterborne diseases: bacterial and protozoal diarrhea, hepatitis A, and typhoid fever
vectorborne diseases: malaria and plague are high risks in some locations
water contact disease: schistosomiasis
animal contact disease: rabies (2008)

Nationality:
noun: Zambian(s)
adjective: Zambian

Ethnic groups: African 98.7%, European 1.1%, other 0.2%

Religions: Christian 50%–75%, Muslim and Hindu 24%–49%, indigenous beliefs 1%

Languages: English (official), major vernaculars—Bemba, Kaonda, Lozi, Lunda, Luvale, Nyanja, Tonga, and about 70 other indigenous languages

Literacy: *definition:* age 15 and over can read and write English
total population: 80.6%
male: 86.8%
female: 74.8% (2003 est.)

GOVERNMENT

Country name:
conventional long form: Republic of Zambia

conventional short form: Zambia
former: Northern Rhodesia
Government type: republic
Capital: *name:* Lusaka
geographic coordinates: 15 25 S, 28 17 E
time difference: UTC+2 (7 hours ahead of Washington, DC during Standard Time)
Administrative divisions: 9 provinces; Central, Copperbelt, Eastern, Luapula, Lusaka, Northern, North-Western, Southern, Western
Independence: 24 October 1964 (from UK)
National holiday: Independence Day, 24 October (1964)
Constitution: 24 August 1991; amended in 1996 to establish presidential term limits
Legal system: based on English common law and customary law; judicial review of legislative acts in an ad hoc constitutional council; has not accepted compulsory ICJ jurisdiction
Suffrage: 18 years of age; universal
Executive branch:
chief of state: President Levy MWANAWASA (since 2 January 2002); Vice President Rupiah BANDA (since 9 October 2006); note—the president is both the chief of state and head of government
head of government: President Levy MWANAWASA (since 2 January 2002); Vice President Rupiah BANDA (since 9 October 2006)
cabinet: Cabinet appointed by the president from among the members of the National Assembly
elections: president elected by popular vote for a five-year term (eligible for a second term); election last held 28 September 2006 (next to be held in 2011); vice president appointed by the president
election results: Levy MWANAWASA reelected president; percent of vote—Levy MWANAWASA 43.0%, Michael SATA 29.4%, Hakainde HICHILEMA 25.3%, Godfrey MIYANDA 1.6%, Winright NGONDO 0.8%
Legislative branch: unicameral National Assembly (158 seats; 150 members are elected by popular vote, 8 members are appointed by the president, to serve five-year terms)
elections: last held 28 September 2006 (next to be held in 2011)
election results: percent of vote by party—NA; seats by party—MMD 72, PF 44, UDA 27, ULP 2, NDF 1, independents 2; seats not determined 2
Judicial branch: Supreme Court (the final court of appeal; justices are appointed by the president); High Court (has unlimited jurisdiction to hear civil and criminal cases)
Political parties and leaders: All Peoples Congress Party [Winright NGONDO]; Forum for Democracy and Development or FDD [Edith NAWAKWI]; Heritage Party or HP [Godfrey MIYANDA]; Liberal Progressive Front or LPF [Roger CHONGWE]; Movement for Multiparty Democracy or MMD [Levy MWANAWASA]; National Democratic Focus or NDF; Patriotic Front or PF [Michael SATA]; Party of Unity for Democracy and Development or PUDD [Dan PULE]; Reform Party [Nevers MUMBA]; United Democratic Alliance or UDA; United Liberal Party or ULP [Sakwiba SIKOTA]; United National Independence Party or UNIP [Tilyenji KAUNDA]; United Party for National Development or UPND [Hakainde HICHILEMA]; Zambia Democratic Congress or ZADECO [Langton SICHONE]; Zambian Republican Party or ZRP [Benjamin MWILA]
Political pressure groups and leaders: NA
International organization participation: ACP, AfDB, AU, C, COMESA, FAO, G-77, IAEA, IBRD, ICAO, ICCt, ICRM, IDA, IFAD, IFC, IFRCS, ILO, IMF, Interpol, IOC, IOM, IPU, ISO (correspondent), ITSO, ITU, ITUC, MIGA, NAM, OPCW, PCA, SADC, UN, UNCTAD, UNESCO, UNHCR, UNIDO, UNMEE, UNMIL, UNMIS, UNOCI, UNWTO, UPU, WCL, WCO, WHO, WIPO, WMO, WTO
Diplomatic representation in the US:
chief of mission: Ambassador Inonge MBIKUSITA-LEWANIKA
chancery: 2419 Massachusetts Avenue NW, Washington, DC 20008
telephone: [1] (202) 265-9717 through 9719
FAX: [1] (202) 332-0826
Diplomatic representation from the US:
chief of mission: Ambassador Carmen M. MARTINEZ
embassy: corner of Independence and United Nations Avenues, Lusaka
mailing address: P. O. Box 31617, Lusaka
telephone: [260] (1) 250-955
FAX: [260] (1) 252-225
Flag description: green field with a panel of three vertical bands of red (hoist side), black, and orange below a soaring orange eagle, on the outer edge of the flag

---

ECONOMY

Economy—overview: Zambia's economy has experienced modest growth in recent years, with real GDP growth in 2005–07 between 5–6% per year. Privatization of government-owned copper mines in the 1990s relieved the government from covering mammoth losses generated by the industry and greatly improved the chances for copper mining to return to profitability and spur economic growth. Copper output has increased steadily since 2004, due to higher copper prices and foreign investment. In 2005, Zambia qualified for debt relief under the Highly Indebted Poor Country Initiative, consisting of approximately USD 6 billion in debt relief. Zambia experienced a bumper harvest in 2007, which helped to boost GDP and agricultural exports and contain inflation. Although poverty continues to be significant problem in Zambia, its economy has strengthened, featuring single-digit inflation, a relatively stable currency, decreasing interest rates, and increasing levels of trade.
GDP (purchasing power parity): $15.92 billion (2007 est.)
GDP (official exchange rate): $11.16 billion (2007 est.)
GDP—real growth rate: 5.3% (2007 est.)
GDP—per capita (PPP): $1,300 (2007 est.)
GDP—composition by sector:
agriculture: 17.3%
industry: 26.2%
services: 56.5% (2007 est.)
Labor force: 4.989 million (2007 est.)
Labor force—by occupation: *agriculture:* 85%
industry: 6%
services: 9% (2004)
Unemployment rate: 50% (2000 est.)
Population below poverty line: 86% (1993)
Household income or consumption by percentage share:
lowest 10%: 1.2%
highest 10%: 38.8% (2004)
Distribution of family income—Gini index: 50.8 (2004)
Inflation rate (consumer prices): 10.7% (2007 est.)
Investment (gross fixed): 26% of GDP (2007 est.)
Budget:
revenues: $2.542 billion
expenditures: $2.678 billion (2007 est.)
Public debt: 28% of GDP (2007 est.)
Agriculture—products: corn, sorghum, rice, peanuts, sunflower seed, vegetables, flowers, tobacco, cotton, sugarcane, cassava (tapioca), coffee; cattle, goats, pigs, poultry, milk, eggs, hides
Industries: copper mining and processing, construction, foodstuffs, beverages, chemicals, textiles, fertilizer, horticulture
Industrial production growth rate: 7.3% (2007 est.)
Electricity—production: 8.85 billion kWh (2005)
Electricity—production by source:
fossil fuel: 0.5%
hydro: 99.5%
nuclear: 0%
other: 0% (2001)
Electricity—consumption: 8.655 billion kWh (2005)

Electricity—exports: 243 million kWh (2005)

Electricity—imports: 465 million kWh (2005)

Oil—production: 150 bbl/day (2005 est.)

Oil—consumption: 14,000 bbl/day (2005 est.)

Oil—exports: 169 bbl/day (2004)

Oil—imports: 13,370 bbl/day (2004)

Oil—proved reserves: NA

Natural gas—production: 0 cu m (2005 est.)

Natural gas—consumption: 0 cu m (2005 est.)

Natural gas—exports: 0 cu m (2005 est.)

Natural gas—imports: 0 cu m (2005)

Natural gas—proved reserves: 0 cu m (1 January 2006 est.)

Current account balance: -$743 million (2007 est.)

Exports: $4.348 billion f.o.b. (2007 est.)

Exports—commodities: copper/cobalt 64%, cobalt, electricity; tobacco, flowers, cotton

Exports—partners: Switzerland 38.4%, South Africa 21.6%, China 10.3%, UK 7.6%, Tanzania 6.4% (2006)

Imports: $3.622 billion f.o.b. (2007 est.)

Imports—commodities: machinery, transportation equipment, petroleum products, electricity, fertilizer; foodstuffs, clothing

Imports—partners: South Africa 47.3%, UAE 10.4%, Zimbabwe 5.7%, Norway 4% (2006)

Economic aid—recipient: $504 million (2007)

Reserves of foreign exchange and gold: $1.09 billion (31 December 2007 est.)

Debt—external: $2.598 billion (31 December 2007 est.)

Stock of direct foreign investment—at home: $NA

Stock of direct foreign investment—abroad: $NA

Market value of publicly traded shares: $4.5 billion (2007)

Currency (code): Zambian kwacha (ZMK)

Currency code: ZMK

Exchange rates: Zambian kwacha per US dollar—3,990.2 (2007), 3,601.5 (2006), 4,463.5 (2005), 4,778.9 (2004), 4,733.3 (2003)

Fiscal year: calendar year

COMMUNICATIONS

Telephones—main lines in use: 93,400 (2006)

Telephones—mobile cellular: 1,663,300 (2006)

Telephone system:

general assessment: facilities are aging but still among the best in Sub-Saharan Africa

domestic: high-capacity microwave radio relay connects most larger towns and cities; several cellular telephone services in operation and network coverage is improving; Internet service is widely available; very small aperture terminal (VSAT) networks are operated by private firms

international: country code—260; satellite earth stations—2 Intelsat (1 Indian Ocean and 1 Atlantic Ocean)

Radio broadcast stations: AM 19, FM 5, shortwave 4 (2001)

Radios: 1.2 million (2001)

Television broadcast stations: 9 (2001)

Televisions: 277,000 (1997)

Internet country code: .zm

Internet hosts: 7,423 (2007)

Internet Service Providers (ISPs): 5 (2001)

Internet users: 334,800 (2005)

TRANSPORTATION

Airports: 107 (2007)

Airports—with paved runways:

total: 9

over 3,047 m: 1

2,438 to 3,047 m: 2

1,524 to 2,437 m: 4

914 to 1,523 m: 2 (2007)

Airports—with unpaved runways:

total: 98

2,438 to 3,047 m: 1

1,524 to 2,437 m: 4

914 to 1,523 m: 64

under 914 m: 29 (2007)

Pipelines: oil 771 km (2007)

Railways:

total: 2,157 km

narrow gauge: 2,157 km 1.067-m gauge

note: includes 891 km of the Tanzania-Zambia Railway Authority (TAZARA) (2006)

Roadways: *total:* 91,440 km

paved: 20,117 km

unpaved: 71,323 km (2001)

Waterways: 2,250 km (includes Lake Tanganyika and the Zambezi and Luapula rivers) (2005)

Ports and terminals: Mpulungu

MILITARY

Military branches: Zambian National Defense Force (ZNDF): Zambian Army, Zambian Air Force, National Service (2008)

Military service age and obligation: 18 years of age for voluntary military service (16 years of age with parental consent); no conscription (2008)

Manpower available for military service: *males age 16–49:* 2,678,668 *females age 16–49:* 2,567,433 (2008 est.)

Manpower fit for military service: *males age 16–49:* 1,329,343 *females age 16–49:* 1,218,114 (2008 est.)

Military expenditures—percent of GDP: 1.8% (2005 est.)

TRANSNATIONAL ISSUES

Disputes—international: in 2004, Zimbabwe dropped objections to plans between Botswana and Zambia to build a bridge over the Zambezi River, thereby de facto recognizing a short, but not clearly delimited, Botswana-Zambia boundary in the river; 42,250 Congolese refugees in Zambia are offered voluntary repatriation in November 2006, most of whom are expected to return in the next two years; Angolan refugees too have been repatriating but 26,450 still remain with 90,000 others from other neighboring states in 2006

Refugees and internally displaced persons: *refugees (country of origin):* 42,565 (Angola); 60,874 (Democratic Republic of the Congo); 4,100 (Rwanda) (2007)

Illicit drugs: transshipment point for moderate amounts of methaqualone, small amounts of heroin, and cocaine bound for southern Africa and possibly Europe; a poorly developed financial infrastructure coupled with a government commitment to combating money laundering make it an unattractive venue for money launderers; major consumer of cannabis

ZIMBABWE

INTRODUCTION

Background: The UK annexed Southern Rhodesia from the [British] South Africa Company in 1923. A 1961 constitution was formulated that favored whites in power. In 1965 the government unilaterally declared its independence, but the UK did not recognize the act and demanded more complete voting rights for the black African majority in the country (then called Rhodesia). UN sanctions and a guerrilla uprising finally led to free elections in 1979 and independence (as Zimbabwe) in 1980. Robert MUGABE, the nation's first prime minister, has been the country's only ruler (as president since 1987) and has dominated the country's political system since independence. His chaotic land redistribution campaign, which began in 2000, caused an exodus of white farmers, crippled the economy, and ushered in widespread shortages of basic

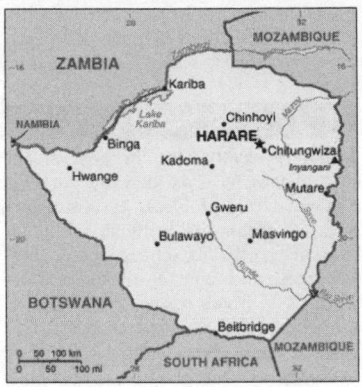

commodities. Ignoring international condemnation, MUGABE rigged the 2002 presidential election to ensure his reelection. The ruling ZANU-PF party used fraud and intimidation to win a two-thirds majority in the March 2005 parliamentary election, allowing it to amend the constitution at will and recreate the Senate, which had been abolished in the late 1980s. In April 2005, Harare embarked on Operation Restore Order, ostensibly an urban rationalization program, which resulted in the destruction of the homes or businesses of 700,000 mostly poor supporters of the opposition, according to UN estimates. President MUGABE in June 2007 instituted price controls on all basic commodities causing panic buying and leaving store shelves empty for months. In October 2007, Constitutional Amendment 18 came into effect allowing for harmonized presidential and parliamentary elections, shortening the length of the presidential term to five years, and moving up the date for parliamentary elections. General elections are expected in March 2008.

GEOGRAPHY

Location: Southern Africa, between South Africa and Zambia
Geographic coordinates: 20 00 S, 30 00 E
Map references: Africa
Area: *total:* 390,580 sq km
land: 386,670 sq km
water: 3,910 sq km
Area—comparative: slightly larger than Montana
Land boundaries: *total:* 3,066 km
border countries: Botswana 813 km, Mozambique 1,231 km, South Africa 225 km, Zambia 797 km
Coastline: 0 km (landlocked)
Maritime claims: none (landlocked)
Climate: tropical; moderated by altitude; rainy season (November to March)
Terrain: mostly high plateau with higher central plateau (high veld); mountains in east

Elevation extremes:
lowest point: junction of the Runde and Save rivers 162 m
highest point: Inyangani 2,592 m
Natural resources: coal, chromium ore, asbestos, gold, nickel, copper, iron ore, vanadium, lithium, tin, platinum group metals
Land use:
arable land: 8.24%
permanent crops: 0.33%
other: 91.43% (2005)
Irrigated land: 1,740 sq km (2003)
Total renewable water resources: 20 cu km (1987)
Freshwater withdrawal (domestic/industrial/agricultural): *total:* 4.21 cu km/yr (14%/7%/79%)
per capita: 324 cu m/yr (2002)
Natural hazards: recurring droughts; floods and severe storms are rare
Environment—current issues: deforestation; soil erosion; land degradation; air and water pollution; the black rhinoceros herd—once the largest concentration of the species in the world—has been significantly reduced by poaching; poor mining practices have led to toxic waste and heavy metal pollution
Environment—international agreements: *party to:* Biodiversity, Climate Change, Desertification, Endangered Species, Law of the Sea, Ozone Layer Protection
signed, but not ratified: none of the selected agreements
Geography—note: landlocked; the Zambezi forms a natural riverine boundary with Zambia; in full flood (February-April) the massive Victoria Falls on the river forms the world's largest curtain of falling water

PEOPLE

Population: 12,382,920
note: estimates for this country explicitly take into account the effects of excess mortality due to AIDS; this can result in lower life expectancy, higher infant mortality, higher death rates, lower population growth rates, and changes in the distribution of population by age and sex than would otherwise be expected (July 2008 est.)
Age structure:
0–14 years: 37% (male 2,314,527/female 2,270,998)
15–64 years: 59.4% (male 3,704,689/female 3,656,173)
65 years and over: 3.5% (male 197,821/female 238,712) (2008 est.)
Median age: *total:* 20.3 years
male: 20.2 years
female: 20.4 years (2008 est.)
Population growth rate: 0.568% (2008 est.)
Birth rate: 27.38 births/1,000 population (2008 est.)

Death rate: 21.7 deaths/1,000 population (2008 est.)
Net migration rate: NA
note: there is an increasing flow of Zimbabweans into South Africa and Botswana in search of better economic opportunities (2008 est.)
Sex ratio:
at birth: 1.03 male(s)/female
under 15 years: 1.02 male(s)/female
15–64 years: 1.01 male(s)/female
65 years and over: 0.83 male(s)/female
total population: 1.01 male(s)/female (2008 est.)
Infant mortality rate:
total: 50.58 deaths/1,000 live births
male: 53.29 deaths/1,000 live births
female: 47.8 deaths/1,000 live births (2008 est.)
Life expectancy at birth:
total population: 39.73 years
male: 40.87 years
female: 38.55 years (2008 est.)
Total fertility rate: 3.03 children born/woman (2008 est.)
HIV/AIDS—adult prevalence rate: 24.6% (2001 est.)
HIV/AIDS—people living with HIV/AIDS: 1.8 million (2001 est.)
HIV/AIDS—deaths: 170,000 (2003 est.)
Major infectious diseases:
degree of risk: high
food or waterborne diseases: bacterial and protozoal diarrhea, hepatitis A, and typhoid fever
vectorborne disease: malaria
water contact disease: schistosomiasis
animal contact disease: rabies (2008)
Nationality:
noun: Zimbabwean(s)
adjective: Zimbabwean
Ethnic groups: African 98% (Shona 82%, Ndebele 14%, other 2%), mixed and Asian 1%, white less than 1%
Religions: syncretic (part Christian, part indigenous beliefs) 50%, Christian 25%, indigenous beliefs 24%, Muslim and other 1%
Languages: English (official), Shona, Sindebele (the language of the Ndebele, sometimes called Ndebele), numerous but minor tribal dialects
Literacy:
definition: age 15 and over can read and write English
total population: 90.7%
male: 94.2%
female: 87.2% (2003 est.)

GOVERNMENT

Country name:
conventional long form: Republic of Zimbabwe
conventional short form: Zimbabwe
former: Southern Rhodesia, Rhodesia
Government type: parliamentary democracy
Capital: *name:* Harare

geographic coordinates: 17 50 S, 31 03 E
time difference: UTC+2 (7 hours ahead of Washington, DC during Standard Time)
Administrative divisions: 8 provinces and 2 cities* with provincial status; Bulawayo*, Harare*, Manicaland, Mashonaland Central, Mashonaland East, Mashonaland West, Masvingo, Matabeleland North, Matabeleland South, Midlands
Independence: 18 April 1980 (from UK)
National holiday: Independence Day, 18 April (1980)
Constitution: 21 December 1979
Legal system: mixture of Roman-Dutch and English common law; has not accepted compulsory ICJ jurisdiction
Suffrage: 18 years of age; universal
Executive branch:
chief of state: Executive President Robert Gabriel MUGABE (since 31 December 1987); Vice President Joseph MSIKA (since December 1999) and Vice President Joyce MUJURU (since 6 December 2004); note—the president is both the chief of state and head of government
head of government: Executive President Robert Gabriel MUGABE (since 31 December 1987); Vice President Joseph MSIKA (since December 1999) and Vice President Joyce MUJURU (since 6 December 2004)
cabinet: Cabinet appointed by the president; responsible to the House of Assembly
elections: presidential candidates nominated with a nomination paper signed by at least 10 registered voters (at least one from each province) and elected by popular vote for a five-year term (no term limits); election last held 9–11 March 2002 (next to be held 28 March 2008); co-vice presidents appointed by the president
election results: Robert Gabriel MUGABE reelected president; percent of vote—Robert Gabriel MUGABE 56.2%, Morgan TSVANGIRAI 41.9%
Legislative branch: bicameral Parliament consists of a Senate (83 seats—50 elected by popular vote for a five-year term, 10 provincial governors nominated by the president, 16 provincial chiefs appointed by the president and deputy president from all provinces except Harare and Bulawayo, and 7 appointed by the president) and a House of Assembly (120 seats—all elected by popular vote for five-year terms)
elections: Senate last held 26 November 2005 (next to be held 28 March 2008; House of Assembly last held 31 March 2005 (next to be held 28 March 2008)
election results: Senate—percent of vote by party—ZANU-PF 73.7%, MDC 20.3%, other 4.4%, independents 1.6%;

seats by party—ZANU-PF 43, MDC 7; House of Assembly—percent of vote by party—ZANU-PF 59.6%, MDC 39.5%, other 0.9%; seats by party—ZANU-PF 78, MDC 41, independents 1
Judicial branch: Supreme Court; High Court
Political parties and leaders: African National Party or ANP [Egypt DZINE-MUNHENZVA]; Movement for Democratic Change or MDC [Morgan TSVANGIRAI, anti-Senate faction; Arthur MUTAMBARA, pro-Senate faction]; Peace Action is Freedom for All or PAFA; United Parties [Abel MUZOREWA]; United People's Party or UPP [Daniel SHUMBA]; Zimbabwe African National Union-Ndonga or ZANU-Ndonga [Wilson KUMBULA]; Zimbabwe African National Union-Patriotic Front or ZANU-PF [Robert Gabriel MUGABE]; Zimbabwe African Peoples Union or ZAPU [Agrippa MADLELA]; Zimbabwe Youth in Alliance or ZIYA
Political pressure groups and leaders: Crisis in Zimbabwe Coalition [Xolani ZITHA]; National Constitutional Assembly or NCA [Lovemore MADHUKU]; Women of Zimbabwe Arise or WOZA [Jenny WILLIAMS]; Zimbabwe Congress of Trade Unions or ZCTU [Wellington CHIBEBE]
International organization participation: ACP, AfDB, AU, COMESA, FAO, G-15, G-77, IAEA, IBRD, ICAO, ICCt (signatory), ICRM, IDA, IFAD, IFC, IFRCS, ILO, IMF, IMO, Interpol, IOC, IOM, IPU, ISO, ITSO, ITU, ITUC, MIGA, NAM, OPCW, PCA, SADC, UN, UNCTAD, UNESCO, UNIDO, UNMIS, UNOCI, UNWTO, UPU, WCL, WCO, WFTU, WHO, WIPO, WMO, WTO
Diplomatic representation in the US:
chief of mission: Ambassador Marina Annette VALERE
chancery: 1608 New Hampshire Avenue NW, Washington, DC 20009
telephone: [1] (202) 332-7100
FAX: [1] (202) 483-9326
Diplomatic representation from the US:
chief of mission: Ambassador James D. MCGEE
embassy: 172 Herbert Chitepo Avenue, Harare
mailing address: P. O. Box 3340, Harare
telephone: [263] (4) 250-593 and 250-594
FAX: [263] (4) 796-488
Flag description: seven equal horizontal bands of green, yellow, red, black, red, yellow, and green with a white isosceles triangle edged in black with its base on the hoist side; a yellow Zimbabwe bird representing the long history of the country is superimposed on a red five-pointed star in the center of the triangle, which symbolizes peace; green symbol-

izes agriculture, yellow—mineral wealth, red—blood shed to achieve independence, and black stands for the native people

ECONOMY

Economy—overview: The government of Zimbabwe faces a wide variety of difficult economic problems as it struggles with an unsustainable fiscal deficit, an overvalued official exchange rate, hyperinflation, and bare store shelves. Its 1998–2002 involvement in the war in the Democratic Republic of the Congo drained hundreds of millions of dollars from the economy. The government's land reform program, characterized by chaos and violence, has badly damaged the commercial farming sector, the traditional source of exports and foreign exchange and the provider of 400,000 jobs, turning Zimbabwe into a net importer of food products. The EU and the US provide food aid on humanitarian grounds. Badly needed support from the IMF has been suspended because of the government's arrears on past loans and the government's unwillingness to enact reforms that would stabilize the economy. The Reserve Bank of Zimbabwe routinely prints money to fund the budget deficit, causing the official annual inflation rate to rise from 32% in 1998, to 133% in 2004, 585% in 2005, passed 1000% in 2006, and 26000% in November 2007. Private sector estimates of inflation in 2007 are well above 100,000%. Meanwhile, the official exchange rate fell from approximately 1 (revalued) Zimbabwean dollar per US dollar in 2003 to 30,000 per US dollar in 2007.
GDP (purchasing power parity): $2.211 billion (2007 est.)
GDP (official exchange rate): $1.726 billion
note: hyperinflation and the plunging value of the Zimbabwean dollar makes Zimbabwe's GDP at the official exchange rate a highly inaccurate statistic (2007 est.)
GDP—real growth rate: -6.1% (2007 est.)
GDP—per capita (PPP): $200 (2007 est.)
GDP—composition by sector:
agriculture: 18.1%
industry: 22.6%
services: 59.3% (2007 est.)
Labor force: 4.032 million (2007 est.)
Labor force—by occupation: *agriculture:* 66%
industry: 10%
services: 24% (1996)
Unemployment rate: 80% (2005 est.)
Population below poverty line: 68% (2004)
Household income or consumption by

percentage share:
lowest 10%: 2%
highest 10%: 40.4% (1995)
Distribution of family income—Gini index: 50.1 (2006)
Inflation rate (consumer prices): 10,453% official data; private sector estimates are much higher (2007 est.)
Investment (gross fixed): 16.9% of GDP (2007 est.)
Budget:
revenues: $2.442 billion
expenditures: $3.017 billion (2007 est.)
Public debt: 211.9% of GDP (2007 est.)
Agriculture—products: corn, cotton, tobacco, wheat, coffee, sugarcane, peanuts; sheep, goats, pigs
Industries: mining (coal, gold, platinum, copper, nickel, tin, clay, numerous metallic and nonmetallic ores), steel; wood products, cement, chemicals, fertilizer, clothing and footwear, foodstuffs, beverages
Industrial production growth rate: 0.2% (2007 est.)
Electricity—production: 9.95 billion kWh (2005)
Electricity—production by source:
fossil fuel: 47%
hydro: 53%
nuclear: 0%
other: 0% (2001)
Electricity—consumption: 12.27 billion kWh (2005)
Electricity—exports: 0 kWh (2005)
Electricity—imports: 3.013 billion kWh (2005)
Oil—production: 0 bbl/day (2005 est.)
Oil—consumption: 16,000 bbl/day (2005 est.)
Oil—exports: 0 bbl/day (2004 est.)
Oil—imports: 13,370 bbl/day (2004 est.)
Oil—proved reserves: 0 bbl (1 January 2006 est.)
Natural gas—production: 0 cu m (2005 est.)
Natural gas—consumption: 0 cu m (2005 est.)
Natural gas—exports: 0 cu m (2005 est.)
Natural gas—imports: 0 cu m (2005)
Natural gas—proved reserves: 0 cu m (1 January 2006 est.)
Current account balance: -$116 million (2005 est.)
Exports: $1.52 billion f.o.b. (2007 est.)
Exports—commodities: platinum, cotton, tobacco, gold, ferroalloys, textiles/clothing
Exports—partners: South Africa 24.8%, Democratic Republic of the Congo 17.6%, Botswana 15.7%, US 10.4% (2006)
Imports: $2.183 billion f.o.b. (2007 est.)
Imports—commodities: machinery and transport equipment, other manufactures, chemicals, fuels
Imports—partners: South Africa 40.8%,

Zambia 29.6%, US 4.9% (2006)
Economic aid—recipient: $367.7 million (2005 est.)
Reserves of foreign exchange and gold: $120 million (31 December 2007 est.)
Debt—external: $5.155 billion (31 December 2007 est.)
Stock of direct foreign investment—at home: $NA
Stock of direct foreign investment—abroad: $NA
Market value of publicly traded shares: $26.56 billion (2006)
Currency (code): Zimbabwean dollar (ZWD)
Currency code: ZWD
Exchange rates: Zimbabwean dollars per US dollar—30,000 (2007), 162.07 (2006), 77.965 (2005), 5.729 (2004), 0.824 (2003)
note: these are official exchange rates; non-official rates vary significantly
Fiscal year: calendar year

COMMUNICATIONS

Telephones—main lines in use: 331,700 (2006)
Telephones—mobile cellular: 832,500 (2006)
Telephone system:
general assessment: system was once one of the best in Africa, but now suffers from poor maintenance; more than 100,000 outstanding requests for connection despite an equally large number of installed but unused main lines
domestic: consists of microwave radio relay links, open-wire lines, radiotelephone communication stations, fixed wireless local loop installations, and a substantial mobile-cellular network; Internet connection is available in Harare and planned for all major towns and for some of the smaller ones
international: country code—263; satellite earth stations—2 Intelsat; 2 international digital gateway exchanges (in Harare and Gweru)
Radio broadcast stations: AM 7, FM 20 (plus 17 repeater stations), shortwave 1 (1998)
Radios: 1.14 million (1997)
Television broadcast stations: 16 (1997)
Televisions: 370,000 (1997)
Internet country code: .zw
Internet hosts: 15,507 (2007)
Internet Service Providers (ISPs): 6 (2000)
Internet users: 1.22 million (2006)

TRANSPORTATION

Airports: 341 (2007)
Airports—with paved runways:
total: 19
over 3,047 m: 3
2,438 to 3,047 m: 2

1,524 to 2,437 m: 4
914 to 1,523 m: 10 (2007)
Airports—with unpaved runways:
total: 322
1,524 to 2,437 m: 4
914 to 1,523 m: 152
under 914 m: 166 (2007)
Pipelines: refined products 270 km (2007)
Railways:
total: 3,077 km
narrow gauge: 3,077 km 1.067-m gauge (313 km electrified) (2006)
Roadways:
total: 97,440 km
paved: 18,514 km
unpaved: 78,926 km (2002)
Waterways: on Lake Kariba (2005)
Ports and terminals: Binga, Kariba

MILITARY

Military branches: Zimbabwe Defense Forces (ZDF): Zimbabwe National Army, Air Force of Zimbabwe (AFZ), Zimbabwe Republic Police (2005)
Military service age and obligation: 18–24 years of age for compulsory military service; women are eligible to serve (2007)
Manpower available for military service:
males age 16–49: 3,264,258
females age 16–49: 3,048,049 (2008 est.)
Manpower fit for military service:
males age 16–49: 1,643,036
females age 16–49: 1,404,663 (2008 est.)
Military expenditures—percent of GDP: 3.8% (2006)

TRANSNATIONAL ISSUES

Disputes—international: Botswana built electric fences and South Africa has placed military along the border to stem the flow of thousands of Zimbabweans fleeing to find work and escape political persecution; Namibia has supported, and in 2004 Zimbabwe dropped objections to, plans between Botswana and Zambia to build a bridge over the Zambezi River, thereby de facto recognizing a short, but not clearly delimited, Botswana-Zambia boundary in the river
Refugees and internally displaced persons: *refugees (country of origin):* 2,500 (Democratic Republic of Congo)
IDPs: 569,685 (MUGABE-led political violence, human rights violations, land reform, and economic collapse) (2007)
Illicit drugs: transit point for cannabis and South Asian heroin, mandrax, and methamphetamines en route to South Africa

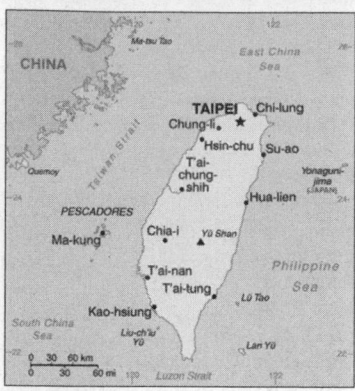

Background: In 1895, military defeat forced China to cede Taiwan to Japan. Taiwan reverted to Chinese control after World War II. Following the Communist victory on the mainland in 1949, 2 million Nationalists fled to Taiwan and established a government using the 1946 constitution drawn up for all of China. Over the next five decades, the ruling authorities gradually democratized and incorporated the local population within the governing structure. In 2000, Taiwan underwent its first peaceful transfer of power from the Nationalist to the Democratic Progressive Party. Throughout this period, the island prospered and became one of East Asia's economic "Tigers." The dominant political issues continue to be the relationship between Taiwan and China—specifically the question of eventual unification—as well as domestic political and economic reform.

GEOGRAPHY

Location: Eastern Asia, islands bordering the East China Sea, Philippine Sea, South China Sea, and Taiwan Strait, north of the Philippines, off the southeastern coast of China
Geographic coordinates: 23 30 N, 121 00 E
Map references: Southeast Asia
Area:
total: 35,980 sq km
land: 32,260 sq km
water: 3,720 sq km
note: includes the Pescadores, Matsu, and Quemoy islands
Area—comparative: slightly smaller than Maryland and Delaware combined
Land boundaries: 0 km
Coastline: 1,566.3 km
Maritime claims:
territorial sea: 12 nm
exclusive economic zone: 200 nm

Climate: tropical; marine; rainy season during southwest monsoon (June to August); cloudiness is persistent and extensive all year
Terrain: eastern two-thirds mostly rugged mountains; flat to gently rolling plains in west
Elevation extremes:
lowest point: South China Sea 0 m
highest point: Yu Shan 3,952 m
Natural resources: small deposits of coal, natural gas, limestone, marble, and asbestos
Land use:
arable land: 24%
permanent crops: 1%
other: 75% (2001)
Irrigated land: NA
Total renewable water resources: 67 cu km (2000)
Natural hazards: earthquakes and typhoons
Environment—current issues: air pollution; water pollution from industrial emissions, raw sewage; contamination of drinking water supplies; trade in endangered species; low-level radioactive waste disposal
Environment—international agreements: *party to:* none of the selected agreements because of Taiwan's international status
signed, but not ratified: none of the selected agreements because of Taiwan's international status
Geography—note: strategic location adjacent to both the Taiwan Strait and the Luzon Strait

PEOPLE

Population: 22,920,946 (July 2008 est.)
Age structure:
0–14 years: 17.3% (male 2,057,458/female 1,900,449)
15–64 years: 72.3% (male 8,362,038/female 8,204,834)
65 years and over: 10.5% (male 1,167,476/female 1,228,691) (2008 est.)
Median age:
total: 36 years
male: 35.5 years
female: 36.6 years (2008 est.)
Population growth rate: 0.238% (2008 est.)
Birth rate: 8.99 births/1,000 population (2008 est.)
Death rate: 6.65 deaths/1,000 population (2008 est.)
Net migration rate: 0.04 migrant(s)/1,000 population (2008 est.)
Sex ratio:
at birth: 1.09 male(s)/female
under 15 years: 1.08 male(s)/female
15–64 years: 1.02 male(s)/female
65 years and over: 0.95 male(s)/female

total population: 1.02 male(s)/female (2008 est.)
Infant mortality rate:
total: 5.45 deaths/1,000 live births
male: 5.75 deaths/1,000 live births
female: 5.11 deaths/1,000 live births (2008 est.)
Life expectancy at birth:
total population: 77.76 years
male: 74.89 years
female: 80.89 years (2008 est.)
Total fertility rate: 1.13 children born/woman (2008 est.)
HIV/AIDS—adult prevalence rate: NA
HIV/AIDS—people living with HIV/AIDS: NA
HIV/AIDS—deaths: NA
Nationality:
noun: Taiwan (singular and plural)
note: example—he or she is from Taiwan; they are from Taiwan
adjective: Taiwan
Ethnic groups: Taiwanese (including Hakka) 84%, mainland Chinese 14%, indigenous 2%
Religions: mixture of Buddhist and Taoist 93%, Christian 4.5%, other 2.5%
Languages: Mandarin Chinese (official), Taiwanese (Min), Hakka dialects
Literacy:
definition: age 15 and over can read and write
total population: 96.1%
male: NA
female: NA (2003)

GOVERNMENT

Country name:
conventional long form: none
conventional short form: Taiwan
local long form: none
local short form: T'ai-wan
former: Formosa
Government type: multiparty democracy
Capital: *name:* Taipei
geographic coordinates: 25 03 N, 121 30 E
time difference: UTC+8 (13 hours ahead of Washington, DC during Standard Time)
Administrative divisions: includes main island of Taiwan plus smaller islands nearby and off coast of China's Fujian Province; Taiwan is divided into 18 counties (hsien, singular and plural), 5 municipalities (shih, singular and plural), and 2 special municipalities (chuan-shih, singular and plural)
note: Taiwan uses a variety of romanization systems; while a modified Wade-Giles system still dominates, the city of Taipei has adopted a Pinyin romanization for street and place names within its boundaries; other local authorities use different romanization systems; names for administrative divisions that follow

are taken from the Taiwan Yearbook 2007 published by the Government Information Office in Taipei.

counties: Changhua, Chiayi [county], Hsinchu, Hualien, Kaohsiung [county], Kinmen, Lienchiang, Miaoli, Nantou, Penghu, Pingtung, Taichung, Tainan, Taipei [county], Taitung, Taoyuan, Yilan, and Yunlin

municipalities: Chiayi [city], Hsinchu, Keelung, Taichung, Tainan

special municipalities: Kaohsiung [city], Taipei [city]

National holiday: Republic Day (Anniversary of the Chinese Revolution), 10 October (1911)

Constitution: 25 December 1947; amended in 1992, 1994, 1997, 1999, 2000, 2005

note: constitution adopted on 25 December 1946; went into effect on 25 December 1947

Legal system: based on civil law system; has not accepted compulsory ICJ jurisdiction

Suffrage: 20 years of age; universal

Executive branch:

chief of state: President MA Ying-jeou (since 20 May 2008); Vice President Vincent SIEW (since 20 May 2008)

head of government: Premier (President of the Executive Yuan) LIO Chao-hsuan (since 20 May 2008); Vice Premier (Vice President of Executive Yuan) Paul CHIU (CHANG-hsiung) (since 20 May 2008)

cabinet: Executive Yuan—(ministers appointed by president on recommendation of premier)

elections: president and vice president elected on the same ticket by popular vote for four-year terms (eligible for a second term); election last held 22 March 2008 (next to be held in March 2012); premier appointed by the president; vice premiers appointed by the president on the recommendation of the premier

election results: MA Ying-jeou elected president on 22 March 2008; percent of vote—MA Ying-jeou 58.45%, Frank HSIEH 41.55%; MA Ying-jeou takes office on 20 May 2008

Legislative branch: unicameral Legislative Yuan (113 seats—73 district members elected by popular vote, 34 at-large members elected on basis of proportion of islandwide votes received by participating political parties, 6 elected by popular vote among aboriginal populations; to serve four-year terms); parties must receive 5% of vote to qualify for at-large seats

elections: Legislative Yuan—last held 12 January 2008 (next to be held in January 2012)

election results: Legislative Yuan—percent of vote by party—KMT 53.5%, DPP 38.2%, NPSU 2.4%, PFP 0.3%, others

1.6%, independents 4%; seats by party—KMT 81, DPP 27, NPSU 3, PFP 1, independent 1

Judicial branch: Judicial Yuan (justices appointed by the president with consent of the Legislative Yuan)

Political parties and leaders: Democratic Progressive Party or DPP [TSAI Ing-wen]; Kuomintang or KMT (Nationalist Party) [WU Po-hsiung]; Non-Partisan Solidarity Union or NPSU [CHANG Po-ya]; People First Party or PFP [James SOONG]

Political pressure groups and leaders: Taiwan independence movement, various business and environmental groups

note: debate on Taiwan independence has become acceptable within the mainstream of domestic politics on Taiwan; political liberalization and the increased representation of opposition parties in Taiwan's legislature have opened public debate on the island's national identity; a broad popular consensus has developed that the island currently enjoys sovereign independence and—whatever the ultimate outcome regarding reunification or independence—that Taiwan's people must have the deciding voice; public opinion polls consistently show a substantial majority of Taiwan people supports maintaining Taiwan's status quo for the foreseeable future; advocates of Taiwan independence oppose the stand that the island will eventually unify with mainland China; goals of the Taiwan independence movement include establishing a sovereign nation on Taiwan and entering the UN; other organizations supporting Taiwan independence include the World United Formosans for Independence and the Organization for Taiwan Nation Building

International organization participation: ADB, APEC, BCIE, ICC, IOC, ITUC, WCL, WTO

Diplomatic representation in the US: none; unofficial commercial and cultural relations with the people of the US are maintained through an unofficial instrumentality, the Taipei Economic and Cultural Representative Office (TECRO), which has its headquarters in Taipei and in the US in Washington, DC; there are also branch offices called Taipei Economic and Cultural Office (TECO) in 12 other US cities

Diplomatic representation from the US: none; unofficial commercial and cultural relations with the people on Taiwan are maintained through an unofficial instrumentality—the American Institute in Taiwan (AIT)—which has offices in the US and Taiwan; US office at 1700 N. Moore St., Suite 1700, Arlington, VA 22209-1996, telephone: [1] (703) 525-8474, FAX: [1] (703) 841-1385); Taiwan offices at #7 Lane 134, Hsin Yi Road,

Section 3, Taipei, Taiwan, telephone: [886] (2) 2162-2000, FAX: [886] (2) 2162-2251; #2 Chung Cheng 3rd Road, 5th Floor, Kaohsiung, Taiwan, telephone: [886] (7) 238-7744, FAX: [886] (7) 238-5237; and the American Trade Center, Room 3208 International Trade Building, Taipei World Trade Center, 333 Keelung Road Section 1, Taipei, Taiwan 10548, telephone: [886] (2) 2720-1550, FAX: [886] (2) 2757-7162

Flag description: red field with a dark blue rectangle in the upper hoist-side corner bearing a white sun with 12 triangular rays

ECONOMY

Economy—overview: Taiwan has a dynamic capitalist economy with gradually decreasing guidance of investment and foreign trade by the authorities. In keeping with this trend, some large, state-owned banks and industrial firms are being privatized. Exports have provided the primary impetus for industrialization. The island runs a large trade surplus, and its foreign reserves are among the world's largest. Despite restrictions on cross-strait links, China has overtaken the US to become Taiwan's largest export market and its second-largest source of imports after Japan. China is also the island's number one destination for foreign direct investment. Strong trade performance in 2007 pushed Taiwan's GDP growth rate above 5%, and unemployment is below 4%.

GDP (purchasing power parity): $695.4 billion (2007 est.)

GDP (official exchange rate): $383.3 billion (2007 est.)

GDP—real growth rate: 5.7% (2007 est.)

GDP—per capita (PPP): $30,100 (2007 est.)

GDP—composition by sector:
agriculture: 1.4%
industry: 27.5%
services: 71.1% (2007 est.)

Labor force: 10.78 million (2007 est.)

Labor force—by occupation: *agriculture:* 5.3%
industry: 36.8%
services: 57.9% (2007 est.)

Unemployment rate: 3.9% (2007 est.)

Population below poverty line: 0.95% (2007 est.)

Household income or consumption by percentage share:
lowest 10%: 6.7%
highest 10%: 41.1% (2002 est.)

Inflation rate (consumer prices): 1.8% (2007 est.)

Investment (gross fixed): 21.2% of GDP (2007 est.)

Budget:
revenues: $76.2 billion
expenditures: $75.65 billion (2007 est.)

Public debt: 27.9% of GDP (2007 est.)
Agriculture—products: rice, corn, vegetables, fruit, tea; pigs, poultry, beef, milk; fish
Industries: electronics, petroleum refining, armaments, chemicals, textiles, iron and steel, machinery, cement, food processing, vehicles, consumer products, pharmaceuticals
Industrial production growth rate: 7.5% (2007 est.)
Electricity—production: 235 billion kWh (2006)
Electricity—production by source:
fossil fuel: 71.4%
hydro: 6%
nuclear: 22.6%
other: 0% (2001)
Electricity—consumption: 221 billion kWh (2006)
Electricity—exports: 0 kWh (2007 est.)
Electricity—imports: 0 kWh (2007)
Oil—production: 406 bbl/day (2006 est.)
Oil—consumption: 816,700 bbl/day (2006 est.)
Oil—exports: 289,200 bbl/day (2006)
Oil—imports: 1.208 million bbl/day (2006)
Oil—proved reserves: 2.24 million bbl (1 January 2007 est.)
Natural gas—production: 462.9 million cu m (2006)
Natural gas—consumption: 10.28 billion cu m (2006)
Natural gas—exports: 0 cu m (2007)
Natural gas—imports: 10.16 billion cu m (2006)
Natural gas—proved reserves: 13.55 billion cu m (2007 est.)
Current account balance: $31.7 billion (2007)
Exports: $246.7 billion f.o.b. (2007 est.)
Exports—commodities: electronic and electrical products, metals, textiles, plastics, chemicals, auto parts (2002)
Exports—partners: China 24%, Hong Kong 15%, US 13.4%, Japan 6.7% (2007 est.)
Imports: $219.3 billion f.o.b. (2007 est.)
Imports—commodities: electronic and electrical products, machinery, petroleum, precision instruments, organic chemicals, metals (2002)
Imports—partners: Japan 21%, China 12.7%, US 12.2%, South Korea 7.1%, Saudi Arabia 4.6% (2007 est.)
Reserves of foreign exchange and gold: $274.7 billion (31 December 2007)
Debt—external: $98.44 billion (31 December 2007)
Stock of direct foreign investment—at home: $92.83 billion (2007)
Stock of direct foreign investment—abroad: $108.9 billion (2007)
Market value of publicly traded shares: $654 billion (28 December 2007)
Currency (code): New Taiwan dollar (TWD)

Currency code: TWD
Exchange rates: New Taiwan dollars per US dollar—32.84 (2007), 32.534 (2006), 31.71 (2005), 34.418 (2004), 34.575 (2003)
Fiscal year: calendar year

COMMUNICATIONS

Telephones—main lines in use: 14.497 million (2006)
Telephones—mobile cellular: 23.249 million (2006)
Telephone system:
general assessment: provides telecommunications service for every business and private need
domestic: thoroughly modern; completely digitalized
international: country code—886; numerous submarine cables provide links throughout Asia, Australia, the Middle East, Europe, and the US; satellite earth stations—2
Radio broadcast stations: AM 140, FM 229, shortwave 49
Radios: 16 million (1994)
Television broadcast stations: 76 (46 digital and 30 analog) (2007)
Televisions: 8.8 million (1998)
Internet country code: .tw
Internet hosts: 5.111 million (2007)
Internet Service Providers (ISPs): 8 (2000)
Internet users: 13.21 million (2005)

TRANSPORTATION

Airports: 41 (2007)
Airports—with paved runways:
total: 38
over 3,047 m: 8
2,438 to 3,047 m: 9
1,524 to 2,437 m: 11
914 to 1,523 m: 7
under 914 m: 3 (2007)
Airports—with unpaved runways:
total: 3
1,524 to 2,437 m: 1
under 914 m: 2 (2007)
Heliports: 4 (2007)
Pipelines: condensate 25 km; gas 661 km (2007)
Railways:
total: 1,588 km
standard gauge: 345 km 1.435-m gauge
narrow gauge: 1,093 km 1.067-m gauge
note: 150 km .762-m gauge (belonging primarily to Taiwan Sugar Corporation and Taiwan Forestry Bureau; some to other entities) (2007)
Roadways:
total: 40,262 km
paved: 38,171 km (includes 976 km of expressways)
unpaved: 2,091 km (2007)
Merchant marine:
total: 102 ships (1000 GRT or over) 2,537,256 GRT/4,203,423 DWT
by type: bulk carrier 33, cargo 20, chem-

ical tanker 2, container 21, passenger/cargo 2, petroleum tanker 15, refrigerated cargo 7, roll on/roll off 2
foreign-owned: 4 (Canada 3, France 1)
registered in other countries: 489 (Bahamas 1, Bolivia 1, Cambodia 1, Honduras 2, Hong Kong 11, Indonesia 2, Italy 11, Liberia 82, Panama 306, Singapore 60, Thailand 1, UK 11, unknown 3) (2007)
Ports and terminals: Chilung (Keelung), Kaohsiung, Taichung

MILITARY

Military branches: Army, Navy (includes Marine Corps), Air Force, Coast Guard Administration, Armed Forces Reserve Command, Combined Service Forces Command, Armed Forces Police Command
Military service age and obligation: 19–35 years of age for male compulsory military service; service obligation 16 months (to be shortened to 14 months as of July 2007 and to 12 months in 2008); women may enlist; women in Air Force service are restricted to noncombat roles; reserve obligation to age 30 (2007)
Manpower available for military service:
males age 16–49: 6,283,134
females age 16–49: 6,098,599 (2008 est.)
Manpower fit for military service:
males age 16–49: 5,112,737
females age 16–49: 5,036,346 (2008 est.)
Manpower reaching militarily significant age annually:
males age 16–49: 164,883
females age 16–49: 152,085 (2008 est.)
Military expenditures—percent of GDP: 2.2% (2006)

TRANSNATIONAL ISSUES

Disputes—international: involved in complex dispute with China, Malaysia, Philippines, Vietnam, and possibly Brunei over the Spratly Islands; the 2002 "Declaration on the Conduct of Parties in the South China Sea" has eased tensions but falls short of a legally binding "code of conduct" desired by several of the disputants; Paracel Islands are occupied by China, but claimed by Taiwan and Vietnam; in 2003, China and Taiwan became more vocal in rejecting both Japan's claims to the uninhabited islands of the Senkaku-shoto (Diaoyu Tai) and Japan's unilaterally declared exclusive economic zone in the East China Sea where all parties engage in hydrocarbon prospecting
Illicit drugs: regional transit point for heroin, methamphetamine, and precursor chemicals; transshipment point for drugs to Japan; major problem with domestic consumption of methamphetamine and heroin; rising problems with use of ketamine and club drugs

INTRODUCTION

Preliminary statement: The evolution of the European Union (EU) from a regional economic agreement among six neighboring states in 1951 to today's supranational organization of 27 countries across the European continent stands as an unprecedented phenomenon in the annals of history. Dynastic unions for territorial consolidation were long the norm in Europe. On a few occasions even country-level unions were arranged—the Polish-Lithuanian Commonwealth and the Austro-Hungarian Empire were examples—but for such a large number of nation-states to cede some of their sovereignty to an overarching entity is truly unique. Although the EU is not a federation in the strict sense, it is far more than a free-trade association such as ASEAN, NAFTA, or Mercosur, and it has many of the attributes associated with independent nations: its own flag, anthem, founding date, and currency, as well as an incipient common foreign and security policy in its dealings with other nations. In the future, many of these nation-like characteristics are likely to be expanded. Thus, inclusion of basic intelligence on the EU has been deemed appropriate as a new, separate entity in The World Factbook. However, because of the EU's special status, this description is placed after the regular country entries.

Background: Following the two devastating World Wars of the first half of the 20th century, a number of European leaders in the late 1940s became convinced that the only way to establish a lasting peace was to unite the two chief belligerent nations—France and Germany—both economically and politically. In 1950, the French Foreign Minister Robert SCHUMAN proposed an eventual union of all Europe, the first step of which would be the integration of the coal and steel industries of Western Europe. The following year the European Coal and Steel Community (ECSC) was set up when six members, Belgium, France, West Germany, Italy, Luxembourg, and the Netherlands, signed the Treaty of Paris. The ECSC was so successful that within a few years the decision was made to integrate other parts of the countries' economies. In 1957, the Treaties of Rome created the European Economic Community (EEC) and the European Atomic Energy Community (Euratom), and the six member states undertook to eliminate trade barriers among themselves by forming a common market. In 1967, the institutions of all three communities were formally merged into the European Community (EC), creating a single Commission, a single Council of Ministers, and the European Parliament. Members of the European Parliament were initially selected by national parliaments, but in 1979 the first direct elections were undertaken and they have been held every five years since. In 1973, the first enlargement of the EC took place with the addition of Denmark, Ireland, and the United Kingdom. The 1980s saw further membership expansion with Greece joining in 1981 and Spain and Portugal in 1986. The 1992 Treaty of Maastricht laid the basis for further forms of cooperation in foreign and defense policy, in judicial and internal affairs, and in the creation of an economic and monetary union—including a common currency. This further integration created the European Union (EU). In 1995, Austria, Finland, and Sweden joined the EU, raising the membership total to 15. A new currency, the euro, was launched in world money markets on 1 January 1999; it became the unit of exchange for all of the EU states except the United Kingdom, Sweden, and Denmark. In 2002, citizens of the 12 euro-area countries began using the euro banknotes and coins. Ten new countries joined the EU in 2004—Cyprus, the Czech Republic, Estonia, Hungary, Latvia, Lithuania, Malta, Poland, Slovakia, and Slovenia—and in 2007 Bulgaria and Romania joined, bringing the current membership to 27. In order to ensure that the EU can continue to function efficiently with an expanded membership, the Treaty of Nice (in force as of 1 February 2003) set forth rules streamlining the size and procedures of EU institutions. An effort to establish an EU constitution, begun in October 2004, failed to attain unanimous ratification. A new effort, undertaken in June 2007, calls for the creation of an Intergovernmental Conference to form a political agreement, known as the Reform Treaty, which is to serve as a constitution. Unlike the constitution, however, the Reform Treaty would amend existing treaties rather than replace them.

GEOGRAPHY

Location: Europe between the North Atlantic Ocean in the west and Russia, Belarus, and Ukraine to the east

Map references: Europe

Area:

total: 4,324,782 sq km

Area—comparative: less than one-half the size of the US

Land boundaries:

total: 12,440.8 km

border countries: Albania 282 km, Andorra 120.3 km, Belarus 1,050 km, Croatia 999 km, Holy See 3.2 km, Liechtenstein 34.9 km, Macedonia 394 km, Moldova 450 km, Monaco 4.4 km, Norway 2,348 km, Russia 2,257 km, San Marino 39 km, Serbia 945 km, Switzerland 1,811 km, Turkey 446 km, Ukraine 1,257 km

note: data for European Continent only

Coastline: 65,992.9 km

Maritime claims: NA

Climate: cold temperate; potentially subarctic in the north to temperate; mild wet winters; hot dry summers in the south

Terrain: fairly flat along the Baltic and Atlantic coast; mountainous in the central and southern areas

Elevation extremes:

lowest point: Lammefjord, Denmark -7 m; Zuidplaspolder, Netherlands -7 m

highest point: Mont Blanc 4,807 m; note—situated on the border between France and Italy

Natural resources: iron ore, natural gas, petroleum, coal, copper, lead, zinc, bauxite, uranium, potash, salt, hydropower, arable land, timber, fish

Land use:

arable land: NA

permanent crops: NA

other: NA

Irrigated land: 168,050 sq km (2003 est.)

Natural hazards: flooding along coasts; avalanches in mountainous area; earthquakes in the south; volcanic eruptions in Italy; periodic droughts in Spain; ice floes in the Baltic

Environment—current issues: NA

Environment—international agreements: *party to:* Air Pollution, Air Pollution-Nitrogen Oxides, Air Pollution-Persistent Organic Pollutants, Air Pollution-Sulphur 94, Antarctic-

Marine Living Resources, Biodiversity, Climate Change, Climate Change-Kyoto Protocol, Desertification, Hazardous Wastes, Law of the Sea, Ozone Layer Protection, Tropical Timber 82, Tropical Timber 94

signed but not ratified: Air Pollution-Volatile Organic Compounds

PEOPLE

Population: 491,018,677 (July 2008 est.)

Age structure:

0–14 years: 15.7% (male 37,208,905/female 35,254,445)

15–64 years: 67.2% (male 155,807,769/female 153,690,235)

65 years and over: 17.1% (male 32,592,595/female 46,273,197) (2008 est.)

Median age: note—see individual country entries of member states

Population growth rate: 0.12% (2008 est.)

Birth rate: 9.97 births/1,000 population (2008 est.)

Death rate: 10.22 deaths/1,000 population (2008 est.)

Net migration rate: 1.46 migrant(s)/1,000 population (2008 est.)

Sex ratio:

at birth: NA

under 15 years: 1.06 male(s)/female

15–64 years: 1.01 male(s)/female

65 years and over: 0.69 male(s)/female

total population: 0.96 male(s)/female (2007 est.)

Infant mortality rate:

total: 5.84 deaths/1,000 live births

male: 6.51 deaths/1,000 live births

female: 5.13 deaths/1,000 live births (2008 est.)

Life expectancy at birth:

total population: 78.51 years

male: 75.39 years

female: 81.82 years (2008 est.)

Total fertility rate: 1.5 children born/woman (2008 est.)

HIV/AIDS—adult prevalence rate: note—see individual country entries of member states

HIV/AIDS—people living with HIV/AIDS: note—see individual country entries of member states

HIV/AIDS—deaths: note—see individual country entries of member states

Religions: Roman Catholic, Protestant, Orthodox, Muslim, Jewish

Languages: Bulgarian, Czech, Danish, Dutch, English, Estonian, Finnish, French, Gaelic, German, Greek, Hungarian, Italian, Latvian, Lithuanian, Maltese, Polish, Portuguese, Romanian, Slovak, Slovene, Spanish, Swedish

note: only official languages are listed; German, the major language of Germany, Austria, and Switzerland, is the most widely spoken mother tongue—over 19% of the EU popula-tion; English is the most widely spoken language—about 49% of the EU popula-tion is conversant with it (2007)

GOVERNMENT

Union name: *conventional long form:* European Union

abbreviation: EU

Political structure: a hybrid intergovern-mental and supranational organization

Capital: *name:* Brussels (Belgium), Strasbourg (France), Luxembourg

geographic coordinates: 50 50 N, 4 20 E

time difference: UTC+1 (6 hours ahead of Washington, DC during Standard Time)

daylight saving time: +1hr, begins last Sunday in March; ends last Sunday in October

note: the Council of the European Union meets in Brussels, Belgium, the European Parliament meets in Brussels and Strasbourg, France, and the Court of Justice of the European Communities meets in Luxembourg

Member states: 27 countries: Austria, Belgium, Bulgaria, Cyprus, Czech Republic, Denmark, Estonia, Finland, France, Germany, Greece, Hungary, Ireland, Italy, Latvia, Lithuania, Luxembourg, Malta, Netherlands, Poland, Portugal, Romania, Slovakia, Slovenia, Spain, Sweden, UK; note—Canary Islands (Spain), Azores and Madeira (Portugal), French Guiana, Guadeloupe, Martinique, and Reunion (France) are sometimes listed separately even though they are legally a part of Spain, Portugal, and France; candidate countries: Croatia, Macedonia, Serbia, Turkey

Independence: 7 February 1992 (Maastricht Treaty signed establishing the EU); 1 November 1993 (Maastricht Treaty entered into force)

National holiday: Europe Day 9 May (1950); note—a Union-wide holiday, the day that Robert SCHUMAN pro-posed the creation of the European Coal and Steel Community to achieve an organized Europe

Constitution: based on a series of treaties: the Treaty of Paris, which set up the European Coal and Steel Community (ECSC) in 1951; the Treaties of Rome, which set up the European Economic Community (EEC) and the European Atomic Energy Community (Euratom) in 1957; the Single European Act in 1986; the Treaty on European Union (Maastricht) in 1992; the Treaty of Amsterdam in 1997; and the Treaty of Nice in 2003; note—a new draft Constitutional Treaty, signed on 29 October 2004 in Rome, gave member states two years for ratification either by parliamentary vote or national referendum before it was scheduled to take effect on 1 November 2006; defeat in French and Dutch referenda in May-June 2005 dealt a severe setback to the ratification process; in June 2007, the European Council agreed on a clear and concise mandate for an Intergovernmental Conference to form a political agreement and put it into legal form; this agreement, known as the Reform Treaty, is to serve as a constitu-tion and will be presented to the European Council in October 2007, in order to begin the ratification process

Legal system: comparable to the legal systems of member states; first suprana-tional law system

Suffrage: 18 years of age; universal

Executive branch:

chief of union: President of the European Commission Jose Manuel DURAO BARROSO (since 22 November 2004)

cabinet: European Commission (com-posed of 27 members, one from each member country; each commissioner responsible for one or more policy areas)

elections: the president of the European Commission is designated by member governments and is confirmed by the European Parliament; working from member state recommendations, the Commission president then assembles a "college" of Commission members; the European Parliament confirms the entire Commission for a five-year term; the last confirmation process was held 18 November 2004 (next to be held in 2009)

election results: European Parliament approved the European Commission by an approval vote of 449 to 149 with 82 abstentions

note: the European Council brings together heads of state and government and the president of the European Commission and meets at least four times a year; its aim is to provide the impetus for the major political issues relating to European integration and to issue general policy guidelines

Legislative branch: two legislative bodies consisting of the Council of the European Union (27 member-state min-isters having 345 votes; the number of votes is roughly proportional to member-states' population; note—the Council is the main decision-making body of the EU) and the European Parliament (785 seats, as of 1 January 2007; seats allo-cated among member states by propor-tion to population; members elected by direct universal suffrage for a five-year term)

elections: last held 10–13 June 2004 (next to be held June 2009)

election results: percent of vote—NA; seats by party—EPP-ED 268, PES 202, ALDE 88, Greens/EFA 42, EUL/NGL 41, IND/DEM 36, UEN 27, independ-ents 28; note—seats by party as of 1

December 2007—EPP-ED 275, PES 217, ALDE 104, UEN 44, Greens/EFA 42, EUL/NGL 41, IND/DEM 24, independents 34, 4 unaccounted for

Judicial branch: Court of Justice of the European Communities (ensures that the treaties are interpreted and applied uniformly throughout the EU; resolve constitutional issues among the EU institutions)—27 justices (one from each member state) appointed for a six-year term; note—for the sake of efficiency, the court can sit with 13 justices known as the "Grand Chamber"; Court of First Instance—27 justices appointed for a six-year term

Political parties and leaders: Confederal Group of the European United Left-Nordic Green Left or EUL/NGL [Francis WURTZ]; European People's Party-European Democrats or EPP-ED [Joseph DAUL]; Group of the Alliance of Liberals and Democrats for Europe or ALDE [Graham R. WATSON]; Group of Greens/European Free Alliance or Greens/EFA [Monica FRASSONI and Daniel Marc COHN-BENDIT]; Identity, Tradition, Sovereignty Group or ITS [Bruno G O L L N I S C H] ; Independence/Democracy Group or IND/DEM [Jens-Peter BONDE and Nigel FARAGE]; Socialist Group in the European Parliament or PES [Martin SCHULZ]; Union for Europe of the Nations Group or UEN [Brian CROWLEY and Cristiana MUSCARDINI]

International organization participation: *European Union:* ARF (dialogue member), ASEAN (dialogue member), IDA, OAS (observer), UN (observer) *European Community:* Australian Group, CBSS, CERN, FAO, EBRD, G-10, NAM (observer), NSG (observer), OECD, UNRWA, WCO, WTO, ZC (observer)
European Central Bank: BIS
European Investment Bank: EBRD, WADB (nonregional member)

Diplomatic representation in the US: *chief of mission:* Ambassador John BRUTON
chancery: 2300 M Street, NW, Washington, DC 20037
telephone: [1] (202) 862-9500
FAX: [1] (202) 429-1766
Diplomatic representation from the US: *chief of mission:* Ambassador C. Boyden GRAY
embassy: 13 Zinnerstraat/Rue Zinner, B-1000 Brussels
mailing address: same as above
telephone: [32] (2) 508-2222
FAX: [32] (2) 512-5720
Flag description: blue field with 12 five-pointed gold stars arranged in a circle in the center, representing the union of the peoples of Europe; the number of stars is fixed

ECONOMY

Economy—overview: Internally, the EU is attempting to lower trade barriers, adopt a common currency, and move toward convergence of living standards. Internationally, the EU aims to bolster Europe's trade position and its political and economic power. Because of the great differences in per capita income among member states (from $7,000 to $69,000) and historic national animosities, the EU faces difficulties in devising and enforcing common policies. For example, since 2003 Germany and France have flouted the member states' treaty obligation to prevent their national budgets from running more than a 3% deficit. In 2004 and 2007, the EU admitted 10 and two countries, respectively, that are, in general, less advanced technologically and economically than the other 15. Eleven established EU member states introduced the euro as their common currency on 1 January 1999 (Greece did so two years later), but the UK, Sweden, and Denmark chose not to participate. Of the 12 most recent member states, only Slovenia (1 January 2007) and Cyprus and Malta (1 January 2008) have adopted the euro; the remaining nine are legally required to adopt the currency upon meeting EU's fiscal and monetary convergence criteria.

GDP (purchasing power parity): $14.38 trillion (2007 est.)
GDP (official exchange rate): $16.62 trillion (2007 est.)
GDP—real growth rate: 3% (2007 est.)
GDP—per capita (PPP): $32,300 (2007 est.)
GDP—composition by sector:
agriculture: 2%
industry: 27.1%
services: 70.7% (2006 est.)
Labor force: 222.5 million (2006 est.)
Labor force—by occupation: *agriculture:* 4.4%
industry: 27.1%
services: 67.1%
note: the remainder is in miscellaneous public and private sector industries and services (2002 est.)
Unemployment rate: 8.5% (2006 est.)
Population below poverty line: note—see individual country entries of member states
Household income or consumption by percentage share:
lowest 10%: 2.8%
highest 10%: 25.1% (2001 est.)
Distribution of family income—Gini index: 30.7 (2003 est.)
Inflation rate (consumer prices): 1.8% (2006 est.)

Investment (gross fixed): 21.6% of GDP (2006 est.)
Agriculture—products: wheat, barley, oilseeds, sugar beets, wine, grapes; dairy products, cattle, sheep, pigs, poultry; fish
Industries: among the world's largest and most technologically advanced, the European Union industrial base includes: ferrous and non-ferrous metal production and processing, metal products, petroleum, coal, cement, chemicals, pharmaceuticals, aerospace, rail transportation equipment, passenger and commercial vehicles, construction equipment, industrial equipment, shipbuilding, electrical power equipment, machine tools and automated manufacturing systems, electronics and telecommunications equipment, fishing, food and beverage processing, furniture, paper, textiles, tourism
Industrial production growth rate: 2.7% (2006 est.)
Electricity—production: 3.02 trillion kWh (2004 est.)
Electricity—consumption: 2.82 trillion kWh (2004 est.)
Electricity—exports: NA kWh
Electricity—imports: NA kWh
Oil—production: 2.874 million bbl/day (2004)
Oil—consumption: 14.55 million bbl/day (2004)
Oil—exports: 6.979 million bbl/day (2001)
Oil—imports: 17.71 million bbl/day (2001)
Oil—proved reserves: 7.072 billion bbl (1 January 2005)
Natural gas—production: 215.4 billion cu m (2005 est.)
Natural gas—consumption: 496.1 billion cu m (2005 est.)
Natural gas—exports: 76.48 billion cu m (2005 est.)
Natural gas—imports: 361.2 billion cu m (2005 est.)
Natural gas—proved reserves: 3.31 trillion cu m (1 January 2006 est.)
Current account balance: $NA
Exports: $1.33 trillion; note—external exports, excluding intra-EU trade (2005)
Exports—commodities: machinery, motor vehicles, aircraft, plastics, pharmaceuticals and other chemicals, fuels, iron and steel, nonferrous metals, wood pulp and paper products, textiles, meat, dairy products, fish, alcoholic beverages.
Exports—partners: US 23.3%, Switzerland 7.6%, Russia 5.2%, China 4.8% (2006)
Imports: $1.466 trillion; note—external imports, excluding intra-EU trade (2005)
Imports—commodities: machinery, vehicles, aircraft, plastics, crude oil, chemicals, textiles, metals, foodstuffs, clothing

Imports—partners: US 13.8%, China 13.4%, Russia 8.2%, Japan 6.2% (2006)
Reserves of foreign exchange and gold: $NA
Currency (code): euro, British pound, Bulgarian lev, Czech koruna, Danish krone, Estonian kroon, Hungarian forint, Latvian lat, Lithuanian litas, Polish zloty, Romanian leu, Slovak koruna, Swedish krona
Currency code: EUR
Exchange rates: euros per US dollar— 0.7345 (2007), 0.7964 (2006), 0.8041 (2005), 0.8054 (2004), 0.886 (2003)
Fiscal year: NA

COMMUNICATIONS

Telephones—main lines in use: 238 million (2005)
Telephones—mobile cellular: 466 million (2005)
Telephone system: note—see individual country entries of member states
Radio broadcast stations: AM 930, FM 13,655, shortwave 71 (1998); note—sum of individual country radio broadcast stations; there is also a European-wide station (Euroradio)
Television broadcast stations: 2,700 (1995); note—sum of individual country television broadcast stations excluding repeaters; there is also a European-wide station (Eurovision)
Internet country code: .eu (effective 2005); note—see country entries of member states for individual country codes
Internet hosts: 50.5 million (2005); note—sum of individual country Internet hosts
Internet users: 247 million (2006)

TRANSPORTATION

Airports: 3,393 (2006)
Airports—with paved runways:
total: 1,991
over 3,047m: 110
2,438 to 3,047m: 347
1,524 to 2,437m: 545
914 to 1,523m: 420

under 914m: 569 (2007)
Airports—with unpaved runways:
total: 1,373
over 3,047m: 2
2,438 to 3,047m: 5
1,524 to 2,437m: 30
914m to 1,523m: 267
under 914m: 1,043 (2007)
Heliports: 100 (2007)
Railways: *total:* 236,436 km
broad gauge: 28,250 km
standard gauge: 200,401 km
narrow gauge: 7,771 km
other: 23 km (2007)
Roadways:
total: 5,269,163 km (includes 56,555 km of expressways)
paved: 4,465,493 km
unpaved: 803,670 km (2006)
Waterways: 52,332 km (2006)
Ports and terminals: Antwerp (Belgium), Barcelona (Spain), Braila (Romania), Bremen (Germany), Burgas (Bulgaria), Constanta (Romania), Copenhagen (Denmark), Galati (Romania), Gdansk (Poland), Hamburg (Germany), Helsinki (Finland), Las Palmas (Canary Islands, Spain), Le Havre (France), Lisbon (Portugal), London (UK), Marseille (France), Naples (Italy), Peiraiefs or Piraeus (Greece), Riga (Latvia), Rotterdam (Netherlands), Stockholm (Sweden), Talinn (Estonia), Tulcea (Romania), Varna (Bulgaria)

MILITARY

Military—note: the five-nation Eurocorps—created in 1992 by France, Germany, Belgium, Spain, and Luxembourg—has deployed troops and police on peacekeeping missions to Bosnia-Herzegovina, Macedonia, and the Democratic Republic of the Congo and assumed command of the ISAF in Afghanistan in August 2004; Eurocorps directly commands the 5,000-man Franco-German Brigade, the Multinational Command Support Brigade, and EUFOR in Bosnia and Herzegovina; in November 2004, the EU Council of Ministers formally committed to creating 13 1,500-man battle groups by the end of 2007, to respond to international crises on a rotating basis; 22 of the EU's 25 nations have agreed to supply troops; France, Italy, and the UK formed the first of three battle groups in 2005; Norway, Sweden, Estonia, and Finland established the Nordic Battle Group effective 1 January 2008; nine other groups are to be formed; a rapid-reaction naval EU Maritime Task Group was stood up in March 2007 (2007)

TRANSNATIONAL ISSUES

Disputes—international: as a political union, the EU has no border disputes with neighboring countries, but Estonia has no land boundary agreements with Russia, Slovenia disputes its land and maritime boundaries with Croatia, and Spain has territorial and maritime disputes with Morocco and with the UK over Gibraltar; the EU has set up a Schengen area—consisting of 22 EU member states that have signed the convention implementing the Schengen agreements or "acquis" (1985 and 1990) on the free movement of persons and the harmonization of border controls in Europe; these agreements became incorporated into EU law with the implementation of the 1997 Treaty of Amsterdam on 1 May 1999; in addition, non-EU states Iceland and Norway (as part of the Nordic Union) have been included in the Schengen area since 1996 (full members in 2001), bringing the total current membership to 24; the UK (since 2000) and Ireland (since 2002) take part in only some aspects of the Schengen area, especially with respect to police and criminal matters; nine of the 12 new member states that joined the EU in 2004 joined Schengen on 21 December 2007; of the three remaining EU states, Cyprus is expected to join by 2009, while Romania and Bulgaria continue to enhance their border security systems

ABBREVIATIONS

ABEDA	Arab Bank for Economic Development in Africa
ACCT	Agency for the French-Speaking Community (see International Organization of the French-speaking World)
ACP Group	African, Caribbean, and Pacific Group of States
AfDB	African Development Bank
AFESD	Arab Fund for Economic and Social Development
Air Pollution	Convention on Long-Range Transboundary Air Pollution
Air Pollution-Nitrogen Oxides	Protocol to the 1979 Convention on Long-Range Transboundary Air Pollution Concerning the Control of Emissions of Nitrogen Oxides or Their Transboundary Fluxes
Air Pollution-Persistent Organic Pollutants	Protocol to the 1979 Convention on Long-Range Transboundary Air Pollution on Persistent Organic Pollutants
Air Pollution-Sulphur 85	Protocol to the 1979 Convention on Long-Range Transboundary Air Pollution on the Reduction of Sulphur Emissions or Their Transboundary Fluxes by at Least 30%
Air Pollution-Sulphur 94	Protocol to the 1979 Convention on Long-Range Transboundary Air Pollution on Further Reduction of Sulphur Emissions
Air Pollution-Volatile Organic Compounds	Protocol to the 1979 Convention on Long-Range Transboundary Air Pollution Concerning the Control of Emissions of Volatile Organic Compounds or Their Transboundary Fluxes
AMF	Arab Monetary Fund
AMU	Arab Maghreb Union
Antarctic-Environmental Protocol	Protocol on Environmental Protection to the Antarctic Treaty
Antarctic Marine Living Resources	Convention on the Conservation of Antarctic Marine Living Resources
Antarctic Seals	Convention for the Conservation of Antarctic Seals
ANZUS	Australia-New Zealand-United States Security Treaty
APEC	Asia-Pacific Economic Cooperation
Arabsat	Arab Satellite Communications Organization
ARF	ASEAN Regional Forum
ADB	Asian Development Bank
ASEAN	Association of Southeast Asian Nations
AU	African Union
Autodin	Automatic Digital Network
BA	Baltic Assembly
bbl/day	barrels per day
BCIE	Central American Bank for Economic Integration
BDEAC	Central African States Development Bank
Benelux	Benelux Economic Union
BIMSTEC	Bay of Bengal Initiative for Multisectoral Technical and Economic Cooperation
Biodiversity	Convention on Biological Diversity
BGN	United States Board on Geographic Names
BIS	Bank for International Settlements
BSEC	Black Sea Economic Cooperation Zone
C	Commonwealth
c.i.f.	cost, insurance, and freight
CACM	Central American Common Market
CAEU	Council of Arab Economic Unity
CAN	Andean Community of Nations
Caricom	Caribbean Community and Common Market
CB	citizen's band mobile radio communications
CBSS	Council of the Baltic Sea States
CCC	Customs Cooperation Council
CDB	Caribbean Development Bank
CE	Council of Europe
CEI	Central European Initiative
CEMAC	Economic and Monetary Community of Central Africa
CEPGL	Economic Community of the Great Lakes Countries
CERN	European Organization for Nuclear Research

CEPT	Conference Europeanne des Poste et Telecommunications
CIS	Commonwealth of Independent States
CITES	see Endangered Species
Climate Change	United Nations Framework Convention on Climate Change
Climate Change-Kyoto Protocol	Kyoto Protocol to the United Nations Framework Convention on Climate Change
COCOM	Coordinating Committee on Export Controls
Comsat	Communications Satellite Corporation
COMESA	Common Market for Eastern and Southern Africa
CP	Colombo Plan
CPLP	Comunidade dos Paises de Lingua Portuguesa
CSTO	Collective Security Treaty Organization
CTBTO	Preparation commission for the Nuclear-Ban-Treaty Operation
CY	calendar year
DC	developed country
DDT	dichloro-diphenyl-trichloro-ethane
Desertification	United Nations Convention to Combat Desertification in Those Countries Experiencing Serious Drought and/or Desertification, Particularly in Africa
DIA	United States Defense Intelligence Agency
DSN	Defense Switched Network
DST	daylight savings time
DWT	deadweight ton
EAC	East African Community
EADB	East African Development Bank
EAEC	Eurasian Economic Community
EAPC	Euro-Atlantic Partnership Council
EBRD	European Bank for Reconstruction and Development
EC	European Community
ECA	Economic Commission for Africa
ECE	Economic Commission for Europe
ECLAC	Economic Commission for Latin America and the Caribbean
ECO	Economic Cooperation Organization
ECOSOC	Economic and Social Council
ECOWAS	Economic Community of West African States
ECSC	European Coal and Steel Community
EEC	European Economic Community
EFTA	European Free Trade Association
EEZ	exclusive economic zone
EIB	European Investment Bank
EMU	European Monetary Union
Endangered Species	Convention on the International Trade in Endangered Species of Wild Flora and Fauna (CITES)
Entente	Council of the Entente
Environmental Modification	Convention on the Prohibition of Military or Any Other Hostile Use of Environmental Modification Techniques
ESA	European Space Agency
ESCAP	Economic and Social Commission for Asia and the Pacific
ESCWA	Economic and Social Commission for Western Asia
est.	estimate
EU	European Union
Euratom	European Atomic Energy Community
Eutelsat	European Telecommunications Satellite Organization
Ex-Im	Export-Import Bank of the United States
f.o.b.	free on board
FAO	Food and Agriculture Organization
FAX	facsimile
FLS	Front Line States
FOC	flags of convenience
FSU	former Soviet Union
FY	fiscal year

FZ	Franc Zone
G-2	Group of 2
G-3	Group of 3
G-5	Group of 5
G-6	Group of 6
G-7	Group of 7
G-8	Group of 8
G-9	Group of 9
G-10	Group of 10
G-15	Group of 15
G-11	Group of 11
G-24	Group of 24
G-77	Group of 77
GATT	General Agreement on Tariffs and Trade; now WTO
GCC	Gulf Cooperation Council
GCTU	General Confederation of Trade Unions
GDP	gross domestic product
GMT	Greenwich Mean Time
GNP	gross national product
GRT	gross register ton
GSM	global system for mobile cellular communications
GUAM	Organization for Democracy and Economic Development; acronym for member states— Georgia, Ukraine, Azerbaijan, Moldova
GWP	gross world product
Hazardous Wastes	Basel Convention on the Control of Transboundary Movements of Hazardous Wastes and Their Disposal
HF	high-frequency
HIV/AIDS	human immunodeficiency virus/acquired immune deficiency syndrome
IADB	Inter-American Development Bank
IAEA	International Atomic Energy Agency
IANA	Internet Assigned Numbers Authority
IBRD	International Bank for Reconstruction and Development (World Bank)
ICAO	International Civil Aviation Organization
ICC	International Chamber of Commerce
ICCt	International Criminal Court
ICJ	International Court of Justice (World Court)
ICRC	International Committee of the Red Cross
ICRM	International Red Cross and Red Crescent Movement
ICSID	International Center for Secretariat of Investment Disputes
ICTR	International Criminal Tribunal for Rwanda
ICTY	International Criminal Tribunal for the former Yugoslavia
IDA	International Development Association
IDB	Islamic Development Bank
IDP	Internally Displaced Person
IEA	International Energy Agency
IFAD	International Fund for Agricultural Development
IFC	International Finance Corporation
IFRCS	International Federation of Red Cross and Red Crescent Societies
IGAD	Inter-Governmental Authority on Development
IHO	International Hydrographic Organization
ILO	International Labor Organization
IMF	International Monetary Fund
IMO	International Maritime Organization
IMSO	International Mobile Satellite Organization
Inmarsat	International Maritime Satellite Organization
InOC	Indian Ocean Commission
INSTRAW	International Research and Training Institute for the Advancement of Women
Intelsat	International Telecommunications Satellite Organization
Interpol	International Criminal Police Organization

Intersputnik	International Organization of Space Communications
IOC	International Olympic Committee
IOM	International Organization for Migration
IPU	Inter-parliamentary Union
ISO	International Organization for Standardization
ISP	Internet Service Provider
ITSO	International Telecommunications Satellites Organization
ITU	International Telecommunication Union
ITUC	International Trade Union Confederation, the successor to ICFTU (International Confederation of Free Trade Unions) and the WCL (World Confederation of Labor)
kHz	kilohertz
km	kilometer
kW	kilowatt
kWh	kilowatt-hour
LAES	Latin American Economic System
LAIA	Latin American Integration Association
LAS	League of Arab States
Law of the Sea	United Nations Convention on the Law of the Sea (LOS)
LDC	less developed country
LLDC	least developed country
London Convention	see Marine Dumping
LOS	see Law of the Sea
m	meter
Marecs	Maritime European Communications Satellite
Marine Dumping	Convention on the Prevention of Marine Pollution by Dumping Wastes and Other Matter
Marine Life Conservation	Convention on Fishing and Conservation of Living Resources of the High Seas
MARPOL	see Ship Pollution
Medarabtel	Middle East Telecommunications Project of the International Telecommunications Union
Mercosur	Southern Cone Common Market
MHz	megahertz
MICAH	International Civilian Support Mission in Haiti
MINURSO	United Nations Mission for the Referendum in Western Sahara
MIGA	Multilateral Investment Geographic Agency
MINUSTAH	United Nations Stabilization Mission in Haiti
MONUC	United Nations Organization Mission in the Democratic Republic of the Congo
NA	not available
NAFTA	North American Free Trade Agreement
NAM	Nonaligned Movement
NATO	North Atlantic Treaty Organization
NC	Nordic Council
NEA	Nuclear Energy Agency
NEGL	negligible
NGA	National Geospatial-Intelligence Agency
NIB	Nordic Investment Bank
NIC	newly industrializing country
NIE	newly industrializing economy
NIS	new independent states
nm	nautical mile
NMT	Nordic Mobile Telephone
NSG	Nuclear Suppliers Group
Nuclear Test Ban	Treaty Banning Nuclear Weapons Tests in the Atmosphere, in Outer Space, and Under Water
NZ	New Zealand
OAPEC	Organization of Arab Petroleum Exporting Countries
OAS	Organization of American States
OAU	Organization of African Unity; see African Union
ODA	official development assistance
OECD	Organization for Economic Cooperation and Development

OECS	Organization of Eastern Caribbean States
OHCHR	Office of the United Nations High Commissioner for Human Rights
OIC	Organization of the Islamic Conference
OIF	International Organization of the French-speaking World
ONUB	United Nations Operation in Burundi
OOF	other official flows
OPANAL	Agency for the Prohibition of Nuclear Weapons in Latin America and the Caribbean
OPCW	Organization for the Prohibition of Chemical Weapons
OPEC	Organization of Petroleum Exporting Countries
OSCE	Organization for Security and Cooperation in Europe
Ozone Layer Protection	Montreal Protocol on Substances That Deplete the Ozone Layer
PCA	Permanent Court of Arbitration
PFP	Partnership for Peace
PIF	Pacific Islands Forum
PPP	purchasing power parity
Ramsar	see Wetlands
RG	Rio Group
SAARC	South Asian Association for Regional Cooperation
SACU	Southern African Customs Union
SACEP	South Asia Co-opeative Environment Programme
SADC	Southern African Development Community
SCO	Shanghai Cooperation Organization
SAFE	South African Far East Cable
SECI	Southeast European Cooperative Initiative
SHF	super-high-frequency
Ship Pollution	Protocol of 1978 Relating to the International Convention for the Prevention of Pollution From Ships, 1973 (MARPOL)
Sparteca	South Pacific Regional Trade and Economic Cooperation Agreement
SPC	Secretariat of the Pacific Communities
SPF	South Pacific Forum
sq km	square kilometer
sq mi	square mile
TAT	Trans-Atlantic Telephone
TEU	Twenty-Foot Equivalent Unit
Tropical Timber 83	International Tropical Timber Agreement, 1983
Tropical Timber 94	International Tropical Timber Agreement, 1994
UAE	United Arab Emirates
UDEAC	Central African Customs and Economic Union
UHF	ultra-high-frequency
UK	United Kingdom
UN	United Nations
UNAMSIL	United Nations Mission in Sierra Leone
UNCLOS	United Nations Convention on the Law of the Sea, also know as LOS
UNCTAD	United Nations Conference on Trade and Development
UNDCP	United Nations Drug Control Program
UNDOF	United Nations Disengagement Observer Force
UNDP	United Nations Development Program
UNEP	United Nations Environment Program
UNESCO	United Nations Educational, Scientific, and Cultural Organization
UNFICYP	United Nations Peace-keeping Force in Cyprus
UNFPA	United Nations Population Fund
UNHCR	United Nations High Commissioner for Refugees
UNICEF	United Nations Children's Fund
UNICRI	United Nations Interregional Crime and Justice Research Institute
UNIDIR	United Nations Institute for Disarmament Research
UNIDO	United Nations Industrial Development Organization
UNIFIL	United Nations Interim Force in Lebanon
UNITAR	United Nations Institute for Training and Research
UNMEE	United Nations Mission in Ethiopia and Eritrea

UNMIK	United Nations Interim Administration Mission in Kosovo
UNMIL	United Nations Mission in Liberia
UNMIT	United Nations Integrated Mission in Timor-Leste
UNMOGIP	United Nations Military Observer Group in India and Pakistan
UNMOVIC	United Nations Monitoring, Verification, and Inspection Commission
UNOCI	United Nations Operation in Cote d'Ivoire
UNOMIG	United Nations Observer Mission in Georgia
UNOPS	United Nations Office of Project Services
UNRISD	United Nations Research Institute for Social Development
UNRWA	United Nations Relief and Works Agency for Palestine Refugees in the Near East
UNSC	United Nations Security Council
UNSSC	Untied Nations System Staff College
UNTSO	United Nations Truce Supervision Organization
UNU	United Nations University
UNWTO	World Tourism Organization
UPU	Universal Postal Union
US	United States
USSR	Union of Soviet Socialist Republics (Soviet Union); used for information dated before 25 December 1991
UTC	Coordinated Universal Time
UV	ultra violet
VHF	very-high-frequency
VSAT	very small aperture terminal
WADB	West African Development Bank
WAEMU	West African Economic and Monetary Union
WCL	World Confederation of Labor
WCO	World Customs Organization
Wetlands	Convention on Wetlands of International Importance Especially As Waterfowl Habitat
WEU	Western European Union
WFP	World Food Program
WFTU	World Federation of Trade Unions
Whaling	International Convention for the Regulation of Whaling
WHO	World Health Organization
WIPO	World Intellectual Property Organization
WMO	World Meteorological Organization
WP	Warsaw Pact
WTO	World Trade Organization
ZC	Zangger Committee

INTERNATIONAL ORGANIZATIONS AND GROUPS

advanced developing countries
another term for those less developed countries (LDCs) with particularly rapid industrial development; see newly industrializing economies (NIEs)

advanced economies
a term used by the International Monetary FUND (IMF) for the top group in its hierarchy of advanced economies, countries in transition, and developing countries; it includes the following 28 advanced economies: Australia, Austria, Belgium, Canada, Denmark, Finland, France, Germany, Greece, Hong Kong, Iceland, Ireland, Israel, Italy, Japan, South Korea, Luxembourg, Netherlands, NZ, Norway, Portugal, Singapore, Spain, Sweden, Switzerland, Taiwan, UK, US; note—this group would presumably also cover the following seven smaller countries of Andorra, Bermuda, Faroe Islands, Holy See, Liechtenstein, Monaco, and San Marino that are included in the more comprehensive group of "developed countries"

African Development Bank Group (AfDB) *note*—regional multilateral development finance institution temporarily located in Tunis, Tunisia; the Bank Group consists of the African Development Bank, the African Development Fund, and the Nigerian Trust Fund
established—10 September 1964
aim—to promote economic development and social progress
regional members—(53) Algeria, Angola, Benin, Botswana, Burkina Faso, Burundi, Cameroon, Cape Verde, Central African Republic, Chad, Comoros, Democratic Republic of the Congo, Republic of the Congo, Cote d'Ivoire, Djibouti, Egypt, Equatorial Guinea, Eritrea, Ethiopia, Gabon, The Gambia, Ghana, Guinea, Guinea-Bissau, Kenya, Lesotho, Liberia, Libya, Madagascar, Malawi, Mali, Mauritania, Mauritius, Morocco, Mozambique, Namibia, Niger, Nigeria, Rwanda, Sao Tome and Principe, Senegal, Seychelles, Sierra Leone, Somalia, South Africa, Sudan, Swaziland, Tanzania, Togo, Tunisia, Uganda, Zambia, Zimbabwe
nonregional members—(24) Argentina, Austria, Belgium, Brazil, Canada, China, Denmark, Finland, France, Germany, India, Italy, Japan, South Korea, Kuwait, Netherlands, Norway, Portugal, Saudi Arabia, Spain, Sweden, Switzerland, UK, US

African Union (AU) *note*—replaces Organization of African Unity (OAU)
established—8 July 2001
aim—to achieve greater unity among African States; to defend states' integrity and independence; to accelerate political, social, and economic integration; to encourage international cooperation; to promote democratic principles and institutions
members—(53) Algeria, Angola, Benin, Botswana, Burkina Faso, Burundi, Cameroon, Cape Verde, Central African Republic, Chad, Comoros, Democratic Republic of the Congo, Republic of the Congo, Cote d'Ivoire, Djibouti, Egypt, Equatorial Guinea, Eritrea, Ethiopia, Gabon, The Gambia, Ghana, Guinea, Guinea-Bissau, Kenya, Lesotho, Liberia, Libya, Madagascar, Malawi, Mali, Mauritania, Mauritius, Mozambique, Namibia, Niger, Nigeria, Rwanda, Sahrawi Arab Democratic Republic (Western Sahara), Sao Tome and Principe, Senegal, Seychelles, Sierra Leone, Somalia, South Africa, Sudan, Swaziland, Tanzania, Togo, Tunisia, Uganda, Zambia, Zimbabwe

African, Caribbean, and Pacific Group of States (ACP Group)
established—6 June 1975
aim—to manage their preferential economic and aid relationship with the EU
members—(79) Angola, Antigua and Barbuda, The Bahamas, Barbados, Belize, Benin, Botswana, Burkina Faso, Burundi, Cameroon, Cape Verde, Central African Republic, Chad, Comoros, Democratic Republic of the Congo, Republic of the Congo, Cook Islands, Cote d'Ivoire, Cuba, Djibouti, Dominica, Dominican Republic, Equatorial Guinea, Eritrea, Ethiopia, Fiji, Gabon, The Gambia, Ghana, Grenada, Guinea, Guinea-Bissau, Guyana, Haiti, Jamaica, Kenya, Kiribati, Lesotho, Liberia, Madagascar, Malawi, Mali, Marshall Islands, Mauritania, Mauritius, Federated States of Micronesia, Mozambique, Namibia, Nauru, Niger, Nigeria, Niue, Palau, Papua New Guinea, Rwanda, Saint Kitts and Nevis, Saint Lucia, Saint Vincent and the Grenadines, Samoa, Sao Tome and Principe, Senegal, Seychelles, Sierra Leone, Solomon Islands, Somalia, South Africa, Sudan, Suriname, Swaziland, Tanzania, Timor-Leste, Togo, Tonga, Trinidad and Tobago, Tuvalu, Uganda, Vanuatu, Zambia, Zimbabwe

Agency for the Prohibition of Nuclear Weapons in Latin America and the Caribbean (OPANAL) *note*—acronym from Organismo para la Proscripcion de las Armas Nucleares en la America Latina y el Caribe (OPANAL)
established—14 February 1967 under the Treaty of Tlatelolco; *effective*—25 April 1969 on the 11th ratification
aim—to encourage the peaceful uses of atomic energy and prohibit nuclear weapons
members—(33) Antigua and Barbuda, Argentina, The Bahamas, Barbados, Belize, Bolivia, Brazil, Chile, Colombia, Costa Rica, Cuba, Dominica, Dominican Republic, Ecuador, El Salvador, Grenada, Guatemala, Guyana, Haiti, Honduras, Jamaica, Mexico, Nicaragua, Panama, Paraguay, Peru, Saint Kitts and Nevis, Saint Lucia, Saint Vincent and the Grenadines, Suriname, Trinidad and Tobago, Uruguay, Venezuela

Andean Community of Nations (CAN) *note*—formerly known as the Andean Group (AG), the Andean Parliament, and most recently as the Andean Common Market (Ancom)
established—26 May 1969; present name established 1 October 1992; *effective*—16 October 1969
aim—to promote harmonious development through economic integration
members—(4) Bolivia, Colombia, Ecuador, Peru
associate members—(5) Argentina, Brazil, Chile, Paraguay, Uruguay
observers—(2) Mexico, Panama

Arab Bank for Economic Development in Africa (ABEDA) *note*—also known as Banque Arabe de Developpement Economique en Afrique (BADEA)
established—18 February 1974; effective—16 September 1974
aim—to promote economic development
members—(17 plus the Palestine Liberation Organization) Algeria, Bahrain, Egypt, Iraq, Jordan, Kuwait, Lebanon, Libya, Mauritania, Morocco, Oman, Qatar, Saudi Arabia, Sudan, Syria, Tunisia, UAE, Palestine Liberation Organization; note—these are all the members of the Arab League excluding Comoros, Djibouti, Somalia, Yemen

Arab Fund for Economic and Social Development (AFESD)
established—16 May 1968
aim—to promote economic and social development
members—(20 plus the Palestine Liberation Organization) Algeria, Bahrain, Djibouti, Egypt, Iraq (suspended 1993), Jordan, Kuwait, Lebanon, Libya, Mauritania, Morocco, Oman, Qatar, Saudi Arabia, Somalia (suspended 1993), Sudan, Syria, Tunisia, UAE, Yemen, Palestine Liberation Organization

Arab Maghreb Union (AMU)
established—17 February 1989
aim—to promote cooperation and integration among the Arab states of northern Africa
members—(5) Algeria, Libya, Mauritania, Morocco, Tunisia

Arab Monetary Fund (AMF)
established—27 April 1976; effective—2 February 1977
aim—to promote Arab cooperation, development, and integration in monetary and economic affairs
members—(21 plus the Palestine Liberation Organization) Algeria, Bahrain, Comoros, Djibouti, Egypt, Iraq, Jordan, Kuwait, Lebanon, Libya, Mauritania, Morocco, Oman, Qatar, Saudi Arabia, Somalia, Sudan, Syria, Tunisia, UAE, Yemen, Palestine Liberation Organization

Arctic Council
established—18 September 1996
aim—to address the common concerns and challenges faced by Arctic governments and the people of the Arctic; to protect the Arctic environment
members—(8) Canada, Denmark (Greenland, Faroe Islands), Finland, Iceland, Norway, Russia, Sweden, US
permanent participants—(6) Aleut International Association, Arctic Athabaskan Council, Gurch'in Council International, Inuit Circumpolar Conference, Russian Association of Indigenous People of the North, Saami Council
observers—(7) China, France, Germany, Italy, Netherlands, Poland, UK

ASEAN Regional Forum (ARF)
established—25 July 1994
aim—to foster constructive dialogue and consultation on political and security issues of common interest and concern
members—(26) Australia, Bangladesh, Brunei, Burma, Cambodia, Canada, China, EU, India, Indonesia, Japan, North Korea, South Korea, Laos, Malaysia, Mongolia, NZ, Pakistan, Papua New Guinea, Philippines, Russia, Singapore, Thailand, Timor-Leste, US, Vietnam

Asian Development Bank (ADB)
established—19 December 1966
aim—to promote regional economic cooperation
members—(48) Afghanistan, Armenia, Australia, Azerbaijan, Bangladesh, Bhutan, Brunei, Burma, Cambodia, China, Cook Islands, Fiji, Georgia, Hong Kong, India, Indonesia, Japan, Kazakhstan, Kiribati, South Korea, Kyrgyzstan, Laos, Malaysia, Maldives, Marshall Islands, Federated States of Micronesia, Mongolia, Nauru, Nepal, NZ, Pakistan, Palau, Papua New Guinea, Philippines, Samoa, Singapore, Solomon Islands, Sri Lanka, Taiwan, Tajikistan, Thailand, Timor-Leste, Tonga, Turkmenistan, Tuvalu, Uzbekistan, Vanuatu, Vietnam
nonregional members—(19) Austria, Belgium, Canada, Denmark, Finland, France, Germany, Ireland, Italy, Luxembourg, Netherlands, Norway, Portugal, Spain, Sweden, Switzerland, Turkey, UK, US

Asia-Pacific Economic Cooperation (APEC)
established—7 November 1989
aim—to promote trade and investment in the Pacific basin
members—(21) Australia, Brunei, Canada, Chile, China, Hong Kong, Indonesia, Japan, South Korea, Malaysia, Mexico, NZ, Papua New Guinea, Peru, Philippines, Russia, Singapore, Taiwan, Thailand, US, Vietnam
observers—(3) Association of Southeast Asian Nations, Pacific Economic Cooperation Council, Pacific Islands Forum Secretariat

Association of Southeast Asian Nations (ASEAN)
established—8 August 1967
aim—to encourage regional economic, social, and cultural cooperation among the non-Communist countries of Southeast Asia
members—(10) Brunei, Burma, Cambodia, Indonesia, Laos, Malaysia, Philippines, Singapore, Thailand, Vietnam
dialogue partners—(11) Australia, Canada, China, EU, India, Japan, South Korea, NZ, Russia, US, UNDP; note—ASEAN promotes cooperation with Pakistan in some areas of mutual interest
observers—(1) Papua New Guinea

Australia Group
established—June 1985
aim—to consult on and coordinate export controls related to chemical and biological weapons
members—(41) Argentina, Australia, Austria, Belgium, Bulgaria, Canada, Croatia, Cyprus, Czech Republic, Denmark, Estonia, European Commission, Finland, France, Germany, Greece, Hungary, Iceland, Ireland, Italy, Japan, South Korea, Latvia, Lithuania, Luxembourg, Malta, Netherlands, NZ, Norway, Poland, Portugal, Romania, Slovakia, Slovenia, Spain, Sweden, Switzerland, Turkey, Ukraine, UK, US

Australia-New Zealand-United States Security Treaty (ANZUS)
established—1 September 1951; *effective*—29 April 1952
aim—to implement a trilateral mutual security agreement, although the US suspended security obligations to NZ on 11 August 1986; Australia and the US continue to hold annual meetings
members—(3) Australia, NZ, US

Baltic Assembly (BA)
established—12 May 1990
aim—to thoroughly discuss various cooperation issues between Baltic states
members—(3) Estonia, Latvia, Lithuania

Bay of Bengal Initiative for Multisectoral Technical and Economic Coopertion (BIMSTEC)
established—June 1997
aim—to foster socio-economic cooperation among members
members—(7) Bangladesh, Bhutan, Burma, India, Nepal, Sri Lanka, Thailand

Bank for International Settlements (BIS)
established—20 January 1930; *effective*—17 March 1930
aim—to promote cooperation among central banks in international financial settlements
members—(54) Algeria, Argentina, Australia, Austria, Belgium, Bosnia and Herzegovina, Brazil, Bulgaria, Canada, Chile, China, Croatia, Czech Republic, Denmark, Estonia, Finland, France, Germany, Greece, Hong Kong, Hungary, Iceland, India, Indonesia, Ireland, Israel, Italy, Japan, South Korea, Latvia, Lithuania, Macedonia, Malaysia, Mexico, Netherlands, NZ, Norway, Philippines, Poland, Portugal, Romania, Russia, Saudi Arabia, Singapore, Slovakia, Slovenia, South Africa, Spain, Sweden, Switzerland, Thailand, Turkey, UK, US; note—Serbia and Montenegro have separate central banks; their links with BIS are currently under review

Benelux Economic Union (Benelux) *note*—acronym from Belgium, Netherlands, and Luxembourg
established—3 February 1958; *effective*—1 November 1960
aim—to develop closer economic cooperation and integration
members—(3) Belgium, Luxembourg, Netherlands

Big Seven *note*—membership is the same as the Group of 7
established—1975
aim—to discuss and coordinate major economic policies
members—(7) Big Six (Canada, France, Germany, Italy, Japan, UK) plus the US

Black Sea Economic Cooperation Zone (BSEC)
established—25 June 1992
aim—to enhance regional stability through economic cooperation
members—(12) Albania, Armenia, Azerbaijan, Bulgaria, Georgia, Greece, Moldova, Romania, Russia, Serbia, Turkey, Ukraine; note—Macedonia is in the process of joining
observers—(17) Austria, Belarus, Black Sea Commission, Commission of the EC, Croatia, Czech Republic, Egypt, Energy Charter Secretariat, France, Germany, International Black Sea Club, Israel, Italy, Poland, Slovakia, Tunisia, US; note—Bosnia and Herzegovina and Slovenia have applied for observer status

Caribbean Community and Common Market (Caricom)
established—4 July 1973; *effective*—1 August 1973
aim—to promote economic integration and development, especially among the less developed countries
members—(15) Antigua and Barbuda, The Bahamas, Barbados, Belize, Dominica, Grenada, Guyana, Haiti, Jamaica, Montserrat, Saint Kitts and Nevis, Saint Lucia, Saint Vincent and the Grenadines, Suriname, Trinidad and Tobago
associate members—(5) Anguilla, Bermuda, British Virgin Islands, Cayman Islands, Turks and Caicos Islands
observers—(7) Aruba, Colombia, Dominican Republic, Mexico, Netherlands Antilles, Puerto Rico, Venezuela

Caribbean Development Bank (CDB)
established—18 October 1969; *effective*—26 January 1970
aim—to promote economic development and cooperation
regional members—(21) Anguilla, Antigua and Barbuda, The Bahamas, Barbados, Belize, British Virgin Islands, Cayman Islands, Colombia, Dominica, Grenada, Guyana, Haiti, Jamaica, Mexico, Montserrat, Saint Kitts and Nevis, Saint Lucia, Saint Vincent and the Grenadines, Trinidad and Tobago, Turks and Caicos Islands, Venezuela
nonregional members—(5) Canada, China, Germany, Italy, UK

Central African Customs and Economic Union (UDEAC)
see Monetary and Economic Community of Central Africa (CEMAC)

Central African States Development Bank (BDEAC) *note*—acronym from Banque de Developpement des Etats de l'Afrique Centrale
established—3 December 1975
aim—to provide loans for economic development
members—(10) African Development Bank (AfDB), Cameroon, Central African States Bank (BEAC), Central African Republic, Chad, Republic of the Congo, Equatorial Guinea, France, Gabon, Kuwait

Central American Bank for Economic Integration (BCIE) *note*—acronym from Banco Centroamericano de Integracion Economico
established—13 December 1960 signature of Articles of Agreement; 31 May 1961 began operations
aim—to promote economic integration and development
members—(5) Costa Rica, El Salvador, Guatemala, Honduras, Nicaragua
nonregional members—(6) Argentina, Colombia, Mexico, Panama, Spain, Taiwan

Central American Common Market (CACM)
established—13 December 1960, collapsed in 1969, reinstated in 1991
aim—to promote establishment of a Central American Common Market
members—(5) Costa Rica, El Salvador, Guatemala, Honduras, Nicaragua; note—Panama, although not a member, pursues full regional cooperation

Central European Initiative (CEI) *note*—evolved from the Quadrilateral Initiative and the Hexagonal Initiative
established—11 November 1989 as the Quadrilateral Initiative, 27 July 1991 became the Hexagonal Initiative, July 1992 its present name was adopted
aim—to form an economic and political cooperation group for the region between the Adriatic and the Baltic Seas
members—(18) Albania, Austria, Belarus, Bosnia and Herzegovina, Bulgaria, Croatia, Czech Republic, Hungary, Italy, Macedonia, Moldova, Montenegro, Poland, Romania, Serbia, Slovakia, Slovenia, Ukraine

centrally planned economies
a term applied mainly to the traditionally Communist states that looked to the former USSR for leadership; most are now evolving toward more democratic and market-oriented systems; also known formerly as the Second World or as as the Communist countries; through the 1980s, this group included Albania, Bulgaria, Cambodia, China, Cuba, Czechoslovakia, German Democratic Republic, Hungary, North Korea, Laos, Mongolia, Montenegro, Poland, Romania, Serbia, USSR, Vietnam

Collective Security Treaty Organization (CSTO)
established—7 October 2002
aim—to coordinate military and political cooperation, to develop multilateral structures and mechanisms of cooperation for ensuring national security of the member states
members—Armenia, Belarus, Kazakhstan, Kyrgyzstan, Russia, Tajikistan, Uzbekistan

Colombo Plan (CP)
established—May 1950 proposal was adopted; 1 July 1951 commenced full operations
aim—to promote economic and social development in Asia and the Pacific
members—(25) Afghanistan, Australia, Bangladesh, Bhutan, Burma, Fiji, India, Indonesia, Iran, Japan, South Korea, Laos, Malaysia, Maldives, Mongolia, Nepal, NZ, Pakistan, Papua New Guinea, Philippines, Singapore, Sri Lanka, Thailand, US, Vietnam

Common Market for Eastern and Southern Africa (COMESA) *note*—formerly known as Preferential Trade Area for Eastern and Southern Africa (PTA)
established—5 November 1993
aim—recognizing, promoting and protecting fundamental human rights, commitment to the principles of liberty and rule of law, maintaining peace and stability through the promotion and strengthening of good neighborliness, commitment to peaceful settlement of disputes among member states
members—(19) Burundi, Comoros, Democratic Republic of the Congo, Djibouti, Egypt, Eritrea, Ethiopia, Kenya, Libya, Madagascar, Malawi, Mauritius, Rwanda, Seychelles, Sudan, Swaziland, Uganda, Zambia, Zimbabwe

Commonwealth (C) *note*—also known as Commonwealth of Nations
established—31 December 1931
aim—to foster multinational cooperation and assistance, as a voluntary association that evolved from the British Empire
members—(53) Antigua and Barbuda, Australia, The Bahamas, Bangladesh, Barbados, Belize, Botswana, Brunei, Cameroon, Canada, Cyprus, Dominica, Fiji (suspended), The Gambia, Ghana, Grenada, Guyana, India, Jamaica, Kenya, Kiribati, Lesotho, Malawi, Malaysia, Maldives, Malta, Mauritius, Mozambique, Namibia, Nauru, NZ, Nigeria, Pakistan (suspended), Papua New Guinea, Saint Kitts and Nevis, Saint Lucia, Saint Vincent and the Grenadines, Samoa, Seychelles, Sierra Leone, Singapore, Solomon Islands, South Africa, Sri Lanka, Swaziland, Tanzania, Tonga, Trinidad and Tobago, Tuvalu, Uganda, UK, Vanuatu, Zambia; note—on 7 December 2003 Zimbabwe withdrew its membership from the Commonwealth

Commonwealth of Independent States (CIS)
established—8 December 1991; effective—21 December 1991
aim—to coordinate intercommonwealth relations and to provide a mechanism for the orderly dissolution of the USSR
members—(12) Armenia, Azerbaijan, Belarus, Georgia, Kazakhstan, Kyrgyzstan, Moldova, Russia, Tajikistan, Turkmenistan, Ukraine, Uzbekistan

Communist countries
traditionally the Marxist-Leninist states with authoritarian governments and command economies based on the Soviet model; most of the original and the successor states are no longer Communist; see centrally planned economies

Comuinidade dos Paises de Lingua Portuguesa (CPLP)
established—1996
aim—to establish a forum for friendship among Portuguese-speaking nations where Portuguese is an official language
members—(8) Angola, Brazil, Cape Verde, Guinea-Bissau, Mozambique, Portugal, Sao Tome and Principe, Timor-Leste
associate observers—(2) Equatorial Guinea, Mauritius

Coordinating Committee on Export Controls (COCOM)
established in 1949 to control the export of strategic products and technical data from member countries to proscribed destinations; members were: Australia, Belgium, Canada, Denmark, France, Germany, Greece, Italy, Japan, Luxembourg, Netherlands, Norway, Portugal, Spain, Turkey, UK, US; abolished 31 March 1994; COCOM members established a new organization, the Wassenaar Arrangement, with expanded membership on 12 July 1996 that focuses on nonproliferation export controls as opposed to East-West control of advanced technology

Council for Mutual Economic Assistance (CEMA) *note*—also known as CMEA or Comecon
established 25 January 1949 to promote the development of socialist economies and abolished 1 January 1991; members included Afghanistan (observer), Albania (had not participated since 1961 break with USSR), Angola (observer), Bulgaria, Cuba, Czechoslovakia, Ethiopia (observer), GDR, Hungary, Laos (observer), Mongolia, Mozambique (observer), Nicaragua (observer), Poland, Romania, USSR, Vietnam, Yemen (observer), Yugoslavia (associate)

Council of Arab Economic Unity (CAEU)
established—3 June 1957; effective—30 May 1964
aim—to promote economic integration among Arab nations
members—(10 plus the Palestine Liberation Organization) Egypt, Iraq, Jordan, Kuwait, Libya, Mauritania, Somalia, Sudan, Syria, Yemen, Palestine Liberation Organization

Council of Europe (CE)
established—5 May 1949; effective—3 August 1949
aim—to promote increased unity and quality of life in Europe
members—(47) Albania, Andorra, Armenia, Austria, Azerbaijan, Belgium, Bosnia and Herzegovina, Bulgaria, Croatia, Cyprus, Czech Republic, Denmark, Estonia, Finland, France, Georgia, Germany, Greece, Hungary, Iceland, Ireland, Italy, Latvia, Liechtenstein, Lithuania, Luxembourg, Macedonia, Malta, Moldova, Monaco, Montenegro, Netherlands, Norway, Poland, Portugal, Romania, Russia, San Marino, Serbia, Slovakia, Slovenia, Spain, Sweden, Switzerland, Turkey, Ukraine, UK
observers—(5) Canada, Holy See, Japan, Mexico, US

Council of the Baltic Sea States (CBSS)
established—6 March 1992
aim—to promote cooperation among the Baltic Sea states in the areas of aid to new democratic institutions, economic development, humanitarian aid, energy and the environment, cultural programs and education, and transportation and communication
members—(12) Denmark, Estonia, EC, Finland, Germany, Iceland, Latvia, Lithuania, Norway, Poland, Russia, Sweden
observers—(7) France, Italy, Netherlands, Slovakia, Ukraine, UK, US

Council of the Entente (Entente)
established—29 May 1959
aim—to promote economic, social, and political coordination
members—(5) Benin, Burkina Faso, Cote d'Ivoire, Niger, Togo

countries in transition
a term used by the International Monetary Fund (IMF) for the middle group in its hierarchy of advanced economies, countries in transition, and developing countries; IMF statistics include the following 28 countries in transition: Albania, Armenia, Azerbaijan, Belarus, Bosnia and Herzegovina, Bulgaria, Croatia, Czech Republic, Estonia, Georgia, Hungary, Kazakhstan, Kyrgyzstan, Latvia, Lithuania, Macedonia, Moldova, Mongolia, Montenegro, Poland, Romania, Russia, Serbia, Slovakia, Slovenia, Tajikistan, Turkmenistan, Ukraine, Uzbekistan; note—this group is identical to the group traditionally referred to as the "former USSR/Eastern Europe" except for the addition of Mongolia

Customs Cooperation Council (CCC) *note*—see World Customs Organization (WCO)

developed countries (DCs)
the top group in the hierarchy of developed countries (DCs), former USSR/Eastern Europe (former USSR/EE), and less developed countries (LDCs); includes the market-oriented economies of the mainly democratic nations in the Organization for Economic Cooperation and Development (OECD), Bermuda, Israel, South Africa, and the European ministates; also known as the First World, high-income countries, the North, industrial countries; generally have a per capita GDP in excess of $10,000 although four OECD countries and South Africa have figures well under $10,000 and two of the excluded OPEC countries have figures of more than $10,000; the 34 DCs are: Andorra, Australia, Austria, Belgium, Bermuda, Canada, Denmark, Faroe Islands, Finland, France, Germany, Greece, Holy See, Iceland, Ireland, Israel, Italy, Japan, Liechtenstein, Luxembourg, Malta, Monaco, Netherlands, NZ, Norway, Portugal, San Marino, South Africa, Spain, Sweden, Switzerland, Turkey, UK, US; note—similar to the new International Monetary Fund (IMF) term "advanced economies" that adds Hong Kong, South Korea, Singapore, and Taiwan but drops Malta, Mexico, South Africa, and Turkey

developing countries

a term used by the International Monetary Fund (IMF) for the bottom group in its hierarchy of advanced economies, countries in transition, and developing countries; IMF statistics include the following 126 developing countries: Afghanistan, Algeria, Angola, Antigua and Barbuda, Argentina, Aruba, The Bahamas, Bahrain, Bangladesh, Barbados, Belize, Benin, Bhutan, Bolivia, Botswana, Brazil, Burkina Faso, Burma, Burundi, Cambodia, Cameroon, Cape Verde, Central African Republic, Chad, Chile, China, Colombia, Comoros, Democratic Republic of the Congo, Republic of the Congo, Costa Rica, Cote d'Ivoire, Cyprus, Djibouti, Dominica, Dominican Republic, Ecuador, Egypt, El Salvador, Equatorial Guinea, Ethiopia, Fiji, Gabon, The Gambia, Ghana, Grenada, Guatemala, Guinea, Guinea-Bissau, Guyana, Haiti, Honduras, India, Indonesia, Iran, Iraq, Jamaica, Jordan, Kenya, Kiribati, Kuwait, Laos, Lebanon, Lesotho, Liberia, Libya, Madagascar, Malawi, Malaysia, Maldives, Mali, Malta, Marshall Islands, Mauritania, Mauritius, Mexico, Federated States of Micronesia, Morocco, Mozambique, Namibia, Nepal, Netherlands Antilles, Nicaragua, Niger, Nigeria, Oman, Pakistan, Panama, Papua New Guinea, Paraguay, Peru, Philippines, Qatar, Rwanda, Saint Kitts and Nevis, Saint Lucia, Saint Vincent and the Grenadines, Samoa, Sao Tome and Principe, Saudi Arabia, Senegal, Seychelles, Sierra Leone, Solomon Islands, Somalia, South Africa, Sri Lanka, Sudan, Suriname, Swaziland, Syria, Tanzania, Thailand, Togo, Trinidad and Tobago, Tunisia, Turkey, UAE, Uganda, Uruguay, Vanuatu, Venezuela, Vietnam, Yemen, Zambia, Zimbabwe; note—this category would presumably also cover the following 46 other countries that are traditionally included in the more comprehensive group of "less developed countries": American Samoa, Anguilla, British Virgin Islands, Brunei, Cayman Islands, Christmas Island, Cocos Islands, Cook Islands, Cuba, Eritrea, Falkland Islands, French Guiana, French Polynesia, Gaza Strip, Gibraltar, Greenland, Grenada, Guadeloupe, Guam, Guernsey, Isle of Man, Jersey, North Korea, Macau, Martinique, Mayotte, Montserrat, Nauru, New Caledonia, Niue, Norfolk Island, Northern Mariana Islands, Palau, Pitcairn Islands, Puerto Rico, Reunion, Saint Helena, Saint Pierre and Miquelon, Tokelau, Tonga, Turks and Caicos Islands, Tuvalu, Virgin Islands, Wallis and Futuna, West Bank, Western Sahara

East African Community (EAC) note—originally established in 1967, it was disbanded in 1977
established—January 2001
aim—to establish a political and economic union among the countries
members—(5) Burundi, Kenya, Rwanda, Tanzania, Uganda

East African Development Bank (EADB)
established—6 June 1967; effective—1 December 1967
aim—to promote economic development
members—(3) Kenya, Tanzania, Uganda

East Asia Summit (EAS)
established—14 December 2005
aim—to promote cooperation in political and security issues; to promote development, financial stability, energy security, economic integration and growth; to eradicate poverty and narrow the development gap in East Asia, and to promote deeper cultural understanding
members—(16) Australia, Brunei, Burma, Cambodia, China, India, Indonesia, Japan, South Korea, Laos, Malaysia, NZ, Philippines, Singapore, Thailand, Vietnam

Economic and Monetary Community of Central Africa (CEMAC) note—was formerly the Central African Customs and Economic Union (UDEAC)
established—8 December 1964; effective—1 January 1966
aim—to promote the establishment of a Central African Common Market
members—(6) Cameroon, Central African Republic, Chad, Republic of the Congo, Equatorial Guinea, Gabon

Economic and Monetary Union (EMU) note—an integral part of the European Union; also known as the European Economic and Monetary Union
established—1–2 December 1969 (proposed at summit conference of heads of government; 7 February 1992 (Maastricht Treaty signed)
aim—to promote a single market by creating a single currency, the euro; timetable—2 May 1998: European exchange rates fixed for 1 January 1999; 1 January 1999: all banks and stock exchanges begin using euros; 1 January 2002: the euro goes into circulation; 1 July 2002 local currencies no longer accepted
members—(15) Austria, Belgium, Cyprus, Finland, France, Germany, Greece, Ireland, Italy, Luxembourg, Malta, Netherlands, Portugal, Slovenia, Spain

Economic and Social Council (ECOSOC)
established—26 June 1945; effective—24 October 1945
aim—to coordinate the economic and social work of the UN; includes five regional commissions (Economic Commission for Africa, Economic Commission for Europe, Economic Commission for Latin America and the Caribbean, Economic and Social Commission for Asia and the Pacific, Economic and Social Commission for Western Asia) and nine functional commissions (Commission for Social Development, Commission on Human Rights, Commission on Narcotic Drugs, Commission on the Status of Women, Commission on Population and Development, Statistical Commission, Commission on Science and Technology for Development, Commission on Sustainable Development, and Commission on Crime Prevention and Criminal Justice)
members—(54) selected on a rotating basis from all regions

Economic Community of the Great Lakes Countries (CEPGL) note—acronym from Communaute Economique des Pays des Grands Lacs
established—20 September 1976
aim—to promote regional economic cooperation and integration
members—(3) Burundi, Democratic Republic of the Congo, Rwanda; note—organization collapsed because of fighting in 1998; reactivated in 2006

Economic Community of West African States (ECOWAS)
established—28 May 1975
aim—to promote regional economic cooperation
members—(15) Benin, Burkina Faso, Cape Verde, Cote d'Ivoire, The Gambia, Ghana, Guinea, Guinea-Bissau, Liberia, Mali, Niger, Nigeria, Senegal, Sierra Leone, Togo

Economic Cooperation Organization (ECO)
established—27–29 January 1985
aim—to promote regional cooperation in trade, transportation, communications, tourism, cultural affairs, and economic development
members—(10) Afghanistan, Azerbaijan, Iran, Kazakhstan, Kyrgyzstan, Pakistan, Tajikistan, Turkey, Turkmenistan, Uzbekistan

Eurasian Economic Community (EAEC or EurasEC) *note*—merged with Central Asian Cooperation Organization (CACO) in 2005
established—May 2001
aim—to create a common economic and energy policy
members—(6) Belarus, Kazakhstan, Kyrgyzstan, Russia, Tajikistan, Uzbekistan
observers—(3) Armenia, Moldova, Ukraine

Euro-Atlantic Partnership Council (EAPC) *note*—began as the North Atlantic Cooperation Council (NACC); an extension of NATO
established—8 November 1991; effective—20 December 1991
aim—to discuss cooperation on mutual political and security issues
members—(49) Albania, Armenia, Austria, Azerbaijan, Belarus, Belgium, Bosnia and Herzegovina, Bulgaria, Canada, Croatia, Czech Republic, Denmark, Estonia, Finland, France, Georgia, Germany, Greece, Hungary, Iceland, Ireland, Italy, Kazakhstan, Kyrgyzstan, Latvia, Lithuania, Luxembourg, Macedonia, Moldova, Montenegro, Netherlands, Norway, Poland, Portugal, Romania, Russia, Serbia, Slovakia, Slovenia, Spain, Sweden, Switzerland, Tajikistan, Turkey, Turkmenistan, Ukraine, UK, US, Uzbekistan

European Bank for Reconstruction and Development (EBRD)
established—8–9 January 1990 (proposals made); 15 April 1991 (bank inaugurated)
aim—to facilitate the transition of seven centrally planned economies in Europe (Bulgaria, former Czechoslovakia, Hungary, Poland, Romania, former USSR, and former Yugoslavia) to market economies by committing 60% of its loans to privatization
members—(63) Albania, Armenia, Australia, Austria, Azerbaijan, Belarus, Belgium, Bosnia and Herzegovina, Bulgaria, Canada, Croatia, Cyprus, Czech Republic, Denmark, Egypt, EC, European Investment Bank (EIB), Estonia, Finland, France, Georgia, Germany, Greece, Hungary, Iceland, Ireland, Israel, Italy, Japan, Kazakhstan, South Korea, Kyrgyzstan, Latvia, Liechtenstein, Lithuania, Luxembourg, Macedonia, Malta, Mexico, Moldova, Mongolia, Montenegro, Morocco, Netherlands, NZ, Norway, Poland, Portugal, Romania, Russia, Serbia, Slovakia, Slovenia, Spain, Sweden, Switzerland, Tajikistan, Turkey, Turkmenistan, Ukraine, UK, US, Uzbekistan

European Community (or European Communities, EC)
established 8 April 1965 to integrate the European Atomic Energy Community (Euratom), the European Coal and Steel Community (ECSC), the European Economic Community (EEC or Common Market), and to establish a completely integrated common market and an eventual federation of Europe; merged into the European Union (EU) on 7 February 1992; member states at the time of merger were Belgium, Denmark, France, Germany, Greece, Ireland, Italy, Luxembourg, Netherlands, Portugal, Spain, UK

European Free Trade Association (EFTA)
established—4 January 1960; effective—3 May 1960
aim—to promote expansion of free trade
members—(4) Iceland, Liechtenstein, Norway, Switzerland

European Investment Bank (EIB)
established—25 March 1957; effective—1 January 1958
aim—to promote economic development of the EU and its predecessors, the EEC and the EC
members—(275) Austria, Belgium, Bulgaria, Cyprus, Czech Republic, Denmark, Estonia, Finland, France, Germany, Greece, Hungary, Ireland, Italy, Latvia, Lithuania, Luxembourg, Malta, Netherlands, Poland, Portugal, Romania, Slovakia, Slovenia, Spain, Sweden, UK

European Organization for Nuclear Research (CERN) *note*—acronym retained from the predecessor organization Conseil
Europeenne pour la Recherche Nucleaire
established—1 July 1953; effective—29 September 1954
aim—to foster nuclear research for peaceful purposes only
members—(20) Austria, Belgium, Bulgaria, Czech Republic, Denmark, Finland, France, Germany, Greece, Hungary, Italy, Netherlands, Norway, Poland, Portugal, Slovakia, Spain, Sweden, Switzerland, UK
observers—(8) EC, India, Israel, Japan, Russia, Turkey, United Nations Educational, Scientific, and Cultural Organization (UNESCO), US

European Space Agency (ESA)
established—31 May 1975
aim—to promote peaceful cooperation in space research and technology
members—(17) Austria, Belgium, Denmark, Finland, France, Germany, Greece, Ireland, Italy, Luxembourg, Netherlands, Norway, Portugal, Spain, Sweden, Switzerland, UK
cooperating states—(5) Canada, Czech Republic, Hungary, Poland, Romania

European Union (EU) *note*—see European Union entry at the end of the "country" listings **First World**
another term for countries with advanced, industrialized economies; this term is fading from use; see developed countries (DCs)

Food and Agriculture Organization (FAO)
established—16 October 1945
aim—to raise living standards and increase availability of agricultural products; a UN specialized agency
members—(191) includes all UN member countries except Brunei, Liechtenstein, and Singapore (189 total); plus Cook Islands, EC, and Niue

former Soviet Union (FSU)
former term often used to identify as a group the successor nations to the Soviet Union or USSR; this group of 15 countries consists of: Armenia, Azerbaijan, Belarus, Estonia, Georgia, Kazakhstan, Kyrgyzstan, Latvia, Lithuania, Moldova, Russia, Tajikistan, Turkmenistan, Ukraine, Uzbekistan

former USSR/Eastern Europe (former USSR/EE)
the middle group in the hierarchy of developed countries (DCs), former USSR/Eastern Europe (former USSR/EE), and less developed countries (LDCs); these countries are in political and economic transition and may well be grouped differently in the near future; this group of 27 countries consists of: Albania, Armenia, Azerbaijan, Belarus, Bosnia and Herzegovina, Bulgaria, Croatia, Czech Republic, Estonia, Georgia, Hungary, Kazakhstan, Kyrgyzstan, Latvia, Lithuania, Macedonia, Moldova, Poland, Romania, Russia, Slovakia, Slovenia, Tajikistan, Turkmenistan, Ukraine, Uzbekistan, Yugoslavia; this group is identical to the IMF group "countries in transition" except for the IMF's inclusion of Mongolia

Four Dragons
the four small Asian less developed countries (LDCs) that have experienced unusually rapid economic growth; also known as the Four Tigers; this group consists of Hong Kong, South Korea, Singapore, Taiwan; these countries are included in the IMF's "advanced economies" group

Franc Zone (FZ) *note*—also known as Conference des Ministres des Finances des Pays de la Zone Franc
established—1964
aim—to form a monetary union among countries whose currencies were linked to the French franc
members—(16) Benin, Burkina Faso, Cameroon, Central African Republic, Chad, Comoros, Republic of the Congo, Cote d'Ivoire, Equatorial Guinea, France, Gabon, Guinea-Bissau, Mali, Niger, Senegal, Togo

Front Line States (FLS)
established to achieve black majority rule in South Africa; has since gone out of existence; members included Angola, Botswana, Mozambique, Namibia, Tanzania, Zambia, Zimbabwe

General Agreement on Tariffs and Trade (GATT)
see the World Trade Organization (WTO)

General Confederation of Trade Unions (GCTU)
established—16 April 1992
aim—to consolidate trade union actions to protect citizens' social and labor rights and interests, to help secure trade unions' rights and guarantees, and to strengthen international trade union solidarity
members—(11) Armenia, Azerbaijan, Belarus, Georgia, Kazakhstan, Kyrgyzstan, Moldova, Russia, Tajikistan, Ukraine, Uzbekistan

Group of 2 (G-2)
informal term that came into use about 1986; to facilitate bilateral economic cooperation between the two most powerful economic giants; members were Japan, US

Group of 3 (G-3)
established—September 1990
aim—mechanism for policy coordination
members—(2) Colombia, Mexico; note—Panama shows interest in joining

Group of 5 (G-5)
established—22 September 1985
aim—to coordinate the economic policies of five major noncommunist economic powers
members—(5) France, Germany, Japan, UK, US

Group of 6 (G-6)
also known as Groupe des Six Sur le Desarmement (not to be confused with the Big Six) was established in 22 May 1984 with the aim of achieving nuclear disarmament; its members were Argentina, Greece, India, Mexico, Sweden, Tanzania

Group of 7 (G-7) *note*—membership is the same as the Big Seven
established—22 September 1985
aim—to facilitate economic cooperation among the seven major noncommunist economic powers
members—(7) Group of 5 (France, Germany, Japan, UK, US) plus Canada and Italy

Group of 8 (G-8)
established—October 1975
aim—to facilitate economic cooperation among the developed countries (DCs) that participated in the Conference on International Economic Cooperation (CIEC), held in several sessions between December 1975 and 3 June 1977
members—(8) Canada, France, Germany, Italy, Japan, Russia, UK, US

Group of 9 (G-9)
established—NA
aim—to discuss matters of mutual interest on an informal basis
members—(9) Austria, Belgium, Bulgaria, Denmark, Finland, Hungary, Romania, Serbia, Sweden

Group of 10 (G-10) *note*—also known as the Paris Club; includes the wealthiest members of the IMF who provide most of the money to be loaned and act as the informal steering committee; name persists despite increased membership
established—October 1962
aim—to coordinate credit policy
members—(11) Belgium, Canada, France, Germany, Italy, Japan, Netherlands, Sweden, Switzerland, UK, US
observers—(4) BIS, EU, IMF, OECD

Group of 11 (G-11) *note*—also known as the Cartagena Group
established in 21–22 June 1984, in Cartagena, Colombia, aim was to provide a forum for largest debtor nations in Latin America; members were: Argentina, Bolivia, Brazil, Chile, Colombia, Dominican Republic, Ecuador, Mexico, Peru, Uruguay, Venezuela

Group of 15 (G-15) *note*—byproduct of the Nonaligned Movement; name persists despite increased membership
established—September 1989
aim—to promote economic cooperation among developing nations; to act as the main political organ for the Nonaligned Movement
members—(17) Algeria, Argentina, Brazil, Chile, Egypt, India, Indonesia, Jamaica, Kenya, Malaysia, Mexico, Nigeria, Peru, Senegal, Sri Lanka, Venezuela, Zimbabwe

Group of 24 (G-24)
established—1 August 1989
aim—to promote the interests of developing countries in Africa, Asia, and Latin America within the IMF
members—(24) Algeria, Argentina, Brazil, Colombia, Democratic Republic of the Congo, Cote d'Ivoire, Egypt, Ethiopia, Gabon, Ghana, Guatemala, India, Iran, Lebanon, Mexico, Nigeria, Pakistan, Peru, Philippines, South Africa, Sri Lanka, Syria, Trinidad and Tobago, Venezuela
observers—(1) China

Group of 77 (G-77)
established—15 June1964; October 1967 first ministerial meeting
aim—to promote economic cooperation among developing countries; name persists in spite of increased membership
members—(129 plus the Palestine Liberation Organization) Afghanistan, Algeria, Angola, Antigua and Barbuda, Argentina, The Bahamas, Bahrain, Bangladesh, Barbados, Belize, Benin, Bhutan, Bolivia, Bosnia and Herzegovina, Botswana, Brazil, Brunei, Burkina Faso, Burma, Burundi, Cambodia, Cameroon, Cape Verde, Central African Republic, Chad, Chile, China, Colombia, Comoros, Democratic Republic of the Congo, Republic of the Congo, Costa Rica, Cote d'Ivoire, Cuba, Djibouti, Dominica, Dominican Republic, Ecuador, Egypt, El Salvador, Equatorial Guinea, Eritrea, Ethiopia, Fiji, Gabon, The Gambia, Ghana, Grenada, Guatemala, Guinea, Guinea-Bissau, Guyana, Haiti, Honduras, India, Indonesia, Iran, Iraq, Jamaica, Jordan, Kenya, North Korea, Kuwait, Laos, Lebanon, Lesotho, Liberia, Libya, Madagascar, Malawi, Malaysia, Maldives, Mali, Marshall Islands, Mauritania, Mauritius, Federated States of Micronesia, Mongolia, Morocco, Mozambique, Namibia, Nepal, Nicaragua, Niger, Nigeria, Oman, Pakistan, Panama, Papua New Guinea, Paraguay, Peru, Philippines, Qatar, Rwanda, Saint Kitts and Nevis, Saint Lucia, Saint Vincent and the Grenadines, Samoa, Sao Tome and Principe, Saudi Arabia, Senegal, Seychelles, Sierra Leone, Singapore, Solomon Islands, Somalia, South Africa, Sri Lanka, Sudan, Suriname, Swaziland, Syria, Tanzania, Thailand, Timor-Leste, Togo, Tonga, Trinidad and Tobago, Tunisia, Turkmenistan, Uganda, UAE, Uruguay, Vanuatu, Venezuela, Vietnam, Yemen, Zambia, Zimbabwe, Palestine Liberation Organization

Gulf Cooperation Council (GCC) *note*—also known as the Cooperation Council for the Arab States of the Gulf
established—25 May 1981
aim—to promote regional cooperation in economic, social, political, and military affairs
members—(6) Bahrain, Kuwait, Oman, Qatar, Saudi Arabia, UAE

Organization for Democracy and Economic Development (GUAM) *note*-acronym standing for the member countries, Georgia, Ukraine, Azerbaijan, Moldova; formerly known as GUUAM before Uzbekistan withdrew in 5 May 2005
established—7 June 2001
aim—commits the countries to cooperation and assistance in social and economic development, the strengthening and broadening of trade and economic relations, and the development and effective use of transport and communications, highways, and related infrastructure crossing the boundaries of the member states
members—(4) Azerbaijan, Georgia, Moldova, Ukraine

high income countries
another term for the industrialized countries with high per capita GDPs; see developed countries (DCs)

Indian Ocean Commission (InOC)
established—21 December 1982
aim—to organize and promote regional cooperation in all sectors, especially economic
members—(5) Comoros, France (for Reunion and Mayotte), Madagascar, Mauritius, Seychelles

industrial countries
another term for the developed countries; see developed countries (DCs)

Inter-American Development Bank (IADB) *note*—also known as Banco Interamericano de Desarrollo (BID)
established—8 April 1959; effective—30 December 1959
aim—to promote economic and social development in Latin America
members—(47) Argentina, Austria, The Bahamas, Barbados, Belgium, Belize, Bolivia, Brazil, Canada, Chile, Colombia, Costa Rica, Croatia, Denmark, Dominican Republic, Ecuador, El Salvador, Finland, France, Germany, Guatemala, Guyana, Haiti, Honduras, Israel, Italy, Jamaica, Japan, South Korea, Mexico, Netherlands, Nicaragua, Norway, Panama, Paraguay, Peru, Portugal, Slovenia, Spain, Suriname, Sweden, Switzerland, Trinidad and Tobago, UK, US, Uruguay, Venezuela

Inter-Governmental Authority on Development (IGAD) *note*—formerly known as Inter-Governmental Authority on Drought and Development (IGADD)
established—15–16 January 1986 as the Inter-Governmental Authority on Drought and Development; revitalized—21 March 1996 as the Inter-Governmental Authority on Development
aim—to promote a social, economic, and scientific community among its members
members—(6) Djibouti, Ethiopia, Kenya, Somalia, Sudan, Uganda; note—Eritrea declared its suspension in 2007

Inter-Parliamentary Union (IPU)
established—1889
aim—fosters contacts among parliamentarians, considers and expresses views of international interest and concern with the purpose of bringing about action by parliaments and parliamentarians, contributes to the defense and promotion of human rights, contributes to better knowledge of representative institutions
members—(146) Afghanistan, Albania, Algeria, Andorra, Angola, Argentina, Armenia, Australia, Austria, Azerbaijan, Bahrain, Bangladesh, Belarus, Belgium, Benin, Bolivia, Bosnia and Herzegovina, Botswana, Brazil, Bulgaria, Burkina Faso, Burundi, Cambodia, Cameroon, Canada, Cape Verde, Chile, China, Colombia, Democratic Republic of the Congo, Republic of the Congo, Costa Rica, Cote d'Ivoire, Croatia, Cuba, Cyprus, Czech Republic, Denmark, Dominican Republic, Ecuador, Egypt, El Salvador, Estonia, Ethiopia, Finland, France, Gabon, The Gambia, Georgia, Germany, Ghana, Greece, Guatemala, Hungary, Iceland, India, Indonesia, Iran, Ireland, Israel, Italy, Japan, Jordan, Kazakhstan, Kenya, North Korea, South Korea, Kuwait, Kyrgyzstan, Laos, Latvia, Lebanon, Liberia, Libya, Lichtenstein, Lithuania, Luxembourg, Macedonia, Madagascar, Malaysia, Maldives, Mali, Malta, Mauritius, Mexico, Moldova, Monaco, Mongolia, Montenegro, Morocco, Mozambique, Namibia, Nepal, Netherlands, NZ, Nicaragua, Niger, Nigeria, Norway, Pakistan, Palau, Panama, Papua New Guinea, Paraguay, Peru, Philippines, Poland, Portugal, Qatar, Romania, Russia, Rwanda, Somalia, Samoa, San Marino, Sao Tome and Principe, Saudi Arabia, Senegal, Serbia, Singapore, Slovakia, Slovenia, South Africa, Spain, Sri Lanka, Sudan, Suriname, Sweden, Switzerland, Syria, Tanzania, Tajikistan, Thailand, Togo, Tunisia, Turkey, Uganda, Ukraine, UAE, UK, Uruguay, Venezuela, Vietnam, Yemen, Zambia, Zimbabwe
associate members—(7) Andean Parliament, Central American Parliament, Community Parliament of the Economic Community of West African States, East African Legislative Assembly, European Parliament, Latin American Parliament, Parliamentary Assembly of the Council of Europe

International Atomic Energy Agency (IAEA)
established—26 October 1956; effective—29 July 1957
aim—to promote peaceful uses of atomic energy
members—(144) Afghanistan, Albania, Algeria, Angola, Argentina, Armenia, Australia, Austria, Azerbaijan, Bangladesh, Belarus, Belgium, Belize, Benin, Bolivia, Bosnia and Herzegovina, Botswana, Brazil, Bulgaria, Burkina Faso, Burma, Cameroon, Canada, Central African Republic, Chad, Chile, China, Colombia, Democratic Republic of the Congo, Costa Rica, Cote d'Ivoire, Croatia, Cuba, Cyprus, Czech Republic, Denmark, Dominican Republic, Ecuador, Egypt, El Salvador, Eritrea, Estonia, Ethiopia, Finland, France, Gabon, Georgia, Germany, Ghana, Greece, Guatemala, Haiti, Holy See, Honduras, Hungary, Iceland, India, Indonesia, Iran, Iraq, Ireland, Israel, Italy, Jamaica, Japan, Jordan, Kazakhstan, Kenya, South Korea, Kuwait, Kyrgyzstan, Latvia, Lebanon, Liberia, Libya, Liechtenstein, Lithuania, Luxembourg, Macedonia, Madagascar, Malawi, Malaysia, Mali, Malta, Marshall Islands, Mauritania, Mauritius, Mexico, Moldova, Monaco, Mongolia, Morocco, Mozambique, Namibia, Netherlands, NZ, Nicaragua, Niger, Nigeria, Norway, Pakistan, Palau, Panama, Paraguay, Peru, Philippines, Poland, Portugal, Qatar, Romania, Russia, Saudi Arabia, Senegal, Serbia, Seychelles, Sierra Leone, Singapore, Slovakia, Slovenia, South Africa, Spain, Sri Lanka, Sudan, Sweden, Switzerland, Syria, Tajikistan, Tanzania, Thailand, Togo, Tunisia, Turkey, Uganda, Ukraine, UAE, UK, US, Uruguay, Uzbekistan, Venezuela, Vietnam, Yemen, Zambia, Zimbabwe; note—Montenegro's membership pending on completion of procedures required to become an IAEA Member State

International Bank for Reconstruction and Development (IBRD) *note*—also known as the World Bank
established—22 July 1944; effective—27 December 1945
aim—to provide economic development loans; a UN specialized agency
members—(185) includes all UN member countries except Andorra, Cuba, North Korea, Liechtenstein, Monaco, Nauru, and Tuvalu

International Chamber of Commerce (ICC)

established—1919

aim—to promote free trade and private enterprise and to represent business interests at national and international levels

members—(91 national committees) Algeria, Argentina, Australia, Austria, Bahrain, Bangladesh, Belgium, Brazil, Bulgaria, Burkina Faso, Cameroon, Canada, Caribbean, Chile, China, Colombia, Costa Rica, Croatia, Cuba, Cyprus, Czech Republic, Denmark, Dominican Republic, Ecuador, Egypt, El Salvador, Finland, France, Georgia, Germany, Ghana, Greece, Guatemala, Hong Kong, Hungary, Iceland, India, Indonesia, Iran, Ireland, Israel, Italy, Japan, Jordan, South Korea, Kuwait, Lebanon, Lithuania, Luxembourg, Madagascar, Malaysia, Mexico, Monaco, Morocco, Nepal, Netherlands, NZ, Nigeria, Norway, Pakistan, Panama, Philippines, Poland, Portugal, Qatar, Romania, Russia, Saudi Arabia, Senegal, Serbia, Singapore, Slovakia, Slovenia, South Africa, Spain, Sri Lanka, Sweden, Switzerland, Syria, Taiwan, Tanzania, Thailand, Togo, Tunisia, Turkey, Ukraine, UAE, UK, US, Uruguay, Venezuela; note—Peru is restructuring

International Civil Aviation Organization (ICAO)

established—7 December 1944; effective—4 April 1947

aim—to promote international cooperation in civil aviation; a UN specialized agency

members—(190) includes all UN member countries except Dominica, Liechtenstein, and Tuvalu (189 total); plus Cook Islands

International Civilian Support Mission in Haiti (MICAH)

established 17 December 1999 to promote respect for human rights; members included Argentina, Benin, Canada, France, India, Mali, Niger, Senegal, Togo, Tunisia, US; closed 2001

International Committee of the Red Cross (ICRC)

established—17 February 1863

aim—to provide humanitarian aid in wartime

members—(15–25 individuals) all Swiss nationals

International Court of Justice (ICJ) *note*—also known as the World Court

established—3 February 1946 superseded Permanent Court of International Justice

aim—primary judicial organ of the UN

members—(15 judges) elected by the UN General Assembly and Security Council to represent all principal legal systems

International Criminal Court (ICCt)

established—11 April 2002

aim—to hold all individuals and countries accountable to international laws of conduct; to specify international standards of conduct; to provide an important mechanism for implementing these standards; to ensure that perpetrators are brought to justice

members (countries that have ratified the treaty)—(105) Afghanistan, Albania, Andorra, Antigua and Barbuda, Argentina, Australia, Austria, Barbados, Belgium, Belize, Benin, Bolivia, Bosnia and Herzegovina, Botswana, Brazil, Bulgaria, Burkina Faso, Burundi, Cambodia, Canada, Central African Republic, Chad, Colombia, Comoros, Democratic Republic of the Congo, Republic of the Congo, Costa Rica, Croatia, Cyprus, Denmark, Djibouti, Dominica, Dominican Republic, Ecuador, Estonia, Fiji, Finland, France, Gabon, The Gambia, Georgia, Germany, Ghana, Greece, Guinea, Guyana, Honduras, Hungary, Iceland, Ireland, Italy, Japan, Jordan, Kenya, South Korea, Latvia, Lesotho, Liberia, Liechtenstein, Lithuania, Luxembourg, Macedonia, Malawi, Mali, Malta, Marshall Islands, Mauritius, Mexico, Mongolia, Montenegro, Namibia, Nauru, Netherlands, NZ, Niger, Nigeria, Norway, Panama, Paraguay, Peru, Poland, Portugal, Romania, Saint Vincent and the Grenadines, Saint Kitts and Nevis, Samoa, San Marino, Senegal, Serbia, Sierra Leone, Slovakia, Slovenia, South Africa, Spain, Sweden, Switzerland, Tajikistan, Tanzania, Timor-Leste, Trinidad and Tobago, Uganda, UK, Uruguay, Venezuela, Zambia

signatory states (countries that have signed, but not ratified, the treaty)—(41) Algeria, Angola, Armenia, The Bahamas, Bahrain, Bangladesh, Cameroon, Cape Verde, Chile, Cote d'Ivoire, Czech Republic, Egypt, Eritrea, Guinea-Bissau, Haiti, Iran, Israel, Jamaica, Kuwait, Kyrgyzstan, Madagascar, Moldova, Monaco, Morocco, Mozambique, Oman, Philippines, Russia, Saint Lucia, Sao Tome and Principe, Seychelles, Solomon Islands, Sudan, Syria, Thailand, Ukraine, UAE, US, Uzbekistan, Yemen, Zimbabwe; note—Israel and US have declared they will not ratify the agreement

International Criminal Police Organization (Interpol)

established—September 1923 set up as the International Criminal Police Commission; 13 June 1956 constitution modified and present name adopted

aim—to promote international cooperation among police authorities in fighting crime

members—(186) Afghanistan, Albania, Algeria, Andorra, Angola, Antigua and Barbuda, Argentina, Armenia, Aruba, Australia, Austria, Azerbaijan, The Bahamas, Bahrain, Bangladesh, Barbados, Belarus, Belgium, Belize, Benin, Bhutan, Bolivia, Bosnia and Herzegovina, Botswana, Brazil, Brunei, Bulgaria, Burkina Faso, Burma, Burundi, Cambodia, Cameroon, Canada, Cape Verde, Central African Republic, Chad, Chile, China, Colombia, Comoros, Democratic Republic of the Congo, Republic of the Congo, Costa Rica, Cote d'Ivoire, Croatia, Cuba, Cyprus, Czech Republic, Denmark, Djibouti, Dominica, Dominican Republic, Ecuador, Egypt, El Salvador, Equatorial Guinea, Eritrea, Estonia, Ethiopia, Fiji, Finland, France, Gabon, The Gambia, Georgia, Germany, Ghana, Greece, Grenada, Guatemala, Guinea, Guinea-Bissau, Guyana, Haiti, Honduras, Hungary, Iceland, India, Indonesia, Iran, Iraq, Ireland, Israel, Italy, Jamaica, Japan, Jordan, Kazakhstan, Kenya, South Korea, Kuwait, Kyrgyzstan, Laos, Latvia, Lebanon, Lesotho, Liberia, Libya, Liechtenstein, Lithuania, Luxembourg, Macedonia, Madagascar, Malawi, Malaysia, Maldives, Mali, Malta, Marshall Islands, Mauritania, Mauritius, Mexico, Moldova, Monaco, Mongolia, Montenegro, Morocco, Mozambique, Namibia, Nauru, Nepal, Netherlands, Netherlands Antilles, NZ, Nicaragua, Niger, Nigeria, Norway, Oman, Pakistan, Panama, Papua New Guinea, Paraguay, Peru, Philippines, Poland, Portugal, Qatar, Romania, Russia, Rwanda, Saint Kitts and Nevis, Saint Lucia, Saint Vincent and the Grenadines, San Marino, Sao Tome and Principe, Saudi Arabia, Senegal,

Serbia, Seychelles, Sierra Leone, Singapore, Slovakia, Slovenia, Somalia, South Africa, Spain, Sri Lanka, Sudan, Suriname, Swaziland, Sweden, Switzerland, Syria, Tajikistan, Tanzania, Thailand, Timor-Leste, Togo, Tonga, Trinidad and Tobago, Tunisia, Turkey, Turkmenistan, Uganda, Ukraine, UAE, UK, US, Uruguay, Uzbekistan, Venezuela, Vietnam, Yemen, Zambia, Zimbabwe *subbureaus*—(11) American Samoa, Anguilla, Bermuda, British Virgin Islands, Cayman Islands, Gibraltar, Hong Kong, Macau, Montserrat, Puerto Rico, Turks and Caicos Islands

International Development Association (IDA)
established—26 January 1960; effective—24 September 1960
aim—to provide economic loans for low-income countries; UN specialized agency and IBRD affiliate
members—(179) Afghanistan, Albania, Algeria, Angola, Argentina, Armenia, Austria, Azerbaijan, Bangladesh, Barbados, Belgium, Belize, Benin, Bhutan, Bolivia, Bosnia and Herzegovina, Botswana, Brazil, Burkina Faso, Burma, Burundi, Cambodia, Cameroon, Canada, Cape Verde, Central African Republic, Chad, Chile, China, Colombia, Comoros, Democratic Republic of the Congo, Republic of the Congo, Costa Rica, Cote d'Ivoire, Croatia, Cyprus, Czech Republic, Denmark, Djibouti, Dominica, Dominican Republic, Ecuador, Egypt, El Salvador, Equatorial Guinea, Eritrea, Ethiopia, EU, Fiji, Finland, France, Gabon, The Gambia, Georgia, Germany, Ghana, Greece, Grenada, Guatemala, Guinea, Guinea-Bissau, Guyana, Haiti, Honduras, Hungary, Iceland, India, Indonesia, Iran, Iraq, Ireland, Israel, Italy, Japan, Jordan, Kazakhstan, Kenya, Kiribati, South Korea, Kuwait, Kyrgyzstan, Laos, Latvia, Lebanon, Lesotho, Liberia, Libya, Luxembourg, Macedonia, Madagascar, Malawi, Malaysia, Maldives, Mali, Marshall Islands, Mauritania, Mauritius, Mexico, Federated States of Micronesia, Moldova, Mongolia, Morocco, Mozambique, Nepal, Netherlands, Nicaragua, Niger, Nigeria, Norway, Oman, Pakistan, Palau, Panama, Papua New Guinea, Paraguay, Peru, Philippines, Poland, Portugal, Russia, Rwanda, Saint Kitts and Nevis, Saint Lucia, Saint Vincent and the Grenadines, Samoa, Sao Tome and Principe, Saudi Arabia, Senegal, Serbia, Sierra Leone, Singapore, Slovakia, Slovenia, Solomon Islands, Somalia, South Africa, Spain, Sri Lanka, Sudan, Swaziland, Sweden, Switzerland, Syria, Tajikistan, Tanzania, Thailand, Timor-Leste, Togo, Tonga, Trinidad and Tobago, Tunisia, Turkey, Uganda, Ukraine, UAE, UK, US, Uzbekistan, Vanuatu, Vietnam, Yemen, Zambia, Zimbabwe

International Energy Agency (IEA)
established—15 November 1974
aim—to promote cooperation on energy matters, especially emergency oil sharing and relations between oil consumers and oil producers; established by the OECD
members—(27) Australia, Austria, Belgium, Canada, Czech Republic, Denmark, Finland, France, Germany, Greece, Hungary, Ireland, Italy, Japan, South Korea, Luxembourg, Netherlands, NZ, Norway, Portugal, Slovakia, Spain, Sweden, Switzerland, Turkey, UK, US

International Federation of Red Cross and Red Crescent Societies (IFRCS) *note*—formerly known as League of Red Cross and Red Crescent Societies (LORCS)
established—5 May 1919
aim—to organize, coordinate, and direct international relief actions; to promote humanitarian activities; to represent and encourage the development of National Societies; to bring help to victims of armed conflicts, refugees, and displaced people; to reduce the vulnerability of people through development programs
members—(185 plus the Palestine Liberation Organization) Afghanistan, Albania, Algeria, Andorra, Angola, Antigua and Barbuda, Argentina, Armenia, Australia, Austria, Azerbaijan, The Bahamas, Bahrain, Bangladesh, Barbados, Belarus, Belgium, Belize, Benin, Bolivia, Bosnia and Herzegovina, Botswana, Brazil, Brunei, Bulgaria, Burkina Faso, Burma, Burundi, Cambodia, Cameroon, Canada, Cape Verde, Central African Republic, Chad, Chile, China, Colombia, Comoros, Democratic Republic of the Congo, Republic of the Congo, Cook Islands, Costa Rica, Cote d'Ivoire, Croatia, Cuba, Czech Republic, Denmark, Djibouti, Dominica, Dominican Republic, Ecuador, Egypt, El Salvador, Equatorial Guinea, Estonia, Ethiopia, Fiji, Finland, France, Gabon, The Gambia, Georgia, Germany, Ghana, Greece, Grenada, Guatemala, Guinea, Guinea-Bissau, Guyana, Haiti, Honduras, Hungary, Iceland, India, Indonesia, Iran, Iraq, Ireland, Israel, Italy, Jamaica, Japan, Jordan, Kazakhstan, Kenya, Kiribati, North Korea, South Korea, Kuwait, Kyrgyzstan, Laos, Latvia, Lebanon, Lesotho, Liberia, Libya, Liechtenstein, Lithuania, Luxembourg, Macedonia, Madagascar, Malawi, Malaysia, Mali, Malta, Mauritania, Mauritius, Mexico, Federated States of Micronesia, Moldova, Monaco, Mongolia, Montenegro, Morocco, Mozambique, Namibia, Nepal, Netherlands, NZ, Nicaragua, Niger, Nigeria, Norway, Pakistan, Palau, Panama, Papua New Guinea, Paraguay, Peru, Philippines, Poland, Portugal, Qatar, Romania, Russia, Rwanda, Saint Kitts and Nevis, Saint Lucia, Saint Vincent and the Grenadines, Samoa, San Marino, Sao Tome and Principe, Saudi Arabia, Senegal, Serbia, Seychelles, Sierra Leone, Singapore, Slovakia, Slovenia, Solomon Islands, Somalia, South Africa, Spain, Sri Lanka, Sudan, Suriname, Swaziland, Sweden, Switzerland, Syria, Tajikistan, Tanzania, Thailand, Timor-Leste, Togo, Tonga, Trinidad and Tobago, Tunisia, Turkey, Turkmenistan, Uganda, Ukraine, UAE, UK, US, Uruguay, Uzbekistan, Vanuatu, Venezuela, Vietnam, Yemen, Zambia, Zimbabwe, Palestine Liberation Organization
observers—(2) Eritrea and Tuvalu

International Finance Corporation (IFC)
established—25 May 1955; effective—24 July 1956
aim—to support private enterprise in international economic development; a UN specialized agency and IBRD affiliate
members—(179) includes all UN member countries except Andorra, Brunei, Cuba, North Korea, Liechtenstein, Monaco, Nauru, Qatar, Saint Vincent and the Grenadines, San Marino, Sao Tome and Principe, Suriname, Tuvalu

International Fund for Agricultural Development (IFAD)
established—November 1974
aim—to promote agricultural development; a UN specialized agency

members—(164)
Category I—(23 industrialized aid contributors) Austria, Belgium, Canada, Denmark, Finland, France, Germany, Greece, Iceland, Ireland, Italy, Japan, Luxembourg, Netherlands, NZ, Norway, Portugal, Spain, Sweden, Switzerland, UK, US
Category II—(12 petroleum-exporting aid contributors) Algeria, Gabon, Indonesia, Iran, Iraq, Kuwait, Libya, Nigeria, Qatar, Saudi Arabia, UAE, Venezuela
Category III—(130 aid recipients) Afghanistan, Albania, Angola, Antigua and Barbuda, Argentina, Armenia, Azerbaijan, Bangladesh, Barbados, Belize, Benin, Bhutan, Bolivia, Bosnia and Herzegovina, Botswana, Brazil, Burkina Faso, Burma, Burundi, Cambodia, Cameroon, Cape Verde, Central African Republic, Chad, Chile, China, Colombia, Comoros, Democratic Republic of the Congo, Republic of the Congo, Cook Islands, Costa Rica, Cote d'Ivoire, Croatia, Cuba, Cyprus, Djibouti, Dominica, Dominican Republic, Ecuador, Egypt, El Salvador, Equatorial Guinea, Eritrea, Ethiopia, Fiji, The Gambia, Georgia, Ghana, Grenada, Guatemala, Guinea, Guinea-Bissau, Guyana, Haiti, Honduras, India, Israel, Jamaica, Jordan, Kazakhstan, Kenya, Kiribati, North Korea, South Korea, Kyrgyzstan, Laos, Lebanon, Lesotho, Liberia, Macedonia, Madagascar, Malawi, Malaysia, Maldives, Mali, Malta, Mauritania, Mauritius, Mexico, Moldova, Mongolia, Morocco, Mozambique, Namibia, Nepal, Nicaragua, Niger, Niue, Oman, Pakistan, Panama, Papua New Guinea, Paraguay, Peru, Philippines, Romania, Rwanda, Saint Kitts and Nevis, Saint Lucia, Saint Vincent and the Grenadines, Samoa, Sao Tome and Principe, Senegal, Serbia (suspended since 1992), Seychelles, Sierra Leone, Solomon Islands, Somalia, South Africa, Sri Lanka, Sudan, Suriname, Swaziland, Syria, Tajikistan, Tanzania, Thailand, Timor-Leste, Togo, Tonga, Trinidad and Tobago, Tunisia, Turkey, Uganda, Uruguay, Vietnam, Yemen, Zambia, Zimbabwe

International Hydrographic Organization (IHO) *note*—name changed from International Hydrographic Bureau on 22 September 1970
established—June 1919; effective—June 1921
aim—to train hydrographic surveyors and nautical cartographers to achieve standardization in nautical charts and electronic chart displays; to provide advice on nautical cartography and hydrography; to develop the sciences in the field of hydrography and techniques used for descriptive oceanograrphy
members—(80) Algeria, Argentina, Australia, Bahrain, Bangladesh, Belgium, Brazil, Burma, Canada, Chile, China (including Hong Kong and Macau), Colombia, Democratic Republic of the Congo (suspended), Croatia, Cuba, Cyprus, Denmark, Dominican Republic (suspended), Ecuador, Egypt, Estonia, Fiji, Finland, France, Germany, Greece, Guatemala, Iceland, India, Indonesia, Iran, Ireland, Italy, Jamaica, Japan, North Korea, South Korea, Kuwait, Latvia, Malaysia, Mauritius, Mexico, Monaco, Morocco, Mozambique, Netherlands, NZ, Nigeria, Norway, Oman, Pakistan, Papua New Guinea, Peru, Philippines, Poland, Portugal, Qatar, Romania, Russia, Saudi Arabia, Serbia, Singapore, Slovenia, South Africa, Spain, Sri Lanka, Suriname (suspended), Sweden, Syria, Thailand, Tonga, Trinidad and Tobago, Tunisia, Turkey, Ukraine, UAE, UK, US, Uruguay, Venezuela

International Labor Organization (ILO)
established—28 June 1919 set up as part of Treaty of Versailles; 11 April 1919 became operative; 14 December 1946 affiliated with the UN
aim—to deal with world labor issues; a UN specialized agency
members—(181) includes all UN member countries except Andorra, Bhutan, North Korea, Liechtenstein, Maldives, Federated States of Micronesia, Monaco, Nauru, Palau, Tonga, and Tuvalu; note—includes the following dependencies: Netherlands (Netherlands Antilles and Aruba)

International Maritime Organization (IMO) *note*—name changed from Intergovernmental Maritime Consultative Organization (IMCO) on 22 May 1982
established—6 March 1948 set up as the Inter-Governmental Maritime Consultative Organization; effective—17 March 1958
aim—to deal with international maritime affairs; a UN specialized agency
members—(167) includes all UN member countries except Afghanistan, Andorra, Armenia, Belarus, Bhutan, Botswana, Burkina Faso, Burundi, Central African Republic, Chad, Kyrgyzstan, Laos, Lesotho, Liechtenstein, Mali, Federated States of Micronesia, Nauru, Niger, Palau, Rwanda, Tajikistan, Uganda, Uzbekistan, Zambia
associate members—(3) Faroe Islands, Hong Kong, Macau

International Monetary Fund (IMF)
established—22 July 1944; effective—27 December 1945
aim—to promote world monetary stability and economic development; a UN specialized agency
members—(185) includes all UN member countries except Andorra, Cuba, North Korea, Liechtenstein, Monaco, Nauru, Tuvalu; note—includes the following dependencies or areas of special interest: China (Hong Kong and Macau), Netherlands (Netherlands Antilles and Aruba)

International Mobile Satellite Organization (IMSO)
established—15 April 1999
aim—acts as watchdog over Inmarsat (International Maritime Satellite Organization), a private company, to make sure it follows ICAO standards and recommended practices; plays an active role in the development of international telecommunications policies
members—(88) Algeria, Argentina, Australia, The Bahamas, Bahrain, Bangladesh, Belarus, Belgium, Bosnia and Herzegovina, Brazil, Brunei, Bulgaria, Cameroon, Canada, Chile, China, Colombia, Costa Rica, Croatia, Cuba, Cyprus, Czech Republic, Denmark, Egypt, Finland, France, Gabon, Germany, Ghana, Greece, Hungary, Iceland, India, Indonesia, Iran, Iraq, Israel, Italy, Japan, Kenya, South Korea, Kuwait, Latvia, Lebanon, Liberia, Libya, Malaysia, Malta, Marshall Islands, Maurtius, Mexico, Monaco, Morocco, Mozambique, Netherlands, NZ, Nigeria, Norway, Oman, Pakistan, Panama, Peru, Philippines, Poland, Portugal, Qatar, Romania, Russia, Saudi Arabia, Senegal, Serbia, Singapore, Slovakia, South Africa, Spain, Sri Lanka, Sweden, Switzerland, Tanzania, Thailand, Timor-Leste, Tunisia, Turkey, Ukraine, UAE, UK, US, Vietnam

International Olympic Committee (IOC)

established—23 June 1894

aim—to promote the Olympic ideals and administer the Olympic games: 2006 Winter Olympics in Turin, Italy; 2008 Summer Olympics in Beijing, China; 2010 Winter Olympics in Vancouver, Canada; 2012 Summer Olympics in London, UK
National Olympic Committees—(204 and the Palestine Liberation Organization) Afghanistan, Albania, Algeria, American Samoa, Andorra, Angola, Antigua and Barbuda, Argentina, Armenia, Aruba, Australia, Austria, Azerbaijan, The Bahamas, Bahrain, Bangladesh, Barbados, Belarus, Belgium, Belize, Benin, Bermuda, Bhutan, Bolivia, Bosnia and Herzegovina, Botswana, Brazil, British Virgin Islands, Brunei, Bulgaria, Burkina Faso, Burma, Burundi, Cambodia, Cameroon, Canada, Cape Verde, Cayman Islands, Central African Republic, Chad, Chile, China, Colombia, Comoros, Democratic Republic of the Congo, Republic of the Congo, Cook Islands, Costa Rica, Cote d'Ivoire, Croatia, Cuba, Cyprus, Czech Republic, Denmark, Djibouti, Dominica, Dominican Republic, Ecuador, Egypt, El Salvador, Equatorial Guinea, Eritrea, Estonia, Ethiopia, Fiji, Finland, France, Gabon, The Gambia, Georgia, Germany, Ghana, Greece, Grenada, Guam, Guatemala, Guinea, Guinea-Bissau, Guyana, Haiti, Honduras, Hong Kong, Hungary, Iceland, India, Indonesia, Iran, Iraq, Ireland, Israel, Italy, Jamaica, Japan, Jordan, Kazakhstan, Kenya, Kiribati, North Korea, South Korea, Kuwait, Kyrgyzstan, Laos, Latvia, Lebanon, Lesotho, Liberia, Libya, Liechtenstein, Lithuania, Luxembourg, Macedonia, Madagascar, Malawi, Malaysia, Maldives, Mali, Malta, Marshall Islands, Mauritania, Mauritius, Mexico, Federated States of Micronesia, Moldova, Monaco, Mongolia, Montenegro, Morocco, Mozambique, Namibia, Nauru, Nepal, Netherlands, Netherlands Antilles, NZ, Nicaragua, Niger, Nigeria, Norway, Oman, Pakistan, Palau, Panama, Papua New Guinea, Paraguay, Peru, Philippines, Poland, Portugal, Puerto Rico, Qatar, Romania, Russia, Rwanda, Saint Kitts and Nevis, Saint Lucia, Saint Vincent and the Grenadines, Samoa, San Marino, Sao Tome and Principe, Saudi Arabia, Senegal, Serbia, Seychelles, Sierra Leone, Singapore, Slovakia, Slovenia, Solomon Islands, Somalia, South Africa, Spain, Sri Lanka, Sudan, Suriname, Swaziland, Sweden, Switzerland, Syria, Taiwan, Tajikistan, Tanzania, Thailand, Timor-Leste, Togo, Tonga, Trinidad and Tobago, Tunisia, Turkey, Turkmenistan, Tuvalu, Uganda, Ukraine, UAE, UK, US, Uruguay, Uzbekistan, Vanuatu, Venezuela, Vietnam, Virgin Islands, Yemen, Zambia, Zimbabwe, Palestine Liberation Organization

International Organization for Migration (IOM)

note—established as Provisional Intergovernmental Committee for the Movement of Migrants from Europe; renamed Intergovernmental Committee for European Migration (ICEM) on 15 November 1952; renamed Intergovernmental Committee for Migration (ICM) in November 1980; current name adopted 14 November 1989

established—5 December 1951

aim—to facilitate orderly international emigration and immigration

members—(122) Afghanistan, Albania, Algeria, Angola, Argentina, Armenia, Australia, Austria, Azerbaijan, The Bahamas, Bangladesh, Belarus, Belgium, Belize, Benin, Bolivia, Bosnia and Herzegovina, Brazil, Bulgaria, Burkina Faso, Burundi, Cambodia, Cameroon, Canada, Cape Verde, Chile, Colombia, Democratic Republic of the Congo, Republic of the Congo, Costa Rica, Cote d'Ivoire, Croatia, Cyprus, Czech Republic, Denmark, Dominican Republic, Ecuador, Egypt, El Salvador, Estonia, Finland, France, Gabon, The Gambia, Georgia, Germany, Ghana, Greece, Guatemala, Guinea, Guinea-Bissau, Haiti, Honduras, Hungary, Iran, Ireland, Israel, Italy, Jamaica, Japan, Jordan, Kazakhstan, Kenya, South Korea, Kyrgyzstan, Latvia, Liberia, Libya, Lithuania, Luxembourg, Madagascar, Mali, Malta, Mauritania, Mauritius, Mexico, Moldova, Montenegro, Morocco, Nepal, Netherlands, NZ, Nicaragua, Niger, Nigeria, Norway, Pakistan, Panama, Paraguay, Peru, Philippines, Poland, Portugal, Romania, Rwanda, Senegal, Serbia, Sierra Leone, Slovakia, Slovenia, South Africa, Spain, Sri Lanka, Sudan, Sweden, Switzerland, Tajikistan, Tanzania, Thailand, Togo, Tunisia, Turkey, Uganda, Ukraine, UK, US, Uruguay, Venezuela, Vietnam, Yemen, Zambia, Zimbabwe

observers—(18) Bahrain, Bhutan, China, Cuba, Ethiopia, Guyana, Holy See, India, Indonesia, Macedonia, Mozambique, Namibia, Papua New Guinea, Russia, San Marino, Sao Tome and Principe, Somalia, Turkmenistan

International Organization for Standardization (ISO)

established—February 1947

aim—to promote the development of international standards with a view to facilitating international exchange of goods and services and to developing cooperation in the sphere of intellectual, scientific, technological and economic activity

members—(105 national standards organizations) Algeria, Argentina, Armenia, Australia, Austria, Azerbaijan, Bahrain, Bangladesh, Barbados, Belarus, Belgium, Bosnia and Herzegovina, Botswana, Brazil, Bulgaria, Canada, Chile, China, Colombia, Democratic Republic of the Congo, Costa Rica, Cote d'Ivoire, Croatia, Cuba, Cyprus, Czech Republic, Denmark, Ecuador, Egypt, Ethiopia, Fiji, Finland, France, Germany, Ghana, Greece, Hungary, Iceland, India, Indonesia, Iran, Iraq, Ireland, Israel, Italy, Jamaica, Japan, Jordan, Kazakhstan, Kenya, North Korea, South Korea, Kuwait, Lebanon, Libya, Lithuania, Luxembourg, Macedonia, Malaysia, Malta, Mauritius, Mexico, Mongolia, Morocco, Netherlands, NZ, Nigeria, Norway, Oman, Pakistan, Panama, Peru, Philippines, Poland, Portugal, Qatar, Romania, Russia, Saint Lucia, Saudi Arabia, Serbia, Singapore, Slovakia, Slovenia, South Africa, Spain, Sri Lanka, Sudan, Sweden, Switzerland, Syria, Tanzania, Thailand, Trinidad and Tobago, Tunisia, Turkey, Ukraine, UAE, UK, US, Uruguay, Uzbekistan, Venezuela, Vietnam, Zimbabwe

correspondent members—(41 plus the Palestine Liberation Organization) Afghanistan, Albania, Angola, Benin, Bhutan, Bolivia, Brunei, Burkina Faso, Burma, Cameroon, Dominican Republic, El Salvador, Eritrea, Estonia, Gabon, Georgia, Guatemala, Hong Kong, Kyrgyzstan, Latvia, Macau, Madagascar, Malawi, Moldova, Montenegro, Mozambique, Namibia, Nepal, Nicaragua, Papua New Guinea, Paraguay, Rwanda, Senegal, Seychelles, Swaziland, Tajikistan, Togo, Turkmenistan, Uganda, Yemen, Zambia, Palestine Liberation Organization

subscriber members—(10) Antigua and Barbuda, Burundi, Cambodia, Dominica, Guyana, Honduras, Laos, Lesotho, Saint Vincent and the Grenadines, Suriname

International Organization of the French-speaking World (OIF) *note*—name changed from Agency of Cultural and Technical Cooperation (ACCT) in 1997; also known as Organisation Internationale de la Francophonie
established—20 March 1970
aim—founded around a common language to promote and spread the cultures of its members and to reinforce cultural and technical cooperation between them
members—(55) Albania, Andorra, Belgium, Benin, Bulgaria, Burkina Faso, Burundi, Cambodia, Cameroon, Canada, Canada—New Brunswick, Canada—Quebec, Cape Verde, Central African Republic, Chad, Comoros, Democratic Republic of Congo, Republic of Congo, Cote d'Ivoire, Cyprus, Djibouti, Dominica, Egypt, Equatorial Guinea, France, French Community of Belgium, Gabon, Ghana, Greece, Guinea, Guinea-Bissau, Haiti, Laos, Lebanon, Luxembourg, Macedonia, Madagascar, Mali, Mauritania, Mauritius, Moldova, Monaco, Morocco, Niger, Romania, Rwanda, Saint Lucia, Sao Tome and Principe, Senegal, Seychelles, Switzerland, Togo, Tunisia, Vanuatu, Vietnam
observers—(13) Armenia, Austria, Croatia, Czech Republic, Georgia, Hungary, Lithuania, Mozambique, Poland, Serbia, Slovakia, Slovenia, Ukraine

International Red Cross and Red Crescent Movement (ICRM)
established—1928
aim—to promote worldwide humanitarian aid through the International Committee of the Red Cross (ICRC) in wartime, and International Federation of Red Cross and Red Crescent Societies (IFRCS; formerly League of Red Cross and Red Crescent Societies or LORCS) in peacetime
National Societies—(185 countries and the Palestine Liberation Organization); note—same as membership for International Federation of Red Cross and Red Crescent Societies (IFRCS)

International Telecommunications Satellites Organization (ITSO)
established—August 1964
aim—to act as a watchdog over Intelsat, Ltd., a private company, to make sure it provides on a global and non-discriminatory basis public telecommunication services
members—(148) Afghanistan, Algeria, Angola, Argentina, Armenia, Australia, Austria, Azerbaijan, The Bahamas, Bahrain, Bangladesh, Barbados, Belgium, Benin, Bhutan, Bolivia, Bosnia and Herzegovina, Botswana, Brazil, Brunei, Bulgaria, Burkina Faso, Cameroon, Canada, Cape Verde, Central African Republic, Chad, China, Colombia, Comoros, Democratic Republic of the Congo, Republic of the Congo, Costa Rica, Cote d'Ivoire, Croatia, Cuba, Cyprus, Czech Republic, Denmark, Dominican Republic, Ecuador, Egypt, El Salvador, Equatorial Guinea, Ethiopia, Fiji, Finland, France, Gabon, The Gambia, Georgia, Germany, Ghana, Greece, Guatemala, Guinea, Guinea-Bissau, Haiti, Holy See, Honduras, Hungary, Iceland, India, Indonesia, Iran, Iraq, Ireland, Israel, Italy, Jamaica, Japan, Jordan, Kazakhstan, Kenya, North Korea, South Korea, Kuwait, Kyrgyzstan, Lebanon, Libya, Liechtenstein, Luxembourg, Madagascar, Malawi, Malaysia, Mali, Malta, Mauritania, Mauritius, Mexico, the Federated States of Micronesia, Monaco, Mongolia, Morocco, Mozambique, Namibia, Nepal, Netherlands, NZ, Nicaragua, Niger, Nigeria, Norway, Oman, Pakistan, Panama, Papua New Guinea, Paraguay, Peru, Philippines, Poland, Portugal, Qatar, Romania, Russia, Rwanda, Saudi Arabia, Senegal, Serbia, Singapore, Somalia, South Africa, Spain, Sri Lanka, Sudan, Swaziland, Sweden, Switzerland, Syria, Tajikistan, Tanzania, Thailand, Togo, Trinidad and Tobago, Tunisia, Turkey, Uganda, UAE, UK, US, Uruguay, Uzbekistan, Venezuela, Vietnam, Yemen, Zambia, Zimbabwe

International Telecommunication Union (ITU)
established—17 May 1865 set up as the International Telegraph Union; 9 December 1932 adopted present name; effective—1 January 1934; affiliated with the UN—15 November 1947
aim—to deal with world telecommunications issues; a UN specialized agency
members—(191) includes all UN member countries except Palau, Timor-Leste (190 total); plus Holy See

International Trade Union Confederation (ITUC) *note*—its predecessors were the Confederation of Free Trade Unions (ICFTU) and the World Confederation of Labor (WCL)
established—3 November 2006
aim—to promote the trade union movement
members—(311 affiliated organizations in the following 154 countries and the Palestine Liberation Organization as of December 2007) Albania, Algeria, Angola, Antigua and Barbuda, Aruba, Argentina, Australia, Austria, Azerbaijan, Bahrain, Bangladesh, Barbados, Belarus, Belgium, Belize, Benin, Bermuda, Bonaire, Bosnia and Herzegovina, Botswana, Brazil, Bulgaria, Burkina Faso, Burundi, Cameroon, Canada, Cape Verde, Central African Republic, Comoros, Chad, Chile, Colombia, Democratic Republic of the Congo, Republic of the Congo, Cook Islands, Costa Rica, Cote d'Ivoire, Curacao, Cyprus, Czech Republic, Denmark, Djibouti, Dominica, Dominican Republic, Ecuador, El Salvador, Eritrea, Estonia, Ethiopia, Fiji, Finland, France, French Polynesia, Gabon, The Gambia, Georgia, Germany, Ghana, Greece, Grenada, Guatemala, Guinea, Guinea-Bissau, Guyana, Haiti, Holy See, Honduras, Hong Kong, Hungary, Iceland, India, Indonesia, Ireland, Israel, Italy, Japan, Jordan, Kenya, Kiribati, South Korea, Kosovo, Kuwait, Latvia, Liberia, Lithuania, Luxembourg, Macedonia, Madagascar, Malawi, Malaysia, Mali, Malta, Mauritania, Mauritius, Mexico, Mongolia, Montenegro, Morocco, Mozambique, Nepal, Netherlands, New Caledonia, NZ, Nicaragua, Niger, Nigeria, Norway, Pakistan, Panama, Paraguay, Peru, Philippines, Poland, Portugal, Puerto Rico, Romania, Russia, Rwanda, Saint Lucia, Saint Vincent and the Grenadines, Samoa, San Marino, Sao Tome and Principe, Senegal, Serbia, Seychelles, Sierra Leone, Singapore, Slovakia, South Africa, Spain, Sri Lanka, Suriname, Swaziland, Sweden, Switzerland, Taiwan, Tanzania, Thailand, Togo, Tonga, Trinidad and Tobago, Tunisia, Turkey, Uganda, Ukraine, UK, US, Vanuatu, Venezuela, Yemen, Zambia, Zimbabwe, and the Palestine Liberation Organization

Islamic Development Bank (IDB)

established—15 December 1973 by declaration of intent; *effective*—12 August 1974
aim—to promote Islamic economic aid and social development
members—(55 plus the Palestine Liberation Organization) Afghanistan, Albania, Algeria, Azerbaijan, Bahrain, Bangladesh, Benin, Brunei, Burkina Faso, Cameroon, Chad, Comoros, Cote d'Ivoire, Djibouti, Egypt, Gabon, The Gambia, Guinea, Guinea-Bissau, Indonesia, Iran, Iraq, Jordan, Kazakhstan, Kuwait, Kyrgyzstan, Lebanon, Libya, Malaysia, Maldives, Mali, Mauritania, Morocco, Mozambique, Niger, Nigeria, Oman, Pakistan, Qatar, Saudi Arabia, Senegal, Sierra Leone, Somalia, Sudan, Suriname, Syria, Tajikistan, Togo, Tunisia, Turkey, Turkmenistan, Uganda, UAE, Uzbekistan, Yemen, Palestine Liberation Organization

Latin American Economic System (LAES) *note*—also known as Sistema Economico Latinoamericana (SELA)

established—17 October 1975
aim—to promote economic and social development through regional cooperation
members—(26) Argentina, the Bahamas, Barbados, Belize, Bolivia, Brazil, Chile, Colombia, Costa Rica, Cuba, Dominican Republic, Ecuador, Guatemala, Guyana, Haiti, Honduras, Jamaica, Mexico, Nicaragua, Panama, Paraguay, Peru, Suriname, Trinidad and Tobago, Uruguay, Venezuela

Latin American Integration Association (LAIA) *note*—also known as Asociacion Latinoamericana de Integracion (ALADI)

established—12 August 1980; *effective*—18 March 1981
aim—to promote freer regional trade
members—(12) Argentina, Bolivia, Brazil, Chile, Colombia, Cuba, Ecuador, Mexico, Paraguay, Peru, Uruguay, Venezuela
observers—(26) China, Corporacion Andina de Fomento, Costa Rica, Dominican Republic, EC, El Salvador, Guatemala, Honduras, Inter-American Development Bank, Inter-American Institute for Cooperation on Agriculture, Italy, Japan, South Korea, Latin America Economic System, Nicaragua, Organization of American States, Panama, Pan-American Health Organization, Portugal, Romania, Russia, Spain, Switzerland, Ukraine, United Nations Development Program, United Nations Economic Commission for Latin America and the Caribbean

League of Arab States (LAS) *note*—also known as Arab League (AL)

established—22 March 1945
aim—aim—to promote economic, social, political, and military cooperation
members—(21 plus the Palestine Liberation Organization) Algeria, Bahrain, Comoros, Djibouti, Egypt, Iraq, Jordan, Kuwait, Lebanon, Libya, Mauritania, Morocco, Oman, Qatar, Saudi Arabia, Somalia, Sudan, Syria, Tunisia, UAE, Yemen, Palestine Liberation Organization
observers—(3) Eritrea, India, Venezuela

least developed countries (LLDCs)

that subgroup of the less developed countries (LDCs) initially identified by the UN General Assembly in 1971 as having no significant economic growth, per capita GDPs normally less than $1,000, and low literacy rates; also known as the undeveloped countries; the 42 LLDCs are: Afghanistan, Bangladesh, Benin, Bhutan, Botswana, Burkina Faso, Burma, Burundi, Cape Verde, Central African Republic, Chad, Comoros, Djibouti, Equatorial Guinea, Eritrea, Ethiopia, The Gambia, Guinea, Guinea-Bissau, Haiti, Kiribati, Laos, Lesotho, Malawi, Maldives, Mali, Mauritania, Mozambique, Nepal, Niger, Rwanda, Samoa, Sao Tome and Principe, Sierra Leone, Somalia, Sudan, Tanzania, Togo, Tuvalu, Uganda, Vanuatu, Yemen

less developed countries (LDCs)

the bottom group in the hierarchy of developed countries (DCs), former USSR/Eastern Europe (former USSR/EE), and less developed countries (LDCs); mainly countries and dependent areas with low levels of output, living standards, and technology; per capita GDPs are generally below $5,000 and often less than $1,500; however, the group also includes a number of countries with high per capita incomes, areas of advanced technology, and rapid rates of growth; includes the advanced developing countries, developing countries, Four Dragons (Four Tigers), least developed countries (LLDCs), low-income countries, middle-income countries, newly industrializing economies (NIEs), the South, Third World, underdeveloped countries, undeveloped countries; the 172 LDCs are: Afghanistan, Algeria, American Samoa, Angola, Anguilla, Antigua and Barbuda, Argentina, Aruba, The Bahamas, Bahrain, Bangladesh, Barbados, Belize, Benin, Bhutan, Bolivia, Botswana, Brazil, British Virgin Islands, Brunei, Burkina Faso, Burma, Burundi, Cambodia, Cameroon, Cape Verde, Cayman Islands, Central African Republic, Chad, Chile, China, Christmas Island, Cocos Islands, Colombia, Comoros, Democratic Republic of the Congo, Republic of the Congo, Cook Islands, Costa Rica, Cote d'Ivoire, Cuba, Cyprus, Djibouti, Dominica, Dominican Republic, Ecuador, Egypt, El Salvador, Equatorial Guinea, Eritrea, Ethiopia, Falkland Islands, Fiji, French Guiana, French Polynesia, Gabon, The Gambia, Gaza Strip, Ghana, Gibraltar, Greenland, Grenada, Guadeloupe, Guam, Guatemala, Guernsey, Guinea, Guinea-Bissau, Guyana, Haiti, Honduras, Hong Kong, India, Indonesia, Iran, Iraq, Isle of Man, Jamaica, Jersey, Jordan, Kenya, Kiribati, North Korea, South Korea, Kuwait, Laos, Lebanon, Lesotho, Liberia, Libya, Macau, Madagascar, Malawi, Malaysia, Maldives, Mali, Marshall Islands, Martinique, Mauritania, Mauritius, Mayotte, Federated States of Micronesia, Mongolia, Montserrat, Morocco, Mozambique, Namibia, Nauru, Nepal, Netherlands Antilles, New Caledonia, Nicaragua, Niger, Nigeria, Niue, Norfolk Island, Northern Mariana Islands, Oman, Palau, Pakistan, Panama, Papua New Guinea, Paraguay, Peru, Philippines, Pitcairn Islands, Puerto Rico, Qatar, Reunion, Rwanda, Saint Helena, Saint Kitts and Nevis, Saint Lucia, Saint Pierre and Miquelon, Saint Vincent and the Grenadines, Samoa, Sao Tome and Principe, Saudi Arabia, Senegal, Seychelles, Sierra Leone, Singapore, Solomon Islands, Somalia, Sri Lanka, Sudan, Suriname, Swaziland, Syria, Taiwan, Tanzania, Thailand, Togo, Tokelau, Tonga, Trinidad and Tobago, Tunisia, Turks and Caicos Islands, Tuvalu, UAE, Uganda, Uruguay, Vanuatu, Venezuela, Vietnam, Virgin Islands, Wallis and Futuna, West Bank, Western Sahara, Yemen, Zambia, Zimbabwe; note—similar to the new International Monetary Fund (IMF) term "developing countries" which adds Malta, Mexico, South Africa, and Turkey but omits in its recently published statistics American Samoa, Anguilla, British Virgin Islands, Brunei, Cayman Islands, Christmas Island, Cocos Islands, Cook Islands, Cuba, Eritrea, Falkland Islands, French Guiana, French Polynesia, Gaza Strip, Gibraltar, Greenland, Grenada, Guadeloupe, Guam, Guernsey, Isle of Man, Jersey, North Korea, Macau, Martinique, Mayotte, Montserrat, Nauru, New Caledonia, Niue, Norfolk Island, Northern Mariana Islands, Palau, Pitcairn Islands, Puerto Rico, Reunion, Saint Helena, Saint Pierre and Miquelon, Tokelau, Tonga, Turks and Caicos Islands, Tuvalu, Virgin Islands, Wallis and Futuna, West Bank, Western Sahara

low-income countries
another term for those less developed countries with below-average per capita GDPs; see less developed countries (LDCs)

middle-income countries
another term for those less developed countries with above-average per capita GDPs; see less developed countries (LDCs)

Multilateral Investment Guarantee Agency (MIGA)
established—12 April 1988
aim—encourages flow of foreign direct investment among member countries by offering investment insurance, consultation, and negotiation on conditions for foreign investment and technical assistance; a UN specialized agency
members—(171) includes all UN member countries except Andorra, Bhutan, Brunei, Burma, Comoros, Cuba, Iraq, Kiribati, North Korea, Liechtenstein, Marshall Islands, Mexico, Monaco, Nauru, NZ, Niger, San Marino, Sao Tome and Principe, Somalia, Tonga, Tuvalu

Near Abroad
Russian term for the 14 non-Russian successor states of the USSR, in which 25 million ethnic Russians live and in which Moscow has expressed a strong national security interest; the 14 countries are Armenia, Azerbaijan, Belarus, Estonia, Georgia, Kazakhstan, Kyrgyzstan, Latvia, Lithuania, Moldova, Tajikistan, Turkmenistan, Ukraine, Uzbekistan

new independent states (NIS)
a term referring to all the countries of the FSU except the Baltic countries (Estonia, Latvia, Lithuania)

newly industrializing countries (NICs)
former term for the newly industrializing economies; see newly industrializing economies (NIEs)

newly industrializing economies (NIEs)
that subgroup of the less developed countries (LDCs) that has experienced particularly rapid industrialization of their economies; formerly known as the newly industrializing countries (NICs); also known as advanced developing countries; usually includes the Four Dragons (Hong Kong, South Korea, Singapore, Taiwan), and Brazil

Nonaligned Movement (NAM)
established—1–6 September 1961
aim—to establish political and military cooperation apart from the traditional East or West blocs
members—(117 plus the Palestine Liberation Organization) Afghanistan, Algeria, Angola, Antigua and Barbuda, The Bahamas, Bahrain, Bangladesh, Barbados, Belarus, Belize, Benin, Bhutan, Bolivia, Botswana, Brunei, Burkina Faso, Burma, Burundi, Cambodia, Cameroon, Cape Verde, Central African Republic, Chad, Chile, Colombia, Comoros, Democratic Republic of the Congo, Republic of the Congo, Cote d'Ivoire, Cuba, Djibouti, Dominica, Dominican Republic, Ecuador, Egypt, Equatorial Guinea, Eritrea, Ethiopia, Gabon, The Gambia, Ghana, Grenada, Guatemala, Guinea, Guinea-Bissau, Guyana, Haiti, Honduras, India, Indonesia, Iran, Iraq, Jamaica, Jordan, Kenya, North Korea, Kuwait, Laos, Lebanon, Lesotho, Liberia, Libya, Madagascar, Malawi, Malaysia, Maldives, Mali, Mauritania, Mauritius, Mongolia, Morocco, Mozambique, Namibia, Nepal, Nicaragua, Niger, Nigeria, Oman, Pakistan, Panama, Papua New Guinea, Peru, Philippines, Qatar, Rwanda, Saint Kitts and Nevis, Saint Lucia, Saint Vincent and the Grenadines, Sao Tome and Principe, Saudi Arabia, Senegal, Seychelles, Sierra Leone, Singapore, Somalia, South Africa, Sri Lanka, Sudan, Suriname, Swaziland, Syria, Tanzania, Thailand, Timor-Leste, Togo, Trinidad and Tobago, Tunisia, Turkmenistan, Uganda, UAE, Uzbekistan, Vanuatu, Venezuela, Vietnam, Yemen, Zambia, Zimbabwe, Palestine Liberation Organization
observers—(15) Armenia, Azerbaijan, Bosnia and Herzegovina, Brazil, China, Costa Rica, Croatia, El Salvadore, Kazakhstan, Kyrgyzstan, Mexico, Paraguay, Serbia, Ukraine, Uruguay
guests—(24) Australia, Austria, Bulgaria, Canada, Cyprus, Czech Republic, Finland, Germany, Greece, Holy See, Hungary, Italy, Netherlands, NZ, Norway, Poland, Portugal, Romania, Russia, Slovakia, Slovenia, Spain, Sweden, Switzerland

Nordic Council (NC)
established—16 March 1952; effective—12 February 1953
aim—to promote regional economic, cultural, and environmental cooperation
members—(5) Denmark (including Faroe Islands and Greenland), Finland (including Aland Islands), Iceland, Norway, Sweden
observers—(3) the Sami (Lapp) local parliaments of Finland, Norway, and Sweden

Nordic Investment Bank (NIB)
established—4 December 1975; effective—1 June 1976
aim—to promote economic cooperation and development
members—(8) Denmark (including Faroe Islands and Greenland), Estonia, Finland (including Aland Islands), Iceland, Latvia, Lithuania, Norway, Sweden

North
a popular term for the rich industrialized countries generally located in the northern portion of the Northern Hemisphere; the counterpart of the South; see developed countries (DCs)

757

North American Free Trade Agreement (NAFTA)
established—17 December 1992
aim—to eliminate trade barriers, promote fair competition, increase investment opportunities, provide protection of intellectual property rights, and create procedures to settle disputes
members—(3) Canada, Mexico, US

North Atlantic Treaty Organization (NATO)
established—4 April 1949
aim—to promote mutual defense and cooperation
members—(26) Belgium, Bulgaria, Canada, Czech Republic, Denmark, Estonia, France, Germany, Greece, Hungary, Iceland, Italy, Latvia, Lithuania, Luxembourg, Netherlands, Norway, Poland, Portugal, Romania, Slovakia, Slovenia, Spain, Turkey, UK, US

Nuclear Energy Agency (NEA) *note*—also known as OECD Nuclear Energy Agency
established—1 February 1958
aim—to promote the peaceful uses of nuclear energy; associated with OECD
members—(28) Australia, Austria, Belgium, Canada, Czech Republic, Denmark, Finland, France, Germany, Greece, Hungary, Iceland, Ireland, Italy, Japan, South Korea, Luxembourg, Mexico, Netherlands, Norway, Portugal, Slovakia, Spain, Sweden, Switzerland, Turkey, UK, US

Nuclear Suppliers Group (NSG) *note*—also known as the London Suppliers Group or the London Group
established—1974; effective—1975
aim—to establish guidelines for exports of nuclear materials, processing equipment for uranium enrichment, and technical information to countries of proliferation concern and regions of conflict and instability
members—(45) Argentina, Australia, Austria, Belarus, Belgium, Brazil, Bulgaria, Canada, China, Croatia, Cyprus, Czech Republic, Denmark, Estonia, Finland, France, Germany, Greece, Hungary, Ireland, Italy, Japan, Kazakhstan, South Korea, Latvia, Lithuania, Luxembourg, Malta, Netherlands, NZ, Norway, Poland, Portugal, Romania, Russia, Slovakia, Slovenia, South Africa, Spain, Sweden, Switzerland, Turkey, Ukraine, UK, US
observer—(1) European Commission (a policy-planning body for the EU)

Organization for Economic Cooperation and Development (OECD)
established—14 December 1960; effective—30 September 1961
aim—to promote economic cooperation and development
members—(30) Australia, Austria, Belgium, Canada, Czech Republic, Denmark, Finland, France, Germany, Greece, Hungary, Iceland, Ireland, Italy, Japan, South Korea, Luxembourg, Mexico, Netherlands, NZ, Norway, Poland, Portugal, Slovakia, Spain, Sweden, Switzerland, Turkey, UK, US
special member—(1) EC

Organization for Security and Cooperation in Europe (OSCE) *note*—formerly the Conference on Security and Cooperation in Europe (CSCE) established 3 July 1975
established—1 January 1995
aim—to foster the implementation of human rights, fundamental freedoms, democracy, and the rule of law; to act as an instrument of early warning, conflict prevention, and crisis management; and to serve as a framework for conventional arms control and confidence building measures
members—(56) Albania, Andorra, Armenia, Austria, Azerbaijan, Belarus, Belgium, Bosnia and Herzegovina, Bulgaria, Canada, Croatia, Cyprus, Czech Republic, Denmark, Estonia, Finland, France, Georgia, Germany, Greece, Holy See, Hungary, Iceland, Ireland, Italy, Kazakhstan, Kyrgyzstan, Latvia, Liechtenstein, Lithuania, Luxembourg, Macedonia, Malta, Moldova, Monaco, Montenegro, Netherlands, Norway, Poland, Portugal, Romania, Russia, San Marino, Serbia, Slovakia, Slovenia, Spain, Sweden, Switzerland, Tajikistan, Turkey, Turkmenistan, Ukraine, UK, US, Uzbekistan
partners for cooperation—(11) Afghanistan, Algeria, Egypt, Israel, Japan, Jordan, South Korea, Mongolia, Morocco, Thailand, Tunisia

Organization for the Prohibition of Chemical Weapons (OPCW)
established—29 April 1997
aim—to enforce the Convention on the Prohibition of the Development, Production, Stockpiling, and Use of Chemical Weapons and on Their Destruction; to provide a forum for consultation and cooperation among the signatories of the Convention
members (countries that have ratified the Convention)—(183) Afghanistan, Albania, Algeria, Andorra, Antigua and Barbuda, Argentina, Armenia, Australia, Austria, Azerbaijan, Bahrain, Bangladesh, Barbadoes, Belarus, Belgium, Belize, Benin, Bhutan, Bolivia, Bosnia and Herzegovina, Botswana, Brazil, Brunei, Bulgaria, Burkina Faso, Burundi, Cambodia, Cameroon, Canada, Cape Verde, Central African Republic, Chad, Chile, China, Colombia, Comoros, Democratic Republic of the Congo, Republic of the Congo, Cook Islands, Costa Rica, Cote d'Ivoire, Croatia, Cuba, Cyprus, Czech Republic, Denmark, Dominica, Djibouti, Ecuador, El Salvador, Equatorial Guinea, Eritrea, Estonia, Ethiopia, Fiji, Finland, France, Gabon, The Gambia, Georgia, Germany, Ghana, Greece, Grenada, Guatemala, Guinea, Guyana, Haiti, Holy See, Honduras, Hungary, Iceland, India, Indonesia, Iran, Ireland, Italy, Jamaica, Japan, Jordan, Kazakhstan, Kenya, Kiribati, South Korea, Kuwait, Kyrgyzstan, Laos, Latvia, Lesotho, Liberia, Libya, Liechtenstein, Lithuania, Luxembourg, Macedonia, Madagascar, Malawi, Malaysia, Maldives, Mali, Malta, Marshall Islands, Mauritania, Mauritius, Mexico, Federated States of Micronesia, Moldova, Monaco, Mongolia, Montenegro, Morocco, Mozambique, Namibia, Nauru, Nepal, Netherlands, NZ, Nicaragua, Niger, Nigeria, Niue, Norway, Oman, Pakistan, Palau, Panama, Papua New Guinea, Paraguay, Peru, Philippines, Poland, Portugal, Qatar, Romania, Russia, Rwanda, Saint Kitts and Nevis, Saint Lucia, Saint Vincent and the Grenadines, Samoa, San Marino, Sao Tome and Principe,

Saudi Arabia, Senegal, Serbia, Seychelles, Sierra Leone, Singapore, Slovakia, Slovenia, Solomon Islands, South Africa, Spain, Sri Lanka, Sudan, Suriname, Swaziland, Sweden, Switzerland, Tajikistan, Tanzania, Thailand, Timor-Leste, Togo, Tonga, Trinidad and Tobago, Tunisia, Turkey, Turkmenistan, Tuvalu, Uganda, Ukraine, UAE, UK, US, Uruguay, Uzbekistan, Vanuatu, Venezuela, Vietnam, Yemen, Zambia, Zimbabwe
signatory states (countries that have signed, but not ratified, the Convention)—(5) The Bahamas, Burma, Dominican Republic, Guinea-Bissau, Israel

Organization of African Unity (OAU)
see African Union

Organization of American States (OAS)
established—14 April 1890 as the International Union of American Republics; 30 April 1948 adopted present charter; effective—13 December 1951
aim—to promote regional peace and security as well as economic and social development
members—(35) Antigua and Barbuda, Argentina, The Bahamas, Barbados, Belize, Bolivia, Brazil, Canada, Chile, Colombia, Costa Rica, Cuba (excluded from formal participation since 1962), Dominica, Dominican Republic, Ecuador, El Salvador, Grenada, Guatemala, Guyana, Haiti, Honduras, Jamaica, Mexico, Nicaragua, Panama, Paraguay, Peru, Saint Kitts and Nevis, Saint Lucia, Saint Vincent and the Grenadines, Suriname, Trinidad and Tobago, US, Uruguay, Venezuela
observers—(60) Algeria, Angola, Armenia, Austria, Azerbaijan, Belgium, Bosnia and Herzegovina, Bulgaria, China, Croatia, Cyprus, Czech Republic, Denmark, Egypt, Equatorial Guinea, Estonia, EU, Finland, France, Georgia, Germany, Ghana, Greece, Holy See, Hungary, India, Ireland, Israel, Italy, Japan, Kazakhstan, South Korea, Latvia, Lebanon, Luxembourg, Morocco, Netherlands, Nigeria, Norway, Pakistan, Philippines, Poland, Portugal, Qatar, Romania, Russia, Saudi Arabia, Serbia, Slovakia, Slovenia, Spain, Sri Lanka, Sweden, Switzerland, Thailand, Tunisia, Turkey, Ukraine, UK, Yemen

Organization of Arab Petroleum Exporting Countries (OAPEC)
established—9 January 1968
aim—to promote cooperation in the petroleum industry
members—(11) Algeria, Bahrain, Egypt, Iraq, Kuwait, Libya, Qatar, Saudi Arabia, Syria, Tunisia (suspended), UAE

Organization of Eastern Caribbean States (OECS)
established—18 June 1981; effective—4 July 1981
aim—to promote political, economic, and defense cooperation
members—(9) Anguilla, Antigua and Barbuda, British Virgin Islands, Dominica, Grenada, Montserrat, Saint Kitts and Nevis, Saint Lucia, Saint Vincent and the Grenadines

Organization of Petroleum Exporting Countries (OPEC)
established—14 September 1960
aim—to coordinate petroleum policies
members—(13) Algeria, Angola, Ecuador, Indonesia, Iran, Iraq, Kuwait, Libya, Nigeria, Qatar, Saudi Arabia, UAE, Venezuela

Organization of the Islamic Conference (OIC)
established—22–25 September 1969
aim—to promote Islamic solidarity in economic, social, cultural, and political affairs
members—(56 plus the Palestine Liberation Organization) Afghanistan, Albania, Algeria, Azerbaijan, Bahrain, Bangladesh, Benin, Brunei, Burkina Faso, Cameroon, Chad, Comoros, Cote d'Ivoire, Djibouti, Egypt, Gabon, The Gambia, Guinea, Guinea-Bissau, Guyana, Indonesia, Iran, Iraq, Jordan, Kazakhstan, Kuwait, Kyrgyzstan, Lebanon, Libya, Malaysia, Maldives, Mali, Mauritania, Morocco, Mozambique, Niger, Nigeria, Oman, Pakistan, Qatar, Saudi Arabia, Senegal, Sierra Leone, Somalia, Sudan, Suriname, Syria, Tajikistan, Togo, Tunisia, Turkey, Turkmenistan, Uganda, UAE, Uzbekistan, Yemen, Palestine Liberation Organization
observers—(14) AU, Bosnia and Herzegovina, Central African Republic, ECO, Islamic Conference Youth Forum for Dialogue and Cooperation, LAS, Moro National Liberation Front, NAM, OAU, Parliamentary Union of the OIC Member States, Russia, Thailand, Turkish Muslim Community of Kibris, UN

Pacific Community (SPC)
local name of the Secretariat of the Pacific Community

Pacific Islands Forum (PIF) *note*—formerly known as South Pacific Forum (SPF)
established—5 August 1971
aim—to promote regional cooperation in political matters
members—(16) Australia, Cook Islands, Fiji, Kiribati, Marshall Islands, Federated States of Micronesia, Nauru, NZ, Niue, Palau, Papua New Guinea, Samoa, Solomon Islands, Tonga, Tuvalu, Vanuatu
associate members—(2) French Polynesia, New Caledonia
observers—(4) Asia Development Bank, The Commonwealth, Timor-Leste, Tokelau

Paris Club
established—1956
aim—to provide a forum for debtor countries to negotiate rescheduling of debt service payments or loans extended by governments or official agencies of participating countries; to help restore normal trade and project finance to debtor countries
members—(19) Australia, Austria, Belgium, Canada, Denmark, Finland, France, Germany, Ireland, Italy, Japan, Netherlands, Norway, Russia, Spain, Sweden, Switzerland, UK, US

Partnership for Peace (PFP)
established—10–11 January 1994
aim—to expand and intensify political and military cooperation throughout Europe, increase stability, diminish threats to peace, and build relationships by promoting the spirit of practical cooperation and commitment to democratic principles that underpin NATO; program under the auspices of NATO
members—(23) Albania, Armenia, Austria, Azerbaijan, Belarus, Bosnia and Herzegovina, Croatia, Finland, Georgia, Ireland, Kazakhstan, Kyrgyzstan, Macedonia, Moldova, Montenegro, Russia, Serbia, Sweden, Switzerland, Tajikistan, Turkmenistan, Ukraine, Uzbekistan; note—a nation that becomes a member of NATO is no longer a member of PFP

Permanent Court of Arbitration (PCA)
established—29 July 1899
aim—to facilitate the settlement of international disputes
members—(107) Argentina, Australia, Austria, Belarus, Belgium, Belize, Benin, Bolivia, Brazil, Bulgaria, Burkina Faso, Cambodia, Cameroon, Canada, Chile, China, Colombia, Democratic Republic of the Congo, Costa Rica, Croatia, Cuba, Cyprus, Czech Republic, Denmark, Dominican Republic, Ecuador, Egypt, El Salvador, Eritrea, Estonia, Ethiopia, Fiji, Finland, France, Germany, Greece, Guatemala, Guyana, Haiti, Honduras, Hungary, Iceland, India, Iran, Iraq, Ireland, Israel, Italy, Japan, Jordan, Kenya, South Korea, Kuwait, Kyrgyzstan, Laos, Latvia, Lebanon, Libya, Liechtenstein, Lithuania, Luxembourg, Macedonia, Malaysia, Malta, Mauritius, Mexico, Montenegro, Morocco, Netherlands, NZ, Nicaragua, Nigeria, Norway, Pakistan, Panama, Paraguay, Peru, Poland, Portugal, Qatar, Romania, Russia, Saudi Arabia, Senegal, Serbia, Singapore, Slovakia, Slovenia, South Africa, Spain, Sri Lanka, Sudan, Suriname, Swaziland, Sweden, Switzerland, Thailand, Togo, Turkey, Uganda, Ukraine, UK, US, Uruguay, Venezuela, Zambia, Zimbabwe

Rio Group (RG)
note—formerly known as Grupo de los Ocho, established NA December 1986; composed of the Contadora Group and the Lima Group
established—1988
aim—to consult on regional Latin American issues
members—(20) Argentina, Bolivia, Brazil, CARICOM, Chile, Colombia, Costa Rica, Dominican Republic, Ecuador, El Salvador, Guatemala, Guyana, Honduras, Mexico, Nicaragua, Panama, Paraguay, Peru, Uruguay, Venezuela

Schengen Convention
established—signed June 1990; effective March 1995
aim—to allow free movement within an area without internal border controls
members—(24) Austria, Belgium, Czech Republic, Denmark, Estonia, Finland, France, Germany, Greece, Hungary, Iceland, Italy, Latvia, Lithuania, Luxembourg, Malta, Netherlands, Norway, Poland, Portugal, Slovakia, Slovenia, Spain, Sweden; note—UK and Ireland have not joined; Switzerland is set to join in the Fall of 2008

Second World
another term for the traditionally Marxist-Leninist states of the USSR and Eastern Europe, with authoritarian governments and command economies based on the Soviet model; the term is fading from use; see centrally planned economies

Secretariat of the Pacific Community (SPC)
established—6 February 1947; effective 29 July 1948
aim—to serve island development in 22 Pacific countries; to develop technical assistance and professional, scientific, and research support; to build planning and management capability
members—(26) America Samoa, Australia, Cook Islands, Fiji, France, French Polynesia, Guam, Kiribati, Marshall Islands, Federated States of Micronesia, Nauru, New Caledonia, Niue, Northern Mariana Islands, NZ, Palau, Papua New Guinea, Pitcairn Islands, Samoa, Solomon Islands, Tokelau, Tonga, Tuvalu, Vanuatu, US, Wallis and Futuna

Shanghai Cooperation Organization (SCO)
established—15 June 2001
aim—to combat terrorism, extremism, and separatism; to safeguard regional security through mutual trust, disarmament, and cooperative security; and to increase cooperation in political, trade, economic, scientific and technological, cultural, and educational fields
members—(6) China, Kazakhstan, Kyrgyzstan, Russia, Tajikistan, Uzbekistan
observer—(4) India, Iran, Mongolia, Pakistan

socialist countries
in general, countries in which the government owns and plans the use of the major factors of production; note—the term is sometimes used incorrectly as a synonym for Communist countries

South
a popular term for the poorer, less industrialized countries generally located south of the developed countries; the counterpart of the North; see less developed countries (LDCs)

South American Community of Nations (CSN)
established—9 December 2004
aim—to coordinate common policies regarding multilateral organizations, to integrate physical infrastructure, and to consolidate the merger of CAN and Mercosur
members—(12) Argentina, Bolivia, Brazil, Chile, Colombia, Ecuador, Guyana, Paraguay, Peru, Surinam, Uruguay, Venezuela
observers—(2) Mexico, Panama

South Asian Association for Regional Cooperation (SAARC)

established—8 December 1985
aim—to promote economic, social, and cultural cooperation
members—(8) Afghanistan. Bangladesh, Bhutan, India, Maldives, Nepal, Pakistan, Sri Lanka
observers—(6) China, EU, Iran, Japan, South Korea, US

South Asia Co-operative Environment Program (SACEP)

established—January 1983
aim—to promote regional cooperation in South Asia in the field of environment, both natural and human, and on issues of economic and social development; to support conservation and management of natural resources of the region
members—(8) Afghanistan, Bangladesh, Bhutan, India, Maldives, Nepal, Pakistan, Sri Lanka

South Pacific Forum (SPF) *note*—see Pacific Island Forum South Pacific Regional Trade and Economic Cooperation Agreement (Sparteca)

established—1981
aim—to redress unequal trade relationships of Australia and New Zealand with small island economies in the Pacific region
members—(16) Australia, Cook Islands, Fiji, Kiribati, Marshall Islands, Federated States of Micronesia, Nauru, NZ, Niue, Palau, Papua New Guinea, Samoa, Solomon Islands, Tonga, Tuvalu, Vanuatu

Southeast European Cooperative Initiative (SECI)

established—6 December 1996
aim—to encourage cooperation among participating states and to facilitate their integration into European structures
members—(12) Albania, Bosnia and Herzegovina, Bulgaria, Croatia, Greece, Hungary, Macedonia, Moldova, Romania, Serbia, Slovenia, Turkey
observers—(18) Austria, Azerbaijan, Belgium, Canada, France, Georgia, Germany, Israel, Italy, Japan, Netherlands, Poland, Portugal, Slovakia, Spain, Ukraine, UK, US

Southeast European Cooperative Initiative (SECI)

established—6 December 1996
aim—to encourage cooperation among participating states and to facilitate their integration into European structures
members—(12) Albania, Bosnia and Herzegovina, Bulgaria, Croatia, Greece, Hungary, Macedonia, Moldova, Romania, Serbia, Slovenia, Turkey
observers—(15) Austria, Azerbaijan, Belgium, Canada, France, Georgia, Germany, Italy, Japan, Netherlands, Portugal, Spain, Ukraine, UK, US

Southern African Customs Union (SACU)

established—11 December 1969
aim—to promote free trade and cooperation in customs matters
members—(5) Botswana, Lesotho, Namibia, South Africa, Swaziland

Southern African Development Community (SADC) *note*—evolved from the Southern African Development Coordination Conference (SADCC)

established—17 August 1992
aim—to promote regional economic development and integration
members—(14) Angola, Botswana, Democratic Republic of the Congo, Lesotho, Madagascar, Malawi, Mauritius, Mozambique, Namibia, South Africa, Swaziland, Tanzania, Zambia, Zimbabwe

Southern Cone Common Market (Mercosur) or Southern Common Market *note*—also known as Mercado Comun del Cono Sur (Mercosur)

established—26 March 1991
aim—to increase regional economic cooperation
members—(4) Argentina, Brazil, Paraguay, Uruguay
associate members—(5) Bolivia, Chile, Colombia, Ecuador, Peru, Venezuela

Third World

another term for the less developed countries; the term is obsolescent; see less developed countries (LDCs)

African Union/United Nations Hybrid Operation in Darfur (UNAMID)

established—31 July 2007
aim—to contribute to the restoration of security conditions which will allow safe humanitarian assistance throughout Darfur, to contribute to the protection of civilian populations under imminent threat of physical attack, to monitor, observe compliance with, and verify the implementation of various ceasefire agreements
members—(18) Bangladesh, Burkina Faso, China, Egypt, Ethiopia, The Gambia, Ghana, Kenya, Malawi, Mali, Nepal, Netherlands, Nigeria, Pakistan, Rwanda, Senegal, South Africa, Thailand

underdeveloped countries

refers to those less developed countries with the potential for above-average economic growth; see less developed countries (LDCs)

undeveloped countries

refers to those extremely poor less developed countries (LDCs) with little prospect for economic growth; see least developed countries (LLDCs)

Union Latina

established—15 May 1954; became functional 1983
aim—to project, protect, and promote the common heritage and unifying idenitites of the Latin, and Latin-influenced, world
members—(37) Andorra, Angola, Bolivia, Brazil, Cape Verde, Chile, Colombia, Cote d'Ivoire, Costa Rica, Cuba, Dominican Republic, Ecuador, El Salvador, France, Guatemala, Guinea-Bissau, Haiti, Honduras, Italy, Mexico, Moldova, Monaco, Mozambique, Nicaragua, Panama, Paraguay, Peru, Philippines, Portugal, Romania, San Marino, Sao Tome and Principe, Senegal, Spain, Timor-Leste, Uruguay, Venezuela
observers—(3) Argentina, Holy See, Order of Malta

United Nations (UN)

established—26 June 1945; effective—24 October 1945
aim—to maintain international peace and security and to promote cooperation involving economic, social, cultural, and humanitarian problems
constituent organizations—the UN is composed of six principal organs and numerous subordinate agencies and bodies as follows:
1) Secretariat
2) General Assembly: Joint United Nations Program on HIV/AIDS (UN-AIDS), International Research and Training Institute for the Advancement of Women (INSTRAW), Office of the United Nations High Commissioner for Human Rights (OHCHR), Organization for the Prohibition of Chemical Weapons (OPCW), Preparation Commission for the Nuclear-Ban-Treaty Operation (CTBTU), United Nations Center for Human Settlements (UN-Habitat), United Nations Children's Fund (UNICEF), United Nations Conference on Trade and Development (UNCTAD), United Nations Democracy Fund (UNDEF), United Nations Development Program (UNDP), United Nations Drug Control Program (UNDCP), United Nations Environment Program (UNEP),United Nations Fund for International Partnerships (UNFIP), United Nations High Commissioner for Refugees (UNHCR), United Nations Institute for Disarmament Research (UNIDIR), United Nations Institute for Training and Research (UNITAR), United Nations Interregional Crime and Justice Research Institute (UNICRI), United Nations Population Fund (UNFPA), United Nations Office of Project Services (UNOPS), United Nations Relief and Works Agency for Palestine Refugees in the Near East (UNRWA), United Nations Research Institute for Social Development (UNRISD), United Nations System Staff College (UNSSC), United Nations University (UNU), World Food Program (WFP)
3) Security Council: International Criminal Tribunal for the Former Yugoslavia (ICTY), International Criminal Tribunal for Rwanda (ICTR), United Nations Compensation Commission, United Nations Disengagement Observer Force (UNDOF), African Union/United Nations Hybrid Operation in Darfur (UNAMID), United Nations Integrated Mission in Timor-Leste (UNMIT), United Nations Interim Administration Mission in Kosovo (UNMIK), United Nations Interim Force in Lebanon (UNIFIL), United Nations Mission in Liberia (UNMIL), United Nations Military Observer Group in India and Pakistan (UNMOGIP), United Nations Operation in Cote d'Ivoire (UNOCI), United Nations Mission for the Referendum in Western Sahara (MINURSO), United Nations Mission in the Central African Republic and Chad (MINURCAT), United Nations Mission in Ethiopia and Eritrea (UNMEE), United Nations Mission in the Sudan (UNMIS), United Nations Observer Mission in Georgia (UNOMIG), United Nations Organization Mission in the Democratic Republic of the Congo (MONUC), United Nations Peace-Keeping Force in Cyprus (UNFICYP), United Nations Stabilization Mission in Haiti (MINUSTAH), and United Nations Truce Supervision Organization (UNTSO)
4) Economic and Social Council (ECOSOC): Commission for Social Development, Commission on Crime Prevention and Criminal Justice, Commission on Narcotics Drugs, Commission on Population and Development, Commission on Science and Technology for Development, Commission on Sustainable Development, Commission on the Status of Women, Economic and Social Commission for Asia and the Pacific (ESCAP), Economic and Social Commission for Western Asia (ESCWA), Economic Commission for Africa (ECA), Economic Commission for Europe (ECE), Economic Commission for Latin America and the Caribbean (ECLAC), Food and Agriculture Organization of the United Nations (FAO), International Atomic Energy Agency (IAEA), International Bank for Reconstruction and Development (IBRD), International Center for Secretariat of Investment Disputes (ICSID), International Civil Aviation Organization (ICAO), International Development Association (IDA), International Finance Corporation (IFC), International Fund for Agricultural Development (IFAD), International Labor Organization (ILO), International Maritime Organization (IMO), International Monetary Fund (IMF), International Telecommunication Union (ITU), Multilateral Investment Geographic Agency (MIGA), Statistical Commission, United Nations Educational, Scientific, and Cultural Organization (UNESCO), United Nations Forum on Forests, United Nations Industrial Development Organization (UNIDO), Universal Postal Union (UPU), World Health Organization (WHO), World Intellectual Property Organization (WIPO), World Meteorological Organization (WMO), World Tourism Organization (UNWTO), and World Trade Organization (WTO)
5) Trusteeship Council (inactive; no trusteeships at this time)
6) International Court of Justice (ICJ)

United Nations Children's Fund (UNICEF) *note*—acronym retained from the predecessor organization, UN International Children's Emergency Fund

established—11 December 1946
aim—to help establish child health and welfare services
members—(36) selected on a rotating basis from all regions

United Nations Conference on Trade and Development (UNCTAD)

established—30 December 1964
aim—to promote international trade
members—(193) all UN members plus Holy See

INTERNATIONAL ORGANIZATIONS AND GROUPS

United Nations Development Program (UNDP)
established—22 November 1965
aim—to provide technical assistance to stimulate economic and social development
members (executive board)—(36) selected on a rotating basis from all regions

United Nations Disengagement Observer Force (UNDOF)
established—31 May 1974
aim—to observe the 1973 Arab-Israeli cease-fire; established by the UN Security Council
members—(6) Austria, Canada, India, Japan, Poland, Slovakia

United Nations Educational, Scientific, and Cultural Organization (UNESCO)
established—16 November 1945; effective—4 November 1946
aim—to promote cooperation in education, science, and culture
members—(193) includes all UN member countries except Liechtenstein (191 total); plus Cook Islands and Niue
associate members—(6) Aruba, British Virgin Islands, Cayman Islands, Macau, Netherlands Antilles, Tokelau

United Nations Environment Program (UNEP)
established—15 December 1972
aim—to promote international cooperation on all environmental matters
members—(58) selected on a rotating basis from all regions

United Nations General Assembly
established—26 June 1945; effective—24 October 1945
aim—to function as the primary deliberative organ of the UN
members—(192) all UN members are represented in the General Assembly

United Nations High Commissioner for Refugees (UNHCR)
established—3 December 1949; effective—1 January 1951
aim—to ensure the humanitarian treatment of refugees and find permanent solutions to refugee problems
members (executive committee)—(72) Algeria, Argentina, Australia, Austria, Bangladesh, Belgium, Brazil, Canada, Chile, China, Colombia, Democratic Republic of the Congo, Costa Rica, Cote d'Ivoire, Cyprus, Denmark, Ecuador, Egypt, Estonia, Ethiopia, Finland, France, Germany, Ghana, Greece, Guinea, Holy See, Hungary, India, Iran, Ireland, Israel, Italy, Japan, Jordan, Kenya, South Korea, Lebanon, Lesotho, Madagascar, Mexico, Morocco, Mozambique, Namibia, Netherlands, NZ, Nicaragua, Nigeria, Norway, Pakistan, Philippines, Poland, Portugal, Romania, Russia, Serbia, Somalia, South Africa, Spain, Sudan, Sweden, Switzerland, Tanzania, Thailand, Tunisia, Turkey, Uganda, UK, US, Venezuela, Yemen, Zambia

United Nations Industrial Development Organization (UNIDO)
established—17 November 1966; effective—1 January 1967
aim—UN specialized agency that promotes industrial development especially among the members
members—(172) includes all UN member countries except Andorra, Antigua and Barbuda, Australia, Brunei, Canada, Estonia, Iceland, Kiribati, Latvia, Liechtenstein, Marshall Islands, Federated States of Micronesia, Nauru, Palau, Samoa, San Marino, Singapore, Solomon Islands, Tuvalu, US

United Nations Institute for Training and Research (UNITAR)
established—11 December 1963 adoption of the resolution establishing the Institute; effective—24 March 1965
aim—to help the UN become more effective through training and research
members (Board of Trustees)—(16) Brazil, Burkino Faso, Egypt, Estonia, France, Ghana, Japan, Kuwait, Mexico, Morocco, Nigeria, Russia, South Africa, Switzerland, Thailand, US; note—the UN Secretary General can appoint up to 30 members

United Nations Integrated Mission in Timor-Leste (UNMIT)
established—25 August 2006
aim—to support the Government, to support the electoral process, to ensure the restoration and maintenance of public security
members—(12) Australia, Bangladesh, Brazil, China, Fiji, Malaysia, NZ, Pakistan, Philippines, Portugal, Sierra Leone, Singapore

United Nations Interim Administration Mission in Kosovo (UNMIK)
established—10 June 1999
aim—to promote the establishment of substantial autonomy and self-government in Kosovo; to perform basic civilian administrative functions; to support the reconstruction of key infrastructure and humanitarian and disaster relief
note—gives civilian support only; works closely with NATO Kosovo Force (KFOR)

United Nations Interim Force in Lebanon (UNIFIL)
established—19 March 1978
aim—to confirm the withdrawal of Israeli forces, and assist in reestablishing Lebanese authority in southern Lebanon; established by the UN Security Council
members—(26) Belgium, China, Croatia, Cyprus, Finland, France, Germany, Ghana, Greece, Guatemala, India, Indonesia, Ireland, Italy, South Korea, Luxembourg, Macedonia, Malaysia, Nepal, Netherlands, Poland, Portugal, Qatar, Spain, Tanzania, Turkey

United Nations Military Observer Group in India and Pakistan (UNMOGIP)
established—24 January 1949
aim—to observe the 1949 India-Pakistan cease-fire; established by the UN Security Council
members—(8) Chile, Croatia, Denmark, Finland, Italy, South Korea, Sweden, Uruguay

United Nations Mission for the Referendum in Western Sahara (MINURSO)

established—29 April 1991

aim—to supervise the cease-fire and conduct a referendum in Western Sahara; established by the UN Security Council

members—(27) Argentina, Austria, Bangladesh, Brazil, China, Croatia, Djibouti, Egypt, El Slavador, France, Ghana, Greece, Guinea, Honduras, Hungary, Ireland, Italy, Kenya, Malaysia, Mongolia, Nigeria, Pakistan, Poland, Russia, Sri Lanka, Uruguay, Yemen

United Nations Mission in the Central African Republic and Chad (MINURCAT)

established—25 September 2007

aim—to create the security and conditions which will to contribute to the protection of refugees, displaced persons, and citizens in danger, to facilitate the provision of humanitarian assistance in eastern Chad and the northeastern Central African Republic, to create favorable conditions for the recontruction and economic and social development of these areas

members—(3) France, Senegal, Sweden

United Nations Mission in Ethiopia and Eritrea (UNMEE)

established—31 July 2000

aim—to monitor the cessation of hostilities

members—(44) Algeria, Austria, Bangladesh, Bolivia, Bosnia and Herzegovina, Brazil, Bulgaria, China, Croatia, Czech Republic, Denmark, Finland, France, The Gambia, Germany, Ghana, Greece, Guatemala, India, Iran, Jordan, Kenya, Kyrgyzstan, Malaysia, Namibia, Nepal, Nigeria, Norway, Pakistan, Paraguay, Peru, Poland, Romania, Russia, South Africa, Spain, Sri Lanka, Sweden, Tanzania, Tunisia, Ukraine, US, Uruguay, Zambia

United Nations Mission in Liberia (UNMIL)

established—19 September 2003

aim—to support the cease-fire agreement and peace process, protect UN facilities and people, support humanitarian activities, and assist in national security reform

note—helps train and organize national police force

United Nations Mission in Sierra Leone (UNAMSIL)

established on 22 October 1999; aim was to to cooperate with the Government of Sierra Leone and the other parties to the Peace Agreement in the implementation of the agreement; to monitor the military and security situation in Sierra Leone; to monitor the disarmament and demobilization of combatants and members of the Civil Defense Forces (CFD); to assist in monitoring respect for international humanitarian law; mandate ended 31 December 2005; members were Bangladesh, Bolivia, China, Croatia, Czech Republic, Egypt, The Gambia, Germany, Ghana, Guinea, Indonesia, Jordan, Kenya, Kyrgyzstan, Malaysia, Nepal, Nigeria, Pakistan, Russia, Slovakia, Sweden, Tanzania, Thailand, Ukraine, UK, Uruguay, Zambia

United Nations Mission in the Sudan (UNMIS)

established—March 2005

aim—to support implementation of the comprehensive Peace Agreement by Monitoring and verifying the implementation of the Cease Fire Agreement, by observing and monitoring movements of armed groups, and by helping disarm, demobilizing and reintegrating armed bands

members—(59) Australia, Bangladesh, Benin, Bolivia, Botswana, Brazil, Burkina Faso, Cambodia, Canada, China, Croatia, Denmark, Ecuador, Egypt, El Salvador, Fiji, Finland, Gabon, Germany, Greece, Guatemala, Guinea, India, Indonesia, Jordan, Kenya, South Korea, Kyrgyzstan, Malawi, Malaysia, Mali, Moldova, Mongolia, Mozambique, Namibia, Nepal, Netherland, NZ, Nigeria, Norway, Pakistan, Paraguay, Peru, Philippines, Poland, Romania, Russia, Rwanda, Sri Lanka, Sweden, Tanzania, Thailand, Turkey, Uganda, Ukraine, UK, Yemen, Zambia, Zimbabwe

United Nations Mission of Support in East Timor (UNMISET)

established on 17 May 2002 to provide assistance to structures critical to public security and to assist in the development of law enforcement agencies; to contribute to extenal security; members were Australia, Bangladesh, Bolivia, Brazil, Denmark, Fiji, Jordan, Malaysia, Mozambique, Nepal, NZ, Pakistan, Philippines, Portugal, Russia, Sweden; completed its mandate 20 May 2005

United Nations Monitoring, Verification, and Inspection Commission (UNMOVIC)

formerly known as United Nations Special Commission for the Elimination of Iraq's Weapons of Mass Destruction (UNSCOM); established December 1999 with the aim to identify, account for, and eliminate Iraq's weapons of mass destruction and the capacity to produce them; commissioners were from Argentina, Brazil, Canada, China, UN Department for Disarmament Affairs, France, Germany, India, Japan, Mexico, Nigeria, Russia, Senegal, Ukraine, UK, US; finished operations 29 June 2007

United Nations Observer Mission in Georgia (UNOMIG)

established—24 August 1993

aim—to verify compliance with the cease-fire agreement, to monitor weapons exclusion zone, and to supervise CIS peacekeeping force for Abkhazia; established by the UN Security Council

members—(32) Albania, Austria, Bangladesh, Croatia, Czech Republic, Denmark, Egypt, France, Germany, Ghana, Greece, Hungary, India, Indonesia, Jordan, South Korea, Lithuania, Moldova, Mongolia, Nepal, Nigeria, Pakistan, Poland, Romania, Russia, Sweden, Switzerland, Turkey, Ukraine, UK, US, Uruguay, Yemen

United Nations Operation in Burundi (ONUB)

was established 21 May 2004 to support and help implement the efforts undertaken by Burundians to restore lasting peace and bring about national reconciliation; members were Algeria, Bangladesh, Belgium, Benin, Bolivia, Burkina Faso, Chad, China,

Egypt, Ethiopia, Gabon, The Gambia, Ghana, Guatemala, Guinea, India, Jordan, Kenya, South Korea, Kyrgyzstan, Malawi, Malaysia, Mali, Mozambique, Nambia, Nepal, Netherlands, Niger, Nigeria, Pakistan, Paraguay, Peru, Philippines, Portugal, Romania, Russia, Senegal, Serbia, South Africa, Spain, Sri Lanka, Thailand, Togo, Tunisia, Uruguay, Yemen, Zambia; mandate was completed 31 December 2006

United Nations Organization Mission in the Democratic Republic of the Congo (MONUC)
established—30 November 1999
aim—to establish contacts with the signatories to the cease-fire agreement and to plan for the observation of the cease-fire and disengagement of forces
members—(18) Bangladesh, Benin, Bolivia, China, Ghana, Guatemala, India, Indonesia, Jordan, Malawi, Morocco, Nepal, Pakistan, Senegal, Serbia, South Africa, Tunisia, Uruguay

United Nations Operation in Cote d'Ivoire (UNOCI)
established—27 February 2004
aim—to facilitate the implementation by the Ivorian parties of the peace agreement signed by them in January 2003
members—(43) Bangladesh, Benin, Bolivia, Brazil, Chad, China, Croatia, Dominican Republic, Ecuador, El Salvador, Ethiopia, France, The Gambia, Ghana, Guatemala, Guinea, India, Ireland, Jordan, Kenya, Moldova, Morocco, Namibia, Nepal, Niger, Nigeria, Pakistan, Paraguay, Peru, Philippines, Poland, Romania, Russia, Senegal, Serbia, Tanzania, Togo, Turkey, Uganda, Uruguay, Yemen, Zambia, Zimbabwe

United Nations Peace-keeping Force in Cyprus (UNFICYP)
established—4 March 1964
aim—to serve as a peacekeeping force between Greek Cypriots and Turkish Cypriots in Cyprus; established by the UN Security Council
members—(16) Argentina, Australia, Austria, Bolivia, Brazil, Canada, Chile, Finland, Hungary, Ireland, South Korea, Paraguay, Peru, Slovakia, UK, Uruguay

United Nations Population Fund (UNFPA) *note*—acronym retained from predecessor organization UN Fund for Population Activities
established—July 1967
aim—to assist both developed and developing countries to deal with their population problems
members (executive board)—(36) selected on a rotating basis from all regions

United Nations Relief and Works Agency for Palestine Refugees in the Near East (UNRWA)
established—8 December 1949
aim—to provide assistance to Palestinian refugees
members (advisory commission)—(22) Australia, Belgium, Canada, Denmark, EC, Egypt, France, Germany, Italy, Japan, Jordan, Lebanon, Netherlands, Norway, Saudi Arabia, Spain, Sweden, Switzerland, Syria, Turkey, UK, US

United Nations Research Institute for Social Development (UNRISD)
established—1963
aim—to conduct research into the problems of economic development during different phases of economic growth
members—no country members, but a Board of Directors consisting of a chairman appointed by the UN secretary general and 10 individual members

United Nations Secretariat
established—26 June 1945; effective—24 October 1945
aim—to serve as the primary administrative organ of the UN; a Secretary General is appointed for a five-year term by the General Assembly on the recommendation of the Security Council
members—the UN Secretary General and staff

United Nations Security Council (UNSC)
established—26 June 1945; effective—24 October 1945
aim—to maintain international peace and security
permanent members—(5) China, France, Russia, UK, US
nonpermanent members—(10) elected for two-year terms by the UN General Assembly; Belgium (2007–08), Burkina Faso (2008–09), Costa Rica (2008–09), Croatia (2008–09), Indonesia (2007–08), Italy (2007–08), Libya (2008–09), Panama (2007–08), South Africa (2007–08), Vietnam (2008–09)

United Nations Stabilization Mission in Haiti (MINUSTAH)
established—30 April 2004
aim—to stabilize Haiti in many areas for at least six months
members—(17) Argentina, Bolivia, Brazil, Canada, Chile, Croatia, Ecuador, France, Guatemala, Jordan, Nepal, Pakistan, Peru, Philippines, Sri Lanka, US, Uruguay

United Nations Truce Supervision Organization (UNTSO)
established—June 1948
aim—to supervise the 1948 Arab-Israeli cease-fire; currently supports timely deployment of reinforcements to other peace-keeping operations in the region as needed; initially established by the UN Security Council

members—(23) Argentina, Australia, Austria, Belgium, Canada, Chile, China, Denmark, Estonia, Finland, France, Ireland, Italy, Nepal, Netherlands, NZ, Norway, Russia, Slovakia, Slovenia, Sweden, Switzerland, US

United Nations Trusteeship Council
established on 26 June 1945, effective on 24 October 1945, to supervise the administration of the 11 UN trust territories; members were China, France, Russia, UK, US; it formally suspended operations 1 November 1995 after the Trust Territory of the Pacific Islands (Palau) became the Republic of Palau, a constitutional government in free association with the US; the Trusteeship Council was not dissolved

United Nations University (UNU)
established—3 December 1973
aim—to conduct research in development, welfare, and human survival and to train scholars
members—(24 members of UNU Council and the Rector are appointed by the Secretary General of the United Nations and the Director General of UNESCO)

Universal Postal Union (UPU)
established—9 October 1874, affiliated with the UN 15 November 1947; *effective*—1 July 1948
aim—to promote international postal cooperation; a UN specialized agency
members—(191) includes all UN member countries except Andorra, Marshall Islands, Federated States of Micronesia, and Palau (189 total); plus Holy See and Overseas Territories of the UK; note—includes the following dependencies or areas of special interest: Australia (Norfolk Island), China (Hong Kong, Macau), Denmark (Faroe Islands, Greenland), France (French Polynesia, French Southern and Antarctic Lands, Mayotte, New Caledonia, Saint Barthelemy, Saint Martin, Saint Pierre and Miquelon, Wallis and Futuna), Netherlands (Aruba, Netherlands Antilles), NZ (Cook Island, Niue, Tokelau), UK (Guernsey, Isle of Man, Jersey; Anguilla, Bermuda, British Indian Ocean Territory, British Virgin Islands, Cayman Islands, Falkland Islands, Gibraltar, Montserrat, Pitcairn Islands, Saint Helena, South Georgia and South Sandwich Islands, Turks and Caicos), US (American Samoa, Guam, Northern Mariana Islands, Puerto Rico, Virgin Islands)

Warsaw Pact (WP)
established 14 May 1955 to promote mutual defense; members met 1 July 1991 to dissolve the alliance; member states at the time of dissolution were: Bulgaria, Czechoslovakia, Hungary, Poland, Romania, and the USSR; earlier members included German Democratic Republic (GDR) and Albania

West African Development Bank (WADB) *note*—also known as Banque Ouest-Africaine de Developpement (BOAD); is a
financial institution of WAEMU
established—14 November 1973
aim—to promote regional economic development and integration
regional members—(9) Central Bank of West African States, Benin, Burkina Faso, Cote d'Ivoire, Guinea-Bissau, Mali, Niger, Senegal, Togo
international/nonregional members—(6) African Development Bank, Belgium, European Investment Bank, France, Germany, People's Bank of China

West African Economic and Monetary Union (WAEMU) *note*—also known as Union Economique et Monetaire Ouest Africaine
(UEMOA)
established—1 August 1994
aim—to increase competitiveness of members' economic markets; to create a common market
members—(8) Benin, Burkina Faso, Cote d'Ivoire, Guinea-Bissau, Mali, Niger, Senegal, Togo

Western European Union (WEU)
established—23 October 1954; *effective*—6 May 1955
aim—to provide mutual defense and to move toward political unification
members—(10) Belgium, France, Germany, Greece, Italy, Luxembourg, Netherlands, Portugal, Spain, UK
associate members—(6) Czech Republic, Hungary, Iceland, Norway, Poland, Turkey
associate partners—(7) Bulgaria, Estonia, Latvia, Lithuania, Romania, Slovakia, Slovenia
observers—(5) Austria, Denmark, Finland, Ireland, Sweden

World Bank Group
includes International Bank for Reconstruction and Development (IBRD), International Development Association (IDA), International Finance Corporation (IFC), and Multilateral Investment Guarantee Agency (MIGA)

World Confederation of Labor (WCL)
established—19 June 1920 as the International Federation of Christian Trade Unions (IFCTU), renamed 4 October 1968
aim—to promote the trade union movement
members—(105 national organizations) Antigua and Barbuda, Argentina, Aruba, Austria, Bangladesh, Belgium, Belize, Benin, Bolivia, Brazil, Bulgaria, Burkina Faso, Cameroon, Canada, Central African Republic, Chad, Chile, Colombia, Democratic Republic of the Congo, Republic of the Congo, Costa Rica, Cote d'Ivoire, Cuba, Cyprus, Czech Republic, Denmark, Dominica, Dominican Republic, Ecuador, El Salvador, France, French Guiana, Gabon, The Gambia, Ghana, Guadeloupe, Guatemala, Guinea, Guyana, Haiti, Honduras, Hong Kong, Hungary, India, Indonesia, Iran, Italy, Japan, Kazakhstan, South Korea, Liberia, Libya, Liechtenstein, Lithuania, Luxembourg, Macedonia, Madagascar, Malawi, Malaysia, Malta, Martinique, Mauritania, Mauritius, Mexico, Morocco, Namibia, Nepal, Netherlands, Netherlands Antilles, Nicaragua, Niger, Pakistan, Panama,

Paraguay, Peru, Philippines, Poland, Portugal, Puerto Rico, Romania, Rwanda, Saint Lucia, Saint Vincent and the Grenadines, Sao Tome and Principe, Senegal, Serbia, Sierra Leone, Singapore, Slovakia, South Africa, Spain, Sri Lanka, Suriname, Switzerland, Taiwan, Thailand, Togo, Trinidad and Tobago, Ukraine, US, Uruguay, Venezuela, Vietnam, Zambia, Zimbabwe

World Customs Organization (WCO) *note*—began as the Customs Cooperation Council (CCC)
established—15 December 1950
aim—to promote international cooperation in customs matters
members—(172) Afghanistan, Albania, Algeria, Andorra, Angola, Argentina, Armenia, Australia, Austria, Azerbaijan, The Bahamas, Bahrain, Bangladesh, Barbados, Belarus, Belgium, Benin, Bermuda, Bhutan, Bolivia, Botswana, Brazil, Brunei, Bulgaria, Burkina Faso, Burma, Burundi, Cambodia, Cameroon, Canada, Cape Verde, Central African Republic, Chad, Chile, China, Colombia, Comoros, Democratic Republic of the Congo, Republic of the Congo, Costa Rica, Cote d'Ivoire, Croatia, Cuba, Cyprus, Czech Republic, Denmark, Dominican Republic, EC, Ecuador, Egypt, El Salvador, Eritrea, Estonia, Ethiopia, Fiji, Finland, France, Gabon, The Gambia, Georgia, Germany, Ghana, Greece, Guatemala, Guinea, Guyana, Haiti, Honduras, Hong Kong, Hungary, Iceland, India, Indonesia, Iran, Iraq, Ireland, Israel, Italy, Jamaica, Japan, Jordan, Kazakhstan, Kenya, South Korea, Kuwait, Kyrgyzstan, Laos, Latvia, Lebanon, Lesotho, Liberia, Libya, Lithuania, Luxembourg, Macau, Macedonia, Madagascar, Malawi, Malaysia, Maldives, Mali, Malta, Mauritania, Mauritius, Mexico, Moldova, Mongolia, Montenegro, Morocco, Mozambique, Namibia, Nepal, Netherlands, Netherlands Antilles, NZ, Nicaragua, Niger, Nigeria, Norway, Oman, Pakistan, Panama, Papua New Guinea, Paraguay, Peru, Philippines, Poland, Portugal, Qatar, Romania, Russia, Rwanda, Saint Lucia, Samoa, Saudi Arabia, Senegal, Serbia, Seychelles, Sierra Leone, Singapore, Slovakia, Slovenia, South Africa, Spain, Sri Lanka, Sudan, Swaziland, Sweden, Switzerland, Syria, Tajikistan, Tanzania, Thailand, Timor-Leste, Togo, Tonga, Trinidad and Tobago, Tunisia, Turkey, Turkmenistan, Uganda, Ukraine, UAE, UK, US, Uruguay, Uzbekistan, Venezuela, Vietnam, Yemen, Zambia, Zimbabwe

World Federation of Trade Unions (WFTU)
established—3 October 1945
aim—to promote the trade union movement
members—(125 and the Palestine Liberation Organization) Afghanistan, Albania, Angola, Antigua and Barbuda, Argentina, Armenia, Australia, Austria, Azerbaijan, Bahrain, Bangladesh, Barbados, Belarus, Benin, Bolivia, Botswana, Brazil, Bulgaria, Burkina Faso, Cambodia, Cameroon, Canada, Chile, Colombia, Democratic Republic of the Congo, Republic of the Congo, Costa Rica, Cote d'Ivoire, Cuba, Cyprus, Czech Republic, Djibouti, Dominican Republic, Ecuador, Egypt, El Salvador, Eritrea, Ethiopia, Fiji, Finland, France, French Guiana, The Gambia, Ghana, Greece, Guadeloupe, Guatemala, Guinea, Guinea-Bissau, Guyana, Haiti, Honduras, Hungary, India, Indonesia, Iran, Iraq, Jamaica, Japan, Jordan, Kazakhstan, North Korea, Kuwait, Kyrgyzstan, Laos, Lebanon, Lesotho, Liberia, Libya, Madagascar, Malawi, Malaysia, Mali, Martinique, Mauritius, Mexico, Mozambique, Nepal, New Caledonia, NZ, Niger, Nigeria, Oman, Pakistan, Panama, Papua New Guinea, Peru, Philippines, Poland, Portugal, Puerto Rico, Reunion, Romania, Russia, Saint Lucia, Saint Pierre and Miquelon, Saint Vincent and the Grenadines, Saudi Arabia, Senegal, Sierra Leone, Slovakia, Solomon Islands, Somalia, South Africa, Sri Lanka, Sudan, Sweden, Syria, Tajikistan, Tanzania, Thailand, Togo, Trinidad and Tobago, Tunisia, Turkey, Turkmenistan, Uganda, Ukraine, Uruguay, Uzbekistan, Vanuatu, Venezuela, Vietnam, Yemen, Zimbabwe, Palestine Liberation Organization

World Food Program (WFP)
established—24 November 1961
aim—to provide food aid in support of economic development or disaster relief; an ECOSOC organization
members—(36) selected on a rotating basis from all regions

World Health Organization (WHO)
established—22 July 1946; effective—7 April 1948
aim—to deal with health matters worldwide; a UN specialized agency
members—(193) includes all UN member countries except Liechtenstein (191 total); plus Cook Islands and Niue

World Intellectual Property Organization (WIPO)
established—14 July 1967; effective—26 April 1970
aim—to furnish protection for literary, artistic, and scientific works; a UN specialized agency
members—(184) includes all UN member countries except Kiribati, Marshall Islands, Federated States of Micronesia, Nauru, Palau, Solomon Islands, Timor-Leste, Tuvalu, Vanuatu (182 total); plus Holy See

World Meteorological Organization (WMO)
established—11 October 1947; effective—4 April 1951
aim—to sponsor meteorological cooperation; a UN specialized agency
members—(188) includes all UN member countries except Andorra, Equatorial Guinea, Grenada, Liechtenstein, Marshall Islands, Nauru, Palau, Saint Kitts and Nevis, Saint Vincent and the Grenadines, San Marino, Tuvalu (181 total); plus Aruba and the Netherlands Antilles, Cook Islands, French Polynesia, Hong Kong, Macau, New Caledonia, and Niue

World Tourism Organization (UNWTO)
established—2 January 1975
aim—to promote tourism as a means of contributing to economic development, international understanding, and peace
members—(153) Afghanistan, Albania, Algeria, Andorra, Angola, Argentina, Armenia, Australia, Austria, Azerbaijan, The Bahamas, Bahrain, Bangladesh, Belarus, Benin, Bhutan, Bolivia, Bosnia and Herzegovina, Botswana, Brazil, Brunei, Bulgaria, Burkina Faso, Burundi, Cambodia, Cameroon, Canada, Cape Verde, Central African Republic, Chad, Chile, China, Colombia,

Democratic Republic of the Congo, Republic of the Congo, Costa Rica, Cote d'Ivoire, Croatia, Cuba, Cyprus, Czech Republic, Djibouti, Dominican Republic, Ecuador, Egypt, El Salvador, Equatorial Guinea, Eritrea, Ethiopia, Fiji, France, Gabon, The Gambia, Georgia, Germany, Ghana, Greece, Guatemala, Guinea, Guinea-Bissau, Haiti, Honduras, Hungary, India, Indonesia, Iran, Iraq, Israel, Italy, Jamaica, Japan, Jordan, Kazakhstan, Kenya, North Korea, South Korea, Kuwait, Kyrgyzstan, Laos, Latvia, Lebanon, Lesotho, Libya, Lithuania, Macedonia, Madagascar, Malawi, Malaysia, Maldives, Mali, Malta, Mauritania, Mauritius, Mexico, Moldova, Monaco, Mongolia, Montenegro, Morocco, Mozambique, Namibia, Nepal, Netherlands, Nicaragua, Niger, Nigeria, Oman, Pakistan, Panama, Papua New Guinea, Paraguay, Peru, Philippines, Poland, Portugal, Qatar, Romania, Russia, Rwanda, San Marino, Sao Tome and Principe, Saudi Arabia, Senegal, Serbia, Seychelles, Sierra Leone, Slovakia, Slovenia, South Africa, Spain, Sri Lanka, Sudan, Swaziland, Switzerland, Syria, Tajikistan, Tanzania, Thailand, Timor-Leste, Togo, Tunisia, Turkey, Turkmenistan, Uganda, Ukraine, UK, Uruguay, Uzbekistan, Venezuela, Vietnam, Yemen, Zambia, Zimbabwe
associate members—(7) Aruba, Flanders, Hong Kong, Macau, Madeira Islands, Netherlands Antilles, Puerto Rico
observers—(1 plus Palestine Liberation Organization) Holy See, Palestine Liberation Organization

World Trade Organization (WTO) *note*—succeeded General Agreement on Tariff and Trade (GATT)
established—15 April 1994; effective—1 January 1995
aim—to provide a forum to resolve trade conflicts between members and to carry on negotiations with the goal of further lowering and/or eliminating tariffs and other trade barriers
members—(151) Albania, Angola, Antigua and Barbuda, Argentina, Armenia, Australia, Austria, Bahrain, Bangladesh, Barbados, Belgium, Belize, Benin, Bolivia, Botswana, Brazil, Brunei, Bulgaria, Burkina Faso, Burma, Burundi, Cambodia, Cameroon, Canada, Central African Republic, Chad, Chile, China, Colombia, Democratic Republic of the Congo, Republic of the Congo, Costa Rica, Cote d'Ivoire, Croatia, Cuba, Cyprus, Czech Republic, Denmark, Djibouti, Dominica, Dominican Republic, Ecuador, Egypt, El Salvador, Estonia, EC, Fiji, Finland, France, Gabon, The Gambia, Georgia, Germany, Ghana, Greece, Grenada, Guatemala, Guinea, Guinea-Bissau, Guyana, Haiti, Honduras, Hong Kong, Hungary, Iceland, India, Indonesia, Ireland, Israel, Italy, Jamaica, Japan, Jordan, Kenya, South Korea, Kuwait, Kyrgyzstan, Latvia, Lesotho, Liechtenstein, Lithuania, Luxembourg, Macau, Macedonia, Madagascar, Malawi, Malaysia, Maldives, Mali, Malta, Mauritania, Mauritius, Mexico, Moldova, Mongolia, Morocco, Mozambique, Namibia, Nepal, Netherlands, NZ, Nicaragua, Niger, Nigeria, Norway, Oman, Pakistan, Panama, Papua New Guinea, Paraguay, Peru, Philippines, Poland, Portugal, Qatar, Romania, Rwanda, Saint Kitts and Nevis, Saint Lucia, Saint Vincent and the Grenadines, Saudi Arabia, Senegal, Sierra Leone, Singapore, Slovakia, Slovenia, Solomon Islands, South Africa, Spain, Sri Lanka, Suriname, Swaziland, Sweden, Switzerland, Taiwan, Tanzania, Thailand, Togo, Tonga, Trinidad and Tobago, Tunisia, Turkey, Uganda, UAE, UK, US, Uruguay, Venezuela, Vietnam, Zambia, Zimbabwe
observers—(31) Afghanistan, Algeria, Andorra, Azerbaijan, The Bahamas, Belarus, Bhutan, Bosnia and Herzegovina, Cape Verde, Comoros, Equatorial Guinea, Ethiopia, Holy See, Iran, Iraq, Kazakhstan, Laos, Lebanon, Libya, Montenegro, Russia, Samoa, Sao Tome and Principe, Serbia, Seychelles, Sudan, Tajikistan, Ukraine, Uzbekistan, Vanuatu, Yemen; note—with the exception of the Holy See, an observer must start accession negotiations within five years of becoming observers

Zangger Committee (ZC)
established—early 1970s
aim—to establish guidelines for the export control provisions of the Nonproliferation of Nuclear Weapons Treaty (NPT)
members—(36) Argentina, Australia, Austria, Belgium, Bulgaria, Canada, China, Croatia, Czech Republic, Denmark, Finland, France, Germany, Greece, Hungary, Ireland, Italy, Japan, South Korea, Luxembourg, Netherlands, Norway, Poland, Portugal, Romania, Russia, Slovakia, Slovenia, South Africa, Spain, Sweden, Switzerland, Turkey, Ukraine, UK, US
observers—(1) EC

Air Pollution
see Convention on Long-Range Transboundary Air Pollution

Air Pollution-Nitrogen Oxides
see Protocol to the 1979 Convention on Long-Range Transboundary Air Pollution Concerning the Control of Emissions of Nitrogen Oxides or Their Transboundary Fluxes

Air Pollution-Persistent Organic Pollutants
see Protocol to the 1979 Convention on Long-Range Transboundary Air Pollution on Persistent Organic Pollutants

Air Pollution-Sulphur 85
see Protocol to the 1979 Convention on Long-Range Transboundary Air Pollution on the Reduction of Sulphur Emissions or Their Transboundary Fluxes by at least 30%

Air Pollution-Sulphur 94
see Protocol to the 1979 Convention on Long-Range Transboundary Air Pollution on Further Reduction of Sulphur Emissions

Air Pollution-Volatile Organic Compounds
see Protocol to the 1979 Convention on Long-Range Transboundary Air Pollution Concerning the Control of Emissions of Volatile Organic Compounds or Their Transboundary Fluxes

Antarctic—Environmental Protocol
see Protocol on Environmental Protection to the Antarctic Treaty

Antarctic Treaty
opened for signature—1 December 1959
entered into force—23 June 1961
objective—to ensure that Antarctica is used for peaceful purposes only (such as international cooperation in scientific research); to defer the question of territorial claims asserted by some nations and not recognized by others; to provide an international forum for management of the region; applies to land and ice shelves south of 60 degrees south latitude
parties—(45) Argentina, Australia, Austria, Belgium, Brazil, Bulgaria, Canada, Chile, China, Colombia, Cuba, Czech Republic, Denmark, Ecuador, Estonia, Finland, France, Germany, Greece, Guatemala, Hungary, India, Italy, Japan, North Korea, South Korea, Netherlands, NZ, Norway, Papua New Guinea, Peru, Poland, Romania, Russia, Slovakia, South Africa, Spain, Sweden, Switzerland, Turkey, Ukraine, UK, US, Uruguay, Venezuela

Basel Convention on the Control of Transboundary Movements of Hazardous Wastes and Their Disposal
note—abbreviated as Hazardous Wastes
opened for signature—22 March 1989
entered into force—5 May 1992
objective—to reduce transboundary movements of wastes subject to the Convention to a minimum consistent with the environmentally sound and efficient management of such wastes; to minimize the amount and toxicity of wastes generated and ensure their environmentally sound management as closely as possible to the source of generation; and to assist LDCs in environmentally sound management of the hazardous and other wastes they generate
parties—(167) Albania, Algeria, Andorra, Antigua and Barbuda, Argentina, Armenia, Australia, Austria, Azerbaijan, The Bahamas, Bahrain, Bangladesh, Barbados, Belarus, Belgium, Belize, Benin, Bhutan, Bolivia, Bosnia and Herzegovina, Botswana, Brazil, Brunei, Bulgaria, Burkina Faso, Burundi, Cambodia, Cameroon, Canada, Cape Verde, Central African Republic, Chad, Chile, China, Colombia, Comoros, Democratic Republic of the Congo, Cook Islands, Costa Rica, Cote d'Ivoire, Croatia, Cuba, Cyprus, Czech Republic, Denmark, Djibouti, Dominica, Dominican Republic, Ecuador, Egypt, El Salvador, Equatorial Guinea, Eritrea, Estonia, Ethiopia, EU, Finland, France, The Gambia, Georgia, Germany, Ghana, Greece, Guatemala, Guinea, Guyana, Honduras, Hungary, Iceland, India, Indonesia, Iran, Ireland, Israel, Italy, Jamaica, Japan, Jordan, Kazakhstan, Kenya, Kiribati, South Korea, Kuwait, Kyrgyzstan, Latvia, Lebanon, Lesotho, Liberia, Libya, Liechtenstein, Lithuania, Luxembourg, Macedonia, Madagascar, Malawi, Malaysia, Maldives, Mali, Malta, Marshall Islands, Mauritania, Mauritius, Mexico, Federated States of Micronesia, Moldova, Monaco, Mongolia, Montenegro, Morocco, Mozambique, Namibia, Nauru, Nepal, Netherlands, NZ, Nicaragua, Niger, Nigeria, Norway, Oman, Pakistan, Panama, Papua New Guinea, Paraguay, Peru, Philippines, Poland, Portugal, Qatar, Romania, Russia, Rwanda, Saint Kitts and Nevis, Saint Lucia, Saint Vincent and the Grenadines, Samoa, Saudi Arabia, Senegal, Serbia, Seychelles, Singapore, Slovakia, Slovenia, South Africa, Spain, Sri Lanka, Sudan, Swaziland, Sweden, Switzerland, Syria, Tanzania, Thailand, Trinidad and Tobago, Tunisia, Turkey, Turkmenistan, Uganda, Ukraine, UAE, UK, Uruguay, Uzbekistan, Venezuela, Vietnam, Yemen, Zambia
countries that have signed, but not yet ratified—(3) Afghanistan, Haiti, US

Biodiversity
see Convention on Biological Diversity

Climate Change
see United Nations Framework Convention on Climate Change

Climate Change-Kyoto Protocol
see Kyoto Protocol to the United Nations Framework Convention on Climate Change

Convention for the Conservation of Antarctic Seals *note*—abbreviated as Antarctic Seals
opened for signature—1 June 1972
entered into force—11 March 1978
objective—to promote and achieve the protection, scientific study, and rational use of Antarctic seals, and to maintain a satisfactory balance within the ecological system of Antarctica
parties—(16) Argentina, Australia, Belgium, Brazil, Canada, Chile, France, Germany, Italy, Japan, Norway, Poland, Russia, South Africa, UK, US
countries that have signed, but not yet ratified—(1) NZ

Convention on Biological Diversity *note*—abbreviated as Biodiversity
opened for signature—5 June 1992
entered into force—29 December 1993
objective—to develop national strategies for the conservation and sustainable use of biological diversity
parties—(186) Afghanistan, Albania, Algeria, Andorra, Angola, Antigua and Barbuda, Argentina, Armenia, Australia, Austria, Azerbaijan, The Bahamas, Bahrain, Bangladesh, Barbados, Belarus, Belgium, Belize, Benin, Bhutan, Bolivia, Bosnia and Herzegovina, Botswana, Brazil, Brunei, Bulgaria, Burkina Faso, Burma, Burundi, Cambodia, Cameroon, Canada, Cape Verde, Central African Republic, Chad, Chile, China, Colombia, Comoros, Democratic Republic of the Congo, Republic of the Congo, Cook Islands, Costa Rica, Cote d'Ivoire, Croatia, Cuba, Cyprus, Czech Republic, Denmark, Djibouti, Dominica, Dominican Republic, Ecuador, Egypt, El Salvador, Equatorial Guinea, Eritrea, Estonia, Ethiopia, EU, Fiji, Finland, France, Gabon, The Gambia, Georgia, Germany, Ghana, Greece, Grenada, Guatemala, Guinea, Guinea-Bissau, Guyana, Haiti, Honduras, Hungary, Iceland, India, Indonesia, Iran, Iraq, Ireland, Israel, Italy, Jamaica, Japan, Jordan, Kazakhstan, Kenya, Kiribati, North Korea, South Korea, Kuwait, Kyrgyzstan, Laos, Latvia, Lebanon, Lesotho, Liberia, Libya, Liechtenstein, Lithuania, Luxembourg, Macedonia, Madagascar, Malawi, Malaysia, Maldives, Mali, Malta, Marshall Islands, Mauritania, Mauritius, Mexico, Federated States of Micronesia, Moldova, Monaco, Mongolia, Morocco, Mozambique, Namibia, Nauru, Nepal, Netherlands, NZ, Nicaragua, Niger, Nigeria, Niue, Norway, Oman, Pakistan, Palau, Panama, Papua New Guinea, Paraguay, Peru, Philippines, Poland, Portugal, Qatar, Romania, Russia, Rwanda, Saint Kitts and Nevis, Saint Lucia, Saint Vincent and the Grenadines, Samoa, San Marino, Sao Tome and Principe, Saudi Arabia, Senegal, Serbia, Seychelles, Sierra Leone, Singapore, Slovakia, Slovenia, Solomon Islands, Somalia, South Africa, Spain, Sri Lanka, Sudan, Suriname, Swaziland, Sweden, Switzerland, Syria, Tajikistan, Tanzania, Thailand, Togo, Tonga, Trinidad and Tobago, Tunisia, Turkey, Turkmenistan, Tuvalu, Uganda, Ukraine, UAE, UK, Uruguay, Uzbekistan, Vanuatu, Venezuela, Vietnam, Yemen, Zambia, Zimbabwe
countries that have signed, but not yet ratified—(1) US

Convention on Fishing and Conservation of Living Resources of the High Seas
note—abbreviated as Marine Life Conservation
opened for signature—29 April 1958
entered into force—20 March 1966
objective—to solve through international cooperation the problems involved in the conservation of living resources of the high seas, considering that because of the development of modern technology some of these resources are in danger of being overexploited
parties—(38) Australia, Belgium, Bosnia and Herzegovina, Burkina Faso, Cambodia, Colombia, Denmark, Dominican Republic, Fiji, Finland, France, Haiti, Indonesia, Jamaica, Kenya, Lesotho, Madagascar, Malawi, Malaysia, Mauritius, Mexico, Netherlands, Nigeria, Portugal, Senegal, Serbia, Sierra Leone, Solomon Islands, South Africa, Spain, Switzerland, Thailand, Tonga, Trinidad and Tobago, Uganda, UK, US, Venezuela
countries that have signed, but not yet ratified—(20) Afghanistan, Argentina, Bolivia, Canada, Costa Rica, Cuba, Ghana, Iceland, Iran, Ireland, Israel, Lebanon, Liberia, Nepal, NZ, Pakistan, Panama, Sri Lanka, Tunisia, Uruguay

Convention on Long-Range Transboundary Air Pollution *note*—abbreviated as Air Pollution
opened for signature—13 November 1979
entered into force—16 March 1983
objective—to protect the human environment against air pollution and to gradually reduce and prevent air pollution, including long-range transboundary air pollution
parties—(49) Armenia, Austria, Azerbaijan, Belarus, Belgium, Bosnia and Herzegovina, Bulgaria, Canada, Croatia, Cyprus, Czech Republic, Denmark, Estonia, EU, Finland, France, Georgia, Germany, Greece, Hungary, Iceland, Ireland, Italy, Kazakhstan, Kyrgyzstan, Latvia, Liechtenstein, Lithuania, Luxembourg, Macedonia, Malta, Moldova, Monaco, Netherlands, Norway, Poland, Portugal, Romania, Russia, Serbia, Slovakia, Slovenia, Spain, Sweden, Switzerland, Turkey, Ukraine, UK, US
countries that have signed, but not yet ratified—(2) Holy See, San Marino

Convention on Wetlands of International Importance Especially as Waterfowl Habitat (Ramsar)
note—abbreviated as Wetlands

opened for signature—2 February 1971
entered into force—21 December 1975
objective—to stem the progressive encroachment on and loss of wetlands now and in the future, recognizing the fundamental ecological functions of wetlands and their economic, cultural, scientific, and recreational value
parties—(153) Albania, Algeria, Antigua and Barbuda, Argentina, Armenia, Australia, Austria, Azerbaijan, The Bahamas, Bahrain, Bangladesh, Barbados, Belarus, Belgium, Belize, Benin, Bolivia, Bosnia and Herzegovina, Botswana, Brazil, Bulgaria, Burkina Faso, Burundi, Cambodia, Cameroon, Canada, Cape Verde, Central African Republic, Chad, Chile, China, Colombia, Comoros, Democratic Republic of the Congo, Republic of the Congo, Costa Rica, Cote d'Ivoire, Croatia, Cuba, Cyprus, Czech Republic, Denmark, Djibouti, Dominican Republic, Ecuador, Egypt, El Salvador, Equatorial Guinea, Estonia, Fiji, Finland, France, Gabon, The Gambia, Georgia, Germany, Ghana, Greece, Guatemala, Guinea, Guinea-Bissau, Honduras, Hungary, Iceland, India, Indonesia, Iran, Ireland, Israel, Italy, Jamaica, Japan, Jordan, Kazakhstan, Kenya, South Korea, Kyrgyzstan, Latvia, Lebanon, Lesotho, Liberia, Libya, Liechtenstein, Lithuania, Luxembourg, Macedonia, Madagascar, Malawi, Malaysia, Mali, Malta, Marshall Islands, Mauritania, Mauritius, Mexico, Moldova, Monaco, Mongolia, Morocco, Mozambique, Namibia, Nepal, Netherlands, NZ, Nicaragua, Niger, Nigeria, Norway, Pakistan, Palau, Panama, Papua New Guinea, Paraguay, Peru, Philippines, Poland, Portugal, Romania, Russia, Rwanda, Saint Lucia, Samoa, Sao Tome and Principe, Senegal, Serbia, Seychelles, Sierra Leone, Slovakia, Slovenia, South Africa, Spain, Sri Lanka, Sudan, Suriname, Sweden, Switzerland, Syria, Tanzania, Tajikistan, Thailand, Togo, Trinidad and Tobago, Tunisia, Turkey, Uganda, Ukraine, UK, US, Uruguay, Uzbekistan, Venezuela, Vietnam, Zambia

Convention on the Conservation of Antarctic Marine Living Resources
note—abbreviated as Antarctic-Marine Living Resources
opened for signature—5 May 1980
entered into force—7 April 1982
objective—to safeguard the environment and protect the integrity of the ecosystem of the seas surrounding Antarctica, and to conserve Antarctic marine living resources
parties—(31) Argentina, Australia, Belgium, Brazil, Bulgaria, Canada, Chile, EU, Finland, France, Germany, Greece, India, Italy, Japan, South Korea, Mauritius, Namibia, Netherlands, NZ, Norway, Peru, Poland, Russia, South Africa, Spain, Sweden, Ukraine, UK, US, Uruguay, Vanuatu

Convention on the International Trade in Endangered Species of Wild Flora and Fauna (CITES)
note—abbreviated as Endangered Species
opened for signature—3 March 1973
entered into force—1 July 1975
objective—to protect certain endangered species from overexploitation by means of a system of import/export permits
parties—(168) Afghanistan, Albania, Algeria, Antigua and Barbuda, Argentina, Australia, Austria, Azerbaijan, The Bahamas, Bangladesh, Barbados, Belarus, Belgium, Belize, Benin, Bhutan, Bolivia, Botswana, Brazil, Brunei, Bulgaria, Burkina Faso, Burma, Burundi, Cambodia, Cameroon, Canada, Cape Verde, Central African Republic, Chad, Chile, China, Colombia, Comoros, Democratic Republic of the Congo, Republic of the Congo, Costa Rica, Cote d'Ivoire, Croatia, Cuba, Cyprus, Czech Republic, Denmark, Djibouti, Dominica, Dominican Republic, Ecuador, Egypt, El Salvador, Equatorial Guinea, Eritrea, Estonia, Ethiopia, Fiji, Finland, France, Gabon, The Gambia, Georgia, Germany, Ghana, Greece, Grenada, Guatemala, Guinea, Guinea-Bissau, Guyana, Honduras, Hungary, Iceland, India, Indonesia, Iran, Ireland, Israel, Italy, Jamaica, Japan, Jordan, Kazakhstan, Kenya, South Korea, Kuwait, Laos, Latvia, Lesotho, Liberia, Libya, Liechtenstein, Lithuania, Luxembourg, Macedonia, Madagascar, Malawi, Malaysia, Mali, Malta, Mauritania, Mauritius, Mexico, Moldova, Monaco, Mongolia, Morocco, Mozambique, Namibia, Nepal, Netherlands, NZ, Nicaragua, Niger, Nigeria, Norway, Pakistan, Panama, Papua New Guinea, Paraguay, Peru, Philippines, Poland, Portugal, Qatar, Romania, Russia, Rwanda, Saint Kitts and Nevis, Palau, Saint Lucia, Saint Vincent and the Grenadines, Sao Tome and Principe, Saudi Arabia, Senegal, Serbia, Seychelles, Sierra Leone, Singapore, Slovakia, Slovenia, Somalia, South Africa, Spain, Sri Lanka, Sudan, Suriname, Swaziland, Sweden, Switzerland, Syria, Tanzania, Thailand, Togo, Trinidad and Tobago, Tunisia, Turkey, Uganda, Ukraine, UAE, UK, US, Uruguay, Uzbekistan, Vanuatu, Venezuela, Vietnam, Yemen, Zambia, Zimbabwe

Convention on the Prevention of Marine Pollution by Dumping Wastes and Other Matter (London Convention)
note—abbreviated as Marine Dumping
opened for signature—29 December 1972
entered into force—30 August 1975
objective—to control pollution of the sea by dumping and to encourage regional agreements supplementary to the Convention; the London Convention came into force in 1996
parties—(88) Afghanistan, Angola, Antigua and Barbuda, Argentina, Australia, Azerbaijan, Barbados, Belarus, Belgium, Bolivia, Brazil, Bulgaria, Canada, Cape Verde, Chile, China, Democratic Republic of the Congo, Costa Rica, Cote d'Ivoire, Croatia, Cuba, Cyprus, Denmark, Dominican Republic, Egypt, Equatorial Guinea, Finland, France, Gabon, Germany, Greece, Guatemala, Haiti, Honduras, Hong Kong (associate member), Hungary, Iceland, Iran, Ireland, Italy, Jamaica, Japan, Jordan, Kenya, Kiribati, South Korea, Libya, Luxembourg, Malta, Mexico, Monaco, Montenegro, Morocco, Nauru, Netherlands, NZ, Nigeria, Norway, Oman, Pakistan, Panama, Papua New Guinea, Peru, Philippines, Poland, Portugal, Russia, Saint Kitts and Nevis, Saint Lucia, Saint Vincent and the Grenadines, Saudi Arabia, Serbia, Seychelles, Slovenia, Solomon Islands, South Africa, Spain, Suriname, Sweden, Switzerland, Tonga, Trinidad and Tobago, Tunisia, Ukraine, UAE, UK, US, Vanuatu
associate members to the London Convention—(2) Faroe Islands, Macau *countries that have signed, but not yet ratified*—(3) Chad, Kuwait, Uruguay

Convention on the Prohibition of Military or Any Other Hostile Use of Environmental Modification Techniques

note—abbreviated as Environmental Modification
opened for signature—10 December 1976
entered into force—5 October 1978
objective—to prohibit the military or other hostile use of environmental modification techniques in order to further world peace and trust among nations
parties—(68) Afghanistan, Algeria, Antigua and Barbuda, Argentina, Australia, Austria, Bangladesh, Belarus, Belgium, Benin, Brazil, Brunei, Bulgaria, Canada, Cape Verde, Chile, Costa Rica, Cuba, Cyprus, Czech Republic, Denmark, Dominica, Egypt, Finland, Germany, Ghana, Greece, Guatemala, Hungary, India, Ireland, Italy, Japan, North Korea, South Korea, Kuwait, Laos, Malawi, Mauritius, Mongolia, Netherlands, NZ, Niger, Norway, Pakistan, Papua New Guinea, Poland, Romania, Russia, Saint Lucia, Saint Vincent and the Grenadines, Sao Tome and Principe, Slovakia, Solomon Islands, Spain, Sri Lanka, Sweden, Switzerland, Tajikistan, Tunisia, Ukraine, UK, US, Uruguay, Uzbekistan, Vietnam, Yemen
countries that have signed, but not yet ratified—(17) Bolivia, Democratic Republic of the Congo, Ethiopia, Holy See, Iceland, Iran, Iraq, Lebanon, Liberia, Luxembourg, Morocco, Nicaragua, Portugal, Sierra Leone, Syria, Turkey, Uganda

Desertification

see United Nations Convention to Combat Desertification in those Countries Experiencing Serious Drought and/or Desertification, Particularly in Africa

Endangered Species

see Convention on the International Trade in Endangered Species of Wild Flora and Fauna (CITES)

Environmental Modification

see Convention on the Prohibition of Military or Any Other Hostile Use of Environmental Modification Techniques

Hazardous Wastes

see Basel Convention on the Control of Transboundary Movements of Hazardous Wastes and Their Disposal

International Convention for the Regulation of Whaling

note—abbreviated as Whaling
opened for signature—2 December 1946
entered into force—10 November 1948
objective—to protect all species of whales from overhunting; to establish a system of international regulation for the whale fisheries to ensure proper conservation and development of whale stocks; and to safeguard for future generations the great natural resources represented by whale stocks
parties—(72) Antigua and Barbuda, Argentina, Australia, Austria, Belgium, Belize, Benin, Brazil, Cambodia, Cameroon, Chile, China, Costa Rica, Cote D'Ivoire, Croatia, Czech Republic, Denmark, Dominica, Finland, France, Gabon, The Gambia, Germany, Grenada, Guatemala, Guinea, Hungary, Iceland, India, Ireland, Israel, Italy, Japan, Kenya, Kiribati, South Korea, Mali, Marshall Islands, Mauritania, Mexico, Monaco, Mongolia, Morocco, Nauru, Netherlands, NZ, Nicaragua, Norway, Oman, Palau, Panama, Peru, Philippines, Portugal, Russia, Saint Kitts and Nevis, Saint Lucia, Saint Vincent and the Grenadines, San Marino, Senegal, Slovakia, Slovenia, Solomon Islands, South Africa, Spain, Suriname, Sweden, Switzerland, Togo, Tuvalu, UK, US

International Tropical Timber Agreement, 1983

note—abbreviated as Tropical Timber 83
opened for signature—18 November 1983
entered into force—1 April 1985; this agreement expired when the International Tropical Timber Agreement, 1994, went into force
objective—to provide an effective framework for cooperation between tropical timber producers and consumers and to encourage the development of national policies aimed at sustainable utilization and conservation of tropical forests and their genetic resources
parties—(54) Australia, Austria, Belgium, Bolivia, Brazil, Burma, Cameroon, Canada, China, Colombia, Democratic Republic of the Congo, Republic of the Congo, Cote d'Ivoire, Denmark, Ecuador, Egypt, EU, Fiji, Finland, France, Gabon, Germany, Ghana, Greece, Guyana, Honduras, India, Indonesia, Ireland, Italy, Japan, South Korea, Liberia, Luxembourg, Malaysia, Nepal, Netherlands, NZ, Norway, Panama, Papua New Guinea, Peru, Philippines, Portugal, Russia, Spain, Sweden, Switzerland, Thailand, Togo, Trinidad and Tobago, UK, US, Venezuela

International Tropical Timber Agreement, 1994 *note*—abbreviated as Tropical Timber 94

opened for signature—26 January 1994
entered into force—1 January 1997
objective—to ensure that by the year 2000 exports of tropical timber originate from sustainably managed sources; to establish a fund to assist tropical timber producers in obtaining the resources necessary to reach this objective
parties—(58) Australia, Austria, Belgium, Bolivia, Brazil, Burma, Cambodia, Cameroon, Canada, Central African Republic, China, Colombia, Democratic Republic of the Congo, Republic of the Congo, Cote d'Ivoire, Denmark, Ecuador, Egypt, EU, Fiji, Finland, France, Gabon, Germany, Ghana, Greece, Guyana, Honduras, India, Indonesia, Ireland, Italy, Japan, South Korea, Liberia, Luxembourg, Malaysia, Nepal, Netherlands, NZ, Norway, Panama, Papua New Guinea, Peru, Philippines, Portugal, Spain, Suriname, Sweden, Switzerland, Thailand, Togo, Trinidad and Tobago, UK, US, Uruguay, Vanuatu, Venezuela

Kyoto Protocol to the United Nations Framework Convention on Climate Change
note—abbreviated as Climate Change-Kyoto Protocol
opened for signature—16 March 1998
entered into force—23 February 2005
objective—to further reduce greenhouse gas emissions by enhancing the national programs of developed countries aimed at this goal and by establishing percentage reduction targets for the developed countries
parties—(181) Albania, Algeria, Angola, Antigua and Barbuda, Argentina, Armenia, Australia, Austria, Azerbaijan, The Bahamas, Bahrain, Bangladesh, Barbados, Belarus, Belgium, Belize, Benin, Bhutan, Bolivia, Bosnia and Herzegovina, Botswana, Brazil, Bulgaria, Burkina Faso, Burma, Burundi, Cambodia, Cameroon, Canada, Cape Verde, Cental African Republic, Chile, China, Colombia, Comoros, Cote d'Ivoire, Democratic Republic of the Congo, Republic of the Congo, Cook Island, Costa Rica, Croatia, Cuba, Cyprus, Czech Republic, Denmark, Djibouti, Dominica, Dominican Republic, Ecuador, Egypt, El Salvador, Equatorial Guinea, Eritrea, Estonia, Ethiopia, EU, Fiji, Finland, France, Gabon, The Gambia, Georgia, Germany, Ghana, Greece, Grenada, Guatemala, Guinea, Guinea-Bissau, Guyana, Haiti, Honduras, Hungary, Iceland, India, Indonesia, Iran, Ireland, Israel, Italy, Jamaica, Japan, Jordan, Kenya, Kiribati, North Korea, South Korea, Kuwait, Kyrgyzstan, Laos, Latvia, Lebanon, Lesotho, Liberia, Libya, Liechtenstein, Lithuania, Luxembourg, Macedonia, Madagascar, Malawi, Malaysia, Maldives, Mali, Malta, Marshall Islands, Mauritania, Mauritius, Mexico, Federated States of Micronesia, Moldova, Monaco, Mongolia, Montenegro, Morocco, Mozambique, Namibia, Nauru, Nepal, Netherlands, NZ, Nicaragua, Niger, Nigeria, Niue, Norway, Oman, Pakistan, Palau, Panama, Papua New Guinea, Paraguay, Peru, Philippines, Poland, Portugal, Qatar, Romania, Russia, Rwanda, Saint Kitts and Nevis, Saint Lucia, Saint Vincent and the Grenadines, Samoa, Saudi Arabia, Senegal, Serbia, Seychelles, Sierra Leone, Singapore, Slovakia, Slovenia, Solomon Islands, South Africa, Spain, Sri Lanka, Sudan, Suriname, Swaziland, Sweden, Switzerland, Syria, Tanzania, Thailand, Togo, Tonga, Trinidad and Tobago, Tunisia, Turkmenistan, Tuvalu, Uganda, Ukraine, UAE, UK, Uruguay, Uzbekistan, Vanuatu, Venezuela, Vietnam, Yemen, Zambia
countries that have signed, but not yet ratified—(4) Kazakhstan, US

Law of the Sea
see United Nations Convention on the Law of the Sea (LOS)

Marine Dumping
see Convention on the Prevention of Marine Pollution by Dumping Wastes and Other Matter (London Convention)

Marine Life Conservation
see Convention on Fishing and Conservation of Living Resources of the High Seas

Montreal Protocol on Substances That Deplete the Ozone Layer *note*—abbreviated as Ozone Layer Protection
opened for signature—16 September 1987
entered into force—1 January 1989
objective—to protect the ozone layer by controlling emissions of substances that deplete it
parties—(189) Afghanistan, Albania, Algeria, Angola, Antigua and Barbuda, Argentina, Armenia, Australia, Austria, Azerbaijan, The Bahamas, Bahrain, Bangladesh, Barbados, Belarus, Belgium, Belize, Benin, Bolivia, Bosnia and Herzegovina, Botswana, Brazil, Brunei, Bulgaria, Burkina Faso, Burma, Burundi, Cambodia, Cameroon, Canada, Cape Verde, Central African Republic, Chad, Chile, China, Colombia, Comoros, Democratic Republic of the Congo, Republic of the Congo, Cook Islands, Costa Rica, Cote d'Ivoire, Croatia, Cuba, Cyprus, Czech Republic, Denmark, Djibouti, Dominica, Dominican Republic, Ecuador, Egypt, El Salvador, Equatorial Guinea, Eritrea, Estonia, Ethiopia, EU, Fiji, Finland, France, Gabon, The Gambia, Georgia, Germany, Ghana, Greece, Grenada, Guatemala, Guinea, Guinea-Bissau, Guyana, Haiti, Honduras, Hungary, Iceland, India, Indonesia, Iran, Ireland, Israel, Italy, Jamaica, Japan, Jordan, Kazakhstan, Kenya, Kiribati, North Korea, South Korea, Kuwait, Kyrgyzstan, Laos, Latvia, Lebanon, Lesotho, Liberia, Libya, Liechtenstein, Lithuania, Luxembourg, Macedonia, Madagascar, Malawi, Malaysia, Maldives, Mali, Malta, Marshall Islands, Mauritania, Mauritius, Mexico, Federated States of Micronesia, Moldova, Monaco, Mongolia, Morocco, Mozambique, Namibia, Nauru, Nepal, Netherlands, NZ, Nicaragua, Niger, Nigeria, Norway, Oman, Pakistan, Palau, Panama, Papua New Guinea, Paraguay, Peru, Philippines, Poland, Portugal, Qatar, Romania, Russia, Rwanda, Saint Kitts and Nevis, Saint Lucia, Saint Vincent and the Grenadines, Samoa, Sao Tome and Principe, Saudi Arabia, Senegal, Serbia, Seychelles, Sierra Leone, Singapore, Slovakia, Slovenia, Solomon Islands, Somalia, South Africa, Spain, Sri Lanka, Sudan, Suriname, Swaziland, Sweden, Switzerland, Syria, Tajikistan, Tanzania, Thailand, Togo, Tonga, Trinidad and Tobago, Tunisia, Turkey, Turkmenistan, Tuvalu, Uganda, Ukraine, UAE, UK, US, Uruguay, Uzbekistan, Vanuatu, Venezuela, Vietnam, Yemen, Zambia, Zimbabwe

Nuclear Test Ban
see Treaty Banning Nuclear Weapons Tests in the Atmosphere, in Outer Space, and Under Water

Ozone Layer Protection
see Montreal Protocol on Substances That Deplete the Ozone Layer

Protocol of 1978 Relating to the International Convention for the Prevention of Pollution From Ships, 1973 (MARPOL)
note—abbreviated as Ship Pollution
opened for signature—17 February 1978
entered into force—2 October 1983

773

objective—to preserve the marine environment through the complete elimination of pollution by oil and other harmful substances and the minimization of accidental discharge of such substances

parties—(139) Algeria, Angola, Antigua and Barbuda, Argentina, Australia, Austria, Azerbaijan, The Bahamas, Bangladesh, Barbados, Belarus, Belgium, Belize, Benin, Bolivia, Brazil, Brunei, Bulgaria, Burma, Cambodia, Canada, Cape Verde, Chile, China, Colombia, Comoros, Republic of Congo, Cote d'Ivoire, Croatia, Cuba, Cyprus, Czech Republic, Denmark, Djibouti, Dominica, Dominican Republic, Ecuador, Egypt, Equatorial Guinea, Estonia, Faroe Islands, Finland, France, Gabon, The Gambia, Georgia, Germany, Ghana, Greece, Guatemala, Guinea, Guyana, Honduras, Hong Kong, Hungary, Iceland, India, Indonesia, Iran, Ireland, Israel, Italy, Jamaica, Japan, Kazakhstan, Kenya, North Korea, South Korea, Latvia, Lebanon, Liberia, Lithuania, Luxembourg, Libya, Macau, Madagascar, Malawi, Malaysia, Maldives, Malta, Marshall Islands, Mauritania, Mauritius, Mexico, Moldova, Monaco, Mongolia, Montenegro, Morocco, Mozambique, Netherlands, NZ, Nicaragua, Nigeria, Norway, Oman, Pakistan, Panama, Papua New Guinea, Peru, Philippines, Poland, Portugal, Qatar Romania, Russia, Saint Kitts and Nevis, Saint Lucia, Saint Vincent and the Grenadines, Samoa, Sao Tome and Principe, Saudi Arabia, Senegal, Serbia, Seychelles, Sierra Leone, Singapore, Slovakia, Slovenia, South Africa, Spain, Sri Lanka, Suriname, Sweden, Switzerland, Syria, Togo, Tonga, Trinidad and Tobago, Tunisia, Turkey, Tuvalu, Ukraine, UK, US, Uruguay, Vanuatu, Venezuela, Vietnam

Protocol on Environmental Protection to the Antarctic Treaty *note*—abbreviated as Antarctic-Environmental Protocol

opened for signature—4 October 1991
entered into force—14 January 1998
objective—to provide for comprehensive protection of the Antarctic environment and dependent and associated ecosystems; applies to the area covered by the Antarctic Treaty
consultative parties—(31) Argentina, Australia, Belgium, Brazil, Bulgaria, Canada, Chile, China, Czech Republic, Ecuador, Finland, France, Germany, India, Italy, Japan, South Korea, Netherlands, NZ, Norway, Peru, Poland, Romania, Russia, South Africa, Spain, Sweden, Ukraine, UK, US, Uruguay
non consultative parties—(12) Austria, Colombia, Cuba, Denmark, Greece, Guatemala, Hungary, North Korea, Papua New Guinea, Slovakia, Switzerland, Turkey

Protocol to the 1979 Convention on Long-Range Transboundary Air Pollution Concerning the Control of Emissions of Nitrogen Oxides or Their Transboundary Fluxes *note*—abbreviated as Air Pollution-Nitrogen Oxides

opened for signature—31 October 1988
entered into force—14 February 1991
objective—to provide for the control or reduction of nitrogen oxides and their transboundary fluxes
parties—(31) Austria, Belarus, Belgium, Bulgaria, Canada, Cyprus, Czech Republic, Denmark, Estonia, EU, Finland, France, Germany, Greece, Hungary, Ireland, Italy, Liechtenstein, Lithuania, Luxembourg, Netherlands, Norway, Russia, Slovakia, Slovenia, Spain, Sweden, Switzerland, Ukraine, UK, US
countries that have signed, but not yet ratified—(1) Poland

Protocol to the 1979 Convention on Long-Range Transboundary Air Pollution Concerning the Control of Emissions of Volatile Organic Compounds or Their Transboundary Fluxes *note*—abbreviated as Air Pollution-Volatile Organic Compounds

opened for signature—18 November 1991
entered into force—29 September 1997
objective—to provide for the control and reduction of emissions of volatile organic compounds in order to reduce their transboundary fluxes so as to protect human health and the environment from adverse effects
parties—(21) Austria, Belgium, Bulgaria, Czech Republic, Denmark, Estonia, Finland, France, Germany, Hungary, Italy, Liechtenstein, Luxembourg, Monaco, Netherlands, Norway, Slovakia, Spain, Sweden, Switzerland, UK
countries that have signed, but not yet ratified—(6) Canada, EU, Greece, Portugal, Ukraine, US

Protocol to the 1979 Convention on Long-Range Transboundary Air Pollution on Further Reduction of Sulphur Emissions

note—abbreviated as Air Pollution-Sulphur 94
opened for signature—14 June 1994
entered into force—5 August 1998
objective—to provide for a further reduction in sulfur emissions or transboundary fluxes
parties—(27) Austria, Belgium, Bulgaria, Canada, Croatia, Cyprus, Czech Republic, Denmark, EU, Finland, France, Germany, Greece, Hungary, Ireland, Italy, Liechtenstein, Luxembourg, Monaco, Netherlands, Norway, Slovakia, Slovenia, Spain, Sweden, Switzerland, UK
countries that have signed, but not yet ratified—(3) Poland, Russia, Ukraine

Protocol to the 1979 Convention on Long-Range Transboundary Air Pollution on Persistent Organic Pollutants

note—abbreviated as Air Pollution-Persistent Organic Pollutants
opened for signature—24 June 1998
entered into force—23 October 2003
objective—to provide for the control and reduction of emissions of persistent organic pollutants in order to reduce their transboundary fluxes so as to protect human health and the environment from adverse effects
parties—(27) Austria, Belgium, Bulgaria, Canada, Cyprus, Czech Republic, Denmark, Estonia, EU, Finland, France, Germany, Hungary, Iceland, Italy, Latvia, Liechtenstein, Lithuania, Luxembourg, Moldova, Netherlands, Norway, Romania, Slovakia, Sweden, Switzerland, UK
countries that have signed, but not yet ratified—(10) Armenia, Croatia, Greece, Ireland, Philippines Poland, Portugal, Spain, Ukraine, US

Protocol to the 1979 Convention on Long-Range Transboundary Air Pollution on the Reduction of Sulphur Emissions or Their Transboundary Fluxes by at Least 30% *note*—abbreviated as Air Pollution-Sulphur 85
opened for signature—8 July 1985
entered into force—2 September 1987
objective—to provide for a 30% reduction in sulfur emissions or transboundary fluxes by 1993
parties—(22) Austria, Belarus, Belgium, Bulgaria, Canada, Czech Republic, Denmark, Estonia, Finland, France, Germany, Hungary, Italy, Liechtenstein, Luxembourg, Netherlands, Norway, Russia, Slovakia, Sweden, Switzerland, Ukraine

Ship Pollution
see Protocol of 1978 Relating to the International Convention for the Prevention of Pollution From Ships, 1973 (MARPOL)

Treaty Banning Nuclear Weapon Tests in the Atmosphere, in Outer Space, and Under Water
note—abbreviated as Nuclear Test Ban
opened for signature—5 August 1963
entered into force—10 October 1963
objective—to obtain an agreement on general and complete disarmament under strict international control in accordance with the objectives of the United Nations; to put an end to the armaments race and eliminate incentives for the production and testing of all kinds of weapons, including nuclear weapons
parties—(113) Afghanistan, Antigua and Barbuda, Argentina, Armenia, Australia, Austria, The Bahamas, Bangladesh, Belgium, Benin, Bhutan, Bolivia, Bosnia and Herzegovina, Botswana, Brazil, Bulgaria, Burma, Canada, Central African Republic, Chad, China, Colombia, Democratic Republic of the Congo, Costa Rica, Cote d'Ivoire, Croatia, Cyprus, Czech Republic, Denmark, Dominican Republic, Ecuador, Egypt, El Salvador, Fiji, Finland, Gabon, The Gambia, Germany, Ghana, Greece, Guatemala, Honduras, Hungary, Iceland, India, Indonesia, Iran, Iraq, Ireland, Israel, Italy, Jamaica, Japan, Jordan, Kenya, South Korea, Kuwait, Laos, Lebanon, Liberia, Luxembourg, Madagascar, Malawi, Malaysia, Malta, Mauritania, Mauritius, Mexico, Morocco, Nepal, Netherlands, NZ, Nicaragua, Niger, Nigeria, Norway, Panama, Papua New Guinea, Peru, Philippines, Poland, Romania, Russia, Rwanda, Samoa, San Marino, Senegal, Serbia, Seychelles, Sierra Leone, Singapore, Slovakia, Slovenia, South Africa, Spain, Sri Lanka, Sudan, Suriname, Swaziland, Sweden, Switzerland, Syria, Thailand, Togo, Tonga, Trinidad and Tobago, Tunisia, Turkey, Uganda, UK, US, Venezuela, Zambia
countries that have signed, but not yet ratified—(17) Algeria, Burkina Faso, Burundi, Cameroon, Chile, Ethiopia, Haiti, Libya, Mali, Pakistan, Paraguay, Portugal, Somalia, Tanzania, Uruguay, Vietnam, Yemen

Tropical Timber 83
see International Tropical Timber Agreement, 1983

Tropical Timber 94
see International Tropical Timber Agreement, 1994

United Nations Convention on the Law of the Sea (LOS) *note*—abbreviated as Law of the Sea
opened for signature—10 December 1982
entered into force—16 November 1994
objective—to set up a comprehensive new legal regime for the sea and oceans; to include rules concerning environmental standards as well as enforcement provisions dealing with pollution of the marine environment
parties—(155) Albania, Algeria, Angola, Antigua and Barbuda, Argentina, Armenia, Australia, Austria, The Bahamas, Bahrain, Bangladesh, Barbados, Belarus, Belgium, Belize, Benin, Bolivia, Bosnia and Herzegovina, Botswana, Brazil, Brunei, Bulgaria, Burkina Faso, Burma, Cameroon, Canada, Cape Verde, Chile, China, Comoros, Democratic Republic of the Congo, Cook Islands, Costa Rica, Cote d'Ivoire, Croatia, Cuba, Cyprus, Czech Republic, Denmark, Djibouti, Dominica, Egypt, Equatorial Guinea, Estonia, EU, Fiji, Finland, France, Gabon, The Gambia, Georgia, Germany, Ghana, Greece, Grenada, Guatemala, Guinea, Guinea-Bissau, Guyana, Haiti, Honduras, Hungary, Iceland, India, Indonesia, Iraq, Ireland, Italy, Jamaica, Japan, Jordan, Kenya, Kiribati, South Korea, Kuwait, Laos, Latvia, Lebanon, Lesotho, Lithuania, Luxembourg, Macedonia, Madagascar, Malaysia, Maldives, Mali, Malta, Marshall Islands, Mauritania, Mauritius, Mexico, Federated States of Micronesia, Moldova, Monaco, Mongolia, Montenegro, Morocco, Mozambique, Namibia, Nauru, Nepal, Netherlands, NZ, Nicaragua, Nigeria, Niue, Norway, Oman, Pakistan, Palau, Panama, Papua New Guinea, Paraguay, Philippines, Poland, Portugal, Qatar, Romania, Russia, Saint Kitts and Nevis, Saint Lucia, Saint Vincent and the Grenadines, Samoa, Sao Tome and Principe, Saudi Arabia, Senegal, Serbia, Seychelles, Sierra Leone, Singapore, Slovakia, Slovenia, Solomon Islands, Somalia, South Africa, Spain, Sri Lanka, Sudan, Suriname, Sweden, Tanzania, Togo, Tonga, Trinidad and Tobago, Tunisia, Tuvalu, Uganda, Ukraine, UK, Uruguay, Vanuatu, Vietnam, Yemen, Zambia, Zimbabwe
countries that have signed, but not yet ratified—(27) Afghanistan, Bangladesh, Bhutan, Burundi, Cambodia, Central African Republic, Chad, Colombia, Republic of the Congo, Dominican Republic, El Salvador, Ethiopia, Iran, North Korea, Lesotho, Liberia, Libya, Liechtenstein, Madagascar, Malawi, Morocco, Niger, Rwanda, Swaziland, Switzerland, Thailand, UAE

United Nations Convention to Combat Desertification in Those Countries Experiencing Serious Drought and/or Desertification, Particularly in Africa *note*—abbreviated as Desertification
opened for signature—14 October 1994
entered into force—26 December 1996
objective—to combat desertification and mitigate the effects of drought through national action programs that incorporate long-term strategies supported by international cooperation and partnership arrangements

parties—(185) Afghanistan, Albania, Algeria, Andorra, Angola, Antigua and Barbuda, Argentina, Armenia, Australia, Austria, Azerbaijan, The Bahamas, Bahrain, Bangladesh, Barbados, Belarus, Belgium, Belize, Benin, Bhutan, Bolivia, Bosnia and Herzegovina, Botswana, Brazil, Brunei, Bulgaria, Burkina Faso, Burma, Burundi, Cambodia, Cameroon, Canada, Cape Verde, Central African Republic, Chad, Chile, China, Colombia, Comoros, Democratic Republic of the Congo, Republic of the Congo, Cook Islands, Costa Rica, Cote d'Ivoire, Croatia, Cuba, Cyprus, Czech Republic, Denmark, Djibouti, Dominica, Dominican Republic, Ecuador, Egypt, El Salvador, Equatorial Guinea, Eritrea, Ethiopia, EU, Fiji, Finland, France, Gabon, The Gambia, Georgia, Germany, Ghana, Greece, Grenada, Guatemala, Guinea, Guinea-Bissau, Guyana, Haiti, Honduras, Hungary, Iceland, India, Indonesia, Iran, Ireland, Israel, Italy, Jamaica, Japan, Jordan, Kazakhstan, Kenya, Kiribati, South Korea, Kuwait, Kyrgyzstan, Laos, Lebanon, Lesotho, Liberia, Libya, Liechtenstein, Luxembourg, Macedonia, Madagascar, Malawi, Malaysia, Maldives, Mali, Malta, Marshall Islands, Mauritania, Mauritius, Mexico, Federated States of Micronesia, Moldova, Monaco, Mongolia, Morocco, Mozambique, Namibia, Nauru, Nepal, Netherlands, NZ, Nicaragua, Niger, Nigeria, Niue, Norway, Oman, Pakistan, Palau, Panama, Papua New Guinea, Paraguay, Peru, Philippines, Poland, Portugal, Qatar, Romania, Rwanda, Saint Kitts and Nevis, Saint Lucia, Saint Vincent and the Grenadines, Samoa, San Marino, Sao Tome and Principe, Saudi Arabia, Senegal, Seychelles, Sierra Leone, Singapore, Slovakia, Slovenia, Solomon Islands, South Africa, Spain, Sri Lanka, Sudan, Suriname, Swaziland, Sweden, Switzerland, Syria, Thailand, Tajikistan, Tanzania, Timor-Leste, Togo, Tonga, Trinidad and Tobago, Tunisia, Turkey, Turkmenistan, Tuvalu, Uganda, UAE, UK, US, Uruguay, Uzbekistan, Vanuatu, Venezuela, Vietnam, Yemen, Zambia, Zimbabwe

United Nations Framework Convention on Climate Change

note—abbreviated as Climate Change
opened for signature—9 May 1992
entered into force—21 March 1994
objective—to achieve stabilization of greenhouse gas concentrations in the atmosphere at a low enough level to prevent dangerous anthropogenic interference with the climate system
parties—(195) Afghanistan, Albania, Algeria, Angola, Antigua and Barbuda, Argentina, Armenia, Australia, Austria, Azerbaijan, The Bahamas, Bahrain, Bangladesh, Barbados, Belarus, Belgium, Belize, Benin, Bhutan, Bolivia, Bosnia and Herzegovina, Botswana, Brazil, Brunei, Bulgaria, Burkina Faso, Burma, Burundi, Cambodia, Cameroon, Canada, Cape Verde, Central African Republic, Chad, Chile, China, Colombia, Comoros, Democratic Republic of the Congo, Republic of the Congo, Cook Islands, Costa Rica, Cote d'Ivoire, Croatia, Cuba, Cyprus, Czech Republic, Denmark, Djibouti, Dominica, Dominican Republic, Ecuador, Egypt, El Salvador, Equatorial Guinea, Eritrea, Estonia, Ethiopia, EU, Fiji, Finland, France, Gabon, The Gambia, Georgia, Germany, Ghana, Greece, Grenada, Guatemala, Guinea, Guinea-Bissau, Guyana, Haiti, Holy See, Honduras, Hungary, Iceland, India, Indonesia, Iraq, Iran, Ireland, Israel, Italy, Jamaica, Japan, Jordan, Kazakhstan, Kenya, Kiribati, North Korea, South Korea, Kuwait, Kyrgyzstan, Laos, Latvia, Lebanon, Lesotho, Liberia, Libya, Liechtenstein, Lithuania, Luxembourg, Macedonia, Madagascar, Malawi, Malaysia, Maldives, Mali, Malta, Marshall Islands, Mauritania, Mauritius, Mexico, Federated States of Micronesia, Moldova, Monaco, Mongolia, Montenegro, Morocco, Mozambique, Namibia, Nauru, Nepal, Netherlands, NZ, Nicaragua, Niger, Nigeria, Niue, Norway, Oman, Pakistan, Palau, Panama, Papua New Guinea, Paraguay, Peru, Philippines, Poland, Portugal, Qatar, Romania, Russia, Rwanda, Saint Kitts and Nevis, Saint Lucia, Saint Vincent and the Grenadines, Samoa, San Marino, Sao Tome and Principe, Saudi Arabia, Senegal, Serbia, Seychelles, Sierra Leone, Singapore, Slovakia, Slovenia, Solomon Islands, Somalia, South Africa, Spain, Sri Lanka, Sudan, Suriname, Swaziland, Sweden, Switzerland, Syria, Tajikistan, Tanzania, Thailand, Timor-Leste, Togo, Tonga, Trinidad and Tobago, Tunisia, Turkey, Turkmenistan, Tuvalu, Uganda, Ukraine, UAE, UK, US, Uruguay, Uzbekistan, Vanuatu, Venezuela, Vietnam, Yemen, Zambia, Zimbabwe

Wetlands
see Convention on Wetlands of International Importance Especially As Waterfowl Habitat (Ramsar)

Whaling
see International Convention for the Regulation of Whaling

APPENDIX D

CROSS-REFERENCE LIST OF COUNTRY DATA CODES

FIPS 10: *Countries, Dependencies, Areas of Special Sovereignty, and Their Principal Administrative Divisions (FIPS 10)* is maintained by the Office of Targeting and Transnational Issues, National Geospatial-Intelligence Agency, and published by the National Institute of Standards and Technology (Department of Commerce). FIPS 10 codes are intended for general use throughout the US Government, especially in activities associated with the mission of the Department of State and national defense programs.

ISO 3166: *Codes for the Representation of Names of Countries (ISO 3166)* is prepared by the International Organization for Standardization. ISO 3166 includes two- and three-character alphabetic codes and three-digit numeric codes that may be needed for activities involving exchange of data with international organizations that have adopted that standard. Except for the numeric codes, ISO 3166 codes have been adopted in the US as FIPS 104-1: *American National Standard Codes for the Representation of Names of Countries, Dependencies, and Areas of Special Sovereignty for Information Interchange.*

STANAG 1059: *Letter Codes for Geographical Entities (8th edition, 2004)* is a Standardization Agreement (STANAG) established and maintained by the North Atlantic Treaty Organization (NATO/OTAN) for the purpose of providing a common set of geospatial identifiers for countries, territories, and possessions. The 8th edition established trigraph codes for each country based upon the ISO 3166-1 alpha-3 character sets. These codes are used throughout NATO.

Internet: The Internet country code is the two-letter digraph maintained by the International Organization for Standardization (ISO) in *the ISO 3166 Alpha-2 list* and used by the Internet Assigned Numbers Authority (IANA) to establish country-coded top-level domains (ccTLDs).

Entity	FIPS 10	ISO 3166			STANAG-1059	Internet	Comment
Afghanistan	AF	AF	AFG	004	AFG	.af	
Albania	AL	AL	ALB	008	ALB	.al	
Algeria	AG	DZ	DZA	012	DZA	.dz	
American Samoa	AQ	AS	ASM	016	ASM	.as	
Andorra	AN	AD	AND	020	AND	.ad	
Angola	AO	AO	AGO	024	AGO	.ao	
Anguilla	AV	AI	AIA	660	AIA	.ai	
Antarctica	AY	AQ	ATA	010	ATA	.aq	ISO defines as the territory south of 60 degrees south latitude
Antigua and Barbuda	AC	AG	ATG	028	ATG	.ag	
Argentina	AR	AR	ARG	032	ARG	.ar	
Armenia	AM	AM	ARM	051	ARM	.am	
Aruba	AA	AW	ABW	533	ABW	.aw	
Ashmore and Cartier Islands	AT	-	-		AUS	-	ISO includes with Australia
Australia	AS	AU	AUS	036	AUS	.au	ISO includes Ashmore and Cartier Islands, Coral Sea Islands
Austria	AU	AT	AUT	040	AUT	.at	
Azerbaijan	AJ	AZ	AZE	031	AZE	.az	
Bahamas, The	BF	BS	BHS	044	BHS	.bs	
Bahrain	BA	BH	BHR	048	BHR	.bh	
Baker Island	FQ	-	-		UMI	-	ISO includes with the US Minor Outlying Islands
Bangladesh	BG	BD	BGD	050	BGD	.bd	
Barbados	BB	BB	BRB	052	BRB	.bb	
Bassas da India	BS	-	-	-	-	-	administered as part of French Southern and Antarctic Lands; no ISO codes assigned
Belarus	BO	BY	BLR	112	BLR	.by	
Belgium	BE	BE	BEL	056	BEL	.be	
Belize	BH	BZ	BLZ	084	BLZ	.bz	
Benin	BN	BJ	BEN	204	BEN	.bj	
Bermuda	BD	BM	BMU	060	BMU	.bm	
Bhutan	BT	BT	BTN	064	BTN	.bt	

Entity	FIPS 10	ISO 3166			STANAG-1059	Internet	Comment
Bolivia	BL	BO	BOL	068	BOL	.bo	
Bosnia and Herzegovina	BK	BA	BIH	070	BIH	.ba	
Botswana	BC	BW	BWA	072	BWA	.bw	
Bouvet Island	BV	BV	BVT	074	BVT	.bv	
Brazil	BR	BR	BRA	076	BRA	.br	
British Indian Ocean Territory	IO	IO	IOT	086	IOT	.io	
British Virgin Islands	VI	VG	VGB	092	VGB	.vg	
Brunei	BX	BN	BRN	096	BRN	.bn	
Bulgaria	BU	BG	BGR	100	BGR	.bg	
Burkina Faso	UV	BF	BFA	854	BFA	.bf	
Burma	BM	MM	MMR	104	MMR	.mm	ISO uses the name Myanmar
Burundi	BY	BI	BDI	108	BDI	.bi	
Cambodia	CB	KH	KHM	116	KHM	.kh	
Cameroon	CM	CM	CMR	120	CMR	.cm	
Canada	CA	CA	CAN	124	CAN	.ca	
Cape Verde	CV	CV	CPV	132	CPV	.cv	
Cayman Islands	CJ	KY	CYM	136	CYM	.ky	
Central African Republic	CT	CF	CAF	140	CAF	.cf	
Chad	CD	TD	TCD	148	TCD	.td	
Chile	CI	CL	CHL	152	CHL	.cl	
China	CH	CN	CHN	156	CHN	.cn	see also Taiwan
Christmas Island	KT	CX	CXR	162	CXR	.cx	
Clipperton Island	IP	-	-		FYP	-	ISO includes with French Polynesia
Cocos (Keeling) Islands	CK	CC	CCK	166	AUS	.cc	
Colombia	CO	CO	COL	170	COL	.co	
Comoros	CN	KM	COM	174	COM	.km	
Congo, Democratic Republic of the	CG	CD	COD	180	COD	.cd	formerly Zaire
Congo, Republic of the	CF	CG	COG	178	COG	.cg	
Cook Islands	CW	CK	COK	184	COK	.ck	
Coral Sea Islands	CR	-	-		AUS	-	ISO includes with Australia
Costa Rica	CS	CR	CRI	188	CRI	.cr	
Cote d'Ivoire	IV	CI	CIV	384	CIV	.ci	
Croatia	HR	HR	HRV	191	HRV	.hr	
Cuba	CU	CU	CUB	192	CUB	.cu	
Cyprus	CY	CY	CYP	196	CYP	.cy	
Czech Republic	EZ	CZ	CZE	203	CZE	.cz	
Denmark	DA	DK	DNK	208	DNK	.dk	
Djibouti	DJ	DJ	DJI	262	DJI	.dj	
Dominica	DO	DM	DMA	212	DMA	.dm	
Dominican Republic	DR	DO	DOM	214	DOM	.do	
Ecuador	EC	EC	ECU	218	ECU	.ec	
Egypt	EG	EG	EGY	818	EGY	.eg	
El Salvador	ES	SV	SLV	222	SLV	.sv	
Equatorial Guinea	EK	GQ	GNQ	226	GNQ	.gq	
Eritrea	ER	ER	ERI	232	ERI	.er	
Estonia	EN	EE	EST	233	EST	.ee	
Ethiopia	ET	ET	ETH	231	ETH	.et	
Europa Island	EU	-	-	-	-	-	administered as part of French Southern and Antarctic Lands; no ISO codes assigned
Falkland Islands (Islas Malvinas)	FK	FK	FLK	238	FLK	.fk	
Faroe Islands	FO	FO	FRO	234	FRO	.fo	

Entity	FIPS 10	ISO 3166			STANAG-1059	Internet	Comment
Fiji	FJ	FJ	FJI	242	FJI	.fj	
Finland	FI	FI	FIN	246	FIN	.fi	
France	FR	FR	FRA	250	FRA	.fr	
France, Metropolitan	-	FX	FXX	249		.fx	ISO limits to the European part of France, excluding French Guiana, French Polynesia, French Southern and Antarctic Lands, Guadeloupe, Martinique, Mayotte, New Caledonia, Reunion, Saint Pierre and Miquelon, Wallis and Futuna
French Guiana	FG	GF	GUF	254	GUF	.gf	
French Polynesia	FP	PF	PYF	258	PYF	.pf	ISO includes Clipperton Island
French Southern and Antarctic Lands	FS	TF	ATF	260	ATF	.tf	FIPS 10-4 does not include the French-claimed portion of Antarctica (Terre Adelie)
Gabon	GB	GA	GAB	266	GAB	.ga	
Gambia, The	GA	GM	GMB	270	GMB	.gm	
Gaza Strip	GZ	PS	PSE	275	PSE	.ps	ISO identifies as Occupied Palestinian Territory
Georgia	GG	GE	GEO	268	GEO	.ge	
Germany	GM	DE	DEU	276	DEU	.de	
Ghana	GH	GH	GHA	288	GHA	.gh	
Gibraltar	GI	GI	GIB	292	GIB	.gi	
Glorioso Islands	GO	-	-	-	-	-	administered as part of French Southern and Antarctic Lands; no ISO codes assigned
Greece	GR	GR	GRC	300	GRC	.gr	
Greenland	GL	GL	GRL	304	GRL	.gl	
Grenada	GJ	GD	GRD	308	GRD	.gd	
Guadeloupe	GP	GP	GLP	312	GLP	.gp	
Guam	GQ	GU	GUM	316	GUM	.gu	
Guatemala	GT	GT	GTM	320	GTM	.gt	
Guernsey	GK	GG	GGY	831	UK	.gg	
Guinea	GV	GN	GIN	324	GIN	.gn	
Guinea-Bissau	PU	GW	GNB	624	GNB	.gw	
Guyana	GY	GY	GUY	328	GUY	.gy	
Haiti	HA	HT	HTI	332	HTI	.ht	
Heard Island and McDonald Islands	HM	HM	HMD	334	HMD	.hm	
Holy See (Vatican City)	VT	VA	VAT	336	VAT	.va	
Honduras	HO	HN	HND	340	HND	.hn	
Hong Kong	HK	HK	HKG	344	HKG	.hk	
Howland Island	HQ	-	-		UMI	-	ISO includes with the US Minor Outlying Islands
Hungary	HU	HU	HUN	348	HUN	.hu	
Iceland	IC	IS	ISL	352	ISL	.is	
India	IN	IN	IND	356	IND	.in	
Indonesia	ID	ID	IDN	360	IDN	.id	
Iran	IR	IR	IRN	364	IRN	.ir	
Iraq	IZ	IQ	IRQ	368	IRQ	.iq	
Ireland	EI	IE	IRL	372	IRL	.ie	
Isle of Man	IM	IM	IMN	833	UK	.im	
Israel	IS	IL	ISR	376	ISR	.il	
Italy	IT	IT	ITA	380	ITA	.it	
Jamaica	JM	JM	JAM	388	JAM	.jm	

Entity	FIPS 10	ISO 3166			STANAG-1059	Internet	Comment
Jan Mayen	JN	-	-		SJM	-	ISO includes with Svalbard
Japan	JA	JP	JPN	392	JPN	.jp	
Jarvis Island	DQ	-	-		UMI	-	ISO includes with the US Minor Outlying Islands
Jersey	JE	JE	JEY	832	UK	.je	
Johnston Atoll	JQ	-	-		UMI	-	ISO includes with the US Minor Outlying Islands
Jordan	JO	JO	JOR	400	JOR	.jo	
Juan de Nova Island	JU	-	-	-	-	-	administered as part of French Southern and Antarctic Lands; no ISO codes assigned
Kazakhstan	KZ	KZ	KAZ	398	KAZ	.kz	
Kenya	KE	KE	KEN	404	KEN	.ke	
Kingman Reef	KQ	-	-		UMI	-	ISO includes with the US Minor Outlying Islands
Kiribati	KR	KI	KIR	296	KIR	.ki	
Korea, North	KN	KP	PRK	408	PRK	.kp	
Korea, South	KS	KR	KOR	410	KOR	.kr	
Kosovo	KV	-	-	-	-	-	ISO codes have not been designated
Kuwait	KU	KW	KWT	414	KWT	.kw	
Kyrgyzstan	KG	KG	KGZ	417	KGZ	.kg	
Laos	LA	LA	LAO	418	LAO	.la	
Latvia	LG	LV	LVA	428	LVA	.lv	
Lebanon	LE	LB	LBN	422	LBN	.lb	
Lesotho	LT	LS	LSO	426	LSO	.ls	
Liberia	LI	LR	LBR	430	LBR	.lr	
Libya	LY	LY	LBY	434	LBY	.ly	
Liechtenstein	LS	LI	LIE	438	LIE	.li	
Lithuania	LH	LT	LTU	440	LTU	.lt	
Luxembourg	LU	LU	LUX	442	LUX	.lu	
Macau	MC	MO	MAC	446	MAC	.mo	
Macedonia	MK	MK	MKD	807	FYR	.mk	
Madagascar	MA	MG	MDG	450	MDG	.mg	
Malawi	MI	MW	MWI	454	MWI	.mw	
Malaysia	MY	MY	MYS	458	MYS	.my	
Maldives	MV	MV	MDV	462	MDV	.mv	
Mali	ML	ML	MLI	466	MLI	.ml	
Malta	MT	MT	MLT	470	MLT	.mt	
Marshall Islands	RM	MH	MHL	584	MHL	.mh	
Martinique	MB	MQ	MTQ	474	MTQ	.mq	
Mauritania	MR	MR	MRT	478	MRT	.mr	
Mauritius	MP	MU	MUS	480	MUS	.mu	
Mayotte	MF	YT	MYT	175	FRA	.yt	
Mexico	MX	MX	MEX	484	MEX	.mx	
Micronesia, Federated States of	FM	FM	FSM	583	FSM	.fm	
Midway Islands	MQ	-	-		UMI	-	ISO includes with the US Minor Outlying Islands
Moldova	MD	MD	MDA	498	MDA	.md	
Monaco	MN	MC	MCO	492	MCO	.mc	
Mongolia	MG	MN	MNG	496	MNG	.mn	
Montenegro	MJ	ME	MNE	499	-	.me	
Montserrat	MH	MS	MSR	500	MSR	.ms	
Morocco	MO	MA	MAR	504	MAR	.ma	
Mozambique	MZ	MZ	MOZ	508	MOZ	.mz	
Myanmar	-	-	-		-	-	see Burma
Namibia	WA	NA	NAM	516	NAM	.na	
Nauru	NR	NR	NRU	520	NRU	.nr	
Navassa Island	BQ	-	-	-	US	-	ISO includes with the US Minor Outlying Islands

Entity	FIPS 10	ISO 3166			STANAG-1059	Internet	Comment
Nepal	NP	NP	NPL	524	NPL	.np	
Netherlands	NL	NL	NLD	528	NLD	.nl	
Netherlands Antilles	NT	AN	ANT	530	ANT	.an	
New Caledonia	NC	NC	NCL	540	NCL	.nc	
New Zealand	NZ	NZ	NZL	554	NZL	.nz	
Nicaragua	NU	NI	NIC	558	NIC	.ni	
Niger	NG	NE	NER	562	NER	.ne	
Nigeria	NI	NG	NGA	566	NGA	.ng	
Niue	NE	NU	NIU	570	NIU	.nu	
Norfolk Island	NF	NF	NFK	574	NFK	.nf	
Northern Mariana Islands	CQ	MP	MNP	580	MNP	.mp	
Norway	NO	NO	NOR	578	NOR	.no	
Oman	MU	OM	OMN	512	OMN	.om	
Pakistan	PK	PK	PAK	586	PAK	.pk	
Palau	PS	PW	PLW	585	PLW	.pw	
Palmyra Atoll	LQ	-	-		UMI	-	ISO includes with the US Minor Outlying Islands
Panama	PM	PA	PAN	591	PAN	.pa	
Papua New Guinea	PP	PG	PNG	598	PNG	.pg	
Paracel Islands	PF	-	-		-	-	
Paraguay	PA	PY	PRY	600	PRY	.py	
Peru	PE	PE	PER	604	PER	.pe	
Philippines	RP	PH	PHL	608	PHL	.ph	
Pitcairn Islands	PC	PN	PCN	612	PCN	.pn	
Poland	PL	PL	POL	616	POL	.pl	
Portugal	PO	PT	PRT	620	PRT	.pt	
Puerto Rico	RQ	PR	PRI	630	PRI	.pr	
Qatar	QA	QA	QAT	634	QAT	.qa	
Reunion	RE	RE	REU	638	REU	.re	
Romania	RO	RO	ROU	642	ROU	.ro	
Russia	RS	RU	RUS	643	RUS	.ru	
Rwanda	RW	RW	RWA	646	RWA	.rw	
Saint Barthelemy	TB	BL	BLM	652	-	.bl	ccTLD .fr and .gp may also be used
Saint Helena	SH	SH	SHN	654	SHN	.sh	
Saint Kitts and Nevis	SC	KN	KNA	659	KNA	.kn	
Saint Lucia	ST	LC	LCA	662	LCA	.lc	
Saint Martin	RN	MF	MAF	663	-	.mf	ccTLD .fr and .gp may also be used
Saint Pierre and Miquelon	SB	PM	SPM	666	SPM	.pm	
Saint Vincent and the Grenadines	VC	VC	VCT	670	VCT	.vc	
Samoa	WS	WS	WSM	882	WSM	.ws	
San Marino	SM	SM	SMR	674	SMR	.sm	
Sao Tome and Principe	TP	ST	STP	678	STP	.st	
Saudi Arabia	SA	SA	SAU	682	SAU	.sa	
Senegal	SG	SN	SEN	686	SEN	.sn	
Serbia	RI	RS	SRB	688	-	.rs	
Seychelles	SE	SC	SYC	690	SYC	.sc	
Sierra Leone	SL	SL	SLE	694	SLE	.sl	
Singapore	SN	SG	SGP	702	SGP	.sg	
Slovakia	LO	SK	SVK	703	SVK	.sk	
Slovenia	SI	SI	SVN	705	SVN	.si	
Solomon Islands	BP	SB	SLB	090	SLB	.sb	
Somalia	SO	SO	SOM	706	SOM	.so	
South Africa	SF	ZA	ZAF	710	ZAF	.za	
South Georgia and the Islands	SX	GS	SGS	239	SGS	.gs	
Spain	SP	ES	ESP	724	ESP	.es	

Entity	FIPS 10	ISO 3166			STANAG-1059	Internet	Comment
Spratly Islands	PG	-	-		-	-	
Sri Lanka	CE	LK	LKA	144	LKA	.lk	
Sudan	SU	SD	SDN	736	SDN	.sd	
Suriname	NS	SR	SUR	740	SUR	.sr	
Svalbard	SV	SJ	SJM	744	SJM	.sj	ISO includes Jan Mayen
Swaziland	WZ	SZ	SWZ	748	SWZ	.sz	
Sweden	SW	SE	SWE	752	SWE	.se	
Switzerland	SZ	CH	CHE	756	CHE	.ch	
Syria	SY	SY	SYR	760	SYR	.sy	
Taiwan	TW	TW	TWN	158	TWN	.tw	
Tajikistan	TI	TJ	TJK	762	TJK	.tj	
Tanzania	TZ	TZ	TZA	834	TZA	.tz	
Thailand	TH	TH	THA	764	THA	.th	
Timor-Leste	TT	TL	TLS	626	TLS	.tl	
Togo	TO	TG	TGO	768	TGO	.tg	
Tokelau	TL	TK	TKL	772	TKL	.tk	
Tonga	TN	TO	TON	776	TON	.to	
Trinidad and Tobago	TD	TT	TTO	780	TTO	.tt	
Tromelin Island	TE	-	-	-	-	-	administered as part of French Southern and Antarctic Lands; no ISO codes assigned
Tunisia	TS	TN	TUN	788	TUN	.tn	
Turkey	TU	TR	TUR	792	TUR	.tr	
Turkmenistan	TX	TM	TKM	795	TKM	.tm	
Turks and Caicos Islands	TK	TC	TCA	796	TCA	.tc	
Tuvalu	TV	TV	TUV	798	TUV	.tv	
Uganda	UG	UG	UGA	800	UGA	.ug	
Ukraine	UP	UA	UKR	804	UKR	.ua	
United Arab Emirates	AE	AE	ARE	784	ARE	.ae	
United Kingdom	UK	GB	GBR	826	GBR	.uk	
United States	US	US	USA	840	USA	.us	
United States Minor Outlying Islands	-	UM	UMI	581		.um	ISO includes Baker Island, Howland Island, Jarvis Island, Johnston Atoll, Kingman Reef, Midway Islands, Navassa Island, Palmyra Atoll, Wake Island
Uruguay	UY	UY	URY	858	URY	.uy	
Uzbekistan	UZ	UZ	UZB	860	UZB	.uz	
Vanuatu	NH	VU	VUT	548	VUT	.vu	
Venezuela	VE	VE	VEN	862	VEN	.ve	
Vietnam	VM	VN	VNM	704	VNM	.vn	
Virgin Islands	VQ	VI	VIR	850	VIR	.vi	
Virgin Islands (UK)	-	-	-			.vg	see British Virgin Islands
Virgin Islands (US)	-	-	-			.vi	see Virgin Islands
Wake Island	WQ	-	-	-	UMI	-	ISO includes with the US Minor Outlying Islands
Wallis and Futuna	WF	WF	WLF	876	WLF	.wf	
West Bank	WE	PS	PSE	275	PSE	.ps	ISO identifies as Occupied Palestinian Territory
Western Sahara	WI	EH	ESH	732	ESH	.eh	
Western Samoa	-	-	-			.ws	see Samoa
World	-	-	-			-	the Factbook uses the W data code from DIAM 65-18 Geopolitical Data Elements and Related Features, Data Standard No. 3, December 1994, published by the Defense Intelligence Agency
Yemen	YM	YE	YEM	887	YEM	.ye	
Zaire	-	-	-			-	see Democratic Republic of the Congo
Zambia	ZA	ZM	ZMB	894	ZMB	.zm	
Zimbabwe	ZI	ZW	ZWE	716	ZWE	.zw	

APPENDIX E
CROSS-REFERENCE LIST OF HYDROGRAPHIC DATA CODES

IHO 23-4th: *Limits of Oceans and Seas*, Special Publication 23, Draft 4th Edition 1986, published by the International Hydrographic Bureau of the International Hydrographic Organization; note—this document has not yet been ratified and only the 3rd Edition (1953) remains in force

IHO 23-3rd: *Limits of Oceans and Seas*, Special Publication 23, 3rd Edition 1953, published by the International Hydrographic Organization

ACIC M 49-1: *Chart of Limits of Seas and Oceans*, revised January 1958, published by the Aeronautical Chart and Information Center (ACIC), United States Air Force; note—ACIC is now part of the National Imagery and Mapping Agency (NIMA)

DIAM 65-18: *Geopolitical Data Elements and Related Features*, Data Standard No. 4, Defense Intelligence Agency Manual 65-18, December 1994, published by the Defense Intelligence Agency

The US Government has not yet adopted a standard for hydrographic codes similar to the Federal Information Processing Standards (FIPS) 10-4 country codes. The names and limits of the following oceans and seas are not always directly comparable because of differences in the customers, needs, and requirements of the individual organizations. Even the number of principal water bodies varies from organization to organization. *Factbook* users, for example, find the Atlantic Ocean and Pacific Ocean entries useful, but none of the following standards include those oceans in their entirety. Nor is there any provision for combining codes or overcodes to aggregate water bodies. The recently delimited Southern Ocean is not included.

Principal Oceans and Seas of the World With Hydrographic Codes by Institution

	IHO 23-4th	IHO 23-3rd*	ACIC M 49-1	DIAM 65-18
Arctic Ocean	9	17	A	5A
Atlantic Ocean	-	-	-	-
Baltic Sea	2	1	B26	7B
Eastern Mediterranean	3.1.2	28 B	-	8E
Indian Ocean	5	45	F	6A
Mediterranean Sea	3.1	28	B11	-
North Atlantic Ocean	1	23	B	1A
North Pacific Ocean	7	57	D	3A
Pacific Ocean	-	-	-	-
South Atlantic Ocean	4	32	C	2A
South China and Eastern Archipelagic Seas	6	49, 48	D18 plus others	3U plus others
South Pacific Ocean	8	61	E	4A
Western Mediterranean	3.1.1	28 A	-	8W

*The letters after the numbers are subdivisions, not footnotes.

CROSS-REFERENCE LIST OF GEOGRAPHIC NAMES

Name	Entry in *The World Factbook*	Latitude (deg min)	Longitude (deg min)
Abidjan (capital)	Cote d'Ivoire	5 19 N	4 02 W
Abkhazia (region)	Georgia	43 00 N	41 00 E
Abu Dhabi (capital)	United Arab Emirates	24 28 N	54 22 E
Abu Musa (island)	Iran	25 52 N	55 03 E
Abuja (capital)	Nigeria	9 12 N	7 11 E
Abyssinia (former name for Ethiopia)	Ethiopia	8 00 N	38 00 E
Acapulco (city)	Mexico	16 51 N	99 55 W
Accra (capital)	Ghana	5 33 N	0 13 W
Adamstown (capital)	Pitcairn Islands	25 04 S	130 05 W
Addis Ababa (capital)	Ethiopia	9 02 N	38 42 E
Adelie Land (claimed by France; also Terre Adelie)	Antarctica	66 30 S	139 00 E
Aden (city)	Yemen	12 46 N	45 01 E
Aden, Gulf of	Indian Ocean	12 30 N	48 00 E
Admiralty Island	United States (Alaska)	57 44 N	134 20 W
Admiralty Islands	Papua New Guinea	2 10 S	147 00 E
Adriatic Sea	Atlantic Ocean	42 30 N	16 00 E
Adygey (region)	Russia	44 30 N	40 10 E
Aegean Islands	Greece	38 00 N	25 00 E
Aegean Sea	Atlantic Ocean	38 30 N	25 00 E
Afars and Issas, French Territory of the (or FTAI; former name for Djibouti)	Djibouti	11 30 N	43 00 E
Afghanestan (local name for Afghanistan)	Afghanistan	33 00 N	65 00 E
Agalega Islands	Mauritius	10 25 S	56 40 E
Agana (city; former name for Hagatna)	Guam	13 28 N	144 45 E
Ajaccio (city)	France (Corsica)	41 55 N	8 44 E
Ajaria (region)	Georgia	41 45 N	42 10 E
Akmola (city; former name for Astana)	Kazakhstan	51 10 N	71 30 E
Aksai Chin (region)	China (de facto), India (claimed)	35 00 N	79 00 E
Al Arabiyah as Suudiyah (local name for Saudi Arabia)	Saudi Arabia	25 00 N	45 00 E
Al Bahrayn (local name for Bahrain)	Bahrain	26 00 N	50 33 E
Al Imarat al Arabiyah al Muttahidah (local name for the United Arab Emirates)	United Arab Emirates	24 00 N	54 00 E
Al Iraq (local name for Iraq)	Iraq	33 00 N	44 00 E
Al Jaza'ir (local name for Algeria)	Algeria	28 00 N	3 00 E
Al Kuwayt (local name for Kuwait)	Kuwait	29 30 N	45 45 E
Al Maghrib (local name for Morocco)	Morocco	32 00 N	5 00 W
Al Urdun (local name for Jordan)	Jordan	31 00 N	36 00 E
Al Yaman (local name for Yemen)	Yemen	15 00 N	48 00 E
Aland Islands	Finland	60 15 N	20 00 E
Alaska (state)	United States	65 00 N	153 00 W
Alaska, Gulf of	Pacific Ocean	58 00 N	145 00 W
Alboran Sea	Atlantic Ocean	36 00 N	2 30 W
Aldabra Islands (Groupe d'Aldabra)	Seychelles	9 25 S	46 22 E
Alderney (island)	Guernsey	49 43 N	2 12 W
Aleutian Islands	United States (Alaska)	52 00 N	176 00 W
Alexander Archipelago (island group)	United States (Alaska)	57 00 N	134 00 W
Alexander Island	Antarctica	71 00 S	70 00 W
Alexandretta (region; former name for Iskenderun)	Turkey	36 34 N	36 08 E
Alexandria (city)	Egypt	31 12 N	29 54 E
Algiers (capital)	Algeria	36 47 N	2 03 E
Alhucemas, Penon de (island group)	Spain	35 13 N	3 53 W
Alma-Ata (city; former name for Almaty)	Kazakhstan	43 15 N	76 57 E
Almaty (former capital)	Kazakhstan	43 15 N	76 57 E
Alofi (capital)	Niue	19 01 S	169 55 W
Alphonse Island	Seychelles	7 01 S	52 45 E
Alsace (region)	France	48 30 N	7 20 E
Amami Strait	Pacific Ocean	28 40 N	129 30 E
Amindivi Islands (former name for Laccadive Islands)	India	11 30 N	72 30 E
Amirante Isles (island group; also Les Amirantes)	Seychelles	6 00 S	53 10 E
Amman (capital)	Jordan	31 57 N	35 56 E
Amsterdam (capital)	Netherlands	52 23 N	4 54 E
Amsterdam Island (Ile Amsterdam)	French Southern and Antarctic Lands	37 52 S	77 32 E

Name	Entry in The World Factbook	Latitude (deg min)	Longitude (deg min)
Amundsen Sea	Southern Ocean	72 30 S	112 00 W
Amur River	China, Russia	52 56 N	141 10 E
Amurskiy Liman (strait)	Pacific Ocean	53 00 N	141 30 E
Anadyrskiy Zaliv (gulf)	Pacific Ocean	64 00 N	177 00 E
Anatolia (region)	Turkey	39 00 N	35 00 E
Andaman Islands	India	12 00 N	92 45 E
Andaman Sea	Indian Ocean	10 00 N	95 00 E
Andorra la Vella (capital)	Andorra	42 30 N	1 30 E
Andros (island)	Greece	37 45 N	24 42 E
Andros Island	The Bahamas	24 26 N	77 57 W
Anegada Passage	Atlantic Ocean	18 30 N	63 40 W
Angkor Wat (ruins)	Cambodia	13 26 N	103 50 E
Anglo-Egyptian Sudan (former name for Sudan)	Sudan	15 00 N	30 00 E
Anjouan (island)	Comoros	12 15 S	44 25 E
Ankara (capital)	Turkey	39 56 N	32 52 E
Annobon (island)	Equatorial Guinea	1 25 S	5 36 E
Antananarivo (capital)	Madagascar	18 52 S	47 30 E
Antigua (island)	Antigua and Barbuda	14 34 N	90 44 W
Antipodes Islands	New Zealand	49 41 S	178 43 E
Antwerp (city)	Belgium	51 13 N	4 25 E
Aomen (local Chinese short-form name for Macau)	Macau	22 10 N	113 33 E
Aozou Strip (region)	Chad	22 00 N	18 00 E
Apia (capital)	Samoa	13 50 S	171 44 W
Aqaba, Gulf of	Indian Ocean	29 00 N	34 30 E
Arab, Shatt al (river)	Iran, Iraq	29 57 N	48 34 E
Arabian Sea	Indian Ocean	15 00 N	65 00 E
Arafura Sea	Pacific Ocean	9 00 S	133 00 E
Aral Sea	Kazakhstan, Uzbekistan	45 00 N	60 00 E
Argun River	China, Russia	53 20 N	121 28 E
Aru Sea	Pacific Ocean	6 15 S	135 00 E
As-Sudan (local name for Sudan)	Sudan	15 00 N	30 00 E
Ascension Island	Saint Helena	7 57 S	14 22 W
Ashgabat, Ashkhabad (capital)	Turkmenistan	37 57 N	58 23 E
Asmara, Asmera (capital)	Eritrea	15 20 N	38 53 E
Assumption Island	Seychelles	9 46 S	46 34 E
Astana (capital; formerly Akmola)	Kazakhstan	51 10 N	71 30 E
Asuncion (capital)	Paraguay	25 16 S	57 40 W
Asuncion Island	Northern Mariana Islands	19 40 N	145 24 E
Atacama (desert)	Chile	23 00 S	70 10 W
Atacama (region)	Chile	24 30 S	69 15 W
Athens (capital)	Greece	37 59 N	23 44 E
Attu Island	United States	52 55 N	172 57 E
Auckland (city)	New Zealand	36 52 S	174 46 E
Auckland Islands	New Zealand	51 00 S	166 30 E
Australes, Iles (island group; also Iles Tubuai)	French Polynesia	23 20 S	151 00 W
Avarua (capital)	Cook Islands	21 12 S	159 46 W
Axel Heiberg Island	Canada	79 30 N	90 00 W
Azad Kashmir (region)	Pakistan	34 30 N	74 00 E
Azarbaycan, Azerbaidzhan (local name for Azerbaijan)	Azerbaijan	40 30 N	47 30 E
Azores (islands)	Portugal	38 30 N	28 00 W
Azov, Sea of	Atlantic Ocean	49 00 N	36 00 E
Bab el Mandeb (strait)	Indian Ocean	12 40 N	43 20 E
Babuyan Channel	Pacific Ocean	18 44 N	121 40 E
Babuyan Islands	Philippines	19 10 N	121 40 E
Baffin Bay	Arctic Ocean	73 00 N	66 00 W
Baffin Island	Canada	68 00 N	70 00 W
Baghdad (capital)	Iraq	33 21 N	44 25 E
Baku (capital; also Baki, Baky)	Azerbaijan	40 23 N	49 51 E
Balabac Strait	Pacific Ocean	7 35 N	117 00 E
Balearic Islands	Spain	39 30 N	3 00 E
Balearic Sea (Iberian Sea)	Atlantic Ocean	40 30 N	2 00 E
Bali (island)	Indonesia	8 20 S	115 00 E
Bali Sea	Indian Ocean	7 45 S	115 30 E
Balintang Channel	Pacific Ocean	19 49 N	121 40 E

Name	Entry in *The World Factbook*	Latitude (deg min)	Longitude (deg min)
Balintang Islands	Philippines	19 55 N	122 10 E
Balkan Peninsula	Albania, Bosnia and Herzegovina, Bulgaria, Croatia, Greece, Macedonia, Montenegro, Romania, Serbia, Slovenia, Turkey (European part)	42 00 N	23 00 E
Balleny Islands	Antarctica	67 00 S	163 00 E
Balochistan (region)	Pakistan	28 00 N	63 00 E
Baltic Sea	Atlantic Ocean	57 00 N	19 00 E
Bamako (capital)	Mali	12 39 N	8 00 W
Banaba (Ocean Island)	Kiribati	0 52 S	169 35 E
Banat (region)	Hungary, Romania, Serbia	45 30 N	21 00 E
Banda Sea	Pacific Ocean	5 00 S	128 00 E
Bandar Seri Begawan (capital)	Brunei	4 53 N	114 56 E
Bangka (island)	Indonesia	2 30 S	106 00 E
Bangkok (capital)	Thailand	13 45 N	100 31 E
Bangui (capital)	Central African Republic	4 22 N	18 35 E
Banjul (capital)	The Gambia	13 28 N	16 39 W
Banks Island	Australia	10 12 S	142 16 E
Banks Island	Canada	75 15 N	121 30 W
Banks Islands (Iles Banks)	Vanuatu	14 00 S	167 30 E
Barbuda (island)	Antigua and Barbuda	17 38 N	61 48 W
Barcelona (city)	Spain	41 25 N	2 13 E
Barents Sea	Arctic Ocean	74 00 N	36 00 E
Barranquilla (city)	Colombia	10 59 N	74 48 W
Bashi Channel	Pacific Ocean	22 00 N	121 00 E
Basilan Strait	Pacific Ocean	6 49 N	122 05 E
Basque Provinces	Spain	43 00 N	2 30 W
Bass Strait	Pacific Ocean	39 20 S	145 30 E
Basse-Terre (capital)	Guadeloupe	16 00 N	61 44 W
Basseterre (capital)	Saint Kitts and Nevis	17 18 N	62 43 W
Bastia (city)	France (Corsica)	42 42 N	9 27 E
Basutoland (former name for Lesotho)	Lesotho	29 30 S	28 30 E
Batan Islands	Philippines	20 30 N	121 50 E
Bavaria (region; also Bayern)	Germany	48 30 N	11 30 E
Beagle Channel	Atlantic Ocean	54 53 S	68 10 W
Bear Island (see Bjornoya)	Svalbard	74 26 N	19 05 E
Beaufort Sea	Arctic Ocean	73 00 N	140 00 W
Bechuanaland (former name for Botswana)	Botswana	22 00 S	24 00 E
Beijing (capital)	China	39 56 N	116 24 E
Beirut (capital)	Lebanon	33 53 N	35 30 E
Bekaa Valley	Lebanon	34 00 N	36 05 E
Belau (Palau Islands)	Palau	7 30 N	134 30 E
Belep Islands (Iles Belep)	New Caledonia	19 45 S	163 40 E
Belfast (city)	United Kingdom	54 36 N	5 55 W
Belgian Congo (former name for Democratic Republic of the Congo)	Democratic Republic of the Congo	0 00 N	25 00 E
Belgie, Belgique (local name for Belgium)	Belgium	50 50 N	4 00 E
Belgrade (capital)	Serbia	44 50 N	20 30 E
Belize City	Belize	17 30 N	88 12 W
Belle Isle, Strait of	Atlantic Ocean	51 35 N	56 30 W
Bellingshausen Sea	Southern Ocean	71 00 S	85 00 W
Belmopan (capital)	Belize	17 15 N	88 46 W
Belorussia (former name for Belarus)	Belarus	53 00 N	28 00 E
Benadir (region; former name of Italian Somaliland)	Somalia	4 00 N	46 00 E
Bengal, Bay of	Indian Ocean	15 00 N	90 00 E
Berau, Gulf of	Pacific Ocean	2 30 S	132 30 E
Bering Island	Russia	55 00 N	166 30 E
Bering Sea	Pacific Ocean	60 00 N	175 00 W
Bering Strait	Pacific Ocean	65 30 N	169 00 W
Berkner Island	Antarctica	79 30 S	49 30 W
Berlin (capital)	Germany	52 31 N	13 24 E
Berlin, East (former name for eastern sector of Berlin)	Germany	52 30 N	13 33 E
Berlin, West (former name for western sector of Berlin)	Germany	52 30 N	13 20 E

Name	Entry in *The World Factbook*	Latitude (deg min)	Longitude (deg min)
Bern (capital)	Switzerland	46 57 N	7 26 E
Bessarabia (region)	Moldova, Romania, Ukraine	47 00 N	28 30 E
Bharat (local name for India)	India	20 00 N	77 00 E
Bhopal (city)	India	23 16 N	77 24 E
Biafra (region)	Nigeria	5 30 N	7 30 E
Big Diomede Island	Russia	65 46 N	169 06 W
Bijagos, Arquipelago dos (island group)	Guinea-Bissau	11 25 N	16 20 W
Bikini Atoll	Marshall Islands	11 35 N	165 23 E
Bilbao (city)	Spain	43 15 N	2 58 W
Bioko (island)	Equatorial Guinea	3 30 N	8 42 E
Biscay, Bay of	Atlantic Ocean	44 00 N	4 00 W
Bishkek (capital)	Kyrgyzstan	42 54 N	74 36 E
Bishop Rock	United Kingdom	49 52 N	6 27 W
Bismarck Archipelago (island group)	Papua New Guinea	5 00 S	150 00 E
Bismarck Sea	Pacific Ocean	4 00 S	148 00 E
Bissau (capital)	Guinea-Bissau	11 51 N	15 35 W
Bjornoya (Bear Island)	Svalbard	74 26 N	19 05 E
Black Forest (region)	Germany	48 00 N	8 15 E
Black Rock (island)	South Georgia and the South Sandwich Islands	53 39 S	41 48 W
Black Sea	Atlantic Ocean	43 00 N	35 00 E
Bloemfontein (judicial capital)	South Africa	29 12 S	26 07 E
Bo Hai (gulf)	Pacific Ocean	38 00 N	120 00 E
Boa Vista (island)	Cape Verde	16 05 N	22 50 W
Bogota (capital)	Colombia	4 36 N	74 05 W
Bohemia (region)	Czech Republic	50 00 N	14 30 E
Bombay (city; see Mumbai)	India	18 58 N	72 50 E
Bonaire (island)	Netherlands Antilles	12 10 N	68 15 W
Bonifacio, Strait of	Atlantic Ocean	41 01 N	14 00 E
Bonin Islands	Japan	27 00 N	142 10 E
Bonn (former capital)	Germany	50 44 N	7 05 E
Bophuthatswana (region; enclave)	South Africa	26 30 S	25 30 E
Bora-Bora (island)	French Polynesia	16 30 S	151 45 W
Bordeaux (city)	France	44 50 N	0 34 W
Borneo (island)	Brunei, Indonesia, Malaysia	0 30 N	114 00 E
Bornholm (island)	Denmark	55 10 N	15 00 E
Bosna i Hercegovina (local name for Bosnia and Herzegovina)	Bosnia and Herzegovina	44 00 N	18 00 E
Bosnia (political region)	Bosnia and Herzegovina	44 00 N	18 00 E
Bosporus (strait)	Atlantic Ocean	41 00 N	29 00 E
Bothnia, Gulf of	Atlantic Ocean	63 00 N	20 00 E
Bougainville (island)	Papua New Guinea	6 00 S	155 00 E
Bougainville Strait	Pacific Ocean	6 40 S	156 10 E
Bounty Islands	New Zealand	47 43 S	174 00 E
Bourbon Island (former name of Reunion)	Reunion	21 06 S	55 36 E
Brasilia (capital)	Brazil	15 47 S	47 55 W
Bratislava (capital)	Slovakia	48 09 N	17 07 E
Brazzaville (capital)	Republic of the Congo	4 16 S	15 17 E
Bridgetown (capital)	Barbados	13 06 N	59 37 W
Brisbane (city)	Australia	27 28 S	153 02 E
Bristol Bay	Pacific Ocean	57 00 N	160 00 W
Bristol Channel	Atlantic Ocean	51 18 N	3 30 W
Britain (see Great Britain)	United Kingdom	54 00 N	2 00 W
British Bechuanaland (region; former name for northwest South Africa)	South Africa	27 30 S	23 30 E
British Central African Protectorate (former name of Nyasaland)	Malawi	13 30 S	34 00 E
British East Africa (former name for British possessions in eastern Africa)	Kenya, Tanzania, Uganda	1 00 N	38 00 E
British Guiana (former name for Guyana)	Guyana	5 00 N	59 00 W
British Honduras (former name for Belize)	Belize	17 15 N	88 45 W
British Solomon Islands (former name for Solomon Islands)	Solomon Islands	8 00 S	159 00 E
British Somaliland (former name for northern Somalia)	Somalia	10 00 N	49 00 E
Brussels (capital)	Belgium	50 50 N	4 20 E
Bubiyan (island)	Kuwait	29 47 N	48 10 E
Bucharest (capital)	Romania	44 26 N	26 06 E

Name	Entry in The World Factbook	Latitude (deg min)	Longitude (deg min)
Budapest (capital)	Hungary	47 30 N	19 05 E
Buenos Aires (capital)	Argentina	34 36 S	58 27 W
Bujumbura (capital)	Burundi	3 23 S	29 22 E
Bukovina (region)	Romania, Ukraine	48 00 N	26 00 E
Byelarus (local name for Belarus)	Belarus	53 00 N	28 00 E
Byelorussia (former name for Belarus)	Belarus	53 00 N	28 00 E
Cabinda (province)	Angola	5 33 S	12 12 E
Cabo Verde (local name for Cape Verde)	Cape Verde	16 00 N	24 00 W
Cabot Strait	Atlantic Ocean	47 20 N	59 30 W
Caicos Islands	Turks and Caicos Islands	21 56 N	71 58 W
Cairo (capital)	Egypt	30 03 N	31 15 E
Calcutta (city)	India	22 32 N	88 21 E
Calgary (city)	Canada	51 02 N	114 04 W
California, Gulf of	Pacific Ocean	28 00 N	112 00 W
Cameroun (local name for Cameroon)	Cameroon	6 00 N	12 00 E
Campbell Island	New Zealand	52 33 S	169 09 E
Campeche, Bay of	Atlantic Ocean	20 00 N	94 00 W
Canal Zone (former name for US possessions in Panama)	Panama	9 00 N	79 45 W
Canarias Sea	Atlantic Ocean	28 00 N	16 00 W
Canary Islands	Spain	28 00 N	15 30 W
Canberra (capital)	Australia	35 17 S	149 08 E
Cancun (city)	Mexico	21 10 N	86 50 W
Canton (city; now Guangzhou)	China	23 06 N	113 16 E
Canton Island (Kanton Island)	Kiribati	2 49 S	171 40 W
Cape Juby (region; former name for Southern Morocco)	Morocco	27 53 N	12 58 W
Cape Province (region; former name for Northern, Western, and Eastern Cape Provinces of South Africa)	South Africa	31 30 S	22 30 E
Cape Town (legislative capital)	South Africa	33 57 S	18 25 E
Cape of Good Hope (cape; also alternate name for Cape Province of South Africa)	South Africa	34 15 S	18 20 E
Caracas (capital)	Venezuela	10 30 N	66 56 W
Cargados Carajos Shoals	Mauritius	16 25 S	59 38 E
Caribbean Sea	Atlantic Ocean	15 00 N	73 00 W
Caroline Islands	Federated States of Micronesia, Palau	7 30 N	148 00 E
Carpatho-Ukraine (region; former name for Zakarpats'ka oblast')	Ukraine	48 22 N	23 32 E
Carpentaria, Gulf of	Pacific Ocean	14 00 S	139 00 E
Casablanca (city)	Morocco	33 35 N	7 34 W
Castries (capital)	Saint Lucia	14 01 N	61 00 W
Catalonia (region)	Spain	42 00 N	2 00 E
Cato Island	Australia	23 15 S	155 32 E
Caucasus (region)	Russia	42 00 N	45 00 E
Cayenne (capital)	French Guiana	4 56 N	52 20 W
Celebes (island)	Indonesia	2 00 S	121 00 E
Celebes Sea	Pacific Ocean	3 00 N	122 00 E
Celtic Sea	Atlantic Ocean	51 00 N	6 30 W
Central African Empire (former name for Central African Republic)	Central African Republic	7 00 N	21 00 E
Ceram (Seram) Sea	Pacific Ocean	2 30 S	129 30 E
Ceska Republika (local name for Czech Republic)	Czech Republic	49 45 N	15 30 E
Ceskoslovensko (former local name for Czechoslovakia)	Czech Republic, Slovakia	49 00 N	17 30 E
Cetinje (capital city)	Montenegro	42 24 N	18 55 E
Ceuta (city)	Spain	35 53 N	5 19 W
Ceylon (former name for Sri Lanka)	Sri Lanka	7 00 N	81 00 E
Chafarinas, Islas (island)	Spain	35 12 N	2 26 W
Chagos Archipelago (Oil Islands)	British Indian Ocean Territory	6 00 S	71 30 E
Challenger Deep (Mariana Trench)	Pacific Ocean	11 22 N	142 36 E
Channel Islands	Guernsey, Jersey	49 20 N	2 20 W
Charlotte Amalie (capital)	Virgin Islands	18 21 N	64 56 W
Chatham Islands	New Zealand	44 00 S	176 30 W
Chechnya (region; also Chechnia)	Russia	43 15 N	45 40 E
Cheju Strait	Pacific Ocean	34 00 N	126 30 E
Cheju-do (island)	Korea, South	33 20 N	126 30 E
Chengdu (city)	China	30 43 N	104 04 E

Name	Entry in The World Factbook	Latitude (deg min)	Longitude (deg min)
Chennai (city; also Madras)	India	13 04 N	80 16 E
Chesterfield Islands (Iles Chesterfield)	New Caledonia	19 52 S	158 15 E
Chihli, Gulf of (see Bo Hai)	Pacific Ocean	38 30 N	120 00 E
Chiloe (island)	Chile	42 50 S	74 00 W
China, People's Republic of	China	35 00 N	105 00 E
China, Republic of	Taiwan	23 30 N	121 00 E
Chisinau (capital; also Kishinev)	Moldova	47 00 N	28 50 E
Choiseul (island)	Solomon Islands	7 05 S	121 00 E
Choson (local name for North Korea)	North Korea	40 00 N	127 00 E
Christmas Island (Indian Ocean)	Australia	10 25 S	105 39 E
Christmas Island (Pacific Ocean; also Kiritimati)	Kiribati	1 52 N	157 20 W
Chukchi Sea	Arctic Ocean	69 00 N	171 00 W
Chuuk Islands (Truk Islands)	Federated States of Micronesia	7 25 N	151 47 W
Cilicia (region)	Turkey	36 50 N	34 30 E
Ciskei (enclave)	South Africa	33 00 S	27 00 E
Citta del Vaticano (local name for Vatican City)	Holy See	41 54 N	12 27 E
Cochin China (region)	Vietnam	11 00 N	107 00 E
Coco, Isla del (island)	Costa Rica	5 32 N	87 04 W
Cocos Islands	Cocos (Keeling) Islands	12 30 S	96 50 E
Colombo (capital)	Sri Lanka	6 56 N	79 51 E
Colon, Archipielago de (Galapagos Islands)	Ecuador	0 00 N	90 30 W
Commander Islands (Komandorskiye Ostrova)	Russia	55 00 N	167 00 E
Comores (local name for Comoros)	Comoros	12 10 S	44 15 E
Con Son (islands)	Vietnam	8 43 N	106 36 E
Conakry (capital)	Guinea	9 31 N	13 43 W
Confederatio Helvetica (local name for Switzerland)	Switzerland	47 00 N	8 00 E
Congo (Brazzaville) (former name for Republic of the Congo)	Republic of the Congo	1 00 S	15 00 E
Congo (Leopoldville) (former name for the Democratic Republic of the Congo)	Democratic Republic of the Congo	0 00 N	25 00 E
Constantinople (city; former name for Istanbul)	Turkey	41 01 N	28 58 E
Cook Strait	Pacific Ocean	41 15 S	174 30 E
Copenhagen (capital)	Denmark	55 40 N	12 35 E
Coral Sea	Pacific Ocean	15 00 S	150 00 E
Corfu (island)	Greece	39 40 N	19 45 E
Corinth (region)	Greece	37 56 N	22 56 E
Corisco (island)	Equatorial Guinea	0 55 N	9 19 E
Corn Islands (Islas del Maiz)	Nicaragua	12 15 N	83 00 W
Corocoro Island	Guyana, Venezuela	3 38 N	66 50 W
Corsica (island; also Corse)	France	42 00 N	9 00 E
Cosmoledo Group (island group; also Atoll de Cosmoledo)	Seychelles	9 43 S	47 35 E
Cotonou (former capital)	Benin	6 21 N	2 26 E
Cotopaxi (volcano)	Ecuador	0 39 S	78 26 W
Courantyne River	Guyana, Suriname	5 57 N	57 06 W
Cozumel (island)	Mexico	20 30 N	86 55 W
Crete (island)	Greece	35 15 N	24 45 E
Crimea (region)	Ukraine	45 00 N	34 00 E
Crimean Peninsula	Ukraine	45 00 N	34 00 E
Crooked Island Passage	Atlantic Ocean	22 55 N	74 35 W
Crozet Islands (Iles Crozet)	French Southern and Antarctic Lands	46 30 S	51 00 E
Cyclades (island group)	Greece	37 00 N	25 10 E
Cyrenaica (region)	Libya	31 00 N	22 00 E
Czechoslovakia (former name for the entity that subsequently split into the Czech Republic and Slovakia)	Czech Republic, Slovakia	49 00 N	18 00 E
D'Entrecasteaux Islands	Papua New Guinea	9 30 S	150 40 E
Dagestan (region)	Russia	43 00 N	47 00 E
Dahomey (former name for Benin)	Benin	9 30 N	2 15 E
Daito Islands	Japan	43 00 N	17 00 E
Dakar (capital)	Senegal	14 40 N	17 26 W
Dalmatia (region)	Croatia	43 00 N	17 00 E
Daman (city; also Damao)	India	20 10 N	73 00 E
Damascus (capital)	Syria	33 30 N	36 18 E
Danger Islands (see Pukapuka Atoll)	Cook Islands	10 53 S	165 49 W
Danish Straits	Atlantic Ocean	58 00 N	11 00 E
Danish West Indies (former name for the Virgin Islands)	Virgin Islands	18 20 N	64 50 W
Danmark (local name)	Denmark	56 00 N	10 00 E
Danzig (city; former name for Gdansk)	Poland	54 23 N	18 40 E

Name	Entry in *The World Factbook*	Latitude (deg min)	Longitude (deg min)
Dao Bach Long Vi (island)	Vietnam	20 08 N	107 44 E
Dar es Salaam (capital)	Tanzania	6 48 S	39 17 E
Dardanelles (strait)	Atlantic Ocean	40 15 N	26 25 E
Davis Strait	Atlantic Ocean	67 00 N	57 00 W
Dead Sea	Israel, Jordan, West Bank	32 30 N	35 30 E
Deception Island	Antarctica	62 56 S	60 34 W
Denmark Strait	Atlantic Ocean	67 00 N	24 00 W
Desolation Islands (Isles Kerguelen)	French Southern and Antarctic Lands	49 30 S	69 30 E
Deutschland (local name for Germany)	Germany	51 00 N	9 00 E
Devils Island (Ile du Diable)	French Guiana	5 17 N	52 35 W
Devon Island	Canada	76 00 N	87 00 W
Dhaka (capital)	Bangladesh	23 43 N	90 25 E
Dhivehi Raajje (local name for Maldives)	Maldives	3 15 N	73 00 E
Dhofar (region)	Oman	17 00 N	54 10 E
Diego Garcia (island)	British Indian Ocean Territory	7 20 S	72 25 E
Diego Ramirez (islands)	Chile	56 30 S	68 43 W
Dili (capital)	Timor-Leste	8 35 S	125 36 E
Dilmun (former name for Bahrain)	Bahrain	7 00 N	81 00 E
Diomede Islands	Russia (Big Diomede), United States (Little Diomede)	65 47 N	169 00 W
Diu (region)	India	20 42 N	70 59 E
Djibouti (capital)	Djibouti	11 30 N	43 15 E
Dnieper (river)	Belarus, Russia, Ukraine (Dnyapro, Dnepr, Dnipro)	46 30 N	32 18 E
Dniester (river)	Moldova, Ukraine (Nistru, Dnister)	46 18 N	30 17 E
Dobruja (region)	Bulgaria, Romania	43 30 N	28 00 E
Dodecanese (island group)	Greece	36 00 N	27 05 E
Dodoma (city)	Tanzania	6 11 S	35 45 E
Doha (capital)	Qatar	25 17 N	51 32 E
Donets Basin	Russia, Ukraine	48 15 N	38 30 E
Douala (city)	Cameroon	4 03 N	9 42 E
Douglas (capital)	Man, Isle of	54 09 N	4 28 W
Dover, Strait of	Atlantic Ocean	51 00 N	1 30 E
Drake Passage	Atlantic Ocean, Southern Ocean	60 00 S	60 00 W
Druk Yul (local name for Bhutan)	Bhutan	27 30 N	90 30 E
Dubai, Dubayy (city)	United Arab Emirates	25 18 N	55 18 E
Dublin (capital)	Ireland	53 20 N	6 15 W
Duesseldorf (city)	Germany	51 13 N	6 47 E
Durban (city)	South Africa	29 51 S	31 02 E
Dushanbe (capital)	Tajikistan	38 35 N	68 48 E
Dutch Antilles (former name for the Netherlands Antilles)	Netherlands Antilles	12 10 N	68 30 W
Dutch East Indies (former name for Indonesia)	Indonesia	5 00 S	120 00 E
Dutch Guiana (former name for Suriname)	Suriname	4 00 N	56 00 W
Dutch West Indies (former name for the Netherlands Antilles)	Netherlands Antilles	12 10 N	68 30 W
Dzungarian Gate (valley)	China, Kazakhstan	45 25 N	82 25 E
East China Sea	Pacific Ocean	30 00 N	126 00 E
East Frisian Islands	Germany	53 44 N	7 25 E
East Germany (German Democratic Republic; former name for eastern portion of Germany)	Germany	52 00 N	13 00 E
East Korea Strait (Eastern Channel or Tsushima Strait)	Pacific Ocean	34 00 N	129 00 E
East Pakistan (former name for Bangladesh)	Bangladesh	24 00 N	90 00 E
East Siberian Sea	Arctic Ocean	74 00 N	166 00 E
Easter Island (Isla de Pascua)	Chile	27 07 S	109 22 W
Eastern Channel (East Korea Strait or Tsushima Strait)	Pacific Ocean	34 00 N	129 00 E
Eastern Samoa (former name for American Samoa)	American Samoa	14 20 S	170 00 W
Edinburgh (city)	United Kingdom	55 57 N	3 11 W
Eesti (local name for Estonia)	Estonia	59 00 N	26 00 E
Eire (local name for Ireland)	Ireland	53 00 N	8 00 W
Elba (island)	Italy	42 46 N	10 17 E
Elemi Triangle (region)	Ethiopia (claimed), Kenya (de facto), Sudan (claimed)	5 00 N	35 30 E
Ellada, Ellas (local name for Greece)	Greece	39 00 N	22 00 E
Ellef Ringnes Island	Canada	78 00 N	103 00 W

791

Name	Entry in The World Factbook	Latitude (deg min)	Longitude (deg min)
Ellesmere Island	Canada	81 00 N	80 00 W
Ellice Islands	Tuvalu	8 00 S	178 00 E
Ellsworth Land (region)	Antarctica	75 00 S	92 00 W
Elobey, Islas de (island group)	Equatorial Guinea	0 59 N	9 33 E
Enderbury Island	Kiribati	3 08 S	171 05 W
Enewetak Atoll (Eniwetok Atoll)	Marshall Islands	11 30 N	162 15 E
England (region)	United Kingdom	52 30 N	1 30 W
English Channel	Atlantic Ocean	50 20 N	1 00 W
Eniwetok Atoll (see Enewetak Atoll)	Marshall Islands	11 30 N	162 15 E
Eolie, Isole (island group)	Italy	38 30 N	15 00 E
Epirus, Northern (region)	Albania, Greece	40 00 N	20 30 E
Episkopi Cantonment (capital)	Akrotiri, Dhekelia	34 40 N	32 51 E
Ertra (local name for Eritrea)	Eritrea	15 00 N	39 00 E
Espana	Spain	40 00 N	4 00 W
Essequibo (region; claimed by Venezuela)	Guyana	6 59 N	58 23 W
Etorofu (island; also Iturup)	Russia (de facto)	44 55 N	147 40 E
Farquhar Group (island group; also Atoll de Farquhar)	Seychelles	10 10 S	51 10 E
Fergana Valley	Kyrgyzstan, Tajikistan, Uzbekistan	41 00 N	72 00 E
Fernando Po (island; see Bioko)	Equatorial Guinea	3 30 N	8 42 E
Fernando de Noronha (island group)	Brazil	3 51 S	32 25 W
Filipinas (local name for the Philippines; also Pilipinas)	Philippines	13 00 N	122 00 E
Finland, Gulf of	Atlantic Ocean	60 00 N	27 00 E
Florence (city)	Italy	43 46 N	11 16 E
Flores (island)	Indonesia	8 45 S	121 00 E
Flores Sea	Pacific Ocean	7 40 S	119 45 E
Florida, Straits of	Atlantic Ocean	25 00 N	79 45 W
Fongafale (largest island of Funafuti)	Tuvalu	8 30 S	179 12 E
Former Soviet Union (FSU)	Armenia, Azerbaijan, Belarus, Estonia, Georgia, Kazakhstan, Kyrgyzstan, Latvia, Lithuania, Moldova, Russia, Tajikistan, Turkmenistan, Ukraine, Uzbekistan		
Formosa (island)	Taiwan	23 30 N	121 00 E
Formosa Strait (see Taiwan Strait)	Pacific Ocean	24 00 N	119 00 E
Foroyar (local name for Faroe Islands)	Faroe Islands	62 00 N	7 00 W
Fort-de-France (capital)	Martinique	14 36 N	61 05 W
Frankfurt am Main (city)	Germany	50 07 N	8 41 E
Franz Josef Land (island group)	Russia	81 00 N	55 00 E
Freetown (capital)	Sierra Leone	8 30 N	13 15 W
French Cameroon (former name for Cameroon)	Cameroon	6 00 N	12 00 E
French Guinea (former name for Guinea)	Guinea	11 00 N	10 00 W
French Indochina (former name for French possessions in southeast Asia)	Cambodia, Laos, Vietnam	15 00 N	107 00 E
French Morocco (former name for Morocco)	Morocco	32 00 N	5 00 W
French Somaliland (former name for Djibouti)	Djibouti	11 30 N	43 00 E
French Sudan (former name for Mali)	Mali	17 00 N	4 00 W
French Territory of the Afars and Issas (or FTAI; former name for Djibouti)	Djibouti	11 30 N	43 00 E
French Togoland (former name for Togo)	Togo	8 00 N	1 10 E
French West Indies (former name for French possessions in the West Indies)	Guadeloupe, Martinique	16 30 N	62 00 W
Friendly Islands	Tonga	20 00 S	175 00 W
Frisian Islands	Denmark, Germany, Netherlands	53 35 N	6 40 E
Frunze (city; former name for Bishkek)	Kyrgyzstan	42 54 N	74 36 E
Funafuti (capital, atoll)	Tuvalu	8 30 S	179 12 E
Fundy, Bay of	Atlantic Ocean	45 00 N	66 00 W
Futuna Islands (Hoorn Islands/Iles de Horne)	Wallis and Futuna	14 19 S	178 05 W
Fyn (island)	Denmark	55 20 N	10 25 E
Gaborone (capital)	Botswana	24 45 S	25 55 E
Galapagos Islands (Archipielago de Colon)	Ecuador	0 00 N	90 30 W
Galicia (region)	Poland, Ukraine	49 30 N	23 00 E
Galicia (region)	Spain	42 45 N	8 10 E
Galilee (region)	Israel	32 54 N	35 20 E
Galleons Passage	Atlantic Ocean	11 00 N	60 55 W

Name	Entry in *The World Factbook*	Latitude (deg min)	Longitude (deg min)
Gambier Islands (Iles Gambier)	French Polynesia	23 09 S	134 58 W
Gaspar Strait	Pacific Ocean	3 00 S	107 00 E
Gdansk (city; formerly Danzig)	Poland	54 23 N	18 40 E
Geneva (city)	Switzerland	46 12 N	6 10 E
Genoa (city)	Italy	44 25 N	8 57 E
George Town (capital)	Cayman Islands	19 20 N	81 23 W
George Town (city)	Malaysia	5 26 N	100 16 E
George Town (city)	The Bahamas	23 30 N	75 46 W
Georgetown (capital)	Guyana	6 48 N	58 10 W
Georgetown (city)	The Gambia	13 30 N	14 47 W
German Democratic Republic (East Germany; former name for eastern portion of Germany)	Germany	52 00 N	13 00 E
German Southwest Africa (former name for Namibia)	Namibia	22 00 S	17 00 E
Germany, Federal Republic of	Germany	51 00 N	9 00 E
Gibraltar (city, peninsula)	Gibraltar	36 11 N	5 22 W
Gibraltar, Strait of	Atlantic Ocean	35 57 N	5 36 W
Gidi Pass	Egypt	30 13 N	33 09 E
Gilbert Islands	Kiribati	1 25 N	173 00 E
Goa (state)	India	15 20 N	74 00 E
Gobi (desert)	China, Mongolia	42 30 N	107 00 E
Godthab (capital; also Nuuk)	Greenland	64 11 N	51 44 W
Golan Heights (region)	Syria	33 00 N	35 45 E
Gold Coast (former name for Ghana)	Ghana	8 00 N	2 00 W
Golfo San Jorge (gulf)	Atlantic Ocean	46 00 S	66 00 W
Golfo San Matias (gulf)	Atlantic Ocean	41 30 S	64 00 W
Good Hope, Cape of	South Africa	34 24 S	18 30 E
Goteborg (city)	Sweden	57 43 N	11 58 E
Gotland (island)	Sweden	57 30 N	18 33 E
Gough Island	Saint Helena	40 20 S	9 55 W
Graham Land (region)	Antarctica	65 00 S	64 00 W
Gran Chaco (region)	Argentina, Paraguay	24 00 S	60 00 W
Grand Bahama (island)	The Bahamas	26 40 N	78 35 W
Grand Banks (fishing ground)	Atlantic Ocean	47 06 N	55 48 W
Grand Cayman (island)	Cayman Islands	19 20 N	81 20 W
Grand Turk (capital; also Cockburn Town)	Turks and Caicos Islands	21 28 N	71 08 W
Great Australian Bight	Indian Ocean	35 00 S	130 00 E
Great Belt (strait; also Store Baelt)	Atlantic Ocean	55 30 N	11 00 E
Great Bitter Lake	Egypt	30 20 N	32 23 E
Great Britain (island)	United Kingdom	54 00 N	2 00 W
Great Channel	Indian Ocean	6 25 N	94 20 E
Great Inagua (island)	The Bahamas	21 00 N	73 20 W
Great Rift Valley	Ethiopia, Kenya	0 30 N	36 00 E
Greater Sunda Islands	Brunei, Indonesia, Malaysia	2 00 S	110 00 E
Green Islands	Papua New Guinea	4 30 S	154 10 E
Greenland Sea	Arctic Ocean	79 00 N	5 00 W
Grenadines, Northern (island group)	Saint Vincent and the Grenadines	13 15 N	61 12 W
Grenadines, Southern (island group)	Grenada	12 07 N	61 40 W
Grytviken (town; on South Georgia)	South Georgia and the South Sandwich Islands	54 15 S	36 45 W
Guadalahara (city)	Mexico	20 40 N	103 24 W
Guadalcanal (island)	Solomon Islands	9 32 S	160 12 E
Guadalupe, Isla de (island)	Mexico	29 11 N	118 17 W
Guangzhou (city; also Canton)	China	23 09 N	113 21 E
Guantanamo Bay (US Naval Base)	Cuba	20 00 N	75 08 W
Guatemala (capital)	Guatemala	14 38 N	90 31 W
Guine-Bissau (local name for Guinea-Bissau)	Guinea-Bissau	12 00 N	15 00 W
Guinea Ecuatorial (local name for Equatorial Guinea)	Equatorial Guinea	2 00 N	10 00 E
Guinea, Gulf of	Atlantic Ocean	3 00 N	2 30 E
Guinee (local name for Guinea)	Guinea	11 00 N	10 00 W
Gustavia (capital)	Saint Barthelemy	17 53 N	62 51 W
Guyane Francaise (local name for French Guiana)	French Guiana	4 00 N	53 00 W
Ha'apai Group (island group)	Tonga	19 42 S	174 29 W
Habomai Islands	Russia (de facto)	43 30 N	146 10 E
Hadhramaut (region)	Yemen	15 00 N	50 00 E
Hagatna (capital; formerly Agana)	Guam	13 28 N	144 45 E
Hague, The (seat of government)	Netherlands	52 05 N	4 18 E

Name	Entry in The World Factbook	Latitude (deg min)	Longitude (deg min)
Haifa (city)	Israel	32 50 N	35 00 E
Hainan Dao (island)	China	19 00 N	109 30 E
Haiphong (city)	Vietnam	20 52 N	106 41 E
Hala'ib Triangle (region)	Egypt (claimed), Sudan (de facto)	22 30 N	35 00 E
Halifax (city)	Canada	44 39 N	63 36 W
Halmahera (island)	Indonesia	1 00 N	128 00 E
Halmahera Sea	Pacific Ocean	0 30 S	129 00 E
Hamburg (city)	Germany	53 34 N	9 59 E
Hamilton (capital)	Bermuda	32 17 N	64 46 W
Han-guk (local name for South Korea	South Korea	37 00 N	127 30 E
Hanoi (capital)	Vietnam	21 02 N	105 51 E
Harare (capital)	Zimbabwe	17 50 S	31 03 E
Harvey Islands (former name for Cook Islands)	Cook Islands	21 14 S	159 46 W
Hatay (province)	Turkey	36 30 N	36 15 E
Havana (capital)	Cuba	23 08 N	82 22 W
Hawaii (island)	United States	19 45 N	155 45 W
Hawaiian Islands	United States	21 00 N	157 45 W
Hawar (island)	Bahrain	25 40 N	50 47 E
Hayastan (local name for Armenia)	Armenia	40 00 N	45 00 E
Heard Island	Heard Island and McDonald Islands	53 06 S	73 30 E
Hejaz (region)	Saudi Arabia	24 30 N	38 30 E
Helsinki (capital)	Finland	60 10 N	24 58 E
Herzegovina (political region)	Bosnia and Herzegovina	44 00 N	18 00 E
Hiiumaa (island)	Estonia	58 50 N	22 30 E
Hispaniola (island)	Dominican Republic, Haiti	18 45 N	71 00 W
Ho Chi Minh City (formerly Saigon)	Vietnam	10 45 N	106 40 E
Hokkaido (island)	Japan	44 00 N	143 00 E
Holland (region)	Netherlands	52 30 N	5 45 E
Hong Kong (special administrative region)	Hong Kong	22 15 N	114 10 E
Honiara (capital)	Solomon Islands	9 26 S	159 57 E
Honshu (island)	Japan	36 00 N	138 00 E
Hormuz, Strait of	Indian Ocean	26 34 N	56 15 E
Horn of Africa (region)	Djibouti, Eritrea, Ethiopia, Somalia	8 00 N	48 00 E
Horn, Cape (Cabo de Hornos)	Chile	55 59 S	67 16 W
Horne, Iles de (island group)	Wallis and Futuna	14 19 S	178 05 W
Hrvatska (local name for Croatia)	Croatia	45 10 N	15 30 E
Hudson Bay	Arctic Ocean	60 00 N	86 00 W
Hudson Strait	Arctic Ocean	62 00 N	71 00 W
Hunter Island	New Caledonia, Vanuatu	22 24 S	172 06 E
Iberian Peninsula	Portugal, Spain	40 00 N	5 00 W
Iceland Sea	Arctic Ocean	68 00 N	20 00 W
Ifni (region; former name of part of Spanish West Africa)	Morocco	29 22 N	10 09 W
Inaccessible Island	Saint Helena	37 17 S	12 40 W
Indochina (region)	Cambodia, Laos, Vietnam	15 00 N	107 00 E
Ingushetia (region)	Russia	43 15 N	45 00 E
Inhambane (region)	Mozambique	22 30 S	34 30 E
Inini (former name for French Guiana)	French Guiana	4 00 N	53 00 W
Inland Sea	Japan	34 20 N	133 30 E
Inner Hebrides (islands)	United Kingdom	56 30 N	6 20 W
Inner Mongolia (region; also Nei Mongol)	China	42 00 N	113 00 E
Ionian Islands	Greece	38 30 N	20 30 E
Ionian Sea	Atlantic Ocean	38 30 N	18 00 E
Irian Jaya (province)	Indonesia	5 00 S	138 00 E
Irish Sea	Atlantic Ocean	53 30 N	5 20 W
Iron Gate (river gorge)	Romania, Serbia	44 41 N	22 31 E
Iskenderun (region; formerly Alexandretta)	Turkey	36 34 N	36 08 E
Islamabad (capital)	Pakistan	33 42 N	73 10 E
Island (local name for Iceland)	Iceland	65 00 N	18 00 W
Islas Malvinas (island group)	Falkland Islands (Islas Malvinas)	51 45 S	59 00 W
Istanbul (city)	Turkey	41 01 N	28 58 E
Istrian Peninsula	Croatia, Slovenia	45 00 N	14 00 E
Italia (local name for Italy)	Italy	42 50 N	12 50 E
Italian East Africa (former name for Italian possessions in eastern Africa)	Eritrea, Ethiopia, Somalia	8 00 N	38 00 E

Name	Entry in *The World Factbook*	Latitude (deg min)	Longitude (deg min)
Italian Somaliland (former name for southern Somalia)	Somalia	10 00 N	49 00 E
Ittihad al-Imarat al-Arabiyah (local name for the United Arab Emirates)	United Arab Emirates	24 00 N	54 00 E
Iturup (island; see Etorofu)	Russia (de facto)	44 55 N	147 40 E
Ityop'iya (local name for Ethiopia)	Ethiopia	8 00 N	38 00 E
Ivory Coast (former name for Cote d'Ivoire)	Cote d'Ivoire	8 00 N	5 00 W
Iwo Jima (island)	Japan	24 47 N	141 20 E
Izmir (region)	Turkey	38 25 N	27 10 E
Jakarta (capital)	Indonesia	6 10 S	106 48 E
James Bay	Arctic Ocean	54 00 N	80 00 W
Jamestown (capital)	Saint Helena	15 56 S	5 44 W
Jammu (city)	India	32 42 N	74 52 E
Jammu and Kashmir (region)	India, Pakistan	34 00 N	76 00 E
Japan, Sea of	Pacific Ocean	40 00 N	135 00 E
Jars, Plain of	Laos	19 27 N	103 10 E
Java (island)	Indonesia	7 30 S	110 00 E
Java Sea	Pacific Ocean	5 00 S	110 00 E
Jerusalem (capital, proclaimed)	Israel, West Bank	31 47 N	35 14 E
Jiddah, Jeddah (city)	Saudi Arabia	21 30 N	39 12 E
Johannesburg (city)	South Africa	26 15 S	28 00 E
Joseph Bonaparte Gulf	Pacific Ocean	14 00 S	128 45 E
Juan Fernandez, Islas de (island group)	Chile	33 00 S	80 00 W
Juan de Fuca, Strait of	Pacific Ocean	48 18 N	124 00 W
Jubal, Strait of	Indian Ocean	27 40 N	33 55 E
Judaea (region)	Israel, West Bank	31 35 N	35 00 E

Jugoslavia, Jugoslavija (local names for Yugoslavia,
    a former Balkan federation)
	Bosnia and Herzegovina,
	    Croatia, Macedonia,
	    Montenegro, Serbia, Slovenia
		43 00 N	21 00 E

Jutland (region)	Denmark	56 00 N	9 15 E
Juventud, Isla de la (Isle of Youth)	Cuba	21 40 N	82 50 W
Kabardino-Balkaria (region)	Russia	43 30 N	43 30 E
Kabul (capital)	Afghanistan	34 31 N	69 12 E
Kaduna (city)	Nigeria	10 33 N	7 27 E
Kailas Range	China, India	30 00 N	82 00 E
Kalaallit Nunaat (local name for Greenland)	Greenland	72 00 N	40 00 W
Kalahari (desert)	Botswana, Namibia	24 30 S	21 00 E
Kalimantan (region)	Indonesia	0 00 N	115 00 E
Kaliningrad (region; formerly part of East Prussia)	Russia	54 30 N	21 00 E
Kamaran (island)	Yemen	15 21 N	42 34 E
Kamchatka Peninsula (Poluostrov Kamchatka)	Russia	56 00 N	160 00 E
Kampala (capital)	Uganda	0 19 N	32 25 E
Kampuchea (former name for Cambodia)	Cambodia	13 00 N	105 00 E
Kane Basin (portion of channel)	Arctic Ocean	79 30 N	68 00 W
Kanton Island	Kiribati	2 49 S	171 40 W
Kara Sea	Arctic Ocean	76 00 N	80 00 E
Karachevo-Cherkessia (region)	Russia	43 40 N	41 50 E
Karachi (city)	Pakistan	24 51 N	67 03 E
Karafuto (island; former name for southern Sakhalin Island)	Russia	50 00 N	143 00 E
Karakoram Pass	China, India	35 30 N	77 50 E
Karelia, Kareliya (region)	Finland, Russia	63 15 N	30 48 E
Karelian Isthmus	Russia	60 25 N	30 00 E
Karimata Strait	Pacific Ocean	2 05 S	108 40 E
Kashmir (region)	India, Pakistan	34 00 N	76 00 E
Katanga (region)	Democratic Republic of the Congo	10 00 S	26 00 E
Kathmandu (capital)	Nepal	27 43 N	85 19 E
Kattegat (strait)	Atlantic Ocean	57 00 N	11 00 E
Kauai Channel	Pacific Ocean	21 45 N	158 50 W
Kazakstan (former name for Kazakhstan)	Kazakhstan	48 00 N	68 00 E
Keeling Islands	Cocos (Keeling) Islands	12 30 S	96 50 E
Kerguelen, Iles (island group)	French Southern and Antarctic Lands	49 30 S	69 30 E
Kermadec Islands	New Zealand	29 50 S	178 15 W
Kerulen River	China, Mongolia	48 48 N	117 00 E
Khabarovsk (city)	Russia	48 27 N	135 06 E
Khanka, Lake	China, Russia	45 00 N	132 24 E
Khartoum (capital)	Sudan	15 36 N	32 32 E

Name	Entry in *The World Factbook*	Latitude (deg min)	Longitude (deg min)
Khios (island)	Greece	38 22 N	26 04 E
Khmer Republic (former name for Cambodia)	Cambodia	13 00 N	105 00 E
Khuriya Muriya Islands (Kuria Muria Islands)	Oman	17 30 N	56 00 E
Khyber Pass	Afghanistan, Pakistan	34 05 N	71 10 E
Kibris (Turkish local name for Cyprus)	Cyprus	35 00 N	33 00 E
Kiel Canal (Nord-Ostsee Kanal)	Atlantic Ocean	53 53 N	9 08 E
Kiev (city; former name for Kyiv)	Ukraine	50 26 N	30 31 E
Kigali (capital)	Rwanda	1 57 S	30 04 E
Kingston (capital)	Jamaica	18 00 N	76 48 W
Kingston (capital)	Norfolk Island	29 03 S	167 58 E
Kingstown (capital)	Saint Vincent and the Grenadines	13 09 N	61 14 W
Kinshasa (capital)	Democratic Republic of the Congo	4 18 S	15 18 E
Kipros (Greek local name for Cyprus)	Cyprus	35 00 N	33 00 E
Kirghiziya, Kirgizia (former name for Kyrgyzstan)	Kyrgyzstan	41 00 N	75 00 E
Kirguizstan (local name for Kyrgyzstan)	Kyrgyzstan	41 00 N	75 00 E
Kiritimati (Christmas Island)	Kiribati	1 52 N	157 20 W
Kishinev (see Chisinau)	Moldova	47 00 N	28 50 E
Kithira Strait	Atlantic Ocean	36 00 N	23 00 E
Kobe (city)	Japan	34 41 N	135 10 E
Kodiak Island	United States	57 49 N	152 23 W
Kola Peninsula (Kol'skiy Poluostrov)	Russia	67 20 N	37 00 E
Kolonia (town; former capital; changed to Palikir)	Federated States of Micronesia	6 58 N	158 13 E
Korea Bay	Pacific Ocean	39 00 N	124 00 E
Korea Strait	Pacific Ocean	34 00 N	129 00 E
Korea, Democratic People's Republic of	North Korea	40 00 N	127 00 E
Korea, Republic of	South Korea	37 00 N	127 30 E
Koror (capital)	Palau	7 20 N	134 29 E
Kosovo (region)	Serbia	42 30 N	21 00 E
Kosrae (island)	Federated States of Micronesia	5 20 N	163 00 E
Kowloon (city)	Hong Kong	22 18 N	114 10 E
Kra, isthmus of	Burma, Thailand	10 20 N	99 00 E
Krakatoa (volcano)	Indonesia	6 07 S	105 24 E
Krakow (city)	Poland	50 03 N	19 56 E
Kuala Lumpur (capital)	Malaysia	3 10 N	101 42 E
Kunashiri (island; also Kunashir)	Russia (de facto)	44 20 N	146 00 E
Kunlun Mountains	China	36 00 N	84 00 E
Kuril Islands	Russia (de facto)	46 10 N	152 00 E
Kuwait (capital)	Kuwait	29 20 N	47 59 E
Kuznetsk Basin	Russia	54 00 N	86 00 E
Kwajalein Atoll	Marshall Islands	9 05 N	167 20 E
Kyiv (capital)	Ukraine	50 26 N	30 31 E
Kyushu (island)	Japan	33 00 N	131 00 E
La Paz (administrative capital)	Bolivia	16 30 S	68 09 W
La Perouse Strait	Pacific Ocean	45 45 N	142 00 E
Labrador (peninsula, region)	Canada	54 00 N	62 00 W
Labrador Sea	Atlantic Ocean	60 00 N	55 00 W
Laccadive Islands	India	10 00 N	73 00 E
Laccadive Sea	Indian Ocean	7 00 N	76 00 E
Lagos (former capital)	Nigeria	6 27 N	3 24 E
Lahore (city)	Pakistan	31 33 N	74 23 E
Lake Erie	Atlantic Ocean	42 30 N	81 00 W
Lake Huron	Atlantic Ocean	45 00 N	83 00 W
Lake Michigan	Atlantic Ocean	43 30 N	87 30 W
Lake Ontario	Atlantic Ocean	43 30 N	78 00 W
Lake Superior	Atlantic Ocean	48 00 N	88 00 W
Lakshadweep (Laccadive Islands)	India	10 00 N	73 00 E
Lantau Island	Hong Kong	22 15 N	113 55 E
Lao (local name for Laos)	Laos	18 00 N	105 00 E
Laptev Sea	Arctic Ocean	76 00 N	126 00 E
Las Palmas (city)	Spain (Canary Islands)	28 06 N	15 24 W
Latakia (region)	Syria	36 00 N	35 50 E
Latvija (local name for Latvia)	Latvia	57 00 N	25 00 E
Lau Group (island group)	Fiji	18 20 S	178 30 E
Lefkosa (see Nicosia)	Cyprus	35 10 N	33 22 E

Name	Entry in *The World Factbook*	Latitude (deg min)	Longitude (deg min)
Leipzig (city)	Germany	51 21 N	12 23 E
Lemnos (island)	Greece	39 54 N	25 21 E
Leningrad (city; former name for Saint Petersburg)	Russia	59 55 N	30 15 E
Lesser Sunda Islands	Indonesia	9 00 S	120 00 E
Lesvos (island)	Greece	39 15 N	26 15 E
Leyte (island)	Philippines	10 50 N	124 50 E
Liancourt Rocks (claimed by Japan)	South Korea	37 15 N	131 50 E
Liaodong Wan (gulf)	Pacific Ocean	40 30 N	121 20 E
Liban (local name for Lebanon)	Lebanon	33 50 N	36 50 E
Libreville (capital)	Gabon	0 23 N	9 27 E
Lietuva (local name for Lithuania)	Lithuania	56 00 N	24 00 E
Ligurian Sea	Atlantic Ocean	43 30 N	9 00 E
Lilongwe (capital)	Malawi	13 59 S	33 44 E
Lima (capital)	Peru	12 03 S	77 03 W
Lincoln Sea	Arctic Ocean	83 00 N	56 00 W
Line Islands	Jarvis Island, Kingman Reef, Kiribati, Palmyra Atoll	0 05 N	157 00 W
Lion, Gulf of	Atlantic Ocean	43 20 N	4 00 E
Lisbon (capital)	Portugal	38 43 N	9 08 W
Little Belt (strait; also Lille Baelt)	Atlantic Ocean	55 05 N	9 55 E
Ljubljana (capital)	Slovenia	46 03 N	14 31 E
Llanos (region)	Venezuela	8 00 N	68 00 W
Lobamba (city)	Swaziland	26 27 S	31 12 E
Lombok (island)	Indonesia	8 28 S	116 40 E
Lombok Strait	Indian Ocean	8 30 S	115 50 E
Lome (capital)	Togo	6 08 N	1 13 E
London (capital)	United Kingdom	51 30 N	0 10 W
Longyearbyen (capital)	Svalbard	78 13 N	15 33 E
Lord Howe Island	Australia	31 30 S	159 00 E
Lorraine (region)	France	48 42 N	6 11 E
Louisiade Archipelago	Papua New Guinea	11 00 S	153 00 E
Lourenco Marques (city; former name for Maputo)	Mozambique	25 56 S	32 34 E
Loyalty Islands (Iles Loyaute)	New Caledonia	21 00 S	167 00 E
Luanda (capital)	Angola	8 48 S	13 14 E
Lubnan (local name for Lebanon)	Lebanon	33 50 N	36 50 E
Lubumbashi (city)	Democratic Republic of the Congo	11 40 S	27 28 E
Lusaka (capital)	Zambia	15 25 S	28 17 E
Luxembourg (capital)	Luxembourg	49 45 N	6 10 E
Luzon (island)	Philippines	16 00 N	121 00 E
Luzon Strait	Pacific Ocean	20 30 N	121 00 E
Lyakhov Islands	Russia	73 45 N	138 00 E
Macao	Macau	22 10 N	113 33 E
Macau (special administrative region)	China	22 10 N	113 33 E
Macquarie Island	Australia	54 36 S	158 54 E
Madagasikara (local name for Madagascar)	Madagascar	20 00 S	47 00 E
Maddalena, Isola	Italy	41 13 N	09 24 E
Madeira Islands	Portugal	32 40 N	16 45 W
Madras (city; see Chennai)	India	13 04 N	80 16 E
Madrid (capital)	Spain	40 24 N	3 41 W
Magellan, Strait of	Atlantic Ocean	54 00 S	71 00 W
Maghreb (region)	Algeria, Libya, Mauritania, Morocco, Tunisia	34 00 N	3 00 E
Magreb (local name for Morocco)	Morocco	32 00 N	5 00 W
Magyarorszag (local name for Hungary)	Hungary	47 00 N	20 00 E
Mahe Island	Seychelles	4 41 S	55 30 E
Maiz, Islas del (Corn Islands)	Nicaragua	12 15 N	83 00 W
Majorca Island (Isla de Mallorca)	Spain	39 30 N	3 00 E
Majuro (capital)	Marshall Islands	7 05 N	171 08 E
Makassar Strait	Pacific Ocean	2 00 S	117 30 E
Makedonija (local name for Macedonia)	Macedonia	41 50 N	22 00 E
Malabo (capital)	Equatorial Guinea	3 45 N	8 47 E
Malacca, Strait of	Indian Ocean	2 30 N	101 20 E
Malagasy Republic	Madagascar	20 00 S	47 00 E
Malay Archipelago	Brunei, Indonesia, Malaysia, Papua New Guinea, Philippines	2 30 N	120 00 E

Name	Entry in *The World Factbook*	Latitude (deg min)	Longitude (deg min)
Malay Peninsula	Malaysia, Thailand	7 10 N	100 35 E
Male (capital)	Maldives	4 10 N	73 31 E
Mallorca, Isla de (island; also Majorca)	Spain	39 30 N	3 00 E
Malmady (region)	Belgium	50 26 N	6 02 E
Malpelo, Isla de (island)	Colombia	4 00 N	90 30 W
Malta Channel	Atlantic Ocean	56 44 N	26 53 E
Malvinas, Islas (island group)	Falkland Islands (Islas Malvinas)	51 45 S	59 00 W
Mamoutzou (capital)	Mayotte	12 47 S	45 14 E
Managua (capital)	Nicaragua	12 09 N	86 17 W
Manama (capital)	Bahrain	26 13 N	50 35 E
Manchukuo (former state)	China	44 00 N	124 00 E
Manchuria (region)	China	44 00 N	124 00 E
Manila (capital)	Philippines	14 35 N	121 00 E
Manipa Strait	Pacific Ocean	3 20 S	127 23 E
Mannar, Gulf of	Indian Ocean	8 30 N	79 00 E
Manua Islands	American Samoa	14 13 S	169 35 W
Maputo (capital)	Mozambique	25 58 S	32 35 E
Marcus Island (Minami-tori-shima)	Japan	24 16 N	154 00 E
Margarita, Isla (island)	Venezuela	10 00 N	64 00 W
Mariana Islands	Guam, Northern Mariana Islands	16 00 N	145 30 E
Marie Byrd Land (region)	Antarctica	77 00 S	130 00 W
Marigot (capital)	Saint Martin	18 04 N	63 05 W
Marion Island	South Africa	46 51 S	37 52 E
Marmara, Sea of	Atlantic Ocean	40 40 N	28 15 E
Marquesas Islands (Iles Marquises)	French Polynesia	9 00 S	139 30 W
Marseille (city)	France	43 18 N	5 23 E
Martin Vaz, Ilhas (island group)	Brazil	20 30 S	28 51 W
Mas a Tierra (Robinson Crusoe Island)	Chile	33 38 S	78 52 W
Mascarene Islands	Mauritius, Reunion	21 00 S	57 00 E
Maseru (capital)	Lesotho	29 28 S	27 30 E
Mata-Utu (capital)	Wallis and Futuna	13 57 S	171 56 W
Matsu (island)	Taiwan	26 13 N	119 56 E
Matthew Island	New Caledonia, Vanuatu	22 20 S	171 20 E
Mauritanie (local name for Mauritania)	Mauritania	20 00 N	12 00 W
Mazatlan (city)	Mexico	23 13 N	106 25 W
Mbabane (capital)	Swaziland	26 18 S	31 06 E
McDonald Islands	Heard Island and McDonald Islands	53 06 S	73 30 E
Mecca (city)	Saudi Arabia	21 27 N	39 49 E
Mediterranean Sea	Atlantic Ocean	36 00 N	15 00 E
Melbourne (city)	Australia	37 49 S	144 58 E
Melilla (exclave)	Spain	35 19 N	2 58 W
Memel (region)	Lithuania	55 43 N	21 30 E
Mesopotamia (region)	Iraq	33 00 N	44 00 E
Messina, Strait of	Atlantic Ocean	38 15 N	15 35 E
Mexico City (capital)	Mexico	19 24 N	99 09 W
Mexico, Gulf of	Atlantic Ocean	25 00 N	90 00 W
Middle Congo (former name for Republic of the Congo)	Republic of the Congo	1 00 S	15 00 E
Milan (city)	Italy	45 28 N	9 11 E
Milwaukee Deep (Puerto Rico Trench)	Atlantic Ocean	19 55 N	65 27 W
Minami-tori-shima (Marcus Island)	Japan	24 16 N	154 00 E
Mindanao (island)	Philippines	8 00 N	125 00 E
Mindanao Sea	Pacific Ocean	9 15 N	124 30 E
Mindoro (island)	Philippines	12 50 N	121 05 E
Mindoro Strait	Pacific Ocean	12 20 N	120 40 E
Mingrelia (region)	Georgia	42 30 N	41 52 E
Minicoy Island	India	8 17 N	73 02 E
Minorca Island (Isla de Menorca)	Spain	40 00 N	4 00 E
Minsk (capital)	Belarus	53 54 N	27 34 E
Misr (local name for Egypt)	Egypt	27 00 N	30 00 E
Mitla Pass	Egypt	30 02 N	32 54 E
Mocambique (local name for Mozambique)	Mozambique	18 15 S	35 00 E
Mogadishu (capital)	Somalia	2 04 N	45 22 E
Moldavia (region)	Moldova, Romania	47 00 N	29 00 E
Molucca Sea	Pacific Ocean	2 00 N	127 00 E
Moluccas (Spice Islands)	Indonesia	2 00 S	128 00 E

Name	Entry in *The World Factbook*	Latitude (deg min)	Longitude (deg min)
Mombasa (city)	Kenya	4 03 S	39 40 E
Mona Passage	Atlantic Ocean	18 30 N	67 45 W
Monaco (capital)	Monaco	43 44 N	7 25 E
Mongol Uls (local name for Mongolia)	Mongolia	46 00 N	105 00 E
Monrovia (capital)	Liberia	6 18 N	10 47 W
Monterrey (city)	Mexico	25 40 N	100 19 W
Montevideo (capital)	Uruguay	34 53 S	56 11 W
Montreal (city)	Canada	45 31 N	73 34 W
Moravia (region)	Czech Republic	49 30 N	17 00 E
Moravian Gate (pass)	Czech Republic	49 35 N	17 50 E
Moroni (capital)	Comoros	11 41 S	43 16 E
Mortlock Islands (Nomoi Islands)	Federated States of Micronesia	5 30 N	153 40 E
Moscow (capital)	Russia	55 45 N	37 35 E
Mount Pinatubo (volcano)	Philippines	15 08 N	120 21 E
Mozambique Channel	Indian Ocean	19 00 S	41 00 E
Mumbai (city; also Bombay)	India	18 58 N	72 50 E
Munich, Muenchen (city)	Germany	48 08 N	11 35 E
Muritaniyah (local name for Mauritania)	Mauritania	20 00 N	12 00 W
Musandam Peninsula	Oman, United Arab Emirates	26 18 N	56 24 E
Muscat (capital)	Oman	23 37 N	58 35 E
Muscat and Oman (former name for Oman)	Oman	21 00 N	57 00 E
Myanma, Myanmar	Burma	22 00 N	98 00 E
N'Djamena (capital)	Chad	12 07 N	15 03 E
Nagorno-Karabakh (region)	Azerbaijan	40 00 N	46 40 E
Nairobi (capital)	Kenya	1 17 S	36 49 E
Namib (desert)	Namibia	24 00 S	15 00 E
Nampo-shoto (island group)	Japan	30 00 N	140 00 E
Nan Madol (ruins)	Federated States of Micronesia	6 85 N	158 35 E
Naples (city)	Italy	40 51 N	14 15 E
Nassau (capital)	The Bahamas	25 05 N	77 21 W
Natal (region)	South Africa	29 00 S	30 25 E
Natuna Besar Islands	Indonesia	3 30 N	102 30 E
Natuna Sea	Pacific Ocean	3 30 N	108 00 E
Naxcivan (region)	Azerbaijan	39 20 N	45 20 E
Naxos (island)	Greece	37 05 N	25 30 E
Nederland (local name for the Netherlands)	Netherlands	52 30 N	5 45 E
Nederlandse Antillen (local name for the Netherlands Antilles)	Netherlands Antilles	12 15 N	68 45 W
Negev (region)	Israel	30 30 N	34 55 E
Negros (island)	Philippines	10 00 N	123 00 E
Nejd (region)	Saudi Arabia	24 05 N	45 15 E
Netherlands East Indies (former name for Indonesia)	Indonesia	5 00 S	120 00 E
Netherlands Guiana (former name for Suriname)	Suriname	4 00 N	56 00 W
Nevis (island)	Saint Kitts and Nevis	17 09 N	62 35 W
New Britain (island)	Papua New Guinea	6 00 S	150 00 E
New Delhi (capital)	India	28 36 N	77 12 E
New Guinea (island)	Indonesia, Papua New Guinea	5 00 S	140 00 E
New Hebrides (island group)	Vanuatu	16 00 S	167 00 E
New Ireland (island)	Papua New Guinea	3 20 N	152 00 E
New Siberian Islands	Russia	75 00 N	142 00 E
New Territories (mainland region)	Hong Kong	22 24 N	114 10 E
Newfoundland (island, with mainland area, and a province)	Canada	52 00 N	56 00 W
Niamey (capital)	Niger	13 31 N	2 07 E
Nicobar Islands	India	8 00 N	93 30 E
Nicosia (capital; also Lefkosia)	Cyprus	35 10 N	33 22 E
Nightingale Island	Saint Helena	37 25 S	12 30 W
Nihon, Nippon (local name for Japan)	Japan	36 00 N	138 00 E
Nomoi Islands (Mortlock Islands)	Federated States of Micronesia	5 30 N	153 40 E
Norge (local name for Norway)	Norway	62 00 N	10 00 E
Norman Isles (Channel Islands)	Guernsey, Jersey	49 20 N	2 20 W
North Atlantic Ocean	Atlantic Ocean	30 00 N	45 00 W
North Channel	Atlantic Ocean	55 10 N	5 40 W
North Frisian Islands	Denmark, Germany	54 50 N	8 12 E
North Greenland Sea	Arctic Ocean	78 00 N	5 00 W
North Island	New Zealand	39 00 S	176 00 E
North Ossetia (region)	Russia	43 00 N	44 10 E
North Pacific Ocean	Pacific Ocean	30 00 N	165 00 W

Name	Entry in The World Factbook	Latitude (deg min)	Longitude (deg min)
North Sea	Atlantic Ocean	56 00 N	4 00 E
North Vietnam (former name for northern portion of Vietnam)	Vietnam	23 00 N	106 00 E
North Yemen (Yemen Arab Republic; now part of Yemen)	Yemen	15 00 N	44 00 E
Northeast Providence Channel	Atlantic Ocean	25 40 N	77 09 W
Northern Areas	Pakistan	36 0 N	75 0 E
Northern Cyprus (region)	Cyprus	35 15 N	33 44 E
Northern Epirus (region)	Albania, Greece	40 00 N	20 30 E
Northern Grenadines (political region)	Saint Vincent and the Grenadines	12 45 N	61 15 W
Northern Ireland	United Kingdom	54 40 N	6 45 W
Northern Rhodesia (former name for Zambia)	Zambia	15 00 S	30 00 E
Northwest Passages	Arctic Ocean	74 40 N	100 00 W
Norwegian Sea	Atlantic Ocean	66 00 N	6 00 E
Nouakchott (capital)	Mauritania	18 06 N	15 57 W
Noumea (capital)	New Caledonia	22 16 S	166 27 E
Nouvelle-Caledonie (local name for New Caledonia)	New Caledonia	21 30 S	165 30 E
Nouvelles Hebrides (former name for Vanuatu)	Vanuatu	16 00 S	167 00 E
Novaya Zemlya (islands)	Russia	74 00 N	57 00 E
Nubia (region)	Egypt, Sudan	20 30 N	33 00 E
Nuku'alofa (capital)	Tonga	21 08 S	175 12 W
Nunavut (region)	Canada	72 00 N	90 00 W
Nuuk (capital; also Godthab)	Greenland	64 11 N	51 44 W
Nyasaland (former name for Malawi)	Malawi	13 30 S	34 00 E
Nyassa (region)	Mozambique	13 30 S	37 00 E
Oahu (island)	United States (Hawaii)	21 30 N	158 00 W
Ocean Island (Banaba)	Kiribati	0 52 S	169 35 E
Ocean Island (Kure Island)	United States	28 25 N	178 20 W
Oesterreich (local name for Austria)	Austria	47 20 N	13 20 E
Ogaden (region)	Ethiopia, Somalia	7 00 N	46 00 E
Oil Islands (Chagos Archipelago)	British Indian Ocean Territory	6 00 S	71 30 E
Okhotsk, Sea of	Pacific Ocean	53 00 N	150 00 E
Okinawa (island group)	Japan	26 30 N	128 00 E
Oland (island)	Sweden	56 45 N	16 40 E
Oman, Gulf of	Indian Ocean	24 30 N	58 30 E
Ombai Strait	Pacific Ocean	8 30 S	125 00 E
Oran (city)	Algeria	35 43 N	0 43 W
Orange River Colony (region; former name of Free State Province of South Africa)	South Africa	28 20 S	26 40 E
Oranjestad (capital)	Aruba	12 33 N	70 06 W
Oresund (The Sound) (strait)	Atlantic Ocean	55 50 N	12 40 E
Orkney Islands	United Kingdom	59 00 N	3 00 W
Osaka (city)	Japan	34 42 N	135 30 E
Oslo (capital)	Norway	59 55 N	10 45 E
Osumi Strait (Van Diemen Strait)	Pacific Ocean	31 00 N	131 00 E
Otranto, Strait of	Atlantic Ocean	40 00 N	19 00 E
Ottawa (capital)	Canada	45 25 N	75 40 W
Ouagadougou (capital)	Burkina Faso	12 22 N	1 31 W
Outer Hebrides (islands)	United Kingdom	57 45 N	7 00 W
Outer Mongolia (region)	Mongolia	46 00 N	105 00 E
P'yongyang (capital)	North Korea	39 01 N	125 45 E
Pacific Islands, Trust Territory of the (former name of a large area of the western North Pacific Ocean)	Marshall Islands, Federated States of Micronesia, Northern Mariana Islands, Palau	10 00 N	155 00 E
Pagan (island)	Northern Mariana Islands	18 08 N	145 47 E
Pago Pago (capital)	American Samoa	14 16 S	170 42 W
Palawan (island)	Philippines	9 30 N	118 30 E
Palermo (city)	Italy	38 07 N	13 21 E
Palestine (region)	Israel, West Bank	32 00 N	35 15 E
Palikir (capital)	Federated States of Micronesia	6 55 N	158 08 E
Palk Strait	Indian Ocean	10 00 N	79 45 E
Pamirs (mountains)	China, Tajikistan	38 00 N	73 00 E
Pampas (region)	Argentina	35 00 S	63 00 W
Panama (capital)	Panama	8 58 N	79 32 W
Panama Canal	Panama	9 00 N	79 45 W
Panama, Gulf of	Pacific Ocean	8 00 N	79 30 W
Panay (island)	Philippines	11 15 N	122 30 E

Name	Entry in *The World Factbook*	Latitude (deg min)	Longitude (deg min)
Pantelleria, Isola di (island)	Italy	36 47 N	12 00 E
Papeete (capital)	French Polynesia	17 32 S	149 34 W
Paramaribo (capital)	Suriname	5 50 N	55 10 W
Parece Vela (island)	Japan	20 20 N	136 00 E
Paris (capital)	France	48 52 N	2 20 E
Pascua, Isla de (Easter Island)	Chile	27 07 S	109 22 W
Pashtunistan (region)	Afghanistan, Pakistan	32 00 N	69 00 E
Passion, Ile de la (island)	Clipperton Island	10 17 N	109 13 W
Patagonia (region)	Argentina	48 00 S	61 00 W
Peking (see Beijing)	China	39 56 N	116 24 E
Pelagian Islands (Isole Pelagie)	Italy	35 40 N	12 40 E
Peleliu (Beliliou) (island)	Palau	7 01 N	134 15 E
Peloponnese (peninsula)	Greece	37 30 N	22 25 E
Pemba Island	Tanzania	5 20 S	39 45 E
Penang Island	Malaysia	5 23 N	100 15 E
Pentland Firth (channel)	Atlantic Ocean	58 44 N	3 13 W
Perim (island)	Yemen	12 39 N	43 25 E
Perouse Strait, La	Pacific Ocean	44 45 N	142 00 E
Persia (former name for Iran)	Iran	32 00 N	53 00 E
Persian Gulf	Indian Ocean	27 00 N	51 00 E
Perth (city)	Australia	31 56 S	115 50 E
Pescadores (islands)	Taiwan	23 30 N	119 30 E
Peshawar (city)	Pakistan	34 01 N	71 40 E
Peter I Island	Antarctica	68 48 S	90 35 W
Petrograd (city; former name for Saint Petersburg)	Russia	59 55 N	30 15 E
Philip Island	Norfolk Island	29 08 S	167 57 E
Philippine Sea	Pacific Ocean	20 00 N	134 00 E
Phnom Penh (capital)	Cambodia	11 33 N	104 55 E
Phoenix Islands	Kiribati	3 30 S	172 00 W
Pinatubo, Mount (volcano)	Philippines	15 08 N	120 21 E
Pines, Isle of (island; former name for Isla de la Juventud)	Cuba	21 40 N	82 50 W
Pleasant Island	Nauru	0 32 S	166 55 E
Plymouth (capital)	Montserrat	16 44 N	62 14 W
Podgorica (administrative capital)	Montenegro	42 26 N	19 16 E
Polska (local name)	Poland	52 00 N	20 00 E
Polynesie Francaise (local name for French Polynesia)	French Polynesia	15 00 S	140 00 W
Pomerania (region)	Germany, Poland	53 40 N	15 35 E
Ponape (Pohnpei) (island)	Federated States of Micronesia	6 55 N	158 15 E
Port Louis (capital)	Mauritius	20 10 S	57 30 E
Port Moresby (capital)	Papua New Guinea	9 30 S	147 10 E
Port-Vila (capital)	Vanuatu	17 44 S	168 19 E
Port-au-Prince (capital)	Haiti	18 32 N	72 20 W
Port-of-Spain (capital)	Trinidad and Tobago	10 39 N	61 31 W
Porto-Novo (capital)	Benin	6 29 N	2 37 E
Portuguese East Africa (former name for Mozambique)	Mozambique	18 15 S	35 00 E
Portuguese Guinea (former name for Guinea-Bissau)	Guinea-Bissau	12 00 N	15 00 W
Portuguese Timor (former name for Timor-Leste)	Timor-Leste	9 00 S	126 00 E
Poznan (city)	Poland	52 25 N	16 55 E
Prague (capital)	Czech Republic	50 05 N	14 28 E
Praia (capital)	Cape Verde	14 55 N	23 31 W
Prathet Thai (local name for Thailand)	Thailand	15 00 N	100 00 E
Pretoria (administrative capital)	South Africa	25 42 S	28 13 E
Prevlaka peninsula	Croatia	42 24 N	18 31 E
Pribilof Islands	United States	57 00 N	170 00 W
Prince Edward Island	Canada	46 20 N	63 20 W
Prince Edward Islands	South Africa	46 35 S	38 00 E
Prince Patrick Island	Canada	76 30 N	119 00 W
Principe (island)	Sao Tome and Principe	1 38 N	7 25 E
Prussia (region)	Germany, Poland, Russia	53 00 N	14 00 E
Pukapuka Atoll	Cook Islands	10 53 S	165 49 W
Punjab (region)	India, Pakistan	30 50 N	73 30 E
Puntland (region)	Somalia	8 21 N	49 08 E
Qazaqstan (local name for Kazakhstan)	Kazakhstan	48 00 N	68 00 E
Qita Ghazzah (local name Gaza Strip)	Gaza Strip	31 25 N	34 20 E
Quebec (city)	Canada	46 48 N	71 15 W
Queen Charlotte Islands	Canada	53 00 N	132 00 W

Name	Entry in *The World Factbook*	Latitude (deg min)	Longitude (deg min)
Queen Elizabeth Islands	Canada	78 00 N	95 00 W
Queen Maud Land (claimed by Norway)	Antarctica	73 30 S	12 00 E
Quemoy (island)	Taiwan	24 27 N	118 23 E
Quito (capital)	Ecuador	0 13 S	78 30 W
Rabat (capital)	Morocco	34 02 N	6 51 W
Ralik Chain (island group)	Marshall Islands	8 00 N	167 00 E
Rangoon (capital; also Yangon)	Burma	16 47 N	96 10 E
Rapa Nui (Easter Island)	Chile	27 07 S	109 22 W
Ratak Chain (island group)	Marshall Islands	9 00 N	171 00 E
Red Sea	Indian Ocean	20 00 N	38 00 E
Redonda (island)	Antigua and Barbuda	16 55 N	62 19 W
Republica Dominicana (local name for Dominican Republic)	Dominican Republic	19 00 N	70 40 W
Republique Centrafricain (local name for Central African Republic)	Central African Republic	7 00 N	21 00 E
Republique Francaise (local name for France)	France	46 00 N	2 00 E
Republique Gabonaise (local name for Gabon)	Gabon	1 00 S	11 45 E
Republique Rwandaise (local name for Rwanda)	Rwanda	2 00 S	30 00 E
Republique Togolaise (local name for Togo)	Togo	8 00 N	1 10 E
Revillagigedo Island	United States (Alaska)	55 35 N	131 06 W
Revillagigedo Islands	Mexico	19 00 N	112 45 W
Reykjavik (capital)	Iceland	64 09 N	21 57 W
Rhodes (island)	Greece	36 10 N	28 00 E
Rhodesia, Northern (former name for Zambia)	Zambia	15 00 S	30 00 E
Rhodesia, Southern (former name for Zimbabwe)	Zimbabwe	20 00 S	30 00 E
Riga (capital)	Latvia	56 57 N	24 06 E
Riga, Gulf of	Atlantic Ocean	57 30 N	23 30 E
Rio Muni (mainland region)	Equatorial Guinea	1 30 N	10 00 E
Rio de Janiero (city)	Brazil	22 55 S	43 17 W
Rio de Oro (region)	Western Sahara	23 45 N	15 45 W
Rio de la Plata (gulf)	Atlantic Ocean	35 00 S	59 00 W
Riyadh (capital)	Saudi Arabia	24 38 N	46 43 E
Road Town (capital)	British Virgin Islands	18 27 N	64 37 W
Robinson Crusoe Island (Mas a Tierra)	Chile	33 38 S	78 52 W
Rocas, Atol das (island)	Brazil	3 51 S	33 49 W
Rockall (island)	United Kingdom	57 35 N	13 48 W
Rodrigues (island)	Mauritius	19 42 S	63 25 E
Rome (capital)	Italy	41 54 N	12 29 E
Roncador Cay (island)	Colombia	13 32 N	80 03 W
Roosevelt Island	Antarctica	79 30 S	162 00 W
Roseau (capital)	Dominica	15 18 N	61 24 W
Ross Dependency (claimed by New Zealand)	Antarctica	80 00 S	180 00 E
Ross Island	Antarctica	81 30 S	175 00 W
Ross Sea	Antarctica, Southern Ocean	76 00 S	175 00 W
Rossiya (local name for Russia)	Russia	60 00 N	100 00 E
Rota (island)	Northern Mariana Islands	14 10 N	145 12 E
Rotuma (island)	Fiji	12 30 S	177 05 E
Ruanda (former name for Rwanda)	Rwanda	2 00 S	30 00 E
Rub al Khali (desert)	Saudi Arabia	19 30 N	49 00 E
Rumelia (region)	Albania, Bulgaria, Macedonia	42 00 N	22 30 E
Ruthenia (region; former name for Carpatho-Ukraine)	Ukraine	48 22 N	23 32 E
Ryukyu Islands	Japan	26 30 N	128 00 E
Saar (region)	Germany	49 25 N	7 00 E
Saaremaa (island)	Estonia	58 25 N	22 30 E
Saba (island)	Netherlands Antilles	17 38 N	63 10 W
Sabah (state)	Malaysia	5 20 N	117 10 E
Sable Island	Canada	43 55 N	59 50 W
Safety Islands (Iles du Salut)	French Guiana	5 20 N	52 37 W
Sahara Occidental (former name for Western Sahara)	Western Sahara	24 30 N	13 00 W
Sahel (region)	Burkina Faso, Chad, The Gambia, Guinea- Bissau, Mali, Mauritania, Niger, Senegal	15 00 N	8 00 W
Saigon (city; former name for Ho Chi Minh City)	Vietnam	10 45 N	106 40 E
Saint Barthelemy (island; also Saint Bart's)	Guadeloupe	17 55 N	62 52 W
Saint Brandon (Cargados Carajos Shoals)	Mauritius	16 25 S	59 38 E
Saint Christopher (island)	Saint Kitts and Nevis	17 20 N	62 45 W
Saint Christopher and Nevis	Saint Kitts and Nevis	17 20 N	62 45 W

Name	Entry in *The World Factbook*	Latitude (deg min)	Longitude (deg min)
Saint Eustatius (island)	Netherlands Antilles	17 30 N	63 00 W
Saint George's (capital)	Grenada	12 03 N	61 45 W
Saint George's Channel	Atlantic Ocean	52 00 N	6 00 W
Saint Helena Island	Saint Helena	15 57 S	5 42 W
Saint Helens, Mount (volcano)	United States	46 15 N	122 12 W
Saint Helier (capital)	Jersey	49 12 N	2 07 W
Saint John (city)	Canada (New Brunswick)	45 16 N	66 04 W
Saint John's (capital)	Antigua and Barbuda	17 06 N	61 51 W
Saint Lawrence Island	United States	49 30 N	67 00 W
Saint Lawrence Seaway	Atlantic Ocean	49 15 N	67 00 W
Saint Lawrence, Gulf of	Atlantic Ocean	48 00 N	62 00 W
Saint Paul Island	Canada	47 12 N	60 09 W
Saint Paul Island	United States	57 11 N	170 16 W
Saint Paul Island (Ile Saint-Paul)	French Southern and Antarctic Lands	38 43 S	77 29 E
Saint Peter Port (capital)	Guernsey	49 27 N	2 32 W
Saint Peter and Saint Paul Rocks (Penedos de Sao Pedro e Sao Paulo)	Brazil	0 23 N	29 23 W
Saint Petersburg (city; former capital)	Russia	59 55 N	30 15 E
Saint Thomas (island)	Virgin Islands	18 21 N	64 55 W
Saint Vincent Passage	Atlantic Ocean	13 30 N	61 00 W
Saint-Denis (capital)	Reunion	20 52 S	55 28 E
Saint-Martin (island; also Sint Maarten)	Guadeloupe	18 04 N	63 04 W
Saint-Pierre (capital)	Saint Pierre and Miquelon	46 46 N	56 11 W
Saipan (island)	Northern Mariana Islands	15 12 N	145 45 E
Sak'art'velo (local name for Georgia)	Georgia	42 00 N	43 30 E
Sakhalin Island (Ostrov Sakhalin)	Russia	51 00 N	143 00 E
Sakishima Islands	Japan	24 30 N	124 00 E
Sala y Gomez, Isla (island)	Chile	26 28 S	105 00 W
Salisbury (city; former name for Harare)	Zimbabwe	17 50 S	105 00 W
Salzburg (city)	Austria	47 48 N	13 02 E
Samar (island)	Philippines	12 00 N	125 00 E
Samaria (region)	West Bank	32 15 N	35 10 E
Samoa Islands	American Samoa, Samoa	14 00 S	171 00 W
Samos (island)	Greece	37 48 N	26 44 E
San Ambrosio, Isla (island)	Chile	26 21 S	79 52 W
San Andres y Providencia, Archipielago (island group)	Colombia	13 00 N	81 30 W
San Bernardino Strait	Pacific Ocean	12 32 N	124 10 E
San Felix, Isla (island)	Chile	26 17 S	80 05 W
San Jose (capital)	Costa Rica	9 56 N	84 05 W
San Juan (capital)	Puerto Rico	18 28 N	66 07 W
San Marino (capital)	San Marino	43 56 N	12 25 E
San Salvador (capital)	El Salvador	13 42 N	89 12 W
Sanaa (capital)	Yemen	15 21 N	44 12 E
Sandzak (region)	Montenegro, Serbia	43 05 N	19 45 E
Santa Cruz (city)	Bolivia	17 48 S	63 10 W
Santa Cruz Islands	Solomon Islands	11 00 S	166 15 E
Santa Sede (local name for the Holy See)	Holy See	41 54 N	12 27 E
Santiago (capital)	Chile	33 27 S	70 40 W
Santo Antao (island)	Cape Verde	17 05 N	25 10 W
Santo Domingo (capital)	Dominican Republic	18 28 N	69 54 W
Sao Paulo (city)	Brazil	23 35 S	46 43 W
Sao Pedro e Sao Paulo, Penedos de (rocks)	Brazil	0 23 N	29 23 W
Sao Tiago (island)	Cape Verde	15 05 N	23 40 W
Sao Tome (island)	Sao Tome and Principe	0 12 N	6 39 E
Sapporo (city)	Japan	43 04 N	141 20 E
Sapudi Strait	Pacific Ocean	7 05 S	114 10 E
Sarajevo (capital)	Bosnia and Herzegovina	43 52 N	18 25 E
Sarawak (state)	Malaysia	2 30 N	113 30 E
Sardinia (island)	Italy	40 00 N	9 00 E
Sargasso Sea (region)	Atlantic Ocean	30 00 N	55 00 W
Sark (island)	Guernsey	49 26 N	2 21 W
Savage Island (former name for Niue)	Niue	19 02 S	169 52 W
Savu Sea	Pacific Ocean	9 30 S	122 00 E
Saxony (region)	Germany	51 00 N	13 00 E
Schleswig-Holstein (region)	Germany	54 31 N	9 33 E

Name	Entry in *The World Factbook*	Latitude (deg min)	Longitude (deg min)
Schweiz (local German name for Switzerland)	Switzerland	47 00 N	8 00 E
Scopus, Mount	Israel, West Bank	31 48 N	35 14 E
Scotia Sea	Atlantic Ocean, Southern Ocean	56 00 S	40 00 W
Scotland (region)	United Kingdom	57 00 N	4 00 W
Scott Island	Antarctica	67 24 S	179 55 W
Senegambia (region; former name of confederation of Senegal and The Gambia)	The Gambia, Senegal	13 50 N	15 25 W
Senyavin Islands	Federated States of Micronesia	6 55 N	158 00 E
Seoul (capital)	South Korea	37 34 N	127 00 E
Serendib (former name for Sri Lanka)	Sri Lanka	7 00 N	81 00 E
Serrana Bank (shoal)	Colombia	14 25 N	80 16 W
Serranilla Bank (shoal)	Colombia	15 51 N	79 46 W
Settlement, The (capital)	Christmas Island	10 25 S	105 43 E
Severnaya Zemlya (island group; also Northland)	Russia	79 30 N	98 00 E
Shaba (region)	Democratic Republic of the Congo	8 00 S	27 00 E
Shag Island	Heard Island and McDonald Islands	53 00 S	72 30 E
Shag Rocks	South Georgia and the South Sandwich Islands	53 33 S	42 02 W
Shanghai (city)	China	31 14 N	121 30 E
Shenyang (city; also Mukden)	China	41 46 N	123 24 E
Shetland Islands	United Kingdom	60 30 N	1 30 W
Shikoku (island)	Japan	33 45 N	133 30 E
Shikotan (island)	Russia (de facto)	43 47 N	146 45 E
Shqiperia (local name for Albania)	Albania	41 00 N	20 00 E
Siam (former name for Thailand)	Thailand	15 00 N	100 00 E
Siberia (region)	Russia	60 00 N	100 00 E
Sibutu Passage	Pacific Ocean	4 50 N	119 35 E
Sicily (island)	Italy	37 30 N	14 00 E
Sicily, Strait of	Atlantic Ocean	37 20 N	11 20 E
Sidra, Gulf of	Atlantic Ocean	31 30 N	18 00 E
Sikkim (state)	India	27 50 N	88 30 E
Silesia (region)	Czech Republic, Germany, Poland	51 00 N	17 00 E
Sinai Peninsula	Egypt	29 30 N	34 00 E
Singapore (capital)	Singapore	1 17 N	103 51 E
Singapore Strait	Pacific Ocean	1 15 N	104 00 E
Sinkiang (autonomous region; also Xinjiang)	China	42 00 N	86 00 E
Sint Eustatius (island)	Netherlands Antilles	17 29 N	62 58 W
Sint Maarten (island; also Saint-Martin)	Netherlands Antilles	18 04 N	63 04 W
Sjaelland (island)	Denmark	55 30 N	12 00 E
Skagerrak (strait)	Atlantic Ocean	57 45 N	9 00 E
Skopje (capital)	Macedonia	41 59 N	21 26 E
Slavonia (region)	Croatia	45 27 N	18 00 E
Slovenija (local name for Slovenia)	Slovenia	46 00 N	15 00 E
Slovensko (local name for Slovakia)	Slovakia	48 40 N	19 30 E
Smyrna (region; former name for Izmir)	Turkey	38 25 N	27 10 E
Society Islands (Iles de la Societe)	French Polynesia	17 00 S	150 00 W
Socotra (island)	Yemen	12 30 N	54 00 E
Sofia (capital)	Bulgaria	42 41 N	23 19 E
Solomon Islands, northern	Papua New Guinea	6 00 S	155 00 E
Solomon Islands, southern	Solomon Islands	8 00 S	159 00 E
Solomon Sea	Pacific Ocean	8 00 S	153 00 E
Somaliland (region)	Somalia	9 30 N	46 00 E
Somers Islands (former name for Bermuda)	Bermuda	32 20 N	64 45 W
Songkhla (city)	Thailand	7 12 N	100 36 E
Sound, The (strait; also Oresund)	Atlantic Ocean	55 50 N	12 40 E
South Atlantic Ocean	Atlantic Ocean	30 00 S	15 00 W
South China Sea	Pacific Ocean	10 00 N	113 00 E
South Georgia (island)	South Georgia and the South Sandwich Islands	54 15 S	36 45 W
South Island	New Zealand	43 00 S	171 00 E
South Korea	South Korea	37 00 N	127 30 E
South Orkney Islands	Antarctica	61 00 S	45 00 W
South Ossetia (region)	Georgia	42 20 N	44 00 E

Name	Entry in *The World Factbook*	Latitude (deg min)	Longitude (deg min)
South Pacific Ocean	Pacific Ocean	30 00 S	130 00 W
South Sandwich Islands	South Georgia and the South Sandwich Islands	57 45 S	26 30 W
South Shetland Islands	Antarctica	62 00 S	59 00 W
South Tyrol (region)	Italy	46 30 N	10 30 E
South Vietnam (former name for the southern portion of Vietnam)	Vietnam	12 00 N	108 00 E
South Yemen (People's Democratic Republic of Yemen; now part of Yemen)	Yemen	14 00 N	48 00 E
South-West Africa (former name for Namibia)	Namibia	22 00 S	17 00 E
Southern Grenadines (island group)	Grenada	12 20 N	61 30 W
Southern Rhodesia (former name for Zimbabwe)	Zimbabwe	20 00 S	30 00 E
Soviet Union (former name of a large Eurasian empire, roughly coequal with the former Russian Empire)	Armenia, Azerbaijan, Belarus, Estonia, Georgia, Kazakhstan, Kyrgyzstan, Latvia, Lithuania, Moldova, Russia, Tajikistan, Turkmenistan, Ukraine, Uzbekistan		
Spanish Guinea (former name for Equatorial Guinea)	Equatorial Guinea	2 00 N	10 00 E
Spanish Morocco (former name for northern Morocco)	Morocco	32 00 N	7 00 W
Spanish North Africa (exclaves)	Spain (Ceuta, Islas Chafarinas, Melilla, Penon de Alhucemas, Penon de Velez de la Gomera)	35 15 N	4 00 W
Spanish Sahara (former name)	Western Sahara	24 30 N	13 00 W
Spanish West Africa (former name for Ifni and Spanish Sahara)	Morocco, Western Sahara	25 00 N	13 00 W
Spice Islands (Moluccas)	Indonesia	2 00 S	28 00 E
Spitsbergen (island)	Svalbard	78 00 N	20 00 E
Srbija (local name for Serbia)	Serbia	44 00 N	21 00 E
St. John's (city)	Canada (Newfoundland)	47 34 N	52 43 W
Stanley (capital)	Falkland Islands (Islas Malvinas)	51 42 S	57 41 W
Stockholm (capital)	Sweden	59 20 N	18 03 E
Strasbourg (city)	France	48 35 N	7 44 E
Stuttgart (city)	Germany	48 46 N	9 11 E
Sucre (constitutional capital)	Bolivia	19 02 S	65 17 W
Suez Canal	Egypt	29 55 N	32 33 E
Suez, Gulf of	Indian Ocean	28 10 N	33 27 E
Suisse (local French name for Switzerland)	Switzerland	47 00 N	8 00 E
Sulawesi (island; Celebes)	Indonesia	2 00 S	121 00 E
Sulawesi Sea	Pacific Ocean	3 00 N	122 00 E
Sulu Archipelago (island group)	Philippines	6 00 N	121 00 E
Sulu Sea	Pacific Ocean	8 00 N	120 00 E
Sumatra (island)	Indonesia	0 00 N	102 00 E
Sumba (island)	Indonesia	10 00 S	120 00 E
Sumba Strait	Pacific Ocean	9 10 S	120 00 E
Sumbawa (island)	Indonesia	8 30 S	118 00 E
Sunda Islands (Soenda Isles)	Indonesia, Malaysia	2 00 S	110 00 E
Sunda Strait	Indian Ocean	6 00 S	105 45 E
Suomi (local name for Finland)	Finland	64 00 N	26 00 E
Surabaya (city)	Indonesia	7 13 S	112 45 E
Surigao Strait	Pacific Ocean	10 15 N	125 23 E
Surinam (former name for Suriname)	Suriname	4 00 N	56 00 W
Suriyah (local name for Syria)	Syria	35 00 N	38 00 E
Surtsey (volcanic island)	Iceland	63 17 N	20 40 W
Suva (capital)	Fiji	18 08 S	178 25 E
Sverdlovsk (city; also Yekaterinburg)	Russia	56 50 N	60 39 E
Sverige (local name for Sweden)	Sweden	62 00 N	15 00 E
Svizzera (local Italian name for Switzerland)	Switzerland	47 00 N	8 00 E
Swains Island	American Samoa	11 03 S	171 15 W
Swan Islands	Honduras	17 25 S	83 56 W
Sydney (city)	Australia	33 53 S	151 13 E
T'bilisi (capital)	Georgia	41 43 N	44 49 E
Tadzhikistan (former name for Tajikistan)	Tajikistan	39 00 N	71 00 E
Tahiti (island)	French Polynesia	17 37 S	149 27 W
Taipei (capital)	Taiwan	25 03 N	121 30 E
Taiwan Strait	Pacific Ocean	24 00 N	119 00 E
Tallinn (capital)	Estonia	59 25 N	24 45 E

Name	Entry in The World Factbook	Latitude (deg min)	Longitude (deg min)
Tanganyika (former name for the mainland portion of Tanzania)	Tanzania	6 00 S	35 00 E
Tangier (city)	Morocco	35 48 N	5 45 W
Tannu-Tuva (region)	Russia	51 25 N	94 45 E
Tarawa (island)	Kiribati	1 25 N	173 00 E
Tartary, Gulf of	Pacific Ocean	50 00 N	141 00 E
Tashkent (capital)	Uzbekistan	41 20 N	69 18 E
Tasman Sea	Pacific Ocean	4 30 S	168 00 E
Tasmania (island)	Australia	43 00 S	147 00 E
Tatar Strait	Pacific Ocean	50 00 N	141 00 E
Taymyr Peninsula (Poluostrov Taymyr)	Russia	76 00 N	104 00 E
Tchad (local name for Chad)	Chad	15 00 N	19 00 E
Tegucigalpa (capital)	Honduras	14 06 N	87 13 W
Tehran (capital)	Iran	35 40 N	51 26 E
Tel Aviv (capital, de facto)	Israel	32 05 N	34 48 E
Teluk Bone (gulf)	Pacific Ocean	4 00 S	120 45 E
Teluk Tomini (gulf)	Pacific Ocean	0 30 S	121 00 E
Terre Adelie (claimed by France; also Adelie Land)	Antarctica	66 30 S	139 00 E
Terres Australes et Antarctiques Francaises (local name for the French Southern and Antarctic Lands)	French Southern and Antarctic Lands	43 00 S	67 00 E
Thailand, Gulf of	Pacific Ocean	10 00 N	101 00 E
The Former Yugoslav Republic of Macedonia	Macedonia	41 50 N	22 00 E
Thessaloniki (city; also Salonika)	Greece	40 38 N	22 57 E
Thimphu (capital)	Bhutan	27 28 N	89 39 E
Thuringia (region)	Germany	51 00 N	11 00 E
Thurston Island	Antarctica	72 20 S	99 00 W
Tiberias, Lake	Israel	32 48 N	35 35 E
Tibet (autonomous region; also Xizang)	China	32 00 N	90 00 E
Tibilisi (see T'bilisi)	Georgia	41 43 N	44 49 E
Tien Shan (mountains)	China, Kyrgyzstan	42 00 N	80 00 E
Tierra del Fuego (island, island group)	Argentina, Chile	54 00 S	69 00 W
Timor (island)	Timor-Leste, Indonesia	9 00 S	125 00 E
Timor Lorosa'e (local name for Timor-Leste)	Timor-Leste	9 00 N	126 00 E
Timor Sea	Pacific Ocean	11 00 S	128 00 E
Tinian (island)	Northern Mariana Islands	15 00 N	145 38 E
Tiran, Strait of	Indian Ocean	28 00 N	34 27 E
Tirana, Tirane (capital)	Albania	41 20 N	19 50 E
Tirol, Tyrol (region)	Austria, Italy	47 00 N	11 00 E
Tobago (island)	Trinidad and Tobago	11 15 N	60 40 W
Tokyo (capital)	Japan	35 42 N	139 46 E
Tonkin, Gulf of	Pacific Ocean	20 00 N	108 00 E
Toronto (city)	Canada	43 40 N	79 23 W
Torres Strait	Pacific Ocean	10 25 S	142 10 E
Torshavn (capital)	Faroe Islands	62 01 N	6 46 W
Toshkent (see Tashkent)	Uzbekistan	41 20 N	69 18 E
Transcarpathia (region; alternate name for Carpatho-Ukraine)	Ukraine	48 22 N	23 32 E
Transjordan (former name for Jordan)	Jordan	31 00 N	36 00 E
Transkei (enclave)	South Africa	32 15 S	28 15 E
Transvaal (region; former name for northeastern South Africa)	South Africa	25 10 S	29 25 E
Transylvania (region)	Romania	46 30 N	24 00 E
Trindade, Ilha de (island)	Brazil	20 31 S	29 20 W
Trinidad (island)	Trinidad and Tobago	10 22 N	61 15 W
Tripoli (capital)	Libya	32 54 N	13 11 E
Tripoli (city)	Lebanon	34 26 N	35 51 E
Tripolitania (region)	Libya	31 00 N	14 00 E
Tristan da Cunha Group (island group)	Saint Helena	37 15 S	12 30 W
Trobriand Islands	Papua New Guinea	8 38 S	151 04 E
Trucial Coast (former name for the United Arab Emirates)	United Arab Emirates	24 00 N	54 00 E
Trucial Oman (former name for the United Arab Emirates)	United Arab Emirates	24 00 N	54 00 E
Trucial States (former name for the United Arab Emirates)	United Arab Emirates	24 00 N	54 00 E
Truk Islands (former name for the Chuuk Islands)	Federated States of Micronesia	7 25 N	151 47 E
Tsugaru Strait	Pacific Ocean	41 35 N	141 00 E
Tuamotu Islands (Iles Tuamotu)	French Polynesia	19 00 S	142 00 W
Tubuai Islands (Iles Tubuai)	French Polynesia	23 00 S	150 00 W
Tunb al Kubra (island)	Iran	26 14 N	55 19 E
Tunb as Sughra (island)	Iran	26 14 N	55 09 E
Tunis (capital)	Tunisia	36 48 N	10 11 E

Name	Entry in The World Factbook	Latitude (deg min)	Longitude (deg min)
Turin (city)	Italy	45 04 N	7 40 E
Turkish Straits (see Bosporus and Dardanelles)	Atlantic Ocean	40 40 N	28 00 E
Turkiye (local name for Turkey)	Turkey	39 00 N	35 00 E
Turkmenia, Turkmeniya (former name for Turkmenistan)	Turkmenistan	40 00 N	60 00 E
Turks Island Passage	Atlantic Ocean	21 40 N	71 00 W
Tuscany (region)	Italy	43 25 N	11 00 E
Tutuila (island)	American Samoa	14 18 S	170 42 W
Tyrrhenian Sea	Atlantic Ocean	40 00 N	12 00 E
Ubangi-Shari (former name for the Central African Republic	Central African Republic	6 38 N	20 33 E
Ukrayina (local name for Ukraine)	Ukraine	49 00 N	32 00 E
Ulaanbaatar (capital)	Mongolia	47 55 N	106 53 E
Ullung-do (island)	South Korea	37 29 N	130 52 E
Ulster (region)	Ireland, United Kingdom	54 35 N	7 00 W
Uman (local name for Oman)	Oman	21 00 N	57 00 E
Unimak Pass (strait)	Pacific Ocean	54 20 N	164 50 W
Union of Soviet Socialist Republics or USSR (former name of a large Eurasian empire, roughly coequal with the former Russian Empire)	Armenia, Azerbaijan, Belarus, Estonia, Georgia, Kazakhstan, Kyrgyzstan, Latvia, Lithuania, Moldova, Russia, Tajikistan, Turkmenistan, Ukraine, Uzbekistan		
United Arab Republic or UAR (former name for a federation between Egypt and Syria)	Egypt, Syria		
Upper Volta (former name for Burkina Faso)	Burkina Faso	13 00 N	2 00 W
Ural Mountains	Kazakhstan, Russia	60 00 N	60 00 E
Urdunn (local name for Jordan)	Jordan	31 00 N	36 00 E
Urundi (former name for Burundi)	Burundi	3 30 S	30 00 E
Ussuri River	China, Russia	48 28 N	135 02 E
Vaduz (capital)	Liechtenstein	47 09 N	9 31 E
Vakhan (Wakhan Corridor)	Afghanistan	37 00 N	73 00 E
Valletta (capital)	Malta	35 54 N	14 31 E
Valley, The (capital)	Anguilla	18 13 N	63 04 W
Van Diemen Strait (Osumi Strait)	Pacific Ocean	31 00 N	131 00 E
Vancouver (city)	Canada	49 16 N	123 08 W
Vancouver Island	Canada	49 45 N	126 00 W
Vatican City (capital)	Holy See	41 54 N	12 27 E
Velez de la Gomera, Penon de (island)	Spain	35 11 N	4 18 W
Venda (enclave)	South Africa	23 00 S	31 00 E
Verde Island Passage	Pacific Ocean	13 34 N	120 51 E
Victoria (capital)	Seychelles	4 38 S	55 27 E
Victoria (island)	Canada	71 00 N	110 00 W
Victoria Land (region)	Antarctica	72 00 S	155 00 E
Vienna (capital)	Austria	48 12 N	16 22 E
Vientiane (capital)	Laos	17 58 N	102 36 E
Vilnius (capital)	Lithuania	54 41 N	25 19 E
Viti Levu (island)	Fiji	18 00 S	178 00 E
Vladivostok (city)	Russia	43 10 N	131 56 E
Vojvodina (region)	Serbia	45 35 N	20 00 E
Volcano Islands	Japan	25 00 N	141 00 E
Vostok Island	Kiribati	10 06 S	152 23 W
Wake Atoll	Wake Island	19 17 N	166 39 E
Wakhan Corridor (see Vakhan)	Afghanistan	37 00 N	73 00 E
Walachia (region)	Romania	44 45 N	26 05 E
Wales (region)	United Kingdom	52 30 N	3 30 W
Wallis Islands	Wallis and Futuna	13 17 S	176 10 W
Walvis Bay (city; former exclave)	Namibia	22 59 S	14 31 E
Warsaw (capital)	Poland	52 15 N	21 00 E
Washington, DC (capital)	United States	38 53 N	77 02 W
Weddell Sea	Southern Ocean	72 00 S	45 00 W
Wellington (capital)	New Zealand	41 28 S	174 51 E
West Frisian Islands	Netherlands	53 26 N	5 30 E
West Germany (Federal Republic of Germany; former name for western portion of Germany)	Germany	53 22 N	5 20 E
West Island (capital)	Cocos (Keeling) Islands	12 10 S	96 55 E
West Korea Strait (Western Channel)	Pacific Ocean	34 40 N	129 00 E
West Pakistan (former name for present-day Pakistan)	Pakistan	30 00 N	70 00 E

Name	Entry in The World Factbook	Latitude (deg min)	Longitude (deg min)
West Siberian Plain	Russia	60 00 N	75 00 E
Western Channel (West Korea Strait)	Pacific Ocean	34 40 N	129 00 E
Western Samoa (former name for Samoa)	Samoa	13 35 S	172 20 W
Wetar Strait	Pacific Ocean	8 20 S	126 30 E
White Sea	Arctic Ocean	65 30 N	38 00 E
Wilkes Land (region)	Antarctica	71 00 S	120 00 E
Willemstad (capital)	Netherlands Antilles	12 06 N	68 56 W
Windhoek (capital)	Namibia	22 34 S	17 06 E
Windward Passage	Atlantic Ocean	20 00 N	73 50 W
Winnipeg (city)	Canada	49 53 N	97 10 W
Wrangel Island (Ostrov Vrangelya)	Russia	71 14 N	179 36 W
Xianggang (local name for Hong Kong)	Hong Kong	22 15 N	114 10 E
Y'israel (local name for Israel)	Israel	31 30 N	34 45 E
Yaitopya (local name for Ethiopia)	Ethiopia	8 00 N	38 00 E
Yalu River	China, North Korea	39 55 N	124 20 E
Yamoussoukro (capital)	Cote d'Ivoire	6 49 N	5 17 W
Yangon (see Rangoon)	Burma	16 47 N	96 10 E
Yaounde (capital)	Cameroon	3 52 N	11 31 E
Yap Islands	Federated States of Micronesia	9 30 N	138 00 E
Yaren (governmental center)	Nauru	0 32 S	166 55 E
Yekaterinburg (city; formerly Sverdlovsk)	Russia	56 50 N	60 39 E
Yellow Sea	Pacific Ocean	36 00 N	123 00 E
Yemen Arab Republic (also Yemen (Sanaa); former name for northern portion of Yemen)	Yemen	15 00 N	44 00 E
Yemen, People's Democratic Republic of (also Yemen (Aden); former name for southern portion of Yemen)	Yemen	14 00 N	46 00 E
Yerevan (capital)	Armenia	40 11 N	44 30 E
Yokohama (city)	Japan	35 26 N	139 37 E
Youth, Isle of (Isla de la Juventud)	Cuba	21 40 N	82 50 W
Yucatan Channel	Atlantic Ocean	21 45 N	85 45 W
Yucatan Peninsula	Mexico	19 30 N	89 00 W
Yugoslavia (former name for a federation of Serbia and Montenegro)	Montenegro, Serbia	43 00 N	21 00 E

Yugoslavia, Kingdom of (former name for a Balkan federation)	Bosnia and Herzegovina,
	  Croatia, Macedonia,
	  Montenegro, Serbia, Slovenia
		43 00 N	19 00 E

Yugoslavia, Socialist Federal Republic of
  (former name for a Balkan federation)
	Bosnia and Herzegovina,
	  Croatia, Macedonia,
	  Montenegro, Serbia, Slovenia
		43 00 N	19 00 E

Zagreb (capital)	Croatia	45 48 N	15 58 E

Zaire (former name for the Democratic
  Republic of the Congo)
	Democratic Republic of
	  the Congo 15 00 S
		30 00 E	

Zakhalinskiy Zaliv (bay)	Pacific Ocean	54 00 N	142 00 E
Zaliv Shelikhova (bay)	Pacific Ocean	60 00 N	157 30 E
Zambezia (region)	Mozambique	16 00 S	37 00 E
Zanzibar (island)	Tanzania	6 10 S	39 11 E
Zhong Guo, Zhonghua (local name for China)	China	35 00 N	105 00 E
Zion, Mount (locale in Jerusalem)	Israel, West Bank	31 46 N	35 14 E
Zurich (city)	Switzerland	47 23 N	8 32 E

APPENDIX G

WEIGHTS AND MEASURES

Note: At this time, only three countries—Burma, Liberia, and the US—have not adopted the International System of Units (SI, or metric system) as their official system of weights and measures. Although use of the metric system has been sanctioned by law in the US since 1866, it has been slow in displacing the American adaptation of the British Imperial System known as the US Customary System. The US is the only industrialized nation that does not mainly use the metric system in its commercial and standards activities, but there is increasing acceptance in science, medicine, government, and many sectors of industry.

Mathematical Notation

Mathematical Power	Name
10^{18} or 1,000,000,000,000,000,000	one quintillion
10^{15} or 1,000,000,000,000,000	one quadrillion
10^{12} or 1,000,000,000,000	one trillion
10^{9} or 1,000,000,000	one billion
10^{6} or 1,000,000	one million
10^{3} or 1,000	one thousand
10^{2} or 100	one hundred
10^{1} or 10	ten
10^{0} or 1	one
10^{-1} or 0.1	one-tenth
10^{-2} or 0.01	one-hundredth
10^{-3} or 0.001	one-thousandth
10^{-6} or 0.000 001	one-millionth
10^{-9} or 0.000 000 001	one-billionth
10^{-12} or 0.000 000 000 001	one-trillionth
10^{-15} or 0.000 000 000 000 001	one-quadrillionth
10^{-18} or 0.000 000 000 000 000 001	one-quintillionth

Metric Interrelationships

Prefix	Symbol	Length, weight, or capacity
yotta	Y	10^{24}
zetta	Z	10^{21}
exa	E	10^{18}
peta	P	10^{15}
tera	T	10^{12}
giga	G	10^{9}
mega	M	10^{6}
kilo	k	10^{3}
hecto	h	10^{2}
deka	da	10^{1}
basic unit	-	1 meter, 1 gram, 1 liter
deci	d	10^{-1}
centi	c	10^{-2}
milli	m	10^{-3}
micro	u	10^{-6}
nano	n	10^{-9}
pico	p	10^{-12}
femto	f	10^{-15}
atto	a	10^{-18}
zepto	z	10^{-21}
yocto	y	10^{-24}

Conversion Factors

To Convert From	To	Multiply By
acres	ares	40.468 564 224
acres	hectares	0.404 685 642 24
acres	square feet	43,560
acres	square kilometers	0.004 046 856 422 4
acres	square meters	4,046.856 422 4
acres	square miles (statute)	0.001 562 50
acres	square yards	4,840
ares	square meters	100
ares	square yards	119.599
barrels, US beer	gallons	31
barrels, US beer	liters	117.347 77
barrels, US petroleum	gallons (British)	34.97

Conversion Factors

To Convert From	To	Multiply By
barrels, US petroleum	gallons (US)	42
barrels, US petroleum	liters	158.987 29
barrels, US proof spirits	gallons	40
barrels, US proof spirits	liters	151.416 47
bushels (US)	bushels (British)	0.968 9
bushels (US)	cubic feet	1.244 456
bushels (US)	cubic inches	2,150.42
bushels (US)	cubic meters	0.035 239 07
bushels (US)	cubic yards	0.046 090 96
bushels (US)	dekaliters	3.523 907
bushels (US)	dry pints	64
bushels (US)	dry quarts	32
bushels (US)	liters	35.239 070 17
bushels (US)	pecks	4
cables	fathoms	120
cables	meters	219.456
cables	yards	240
carat	milligrams	200
centimeters	feet	0.032 808 40
centimeters	inches	0.393 700 8
centimeters	meters	0.01
centimeters	yards	0.010 936 13
centimeters, cubic	cubic inches	0.061 023 744
centimeters, square	square feet	0.001 076 39
centimeters, square	square inches	0.155 000 31
centimeters, square	square meters	0.000 1
centimeters, square	square yards	0.000 119 599
chains, square surveyor's	ares	4.046 86
chains, square surveyor's	square feet	4,356
chains, surveyor's	feet	66
chains, surveyor's	meters	20.116 8
chains, surveyor's	rods	4
cords of wood	cubic feet	128
cords of wood	cubic meters	3.624 556
cords of wood	cubic yards	4.740 7
cups	liquid ounces (US)	8
cups	liters	0.236 588 2
degrees Celsius	degrees Fahrenheit	multiply by 1.8 and add 32
degrees Fahrenheit	degrees Celsius	subtract 32 and divide by 1.8
dekaliters	bushels	0.283 775 9
dekaliters	cubic feet	0.353 146 7
dekaliters	cubic inches	610.237 4
dekaliters	dry pints	18.161 66
dekaliters	dry quarts	9.080 829 8
dekaliters	liters	10
dekaliters	pecks	1.135 104
drams, avoirdupois	avoirdupois ounces	0.062 55
drams, avoirdupois	grains	27.344
drams, avoirdupois	grams	1.771 845 2
drams, troy	grains	60
drams, troy	grams	3.887 934 6
drams, troy	scruples	3
drams, troy	troy ounces	0.125
drams, liquid (US)	cubic inches	0.226
drams, liquid (US)	liquid drams (British)	1.041
drams, liquid (US)	liquid ounces	0.125
drams, liquid (US)	milliliters	3.696 69
drams, liquid (US)	minims	60
fathoms	feet	6
fathoms	meters	1.828 8
feet	centimeters	30.48
feet	inches	12
feet	kilometers	0.000 304 8
feet	meters	0.304 8
feet	statute miles	0.000 189 39
feet	yards	0.333 333 3

Conversion Factors

To Convert From	To	Multiply By
feet, cubic	bushels	0.803 563 95
feet, cubic	cubic decimeters	28.316 847
feet, cubic	cubic inches	1,728
feet, cubic	cubic meters	0.028 316 846 592
feet, cubic	cubic yards	0.037 037 04
feet, cubic	dry pints	51.428 09
feet, cubic	dry quarts	25.714 05
feet, cubic	gallons	7.480 519
feet, cubic	gills	239.376 6
feet, cubic	liquid ounces	957.506 5
feet, cubic	liquid pints	59.844 16
feet, cubic	liquid quarts	29.922 08
feet, cubic	liters	28.316 846 592
feet, cubic	pecks	3.214 256
feet, square	acres	0.000 022 956 8
feet, square	square centimeters	929.030 4
feet, square	square decimeters	9.290 304
feet, square	square inches	144
feet, square	square meters	0.092 903 04
feet, square	square yards	0.111 111 1
furlongs	feet	660
furlongs	inches	7,920
furlongs	meters	201.168
furlongs	statute miles	0.125
furlongs	yards	220
gallons, liquid (US)	cubic feet	0.133 680 6
gallons, liquid (US)	cubic inches	231
gallons, liquid (US)	cubic meters	0.003 785 411 784
gallons, liquid (US)	cubic yards	0.004 951 13
gallons, liquid (US)	gills (US)	32
gallons, liquid (US)	liquid gallons (British)	0.832 67
gallons, liquid (US)	liquid ounces	128
gallons, liquid (US)	liquid pints	8
gallons, liquid (US)	liquid quarts	4
gallons, liquid (US)	liters	3.785 411 784
gallons, liquid (US)	milliliters	3,785.411 784
gallons, liquid (US)	minims	61,440
gills (US)	centiliters	11.829 4
gills (US)	cubic feet	0.004 177 517
gills (US)	cubic inches	7.218 75
gills (US)	gallons	0.031 25
gills (US)	gills (British)	0.832 67
gills (US)	liquid ounces	4
gills (US)	liquid pints	0.25
gills (US)	liquid quarts	0.125
gills (US)	liters	0.118 294 118 25
gills (US)	milliliters	118.294 118 25
gills (US)	minims	1,920
grains	avoirdupois drams	0.036 571 43
grains	avoirdupois ounces	0.002 285 71
grains	avoirdupois pounds	0.000 142 86
grains	grams	0.064 798 91
grains	kilograms	0.000 064 798 91
grains	milligrams	64.798 910
grains	pennyweights	0.042
grains	scruples	0.05
grains	troy drams	0.016 6
grains	troy ounces	0.002 083 33
grains	troy pounds	0.000 173 61
grams	avoirdupois drams	0.564 383 39
grams	avoirdupois ounces	0.035 273 961
grams	avoirdupois pounds	0.002 204 622 6
grams	grains	15.432 361
grams	kilograms	0.001
grams	milligrams	1,000
grams	troy ounces	0.032 150 746 6

Conversion Factors

To Convert From	To	Multiply By
grams	troy pounds	0.002 679 23
hands (height of horse)	centimeters	10.16
hands (height of horse)	inches	4
hectares	acres	2.471 053 8
hectares	square feet	107,639.1
hectares	square kilometers	0.01
hectares	square meters	10,000
hectares	square miles	0.003 861 02
hectares	square yards	11,959.90
hundredweights, long	avoirdupois pounds	112
hundredweights, long	kilograms	50.802 345
hundredweights, long	long tons	0.05
hundredweights, long	metric tons	0.050 802 345
hundredweights, long	short tons	0.056
hundredweights, short	avoirdupois pounds	100
hundredweights, short	kilograms	45.359 237
hundredweights, short	long tons	0.044 642 86
hundredweights, short	metric tons	0.045 359 237
hundredweights, short	short tons	0.05
inches	centimeters	2.54
inches	feet	0.083 333 33
inches	meters	0.025 4
inches	millimeters	25.4
inches	yards	0.027 777 78
inches, cubic	bushels	0.000 465 025
inches, cubic	cubic centimeters	16.387 064
inches, cubic	cubic feet	0.000 578 703 7
inches, cubic	cubic meters	0.000 016 387 064
inches, cubic	cubic yards	0.000 021 433 47
inches, cubic	dry pints	0.029 761 6
inches, cubic	dry quarts	0.014 880 8
inches, cubic	gallons	0.004 329 0
inches, cubic	gills	0.138 528 1
inches, cubic	liquid ounces	0.554 112 6
inches, cubic	liquid pints	0.034 632 03
inches, cubic	liquid quarts	0.017 316 02
inches, cubic	liters	0.016 387 064
inches, cubic	milliliters	16.387 064
inches, cubic	minims (US)	265.974 0
inches, cubic	pecks	0.001 860 10
inches, square	square centimeters	6.451 600
inches, square	square feet	0.006 944 44
inches, square	square meters	0.000 645 16
inches, square	square yards	0.000 771 605
kilograms	avoirdupois drams	564.383 4
kilograms	avoirdupois ounces	35.273 962
kilograms	avoirdupois pounds	2.204 622 622
kilograms	grains	15,432.36
kilograms	grams	1,000
kilograms	long tons	0.000 984 2
kilograms	metric tons	0.001
kilograms	short hundredweights	0.022 046 23
kilograms	short tons	0.001 102 31
kilograms	troy ounces	32.150 75
kilograms	troy pounds	2.679 229
kilometers	meters	1,000
kilometers	statute miles	0.621 371 192
kilometers, square	acres	247.105 38
kilometers, square	hectares	100
kilometers, square	square meters	1,000,000
kilometers, square	statute miles	0.386 102 16
knots (nautical mi/hr)	kilometers/hour	1.852
knots (nautical mi/hr)	statute miles/hour	1.151
leagues, nautical	kilometers	5.556
leagues, nautical	nautical miles	3
leagues, statute	kilometers	4.828 032

Conversion Factors

To Convert From	To	Multiply By
leagues, statute	statute miles	3
links, square surveyor's	square centimeters	404.686
links, square surveyor's	square inches	62.726 4
links, surveyor's	centimeters	20.116 8
links, surveyor's	chains	0.01
links, surveyor's	inches	7.92
liters	bushels	0.028 377 59
liters	cubic feet	0.035 314 67
liters	cubic inches	61.023 74
liters	cubic meters	0.001
liters	cubic yards	0.001 307 95
liters	dekaliters	0.1
liters	dry pints	1.816 166
liters	dry quarts	0.908 082 98
liters	gallons	0.264 172 052
liters	gills (US)	8.453 506
liters	liquid ounces	33.814 02
liters	liquid pints	2.113 376
liters	liquid quarts	1.056 688 2
liters	milliliters	1,000
liters	pecks	0.113 510 4
meters	centimeters	100
meters	feet	3.280 839 895
meters	inches	39.370 079
meters	kilometers	0.001
meters	millimeters	1,000
meters	statute miles	0.000 621 371
meters	yards	1.093 613 298
meters, cubic	bushels	28.377 59
meters, cubic	cubic feet	35.314 666 7
meters, cubic	cubic inches	61,023.744
meters, cubic	cubic yards	1.307 950 619
meters, cubic	gallons	264.172 05
meters, cubic	liters	1,000
meters, cubic	pecks	113.510 4
meters, square	acres	0.000 247 105 38
meters, square	hectares	0.000 1
meters, square	square centimeters	10,000
meters, square	square feet	10.763 910 4
meters, square	square inches	1,550.003 1
meters, square	square yards	1.195 990 046
microns	meters	0.000 001
microns	inches	0.000 039 4
mils	inches	0.001
mils	millimeters	0.025 4
miles, nautical	kilometers	1.852 0
miles, nautical	statute miles	1.150 779 4
miles, statute centimeters 160,934.4		
miles, statute	feet	5,280
miles, statute	furlongs	8
miles, statute	inches	63,360
miles, statute	kilometers	1.609 344
miles, statute	meters	1,609.344
miles, statute	rods	320
miles, statute	yards	1,760
miles, square nautical	square kilometers	3.429 904
miles, square nautical	square statute miles	1.325
miles, square statute	acres	640
miles, square statute	hectares	258.998 811 033 6
miles, square statute	sections	1
miles, square statute	square kilometers	2.589 988 110 336
miles, square statute	square nautical miles	0.755 miles
miles, square statute	square rods	102,400
milligrams	grains	0.015 432 358 35
milliliters	cubic inches	0.061 023 744
milliliters	gallons	0.000 264 17

Conversion Factors

To Convert From	To	Multiply By
milliliters	gills (US)	0.008 453 5
milliliters	liquid ounces	0.033 814 02
milliliters	liquid pints	0.002 113 4
milliliters	liquid quarts	0.001 056 7
milliliters	liters	0.001
milliliters	minims	16.230 73
millimeters inches 0.039 370 078 7		
minims (US)	cubic inches	0.003 759 77
minims (US)	gills (US)	0.000 520 83
minims (US)	liquid ounces	0.002 083 33
minims (US)	milliliters	0.061 611 52
minims (US)	minims (British)	1.041
ounces, avoirdupois	avoirdupois drams	16
ounces, avoirdupois	avoirdupois pounds	0.062 5
ounces, avoirdupois	grains	437.5
ounces, avoirdupois	grams	28.349 523 125
ounces, avoirdupois	kilograms	0.028 349 523 125
ounces, avoirdupois	troy ounces	0.911 458 3
ounces, avoirdupois	troy pounds	0.075 954 86
ounces, liquid (US)	cubic feet	0.001 044 38
ounces, liquid (US)	centiliters	2.957 35
ounces, liquid (US)	cubic inches	1.804 687 5
ounces, liquid (US)	gallons	0.007 812 5
ounces, liquid (US)	gills (US)	0.25
ounces, liquid (US)	liquid drams	8
ounces, liquid (US)	liquid ounces (British)	1.041
ounces, liquid (US)	liquid pints	0.062 5
ounces, liquid (US)	liquid quarts	0.031 25
ounces, liquid (US)	liters	0.029 573 53
ounces, liquid (US)	milliliters	29.573 529 6
ounces, liquid (US)	minims	480
ounces, troy	avoirdupois drams	17.554 29
ounces, troy	avoirdupois ounces	1.097 143
ounces, troy	avoirdupois pounds	0.068 571 43
ounces, troy	grains	480
ounces, troy	grams	31.103 476 8
ounces, troy	pennyweights	20
ounces, troy	troy drams	8
ounces, troy	troy pounds	0.083 333 3
paces (US)	centimeters	76.2
paces (US)	inches	30
pecks (US)	bushels	0.25
pecks (US)	cubic feet	0.311 114
pecks (US)	cubic inches	537.605
pecks (US)	cubic meters	0.008 809 77
pecks (US)	cubic yards	0.011 522 74
pecks (US)	dekaliters	0.880 976 75
pecks (US)	dry pints	16
pecks (US)	dry quarts	8
pecks (US)	liters	8.809 767 5
pecks (US)	pecks (British)	0.968 9
pennyweights	grains	24
pennyweights	grams	1.555 173 84
pennyweights	troy ounces	0.05
pints, dry (US)	bushels	0.015 625
pints, dry (US)	cubic feet	0.019 444 63
pints, dry (US)	cubic inches	33.600 312 5
pints, dry (US)	dekaliters	0.055 061 05
pints, dry (US)	dry pints (British)	0.968 9
pints, dry (US)	dry quarts	0.5
pints, dry (US)	liters	0.550 610 47
pints, liquid (US)	cubic feet	0.016 710 07
pints, liquid (US)	cubic inches	28.875
pints, liquid (US)	deciliters	4.731 76
pints, liquid (US)	gallons	0.125
pints, liquid (US)	gills (US)	4

Conversion Factors

To Convert From	To	Multiply By
pints, liquid (US)	liquid ounces	16
pints, liquid (US)	liquid pints (British)	0.832 67
pints, liquid (US)	liquid quarts	0.5
pints, liquid (US)	liters	0.473 176 473
pints, liquid (US)	milliliters	473.176 473
pints, liquid (US)	minims	7,680
points (typographical)	inches	0.013 837
points (typographical)	millimeters	0.351 459 8
pounds, avoirdupois	avoirdupois drams	256
pounds, avoirdupois	avoirdupois ounces	16
pounds, avoirdupois	grains	7,000
pounds, avoirdupois	grams	453.592 37
pounds, avoirdupois	kilograms	0.453 592 37
pounds, avoirdupois	long tons	0.000 446 428 6
pounds, avoirdupois	metric tons	0.000 453 592 37
pounds, avoirdupois	quintals	0.004 535 92
pounds, avoirdupois	short tons	0.000 5
pounds, avoirdupois	troy ounces	14.583 33
pounds, avoirdupois	troy pounds	1.215 278
pounds, troy	avoirdupois drams	210.651 4
pounds, troy	avoirdupois ounces	13.165 71
pounds, troy	avoirdupois pounds	0.822 857 1
pounds, troy	grains	5,760
pounds, troy	grams	373.241 721 6
pounds, troy	kilograms	0.373 241 721 6
pounds, troy	pennyweights	240
pounds, troy	troy ounces	12
quarts, dry (US)	bushels	0.031 25
quarts, dry (US)	cubic feet	0.038 889 25
quarts, dry (US)	cubic inches	67.200 625
quarts, dry (US)	dekaliters	0.110 122 1
quarts, dry (US)	dry pints	2
quarts, dry (US)	dry quarts (British)	0.968 9
quarts, dry (US)	liters	1.101 221
quarts, dry (US)	pecks	0.125
quarts, dry (US)	pints, dry (US)	2
quarts, liquid (US)	cubic feet	0.033 420 14
quarts, liquid (US)	cubic inches	57.75
quarts, liquid (US)	deciliters	9.463 53
quarts, liquid (US)	gallons	0.25
quarts, liquid (US)	gills (US)	8
quarts, liquid (US)	liquid ounces	32
quarts, liquid (US)	liquid pints (US)	2
quarts, liquid (US)	liquid quarts (British)	0.832 67
quarts, liquid (US)	liters	0.946 352 946
quarts, liquid (US)	milliliters	946.352 946
quarts, liquid (US)	minims	15,360
quintals	avoirdupois pounds	220.462 26
quintals	kilograms	100
quintals	metric tons	0.1
rods	feet	16.5
rods	meters	5.029 2
rods	yards	5.5
rods, square	acres	0.006 25
rods, square	square meters	25.292 85
rods, square	square yards	30.25
scruples	grains	20
scruples	grams	1.295 978 2
scruples	troy drams	0.333
sections (US)	square kilometers	2.589 988 1
sections (US)	square statute miles	1
spans	centimeters	22.86
spans	inches	9
steres	cubic meters	1
steres	cubic yards	1.307 95
tablespoons	milliliters	14.786 76

815

Conversion Factors

To Convert From	To	Multiply By
tablespoons	teaspoons	3
teaspoons	milliliters	4.928 922
teaspoons	tablespoons	0.333 333
ton-miles, long	metric ton-kilometers	1.635 169
ton-miles, short	metric ton-kilometers	1.459 972
tons, gross register	cubic feet of permanently enclosed space	100
tons, gross register	cubic meters of permanently enclosed space	2.831 684 7
tons, long (deadweight)	avoirdupois ounces	35,840
tons, long (deadweight)	avoirdupois pounds	2,240
tons, long (deadweight)	kilograms	1,016.046 909 8
tons, long (deadweight)	long hundredweights	20
tons, long (deadweight)	metric tons	1.016 046 908 8
tons, long (deadweight)	short hundredweights	22.4
tons, long (deadweight)	short tons	1.12
tons, metric	avoirdupois pounds	2,204.623
tons, metric	kilograms	1,000
tons, metric	long hundredweights	19.684 130 3
tons, metric	long tons	0.984 206 5
tons, metric	quintals	10
tons, metric	short hundredweights	22.046 23
tons, metric	short tons	1.102 311 3
tons, metric	troy ounces	32,150.75
tons, net register	cubic feet of permanently enclosed space for cargo and passengers	100
tons, net register	cubic meters of permanently enclosed space for cargo and passengers	2.831 684 7
tons, shipping	cubic feet of permanently enclosed cargo space	42
tons, shipping	cubic meters of permanently enclosed cargo space	1.189 307 574
tons, short	avoirdupois pounds	2,000
tons, short	kilograms	907.184 74
tons, short	long hundredweights	17.857 14
tons, short	long tons	0.892 857 1
tons, short	metric tons	0.907 184 74
tons, short	short hundredweights	20
townships (US)	sections	36
townships (US)	square kilometers	93.239 572
townships (US)	square statute miles	36
miles, square statute	acres	640
miles, square statute	hectares	258.998 811 033 6
miles, square statute	square feet	27,878,400
miles, square statute	square meters	2,589,988.110 336
miles, square statute	square yards	3,097,600
yards	centimeters	91.44
yards	feet	3
yards	inches	36
yards	meters	0.914 4
yards	miles	0.000 568 18
yards, cubic	bushels	21.696 227
yards, cubic	cubic feet	27
yards, cubic	cubic inches	46,656
yards, cubic	cubic meters	0.764 554 857 984
yards, cubic	gallons	201.974 0
yards, cubic	liters	764.554 857 984
yards, cubic	pecks	86.784 91
yards, square	acres	0.000 206 611 6
yards, square	hectares	0.000 083 612 736
yards, square	square centimeters	8,361.273 6
yards, square	square feet	9
yards, square	square inches	1,296
yards, square	square meters	0.836 127 36
yards, square	square miles	0.000 000 322 830 6

REFERENCE MAPS

AFRICA

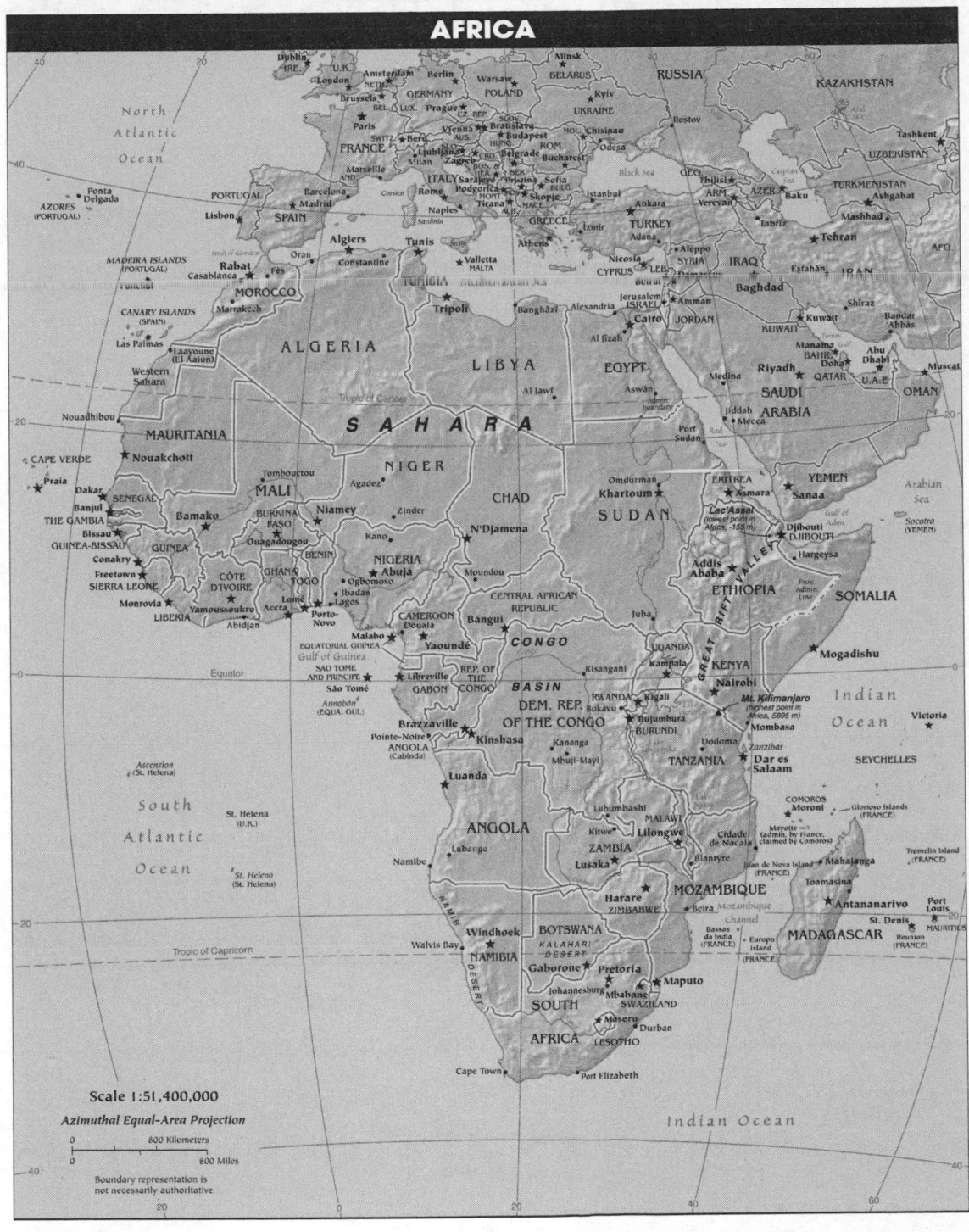

Scale 1:51,400,000

Azimuthal Equal-Area Projection

0   800 Kilometers

0   800 Miles

Boundary representation is
not necessarily authoritative.

ANTARCTIC REGION

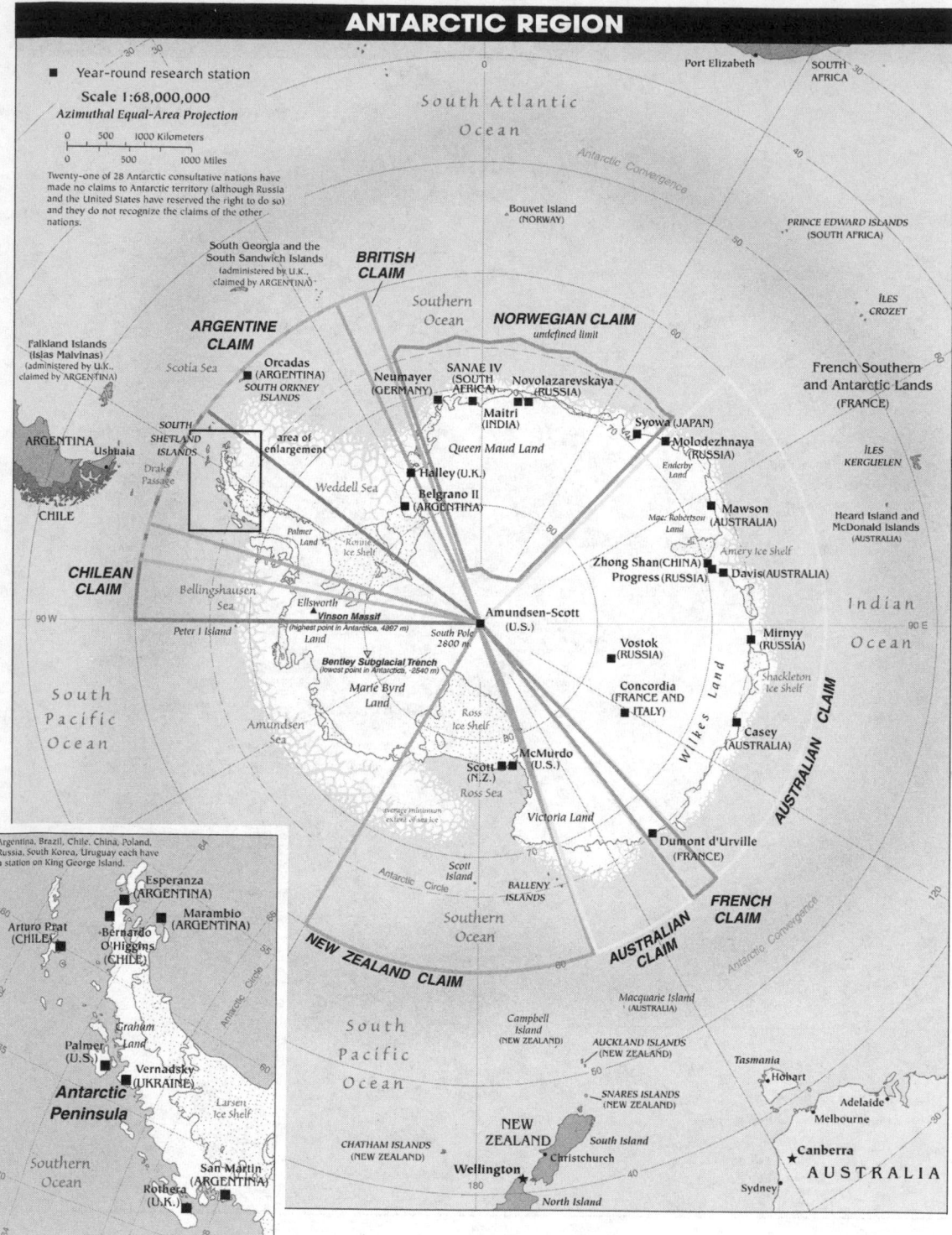

■ Year-round research station

Scale 1:68,000,000
Azimuthal Equal-Area Projection

0   500   1000 Kilometers

0   500   1000 Miles

Twenty-one of 28 Antarctic consultative nations have
made no claims to Antarctic territory (although Russia
and the United States have reserved the right to do so)
and they do not recognize the claims of the other
nations.

South Atlantic
Ocean

Port Elizabeth

SOUTH
AFRICA

Bouvet Island
(NORWAY)

PRINCE EDWARD ISLANDS
(SOUTH AFRICA)

ÎLES
CROZET

South Georgia and the
South Sandwich Islands
(administered by U.K.,
claimed by ARGENTINA)

**BRITISH
CLAIM**

Southern
Ocean

NORWEGIAN CLAIM
undefined limit

French Southern
and Antarctic Lands
(FRANCE)

**ARGENTINE
CLAIM**

Falkland Islands
(Islas Malvinas)
(administered by U.K.,
claimed by ARGENTINA)

Scotia Sea

Orcadas
(ARGENTINA)
SOUTH ORKNEY
ISLANDS

Neumayer
(GERMANY)

SANAE IV
(SOUTH
AFRICA)

Novolazarevskaya
(RUSSIA)

Syowa (JAPAN)

ÎLES
KERGUELEN

ARGENTINA

Ushuaia

SOUTH
SHETLAND
ISLANDS

Drake
Passage

area of
enlargement

Maitri
(INDIA)

Molodezhnaya
(RUSSIA)

Queen Maud Land

Enderby
Land

Heard Island and
McDonald Islands
(AUSTRALIA)

CHILE

Weddell Sea

Halley (U.K.)

Belgrano II
(ARGENTINA)

Mac. Robertson
Land

Mawson
(AUSTRALIA)

Palmer
Land

Ronne
Ice Shelf

Amery Ice Shelf

**CHILEAN
CLAIM**

Bellingshausen
Sea

Ellsworth
Land

Vinson Massif
(highest point in Antarctica, 4897 m)

Zhong Shan(CHINA)
Progress (RUSSIA)

Davis(AUSTRALIA)

Indian
Ocean

90 W

Peter I Island

Bentley Subglacial Trench
(lowest point in Antarctica, -2540 m)

South Pole
2800 m

Amundsen-Scott
(U.S.)

Vostok
(RUSSIA)

Mirnyy
(RUSSIA)

Shackleton
Ice Shelf

90 E

South
Pacific
Ocean

Amundsen
Sea

Marie Byrd
Land

Ross
Ice Shelf

Concordia
(FRANCE AND
ITALY)

Casey
(AUSTRALIA)

AUSTRALIAN CLAIM

McMurdo
(U.S.)

Scott
(N.Z.)

Ross Sea

Victoria Land

average minimum
extent of sea ice

Scott
Island

Dumont d'Urville
(FRANCE)

Antarctic Circle

BALLENY
ISLANDS

NEW ZEALAND CLAIM

Southern
Ocean

**FRENCH
CLAIM**

**AUSTRALIAN
CLAIM**

Macquarie Island
(AUSTRALIA)

Campbell
Island
(NEW ZEALAND)

AUCKLAND ISLANDS
(NEW ZEALAND)

Tasmania

Hobart

Argentina, Brazil, Chile, China, Poland,
Russia, South Korea, Uruguay each have
a station on King George Island.

Esperanza
(ARGENTINA)

Marambio
(ARGENTINA)

Arturo Prat
(CHILE)

Bernardo
O'Higgins
(CHILE)

Palmer
(U.S.)

Graham
Land

Vernadsky
(UKRAINE)

**Antarctic
Peninsula**

Larsen
Ice Shelf

Antarctic Circle

SNARES ISLANDS
(NEW ZEALAND)

Adelaide

Melbourne

Canberra

AUSTRALIA

South
Pacific
Ocean

CHATHAM ISLANDS
(NEW ZEALAND)

**NEW
ZEALAND**

South Island

Christchurch

Sydney

Southern
Ocean

San Martin
(ARGENTINA)

Rothera
(U.K.)

Wellington

North Island

ARCTIC REGION

North Pacific Ocean

Bering Sea

ALEUTIAN ISLANDS

KURIL ISLANDS
occupied by the Soviet Union in 1945,
administered by Russia, claimed by Japan.

JAPAN

Petropavlovsk-
Kamchatskiy

Sea of
Okhotsk

Sakhalin

Khabarovsk

Kodiak

Bethel

Gulf of
Alaska

Anchorage
Valdez

Juneau

Whitehorse

Dawson

Fairbanks

UNITED STATES

Nome

Bering
Strait

Providniya Anadyr'

Magadan

Okhotsk

Arctic Circle

Amur

Oymyakon

Yakutsk

Pevek   Cherskiy

Chukchi
Sea

Wrangel
Island

Verkhoyansk

Watson
Lake

Prudhoe
Bay

Barrow
average minimum
extent of sea ice
(as of 1975)

Beaufort
Sea

East
Siberian
Sea

Tiksi

Hay
River

Echo Bay

Yellowknife

Inuvik

Banks
Island

Laptev
Sea

NEW
SIBERIAN
ISLANDS

Great Slave
Lake

Victoria
Island

Lake
Athabasca

Cambridge
Bay

Arctic
Ocean

North
Pole

SEVERNAYA
ZEMLYA

10°C (50°F) isotherm,
July

RUSSIA

CANADA

QUEEN
ELIZABETH
ISLANDS

Ellesmere
Island

Noril'sk

Rankin Inlet

Resolute

Repulse Bay

Alert

Qaanaaq
(Thule)

FRANZ
JOSEF
LAND

Kara
Sea

Dikson

Baffin
Island

Baffin
Bay

Nord

NOVAYA
ZEMLYA

Hudson
Bay

Iqaluit

Longyearbyen

Svalbard
(NORWAY)

Barents Sea

Davis Strait

Greenland
(DENMARK)

Greenland
Sea

Bjørnøya
(NORWAY)

Murmansk

Arkhangel'sk

Perm'

Kangerlussuaq
(Søndre Strømfjord)

Ittoqqortoormiit
(Scoresbysund)

Severnaya Dvina

Kazan'

Paamiut
(Frederikshåb)

Nuuk
(Godthåb)

Tasiilaq

Narsarsuaq

Jan Mayen
(NORWAY)

Lake
Onega

Samara

Labrador
Sea

Norwegian
Sea

Tromsø

Lake
Ladoga

Nizhniy
Novgorod

KAZ.

Denmark Strait

Arctic Circle

FINLAND

Moscow

Saratov

ICELAND

Reykjavík

Tórshavn

Faroe
Islands
(DENMARK)

NORWAY

SWEDEN

Helsinki

Oslo Stockholm

Saint
Petersburg

Tallinn

EST.

Volgograd

North Atlantic Ocean

Scale 1:39,000,000

Azimuthal Equal-Area Projection

0   500 Kilometers

0   500 Miles

The Arctic region is often defined as that area where the
average temperature for the warmest month is below 10°C.

SHETLAND
ISLANDS

Copenhagen

DENMARK

Belfast

IRE.
Dublin

U.K.

North
Sea

Baltic
Sea

RUS.

Riga

Vilnius

LATVIA

LITH.

Minsk

BELARUS

Warsaw

Berlin

GERMANY

POLAND

Kyiv

UKRAINE

Kharkiv

Rostov

Black Sea

ASIA

Scale 1:48,000,000
Azimuthal Equal-Area Projection

0 ————— 800 Kilometers
0 ————— 800 Miles

Boundary representation is
not necessarily authoritative.

CENTRAL AMERICA AND THE CARIBBEAN

EUROPE

Scale 1: 19,500,000
Lambert Conformal Conic Projection,
standard parallels 40°N and 56°N

Boundary representation is not
necessarily authoritative

MIDDLE EAST

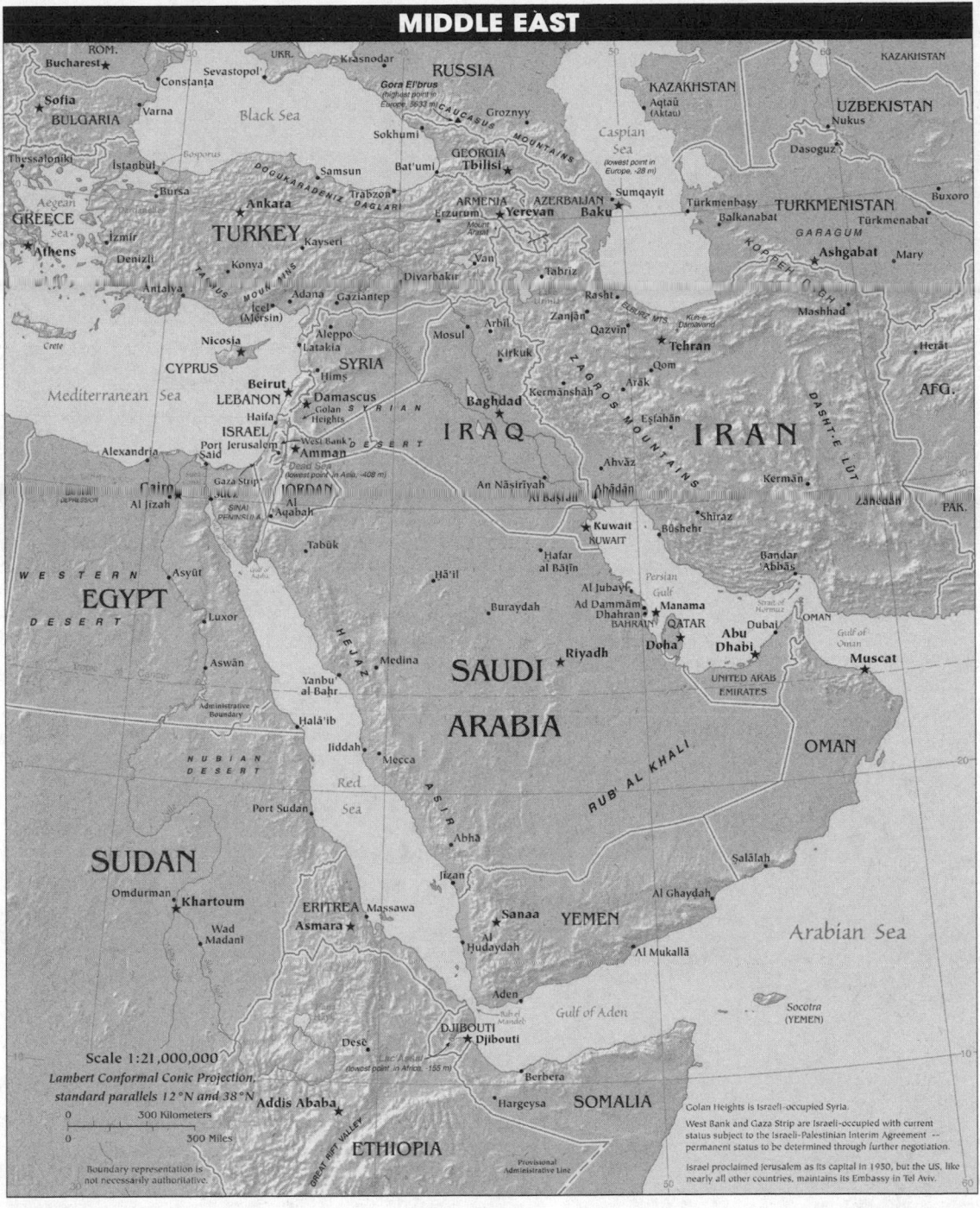

ROM.
Bucharest ★
Constanța
UKR.
Krasnodar
Sevastopol'
RUSSIA
Gora El'brus
(highest point in
Europe, 5633 m)
CAUCASUS MOUNTAINS
Grozny
KAZAKHSTAN
Aqtaū
(Aktau)
UZBEKISTAN
Nukus
Sofia ★
BULGARIA
Varna
Black Sea
Sokhumi
Caspian
Sea
(lowest point in
Europe, -28 m)
Dasoguz
Thessaloníki
İstanbul
Bosporus
Samsun
Bat'umi
GEORGIA
Tbilisi
Sumqayit
Türkmenbasy
TURKMENISTAN
Buxoro
Aegean
Sea
GREECE
Athens ★
İzmir
Bursa
DOĞUKARADENİZ
DAĞLARI
Trabzon
ARMENIA
Erzurum
Yerevan ★
Mount
Ararat
AZERBAIJAN
Baku ★
Balkanabat
GARAGUM
KOPPEH
Ashgabat ★
Türkmenabat
Mary
Denizli
Ankara ★
TURKEY
Kayseri
Konya
Van
Tabriz
Rasht
Zanjān
Qazvin
ELBURZ MTS
Küh-e
Damāvand
DAG
Mashhad
Antalya
TAURUS MOUNTAINS
Adana
Gaziantep
Diyarbakır
Urmia
ZAGROS
Tehran ★
Qom
Herāt
Crete
İçel
(Mersin)
Aleppo
Latakia
Mosul
Arbīl
Kirkuk
MOUNTAINS
Arāk
Eşfahān
IRAN
AFG.
Nicosia ★
CYPRUS
SYRIA
Hims
Kermānshāh
DASHT-E LŪT
Mediterranean Sea
Beirut ★
LEBANON
Damascus ★
Golan
Heights
SYRIAN
Baghdad ★
IRAQ
Ahvāz
Kerman
Haifa
ISRAEL
Port
Said
Alexandria
Jerusalem ★
West Bank
Amman ★
Dead Sea
(lowest point in Asia, -408 m)
JORDAN
DESERT
An Nāṣirīyah
Al Baṣrah
Abādān
Shīrāz
Zāhedān
PAK.
Cairo ★
Al Jīzah
Gaza Strip
Suez
SINAI
PENINSULA
Al 'Aqabah
Tabūk
Hafar
al Bāṭin
Kuwait ★
KUWAIT
Būshehr
Bandar
'Abbās
OMAN
Gulf of
Oman
DEPRESSION
Asyūṭ
WESTERN
EGYPT
DESERT
Luxor
Aswān
HEJAZ
Medina
Ḥā'il
Buraydah
Al Jubayl
Ad Dammām
Dhahran
BAHRAIN
Al Jubayl
Manama ★
Doha ★
QATAR
Persian
Gulf
Strait of
Hormuz
Abu
Dhabi ★
Dubai
Muscat ★
Yanbu'
al Bahr
SAUDI
Riyadh ★
UNITED ARAB
EMIRATES
Halā'ib
ARABIA
Jiddah
Mecca
ASIR
RUB' AL KHALI
OMAN
NUBIAN
DESERT
Port Sudan
Red
Sea
Abhā
Administrative
Boundary
SUDAN
Omdurman
Khartoum ★
Wad
Madanī
ERITREA
Asmara ★
Massawa
Jīzan
Sanaa ★
Al Ḥudaydah
YEMEN
Al Ghaydah
Ṣalālah
Al Mukallā
Arabian Sea
Socotra
(YEMEN)
Dese
DJIBOUTI
★ Djibouti
Lac 'Assal
(lowest point in Africa, 155 m)
Aden
Bab el
Mandeb
Gulf of Aden
Berbera

Scale 1:21,000,000
Lambert Conformal Conic Projection,
standard parallels 12°N and 38°N

0   300 Kilometers
0   300 Miles

Boundary representation is
not necessarily authoritative.

Addis Ababa ★
GREAT RIFT VALLEY
Hargeysa
SOMALIA
Provisional
Administrative Line
ETHIOPIA

Golan Heights is Israeli-occupied Syria.

West Bank and Gaza Strip are Israeli-occupied with current
status subject to the Israeli-Palestinian Interim Agreement —
permanent status to be determined through further negotiation.

Israel proclaimed Jerusalem as its capital in 1950, but the US, like
nearly all other countries, maintains its Embassy in Tel Aviv.

NORTH AMERICA

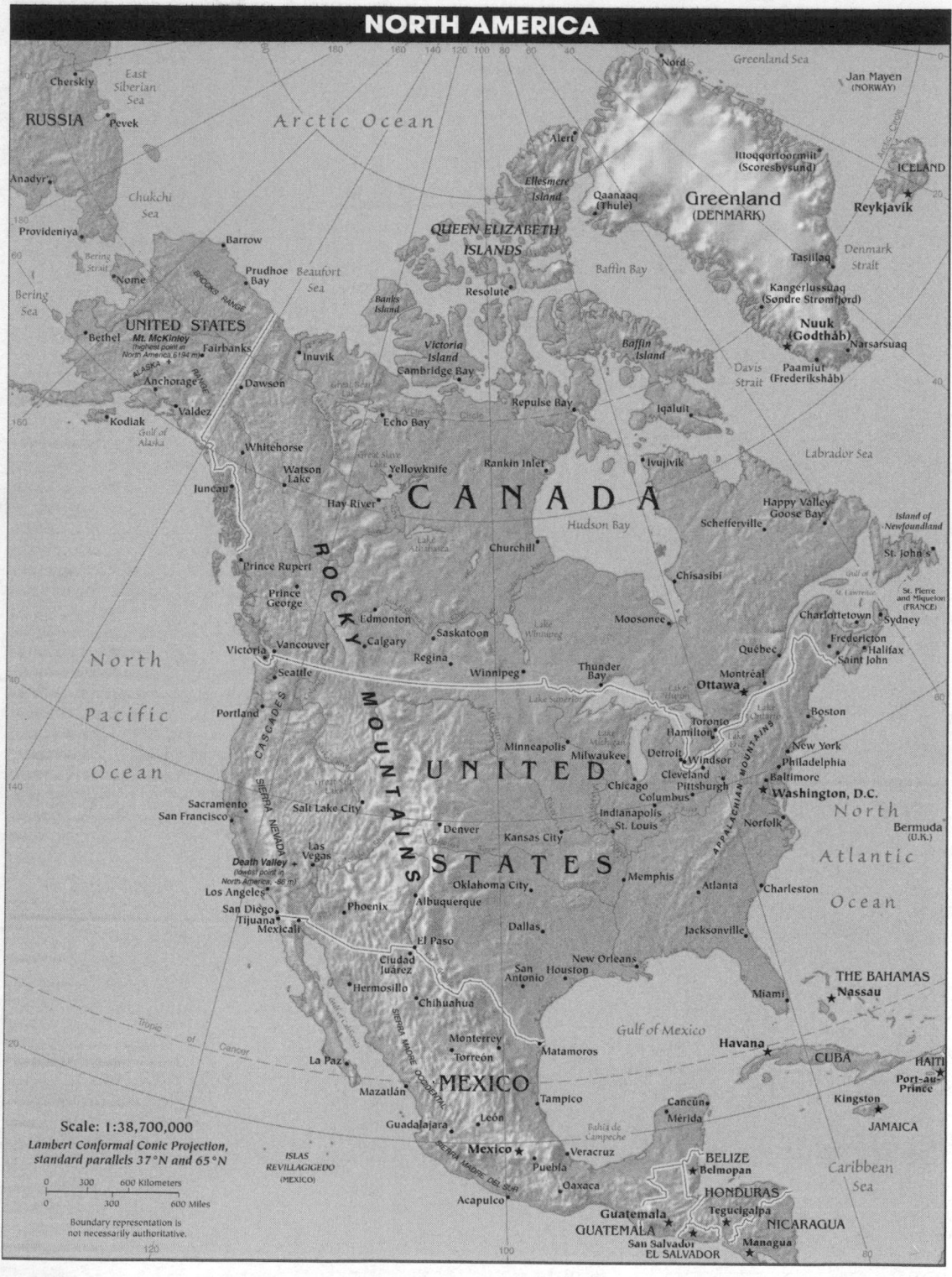

OCEANIA

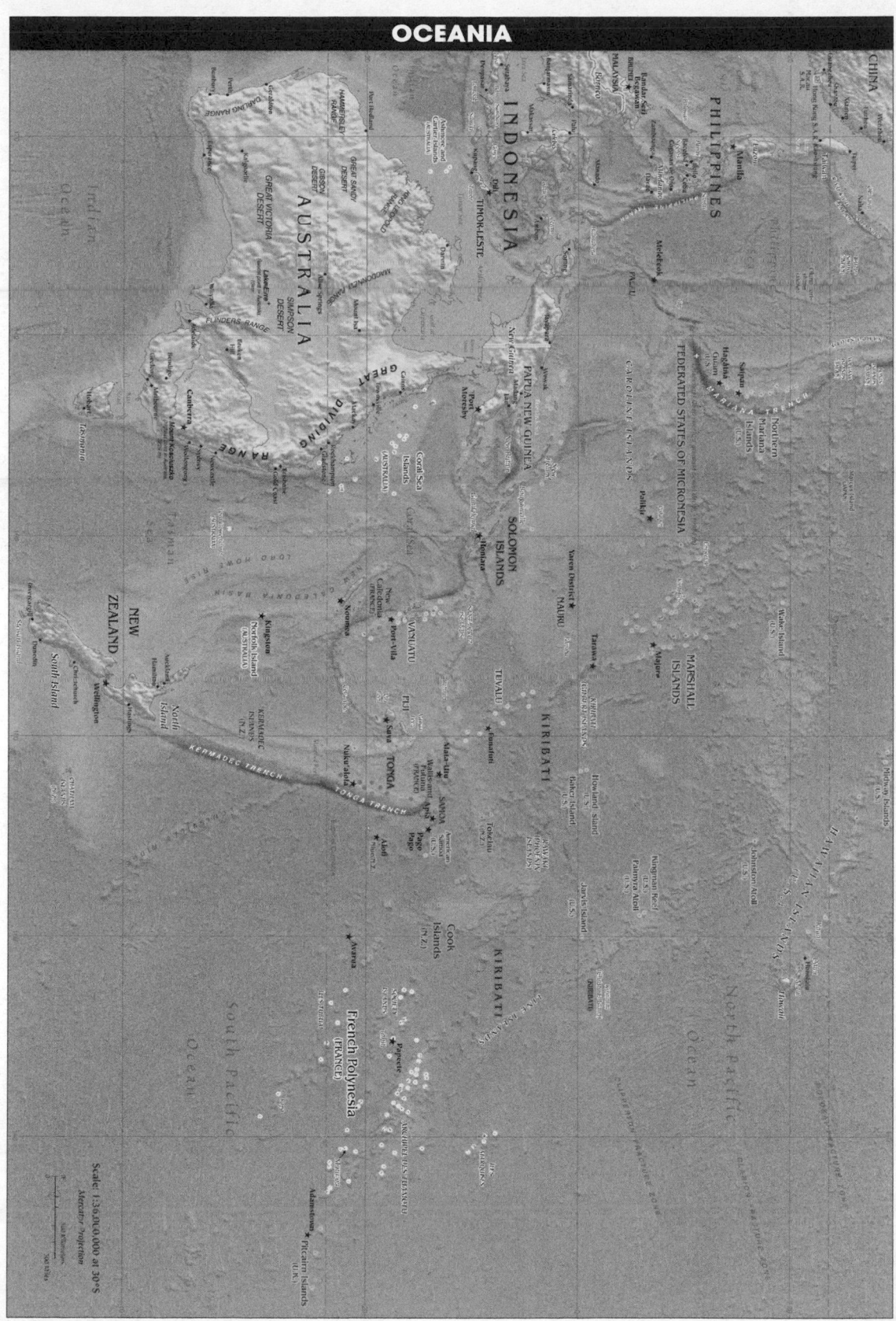

PHYSICAL MAP OF THE WORLD

POLITICAL MAP OF THE WORLD

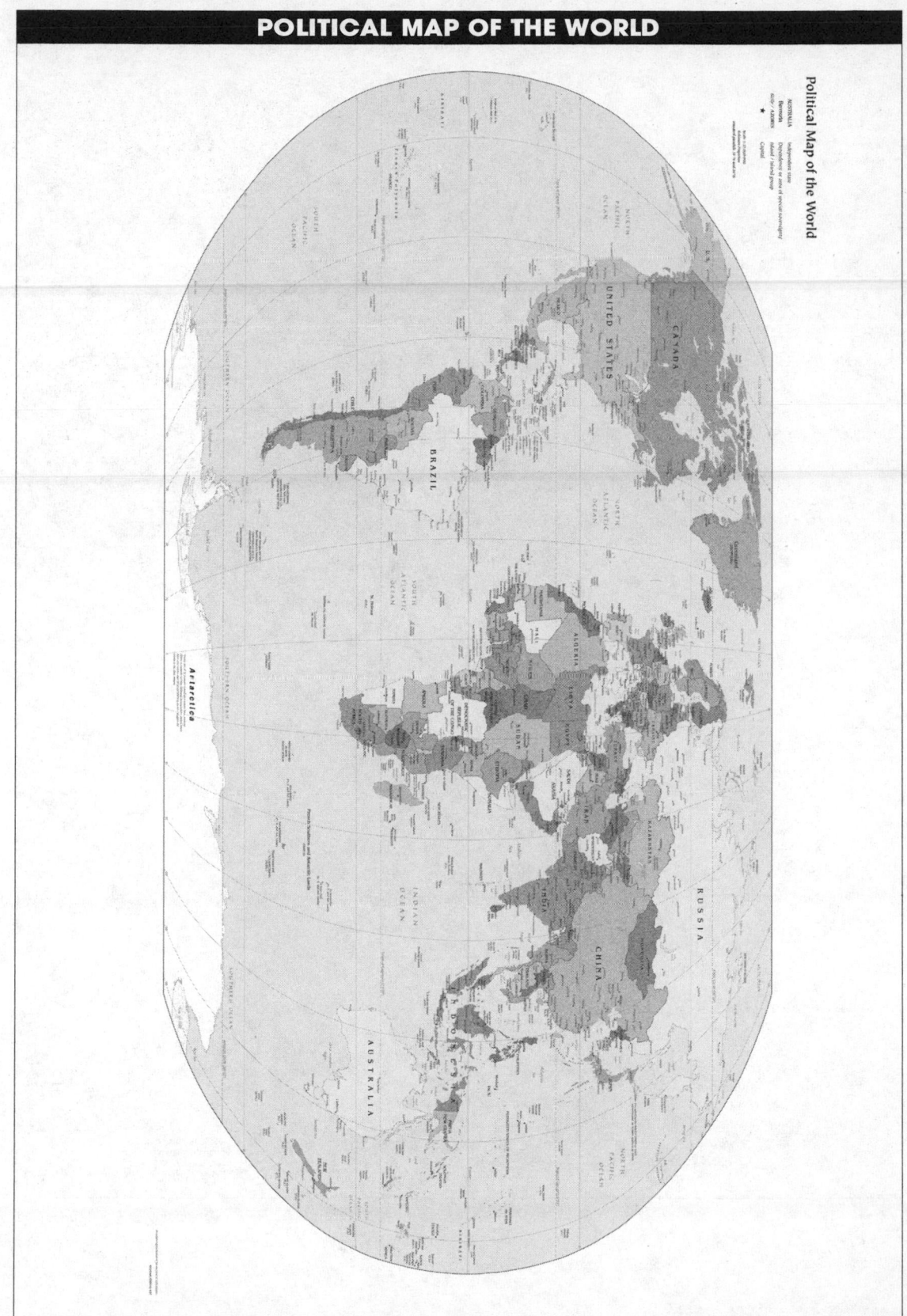

Political Map of the World

SOUTH AMERICA

Caribbean Sea

Martinique (FRANCE)
ST. LUCIA
St. VINCENT AND
THE GRENADINES
GRENADA

BARBADOS

HONDURAS
Tegucigalpa

Isla de
San Andrés
(COLOMBIA)

Aruba
(NETH.)
Netherlands
Antilles
(NETH.)

NICARAGUA
Managua

Barranquilla
Cartagena

Maracaibo

Caracas

Port-of-Spain
TRINIDAD AND
TOBAGO

North
Atlantic
Ocean

San José

Panama

Valencia
Barquisimeto

Ciudad
Guayana

Georgetown

Paramaribo

COSTA RICA

PANAMA

Cúcuta

San Cristóbal

VENEZUELA

GUYANA

SURINAME

Cayenne

French
Guiana
(FRANCE)

Medellín

Bogotá

GUIANA

HIGHLANDS

Isla de Malpelo
(COLOMBIA)

Cali

COLOMBIA

Boa Vista

Macapá

Quito

Equator

ECUADOR
Guayaquil

A M A Z O N

Manaus

Santarém

Belém

São Luís

Fortaleza

Iquitos

B A S I N

Piura

Trujillo

Huánuco

Río
Branco

Pôrto
Velho

BRAZIL

Natal

Recife

Maceió

South
Pacific
Ocean

PERU

Lima

Cusco

Trinidad

MATO GROSSO
PLATEAU

Cuiabá

BRAZILIAN

Brasília
Goiânia

HIGHLANDS

Salvador

Arequipa

La Paz

BOLIVIA

Cochabamba
Sucre

Santa Cruz

Potosí

Uberlândia

Belo
Horizonte

Arica

ALTIPLANO

Vitória

Iquique

Campo
Grande

Rio de Janeiro

Antofagasta

Tropic of Capricorn

Isla San Ambrosio
(CHILE)

Isla San Félix
(CHILE)

PARAGUAY

Salta

San Miguel
de Tucumán

Asunción

São Paulo

Curitiba

Santos

Resistencia

Florianópolis

CHILE

ARCHIPIÉLAGO
JUAN FERNÁNDEZ
(CHILE)

Cerro Aconcagua
(highest point in
South America, 6962 m)

Córdoba

Santa Fe

Pôrto Alegre

Valparaíso

Mendoza

Rosario

Salto

URUGUAY

South
Atlantic
Ocean

Santiago

PAMPAS

Buenos Aires

La Plata

Montevideo

Concepción

ANDES

ARGENTINA

Bahía Blanca

Mar del Plata

San Carlos de
Bariloche

Puerto Montt

PATAGONIA

Comodoro Rivadavia

Laguna del Carbón
(lowest point in South America and
the Western Hemisphere, -105 m)

Scale 1:35,000,000

Azimuthal Equal-Area Projection

500 Kilometers

500 Miles

Rio
Gallegos

Strait of
Magellan

Stanley

Falkland Islands
(Islas Malvinas)
(administered by U.K.,
claimed by ARGENTINA)

Boundary representation is
not necessarily authoritative.

Punta Arenas

Ushuaia

Cape
Horn

South Georgia and the
South Sandwich Islands
(administered by U.K.,
claimed by ARGENTINA)

ANDES

SOUTHEAST ASIA

Scale 1:32,000,000 at 5°N

Mercator Projection

| 0 | 500 Kilometers |
| 0 | 500 Miles |

Boundary representation is not necessarily authoritative.
Names in Vietnam are shown without diacritical marks.

STANDARD TIME ZONES OF THE WORLD

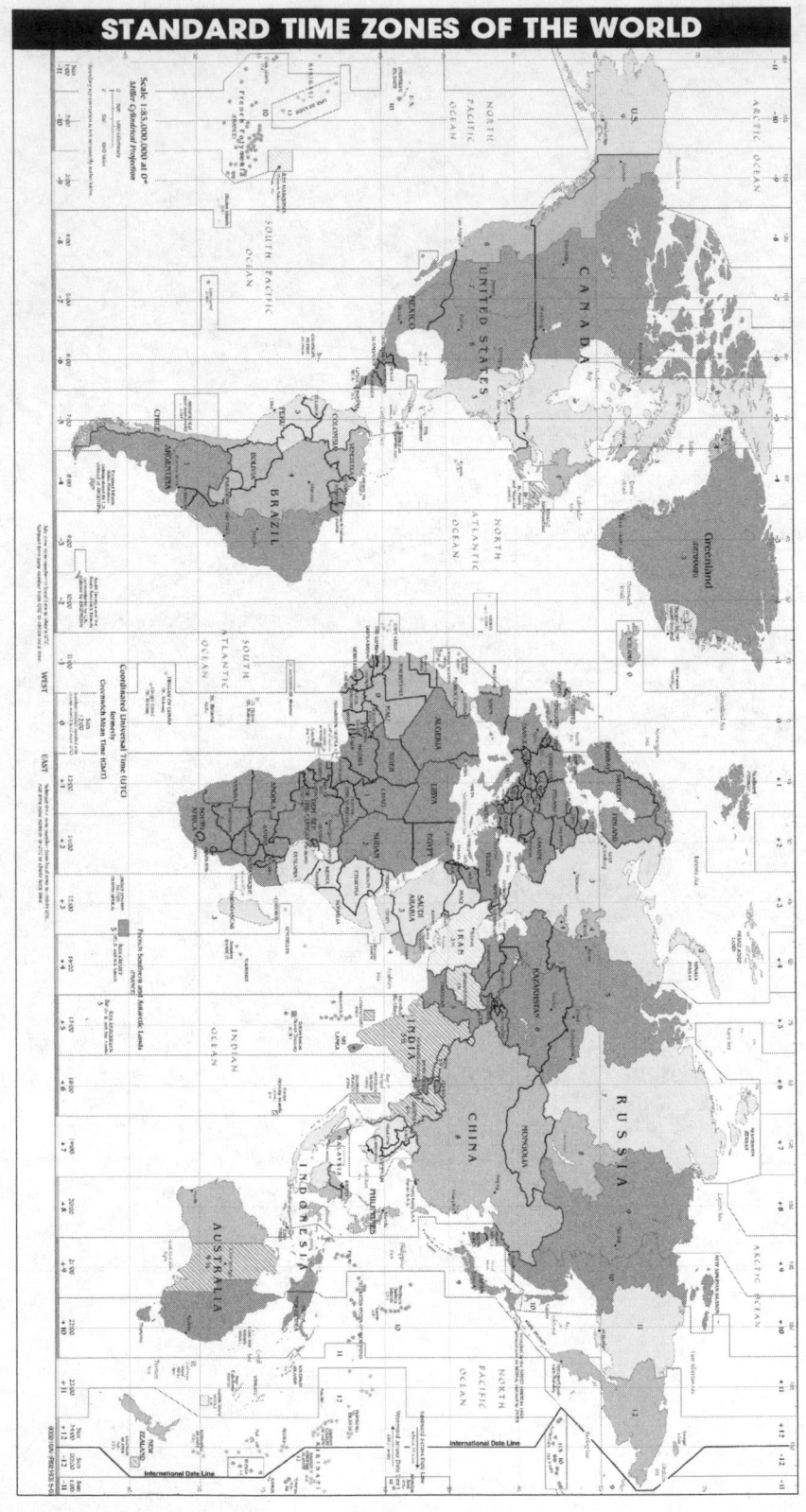

UNITED STATES

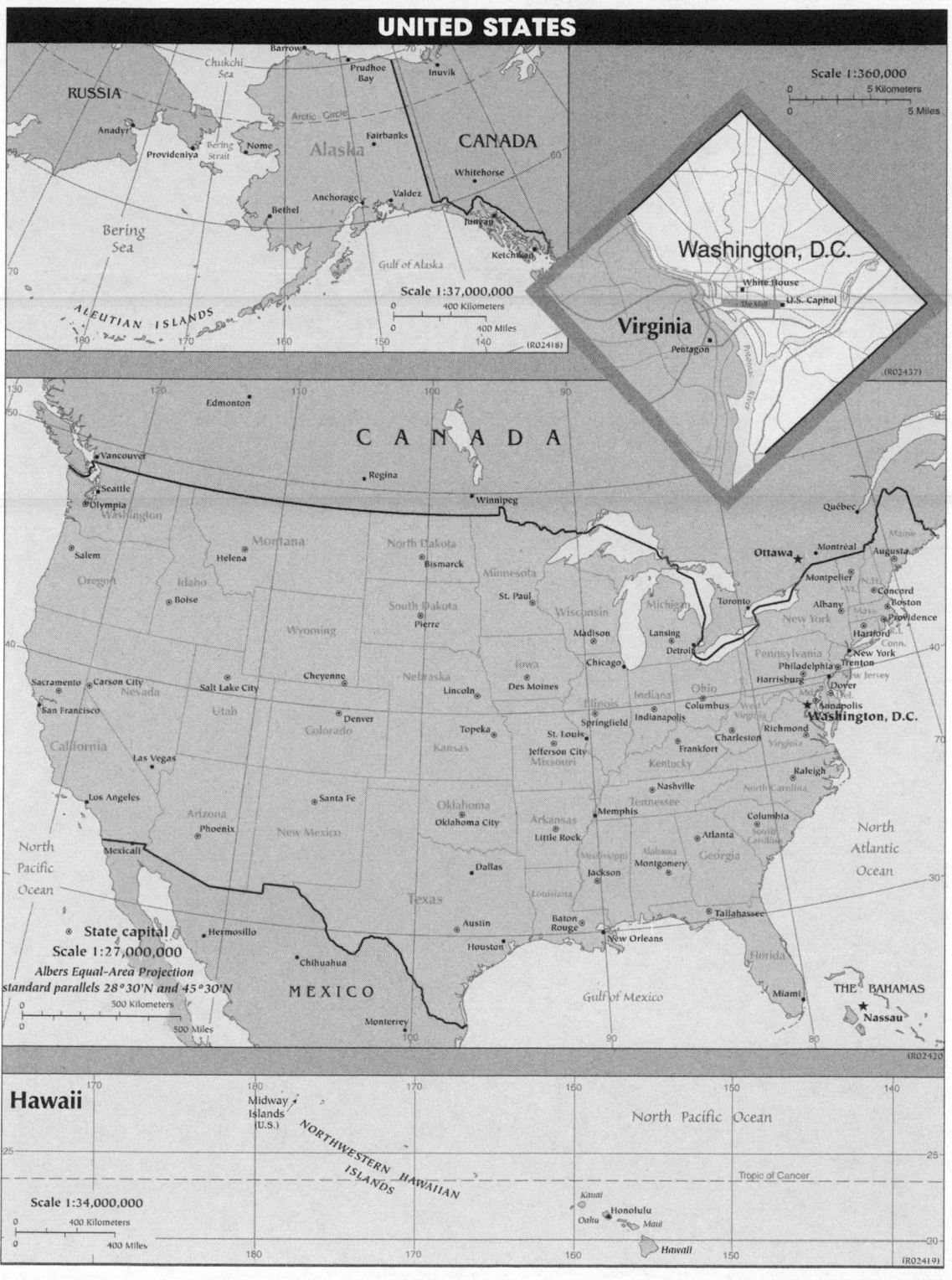

Scale 1:360,000

RUSSIA

Chukchi Sea

Barrow

Prudhoe Bay

Inuvik

Anadyr

Providéniya

Nome

Bering Strait

Fairbanks

CANADA

Alaska

Whitehorse

Anchorage

Valdez

Bethel

Juneau

Gulf of Alaska

Ketchikan

Bering Sea

ALEUTIAN ISLANDS

Scale 1:37,000,000

400 Kilometers
400 Miles

(R02418)

Washington, D.C.

White House

The Mall

U.S. Capitol

Virginia

Pentagon

(R02437)

CANADA

Edmonton

Vancouver

Regina

Winnipeg

Seattle

Olympia

Washington

Montana

Québec

Salem

Oregon

Idaho

Helena

North Dakota

Bismarck

Minnesota

St. Paul

Wisconsin

Michigan

Lansing

Toronto

Ottawa

Montréal

Montpelier

Maine

Augusta

Boise

South Dakota

Pierre

Madison

Concord

Boston

Albany

New York

Providence

Hartford

Conn.

Sacramento

Carson City

Nevada

Salt Lake City

Cheyenne

Nebraska

Iowa

Des Moines

Chicago

Detroit

Lincoln

Indiana

Ohio

Columbus

Pennsylvania

Harrisburg

Philadelphia

New Jersey

Dover

New York

Trenton

San Francisco

California

Utah

Denver

Colorado

Topeka

Kansas

St. Louis

Springfield

Illinois

Indianapolis

Frankfort

West Virginia

Charleston

Richmond

Virginia

Washington, D.C.

Annapolis

Las Vegas

Jefferson City

Missouri

Kentucky

Nashville

North Carolina

Raleigh

Los Angeles

Santa Fe

Oklahoma

Oklahoma City

Arkansas

Little Rock

Memphis

Tennessee

Columbia

South Carolina

Atlanta

Georgia

North Atlantic Ocean

Mexicali

Arizona

Phoenix

New Mexico

Dallas

Mississippi

Jackson

Alabama

Montgomery

North Pacific Ocean

● State capital

Scale 1:27,000,000

Albers Equal-Area Projection
standard parallels 28°30'N and 45°30'N

Hermosillo

Austin

Texas

Houston

Baton Rouge

New Orleans

Louisiana

Tallahassee

Florida

Miami

THE BAHAMAS

500 Kilometers
500 Miles

Chihuahua

MEXICO

Monterrey

Gulf of Mexico

Nassau

(R02420)

Hawaii

Midway Islands (U.S.)

NORTHWESTERN HAWAIIAN ISLANDS

North Pacific Ocean

Tropic of Cancer

Scale 1:34,000,000

Kauai

Oahu

Honolulu

Maui

400 Kilometers
400 Miles

Hawaii

(R02419)